3. Solve the related inequality analytically and graphically.

Illustration: Solve the inequality $2x^2 - 4x - 6 \le 0$.

Solution Divide a number line into intervals determined by the zeros of $f(x) = 2x^2 - 4x - 6$, which are -1 and 3. Choose a test value from each interval to identify values for which $f(x) \le 0$.

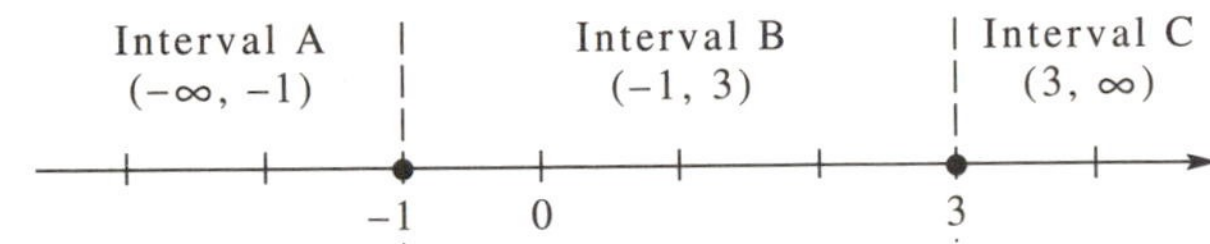

Interval	Test Value	Is $f(x) = 2x^2 - 4x - 6 \le 0$ True or False?
A: $(-\infty, -1)$	-2	$f(-2) = 2(-2)^2 - 4(-2) - 6 \le 0$? $10 \le 0$ False
B: $(-1, 3)$	0	$f(0) = 2(0)^2 - 4(0) - 6 \le 0$? $-6 \le 0$ **True**
C: $(3, \infty)$	4	$f(4) = 2(4)^2 - 4(4) - 6 \le 0$? $10 \le 0$ False

The polynomial is negative or zero on the interval $[-1, 3]$. The calculator graph in Illustration 2 supports this solution, since the graph lies *on* or *below* the x-axis on this interval.

4. Apply analytic and graphical methods to solve an application of that class of function.

Illustration: If an object is projected directly upward from the ground with an initial velocity of 64 feet per second, then (neglecting air resistance) the height of the object t seconds after it is projected is modeled by $s(t) = -16t^2 + 64t$, where $s(t)$ is in feet. After how many seconds does it reach a height of 28 feet?

Analytic Solution

We must solve the equation $s(t) = 28$.

$$s(t) = -16t^2 + 64t$$

$$28 = -16t^2 + 64t \qquad \text{Let } s(t) = 28.$$

$$16t^2 - 64t + 28 = 0 \qquad \text{Standard form}$$

$$4t^2 - 16t + 7 = 0 \qquad \text{Divide by 4.}$$

$$(2t - 1)(2t - 7) = 0 \qquad \text{Factor.}$$

$$t = .5 \quad \text{or} \quad t = 3.5 \qquad \text{Zero-product property}$$

The object reaches a height of 28 feet twice, at .5 second (on its way up) and at 3.5 seconds (on its way down).

Graphing Calculator Solution

Using the intersection-of-graphs method, we see that the graphs of $y = -16x^2 + 64x$ and $y = 28$ intersect at points whose coordinates are $(.5, 28)$ and $(3.5, 28)$, confirming our analytic answer.

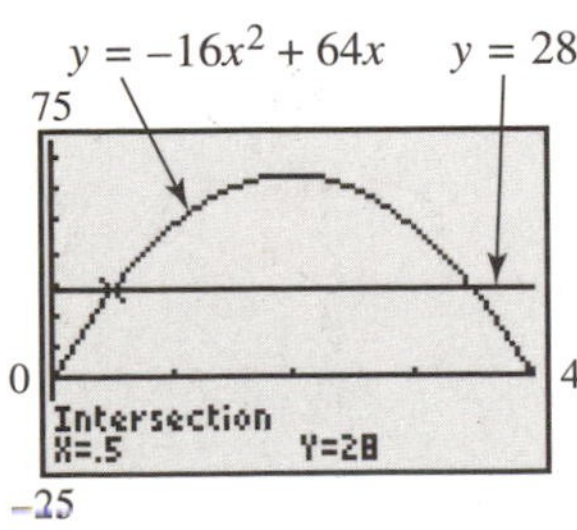

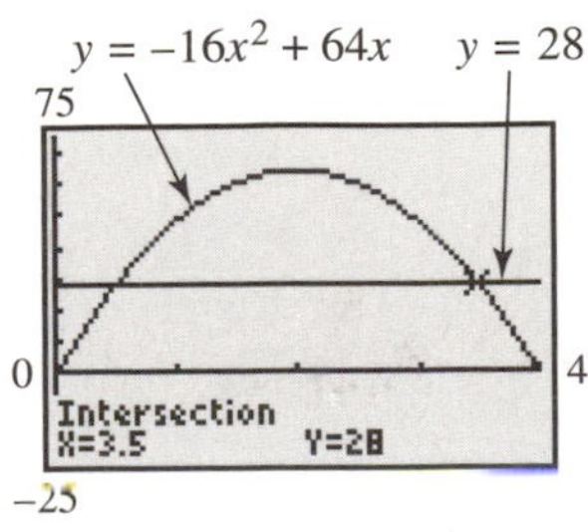

Fourth Edition

A Graphical Approach to Precalculus with Limits: A Unit Circle Approach

John Hornsby
University of New Orleans

Margaret L. Lial
American River College

Gary K. Rockswold
Minnesota State University, Mankato

BOSTON SAN FRANCISCO NEW YORK
LONDON TORONTO SYDNEY TOKYO SINGAPORE MADRID
MEXICO CITY MUNICH PARIS CAPE TOWN HONG KONG MONTREAL

Publisher: **Greg Tobin**
Executive Editor: **Anne Kelly**
Executive Project Manager: **Christine O'Brien**
Editorial Assistant: **Ashley O'Shaughnessy**
Managing Editor: **Karen Wernholm**
Senior Designer: **Barbara T. Atkinson**
Supplements Supervisor: **Emily Portwood**
Media Producer: **Christine Stavrou**
Software Development: **Malcolm Litowitz and Ted Hartman**
Senior Marketing Manager: **Becky Anderson**
Marketing Assistant: **Maureen McLaughlin**
Senior Author Support/Technology Specialist: **Joe Vetere**
Senior Prepress Supervisor: **Caroline Fell**
Rights and Permissions Advisor: **Shannon Barbe**
Senior Manufacturing Buyer: **Evelyn Beaton**
Cover and Text Design: **Dutton & Sherman Design**
Production Coordination: **WestWords, Inc.**
Composition: **Beacon Publishing Services**
Illustrations: **Techsetters, Inc.**
Cover photo: **Keys on flute, close-up © Comstock Images**

In memory of the victims
of Hurricane Katrina
E.J.H.
M.L.L.
G.K.R.

Photo Credits: For permission to use copyrighted material, grateful acknowledgment is made to the copyright holders on page I-18, which is hereby made part of this copyright page.

Library of Congress Cataloging-in-Publication Data

Hornsby, E. John.
A graphical approach to precalculus with limits.—4th ed./John Hornsby, Margaret Lial, Gary Rockswold.
p. cm.
Includes index.
ISBN 0-321-35696-9
1. Functions. I. Lial, Margaret L. II. Rockswold, Gary K. III. Title.
QA331.3.H67 2007
512′.1—dc22 2005047685

4 5 6 7 8 9 10—DOW—10 09 08

Contents

9 Trigonometric Identities and Equations 676

10 Applications of Trigonometry; Vectors 743

Preface

In 1996, *A Graphical Approach to Precalculus with Limits: A Unit Circle Approach* was one of the first texts to tailor the traditional precalculus table of contents to maximize the impact of the graphing calculator. Now, a decade later, we have adapted this fourth edition in an attempt to address the changing needs of today's students. The text maintains its unique table of contents and functions-based approach (as outlined on the inside front cover), but also includes additional components to build skill, address critical thinking, solve applications, and apply technology to support traditional algebraic solutions—areas where we feel today's students need more assistance than in previous years.

While the text has evolved significantly from earlier editions, it retains the strengths of those editions and provides new and relevant opportunities for students and instructors alike. It incorporates an open design, helpful features, careful explanations of topics, and a comprehensive package of supplements and study aids. We offer for the first time an *Annotated Instructor's Edition,* which provides answers to most nongraphing exercises beside or below each problem. As in earlier editions, answers to both even- and odd-numbered exercises are provided in the back of the text for the instructor.

This text is part of a series that also includes the following books:

- *A Graphical Approach to College Algebra, Fourth Edition,* by Hornsby, Lial, and Rockswold
- *A Graphical Approach to Algebra and Trigonometry, Fourth Edition,* by Hornsby, Lial, and Rockswold
- *A Graphical Approach to Precalculus, Fourth Edition,* by Hornsby, Lial, and Rockswold

Our Approach

It is our desire that this text allow students to easily make connections between graphs of functions and their associated equations and inequalities. Using linear functions as the basis for the presentation in Chapter 1, we introduce a four-step process for the study of functions in subsequent chapters. After introducing a class of functions,

1. We examine the nature of its graph, using hand-drawn and calculator graphs.
2. We solve equations analytically and use graphing calculators to find and support solutions obtained with the x-intercept method or the intersection-of-graphs method.
3. We solve the associated inequalities analytically and graphically.
4. We apply analytic and graphical methods to modeling and traditional applications.

This unifying approach is illustrated on the inside front cover of the text. By using this approach consistently with each class of functions, we hope that students will better understand, connect, and apply concepts and skills.

The book is written with the assumption that students have access to graphing calculators. Although we have chosen to use screens from the TI-83/84 Plus models, we do not include specific keystroke instructions because of the wide variety of models available. Students should refer to the manuals provided with their calculators for specific information. Keystroke instructions for selected examples are included in the *Graphing Calculator Manual* that accompanies the text.

Content Changes

You will find many places in the text where we have polished individual presentations and added examples, exercises, and applications based on reviewer feedback. Some of the changes you may notice include the following:

- There are approximately 1200 additional exercises, many of which are devoted to skill development.
- Graphing by hand has been expanded throughout the text. Over 25 graphs featuring a new "hand-drawn" style that simulates what a student might actually sketch on graph paper have been incorporated into discussions and examples.
- Real-world data in over 75 applications have been updated.
- Section 2.3 includes a new subsection on horizontal stretching and shrinking.
- Quadratic and rational inequalities are solved analytically by using test values in Sections 3.3 and 4.3.
- Section 3.7 includes new subsections on Descartes' Rule of Signs and the Boundedness Theorem.
- Sections 6.1 and 6.2 provide new subsections on translations of parabolas, ellipses, and hyperbolas. New figures have been included to enhance presentations on forms of conic sections.
- Systems of three equations in three variables in Section 7.2 are solved by using traditional elimination and substitution.
- Section 8.5 has a new subsection on applications of trigonometric functions.
- Section 9.4 includes a new subsection that reviews inverse functions.
- Chapter R now ends with a test over the material covered in the chapter.
- Appendix A includes the geometry formulas formerly covered in Section R.6. Appendix B: Deciding Which Model Best Fits a Set of Data is new.
- In addition, the following topics have been expanded:
  Using slope–intercept form to write and graph equations of lines
  Discussion of composition of functions
  Graphing rational functions
  Reviewing rational exponents
  Finding inverses of functions
  Finding matrix inverses
  Graphing trigonometric functions
  Graphing polar equations of lines and circles

Features

We are pleased to offer the following new or enhanced features:

Chapter Openers These provide a motivating application topic that is tied to the chapter content, plus a list of the sections in the chapter. Most openers have been updated or are entirely new to this edition.

Enhanced Examples We have added even more examples in this edition and have carefully polished all solutions and incorporated more side comments. New red comments with pointers provide students with on-the-spot reminders and warnings about common pitfalls.

Dual-Solution Format Selected examples continue to provide side-by-side analytic and graphing calculator solutions, to connect traditional analytic methods for solving problems with graphical methods of solution or support. Some of these examples are now marked with GCM to indicate that additional information on the graphing calculator solution is included in the *Graphing Calculator Manual* that accompanies the text.

Figures and Photos Today's students are more visually oriented than ever. As a result, we have made a concerted effort to provide more figures, diagrams, tables, and graphs, including the new "hand-drawn" style of graphs, whenever possible. The number of photos accompanying applications in examples and exercises is almost double the number in the previous edition.

Function Capsules These special boxes offer a comprehensive, visual introduction to each class of functions and serve as an excellent resource for reference and review. Each capsule includes traditional and calculator graphs and a calculator table of values, as well as the domain, range, and other specific information about the function. Abbreviated versions of all function capsules are provided on the inside back cover of the text.

What Went Wrong? This popular feature anticipates typical errors that students make when using graphing technology and provides an avenue for instructors to highlight and discuss such errors. Answers are included on the same page as the "What Went Wrong?" boxes.

Cautions and Notes These warn students of common errors and emphasize important ideas throughout the exposition.

Looking Ahead to Calculus These margin notes provide glimpses of how the precalculus topics currently being studied are used in calculus.

Technology Notes Also appearing in the margin, these notes provide tips to students on how to use graphing calculators more effectively. Some notes are now marked with GCM, to indicate that additional related information on graphing technology is included in the *Graphing Calculator Manual* that accompanies the text.

For Discussion These activities appear within the exposition or in the margins and offer material on important concepts for instructors and students to investigate or discuss in class.

Exercise Sets We have taken special care to respond to the suggestions of users and reviewers and have added about 1200 new exercises to this edition on the basis of their feedback. As a result, the text includes more problems than ever to provide students with ample opportunities to practice, apply, connect, and extend concepts and skills. We have included writing exercises, as well as multiple-choice, matching, true/false, and completion problems. Exercises marked ***Concept Check*** focus on mathematical thinking and conceptual understanding.

Relating Concepts These groups of exercises appear in selected exercise sets. They tie together topics and highlight relationships among various concepts and skills. All answers to these problems appear in the answer section at the back of the book.

Reviewing Basic Concepts These sets of exercises appear every two or three sections and allow students to review and check their understanding of the material in preceding sections. All answers to these problems are included in the answer section.

Chapter Review Material One of the most popular features of the text, each end-of-chapter Summary features a section-by-section list of Key Terms and Symbols, in addition to Key Concepts. A comprehensive set of Chapter Review Exercises and a Chapter Test are also included.

Chapter Projects Each chapter concludes with a project that students can complete individually or collaboratively, using the material from the chapter.

Supplements

For a comprehensive list of the supplements and study aids that accompany *A Graphical Approach to Precalculus with Limits: A Unit Circle Approach, Fourth Edition,* see pages xviii and xix.

Acknowledgments

Previous editions of this text were published after thousands of hours of work, not only by the authors, but also by reviewers, instructors, students, answer checkers, and editors. To these individuals and to all those who have worked in some way on this text over the years, we are most grateful for your contributions. We could not have done it without you. We especially wish to thank the following individuals who provided valuable input into this and previous editions of the text.

Judy Ahrens	Pellissippi State Technical College
Randall Allbritton	Daytona Beach Community College
Jamie Ashby	Texarkana College
Gloria Bass	Mercer University
Pat Bassett	Palm Beach Atlantic University
Daniel Biles	Western Kentucky University
Linda Buchanan	Howard College
Sylvia Calcano	Lake City Community College
Faye Childress	Central Piedmont Community College
Bettyann Daley	University of Delaware
Jacqueline Donofrio	Monroe Community College
Patricia Dueck	Scottsdale Community College
Mickle Duggan	East Central University
Douglas Dunbar	Okaloosa-Walton Community College
Nancy Eschen	Florida Community College at Jacksonville
Donna Fatheree	University of Louisiana at Lafayette
Linda Fosnaugh	Midwestern State University
William Frederick	Indiana Purdue University, Fort Wayne
Henry Graves	Trident Technical College
Kim Gregor	Delaware Technical and Community College
Sandee House	Georgia Perimeter College
W. H. Howland	University of St. Thomas (Houston)
Tuesday J. Johnson	New Mexico State University
Cheryl Kane	University of Nebraska
Mike Keller	St. John's River Community College

Rosemary Kradel	Lehigh Carbon Community College
Nancy Matthews	University of Oklahoma
Peggy Miller	University of Nebraska at Kearney
Phillip Miller	Indiana University Southeast
Stacey McNiel	Lake City Community College
Richard Montgomery	The University of Connecticut
Michael Nasab	Long Beach City College
Jon Odell	Richland Community College
Karen Pender	Chaffey College
Zikica Perovic	Normandale Community College
Mary Anne Petruska	Pensacola Junior College
Susan Pfeifer	Butler County Community College, Andover
John Putnam	University of Northern Colorado
Charles Roberts	Mercer University
Donna Saye	Georgia Southern University
Alicia Schlintz	Meredith College
Linda K. Schmidt	Greenville Technical College
Mike Shirazi	Germanna Community College
Jed Soifer	Atlantic Cape Community College
Betty Swift	Cerritos College
Jennifer Walsh	Daytona Beach Community College
Robert Woods	Broome Community College
Fred Worth	Henderson State University
Kevin Yokoyama	College of the Redwoods

Over the years, we have come to rely on an extensive team of experienced professionals. Our sincere thanks go to these dedicated individuals at Addison-Wesley who worked long and hard to make this revision a success: Greg Tobin, Anne Kelly, Becky Anderson, Christine O'Brien, Karen Wernholm, Barbara Atkinson, and Ashley O'Shaughnessy.

Terry McGinnis continues to provide invaluable behind-the-scenes guidance—we have come to rely on her expertise during all phases of the revision process. Terry Krieger did an outstanding job preparing the student answer section, including the answer art, the answers for the *Instructor's Edition,* and, with Randy Krieger, the solutions manuals. Thanks are also due Jami Darby, Melena Fenn, and WestWords, Inc., for their excellent production work; Virginia Starkenburg, who researched and wrote the new chapter openers and updated application data in examples and exercises; Sandee House for writing new Appendix B on modeling; Perian Herring for checking the answers to all exercises and accuracy checking portions of the text; and Paul Lorczak, John Samons, and Janis Cimperman for accuracy checking page proofs.

Production of this edition of the text was accomplished during the hurricane season of 2005. A portion of the royalties will be donated to the victims of Hurricane Katrina.

As an author team, we are committed to providing the best possible text to help instructors teach and students succeed using graphing technology. As we continue to work toward this goal, we would welcome any comments or suggestions you might have via e-mail to *math@aw.com.*

John Hornsby
Marge Lial
Gary Rockswold

Student Supplements

Student's Solutions Manual

- By Randy Krieger and Terry A. Krieger, *Rochester Community and Technical College*
- Provides detailed solutions to odd-numbered Section and Chapter Review Exercises and to all Relating Concepts, Reviewing Basic Concepts, and Chapter Test problems
  ISBN: 0-321-35819-8

Graphing Calculator Manual

- By Darryl Nester, *Bluffton University*
- Provides instructions and keystroke operations for the TI-83/84 Plus, TI-85, TI-86, and TI-89
- Keyed directly to text Examples and Technology Notes marked GCM
  ISBN: 0-321-35820-1

Videotape Series

- Provides comprehensive coverage of each section and topic in the text in an engaging format that stresses student interaction
- Features examples and problems taken from the text
- ISBN: 0-321-35815-5

Digital Video Tutor

- Complete set of digitized videos for student use at home or on campus
- Ideal for distance learning or supplemental instruction
  ISBN: 0-321-35814-7

Addison-Wesley Math Tutor Center

- Provides free tutoring through a registration number that can be packaged with a new textbook or purchased separately
- Staffed by qualified college mathematics instructors
- Accessible via toll-free telephone, toll-free fax, e-mail, and the Internet at www.aw-bc.com/tutorcenter

Instructor Supplements

New! Annotated Instructor's Edition

- Includes answers beside or below exercises for quick reference where possible
- Provides answers to both even- and odd-numbered exercises at the back of the text
  ISBN: 0-321-36153-9

Instructor's Solutions Manual

- By Randy Krieger and Terry A. Krieger, *Rochester Community and Technical College*
- Provides complete solutions to all text exercises
  ISBN: 0-321-35813-9

Instructor's Testing Manual

- By Paul Lorczak
- Contains four alternative tests per chapter
- Answer keys are included
  ISBN: 0-321-35816-3

TestGen®

- Enables instructors to build, edit, print, and administer tests
- Features a computerized bank of questions developed to cover all text objectives
- Available on a dual-platform Windows/Macintosh CD-ROM
  ISBN: 0-321-35818-X

PowerPoint Lecture Presentation

- Features presentations written and designed specifically for this text, including figures and examples from the text
- Available within MyMathLab or from the Instructor Resource Center at www.aw-bc.com/irc

New! Insider's Guide

- Includes resources to help faculty with course preparation and classroom management
- Provides helpful teaching tips, correlated to each section of the text, as well as general teaching advice
  ISBN: 0-321-35817-1

New! Adjunct Support Center

- Offers consultation on suggested syllabi, helpful tips on using the textbook support package, assistance with content, and advice on classroom strategies
- Available Sunday–Thursday evenings from 5 P.M. to midnight EST; telephone: 1-800-435-4084; e-mail: AdjunctSupport@aw.com; fax: 1-877-262-9774

MathXL®

MathXL is a powerful online homework, tutorial, and assessment system that accompanies this textbook. With MathXL, instructors can create, edit, and assign online homework and tests using algorithmically generated exercises correlated at the objective level to the text. Instructors can also create and assign their own online exercises and import TestGen tests for added flexibility. All student work is tracked in MathXL's online gradebook. Students can take chapter tests in MathXL and receive personalized study plans based on their test results. The study plan diagnoses weaknesses and links students directly to tutorial exercises for the objectives they need to study and on which they need to be retested. Students can also access supplemental animations and video clips directly from selected exercises. MathXL is available to qualified adopters. For more information, visit our Web site at www.mathxl.com or contact your local sales representative.

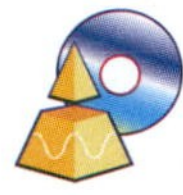

MathXL® Tutorials on CD ISBN: 0-321-35821-X

This interactive tutorial CD-ROM provides algorithmically generated practice exercises that are correlated at the objective level to the exercises in the textbook. Every practice exercise is accompanied by an example and a guided solution designed to involve students in the solution process. Selected exercises may also include a video clip to help students visualize concepts. The software provides helpful feedback for incorrect answers and can generate printed summaries of students' progress.

MyMathLab MyMathLab is a series of text-specific, easily customizable online courses for textbooks in mathematics and statistics. MyMathLab is powered by CourseCompass™—Pearson Education's online teaching and learning environment—and by MathXL—our online homework, tutorial, and assessment system. MyMathLab gives instructors the tools needed to deliver all or a portion of their course online, whether students are in a lab setting or working from home. MyMathLab provides a rich and flexible set of course materials, featuring free-response exercises that are algorithmically generated for unlimited practice and mastery. Students can also use online tools, such as video lectures, animations, and a multimedia textbook, to independently improve their understanding and performance. Instructors can use MyMathLab's homework and test managers to select and assign online exercises correlated directly to the textbook, and they can also create and assign their own online exercises and import TestGen tests for added flexibility. MyMathLab's online gradebook—designed specifically for mathematics and statistics—automatically tracks students' homework and test results and gives the instructor control over how to calculate final grades. Instructors can also add off-line (paper-and-pencil) grades to the gradebook. MyMathLab is available to qualified adopters. For more information, visit our Web site at www.mymathlab.com or contact your local sales representative.

InterAct Math Tutorial Web site www.interactmath.com

Get practice and tutorial help online! This interactive tutorial Web site provides algorithmically generated practice exercises that correlate directly to the exercises in the textbook. Students can retry an exercise as many times as they like, with new values each time, for unlimited practice and mastery. Every exercise is accompanied by an interactive guided solution that provides helpful feedback for incorrect answers, and students can also view a worked-out sample problem that steps them through an exercise similar to the one they are working on.

1

Linear Functions, Equations, and Inequalities

DURING THE PAST 20 years, average temperatures have risen 7°F over much of the Arctic and ice has melted at an alarming rate. Although a 7° temperature change may not seem like it would have much effect on Earth's climate, it took only an 11° decrease in average global temperatures to cause the last ice age. Calculations show that if all the ice in the Arctic were to melt, sea levels would rise 25 feet.

This melting phenomenon has also occurred in the Antarctic, where temperatures have risen approximately 5° in the last 50 years, or about $\frac{1}{10}^{\circ}$ per year. If this trend were to continue for the next hundred years, the average temperature in the Antarctic would rise an additional 10°F, causing major climatic change.

What will the future bring? *Linear functions* can help make these types of estimates. In this chapter, we learn how to model many different phenomena with linear functions and equations.

Source: Hudson, D., "Clear and Present Danger," *Discover*, January 2001.

Chapter Outline

1.1 Real Numbers and the Rectangular Coordinate System

Sets of Real Numbers ■ The Rectangular Coordinate System ■ Viewing Windows ■ Roots ■ Distance and Midpoint Formulas

Sets of Real Numbers

The idea of counting goes back to the beginning of civilization. When people first counted, they used only the **natural numbers,** written in set notation as

$$\{1, 2, 3, 4, 5, \ldots\}. \quad \text{Natural numbers}$$

Including 0 in the set of natural numbers gives the set of **whole numbers,**

$$\{0, 1, 2, 3, 4, 5, \ldots\}. \quad \text{Whole numbers}$$

The negatives of the natural numbers, included in the set of whole numbers, give the set of **integers,**

$$\{\ldots, -3, -2, -1, 0, 1, 2, 3, \ldots\}. \quad \text{Integers}$$

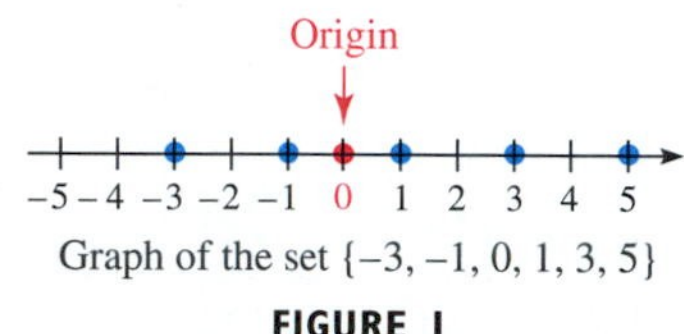

Graph of the set {–3, –1, 0, 1, 3, 5}

FIGURE 1

Integers can be shown pictorially—that is, **graphed**—on a **number line.** See Figure 1. The point corresponding to 0 is called the **origin.** Every number corresponds to one and only one point on the number line, and each point corresponds to one and only one number. This correspondence is called a **coordinate system.** The number associated with a given point is called the **coordinate** of the point.

The result of dividing two integers (with a nonzero divisor) is called a *rational number,* or *fraction*. A **rational number** is an element of the set

$$\left\{\frac{p}{q}\,\middle|\, p, q \text{ are integers}, q \neq 0\right\}. \quad \text{Rational numbers}$$

Rational numbers include the natural numbers, whole numbers, and integers. For example, the integer -3 is a rational number because it can be written as $\frac{-3}{1}$. Numbers that can be written as repeating or terminating decimals are also rational numbers. For example, $.\overline{6} = .66666\ldots$ represents a rational number that can be expressed as the fraction $\frac{2}{3}$.

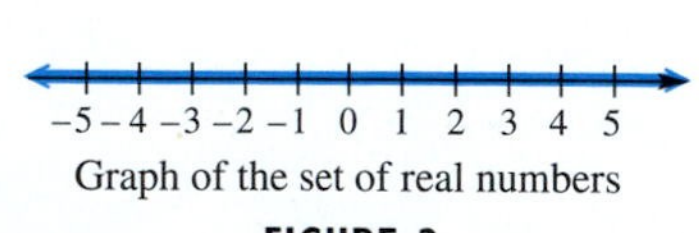

Graph of the set of real numbers

FIGURE 2

The set of all numbers that correspond to points on a number line is called the **real numbers,** as shown in Figure 2. Real numbers can be represented by decimals. Since every fraction has a decimal form—for example, $\frac{1}{4} = .25$—real numbers include rational numbers.

Some real numbers cannot be represented by quotients of integers. These numbers are called **irrational numbers.** Examples of irrational numbers include $\sqrt{3}$, $\sqrt{5}$, $\sqrt[3]{10}$, and $\sqrt[5]{20}$, but not $\sqrt{1}$, $\sqrt{4}$, $\sqrt{9}, \ldots$, which equal $1, 2, 3, \ldots$, and hence are rational numbers. Another irrational number is π, which is approximately equal to 3.14159. The numbers in the set $\left\{-\frac{2}{3}, 0, \sqrt{2}, \sqrt{5}, \pi, 4\right\}$ can be located on a number line, as shown in Figure 3. (Only $\sqrt{2}$, $\sqrt{5}$, and π are irrational here.) Note that $\sqrt{2}$ is approximately equal to 1.41, so it is located between 1 and 2, slightly closer to 1.

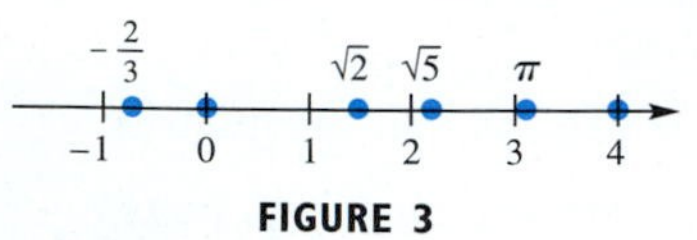

FIGURE 3

The Rectangular Coordinate System

If we place two number lines at right angles, intersecting at their origins, we obtain a two-dimensional **rectangular coordinate system.** The number lines intersect at the *origin* of the system, designated 0. The horizontal line is called the ***x*-axis,** and the

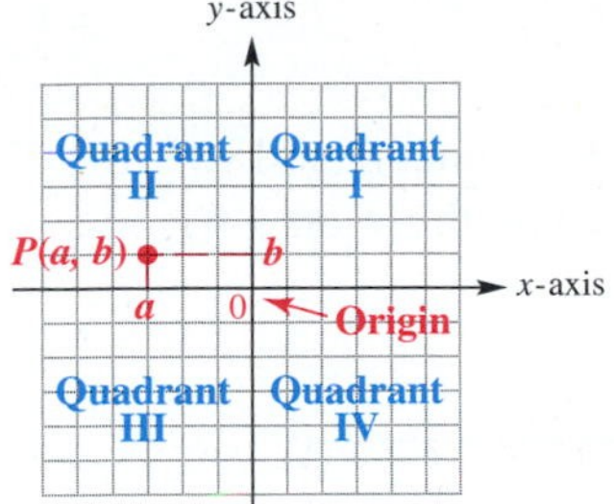

Rectangular coordinate system

FIGURE 4

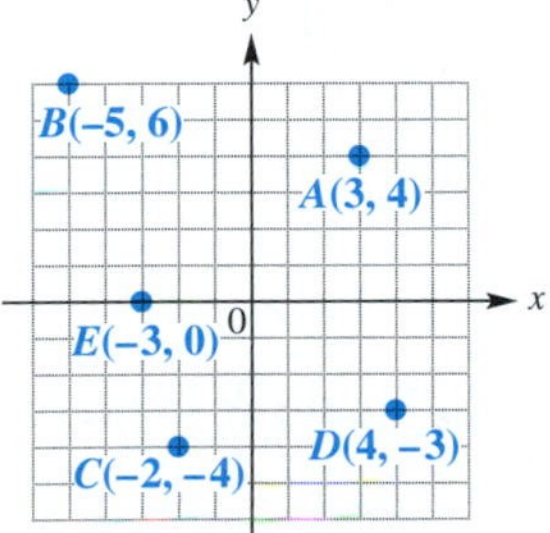

FIGURE 5

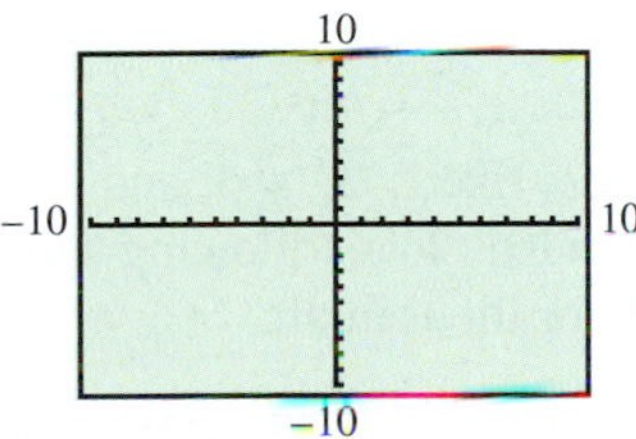

Standard viewing window

FIGURE 6

GCM TECHNOLOGY NOTE

You should consult the graphing calculator manual that accompanies this text or your owner's manual to see how to set the viewing window on your screen. Remember that different settings will result in different views of graphs.

vertical line is called the ***y*-axis.** On the x-axis, positive numbers are located to the right of the origin, and negative numbers are located to the left. On the y-axis, positive numbers are located above the origin, negative numbers below.

This rectangular coordinate system is also called the **Cartesian coordinate system,** named after René Descartes (1596–1650). The plane into which the coordinate system is introduced is the **coordinate plane,** or ***xy*-plane.** The x-axis and y-axis divide the plane into four regions, or **quadrants,** as shown in Figure 4. The points on the x-axis or y-axis belong to no quadrant.

Each point P in the xy-plane corresponds to a unique ordered pair (a, b) of real numbers. The numbers a and b are the *coordinates* of point P. We call a the ***x*-coordinate** and b the ***y*-coordinate.** To locate, for example, the point corresponding to the ordered pair $(3, 4)$ on the xy-plane, draw a vertical line through 3 on the x-axis and a horizontal line through 4 on the y-axis. These two lines intersect at point A in Figure 5. Point A corresponds to the ordered pair $(3, 4)$. Additional points are labeled B–E. The coordinates of the origin are $(0, 0)$. The point P corresponding to the ordered pair (a, b) is often written as $P(a, b)$, as in Figure 4, and referred to as "the point (a, b)."

Viewing Windows

The rectangular (Cartesian) coordinate system extends indefinitely in all directions. We can show only a portion of such a system in a text figure. Similar limitations occur with the viewing "window" on a calculator screen. Figure 6 shows a calculator screen that has been set to have a minimum x-value of -10, a maximum x-value of 10, a minimum y-value of -10, and a maximum y-value of 10. The tick marks on the axes have been set to be 1 unit apart. Thus, there are 10 tick marks on the positive x-axis. This window is called the **standard viewing window.**

To convey information about a viewing window, we use the following abbreviations:

Xmin: minimum value of x	**Ymin:** minimum value of y
Xmax: maximum value of x	**Ymax:** maximum value of y
Xscl: scale (distance between tick marks) on the x-axis	**Yscl:** scale (distance between tick marks) on the y-axis.

To further condense this information, we use the following symbolism, which gives viewing information for the window in Figure 6:

Xmin, Xmax, Ymin, Ymax

$$[-10, 10] \text{ by } [-10, 10].$$

Xscl = 1 Yscl = 1

Figure 7 shows several other viewing windows. Notice that Figures 7(b) and 7(c) look exactly alike, and unless we are told what the settings are, we have no way of distinguishing between them. What are Xscl and Yscl in each figure?

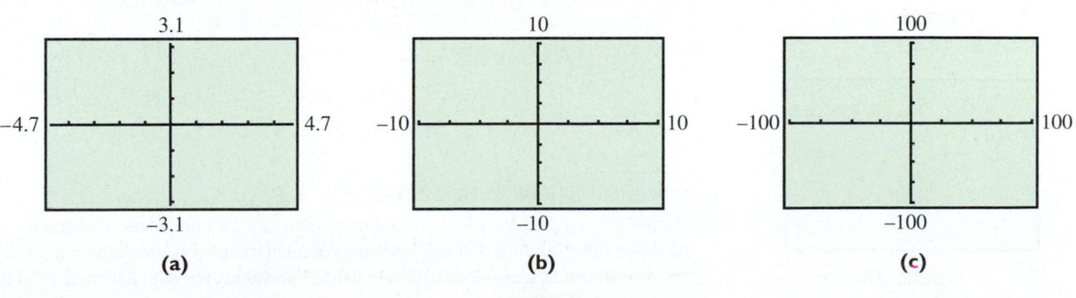

FIGURE 7

What Went Wrong?

A student learning how to use a graphing calculator could not understand why the axes on the graph were so "thick," as seen in Figure A, while those on a friend's calculator were not, as seen in Figure B.

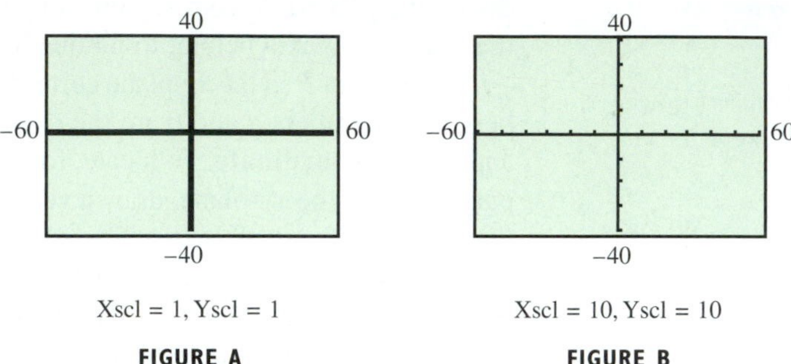

FIGURE A FIGURE B

What Went Wrong? How can the student correct the problem in Figure A so that the axes look like those in Figure B?

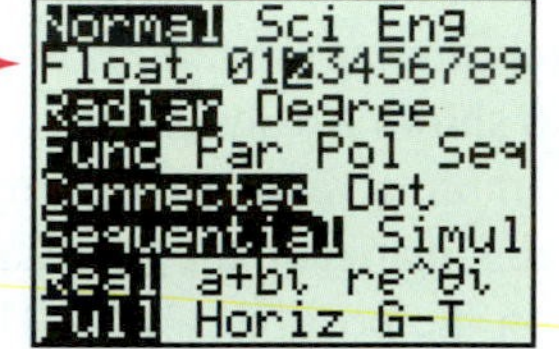

TI-83 Plus

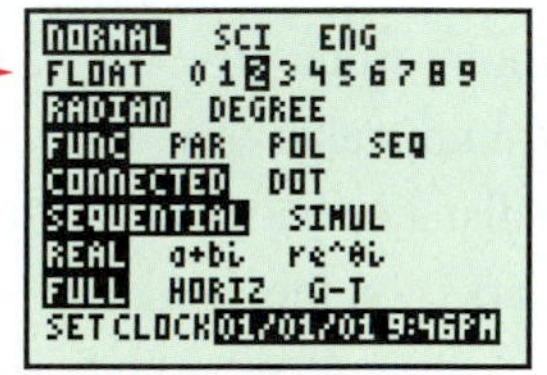

TI-84 Plus

FIGURE 8

Roots

Although calculators have the capability to express numbers like $\sqrt{2}$, $\sqrt[3]{5}$, and π to many decimal places, we often ask that answers be rounded. The following table reviews rounding numbers to the nearest tenth, hundredth, or thousandth.

Rounding Numbers

Number	Nearest Tenth	Nearest Hundredth	Nearest Thousandth
1.3782	1.4	1.38	1.378
201.6666	201.7	201.67	201.667
.0819	.1	.08	.082

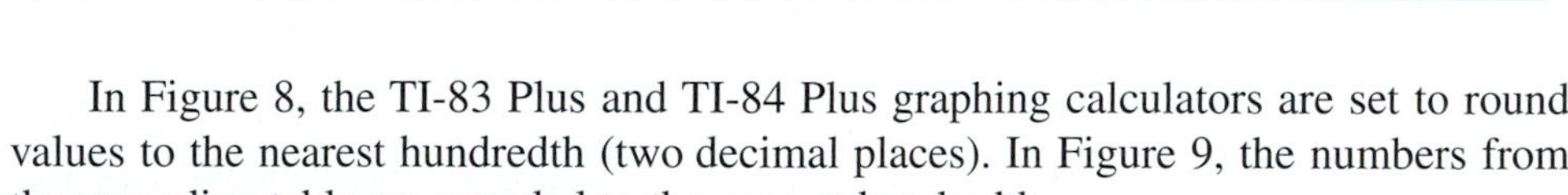

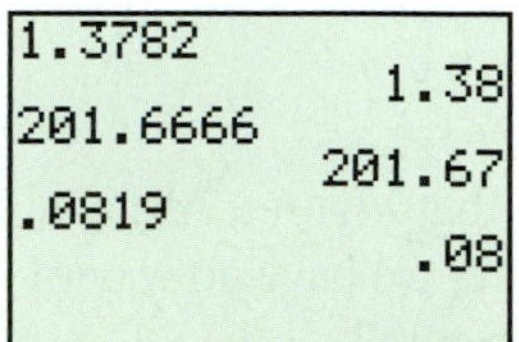

FIGURE 9

In Figure 8, the TI-83 Plus and TI-84 Plus graphing calculators are set to round values to the nearest hundredth (two decimal places). In Figure 9, the numbers from the preceding table are rounded to the nearest hundredth.

The symbol $\approx$ indicates that two expressions are *approximately equal*. For example, $\pi \approx 3.14$, but $\pi \neq 3.14$, since $\pi = 3.141592654\ldots$. See Figure 10.

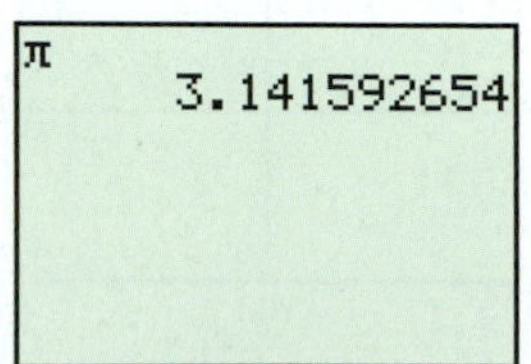

FIGURE 10

Answer to What Went Wrong?

Since Xscl = 1 and Yscl = 1 in Figure A, there are 120 tick marks along the x-axis and 80 tick marks along the y-axis. The resolution of the graphing calculator screen is not high enough to show all these tick marks, so the axes appear as heavy black lines instead. The values for Xscl and Yscl need to be larger, as in Figure B.

GCM **EXAMPLE 1 Finding Roots on a Calculator**

Approximate each root to the nearest thousandth. (*Note:* You can use the fact that $\sqrt[n]{a} = a^{1/n}$ to find roots.)

(a) $\sqrt{23}$ **(b)** $\sqrt[3]{87}$ **(c)** $\sqrt[4]{12}$

Solution

(a) The screen in Figure 11(a) shows that an approximation for $\sqrt{23}$, to the nearest thousandth, is 4.796. It is displayed twice, once for $\sqrt{23}$ and once for $23^{1/2}$.

(b) To the nearest thousandth, $\sqrt[3]{87} \approx 4.431$. See Figure 11(b).

(c) Figure 11(c) indicates $\sqrt[4]{12} \approx 1.861$ in three different ways.

In all the screens, note the inclusion of parentheses.

TECHNOLOGY NOTE

Many graphing calculators have built-in keys for calculating square roots and menus for calculating other types of roots. On some models, roots are located under the MATH menu.

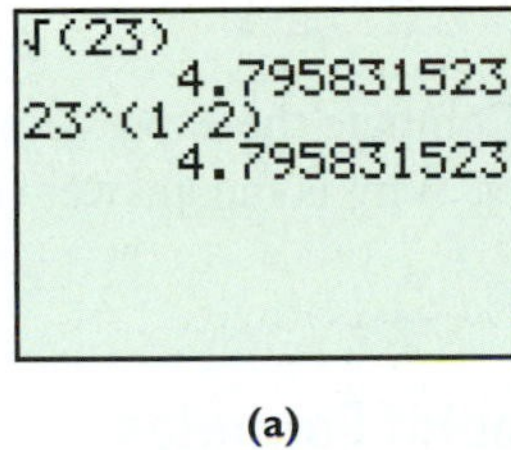

(a)

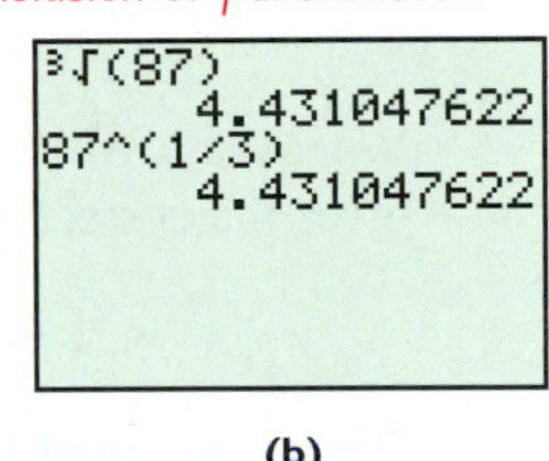

(b)

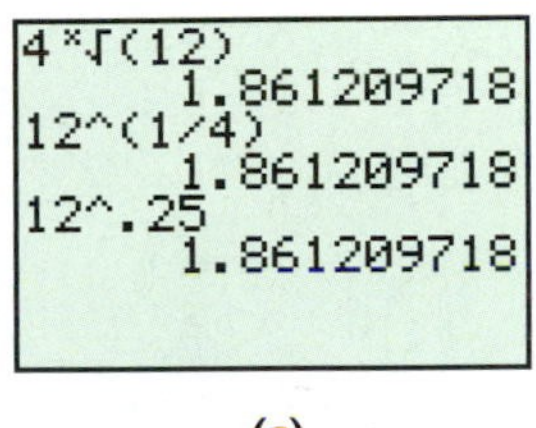

(c)

FIGURE 11

EXAMPLE 2 Approximating Expressions with a Calculator

Approximate each expression to the nearest hundredth.

(a) $\dfrac{3.8 - 1.4}{5.4 + 3.5}$ **(b)** $3\pi^4 - 9^2$ **(c)** $\sqrt{(4 - 1)^2 + (-3 - 2)^2}$

Solution

(a) See Figure 12(a). To the nearest hundredth,

$$\frac{3.8 - 1.4}{5.4 + 3.5} \approx .27.$$

(b) When entering π, use the built-in key for π rather than 3.14. Many calculators also have a special key to calculate the square of a number. To the nearest hundredth, $3\pi^4 - 9^2 \approx 211.23$. See Figure 12(b).

(c) From Figure 12(c), $\sqrt{(4 - 1)^2 + (-3 - 2)^2} \approx 5.83$.

Insert parentheses around both the numerator and the denominator.

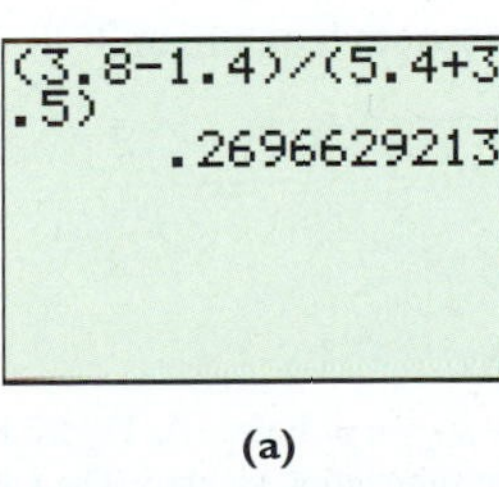

(a)

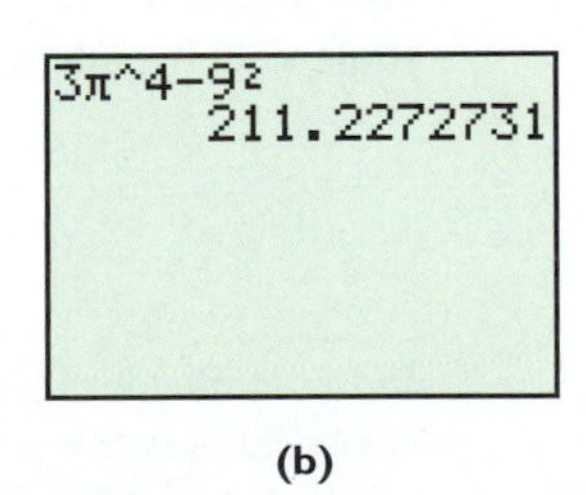

(b)

Do not confuse the negation and subtraction symbols.

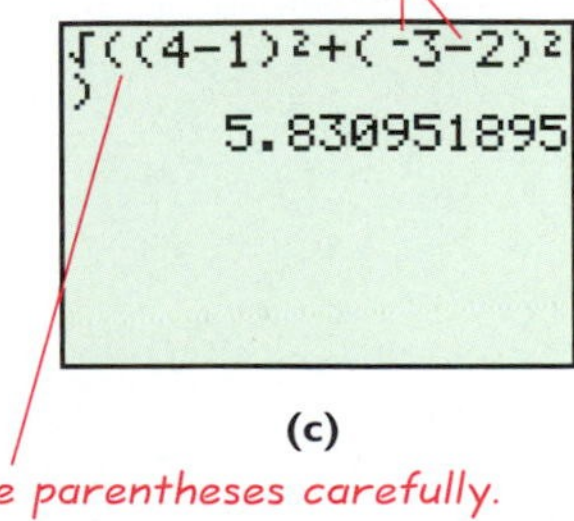

(c)

Use parentheses carefully.

FIGURE 12

What Went Wrong?

Two students were asked to compute the expression $(2 + 9) - (8 + 13)$ on a TI-83/84 Plus calculator. One student obtained the answer -10, as seen in Figure A, while the other obtained -231, as seen in Figure B.

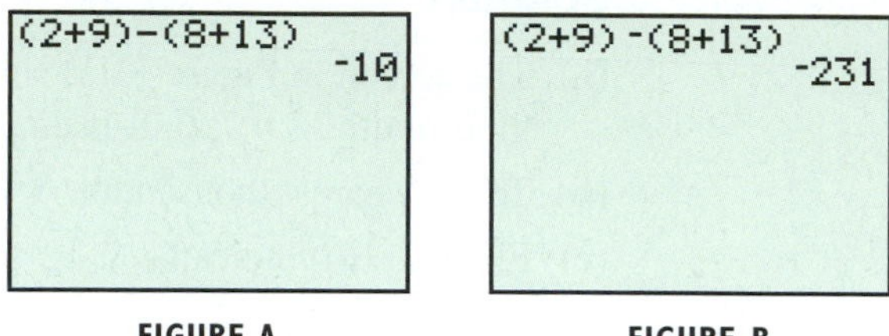

FIGURE A FIGURE B

What Went Wrong? Compute the expression by hand to determine which screen gives the correct answer. Why is the answer on the other screen incorrect?

Distance and Midpoint Formulas

The Pythagorean theorem can be used to calculate the lengths of the sides of a right triangle. In a right triangle, the sides that form the right angle are called **legs** and the side opposite the right angle (the longest side) is called the **hypotenuse.**

Pythagorean Theorem

In a right triangle, the sum of the squares of the lengths of the legs is equal to the square of the length of the hypotenuse.

$$a^2 + b^2 = c^2$$

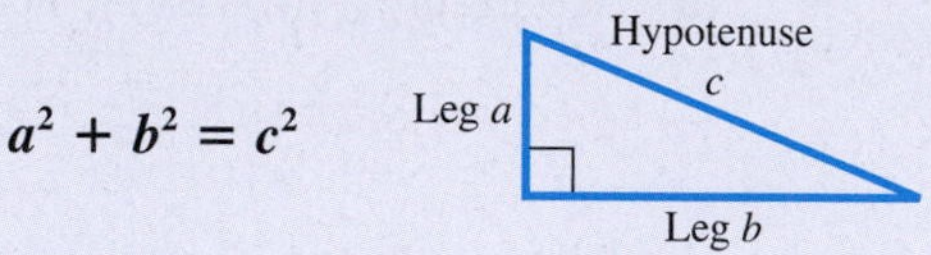

NOTE The *converse* of the Pythagorean theorem is also true. That is, if a, b, and c are lengths of the sides of a triangle and $a^2 + b^2 = c^2$, then the triangle is a right triangle with hypotenuse c. For example, if a triangle has sides with lengths 3, 4, and 5, then it is a right triangle with hypotenuse of length 5 because $3^2 + 4^2 = 5^2$.

Answer to What Went Wrong?

The correct answer is -10, as shown in Figure A. Figure B gives an incorrect answer because the negation symbol is used, rather than the subtraction symbol. The calculator computed $2 + 9 = 11$ and then *multiplied* by the negative of $8 + 13$ (that is, -21), to obtain the incorrect answer, -231.

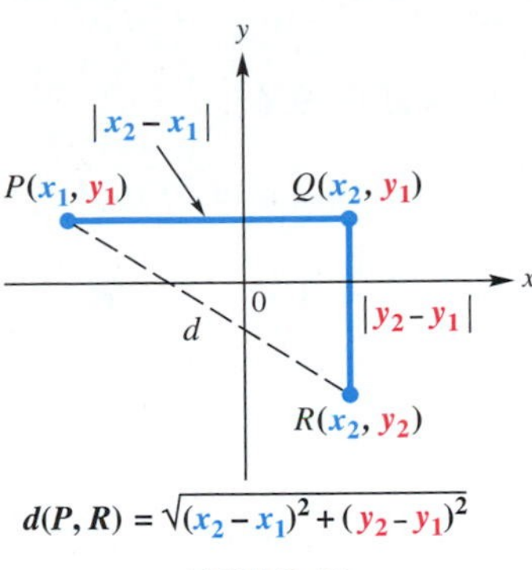

$d(P, R) = \sqrt{(x_2 - x_1)^2 + (y_2 - y_1)^2}$

FIGURE 13

To derive a formula to find the distance between two points in the *xy*-plane, let $P(x_1, y_1)$ and $R(x_2, y_2)$ be any two distinct points in the plane, as shown in Figure 13. Complete a triangle by locating point Q with coordinates (x_2, y_1). The Pythagorean theorem gives the distance between P and R as

$$d(P, R) = \sqrt{(x_2 - x_1)^2 + (y_2 - y_1)^2}.$$

NOTE Absolute value bars are not necessary in this formula, since, for all real numbers a and b, $|a - b|^2 = (a - b)^2$.

Distance Formula

Suppose that $P(x_1, y_1)$ and $R(x_2, y_2)$ are two points in a coordinate plane. Then the distance between P and R, written $d(P, R)$, is given by the **distance formula**

$$\mathbf{d(P, R) = \sqrt{(x_2 - x_1)^2 + (y_2 - y_1)^2}.}$$

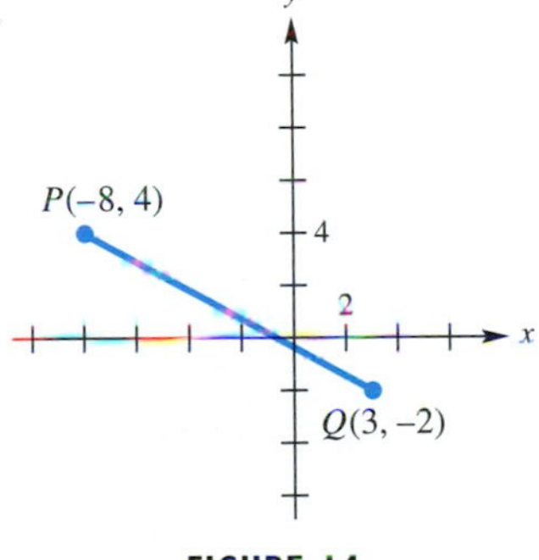

FIGURE 14

EXAMPLE 3 Using the Distance Formula

Use the distance formula to find $d(P, Q)$ in Figure 14.

Solution

$$\begin{aligned} d(P, Q) &= \sqrt{(x_2 - x_1)^2 + (y_2 - y_1)^2} && \text{Distance formula} \\ &= \sqrt{[3 - (-8)]^2 + (-2 - 4)^2} && x_1 = -8,\ y_1 = 4,\ x_2 = 3,\ y_2 = -2 \\ &= \sqrt{11^2 + (-6)^2} \\ &= \sqrt{121 + 36} \\ &= \sqrt{157} && \text{Leave in radical form.} \end{aligned}$$

Be careful when subtracting a negative number.

The *midpoint* of a line segment is the point on the segment that lies the same distance from both endpoints. See Figure 15. The midpoint is found by calculating the *average* of the *x*-coordinates and the *average* of the *y*-coordinates of the endpoints of the segment.

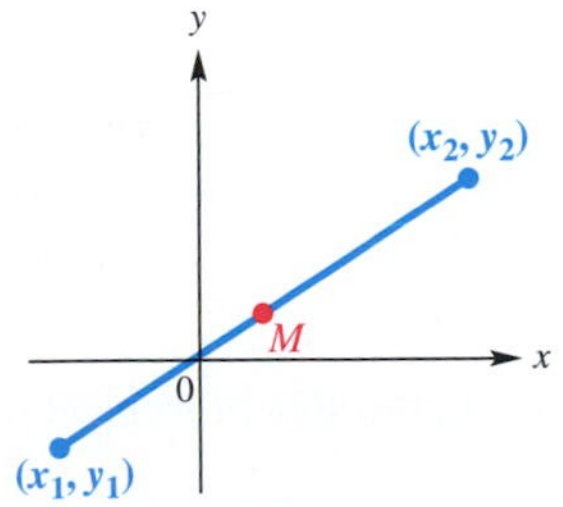

Point M is the midpoint of the segment joining (x_1, y_1) and (x_2, y_2).

FIGURE 15

Midpoint Formula

The **midpoint** of the line segment with endpoints (x_1, y_1) and (x_2, y_2) is

$$\left(\frac{x_1 + x_2}{2}, \frac{y_1 + y_2}{2}\right).$$

EXAMPLE 4 **Using the Midpoint Formula**

Find the midpoint M of the segment with endpoints $(8, -4)$ and $(-9, 6)$.

Solution Let $(x_1, y_1) = (8, -4)$ and $(x_2, y_2) = (-9, 6)$ in the midpoint formula.

$$M = \left(\frac{x_1 + x_2}{2}, \frac{y_1 + y_2}{2}\right) = \left(\frac{8 + (-9)}{2}, \frac{-4 + 6}{2}\right) = \left(-\frac{1}{2}, 1\right)$$ ■

EXAMPLE 5 **Estimating Tuition and Fees**

In 1998, average tuition and fees at public colleges and universities were \$3293, whereas they were \$5132 in 2004. Use the midpoint formula to estimate tuition and fees in 2001. Compare your estimate with the actual value of \$4221. (*Source:* The College Board.)

Solution Notice that the year 2001 lies midway between 1998 and 2004. Therefore, we can use the midpoint formula to find the midpoint of the line segment joining (1998, 3293) and (2004, 5132).

$$\left(\frac{1998 + 2004}{2}, \frac{3293 + 5132}{2}\right) = (2001, 4212.50)$$

The midpoint formula estimates tuition and fees at public colleges and universities to be \$4212.50 in 2001. This is within \$10 of the actual value. ■

1.1 Exercises

For each set, list all elements that belong to the **(a)** *natural numbers,* **(b)** *whole numbers,* **(c)** *integers,* **(d)** *rational numbers,* **(e)** *irrational numbers, and* **(f)** *real numbers.*

1. $\left\{-6, -\frac{12}{4}, -\frac{5}{8}, -\sqrt{3}, 0, .31, .\overline{3}, 2\pi, 10, \sqrt{17}\right\}$

2. $\left\{-8, -\frac{14}{7}, -.245, 0, \frac{6}{2}, 8, \sqrt{81}, \sqrt{12}\right\}$

3. $\left\{-\sqrt{100}, -\frac{13}{6}, -1, 5.23, 9.\overline{14}, 3.14, \frac{22}{7}\right\}$

4. $\{-\sqrt{49}, -.405, -.\overline{3}, .1, 3, 18, 6\pi, 56\}$

Classify each number as one or more of the following: natural number, integer, rational number, *or* real number.

5. 4,000,000,000 (The amount in dollars that drugmaker Pfizer plans to cut costs by the year 2008)

6. 92,000,000,000 (The amount in dollars spent on diabetes care in the year 2002)

7. -250 (The amount in billions of dollars of U.S. public school debt)

8. -3 (The annual percent change in the area of tropical rain forests)

9. $\frac{2}{9}$ (The fraction of plants that drugmaker Pfizer plans to close in order to cut costs)

10. -3.5 (The amount in billions of dollars that the Motion Picture Association of America estimates is lost annually due to piracy)

11. $5\sqrt{2}$ (The length of the diagonal of a square measuring 5 units on each side)

12. π (The ratio of the circumference of a circle to its diameter)

Concept Check *For each measured quantity, state the set of numbers that is most appropriate to describe it. Choose from the* natural numbers, integers, *and* rational numbers.

13. Populations of cities

14. Distances to nearby cities on road signs

15. Shoe sizes

16. Prices paid (in dollars and cents) for gasoline tank fill-ups

17. Daily low winter temperatures in U.S. cities

18. Golf scores relative to par

Graph each set of numbers on a number line.

19. $\{-4, -3, -2, -1, 0, 1\}$

20. $\{-6, -5, -4, -3, -2\}$

21. $\left\{-.5, .75, \frac{5}{3}, 3.5\right\}$

22. $\left\{-.6, \frac{9}{8}, 2.5, \frac{13}{4}\right\}$

23. Explain the distinction between a rational number and an irrational number.

24. ***Concept Check*** Using his calculator, a student found the decimal 1.414213562 when he evaluated $\sqrt{2}$. Is this decimal the exact value of $\sqrt{2}$ or just an approximation of $\sqrt{2}$? Should he write $\sqrt{2} = 1.414213562$ or $\sqrt{2} \approx 1.414213562$?

Locate each point on a rectangular coordinate system. Identify the quadrant, if any, in which each point lies.

25. $(2, 3)$

26. $(-1, 2)$

27. $(-3, -2)$

28. $(1, -4)$

29. $(0, 5)$

30. $(-2, -4)$

31. $(-2, 4)$

32. $(3, 0)$

33. $(-2, 0)$

34. $(3, -3)$

Name the possible quadrants in which the point (x, y) can lie if the given condition is true.

35. $xy > 0$

36. $xy < 0$

37. $\frac{x}{y} < 0$

38. $\frac{x}{y} > 0$

39. ***Concept Check*** If the x-coordinate of a point is 0, the point must lie on which axis?

40. ***Concept Check*** If the y-coordinate of a point is 0, the point must lie on which axis?

Give the values of Xmin, Xmax, Ymin, *and* Ymax *for each screen, given the values for* Xscl *and* Yscl. *Use the notation described in this section.*

41.
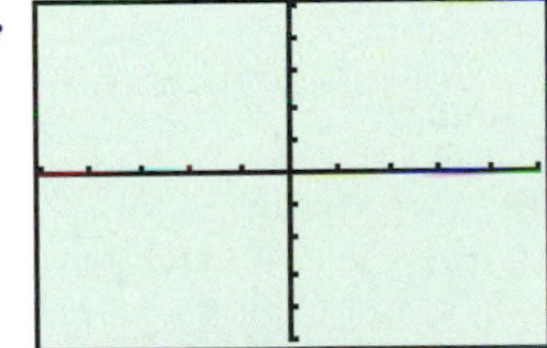
Xscl = 1, Yscl = 5

42.
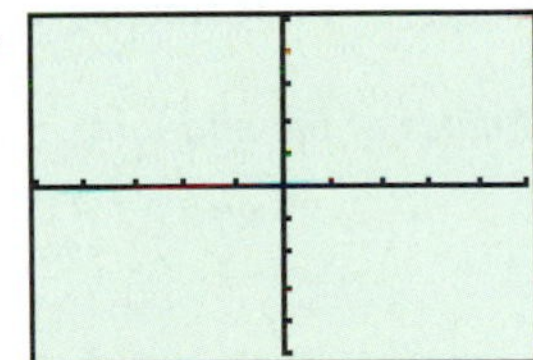
Xscl = 5, Yscl = 1

43.
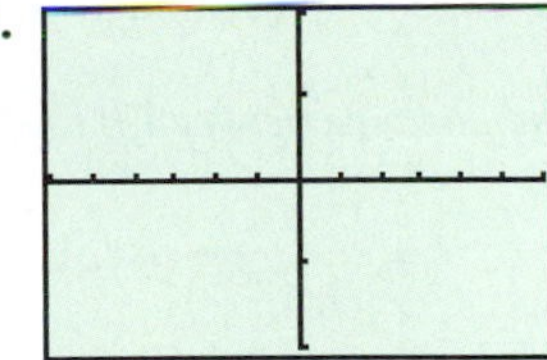
Xscl = 10, Yscl = 50

44.
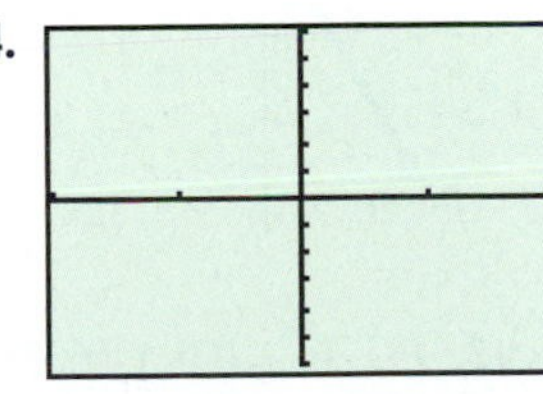
Xscl = 50, Yscl = 10

45.
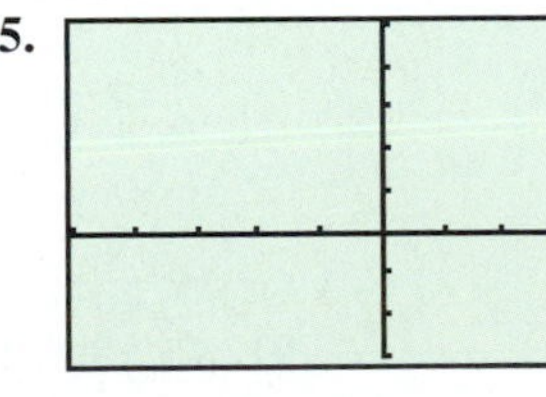
Xscl = 100, Yscl = 100

46.
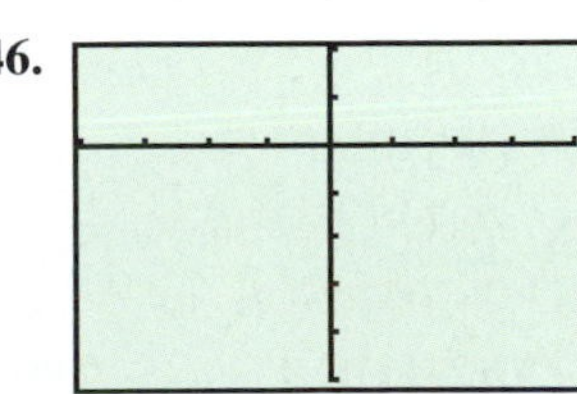
Xscl = 75, Yscl = 75

Set the viewing window of your calculator to the given specifications. Make a sketch of your window.

47. $[-10, 10]$ by $[-10, 10]$
Xscl = 1 Yscl = 1

48. $[-40, 40]$ by $[-30, 30]$
Xscl = 5 Yscl = 5

49. $[-5, 10]$ by $[-5, 10]$
Xscl = 3 Yscl = 3

50. $[-3.5, 3.5]$ by $[-4, 10]$
Xscl = 1 Yscl = 1

51. $[-100, 100]$ by $[-50, 50]$
Xscl = 20 Yscl = 25

52. $[-4.7, 4.7]$ by $[-3.1, 3.1]$
Xscl = 1 Yscl = 1

 **53.** Set your viewing window to $[-10, 10]$ by $[-10, 10]$, and then set Xscl to 0 and Yscl to 0. What do you notice? Make a conjecture as to how to set the screen with no tick marks on the axes.

 **54.** Set your viewing window to $[-50, 50]$ by $[-50, 50]$, and then set Xscl to 1 and Yscl to 1. Observe this screen and describe the appearance of the axes compared with those seen in the standard window. Why do you think they appear this way? How can you change your scale settings so that this "problem" is alleviated?

Find a decimal approximation of each root or power. Round answers to the nearest thousandth.

55. $\sqrt{58}$ **56.** $\sqrt{97}$ **57.** $\sqrt[3]{33}$ **58.** $\sqrt[3]{91}$ **59.** $\sqrt[4]{86}$

60. $\sqrt[4]{123}$ **61.** $19^{1/2}$ **62.** $29^{1/3}$ **63.** $46^{1.5}$ **64.** $23^{2.75}$

Approximate each expression to the nearest hundredth.

65. $\dfrac{5.6 - 3.1}{8.9 + 1.3}$ **66.** $\dfrac{34 + 25}{23}$ **67.** $\sqrt{\pi^3 + 1}$

68. $\sqrt[3]{2.1 - 6^2}$ **69.** $3(5.9)^2 - 2(5.9) + 6$ **70.** $2\pi^3 - 5\pi^2 - 3$

71. $\sqrt{(4 - 6)^2 + (7 + 1)^2}$ **72.** $\sqrt{[-1 - (-3)]^2 + (-5 - 3)^2}$ **73.** $\dfrac{\sqrt{\pi - 1}}{\sqrt{1 + \pi}}$

74. $\sqrt[3]{4.5 \times 10^5 + 3.7 \times 10^2}$ **75.** $\dfrac{2}{1 - \sqrt[3]{5}}$ **76.** $1 - \dfrac{4.5}{3 - \sqrt{2}}$

Find the length of the unknown side of the right triangle. In each case, a and b represent the lengths of the legs and c represents the length of the hypotenuse.

77. $a = 8, b = 15$; find c **78.** $a = 7, b = 24$; find c

79. $a = 13, c = 85$; find b **80.** $a = 14, c = 50$; find b

81. $a = 5, b = 8$; find c **82.** $a = 9, b = 10$; find c

83. $b = \sqrt{13}, c = \sqrt{29}$; find a **84.** $b = \sqrt{7}, c = \sqrt{11}$; find a

a
c
b
Typical labeling

Find **(a)** *the distance between P and Q and* **(b)** *the coordinates of the midpoint of the segment joining P and Q.*

85.

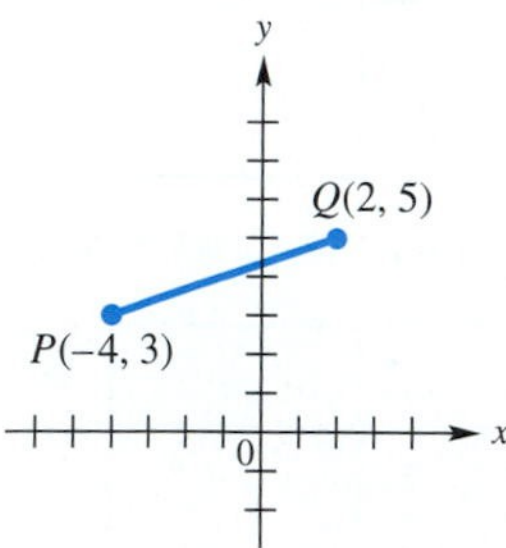

86.

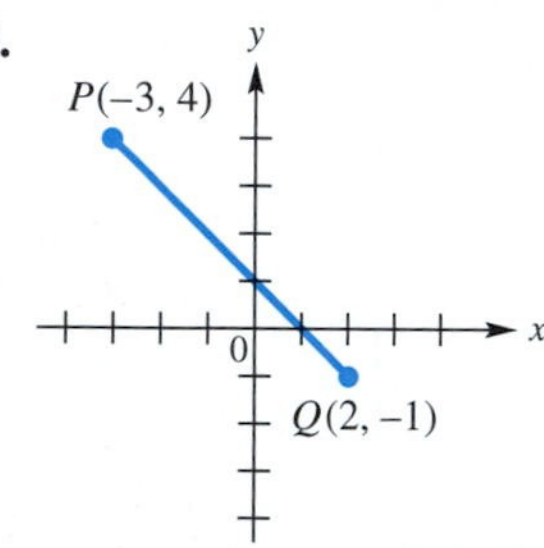

87.

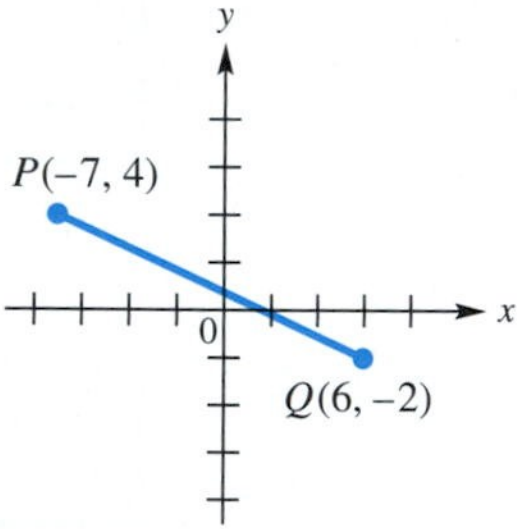

88.

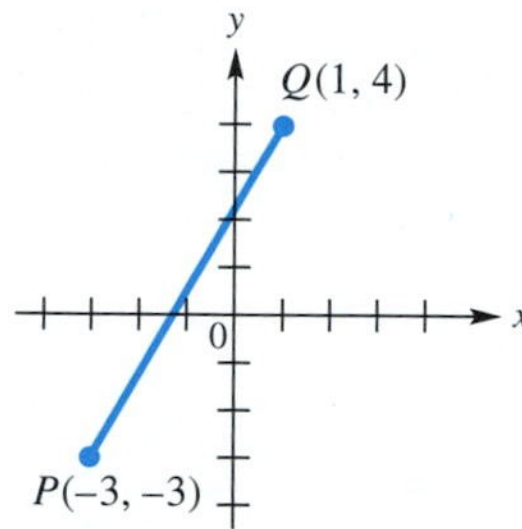

89. $P(5, 7), Q(2, 11)$ **90.** $P(-2, 5), Q(4, -3)$ **91.** $P(-8, -2), Q(-3, -5)$ **92.** $P(-6, -10), Q(6, 5)$

93. $P(9.2, 3.4), Q(6.2, 7.4)$ **94.** $P(8.9, 1.6), Q(3.9, 13.6)$ **95.** $P(13x, -23x), Q(6x, x), x > 0$ **96.** $P(12y, -3y), Q(20y, 12y), y > 0$

Suppose that P is the endpoint of a segment PQ and M is the midpoint of PQ. Find the coordinates of endpoint Q.

97. $P(7, -4), M(8, 5)$ **98.** $P(13, 5), M(-2, -4)$

99. $P(5.64, 8.21), M(-4.04, 1.60)$ **100.** $P(-10.32, 8.55), M(1.55, -2.75)$

Solve each problem.

101. *(Modeling) Tuition and Fees* From 1998–2004, the average annual cost (in dollars) of tuition and fees at private four-year colleges rose in an approximately linear fashion. The graph depicts this growth with a line segment. Use the midpoint formula to approximate the cost during the year 2001.

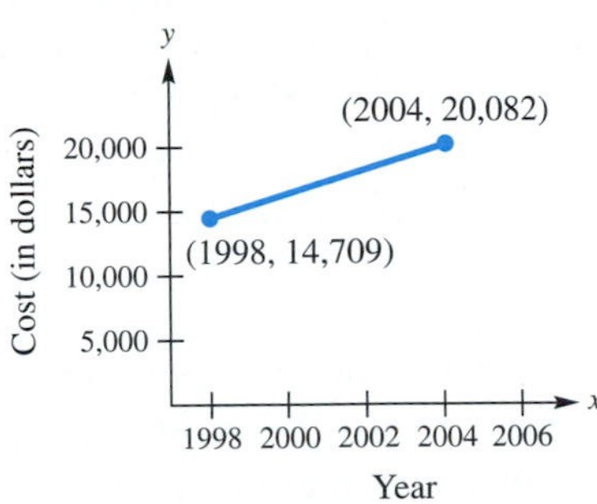

Source: The College Board.

102. *(Modeling) Two-Year College Enrollment* Projected enrollments in two-year colleges for 2005, 2007, and 2009 are shown in the table. Use the midpoint formula to estimate the enrollments for 2006 and 2008.

Year	Enrollment (in thousands)
2005	6038
2007	6146
2009	6257

Source: Statistical Abstract of the United States.

103. *(Modeling) Poverty-Level Income Cutoffs* The table lists poverty-level income cutoffs for a family of four since 1983. Use the midpoint formula to estimate the poverty-level cutoffs (rounded to the nearest dollar) in 1988 and 1998.

Year	Income (in dollars)
1983	10,178
1993	14,763
2003	18,810

Source: U.S. Census Bureau.

104. *Geometry* Triangles can be classified by their sides.

(a) An **isosceles triangle** has at least two sides of equal length. Determine whether the triangle with vertices $(0, 0)$, $(3, 4)$, and $(7, 1)$ is isosceles.

(b) An **equilateral triangle** has all sides of equal length. Determine whether the triangle with vertices $(-1, -1)$, $(2, 3)$, and $(-4, 3)$ is equilateral.

105. *Distance between Cars* At 9:00 A.M., Car A is traveling north at 50 mph and is located 50 miles south of Car B. Car B is traveling west at 20 mph.

(a) Let $(0, 0)$ be the initial coordinates of Car B in the xy-plane, where units are in miles. Plot the locations of each car at 9:00 A.M. and at 11:00 A.M.

(b) Find the distance between the cars at 11:00 A.M.

106. *Distance between Ships* Two ships leave the same harbor at the same time. The first ship heads north at 20 mph and the second ship heads west at 15 mph.

(a) Draw a sketch depicting their positions after t hours.

(b) Write an expression that gives the distance between the ships after t hours.

107. One of the most popular proofs of the Pythagorean theorem uses the figure shown here. Determine the area of the figure in two ways. First, find the area of the large square, using the formula for the area of a square. Then, find its area as the sum of the areas of the smaller square and the four right triangles. Set these expressions equal to each other and simplify to obtain $a^2 + b^2 = c^2$.

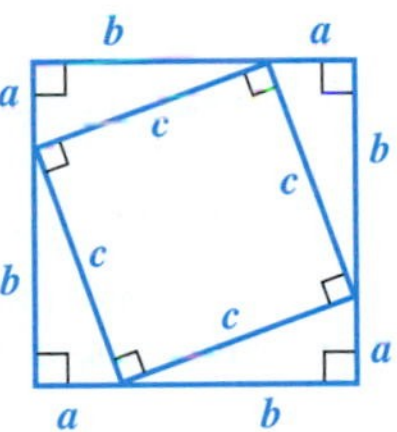

108. Prove that the midpoint of the line segment joining endpoints $P(x_1, y_1)$ and $Q(x_2, y_2)$ is

$$M\left(\frac{x_1 + x_2}{2}, \frac{y_1 + y_2}{2}\right)$$

by showing that the distance between P and M is equal to the distance between M and Q and that the sum of these distances is equal to the distance between P and Q.

1.2 Introduction to Relations and Functions

Set-Builder Notation and Interval Notation ■ Relations, Domain, and Range ■ Functions ■ Tables ■ Function Notation

Set-Builder Notation and Interval Notation

Suppose we wish to symbolize the set of real numbers greater than -2. One way is to write it as $\{x \mid x > -2\}$, read "the set of all x such that x is greater than -2." This is called **set-builder notation,** since the variable x is used to "build" the set. On a number line, we graph the elements of this set by drawing a line from -2 to the right. We use a parenthesis at -2 to indicate that -2 is not an element of the given set. See Figure 16.

FIGURE 16

The set of numbers greater than -2 is an example of an **interval** on the number line. A simplified notation, called **interval notation,** is used for writing intervals. With this notation, the interval of all numbers greater than -2 is written $(-2, \infty)$. The **infinity symbol** ∞ does not indicate a number; it shows that the interval includes all real numbers greater than -2. The left parenthesis indicates that -2 is *not* included. A square bracket, **[**, would be used if -2 were included. ***A parenthesis is always used next to the infinity symbol in interval notation.*** The set of all real numbers is written $(-\infty, \infty)$ in interval notation.

The chart summarizes the various types of intervals. (In each case, we assume that $a < b$.)

Type of Interval	Set-Builder Notation	Corresponding Interval Notation	Corresponding Graph
Open interval	$\{x \mid a < x < b\}$	(a, b)	a b
Closed interval	$\{x \mid a \le x \le b\}$	$[a, b]$	a b
Half-open (or half-closed) interval	$\{x \mid a < x \le b\}$	$(a, b]$	a b
	$\{x \mid a \le x < b\}$	$[a, b)$	a b
Unbounded interval	$\{x \mid x > a\}$	(a, ∞)	a
	$\{x \mid x \ge a\}$	$[a, \infty)$	a
	$\{x \mid x < a\}$	$(-\infty, a)$	a
	$\{x \mid x \le a\}$	$(-\infty, a]$	a
All real numbers	$\{x \mid x \text{ is real}\}$	$(-\infty, \infty)$	a b

CAUTION Interval notation for the open interval (a, b) looks exactly like the notation for the ordered pair (a, b). When the need arises, we will distinguish between them by referring to "the interval (a, b)" or "the point (a, b)."

Relations, Domain, and Range

The bar graph shown in Figure 17 shows the number of recreational visitors, in millions, to national parks in the United States for selected years between 1994 and 2002.

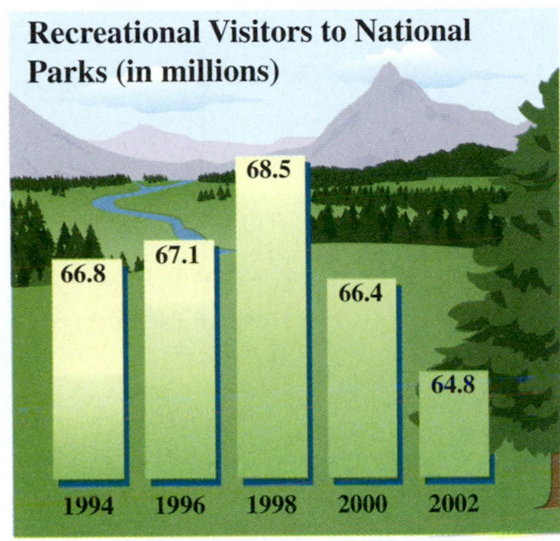

Source: National Park Service.

FIGURE 17

Since each year in the graph is paired with a number of visitors, we may depict this information as a set of ordered pairs. The first component represents the year, and the second component represents the number of visitors in millions.

$$\{(1994, 66.8), (1996, 67.1), (1998, 68.5), (2000, 66.4), (2002, 64.8)\}$$

Such a set of ordered pairs is called a *relation.*

Relation

A **relation** is a set of ordered pairs.

If we denote the ordered pairs of a relation by (x, y), then the set of all x-values is called the **domain** of the relation and the set of all y-values is called the **range** of the relation. For the relation represented by the bar graph in Figure 17,

$$\text{Domain} = \{1994, 1996, 1998, 2000, 2002\}$$

and

$$\text{Range} = \{66.8, 67.1, 68.5, 66.4, 64.8\}.$$

Here are three other examples of relations:

$$F = \{(1, 2), (-2, 5), (3, -1)\},$$

$$G = \{(-2, 1), (-1, 0), (0, 1), (1, 2), (2, 2)\},$$

$$H = \{(-4, 1), (-2, 1), (-2, 0)\}$$

$F = \{(1,2), (-2,5), (3,-1)\}$

$G = \{(-2,1), (-1,0), (0,1), (1,2), (2,2)\}$

$H = \{(-4,1), (-2,1), (-2,0)\}$

For the relations F, G, and H, repeated in the margin,

$$\text{Domain of } F = \{1, -2, 3\}, \qquad \text{Range of } F = \{2, 5, -1\};$$
$$\text{Domain of } G = \{-2, -1, 0, 1, 2\}, \qquad \text{Range of } G = \{1, 0, 2\};$$
$$\text{Domain of } H = \{-4, -2\}, \qquad \text{Range of } H = \{1, 0\}.$$

Since a relation is a set of ordered pairs, it may be represented graphically in the rectangular coordinate system. See Figure 18.

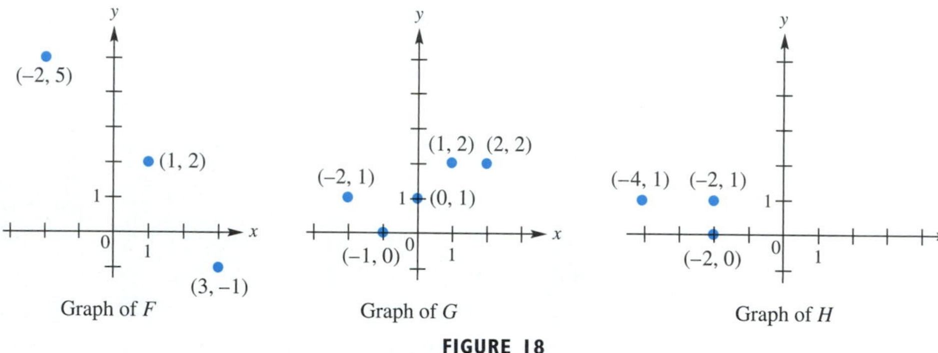

FIGURE 18

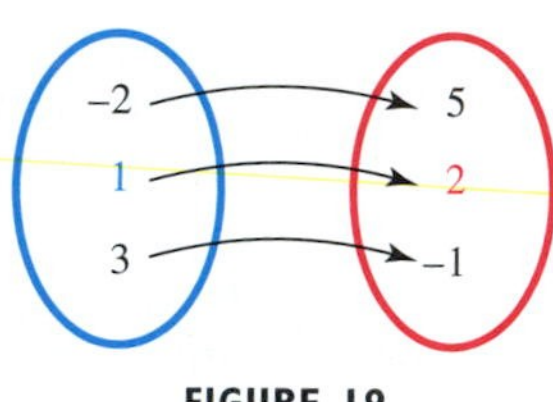

FIGURE 19

A relation can be illustrated with a "mapping" diagram, as shown in Figure 19 for relation F. The arrow from 1 to 2 indicates that the ordered pair $(1, 2)$ belongs to F.

Some relations contain infinitely many ordered pairs. For example, let F represent a relation consisting of all ordered pairs having the form $(x, 2x)$, where x is a real number. Since there are infinitely many values for x, there are infinitely many ordered pairs in F. Five such ordered pairs are plotted in Figure 20(a) and suggest that the graph of F is a line. The graph of F includes all points (x, y), such that $y = 2x$, as shown in Figure 20(b).

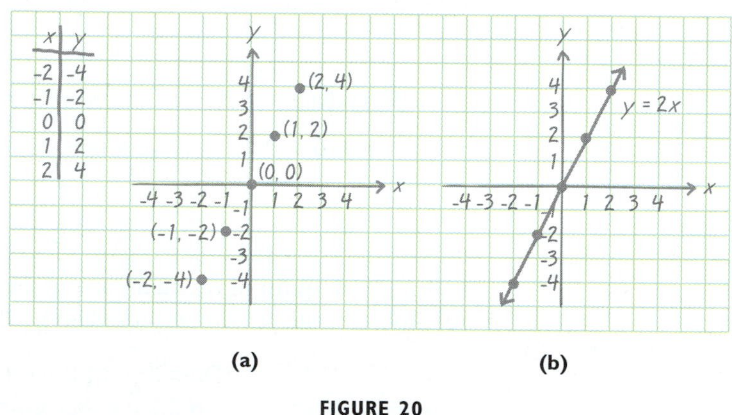

(a) **(b)**

FIGURE 20

A graph of a line or curve in the xy-plane represents a relation. The x-values of the points in the graph correspond to the domain of the relation, and the y-values correspond to the range of the relation.

EXAMPLE 1 **Determining Domains and Ranges from Graphs**

Give the domain and range of each relation from its graph.

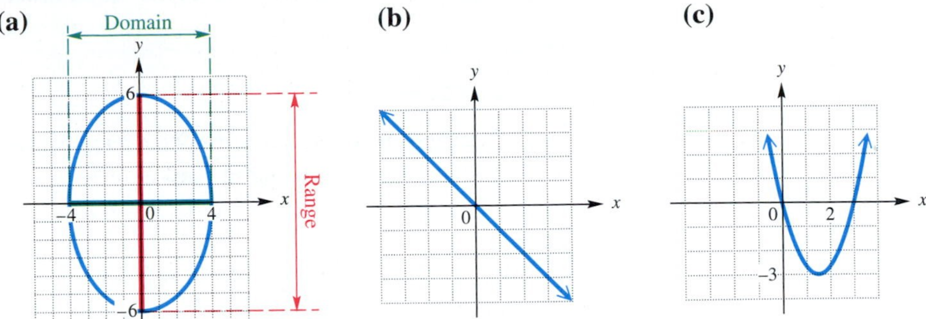

Solution

(a) The x-values of the points on the graph include all numbers between -4 and 4 inclusive. The y-values include all numbers between -6 and 6 inclusive. Using interval notation, the domain is $[-4, 4]$, and the range is $[-6, 6]$.

(b) The arrowheads indicate that the line extends indefinitely left and right, as well as upward and downward. Therefore, both the domain and range are the set of all real numbers, written $(-\infty, \infty)$.

(c) The arrowheads indicate that the graph extends indefinitely left and right, as well as upward. The domain is $(-\infty, \infty)$. There is a least y-value, -3, so the range includes all numbers greater than or equal to -3, written $[-3, \infty)$. ■

EXAMPLE 2 **Finding Domain and Range from a Calculator Graph**

Figure 21 shows a graph on a screen with viewing window $[-5, 5]$ by $[-5, 5]$, Xscl $= 1$, Yscl $= 1$. Give the domain and range of this relation.

TECHNOLOGY NOTE

In Figure 21, we see a calculator graph that is formed by a rather jagged curve. These representations are sometimes called *jaggies* and are typically found on low-resolution graphers, such as graphing calculators. In general, most curves in this book are smooth, and jaggies are just a part of the limitations of technology.

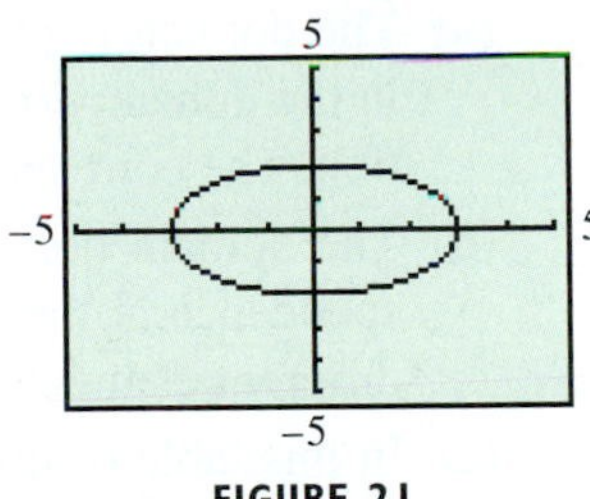

FIGURE 21

Solution Since the scales on both axes are 1, the graph *appears* to have minimum x-value -3, maximum x-value 3, minimum y-value -2, and maximum y-value 2. Observation leads us to conclude that the domain is $[-3, 3]$ and the range is $[-2, 2]$. ■

Functions

Suppose that a sales tax rate is 6%. Then a purchase of \$200 results in a sales tax of $.06 \times 200 = \$12$. This calculation can be summarized by the ordered pair $(200, 12)$. The ordered pair $(50, 3)$ indicates that a purchase of \$50 results in a sales tax of \$3. For each purchase of x dollars, there is exactly one sales tax amount.

Calculating sales tax y on a purchase of x dollars results in a set of ordered pairs (x, y), where $y = .06x$. This set of ordered pairs represents a special type of relation called a *function*. ***In a function, each x-value must correspond to exactly one y-value.***

Function

A **function** is a relation in which each element in the domain corresponds to exactly one element in the range.*

If x represents any element in the domain, x is called the **independent variable.** If y represents any element in the range, y is called the **dependent variable,** because the value of y *depends on* the value of x. For example, sales tax depends on purchase price.

In most applications of functions, the correspondence between domain and range elements is defined with an equation, such as $y = 9x - 5$. The equation is usually solved for y because y is the dependent variable. As we choose values from the domain for x, we can determine corresponding y-values of the ordered pairs of the function. (These equations can use other variables besides x and y.)

EXAMPLE 3 Deciding whether Relations are Functions

Give the domain and range of each relation. Decide whether each relation is a function.

(a) $\{(1,2),(3,4),(5,6),(7,8),(9,10)\}$ **(b)** $\{(1,1),(1,2),(1,3),(2,4)\}$

(c)

x	-5	-4	-3	-2	-1	0	1
y	2	2	2	2	2	2	2

(d) $y = x - 2$

Solution

$D = \{1, 3, 5, 7, 9\}$ Domain

$R = \{2, 4, 6, 8, 10\}$ Range

$D = \{1, 2\}$

$R = \{1, 2, 3, 4\}$

(a) The domain is $\{1, 3, 5, 7, 9\}$, and the range is $\{2, 4, 6, 8, 10\}$. Since each element in the domain corresponds to exactly one element in the range, this set is a function. The correspondence is shown in the margin.

(b) The domain here is $\{1, 2\}$, and the range is $\{1, 2, 3, 4\}$. As shown in the correspondence in the margin, one element in the domain, 1, has been assigned three different elements from the range, so this relation is *not* a function.

(c) In this table of ordered pairs, the domain is $\{-5, -4, -3, -2, -1, 0, 1\}$ and the range is $\{2\}$. Although every element in the domain corresponds to the same element in the range, this is a function because each element in the domain has exactly one range element assigned to it.

(d) Since y is always found by subtracting 2 from x, each x corresponds to just one value of y, so this relation is a function. Any number can be used for x, and each x will give a number that is 2 smaller for y; thus, both the domain and range are the set of real numbers $(-\infty, \infty)$. ■

*An alternative definition of *function* based on the idea of correspondence is given later in the section.

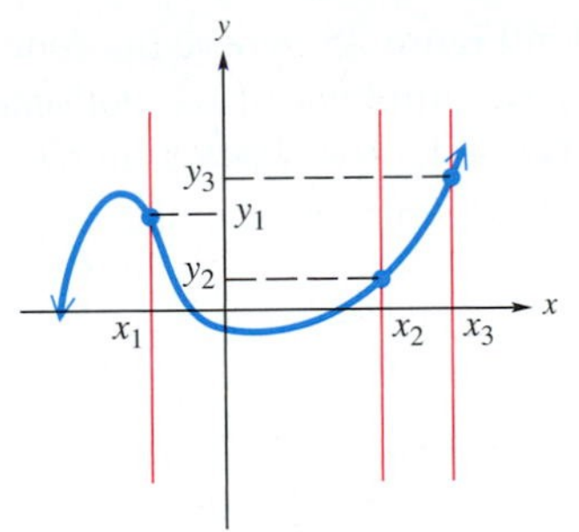

This *is* the graph of a function.

(a)

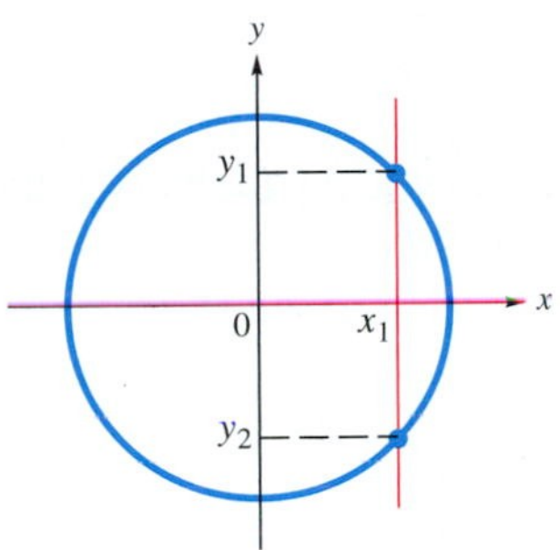

This *is not* the graph of a function.

(b)

FIGURE 22

There is a quick way to tell whether a given graph is the graph of a function. In the graph in Figure 22(a), each value of x corresponds to only one value of y, so this is the graph of a function. By contrast, the graph in Figure 22(b) is *not* the graph of a function. If $x = x_1$, the vertical line through x_1 intersects the graph at two points, showing that there are two values of y that correspond to this x-value. This concept is known as the *vertical line test* for a function.

Vertical Line Test

If every vertical line intersects a graph in no more than one point, then the graph is the graph of a function.

EXAMPLE 4 Using the Vertical Line Test

(a) Is the graph in Figure 23 the graph of a function?

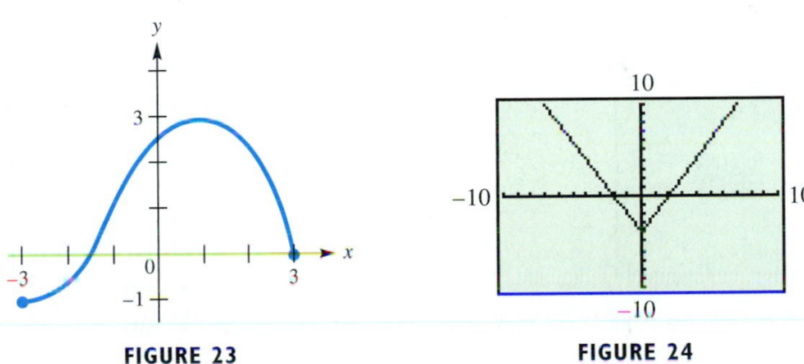

FIGURE 23 **FIGURE 24**

(b) Assuming that the graph in Figure 24 extends left and right indefinitely and upward indefinitely, does it appear to be the graph of a function?

Solution

(a) Any vertical line will intersect the graph at most once. Therefore, the graph satisfies the vertical line test and is the graph of a function.

(b) It appears that no vertical line will intersect the graph more than once, so we may conclude that it is the graph of a function. ■

While the concept of a function is crucial to the study of mathematics, the definition of the term may vary from text to text. We now give an alternative definition.

Function (Alternative Definition)

A function is a correspondence in which each element x from a set called the domain is paired with one and only one element y from a set called the range.

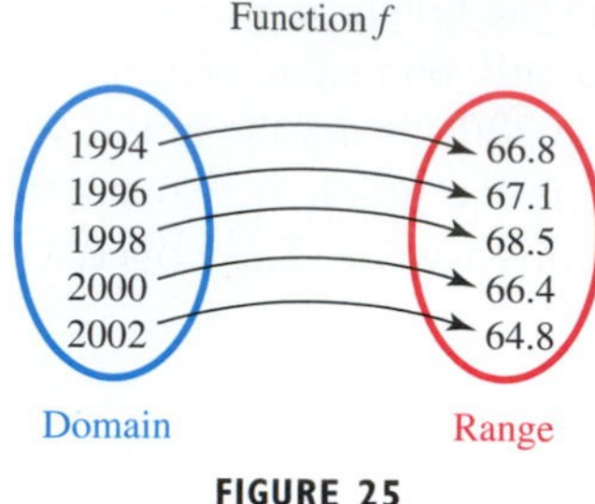

FIGURE 25

This idea of a correspondence, or mapping, is illustrated in Figure 25, where the function f consists of the ordered pairs in Figure 17, which gave numbers of recreational visitors to national parks (in millions). Function f pairs 1994 with 66.8, 1996 with 67.1, and so on. Each value in the domain is paired with one and only one value in the range.

Tables

A convenient way to display ordered pairs in a function is by using a table. An equation such as $y = 9x - 5$ describes a function. If we choose x-values to be $0, 1, 2, \ldots, 6$, then the corresponding y-values are

Use parentheses around substituted values to avoid errors.

$$y = 9(0) - 5 = -5 \qquad y = 9(1) - 5 = 4$$
$$y = 9(2) - 5 = 13 \qquad y = 9(3) - 5 = 22$$
$$y = 9(4) - 5 = 31 \qquad y = 9(5) - 5 = 40$$
$$y = 9(6) - 5 = 49.$$

These ordered pairs, $(0, -5)$, $(1, 4)$, …, $(6, 49)$, can be organized in a table. A graphing calculator can also generate this table, as shown in Figure 26.

GCM TECHNOLOGY NOTE

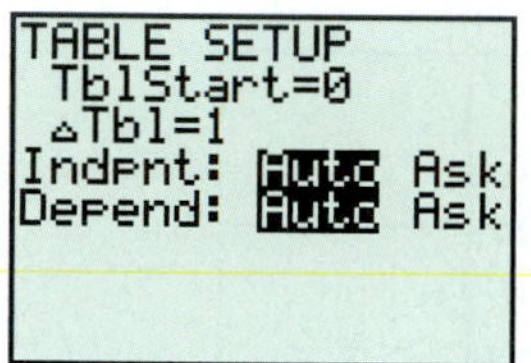
TABLE SETUP
TblStart=0
ΔTbl=1
Indpnt: Auto Ask
Depend: Auto Ask

This screen indicates that the table will *start* with 0 and have an *increment* of 1. Both variables appear *automatically.* TblStart represents the initial value of the independent variable (0 in the table in Figure 26) and ΔTbl represents the difference, or increment, between successive values of the independent variable (1 in Figure 26 because $1 - 0 = 1$, $2 - 1 = 1$, $3 - 2 = 1$, and so on).

x	y
0	−5
1	4
2	13
3	22
4	31
5	40
6	49

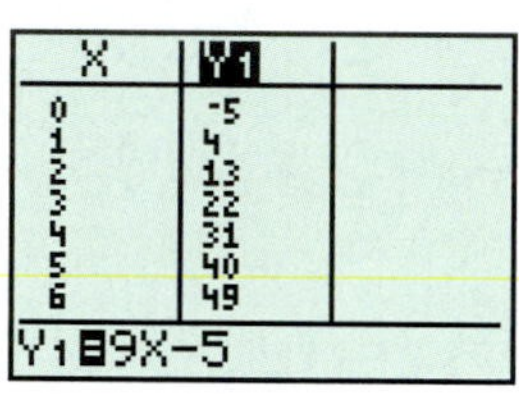

X	Y1
0	-5
1	4
2	13
3	22
4	31
5	40
6	49

Y1■9X−5

FIGURE 26

Function Notation

To say that y is a function of x means that for each value of x from the domain of the function, there is exactly one value of y. To emphasize that *y is a function of x*, or that *y depends on x*, it is common to write

$$y = f(x),$$

with $f(x)$ read **"f of x."** This notation is called **function notation.**

Function notation is used frequently when functions are defined by equations. For the function defined by $y = 9x - 5$, we may name this function f and write $f(x) = 9x - 5$. ***Note that $f(x)$ is simply another name for y.*** If $x = 2$, then we find y, or $f(2)$, by replacing x with 2.

$$f(2) = 9 \cdot 2 - 5 = 13$$

The statement "if $x = 2$, then $y = 13$" represents the ordered pair $(2, 13)$ and is abbreviated with function notation as

$$f(2) = 13.$$

Also, $f(0) = 9 \cdot 0 - 5 = -5$, and $f(-3) = 9(-3) - 5 = -32$.

These ideas can be explained as follows:

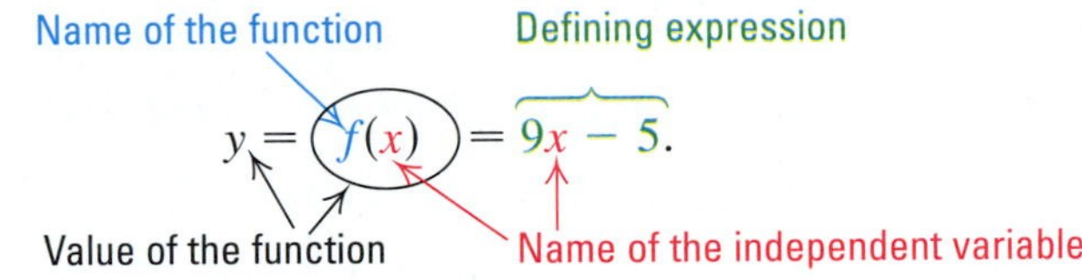

CAUTION ***The symbol f(x) does not indicate "f times x,"*** but represents the y-value for the indicated x-value. For example, $f(2)$ is the y-value that corresponds to the x-value 2.

We can use variables other than f, such as g and h, to represent functions.

Looking Ahead to Calculus

Functions are evaluated frequently in calculus. They can be evaluated in a variety of ways, as illustrated in Example 5.

EXAMPLE 5 Using Function Notation

For each function, find $f(3)$.

(a) $f(x) = 3x - 7$

(b) The function f depicted in Figure 27

(c) The function f graphed in Figure 28

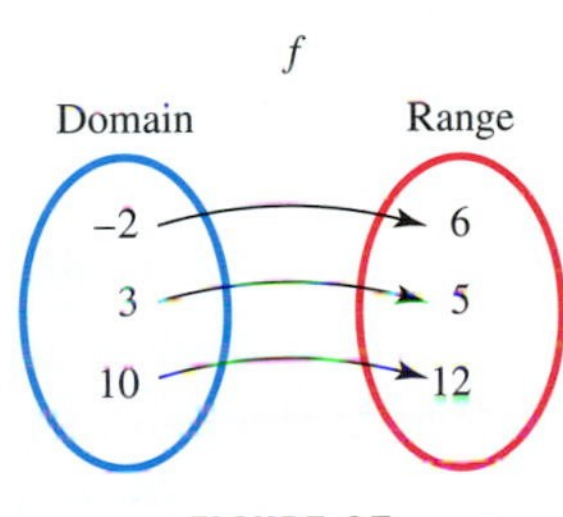

FIGURE 27

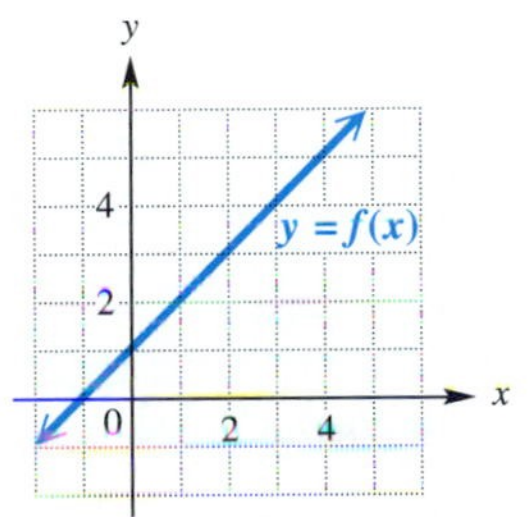

FIGURE 28

(d) The function f defined by the table

x	1	2	3	4
$f(x)$	−15	−12	−9	−6

Solution

(a) Replace x with 3 to get $f(3) = 3(3) - 7 = 2$.

(b) In Figure 27, 3 in the domain is paired with 5 in the range, so $f(3) = 5$.

(c) To evaluate $f(3)$, begin by finding 3 on the x-axis. See Figure 29. Then move upward until the graph of f is reached. Moving horizontally to the y-axis gives 4 for the corresponding y-value. Thus, $f(3) = 4$.

(d) From the table, $f(3) = -9$. ■

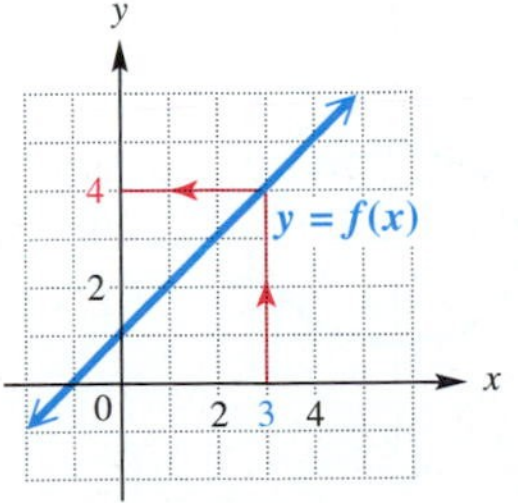

FIGURE 29

1.2 Exercises

Using interval notation, write each set. Then graph it on a number line.

1. $\{x \mid -1 < x < 4\}$ **2.** $\{x \mid x \geq -3\}$ **3.** $\{x \mid x < 0\}$

4. $\{x \mid 8 > x > 3\}$ **5.** $\{x \mid 1 \leq x < 2\}$ **6.** $\{x \mid -5 < x \leq -4\}$

Using the variable x, write each interval using set-builder notation.

7. $(-4, 3)$ **8.** $[2, 7)$ **9.** $(-\infty, -1]$ **10.** $(3, \infty)$

11. **13.**

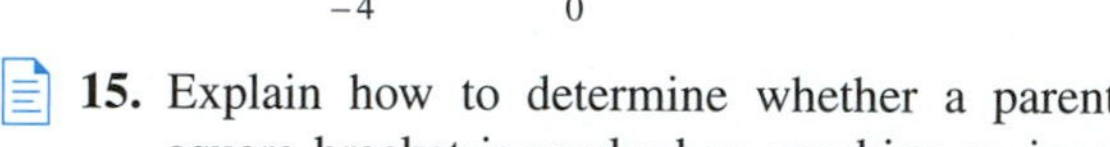

12. **14.**

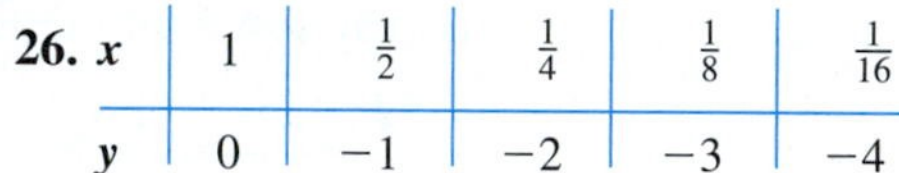

15. Explain how to determine whether a parenthesis or a square bracket is used when graphing an inequality on a number line.

16. ***Concept Check*** The three-part inequality $a < x < b$ means "a is less than x, *and* x is less than b." Which one of the following inequalities is not satisfied by some real number x?

A. $-3 < x < 5$ **B.** $0 < x < 4$
C. $-3 < x < -2$ **D.** $-7 < x < -10$

Determine the domain and range of each relation, and tell whether the relation is a function. Assume that a calculator graph extends indefinitely and a table includes only the points shown.

17. $\{(5, 1), (3, 2), (4, 9), (7, 6)\}$ **18.** $\{(8, 0), (5, 4), (9, 3), (3, 8)\}$ **19.** $\{(1, 6), (2, 6), (3, 6)\}$

20. $\{(-10, 5), (-20, 5), (-30, 5)\}$ **21.** $\{(4, 1), (3, -5), (-2, 3), (3, 7)\}$ **22.** $\{(0, 5), (1, 3), (0, -4)\}$

23.

x	11	12	13	14
y	−6	−6	−7	−6

24.

x	1	1	1	1
y	12	13	14	15

25.

x	0	1	2	3	4
y	$\sqrt{2}$	$\sqrt{3}$	$\sqrt{5}$	$\sqrt{6}$	$\sqrt{7}$

26.

x	1	$\frac{1}{2}$	$\frac{1}{4}$	$\frac{1}{8}$	$\frac{1}{16}$
y	0	−1	−2	−3	−4

27.

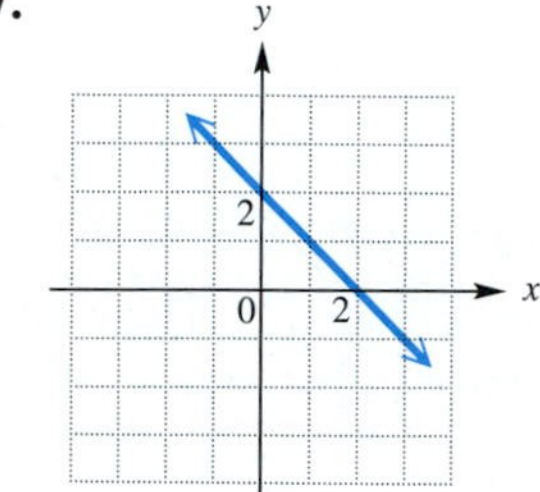

28.

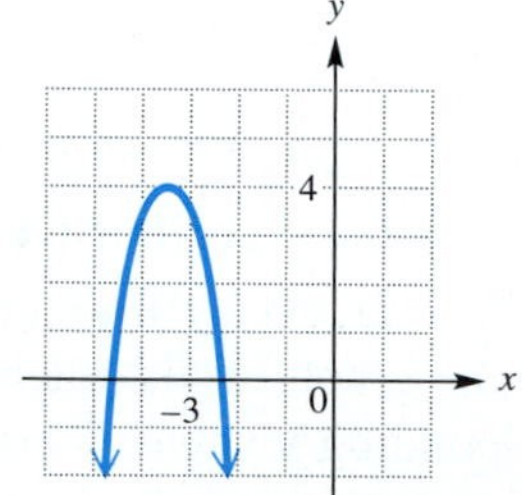

29.

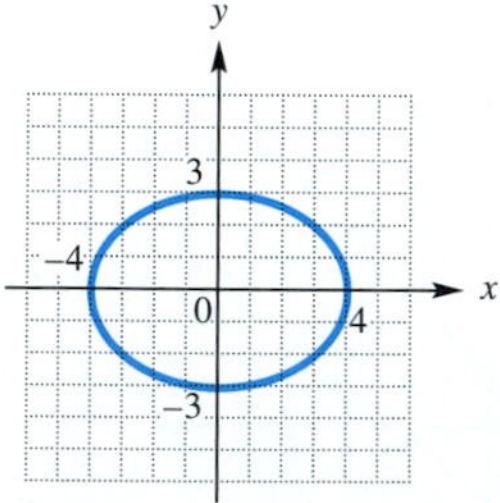

30.

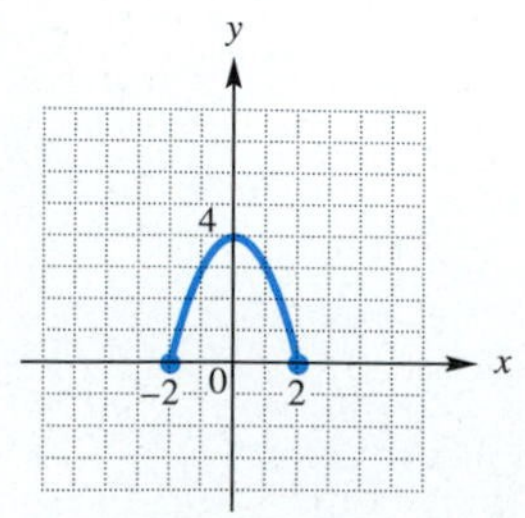

31.

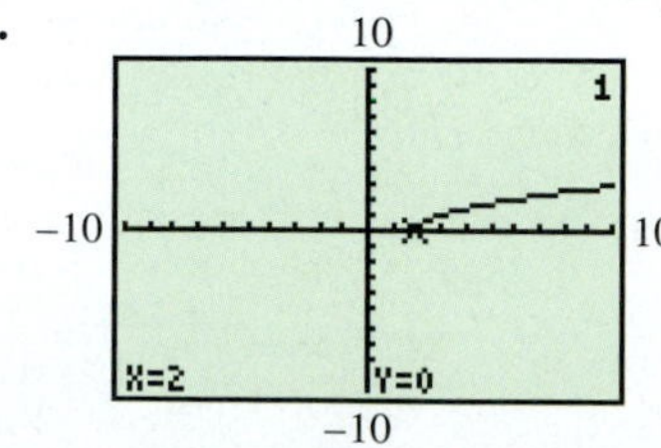

32.

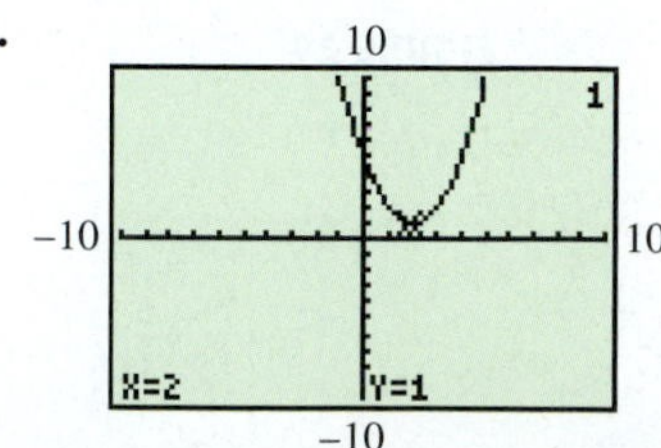

33.

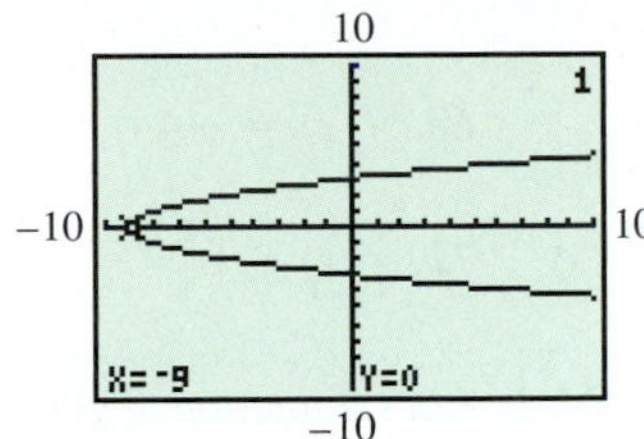

34.

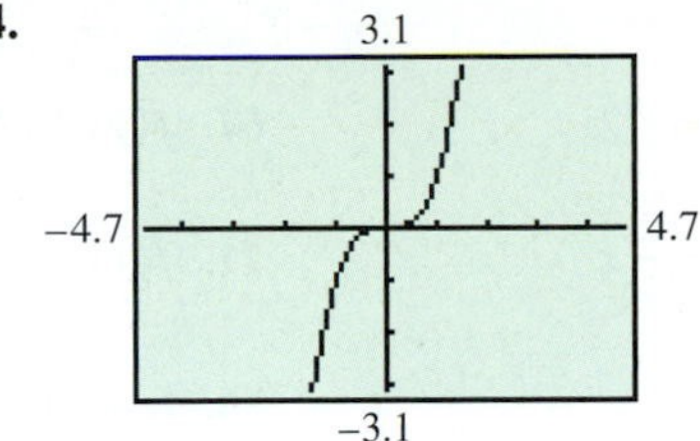

35.

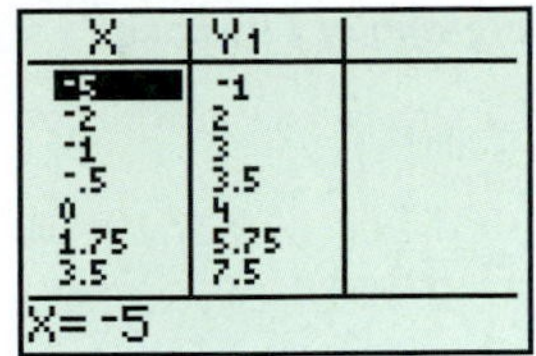

X	Y1
-5	-1
-2	2
-1	3
-.5	3.5
0	4
1.75	5.75
3.5	7.5

X= -5

36.

X	Y1
-2	5
-1	0
0	-3
5	12
9	60
10	77
13	140

X= -2

37.

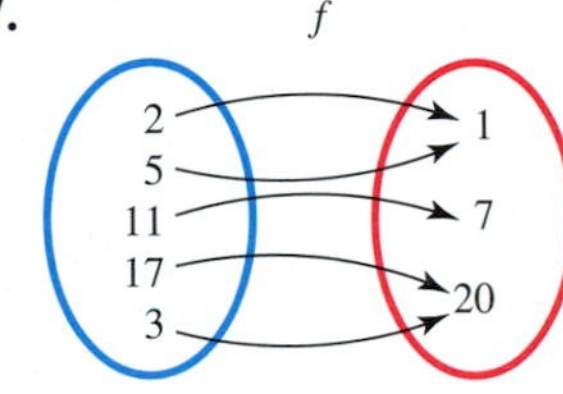

38.

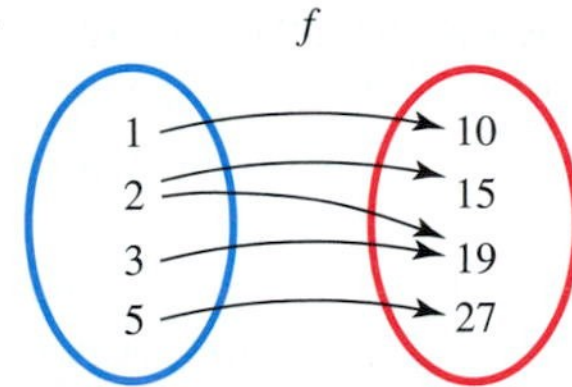

Find each function value.

39. $f(-2)$ if $f(x) = Y_1$ as defined in Exercise 35

40. $f(5)$ if $f(x) = Y_1$ as defined in Exercise 36

41. $f(11)$ for the function f in Exercise 37

42. $f(5)$ for the function f in Exercise 37

Find $f(x)$ at the indicated value of x.

43. $f(x) = 3x - 4, x = -2$

44. $f(x) = 5x + 6, x = -5$

45. $f(x) = 2x^2 - x + 3, x = 1$

46. $f(x) = 3x^2 + 2x - 5, x = 2$

47. $f(x) = -x^2 + x + 2, x = 4$

48. $f(x) = -x^2 - x - 6, x = 3$

49. $f(x) = 5, x = 9$

50. $f(x) = -4, x = 12$

51. $f(x) = \sqrt{x^3 + 12}, x = -2$

52. $f(x) = \sqrt[3]{x^2 - x + 6}, x = 2$

53. $f(x) = |5 - 2x|, x = 8$

54. $f(x) = \left|6 - \frac{1}{2}x\right|, x = 20$

Concept Check *Work each problem.*

55. If $f(-2) = 3$, identify a point on the graph of f.

56. If $f(3) = -9.7$, identify a point on the graph of f.

57. If the point $(7, 8)$ lies on the graph of f, then $f(\_\_\_\_) = \_\_\_\_$.

58. If the point $(-3, 2)$ lies on the graph of f, then $f(\_\_\_\_) = \_\_\_\_$.

Use the graph of $y = f(x)$ to find each function value: **(a)** $f(-2)$, **(b)** $f(0)$, **(c)** $f(1)$, *and* **(d)** $f(4)$.

59.

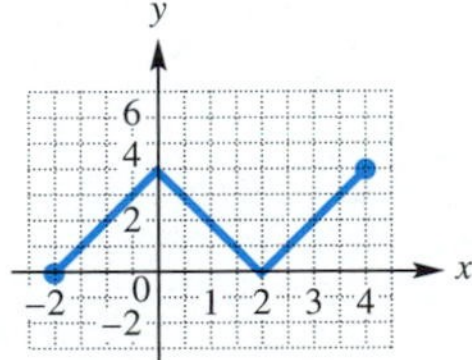

60.

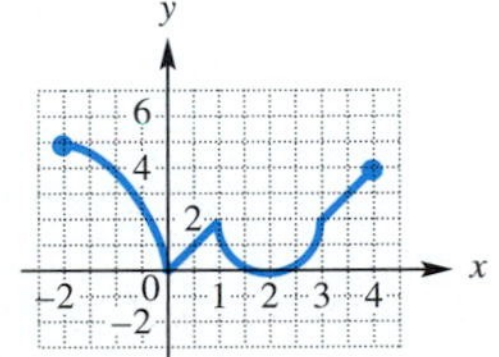

61.

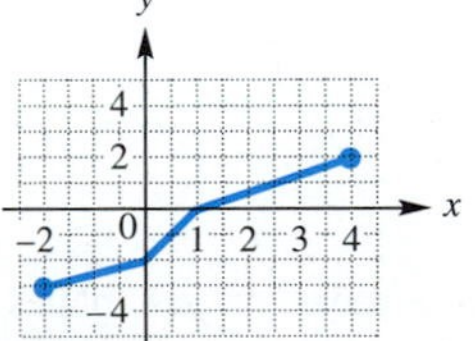

62.

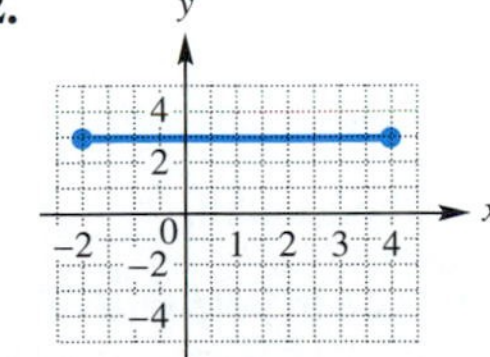

63. Explain each term in your own words.
(a) Relation
(b) Function
(c) Domain of a function
(d) Range of a function

64. *Radio Stations* The function f gives the number y in thousands of radio stations on the air during year x:

$$f = \{(1950, 2.8), (1975, 7.7), (1997, 12.8), (2000, 12.8), (2003, 13.4)\}.$$

(*Source:* M. Street Corporation.)
(a) Use a mapping diagram (see Figure 25) to represent f.
(b) Evaluate $f(2000)$ and explain what it means.
(c) Identify the domain and range of f.

Sketch the graph of f by hand.

65. $f(x) = x - 3$

66. $f(x) = 1 - 2x$

67. $f(x) = 3$

68. $f(x) = -4$

69. $f(x) = \frac{1}{2}x$

70. $f(x) = -\frac{2}{3}x$

71. $f(x) = x^2$

72. $f(x) = |x|$

(Modeling) *Solve each problem.*

73. ***Distance to Lightning*** When a bolt of lightning strikes in the distance, there is often a delay between seeing the lightning and hearing the thunder. The function defined by $f(x) = \frac{x}{5}$ computes the approximate distance in miles between an observer and a bolt of lightning when the delay is x seconds. (*Source:* Weidner, R., and R. Sells, *Elementary Classical Physics,* Allyn and Bacon, Inc., 1965.)

(a) Find $f(15)$ and interpret the result.

(b) Graph $y = f(x)$. Let the domain of f be $0 \le x \le 20$.

74. ***Air Temperature*** When the relative humidity is 100%, air cools 5.8°F for every 1-mile increase in altitude. If the temperature is 80°F on the ground, then $f(x) = 80 - 5.8x$ calculates the air temperature x miles above the ground. Find $f(3)$ and interpret the result. (*Source:* Battan, L., *Weather in Your Life,* W. H. Freeman, 1983.)

75. ***Sales Tax*** If the sales tax rate is 7.5%, write a function f that calculates the sales tax on a purchase of x dollars. What is the sales tax on a purchase of $86?

76. ***Income and Level of Education*** Function f gives the median 2002 individual income (in dollars) by educational attainment for people 25 years old and over. This function is defined by $f(N) = 23{,}267$, $f(H) = 35{,}646$, $f(B) = 69{,}156$, and $f(M) = 76{,}470$, where N denotes no high school diploma, H a high school diploma, B a bachelor's degree, and M a master's degree. (*Source:* U.S. Census Bureau.)

(a) Write f as a set of ordered pairs.

(b) Give the domain and range of f.

(c) Discuss the relationship between education and income.

77. ***Tuition and Fees*** If college tuition costs $92 per credit and fees are fixed at $75, write a formula for a function f that calculates the tuition and fees for taking x credits. What is the total cost of taking 11 credits?

78. ***Converting Units of Measure*** Write a formula for a function f that converts x gallons to quarts. How many quarts are there in 19 gallons?

Reviewing Basic Concepts (Sections 1.1 and 1.2)

1. Plot the points $(-3,1)$, $(-2,-1)$, $(2,-3)$, $(1,1)$, and $(0,2)$. Label each point.

2. Find the length and midpoint of the line segment that connects the points $P(-4,5)$ and $Q(6,-2)$.

3. Use a calculator to approximate $\dfrac{\sqrt{5}+\pi}{\sqrt[3]{3}+1}$ to the nearest thousandth.

4. The hypotenuse of a right triangle measures 61 inches, and one of its legs measures 11 inches. Find the length of the other leg.

5. Use interval notation to write the sets $\{x \mid -2 < x \le 5\}$ and $\{x \mid x \ge 4\}$.

6. Determine whether the relation shown in the graph is a function. What is its domain and range?

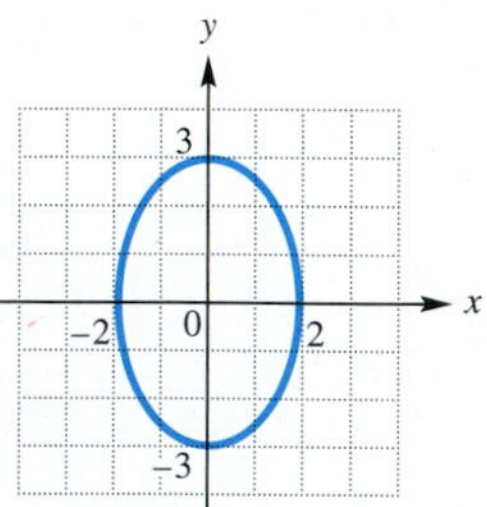

7. Find $f(-5)$ if $f(x) = 3 - 4x$.

8. Use the graph to find $f(2)$ and $f(-1)$.

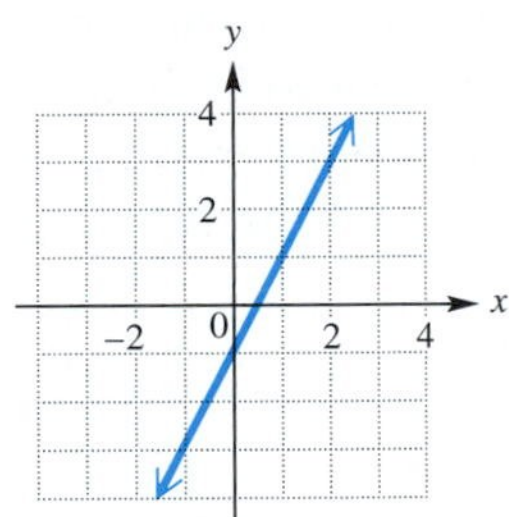

9. Graph $f(x) = \frac{1}{2}x - 1$ by hand.

10. Find the distance between the points $(12, -3)$ and $(-4, 27)$.

1.3 Linear Functions

Basic Concepts about Linear Functions ■ Slope of a Line ■ Slope–Intercept Form of the Equation of a Line

Basic Concepts about Linear Functions

Suppose that by noon 2 inches of rain had fallen during a storm. Rain continued to fall at the rate of $\frac{1}{2}$ inch per hour in the afternoon. Then the total rainfall x hours past noon is given by

$$f(x) = \frac{1}{2}x + 2.$$

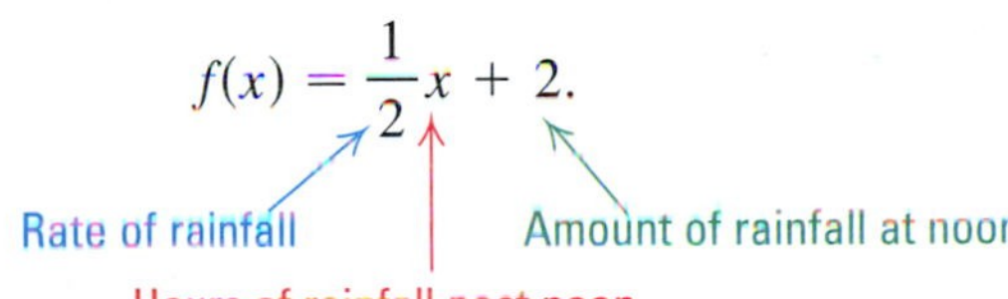

At 3:00 P.M., the total rainfall equaled

$$f(3) = \frac{1}{2}(3) + 2 = 3.5 \text{ inches.}$$

This function f satisfies the general form $f(x) = ax + b$ (where $a = \frac{1}{2}$ and $b = 2$) and is called a *linear function.*

Linear Function

A function f defined by $f(x) = ax + b$, where a and b are real numbers, is called a **linear function.**

The graph of $f(x) = ax + b$ is a line and can be found by graphing $y = ax + b$. For example, the graph of $f(x) = 3x + 6$ is the line determined by $y = 3x + 6$. An equation such as $y = 3x + 6$ is called a **linear equation in two variables.** A **solution** is an ordered pair (x, y) that makes the equation true. Verify that $(-2, 0)$, $(-1, 3)$, $(0, 6)$, and $(1, 9)$ are all solutions of $y = 3x + 6$.

Graphing linear equations by hand involves plotting points whose coordinates are solutions of the equation and then connecting them with a straight line. Figure 30(a) shows the ordered pairs just mentioned for the linear equation $y = 3x + 6$, accompanied by a *table of values.* Notice that the points appear to lie in a straight line; that is indeed the case. Since we may substitute *any* real number for x, we connect these points with a line to obtain the graph of $f(x) = 3x + 6$, as shown in Figure 30(b).

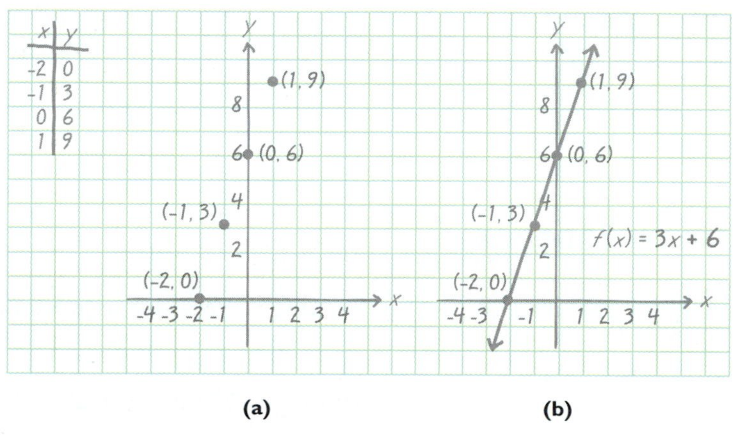

(a) (b)

FIGURE 30

To graph the linear function defined by $f(x) = 3x + 6$ on a calculator, we enter $3x + 6$ for one of the y-variables. Using the standard viewing window, we get the graph shown in Figure 31(a). A graphing calculator will also give a table of selected points, as shown in Figure 31(b).

TECHNOLOGY NOTE

To graph a function f on a graphing calculator, be sure to enter the formula for f and then set an appropriate viewing window.

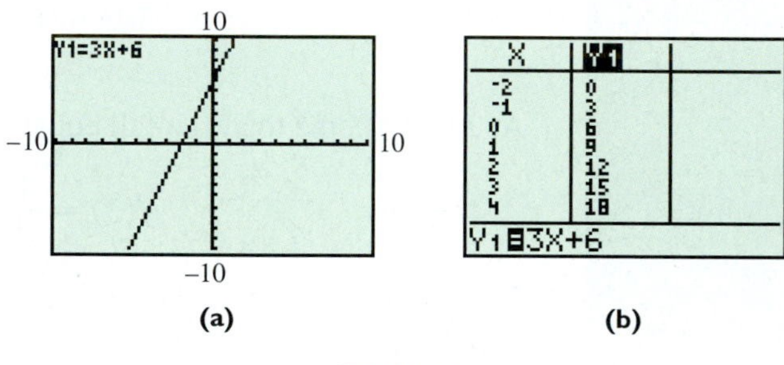

(a) (b)

FIGURE 31

From geometry, we know that two distinct points determine a line. Therefore, if we know the coordinates of two points, we can graph the line. For the equation $y = 3x + 6$, suppose we let $x = 0$. Then $y = 3(0) + 6 = 6$.

Now, suppose we let $y = 0$ and find x.

$$
\begin{aligned}
y &= 3x + 6 & \\
0 &= 3x + 6 & \text{Let } y = 0. \\
-6 &= 3x & \text{Subtract 6.} \\
x &= -2 & \text{Divide by 3; rewrite.}
\end{aligned}
$$

The points $(0, 6)$ and $(-2, 0)$ lie on the graph of $y = 3x + 6$ and are sufficient for obtaining the graph in Figure 30(b). The numbers 6 and -2 are called the ***y*- and *x*-intercepts** of the line respectively.

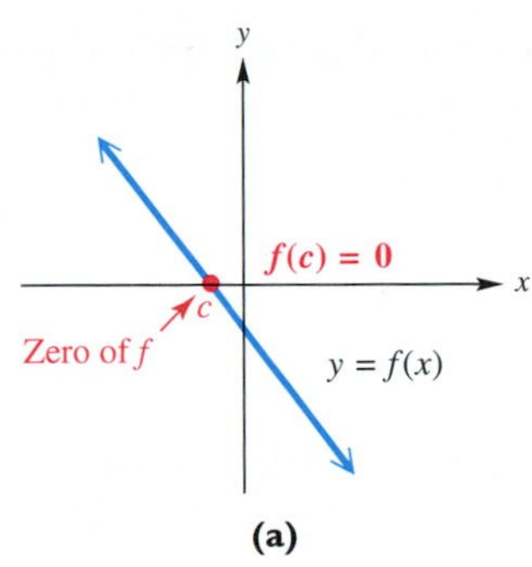

(a)

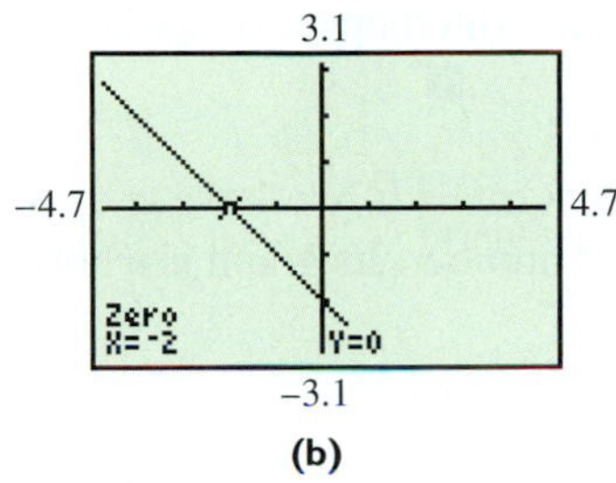

(b)

FIGURE 32

Locating x- and y-Intercepts

To find the x-intercept of the graph of $y = ax + b$, let $y = 0$ and solve for x (assuming that $a \neq 0$). To find the y-intercept, let $x = 0$ and solve for y.

The x-intercept of the graph of a linear function is a value that makes $f(x) = 0$ a true statement; that is, it causes the function value to equal zero. In general, such a number is called a *zero* of the function.

Zero of a Function

Let f be a function. Then any number c for which $f(c) = 0$ is called a **zero** of the function f.

NOTE ***If c is a zero of f, then f(c) = 0 and c is an x-intercept of the graph of f.*** See Figure 32(a). A calculator can also be directed to find a zero, as in Figure 32(b).

EXAMPLE 1 Graphing a Line

Graph the function defined by $f(x) = -2x + 5$. What is the zero of f?

Analytic Solution

The graph of $f(x) = -2x + 5$ and its intercepts are shown in Figure 33. The zero of f is 2.5.

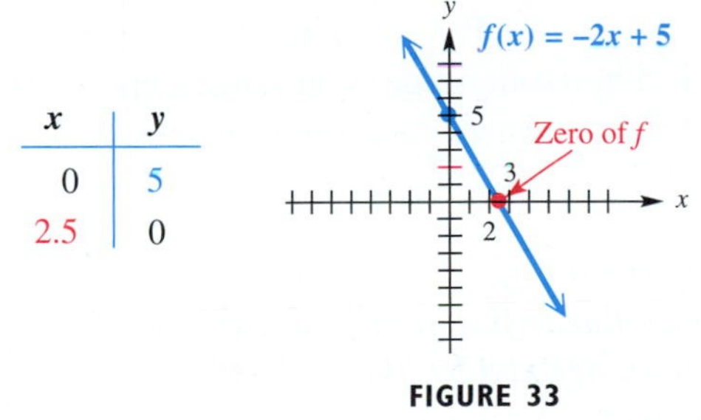

x	y
0	5
2.5	0

FIGURE 33

Graphing Calculator Solution

A calculator graph is shown in Figure 34. The x-intercept is 2.5, so 2.5 is the zero of f.

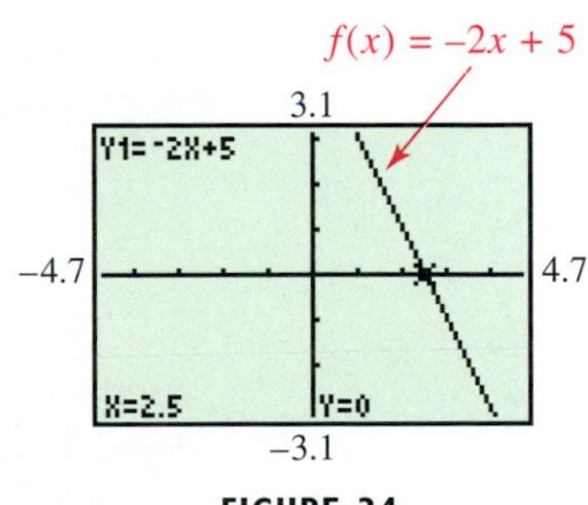

FIGURE 34

EXAMPLE 2 Finding a Formula for a Function

A 100-gallon tank is initially full of water and is being drained at a rate of 5 gallons per minute.

(a) Write a formula for a linear function f that models the number of gallons of water in the tank after x minutes.

(b) How much water is in the tank after 4 minutes?

(c) Use the x- and y-intercepts to graph f. Interpret each intercept.

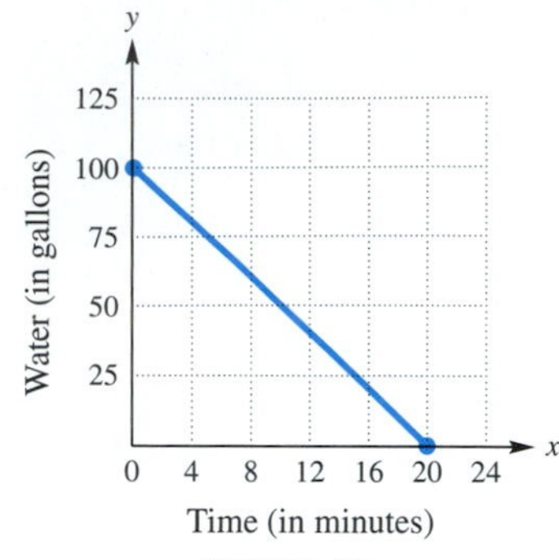

FIGURE 35

Solution

(a) The amount of water in the tank is *decreasing* at a constant rate of 5 gallons per minute, so the constant rate of change is -5. The initial amount of water is equal to 100 gallons, so

$$f(x) = (\text{constant rate of change})x + (\text{initial amount})$$
$$f(x) = -5x + 100.$$

(b) After 4 minutes, the tank held $f(4) = -5(4) + 100 = 80$ gallons.

(c) The graph of $y = -5x + 100$ has y-intercept 100 because $y = 100$ when $x = 0$. To find the x-intercept, let $y = 0$ and solve the equation $0 = -5x + 100$, obtaining $x = 20$. The graph of f is shown in Figure 35. The x-intercept corresponds to the time in minutes that it takes to empty the tank. The y-intercept corresponds to the number of gallons of water in the tank initially. ■

Suppose that for a linear function defined by $f(x) = ax + b$, we have $a = 0$. Then the function becomes $f(x) = b$, where b is some real number. Its graph is a horizontal line.

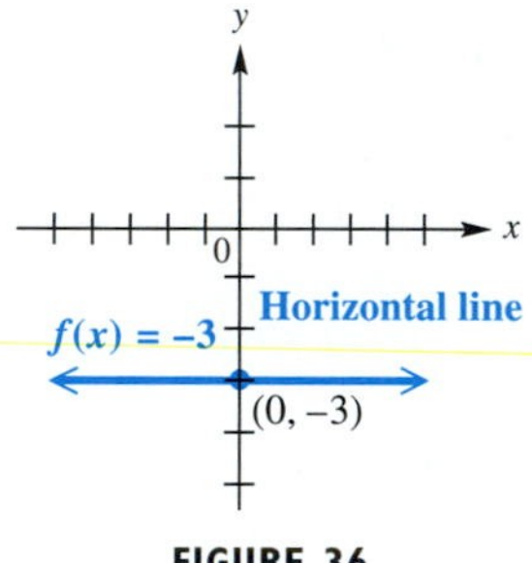

FIGURE 36

EXAMPLE 3 Sketching the Graph of $f(x) = b$

Graph the function defined by $f(x) = -3$.

Solution Since y always equals -3, the y-intercept is -3. Since the value of y can never be 0, the graph has no x-intercept. The only way that a straight line can have no x-intercept is for it to be parallel to the x-axis, as shown in Figure 36. ■

The function discussed in Example 3 is a *constant function.*

Constant Function

A function defined by $\boldsymbol{f(x) = b}$, where b is a real number, is called a **constant function.** Its graph is a horizontal line with y-intercept b. For $b \neq 0$, it has no x-intercept. (Every constant function is also linear.)

Unless otherwise specified, the domain of a linear function is the set of all real numbers. The range of a nonconstant linear function is also the set of all real numbers. The range of a constant function defined by $f(x) = b$ is $\{b\}$.

The choice of viewing window may give drastically different views of a calculator graph, as shown in Figure 37 for the graph of $f(x) = 3x + 6$.

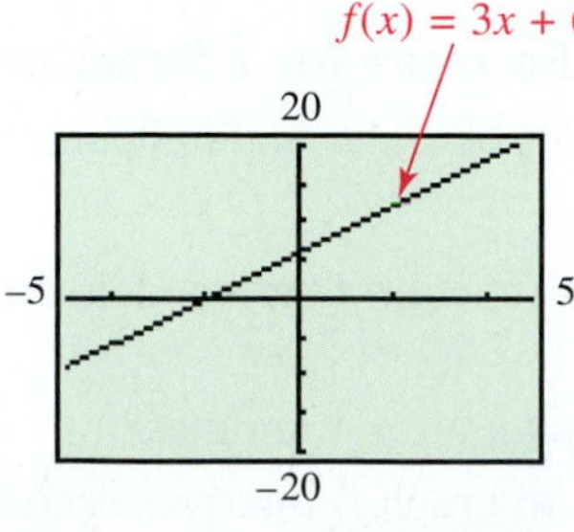

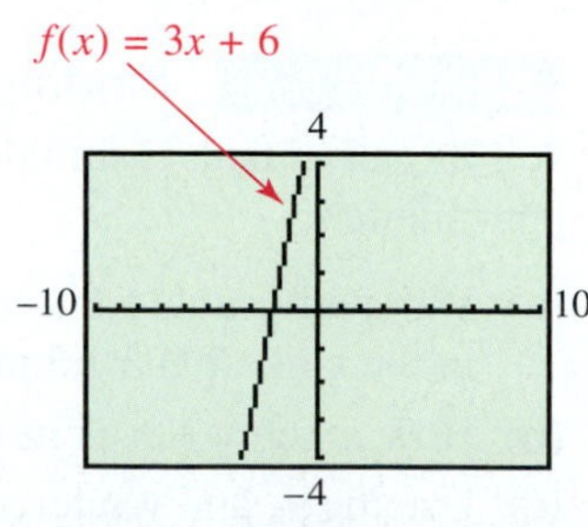

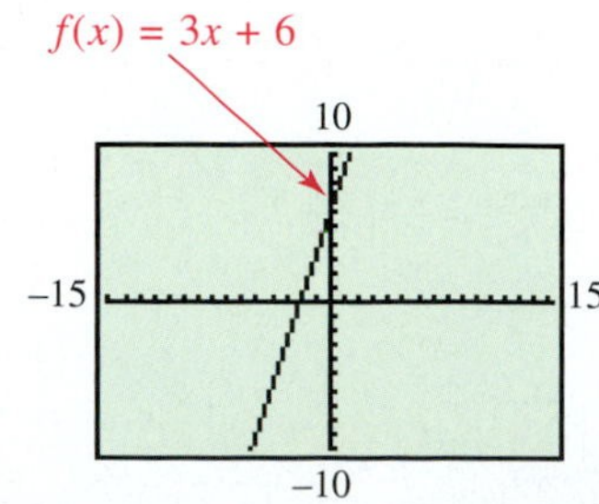

FIGURE 37

We usually want a window that shows the most important features of a particular graph. Such a graph is called a **comprehensive graph.*** The choice of window for a comprehensive graph is not unique—there are many acceptable ones. ***For a line, a comprehensive graph shows all intercepts of the line.***

EXAMPLE 4 Finding a Comprehensive Graph of a Line

Find a comprehensive graph of $g(x) = -.75x + 12.5$.

Solution The window $[-10, 10]$ by $[-10, 10]$ of Figure 38(a) does not show either intercept, so it will not work. We can alter the window size to, for example, $[-10, 20]$ by $[-10, 20]$ to obtain a comprehensive graph. See Figure 38(b).

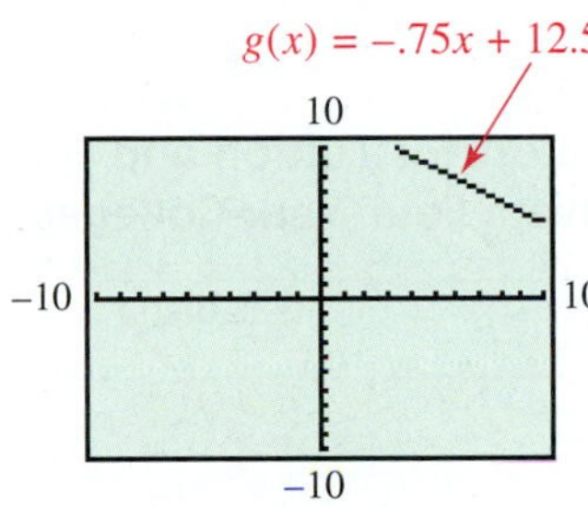

This *is not* a comprehensive graph of the line, since no intercepts are visible.

(a)

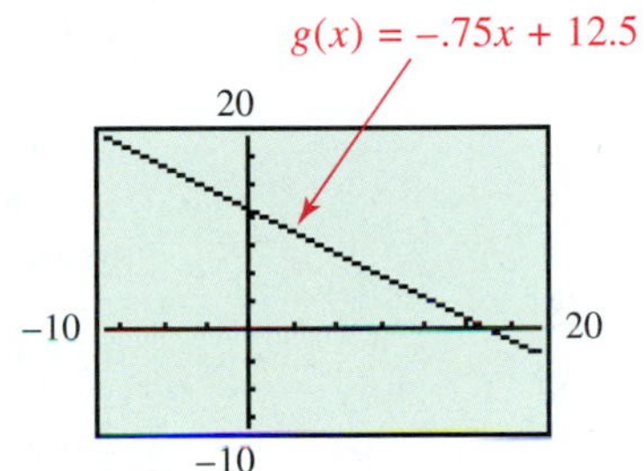

This *is* a comprehensive graph of the line, since both intercepts are visible.

(b)

FIGURE 38

What Went Wrong?

A student learning to use a graphing calculator attempted to graph $y = \frac{1}{2}x + 15$. However, she obtained the blank screen shown here:

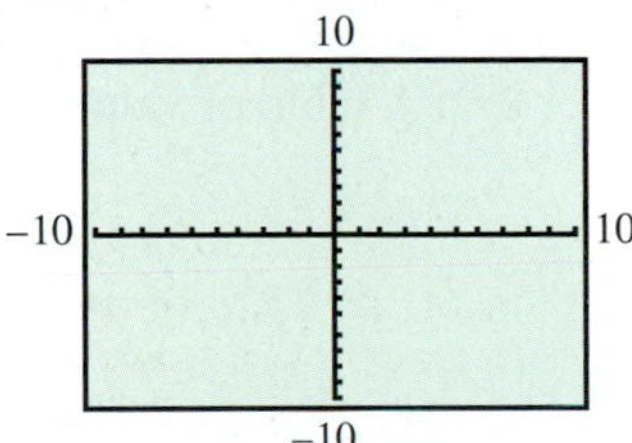

What Went Wrong? How can she obtain a comprehensive graph of this linear function?

Answer to What Went Wrong?

The y-intercept is 15 and the x-intercept is -30, so the window size must be increased to show the intercepts. For example, a window of $[-40, 40]$ by $[-30, 30]$ would work.

---

*The term *comprehensive graph* was coined by Shoko Aogaichi Brant and Edward A. Zeidman in the text *Intermediate Algebra: A Functional Approach* (HarperCollins College Publishers, 1996), with the assistance of Professor Brant's daughter Jennifer. The authors thank them for permission to use the terminology in this text.

Looking Ahead to Calculus

The concept of slope of a line is extended in calculus to general curves. The *slope of a curve* at a point is understood to mean the slope of the line tangent to the curve at that point.

Slope of a Line

In 1984, the average annual cost for tuition and fees at private four-year colleges was \$5991. By 2004, this cost had increased to \$20,082. The line graphed in Figure 39 is actually somewhat misleading, since it indicates that the increase in cost was the same from year to year. However, we can use the graph to determine the *average* yearly increase in cost. Over the 20-year span, the cost increased \$14,091. Therefore, the average yearly increase was

$$\frac{\$20{,}082 - \$5991}{2004 - 1984} = \frac{\$14{,}091}{20} \approx \$705.$$

This quotient is an illustration of the *slope* of the line joining $(1984, 5991)$ and $(2004, 20{,}082)$.

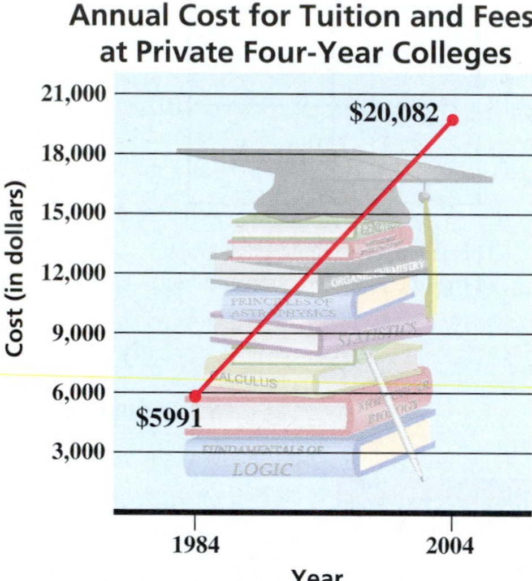

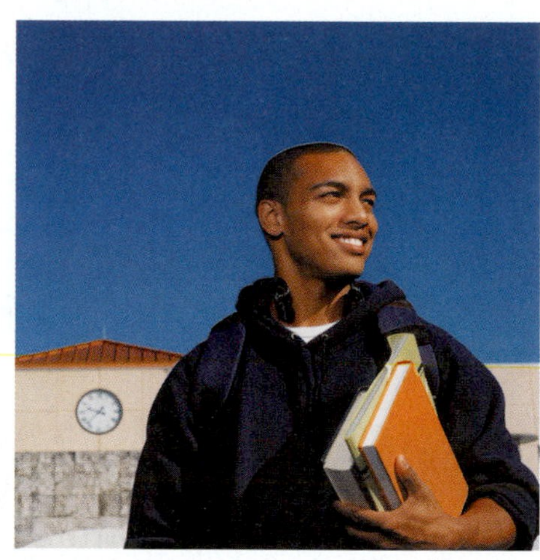

Source: The College Board.

FIGURE 39

A table of values and the graph of the line $y = 3x + 1$ are shown in Figure 40.

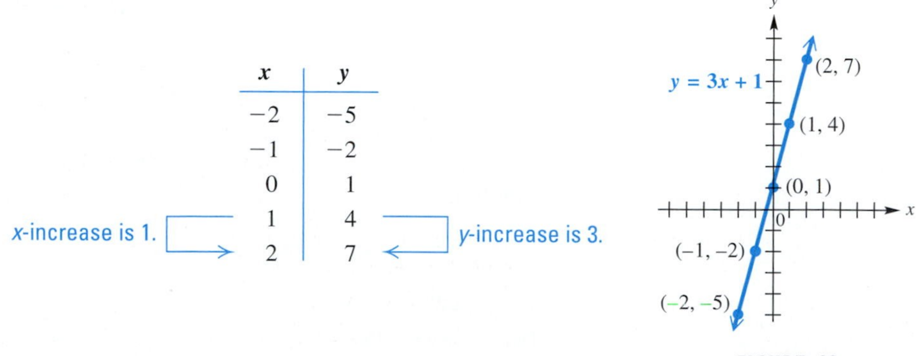

x	y
−2	−5
−1	−2
0	1
1	4
2	7

FIGURE 40

For each increase of 1 for the x-value, the y-value increases by 3. The slope of the line is 3.

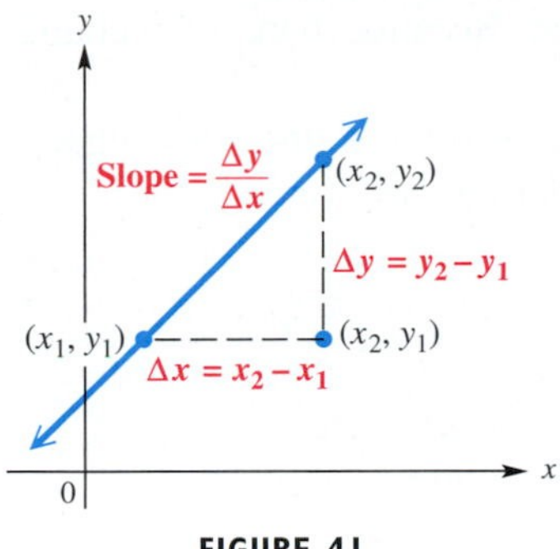

FIGURE 41

Geometrically, slope is a numerical measure of the steepness of a line and may be interpreted as the ratio of *rise* to *run*. To calculate slope, start with the line through the two distinct points (x_1, y_1) and (x_2, y_2), where $x_1 \neq x_2$. See Figure 41. The difference $x_2 - x_1$ is called the **change in *x***, denoted $\mathbf{\Delta x}$ (read **"delta *x*"**), where $\mathbf{\Delta}$ is the Greek letter **delta.** In the same way, the **change in *y*** is denoted $\mathbf{\Delta y} = y_2 - y_1$. The *slope* of a nonvertical line is defined as the quotient of the change in *y* and the change in *x*.

Slope

The **slope** *m* of the line passing through the points (x_1, y_1) and (x_2, y_2) is

$$m = \frac{\Delta y}{\Delta x} = \frac{y_2 - y_1}{x_2 - x_1}, \quad \text{where } \Delta x = x_2 - x_1 \neq 0.$$

CAUTION ***When using the slope formula, it makes no difference which point is (x_1, y_1) or (x_2, y_2); however, be consistent.*** Start with the *x*- and *y*-values of one point (either one), and subtract the corresponding values of the *other* point.

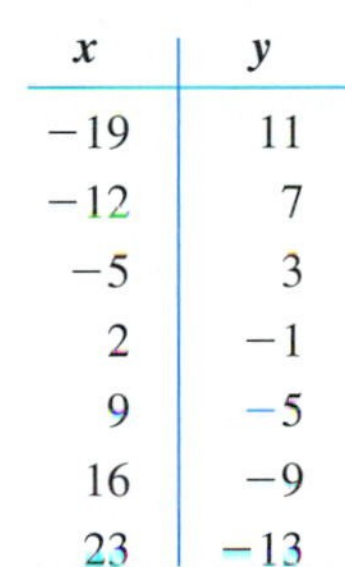

x	y
−19	11
−12	7
−5	3
2	−1
9	−5
16	−9
23	−13

EXAMPLE 5 **Using the Slope Formula to Find Slope**

A table of values for a linear function is shown in the margin. Determine the slope of the graph of the line. Sketch the graph.

Solution Because the slope of a line is the same regardless of the two points chosen, we can choose any two points from the table. If we let

$$(2, -1) = (x_1, y_1) \quad \text{and} \quad (-5, 3) = (x_2, y_2)$$

then

$$m = \frac{y_2 - y_1}{x_2 - x_1} = \frac{3 - (-1)}{-5 - 2} = -\frac{4}{7}.$$

Start with the x- and y-values of the same point.

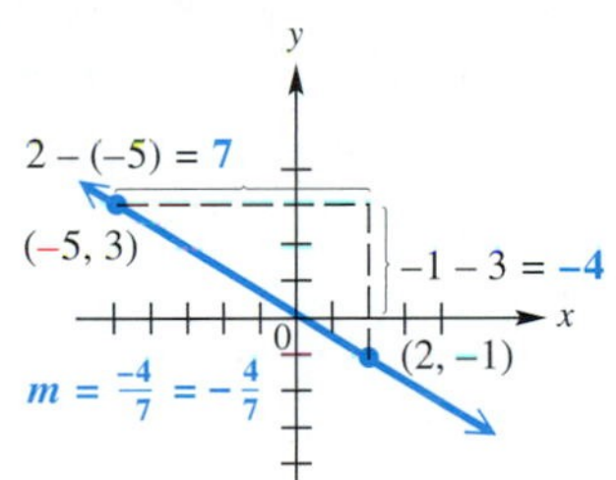

FIGURE 42

See Figure 42. By contrast, if $(-5, 3) = (x_1, y_1)$ and $(2, -1) = (x_2, y_2)$, then

$$m = \frac{-1 - 3}{2 - (-5)} = -\frac{4}{7}.$$

The slope is $-\frac{4}{7}$ no matter which point is considered first. ■

EXAMPLE 6 **Using the Slope and a Point to Graph a Line**

Graph the line that passes through $(2, 1)$ and has slope $-\frac{4}{3}$.

Solution Start by locating the point $(2, 1)$ on the graph. Find a second point on the line by using the definition of slope.

$$\text{slope} = \frac{\text{change in } y}{\text{change in } x} = \frac{-4}{3}$$

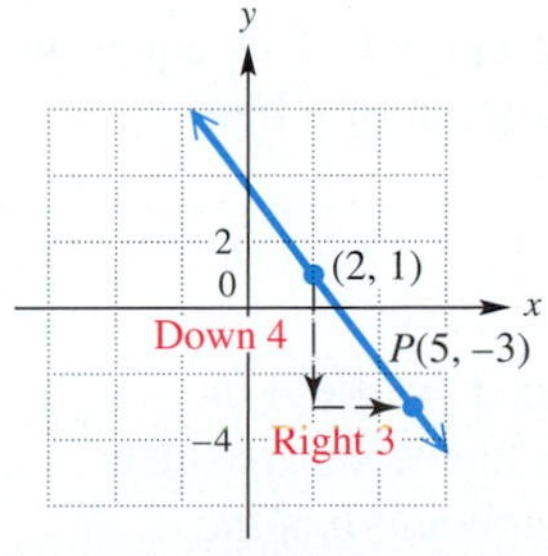

FIGURE 43

Move *down* 4 units from $(2, 1)$ and then 3 units to the right to obtain $P(5, -3)$. Draw a line through this second point *P* and $(2, 1)$, as shown in Figure 43. ■

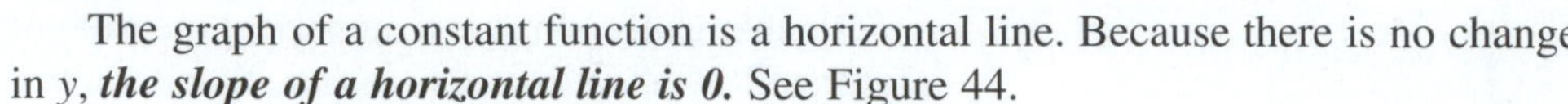

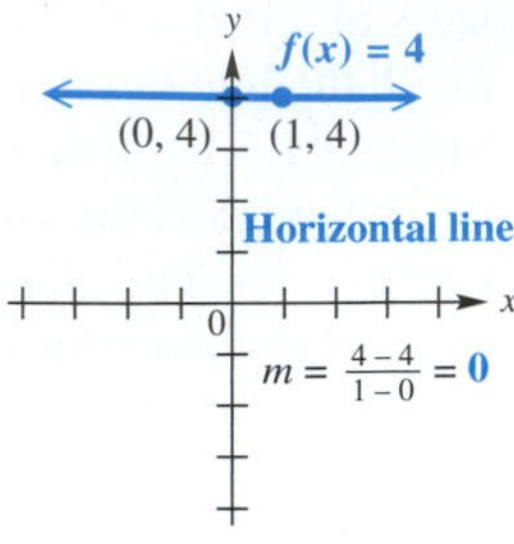

FIGURE 44

The graph of a constant function is a horizontal line. Because there is no change in y, ***the slope of a horizontal line is 0.*** See Figure 44.

In general, a line with positive slope *rises* from left to right, a line with negative slope *falls* from left to right, and a line with slope 0 is *horizontal.*

Geometric Orientation Based on Slope

For a line with slope m, if $m > 0$, the line rises from left to right. If $m < 0$, it falls from left to right. If $m = 0$, the line is horizontal.

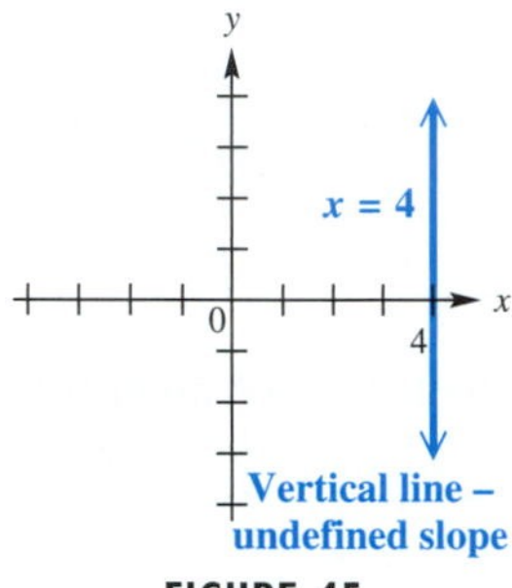

FIGURE 45

In the slope formula, we have the condition $\Delta x = x_2 - x_1 \neq 0$. This means that $x_2 \neq x_1$. If we graph a line with two points having the same x-values, we get a vertical line. See Figure 45. Notice that this is *not* the graph of a function, since 4 appears as the first number in more than one ordered pair. If we were to apply the slope formula, the denominator would be 0. As a result, ***the slope of a vertical line is undefined.***

Vertical Line

A vertical line with x-intercept a has an equation of the form $\boldsymbol{x = a}$. Its slope is undefined.

Slope–Intercept Form of the Equation of a Line

FIGURE 46

In Figure 46, the slope m of the line $y = 3x + 1$ is 3 and the y-intercept is 1. In general, if $f(x) = ax + b$, then the slope of the graph of $f(x)$ is a and the y-intercept is b. To verify this fact, notice that $f(0) = a(0) + b = b$. Thus, the graph of f passes through the point $(0, b)$ and b is the y-intercept. Since $f(1) = a(1) + b = a + b$, the graph of f also passes through the point $(1, a + b)$. The slope of the line that passes through the points $(0, b)$ and $(1, a + b)$ is

$$m = \frac{a + b - b}{1 - 0} = a.$$

See Figure 47. Because the slope of the graph of $f(x) = ax + b$ is a, it is often convenient to use m rather than a in the general form of the equation. Therefore, we write either

$$\boldsymbol{f(x) = mx + b} \quad \text{or} \quad \boldsymbol{y = mx + b}$$

to indicate a linear function. The slope is m, and the y-intercept is b. This equation for $f(x)$ is generally called the *slope–intercept form* of the equation of a line.

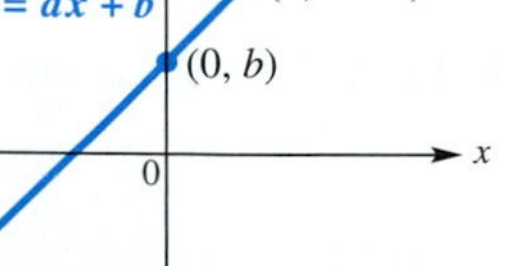

FIGURE 47

Slope–Intercept Form

The **slope–intercept form** of the equation of a line is $\boldsymbol{y = mx + b}$, where m is the slope and b is the y-intercept. (Linear functions are often written in the form $f(x) = ax + b$, where a is the slope and b is the y-intercept of the graph.)

EXAMPLE 7 Matching Graphs with Equations

Figure 48 shows four graphs of lines. Their equations are

$$y = 2x + 3, \quad y = -2x + 3, \quad y = 2x - 3, \quad \text{and} \quad y = -2x - 3,$$

but not necessarily in that order. Match each equation with its graph.

A.

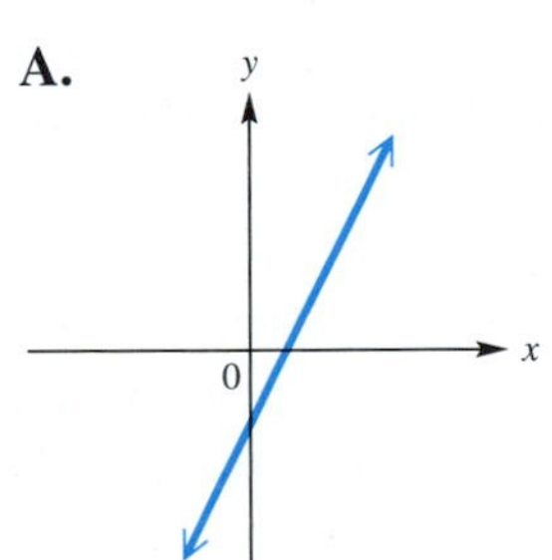

B.

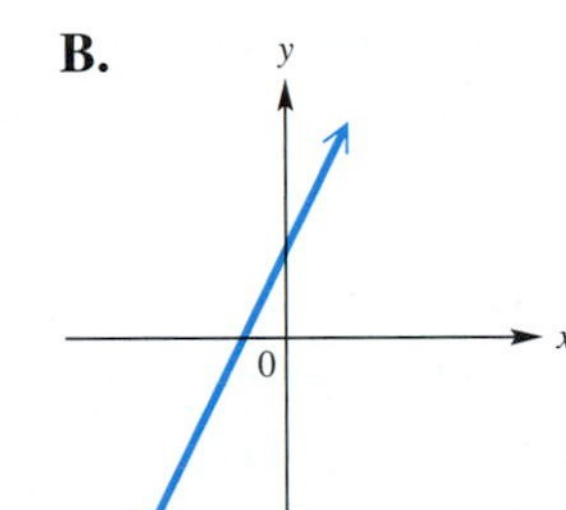

C.

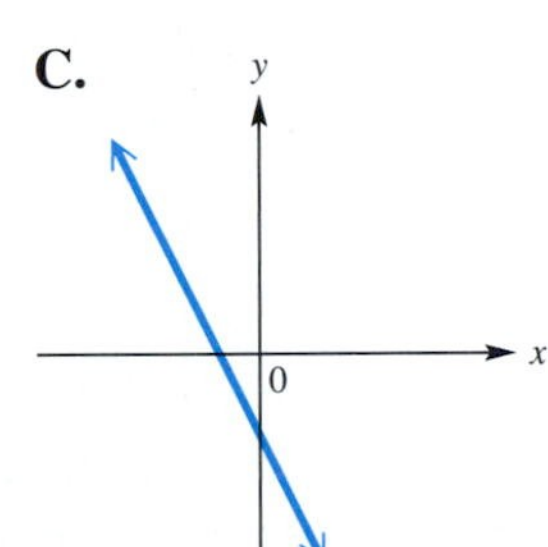

D.

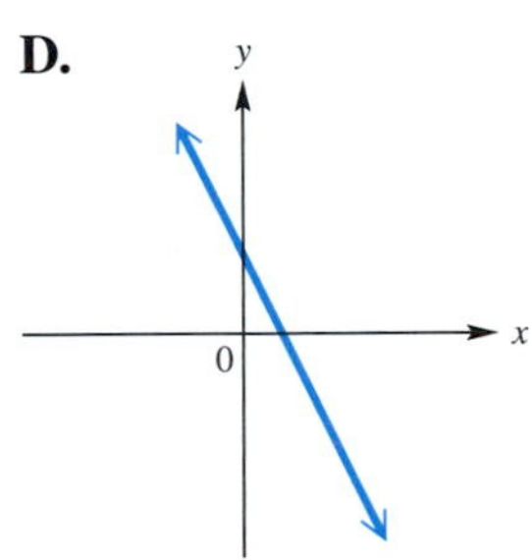

FIGURE 48

Looking Ahead to Calculus

Slope represents a rate of change. In calculus, rates of change of nonlinear functions are studied extensively by using the *derivative*.

Solution The sign of m determines whether the graph rises or falls from left to right. Also, if $b > 0$, the y-intercept is *above* the x-axis, and if $b < 0$, the y-intercept is *below* the x-axis. Therefore,

$y = 2x + 3$ is shown in B, since the graph rises from left to right and the y-intercept is positive;

$y = -2x + 3$ is shown in D, since the graph falls from left to right and the y-intercept is positive;

$y = 2x - 3$ is shown in A, since the graph rises from left to right and the y-intercept is negative;

$y = -2x - 3$ is shown in C, since the graph falls from left to right and the y-intercept is negative. ■

EXAMPLE 8 Interpreting Slope

In 1980, passengers traveled a total of 4.5 billion miles on Amtrak, and in 2000 they traveled 5.5 billion miles. (*Source:* U.S. Department of Transportation.)

(a) Find the slope m of the line passing through the points $(1980, 4.5)$ and $(2000, 5.5)$.

(b) Interpret the slope.

Solution

(a) $m = \dfrac{5.5 - 4.5}{2000 - 1980} = \dfrac{1}{20} = .05$

(b) Because the slope is positive, the number of passenger-miles traveled on Amtrak increased, on average, by about .05 billion, or 50 million miles per year, between 1980 and 2000. The number of passenger-miles did not increase by exactly 50 million each year. ***Thus, slope gives an average rate of change.*** ■

EXAMPLE 9 **Interpreting Slope–Intercept Form**

In 2004, there were approximately 39 million people worldwide living with HIV/AIDS.* At that time, the infection rate was 4.9 million people per year. (*Source:* Joint United Nations Programme on HIV/AIDS (UNAIDS.org).)

(a) Find values for m and b so that $y = mx + b$ models the total number of people y in millions who were living with HIV/AIDS in year x, where $x = 0$ corresponds to 2004, $x = 1$ to 2005, and so on.

(b) Estimate the number of people who could be living with HIV/AIDS in 2007.

Solution

(a) Since there were 39 million people living with HIV/AIDS in 2004 and $x = 0$ corresponds to 2004, the line determined by $y = mx + b$ must pass through the point $(0, 39)$. Thus, the y-intercept is $b = 39$. The infection rate is 4.9 million per year, so $m = 4.9$. The equation is

$$y = 4.9x + 39.$$

(b) The year 2007 corresponds to $x = 3$, so

$$y = 4.9(3) + 39 = 53.7.$$

Thus, according to this model, approximately 53.7 million people could be living with HIV/AIDS in 2007. ■

EXAMPLE 10 **Finding an Equation from a Graph**

Use the graph of the linear function f in Figure 49 to complete the following.

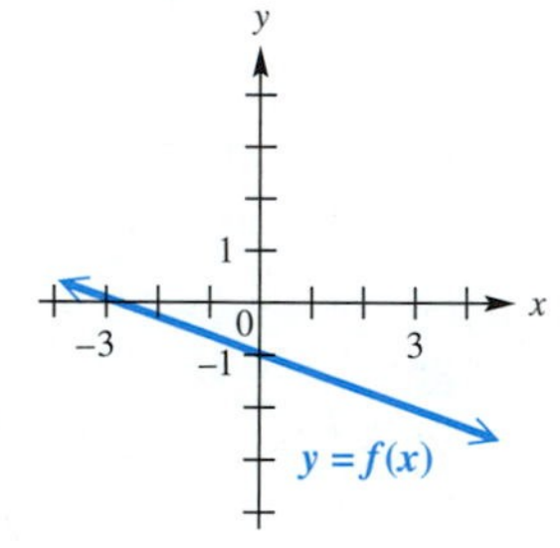

FIGURE 49

(a) Find the slope, y-intercept, and x-intercept.

(b) Write the equation defining f.

(c) Find any zeros of f.

Solution

(a) The line falls 1 unit each time the x-value increases by 3 units. Therefore, the slope is $\frac{-1}{3} = -\frac{1}{3}$. The graph intersects the y-axis at the point $(0, -1)$ and intersects the x-axis at the point $(-3, 0)$. Therefore, the y-intercept is -1 and the x-intercept is -3.

(b) Because the slope is $-\frac{1}{3}$ and the y-intercept is -1, it follows that the equation defining f is

$$f(x) = -\frac{1}{3}x - 1.$$

(c) Zeros of f correspond to x-intercepts, so the only zero is -3. ■

*"Living with HIV/AIDS" is the terminology used by the Joint United Nations Programme on HIV/AIDS.

1.3 Exercises

Work each problem related to linear functions.

(a) *Evaluate* $f(-2)$ *and* $f(4)$. **(b)** *Find the zero of* f.

(c) *Graph* f. *How can the graph of* f *be used to determine the zero of* f?

1. $f(x) = x + 2$ **2.** $f(x) = -3x + 2$ **3.** $f(x) = 2 - \frac{1}{2}x$ **4.** $f(x) = \frac{1}{4}x + \frac{1}{2}$

5. $f(x) = \frac{1}{3}x$ **6.** $f(x) = -3x$ **7.** $f(x) = .4x + .15$ **8.** $f(x) = x + .5$

9. $f(x) = \frac{2 - x}{4}$ **10.** $f(x) = \frac{3 - 3x}{6}$ **11.** $f(x) = \frac{3x + \pi}{2}$ **12.** $f(x) = \frac{4x + \pi}{3}$

Graph each linear function by hand. Give the **(a)** *x-intercept,* **(b)** *y-intercept,* **(c)** *domain,* **(d)** *range, and* **(e)** *slope of the line.*

13. $f(x) = x - 4$ **14.** $f(x) = -x + 4$ **15.** $f(x) = 3x - 6$ **16.** $f(x) = \frac{2}{3}x - 2$

17. $f(x) = -\frac{2}{5}x + 2$ **18.** $f(x) = \frac{4}{3}x - 3$ **19.** $f(x) = 3x$ **20.** $f(x) = -.5x$

21. ***Concept Check*** Based on the graphs of the functions in Exercises 19 and 20, what conclusion can you make about one particular point that *must* lie on the graph of the line $y = ax$ (where $b = 0$)?

22. ***Concept Check*** Using the concept of slope and your answer in Exercise 21, give the equation of the line whose graph is shown at the right.

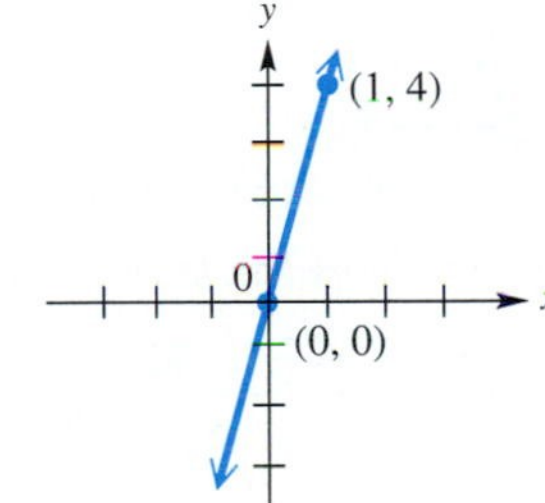

Graph each line by hand. Also, give the **(a)** *x-intercept (if any),* **(b)** *y-intercept (if any),* **(c)** *domain,* **(d)** *range, and* **(e)** *slope of the line (if defined).*

23. $f(x) = -3$ **24.** $f(x) = 5$ **25.** $x = -1.5$ **26.** $f(x) = \frac{5}{4}$ **27.** $x = 2$ **28.** $x = -3$

29. ***Concept Check*** What special name is given to the functions found in Exercises 23, 24, and 26?

Give the equation of the line illustrated.

30.

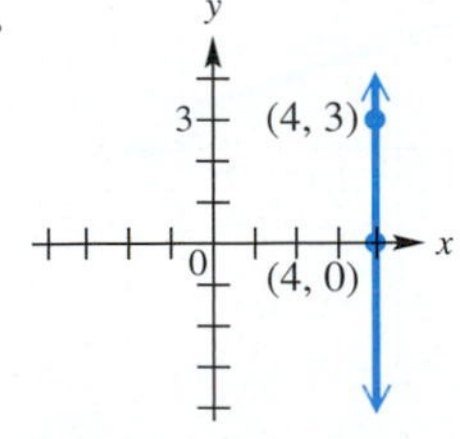

31.

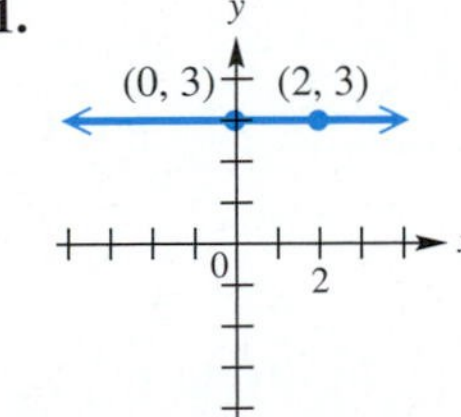

32.

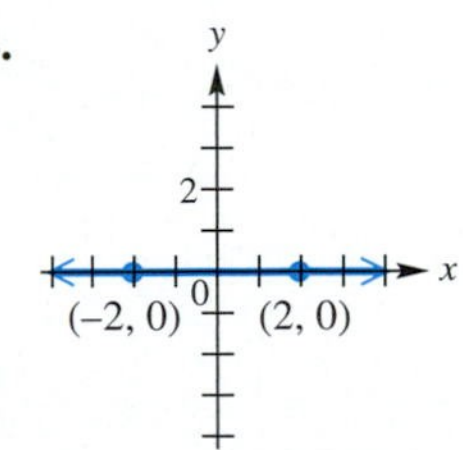

33.

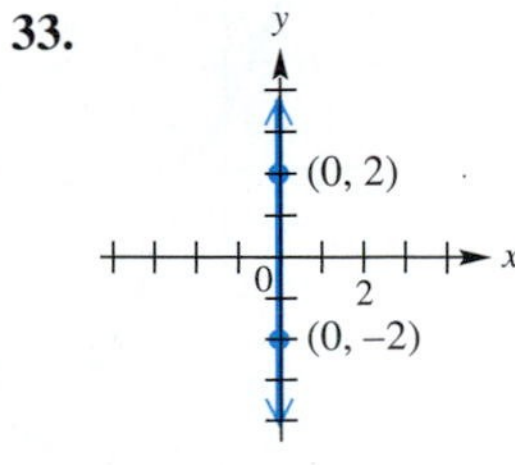

34. *Concept Check* Answer each question.
(a) What is the equation of the x-axis? **(b)** What is the equation of the y-axis?

Graph each linear function on a graphing calculator, using the two different windows given. State which window gives the comprehensive graph.

35. $f(x) = 4x + 20$
Window A: $[-10, 10]$ by $[-10, 10]$
Window B: $[-10, 10]$ by $[-5, 25]$

36. $f(x) = -5x + 30$
Window A: $[-10, 10]$ by $[-10, 40]$
Window B: $[-5, 5]$ by $[-5, 40]$

37. $f(x) = 3x + 10$
Window A: $[-3, 3]$ by $[-5, 5]$
Window B: $[-5, 5]$ by $[-10, 14]$

38. $f(x) = -6$
Window A: $[-5, 5]$ by $[-5, 5]$
Window B: $[-10, 10]$ by $[-10, 10]$

Find the slope (if defined) of the line that passes through the given points.

39. $(-2, 1)$ and $(3, 6)$
40. $(-2, 3)$ and $(-1, 2)$
41. $(8, 4)$ and $(-1, -3)$
42. $(-4, -3)$ and $(5, 0)$
43. $(-11, 3)$ and $(-11, 5)$
44. $(-8, 2)$ and $(-8, 1)$
45. $\left(\frac{2}{3}, 9\right)$ and $\left(\frac{1}{2}, 9\right)$
46. $(.12, .36)$ and $(.18, .36)$

Concept Check *Match each equation with the graph that it most closely resembles.*

47. $y = 3x + 6$
48. $y = -3x + 6$
49. $y = -3x - 6$
50. $y = 3x - 6$
51. $y = 3x$
52. $y = -3x$
53. $y = 3$
54. $y = -3$

A. **B.** **C.** **D.**

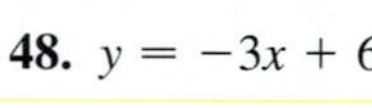
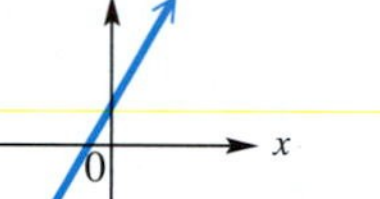
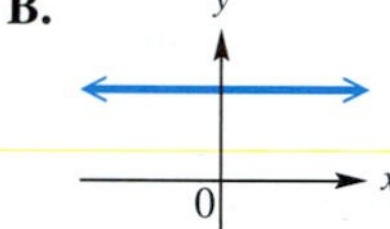
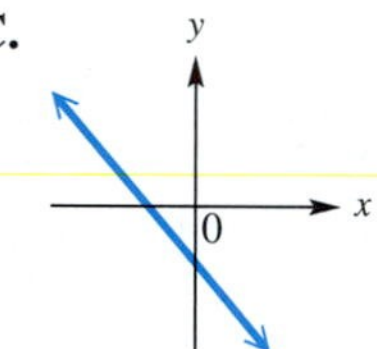
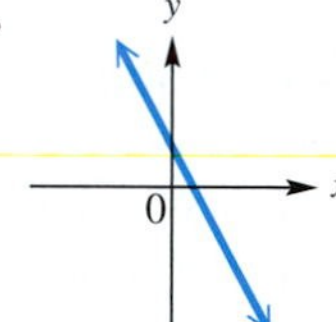

E. **F.** **G.** **H.**

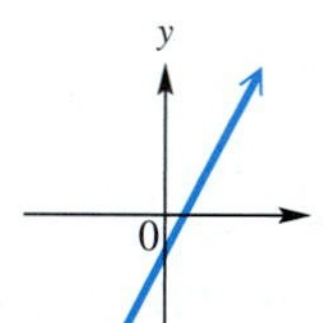
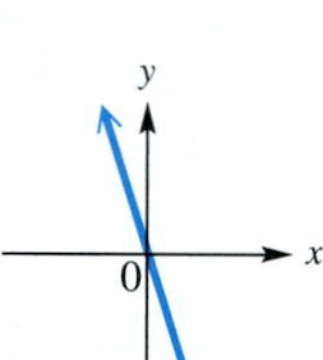
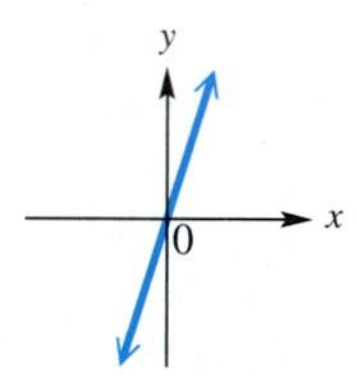

The graph of a linear function f is shown. **(a)** *Identify the slope, y-intercept, and x-intercept.* **(b)** *Write a formula for f.* **(c)** *Estimate the zero of f.*

55.

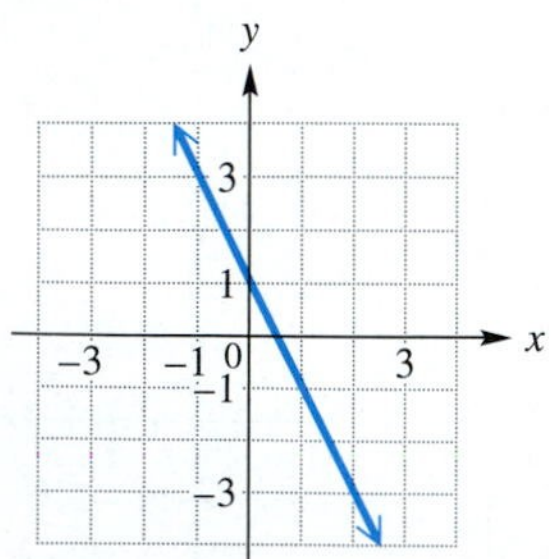

56.

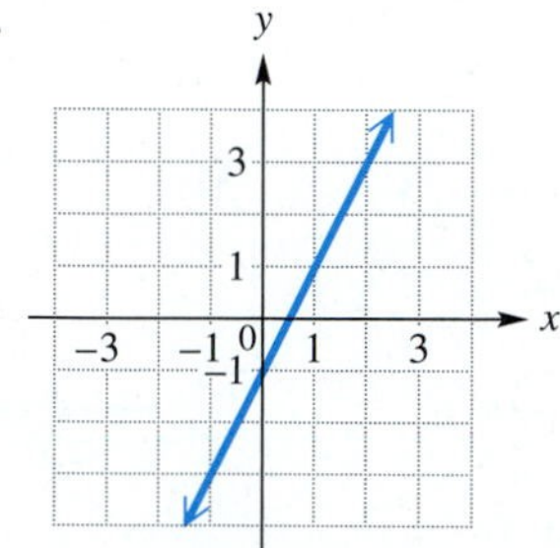

57.

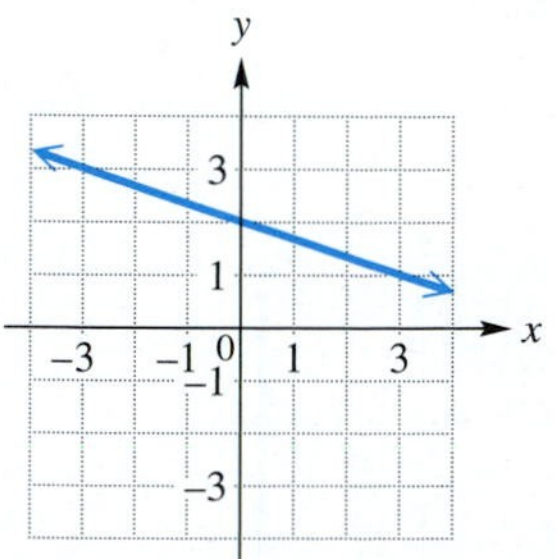

58.

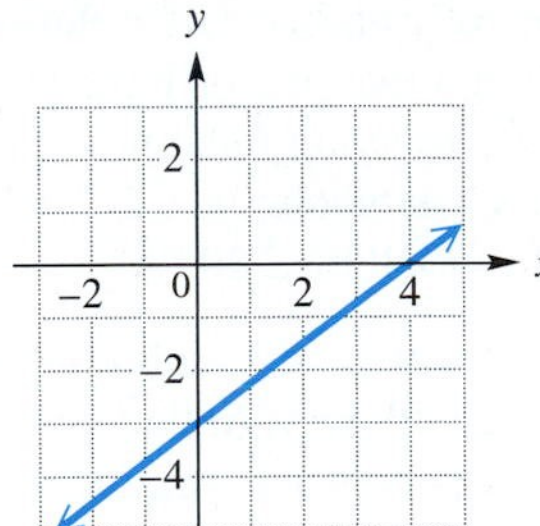

59.

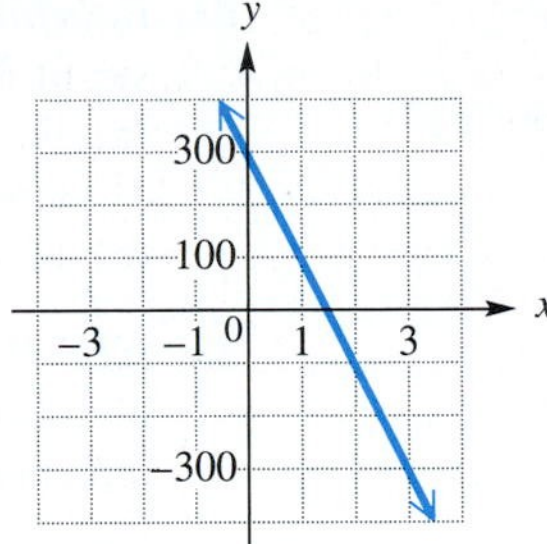

60.

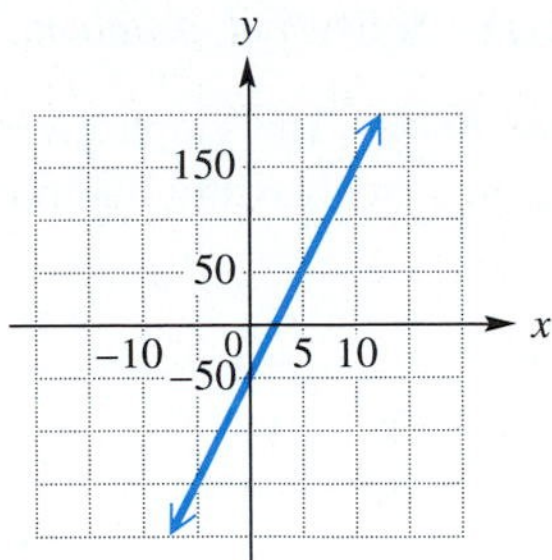

Concept Check *A linear function f has the ordered pairs listed in the table. Find the slope m of the graph of f, use the table to find the y-intercept of the line, and give the equation that defines f.*

61.

x	$f(x)$
-3	-10
-2	-6
-1	-2
0	2
1	6

62.

x	$f(x)$
-2	-11
-1	-8
0	-5
1	-2
2	1

63.

x	$f(x)$
$-.4$	-2.54
$-.2$	-2.82
0	-3.1
$.2$	-3.38
$.4$	-3.66

64.

x	$f(x)$
-100	-4
-50	-4
0	-4
50	-4
100	-4

Concept Check *Match each equation in Exercises 65–68 with the line in choices A–D that would most closely resemble its graph, where $k > 0$.*

65. $y = k$ **66.** $y = -k$ **67.** $x = k$ **68.** $x = -k$

A.

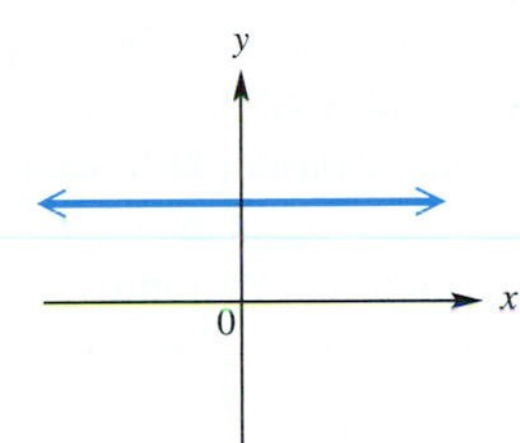

B.

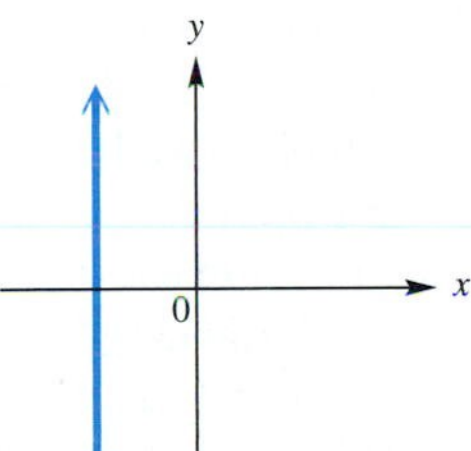

C.

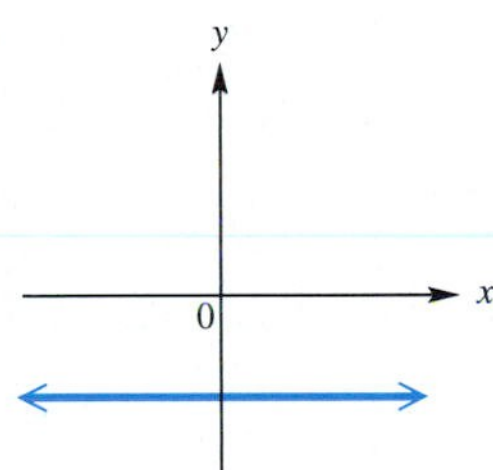

D.

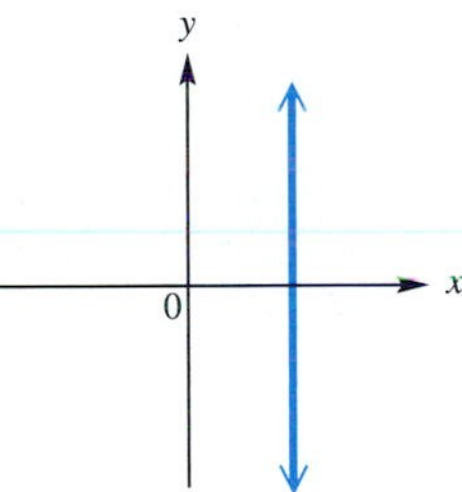

Sketch by hand the graph of the line passing through the given point and having the given slope. Label two points on the line.

69. Through $(-1, 3)$, $m = \frac{3}{2}$

70. Through $(-2, 8)$, $m = -1$

71. Through $(3, -4)$, $m = -\frac{1}{3}$

72. Through $(-2, -3)$, $m = -\frac{3}{4}$

73. Through $(-1, 4)$, $m = 0$

74. Through $\left(\frac{9}{4}, 2\right)$, undefined slope

75. Through $(0, -4)$, $m = \frac{3}{4}$

76. Through $(0, 5)$, $m = -2.5$

77. Through $(-3, 0)$, undefined slope

78. *Concept Check* Refer to Exercises 75 and 76.

(a) Give the equation of the line described in Exercise 75.

(b) Give the equation of the line described in Exercise 76.

(Modeling) *Solve each problem.*

79. ***Water Flow*** The graph gives the number of gallons of water in a small swimming pool after x hours.

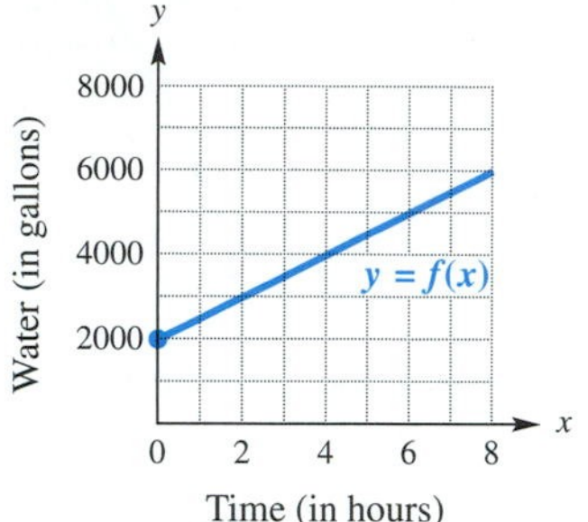

(a) Write a formula for $f(x)$.

(b) Interpret both the slope and the y-intercept.

(c) Use the graph to estimate how much water was in the pool after 7 hours. Verify your answer by evaluating $f(x)$.

80. ***Fuel Consumption*** The table shows the distance y traveled in miles by a car burning x gallons of gasoline.

x (gallons)	5	10	12	16
y (miles)	115	230	276	368

(a) Find values for a and b so that $f(x) = ax + b$ models the data exactly. That is, find values for a and b so that the graph of f passes through the data points in the table.

(b) Interpret the slope of the graph of f.

(c) How many miles could be driven using 20 gallons of gasoline?

81. ***Rainfall*** By noon, 3 inches of rain had fallen during a storm. Rain continued to fall at a rate of $\frac{1}{4}$ inch per hour.

(a) Find a formula for a linear function f that models the amount of rainfall x hours past noon.

(b) Find the total amount of rainfall by 2:30 P.M.

82. ***HIV Infections*** In 1998, there were 47 million people worldwide who had been living with HIV. At that time, the infection rate was 5.8 million people per year. (*Source:* United Nations AIDS and World Health Organization.)

(a) Write a formula for a linear function f that models the total number of people in millions who were living with HIV x years after 1998.

(b) Estimate the number of people who may have been living with HIV by the year 2006.

83. ***Birthrate*** In 1995, the number of births per 1000 people in the United States was 14.8 and was decreasing at .096 birth per 1000 people each year. (*Source:* U.S. Census Bureau.)

(a) Write a formula for a linear function f that models the birthrate in year x, where $x = 0$ corresponds to 1995, $x = 1$ to 1996, and so on.

(b) Estimate the birthrate in 2001, and compare it with the actual value of 14.1.

84. ***Birthrate*** In 1998, the number of births per 1000 people in the United States was 14.6 and was decreasing at .16 birth per 1000 people each year. (*Source:* U.S. Census Bureau.)

(a) Write a formula for a linear function f that models the birthrate in year x, where $x = 0$ corresponds to 1998, $x = 1$ to 1999, and so on.

(b) Estimate the birthrate in 2001, and compare it with the actual value of 14.1.

1.4 Equations of Lines and Linear Models

Point–Slope Form of the Equation of a Line ■ Standard Form of the Equation of a Line ■ Parallel and Perpendicular Lines ■ Linear Models and Regression

Point–Slope Form of the Equation of a Line

Figure 50 shows a line passing through the fixed point (x_1, y_1) with slope m. Let (x, y) be any other point on the line. Then, by the slope formula,

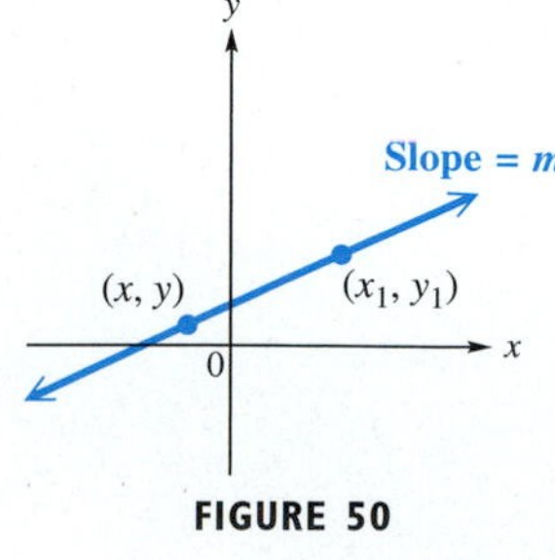

FIGURE 50

$$m = \frac{y - y_1}{x - x_1}$$

$$m(x - x_1) = y - y_1 \quad \text{Multiply both sides by } x - x_1.$$

$$y - y_1 = m(x - x_1). \quad \text{Rewrite.}$$

This result is called the **point–slope form** of the equation of a line.

Looking Ahead to Calculus

In calculus, it is often necessary to find the equation of a line, given its slope and a point on the line. The point–slope form is a valuable tool in these situations.

Point–Slope Form

The line with slope m passing through the point (x_1, y_1) has equation

$$y - y_1 = m(x - x_1).$$

EXAMPLE 1 Using Point–Slope Form

Find the slope–intercept form of the line passing through the two points shown in Figure 51.

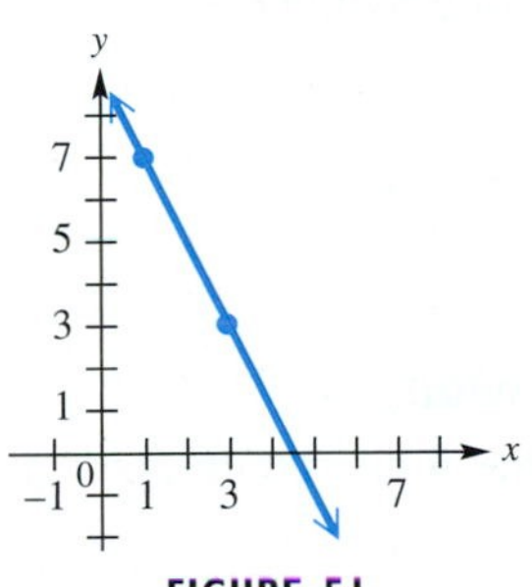

FIGURE 51

Solution The points are $(1, 7)$ and $(3, 3)$. Find the slope of the line.

Start with the x- and y-values of the same point.

$$m = \frac{7 - 3}{1 - 3} = \frac{4}{-2} = -2$$

Now, use either point—say, $(1, 7)$—with $m = -2$ in the point–slope form.

$$y - y_1 = m(x - x_1) \quad \text{Point–slope form}$$

$$y - 7 = -2(x - 1) \quad y_1 = 7,\ m = -2,\ x_1 = 1$$

Be careful with signs.

$$y - 7 = -2x + 2 \quad \text{Distributive property}$$

$$y = -2x + 9 \quad \text{Add 7 to obtain slope–intercept form.}$$

■

EXAMPLE 2 Using Point–Slope Form

The table in the margin lists points found on the line $y = mx + b$. Find the slope–intercept form of the equation of the line.

x	y
2	3
3	5
4	7
5	9
6	11
7	13
8	15

Solution Choose any two points to find the slope of the line.

$$m = \frac{11 - 3}{6 - 2} = \frac{8}{4} = 2 \quad \text{Use (2, 3) and (6, 11).}$$

Now, use a point on the line—say, $(2, 3)$—with $m = 2$ in the point–slope form.

$$y - y_1 = m(x - x_1) \quad \text{Point–slope form}$$

$$y - 3 = 2(x - 2) \quad y_1 = 3,\ m = 2,\ x_1 = 2$$

$$y - 3 = 2x - 4 \quad \text{Distributive property}$$

$$y = 2x - 1 \quad \text{Add 3 to obtain slope–intercept form.}$$

■

Standard Form of the Equation of a Line

Another form of the equation of a line is *standard form*.

Standard Form

A linear equation written in the form

$$Ax + By = C,$$

where A, B, and C are real numbers (A and B not both 0), is said to be in **standard form.**

One advantage of standard form is that it allows the quick calculation of both intercepts. For example, given $3x + 2y = 6$, we can find the x-intercept by letting $y = 0$ and the y-intercept by letting $x = 0$.

x-intercept: $3x + 2(0) = 6$	y-intercept: $3(0) + 2y = 6$
$3x = 6$	$2y = 6$
$x = 2$	$y = 3$

This information is useful when sketching the graph of the line by hand.

EXAMPLE 3 **Graphing an Equation in Standard Form**

Graph $3x + 2y = 6$.

Analytic Solution

As just calculated, the points $(2, 0)$ and $(0, 3)$ correspond to the x- and y-intercepts respectively. Plot these two points and connect them with a straight line. See Figure 52.

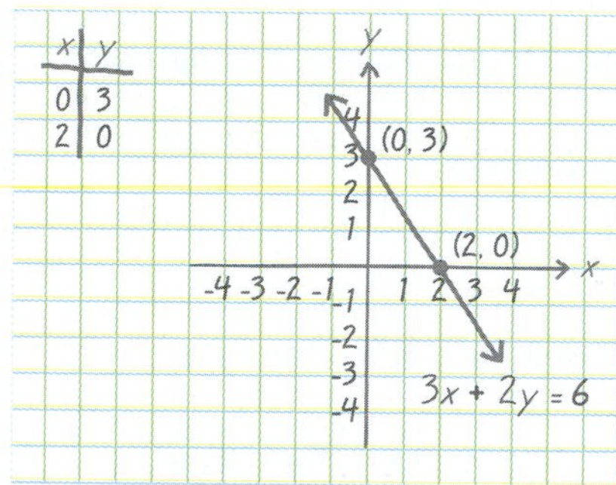

FIGURE 52

Graphing Calculator Solution

Solve the equation for y so that it can be entered into a calculator.

$$3x + 2y = 6 \quad \text{Given equation}$$
$$2y = -3x + 6 \quad \text{Subtract } 3x.$$
$$y = -1.5x + 3 \quad \text{Divide by 2.}$$

The desired graph is shown in Figure 53.

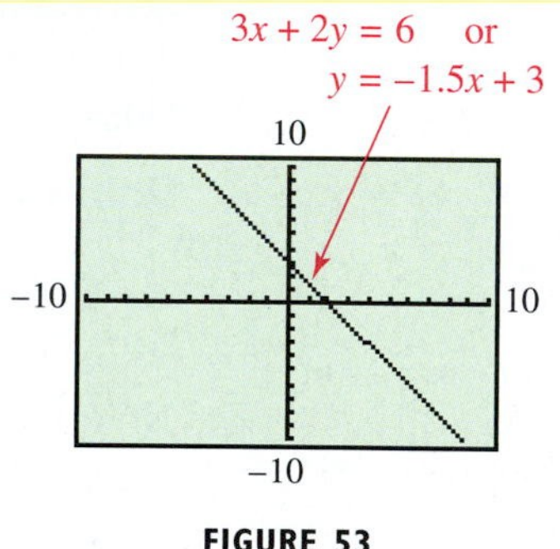

FIGURE 53

NOTE Because of the usefulness of the slope–intercept form when graphing with a graphing calculator, we emphasize that form in much of our work.

Parallel and Perpendicular Lines

FOR DISCUSSION

In the standard viewing window of your calculator, graph the following four lines:

$$y_1 = 2x - 6, \quad y_2 = 2x - 2, \quad y_3 = 2x, \quad y_4 = 2x + 4.$$

What is the slope of each line? What geometric term seems to describe the set of lines?

Two lines in a plane are *parallel* if they do not intersect. Although the exercise in the "For Discussion" box does not actually prove the result that follows, it provides visual support.

Parallel Lines

Two distinct nonvertical lines are **parallel** if and only if they have the same slope.

EXAMPLE 4 Using the Slope Relationship for Parallel Lines

Find the equation of the line that passes through the point $(3, 5)$ and is parallel to the line with equation $2x + 5y = 4$. Graph both lines in the standard viewing window.

Solution Since the point $(3, 5)$ is on the line, we need only find the slope by writing the given equation in slope–intercept form. (That is, solve for y.)

$$2x + 5y = 4 \qquad \text{Given equation}$$

$$y = -\frac{2}{5}x + \frac{4}{5} \qquad \text{Subtract } 2x\text{; divide by 5.}$$

The slope is $-\frac{2}{5}$. Since the lines are parallel, $-\frac{2}{5}$ is also the slope of the line whose equation we must find. Substitute into the point–slope form.

$$y - y_1 = m(x - x_1) \qquad \text{Point–slope form}$$

$$y - 5 = -\frac{2}{5}(x - 3) \qquad y_1 = 5,\ m = -\tfrac{2}{5},\ x_1 = 3$$

$$5(y - 5) = -2(x - 3) \qquad \text{Multiply by 5.}$$

$$5y - 25 = -2x + 6 \qquad \text{Distributive property}$$

Be careful with signs.

$$5y = -2x + 31 \qquad \text{Add 25.}$$

$$y = -\frac{2}{5}x + \frac{31}{5} \qquad \text{Divide by 5 to obtain slope–intercept form of the desired line.}$$

An alternative method for finding this equation involves using the slope $-\frac{2}{5}$ and the point $(3, 5)$ in the slope–intercept form to find b.

$$y = mx + b \qquad \text{Slope–intercept form}$$

$$5 = -\frac{2}{5}(3) + b \qquad y = 5,\ m = -\tfrac{2}{5},\ x = 3$$

$$5 = -\frac{6}{5} + b \qquad \text{Multiply.}$$

$$b = \frac{31}{5} \qquad \text{Solve for } b.$$

Therefore, the equation is $y = -\frac{2}{5}x + \frac{31}{5}$, which agrees with our earlier result. Figure 54 provides support, as the lines *appear* to be parallel. ■

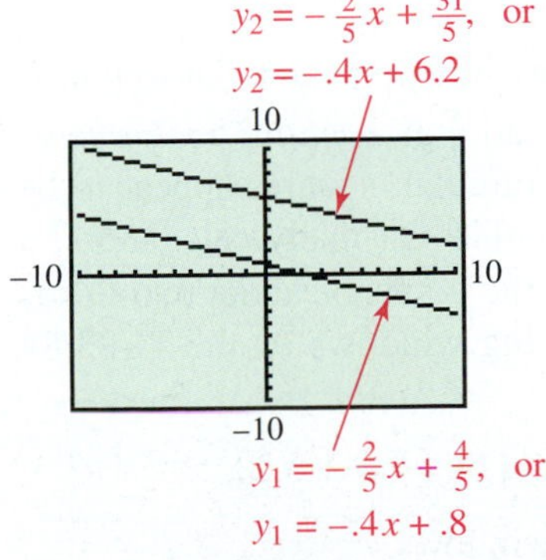

FIGURE 54

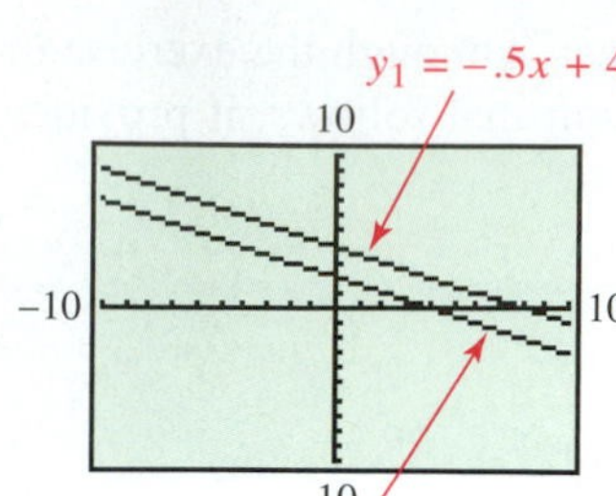

FIGURE 55

When using a graphing calculator, be aware that visual support (as seen in Example 4) does not necessarily prove the result. For example, Figure 55 shows the graphs of $y_1 = -.5x + 4$ and $y_2 = -.5001x + 2$. Although they *appear* to be parallel by visual inspection, they are *not* parallel, because the slope of y_1 is $-.5$ and the slope of y_2 is $-.5001$.

A CAUTION AGAINST RELYING TOO HEAVILY ON TECHNOLOGY

Analysis of a calculator graph is often not sufficient to draw correct conclusions. While this technology is incredibly powerful and pedagogically useful, we cannot rely on it alone in our work. We must understand the basic concepts of algebra as well.

FOR DISCUSSION

Using a "square" viewing window, such as $[-9.4, 9.4]$ by $[-6.2, 6.2]$, graph each pair of lines. Graph each group separately.

I	II	III	IV
$y_1 = 4x + 1$	$y_1 = -\frac{2}{3}x + 3$	$y_1 = 6x - 3$	$y_1 = \frac{13}{7}x - 3$
$y_2 = -\frac{1}{4}x + 3$	$y_2 = \frac{3}{2}x - 4$	$y_2 = -\frac{1}{6}x + 4$	$y_2 = -\frac{7}{13}x + 4$

What geometric term applies to each pair of lines? What is the product of the slopes for each pair of lines?

As in the earlier "For Discussion" box, we have not proved the result that follows; rather, we have provided visual support for it.

Perpendicular Lines

Two lines, neither of which is vertical, are **perpendicular** if and only if their slopes have product -1.

For example, if the slope of a line is $-\frac{3}{4}$, the slope of any line perpendicular to it is $\frac{4}{3}$, since $-\frac{3}{4}\left(\frac{4}{3}\right) = -1$. We refer to numbers such as $-\frac{3}{4}$ and $\frac{4}{3}$ as *negative reciprocals.*

In a **square viewing window,** circles appear to be circular, squares appear to be square, and perpendicular lines appear to be perpendicular. On many calculators, a square viewing window requires that the distance along the y-axis be about two-thirds the distance along the x-axis. Examples of square viewing windows on the TI-83/84 Plus calculators are

$$[-4.7, 4.7] \text{ by } [-3.1, 3.1] \quad \text{and} \quad [-9.4, 9.4] \text{ by } [-6.2, 6.2].$$

Figure 56 illustrates the importance of square viewing windows.

TECHNOLOGY NOTE

Many calculators can set a square viewing window automatically. Check the graphing calculator manual that accompanies this text or your owner's manual, or look under the ZOOM menu.

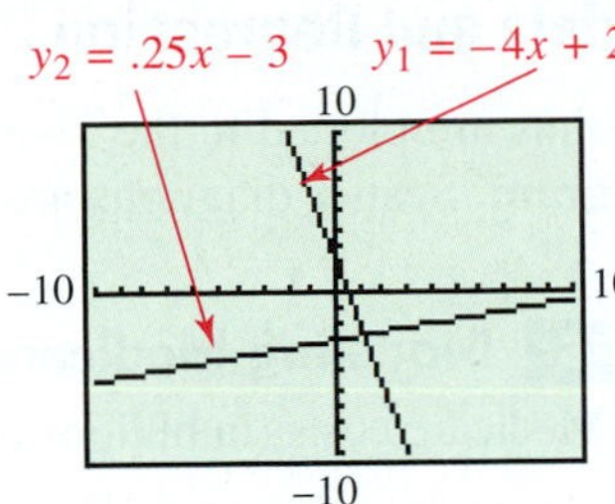

Although the graphs of these lines are perpendicular, they do not appear to be when graphed in a standard viewing window.

(a)

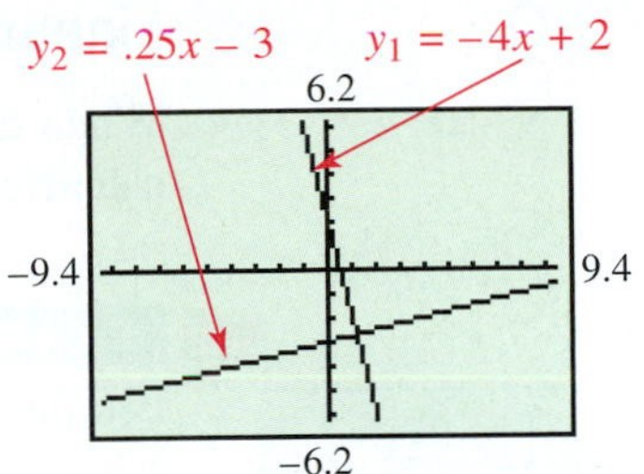

Visual support for perpendicularity is more obvious using a square viewing window.

(b)

FIGURE 56

GCM **EXAMPLE 5** **Using the Slope Relationship for Perpendicular Lines**

Find the equation of the line that passes through the point $(3, 5)$ and is perpendicular to the line with equation $2x + 5y = 4$. Graph both lines in a square viewing window.

Solution In Example 4, we found that the slope of the given line is $-\frac{2}{5}$, so the slope of any line perpendicular to it is $\frac{5}{2}$. We can use either method shown there to find the equation of the line. The second method yields the following:

$$y = mx + b \qquad \text{Slope–intercept form}$$

$$5 = \frac{5}{2}(3) + b \qquad y = 5,\ m = \tfrac{5}{2},\ x = 3$$

$$5 = \frac{15}{2} + b \qquad \text{Multiply.}$$

$$b = -\frac{5}{2}. \qquad \text{Solve for } b.$$

Thus, the equation is $y = \frac{5}{2}x - \frac{5}{2}$.

Graphing both equations in slope–intercept form and using a square viewing window provides visual support for our answer. See Figure 57.

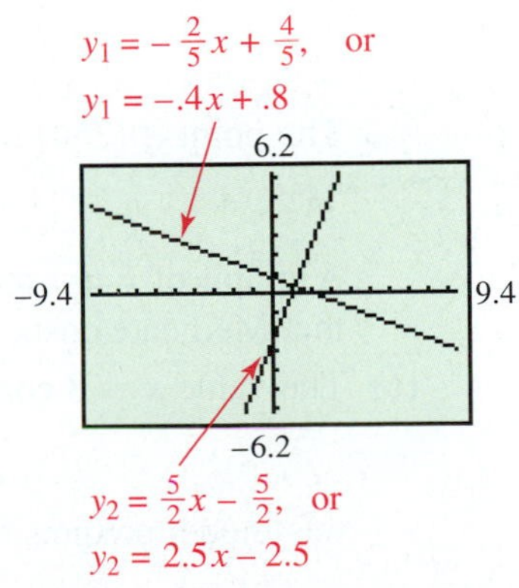

FIGURE 57

Linear Models and Regression

When data points are plotted in the xy-plane, the resulting graph is sometimes called a **scatter diagram.** Scatter diagrams are often helpful for analyzing trends in data.

GCM **EXAMPLE 6** **Modeling Medicare Costs with a Linear Function**

Estimates for Medicare costs (in billions of dollars) are shown in the table in the margin.

x (year)	y (cost)
2002	264
2003	281
2004	299
2005	318
2006	336
2007	354

Source: U.S. Center for Medicare and Medicaid Services.

(a) Make a scatter diagram of the data. Let $x = 0$ correspond to 2002, $x = 1$ to 2003, and so on. What type of function might model the data?

(b) Find a linear function f that models the data. Graph f and the data in the same viewing window. Interpret the slope m.

(c) Use $f(x)$ to predict Medicare costs in 2010.

Solution

(a) Since $x = 0$ corresponds to 2002, $x = 1$ corresponds to 2003, and so on, the data points can be expressed as the ordered pairs

$$(0, 264), \quad (1, 281), \quad (2, 299), \quad (3, 318), \quad (4, 336), \quad \text{and} \quad (5, 354).$$

Scatter diagrams are shown in Figure 58. The data appear to be approximately linear, so a linear function might be appropriate.

TECHNOLOGY NOTE

To make a scatter diagram with a graphing calculator, you may need to use the *list* feature by entering the x-values in list L_1 and the y-values in list L_2. See Figure 59.

L1	L2	L3 1
0	264	
1	281	
2	299	
3	318	
4	336	
5	354	

L1 ={0,1,2,3,4,5}

FIGURE 59

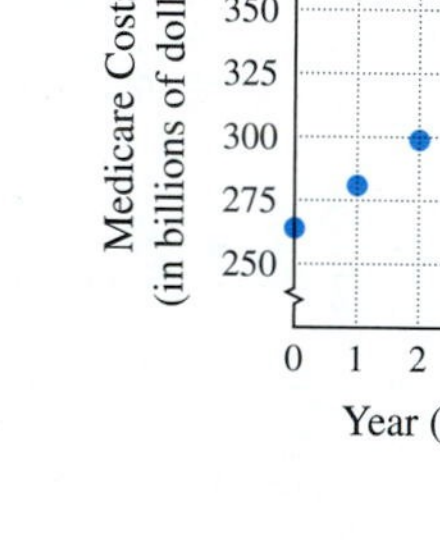

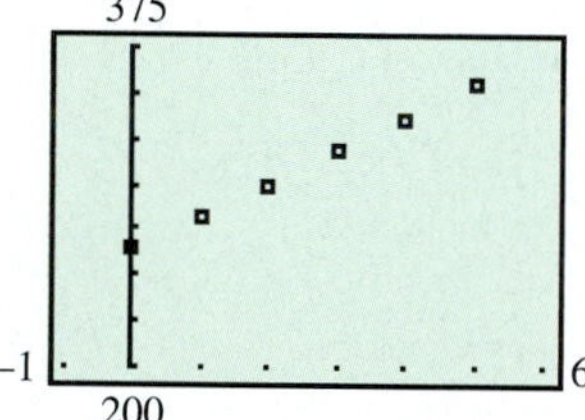

FIGURE 58

(b) We start by choosing two data points that the line should pass through. For example, if we use (0, 264) and (3, 318), then the slope of the line is

$$m = \frac{318 - 264}{3 - 0} = 18.$$

Start with the x- and y-values of the same point.

The point (0, 264) indicates that the y-intercept b is 264. Thus,

$$f(x) = 18x + 264.$$

A graph of f and the data are shown in Figure 60. The slope $m = 18$ indicates that Medicare costs might increase, on average, by $18 billion per year.

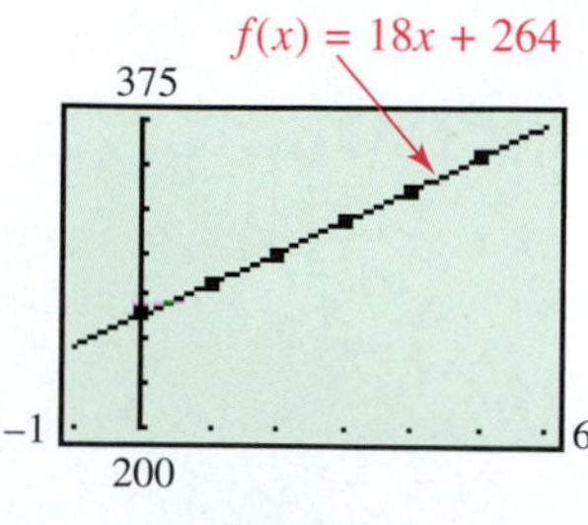

FIGURE 60

(c) The value $x = 8$ corresponds to the year 2010. Since

$$f(8) = 18(8) + 264 = 408,$$

this model predicts that Medicare costs will reach $408 billion in 2010. ■

NOTE The formula $f(x) = 18x + 264$ found in Example 6 is not unique. If two different points are chosen, a different formula for $f(x)$ will result. However, all such formulas for $f(x)$ should be in approximate agreement.

The method for finding the model in Example 6 used algebraic concepts that do not always give a unique line. Graphing calculators are capable of finding the line of "best fit," called the **least-squares regression line,** by using a technique from statistics known as **least-squares regression.**

GCM **EXAMPLE 7** **Finding the Least-Squares Regression Line**

Use a graphing calculator to find the least-squares regression line that models the Medicare costs presented in Example 6. Graph the data and the line in the same viewing window.

TECHNOLOGY NOTE

To find the equation of a least-squares regression line, refer to the graphing calculator manual that accompanies this text or your owner's manual.

Solution Figure 61 shows how a TI-83/84 Plus graphing calculator finds the regression line for the data in Example 6. In Figure 61(a), the years are entered into list L_1, and Medicare costs are entered into list L_2. In Figure 61(b), the formula for the regression line is calculated to be

$$y \approx 18.11x + 263.38.$$

Notice that this equation is not exactly the same as the one found for $f(x)$ in Example 6. In Figure 61(c), both the data and the regression line are graphed.

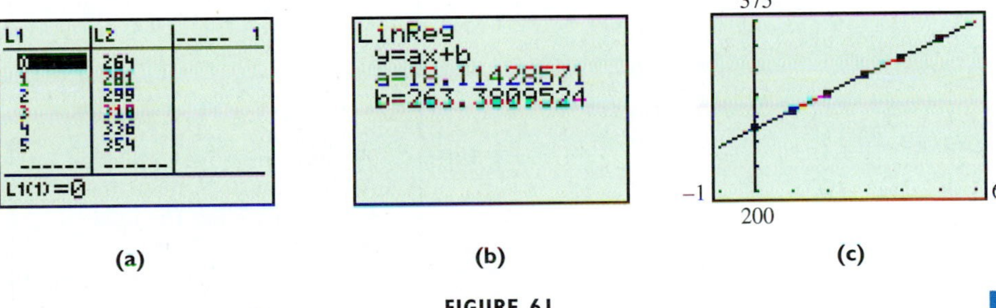

(a) (b) (c)

FIGURE 61

TECHNOLOGY NOTE

By choosing the DiagnosticOn option on the TI-83/84 Plus, the correlation coefficient r and its square r^2 are displayed. See Figure 62, and compare with Figure 61(b).

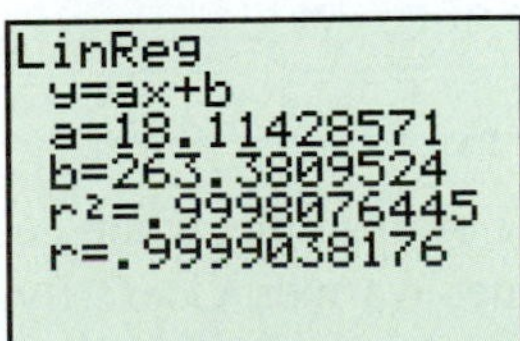

FIGURE 62

Once an equation for the least-squares regression line has been found, it is reasonable to ask, "Just how good is this line for predictive purposes?" If the line *fits* the observed data points, then future pairs of points might be expected to do so also.

One common measure of the strength of the linear relationship in a data set is called the **correlation coefficient,** denoted r, where $-1 \le r \le 1$. When r is positive and near 1, low x-values correspond to low y-values and high x-values correspond to high y-values. For example, there is a positive correlation between years of education x and income y. More years of education correlate with higher income. When r is near -1, the reverse is true: Low x-values correspond to high y-values and high x-values correspond to low y-values. For example, as latitude increases (moving toward either pole), the average yearly temperature decreases. Therefore, there will be a negative correlation between latitude and average annual temperature. If $r \approx 0$, then there is little or no correlation between the data points. In this case, a least-squares regression line does not provide a suitable model.

A summary of these concepts is shown in the table on the next page.

Correlation Coefficient r $(-1 \leq r \leq 1)$

Value of r	Comments	Sample Scatter Diagram
$r = 1$	There is an exact linear fit. The line passes through all data points and has a positive slope.	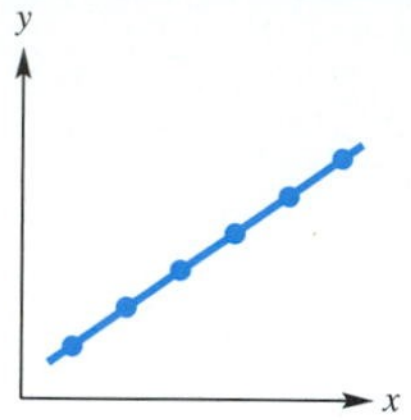
$r = -1$	There is an exact linear fit. The line passes through all data points and has a negative slope.	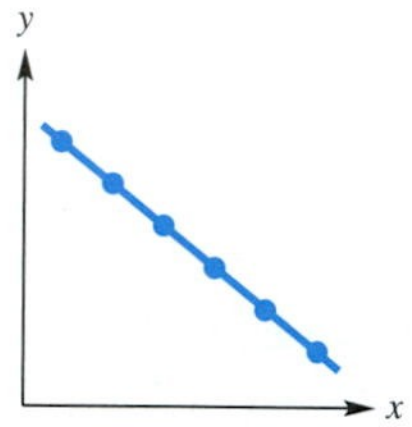
$0 < r < 1$	There is a positive correlation. As the x-values increase, so do the y-values. The fit is not exact.	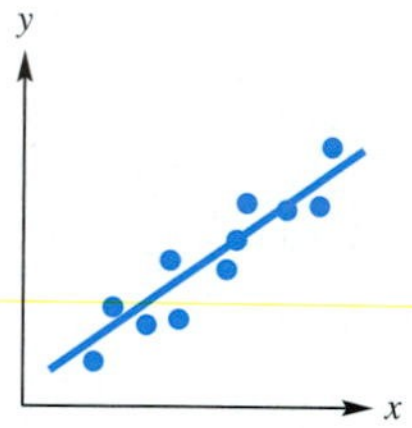
$-1 < r < 0$	There is a negative correlation. As the x-values increase, the y-values decrease. The fit is not exact.	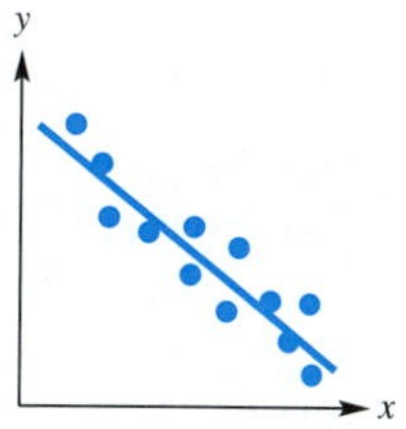
$r = 0$	There is no correlation. The data have no tendency toward being linear. A regression line predicts poorly.	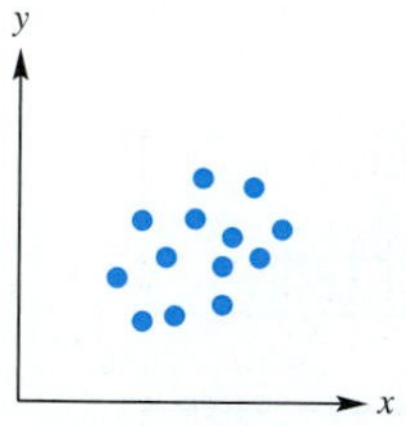

There is a difference between correlation and causation. For example, when geese begin to fly north, summer is coming and the weather becomes warmer. Geese flying north correlate with warmer weather. However, geese flying north clearly do not *cause* warmer weather. Correlation does not always indicate causation.

EXAMPLE 8 **Predicting Airline Passenger Growth**

The table lists the estimated numbers in millions of airline passengers at some of the fastest-growing airports in 1992 and 2005.

Airline Passengers (in millions)

Airport	1992	2005
Harrisburg International	.7	1.4
Dayton International	1.1	2.4
Austin Robert Mueller	2.2	4.7
Milwaukee General Mitchell	2.2	4.4
Sacramento Metropolitan	2.6	5.0
Fort Lauderdale–Hollywood	4.1	8.1
Washington Dulles	5.3	10.9
Greater Cincinnati	5.8	12.3

Source: Federal Aviation Administration.

(a) Make a scatter diagram of the data, using the 1992 data for x-values and the corresponding 2005 data for y-values. Predict whether the correlation coefficient will be positive or negative.

(b) Use a calculator to find the least-squares regression line.

(c) Raleigh–Durham International Airport had 4.9 million passengers in 1992. Assuming that Raleigh–Durham International follows the same trend in growth as the other airports, estimate the number of passengers it may have in 2005. Compare this result with the Federal Aviation Administration (FAA) estimate of 10.3 million passengers.

14

0 6

0

(a)

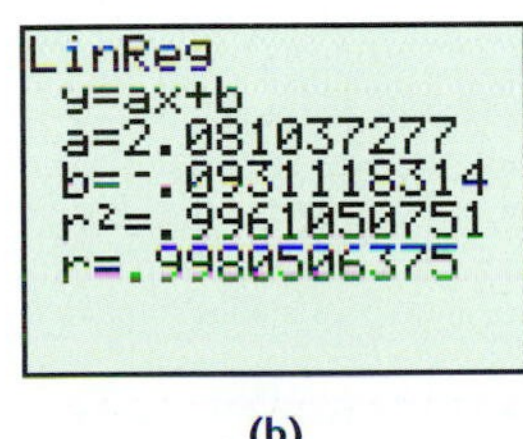

(b)

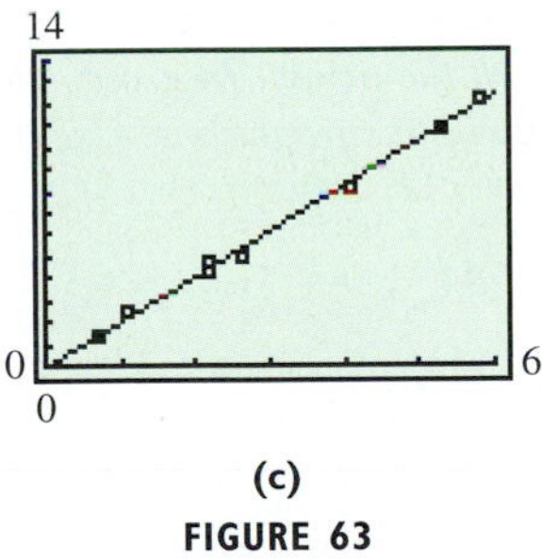

(c)

FIGURE 63

Solution

(a) Plot the data as shown in Figure 63(a). Since increasing x-values correspond to increasing y-values, the correlation coefficient will be positive.

(b) The equation of the regression line is shown in Figure 63(b). The line

$$y \approx 2.081x - .0931$$

and the data are graphed together in Figure 63(c). Note that $r \approx .998$, which is positive as predicted.

(c) When $x = 4.9$, $y = 2.081(4.9) - .0931 \approx 10.1$ million passengers. This is quite close to the FAA prediction of 10.3 million. ■

1.4 *Exercises*

Write the slope–intercept form of the line that passes through the given point with slope m.

1. Through $(1, 3)$, $m = -2$
2. Through $(2, 4)$, $m = -1$
3. Through $(-5, 4)$, $m = 1.5$
4. Through $(-4, 3)$, $m = .75$
5. Through $(-8, 1)$, $m = -.5$
6. Through $(-5, 9)$, $m = -.75$
7. Through $\left(\frac{1}{2}, -4\right)$, $m = 2$
8. Through $\left(5, -\frac{1}{3}\right)$, $m = 3$
9. Through $\left(\frac{1}{4}, \frac{2}{3}\right)$, $m = \frac{1}{2}$

10. Why is it not possible to write a slope–intercept form of the equation of the line through the points $(12, 6)$ and $(12, -2)$?

Concept Check *Find the slope–intercept form of the line shown in each graph.*

11.
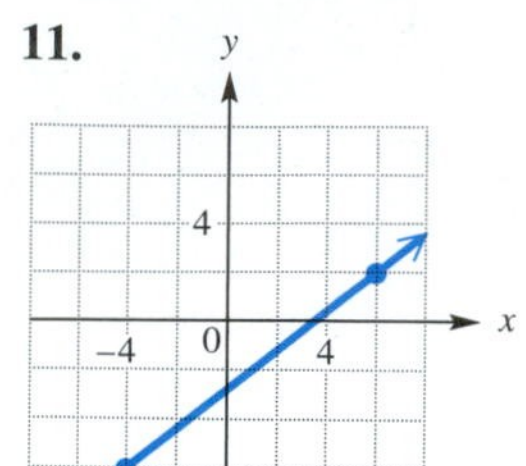

12.
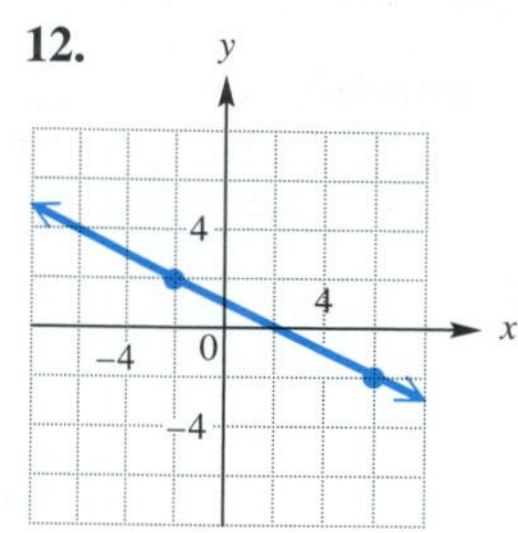

13.
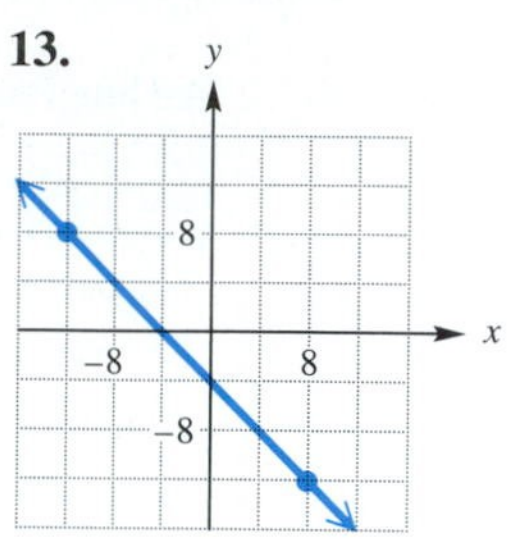

14.
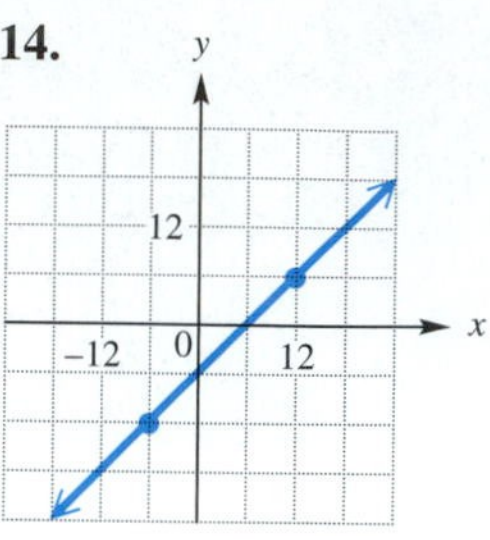

Find the slope–intercept form of the line satisfying the given conditions.

15. Through $(4, 8)$ and $(0, 4)$

16. Through $(3, 6)$ and $(0, 10)$

17. Through $(3, -8)$ and $(5, -3)$

18. Through $(-5, 4)$ and $(-3, 2)$

19. Through $(2, 3.5)$ and $(6, -2.5)$

20. Through $(-1, 6.25)$ and $(2, -4.25)$

21. Through $(0, 5)$ and $(10, 0)$

22. Through $(0, -8)$ and $(4, 0)$

23.

x	y
-7	-44
-6	-36
-5	-28
-4	-20

24.

x	y
-2.4	5.2
1.3	-24.4
1.75	-28
2.98	-37.84

Graph each line by hand, finding intercepts to determine two points on the line.

25. $x - y = 4$

26. $x + y = 4$

27. $3x - y = 6$

28. $2x - 3y = 6$

29. $2x + 5y = 10$

30. $4x - 3y = 9$

A line having an equation of the form $y = kx$, where k is a real number, $k \neq 0$, will always pass through the origin. To graph such an equation by hand, we must determine a second point and then join the origin and that second point with a straight line. Use this method to graph each line.

31. $y = 3x$

32. $y = -2x$

33. $y = -.75x$

34. $y = 1.5x$

Write each equation in the form $y = mx + b$. Then, using your calculator, graph the line in the window indicated.

35. $5x + 3y = 15$
$[-10, 10]$ by $[-10, 10]$

36. $6x + 5y = 9$
$[-10, 10]$ by $[-10, 10]$

37. $-2x + 7y = 4$
$[-5, 5]$ by $[-5, 5]$

38. $-.23x - .46y = .82$
$[-5, 5]$ by $[-5, 5]$

39. $1.2x + 1.6y = 5.0$
$[-6, 6]$ by $[-4, 4]$

40. $2y - 5x = 0$
$[-10, 10]$ by $[-10, 10]$

Find the equation of the line satisfying the given conditions, giving it in slope–intercept form if possible.

41. Through $(-1, 4)$, parallel to $x + 3y = 5$

42. Through $(3, -2)$, parallel to $2x - y = 5$

43. Through $(1, 6)$, perpendicular to $3x + 5y = 1$

44. Through $(-2, 0)$, perpendicular to $8x - 3y = 7$

45. Through $(-5, 7)$, perpendicular to $y = -2$

46. Through $(1, -4)$, perpendicular to $x = 4$

47. Through $(-5, 8)$, parallel to $y = -.2x + 6$

48. Through $(-4, -7)$, parallel to $x + y = 5$

49. Through the origin, perpendicular to $2x + y = 6$

50. Through the origin, parallel to $y = -3.5x + 7.4$

51. Refer to the given figure and complete parts (a)–(h) to prove that if two lines are perpendicular and neither line is parallel to an axis, then the lines have slopes whose product is -1.

(a) In triangle OPQ, angle POQ is a right angle if and only if

$$[d(O,P)]^2 + [d(O,Q)]^2 = [d(P,Q)]^2.$$

What theorem from geometry is this?

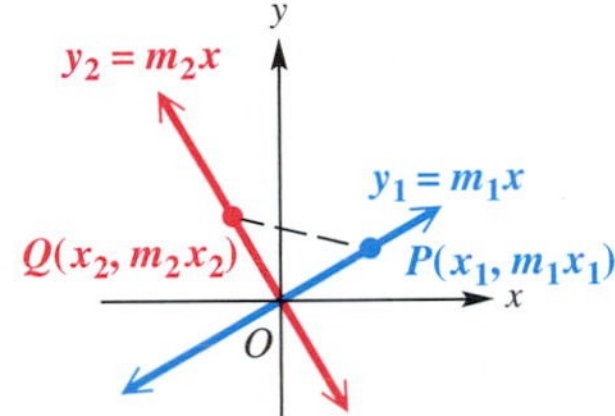

(b) Find an expression for the distance $d(O,P)$.

(c) Find an expression for the distance $d(O,Q)$.

(d) Find an expression for the distance $d(P,Q)$.

(e) Use your results from parts (b)–(d) and substitute into the equation in part (a). Simplify to show that the resulting equation is $-2m_1m_2x_1x_2 - 2x_1x_2 = 0$.

(f) Factor $-2x_1x_2$ from the final form of the equation in part (e).

(g) Use the zero-product property from intermediate algebra to solve the equation in part (f) to show that $m_1m_2 = -1$.

(h) State your conclusion on the basis of parts (a)–(g).

52. *Concept Check* The following tables show ordered pairs for linear functions defined in Y_1 and Y_2. Determine whether the lines are parallel, perpendicular, or neither parallel nor perpendicular.

(a)

X	Y1	Y2
0	-3	4
1	1	3.75
2	5	3.5
3	9	3.25
4	13	3
5	17	2.75
6	21	2.5

X=0

(b)

X	Y1	Y2
0	-3	5
1	2	5.2
2	7	5.4
3	12	5.6
4	17	5.8
5	22	6
6	27	6.2

X=0

(c)

X	Y1	Y2
0	-3	12
1	2	17
2	7	22
3	12	27
4	17	32
5	22	37
6	27	42

X=0

(d)

X	Y1	Y2
0	2	-2
1	-2	2
2	-6	6
3	-10	10
4	-14	14
5	-18	18
6	-22	22

X=0

Relating Concepts

For individual or group investigation (Exercises 53–62)

The table at the side shows ordered pairs that lie on the graph of a linear function. ***Work Exercises 53–62 in order,*** *to investigate connections between the slope formula, distance formula, midpoint formula, and linear functions.*

x	y
0	-6
1	-3
2	0
3	3
4	6
5	9
6	12

53. Use the first two points in the table to find the slope of the line.

54. Use the second and third points in the table to find the slope of the line.

55. Make a conjecture by filling in the blank: If we use any two points on a line to find its slope, we find that the slope is ________ in all cases.

56. Use the distance formula to find the distance between the first two points in the table.

57. Use the distance formula to find the distance between the second and fourth points in the table.

58. Use the distance formula to find the distance between the first and fourth points in the table.

59. Add the results in Exercises 56 and 57, and compare the sum with the answer you found in Exercise 58. What do you notice?

60. Fill in the blanks, basing your answers on your observations in Exercises 56–59. If points A, B, and C lie on a line in that order, then the distance between A and B added to the distance between ________ and ________ is equal to the distance between ________ and ________.

61. Use the midpoint formula to find the midpoint of the segment joining $(0, -6)$ and $(6, 12)$. Compare your answer with the middle entry in the table. What do you notice?

62. If the table were set up to show an x-value of 4.5, what would be the corresponding y-value?

(Modeling) Solve each problem.

63. ***Distance*** A person is riding a bicycle along a straight highway. The graph shows the rider's distance y in miles from an interstate highway after x hours.

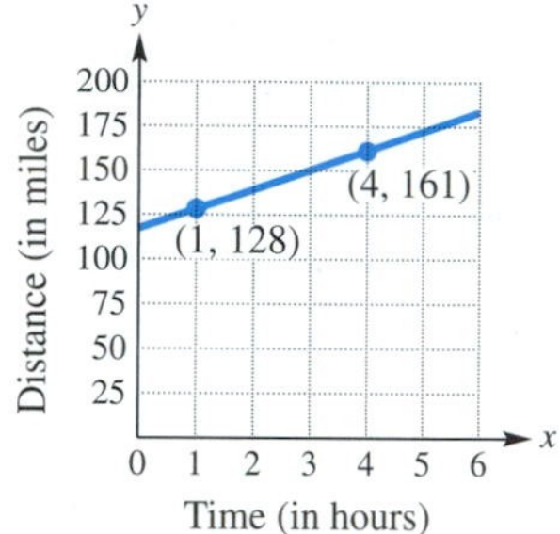

(a) Find the slope–intercept form of the line.
(b) How fast is the biker traveling?
(c) How far was the biker from the interstate highway initially?
(d) How far was the biker from the interstate highway after 1 hour and 15 minutes?

64. ***Water in a Tank*** The graph shows the amount of water y in a 100-gallon tank after x minutes have elapsed.

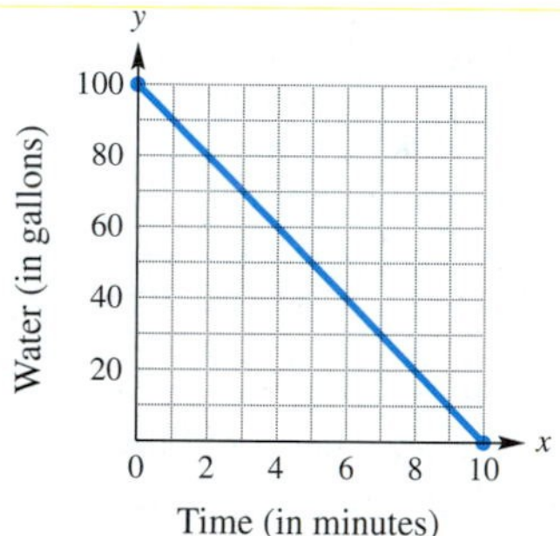

(a) Is water entering or leaving the tank? How much water is in the tank after 3 minutes?
(b) Find both the x- and y-intercepts. Interpret their meanings.
(c) Find the slope–intercept form of the equation of the line. Interpret the slope.
(d) Estimate the x-coordinate of the point $(x, 50)$ that lies on the line.

65. ***Farm Pollution*** In 1979, the number of farm pollution incidents reported in England and Wales was 1480. This number increased roughly at a rate of 280 per year after 1979. (*Source:* Mason, C., *Biology of Freshwater Pollution,* John Wiley and Sons, 1991.)

(a) Find an equation of a line $y = mx + b$ that models the information, where y represents the number of farm pollution incidents during the year x. Let $x = 0$ correspond to 1979, $x = 1$ to 1980, and so on.
(b) Estimate the number of incidents in 1988.

66. ***Average Wages*** The average hourly wage (adjusted to 1982 dollars) was \$7.66 in 1990 and \$8.24 in 2002. (*Source:* U.S. Census Bureau.)

(a) Find an equation of a line that passes through the points $(1990, 7.66)$ and $(2002, 8.24)$.
(b) Interpret the slope.
(c) Use the equation from part (a) to approximate the hourly wage in 1995. Compare it with the actual value of \$7.53.

67. ***Temperature Scales*** The table shows equivalent temperatures in degrees Celsius and degrees Fahrenheit.

°F	−40	32	59	95	212
°C	−40	0	15	35	100

(a) Plot the data by having the x-axis correspond to Fahrenheit temperature and the y-axis to Celsius temperature. What type of relation exists between the data?
(b) Find a function C that uses the Fahrenheit temperature x to calculate the corresponding Celsius temperature. Interpret the slope.
(c) What is a temperature of 83°F in degrees Celsius?

68. ***Population*** Asian-American populations (in millions) are shown in the table.

Year	1996	1998	2000	2002
Population (in millions)	9.7	10.5	11.2	11.6

Source: U.S. Census Bureau.

(a) Use the points $(1996, 9.7)$ and $(2002, 11.6)$ to find the point–slope form of a line that models the data. Let $(x_1, y_1) = (1996, 9.7)$.
(b) Estimate the Asian-American population in 2006 to the nearest tenth of a million.

69. ***Civilian Labor*** The table lists the percent of the female population that worked in the civilian labor force from 1980 to 2002.

Year	1980	1990	1995	2000	2002
Percent (%)	51.5	57.5	58.9	59.9	59.6

Source: U.S. Census Bureau.

(a) Find the point–slope form of the line that passes through the points (1980, 51.5) and (2002, 59.6). Let $(x_1, y_1) = (1980, 51.5)$.

(b) Use this equation to estimate the percents in 1990, 1995, and 2000 to the nearest tenth of a percent. Compare these estimates with the actual values shown in the table.

70. ***Civilian Labor*** The table lists the percent of the male population that worked in the civilian labor force from 1980 to 2002.

Year	1980	1990	1995	2000	2002
Percent (%)	77.4	76.4	75.0	74.8	74.1

Source: U.S. Census Bureau.

(a) Find the point–slope form of the line that passes through (1980, 77.4) and (2000, 74.8).

(b) Find the point–slope form of the line that passes through (1990, 76.4) and (2002, 74.1).

(c) Use the equations from parts (a) and (b) to predict the percent for 2003.

(d) Discuss how well each line models the data.

71. ***Tuition and Fees*** The table lists the average tuition and fees at private colleges and universities for selected years.

Year	1985	1990	1995	1999	2004
Tuition and Fees (in dollars)	5418	9391	12,432	15,380	20,082

Source: The College Board.

(a) Find the equation of the least-squares regression line that models the data.

(b) Graph the data and the regression line in the same viewing window.

(c) Estimate tuition and fees in 1992, and compare it with the actual value of $10,498.

72. ***Tuition and Fees*** The table lists the average tuition and fees at public colleges and universities for selected years.

Year	1985	1990	1995	1999	2004
Tuition and Fees (in dollars)	1242	1809	2860	3356	5132

Source: The College Board.

(a) Find the equation of the least-squares regression line that models the data.

(b) Graph the data and the regression line in the same viewing window.

(c) Estimate tuition and fees in 2006.

(d) Use the model to predict tuition and fees in 2010.

73. ***Distant Galaxies*** In the late 1920s, the famous observational astronomer Edwin P. Hubble (1889–1953) determined the distances to several galaxies and the velocities at which they were receding from Earth. Four galaxies with their distances in light-years and velocities in miles per second are listed in the table.

Galaxy	**Distance**	**Velocity**
Virgo	50	990
Ursa Minor	650	9,300
Corona Borealis	950	15,000
Bootes	1,700	25,000

Source: Sharov, A., and I. Novikov, *Edwin Hubble, the Discoverer of the Big Bang Universe,* Cambridge University Press, 1993.

(a) Let x represent distance and y represent velocity. Find the equation of the least-squares regression line that models the data.

(b) If the galaxy Hydra is receding at a speed of 37,000 miles per second, estimate its distance from Earth.

74. ***Heights and Weights of Men*** A sample of 10 adult men gave the following data on their heights and weights:

Height, x (in inches)	**Weight, y (in pounds)**
62	120
62	140
63	130
65	150
66	142
67	130
68	135
68	175
70	149
72	168

(a) Find the equation of the least-squares regression line.

(b) Using the result of part (a), predict the weight of a man whose height is 60 inches.

(c) What would be the predicted weight of a man whose height is 70 inches?

(d) Find the correlation coefficient. Interpret your answer.

(Modeling) *In Exercises 75 and 76, obtain the least-squares regression line and the correlation coefficient. Make a statement about the correlation.*

75. ***Gestation Period and Life Span of Animals***

Animal	Average Gestation or Incubation Period, x (in days)	Record Life Span, y (in years)
Cat	63	26
Dog	63	24
Duck	28	15
Elephant	624	71
Goat	151	17
Guinea pig	68	6
Hippopotamus	240	49
Horse	336	50
Lion	108	29
Parakeet	18	12
Pig	115	22
Rabbit	31	15
Sheep	151	16

Source: 1995 *Information Please Almanac.*

76. ***City Sizes in the World (Population and Area)***

Rank	City	Population, x (in thousands)	Area, y (in square miles)
1	Tokyo–Yokohama, Japan	34,450	1,089
2	Mexico City, Mexico	18,066	522
3	New York, United States	17,846	1,274
4	Saõ Paulo, Brazil	17,099	451
5	Mumbai (Bombay), India	16,086	95
6	Kolkata (Calcutta), India	13,058	209
7	Shanghai, China	12,887	2,232
8	Buenos Aires, Argentina	12,583	535
9	Delhi, India	12,441	535
10	Los Angeles, United States	11,814	472

Source: *World Almanac and Book of Facts*, 2005.

Reviewing Basic Concepts (Sections 1.3 and 1.4)

1. Write the formula for a linear function f with slope 1.4 and y-intercept -3.1. Find $f(1.3)$.

2. Graph $f(x) = -2x + 1$ by hand. State the x-intercept, y-intercept, domain, range, and slope.

3. Find the slope of the line passing through $(-2,4)$ and $(5,6)$.

4. Give the equations of a vertical line and a horizontal line passing through $(-2,10)$.

5. Let $f(x) = .5x - 1.4$. Graph f in the standard viewing window of a graphing calculator, and make a table of values for f at $x = -3, -2, -1, \ldots, 3$.

6. Give the slope–intercept form of the line shown in the figure.

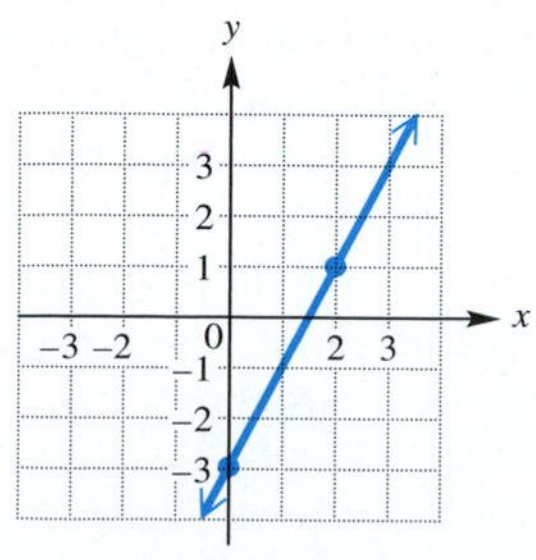

7. Find the slope–intercept form of the line passing through $(-2,4)$ and $(5,2)$.

8. Find the slope–intercept form of the line perpendicular to $3x - 2y = 5$ and passing through $(-1,3)$.

9. ***Average Household Size*** The table lists the average number y of people per household for various years x.

x	1940	1950	1960	1970
y	3.67	3.37	3.33	3.14

x	1980	1990	1998
y	2.76	2.63	2.62

Source: U.S. Census Bureau.

(a) Make a scatter diagram of the data.
(b) Decide whether the correlation coefficient is positive, negative, or zero.
(c) Find a least-squares regression line that models these data. What is the correlation coefficient?
(d) Estimate the number of people per household in 1975, and compare it with the actual value of 2.94.

1.5 Linear Equations and Inequalities

Solving Linear Equations ■ Graphical Approaches to Solving Linear Equations ■ Identities and Contradictions ■ Solving Linear Inequalities ■ Graphical Approaches to Solving Linear Inequalities ■ Three-Part Inequalities

Solving Linear Equations

In this text, we use two distinct approaches to solving equations: the *analytic approach,* in which we use paper and pencil to transform complicated equations into simpler ones, and the *graphical approach,* in which we often support our analytic solutions by using graphs or tables.

An **equation** is a statement that two expressions are equal. To *solve* an equation means to find all numbers that make the equation a true statement. Such numbers are called **solutions** or **roots** of the equation. A number that is a solution of an equation is said to *satisfy* the equation, and the solutions of an equation make up its **solution set.**

In this section, we solve *linear equations in one variable.* A linear equation in one variable has one solution.

Linear Equation in One Variable

A **linear equation in one variable** is an equation that can be written in the form

$$ax + b = 0, \quad a \neq 0.$$

One way to solve an equation is to rewrite it as a series of simpler **equivalent equations,** each of which has the same solution set as the original one. Equivalent equations are obtained by using the properties of equality.

Addition and Multiplication Properties of Equality

For real numbers a, b, and c,

$a = b$ and $a + c = b + c$ are equivalent.
(*The same number may be added to each side of an equation without changing the solution set.*)

If $c \neq 0$, then $a = b$ and $ac = bc$ are equivalent.
(*Each side of an equation may be multiplied by the same nonzero number without changing the solution set.*)

Extending these two properties allows us to subtract the same number from each side of an equation and to divide each side by the same nonzero number.

EXAMPLE 1 Solving a Linear Equation

Solve $10 + 3(2x - 4) = 17 - (x + 5)$.

Solution

$$10 + 3(2x - 4) = 17 - (x + 5)$$

$$10 + 6x - 12 = 17 - x - 5 \quad \text{Distributive property}$$

Be careful with signs.

$$-2 + 7x = 12 \quad \text{Add } x \text{ to each side; combine terms.}$$

$$7x = 14 \quad \text{Add 2 to each side.}$$

$$x = 2 \quad \text{Divide each side by 7.}$$

Check:

$$10 + 3(2x - 4) = 17 - (x + 5) \quad \text{Original equation}$$

$$10 + 3(2 \cdot 2 - 4) = 17 - (2 + 5) \quad ? \quad \text{Let } x = 2.$$

$$10 + 3(4 - 4) = 17 - 7 \quad ?$$

$$10 = 10 \quad \text{True; the answer checks.}$$

The solution set is $\{2\}$. ■

EXAMPLE 2 Solving a Linear Equation with Fractional Coefficients

Solve $\frac{x + 7}{6} + \frac{2x - 8}{2} = -4$.

Solution To eliminate fractions, multiply each side of the equation by the least common denominator.

$$\frac{x + 7}{6} + \frac{2x - 8}{2} = -4$$

$$6\left(\frac{x + 7}{6} + \frac{2x - 8}{2}\right) = 6(-4) \quad \text{Multiply by the LCD, 6.}$$

$$6\left(\frac{x + 7}{6}\right) + 6\left(\frac{2x - 8}{2}\right) = 6(-4) \quad \text{Distributive property}$$

$$x + 7 + 3(2x - 8) = -24 \quad \text{Simplify.}$$

$$x + 7 + 6x - 24 = -24 \quad \text{Distributive property}$$

$$7x - 17 = -24 \quad \text{Combine terms.}$$

$$7x = -7 \quad \text{Add 17.}$$

$$x = -1 \quad \text{Divide by 7.}$$

An analytic check will verify that the solution set is $\{-1\}$. ■

EXAMPLE 3 Solving a Linear Equation with Decimal Coefficients

Solve $.06x + .09(15 - x) = .07(15)$.

Solution Since each decimal number is given in hundredths, multiply each side of the equation by 100. (To multiply a number by 100, move the decimal point two places to the right.)

$$\begin{aligned} .06x + .09(15 - x) &= .07(15) & \\ 6x + 9(15 - x) &= 7(15) & \text{Multiply by 100.} \\ 6x + 135 - 9x &= 105 & \text{Distributive property; multiply.} \\ -3x + 135 &= 105 & \text{Combine terms.} \\ -3x &= -30 & \text{Subtract 135.} \\ x &= 10 & \text{Divide by } -3. \end{aligned}$$

An analytic check will verify that the solution set is $\{10\}$. ■

The equations solved in Examples 1–3 each have one solution and are called **conditional equations.**

Graphical Approaches to Solving Linear Equations

Since an equation always contains an equals sign, we can think of a linear equation as being in the form

$$f(x) = g(x),$$

where $f(x)$ and $g(x)$ are formulas for linear functions. The equation

$$\underbrace{10 + 3(2x - 4)}_{f(x)} = \underbrace{17 - (x + 5)}_{g(x)}$$

from Example 1 is in this form. To find the solution set graphically, we graph $y_1 = f(x)$ and $y_2 = g(x)$ and locate their point of intersection. In general, if f and g are linear functions, then their graphs are lines that intersect at a single point, no point, or infinitely many points, as illustrated in Figure 64.

Figure 65(a) shows graphs of $Y_1 = 10 + 3(2X - 4)$ and $Y_2 = 17 - (X + 5)$ intersecting at the point $(2, 10)$. The x-coordinate 2 is the solution of the equation from Example 1, and the y-coordinate 10 is the value obtained when either Y_1 or Y_2 is evaluated at $X = 2$. In Figure 65(b), a table of values for Y_1 and Y_2 shows that $Y_1 = Y_2 = 10$ when $X = 2$. Creating a table to find a solution of an equation gives a *numerical solution*. Locating a numerical solution involves searching a table to find an x-value at which $Y_1 = Y_2$. A table is less practical when the solution is not an integer.

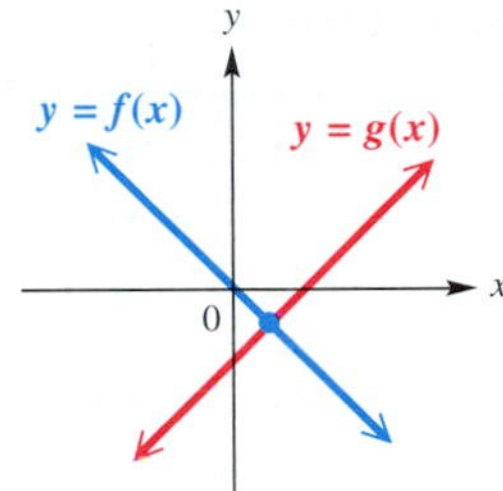

One point of intersection

(a)

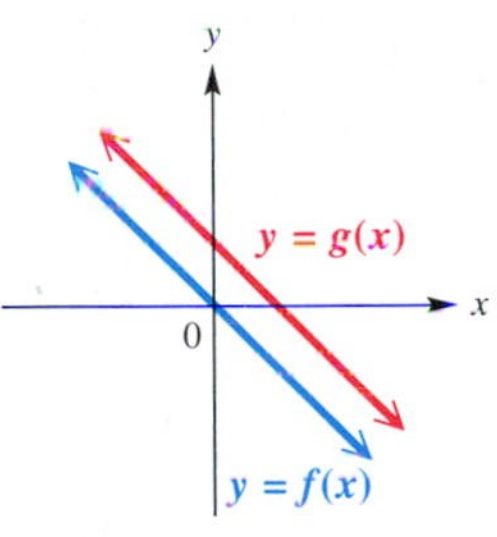

No point of intersection

(b)

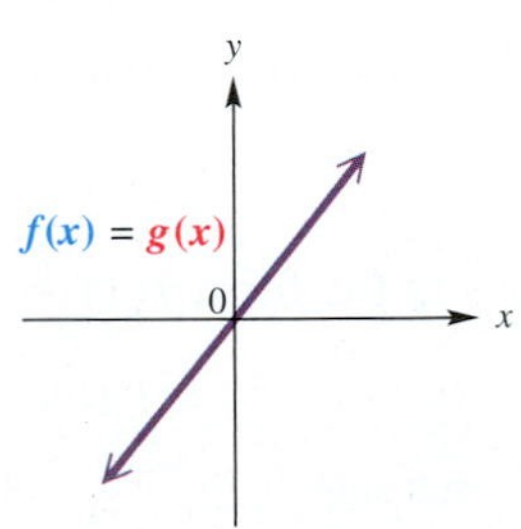

Infinitely many points of intersection (the lines coincide)

(c)

FIGURE 64

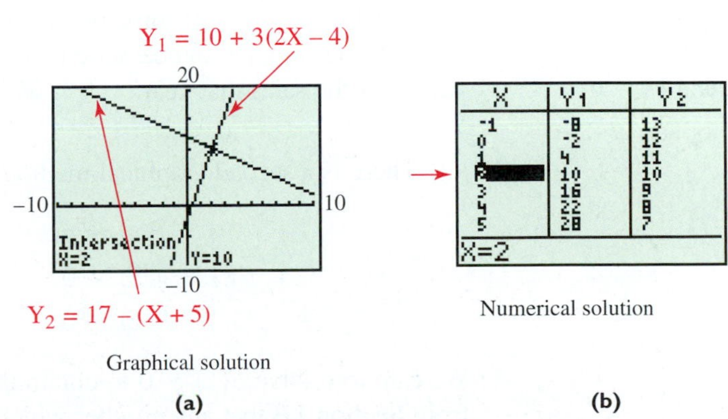

Graphical solution

(a)

Numerical solution

(b)

FIGURE 65

TECHNOLOGY NOTE

Graphing calculators often have built-in routines to find the intersection of two graphs. Sometimes these routines are located under the CALC menu.

The method shown in Figure 65(a) is called the *intersection-of-graphs method.*

Intersection-of-Graphs Method of Graphical Solution

To solve the equation $f(x) = g(x)$ graphically, graph

$$y_1 = f(x) \quad \text{and} \quad y_2 = g(x).$$

The x-coordinate of any point of intersection of the two graphs is a solution of the equation.

GCM EXAMPLE 4 Applying the Intersection-of-Graphs Method

The percent share of music sales (in dollars) that compact discs (CDs) held from 1987 to 1998 can be modeled by

$$f(x) = 5.91x + 13.7.$$

During the same period, the percent share of music sales that cassette tapes held can be modeled by

$$g(x) = -4.71x + 64.7.$$

In these formulas, $x = 0$ corresponds to 1987, $x = 1$ to 1988, and so on. Use the intersection-of-graphs method to estimate the year when sales of CDs equaled sales of cassettes. (*Source:* Recording Industry Association of America.)

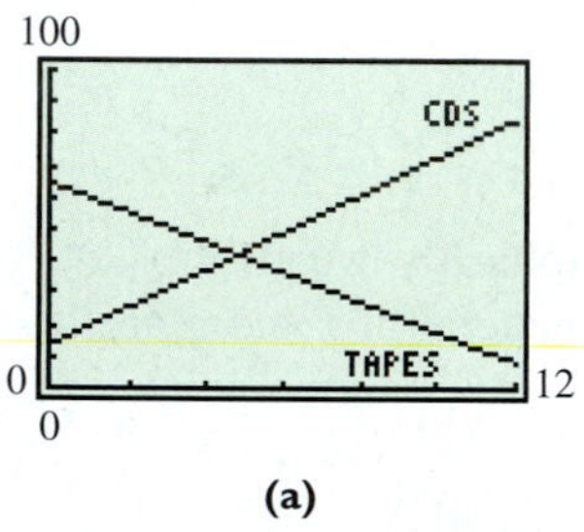

(a)

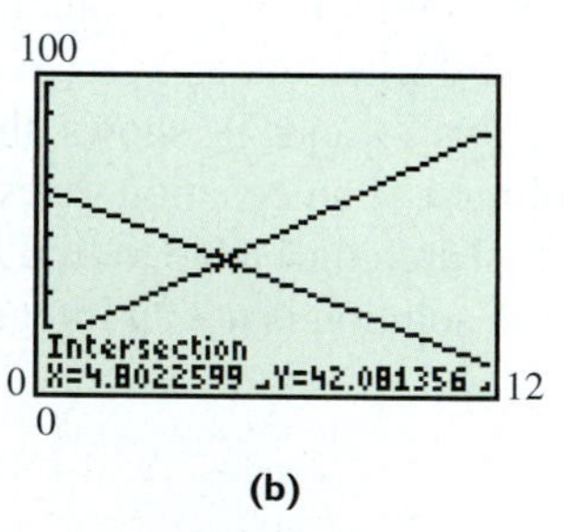

(b)

FIGURE 66

Solution We solve the linear equation $f(x) = g(x)$, that is,

$$5.91x + 13.7 = -4.71x + 64.7, \qquad \text{Let } f(x) = g(x).$$

by graphing

$$y_1 = 5.91x + 13.7 \quad \text{and} \quad y_2 = -4.71x + 64.7,$$

as shown in Figure 66(a). In Figure 66(b), the two graphs intersect near the point $(4.8, 42.1)$. Since $x = 0$ corresponds to 1987, and $1987 + 4.8 \approx 1992$, it follows that in 1992 sales of CDs and cassette tapes were approximately equal. Both shared about 42.1% of the sales that year. ■

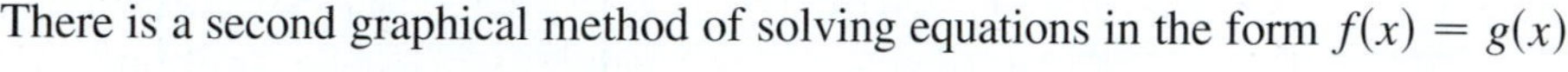

There is a second graphical method of solving equations in the form $f(x) = g(x)$.

$$\begin{aligned} f(x) &= g(x) \\ f(x) - g(x) &= 0 \qquad \text{Subtract } g(x) \text{ from each side.} \\ F(x) &= 0 \qquad \text{Let } F(x) = f(x) - g(x). \end{aligned}$$

We can now solve $F(x) = 0$ to obtain the solution set of the original equation. Recall from Section 1.3 that any number which satisfies this equation is an x-intercept of the graph of $y = F(x)$. It is also called a *zero* of F. The *x-intercept method* of graphical solution uses this idea.

x-Intercept Method of Graphical Solution

To solve the equation $f(x) = g(x)$ graphically, graph

$$y = f(x) - g(x) = F(x).$$

Any x-intercept of the graph of $y = F(x)$ (or zero of the function F) is a solution of the equation.

TECHNOLOGY NOTE

Graphing calculators often have built-in programs to locate x-intercepts (roots, zeros).

The words ***root, solution,*** and ***zero*** all refer to the same basic concept. When solving equations, remember this: ***The real solutions (or roots or zeros) of an equation of the form $f(x) = 0$ correspond to the x-intercepts of the graph of $y = f(x)$.***

GCM **EXAMPLE 5 Using the x-Intercept Method**

Solving analytically, the solution set of

$$6x - 4(3 - 2x) = 5(x - 4) - 10$$

is $\{-2\}$. Use the x-intercept method to solve this equation. Then use a table to obtain a numerical solution.

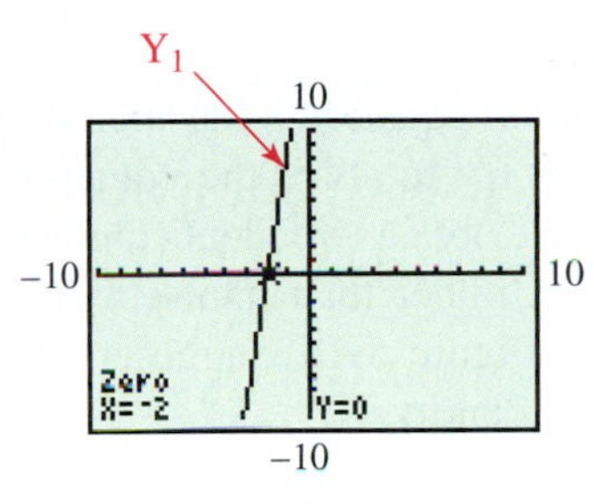

(a)

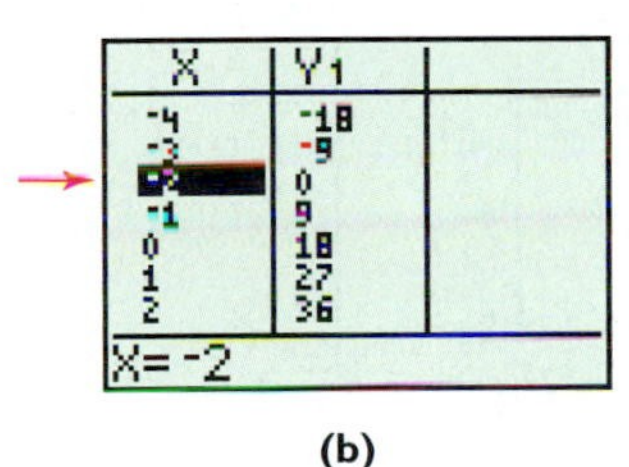

X	Y1
-4	-18
-3	-9
-2	0
-1	9
0	18
1	27
2	36

X=-2

(b)

FIGURE 67

Solution Begin by writing the equation in the form $F(x) = 0$.

$$6x - 4(3 - 2x) - (5(x - 4) - 10) = 0 \quad \text{Subtract } 5(x - 4) - 10.$$

Use parentheses.

Graph $Y_1 = 6X - 4(3 - 2X) - (5(X - 4) - 10)$, and locate any x-intercepts or zeros of Y_1. Figure 67(a) shows x-intercept -2, confirming the solution given in the statement of the problem.

Figure 67(b) shows a table of values for the equation $Y_1 = 6X - 4(3 - 2X) - (5(X - 4) - 10)$. To find a solution, look for an x-value at which $Y_1 = 0$. Notice that $Y_1 = 0$ when $X = -2$. ■

FOR DISCUSSION

In Example 5, we solved a linear equation by letting

$$f(x) = 6x - 4(3 - 2x) \quad \text{and} \quad g(x) = 5(x - 4) - 10$$

and then graphing $F(x) = f(x) - g(x)$. This time, graph

$$y_2 = g(x) - f(x), \quad \text{rather than} \quad y_1 = f(x) - g(x).$$

Does the graph of y_2 have the same x-intercept as the graph of y_1? Make a conjecture concerning the order in which f and g are subtracted when the x-intercept method is used.

Identities and Contradictions

Every equation solved thus far in this section has been a conditional equation with one solution. However, since the graphs of two linear functions may not intersect or may coincide (as shown in Figures 64(b) and (c), respectively, on page 53), there are two other situations that may arise in solving equations.

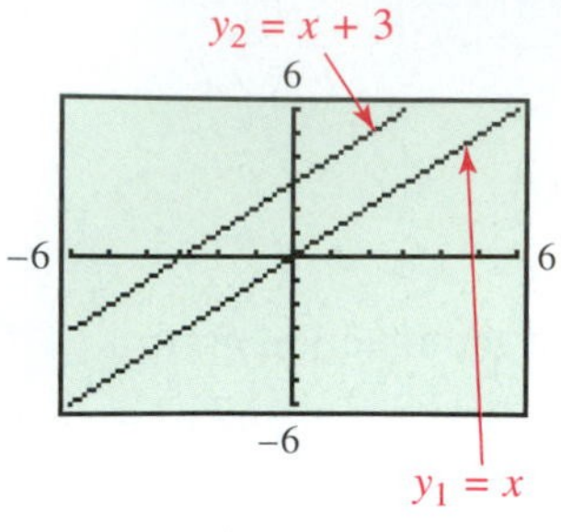

FIGURE 68

A **contradiction** is an equation that has no solution. For example, no values for x satisfy the equation $x = x + 3$, because there is no real number equal to itself plus 3. The equation is a contradiction. If we try to solve this equation analytically by subtracting x from each side, we obtain $0 = 3$, a false statement. If we try to solve $x = x + 3$ graphically by letting $y_1 = x$ and $y_2 = x + 3$, we obtain two parallel lines, as shown in Figure 68. Since the lines do not intersect, the solution set is the **empty set** or **null set,** denoted **$\emptyset$.**

An **identity** is an equation that is true for all values in the domain of its variables. For example, the equation $2(x - 3) = 2x - 6$ is true for all real numbers. If we try to solve this equation analytically, we obtain

$$2(x - 3) = 2x - 6$$
$$2x - 6 = 2x - 6 \quad \text{Distributive property}$$
$$2x = 2x \quad \text{Add 6.}$$
$$0 = 0. \quad \text{Subtract } 2x.$$

The equation $0 = 0$ is always true, indicating that the given equation is an identity. The solution set is {all real numbers}, or $(-\infty, \infty)$. If we try to solve the equation graphically by letting $Y_1 = 2(X - 3)$ and $Y_2 = 2X - 6$, we obtain two identical lines as shown in Figure 69(a). Since these two lines coincide, rather than intersect at a point, the solution set is {all real numbers}. The table in Figure 69(b) suggests that $Y_1 = Y_2$ for all X-values and that the given equation is an identity.

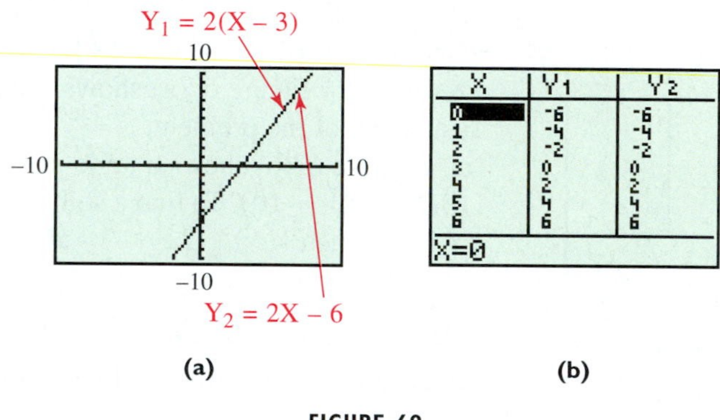

(a) (b)

FIGURE 69

NOTE Contradictions and identities are *not* linear equations, because they cannot be written as $ax + b = 0$ with $a \neq 0$. Contradictions and identities simplify to equations that are either always false or always true. ***Linear equations always have exactly one solution.***

Solving Linear Inequalities

An equation says that two expressions are equal; an **inequality** says that one expression is greater than, greater than or equal to, less than, or less than or equal to another. As with equations, a value of the variable for which the inequality is true is a solution of the inequality, and the set of all such solutions is the solution set of the inequality. Two inequalities with the same solution set are **equivalent inequalities.**

Inequalities are solved using the properties of inequality.

Addition and Multiplication Properties of Inequality

For real numbers a, b, and c,

(a) $a < b$ **and** $a + c < b + c$ **are equivalent.**

(The same number may be added to each side of an inequality without changing the solution set.)

(b) If $c > 0$**, then** $a < b$ **and** $ac < bc$ **are equivalent.**

(Each side of an inequality may be multiplied by the same positive number without changing the solution set.)

(c) If $c < 0$**, then** $a < b$ **and** $ac > bc$ **are equivalent.**

(Each side of an inequality may be multiplied by the same negative number without changing the solution set, as long as the direction of the inequality symbol is reversed.)

Similar properties exist for $>$, $\leq$, and $\geq$.

NOTE Because division is defined in terms of multiplication, the word "multiplied" may be replaced by "divided" in parts (b) and (c) of the properties of inequality. Similarly, in part (a), the words "added to" may be replaced by "subtracted from."

Pay careful attention to part (c). ***If each side of an inequality is multiplied by a negative number, the direction of the inequality symbol must be reversed.*** For example, starting with $-3 < 5$ and multiplying each side by the *negative* number -2 gives a true result only if the direction of the inequality symbol is reversed.

$$-3 < 5$$

$$-3(-2) > 5(-2)$$

$$6 > -10$$

A similar situation exists when each side is divided by a negative number.

A *linear inequality in one variable* is defined as follows.

Linear Inequality in One Variable

A **linear inequality in one variable** is an inequality that can be written in one of the following forms, where $a \neq 0$:

$$ax + b > 0, \qquad ax + b < 0, \qquad ax + b \geq 0, \qquad ax + b \leq 0.$$

We solve a linear inequality analytically using essentially the same steps as those used to solve a linear equation. The solution set of a linear inequality is typically an interval of the real number line.

FOR DISCUSSION

Solve each equation or inequality:

$$2(x - 3) + x = 9,$$
$$2(x - 3) + x < 9,$$
$$2(x - 3) + x > 9.$$

How can the solution of the equation be used to help solve the two inequalities?

EXAMPLE 6 **Solving a Linear Inequality**

Solve $3x - 2(2x + 4) \leq 2x + 1$.

Solution

$$3x - 2(2x + 4) \leq 2x + 1$$
$$3x - 4x - 8 \leq 2x + 1 \quad \text{Distributive property}$$
$$-x - 8 \leq 2x + 1 \quad \text{Combine terms.}$$
$$-3x \leq 9 \quad \text{Subtract } 2x\text{; add 8.}$$
$$x \geq -3 \quad \text{Divide by } -3\text{; reverse the direction of the inequality symbol.}$$

Be careful with signs.

In interval notation, the solution set is $[-3, \infty)$. ■

EXAMPLE 7 **Solving Linear Inequalities**

Solve each inequality.

(a) $2x - 3 < \dfrac{x + 2}{-3}$ **(b)** $-3(4x - 4) \geq 4 - (x - 1)$

Solution

(a)
$$2x - 3 < \frac{x + 2}{-3}$$
$$-3(2x - 3) > -3\left(\frac{x + 2}{-3}\right) \quad \text{Multiply by } -3\text{; reverse the inequality symbol.}$$
$$-6x + 9 > x + 2 \quad \text{Distributive property; simplify.}$$
$$7 > 7x \quad \text{Add } 6x\text{; subtract 2.}$$
$$1 > x \quad \text{Divide by 7.}$$
$$x < 1 \quad \text{Rewrite.}$$

Be careful with signs.

The solution set is $(-\infty, 1)$.

(b)
$$-3(4x - 4) \geq 4 - (x - 1)$$
$$-12x + 12 \geq 4 - x + 1 \quad \text{Distributive property}$$
$$-12x + 12 \geq -x + 5 \quad \text{Simplify.}$$
$$-11x \geq -7 \quad \text{Add } x\text{; subtract 12.}$$
$$x \leq \frac{7}{11} \quad \text{Divide by } -11\text{; reverse the inequality symbol.}$$

Be careful with signs.

The solution set is $\left(-\infty, \frac{7}{11}\right]$. ■

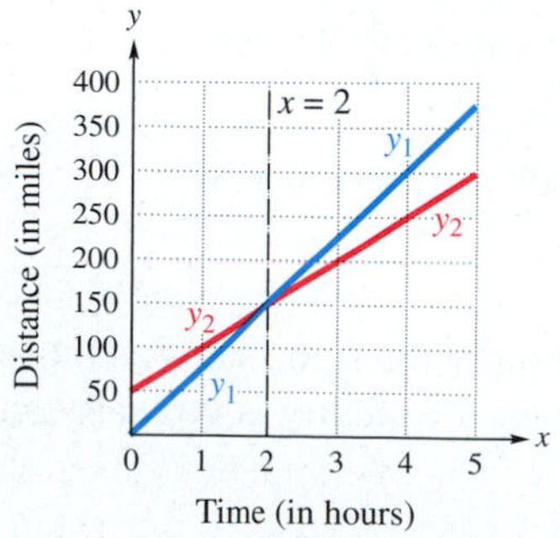

FIGURE 70

Graphical Approaches to Solving Linear Inequalities

Figure 70 shows the distances of two cars from Chicago, Illinois, x hours after traveling in the same direction on a freeway. The distance for Car 1 is denoted y_1, and the distance for Car 2 is denoted y_2. After $x = 2$ hours, $y_1 = y_2$ and both cars are 150 miles from Chicago. To the left of the dashed vertical line $x = 2$, the graph of y_1 is below the graph of y_2, so Car 1 is closer to Chicago than Car 2. Thus,

$$y_1 < y_2 \quad \text{when} \quad x < 2.$$

To the right of the dashed vertical line $x = 2$, the graph of y_1 is above the graph of y_2, so Car 1 is farther from Chicago than Car 2. Thus,

$$y_1 > y_2 \quad \text{when} \quad x > 2.$$

This discussion leads to an extension of the intersection-of-graphs method.

TECHNOLOGY NOTE

When solving linear inequalities graphically, the calculator will not determine whether the endpoint of the interval is included or excluded. This must be done by looking at the inequality symbol in the given inequality.

Intersection-of-Graphs Method of Solution of a Linear Inequality

Suppose that f and g are linear functions. The solution set of $f(x) > g(x)$ is the set of all real numbers x such that the graph of f is **above** the graph of g. The solution set of $f(x) < g(x)$ is the set of all real numbers x such that the graph of f is **below** the graph of g.

If an inequality involves one of the symbols $\geq$ or $\leq$, the same method applies, with the solution of the corresponding equation included in the solution set.

GCM EXAMPLE 8 Using the Intersection-of-Graphs Method

The solution set of the inequality

$$3x - 2(2x + 4) \leq 2x + 1,$$

solved in Example 6, is $[-3, \infty)$. Solve this inequality graphically.

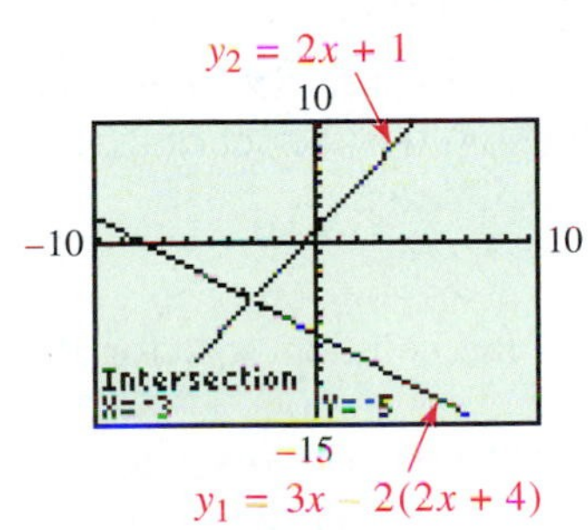

FIGURE 71

Solution Graph

$$y_1 = 3x - 2(2x + 4) \quad \text{and} \quad y_2 = 2x + 1.$$

Figure 71 indicates that the point of intersection of the two lines is $(-3, -5)$. Because the graph of y_1 is *below* the graph of y_2 when x is *greater than* -3, the solution set of $y_1 \leq y_2$ is $[-3, \infty)$, agreeing with our analytic solution. ■

EXAMPLE 9 Applying the Intersection-of-Graphs Method

When the air temperature reaches the dew point, fog may form. This phenomenon also causes clouds to form at higher altitudes. Both the air temperature and the dew point decrease at a constant rate as the altitude above ground level increases. If the ground-level temperature and dew point are T_0 and D_0, respectively, the air temperature at an altitude of x miles can be approximated by

$$T(x) = T_0 - 29x$$

and the dew point by

$$D(x) = D_0 - 5.8x.$$

If $T_0 = 75°\text{F}$ and $D_0 = 55°\text{F}$, determine the altitudes where clouds will not form.

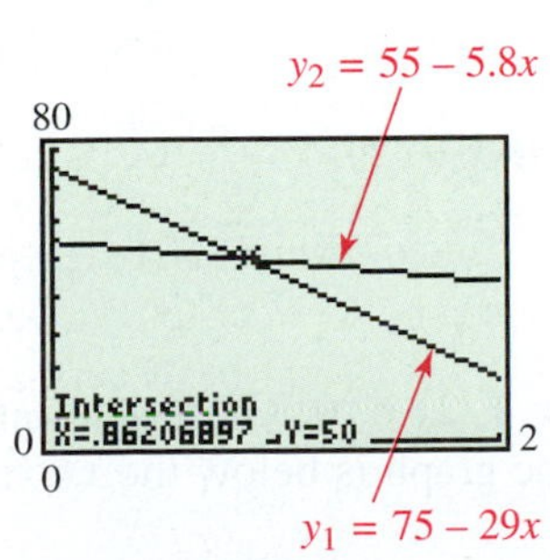

FIGURE 72

Solution Since $T_0 = 75$ and $D_0 = 55$, let $T(x) = 75 - 29x$ and $D(x) = 55 - 5.8x$. Graph $y_1 = 75 - 29x$ and $y_2 = 55 - 5.8x$, as shown in Figure 72. The graphs intersect near $(.86, 50)$. This means that the air temperature and dew point are both 50°F at about .86 mile above ground level. Clouds will not form below this altitude—that is, when the graph of y_1 is above the graph of y_2. The solution set is $[0, .86)$, where the endpoint .86 is approximate. ■

In Example 9, the endpoint, .86 mile, is an approximation to the nearest hundredth. In solving inequalities graphically, the appropriate symbol for the endpoint, a parenthesis or a bracket, may not actually be valid because of rounding. As a result, we state an agreement to be used throughout this text.

Agreement on Inclusion or Exclusion of Endpoints for Approximations

When an approximation is used for an endpoint in specifying an interval, we continue to use parentheses in specifying inequalities involving $<$ or $>$ and square brackets in specifying inequalities involving $\leq$ or $\geq$.

The x-intercept method can also be used to solve inequalities. For example, to solve $f(x) > g(x)$, we can rewrite the inequality as $f(x) - g(x) > 0$ or $F(x) > 0$, where $F(x) = f(x) - g(x)$. All solutions of the given inequality correspond to the x-values where the graph of $y = F(x)$ is above the x-axis.

GCM TECHNOLOGY NOTE

If two functions defined by Y_1 and Y_2 are already entered into your calculator, you can enter Y_3 as $Y_2 - Y_1$. Then, if you direct the calculator to graph Y_3 only, you can solve the inequality $Y_1 > Y_2$ by first finding the x-intercept of Y_3.

x-Intercept Method of Solution of a Linear Inequality

The solution set of $\boldsymbol{F(x) > 0}$ is the set of all real numbers x such that the graph of F is **above** the x-axis. The solution set of $\boldsymbol{F(x) < 0}$ is the set of all real numbers x such that the graph of F is **below** the x-axis.

Figure 73 illustrates this discussion and summarizes the solution sets for the appropriate inequalities.

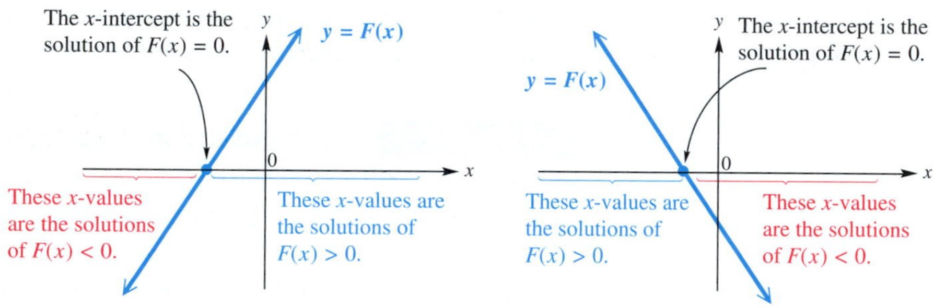

FIGURE 73

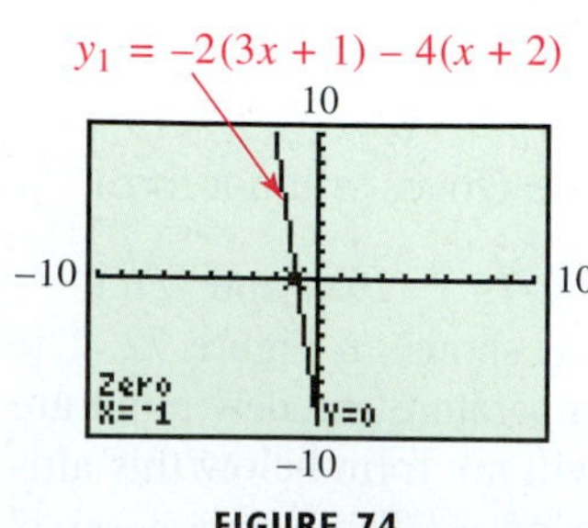

FIGURE 74

EXAMPLE 10 Using the x-Intercept Method

Solve each inequality using the x-intercept method.

(a) $-2(3x + 1) < 4(x + 2)$ **(b)** $-2(3x + 1) \leq 4(x + 2)$

(c) $-2(3x + 1) > 4(x + 2)$ **(d)** $-2(3x + 1) \geq 4(x + 2)$

Solution

(a) The inequality can be written as $-2(3x + 1) - 4(x + 2) < 0$. Graph the left side of the inequality as y_1, as shown in Figure 74. The graph is below the x-axis when $x > -1$, so the solution set is $(-1, \infty)$.

(b) The solution set is $[-1, \infty)$, because the endpoint at $x = -1$ is included.

(c) The graph in Figure 74 on the previous page is above the x-axis when $x < -1$, so the solution set is $(-\infty, -1)$.

(d) The solution set is $(-\infty, -1]$, because the endpoint at $x = -1$ is included.

Notice that the same graph was used to solve all four inequalities. ■

Three-Part Inequalities

Sometimes a mathematical expression is limited to an interval of values. For example, suppose that $2x + 1$ must satisfy both

$$2x + 1 \geq -3 \quad \text{and} \quad 2x + 1 \leq 2.$$

These two inequalities can be combined into the **three-part inequality**

$$-3 \leq 2x + 1 \leq 2. \qquad \text{Three-part inequality}$$

Three-part inequalities occur in manufacturing, where error tolerances are often maintained. Suppose that aluminum cans are to be constructed with radius 1.4 inches. However, because the manufacturing process is not exact, the radius r actually varies from 1.38 inches to 1.42 inches. We can express possible values for r by using the three-part inequality

$$1.38 \leq r \leq 1.42.$$

$C = 2\pi r$

See the figure in the margin. Then the circumference ($C = 2\pi r$) of the can varies between $2\pi(1.38) \approx 8.67$ inches and $2\pi(1.42) \approx 8.92$ inches. This result is given (approximately) by

$$8.67 \leq C \leq 8.92.$$

GCM **EXAMPLE 11** **Solving a Three-Part Inequality**

Solve $-2 < 5 + 3x < 20$.

Analytic Solution

Work with all three parts of the inequality at the same time.

$$-2 < 5 + 3x < 20$$

$$-7 < 3x < 15 \qquad \text{Subtract 5.}$$

$$-\frac{7}{3} < x < 5 \qquad \text{Divide by 3.}$$

The open interval $\left(-\frac{7}{3}, 5\right)$ is the solution set. Notice that an open interval is used because equality is *not* included in the three-part inequality.

Graphing Calculator Solution

Graph $y_1 = -2$, $y_2 = 5 + 3x$, and $y_3 = 20$, as shown in Figure 75. The x-values of the points of intersection are $-\frac{7}{3} = -2.\overline{3}$, and 5, confirming that our analytic work is correct. Notice how the slanted line, y_2, lies *between* the graphs of $y_1 = -2$ and $y_3 = 20$ for x-values between $-\frac{7}{3}$ and 5.

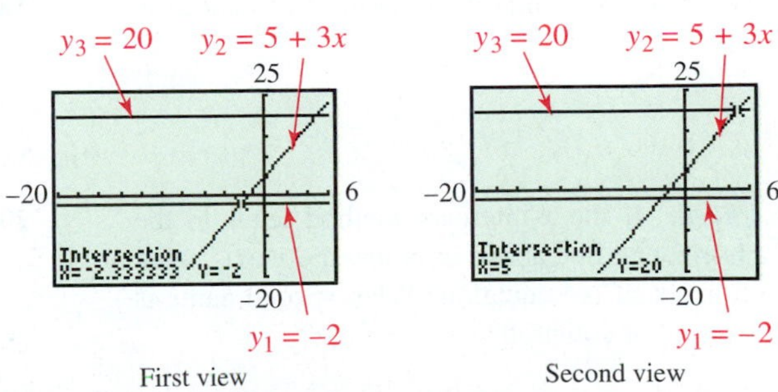

FIGURE 75 ■

1.5 Exercises

Find the zero of the function f.

1. $f(x) = -3x - 12$
2. $f(x) = 5x - 30$
3. $f(x) = 5x$
4. $f(x) = -2x$
5. $f(x) = 2(3x - 5) + 8(4x + 7)$
6. $f(x) = -4(2x - 3) + 8(2x + 1)$
7. $f(x) = 3x + 6(x - 4)$
8. $f(x) = -8x + .5(2x + 8)$
9. $f(x) = 1.5x + 2(x - 3) + 5.5(x + 9)$
10. ***Concept Check*** If c is a zero of the linear function defined by $f(x) = mx + b$, $m \neq 0$, then the point at which the graph intersects the x-axis has coordinates (______, ______).

In Exercises 11–13, two linear functions, y_1 and y_2, are graphed with their point of intersection indicated. Give the solution set of $y_1 = y_2$.

11.

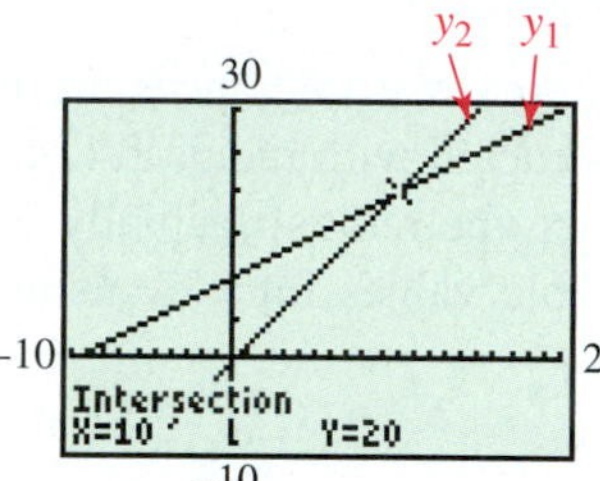

12.

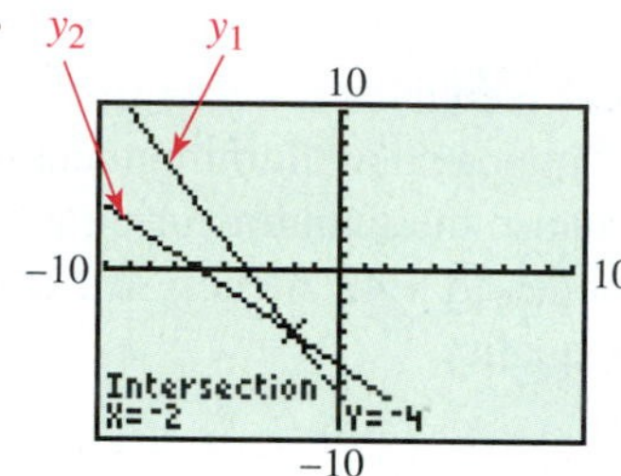

13.

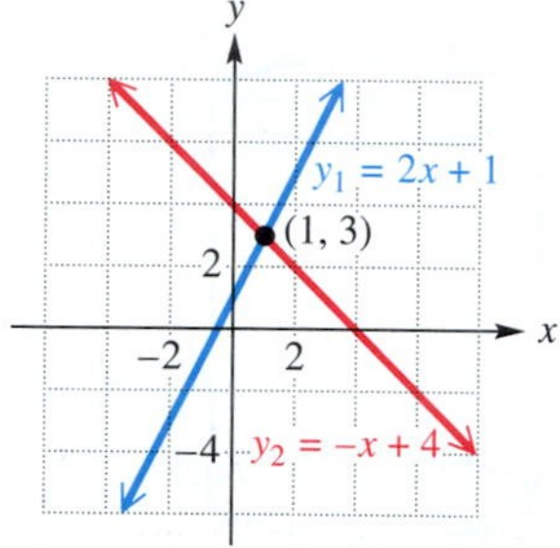

In Exercises 14–16, use the graph of $y = y_1 - y_2$ to solve the equation $y_1 = y_2$, where y_1 and y_2 represent linear functions.

14.

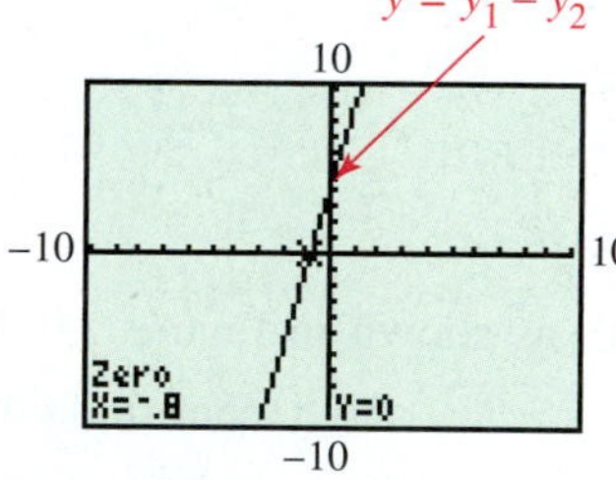

15.

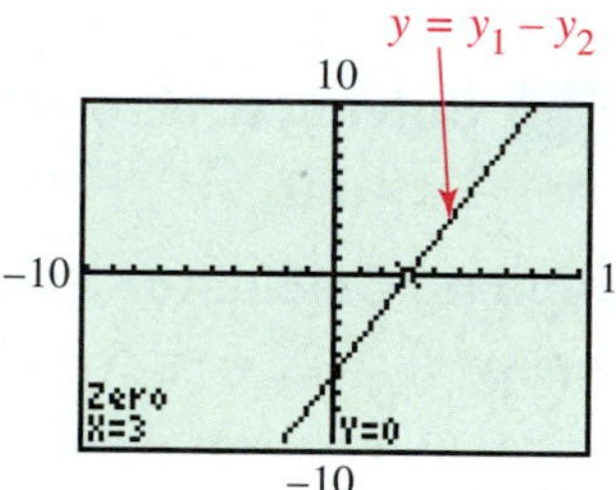

16.

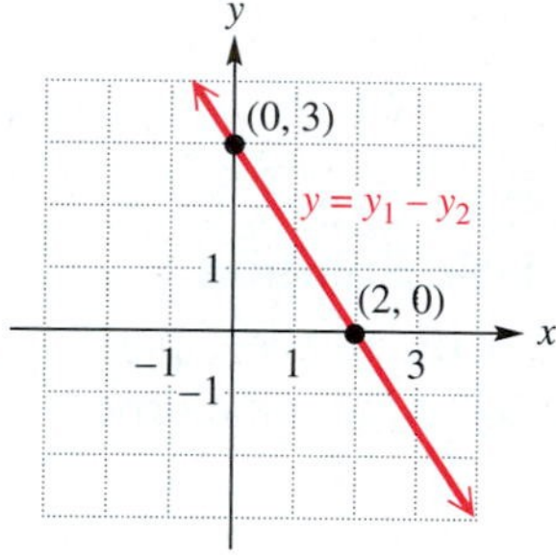

17. Interpret the y-value shown at the bottom of the screen in Exercise 11.
18. If you were asked to solve

$$2x + 3 = 4x - 12$$

by the x-intercept method, why would you *not* get the correct answer by graphing $y_1 = 2x + 3 - 4x - 12$?

19. ***Concept Check*** If the x-intercept method leads to the graph of a horizontal line above or below the x-axis, what is the solution set of the equation? What special name is given to this kind of equation?
20. ***Concept Check*** If the x-intercept method leads to a horizontal line that coincides with the x-axis, what is the solution set of the equation? What special name is given to this kind of equation?

Solve each equation analytically. Check it analytically by direct substitution, and then support your solution graphically.

21. $2x - 5 = x + 7$
22. $9x - 17 = 2x + 4$
23. $.01x + 3.1 = 2.03x - 2.96$
24. $.04x + 2.1 = .02x + 1.92$

25. $-(x + 5) - (2 + 5x) + 8x = 3x - 5$

26. $-(8 + 3x) + 5 = 2x + 3$

27. $\dfrac{2x + 1}{3} + \dfrac{x - 1}{4} = \dfrac{13}{2}$

28. $\dfrac{x - 2}{4} + \dfrac{x + 1}{2} = 1$

29. $\dfrac{1}{2}(x - 3) = \dfrac{5}{12} + \dfrac{2}{3}(2x - 5)$

30. $\dfrac{7}{3}(2x - 1) = \dfrac{1}{5}x + \dfrac{2}{5}(4 - 3x)$

31. $.1x - .05 = -.07x$

32. $1.1x - 2.5 = .3(x - 2)$

33. $.40x + .60(100 - x) = .45(100)$

34. $1.30x + .90(.50 - x) = 1.00(50)$

35. $2[x - (4 + 2x) + 3] = 2x + 2$

36. $6[x - (2 - 3x) + 1] = 4x - 6$

37. $\dfrac{5}{6}x - 2x + \dfrac{1}{3} = \dfrac{1}{3}$

38. $\dfrac{3}{4} + \dfrac{1}{5}x - \dfrac{1}{2} = \dfrac{4}{5}x$

39. $5x - (8 - x) = 2[-4 - (3 + 5x - 13)]$

40. $-[x - (4x + 2)] = 2 + (2x + 7)$

Each table shows selected ordered pairs for two linear functions Y_1 *and* Y_2. *Use the table to solve the given equation.*

41. $Y_1 = Y_2$

X	Y1	Y2
0	4	16
1	5	14
2	6	12
3	7	10
4	8	8
5	9	6
6	10	4

X=0

42. $Y_1 = Y_2$

X	Y1	Y2
-9	-35	-23
-8	-32	-22
-7	-29	-21
-6	-26	-20
-5	-23	-19
-4	-20	-18
-3	-17	-17

X= -9

43. $Y_1 - Y_2 = 0$

X	Y1	Y2
0	0	3
.5	1.5	3.5
1	3	4
1.5	4.5	4.5
2	6	5
2.5	7.5	5.5
3	9	6

X=0

44. $Y_1 - Y_2 = 0$

X	Y1	Y2
5.5	8	8
6	9	8.5
6.5	10	9
7	11	9.5
7.5	12	10
8	13	10.5
8.5	14	11

X=5.5

Use the intersection-of-graphs method to approximate each solution to the nearest hundredth.

45. $4(.23x + \sqrt{5}) = \sqrt{2}x + 1$

46. $9(-.84x + \sqrt{17}) = \sqrt{6}x - 4$

47. $2\pi x + \sqrt[3]{4} = .5\pi x - \sqrt{28}$

48. $3\pi x - \sqrt[4]{3} = .75\pi x + \sqrt{19}$

49. $.23(\sqrt{3} + 4x) - .82(\pi x + 2.3) = 5$

50. $-.15(6 + \sqrt{2}x) + 1.4(2\pi x - 6.1) = 10$

Determine analytically whether each equation is an identity or a contradiction. Give the solution set. Use a graph or table to support your answer.

51. $5x + 5 = 5(x + 3) - 10$

52. $5 - 4x = 5x - (9 + 9x)$

53. $6(2x + 1) = 4x + 8\left(x + \dfrac{3}{4}\right)$

54. $3(x + 2) - 5(x + 2) = -2x - 4$

55. $-4[6 - (-2 + 3x)] = 21 + 12x$

56. $-3[-5 - (-9 + 2x)] = 2(3x - 1)$

57. $\dfrac{1}{2}x - 2(x - 1) = 2 - \dfrac{3}{2}x$

58. $.5(x - 2) + 12 = .5x + 11$

59. $\dfrac{x - 1}{2} = \dfrac{3x - 2}{6}$

60. $\dfrac{2x - 1}{3} = \dfrac{2x + 1}{3}$

Use the graph at the right to solve each equation or inequality.

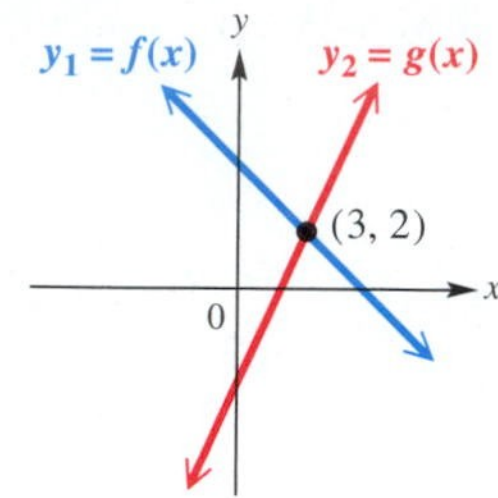

61. $f(x) = g(x)$

62. $f(x) > g(x)$

63. $f(x) < g(x)$

64. $g(x) - f(x) \geq 0$

65. $y_1 - y_2 \geq 0$

66. $y_2 > y_1$

67. $f(x) \leq g(x)$

68. $f(x) \geq g(x)$

69. $f(x) \leq 2$

70. $g(x) \leq 2$

Refer to the graph of the linear function defined by $y = f(x)$ to solve each inequality. Express solution sets in interval notation.

71. **(a)** $f(x) > 0$
(b) $f(x) < 0$
(c) $f(x) \geq 0$
(d) $f(x) \leq 0$

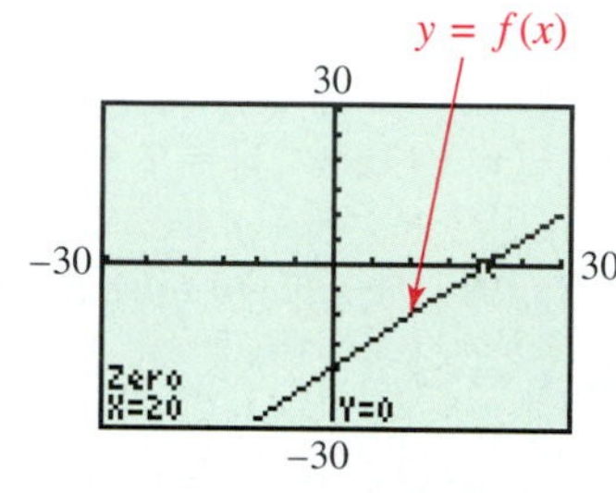

72. **(a)** $f(x) < 0$
(b) $f(x) \leq 0$
(c) $f(x) \geq 0$
(d) $f(x) > 0$

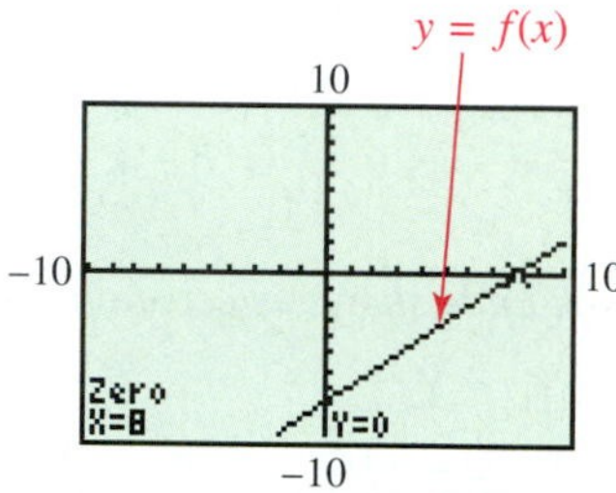

Concept Check *In Exercises 73 and 74, f and g are linear functions.*

73. If the solution set of $f(x) \geq g(x)$ is $[4, \infty)$, what is the solution set of each equation or inequality?
(a) $f(x) = g(x)$ **(b)** $f(x) > g(x)$ **(c)** $f(x) < g(x)$

74. If the solution set of $f(x) < g(x)$ is $(-\infty, 3)$, what is the solution set of each equation or inequality?
(a) $f(x) = g(x)$ **(b)** $f(x) \geq g(x)$ **(c)** $f(x) \leq g(x)$

Solve each inequality analytically, writing the solution set in interval notation. Support your answer graphically. (Hint: Once part (a) is done, the answer to part (b) follows from the answer to part (a).)

75. **(a)** $9 - (x + 1) < 0$
(b) $9 - (x + 1) \geq 0$

76. **(a)** $5 + 3(1 - x) \geq 0$
(b) $5 + 3(1 - x) < 0$

77. **(a)** $2x - 3 > x + 2$
(b) $2x - 3 \leq x + 2$

78. **(a)** $5 - 3x \leq 7 + x$
(b) $5 - 3x > 7 + x$

79. **(a)** $10x + 5 - 7x \geq 8(x + 2) + 4$
(b) $10x + 5 - 7x < 8(x + 2) + 4$

80. **(a)** $6x + 2 + 10x > -2(2x + 4) + 10$
(b) $6x + 2 + 10x \leq -2(2x + 4) + 10$

81. **(a)** $x + 2(-x + 4) - 3(x + 5) < -4$
(b) $x + 2(-x + 4) - 3(x + 5) \geq -4$

82. **(a)** $-11x - (6x - 4) + 5 - 3x \leq 1$
(b) $-11x - (6x - 4) + 5 - 3x > 1$

Solve each inequality. Write the solution set in interval notation. Support your answer graphically.

83. $\frac{1}{3}x - \frac{1}{5}x \leq 2$

84. $\frac{3x}{2} + \frac{4x}{7} \geq -5$

85. $\frac{x - 2}{2} - \frac{x + 6}{3} > -4$

86. $\frac{2x + 3}{5} - \frac{3x - 1}{2} < \frac{4x + 7}{2}$

87. $.6x - 2(.5x + .2) \leq .4 - .3x$

88. $-.9x - (.5 + .1x) > -.3x - .5$

89. $-\frac{1}{2}x + .7x - 5 > 0$

90. $\frac{3}{4}x - .2x - 6 \leq 0$

91. $-4(3x + 2) \geq -2(6x + 1)$

92. **Concept Check** Suppose that in solving a linear inequality, the final statement in the solution process is $5 > 0$. What is the solution set?

93. ***Distance*** The linear function f computes the distance y in miles between a car and the city of Omaha after x hours, where $0 \leq x \leq 6$. The graphs of f and the horizontal lines $y = 100$ and $y = 200$ are shown in the figure. Use the graphs to answer the questions that follow.

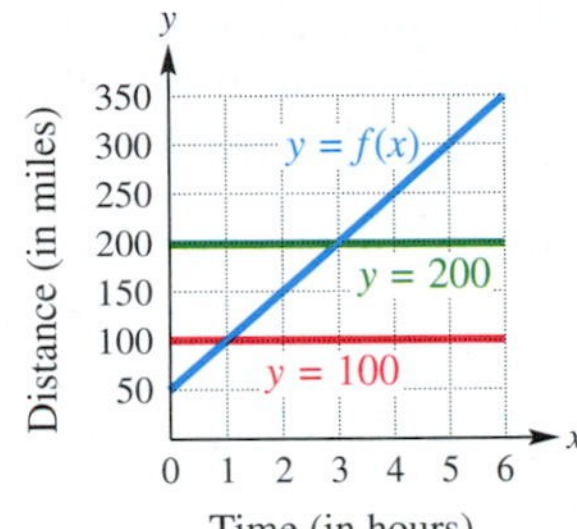

(a) Is the car moving toward or away from Omaha? Explain.
(b) At what time is the car 100 miles from Omaha? 200 miles?
(c) When is the car from 100 to 200 miles (inclusive) from Omaha?
(d) When is the car's distance from Omaha greater than 100 miles?

94. ***Concept Check*** Use the figure to solve each equation or inequality.

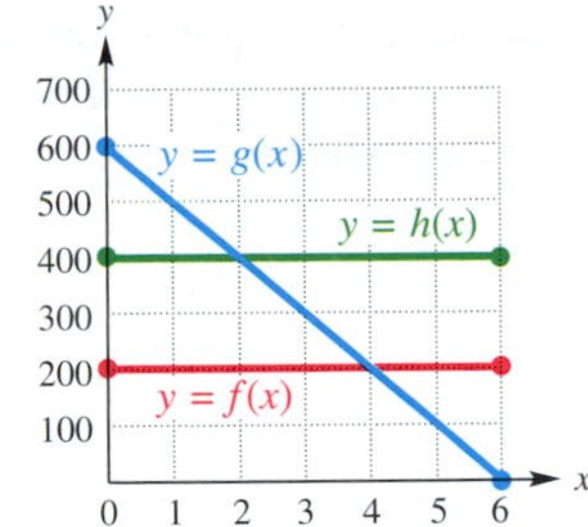

(a) $f(x) = g(x)$ **(b)** $g(x) = h(x)$
(c) $f(x) < g(x) < h(x)$ **(d)** $g(x) > h(x)$
(e) $f(x) = h(x)$ **(f)** $f(x) \leq h(x)$
(g) $h(x) > g(x)$ **(h)** $h(x) < 500$
(i) $g(x) < 200$ **(j)** $f(x) = 200$

Solve each three-part inequality analytically. Support your answer graphically.

95. $4 \leq 2x + 2 \leq 10$ **96.** $-4 \leq 2x - 1 \leq 5$ **97.** $-10 > 3x + 2 > -16$ **98.** $4 > 6x + 5 > -1$

99. $-3 \leq \dfrac{x-4}{-5} < 4$ **100.** $1 < \dfrac{4x-5}{-2} < 9$ **101.** $\sqrt{2} \leq \dfrac{2x+1}{3} \leq \sqrt{5}$ **102.** $\pi \leq 5 - 4x < 7\pi$

Relating Concepts

For individual or group discussion (Exercises 103–106)

The solution set of a linear equation is closely related to the solution set of a linear inequality. ***Work Exercises 103–106 in order,*** *to investigate this connection.*

103. Use the x-intercept method to find the solution set of $3.7x - 11.1 = 0$. How many solutions are there? How many solutions are there to any linear equation in one variable?

104. The solution from Exercise 103 is sometimes called a *boundary number.* Find the solution sets of $3.7x - 11.1 < 0$ and $3.7x - 11.1 > 0$. Write your answers in interval notation. Explain why the *boundary number* is appropriate for the solution you found in Exercise 103.

105. Use the x-intercept method to find the solution set of

$$-4x + 6 = 0.$$

Then find the solution sets of

$$-4x + 6 < 0 \quad \text{and} \quad -4x + 6 > 0.$$

106. Generalize your results from Exercises 103–105 by answering the questions that follow.
(a) What is the solution set of $ax + b = 0$ if $a \neq 0$?
(b) Suppose $a > 0$. What are the solution sets of $ax + b < 0$ and $ax + b > 0$? Write your answers in interval notation.
(c) Suppose $a < 0$. What are the solution sets of $ax + b < 0$ and $ax + b > 0$? Write your answers in interval notation.

(Modeling) *Solve each problem.*

107. *Clouds and Temperature* Refer to Example 9. Suppose the ground-level temperature is 65°F and the dew point is 50°F.
(a) Use the intersection-of-graphs method to estimate the altitudes at which clouds will not form.
(b) Solve part (a) analytically.

108. *Prices of Homes* The median price of a single-family home from 2001 to 2004 can be approximated by

$$f(x) = 44{,}500x + 129{,}650,$$

where $x = 0$ corresponds to 2001 and $x = 3$ to 2004. (*Source:* U.S. Department of Commerce.)
(a) Interpret the slope of the graph of f.
(b) Estimate the years when the median price range was from \$174,000 to \$219,000.

109. *Error Tolerances* Suppose that an aluminum can is manufactured so that its radius r can vary from 1.99 inches to 2.01 inches. What range of values is possible for the circumference C of the can? Express your answer by using a three-part inequality.

110. *Error Tolerances* Suppose that a square picture frame has sides that vary between 9.9 inches and 10.1 inches. What range of values is possible for the perimeter P of the picture frame? Express your answer by using a three-part inequality.

1.6 Applications of Linear Functions

Problem-Solving Strategies ■ Applications of Linear Equations ■ Break-Even Analysis ■ Direct Variation ■ Formulas

Problem-Solving Strategies

To become proficient at solving problems, we need to establish a procedure to guide our thinking. These steps may be helpful in solving application problems.

Solving Application Problems

Step 1 **Read the problem** and make sure you understand it. Assign a variable to what you are being asked to find. If necessary, write other quantities in terms of this variable.

Step 2 **Write an equation** that relates the quantities described in the problem. You may need to sketch a diagram and refer to known formulas.

Step 3 **Solve the equation** and determine the solution.

Step 4 **Look back and check** your solution. Does it seem reasonable?

Applications of Linear Equations

EXAMPLE 1 Determining the Dimensions of a Television Screen

A new generation of televisions has a 16:9 *aspect ratio*. This means that the length of the television's rectangular screen is $\frac{16}{9}$ times its width. If the perimeter of the screen is 136 inches, find the length and the width of the screen.

Analytic Solution

Step 1 If we let x represent the width of the screen, then $\frac{16}{9}x$ represents the length.

Step 2 See Figure 76.

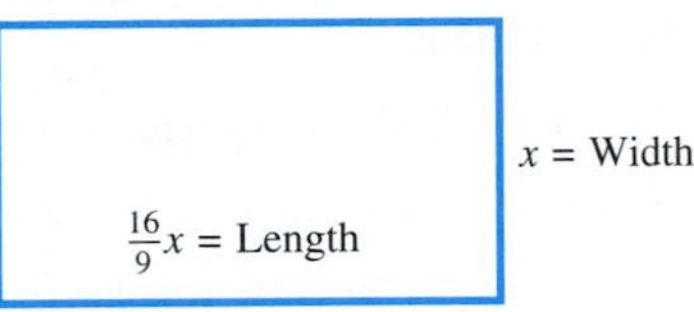

FIGURE 76

We can use the formula for the perimeter of a rectangle to write an equation.

$P = 2L + 2W$ Perimeter formula

Step 3 $136 = 2\left(\frac{16}{9}x\right) + 2x$ $P = 136$, $L = \frac{16}{9}x$, $W = x$

$136 = \frac{32}{9}x + 2x$ Multiply.

$136 = \frac{50}{9}x$ Combine terms.

$x = 24.48$ Multiply by $\frac{9}{50}$.

The width of the screen is 24.48 inches, and the length is $\frac{16}{9}(24.48) = 43.52$ inches.

Step 4 To check this answer, verify that the sum of the lengths of the sides is 136 inches.

Graphing Calculator Solution

We graph

$$y_1 = 2\left(\frac{16}{9}\right)x + 2x$$

and

$$y_2 = 136.$$

As seen in Figure 77, the point of intersection of the graphs is (24.48, 136). The x-coordinate supports our answer of 24.48 inches for the width.

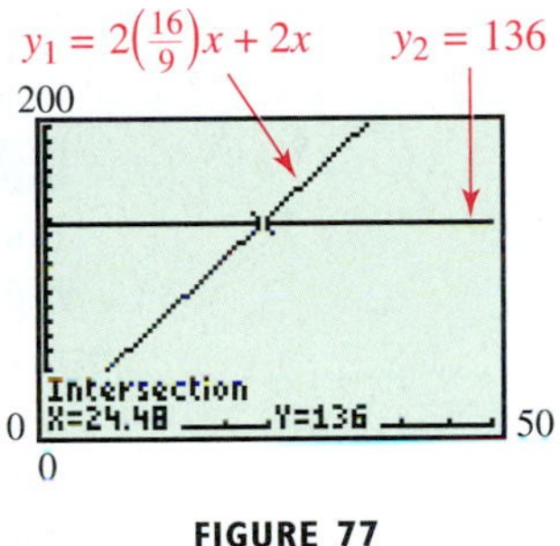

FIGURE 77

FOR DISCUSSION

Observe Figure 77, and disregard the graph of $y_2 = 136$. What remains is the graph of a linear function that gives the perimeter y of all rectangles satisfying the 16-by-9 aspect ratio as a *function* of the rectangle's width x. Trace along the graph, back and forth, and describe what the display at the bottom represents. Why would nonpositive values of x be meaningless here? If such a television has perimeter 117 inches, what would be its approximate width?

EXAMPLE 2 **Solving a Mixture-of-Concentrations Problem**

How much pure alcohol should be added to 20 liters of 40% alcohol to increase the concentration to 50% alcohol?

Solution

Step 1 We let x represent the number of liters of pure alcohol that must be added. The information can be summarized in a "box diagram." See Figure 78.

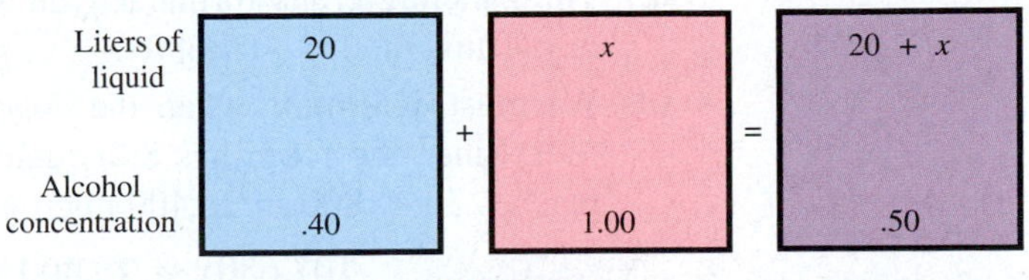

FIGURE 78

Step 2 In each box, we have the number of liters and the alcohol concentration. We multiply the two values in each box to get the amount of pure alcohol.

The amount of pure alcohol on the left must equal the amount of pure alcohol on the right, so the equation to solve is

$$\underbrace{.40(20)}_{\text{Liters of pure alcohol in starting mixture}} + \underbrace{1.00x}_{\text{Liters of pure alcohol added}} = \underbrace{.50(20 + x)}_{\text{Liters of pure alcohol in final mixture}}.$$

Step 3

$$40(20) + 100x = 50(20 + x) \quad \text{Multiply by 100.}$$

$$800 + 100x = 1000 + 50x \quad \text{Distributive property}$$

$$50x = 200 \quad \text{Subtract } 50x\text{; subtract 800.}$$

$$x = 4 \quad \text{Divide by 50.}$$

Move the decimal points two places to the right.

Therefore, 4 liters of pure alcohol must be added.

Step 4 *Check* to see that $.40(20) + 1.00(4) = .50(20 + 4)$ is true. ■

EXAMPLE 3 Examining the Effect of Formaldehyde on the Eyes

Formaldehyde is a volatile organic compound that has come to be recognized as a highly toxic indoor air pollutant. It is found in building materials such as fiberboard, plywood, foam insulation, and carpeting. When concentrations of formaldehyde in the air exceed 33 micrograms per cubic foot (1 microgram $= 10^{-6}$ gram), a strong odor and irritation to the eyes often occur. One square foot of hardwood plywood paneling can emit 3365 micrograms of formaldehyde per day. (*Source:* Hines, A., T. Ghosh, S. Layalka, and R. Warder, *Indoor Air Quality & Control*, Prentice Hall, 1993.)

A 4- by 8-foot sheet of this paneling is attached to an 8-foot wall in a room having dimensions 10 by 10 feet.

(a) How many cubic feet of air are there in the room?

(b) Find the total number of micrograms of formaldehyde that are released into the air by the paneling each day.

(c) If there is no ventilation in the room, write a linear function that models the amount of formaldehyde F in the room after x days.

(d) How long will it take before a person's eyes become irritated in the room?

Solution

(a) The volume of the room is $8 \times 10 \times 10 = 800$ cubic feet.

(b) The paneling releases 3365 micrograms for each square foot of area. The area of the sheet is $4 \times 8 = 32$ square feet, so the paneling will release $32 \times 3365 = 107{,}680$ micrograms of formaldehyde into the air each day.

(c) The paneling emits formaldehyde at a constant rate of 107,680 micrograms per day. Thus, $F(x) = 107{,}680x$.

(d) We must determine when the concentration exceeds 33 micrograms per cubic foot. Since the room has 800 cubic feet, this will occur when the total amount reaches $33 \times 800 = 26{,}400$ micrograms.

$$107{,}680x = 26{,}400 \quad \text{Solve } F(x) = 26{,}400.$$

$$x = \frac{26{,}400}{107{,}680} \quad \text{Divide by 107,680.}$$

$$x \approx .25 \quad \text{Approximate.}$$

It will take approximately $\frac{1}{4}$ day, or 6 hours. ■

Break-Even Analysis

By expressing a company's cost of producing a product and the revenue the company receives from selling the product as functions, the company can determine at what point it will break even. In other words, we try to answer the question, "For what number of items sold will the revenue collected equal the cost of producing those items?"

EXAMPLE 4 Determining the Break-Even Point

Peripheral Visions, Inc., produces studio-quality compact discs (CDs) of live concerts. The company places an ad in a trade newsletter. The cost of the ad is \$100. Each CD costs \$20 to produce, and the company charges \$24 per CD.

(a) Express the cost C as a function of x, the number of CDs produced.

(b) Express the revenue R as a function of x, the number of CDs sold.

(c) For what value of x does revenue equal cost?

(d) Graph $y_1 = C(x)$ and $y_2 = R(x)$ in an appropriate window to support the answer in part (c).

(e) Use a table to support the answer in part (c).

Solution

(a) The *fixed cost* is \$100, and for each CD produced, the *variable cost* is \$20. The total cost C in dollars can be expressed as a function of x, the number of CDs produced:

$$C(x) = 20x + 100.$$

(b) Since each CD sells for \$24, the revenue R in dollars is given by $R(x) = 24x$.

(c) The company will break even (earn no profit and suffer no loss) when revenue equals cost.

$$R(x) = C(x)$$
$$24x = 20x + 100 \quad \text{Substitute for } R(x) \text{ and } C(x).$$
$$4x = 100 \quad \text{Subtract } 20x.$$
$$x = 25 \quad \text{Divide by 4.}$$

When 25 CDs are sold, the company will break even.

(d) The graphs shown in Figure 79 confirm our solution of X = 25. The y-value of 600 indicates that when 25 CDs are sold, both the cost and the revenue are \$600.

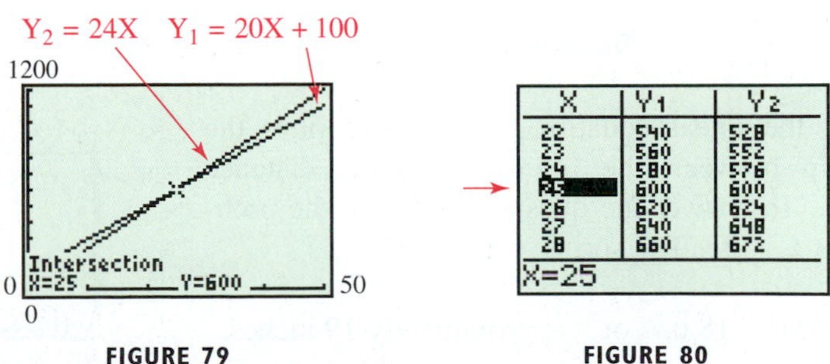

FIGURE 79

X	Y1	Y2
22	540	528
23	560	552
24	580	576
25	600	600
26	620	624
27	640	648
28	660	672

X=25

FIGURE 80

(e) The table in Figure 80 shows that when the number of CDs is 25, both function values are 600, numerically supporting our answer in part (c). ■

Direct Variation

A common application involving linear functions deals with quantities that *vary directly* (or are in direct proportion).

Direct Variation

A number y **varies directly** with x if there exists a nonzero number k such that

$$y = kx.$$

The number k is called the **constant of variation.**

Notice that the graph of $y = kx$ is simply a straight line with slope k, passing through the origin. See Figure 81.

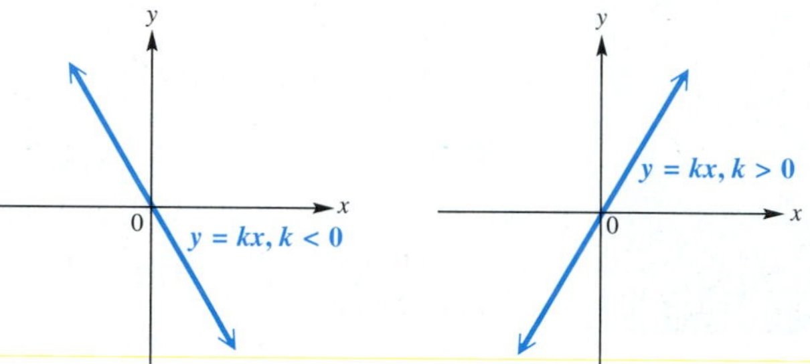

FIGURE 81

If we divide both sides of $y = kx$ by x, we get $\frac{y}{x} = k$, indicating that in a direct variation, the quotient (or proportion) of the two quantities is constant.

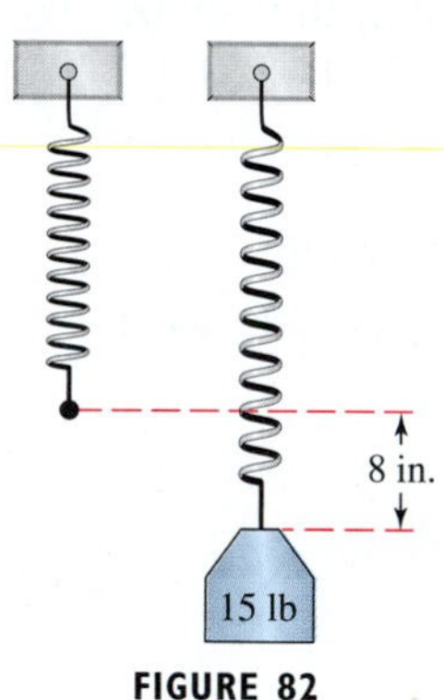

FIGURE 82

EXAMPLE 5 Solving a Direct Variation Problem (Hooke's Law)

Hooke's law states that the distance y a spring stretches varies directly with the force x applied. If a force of 15 pounds stretches a spring 8 inches, how much will a force of 35 pounds stretch the spring? See Figure 82.

Analytic Solution

Using the fact that when $x = 15$, $y = 8$, we can find k.

$$y = kx \qquad \text{Direct variation}$$

$$8 = k \cdot 15 \qquad \text{Let } x = 15,\ y = 8.$$

$$k = \frac{8}{15} \qquad \text{Divide by 15.}$$

Therefore, the linear equation $y = \frac{8}{15}x$ describes the relationship between the force x and the distance stretched y. To answer the question stated in the problem, we let $x = 35$. The spring will stretch

$$y = \frac{8}{15}(35) = 18.\overline{6}, \quad \text{or} \quad \text{approximately 19 inches.}$$

Graphing Calculator Solution

Locating the point where $X = 35$, we see in Figure 83 that $Y_1 = 18.\overline{6}$. The table of values in Figure 84 numerically supports our analytic solution. The table also allows us to find the distance in inches the spring will stretch for other values of X, such as 32, 33, and so on.

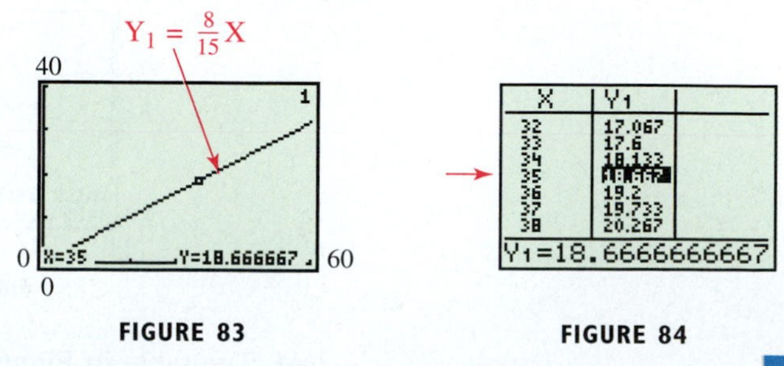

FIGURE 83

FIGURE 84

EXAMPLE 6 **Using Similar Triangles**

A grain bin in the shape of an inverted cone has height 11 feet and radius 3.5 feet. See Figure 85. If the grain in the bin is 7 feet high, calculate the volume of grain.

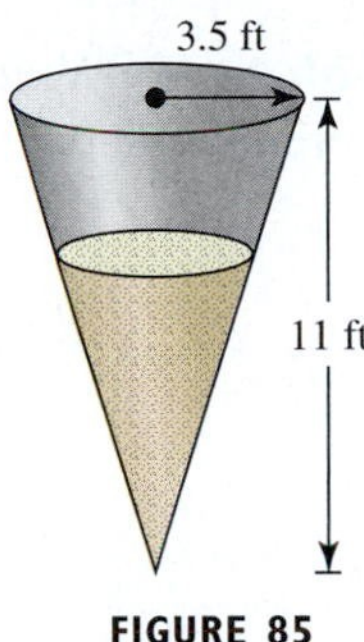

FIGURE 85

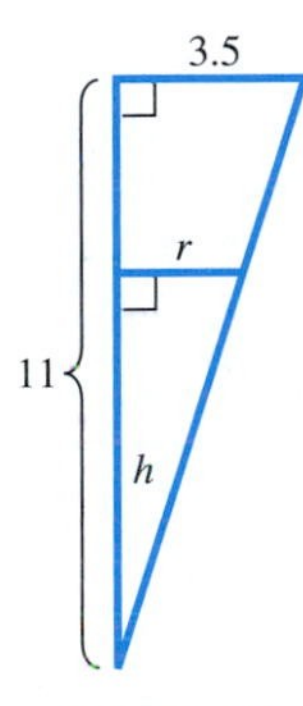

FIGURE 86

Solution The volume of a cone is $V = \frac{1}{3}\pi r^2 h$. The height h of the grain in the bin is 7 feet, but we are not given the radius r of the bin at that height. We can find the radius r corresponding to a height of 7 feet by using similar triangles, as illustrated in Figure 86. The sides of similar triangles are proportional, so

$$\frac{r}{h} = \frac{3.5}{11}$$

$$r = \frac{3.5}{11}h. \quad \text{Multiply by } h.$$

Letting $h = 7$ gives

$$r = \frac{3.5}{11}(7) \approx 2.227.$$

The volume of the grain is

$$V = \frac{1}{3}\pi(2.227)^2(7) \approx 36.4 \text{ cubic feet.}$$

Formulas

The formula $A = \frac{1}{2}h(b_1 + b_2)$ gives the area A of a trapezoid in terms of its height h and its two parallel bases b_1 and b_2. See Figure 87. Notice that A is alone on one side of the formula. Suppose, however, that we want to write the formula so that it is solved for b_1. We can use the methods of solving linear equations analytically to do this.

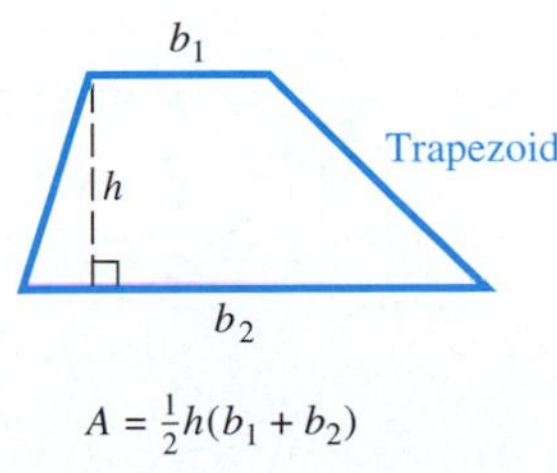

$A = \frac{1}{2}h(b_1 + b_2)$

FIGURE 87

EXAMPLE 7 **Solving a Formula for a Specified Variable**

A trapezoid has area 169 square inches, height 13 inches, and base 19 inches. Find the length of the other base by solving the formula $A = \frac{1}{2}h(b_1 + b_2)$ for b_1 and substituting.

Solution We treat the formula as if b_1 were the only variable and all other variables were constants.

Multiply both sides by 2.

$$A = \frac{1}{2}h(b_1 + b_2) \quad \text{Area formula}$$

$$2A = h(b_1 + b_2) \quad \text{Multiply by 2.}$$

$$\frac{2A}{h} = b_1 + b_2 \quad \text{Divide by } h.$$

$$\frac{2A}{h} - b_2 = b_1 \quad \text{Subtract } b_2.$$

$$b_1 = \frac{2A}{h} - b_2 \quad \text{Rewrite.}$$

To determine b_1, we substitute $A = 169$, $h = 13$, and $b_2 = 19$.

$$b_1 = \frac{2(169)}{13} - 19 = 7 \text{ inches}$$

■

EXAMPLE 8 **Solving Formulas for Specified Variables**

Solve each formula for the specified variable.

(a) $P = 2W + 2L$ for L (Perimeter of a rectangle)

(b) $C = \frac{5}{9}(F - 32)$ for F (Fahrenheit to Celsius)

Solution

(a)

$$P = 2W + 2L$$

$$P - 2W = 2L \quad \text{Subtract } 2W.$$

$$\frac{P - 2W}{2} = L \quad \text{Divide by 2.}$$

$$L = \frac{P}{2} - W \quad \text{Rewrite; simplify.}$$

(b)

$$C = \frac{5}{9}(F - 32)$$

$$\frac{9}{5}C = F - 32 \quad \text{Multiply by } \tfrac{9}{5}.$$

$$F = \frac{9}{5}C + 32 \quad \text{Add 32; rewrite.}$$

■

1.6 Exercises

Concept Check *Work Exercises 1–6 mentally.*

1. ***Acid Solution*** If 40 L of an acid solution is 75% acid, how much pure acid is there in the mixture?

2. ***Direct Variation*** If y varies directly as x, and $y = 2$ when $x = 4$, what is the value of y when $x = 12$?

3. ***Acid Mixture*** Suppose two acid solutions are mixed. One is 26% acid and the other is 32% acid. Which one of the following concentrations cannot possibly be the concentration of the mixture?
   A. 36% **B.** 28% **C.** 30% **D.** 31%

4. ***Sale Price*** Suppose that a computer that originally sold for x dollars has been discounted 30%. Which one of the following expressions does not represent the sale price of the computer?
   A. $x - .30x$ **B.** $.70x$ **C.** $\frac{7}{10}x$ **D.** $x - .30$

5. ***Unknown Numbers*** Consider the following problem:

   One number is three less than six times a second number. Their sum is 32. Find the numbers.

   If x represents the second number, which equation is correct for solving this problem?
   A. $32 - (x + 3) = 6x$ **B.** $(3 - 6x) + x = 32$
   C. $32 - (3 - 6x) = x$ **D.** $(6x - 3) + x = 32$

6. ***Unknown Numbers*** Consider the following problem:

   The difference between six times a number and 9 is equal to five times the sum of the number and 2. Find the number.

   If x represents the number, which equation is correct for solving this problem?
   A. $6x - 9 = 5(x + 2)$ **B.** $9 - 6x = 5(x + 2)$
   C. $6x - 9 = 5x + 2$ **D.** $9 - 6x = 5x + 2$

Solve each problem analytically, and support your solution graphically.

7. ***Perimeter of a Rectangle*** The perimeter of a rectangle is 98 centimeters. The width is 19 centimeters. Find the length.

8. ***Perimeter of a Storage Shed*** Lee Guidroz must build a rectangular storage shed. He wants the length to be 3 feet greater than the width, and the perimeter must be 22 feet. Find the length and the width of the shed.

9. ***Dimensions of a Label*** The length of a rectangular label is 2.5 centimeters less than twice the width. The perimeter is 40.6 centimeters. Find the width.

10. ***Dimensions of a Label*** The length of a rectangular mailing label is 3 centimeters less than twice the width. The perimeter is 54 centimeters. Find the dimensions of the label.

11. ***Dimensions of a Square*** If the length of a side of a square is increased by 3 centimeters, the perimeter of the new square is 40 centimeters more than twice the length of the side of the original square. Find the length of the side of the original square.

12. ***World's Largest Needlepoint*** The world's largest needlepoint has a perimeter of 940 inches. It was made by Vahid Saadati and 650 volunteers in Canada beginning in August 2000. Its length is 120 inches more than its width. What is its length in yards? (*Source: Guiness Book of Records, 2005.*)

13. ***Aspect Ratio of a Television Monitor*** The aspect ratio of a conventional television monitor is 4:3. One such model Toshiba television has a rectangular viewing screen with perimeter 98 inches. What are the length and width of the screen? Since televisions are advertised by the diagonal measure of their screens, how would this monitor be advertised?

14. ***Aspect Ratio of a Television Monitor*** Repeat Exercise 13 for a Sony television screen having perimeter 126 inches.

15. ***Dimensions of a Puzzle Piece*** A puzzle piece in the shape of a triangle has perimeter 30 centimeters. Two sides of the triangle are each twice as long as the shortest side. Find the length of the shortest side.

16. ***Perimeter of a Plot of Land*** The perimeter of a triangular plot of land is 2400 feet. The longest side is 200 feet less than twice the shortest. The middle side is 200 feet less than the longest side. Find the lengths of the three sides of the triangular plot.

17. ***World's Largest Tablecloth*** The world's largest rectangular tablecloth has perimeter 44,252 inches. It was manufactured by Döhler, S.A., of Joinville, Santa Catarina, Brazil, on December 7, 1999. Its length is 22,000 inches more than its width. What is its length, to the nearest hundredth, in yards? (*Source*: www.guinnessworldrecords.com/index.asp)

18. ***Smallest Ticket Size*** The smallest ticket ever produced was for admission to the Asian-Pacific Exposition Fukuoka '89 in Japan. The ticket was rectangular, with length 4 millimeters more than its width. If the width had been increased by 1 millimeter and the length had been increased by 4 millimeters, the ticket would have had perimeter equal to 30 millimeters. What were the actual dimensions of the ticket? (*Source: The Guinness Book of World Records.*)

19. ***Acid Mixture*** How many gallons of a 5% acid solution must be mixed with 5 gallons of a 10% solution to obtain a 7% solution?

20. ***Acid Mixture*** Bill Charlson needs 10% hydrochloric acid for a chemistry experiment. How much 5% acid should he mix with 60 milliliters of 20% acid to obtain a 10% solution?

21. ***Alcohol Mixture*** How many gallons of pure alcohol should be mixed with 20 gallons of a 15% alcohol solution to obtain a mixture that is 25% alcohol?

22. ***Alcohol Mixture*** A chemist wishes to strengthen a mixture from 10% alcohol to 30% alcohol. How much pure alcohol should be added to 7 liters of the 10% mixture?

23. ***Saline Solution Mixture*** How much water should be added to 8 milliliters of 6% saline solution to reduce the concentration to 4% saline?

24. ***Acid Mixture*** How much water should be added to 20 liters of an 18% acid solution to reduce the concentration to 15% acid?

25. ***Antifreeze Mixture*** An automobile radiator holds 16 liters of fluid. There is currently a mixture in the radiator that is 80% antifreeze and 20% water. How much of this mixture should be drained and replaced by pure antifreeze so that the resulting mixture is 90% antifreeze?

26. ***Antifreeze Mixture*** An automobile radiator contains a 10-quart mixture of water and antifreeze that is 40% antifreeze. How much should the owner drain from the radiator and replace with pure antifreeze so that the liquid in the radiator will be 80% antifreeze?

Exercises 27 and 28 involve octane rating of gasoline, a measure of its antiknock qualities. In one measure of octane, a standard fuel is made with only two ingredients: heptane and isooctane. For this fuel, the octane rating is the percent of isooctane. An actual gasoline blend is then compared with a standard fuel. For example, a gasoline with an octane rating of 98 has the same antiknock properties as a standard fuel that is 98% isooctane.

27. ***Octane Rating of Gasoline*** How many gallons of 94-octane gasoline should be mixed with 400 gallons of 99-octane gasoline to obtain a mixture that is 97-octane?

28. ***Octane Rating of Gasoline*** How many gallons of 92-octane and 98-octane gasoline should be mixed together to provide 120 gallons of 96-octane gasoline?

(Modeling) *Solve each problem.*

29. ***Ventilation in a Classroom*** According to the American Society of Heating, Refrigerating and Air-Conditioning Engineers, Inc. (ASHRAE), a nonsmoking classroom should have a ventilation rate of 15 cubic feet per minute for each person in the room. (*Source:* ASHRAE, 1989.)
    (a) Write a linear function that describes the total ventilation V (in cubic feet per hour) necessary for a classroom with x people.
    (b) A common unit of ventilation is an *air change per hour* (ach). One ach is equivalent to exchanging all of the air in a room every hour. If x people are in a classroom having a volume of 15,000 cubic feet, determine how many air exchanges per hour are necessary to keep the room properly ventilated.
    (c) Find the necessary number of ach A if the classroom has 40 people in it.
    (d) In areas like bars and lounges that allow smoking, the ventilation rate should be increased to 50 cubic feet per minute per person. Compared with classrooms, ventilation should be increased by what factor in heavy-smoking areas?

30. ***Indoor Air Quality and Control*** The excess lifetime cancer risk R is a measure of the likelihood that an individual will develop cancer from a particular pollutant. For example, if $R = .01$, then a person has a 1% increased chance of developing cancer during a lifetime. (This would translate into 1 case of cancer for every 100 people during an average lifetime.) The value of R for formaldehyde, a highly toxic indoor air pollutant, can be calculated from the linear model $R = kd$, where k is a constant and d is the daily dose in parts per million. The constant k for formaldehyde can be calculated with the formula

$$k = \frac{.132B}{W},$$

where B is the total number of cubic meters of air a person breathes in one day and W is a person's weight in kilograms. (*Source:* Hines, A., T. Ghosh, S. Loyalka, and R. Warder, *Indoor Air: Quality & Control,* Prentice Hall, 1993; Ritchie, I., and R. Lehnen, "An Analysis of Formaldehyde Concentration in Mobile and Conventional Homes," *J. Env. Health* 47: 300–305.)

(a) Find k for a person who breathes in 20 cubic meters of air per day and weighs 75 kilograms.

(b) Mobile homes in Minnesota were found to have a mean daily dose d of .42 part per million. Calculate R, using the value of k found in part (a).

(c) For every 5000 people, how many cases of cancer could be expected each year from these levels of formaldehyde? Assume an average life expectancy of 72 years.

31. ***Indoor Air Quality and Control*** (See Exercise 30.) For nonsmokers exposed to environmental tobacco smoke (passive smokers), $R = 1.5 \times 10^{-3}$. (*Source:* Hines A., T. Ghosh, S. Loyalka, and R. Warder, *Indoor Air: Quality & Control,* Prentice Hall, 1993.)

(a) If the average life expectancy is 72 years, what is the excess lifetime cancer risk from secondhand tobacco smoke per year?

(b) Write a linear equation that will model the expected number of cancer cases C per year if there are x passive smokers.

(c) Estimate the number of cancer cases each year per 100,000 passive smokers.

(d) The excess lifetime risk of death from smoking is $R = .44$. Currently 26% of the U.S. population smoke. If the U.S. population is 260 million, approximate the excess number of deaths caused by smoking each year.

32. ***Bicycle Safety*** A survey found that 76% of bicycle riders do not wear helmets. (*Source:* Opinion Research Corporation for Glaxo Wellcome, Inc.)

(a) Write a linear function f that computes the number of people who do not wear helmets among x bicycle riders.

(b) There are approximately 38.7 million riders of all ages who do not wear helmets. Write a linear equation whose solution gives the total number of bicycle riders. Find this number of riders.

(Modeling) In Exercises 33–36,

(a) *Express the cost C as a function of x, where x represents the number of items as described.*

(b) *Express the revenue R as a function of x.*

(c) *Determine analytically the value of x for which revenue equals cost.*

(d) *Graph $y_1 = C(x)$ and $y_2 = R(x)$ on the same xy-axes and interpret the graphs.*

33. ***Stuffing Envelopes*** Bonnie Robillard stuffs envelopes for extra income during her spare time. Her initial cost to obtain the necessary information for the job was \$200.00. Each envelope costs \$.02, and she gets paid \$.04 per envelope stuffed. Let x represent the number of envelopes stuffed.

34. ***Copier Service*** B. K. Tomlin runs a copying service in his home. He paid \$3500 for the copier and a lifetime service contract. Each sheet of paper he uses costs \$.01, and he gets paid \$.05 per copy. Let x represent the number of copies he makes.

35. ***Delivery Service*** Wayne Pourciau operates a delivery service in a southern city. His start-up costs amounted to \$2300. He estimates that it costs him (in terms of gasoline, wear and tear on his car, etc.) \$3.00 per delivery. He charges \$5.50 per delivery. Let x represent the number of deliveries he makes.

36. ***Baking and Selling Cakes*** Vangie Tomlin bakes cakes and sells them at county fairs. Her initial cost for the Pointe Coupee parish fair was \$40.00. She figures that each cake costs \$2.50 to make, and she charges \$6.50 per cake. Let x represent the number of cakes sold. (Assume that there were no cakes left over.)

In Exercises 37–40, find the constant of variation k and the undetermined value in the table if y is directly proportional to x.

37.

x	3	5	6	8
y	7.5	12.5	15	?

38.

x	1.2	4.3	5.7	?
y	3.96	14.19	18.81	23.43

39. Sales tax y on a purchase of x dollars

x	\$25	\$55	?
y	\$1.50	\$3.30	\$5.10

40. Cost y of buying x compact discs having the same price

x	3	4	5
y	\$41.97	\$55.96	?

Solve each problem.

41. ***Pressure of a Liquid*** The pressure exerted by a certain liquid at a given point is directly proportional to the depth of the point beneath the surface of the liquid. If the pressure exerted at 30 feet is 13 pounds per square inch, what is the pressure exerted at 70 feet?

42. ***Rate of Nerve Impulses*** The rate at which impulses are transmitted along a nerve fiber is directly proportional to the diameter of the fiber. If the rate for a certain fiber is 40 meters per second when the diameter is 6 micrometers, what is the rate if the diameter is 8 micrometers?

43. ***Cost of Tuition*** The cost of tuition is directly proportional to the number of credits taken. If 11 credits cost \$720.50, find the cost of taking 16 credits. What is the constant of variation?

44. ***Strength of a Beam*** The maximum load that a horizontal beam can carry is directly proportional to its width. If a beam 1.5 inches wide can support a load of 250 pounds, find the load that a beam of the same type can support if its width is 3.5 inches. What is the constant of variation?

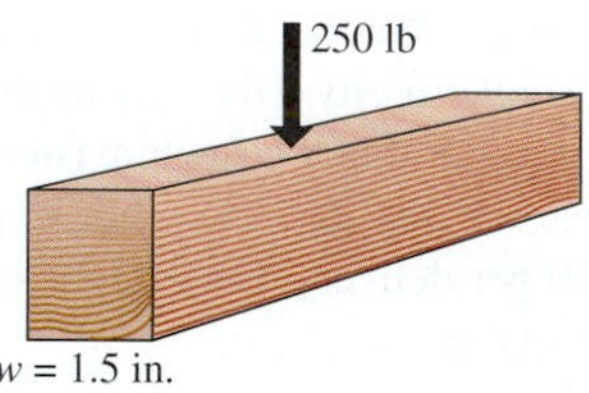

45. ***Volume of Water*** A water tank in the shape of an inverted cone has height 11 feet and radius 5.0 feet, as illustrated in the accompanying figure. Find the volume of the water in the tank when the water is 6 feet deep.

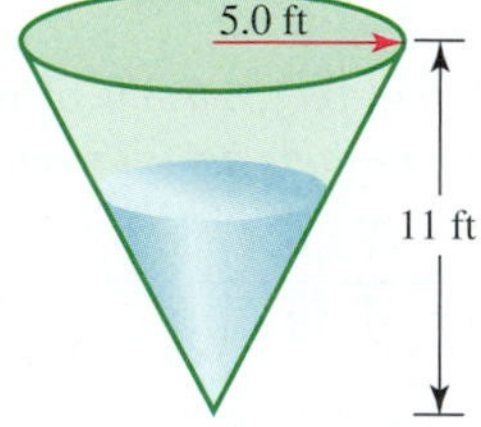

46. ***Volume of Water*** A water tank in the shape of an inverted cone has height 6 feet and radius 2 feet. If the water level in the tank is 3.5 feet, calculate the volume of the water.

47. ***Height of a Tree*** A certain tree casts a shadow 45 feet long. At the same time, the shadow cast by a vertical stick 2 feet high is 1.75 feet long. How tall is the tree? (*Hint:* Use similar triangles.)

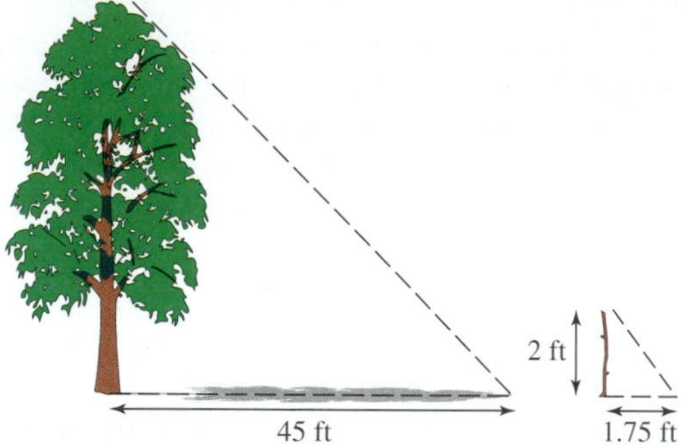

48. ***Height of a Streetlight*** A person 66 inches tall is standing 15 feet from a streetlight. If the person casts a shadow 84 inches long, how tall is the streetlight?

49. ***Hooke's Law*** If a 3-pound weight stretches a spring 2.5 inches, how far will a 17-pound weight stretch the spring?

50. ***Hooke's Law*** If a 9.8-pound weight stretches a spring .75 inch, how much weight would be needed to stretch the spring 3.1 inches?

Biologists use direct variation to estimate the number of individuals of a species in a particular area. They first capture a sample of individuals from the area and mark each specimen with a harmless tag. Later, they return and capture another sample from the same area. They base their estimate on the theory that the proportion of tagged specimens in the new sample is the same as the proportion of tagged individuals in the entire area. Use this idea to work Exercises 51 and 52.

51. ***Estimating Fish in a Lake*** Biologists tagged 250 fish in City Park Lake on October 12. On a later date, they found 7 tagged fish in a sample of 350. Estimate, to the nearest hundred, the total number of fish in the lake.

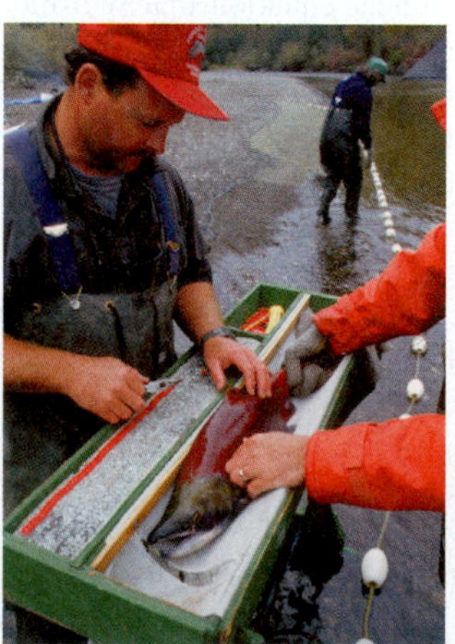

52. ***Estimating Seal Pups in a Breeding Area*** According to an actual survey in 1961, to estimate the number of seal pups in a certain breeding area in Alaska, 4963 pups were tagged in early August. In late August, a sample of 900 pups was examined and 218 of these were found to have been tagged. Use this information to estimate, to the nearest hundred, the total number of seal pups in this breeding area. (*Source:* "Estimating the Size of Wildlife Populations," Chatterjee, S., in *Statistics by Example,* 1973, obtained from data in *Transactions of the American Fisheries Society,* July 1968.)

(Modeling) *In Exercises 53–56, assume that a linear relationship exists between the two quantities.*

53. ***Solar Heater Production*** A company finds that it can produce 10 solar heaters for \$7500, whereas producing 20 heaters costs \$13,900.

(a) Express the cost y as a linear function of the number of heaters, x.

(b) Determine analytically the cost to produce 25 heaters.

(c) Support the result of part (b) graphically.

54. ***Cricket Chirping*** At 68°F, a certain species of cricket chirps 24 times per minute. At 40°F, the same cricket chirps 86 times per minute. (*Source:* Bushaw, D., et al., *A Sourcebook of Applications of School Mathematics,* MAA, 1980. Reprinted with permission.)

(a) Express the number of chirps, y, as a linear function of the Fahrenheit temperature.

(b) If the temperature is 60°F, how many times will the cricket chirp per minute?

(c) If you count the number of cricket chirps in one-half minute and hear 40 chirps, what is the temperature?

55. ***Appraised Value of a Home*** In 1990, a house was purchased for \$120,000. In 2000, it was appraised for \$146,000.

(a) If $x = 0$ represents 1990 and $x = 10$ represents 2000, express the appraised value of the house, y, as a linear function of the number of years, x, after 1990.

(b) What will the house be worth in the year 2004?

(c) What does the slope of the line represent (in your own words)?

56. ***Depreciation of a Photocopier*** A photocopier sold for \$3000 in 1994 when it was purchased. Its value in 2002 had depreciated to \$600.

(a) If $x = 0$ represents 1994 and $x = 8$ represents 2002, express the value of the machine, y, as a linear function of the number of years, x, after 1994.

(b) Graph the function from part (a) in a window [0, 10] by [0, 4000]. How would you interpret the y-intercept in terms of this particular situation?

(c) Use your calculator to determine the value of the machine in 1998, and verify your result analytically.

57. ***Snowmaking and Water Consumption*** Ski resorts require large amounts of water in order to make snow. Snowmass Ski Area in Colorado plans to pump between 1120 and 1900 gallons of water per minute at least 12 hours per day from Snowmass Creek between mid-October and late December. (*Source:* York Snow, Incorporated.)

(a) Determine a linear function that calculates the *minimum* amount of water A (in gallons) pumped after x days from mid-October to late December.

(b) Find the minimum amount of water pumped in 30 days.

(c) Suppose the water being pumped from Snowmass Creek was used to fill swimming pools. The average backyard swimming pool holds 20,000 gallons of water. Determine a linear function that will give the minimum number of pools P that could be filled after x days. How many pools could be filled each day?

(d) In how many days could a minimum of 1000 pools be filled?

58. ***Speeding Fines*** Suppose that speeding fines are determined by the linear function

$$y = 10(x - 65) + 50, \quad x > 65,$$

where y is the cost in dollars of the fine if a person is caught driving x miles per hour.

(a) Radar clocked a driver at 76 mph. How much was the fine?

(b) While balancing his checkbook, Johnny ran across a canceled check that his wife Gwen had written to the Department of Motor Vehicles for a speeding fine. The check was written for \$100. How fast was Gwen driving?

(c) At what whole-number speed are tickets first given?

(d) For what speeds is the fine greater than \$200?

59. ***Expansion and Contraction of Gases*** In 1787, Jacques Charles noticed that gases expand when heated and contract when cooled. Suppose that a particular gas follows the model

$$y = \frac{5}{3}x + 455,$$

where x is the temperature in Celsius and y is the volume in cubic centimeters. (*Source:* Bushaw, D., et al., *A Sourcebook of Applications of School Mathematics,* MAA, 1980. Reprinted with permission.)

(a) What is the volume when the temperature is 27°C?

(b) What is the temperature when the volume is 605 cubic centimeters?

(c) Determine what temperature gives a volume of 0 cubic centimeters (that is, absolute zero, or the coldest possible temperature).

60. ***Tail Length of a Snake*** It has been reported that the total length x and the tail length y of females of the snake species *Lampropeltis polyzona* are nearly linearly related by the model

$$y = .134x - 1.18,$$

where x is the length of the snake in millimeters. If a snake of this species measures 1000 millimeters, what is its tail length to the nearest millimeter? (*Source:* Bushaw, D., et al., *A Sourcebook of Applications of School Mathematics*, MAA, 1980. Reprinted with permission.)

Solve each formula for the specified variable.

61. $I = PRT$ for P (Simple interest)

62. $V = LWH$ for L (Volume of a box)

63. $P = 2L + 2W$ for W (Perimeter of a rectangle)

64. $P = a + b + c$ for c (Perimeter of a triangle)

65. $A = \frac{1}{2}h(b_1 + b_2)$ for h (Area of a trapezoid)

66. $A = \frac{1}{2}h(b_1 + b_2)$ for b_1 (Area of a trapezoid)

67. $S = 2LW + 2WH + 2HL$ for H (Surface area of a rectangular box)

68. $S = 2\pi rh + 2\pi r^2$ for h (Surface area of a cylinder)

69. $V = \frac{1}{3}\pi r^2 h$ for h (Volume of a cone)

70. $y = a(x - h)^2 + k$ for a (Mathematics)

71. $S = \frac{n}{2}(a_1 + a_n)$ for n (Mathematics)

72. $S = \frac{n}{2}[2a_1 + (n - 1)d]$ for a_1 (Mathematics)

73. $s = \frac{1}{2}gt^2$ for g (Distance traveled by a falling object)

74. $A = \frac{24f}{B(p + 1)}$ for p (Approximate annual interest rate)

Investment problems such as those in Exercises 75–80 can be solved by using a method similar to the one explained in Example 2, along with the simple-interest formula $I = PRT$. Solve each problem.

75. ***Real-Estate Financing*** Cody Westmoreland wishes to sell a piece of property for $240,000. He wants the money to be paid off in two ways: a short-term note at 6% interest and a long-term note at 5%. Find the amount of each note if the total annual interest paid is $13,000.

76. ***Buying and Selling Land*** Ms. Gay Aguillard bought two plots of land for a total of $120,000. When she sold the first plot, she made a profit of 15%. When she sold the second, she lost 10%. Her total profit was $5500. How much did she pay for each piece of land?

77. ***Retirement Planning*** In planning her retirement, Mary Lynn Ellis deposits some money at 2.5% interest with twice as much deposited at 3%. Find the amount deposited at each rate if the total annual interest income is $850.

78. ***Investing a Building Fund*** A church building fund has invested some money in two ways: part of the money at 4% interest and four times as much at 3.5%. Find the amount invested at each rate if the total annual income from interest is $3600.

79. ***Lottery Winnings*** Nancy B. Kindy won $200,000 in a state lottery. She first paid income tax of 30% on the winnings. Of the rest, she invested some at 1.5% and some at 4%, earning $4350 interest per year. How much did she invest at each rate?

80. ***Cookbook Royalties*** Latasha Williams earned $48,000 from royalties on her cookbook. She paid a 28% income tax on these royalties. The balance was invested in two ways, some of it at 3.25% interest and some at 1.75%. The investments produced $904.80 interest per year. Find the amount invested at each rate.

Reviewing Basic Concepts (Sections 1.5 and 1.6)

1. Solve $3(x - 5) + 2 = 1 - (4 + 2x)$ analytically. Support your result graphically or numerically.

2. Solve $\pi(1 - x) = .6(3x - 1)$ with the intersection-of-graphs method. Round your answer to the nearest thousandth.

3. Find the zero of $f(x) = \frac{1}{3}(4x - 2) + 1$ analytically. Use the x-intercept method to support your analytic result graphically.

4. Determine whether each equation is an identity, a contradiction, or a conditional equation. Give the solution set.
(a) $4x - 5 = -2(3 - 2x) + 3$
(b) $5x - 9 = 5(-2 + x) + 1$
(c) $5x - 4 = 3(6 - x)$

5. Solve $2x + 3(x + 2) < 1 - 2x$ analytically. Express the solution set in interval notation. Support your result graphically.

6. Solve $-5 \leq 1 - 2x < 6$ analytically.

7. Use the figure to solve **(a)** $f(x) = g(x)$ **(b)** $f(x) \leq g(x)$.

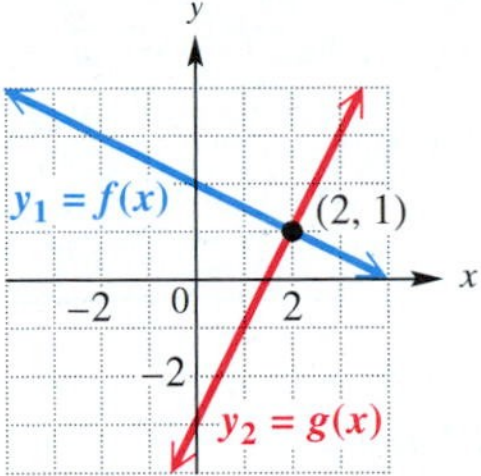

8. ***Height of a Tree*** A tree casts a 27-foot shadow, and a 6-foot person casts a 4-foot shadow. How tall is the tree?

9. ***(Modeling) Production of Compact Discs*** A student is planning to produce and sell compact discs (CDs) with recorded music for \$5.50 each. A computer with a CD burner costs \$2500, and each blank disc costs \$1.50.
   (a) Find a formula for a function R that calculates the revenue from selling x discs.
   (b) Find a formula for a function C that calculates the cost of recording x discs. Be sure to include the fixed cost of the computer.
   (c) Find the break-even point. That is, determine how many discs must be sold for revenue to equal cost.

10. Solve $V = \pi r^2 h$ for h.

CHAPTER 1 *SUMMARY*

KEY TERMS & SYMBOLS

1.1 Real Numbers and the Rectangular Coordinate System

natural numbers
whole numbers
integers
number line
origin
graph
coordinate system
coordinate
rational numbers
real numbers
irrational numbers
rectangular (Cartesian) coordinate system
x-axis
y-axis
coordinate (xy-) plane
quadrants
x-coordinate
y-coordinate
standard viewing window
Xmin
Xmax
Ymin
Ymax
Xscl
Yscl

KEY CONCEPTS

Natural numbers $\{1, 2, 3, 4, 5, \ldots\}$
Whole numbers $\{0, 1, 2, 3, 4, 5, \ldots\}$
Integers $\{\ldots, -3, -2, -1, 0, 1, 2, 3, \ldots\}$
Rational numbers $\left\{\frac{p}{q} \,\middle|\, p, q \text{ are integers}, q \neq 0\right\}$
Real numbers $\{x \mid x \text{ corresponds to a point on a number line}\}$
(The real numbers include both the rational and the irrational numbers.)

RECTANGULAR COORDINATE SYSTEM
Points are located by using ordered pairs in the xy-plane.

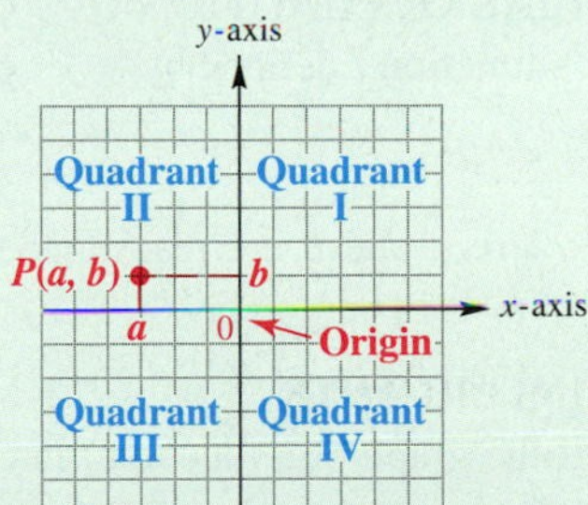

ROOTS
The nth root of a real number a can be expressed as $\sqrt[n]{a}$ or $a^{1/n}$. Calculators are often used to approximate roots of real numbers.

DISTANCE FORMULA
Suppose that $P(x_1, y_1)$ and $R(x_2, y_2)$ are two points in a coordinate plane. Then the distance between P and R is given by the distance formula

$$d(P, R) = \sqrt{(x_2 - x_1)^2 + (y_2 - y_1)^2}.$$

(continued)

KEY TERMS & SYMBOLS	KEY CONCEPTS
	MIDPOINT FORMULA The midpoint of the line segment with endpoints (x_1, y_1) and (x_2, y_2) is $$\left(\frac{x_1 + x_2}{2}, \frac{y_1 + y_2}{2}\right).$$
1.2 Introduction to Relations and Functions set-builder notation interval interval notation infinity symbol, ∞ relation domain range function independent variable dependent variable vertical line test function notation, $f(x)$	**INTERVAL NOTATION** Interval notation can be used to denote sets of real numbers: $\{x \mid x \leq a\}$ is equivalent to $(-\infty, a]$. $\{x \mid a \leq x < b\}$ is equivalent to $[a, b)$. $\{x \mid x > b\}$ is equivalent to (b, ∞). **RELATION** A relation is a set of ordered pairs. **DOMAIN AND RANGE** If we denote the ordered pairs of a relation by (x, y), then the set of all x-values is called the domain of the relation and the set of all y-values is called the range of the relation. **FUNCTION** A function is a relation in which each element in the domain corresponds to exactly one element in the range. **VERTICAL LINE TEST** If every vertical line intersects a graph in no more than one point, then the graph is the graph of a function. **FUNCTION NOTATION** To denote that y is a function of x, we write $\boldsymbol{y = f(x)}$. The independent variable is x and the dependent variable is y.
1.3 Linear Functions linear function linear equation in two variables solution x-intercept y-intercept zero of a function constant function comprehensive graph change in x, Δx change in y, Δy slope, m slope–intercept form	**LINEAR FUNCTION** A function f defined by $$f(x) = ax + b,$$ where a and b are real numbers, is called a linear function. The graph of a linear function is a line. **INTERCEPTS** If the graph of a function f intersects the x-axis at $(a, 0)$, then a is an x-intercept. If the graph of a function f intersects the y-axis at $(0, b)$, then b is the y-intercept. **ZERO OF A FUNCTION f** Any number c for which $\boldsymbol{f(c) = 0}$ is called a zero of f. The real zeros of a function f correspond to the x-intercepts of its graph. **CONSTANT FUNCTION** A function defined by $\boldsymbol{f(x) = b}$, where b is a real number, is called a constant function. A constant function is a special type of linear function whose graph is a horizontal line. *(continued)*

KEY TERMS & SYMBOLS	KEY CONCEPTS
	SLOPE The slope m of the line passing through the points (x_1, y_1) and (x_2, y_2) is $$m = \frac{\Delta y}{\Delta x} = \frac{y_2 - y_1}{x_2 - x_1}, \quad \Delta x \neq 0.$$ **GEOMETRIC ORIENTATION BASED ON SLOPE** For a line with slope m, if $m > 0$, the line rises from left to right. If $m < 0$, it falls from left to right. If $m = 0$, the line is horizontal. **VERTICAL LINE** A vertical line with x-intercept a has an equation of the form $\boldsymbol{x = a}$. Its slope is undefined. **SLOPE–INTERCEPT FORM** The slope–intercept form of the equation of a line with slope m and y-intercept b is $$\boldsymbol{y = mx + b}.$$
1.4 Equations of Lines and Linear Models point–slope form standard form parallel lines perpendicular lines square viewing window scatter diagram least-squares regression line least-squares regression correlation coefficient r	**POINT–SLOPE FORM** The line with slope m passing through the point (x_1, y_1) has equation $$\boldsymbol{y - y_1 = m(x - x_1)}.$$ **STANDARD FORM** A linear equation written in the form $$\boldsymbol{Ax + By = C},$$ where A, B, and C are real numbers (A and B not both 0), is said to be in standard form. **PARALLEL AND PERPENDICULAR LINES** Two distinct nonvertical lines are parallel if and only if they have the same slope. Two lines, neither of which is vertical, are perpendicular if and only if their slopes have product -1. **LINEAR MODELS AND REGRESSION** If a collection of data points approximates a straight line, then we can find the equation of such a line. Choosing two data points, we apply the methods of Sections 1.3 and 1.4 to find this equation. The equation will vary with the two points chosen. The line of best fit can be found using the linear regression feature of a graphing calculator.
1.5 Linear Equations and Inequalities equation solution (root) solution set linear equation in one variable equivalent equations conditional equation	**LINEAR EQUATIONS AND INEQUALITIES** A linear equation can be written in the form $\boldsymbol{ax + b = 0}$, where $a \neq 0$. A linear inequality can be written in one of the following forms: $$\boldsymbol{ax + b < 0}, \quad \boldsymbol{ax + b > 0}, \quad \boldsymbol{ax + b \leq 0}, \quad \text{and} \quad \boldsymbol{ax + b \geq 0}.$$ **ANALYTIC SOLUTIONS** To solve linear equations, use the addition and multiplication properties of equality. To solve linear inequalities, use the properties of inequality. ***When multiplying or dividing each side of an inequality by a negative number, be sure to reverse the direction of the inequality symbol.*** *(continued)*

KEY TERMS & SYMBOLS	KEY CONCEPTS
contradiction empty (null) set $\emptyset$ identity inequality equivalent inequalities linear inequality in one variable three-part inequality	**INTERSECTION-OF-GRAPHS METHOD OF GRAPHICAL SOLUTION** To solve the equation $f(x) = g(x)$, graph $$y_1 = f(x) \quad \text{and} \quad y_2 = g(x).$$ The x-coordinate of any point of intersection of the two graphs is a solution of the equation. The solution set of $f(x) > g(x)$ is the set of all real numbers x such that the graph of f is **above** the graph of g. The solution set of $f(x) < g(x)$ is the set of all real numbers x such that the graph of f is **below** the graph of g. **x-INTERCEPT METHOD OF GRAPHICAL SOLUTION** To solve the equation $f(x) = g(x)$, graph $$y = f(x) - g(x) = F(x).$$ Any x-intercept of the graph of $y = F(x)$ is a solution of the equation. The solution set of $F(x) > 0$ is the set of all real numbers x such that the graph of F is **above** the x-axis. The solution set of $F(x) < 0$ is the set of all real numbers x such that the graph of F is **below** the x-axis.
1.6 Applications of Linear Functions varies directly constant of variation	**PROBLEM-SOLVING STRATEGY** **1.** Read the problem. **2.** Write an equation. **3.** Solve the equation. **4.** Look back and check. **DIRECT VARIATION** A number y varies directly with x if there exists a nonzero number k such that $y = kx$. The number k is called the constant of variation.

CHAPTER 1 Review Exercises

Let A represent the point with coordinates $(-1, 16)$, *and let B represent the point with coordinates* $(5, -8)$.

1. Find the exact distance between points A and B.

2. Find the coordinates of the midpoint of the line segment connecting points A and B.

3. Find the slope of the line segment AB.

4. Find the equation of the line passing through points A and B. Write it in $y = mx + b$ form.

Consider the graph of $3x + 4y = 144$ *in Exercises 5–8.*

5. What is the slope of the line?

6. What is the x-intercept of the line?

7. What is the y-intercept of the line?

8. Give a viewing window that will show a comprehensive graph. (There are many possible such windows.)

9. Suppose that f is a linear function such that $f(3) = 6$ and $f(-2) = 1$. Find $f(8)$. (*Hint:* Find a formula for $f(x)$.)

10. Find the equation of the line perpendicular to the graph of $y = -4x + 3$ and passing through the point $(-2, 4)$. Write it in $y = mx + b$ form.

For each line shown in Exercises 11 and 12, do the following.

(a) *Find the slope.*

(b) *Find the slope–intercept form of the equation.*

(c) *Find the midpoint of the segment connecting the two points identified on the graph.*

(d) *Find the distance between the two points identified on the graph.*

11.

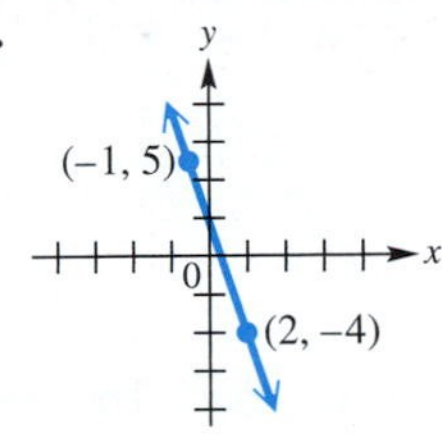

12.

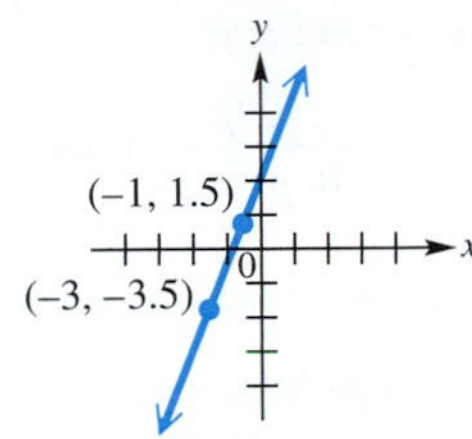

Concept Check *Choose the letter of the graph that would most closely resemble the graph of $f(x) = mx + b$, given the conditions on m and b.*

13. $m < 0, b < 0$

14. $m > 0, b < 0$

15. $m < 0, b > 0$

16. $m > 0, b > 0$

17. $m = 0$

18. $b = 0$

A.

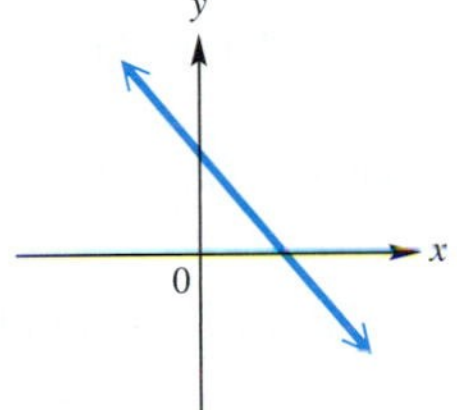

B.

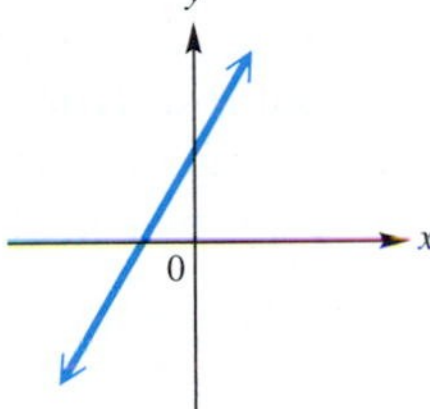

C.

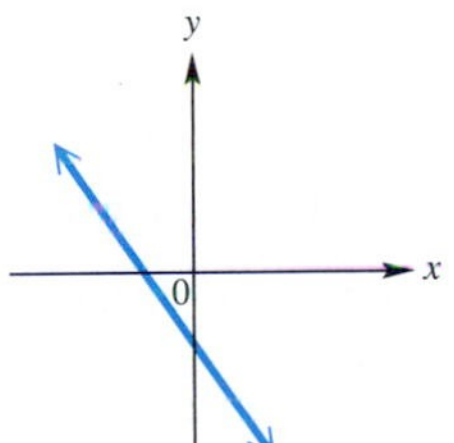

D.

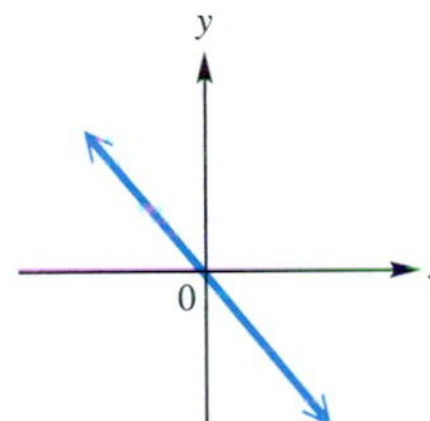

E.

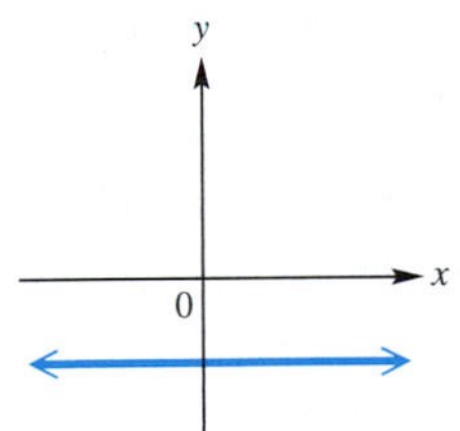

F.

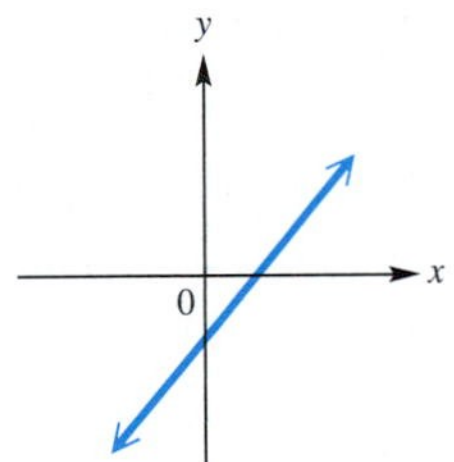

19. What are the domain and range of the relation graphed on the screen? (*Hint:* Pay attention to scale.)

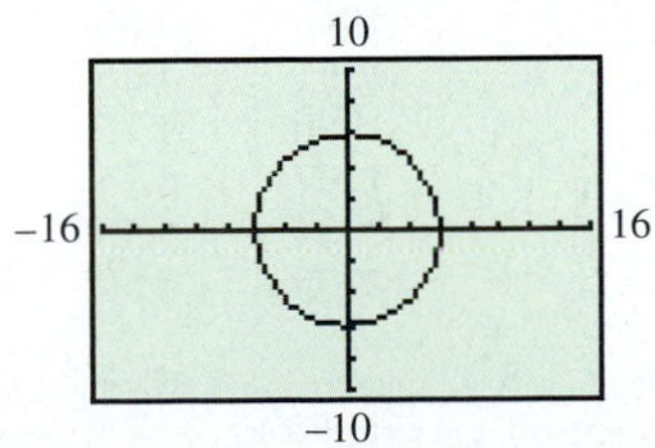

20. ***Concept Check*** *True* or *false*? The graphs of the equations $y_1 = 5.001x - 3$ and $y_2 = 5x + 6$ are shown in the screen. From this view, we may correctly conclude that these lines are parallel.

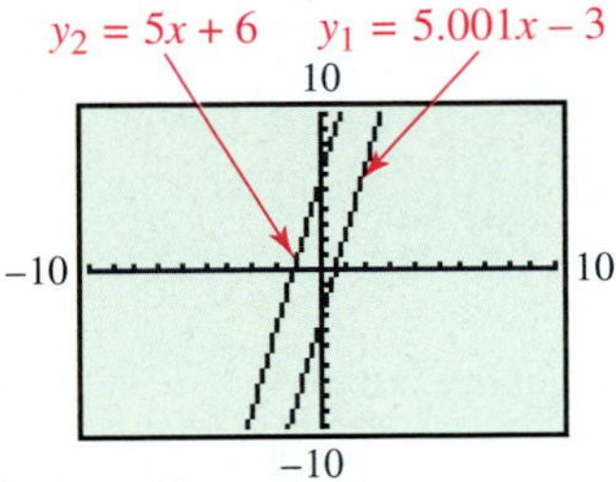

Refer to the graphs of the linear functions defined by $y_1 = f(x)$ and $y_2 = g(x)$ in the figure. Match the solution set in Column II with the equation or inequality in Column I. Choices may be used once, more than once, or not at all.

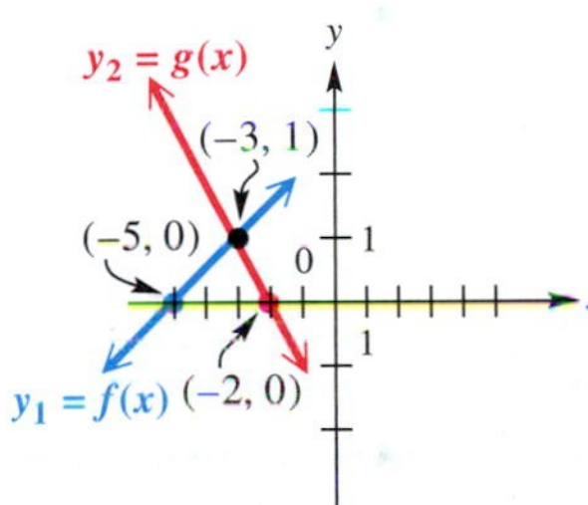

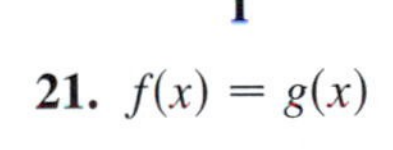

I	II	
21. $f(x) = g(x)$	**A.** $(-\infty, -3]$	**I.** $\{-3\}$
22. $f(x) > g(x)$	**B.** $(-\infty, -3)$	**J.** $[-3, \infty)$
23. $f(x) < g(x)$	**C.** $\{3\}$	**K.** $(-3, \infty)$
24. $g(x) \geq f(x)$	**D.** $\{2\}$	**L.** $\{1\}$
25. $y_2 - y_1 = 0$	**E.** $\{(3, 2)\}$	**M.** $(-\infty, -5)$
26. $f(x) < 0$	**F.** $\{-5\}$	**N.** $(-5, \infty)$
27. $g(x) > 0$	**G.** $\{-2\}$	**O.** $(-\infty, -2)$
28. $y_2 - y_1 < 0$	**H.** $\{0\}$	**P.** $(-2, \infty)$

Solve each equation or inequality using analytic methods. Support your answer graphically. Identify any identities or contradictions.

29. $5[3 + 2(x - 6)] = 3x + 1$

30. $\frac{x}{4} - \frac{x + 4}{3} = -2$

31. $-3x - (4x + 2) = 3$

32. $-2x + 9 + 4x = 2(x - 5) - 3$

33. $.5x + .7(4 - 3x) = .4x$

34. $\frac{x}{4} - \frac{5x - 3}{6} = 2 - \frac{7x + 18}{12}$

35. $x - 8 < 1 - 2x$

36. $\frac{4x - 1}{3} \geq \frac{x}{5} - 1$

37. Solve the inequality $-6 \leq \frac{4 - 3x}{7} < 2$ analytically.

38. Consider the linear function defined by

$$f(x) = 5\pi x + (\sqrt{3})x - 6.24(x - 8.1) + (\sqrt[3]{9})x.$$

(a) Solve the equation $f(x) = 0$ graphically. Give the solution to the nearest hundredth. Then explain how you went about solving the equation graphically.

(b) Refer to your graph, and give the solution set of $f(x) < 0$.

(c) Refer to your graph, and give the solution set of $f(x) \geq 0$.

(Modeling) Production of Compact Discs *A company produces studio-quality compact discs (CDs) of live concerts. The company places an ad in a trade newsletter. The cost of the ad is* \$150. *Each CD costs* \$30 *to produce, and the company charges* \$37.50 *per CD. Use this information to work Exercises 39–42.*

39. Express the company's cost C as a function of x, the number of CDs produced and sold.

40. Assuming that the company sells x CDs, express the revenue as a function of x.

41. Determine analytically the value of x for which revenue equals cost.

42. The graph shows $y = C(x)$ and $y = R(x)$. Discuss how the graph illustrates when the company is losing money, when it is breaking even, and when it is making a profit.

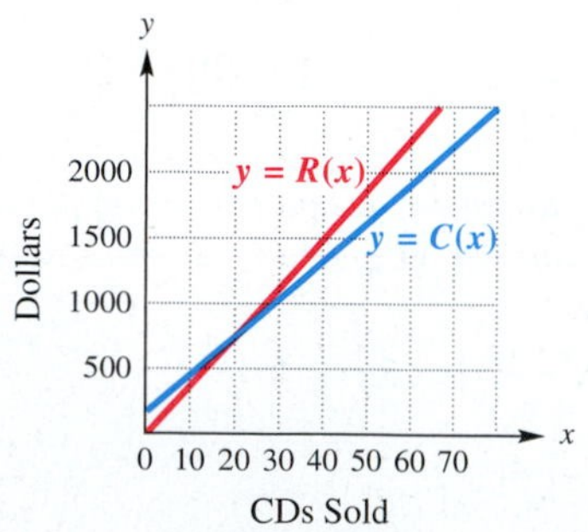

Solve the formula for approximate annual interest rate for the variable indicated.

43. $A = \frac{24f}{B(p + 1)}$ for f

44. $A = \frac{24f}{B(p + 1)}$ for B

(Modeling) Solve each problem.

45. ***Temperature Levels above Earth's Surface*** The linear function defined by

$$f(x) = -3.52x + 58.6$$

gives an approximation for the temperature in degrees Fahrenheit above the surface of Earth, were x is in thousands of feet and y is the temperature. (*Source:* Schwartz, Richard H., *Mathematics and Global Survival,* Ginn Press, 1991.)

(a) When the height is 5000 feet, what is the temperature? Solve analytically.

(b) When the temperature is $-15°$F, what is the height? Solve analytically.

(c) Explain how the answers in parts (a) and (b) can be supported graphically.

46. ***Prevention of Indoor Pollutants*** One of the most effective ways of removing contaminants from the air while cooking is to use a *vented* range hood. If a range hood removes F liters of air per second, then the percent P of contaminants that are also removed from the surrounding air can be expressed by the linear function defined by

$$P = 1.06F + 7.18,$$

where $10 \leq F \leq 75$. What flow rate must a range hood have to remove 50% of the contaminants from the air? (*Source:* Rezvan, R. L., "Effectiveness of Local Ventilation in Removing Simulated Pollutants from Point Sources," in *Proceedings of the Third International Conference on Indoor Air Quality and Climate,* 1984.)

47. ***Minimum Hourly Wage*** The table shows the changes in the minimum hourly wage in the United States for selected years from 1978 to 2005.

Year	Wage
1978	\$2.65
1979	2.90
1980	3.10
1981	3.35
1990	3.80
1991	4.25
1996	4.75
1997	5.15
2004	5.50
2005	6.50

Source: U.S. Labor Department.

(a) Find a least-squares regression linear model for the minimum hourly wage.

(b) Use a graphing calculator to plot the data and your linear model. Comment on how well the data fit a linear model.

(c) The 1999 minimum wage was $5.15. Use the linear model from part (a) to see how close an approximation it gives for 1999.

48. *Indianapolis 500 Pole Speeds* The pole speeds for selected drivers in the Indianapolis 500 from 1991 to 2003 are given in the table. Find the correlation coefficient for the least-squares regression linear model, and explain why this model would not be a good one to predict future pole speeds.*

Year	Driver	Pole Speed
1991	Rick Mears	224.113 mph
1992	Roberto Guerrero	232.482 mph
1993	Arie Luyendyk	223.967 mph
1994	Al Unser, Jr.	228.011 mph
1995	Scott Brayton	231.604 mph
1996	Scott Brayton	233.718 mph
1997	Arie Luyendyk	218.263 mph
1998	Billy Boat	223.503 mph
1999	Arie Luyendyk	225.179 mph
2000	Greg Ray	223.471 mph
2001	Scott Sharp	226.037 mph
2002	Bruno Junqueira	231.342 mph
2003	Helio Castroneves	231.725 mph

Source: www.indy500.com

49. *Speed of a Batted Baseball* Suppose a baseball is thrown at 85 mph. The ball will travel 320 feet when hit by a bat swung at 50 mph and 440 feet when hit by a bat swung at 80 mph. Let y be the number of feet traveled by the ball when hit by a bat swung at x mph. Find the equation of the line given by $y = mx + b$ that models the data. (*Note*: This function is valid for $50 \leq x \leq 90$, where the bat is 35 inches long, weighs 32 ounces, and is swung slightly upward to drive the ball at an angle of 35°.) How much farther will a ball travel for each 1-mph increase in the speed of the bat? (*Source*: Adair, Robert K., *The Physics of Baseball,* HarperCollins Publishers, 1990.)

*The authors wish to thank Randall Leigh of the University of Southern Indiana for his input into this exercise.

Solve each application of linear equations.

50. *Dimensions of a Recycling Bin* A recycling bin is in the shape of a rectangular box. Find the height of the box if its length is 18 feet, its width is 8 feet, and its surface area is 496 square feet. (In the figure, let h = height. Assume that the given surface area includes the top lid of the box.)

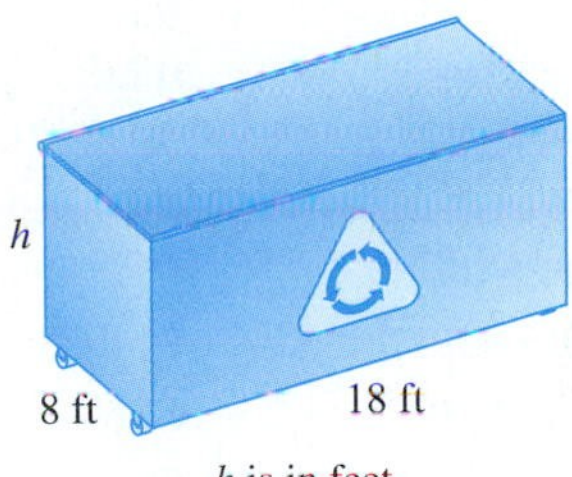

h is in feet.

51. *Running Speeds in Track Events* In 2002, Tim Montgomery (USA) set a world record in the 100-meter dash with a time of 9.78 seconds. If this pace could be maintained for an entire 26.2-mile marathon, how would that time compare with the fastest marathon time of 2 hours, 4 minutes, and 55 seconds, by Paul Tergat of Kenya in 2003? (*Hint*: 1 meter ≈ 3.281 feet.) (*Source*: International Association of Athletics Federations.)

52. *Temperature of Venus* Venus is our solar system's hottest planet, with a surface temperature of 864°F. What is this temperature in Celsius? (*Hint:* Use $C = \frac{5}{9}(F - 32)$.) (*Source: Guinness Book of Records*, 1995.)

53. *Pressure in a Liquid* The pressure on a point in a liquid is directly proportional to the distance from the surface to the point. In a certain liquid, the pressure at a depth of 4 meters is 3000 kilograms per square meter. Find the pressure at a depth of 10 meters.

54. ***(Modeling) Data Analysis*** The table lists data that are exactly linear.

x	-3	-2	-1	0	1	2	3
y	6.6	5.4	4.2	3	1.8	.6	$-.6$

(a) Determine the slope–intercept form of the line that passes through these data points.
(b) Predict y when $x = -1.5$ and when $x = 3.5$.

55. ***(Modeling) Math SAT Scores*** The table lists average mathematics SAT scores for selected years.

Year	1996	1998	2000	2002	2004
Score	508	512	514	516	518

Source: The College Board.

(a) Use regression to find a linear function f that models these data.
(b) Use f to approximate the average mathematics SAT score in 2003.
(c) Use f to predict the score in 2008.

56. ***(Modeling) Airline Passenger Totals*** The table lists the numbers in millions of airline passengers at 10 of the world's busiest airports in 2000 and 2003.

Rank (as of 2004)	City (Airport)	2000 Total Passengers (in millions)	2003 Total Passengers (in millions)
1	Atlanta (ATL)	80.2	79.1
2	Chicago (ORD)	72.1	69.5
3	London (LHR)	64.6	63.5
4	Tokyo (HND)	56.4	62.9
5	Los Angeles (LAX)	66.4	55.0
6	Dallas–Ft. Worth (DFW)	60.7	53.3
7	Frankfurt/Main (FRA)	49.4	48.4
8	Paris (CDG)	48.2	48.2
9	Amsterdam (AMS)	39.6	40.0
10	Denver (DEN)	38.8	37.5

Source: Airports Council International (ACI).

(a) Use a calculator to find the least-squares linear regression model for these data.
(b) Phoenix Sky Harbor Airport had 36 million passengers in 2000. Assuming that it follows the same trend as the other airports in the table, estimate the number of passengers it had in 2003. Compare this result with the Airports Council International (ACI) figure of 37.4 million passengers.

57. ***Alcohol Mixture*** A chemist wishes to strengthen a mixture that is 10% alcohol to one that is 30% alcohol. How much pure alcohol should be added to 12 liters of the 10% mixture?

58. ***Acid Mixture*** A student needs 10% hydrochloric acid for a chemistry experiment. How much 5% acid should be mixed with 120 milliliters of 20% acid to get a 10% solution?

59. ***(Modeling) DVD Production*** A company produces DVDs. The revenue from the sale of x DVDs is $R(x) = 8x$. The cost to produce x DVDs is $C(x) = 3x + 1500$. Revenue and cost are in dollars. In what interval will the company at least break even?

60. ***Intelligence Quotient*** A person's intelligence quotient (IQ) is found by multiplying the mental age by 100 and dividing by the chronological age.
(a) Jack is 7 years old. His IQ is 130. Find his mental age.
(b) If a person is 16 years old with a mental age of 20, what is the person's IQ?

CHAPTER 1 Test

1. For each function, determine the
   (i) domain. **(ii)** range.
   (iii) x-intercept(s). **(iv)** y-intercept.

(a)

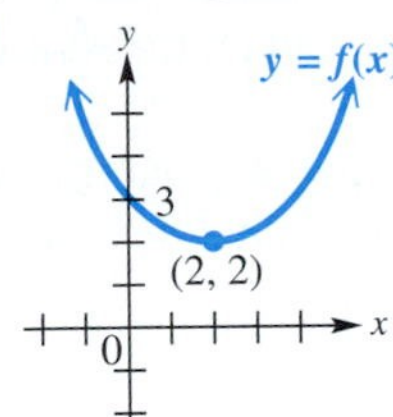

(b)

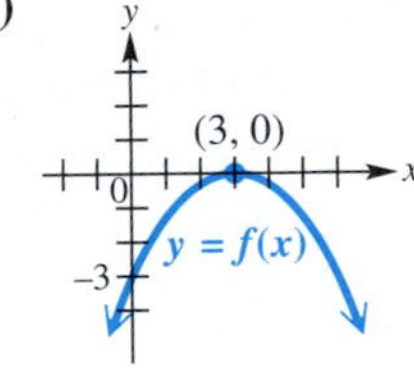

(c)

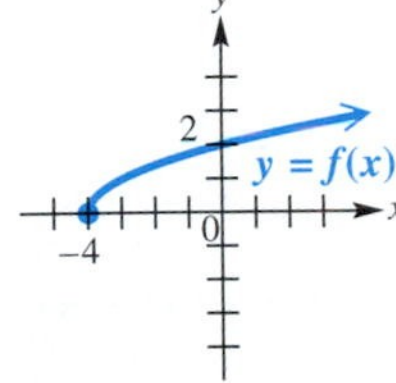

2. Use the figure to solve each equation or inequality.
   (a) $f(x) = g(x)$
   (b) $f(x) < g(x)$
   (c) $f(x) \geq g(x)$
   (d) $y_2 - y_1 = 0$

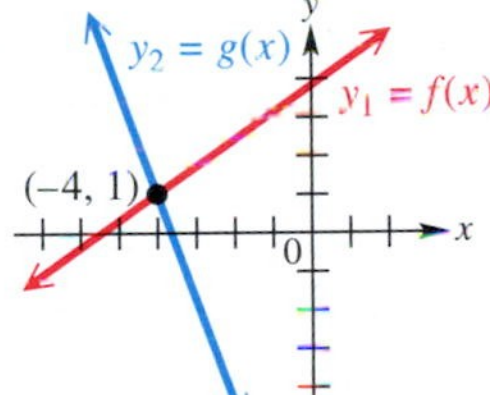

3. Use the screen to solve the equation or inequality. Here, f is a linear function defined over the domain of real numbers.
   (a) $y_1 = 0$
   (b) $y_1 < 0$
   (c) $y_1 > 0$
   (d) $y_1 \leq 0$

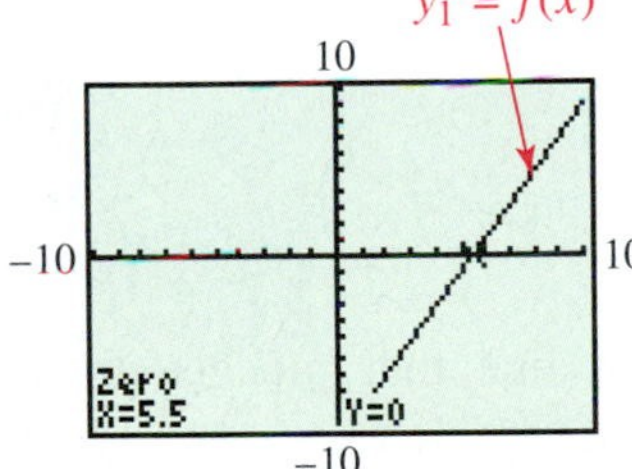

4. Consider the linear functions defined by

$$f(x) = 3(x - 4) - 2(x - 5)$$

and

$$g(x) = -2(x + 1) - 3.$$

   (a) Solve $f(x) = g(x)$ analytically, showing all steps. Also, check analytically.
   (b) Graph $y_1 = f(x)$ and $y_2 = g(x)$. Use your result in part (a) to find the solution set of $f(x) > g(x)$. Explain your answer.
   (c) Repeat part (b) for $f(x) < g(x)$.

5. Consider the linear function defined by

$$f(x) = -\frac{1}{2}(8x + 4) + 3(x - 2).$$

   (a) Solve the equation $f(x) = 0$ analytically.
   (b) Solve the inequality $f(x) \leq 0$ analytically.
   (c) Graph $y = f(x)$ in an appropriate viewing window, and explain how the graph supports your answers in parts (a) and (b).

6. ***(Modeling) Cable Television Rates*** The graph depicts how the average monthly rates for cable television increased during the years 1982 through 2002.

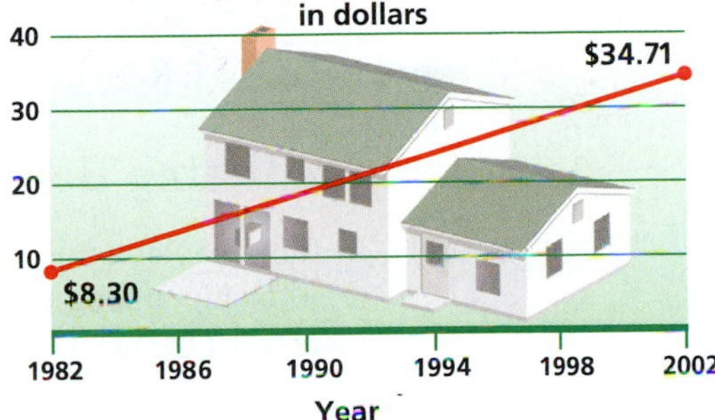

Source: U.S. Census Bureau.

   (a) Use the midpoint formula to approximate the average monthly rate in 1992.
   (b) Find the slope of the line, and explain its meaning in the given context.

7. Find the equation of the line passing through the point $(-3, 5)$ and
   (a) parallel to the line with equation $y = -2x + 4$.
   (b) perpendicular to the line with equation $-2x + y = 0$.

8. Find the x- and y-intercepts of the line whose standard form is $3x - 4y = 6$. What is the slope of this line?

9. Give the equations of both the horizontal and vertical lines passing through the point $(-3, 7)$.

10. ***(Modeling) Windchill Factor*** The table shows the windchill factor for various wind speeds when the Fahrenheit temperature is 40°.

Wind Speed (mph)	Degrees
10	34
15	32
20	30
25	29
30	28
35	28

(continued)

10. (a) Find the least-squares regression line for the data. Give the correlation coefficient.
    (b) Use this line to predict the windchill factor when the wind speed is 40 mph.

11. ***(Modeling) Rainfall*** Suppose that during a storm, rain is falling at a rate of 1 inch per hour. The water coming from a circular roof with a radius of 20 feet is running down a spout that can accommodate 400 gallons of water per hour.

(a) Determine the number of cubic inches of water falling on the roof in one hour. (*Hint:* Each hour, a layer of water 1 inch thick falls on the roof.)
(b) One gallon equals about 231 cubic inches. Write a formula for a function g that computes the gallons of water landing on the roof in x hours.
(c) How many gallons of water land on the roof during a 2.5-hour rainstorm?
(d) Will one downspout be sufficient to handle this type of rainfall? If not, how many should there be?

CHAPTER 1 PROJECT

Predicting Heights and Weights of Athletes

In this chapter project, we investigate the height–weight relationship among professional athletes. The tables in the margin list the heights and weights of 10 female and 10 male professional basketball players.

Notice that taller athletes tend to be heavier, but not always. For example, there is a 64-inch female athlete who weighs more than a 68-inch female athlete.

Female Basketball Players	
Height (in.)	**Weight (lb)**
74	169
74	181
77	199
64	139
68	135
71	161
78	187
68	150
71	181
73	194

Male Basketball Players	
Height (in.)	**Weight (lb)**
74	197
75	202
71	173
77	220
83	220
83	250
81	225
76	219
79	215
85	246

Source: WNBA; NBA.

Activities

1. Find the least-squares regression line that models the height–weight relationship among the 10 female basketball players. Let $y = mx + b$, where x represents height in inches and y represents weight in pounds. What is the correlation coefficient, and what does it indicate?
2. Use this regression line to predict the weight of a female basketball player who is 75 inches tall.
3. For each 1-inch increase in height, predict the corresponding increase in weight for female basketball players.
4. Now find the least-squares regression line that models the height–weight relationship among the 10 male basketball players. What is the correlation coefficient, and what does it indicate?
5. Predict the weight of a male basketball player who is 80 inches tall.
6. For each 1-inch increase in height, predict the corresponding increase in weight for male basketball players.

2

Analysis of Graphs of Functions

IS THERE A connection between the seemingly diverse subjects of mathematics and art? The answer is yes, and *symmetry* is one of the bridges. The graph of a mathematical equation can be replicated multiple times by using three types of symmetry—*reflection, rotation,* and *translation*—which transforms the plane into a work of art. Each replication has a relationship to the original graph and is represented by a new equation that indicates its change in location and orientation from the original position.

Symmetry is used in mandala (Sanskrit for "circle") and kaleidoscope art. Mandalas generally exhibit a center, radial symmetry, and coordinate points that are reflected across the x-axis, y-axis, or myriad diagonal lines through the origin. Kaleidoscopes use internal mirrors to reflect images multiple times, resulting in the formation of symmetrical patterns.

In this chapter, we learn to recognize symmetry with respect to the x-axis, y-axis, and origin and to reflect, translate, and rotate graphs of equations.

Source: Virginia Starkenburg, San Diego City College.

Chapter Outline

2.1 Graphs of Basic Functions and Relations; Symmetry

Continuity ■ Increasing and Decreasing Functions ■ The Identity Function ■ The Squaring Function and Symmetry with Respect to the y-Axis ■ The Cubing Function and Symmetry with Respect to the Origin ■ The Square Root and Cube Root Functions ■ The Absolute Value Function ■ The Relation $x = y^2$ and Symmetry with Respect to the x-Axis ■ Even and Odd Functions

Continuity

Looking Ahead to Calculus

Many calculus theorems apply only to continuous functions.

The graph of a linear function, a straight line, may be drawn by hand over any interval of its domain without picking the pencil up from the paper. We say that a function with this property is *continuous* over any interval. Intuitively, the graph of a continuous function has no breaks or holes in it. The formal definition of continuity requires concepts from calculus, but we can give an informal definition.

Continuity (Informal Definition)

A function is **continuous** over an interval of its domain if its hand-drawn graph over that interval can be sketched without lifting the pencil from the paper.

If a function is not continuous at a point, then it may have a point of *discontinuity* (Figure 1(a)), or it may have a vertical *asymptote* (a vertical line that the graph does not intersect, as in Figure 1(b)). Asymptotes will be discussed further in Chapter 4.

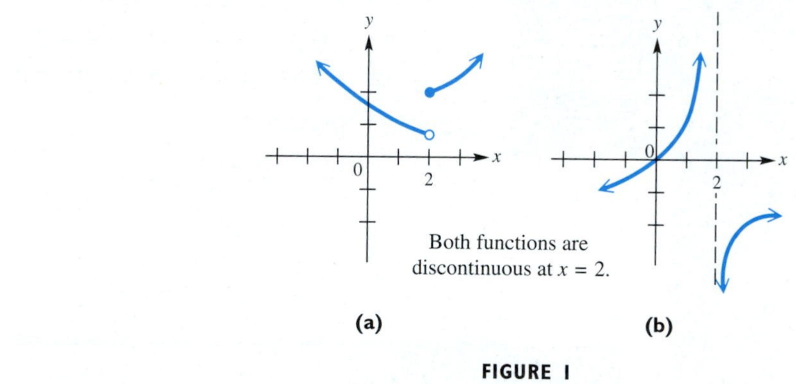

FIGURE 1

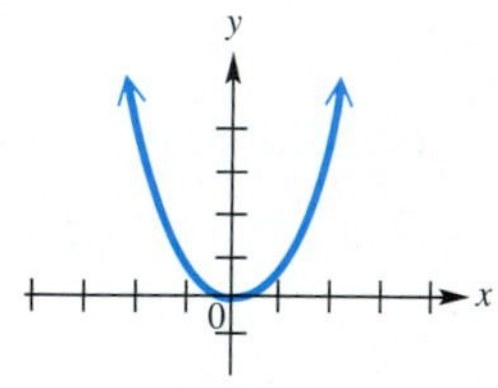

FIGURE 2

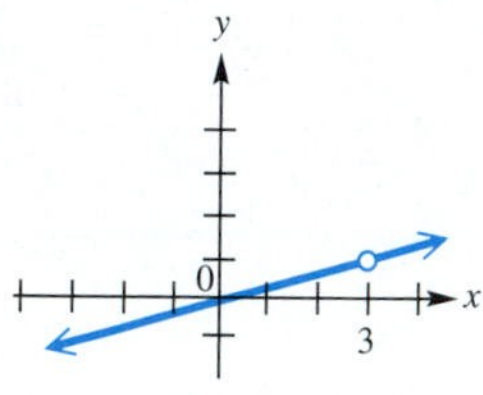

FIGURE 3

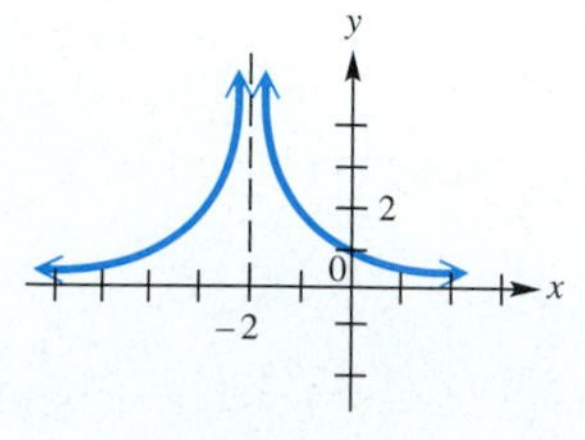

FIGURE 4

EXAMPLE 1 Determining Intervals of Continuity

Figures 2, 3, and 4 show graphs of functions. Indicate the intervals of the domain over which they are continuous.

Solution The function in Figure 2 is continuous over the entire domain of real numbers, $(-\infty, \infty)$. The function in Figure 3 has a point of discontinuity at $x = 3$. It is continuous over the interval $(-\infty, 3)$ and the interval $(3, \infty)$. Finally, the function in Figure 4 has a vertical asymptote at $x = -2$, as indicated by the dashed line. It is continuous over the intervals $(-\infty, -2)$ and $(-2, \infty)$. ■

Increasing and Decreasing Functions

If a continuous function does not take on the same value throughout an interval, then its graph will either rise from left to right or fall from left to right or some combination of both. We use the words *increasing* and *decreasing* to describe this behavior. For example, a linear function with positive slope is increasing over its entire domain, while one with negative slope is decreasing. See Figure 5.

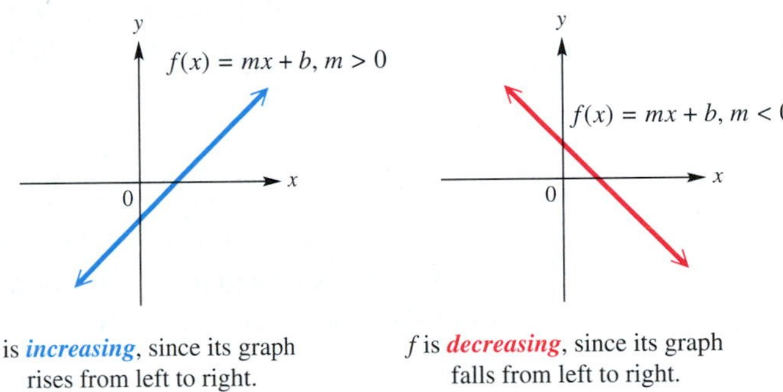

f is ***increasing***, since its graph rises from left to right.

f is ***decreasing***, since its graph falls from left to right.

FIGURE 5

Informally speaking, a function *increases* on an interval of its domain if its graph *rises from left to right* on the interval. A function *decreases* on an interval of its domain if its graph *falls from left to right* on the interval. A function is *constant* on an interval of its domain if its graph is *horizontal* on the interval. (While some functions are increasing, decreasing, or constant over their entire domains, many are not. Hence, we speak of increasing, decreasing, or constant over an interval of the domain.)

Criteria for Increasing, Decreasing, and Constant on an Interval

Suppose that a function f is defined over an interval I.

(a) f **increases** on I if, whenever $x_1 < x_2$, $f(x_1) < f(x_2)$;

(b) f **decreases** on I if, whenever $x_1 < x_2$, $f(x_1) > f(x_2)$;

(c) f is **constant** on I if, for every x_1 and x_2, $f(x_1) = f(x_2)$.

Looking Ahead to Calculus

Determining where a function increases, decreases, or is constant is important in calculus. The *derivative* of a function provides a formula for determining the slope of a line tangent to the curve. If the slope of the tangent line is positive over an interval, the function is increasing on that interval; similarly, if the slope is negative, the function is decreasing; and if the slope is 0, the function is constant. Apply these concepts to Figure 6.

Figure 6 illustrates these ideas.

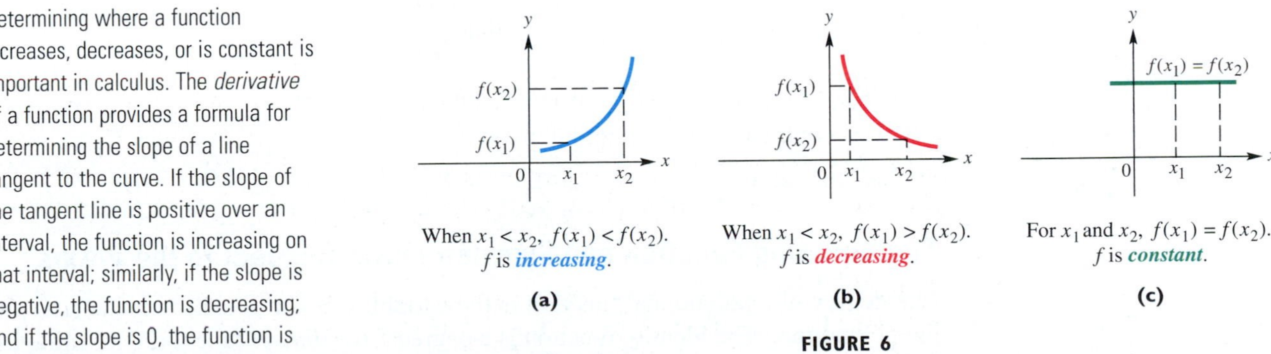

When $x_1 < x_2$, $f(x_1) < f(x_2)$. f is ***increasing***.

(a)

When $x_1 < x_2$, $f(x_1) > f(x_2)$. f is ***decreasing***.

(b)

For x_1 and x_2, $f(x_1) = f(x_2)$. f is ***constant***.

(c)

FIGURE 6

NOTE To decide whether a function is increasing, decreasing, or constant on an interval, ask yourself, **"What does *y* do as *x* goes from left to right?"**

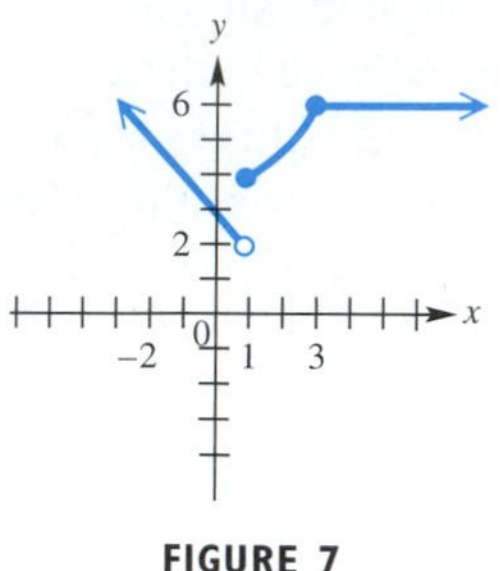

FIGURE 7

EXAMPLE 2 Determining Intervals over Which a Function Is Increasing, Decreasing, or Constant

Figure 7 shows the graph of a function. Determine the intervals over which the function is increasing, decreasing, or constant.

Solution We should ask, "What is happening to the y-values as the x-values are getting larger?" Moving from left to right on the graph, we see that on the interval $(-\infty, 1)$, the y-values are *decreasing*; on the interval $[1, 3]$, the y-values are *increasing*; and on the interval $[3, \infty)$, the y-values are *constant* (and equal to 6). Therefore, the function is decreasing on $(-\infty, 1)$, increasing on $[1, 3]$, and constant on $[3, \infty)$. ■

CAUTION A common error involves writing range values (y-values) when determining intervals like those in Example 2. ***Remember that we are determining intervals of the domain, and thus are interested in x-values for our interval designations. Range values are* not *written when designating such intervals.***

The Identity Function

If we let $m = 1$ and $b = 0$ in the general form of the linear function defined by $f(x) = mx + b$ from Section 1.3, we get the **identity function,** or $\boldsymbol{f(x) = x}$. This function pairs every real number with itself. See Figure 8.

FUNCTION CAPSULE

IDENTITY FUNCTION $f(x) = x$

Domain: $(-\infty, \infty)$ Range: $(-\infty, \infty)$

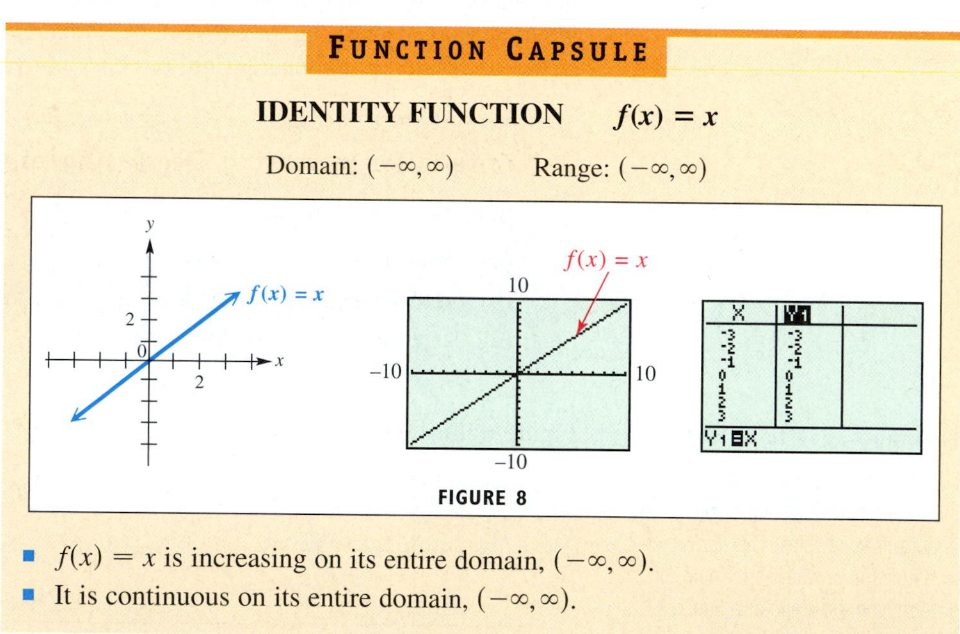

FIGURE 8

- $f(x) = x$ is increasing on its entire domain, $(-\infty, \infty)$.
- It is continuous on its entire domain, $(-\infty, \infty)$.

The Squaring Function and Symmetry with Respect to the y-Axis

The **degree** of a polynomial function in the variable x is the greatest exponent on x in the polynomial. The identity function is a degree 1 (or *linear*) function. The **squaring function,** defined by $\boldsymbol{f(x) = x^2}$, is the simplest degree 2 (or *quadratic*) function. (The word *quadratic* refers to degree 2.) See Figure 9 on the next page.

The squaring function pairs every real number with its square. Its graph is called a **parabola.** The point at which the graph changes from decreasing to increasing (the point $(0, 0)$) is called the **vertex** of the graph. (For a parabola that opens downward, the vertex is the point at which the graph changes from increasing to decreasing.)

FUNCTION CAPSULE

SQUARING FUNCTION $f(x) = x^2$

Domain: $(-\infty, \infty)$ Range: $[0, \infty)$

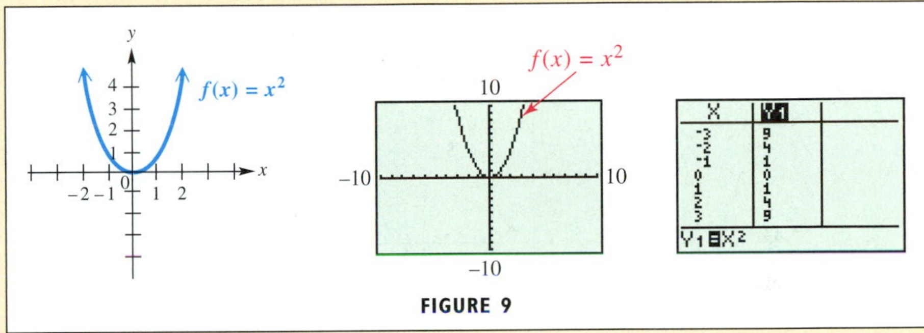

FIGURE 9

- $f(x) = x^2$ decreases on the interval $(-\infty, 0]$ and increases on the interval $[0, \infty)$.
- It is continuous on its entire domain, $(-\infty, \infty)$.

Looking Ahead to Calculus

Calculus allows us to find areas of regions in the plane. For example, we can find the area of the region below the graph of $y = x^2$, above the x-axis, bounded on the left by the line $x = -2$, and bounded on the right by $x = 2$. Draw a sketch of this region, and notice that, due to the symmetry of the graph of $y = x^2$, the desired area is twice that of the area to the right of the y-axis. Thus, symmetry can be used to reduce the original problem to an easier one by simply finding the area to the right of the y-axis, and then doubling the answer.

If we were able to "fold" the graph of $f(x) = x^2$ along the y-axis, the two halves would coincide exactly. In mathematics we refer to this property as symmetry, and we say that the graph of $f(x) = x^2$ is *symmetric with respect to the y-axis.*

Symmetry with Respect to the y-Axis

If a function f is defined so that

$$f(-x) = f(x)$$

for all x in its domain, then the graph of f is **symmetric with respect to the y-axis.**

To illustrate that the graph of $f(x) = x^2$ is symmetric with respect to the y-axis, note that

$$f(-4) = f(4) = 16, \qquad f(-3) = f(3) = 9,$$
$$f(-2) = f(2) = 4, \qquad f(-1) = f(1) = 1.$$

This pattern holds for any real number x, since

$$f(-x) = (-x)^2 = (-1)^2x^2 = x^2 = f(x).$$

In general, if a graph is symmetric with respect to the y-axis, the following is true:

If (a, b) is on the graph, so is $(-a, b)$.

The Cubing Function and Symmetry with Respect to the Origin

The **cubing function,** defined by $f(x) = x^3$, is the simplest degree 3 function. It pairs with each real number the third power, or cube, of the number. See Figure 10. The point at which the graph of the cubing function changes from "opening downward" to "opening upward" (the point $(0, 0)$) is called an **inflection point.**

FUNCTION CAPSULE

CUBING FUNCTION $f(x) = x^3$

Domain: $(-\infty, \infty)$ Range: $(-\infty, \infty)$

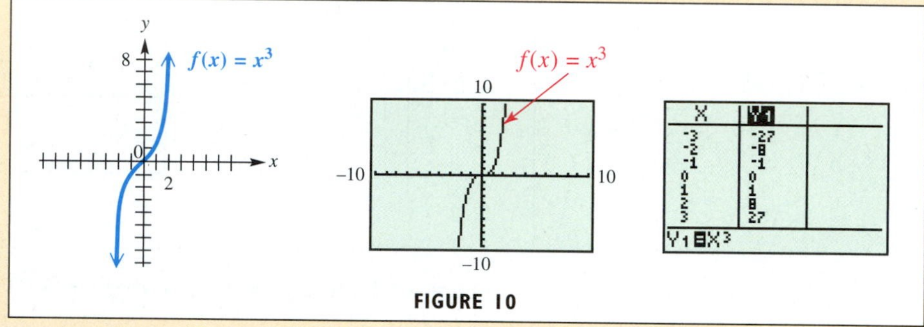

FIGURE 10

- $f(x) = x^3$ increases on its entire domain, $(-\infty, \infty)$.
- It is continuous on its entire domain, $(-\infty, \infty)$.

If we were able to "fold" the graph of $f(x) = x^3$ along the y-axis and then along the x-axis, forming a "corner" at the origin, the two parts of the graph would coincide exactly. We say that the graph of $f(x) = x^3$ is *symmetric with respect to the origin.*

Symmetry with Respect to the Origin

If a function f is defined so that

$$f(-x) = -f(x)$$

for all x in its domain, then the graph of f is **symmetric with respect to the origin.**

FOR DISCUSSION

Suppose the graph of $f(x) = x^3$ is rotated about the origin half a turn, or 180°.

1. How does this graph compare with the original graph?
2. Explain how you can use rotation to determine whether a graph is symmetric with respect to the origin.

To illustrate that the graph of $f(x) = x^3$ is symmetric with respect to the origin, note that

$$f(-2) = -f(2) = -8, \qquad f(-1) = -f(1) = -1.$$

This pattern holds for any real number x, since

$$f(-x) = (-x)^3 = (-1)^3x^3 = -x^3 = -f(x).$$

In general, if a graph is symmetric with respect to the origin, the following is true:

If (a, b) is on the graph, so is $(-a, -b)$.

EXAMPLE 3 Determining Symmetry

Show that **(a)** $f(x) = x^4 - 3x^2 - 8$ has a graph that is symmetric with respect to the y-axis; **(b)** $f(x) = x^3 - 4x$ has a graph that is symmetric with respect to the origin.

Analytic Solution

(a) We must show that $f(-x) = f(x)$ for any x.

$$f(-x) = (-x)^4 - 3(-x)^2 - 8 \quad \text{Substitute } -x \text{ for } x.$$

$(ab)^m = a^m b^m$

$$= (-1)^4x^4 - 3(-1)^2x^2 - 8$$
$$= x^4 - 3x^2 - 8$$
$$= f(x)$$

This *proves* that the graph is symmetric with respect to the y-axis.

(b) We must show that $f(-x) = -f(x)$ for any x.

$$f(-x) = (-x)^3 - 4(-x) \quad \text{Substitute } -x \text{ for } x.$$

Be careful with signs.

$$= (-1)^3x^3 + 4x$$
$$= -x^3 + 4x \quad *$$
$$= -(x^3 - 4x) \quad \text{Factor out } -1.$$
$$= -f(x)$$

In the line denoted *, the signs of the coefficients are *opposites* of those in $f(x)$. We factored out -1 to show that the final result is $-f(x)$.

Graphing Calculator Solution

Figures 11 and 12 support the analytic proofs.

(a)

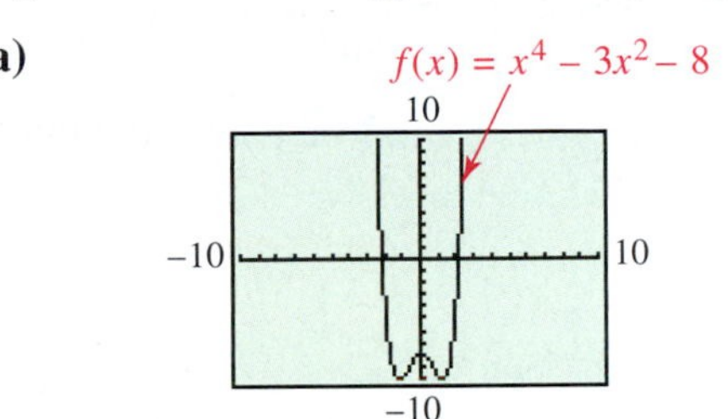

This graph is symmetric with respect to the y-axis.

FIGURE 11

(b)

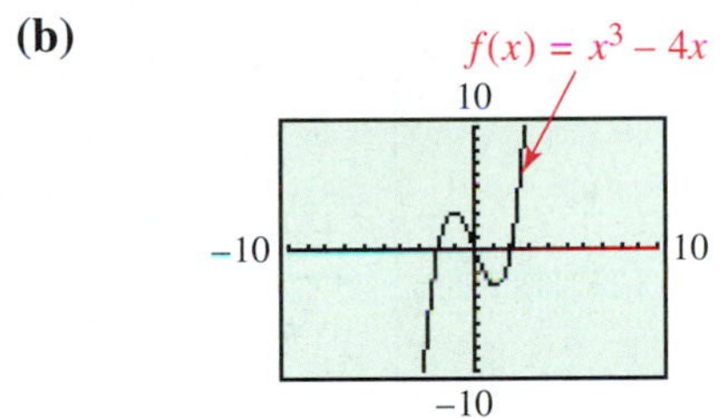

This graph is symmetric with respect to the origin.

FIGURE 12

The Square Root and Cube Root Functions

The **square root function** is defined by $f(x) = \sqrt{x}$. See Figure 13. For the function value to be a real number, the domain must be restricted to $[0, \infty)$.

FUNCTION CAPSULE

SQUARE ROOT FUNCTION $\quad f(x) = \sqrt{x}$

Domain: $[0, \infty)$ $\qquad$ Range: $[0, \infty)$

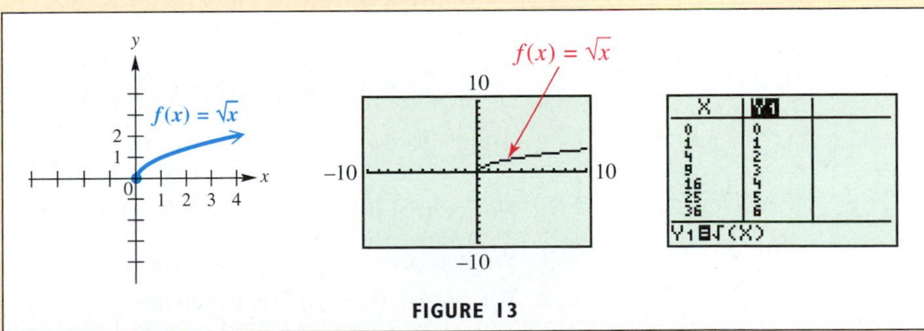

FIGURE 13

- $f(x) = \sqrt{x}$ increases on its entire domain, $[0, \infty)$.
- It is continuous on its entire domain, $[0, \infty)$.

GCM TECHNOLOGY NOTE

The definition of a rational exponent allows us to enter $\sqrt{x}$ as $x^{1/2}$ and $\sqrt[3]{x}$ as $x^{1/3}$ when using a calculator. Further discussion of rational exponents can be found in Chapter R.

The **cube root function,** defined by $f(x) = \sqrt[3]{x}$ (Figure 14), differs from the square root function in that *any* real number—positive, 0, or *negative*—has a real cube root. Thus, the domain is $(-\infty, \infty)$.

$$\text{When} \quad x > 0, \quad \sqrt[3]{x} > 0,$$
$$\text{when} \quad x = 0, \quad \sqrt[3]{x} = 0,$$
and
$$\text{when} \quad x < 0, \quad \sqrt[3]{x} < 0.$$

As a result, the range is also $(-\infty, \infty)$.

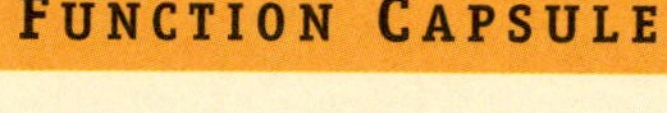

CUBE ROOT FUNCTION $\quad f(x) = \sqrt[3]{x}$

Domain: $(-\infty, \infty)$ Range: $(-\infty, \infty)$

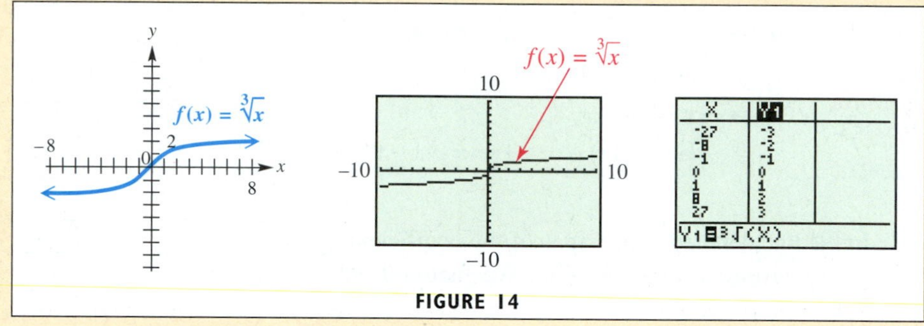

FIGURE 14

- $f(x) = \sqrt[3]{x}$ increases on its entire domain, $(-\infty, \infty)$.
- It is continuous on its entire domain, $(-\infty, \infty)$.

The Absolute Value Function

GCM TECHNOLOGY NOTE

You should become familiar with the command on your particular calculator that allows you to graph the absolute value function.

On a number line, the absolute value of a real number x, denoted $|x|$, represents its undirected distance from the origin, 0. The **absolute value function,** denoted by $f(x) = |x|$**,** pairs every real number with its absolute value and is defined as follows.

Absolute Value $|x|$

$$f(x) = |x| = \begin{cases} x & \text{if } x \geq 0 \\ -x & \text{if } x < 0 \end{cases}$$

Notice that this function is defined in two parts.

1. We use $|x| = x$ if x is positive or 0.
2. We use $|x| = -x$ if x is negative.

Since x can be any real number, the domain of the absolute value function is $(-\infty, \infty)$, but since $|x|$ cannot be negative, the range is $[0, \infty)$. See Figure 15.

FUNCTION CAPSULE

ABSOLUTE VALUE FUNCTION $f(x) = |x|$

Domain: $(-\infty, \infty)$ Range: $[0, \infty)$

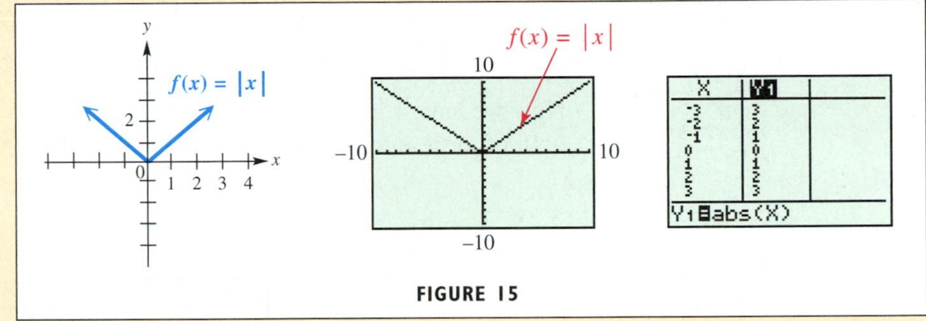

FIGURE 15

- $f(x) = |x|$ decreases on the interval $(-\infty, 0]$ and increases on the interval $[0, \infty)$.
- It is continuous on its entire domain, $(-\infty, \infty)$.

FOR DISCUSSION

Based on the functions discussed so far in this section, answer each question.

1. Which functions have graphs that are symmetric with respect to the y-axis?
2. Which functions have graphs that are symmetric with respect to the origin?
3. Which functions have graphs with neither of these symmetries?
4. Why is it not possible for the graph of a nonzero function to be symmetric with respect to the x-axis?

The Relation $x = y^2$ and Symmetry with Respect to the x-Axis

Recall that a function is a relation where every domain value is paired with one and only one range value. There are cases where we are interested in graphing relations that are *not* functions. Consider the relation defined by the equation $x = y^2$. The table of selected ordered pairs indicates that this relation has two y-values for each positive value of x. If we plot these points and join them with a smooth curve, we find that the graph of $x = y^2$ is a parabola opening to the right. See Figure 16.

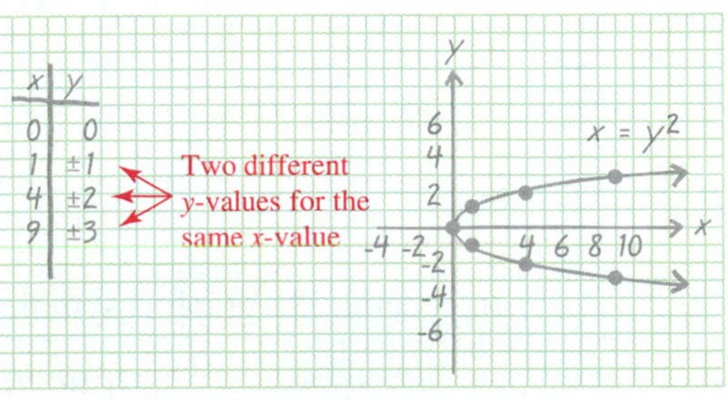

FIGURE 16

GCM TECHNOLOGY NOTE

You should be aware of how function mode (as opposed to parametric mode) is activated for your model.

If a graphing calculator is set in function mode, it is not possible to graph $x = y^2$ directly. (However, if it is set in *parametric* mode, such a curve is possible with direct graphing.) To overcome this problem, we begin with $x = y^2$ and take the square root on each side.

Choose both the positive and negative square roots of x.

$$x = y^2 \quad \text{Given equation}$$
$$y^2 = x \quad \text{Transform so that } y \text{ is on the left.}$$
$$y = \pm\sqrt{x} \quad \text{Take square roots.}$$

Now we have $x = y^2$ defined by two *functions*: $Y_1 = \sqrt{X}$ and $Y_2 = -\sqrt{X}$. Entering both of these into a calculator gives the graph shown in Figure 17(a).

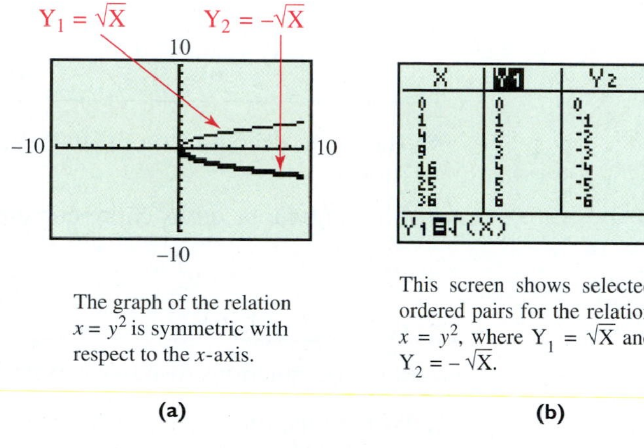

The graph of the relation $x = y^2$ is symmetric with respect to the x-axis.

(a)

This screen shows selected ordered pairs for the relation $x = y^2$, where $Y_1 = \sqrt{X}$ and $Y_2 = -\sqrt{X}$.

(b)

FIGURE 17

If we were to "fold" the graph of $x = y^2$ along the x-axis, the two halves of the parabola would coincide. We say that the graph of $x = y^2$ exhibits *symmetry with respect to the x-axis.* A table supports this symmetry, as seen in Figure 17(b).

Symmetry with Respect to the x-Axis

If replacing y with $-y$ in an equation results in the same equation, then the graph is **symmetric with respect to the x-axis.**

To illustrate this property, begin with $x = y^2$ and replace y with $-y$.

$(ab)^m = a^m b^m$

$$x = y^2$$
$$x = (-y)^2 \quad \text{Substitute } -y \text{ for } y.$$
$$x = (-1)^2 y^2$$
$$x = y^2 \quad \text{The result is the same equation.}$$

In general, if a graph is symmetric with respect to the x-axis, the following is true:

If (a, b) is on the graph, so is $(a, -b)$.

A summary of the types of symmetry just discussed follows.

Type of Symmetry	Example	Basic Fact about Points on the Graph
y-axis symmetry	y; $(-a, b)$; (a, b); 0; x	If (a, b) is on the graph, so is $(-a, b)$.
Origin symmetry	y; (a, b); 0; x; $(-a, -b)$	If (a, b) is on the graph, so is $(-a, -b)$.
x-axis symmetry (not possible for a nonzero function)	y; (a, b); 0; x; $(a, -b)$	If (a, b) is on the graph, so is $(a, -b)$.

Even and Odd Functions

The concepts of symmetry with respect to the y-axis and symmetry with respect to the origin are closely associated with the concepts of *even* and *odd functions.*

Even and Odd Functions

A function f is called an **even function** if $f(-x) = f(x)$ for all x in the domain of f. (Its graph is symmetric with respect to the y-axis.)

A function f is called an **odd function** if $f(-x) = -f(x)$ for all x in the domain of f. (Its graph is symmetric with respect to the origin.)

As an illustration, observe the following:

$f(x) = x^2$ is an even function because

$$\begin{aligned} f(-x) &= (-x)^2 \\ &= (-1)^2x^2 \\ &= x^2 \\ &= f(x). \end{aligned}$$

$f(x) = x^3$ is an odd function because

$$\begin{aligned} f(-x) &= (-x)^3 \\ &= (-1)^3x^3 \\ &= -x^3 \\ &= -f(x). \end{aligned}$$

A function may be neither even nor odd; for example, $f(x) = \sqrt{x}$ is neither even nor odd.

EXAMPLE 4 Determining whether Functions Are Even, Odd, or Neither

Decide whether each function defined is *even, odd,* or *neither.*

(a) $f(x) = 8x^4 - 3x^2$ **(b)** $f(x) = 6x^3 - 9x$ **(c)** $f(x) = 3x^2 + 5x$

Solution

(a) Replacing x with $-x$ gives

$$f(-x) = 8(-x)^4 - 3(-x)^2 = 8x^4 - 3x^2 = f(x).$$

Since $f(-x) = f(x)$ for each x in the domain of the function, f is even.

(b) $f(-x) = 6(-x)^3 - 9(-x) = -6x^3 + 9x = -(6x^3 - 9x) = -f(x)$

The function f is odd. *Be careful with signs.*

(c) $f(-x) = 3(-x)^2 + 5(-x) = 3x^2 - 5x$

Since $f(-x) \neq f(x)$ and $f(-x) \neq -f(x)$, f is neither even nor odd. ■

FOR DISCUSSION

The three functions discussed in Example 4 are graphed in Figure 18, but not necessarily in the same order as in the example. Without actually using your calculator, identify each function.

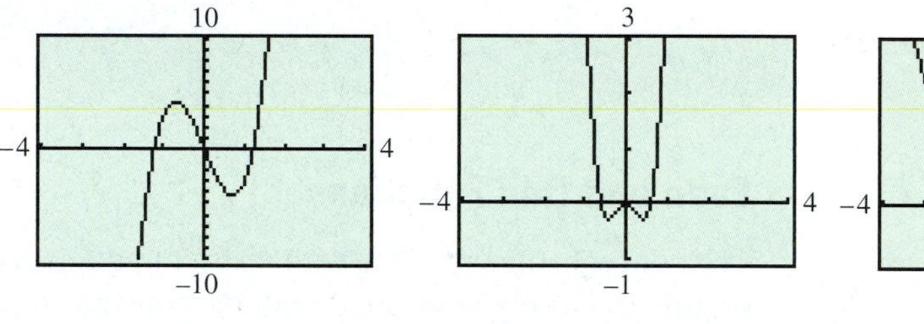

FIGURE 18

2.1 Exercises

Concept Check *Fill in each blank with the correct response.*

1. The domain and the range of the identity function are both ________.

2. The domain of the squaring function is ________, and its range is ________.

3. The graph of the cubing function changes from "opening downward" to "opening upward" at the point ________.

4. The domain of the square root function is ________, and its range is ________.

5. The cube root function ________ (increases/decreases) on its entire domain.

6. The absolute value function decreases on the interval ________ and increases on the interval ________.

7. The graph of the relation given by $x = y^2$ is symmetric with respect to the ________.

8. The function defined by $f(x) = x^4 + x^2$ is an ________ (even/odd) function.

9. The function defined by $f(x) = x^3 + x$ is an ________ (even/odd) function.

10. If a function is even, its graph is symmetric with respect to the ________; if it is odd, its graph is symmetric with respect to the ________.

Determine the intervals of the domain over which each function is continuous.

11.

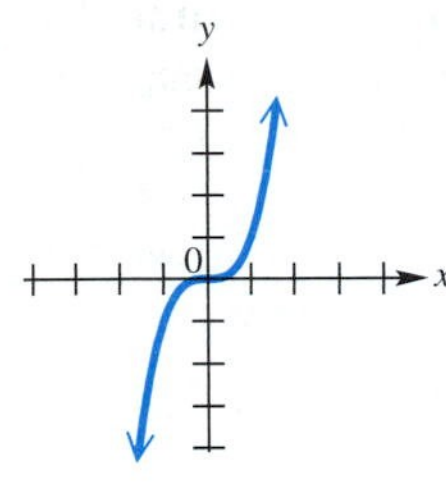

12.

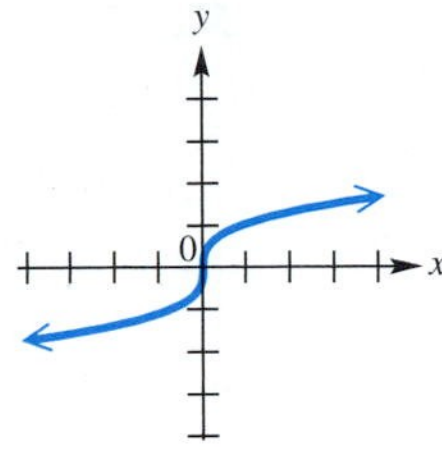

13.

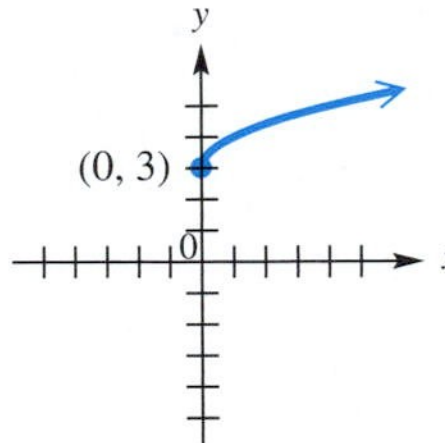

14.

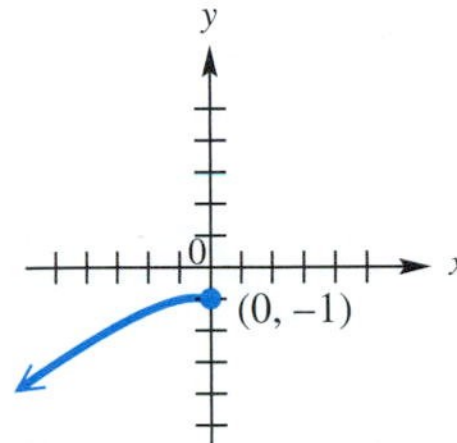

15.

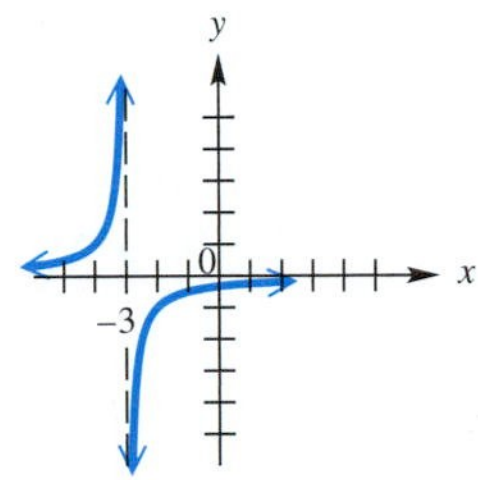

16.

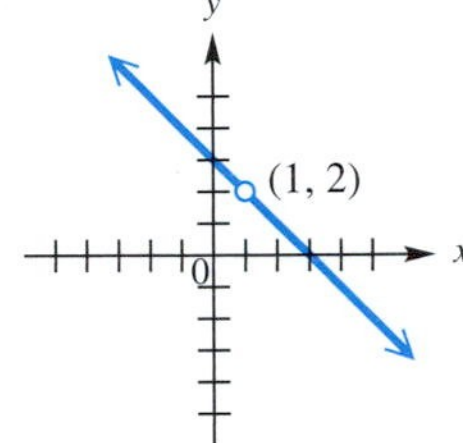

Determine the intervals of the domain over which each function is **(a)** *increasing,* **(b)** *decreasing, and* **(c)** *constant. Then give the* **(d)** *domain and* **(e)** *range.*

17.

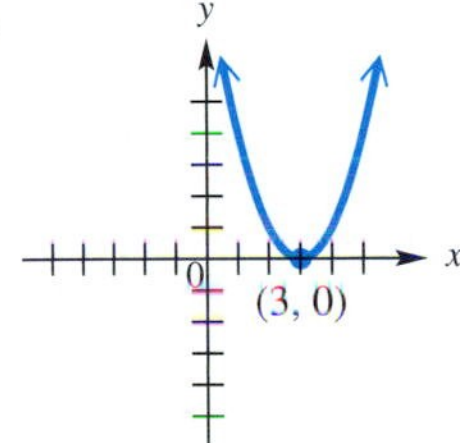

18.

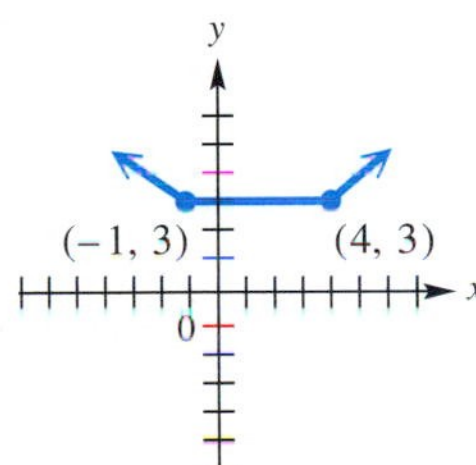

19.

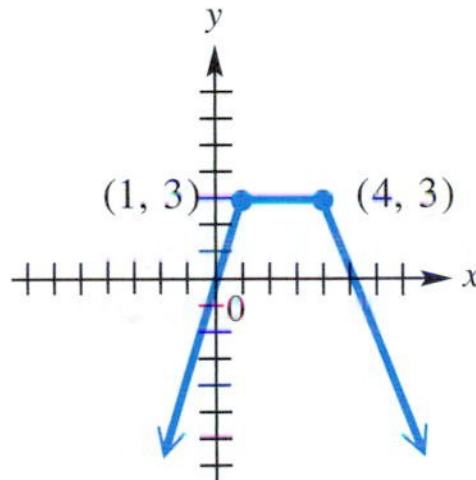

20.

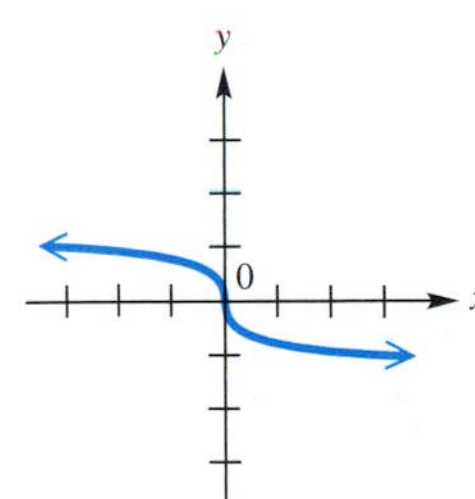

21.

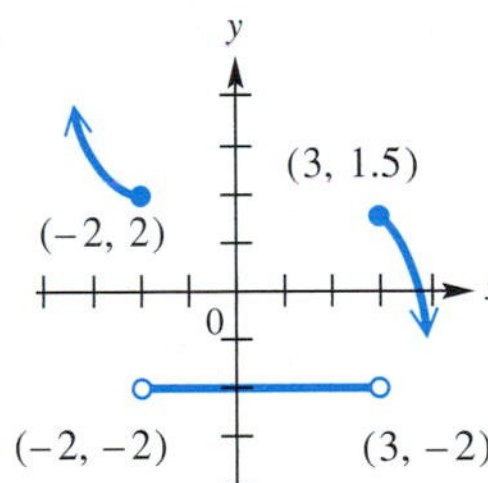

22.

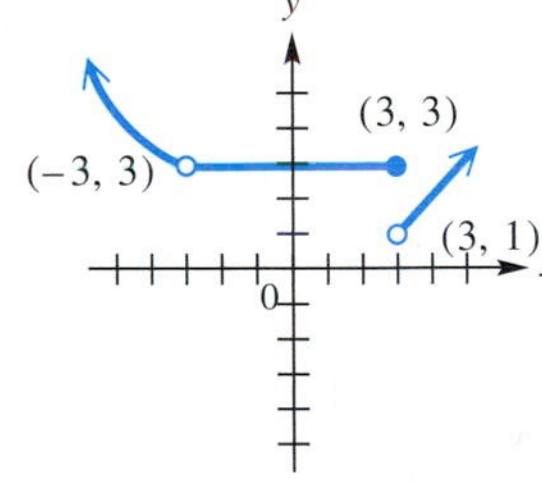

In Exercises 23–34, graph each function in the standard viewing window of your calculator, and trace from left to right along a representative portion of the specified interval. Then fill in the blank of the following sentence with either **increasing** *or* **decreasing**:

OVER THE INTERVAL SPECIFIED, THIS FUNCTION IS ________.

23. $f(x) = x^5$; $(-\infty, \infty)$

24. $f(x) = -x^3$; $(-\infty, \infty)$

25. $f(x) = x^4$; $(-\infty, 0]$

26. $f(x) = x^4$; $[0, \infty)$

27. $f(x) = -|x|$; $(-\infty, 0]$

28. $f(x) = -|x|$; $[0, \infty)$

29. $f(x) = -\sqrt[3]{x}$; $(-\infty, \infty)$

30. $f(x) = -\sqrt{x}$; $[0, \infty)$

31. $f(x) = 1 - x^3$; $(-\infty, \infty)$

32. $f(x) = x^2 - 2x$; $[1, \infty)$

33. $f(x) = 2 - x^2$; $(-\infty, 0]$

34. $f(x) = |x + 1|$; $(-\infty, -1]$

Using visual observation, determine whether each graph is symmetric with respect to the **(a)** *x-axis,* **(b)** *y-axis, or* **(c)** *origin.*

35.

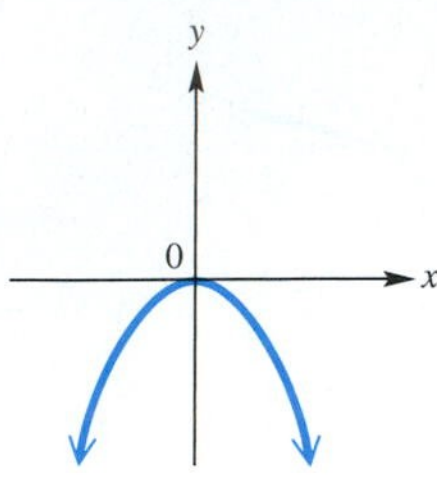

36.

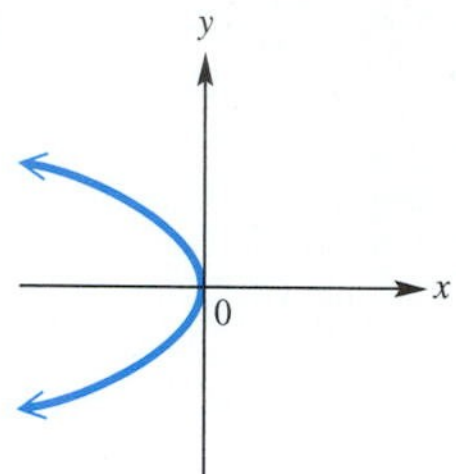

37.

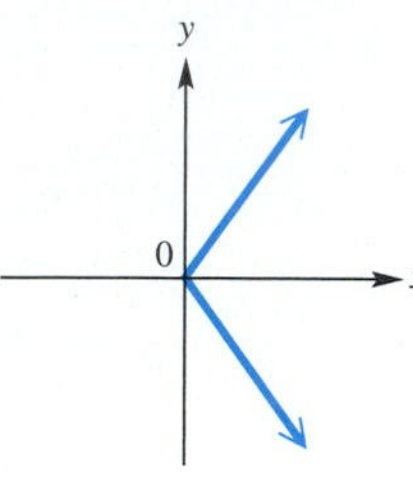

38.

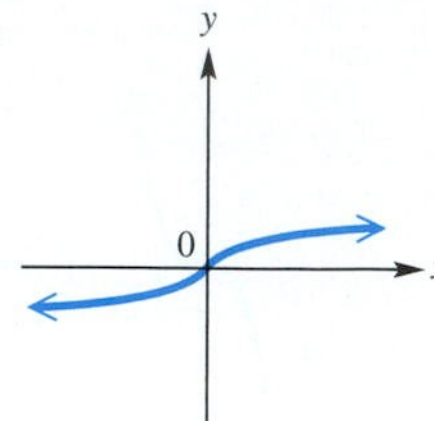

39.

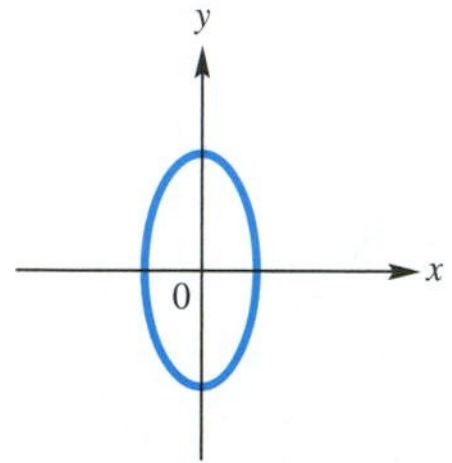

40.

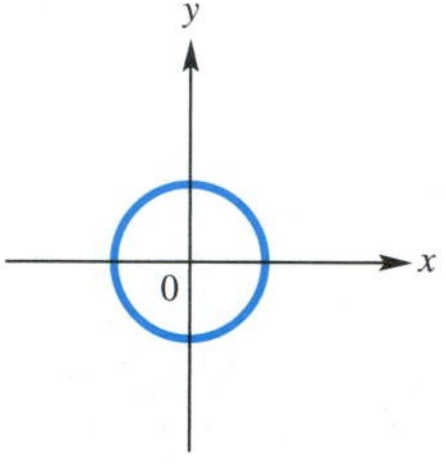

41.

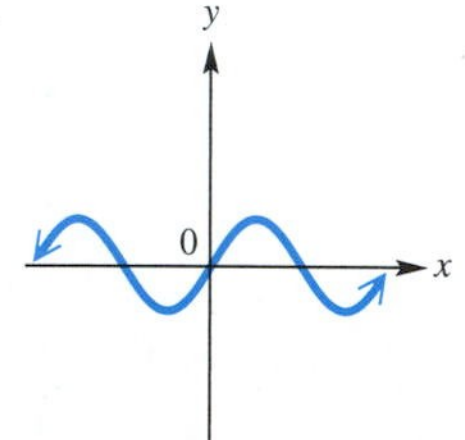

42.

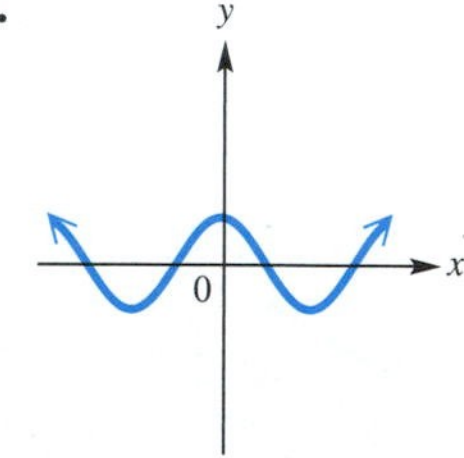

43. *Concept Check* Complete the table if f is an even function.

x	-3	-2	-1	1	2	3
$f(x)$	21		-25		-12	

44. *Concept Check* Complete the table if g is an odd function.

x	-5	-3	-2	0	2	3	5
$g(x)$	13		-5			-1	

45. *Concept Check* Complete the left half of the graph of $y = f(x)$ in the figure for each of the following conditions:

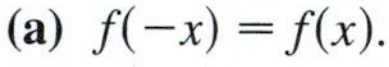

(a) $f(-x) = f(x)$.

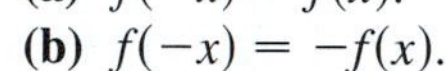

(b) $f(-x) = -f(x)$.

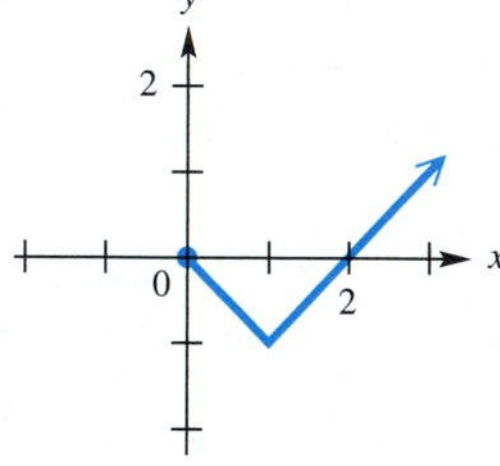

46. *Concept Check* Complete the right half of the graph of $y = f(x)$ in the figure for each of the following conditions:

(a) f is odd.

(b) f is even.

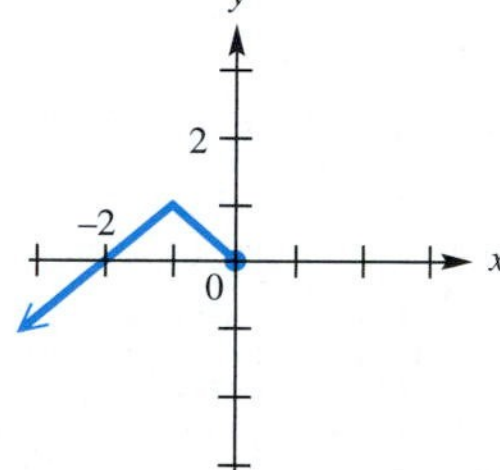

Based on the ordered pairs seen in each pair of tables, make a conjecture about whether the function defined in Y_1 *is* even, odd, *or* neither even nor odd.

47.

X	Y_1
0	0
1	1
2	4
3	9
4	16
5	25
6	36

X=0

X	Y_1
0	0
-1	1
-2	4
-3	9
-4	16
-5	25
-6	36

X=0

48.

X	Y_1
0	0
1	1
2	8
3	27
4	64
5	125
6	216

X=0

X	Y_1
0	0
-1	-1
-2	-8
-3	-27
-4	-64
-5	-125
-6	-216

X=0

Prove analytically that each function is even.

49. $f(x) = x^4 - 7x^2 + 6$

50. $f(x) = -2x^6 - 8x^2$

51. $f(x) = x^6 - 4x^4 + 5$

52. $f(x) = 8$

53. $f(x) = |5x|$

54. $f(x) = \sqrt{x^2 + 1}$

Prove analytically that each function is odd.

55. $f(x) = 3x^3 - x$

56. $f(x) = -x^5 + 2x^3 - 3x$

57. $f(x) = 3x^5 - x^3 + 7x$

58. $f(x) = x^3 - 4x$

59. $f(x) = \frac{1}{2x}$

60. $f(x) = 4x - \frac{1}{x}$

Use the analytic method of Example 3 to determine whether the graph of the given function is symmetric with respect to the y-axis, symmetric with respect to the origin, *or* neither. *Use your calculator and the standard window to support your conclusion.*

61. $f(x) = -x^3 + 2x$

62. $f(x) = x^5 - 2x^3$

63. $f(x) = .5x^4 - 2x^2 + 1$

64. $f(x) = .75x^2 + |x| + 1$

65. $f(x) = x^3 - x + 3$

66. $f(x) = x^4 - 5x + 2$

67. $f(x) = x^6 - 4x^3$

68. $f(x) = x^3 - 3x$

69. $f(x) = -6$

70. $f(x) = |-x|$

71. $f(x) = \frac{1}{4x^3}$

72. $f(x) = \sqrt{x^2}$

73. Refer to Exercises 61–72, and determine which exercises have functions that are
(a) even **(b)** odd **(c)** neither even nor odd.

74. Write an explanation of the connections between symmetry with respect to the y-axis, symmetry with respect to the origin, even functions, and odd functions.

2.2 Vertical and Horizontal Shifts of Graphs

Vertical Shifts ■ Horizontal Shifts ■ Combinations of Vertical and Horizontal Shifts ■ Effects of Shifts on Domain and Range ■ Horizontal Shifts Applied to Equations for Modeling

In this section we examine how the graphs of some basic functions can be *shifted,* or **translated,** vertically and horizontally in the plane.

Vertical Shifts

FOR DISCUSSION

In each group that follows, we give four related functions. Graph the functions in the first group (Group A) in the standard viewing window, and then answer the questions. Repeat the process for Group B, Group C, and Group D.

A	B	C	D
$y_1 = x^2$	$y_1 = x^3$	$y_1 = \sqrt{x}$	$y_1 = \sqrt[3]{x}$
$y_2 = x^2 + 3$	$y_2 = x^3 + 3$	$y_2 = \sqrt{x} + 3$	$y_2 = \sqrt[3]{x} + 3$
$y_3 = x^2 - 2$	$y_3 = x^3 - 2$	$y_3 = \sqrt{x} - 2$	$y_3 = \sqrt[3]{x} - 2$
$y_4 = x^2 + 5$	$y_4 = x^3 + 5$	$y_4 = \sqrt{x} + 5$	$y_4 = \sqrt[3]{x} + 5$

1. How does the graph of y_2 compare with the graph of y_1?
2. How does the graph of y_3 compare with the graph of y_1?
3. How does the graph of y_4 compare with the graph of y_1?

(continued)

GCM TECHNOLOGY NOTE

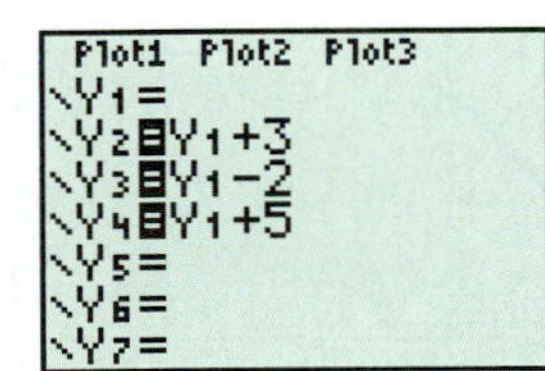

This screen shows how to minimize keystrokes in the activity in the "For Discussion" box. The entry for Y_1 can be altered as needed.

4. If $c > 0$, how do you think the graph of $y = f(x) + c$ would compare with the graph of $y = f(x)$?
5. If $c > 0$, how do you think the graph of $y = f(x) - c$ would compare with the graph of $y = f(x)$?

Choosing your own value of c, support your answers to Items 4 and 5 graphically. (Be sure that your choice is appropriate for the standard window.)

In each group of functions in the preceding activity, we obtained a *vertical shift* of the graph of the basic function with which we started. Our observations can be generalized to any function.

Vertical Shifting of the Graph of a Function

If $c > 0$, then the graph of $y = f(x) + c$ is obtained by shifting the graph of $y = f(x)$ *upward* a distance of c units. The graph of $y = f(x) - c$ is obtained by shifting the graph of $y = f(x)$ *downward* a distance of c units.

In Figure 19, we graphically interpret the preceding statements.

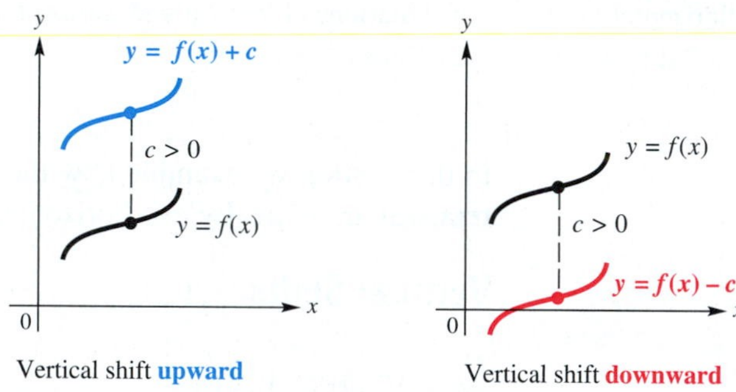

FIGURE 19

EXAMPLE 1 Recognizing Vertical Shifts

Give the equation of each function graphed in Figure 20.

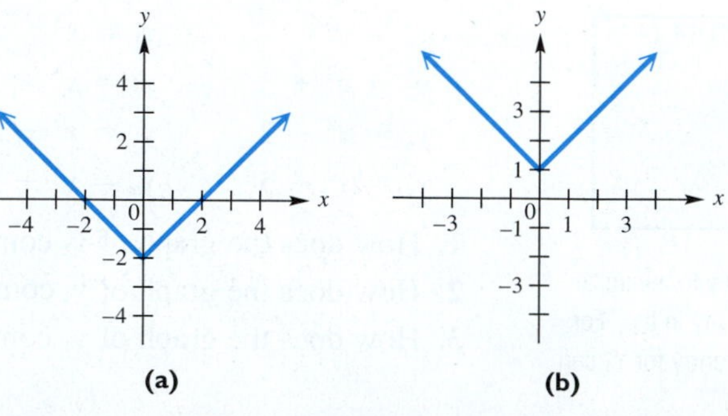

FIGURE 20

Solution Both graphs are vertical translations of the graph of $y = |x|$. In Figure 20(a), the graph has been shifted 2 units downward, so the equation is $y = |x| - 2$. In Figure 20(b), the graph has been shifted 1 unit upward, so the equation is $y = |x| + 1$. ■

Horizontal Shifts

FOR DISCUSSION

This discussion parallels the one earlier in this section. Follow the same general directions.

A	B	C	D
$y_1 = x^2$	$y_1 = x^3$	$y_1 = \sqrt{x}$	$y_1 = \sqrt[3]{x}$
$y_2 = (x - 3)^2$	$y_2 = (x - 3)^3$	$y_2 = \sqrt{x - 3}$	$y_2 = \sqrt[3]{x - 3}$
$y_3 = (x - 5)^2$	$y_3 = (x - 5)^3$	$y_3 = \sqrt{x - 5}$	$y_3 = \sqrt[3]{x - 5}$
$y_4 = (x + 4)^2$	$y_4 = (x + 4)^3$	$y_4 = \sqrt{x + 4}$	$y_4 = \sqrt[3]{x + 4}$

1. How does the graph of y_2 compare with the graph of y_1?
2. How does the graph of y_3 compare with the graph of y_1?
3. How does the graph of y_4 compare with the graph of y_1?
4. If $c > 0$, how do you think the graph of $y = f(x - c)$ would compare with the graph of $y = f(x)$?
5. If $c > 0$, how do you think the graph of $y = f(x + c)$ would compare with the graph of $y = f(x)$?

Choosing your own value of c, support your answers to Items 4 and 5 graphically. (Again, be sure that your choice is appropriate for the standard window.)

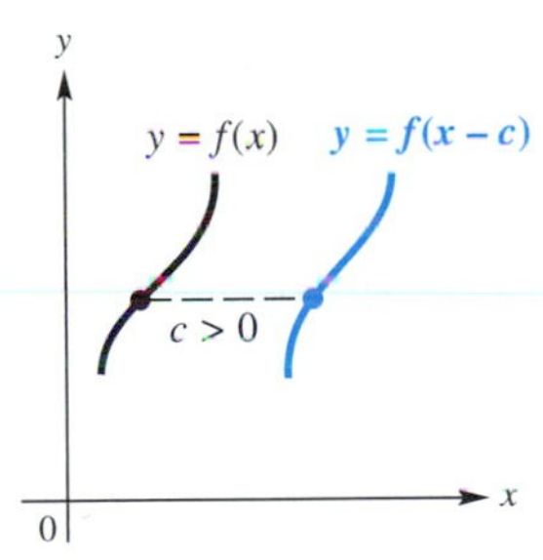

Horizontal shift to the **right**

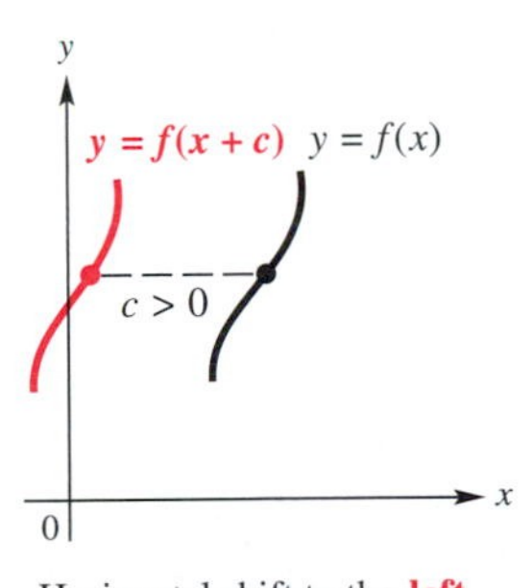

Horizontal shift to the **left**

FIGURE 21

The results of the preceding discussion should remind you of the results found earlier when we shifted graphs of functions vertically. Now we see how such graphs can be shifted *horizontally*. The observations can be generalized as follows.

Horizontal Shifting of the Graph of a Function

If $c > 0$, the graph of $y = f(x - c)$ is obtained by shifting the graph of $y = f(x)$ to the *right* a distance of c units. The graph of $y = f(x + c)$ is obtained by shifting the graph of $y = f(x)$ to the *left* a distance of c units.

In Figure 21, we graphically interpret the preceding statements.

CAUTION ***Be careful when translating graphs horizontally.*** To determine the direction and magnitude of horizontal shifts, find the value of x that would cause the expression in parentheses to equal 0. For example, the graph of $f(x) = (x - 5)^2$ would be shifted 5 units to the *right*, because $+5$ would cause $x - 5$ to equal 0. By contrast, the graph of $f(x) = (x + 5)^2$ would be shifted 5 units to the *left*, because -5 would cause $x + 5$ to equal 0.

EXAMPLE 2 **Recognizing Horizontal Shifts**

Give the equation of each function graphed in Figure 22.

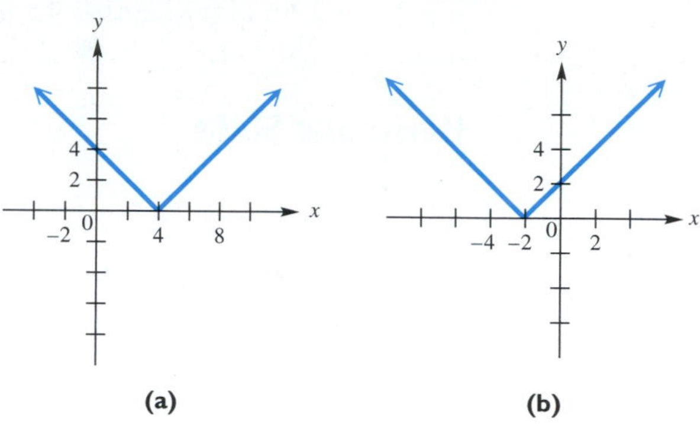

FIGURE 22

Solution Both graphs are horizontal translations of the graph of $y = |x|$. In Figure 22(a), the graph has been shifted 4 units to the right, so the equation is $y = |x - 4|$. In Figure 22(b), the graph has been shifted 2 units to the left, so the equation is $y = |x + 2|$. (Be sure that you understand why the signs are as they appear here.) ■

Combinations of Vertical and Horizontal Shifts

Now that we have seen how graphs of functions can be shifted vertically or horizontally, we extend these ideas to graphs that are obtained by applying *both* types of translations.

EXAMPLE 3 **Applying Both Vertical and Horizontal Shifts**

Describe how the graph of $y_2 = |x + 2| - 6$ would be obtained by translating the graph of $y_1 = |x|$. Sketch the graphs of y_1 and y_2 on the same xy-plane by hand.

Solution The function defined by $y_2 = |x + 2| - 6$ is translated 2 units to the *left* (because of the $|x + 2|$) and 6 units *downward* (because of -6) compared with the graph of $y_1 = |x|$. See Figure 23. Note that at the point $(-2, -6)$, the graph of y_2 changes from decreasing to increasing.

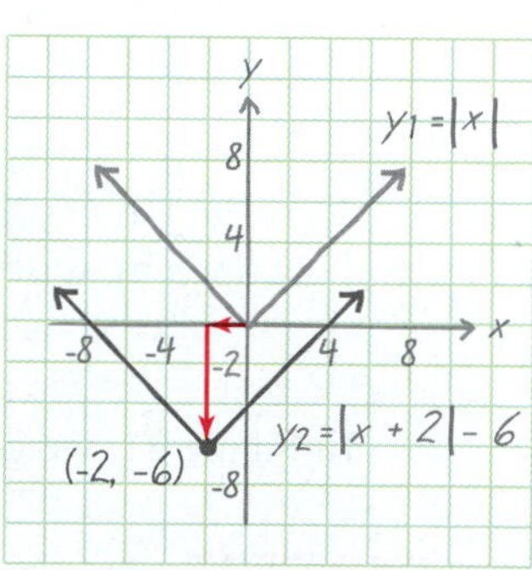

FIGURE 23 ■

Effects of Shifts on Domain and Range

The domains and ranges of functions may or may not be affected by vertical and horizontal shifts, as the next example illustrates.

EXAMPLE 4 Determining Domains and Ranges of Shifted Graphs

Give the domain and range of each function graphed in Figure 24.

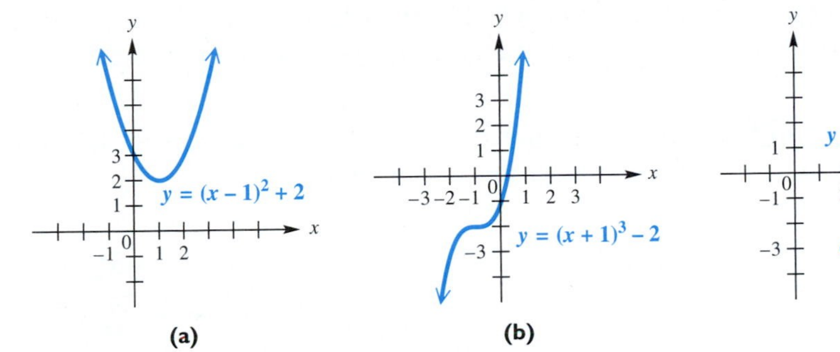

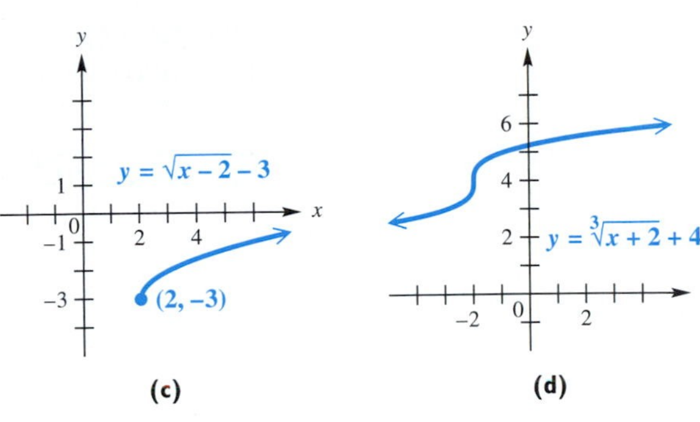

FIGURE 24

Solution

(a) To obtain the graph of $y = (x - 1)^2 + 2$ in Figure 24(a), the graph of $y = x^2$ was translated 1 unit to the right and 2 units upward. The original domain $(-\infty, \infty)$ is not affected. However, the range of this function is $[2, \infty)$, because of the vertical translation.

(b) The graph of $y = (x + 1)^3 - 2$ in Figure 24(b) is obtained by vertical and horizontal shifts of the graph of $y = x^3$, a function that has both domain and range $(-\infty, \infty)$. Neither is affected here, so the domain and range of $y = (x + 1)^3 - 2$ are also $(-\infty, \infty)$.

(c) The function defined by $y = \sqrt{x}$ has domain $[0, \infty)$. The graph of $y = \sqrt{x - 2} - 3$ in Figure 24(c) is obtained by shifting the basic graph 2 units to the right, so the new domain is $[2, \infty)$. The original range, $[0, \infty)$, has also been affected by the shift of the graph 3 units downward, so the new range is $[-3, \infty)$.

(d) The situation in Figure 24(d) is similar to that of Figure 24(b). The graph of $y = \sqrt[3]{x}$, with domain and range $(-\infty, \infty)$, is shifted 2 units to the left and 4 units upward. Regardless of the direction and size of these shifts, the domain and range are unaffected and remain $(-\infty, \infty)$. ■

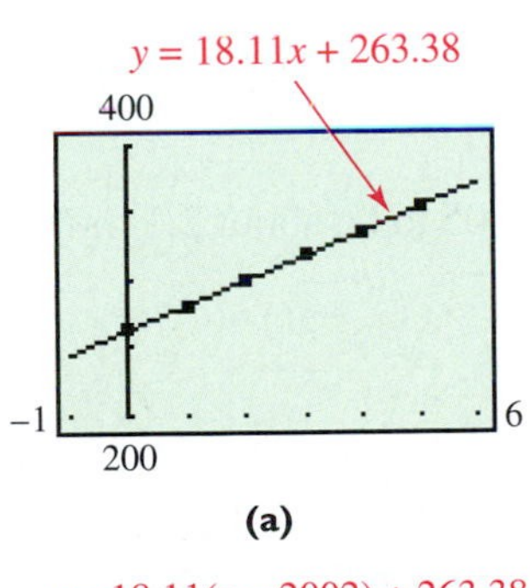

(a)

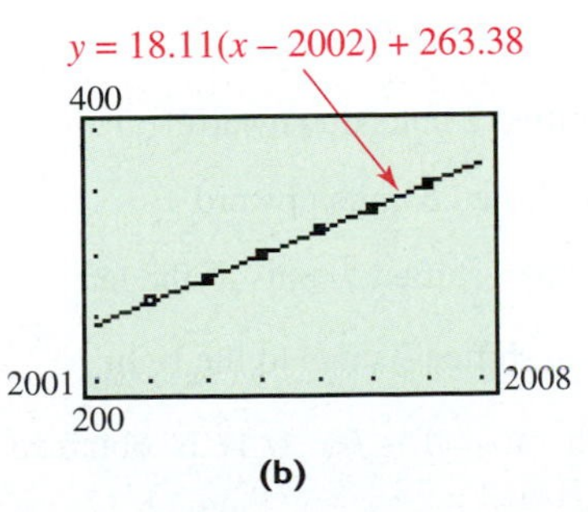

(b)

FIGURE 25

Horizontal Shifts Applied to Equations for Modeling

In Examples 6 and 7 of Section 1.4, we examined data that showed estimates for Medicare costs (in billions of dollars) between the years 2002 and 2007. To simplify our work, $x = 0$ corresponded to the year 2002, $x = 1$ to 2003, and so on. We determined that the least-squares regression line has equation

$$y \approx 18.11x + 263.38.$$

The graph of this line is shown in Figure 25(a). To use the values of the years 2002 through 2007 directly in the regression equation, we must shift the graph of the line 2002 units to the right. The equation of this new line is

$$y \approx 18.11(x - 2002) + 263.38.$$

Using a graphing calculator to plot the set of points

$$\{(2002, 264), (2003, 281), (2004, 299), (2005, 318), (2006, 336), (2007, 354)\}$$

and graphing $y = 18.11(x - 2002) + 263.38$ in the viewing window $[2001, 2008]$ by $[200, 400]$ gives the graph shown in Figure 25(b) on the previous page.

EXAMPLE 5 Applying a Horizontal Shift to an Equation Model

The table lists U.S. sales of Toyota vehicles in millions.

Year	Vehicles
1998	1.4
1999	1.5
2000	1.6
2001	1.7
2002	1.8

Source: Autodata.

These data can be modeled by the equation $y = .1x + 1.4$, where $x = 0$ corresponds to 1998, $x = 1$ to 1999, and so on.

(a) Find the corresponding equation that allows direct input of the year.

(b) Predict U.S. sales of Toyota vehicles in 2005.

Solution

(a) Because 1998 corresponds to 0, the graph of $y = .1x + 1.4$ would have to be shifted 1998 units to the right. Thus, the equation becomes

$$y = .1(x - 1998) + 1.4.$$

(b) To predict sales in 2005, substitute 2005 for x and evaluate the expression.

$$y = .1(2005 - 1998) + 1.4 = 2.1$$

This model predicts U.S. sales of Toyota vehicles in 2005 to be about 2.1 million. ■

2.2 Exercises

Write the equation that results in the desired translation.

1. The squaring function, shifted 3 units upward
2. The cubing function, shifted 2 units downward
3. The square root function, shifted 4 units downward
4. The cube root function, shifted 6 units upward
5. The absolute value function, shifted 4 units to the right
6. The absolute value function, shifted 3 units to the left
7. The cubing function, shifted 7 units to the left
8. The square root function, shifted 9 units to the right
9. Explain how the graph of $g(x) = f(x) + 4$ is obtained from the graph of $y = f(x)$.
10. Explain how the graph of $g(x) = f(x + 4)$ is obtained from the graph of $y = f(x)$.

Exercises 11–19 are grouped in threes. For each group, match the correct graph, A, B, or C, to the function. Confirm your answers with your calculator.

11. $y = x^2 - 3$ | **12.** $y = (x - 3)^2$ | **13.** $y = (x + 3)^2$

A.

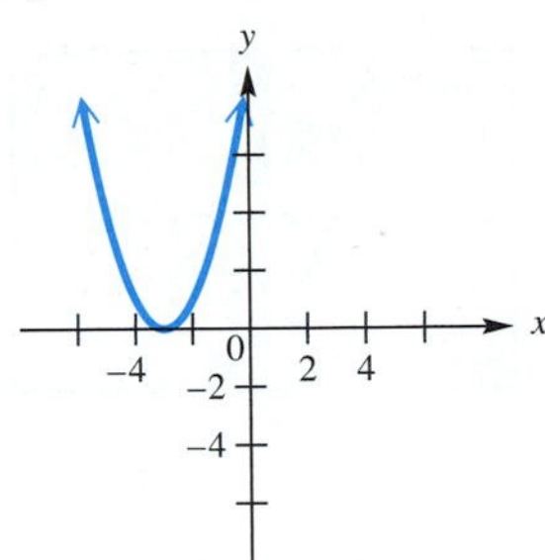

B.

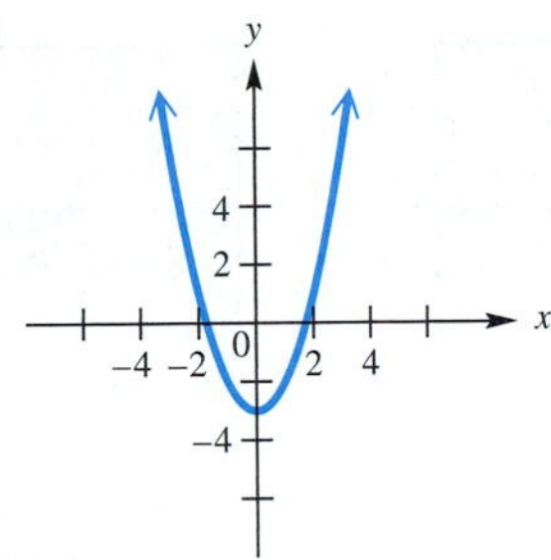

C.

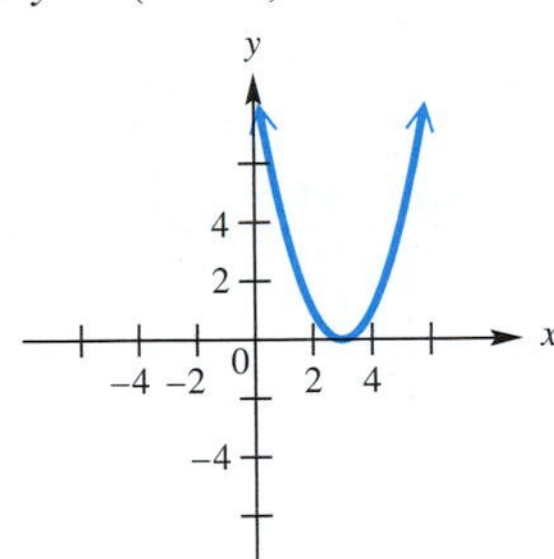

14. $y = |x| + 4$ | **15.** $y = |x + 4| - 3$ | **16.** $y = |x - 4| - 3$

A.

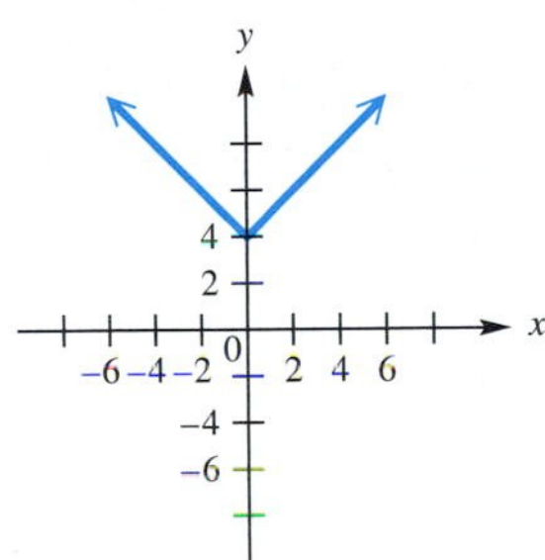

B.

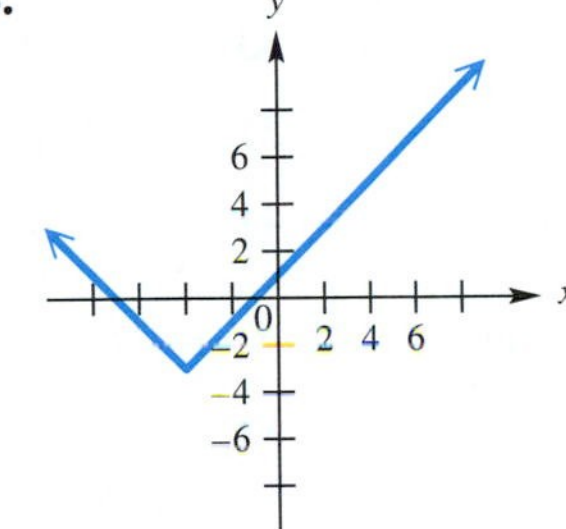

C.

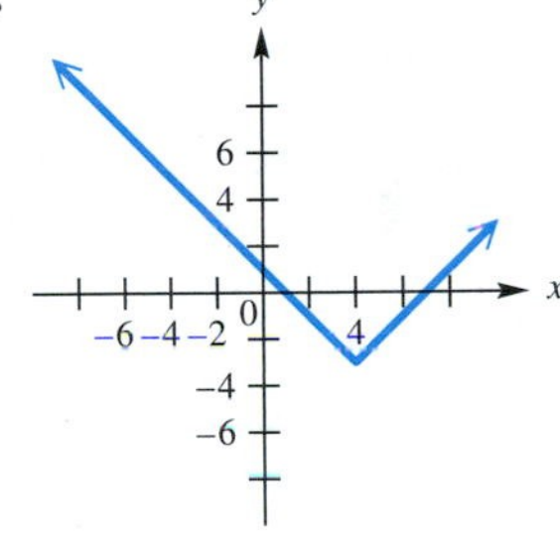

17. $y = (x - 3)^3$ | **18.** $y = (x - 2)^3 - 4$ | **19.** $y = (x + 2)^3 - 4$

A.

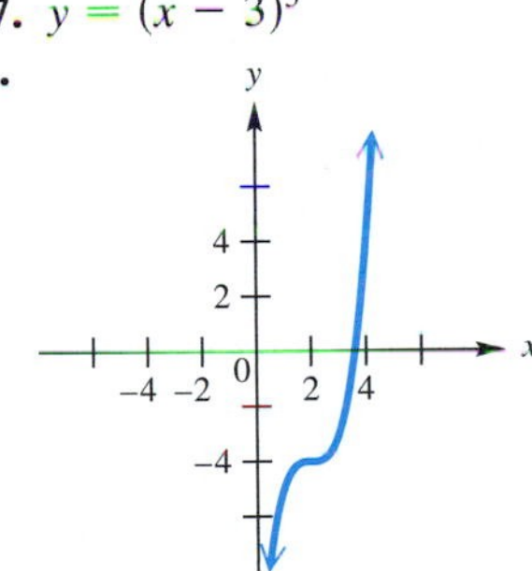

B.

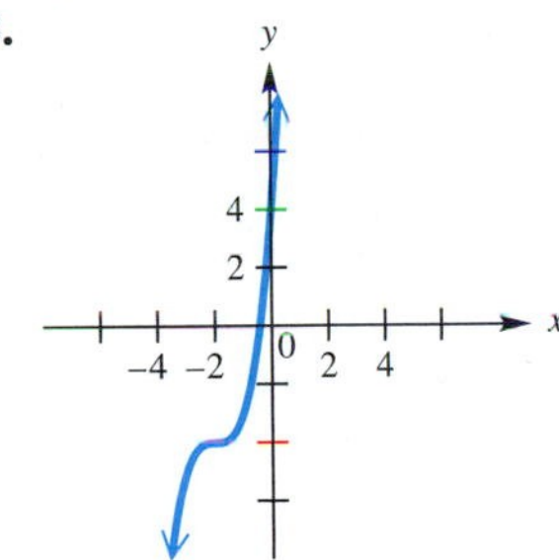

C.

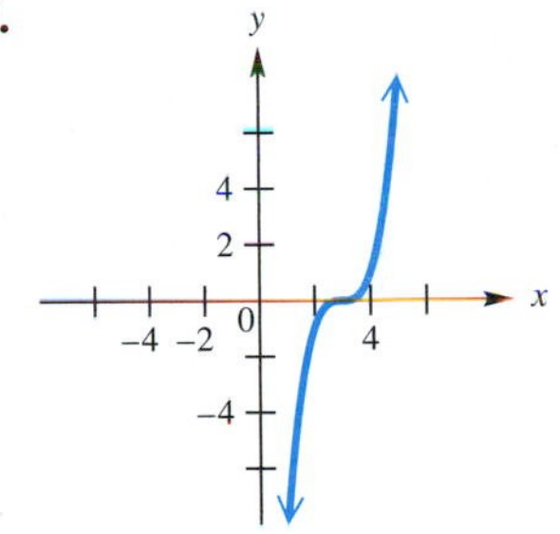

20. ***Concept Check*** In which quadrant does the lowest point on the graph of $y = |x - h|^2 + k$ lie if $h < 0$ and $k < 0$?

Use the results of the specified exercises to determine **(a)** *the domain and* **(b)** *the range of each function.*

21. $y = x^2 - 3$ (Exercise 11)

22. $y = (x - 3)^2$ (Exercise 12)

23. $y = |x + 4| - 3$ (Exercise 15)

24. $y = |x - 4| - 3$ (Exercise 16)

25. $y = (x - 3)^3$ (Exercise 17)

26. $y = (x - 2)^3 - 4$ (Exercise 18)

Concept Check *The function* Y_2 *is defined as* $Y_1 + k$ *for some real number k. Based on the table shown, what is the value of k?*

27.

X	Y_1	Y_2
0	15	19
1	10	14
2	7	11
3	6	10
4	7	11
5	10	14
6	15	19

X=0

28.

X	Y_1	Y_2
0	−3	−5
1	2	0
2	5	3
3	6	4
4	5	3
5	2	0
6	−3	−5

X=0

Concept Check *The function* Y_2 *is defined as* $Y_1 + k$ *for some real number k. Based on the two views of the graphs of* Y_1 *and* Y_2 *and the displays at the bottoms of the screens, what is the value of k?*

29.

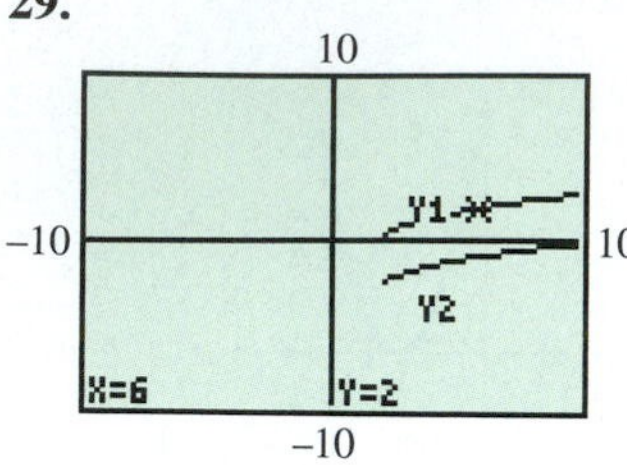

(6, 2) lies on the graph of Y_1.
First view

(6, −1) lies on the graph of Y_2.
Second view

30.

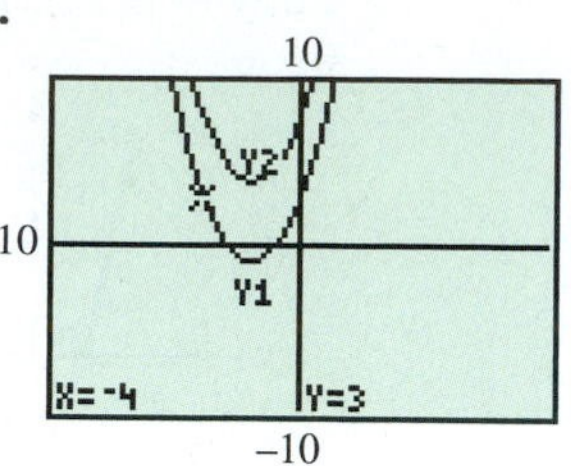

(−4, 3) lies on the graph of Y_1.
First view

(−4, 8) lies on the graph of Y_2.
Second view

Use translations of one of the basic functions defined by $y = x^2$, $y = x^3$, $y = \sqrt{x}$, *or* $y = |x|$ *to sketch a graph of* $y = f(x)$ *by hand.*

31. $y = (x - 1)^2$

32. $y = \sqrt{x + 2}$

33. $y = x^3 + 1$

34. $y = |x + 2|$

35. $y = (x - 1)^3$

36. $y = |x| - 3$

37. $y = \sqrt{x - 2} - 1$

38. $y = \sqrt{x + 3} - 4$

39. $y = (x + 2)^2 + 3$

40. $y = (x - 4)^2 - 4$

41. $y = |x + 4| - 2$

42. $y = (x + 3)^3 - 1$

Concept Check *Suppose that h and k are both positive numbers. Match each equation with the correct graph.*

43. $y = (x - h)^2 - k$

44. $y = (x + h)^2 - k$

45. $y = (x + h)^2 + k$

46. $y = (x - h)^2 + k$

A.

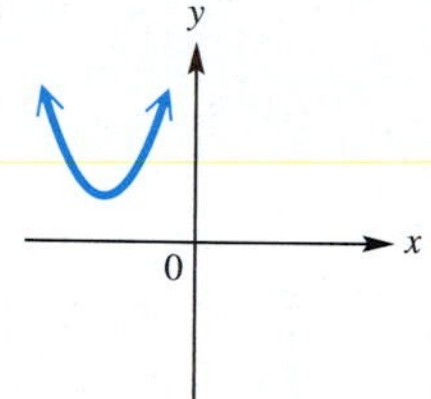

B.

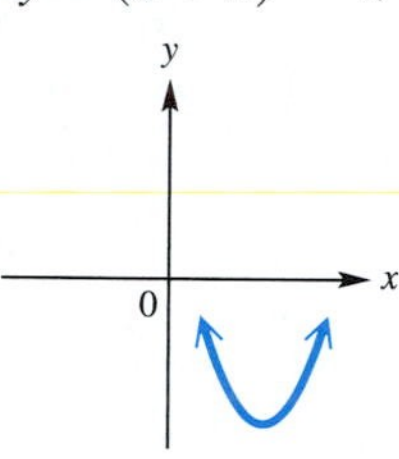

C.

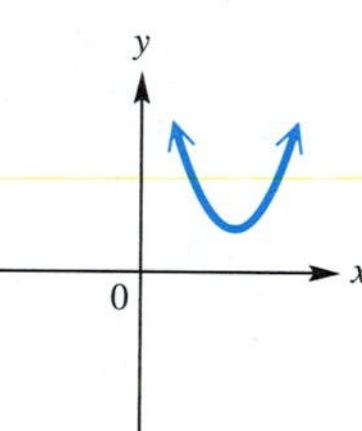

D.

Concept Check *Given the graph shown, sketch by hand the graph of each function described, indicating how the three points labeled on the original graph have been translated.*

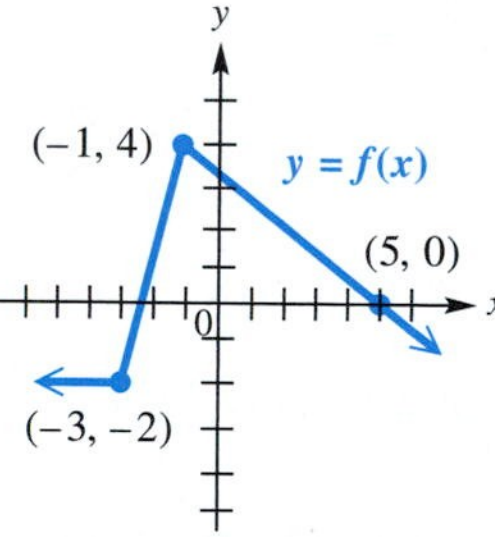

47. $y = f(x) + 2$

48. $y = f(x) - 2$

49. $y = f(x + 2)$

50. $y = f(x - 2)$

Each graph is a translation of the graph of one of the basic functions defined by $y = x^2$, $y = x^3$, $y = \sqrt{x}$, *or* $y = |x|$. *Find the equation that defines each function. Then, using the concepts of increasing and decreasing functions discussed in Section 2.1, determine the interval of the domain over which the function is* **(a)** *increasing and* **(b)** *decreasing.*

51.

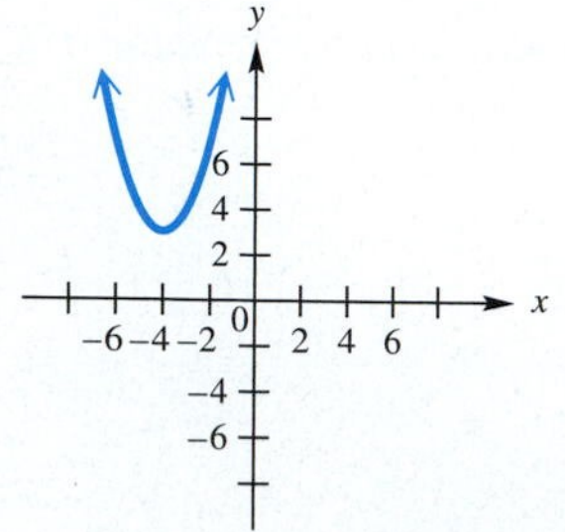

52.

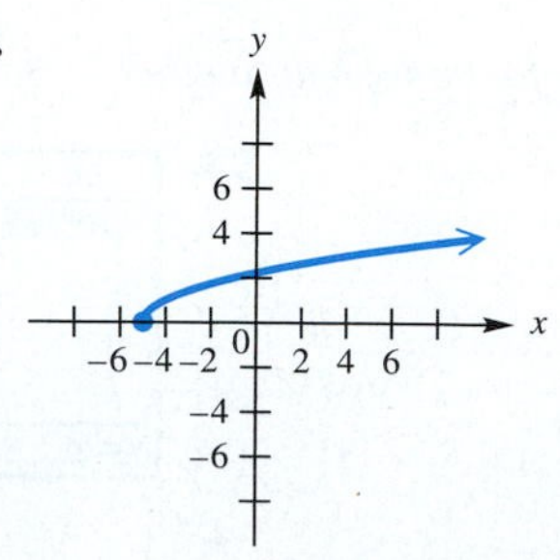

53.

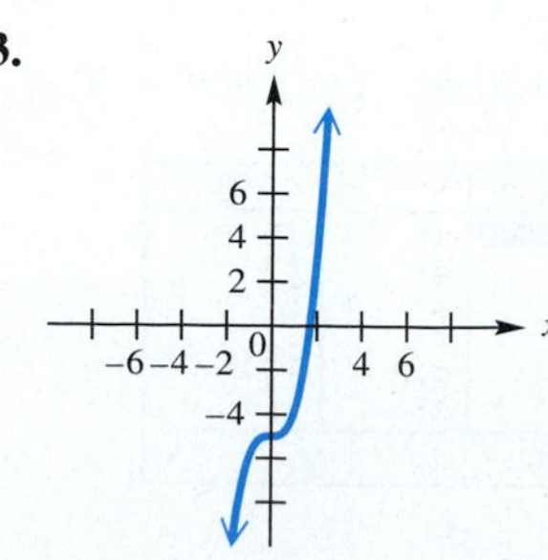

54.

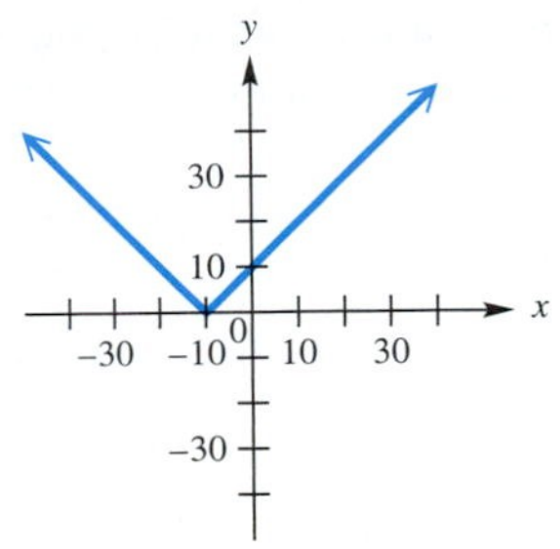

55.

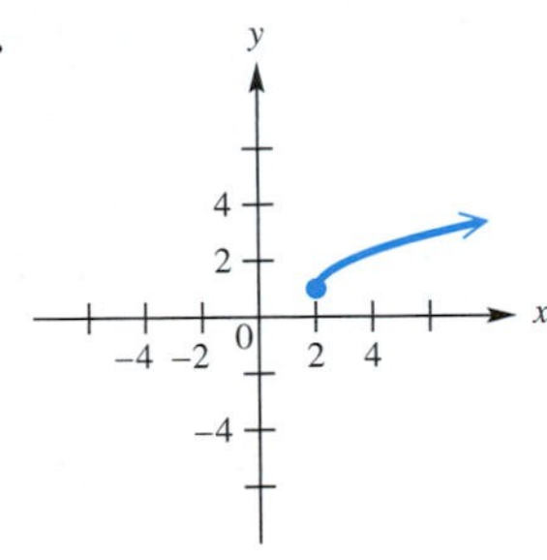

56.

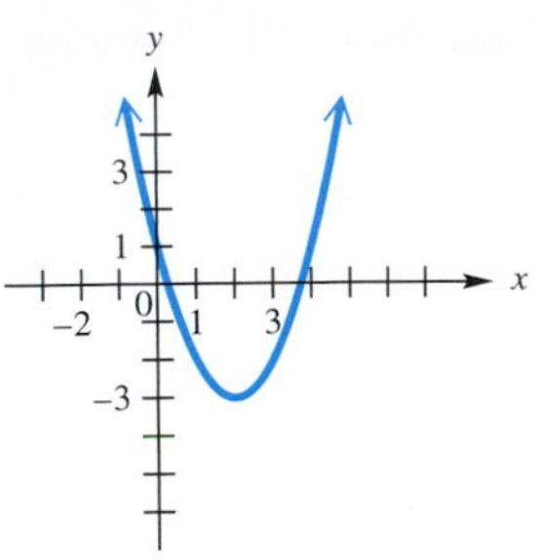

Relating Concepts

For individual or group investigation (Exercises 57–60)

Use the x-intercept method and the given graph to solve each equation or inequality.

57. **(a)** $f(x) = 0$
(b) $f(x) > 0$
(c) $f(x) < 0$

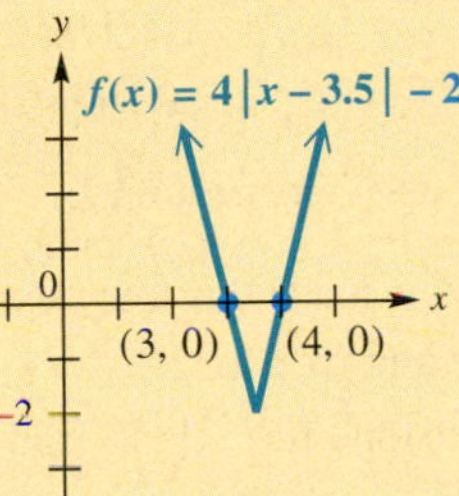

58. **(a)** $f(x) = 0$
(b) $f(x) > 0$
(c) $f(x) < 0$

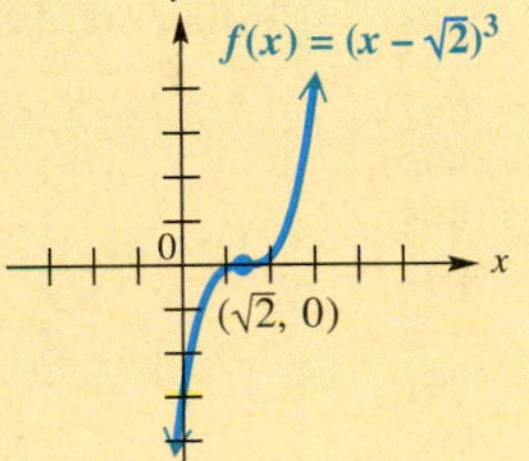

59. **(a)** $f(x) = 0$
(b) $f(x) \geq 0$
(c) $f(x) \leq 0$

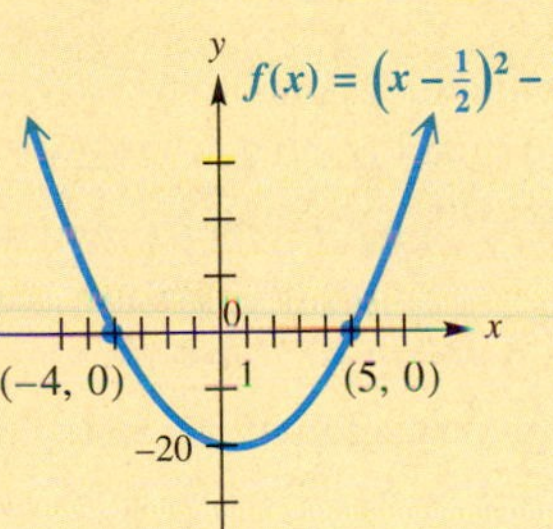

60. **(a)** $f(x) = 0$
(b) $f(x) \geq 0$
(c) $f(x) \leq 0$

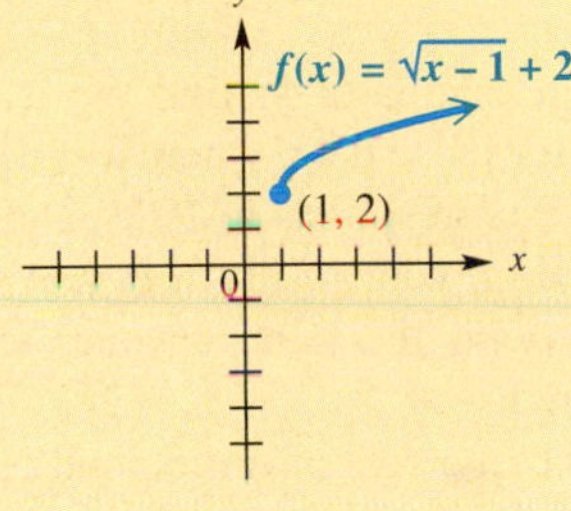

61. ***Concept Check*** The graph is a translation of $y = |x|$. What are the values of h and k if the equation is of the form $y = |x - h| + k$?

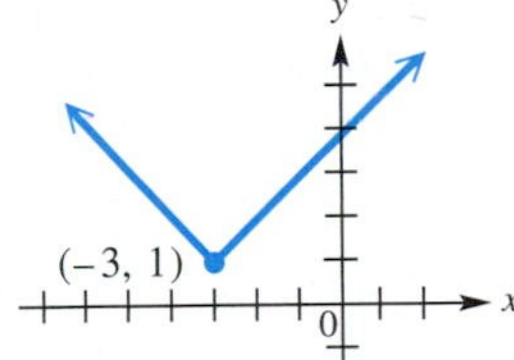

62. ***Concept Check*** Suppose that the graph of $y = x^2$ is translated in such a way that its domain is $(-\infty, \infty)$ and its range is $[38, \infty)$. What values of h and k can be used if the new function is of the form $y = (x - h)^2 + k$?

(Modeling) *Solve each problem.*

63. ***Cost of Private College Education*** The linear equation

$$y = 895.5x + 14{,}709$$

gives an approximation of the annual cost y (in dollars) for tuition and fees at private universities, where $x = 0$ corresponds to 1998, $x = 1$ corresponds to 1999, and so on. (*Source*: The College Board.)

(a) Write an equation that yields the same y-values when the exact year is entered.

(b) Predict the cost of tuition and fees in 2006.

64. *Federal Debt* The linear equation

$$y = 299.8x + 5249.1$$

gives an approximation of the total federal debt y (in billions of dollars), where $x = 0$ corresponds to 1998, $x = 1$ corresponds to 1999, and so on. (*Source:* U.S. Bureau of the Public Debt.)

(a) Write an equation that yields the same y-values when the exact year is entered.

(b) Predict the total federal debt in 2007.

65. *Cost of Public College Education* The table lists the average annual costs (in dollars) of tuition and fees at public four-year colleges for selected years.

Year	Tuition and Fees (in dollars)
1991	2137
1993	2527
1995	2860
1997	3111
1999	3356
2001	3857
2003	4632

Source: The College Board.

(a) Use a calculator to find the least-squares regression line for these data, where $x = 0$ corresponds to 1991, $x = 2$ corresponds to 1993, and so on.

(b) Based on your result from part (a), write an equation that yields the same y-values when the exact year is entered.

(c) Predict the cost of tuition and fees in 2008.

66. *Women in the Workforce* The table shows how the percent of women in the civilian workforce has changed from 1965 to 2000.

Year	Percent of Women in the Workforce
1965	39.3
1970	43.3
1975	46.3
1980	51.5
1985	54.5
1990	57.5
1995	58.9
2000	59.9

Source: U.S. Bureau of Labor Statistics.

(a) Use a calculator to find the least-squares regression line for these data, where $x = 0$ corresponds to 1965, $x = 5$ corresponds to 1970, and so on.

(b) Based on your result from part (a), write an equation that yields the same y-values when the exact year is entered.

(c) Predict the percent of population of women in the civilian workforce in 2010.

Relating Concepts

For individual or group investigation (Exercises 67–74)

Recall from Chapter 1 that a unique line is determined by two different points on the line and that the values of m and b can then be determined for the general form of the linear function defined by

$$f(x) = mx + b.$$

Work these exercises in order.

67. Sketch by hand the line that passes through the points $(1, -2)$ and $(3, 2)$.

68. Use the slope formula to find the slope of this line.

69. Find the equation of the line, and write it in the form $y_1 = mx + b$.

70. Keeping the same two x-values as indicated in Exercise 67, add 6 to each y-value. What are the coordinates of the two new points?

71. Find the slope of the line passing through the points determined in Exercise 70.

72. Find the equation of this new line, and write it in the form $y_2 = mx + b$.

73. Graph both y_1 and y_2 in the standard viewing window of your calculator, and describe how the graph of y_2 can be obtained by vertically translating the graph of y_1. What is the value of the constant in this vertical translation? Where do you think it comes from?

74. Fill in the blanks with the correct responses, based on your work in Exercises 67–73.

If the points (x_1, y_1) and (x_2, y_2) lie on a line, then when we add the positive constant c to each y-value, we obtain the points $(x_1, y_1 +$ ______$)$ and $(x_2, y_2 +$ ______$)$. The slope of the new line is ______________ (the same as/different from) the slope of the original line. The graph of the new line can be obtained by shifting the graph of the original line ______ units in the ____________ direction.

2.3 Stretching, Shrinking, and Reflecting Graphs

Vertical Stretching ■ Vertical Shrinking ■ Horizontal Stretching and Shrinking ■ Reflecting across an Axis ■ Combining Transformations of Graphs

In the previous section, we saw how adding or subtracting a constant can cause a vertical or horizontal shift. Now we will see how multiplying by a constant alters the graph of a function.

Vertical Stretching

FOR DISCUSSION

In each group that follows, we give four related functions. Graph the functions in the first group (Group A), and then answer the questions. Repeat the process for Group B and Group C. Use the window specified for each group.

A	B	C
$[-5, 5]$ by $[-5, 20]$	$[-5, 15]$ by $[-5, 10]$	$[-20, 20]$ by $[-10, 10]$
$y_1 = x^2$	$y_1 = \sqrt{x}$	$y_1 = \sqrt[3]{x}$
$y_2 = 2x^2$	$y_2 = 2\sqrt{x}$	$y_2 = 2\sqrt[3]{x}$
$y_3 = 3x^2$	$y_3 = 3\sqrt{x}$	$y_3 = 3\sqrt[3]{x}$
$y_4 = 4x^2$	$y_4 = 4\sqrt{x}$	$y_4 = 4\sqrt[3]{x}$

1. How does the graph of y_2 compare with the graph of y_1?
2. How does the graph of y_3 compare with the graph of y_1?
3. How does the graph of y_4 compare with the graph of y_1?
4. If we choose $c > 4$, how do you think the graph of $y_5 = c \cdot y_1$ would compare with the graph of y_4? Provide graphical support by choosing such a value of c.

TECHNOLOGY NOTE

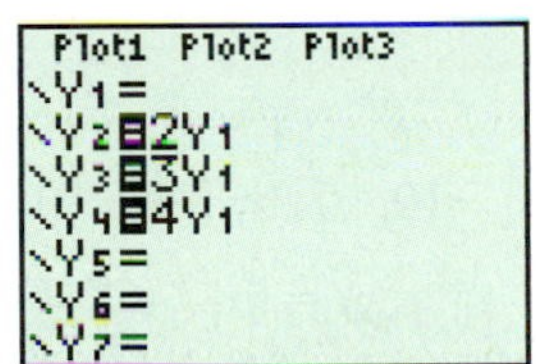

By defining Y_1 as directed in parts A, B, and C, and defining Y_2, Y_3, and Y_4 as shown here, you can minimize your keystrokes. (These graphs will *not* appear unless Y_1 is defined.)

In each group of functions in the preceding activity, we obtained a *vertical stretch* of the graph of the basic function with which we started. These observations can be generalized to any function.

Vertical Stretching of the Graph of a Function

If a point (x, y) lies on the graph of $y = f(x)$, then the point (x, cy) lies on the graph of $y = cf(x)$. If $c > 1$, then the graph of $y = cf(x)$ is a *vertical stretching* of the graph of $y = f(x)$ by a factor of c.

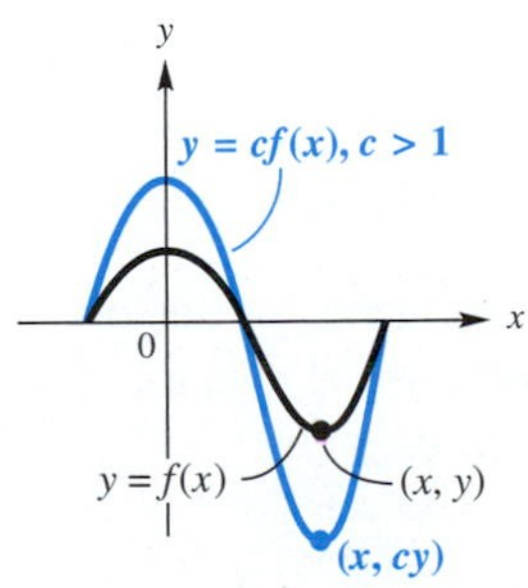

FIGURE 26

In Figure 26, we graphically interpret the preceding statement. Notice that the graphs have the same x-intercepts.

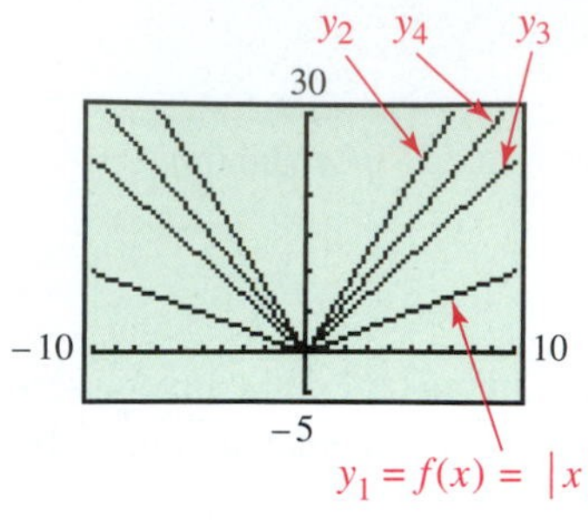

FIGURE 27

EXAMPLE 1 Recognizing Vertical Stretches

In Figure 27, the graph labeled y_1 is that of the function defined by $f(x) = |x|$. The other three functions, y_2, y_3, and y_4, are defined as follows, but not necessarily in the given order: $2.4|x|$, $3.2|x|$, and $4.3|x|$. Determine the correct equation for each graph.

Solution The values of c here are 2.4, 3.2, and 4.3. The vertical heights of the points with the same x-coordinates of the three graphs will correspond to the magnitudes of these c values. Thus, the graph just above $y_1 = |x|$ will be that of $y = 2.4|x|$, the "highest" graph will be that of $y = 4.3|x|$, and the graph of $y = 3.2|x|$ will lie "between" the others. Therefore, $y_2 = 4.3|x|$, $y_3 = 2.4|x|$, and $y_4 = 3.2|x|$.

If we traced to any point on the graph of y_1 and then moved to the other graphs one by one, the y-values of the points would be multiplied by the appropriate values of c. You may wish to experiment with your calculator in this way. ■

Vertical Shrinking

FOR DISCUSSION

This discussion parallels the earlier one in this section. Follow the same general directions. (*Note:* The fractions $\frac{3}{4}$, $\frac{1}{2}$, and $\frac{1}{4}$ may be entered as their decimal equivalents when plotting the graphs.)

A	B	C
$[-5, 5]$ by $[-5, 20]$	$[-5, 15]$ by $[-2, 5]$	$[-10, 10]$ by $[-2, 10]$
$y_1 = x^2$	$y_1 = \sqrt{x}$	$y_1 = \|x\|$
$y_2 = \frac{3}{4}x^2$	$y_2 = \frac{3}{4}\sqrt{x}$	$y_2 = \frac{3}{4}\|x\|$
$y_3 = \frac{1}{2}x^2$	$y_3 = \frac{1}{2}\sqrt{x}$	$y_3 = \frac{1}{2}\|x\|$
$y_4 = \frac{1}{4}x^2$	$y_4 = \frac{1}{4}\sqrt{x}$	$y_4 = \frac{1}{4}\|x\|$

1. How does the graph of y_2 compare with the graph of y_1?
2. How does the graph of y_3 compare with the graph of y_1?
3. How does the graph of y_4 compare with the graph of y_1?
4. If we choose $0 < c < \frac{1}{4}$, how do you think the graph of $y_5 = c \cdot y_1$ would compare with the graph of y_4? Provide support by choosing such a value of c.

TECHNOLOGY NOTE

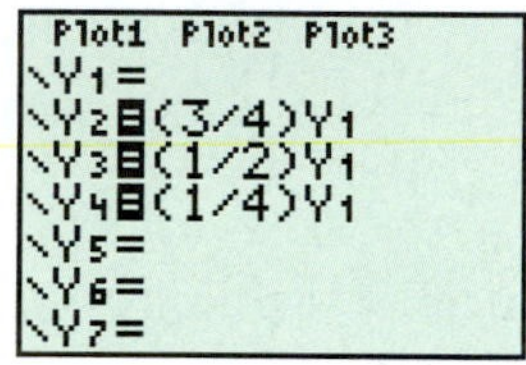

You can use a screen such as this to minimize your keystrokes in parts A, B, and C. Again, Y_1 must be defined in order to obtain the other graphs.

In this "For Discussion" activity, we began with a basic function y_1, and in each case the graph of y_1 was *vertically shrunk* (or *compressed*). These observations can also be generalized to any function.

Vertical Shrinking of the Graph of a Function

If a point (x, y) lies on the graph of $y = f(x)$, then the point (x, cy) lies on the graph of $y = cf(x)$. If $0 < c < 1$, then the graph of $y = cf(x)$ is a *vertical shrinking* of the graph of $y = f(x)$ by a factor of c.

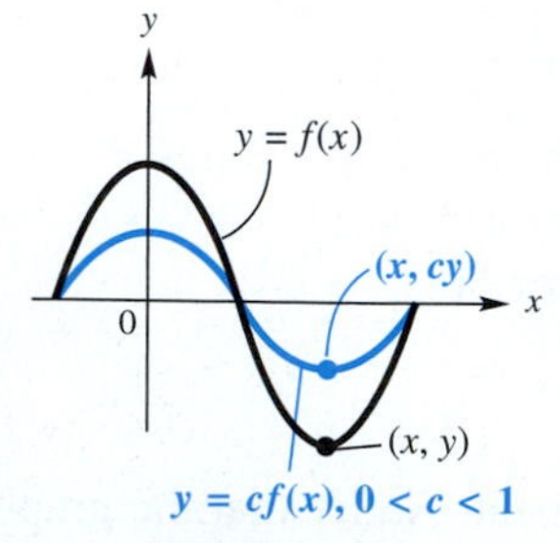

FIGURE 28

Figure 28 shows a graphical interpretation of vertical shrinking.

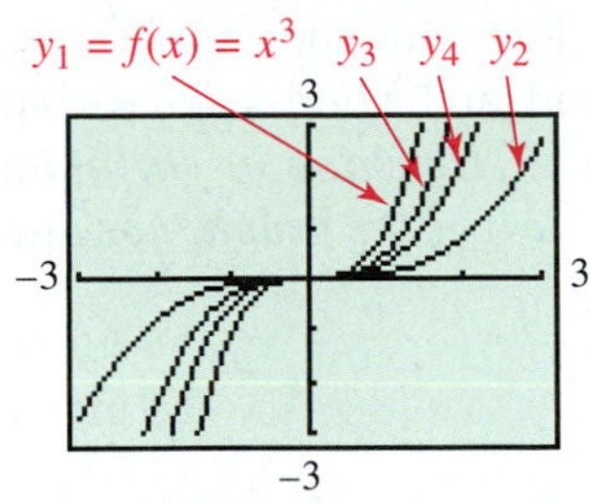

FIGURE 29

TECHNOLOGY NOTE

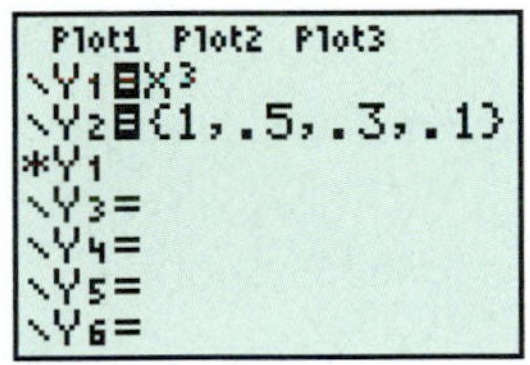

This method of defining Y_1 and Y_2—using a list of coefficients in Y_2—will allow you to duplicate Figure 29.

TECHNOLOGY NOTE

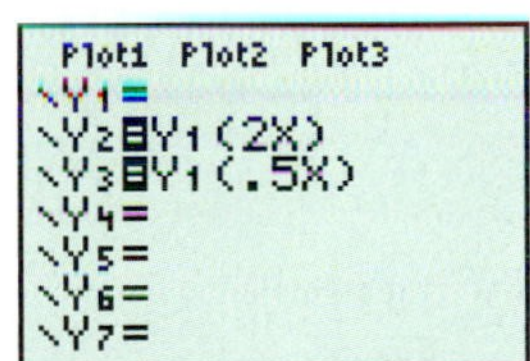

To minimize keystrokes when graphing Y_2 and Y_3, you can use a screen such as this.

EXAMPLE 2 **Recognizing Vertical Shrinks**

In Figure 29, the graph labeled y_1 is that of the function defined by $f(x) = x^3$. The other three functions, y_2, y_3, and y_4, are defined as follows, but not necessarily in the given order: $.5x^3$, $.3x^3$, and $.1x^3$. Determine the correct equation for each graph.

Solution The smaller the positive value of c, where $0 < c < 1$, the more compressed toward the x-axis the graph will be. Since we have $c = .5, .3$, and $.1$, the function rules must be as follows: $y_2 = .1x^3$, $y_3 = .5x^3$, and $y_4 = .3x^3$. ■

Horizontal Stretching and Shrinking

FOR DISCUSSION

This discussion parallels the two earlier ones in this section. Follow the same general directions. (*Note:* We give three related functions, defined by $y_1 = f(x)$, $y_2 = f(2x)$, and $y_3 = f\left(\frac{1}{2}x\right)$, in each group.) Use the standard viewing window to graph the equations.

A	B	C
$y_1 = 5 - x^2$	$y_1 = x^3 - 4x$	$y_1 = \lvert x\rvert - 3$
$y_2 = 5 - (2x)^2$	$y_2 = (2x)^3 - 4(2x)$	$y_2 = \lvert 2x\rvert - 3$
$y_3 = 5 - \left(\frac{1}{2}x\right)^2$	$y_3 = \left(\frac{1}{2}x\right)^3 - 4\left(\frac{1}{2}x\right)$	$y_3 = \left\lvert\frac{1}{2}x\right\rvert - 3$

1. How does the graph of y_2 compare with the graph of y_1?
2. How does the graph of y_3 compare with the graph of y_1?
3. How does the graph of $y = f(cx)$ compare with the graph of $y = f(x)$ when $c > 1$ and when $0 < c < 1$. Confirm your answer by graphing.

In each group of functions in the preceding activity, we compared the graph of $y_1 = f(x)$ with the graphs of functions of the form $y = f(cx)$. In each case, we obtained a graph that was either a *horizontal stretch* or *shrink* of $y = f(x)$.

Consider the line graph in Figure 30. The graph can be stretched or shrunk horizontally. On one hand, if the line graph represents the graph of a function f, then the graph of $y = f\left(\frac{1}{2}x\right)$ in Figure 31 is a horizontal stretching of the graph of $y = f(x)$. On the other hand, the graph of $y = f(2x)$ in Figure 32 represents a horizontal shrinking of the graph of $y = f(x)$.

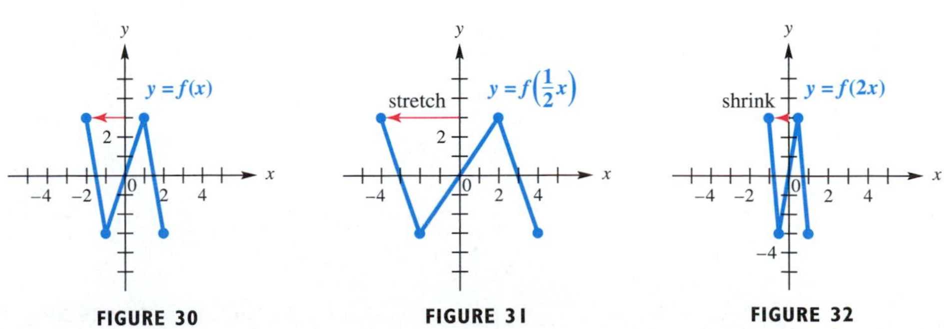

FIGURE 30 **FIGURE 31** **FIGURE 32**

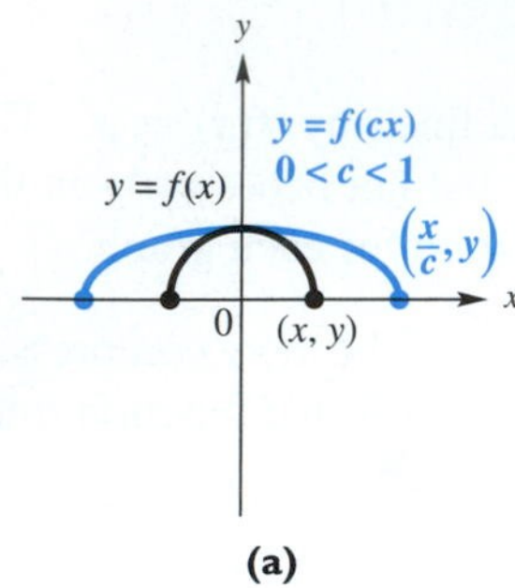

(a)

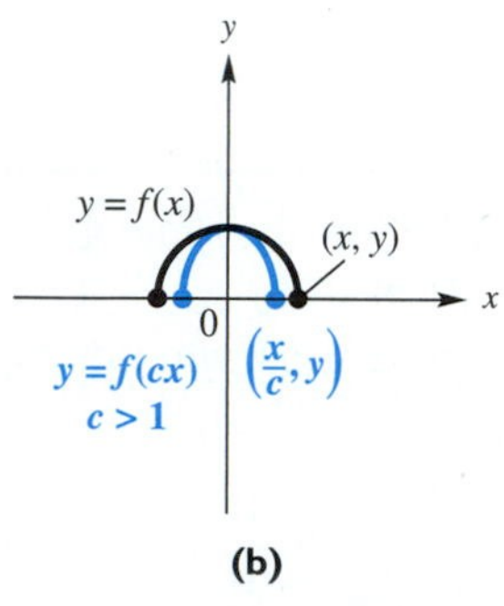

(b)

FIGURE 33

If one were to imagine the graph of $y = f(x)$ as a flexible wire, then a horizontal stretching would happen if the wire were pulled on each end, and a horizontal shrinking would happen if the wire were compressed. ***Horizontal stretching or shrinking does not change the height (maximum or minimum y-values) of the graph, nor does it change the y-intercept.***

These observations can be generalized to any function.

Horizontal Stretching or Shrinking of the Graph of a Function

If a point (x, y) lies on the graph of $y = f(x)$, then the point $(\frac{x}{c}, y)$ lies on the graph of $y = f(cx)$.

(a) If $0 < c < 1$, then the graph of $y = f(cx)$ is a *horizontal stretching* of the graph of $y = f(x)$.

(b) If $c > 1$, then the graph of $y = f(cx)$ is a *horizontal shrinking* of the graph of $y = f(x)$.

In Figure 33, we interpret these statements graphically. For example, if the point $(2, 0)$ is on the graph of $y = f(x)$, then the point $\left(\frac{2}{2}, 0\right) = (1, 0)$ is on the graph of $y = f(2x)$ and the point $\left(\frac{2}{\frac{1}{2}}, 0\right) = (4, 0)$ is on the graph of $y = f\left(\frac{1}{2}x\right)$. Notice in Figure 33, and also in the next example, that horizontal stretching or shrinking changes the x-intercepts, but *not* the y-intercept.

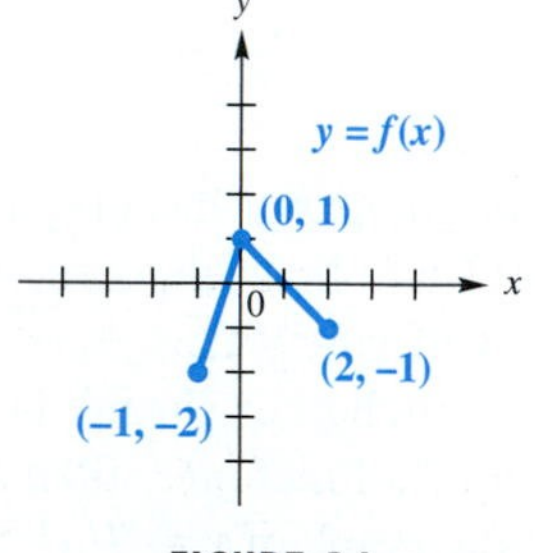

FIGURE 34

EXAMPLE 3 Stretching a Graph Horizontally

Use the graph of $y = f(x)$ in Figure 34 to sketch a graph of $y = f\left(\frac{1}{2}x\right)$.

Solution The graph of $y = f\left(\frac{1}{2}x\right)$ is a horizontal stretching of the graph of $y = f(x)$ and can be obtained by dividing each x-coordinate on the graph of $y = f(x)$ by $\frac{1}{2} = .5$. For the points $(-1, -2)$, $(0, 1)$, and $(2, -1)$, we obtain the following:

$$\left(\frac{-1}{.5}, -2\right) = (-2, -2), \quad \left(\frac{0}{.5}, 1\right) = (0, 1), \quad \left(\frac{2}{.5}, -1\right) = (4, -1). \qquad \text{Points on } y = f\left(\tfrac{1}{2}x\right)$$

Next, we sketch a line graph with these points. See Figure 35.

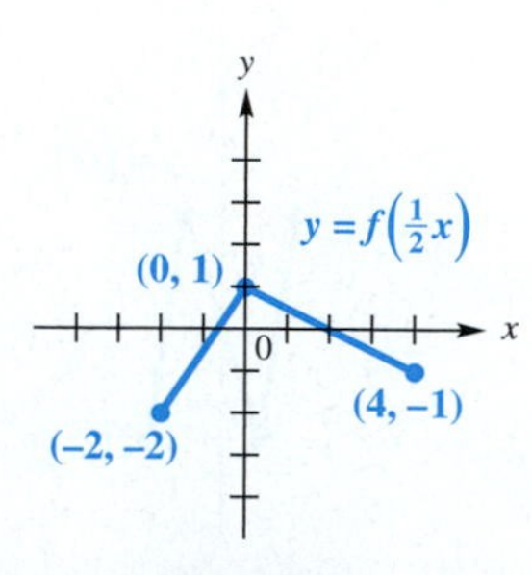

FIGURE 35

Reflecting across an Axis

We have seen how graphs can be transformed by shifting, stretching, and shrinking. We now examine how graphs can be reflected across an axis.

FOR DISCUSSION

In each pair that follows, we give two related functions. Graph $y_1 = f(x)$ and $y_2 = -f(x)$ in the standard viewing window, and then answer the questions about each pair.

A	B	C	D
$y_1 = x^2$	$y_1 = \lvert x \rvert$	$y_1 = \sqrt{x}$	$y_1 = x^3$
$y_2 = -x^2$	$y_2 = -\lvert x \rvert$	$y_2 = -\sqrt{x}$	$y_2 = -x^3$

With respect to the x-axis,

1. how does the graph of y_2 compare with the graph of y_1?
2. how would the graph of $y = -\sqrt[3]{x}$ compare with the graph of $y = \sqrt[3]{x}$, based on your answer to Item 1? Confirm your answer by graphing each equation.

In each pair that follows, we give two related functions. Graph $y_1 = f(x)$ and $y_2 = f(-x)$ in the standard viewing window, and then answer the questions about each pair.

E	F	G
$y_1 = \sqrt{x}$	$y_1 = \sqrt{x - 3}$	$y_1 = \sqrt[3]{x + 4}$
$y_2 = \sqrt{-x}$	$y_2 = \sqrt{-x - 3}$	$y_2 = \sqrt[3]{-x + 4}$

With respect to the y-axis,

3. how does the graph of y_2 compare with the graph of y_1?
4. how would the graph of $y = \sqrt[3]{-x}$ compare with the graph of $y = \sqrt[3]{x}$, based on your answer to Item 3? Confirm your answer by graphing each equation.

TECHNOLOGY NOTE

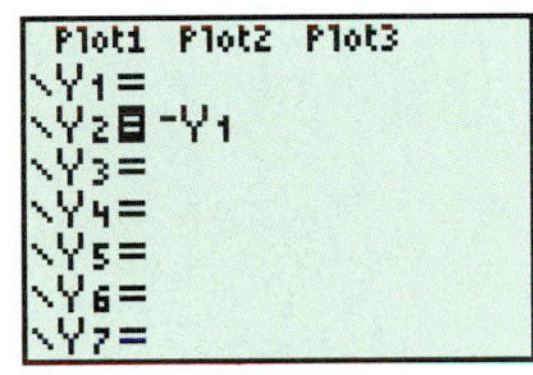

By defining Y_1 as directed in parts A, B, C, and D, and defining Y_2 as shown here, you can minimize your keystrokes.

GCM TECHNOLOGY NOTE

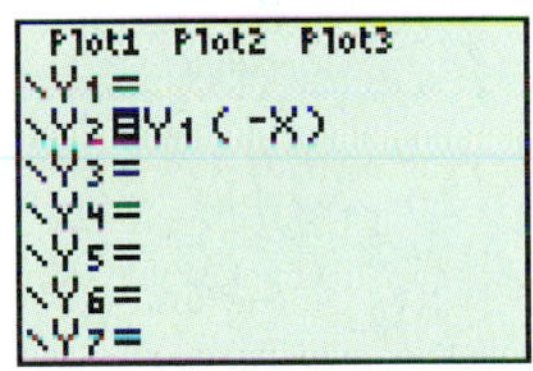

By defining Y_1 as directed in parts E, F, and G, and defining Y_2 as shown here (using function notation), you can minimize your keystrokes.

In this "For Discussion" activity, we began with a basic function y_1. In the first group of functions, we reflected y_1 across the x-axis; in the second group of functions, we reflected y_1 across the y-axis. The results of the preceding discussion can be formally summarized.

Reflecting the Graph of a Function across an Axis

For a function defined by $y = f(x)$,

(a) the graph of $y = -f(x)$ is a reflection of the graph of f across the x-axis.
(b) the graph of $y = f(-x)$ is a reflection of the graph of f across the y-axis.

Figure 36 shows how reflections affect the graph of a function in general.

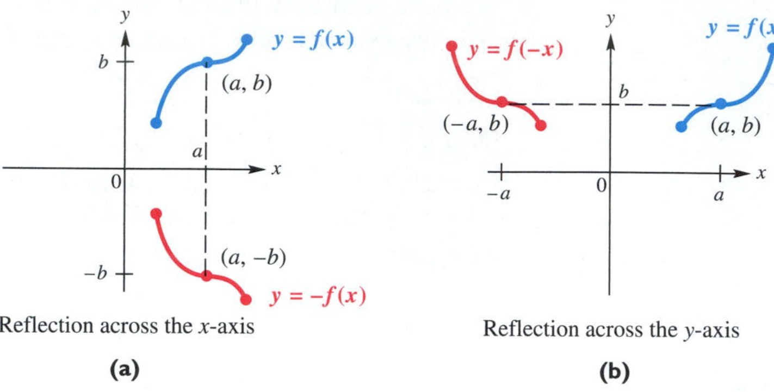

FIGURE 36

EXAMPLE 4 Applying Reflections across Axes

Figure 37 shows the graph of a function $y = f(x)$.

(a) Sketch the graph of $y = -f(x)$. **(b)** Sketch the graph of $y = f(-x)$.

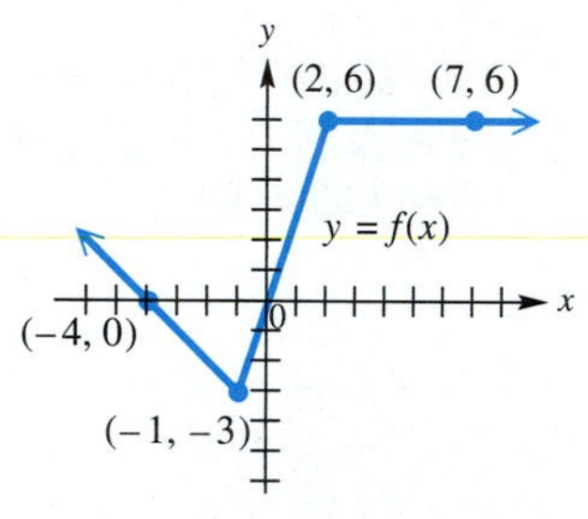

FIGURE 37

Solution

(a) We must reflect the graph across the x-axis. This means that if a point (a, b) lies on the graph of $y = f(x)$, then the point $(a, -b)$ must lie on the graph of $y = -f(x)$. Using the labeled points, we find the graph of $y = -f(x)$ in Figure 38.

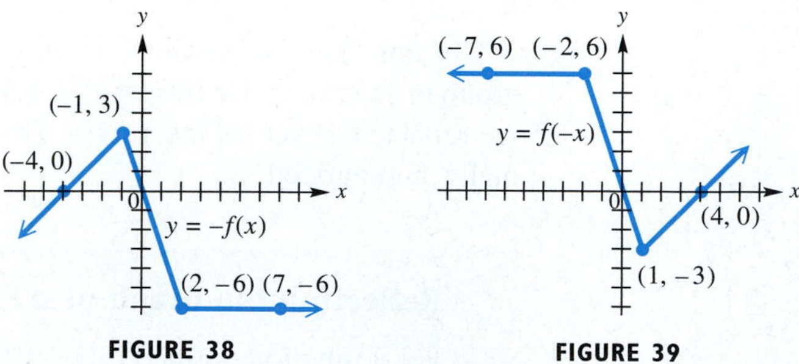

FIGURE 38 **FIGURE 39**

(b) Here we must reflect the graph across the y-axis, meaning that if a point (a, b) lies on the graph of $y = f(x)$, then the point $(-a, b)$ must lie on the graph of $y = f(-x)$. Thus, we obtain the graph of $y = f(-x)$, as shown in Figure 39. ■

What Went Wrong?

A student entered three functions Y_1, Y_2, and Y_3. The calculator graphed the first two as shown, but gave a syntax error when it attempted to graph the third.

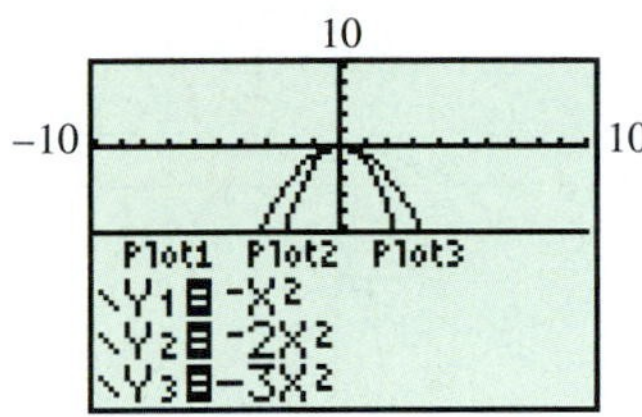

What Went Wrong? What must the student do in order to obtain the desired graph for $y = -3x^2$?

Combining Transformations of Graphs

We can create an infinite number of functions by stretching or shrinking, shifting upward, downward, left, or right, and reflecting across an axis. ***To determine the order in which the transformations are made, use the order of operations.***

EXAMPLE 5 Describing Combinations of Transformations of Graphs

(a) Describe how the graph of $y = -3(x - 4)^2 + 5$ can be obtained by transforming the graph of $y = x^2$. Sketch its graph.

(b) Give the equation of the function that would be obtained by shifting the graph of $y = |x|$ to the left 3 units, vertically shrinking the graph by a factor of $\frac{2}{3}$, reflecting across the x-axis, and shifting the graph 4 units downward, in that order. Sketch its graph.

Solution

(a) In the definition of the function, $(x - 4)^2$ indicates that the graph of $y = x^2$ must be shifted 4 units to the right. Since the coefficient of $(x - 4)^2$ is -3 (a negative number with absolute value greater than 1), the graph is stretched vertically by a factor of 3 and then reflected across the x-axis. The constant $+5$ indicates that the graph is shifted upward 5 units. These ideas are summarized here:

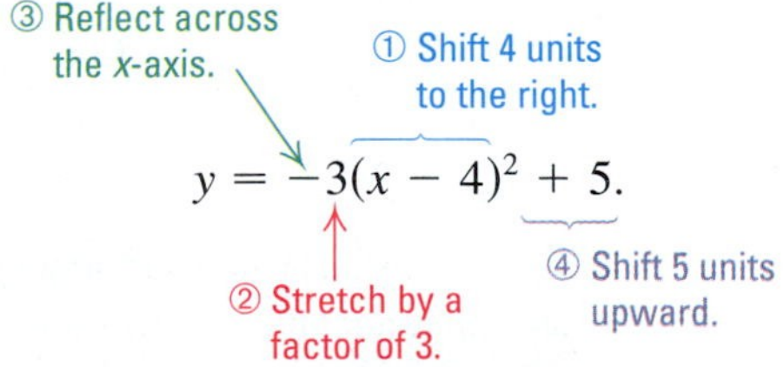

Answer to What Went Wrong?

The student used a subtraction sign rather than a negative sign to define Y_3. Notice the difference between the signs in Y_1 and Y_2 compared with that in Y_3. The student must reenter Y_3 with a negative sign.

The graphs in Figure 40 demonstrate the preceding four steps to graph the transformation of $y = x^2$ by hand. It is not necessary to sketch all four graphs. Only the final graph is necessary.

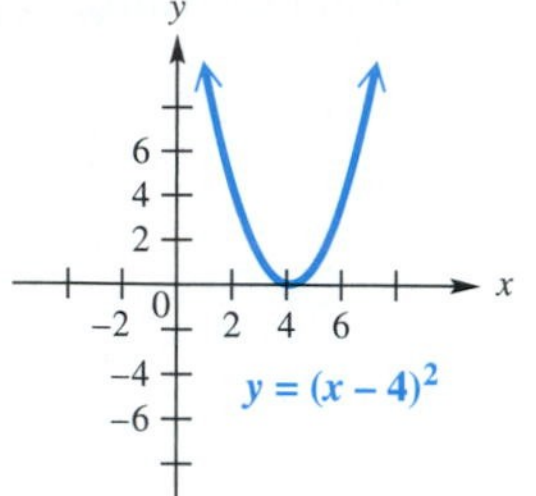

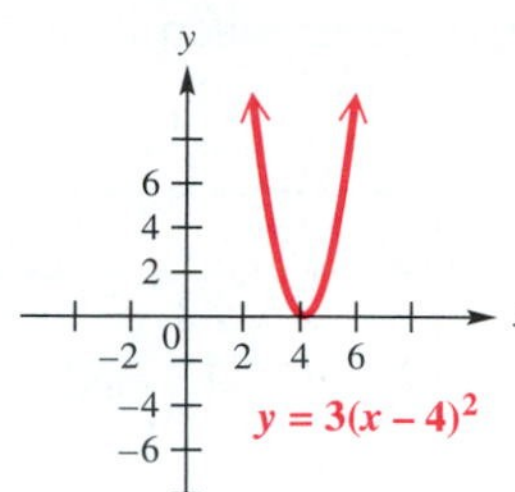

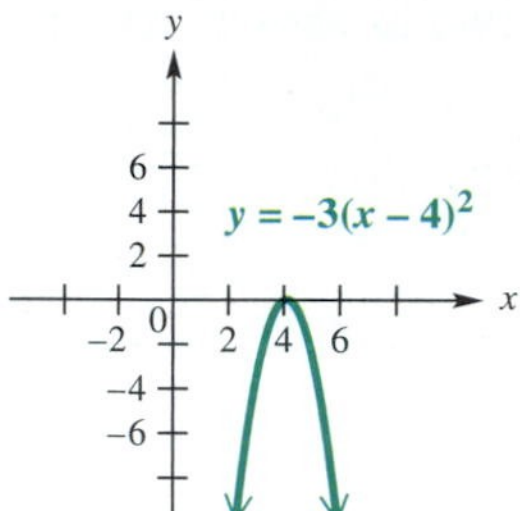

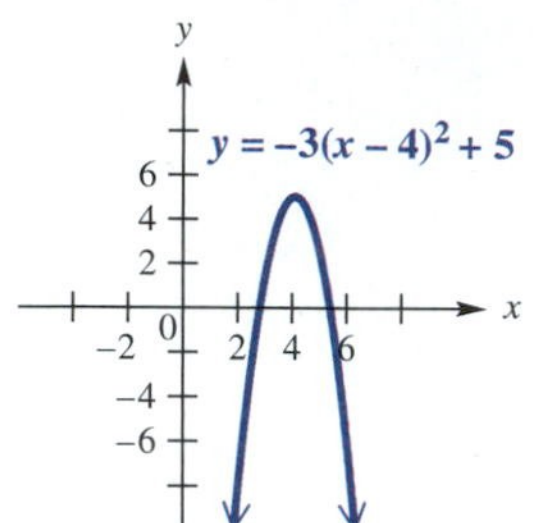

FIGURE 40

(b) Shifting the graph of $y = |x|$ to the left 3 units means that $|x|$ is transformed to $|x + 3|$. Vertically shrinking the graph by a factor of $\frac{2}{3}$ means multiplying by $\frac{2}{3}$, and reflecting across the x-axis changes $\frac{2}{3}$ to $-\frac{2}{3}$. Finally, shifting the graph 4 units downward means subtracting 4. Putting this all together leads to the equation

$$y = -\frac{2}{3}|x + 3| - 4.$$

The final graph is shown is Figure 41.

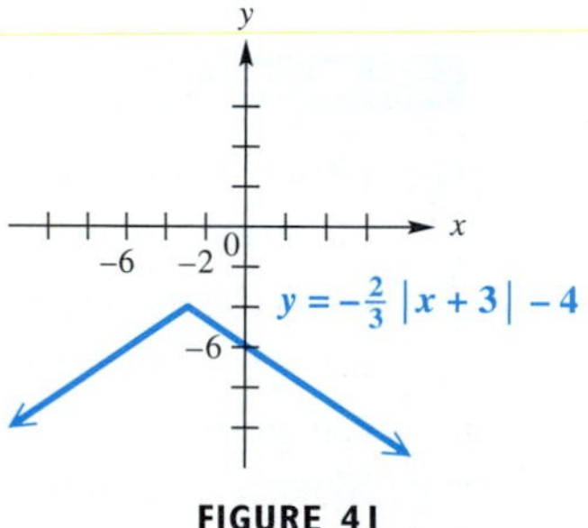

FIGURE 41

CAUTION ***The order in which the transformations are made is important.*** If they are made in a different order, a different equation can result. For example, changing the order of a stretch and shift can result in a different equation and graph.

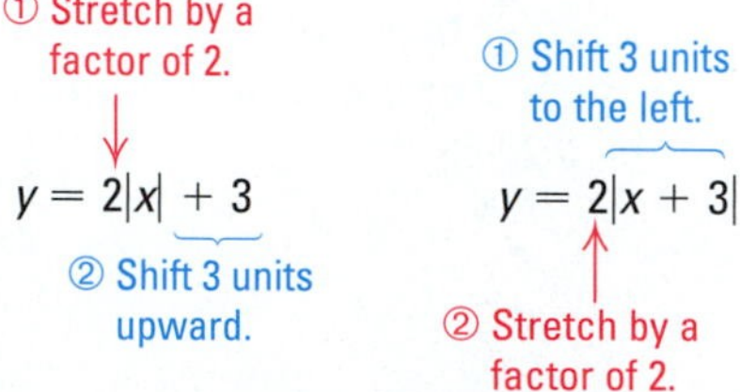

Also, be careful when performing reflections and shifts. For example, to graph $y = \sqrt{-(x - 2)}$, we can reflect the graph of $y = \sqrt{x}$ across the y-axis to obtain $y = \sqrt{-x}$, then shift the graph to the right 2 units to obtain $y = \sqrt{-(x - 2)}$. Or, we can shift the graph of $y = \sqrt{x}$ to the left 2 units to obtain $y = \sqrt{x + 2}$, then reflect it across the y-axis to obtain $y = \sqrt{-x + 2}$, or $y = \sqrt{-(x - 2)}$. Either approach is correct.

EXAMPLE 6 **Recognizing a Combination of Transformations**

Figure 42 shows two views of the graph of $y = |x|$ and another graph illustrating a combination of transformations. Find the equation of the transformed graph.

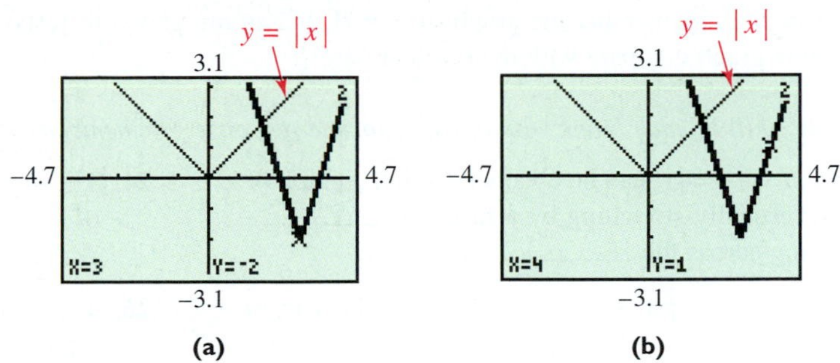

FIGURE 42

Solution Figure 42(a) shows that the lowest point on the transformed graph has coordinates $(3, -2)$, indicating that the graph has been shifted 3 units to the right and 2 units downward. Figure 42(b) shows that a point on the right side of the transformed graph has coordinates $(4, 1)$; thus, the slope of this ray is

Start with the x- and y-values of the same point.

$$m = \frac{-2 - 1}{3 - 4} = \frac{-3}{-1} = 3.$$

The stretch factor is 3, and the equation of the transformed graph is

$$y = 3|x - 3| - 2.$$

2.3 Exercises

Write the equation that results in the desired transformation.

1. The squaring function, stretched by a factor of 2
2. The cubing function, vertically shrunk by a factor of $\frac{1}{2}$
3. The square root function, reflected across the y-axis
4. The cube root function, reflected across the x-axis
5. The absolute value function, vertically stretched by a factor of 3 and reflected across the x-axis
6. The absolute value function, vertically shrunk by a factor of $\frac{1}{3}$ and reflected across the y-axis
7. The cubing function, vertically shrunk by a factor of .25 and reflected across the y-axis
8. The square root function, vertically shrunk by a factor of .2 and reflected across the x-axis

Use transformations of graphs to sketch the graphs of y_1, y_2, and y_3 *by hand. Check by graphing in an appropriate viewing window of your calculator.*

9. $y_1 = x, y_2 = x + 3, y_3 = x - 3$
10. $y_1 = x^3, y_2 = x^3 + 4, y_3 = x^3 - 4$
11. $y_1 = |x|, y_2 = |x - 3|, y_3 = |x + 3|$
12. $y_1 = |x|, y_2 = |x| - 3, y_3 = |x| + 3$
13. $y_1 = \sqrt{x}, y_2 = \sqrt{x + 6}, y_3 = \sqrt{x - 6}$
14. $y_1 = |x|, y_2 = 2|x|, y_3 = 2.5|x|$
15. $y_1 = \sqrt[3]{x}, y_2 = -\sqrt[3]{x}, y_3 = -2\sqrt[3]{x}$
16. $y_1 = x^2, y_2 = (x - 2)^2 + 1, y_3 = -(x + 2)^2$
17. $y_1 = |x|, y_2 = -2|x - 1| + 1, y_3 = -\frac{1}{2}|x| - 4$
18. $y_1 = \sqrt{x}, y_2 = -\sqrt{x}, y_3 = \sqrt{-x}$

19. $y_1 = x^2 - 1,\ y_2 = \left(\frac{1}{2}x\right)^2 - 1,\ y_3 = (2x)^2 - 1$

20. $y_1 = 3 - |x|,\ y_2 = 3 - |3x|,\ y_3 = 3 - \left|\frac{1}{3}x\right|$

21. $y_1 = \sqrt[3]{x},\ y_2 = \sqrt[3]{-x},\ y_3 = \sqrt[3]{-(x-1)}$

22. ***Concept Check*** Suppose that the graph of $y = f(x)$ is symmetric with respect to the y-axis and is reflected across the y-axis. How will the new graph compare with the original one?

Concept Check *Fill in each blank with the appropriate response. (Remember that the vertical stretch or shrink factor is positive.)*

23. The graph of $y = -4x^2$ can be obtained from the graph of $y = x^2$ by vertically stretching by a factor of ________ and reflecting across the ________-axis.

24. The graph of $y = -6\sqrt{x}$ can be obtained from the graph of $y = \sqrt{x}$ by vertically stretching by a factor of ________ and reflecting across the ________-axis.

25. The graph of $y = -\frac{1}{4}|x + 2| - 3$ can be obtained from the graph of $y = |x|$ by shifting horizontally ________ units to the ________, vertically shrinking by a factor of ________, reflecting across the ________-axis, and shifting vertically ________ units in the ________ direction.

26. The graph of $y = -\frac{2}{5}|-x| + 6$ can be obtained from the graph of $y = |x|$ by reflecting across the ________-axis, vertically shrinking by a factor of ________, reflecting across the ________-axis, and shifting vertically ________ units in the ________ direction.

27. The graph of $y = 6\sqrt[3]{x - 3}$ can be obtained from the graph of $y = \sqrt[3]{x}$ by shifting horizontally ________ units to the ________ and stretching vertically by a factor of ________.

28. The graph of $y = .5\sqrt[3]{x + 2}$ can be obtained from the graph of $y = \sqrt[3]{x}$ by shifting horizontally ________ units to the ________ and shrinking vertically by a factor of ________.

Give the equation of each function whose graph is described.

29. The graph of $y = x^2$ is vertically shrunk by a factor of $\frac{1}{2}$, and the resulting graph is shifted 7 units downward.

30. The graph of $y = x^3$ is vertically stretched by a factor of 3. This graph is then reflected across the x-axis. Finally, the graph is shifted 8 units upward.

31. The graph of $y = \sqrt{x}$ is shifted 3 units to the right. This graph is then vertically stretched by a factor of 4.5. Finally, the graph is shifted 6 units downward.

32. The graph of $y = \sqrt[3]{x}$ is shifted 2 units to the left. This graph is then vertically stretched by a factor of 1.5. Finally, the graph is shifted 8 units upward.

In Exercises 33 and 34, the graph of $y = f(x)$ has been transformed to the graph of $y = g(x)$. No shrinking or stretching is involved. Give the equation of $y = g(x)$.

33.

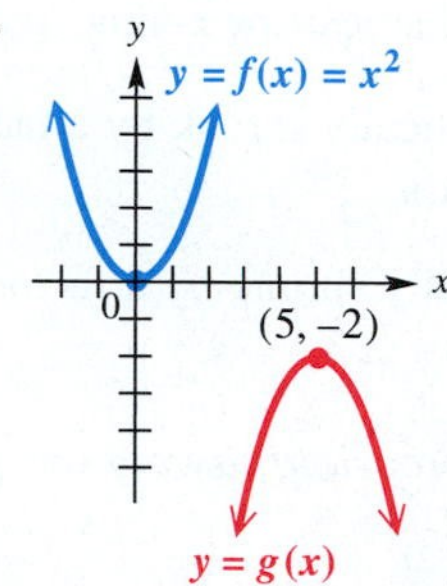

34.

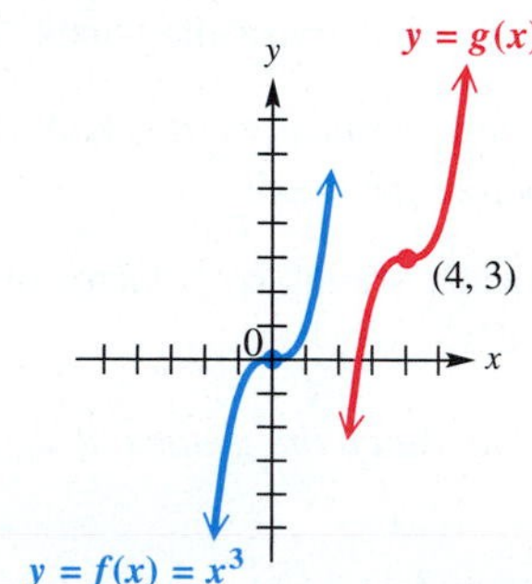

Use transformations of graphs to sketch a graph of $y = f(x)$ by hand.

35. $f(x) = \sqrt{x - 3} + 2$

36. $f(x) = |x + 2| - 3$

37. $f(x) = \sqrt{2x}$

38. $f(x) = \frac{1}{2}(x + 2)^2$

39. $f(x) = |2x|$

40. $f(x) = \frac{1}{2}|x|$

41. $f(x) = 1 - \sqrt{x}$

42. $f(x) = 2\sqrt{x - 2} - 1$

43. $f(x) = -\sqrt{1 - x}$

44. $f(x) = \sqrt{-x} - 1$

45. $f(x) = \sqrt{-(x + 1)}$

46. $f(x) = 2 + \sqrt{-(x - 3)}$

47. $f(x) = (x - 1)^3$

48. $f(x) = (x + 2)^3$

49. $f(x) = -x^3$

50. $f(x) = (-x)^3 + 1$

In Exercises 51–56, each figure shows the graph of a function defined by $y = f(x)$. Sketch by hand the graphs of the functions in parts (a), (b), and (c), and answer the question of part (d).

51. **(a)** $y = -f(x)$ **(b)** $y = f(-x)$ **(c)** $y = 2f(x)$
(d) What is $f(0)$?

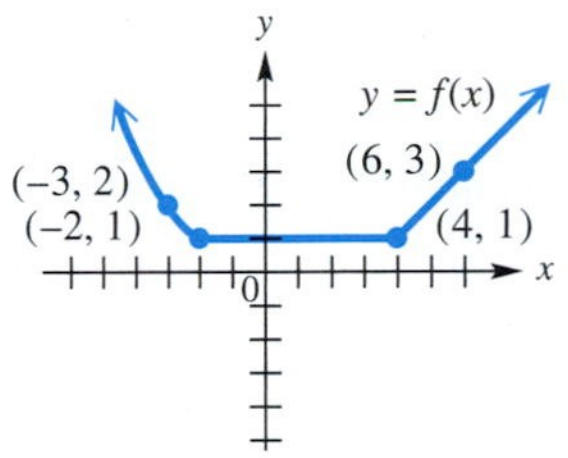

52. **(a)** $y = -f(x)$ **(b)** $y = f(-x)$ **(c)** $y = 3f(x)$
(d) What is $f(4)$?

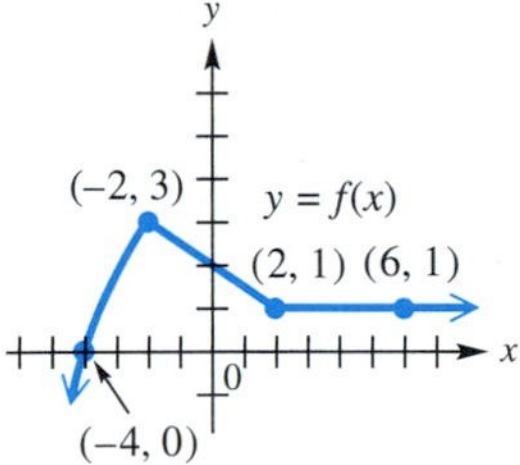

53. **(a)** $y = -f(x)$ **(b)** $y = f(-x)$ **(c)** $y = f(x + 1)$
(d) What are the x-intercepts of $y = f(x - 1)$?

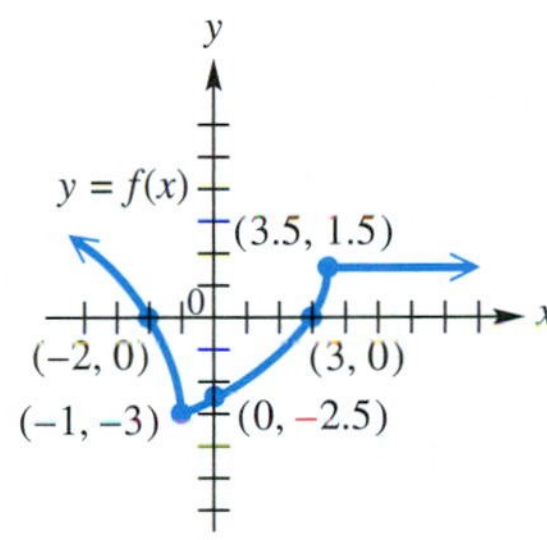

54. **(a)** $y = -f(x)$ **(b)** $y = f(-x)$ **(c)** $y = \frac{1}{2}f(x)$
(d) On what interval of the domain is $f(x) < 0$?

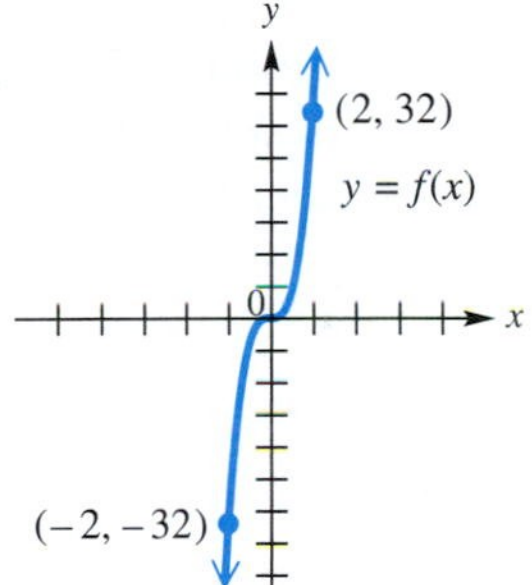

55. **(a)** $y = -f(x)$ **(b)** $y = f\left(\frac{1}{3}x\right)$ **(c)** $y = .5f(x)$
(d) What symmetry does the graph of $y = f(x)$ exhibit?

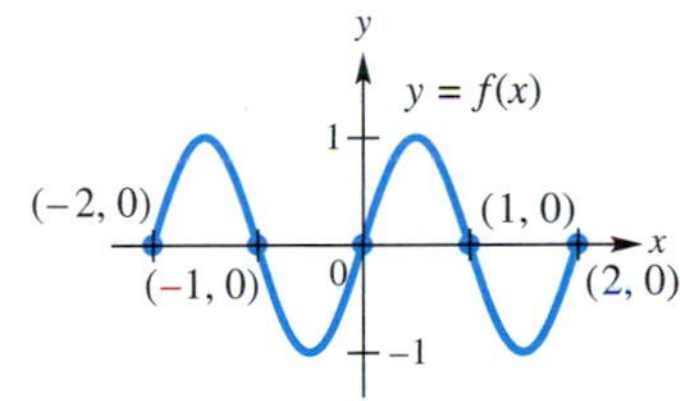

56. **(a)** $y = f(2x)$ **(b)** $y = f(-x)$ **(c)** $y = 3f(x)$
(d) What symmetry does the graph of $y = f(x)$ exhibit?

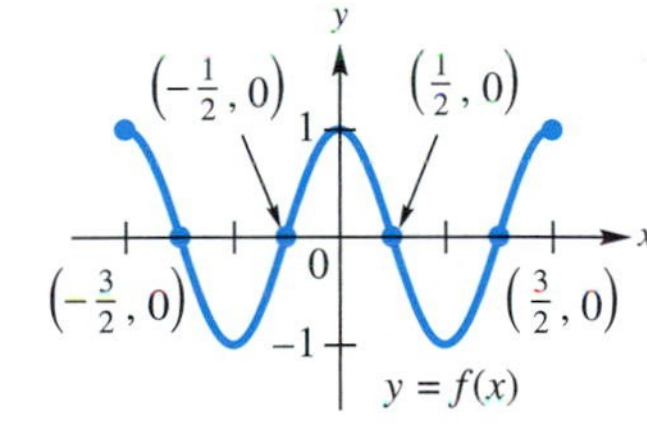

In Exercises 57–60, use the accompanying graph of $y = f(x)$ to sketch a graph of each equation.

57. **(a)** $y = f(x) + 1$
(b) $y = -f(x) - 1$
(c) $y = 2f\left(\frac{1}{2}x\right)$

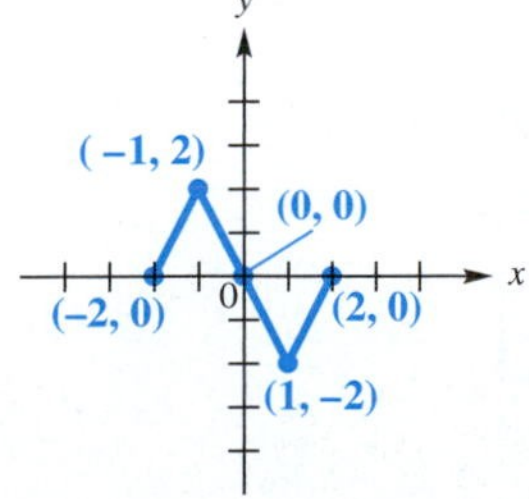

58. **(a)** $y = f(x) - 2$
(b) $y = f(x - 1) + 2$
(c) $y = 2f(x)$

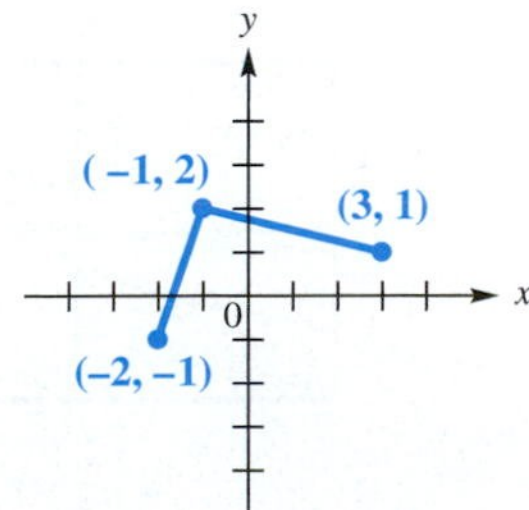

59. **(a)** $y = f(2x) + 1$
(b) $y = 2f\left(\frac{1}{2}x\right) + 1$
(c) $y = \frac{1}{2}f(x - 2)$

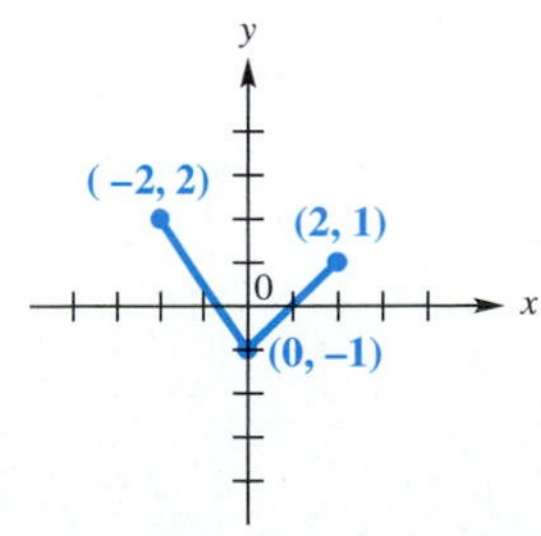

60. **(a)** $y = f(2x)$
(b) $y = f\left(\frac{1}{2}x\right) - 1$
(c) $y = 2f(x) - 1$

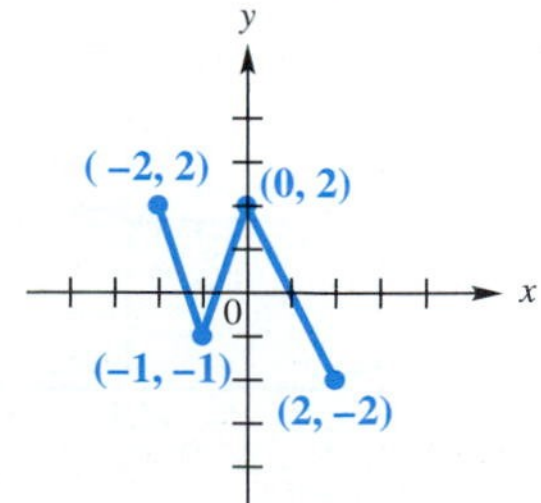

61. **Concept Check** If r is an x-intercept of the graph of $y = f(x)$, what statement can be made about the x-intercept of the graph of each function? (*Hint*: Make a sketch.)
(a) $y = -f(x)$
(b) $y = f(-x)$
(c) $y = -f(-x)$

62. **Concept Check** If b is the y-intercept of the graph of $y = f(x)$, what statement can be made about the y-intercept of the graph of each function? (*Hint:* Make a sketch.)
(a) $y = -f(x)$
(b) $y = f(-x)$
(c) $y = 5f(x)$
(d) $y = -3f(x)$

Let the domain of $f(x)$ be $[-1, 2]$ and the range be $[0, 3]$. Find the domain and range of each of the following.

63. $f(x - 2)$
64. $5f(x + 1)$
65. $-f(x)$
66. $f(x - 3) + 1$
67. $f(2x)$
68. $2f(x - 1)$
69. $3f\left(\frac{1}{4}x\right)$
70. $-2f(4x)$
71. $f(-x)$
72. $-2f(-x)$
73. $f(-3x)$
74. $\frac{1}{3}f(x - 3)$

In Exercises 75–77, each function has a graph with an endpoint (a translation of the point $(0, 0)$.) Enter each into your calculator in an appropriate viewing window, and using your knowledge of the graph of $y = \sqrt{x}$, determine the domain and range of the function. (Hint: Locate the endpoint.)

75. $y = 10\sqrt{x - 20} + 5$
76. $y = -2\sqrt{x + 15} - 18$
77. $y = -.5\sqrt{x + 10} + 5$

78. **Concept Check** Based on your observations in Exercise 75, what are the domain and range of $f(x) = a\sqrt{x - h} + k$ if $a > 0$, $h > 0$, and $k > 0$?

Concept Check *The sketch shows an example of a function defined by $y = f(x)$ that increases on the interval $[a, b]$. Use this graph as a visual aid, and apply the concepts of reflection introduced in this section to answer each question.*

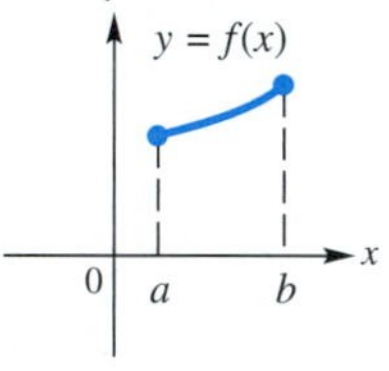

79. Does the graph of $y = -f(x)$ increase or decrease on the interval $[a, b]$?

80. Does the graph of $y = f(-x)$ increase or decrease on the interval $[-b, -a]$?

81. Does the graph of $y = -f(-x)$ increase or decrease on the interval $[-b, -a]$?

82. If $c > 0$, does the graph of $y = -cf(x)$ increase or decrease on the interval $[a, b]$?

State the intervals over which each function is **(a)** *increasing,* **(b)** *decreasing, and* **(c)** *constant.*

83. The function graphed in Figure 37
84. The function graphed in Figure 38
85. The function graphed in Figure 39
86. $y = -\frac{2}{3}|x + 3| - 4$ (See Figure 41.)

Concept Check *Shown here are the graphs of $y_1 = \sqrt[3]{x}$ and $y_2 = 5\sqrt[3]{x}$. The point whose coordinates are given at the bottom of the screen lies on the graph of y_1. Use this graph, not your calculator, to find the value of y_2 for the same value of x shown.*

87.

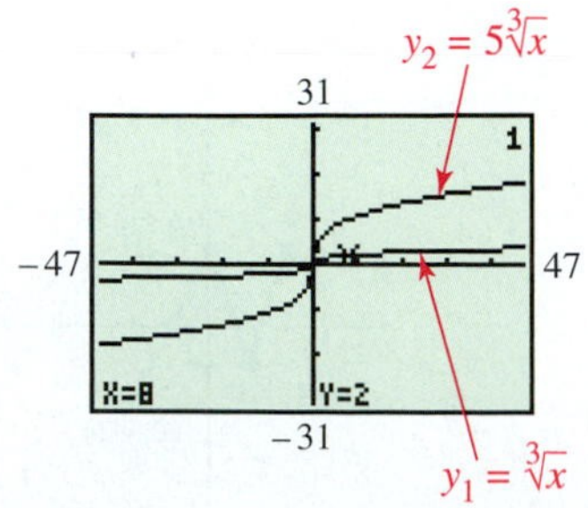

88.

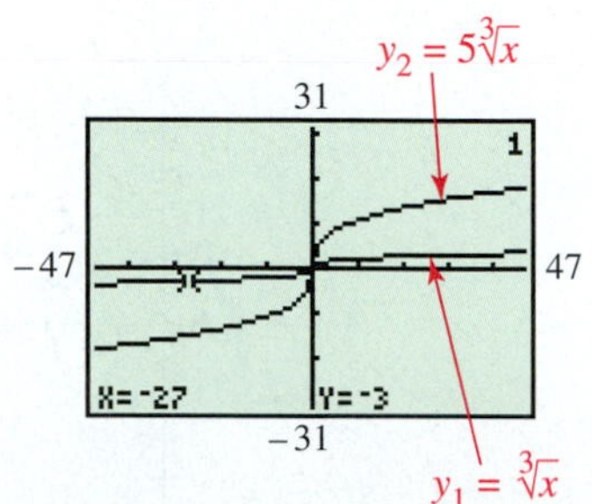

The figure shows a transformation of the graph of $y = |x|$. Write the equation for the graph. Refer to Example 6 as needed.

89.

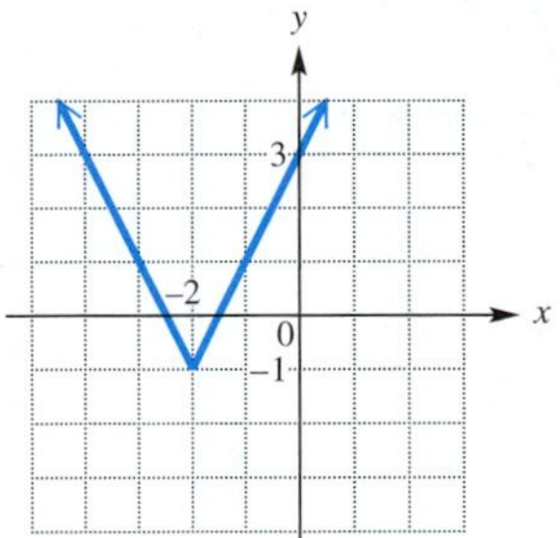

90.

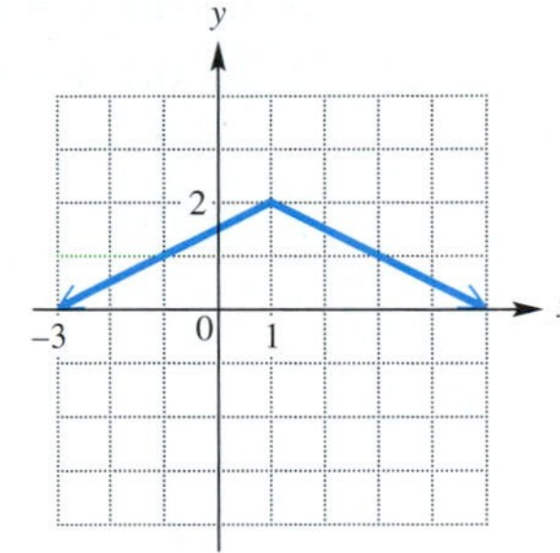

91.

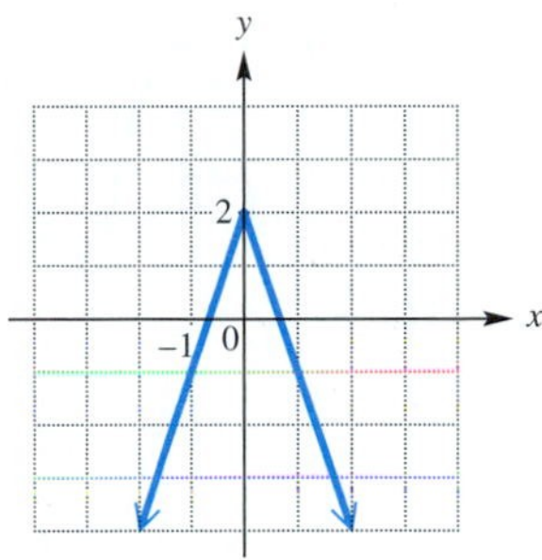

92.

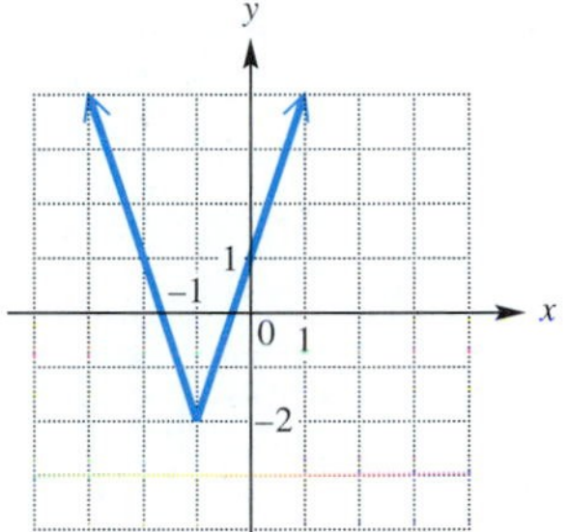

Reviewing Basic Concepts (Sections 2.1–2.3)

1. Suppose that f is defined for all real numbers and $f(3) = 6$. For each of the given assumptions, find another function value.

(a) The graph of $y = f(x)$ is symmetric with respect to the origin.
(b) The graph of $y = f(x)$ is symmetric with respect to the y-axis.
(c) For all x, $f(-x) = -f(x)$.
(d) For all x, $f(-x) = f(x)$.

2. Match each equation in Column I with a description of its graph from Column II as it relates to the graph of $y = x^2$.

I	II
(a) $y = (x - 7)^2$	**A.** A shift of 7 units to the left
(b) $y = x^2 - 7$	**B.** A shift of 7 units to the right
(c) $y = 7x^2$	**C.** A horizonal stretch
(d) $y = (x + 7)^2$	**D.** A shift of 7 units downward
(e) $y = \left(\frac{1}{3}x\right)^2$	**E.** A vertical stretch by a factor of 7

3. Match each equation in parts (a)–(h) with the sketch of its graph in choices A–H below. The graph of $y = x^2$ is shown here.

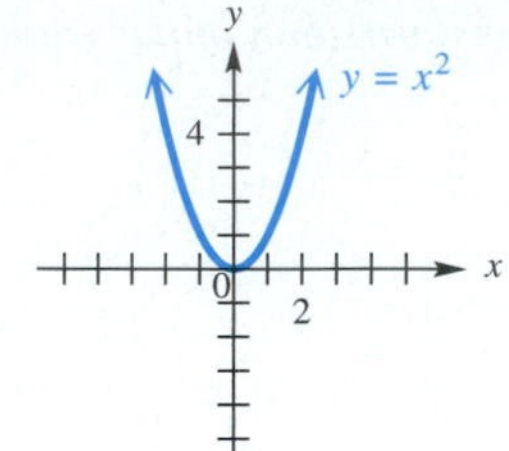

(a) $y = x^2 + 2$ **(b)** $y = x^2 - 2$

(c) $y = (x + 2)^2$ **(d)** $y = (x - 2)^2$

(e) $y = 2x^2$ **(f)** $y = -x^2$

(g) $y = (x - 2)^2 + 1$ **(h)** $y = (x + 2)^2 + 1$

A.

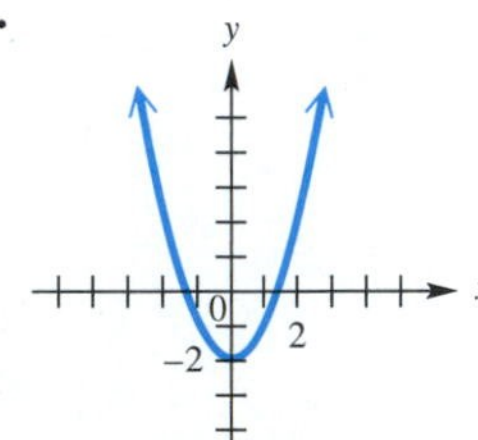

B.

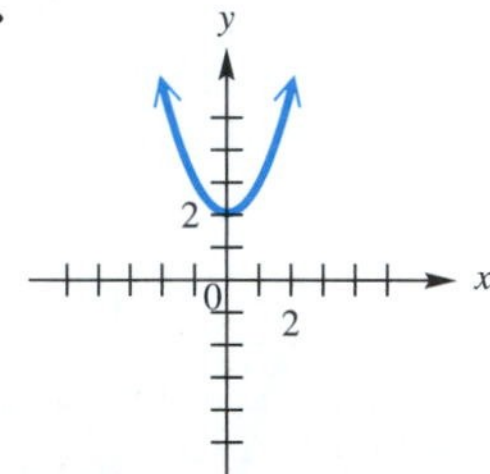

C.

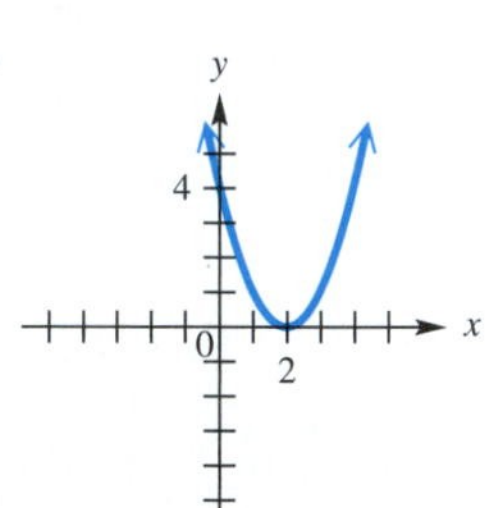

D.

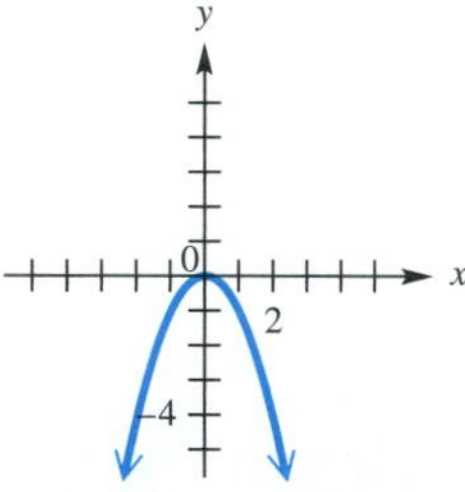

E.

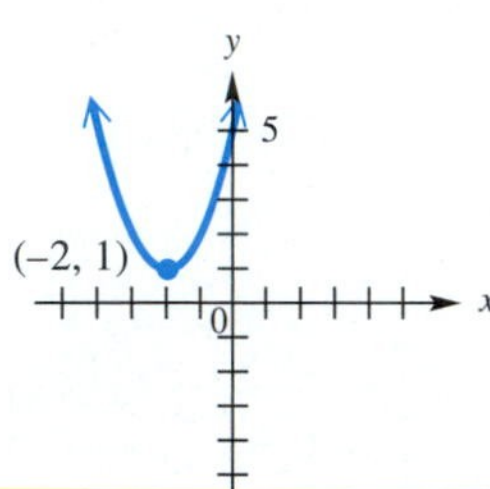

F.

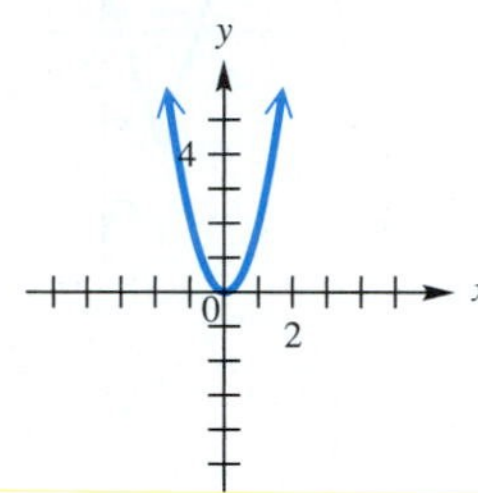

G.

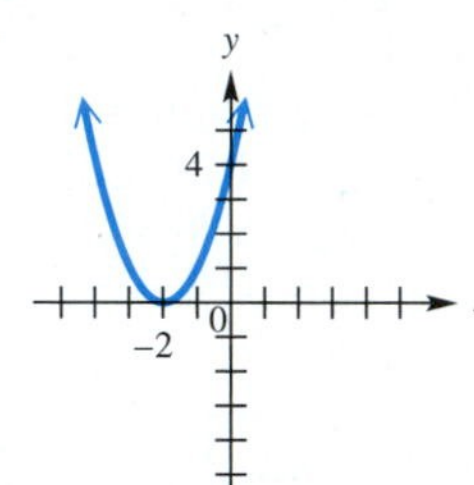

H.

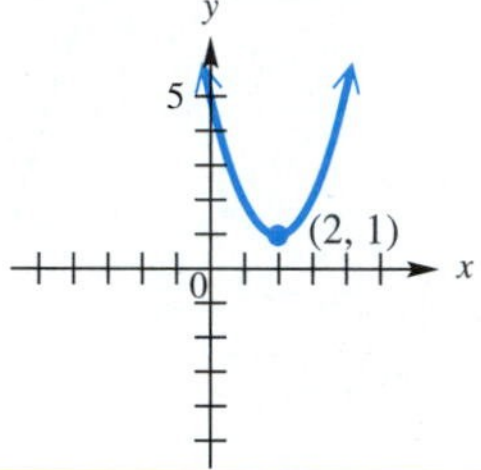

4. Sketch each graph of $y = f(x)$ by hand.

(a) $y = |x| + 4$ **(b)** $y = |x + 4|$ **(c)** $y = |x - 4|$ **(d)** $y = |x + 2| - 4$ **(e)** $y = -|x - 2| + 4$

5. Describe the transformation of each graph of $f(x) = |x|$ or $g(x) = \sqrt{x}$, and then give the equation of the graph.

(a)

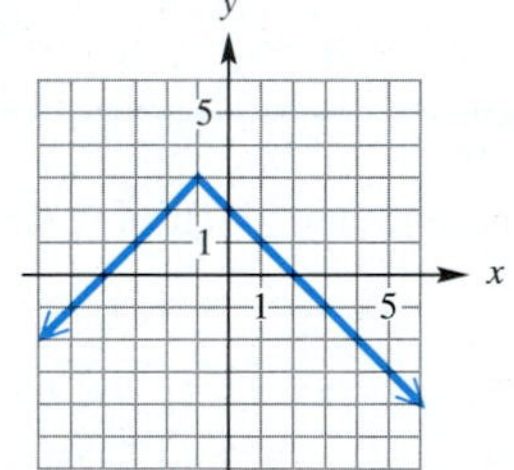

(b)

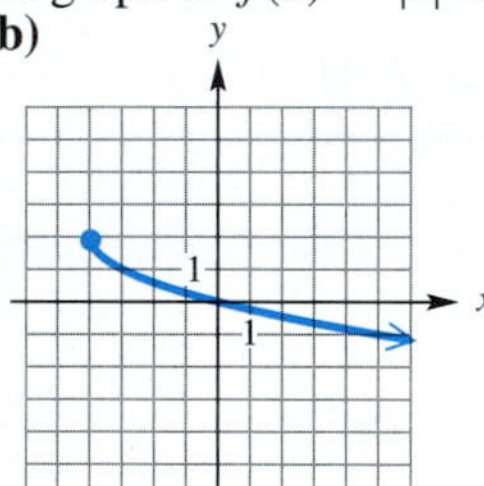

(c)

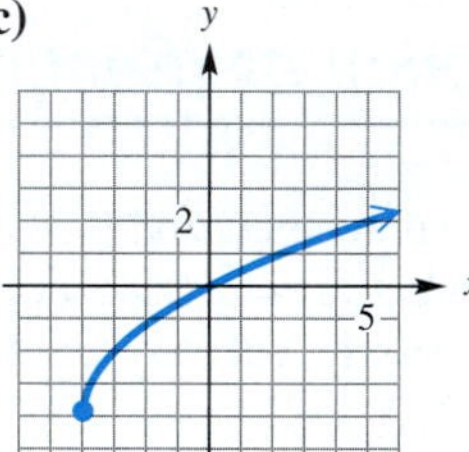

(d)

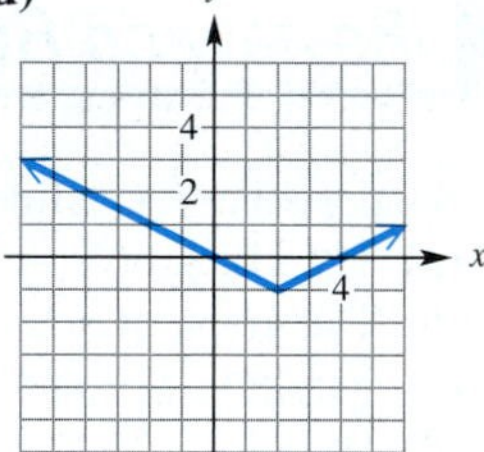

6. Consider the two functions in the figure.

(a) Find a value of c for which $g(x) = f(x) + c$.

(b) Find a value of c for which $g(x) = f(x + c)$.

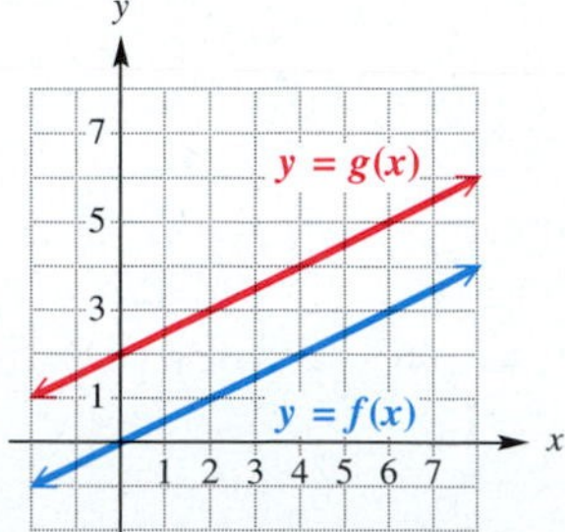

7. Suppose the equation $y = F(x)$ is changed to the equation $y = F(x + h)$. How are the graphs of these equations related? Is the graph of $y = F(x) + h$ the same as the graph of $y = F(x + h)$? If not, how do they differ?

8. Suppose the equation $y = F(x)$ is changed to $y = cF(x)$, for some constant c. What is the effect on the graph of $y = F(x)$? Discuss the effect depending on whether $c > 0$ or $c < 0$, and $|c| > 1$ or $|c| < 1$.

9. Complete the table if **(a)** f is an even function and **(b)** f is an odd function.

x	$f(x)$
-3	4
-2	-6
-1	5
1	
2	
3	

10. ***(Modeling) Carbon Monoxide Levels*** The 8-hour maximum carbon monoxide levels (in parts per million) for the United States from 1992 to 2002 can be modeled by

$$f(x) = -.279x + 5.532,$$

where $x = 0$ corresponds to 1992. (*Source:* U.S. Environmental Protection Agency, 2004.)

(a) Find a function represented by $g(x)$ that models the same carbon monoxide levels, except that x is the actual year between 1992 and 2002.

(b) Use $g(x)$ to predict the 8-hour maximum carbon monoxide levels for the United States in 2006.

2.4 Absolute Value Functions: Graphs, Equations, Inequalities, and Applications

The Graph of $y = |f(x)|$ ■ Properties of Absolute Value ■ Equations and Inequalities Involving Absolute Value ■ An Application Involving Absolute Value

TECHNOLOGY NOTE

```
abs(-3)
                3
abs(0)
                0
abs(3)
                3
```

The command abs(x) is used by some graphing calculators to find absolute value.

In this section we investigate the absolute value function in detail.

The Graph of $y = |f(x)|$

Geometrically, the absolute value of a real number is its undirected distance from 0 on the number line. ***As a result, the absolute value of a real number is never negative; it is always greater than or equal to 0.*** Thus, the function defined by

$$f(x) = |x| = \begin{cases} x & \text{if } x \geq 0 \\ -x & \text{if } x < 0 \end{cases}$$

in Section 2.1 is just an extension of the definition of absolute value. The expression within the absolute value bars, x, is the defining expression for the identity function $y = x$.

Now we extend the concept and consider the definition of a function defined by the *absolute value of a function f*:

$$|f(x)| = \begin{cases} f(x) & \text{if } f(x) \geq 0 \\ -f(x) & \text{if } f(x) < 0 \end{cases}.$$

To graph a function of the form $y = |f(x)|$, the first part of the definition indicates that the graph is the same as that of $y = f(x)$ for values of $f(x)$ (that is, range values) that are nonnegative. The second part of the definition indicates that for range values that are negative, the graph of $y = f(x)$ is reflected across the x-axis. The domain of $y = |f(x)|$ is the same as the domain of f, while the range of $y = |f(x)|$ will be a subset of $[0, \infty)$.

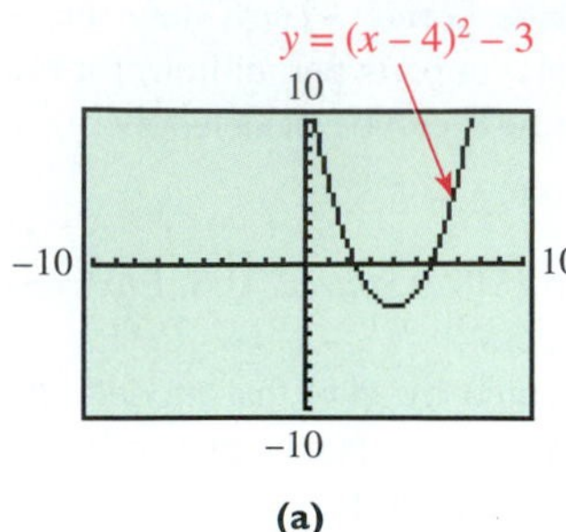

(a)

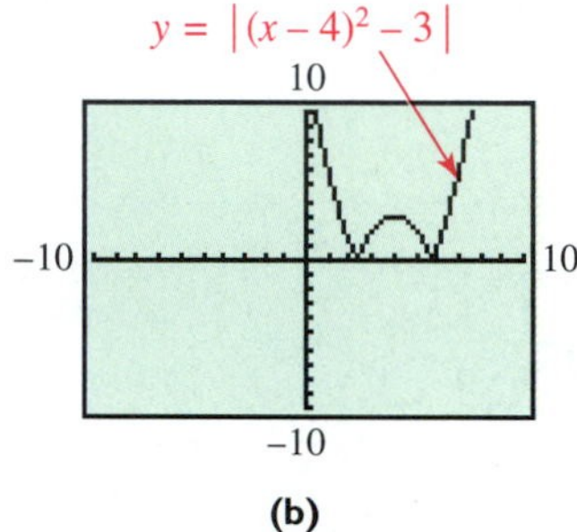

(b)

FIGURE 43

EXAMPLE 1 Finding Domains and Ranges of $y = f(x)$ and $y = |f(x)|$

Figure 43(a) shows the graph of

$$y = (x - 4)^2 - 3,$$

which is the graph of $y = x^2$ shifted 4 units to the right and 3 units downward. Figure 43(b) shows the graph of

$$y = |(x - 4)^2 - 3|.$$

Give the domain and range of each function.

Solution Notice that all points with negative y-values in the first graph have been reflected across the x-axis in the second graph, while all points with nonnegative y-values are the same for both graphs. The domain of each function is $(-\infty, \infty)$. The range of $y = (x - 4)^2 - 3$ is $[-3, \infty)$, while the range of $y = |(x - 4)^2 - 3|$ is $[0, \infty)$. ■

EXAMPLE 2 Graphing $y = |f(x)|$, Given the Graph of $y = f(x)$

Figure 44 shows the graph of a function $y = f(x)$. Use the figure to sketch the graph of $y = |f(x)|$. Give the domain and range of each function.

Solution The graph will remain the same for points whose y-values are nonnegative, while it will be reflected across the x-axis for all other points. Figure 45 shows the graph of $y = |f(x)|$. The domain of both functions is $(-\infty, \infty)$. The range of $y = f(x)$ is $[-3, \infty)$, while the range of $y = |f(x)|$ is $[0, \infty)$. ■

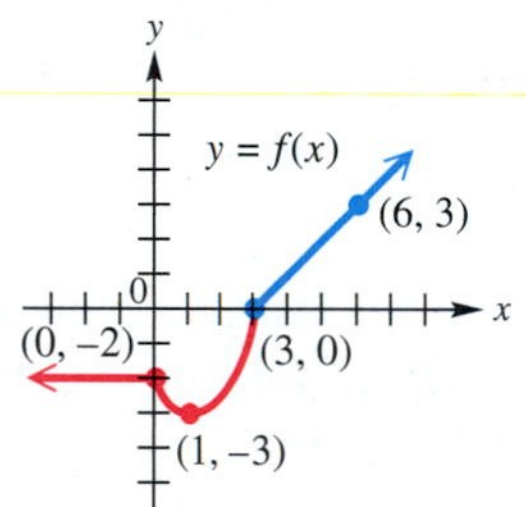

FIGURE 44

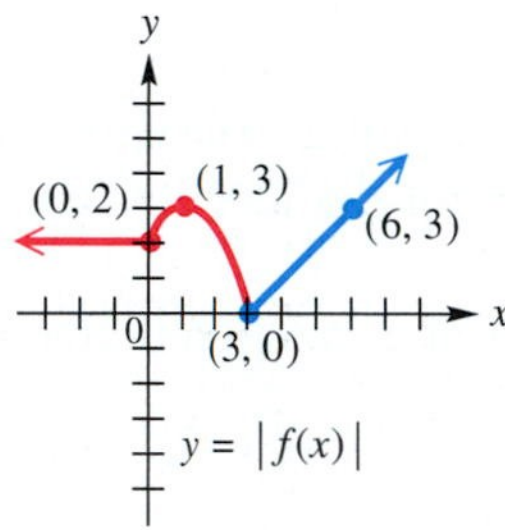

FIGURE 45

Properties of Absolute Value

Several properties of absolute value will be useful in our work.

Properties of Absolute Value

For all real numbers a and b,

1. $|ab| = |a| \cdot |b|$
   (*The absolute value of a product is equal to the product of the absolute values.*)
2. $\left|\frac{a}{b}\right| = \frac{|a|}{|b|}$ $(b \neq 0)$
   (*The absolute value of a quotient is equal to the quotient of the absolute values.*)
3. $|a| = |-a|$
   (*The absolute value of a number is equal to the absolute value of its additive inverse.*)
4. $|a| + |b| \geq |a + b|$ **(triangle inequality)**
   (*The sum of the absolute values of two numbers is greater than or equal to the absolute value of their sum.*)

Consider the absolute value function $y = |2x + 11|$, and observe the following sequence of transformations:

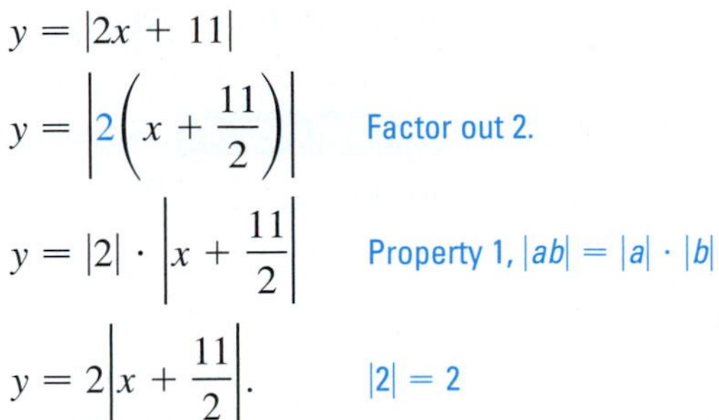

$$y = |2x + 11|$$

$$y = \left|2\left(x + \frac{11}{2}\right)\right| \qquad \text{Factor out 2.}$$

$$y = |2| \cdot \left|x + \frac{11}{2}\right| \qquad \text{Property 1, } |ab| = |a| \cdot |b|$$

$$y = 2\left|x + \frac{11}{2}\right|. \qquad |2| = 2$$

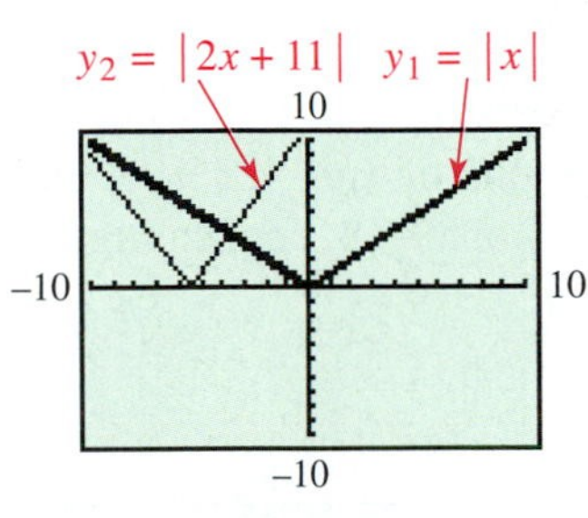

FIGURE 46

Using the concepts of the previous two sections, we can find the graph of this function by starting with the graph of $y = |x|$, shifting $\frac{11}{2}$ units to the left, and then vertically stretching by a factor of 2. The graphs of $y_1 = |x|$ and $y_2 = |2x + 11|$ in Figure 46 support this statement.

We are often interested in absolute value functions of the form $f(x) = |ax + b|$, where the expression inside the absolute value bars is linear. ***A comprehensive graph of $f(x) = |ax + b|$ will include all intercepts and the lowest point on the "V-shaped" graph.***

EXAMPLE 3 Graphing $y = |ax + b|$ by Hand

Graph each equation. Then comment on how the graphs are related.

(a) $y = 2x - 1$ **(b)** $y = |2x - 1|$

Solution

(a) The graph of $y = 2x - 1$ is a line with slope 2 and y-intercept -1, as shown in Figure 47.

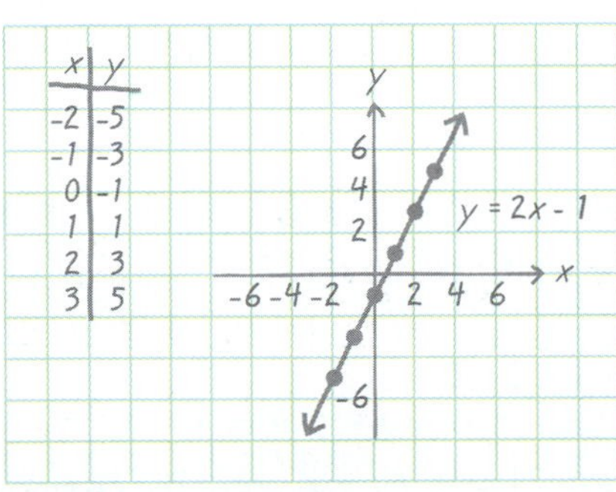

FIGURE 47

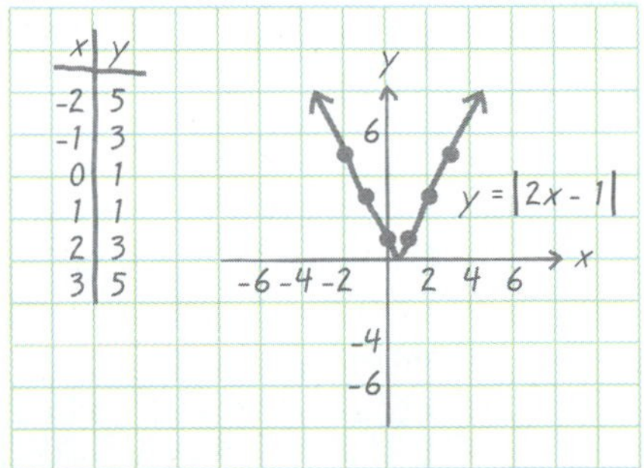

FIGURE 48

(b) The graph of $y = |2x - 1|$ is V-shaped, as shown in Figure 48. One way to obtain the graph of $y = |2x - 1|$ is to reflect the graph of $y = 2x - 1$ across the x-axis whenever the graph is below that axis. ■

In general, the graph of $y = ax + b$ $(a \neq 0)$ is a line that crosses the x-axis, whereas the graph of $y = |ax + b|$ is V-shaped. Both graphs have the same x-intercept.

Equations and Inequalities Involving Absolute Value

We first investigate equations (and related inequalities) that involve absolute values of linear functions and constants.

EXAMPLE 4 Solving an Absolute Value Equation

Solve $|2x + 1| = 7$.

Analytic Solution

For $|2x + 1|$ to equal 7, $2x + 1$ must be 7 units from 0 on the number line. This can happen only when $2x + 1 = 7$ or $2x + 1 = -7$. Solve this compound equation as follows:

$$\begin{aligned} 2x + 1 &= 7 \quad \text{or} \quad 2x + 1 = -7 \\ 2x &= 6 \quad \text{or} \quad 2x = -8 \qquad \text{Subtract 1.} \\ x &= 3 \quad \text{or} \quad x = -4. \qquad \text{Divide by 2.} \end{aligned}$$

Check by substitution in the original equation that the solution set is $\{-4, 3\}$.

Graphing Calculator Solution

Figure 49 shows that the graphs of $y_1 = |2x + 1|$ and $y_2 = 7$ intersect when $x = -4$ or $x = 3$, confirming the analytic solution.

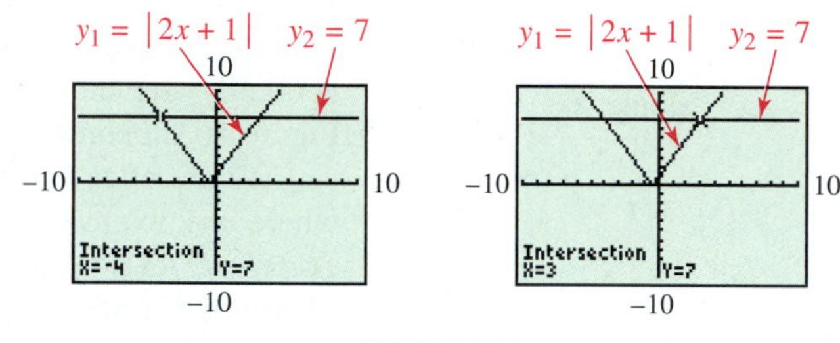

FIGURE 49

CAUTION A common error when solving $|2x + 1| = 7$ analytically is to consider only one part of the "or" statement, $2x + 1 = 7$. Remember that $2x + 1 = -7$ must also be considered.

In Example 4, we used the property that if $|ax + b| = k$, where $k > 0$, then

$$ax + b = k \quad \text{or} \quad ax + b = -k.$$

The following summary indicates the general methods for solving absolute value equations and inequalities analytically.

Looking Ahead to Calculus

The formal definition of *limit*, one of the most important concepts in calculus, is given here. It uses two absolute value inequalities.

Suppose that a function f is defined at every number in an open interval I containing a, except perhaps at a itself. Then the limit of $f(x)$ as x approaches a is L, written

$$\lim_{x \to a} f(x) = L,$$

if, for every $\epsilon > 0$, there exists a $\delta > 0$ such that

$$|f(x) - L| < \epsilon$$

whenever

$$0 < |x - a| < \delta.$$

Solving Absolute Value Equations and Inequalities

Let k be a positive number.

1. To solve $|ax + b| = k$, solve the compound equation
$$\boldsymbol{ax + b = k} \quad \text{or} \quad \boldsymbol{ax + b = -k}.$$
2. To solve $|ax + b| > k$, solve the compound inequality
$$\boldsymbol{ax + b > k} \quad \text{or} \quad \boldsymbol{ax + b < -k}.$$
3. To solve $|ax + b| < k$, solve the three-part inequality
$$\boldsymbol{-k < ax + b < k}.$$

Inequalities involving $\le$ or $\ge$ are solved similarly, by using the equality part of the symbol as well.

EXAMPLE 5 Solving Absolute Value Inequalities

Solve each inequality.

(a) $|2x + 1| < 7$ **(b)** $|2x + 1| > 7$

Analytic Solution

(a) The expression $2x + 1$ must represent a number that is less than 7 units from 0 on the number line. This means that $2x + 1$ must be *between* -7 and 7. This statement is written as a three-part inequality.

$$-7 < 2x + 1 < 7$$

$$-8 < 2x < 6 \quad \text{Subtract 1 from each part.}$$

$$-4 < x < 3 \quad \text{Divide each part by 2.}$$

The solution set is the open interval $(-4, 3)$.

(b) This absolute value inequality must be rewritten as

$$2x + 1 > 7 \quad \text{or} \quad 2x + 1 < -7,$$

because $2x + 1$ must represent a number that is *more* than 7 units from 0 on either side of the number line. Now, solve this compound inequality.

$$2x + 1 > 7 \quad \text{or} \quad 2x + 1 < -7$$

$$2x > 6 \quad \text{or} \quad 2x < -8 \quad \text{Subtract 1.}$$

$$x > 3 \quad \text{or} \quad x < -4 \quad \text{Divide by 2.}$$

The solution set is $(-\infty, -4) \cup (3, \infty)$.

Graphing Calculator Solution

(a) As seen in Figure 50, the graph of $y_1 = |2x + 1|$ lies *below* the graph of $y_2 = 7$ for x-values between -4 and 3, so the solution set is $(-4, 3)$.

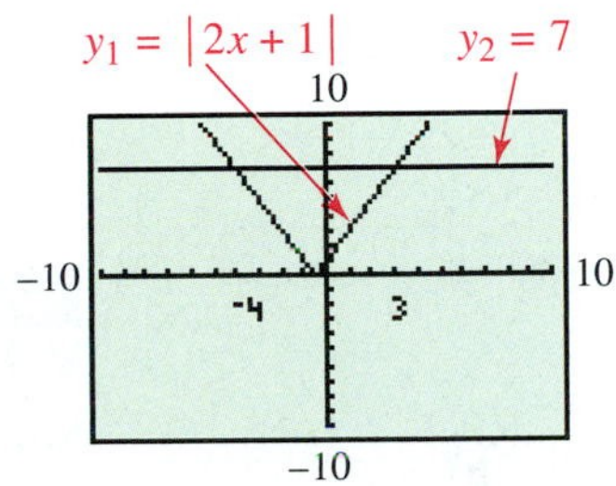

FIGURE 50

(b) The graph of $y_1 = |2x + 1|$ lies *above* the graph of $y_2 = 7$ for x-values less than -4 or greater than 3, confirming the analytic result. ■

An absolute value equation of the form $|ax + b| = k$, where $k < 0$, has solution set $\emptyset$, since the absolute value of a real number cannot be negative. The related inequalities, $|ax + b| > k$ and $|ax + b| < k$, have solution sets $(-\infty, \infty)$ and $\emptyset$, respectively. (What is the solution set of $|ax + b| = 0$?)

EXAMPLE 6 Solving Special Cases

Solve each equation or inequality.

(a) $|3x + 5| = -5$ **(b)** $|3x + 5| < -5$ **(c)** $|3x + 5| > -5$

Solution

(a) Because the absolute value of an expression will never be -5, this equation has no solution. The solution set is $\emptyset$.

(b) Using reasoning similar to that in part (a), the absolute value of an expression will never be less than -5. Once again, the solution set is $\emptyset$.

(c) Because the absolute value of an expression will always be greater than or equal to 0, the absolute value of an expression will always be greater than -5. The solution set is $(-\infty, \infty)$. ■

If two quantities have the same absolute value, they must either be equal to each other or be negatives of each other. This fact allows us to solve absolute value equations (and related inequalities) of the form $|ax + b| = |cx + d|$.

Solving $|ax + b| = |cx + d|$

To solve the equation $|ax + b| = |cx + d|$ analytically, solve the compound equation

$$ax + b = cx + d \quad \text{or} \quad ax + b = -(cx + d).$$

EXAMPLE 7 Solving an Equation Involving Two Absolute Value Expressions

Solve $|x + 6| = |2x - 3|$.

Analytic Solution

This equation is satisfied if

$$x + 6 = 2x - 3 \quad \text{or} \quad x + 6 = -(2x - 3).$$

Solve each equation.

$$x + 6 = 2x - 3 \quad \text{or} \quad x + 6 = -(2x - 3)$$
$$9 = x \qquad x + 6 = -2x + 3$$
$$3x = -3$$
$$x = -1$$

Be careful with signs.

Check each solution by substitution in the original equation. The solution set is $\{-1, 9\}$.

Graphing Calculator Solution

Let $y_1 = |x + 6|$ and $y_2 = |2x - 3|$. The equation $y_1 = y_2$ is equivalent to $y_1 - y_2 = 0$, so we graph

$$y_3 = |x + 6| - |2x - 3|$$

and find the x-intercepts, -1 and 9. See Figure 51.

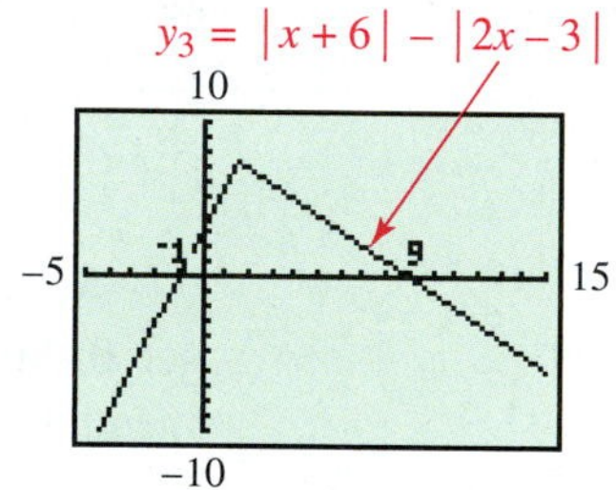

FIGURE 51

EXAMPLE 8 Solving Inequalities Involving Two Absolute Value Expressions

Solve each inequality graphically by referring to Example 7 and Figure 51.

(a) $|x + 6| < |2x - 3|$ **(b)** $|x + 6| \geq |2x - 3|$

Solution

(a) The inequality $y_1 < y_2$ is equivalent to $y_1 - y_2 < 0$, or $y_3 < 0$. In Figure 51, note that the graph of y_3 is below the x-axis on the interval

$$(-\infty, -1) \cup (9, \infty).$$

(b) The inequality $y_3 \geq 0$ is satisfied by the closed interval $[-1, 9]$.

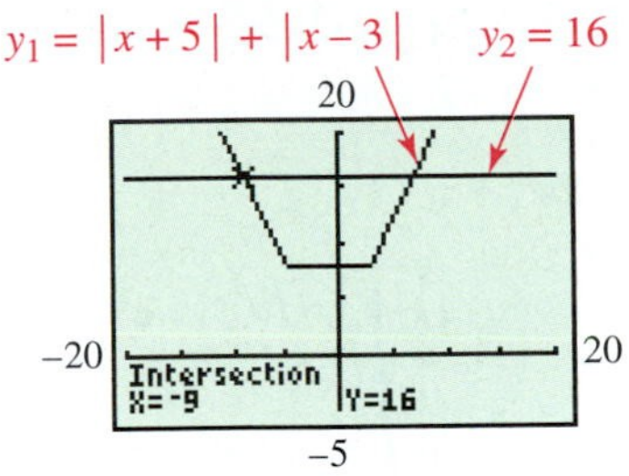

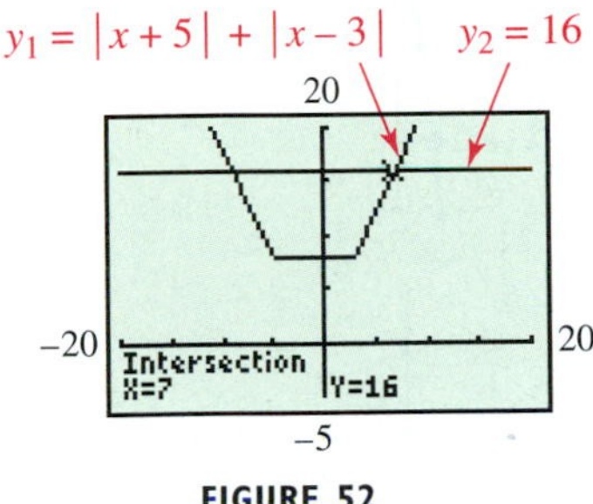

FIGURE 52

EXAMPLE 9 Solving an Equation Involving a Sum of Absolute Values

Solve $|x + 5| + |x - 3| = 16$ graphically by the intersection-of-graphs method. Check the solutions by substitution.

Solution Let $y_1 = |x + 5| + |x - 3|$ and $y_2 = 16$. Their graphs are shown in Figure 52. Locate the points of intersection of the graphs to find that the x-coordinates of the points are -9 and 7. To check, substitute each solution into the original equation.

$$\text{Let } x = -9: \quad |(-9) + 5| + |(-9) - 3| = |-4| + |-12| = 4 + 12 = 16. \checkmark$$

$$\text{Let } x = 7: \quad |7 + 5| + |7 - 3| = |12| + |4| = 12 + 4 = 16. \checkmark$$

Therefore, the solution set is $\{-9, 7\}$. ■

An Application Involving Absolute Value

EXAMPLE 10 Analyzing the Temperature in Spokane

The inequality $|x - 48| \le 21$ describes the range of average monthly temperatures x in degrees Fahrenheit for Spokane, Washington. Solve this inequality and interpret the result. (*Source:* yahoo.com weather.)

Analytic Solution

To solve $|x - 48| \le 21$, we must solve this compound inequality.

$$-21 \le x - 48 \le 21$$
$$-21 + 48 \le x - 48 + 48 \le 21 + 48$$
$$27 \le x \le 69$$

This means that the average monthly temperatures in Spokane, Washington, range from 27°F through 69°F. The average monthly temperatures are always within 21°F of 48°F.

Graphing Calculator Solution

Figure 53 shows the graphs of $y_1 = |x - 48|$ and $y_2 = 21$. The graph of y_1 lies below the graph of y_2 when x-values are between 27 and 69. This confirms the analytic solution. The endpoints are included because the original inequality includes equality.

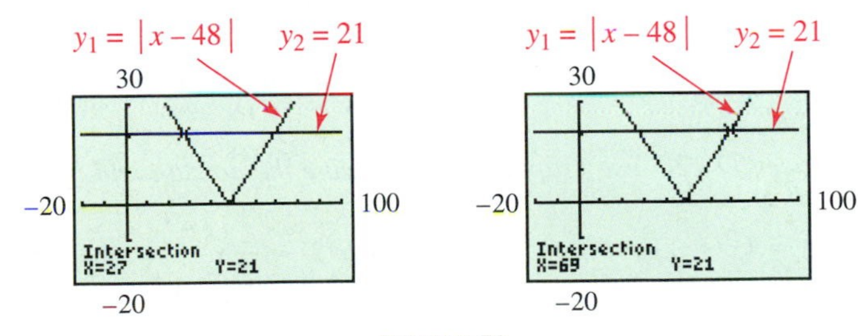

FIGURE 53

■

2.4 Exercises

Concept Check *Give a short answer to each question.*

1. If $f(a) = -5$, what is the value of $|f(a)|$?

2. How does the graph of $f(x) = x^2$ compare with the graph of $f(x) = |x^2|$?

3. What is the range of $y = |f(x)|$ if $f(x) = -x^2$?

4. If the range of $y = f(x)$ is $[-2, \infty)$, what is the range of $y = |f(x)|$?

5. If the range of $y = f(x)$ is $(-\infty, -2]$, what is the range of $y = |f(x)|$?

6. Why can't the range of $y = |f(x)|$ include -1, for any function f?

In Exercises 7–15, you are given graphs of functions $y = f(x)$. Sketch the graph of $y = |f(x)|$ by hand.

7.

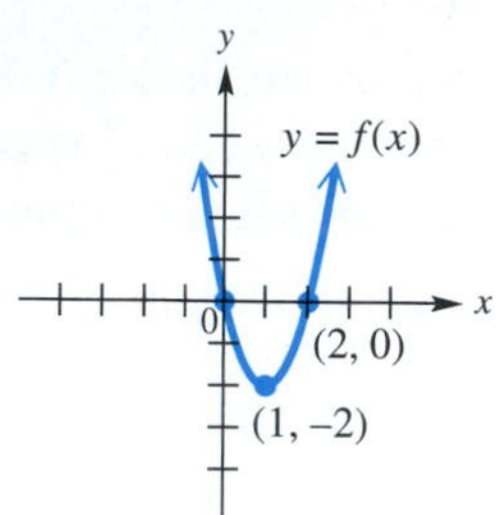

8.

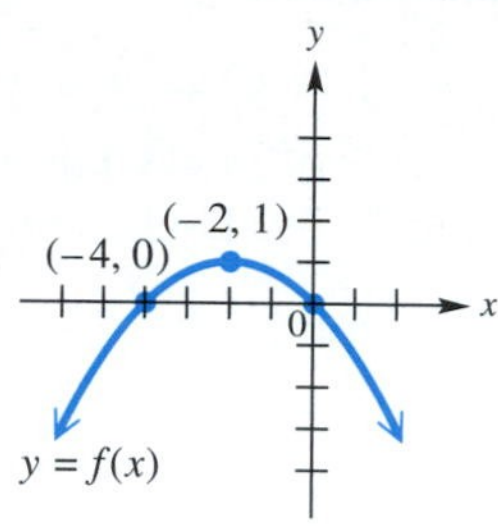

9.

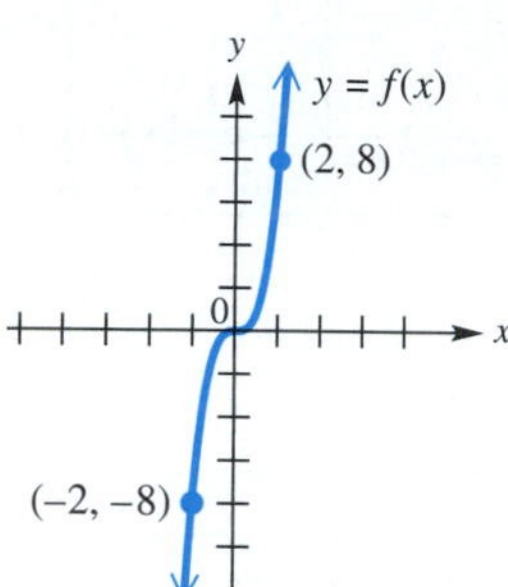

10.

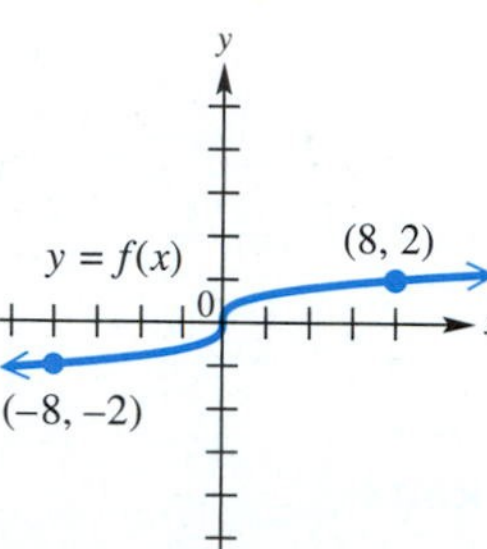

11.

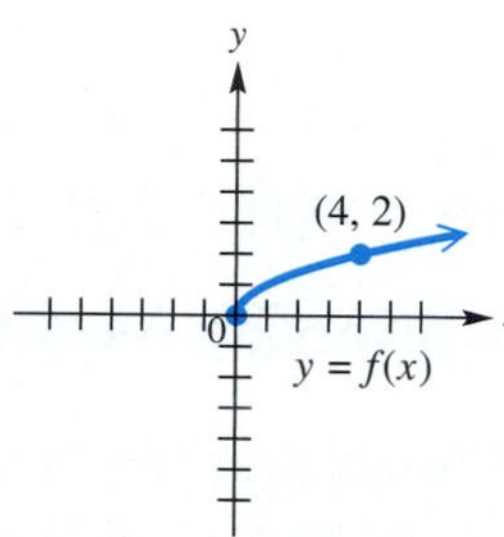

12.

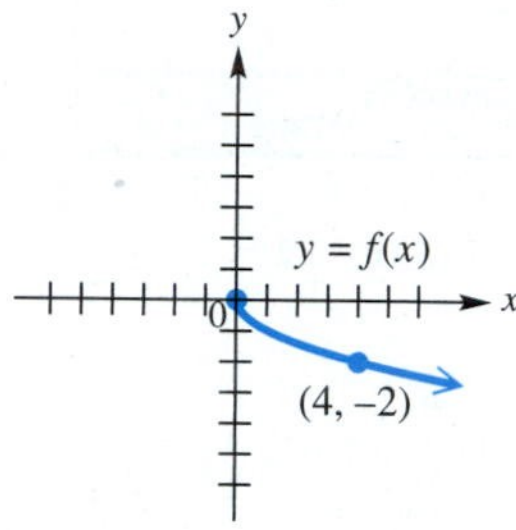

13.

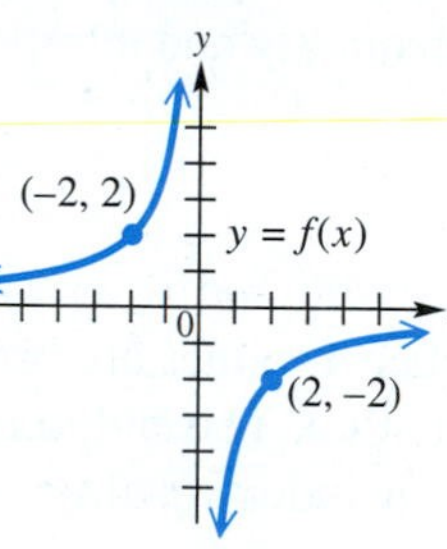

14.

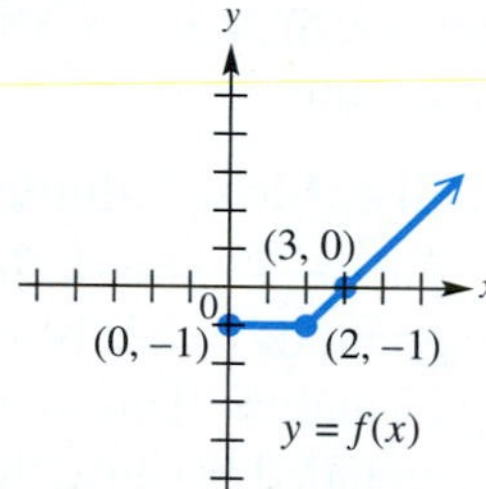

15.

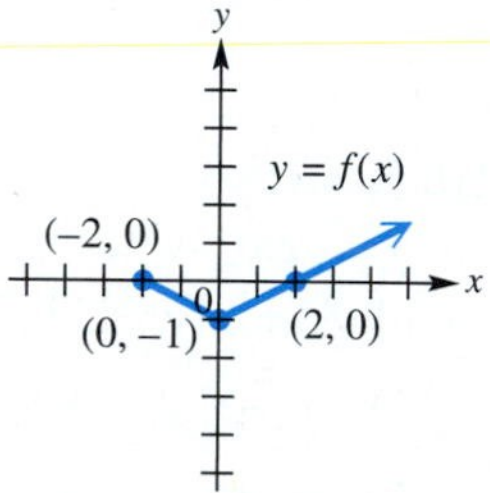

16. Explain in your own words the procedure you used to find the graphs of $y = |f(x)|$ in Exercises 7–15.

In Exercises 17–20, use graphing to determine the domain and range of $y = f(x)$ and of $y = |f(x)|$.

17. $f(x) = (x + 1)^2 - 2$

18. $f(x) = 2 - \frac{1}{2}x$

19. $f(x) = -1 - (x - 2)^2$

20. $f(x) = -|x + 2| - 2$

In Exercises 21–24, a line graph of $y = f(x)$ is given. Find the domain and range of $y = f(x)$ and of $y = |f(x)|$.

21.

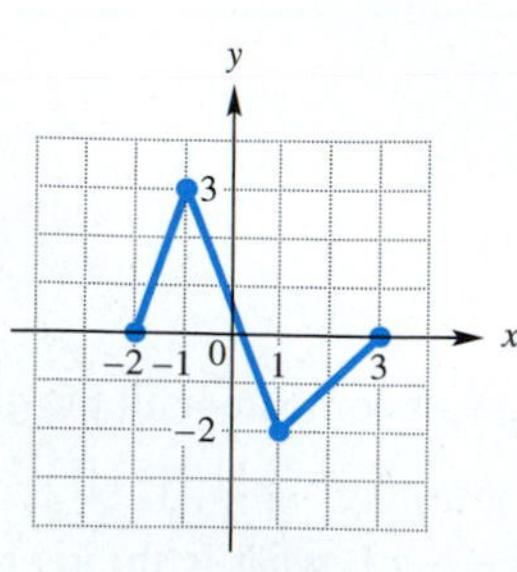

22.

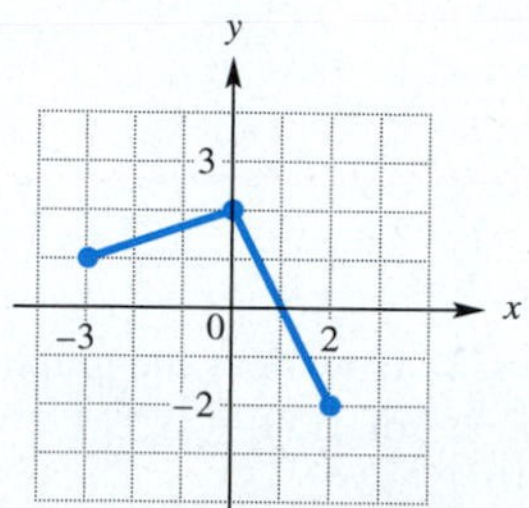

23.

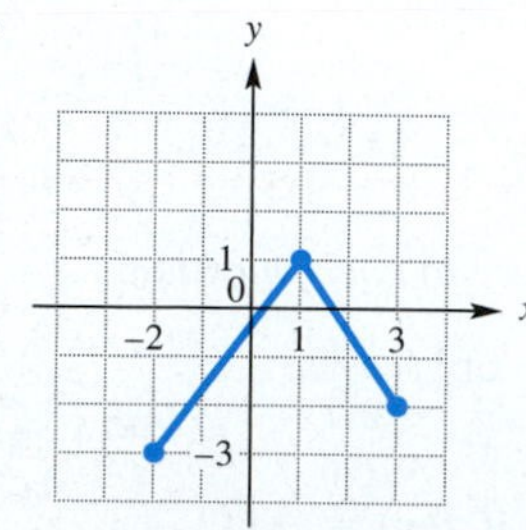

24.

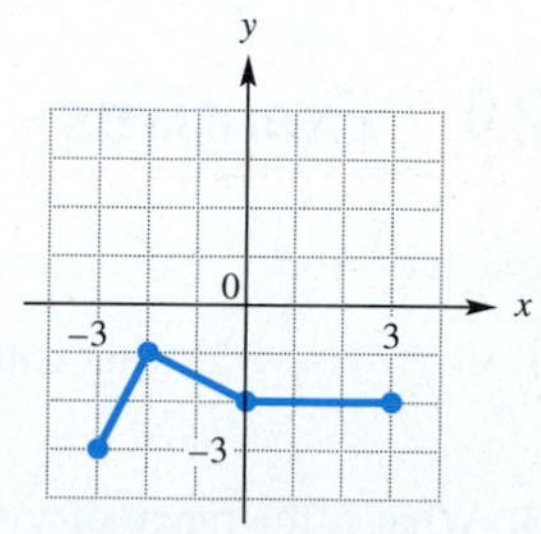

25. The graph of a function $y = f(x)$ is shown. Sketch by hand, in order, the graph of each of the functions that follow. Use the concept of reflecting introduced in Section 2.3 and the concept of graphing $y = |f(x)|$ introduced in this section.

(a) $y = f(-x)$

(b) $y = -f(-x)$

(c) $y = |-f(-x)|$

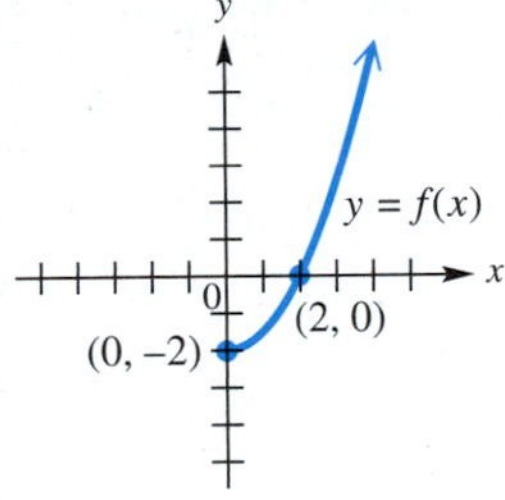

26. Repeat Exercise 25 for the graph of $y = f(x)$ shown here.

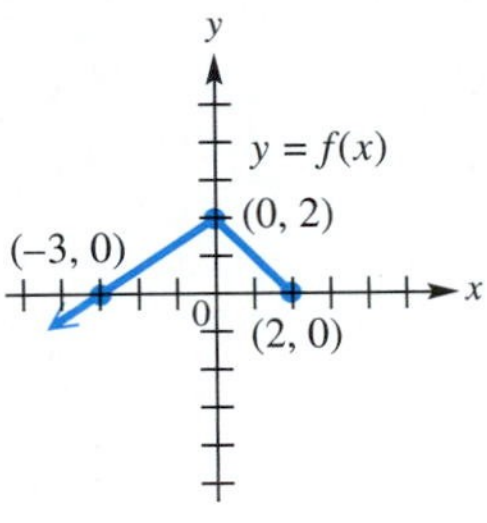

In Exercises 27 and 28, one graph is that of $y = f(x)$ and the other is that of $y = |f(x)|$. State which is which.

27. A.

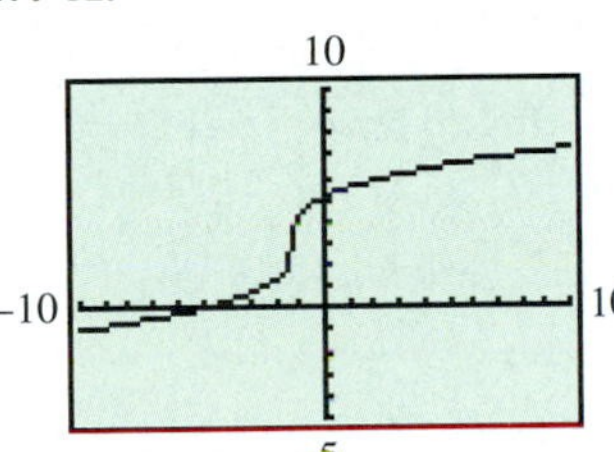

B.

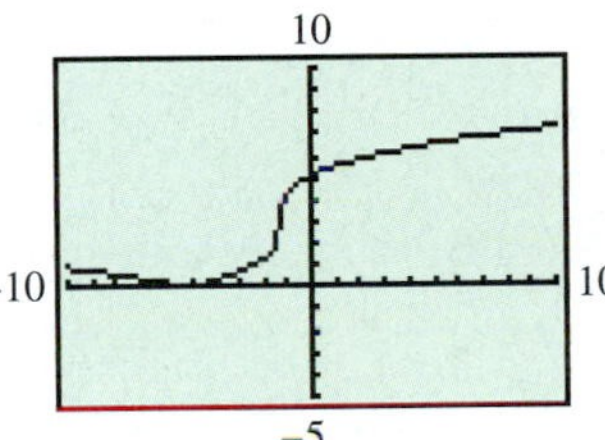

28. A.

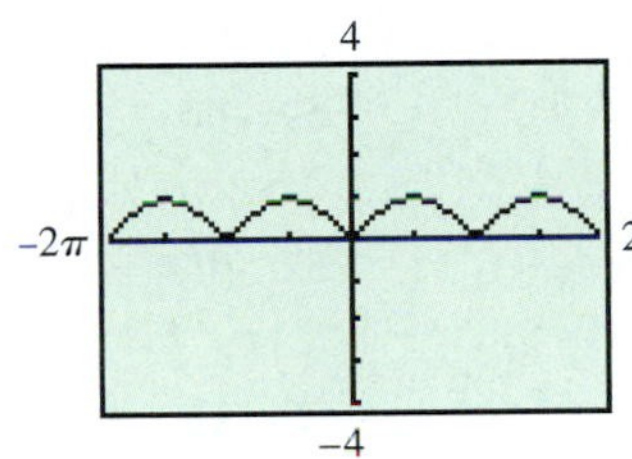

B.

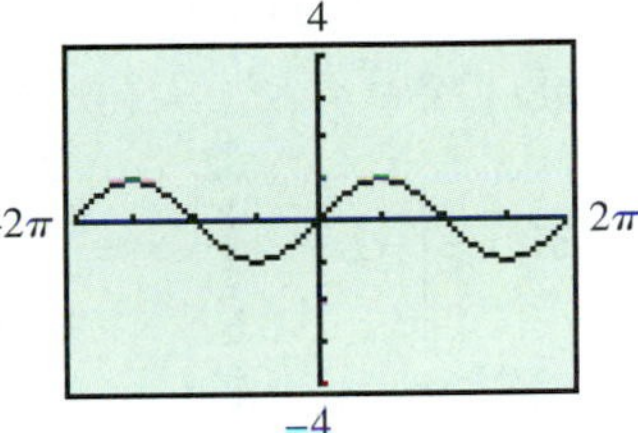

In Exercises 29–32, use the graph, along with the indicated points, to give the solution set of

(a) $y_1 = y_2$. **(b)** $y_1 < y_2$. **(c)** $y_1 > y_2$.

29.

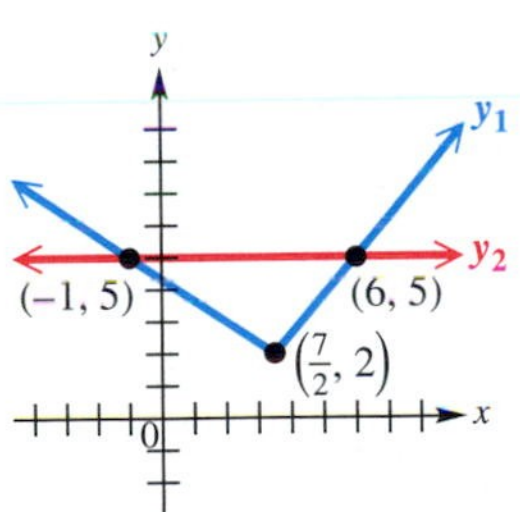

30.

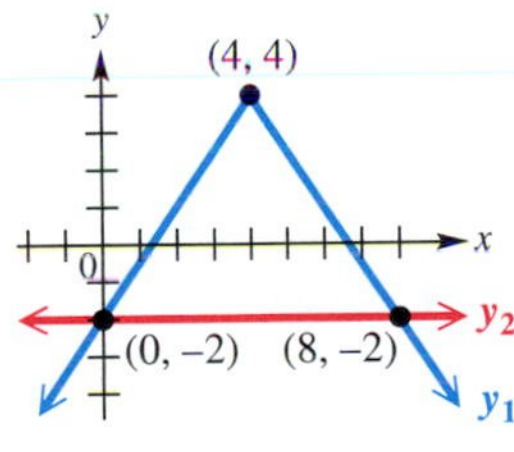

31.

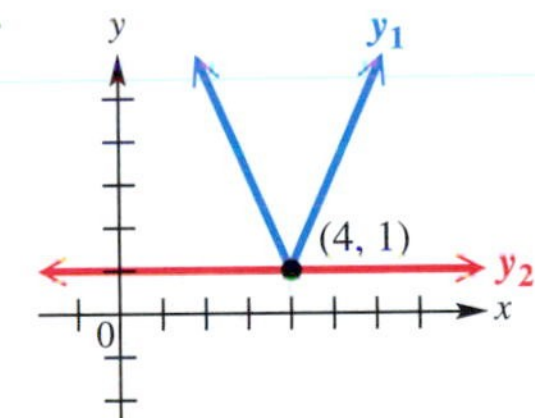

32.

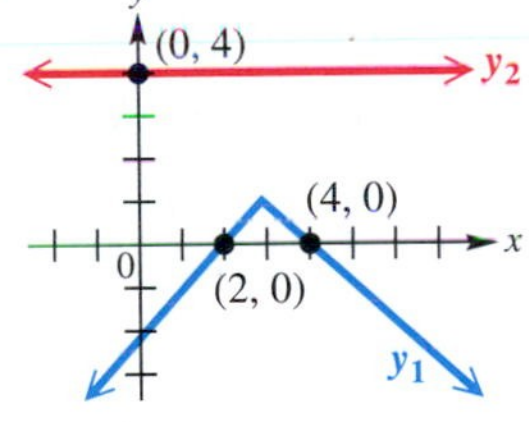

Relating Concepts

For individual or group investigation (Exercises 33–38)

The figure shows the graphs of $f(x) = |.5x + 6|$ and $g(x) = 3x - 14$. ***Work Exercises 33–38 in order,*** *without actually using your graphing calculator.*

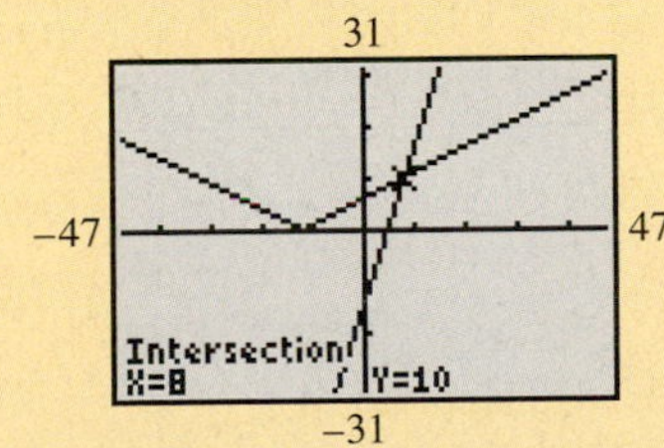

33. Which graph is that of $y = f(x)$? How do you know?

34. Which graph is that of $y = g(x)$? How do you know?

35. Solve $f(x) = g(x)$ based on the display at the bottom of the screen.

36. Solve $f(x) > g(x)$ based on the graphs and the display.

37. Solve $f(x) < g(x)$ based on the graphs and the display.

38. What is the solution set of $|.5x + 6| - (3x - 14) = 0$?

Solve each group of equations and inequalities analytically. Support your solutions graphically.

39. **(a)** $|x + 4| = 9$
(b) $|x + 4| > 9$
(c) $|x + 4| < 9$

40. **(a)** $|x - 3| = 5$
(b) $|x - 3| > 5$
(c) $|x - 3| < 5$

41. **(a)** $|-2x + 7| = 3$
(b) $|-2x + 7| \geq 3$
(c) $|-2x + 7| \leq 3$

42. **(a)** $|-3x - 9| = 6$
(b) $|-3x - 9| \geq 6$
(c) $|-3x - 9| \leq 6$

43. **(a)** $|2x + 1| + 3 = 5$
(b) $|2x + 1| + 3 \leq 5$
(c) $|2x + 1| + 3 \geq 5$

44. **(a)** $|4x + 7| = 0$
(b) $|4x + 7| > 0$
(c) $|4x + 7| < 0$

45. **(a)** $|7x - 5| = 0$
(b) $|7x - 5| \geq 0$
(c) $|7x - 5| \leq 0$

46. **(a)** $|\pi x + 8| = -4$
(b) $|\pi x + 8| < -4$
(c) $|\pi x + 8| > -4$

47. **(a)** $|\sqrt{2}x - 3.6| = -1$
(b) $|\sqrt{2}x - 3.6| \leq -1$
(c) $|\sqrt{2}x - 3.6| \geq -1$

Solve each equation or inequality.

48. $|2x + 4| + 2 = 10$

49. $3|4 - 3x| - 4 = 8$

50. $5|x + 3| - 2 = 18$

51. $\frac{1}{2}\left|-2x + \frac{1}{2}\right| = \frac{3}{4}$

52. $|3(x - 5) + 2| + 3 = 9$

53. $4.2|.5 - x| + 1 = 3.1$

54. $|3x - 1| < 8$

55. $|15 - x| < 7$

56. $|7 - 4x| \leq 11$

57. $|2x - 3| > 1$

58. $|5x - 7| > -2$

59. $|-3x + 8| \geq 3$

60. $\left|\frac{1}{2}x - 2\right| > -1$

61. $\left|6 - \frac{1}{3}x\right| > 0$

62. $|8x - 4| < 0$

63. $|-2x + 7| \leq -6$

64. $\left|\frac{3}{4}x - \frac{1}{2}\right| < -4$

65. $|7x - 5| \geq -5$

66. Explain how to solve an equation of the form $|ax + b| = |cx + d|$ analytically.

An equation of the form $|f(x)| = |g(x)|$ is given.
(a) *Solve the equation analytically and support the solution graphically.*
(b) *Solve $|f(x)| > |g(x)|$ graphically.*
(c) *Solve $|f(x)| < |g(x)|$ graphically.*

67. $|3x + 1| = |2x - 7|$

68. $|x - 4| = |7x + 12|$

69. $|-2x + 5| = |x + 3|$

70. $|-5x + 1| = |3x - 4|$

71. $\left|x - \frac{1}{2}\right| = \left|\frac{1}{2}x - 2\right|$

72. $|x + 3| = \left|\frac{1}{3}x + 8\right|$

73. $|4x + 1| = |4x + 6|$

74. $|6x + 9| = |6x - 3|$

75. $|.25x + 1| = |.75x - 3|$

76. $|.40x + 2| = |.60x - 5|$

Solve each equation graphically.

77. $|x + 1| + |x - 6| = 11$

78. $|2x + 2| + |x + 1| = 9$

79. $|x| + |x - 4| = 8$

80. $|.5x + 2| + |.25x + 4| = 9$

(Modeling) Average Temperature *Each inequality describes the range of average monthly temperatures T in degrees Fahrenheit at a certain location.*

(a) *Solve the inequality.* **(b)** *Interpret the result.*

81. $|T - 43| \le 24$, Marquette, Michigan

82. $|T - 62| \le 19$, Memphis, Tennessee

83. $|T - 50| \le 22$, Boston, Massachusetts

84. $|T - 10| \le 36$, Chesterfield, Canada

85. $|T - 61.5| \le 12.5$, Buenos Aires, Argentina

86. $|T - 43.5| \le 8.5$, Punta Arenas, Chile

(Modeling) *Solve each problem.*

87. *Weights of Babies* Dr. Cazayoux has found that, over the years, 95% of the babies he delivered weighed x pounds, where $|x - 8.0| \le 1.5$. What range of weights corresponds to this inequality?

88. *Conversion of Methanol to Gasoline* The industrial process that is used to convert methanol to gasoline is carried out at a temperature range of 680°F to 780°F. Using F as the variable, write an absolute value inequality that corresponds to this range.

89. *Blood Pressure* Systolic blood pressure is the maximum pressure produced by each heartbeat. Both low blood pressure and high blood pressure are cause for medical concern. Therefore, health care professionals are interested in a patient's "pressure difference from normal," or P_d. If 120 is considered a normal systolic pressure, $P_d = |P - 120|$, where P is the patient's recorded systolic pressure. For example, a patient with a systolic pressure P of 113 would have a pressure difference from normal of $P_d = |P - 120| = |113 - 120| = |-7| = 7$.

(a) Calculate the P_d value for a woman whose actual systolic pressure is 116 and whose normal value should be 125.

(b) If a patient's P_d value is 17 and the normal pressure for his sex and age should be 130, what are the two possible values for his systolic blood pressure?

90. *Kite Flying* When a model kite was flown in crosswinds in tests to determine its limits of power extraction, it attained speeds of 98 to 148 feet per second in winds of 16 to 26 feet per second. Using x as the variable in each case, write absolute value inequalities that correspond to these ranges.

Tolerances *In quality control and other applications, we often wish to keep the difference between two quantities within some predetermined amount, or tolerance. For example, suppose* $y = 2x + 1$ *and we want* y *to be within .01 unit of 4. This criterion can be written as* $|y - 4| < .01$. *To find the values of* x *that satisfy this condition on* y, *we use properties of absolute value as follows.*

$$|y - 4| < .01$$

$$|2x + 1 - 4| < .01 \quad \text{Substitute } 2x + 1 \text{ for } y.$$

$$|2x - 3| < .01$$

$$-.01 < 2x - 3 < .01 \quad \text{Write a three-part inequality.}$$

$$2.99 < 2x < 3.01 \quad \text{Add 3 to each part.}$$

$$1.495 < x < 1.505 \quad \text{Divide each part by 2.}$$

Keeping x *between* 1.495 *and* 1.505 *will ensure that the difference between* y *and* 4 *is less than* .01.

In Exercises 91–94, find the open interval in which x *must lie in order for the given condition to hold.*

91. $y = 2x + 1$, and the difference between y and 1 is less than .1.

92. $y = 3x - 6$, and the difference between y and 2 is less than .01.

93. $y = 4x - 8$, and the difference between y and 3 is less than .001.

94. $y = 5x + 12$, and the difference between y and 4 is less than .0001.

Windchill Factor *The windchill factor is a measure of the cooling effect that the wind has on a person's skin. It calculates the equivalent cooling temperature if there were no wind. The table shown here can be used to determine the windchill factor for various wind speeds and temperatures.*

Wind/°F	40°	30°	20°	10°	0°	−10°	−20°	−30°	−40°
5 mph	36	25	13	1	−11	−22	−34	−46	−57
10 mph	34	21	9	−4	−16	−28	−41	−53	−66
15 mph	32	19	6	−7	−19	−32	−45	−58	−71
20 mph	30	17	4	−9	−22	−35	−48	−61	−74
25 mph	29	16	3	−11	−24	−37	−51	−64	−78
30 mph	28	15	1	−12	−26	−39	−53	−67	−80
35 mph	28	14	0	−14	−27	−41	−55	−69	−82
40 mph	27	13	−1	−15	−29	−43	−57	−71	−84

Source: National Weather Service, 2005.

Suppose that we wish to determine the difference between two of these entries and are interested only in the magnitude, or absolute value, of this difference. We subtract the two entries and find the absolute value. For example, the difference in windchill factors for wind at 20 mph with a 20°F temperature and wind at 30 mph with a 40°F temperature is $|4° - 28°| = 24°\text{F}$*, or equivalently,* $|28° - 4°| = 24°\text{F}$.

Find the absolute value of the difference of the two indicated windchill factors.

95. Wind at 15 mph with a 30°F temperature and wind at 10 mph with a −10°F temperature

96. Wind at 20 mph with a −20°F temperature and wind at 5 mph with a 30°F temperature

97. Wind at 30 mph with a −30°F temperature and wind at 15 mph with a −20°F temperature

98. Wind at 40 mph with a 40°F temperature and wind at 25 mph with a −30°F temperature

Solve each equation or inequality graphically. Express solutions or endpoints of intervals rounded to the nearest hundredth if necessary.

99. $|2x + 7| = 6x - 1$

100. $-|3x - 12| \geq -x - 1$

101. $|x - 4| > .5x - 6$

102. $2x + 8 > -|3x + 4|$

103. $|3x + 4| < -3x - 14$

104. $|x - \sqrt{13}| + \sqrt{6} \leq -x - \sqrt{10}$

2.5 Piecewise-Defined Functions

Graphing Piecewise-Defined Functions ■ The Greatest Integer Function ■ Applications of Piecewise-Defined Functions

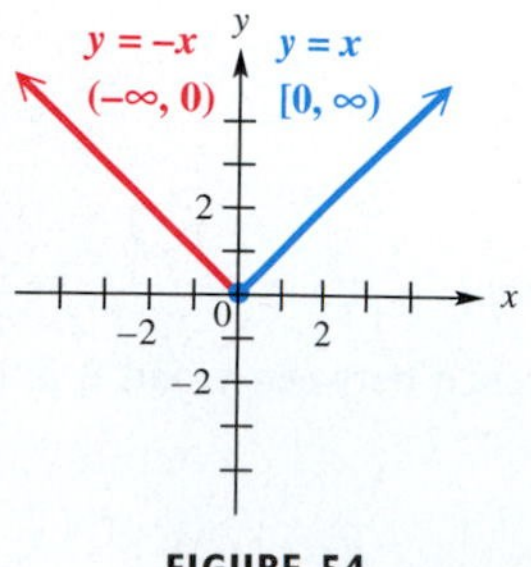

FIGURE 54

Graphing Piecewise-Defined Functions

The absolute value function is a simple example of a function defined by different rules over different subsets of its domain. Such a function is called a **piecewise-defined function.** Recall that the domain of $f(x) = |x|$ is $(-\infty, \infty)$. For the interval $[0, \infty)$ of the domain, we use the rule $f(x) = x$. For the interval $(-\infty, 0)$, we use the rule $f(x) = -x$. See Figure 54. Thus, the graph of $f(x) = |x|$ is composed of two "pieces." One piece comes from the graph of $y = x$ and the other from $y = -x$.

GCM **EXAMPLE 1 Finding Function Values for a Piecewise-Defined Function**

Consider the piecewise-defined function

$$f(x) = \begin{cases} x + 2 & \text{if } x \le 0 \\ \dfrac{1}{2}x^2 & \text{if } x > 0 \end{cases}.$$

Find **(a)** $f(-3)$. **(b)** $f(0)$. **(c)** $f(3)$. **(d)** Sketch the graph of f.

Solution

(a) Since $-3 \le 0$ (specifically, $-3 < 0$), we use the rule $f(x) = x + 2$. Thus,

$$f(-3) = -3 + 2 = -1. \quad \text{Let } x = -3.$$

This means that the graph of f will contain the point $(-3, -1)$.

(b) Since $0 \le 0$ (because $0 = 0$), we again use the rule $f(x) = x + 2$.

$$f(0) = 0 + 2 = 2 \quad \text{Let } x = 0.$$

The point $(0, 2)$ lies on the graph of f, meaning that the y-intercept of the graph is 2.

(c) The number 3 is greater than 0, and the second part of the rule f indicates that we must use the rule $f(x) = \frac{1}{2}x^2$.

$$f(3) = \frac{1}{2}(3)^2 = \frac{1}{2}(9) = 4.5 \quad \text{Let } x = 3.$$

(d) We graph the ray defined by $y = x + 2$, choosing x so that $x \le 0$, with a solid endpoint at $(0, 2)$. The ray has slope 1 and y-intercept 2. Then we graph $y = \frac{1}{2}x^2$ for $x > 0$. This graph will be half of a parabola with an open endpoint at $(0, 0)$. Figure 55 shows a table of values and a hand-drawn graph of the piecewise-defined function given by $y = f(x)$.

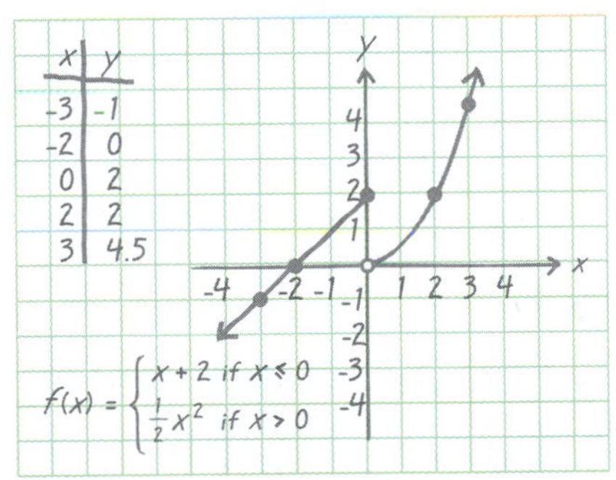

FIGURE 55

TECHNOLOGY NOTE

In general, it is best to graph piecewise-defined functions with the calculator in dot mode, especially when the graph exhibits discontinuities. Otherwise, the calculator may attempt to connect portions of the graph that are actually separate from one another.

EXAMPLE 2 Graphing a Piecewise-Defined Function

Graph the function defined by

$$f(x) = \begin{cases} x + 1 & \text{if } x \le 2 \\ -2x + 7 & \text{if } x > 2 \end{cases}.$$

Give the domain and range.

Analytic Solution

To graph

$$f(x) = \begin{cases} x + 1 & \text{if } x \le 2 \\ -2x + 7 & \text{if } x > 2 \end{cases}$$

for x-values less than or equal to 2, we graph the portion of the line $y = x + 1$ to the left of, and including, the point $(2, 3)$. For x-values greater than 2, we graph the portion of the line $y = -2x + 7$ to the right of $(2, 3)$. This is a continuous function. See Figure 56.

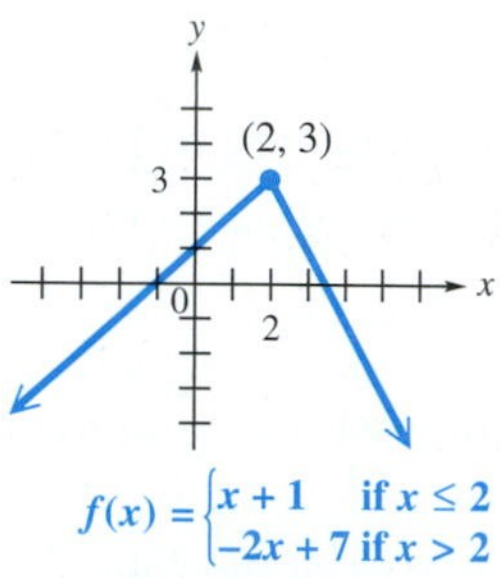

$f(x) = \begin{cases} x + 1 & \text{if } x \le 2 \\ -2x + 7 & \text{if } x > 2 \end{cases}$

FIGURE 56

The domain is $(-\infty, \infty)$, and the range is $(-\infty, 3]$.

Graphing Calculator Solution

To graph a piecewise-defined function with a graphing calculator, we use the *test* feature, which returns a 1 for a true statement and a 0 for a false statement. Figure 57(a) shows how $f(x)$ can be entered as Y_1.* Notice that the two defining expressions are multiplied by an expression composed of an inequality involving X (which will be 1 or 0, depending on the value of X as the graphing occurs). When $X \le 2$, the expression is

$$(X + 1)(1) + (-2X + 7)(0) = X + 1;$$

when $X > 2$, it is

$$(X + 1)(0) + (-2X + 7)(1) = -2X + 7.$$

The graph of Y_1 is shown in dot mode in Figure 57(b).

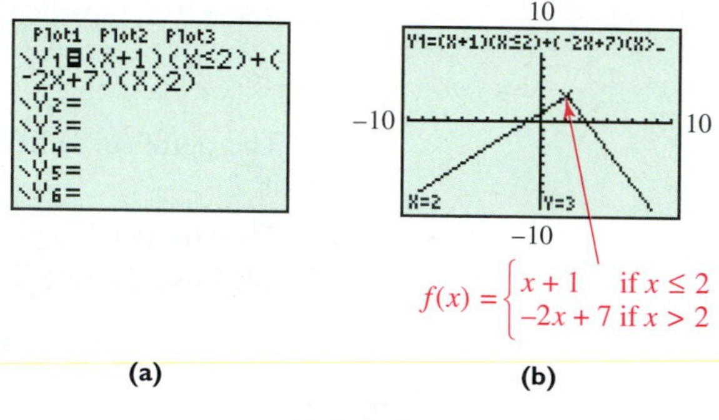

$f(x) = \begin{cases} x + 1 & \text{if } x \le 2 \\ -2x + 7 & \text{if } x > 2 \end{cases}$

(a) (b)

FIGURE 57

EXAMPLE 3 Finding the Formula for a Piecewise-Defined Function

Write a formula for the piecewise-defined function f shown in Figure 58.

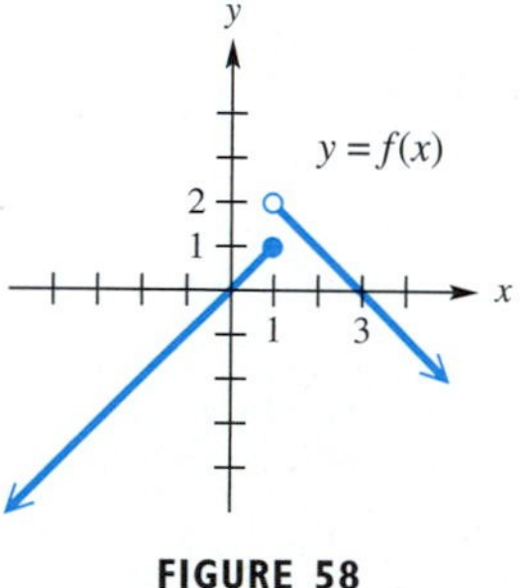

FIGURE 58

Solution The graph of f consists of two parts. When $x \le 1$, the graph is a line with slope 1 and y-intercept 0. Thus, the equation of this line is $y = x$. When $x > 1$, the graph is a line with slope -1 and x-intercept 3. Using the point–slope form, we find that the equation of the line passing through the point $(3, 0)$ with slope -1 is

$$y - 0 = -1(x - 3) \quad \text{or} \quad y = -x + 3.$$

Thus, we can write a formula for the piecewise-defined function f as follows:

$$f(x) = \begin{cases} x & \text{if } x \le 1 \\ -x + 3 & \text{if } x > 1 \end{cases}.$$

NOTE Other acceptable forms exist for defining piecewise-defined functions. We give most obvious ones.

*In defining the function as Y_1, the user has a choice whether to use the multiplication symbol, *.

The Greatest Integer Function

An important example of a piecewise-defined function is the **greatest integer function.** ***The notation*** $[\![x]\!]$ ***represents the greatest integer less than or equal to x;*** thus,

$$f(x) = [\![x]\!] = \begin{cases} x & \text{if } x \text{ is an integer} \\ \text{the greatest integer less than } x & \text{if } x \text{ is not an integer} \end{cases}.$$

GCM **EXAMPLE 4 Evaluating** $[\![x]\!]$

Evaluate $[\![x]\!]$ for each value of x.

(a) 4 **(b)** -5 **(c)** 2.46 **(d)** π **(e)** $-6\frac{1}{2}$

Analytic Solution

(a) $[\![4]\!] = 4$, since 4 is an integer.

(b) $[\![-5]\!] = -5$, since -5 is an integer.

(c) $[\![2.46]\!] = 2$, since 2.46 is not an integer and 2 is the greatest integer less than 2.46.

(d) $[\![\pi]\!] = 3$, since $\pi \approx 3.14$.

(e) $[\![-6\frac{1}{2}]\!] = -7$, since -7 is the greatest integer less than $-6\frac{1}{2}$.

Graphing Calculator Solution

(a)–(e) The command "int" is used by some graphing calculators for the greatest integer function. See the list in Figure 59.

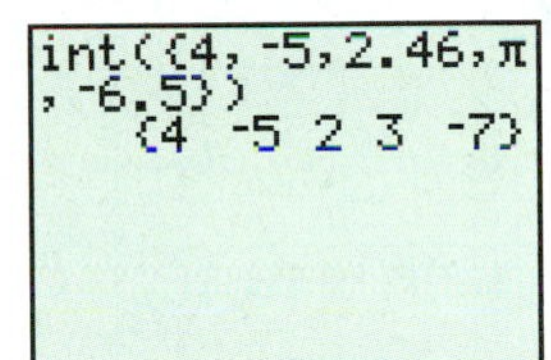

FIGURE 59

The greatest integer function (shown in Figure 60) is called a **step function.** (Why?)

TECHNOLOGY NOTE

You should learn how to activate the greatest integer function command on your particular model. Keep in mind that to obtain an accurate graph of a step function, your calculator must be in dot mode.

FUNCTION CAPSULE

GREATEST INTEGER FUNCTION $f(x) = [\![x]\!]$

Domain: $(-\infty, \infty)$ Range: $\{x \mid x \text{ is an integer}\} = \{\ldots, -3, -2, -1, 0, 1, 2, 3, \ldots\}$

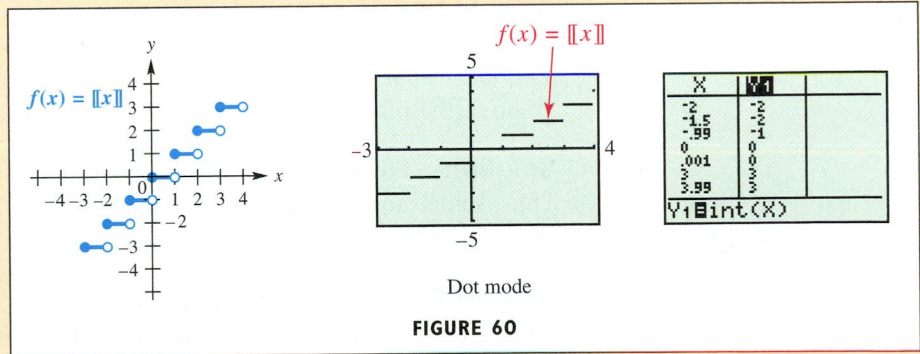

FIGURE 60

- $f(x) = [\![x]\!]$ is discontinuous at integer values of its domain, $(-\infty, \infty)$.
- The x-intercepts are all real numbers in the interval $[0, 1)$, and the y-intercept is 0.

EXAMPLE 5 Graphing a Step Function

Graph the function defined by $y = [\![\frac{1}{2}x + 1]\!]$. Give the domain and range.

Analytic Solution

Try some values of x in the equation to see how the values of y behave. Some sample ordered pairs are given in the accompanying table. These ordered pairs suggest that if x is in the interval $[0, 2)$, then $y = 1$. For x in $[2, 4)$, $y = 2$, and so on. The graph is shown in Figure 61. The domain is $(-\infty, \infty)$, and the range is $\{\ldots, -2, -1, 0, 1, 2, \ldots\}$.

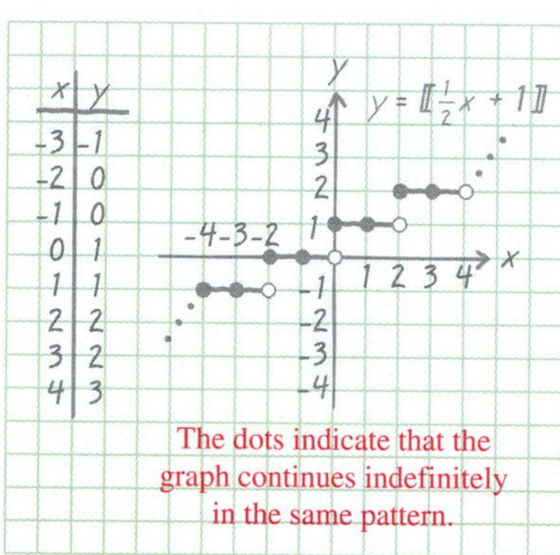

FIGURE 61

Graphing Calculator Solution

Figure 62(a) shows the graph of $Y_1 = [\![\frac{1}{2}X + 1]\!]$ in the window $[-4, 4]$ by $[-4, 4]$. Scroll up and down, using the table feature to confirm that the range is $\{\ldots, -2, -1, 0, 1, 2, \ldots\}$. See Figure 62(b).

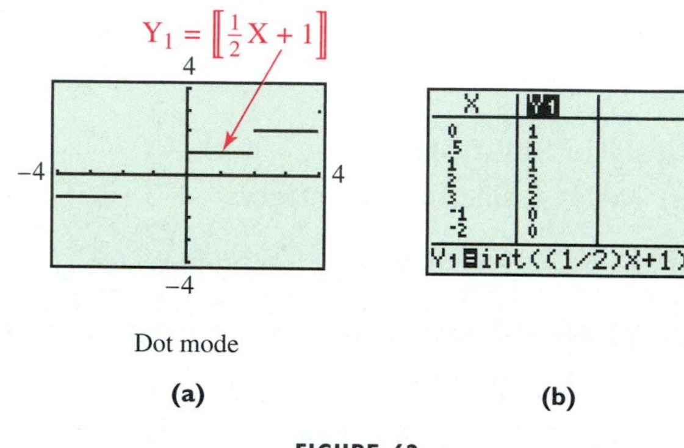

(a) (b)

FIGURE 62

In Figure 62(a), the inclusion or exclusion of the endpoints for each segment is not readily determined from the calculator graph. However, the traditional graph in Figure 61 *does* show whether endpoints are included or excluded. ***While technology is incredibly powerful and useful, we cannot rely on it alone in our study of graphs of functions.***

Applications of Piecewise-Defined Functions

EXAMPLE 6 Applying the Greatest Integer Function to Parking Rates

Downtown Parking charges a \$5 base fee for parking through 1 hour, and \$1 for each additional hour or fraction thereof. The maximum fee for 24 hours is \$15. Sketch a graph of the function that describes this pricing scheme.

Solution For any amount of time during and up to the first hour, the rate is \$5. Thus, some sample ordered pairs for the function in the interval $(0, 1]$ would be

$$(.25, 5), (.5, 5), (.75, 5), \text{ and } (1, 5).$$

After the first hour, the price immediately jumps (or steps up) to \$6 and remains \$6 until the time equals 2 hours. It then jumps to \$7 during the third hour, and so on. During the 11th hour, it will have jumped to \$15 and will remain at \$15 for the rest of the 24-hour period. Figure 63 on the next page shows the graph of this function on the interval $(0, 24]$. (A closed endpoint is found at the *right* extreme value rather than the left, because the increase does not take effect until after the first hour of use.) The range of the function is $\{5, 6, 7, \ldots, 15\}$.

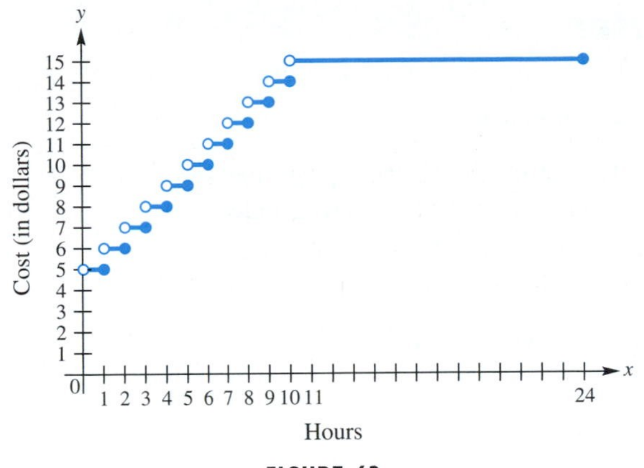

FIGURE 63

EXAMPLE 7 Using a Piecewise-Defined Function to Analyze Pollution

Due to acid rain, the percentage of lakes in Scandinavia that had lost their population of brown trout increased dramatically between 1940 and 1975. Based on a sample of 2850 lakes, this percentage can be approximated by the piecewise-defined function

$$f(x) = \begin{cases} \frac{11}{20}(x - 1940) + 7 & \text{if } 1940 \le x < 1960 \\ \frac{32}{15}(x - 1960) + 18 & \text{if } 1960 \le x \le 1975 \end{cases}.$$

(*Source:* C. Mason, *Biology of Freshwater Pollution*, John Wiley and Sons, 1991.) Determine the percent of lakes that had lost brown trout by 1950 and by 1972. Interpret your results.

Solution Because $1940 \le 1950 < 1960$, use the first formula to calculate $f(1950)$.

$$f(1950) = \frac{11}{20}(1950 - 1940) + 7 = 12.5$$

By 1950, about 12.5% of the lakes had lost their population of brown trout. In a similar manner, $f(1972)$ can be evaluated by using the second formula.

$$f(1972) = \frac{32}{15}(1972 - 1960) + 18 = 43.6$$

By 1972, about 43.6% of the lakes had lost their population of brown trout.

2.5 Exercises

(Modeling) *Solve each problem.*

1. ***Speed Limits*** The graph of $y = f(x)$ gives the speed limit y along a rural highway after traveling x miles.

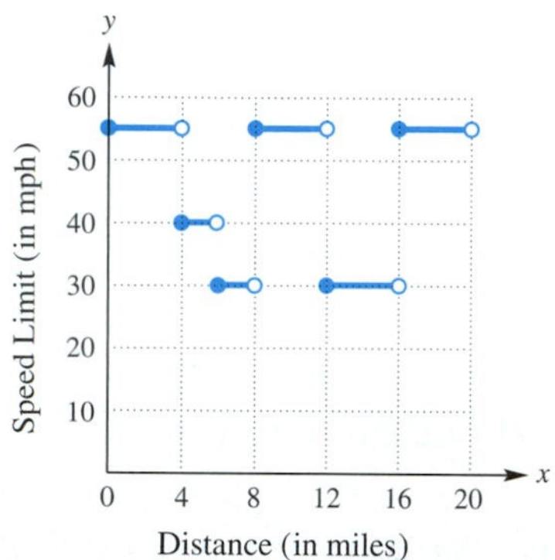

(a) What are the highest and lowest speed limits along this stretch of highway?
(b) Estimate the miles of highway with a speed limit of 55 mph.
(c) Evaluate $f(4)$, $f(12)$, and $f(18)$.
(d) At what x-values is the graph discontinuous? Interpret each discontinuity.

2. ***ATM*** The graph of $y = f(x)$ depicts the amount of money y in dollars in an automatic teller machine (ATM) after x minutes.

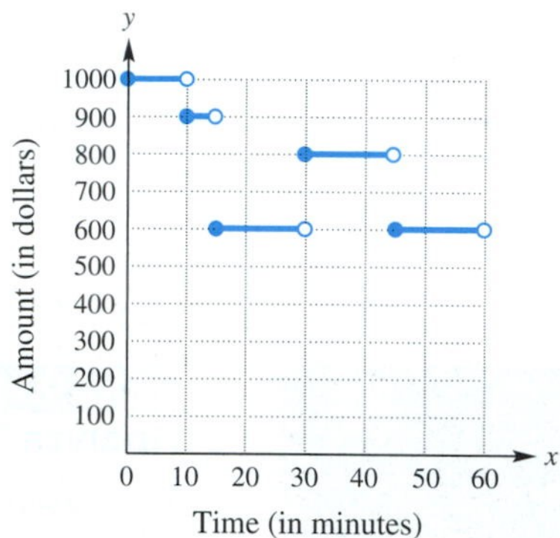

(a) Determine the initial and final amounts of money in the ATM.
(b) Evaluate $f(10)$ and $f(50)$. Is f continuous?
(c) How many withdrawals occurred during this period?
(d) When did the largest withdrawal occur? How much was it?
(e) How much was deposited into the machine?

3. ***Swimming Pool Levels*** The graph of $y = f(x)$ represents the amount of water in thousands of gallons remaining in a swimming pool after x days.

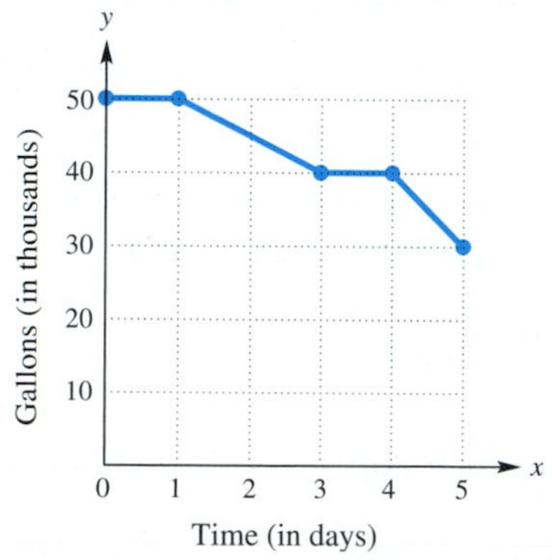

(a) Estimate the initial and final amounts of water contained in the pool.
(b) When did the amount of water in the pool remain constant?
(c) Approximate $f(2)$ and $f(4)$.
(d) At what rate was water being drained from the pool when $1 \le x \le 3$?

4. ***Gasoline Usage*** The graph shows the gallons of gasoline y in the gas tank of a car after x hours.

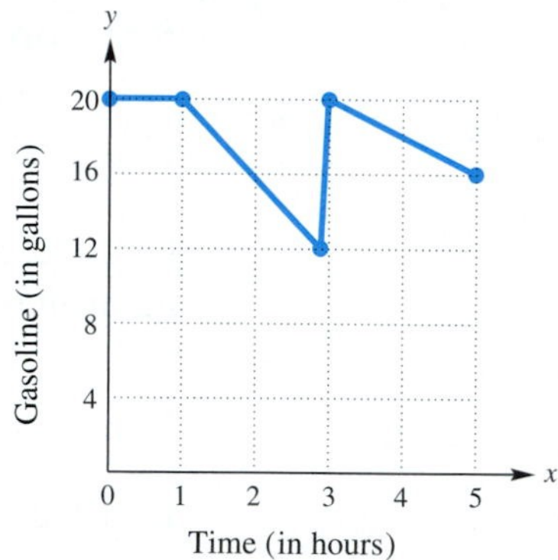

(a) Estimate how much gasoline was in the gas tank when $x = 3$.
(b) Interpret the graph.
(c) When did the car burn gasoline at the fastest rate?

For each piecewise-defined function, find **(a)** $f(-5)$, **(b)** $f(-1)$, **(c)** $f(0)$, *and* **(d)** $f(3)$.

5. $f(x) = \begin{cases} 2x & \text{if } x \le -1 \\ x - 1 & \text{if } x > -1 \end{cases}$

6. $f(x) = \begin{cases} x - 2 & \text{if } x < 3 \\ 5 - x & \text{if } x \ge 3 \end{cases}$

7. $f(x) = \begin{cases} 2 + x & \text{if } x < -4 \\ -x & \text{if } -4 \le x \le 2 \\ 3x & \text{if } x > 2 \end{cases}$

8. $f(x) = \begin{cases} -2x & \text{if } x < -3 \\ 3x - 1 & \text{if } -3 \le x \le 2 \\ -4x & \text{if } x > 2 \end{cases}$

Graph each piecewise-defined function by hand.

9. $f(x) = \begin{cases} x - 1 & \text{if } x \le 3 \\ 2 & \text{if } x > 3 \end{cases}$

10. $f(x) = \begin{cases} 6 - x & \text{if } x \le 3 \\ 3x - 6 & \text{if } x > 3 \end{cases}$

11. $f(x) = \begin{cases} 4 - x & \text{if } x < 2 \\ 1 + 2x & \text{if } x \ge 2 \end{cases}$

12. $f(x) = \begin{cases} 2x + 1 & \text{if } x \ge 0 \\ x & \text{if } x < 0 \end{cases}$

13. $f(x) = \begin{cases} 2 + x & \text{if } x < -4 \\ -x & \text{if } -4 \le x \le 5 \\ 3x & \text{if } x > 5 \end{cases}$

14. $f(x) = \begin{cases} -2x & \text{if } x < -3 \\ 3x - 1 & \text{if } -3 \le x \le 2 \\ -4x & \text{if } x > 2 \end{cases}$

15. $f(x) = \begin{cases} -\frac{1}{2}x^2 + 2 & \text{if } x \le 2 \\ \frac{1}{2}x & \text{if } x > 2 \end{cases}$

16. $f(x) = \begin{cases} x^3 + 5 & \text{if } x \le 0 \\ -x^2 & \text{if } x > 0 \end{cases}$

17. $f(x) = \begin{cases} 2x & \text{if } -5 \le x < -1 \\ -2 & \text{if } -1 \le x < 0 \\ x^2 - 2 & \text{if } 0 \le x \le 2 \end{cases}$

18. $f(x) = \begin{cases} .5x^2 & \text{if } -4 \le x \le -2 \\ x & \text{if } -2 < x < 2 \\ x^2 - 4 & \text{if } 2 \le x \le 4 \end{cases}$

19. $f(x) = \begin{cases} x^3 + 3 & \text{if } -2 \le x \le 0 \\ x + 3 & \text{if } 0 < x < 1 \\ 4 + x - x^2 & \text{if } 1 \le x \le 3 \end{cases}$

20. $f(x) = \begin{cases} -2x & \text{if } -3 \le x < -1 \\ x^2 + 1 & \text{if } -1 \le x \le 2 \\ \frac{1}{2}x^3 + 1 & \text{if } 2 < x \le 3 \end{cases}$

Match each piecewise-defined function with its calculator graph in choices A–D.

21. $f(x) = \begin{cases} x^2 - 4 & \text{if } x \ge 0 \\ -x + 5 & \text{if } x < 0 \end{cases}$

22. $g(x) = \begin{cases} |x - 4| & \text{if } x \ge -1 \\ -x^2 & \text{if } x < -1 \end{cases}$

23. $h(x) = \begin{cases} 6 & \text{if } x \ge 0 \\ -6 & \text{if } x < 0 \end{cases}$

24. $k(x) = \begin{cases} \sqrt{x} & \text{if } x \ge 0 \\ -x^2 & \text{if } x < 0 \end{cases}$

A.

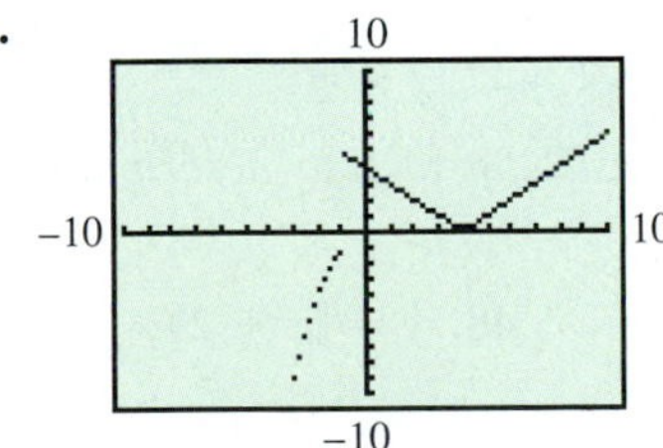

Dot mode

B.

10
−10
10
−10

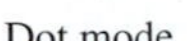

Dot mode

C.

10
−10
10
−10

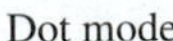

Dot mode

D.

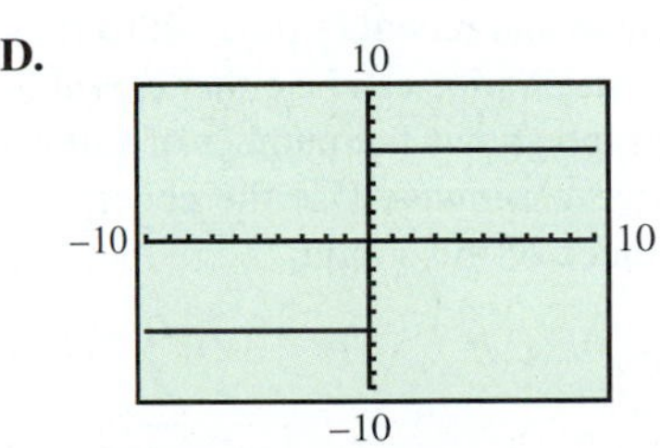

Dot mode

Use a graphing calculator in dot mode and the window indicated to graph each piecewise-defined function.

25. Exercise 9, window $[-4, 6]$ by $[-2, 4]$

26. Exercise 10, window $[-2, 8]$ by $[-2, 10]$

27. Exercise 11, window $[-4, 6]$ by $[-2, 8]$

28. Exercise 12, window $[-5, 4]$ by $[-3, 8]$

29. Exercise 13, window $[-12, 12]$ by $[-6, 20]$

30. Exercise 14, window $[-6, 6]$ by $[-10, 8]$

31. Exercise 15, window $[-5, 6]$ by $[-2, 4]$

32. Exercise 16, window $[-3, 4]$ by $[-3, 6]$

Write a formula for a piecewise-defined function f for each graph shown. Give the domain and range.

33.

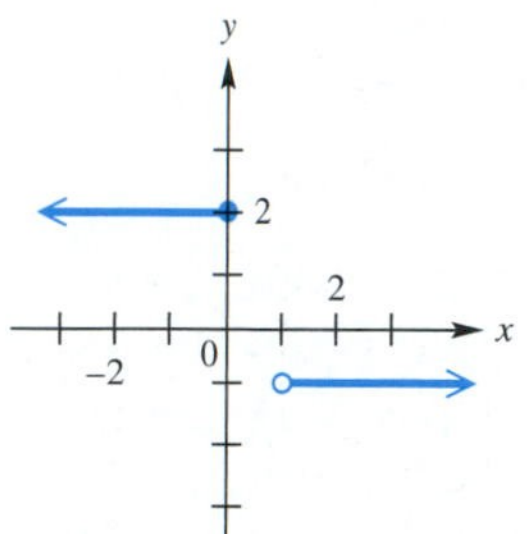

34.

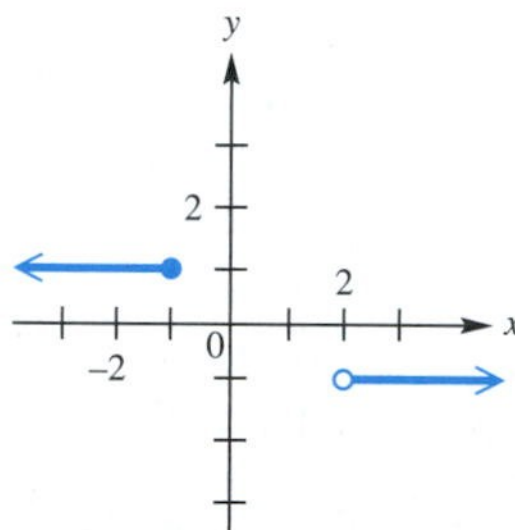

35.

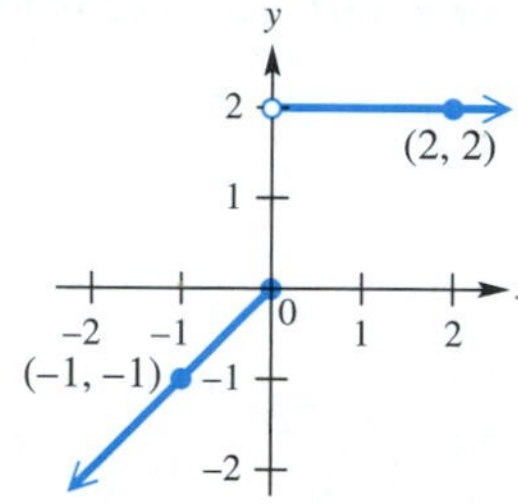

36.

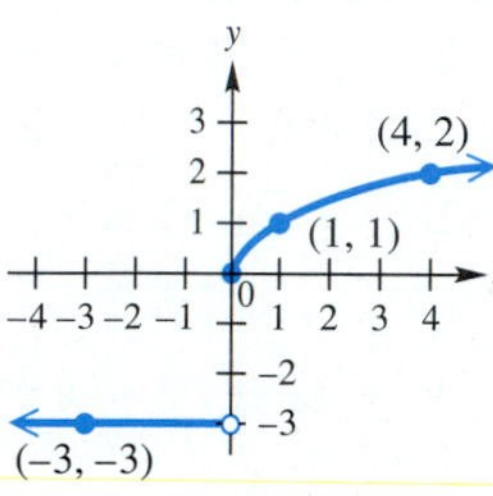

37.

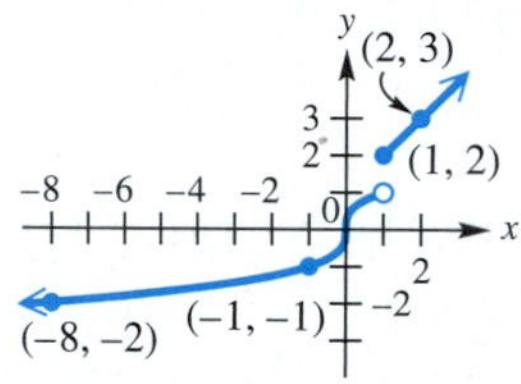

38.

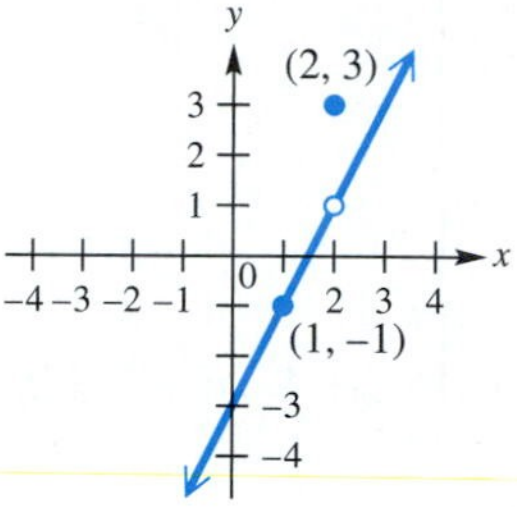

39. Why is the following not a piecewise-defined function?

$$f(x) = \begin{cases} x + 7 & \text{if } x \leq 4 \\ x^2 & \text{if } x \geq 4 \end{cases}$$

40. Consider the "function" defined in Exercise 39. Suppose you were asked to find $f(4)$. How would you respond?

Concept Check *Describe how the graph of the given function can be obtained from the graph of* $y = [\![x]\!]$.

41. $y = [\![x]\!] - 1.5$

42. $y = [\![-x]\!]$

43. $y = -[\![x]\!]$

44. $y = [\![x + 2]\!]$

Use a graphing calculator in dot mode with window $[-5, 5]$ *by* $[-3, 3]$ *to graph each equation. (Refer to your descriptions in Exercises 41–44.)*

45. $y = [\![x]\!] - 1.5$

46. $y = [\![-x]\!]$

47. $y = -[\![x]\!]$

48. $y = [\![x + 2]\!]$

(Modeling) *Solve each problem.*

49. ***Flow Rates*** A water tank has an inlet pipe with a flow rate of 5 gallons per minute and an outlet pipe with a flow rate of 3 gallons per minute. A pipe can be either closed or completely open. The graph shows the number of gallons of water in the tank after x minutes. Use the concept of slope to interpret each piece of this graph.

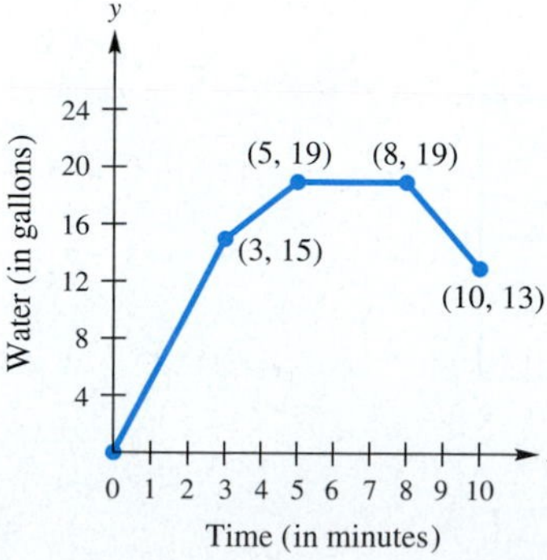

50. ***Shoe Size*** Professional basketball player Shaquille O'Neal is 7 feet, 1 inch, tall and weighs 325 pounds. The table lists his shoe sizes at certain ages.

Age	20	21	22	23
Shoe Size	19	20	21	22

Source: USA Today.

(a) Write a formula that gives his shoe size y at age $x = 20$, 21, 22, and 23.

(b) Suppose that after age 23 his shoe size did not change. Sketch a graph of a continuous, piecewise-defined function f that models his shoe size from ages 20 to 26.

51. ***Insulin Level*** When a diabetic takes long-acting insulin, it reaches its peak effect on the blood sugar level in about 3 hours. This effect remains fairly constant for 5 hours, then declines, and is very low until the next injection. In a typical patient, the level of insulin might be given by the function defined by

$$f(t) = \begin{cases} 40t + 100 & \text{if } 0 \le t \le 3 \\ 220 & \text{if } 3 < t \le 8 \\ -80t + 860 & \text{if } 8 < t \le 10 \\ 60 & \text{if } 10 < t \le 24 \end{cases},$$

where $f(t)$ is the blood sugar level, in appropriate units, at time t measured in hours from the injection time. Chuck takes his insulin at 6 A.M. Find his blood sugar level at

(a) 7 A.M. **(b)** 9 A.M.
(c) 10 A.M. **(d)** noon
(e) 2 P.M. **(f)** 5 P.M.
(g) midnight.
(h) Graph $y = f(t)$ for $0 \le t \le 18$.
(i) Why is it reasonable that f is continuous on its domain?

52. ***Snow Depth*** The snow depth in Michigan's Isle Royale National Park varies throughout the winter. In a typical winter, the snow depth in inches is approximated by the function defined by

$$f(x) = \begin{cases} 6.5x & \text{if } 0 \le x \le 4 \\ -5.5x + 48 & \text{if } 4 < x \le 6 \\ -30x + 195 & \text{if } 6 < x \le 6.5 \end{cases},$$

where x represents the time in months, with $x = 0$ corresponding to the beginning of October, $x = 1$ to the beginning of November, and so on.

(a) Graph the function for $0 \le x \le 6.5$.
(b) In what month is the snow the deepest? What is the greatest snow depth?
(c) In what months does the snow begin and end?

53. ***Rabies Cases*** The table lists the approximate number of animal rabies cases in the United States from 1997 to 2001.

Year	1997	1998	1999	2000	2001
Cases	8100	7400	6700	6900	7100

Source: U.S. Census Bureau.

(a) Describe the change in the data from one year to the next.

(b) Determine a continuous piecewise-defined function f that approximates the data. Let $x = 0$ correspond to the year 1997.

54. ***Minimum Wage*** The table lists the federal minimum wage rates in current dollars for the years 1981–2005. Sketch a graph of the data as a piecewise-defined function. (Assume that wages take effect on Jan. 1 of the year in the first year of the interval.)

Year(s)	Wage
1981–89	\$3.35
1990	\$3.80
1991–95	\$4.25
1996	\$4.75
1997–2003	\$5.15
2004	\$5.50
2005	\$6.50

Source: U.S. Bureau of Labor Statistics.

55. ***Violent Crimes*** Using interval notation, the table lists the numbers of victims of violent crime per 1000 people in 2003 by age group.

Age	Crime Rate
[12, 16)	52
[16, 20)	53
[20, 25)	44
[25, 35)	27
[35, 50)	19
[50, 65)	10
[65, 90)	2

Source: U.S. Department of Justice.

(a) Sketch the graph of a piecewise-defined function that models the data, where x represents age.

(b) Discuss the impact that age has on the likelihood of being a victim of a violent crime.

56. ***New Homes*** Using interval notation, the table on the next page lists the numbers in millions of houses built for various time intervals from 1950 to 2000.

Year	Houses
[1950, 1960)	13.6
[1960, 1970)	16.1
[1970, 1980)	11.6
[1980, 1990)	9.9
[1990, 2000)	11.0

Source: U.S. Census Bureau, *American Housing Survey.*

(a) Sketch the graph of a piecewise-defined function that models the data, where x represents the year.

(b) Discuss the trends in housing starts between 1950 and 2000.

57. ***Cellular Phone Bills*** Suppose that the charges for a cellular phone call are \$.50 for the first minute and \$.25 for each additional minute. Assume that a fraction of a minute is rounded up.

(a) Determine the cost of a phone call lasting 3.5 minutes.

(b) Find a formula for a function f that computes the cost of a telephone call x minutes long, where $0 < x \leq 5$. (*Hint:* Express f as a piecewise-defined function.)

58. ***Lumber Costs*** Lumber that is used to frame walls of houses is frequently sold in multiples of 2 feet. If the length of a board is not exactly a multiple of 2 feet, there is often no charge for the additional length. For example, if a board measures at least 8 feet, but less than 10 feet, then the consumer is charged for only 8 feet.

(a) Suppose that the cost of lumber is \$.80 every 2 feet. Find a formula for a function f that computes the cost of a board x feet long for $6 \leq x \leq 18$.

(b) Use a graphing calculator to graph f.

(c) Determine the costs of boards with lengths of 8.5 feet and 15.2 feet.

59. ***Water in a Tank*** Sketch a graph that depicts the amount of water in a 100-gallon tank. The tank is initially empty and then filled at a rate of 5 gallons per minute. Immediately after it is full, a pump is used to empty the tank at 2 gallons per minute.

60. ***Distance from Home*** Sketch a graph showing the distance a person is from home after x hours if he or she drives on a straight road at 40 mph to a park 20 miles away, remains at the park for 2 hours, and then returns home at a speed of 20 mph.

61. ***Cost to Send a Package*** An express-mail company charges \$25 for a package weighing up to 2 pounds. For each additional pound or fraction of a pound, there is an additional charge of \$3. Let $D(x)$ represent the cost to send a package weighing x pounds. Graph $D(x)$ for x in the interval $(0, 6]$.

62. ***Light-Vehicle Market Share*** The light-vehicle market share (in percent) in the United States for domestic cars is shown in the graph. Let $x = 3$ represent 1993, $x = 6$ represent 1996, and so on.

Light-Vehicle Market Share

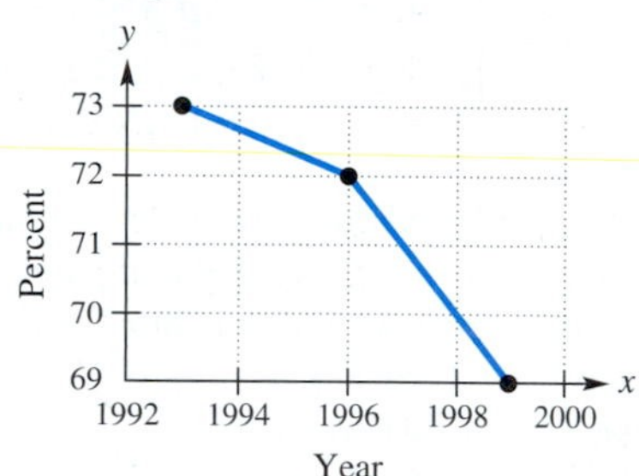

Source: J.D. Power & Associates.

(a) Use the points on the graph to write equations for the line segments in the intervals $[3, 6]$ and $(6, 9]$.

(b) Define $f(x)$ for the piecewise-defined function.

2.6 Operations and Composition

Operations on Functions ■ The Difference Quotient ■ Composition of Functions ■ Applications of Operations and Composition

Operations on Functions

Just as we add, subtract, multiply, and divide real numbers, we can also perform these operations on functions.

Operations on Functions

Given two functions f and g, for all values of x for which both $f(x)$ and $g(x)$ are defined, the functions $f + g$, $f - g$, fg, and $\frac{f}{g}$ are defined as follows:

Sum	$(f + g)(x) = f(x) + g(x)$
Difference	$(f - g)(x) = f(x) - g(x)$
Product	$(fg)(x) = f(x) \cdot g(x)$
Quotient	$\left(\frac{f}{g}\right)(x) = \frac{f(x)}{g(x)}, \quad g(x) \neq 0.$

The domains of $f + g$, $f - g$, and fg include all real numbers in the intersection of the domains of f and g, while the domain of $\frac{f}{g}$ includes those real numbers in the intersection of the domains of f and g for which $g(x) \neq 0$.

GCM **EXAMPLE 1** **Using the Operations on Functions**

Let $f(x) = x^2 + 1$ and $g(x) = 3x + 5$. Perform the operations.

(a) $(f + g)(1)$ **(b)** $(f - g)(-3)$ **(c)** $(fg)(5)$ **(d)** $\left(\frac{f}{g}\right)(0)$

Analytic Solution

(a) Since $f(1) = 2$ and $g(1) = 8$, use the preceding definition to get

$$\begin{aligned}(f + g)(1) &= f(1) + g(1) \\ &= 2 + 8 \\ &= 10.\end{aligned}$$

(b) $$\begin{aligned}(f - g)(-3) &= f(-3) - g(-3) \\ &= 10 - (-4) \\ &= 14\end{aligned}$$

(c) $(fg)(5) = f(5) \cdot g(5) = 26 \cdot 20 = 520$

(d) $\left(\frac{f}{g}\right)(0) = \frac{f(0)}{g(0)} = \frac{1}{5}$

Graphing Calculator Solution

Figure 64(a) shows f entered as Y_1 and g as Y_2.

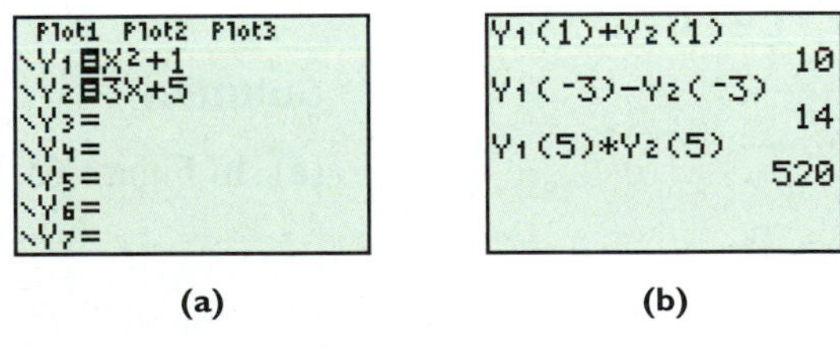

(a) (b)

FIGURE 64

Because the calculator is capable of function notation, we can calculate parts (a)–(c) as shown in Figure 64(b). Part (d) can be found similarly.

GCM EXAMPLE 2 Using the Operations on Functions

Let $f(x) = 8x - 9$ and $g(x) = \sqrt{2x - 1}$. Perform the operations.

(a) $(f + g)(x)$ **(b)** $(f - g)(x)$ **(c)** $(fg)(x)$ **(d)** $\left(\frac{f}{g}\right)(x)$

(e) Give the domains of f, g, $f + g$, $f - g$, fg, and $\frac{f}{g}$.

Solution

(a) $(f + g)(x) = f(x) + g(x) = 8x - 9 + \sqrt{2x - 1}$

(b) $(f - g)(x) = f(x) - g(x) = 8x - 9 - \sqrt{2x - 1}$

(c) $(fg)(x) = f(x) \cdot g(x) = (8x - 9)\sqrt{2x - 1}$

(d) $\left(\frac{f}{g}\right)(x) = \frac{f(x)}{g(x)} = \frac{8x - 9}{\sqrt{2x - 1}}$

(e) The domain of f is $(-\infty, \infty)$, while the domain of $g(x) = \sqrt{2x - 1}$ includes all real numbers that satisfy $2x - 1 \geq 0$; the domain of g is the interval $\left[\frac{1}{2}, \infty\right)$. The domain of $f + g$, $f - g$, and fg is thus $\left[\frac{1}{2}, \infty\right)$. With $\frac{f}{g}$, the denominator cannot be 0, so the value $\frac{1}{2}$ is excluded from the domain. The domain of $\frac{f}{g}$ is $\left(\frac{1}{2}, \infty\right)$. ■

EXAMPLE 3 Evaluating Combinations of Functions

If possible, use the given representations of functions f and g to evaluate

$$(f + g)(4), \quad (f - g)(-2), \quad (fg)(1), \quad \text{and} \quad \left(\frac{f}{g}\right)(0).$$

(a) 

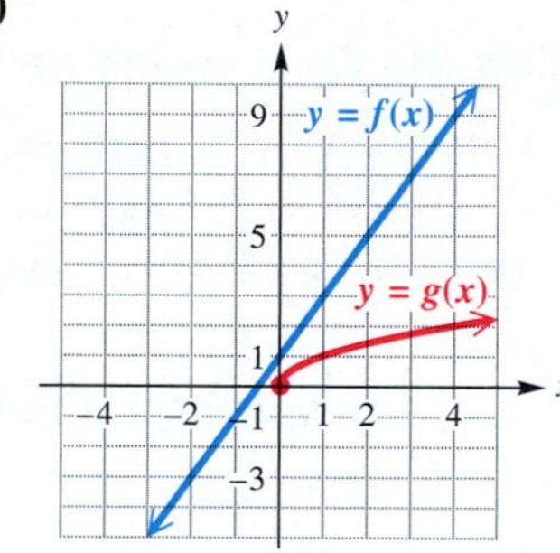

FIGURE 65

(b)

x	$f(x)$	$g(x)$
−2	−3	undefined
0	1	0
1	3	1
4	9	2

(c) $f(x) = 2x + 1, \quad g(x) = \sqrt{x}$

Solution

(a) In Figure 65, $f(4) = 9$ and $g(4) = 2$. Thus,

$$(f + g)(4) = f(4) + g(4) = 9 + 2 = 11.$$

Although $f(-2) = -3$ in Figure 65, $g(-2)$ is undefined because -2 is not in the domain of g. Thus $(f - g)(-2)$ is undefined.

The domains of f and g include 1, so

$$(fg)(1) = f(1) \cdot g(1) = 3 \cdot 1 = 3.$$

The graph of g intersects the origin, so $g(0) = 0$. Thus, $\left(\frac{f}{g}\right)(0)$ is undefined.

(b) From the table, $f(4) = 9$ and $g(4) = 2$. As in part (a),

$$(f + g)(4) = f(4) + g(4) = 9 + 2 = 11.$$

In the table $g(-2)$ is undefined, so $(f - g)(-2)$ is also undefined. Similarly,

$$(fg)(1) = f(1) \cdot g(1) = 3 \cdot 1 = 3,$$

and $$\left(\frac{f}{g}\right)(0) = \frac{f(0)}{g(0)} \text{ is undefined, since } g(0) = 0.$$

(c) Use the formulas $f(x) = 2x + 1$ and $g(x) = \sqrt{x}$.

Use the order of operations.

$$(f + g)(4) = f(4) + g(4) = (2 \cdot 4 + 1) + \sqrt{4} = 9 + 2 = 11$$

$$(f - g)(-2) = f(-2) - g(-2) = [2(-2) + 1] - \sqrt{-2} \text{ is undefined.}$$

$$(fg)(1) = f(1) \cdot g(1) = (2 \cdot 1 + 1)\sqrt{1} = 3(1) = 3$$

$$\left(\frac{f}{g}\right)(0) = \frac{f(0)}{g(0)} \text{ is undefined, since } g(0) = 0.$$ ■

The Difference Quotient

Suppose that the point P lies on the graph of $y = f(x)$ as in Figure 66, and suppose that h is a positive number. If we let $(x, f(x))$ denote the coordinates of P and $(x + h, f(x + h))$ denote the coordinates of Q, then the line joining P and Q has slope

$$m = \frac{f(x + h) - f(x)}{(x + h) - x} = \frac{\mathbf{f(x + h) - f(x)}}{\mathbf{h}}.$$

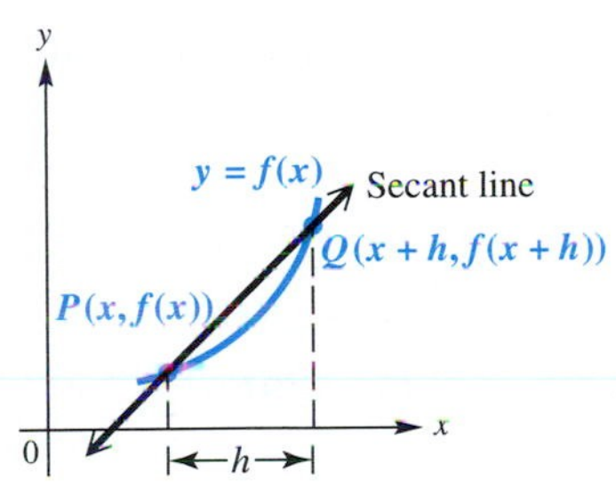
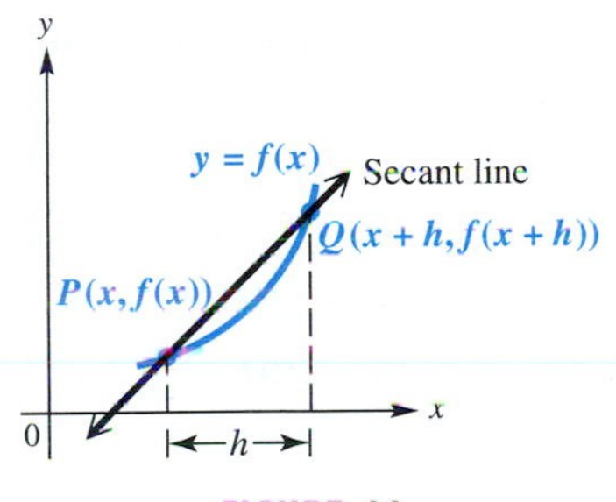

FIGURE 66

This expression, called the **difference quotient,** is important in the study of calculus.

Figure 66 shows the graph of the line PQ (called a *secant line*). As h approaches 0, the slope of this secant line approaches the slope of the line tangent to the curve at P. Important applications of this idea are developed in calculus, where the concepts of *limit* and *derivative* are investigated.

Looking Ahead to Calculus

The difference quotient is essential in the definition of the *derivative of a function* in calculus. The derivative provides a formula, in function form, for finding the slope of the tangent line to the graph of the function at a given point. Observe Figure 66 and notice that as h approaches 0, the line PQ approaches tangency at point P. Thus, the slope of PQ approaches the slope of the tangent line at P.

EXAMPLE 4 Finding the Difference Quotient

Let $f(x) = 2x^2 - 3x$. Find the difference quotient and simplify the expression.

Solution To find $f(x + h)$, replace x in $f(x)$ with $x + h$ to get

$$f(x + h) = 2(x + h)^2 - 3(x + h).$$

$$\frac{f(x + h) - f(x)}{h} = \frac{2(x + h)^2 - 3(x + h) - (2x^2 - 3x)}{h}$$

Be careful with signs.

$$= \frac{2(x^2 + 2xh + h^2) - 3x - 3h - 2x^2 + 3x}{h}$$ Square $x + h$; distributive property.

Remember the middle term when squaring $x + h$.

$$= \frac{2x^2 + 4xh + 2h^2 - 3x - 3h - 2x^2 + 3x}{h}$$ Distributive property

$$= \frac{4xh + 2h^2 - 3h}{h}$$ Combine terms.

$$= \frac{h(4x + 2h - 3)}{h}$$ Factor out h.

$$= 4x + 2h - 3$$ Divide. ■

CAUTION ***Notice that the function notation $f(x + h)$ is not the same as $f(x) + f(h)$.*** For $f(x) = 2x^2 - 3x$, as shown in Example 4,

$$f(x + h) = 2(x + h)^2 - 3(x + h) = 2x^2 + 4xh + 2h^2 - 3x - 3h,$$

but $$f(x) + f(h) = (2x^2 - 3x) + (2h^2 - 3h) = 2x^2 - 3x + 2h^2 - 3h.$$

These expressions differ by $4xh$.

Composition of Functions

The diagram in Figure 67 shows a function f that assigns, to each x in its domain, a value $f(x)$. Then another function g assigns, to each $f(x)$ in its domain, a value $g[f(x)]$. This two-step process takes an element x and produces a corresponding element $g[f(x)]$.

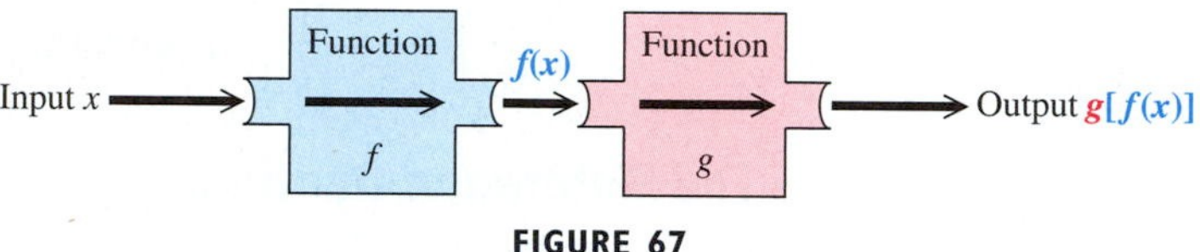

FIGURE 67

The function with y-values $g[f(x)]$ is called the *composition* of function g and f, written $\boldsymbol{g \circ f}$ and read "**g of f.**"

Composition of Functions

If f and g are functions, then the **composite function,** or **composition,** of g and f is

$$(g \circ f)(x) = g[f(x)]$$

for all x in the domain of f such that $f(x)$ is in the domain of g.

As a real-life example of how composite functions occur, consider the following retail situation:

> *A* \$40 *pair of blue jeans is on sale for* 25% *off. If you purchase the jeans before noon, the retailer offers an additional* 10% *off. What is the final sale price of the blue jeans?*

You might be tempted to say that the jeans are 35% off and calculate \$40(.35)=\$14, giving a final sale price of \$40 − \$14 = \$26 for the jeans. ***This is not correct.*** To find the final sale price, we must first find the price after taking 25% off and then take an additional 10% off *that* price.

\$40(.25) = \$10, giving a sale price of \$40 − \$10 = **\$30**. — Take 25% off original price.

\$30(.10) = \$3, giving a ***final sale price*** of \$30 − \$3 = **\$27**. — Take additional 10% off.

This is the idea behind composition of functions.

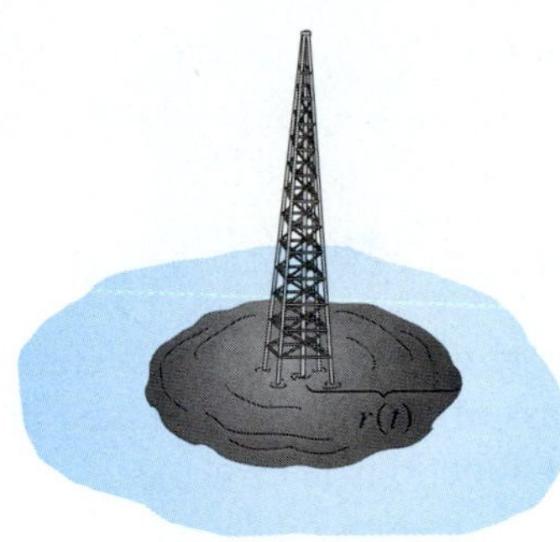

As another example, suppose an oil well off the California coast is leaking, spreading oil in a circular layer over the water's surface. (See the figure and the accompanying table in the margin.) At any time t, in minutes, after the beginning of the leak, the radius of the circular oil slick is $r(t) = 5t$ feet. Since $A(r) = \pi r^2$ gives the area of a circle of radius r, the area can be expressed as a function of time by substituting $5t$ for r in $A(r) = \pi r^2$ to get

$$A(r) = \pi r^2$$
$$A[r(t)] = \pi(5t)^2 = 25\pi t^2.$$

The function $A[r(t)]$ is a composite function of the functions A and r.

t	$r(t)$	$A[r(t)]$
1	5	$\pi(5)^2 = 25\pi$
2	10	$\pi(10)^2 = 100\pi$
3	15	$\pi(15)^2 = 225\pi$
t	$5t$	$\pi(5t)^2 = 25\pi t^2$

GCM **EXAMPLE 5** **Evaluating Composite Functions**

Let $f(x) = 2x - 1$ and $g(x) = \dfrac{4}{x-1}$. Perform the compositions.

(a) $(f \circ g)(2)$ **(b)** $(g \circ f)(-3)$

Analytic Solution

(a) First find $g(2)$. Since $g(x) = \frac{4}{x-1}$,

$$g(2) = \frac{4}{2-1} = \frac{4}{1} = 4.$$

Now find $(f \circ g)(2) = f[g(2)]$:

$$f[g(2)] = f(4) = 2(4) - 1 = 7.$$

(b) Since $f(-3) = 2(-3) - 1 = -7$,

$$(g \circ f)(-3) = g[f(-3)] = g(-7)$$
$$= \frac{4}{-7-1} = \frac{4}{-8}$$
$$= -\frac{1}{2}.$$

Don't confuse composition with multiplication.

Graphing Calculator Solution

In Figure 68, $Y_1 = f(x)$ and $Y_2 = g(x)$.

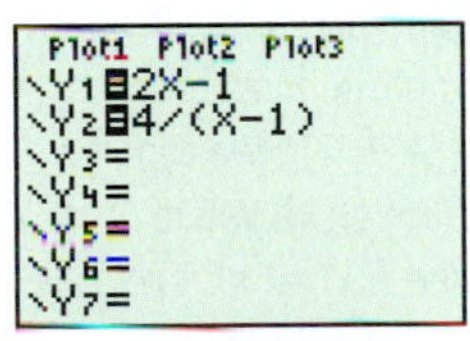

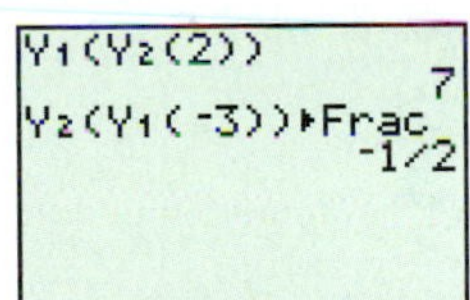

FIGURE 68

GCM **EXAMPLE 6** **Finding Composite Functions**

Let $f(x) = 4x + 1$ and $g(x) = 2x^2 + 5x$. Perform the compositions.

(a) $(g \circ f)(x)$ **(b)** $(f \circ g)(x)$

Solution

(a)

$$(g \circ f)(x) = g[f(x)] = g(4x+1) \qquad f(x) = 4x + 1$$
$$= 2(4x+1)^2 + 5(4x+1) \qquad g(x) = 2x^2 + 5x$$
$$= 2(16x^2 + 8x + 1) + 20x + 5 \qquad \text{Square } 4x+1\text{; distributive property.}$$
$$= 32x^2 + 16x + 2 + 20x + 5 \qquad \text{Distributive property}$$
$$= 32x^2 + 36x + 7 \qquad \text{Combine terms.}$$

$(x + y)^2 = x^2 + 2xy + y^2$

(b) $(f \circ g)(x) = f[g(x)]$ Definition of composition

$= f(2x^2 + 5x)$ $g(x) = 2x^2 + 5x$

$= 4(2x^2 + 5x) + 1$ $f(x) = 4x + 1$

$= 8x^2 + 20x + 1$ Distributive property ■

As Example 6 shows, it is not always true that $f \circ g = g \circ f$. However, the composite functions $f \circ g$ and $g \circ f$ are equal for a special class of functions, discussed in Section 5.1. In Example 6, the domain of both composite functions is $(-\infty, \infty)$.

CAUTION ***In general, the composite function $f \circ g$ is not the same as the product fg.*** For example, if $f(x) = 4x + 1$ and $g(x) = 2x^2 + 5x$ as defined in Example 6,

$$(f \circ g)(x) = 8x^2 + 20x + 1,$$

but

$$(fg)(x) = (4x + 1)(2x^2 + 5x) = 8x^3 + 22x^2 + 5x.$$

EXAMPLE 7 Finding Functions That Form a Given Composite Function

Suppose that $h(x) = (x^2 - 5)^3 - 4(x^2 - 5) + 3$. Find functions f and g so that $(f \circ g)(x) = h(x)$.

Solution Note the repeated quantity $x^2 - 5$ in $h(x)$. If we let $g(x) = x^2 - 5$ and $f(x) = x^3 - 4x + 3$, then

$$\begin{aligned}(f \circ g)(x) &= f[g(x)] \\ &= f(x^2 - 5) \\ &= (x^2 - 5)^3 - 4(x^2 - 5) + 3 \\ &= h(x).\end{aligned}$$

Other pairs of functions f and g also work. For instance, we can use

$$f(x) = (x - 5)^3 - 4(x - 5) + 3 \quad \text{and} \quad g(x) = x^2.$$ ■

Looking Ahead to Calculus

Finding the derivative of a function in calculus is called *differentiation.* To differentiate a composite function such as $h(x) = (3x + 2)^4$, we can interpret $h(x)$ as $(f \circ g)(x)$, where $g(x) = 3x + 2$ and $f(x) = x^4$. The *chain rule* allows us to differentiate composite functions.

Applications of Operations and Composition

There are many applications in business, economics, physics, and other fields that combine functions. For example, in manufacturing, the cost of producing a product usually consists of two parts: a *fixed cost* for designing the product, setting up a factory, training workers, and so on, and a *variable cost* per item for labor, materials, packaging, shipping, and so on.

A *linear cost function* has the form

$$C(x) = mx + b,$$ Cost function

where m represents the variable cost per item and b represents the fixed cost. The revenue from selling a product depends on the price per item and the number of items sold, as given by the *revenue function*,

$$R(x) = px,$$ Revenue function

where p is the price per item and $R(x)$ is the revenue from the sale of x items. The profit is described by the *profit function*, given by

$$P(x) = R(x) - C(x).$$ Profit function

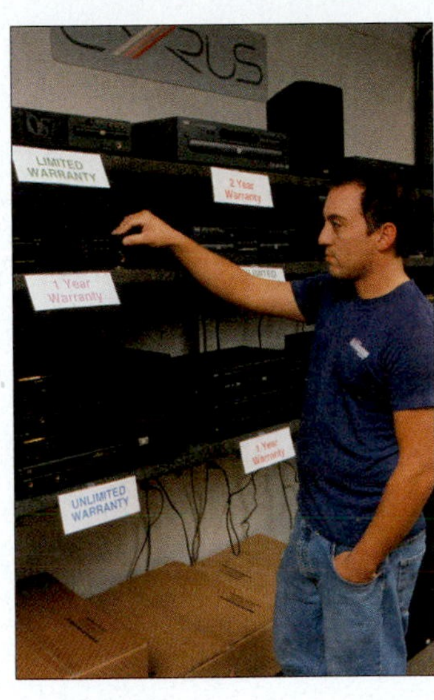

EXAMPLE 8 **Finding and Analyzing Cost, Revenue, and Profit**

Suppose that a businessman invests \$1500 as his fixed cost in a new venture that produces and sells a device that makes programming a VCR easier. Each such device costs \$100 to manufacture.

(a) Write a cost function for the product if x represents the number of devices produced. Assume that the function is linear.

(b) Find the revenue function if each device in part (a) sells for \$125.

(c) Give the profit function for the item in parts (a) and (b).

(d) How many items must be produced and sold before the company makes a profit?

Solution

(a) Since the cost function is linear, it will have the form $C(x) = mx + b$, with $m = 100$ and $b = 1500$. That is,

$$C(x) = 100x + 1500.$$

(b) If we let $p = 125$, then the revenue function is $R(x) = px = 125x$.

(c) Using the results of parts (a) and (b), we obtain the profit function

$$\begin{aligned} P(x) &= R(x) - C(x) \\ &= 125x - (100x + 1500) \\ &= 125x - 100x - 1500 \\ &= 25x - 1500. \end{aligned}$$

Be careful with signs.

(d) To make a profit, $P(x)$ must be positive. Set $P(x) = 25x - 1500 > 0$ and solve for x.

$$\begin{aligned} 25x - 1500 &> 0 \\ 25x &> 1500 && \text{Add 1500.} \\ x &> 60 && \text{Divide by 25.} \end{aligned}$$

At least 61 devices must be sold for the company to make a profit. ■

EXAMPLE 9 **Modeling the Surface Area of a Ball**

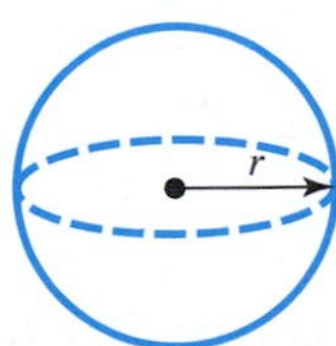

The formula for the surface area S of a sphere is $S = 4\pi r^2$, where r is the radius of the sphere. See the figure.

(a) Construct a model $S(r)$ that describes the amount of surface area gained if the radius r of a ball is increased by 2 inches.

(b) Determine the amount of extra material needed to manufacture a ball of radius 22 inches compared with a ball of radius 20 inches.

Solution

(a) In words, we have

surface area gained = larger surface area − smaller surface area.

This translates as a difference of functions: $S(r) = 4\pi(r + 2)^2 - 4\pi r^2$.

(b) $S(r) = 4\pi(r + 2)^2 - 4\pi r^2$ — Model from part (a)

$S(20) = 4\pi(20 + 2)^2 - 4\pi(20)^2$ — Let $r = 20$.

$= 1936\pi - 1600\pi$

$= 336\pi$

≈ 1056

Thus, it would take about 1056 square inches of extra material. A graph of

$$Y_1 = 4\pi(X + 2)^2 - 4\pi X^2$$

is shown in Figure 69. The display at the bottom agrees with the analytic solution.

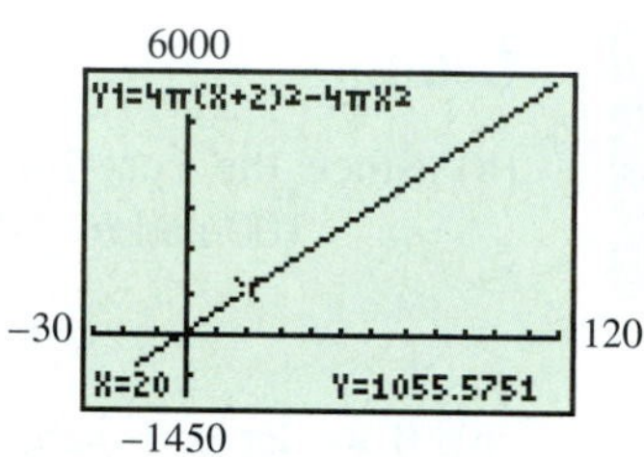

FIGURE 69

2.6 Exercises

Concept Check *Let $f(x) = x^2$ and $g(x) = 2x - 5$. Match each function in Group I with the correct expression in Group II.*

I		II	
1. $(f + g)(x)$	**2.** $(f - g)(x)$	**A.** $4x^2 - 20x + 25$	**B.** $x^2 - 2x + 5$
3. $(fg)(x)$	**4.** $\left(\frac{f}{g}\right)(x)$	**C.** $2x^2 - 5$	**D.** $\frac{x^2}{2x - 5}$
5. $(f \circ g)(x)$	**6.** $(g \circ f)(x)$	**E.** $x^2 + 2x - 5$	**F.** $2x^3 - 5x^2$

Let $f(x) = 4x^2 - 2x$ and $g(x) = 8x + 1$. Perform the composition or operation indicated.

7. $(f \circ g)(3)$ **8.** $(g \circ f)(-2)$ **9.** $(f \circ g)(x)$ **10.** $(g \circ f)(x)$ **11.** $(f + g)(3)$

12. $(f + g)(-5)$ **13.** $(fg)(4)$ **14.** $(fg)(-3)$ **15.** $\left(\frac{f}{g}\right)(-1)$ **16.** $\left(\frac{f}{g}\right)(4)$

17. $(f \circ g)(2)$ **18.** $(f \circ g)(-5)$ **19.** $(g \circ f)(2)$ **20.** $(g \circ f)(-5)$

For each pair of functions, **(a)** *find $(f + g)(x)$, $(f - g)(x)$, and $(fg)(x)$;* **(b)** *give the domains of the functions in part (a);* **(c)** *find $\frac{f}{g}$ and give its domain;* **(d)** *find $f \circ g$ and give its domain; and* **(e)** *find $g \circ f$ and give its domain.*

21. $f(x) = 4x - 1, g(x) = 6x + 3$

22. $f(x) = 9 - 2x, g(x) = -5x + 2$

23. $f(x) = |x + 3|, g(x) = 2x$

24. $f(x) = |2x - 4|, g(x) = x + 1$

25. $f(x) = \sqrt[3]{x + 4}, g(x) = x^3 + 5$

26. $f(x) = \sqrt[3]{6 - 3x}, g(x) = 2x^3 + 1$

27. $f(x) = \sqrt{x^2 + 3}, g(x) = x + 1$

28. $f(x) = \sqrt{2 + 4x^2}, g(x) = x$

Use the graph to evaluate each expression.

29. **(a)** $(f+g)(2)$
(b) $(f-g)(1)$
(c) $(fg)(0)$
(d) $\left(\frac{f}{g}\right)(1)$

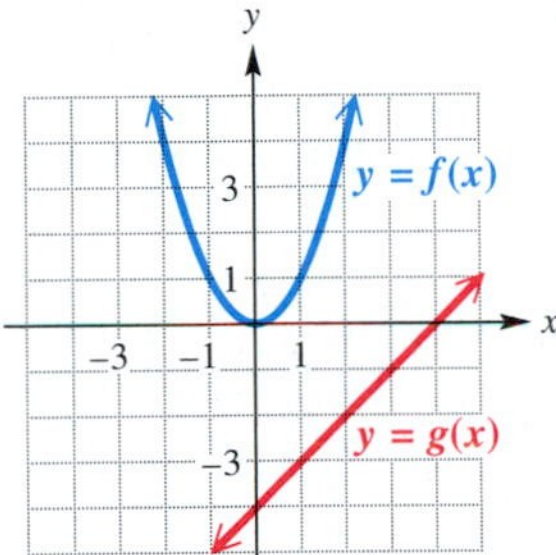

30. **(a)** $(f+g)(0)$
(b) $(f-g)(-1)$
(c) $(fg)(1)$
(d) $\left(\frac{f}{g}\right)(2)$

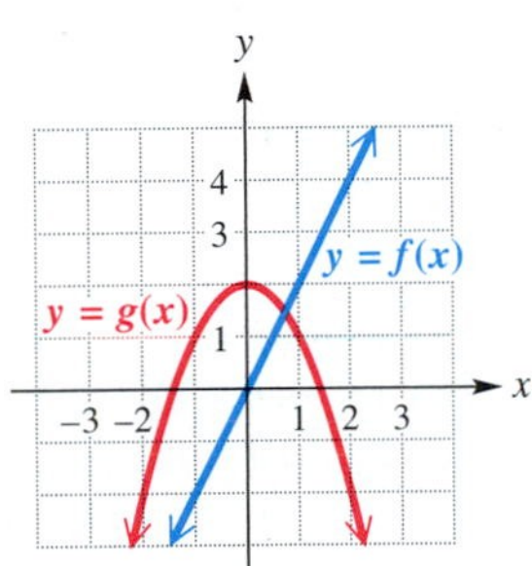

31. **(a)** $(f+g)(-1)$
(b) $(f-g)(-2)$
(c) $(fg)(0)$
(d) $\left(\frac{f}{g}\right)(2)$

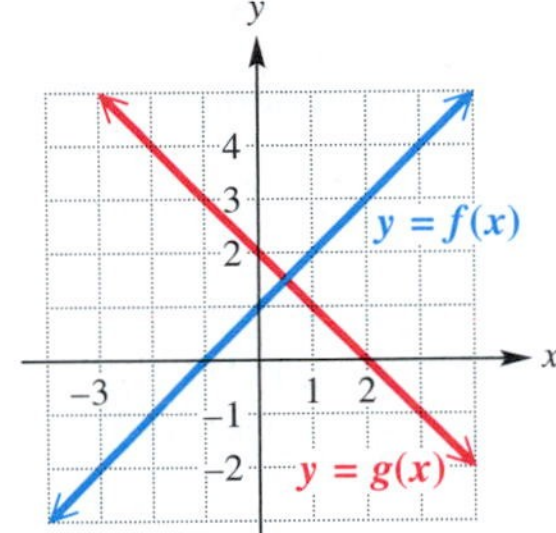

32. **(a)** $(f+g)(1)$
(b) $(f-g)(0)$
(c) $(fg)(-1)$
(d) $\left(\frac{f}{g}\right)(1)$

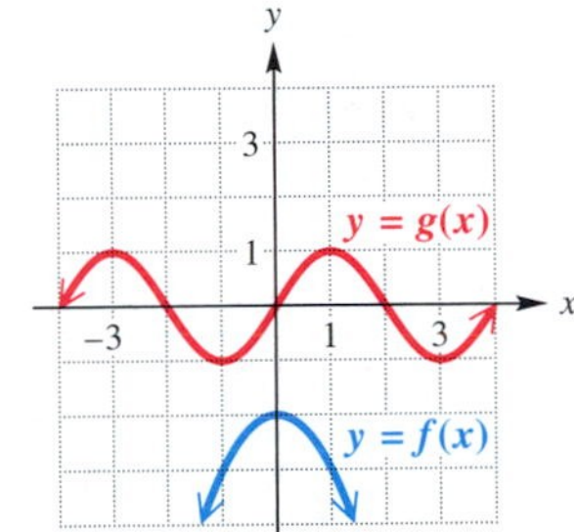

Use the table to evaluate each expression, if possible.

(a) $(f+g)(2)$ **(b)** $(f-g)(4)$ **(c)** $(fg)(-2)$ **(d)** $\left(\frac{f}{g}\right)(0)$

33.

x	$f(x)$	$g(x)$
−2	0	6
0	5	0
2	7	−2
4	10	5

34.

x	$f(x)$	$g(x)$
−2	−4	2
0	8	−1
2	5	4
4	0	0

35. Use the table in Exercise 33 to complete the following table:

x	$(f+g)(x)$	$(f-g)(x)$	$(fg)(x)$	$\left(\frac{f}{g}\right)(x)$
−2				
0				
2				
4				

36. Use the table in Exercise 34 to complete the following table:

x	$(f+g)(x)$	$(f-g)(x)$	$(fg)(x)$	$\left(\frac{f}{g}\right)(x)$
−2				
0				
2				
4				

Sodas Consumed by Adolescents *The graph shows the number of sodas (soft drinks) adolescents (age 12–19) drank per week from 1978 to 1996. $G(x)$ gives the number of sodas for girls, $B(x)$ gives the number of sodas for boys, and $T(x)$ gives the total number for both groups. Use the graph to do the following.*

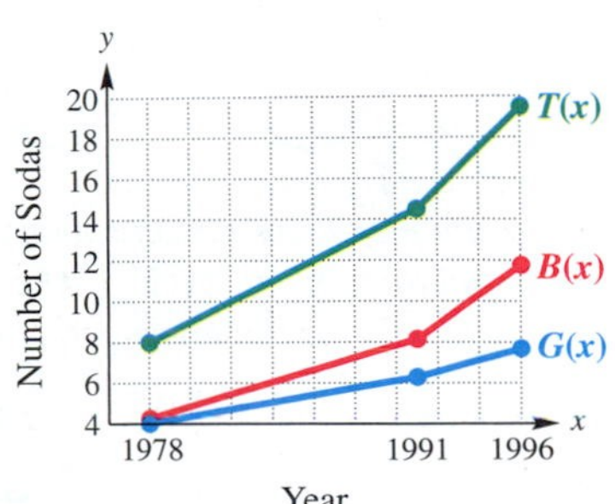

Source: U.S. Department of Agriculture.

37. Estimate $G(1996)$ and $B(1996)$, and use your results to estimate $T(1996)$.

38. Estimate $G(1991)$ and $B(1991)$, and use your results to estimate $T(1991)$.

39. Use the slopes of the line segments to decide in which period (1978–1991 or 1991–1996) the number of sodas per week increased more rapidly.

40. Explain how to obtain the graph of $T(x)$ by using the graphs of $B(x)$ and $G(x)$.

Science and Space/Technology Spending *The graph shows dollars (in billions) spent for general science and for space and other technologies in selected years. $G(x)$ represents the dollars spent for general science, and $S(x)$ the dollars spent for space and other technologies. $T(x)$ represents total expenditures for the two categories. Use the graph to answer the following.*

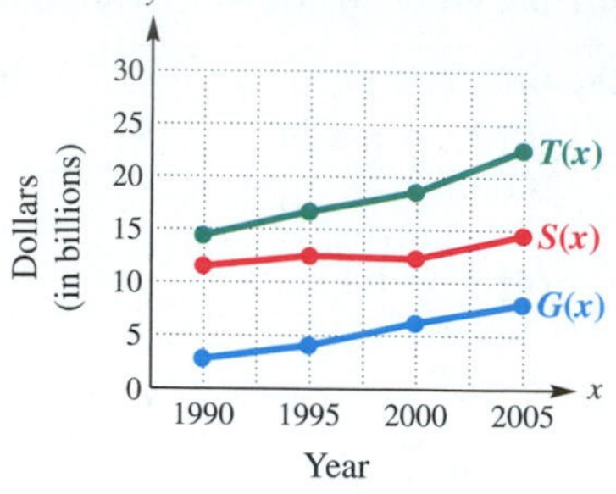

Source: U.S. Office of Management and Budget.

41. Estimate $(T-S)(2000)$. What does this result represent?

42. Estimate $(T-G)(2005)$. What does this result represent?

43. In which of the categories was spending almost static for several years? In which years did this occur?

44. In which period and which category did spending for $G(x)$ or $S(x)$ increase most?

Use the graph to evaluate each expression. (Hint: *Extend the ideas of Example 3.*)

45. (a) $(f \circ g)(4)$ **(b)** $(g \circ f)(3)$ **(c)** $(f \circ f)(2)$

46. (a) $(f \circ g)(2)$ **(b)** $(g \circ g)(0)$ **(c)** $(g \circ f)(4)$

47. (a) $(f \circ g)(1)$ **(b)** $(g \circ f)(-2)$ **(c)** $(g \circ g)(-2)$

48. (a) $(f \circ g)(-2)$ **(b)** $(g \circ f)(1)$ **(c)** $(f \circ f)(0)$

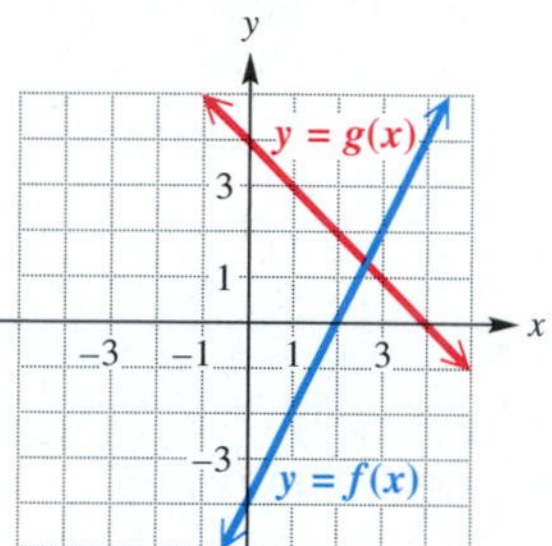

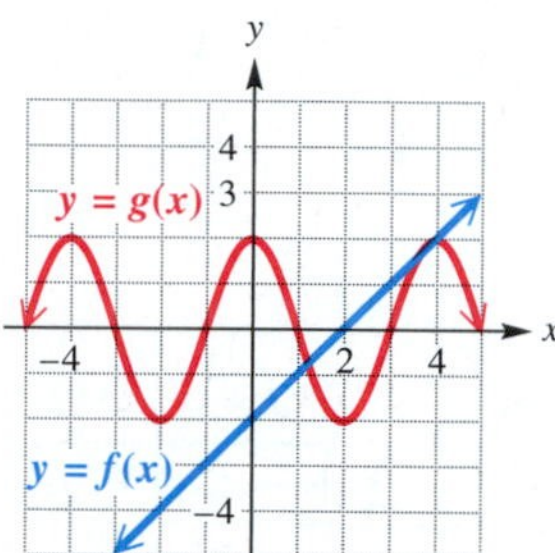

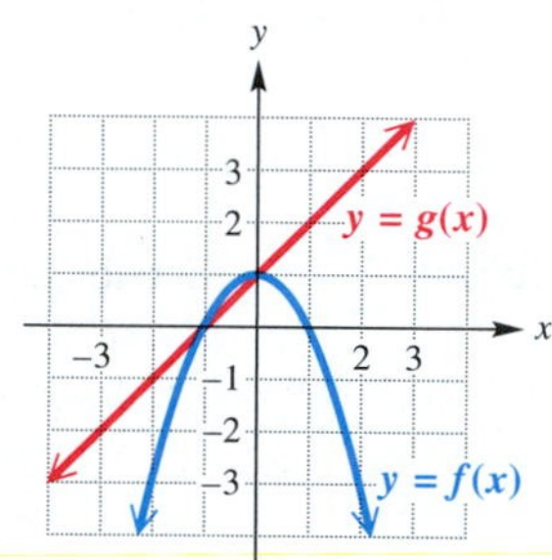

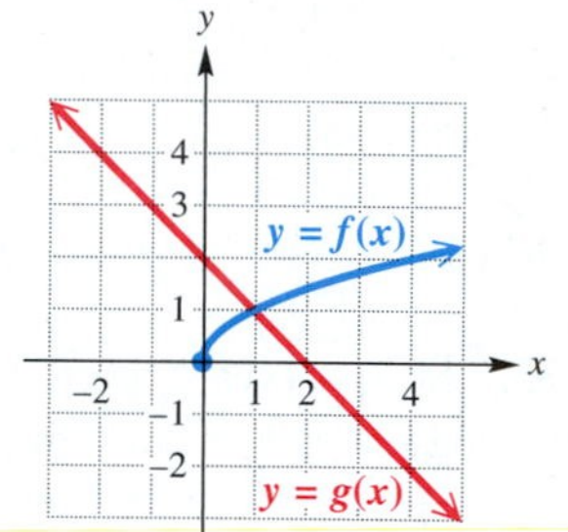

In Exercises 49 and 50, tables for functions f and g are given. Evaluate each expression, if possible.
(a) $(g \circ f)(1)$ **(b)** $(f \circ g)(4)$ **(c)** $(f \circ f)(3)$

49.

x	$f(x)$
1	4
2	3
3	1
4	2

x	$g(x)$
1	2
2	3
3	4
4	5

50.

x	$f(x)$
1	2
3	6
4	5
6	7

x	$g(x)$
2	4
3	2
5	6
7	0

51. Use the tables for f and g in Exercise 49 to complete the composition shown in the diagram.

3 —g→ ? —f→ ?

52. Use the tables for f and g in Exercise 50 to complete the composition shown in the diagram.

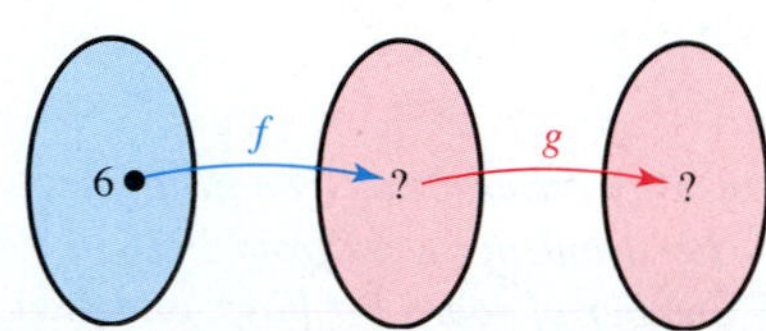

The graphing calculator screen on the left shows three functions: Y_1, Y_2, *and* Y_3. *The last of these,* Y_3, *is defined as* $Y_1 \circ Y_2$, *indicated by the notation* $Y_3 = Y_1(Y_2)$. *The table on the right shows selected values of* X, *along with the calculated values of* Y_3. *Predict the display for* Y_3 *for the given value of* X.

53. $X = -1$

54. $X = -2$

55. $X = 7$

56. $X = 8$

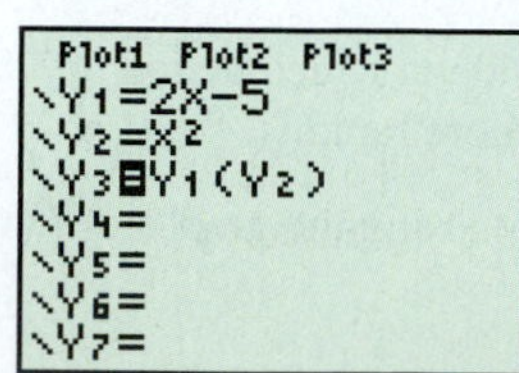

X	Y3
0	-5
1	-3
2	3
3	13
4	27
5	45
6	67

X=0

In Exercises 57–64, use $f(x)$ and $g(x)$ to find each composition. Identify its domain.
(a) $(f \circ g)(x)$ **(b)** $(g \circ f)(x)$ **(c)** $(f \circ f)(x)$

57. $f(x) = x^3,\ g(x) = x^2 + 3x - 1$

58. $f(x) = 2 - x,\ g(x) = \dfrac{1}{x^2}$

59. $f(x) = x^2,\ g(x) = \sqrt{1 - x}$

60. $f(x) = x + 2,\ g(x) = x^4 + x^2 - 3x - 4$

61. $f(x) = \dfrac{1}{x + 1},\ g(x) = 5x$

62. $f(x) = x + 4,\ g(x) = \sqrt{4 - x^2}$

63. $f(x) = 2x + 1,\ g(x) = 4x^3 - 5x^2$

64. $f(x) = \dfrac{x - 3}{2},\ g(x) = 2x + 3$

For certain pairs of functions f and g, $(f \circ g)(x) = x$ and $(g \circ f)(x) = x$. Show that this is true for the pairs in Exercises 65–68.

65. $f(x) = 4x + 2, g(x) = \dfrac{1}{4}(x - 2)$

66. $f(x) = -3x, g(x) = -\dfrac{1}{3}x$

67. $f(x) = \sqrt[3]{5x + 4}, g(x) = \dfrac{1}{5}x^3 - \dfrac{4}{5}$

68. $f(x) = \sqrt[3]{x + 1}, g(x) = x^3 - 1$

Functions such as the pairs in Exercises 65–68 are called inverse functions, *because the result of composition in both directions is the identity function. (Inverse functions will be discussed in detail in Section 5.1.)*

69. In a square viewing window, graph $y_1 = \sqrt[3]{x - 6}$ and $y_2 = x^3 + 6$, an example of a pair of inverse functions. Now graph $y_3 = x$. Describe how the graph of y_2 can be obtained from the graph of y_1, using the graph $y_3 = x$ as a basis for your description.

70. Repeat Exercise 69 for $y_1 = 5x - 3$ and $y_2 = \frac{1}{5}(x + 3)$.

Determine the difference quotient $\dfrac{f(x + h) - f(x)}{h}$ *(where $h \neq 0$) for each function f. Simplify completely.*

71. $f(x) = 4x + 3$

72. $f(x) = 5x - 6$

73. $f(x) = -6x^2 - x + 4$

74. $f(x) = \dfrac{1}{2}x^2 + 4x$

75. $f(x) = x^3$

76. $f(x) = -2x^3$

Consider the function h as defined. Find functions f and g such that $(f \circ g)(x) = h(x)$. (There are many possible ways to do this.)

77. $h(x) = (6x - 2)^2$

78. $h(x) = (11x^2 + 12x)^2$

79. $h(x) = \sqrt{x^2 - 1}$

80. $h(x) = (2x - 3)^3$

81. $h(x) = \sqrt{6x} + 12$

82. $h(x) = \sqrt[3]{2x + 3} - 4$

(Modeling) Cost/Revenue/Profit Analysis *For each situation, if x represents the number of items produced,* **(a)** *write a cost function,* **(b)** *find a revenue function if each item sells for the price given,* **(c)** *state the profit function,* **(d)** *determine analytically how many items must be produced before a profit is realized (assume whole numbers of items), and* **(e)** *support the result of part (d) graphically.*

83. The fixed cost is \$500, the cost to produce an item is \$10, and the selling price of the item is \$35.

84. The fixed cost is \$180, the cost to produce an item is \$11, and the selling price of the item is \$20.

85. The fixed cost is \$2700, the cost to produce an item is \$100, and the selling price of the item is \$280.

86. The fixed cost is \$1000, the cost to produce an item is \$200, and the selling price of the item is \$240.

(Modeling) *Solve each application of operations and composition of functions.*

87. *Volume of a Sphere* The formula for the volume of a sphere is $V = \frac{4}{3}\pi r^3$, where r represents the radius of the sphere.

(a) Construct a model representing the volume gained when the radius of a sphere of r inches is increased by 3 inches.

(b) Graph the model found in part (a), using y for V and x for r, in the window $[0, 10]$ by $[0, 1500]$.

(c) Use your calculator to graphically find the amount of volume gained when a sphere of 4-inch radius is increased to a 7-inch radius.

(d) Verify your result in part (c) analytically.

88. ***Surface Area of a Sphere*** Rework Example 9(a), but consider what happens when the radius is doubled (rather than increased by 2 inches).

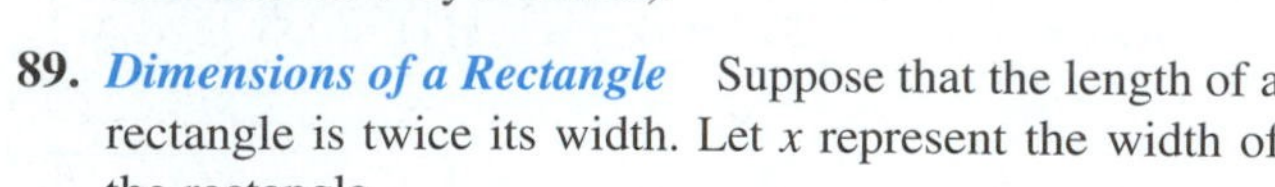

89. ***Dimensions of a Rectangle*** Suppose that the length of a rectangle is twice its width. Let x represent the width of the rectangle.

(a) Write a formula for the perimeter P of the rectangle in terms of x alone. Then use $P(x)$ notation to describe it as a function. What type of function is this?

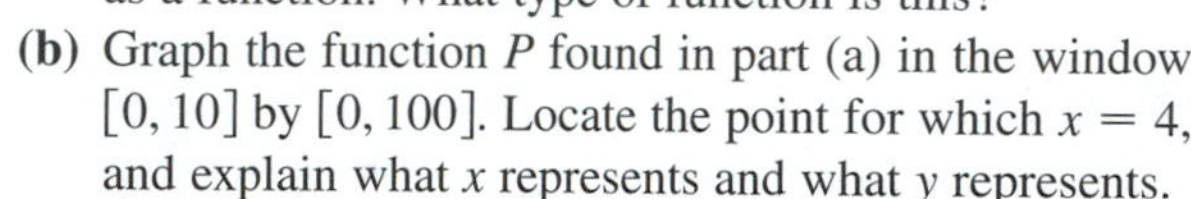

(b) Graph the function P found in part (a) in the window $[0, 10]$ by $[0, 100]$. Locate the point for which $x = 4$, and explain what x represents and what y represents.

(c) On the graph of P, locate the point with x-value 4. Then sketch a rectangle satisfying the conditions described earlier, and evaluate its perimeter if its width is this x-value. Use the standard perimeter formula. How does the result compare with the y-value shown on your screen?

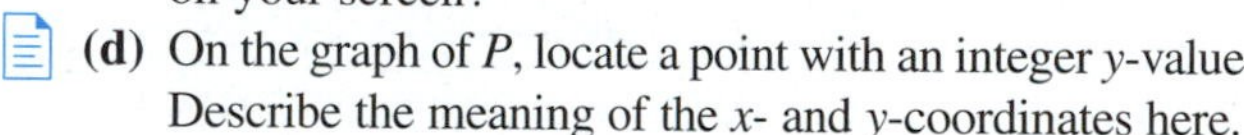

(d) On the graph of P, locate a point with an integer y-value. Describe the meaning of the x- and y-coordinates here.

90. ***Perimeter of a Square*** The perimeter x of a square with side length s is given by the formula

$$x = 4s.$$

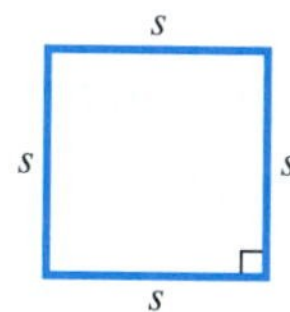

(a) Solve for s in terms of x.

(b) If y represents the area of this square, write y as a function of the perimeter x.

(c) Use the composite function of part (b) to analytically find the area of a square with perimeter 6.

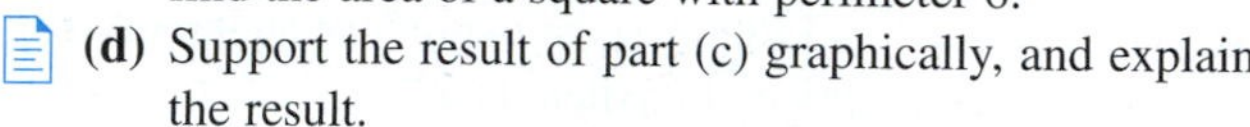

(d) Support the result of part (c) graphically, and explain the result.

91. ***Area of an Equilateral Triangle*** The area A of an equilateral triangle with sides of length x is given by

$$A(x) = \frac{\sqrt{3}}{4}x^2.$$

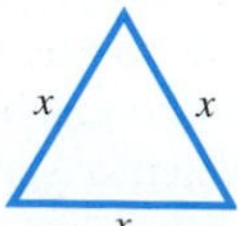

(a) Find $A(2x)$, the function representing the area of an equilateral triangle with sides of length twice the original length.

(b) Find analytically the area of an equilateral triangle with side length 16. Use the given formula for $A(x)$.

(c) Support the result of part (b) graphically.

92. ***Emission of Pollutants*** When a thermal inversion layer is over a city, pollutants cannot rise vertically, but are trapped below the layer and must disperse horizontally. Assume that a factory smokestack begins emitting a pollutant at 8 A.M. and that the pollutant disperses horizontally over a circular area. Let t represent the time in hours since the factory began emitting pollutants ($t = 0$ represents 8 A.M.) and assume that the radius of the circle of pollution is $r(t) = 2t$ miles. $A(r) = \pi r^2$ represents the area of a circle of radius r.

(a) Find $(A \circ r)(t)$.

(b) Interpret $(A \circ r)(t)$.

(c) What is the area of the circular region covered by the layer at noon?

(d) Support your result graphically.

93. ***Hollywood Movies*** From 1999 to 2003, the cost to produce and market Hollywood movies increased. In the table, f computes the average cost to produce a movie and g computes the average cost to market a movie, both in millions of dollars, in the year x.

x	1999	2000	2001	2002	2003
$f(x)$	52	55	48	59	64
$g(x)$	24	27	31	30	39

Source: Motion Picture Association of America.

(a) Make a table for a function h that computes the average cost to produce *and* market a movie in the year x.

(b) Write an equation that relates $f(x)$, $g(x)$, and $h(x)$.

94. ***Agriculture*** Crops and animals produced in the United States and sold domestically or abroad pump billions of dollars into the U.S. economy each year. In the table, x represents the year and f computes the total value of crops and animals produced in the United States and sold. The function g computes the total value of crops only. Values are given in billions of dollars.

x	1998	1999	2000	2001	2002
$f(x)$	195.9	187.9	194.1	201.5	190.4
$g(x)$	101.7	92.6	94.8	95.1	97.2

Source: Statistical Abstract of the United States, 2003.

(a) Make a table for a function h defined by $h(x) = f(x) - g(x)$.
(b) Interpret what h computes.

95. ***Acid Rain*** A common air pollutant responsible for acid rain is sulfur dioxide (SO_2). Emissions of SO_2 from burning coal during year x are computed by $f(x)$ in the table. Emissions of SO_2 from burning oil are computed by $g(x)$. Amounts are given in millions of tons.

x	1860	1900	1940	1970	2000
$f(x)$	2.4	12.6	24.2	32.4	55.0
$g(x)$	0	.2	2.3	17.6	23.0

Source: Freedman, B., *Environmental Ecology*, Second Edition, Academic Press, 1995.

(a) Evaluate $(f + g)(1970)$.
(b) Interpret $(f + g)(x)$.
(c) Make a table for $(f + g)(x)$.

96. ***Methane Emissions*** The greenhouse gas methane lets sunlight into the atmosphere, but blocks heat from escaping the earth's atmosphere. Methane is a by-product of burning fossil fuels. In the table, f models the predicted methane emissions in millions of tons produced by developed countries during year x. The function g models the same emissions for developing countries.

x	1990	2000	2010	2020	2030
$f(x)$	27	28	29	30	31
$g(x)$	5	7.5	10	12.5	15

Source: Nilsson, A., *Greenhouse Earth,* John Wiley and Sons, 1992.

(a) Make a table for a function h that models the total predicted methane emissions for developed *and* developing countries.
(b) Write an equation that relates $f(x)$, $g(x)$, and $h(x)$.

97. ***China's Energy Production*** Predicted energy production in China is shown in the figure. The function f computes total coal production, the function g total coal *and* oil production. Energy units are in million metric tons of oil equivalent (Mtoe). Let the function h compute China's oil production. (*Source:* Priddle, R., *Coal in the Energy Supply of China*, Coal Energy Advisory Board/International Energy Agency, 1999.)

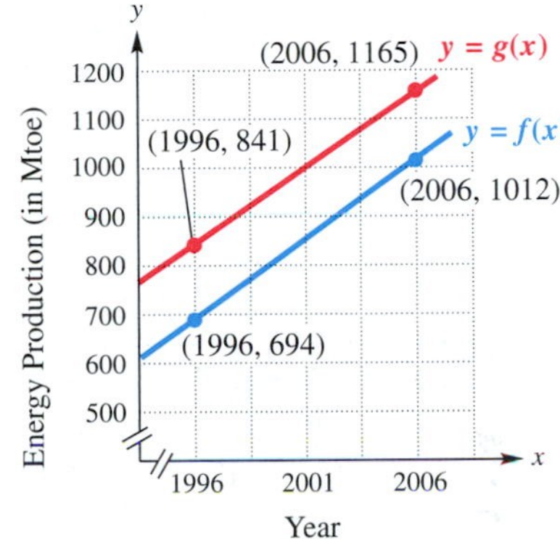

(a) Write an equation that relates $f(x)$, $g(x)$, and $h(x)$.
(b) Evaluate $h(1996)$ and $h(2006)$.
(c) Determine a formula for h. (*Hint:* h is linear.)

98. ***AIDS*** During the early years of the AIDS epidemic, cases and cumulative deaths reported for selected years x were modeled quite well by quadratic functions. For 1982–1994, the numbers of AIDS cases are modeled by

$$f(x) = 3200(x - 1982)^2 + 1586$$

and the numbers of deaths are modeled by

$$g(x) = 1900(x - 1982)^2 + 619.$$

Year	Cases	Deaths
1982	1,586	619
1984	10,927	5,605
1986	41,910	24,593
1988	106,304	61,911
1990	196,576	120,811
1992	329,205	196,283
1994	441,528	270,533

Source: U.S. Department of Health and Human Services.

(a) Graph $h(x) = \frac{g(x)}{f(x)}$ in the window $[1982, 1994]$ by $[0, 1]$. Interpret the graph.
(b) Compute the ratio $\frac{\text{deaths}}{\text{cases}}$ for each year. Compare the results with those from part (a).

Reviewing Basic Concepts (Sections 2.4–2.6)

1. Solve the equation in part (a) and the related inequalities in parts (b) and (c).

   (a) $\left|\frac{1}{2}x + 2\right| = 4$ (b) $\left|\frac{1}{2}x + 2\right| > 4$

   (c) $\left|\frac{1}{2}x + 2\right| \leq 4$

2. Given the graph of $y = f(x)$ below, sketch the graph of $y = |f(x)|$.

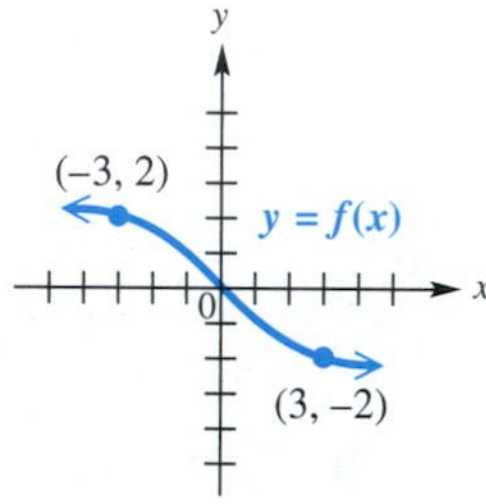

3. *(Modeling) Carbon Dioxide Emissions* When humans breathe, carbon dioxide is emitted. In one study, the emission rates of carbon dioxide by college students were measured during both lectures and exams. The average individual rate R_L (in grams per hour) during a lecture class satisfied the inequality

$$|R_L - 26.75| \leq 1.42,$$

whereas during an exam the rate R_E satisfied the inequality

$$|R_E - 38.75| \leq 2.17.$$

(*Source:* Wang, T. C., ASHRAE Trans., 81 (Part 1), 32 (1975).)

   (a) Find the range of values for R_L and R_E.

   (b) The class had 225 students. If T_L and T_E represent the total amounts of carbon dioxide in grams emitted during a 1-hour lecture and exam, respectively, write inequalities that describe the ranges for T_L and T_E.

4. For $f(x) = \begin{cases} 2x + 3 & \text{if } -3 \leq x < 0 \\ x^2 + 4 & \text{if } x \geq 0 \end{cases}$, find each value.

   (a) $f(-3)$ (b) $f(0)$ (c) $f(2)$

5. Consider the piecewise-defined function given by

$$f(x) = \begin{cases} -x^2 & \text{if } x \leq 0 \\ x - 4 & \text{if } x > 0 \end{cases}.$$

   (a) Sketch its graph.

   (b) Use a graphing calculator to obtain a graph in the window $[-10, 10]$ by $[-10, 10]$.

6. Given $f(x) = -3x - 4$ and $g(x) = x^2$, perform the composition or operation indicated.

   (a) $(f + g)(1)$ (b) $(f - g)(3)$

   (c) $(fg)(-2)$ (d) $\left(\frac{f}{g}\right)(-3)$

   (e) $(f \circ g)(x)$ (f) $(g \circ f)(x)$

7. Find functions f and g so that $h(x) = (f \circ g)(x)$, if $h(x) = (x + 2)^4$.

8. Find the difference quotient

$$\frac{f(x + h) - f(x)}{h}, \quad h \neq 0,$$

for $f(x) = -2x^2 + 3x - 5$.

9. *(Modeling) Author Royalties* An author invests her royalties in two accounts for 1 year.

   (a) The first account pays 4% simple interest. If she invests x dollars in this account, write an expression for y_1 in terms of x, where y_1 represents the amount of interest earned.

   (b) In a second account, she invests $500 more than she invested in the first account. This second account pays 2.5% simple interest. Write an expression for y_2, the amount of interest earned.

   (c) What does $y_1 + y_2$ represent?

   (d) Graph $y_1 + y_2$ in the window $[0, 1000]$ by $[0, 100]$. Use the graph to find the amount of interest she will receive if she invests $250 in the first account.

   (e) Support the result of part (d) analytically.

10. *Geometry* The surface area of a cone (excluding the bottom) is given by

$$S = \pi r\sqrt{r^2 + h^2},$$

where r is its radius and h is its height, as shown in the figure. If the height is twice the radius, write a formula for S in terms of r.

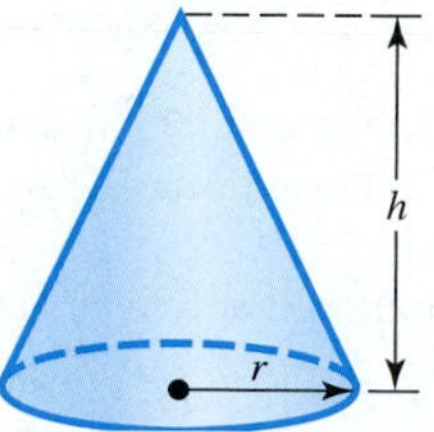

CHAPTER 2 SUMMARY

KEY TERMS & SYMBOLS

2.1 Graphs of Basic Functions and Relations; Symmetry

continuity
increasing function
decreasing function
constant function
identity function
degree
squaring function
parabola
vertex
symmetry
cubing function
inflection point
square root function
cube root function
absolute value function
even function
odd function

KEY CONCEPTS

CONTINUITY (Informal Definition)

A function is continuous over an interval of its domain if its hand-drawn graph over that interval can be sketched without lifting the pencil from the paper.

INCREASING, DECREASING, AND CONSTANT FUNCTIONS

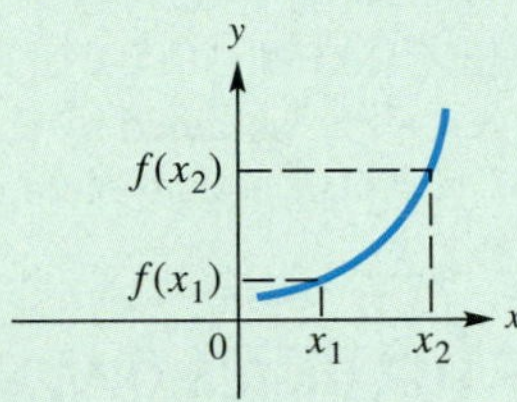

When $x_1 < x_2$, $f(x_1) < f(x_2)$. f is **increasing**.

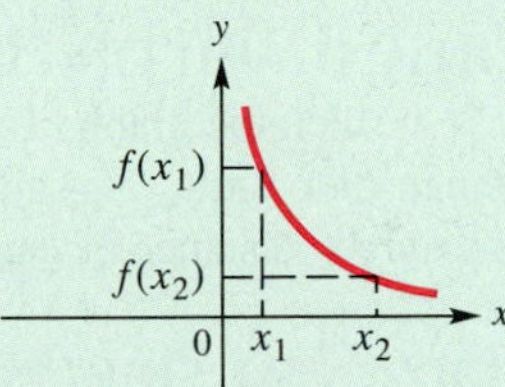

When $x_1 < x_2$, $f(x_1) > f(x_2)$. f is **decreasing**.

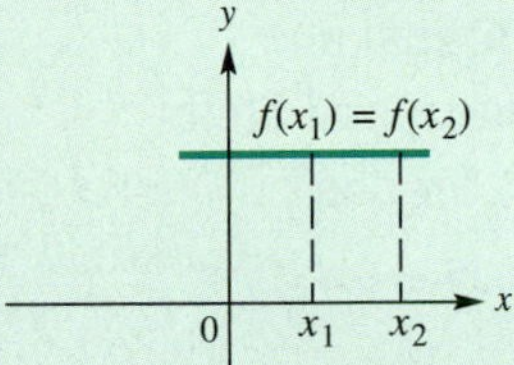

For x_1 and x_2, $f(x_1) = f(x_2)$. f is **constant**.

TYPES OF SYMMETRY

y-Axis Symmetry $f(-x) = f(x)$	Origin Symmetry $f(-x) = -f(x)$	x-Axis Symmetry (not possible for a nonzero function)
y, x, 0, $(-a, b)$, (a, b)	y, x, 0, (a, b), $(-a, -b)$	y, x, 0, (a, b), $(a, -b)$

EVEN AND ODD FUNCTIONS

A function f is called an **even function** if $f(-x) = f(x)$ for all x in the domain of f. (Its graph is symmetric with respect to the y-axis.)

A function f is called an **odd function** if $f(-x) = -f(x)$ for all x in the domain of f. (Its graph is symmetric with respect to the origin.)

BASIC FUNCTIONS

- The **identity function,** defined by $f(x) = x$, is increasing and continuous on its entire domain, $(-\infty, \infty)$.
- The **squaring function,** defined by $f(x) = x^2$, decreases on the interval $(-\infty, 0]$, increases on the interval $[0, \infty)$, and is continuous on its entire domain, $(-\infty, \infty)$.
- The **cubing function,** defined by $f(x) = x^3$, increases and is continuous on its entire domain, $(-\infty, \infty)$.

(continued)

KEY TERMS & SYMBOLS	KEY CONCEPTS
	• The **square root function,** defined by $f(x) = \sqrt{x}$, increases and is continuous on its entire domain, $[0, \infty)$. • The **cube root function,** defined by $f(x) = \sqrt[3]{x}$, increases and is continuous on its entire domain, $(-\infty, \infty)$. • The **absolute value function,** defined by $f(x) = \|x\|$, decreases on the interval $(-\infty, 0]$, increases on the interval $[0, \infty)$, and is continuous on its entire domain, $(-\infty, \infty)$.
2.2 Vertical and Horizontal Shifts of Graphs translation	**VERTICAL SHIFTING OF THE GRAPH OF A FUNCTION** If $c > 0$, then the graph of $y = f(x) + c$ is obtained by shifting the graph of $y = f(x)$ ***upward*** a distance of c units. The graph of $y = f(x) - c$ is obtained by shifting the graph of $y = f(x)$ ***downward*** a distance of c units. **HORIZONTAL SHIFTING OF THE GRAPH OF A FUNCTION** If $c > 0$, then the graph of $y = f(x - c)$ is obtained by shifting the graph of $y = f(x)$ to the ***right*** a distance of c units. The graph of $y = f(x + c)$ is obtained by shifting the graph of $y = f(x)$ to the ***left*** a distance of c units.
2.3 Stretching, Shrinking, and Reflecting Graphs	**VERTICAL STRETCHING OF THE GRAPH OF A FUNCTION** If a point (x, y) lies on the graph of $y = f(x)$, then the point (x, cy) lies on the graph of $y = cf(x)$. If $c > 1$, then the graph of $y = cf(x)$ is a ***vertical stretching*** of the graph of $y = f(x)$ by a factor of c. **VERTICAL SHRINKING OF THE GRAPH OF A FUNCTION** If a point (x, y) lies on the graph of $y = f(x)$, then the point (x, cy) lies on the graph of $y = cf(x)$. If $0 < c < 1$, then the graph of $y = cf(x)$ is a ***vertical shrinking*** of the graph of $y = f(x)$ by a factor of c. **HORIZONTAL STRETCHING OF THE GRAPH OF A FUNCTION** If a point (x, y) lies on the graph of $y = f(x)$, then the point $\left(\frac{x}{c}, y\right)$ lies on the graph of $y = f(cx)$. If $0 < c < 1$, then the graph of $y = f(cx)$ is a ***horizontal stretching*** of the graph of $y = f(x)$. **HORIZONTAL SHRINKING OF THE GRAPH OF A FUNCTION** If a point (x, y) lies on the graph of $y = f(x)$, then the point $\left(\frac{x}{c}, y\right)$ lies on the graph of $y = f(cx)$. If $c > 1$, then the graph of $y = f(cx)$ is a ***horizontal shrinking*** of the graph of $y = f(x)$. **REFLECTING THE GRAPH OF A FUNCTION ACROSS AN AXIS** For a function defined by $y = f(x)$, **(a)** the graph of $y = -f(x)$ is a reflection of the graph of f across the **x-axis.** **(b)** the graph of $y = f(-x)$ is a reflection of the graph of f across the **y-axis.**

(continued)

KEY TERMS & SYMBOLS	KEY CONCEPTS
2.4 Absolute Value Functions: Graphs, Equations, Inequalities, and Applications	**PROPERTIES OF ABSOLUTE VALUE** For all real numbers a and b, **1.** $\lvert ab\rvert = \lvert a\rvert \cdot \lvert b\rvert$ **2.** $\left\lvert \frac{a}{b} \right\rvert = \frac{\lvert a\rvert}{\lvert b\rvert}$ $(b \neq 0)$ **3.** $\lvert a\rvert = \lvert -a\rvert$ **4.** $\lvert a\rvert + \lvert b\rvert \geq \lvert a + b\rvert$ (triangle inequality). **GRAPH OF $y = \lvert f(x)\rvert$** The graph of $y = \lvert f(x)\rvert$ is obtained from the graph of $y = f(x)$ by reflecting the portion of the graph below the x-axis across the x-axis and leaving the graph unchanged for the portion on or above the x-axis. **SOLVING ABSOLUTE VALUE EQUATIONS AND INEQUALITIES** **1.** To solve $\lvert ax + b\rvert = k, k > 0$, solve the compound equation $ax + b = k$ **or** $ax + b = -k$. **2.** To solve $\lvert ax + b\rvert > k,\ k > 0$, solve the compound inequality $ax + b > k$ **or** $ax + b < -k$. **3.** To solve $\lvert ax + b\rvert < k,\ k > 0$, solve the three-part inequality $-k < ax + b < k$.
2.5 Piecewise-Defined Functions piecewise-defined function greatest integer function step function	**PIECEWISE-DEFINED FUNCTION** A piecewise-defined function is defined by different rules over different subsets of its domain. **GREATEST INTEGER FUNCTION** $f(x) = [\![x]\!] = \begin{cases} x & \text{if } x \text{ is an integer} \\ \text{the greatest integer less than } x & \text{if } x \text{ is not an integer} \end{cases}$
2.6 Operations and Composition difference quotient composite function, $g \circ f$	**OPERATIONS ON FUNCTIONS** Given two functions f and g, for all values for which both $f(x)$ and $g(x)$ are defined, the functions $f + g$, $f - g$, fg, and $\frac{f}{g}$ are defined as follows: **Sum** $(f + g)(x) = f(x) + g(x)$ **Difference** $(f - g)(x) = f(x) - g(x)$ **Product** $(fg)(x) = f(x) \cdot g(x)$ **Quotient** $\left(\frac{f}{g}\right)(x) = \frac{f(x)}{g(x)},\ g(x) \neq 0.$ The domains of $f + g$, $f - g$, and fg include all real numbers in the intersection of the domains of f and g, while the domain of $\frac{f}{g}$ includes those real numbers in the intersection of the domains of f and g for which $g(x) \neq 0$. *(continued)*

KEY TERMS & SYMBOLS	KEY CONCEPTS
	DIFFERENCE QUOTIENT $$\frac{f(x+h)-f(x)}{h}, \quad h \neq 0$$ **COMPOSITION OF FUNCTIONS** If f and g are functions, then the composite function, or composition, of g and f is $$(g \circ f)(x) = g[f(x)]$$ for all x in the domain of f such that $f(x)$ is in the domain of g.

CHAPTER 2 Review Exercises

Concept Check *Draw sketches of the graphs of the basic functions introduced in Section 2.1, namely,*

$$f(x)=x, \quad f(x)=x^2, \quad f(x)=x^3,$$
$$f(x)=\sqrt{x}, \quad f(x)=\sqrt[3]{x}, \quad f(x)=|x|.$$

Use your sketches to determine whether each statement in Exercises 1–10 is true *or* false. *If* false, *tell why.*

1. The range of $f(x)=x^2$ is the same as the range of $f(x)=|x|$.
2. $f(x)=x^2$ and $f(x)=|x|$ increase on the same interval.
3. $f(x)=\sqrt{x}$ and $f(x)=\sqrt[3]{x}$ have the same domain.
4. $f(x)=\sqrt[3]{x}$ decreases on its entire domain.
5. $f(x)=x$ has its domain equal to its range.
6. $f(x)=\sqrt{x}$ is continuous on the interval $(-\infty, 0)$.
7. None of the basic functions decrease on the interval $[0, \infty)$.
8. Both $f(x)=x$ and $f(x)=x^3$ have graphs that are symmetric with respect to the origin.
9. Both $f(x)=x^2$ and $f(x)=|x|$ have graphs that are symmetric with respect to the y-axis.
10. None of the graphs shown are symmetric with respect to the x-axis.

In Exercises 11–18, give the interval described.

11. Domain of $f(x)=\sqrt{x}$ **12.** Range of $f(x)=|x|$

13. Range of $f(x)=\sqrt[3]{x}$ **14.** Domain of $f(x)=x^2$

15. The largest interval over which $f(x)=\sqrt[3]{x}$ is increasing

16. The largest interval over which $f(x)=|x|$ is increasing

17. Domain of $x=y^2$ **18.** Range of $x=y^2$

In Exercises 19–30, graph $y=f(x)$ by hand.

19. $f(x)=(x+3)-1$ **20.** $f(x)=-\frac{1}{2}x+1$

21. $f(x)=(x+1)^2-2$ **22.** $f(x)=-2x^2+3$

23. $f(x)=-x^3+2$ **24.** $f(x)=(x-3)^3$

25. $f(x)=\sqrt{\frac{1}{2}x}$ **26.** $f(x)=\sqrt{x-2}+1$

27. $f(x)=2\sqrt[3]{x}$ **28.** $f(x)=\sqrt[3]{x}-2$

29. $f(x)=|x-2|+1$ **30.** $f(x)=|-2x+3|$

31. Consider the function whose graph is shown here.

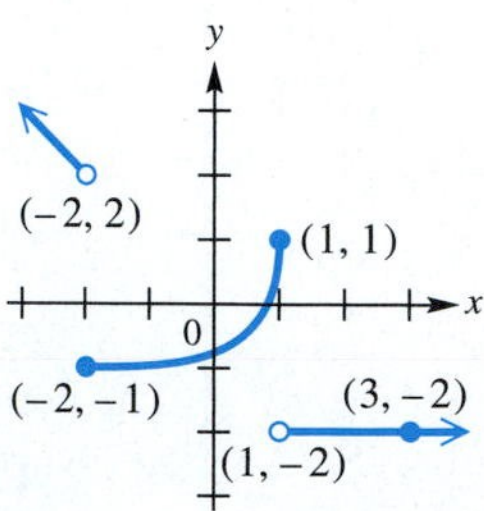

Give the interval(s) over which the function
(a) is continuous. **(b)** increases.
(c) decreases. **(d)** is constant.
(e) What is the domain of the function?
(f) What is the range of the function?

32. The screen shows the graph of $x = y^2 - 4$. Give the two functions that must be used to graph this relation if the calculator is in function mode.

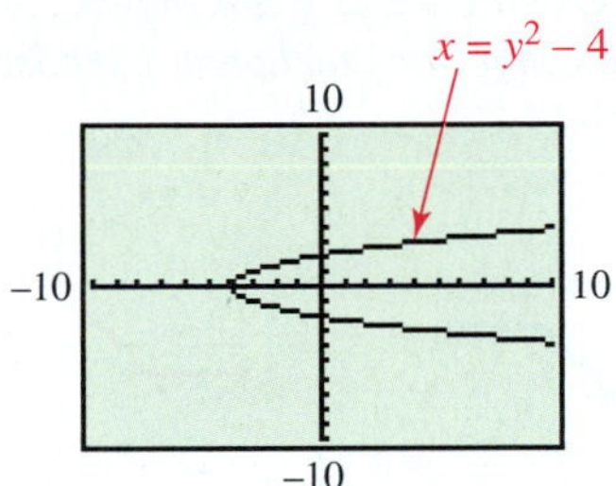

In Exercises 33–38, determine whether the given relation has x-axis symmetry, y-axis symmetry, origin symmetry, *or* none of these *symmetries. (More than one choice is possible.) Also, if the relation is a function, determine whether it is an* even function, *an* odd function, *or* neither.

33.

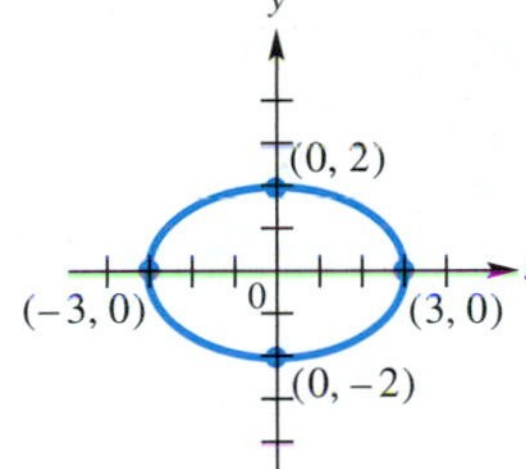

34. $F(x) = x^3 - 6$

35. $y = |x| + 4$

36. $f(x) = \sqrt{x - 5}$

37. $y^2 = x - 5$

38. $f(x) = 3x^4 + 2x^2 + 1$

Concept Check *Decide whether each statement in Exercises 39–44 is* true *or* false. *If false, tell why.*

39. The graph of a function (except for the constant function defined by $f(x) = 0$) cannot be symmetric with respect to the x-axis.

40. The graph of an even function is symmetric with respect to the y-axis.

41. The graph of an odd function is symmetric with respect to the origin.

42. If (a, b) is on the graph of an even function, so is $(a, -b)$.

43. If (a, b) is on the graph of an odd function, so is $(-a, b)$.

44. The constant function defined by $f(x) = 0$ is both even and odd.

45. Use the terminology of Sections 2.2 and 2.3 to describe how the graph of $y = -3(x + 4)^2 - 8$ can be obtained from the graph of $y = x^2$.

46. Give the equation of the function whose graph is obtained by reflecting the graph of $y = \sqrt{x}$ across the y-axis, reflecting across the x-axis, shrinking vertically by a factor of $\frac{2}{3}$, and, finally, translating 4 units upward.

The graph of a function f is shown in the figure. Sketch the graph of each function as defined in Exercises 47–52.

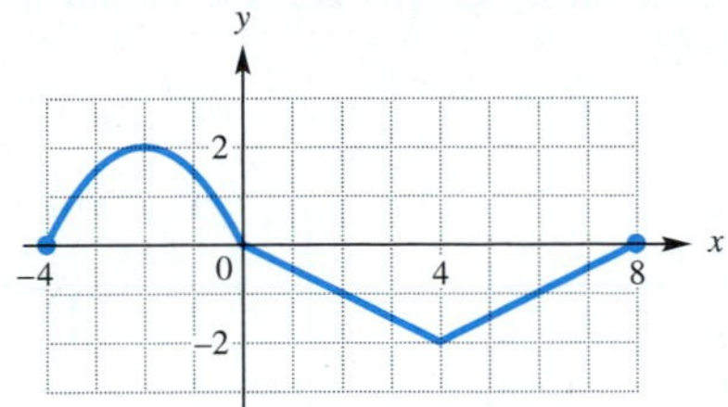

47. $y = f(x) + 3$

48. $y = f(x - 2)$

49. $y = f(x + 3) - 2$

50. $y = |f(x)|$

51. $y = f(4x)$

52. $y = f\left(\frac{1}{2}x\right)$

Let the domain of $f(x)$ be $[-3, 4]$ and the range be $[-2, 5]$. Find the domain and range of each of the following.

53. $f(x) + 4$

54. $5f(x + 10)$

55. $-f(2x)$

56. $f(x - 1) + 3$

The graph of a function defined by $y = f(x)$ is given. Sketch the graph of $y = |f(x)|$.

57.

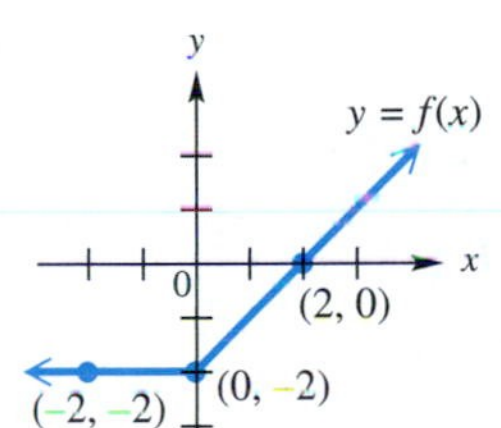

58.

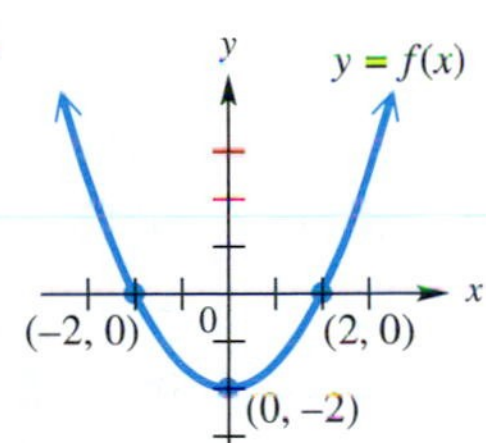

59.

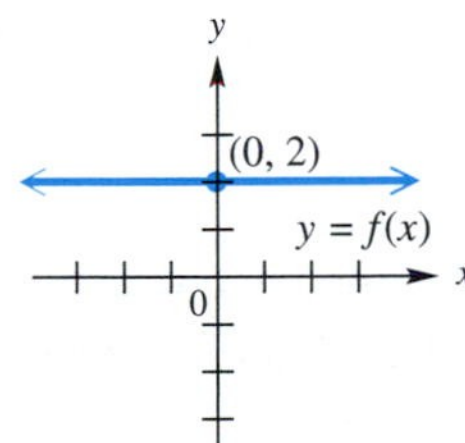

60.

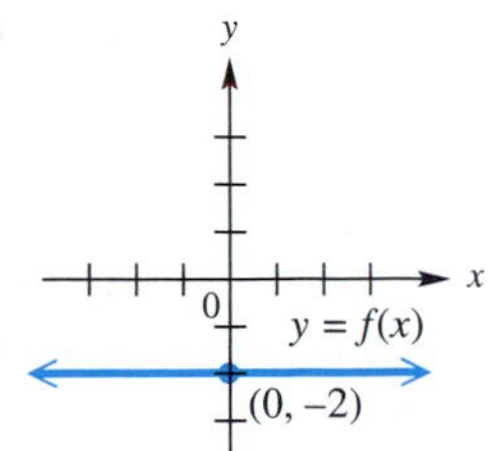

Solve each equation or inequality analytically.

61. $|4x + 3| = 12$

62. $|-2x - 6| + 4 = 1$

63. $|5x + 3| = |x + 11|$

64. $|2x + 5| = 7$

65. $|2x + 5| \leq 7$

66. $|2x + 5| \geq 7$

67. $|5x - 12| > 0$

68. $|6 - x| \leq -4$

69. $2|3x - 1| + 1 = 21$

70. $|2x + 1| = |-3x + 1|$

71. The graphs of $y_1 = |2x + 5|$ and $y_2 = 7$ are shown, along with the two points of intersection of the graphs. Explain how these screens support the answers to Exercises 64–66.

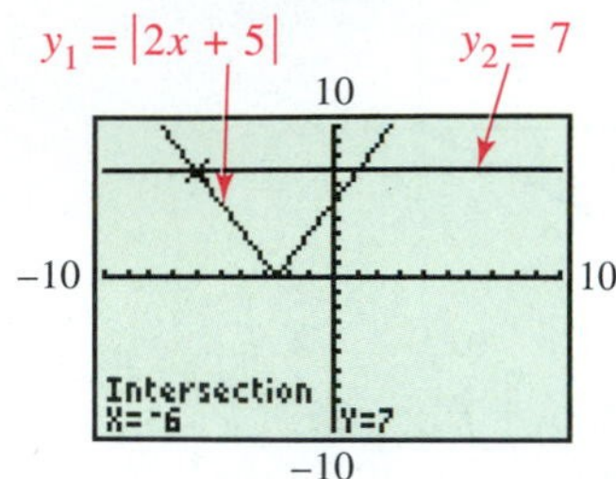

First view

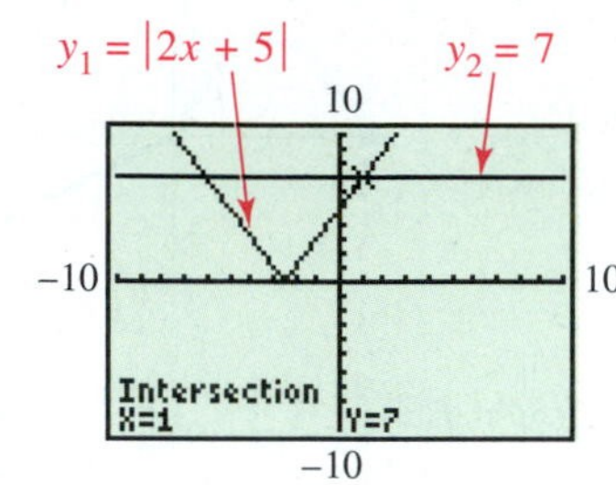

Second view

72. Solve the equation $|x + 1| + |x - 3| = 8$ graphically. Then, give an analytic check by substituting the values in the solution set directly into the left-hand side of the equation.

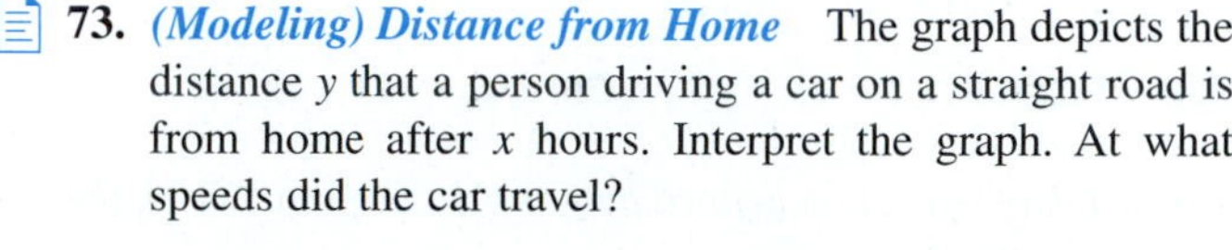

73. ***(Modeling) Distance from Home*** The graph depicts the distance y that a person driving a car on a straight road is from home after x hours. Interpret the graph. At what speeds did the car travel?

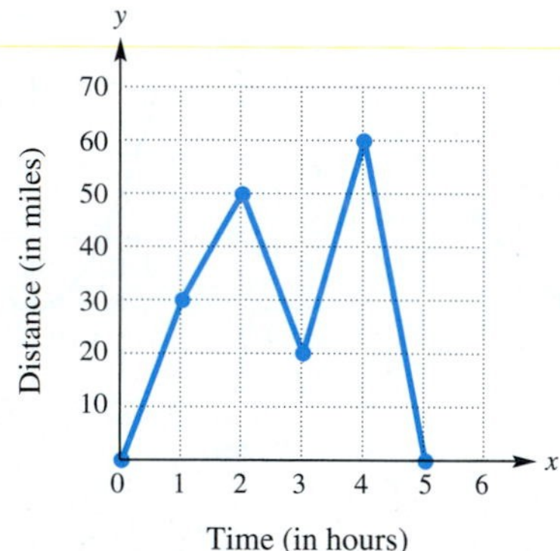

74. ***(Modeling) Water in a Tank*** An initially full 500-gallon water tank is emptied at a constant rate of 50 gallons per minute. Then the tank is filled by a pump that outputs 25 gallons of water per minute. Sketch a graph that depicts the amount of water in the tank after x minutes.

Sketch the graph of each function by hand.

75. $f(x) = \begin{cases} 3x + 1 & \text{if } x < 2 \\ -x + 4 & \text{if } x \geq 2 \end{cases}$

76. $f(x) = \begin{cases} |x| & \text{if } x < 3 \\ 6 - x & \text{if } x \geq 3 \end{cases}$

77. Graph the function in Exercise 75, using a graphing calculator with the window $[-10, 10]$ by $[-10, 10]$.

78. Use a graphing calculator to graph $f(x) = [\![x - 3]\!]$ in the window $[-5, 5]$ by $[-5, 5]$.

The graphs of functions f and g are shown. Use these graphs to evaluate each composite function in Exercises 79–86.

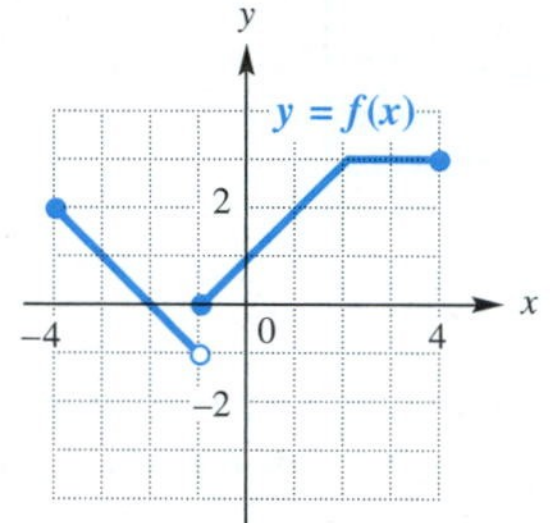

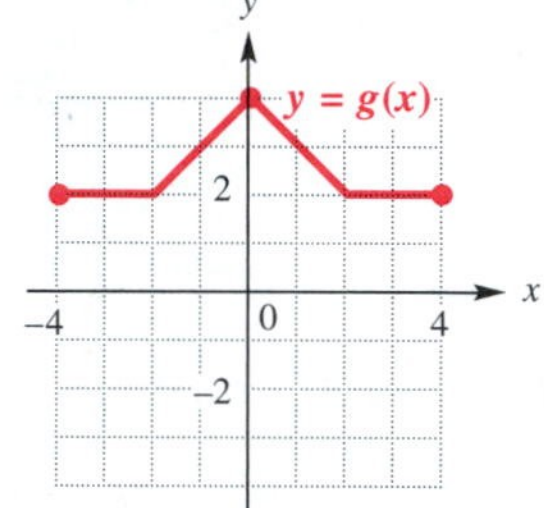

79. $(f + g)(1)$

80. $(f - g)(0)$

81. $(fg)(-1)$

82. $\left(\dfrac{f}{g}\right)(2)$

83. $(f \circ g)(2)$

84. $(g \circ f)(2)$

85. $(g \circ f)(-4)$

86. $(f \circ g)(-2)$

Use the table to evaluate each expression in Exercises 87–90, if possible.

87. $(f + g)(1)$

88. $(f - g)(3)$

89. $(fg)(-1)$

90. $\left(\dfrac{f}{g}\right)(0)$

x	$f(x)$	$g(x)$
−1	3	−2
0	5	0
1	7	1
3	9	9

Use the given tables for f and g to evaluate each expression in Exercises 91 and 92.

91. $(g \circ f)(-2)$

92. $(f \circ g)(3)$

x	$f(x)$
−2	1
0	4
2	3
4	2

x	$g(x)$
1	2
2	4
3	−2
4	0

For the given function, find and simplify $\dfrac{f(x + h) - f(x)}{h}$.

93. $f(x) = 2x + 9$

94. $f(x) = x^2 - 5x + 3$

Find functions f and g such that $(f \circ g)(x) = h(x)$.

95. $h(x) = (x^3 - 3x)^2$

96. $h(x) = \dfrac{1}{x - 5}$

(Modeling) Solve each problem.

97. *Volume of a Sphere* The formula for the volume of a sphere is $V(r) = \frac{4}{3}\pi r^3$, where r represents the radius of the sphere. Construct a model function D representing the volume gained when the radius of a sphere of r inches is increased by 4 inches.

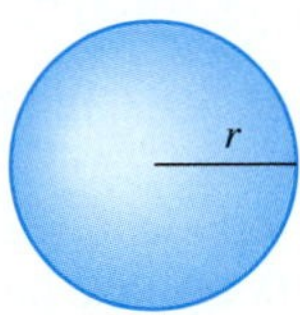

98. *Dimensions of a Cylinder* A cylindrical can with a top and bottom makes the most efficient use of materials when its height is the same as the diameter of its top.

(a) Express the volume V of such a can as a function of the diameter d of its top.

(b) Express the surface area S of such a can as a function of the diameter d of its top. (*Hint:* The curved side is made from a rectangle whose length is the circumference of the top of the can.)

99. *Relationship of Measurement Units* There are 36 inches in 1 yard, and there are 1760 yards in 1 mile. Express the number of inches in x miles by forming two functions and then finding their composition.

100. *Perimeter of a Rectangle* Suppose the length of a rectangle is twice its width. Let x represent the width of the rectangle. Write a formula for the perimeter P of the rectangle in terms of x alone. Then use $P(x)$ notation to describe it as a function. What type of function is this?

CHAPTER 2 Test

1. Match the set described in Column I with the correct interval notation from Column II. Choices in Column II may be used once, more than once, or not at all.

I	II
(a) Domain of $f(x) = \sqrt{x} + 3$	**A.** $[-3, \infty)$
(b) Range of $f(x) = \sqrt{x - 3}$	**B.** $[3, \infty)$
(c) Domain of $f(x) = x^2 - 3$	**C.** $(-\infty, \infty)$
(d) Range of $f(x) = x^2 + 3$	**D.** $[0, \infty)$
(e) Domain of $f(x) = \sqrt[3]{x - 3}$	**E.** $(-\infty, 3)$
(f) Range of $f(x) = \sqrt[3]{x} + 3$	**F.** $(-\infty, 3]$
(g) Domain of $f(x) = \lvert x\rvert - 3$	**G.** $(3, \infty)$
(h) Range of $f(x) = \lvert x + 3\rvert$	**H.** $(-\infty, 0]$
(i) Domain of $x = y^2$	
(j) Range of $x = y^2$	

2. The graph of $y = f(x)$ is shown here. Sketch the graph of each equation in parts (a)–(f). Label three points on the graph.

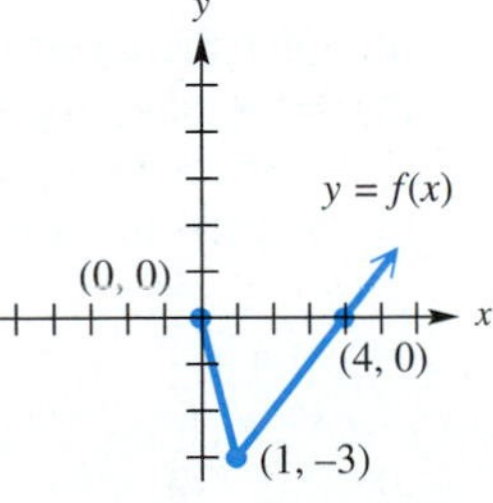

(a) $y = f(x) + 2$
(b) $y = f(x + 2)$
(c) $y = -f(x)$
(d) $y = f(-x)$
(e) $y = 2f(x)$
(f) $y = \lvert f(x)\rvert$

3. If the point $(-2, 4)$ lies on the graph of $y = f(x)$, determine a point on the graph of each equation.

(a) $y = f(2x)$ **(b)** $y = f\left(\frac{1}{2}x\right)$

4. Graph $y = f(x)$ by hand.

(a) $f(x) = -(x - 2)^2 + 4$ **(b)** $y = 2\sqrt{-x}$

5. Observe the coordinates displayed at the bottom of the screen showing only the right half of the graph of $y = f(x)$. Answer the following based on your observation.

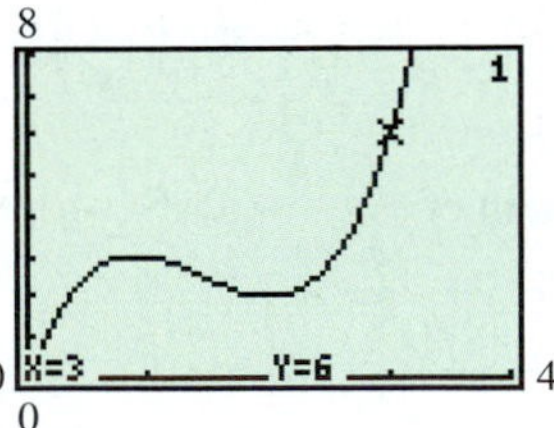

(a) If the graph is symmetric with respect to the y-axis, what are the coordinates of another point on the graph?

(b) If the graph is symmetric with respect to the origin, what are the coordinates of another point on the graph?

(c) Suppose the graph is symmetric with respect to the y-axis. Sketch a typical viewing window with dimensions $[-4, 4]$ by $[0, 8]$. Then draw the graph you would expect to see in this window.

6. (a) Write a brief description that explains how the graph of $y = 4\sqrt[3]{x+2} - 5$ can be obtained by transforming the graph of $y = \sqrt[3]{x}$.

(b) Sketch by hand the graph of $y = -\frac{1}{2}|x - 3| + 2$. State the domain and range.

7. Consider the graph of the function shown here.

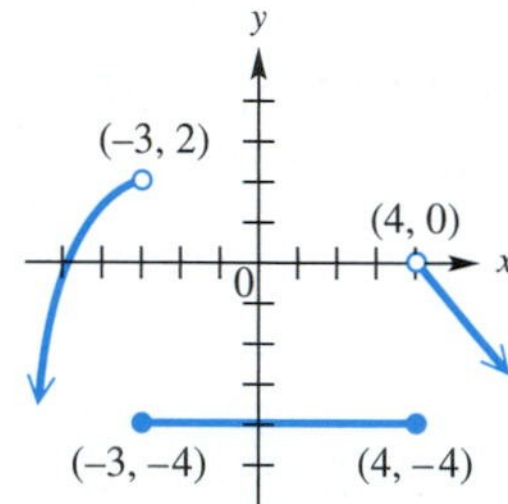

(a) State the interval over which the function is increasing.

(b) State the interval over which the function is decreasing.

(c) State the interval over which the function is constant.

(d) State the intervals over which the function is continuous.

(e) What is the domain of the function?

(f) What is the range of the function?

8. Solve each equation or inequality analytically. Then graph $y_1 = |4x + 8|$ and $y_2 = 4$ in the standard viewing window of a graphing calculator, and state how the graphs support your solution in each case.

(a) $|4x + 8| = 4$

(b) $|4x + 8| < 4$

(c) $|4x + 8| > 4$

9. Given $f(x) = 2x^2 - 3x + 2$ and $g(x) = -2x + 1$, find the following.

(a) $(f - g)(x)$ (b) $\left(\frac{f}{g}\right)(x)$

(c) The domain of $\frac{f}{g}$ (d) $(f \circ g)(x)$

(e) $\frac{f(x+h) - f(x)}{h}$ $(h \neq 0)$

10. Consider the piecewise-defined function given by

$$f(x) = \begin{cases} -x^2 + 3 & \text{if } x \leq 1 \\ \sqrt[3]{x} + 2 & \text{if } x > 1 \end{cases}.$$

(a) Graph f by hand.

(b) Use a graphing calculator to obtain an accurate graph in the window $[-4.7, 4.7]$ by $[-5.1, 5.1]$.

11. *(Modeling) Long-Distance Call Charges* A certain long-distance carrier provides service between Podunk and Nowheresville. If x represents the number of minutes for the call, where $x > 0$, then the function defined by

$$f(x) = .40[\![x]\!] + .75$$

gives the total cost of the call in dollars.

(a) Using dot mode and window $[0, 10]$ by $[0, 6]$, graph this function on a graphing calculator.

(b) Use the graph to find the cost of a call that is 5.5 minutes long.

12. *(Modeling) Cost, Revenue, and Profit Analysis* Tyler McGinnis starts up a small business manufacturing bobble-head figures of famous baseball players. His initial cost is \$3300. Each figure costs \$4.50 to manufacture.

(a) Write a cost function C, where x represents the number of figures manufactured.

(b) Find the revenue function R if each figure in part (a) sells for \$10.50.

(c) Give the profit function P.

(d) How many figures must be produced and sold before Tyler earns a profit?

(e) Support the result of part (d) graphically.

CHAPTER 2 PROJECT

Modeling the Movement of a Cold Front

A weather map of the United States is shown in Figure A. A cold front was traveling in a southeasterly direction, roughly in the shape of a circular arc passing north of Dallas and west of Detroit. The center of the arc was located near Pierre, South Dakota, and had a radius of about 750 miles. Rectangular coordinate axes have been superimposed on the map, with Pierre at the origin. Thus, Pierre has coordinates $(0, 0)$, and the equation of the front can be modeled by

$$f(x) = -\sqrt{750^2 - x^2},$$

where x is in miles, and $0 \le x \le 750$. (*Source:* AccuWeather, Inc.)

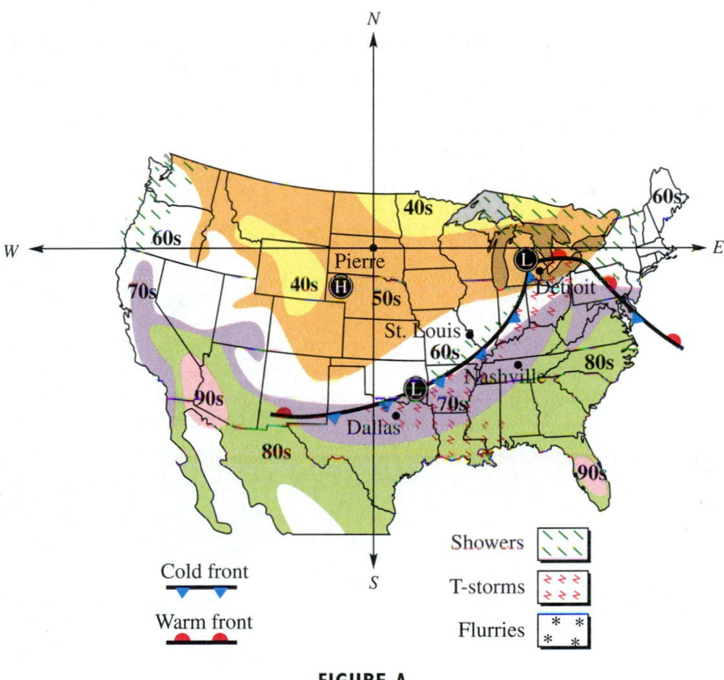

FIGURE A

St. Louis is located at $(535, -400)$, and Nashville is at $(730, -570)$. Figure B shows the graph of f, along with these two cities plotted as data points. Notice that the cold front had passed through St. Louis but had not yet reached Nashville.

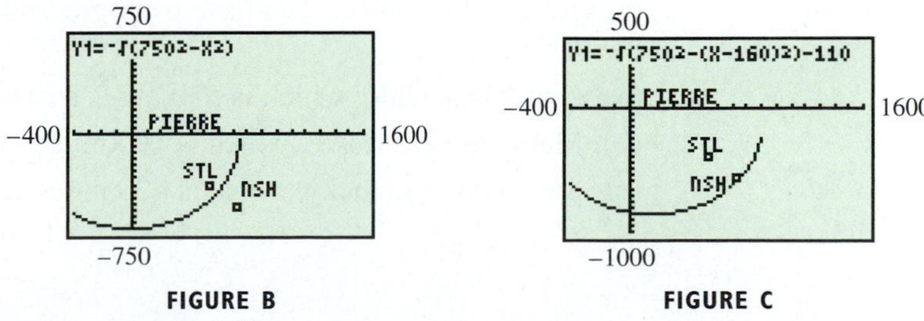

FIGURE B

FIGURE C

During the next 12 hours, the center of the front moved approximately 110 miles south and 160 miles east. If we assume that the front did not change shape, its new position can be modeled by

$$g(x) = -\sqrt{750^2 - (x - 160)^2} - 110.$$

We can see from Figure C on the previous page that the cold front did indeed reach Nashville after 12 hours.

Activity

Suppose that a cold front is passing through the United States at noon with a shape described by

$$f(x) = \frac{1}{20}x^2.$$

Each unit represents 100 miles. Des Moines, Iowa, is located at $(0, 0)$, and the positive y-axis points north. See Figure D.

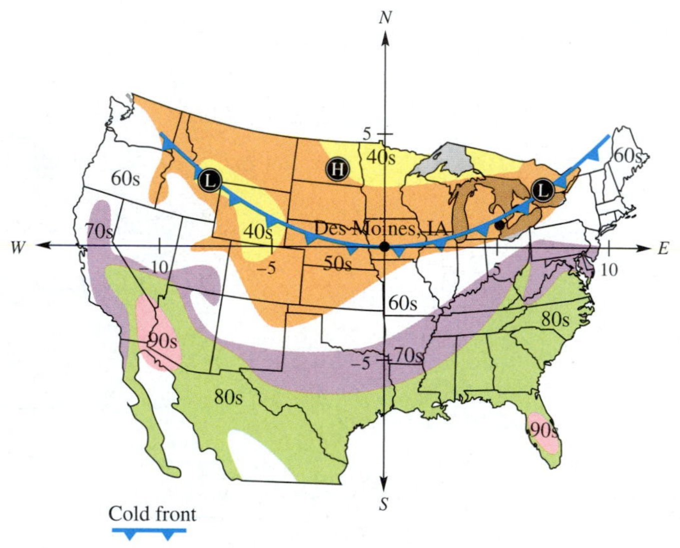

FIGURE D

1. If the cold front is moving south at 40 mph for 4 hours and retains its present shape, what would be the equation of its graph at that time?
2. Suppose that by midnight the vertex of the front has moved 250 miles south and 210 miles east of Des Moines, maintaining the same shape. Divide the class into three groups (a), (b), and (c), and have each group determine whether the front has reached the designated city.
   (a) Columbus, Ohio, which is 550 miles east and 80 miles south of Des Moines
   (b) Memphis, Tennessee, which is 190 miles east and 430 miles south of Des Moines
   (c) Louisville, Kentucky, which is 420 miles east and 230 miles south of Des Moines

3 Polynomial Functions

THE LAST basketball player in the NBA to shoot foul shots underhand was Rick Barry, who retired in 1980. On average, he was able to make about 9 out of 10 shots. Since then, every NBA player has used the overhand style of shooting foul shots—even though this style has often resulted in lower free-throw percentages.

According to Peter Brancazio, a physics professor emeritus from Brooklyn College and author of *Sports Science,* an underhand shot obtains a higher arc and has a better chance of going through the hoop than does a ball with a flatter arc. If a basketball is tossed at an angle of 32 degrees or less, it will likely hit the back of the rim and bounce out. The optimal angle is greater than 45 degrees and depends on the height at which the ball is released. Lower release points require steeper arcs and increase the chances of the ball passing through the hoop.

In this chapter, we use *polynomial functions* to analyze applied problems such as shooting foul shots.

Source: Rist, C., "The Physics of Foul Shots," *Discover*, October 2000. (Photograph reprinted with permission.)

Chapter Outline

3.1 Complex Numbers

The Number i ■ Operations with Complex Numbers

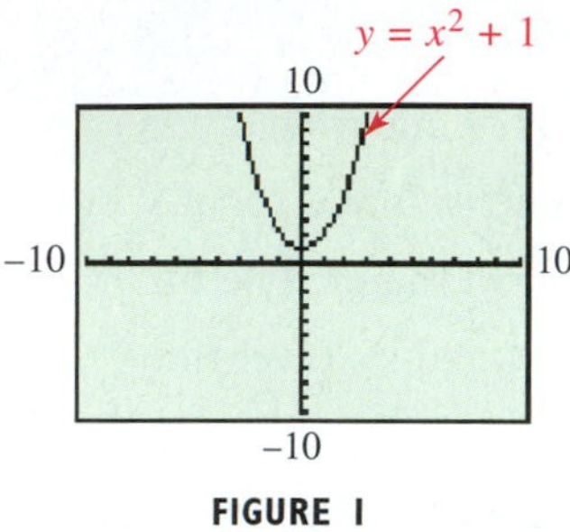

FIGURE 1

The Number i

The graph of $y = x^2 + 1$ in Figure 1 does not intersect the x-axis; therefore, there are no *real* solutions of the equation $x^2 + 1 = 0$. This equation is equivalent to $x^2 = -1$, and we know from experience that no real number has a square of -1.

The **complex number system** is an extended number system that includes the set of real numbers as a subset. A basic unit of this new system is i, which is defined as the principal square root of -1. Thus, $i^2 = -1$.

Imaginary Unit i

$$i = \sqrt{-1} \quad \text{or} \quad i^2 = -1$$

GCM TECHNOLOGY NOTE

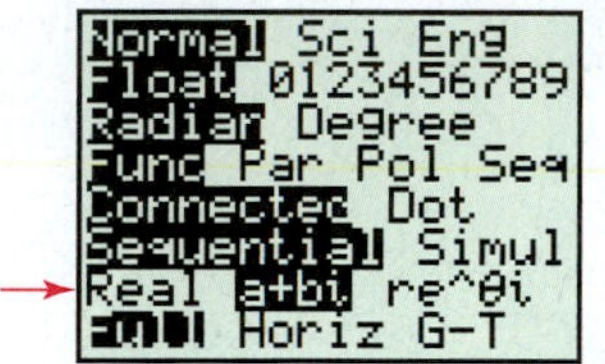

Some graphing calculators such as the TI-83/84 Plus are capable of complex number operations, as indicated by $a + bi$ here.

Numbers of the form $a + bi$, where a and b are real numbers, are called **complex numbers.** In the complex number $a + bi$, a is called the **real part** and b is called the **imaginary part.***

The relationships among the various sets of numbers are shown in Figure 2.

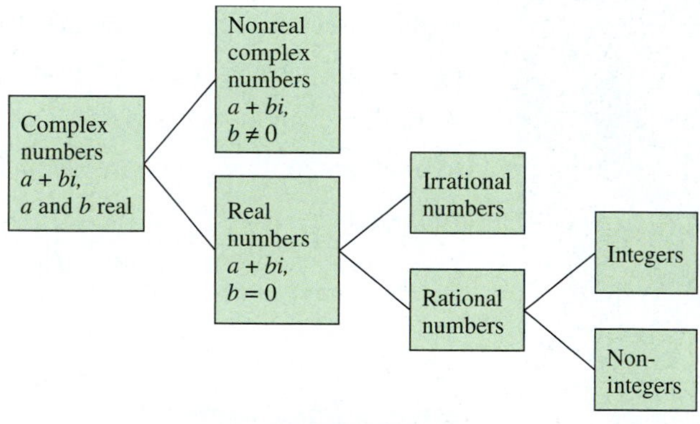

FIGURE 2

Two complex numbers $a + bi$ and $c + di$ are *equal* provided that their real parts are equal and their imaginary parts are equal.

Equality of Complex Numbers

$$a + bi = c + di \quad \text{if and only if} \quad a = c \quad \text{and} \quad b = d$$

*In some texts, the term bi is defined to be the imaginary part.

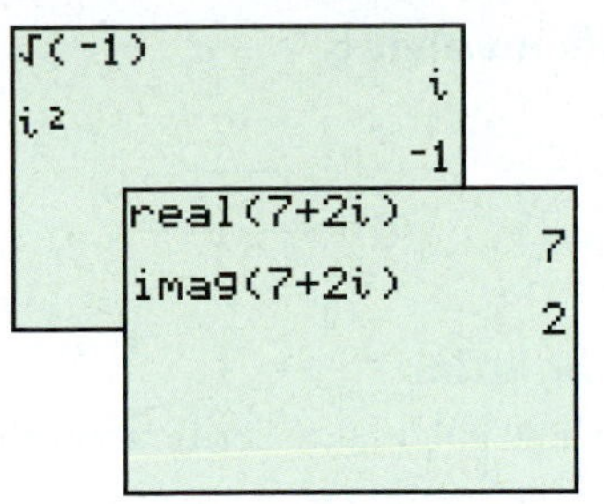

Some graphing calculators are capable of working with complex numbers, as seen here. The top screen supports the definition of i. The bottom screen shows how the calculator returns the real and imaginary parts of $7 + 2i$.

For a complex number $a + bi$, if $b = 0$, then $a + bi = a$, which is a real number. ***Thus, the set of real numbers is a subset of the set of complex numbers.*** If $a = 0$ and $b \neq 0$, the complex number is said to be a **pure imaginary number.** For example, $3i$ is a pure imaginary number. A number such as $7 + 2i$ is a **nonreal complex number.** A complex number written in the form $a + bi$ (or $a + ib$) is in **standard form.** (The form $a + ib$ is used to write expressions such as $i\sqrt{5}$, since $\sqrt{5}i$ could be mistaken for $\sqrt{5i}$.)

Many of the solutions of quadratic equations in Section 3.3 will involve expressions such as $\sqrt{-a}$, for a positive real number a, defined as follows.

The Expression $\sqrt{-a}$

If $a > 0$, then

$$\sqrt{-a} = i\sqrt{a}.$$

GCM **EXAMPLE 1** **Writing $\sqrt{-a}$ as $i\sqrt{a}$**

Write each expression as the product of i and a real number.

(a) $\sqrt{-16}$ **(b)** $\sqrt{-75}$

Analytic Solution

(a)
$$\begin{aligned}\sqrt{-16} &= \sqrt{-1 \cdot 16} \\ &= i\sqrt{16} && \sqrt{-1} = i \\ &= 4i && \sqrt{16} = 4\end{aligned}$$

(b)
$$\begin{aligned}\sqrt{-75} &= \sqrt{-1 \cdot 75} \\ &= i\sqrt{75} && \sqrt{-1} = i \\ &= i\sqrt{25 \cdot 3} \\ &= 5i\sqrt{3} && \sqrt{25} = 5\end{aligned}$$

Graphing Calculator Solution

The calculator in Figure 3 is in complex number mode.

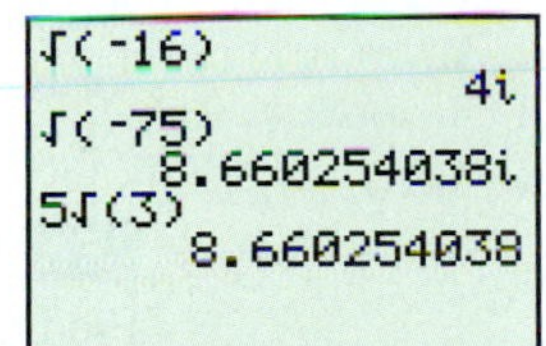

FIGURE 3

A product or quotient with a negative radicand is simplified by first rewriting $\sqrt{-a}$ as $i\sqrt{a}$ for positive a. Then the properties of real numbers are applied, together with the fact that $i^2 = -1$.

CAUTION ***When working with negative radicands, use the definition $\sqrt{-a} = i\sqrt{a}$ before using any of the other rules for radicals.*** In particular, the rule $\sqrt{c} \cdot \sqrt{d} = \sqrt{cd}$ is valid only when c and d are *not* both negative. For example,

$$\sqrt{(-4)(-9)} = \sqrt{36} = 6,$$

while
$$\sqrt{-4} \cdot \sqrt{-9} = 2i(3i) = 6i^2 = -6,$$

so
$$\sqrt{-4} \cdot \sqrt{-9} \neq \sqrt{(-4)(-9)}.$$

GCM EXAMPLE 2 Finding Products and Quotients Involving $\sqrt{-a}$

Multiply or divide as indicated.

(a) $\sqrt{-7} \cdot \sqrt{-7}$ **(b)** $\sqrt{-6} \cdot \sqrt{-10}$ **(c)** $\dfrac{\sqrt{-20}}{\sqrt{-2}}$ **(d)** $\dfrac{\sqrt{-48}}{\sqrt{24}}$

Analytic Solution

(a) $\sqrt{-7} \cdot \sqrt{-7} = i\sqrt{7} \cdot i\sqrt{7}$ $\quad \sqrt{-1} = i$

$= i^2 \cdot (\sqrt{7})^2$

$= -1 \cdot 7$ $\quad i^2 = -1$

$= -7$

(b) $\sqrt{-6} \cdot \sqrt{-10} = i\sqrt{6} \cdot i\sqrt{10}$ $\quad \sqrt{-1} = i$

$= i^2 \cdot \sqrt{60}$

$= -1 \cdot 2\sqrt{15}$

$= -2\sqrt{15}$

(c) $\dfrac{\sqrt{-20}}{\sqrt{-2}} = \dfrac{i\sqrt{20}}{i\sqrt{2}} = \sqrt{\dfrac{20}{2}} = \sqrt{10}$ $\quad \dfrac{\sqrt{a}}{\sqrt{b}} = \sqrt{\dfrac{a}{b}}$

(d) $\dfrac{\sqrt{-48}}{\sqrt{24}} = \dfrac{i\sqrt{48}}{\sqrt{24}} = i\sqrt{\dfrac{48}{24}} = i\sqrt{2}$

Graphing Calculator Solution

The screens in Figure 4 show calculator verification for parts (b) and (d).

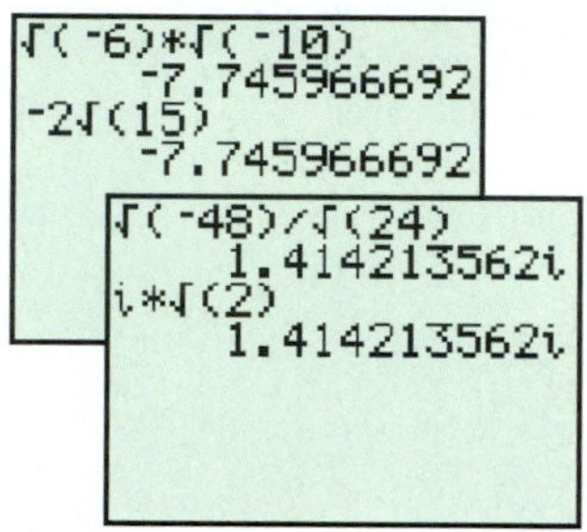

FIGURE 4

Operations with Complex Numbers

We find the sum, difference, and product of two complex numbers just as we would for two binomials.

GCM EXAMPLE 3 Adding and Subtracting Complex Numbers

Find each sum or difference.

(a) $(3 - 4i) + (-2 + 6i)$ **(b)** $(-9 + 7i) + (3 - 15i)$

(c) $(-4 + 3i) - (6 - 7i)$ **(d)** $(12 - 5i) - (8 - 3i)$

Analytic Solution

(a) $(3 - 4i) + (-2 + 6i) = \underbrace{[3 + (-2)]}_{\text{Add real parts.}} + \underbrace{(-4 + 6)}_{\text{Add imaginary parts.}}i$

$= 1 + 2i$

(b) $(-9 + 7i) + (3 - 15i) = -6 - 8i$

(c) $(-4 + 3i) - (6 - 7i) = \underbrace{(-4 - 6)}_{\text{Subtract real parts.}} + \underbrace{[3 - (-7)]}_{\text{Subtract imaginary parts.}}i$

$= -10 + 10i$

(d) $(12 - 5i) - (8 - 3i) = 4 - 2i$ $\quad -5i - (-3i) = -2i$

Graphing Calculator Solution

Figure 5 shows the calculator results.

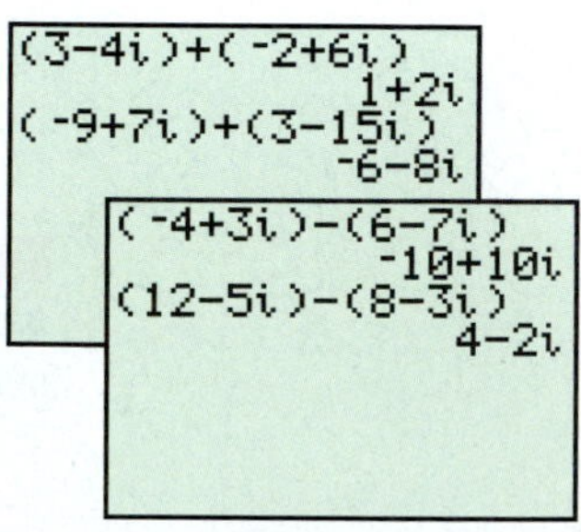

FIGURE 5

The product of two complex numbers is found by multiplying as if the numbers were binomials and using the fact that $i^2 = -1$.

GCM **EXAMPLE 4 Multiplying Complex Numbers**

Find each product.

(a) $(2 - 3i)(3 + 4i)$ **(b)** $(5 - 4i)(7 - 2i)$

(c) $(6 + 5i)(6 - 5i)$ **(d)** $(4 + 3i)^2$

Analytic Solution

(a) $(2 - 3i)(3 + 4i) = 2(3) + 2(4i) - 3i(3) - 3i(4i)$ FOIL

$= 6 + 8i - 9i - 12i^2$

$= 6 - i - 12(-1)$ $i^2 = -1$

$= 18 - i$

(b) $(5 - 4i)(7 - 2i) = 5(7) + 5(-2i) - 4i(7) - 4i(-2i)$ FOIL

$= 35 - 10i - 28i + 8i^2$

$= 35 - 38i + 8(-1)$

$= 27 - 38i$

Be careful with signs.

(c) $(6 + 5i)(6 - 5i) = 6^2 - 25i^2$ Product of the sum and difference of two terms

$= 36 - 25(-1)$ $i^2 = -1$

$= 61$

(d) $(4 + 3i)^2 = 4^2 + 2(4)(3i) + (3i)^2$ Square of a binomial

$= 16 + 24i + (-9)$ $(3i)^2 = 3^2 i^2 = 9(-1) = -9$

$= 7 + 24i$

Remember the middle term when squaring a binomial.

Graphing Calculator Solution

Calculator solutions are shown in Figure 6.

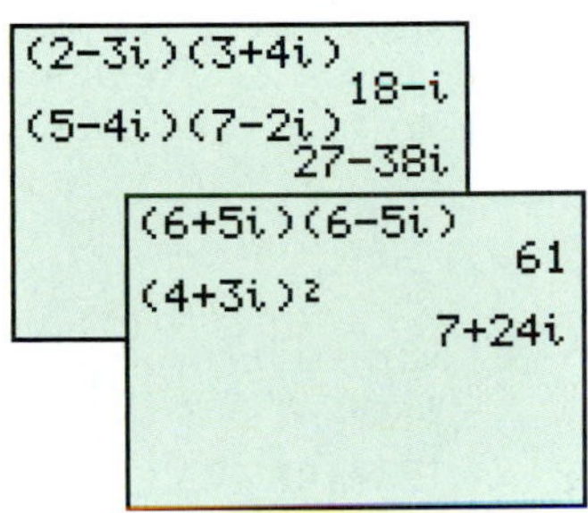

FIGURE 6

The TI-83/84 Plus calculator does *not* need to be in complex number mode to perform the calculations shown in the screens of Figure 6. ■

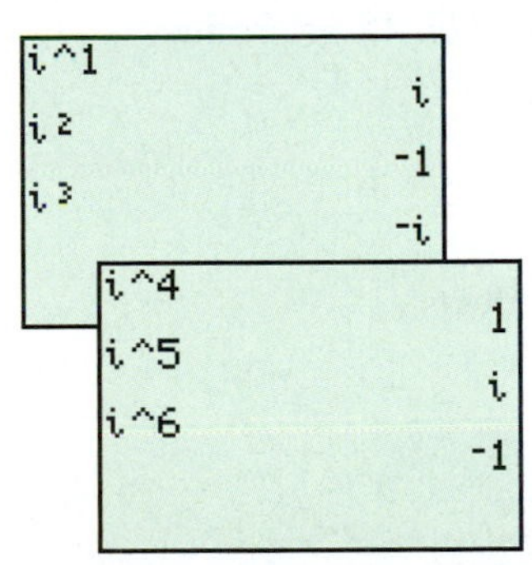

These screens show how powers of i follow a cycle.

By definition, $i^1 = i$ and $i^2 = -1$. Now, observe the following pattern.

$$
\begin{aligned}
i^1 &= i \\
i^2 &= -1 \\
i^3 &= i^2 \cdot i = -1 \cdot i = -i \\
i^4 &= i^3 \cdot i = -i \cdot i = -i^2 = -(-1) = 1 \\
i^5 &= i^4 \cdot i = 1 \cdot i = i \\
i^6 &= i^4 \cdot i^2 = 1 \cdot (-1) = -1 \\
i^7 &= i^4 \cdot i^3 = 1 \cdot i^3 = -i \\
i^8 &= (i^4)^2 = 1^2 = 1
\end{aligned}
$$

As seen for i^5, i^6, i^7, and i^8, any larger power of i may be found by writing the power as a product of two powers of i, with one exponent a multiple of 4, and then simplifying. Since $(i^4)^n = 1$ for all natural numbers n, we can then simplify by considering the other factor. For example,

$$i^{53} = (i^4)^{13} \cdot i^1 = i.$$

EXAMPLE 5 **Simplifying Powers of *i***

Simplify each power of i.

(a) i^{13} **(b)** i^{56} **(c)** i^{-3}

Analytic Solution

In each case, we use the fact that $i^4 = 1$.

(a)
$$\begin{aligned} i^{13} &= i^{12} \cdot i \\ &= (i^4)^3 \cdot i \\ &= 1^3 \cdot i \\ &= i \end{aligned}$$

(b) $i^{56} = (i^4)^{14} = 1^{14} = 1$

(c)
$$\begin{aligned} i^{-3} &= i^{-4} \cdot i \\ &= (i^4)^{-1} \cdot i \\ &= (1)^{-1} \cdot i \\ &= i \end{aligned}$$

Graphing Calculator Solution

The screen in Figure 7 agrees with the analytic results for parts (a) and (b). In the first case, the real part 0 is approximated as -3×10^{-13}, and in the second case, the imaginary part 0 is approximated as -4×10^{-13}.

This screen provides another example of the limitations of technology. The results may be misinterpreted if the mathematical concepts are not understood.

```
i^13
                -3E-13+i
i^56
                1-4E-13i
```

FIGURE 7

In Example 4(c), the factors $6 + 5i$ and $6 - 5i$ have the same real parts but *opposite* imaginary parts. Pairs of complex numbers satisfying these conditions are called *complex conjugates,* or simply *conjugates.* ***The product of a complex number and its conjugate is always a real number.***

Complex Conjugates

The **conjugate** of the complex number $a + bi$ is $a - bi$. Their product is

$$(a + bi)(a - bi) = a^2 + b^2.$$

The table shows several pairs of conjugates and their products.

Number	Conjugate	Product
$3 - i$	$3 + i$	$(3 - i)(3 + i) = 9 + 1 = 10$
$2 + 7i$	$2 - 7i$	$(2 + 7i)(2 - 7i) = 53$
$-6i$	$6i$	$(-6i)(6i) = 36$

All products are real numbers.

To find the quotient of two complex numbers in standard form, we multiply numerator and denominator by the conjugate of the denominator.

GCM **EXAMPLE 6** **Dividing Complex Numbers**

Find each quotient.

(a) $\frac{7+i}{3-i}$ **(b)** $\frac{3+2i}{5-i}$ **(c)** $\frac{3}{i}$

Analytic Solution

(a) $\frac{7+i}{3-i} = \frac{(7+i)(3+i)}{(3-i)(3+i)}$ Multiply numerator and denominator by $3+i$.

$= \frac{21+7i+3i+i^2}{9-i^2}$ Multiply.

$= \frac{21+10i-1}{9-(-1)}$ $i^2 = -1$

$= \frac{20+10i}{10}$ Use parentheses to avoid sign errors.

$= \frac{20}{10} + \frac{10}{10}i = 2 + i$ $\frac{a+bi}{c} = \frac{a}{c} + \frac{b}{c}i$

(b) $\frac{3+2i}{5-i} = \frac{(3+2i)(5+i)}{(5-i)(5+i)}$ Multiply numerator and denominator by $5+i$.

$= \frac{15+3i+10i+2i^2}{25-i^2}$ Multiply.

$= \frac{13+13i}{26}$ $i^2 = -1$; simplify.

$= \frac{13}{26} + \frac{13}{26}i = \frac{1}{2} + \frac{1}{2}i$

(c) $\frac{3}{i} = \frac{3(-i)}{i(-i)} = \frac{-3i}{-i^2} = \frac{-3i}{-(-1)} = -3i$

In each case, check by multiplying quotient by divisor to obtain the dividend.

Graphing Calculator Solution

A calculator gives the same quotients. See Figure 8.

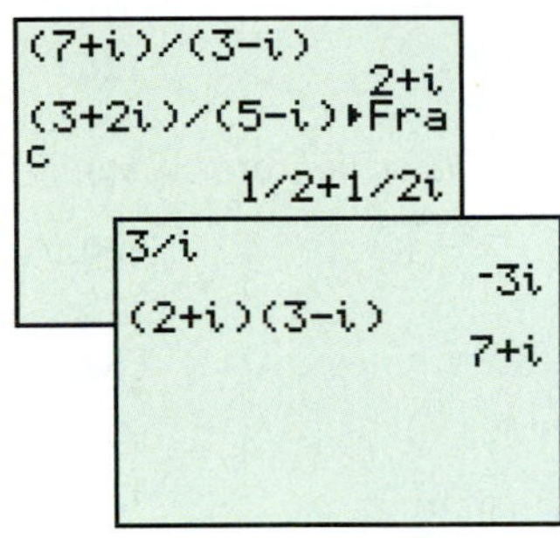

FIGURE 8

Notice in the top screen that $1/2i$ is actually $\frac{1}{2}i$.

As with division of real numbers, a quotient of complex numbers can be checked by multiplying the quotient by the divisor to obtain the dividend. For example, the final entry in the bottom screen indicates a check for the result in part (a). ■

3.1 Exercises

Note: For the exercises in this section, use your graphing calculator with complex number capability to confirm your answers when possible.

Concept Check *For each complex number,* **(a)** *state the real part,* **(b)** *state the imaginary part, and* **(c)** *identify the number as one of the following:* real, pure imaginary, *or* nonreal complex.

1. $-9i$ **2.** $3i$ **3.** π **4.** $\sqrt{2}$ **5.** $3 + 7i$

6. $-8 + 4i$ **7.** $i\sqrt{7}$ **8.** $-i\sqrt{3}$ **9.** $\sqrt{-7}$ **10.** $\sqrt{-10}$

Concept Check *Determine whether each statement is* true *or* false. *If* false, *tell why.*

11. Every real number is a complex number.

12. No real number is a pure imaginary number.

13. Every pure imaginary number is a complex number.

14. A number can be both real and complex.

15. There is no real number that is a complex number.

16. A complex number might not be a pure imaginary number.

Write each number in simplest form, without a negative radicand.

17. $\sqrt{-100}$ **18.** $\sqrt{-169}$ **19.** $-\sqrt{-400}$ **20.** $-\sqrt{-225}$

21. $-\sqrt{-39}$ **22.** $-\sqrt{-95}$ **23.** $5 + \sqrt{-4}$ **24.** $-7 + \sqrt{-100}$

25. $9 - \sqrt{-50}$ **26.** $-11 - \sqrt{-24}$ **27.** $i\sqrt{-9}$ **28.** $i\sqrt{-16}$

Multiply or divide as indicated. Simplify each answer.

29. $\sqrt{-13} \cdot \sqrt{-13}$ **30.** $\sqrt{-17} \cdot \sqrt{-17}$ **31.** $\sqrt{-3} \cdot \sqrt{-8}$ **32.** $\sqrt{-5} \cdot \sqrt{-15}$

33. $\frac{\sqrt{-30}}{\sqrt{-10}}$ **34.** $\frac{\sqrt{-70}}{\sqrt{-7}}$ **35.** $\frac{\sqrt{-24}}{\sqrt{8}}$ **36.** $\frac{\sqrt{-54}}{\sqrt{27}}$

37. $\frac{\sqrt{-10}}{\sqrt{-40}}$ **38.** $\frac{\sqrt{-40}}{\sqrt{20}}$ **39.** $\frac{\sqrt{-6} \cdot \sqrt{-2}}{\sqrt{3}}$ **40.** $\frac{\sqrt{-12} \cdot \sqrt{-6}}{\sqrt{8}}$

Add or subtract as indicated. Write each sum or difference in standard form.

41. $(3 + 2i) + (4 - 3i)$ **42.** $(4 - i) + (2 + 5i)$ **43.** $(-2 + 3i) - (-4 + 3i)$

44. $(-3 + 5i) - (-4 + 5i)$ **45.** $(3 - 8i) + (2i + 4)$ **46.** $(9 - 5i) - (3i - 6)$

47. $(2 - 5i) - (3 + 4i) - (-2 + i)$ **48.** $(-4 - i) - (2 + 3i) + (-4 + 5i)$

49. $(-6 + 5i) + (4 - 4i) + (2 - i)$ **50.** $(7 + 9i) + (1 - 2i) + (-8 - 7i)$

Multiply as indicated. Write each product in standard form.

51. $(2 + i)(3 - 2i)$ **52.** $(-2 + 3i)(4 - 2i)$ **53.** $(2 + 4i)(-1 + 3i)$

54. $(1 + 3i)(2 - 5i)$ **55.** $(-3 + 2i)^2$ **56.** $(2 + i)^2$

57. $(3 + i)(-3 - i)$ **58.** $(-5 - i)(5 + i)$ **59.** $(2 + 3i)(2 - 3i)$

60. $(6 - 4i)(6 + 4i)$ **61.** $(\sqrt{6} + i)(\sqrt{6} - i)$ **62.** $(\sqrt{2} - 4i)(\sqrt{2} + 4i)$

63. $i(3 - 4i)(3 + 4i)$ **64.** $i(2 + 7i)(2 - 7i)$ **65.** $3i(2 - i)^2$

66. $-5i(4 - 3i)^2$ **67.** $(2 + i)(2 - i)(4 + 3i)$ **68.** $(3 - i)(3 + i)(2 - 6i)$

Simplify each power of i to i, 1, $-i$, or -1.

69. i^5 **70.** i^8 **71.** i^{15} **72.** i^{19} **73.** i^{64}

74. i^{102} **75.** i^{-6} **76.** i^{-15} **77.** $\frac{1}{i^9}$ **78.** $\frac{1}{i^{12}}$

79. $\frac{1}{i^{-51}}$ **80.** $\frac{1}{i^{-46}}$ **81.** $\frac{-1}{-i^{12}}$ **82.** $\frac{-1}{-i^{15}}$

83. **Concept Check** Show that $\frac{\sqrt{2}}{2} + \frac{\sqrt{2}}{2}i$ is a square root of i.

84. **Concept Check** Show that $\frac{\sqrt{3}}{2} + \frac{1}{2}i$ is a cube root of i.

Find the conjugate of each number.

85. $5 - 3i$

86. $-3 + i$

87. $-18i$

88. $\sqrt{7}$

Divide as indicated. Write each quotient in standard form.

89. $\frac{-19 - 9i}{4 + i}$

90. $\frac{-12 - 5i}{3 - 2i}$

91. $\frac{1 - 3i}{1 + i}$

92. $\frac{-3 + 4i}{2 - i}$

93. $\frac{-6 + 8i}{4 + 3i}$

94. $\frac{2 - i}{2 + i}$

95. $\frac{4 - 3i}{4 + 3i}$

96. $\frac{3}{-i}$

97. $\frac{-7}{3i}$

98. $\frac{-10}{i}$

99. Explain why the method of dividing complex numbers (that is, multiplying both the numerator and the denominator by the conjugate of the denominator) works. What property justifies this process?

100. Suppose that your friend Virginia Starkenburg tells you that she has discovered a method of simplifying a positive power of i. "Just divide the exponent by 4," she says, "and then look at the remainder. Then, refer to the short table of powers of i in this section. The given power of i is equal to i to the power indicated by the remainder. And if the remainder is 0, the result is $i^0 = 1$." Explain why Virginia's method works.

3.2 Quadratic Functions and Graphs

Completing the Square ■ Graphs of Quadratic Functions ■ Vertex Formula ■ Extreme Values ■ Applications and Quadratic Models

In Chapter 2 we saw that the graph of $y = x^2$ is a parabola. See Figure 9. The function defined by $P(x) = x^2$ is the simplest example of a *quadratic function.*

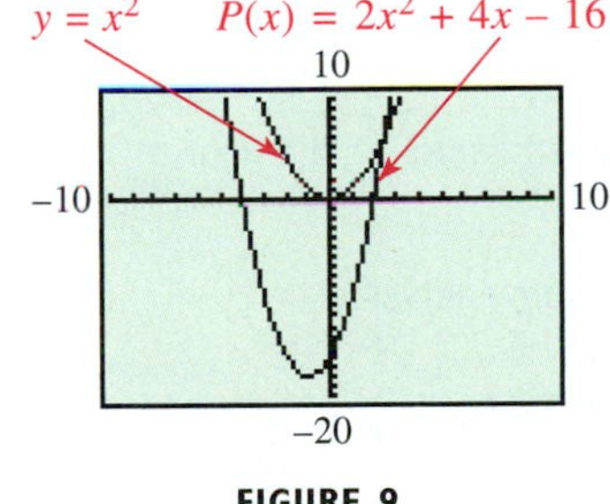

FIGURE 9

Quadratic Function

A function defined by

$$P(x) = ax^2 + bx + c,$$

with $a \neq 0$, is called a **quadratic function.**

Quadratic functions, as well as linear functions, are examples of *polynomial functions.* We often use P to name a polynomial function, and we do so with quadratic functions throughout this chapter.

Completing the Square

Recall from Chapter 2 that the graph of

$$g(x) = a(x - h)^2 + k$$

has the same general shape as the graph of $f(x) = x^2$, but may be stretched, shrunk, reflected, or shifted. Consider $P(x) = 2x^2 + 4x - 16$, graphed in Figure 9. It appears that the graph of P can be obtained from the graph of $y = x^2$, also shown in the figure, by a vertical stretch with a factor greater than 1, a shift to the left, and a shift downward. We can determine the sizes of these transformations by writing the function in the form $P(x) = a(x - h)^2 + k$. This process is called *completing the square.*

Completing the Square

To transform the equation $P(x) = ax^2 + bx + c$ into the form $P(x) = a(x - h)^2 + k$, follow these steps:

Step 1 Divide each side of the equation by a so that the coefficient of x^2 is 1.

Step 2 Add $-\frac{c}{a}$ to each side.

Step 3 Add to each side the square of half the coefficient of x—that is, $\left(\frac{b}{2a}\right)^2$.

Step 4 Factor the right side as the square of a binomial and combine terms on the left.

Step 5 Isolate the term involving $P(x)$ on the left.

Step 6 Multiply each side by a.

We can apply this procedure to $P(x) = 2x^2 + 4x - 16$.

$$P(x) = 2x^2 + 4x - 16$$

$$\frac{P(x)}{2} = x^2 + 2x - 8$$ Divide by 2 so that the coefficient of x^2 is 1.

$$\frac{P(x)}{2} + 8 = x^2 + 2x$$ Add 8 to each side.

$$\frac{P(x)}{2} + 8 + 1 = x^2 + 2x + 1$$ Add $\left[\frac{1}{2}(\mathbf{2})\right]^2 = 1$ to each side to complete the square on the right.

Be sure to add 1 to each side.

$$\frac{P(x)}{2} + 9 = (x + 1)^2$$ Add on the left; factor on the right.

$$\frac{P(x)}{2} = (x + 1)^2 - 9$$ Subtract 9 from each side.

$$P(x) = 2(x + 1)^2 - 18$$ Multiply each side by 2.

$$P(x) = \underset{a}{2}[x - (\underset{h}{-1})]^2 - \underset{k}{18}$$

From this form of the equation, we see that $a = 2$, $h = -1$, and $k = -18$. Thus, the graph of $P(x) = 2x^2 + 4x - 16$ is the graph of $y = x^2$ vertically stretched by a factor of 2 and shifted 1 unit left and 18 units downward. The vertex of the parabola has coordinates $(-1, -18)$, the domain is $(-\infty, \infty)$, and the range is $[-18, \infty)$. The function decreases on the interval $(-\infty, -1]$ and increases on the interval $[-1, \infty)$. (See Figure 9 on the previous page.)

EXAMPLE 1 Completing the Square

Complete the square on $P(x) = -x^2 - 6x - 8$.

Looking Ahead to Calculus

An important concept in calculus is the *definite integral*. If $P(x) \geq 0$ on the interval $[c, d]$, then the symbol

$$\int_c^d P(x)\,dx$$

represents the area of the region above the x-axis and below the graph of P from $x = c$ to $x = d$. For example, in Figure 10 on the next page, with

$$P(x) = -x^2 - 6x - 8,$$

$c = -4$, and $d = -2$, calculus provides the tools for determining that the area enclosed by the parabola and the x-axis is $\frac{4}{3}$ (square units).

Solution

$$P(x) = -x^2 - 6x - 8$$

$$-P(x) = x^2 + 6x + 8 \quad \text{Multiply by } -1.$$

$$-P(x) - 8 = x^2 + 6x \quad \text{Subtract 8.}$$

$$-P(x) - 8 + 9 = x^2 + 6x + 9 \quad \text{Add } \left[\tfrac{1}{2}(6)\right]^2 = 9 \text{ to each side.}$$

Be sure to add 9 to each side.

$$-P(x) + 1 = (x + 3)^2 \quad \text{Add on the left; factor on the right.}$$

$$-P(x) = (x + 3)^2 - 1 \quad \text{Subtract 1.}$$

$$P(x) = -(x + 3)^2 + 1 \quad \text{Multiply by } -1.$$

■

Graphs of Quadratic Functions

The quadratic function defined by $P(x) = -x^2 - 6x - 8 = -(x + 3)^2 + 1$ from Example 1 can now be graphed easily because it is written in the form $P(x) = a(x - h)^2 + k$. Also, recall from Chapter 1 that the y-intercept of the graph of an equation is the y-value that corresponds to $x = 0$. For a parabola given in the form $P(x) = ax^2 + bx + c$, the y-intercept is $P(0) = c$. ***A comprehensive graph of a quadratic function shows all intercepts and the vertex of the parabola.***

Characteristics of the Graph of $P(x) = a(x - h)^2 + k$

The graph of $P(x) = a(x - h)^2 + k$ $(a \neq 0)$

(a) is a parabola with vertex (h, k) and vertical line $x = h$ as axis of symmetry;

(b) opens upward if $a > 0$ and downward if $a < 0$;

(c) is wider than the graph of $y = x^2$ if $0 < |a| < 1$ and narrower than the graph of $y = x^2$ if $|a| > 1$.

EXAMPLE 2 Graphing a Quadratic Function

Graph the function defined by $P(x) = -(x + 3)^2 + 1$ from Example 1. Give the domain and range and the intervals over which the function is increasing or decreasing.

Analytic Solution

The vertex (h, k) is $(-3, 1)$, so the graph is that of $f(x) = x^2$ shifted 3 units to the left and 1 unit upward. Since

$$P(0) = -(0 + 3)^2 + 1 = -8,$$

the y-intercept is -8. We plot the vertex and y-intercept. To locate another point on the graph, we choose a value of x on the same side of the vertex as the y-intercept. If $x = -1$, for example, then

$$P(-1) = -(-1 + 3)^2 + 1 = -3.$$

Graphing Calculator Solution

A calculator graph is shown in Figure 11 on the next page. Since the vertex is $(-3, 1)$ and the y-intercept is -8, choose the viewing window $[-8, 2]$ by $[-10, 2]$, which gives a comprehensive graph of the function.

(continued)

Thus the point $(-1, -3)$ lies on the graph. The line $x = -3$ through the vertex of the parabola is called the **axis of symmetry:** If the graph were folded along this line, the two halves would coincide. Using symmetry about $x = -3$, we plot the points $(-5, -3)$ and $(-6, -8)$, which also lie on the graph.

Because the coefficient of $(x + 3)^2$ is -1, the graph opens downward and has the same shape as that of $f(x) = x^2$. See Figure 10.

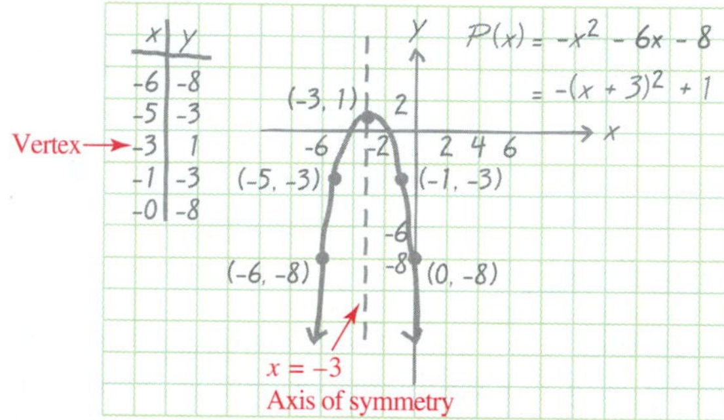

FIGURE 10

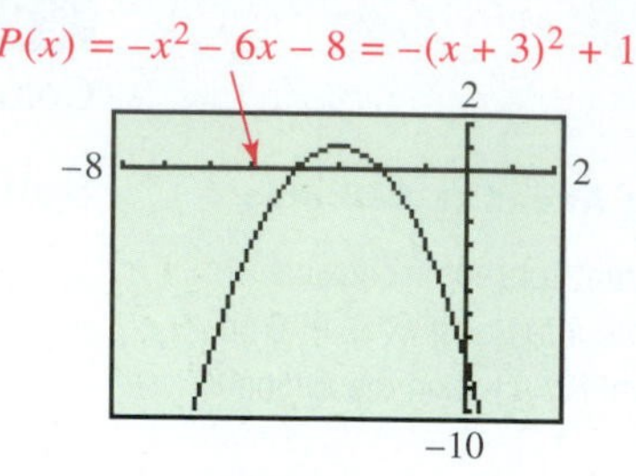

FIGURE 11

The domain is $(-\infty, \infty)$. Because the graph opens downward, the vertex $(-3, 1)$ is the highest point on the graph. Thus, the range is $(-\infty, 1]$. The function increases on the interval $(-\infty, -3]$ and decreases on $[-3, \infty)$. ■

Looking Ahead to Calculus

The derivative of a function provides a formula for finding the slope of a line tangent to the graph of the function. Using the methods of calculus, we can show that the function of Examples 1 and 2,

$$P(x) = -x^2 - 6x - 8,$$

has derivative

$$P'(x) = -2x - 6.$$

If we solve $P'(x) = 0$, we find the x-value for which the graph of P has a horizontal tangent. Solve this equation and show that its solution gives the x-value of the vertex. Notice that if you draw a tangent line at the vertex, it is a horizontal line with slope 0.

Vertex Formula

We can determine the coordinates of the vertex of the graph of a quadratic function by completing the square, as shown earlier. Rather than go through that procedure for each individual function, we generalize the result for the standard form of the quadratic function defined by $P(x) = ax^2 + bx + c$, where $a \neq 0$.

$$P(x) = ax^2 + bx + c \qquad \text{Standard form}$$

$$y = ax^2 + bx + c \qquad \text{Replace } P(x) \text{ with } y.$$

$$\frac{y}{a} = x^2 + \frac{b}{a}x + \frac{c}{a} \qquad \text{Divide by } a.$$

$$\frac{y}{a} - \frac{c}{a} = x^2 + \frac{b}{a}x \qquad \text{Subtract } \tfrac{c}{a}.$$

$$\frac{y}{a} - \frac{c}{a} + \frac{b^2}{4a^2} = x^2 + \frac{b}{a}x + \frac{b^2}{4a^2} \qquad \text{Add } \left[\tfrac{1}{2}\left(\tfrac{b}{a}\right)\right]^2 = \tfrac{b^2}{4a^2}.$$

$$\frac{y}{a} + \frac{b^2 - 4ac}{4a^2} = \left(x + \frac{b}{2a}\right)^2 \qquad \text{Combine terms on the left; factor on the right.}$$

$$\frac{y}{a} = \left(x + \frac{b}{2a}\right)^2 - \frac{b^2 - 4ac}{4a^2} \qquad \text{Isolate } y\text{-term on the left.}$$

$$y = a\left(x + \frac{b}{2a}\right)^2 + \frac{4ac - b^2}{4a} \qquad \text{Multiply by } a.$$

$$P(x) = a\Big[x - \underbrace{\left(-\frac{b}{2a}\right)}_{h}\Big]^2 + \underbrace{\frac{4ac - b^2}{4a}}_{k} \qquad \text{Write in the form } P(x) = a(x - h)^2 + k.$$

TECHNOLOGY NOTE

Most current models of graphing calculators are capable of determining the coordinates of the "highest point" or "lowest point" in a designated interval of a graph, usually with commands like "maximum" or "minimum." With this capability, you can find the coordinates of the vertex of a parabola graphically.

The final equation shows that the vertex (h, k) can be expressed in terms of a, b, and c. It is not necessary to memorize the expression for k, since it is equal to $P(h) = P\left(-\frac{b}{2a}\right)$.

Vertex Formula

The vertex of the graph of $P(x) = ax^2 + bx + c$ $(a \neq 0)$ is the point

$$\left(-\frac{b}{2a}, P\left(-\frac{b}{2a}\right)\right).$$

GCM **EXAMPLE 3** **Using the Vertex Formula**

Find the coordinates of the vertex of the graph of $P(x) = -.65x^2 + \sqrt{2}x + 4$.

(a) Give the exact values of the coordinates.

(b) Give the approximate values of the coordinates to the nearest hundredth.

Analytic Solution

(a) Apply the vertex formula, with $a = -.65$ and $b = \sqrt{2}$.

$$x = -\frac{b}{2a} = -\frac{\sqrt{2}}{2(-.65)} = \frac{\sqrt{2}}{2(.65)}$$

$$y = P\left(-\frac{b}{2a}\right) = -.65\left(\frac{\sqrt{2}}{2(.65)}\right)^2 + \sqrt{2}\left(\frac{\sqrt{2}}{2(.65)}\right) + 4$$

These are the *exact* values of x and y.

(b) Using a calculator to approximate these values, we find that, to the nearest hundredth,

$$x \approx 1.09 \quad \text{and} \quad y \approx 4.77.$$

Thus, the vertex is $(1.09, 4.77)$.

Graphing Calculator Solution

Graphing the function and using the calculator to locate the highest point on the graph, we see a display of X ≈ 1.0878556 and Y ≈ 4.7692308 in Figure 12. These values agree with the approximate values in our analytic solution.

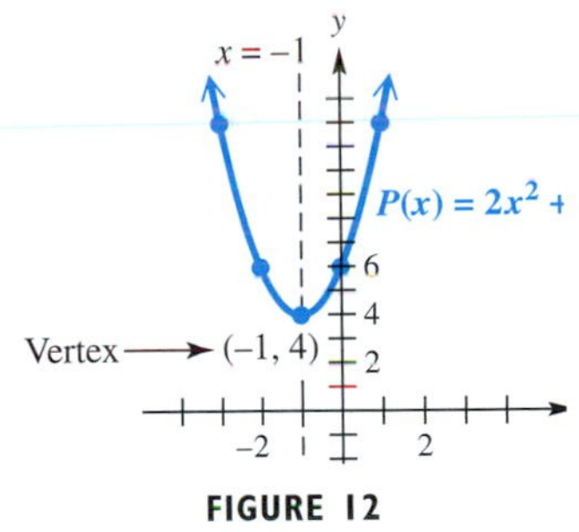

FIGURE 12

EXAMPLE 4 **Graphing a Quadratic Function**

Graph the function defined by $P(x) = 2x^2 + 4x + 6$.

Solution For this quadratic function, $a = 2$, $b = 4$, and $c = 6$. Find the coordinates of the vertex by using the vertex formula.

Use parentheses around substituted values to avoid errors.

$$x = -\frac{b}{2a} = -\frac{4}{2(2)} = -1$$

$$y = 2(-1)^2 + 4(-1) + 6 = 4$$

Thus, the vertex is $(-1, 4)$. The y-intercept is $P(0) = 6$. The table shows five key points on the graph, seen in Figure 13 on the next page.

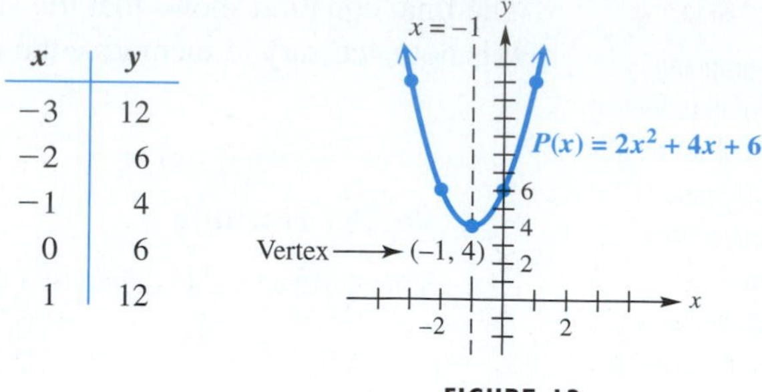

x	y
-3	12
-2	6
-1	4
0	6
1	12

FIGURE 13

Extreme Values

TECHNOLOGY NOTE

Graphing calculators have the capabilities of locating extrema to great accuracy. The TI-83/84 Plus, for example, has fMax(and fMin(capabilities, which allow the user to locate extrema without graphing. See Figure 15 in Example 5(a).

The vertex of the graph of $P(x) = ax^2 + bx + c$ is the lowest point on the graph of the function if $a > 0$ and the highest point if $a < 0$. Such points are called **extreme points** (also **extrema;** singular, **extremum**). For the quadratic function defined by $P(x) = ax^2 + bx + c$,

(a) if $a > 0$, then the vertex (h, k) is called the **minimum point** of the graph. The **minimum value** of the function is $P(h) = k$.

(b) if $a < 0$, then the vertex (h, k) is called the **maximum point** of the graph. The **maximum value** of the function is $P(h) = k$.

Figure 14 illustrates these ideas.

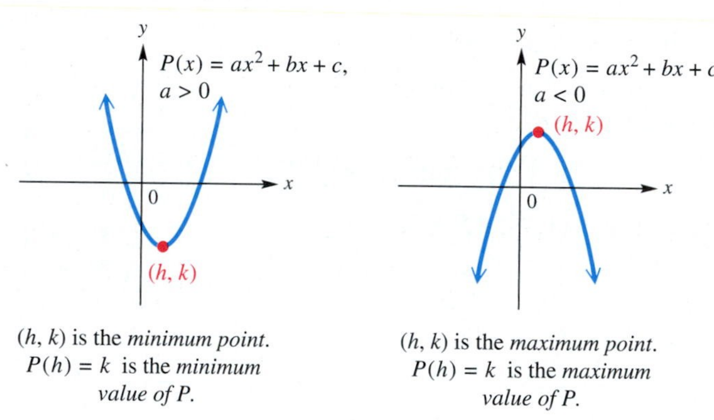

FIGURE 14

NOTE For a quadratic function whose graph has two x-intercepts, the x-coordinate of the extreme point (vertex) will always be exactly halfway between those intercepts.

GCM **EXAMPLE 5** **Identifying Extreme Points and Extreme Values**

Give the coordinates of the extreme point of the graph of each function and the corresponding maximum or minimum value of the function.

(a) $P(x) = 2x^2 + 4x - 16$ **(b)** $P(x) = -x^2 - 6x - 8$

Analytic Solution

(a) In the discussion preceding Example 1, we found that the vertex of the graph of

$$\begin{aligned} P(x) &= 2x^2 + 4x - 16 \\ &= 2(x + 1)^2 - 18 \end{aligned}$$

is $(-1, -18)$. The graph opens upward, since $a = 2 > 0$ (as seen in Figure 9 on page 181), so the vertex $(-1, -18)$ is the minimum point and -18 is the minimum value of the function.

(b) We found the vertex $(-3, 1)$ of this parabola in Example 1 and graphed the function in Example 2. Since $a = -1 < 0$ in

$$\begin{aligned} P(x) &= -x^2 - 6x - 8 \\ &= -(x + 3)^2 + 1, \end{aligned}$$

the graph of the parabola opens downward (as seen in Figure 10 on page 184), so $(-3, 1)$ is the maximum point and 1 is the maximum function value.

Graphing Calculator Solution

(a) The screens in Figure 15 use the fMin calculator function, which gives the x-value where the minimum value occurs on the graph of $P(x) = 2x^2 + 4x - 16$. The y-value, found by substitution, is the minimum value of the function.

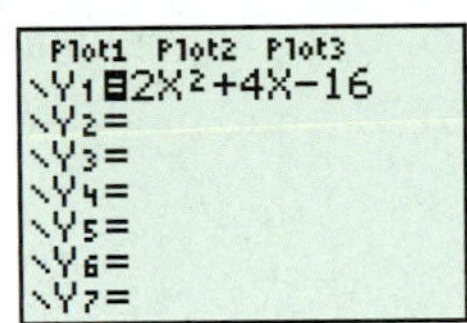

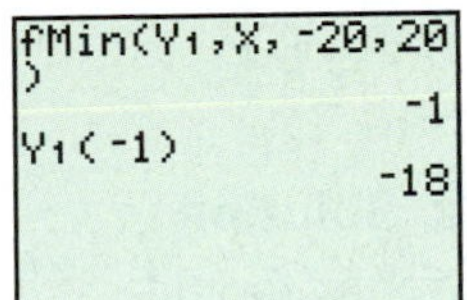

FIGURE 15

(b) The screen in Figure 16 shows an alternative way to find the coordinates of an extreme point with a graphing calculator. The results agree with our analytic results in part (b).

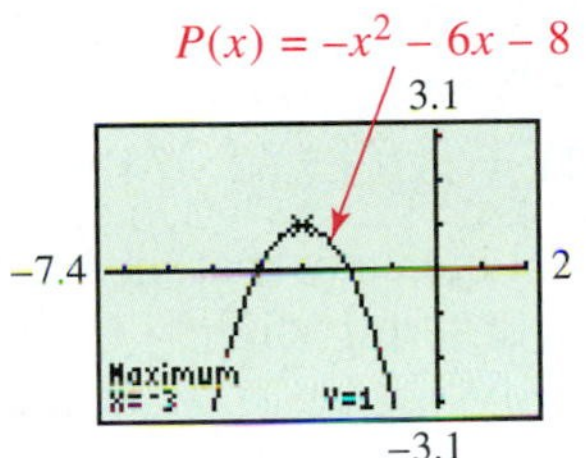

FIGURE 16

Applications and Quadratic Models

Quadratic functions make good models for many applications.

GCM **EXAMPLE 6** **Modeling Hospital Spending with a Quadratic Function**

The table gives data for annual percent increase y in spending on hospital services in the years 1994–2001, where x is the number of years since 1990. The data are plotted in the scatter diagram in Figure 17.

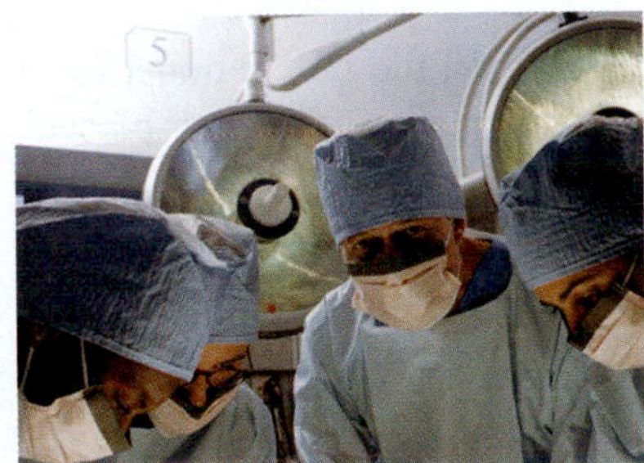

Year x	Percent Increase y
4	1.8
5	.8
6	.5
7	1.3
8	3.4
9	5.8
10	7.1
11	12.0

Source: Center for Studying Health System Change.

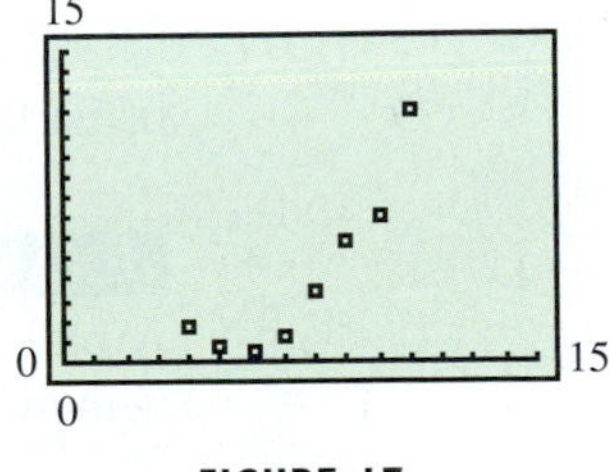

FIGURE 17

The scatter diagram suggests that a quadratic function with a positive value of a (so that the graph opens upward) would be a good model for these data. The function defined by

$$f(x) = .37x^2 - 4.1x + 12$$

gives a good approximation of the data.

(a) Use $f(x)$ to approximate the year when the percent increase was a minimum.

(b) Use the model to find the minimum percent increase.

Analytic Solution

Use the vertex formula.

(a) The x-value of the minimum point is

$$-\frac{b}{2a} = -\frac{-4.1}{2(.37)} \approx 5.54 \approx 6.$$

Since $x = 6$ corresponds to 1996, the percent increase was at a minimum in 1996.

(b) The minimum value is

$$f(5.54) = .37(5.54)^2 - 4.1(5.54) + 12 \approx .64$$

Thus, the model indicates that the minimum percent increase was about .64, differing slightly from the data value of .5 in the table.

Graphing Calculator Solution

Use a calculator to graph the function, and then use the capabilities of the calculator to get the coordinates of the minimum point. See Figure 18.

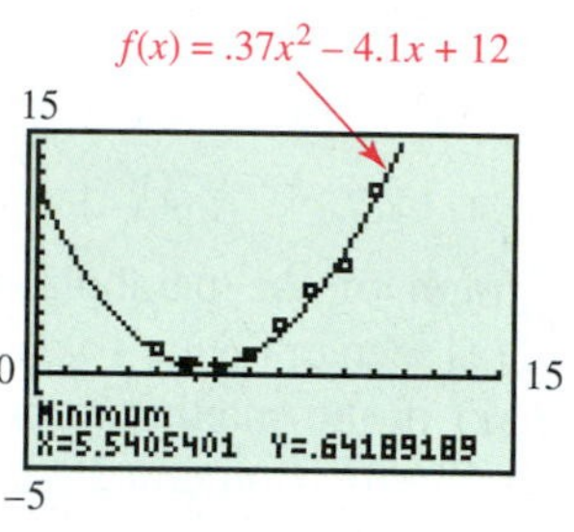

FIGURE 18

An important application of quadratic functions deals with the height of a projected object as a function of the time elapsed after it is launched.

TECHNOLOGY NOTE

In the formula for the height of a projected object, height is a function of time, and we use t to represent the independent variable. However, when graphing this type of function on a calculator, we use x, since graphing calculators use x for the independent variable. There is no difference between the formulas

$$s(t) = -16t^2 + v_0t + s_0$$

and

$$s(x) = -16x^2 + v_0x + s_0.$$

Height of a Projected Object

If air resistance is neglected, the height s (in feet) of an object projected directly upward from an initial height s_0 feet with initial velocity v_0 feet per second is

$$\mathbf{s(t) = -16t^2 + v_0t + s_0,}$$

where t is the number of seconds after the object is projected.

The coefficient of t^2 (that is, -16) is a constant based on the gravitational force of Earth. This constant varies on other surfaces, such as the moon or the other planets.

EXAMPLE 7 Solving a Problem Involving Projectile Motion

A ball is thrown directly upward from an initial height of 100 feet with an initial velocity of 80 feet per second.

(a) Give the function that describes the height of the ball in terms of time t.

(b) Graph this function so that the y-intercept, the positive x-intercept, and the vertex are visible.

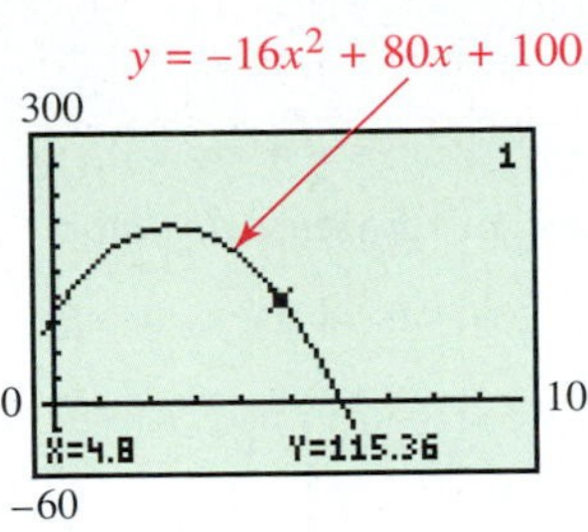

FIGURE 19

(c) Figure 19 shows that the point (4.8, 115.36) lies on the graph of the function. What does this mean for this particular situation?

(d) After how many seconds does the projectile reach its maximum height? What is the maximum height? Solve analytically and graphically.

(e) For what interval of time is the height of the ball greater than 160 feet? Determine the answer graphically.

(f) After how many seconds will the ball fall to the ground? Determine the answer graphically.

Solution

(a) Use the projectile height function with $v_0 = 80$ and $s_0 = 100$.

$$s(t) = -16t^2 + 80t + 100$$

(b) There are many suitable choices for such a window. One choice is $[0, 10]$ by $[-60, 300]$, as in Figure 19, which shows the graph of $y = -16x^2 + 80x + 100$. (Here, $x = t$.)

(c) In Figure 19, when $x = 4.8$, $y = 115.36$. Therefore, when 4.8 seconds have elapsed, the projectile is at a height of 115.36 feet.

(d) Use the vertex formula to find the coordinates of the vertex of the parabola.

$$x = -\frac{b}{2a} = -\frac{80}{2(-16)} = 2.5$$

$$y = -16(2.5)^2 + 80(2.5) + 100 = 200$$

After 2.5 seconds, the ball reaches its maximum height of 200 feet.

Using the capabilities of the calculator, we see that the vertex coordinates are indeed (2.5, 200). See Figure 20.

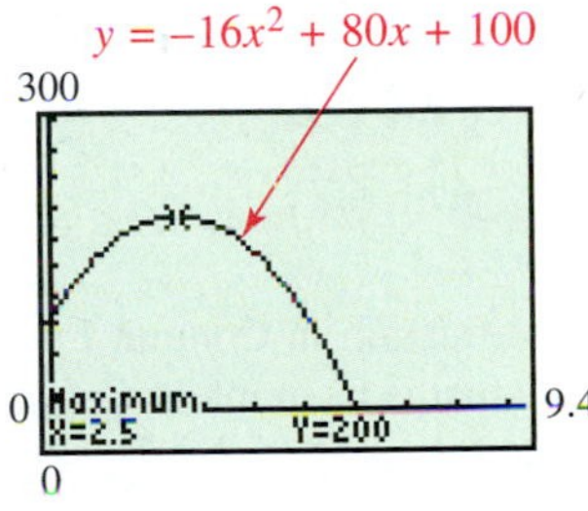

FIGURE 20

CAUTION It is easy to misinterpret the graph in Figure 20. ***This graph does not depict the path followed by the ball; it depicts height as a function of time.***

(e) With $y_1 = -16x^2 + 80x + 100$ and $y_2 = 160$ graphed, as shown in Figure 21, locate the two points of intersection. The x-coordinates of these points are approximately .92 and 4.08, respectively. Thus, between .92 and 4.08 seconds, the ball is more than 160 feet above the ground; that is, $y_1 > y_2$.

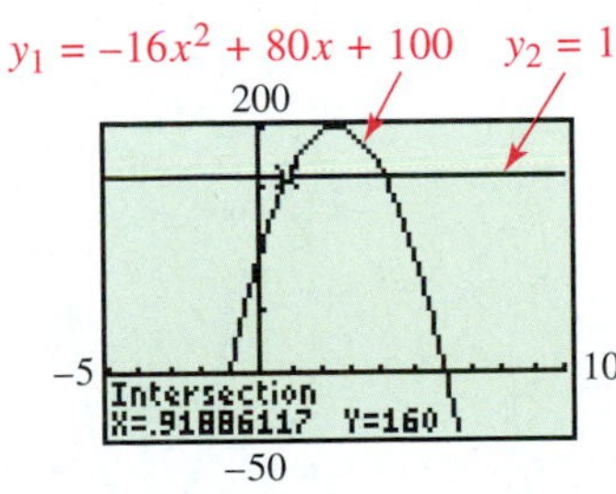

FIGURE 21

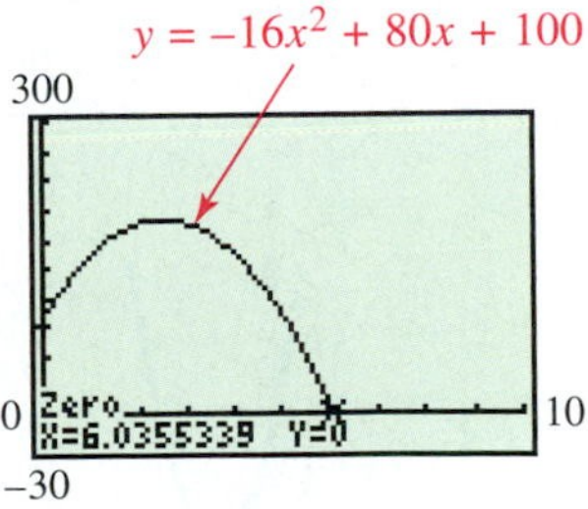

FIGURE 22

(f) When the ball falls to the ground, its height will be 0 feet, so find the positive x-intercept of the graph. As shown in Figure 22, this intercept is about 6.04, which means that the ball reaches the ground about 6.04 seconds after it is thrown. ■

3.2 Exercises

Concept Check *Match each function in Exercises 1–4 with its graph in A–D. Then check your answers with your calculator.*

1. $y = (x - 4)^2 - 3$

2. $y = -(x - 4)^2 + 3$

3. $y = (x + 4)^2 - 3$

4. $y = -(x + 4)^2 + 3$

A.

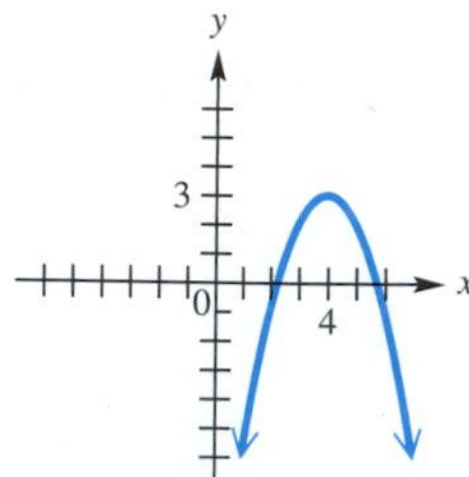

B.

C.

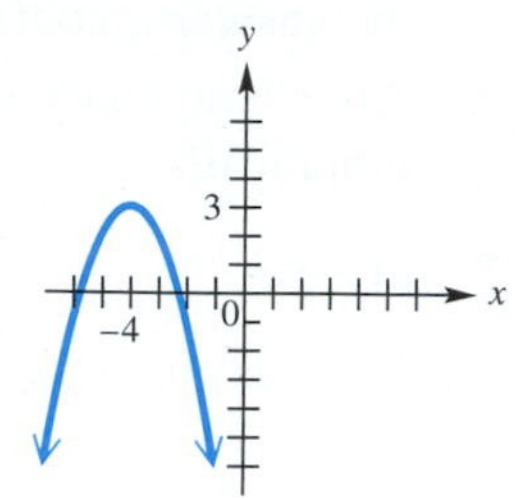

D.

For each quadratic function defined in Exercises 5–16, **(a)** *write the function in the form $P(x) = a(x - h)^2 + k$;* **(b)** *give the vertex of the parabola; and* **(c)** *graph the function by hand. Then support your answers with a calculator graph in a viewing window that shows a comprehensive graph of the function.*

5. $P(x) = x^2 - 2x - 15$

6. $P(x) = x^2 + 2x - 15$

7. $P(x) = -x^2 - 3x + 10$

8. $P(x) = -x^2 + 3x + 10$

9. $P(x) = x^2 - 6x$

10. $P(x) = x^2 + 4x$

11. $P(x) = 2x^2 - 2x - 24$

12. $P(x) = 3x^2 + 3x - 6$

13. $P(x) = -2x^2 + 6x$

14. $P(x) = -4x^2 + 4x$

15. $P(x) = 3x^2 + 4x - 1$

16. $P(x) = 4x^2 + 3x - 1$

17. *Concept Check* Match each equation in Column I with the description of the parabola that is its graph in Column II.

I	II
(a) $y = (x - 4)^2 - 2$	**A.** Vertex $(2, -4)$, opens downward
(b) $y = (x - 2)^2 - 4$	**B.** Vertex $(2, -4)$, opens upward
(c) $y = -(x - 4)^2 - 2$	**C.** Vertex $(4, -2)$, opens downward
(d) $y = -(x - 2)^2 - 4$	**D.** Vertex $(4, -2)$, opens upward

18. *Concept Check* Match each equation in Column I with the description of the parabola that is its graph in Column II, assuming $a > 0$, $h > 0$, and $k > 0$.

I	II
(a) $y = -a(x + h)^2 + k$	**A.** Vertex in quadrant I, two x-intercepts
(b) $y = a(x - h)^2 + k$	**B.** Vertex in quadrant I, no x-intercepts
(c) $y = a(x + h)^2 + k$	**C.** Vertex in quadrant II, two x-intercepts
(d) $y = -a(x - h)^2 + k$	**D.** Vertex in quadrant II, no x-intercepts

Use the graphs of the functions in Exercises 19–24 to do each of the following: **(a)** *Give the coordinates of the vertex.* **(b)** *Give the domain and range.* **(c)** *Give the equation of the axis of symmetry.* **(d)** *Give the interval over which the function is increasing.* **(e)** *Give the interval over which the function is decreasing.* **(f)** *State whether the vertex is a maximum or minimum point, and give the corresponding maximum or minimum value of the function.*

19.

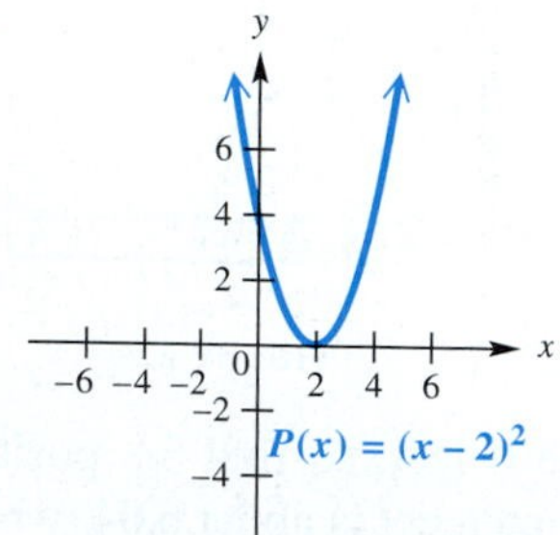

20.

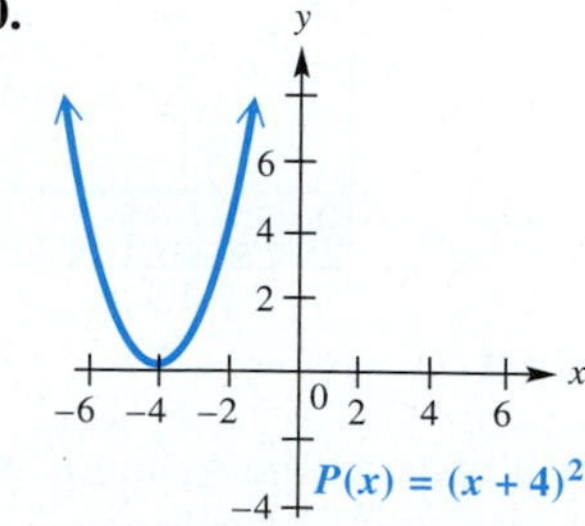

21.

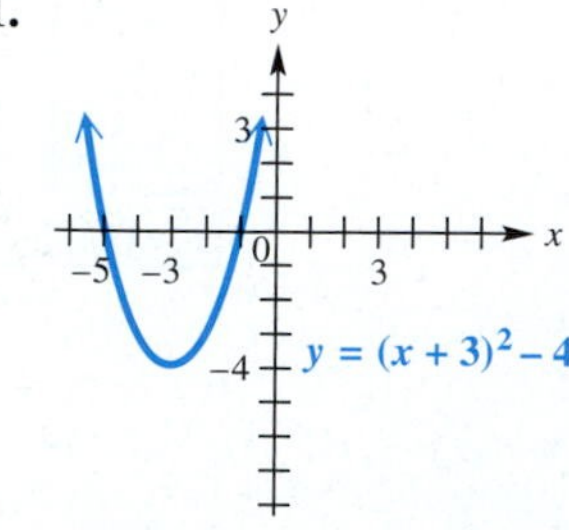

22.

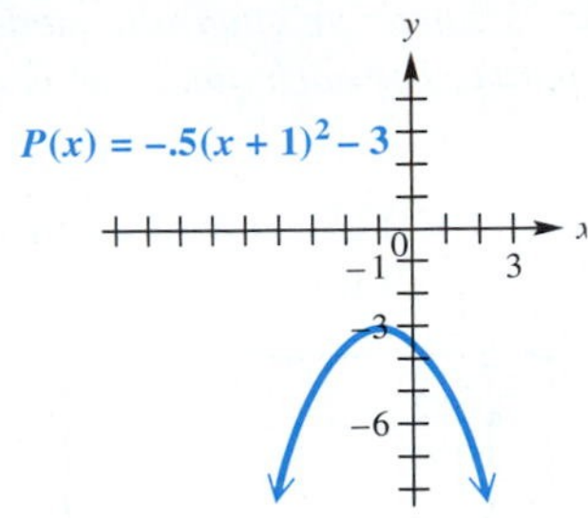

23.

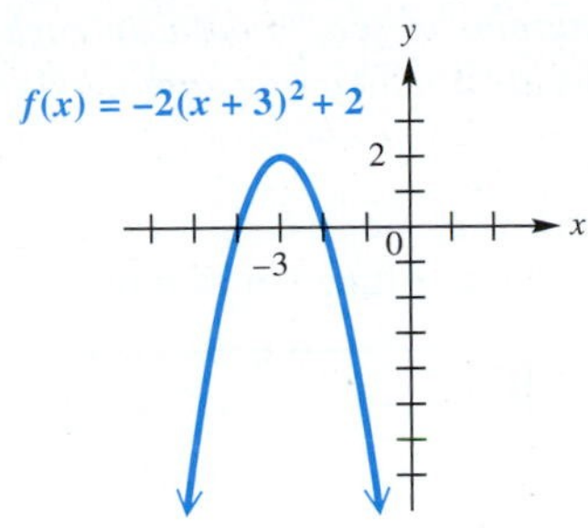

24.

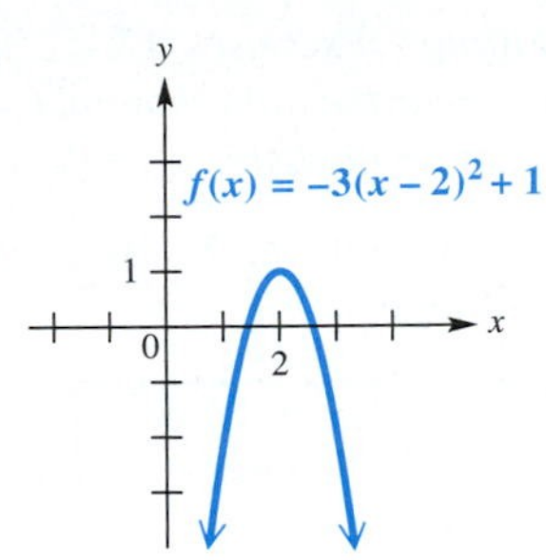

For each quadratic function defined in Exercises 25–32, **(a)** *use the vertex formula to find the coordinates of the vertex and* **(b)** *graph the function by hand.*

25. $P(x) = x^2 - 10x + 21$

26. $P(x) = x^2 - 2x + 3$

27. $y = -x^2 + 4x - 2$

28. $y = -x^2 + 2x + 1$

29. $P(x) = 2x^2 - 4x + 5$

30. $P(x) = 2x^2 - 8x + 9$

31. $P(x) = -3x^2 + 24x - 46$

32. $P(x) = -3x^2 - 18x - 23$

Graph each function in a viewing window that will allow you to use your calculator to approximate **(a)** *the coordinates of the vertex and* **(b)** *the x-intercepts. Give values to the nearest hundredth.*

33. $P(x) = -.32x^2 + \sqrt{3}x + 2.86$

34. $P(x) = -\sqrt{2}x^2 + .45x + 1.39$

35. $y = 1.34x^2 - 3x + \sqrt{5}$

36. $y = -.55x^2 + 3.21x$

37. $y = -1.24x^2 + 1.68x$

38. $y = 2.96x^2 + 1.31$

Concept Check *The figure shows the graph of a quadratic function $y = f(x)$. Use it to work Exercises 39–42.*

39. What is the minimum value of $f(x)$?

40. For what value of x is $f(x)$ as small as possible?

41. How many real solutions are there of the equation $f(x) = 1$?

42. How many real solutions are there of the equation $f(x) = 3$?

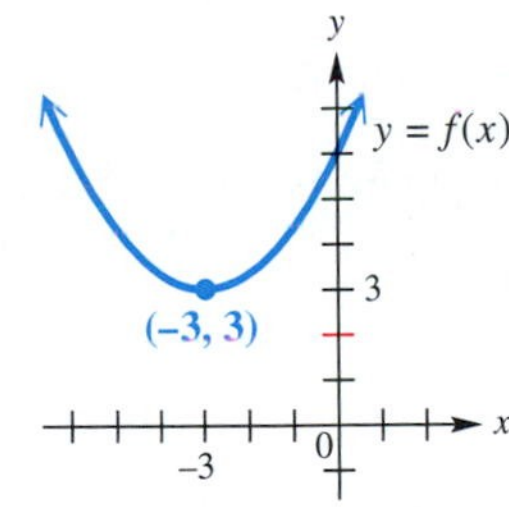

In each table, Y_1 is defined as a quadratic function. Use the table to do the following: **(a)** *Determine the coordinates of the vertex of the graph.* **(b)** *Determine whether the vertex is a minimum point or a maximum point.* **(c)** *Find the minimum or maximum value of the function.* **(d)** *Determine the range of the function.*

43.

X	Y1
1	15
2	0
3	-9
4	-12
5	-9
6	0
7	15

X=1

44.

X	Y1
-6	27
-5	17
-4	11
-3	9
-2	11
-1	17
0	27

X= -6

45.

X	Y1
0	-7
.5	-2
1	1
1.5	2
2	1
2.5	-2
3	-7

X=0

46.

X	Y1
-4	-9.75
-3.5	-6
-3	-3.75
-2.5	-3
-2	-3.75
-1.5	-6
-1	-9.75

X= -4

Curve Fitting *Exercises 47–52 show scatter diagrams of sets of data. In each case, tell whether a linear or quadratic model is appropriate for the data. For each linear model, decide whether the slope should be positive or negative. For each quadratic model, decide whether the leading coefficient a should be positive or negative.*

47. Social Security assets as a function of time

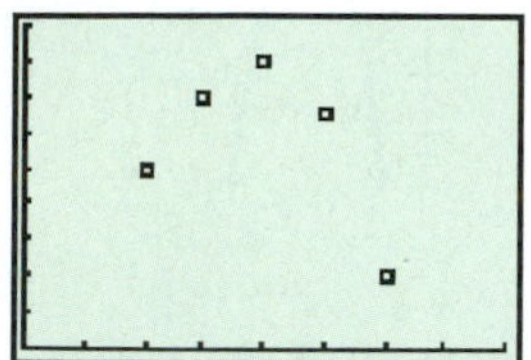

48. Growth in science centers/museums as a function of time

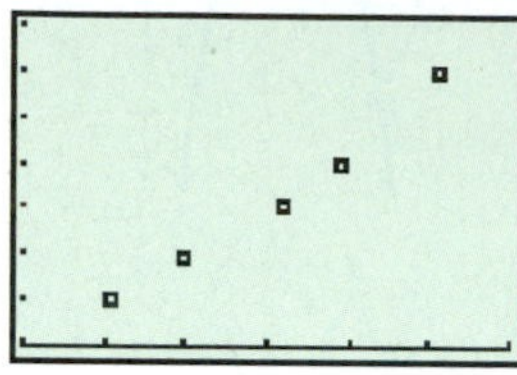

49. Value of U.S. salmon catch as a function of time

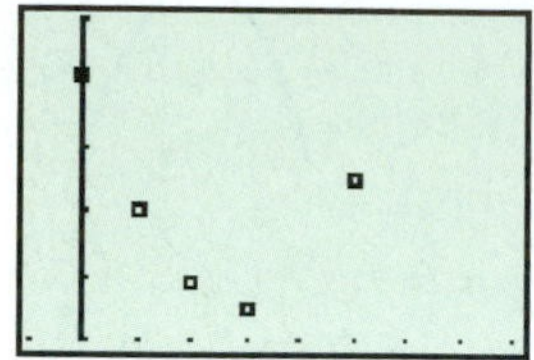

50. Height of an object thrown upward from a building as a function of time

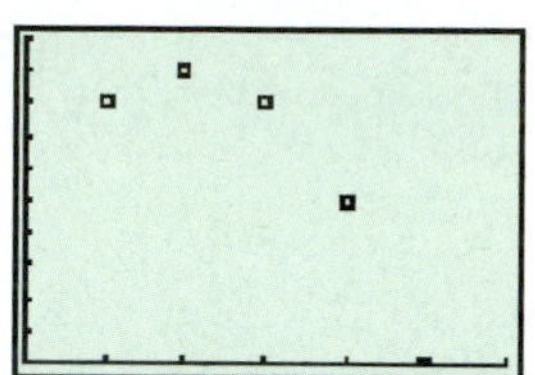

51. Number of shopping centers as a function of time

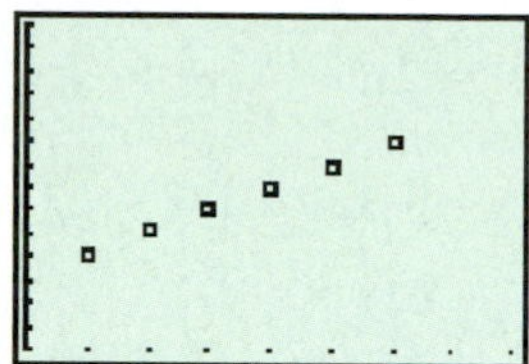

52. Newborns with AIDS as a function of time

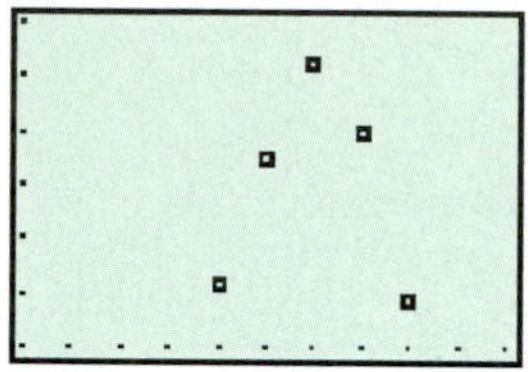

(Modeling) *Solve each problem.*

53. ***Business Mergers and Acquisitions*** The approximate value of U.S. business mergers and acquisitions for selected years between 1995 and 2002 is shown in the table. In the year column, 5 represents 1995, 7 represents 1997, and so on.

Year	Value of Business Mergers and Acquisitions (in trillions of dollars)
5	5.0
7	8.8
8	9.6
9	9.6
11	7.7
12	7.0

Source: U.S. Census Bureau.

(a) Use a scatter diagram to decide whether a linear or quadratic function would model the data better. If quadratic, should the coefficient a of x^2 be positive or negative?

(b) The data are approximated by

$$f(x) = -.3053x^2 + 5.403x - 14.256,$$

where $x = 5$ represents 1995, $x = 7$ represents 1997, and so on. Analytically determine the coordinates of the vertex of the graph.

(c) Interpret the answer to part (b) as it relates to this application. How does your answer compare with the data given in the table?

54. ***Social Security Assets*** The graph shows how Social Security assets are expected to change as the number of retirees receiving benefits increases.

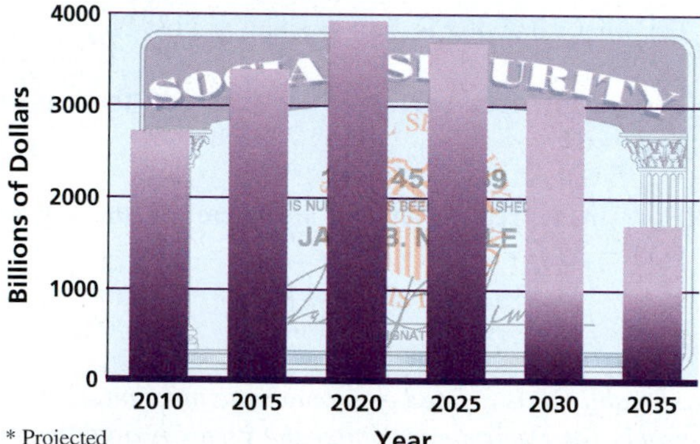

* Projected

Source: Social Security Administration.

The graph suggests that a quadratic function would be a good fit to the data, which are approximated by

$$f(x) = -10.36x^2 + 431.8x - 650.$$

In the model, $x = 10$ represents 2010, $x = 15$ represents 2015, and so on, and $f(x)$ is in billions of dollars.

(a) Explain why the coefficient of x^2 in the model is negative, based on the graph.
(b) Analytically determine the vertex of the graph.
(c) Interpret the answer to part (b) as it relates to this application.

55. ***Domestic Catfish Harvest*** The U.S. domestic catfish harvest (in millions of dollars) for the years 1995–2002 can be approximated by the quadratic function defined by

$$f(x) = -6.15x^2 + 44.43x + 400,$$

where $x = 0$ represents 1995, $x = 1$ represents 1996, and so on. (*Source:* U.S. Census Bureau.)

(a) Since the coefficient of x^2 in the model is negative, the graph of this quadratic function is a parabola that opens downward. Will the y-value of the vertex of the graph be a maximum or minimum?
(b) In what year was the maximum domestic catfish harvest? (Round down to the nearest year.) Use the actual x-value of the vertex, to the nearest tenth, to find this catch.

56. ***Tuition Increase*** The percent increase for in-state tuition at Iowa public universities during the years 1992–2002 can be modeled by

$$f(x) = .156x^2 - 2.05x + 10.2,$$

where $x = 2$ represents 1992, $x = 3$ represents 1993, and so on. (*Source:* Iowa Board of Regents.)
(a) Based on this model, by what percent (to the nearest tenth) did tuition increase in 2001?
(b) In what year was the minimum tuition increase? (Round down to the nearest year.) To the nearest tenth, by what percent did tuition increase that year?

(Modeling) *The formula for height of a projectile is*

$$s(t) = -16t^2 + v_0t + s_0,$$

where t is time in seconds, s_0 is the initial height in feet, v_0 is the initial velocity in feet per second, and $s(t)$ is in feet. Use this formula to solve each problem.

57. ***Height of a Projected Rock*** A rock is projected upward from ground level with an initial velocity of 90 feet per second. Let t represent the amount of time elapsed after it is thrown.

(a) Explain why t cannot be a negative number in this situation.
(b) Explain why $s_0 = 0$ in this problem.
(c) Give the function s that describes the height of the rock as a function of t.
(d) How high will the rock be 1.5 seconds after it is thrown?
(e) What is the maximum height attained by the rock? After how many seconds will this happen? Determine the answer analytically and graphically.
(f) After how many seconds will the rock hit the ground? Determine the answer graphically.

58. ***Height of a Toy Rocket*** A toy rocket is launched from the top of a building 50 feet tall at an initial velocity of 200 feet per second. Let t represent the amount of time elapsed after the launch.

(a) Express the height s as a function of the time t.
(b) Determine both analytically and graphically the time at which the rocket reaches its highest point. How high will it be at that time?
(c) For what time interval will the rocket be more than 300 feet above ground level? Determine the answer graphically, and give times to the nearest tenth of a second.
(d) After how many seconds will the rocket hit the ground? Determine the answer graphically.

59. ***Height of a Projected Ball*** A ball is launched upward from ground level with an initial velocity of 150 feet per second.
   (a) Determine graphically whether the ball will reach a height of 355 feet. If it will, determine the time(s) when this happens. If it will not, explain why, using a graphical interpretation.
   (b) Repeat part (a) for a ball launched from a height of 30 feet with an initial velocity of 250 feet per second.

60. ***Height of a Projected Ball on the Moon*** An astronaut on the moon throws a baseball upward. The astronaut is 6 feet, 6 inches, tall and the initial velocity of the ball is 30 feet per second. The height of the ball is approximated by the function defined by

$$s(t) = -2.7t^2 + 30t + 6.5,$$

where t is the number of seconds after the ball was thrown.
   (a) After how many seconds is the ball 12 feet above the moon's surface?
   (b) How many seconds will it take for the ball to return to the surface?
   (c) The ball will never reach a height of 100 feet. How can this be determined analytically?

Find the equation of the quadratic function satisfying the given conditions. (Hint: Determine values of a, h, and k that satisfy $P(x) = a(x - h)^2 + k$.) Express your answer in the form $P(x) = ax^2 + bx + c$. Use your calculator to support your results.

61. Vertex $(-1, -4)$; through $(5, 104)$

62. Vertex $(-2, -3)$; through $(0, -19)$

63. Vertex $(8, 3)$; through $(10, 5)$

64. Vertex $(-6, -12)$; through $(6, 24)$

65. Vertex $(-4, -2)$; through $(2, -26)$

66. Vertex $(5, 6)$; through $(1, -6)$

Concept Check *Sketch a graph that satisfies each set of given conditions.*

67. Vertex $(-2, -3)$; through $(1, 4)$

68. Vertex $(5, 6)$; through $(1, -6)$

69. Maximum value of 1 at $x = 3$; y-intercept is -4

70. Minimum value of -4 at $x = -3$; y-intercept is 3

3.3 Quadratic Equations and Inequalities

Zero-Product Property ■ Solving $x^2 = k$ ■ Quadratic Formula and the Discriminant ■ Solving Quadratic Equations ■ Solving Quadratic Inequalities ■ Formulas Involving Quadratics ■ Another Quadratic Model

A *quadratic equation* is defined as follows.

Quadratic Equation in One Variable

An equation that can be written in the form

$$ax^2 + bx + c = 0,$$

where a, b, and c are real numbers with $a \neq 0$, is a **quadratic equation in standard form.**

Zero-Product Property

Consider the following problems:

1. Find the ***zeros of the quadratic function*** defined by $P(x) = 2x^2 + 4x - 16$.
2. Find the ***x-intercepts of the graph*** of $P(x) = 2x^2 + 4x - 16$.
3. Find the ***solution set of the equation*** $2x^2 + 4x - 16 = 0$.

Each of these problems is solved by finding the numbers that make $2x^2 + 4x - 16$ equal 0.

One way to solve a quadratic equation $P(x) = 0$ is to use the zero-product property.

> **Zero-Product Property**
>
> If a and b are complex numbers and $ab = 0$, then $a = 0$ or $b = 0$ or both.

EXAMPLE 1 **Using the Zero-Product Property (Two Solutions)**

Solve $2x^2 + 4x - 16 = 0$.

Analytic Solution

$2x^2 + 4x - 16 = 0$	Solve $P(x) = 0$.
$x^2 + 2x - 8 = 0$	Divide each term by 2.
$(x + 4)(x - 2) = 0$	Factor.
$x + 4 = 0$ or $x - 2 = 0$	Zero-product property
$x = -4$ or $x = 2$	Solve each equation.

The solution set of the equation is $\{-4, 2\}$. We can check each solution by substituting it into the given equation. These results also tell us that the zeros of $P(x)$ and the x-intercepts of the graph of $P(x)$ are -4 and 2.

Graphing Calculator Solution

The screens in Figure 23 show the graph of

$$P(x) = 2x^2 + 4x - 16$$

with zeros -4 and 2. That is, $P(-4) = 0$ and $P(2) = 0$. Since the zeros satisfy $P(x) = 0$, they are also the solutions of the equation and the x-intercepts of the graph.

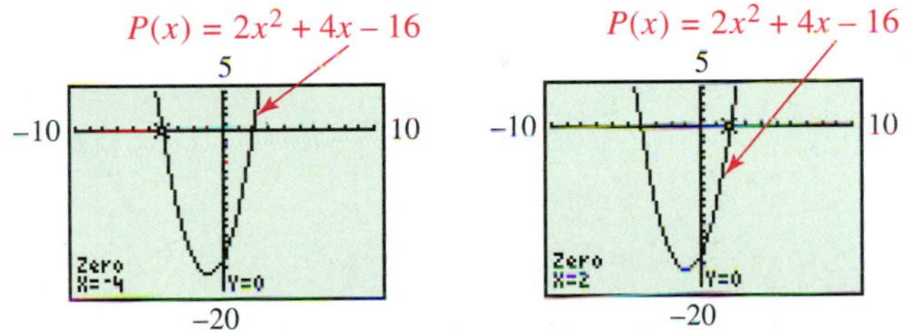

FIGURE 23

EXAMPLE 2 **Using the Zero-Product Property (One Solution)**

Find all zeros of the quadratic function defined by $P(x) = x^2 - 6x + 9$.

Analytic Solution

$x^2 - 6x + 9 = 0$	Solve $P(x) = 0$.
$(x - 3)^2 = 0$	Factor.
$x - 3 = 0$ or $x - 3 = 0$	Zero-product property
$x = 3$ or $x = 3$	Solve each equation.

There is only one *distinct* zero, 3. It is sometimes called a *double zero,* or *double solution* (*root*) of the equation.

(Notice that finding the zeros of a function defined by $P(x)$ and solving the equation $P(x) = 0$ are equivalent procedures.)

Check:

$$3^2 - 6(3) + 9 = 0 \quad ?$$
$$9 - 18 + 9 = 0 \quad ?$$
$$0 = 0 \quad \text{True}$$

Graphing Calculator Solution

The graph of the function in Figure 24 shows why there is only one zero: The graph is tangent to the x-axis at the vertex, (3, 0). The table also indicates that a zero occurs at $x = 3$, which must be a minimum value. Because of the symmetry shown in the table, 3 is the only zero, which numerically supports our analytic result.

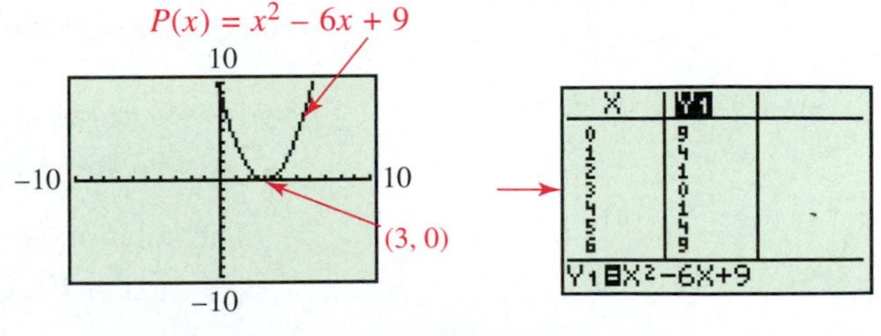

FIGURE 24

CAUTION A purely graphical approach may not prove to be as useful with a graph like the one in Figure 24, since the vertex may be slightly above or slightly below the x-axis. ***This is why we need to understand the algebraic concepts presented in this section—only when we know the mathematics can we use the technology to its utmost.***

Figure 25 shows possible numbers of x-intercepts of the graph of a quadratic function that opens upward $(a > 0)$.

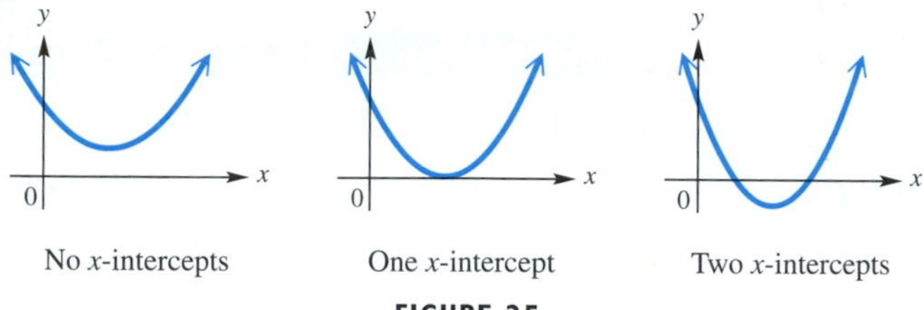

FIGURE 25

Figure 26 shows possible numbers of x-intercepts of the graph of a quadratic function that opens downward $(a < 0)$. Thus, a quadratic equation can have zero, one, or two real solutions.

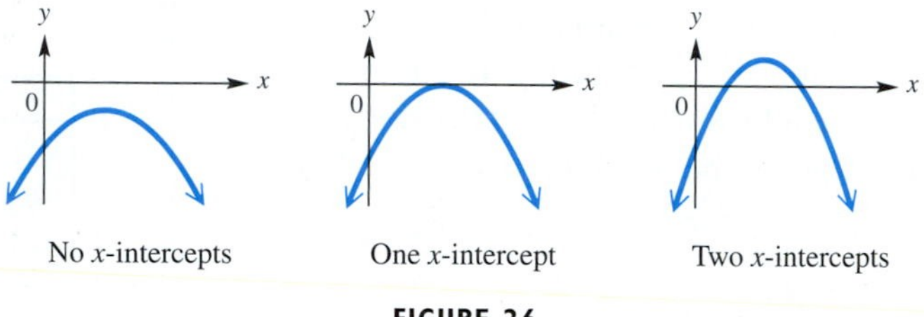

FIGURE 26

The zero-product property is useful only if the quadratic expression can be factored easily. We now develop more general methods for solving quadratic equations.

Solving $x^2 = k$

We can solve quadratic equations of the form $x^2 = k$, where k is a real number, by factoring, using the following sequence of equivalent equations.

$$x^2 = k$$

$$x^2 - k = 0 \quad \text{Subtract } k.$$

$$\left(x - \sqrt{k}\right)\left(x + \sqrt{k}\right) = 0 \quad \text{Factor.}$$

$$x - \sqrt{k} = 0 \quad \text{or} \quad x + \sqrt{k} = 0 \quad \text{Zero-product property}$$

$$x = \sqrt{k} \quad \text{or} \quad x = -\sqrt{k} \quad \text{Solve each equation.}$$

We call this result the *square root property* for solving quadratic equations.

Square Root Property

The solution set of $x^2 = k$ is

(a) $\left\{\pm\sqrt{k}\right\}$ **if** $k > 0$ **(b)** $\{0\}$ **if** $k = 0$ **(c)** $\left\{\pm i\sqrt{|k|}\right\}$ **if** $k < 0$.

As shown in Figure 27, the graph of $y_1 = x^2$ intersects the graph of $y_2 = k$ twice if $k > 0$, once if $k = 0$, and not at all if $k < 0$.

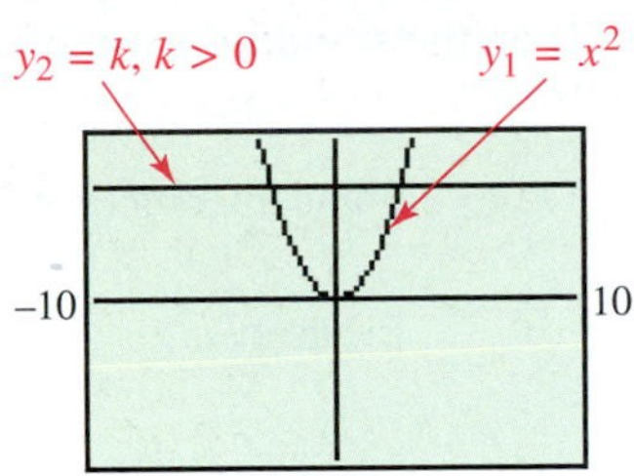

There are two points of intersection if $k > 0$.

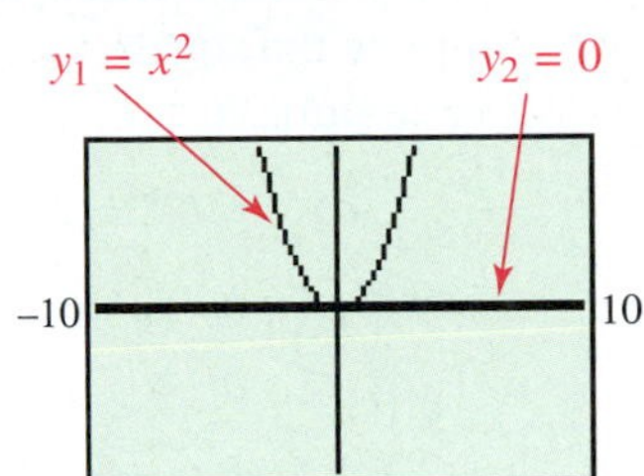

There is only one point of intersection (the origin) if $k = 0$.

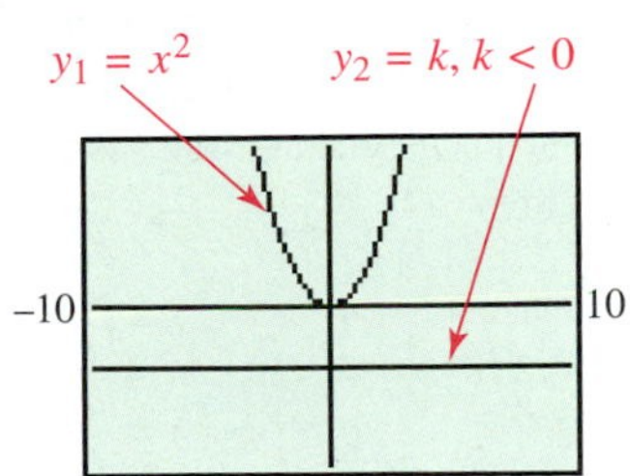

There are no points of intersection if $k < 0$.

FIGURE 27

EXAMPLE 3 Using the Square Root Property

Solve each equation.

(a) $2x^2 = -10$ **(b)** $2(x-1)^2 = 14$

Analytic Solution

(a) $2x^2 = -10$

$x^2 = -5$ Divide by 2.

There is no real number whose square is -5, so this equation has two pure imaginary solutions. Apply the square root property.

$$x = \sqrt{-5} \quad \text{or} \quad x = -\sqrt{-5}$$

$$x = i\sqrt{5} \quad \text{or} \quad x = -i\sqrt{5}$$

The solution set is $\{\pm i\sqrt{5}\}$. These solutions can be checked by substituting $i\sqrt{5}$ and $-i\sqrt{5}$ in the original equation.

(b) $2(x-1)^2 = 14$

$(x-1)^2 = 7$ Divide by 2.

$x - 1 = \pm\sqrt{7}$ Square root property

$x = 1 \pm \sqrt{7}$ Add 1.

Check: $2\left[\left(1 \pm \sqrt{7}\right) - 1\right]^2 = 14$?

$2(\pm\sqrt{7})^2 = 14$?

$2(7) = 14$?

$14 = 14$ True

The solution set is $\{1 \pm \sqrt{7}\}$.

Graphing Calculator Solution

(a) Notice that the graphs of $y_1 = 2x^2$ and $y_2 = -10$ in Figure 28 do not intersect. Thus, there are no *real* solutions.

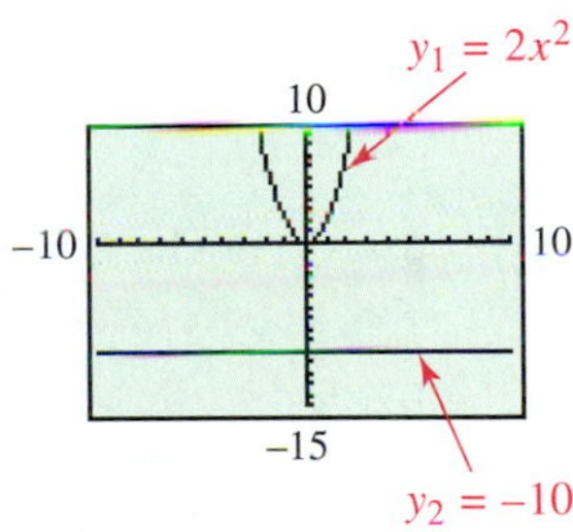

FIGURE 28

(b) Graph $y_1 = 2(x-1)^2$ and $y_2 = 14$, and locate the points of intersection. The screens in Figure 29 show the x-coordinates of these points, which are approximations of $1 \pm \sqrt{7}$, confirming the solution set $\{1 \pm \sqrt{7}\}$.

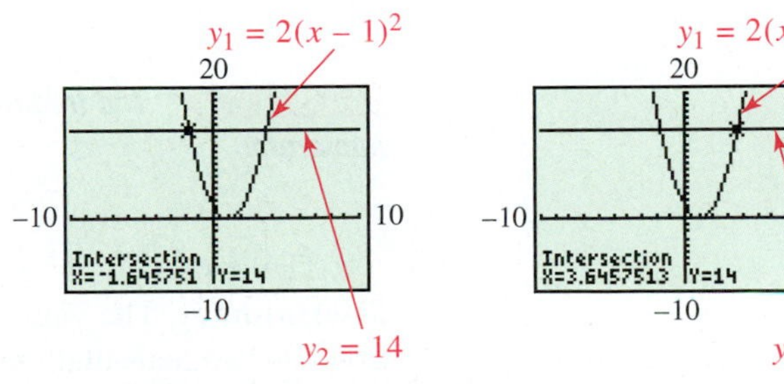

FIGURE 29

GCM TECHNOLOGY NOTE

Programs for the various makes and models of graphing calculators are available from the manufacturers, users' groups, and the Web site for this text.

Quadratic Formula and the Discriminant

There is a general formula that can be used to solve any quadratic equation. To find it, we complete the square (Section 3.2) on the standard form of a quadratic equation. For now, we assume $a > 0$.

$$ax^2 + bx + c = 0 \qquad \text{Standard form } (a \neq 0)$$

$$x^2 + \frac{b}{a}x + \frac{c}{a} = 0 \qquad \text{Divide by } a.$$

$$x^2 + \frac{b}{a}x = -\frac{c}{a} \qquad \text{Add } -\tfrac{c}{a}.$$

$$x^2 + \frac{b}{a}x + \frac{b^2}{4a^2} = \frac{b^2}{4a^2} - \frac{c}{a} \qquad \text{Add } \left[\tfrac{1}{2}\left(\tfrac{b}{a}\right)\right]^2 = \tfrac{b^2}{4a^2}.$$

Be sure to add $\frac{b^2}{4a^2}$ to *each* side.

$$\left(x + \frac{b}{2a}\right)^2 = \frac{b^2 - 4ac}{4a^2} \qquad \text{Factor on the left; simplify on the right.}$$

$$x + \frac{b}{2a} = \sqrt{\frac{b^2 - 4ac}{4a^2}} \quad \text{or} \quad x + \frac{b}{2a} = -\sqrt{\frac{b^2 - 4ac}{4a^2}} \qquad \text{Square root property}$$

$$x + \frac{b}{2a} = \frac{\sqrt{b^2 - 4ac}}{2a} \quad \text{or} \quad x + \frac{b}{2a} = \frac{-\sqrt{b^2 - 4ac}}{2a} \qquad \sqrt{4a^2} = 2a \ (a > 0)$$

$$x = \frac{-b + \sqrt{b^2 - 4ac}}{2a} \quad \text{or} \quad x = \frac{-b - \sqrt{b^2 - 4ac}}{2a} \qquad \text{Add } -\tfrac{b}{2a}.$$

These two results are also valid if $a < 0$. A compact form, called the *quadratic formula*, follows.

Quadratic Formula

The solutions of the quadratic equation $\boldsymbol{ax^2 + bx + c = 0}$, where $a \neq 0$, are given by

$$\frac{-b \pm \sqrt{b^2 - 4ac}}{2a}.$$

CAUTION ***The fraction bar in the quadratic formula extends under the $-b$ term in the numerator.***

The expression $\boldsymbol{b^2 - 4ac}$ under the radical in the quadratic formula is called the **discriminant.** The value of the discriminant determines whether the quadratic equation has two real solutions, one real solution, or no real solutions. In the last case, there will be two nonreal complex solutions.

Effect of the Discriminant

If a, b, and c are real numbers, $a \neq 0$, then the complex solutions of $ax^2 + bx + c = 0$ are described as follows, based on the value of the discriminant, $b^2 - 4ac$.

Value of $b^2 - 4ac$	Number of Solutions	Type of Solutions
Positive	Two	Real
Zero	One (a double solution)	Real
Negative	Two	Nonreal complex

Furthermore, if a, b, and c are *integers*, $a \neq 0$, the real solutions are *rational* if $b^2 - 4ac$ is the square of a positive integer.

NOTE The final sentence in the preceding box suggests that a quadratic equation can be solved by factoring if $b^2 - 4ac$ is a perfect square.

Solving Quadratic Equations

EXAMPLE 4 **Using the Quadratic Formula**

Solve $x(x - 2) = 2x - 2$ with the quadratic formula.

Analytic Solution

We must write the equation in standard form.

$$x(x - 2) = 2x - 2$$

$$x^2 - 2x = 2x - 2 \quad \text{Distributive property}$$

$$x^2 - 4x + 2 = 0 \quad \text{Subtract } 2x\text{; add 2.}$$

Substitute the values of a, b, and c into the quadratic formula.

$$x = \frac{-b \pm \sqrt{b^2 - 4ac}}{2a} \quad \text{Quadratic formula}$$

$$= \frac{-(-4) \pm \sqrt{(-4)^2 - 4(1)2}}{2(1)} \quad a = 1, b = -4, c = 2$$

$$= \frac{4 \pm \sqrt{16 - 8}}{2} = \frac{4 \pm \sqrt{8}}{2}$$

$$= \frac{4 \pm 2\sqrt{2}}{2} = \frac{2(2 \pm \sqrt{2})}{2} \quad \text{Factor first. Then divide out the common factor.}$$

$$x = 2 \pm \sqrt{2} \quad \text{Lowest terms}$$

The solution set is $\{2 + \sqrt{2}, 2 - \sqrt{2}\}$, abbreviated $\{2 \pm \sqrt{2}\}$.

Graphing Calculator Solution

Using the intersection-of-graphs method, we graph

$$y_1 = x(x - 2)$$

and

$$y_2 = 2x - 2.$$

(Note that the original form of the equation is $y_1 = y_2$.) Verify that the x-coordinates of the points of intersection in Figure 30 are approximations of $2 - \sqrt{2}$ and $2 + \sqrt{2}$.

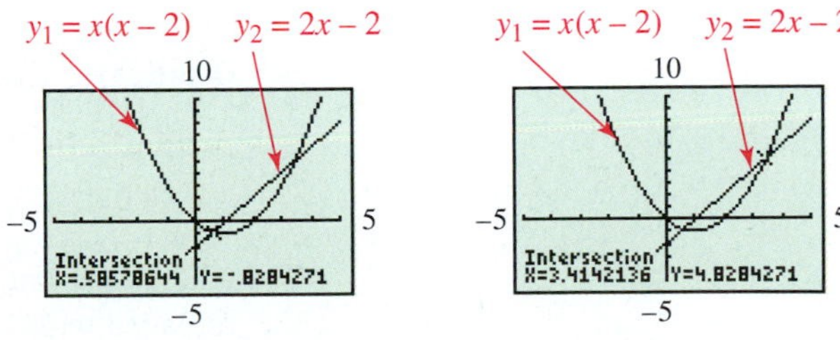

FIGURE 30

EXAMPLE 5 **Using the Quadratic Formula**

Solve $2x^2 - x + 4 = 0$ with the quadratic formula.

Solution

$$x = \frac{-b \pm \sqrt{b^2 - 4ac}}{2a}$$ Quadratic formula

Use parentheses around substituted values to avoid errors.

$$x = \frac{-(-1) \pm \sqrt{(-1)^2 - 4(2)(4)}}{2(2)}$$ $a = 2, b = -1, c = 4$

$$x = \frac{1 \pm \sqrt{1 - 32}}{4}$$

$$x = \frac{1 \pm \sqrt{-31}}{4}$$

Because the discriminant is negative, there are two nonreal complex solutions.

$$x = \frac{1 \pm i\sqrt{31}}{4} = \frac{1}{4} \pm i\frac{\sqrt{31}}{4}$$ Write in $a + bi$ form.

The solution set is $\left\{\frac{1}{4} \pm i\frac{\sqrt{31}}{4}\right\}$. ■

NOTE If we were to graph $y = 2x^2 - x + 4$, we would see that there are no x-intercepts, supporting our result in Example 5. There are no *real* solutions. (See Example 3(a) as well.)

FOR DISCUSSION

Solve $x^2 = 4x - 4$ analytically. Then, graph

$$y_1 = x^2 \quad \text{and} \quad y_2 = 4x - 4,$$

and use the intersection-of-graphs method to support your result.

1. Do you encounter a problem if you use the standard viewing window?
2. Why do you think that analytic methods of solution are essential for understanding graphical methods?

Solving Quadratic Inequalities

Quadratic Inequality in One Variable

A **quadratic inequality** is an inequality that can be written in the form

$$ax^2 + bx + c < 0,$$

where a, b, and c are real numbers with $a \neq 0$. (The symbol $<$ can be replaced with $>$, $\leq$, or $\geq$.)

To solve a quadratic inequality analytically, we find the solutions of the corresponding quadratic equation, and then test values in the intervals on a number line determined by those solutions.

Graphical Interpretations for Quadratic Equations and Inequalities

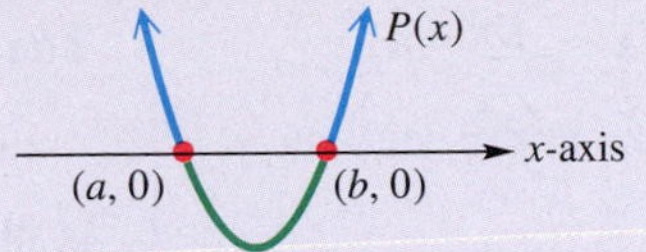

Solution Set of	Is
$P(x) = 0$	$\{a, b\}$
$P(x) < 0$	The interval (a, b)
$P(x) > 0$	$(-\infty, a) \cup (b, \infty)$

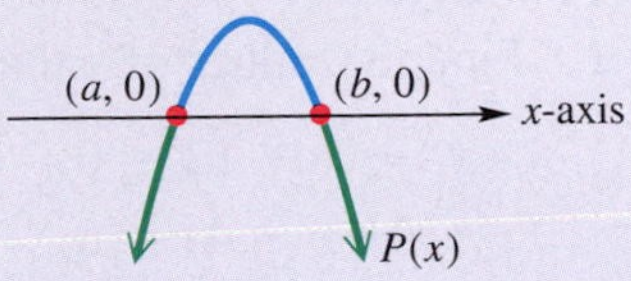

Solution Set of	Is
$P(x) = 0$	$\{a, b\}$
$P(x) < 0$	$(-\infty, a) \cup (b, \infty)$
$P(x) > 0$	The interval (a, b)

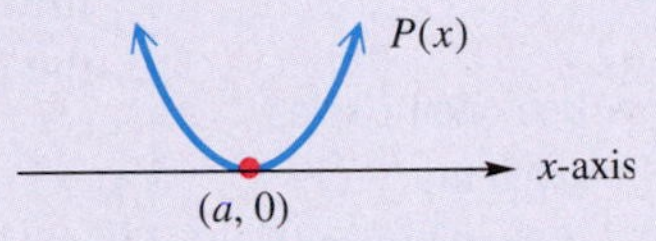

Solution Set of	Is
$P(x) = 0$	$\{a\}$
$P(x) < 0$	$\emptyset$
$P(x) > 0$	$(-\infty, a) \cup (a, \infty)$

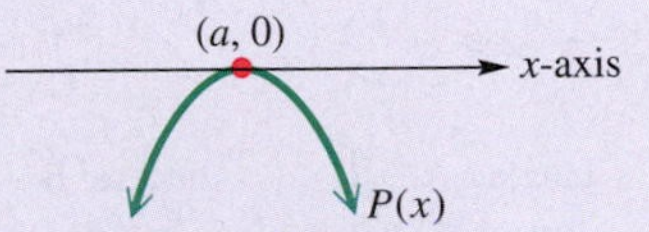

Solution Set of	Is
$P(x) = 0$	$\{a\}$
$P(x) < 0$	$(-\infty, a) \cup (a, \infty)$
$P(x) > 0$	$\emptyset$

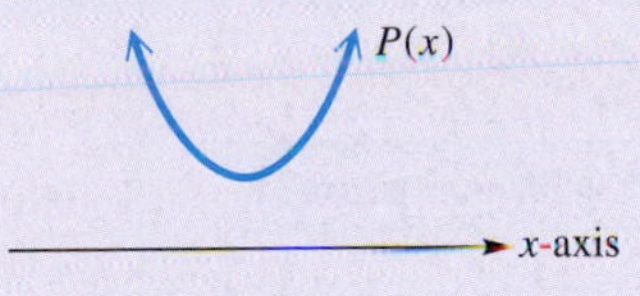

Solution Set of	Is
$P(x) = 0$	$\emptyset$
$P(x) < 0$	$\emptyset$
$P(x) > 0$	$(-\infty, \infty)$

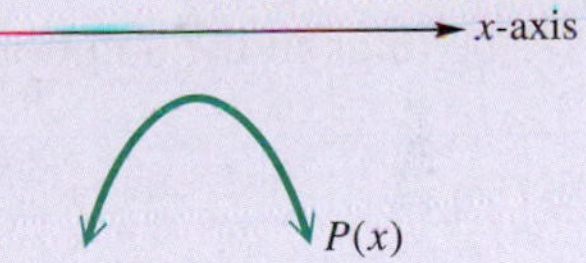

Solution Set of	Is
$P(x) = 0$	$\emptyset$
$P(x) < 0$	$(-\infty, \infty)$
$P(x) > 0$	$\emptyset$

To solve a quadratic inequality, follow these steps.

Solving a Quadratic Inequality

Step 1 Solve the corresponding quadratic equation.

Step 2 Identify the intervals determined by the solutions of the equation.

Step 3 Use a test value from each interval to determine which intervals form the solution set.

EXAMPLE 6 Solving a Quadratic Inequality

Solve $x^2 - x - 12 < 0$.

Analytic Solution

Step 1 Find the values of x that satisfy $x^2 - x - 12 = 0$.

$x^2 - x - 12 = 0$	Corresponding quadratic equation
$(x + 3)(x - 4) = 0$	Factor.
$x + 3 = 0$ or $x - 4 = 0$	Zero-product property
$x = -3$ or $x = 4$	Solve each equation.

Step 2 The two numbers -3 and 4 divide a number line into the three intervals shown in Figure 31. If a value in Interval A, for example, makes the polynomial $x^2 - x - 12$ negative, then all values in Interval A will make the polynomial negative.

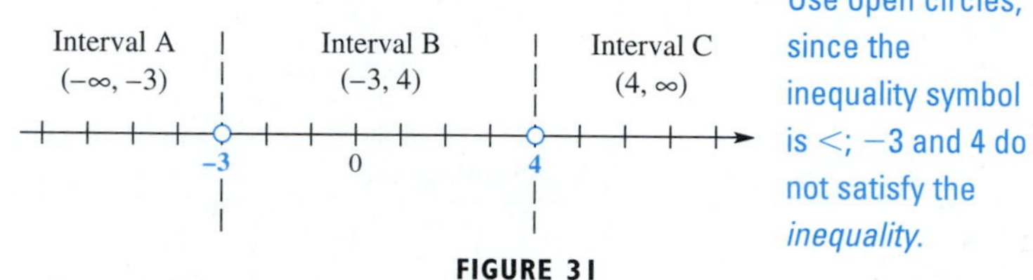

FIGURE 31

Use open circles, since the inequality symbol is $<$; -3 and 4 do not satisfy the *inequality.*

Step 3 Choose a test value in each interval to see if it satisfies the original inequality, $x^2 - x - 12 < 0$. If the test value makes the statement true, then the entire interval belongs to the solution set.

Interval	Test Value	Is $x^2 - x - 12 < 0$ True or False?
A: $(-\infty, -3)$	-4	$(-4)^2 - (-4) - 12 < 0$? $8 < 0$ False
B: $(-3, 4)$	0	$0^2 - 0 - 12 < 0$? $-12 < 0$ True
C: $(4, \infty)$	5	$5^2 - 5 - 12 < 0$? $8 < 0$ False

Since the values in Interval B make the inequality true, the solution set is the interval $(-3, 4)$.

Graphing Calculator Solution

The graph of

$$Y_1 = X^2 - X - 12$$

is shown in Figure 32. When the graph lies *below* the x-axis, the function values are negative. This happens in the interval $(-3, 4)$, which is the same as the analytic result.

The graph of the function is not sufficient to determine whether the endpoints should be included or excluded from the solution set. We make this decision based on the symbol $<$ in the original inequality. Thus, the endpoints are excluded.

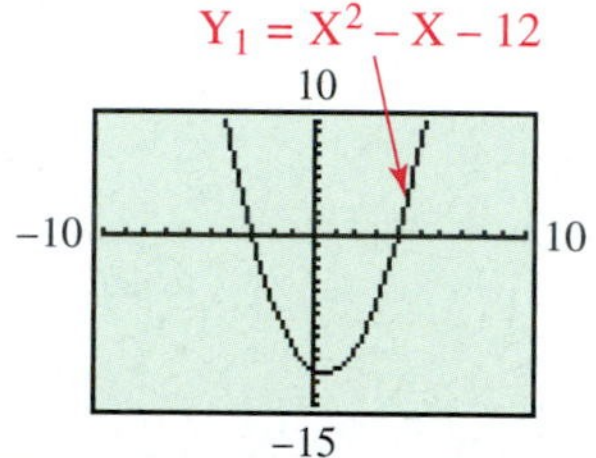

X	Y1	
-4	8	
-3	0	
-2	-6	
0	-12	
3	-6	
4	0	
5	8	

Y1■X²−X−12

FIGURE 32

The table in Figure 32 numerically supports the analytic result. Values of X in the interval $(-3, 4)$ make $Y_1 < 0$, while values outside that interval make $Y_1 \geq 0$. ■

NOTE Inequalities that use the symbols $<$ and $>$ are called **strict inequalities;** $\leq$ and $\geq$ are used in **nonstrict inequalities.** The solutions of the equation $x^2 - x - 12 = 0$ in Example 6 were not included in the solution set since the inequality was *strict.* In Example 7 which follows, the solutions of the equation will be included since the inequality there is *nonstrict.*

EXAMPLE 7 Solving a Quadratic Inequality

Solve $2x^2 \geq -5x + 12$.

Analytic Solution

Step 1 Find the values of x that satisfy $2x^2 + 5x - 12 = 0$.

$$2x^2 + 5x - 12 = 0 \quad \text{Corresponding quadratic equation}$$

$$(2x - 3)(x + 4) = 0 \quad \text{Factor.}$$

$$2x - 3 = 0 \quad \text{or} \quad x + 4 = 0 \quad \text{Zero-product property}$$

$$x = \frac{3}{2} \quad \text{or} \quad x = -4 \quad \text{Solve each equation.}$$

Step 2 The values $\frac{3}{2}$ and -4 form the intervals $(-\infty, -4)$, $\left(-4, \frac{3}{2}\right)$, and $\left(\frac{3}{2}, \infty\right)$ on the number line. See Figure 33.

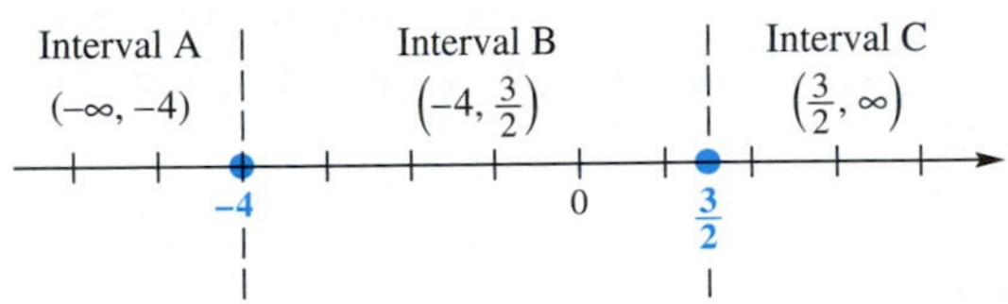

Use closed circles, since the inequality symbol is $\geq$; -4 and $\frac{3}{2}$ satisfy the *inequality.*

FIGURE 33

Step 3 Choose a test value from each interval.

Interval	Test Value	Is $2x^2 \geq -5x + 12$ True or False?
A: $(-\infty, -4)$	-5	$2(-5)^2 \geq -5(-5) + 12$? $50 \geq 37$ True
B: $\left(-4, \frac{3}{2}\right)$	0	$2(0)^2 \geq -5(0) + 12$? $0 \geq 12$ False
C: $\left(\frac{3}{2}, \infty\right)$	2	$2(2)^2 \geq -5(2) + 12$? $8 \geq 2$ True

Intervals A and C make the inequality true, so the solution set is the union of the intervals with endpoints, written $(-\infty, -4] \cup \left[\frac{3}{2}, \infty\right)$.

Graphing Calculator Solution

Use the x-intercept method here to see the intervals where $2x^2 + 5x - 12 \geq 0$. Write the original inequality in the form

$$Y_1 - Y_2 = 2X^2 + 5X - 12 \geq 0,$$

and then find the domain values where the graph of $Y = Y_1 - Y_2$ *lies above or on* the x-axis. From the graph, the intervals $(-\infty, -4]$ and $[1.5, \infty)$ give these values, supporting our analytic result. See Figure 34.

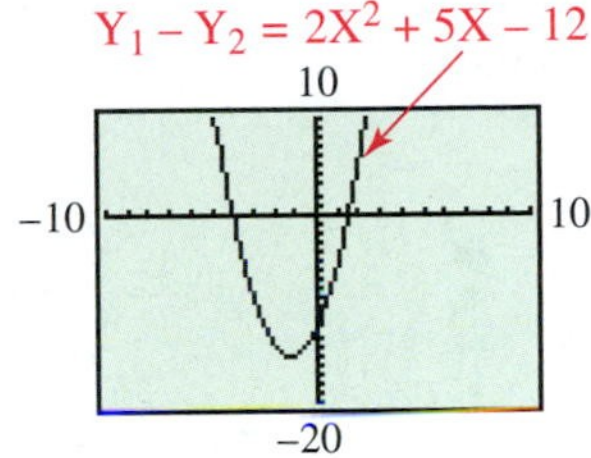

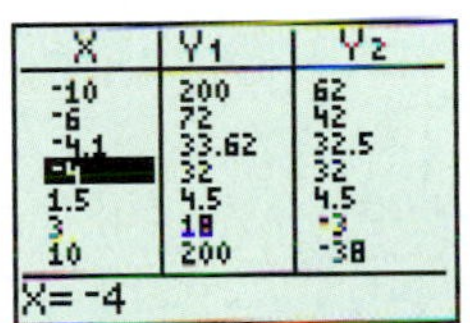

X	Y1	Y2
-10	200	62
-6	72	42
-4.1	33.62	32.5
-4	32	32
1.5	4.5	4.5
3	18	-3
10	200	-38

X= -4

FIGURE 34

The table numerically supports the analytic result because choosing values of X less than or equal to -4 *or* greater than or equal to 1.5 gives $Y_1 \geq Y_2$.

FOR DISCUSSION

1. Graph $P(x) = x^2 - x - 20$ in a window that shows both x-intercepts, and, without any analytic work, solve each equation or inequality.

   (a) $x^2 - x - 20 = 0$ **(b)** $x^2 - x - 20 < 0$ **(c)** $x^2 - x - 20 > 0$

2. Graph $P(x) = x^2 + x + 20$ in a window that shows the vertex and the y-intercept. (Why won't the standard window work?) Then, without any analytic work, solve each equation or inequality for real solutions.

   (a) $x^2 + x + 20 = 0$ **(b)** $x^2 + x + 20 < 0$ **(c)** $x^2 + x + 20 > 0$

From the "For Discussion" box, we see that solving a quadratic equation leads easily to the solution sets of the corresponding inequalities.

EXAMPLE 8 **Solving Quadratic Inequalities**

Use the graph in Figure 35 and the result of Example 4 to solve each inequality.

(a) $x^2 - 4x + 2 \le 0$ **(b)** $x^2 - 4x + 2 \ge 0$

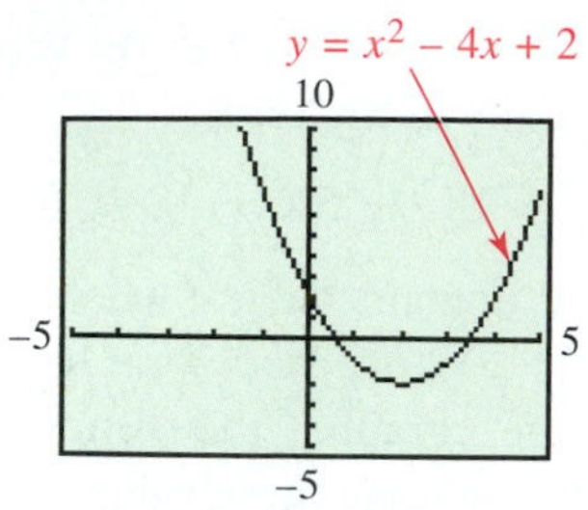

The intercepts are $2 - \sqrt{2} \approx .58578644$ and $2 + \sqrt{2} \approx 3.4142136$.

FIGURE 35

Solution

(a) In Example 4, we found that the exact solutions for $x^2 - 4x + 2 = 0$ are $2 - \sqrt{2}$ and $2 + \sqrt{2}$. Since the graph of $y = x^2 - 4x + 2$ lies below or intersects the x-axis (that is, $y \le 0$) between or at these values, as shown in Figure 35, the solution set for the inequality $x^2 - 4x + 2 \le 0$ is the closed interval $\left[2 - \sqrt{2}, 2 + \sqrt{2}\right]$.

(b) Figure 35 shows that the graph of $y = x^2 - 4x + 2$ is above or intersects the x-axis (that is, $y \ge 0$) on $\left(-\infty, 2 - \sqrt{2}\right] \cup \left[2 + \sqrt{2}, \infty\right)$, which is the solution set of $x^2 - 4x + 2 \ge 0$. ■

Formulas Involving Quadratics

EXAMPLE 9 **Solving for a Squared Variable**

Solve each equation for the indicated variable.

(a) $A = \dfrac{\pi d^2}{4}$ for d **(b)** $rt^2 - st = k$ for t $(r \neq 0)$

Solution

(a)

$$A = \frac{\pi d^2}{4}$$

$$4A = \pi d^2 \qquad \text{Multiply by 4.}$$

$$d^2 = \frac{4A}{\pi} \qquad \text{Divide by } \pi.$$

$$d = \pm\sqrt{\frac{4A}{\pi}} \qquad \text{Square root property}$$

Remember both the positive and negative square roots.

$$d = \frac{\pm 2\sqrt{A}}{\sqrt{\pi}} \cdot \frac{\sqrt{\pi}}{\sqrt{\pi}} = \frac{\pm 2\sqrt{A\pi}}{\pi} \qquad \text{Rationalize the denominator.}$$

(b) Because this equation has a term with t as well as one with t^2, use the quadratic formula. Subtract k from each side to get $rt^2 - st - k = 0$.

$$t = \frac{-b \pm \sqrt{b^2 - 4ac}}{2a} \qquad \text{Quadratic formula}$$

Take extra care when substituting variables in formulas.

$$t = \frac{-(-s) \pm \sqrt{(-s)^2 - 4(r)(-k)}}{2(r)} \qquad a = r, b = -s, c = -k$$

$$t = \frac{s \pm \sqrt{s^2 + 4rk}}{2r}$$ ■

Another Quadratic Model

The intersection-of-graphs method can be used to find specific function values with a quadratic model.

EXAMPLE 10 Modeling Length of Life

According to recent data from the Teachers Insurance and Annuity Association (TIAA), the survival rate after age 65 can be approximated by

$$S(x) = 1 - .058x - .076x^2,$$

where x is measured in decades. This function gives the probability that an individual who reaches age 65 will live at least x decades ($10x$ years) longer. (*Source:* Lial, M., R. Greenwell, and N. Ritchey, *Calculus with Applications,* 7th ed., Addison-Wesley, 2002.) For example,

$$S(1) = 1 - .058(1) - .076(1)^2 = .866.$$

This means that a 65-year old person has a .866 probability of living at least 1 more decade, to age 75.

Find the age for which the survival rate is .5 for people who reach 65. Interpret the result.

Solution Let $y_1 = 1 - .058x - .076x^2$ and $y_2 = .5$. Graph both functions and find the positive x-value of an intersection point. See Figure 36.

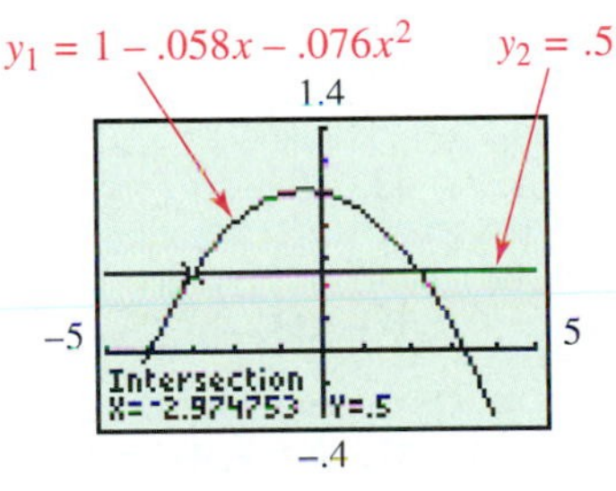

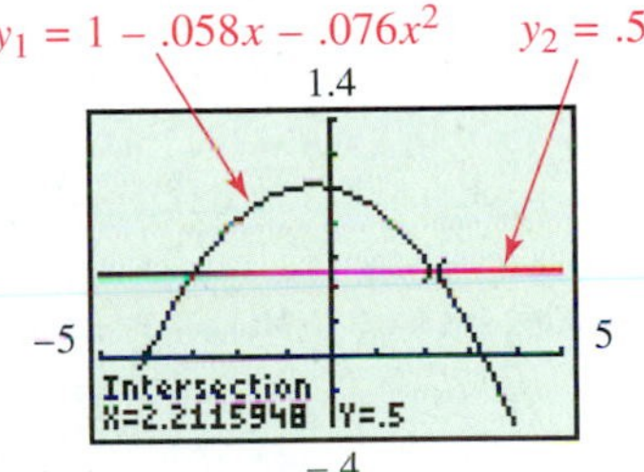

FIGURE 36

Figure 36 shows that the graphs intersect at $x \approx -3.0$ and $x \approx 2.2$. This means that half the people who reach age 65 will live about 2.2 decades or 22 years longer, to age 87. (The negative value must be discarded, since it does not make sense in the problem.) ■

3.3 Exercises

Concept Check *Match the equation in Column I with its solution(s) in Column II.*

I		II	
1. $x^2 = 4$	**2.** $x^2 = -4$	**A.** $\pm 2i$	**B.** $\pm 2\sqrt{2}$
3. $x^2 + 2 = 0$	**4.** $x^2 - 2 = 0$	**C.** $\pm i\sqrt{2}$	**D.** 2
5. $x^2 = -8$	**6.** $x^2 = 8$	**E.** $\pm\sqrt{2}$	**F.** -2
7. $x - 2 = 0$	**8.** $x + 2 = 0$	**G.** ± 2	**H.** $\pm 2i\sqrt{2}$

Concept Check *Answer each question.*

9. Which one of the following equations is set up for direct use of the zero-product property? Solve it.
A. $3x^2 - 17x - 6 = 0$ **B.** $(2x + 5)^2 = 7$
C. $x^2 + x = 12$ **D.** $(3x + 1)(x - 7) = 0$

10. Which one of the following equations is set up for direct use of the square root property? Solve it.
A. $3x^2 - 17x - 6 = 0$ **B.** $(2x + 5)^2 = 7$
C. $x^2 + x = 12$ **D.** $(3x + 1)(x - 7) = 0$

11. Only one of the following equations does not require Step 1 of the method for completing the square described in Section 3.2. Which one is it? Solve it.
A. $3x^2 - 17x - 6 = 0$ **B.** $(2x + 5)^2 = 7$
C. $x^2 + x = 12$ **D.** $(3x + 1)(x - 7) = 0$

12. Only one of the following equations is set up so that the values of a, b, and c can be determined immediately. Which one is it? Solve it.
A. $3x^2 - 17x - 6 = 0$ **B.** $(2x + 5)^2 = 7$
C. $x^2 + x = 12$ **D.** $(3x + 1)(x - 7) = 0$

Solve each equation. For equations with real solutions, support your answers graphically.

13. $x^2 = 16$ **14.** $x^2 = 144$ **15.** $2x^2 = 90$ **16.** $2x^2 = 48$

17. $x^2 = -16$ **18.** $x^2 = -100$ **19.** $x^2 = -18$ **20.** $x^2 = -32$

21. $(3x - 1)^2 = 12$ **22.** $(4x + 1)^2 = 20$ **23.** $(5x - 3)^2 = -3$ **24.** $(-2x + 5)^2 = -8$

25. $x^2 = 2x + 24$ **26.** $x^2 = 3x + 18$ **27.** $3x^2 - 2x = 0$ **28.** $5x^2 + 3x = 0$

29. $x(14x + 1) = 3$ **30.** $x(12x + 11) = -2$ **31.** $-4 + 9x - 2x^2 = 0$ **32.** $-5 + 16x - 3x^2 = 0$

33. $\frac{1}{3}x^2 - \frac{1}{3}x = 24$ **34.** $\frac{1}{6}x^2 + \frac{1}{6}x = 5$ **35.** $(x + 2)(x - 1) = 7x + 5$ **36.** $(x + 4)(x - 1) = -5x - 4$

37. $x^2 - 2x - 4 = 0$ **38.** $x^2 + 8x + 13 = 0$ **39.** $2x^2 + 2x = -1$ **40.** $9x^2 - 12x = -8$

41. $x(x - 1) = 1$ **42.** $x(x - 3) = 2$ **43.** $x^2 - 5x = x - 7$ **44.** $11x^2 - 3x + 2 = 4x + 1$

45. $4x^2 - 12x = -11$ **46.** $x^2 = 2x - 5$ **47.** $\frac{1}{3}x^2 + \frac{1}{4}x - 3 = 0$ **48.** $\frac{2}{3}x^2 + \frac{1}{4}x = 3$

49. $(3 - x)^2 = 25$ **50.** $(2 + x)^2 = 49$ **51.** $2x^2 - 4x = 1$ **52.** $3x^2 - 6x = 4$

53. $x^2 = -1 - x$ **54.** $x^2 = -3 - 3x$ **55.** $4x^2 - 20x + 25 = 0$ **56.** $9x^2 + 12x + 4 = 0$

57. $-3x^2 + 4x + 4 = 0$ **58.** $-5x^2 + 28x + 12 = 0$

59. $(x + 5)(x - 6) = (2x - 1)(x - 4)$ **60.** $(x + 2)(3x - 4) = (x + 5)(2x - 5)$

Evaluate the discriminant, and use it to determine the number of real solutions of the equation. If the equation does have real solutions, tell whether they are rational *or* irrational. *Do not actually solve the equation.*

61. $x^2 + 8x + 16 = 0$ **62.** $8x^2 = 14x - 3$ **63.** $4x^2 = 6x + 3$

64. $2x^2 - 4x + 1 = 0$ **65.** $9x^2 + 11x + 4 = 0$ **66.** $3x^2 = 4x - 5$

Concept Check *For each pair of numbers, find the values of a, b, and c for which the quadratic equation $ax^2 + bx + c = 0$ has the given numbers as solutions. Answers may vary.* (Hint: *Use the zero-product property in reverse.*)

67. 4, 5 **68.** $-3, 2$ **69.** $1 + \sqrt{2}, 1 - \sqrt{2}$ **70.** $i, -i$

Concept Check *Sketch a graph of $f(x) = ax^2 + bx + c$ that satisfies each set of conditions.*

71. $a < 0, b^2 - 4ac = 0$ **72.** $a > 0, b^2 - 4ac < 0$ **73.** $a < 0, b^2 - 4ac < 0$

74. $a < 0, b^2 - 4ac > 0$ **75.** $a > 0, b^2 - 4ac > 0$ **76.** $a > 0, b^2 - 4ac = 0$

Concept Check Exercises 77–90 refer to the graphs of the quadratic functions f, g, and h shown here.

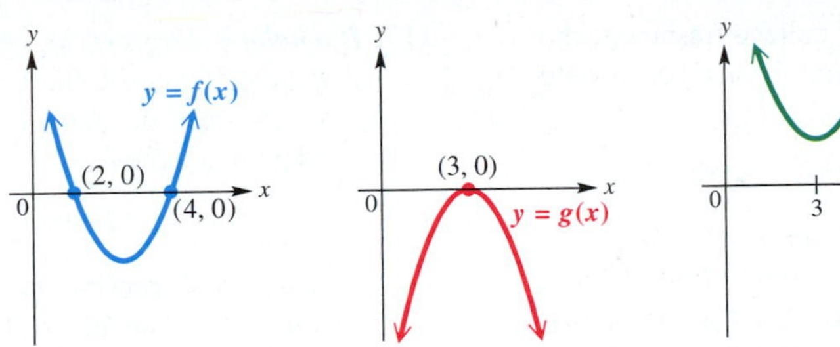

77. What is the solution set of $f(x) = 0$?

78. What is the solution set of $f(x) < 0$?

79. What is the solution set of $f(x) > 0$?

80. What is the solution set of $g(x) = 0$?

81. What is the solution set of $g(x) < 0$?

82. What is the solution set of $g(x) > 0$?

83. Solve $h(x) > 0$.

84. Solve $h(x) < 0$.

85. How many real solutions does $h(x) = 0$ have? How many nonreal complex solutions does it have?

86. What is the value of the discriminant of $g(x)$?

87. What is the x-coordinate of the vertex of the graph of $y = f(x)$?

88. What is the equation of the axis of symmetry of the graph of $y = g(x)$?

89. Does the graph of $y = g(x)$ have a y-intercept? If so, is it positive or negative?

90. Is the minimum value of h positive or negative?

Solve each inequality analytically. Support your answers graphically. Give exact values for endpoints.

91. **(a)** $x^2 + 4x + 3 \geq 0$
(b) $x^2 + 4x + 3 < 0$

92. **(a)** $x^2 + 6x + 8 < 0$
(b) $x^2 + 6x + 8 \geq 0$

93. **(a)** $2x^2 - 9x > -4$
(b) $2x^2 - 9x \leq -4$

94. **(a)** $3x^2 + 13x + 10 \leq 0$
(b) $3x^2 + 13x + 10 > 0$

95. **(a)** $-x^2 - x \leq 0$
(b) $-x^2 - x > 0$

96. **(a)** $-x^2 + 2x \leq 0$
(b) $-x^2 + 2x > 0$

97. **(a)** $x^2 - x + 1 < 0$
(b) $x^2 - x + 1 \geq 0$

98. **(a)** $2x^2 - x + 3 < 0$
(b) $2x^2 - x + 3 \geq 0$

99. **(a)** $2x + 1 \geq x^2$
(b) $2x + 1 < x^2$

100. **(a)** $x^2 + 5x < 2$
(b) $x^2 + 5x \geq 2$

Solve each formula for the indicated variable. Assume that no denominators are 0.

101. $s = \frac{1}{2}gt^2$ for t

102. $A = \pi r^2$ for r

103. $a^2 + b^2 = c^2$ for a

104. $A = s^2$ for s

105. $S = 4\pi r^2$ for r

106. $V = \frac{1}{3}\pi r^2 h$ for r

107. $V = e^3$ for e

108. $V = \frac{4}{3}\pi r^3$ for r

109. $F = \frac{kMv^4}{r}$ for v

110. $s = s_0 + gt^2 + k$ for t

111. $P = \frac{E^2R}{(r + R)^2}$ for R

112. $S = 2\pi rh + 2\pi r^2$ for r

Solve each equation for x and then for y.

113. $x^2 + xy + y^2 = 0 \quad (x > 0, y > 0)$

114. $4x^2 - 2xy + 3y^2 = 2$

115. $3y^2 + 4xy - 9x^2 = -1$

(Modeling) *Solve each problem.*

116. *Medical Degrees* The percent of college freshmen who planned to get a professional degree in a medical field between 1992 and 1998 can be modeled by

$$f(x) = -.2369x^2 + 1.425x + 6.905,$$

where $x = 0$ represents 1992. (*Source: The American Freshman: National Norms for Fall 1992–1998,* Higher Education Research Institute, UCLA.) Based on this model, in what year did the percent of freshmen planning to get a medical degree reach its maximum? What was the maximum percent?

117. *Bachelor's Degrees in Mathematics* The number of college graduates in the United States who earned bachelor's degrees in mathematics between 1980 and 2002 can be modeled by

$$f(x) = -21.15x^2 + 476.2x + 11{,}973,$$

where $x = 0$ represents 1980. (*Source:* U.S. Census Bureau.) Based on this model, in what year did the number of graduates receiving bachelor's degrees in mathematics reach its maximum? What was the maximum number?

 **118.** *Length of Life* Refer to Example 10. Solve $S(x) = 0$ and interpret the result.

Reviewing Basic Concepts (Sections 3.1–3.3)

Perform each operation. Write answers in standard form.

1. $(5 + 6i) - (2 - 4i) - 3i$

2. $i(5 + i)(5 - i)$

3. $\dfrac{-10 - 10i}{2 + 3i}$

Give the following for the function defined by $P(x) = 2x^2 + 8x + 5$.

4. The graph of P

5. The vertex of the graph and whether it is a maximum or minimum point

6. The equation of the axis of symmetry of the graph

7. The domain and range of P

Solve each equation or inequality analytically, and support your result graphically.

8. $9x^2 = 25$

9. $3x^2 - 5x = 2$

10. $-x^2 + x + 3 = 0$

11. $3x^2 - 5x - 2 \le 0$

12. $x^2 + x - 3 > 0$

13. $x(3x - 1) \le -4$

3.4 Further Applications of Quadratic Functions and Models

Applications of Quadratic Functions ■ Quadratic Models

Applications of Quadratic Functions

EXAMPLE 1 Finding the Area of a Rectangular Region

A farmer wishes to enclose a rectangular region. He has 120 feet of fencing and plans to use his barn as one side of the enclosure. See Figure 37. Let x represent the length of one of the parallel sides of the fencing.

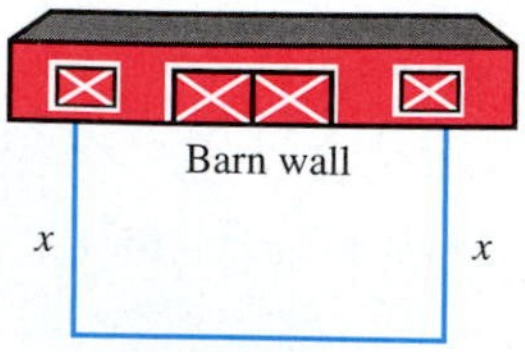

(a) Determine a function A that represents the area of the region in terms of x.

(b) What are the restrictions on x?

(c) Graph the function in a viewing window that shows both x-intercepts and the vertex of the graph.

FIGURE 37

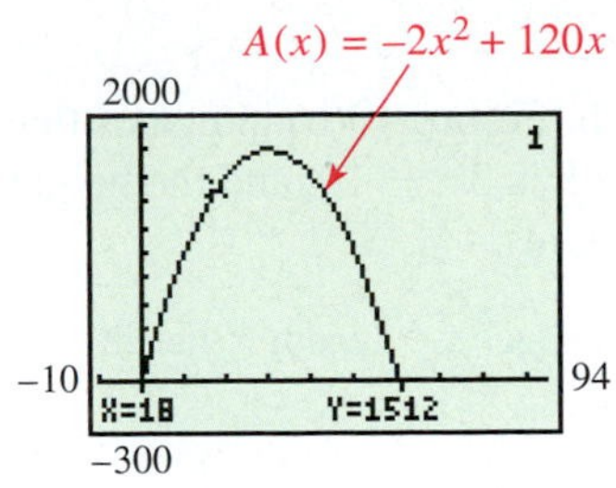

FIGURE 38

(d) Figure 38 shows the cursor at (18, 1512). Interpret this information.

(e) What is the maximum area the farmer can enclose?

Solution

(a) The lengths of the sides of the region bordered by the fencing are x, x, and $120 - 2x$, as shown in Figure 37. Area = width × length, so the function is

$$A(x) = x(120 - 2x) \quad \text{or} \quad A(x) = -2x^2 + 120x.$$

(b) Since x represents a length, $x > 0$. Furthermore, the side of length $120 - 2x > 0$, or $x < 60$. Putting these two restrictions together gives $0 < x < 60$ or the interval $(0, 60)$ as the *theoretical* domain.

(c) See Figure 38.

(d) If each parallel side of fencing measures 18 feet, the area of the enclosure is 1512 square feet. This can be written $A(18) = 1512$ and checked as follows: If the width is 18 feet, the length is $120 - 2(18) = 84$ feet, and $18 \times 84 = 1512$.

(e) The maximum value of the quadratic function occurs at the vertex. For $A(x) = -2x^2 + 120x$, use $a = -2$ and $b = 120$ in the vertex formula from Section 3.2.

$$x = -\frac{b}{2a} = -\frac{120}{2(-2)} = 30$$

$$A(30) = -2(30)^2 + 120(30) = 1800$$

Therefore, the farmer can enclose a maximum of 1800 square feet if the parallel sides of fencing measure 30 feet.

Alternatively, use a calculator to locate the vertex of the parabola, and observe the x- and y-values there. The graph in Figure 39(a) confirms the analytic results. The table in Figure 39(b) provides numerical support. ■

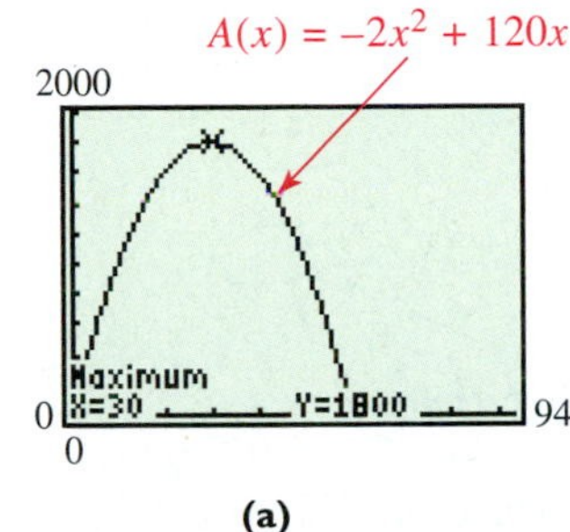

(a)

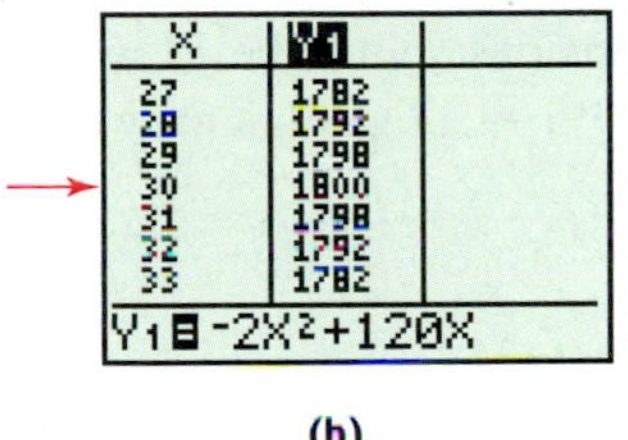

(b)

FIGURE 39

TECHNOLOGY NOTE

In the calculator graph in Figure 39(a), the display at the bottom obscures the view of the intercepts. To overcome this problem, lower the minimum values of x and y enough so that when the display appears, the x- and y-axes are visible, as in Figure 38.

CAUTION ***Be careful when interpreting the meanings of the coordinates of the vertex.*** The first coordinate, x, gives the *domain value* for which the *function value* is a maximum or minimum. Read the problem carefully to determine whether you are asked to find the value of the independent variable x, the function value y, or both.

EXAMPLE 2 Finding the Volume of a Box

A machine produces rectangular sheets of metal satisfying the condition that the length is three times the width. Furthermore, equal-sized squares measuring 5 inches on a side can be cut from the corners so that the resulting piece of metal can be shaped into an open box by folding up the flaps. See Figure 40(a) on the next page.

(a) Determine a function V that expresses the volume of the box in terms of the width x of the original sheet of metal.

(b) What restrictions must be placed on x?

(c) If specifications call for the volume of such a box to be 1435 cubic inches, what should the dimensions of the original piece of metal be?

(d) What dimensions of the original piece of metal will assure a volume greater than 2000, but less than 3000, cubic inches? Solve graphically.

5
$x - 10$
$3x - 10$

(a)

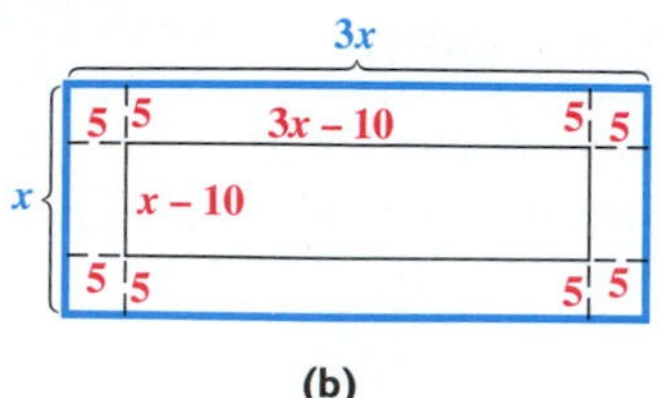

(b)

FIGURE 40

Solution

(a) If x represents the width, then $3x$ represents the length. Figure 40(b) indicates that the width of the bottom of the box is $x - 10$, the length is $3x - 10$, and the height is 5 inches (the length of the side of each cut-out square).

$$V(x) = (3x - 10)(x - 10)(5) \qquad \text{Volume} = \text{length} \times \text{width} \times \text{height}$$
$$= 15x^2 - 200x + 500 \qquad \text{Multiply.}$$

(b) Since the dimensions of the box represent positive numbers, $3x - 10 > 0$ and $x - 10 > 0$, or $x > \frac{10}{3}$ and $x > 10$. Both conditions are satisfied when $x > 10$, so the theoretical domain of x is $(10, \infty)$.

(c)

$$15x^2 - 200x + 500 = 1435 \qquad \text{Let } V(x) = 1435.$$
$$15x^2 - 200x - 935 = 0 \qquad \text{Subtract 1435.}$$
$$3x^2 - 40x - 187 = 0 \qquad \text{Divide by 5.}$$
$$(3x + 11)(x - 17) = 0 \qquad \text{Factor.}$$
$$3x + 11 = 0 \quad \text{or} \quad x - 17 = 0 \qquad \text{Zero-product property}$$
$$x = -\frac{11}{3} \quad \text{or} \quad x = 17 \qquad \text{Solve each equation.}$$

Only 17 satisfies the condition that $x > 10$. The dimensions should be 17 inches by $3(17) = 51$ inches. Since $(51 - 10) \cdot (17 - 10) \cdot 5 = 1435$, this answer is correct.

In the same window, we can also graph

$$y_1 = 15x^2 - 200x + 500 \quad \text{and} \quad y_2 = 1435.$$

As shown in Figure 41, the graphs intersect at the point with x-value 17, supporting the analytic result.

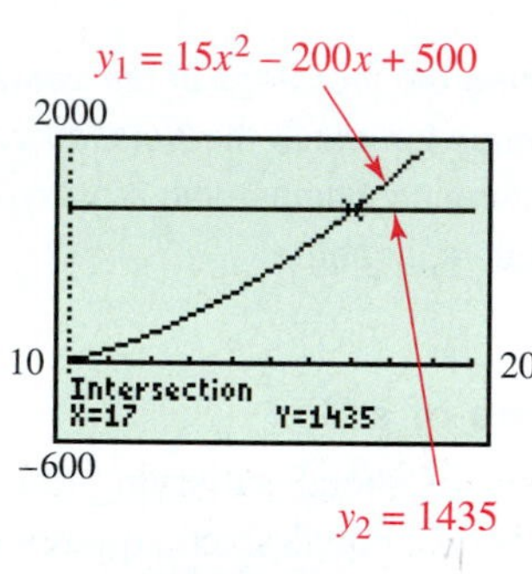

FIGURE 41

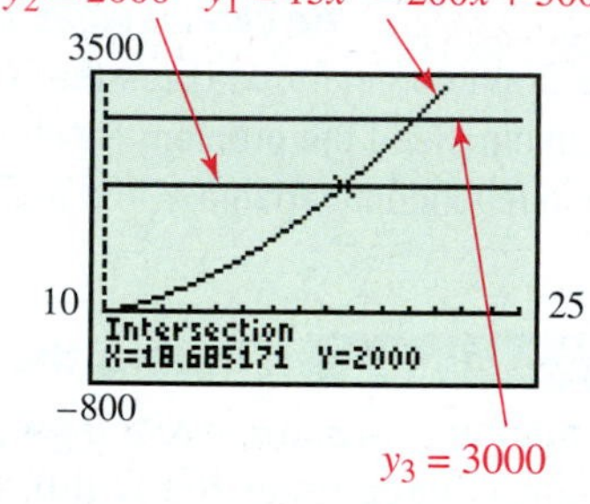

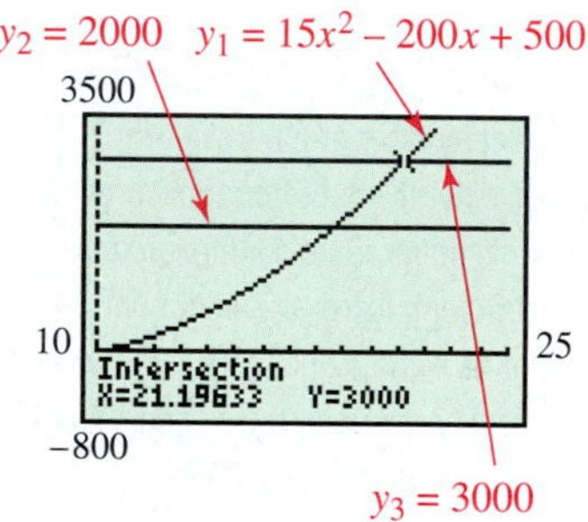

FIGURE 42

(d) Using the graphs of the functions

$$y_1 = 15x^2 - 200x + 500, \quad y_2 = 2000, \quad \text{and} \quad y_3 = 3000,$$

as shown in Figure 42, we find that the points of intersection of the graphs are *approximately* (18.7, 2000) and (21.2, 3000) for $x > 10$. Therefore, the width of the rectangle should be between 18.7 and 21.2 inches, with the corresponding length three times these values (that is, between $3(18.7) \approx 56.1$ and $3(21.2) \approx 63.6$ inches). ■

EXAMPLE 3 Solving a Problem Requiring the Pythagorean Theorem

Juan Pérez finds a piece of property in the shape of a right triangle. To get some idea of its dimensions, he measures the three sides, starting with the shortest one. He finds that the longer leg is approximately 20 meters longer than twice the length of the shorter leg. The length of the hypotenuse is approximately 10 meters longer than the length of the longer leg. Find the lengths of the sides of the triangular lot.

Analytic Solution

Let s = the length of the shorter leg in meters, so $2s + 20$ meters is the length of the longer leg, and $(2s + 20) + 10 = 2s + 30$ meters is the length of the hypotenuse. See Figure 43.

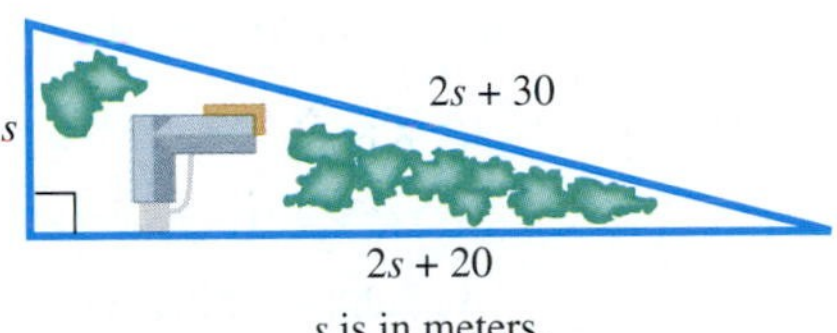

s is in meters.

FIGURE 43

The domain of s is $(0, \infty)$. Write and solve an equation.

$$s^2 + (2s + 20)^2 = (2s + 30)^2 \quad \text{Pythagorean theorem}$$

$$s^2 + 4s^2 + 80s + 400 = 4s^2 + 120s + 900 \quad (x + y)^2 = x^2 + 2xy + y^2$$

$$s^2 - 40s - 500 = 0 \quad \text{Combine terms.}$$

$$(s - 50)(s + 10) = 0 \quad \text{Factor.}$$

$$s - 50 = 0 \quad \text{or} \quad s + 10 = 0 \quad \text{Zero-product property}$$

$$s = 50 \quad \text{or} \quad s = -10 \quad \text{Solve each equation.}$$

The solution -10 is not in the domain. The lengths of the sides of the triangular lot are 50 meters, $2(50) + 20 = 120$ meters, and $2(50) + 30 = 130$ meters.

Graphing Calculator Solution

Replace s with x, and using

$$y_1 = x^2 + (2x + 20)^2$$

and $$y_2 = (2x + 30)^2,$$

graph $y_1 - y_2$. The x-intercept is 50, as shown in Figure 44. We use the x-intercept method with $y_1 - y_2$ here because the y-values for y_1 and y_2 are large.

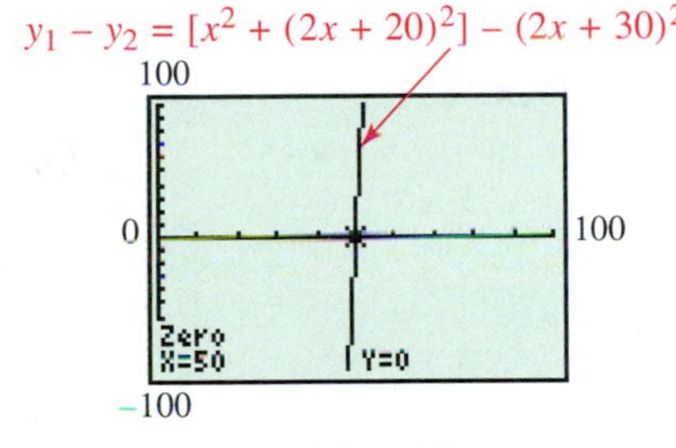

FIGURE 44

Quadratic Models

By extending the concept of regression from Section 1.4 to polynomials, we can sometimes fit data to a quadratic function to provide a quadratic model.

Year, x	Percent 65 and older, y
1900	4.1
1920	4.7
1940	6.8
1960	9.3
1980	11.3
2000	12.4
2020*	16.5
2040*	20.6

*Projected

Source: U.S. Census Bureau.

EXAMPLE 4 Modeling Population Data

The percent of Americans 65 and older for selected years is shown in the table.

(a) Plot the data, letting $x = 0$ correspond to 1900.

(b) Find a quadratic function of the form $f(x) = a(x - h)^2 + k$ that models the data by using $(0, 4.1)$ as the vertex and a second point, such as $(140, 20.6)$, to determine a.

(c) Graph f in the same viewing window as the data. How well does f model the percent of Americans 65 and older?

(d) Use f to predict the number of Americans 65 and older in 2010.

(e) Use the quadratic regression feature of a graphing calculator to determine the quadratic function g that provides the best fit for the data. Graph g with the data.

Solution

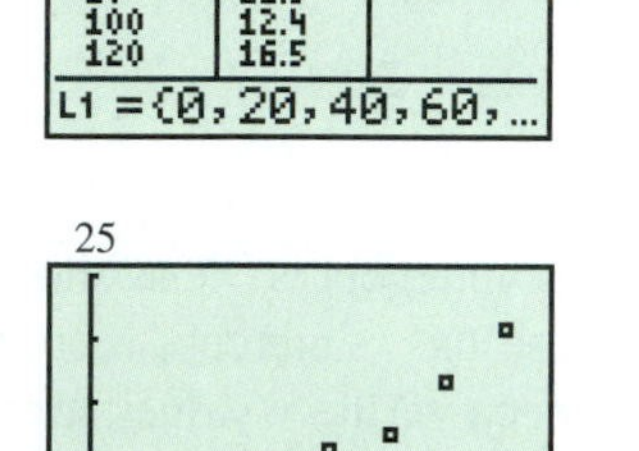

FIGURE 45

(a) Using the statistical feature of a graphing calculator and letting the x-list L_1 be $\{0, 20, 40, \ldots, 140\}$ and the y-list L_2 be $\{4.1, 4.7, 6.8, \ldots, 20.6\}$ gives the scatter diagram shown in Figure 45.

(b) Since $(0, 4.1)$ is the lowest point in the scatter diagram of the data, let it correspond to the vertex on the graph of f.

$$f(x) = a(x - h)^2 + k \qquad \text{Given form of quadratic function}$$

$$f(x) = a(x - 0)^2 + 4.1 \qquad \text{Let } h = 0 \text{ and } k = 4.1.$$

To determine a, use the given ordered pair $(140, 20.6)$.

$$20.6 = a(140)^2 + 4.1 \qquad \text{Let } x = 140 \text{ and } f(140) = 20.6.$$

$$16.5 = 19{,}600a \qquad \text{Subtract 4.1; square 140.}$$

$$a = .0008418 \qquad \text{Rounded}$$

Thus, $$f(x) = .0008418(x - 0)^2 + 4.1$$

$$= .0008418x^2 + 4.1. \qquad \text{Quadratic model}$$

(Note that choosing other second points would produce different models.)

(c) Figure 46(a) shows the graph of f plotted with the data points. There is a fairly good fit, especially for the later years.

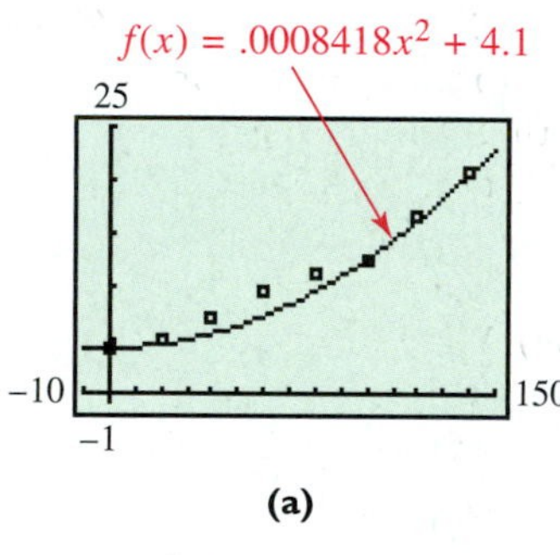

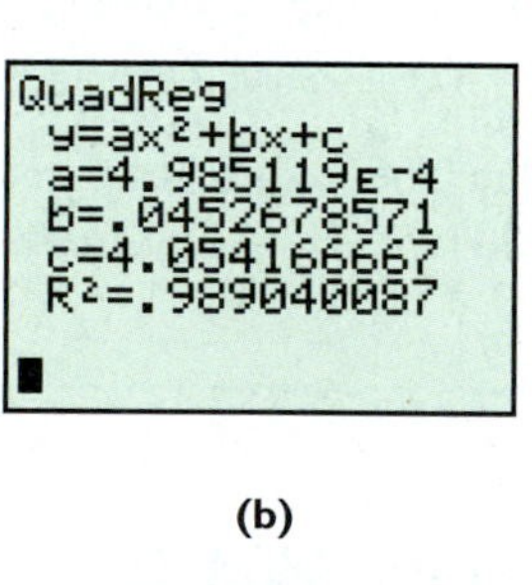

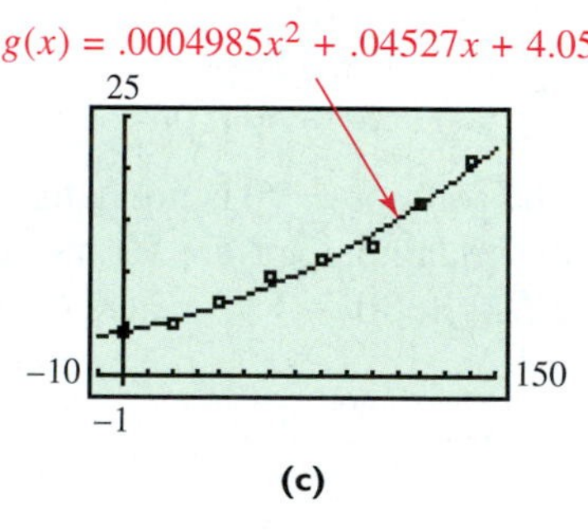

(a) (b) (c)

FIGURE 46

TECHNOLOGY NOTE

The TI-83/84 Plus computes r (the correlation coefficient) and r^2 (the coefficient of determination) for a regression that yields two estimates (a and b), such as linear regression. For a regression that yields three or more estimates (such as a, b, and c for quadratic regression, as in Figure 46(b)), the calculator computes R^2, called the *coefficient of multiple determination.*

(d) For the year 2010, $x = 2010 - 1900 = 110$.

$$f(110) = .0008418(110)^2 + 4.1$$

$$\approx 14.3$$

This model predicts 14.3 million Americans will be 65 or older in 2010.

(e) Figure 46(b) shows the calculator quadratic regression model for the data,

$$g(x) \approx .0004985x^2 + .04527x + 4.054,$$

graphed in the same window as the data in Figure 46(c). Comparing this graph with the one in Figure 46(a), we see that it is a closer fit to the data. This is because all data are used in determining function g, while only the first and last data points were used to determine function f. ■

3.4 Exercises

(Modeling) Solve each problem.

1. ***Area of a Parking Lot*** For the rectangular parking area of the shopping center shown, which equation says that the area is 40,000 square yards?

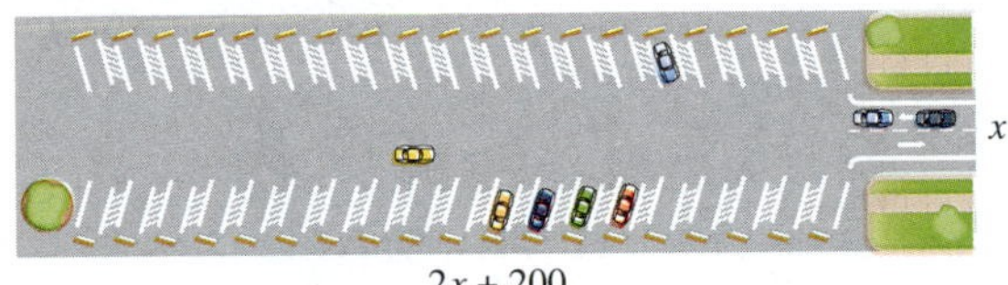

A. $x(2x + 200) = 40{,}000$
B. $2x + 2(2x + 200) = 40{,}000$
C. $x + (2x + 200) = 40{,}000$
D. None of the above

2. ***Area of a Picture*** The mat around the picture shown measures x inches across. Which equation says that the area of the picture itself is 600 square inches?

A. $2(34 - 2x) + 2(21 - 2x) = 600$
B. $(34 - 2x)(21 - 2x) = 600$
C. $(34 - x)(21 - x) = 600$
D. $x(34)(21) = 600$

3. ***Sum of Two Numbers*** Suppose that x represents one of two *positive* numbers whose sum is 30.
   (a) Represent the other of the two numbers in terms of x.
   (b) What are the restrictions on x?
   (c) Determine a function P that represents the product of the two numbers.
   (d) Determine analytically and support graphically the two such numbers whose product is a maximum. What is this maximum product?

4. ***Sum of Two Numbers*** Suppose that x represents one of two *positive* numbers whose sum is 45.
   (a) Represent the other of the two numbers in terms of x.
   (b) What are the restrictions on x?
   (c) Determine a function P that represents the product of the two numbers.
   (d) For what two such numbers is the product equal to 504? Determine analytically.
   (e) Determine analytically and support graphically the two such numbers whose product is a maximum.

5. ***Area of a Parking Lot*** American River College has plans to construct a rectangular parking lot on land bordered on one side by a highway. There are 640 feet of fencing available to fence the other three sides. Let x represent the length of each of the two parallel sides of fencing.

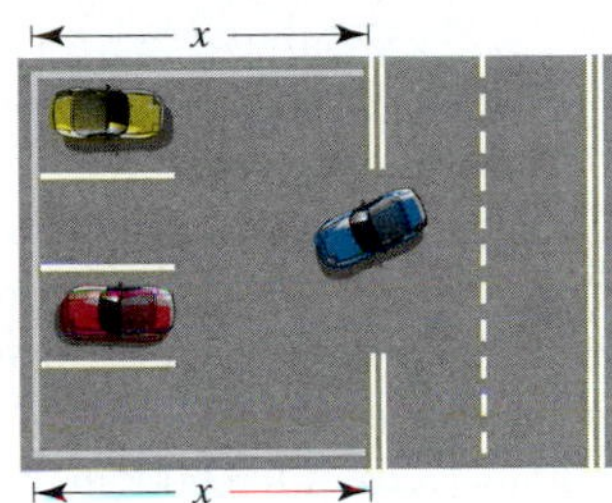

   (a) Express the length of the remaining side to be fenced in terms of x.
   (b) What are the restrictions on x?
   (c) Determine a function A that represents the area of the parking lot in terms of x.
   (d) Graph the function A from part (c) in a viewing window of $[0, 320]$ by $[0, 55{,}000]$. Determine graphically the values of x that will give an area between 30,000 and 40,000 square feet.
   (e) What dimensions will give a maximum area, and what will this area be? Determine analytically and support graphically.

6. ***Area of a Rectangular Region*** A farmer wishes to enclose a rectangular region bordering a river with fencing, as shown in the diagram. Suppose that x represents the length of each of the three parallel pieces of fencing. She has 600 feet of fencing available.

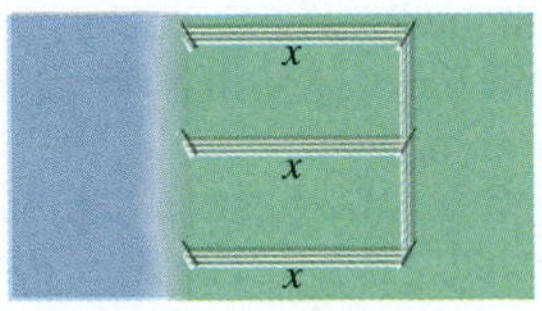

   (a) What is the length of the remaining piece of fencing in terms of x?
   (b) Determine a function A that represents the total area of the enclosed region. Give any restrictions on x.

(c) What dimensions for the total enclosed region would give an area of 22,500 square feet? Determine the answer analytically.

(d) Use a graph to find the maximum area that can be enclosed.

7. ***Volume of a Box*** A piece of cardboard is twice as long as it is wide. It is to be made into a box with an open top by cutting 2-inch squares from each corner and folding up the sides. Let x represent the width of the original piece of cardboard.

(a) Represent the length of the original piece of cardboard in terms of x.

(b) What will be the dimensions of the bottom rectangular base of the box? Give the restrictions on x.

(c) Determine a function V that represents the volume of the box in terms of x.

(d) For what dimensions of the bottom of the box will the volume be 320 cubic inches? Determine analytically and support graphically.

(e) Determine graphically (to the nearest tenth of an inch) the values of x if the box is to have a volume between 400 and 500 cubic inches.

8. ***Volume of a Box*** A piece of sheet metal is 2.5 times as long as it is wide. It is to be made into a box with an open top by cutting 3-inch squares from each corner and folding up the sides. Let x represent the width of the original piece of sheet metal.

(a) Represent the length of the original piece of sheet metal in terms of x.

(b) What are the restrictions on x?

(c) Determine a function V that represents the volume of the box in terms of x.

(d) For what values of x (that is, original widths) will the volume of the box be between 600 and 800 cubic inches? Determine the answer graphically, and give values to the nearest tenth of an inch.

9. ***Radius of a Can*** A can of garbanzo beans has surface area 54.19 square inches. Its height is 4.25 inches. What is the radius of the circular top? See the figure.

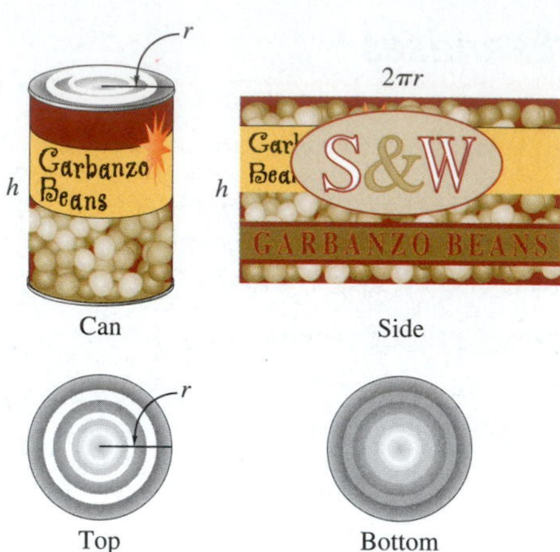

(*Hint:* The surface area consists of the circular top and bottom and a rectangle that represents the side cut open vertically and unrolled.)

10. ***Dimensions of a Cereal Box*** The volume of a 10-ounce box of cereal is 182.742 cubic inches. The width of the box is 3.1875 inches less than the length, and its depth is 2.3125 inches. Find the length and width of the box to the nearest thousandth.

11. ***Radius Covered by a Circular Lawn Sprinkler*** A square lawn has area 800 square feet. A sprinkler placed at the center of the lawn sprays water in a circular pattern that just covers the lawn. What is the radius of the circle?

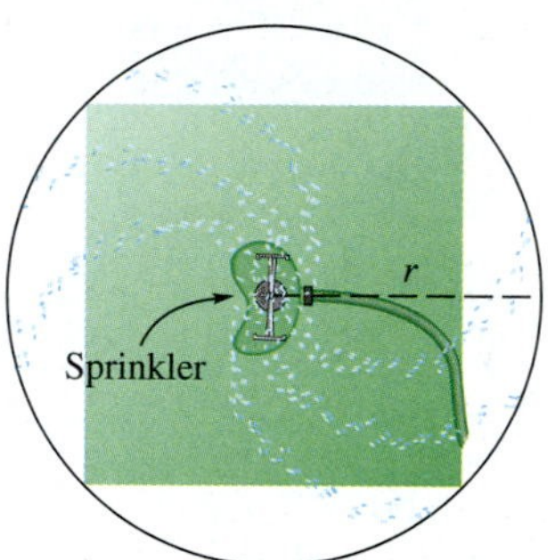

12. ***Height of a Kite*** A kite is flying on 50 feet of string. How high is it above the ground if its height is 10 feet more than the horizontal distance from the person flying it? Assume that the string is being held at ground level.

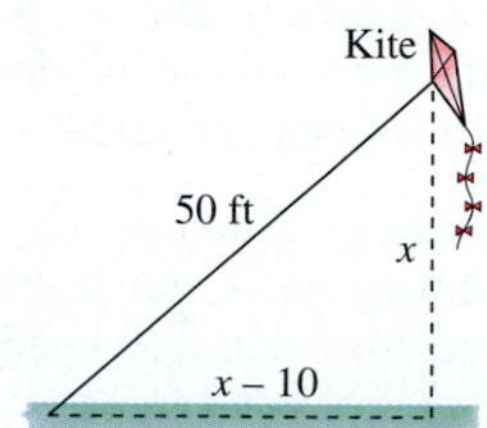

13. *Height of a Dock* A boat with a rope attached at water level is being pulled into a dock. When the boat is 12 feet from the dock, the length of the rope is 3 feet more than twice the height of the dock above the water. Find the height of the dock.

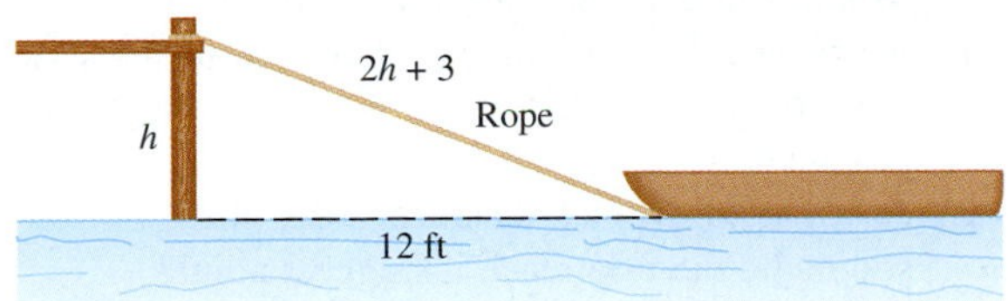

14. *Length of a Walkway* A nature conservancy group has purchased some land that includes a wetland. To make the wetland available for the public to see and learn about, the group will construct a raised wooden walkway through it. The walkway will be a loop that begins and ends at a nature center. To enclose the most interesting part of the wetlands, the walkway will have the shape of a right triangle with one leg 700 yards longer than the other and the hypotenuse 100 yards longer than the longer leg. Find the total length of the walkway.

15. *Length of a Ladder* A building is 2 feet from a 9-foot fence that surrounds the property. A worker wants to wash a window in the building 13 feet from the ground. He plans to place a ladder over the fence so that it rests against the building. He decides he should place the ladder at least 8 feet from the fence for stability. To the nearest foot, how long a ladder will he need?

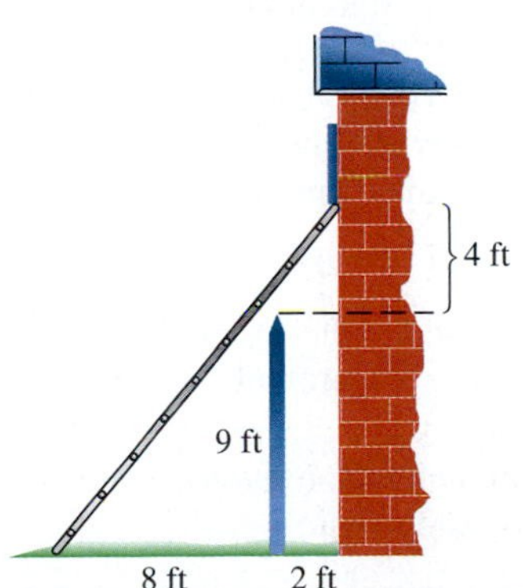

16. *Dimensions of a Solar Panel Frame* Christine has a solar panel with a width of 26 inches, as shown in the figure at the top of the next column. To get the proper inclination for her climate, she needs a right triangular support frame with one leg twice as long as the other. To the nearest tenth of an inch, what dimensions should the frame have?

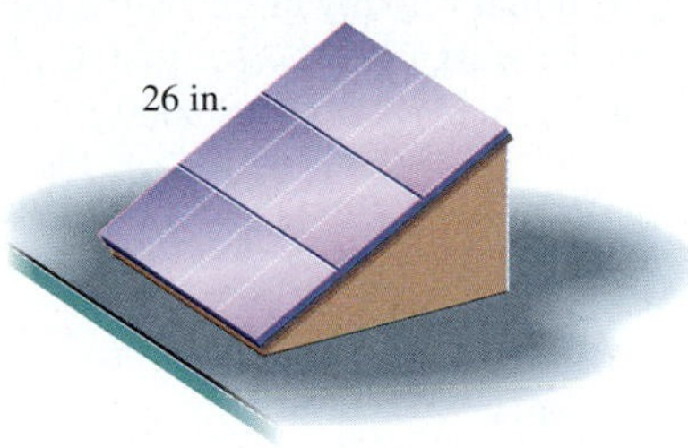

17. *Apartment Rental* The manager of an 80-unit apartment complex knows from experience that at a rent of \$400 per month, all units will be rented. However, for each increase of \$20 in rent, he can expect one unit to be vacated. Let x represent the number of \$20 increases over \$400.
   (a) Express, in terms of x, the number of apartments that will be rented if x increases of \$20 are made. (For example, with three such increases, the number of apartments rented will be $80 - 3 = 77$.)
   (b) Express the rent per apartment if x increases of \$20 are made. (For example, if he increases rent $\$60 = 3 \times \$20$, the rent per apartment is $400 + 3(20) = \$460$.)
   (c) Determine a revenue function R in terms of x that will give the revenue generated as a function of the number of \$20 increases.
   (d) For what number of increases will the revenue be \$37,500?
   (e) What rent should he charge in order to achieve the maximum revenue?

18. *Seminar Fee* When Respect Brings Success charges \$600 for a seminar on management techniques, it attracts 1000 people. For each decrease of \$20 in the charge, an additional 100 people will attend the seminar. Let x represent the number of \$20 decreases in the charge.
   (a) Determine a revenue function R that will give revenue generated as a function of the number of \$20 decreases.
   (b) Find the value of x that maximizes the revenue. What should the company charge to maximize the revenue?
   (c) What is the maximum revenue the company can generate?

19. *Shooting a Foul Shot* Refer to the chapter introduction. To make a foul shot in basketball, the ball must follow a parabolic arc that depends on both the angle and velocity with which the basketball is released. If a person shoots the basketball overhand from a position 8 feet above the floor, then the path can sometimes be modeled by the quadratic function defined by

$$f(x) = \frac{-16x^2}{.434v^2} + 1.15x + 8,$$

where v is the initial velocity of the ball in feet per second, as illustrated in the figure. (*Source:* Rist, C., "The Physics of Foul Shots," *Discover,* October 2000.)

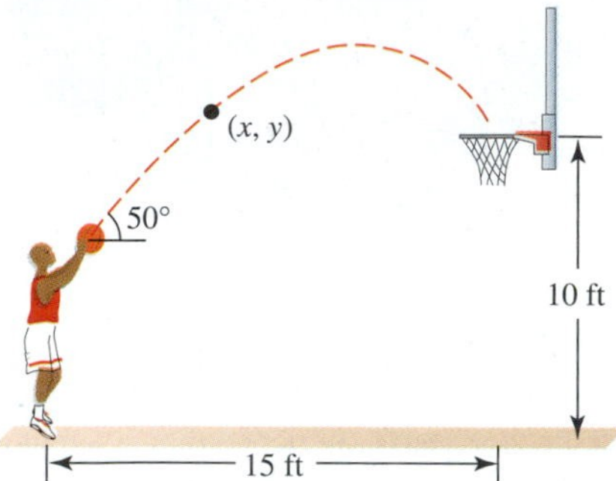

(a) If the basketball hoop is 10 feet high and located 15 feet away, what initial velocity v should the basketball have?

(b) Check your answer from part (a) graphically. Plot the point $(0, 8)$ where the ball is released and the point $(15, 10)$ where the basketball hoop is. Does your graph pass through both points?

(c) What is the maximum height of the basketball?

20. ***Shooting a Foul Shot*** Refer to Exercise 19. If a person releases a basketball underhand from a position 3 feet above the floor, it often has a steeper arc than if it is released overhand, and its path can sometimes be modeled by

$$f(x) = \frac{-16x^2}{.117v^2} + 2.75x + 3.$$

Complete parts (a), (b), and (c) of Exercise 19. Then compare the paths for the overhand shot and the underhand shot.

21. ***Path of a Frog's Leap*** A frog leaps from a stump 3 feet high and lands 4 feet from the base of the stump. We can consider the initial position of the frog to be at $(0, 3)$ and its landing position to be at $(4, 0)$.

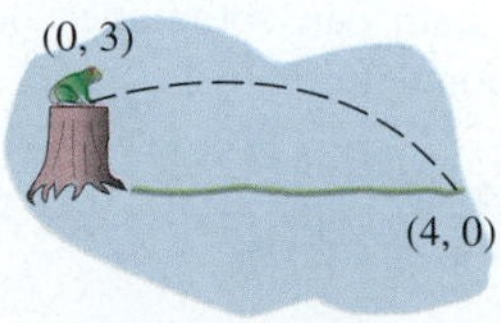

It is determined that the height h in feet of the frog as a function of its distance x from the base of the stump is given by

$$h(x) = -.5x^2 + 1.25x + 3.$$

(a) How high was the frog when its horizontal distance x from the base of the stump was 2 feet?

(b) What was the horizontal distance from the base of the stump when the frog was 3.25 feet above the ground?

(c) At what horizontal distance from the base of the stump did the frog reach its highest point?

(d) What was the maximum height reached by the frog?

22. ***Path of a Frog's Leap*** Refer to Exercise 21. Suppose that the initial position of the frog is $(0, 4)$ and its landing position is $(6, 0)$. The height of the frog in feet is given by

$$h(x) = -\frac{1}{3}x^2 + \frac{4}{3}x + 4.$$

(a) What was the horizontal distance x from the base of the stump when the frog reached maximum height?

(b) What was the maximum height?

23. ***Airplane Landing Speed*** To determine the appropriate landing speed of Vangie's airplane, we might use

$$f(x) = \frac{1}{10}x^2 - 3x + 22,$$

where x is the initial landing speed in feet per second and $f(x)$ is the length of the runway in feet. If the landing speed is too fast, she may run out of runway; if the speed is too slow, the plane may stall. If the runway is 800 feet long, what is the appropriate landing speed? What is the landing speed in mph? (*Hint:* 5280 feet = 1 mile)

24. ***Fatality Rate*** As a function of age group x, the fatality rate (per 100,000 population) for males killed in automobile accidents can be approximated by

$$f(x) = 1.8x^2 - 12x + 37.4,$$

where $x = 0$ represents ages 21–24, $x = 1$ represents ages 25–34, $x = 2$ represents ages 35–44, and so on. Find the age group at which the accident rate is a minimum, and find the minimum rate. (*Source:* National Highway Traffic Safety Administration.)

25. ***Carbon Monoxide Exposure*** Carbon monoxide (CO) combines with the hemoglobin of the blood to form carboxyhemoglobin (COHb), which reduces the transport of oxygen to tissues. Smokers routinely have a 4% to 6% COHb level in their blood, which can cause symptoms such as blood flow alterations, visual impairment, and poorer vigilance. The quadratic function defined by

$$T(x) = .00787x^2 - 1.528x + 75.89$$

approximates the exposure time in hours necessary to reach this 4% to 6% level, where $50 \leq x \leq 100$ is the amount of carbon monoxide present in the air in parts per million (ppm). (*Source: Indoor Air Quality Environmental Information Handbook: Combustion Sources,* U.S. Department of Energy, 1985.)

(a) A kerosene heater or a room full of smokers is capable of producing 50 ppm of carbon monoxide. How long would it take for a nonsmoking person to start feeling the symptoms mentioned?

(b) Find the carbon monoxide concentration necessary for a person to reach the 4% to 6% COHb level in 3 hours.

26. ***Carbon Monoxide Exposure*** Refer to Exercise 25. High concentrations of carbon monoxide (CO) can cause coma and death. The time required for a person to reach a COHb level capable of causing a coma can be approximated by

$$T(x) = .0002x^2 - .316x + 127.9,$$

where T is the exposure time in hours necessary to reach that level and $500 \leq x \leq 800$ is the amount of carbon monoxide in parts per million (ppm). (*Source: Indoor Air Quality Environmental Information Handbook: Combustion Sources,* U.S. Department of Energy, 1985.)

(a) What is the exposure time when $x = 600$ ppm?

(b) Estimate the concentration of CO necessary to produce a coma in 4 hours.

27. ***Americans Over 100 Years of Age*** The table lists the number of Americans (in thousands) who were predicted to be over 100 years old for selected years.

Year	Number (in thousands)
1994	50
1996	56
1998	65
2000	75
2002	94
2004	110

Source: U.S. Census Bureau.

(a) Plot the data. Let $x = 4$ correspond to the year 1994, $x = 6$ correspond to 1996, and so on.

(b) Find a quadratic function defined by

$$f(x) = a(x - h)^2 + k$$

that models the data. Use $(4, 50)$ as the vertex and $(14, 110)$ as the other point to determine a.

(c) Plot the data together with f in the same window. How well does f model the number of Americans (in thousands) who are expected to be over 100 years old?

(d) Use the quadratic regression feature of a graphing calculator to determine the quadratic function g that provides the best fit for the data.

(e) Use the functions f and g to predict the number of Americans, in thousands, who will be over 100 years old in the following years:

(i) 2006
(ii) 2010
(iii) 2015.

28. ***Hospital Spending*** The data for annual percent increase in spending on hospital services from Section 3.2, Example 6, are repeated in the table here.

Year	Percent Increase
1994	1.8
1995	.8
1996	.5
1997	1.3
1998	3.4
1999	5.8
2000	7.1
2001	12.0

Source: Center for Studying Health System Change.

(a) Plot the data, where x is the number of years since 1990. Then graph the function defined by

$$f(x) = .37x^2 - 4.1x + 12$$

given in Example 6 in the same window. Does the function appear to fit the data well?

(b) Use the quadratic regression feature of a graphing calculator to find the quadratic function g that best fits the data when the points $(4, 1.8)$ and $(5, .8)$ are not used.

(c) Use f and g to predict the percent increases for the following years:

(i) 2002
(ii) 2003
(iii) 2004.

29. ***Automobile Stopping Distance*** Selected values of the stopping distance y in feet of a car traveling x mph are given in the table.

Speed (in mph)	Stopping Distance (in feet)
20	46
30	87
40	140
50	240
60	282
70	371

Source: National Safety Institute Student Workbook, 1993, p. 7.

(a) Plot the data.

(b) The quadratic function defined by

$$f(x) = .056057x^2 + 1.06657x$$

is one model of the data. Find and interpret $f(45)$.

(c) Graph the function in the same window as the data. How well does f model the stopping distance?

30. ***Coast-Down Time*** The coast-down time y for a typical car as it drops 10 mph from an initial speed x depends on several factors, such as average drag, tire pressure, and whether the transmission is in neutral. The table gives the coast-down time in seconds for a car under standard conditions for selected speeds in miles per hour.

Initial Speed (in mph)	Coast-Down Time (in seconds)
30	30
35	27
40	23
45	21
50	18
55	16
60	15
65	13

Source: Scientific American, December 1994.

(a) Plot the data.

(b) Use the quadratic regression feature of a graphing calculator to find the quadratic function g that best fits the data. Graph this function in the same window as the data. Is g a good model for the data?

(c) Use g to predict the coast-down time at an initial speed of 70 mph.

(d) Use the graph to find the speed that corresponds to a coast-down time of 24 seconds.

3.5 Higher-Degree Polynomial Functions and Graphs

Cubic Functions ■ Quartic Functions ■ Extrema ■ End Behavior ■ x-Intercepts (Real Zeros) ■ Comprehensive Graphs ■ Curve Fitting and Polynomial Models

Looking Ahead to Calculus

In calculus, polynomial functions are used to approximate more complicated functions, such as trigonometric, exponential, or logarithmic functions. For example, the trigonometric function $\sin x$ is approximated by the polynomial

$$x - \frac{x^3}{3!} + \frac{x^5}{5!} - \frac{x^7}{7!}$$

for values of x near 0. (See Chapter 11 for an explanation of factorial (!) notation.)

Linear and quadratic functions are examples of *polynomial functions.*

Polynomial Function

A **polynomial function of degree n in the variable x** is a function defined by

$$P(x) = a_n x^n + a_{n-1}x^{n-1} + \cdots + a_1 x + a_0,$$

where each a_i is a real number, $a_n \neq 0$, and n is a whole number.*

* While our definition requires real coefficients, the definition of a polynomial function can be extended to include nonreal complex numbers as coefficients.

The behavior of the graph of a polynomial function is due largely to the value of the coefficient a_n and the *parity* (that is, "evenness" or "oddness") of the exponent n on the term of greatest degree. For this reason, we will refer to a_n as the **leading coefficient** and to a_nx^n as the **dominating term.** The term a_0 is the constant term of the polynomial function, and since $P(0) = a_0$, it is the y-intercept of the graph.

As we study the graphs of polynomial functions, we use the following properties:

1. A polynomial function (unless otherwise specified) has domain $(-\infty, \infty)$.
2. The graph of a polynomial function is a smooth, continuous curve with no sharp corners.

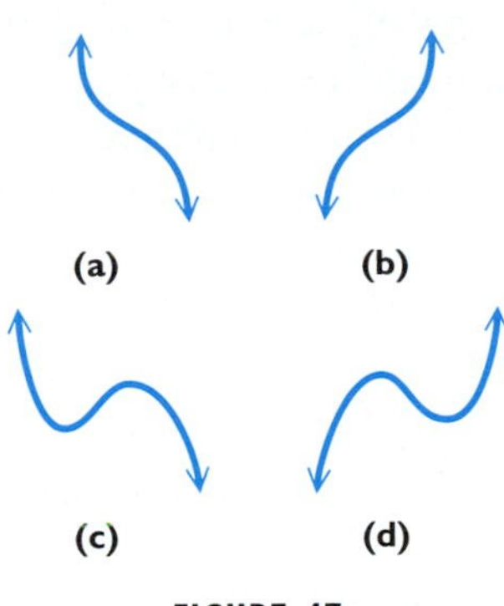

FIGURE 47

Cubic Functions

A polynomial function of the form $\boldsymbol{P(x) = ax^3 + bx^2 + cx + d,\ a \neq 0,}$ is a third-degree function, or **cubic function.** The simplest cubic function is $f(x) = x^3$. The graph of a cubic function generally resembles one of the shapes shown in Figure 47.

FOR DISCUSSION

1. Without graphing, which of the shapes in Figure 47 depicts the general form of the graph of $f(x) = x^3$?
2. Using the concepts of reflection of graphs from Chapter 2, which shape in Figure 47 most closely resembles the graph of $g(x) = -x^3$?
3. Graph each function in the standard window, and determine which shape in Figure 47 the graph most closely resembles.
   (a) $P(x) = x^3 + 5x^2 + 5x - 2$ **(b)** $f(x) = -x^3 + 5x - 1$
   (c) $g(x) = x^3 + 3x^2 - 3x + 1$

Quartic Functions

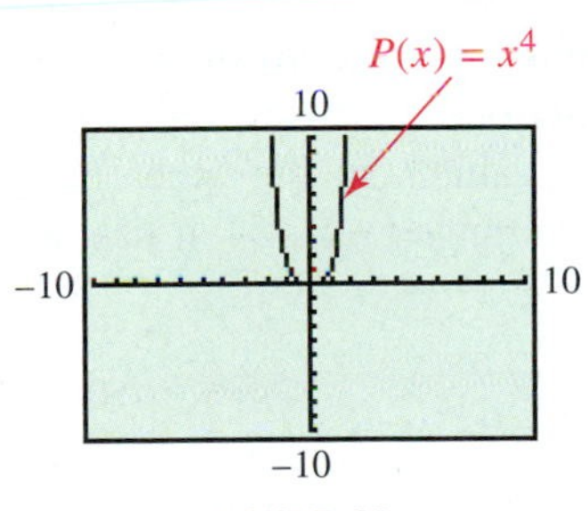

FIGURE 48

A polynomial function of the form $\boldsymbol{P(x) = ax^4 + bx^3 + cx^2 + dx + e,\ a \neq 0,}$ is a fourth-degree function, or **quartic function.** The simplest quartic function, $P(x) = x^4$, is graphed in Figure 48. Notice that it resembles the graph of the squaring function; however, it is *not* actually a parabola.

If we graph a quartic function in an appropriate window, the graph will generally resemble one of the shapes shown in Figure 49. The dashed portions in (c) and (d) indicate that there may be irregular, but smooth, behavior in those intervals.

1. Which of the shapes in Figure 49 depicts the general form of the graphs of $f(x) = x^2$ and $g(x) = x^4$?
2. Using the concepts of reflection of graphs from Chapter 2, which shape in Figure 49 most closely resembles the graph of $h(x) = -x^4$?
3. Graph each function in the specified window, and determine which shape in Figure 49 the graph most closely resembles.
   (a) $y = 3x^4 + x - 2$; $[-10, 10]$ by $[-10, 10]$
   (b) $y = -2x^4 - x^3 + x - 3$; $[-10, 10]$ by $[-10, 10]$
   (c) $y = -x^4 + 12x^3$; $[-15, 15]$ by $[-5000, 2500]$

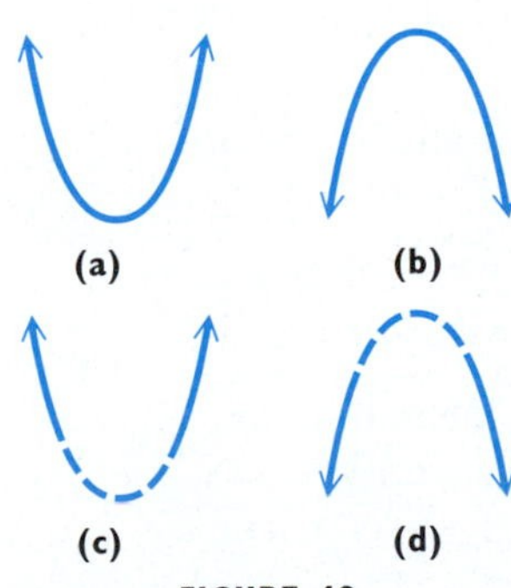

FIGURE 49

TECHNOLOGY NOTE

The feature described in the Technology Note in Section 3.2 that refers to maxima and minima also applies to polynomial functions of higher degree, provided that an appropriate interval is designated.

Looking Ahead to Calculus

Suppose we need to find the x-coordinates of the two turning points of the graph of

$$f(x) = 2x^3 - 8x^2 + 9.$$

We could use the "maximum" and "minimum" capabilities of a graphing calculator and determine that, to the nearest thousandth, they are 0 and 2.667, respectively. In calculus, their exact values can be found by determining the zeros of the derivative function of $f(x)$,

$$f'(x) = 6x^2 - 16x.$$

Factoring would show that the two zeros are 0 and $\frac{8}{3}$, which agree with the approximations found with a graphing calculator.

Extrema

In Figures 47–49, several graphs have **turning points** where the function changes from increasing to decreasing or vice versa. We first saw this behavior with quadratic functions, where the vertex was a maximum or minimum point on the graph. In general, the highest point at a "peak" is known as a **local maximum point,** and the lowest point at a "valley" is known as a **local minimum point.** Function values at such points are called **local maxima** (plural of *maximum*) and **local minima** (plural of *minimum*). Collectively, these values are called *extrema* (plural of *extremum*), as mentioned in Section 3.2. Figure 50 and the accompanying chart illustrate these ideas for typical graphs.

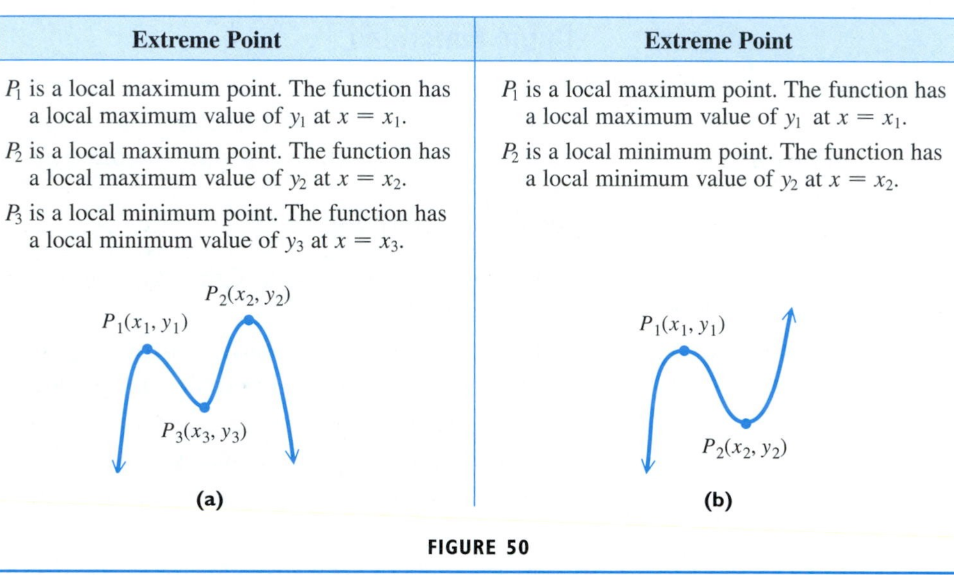

Extreme Point	Extreme Point
P_1 is a local maximum point. The function has a local maximum value of y_1 at $x = x_1$. P_2 is a local maximum point. The function has a local maximum value of y_2 at $x = x_2$. P_3 is a local minimum point. The function has a local minimum value of y_3 at $x = x_3$.	P_1 is a local maximum point. The function has a local maximum value of y_1 at $x = x_1$. P_2 is a local minimum point. The function has a local minimum value of y_2 at $x = x_2$.

FIGURE 50

Notice in Figure 50(a) that point P_2 is the absolute highest point on the graph, and the range of the function is $(-\infty, y_2]$. We call P_2 the **absolute maximum point** on the graph and y_2 the **absolute maximum value** of the function. Because the y-values approach $-\infty$, this function has no absolute minimum value. On the other hand, because the graph in Figure 50(b) is that of a function with range $(-\infty, \infty)$, it has neither an absolute maximum nor an absolute minimum.

Absolute and Local Extrema

Let c be in the domain of P. Then

(a) $P(c)$ is an **absolute maximum** if $P(c) \geq P(x)$ for all x in the domain of P.

(b) $P(c)$ is an **absolute minimum** if $P(c) \leq P(x)$ for all x in the domain of P.

(c) $P(c)$ is a **local maximum** if $P(c) \geq P(x)$ when x is *near* c.

(d) $P(c)$ is a **local minimum** if $P(c) \leq P(x)$ when x is *near* c.

The expression "near c" means that there is an open interval in the domain of P containing c, where $P(c)$ satisfies the inequality.

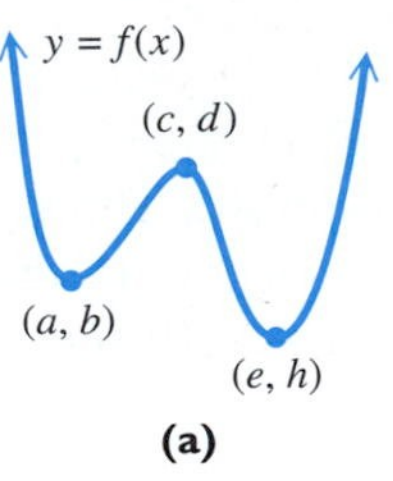

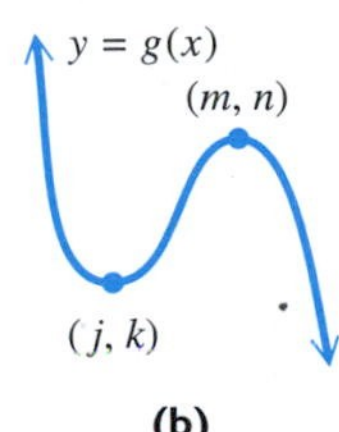

FIGURE 51

EXAMPLE 1 Identifying Local and Absolute Extreme Points

Consider the graphs in Figure 51.

(a) Identify and classify the local extreme points of f.

(b) Identify and classify the local extreme points of g.

(c) Describe the absolute extreme points for f and g.

Solution

(a) The points (a, b) and (e, h) are local minimum points. The point (c, d) is a local maximum point.

(b) The point (j, k) is a local minimum point and the point (m, n) is a local maximum point.

(c) The absolute minimum point of function f is (e, h), and the absolute minimum value is h, since the range of f is $[h, \infty)$. Function f has no absolute maximum value. Function g has no absolute extrema, since its range is $(-\infty, \infty)$. ■

NOTE A function may have more than one absolute maximum or minimum *point,* but only one absolute maximum or minimum *value.*

It is possible for the graph of a polynomial function to have a maximum or a minimum point that is not apparent in a particular window. This is an example of **hidden behavior.**

$P(x) = x^3 - 2x^2 + x - 2$

10, −10, 10, −10

FIGURE 52

EXAMPLE 2 Examining Hidden Behavior

Figure 52 shows the graph of $P(x) = x^3 - 2x^2 + x - 2$ in the standard viewing window. Make a conjecture concerning possible hidden behavior, and verify it.

Solution In Figure 52, because the graph levels off in the domain interval $[0, 2]$, there may be behavior that is not apparent in the given window. By changing the window to $[-2.5, 2.5]$ by $[-4.5, .5]$, we see that there are two extrema there. The local maximum point, as seen in Figure 53, has approximate coordinates $(.33, -1.85)$. There is also a local minimum point at $(1, -2)$. ■

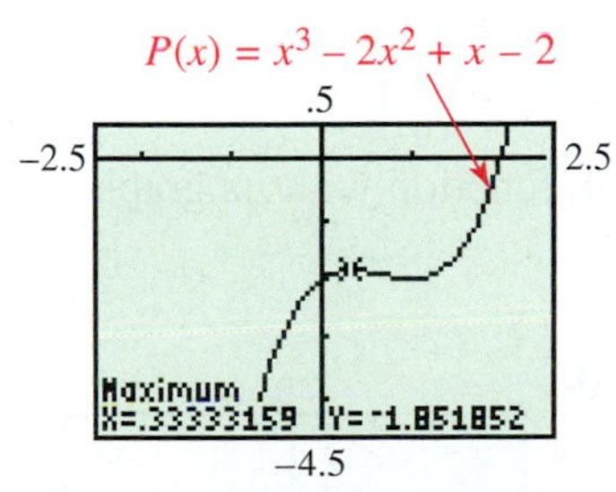

FIGURE 53

A quadratic function has degree 2 and has only one turning point (its vertex). Extending this idea, we would expect a third-degree polynomial function to have at most two turning points, a fourth-degree polynomial function to have at most three turning points, and so on. This is actually the case.

Number of Turning Points

The number of turning points of the graph of a polynomial function of degree $n \geq 1$ is at most $n - 1$.

NOTE The above property implies that the number of local extrema is *at most* $n - 1$ for a *polynomial function* of degree $n \geq 1$. The graph may have fewer than $n - 1$ local extrema.

End Behavior

If the value of a is positive for the quadratic function defined by $P(x) = ax^2 + bx + c$, the graph opens upward; if a is negative, the graph opens downward. The sign of a determines the **end behavior** of the graph. In general, the end behavior of the graph of a polynomial function is determined by the sign of the leading coefficient and the parity of the degree.

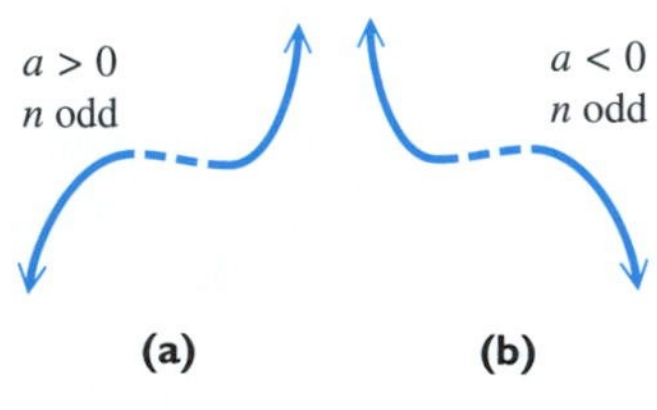

FIGURE 54

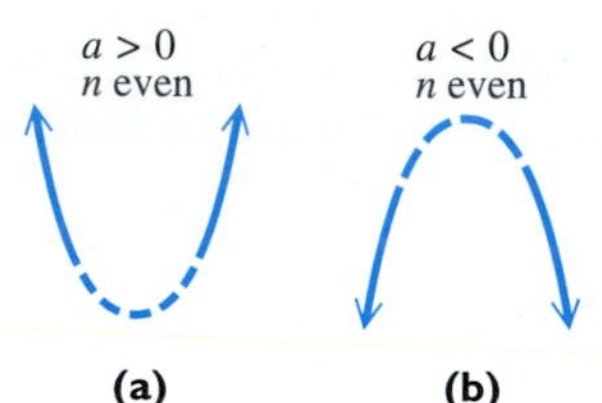

FIGURE 55

End Behavior of Graphs of Polynomial Functions

Suppose that ax^n is the dominating term of a polynomial function P of ***odd degree.*** (The dominating term is the term of greatest degree.)

1. If $a > 0$, then as $x \to \infty$, $P(x) \to \infty$, and as $x \to -\infty$, $P(x) \to -\infty$. Therefore, the end behavior of the graph is of the type shown in Figure 54(a). We symbolize it as $\swarrow\!\nearrow$.
2. If $a < 0$, then as $x \to \infty$, $P(x) \to -\infty$, and as $x \to -\infty$, $P(x) \to \infty$. Therefore, the end behavior of the graph is of the type shown in Figure 54(b). We symbolize it as $\nwarrow\!\searrow$.

Suppose that ax^n is the dominating term of a polynomial function P of ***even degree.***

1. If $a > 0$, then as $|x| \to \infty$, $P(x) \to \infty$. Therefore, the end behavior of the graph is of the type shown in Figure 55(a). We symbolize it as $\nwarrow\!\nearrow$.
2. If $a < 0$, then as $|x| \to \infty$, $P(x) \to -\infty$. Therefore, the end behavior of the graph is of the type shown in Figure 55(b). We symbolize it as $\swarrow\!\searrow$.

EXAMPLE 3 Determining End Behavior Given the Defining Polynomial

The graphs of the functions defined as follows are shown in Figure 56.

$$f(x) = x^4 - x^2 + 5x - 4, \qquad g(x) = -x^6 + x^2 - 3x - 4,$$

$$h(x) = 3x^3 - x^2 + 2x - 4, \quad \text{and} \quad k(x) = -x^7 + x - 4.$$

Based on the discussion in the preceding box, match each function with its graph.

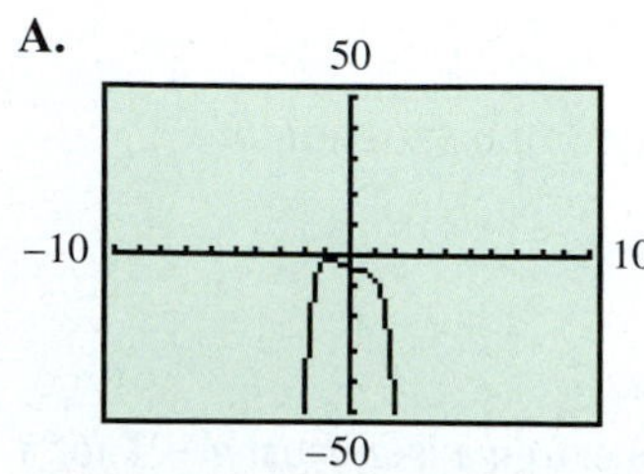

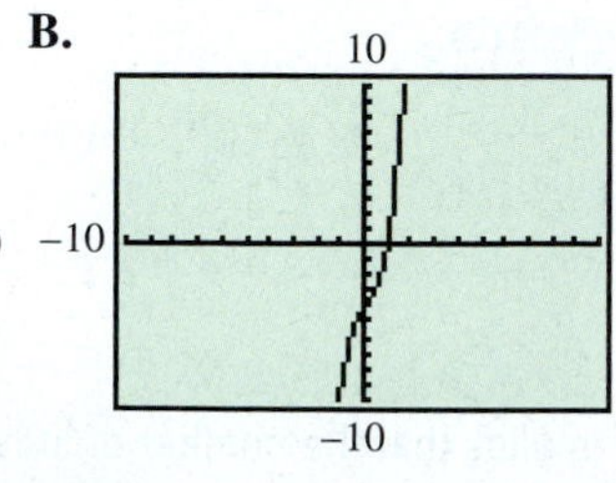

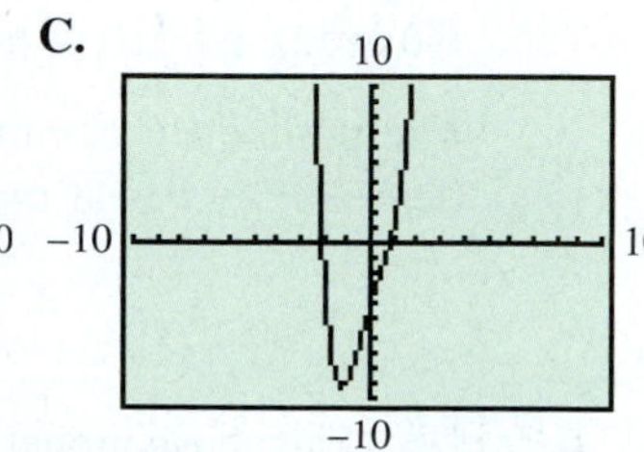

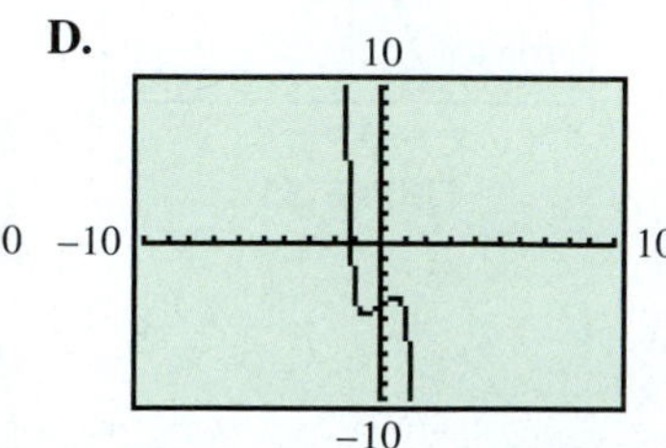

FIGURE 56

Solution Because f is of even degree with positive leading coefficient, its graph is in C. Because g is of even degree with negative leading coefficient, its graph is in A. Function h has odd degree and the coefficient of the dominating term is positive, so its graph is in B. Because function k has odd degree and a negative coefficient of the dominating term, its graph is in D. ■

x-Intercepts (Real Zeros)

A linear function can have no more than one x-intercept, and a quadratic function can have no more than two x-intercepts. Figure 57 shows how a cubic function may have one, two, or three x-intercepts. These observations suggest the following important property of polynomial functions.

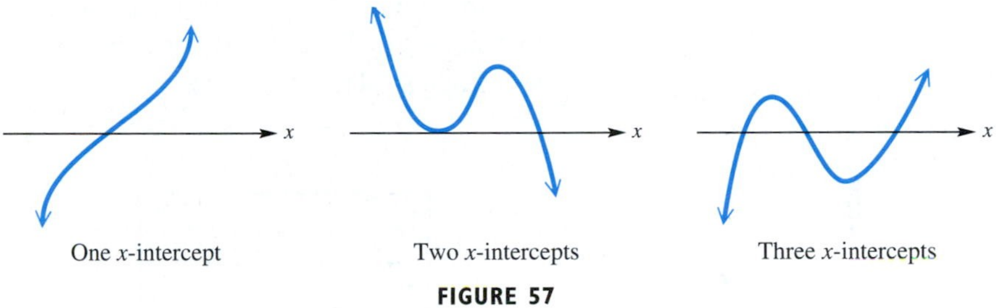

FIGURE 57

Number of x-Intercepts (Real Zeros) of a Polynomial Function

The graph of a polynomial function of degree n will have at most n x-intercepts (real zeros).

EXAMPLE 4 **Determining x-Intercepts Graphically**

Graphically find the x-intercepts (real zeros) of the polynomial function defined by

$$P(x) = x^3 + 5x^2 + 5x - 2.$$

Solution Using the standard window, we get the graph shown in Figure 58, which has three x-intercepts. We find that the x-intercepts (real zeros) are -2, approximately -3.30 (shown in the display), and approximately .30.

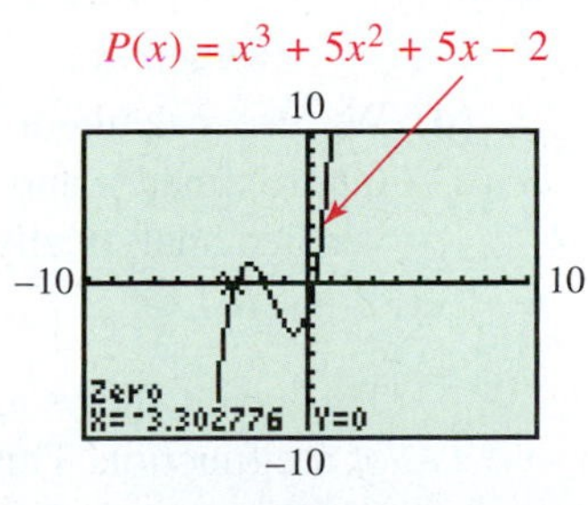

FIGURE 58 ■

EXAMPLE 5 **Analyzing a Polynomial Function**

Perform the following for the fifth-degree polynomial function defined by

$$P(x) = x^5 + 2x^4 - x^3 + x^2 - x - 4.$$

(a) Determine its domain. **(b)** Determine its range.

(c) Use its graph to find approximations of its local extreme points.

(d) Use its graph to find its approximate and/or exact x-intercepts.

Looking Ahead to Calculus

In calculus, the derivative of a function f is a function f' that gives the slope of the tangent line at any value in the domain. At turning points, the slope—and thus the derivative—is 0. By solving the equation $f'(x) = 0$, the number and location of the extrema of f can be identified. In Example 5(c), this method would verify that there are only two extrema.

Solution

(a) This is a polynomial function, so its domain is $(-\infty, \infty)$.

(b) This function is of odd degree. Its range is $(-\infty, \infty)$.

(c) From the graph of P in Figures 59 and 60, it appears that there are only two extreme points. We find that the local maximum in Figure 59 has approximate coordinates $(-2.02, 10.01)$ and the local minimum in Figure 60 has approximate coordinates $(.41, -4.24)$.

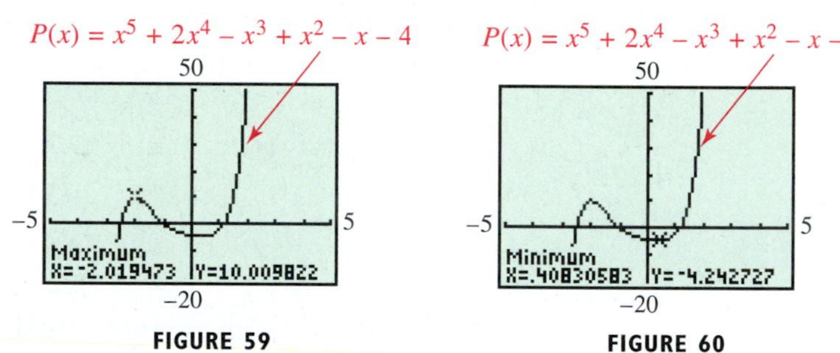

FIGURE 59 **FIGURE 60**

The preceding graphical results can also be found by using a built-in utility, as shown in the two screens in Figure 61. Here Y_1 is defined as

$$X^5 + 2X^4 - X^3 + X^2 - X - 4,$$

so the local maximum is in the interval $[-3, -1]$ and the local minimum is in the interval $[0, 1]$.

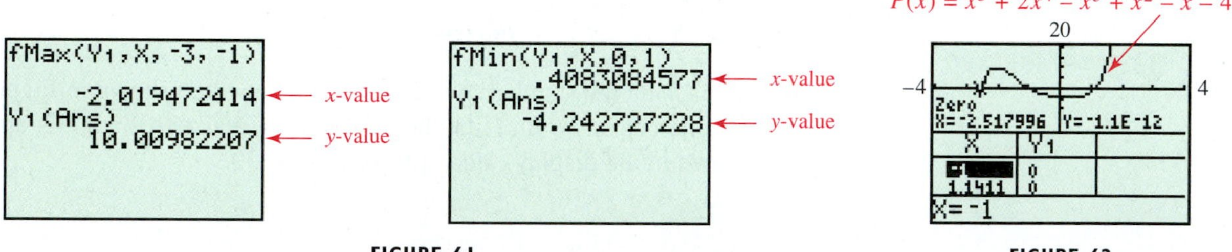

FIGURE 61 **FIGURE 62**

(d) We use calculator methods to find that the x-intercepts are -1 (exact), 1.14 (approximate), and -2.52 (approximate). See Figure 62. The first of these can be verified analytically by evaluating $P(-1)$.

$$P(-1) = (-1)^5 + 2(-1)^4 - (-1)^3 + (-1)^2 - (-1) - 4 = 0$$

This result shows again that an x-intercept of the graph of a function is a real zero of the function. This function has only three x-intercepts and thus three real zeros, which supports the fact that a polynomial function of degree n will have *at most* n x-intercepts. It may have fewer, as in this case. ■

EXAMPLE 6 Analyzing a Polynomial Function

Perform the following for the fourth degree polynomial function defined by

$$P(x) = x^4 + 2x^3 - 15x^2 - 12x + 36.$$

(a) State the domain.

(b) Use the graph of $P(x)$ to find approximations of its local extreme points. Does it have an absolute minimum? What is the range of the function?

(c) Use its graph to find its x-intercepts.

Solution

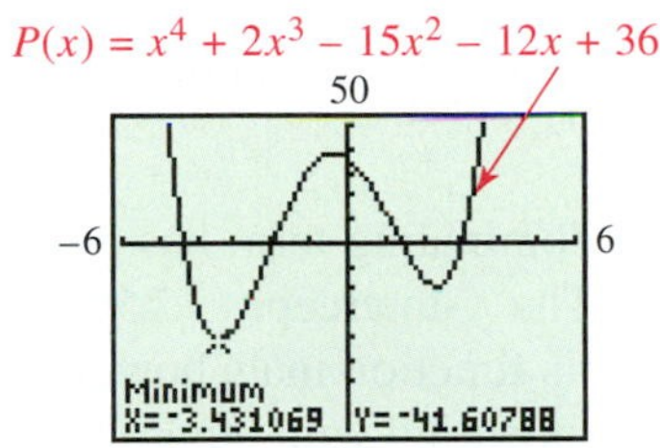

−80

The other two extreme points are (−.38, 38.31) and (2.31, −18.63).

FIGURE 63

(a) Because $P(x)$ is a polynomial function, its domain is $(-\infty, \infty)$.

(b) A window of $[-6, 6]$ by $[-80, 50]$ provides a view of the extreme points, as well as all intercepts. See Figure 63. Using the capabilities of a calculator, we find that the two local minimum points have approximate coordinates $(-3.43, -41.61)$ and $(2.31, -18.63)$, and the local maximum point has approximate coordinates $(-.38, 38.31)$. Because the end behavior is ↖ ↗ and the point $(-3.43, -41.61)$ is the lowest point on the graph, the absolute minimum value of the function is approximately -41.61; therefore, the range is approximately $[-41.61, \infty)$.

(c) This function has the maximum number of x-intercepts possible, four. Using the calculator capabilities, we find that two exact values for the x-intercepts are -2 and 3, while, to the nearest hundredth, the other two are -4.37 and 1.37. ■

What Went Wrong?

A student graphed $y = .045x^4 - 2x^2 + 2$ in the decimal window of a popular model of graphing calculator. Because the polynomial has even degree and positive leading coefficient, she expected to find end behavior ↖ ↗. However, this graph indicates ↙ ↘ as end behavior.

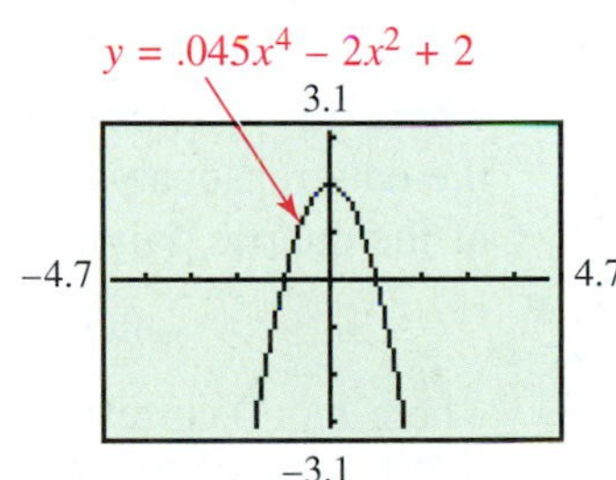

What Went Wrong? Is there a way to graph this function so that the correct end behavior is apparent?

Answer to What Went Wrong?

The window must be enlarged. Scrolling through a table should indicate a suitable window. One example is $[-10, 10]$ by $[-25, 10]$.

$P(x) = x^6 - 36x^4 + 288x^2 - 256$

50

−1.25 1.25

−400

FIGURE 64

Comprehensive Graphs

The most important features of the graph of a polynomial function are its intercepts, its extrema, and its end behavior. For this reason, ***a comprehensive graph of a polynomial function*** will exhibit the following features:

1. all x-intercepts (if any),
2. the y-intercept,
3. all extreme points (if any),
4. enough of the graph to reveal the correct end behavior.

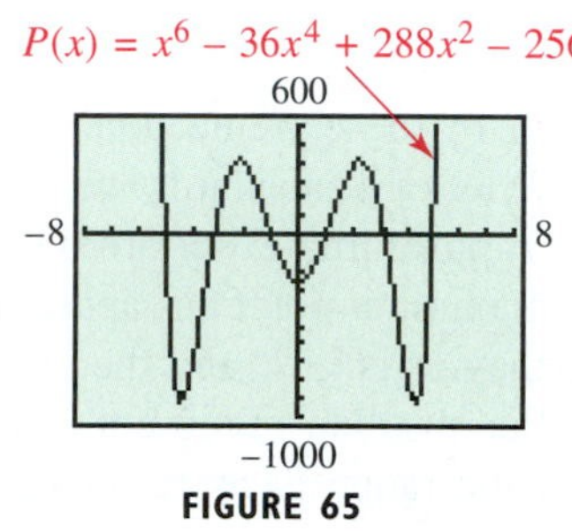

FIGURE 65

EXAMPLE 7 Determining an Appropriate Window

Is the graph of $P(x) = x^6 - 36x^4 + 288x^2 - 256$ in Figure 64 a comprehensive graph? If not, provide one.

Solution Since the function is of even degree and the dominating term has positive coefficient, the end behavior seems to be correct. The y-intercept, -256, is shown, as are two x-intercepts and one local minimum. This function may, however, have up to six x-intercepts, since it is of degree 6. By experimenting with other viewing windows, we see that a window of $[-8, 8]$ by $[-1000, 600]$ shows a total of five local extrema, and four x-intercepts that were not apparent in the earlier figure. See Figure 65. Since there can be no more than five local extrema, this second view (*not* the first view in Figure 64) gives a comprehensive graph. ■

Curve Fitting and Polynomial Models

We have used graphing calculators to find linear and quadratic regression equations to model data. Some data can be modeled best with cubic (degree 3) or quartic (degree 4) regression equations.

For example, the table gives the average monthly sales for manufacturing in the United States during 1992–2003. The data are graphed in Figure 66(a). The cubic polynomial function defined by

$$f(x) = -.0956x^3 + .2557x^2 + 16.61x + 239,$$

where $x = 0$ corresponds to 1992, $x = 5$ corresponds to 1997, and so on, models the data reasonably well, as shown in Figure 66(b). Figure 66(c) shows the graph of the quartic polynomial function defined by

$$g(x) = .0346x^4 - .856x^3 + 5.495x^2 + 4.99x + 243,$$

where $x = 0$ corresponds to 1992, $x = 5$ to 1997, and so on.

Year	Average Monthly Manufacturing Sales (in billions of dollars)
1992	242
1993	252
1994	270
1995	290
1996	300
1997	320
1998	325
1999	336
2000	351
2001	331
2002	324
2003	333

Source: U.S. Census Bureau.

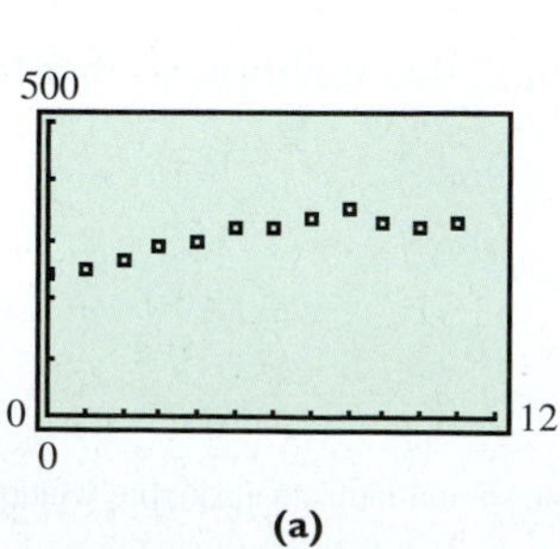

(a)

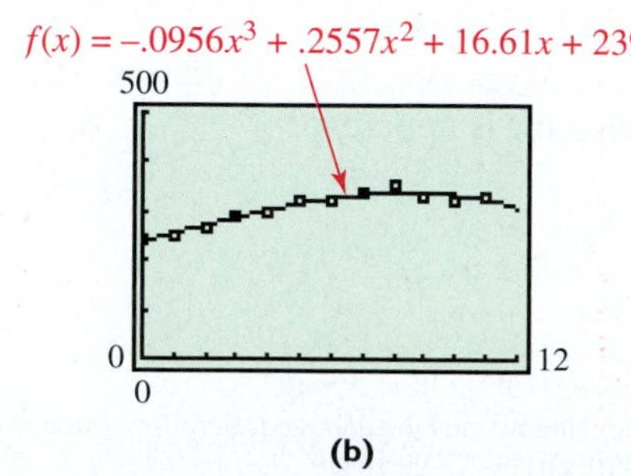

(b)

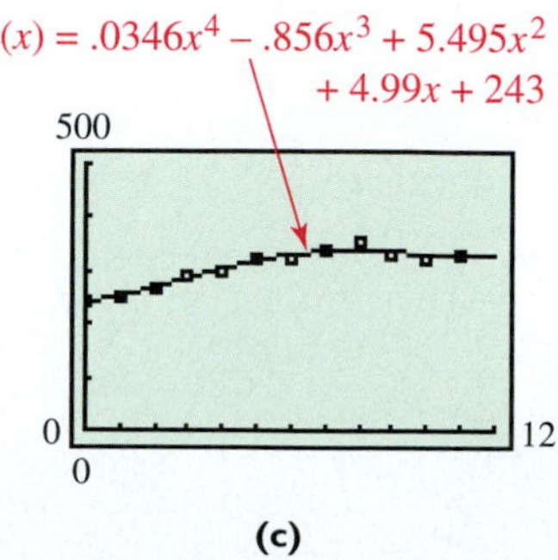

(c)

FIGURE 66

The quartic function g provides a better model for two reasons. The value of R^2 is .975, compared with .967 for the cubic model, which indicates a closer correlation with the data. The quartic model also predicts an increase in manufacturing for years beyond 2003, while the cubic model predicts a decrease. Historically, except for short periods, manufacturing sales have increased steadily over time, making the quartic model more consistent with historical trends than the cubic model.

3.5 Exercises

Relating Concepts

For individual or group investigation (Exercises 1–4)

The concepts of stretching, shrinking, translating, and reflecting graphs presented in Sections 2.2 and 2.3 can be applied to polynomial functions of the form $P(x) = x^n$. *For example, the graph of* $y = -2(x + 4)^4 - 6$ *can be obtained from the graph of* $y = x^4$ *by shifting 4 units to the left, stretching vertically by a factor of 2, reflecting across the x-axis, and shifting downward 6 units, so the graph should resemble the graph at the right.*

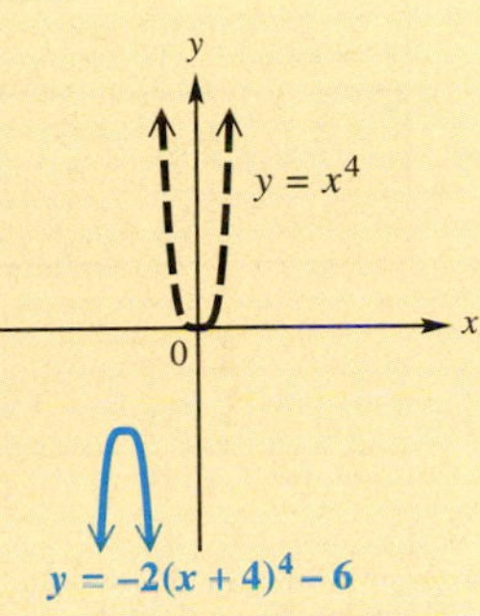

If we expand the expression $-2(x + 4)^4 - 6$ *algebraically, we get*

$$-2x^4 - 32x^3 - 192x^2 - 512x - 518.$$

Thus, the graph of $y = -2(x + 4)^4 - 6$ *is the same as that of*

$$y = -2x^4 - 32x^3 - 192x^2 - 512x - 518.$$

In Exercises 1–4, two forms of the same polynomial function are given. Sketch by hand the general shape of the graph of the function, using the concepts of Chapter 2, and describe the transformations. Then, support your answer by graphing it on your calculator in a suitable window.

1. $y = 2(x + 3)^4 - 7$
$y = 2x^4 + 24x^3 + 108x^2 + 216x + 155$

2. $y = -3(x + 1)^4 + 12$
$y = -3x^4 - 12x^3 - 18x^2 - 12x + 9$

3. $y = -3(x - 1)^3 + 12$
$y = -3x^3 + 9x^2 - 9x + 15$

4. $y = .5(x - 1)^5 + 13$
$y = .5x^5 - 2.5x^4 + 5x^3 - 5x^2 + 2.5x + 12.5$

Concept Check *Use the graphs shown here, which include all extrema, for Exercises 5–12.*

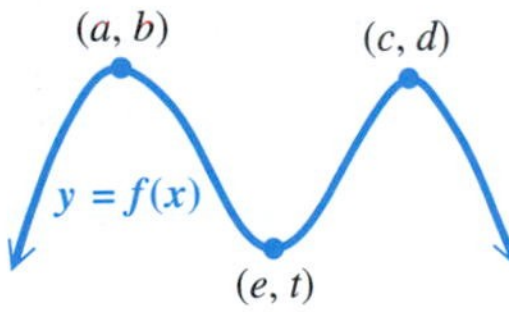

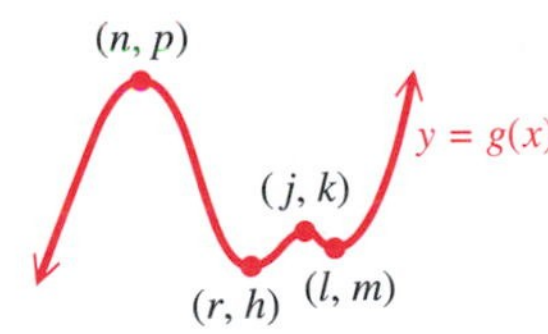

5. Use the extrema to determine the minimum degree of f.

6. Use the extrema to determine the minimum degree of g.

7. Give all local extreme points of f. Tell whether each is a maximum or minimum.

8. Give all local extreme points of g. Tell whether each is a maximum or minimum.

9. Describe all absolute extreme points of f.

10. Describe all absolute extreme points of g.

11. Give the local and absolute extreme values of f.

12. Give the local and absolute extreme values of g.

Give a short written answer in Exercises 13–16.

13. The graphs of $f(x) = x^n$ for $n = 3, 5, 7, \ldots$ resemble each other. Describe how they are the same. As n gets larger, what happens to the graph?

14. Repeat Exercise 13 for $f(x) = x^n$, where $n = 2, 4, 6, \ldots$.

15. Using a window of $[-1,1]$ by $[-1,1]$, graph the odd-degree polynomial functions

$$y=x, \quad y=x^3, \quad \text{and} \quad y=x^5.$$

Describe the behavior of these functions relative to each other. Predict the behavior of the graph of $y=x^7$ in the same window, and then graph it to support your prediction.

16. Repeat Exercise 15 for the even-degree polynomial functions

$$y=x^2, \quad y=x^4, \quad \text{and} \quad y=x^6.$$

Predict the behavior of the graph of $y=x^8$ in the same window, and then graph it to support your prediction.

Each function is graphed in a window that results in hidden behavior. Experiment with various windows to locate the extreme points on the graph of the function.

17.

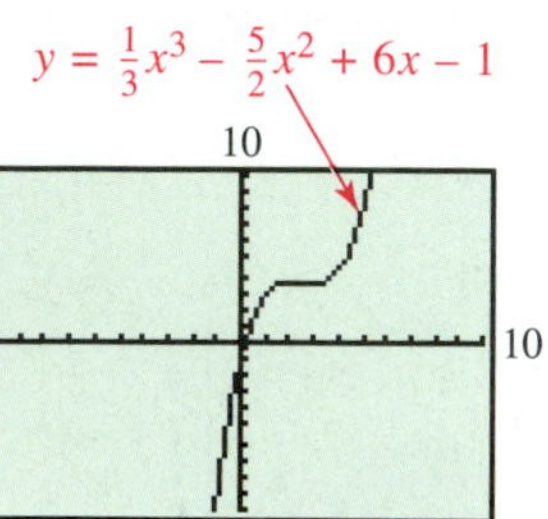

18.

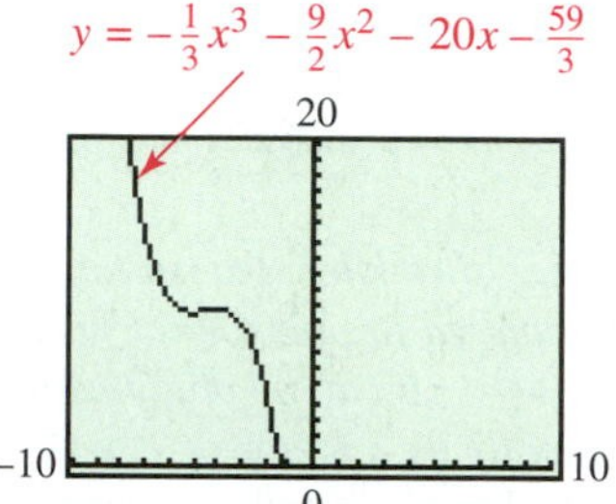

19.

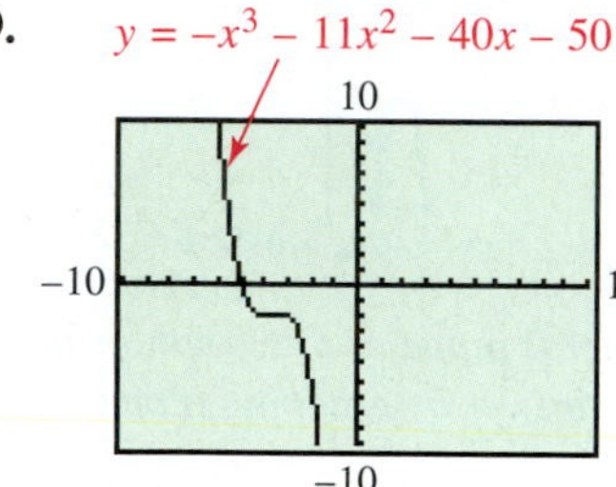

20.

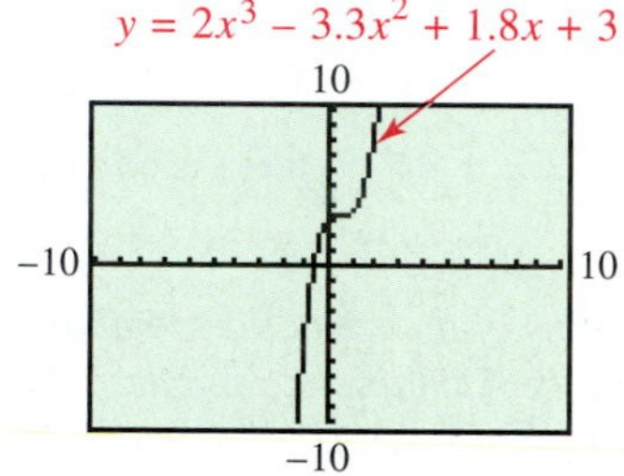

In Exercises 21–24, it is not apparent from the standard viewing window whether the graph of the quadratic function intersects the x-axis once, twice, or not at all. This is another example of hidden behavior. Experiment with various windows to determine the number of x-intercepts. If there are x-intercepts, give their values to the nearest hundredth.

21.

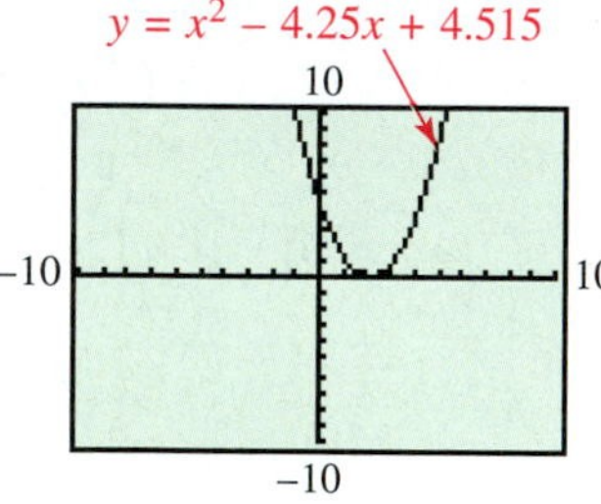

22.

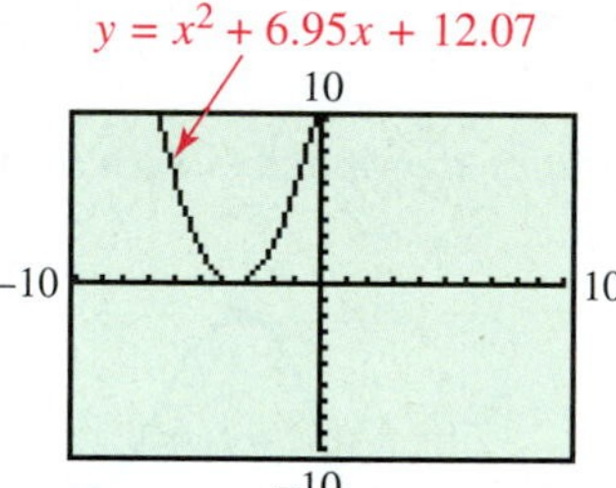

23.

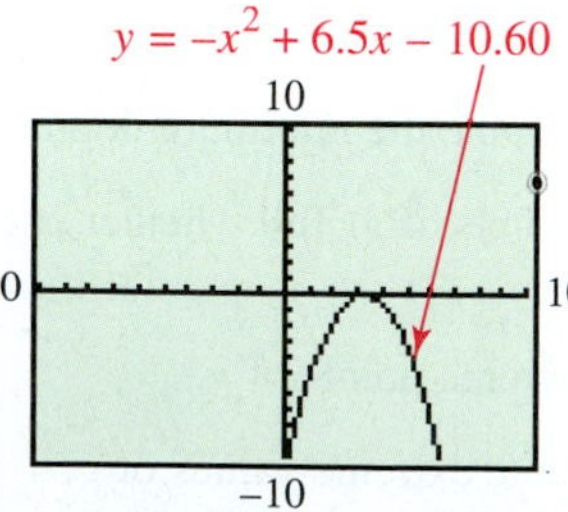

24.

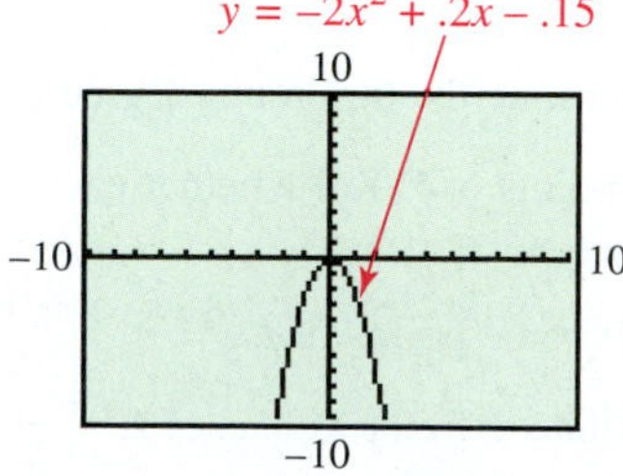

Use an end behavior diagram (↖ ↗, ↙ ↘, ↖ ↘, or ↙ ↗) to describe the end behavior of the graph of each function. Then, verify your answer by graphing the function on your calculator.

25. $P(x)=\sqrt{5}x^3+2x^2-3x+4$

26. $P(x)=-\sqrt{7}x^3-4x^2+2x-1$

27. $P(x)=-\pi x^5+3x^2-1$

28. $P(x) = \pi x^7 - x^5 + x - 1$

29. $P(x) = 2.74x^4 - 3x^2 + x - 2$

30. $P(x) = \sqrt{6}x^6 - x^5 + 2x - 2$

31. $P(x) = -x^4 + x^5 - \pi x^6 - x + 3$

32. $P(x) = -x - 3.2x^3 + x^2 - 2.84x^4$

33. $P(x) = x^{10,000}$

34. $P(x) = -x^{104,266}$

35. $P(x) = -3x^{15,297}$

36. $P(x) = 12x^{107,499}$

Without using a calculator, match each function in Exercises 37–40 with the correct graph in choices A–D.

37. $f(x) = 2x^3 + x^2 - x + 3$

38. $g(x) = -2x^3 - x + 3$

39. $h(x) = -2x^4 + x^3 - 2x^2 + x + 3$

40. $k(x) = 2x^4 - x^3 - 2x^2 + 3x + 3$

A.

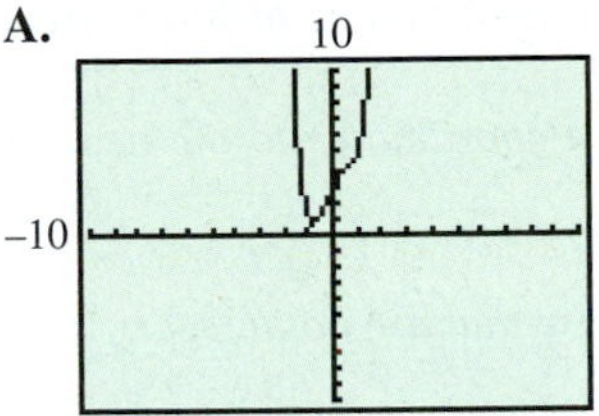

B.

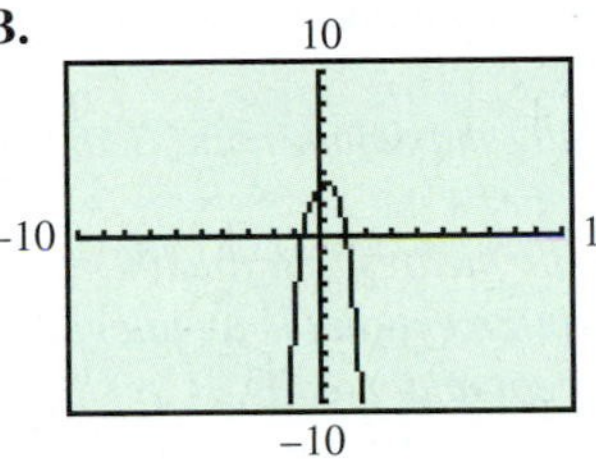

C.

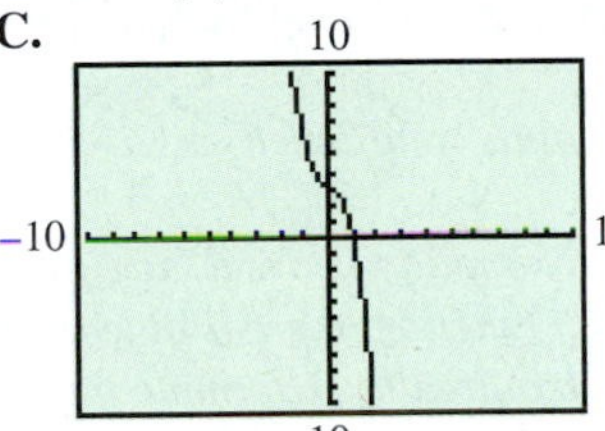

D.

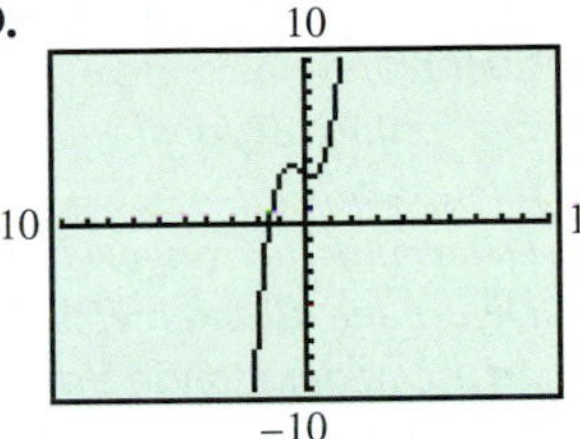

Concept Check *The graphs below show*

$$y = x^3 - 3x^2 - 6x + 8, \qquad y = x^4 + 7x^3 - 5x^2 - 75x,$$
$$y = -x^3 + 9x^2 - 27x + 17, \quad \text{and} \quad y = -x^5 + 36x^3 - 22x^2 - 147x - 90,$$

but not necessarily in that order. Assuming that each is a comprehensive graph, answer each question in Exercises 41–50.

A.

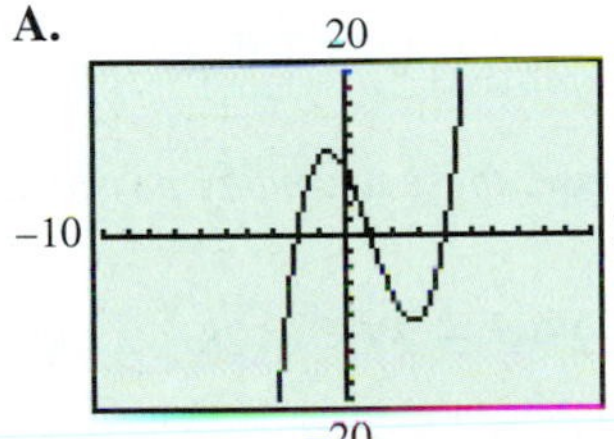

B.

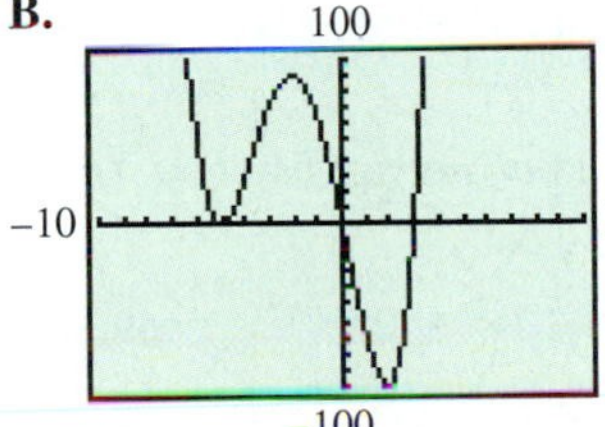

C.

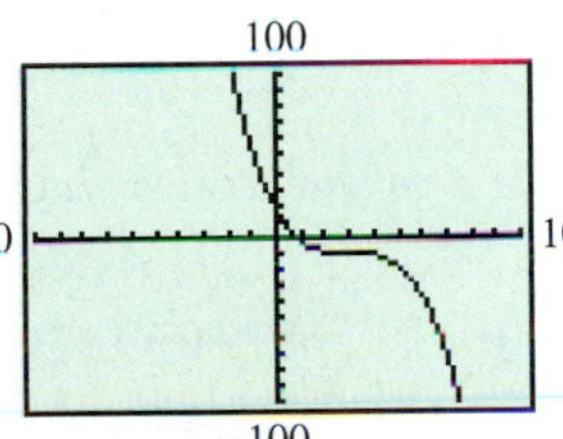

D.

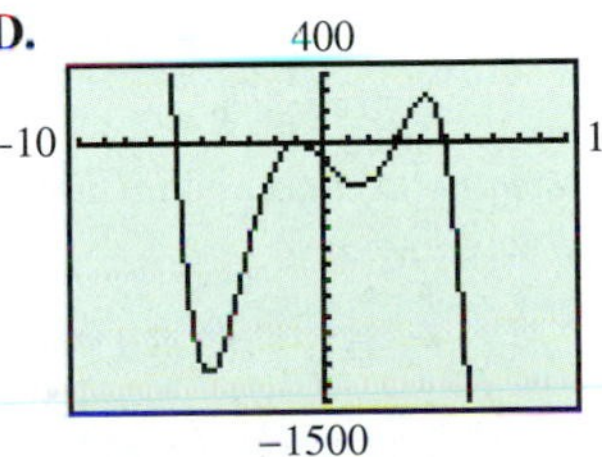

41. Which graph is that of $y = x^3 - 3x^2 - 6x + 8$?

42. Which graph is that of $y = x^4 + 7x^3 - 5x^2 - 75x$?

43. How many real zeros does the graph in C have?

44. The graph of $y = -x^3 + 9x^2 - 27x + 17$ is either C or D. Which is it?

45. Which of the graphs cannot be that of a cubic polynomial function?

46. How many positive real zeros does the function graphed in D have?

47. How many negative real zeros does the function graphed in A have?

48. Is the absolute minimum value of the function graphed in B a positive number or a negative number?

49. Which one of the graphs is that of a function whose range is *not* $(-\infty, \infty)$?

50. One of the following is an approximation for the local maximum point of the graph in A. Which one is it?
A. $(.73, 10.39)$ **B.** $(-.73, 10.39)$
C. $(-.73, -10.39)$ **D.** $(.73, -10.39)$

Concept Check *Without graphing, answer* true *or* false *to each statement. Then, support your answer by graphing.*

51. The function defined by $f(x) = x^3 + 2x^2 - 4x + 3$ has four real zeros.

52. The function defined by $f(x) = x^3 + 3x^2 + 3x + 1$ must have at least one real zero.

53. If a polynomial function of even degree has a negative leading coefficient and a positive y-intercept, it must have at least two real zeros.

54. The function defined by $f(x) = 3x^4 + 5$ has no real zeros.

55. The function defined by $f(x) = -3x^4 + 5$ has two real zeros.

56. The graph of $f(x) = x^3 - 3x^2 + 3x - 1 = (x - 1)^3$ has exactly one x-intercept.

57. A fifth-degree polynomial function cannot have a single real zero.

58. An even-degree polynomial function must have at least one real zero.

For the functions defined in Exercises 59–66, use your graphing calculator to find a comprehensive graph and answer each of the following.

(a) *Determine the domain.*

(b) *Determine all local minimum points, and tell if any is an absolute minimum point. (Approximate coordinates to the nearest hundredth.)*

(c) *Determine all local maximum points, and tell if any is an absolute maximum point. (Approximate coordinates to the nearest hundredth.)*

(d) *Determine the range. (If an approximation is necessary, give it to the nearest hundredth.)*

(e) *Determine all intercepts. For each function, there is at least one x-intercept that is an integer. For those that are not integers, give an approximation to the nearest hundredth. Determine the y-intercept analytically.*

(f) *Give the interval(s) over which the function is increasing.*

(g) *Give the interval(s) over which the function is decreasing.*

59. $P(x) = -2x^3 - 14x^2 + 2x + 84$

60. $P(x) = -3x^3 + 6x^2 + 39x - 60$

61. $P(x) = x^5 + 4x^4 - 3x^3 - 17x^2 + 6x + 9$

62. $P(x) = -2x^5 + 7x^4 + x^3 - 20x^2 + 4x + 16$

63. $P(x) = 2x^4 + 3x^3 - 17x^2 - 6x - 72$

64. $P(x) = 3x^4 - 33x^2 + 54$

65. $P(x) = -x^6 + 24x^4 - 144x^2 + 256$

66. $P(x) = -3x^6 + 2x^5 + 9x^4 - 8x^3 + 11x^2 + 4$

Determine a window that will provide a comprehensive graph of each polynomial function. (In each case, there are many possible such windows.)

67. $P(x) = 4x^5 - x^3 + x^2 + 3x - 16$

68. $P(x) = 3x^5 - x^4 + 12x^2 - 25$

69. $P(x) = 2.9x^3 - 37x^2 + 28x - 143$

70. $P(x) = -5.9x^3 + 16x^2 - 120$

71. $P(x) = \pi x^4 - 13x^2 + 84$

72. $P(x) = 2\pi x^4 - 12x^2 + 100$

(Modeling) *Solve each problem.*

73. ***University Spending on Research*** The table lists the annual amount (in billions of dollars) spent by degree-granting public universities on research programs.

Year	Amount	Year	Amount
1990	6.6	1996	8.5
1991	6.9	1997	9.0
1992	7.3	1998	9.7
1993	7.7	1999	10.6
1994	8.1	2000	11.4
1995	8.2		

Source: National Center for Education Statistics.

(a) Graph the data with the following three function definitions, where x represents the year, and $x = 0$ corresponds to 1990.

(i) $f(x) = .03x^2 + .14x + 6.78$

(ii) $g(x) = .45x + 6.3$

(iii) $h(x) = .0065x^3 - .067x^2 + .51x + 6.55$

(b) Which function definition best models the data?

74. ***Life Expectancy of Americans*** One result of improved health care is that people are living longer. The table lists the number of Americans (in thousands) who are expected to be over 100 years old for selected years.

Year	Number	Year	Number
1998	65	2004	110
2000	75	2006	132
2002	94	2008	153

Source: U.S. Census Bureau.

(a) Use graphing to determine which polynomial best models the number of Americans over 100 years old, where $x = 0$ corresponds to 1998.

(i) $f(x) = 8.957x + 60.048$

(ii) $g(x) = .2991x^2 + 5.966x + 64.036$

(iii) $h(x) = -.03x^3 + .698x^2 + 4.507x + 64.675$

(b) Use your choice from part (a) to predict the number of Americans who will be over 100 years old in 2012.

Reviewing Basic Concepts (Sections 3.4 and 3.5)

1. ***Dimensions of a Garden*** An ecology center wants to set up an experimental garden using 300 meters of fencing to enclose a rectangular area of 5000 square meters.

(a) Let x meters represent the width of the garden. Why must the length be $150 - x$ meters?

(b) Define an expression $A(x)$ that represents the area of the garden.

(c) What are the restrictions on x?

(d) Use the given values for area and length of fencing to find the dimensions of the garden.

2. ***Research Funding*** National Science Foundation funding for research increased in the decades of the 1980s and 1990s. The table gives funding at the beginning of each decade since 1951. In the table, years are given as the number of years since 1900. Thus, $x = 51$ corresponds to 1951, $x = 61$ corresponds to 1961, and so on.

Year x	51	61	71	81	91	101
Dollars y (in billions)	.1	.2	.6	1.1	2.3	4.7

Source: U.S. National Science Foundation.

(a) Plot the data.

(b) Find a function defined by $f(x) = a(x - h)^2 + k$ that models the data by using $(51, .1)$ as the vertex and $(101, 4.7)$ as the second point.

(c) Use the quadratic regression feature of a graphing calculator to determine the quadratic function g that best fits the data.

(d) Plot f and g in the same window as the data. Discuss how well the models fit the data.

For $P(x) = 2x^3 - 9x^2 + 4x + 15$, *answer the following.*

3. Predict the end behavior.

4. What is the maximum number of extrema the graph of P can have? What is the maximum number of zeros?

For $P(x) = x^4 + 4x^3 - 20$, *answer the following.*

5. Predict the end behavior.

6. Give a comprehensive graph of function P.

7. Find all extreme points. Tell whether each one is a maximum or minimum and a local or absolute extremum.

8. Find all intercepts to the nearest hundredth.

3.6 Topics in the Theory of Polynomial Functions (I)

Intermediate Value Theorem ■ Division of Polynomials and Synthetic Division ■ Remainder and Factor Theorems

The topics in this section and the next complement the graphical work done in Section 3.5 and pave the way for the work in Section 3.8 on equations, inequalities, and applications of polynomial functions.

Intermediate Value Theorem

The intermediate value theorem applies to the zeros of every polynomial function with *real coefficients*. It uses the fact that graphs of polynomial functions are continuous curves, with no gaps or sudden jumps. The proof requires advanced methods.

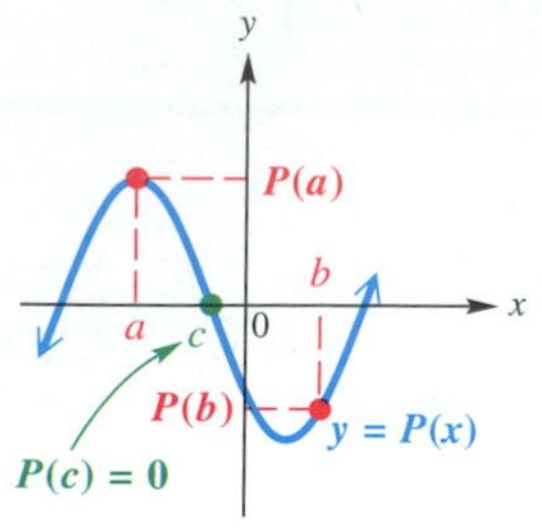

FIGURE 67

Intermediate Value Theorem

If $P(x)$ defines a polynomial function with only real coefficients, and if, for real numbers a and b, the values $P(a)$ and $P(b)$ are opposite in sign, then there exists at least one real zero between a and b.

To see how the intermediate value theorem is applied, note that, in Figure 67, $P(a)$ and $P(b)$ are opposite in sign, so 0 is between $P(a)$ and $P(b)$. Then, by this theorem, there must be a number c in $[a, b]$ such that $P(c) = 0$.

EXAMPLE 1 Applying the Intermediate Value Theorem

Show that the polynomial function defined by $P(x) = x^3 - 2x^2 - x + 1$ has a real zero between 2 and 3.

Analytic Solution

Using

$$P(x) = x^3 - 2x^2 - x + 1,$$

we evaluate $P(2)$ and $P(3)$.

$$P(2) = 2^3 - 2(2)^2 - 2 + 1 = -1$$
$$P(3) = 3^3 - 2(3)^2 - 3 + 1 = 7$$

Since $P(2) = -1$ and $P(3) = 7$ differ in sign, the intermediate value theorem assures us that there is a real zero between 2 and 3.

Graphing Calculator Solution

The graph of $P(x) = x^3 - 2x^2 - x + 1$ in Figure 68 shows that there is an x-intercept between 2 and 3, confirming our analytic result. Using the table, we see that the zero lies between 2.246 and 2.247, since there is a sign change in the function values there.

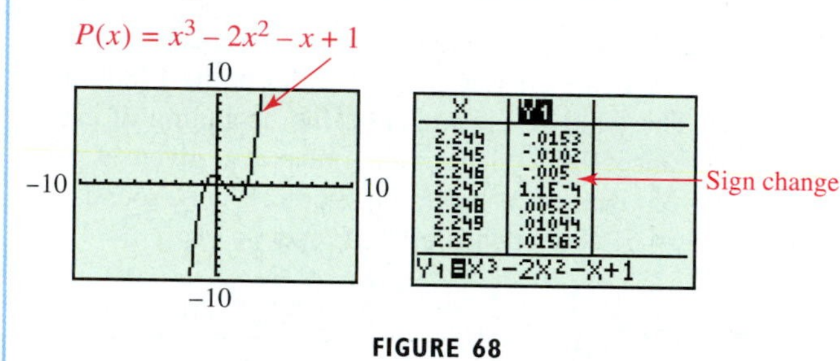

FIGURE 68

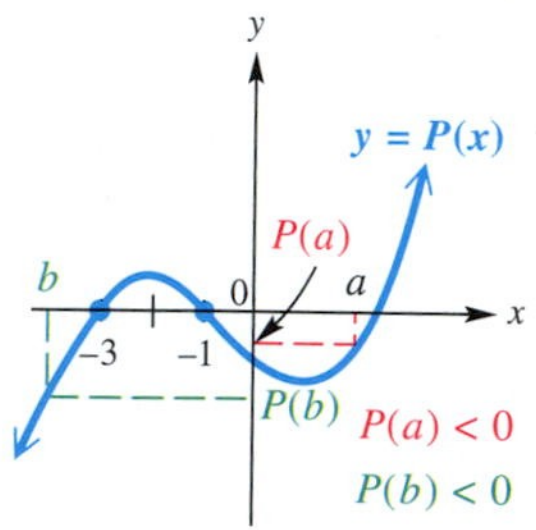

FIGURE 69

CAUTION ***Be careful how you interpret the intermediate value theorem.*** If $P(a)$ and $P(b)$ are *not* opposite in sign, it does not necessarily mean that there is no zero between a and b. For example, in Figure 69, $P(a)$ and $P(b)$ are both negative, but -3 and -1, which are between a and b, are zeros of P.

Division of Polynomials and Synthetic Division

We can use long division to determine whether one whole number is a factor of another. We can also use it to determine whether one polynomial is a factor of another.

EXAMPLE 2 Dividing a Polynomial by a Binomial

Divide $3x^3 - 2x + 5$ by $x - 3$. Determine the quotient and the remainder.

Solution The powers of the variable in the dividend $(3x^3 - 2x + 5)$ must be descending, which they are. Insert the term $0x^2$ to act as a placeholder.

Missing term

$$x - 3\overline{)3x^3 + 0x^2 - 2x + 5}$$

Divide as with whole numbers. Start with $\frac{3x^3}{x} = 3x^2$.

$$
\begin{array}{r}
3x^2 \phantom{{}+0x^2-2x+5} \\
x-3\overline{)3x^3+0x^2-2x+5} \\
\underline{3x^3-9x^2} \phantom{{}-2x+5} \\
9x^2 \phantom{{}-2x+5}
\end{array}
$$

$\leftarrow \frac{3x^3}{x} = 3x^2$

$\leftarrow 3x^2(x-3)$

$\leftarrow$ Subtract.

To subtract, think $3x^3 - 3x^3$ and $0x^2 - (-9x^2)$.

Bring down the next term.

$$
\begin{array}{r}
3x^2 \phantom{{}+0x^2-2x+5} \\
x-3\overline{)3x^3+0x^2-2x+5} \\
\underline{3x^3-9x^2} \phantom{{}-2x+5} \\
9x^2-2x \phantom{{}+5}
\end{array}
$$

$\leftarrow$ Bring down $-2x$.

In the next step, divide: $\frac{9x^2}{x} = 9x$.

$$
\begin{array}{r}
3x^2+9x \phantom{{}+5} \\
x-3\overline{)3x^3+0x^2-2x+5} \\
\underline{3x^3-9x^2} \phantom{{}-2x+5} \\
9x^2-2x \phantom{{}+5} \\
\underline{9x^2-27x} \phantom{{}+5} \\
25x+5
\end{array}
$$

$\leftarrow \frac{9x^2}{x} = 9x$

$\leftarrow 9x(x-3)$

$\leftarrow$ Subtract and bring down 5.

Divide: $\frac{25x}{x} = 25$.

$$
\begin{array}{r}
3x^2+9x+25 \\
x-3\overline{)3x^3+0x^2-2x+5} \\
\underline{3x^3-9x^2} \phantom{{}-2x+5} \\
9x^2-2x \phantom{{}+5} \\
\underline{9x^2-27x} \phantom{{}+5} \\
25x+5 \\
\underline{25x-75} \\
80
\end{array}
$$

$\leftarrow \frac{25x}{x} = 25$

$\leftarrow 25(x-3)$

$\leftarrow$ Subtract.

The quotient is $3x^2 + 9x + 25$ with a remainder of 80. ■

Notice in Example 2 that we divided a cubic polynomial (degree 3) by a linear polynomial (degree 1) and obtained a quadratic polynomial quotient (degree 2). Notice also that $3 - 1 = 2$, so the degree of the quotient polynomial is found by subtracting the degree of the divisor from the degree of the dividend. Also, since the remainder is a nonzero constant 80, we can write it as the numerator of a fraction with denominator $x - 3$ to express the fractional part of the quotient.

$$\frac{3x^3 - 2x + 5}{x - 3} = \underbrace{3x^2 + 9x + 25}_{\text{Quotient polynomial}} + \underbrace{\frac{80}{x-3}}_{\text{Fractional part of the quotient}}$$

Dividend → $3x^3 - 2x + 5$; Divisor → $x - 3$; Remainder ← 80; Divisor ← $x - 3$

The following rules apply when dividing a polynomial by a binomial of the form $x - k$.

Division of a Polynomial by $x - k$

1. If the degree n polynomial $P(x)$ is divided by $x - k$, then the quotient polynomial, $Q(x)$, has degree $n - 1$.
2. The remainder R is a constant (and may be 0). The complete quotient for $\frac{P(x)}{x - k}$ may be written as

$$\frac{P(x)}{x - k} = Q(x) + \frac{R}{x - k}.$$

Long division of a polynomial by a binomial of the form $x - k$ can be condensed. Using the division performed in Example 2, observe the following:

$$\begin{array}{r}
3x^2 + 9x + 25 \\
x - 3\overline{)3x^3 + 0x^2 - 2x + 5} \\
\underline{3x^3 - 9x^2} \qquad\qquad \\
9x^2 - 2x \qquad \\
\underline{9x^2 - 27x} \qquad \\
25x + 5 \\
\underline{25x - 75} \\
80
\end{array}
\qquad
\begin{array}{rrrrr}
 & 3 & 9 & 25 & \\
1 - 3\overline{)} & 3 & 0 & -2 & 5 \\
 & \underline{3} & \underline{-9} & & \\
 & & 9 & -2 & \\
 & & \underline{9} & \underline{-27} & \\
 & & & 25 & 5 \\
 & & & \underline{25} & \underline{-75} \\
 & & & & 80
\end{array}$$

On the right, exactly the same division is shown without the variables. All the numbers in color on the right are repetitions of the numbers directly above them, so they can be omitted, as shown below on the left. Since the coefficient of x in the divisor is always 1, it can be omitted, too.

$$\begin{array}{rrrrr}
 & 3 & 9 & 25 & \\
-3\overline{)} & 3 & 0 & -2 & 5 \\
 & & \underline{-9} & & \\
 & & 9 & -2 & \\
 & & & \underline{-27} & \\
 & & & 25 & 5 \\
 & & & & \underline{-75} \\
 & & & & 80
\end{array}
\qquad
\begin{array}{rrrrr}
 & 3 & 9 & 25 & \\
-3\overline{)} & 3 & 0 & -2 & 5 \\
 & & \underline{-9} & & \\
 & & 9 & & \\
 & & & \underline{-27} & \\
 & & & 25 & \\
 & & & & \underline{-75} \\
 & & & & 80
\end{array}$$

The numbers in color on the left are again repetitions of the numbers directly above them; they may be omitted, as shown on the right.

Now the problem can be condensed. If the 3 in the dividend is brought down to the beginning of the bottom row, the top row can be omitted, since it duplicates the bottom row.

$$\begin{array}{rrrrr}
-3\overline{)} & 3 & 0 & -2 & 5 \\
 & & -9 & -27 & -75 \\
\hline
 & 3 & 9 & 25 & 80
\end{array}$$

To simplify the arithmetic, we replace subtraction in the second row by addition and compensate by changing the -3 at the upper left to its additive inverse, 3.

$$\begin{array}{r|rrrr} 3 & 3 & 0 & -2 & 5 \\ & & 9 & 27 & 75 \\ \hline & 3 & 9 & 25 & 80 \end{array}$$

Additive inverse → 3; Signs changed ← 9, 27, 75

Quotient → $3x^2 + 9x + 25 + \dfrac{80}{x-3}$ ← Remainder

This abbreviated form of long division of polynomials is called **synthetic division.**

EXAMPLE 3 Using Synthetic Division

Use synthetic division to divide $5x^3 - 6x^2 - 28x + 8$ by $x + 2$.

Solution Express $x + 2$ in the form $x - k$ by writing it as $x - (-2)$. Use this and the coefficients of the polynomial to obtain

$$-2\overline{)5 \quad -6 \quad -28 \quad 8.}$$

$x + 2$ leads to -2. → ; ← Coefficients

Bring down the 5, and multiply: $-2(5) = -10$.

$$\begin{array}{r|rrrr} -2 & 5 & -6 & -28 & 8 \\ & \downarrow & -10 & & \\ \hline & 5 & & & \end{array}$$

Add -6 and -10 to obtain -16. Multiply: $-2(-16) = 32$.

$$\begin{array}{r|rrrr} -2 & 5 & -6 & -28 & 8 \\ & & -10 & 32 & \\ \hline & 5 & -16 & & \end{array}$$

Add -28 and 32, obtaining 4. Finally, $-2(4) = -8$.

$$\begin{array}{r|rrrr} -2 & 5 & -6 & -28 & 8 \\ & & -10 & 32 & -8 \\ \hline & 5 & -16 & 4 & \end{array}$$

Add 8 and -8 to obtain 0.

$$\begin{array}{r|rrrr} -2 & 5 & -6 & -28 & 8 \\ & & -10 & 32 & -8 \\ \hline & 5 & -16 & 4 & 0 \end{array}$$

Quotient: 5, -16, 4; ← Remainder: 0

Since the divisor $x - k$ has degree 1, the degree of the quotient will always be one less than the degree of the polynomial to be divided. Thus,

$$\frac{5x^3 - 6x^2 - 28x + 8}{x + 2} = 5x^2 - 16x + 4.$$

Notice that the divisor $x + 2$ is a *factor* of $5x^3 - 6x^2 - 28x + 8$ because the remainder is 0, so $5x^3 - 6x^2 - 28x + 8 = (x + 2)(5x^2 - 16x + 4)$. ■

Remainder and Factor Theorems

In Example 2, we divided $3x^3 - 2x + 5$ by $x - 3$ and obtained a remainder of 80. If we evaluate $P(x) = 3x^3 - 2x + 5$ at $x = 3$, we get

$$P(3) = 3(3)^3 - 2(3) + 5 = 81 - 6 + 5 = 80.$$

Notice that the remainder is equal to $P(3)$. In Example 3, we divided the polynomial $5x^3 - 6x^2 - 28x + 8$ by $x - (-2)$ and obtained a remainder of 0. If we evaluate $P(x) = 5x^3 - 6x^2 - 28x + 8$ at $x = -2$, we get

$$\begin{aligned} P(-2) &= 5(-2)^3 - 6(-2)^2 - 28(-2) + 8 \\ &= -40 - 24 + 56 + 8 \\ &= 0. \end{aligned}$$

Use parentheses around substituted values to avoid errors.

The remainder is equal to $P(-2)$. These examples illustrate the *remainder theorem.*

Remainder Theorem

If a polynomial $P(x)$ is divided by $x - k$, the remainder is equal to $P(k)$.

EXAMPLE 4 Using the Remainder Theorem

Use the remainder theorem and synthetic division to find $P(-2)$ if

$$P(x) = -x^4 + 3x^2 - 4x - 5.$$

Analytic Solution

Use synthetic division to find the remainder when $P(x)$ is divided by $x - (-2)$.

$$\begin{array}{r|rrrrr} -2 & -1 & 0 & 3 & -4 & -5 \\ & & 2 & -4 & 2 & 4 \\ \hline & -1 & 2 & -1 & -2 & -1 \end{array}$$

Remember to insert 0 for the missing x^3-term.

Remainder

Since the remainder is -1, $P(-2) = -1$ by the remainder theorem.

Graphing Calculator Solution

The graph of $P(x) = -x^4 + 3x^2 - 4x - 5$ in Figure 70 indicates that the point $(-2, -1)$ lies on the graph, so $P(-2) = -1$. Alternatively, the table in Figure 70 shows that $P(-2) = -1$.

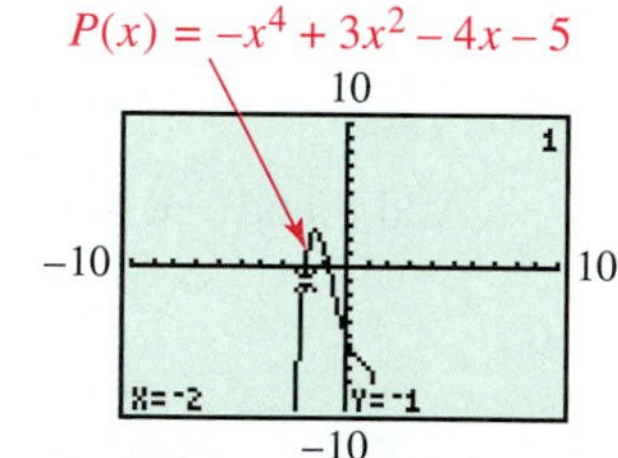

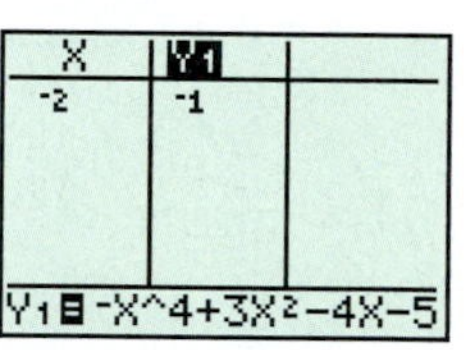

FIGURE 70

EXAMPLE 5 Deciding whether a Number Is a Zero of a Polynomial Function

Decide whether the given number is a zero of the given polynomial function.

(a) 2; $P(x) = x^3 - 4x^2 + 9x - 10$ **(b)** -2; $P(x) = \frac{3}{2}x^3 - x^2 + \frac{3}{2}x$

Analytic Solution

(a) Use synthetic division.

Proposed zero → 2)	1	−4	9	−10	
		2	−4	10	
	1	−2	5	0	← Remainder

Since the remainder is 0, $P(2) = 0$, and 2 is a zero of the given polynomial function.

(b)

Proposed zero → −2)	$\frac{3}{2}$	−1	$\frac{3}{2}$	0	Use 0 for the missing constant term.
		−3	8	−19	
	$\frac{3}{2}$	−4	$\frac{19}{2}$	−19	← Remainder

The remainder is not 0, so -2 is not a zero of P. In fact, $P(-2) = -19$. From this, we know that the point $(-2, -19)$ lies on the graph of P.

Graphing Calculator Solution

The first screen in Figure 71 shows the polynomials entered as Y_1 and Y_2. The second screen shows the results of finding $Y_1(2)$ and $Y_2(-2)$ for parts (a) and (b).

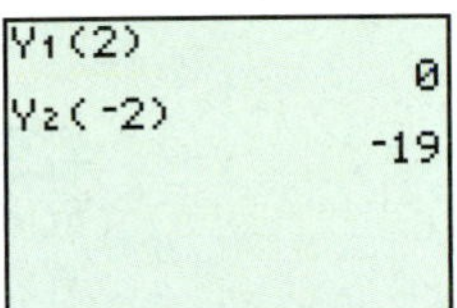

FIGURE 71

In Example 5(a), we showed that 2 is a zero of the polynomial function defined by $P(x) = x^3 - 4x^2 + 9x - 10$. The first three numbers in the bottom row of the synthetic division indicate the coefficients of the quotient polynomial. Thus,

$$\frac{P(x)}{x-2} = x^2 - 2x + 5$$

$$P(x) = (x-2)(x^2 - 2x + 5), \qquad \text{Multiply by } x - 2.$$

indicating that $x - 2$ is a *factor* of $P(x)$.

By the remainder theorem, if $P(k) = 0$, then the remainder when $P(x)$ is divided by $x - k$ is 0. This means that $x - k$ is a factor of $P(x)$. Conversely, if $x - k$ is a factor of $P(x)$, then $P(k)$ must equal 0. This is summarized in the *factor theorem.*

Factor Theorem

The polynomial $x - k$ is a factor of the polynomial $P(x)$ if and only if $P(k) = 0$.

EXAMPLE 6 Using the Factor Theorem

Determine whether the second polynomial is a factor of the first.

(a) $P(x) = 4x^3 + 24x^2 + 48x + 32$; $x + 2$

(b) $P(x) = 2x^4 + 3x^2 - 5x + 7$; $x - 1$

Solution

(a)

−2)	4	24	48	32	
		−8	−32	−32	
	4	16	16	0	← Remainder

Use synthetic division to divide $P(x)$ by $x + 2 = x - (-2)$.

Since the remainder is 0, $x + 2$ is a factor of $P(x) = 4x^3 + 24x^2 + 48x + 32$. A factored form (but not necessarily *completely* factored form) of the polynomial is $(x + 2)(4x^2 + 16x + 16)$.

(b) By the factor theorem, $x - 1$ will be a factor of $P(x) = 2x^4 + 3x^2 - 5x + 7$ if $P(1) = 0$.

$$\begin{array}{r|rrrrr} 1 & 2 & 0 & 3 & -5 & 7 \\ & & 2 & 2 & 5 & 0 \\ \hline & 2 & 2 & 5 & 0 & 7 \end{array}$$

Use synthetic division to divide $P(x)$ by $x - 1$. ← $P(1) = 7$

Since the remainder is 7, $P(1) = 7$, not 0, so $x - 1$ is *not* a factor of $P(x)$. ■

NOTE An easy way to determine $P(1)$ for a polynomial function P is simply to add the coefficients of $P(x)$. This method works because every power of 1 is equal to 1. For example, using $P(x) = 2x^4 + 3x^2 - 5x + 7$ as shown in Example 6(b), we have $P(1) = 2 + 3 - 5 + 7 = 7$, confirming our result found by synthetic division.

EXAMPLE 7 Examining x-Intercepts, Zeros, and Solutions

Consider the polynomial function defined by $P(x) = 2x^3 + 5x^2 - x - 6$.

(a) Show that -2, $-\frac{3}{2}$, and 1 are zeros of P, and write $P(x)$ in factored form with all factors linear.

(b) Graph P in a suitable viewing window and locate the x-intercepts.

(c) Solve the polynomial equation $2x^3 + 5x^2 - x - 6 = 0$.

Solution

(a)
$$\begin{array}{r|rrrr} -2 & 2 & 5 & -1 & -6 \\ & & -4 & -2 & 6 \\ \hline & 2 & 1 & -3 & 0 \end{array}$$

Use synthetic division to divide $P(x)$ by $x - (-2)$. ← $P(-2) = 0$

Since $P(-2) = 0$, $x + 2$ is a factor; thus, $P(x) = (x + 2)(2x^2 + x - 3)$. Rather than show that $-\frac{3}{2}$ and 1 are zeros of $P(x)$, we need only show that they are zeros of $2x^2 + x - 3$ by factoring directly.

$$2x^2 + x - 3 = (2x + 3)(x - 1) \quad \text{Factored form}$$

The solutions of $(2x + 3)(x - 1) = 0$ are, as predicted, $-\frac{3}{2}$ and 1. The completely factored form of $P(x)$ is

$$P(x) = (x + 2)(2x + 3)(x - 1).$$

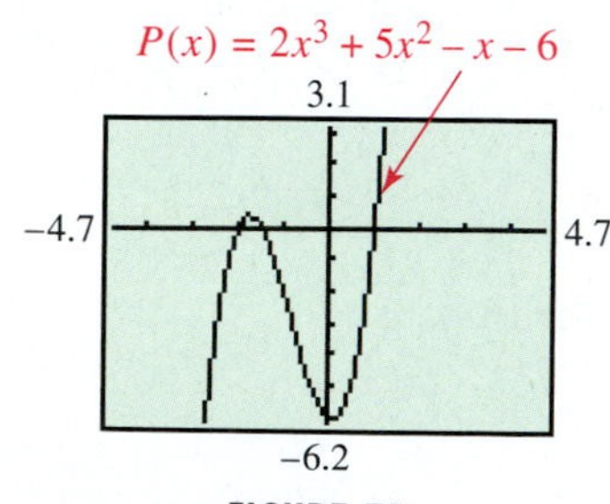

FIGURE 72

(b) Figure 72 shows the graph of this function. The calculator can be used to determine the x-intercepts: -2, $-\frac{3}{2}$, and 1.

(c) From part (a), the zeros of P are -2, $-\frac{3}{2}$, and 1. Because the zeros of P are the solutions of $P(x) = 0$, the solution set is $\{-2, -\frac{3}{2}, 1\}$. ■

NOTE In Example 7(a), it was easy to use factoring to find the zeros generated by the factor $2x^2 + x - 3$. It is *always* possible to use the quadratic formula at this stage of the procedure and, in fact, often necessary if the quadratic factor cannot be factored further using integer coefficients.

3.6 Exercises

Use the intermediate value theorem to show that each function has a real zero between the two numbers given. Then, use your calculator to approximate the zero to the nearest hundredth.

1. $P(x) = 3x^2 - 2x - 6$; 1 and 2

2. $P(x) = x^3 + x^2 - 5x - 5$; 2 and 3

3. $P(x) = 2x^3 - 8x^2 + x + 16$; 2 and 2.5

4. $P(x) = 3x^3 + 7x^2 - 4$; $\frac{1}{2}$ and 1

5. $P(x) = 2x^4 - 4x^2 + 3x - 6$; 1.5 and 2

6. $P(x) = x^4 - 4x^3 - x + 1$; .3 and 1

7. $P(x) = -x^4 + 2x^3 + x + 12$; 2.7 and 2.8

8. $P(x) = -2x^4 + x^3 - x^2 + 3$; -1 and $-.9$

9. $P(x) = x^5 - 2x^3 + 1$; -1.6 and -1.5

10. $P(x) = 2x^7 - x^4 + x - 4$; 1.1 and 1.2

11. *Concept Check* Suppose that a polynomial function P is defined in such a way that $P(2) = -4$ and $P(2.5) = 2$. What conclusion does the intermediate value theorem allow you to make?

12. Suppose that a polynomial function P is defined in such a way that $P(3) = -4$ and $P(4) = -10$. Can we be certain that there is no zero between 3 and 4? Explain, using a graph.

Find each quotient when $P(x)$ is divided by the binomial following it.

13. $P(x) = x^3 + 2x^2 - 17x - 10$; $x + 5$

14. $P(x) = x^4 + 4x^3 + 2x^2 + 9x + 4$; $x + 4$

15. $P(x) = 3x^3 - 11x^2 - 20x + 3$; $x - 5$

16. $P(x) = x^4 - 3x^3 - 5x^2 + 2x - 16$; $x - 3$

17. $P(x) = x^4 - 3x^3 - 4x^2 + 12x$; $x - 2$

18. $P(x) = 2x^4 + 3x^3 - 5x^2 - 18x$; $x - 2$

19. $P(x) = x^3 + 2x^2 - 3$; $x - 1$

20. $P(x) = x^3 - 2x^2 - 9$; $x - 3$

21. $P(x) = -2x^3 - x - 2$; $x + 1$

22. $P(x) = -3x^3 - x - 5$; $x + 1$

23. $P(x) = x^5 - 1$; $x - 1$

24. $P(x) = x^7 + 1$; $x + 1$

Use synthetic division to find $P(k)$.

25. $k = 3$; $P(x) = x^2 - 4x + 3$

26. $k = -2$; $P(x) = x^2 + 5x + 6$

27. $k = -2$; $P(x) = 5x^3 + 2x^2 - x + 5$

28. $k = 2$; $P(x) = 2x^3 - 3x^2 - 5x + 4$

29. $k = 2$; $P(x) = x^2 - 5x + 1$

30. $k = 3$; $P(x) = x^2 - x + 3$

31. $k = .5$; $P(x) = x^3 - x + 4$

32. $k = 1.5$; $P(x) = x^3 + x - 3$

33. $k = \sqrt{2}$; $P(x) = x^4 - x^2 - 3$

34. $k = \sqrt{3}$; $P(x) = x^4 + 2x^2 - 10$

35. $k = \sqrt[3]{4}$; $P(x) = -x^3 + x + 4$

36. $k = \sqrt[5]{3}$; $P(x) = -x^5 + 2x + 3$

Use synthetic division to determine whether the given number is a zero of the polynomial.

37. 2; $P(x) = x^2 + 2x - 8$

38. -1; $P(x) = x^2 + 4x - 5$

39. 4; $P(x) = 2x^3 - 6x^2 - 9x + 6$

40. -4; $P(x) = 9x^3 + 39x^2 + 12x$

41. $-.5$; $P(x) = 4x^3 + 12x^2 + 7x + 1$

42. $-.25$; $P(x) = 8x^3 + 6x^2 - 3x - 1$

43. -5; $P(x) = 8x^3 + 50x^2 + 47x + 15$

44. -4; $P(x) = 6x^3 + 25x^2 + 3x - 3$

45. $\sqrt{6}$; $P(x) = -2x^6 + 5x^4 - 3x^2 + 270$

46. $\sqrt{7}$; $P(x) = -3x^6 + 7x^4 - 5x^2 + 721$

Relating Concepts

For individual or group investigation (Exercises 47–52)

The close relationships among ***x-intercepts of a graph of a function, real zeros of the function,*** *and* ***real solutions of the associated equation*** *should, by now, be apparent to you. Consider the graph of the polynomial function shown here defined by* $P(x) = x^3 - 2x^2 - 11x + 12$, *and* ***work Exercises 47–52 in order.***

47. What are the linear factors of $P(x)$?

48. What are the solutions of the equation $P(x) = 0$?

49. What are the zeros of the function P?

50. If $P(x)$ is divided by $x - 2$, what is the remainder? What is $P(2)$?

51. Give the solution set of $P(x) > 0$, using interval notation.

52. Give the solution set of $P(x) < 0$, using interval notation.

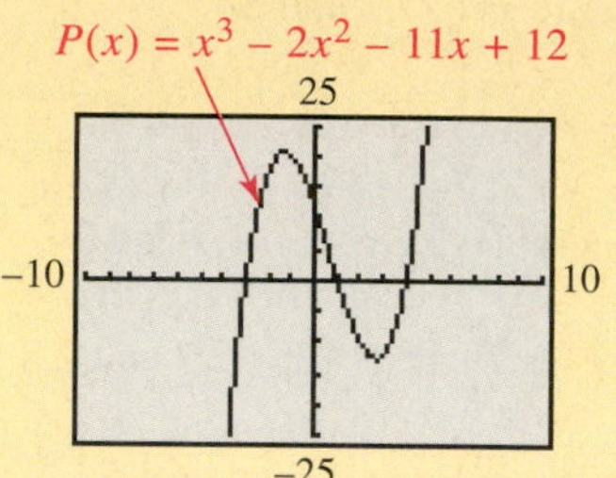

The x-intercepts are –3, 1, and 4.

For each polynomial, at least one zero is given. Find all others analytically.

53. $P(x) = x^3 - 2x^2 - 5x + 6$; 3

54. $P(x) = x^3 + 2x^2 - 11x - 12$; 3

55. $P(x) = x^3 - 2x + 1$; 1

56. $P(x) = 2x^3 + 8x^2 - 11x - 5$; -5

57. $P(x) = 3x^3 + 5x^2 - 3x - 2$; -2

58. $P(x) = x^3 - 7x^2 + 13x - 3$; 3

59. $P(x) = x^4 - 41x^2 + 180$; -6 and 6

60. $P(x) = x^4 - 52x^2 + 147$; -7 and 7

61. $P(x) = -x^3 + 8x^2 + 3x - 24$; 8

62. $P(x) = -x^3 + 4x^2 + 7x - 28$; 4

Factor $P(x)$ *into linear factors given that* k *is a zero of* P.

63. $P(x) = 2x^3 - 3x^2 - 17x + 30$; $k = 2$

64. $P(x) = 2x^3 - 3x^2 - 5x + 6$; $k = 1$

65. $P(x) = 6x^3 + 25x^2 + 3x - 4$; $k = -4$

66. $P(x) = 8x^3 + 50x^2 + 47x - 15$; $k = -5$

67. $P(x) = -6x^3 - 13x^2 + 14x - 3$; $k = -3$

68. $P(x) = -6x^3 - 17x^2 + 63x - 10$; $k = -5$

69. $P(x) = x^3 + 5x^2 - 3x - 15$; $k = -5$

70. $P(x) = x^3 + 9x^2 - 7x - 63$; $k = -9$

71. $P(x) = x^3 - 2x^2 - 7x - 4$; $k = -1$

72. $P(x) = x^3 + x^2 - 21x - 45$; $k = -3$

3.7 Topics in the Theory of Polynomial Functions (II)

Complex Zeros and the Fundamental Theorem of Algebra ■ Number of Zeros ■ Rational Zeros Theorem ■ Descartes' Rule of Signs ■ Boundedness Theorem

Complex Zeros and the Fundamental Theorem of Algebra

In Example 5 of Section 3.3, we found that the nonreal complex solutions of the equation $2x^2 - x + 4 = 0$ are $\frac{1}{4} + i\frac{\sqrt{31}}{4}$ and $\frac{1}{4} - i\frac{\sqrt{31}}{4}$. These two solutions are complex conjugates. This is not a coincidence, as given in the *conjugate zeros theorem.*

Conjugate Zeros Theorem

If $P(x)$ defines a polynomial function having only *real* coefficients, and if $a + bi$ is a zero of $P(x)$, then the conjugate $a - bi$ is also a zero of $P(x)$.

EXAMPLE 1 Defining a Polynomial Function Satisfying Given Conditions

(a) Find a cubic polynomial function P in standard form with real coefficients having zeros 3 and $2 + i$.

(b) Find a polynomial function P satisfying the conditions of part (a), with the additional requirement $P(-2) = 4$. Support the result graphically.

Solution

(a) By the conjugate zeros theorem, $2 - i$ must also be a zero of the function. Since the defining polynomial will be cubic, it will have three linear factors, and by the factor theorem they must be $x - 3$, $x - (2 + i)$, and $x - (2 - i)$.

$$\begin{aligned} P(x) &= (x - 3)[x - (2 + i)][x - (2 - i)] \\ &= (x - 3)(x - 2 - i)(x - 2 + i) \\ &= (x - 3)(x^2 - 4x + 5) \\ &= x^3 - 7x^2 + 17x - 15 \end{aligned}$$

Multiplying the polynomial by any real nonzero constant a will also yield a function satisfying the given conditions, so a more general form is

$$P(x) = a(x^3 - 7x^2 + 17x - 15).$$

(b) We must define $P(x) = a(x^3 - 7x^2 + 17x - 15)$ in such a way that $P(-2) = 4$. To find a, let $x = -2$, and set the result equal to 4. Then solve for a.

$$a[(-2)^3 - 7(-2)^2 + 17(-2) - 15] = 4 \quad \text{Substitute.}$$

Use parentheses around substituted values to avoid errors.

$$a(-8 - 28 - 34 - 15) = 4$$

$$-85a = 4$$

$$a = -\frac{4}{85} \quad \text{Divide by } -85.$$

Therefore, the desired function is defined by

$$\begin{aligned} P(x) &= -\frac{4}{85}(x^3 - 7x^2 + 17x - 15) \\ &= -\frac{4}{85}x^3 + \frac{28}{85}x^2 - \frac{4}{5}x + \frac{12}{17}. \quad \text{Distributive property} \end{aligned}$$

Be careful with signs.

We can support this result by graphing $P(x) = -\frac{4}{85}x^3 + \frac{28}{85}x^2 - \frac{4}{5}x + \frac{12}{17}$ and showing that the point $(-2, 4)$ lies on the graph. See Figure 73. ■

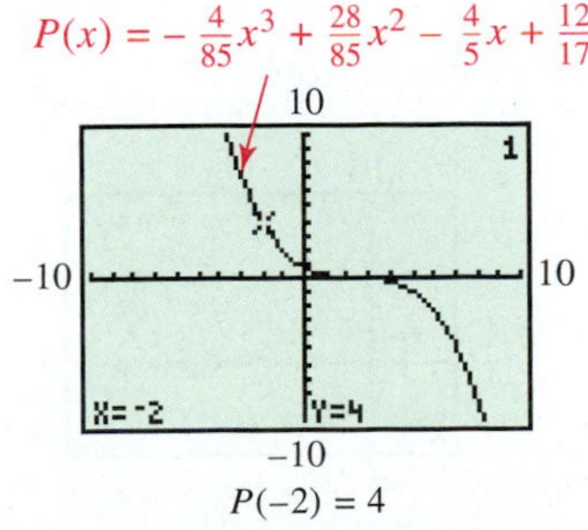

$P(-2) = 4$

FIGURE 73

Number of Zeros

Carl Friedrich Gauss (1777–1855)

The *fundamental theorem of algebra* was first proved by Carl Friedrich Gauss in his doctoral thesis in 1799, when he was 22 years old. Although many proofs of this result have been given, all involve mathematics beyond this book.

> **Fundamental Theorem of Algebra**
>
> Every function defined by a polynomial of degree 1 or more has at least one complex zero.

From the fundamental theorem, if $P(x)$ is of degree 1 or more, then there is some number k such that $P(k) = 0$. Thus, by the factor theorem, $P(x) = (x - k) \cdot Q(x)$ for some polynomial $Q(x)$. The fundamental theorem and the factor theorem can be used to factor $Q(x)$ in the same way. Assuming that $P(x)$ has degree n, repeating this process n times gives

$$P(x) = a(x - k_1)(x - k_2) \cdots (x - k_n),$$

where a is the leading coefficient of $P(x)$. Each factor leads to a zero of $P(x)$, so $P(x)$ has n zeros $k_1, k_2, k_3, \ldots, k_n$. This suggests the *number of zeros theorem.*

> **Number of Zeros Theorem**
>
> A function defined by a polynomial of degree n has at most n distinct complex zeros.

EXAMPLE 2 Finding All Zeros of a Polynomial Function

Find all complex zeros of $P(x) = x^4 - 7x^3 + 18x^2 - 22x + 12$, given that $1 - i$ is a zero.

Analytic Solution

This quartic function will have at most four complex zeros. Since $1 - i$ is a zero and the coefficients are real numbers, by the conjugate zeros theorem $1 + i$ is also a zero. The remaining zeros are found by first dividing the original polynomial by $x - (1 - i)$.

$$\begin{array}{r|rrrrr} 1 - i & 1 & -7 & 18 & -22 & 12 \\ & & 1 - i & -7 + 5i & 16 - 6i & -12 \\ \hline & 1 & -6 - i & 11 + 5i & -6 - 6i & 0 \end{array}$$

Next, divide the quotient from the first division by $x - (1 + i)$.

$$\begin{array}{r|rrrr} 1 + i & 1 & -6 - i & 11 + 5i & -6 - 6i \\ & & 1 + i & -5 - 5i & 6 + 6i \\ \hline & 1 & -5 & 6 & 0 \end{array}$$

Graphing Calculator Solution

By the conjugate zeros theorem, since $1 - i$ is a zero, $1 + i$ is also a zero. We can use a graphing calculator to find the real zeros as in Figure 74, which shows the graph of

$$P(x) = x^4 - 7x^3 + 18x^2 - 22x + 12,$$

with the real zeros identified at the bottom.

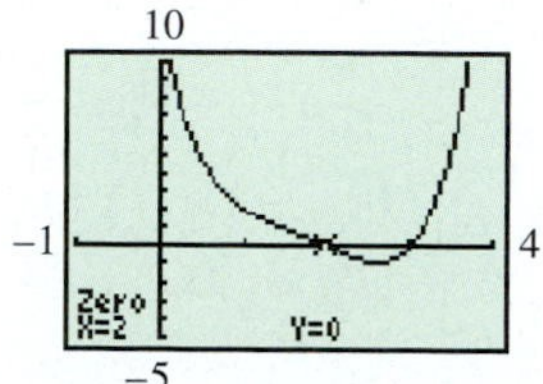

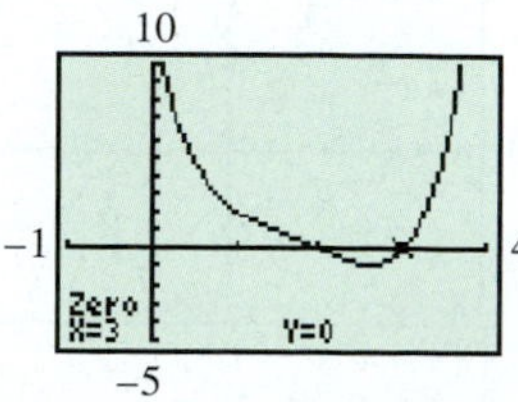

FIGURE 74

(continued)

Now find the zeros of the function defined by the quadratic polynomial $x^2 - 5x + 6$ by solving the equation $x^2 - 5x + 6 = 0$. By factoring, we see that the other zeros are 2 and 3. Thus, this function has four complex zeros: $1 - i$, $1 + i$, 2, and 3.

If no zero had been given for this function, one approach would be to find the real zeros using the graph, and then using synthetic division twice. Once a quadratic factor is determined, the quadratic formula can then be used. ■

The number of zeros theorem says that a polynomial function of degree n has *at most n* distinct zeros. In the polynomial function defined by

$$\begin{aligned} P(x) &= x^6 + x^5 - 5x^4 - x^3 + 8x^2 - 4x \\ &= x(x + 2)^2(x - 1)^3, \end{aligned}$$

each factor leads to a zero of the function. The factor x leads to a *single* zero of 0, the factor $(x + 2)^2$ leads to a zero of -2 appearing *twice,* and the factor $(x - 1)^3$ leads to a zero of 1 appearing *three* times. The number of times a zero appears is referred to as the **multiplicity of the zero.**

EXAMPLE 3 Defining a Polynomial Function Satisfying Given Conditions

Find a polynomial function with real coefficients of least possible degree having a zero 2 of multiplicity 3, a zero 0 of multiplicity 2, and a zero i of single multiplicity.

Solution By the conjugate zeros theorem, this polynomial function must also have a zero of $-i$. This means there are seven zeros, so the least possible degree of the polynomial is 7. Therefore,

$$P(x) = x^2(x - 2)^3(x - i)(x + i) \quad \text{Factor theorem}$$

$$P(x) = x^7 - 6x^6 + 13x^5 - 14x^4 + 12x^3 - 8x^2. \quad \text{Multiply.}$$

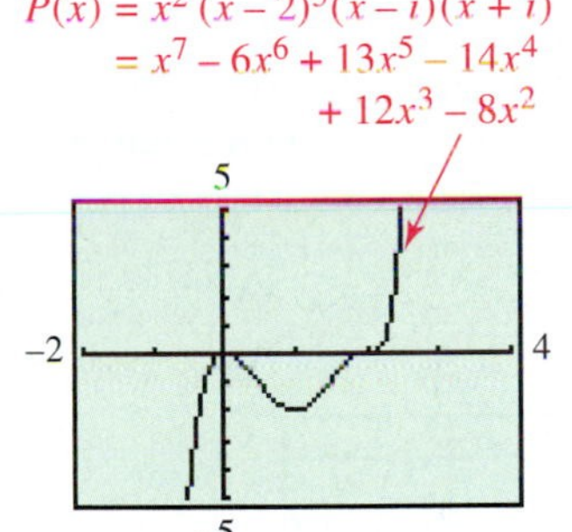

The graph is tangent to the x-axis at $x = 0$, and crosses the x-axis at $x = 2$.

FIGURE 75

This is one of infinitely many such functions. Multiplying $P(x)$ by a nonzero constant will yield another polynomial function satisfying these conditions.

The graph of this function in Figure 75 shows that it has only two distinct x-intercepts, corresponding to real zeros of 0 and 2. ■

FOR DISCUSSION

Graph each function in the indicated window. Then respond to the following items.

$P(x) = (x + 3)(x - 2)^2$; $[-10, 10]$ by $[-30, 30]$

$P(x) = (x + 3)^2(x - 2)^3$; $[-4, 4]$ by $[-125, 50]$

$P(x) = x^2(x - 1)(x + 2)^2$; $[-4, 4]$ by $[-5, 5]$

1. Describe the behavior of the graph at each x-intercept that corresponds to a zero of odd multiplicity.
2. Describe the behavior of the graph at each x-intercept that corresponds to a zero of even multiplicity.

The observations in the "For Discussion" box indicate that the behavior of the graph of a polynomial function near an x-intercept depends on the *parity* of multiplicity of the zero that leads to the x-intercept.

If the zero of the polynomial function is of *odd* multiplicity, then the graph will cross the x-axis at the corresponding x-intercept. If the zero is of *even* multiplicity, then the graph will be tangent to the x-axis at the corresponding x-intercept. (That is, it will touch but not cross the x-axis.) See Figure 76.

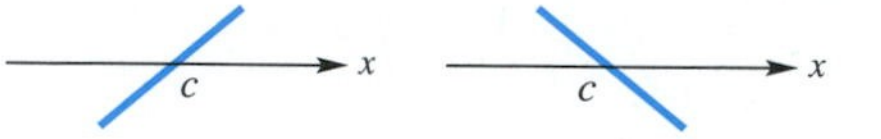

The graph crosses the x-axis at $(c, 0)$ if c is a zero of odd multiplicity.

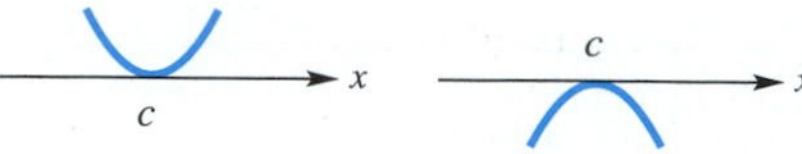

The graph is tangent to the x-axis at $(c, 0)$ if c is a zero of even multiplicity.

FIGURE 76

By observing the dominating term and noting the parity of multiplicities of zeros of a polynomial function in factored form, we can sketch a rough graph.

EXAMPLE 4 Sketching a Graph of a Polynomial Function by Hand

Consider the polynomial function defined by

$$\begin{aligned} P(x) &= -2x^5 - 18x^4 - 38x^3 + 42x^2 + 112x - 96 \\ &= -2(x+4)^2(x+3)(x-1)^2. \end{aligned}$$

Factored form

Sketch the graph of P by hand. Confirm the result with a calculator.

Solution Because the dominating term is $-2x^5$, the end behavior of the graph will be ↖↘. Since -4 and 1 are both x-intercepts determined by zeros of even multiplicity, the graph will be tangent to the x-axis at these intercepts. Because -3 is a zero of multiplicity 1, the graph will cross the x-axis at $x = -3$. The y-intercept is -96. This information leads to the rough sketch in Figure 77(a).

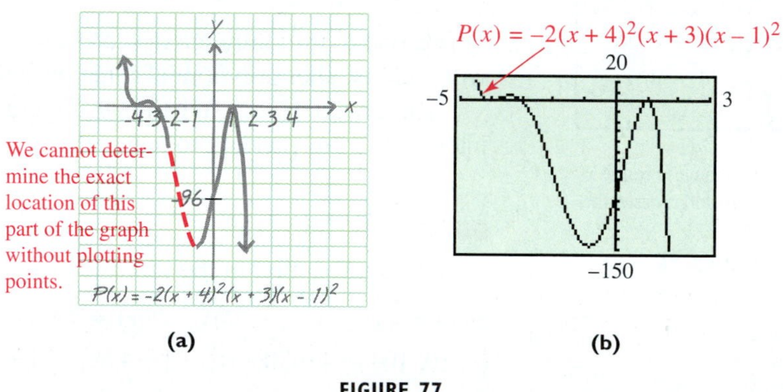

(a) **(b)**

FIGURE 77

The hand-drawn graph does not necessarily give a good indication of local extrema. The calculator graph shown in Figure 77(b) fills in the details. ■

Rational Zeros Theorem

The *rational zeros theorem* gives a method to determine all possible candidates for rational zeros of a polynomial function with integer coefficients.

Rational Zeros Theorem

Let $P(x) = a_nx^n + a_{n-1}x^{n-1} + \cdots + a_1x + a_0$, where $a_n \neq 0$, define a polynomial function with integer coefficients. If $\frac{p}{q}$ is a rational number written in lowest terms, and if $\frac{p}{q}$ is a zero of $P(x)$, then p is a factor of the constant term a_0, and q is a factor of the leading coefficient a_n.

Proof $P(\frac{p}{q}) = 0$, since $\frac{p}{q}$ is a zero of $P(x)$, so

$$a_n\left(\frac{p}{q}\right)^n + a_{n-1}\left(\frac{p}{q}\right)^{n-1} + \cdots + a_1\left(\frac{p}{q}\right) + a_0 = 0$$

$$a_n\left(\frac{p^n}{q^n}\right) + a_{n-1}\left(\frac{p^{n-1}}{q^{n-1}}\right) + \cdots + a_1\left(\frac{p}{q}\right) + a_0 = 0$$

$$a_np^n + a_{n-1}p^{n-1}q + \cdots + a_1pq^{n-1} = -a_0q^n \quad \text{Multiply by } q^n\text{; add } -a_0q^n.$$

$$p(a_np^{n-1} + a_{n-1}p^{n-2}q + \cdots + a_1q^{n-1}) = -a_0q^n. \quad \text{Factor out } p.$$

Thus, $-a_0q^n$ equals the product of the two factors, p and $(a_np^{n-1} + \cdots + a_1q^{n-1})$. For this reason, p must be a factor of $-a_0q^n$. Since it was assumed that $\frac{p}{q}$ is written in lowest terms, p and q have no common factor other than 1, so p is not a factor of q^n. Thus, p must be a factor of a_0. In a similar way, it can be shown that q is a factor of a_n.

EXAMPLE 5 Using the Rational Zeros Theorem

Perform the following for the polynomial function defined by

$$P(x) = 6x^4 + 7x^3 - 12x^2 - 3x + 2.$$

(a) List all possible rational zeros.

(b) Use a graph to eliminate some of the possible zeros listed in part (a).

(c) Find all rational zeros and factor $P(x)$.

Solution

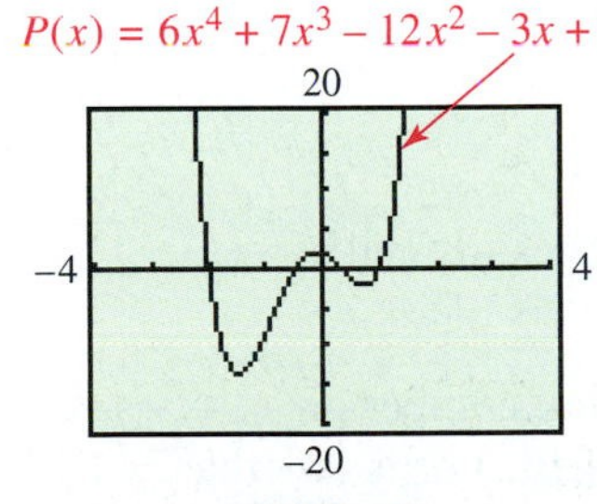

FIGURE 78

(a) For a rational number $\frac{p}{q}$ to be a zero, p must be a factor of $a_0 = 2$ and q must be a factor of $a_4 = 6$. Thus, p can be ±1 or ±2, and q can be ±1, ±2, ±3, or ±6. The possible rational zeros, $\frac{p}{q}$, are $\pm1, \pm2, \pm\frac{1}{2}, \pm\frac{1}{3}, \pm\frac{1}{6}, \pm\frac{2}{3}$.

(b) From Figure 78, we see that the zeros are no less than -2 and no greater than 1, so we can eliminate 2. Furthermore, -1 is not a zero, since the graph does not intersect the x-axis at $(-1, 0)$. At this point, we have no way of knowing whether the zeros indicated on the graph are rational numbers. They may be irrational.

(c) We use the remainder theorem to show that 1 and -2 are zeros.

$$\begin{array}{r|rrrrr} 1 & 6 & 7 & -12 & -3 & 2 \\ & & 6 & 13 & 1 & -2 \\ \hline & 6 & 13 & 1 & -2 & 0 \end{array}$$

The 0 remainder shows that 1 is a zero. Now we use the quotient polynomial $6x^3 + 13x^2 + x - 2$ and synthetic division to find that -2 is also a zero.

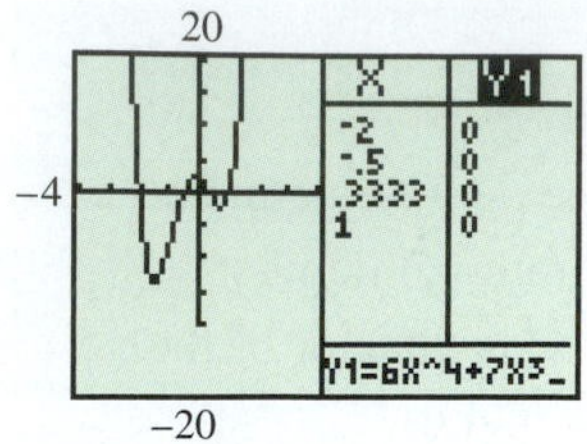
This graph/table verifies that the four zeros of the function of Example 5 are $-2, -\frac{1}{2}, \frac{1}{3}$, and 1.

$$\begin{array}{r|rrrr} -2 & 6 & 13 & 1 & -2 \\ & & -12 & -2 & 2 \\ \hline & 6 & 1 & -1 & 0 \end{array}$$

The new quotient polynomial is $6x^2 + x - 1$, which is easily factored as $(3x - 1)(2x + 1)$. Thus, the remaining two zeros are $\frac{1}{3}$ and $-\frac{1}{2}$.

Since the four zeros of $P(x) = 6x^4 + 7x^3 - 12x^2 - 3x + 2$ are $1, -2, \frac{1}{3}$, and $-\frac{1}{2}$, the factors are $x - 1$, $x + 2$, $x - \frac{1}{3}$, and $x + \frac{1}{2}$, and it follows that

$$P(x) = a(x-1)(x+2)\left(x - \frac{1}{3}\right)\left(x + \frac{1}{2}\right)$$

The polynomial can be left in this form.

$$= 6(x-1)(x+2)\left(x - \frac{1}{3}\right)\left(x + \frac{1}{2}\right) \quad \text{Leading coefficient of } P(x) \text{ is 6; let } a = 6.$$

$$= (x-1)(x+2)(3)\left(x - \frac{1}{3}\right)(2)\left(x + \frac{1}{2}\right) \quad \text{Simplify.}$$

$$= (x-1)(x+2)(3x-1)(2x+1).$$ ■

CAUTION The rational zeros theorem gives only *possible* rational zeros; it does not tell us whether these rational numbers are *actual* zeros. We must rely on other methods to determine whether they are indeed zeros. Furthermore, the function must have integer coefficients. To apply the rational zeros theorem to a polynomial with fractional coefficients, multiply by the least common denominator of all the fractions. For example, any rational zeros of $P(x)$ will also be rational zeros of $Q(x)$.

$$P(x) = x^4 - \frac{1}{6}x^3 + \frac{2}{3}x^2 - \frac{1}{6}x - \frac{1}{3}$$

$$Q(x) = 6x^4 - x^3 + 4x^2 - x - 2 \quad \text{Multiply the terms of } P(x) \text{ by 6.}$$

Descartes' Rule of Sign

Descartes' rule of signs helps to determine the number of positive and negative real zeros of a polynomial function.

Descartes' Rule of Signs

Let $P(x)$ define a polynomial function with real coefficients and a nonzero constant term, with terms in descending powers of x.

(a) The number of positive real zeros either equals the number of variations in sign occurring in the coefficients of $P(x)$ or is less than the number of variations by a positive even integer.

(b) The number of negative real zeros either equals the number of variations in sign occurring in the coefficients of $P(-x)$ or is less than the number of variations by a positive even integer.

A *variation in sign* is a change from positive to negative or negative to positive in successive terms of the polynomial when written in descending powers of the variable. Missing terms (those with 0 coefficients) can be ignored.

EXAMPLE 6 Applying Descartes' Rule of Signs

Determine the possible number of positive real zeros and negative real zeros of $P(x) = x^4 - 6x^3 + 8x^2 + 2x - 1$.

Solution We first consider the possible number of positive zeros by observing that $P(x)$ has three variations in signs.

$$+x^4 \underbrace{-}_{1} 6x^3 \underbrace{+}_{2} 8x^2 + 2x \underbrace{-}_{3} 1$$

Thus, by Descartes' rule of signs, $P(x)$ has either 3 or $3 - 2 = 1$ positive real zeros.

For negative zeros, consider the variations in signs for $P(-x)$.

$$\begin{aligned} P(-x) &= (-x)^4 - 6(-x)^3 + 8(-x)^2 + 2(-x) - 1 \\ &= x^4 + 6x^3 + 8x^2 \underbrace{-}_{1} 2x - 1 \end{aligned}$$

Since there is only one variation in sign, $P(x)$ has only one negative real zero. ■

Boundedness Theorem

The *boundedness theorem* shows how the bottom row of a synthetic division is used to place upper and lower bounds on possible real zeros of a polynomial function.

Boundedness Theorem

Let $P(x)$ define a polynomial function of degree $n \geq 1$ with real coefficients and with a positive leading coefficient. If $P(x)$ is divided synthetically by $x - c$, and

(a) if $c > 0$ and all numbers in the bottom row of the synthetic division are nonnegative, then $P(x)$ has no zero greater than c;

(b) if $c < 0$ and the numbers in the bottom row of the synthetic division alternate in sign (with 0 considered positive or negative, as needed), then $P(x)$ has no zero less than c.

EXAMPLE 7 Using the Boundedness Theorem

Show that the real zeros of the polynomial function $P(x) = 2x^4 - 5x^3 + 3x + 1$ satisfy the following conditions.

(a) No real zero is greater than 3. **(b)** No real zero is less than -1.

Solution

(a) Since $P(x)$ has real coefficients and the leading coefficient, 2, is positive, use the boundedness theorem. Divide $P(x)$ synthetically by $x - 3$.

$$\begin{array}{r|rrrrr} c > 0 \longrightarrow 3 & 2 & -5 & 0 & 3 & 1 \\ & & 6 & 3 & 9 & 36 \\ \hline & 2 & 1 & 3 & 12 & 37 \end{array} \longleftarrow \text{All are nonnegative.}$$

Thus, $P(x)$ has no real zero greater than 3.

(b) Divide $P(x) = 2x^4 - 5x^3 + 3x + 1$ synthetically by $x + 1$.

$$c < 0 \longrightarrow -1\overline{)\begin{array}{rrrrr} 2 & -5 & 0 & 3 & 1 \\ & -2 & 7 & -7 & 4 \\ \hline 2 & -7 & 7 & -4 & 5 \end{array}}$$

Divide $P(x)$ by $x+1$.

$\leftarrow$ The numbers alternate in sign.

Thus, $P(x)$ has no zero less than -1. ■

3.7 Exercises

Find a cubic polynomial in standard form with real coefficients, having the given zeros.

1. 4 and $2 + i$

2. -3 and $6 + 2i$

3. 5 and i

4. -9 and $-i$

5. 0 and $3 + i$

6. 0 and $4 - 3i$

For Exercises 7–12, find a function $P(x)$ defined by a polynomial of degree 3 with real coefficients that satisfies the given conditions.

7. Zeros of -3, -1, and 4; $P(2) = 5$

8. Zeros of 1, -1, and 0; $P(2) = -3$

9. Zeros of -2, 1, and 0; $P(-1) = -1$

10. Zeros of 2, 5, and -3; $P(1) = -4$

11. Zeros of 4 and $1 + i$; $P(2) = 4$

12. Zeros of -7 and $2 - i$; $P(1) = 9$

For each polynomial, one or more zeros are given. Find all remaining zeros.

13. $P(x) = x^3 - x^2 - 4x - 6$; 3 is a zero.

14. $P(x) = x^3 - 5x^2 + 17x - 13$; 1 is a zero.

15. $P(x) = x^4 + 2x^3 - 10x^2 - 18x + 9$; -3 and 3 are zeros.

16. $P(x) = 2x^4 - x^3 - 27x^2 + 16x - 80$; -4 and 4 are zeros.

17. $P(x) = x^4 - x^3 + 10x^2 - 9x + 9$; $3i$ is a zero.

18. $P(x) = 2x^4 - 2x^3 + 55x^2 - 50x + 125$; $-5i$ is a zero.

For Exercises 19–30, find a polynomial function $P(x)$ of least possible degree, having real coefficients, with the given zeros.

19. 5 and -4

20. 6 and -2

21. -3, 2, and i

22. $1 + \sqrt{2}$, $1 - \sqrt{2}$, and 3

23. $1 - \sqrt{3}$, $1 + \sqrt{3}$, and 1

24. $-2 + i$, $-2 - i$, 3, and -3

25. $3 + 2i$, -1, and 2

26. 2 and $3i$

27. -1 and $6 - 3i$

28. $1 + 2i$ and 2 (multiplicity 2)

29. $2 + i$ and -3 (multiplicity 2)

30. 5 (multiplicity 2) and $-2i$

Concept Check *Use the concepts of this section for Exercises 31–36.*

31. Show analytically that -2 is a zero of multiplicity 2 of $P(x) = x^4 + 2x^3 - 7x^2 - 20x - 12$, and find all other complex zeros. Then, write $P(x)$ in factored form.

32. Show analytically that -1 is a zero of multiplicity 3 of $P(x) = x^5 + 9x^4 + 33x^3 + 55x^2 + 42x + 12$, and find all other complex zeros. Then, write $P(x)$ in factored form.

33. What are the possible numbers of real zeros (counting multiplicities) for a polynomial function with real coefficients of degree 5?

34. Explain why a function defined by a polynomial of degree 4 with real coefficients has either zero, two, or four real zeros (counting multiplicities).

35. Determine whether the description of the polynomial function defined by $P(x)$ with real coefficients is *possible* or *not possible*.

(a) $P(x)$ is of degree 3 and has zeros of 1, 2, and $1 + i$.
(b) $P(x)$ is of degree 4 and has four nonreal complex zeros.
(c) $P(x)$ is of degree 5 and -6 is a zero of multiplicity 6.
(d) $P(x)$ has $1 + 2i$ as a zero of multiplicity 2.

36. Suppose that k, a, b, and c are real numbers, $a \neq 0$, and a polynomial function defined by $P(x)$ may be expressed in factored form as $(x - k)(ax^2 + bx + c)$.

(a) What is the degree of P?
(b) What are the possible numbers of distinct *real* zeros of P?
(c) What are the possible numbers of *nonreal complex* zeros of P?

(d) Use the discriminant to explain how to determine the number and type of zeros P has.

In Exercises 37–46, draw by hand a rough sketch of the graph of each function. (You may wish to support your answer with a calculator graph.)

37. $P(x) = 2x^3 - 5x^2 - x + 6 = (x + 1)(2x - 3)(x - 2)$

38. $P(x) = x^3 + x^2 - 8x - 12 = (x + 2)^2(x - 3)$

39. $P(x) = x^4 - 18x^2 + 81 = (x - 3)^2(x + 3)^2$

40. $P(x) = x^4 - 8x^2 + 16 = (x + 2)^2(x - 2)^2$

41. $P(x) = 2x^4 + x^3 - 6x^2 - 7x - 2 = (2x + 1)(x - 2)(x + 1)^2$

42. $P(x) = 3x^4 - 7x^3 - 6x^2 + 12x + 8 = (3x + 2)(x + 1)(x - 2)^2$

43. $P(x) = x^4 + 3x^3 - 3x^2 - 11x - 6 = (x + 3)(x + 1)^2(x - 2)$

44. $P(x) = -2x^5 + 5x^4 + 34x^3 - 30x^2 - 84x + 45 = (x + 3)(2x - 1)(x - 5)(3 - x^2)$

45. $P(x) = 2x^5 - 10x^4 + x^3 - 5x^2 - x + 5 = (x - 5)(x^2 + 1)(2x^2 - 1)$

46. $P(x) = 3x^4 - 4x^3 - 22x^2 + 15x + 18 = (3x + 2)(x - 3)(x^2 + x - 3)$

For each polynomial function, **(a)** *list all possible rational zeros,* **(b)** *use a graph to eliminate some of the possible zeros listed in part (a),* **(c)** *find all rational zeros, and* **(d)** *factor $P(x)$.*

47. $P(x) = x^3 - 2x^2 - 13x - 10$

48. $P(x) = x^3 + 5x^2 + 2x - 8$

49. $P(x) = x^3 + 6x^2 - x - 30$

50. $P(x) = x^3 - x^2 - 10x - 8$

51. $P(x) = 6x^3 + 17x^2 - 31x - 12$

52. $P(x) = 15x^3 + 61x^2 + 2x - 8$

53. $P(x) = 12x^3 + 20x^2 - x - 6$

54. $P(x) = 12x^3 + 40x^2 + 41x + 12$

55. $P(x) = 24x^3 + 40x^2 - 2x - 12$

56. $P(x) = 24x^3 + 80x^2 + 82x + 24$

Find all rational zeros of each polynomial function.

57. $P(x) = x^3 + \frac{1}{2}x^2 - \frac{11}{2}x - 5$

58. $P(x) = \frac{10}{7}x^4 - x^3 - 7x^2 + 5x - \frac{5}{7}$

59. $P(x) = \frac{1}{6}x^4 - \frac{11}{12}x^3 + \frac{7}{6}x^2 - \frac{11}{12}x + 1$

60. $P(x) = x^4 - \frac{1}{6}x^3 + \frac{2}{3}x^2 - \frac{1}{6}x - \frac{1}{3}$

Use Descartes' rule of signs to determine the possible number of positive and negative real zeros for each function. Then, use a graph to determine the actual numbers of positive and negative real zeros.

61. $P(x) = 2x^3 - 4x^2 + 2x + 7$

62. $P(x) = x^3 + 2x^2 + x - 10$

63. $P(x) = 5x^4 + 3x^2 + 2x - 9$

64. $P(x) = 3x^4 + 2x^3 - 8x^2 - 10x - 1$

65. $P(x) = x^5 + 3x^4 - x^3 + 2x + 3$

66. $P(x) = 2x^5 - x^4 + x^3 - x^2 + x + 5$

Use the boundedness theorem to show that the real zeros of each polynomial function satisfy the given conditions.

67. $P(x) = x^4 - x^3 + 3x^2 - 8x + 8$; no real zero greater than 2

68. $P(x) = 2x^5 - x^4 + 2x^3 - 2x^2 + 4x - 4$; no real zero greater than 1

69. $P(x) = x^4 + x^3 - x^2 + 3$; no real zero less than -2

70. $P(x) = x^5 + 2x^3 - 2x^2 + 5x + 5$; no real zero less than -1

71. $P(x) = 3x^4 + 2x^3 - 4x^2 + x - 1$; no real zero greater than 1

72. $P(x) = 3x^4 + 2x^3 - 4x^2 + x - 1$; no real zero less than -2

73. $P(x) = x^5 - 3x^3 + x + 2$; no real zero greater than 2

74. $P(x) = x^5 - 3x^3 + x + 2$; no real zero less than -3

Concept Check *In Exercises 75 and 76, find a cubic polynomial function having the graph shown.*

75.

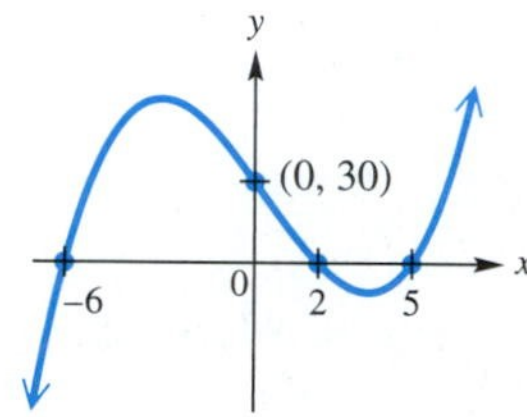

76.

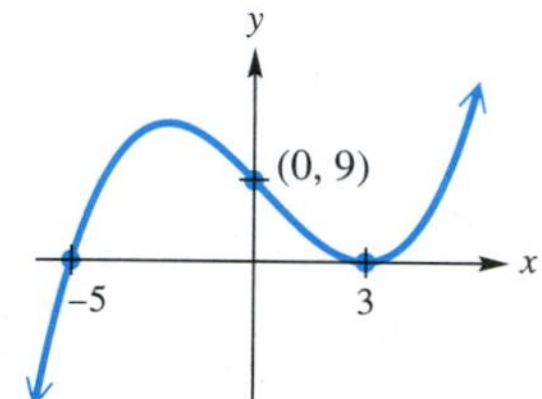

Relating Concepts

For individual or group investigation (Exercises 77–82)

In this chapter, we have studied many characteristics of polynomial functions. This set of exercises is designed to put these ideas together.

For each polynomial function in Exercises 77–81, ***do the following in order.***

(a) *Use Descartes' rule of signs to find the possible number of positive and negative real zeros.*
(b) *Use the rational zeros theorem to determine the possible rational zeros of the function.*
(c) *Find the rational zeros, if any.*
(d) *Find all other real zeros, if any.*
(e) *Find any other nonreal complex zeros, if any.*
(f) *Find the x-intercepts of the graph, if any.*
(g) *Find the y-intercept of the graph.*
(h) *Use synthetic division to find P(4), and give the coordinates of the corresponding point on the graph.*
(i) *Determine the end behavior of the graph.*
(j) *Sketch the graph. (You may wish to support your answer with a calculator graph.)*

77. $P(x) = -2x^4 - x^3 + x + 2$

78. $P(x) = 4x^5 + 8x^4 + 9x^3 + 27x^2 + 27x$ (*Hint:* Factor out x first.)

79. $P(x) = 3x^4 - 14x^2 - 5$ (*Hint:* Factor the polynomial.)

80. $P(x) = -x^5 - x^4 + 10x^3 + 10x^2 - 9x - 9$

81. $P(x) = -3x^4 + 22x^3 - 55x^2 + 52x - 12$

82. For the polynomial functions in Exercises 77–81 that have irrational zeros, find approximations to the nearest thousandth.

3.8 Polynomial Equations and Inequalities; Further Applications and Models

Polynomial Equations and Inequalities ■ Complex *n*th Roots ■ Applications and Polynomial Models

FOR DISCUSSION

Although methods of solving linear and quadratic equations were known since the Babylonians, mathematicians struggled for centuries to find a formula that solved cubic equations to find the zeros of cubic functions. In 1545, a method of solving a cubic equation of the form $x^3 + mx = n$, developed by Niccolò Tartaglia, was published in the *Ars Magna*, a work by Girolamo Cardano. The formula for finding the one real solution of the equation is

$$x = \sqrt[3]{\frac{n}{2} + \sqrt{\left(\frac{n}{2}\right)^2 + \left(\frac{m}{3}\right)^3}} - \sqrt[3]{\frac{-n}{2} + \sqrt{\left(\frac{n}{2}\right)^2 + \left(\frac{m}{3}\right)^3}}.$$

(*Source:* Gullberg, J., *Mathematics from the Birth of Numbers*, W.W. Norton & Company, 1997.) Use this formula to show that the equation $x^3 + 9x = 26$ has 2 as its one real solution.

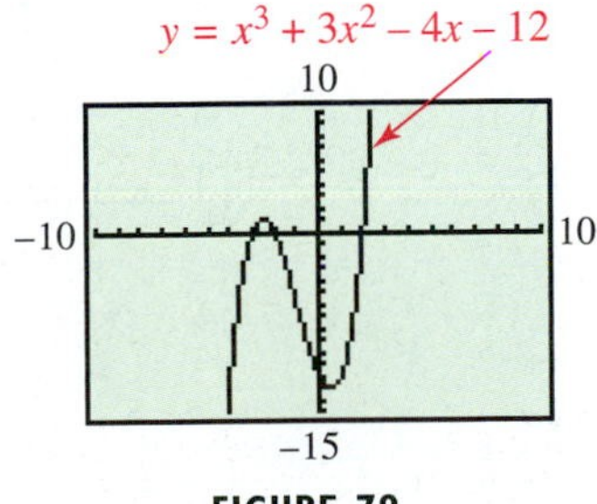

FIGURE 79

In 1824, Norwegian mathematician Niels Henrik Abel proved that it is impossible to find a formula that will yield solutions of the general *quintic* (fifth-degree) equation. A similar result holds for polynomial equations of degree greater than 5.

We use elementary methods to solve *some* higher-degree polynomial equations analytically in this section. Graphing calculators support the analytic work and enable us to find accurate approximations of solutions of polynomial equations that cannot be solved easily, or at all, by analytic methods.

Polynomial Equations and Inequalities

EXAMPLE 1 Solving a Polynomial Equation and Associated Inequalities

(a) Solve $x^3 + 3x^2 - 4x - 12 = 0$ using the zero-product property.

(b) Graph $y = x^3 + 3x^2 - 4x - 12$.

(c) Use the graph from part (b) to solve the inequalities

$$x^3 + 3x^2 - 4x - 12 > 0 \quad \text{and} \quad x^3 + 3x^2 - 4x - 12 \le 0.$$

Solution

(a)

$x^3 + 3x^2 - 4x - 12 = 0$	
$(x^3 + 3x^2) + (-4x - 12) = 0$	Group terms with common factors.
Be careful with signs. $x^2(x + 3) - 4(x + 3) = 0$	Factor out common factors in each group.
$(x + 3)(x^2 - 4) = 0$	Factor out $x + 3$.
$(x + 3)(x - 2)(x + 2) = 0$	Factor the difference of squares.
$x = -3$ or $x = 2$ or $x = -2$	Zero-product property

The solution set is $\{-3, -2, 2\}$.

(b) Graph $y = f(x) = x^3 + 3x^2 - 4x - 12$ and verify that the x-intercepts are -3, -2, and 2, supporting the analytic solution. See Figure 79.

(c) Recall that the solution set of $f(x) > 0$ consists of all numbers in the domain of f for which the graph lies *above* the x-axis. In Figure 79, this occurs in the intervals $(-3, -2) \cup (2, \infty)$, the solution set of this inequality. To solve $x^3 + 3x^2 - 4x - 12 \le 0$, locate the intervals where the graph *lies below or intersects* the x-axis. From the figure, the solution set is $(-\infty, -3] \cup [-2, 2]$. The endpoints are included here because this is a nonstrict inequality. ■

An equation is **quadratic in form** if it can be written as

$$au^2 + bu + c = 0,$$

where $a \neq 0$ and u is an algebraic expression.

EXAMPLE 2 Solving an Equation Quadratic in Form and Associated Inequalities

(a) Solve $x^4 - 6x^2 - 40 = 0$ analytically. Find all complex solutions.

(b) Graph $y = x^4 - 6x^2 - 40$, and use the graph to solve the inequalities

$$x^4 - 6x^2 - 40 \geq 0 \quad \text{and} \quad x^4 - 6x^2 - 40 < 0.$$

Give endpoints of intervals in both exact and approximate form.

Solution

(a)

$$x^4 - 6x^2 - 40 = 0$$

$$(x^2)^2 - 6x^2 - 40 = 0 \qquad x^4 = (x^2)^2$$

$$u^2 - 6u - 40 = 0 \qquad \text{Let } u = x^2.$$

$$(u - 10)(u + 4) = 0 \qquad \text{Factor.}$$

$$u = 10 \quad \text{or} \quad u = -4 \qquad \text{Zero-product property}$$

$$x^2 = 10 \quad \text{or} \quad x^2 = -4 \qquad \text{Replace } u \text{ with } x^2.$$

$$x = \pm\sqrt{10} \quad \text{or} \quad x = \pm 2i \qquad \text{Square root property}$$

Remember both the positive and negative square roots.

The solution set is $\{-\sqrt{10}, \sqrt{10}, -2i, 2i\}$.

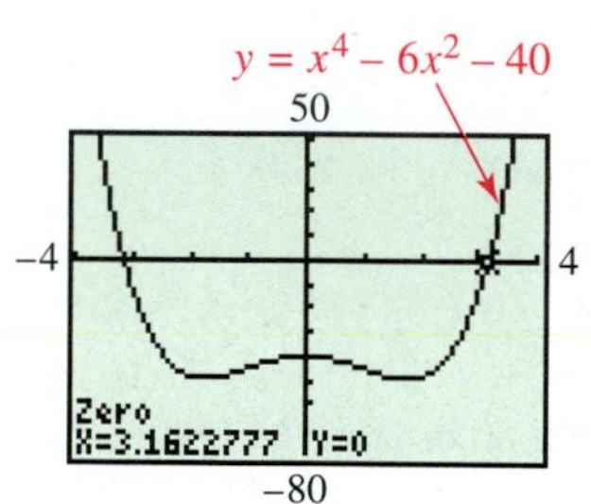

FIGURE 80

(b) The graph of $y = x^4 - 6x^2 - 40$ is shown in Figure 80. The x-intercepts are approximately -3.16 and 3.16 (approximations of $-\sqrt{10}$ and $\sqrt{10}$, respectively). The graph cannot support the imaginary solutions.

Since the graph lies above or intersects the x-axis for real numbers less than or equal to $-\sqrt{10}$ and for real numbers greater than or equal to $\sqrt{10}$, the solution set of $x^4 - 6x^2 - 40 \geq 0$ is

$$(-\infty, -\sqrt{10}] \cup [\sqrt{10}, \infty) \qquad \text{Exact form}$$

or

$$(-\infty, -3.16] \cup [3.16, \infty). \qquad \text{Approximate form}$$

By similar reasoning, the solution set of $x^4 - 6x^2 - 40 < 0$ is

$$(-\sqrt{10}, \sqrt{10}) \qquad \text{Exact form}$$

or

$$(-3.16, 3.16). \qquad \text{Approximate form}$$

The nonreal complex solutions do not affect the solution sets of the inequalities. ■

EXAMPLE 3 Solving a Polynomial Equation

(a) Show that 2 is a real solution of $P(x) = x^3 + 3x^2 - 11x + 2 = 0$, and then find all solutions of this equation.

(b) Support the result of part (a) graphically.

Solution

(a)

$$\begin{array}{r|rrrr} 2 & 1 & 3 & -11 & 2 \\ & & 2 & 10 & -2 \\ \hline & 1 & 5 & -1 & 0 \end{array}$$

Use synthetic division.

$0 \leftarrow P(2) = 0$ by the remainder theorem.

Coefficients of the quotient polynomial

By the factor theorem, $x - 2$ is a factor of $P(x)$, so

$$P(x) = (x - 2)(x^2 + 5x - 1).$$

To find the other zeros of P, solve $x^2 + 5x - 1 = 0$.

$$x = \frac{-b \pm \sqrt{b^2 - 4ac}}{2a} = \frac{-5 \pm \sqrt{5^2 - 4(1)(-1)}}{2(1)} \quad \text{Quadratic formula; } a = 1, b = 5, c = -1$$

$$= \frac{-5 \pm \sqrt{29}}{2}$$

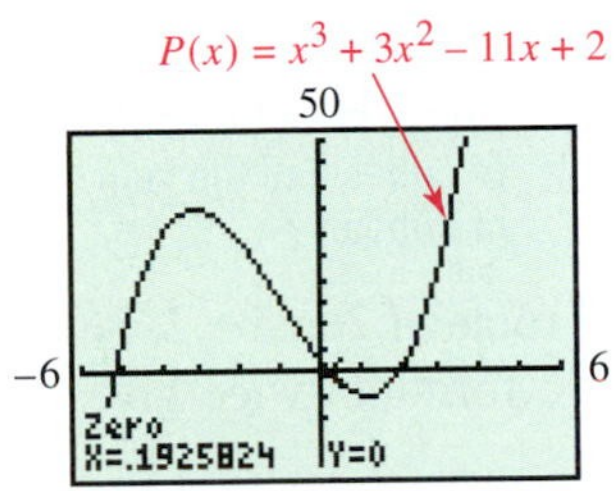

FIGURE 81

The solution set is $\left\{\frac{-5 - \sqrt{29}}{2}, \frac{-5 + \sqrt{29}}{2}, 2\right\}$.

(b) The graph of $P(x) = x^3 + 3x^2 - 11x + 2$ is shown in Figure 81. The x-intercepts are 2, approximately -5.19, and approximately .19. The latter two approximations are for $\frac{-5 - \sqrt{29}}{2}$ and $\frac{-5 + \sqrt{29}}{2}$, supporting the analytic results. ■

The associated inequalities for the equation in Example 3 could be solved by using the solutions and graph as in Examples 1 and 2.

EXAMPLE 4 Solving an Equation and Associated Inequalities Graphically

Let $P(x) = 2.45x^3 - 3.14x^2 - 6.99x + 2.58$. Use a graph to solve $P(x) = 0$, $P(x) > 0$, and $P(x) < 0$. Express solutions of the equation and endpoints of the intervals for the inequalities to the nearest hundredth.

Solution The graph of P is shown in Figure 82. Using a calculator, we find that the approximate x-intercepts are -1.37, .33, and 2.32. Therefore, the solution set of the equation $P(x) = 0$ is $\{-1.37, .33, 2.32\}$. Based on the graph, the solution sets of $P(x) > 0$ and $P(x) < 0$ are, respectively,

$$(-1.37, .33) \cup (2.32, \infty) \quad \text{and} \quad (-\infty, -1.37) \cup (.33, 2.32). \quad ■$$

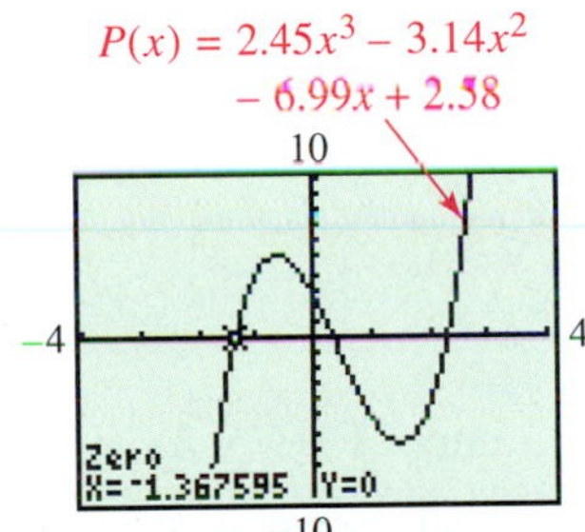

The other two x-intercepts are approximately .33 and 2.32.

FIGURE 82

NOTE The graphical method of solving $P(x) = 0$ in Example 4 would not have yielded nonreal complex solutions had there been any. Only real solutions are obtained with this method.

Complex nth Roots

If n is a positive integer and k is a nonzero complex number, then a solution of $x^n = k$ is called an ***n*th root of *k*.** For example, since -1 and 1 are solutions of $x^2 = 1$, they are called second, or square, roots of 1. Similarly, the pure imaginary numbers $-2i$ and $2i$ are called square roots of -4, since $(\pm 2i)^2 = -4$.

The real numbers -2 and 2 are sixth roots of 64, since $(\pm 2)^6 = 64$. However, 64 has four more complex sixth roots. While a complete discussion of the next theorem requires concepts from trigonometry, we state it and use it to solve particular problems involving nth roots.

Complex nth Roots Theorem

If n is a positive integer and k is a nonzero complex number, then the equation $x^n = k$ has *exactly* n complex roots.

EXAMPLE 5 Finding *n*th Roots of a Number

Find all six complex sixth roots of 64.

Solution The roots must be solutions of $x^6 = 64$.

$$x^6 = 64$$

$$x^6 - 64 = 0 \qquad \text{Subtract 64.}$$

$$(x^3 - 8)(x^3 + 8) = 0 \qquad \text{Factor the difference of squares.}$$

$$(x - 2)(x^2 + 2x + 4)(x + 2)(x^2 - 2x + 4) = 0 \qquad \text{Factor the difference of cubes and the sum of cubes.}$$

Now apply the zero-product property to obtain the real roots of 2 and -2. Setting the quadratic factors equal to 0 and applying the quadratic formula twice gives the remaining four complex roots, none of which are real.

$$x^2 + 2x + 4 = 0 \qquad\qquad x^2 - 2x + 4 = 0$$

$$x = \frac{-2 \pm \sqrt{2^2 - 4(1)(4)}}{2(1)} \qquad\qquad x = \frac{2 \pm \sqrt{(-2)^2 - 4(1)(4)}}{2(1)}$$

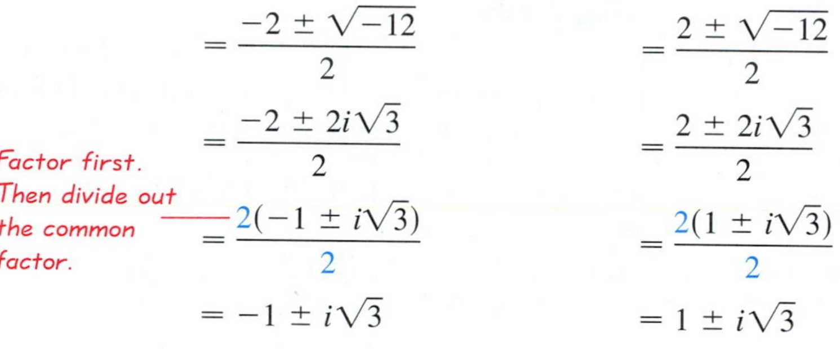

$$= \frac{-2 \pm \sqrt{-12}}{2} \qquad\qquad = \frac{2 \pm \sqrt{-12}}{2}$$

$$= \frac{-2 \pm 2i\sqrt{3}}{2} \qquad\qquad = \frac{2 \pm 2i\sqrt{3}}{2}$$

$$= \frac{2(-1 \pm i\sqrt{3})}{2} \qquad\qquad = \frac{2(1 \pm i\sqrt{3})}{2}$$

$$= -1 \pm i\sqrt{3} \qquad\qquad = 1 \pm i\sqrt{3}$$

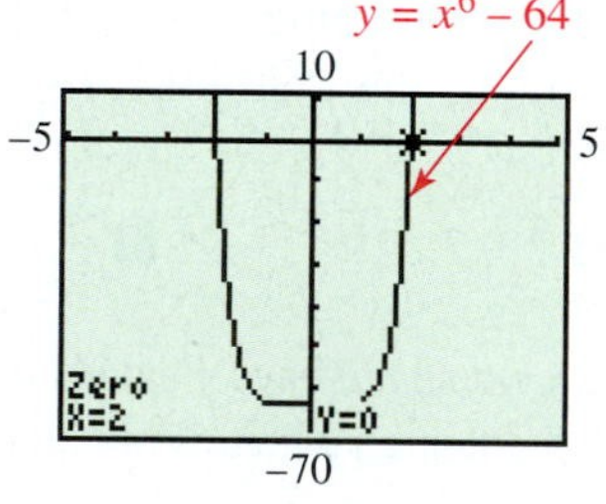

FIGURE 83

Therefore, the six complex sixth roots of 64 are

$$2, \quad -2, \quad -1 + i\sqrt{3}, \quad -1 - i\sqrt{3}, \quad 1 + i\sqrt{3}, \quad \text{and} \quad 1 - i\sqrt{3}.$$

The graph of $y = x^6 - 64$ confirms the two real sixth roots of 64. See Figure 83. The x-intercepts are -2 and 2. ■

Applications and Polynomial Models

EXAMPLE 6 Using a Polynomial Function to Model the Volume of a Box

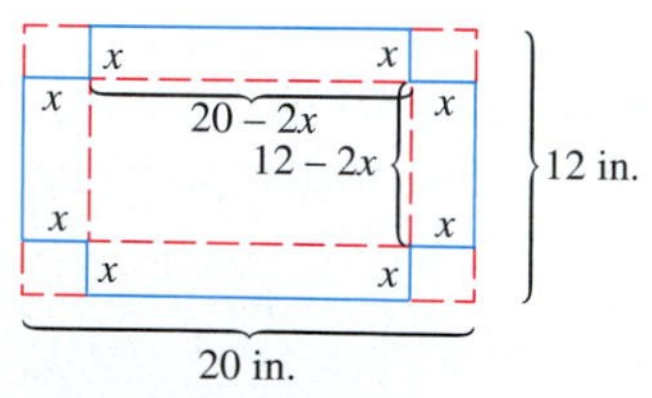

A box with an open top is to be constructed from a rectangular 12-inch by 20-inch piece of cardboard by cutting equal-sized squares from each corner and folding up the sides.

(a) If x represents the length of the side of each such square, determine a function V that describes the volume of the box in terms of x.

(b) Graph V in the window $[0, 6]$ by $[0, 300]$, and locate a point on the graph. Interpret the displayed values of x and y.

(c) Determine the value of x for which the volume of the box is maximized. What is this volume?

(d) For what values of x is the volume equal to 200 cubic inches? greater than 200 cubic inches? less than 200 cubic inches?

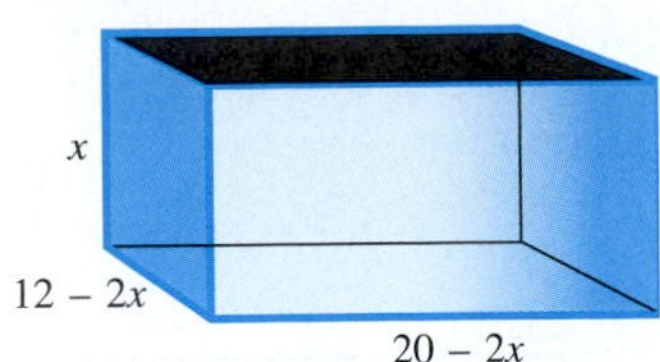

FIGURE 84

Solution

(a) As shown in Figure 84, the dimensions (in inches) of the box to be formed will be

$$\text{length} = 20 - 2x, \quad \text{width} = 12 - 2x, \quad \text{height} = x.$$

Furthermore, x must be positive, and both $20 - 2x$ and $12 - 2x$ must be positive, implying that $0 < x < 6$. The desired function is

$$V(x) = (20 - 2x)(12 - 2x)x \qquad \text{Volume = length} \times \text{width} \times \text{height}$$
$$= 4x^3 - 64x^2 + 240x.$$

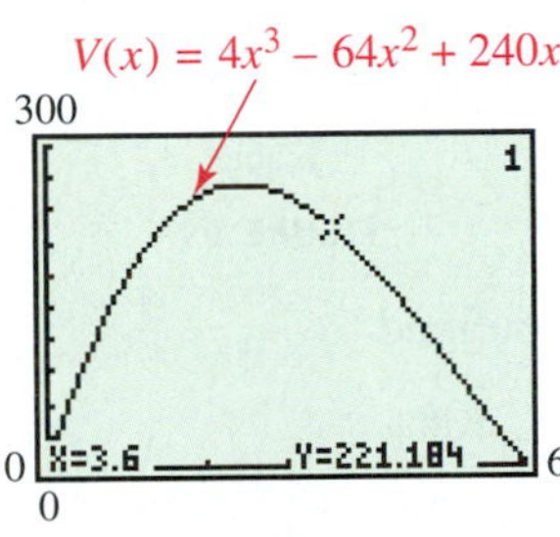

FIGURE 85

(b) Figure 85 shows the graph of V with the cursor at the arbitrarily chosen point (3.6, 221.184). This means that when the side of each cut-out square measures 3.6 inches, the volume of the resulting box is 221.184 cubic inches.

(c) Use a calculator to find the local maximum point on the graph of V. To the nearest hundredth, the coordinates of this point are (2.43, 262.68). See Figure 86. Therefore, when $x \approx 2.43$ is the measure of the side of each cut-out square, the volume of the box is at its maximum, approximately 262.68 cubic inches.

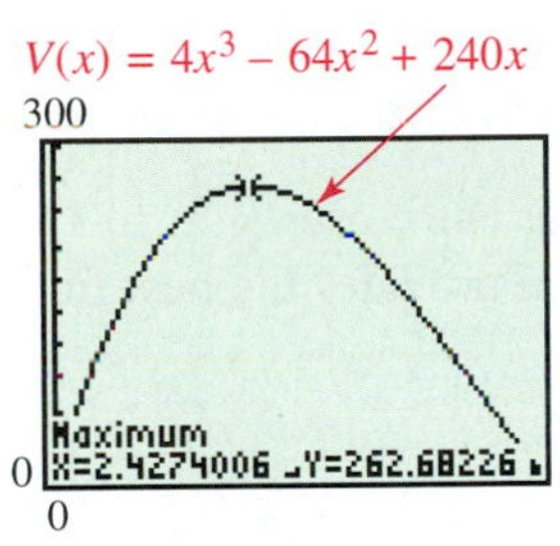

FIGURE 86

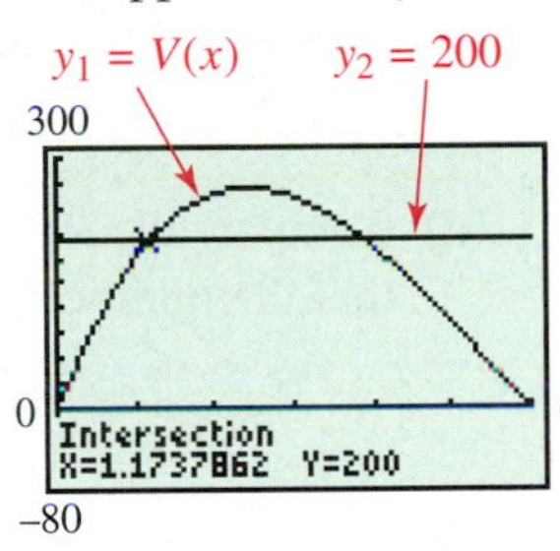

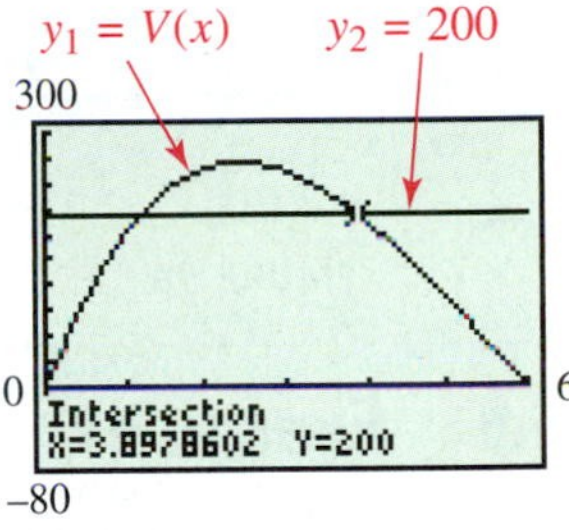

FIGURE 87

(d) The graphs of $y_1 = V(x)$ and $y_2 = 200$ are shown in Figure 87. The points of intersection of the line and the cubic curve are approximately (1.17, 200) and (3.90, 200), so the volume is equal to 200 cubic inches for $x \approx 1.17$ or 3.90, is greater than 200 cubic inches for $1.17 < x < 3.90$, and is less than 200 cubic inches for $0 < x < 1.17$ or $3.90 < x < 6$. ■

EXAMPLE 7 Examining Polynomial Models for Debit Card Use

The table shows the number of transactions, in millions, by users of bank debit cards for selected years.

Year	1995	1998	2000	2002	2005*
Transactions (in millions)	829	3765	8300	13,300	22,700

*Projected
Source: Statistical Abstract of the United States, 2003.

(a) With $x = 0$ representing 1995, $x = 3$ representing 1998, and so on, use the regression feature of a calculator to determine the quadratic function that best fits the data. Plot the data and the graph.

(b) Repeat part (a) for a cubic function.

(c) Repeat part (a) for a quartic function.

(d) Compare R^2 for the three functions found in parts (a)–(c) to decide which function best fits the data.

Solution

(a) As shown in Figure 88, the best-fitting regression equation is

$$y \approx 146.61x^2 + 748.1x + 672.2.$$

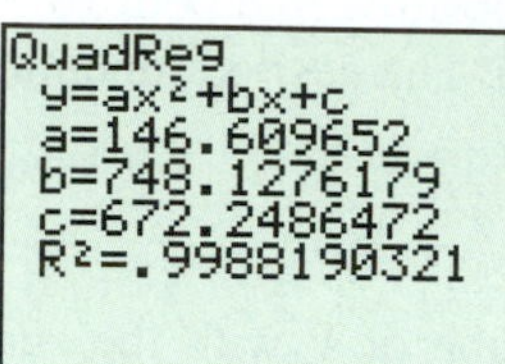

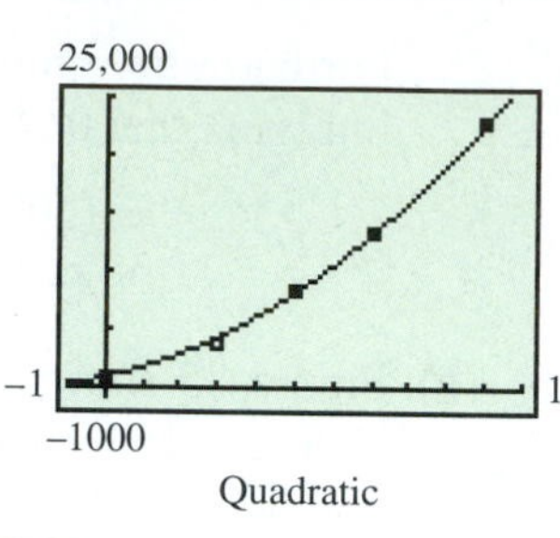

Quadratic

FIGURE 88

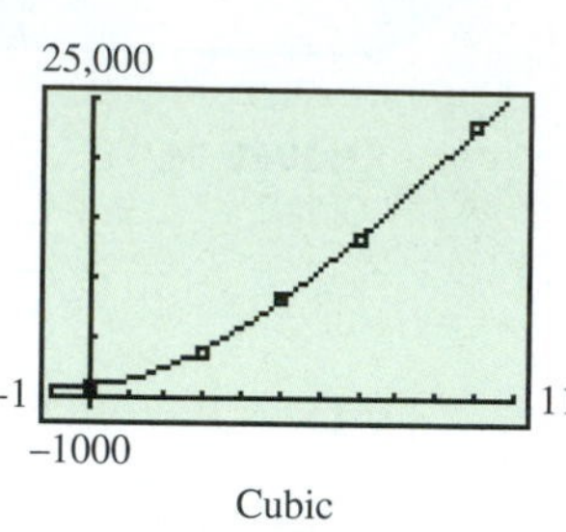

Cubic

FIGURE 89

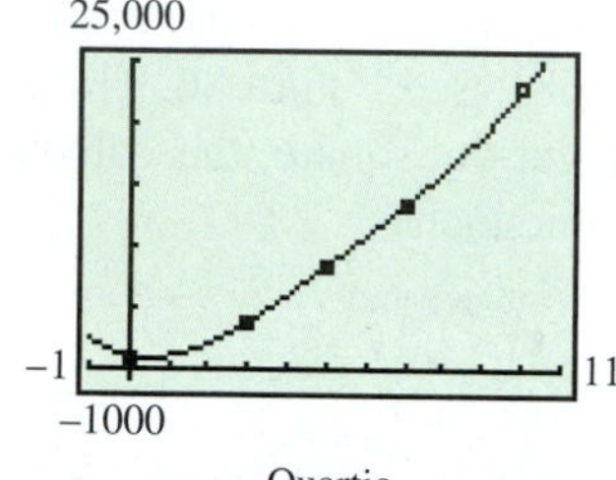

Quartic

FIGURE 90

(b) The best-fitting cubic function, shown in Figure 89, is defined by

$$y \approx -9.364x^3 + 287.07x^2 + 252.8x + 808.$$

(c) The best-fitting quartic function, shown in Figure 90, is defined by

$$y \approx 3.831x^4 - 85.99x^3 + 757.9x^2 - 624.7x + 829.$$

(d) Compare the R^2 values. The closer R^2 is to 1, the better the approximation. Verify that the quartic function provides the best fit. ■

3.8 Exercises

Find all complex solutions of each equation.

1. $7x^3 + x = 0$
2. $2x^3 - 4x = 0$
3. $3x^3 + 2x^2 - 3x - 2 = 0$
4. $5x^3 - x^2 + 10x - 2 = 0$
5. $x^4 - 11x^2 + 10 = 0$
6. $x^4 + x^2 - 6 = 0$

Solve each equation analytically for all complex solutions, giving exact *forms in your solution set. Then, graph the left side of the equation as y_1 in the suggested viewing window and, using the capabilities of your calculator, support the real solutions.*

7. $4x^4 - 25x^2 + 36 = 0$; $[-5, 5]$ by $[-5, 100]$
8. $4x^4 - 29x^2 + 25 = 0$; $[-5, 5]$ by $[-50, 100]$
9. $x^4 - 15x^2 - 16 = 0$; $[-5, 5]$ by $[-100, 100]$
10. $9x^4 + 35x^2 - 4 = 0$; $[-3, 3]$ by $[-10, 100]$
11. $x^3 - x^2 - 64x + 64 = 0$; $[-10, 10]$ by $[-300, 300]$
12. $x^3 + 6x^2 - 100x - 600 = 0$; $[-15, 15]$ by $[-1000, 300]$
13. $-2x^3 - x^2 + 3x = 0$; $[-4, 4]$ by $[-10, 10]$
14. $-5x^3 + 13x^2 + 6x = 0$; $[-4, 4]$ by $[-2, 30]$
15. $x^3 + x^2 - 7x - 7 = 0$; $[-10, 10]$ by $[-20, 20]$
16. $x^3 + 3x^2 - 19x - 57 = 0$; $[-10, 10]$ by $[-100, 50]$
17. $-3x^3 - x^2 + 6x = 0$; $[-4, 4]$ by $[-10, 10]$
18. $-4x^3 - x^2 + 4x = 0$; $[-4, 4]$ by $[-10, 10]$
19. $3x^3 + 3x^2 + 3x = 0$; $[-5, 5]$ by $[-5, 5]$
20. $2x^3 + 2x^2 + 12x = 0$; $[-10, 10]$ by $[-20, 20]$
21. $x^4 + 17x^2 + 16 = 0$; $[-4, 4]$ by $[-10, 40]$
22. $36x^4 + 85x^2 + 9 = 0$; $[-4, 4]$ by $[-10, 40]$
23. $x^6 + 19x^3 - 216 = 0$; $[-4, 4]$ by $[-350, 200]$
24. $8x^6 + 7x^3 - 1 = 0$; $[-4, 4]$ by $[-5, 100]$

Graph each polynomial function by hand, as shown in Section 3.7. Then solve each equation and inequality.

25. $P(x) = x^3 - 3x^2 - 6x + 8$
$= (x - 4)(x - 1)(x + 2)$
(a) $P(x) = 0$ **(b)** $P(x) < 0$ **(c)** $P(x) > 0$

26. $P(x) = x^3 + 4x^2 - 11x - 30$
$= (x - 3)(x + 2)(x + 5)$
(a) $P(x) = 0$ **(b)** $P(x) < 0$ **(c)** $P(x) > 0$

27. $P(x) = 2x^4 - 9x^3 - 5x^2 + 57x - 45$
$= (x - 3)^2(2x + 5)(x - 1)$
(a) $P(x) = 0$ **(b)** $P(x) < 0$ **(c)** $P(x) > 0$

28. $P(x) = 4x^4 + 27x^3 - 42x^2 - 445x - 300$
$= (x + 5)^2(4x + 3)(x - 4)$
(a) $P(x) = 0$ **(b)** $P(x) < 0$ **(c)** $P(x) > 0$

29. $P(x) = -x^4 - 4x^3 + 3x^2 + 18x$
$= x(2 - x)(x + 3)^2$
(a) $P(x) = 0$ **(b)** $P(x) \geq 0$ **(c)** $P(x) \leq 0$

30. $P(x) = -x^4 + 2x^3 + 8x^2$
$= x^2(4 - x)(x + 2)$
(a) $P(x) = 0$ **(b)** $P(x) \geq 0$ **(c)** $P(x) \leq 0$

Use a graphical method to find all real *solutions of each equation. Express solutions to the nearest hundredth.*

31. $.86x^3 - 5.24x^2 + 3.55x + 7.84 = 0$

32. $-2.47x^3 - 6.58x^2 - 3.33x + .14 = 0$

33. $-\sqrt{7}x^3 + \sqrt{5}x^2 + \sqrt{17} = 0$

34. $\sqrt{10}x^3 - \sqrt{11}x - \sqrt{8} = 0$

35. $2.45x^4 - 3.22x^3 = -.47x^2 + 6.54x + 3$

36. $\sqrt{17}x^4 - \sqrt{22}x^2 = -1$

Find all n complex solutions of each equation of the form $x^n = k$.

37. $x^2 = -1$ **38.** $x^2 = -4$ **39.** $x^3 = -1$ **40.** $x^3 = -8$

41. $x^3 = 27$ **42.** $x^3 = 64$ **43.** $x^4 = 16$ **44.** $x^4 = 81$

45. $x^3 = -64$ **46.** $x^3 = -27$ **47.** $x^2 = -18$ **48.** $x^2 = -52$

(Modeling) Solve each problem.

49. ***Floating Ball*** The polynomial function defined by

$$f(x) = \frac{\pi}{3}x^3 - 5\pi x^2 + \frac{500\pi d}{3}$$

can be used to find the depth that a ball 10 centimeters in diameter sinks in water. The constant d is the density of the ball, where the density of water is 1. The smallest *positive* zero of $f(x)$ equals the depth that the ball sinks. Approximate this depth for each material and interpret the results.
(a) A wooden ball with $d = .8$
(b) A solid aluminum ball with $d = 2.7$
(c) A spherical water balloon with $d = 1$

50. ***Concentration of Toxin*** A survey team measures the concentration (in parts per million) of a particular toxin in a local river. On a normal day, the concentration of the toxin at time x (in hours) after the factory upstream dumps its waste is given by

$$g(x) = -.006x^4 + .14x^3 - .05x^2 + .02x,$$

where $0 \leq x \leq 24$.
(a) Graph $y = g(x)$ in the window $[0, 24]$ by $[0, 200]$.
(b) Estimate the time at which the concentration is greatest.
(c) A concentration greater than 100 parts per million is considered polluted. Using the graph from part (a), estimate the period during which the river is polluted.

51. ***Floating Ball*** Refer to Exercise 49. If a ball has a 20-centimeter diameter, then

$$f(x) = \frac{\pi}{3}x^3 - 10\pi x^2 + \frac{4000\pi d}{3}$$

can be used to determine the depth that it sinks to in water. Find the depth that this size ball sinks when $d = .6$.

52. ***Floating Ball*** Refer to Exercise 49. Determine the depth to which a pine ball with a 10-centimeter diameter sinks in water if $d = .55$.

53. ***Volume of a Box*** A rectangular piece of cardboard measuring 12 inches by 18 inches is to be made into a box with an open top by cutting equal-sized squares from each corner and folding up the sides. Let x represent the length of a side of each such square in inches.

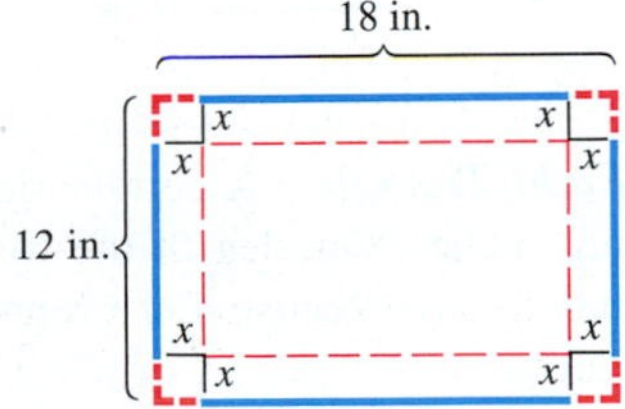

(a) Give the restrictions on x.
(b) Determine a function V that gives the volume of the box as a function of x.

(c) For what value of x will the volume be a maximum? What is this maximum volume?
(d) For what values of x will the volume be greater than 80 cubic inches?

54. ***Construction of a Rain Gutter*** A rectangular piece of sheet metal is 20 inches wide. It is to be made into a rain gutter by turning up the edges to form parallel sides. Let x represent the length of each of the parallel sides in inches. See the figure.

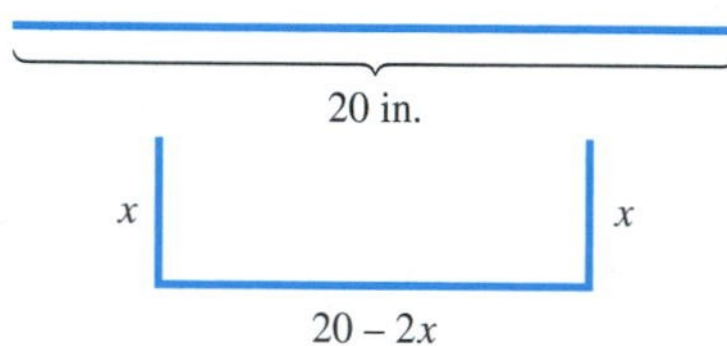

(a) Give the restrictions on x.
(b) Determine a function A that gives the area of a cross section of the gutter.
(c) For what value of x will A be a maximum (and thus maximize the amount of water that the gutter will hold)? What is this maximum area?
(d) For what values of x will the area of a cross section be less than 40 square inches?

55. ***Buoyancy of a Spherical Object*** It has been determined that a spherical object of radius 4 inches with specific gravity .25 will sink in water to a depth of x inches, where x is the least positive root of the equation

$$x^3 - 12x^2 + 64 = 0.$$

To what depth will this object sink if $x < 10$?

56. ***Area of a Rectangle*** Find the value of x in the figure that will maximize the area of rectangle $ABCD$.

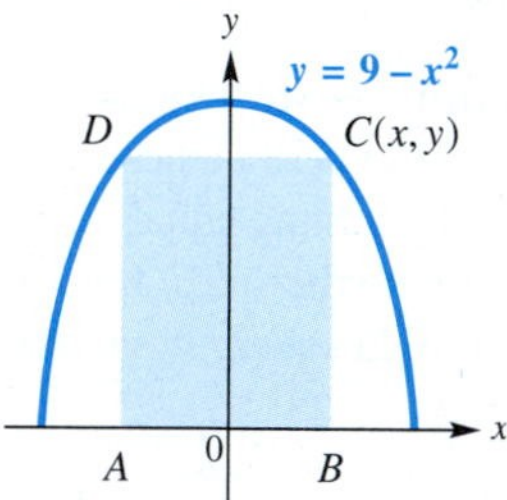

57. ***Sides of a Right Triangle*** A certain right triangle has area 84 square inches. One leg of the triangle measures 1 inch less than the hypotenuse. Let x represent the length of the hypotenuse.
(a) Express the length of the leg described in terms of x.
(b) Express the length of the other leg in terms of x.
(c) Write an equation based on the information determined thus far. Square each side, and then write the equation with one side as a polynomial with integer coefficients, in descending powers, and the other side equal to 0.
(d) Solve the equation in part (c) graphically. Find the lengths of the three sides of the triangle.

58. ***Butane Gas Storage*** A storage tank for butane gas is to be built in the shape of a right circular cylinder having altitude 12 feet, with a half sphere attached to each end. If x represents the radius of each half sphere, what radius should be used to cause the volume of the tank to be 144π cubic feet?

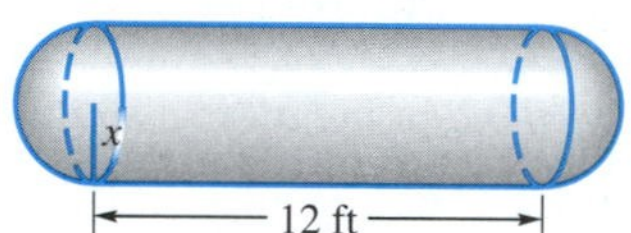

59. ***Volume of a Box*** A standard piece of notebook paper measuring 8.5 inches by 11 inches is to be made into a box with an open top by cutting equal-sized squares from each corner and folding up the sides. Let x represent the length of a side of each such square in inches.
(a) Use the table feature of your graphing calculator to find the maximum volume of the box.
(b) Use the table feature to determine when the volume of the box will be greater than 40 cubic inches.

60. ***Highway Design*** To allow enough distance for cars to pass on two-lane highways, engineers calculate minimum sight distances between curves and hills. The table shows the minimum sight distance y in feet for a car traveling at x mph.

x (in mph)	20	30	40	50
y (in feet)	810	1090	1480	1840

x (in mph)	60	65	70
y (in feet)	2140	2310	2490

Source: Haefner, L., *Introduction to Transportation Systems,* Holt, Rinehart and Winston, 1986.

(a) Make a scatter diagram of the data.
(b) Use the regression feature of a calculator to find the best-fitting linear function for the data. Graph the function with the data.

(c) Repeat part (b) for a cubic function.

(d) Use both functions from parts (b) and (c) to estimate the minimum sight distance for a car traveling 43 mph.

(e) Which function fits the data better? Why?

61. ***Water Pollution*** In one study, freshwater mussels were used to monitor copper discharge into a river from an electroplating works. Copper in high doses can be lethal to aquatic life. The table lists copper concentrations in mussels after 45 days at various distances downstream from the plant. The concentration C is measured in micrograms of copper per gram of mussel x kilometers downstream.

x	5	21	37	53	59
C	20	13	9	6	5

Sources: Foster, R., and J. Bates, "Use of mussels to monitor point source industrial discharges," *Eviron. Sci. Technol.*; Mason, C., *Biology of Freshwater Pollution*, John Wiley and Sons, 1991.

(a) Make a scatter diagram of the data.

(b) Use the regression feature of a calculator to find the best-fitting quadratic function C for the data. Graph the function with the data.

(c) Repeat part (b) for a cubic function.

(d) By comparing graphs of the functions in parts (b) and (c) with the data, decide which function best fits the given data.

(e) Concentrations above 10 are lethal to mussels. Find the values of x for which this is the case.

62. ***Temperature*** The cubic polynomial function defined by

$$f(x) = -.184x^3 + 1.45x^2 + 10.7x - 27.9$$

models monthly average temperature at Trout Lake, Canada, in degrees Fahrenheit, where $x = 1$ corresponds to January and $x = 12$ represents December. Find and interpret the zeros of f.

63. ***Cardiac Output*** A technique for measuring cardiac output depends on the concentration of a dye in the bloodstream after a known amount is injected into a vein near the heart. For a normal heart, the concentration of dye in the bloodstream at time x (in seconds) is modeled by

$$g(x) = -.006x^4 + .140x^3 - .053x^2 + 1.79x.$$

When does the concentration reach 10 units?

64. ***Roots of Unity*** Consider the equation $x^8 = 1$. This equation has eight distinct solutions, each of which is an eighth root of 1. (Roots of 1 are often called *roots of unity.*)

(a) Graph $y = x^8 - 1$ to determine the two real eighth roots of unity.

(b) Show analytically that i is also an eighth root of unity.

(c) Based on the conjugate zeros theorem and the result of part (b), what must be another eighth root of unity? Verify analytically that this is a root.

(d) Using concepts from trigonometry, one can show that

$$\frac{\sqrt{2}}{2} + \frac{\sqrt{2}}{2}i \quad \text{and} \quad -\frac{\sqrt{2}}{2} + \frac{\sqrt{2}}{2}i$$

are nonreal complex eighth roots of unity. Based on the conjugate zeros theorem, what two other imaginary numbers must be eighth roots of unity?

(e) List all eight eighth roots of unity.

Reviewing Basic Concepts (Sections 3.6–3.8)

For Exercises 1–3, use $P(x) = 2x^4 - 7x^3 + 29x - 30$.

1. Find $P(3)$.

2. Is 2 a zero of P?

3. One zero of P is $2 + i$. Factor $P(x)$ into linear factors.

4. Find a polynomial $P(x)$ with real coefficients that has zeros of $\frac{3}{2}$ and i and for which $P(3) = 15$.

5. Find a polynomial $P(x)$ of least possible degree with real coefficients and zeros -4 (multiplicity 2) and $1 + 2i$.

6. Given $P(x) = 2x^3 + x^2 - 11x - 10$, list all zeros of P.

7. Find all complex solutions of $3x^4 - 12x^2 + 1 = 0$.

8. ***(Modeling) Bank Debit Cards*** The table shows the number of bank debit cards issued, in millions.

Year	1998	2000	2002	2005*
Cards (in millions)	100	137	175	205

*Projected
Source: Statistical Abstract of the United States, 2003.

Let $x = 0$ correspond to 1998, $x = 2$ correspond to 2000, and so on. Determine a quadratic model $P(x)$ for the data, using the regression feature of your calculator. Use the model to estimate $P(x)$ when $x = 9$, and interpret the result.

CHAPTER 3 SUMMARY

KEY TERMS & SYMBOLS	KEY CONCEPTS

3.1 Complex Numbers

Key Terms & Symbols

complex number system
complex number
real part
imaginary part
pure imaginary number
nonreal complex number
standard form of a complex number
complex conjugates (conjugates)

Key Concepts

IMAGINARY UNIT *i*

$$i = \sqrt{-i} \quad \text{or} \quad i^2 = -1$$

COMPLEX NUMBER

A number in the form $a + bi$, where a and b are real numbers and i is the imaginary unit, is called a complex number.

THE EXPRESSION $\sqrt{-a}$

If $a > 0$, then $\sqrt{-a} = i\sqrt{a}$.

COMPLEX CONJUGATE

The complex conjugate of $a + bi$ is $a - bi$.

OPERATIONS WITH COMPLEX NUMBERS

Addition: $(a + bi) + (c + di) = (a + c) + (b + d)i$

Subtraction: $(a + bi) - (c + di) = (a - c) + (b - d)i$

Multiplication: $(a + bi)(c + di) = (ac - bd) + (ad + bc)i$

Division: To divide complex numbers, multiply both the numerator and the denominator by the conjugate of the denominator.

3.2 Quadratic Functions and Graphs

Key Terms & Symbols

quadratic function
axis of symmetry
extreme points (extrema, extremum)
minimum point
minimum value
maximum point
maximum value

Key Concepts

CHARACTERISTICS OF THE GRAPH OF $P(x) = a(x - h)^2 + k$

The graph of $P(x) = a(x - h)^2 + k$ $(a \neq 0)$

(a) is a parabola with vertex (h, k) and vertical line $x = h$ as axis of symmetry;

(b) opens upward if $a > 0$ and downward if $a < 0$;

(c) is wider than the graph of $y = x^2$ if $0 < |a| < 1$ and narrower than the graph of $y = x^2$ if $|a| > 1$.

VERTEX FORMULA

The vertex of the graph of $P(x) = ax^2 + bx + c$ $(a \neq 0)$ is the point $\left(-\frac{b}{2a}, P\left(-\frac{b}{2a}\right)\right)$.

EXTREME VALUES

For the quadratic function defined by $P(x) = ax^2 + bx + c$,

(a) if $a > 0$, then the vertex (h, k) is called the **minimum point** of the graph. The **minimum value** of the function is $P(h) = k$.

(b) if $a < 0$, then the vertex (h, k) is called the **maximum point** of the graph. The **maximum value** of the function is $P(h) = k$.

3.3 Quadratic Equations and Inequalities

Key Terms & Symbols

quadratic equation in one variable
quadratic equation in standard form

Key Concepts

ZERO-PRODUCT PROPERTY

If a and b are complex numbers and $ab = 0$, then $a = 0$ or $b = 0$ or both.

SQUARE ROOT PROPERTY

The solution set of $x^2 = k$ is

(a) $\{\pm\sqrt{k}\}$ if $k > 0$ **(b)** $\{0\}$ if $k = 0$ **(c)** $\{\pm i\sqrt{|k|}\}$ if $k < 0$.

(continued)

KEY TERMS & SYMBOLS	KEY CONCEPTS
discriminant quadratic inequality in one variable	**QUADRATIC FORMULA** The solutions of the quadratic equation $ax^2 + bx + c = 0$, where $a \neq 0$, are given by $$\frac{-b \pm \sqrt{b^2 - 4ac}}{2a}.$$ **EFFECT OF THE DISCRIMINANT** If a, b, and c are real numbers, $a \neq 0$, then the complex solutions of $ax^2 + bx + c = 0$ are described as follows, based on the value of the discriminant, $b^2 - 4ac$. (see table below) **SOLVING A QUADRATIC INEQUALITY** ***Step 1*** Solve the corresponding quadratic equation. ***Step 2*** Identify the intervals determined by the solutions of the equation. ***Step 3*** Use a test value from each interval to determine which intervals are in the solution set.
3.4 Further Applications of Quadratic Functions and Models	**QUADRATIC MODEL** A quadratic function can be used to model data. A model can be determined by choosing an appropriate data point (h, k) as vertex and using another point to find the value of a in $f(x) = a(x - h)^2 + k$.
3.5 Higher-Degree Polynomial Functions and Graphs polynomial function leading coefficient dominating term cubic function quartic function turning points local maximum point local minimum point local maximum local minimum absolute maximum (minimum) point absolute maximum (minimum) value $x \rightarrow -\infty$ $x \rightarrow \infty$	**EXTREMA OF POLYNOMIAL FUNCTIONS** $P_1(x_1, y_1)$ $P_2(x_2, y_2)$ $P_3(x_3, y_3)$ P_1 is a local maximum point. P_2 is a local and absolute maximum point. P_3 is a local minimum point. **NUMBER OF TURNING POINTS** The number of turning points of the graph of a polynomial function of degree $n \geq 1$ is at most $n - 1$. **END BEHAVIOR OF GRAPHS OF POLYNOMIAL FUNCTIONS** $a > 0$, n odd; $a < 0$, n odd; $a > 0$, n even; $a < 0$, n even *(continued)*

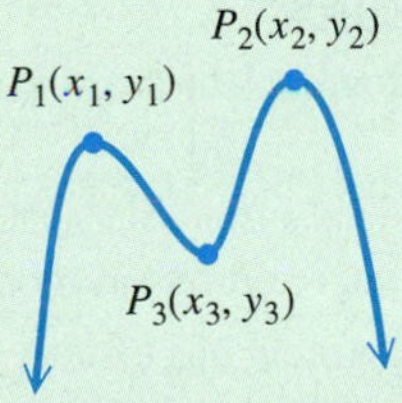

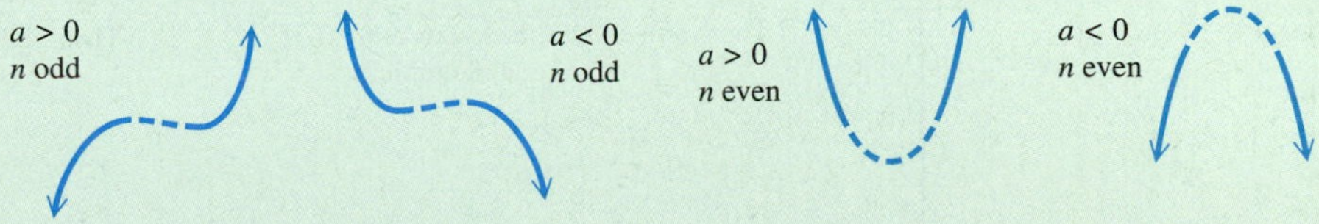

Value of $b^2 - 4ac$	Number of Solutions	Type of Solutions
Positive	Two	Real
Zero	One (a double solution)	Real
Negative	Two	Nonreal complex

KEY TERMS & SYMBOLS	KEY CONCEPTS
	NUMBER OF x-INTERCEPTS (REAL ZEROS) OF A POLYNOMIAL FUNCTION The graph of a polynomial function of degree n will have at most n x-intercepts (real zeros).
3.6 Topics in the Theory of Polynomial Functions (I) synthetic division	**INTERMEDIATE VALUE THEOREM** If $P(x)$ defines a polynomial function with only real coefficients, and if, for real numbers a and b, the values $P(a)$ and $P(b)$ are opposite in sign, then there exists at least one real zero between a and b. **DIVISION OF A POLYNOMIAL BY $x - k$** 1. If the degree n polynomial $P(x)$ is divided by $x - k$, then the quotient polynomial, $Q(x)$, has degree $n - 1$. 2. The remainder R is a constant (and may be 0). The complete quotient for $\frac{P(x)}{x-k}$ may be written as $$\frac{P(x)}{x-k} = Q(x) + \frac{R}{x-k}.$$ **REMAINDER THEOREM** If a polynomial $P(x)$ is divided by $x - k$, the remainder is equal to $P(k)$. **FACTOR THEOREM** The polynomial $x - k$ is a factor of the polynomial $P(x)$ if and only if $P(k) = 0$.
3.7 Topics in the Theory of Polynomial Functions (II) multiplicity of a zero variation in sign	**CONJUGATE ZEROS THEOREM** If $P(x)$ defines a polynomial function having only real coefficients, and if $a + bi$ is a zero of $P(x)$, then the conjugate $a - bi$ is also a zero of $P(x)$. **FUNDAMENTAL THEOREM OF ALGEBRA** Every function defined by a polynomial of degree 1 or more has at least one complex zero. **NUMBER OF ZEROS THEOREM** 1. A function defined by a polynomial of degree n has at most n distinct complex zeros. 2. A function defined by a polynomial of degree n has exactly n complex zeros if zeros of multiplicity m are counted m times. **BEHAVIOR OF THE GRAPH OF A POLYNOMIAL FUNCTION NEAR THE x-AXIS** The graph crosses the x-axis at $(c, 0)$ if c is a zero of odd multiplicity. The graph is tangent to the x-axis at $(c, 0)$ if c is a zero of even multiplicity. *(continued)*

KEY TERMS & SYMBOLS	KEY CONCEPTS
	RATIONAL ZEROS THEOREM Let $P(x) = a_nx^n + a_{n-1}x^{n-1} + \cdots + a_1x + a_0$, where $a_n \neq 0$, define a polynomial function with integer coefficients. If $\frac{p}{q}$ is a rational number written in lowest terms, and if $\frac{p}{q}$ is a zero of $P(x)$, then p is a factor of the constant term a_0 and q is a factor of the leading coefficient a_n. **DESCARTES' RULE OF SIGNS** Let $P(x)$ define a polynomial function with real coefficients and a nonzero constant term, with terms in descending powers of x. **(a)** The number of positive real zeros either equals the number of variations in sign occurring in the coefficients of $P(x)$ or is less than the number of variations by a positive even integer. **(b)** The number of negative real zeros either equals the number of variations in sign occurring in the coefficients of $P(-x)$ or is less than the number of variations by a positive even integer. **BOUNDEDNESS THEOREM** Let $P(x)$ define a polynomial function of degree $n \geq 1$ with real coefficients and with a positive leading coefficient. If $P(x)$ is divided synthetically by $x - c$, and **(a)** if $c > 0$ and all numbers in the bottom row of the synthetic division are nonnegative, then $P(x)$ has no zero greater than c. **(b)** if $c < 0$ and the numbers in the bottom row of the synthetic division alternate in sign (with 0 considered positive or negative, as needed), then $P(x)$ has no zero less than c.
3.8 Polynomial Equations and Inequalities: Further Applications and Models quadratic in form	**COMPLEX nTH ROOTS THEOREM** If n is a positive integer and k is a nonzero complex number, then the equation $x^n = k$ has *exactly* n complex roots. **POLYNOMIAL MODELS** Graphing calculators can be used to find the best-fitting polynomial function models for a set of data points.

CHAPTER 3 Review Exercises

Let $w = 17 - i$ and let $z = 1 - 3i$. Write each complex number in standard form.

1. $w + z$ **2.** $w - z$ **3.** wz

4. w^2 **5.** $\frac{1}{z}$ **6.** $\frac{w}{z}$

Consider the function defined by $P(x) = 2x^2 - 6x - 8$ for Exercises 7–16.

7. What is the domain of P?

8. Determine analytically the coordinates of the vertex of the graph.

9. Use an end-behavior diagram to describe the end behavior of the graph of P.

10. Determine analytically the x-intercepts, if any, of the graph of P.

11. Determine analytically the y-intercept of the graph of P.

12. What is the range of P?

13. Over what interval is the function increasing? Over what interval is it decreasing?

14. Give the solution set of each equation or inequality.
 (a) $2x^2 - 6x - 8 = 0$
 (b) $2x^2 - 6x - 8 > 0$
 (c) $2x^2 - 6x - 8 \leq 0$

15. The graph of P is shown here. Explain how the graph supports your solution sets in Exercise 14.

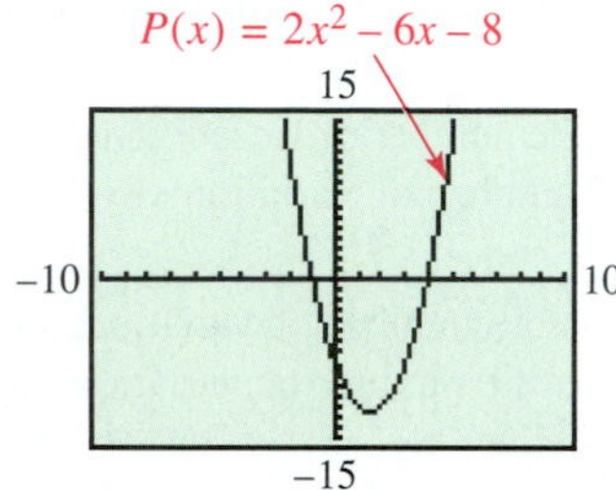

16. What is the equation of the axis of symmetry of the graph in Exercise 15?

In Exercises 17–20, consider the function defined by

$$P(x) = -2.64x^2 + 5.47x + 3.54.$$

17. Use the discriminant to explain how you can determine the number of x-intercepts the graph of P will have even before graphing it on your calculator.

18. Graph the function in the standard window of your calculator, and use the root-finding capabilities to solve the equation $P(x) = 0$. Express solutions as approximations to the nearest hundredth.

19. Use the capabilities of your calculator to find the coordinates of the vertex of the graph. Express coordinates to the nearest hundredth.

20. Verify analytically that your answer in Exercise 19 is correct.

21. The figure shows the graph of a quadratic function $y = f(x)$.

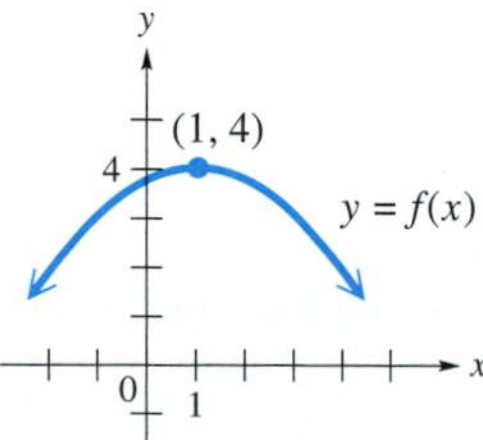

 (a) What is the maximum value of $f(x)$?
 (b) For what value of x is $f(x)$ a maximum?
 (c) How many real solutions are there of the equation $f(x) = 2$?
 (d) How many real solutions are there of the equation $f(x) = 6$?

(Modeling) Height of a Projectile *A projectile is fired vertically upward, and its height in feet after t seconds is given by*

$$s(t) = -16t^2 + 800t + 600.$$

Graph this function in the window $[0, 60]$ by $[-1000, 11{,}000]$, and use either analytic or graphical methods to answer Exercises 22–26.

22. From what height was the projectile fired?

23. After how many seconds will it reach its maximum height?

24. What is the maximum height it will reach?

25. Between what two times (in seconds, to the nearest tenth) will it be more than 5000 feet above the ground?

26. How long will the projectile be in the air? Give your answer to the nearest tenth of a second.

(Modeling) *Solve each problem.*

27. ***Volume of a Box*** A piece of cardboard is 3 times as long as it is wide. Equal-sized squares measuring 4 inches on each side are to be cut from the corners of the piece of cardboard, and the flaps will be folded up to form a box with an open top.
 (a) Determine a function V that describes the volume of the box as a function of x, where x is the original width in inches.
 (b) What should be the original dimensions of the piece of cardboard if the box is to have a volume of 2496 cubic inches? Solve this problem analytically.
 (c) Support the answer in part (b) graphically.

28. ***Concentration of Atmospheric CO_2*** In 1990, the International Panel on Climate Change (IPCC) stated that if current trends of burning fossil fuel and deforestation continue, then future amounts of atmospheric carbon dioxide in parts per million (ppm) will increase, as shown in the table.

Year	Carbon Dioxide
1990	353
2000	375
2075	590
2175	1090
2275	2000

Source: IPCC.

 (a) Plot the data. Let $x = 0$ correspond to 1990.
 (b) Find a function of the form $P(x) = a(x - h)^2 + k$ that models the data. Use $(0, 353)$ as the vertex and $(285, 2000)$ as another point to determine a.
 (c) Use the regression feature of a graphing calculator to find the best-fitting quadratic function Q for the data.

29. Use the intermediate value theorem to show that the polynomial function defined by $P(x) = -3x^3 - x^2 + 2x - 4$ has a real zero between -2 and -1.

30. Use synthetic division to find the quotient $Q(x)$ and the remainder R.

(a) $\dfrac{x^3 + x^2 - 11x - 10}{x - 3}$ **(b)** $\dfrac{3x^3 + 8x^2 + 5x + 10}{x + 2}$

In Exercises 31–34, use synthetic division to find P(2).

31. $P(x) = -x^3 + 5x^2 - 7x + 1$

32. $P(x) = 2x^3 - 3x^2 + 7x - 12$

33. $P(x) = 5x^4 - 12x^2 + 2x - 8$

34. $P(x) = x^5 + 4x^2 - 2x - 4$

35. If $P(x)$ is defined by a polynomial with real coefficients, and $7 + 2i$ is a zero of the function, what other complex number must be a zero?

In Exercises 36–39, find a polynomial function with real coefficients and of least possible degree having the given zeros.

36. $-1, 4, 7$

37. $8, 2, 3$

38. $\sqrt{3}, -\sqrt{3}, 2, 3$

39. $-2 + \sqrt{5}, -2 - \sqrt{5}, -2, 1$

40. Is -1 a zero of $P(x) = 2x^4 + x^3 - 4x^2 + 3x + 1$?

41. Is $x + 1$ a factor of $P(x) = x^3 + 2x^2 + 3x + 2$?

42. Find a polynomial function P with real coefficients of degree 4 with 3, 1, and $-1 - 3i$ as zeros and for which $P(2) = -36$.

43. Find all zeros of $P(x) = x^4 - 3x^3 - 8x^2 + 22x - 24$, given that $1 - i$ is a zero.

44. Find all zeros of $P(x) = 2x^4 - x^3 + 7x^2 - 4x - 4$, given that 1 and $2i$ are zeros.

45. Find all rational zeros of

$$P(x) = 3x^5 - 4x^4 - 26x^3 - 21x^2 - 14x + 8$$

by first listing the possible rational zeros based on the rational zeros theorem.

46. Use Descartes' rule of signs to determine the possible number of positive real zeros and negative real zeros of $P(x) = 3x^4 + x^3 - x^2 - 2x - 1$.

47. Use the boundedness theorem to show that

$$P(x) = 2x^4 + 3x^3 - 5x^2 + 8x - 10$$

has no real zero greater than 2 and no real zero less than -4.

In Exercises 48–52, consider the function defined by

$$P(x) = x^3 - 2x^2 - 4x + 3.$$

48. Graph the function in an appropriate window to show a comprehensive graph. Based on the graph, how many real solutions does the equation $x^3 - 2x^2 - 4x + 3 = 0$ have? Use your calculator to find the real root that is an integer.

49. Use your answer in Exercise 48 along with synthetic division to factor $x^3 - 2x^2 - 4x + 3$ so that one factor is linear and the other is quadratic.

50. Find exact values of any remaining zeros analytically.

51. Use your calculator to support your answer in Exercise 50.

52. Give the solution set of each inequality, using exact values.

(a) $x^3 - 2x^2 - 4x + 3 > 0$

(b) $x^3 - 2x^2 - 4x + 3 \leq 0$

53. Use an analytic method to find all solutions of the equation $x^3 + 2x^2 + 5x = 0$. Then, without graphing, give the exact values of all x-intercepts of the graph.

54. For the graph of $P(x) = x^4 - 5x^3 + x^2 + 21x - 18$, suppose you know that all zeros of P are integers and each zero has multiplicity 1 or 2. Give the factored form of $P(x)$. (*Hint:* Once you have determined your answer, graph the factored form to see if it matches the graph shown here.)

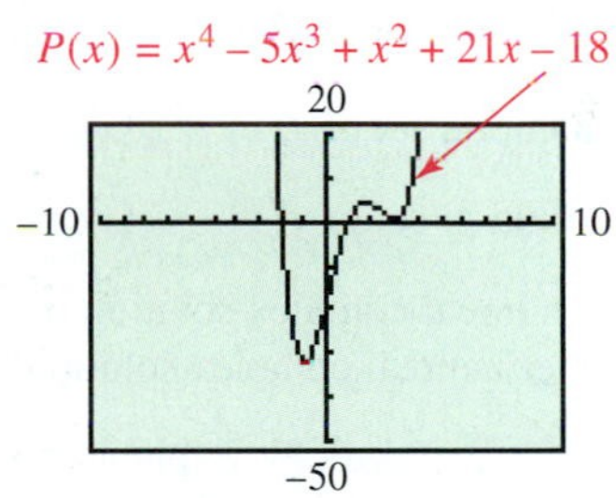

The x-intercepts are −2, 1, and 3.

55. The table shows several ordered pairs that lie on the graph of Y_1. Use one of the ordered pairs to find a factor of Y_1. Then, find the exact values of all zeros of Y_1, and write Y_1 as a product of three linear factors.

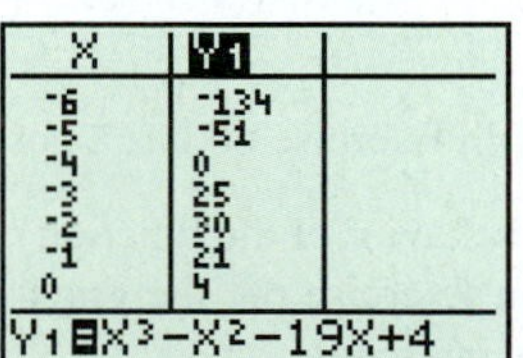

X	Y1
-6	-134
-5	-51
-4	0
-3	25
-2	30
-1	21
0	4

Y1=X³−X²−19X+4

56. Use the intermediate value theorem to determine which two x-values from the table have a zero between them. Explain.

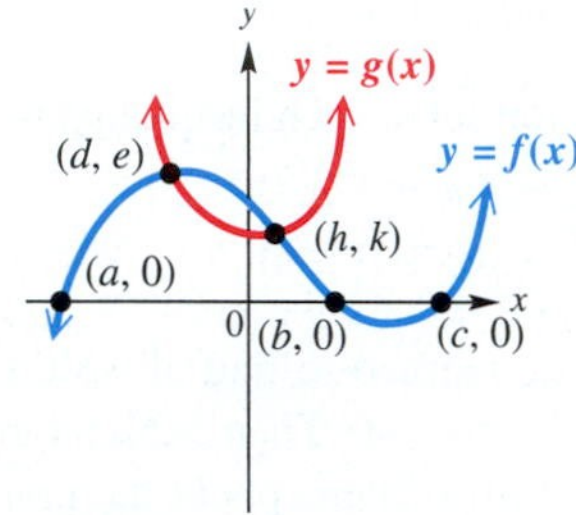

X	Y1
.6	.784
.7	.157
.8	-.512
.9	-1.229
1	-2
1.1	-2.831
1.2	-3.728

Y1 = -X³-5X+4

Concept Check *Comprehensive graphs of polynomial functions f and g are shown here. They have only real coefficients. Answer Exercises 57–65 based on the graphs.*

57. Is the degree of g even or odd?

58. Is the degree of f even or odd?

59. Is the leading coefficient of f positive or negative?

60. How many real solutions does $g(x) = 0$ have?

61. Express the solution set of $f(x) < 0$ in interval form.

62. What is the solution set of $f(x) > g(x)$?

63. What is the solution set of $f(x) - g(x) = 0$?

64. If $r + pi$ is a nonreal complex solution of $g(x) = 0$, what must be another nonreal complex solution?

65. Suppose that f is of degree 3. Explain why f cannot have nonreal complex zeros.

Concept Check *Answer* true *or* false *to each statement in Exercises 66–71.*

66. The function f defined by $f(x) = 3x^7 - 8x^6 + 9x^5 + 12x^4 - 18x^3 + 26x^2 - x + 500$ has eight x-intercepts.

67. The function f in Exercise 66 may have up to six local extrema.

68. The function f in Exercise 66 has a positive y-intercept.

69. Based on end behavior of the function f in Exercise 66 and your answer in Exercise 68, the graph must have at least one negative x-intercept.

70. If a polynomial function of even degree has a positive leading coefficient and a negative y-intercept, it must have at least two real zeros.

71. Because $-\frac{1}{2} + i\frac{\sqrt{3}}{2}$ is a nonreal complex zero of

$$f(x) = x^2 + x + 1,$$

another zero must be $\frac{1}{2} + i\frac{\sqrt{3}}{2}$.

Graph the function defined by

$$P(x) = -2x^5 + 15x^4 - 21x^3 - 32x^2 + 60x$$

in the window $[-8, 8]$ *by* $[-100, 200]$ *to obtain a comprehensive graph. Then, use your calculator and the concepts of this chapter to answer Exercises 72–76.*

72. How many local maxima does this function have?

73. One local minimum point lies on the x-axis and has an integer as its x-value. What are its coordinates?

74. The greatest x-intercept is 5. Therefore, $x - 5$ is a factor of $P(x)$. Use synthetic division to find the quotient polynomial $Q(x)$ obtained when $P(x)$ is divided by $x - 5$.

75. What is the range of P?

76. The graph has a local minimum point with a negative x-value. Use your calculator to find its coordinates. Express them to the nearest hundredth.

77. Solve the equation $3x^3 + 2x^2 - 21x - 14 = 0$ analytically for all complex solutions, giving exact values in your solution set. Then, graph the left side of the equation, and support the real solutions.

78. Consider the polynomial function defined by

$$\begin{aligned} P(x) &= -x^4 + 3x^3 + 3x^2 - 7x - 6 \\ &= (-x + 2)(x - 3)(x + 1)^2. \end{aligned}$$

Graph the polynomial by hand as explained in Section 3.7, and solve each equation or inequality.

(a) $P(x) = 0$ **(b)** $P(x) > 0$

(c) $P(x) < 0$

(Modeling) *Solve each problem.*

79. *Dimensions of a Cube* After a 2-inch slice is cut off the top of a cube, the resulting solid has volume 32 cubic inches. Find the dimensions of the original cube.

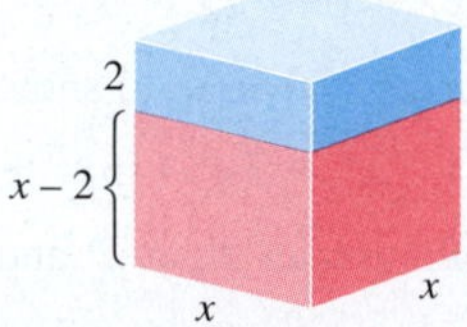

80. ***Height of a Projected Object*** If air resistance is neglected, an object projected straight upward with an initial velocity of 40 meters per second from a height of 50 meters will be at a height s meters after t seconds, where

$$s(t) = -4.9t^2 + 40t + 50.$$

(a) What are the restrictions on t?
(b) Graph the function s in a suitable window.
(c) Use the graph to find the time at which the object reaches its highest point.
(d) Use the graph to find the maximum height reached by the object.
(e) Write the equation that allows you to find the time at which the object reaches the ground. Then, solve the equation, using the method of your choice. Give your answer to the nearest hundredth of a second.

81. ***Military Personnel on Active Duty*** The number of military personnel on active duty in the United States during the period from 1990 through 2004 is approximated by the cubic model

$$y = -.1885x^3 + 10.617x^2 - 157.075x + 2080.8,$$

where $x = 0$ corresponds to 1990 and y is in thousands. Based on this model, how many military personnel were on active duty in 2001? (*Source:* U.S. Department of Defense.)

82. ***Medicare Beneficiary Spending*** Out-of-pocket spending projections for a typical Medicare beneficiary as a share of his or her income are given in the table.

Year	Percent of Income
1998	18.6
2000	19.3
2005	21.7
2010	24.7
2015	27.5
2020	28.3
2025	28.6

Source: Urban Institute's Analysis of 1998 Medicare Trustees' Report.

Let $x = 0$ represent 1990, so that $x = 8$ represents 1998. Use a graphing calculator to do the following.
(a) Plot the data.
(b) Find a quadratic function to model the data.
(c) Find a cubic function to model the data.
(d) Graph each function in the same viewing window as the data.
(e) Compare the two functions. Which is a better fit to the data?
(f) Would a linear model be appropriate for these data? Why or why not?

CHAPTER 3 Test

1. Perform each operation with complex numbers. Give answers in standard form.

(a) $(8 - 7i) - (-12 + 2i)$ **(b)** $\dfrac{11 + 10i}{2 + 3i}$

(c) Simplify i^{65}. **(d)** $2i(3 - i)^2$

2. For the function defined by $P(x) = -2x^2 - 4x + 6$, do the following.
(a) Find the vertex analytically.
(b) Give a comprehensive graph, and use a calculator to support your result in part (a).
(c) Find the zeros of P and support your result, using a graph for one zero and a table for the other.
(d) Find the y-intercept analytically.
(e) State the domain and range of P.
(f) Give the interval over which the function is increasing and the interval over which it is decreasing.

3. (a) Solve the quadratic equation $3x^2 + 3x - 2 = 0$ analytically. Give solutions in exact form.
(b) Graph $P(x) = 3x^2 + 3x - 2$ with a calculator. Use your results in part (a) along with the graph to give the solution set of each inequality. Express endpoints of intervals in exact form.
(i) $P(x) < 0$
(ii) $P(x) \geq 0$

4. ***Dimensions of a Box*** The width of a rectangular box is 3 times its height, and its length is 11 inches more than its height. Find the dimensions of the box if its volume is 720 cubic inches.

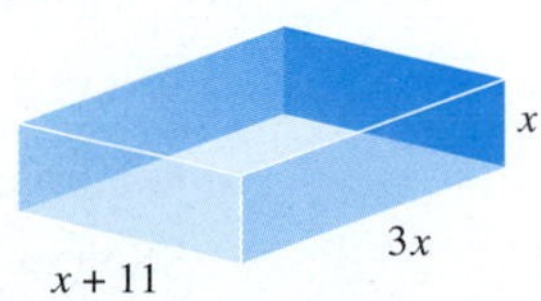

5. ***(Modeling) U.S. Air Passengers*** The table gives the number of air passengers (in millions) in the United States for selected years.

Year	1990	1992	1994	1996	1998	2000
Passengers	470	475	530	580	620	670

Source: U.S. Bureau of Transportation Statistics.

(a) Plot the data. Let $x = 0$ correspond to the year 1990, $x = 2$ correspond to 1992, and so on.

(b) Find a function of the form $f(x) = a(x - h)^2 + k$ that models the data. Let $(0, 470)$ represent the vertex, and use $(10, 670)$ to determine the value of a.

(c) Use the regression feature of your graphing calculator to find the best-fitting quadratic function g for the data. Graph both functions g and f from part (b) in the same window as the data points. Which function is the better fit?

6. **(a)** Given that the function defined by

$$f(x) = x^6 - 5x^5 + 3x^4 + x^3 + 40x^2 - 24x - 72$$

has 3 as a zero of multiplicity 2, 2 as a single zero, and -1 as a single zero, find all other zeros of f.

(b) Give an end-behavior diagram. Use the information from part (a) to sketch the graph of f by hand.

7. Perform the following for the function defined by

$$f(x) = 4x^4 - 21x^2 - 25.$$

(a) Find all zeros analytically.

(b) Find a comprehensive graph of f, and support the real zeros found in part (a).

(c) Discuss the symmetry of the graph of f.

(d) Use the graph and the results of part (a) to find the solution set of each inequality.

(i) $f(x) \geq 0$

(ii) $f(x) < 0$

8. Perform the following for the function defined by

$$f(x) = 3x^4 + 5x^3 - 35x^2 - 55x + 22.$$

(a) List all possible rational zeros.

(b) Find all rational zeros.

(c) Use the intermediate value theorem to show that there must be a zero between 3 and 4.

(d) Use Descartes' rule of signs to determine the possible number of positive zeros and negative zeros.

(e) Use the boundedness theorem to show that there is no zero less than -5 and no zero greater than 4.

9. **(a)** Use only a graphical method to find the real solutions, to the nearest thousandth if necessary, of

$$x^5 - 4x^4 + 2x^3 - 4x^2 + 6x - 1 = 0.$$

(b) Based on the degree and your answer in part (a), how many nonreal complex solutions does the equation have?

10. ***(Modeling) U.S. Air Passengers*** Refer to the data found in Exercise 5. Again, letting $x = 0$ represent 1990, $x = 2$ represent 1992, and so on, do the following.

(a) Use a graphing calculator to find the best-fitting cubic function for the data.

(b) Repeat part (a) for a quartic function.

(c) Graph the functions of parts (a) and (b) over the data points.

(d) Based on the growth through the 1990s, which function gives the best estimate of the number of passengers in 2001?

CHAPTER 3 PROJECT

Creating a Social Security Polynomial

Each Social Security number is unique to one person. In this activity, you can construct a Social Security polynomial. Let's agree that the polynomial function will be defined as follows, where a_i represents the ith digit in the Social Security number that you construct:

$$\text{SSN}(x) = (x - a_1)(x + a_2)(x - a_3)(x + a_4)(x - a_5)(x + a_6)(x - a_7)(x + a_8)(x - a_9).$$

For example, if the Social Security number is 539-58-0954, the polynomial is

$$\text{SSN}(x) = (x - 5)(x + 3)(x - 9)(x + 5)(x - 8)(x + 0)(x - 9)(x + 5)(x - 4).$$

A comprehensive graph of this function is shown in Figure A. In Figure B, we show a screen obtained by zooming in on the positive zeros, as the comprehensive graph does not show the local behavior well in this region.

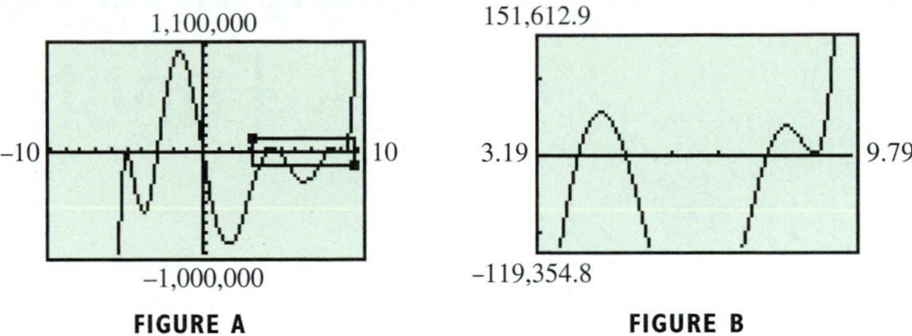

FIGURE A **FIGURE B**

Activities

(These activities can be performed by individual students, or by students in groups, where the group chooses digits for a Social Security number to be used.)

1. Construct your Social Security polynomial SSN(x).
2. What is the degree of this polynomial? What is the dominating term?
3. Construct an end-behavior diagram.
4. List all zeros of multiplicity 1.
5. List all zeros of multiplicity 2.
6. List all zeros of multiplicity 3 or higher.
7. Using your graphing calculator, find a comprehensive graph of SSN(x). It should include all local extreme points and all intercepts, and should illustrate end behavior. Sketch the graph on paper, or print it out, using the appropriate computer software. Regardless of which type of graph you submit, give the values of Xmin, Xmax, Xscl, Ymin, Ymax, and Yscl. (*Hint:* Ymin and Ymax will be quite large in comparison to values you might be accustomed to using.) You may need to include "enlargements" of certain parts of the graph, because they may not show enough detail in one window.
8. List the coordinates of each local maximum.
9. List the coordinates of each local minimum.
10. What are the domain and the range of SSN(x)?
11. List the intervals of the domain over which SSN(x) is increasing.
12. List the intervals of the domain over which SSN(x) is decreasing.
13. What must be true of the Social Security number that leads to a polynomial whose graph passes through the origin?

4

Rational, Power, and Root Functions

TRAFFIC CONGESTION IS one of the biggest hassles in modern life, and it's getting worse each year. Although the population of the United States has increased by about 26% since 1982, the time spent waiting in traffic has increased by 236%. Today, the average driver spends over 51 hours a year stuck in traffic. Traffic congestion costs Americans $70 billion a year due to wasted fuel and lost time from work. The following table lists the five cities with the longest commutes in terms of annual amount of extra travel time to and from work because of traffic.

City	Hours
Los Angeles	90
San Francisco	68
Denver	64
Miami	63
Chicago	61

Source: Texas Transportation Institute.

Traffic is subject to a *nonlinear effect.* According to Joe Sussman, an engineer from the Massachusetts Institute of Technology, you can put more cars on the road up to a point. Then, if traffic intensity increases even slightly, congestion and waiting time increase dramatically. This nonlinear effect can also occur when you are waiting to enter a parking ramp. In Section 4.3, we use a rational function to model this phenomenon.

Source: Longman, Phillip J., "American Gridlock," *U.S. News & World Report,* May 28, 2001.

Chapter Outline

4.1 Rational Functions and Graphs

The Reciprocal Function ■ The Rational Function Defined by $f(x) = \frac{1}{x^2}$

A **rational expression** is a fraction that is the quotient of two polynomials. A function defined by a rational expression is called a *rational function.*

X	Y1
-1	-1
-.1	-10
-.01	-100
-.001	-1000
-1E-4	-10000
-1E-5	-1E5
-1E-6	-1E6

$Y_1 = 1/X$

As X approaches 0 from the left, $Y_1 = \frac{1}{X}$ approaches $-\infty$.

X	Y1
1	1
.1	10
.01	100
.001	1000
1E-4	10000
1E-5	100000
1E-6	1E6

$Y_1 = 1/X$

As X approaches 0 from the right, $Y_1 = \frac{1}{X}$ approaches ∞.

FIGURE 1

Rational Function

A function f of the form $\frac{p}{q}$ defined by

$$f(x) = \frac{p(x)}{q(x)},$$

where $p(x)$ and $q(x)$ are polynomials, with $q(x) \neq 0$, is called a **rational function.**

Some examples of rational functions are

$$f(x) = \frac{1}{x}, \quad f(x) = \frac{x+1}{2x^2+5x-3}, \quad \text{and} \quad f(x) = \frac{3x^2-3x-6}{x^2+8x+16}.$$

Rational functions

Since any values of x such that $q(x) = 0$ are excluded from the domain of a rational function, this type of function often has a *discontinuous* graph (a graph that has one or more breaks in it).

The Reciprocal Function

The simplest rational function with a variable denominator is the **reciprocal function,** defined by

$$f(x) = \frac{1}{x}.$$

Reciprocal function

The domain of this function is the set of all real numbers except 0. The number 0 cannot be used as a value of x, but it is helpful to find values of $f(x)$ for some values of x close to 0. We can use the table feature of a graphing calculator to do this. The tables in Figure 1 suggest that $|f(x)|$ gets larger and larger as x gets closer and closer to 0. Symbolically,

$$|f(x)| \to \infty \quad \text{as} \quad x \to 0.$$

(The symbol $x \to 0$ means that x approaches 0, without necessarily ever being equal to 0.) Since x cannot equal 0, the graph of $f(x) = \frac{1}{x}$ will never intersect the vertical line $x = 0$. This line is called a **vertical asymptote.**

On the other hand, as $|x|$ gets larger and larger, the values of $f(x) = \frac{1}{x}$ get closer and closer to 0, as shown in the tables in Figure 2. Letting $|x|$ get larger and larger without bound (written $|x| \to \infty$) causes the graph of $f(x) = \frac{1}{x}$ to move closer and closer to the horizontal line $y = 0$. This line is called a **horizontal asymptote.**

X	Y1
1	1
10	.1
100	.01
1000	.001
10000	1E-4
100000	1E-5
1E6	1E-6

$Y_1 = 1/X$

As X approaches ∞, $Y_1 = \frac{1}{X}$ approaches 0 through positive values.

X	Y1
-1	-1
-10	-.1
-100	-.01
-1000	-.001
-10000	-1E-4
-1E5	-1E-5
-1E6	-1E-6

$Y_1 = 1/X$

As X approaches $-\infty$, $Y_1 = \frac{1}{X}$ approaches 0 through negative values.

FIGURE 2

The graph and important features of $f(x) = \frac{1}{x}$ are shown in Figure 3.

FUNCTION CAPSULE

RECIPROCAL FUNCTION $f(x) = \dfrac{1}{x}$

Domain: $(-\infty, 0) \cup (0, \infty)$ Range: $(-\infty, 0) \cup (0, \infty)$

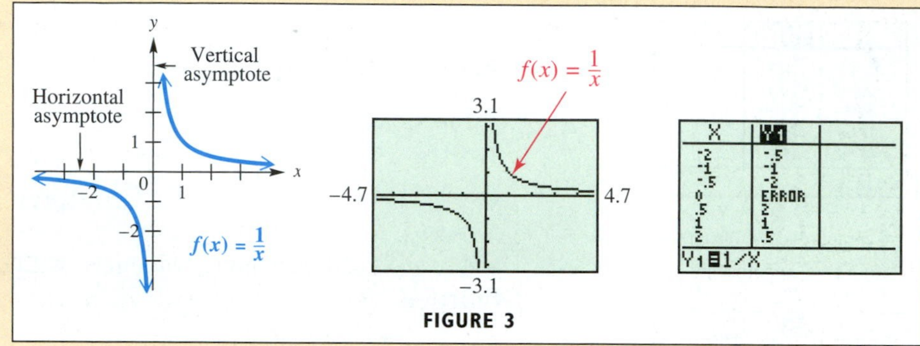

FIGURE 3

- $f(x) = \frac{1}{x}$ decreases on the intervals $(-\infty, 0)$ and $(0, \infty)$.
- It is discontinuous at $x = 0$.
- The y-axis is a vertical asymptote, and the x-axis is a horizontal asymptote.
- It is an odd function, and its graph is symmetric with respect to the origin.

EXAMPLE 1 Graphing a Rational Function

Graph $y = -\dfrac{2}{x}$. Give the domain and range.

Solution The expression $-\frac{2}{x}$ can be written as either $2\left(\frac{1}{-x}\right)$ or $-2\left(\frac{1}{x}\right)$, indicating that the graph may be obtained by stretching the graph of $y = \frac{1}{x}$ vertically by a factor of 2 and reflecting it across either the y-axis or x-axis. The x- and y-axes remain the horizontal and vertical asymptotes. The domain and range are still $(-\infty, 0) \cup (0, \infty)$. See Figure 4.

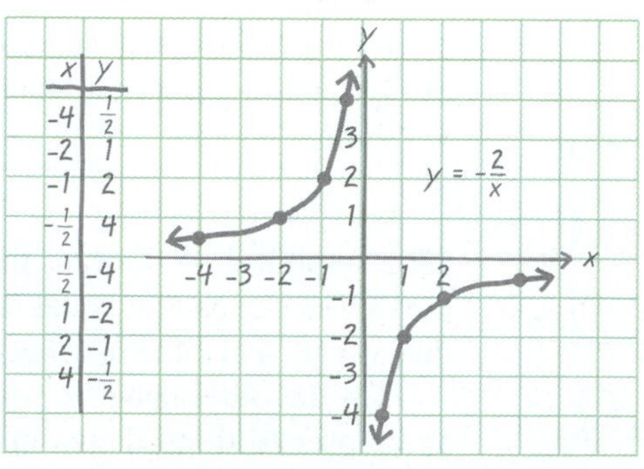

FIGURE 4

GCM **EXAMPLE 2** **Graphing a Rational Function**

Graph $y = \frac{2}{x + 1}$. Give the domain and range.

Analytic Solution

The expression $\frac{2}{x + 1}$ can be written as $2\left(\frac{1}{x + 1}\right)$, indicating that the graph may be obtained by shifting the graph of $y = \frac{1}{x}$ to the left 1 unit and stretching it vertically by a factor of 2. The graph is shown in Figure 5.

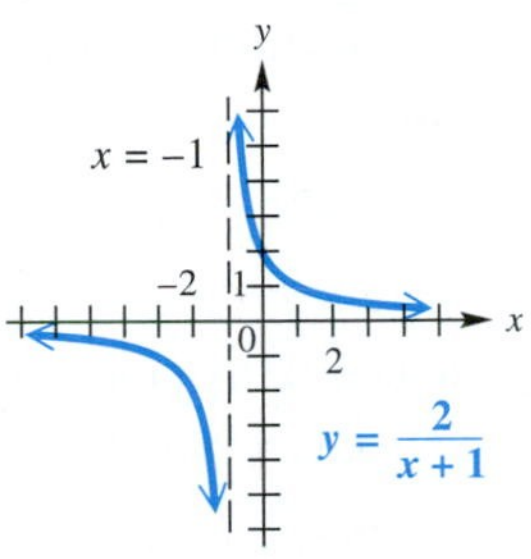

FIGURE 5

The horizontal shift affects the domain, which is now $(-\infty, -1) \cup (-1, \infty)$. The line $x = -1$ is the vertical asymptote, and the line $y = 0$ (the x-axis) remains the horizontal asymptote. The range is still $(-\infty, 0) \cup (0, \infty)$.

Graphing Calculator Solution

If a calculator is in *connected mode,* the graph it generates of a rational function may show a vertical line for each vertical asymptote. While this may be interpreted as a graph of the asymptote, *dot mode* produces a more realistic graph. See Figure 6.

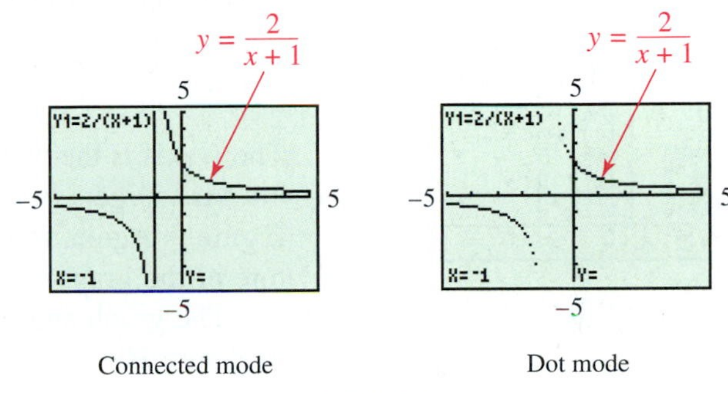

FIGURE 6

We can eliminate the vertical line in the connected mode graph by changing the viewing window for x slightly to $[-5, 3]$. This works because it places the asymptote exactly in the center of the screen. You may need to try various modes and windows to obtain an accurate calculator graph of a rational function. ■

EXAMPLE 3 **Graphing a Rational Function**

Use division and translations to graph $y = \frac{x + 2}{x + 1}$. Give the domain and range.

Solution First, we rewrite the given expression by dividing $x + 1$ into $x + 2$.

$$\begin{array}{r} 1 \\ x + 1 \overline{)\, x + 2} \\ \underline{x + 1} \\ 1 \end{array}$$

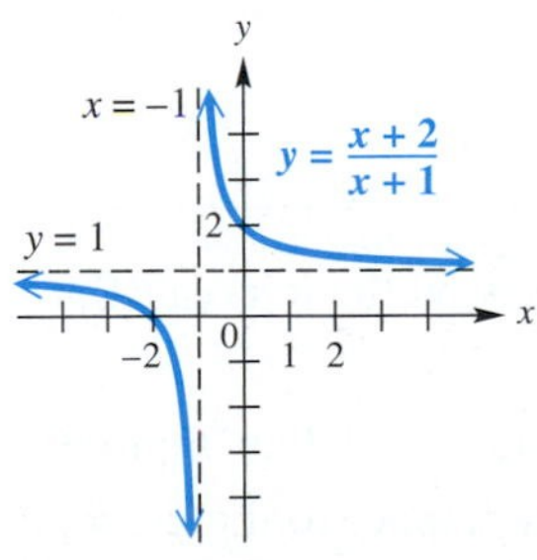

FIGURE 7

The rational expression $\frac{x + 2}{x + 1}$ is equal to $1 + \frac{1}{x + 1}$. Note that the expression $1 + \frac{1}{x + 1}$ can be obtained from the expression $\frac{1}{x}$ by replacing x with $x + 1$ and then adding 1. Thus, the graph of $y = 1 + \frac{1}{x + 1}$ can be obtained by shifting the graph of $y = \frac{1}{x}$ left 1 unit and upward 1 unit. The resulting graph is shown in Figure 7. The domain is $(-\infty, -1) \cup (-1, \infty)$ and the range is $(-\infty, 1) \cup (1, \infty)$. ■

The Rational Function Defined by $f(x) = \frac{1}{x^2}$

X	Y1
-1	1
-.1	100
-.01	10000
-.001	1E6
-1E-4	1E8
-1E-5	1E10
-1E-6	1E12

Y1=1/X²

As X approaches 0 from the left, $Y_1 = \frac{1}{X^2}$ approaches ∞.

X	Y1
1	1
.1	100
.01	10000
.001	1E6
1E-4	1E8
1E-5	1E10
1E-6	1E12

Y1=1/X²

As X approaches 0 from the right, $Y_1 = \frac{1}{X^2}$ approaches ∞.

FIGURE 8

The rational function defined by

$$f(x) = \frac{1}{x^2} \qquad \text{Rational function}$$

has domain $(-\infty, 0) \cup (0, \infty)$. We can use the table feature of a graphing calculator to examine values of $f(x)$ for some x-values close to 0. See Figure 8.

The tables suggest that $f(x)$ gets larger and larger as x gets closer and closer to 0. Notice that as x approaches 0 from *either* side, function values are all positive and there is symmetry with respect to the y-axis. Thus,

$$f(x) \rightarrow \infty \quad \text{as} \quad x \rightarrow 0.$$

The y-axis is the vertical asymptote.

As $|x|$ gets larger and larger, $f(x)$ approaches 0, as suggested by the tables in Figure 9. Again, function values are all positive. The x-axis is the horizontal asymptote of the graph.

The graph and important features of $f(x) = \frac{1}{x^2}$ are summarized here and shown in Figure 10.

X	Y1
1	1
10	.01
100	1E-4
1000	1E-6
10000	1E-8
100000	1E-10
1E6	1E-12

Y1=1/X²

As X approaches ∞, $Y_1 = \frac{1}{X^2}$ approaches 0 through positive values.

X	Y1
-1	1
-10	.01
-100	1E-4
-1000	1E-6
-10000	1E-8
-1E5	1E-10
-1E6	1E-12

Y1=1/X²

As X approaches $-\infty$, $Y_1 = \frac{1}{X^2}$ approaches 0 through positive values.

FIGURE 9

FUNCTION CAPSULE

RATIONAL FUNCTION $\quad f(x) = \frac{1}{x^2}$

Domain: $(-\infty, 0) \cup (0, \infty)$ $\qquad$ Range: $(0, \infty)$

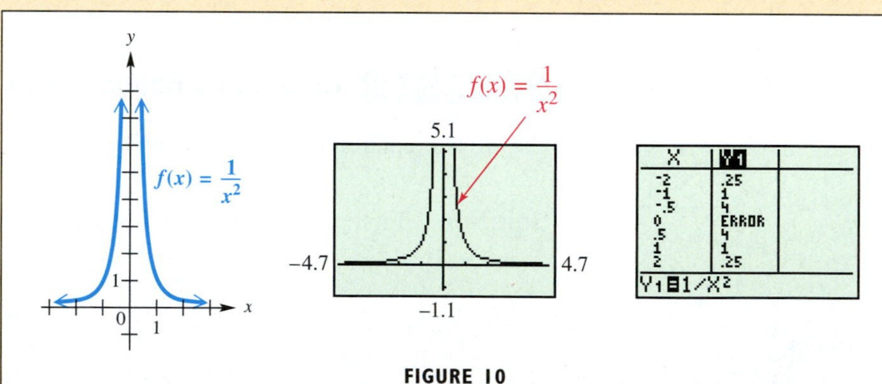

FIGURE 10

- $f(x) = \frac{1}{x^2}$ increases on the interval $(-\infty, 0)$ and decreases on the interval $(0, \infty)$.
- It is discontinuous at $x = 0$.
- The y-axis is a vertical asymptote, and the x-axis is a horizontal asymptote.
- It is an even function, and its graph is symmetric with respect to the y-axis.

EXAMPLE 4 **Graphing a Rational Function**

Graph $y = \frac{1}{(x-1)^2}$. Give the domain and range.

Analytic Solution

The equation $y = \frac{1}{(x-1)^2}$ is equivalent to

$$y = f(x-1), \quad \text{where} \quad f(x) = \frac{1}{x^2}.$$

This indicates that the graph will be shifted right 1 unit. The vertical asymptote is now $x = 1$, but the horizontal asymptote remains $y = 0$. See Figure 11. The horizontal shift affects the domain, which is now $(-\infty, 1) \cup (1, \infty)$. The range remains $(0, \infty)$.

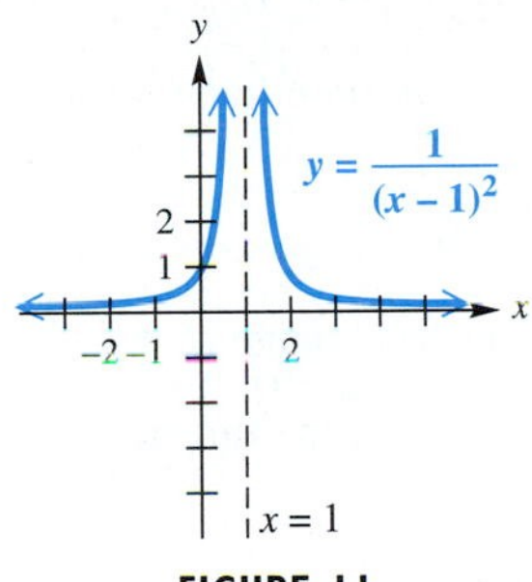

FIGURE 11

Graphing Calculator Solution

The graph of

$$y = \frac{1}{(x-1)^2}$$

is shown in a *decimal window,* or *friendly window,* in Figure 12. In a decimal window, the calculator does not produce an undesired vertical line at $x = 1$. Note that asymptotes are not part of the graph of a rational function. They simply aid us in graphing. You may need to try various modes and windows to obtain an accurate calculator graph of a rational function.

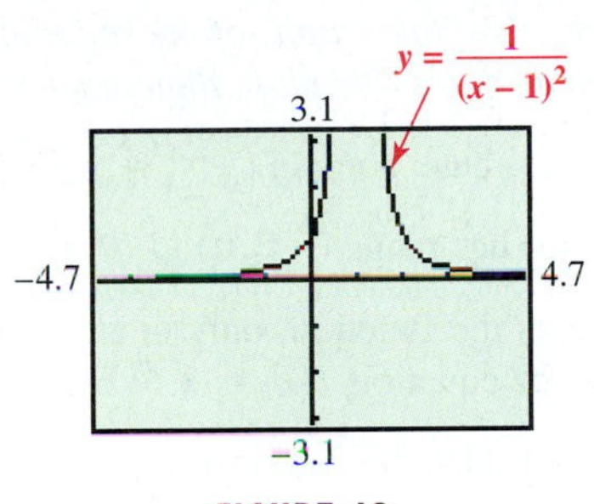

FIGURE 12

EXAMPLE 5 **Graphing a Rational Function**

Graph $y = \frac{1}{(x+2)^2} - 1$. Give the domain and range.

Solution The equation $y = \frac{1}{(x+2)^2} - 1$ is equivalent to

$$y = f(x+2) - 1, \quad \text{where} \quad f(x) = \frac{1}{x^2}.$$

This indicates that the graph will be shifted left 2 units and downward 1 unit. The horizontal shift affects the domain, which is now $(-\infty, -2) \cup (-2, \infty)$, while the vertical shift affects the range, now $(-1, \infty)$. The vertical asymptote has equation $x = -2$, and the horizontal asymptote has equation $y = -1$. The graph is shown in Figure 13 and is a translation of $y = \frac{1}{x^2}$.

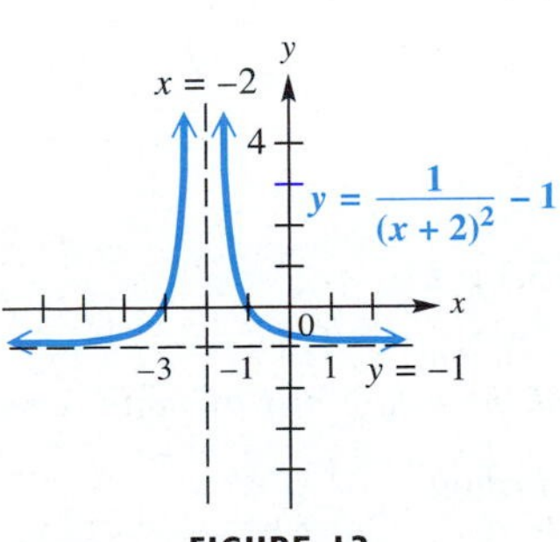

FIGURE 13

4.1 Exercises

Concept Check *Provide a short answer to each question.*

1. What is the domain of $f(x) = \frac{1}{x}$? What is its range?

2. What is the domain of $f(x) = \frac{1}{x^2}$? What is its range?

3. What is the interval over which $f(x) = \frac{1}{x}$ increases? decreases? is constant?

4. What is the interval over which $f(x) = \frac{1}{x^2}$ increases? decreases? is constant?

5. What is the equation of the vertical asymptote of the graph of $y = \frac{1}{x-3} + 2$? of the horizontal asymptote?

6. What is the equation of the vertical asymptote of the graph of $y = \frac{1}{(x+2)^2} - 4$? of the horizontal asymptote?

7. Is $f(x) = \frac{1}{x^2}$ an even or odd function? What symmetry does its graph exhibit?

8. Is $f(x) = \frac{1}{x}$ an even or odd function? What symmetry does its graph exhibit?

Concept Check *Use the graphs of the rational functions in A–D below to answer each question in Exercises 9–16. Give all possible answers, as there may be more than one correct choice.*

9. Which choices have domain $(-\infty, 3) \cup (3, \infty)$?

10. Which choice has range $(-\infty, 3) \cup (3, \infty)$?

11. Which choice has range $(-\infty, 0) \cup (0, \infty)$?

12. Which choices have range $(0, \infty)$?

13. If f represents the function, only one choice has a single solution of the equation $f(x) = 3$. Which one is it?

14. What is the range of the function in A?

15. Which choices have the x-axis as a horizontal asymptote?

16. Which choices are symmetric with respect to a vertical line?

A.

B.

C.

D.

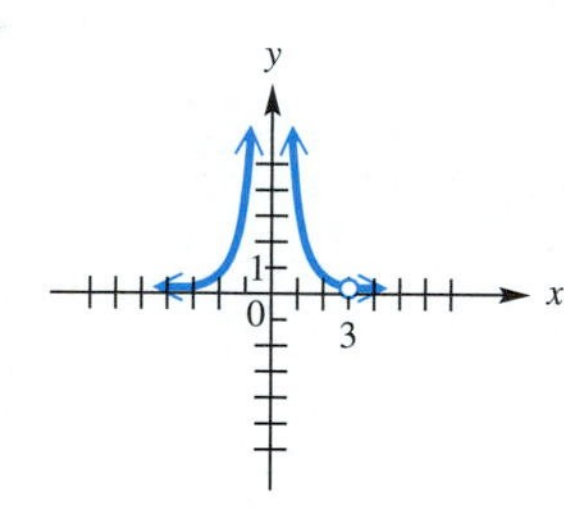

Graph each rational function in the given windows, and determine which window gives the most accurate depiction of the graph.

17. $f(x) = \frac{1}{x+5}$
   A. $[-4.7, 4.7]$ by $[-3.1, 3.1]$
   B. $[-14.4, 4.4]$ by $[0, 5]$
   C. $[-9.4, 9.4]$ by $[-3.1, 3.1]$

18. $f(x) = \frac{1}{x^2} + 6$
   A. $[-4.7, 4.7]$ by $[0, 12.4]$
   B. $[-4.7, 4.7]$ by $[-6.2, 6.2]$
   C. $[-4.7, 4.7]$ by $[-3.1, 3.1]$

Explain how the graph of f can be obtained from the graph of $y = \frac{1}{x}$ or $y = \frac{1}{x^2}$. Draw a sketch of the graph of f by hand. Then generate an accurate depiction of the graph with a graphing calculator. Finally, give the domain and range.

19. $f(x) = \frac{2}{x}$

20. $f(x) = -\frac{3}{x}$

21. $f(x) = \frac{1}{x+2}$

22. $f(x) = \frac{1}{x-3}$

23. $f(x) = \frac{1}{x} + 1$ **24.** $f(x) = \frac{1}{x} - 2$ **25.** $f(x) = \frac{1}{x-1} + 1$ **26.** $f(x) = \frac{2}{x+2} - 1$

27. $f(x) = \frac{1}{x^2} - 2$ **28.** $f(x) = \frac{1}{x^2} + 3$ **29.** $f(x) = -\frac{2}{x^2}$ **30.** $f(x) = \frac{.5}{x^2}$

31. $f(x) = \frac{1}{(x-3)^2}$ **32.** $f(x) = \frac{-2}{(x-3)^2}$ **33.** $f(x) = \frac{-1}{(x+2)^2} - 3$ **34.** $f(x) = \frac{-1}{(x-4)^2} + 2$

Concept Check *The figures in A–D show the four ways that the graph of a rational function can approach the vertical line $x = 2$ as an asymptote. Identify the graph of each rational function in Exercises 35–38.*

35. $f(x) = \frac{1}{(x-2)^2}$ **36.** $f(x) = \frac{1}{x-2}$ **37.** $f(x) = \frac{-1}{x-2}$ **38.** $f(x) = \frac{-1}{(x-2)^2}$

A.

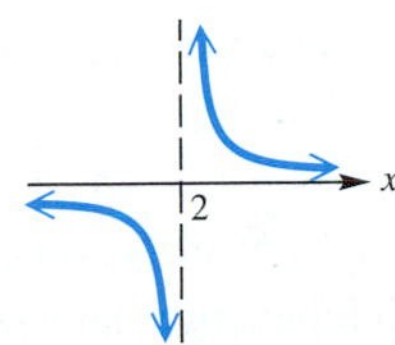

B.

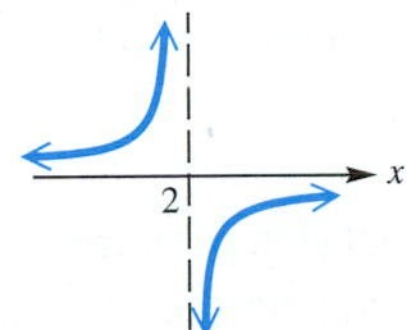

C.

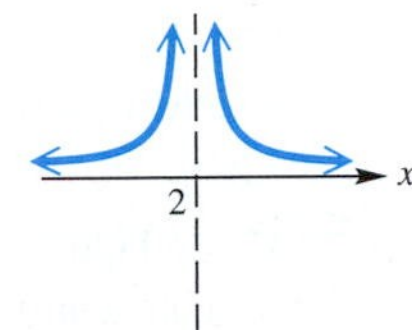

D.

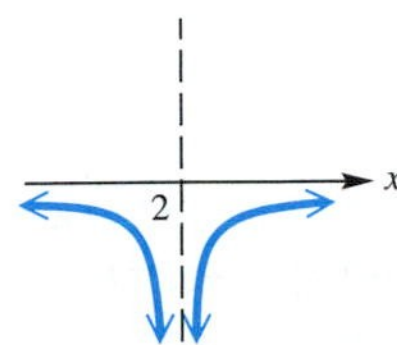

Concept Check *Suppose that the graph of a rational function f has vertical asymptote $x = 1$, horizontal asymptote $y = 2$, domain $(-\infty, 1) \cup (1, \infty)$, and range $(-\infty, 2) \cup (2, \infty)$. Give the vertical asymptote, horizontal asymptote, domain, and range for the graph of each shifted function.*

39. $y = f(x + 2) - 1$ **40.** $y = f(x - 1) + 3$

Use the method described in Example 3 to rewrite the rational expression, sketch the graph by hand, and generate an accurate depiction of each function with a graphing calculator.

41. $f(x) = \frac{x-1}{x-2}$ **42.** $f(x) = \frac{x-2}{x-3}$ **43.** $f(x) = \frac{-2x-5}{x+3}$

44. $f(x) = \frac{-2x-1}{x+1}$ **45.** $f(x) = \frac{2x-5}{x-3}$ **46.** $f(x) = \frac{2x-1}{x-1}$

4.2 More on Graphs of Rational Functions

Vertical and Horizontal Asymptotes ■ Graphing Techniques ■ Oblique Asymptotes ■ Graphs with Points of Discontinuity

Vertical and Horizontal Asymptotes

In this section, we discuss more general rational functions, such as

$$f(x) = \frac{x+1}{2x^2 + 5x - 3}, \quad f(x) = \frac{2x+1}{x-3}, \quad \text{and} \quad f(x) = \frac{x^2+1}{x-2}.$$

Rational functions

Thus far, we have seen how graphs of rational functions may approach vertical or horizontal asymptotes as $|x| \to \infty$, or as $|x| \to a$ if the function is undefined at $x = a$. We now give formal definitions for vertical and horizontal asymptotes.

Looking Ahead to Calculus

The rational function defined by

$$f(x) = \frac{2}{x+1}$$

has a vertical asymptote at $x = -1$. In calculus, the behavior of the graph of this function for values close to -1 is described by using *one-sided limits.* As x approaches -1 from the *left,* the function values decrease without bound. This is written

$$\lim_{x \to -1^-} f(x) = -\infty.$$

As x approaches -1 from the *right,* the function values increase without bound. This is written

$$\lim_{x \to -1^+} f(x) = \infty.$$

Asymptotes

Let $p(x)$ and $q(x)$ define polynomials. For the rational function defined by $f(x) = \frac{p(x)}{q(x)}$, written in *lowest terms,* and for real numbers a and b,

1. If $|f(x)| \to \infty$ as $x \to a$, then the line $x = a$ is a **vertical asymptote.**
2. If $f(x) \to b$ as $|x| \to \infty$, then the line $y = b$ is a **horizontal asymptote.**

Locating asymptotes is important in graphing rational functions. We find vertical asymptotes by determining the values of x that make the denominator, but *not* the numerator, equal to 0. To find horizontal asymptotes (and, in some cases, *oblique asymptotes*), we consider what happens to $f(x)$ as $|x| \to \infty$. These asymptotes determine the end behavior of the graph.

NOTE In Figures 14–16 accompanying Examples 1–3, we show the graphs of the rational functions to illustrate asymptotes. Techniques for actually graphing such functions follow later in the section.

EXAMPLE 1 Finding Asymptotes

Find the asymptotes of the graph of $f(x) = \frac{x+1}{2x^2 + 5x - 3}$.

Use this procedure in general to find vertical asymptotes.

Solution To find the vertical asymptotes, we set the denominator equal to 0 and solve.

$$2x^2 + 5x - 3 = 0$$

$$(2x - 1)(x + 3) = 0 \quad \text{Factor.}$$

$$2x - 1 = 0 \quad \text{or} \quad x + 3 = 0 \quad \text{Zero-product property}$$

$$x = \frac{1}{2} \quad \text{or} \quad x = -3 \quad \text{Solve each equation.}$$

Because the numerator is not equal to 0 at these values of x, the equations of the vertical asymptotes are $x = \frac{1}{2}$ and $x = -3$. See Figure 14.

Use this procedure in general to find horizontal asymptotes.

To find the equation of the horizontal asymptote, we divide each term by the variable factor of greatest degree in the expression for $f(x)$. In this case, we divide each term by x^2.

$$f(x) = \frac{\frac{x}{x^2} + \frac{1}{x^2}}{\frac{2x^2}{x^2} + \frac{5x}{x^2} - \frac{3}{x^2}} = \frac{\frac{1}{x} + \frac{1}{x^2}}{2 + \frac{5}{x} - \frac{3}{x^2}}$$

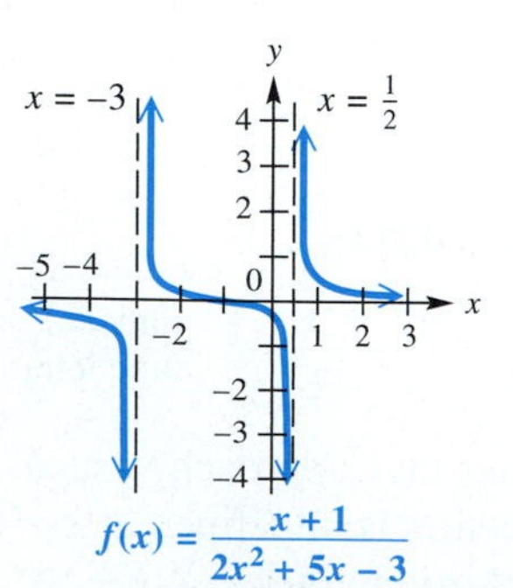

$f(x) = \frac{x+1}{2x^2 + 5x - 3}$

FIGURE 14

As $|x|$ gets larger and larger, the quotients $\frac{1}{x}$, $\frac{1}{x^2}$, $\frac{5}{x}$, and $\frac{3}{x^2}$ all approach 0, and the value of $f(x)$ approaches

$$\frac{0 + 0}{2 + 0 - 0} = \frac{0}{2} = 0.$$

The line $y = 0$ (that is, the x-axis) is therefore the horizontal asymptote. See Figure 14. ■

EXAMPLE 2 **Finding Asymptotes**

Find the asymptotes of the graph of $f(x) = \dfrac{2x + 1}{x - 3}$.

Solution Set the denominator, $x - 3$, equal to 0 to find that the vertical asymptote has equation $x = 3$. To find the horizontal asymptote, divide each term in the rational expression by x, since the greatest exponent on x in the expression is 1.

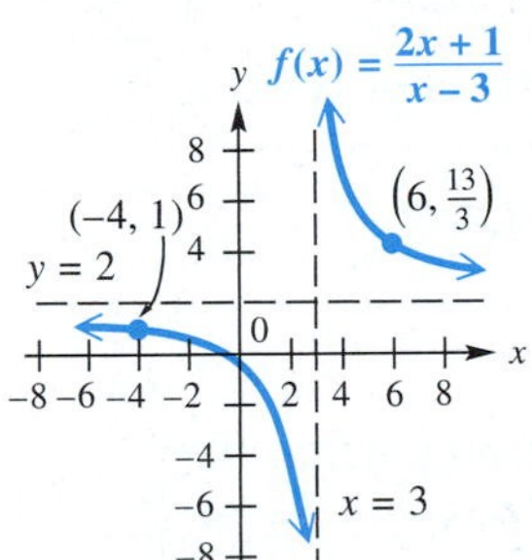

FIGURE 15

$$f(x) = \frac{2x + 1}{x - 3} = \frac{\frac{2x}{x} + \frac{1}{x}}{\frac{x}{x} - \frac{3}{x}} = \frac{2 + \frac{1}{x}}{1 - \frac{3}{x}}$$

As $|x|$ gets larger and larger, both $\frac{1}{x}$ and $\frac{3}{x}$ approach 0, and $f(x)$ approaches

$$\frac{2 + 0}{1 - 0} = \frac{2}{1} = 2.$$

Thus, the line $y = 2$ is the horizontal asymptote. See Figure 15. ■

EXAMPLE 3 **Finding Asymptotes**

Find the asymptotes of the graph of $f(x) = \dfrac{x^2 + 1}{x - 2}$.

Solution Setting the denominator, $x - 2$, equal to 0 shows that the vertical asymptote has equation $x = 2$. If we divide by the variable factor of greatest degree of x as before (x^2 in this case), we see that there is no horizontal asymptote, because

$$f(x) = \frac{\frac{x^2}{x^2} + \frac{1}{x^2}}{\frac{x}{x^2} - \frac{2}{x^2}} = \frac{1 + \frac{1}{x^2}}{\frac{1}{x} - \frac{2}{x^2}}$$

does not approach any real number as $|x| \to \infty$, since $\frac{1}{0}$ is undefined. This happens whenever the degree of the numerator is greater than the degree of the denominator. In such cases, we divide the denominator into the numerator to write the expression in another form.

$$\begin{array}{r l}
& \quad\; x + 2 \quad \leftarrow \frac{x^2}{x},\ \frac{2x}{x} \\
x - 2\,\big) & x^2 + 0x + 1 \\
& x^2 - 2x \qquad \text{Multiply } x(x - 2). \\
& \qquad 2x + 1 \qquad \text{Subtract; bring down.} \\
& \qquad 2x - 4 \qquad \text{Multiply } 2(x - 2). \\
& \qquad\quad\; 5 \qquad \text{Subtract.}
\end{array}$$

Include 0x for the missing x-term.

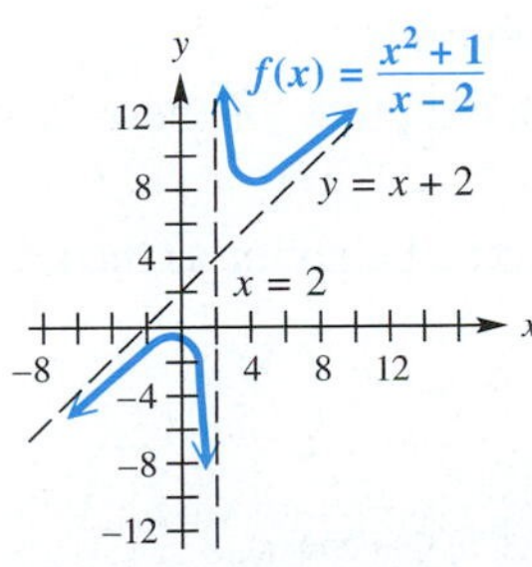

FIGURE 16

(We could have used synthetic division.) We can now write the function as

$$f(x) = \frac{x^2 + 1}{x - 2} = x + 2 + \frac{5}{x - 2}.$$

For very large values of $|x|$, $\frac{5}{x - 2}$ is close to 0, and the graph approaches the line $y = x + 2$. This line is an **oblique asymptote** (slanted, neither vertical nor horizontal) for the graph of the function. See Figure 16. ■

The results of Examples 1–3 can be summarized as follows.

Determining Asymptotes

To find asymptotes of a rational function defined by a rational expression *in lowest terms,* use the following procedures:

1. **Vertical Asymptotes**
   Find any vertical asymptotes by setting the denominator equal to 0 and solving for x. If a is a zero of the denominator but not the numerator, then the line $x = a$ **is a vertical asymptote.**
2. **Other Asymptotes**
   Determine any other asymptotes. Consider three possibilities:
   (a) If the numerator has lesser degree than the denominator, then there is a **horizontal asymptote** $y = 0$ (the x-axis).
   (b) If the numerator and denominator have the same degree and the function is of the form

$$f(x) = \frac{a_n x^n + \cdots + a_0}{b_n x^n + \cdots + b_0}, \quad \text{where } b_n \neq 0,$$

   then dividing by x^n in the numerator and denominator produces the **horizontal asymptote** $y = \frac{a_n}{b_n}$.
   (c) If the numerator is of degree exactly one greater than the denominator, then there may be an **oblique asymptote.*** To find it, divide the numerator by the denominator and disregard any remainder. Set the rest of the quotient equal to y to get the equation of the asymptote.

NOTE The graph of a rational function may have more than one vertical asymptote, or it may have none at all. ***The graph cannot intersect any vertical asymptote. There can be only one other (nonvertical) asymptote, and the graph may intersect that asymptote.*** This will be seen in Examples 4 and 6.

Graphing Techniques

TECHNOLOGY NOTE

Determining intercepts and asymptotes analytically will help to obtain a realistic comprehensive graph of a rational function. Choosing Xmin and Xmax so that the calculator will attempt to compute a value for which the rational function is undefined will eliminate the misleading vertical line that occasionally appears when the calculator is in connected mode.

A comprehensive graph of a rational function exhibits these features:

1. all intercepts, both x- and y-;
2. location of all asymptotes: vertical, horizontal, and/or oblique;
3. the point at which the graph intersects its nonvertical asymptote (if there is any such point);
4. enough of the graph to exhibit the correct end behavior (e.g., behavior as the graph approaches its nonvertical asymptote).

*More involved rational functions, such as $f(x) = \frac{8x^3 - 1}{x}$, are not covered in this book.

Graphing a Rational Function

Let $f(x) = \frac{p(x)}{q(x)}$ define a function with rational expression in lowest terms. To sketch its graph, follow these steps:

Step 1 Find all vertical asymptotes.

Step 2 Find all horizontal or oblique asymptotes.

Step 3 Find the y-intercept, if possible, by evaluating $f(0)$.

Step 4 Find the x-intercepts, if any, by solving $f(x) = 0$. (These will be the zeros of the numerator $p(x)$.)

Step 5 Determine whether the graph will intersect its nonvertical asymptote $y = b$ by solving $f(x) = b$, where b is the y-value of the horizontal asymptote, or by solving $f(x) = mx + b$, where $y = mx + b$ is the equation of the oblique asymptote.

Step 6 Plot selected points as necessary. Choose an x-value in each interval of the domain determined by the vertical asymptotes and x-intercepts.

Step 7 Complete the sketch.

EXAMPLE 4 Graphing a Rational Function with the *x*-Axis as Horizontal Asymptote

Graph $f(x) = \dfrac{x+1}{2x^2 + 5x - 3}$.

Solution

Step 1 From Example 1, the vertical asymptotes have equations $x = \frac{1}{2}$ and $x = -3$.

Step 2 Again, as shown in Example 1, the horizontal asymptote is the x-axis.

Step 3 The y-intercept is $-\frac{1}{3}$, since

$$f(0) = \frac{0+1}{2(0)^2 + 5(0) - 3} = -\frac{1}{3}.$$

Step 4 The x-intercept is found by solving $f(x) = 0$.

$$\frac{x+1}{2x^2 + 5x - 3} = 0$$

$$x + 1 = 0 \quad \text{If a fraction equals 0, then its numerator must equal 0.}$$

$$x = -1 \quad \text{The } x\text{-intercept is } -1.$$

Step 5 To determine whether the graph intersects its horizontal asymptote, solve

$$f(x) = 0. \leftarrow y\text{-value of horizontal asymptote}$$

The horizontal asymptote is the x-axis; the solution of this equation was found in Step 4. The graph intersects its horizontal asymptote at $(-1, 0)$.

Step 6 Plot a point in each of the intervals determined by the x-intercepts and vertical asymptotes, $(-\infty, -3)$, $(-3, -1)$, $\left(-1, \frac{1}{2}\right)$ and $\left(\frac{1}{2}, \infty\right)$, to get an idea of how the graph behaves in each region. See the table and the graph on the next page.

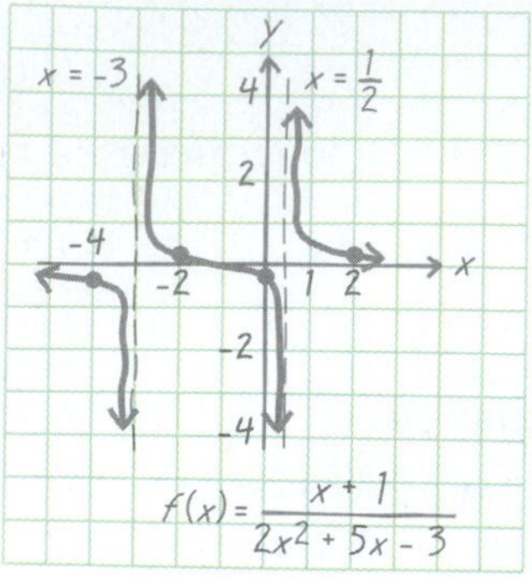

FIGURE 17

Interval	Test Value	Value of $f(x)$	Sign of $f(x)$	Graph Above or Below x-Axis
$(-\infty, -3)$	-4	$-\frac{1}{3}$	Negative	Below
$(-3, -1)$	-2	$\frac{1}{5}$	Positive	Above
$\left(-1, \frac{1}{2}\right)$	0	$-\frac{1}{3}$	Negative	Below
$\left(\frac{1}{2}, \infty\right)$	2	$\frac{1}{5}$	Positive	Above

Step 7 Complete the sketch as shown in Figure 17. ■

EXAMPLE 5 Graphing a Rational Function That Does Not Intersect Its Horizontal Asymptote

Graph $f(x) = \dfrac{2x + 1}{x - 3}$.

Solution

Steps 1 and 2 As determined in Example 2, the equation of the vertical asymptote is $x = 3$. The horizontal asymptote has equation $y = 2$.

Step 3 $f(0) = -\frac{1}{3}$, so the y-intercept is $-\frac{1}{3}$.

Step 4 Solve $f(x) = 0$ to find the x-intercept(s).

$$\frac{2x + 1}{x - 3} = 0$$

$$2x + 1 = 0 \qquad \text{If a fraction equals 0, then its numerator must equal 0.}$$

$$x = -\frac{1}{2} \qquad \text{The x-intercept is } -\tfrac{1}{2}.$$

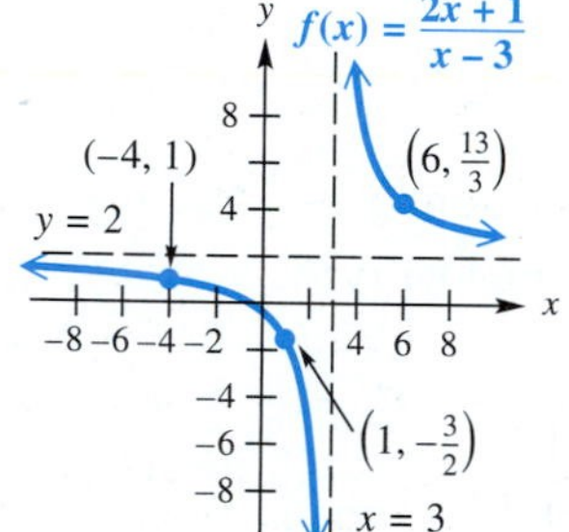

FIGURE 18

Step 5 The graph does not intersect its horizontal asymptote, since $f(x) = 2$ has no solution. (Verify this.)

Steps 6 and 7 The points $(-4, 1)$, $\left(1, -\frac{3}{2}\right)$, and $\left(6, \frac{13}{3}\right)$ are on the graph and can be used to complete the sketch, as seen in Figure 18. ■

EXAMPLE 6 Graphing a Rational Function That Intersects Its Horizontal Asymptote

Graph $f(x) = \dfrac{3x^2 - 3x - 6}{x^2 + 8x + 16}$.

Solution

Step 1 To find the vertical asymptote(s), solve $x^2 + 8x + 16 = 0$.

$$x^2 + 8x + 16 = 0 \qquad \text{Set the denominator equal to 0.}$$

$$(x + 4)^2 = 0 \qquad \text{Factor as a perfect square trinomial.}$$

$$x = -4 \qquad \text{Solve.}$$

Since the numerator is not 0 when $x = -4$, the only vertical asymptote has equation $x = -4$.

Looking Ahead to Calculus

The rational function defined by

$$f(x) = \frac{2x + 1}{x - 3}$$

in Example 5 has horizontal asymptote $y = 2$. In calculus, the behavior of the graph of this function as x approaches $-\infty$ and as x approaches ∞ is described by using *limits at infinity*. As x approaches $-\infty$, $f(x)$ approaches 2 (from below). This is written

$$\lim_{x \to -\infty} f(x) = 2.$$

As x approaches ∞, $f(x)$ approaches 2 (from above). This is written

$$\lim_{x \to \infty} f(x) = 2.$$

Step 2 We divide all terms by x^2 to get the equation of the horizontal asymptote,

$$y = \frac{3}{1}, \quad \text{or} \quad y = 3.$$

(3 ← Leading coefficient of numerator; 1 ← Leading coefficient of denominator)

Step 3 The y-intercept is $f(0) = -\frac{3}{8}$.

Step 4 To find the x-intercept(s), if any, we solve $f(x) = 0$.

$$\frac{3x^2 - 3x - 6}{x^2 + 8x + 16} = 0$$

$3x^2 - 3x - 6 = 0$ — Set the numerator equal to 0.

$x^2 - x - 2 = 0$ — Divide by 3.

$(x - 2)(x + 1) = 0$ — Factor.

$x = 2 \quad \text{or} \quad x = -1$ — Zero-product property

The x-intercepts are -1 and 2.

Step 5 We set $f(x) = 3$ and solve to locate the point where the graph intersects the horizontal asymptote.

$\dfrac{3x^2 - 3x - 6}{x^2 + 8x + 16} = 3$ — Let $f(x) = 3$.

$3x^2 - 3x - 6 = 3x^2 + 24x + 48$ — Multiply by $x^2 + 8x + 16$.

$-3x - 6 = 24x + 48$ — Subtract $3x^2$.

$-27x = 54$ — Subtract $24x$; add 6.

$x = -2$ — Divide by -27.

The graph intersects its horizontal asymptote at $(-2, 3)$.

Steps 6 and 7 Some other points that lie on the graph are $(-10, 9)$, $\left(-8, 13\frac{1}{8}\right)$, and $\left(5, \frac{2}{3}\right)$. These can be used to complete the sketch, as shown in Figure 19. ■

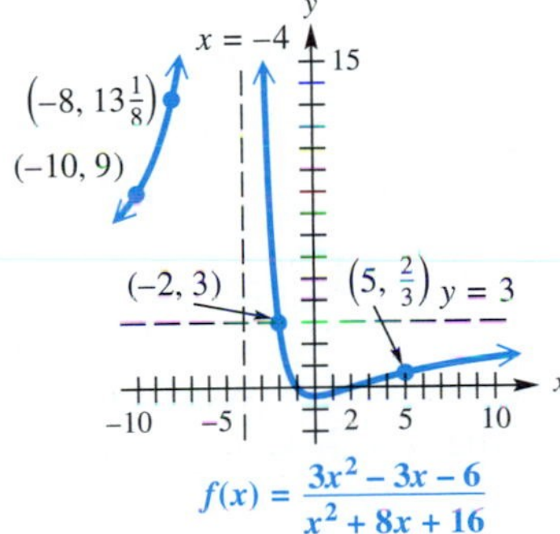

FIGURE 19

Notice the behavior of the graph of the function in Figure 19 near the line $x = -4$. As $x \to -4$ from either side, $f(x) \to \infty$. On the other hand, if we examine the behavior of the graph in Figure 18 near the line $x = 3$, $f(x) \to -\infty$ as x approaches 3 from the left, while $f(x) \to \infty$ as x approaches 3 from the right. The behavior of the graph of a rational function near a vertical asymptote $x = a$ will partially depend on the exponent on $x - a$ in the denominator.

Behavior of Graphs of Rational Functions Near Vertical Asymptotes

Suppose that $f(x)$ is defined by a rational expression in lowest terms. If n is the largest positive integer such that $(x - a)^n$ is a factor of the denominator of $f(x)$, the graph will behave in the manner illustrated.

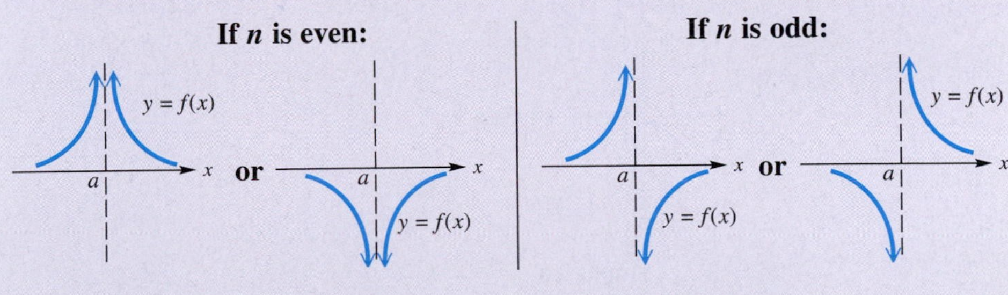

In Section 3.7, we observed that the behavior of the graph of a polynomial function near its zeros is dependent on whether the multiplicity of the zero is even or odd. The same statement can be made for rational functions.

Behavior of Graphs of Rational Functions Near x-Intercepts

Suppose that $f(x)$ is defined by a rational expression in lowest terms. If n is the largest positive integer such that $(x - c)^n$ is a factor of the numerator of $f(x)$, the graph will behave in the manner illustrated.

If n is even: | **If n is odd:**

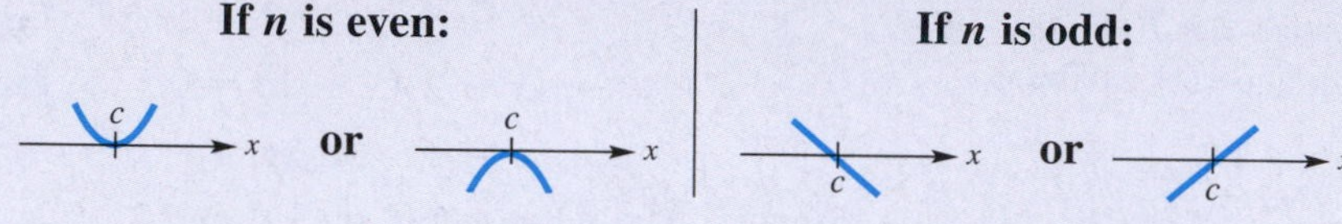

Oblique Asymptotes

EXAMPLE 7 Graphing a Rational Function with an Oblique Asymptote

Graph $f(x) = \dfrac{x^2 + 1}{x - 2}$.

Analytic Solution

As shown in Example 3, the vertical asymptote has equation $x = 2$, and the graph has an oblique asymptote with equation $y = x + 2$. Refer to the box on the preceding page to determine the behavior of the graph near the asymptote $x = 2$. The y-intercept is $-\frac{1}{2}$, and since the numerator, $x^2 + 1$, has no real zeros, the graph has no x-intercepts. The graph does not intersect its oblique asymptote, because

$$\frac{x^2 + 1}{x - 2} = x + 2$$

has no solution. (Verify this.) Using the y-intercept, asymptotes, the points $\left(4, \frac{17}{2}\right)$ and $\left(-1, -\frac{2}{3}\right)$, and the general behavior of the graph near its asymptotes leads to the graph in Figure 20.

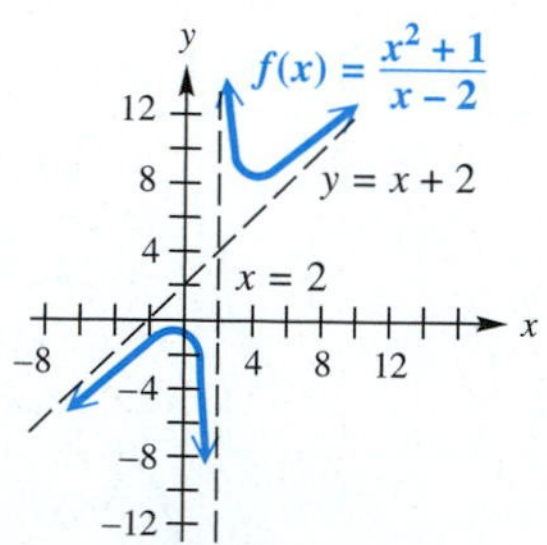

FIGURE 20

Graphing Calculator Solution

A calculator graph is shown in Figure 21. If we define

$$Y_1 = \frac{X^2 + 1}{X - 2} \quad \text{and} \quad Y_2 = X + 2$$

and observe a table of values as $|X|$ gets larger and larger, we see that for these large values, $Y_1 \approx Y_2$.

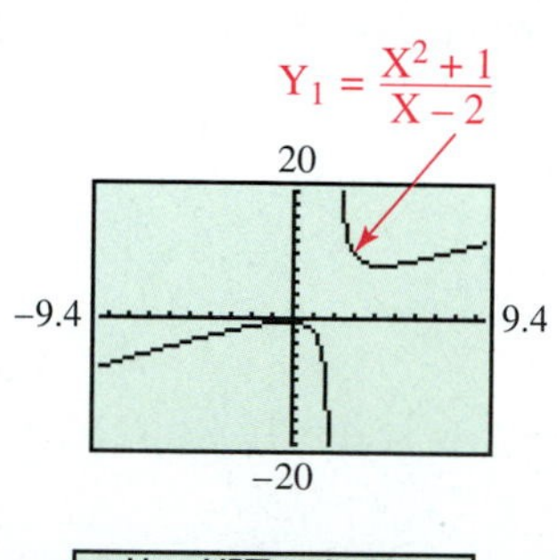

X	Y1	Y2
25	27.217	27
-25	-23.19	-23
50	52.104	52
-50	-48.1	-48
100	102.05	102
-100	-98.05	-98

Y1=(X²+1)/(X−2)

FIGURE 21

Graphs with Points of Discontinuity

As mentioned earlier, a rational function must be defined by an expression in lowest terms before we can use the methods discussed thus far to hand-sketch the graph. A rational function that is not in lowest terms usually has a "hole," or point of discontinuity, in its graph.

GCM **EXAMPLE 8** **Graphing a Rational Function Defined by an Expression That Is Not in Lowest Terms**

Graph $f(x) = \frac{x^2 - 4}{x - 2}$.

Analytic Solution

The domain of this function cannot include 2. The expression $\frac{x^2-4}{x-2}$ should be written in lowest terms.

$$\begin{aligned} f(x) &= \frac{x^2 - 4}{x - 2} \\ &= \frac{(x + 2)(x - 2)}{x - 2} && \text{Factor.} \\ &= x + 2 \quad (x \neq 2) && \text{Lowest terms} \end{aligned}$$

Therefore, the graph of this function will be the same as the graph of

$$y = x + 2$$

(a straight line), with the exception of the point with x-value 2. A "hole" appears in the graph at $(2, 4)$. See Figure 22.

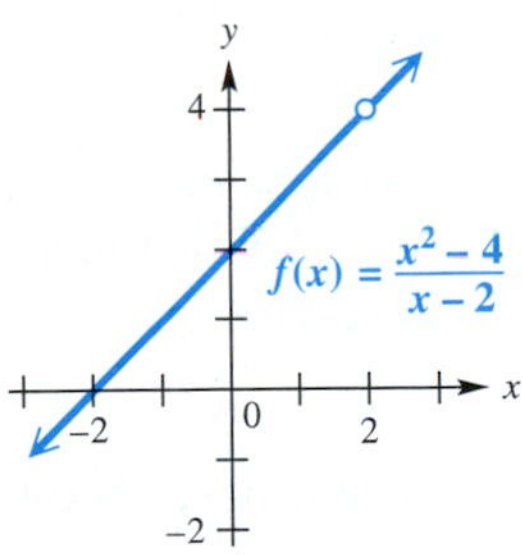

FIGURE 22

Graphing Calculator Solution

If we set the window of a graphing calculator so that an x-value for the location of the tracing cursor is 2, then we can see from the display that the calculator cannot determine a value for Y. We define

$$Y_1 = \frac{X^2 - 4}{X - 2}$$

and graph it in such a window, as in Figure 23. The error message in the table further supports the existence of a discontinuity at $X = 2$. (For the table, $Y_2 = X + 2$.)

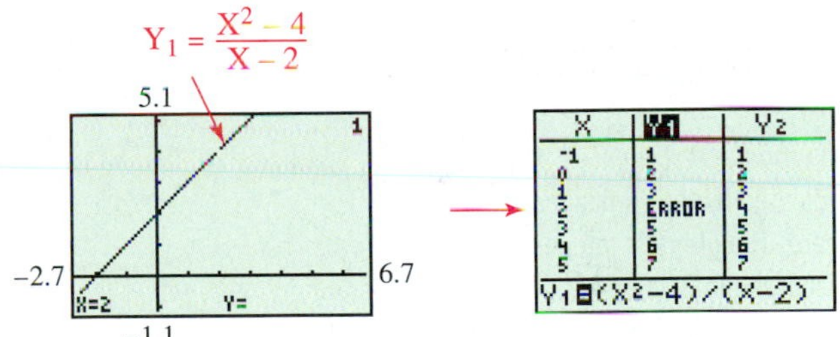

FIGURE 23

Notice the visible discontinuity at $X = 2$ in the graph. The window was chosen so that the "hole" would be visible. This requires a *decimal viewing window* or a window with x-values centered at 2. Other choices may not show this discontinuity.

FOR DISCUSSION

Let a rational function f be defined by $f(x) = \frac{p(x)}{q(x)}$, where $p(x)$ and $q(x)$ are polynomials. Suppose that $p(k) = 0$ and $q(k) = 0$. Discuss how you could determine whether the graph of f has a "hole" in its graph at $x = k$. Try your method on $f(x) = \frac{x^2 + x - 6}{2x^2 + 5x - 3}$ and $k = -3$.

To summarize, rational functions presented in Sections 4.1 and 4.2 fall into these categories:

1. $y = \frac{1}{x}$ and $y = \frac{1}{x^2}$ and their transformations (Section 4.1)
2. Degree of numerator $\leq$ degree of denominator (Examples 1, 2, 4, 5, and 6 of this section)
3. Degree of numerator $>$ degree of denominator (Examples 3 and 7 of this section)
4. Those with common variable factors in numerator and denominator (Example 8 of this section)

Looking Ahead to Calculus

Different types of discontinuity are discussed in calculus. The function in Example 8,

$$f(x) = \frac{x^2 - 4}{x - 2},$$

is said to have *removable discontinuity* at $x = 2$, since the discontinuity can be removed by redefining f at 2. The function in Example 7,

$$f(x) = \frac{x^2 + 1}{x - 2},$$

has *infinite discontinuity* at $x = 2$, as indicated by the vertical asymptote there. The greatest integer function, discussed in Section 2.5, has *jump discontinuities*, because the function values "jump" from one value to another for integer domain values.

What Went Wrong?

A student was asked to determine the domain of the rational function defined by

$$f(x) = \frac{x^2 - 40}{x^2 - 4}.$$

Using the techniques of this section, the student determined that -2 and 2 are excluded, and gave her answer as $(-\infty, -2) \cup (-2, 2) \cup (2, \infty)$. She then graphed the function in the window $[-9.4, 9.4]$ by $[-6.2, 6.2]$ and obtained the graph shown in this screen.

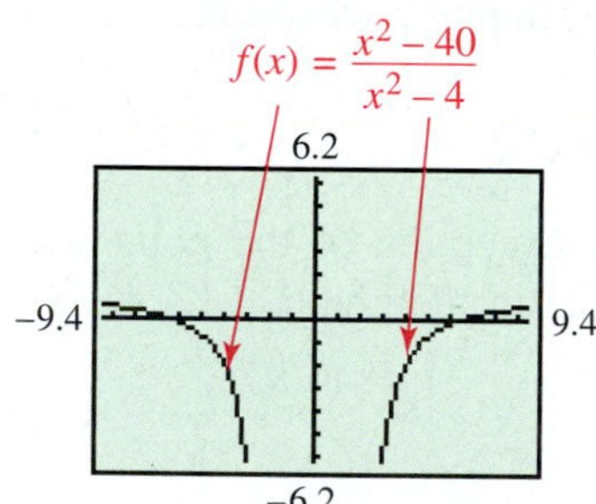

She became puzzled because no portion of the graph appeared between -2 and 2. So she changed her answer to $(-\infty, -2) \cup (2, \infty)$. Her instructor counted her second answer incorrect and told her that she should have given her first answer, which was correct.

What Went Wrong? Why did her graph not indicate any points between -2 and 2?

Answer to What Went Wrong?

There is a U-shaped portion of the graph between -2 and 2, but it appears above the maximum y-value shown in the screen (6.2). She should increase the y-maximum value. (The window $[-9.4, 9.4]$ by $[-6.2, 18.6]$ works well.)

4.2 Exercises

Concept Check *Match the rational function in Column I with the appropriate description in Column II. Choices in Column II can be used only once.*

I	II
1. $f(x) = \dfrac{x+7}{x+1}$	**A.** The x-intercept is -3.
2. $f(x) = \dfrac{x+10}{x+2}$	**B.** The y-intercept is 5.
3. $f(x) = \dfrac{1}{x+12}$	**C.** The horizontal asymptote is $y = 4$.
4. $f(x) = \dfrac{-3+x^2}{x^2}$	**D.** The vertical asymptote is $x = -1$.
5. $f(x) = \dfrac{x^2-16}{x+4}$	**E.** There is a "hole" in its graph at $x = -4$.
6. $f(x) = \dfrac{4x+3}{x-7}$	**F.** The graph has an oblique asymptote.
7. $f(x) = \dfrac{x^2+3x+4}{x-5}$	**G.** The x-axis is its horizontal asymptote.
8. $f(x) = \dfrac{x+3}{x-6}$	**H.** The y-axis is its vertical asymptote.

Give the equations of any vertical, horizontal, or oblique asymptotes for the graph of each rational function.

9. $f(x) = \dfrac{3}{x-5}$ **10.** $f(x) = \dfrac{-6}{x+9}$ **11.** $f(x) = \dfrac{4-3x}{2x+1}$ **12.** $f(x) = \dfrac{2x+6}{x-4}$

13. $f(x) = \dfrac{x^2-1}{x+3}$ **14.** $f(x) = \dfrac{x^2+4}{x-1}$ **15.** $f(x) = \dfrac{x^2-2x-3}{2x^2-x-10}$ **16.** $f(x) = \dfrac{3x^2-6x-24}{5x^2-26x+5}$

17. *Concept Check* Which function has a graph that does not have a vertical asymptote?

A. $f(x) = \dfrac{1}{x^2+2}$ **B.** $f(x) = \dfrac{1}{x^2-2}$ **C.** $f(x) = \dfrac{3}{x^2}$ **D.** $f(x) = \dfrac{2x+1}{x-8}$

18. *Concept Check* Which function has a graph that does not have a horizontal asymptote?

A. $f(x) = \dfrac{2x-7}{x+3}$ **B.** $f(x) = \dfrac{3x}{x^2-9}$ **C.** $f(x) = \dfrac{x^2-9}{x+3}$ **D.** $f(x) = \dfrac{x+5}{(x+2)(x-3)}$

Identify any vertical, horizontal, or oblique asymptotes in the graph of $y = f(x)$. State the domain of f.

19.

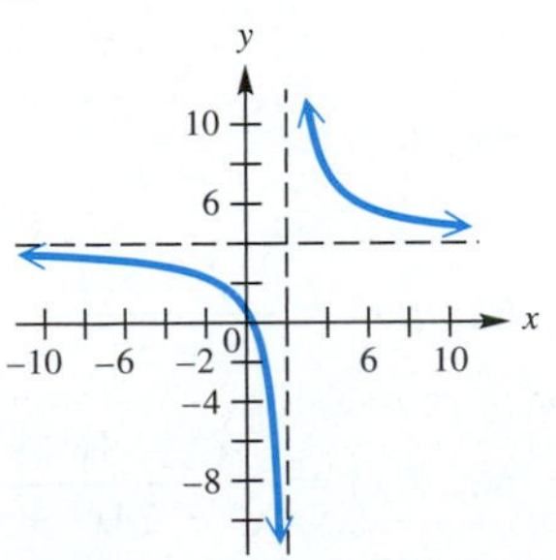

20.

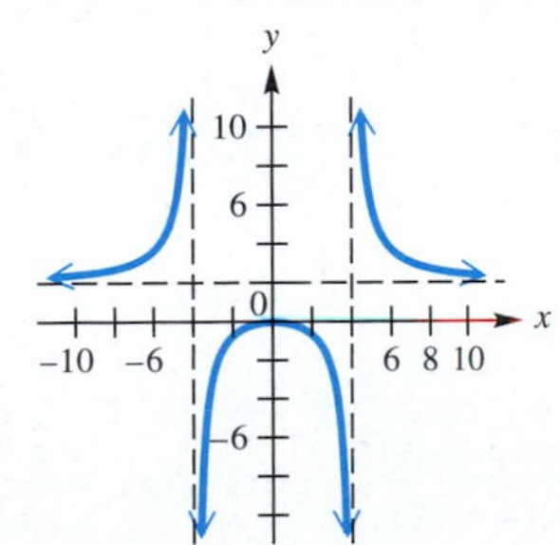

21.

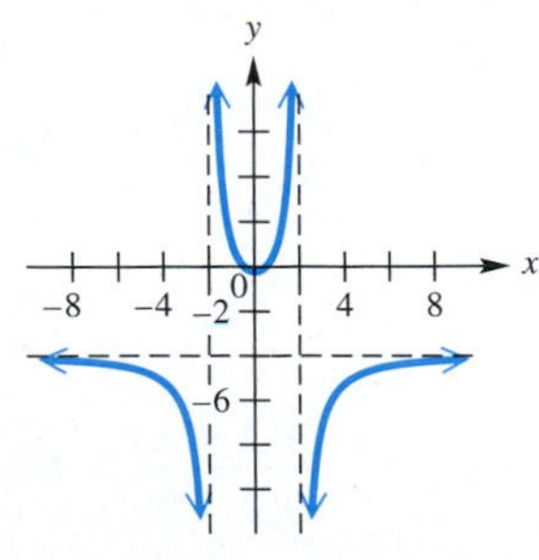

22.

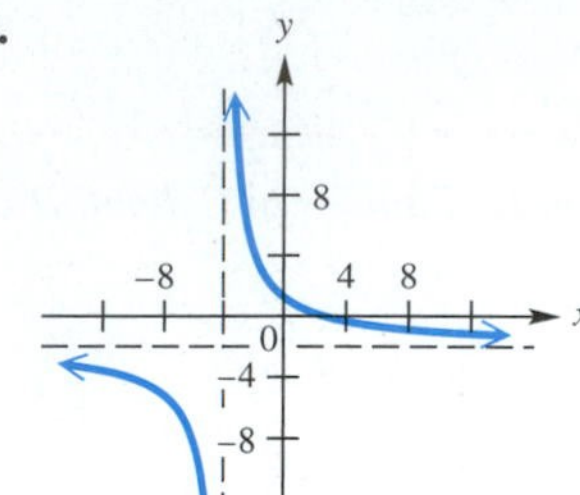

23.

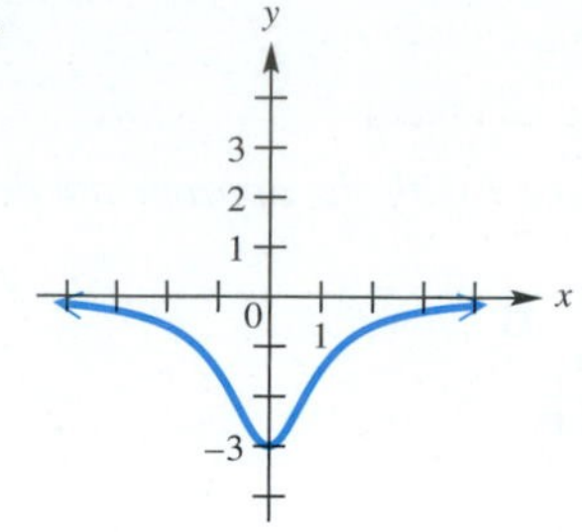

24.

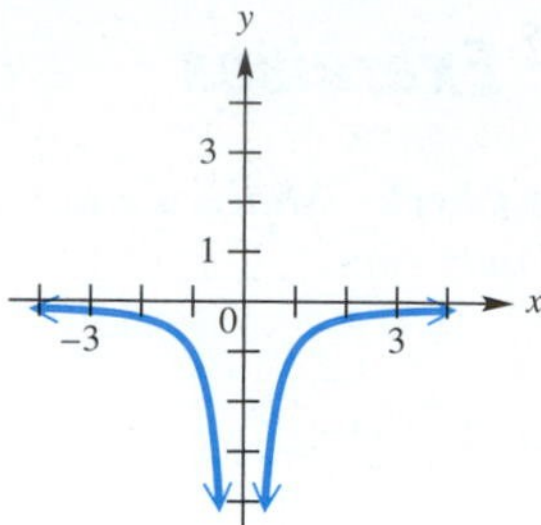

25.

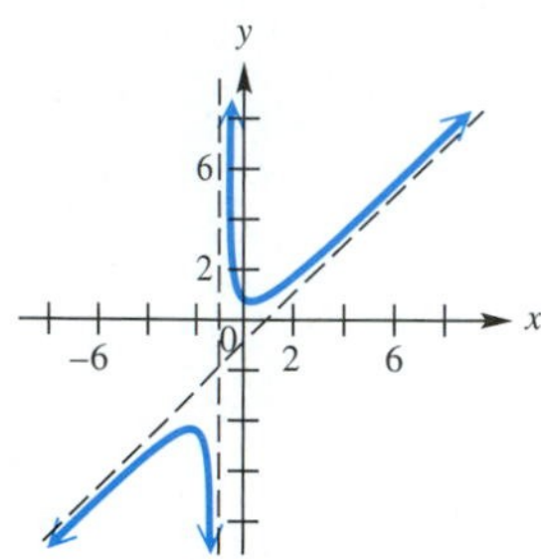

26.

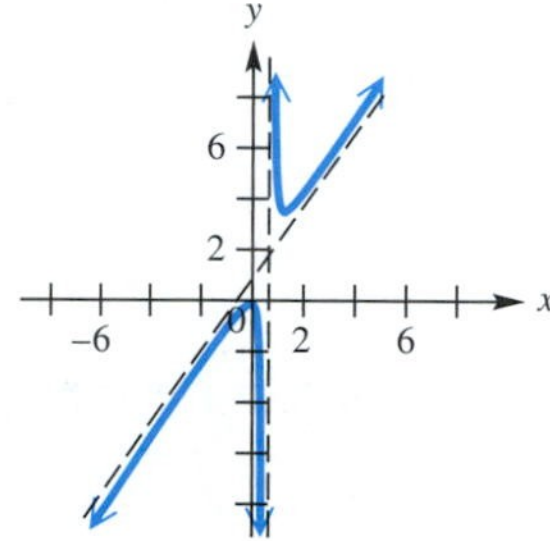

Sketch a graph of each rational function. Your graph should include all asymptotes and be comprehensive. You may want to use your graphing calculator as an aid.

27. $f(x) = \dfrac{x+1}{x-4}$

28. $f(x) = \dfrac{x-5}{x+3}$

29. $f(x) = \dfrac{x+2}{x-3}$

30. $f(x) = \dfrac{x-3}{x+4}$

31. $f(x) = \dfrac{4-2x}{8-x}$

32. $f(x) = \dfrac{6-3x}{4-x}$

33. $f(x) = \dfrac{3x}{(x+1)(x-2)}$

34. $f(x) = \dfrac{2x+1}{(x+2)(x+4)}$

35. $f(x) = \dfrac{5x}{x^2-1}$

36. $f(x) = \dfrac{x}{4-x^2}$

37. $f(x) = \dfrac{(x+6)(x-2)}{(x+3)(x-4)}$

38. $f(x) = \dfrac{(x+3)(x-5)}{(x+1)(x-4)}$

39. $f(x) = \dfrac{3x^2+3x-6}{x^2-x-12}$

40. $f(x) = \dfrac{4x^2+4x-24}{x^2-3x-10}$

41. $f(x) = \dfrac{9x^2-1}{x^2-4}$

42. $f(x) = \dfrac{16x^2-9}{x^2-9}$

43. $f(x) = \dfrac{(x-3)(x+1)}{(x-1)^2}$

44. $f(x) = \dfrac{x(x-2)}{(x+3)^2}$

45. $f(x) = \dfrac{x}{x^2-9}$

46. $f(x) = \dfrac{-5}{2x+4}$

47. $f(x) = \dfrac{1}{x^2+1}$

48. $f(x) = \dfrac{(x-5)(x-2)}{x^2+9}$

49. $f(x) = \dfrac{(x+4)^2}{(x-1)(x+5)}$

50. $f(x) = \dfrac{(x+1)^2}{(x+2)(x-3)}$

51. $f(x) = \dfrac{20+6x-2x^2}{8+6x-2x^2}$

52. $f(x) = \dfrac{18+6x-4x^2}{4+6x+2x^2}$

53. $f(x) = \dfrac{x^2+1}{x+3}$

54. $f(x) = \dfrac{2x^2+3}{x-4}$

55. $f(x) = \dfrac{x^2+2x}{2x-1}$

56. $f(x) = \dfrac{x^2-x}{x+2}$

57. $f(x) = \dfrac{x^2-9}{x+3}$

58. $f(x) = \dfrac{x^2-16}{x+4}$

59. $f(x) = \dfrac{2x^2-5x-2}{x-2}$

60. $f(x) = \dfrac{x^2-5}{x-3}$

61. $f(x) = \dfrac{x^2-1}{x^2-4x+3}$

62. $f(x) = \dfrac{x^2-4}{x^2+3x+2}$

63. $f(x) = \dfrac{(x^2-9)(2+x)}{(x^2-4)(3+x)}$

64. $f(x) = \dfrac{(x^2-16)(3+x)}{(x^2-9)(4+x)}$

65. $f(x) = \dfrac{x^4-20x^2+64}{x^4-10x^2+9}$

66. $f(x) = \dfrac{x^4-5x^2+4}{x^4-24x^2+108}$

Concept Check *In each table, Y_1 is defined by a rational expression of the form $\frac{X - p}{X - q}$. Use the table to find the values of p and q.*

67.

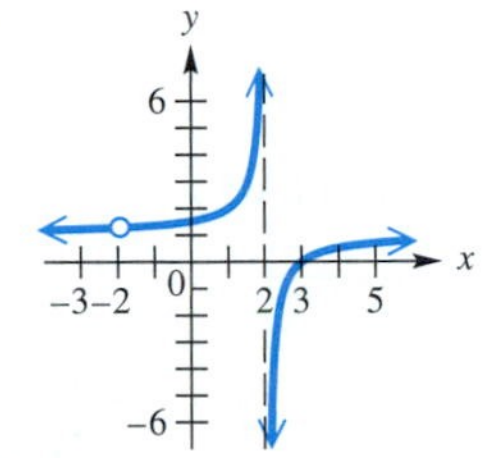

X	Y1
0	2
1	3
2	ERROR
3	-1
4	0
5	.33333
6	.5

X=0

68.

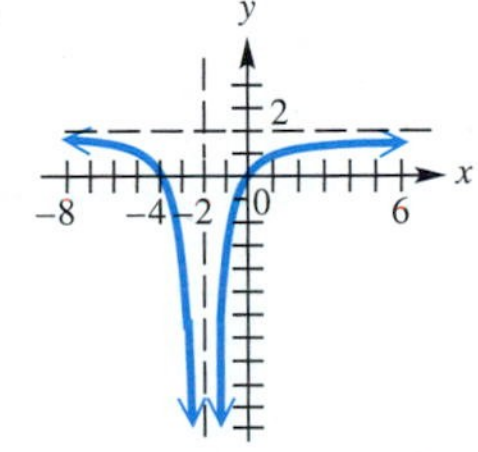

X	Y1
0	3
1	ERROR
2	-1
3	0
4	.33333
5	.5
6	.6

X=0

69.

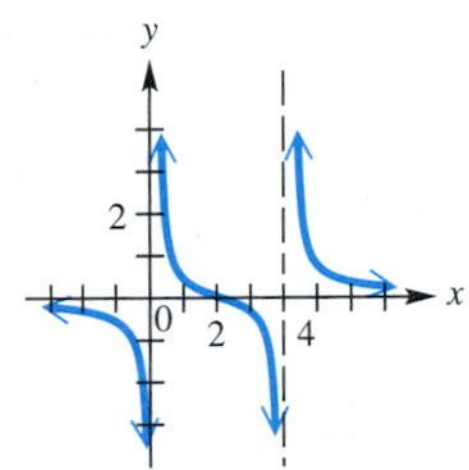

X	Y1
-3	.5
-2	0
-1	ERROR
0	2
1	1.5
2	1.3333
3	1.25

X=-3

70.

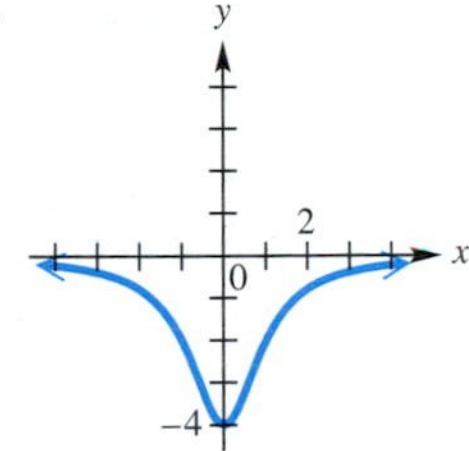

X	Y1
-7	.75
-6	.66667
-5	.5
-4	0
-3	ERROR
-2	2
-1	1.5

X=-7

Concept Check *Find an equation for the rational function graphed.*

71. **72.** **73.** **74.**

Consider $f(x) = \frac{x + 3}{x^2 + x + 4}$, *which is defined by a rational expression in lowest terms and whose denominator is a quadratic polynomial.*

75. To find vertical asymptotes, set the denominator equal to 0 and solve. What are the complex solutions of the equation you solved? What are the real solutions? Does f have any vertical asymptotes?

76. With your calculator in connected mode and using a window of $[-10, 10]$ by $[-1, 3]$, graph the function. Why is connected mode acceptable here to get a realistic view of the graph?

Relating Concepts

For individual or group investigation (Exercises 77–80)

Recall from Section 2.3 that if we are given the graph of $y = f(x)$, we can obtain the graph of $y = -f(x)$ by reflecting across the x-axis, and we can obtain the graph of $y = f(-x)$ by reflecting across the y-axis. In Exercises 77–80, you are given the graph of a rational function $y = f(x)$. Draw a sketch by hand of the graph of **(a)** $y = -f(x)$ *and* **(b)** $y = f(-x)$.

77.

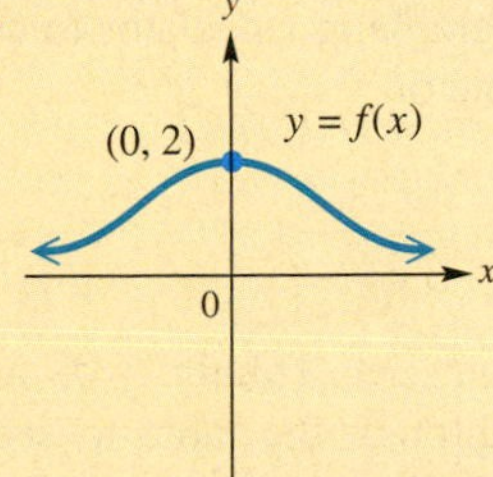

78.

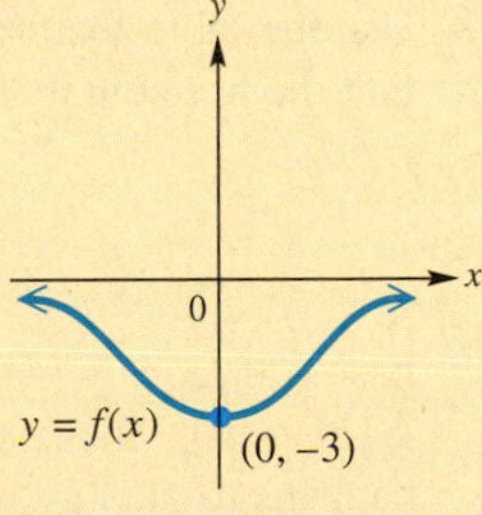

79.

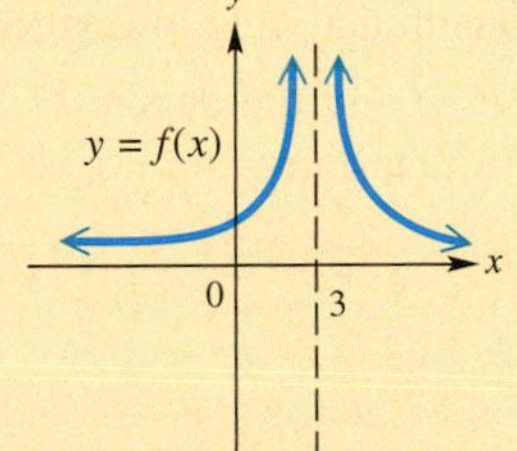

80.

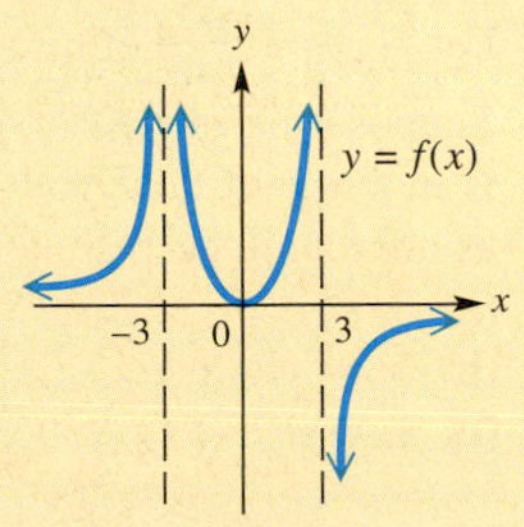

Each rational function in Exercises 81–84 has an oblique asymptote. Determine the equation of this asymptote. Then, use a graphing calculator to graph both the function and the asymptote in the window indicated.

81. $f(x) = \frac{2x^2 + 3}{4 - x}$; window: $[-18.8, 18.8]$ by $[-50, 25]$

82. $f(x) = \frac{x^2 + 9}{x + 3}$; window: $[-9.4, 9.4]$ by $[-25, 25]$

83. $f(x) = \frac{x - x^2}{x + 2}$; window: $[-9.4, 9.4]$ by $[-15, 25]$

84. $f(x) = \frac{x^2 + 2x}{1 - 2x}$; window: $[-4.7, 4.7]$ by $[-5, 5]$

85. *Concept Check* $f(x) = \dfrac{x^5 + x^4 + x^2 + 1}{x^4 + 1}$

becomes $f(x) = x + 1 + \dfrac{x^2 - x}{x^4 + 1}$

after the numerator is divided by the denominator.

(a) What is the equation of the oblique asymptote of the graph of the function?

(b) Where does the graph of the function intersect its asymptote?

(c) As $x \to \infty$, does the graph of the function approach its asymptote from above or below?

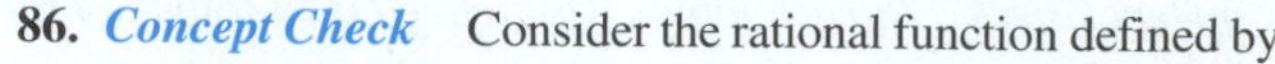

86. *Concept Check* Consider the rational function defined by

$$f(x) = \frac{x^3 - 4x^2 + x + 6}{x^2 + x - 2}.$$

Divide the numerator by the denominator and use the method of Example 3 to determine the equation of the oblique asymptote. Then, determine the coordinates of the point where the graph of f intersects its oblique asymptote. Use a calculator to support your answer.

87. Use long division of polynomials to show that, for the function defined by

$$f(x) = \frac{x^4 - 5x^2 + 4}{x^2 + x - 12},$$

if we divide the numerator by the denominator, then the quotient polynomial is $x^2 - x + 8$, and the remainder is $-20x + 100$. Graph both $f(x)$ and $g(x) = x^2 - x + 8$ in the window $[-50, 50]$ by $[0, 1000]$. Comment on the appearance of the two graphs. Explain how the graph of f approaches that of g as $|x|$ gets very large.

88. Suppose a friend tells you that the graph of

$$f(x) = \frac{x^2 - 25}{x + 5}$$

has a vertical asymptote with equation $x = -5$. Is this correct? If not, describe the behavior of the graph at $x = -5$.

Each rational function in Exercises 89 and 90 has a "hole" in its graph.

(a) *Write the fraction in lowest terms, and call this new function* $g(x)$.

(b) *Use a table with* $Y_1 = f(x)$ *and* $Y_2 = g(x)$ *to support the fact that, for all values except those for which* f *is undefined,* $f(x) = g(x)$.

(c) *What are the coordinates of the "hole"? Support your answer with a graph.*

89. $f(x) = \dfrac{x^2 - 9}{x + 3}$

90. $f(x) = \dfrac{x^2 - 36}{6 - x}$

91. Let $f(x) = \frac{p(x)}{q(x)}$ define a rational function in lowest terms. Suppose the degree of $p(x)$ is m and the degree of $q(x)$ is n. Explain how you would determine the nonvertical asymptote in each situation.

(a) $m < n$

(b) $m = n$

(c) $m = n + 1$

92. Use the table feature of a graphing calculator to confirm that the horizontal asymptote of

$$f(x) = \frac{6x^2 + 3}{2x^2}$$

has equation $y = 3$. (*Hint:* Let Tblmin $= 0$, and let ΔTbl $= 10$. Then, scroll through the table to see what happens to $f(x)$ as $x \to \infty$ and $x \to -\infty$.)

4.3 Rational Equations, Inequalities, Applications, and Models

Solving Rational Equations and Inequalities ■ Applications and Models of Rational Functions ■ Inverse Variation ■ Combined and Joint Variation

Solving Rational Equations and Inequalities

A **rational equation** (or **rational inequality**) is an equation (or inequality) with at least one term having a variable expression in a denominator or at least one term having a variable expression raised to a negative integer power. Some examples are

$$\frac{x+2}{2x+1}=1, \quad \frac{x+2}{2x+1}\leq 1, \quad \frac{x}{x-2}+\frac{1}{x+2}=\frac{8}{x^2-4}, \quad \text{and} \quad 7x^{-4}-8x^{-2}+1=0.$$

Rational equations and inequalities

When solving rational equations and inequalities, remember that an expression with a variable denominator or with a negative exponent may be undefined for certain values of the variable. We must identify these values (which are actually values at which the associated rational function will have a vertical asymptote or "hole").

To solve a rational equation, we multiply each side of the equation by the least common denominator (LCD) of all the terms in the equation.

EXAMPLE 1 **Solving a Rational Equation**

Solve $\frac{x+2}{2x+1}=1$.

Analytic Solution

Set the denominator, $2x+1$, equal to 0 to determine that the rational expression is undefined for $x=-\frac{1}{2}$. To clear fractions, multiply each side of the equation by $2x+1$.

$$\left(\frac{x+2}{2x+1}\right)(2x+1)=(1)(2x+1)$$
$$x+2=2x+1$$
$$1=x$$

Check by substituting 1 for x in the original equation.

$$\frac{1+2}{2(1)+1}=1 \quad ? \quad \text{Let } x=1.$$
$$\frac{3}{3}=1 \quad ?$$
$$1=1 \quad \text{True}$$

The solution set is $\{1\}$.

Graphing Calculator Solution

Because graphs of rational functions usually consist of several parts, using the intersection-of-graphs method can be confusing. Rewrite the equation as

$$\frac{x+2}{2x+1}-1=0,$$

and graph
$$y=\frac{x+2}{2x+1}-1.$$

The x-intercept method shows that the zero of the function is 1. See Figure 24.

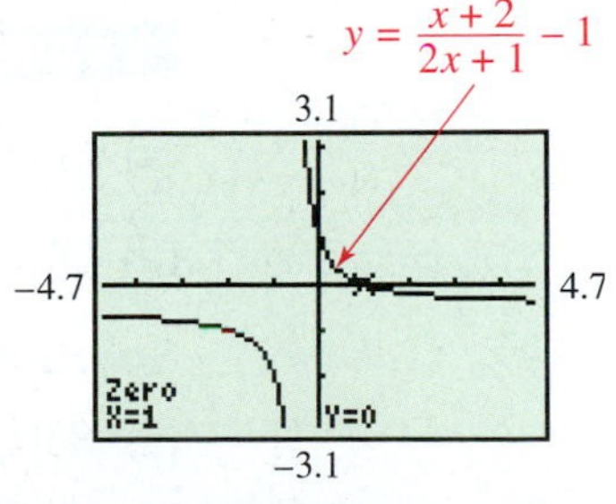

FIGURE 24

EXAMPLE 2 Solving a Rational Equation

Solve $\dfrac{x}{x-2} + \dfrac{1}{x+2} = \dfrac{8}{x^2-4}$.

Analytic Solution

For this equation, $x \neq \pm 2$. (Why?)

$$\frac{x}{x-2} + \frac{1}{x+2} = \frac{8}{x^2-4}$$

$$x(x+2) + 1(x-2) = 8 \qquad \text{Multiply by the LCD, } (x-2)(x+2).$$

$$x^2 + 2x + x - 2 = 8 \qquad \text{Distributive property}$$

$$x^2 + 3x - 10 = 0 \qquad \text{Standard form}$$

$$(x+5)(x-2) = 0 \qquad \text{Factor.}$$

$$x + 5 = 0 \quad \text{or} \quad x - 2 = 0 \qquad \text{Zero-product property}$$

$$x = -5 \quad \text{or} \quad x = 2 \qquad \text{2 is not in the domain.}$$

The numbers -5 and 2 are the *possible* solutions of the equation, but 2 is *not* in the domain of the original equation and, therefore, must be rejected. Such a value is *extraneous*. The solution set is $\{-5\}$.

Graphing Calculator Solution

Rewrite the equation as

$$\frac{x}{x-2} + \frac{1}{x+2} - \frac{8}{x^2-4} = 0,$$

and let y equal the left side of this equation. Figure 25 shows that the zero of the function is -5, supporting the analytic solution.

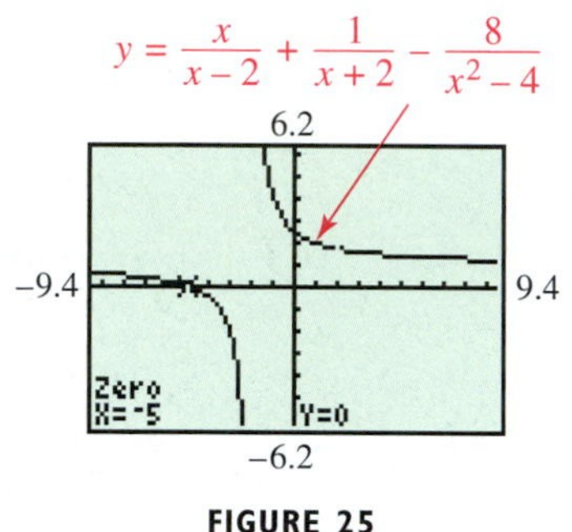

FIGURE 25

Examples 1 and 2 illustrate the general procedure for solving rational equations.

Solving a Rational Equation

Step 1 Determine all values for which the rational equation has undefined expressions.

Step 2 Multiply each side of the equation by the least common denominator of all rational expressions in the equation.

Step 3 Solve the resulting equation.

Step 4 Reject any values found in Step 1.

To solve a rational inequality analytically, we use tables and test values, first introduced in Section 3.3 to solve quadratic inequalities.

EXAMPLE 3 Solving a Rational Inequality

(a) Solve $\dfrac{x+2}{2x+1} \leq 1$.

(b) Discuss how the graph in Figure 24 supports the solution.

Solution

(a) Subtract 1 from each side and combine terms on the left side to get a single rational expression.

$$\frac{x+2}{2x+1} - 1 \le 0 \quad \text{Subtract 1.}$$

$$\frac{x+2}{2x+1} - \frac{2x+1}{2x+1} \le 0 \quad \text{The common denominator is } 2x+1.$$

Be careful with signs.

$$\frac{x+2-(2x+1)}{2x+1} \le 0 \quad \text{Write as a single rational expression.}$$

$$\frac{x+2-2x-1}{2x+1} \le 0 \quad \text{Distributive property}$$

$$\frac{-x+1}{2x+1} \le 0 \quad \text{Combine terms in the numerator.}$$

The quotient possibly changes sign only where the x-values make the numerator or denominator 0. This occurs at

$$-x+1=0 \quad \text{or} \quad 2x+1=0$$

$$x=1 \quad \text{or} \quad x=-\frac{1}{2}.$$

These two x-values divide the real number line into three intervals: $\left(-\infty, -\frac{1}{2}\right)$, $\left(-\frac{1}{2}, 1\right)$, and $(1, \infty)$. See Figure 26.

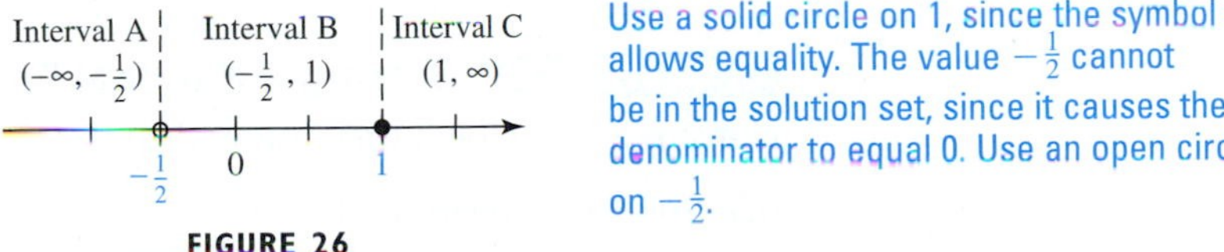

FIGURE 26

Use a solid circle on 1, since the symbol $\le$ allows equality. The value $-\frac{1}{2}$ cannot be in the solution set, since it causes the denominator to equal 0. Use an open circle on $-\frac{1}{2}$.

To determine the sign of the expression $\frac{-x+1}{2x+1}$, choose a test value from each interval and substitute it into the expression, as shown in the following table.

Interval	Test Value	Is $\frac{-x+1}{2x+1} \le 0$ True or False?
A: $\left(-\infty, -\frac{1}{2}\right)$	-1	$\frac{1+1}{2(-1)+1} \le 0$? $-2 \le 0$ True
B: $\left(-\frac{1}{2}, 1\right)$	0	$\frac{0+1}{2(0)+1} \le 0$? $1 \le 0$ False
C: $(1, \infty)$	2	$\frac{-2+1}{2(2)+1} \le 0$? $-\frac{1}{5} \le 0$ True

From the table, we see that values of x in the intervals $\left(-\infty, -\frac{1}{2}\right)$ and $(1, \infty)$ make the quotient negative, as required. Therefore, the solution set is $\left(-\infty, -\frac{1}{2}\right) \cup [1, \infty)$. From our previous discussion, 1 is included in the solution set because it results in equality, whereas $-\frac{1}{2}$ is not included because it results in a denominator that is equal to 0.

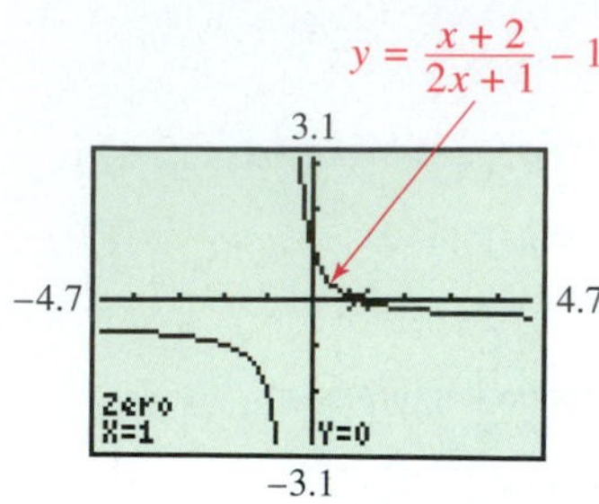

(b) The original inequality is equivalent to $\frac{x+2}{2x+1} - 1 \leq 0$. From Figure 24, repeated in the margin, we see that the graph lies *below* the x-axis for x-values less than the x-value of the vertical asymptote $\left(x = -\frac{1}{2}\right)$ as well as for x-values greater than 1. This agrees with the solution set obtained in part (a). ■

CAUTION ***Be careful with endpoints of intervals when solving rational inequalities.***

Example 3 illustrates the general procedure for solving a rational inequality.

Solving a Rational Inequality

Step 1 Rewrite the inequality, if necessary, so that 0 is on one side and there is a single fraction on the other side.

Step 2 Determine the values that will cause either the numerator or the denominator of the rational expression to equal 0. These values determine the intervals on the number line to consider.

Step 3 Use a test value from each interval to determine which intervals form the solution set.

EXAMPLE 4 Solving a Rational Equation and Associated Inequalities

Consider the rational function defined by $f(x) = 7x^{-4} - 8x^{-2} + 1$.

(a) Solve the equation $f(x) = 0$.

(b) Graph the equation $y = f(x)$. Identify the zeros of $f(x)$.

(c) Use the graph to solve the inequalities $f(x) \leq 0$ and $f(x) \geq 0$.

Solution

(a) The expression for $f(x)$ may be written equivalently as

$$\frac{7}{x^4} - \frac{8}{x^2} + 1.$$

For this expression, $x \neq 0$. Solve the equation $f(x) = 0$.

$$\frac{7}{x^4} - \frac{8}{x^2} + 1 = 0$$

$$7 - 8x^2 + x^4 = 0 \qquad \text{Multiply by } x^4.$$

$$x^4 - 8x^2 + 7 = 0 \qquad \text{Rewrite.}$$

$$(x^2 - 7)(x^2 - 1) = 0 \qquad \text{Factor.}$$

$$(x^2 - 7)(x + 1)(x - 1) = 0 \qquad \text{Factor again.}$$

$$x^2 - 7 = 0 \quad \text{or} \quad x + 1 = 0 \quad \text{or} \quad x - 1 = 0 \qquad \text{Zero-product property}$$

$$x = \pm\sqrt{7} \quad \text{or} \quad x = -1 \quad \text{or} \quad x = 1 \qquad \text{Solve each equation.}$$

The solution set is $\{\pm\sqrt{7}, \pm 1\}$.

(b) Figure 27 shows that the graph of $f(x) = 7x^{-4} - 8x^{-2} + 1$ has four zeros. The display indicates that one zero is 2.6457513, an approximation for $\sqrt{7}$. The other zeros, $-\sqrt{7}$, -1, and 1, can be found similarly. The table provides further support.

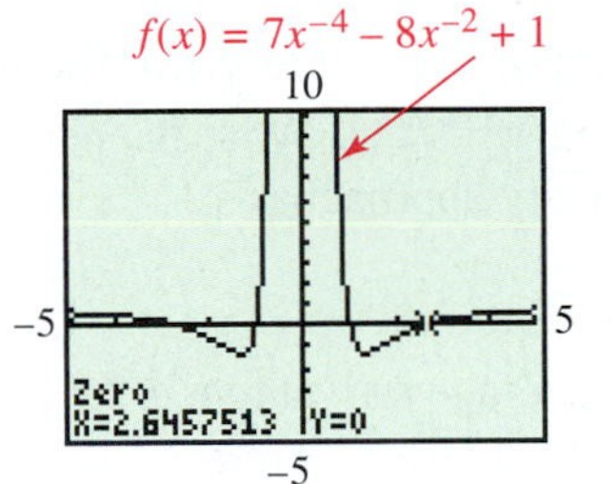

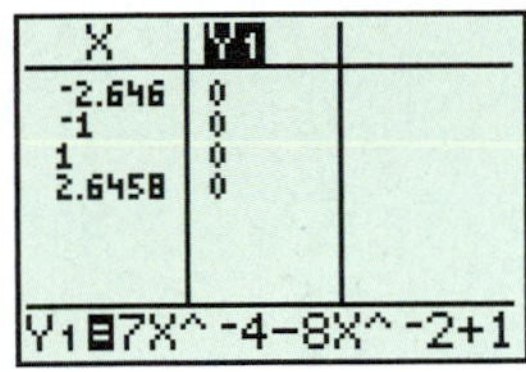

FIGURE 27

(c) The solution set of $f(x) \leq 0$ is $[-\sqrt{7}, -1] \cup [1, \sqrt{7}]$, since these are the values for which the graph lies *on* or *below* the x-axis. Similarly, the solution set of $f(x) \geq 0$ is

$$\left(-\infty, -\sqrt{7}\right] \cup [-1, 0) \cup (0, 1] \cup \left[\sqrt{7}, \infty\right).$$

Notice that 0 is not included, because it is not in the domain of $f(x)$. ■

FOR DISCUSSION

Tables can be used to numerically support analytic solutions of equations. In Example 4, we showed that the equation $7x^{-4} - 8x^{-2} + 1 = 0$ has solutions $-\sqrt{7}$, -1, 1, and $\sqrt{7}$. By creating a table with these four values, as in Figure 27, we can show that the function value in each case is 0. The X-values -2.646 and 2.6458 shown in the table are decimal approximations for $-\sqrt{7}$ and $\sqrt{7}$. We must enter these X-values directly as $-\sqrt{7}$ and $\sqrt{7}$; if we use approximations, we will not get 0 for the corresponding Y_1 values.

1. Use a table in the manner described to support the solution in Example 1.
2. Use a table to support the single solution in Example 2. What happens when you input the extraneous value 2?

Applications and Models of Rational Functions

EXAMPLE 5 Analyzing Traffic Intensity

Vehicles arrive randomly at a parking ramp at an average rate of 2.6 per minute. The parking attendant can admit 3.2 vehicles per minute. However, since arrivals are random, lines form at various times. (*Source:* Mannering, F. and W. Kilareski, *Principles of Highway Engineering and Traffic Control,* 2d ed., John Wiley and Sons, 1998.)

(a) The *traffic intensity* x is defined as the ratio of the average arrival rate to the average admittance rate. Determine x for this parking ramp.

(b) The average number of vehicles waiting in line to enter the ramp is modeled by

$$f(x) = \frac{x^2}{2(1 - x)},$$

where $0 \le x < 1$ is the traffic intensity. Compute $f(x)$ for this parking ramp.

(c) Graph $y = f(x)$. What happens to the number of vehicles waiting as the traffic intensity approaches 1?

Solution

(a) The average arrival rate is 2.6 vehicles per minute and the average admittance rate is 3.2 vehicles per minute, so

$$x = \frac{2.6}{3.2} = .8125.$$

(b) In part (a), we found that $x = .8125$. Thus,

$$f(.8125) = \frac{.8125^2}{2(1 - .8125)} \approx 1.76 \text{ vehicles.}$$

(c) From the graph shown in Figure 28, we see that as x approaches 1, $y = f(x)$ gets very large; that is, the number of waiting vehicles gets very large. A small increase in the traffic intensity can result in a dramatic increase in the number of vehicles waiting in line.

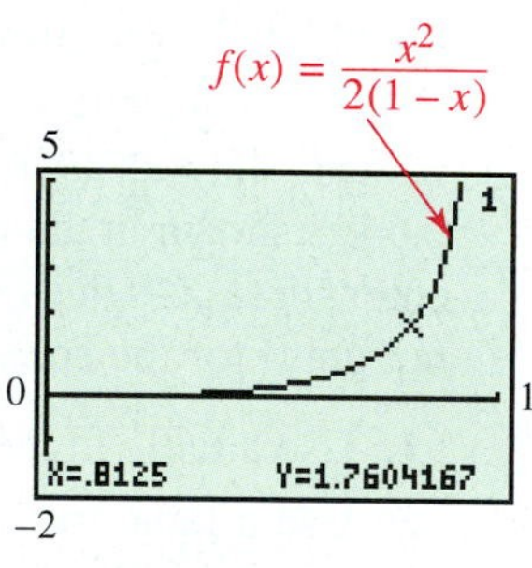

FIGURE 28

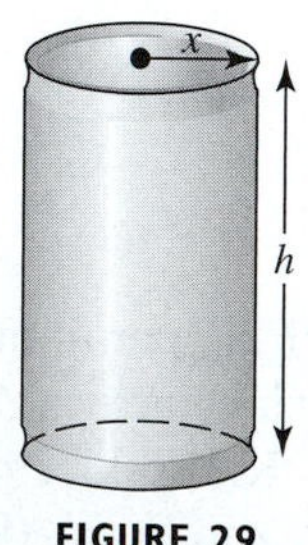

FIGURE 29

EXAMPLE 6 Solving a Problem Involving Manufacturing

A manufacturer wants to construct cylindrical aluminum cans with volume 2000 cubic centimeters (2 liters). What radius and height of the can will minimize the amount of aluminum used? What will this amount be?

Solution The two unknowns are the radius x and the height h of the can, as shown in Figure 29. Minimizing the amount of aluminum used requires minimizing the surface area S of the can, given by the formula

$$S = 2\pi xh + 2\pi x^2. \quad x \text{ is radius; } h \text{ is height.}$$

Looking Ahead to Calculus

The problem in Example 6 cannot be solved in college algebra without the aid of a computer or a graphing calculator. The traditional calculus method of solving this problem would involve (analytically) finding the appropriate zero of the derivative of $S(x)$:

$$S'(x) = -4000x^{-2} + 4\pi x.$$

Since the volume of the can is to be 2000 cubic centimeters and the formula for volume is $V = \pi x^2 h$ (where x is radius and h is height), we have

$$V = \pi x^2 h$$

$$2000 = \pi x^2 h \quad \text{Let } V = 2000.$$

$$h = \frac{2000}{\pi x^2}. \quad \text{Solve for } h.$$

Now we can write the surface area S as a function of x alone.

$$S(x) = 2\pi x\left(\frac{2000}{\pi x^2}\right) + 2\pi x^2 \quad \text{Substitute for } h.$$

$$= \frac{4000}{x} + 2\pi x^2 \quad *$$

$$= \frac{4000 + 2\pi x^3}{x} \quad \text{Combine terms.*}$$

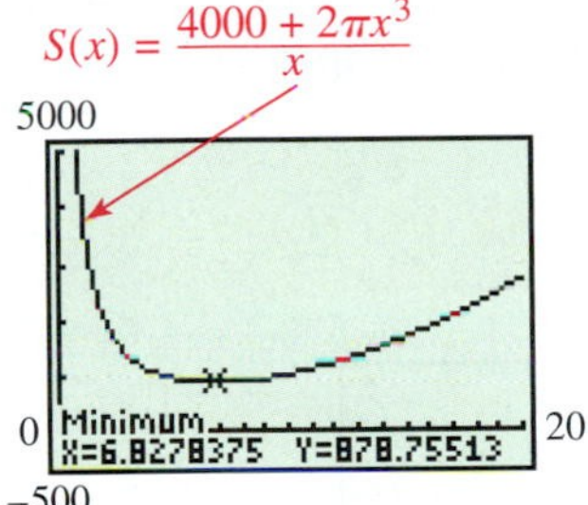

The local minimum point is approximately (6.83, 878.76).

FIGURE 30

Since x represents the radius, it must be a positive number. We graph the function and find that the local minimum point is approximately (6.83, 878.76). See Figure 30. Therefore, rounded to the nearest hundredth, the radius should be 6.83 centimeters and the height should be $\frac{2000}{\pi(6.83)^2}$, or about 13.65 centimeters. These dimensions lead to a minimum amount of 878.76 square centimeters of aluminum used. ■

Inverse Variation

Recall from Section 1.6 that when two quantities vary directly, an increase in one quantity results in an increase in the other. When two quantities *vary inversely,* an *increase* in one quantity results in a *decrease* in the other.

Inverse Variation as the *n*th Power

Let x and y denote two quantities and n be a positive number. Then y is **inversely proportional to the *n*th power** of x, or y **varies inversely as the *n*th power** of x, if there exists a nonzero number k such that

$$y = \frac{k}{x^n}.$$

If $y = \frac{k}{x}$, then y is **inversely proportional** to x, or y **varies inversely** as x.

For example, it takes 4 hours to travel 100 miles at 25 miles per hour, but only 2 hours to travel 100 miles at 50 miles per hour. Greater speed results in less travel time. If s represents the average speed of a car and t is the time to travel 100 miles, then $s \cdot t = 100$ or $t = \frac{100}{s}$. Doubling the speed cuts the time in half. The quantities t and s vary inversely. The constant of variation here is 100.

*These steps are not necessary to obtain the appropriate graph.

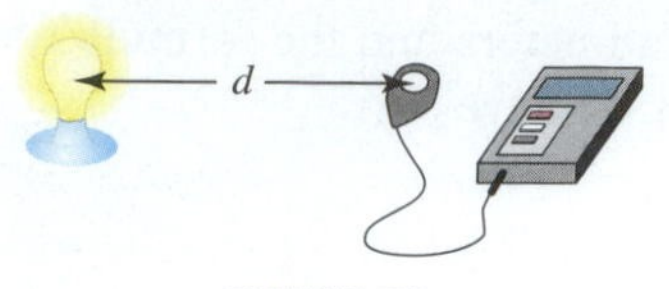

FIGURE 31

EXAMPLE 7 Modeling the Intensity of Light

The intensity of light I is inversely proportional to the second power of the distance d. See Figure 31. The equation

$$I = \frac{k}{d^2}$$

models this phenomenon. At a distance of 3 meters, a 100-watt bulb produces an intensity of .88 watt per square meter. (*Source:* Weidner, R. and R. Sells, *Elementary Classical Physics,* Volume 2, Allyn and Bacon, 1965.)

Find the constant of variation k, and then determine the intensity of the light at a distance of 2 meters.

Analytic Solution

Substitute $d = 3$ and $I = .88$ into the variation equation, and solve for k.

$$I = \frac{k}{d^2}$$

$$.88 = \frac{k}{3^2}$$

$$k = 7.92$$

Since $k = 7.92$, the equation is

$$I = \frac{7.92}{d^2}, \quad \text{and when } d = 2, \quad I = \frac{7.92}{2^2} = 1.98.$$

The intensity at 2 meters is 1.98 watts per square meter.

Graphing Calculator Solution

From the analytic solution, $k = 7.92$. The formula $I = \frac{7.92}{d^2}$ can be written as $f(x) = \frac{7.92}{x^2}$, a rational function. Enter this function into a calculator, and graph it for positive values of x. Calculate $f(2)$ by locating the point having $x = 2$, and determine that $f(2) = 1.98$. See Figure 32.

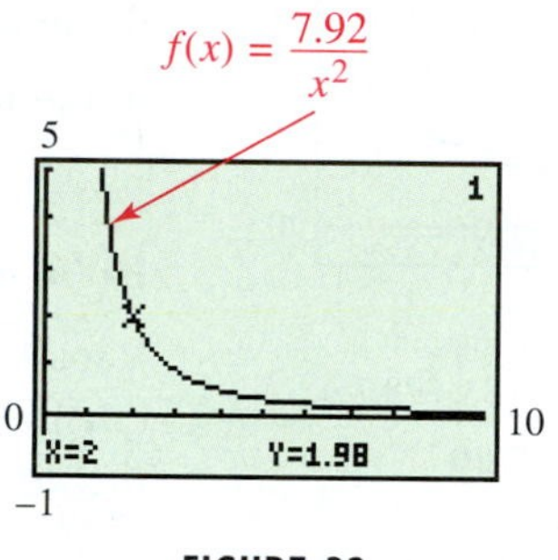

FIGURE 32

Combined and Joint Variation

One variable may depend on more than one other variable. Such variation is called **combined variation.** More specifically, when a variable depends on the *product* of two or more other variables, we refer to it as *joint variation.*

Joint Variation

Let m and n be real numbers. Then z **varies jointly** as the nth power of x and the mth power of y if a nonzero real number k exists such that

$$z = kx^n y^m.$$

For example, the volume of a cylinder is given by $V = \pi r^2 h$. We say that V *varies jointly* as h and the square of r. The constant of variation here is π.

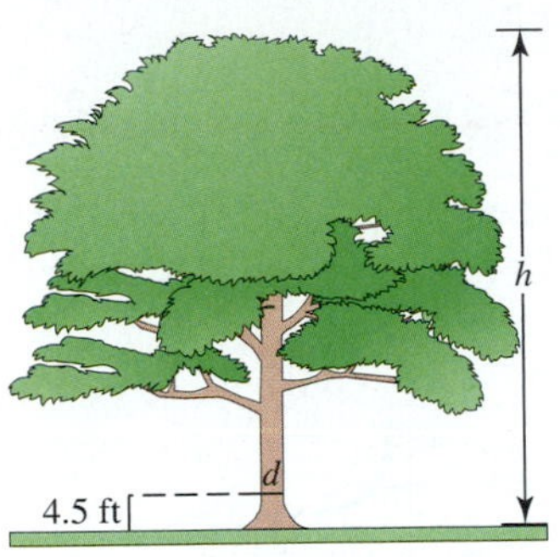

FIGURE 33

EXAMPLE 8 Modeling the Amount of Wood in a Tree

In forestry, formulas are developed to find the amount of wood contained in a tree of height h and diameter d. See Figure 33. The volume V of wood in a tree varies jointly as the 1.12 power of h and the 1.98 power of d. (The diameter is measured 4.5 feet above the ground.) (*Source:* Ryan, B., B. Joiner, and T. Ryan, *Minitab Handbook,* Duxbury Press, 1985.)

(a) Write an equation that relates V, h, and d.

(b) A tree with a 13.8-inch diameter and a 64-foot height has a volume of 25.14 cubic feet. Estimate the constant of variation k.

(c) Estimate the volume of wood in a tree with $d = 11$ inches and $h = 47$ feet.

Solution

(a) Let $V = kh^{1.12}d^{1.98}$, where k is the constant of variation.

(b) Substitute $d = 13.8$, $h = 64$, and $V = 25.14$ into the equation, and solve for k.

$$25.14 = k(64^{1.12})(13.8^{1.98})$$

$$k = \frac{25.14}{(64^{1.12})(13.8^{1.98})} \approx .00132$$

Thus, $V = .00132h^{1.12}d^{1.98}$.

(c) For the given tree, $V = .00132(47^{1.12})(11^{1.98}) \approx 11.4$ cubic feet. ■

This screen illustrates how the computations in Examples 8(b) and (c) are made using a graphing calculator.

EXAMPLE 9 Solving a Combined Variation Problem in Photography

In the photography formula

$$L = \frac{25F^2}{st},$$

the luminance L (in foot-candles) varies directly as the square of the F-stop F and inversely as the product of the film ASA number s and the shutter speed t. The constant of variation is 25.

Suppose we want to use 200 ASA film and a shutter speed of $\frac{1}{250}$ when 500 foot-candles of light are available. What would be an appropriate F-stop?

Solution Substitute the given values for the variables. Then, solve for F.

$$L = \frac{25F^2}{st}$$

$$500 = \frac{25F^2}{200\left(\frac{1}{250}\right)} \qquad L = 500,\ s = 200,\ t = \tfrac{1}{250}$$

$$400 = 25F^2 \qquad \text{Simplify.}$$

$$16 = F^2 \qquad \text{Divide by 25.}$$

$$4 = F \qquad \text{Square root property; } F > 0$$

An F-stop of 4 would be appropriate. ■

4.3 Exercises

In Exercises 1–12, the graph of a rational function $y = f(x)$ is given. Use the graph to give the solution set of **(a)** $f(x) = 0$, **(b)** $f(x) < 0$, *and* **(c)** $f(x) > 0$. *Use set notation for part (a) and interval notation for parts (b) and (c).*

1.

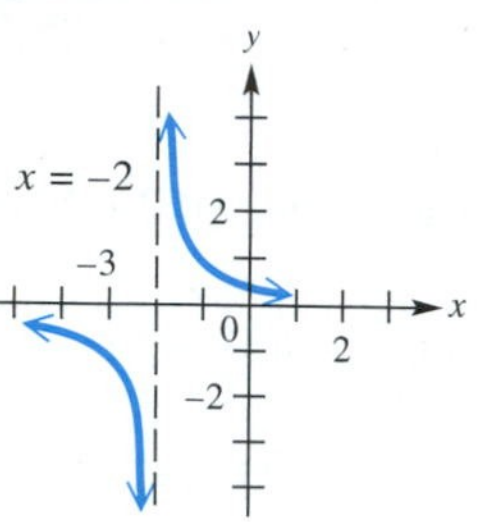

2.

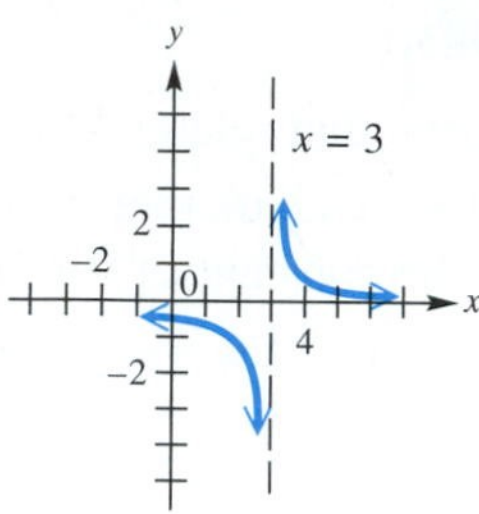

3.

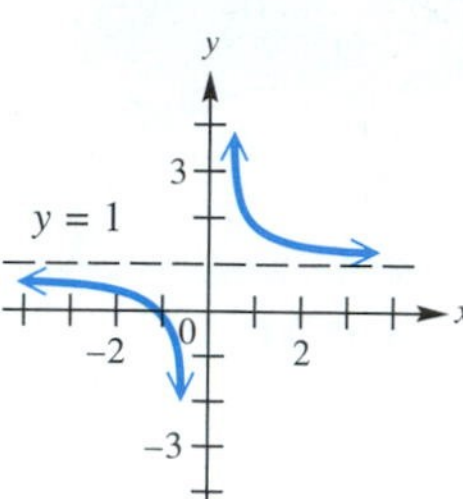

4.

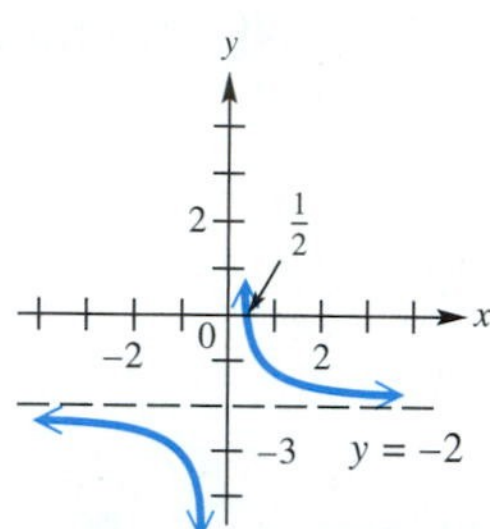

5.

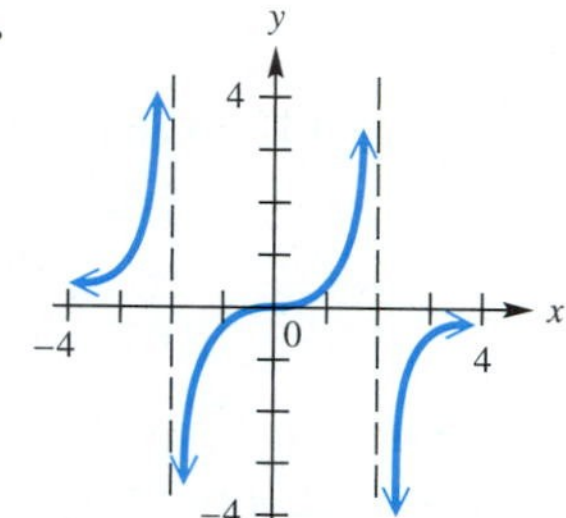

6.

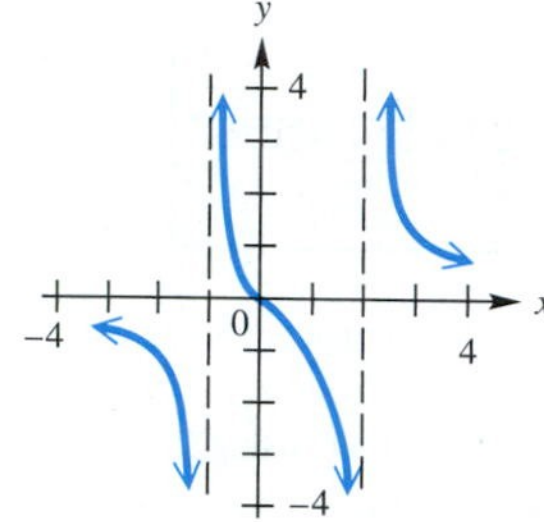

7.

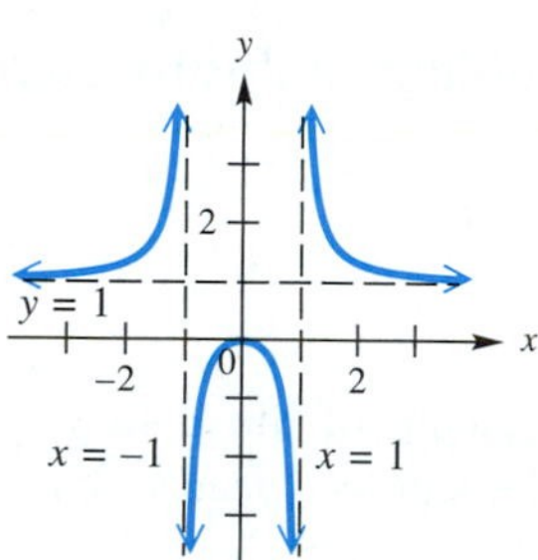

8.

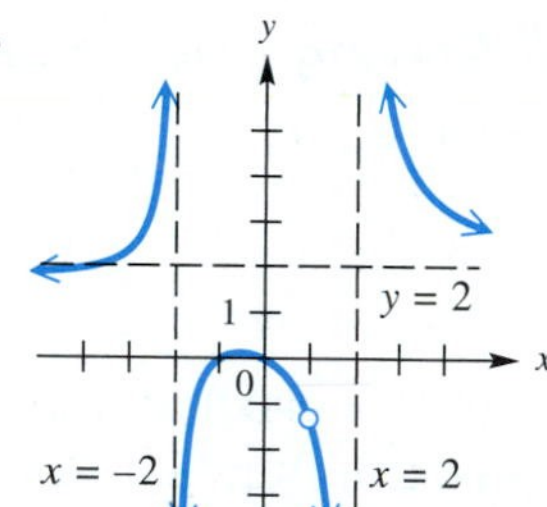

9.

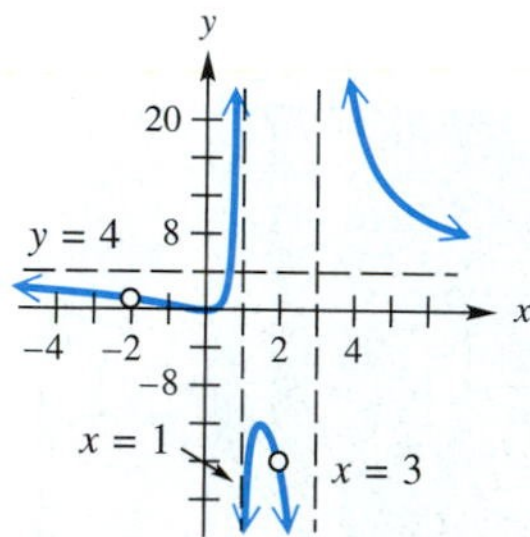

10.

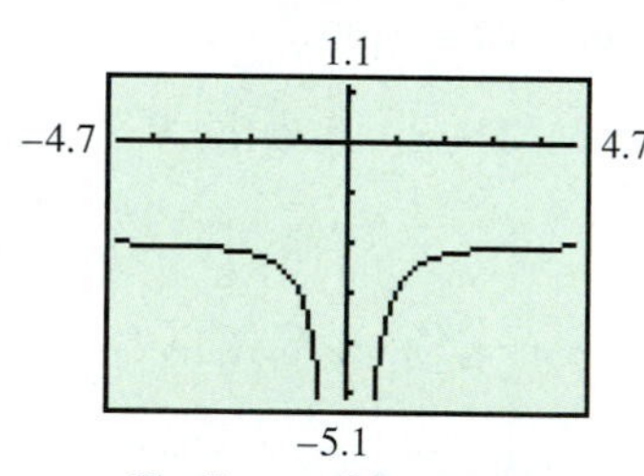

The line $x = 0$ is a vertical asymptote.

11.

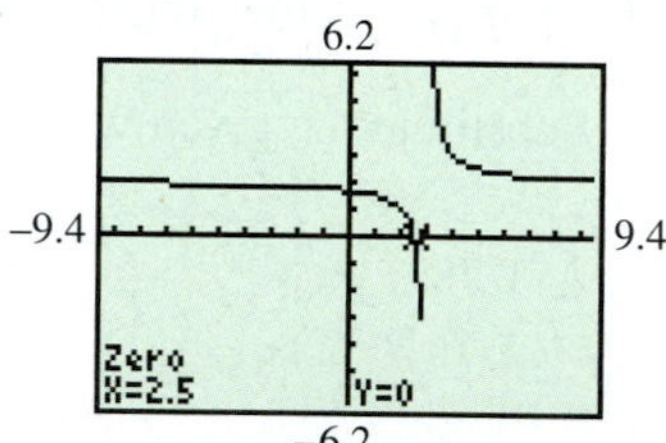

The line $x = 3$ is a vertical asymptote.

12.

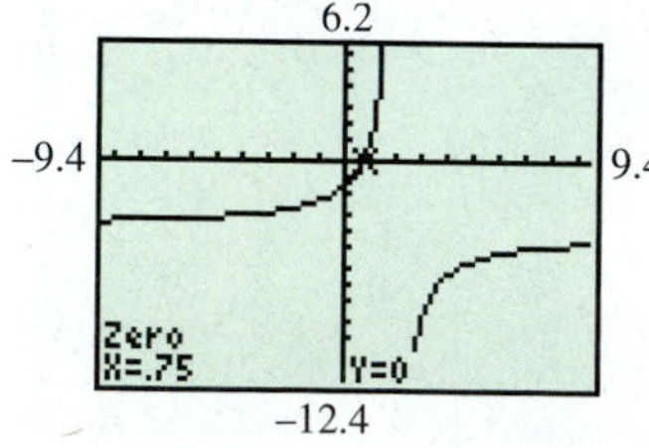

The line $x = 2$ is a vertical asymptote.

13. A student attempted to solve the inequality at the right by multiplying each side by $x + 2$. He wrote the solution set as $\left(-\infty, \frac{1}{2}\right]$. Is his solution correct? Explain.

$$\frac{2x - 1}{x + 2} \le 0$$

$$2x - 1 \le 0$$

$$x \le \frac{1}{2}$$

14. ***Concept Check*** Explain why it is incorrect to solve the inequality $\frac{1}{x} < 1$ by multiplying each side of the inequality by x.

Use the methods of Examples 1 and 3 to solve the rational equation and associated inequalities given in Exercises 15–26. Then, support your answer by using the x-intercept method with a calculator graph in the suggested window.

15. **(a)** $\frac{x-3}{x+5} = 0$ **(b)** $\frac{x-3}{x+5} \le 0$ **(c)** $\frac{x-3}{x+5} \ge 0$

Window: $[-10, 10]$ by $[-5, 8]$

16. **(a)** $\frac{x+1}{x-4} = 0$ **(b)** $\frac{x+1}{x-4} \ge 0$ **(c)** $\frac{x+1}{x-4} \le 0$

Window: $[-10, 10]$ by $[-5, 10]$

17. **(a)** $\frac{x-1}{x+2} = 1$ **(b)** $\frac{x-1}{x+2} > 1$ **(c)** $\frac{x-1}{x+2} < 1$

Window: $[-10, 10]$ by $[-5, 10]$

18. **(a)** $\frac{x-6}{x+2} = -1$ **(b)** $\frac{x-6}{x+2} < -1$ **(c)** $\frac{x-6}{x+2} > -1$

Window: $[-10, 10]$ by $[-10, 10]$

19. **(a)** $\frac{1}{x-1} = \frac{5}{4}$ **(b)** $\frac{1}{x-1} < \frac{5}{4}$ **(c)** $\frac{1}{x-1} > \frac{5}{4}$

Window: $[-5, 5]$ by $[-5, 5]$

20. **(a)** $\frac{6}{5-3x} = 2$ **(b)** $\frac{6}{5-3x} \le 2$ **(c)** $\frac{6}{5-3x} \ge 2$

Window: $[-5, 5]$ by $[-5, 5]$

21. **(a)** $\frac{4}{x-2} = \frac{3}{x-1}$ **(b)** $\frac{4}{x-2} \le \frac{3}{x-1}$ **(c)** $\frac{4}{x-2} \ge \frac{3}{x-1}$

Window: $[-3, 3]$ by $[-20, 20]$

22. **(a)** $\frac{4}{x+1} = \frac{2}{x+3}$ **(b)** $\frac{4}{x+1} < \frac{2}{x+3}$ **(c)** $\frac{4}{x+1} > \frac{2}{x+3}$

Window: $[-8, 5]$ by $[-10, 10]$

23. **(a)** $\frac{1}{(x-2)^2} = 0$ **(b)** $\frac{1}{(x-2)^2} < 0$ **(c)** $\frac{1}{(x-2)^2} > 0$

Window: $[-5, 10]$ by $[-5, 10]$

24. **(a)** $\frac{-2}{(x+3)^2} = 0$ **(b)** $\frac{-2}{(x+3)^2} > 0$ **(c)** $\frac{-2}{(x+3)^2} < 0$

Window: $[-10, 5]$ by $[-10, 5]$

25. **(a)** $\frac{5}{x+1} = \frac{12}{x+1}$ **(b)** $\frac{5}{x+1} > \frac{12}{x+1}$ **(c)** $\frac{5}{x+1} < \frac{12}{x+1}$

Window: $[-10, 10]$ by $[-10, 10]$

26. **(a)** $\frac{7}{x+2} = \frac{1}{x+2}$ **(b)** $\frac{7}{x+2} \ge \frac{1}{x+2}$ **(c)** $\frac{7}{x+2} \le \frac{1}{x+2}$

Window: $[-10, 10]$ by $[-10, 10]$

In some cases, it is possible to solve a rational inequality simply by deciding what sign the numerator and the denominator must have and then using the rules for quotients of positive and negative numbers to determine the solution set. For example, consider the rational inequality

$$\frac{1}{x^2+1} > 0.$$

The numerator of the rational expression, 1, is positive, and the denominator, $x^2 + 1$, must always be positive because it is the sum of a nonnegative number, x^2, and a positive number, 1. Therefore, the rational expression is the quotient of two positive numbers,

(continued)

which is positive. Because the inequality requires that the rational expression be greater than 0, and this will always be true, the solution set is $(-\infty, \infty)$.

Use similar reasoning to solve each inequality.

27. $\dfrac{-1}{x^2+2} < 0$

28. $\dfrac{5}{x^2+2} < 0$

29. $\dfrac{-5}{x^2+2} > 0$

30. $\dfrac{x^4+2}{-6} \le 0$

31. $\dfrac{x^4+2}{-6} \ge 0$

32. $\dfrac{x^4+x^2+3}{x^2+2} < 0$

33. $\dfrac{x^4+x^2+3}{x^2+2} > 0$

34. $\dfrac{(x-1)^2}{x^2+4} > 0$

35. $\dfrac{(x-1)^2}{x^2+4} \le 0$

36. ***Concept Check*** Let $f(x) = \dfrac{x^2-4}{x^2-4}$. Solve each equation or inequality.

(a) $f(x) = 0$ **(b)** $f(x) < 0$ **(c)** $f(x) > 0$ **(d)** $f(x) = 1$

Find all complex solutions for each equation. Support your real solutions with a graph, using an appropriate window.

37. $\dfrac{2x}{x^2-1} = \dfrac{2}{x+1} - \dfrac{1}{x-1}$

38. $\dfrac{8x}{4x^2-1} = \dfrac{3}{2x+1} + \dfrac{3}{2x-1}$

39. $\dfrac{4}{x^2-3x} - \dfrac{1}{x^2-9} = 0$

40. $\dfrac{2}{x^2-2x} - \dfrac{3}{x^2-x} = 0$

41. $1 - \dfrac{13}{x} + \dfrac{36}{x^2} = 0$

42. $1 - \dfrac{3}{x} - \dfrac{10}{x^2} = 0$

43. $1 + \dfrac{3}{x} = \dfrac{5}{x^2}$

44. $4 + \dfrac{7}{x} = -\dfrac{1}{x^2}$

45. $\dfrac{x}{2-x} + \dfrac{2}{x} - 5 = 0$

46. $\dfrac{2x}{x-3} + \dfrac{4}{x} - 6 = 0$

47. $x^{-4} - 3x^{-2} - 4 = 0$

48. $x^{-4} - 5x^{-2} - 36 = 0$

49. $\dfrac{1}{x+2} + \dfrac{3}{x+7} = \dfrac{5}{x^2+9x+14}$

50. $\dfrac{1}{x+3} + \dfrac{4}{x+5} = \dfrac{2}{x^2+8x+15}$

51. $\dfrac{x}{x-3} + \dfrac{4}{x+3} = \dfrac{18}{x^2-9}$

52. $\dfrac{2x}{x-3} + \dfrac{4}{x+3} = \dfrac{24}{9-x^2}$

53. $9x^{-1} + 4(6x-3)^{-1} = 2(6x-3)^{-1}$

54. $x(x-2)^{-1} + x(x+2)^{-1} = 8(x^2-4)^{-1}$

Solve each rational inequality.

55. $\dfrac{(x+1)^2}{x-2} \le 0$

56. $\dfrac{2x}{(x-2)^2} > 0$

57. $\dfrac{3-2x}{1+x} < 0$

58. $\dfrac{x+1}{4-2x} \ge 1$

59. $\dfrac{(x+1)(x-2)}{(x+3)} < 0$

60. $\dfrac{x(x-3)}{x+2} \ge 0$

61. $\dfrac{2x-5}{x^2-1} \ge 0$

62. $\dfrac{5-x}{x^2-x-2} < 0$

63. $\dfrac{1}{x-3} \le \dfrac{5}{x-3}$

64. $\dfrac{3}{2-x} > \dfrac{x}{2+x}$

65. $2 - \dfrac{5}{x} + \dfrac{2}{x^2} \ge 0$

66. $\dfrac{1}{x-1} + \dfrac{1}{x+1} > \dfrac{3}{4}$

Solve the equation in part (a) graphically, expressing solutions to the nearest hundredth. Then, use the graph to solve the associated inequalities in parts (b) and (c), expressing endpoints to the nearest hundredth.

67. (a) $\dfrac{\sqrt{2}x+5}{x^3-\sqrt{3}} = 0$ **(b)** $\dfrac{\sqrt{2}x+5}{x^3-\sqrt{3}} > 0$

(c) $\dfrac{\sqrt{2}x+5}{x^3-\sqrt{3}} < 0$

68. (a) $\dfrac{\sqrt[3]{7}x^3-1}{x^2+2} = 0$ **(b)** $\dfrac{\sqrt[3]{7}x^3-1}{x^2+2} > 0$

(c) $\dfrac{\sqrt[3]{7}x^3-1}{x^2+2} < 0$

(Modeling) Solve each problem.

69. ***Insect Population*** Suppose that an insect population in millions is modeled by

$$f(x) = \frac{10x + 1}{x + 1},$$

where $x \geq 0$ is in months.

(a) Graph f in the window $[0, 14]$ by $[0, 14]$. Find the equation of the horizontal asymptote.

(b) Determine the initial insect population.

(c) What happens to the population after several months?

(d) Interpret the horizontal asymptote.

70. ***Fish Population*** Suppose that the population of a species of fish in thousands is modeled by

$$f(x) = \frac{x + 10}{.5x^2 + 1},$$

where $x \geq 0$ is in years.

(a) Graph f in the window $[0, 12]$ by $[0, 12]$. What is the equation of the horizontal asymptote?

(b) Determine the initial population.

(c) What happens to the population of this fish after many years?

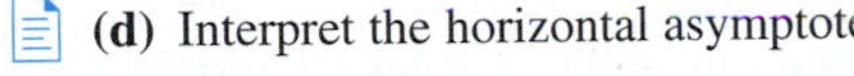

(d) Interpret the horizontal asymptote.

71. ***Time Spent in Line*** Suppose the average number of vehicles arriving at the main gate of an amusement park is equal to 10 per minute, while the average number of vehicles being admitted through the gate per minute is equal to x. Then the average waiting time in minutes for each vehicle at the gate is given by

$$f(x) = \frac{x - 5}{x^2 - 10x},$$

where $x > 10$. (*Source:* Mannering, F. and W. Kilareski, *Principles of Highway Engineering and Traffic Analysis,* 2d. ed., John Wiley and Sons, 1998.)

(a) Estimate the admittance rate x that results in an average wait of 15 seconds.

(b) If one attendant can serve 5 vehicles per minute, how many attendants are needed to keep the average wait to 15 seconds or less?

72. ***Length of Lines*** See Example 5. Determine the traffic intensity x when the average number of vehicles in line equals 3.

73. ***Construction*** Find possible dimensions for a closed box with volume 196 cubic inches, surface area 280 square inches, and length that is twice the width.

74. ***Volume of a Cylindrical Can*** A metal cylindrical can with an *open top* and *closed bottom* is to have volume 4 cubic feet. Find the dimensions that require the least amount of material. What would this amount be? (Compare this problem with Example 6.)

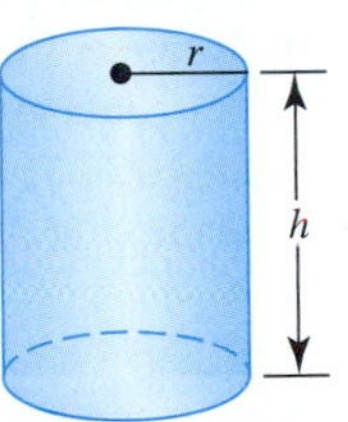

75. ***Train Curves*** When curves are designed for trains, sometimes the outer rail is elevated or *banked* so that a locomotive can safely negotiate the curve at a higher speed. Suppose a circular curve is being designed for a speed of 60 mph. The rational function given by

$$f(x) = \frac{2540}{x}$$

computes the elevation y in inches of the outer track for a curve with a radius of x feet, where $y = f(x)$. (*Source:* Haefner, L., *Introduction to Transportation Systems,* Holt, Rinehart and Winston, 1986.)

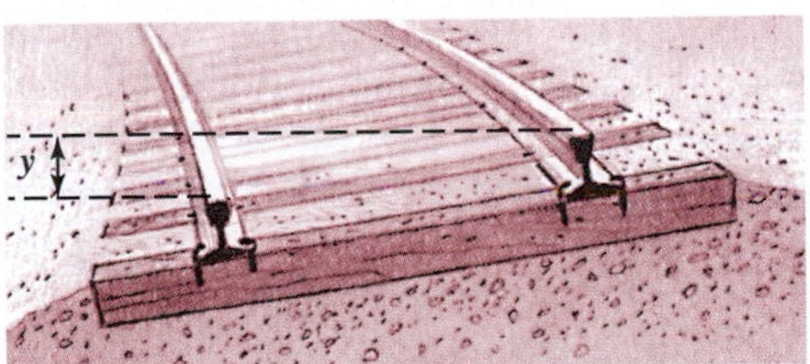

(a) Evaluate $f(400)$ and interpret its meaning.

(b) Graph f in the window $[0, 600]$ by $[0, 50]$. Discuss how the elevation of the outer rail changes with the radius x.

(c) Interpret the horizontal asymptote.

(d) What radius is associated with an elevation of 12.7 inches?

76. ***Recycling*** A cost–benefit function C computes the cost in millions of dollars of implementing a city recycling project when x percent of the citizens participate, where

$$C(x) = \frac{1.2x}{100 - x}.$$

(a) Graph C in the window $[0, 100]$ by $[0, 10]$. Interpret the graph as x approaches 100.

(b) If 75% participation is expected, determine the cost for the city.

(c) The city plans to spend \$5 million on this recycling project. Estimate graphically the percentage of participation that can be expected.

(d) Solve part (c) analytically.

77. ***Braking Distance*** The *grade* x of a hill is a measure of its steepness. For example, if a road rises 10 feet for every 100 feet of horizontal distance, then it has an uphill grade of $x = \frac{10}{100}$, or 10%. Grades are typically kept quite small—usually less than 10%. The braking (or stopping) distance D for a car traveling at 50 mph on a wet, uphill grade is given by

$$D(x) = \frac{2500}{30(.3 + x)}.$$

(*Source:* Haefner, L., *Introduction to Transportation Systems,* Holt, Rinehart and Winston, 1986.)

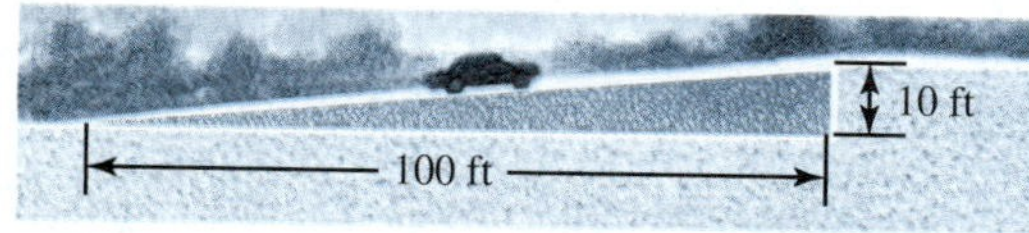

(a) Evaluate $D(.05)$ and interpret the result.

(b) Describe what happens to braking distance as the hill becomes steeper. Does this agree with your driving experience?

(c) Estimate the grade associated with a braking distance of 220 feet.

78. ***Braking Distance*** See Exercise 77. If a car is traveling 50 mph downhill, then its braking distance on wet pavement is given by

$$D(x) = \frac{2500}{30(.3 + x)},$$

where $x < 0$ for a downhill grade.

(a) Evaluate $D(-.1)$ and interpret the result.

(b) What happens to braking distance as the downhill grade becomes steeper? Does this agree with your driving experience?

(c) The graph of D has a vertical asymptote at $x = -.3$. Give the physical significance of this asymptote.

(d) Estimate the grade associated with a braking distance of 350 feet.

Solve each problem.

79. Suppose r varies directly as the square of m and inversely as s. If $r = 12$ when $m = 6$ and $s = 4$, find r when $m = 4$ and $s = 10$.

80. Suppose p varies directly as the square of z and inversely as r. If $p = \frac{32}{5}$ when $z = 4$ and $r = 10$, find p when $z = 2$ and $r = 16$.

81. Let a vary directly as m and n^2 and inversely as y^3. If $a = 9$ when $m = 4$, $n = 9$, and $y = 3$, find a if $m = 6$, $n = 2$, and $y = 5$.

82. If y varies directly as x and inversely as m^2 and r^2, and $y = \frac{5}{3}$ when $x = 1$, $m = 2$, and $r = 3$, find y if $x = 3$, $m = 1$, and $r = 8$.

83. For $k > 0$, if y varies directly as x, when x increases, y ________, and when x decreases, y ________.

84. For $k > 0$, if y varies inversely as x, when x increases, y ________, and when x decreases, y ________.

In Exercises 85–88, assume that the constant of variation is positive.

85. Let y be inversely proportional to x. If x doubles, what happens to y?

86. Let y vary inversely as the second power of x. If x doubles, what happens to y?

87. Suppose y varies directly as the third power of x. If x triples, what happens to y?

88. Suppose y is directly proportional to the second power of x. If x is halved, what happens to y?

(Modeling) Solve each problem.

89. ***Body Mass Index*** In 1998, the federal government developed the *body mass index* (BMI) to determine ideal weights. A person's BMI is directly proportional to his or her weight in pounds and inversely proportional to the square of his or her height in inches. (A BMI of 19 to 25 corresponds to a healthy weight.) A 6-foot-tall person weighing 177 pounds has a BMI of 24. Find the BMI (to the nearest whole number) of a person whose weight is 130 pounds and whose height is 66 inches. *(Source: Washington Post.)*

90. ***Volume of a Gas*** Natural gas provides 35.8% of U.S. energy. (*Source:* U.S. Energy Department.) The volume of a gas varies inversely as the pressure and directly as the temperature. (Temperature must be measured in degrees *Kelvin* (K), a unit of measurement used in physics.) If a certain gas occupies a volume of 1.3 liters at 300 K and a pressure of 18 newtons per square centimeter, find the volume at 340 K and a pressure of 24 newtons per square centimeter.

91. ***Electrical Resistance*** The electrical resistance R of a wire varies inversely as the square of its diameter d. If a 25-foot wire with diameter 2 millimeters has resistance .5 ohm, find the resistance of a wire having the same length and diameter 3 millimeters.

92. ***Poiseuille's Law*** According to Poiseuille's law, the resistance to flow of a blood vessel R is directly proportional to the length l and inversely proportional to the fourth power of the radius r. (*Source:* Hademenos, George J., "The Biophysics of Stroke," *American Scientist,* May–June 1997.) If $R = 25$ when $l = 12$ and $r = .2$, find R as r increases to .3 while l is unchanged.

93. ***Gravity*** The weight of an object varies inversely as the square of its distance from the center of Earth. The radius of Earth is approximately 4000 miles. If a person weighs 160 pounds on Earth's surface, what would this individual weigh 8000 miles above the surface of Earth?

94. ***Hubble Telescope*** The brightness or intensity of starlight varies inversely as the square of its distance from Earth. The Hubble Telescope can see stars whose intensities are $\frac{1}{50}$ of the faintest star now seen by ground-based telescopes. Determine how much farther the Hubble Telescope can see into space than ground-based telescopes. (*Source:* National Aeronautics and Space Administration.)

95. ***Volume of a Cylinder*** The volume of a right circular cylinder is jointly proportional to the square of the radius of the circular base and to the height. If the volume is 300 cubic centimeters when the height is 10.62 centimeters and the radius is 3 centimeters, find the volume of a cylinder with radius 4 centimeters and height 15.92 centimeters.

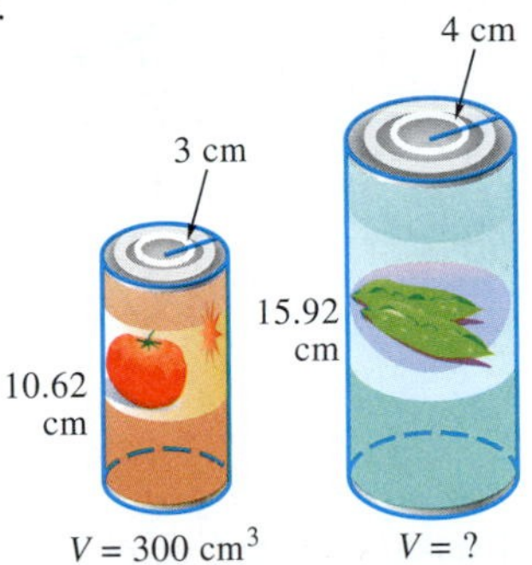

96. ***Photography*** See Example 9. If 125 foot-candles of light are available and an F-stop of 2 is used with 200 ASA film, what shutter speed should be used?

In each table, $Y_1 = \frac{k}{X}$ *for some value of k. Find the value of k.*

97.

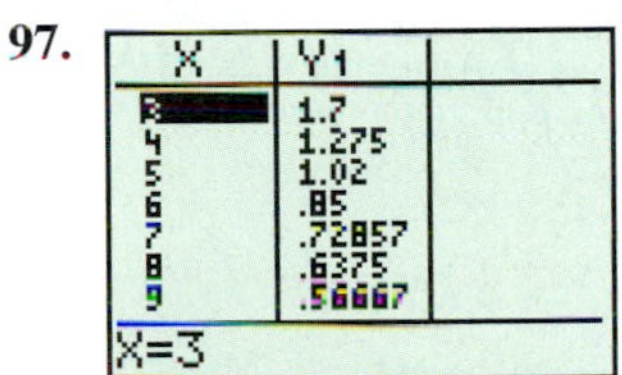

X	Y_1
3	1.7
4	1.275
5	1.02
6	.85
7	.72857
8	.6375
9	.56667

X=3

98.

X	Y_1
5	.58
6	.48333
7	.41429
8	.3625
9	.32222
10	.29
11	.26364

X=5

99.

X	Y_1
-3	.46667
-2	.7
-1	1.4
0	ERROR
1	-1.4
2	-.7
3	-.4667

X= -3

100.

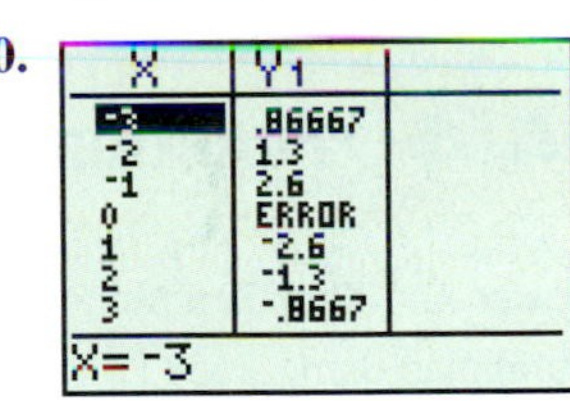

X	Y_1
-3	.86667
-2	1.3
-1	2.6
0	ERROR
1	-2.6
2	-1.3
3	-.8667

X= -3

Reviewing Basic Concepts (Sections 4.1–4.3)

1. Sketch the graph of $y = \frac{1}{x + 2} - 3$. Then obtain an accurate representation of the graph by using a graphing calculator.

2. What is the domain of the rational function defined by $f(x) = \frac{3}{x^2 - 1}$?

3. What is the equation of the vertical asymptote of the graph of $f(x) = \frac{4x + 3}{x - 6}$?

4. What is the equation of the horizontal asymptote of the graph of $f(x) = \frac{x^2 + 3}{x^2 - 4}$?

5. What is the equation of the oblique asymptote of the graph of $f(x) = \frac{x^2 + x + 5}{x + 3}$?

6. Sketch the graph of $f(x) = \frac{3x + 6}{x - 4}$ by hand. Then obtain an accurate representation of the graph by using a graphing calculator.

7. The graph of a rational function f is shown below. Give the solution set of each equation or inequality.
(a) $f(x) = 0$ **(b)** $f(x) > 0$ **(c)** $f(x) < 0$

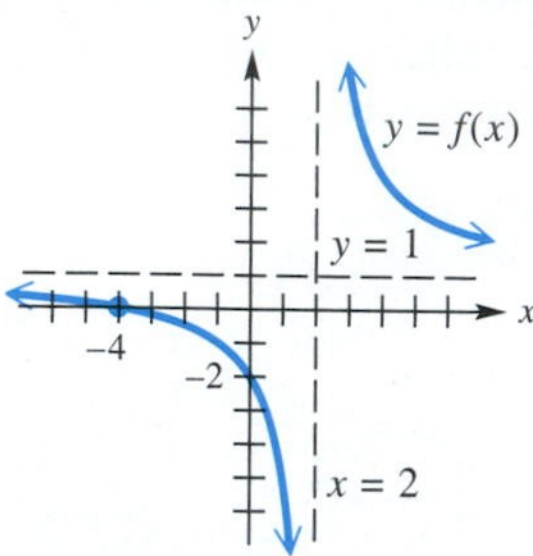

8. Find the solution set of $\dfrac{x + 4}{3x + 1} > 1$.

9. Fill in the blanks with the correct responses: $b = \frac{24}{h}$ is the formula for the base of a parallelogram with area 24. The base of this parallelogram varies ________ as its ________. The constant of variation is ________.

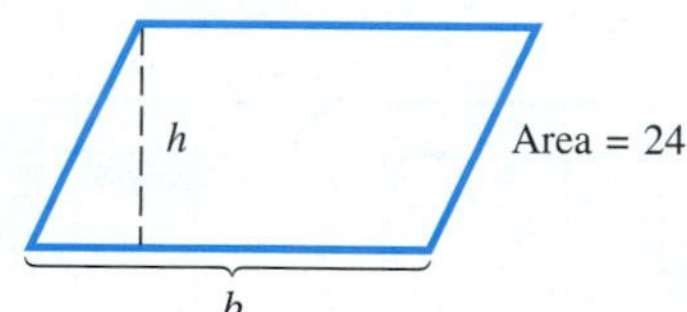

10. *(Modeling) Vibrations of a Guitar String* The number of vibrations per second (the pitch) of a steel guitar string varies directly as the square root of the tension and inversely as the length of the string. If the number of vibrations per second is 5 when the tension is 225 kilograms and the length is .60 meter, find the number of vibrations per second when the tension is 196 kilograms and the length is .65 meter.

4.4 Functions Defined by Powers and Roots

Power and Root Functions ■ Modeling Using Power Functions ■ Graphs of $f(x) = \sqrt[n]{ax + b}$ ■ Graphing Circles and Horizontal Parabolas Using Root Functions

Power and Root Functions

Planet	x	y
Mercury	.387	.241
Venus	.723	.615
Earth	1.00	1.00
Mars	1.52	1.88
Jupiter	5.20	11.9
Saturn	9.54	29.5

Source: Ronan, C., *The Natural History of the Universe*, MacMillan, 1991.

Johannes Kepler (1571–1630) was the first to recognize that the orbits of planets are elliptical, rather than circular. He also found that a *power function* models the relationship between a planet's distance from the sun and its period of revolution. The table lists the average distance x from the sun and the time y in years for several planets to orbit the sun. The distance x has been normalized so that Earth is 1 unit away from the sun. For example, Jupiter is 5.20 times farther from the sun than Earth and requires 11.9 years to orbit the sun.

A scatter diagram of the data in the table is shown in Figure 34. To model the data, we might try a polynomial, such as the linear model $f(x) = x$ or the quadratic model $g(x) = x^2$. Figure 35 shows that $f(x) = x$ increases too slowly and $g(x) = x^2$ increases too quickly. To model the data, we need a function in the form $h(x) = x^b$, where $1 < b < 2$. In a polynomial, the exponent b must be a *nonnegative integer,* whereas in a power function, b can be *any real number.* See Figure 36.

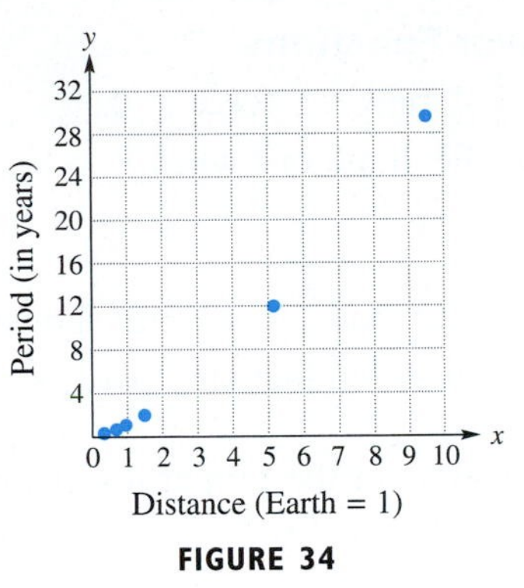

FIGURE 34

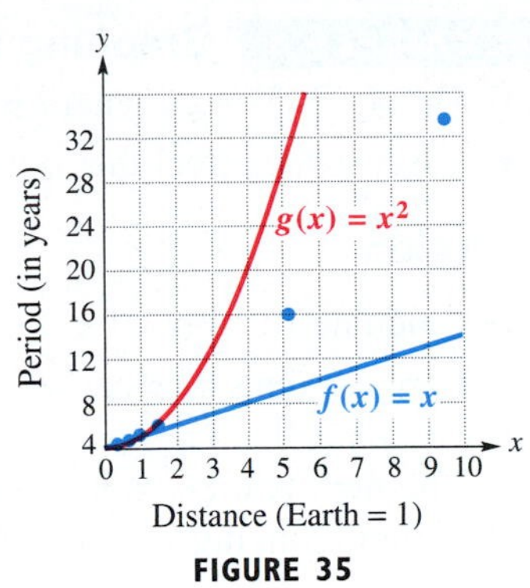

FIGURE 35

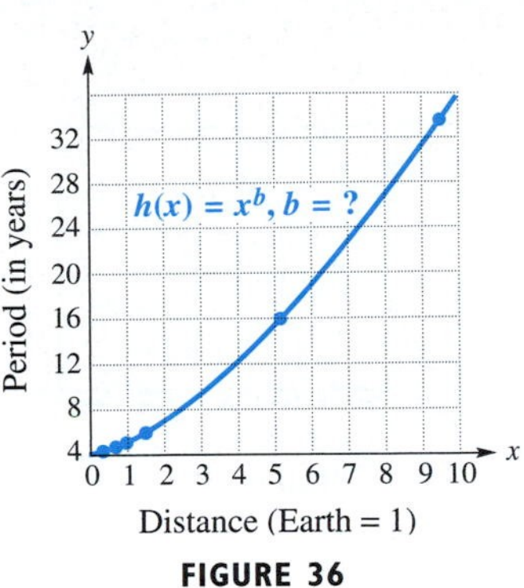

FIGURE 36

A *root function* is a special type of power function. These functions are defined as follows.

Power and Root Functions

A function f given by

$$f(x) = x^b,$$

where b is a constant, is a **power function.** If $b = \frac{1}{n}$ for some integer $n \geq 2$, then f is a **root function** given by

$$f(x) = x^{1/n}, \quad \text{or equivalently,} \quad f(x) = \sqrt[n]{x}.$$

Examples of power functions include

$$f(x) = x^2, \quad f(x) = x^{\pi}, \quad f(x) = x^4, \quad \text{and} \quad f(x) = \sqrt[3]{x^2}.$$ Power functions

Frequently, the domain of a power function f is restricted to nonnegative numbers. Suppose the rational number $\frac{p}{q}$ is written in lowest terms. Then the domain of $f(x) = x^{p/q}$ is all real numbers whenever q is odd and all nonnegative real numbers whenever q is even. If b is an irrational number, the domain of $f(x) = x^b$ is all nonnegative real numbers. For example, the domain of $f(x) = x^{1/3}$ $\left(f(x) = \sqrt[3]{x}\right)$ is all real numbers, whereas the domains of $g(x) = x^{1/2}$ $\left(g(x) = \sqrt{x}\right)$ and $h(x) = x^{\sqrt{2}}$ are all nonnegative numbers.

Graphs of three common power functions are shown in Figures 37–39. We first introduced the square root and cube root functions in Section 2.1.

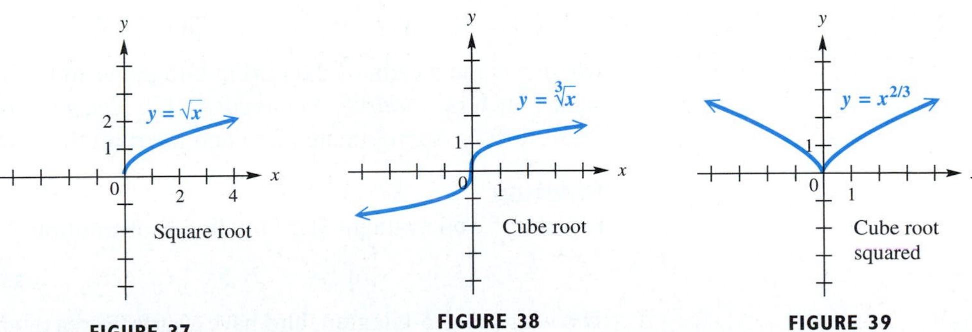

FIGURE 37

FIGURE 38

FIGURE 39

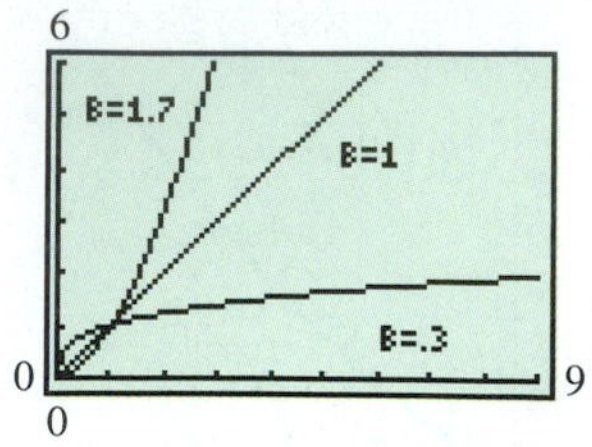

EXAMPLE 1 Graphing Power Functions

(a) Graph $f(x) = x^b$, where $b = .3$, 1, and 1.7, for $x \geq 0$.

(b) Discuss the effect that b has on the graph of f when $b > 0$.

Solution

(a) Calculator graphs of $y = x^{.3}$, $y = x^1$, and $y = x^{1.7}$ are shown in Figure 40.

(b) Larger values of b cause the graph of f to increase faster. ■

6
B=1.7
B=1
B=.3
0 9
0

FIGURE 40

The next two examples review some of the properties of rational exponents and radical notation, discussed in Sections R.4 and R.5.

EXAMPLE 2 Applying Properties of Rational Exponents

Simplify each expression by hand.

(a) $16^{3/4}$ **(b)** $\left(\sqrt[3]{-64}\right)^4$ **(c)** $(-125)^{2/3}$

Solution

(a) $16^{3/4} = \left(\sqrt[4]{16}\right)^3 = (2)^3 = 8$ $\quad a^{m/n} = (\sqrt[n]{a})^m$

(b) $\left(\sqrt[3]{-64}\right)^4 = (-4)^4 = 256$

(c) $(-125)^{2/3} = (\sqrt[3]{-125})^2 = (-5)^2 = 25$ ■

EXAMPLE 3 Writing Radicals with Rational Exponents

Use positive rational exponents to write each expression. Assume variables are positive.

(a) $\sqrt{x}$ **(b)** $\sqrt[3]{x^2}$ **(c)** $(\sqrt[4]{z})^{-5}$ **(d)** $\sqrt{\sqrt[3]{y} \cdot \sqrt[4]{y}}$

Solution

(a) $\sqrt{x} = x^{1/2}$ **(b)** $\sqrt[3]{x^2} = (x^2)^{1/3} = x^{2/3}$

(c) $(\sqrt[4]{z})^{-5} = (z^{1/4})^{-5} = z^{-5/4} = \dfrac{1}{z^{5/4}}$ $\quad a^{-n} = \frac{1}{a^n}$

(d) $\sqrt{\sqrt[3]{y} \cdot \sqrt[4]{y}} = (y^{1/3} \cdot y^{1/4})^{1/2} = (y^{(1/3)+(1/4)})^{1/2}$ $\quad a^m \cdot a^n = a^{m+n}$

$= (y^{7/12})^{1/2} = y^{7/24}$ $\quad (a^m)^n = a^{mn}$ ■

Modeling Using Power Functions

EXAMPLE 4 Modeling Wing Size of a Bird

Heavier birds have larger wings with more surface area than do lighter birds. For some species of birds, this relationship can be modeled by

$$S(x) = .2x^{2/3},$$

where x is the weight of the bird in kilograms and S is the surface area of the wings in square meters. (*Source:* Pennycuick, C., *Newton Rules Biology,* Oxford University Press, 1992.) Approximate $S(.5)$ and interpret the result.

Solution

Let $x = .5$ and evaluate $S(.5)$ by direct substitution.

$$S(.5) = .2(.5)^{2/3} \approx .126. \quad \text{Use a calculator.}$$

The wings of a .5-kilogram bird have a surface area of about .126 square meter. ■

GCM EXAMPLE 5 Modeling the Period of Planetary Orbits

Planet	x	y
Mercury	.387	.241
Venus	.723	.615
Earth	1.00	1.00
Mars	1.52	1.88
Jupiter	5.20	11.9
Saturn	9.54	29.5

Source: Ronan, C., *The Natural History of the Universe*, MacMillan, 1991.

Use the data in the table to complete the following. (Refer to the opening paragraph of this section.)

(a) Make a scatter diagram of the data. Graphically estimate a value for b so that $f(x) = x^b$ models the data.

(b) Numerically check the accuracy of f.

(c) The average distances of Uranus, Neptune, and Pluto from the sun are 19.2, 30.1, and 39.5, respectively. Use f to estimate the periods of revolution for these planets. Compare these answers with the actual values of 84.0, 164.8, and 248.5 years.

Solution

(a) We make a scatter diagram of the data and then graph $y = x^b$ for different values of b. From the calculator graphs of $y = x^{1.4}$, $y = x^{1.5}$, and $y = x^{1.6}$ in Figures 41–43, we see that $b \approx 1.5$.

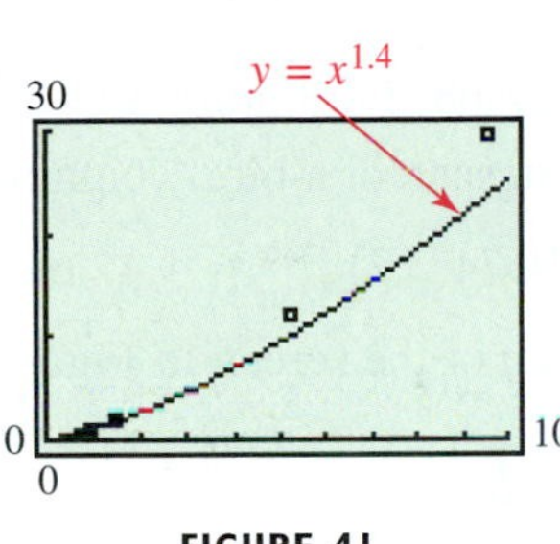

FIGURE 41

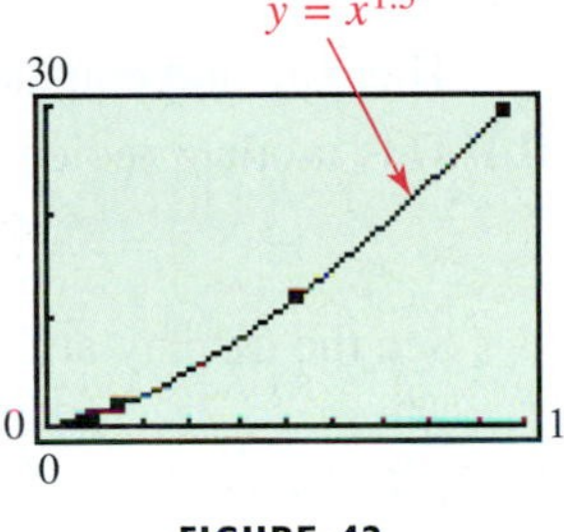

FIGURE 42

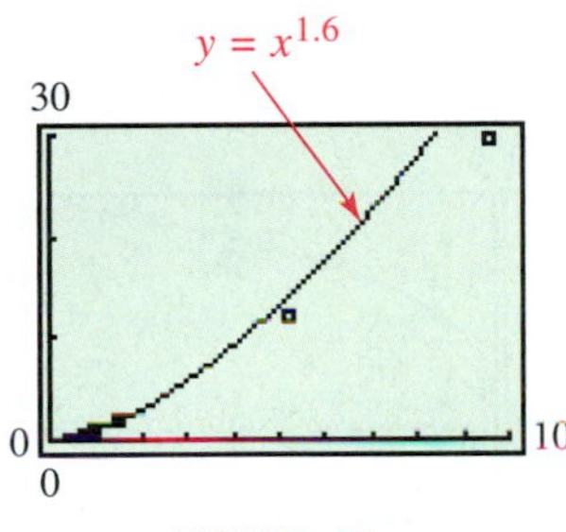

FIGURE 43

(b) The values shown in Figure 44 model the data in the table remarkably well.

X	Y1
.387	.24075
.723	.61476
1	1
1.52	1.874
5.2	11.858
9.54	29.466

Y1 = X^1.5

FIGURE 44

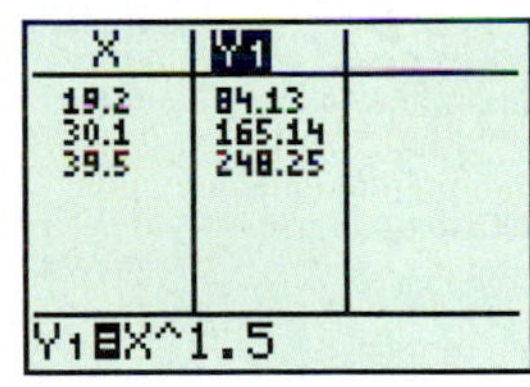

X	Y1
19.2	84.13
30.1	165.14
39.5	248.25

Y1 = X^1.5

FIGURE 45

(c) To approximate the number of years for Uranus, Neptune, and Pluto to orbit the sun, we evaluate $f(x) = x^{1.5}$ at $x = 19.2$, 30.1, and 39.5, as shown in Figure 45. These values are close to the actual values. ■

Rather than visually fit a curve to data as we did in Example 5, we can use least-squares regression, first introduced in Section 1.4, to fit the data.

GCM EXAMPLE 6 Modeling the Length of a Bird's Wing

The table lists the weight W and the wingspan L for birds of a particular species.

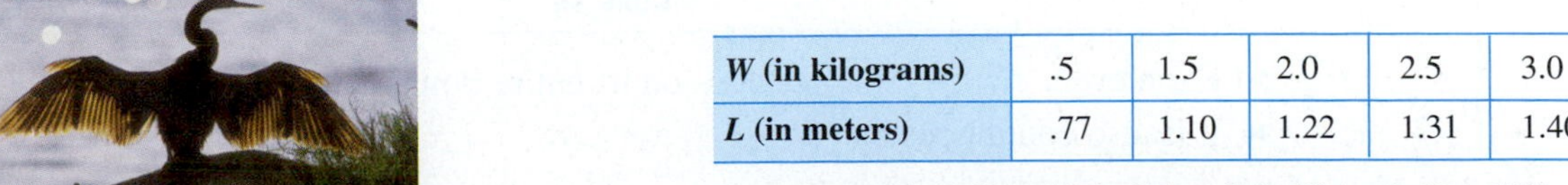

W (in kilograms)	.5	1.5	2.0	2.5	3.0
L (in meters)	.77	1.10	1.22	1.31	1.40

Source: Pennycuick, C., *Newton Rules Biology*, Oxford University Press, 1992.

(a) Use power regression to model the data with $L = aW^b$. Graph the data and the equation.

(b) Approximate the wingspan for a bird weighing 3.2 kilograms.

Solution

(a) Let x be the weight W and y be the length L. Enter the data into a graphing calculator, and then select power regression (PwrReg), as shown in Figures 46 and 47. Based on the results shown in Figure 48, $y \approx .9674x^{.3326}$, or $L \approx .9674W^{.3326}$.

L1	L2	L3
.5	.77	------
1.5	1.1	
2	1.22	
2.5	1.31	
3	1.4	
------	------	

L1(1)=.5

FIGURE 46

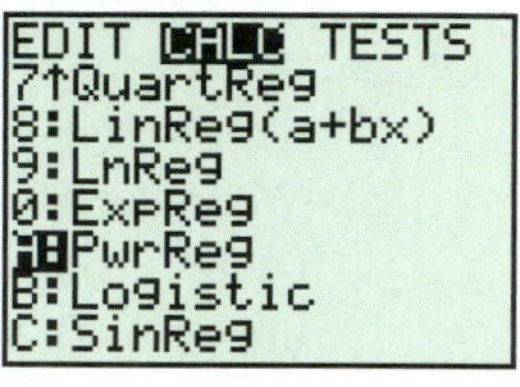

FIGURE 47

FIGURE 48

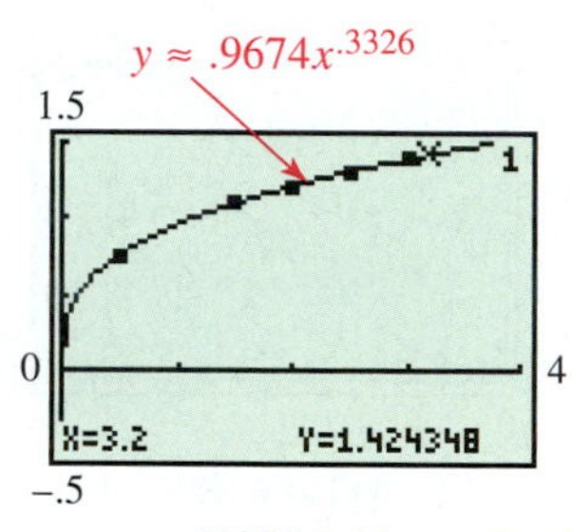

FIGURE 49

The data and equation are graphed in Figure 49.

(b) This model predicts the wingspan of a bird weighing 3.2 kilograms to be

$$L = .9674(3.2)^{.3326} \approx 1.42 \text{ meters.}$$

See the display at the bottom of the screen in Figure 49. ■

Graphs of $f(x) = \sqrt[n]{ax + b}$

In Section 2.1, we introduced the graph of the *square* root function defined by $f(x) = \sqrt{x}$. When n is even, the graph of the **root function** defined by $f(x) = \sqrt[n]{x}$ resembles the graph of the square root function; as n gets larger, the graph lies closer to the x-axis for $x \geq 1$. See Figure 50.

TECHNOLOGY NOTE

The definition of $x^{1/n}$ allows us to use a rational exponent as well as a radical when we graph a root function.

FUNCTION CAPSULE

ROOT FUNCTION, n EVEN $\quad f(x) = \sqrt[n]{x}$

Domain: $[0, \infty)$ $\quad$ Range: $[0, \infty)$

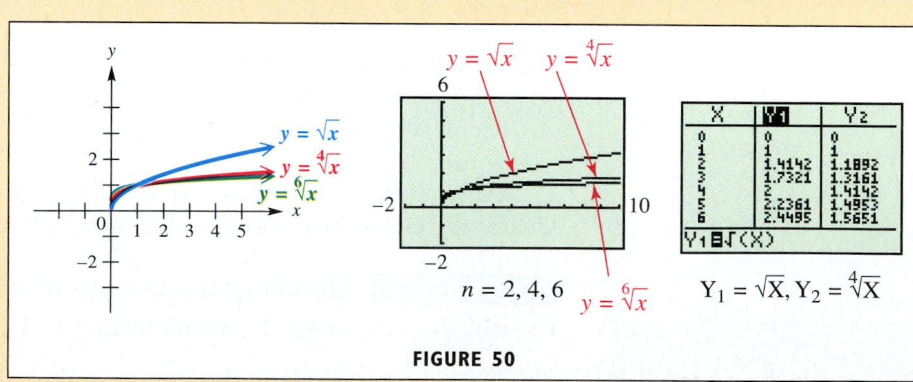

FIGURE 50

- For n even, $f(x) = \sqrt[n]{x}$ increases on its entire domain, $[0, \infty)$.
- It is also continuous on $[0, \infty)$.

A discussion similar to the preceding one pertains to the graph of $f(x) = \sqrt[n]{x}$, where n is odd. We introduced the graph of the *cube* root function defined by $f(x) = \sqrt[3]{x}$ in Section 2.1. When n is odd, the graph of the root function defined by $f(x) = \sqrt[n]{x}$ resembles the graph of the cube root function; for larger n, the graph lies closer to the x-axis for $|x| \geq 1$. See Figure 51.

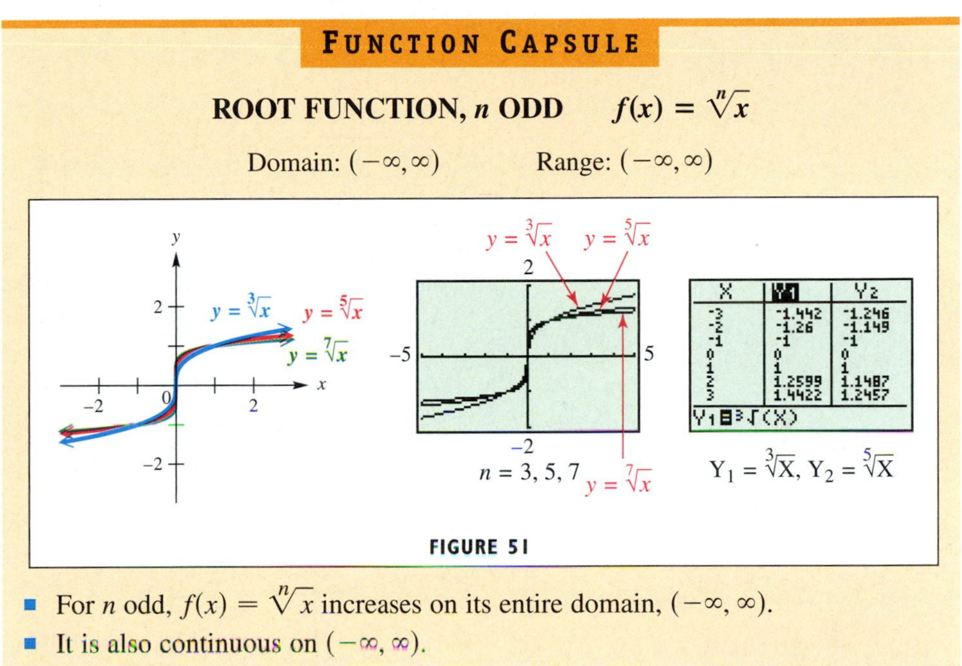

FUNCTION CAPSULE

ROOT FUNCTION, n ODD $\quad f(x) = \sqrt[n]{x}$

Domain: $(-\infty, \infty)$ Range: $(-\infty, \infty)$

X	Y1	Y2
-3	-1.442	-1.246
-2	-1.26	-1.149
-1	-1	-1
0	0	0
1	1	1
2	1.2599	1.1487
3	1.4422	1.2457

Y1 ■ ³√(X)

FIGURE 51

- For n odd, $f(x) = \sqrt[n]{x}$ increases on its entire domain, $(-\infty, \infty)$.
- It is also continuous on $(-\infty, \infty)$.

To determine the domain of a root function of the form

$$f(x) = \sqrt[n]{ax + b},$$

we must note the parity (i.e., whether it is even or odd) of n. If n is even, then $ax + b$ must be greater than or equal to 0; if n is odd, then $ax + b$ can be any real number.

EXAMPLE 7 Finding Domains of Root Functions

Find the domain of each function.

(a) $f(x) = \sqrt{4x + 12}$ **(b)** $g(x) = \sqrt[3]{-8x + 8}$

Solution

(a) For the function to be defined, $4x + 12$ must be greater than or equal to 0, since $f(x)$ contains an even root ($n = 2$).

$$4x + 12 \geq 0$$
$$4x \geq -12$$
$$x \geq -3$$

The domain of f is $[-3, \infty)$.

(b) Because $n = 3$ is an odd number, the domain of g is $(-\infty, \infty)$. ■

EXAMPLE 8 **Transforming Graphs of Root Functions**

(a) Explain how the graph of $y = \sqrt{4x + 12}$ can be obtained from the graph of $y = \sqrt{x}$. Then, graph both equations on the same coordinate axes.

(b) Repeat part (a) for the graph of $y = \sqrt[3]{-8x + 8}$, compared with the graph of $y = \sqrt[3]{x}$.

Solution

(a)

$$y = \sqrt{4x + 12}$$

$$y = \sqrt{4(x + 3)} \qquad \text{Factor.}$$

$$y = \sqrt{4}\sqrt{x + 3} \qquad \sqrt{ab} = \sqrt{a} \cdot \sqrt{b}$$

$$y = 2\sqrt{x + 3} \qquad \sqrt{4} = 2$$

The graph of this function can be obtained from the graph of $y = \sqrt{x}$ by shifting it left 3 units and stretching vertically by a factor of 2. See Figure 52. The domain of $y = \sqrt{4x + 12}$ is $[-3, \infty)$, as determined in Example 7(a). The range is $[0, \infty)$.

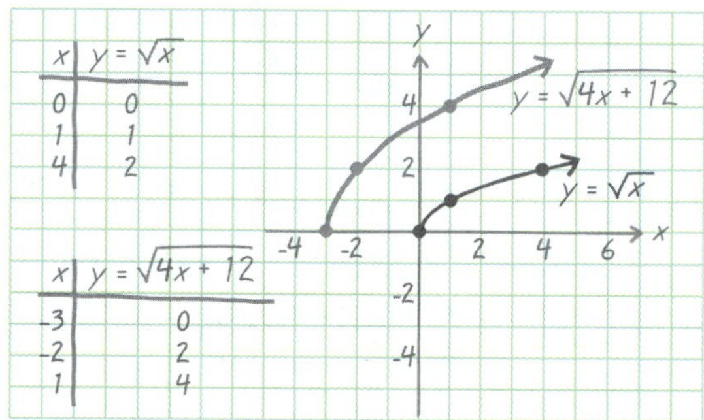

FIGURE 52

(b)

$$y = \sqrt[3]{-8x + 8}$$

$$y = \sqrt[3]{-8(x - 1)} \qquad \text{Factor.}$$

Be careful with signs.

$$y = \sqrt[3]{-8} \cdot \sqrt[3]{x - 1} \qquad \sqrt[3]{ab} = \sqrt[3]{a} \cdot \sqrt[3]{b}$$

$$y = -2\sqrt[3]{x - 1} \qquad \sqrt[3]{-8} = -2$$

The graph can be obtained by shifting the graph of $y = \sqrt[3]{x}$ to the right 1 unit, stretching vertically by a factor of 2, and reflecting across the x-axis (because of the negative sign in -2). Figure 53 shows the graphs and tables. The domain and range are both $(-\infty, \infty)$.

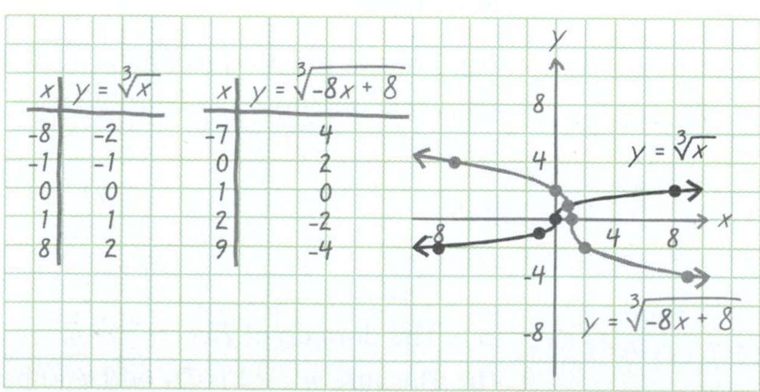

FIGURE 53

Graphing Circles and Horizontal Parabolas Using Root Functions

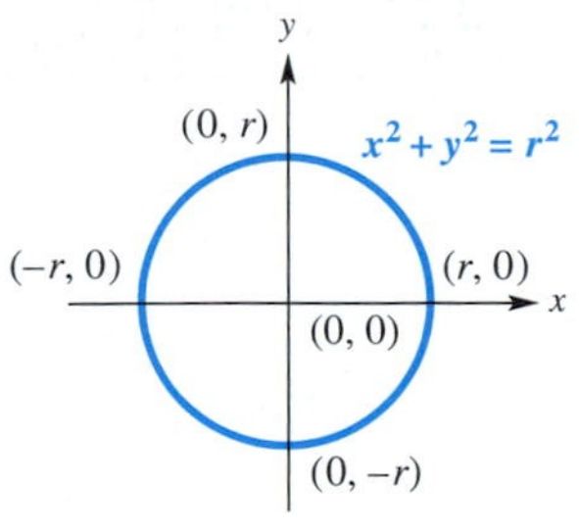

FIGURE 54

If we consider the graph of all points (x, y) that lie a fixed distance r (with $r > 0$) from the origin, then, by the distance formula,

$$\sqrt{(x-0)^2 + (y-0)^2} = r$$

$$x^2 + y^2 = r^2. \quad \text{Square each side.}$$

The graph of this equation is a circle with center at $(0, 0)$ and radius r. See Figure 54.

Because a circle is *not* the graph of a function, a graphing calculator in function mode is not appropriate for graphing a circle. However, if we imagine the circle as being the union of the graphs of two functions—one the top semicircle and the other the bottom semicircle—then we can graph both functions in the same square viewing window to obtain a circle. To accomplish this, we solve for y in $x^2 + y^2 = r^2$.

$$x^2 + y^2 = r^2 \quad \text{Equation of a circle}$$

$$y^2 = r^2 - x^2 \quad \text{Subtract } x^2.$$

$$y = \pm\sqrt{r^2 - x^2} \quad \text{Square root property}$$

Remember both the positive and negative square roots.

The final result yields two equations, both of which define root functions:

$$y_1 = \sqrt{r^2 - x^2} \quad \text{The semicircle above the } x\text{-axis}$$

and

$$y_2 = -\sqrt{r^2 - x^2}. \quad \text{The semicircle below the } x\text{-axis}$$

Since $y_2 = -y_1$, the graph of the "bottom" semicircle is simply a reflection of the graph of the "top" semicircle across the x-axis.

GCM **EXAMPLE 9** **Graphing a Circle**

Use a calculator in function mode to graph the circle $x^2 + y^2 = 4$.

Solution This graph can be obtained by graphing the two root functions

$$y_1 = \sqrt{4 - x^2} \quad \text{and} \quad y_2 = -y_1 = -\sqrt{4 - x^2}$$

in the same window. See Figure 55. To obtain a graph that is visually in correct proportions, we use a square window. Because it is a decimal window, the two semicircles connect at the x-axis. (See the Technology Note.)

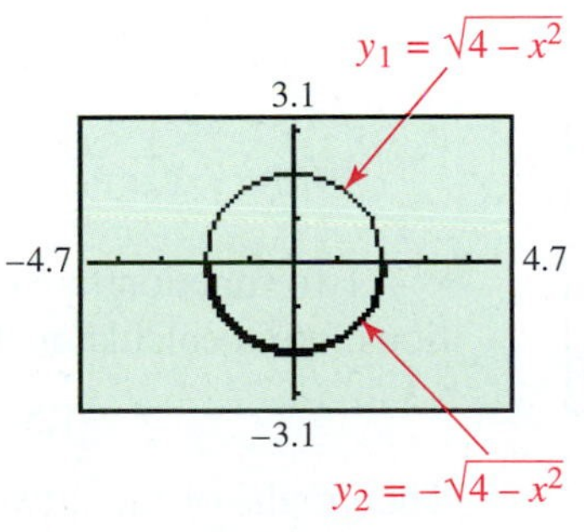

FIGURE 55

From Sections 3.2 and 3.3, we know that the graph of $y = ax^2 + bx + c$ $(a \neq 0)$ is a parabola with a *vertical* axis of symmetry. Such an equation defines a function. If we reverse the roles of x and y, however, we obtain the equation $\boldsymbol{x = ay^2 + by + c}$, whose graph is also a parabola, but with a *horizontal* axis of symmetry.

TECHNOLOGY NOTE

A graph of the circle $x^2 + y^2 = 36$ is shown.

$y_1 = \sqrt{36 - x^2}$
10
−15
15
−10
$y_2 = -y_1 = -\sqrt{36 - x^2}$

Although a square window is used, it is not a decimal window, so the two semicircles that form the graph do not completely "connect." *Mathematically, this is a complete circle.* ***We must understand the concepts to interpret what we see on the screen.***

Figure 56 shows a graph of $x = 2y^2 + 6y + 5$. Notice that this is *not* the graph of a function. However, if we consider it to be the union of the graphs of two functions, one the top half-parabola and the other the bottom, then it can be graphed by using a calculator in function mode.

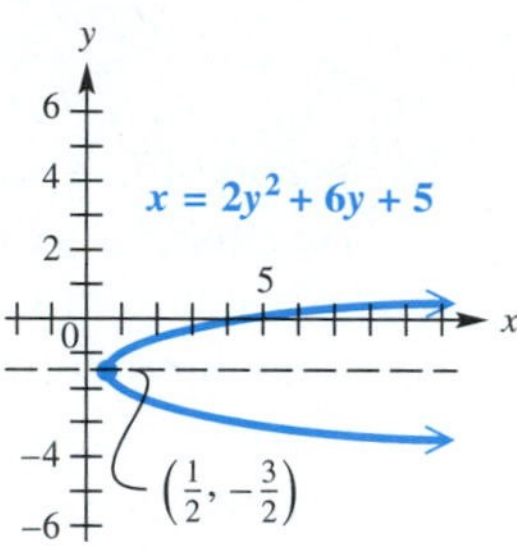

FIGURE 56

EXAMPLE 10 Graphing a Horizontal Parabola

Graph $x = 2y^2 + 6y + 5$ and its axis of symmetry in the window $[-2, 8]$ by $[-4, 2]$.

Solution We must solve for y by completing the square.

$$x = 2y^2 + 6y + 5$$

$$\frac{x}{2} = y^2 + 3y + \frac{5}{2} \qquad \text{Divide by 2.}$$

$$\frac{x}{2} - \frac{5}{2} = y^2 + 3y \qquad \text{Subtract } \tfrac{5}{2}.$$

$$\frac{x}{2} - \frac{5}{2} + \frac{9}{4} = y^2 + 3y + \frac{9}{4} \qquad \text{Add } \left[\tfrac{1}{2}(3)\right]^2 = \tfrac{9}{4}.$$

Be sure to add $\frac{9}{4}$ to *each* side.

$$\frac{x}{2} - \frac{1}{4} = \left(y + \frac{3}{2}\right)^2 \qquad \text{Add on the left; factor on the right.}$$

$$\left(y + \frac{3}{2}\right)^2 = \frac{x}{2} - \frac{1}{4} \qquad \text{Rewrite.}$$

$$y + \frac{3}{2} = \pm\sqrt{\frac{x}{2} - \frac{1}{4}} \qquad \text{Square root property}$$

$$y = -\frac{3}{2} \pm \sqrt{\frac{x}{2} - \frac{1}{4}} \qquad \text{Subtract } \tfrac{3}{2}.$$

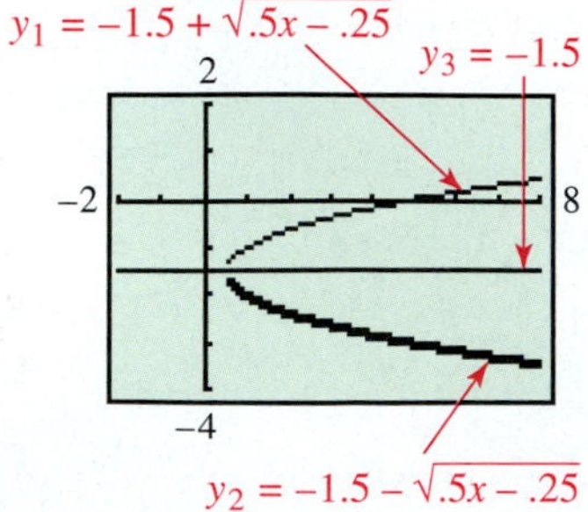

FIGURE 57

Two functions are now defined. It is easier to use decimals when entering the equations into a calculator. Therefore, let

$$y_1 = -1.5 + \sqrt{.5x - .25} \quad \text{and} \quad y_2 = -1.5 - \sqrt{.5x - .25}.$$

The graphs of these two functions together form the parabola with horizontal axis of symmetry $y_3 = -1.5$, shown in Figure 57. ■

4.4 Exercises

Note to student: At this point, you may want to review the material on radicals and rational exponents in Chapter R.

Evaluate each expression without using a calculator.

1. $\sqrt{169}$ **2.** $-\sqrt[3]{64}$ **3.** $\sqrt[5]{-32}$ **4.** $\sqrt[4]{16}$ **5.** $81^{3/2}$

6. $27^{4/3}$ **7.** $125^{-2/3}$ **8.** $(\sqrt[3]{-27})^2$ **9.** $(-1000)^{2/3}$ **10.** $(-125)^{-4/3}$

11. $8^{2/3}$ **12.** $-16^{3/2}$ **13.** $16^{-3/4}$ **14.** $25^{-3/2}$ **15.** $-81^{.5}$

16. $32^{1/5}$ **17.** $64^{1/6}$ **18.** $16^{-.25}$ **19.** $(-9^{3/4})^2$ **20.** $(4^{-1/2})^{-4}$

Use positive rational exponents to rewrite each expression. Assume variables are positive.

21. $\sqrt[3]{2x}$ **22.** $\sqrt{x+1}$ **23.** $\sqrt[3]{z^5}$ **24.** $\sqrt[5]{x^2}$ **25.** $(\sqrt[4]{y})^{-3}$

26. $(\sqrt[3]{y^2})^{-5}$ **27.** $\sqrt{x} \cdot \sqrt[3]{x}$ **28.** $(\sqrt[5]{z})^{-3}$ **29.** $\sqrt{y \cdot \sqrt{y}}$ **30.** $\dfrac{\sqrt[3]{x}}{\sqrt{x}}$

Use a calculator to find each root or power. Give as many digits as your display shows.

31. $\sqrt[3]{-4}$ **32.** $\sqrt[5]{-3}$ **33.** $\sqrt[3]{-125}$ **34.** $\sqrt[5]{-243}$

35. $\sqrt[3]{-17}$ **36.** $\sqrt[5]{-8}$ **37.** $\sqrt[6]{\pi^2}$ **38.** $\sqrt[6]{\pi^{-1}}$

39. $13^{-1/3}$ **40.** $15^{-1/6}$ **41.** $32^{.2}$ **42.** $81^{.25}$

43. $\left(\dfrac{5}{6}\right)^{-1.3}$ **44.** $\left(\dfrac{4}{7}\right)^{-.6}$ **45.** π^{-3} **46.** $(2\pi)^{4/3}$

Evaluate $f(x)$ at the given x. Approximate each result to the nearest hundredth.

47. $f(x) = x^{1.62}$, $x = 1.2$

48. $f(x) = x^{-.71}$, $x = 3.8$

49. $f(x) = x^{3/2} - x^{1/2}$, $x = 50$

50. $f(x) = x^{5/4} - x^{-3/4}$, $x = 7$

Concept Check *Match $f(x)$ with its graph. Assume that a and b are constants with $0 < a < 1 < b$.*

51. $f(x) = x^a$

52. $f(x) = x^b$

A.

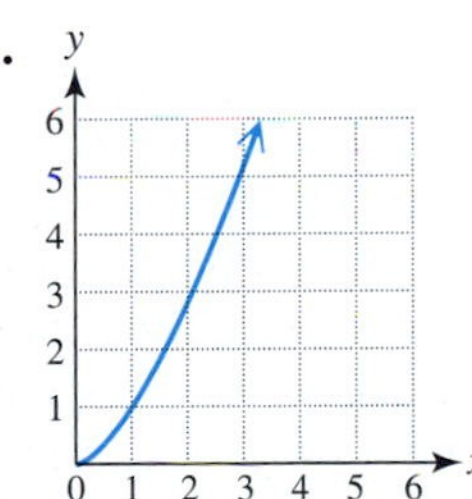

B.

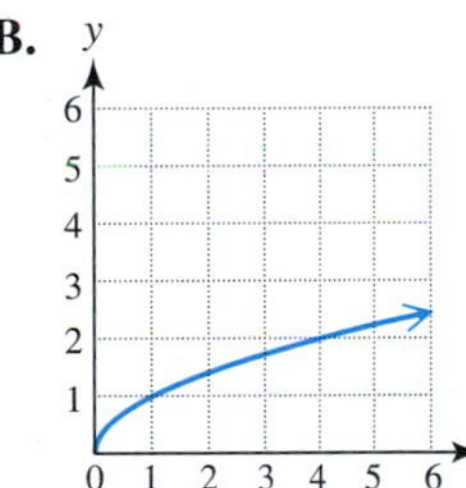

53. Consider the expression $16^{-3/4}$.

(a) Simplify this expression without using a calculator. Give the answer in both decimal and $\frac{a}{b}$ form.

(b) Write two different radical expressions that are equivalent to it, and use your calculator to evaluate them to show that the result is the same as the decimal form you found in part (a).

(c) If your calculator has the capability to convert decimal numbers to fractions, use it to verify your results in part (a).

54. Consider the expression $5^{.47}$.

(a) Use the exponentiation capability of your calculator to find an approximation. Give as many digits as your calculator displays.

(b) Use the fact that $.47 = \frac{47}{100}$ to write the expression as a radical, and then use the root-finding capability of your calculator to find an approximation that agrees with the one found in part (a).

Relating Concepts

For individual or group investigation (Exercises 55–58)

Duplicate each screen on your calculator. The screens show multiple ways of finding an approximation for $\sqrt[6]{9}$. ***Work Exercises 55–58 in order*** *using your calculator.*

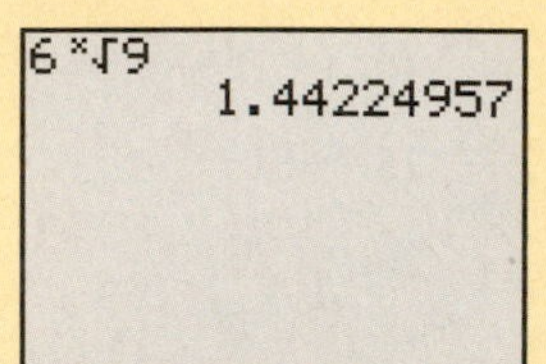

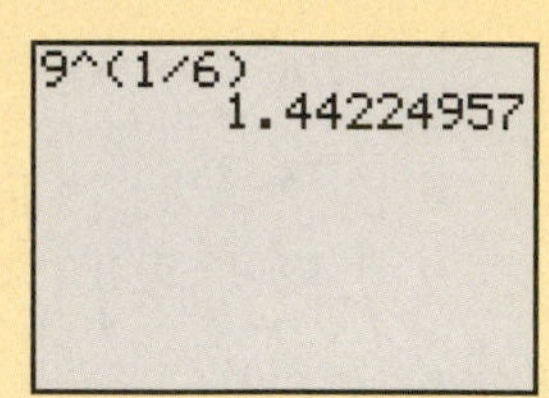

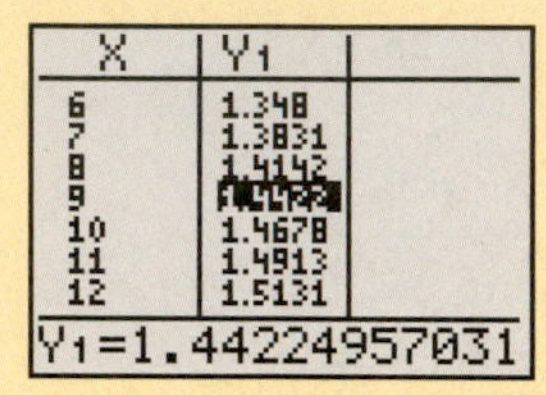

In this table, $Y_1 = \sqrt[6]{X}$.

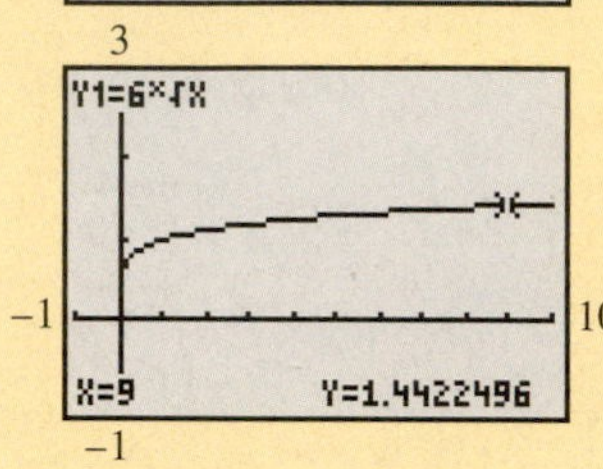

55. Use a radical expression to approximate $\sqrt[6]{81}$ to as many decimal places as the calculator will give.
56. Use a rational exponent to repeat Exercise 55.
57. Use a table to repeat Exercise 55.
58. Use the graph of $y = \sqrt[6]{x}$ to repeat Exercise 55.

(Modeling) Solve each problem.

59. ***Wing Size*** Suppose that the surface area S of a bird's wings, in square feet, can be modeled by

$$S(w) = 1.27w^{2/3},$$

where w is the weight of the bird in pounds. Estimate the surface area of a bird's wings if the bird weighs 4.0 pounds.

60. ***Wingspan*** The wingspan L in feet of a bird weighing W pounds is given by

$$L = 2.43W^{.3326}.$$

Estimate the wingspan of a bird that weighs 5.2 pounds.

61. ***Planetary Orbits*** The formula

$$f(x) = x^{1.5}$$

calculates the number of years it would take for a planet to orbit the sun if its average distance from the sun is x times farther than Earth. If there were a planet located 15 times farther from the sun than Earth, how many years would it take for the planet to orbit the sun?

62. ***Sight Distance*** A formula for calculating the distance one can see from an airplane to the horizon on a clear day is given by

$$f(x) = 1.22x^{.5},$$

where x is the altitude of the plane in feet and $f(x)$ is in miles. If a plane is flying at 30,000 feet, how far can the pilot see?

63. ***Trout and Pollution*** Rainbow trout are sensitive to zinc ions in the water. High concentrations are lethal. The average survival times x in minutes for trout in various concentrations of zinc ions y in milligrams per liter (mg/L) are listed in the table.

x (in minutes)	.5	1	2	3
y (in mg/L)	4500	1960	850	525

Source: Mason, C., *Biology of Freshwater Pollution,* John Wiley and Sons, 1991.

(a) The data can be modeled by $f(x) = ax^b$, where a and b are constants. Determine a. (*Hint:* Let $f(1) = 1960$.)
(b) Estimate b.
(c) Evaluate $f(4)$ and interpret the result.

64. ***Asbestos and Cancer*** Insulation workers who were exposed to asbestos and employed before 1960 experienced an increased likelihood of lung cancer. If a group of insulation workers have a cumulative total of 100,000 years of work experience, with their first date of employment x years ago, then the number of lung cancer cases occurring within the group can be modeled by

$$N(x) = .00437x^{3.2}.$$

(*Source:* Walker, A., *Observation and Inference: An Introduction to the Methods of Epidemiology,* Epidemiology Resources, Inc., 1991.)

(a) Calculate $N(x)$ when $x = 5$, 10, and 20. What happens to the likelihood of cancer as x increases?
(b) If x doubles, does the number of cancer cases also double?

65. ***Fiddler Crab Size*** One study of the male fiddler crab showed a connection between the weight of the large claws and the animal's total body weight. For a crab weighing over .75 gram, the weight of its claws can be estimated by

$$f(x) = .445x^{1.25}.$$

The input x is the weight of the crab in grams, and the output $f(x)$ the weight of the claws in grams. Predict the weight of the claws for a crab that weighs 2 grams. (*Source:* Huxley, J., *Problems of Relative Growth,* Methuen and Co., 1932; Brown, D. and P. Rothery, *Models in Biology: Mathematics, Statistics, and Computing,* John Wiley and Sons, 1993.)

66. ***Weight and Height of Men*** The average weight in pounds for a man can be estimated by

$$f(x) = .117x^{1.7},$$

where x represents the man's height in inches and $f(x)$ is his weight in pounds. What is the average weight of a 68-inch-tall man?

67. ***Animal Pulse Rate and Weight*** According to one model, the rate at which an animal's heart beats varies with its weight. Smaller animals tend to have faster pulses, whereas larger animals tend to have slower pulses. The table in the next column lists average pulse rates in beats per minute (bpm) for animals with various weights in pounds (lb). Use regression (or some other method) to find values for a and b so that $f(x) = ax^b$ models these data.

Weight (in lb)	40	150	400	1000	2000
Pulse (in bpm)	140	72	44	28	20

Source: Pennycuick, C., *Newton Rules Biology*, Oxford University Press, 1992.

68. ***Animal Pulse Rate and Weight*** Use the results of Exercise 67 to calculate the pulse rates for a 60-pound dog and a 2-ton whale.

Determine the domain of each function.

69. $f(x) = \sqrt{5 + 4x}$

70. $f(x) = \sqrt{9x + 18}$

71. $f(x) = -\sqrt{6 - x}$

72. $f(x) = -\sqrt{2 - .5x}$

73. $f(x) = \sqrt[3]{8x - 24}$

74. $f(x) = \sqrt[5]{x + 32}$

75. $f(x) = \sqrt{49 - x^2}$

76. $f(x) = \sqrt{81 - x^2}$

77. $f(x) = \sqrt{x^3 - x}$

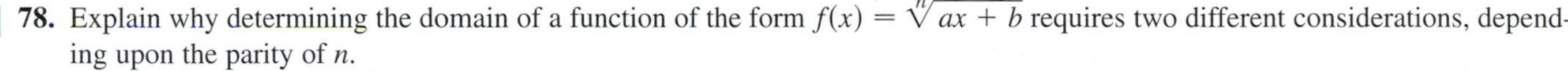

78. Explain why determining the domain of a function of the form $f(x) = \sqrt[n]{ax + b}$ requires two different considerations, depending upon the parity of n.

Graph each function by hand. The domains were determined in Exercises 69–76. Use the graph to **(a)** *find the range,* **(b)** *give the interval over which the function is increasing,* **(c)** *give the interval over which the function is decreasing, and* **(d)** *solve the equation $f(x) = 0$ by observing your graph.*

79. $f(x) = \sqrt{5 + 4x}$

80. $f(x) = \sqrt{9x + 18}$

81. $f(x) = -\sqrt{6 - x}$

82. $f(x) = -\sqrt{2 - .5x}$

83. $f(x) = \sqrt[3]{8x - 24}$

84. $f(x) = \sqrt[5]{x + 32}$

85. $f(x) = \sqrt{49 - x^2}$

86. $f(x) = \sqrt{81 - x^2}$

Concept Check *Use transformations to explain how the graph of the given function can be obtained from the graph of the appropriate root function* $\left(y = \sqrt{x} \text{ or } y = \sqrt[3]{x}\right)$.

87. $y = \sqrt{9x + 27}$

88. $y = \sqrt{16x + 16}$

89. $y = \sqrt{4x + 16} + 4$

90. $y = \sqrt{32 - 4x} - 3$

91. $y = \sqrt[3]{27x + 54} - 5$

92. $y = \sqrt[3]{8x - 8}$

In Exercises 93–100, describe the graph of the equation as either a circle or a parabola with a horizontal axis of symmetry. Then, determine two functions, designated by y_1 *and* y_2*, such that their union will give the graph of the given equation. Finally, graph the equation in the given viewing window.*

93. $x^2 + y^2 = 100$;
$[-15, 15]$ by $[-10, 10]$

94. $x^2 + y^2 = 81$;
$[-15, 15]$ by $[-10, 10]$

95. $(x - 2)^2 + y^2 = 9$;
$[-9.4, 9.4]$ by $[-6.2, 6.2]$

96. $(x + 3)^2 + y^2 = 16$;
$[-9.4, 9.4]$ by $[-6.2, 6.2]$

97. $x = y^2 + 6y + 9$;
$[-10, 10]$ by $[-10, 10]$

98. $x = y^2 - 8y + 16$;
$[-10, 10]$ by $[-10, 10]$

99. $x = 2y^2 + 8y + 1$;
$[-10, 10]$ by $[-10, 10]$

100. $x = -3y^2 - 6y + 2$;
$[-9.4, 9.4]$ by $[-6.2, 6.2]$

4.5 Equations, Inequalities, and Applications Involving Root Functions

Equations and Inequalities ■ An Application of Root Functions

Equations and Inequalities

To solve an equation involving roots, such as

$$\sqrt{11 - x} - x = 1,$$

in which the variable appears in a radicand, we use the following property to eliminate the radical.

Power Property

If P and Q are algebraic expressions, then every solution of the equation $P = Q$ is also a solution of the equation $P^n = Q^n$, for any positive integer n.

CAUTION The power property does *not* say that the equations $P = Q$ and $P^n = Q^n$ are equivalent; it says only that each solution of the original equation $P = Q$ is also a solution of the new equation $P^n = Q^n$.

When the power property is used to solve equations, the new equation may have *more* solutions than the original equation. For example, the solution set of the equation $x = -2$ is $\{-2\}$. If we square each side of the equation $x = -2$, we get the new equation $x^2 = 4$, which has solution set $\{-2, 2\}$. Since the solution sets are not equal, the equations are not equivalent. ***When an equation contains radicals or rational exponents, it is essential to check all proposed solutions in the original equation.***

To solve equations involving roots, follow these steps.

Solving an Equation Involving a Root Function

Step 1 Isolate a term involving a root on one side of the equation.

Step 2 Raise each side of the equation to a positive integer power that will eliminate the radical or rational exponent.

Step 3 Solve the resulting equation. (If a root is still present after Step 2, repeat Steps 1 and 2.)

Step 4 Check each proposed solution in the original equation.

EXAMPLE 1 **Solving an Equation Involving a Square Root**

Solve $\sqrt{11 - x} - x = 1$.

Analytic Solution

$$\sqrt{11 - x} - x = 1$$

$$\sqrt{11 - x} = 1 + x \quad \text{Isolate the radical.}$$

$$(\sqrt{11 - x})^2 = (1 + x)^2 \quad \text{Square each side.}$$

$$11 - x = 1 + 2x + x^2 \quad (x + y)^2 = x^2 + 2xy + y^2$$

$$x^2 + 3x - 10 = 0 \quad \text{Standard form}$$

$$(x + 5)(x - 2) = 0 \quad \text{Factor.}$$

$$x + 5 = 0 \quad \text{or} \quad x - 2 = 0 \quad \text{Zero-product property}$$

$$x = -5 \quad \text{or} \quad x = 2 \quad \text{Solve each equation.}$$

Check the proposed solutions, -5 and 2, in the *original* equation.

$$\sqrt{11 - x} - x = 1 \quad \text{Original equation}$$

Let $x = -5$.	Let $x = 2$.
$\sqrt{11 - (-5)} - (-5) = 1$?	$\sqrt{11 - 2} - 2 = 1$?
$\sqrt{16} + 5 = 1$?	$\sqrt{9} - 2 = 1$?
$4 + 5 = 1$?	$3 - 2 = 1$?
$9 = 1$ False	$1 = 1$ True

Squaring each side of the equation led to the *extraneous* value -5, as indicated by the false statement $9 = 1$. Therefore, the only solution of the equation is 2. The solution set is $\{2\}$.

Graphing Calculator Solution

The equation in the second line of the analytic solution has the same solution set as the original equation, because each side of the equation has not yet been squared. If we graph

$$y_1 = \sqrt{11 - x} \quad \text{and} \quad y_2 = 1 + x$$

and observe the x-coordinate of the only point of intersection of the graphs, we see that it is 2, confirming the analytic solution. See Figure 58. Note that extraneous values do not occur in graphical solutions.

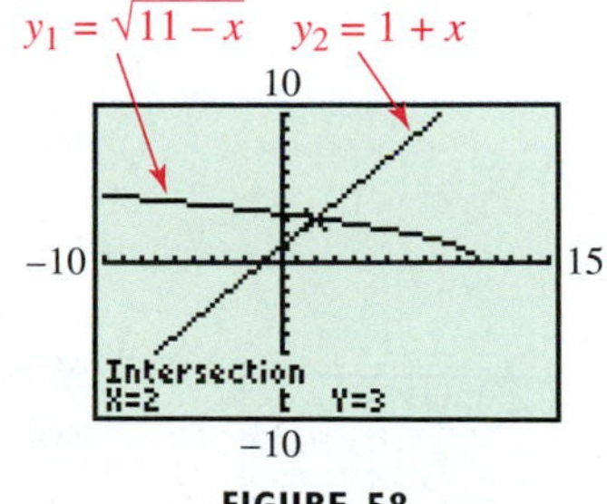

FIGURE 58

EXAMPLE 2 **Solving an Equation Involving a Rational Exponent**

Solve $(5 - 5x)^{1/2} + x = 1$.

Analytic Solution

$$(5 - 5x)^{1/2} + x = 1$$

$$(5 - 5x)^{1/2} = 1 - x \quad \text{Isolate the term with the rational exponent.}$$

$$[(5 - 5x)^{1/2}]^2 = (1 - x)^2 \quad \text{Square each side.}$$

$$5 - 5x = 1 - 2x + x^2 \quad (x - y)^2 = x^2 - 2xy + y^2$$

$$x^2 + 3x - 4 = 0 \quad \text{Standard form}$$

$$(x + 4)(x - 1) = 0 \quad \text{Factor.}$$

$$x + 4 = 0 \quad \text{or} \quad x - 1 = 0 \quad \text{Zero-product property}$$

$$x = -4 \quad \text{or} \quad x = 1 \quad \text{Solve each equation.}$$

Check the proposed solutions, -4 and 1, in the *original* equation.

$$(5 - 5x)^{1/2} + x = 1 \quad \text{Original equation}$$

Let $x = -4$.	Let $x = 1$.
$[5 - 5(-4)]^{1/2} + (-4) = 1$?	$[5 - 5(1)]^{1/2} + (1) = 1$?
$25^{1/2} + (-4) = 1$?	$0^{1/2} + 1 = 1$?
$5 + (-4) = 1$?	$0 + 1 = 1$?
$1 = 1$ True	$1 = 1$ True

Both proposed solutions check, so the solution set is $\{-4, 1\}$.

Graphing Calculator Solution

To use the x-intercept method, we let

$$y_1 = (5 - 5x)^{1/2} + x \quad \text{and} \quad y_2 = 1.$$

We graph $y_3 = y_1 - y_2$ to produce the curve shown in Figure 59. The displays at the bottom of the screens confirm the solutions, -4 and 1.

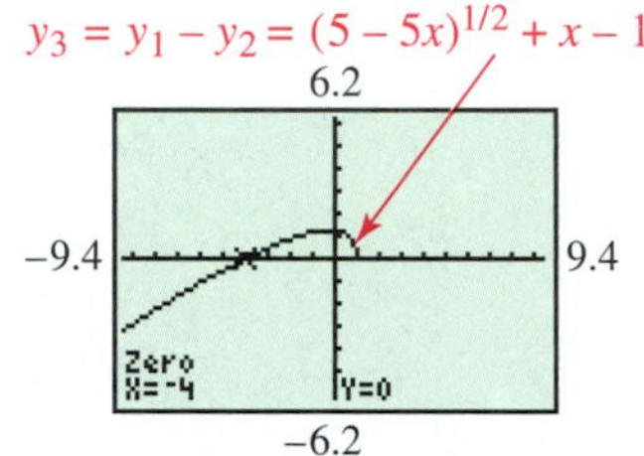

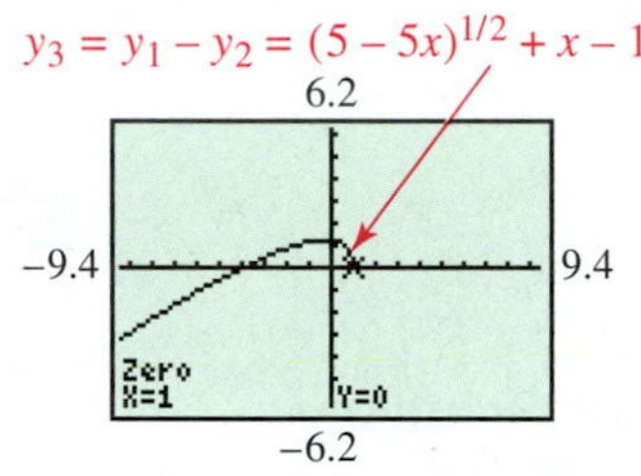

FIGURE 59

EXAMPLE 3 **Solving an Equation Involving Cube Roots**

Solve $\sqrt[3]{x^2 + 3x} = \sqrt[3]{5}$.

Solution

$$\sqrt[3]{x^2 + 3x} = \sqrt[3]{5}$$

$$\left(\sqrt[3]{x^2 + 3x}\right)^3 = \left(\sqrt[3]{5}\right)^3 \quad \text{Cube each side.}$$

$$x^2 + 3x = 5 \quad \text{Simplify.}$$

$$x^2 + 3x - 5 = 0 \quad \text{Standard form}$$

Since this equation cannot be solved by factoring, we use the quadratic formula, with $a = 1$, $b = 3$, and $c = -5$.

$$x = \frac{-b \pm \sqrt{b^2 - 4ac}}{2a} = \frac{-3 \pm \sqrt{3^2 - 4(1)(-5)}}{2(1)} = \frac{-3 \pm \sqrt{29}}{2}$$

While an analytic check is quite involved, it can be shown that both of these values are solutions. The solution set is

$$\left\{\frac{-3 + \sqrt{29}}{2}, \frac{-3 - \sqrt{29}}{2}\right\}.$$

```
(-3+√(29))/2
          1.192582404
³√(Ans²+3Ans)
          1.709975947
³√(5)
          1.709975947
```

FIGURE 60

In Figure 60, a graphing calculator has been used to support the fact that $\frac{-3 + \sqrt{29}}{2}$ is a solution. The second solution can be supported similarly.

EXAMPLE 4 **Solving an Equation Involving Roots (Squaring Twice)**

Solve $\sqrt{2x+3} - \sqrt{x+1} = 1$.

Solution Isolate the more complicated radical.

$$\sqrt{2x+3} - \sqrt{x+1} = 1$$

$$\sqrt{2x+3} = 1 + \sqrt{x+1} \quad \text{Isolate } \sqrt{2x+3}.$$

$$\left(\sqrt{2x+3}\right)^2 = \left(1 + \sqrt{x+1}\right)^2 \quad \text{Square each side.}$$

$$2x + 3 = 1 + 2\sqrt{x+1} + x + 1 \quad \text{Simplify.}$$

Remember the middle term.

$$x + 1 = 2\sqrt{x+1} \quad \text{Isolate the remaining radical.}$$

$$(x+1)^2 = \left(2\sqrt{x+1}\right)^2 \quad \text{Square each side again.}$$

$$x^2 + 2x + 1 = 4(x+1) \quad \text{Simplify.}$$

$(ab)^m = a^m b^m$

$$x^2 + 2x + 1 = 4x + 4 \quad \text{Distributive property}$$

$$x^2 - 2x - 3 = 0 \quad \text{Standard form}$$

$$(x-3)(x+1) = 0 \quad \text{Factor.}$$

$$x - 3 = 0 \quad \text{or} \quad x + 1 = 0 \quad \text{Zero-product property}$$

$$x = 3 \quad \text{or} \quad x = -1 \quad \text{Solve each equation.}$$

Check that both proposed solutions, 3 and -1, are solutions of the original equation, giving $\{3, -1\}$ as the solution set. ■

What Went Wrong?

To solve the equation $x + 4 = \sqrt{x+6}$, a student squared each side to get $x^2 + 8x + 16 = x + 6$ and then subtracted x and 6 from each side to obtain $x^2 + 7x + 10 = 0$. Factoring led to $(x+5)(x+2) = 0$ and the possible solutions -5 and -2. He found that, of these, only -2 is an actual solution. To support his solution, he entered $Y_1 = X + 4$ and $Y_2 = \sqrt{X+6}$ into his calculator. Using the intersection-of-graphs method, he obtained the screen on the left.

Next, he decided to use the x-intercept method to show that -2 is a solution. He got the screen on the right. Now he knew there was a problem.

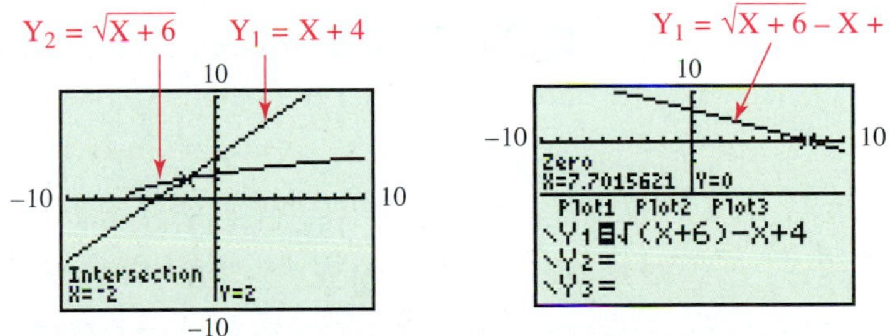

What Went Wrong? How can he correct his work so that the screen on the right shows a zero of -2?

Answer to What Went Wrong?

He entered $\sqrt{\ }(X + 6) - X + 4$ rather than $\sqrt{\ }(X + 6) - (X + 4)$. He should insert parentheses around $X + 4$ in Y_1.

EXAMPLE 5 **Solving Inequalities Involving Roots**

Solve each inequality.

(a) $\sqrt[3]{x^2 + 3x} \le \sqrt[3]{5}$ **(b)** $\sqrt{2x + 3} - \sqrt{x + 1} > 1$

Solution

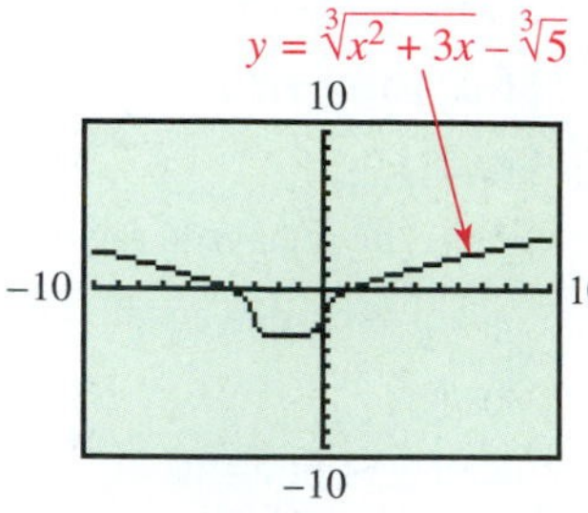

$y = \sqrt[3]{x^2 + 3x} - \sqrt[3]{5}$

10

−10 10

−10

FIGURE 61

(a) The associated equation was solved in Example 3. Its solutions are

$$\frac{-3 + \sqrt{29}}{2} \quad \text{and} \quad \frac{-3 - \sqrt{29}}{2}.$$

We use the x-intercept method to solve this inequality. As seen in Figure 61, the graph of $y = \sqrt[3]{x^2 + 3x} - \sqrt[3]{5}$ lies on or below the x-axis in the interval between the two x-intercepts, including the endpoints. Therefore, the solution set of the given inequality is

$$\left[\frac{-3 - \sqrt{29}}{2}, \frac{-3 + \sqrt{29}}{2}\right].$$

(b) This inequality is equivalent to $\sqrt{2x + 3} - \sqrt{x + 1} - 1 > 0$. The associated equation was solved in Example 4; its solutions are 3 and -1. The graph of $y_3 = \sqrt{2x + 3} - \sqrt{x + 1} - 1$ lies above the x-axis for real numbers greater than 3, as we see in Figure 62. The solution set is $(3, \infty)$.

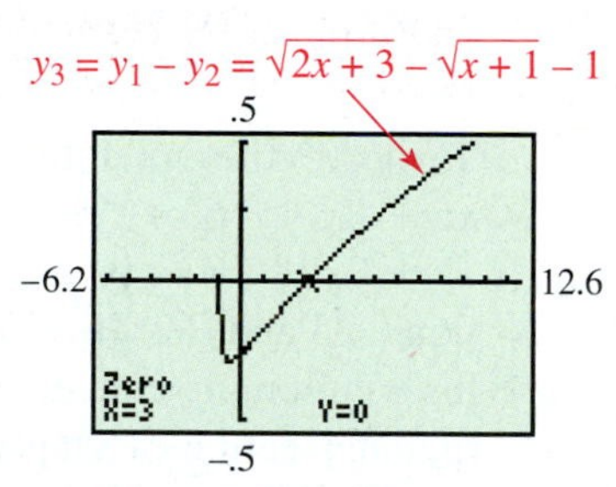

FIGURE 62

EXAMPLE 6 **Solving an Equation Having Negative Exponents**

Solve $15x^{-2} - 19x^{-1} + 6 = 0$.

Solution

$$\begin{aligned}
15x^{-2} - 19x^{-1} + 6 &= 0 \\
x^2(15x^{-2} - 19x^{-1} + 6) &= x^2(0) && \text{Multiply each side by } x^2. \\
15x^2x^{-2} - 19x^2x^{-1} + 6x^2 &= 0 && \text{Distributive property} \\
15 - 19x + 6x^2 &= 0 && \text{Properties of exponents} \\
6x^2 - 19x + 15 &= 0 && \text{Rewrite.} \\
(2x - 3)(3x - 5) &= 0 && \text{Factor.} \\
x = \frac{3}{2} \quad \text{or} \quad x &= \frac{5}{3} && \text{Zero-product property}
\end{aligned}$$

The solutions check. The solution set is $\{\frac{3}{2}, \frac{5}{3}\}$.

EXAMPLE 7 **Solving an Equation Having Fractional Exponents**

Solve $2x^{2/3} + 5x^{1/3} - 3 = 0$.

Solution

$$2x^{2/3} + 5x^{1/3} - 3 = 0$$

$$2(x^{1/3})^2 + 5(x^{1/3}) - 3 = 0 \qquad \text{Properties of exponents}$$

$$2u^2 + 5u - 3 = 0 \qquad \text{Let } u = x^{1/3} \text{ and substitute.}$$

$$(2u - 1)(u + 3) = 0 \qquad \text{Factor.}$$

$$u = \frac{1}{2} \quad \text{or} \quad u = -3 \qquad \text{Zero-product property}$$

$$x^{1/3} = \frac{1}{2} \quad \text{or} \quad x^{1/3} = -3 \qquad \text{Substitute } x^{1/3} \text{ for } u.$$

$$(x^{1/3})^3 = \left(\frac{1}{2}\right)^3 \quad \text{or} \quad (x^{1/3})^3 = (-3)^3 \qquad \text{Cube each side.}$$

$$x = \frac{1}{8} \quad \text{or} \quad x = -27 \qquad \text{Simplify.}$$

The solutions check, so the solution set is $\left\{\frac{1}{8}, -27\right\}$. ■

An Application of Root Functions

EXAMPLE 8 **Solving a Cable Installation Problem Involving a Root Function**

A company wishes to run a utility cable from point A on the shore (as shown in Figure 63) to an installation at point B on the island. The island is 6 miles from the shore. It costs \$400 per mile to run the cable on land and \$500 per mile under water. Assume that the cable starts at A, runs along the shoreline, and then angles and runs under water to the island. Let x represent the distance from C at which the under water portion of the cable run begins, and let the distance between A and C be 9 miles.

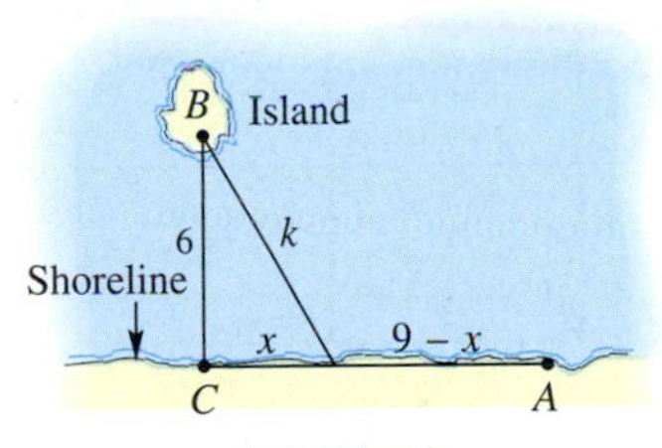

FIGURE 63

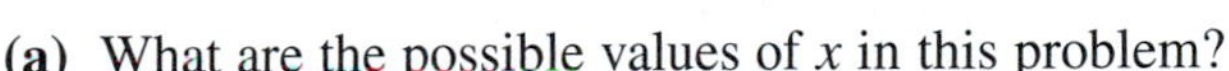

(a) What are the possible values of x in this problem?

(b) Express the cost of laying the cable as a function of x.

(c) Find the total cost if 3 miles of cable are on land.

(d) Find the point at which the line should begin to angle to minimize the total cost. What is this total cost? Solve graphically.

Solution

(a) The value of x must be a real number greater than or equal to 0 and less than or equal to 9, meaning that x must be in the interval $[0, 9]$.

(b) The total cost is found by adding the cost of the cable on land to the cost of the cable under water. If we let k represent the length of the cable under water, then

$$k^2 = 6^2 + x^2 \qquad \text{Use the Pythagorean theorem.}$$

$$k^2 = 36 + x^2$$

$$k = \sqrt{36 + x^2}. \qquad \text{Square root property; } k > 0$$

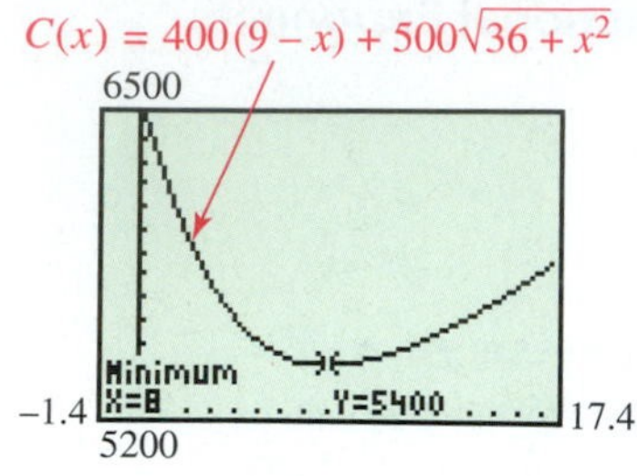

FIGURE 64

The cost of the cable on land is $400(9 - x)$ dollars, while the cost of the cable under water is $500k$ or $500\sqrt{36 + x^2}$ dollars. The total cost $C(x)$ is given by

$$C(x) = 400(9 - x) + 500\sqrt{36 + x^2}. \quad \text{Total cost in dollars}$$

(c) According to Figure 63, if 3 miles of cable are on land, then $3 = 9 - x$, giving $x = 6$. We must evaluate $C(6)$.

$$\begin{aligned} C(6) &= 400(9 - 6) + 500\sqrt{36 + 6^2} \\ &\approx 5442.64 \text{ dollars} \end{aligned}$$

(d) From Figure 64, the absolute minimum value of the function on the interval $[0, 9]$ is found when $x = 8$, meaning that $9 - 8 = 1$ mile should be along land and $\sqrt{36 + 8^2} = 10$ miles should be under water. We must evaluate $C(8)$.

$$\begin{aligned} C(8) &= 400(9 - 8) + 500\sqrt{36 + 8^2} \\ &= 5400 \text{ dollars} \quad \text{Minimum total cost} \end{aligned}$$

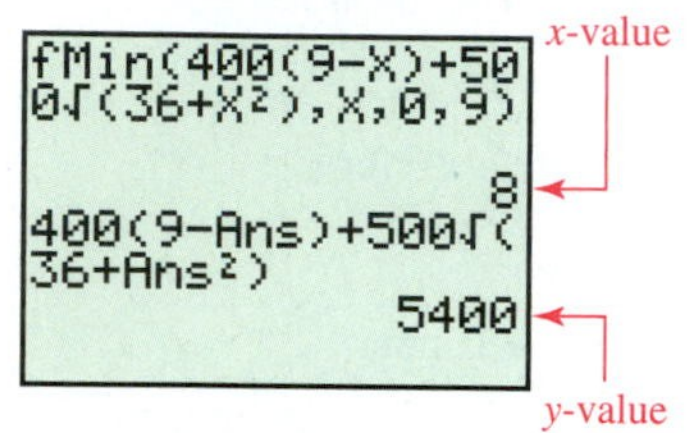

FIGURE 65

Figure 65 shows how the coordinates of the minimum point can be determined without actually graphing the function. ■

4.5 Exercises

Use the calculator graph to find the solution set of the given equation and inequalities.

1. (a) $\sqrt{x + 5} = x - 1$
(b) $\sqrt{x + 5} \leq x - 1$
(c) $\sqrt{x + 5} \geq x - 1$

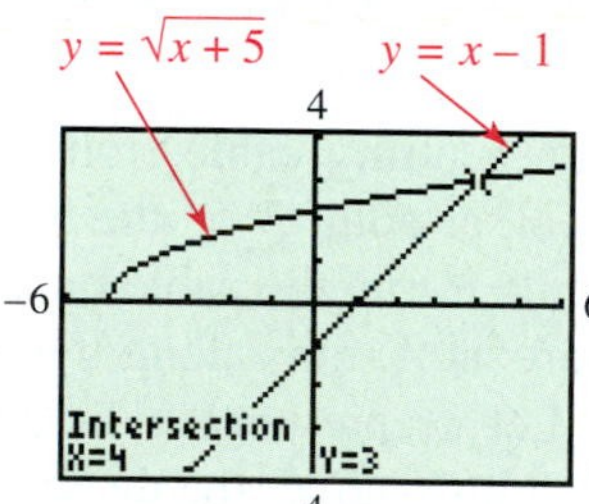

2. (a) $-\sqrt{x + 12} = x$
(b) $-\sqrt{x + 12} \leq x$
(c) $-\sqrt{x + 12} \geq x$

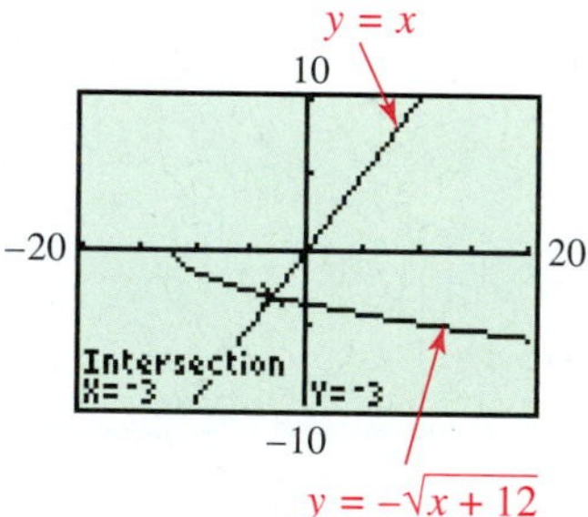

3. (a) $\sqrt[3]{2 - 2x} = \sqrt[3]{2x + 14}$
(b) $\sqrt[3]{2 - 2x} > \sqrt[3]{2x + 14}$
(c) $\sqrt[3]{2 - 2x} < \sqrt[3]{2x + 14}$

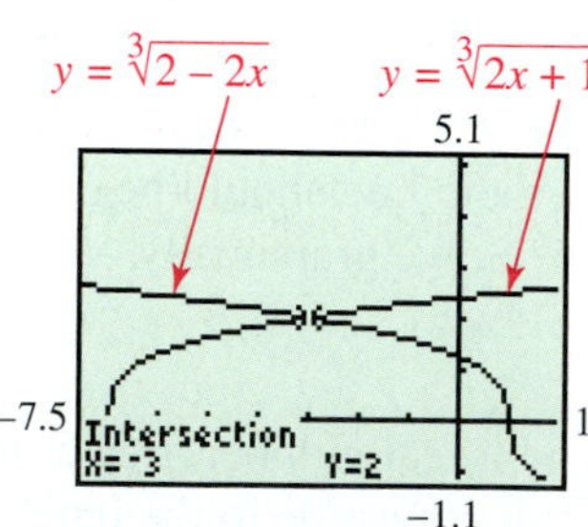

4. (a) $\sqrt[3]{3x} = \sqrt[3]{7 - 4x}$
(b) $\sqrt[3]{3x} > \sqrt[3]{7 - 4x}$
(c) $\sqrt[3]{3x} < \sqrt[3]{7 - 4x}$

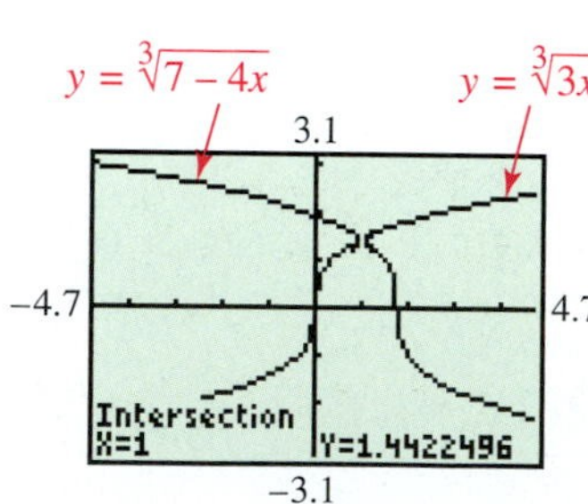

Consider the rough sketch at the right of the graphs of

$$y_1 = \sqrt{x} \quad \textit{and} \quad y_2 = -x + 6.$$

There is only one point of intersection of the graphs and thus only one real solution of the equation $y_1 = y_2$. Using the methods of Chapter 3, we can show that the solution set of

$$\sqrt{x} = -x + 6$$

is $\{4\}$, with extraneous value 9.

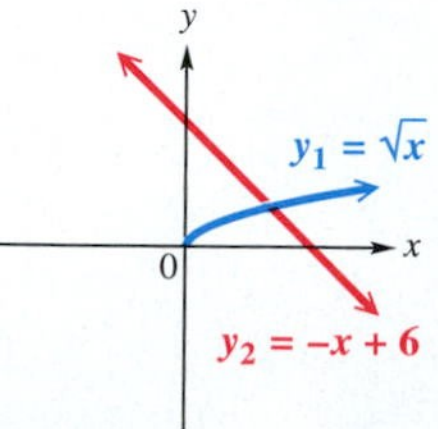

For Exercises 5–9, begin by drawing a rough sketch to determine the number of real solutions for the equation $y_1 = y_2$, as explained on the preceding page. Then use an analytic method to confirm your answer. Give the solution set and any extraneous values that may occur.

5. $y_1 = \sqrt{x}$, $y_2 = 2x - 1$
6. $y_1 = \sqrt{x}$, $y_2 = x - 6$
7. $y_1 = \sqrt{x}$, $y_2 = -x + 3$
8. $y_1 = \sqrt{x}$, $y_2 = 3x$
9. $y_1 = \sqrt[3]{x}$, $y_2 = x^2$

 **10.** Use a hand-drawn graph to explain why $\sqrt{x} = -x - 5$ has no real solutions.

Use an analytic method to solve each equation in part (a). Support the solution with a graph. Then use the graph to solve the inequalities in parts (b) and (c).

11. **(a)** $\sqrt{3x + 7} = 2$ **(b)** $\sqrt{3x + 7} > 2$ **(c)** $\sqrt{3x + 7} < 2$

12. **(a)** $\sqrt{2x + 13} = 3$ **(b)** $\sqrt{2x + 13} > 3$ **(c)** $\sqrt{2x + 13} < 3$

13. **(a)** $\sqrt{4x + 13} = 2x - 1$ **(b)** $\sqrt{4x + 13} > 2x - 1$ **(c)** $\sqrt{4x + 13} < 2x - 1$

14. **(a)** $\sqrt{3x + 7} = 3x + 5$ **(b)** $\sqrt{3x + 7} > 3x + 5$ **(c)** $\sqrt{3x + 7} < 3x + 5$

15. **(a)** $\sqrt{5x + 1} + 2 = 2x$ **(b)** $\sqrt{5x + 1} + 2 > 2x$ **(c)** $\sqrt{5x + 1} + 2 < 2x$

16. **(a)** $\sqrt{3x + 4} + x = 8$ **(b)** $\sqrt{3x + 4} + x > 8$ **(c)** $\sqrt{3x + 4} + x < 8$

17. **(a)** $\sqrt{3x - 6} + 2 = \sqrt{5x - 6}$ **(b)** $\sqrt{3x - 6} + 2 > \sqrt{5x - 6}$ **(c)** $\sqrt{3x - 6} + 2 < \sqrt{5x - 6}$

18. **(a)** $\sqrt{2x - 4} + 2 = \sqrt{3x + 4}$ **(b)** $\sqrt{2x - 4} + 2 > \sqrt{3x + 4}$ **(c)** $\sqrt{2x - 4} + 2 < \sqrt{3x + 4}$

19. **(a)** $\sqrt[3]{x^2 - 2x} = \sqrt[3]{x}$ **(b)** $\sqrt[3]{x^2 - 2x} > \sqrt[3]{x}$ **(c)** $\sqrt[3]{x^2 - 2x} < \sqrt[3]{x}$

20. **(a)** $\sqrt[3]{4x^2 - 4x + 1} = \sqrt[3]{x}$ **(b)** $\sqrt[3]{4x^2 - 4x + 1} > \sqrt[3]{x}$ **(c)** $\sqrt[3]{4x^2 - 4x + 1} < \sqrt[3]{x}$

21. **(a)** $\sqrt[4]{3x + 1} = 1$ **(b)** $\sqrt[4]{3x + 1} > 1$ **(c)** $\sqrt[4]{3x + 1} < 1$

22. **(a)** $\sqrt[4]{x - 15} = 2$ **(b)** $\sqrt[4]{x - 15} > 2$ **(c)** $\sqrt[4]{x - 15} < 2$

23. **(a)** $(2x - 5)^{1/2} - 2 = (x - 2)^{1/2}$ **(b)** $(2x - 5)^{1/2} - 2 \geq (x - 2)^{1/2}$ **(c)** $(2x - 5)^{1/2} - 2 \leq (x - 2)^{1/2}$

24. **(a)** $(x + 5)^{1/2} - 2 = (x - 1)^{1/2}$ **(b)** $(x + 5)^{1/2} - 2 \geq (x - 1)^{1/2}$ **(c)** $(x + 5)^{1/2} - 2 \leq (x - 1)^{1/2}$

25. **(a)** $(x^2 + 6x)^{1/4} = 2$ **(b)** $(x^2 + 6x)^{1/4} > 2$ **(c)** $(x^2 + 6x)^{1/4} < 2$

26. **(a)** $(x^2 + 2x)^{1/4} = 3^{1/4}$ **(b)** $(x^2 + 2x)^{1/4} > 3^{1/4}$ **(c)** $(x^2 + 2x)^{1/4} < 3^{1/4}$

27. **(a)** $(2x - 1)^{2/3} = x^{1/3}$ **(b)** $(2x - 1)^{2/3} > x^{1/3}$ **(c)** $(2x - 1)^{2/3} < x^{1/3}$

28. **(a)** $(x - 3)^{2/5} = (4x)^{1/5}$ **(b)** $(x - 3)^{2/5} > (4x)^{1/5}$ **(c)** $(x - 3)^{2/5} < (4x)^{1/5}$

Solve each equation. Check your answers.

29. $\sqrt{5x - 6} = x$
30. $x - 5 = \sqrt{5x - 1}$
31. $\sqrt{x + 5} + 1 = x$
32. $\sqrt{4 - 3x} = x + 8$
33. $\sqrt{2x + 3} - \sqrt{x + 1} = 1$
34. $\sqrt{3x + 4} - \sqrt{2x - 4} = 2$
35. $\sqrt[3]{x + 1} = -3$
36. $\sqrt[3]{x} + 5 = 4$
37. $\sqrt[3]{x + 1} = \sqrt[3]{2x - 1}$
38. $\sqrt[3]{2x^2 + 1} = \sqrt[3]{1 - x}$
39. $\sqrt[4]{x - 2} + 4 = 20$
40. $\sqrt[4]{2x + 3} = \sqrt{x + 1}$
41. $x^{2/5} = 4$
42. $x^{2/3} = 16$
43. $2x^{1/3} - 5 = 1$
44. $4x^{3/2} + 5 = 21$
45. $x^{-2} + 3x^{-1} + 2 = 0$
46. $2x^{-2} - x^{-1} = 3$
47. $5x^{-2} + 13x^{-1} = 28$
48. $3x^{-2} - 19x^{-1} + 20 = 0$
49. $x^{2/3} - x^{1/3} - 6 = 0$
50. $x^{2/3} + 9x^{1/3} + 14 = 0$
51. $x^{3/4} - x^{1/2} - x^{1/4} + 1 = 0$
52. $x^{3/4} - 2x^{1/2} - 4x^{1/4} + 8 = 0$

Relating Concepts

For individual or group investigation (Exercises 53–64)

Exercises 53–64 incorporate many concepts from Chapter 3 with the method of solving equations involving roots. ***Work them in order.*** *Consider the equation* $\sqrt[3]{4x-4}=\sqrt{x+1}$.

53. Rewrite the equation, using rational exponents.

54. What is the least common denominator of the rational exponents found in Exercise 53?

55. Raise each side of the equation in Exercise 53 to the power indicated by your answer in Exercise 54.

56. Show that the equation in Exercise 55 is equivalent to $x^3-13x^2+35x-15=0$.

57. Graph the cubic function defined by the polynomial on the left side of the equation in Exercise 56 in the window $[-5,10]$ by $[-100,100]$. How many real solutions does the equation have?

58. Use synthetic division to show that 3 is a zero of the polynomial $P(x)=x^3-13x^2+35x-15$.

59. Use the result of Exercise 58 to factor $P(x)$ so that one factor is linear and the other is quadratic.

60. Set the quadratic factor of $P(x)$ from Exercise 59 equal to 0, and solve the equation, using the quadratic formula.

61. What are the three proposed solutions of the original equation, $\sqrt[3]{4x-4}=\sqrt{x+1}$?

62. Let $y_1=\sqrt[3]{4x-4}$ and $y_2=\sqrt{x+1}$. Graph $y_3=y_1-y_2$ in the window $[-2,20]$ by $[-.5,.5]$ to determine the number of real solutions of the original equation.

63. Use both an analytic method and your calculator to solve the original equation.

64. Write an explanation of how the solutions of the equation in Exercise 56 relate to the solutions of the original equation. Discuss any extraneous solutions that may be involved.

Solve each equation involving "nested" radicals for all real solutions analytically. Support your solutions with a graph.

65. $\sqrt{\sqrt{x}}=x$

66. $\sqrt[3]{\sqrt[3]{x}}=x$

67. $\sqrt{\sqrt{28x+8}}=\sqrt{3x+2}$

68. $\sqrt{\sqrt{2x+10}}=\sqrt{2x-2}$

69. $\sqrt[3]{\sqrt{32x}}=\sqrt[3]{x+6}$

70. $\sqrt[3]{\sqrt{x+63}}=\sqrt[3]{2x+6}$

(Modeling) Solve each problem.

71. *Velocity of a Meteorite* The velocity v of a meteorite approaching Earth is given by

$$v=\frac{k}{\sqrt{d}},$$

measured in kilometers per second, where d is its distance from the center of Earth and k is a constant. If $k=350$, what is the velocity of a meteorite that is 6000 kilometers away from the center of Earth? Round to the nearest tenth.

72. *Illumination* The illumination I in foot-candles produced by a light source is related to the distance d in feet from the light source by the equation

$$d=\sqrt{\frac{k}{I}},$$

where k is a constant. If $k=400$, how far from the source will the illumination be 14 foot-candles? Round to the nearest hundredth of a foot.

73. *Period of a Pendulum* The period P of a pendulum in seconds depends on its length L in feet and is given by

$$P=2\pi\sqrt{\frac{L}{32}}.$$

If the length of a pendulum is 5 feet, what is its period? Round to the nearest tenth.

74. *Visibility from an Airplane* A formula for calculating the distance d one can see from an airplane to the horizon on a clear day is

$$d=1.22\sqrt{x},$$

where x is the altitude of the plane in feet and d is given in miles. How far can one see to the horizon in a plane flying at each altitude? Give answers to the nearest mile.

(a) 15,000 feet **(b)** 20,000 feet

75. ***Speed of a Car in an Accident*** To estimate the speed s at which a car was traveling at the time of an accident, a police officer drives a car like the one involved in the accident under conditions similar to those during which the accident took place and then skids to a stop. If the car is driven at 30 miles per hour, the speed at the time of the accident is given by

$$s = 30\sqrt{\frac{a}{p}},$$

where a is the length of the skid marks and p is the length of the marks in the police test. Find s if $a = 900$ feet and $p = 97$ feet.

76. ***Plant Species and Land Area*** A research biologist has shown that the number S of different plant species on a Galápagos Island is related to the area A of the island by

$$S = 28.6\sqrt[3]{A}.$$

How many plant species would exist on such an island with each area?

(a) 100 square miles **(b)** 1500 square miles

77. ***Wire between Two Poles*** Two vertical poles of lengths 12 feet and 16 feet are situated on level ground, 20 feet apart, as shown in the figure. A piece of wire is to be strung from the top of the 12-foot pole to the top of the 16-foot pole, attached to a stake in the ground at a point P on a line formed by the vertical poles. Let x represent the distance from P to D, the base of the 12-foot pole.

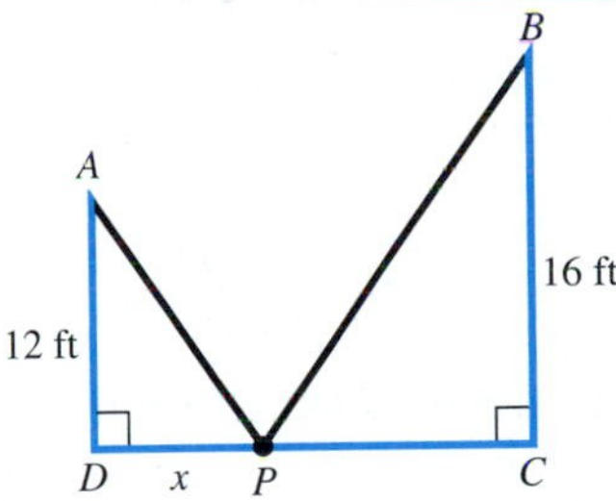

(a) Express the distance from P to C in terms of x.

(b) What are the restrictions on the value of x in this problem?

(c) Use the Pythagorean theorem to express the lengths AP and BP in terms of x.

(d) Give a function f that expresses the total length of the wire used.

(e) Graph f in the window $[0, 20]$ by $[0, 50]$. Use your calculator to find $f(4)$, and interpret your result.

(f) Find the value of x that will minimize the amount of wire used.

(g) Write a short paragraph summarizing what this problem has examined and the results you have obtained.

78. ***Wire between Two Poles*** Repeat Exercise 77 if the heights of the poles are 9 feet and 12 feet and the distance between the poles is 16 feet. Let P be x feet from the 9-foot pole. In part (e), use the window $[0, 16]$ by $[0, 50]$.

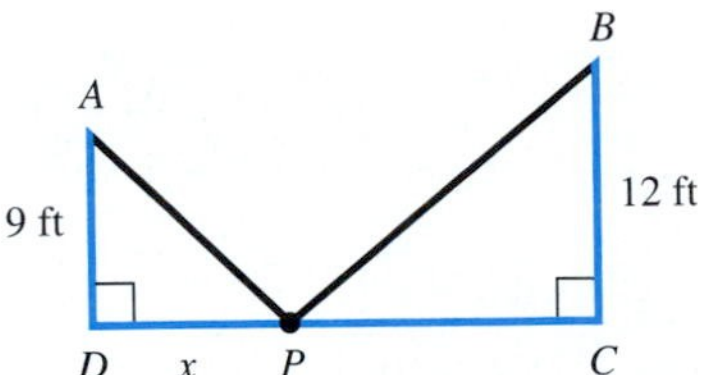

79. ***Hunter Returns to His Cabin*** A hunter is at a point on a riverbank. He wants to get to his cabin, located 3 miles north and 8 miles west. He can travel 5 mph on the river but only 2 mph on this very rocky land. How far upriver should he go in order to reach the cabin in a minimum amount of time? (*Hint:* distance = rate × time.)

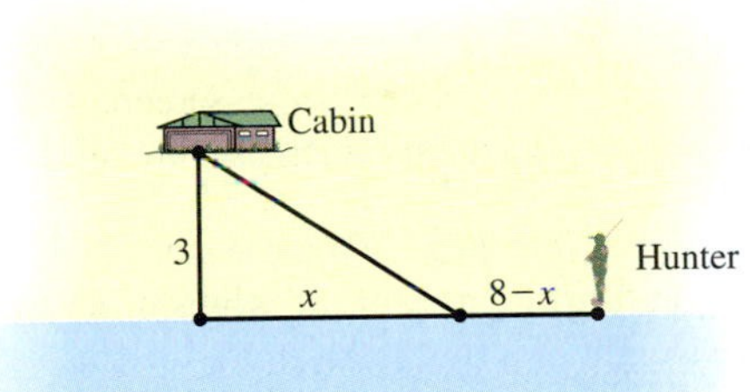

80. ***Homing Pigeon Flight*** Homing pigeons avoid flying over large bodies of water, preferring to fly around them instead. (One possible explanation is the fact that extra energy is required to fly over water because air pressure drops over water in the daytime.) Assume that a pigeon released from a boat 1 mile from the shore of a lake (point B in the figure) flies first to point P on the shore and then along the straight edge of the lake to reach its home at L. If L is 2 miles from point A, the point on the shore closest to the boat, and if a pigeon needs $\frac{4}{3}$ as much energy to fly over water as over land, find the location of point P that minimizes the pigeon's energy use.

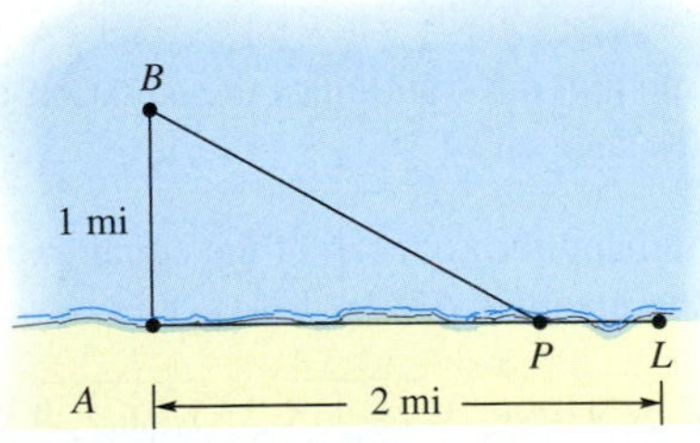

81. ***Cruise Ship Travel*** At noon, the cruise ship *Celebration* is 60 miles due south of the cruise ship *Inspiration* and is sailing north at a rate of 30 mph. If the *Inspiration* is sailing west at a rate of 20 mph, find the time at which the distance d between the ships is a minimum. What is this distance?

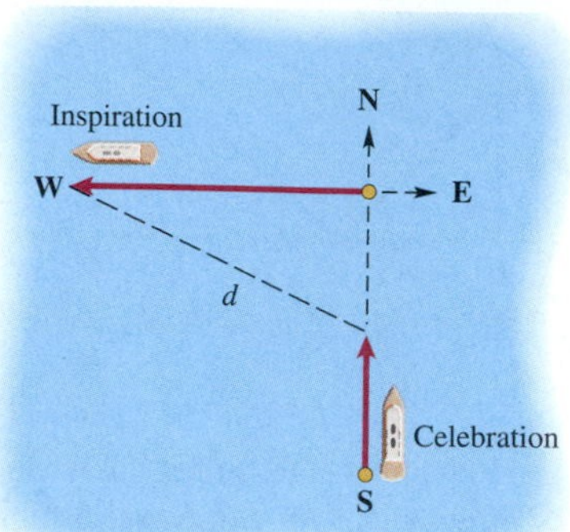

82. ***Fisherman Returns to Camp*** "Mido" Simon is in his bass boat, the *Lesley,* 3 miles from the nearest point on the shore. He wishes to reach his camp at Katie Point, 6 miles farther down the shoreline. If Mido's motor is disabled, and he can row his boat at a rate of 4 mph and walk at a rate of 5 mph, find the least amount of time that he will need to reach the camp.

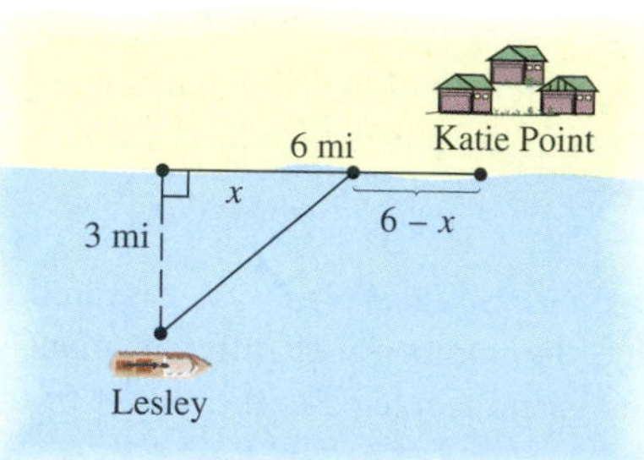

Reviewing Basic Concepts (Sections 4.4 and 4.5)

1. Use a graphing calculator to graph $y = x^{.7}$, $y = x^{1.2}$, and $y = x^{2.4}$ for $x \geq 0$ on the same screen. Describe what happens as the exponent n increases in value.

2. ***(Modeling) Wing Size*** Suppose that the relationship between the surface area of the wings of a species of bird and the weight of a bird is given by

$$S(w) = .3w^{3/4},$$

where w is the weight of the bird in kilograms and S is the surface area of the wings in square meters. If a bird weighs .75 kilogram, what is the surface area of the wings?

3. To graph the circle with equation $x^2 + y^2 = 16$ with a graphing calculator in function mode, what two expressions must be used for y_1 and y_2? Graph the circle in a decimal window.

4. To graph the horizontal parabola $x = y^2 + 4y + 6$ with a graphing calculator in function mode, what two expressions must be used for y_1 and y_2? Graph the parabola in a square window.

5. Solve the equation $\sqrt{3x + 4} = 8 - x$ analytically.

6. Use a graph and the solution of the equation in Exercise 5 to solve the inequality $\sqrt{3x + 4} > 8 - x$.

7. Use a graph and the solution of the equation in Exercise 5 to solve the inequality $\sqrt{3x + 4} < 8 - x$.

8. Solve the equation $\sqrt{3x + 4} + \sqrt{5x + 6} = 2$ analytically, and support the solution with a graph.

9. Explain why the table shows error messages for all values of X less than 15.

X	Y1
12	ERROR
13	ERROR
14	ERROR
15	0
16	1
17	1.4142
18	1.7321

Y1=√(X−15)

10. ***(Modeling) Animal Pulse Rates*** An 18th-century study found that the pulse rate of an animal could be approximated by

$$f(x) = \frac{1607}{\sqrt[4]{x^3}},$$

where x is the length of the animal in inches and $f(x)$ is the approximate number of heartbeats per minute. (*Source:* Lancaster, H., *Quantitative Methods in Biology and Medical Sciences,* Springer-Verlag, 1994.)

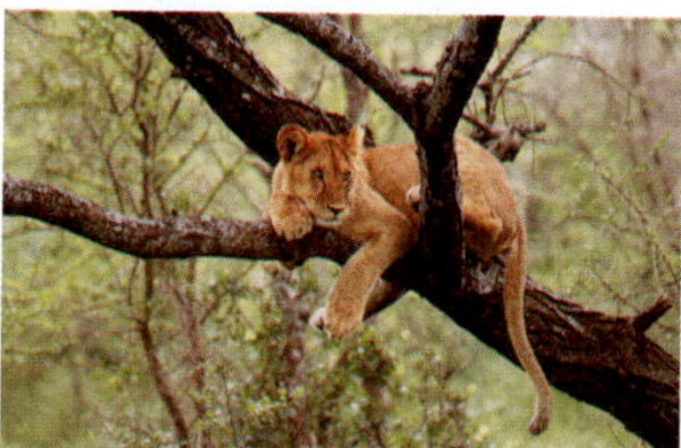

(a) Use f to estimate the pulse rates of a 2-foot dog and a 5.5-foot person.

(b) What length corresponds to a pulse rate of 400 beats per minute?

CHAPTER 4 SUMMARY

KEY TERMS & SYMBOLS	KEY CONCEPTS
4.1 Rational Functions and Graphs rational expression rational function discontinuous reciprocal function vertical asymptote horizontal asymptote	**RATIONAL FUNCTION** A function f of the form $\frac{p}{q}$ defined by $$f(x) = \frac{p(x)}{q(x)},$$ where $p(x)$ and $q(x)$ are polynomials, with $q(x) \neq 0$, is called a rational function. $f(x) = \frac{1}{x}$ (the reciprocal function) decreases on the intervals $(-\infty, 0)$ and $(0, \infty)$. ■ It is discontinuous at $x = 0$. ■ The y-axis is a vertical asymptote, and the x-axis is a horizontal asymptote. ■ It is an odd function, and its graph is symmetric with respect to the origin. $f(x) = \frac{1}{x^2}$ increases on the interval $(-\infty, 0)$ and decreases on the interval $(0, \infty)$. ■ It is discontinuous at $x = 0$. ■ The y-axis is a vertical asymptote, and the x-axis is a horizontal asymptote. ■ It is an even function, and its graph is symmetric with respect to the y-axis.
4.2 More on Graphs of Rational Functions oblique asymptote	**GRAPHING A RATIONAL FUNCTION** Let $f(x) = \frac{p(x)}{q(x)}$ define a function where the rational expression is in lowest terms. To sketch its graph, follow these steps: ***Step 1*** Find all vertical asymptotes. (See page 280.) ***Step 2*** Find all horizontal or oblique asymptotes. (See page 280.) ***Step 3*** Find the y-intercept, if possible, by evaluating $f(0)$. ***Step 4*** Find the x-intercepts, if any, by solving $f(x) = 0$. (These will be the zeros of the numerator $p(x)$.) ***Step 5*** Determine whether the graph will intersect its nonvertical asymptote by solving $f(x) = b$, where b is the y-value of the horizontal asymptote, or by solving $f(x) = mx + b$, where $y = mx + b$ is the equation of the oblique asymptote. ***Step 6*** Plot selected points as necessary. Choose an x-value in each interval of the domain determined by the vertical asymptotes and x-intercepts. ***Step 7*** Complete the sketch.
4.3 Rational Equations, Inequalities, Applications, and Models rational equation rational inequality inverse variation combined variation joint variation	**SOLVING A RATIONAL EQUATION** ***Step 1*** Determine all values for which the rational equation has undefined expressions. ***Step 2*** Multiply each side of the equation by the LCD of all rational expressions in the equation. ***Step 3*** Solve the resulting equation. ***Step 4*** Reject any values found in Step 1. *(continued)*

KEY TERMS & SYMBOLS	KEY CONCEPTS
	SOLVING A RATIONAL INEQUALITY ***Step 1*** Rewrite the inequality, if necessary, so that 0 is on one side and there is a single fraction on the other side. ***Step 2*** Determine the values that will cause either the numerator or the denominator of the rational expression to equal 0. These values determine the intervals on the number line to consider. ***Step 3*** Use a test value from each interval to determine which intervals form the solution set. **INVERSE VARIATION AS THE *n*TH POWER** Let x and y denote two quantities and n be a positive number. Then y is inversely proportional to the nth power of x, or y varies inversely as the nth power of x, if there exists a nonzero number k such that $$y = \frac{k}{x^n}.$$ If $y = \frac{k}{x}$, then y is inversely proportional to x, or y varies inversely as x. **JOINT VARIATION** Let m and n be real numbers. Then z varies jointly as the nth power of x and the mth power of y if a nonzero real number k exists such that $$z = kx^n y^m.$$
4.4 Functions Defined by Powers and Roots power function root function	**POWER AND ROOT FUNCTIONS** A function f given by $$f(x) = x^b,$$ where b is a constant, is a power function. If $b = \frac{1}{n}$ for some integer $n \geq 2$, then f is a root function given by $$f(x) = x^{1/n}, \quad \text{or equivalently,} \quad f(x) = \sqrt[n]{x}.$$ ■ For n even, the root function defined by $f(x) = \sqrt[n]{x}$ increases and is continuous on its entire domain, $[0, \infty)$. ■ For n odd, the root function defined by $f(x) = \sqrt[n]{x}$ increases and is continuous on its entire domain, $(-\infty, \infty)$.
4.5 Equations, Inequalities, and Applications Involving Root Functions	**SOLVING AN EQUATION INVOLVING A ROOT FUNCTION** ***Step 1*** Isolate a term involving a root on one side of the equation. ***Step 2*** Raise each side of the equation to a positive integer power that will eliminate the radical or rational exponent. ***Step 3*** Solve the resulting equation. (If a root is still present after Step 2, repeat Steps 1 and 2.) ***Step 4*** Check each proposed solution in the original equation.

CHAPTER 4 Review Exercises

For each rational function, do the following.

(a) *Explain how the graph of the function can be obtained from the graph of $y = \frac{1}{x}$ or $y = \frac{1}{x^2}$.*

(b) *Sketch the graph by hand.*

(c) *Use a graphing calculator to obtain an accurate depiction of the graph.*

1. $y = -\dfrac{1}{x} + 6$

2. $y = \dfrac{4}{x} - 3$

3. $y = -\dfrac{1}{(x-2)^2}$

4. $y = \dfrac{2}{x^2} + 1$

5. Under what conditions will the graph of a rational function defined by an expression written in lowest terms have an oblique asymptote?

Sketch a graph of each rational function. Your graph should include all asymptotes and be comprehensive. You may want to use your graphing calculator as an aid.

6. $f(x) = \dfrac{4x - 3}{2x - 1}$

7. $f(x) = \dfrac{6x}{(x-1)(x+2)}$

8. $f(x) = \dfrac{2x}{x^2 - 1}$

9. $f(x) = \dfrac{x^2 + 4}{x + 2}$

10. $f(x) = \dfrac{x^2 - 1}{x}$

11. $f(x) = \dfrac{-2}{x^2 + 1}$

12. $f(x) = \dfrac{x^2 - 1}{x + 1}$

13. $f(x) = \dfrac{x^2 - x - 2}{x^2 + 3x + 2}$

Find an equation for each rational function graphed. (Answers may vary.)

14.

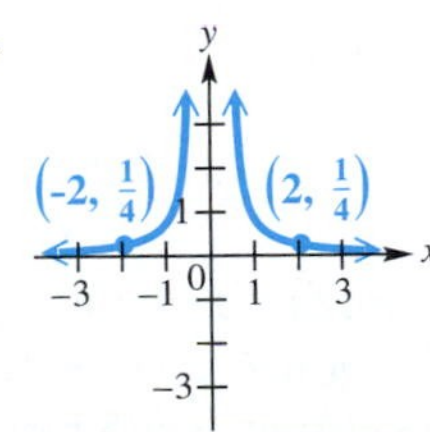

15.

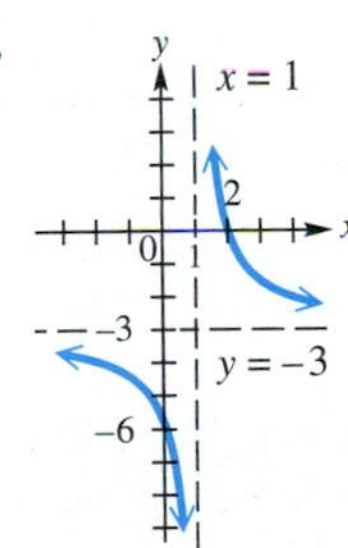

16.

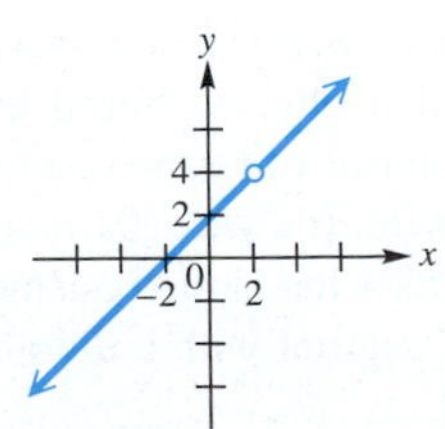

Solve the rational equation in part (a) analytically. Then, use a graph to determine the solution sets of the associated inequalities in parts (b) and (c).

17. (a) $\dfrac{3x - 2}{x + 1} = 0$

(b) $\dfrac{3x - 2}{x + 1} < 0$

(c) $\dfrac{3x - 2}{x + 1} > 0$

18. (a) $\dfrac{5}{2x + 5} = \dfrac{3}{x + 2}$

(b) $\dfrac{5}{2x + 5} < \dfrac{3}{x + 2}$

(c) $\dfrac{5}{2x + 5} > \dfrac{3}{x + 2}$

19. (a) $\dfrac{3}{x - 2} + \dfrac{1}{x + 1} = \dfrac{1}{x^2 - x - 2}$

(b) $\dfrac{3}{x - 2} + \dfrac{1}{x + 1} \le \dfrac{1}{x^2 - x - 2}$

(c) $\dfrac{3}{x - 2} + \dfrac{1}{x + 1} \ge \dfrac{1}{x^2 - x - 2}$

20. (a) $1 - \dfrac{5}{x} + \dfrac{6}{x^2} = 0$

(b) $1 - \dfrac{5}{x} + \dfrac{6}{x^2} \le 0$

(c) $1 - \dfrac{5}{x} + \dfrac{6}{x^2} \ge 0$

A comprehensive graph of a rational function f is shown. Use the graph to find each solution set.

21. $f(x) = 0$

22. $f(x) > 0$

23. $f(x) < 0$

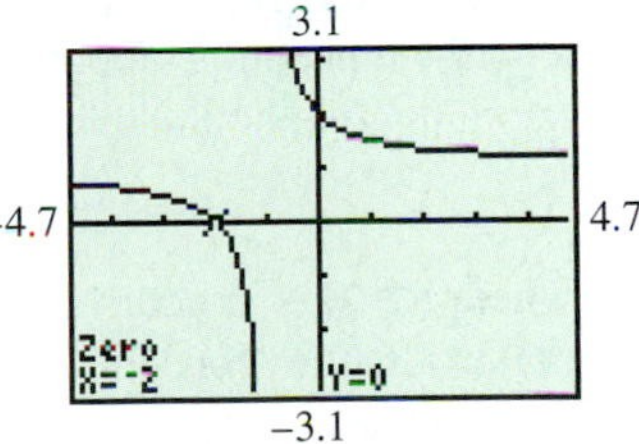

The graph has vertical asymptote $x = -1$.

(Modeling) Solve each problem.

24. *Environmental Pollution* In situations involving environmental pollution, a cost–benefit model expresses cost as a function of the percentage of pollutant removed from the environment. Suppose a cost–benefit model is expressed as

$$C(x) = \frac{6.7x}{100 - x},$$

where $C(x)$ is the cost, in thousands of dollars, of removing x percent of a certain pollutant.

(a) Use a graphing calculator to graph the function in the window $[0, 100]$ by $[0, 150]$.

(b) How much would it cost to remove 95% of the pollutant?

25. ***Antique-Car Competition*** Antique-car owners often enter their cars in a *concours d'elegance* in which a maximum of 100 points can be awarded to a particular car. Points are awarded for the general attractiveness of the car. The function defined by

$$C(x) = \frac{10x}{49(101 - x)}$$

expresses the cost, in thousands of dollars, of restoring a car so that it will win x points.

(a) Use a graphing calculator to graph the function in the window $[0, 101]$ by $[0, 5]$.

(b) How much would an owner expect to pay to restore a car in order to earn 95 points?

26. ***Concentration of a Drug*** The concentration of a drug in a medical patient's bloodstream is given by

$$f(t) = \frac{5}{t^2 + 1},$$

where $t \geq 0$ is in hours and $f(t)$ is in milligrams per liter.

(a) Does the concentration of the drug increase or decrease?

(b) The patient should not take a second dose until the concentration is below 1.5 milligrams per liter. How long should the patient wait before taking a second dose?

27. ***Waiting in Line*** A parking lot attendant can wait on 40 cars per hour. If cars arrive randomly at a rate of x cars per hour, then the average line length to enter the ramp is given by

$$f(x) = \frac{x^2}{1600 - 40x}, \quad \text{where } 0 \leq x < 40.$$

(a) Solve the inequality $f(x) \leq 8$.

(b) Interpret your answer from part (a).

28. ***Slippery Roads*** The coefficient of friction x measures the friction between the tires of a car and the road, where $0 < x \leq 1$. A smaller value of x indicates that the road is more slippery. If a car is traveling at 60 mph, then braking distance D in feet is given by

$$D(x) = \frac{120}{x}.$$

(a) What happens to the braking distance as the coefficient of friction becomes smaller?

(b) Find values for the coefficient of friction x that correspond to a braking distance of 400 feet or more.

29. Let y be inversely proportional to x. When $x = 6$, $y = 5$. Find y when $x = 15$.

30. Let z be inversely proportional to the third power of t. When $t = 5$, $z = .08$. Find z when $t = 2$.

31. Suppose m varies jointly as n and the square of p, and inversely as q. If m is 20 when n is 5, p is 6, and q is 18, find m when n is 7, p is 11, and q is 2.

32. The formula for the height of a right circular cone with volume 100 is

$$h = \frac{\frac{300}{\pi}}{r^2}.$$

The height of this cone varies ________ as the ________ of the ________ of its base. The constant of variation is ________.

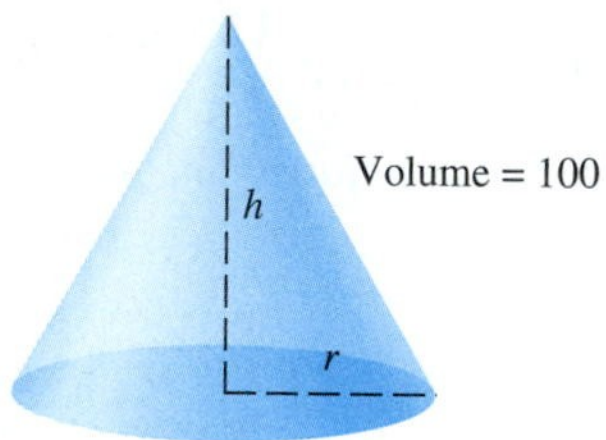

33. ***Illumination*** The illumination produced by a light source varies inversely as the square of the distance from the source. The illumination of a light source at 5 meters is 70 candela. What is the illumination 12 meters from the source?

34. ***Current Flow*** In electric current flow, it is found that the resistance (measured in units called ohms) offered by a fixed length of wire of a given material varies inversely as the square of the diameter of the wire. If a wire .01 inch in diameter has a resistance of .4 ohm, what is the resistance of a wire of the same length and material with a diameter of .03 inch?

35. ***Simple Interest*** Simple interest varies jointly as principal and time. If \$1000 left at interest for 2 years earned \$110, find the amount of interest earned by \$5000 for 5 years.

36. ***Car Skidding*** The force needed to keep a car from skidding on a curve varies inversely as the radius of the curve and jointly as the weight of the car and the square of the speed. It takes 3000 pounds of force to keep a 2000-pound car from skidding on a curve of radius 500 feet at 30 mph. What force is needed to keep the same car from skidding on a curve of radius 800 feet at 60 mph?

37. ***Sports Arena Construction*** The roof of a new sports arena rests on round concrete pillars. The maximum load a cylindrical column of circular cross section can hold varies directly as the fourth power of the diameter and inversely as the square of the height. The arena has 9-meter-tall columns that are 1 meter in diameter and will support a load of 8 metric tons. How many metric tons will be supported by a column 12 meters high and $\frac{2}{3}$ meter in diameter?

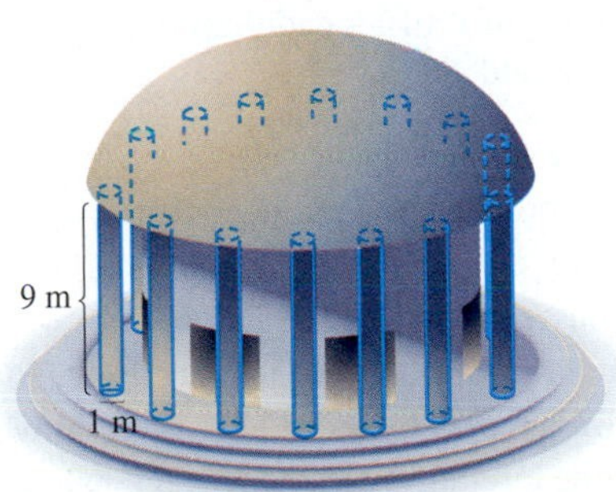

Load = 8 metric tons

38. ***Sports Arena Construction*** A sports arena requires a beam 16 meters long, 24 centimeters wide, and 8 centimeters high. The maximum load of a horizontal beam that is supported at both ends varies directly as the width and square of the height of the beam and inversely as the length between supports. If a beam of the same material 8 meters long, 12 centimeters wide, and 15 centimeters high can support a maximum of 400 kilograms, what is the maximum load the beam in the arena will support?

39. ***Weight of an Object*** The weight w of an object varies inversely as the square of the distance d between the object and the center of Earth. If a man weighs 90 kilograms on the surface of Earth, how much would he weigh 800 kilometers above the surface? (The radius of Earth is about 6400 kilometers.)

Suppose that a and b are positive numbers. Sketch the general shape of the graph of each function.

40. $y = -a\sqrt{x}$

41. $y = \sqrt[3]{x + a}$

42. $y = \sqrt[3]{x - a}$

43. $y = -a\sqrt[3]{x} - b$

44. $y = \sqrt{x + a} + b$

Evaluate each expression without *the use of a calculator. Then, check your results with a calculator.*

45. $\sqrt{144}$

46. $\sqrt[3]{-64}$

47. $\sqrt[3]{\dfrac{1}{27}}$

48. $\sqrt[4]{81}$

49. $\sqrt[5]{-\dfrac{32}{243}}$

50. $-(-32)^{1/5}$

51. $36^{-3/2}$

52. $-1000^{2/3}$

53. $(-27)^{-4/3}$

54. $16^{3/4}$

Use your calculator to find each root or power. Give as many digits as your display shows.

55. $\sqrt[5]{84.6}$

56. $\sqrt[4]{\dfrac{1}{16}}$

57. $\left(\dfrac{1}{8}\right)^{4/3}$

58. $12^{1/3}$

Consider the function defined by $f(x) = -\sqrt{2x - 4}$ *for Exercises 59–61.*

59. Find the domain of f analytically.

60. Use a graph to determine the range of f.

61. Give the interval (if any) over which the function is
(a) increasing. **(b)** decreasing.

62. Determine the expressions for y_1 and y_2 that are required to graph the circle $x^2 + y^2 = 25$ with a graphing calculator in function mode. Then graph the circle in a square window.

Solve each equation involving radicals or rational exponents in part (a) analytically. Then, use a graph to determine the solution sets of the associated inequalities in parts (b) and (c).

63. **(a)** $\sqrt{5 + 2x} = x + 1$
(b) $\sqrt{5 + 2x} > x + 1$
(c) $\sqrt{5 + 2x} < x + 1$

64. **(a)** $\sqrt{2x + 1} - \sqrt{x} = 1$
(b) $\sqrt{2x + 1} - \sqrt{x} > 1$
(c) $\sqrt{2x + 1} - \sqrt{x} < 1$

65. **(a)** $\sqrt[3]{6x + 2} = \sqrt[3]{4x}$
(b) $\sqrt[3]{6x + 2} \geq \sqrt[3]{4x}$
(c) $\sqrt[3]{6x + 2} \leq \sqrt[3]{4x}$

66. **(a)** $(x - 2)^{2/3} - x^{1/3} = 0$
(b) $(x - 2)^{2/3} - x^{1/3} \geq 0$
(c) $(x - 2)^{2/3} - x^{1/3} \leq 0$

Relating Concepts

For individual or group investigation (Exercises 67–70)

Exercises 67–70 refer to the functions defined by

$$y_1 = \sqrt{3x + 12} - 4 \quad \textit{and} \quad y_2 = \sqrt[3]{3x + 12} - 6.$$

Use a graph to help answer each exercise.

67. How many solutions does the equation $y_1 = y_2$ have?

68. Based on your answer to Exercise 67, how many x-intercepts does the graph of $f(x) = y_1 - y_2$ have?

69. Describe the appearance of the graph of $y = -f(x)$.

70. Describe the appearance of the graph of $y = f(-x)$.

Solve each equation. Check your results.

71. $x^5 = 1024$

72. $x^{1/3} = 4$

73. $\sqrt{x - 2} = x - 4$

74. $x^{3/2} = 27$

75. $2x^{1/4} + 3 = 6$

76. $\sqrt{x - 2} = 14 - x$

77. $\sqrt[3]{2x - 3} + 1 = 4$

78. $x^{1/3} + 3x^{1/3} = -2$

79. $2x^{-2} - 5x^{-1} = 3$

80. $x^{-3} + 2x^{-2} + x^{-1} = 0$

81. $x^{2/3} - 4x^{1/3} - 5 = 0$

82. $x^{3/4} - 16x^{1/4} = 0$

83. $\sqrt{x + 1} + 1 = \sqrt{2x}$

84. $\sqrt{x - 2} = 5 - \sqrt{x + 3}$

(Modeling) *Solve each problem.*

85. ***Swing of a Pendulum*** A simple pendulum will swing back and forth in regular time intervals (periods). Grandfather clocks use pendulums to keep accurate time. The relationship between the length L of a pendulum and the time T for one complete oscillation can be expressed by the equation $L = kT^n$, where k is a constant and n is a positive integer to be determined. The following data were taken for different lengths of pendulums.

T (in seconds)	L (in feet)
1.11	1.0
1.36	1.5
1.57	2.0
1.76	2.5
1.92	3.0
2.08	3.5
2.22	4.0

(a) As L increases, what happens to T?
(b) Discuss how n and k can be found.
(c) Use the data to approximate k, and determine the best value for n.
(d) Using the values of k and n from part (c), predict T for a pendulum having a length of 5 feet.
(e) If the length L of a pendulum doubles, what happens to the period T?

86. ***Volume of a Cylindrical Package*** A company plans to package its product in a cylinder that is open at one end. The cylinder is to have a volume of 27π cubic inches. What radius should the circular bottom of the cylinder have to minimize the cost of the material? (*Hint:* The volume V of a circular cylinder is $V = \pi r^2 h$, where r is the radius of the circular base and h is the height; the surface area S of a cylinder open at one end is $S = 2\pi rh + \pi r^2$.)

CHAPTER 4 Test

For each rational function, do the following.

(a) *Sketch its graph.*

(b) *Explain how the graph is obtained from the graph of* $y = \frac{1}{x}$ *or* $y = \frac{1}{x^2}$.

(c) *Use a graphing calculator to obtain an accurate depiction of the graph.*

1. $f(x) = -\frac{1}{x}$ **2.** $f(x) = \frac{-1}{x^2} - 3$

3. Consider the rational function defined by

$$f(x) = \frac{x^2 + x - 6}{x^2 - 3x - 4} = \frac{(x + 3)(x - 2)}{(x - 4)(x + 1)}.$$

Determine the answers to (a)–(e) analytically.

(a) Equations of the vertical asymptotes
(b) Equation of the horizontal asymptote
(c) y-intercept
(d) x-intercepts, if any
(e) Coordinates of the point where the graph of f intersects its horizontal asymptote

Now sketch a comprehensive graph of f.

4. Find the equation of the oblique asymptote of the graph of the rational function defined by

$$f(x) = \frac{2x^2 + x - 3}{x - 2}.$$

Then, graph the function and its asymptotes.

5. Consider the rational function defined by

$$f(x) = \frac{x^2 - 16}{x + 4}.$$

(a) For what value of x does the graph exhibit a "hole"?
(b) Graph the function and show the "hole" in the graph.

6. (a) Solve the following rational equation analytically.

$$\frac{3}{x - 2} + \frac{21}{x^2 - 4} = \frac{14}{x + 2}$$

(b) Use the result of part (a) and a graph to find the solution set of

$$\frac{3}{x - 2} + \frac{21}{x^2 - 4} \geq \frac{14}{x + 2}.$$

7. ***(Modeling) Waiting in Line*** A small section of highway that is under construction can accommodate at most 40 cars per minute. If cars arrive randomly at an average rate of x vehicles per minute, then the average wait W in minutes for a car to pass through this section of highway is approximated by

$$W(x) = \frac{1}{40 - x}, \quad \text{where } 0 \leq x < 40.$$

(a) Evaluate $W(30)$, $W(39)$, and $W(39.9)$. Interpret the results.
(b) Graph W, using the window $[0, 40]$ by $[-.5, 1]$. Identify the vertical asymptote. What happens to W as x approaches 40?
(c) Find x when the wait is 5 minutes.

8. ***(Modeling) Measure of Malnutrition*** The *pelidisi,* a measure of malnutrition, varies directly as the cube root of a person's weight in grams and inversely as the person's sitting height in centimeters. A person with a pelidisi below 100 is considered to be undernourished, while a pelidisi greater than 100 indicates overfeeding. A person who weighs 48,820 grams and has a sitting height of 78.7 centimeters has a pelidisi of 100. Find the pelidisi (to the nearest whole number) of a person whose weight is 54,430 grams and whose sitting height is 88.9 centimeters. Is this individual undernourished or overfed?

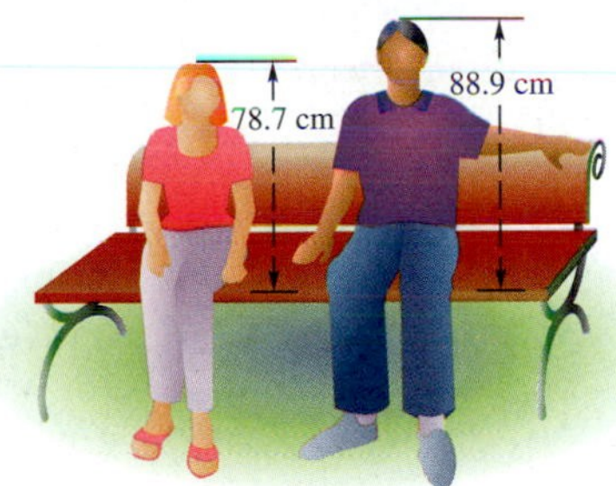

9. ***(Modeling) Volume of a Cylindrical Can*** A manufacturer wants to construct a cylindrical aluminum can with a volume of 4000 cubic centimeters. If x represents the radius of the circular top and bottom of the can, the surface area S as a function of x is given by

$$S(x) = \frac{8000 + 2\pi x^3}{x}.$$

Use a graph to find the radius that will minimize the amount of aluminum needed, and determine what this amount will be. (*Hint:* Use the window $[-5, 42]$ by $[-1000, 10{,}000]$.)

10. Graph the function defined by $f(x) = -\sqrt{5-x}$ in the standard viewing window. Then, do the following.

(a) Determine the domain analytically.

(b) Use the graph to find the range.

(c) This function ______________ over its entire domain. (increases/decreases)

(d) Solve the equation $f(x) = 0$ graphically.

(e) Solve the inequality $f(x) < 0$ graphically.

11. **(a)** Solve the equation $\sqrt{4-x} = x + 2$ analytically, and support the solution(s) with a graph.

(b) Use the graph in part (a) to find the solution set of $\sqrt{4-x} > x + 2$.

(c) Use the graph in part (a) to find the solution set of $\sqrt{4-x} \leq x + 2$.

12. *(Modeling) Laying a Telephone Cable* A telephone company wishes to minimize the cost of laying a cable from point P to point R. Points P and Q are directly opposite each other along the banks of a straight river 300 yards wide. Point R lies on the same side of the river as point Q, but 600 yards away. If the cost per yard for the cable is \$125 per yard under the water and \$100 per yard on land, how should the company lay the cable to minimize the cost?

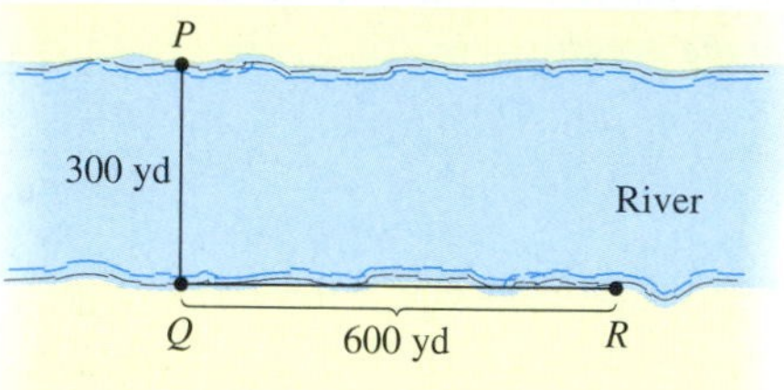

CHAPTER 4 PROJECT

How Rugged Is Your Coastline?*

An interesting feature of coastlines is that their ruggedness is independent of the distance from which they are viewed. From an airplane, we see irregularities as bays, peninsulas, river mouths, and so on. On foot, we see each rock outcropping and creek that makes the coastline appear more rugged. An ant sees every pebble as a mountain to be scaled. An interesting result of this phenomenon is that the total distance you travel along a coastline is dependent on the size of the steps you take. The closer (or smaller) you are, the smaller your steps will be. This means you will have more obstacles in your way, which results in a longer distance to travel. The more rugged the coastline, the longer it will be. In theory, then, if you could take small enough steps on a rugged enough coastline, the length of the coastline would approach infinity.

From a mathematical perspective, we can say that the number of steps needed (y) varies inversely with some power of the size of the steps taken (x):

$$y = \frac{k}{x^n}.$$

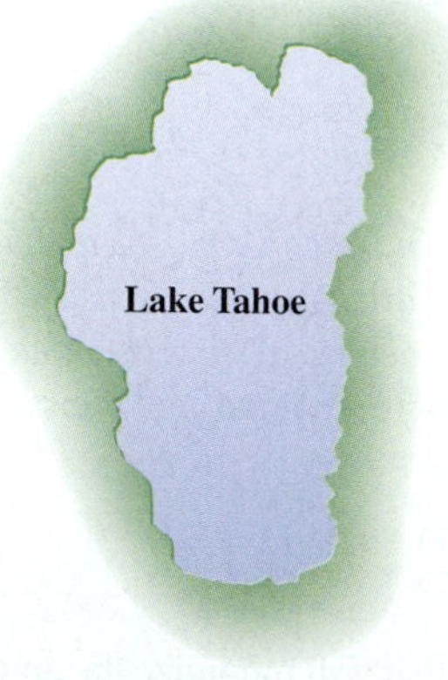

Each coastline will have different values for k and n, depending on its ruggedness. To find these values, you must walk the coastline. Gathering such data from a real coastline is, of course, too difficult as a classroom activity, but we can obtain similar data by using maps and "walking" the coastline with compasses (the kind used in geometry for drawing circles).

The data on the next page were gathered by "walking" compasses along the coastline of a map of Lake Tahoe (on the California–Nevada border). For each walk, the compasses were opened to the given step size, and the legs of the compasses were walked around the lake, counting the number of steps until the journey was completed.

*This project is based on an idea presented by Lori Lambertson of the Nueva School and the Exploratorium in San Francisco.

SIZE	NUMBR	----- 1
4	4.75	
2	10.75	
1	23.5	
.5	51	
.25	104	
------	------	

SIZE={4,2,1,.5,...

FIGURE A

Step Size	Number of Steps	Total Length (step size times number of steps)
4 inches	4.75	19 inches
2 inches	10.75	21.5 inches
1 inch	23.5	23.5 inches
$\frac{1}{2}$ inch	51	25.5 inches
$\frac{1}{4}$ inch	104	26 inches

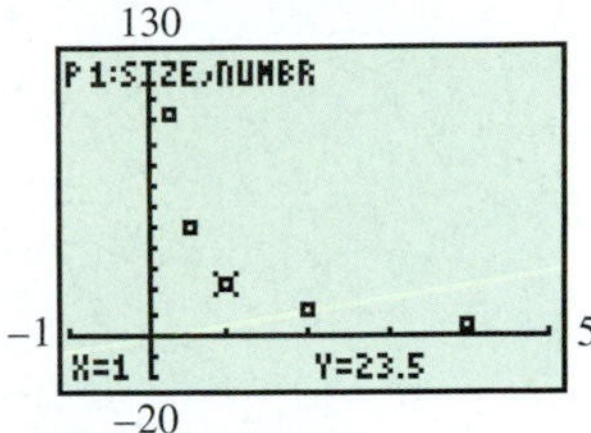

FIGURE B

Now we can analyze these data. First, enter them into the list editor of a graphing calculator, as shown in Figure A. Next, draw a scatter diagram of the data in an appropriate viewing window, as seen in Figure B. To find an equation to model the data, we solve for k by substituting the coordinates $(1, 23.5)$ for x and y in $y = \frac{k}{x^n}$. (We choose this point because any power of 1 is 1.) We find that $k = 23.5$.

We now know that the equation will be of the form $y = \frac{23.5}{x^n}$ for some value of n. Using the data point $(4, 4.75)$ leads to $4.75 = \frac{23.5}{4^n}$.

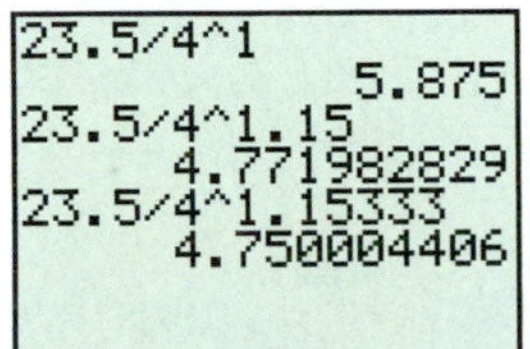

FIGURE C

Solving this equation for n poses a problem: We have not yet discussed how to solve such an equation analytically. (This is discussed in Chapter 5.) However, the multiline display and editing capabilities of a graphing calculator make it possible to estimate n. To do this, enter $23.5/4^x$ onto the home screen, and use guessing to come up with a value of x that leads to approximately 4.75. As seen in the final display of Figure C, a guess of 1.15333 works. Finally, with $n \approx 1.153$, the equation we are looking for is

$$Y_1 = \frac{23.5}{X^{1.153}}.$$

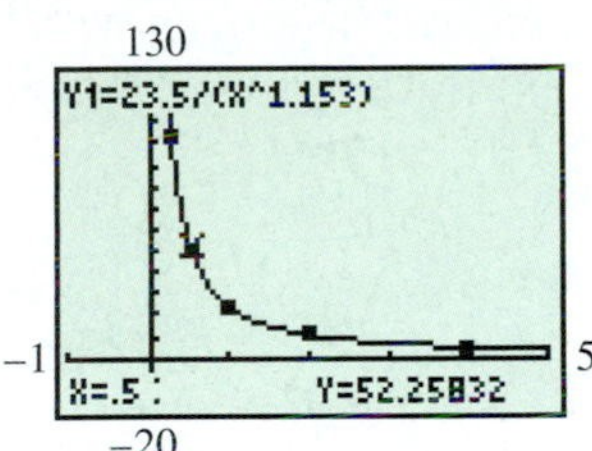

FIGURE D

Figure D shows the graph of this equation with the scatter diagram and illustrates a remarkably good fit. The display at the bottom of the screen indicates that when $X = .5$, $Y \approx 52.26$, which is very close to the experimental value of 51 seen in the original table.

Activities

1. Use the equation to determine how long the coastline would be if you "walked" with step sizes of 6 inches, .1 inch, and .01 inch, respectively. Remember that the equation gives you the number of steps, so multiply that value by the step size to get the total length.
2. Gather data from a map of your choice. Using compasses, walk the coastline of your map and count the number of steps. Change the size of the opening of the compasses, and walk the coastline again. After you have done this with four or five different step sizes, find a mathematical model (equation) for your data, using the procedure described.

Inverse, Exponential, and Logarithmic Functions

IN DECEMBER 2004, a massive earthquake struck the ocean floor off the west coast of Sumatra. Registering magnitude 9.0 on the Richter scale, the earthquake generated a tsunami in the Indian Ocean that killed 283,000 people in several coastal countries.

The Sumatra earthquake occurred along the interface of the India and Burma tectonic plates. Pressure gradually built up due to friction as the India plate subducted under the Burma plate at a rate of 2.5 inches per year. When the pressure reached a high enough level, it overcame friction and generated the earthquake as the India plate thrust forward. The Burma plate displaced an enormous volume of water as it pushed upward 30 feet during the earthquake, resulting in the devastating tsunami.

The magnitude of an earthquake is an indication of its intensity and is based on the amplitude, or height, of seismic waves measured at seismograph sites. Magnitude is measured with the Richter scale, an application of *logarithmic functions*. Logarithmic functions and their inverses, *exponential functions,* have many other applications in science and business, as we will see in this chapter.

Source: U.S. Geological Survey, British Geological Survey, and Canada Geological Survey, 2005.

Chapter Outline

5.1 Inverse Functions

Inverse Operations

Addition and subtraction are inverse operations: Starting with a number x, adding 5, and subtracting 5 gives x back as the result. Similarly, some functions are *inverses* of each other. For example, the functions defined by

$$f(x) = 8x \quad \text{and} \quad g(x) = \frac{1}{8}x$$

are inverses of each other with respect to function composition. This means that if a value of x such as $x = 12$ is chosen, then

$$f(12) = 8 \cdot 12 = 96.$$

Calculating $g(96)$ gives

$$g(96) = \frac{1}{8} \cdot 96 = 12.$$

Thus, $g[f(12)] = 12$. Also, $f[g(12)] = 12$. For any value of x, it can be shown that

$$f[g(x)] = x \quad \text{and} \quad g[f(x)] = x.$$

One-to-One Functions

For the function $y = f(x) = 5x - 8$, any two different values of x produce two different values of y. On the other hand, for the function $y = g(x) = x^2$, two different values of x can lead to the *same* value of y; for example, both $x = 4$ and $x = -4$ give $y = 4^2 = (-4)^2 = 16$. A function such as $y = f(x) = 5x - 8$, for which different elements from the domain always lead to different elements from the range, is called a *one-to-one function.*

One-to-One Function

A function f is a **one-to-one function** if, for elements a and b from the domain of f,

$$\boldsymbol{a \neq b} \quad \text{implies} \quad \boldsymbol{f(a) \neq f(b).}$$

Using the concept of the *contrapositive* from the study of logic, the last line in the preceding box is equivalent to

$$\boldsymbol{f(a) = f(b)} \quad \text{implies} \quad \boldsymbol{a = b.}$$

We use this statement to show that a function f is one-to-one in the next example.

EXAMPLE 1 Deciding whether Functions Are One-to-One

Decide whether each function is one-to-one.

(a) $f(x) = -4x + 12$ **(b)** $f(x) = \sqrt{25 - x^2}$

Solution

(a) We show that $f(a) = f(b)$ leads to the result $a = b$.

$$f(a) = f(b)$$

$$-4a + 12 = -4b + 12 \qquad f(x) = -4x + 12$$

$$-4a = -4b \qquad \text{Subtract 12.}$$

$$a = b \qquad \text{Divide by } -4.$$

By the definition, $f(x) = -4x + 12$ is one-to-one.

(b) If we choose $a = 3$ and $b = -3$, then $3 \neq -3$, but

$$f(3) = \sqrt{25 - 3^2} = \sqrt{25 - 9} = \sqrt{16} = 4$$

and $$f(-3) = \sqrt{25 - (-3)^2} = \sqrt{25 - 9} = 4.$$

Here, even though $3 \neq -3$, $f(3) = f(-3) = 4$. By definition, this is *not* a one-to-one function. ■

As shown in Example 1(b), a way to demonstrate that a function is *not* one-to-one is to produce a pair of unequal numbers that lead to the same function value. The *horizontal line test* also tells us whether a function is one-to-one.

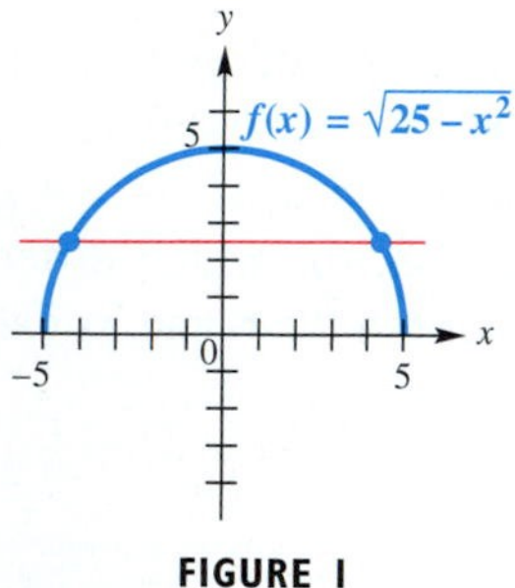

FIGURE 1

> **Horizontal Line Test**
>
> If every horizontal line intersects the graph of a function at no more than one point, then the function is one-to-one.

NOTE In Example 1(b), the graph of the function is a semicircle, as shown in Figure 1. There are infinitely many horizontal lines that intersect the graph of a semicircle in two points, so the horizontal line test shows that the function is not one-to-one.

EXAMPLE 2 Using the Horizontal Line Test

Determine whether each graph is the graph of a one-to-one function.

(a) **(b)**

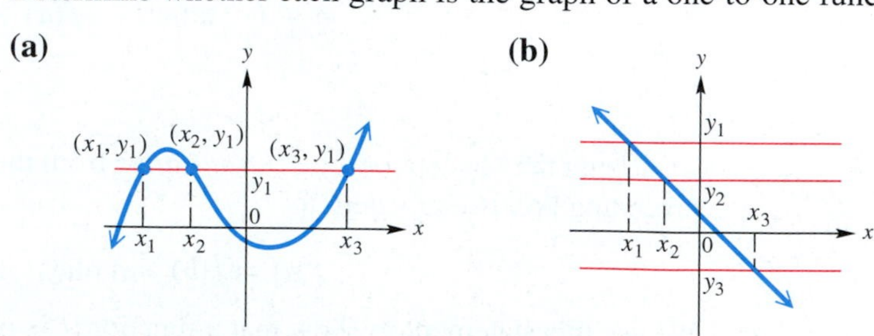

FOR DISCUSSION

Based on your knowledge of the basic functions studied so far in this text, answer each question. In each case, assume that the function has the largest possible domain.

1. Is a nonconstant linear function always one-to-one?
2. Is an odd-degree polynomial function always one-to-one? Why or why not?
3. Is an even-degree polynomial function ever one-to-one? Why or why not?

Solution

(a) Each point where the horizontal line intersects the graph has the same value of y but a different value of x. Since more than one (here, three) different values of x lead to the same value of y, the function is *not* one-to-one.

(b) Every horizontal line will intersect the graph at exactly one point. This function is one-to-one. ■

NOTE A function that is either increasing or decreasing on its entire domain, such as $f(x) = x^3$ or $g(x) = \sqrt{x}$, must be one-to-one.

Inverse Functions and Their Graphs

Certain pairs of one-to-one functions "undo" one another. For example, if

$$f(x) = 8x + 5 \quad \text{and} \quad g(x) = \frac{x - 5}{8},$$

then

$$f(10) = 8 \cdot 10 + 5 = 85 \quad \text{and} \quad g(85) = \frac{85 - 5}{8} = 10.$$

Starting with 10, we "applied" function f and then "applied" function g to the result, which gave back the number 10. See Figure 2.

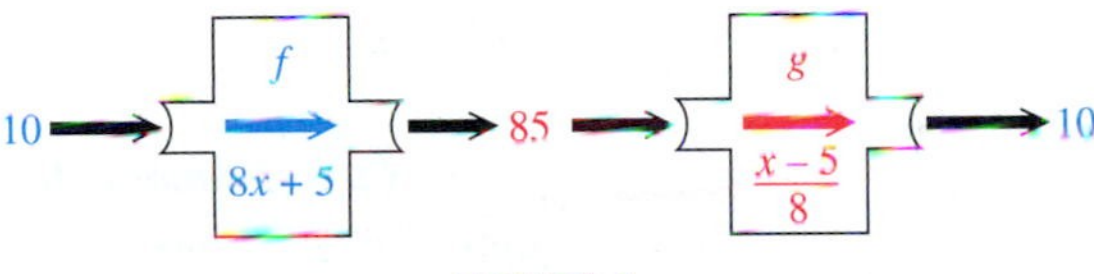

FIGURE 2

Similarly, for these same functions, check that

$$f(3) = 29 \quad \text{and} \quad g(29) = 3,$$
$$f(-5) = -35 \quad \text{and} \quad g(-35) = -5,$$
$$g(2) = -\frac{3}{8} \quad \text{and} \quad f\left(-\frac{3}{8}\right) = 2.$$

In particular, for these functions,

$$f[g(2)] = 2 \quad \text{and} \quad g[f(2)] = 2.$$

In fact for *any* value of x,

$$f[g(x)] = x \quad \text{and} \quad g[f(x)] = x,$$

or

$$(f \circ g)(x) = x \quad \text{and} \quad (g \circ f)(x) = x.$$

Because of this property, g is called the *inverse* of f.

Inverse Function

Let f be a one-to-one function. Then g is the **inverse function** of f and f is the inverse function of g if

$$(f \circ g)(x) = x \quad \text{for every } x \text{ in the domain of } g,$$

and

$$(g \circ f)(x) = x \quad \text{for every } x \text{ in the domain of } f.$$

EXAMPLE 3 Deciding whether Two Functions Are Inverses

Let f be defined by $f(x) = x^3 - 1$ and g by $g(x) = \sqrt[3]{x + 1}$. Is g the inverse function of f?

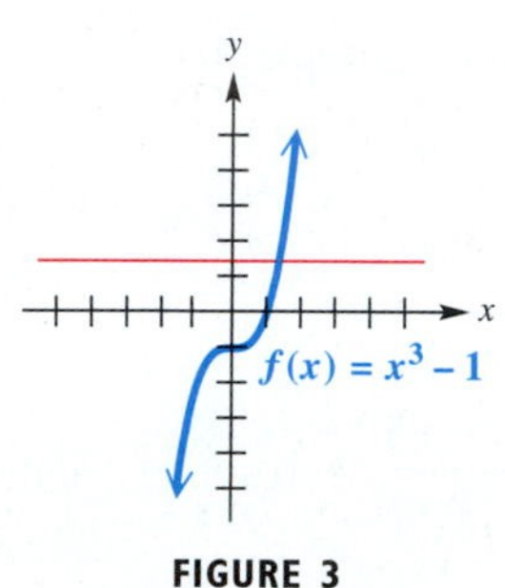

FIGURE 3

Solution As shown in Figure 3, the horizontal line test applied to the graph indicates that f is one-to-one, so it does have an inverse. Now find $(f \circ g)(x)$ and $(g \circ f)(x)$.

$$\begin{aligned}(f \circ g)(x) = f[g(x)] &= \left(\sqrt[3]{x+1}\right)^3 - 1\\ &= x + 1 - 1\\ &= x\end{aligned}$$

$$\begin{aligned}(g \circ f)(x) = g[f(x)] &= \sqrt[3]{(x^3 - 1) + 1}\\ &= \sqrt[3]{x^3}\\ &= x\end{aligned}$$

Since $(f \circ g)(x) = x$ and $(g \circ f)(x) = x$, function g is the inverse of function f. Also, note that f is the inverse of g. ■

A special notation is often used for inverse functions: If g is the inverse function of f, then g can be written as f^{-1} (read **"f-inverse"**). In Example 3,

$$f^{-1}(x) = \sqrt[3]{x + 1}.$$

CAUTION ***Do not confuse the −1 in f^{-1} with a negative exponent.*** The symbol $f^{-1}(x)$ does not represent $\frac{1}{f(x)}$; rather, it represents the inverse function of f.

The definition of inverse function can be used to show that the domain of f equals the range of f^{-1}, and the range of f equals the domain of f^{-1}. See Figure 4.

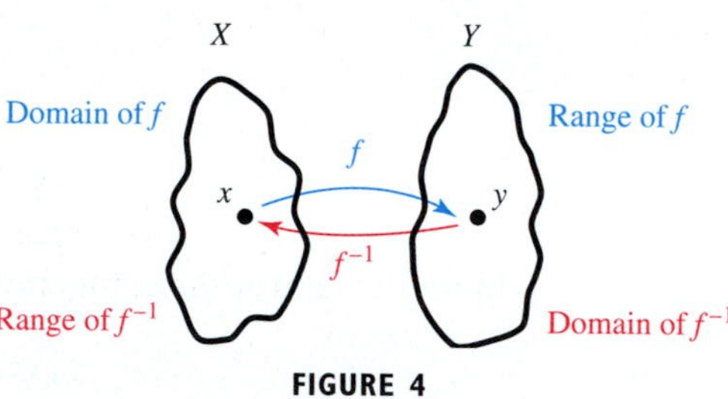

FIGURE 4

EXAMPLE 4 Finding Inverses of One-to-One Functions

Determine whether each function is one-to-one. If so, find its inverse.

(a) $f = \{(-2, 1), (-1, 0), (0, 1), (1, 2), (2, 2)\}$

(b) $g = \{(3, 1), (0, 2), (2, 3), (4, 0)\}$

Solution

(a) Each x-value in f corresponds to just one y-value. However, the y-value 2 corresponds to two x-values: 1 and 2. Also, the y-value 1 corresponds to both -2 and 0. Because some y-values correspond to more than one x-value, f is not one-to-one and does not have an inverse.

(b) Every x-value in g corresponds to only one y-value, and every y-value corresponds to only one x-value, so g is a one-to-one function. The inverse function is found by interchanging the x- and y-values in each ordered pair.

$$g^{-1} = \{(1, 3), (2, 0), (3, 2), (0, 4)\}$$

Notice how the domain and range of g become the range and domain, respectively, of g^{-1}. ■

Equations of Inverse Functions

The inverse of a one-to-one function is found by interchanging the x- and y-values of each of its ordered pairs. The equation of the inverse of a function defined by $y = f(x)$ is found in the same way.

Finding the Equation of the Inverse of $y = f(x)$

For a one-to-one function f defined by an equation $y = f(x)$, find the defining equation of the inverse as follows:

Step 1 Interchange x and y.

Step 2 Solve for y.

Step 3 Replace y with $f^{-1}(x)$.

(Any restrictions on x and y should be considered.)

X	Y1
1	2
6	6
11	10
-4	-2
-9	-6
51	42
181	146

Y1■(4X+6)/5

X	Y2
2	1
6	6
10	11
-2	-4
-6	-9
42	51
146	181

Y2■(5X−6)/4

Using the method of Example 5, we see that the inverse of $Y_1 = (4X + 6)/5$ is $Y_2 = (5X - 6)/4$. In the tables of selected points for these functions, notice that the X- and Y-values are interchanged. This is typical for inverse functions.

EXAMPLE 5 Finding Equations of Inverses

Decide whether each equation defines a one-to-one function. If so, find the equation of the inverse.

(a) $f(x) = 2x + 5$ **(b)** $y = x^2 + 2$ **(c)** $f(x) = (x - 2)^3$

Solution

(a) The graph of $y = 2x + 5$ is a nonhorizontal line, so by the horizontal line test, f is a one-to-one function. To find the equation of the inverse, follow the steps in the preceding box, first replacing $f(x)$ with y.

$$y = 2x + 5 \qquad y = f(x)$$

$$x = 2y + 5 \qquad \text{Interchange } x \text{ and } y. \text{ (Step 1)}$$

$$2y = x - 5 \qquad \text{Solve for } y. \text{ (Step 2)}$$

$$y = \frac{x - 5}{2}$$

$$f^{-1}(x) = \frac{x - 5}{2} \qquad \text{Replace } y \text{ with } f^{-1}(x). \text{ (Step 3)}$$

(b) The equation $y = x^2 + 2$ has a parabola opening up as its graph, so some horizontal lines will intersect the graph at two points. For example, both $x = 3$ and $x = -3$ correspond to $y = 11$. Because of the x^2-term, there are many pairs of x-values that correspond to the same y-value. Thus, the function defined by $y = x^2 + 2$ is not one-to-one and does not have an inverse.

(c) Refer to Sections 2.1 and 2.2 to see that translations of the graph of the cubing function are one-to-one.

$$y = (x - 2)^3 \quad \text{Replace } f(x) \text{ with } y.$$

$$x = (y - 2)^3 \quad \text{Interchange } x \text{ and } y.$$

$$\sqrt[3]{x} = \sqrt[3]{(y - 2)^3} \quad \text{Take the cube root of each side.}$$

$$\sqrt[3]{x} = y - 2$$

$$\sqrt[3]{x} + 2 = y \quad \text{Solve for } y.$$

$$f^{-1}(x) = \sqrt[3]{x} + 2 \quad \text{Replace } y \text{ with } f^{-1}(x).$$

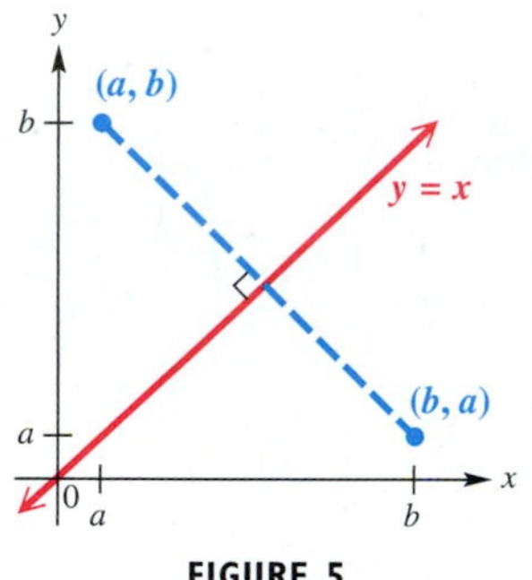

FIGURE 5

Suppose f and f^{-1} are inverse functions and $f(a) = b$ for real numbers a and b. Then, by the definition of inverse function, $f^{-1}(b) = a$. This means that if a point (a, b) is on the graph of f, then (b, a) will belong to the graph of f^{-1}. As shown in Figure 5, the points (a, b) and (b, a) are reflections of one another across the line $y = x$.

Geometric Relationship between the Graphs of f and f^{-1}

If a function f is one-to-one, then the graph of its inverse f^{-1} is a reflection of the graph of f across the line $y = x$.

Figure 6 illustrates this idea for each function f (in blue) and its inverse f^{-1} (in red).

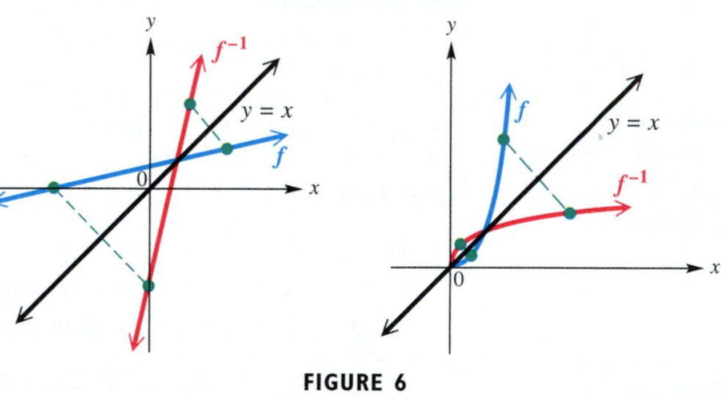

FIGURE 6

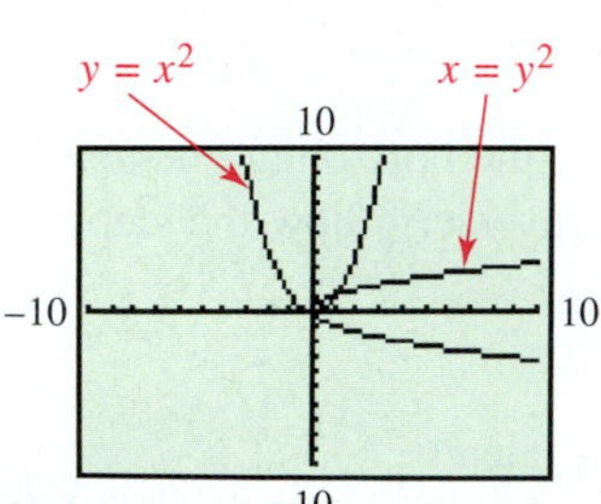

Despite the fact that $y = x^2$ is not one-to-one, the calculator will draw its "inverse," $x = y^2$.

FIGURE 7

CAUTION Some graphing calculators have the capability of "drawing the inverse" of a function. This feature does not require that the function be one-to-one, however, so the resulting figure may not be the graph of a function. See Figure 7. ***Again, it is necessary to understand the mathematics to interpret results correctly.***

GCM **EXAMPLE 6 Finding the Inverse of a Function with a Restricted Domain**

Let $f(x) = \sqrt{x + 5}$. Find $f^{-1}(x)$.

Solution The domain of f is restricted to the interval $[-5, \infty)$. Since f is an increasing function, as shown in Figure 8, f is one-to-one and thus has an inverse function.

$$y = \sqrt{x + 5}, \quad x \geq -5 \qquad \text{Let } y = f(x).$$

$$x = \sqrt{y + 5}, \quad y \geq -5 \qquad \text{Interchange } x \text{ and } y.$$

$$x^2 = y + 5 \qquad \text{Square both sides.}$$

$$y = x^2 - 5 \qquad \text{Solve for } y.$$

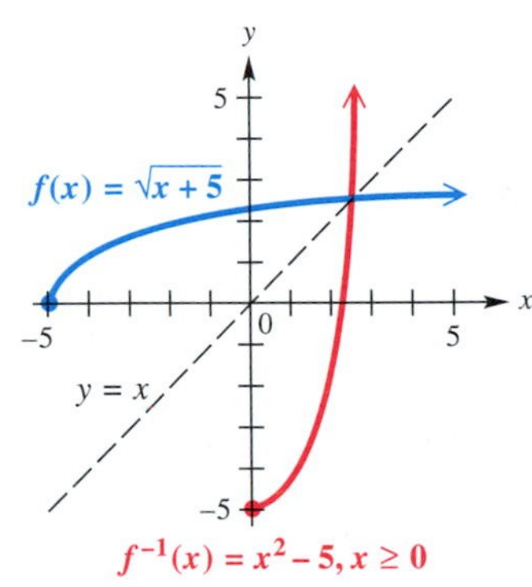

FIGURE 8

We cannot simply give $f^{-1}(x)$ as $x^2 - 5$. The domain of f is $[-5, \infty)$ and its range is $[0, \infty)$. The range of f is the domain of f^{-1}, so f^{-1} must be defined as

$$f^{-1}(x) = x^2 - 5, \quad x \geq 0.$$

As a check, the range of f^{-1}, $[-5, \infty)$, is the domain of f. Graphs of f and f^{-1} are shown in Figure 8. The line $y = x$ is included on the graphs to show that the graphs of f and f^{-1} are reflections of one another—that is, mirror images with respect to this line. ■

TECHNOLOGY NOTE

When graphing inverse functions on a calculator, square windows work best.

Important Facts about Inverses

1. If f is one-to-one, then f^{-1} exists.
2. The domain of f is equal to the range of f^{-1}, and the range of f is equal to the domain of f^{-1}.
3. If the point (a, b) lies on the graph of f, then (b, a) lies on the graph of f^{-1}. The graphs of f and f^{-1} are reflections of each other across the line $y = x$.

An Application of Inverse Functions

FOR DISCUSSION

Why is a one-to-one function essential in the encoding–decoding process described here?

Inverse functions are used to send and receive coded information. A simple example might use the function defined by $f(x) = 3x + 1$, which is one-to-one. Suppose that each letter of the alphabet is assigned a numerical value according to its position as follows:

A	1	H	8	O	15	V	22
B	2	I	9	P	16	W	23
C	3	J	10	Q	17	X	24
D	4	K	11	R	18	Y	25
E	5	L	12	S	19	Z	26.
F	6	M	13	T	20		
G	7	N	14	U	21		

A	1	N	14
B	2	O	15
C	3	P	16
D	4	Q	17
E	5	R	18
F	6	S	19
G	7	T	20
H	8	U	21
I	9	V	22
J	10	W	23
K	11	X	24
L	12	Y	25
M	13	Z	26

EXAMPLE 7 Using Functions to Encode and Decode a Message

Use the one-to-one function defined by $f(x) = 3x + 1$ and the numerical values repeated in the margin to encode and decode the message BE VERY CAREFUL.

Solution The message BE VERY CAREFUL would be encoded as

7 16 67 16 55 76 10 4 55 16 19 64 37

because B corresponds to 2 and

$$f(2) = 3(2) + 1 = 7,$$

E corresponds to 5 and

$$f(5) = 3(5) + 1 = 16,$$

and so on. Using the inverse $f^{-1}(x) = \frac{x-1}{3}$ to decode yields

$$f^{-1}(7) = \frac{7-1}{3} = 2,$$

which corresponds to B,

$$f^{-1}(16) = \frac{16-1}{3} = 5,$$

which corresponds to E, and so on. The table feature of a graphing calculator can be useful in this procedure. Figure 9 shows how decoding can be accomplished for the words BE VERY. ■

X	Y_1
7	2
16	5
67	22
16	5
55	18
76	25

$Y_1 = (X-1)/3$

FIGURE 9

5.1 Exercises

Decide whether each function graphed or defined is one-to-one.

1. $f(x) = -3x + 5$
2. $f(x) = -5x + 2$
3. $f(x) = x^2$
4. $f(x) = -x^2$
5. $f(x) = \sqrt{36 - x^2}$
6. $f(x) = -\sqrt{100 - x^2}$

7.

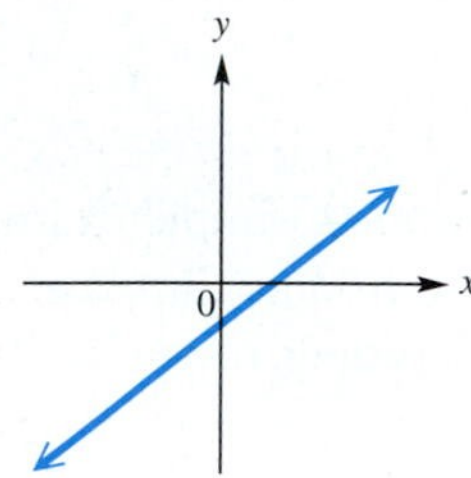

8.

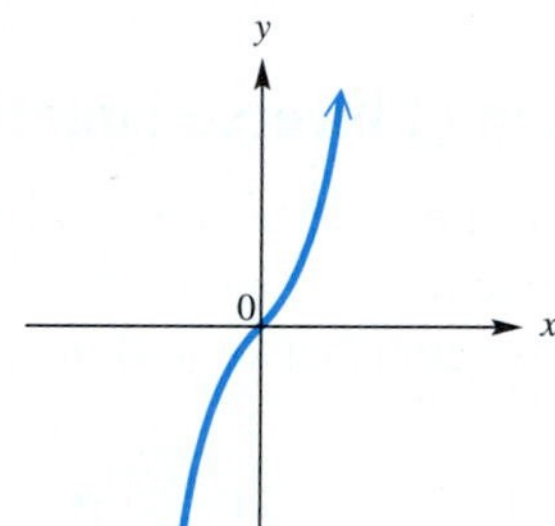

9.

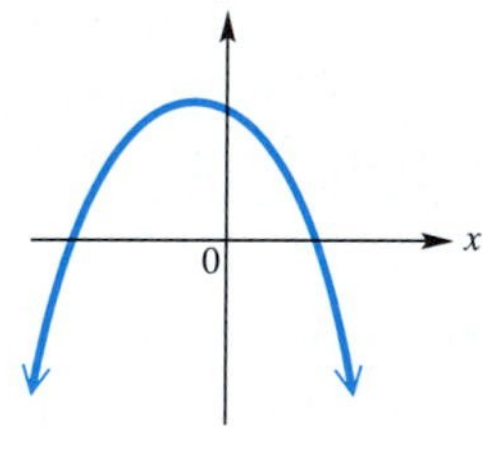

10.

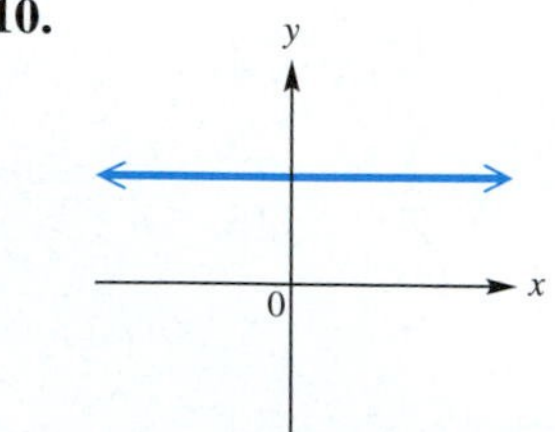

11.

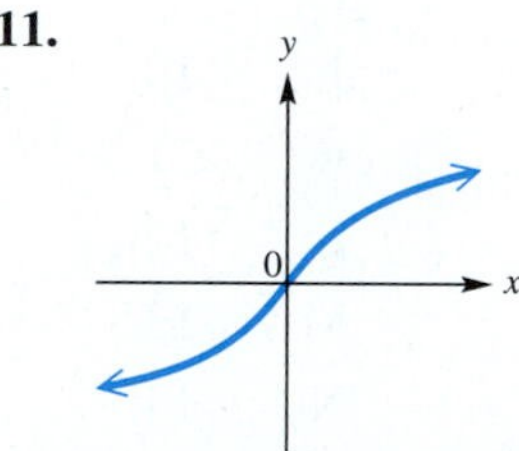

12.

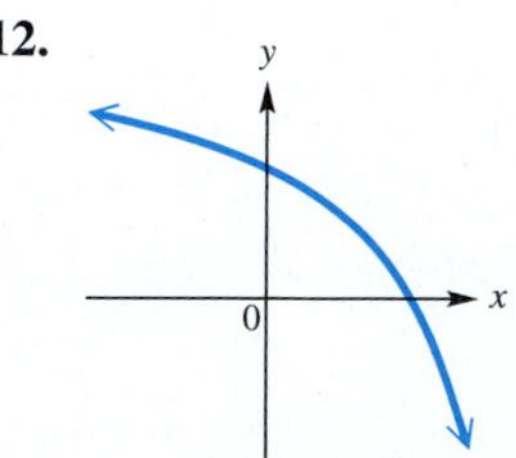

13.

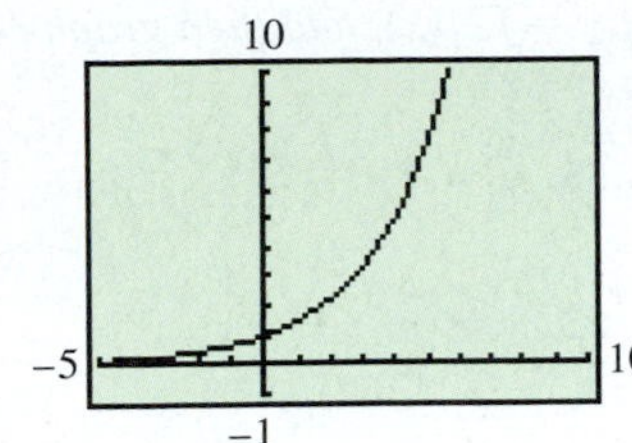

14.

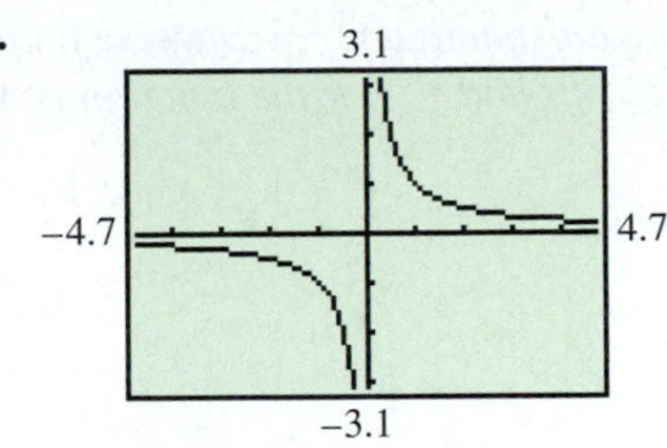

15.

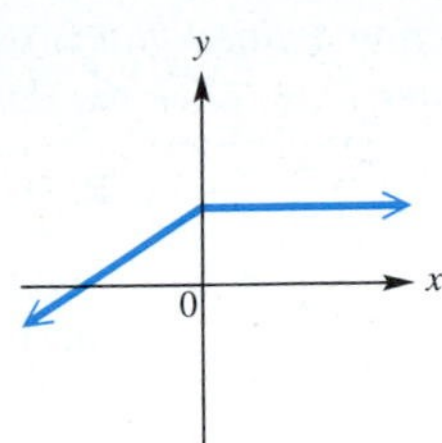

16. $f(x) = \begin{cases} 3 & \text{if } x \geq 0 \\ -x & \text{if } x < 0 \end{cases}$

17. $y = (x - 2)^2$

18. $y = -(x + 3)^2 - 8$

19. $y = 2x^3 + 1$

20. $y = -2x^5 - 4$

21. $y = -\sqrt[3]{x + 5}$

22. $y = \dfrac{1}{x + 2}$

23. $y = \dfrac{-4}{x - 8}$

24. $f(x) = -7$

25. Explain why a polynomial function of even degree cannot be one-to-one.

26. Explain why in some cases a polynomial function of odd degree is not one-to-one.

Concept Check *Answer the following.*

27. For a function to have an inverse, it must be __________.

28. For a function f to be of the type mentioned in Exercise 27, if $a \neq b$, then __________.

29. If f and g are inverses, then $(f \circ g)(x) =$ __________ and __________ $= x$.

30. The domain of f is equal to the __________ of f^{-1}, and the range of f is equal to the __________ of f^{-1}.

31. If the point (a, b) lies on the graph of f and f has an inverse, then the point __________ lies on the graph of f^{-1}.

32. If $f(x) = x$, then for any function g, $(f \circ g)(x) = (g \circ f)(x) =$ __________.

33. If a function f has an inverse, then the graph of f^{-1} may be obtained by reflecting the graph of f across the line with equation __________.

34. If a function f has an inverse and $f(-3) = 6$, then $f^{-1}(6) =$ __________.

35. If $f(-4) = 16$ and $f(4) = 16$, then f __________ (does/does not) have an inverse because __________.

36. If f is a function that has an inverse and the graph of f lies completely within the second quadrant, then the graph of f^{-1} lies completely within the __________ quadrant.

Use the definition of inverse functions to show analytically that f and g are inverses.

37. $f(x) = 3x - 7, \quad g(x) = \dfrac{x + 7}{3}$

38. $f(x) = 4x + 3, \quad g(x) = \dfrac{x - 3}{4}$

39. $f(x) = x^3 + 4, \quad g(x) = \sqrt[3]{x - 4}$

40. $f(x) = x^3 - 7, \quad g(x) = \sqrt[3]{x + 7}$

41. $f(x) = -x^5, \quad g(x) = -\sqrt[5]{x}$

42. $f(x) = -x^7, \quad g(x) = -\sqrt[7]{x}$

Determine whether each function is one-to-one. If so, find its inverse.

43. $f = \{(10, 4), (20, 5), (30, 6), (40, 7)\}$

44. $g = \{(5, 12), (10, 22), (15, 32), (20, 42)\}$

45. $f = \{(1, 5), (2, 6), (3, 5), (4, 8)\}$

46. $g = \{(0, 10), (1, 20), (2, 10), (3, 40)\}$

47. $f = \{(0, 0^2), (1, 1^2), (2, 2^2), (3, 3^2), (4, 4^2)\}$

48. $g = \{(0, 0^4), (-1, (-1)^4), (-2, (-2)^4), (-3, (-3)^4), (-4, (-4)^4)\}$

Concept Check *In Exercises 49–54, an everyday activity is described. Keeping in mind that an inverse operation "undoes" what an operation does, describe the inverse activity.*

49. Tying your shoelaces

50. Pressing a car's accelerator

51. Entering a room

52. Climbing the stairs

53. Wrapping a package

54. Putting on a coat

For each function defined that is one-to-one, write an equation for the inverse function in the form $y = f^{-1}(x)$, and then graph f and f^{-1} on the same axes. Give the domain and range of f and f^{-1}. If the function is not one-to-one, say so.

55. $y = 3x - 4$ **56.** $y = 4x - 5$ **57.** $y = x^3 + 1$ **58.** $y = -x^3 - 2$

59. $y = x^2$ **60.** $y = -x^2 + 2$ **61.** $y = \frac{1}{x}$ **62.** $y = \frac{4}{x}$

63. $y = \frac{2}{x + 3}$ **64.** $y = \frac{3}{x - 4}$ **65.** $f(x) = \sqrt{6 + x}, x \geq -6$ **66.** $f(x) = -\sqrt{x^2 - 16}, x \geq 4$

Concept Check *Let $f(x) = x^3$. Evaluate each expression.*

67. $f(2)$ **68.** $f(0)$ **69.** $f(-2)$ **70.** $f^{-1}(8)$ **71.** $f^{-1}(0)$ **72.** $f^{-1}(-8)$

Concept Check *The graph of a function f is shown in the figure. Use the graph to find each value.*

73. $f^{-1}(4)$ **74.** $f^{-1}(2)$

75. $f^{-1}(0)$ **76.** $f^{-1}(-2)$

77. $f^{-1}(-3)$ **78.** $f^{-1}(-4)$

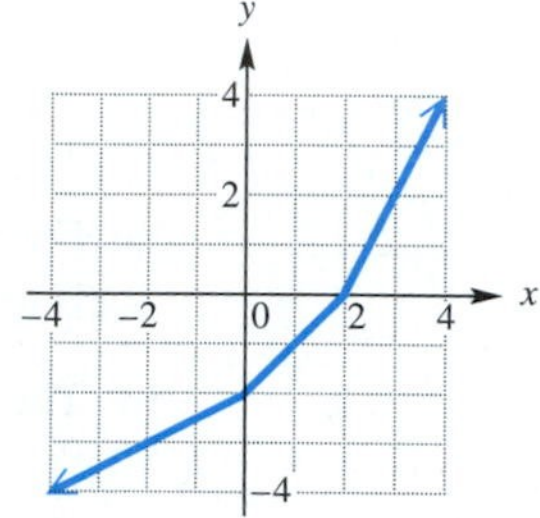

Decide whether the pair of functions shown in the graph appear to be inverses.

79.

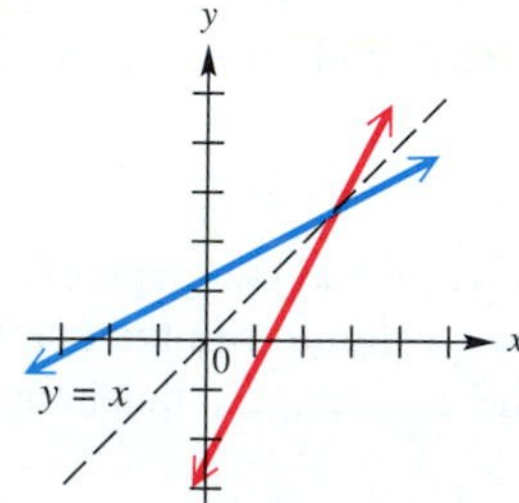

80.

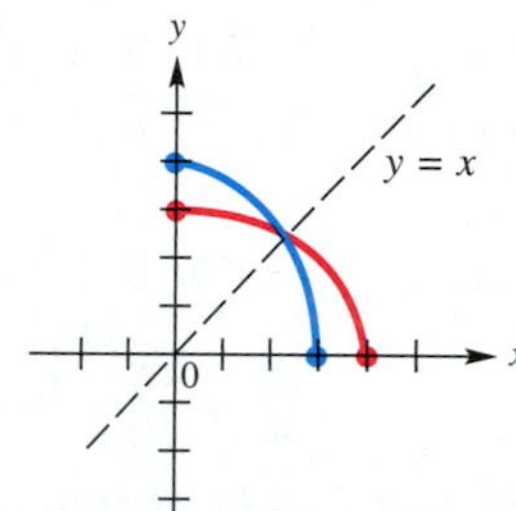

81.

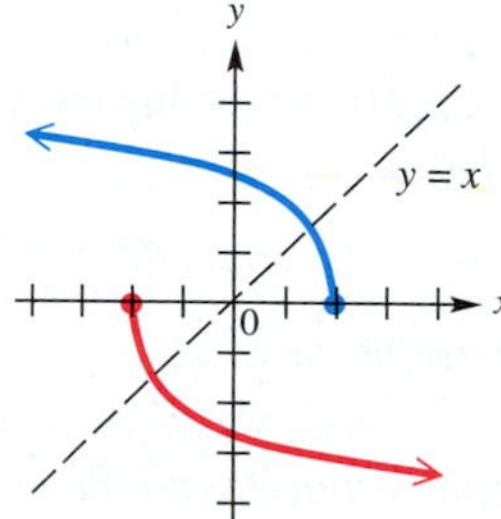

82.

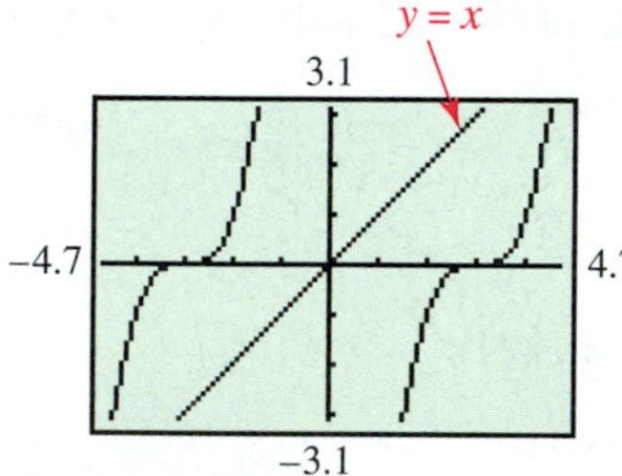

83.

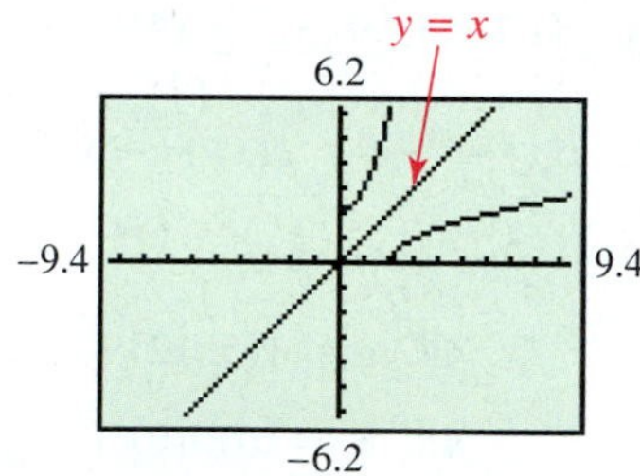

84.

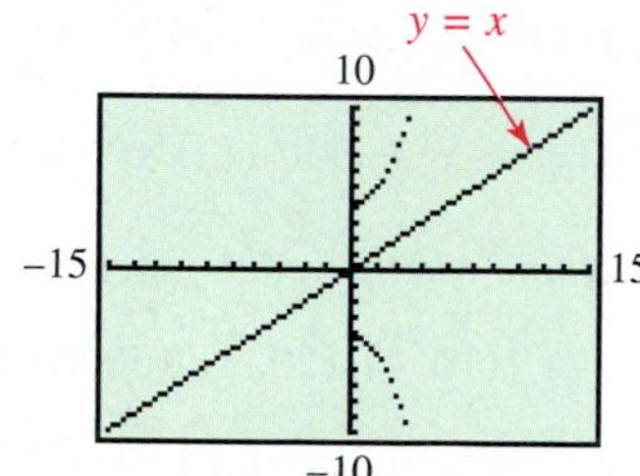

Dot mode

85. $f(x) = -\frac{3}{11}x, \quad g(x) = -\frac{11}{3}x$ **86.** $f(x) = 2x + 4, \quad g(x) = \frac{1}{2}x - 2$ **87.** $f(x) = 5x - 5, \quad g(x) = \frac{1}{5}x + 1$

88. $f(x) = 8x - 7, \quad g(x) = \frac{x + 8}{7}$ **89.** $f(x) = \frac{1}{x}, \quad g(x) = \frac{1}{x}$ **90.** $f(x) = \frac{2x + 3}{x - 1}, \quad g(x) = \frac{x + 3}{x - 2}$

Concept Check *Indicate whether the screens suggest that Y_1 and Y_2 are inverse functions.*

91.

X	Y_1
1	6
3	8
-4	1
9	14
3.2	8.2
0	5
-10	-5

X=1

X	Y_2
6	1
8	3
1	-4
14	9
8.2	3.2
5	0
-5	-10

X=6

92.

X	Y_1
1	7
3	9
-4	2
9	15
3.2	9.2
0	6
-10	-4

X=1

X	Y_2
7	-1
9	-3
2	4
15	-9
9.2	-3.2
6	0
-4	10

X=7

Graph the inverse of each one-to-one function.

93.

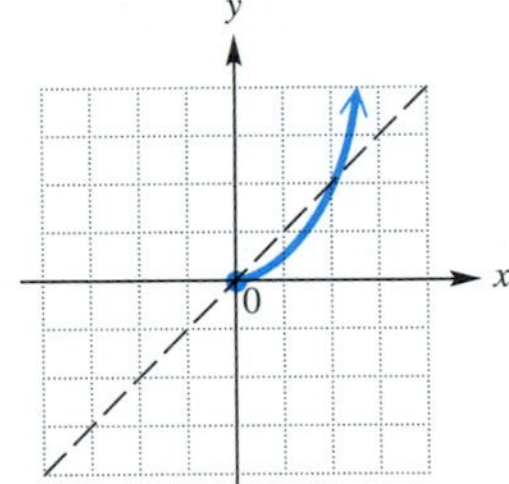

94.

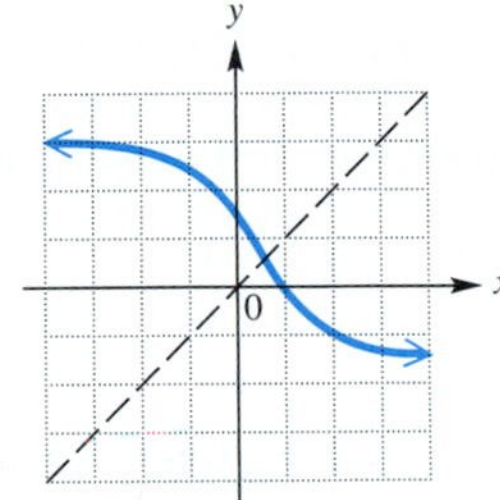

95.

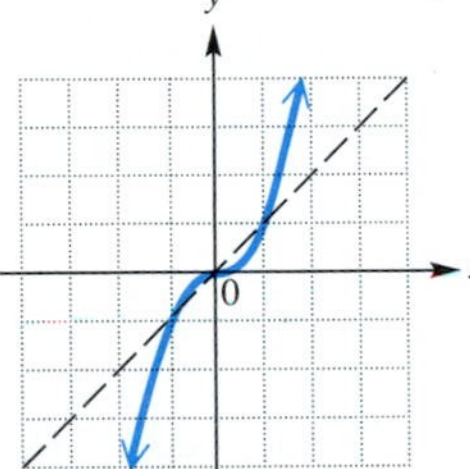

96.

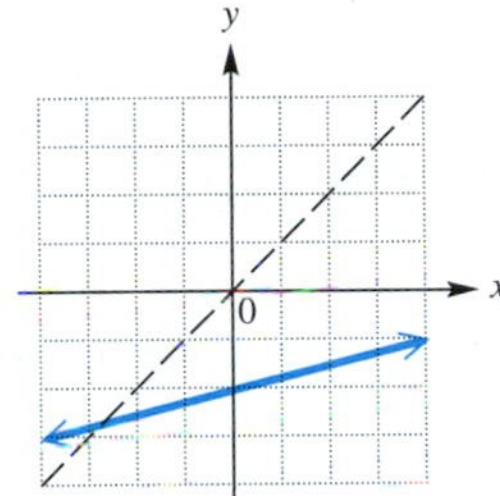

97.

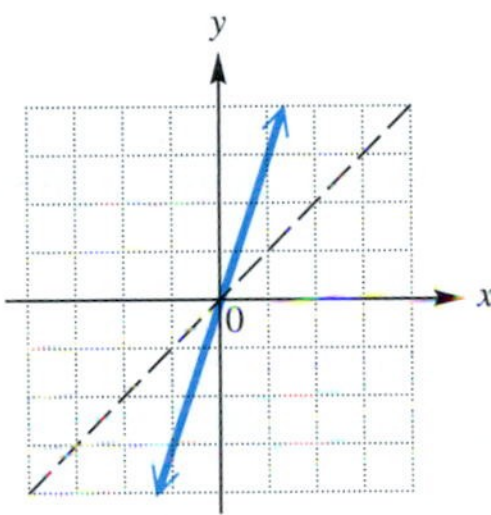

98.

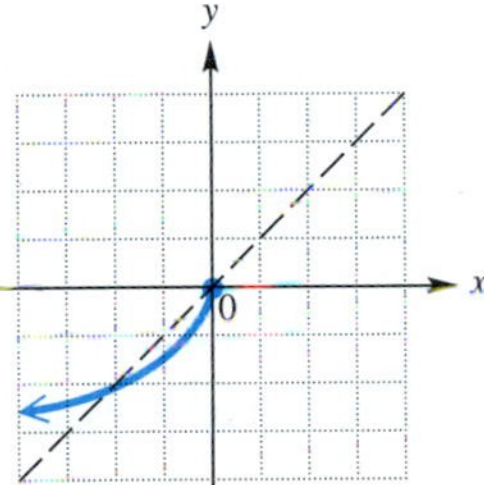

Concept Check *Answer the following.*

99. Suppose $f(x)$ is the number of cars that can be built for x dollars. What does $f^{-1}(1000)$ represent?

100. Suppose $f(r)$ is the volume (in cubic inches) of a sphere of radius r inches. What does $f^{-1}(5)$ represent?

101. If a line has nonzero slope a, what is the slope of its reflection across the line $y = x$?

102. Find $f^{-1}[f(2)]$, where $f(2) = 3$.

On a graphing calculator, graph each function defined, using the given viewing window. Use the graph to decide which functions are one-to-one. If a function is one-to-one, give the equation of its inverse function and graph the inverse function on the same coordinate axes.

103. $f(x) = 6x^3 + 11x^2 - x - 6$; $[-3, 2]$ by $[-10, 10]$

104. $f(x) = x^4 - 5x^2 + 6$; $[-3, 3]$ by $[-1, 8]$

105. $f(x) = \dfrac{x - 5}{x + 3}$; $[-9.4, 9.4]$ by $[-6.2, 6.2]$

106. $f(x) = \dfrac{-x}{x - 4}$; $[-4.7, 9.4]$ by $[-6.2, 6.2]$

While a function may not be one-to-one when defined over its "natural" domain, it may be possible to restrict the domain in such a way that it is one-to-one and the range of the function is unchanged. For example, if we restrict the domain of the function $f(x) = x^2$ (which is not one-to-one over $(-\infty, \infty)$) to $[0, \infty)$, we obtain a one-to-one function whose range is still $[0, \infty)$. See the figure at the right. Notice that we could choose to restrict the domain of $f(x) = x^2$ to $(-\infty, 0]$ and also obtain the graph of a one-to-one function, except that it would be the left half of the parabola.

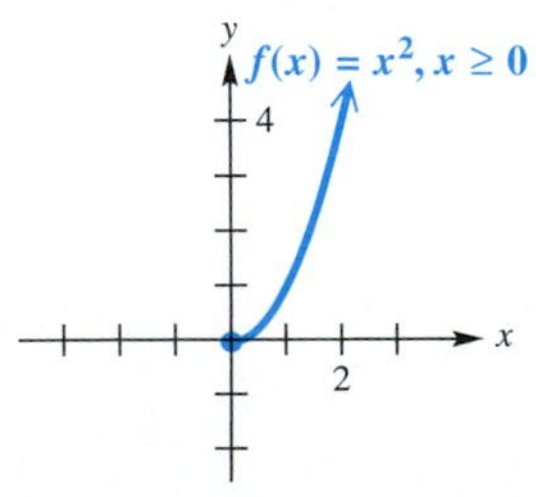

(continued)

For each function defined in Exercises 107–112, decide on a suitable restriction on the domain so that the function is one-to-one and the range is not changed. You may wish to use a graph to help you decide.

107. $f(x) = -x^2 + 4$

108. $f(x) = (x - 1)^2$

109. $f(x) = |x - 6|$

110. $f(x) = x^4$

111. $f(x) = x^4 + x^2 - 6$

112. $f(x) = -\sqrt{x^2 - 16}$

Using the restrictions on the functions defined in Exercises 113–116, find a formula for f^{-1}.

113. $f(x) = -x^2 + 4, x \geq 0$

114. $f(x) = (x - 1)^2, x \geq 1$

115. $f(x) = |x - 6|, x \geq 6$

116. $f(x) = x^4, x \geq 0$

Use the alphabet coding assignment given in Example 7 for Exercises 117–120.

117. The function defined by $f(x) = 4x - 1$ was used to encode the following message:

79 71 19 3 75 83 71 19 31 83 55 79 35 75 59 55.

Find the inverse function and decode the message.

118. The function defined by $f(x) = 3x - 3$ was used to encode the following message:

54 12 57 54 0 24 33 57 42 9 0 72.

Find the inverse function and decode the message.

119. Encode the message NO PROBLEM, using the one-to-one function defined by $f(x) = x^3 + 1$. Give the inverse function the decoder would need when the message is received.

120. Encode the message BEGIN OPERATIONS, using the one-to-one function defined by $f(x) = (x + 2)^3$. Give the inverse function the decoder would need when the message is received.

5.2 Exponential Functions

Real-Number Exponents ■ Graphs of Exponential Functions ■ Exponential Equations (Type 1) ■ Compound Interest ■ The Number *e* and Continuous Compounding ■ An Application of Exponential Functions

Real-Number Exponents

Recall that if $r = \frac{m}{n}$ is a rational number, then, for appropriate integer values of m and n,

$$a^{m/n} = \left(\sqrt[n]{a}\right)^m.$$

For example,

$$16^{3/4} = \left(\sqrt[4]{16}\right)^3 = 2^3 = 8,$$

$$27^{-1/3} = \frac{1}{27^{1/3}} = \frac{1}{\sqrt[3]{27}} = \frac{1}{3}, \quad \text{and} \quad 64^{-1/2} = \frac{1}{64^{1/2}} = \frac{1}{\sqrt{64}} = \frac{1}{8}.$$

In this section, we extend the definition of a^r to include all real (not just rational) values of the exponent r. For example, the symbol $2^{\sqrt{3}}$ might be evaluated by approximating the exponent $\sqrt{3}$ by the numbers 1.7, 1.73, 1.732, and so on. Since these decimals approach $\sqrt{3}$ more and more closely, it seems reasonable that $2^{\sqrt{3}}$ should be approximated more and more closely by the numbers $2^{1.7}$, $2^{1.73}$, $2^{1.732}$, and so on. (Recall, for example, that $2^{1.7} = 2^{17/10} = \left(\sqrt[10]{2}\right)^{17}$.) In fact, this is exactly how $2^{\sqrt{3}}$ is defined (in a more advanced course).

With this interpretation of real exponents, all rules and theorems for exponents are valid for real number exponents as well as rational ones. In addition to the usual rules for exponents (see Chapter R), we use several new properties in this chapter. For example, if $y = 2^x$, then each value of x leads to exactly one value of y; therefore, $y = 2^x$ defines a function. Furthermore,

$$\text{if} \quad 3^x = 3^4, \quad \text{then} \quad x = 4,$$

and for $p > 0$,

$$\text{if} \quad p^2 = 3^2, \quad \text{then} \quad p = 3.$$

Also,

$$4^2 < 4^3, \quad \text{but} \quad \left(\frac{1}{4}\right)^2 > \left(\frac{1}{4}\right)^3.$$

In general, when $a > 1$, increasing the exponent on a leads to a *larger* number, but when $0 < a < 1$, increasing the exponent on a leads to a *smaller* number.

These properties are generalized below. Proofs of the properties are not given here, as they require more advanced mathematics.

Additional Properties of Exponents

For any real number $a > 0$, $a \neq 1$, the following statements are true:

(a) a^x **is a unique real number for all real numbers** x**.**

(b) $a^b = a^c$ **if and only if** $b = c$**.**

(c) If $a > 1$ **and** $m < n$**, then** $a^m < a^n$**.**

(d) If $0 < a < 1$ **and** $m < n$**, then** $a^m > a^n$**.**

Properties (a) and (b) require $a > 0$ so that a^x is always defined. For example, $(-6)^x$ is not a real number if $x = \frac{1}{2}$. This means that a^x will always be positive, since a is positive. In Property (a), $a \neq 1$ because $1^x = 1$ for every real-number value of x, and thus each value of x leads to the same real number, 1. For Property (b) to hold, a must not equal 1. For example, $1^4 = 1^5$, even though $4 \neq 5$.

Graphs of Exponential Functions

A graphing calculator can easily find approximations of numbers raised to irrational powers. For example,

$$2^{\sqrt{6}} \approx 5.462228786, \quad \left(\frac{1}{2}\right)^{\sqrt{3}} \approx .3010237439, \quad \text{and} \quad .5^{-\sqrt{2}} \approx 2.665144143.$$

Later we find these approximations by using graphs of *exponential functions.*

Exponential Function

If $a > 0$, $a \neq 1$, then

$$f(x) = a^x$$

is the **exponential function** with base a.

NOTE ***We do not allow 1 as a base for the exponential function*** because $1^x = 1$ for all real x, leading to the constant function defined by $f(x) = 1$.

EXAMPLE 1 Graphing an Exponential Function ($a > 1$)

Graph $f(x) = 2^x$. Determine the domain and range of f.

Solution Make a table of values, plot the points, and connect them with a smooth curve, as shown in Figure 10. As the graph suggests, the domain is $(-\infty, \infty)$ and the range is $(0, \infty)$.

It is difficult to graph values of a^x for $x < -2$. The calculator table in Figure 11 shows that a^x continues to decrease (but remains positive) as x decreases.

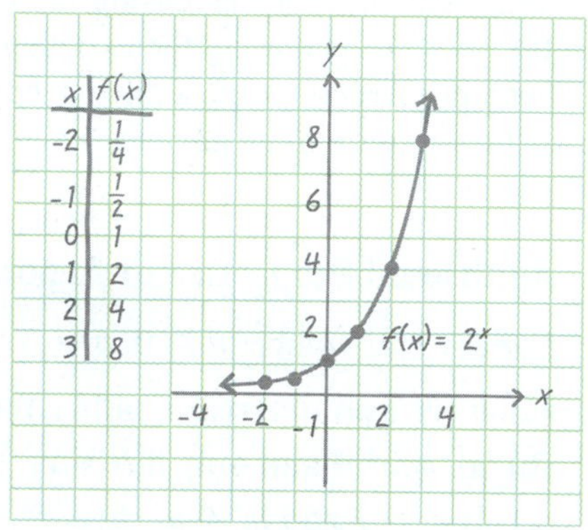

FIGURE 10

X	Y1
-2	.25
-2.5	.17678
-3	.125
-3.5	.08839
-4	.0625
-4.5	.04419
-5	.03125

Y1=2^X

FIGURE 11

TECHNOLOGY NOTE

Because of the limited resolution of the graphing calculator screen, it is difficult to interpret how the graph of the exponential function defined by $f(x) = a^x$ behaves when the curve is close to the x-axis, as seen in Figure 14. Remember that there is no endpoint and that the curve approaches, but never touches, the x-axis. Tracing along the curve and observing the y-coordinates as $x \to -\infty$ when $a > 1$ and as $x \to \infty$ when $0 < a < 1$ or looking at a table helps support this fact.

EXAMPLE 2 Graphing an Exponential Function ($0 < a < 1$)

Graph $g(x) = \left(\frac{1}{2}\right)^x$. Determine the domain and range of g.

Solution Use the same procedure as in Example 1. See Figures 12 and 13. Again, the domain is $(-\infty, \infty)$ and the range is $(0, \infty)$, but here the function *decreases* on its domain. (In Example 1, the function *increases* on its domain.)

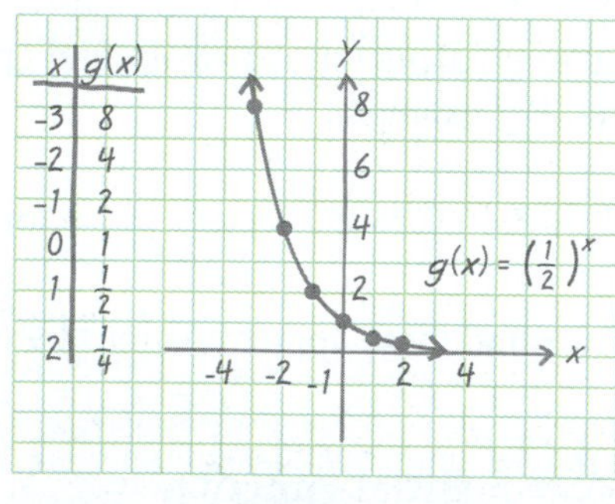

FIGURE 12

X	Y1
2	.25
2.5	.17678
3	.125
3.5	.08839
4	.0625
4.5	.04419
5	.03125

Y1=(1/2)^X

FIGURE 13

The behavior of the graph of an exponential function depends, in general, on the magnitude of a. Figure 14(a) shows the graphs of $y = a^x$ for $a = 2$, 3, and 4. The function is *increasing* for these values of a. As a becomes larger, the graph becomes "steeper" to the right of the y-axis.

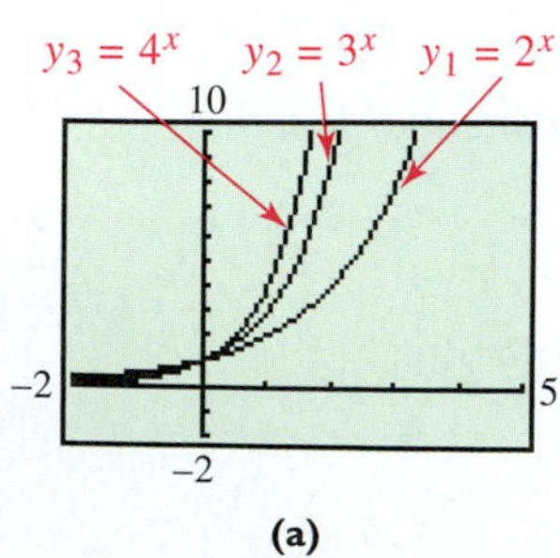

(a)

FIGURE 14

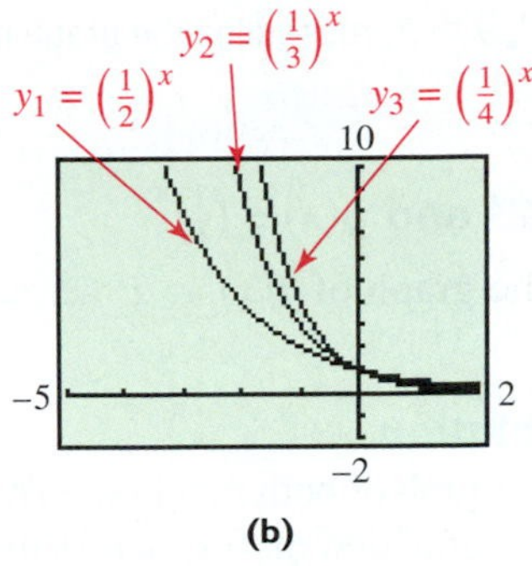

(b)

FIGURE 14

Figure 14(b) shows the graphs of $y = a^x$ for $a = \frac{1}{2}, \frac{1}{3}$, and $\frac{1}{4}$. If the base a is between 0 and 1, as a gets closer to 0, the graph becomes "steeper" to the left of the y-axis. The function is *decreasing* for these values of a.

Figure 14 leads to generalizations about the graphs of exponential functions.

FUNCTION CAPSULE

EXPONENTIAL FUNCTION $\quad f(x) = a^x, \quad a > 1$

Domain: $(-\infty, \infty)$ Range: $(0, \infty)$

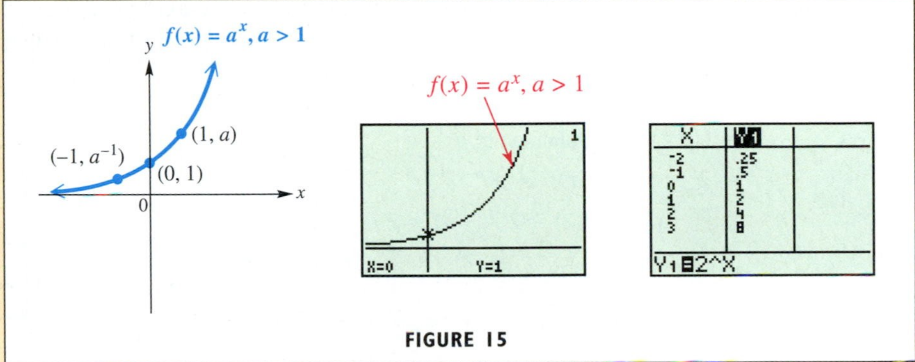

FIGURE 15

- $f(x) = a^x, a > 1$, is increasing and continuous on its entire domain, $(-\infty, \infty)$.
- The x-axis is the horizontal asymptote as $x \rightarrow -\infty$.
- The graph passes through the points $(-1, a^{-1})$, $(0, 1)$, and $(1, a)$.

FUNCTION CAPSULE

EXPONENTIAL FUNCTION $\quad f(x) = a^x, \quad 0 < a < 1$

Domain: $(-\infty, \infty)$ Range: $(0, \infty)$

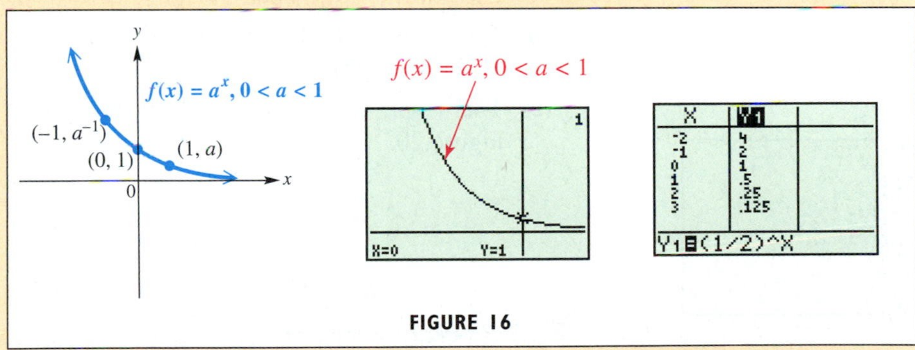

FIGURE 16

- $f(x) = a^x, 0 < a < 1$, is decreasing and continuous on its entire domain, $(-\infty, \infty)$.
- The x-axis is the horizontal asymptote as $x \rightarrow \infty$.
- The graph passes through the points $(-1, a^{-1})$, $(0, 1)$, and $(1, a)$.

FOR DISCUSSION

1. Using a standard viewing window, we cannot observe how the graph of the exponential function defined by $f(x) = 2^x$ behaves for values of x less than about -3. Use the window $[-10, 0]$ by $[-.5, .5]$, and then trace to the left to see what happens.

2. Repeat Item 1 for the exponential function defined by $f(x) = \left(\frac{1}{2}\right)^x$. Use the window $[0, 10]$ by $[-.5, .5]$, and trace to the right.

3. Complete the following statement: For the graph of an exponential function defined by $f(x) = a^x, a > 0, a \neq 1$, every range value appears exactly once; thus, f is a(n)________ function. Because of this, f has a(n) ________. (*Hint:* Recall the concepts that were studied in Section 5.1.)

NOTE Since $a^{-x} = (a^{-1})^x = \left(\frac{1}{a}\right)^x$, the graph of $f(x) = a^{-x}$, $a > 1$, resembles the graph in Figure 16 on the preceding page.

GCM EXAMPLE 3 Comparing the Graphs of $f(x) = 2^x$ and $g(x) = \left(\frac{1}{2}\right)^x$

Explain why the graph of $g(x) = \left(\frac{1}{2}\right)^x$ is a reflection of the graph of $f(x) = 2^x$ across the y-axis.

Analytic Solution

We must show that $g(x) = f(-x)$ to prove analytically that the graph of g is a reflection of the graph of f across the y-axis.

$$
\begin{aligned}
g(x) &= \left(\frac{1}{2}\right)^x \\
&= (2^{-1})^x && \text{Definition of } a^{-1} \\
&= 2^{-x} && (a^m)^n = a^{mn} \\
&= f(-x) && f(x) = 2^x;\ f(-x) = 2^{-x}
\end{aligned}
$$

Graphing Calculator Solution

Figure 17 shows a calculator graph of both functions. The graph indicates that $g(x)$ is a reflection of $f(x)$ across the y-axis. Note that if (a, b) lies on the graph of $f(x)$, then $(-a, b)$ lies on that of $g(x)$.

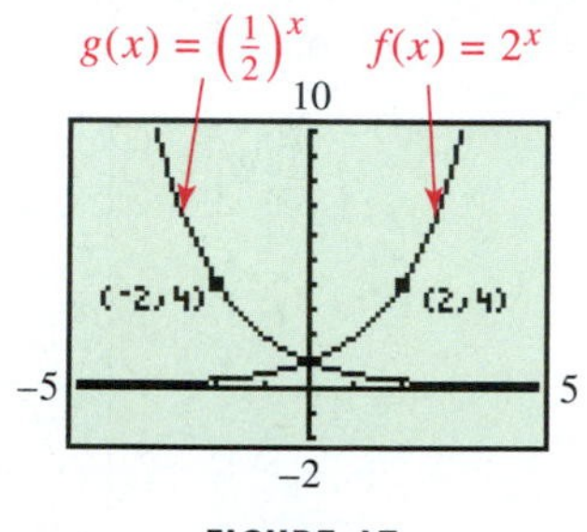

FIGURE 17

EXAMPLE 4 Graphing Reflections and Translations

Graph each function.

(a) $f(x) = -2^x$ **(b)** $f(x) = 2^{x+3}$ **(c)** $f(x) = 2^x + 3$

Solution

(a) The graph of $f(x) = -2^x$ is that of $f(x) = 2^x$ reflected across the x-axis. The domain is $(-\infty, \infty)$ and the range is $(-\infty, 0)$. See Figure 18.

(b) The graph of $f(x) = 2^{x+3}$ is the graph of $f(x) = 2^x$ translated 3 units to the left, as shown in Figure 19.

(c) The graph of $f(x) = 2^x + 3$ is that of $f(x) = 2^x$ translated 3 units up. See Figure 20.

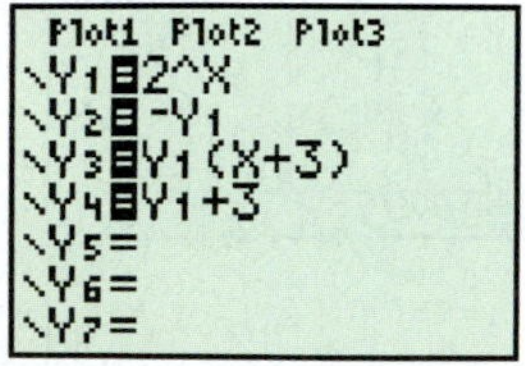

The screen shows how a graphing calculator can be directed to graph the three functions in Example 4. Y_1 is defined as 2^X, and Y_2, Y_3, and Y_4 are defined as reflections and/or translations of Y_1.

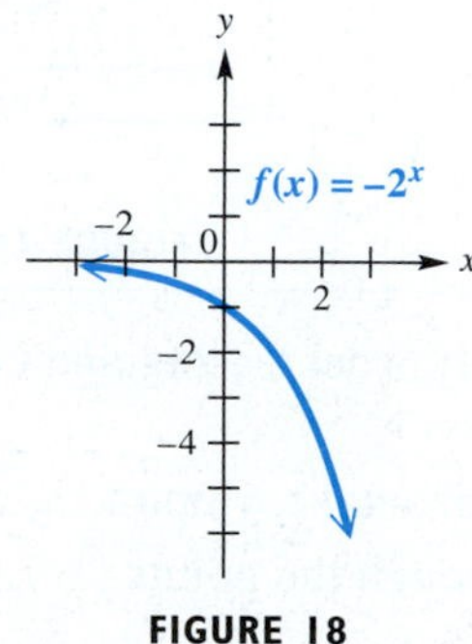

FIGURE 18

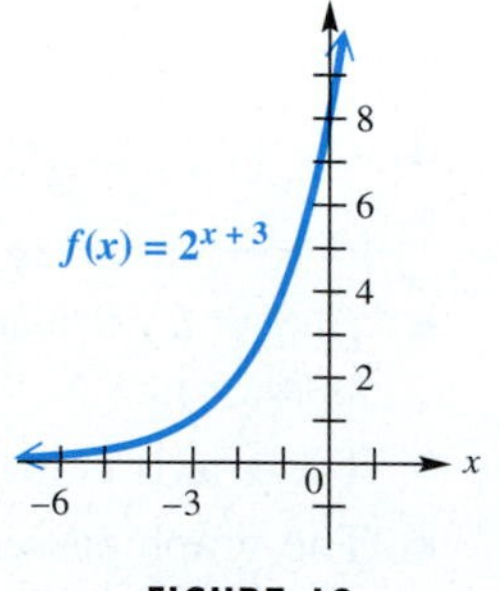

FIGURE 19

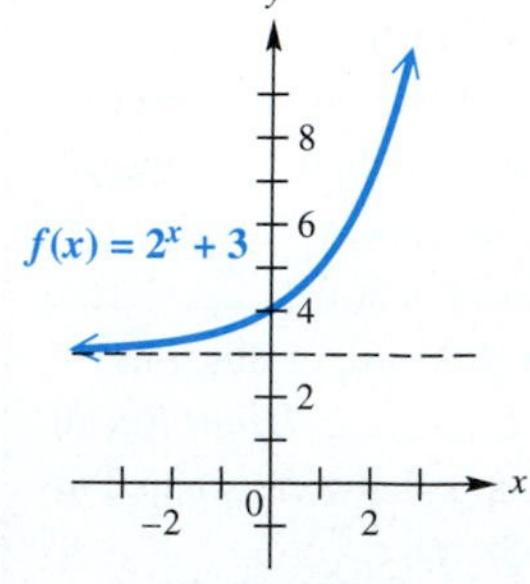

FIGURE 20

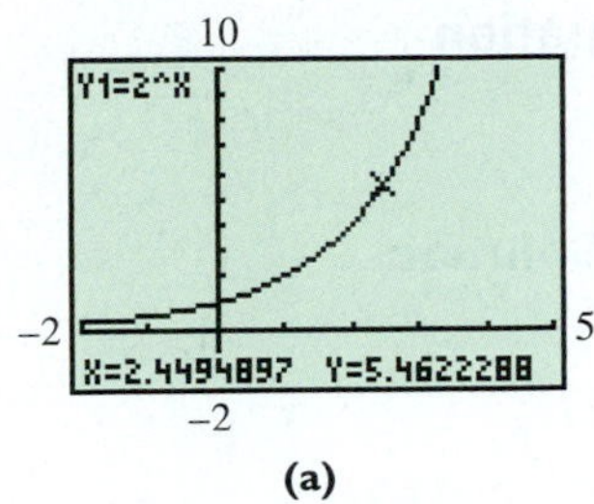

(a)

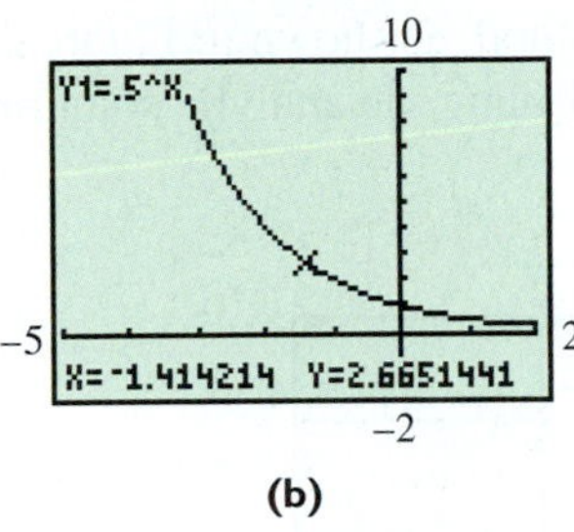

(b)

FIGURE 21

GCM **EXAMPLE 5 Using Graphs to Evaluate Exponential Expressions**

Use a graph to evaluate each expression.

(a) $2^{\sqrt{6}}$ **(b)** $.5^{-\sqrt{2}}$

Solution

(a) Figure 21(a) shows the graph of $Y_1 = 2^X$, with Y_1 evaluated for $X = \sqrt{6} \approx 2.4494897$. From the bottom of the screen, we see that $Y = 2^{\sqrt{6}} \approx 5.4622288$, which confirms our earlier result when we raised numbers to irrational powers.

(b) Using the graph of $Y_1 = .5^X$ with $X = -\sqrt{2} \approx -1.414214$, we find that $Y \approx 2.6651441$, again supporting our earlier result. See Figure 21(b). ■

Exponential Equations (Type 1)

The equation $25^x = 125$ is different from any equation studied so far in this book, because the variable appears in the exponent. The base on the left side (25) and the constant on the right side (125) can both be written as powers of a common base, 5. We refer to such an equation as a **Type 1 exponential equation.**

EXAMPLE 6 Solving Type 1 Exponential Equations

Solve each equation.

(a) $25^x = 125$ **(b)** $\left(\frac{1}{3}\right)^x = 81$

Solution

(a)

$$\begin{aligned} 25^x &= 125 \\ (5^2)^x &= 5^3 && \text{Write with the same base.} \\ 5^{2x} &= 5^3 && (a^m)^n = a^{mn} \\ 2x &= 3 && \text{Set exponents equal (Property (b)).} \\ x &= \frac{3}{2} && \text{Divide by 2.} \end{aligned}$$

Check:

$$\begin{aligned} 25^{3/2} &= 125 \ ? && \text{Let } x = \tfrac{3}{2} \text{ in the original equation.} \\ (\sqrt{25})^3 &= 125 \ ? \\ 5^3 &= 125 \ ? \\ 125 &= 125 && \text{True} \end{aligned}$$

The solution set is $\left\{\frac{3}{2}\right\}$.

(b)

$$\begin{aligned} \left(\frac{1}{3}\right)^x &= 81 \\ (3^{-1})^x &= 3^4 && \text{Write with the same base; } \tfrac{1}{3} = 3^{-1} \text{ and } 81 = 3^4. \\ 3^{-x} &= 3^4 && (a^m)^n = a^{mn} \\ -x &= 4 && \text{Set exponents equal (Property (b)).} \\ x &= -4 && \text{Multiply by } -1. \end{aligned}$$

Check that the solution set is $\{-4\}$. ■

EXAMPLE 7 **Solving a Type 1 Exponential Equation**

Solve $1.5^{x+1} = \left(\frac{27}{8}\right)^x$.

Analytic Solution

Here, $1.5 = \frac{3}{2}$ and $\frac{27}{8} = \left(\frac{3}{2}\right)^3$, so each base can be written as a power of $\frac{3}{2}$.

$$1.5^{x+1} = \left(\frac{27}{8}\right)^x$$

$$\left(\frac{3}{2}\right)^{x+1} = \left[\left(\frac{3}{2}\right)^3\right]^x \qquad \text{Write with the same base.}$$

$$\left(\frac{3}{2}\right)^{x+1} = \left(\frac{3}{2}\right)^{3x} \qquad (a^m)^n = a^{mn}$$

$$x + 1 = 3x \qquad \text{Set exponents equal (Property (b)).}$$

$$1 = 2x \qquad \text{Subtract } x.$$

$$x = \frac{1}{2} \quad \text{or} \quad .5 \qquad \text{Divide by 2; rewrite.}$$

Check that the solution set is $\{.5\}$.

Graphing Calculator Solution

Graph

$$y = 1.5^{x+1} - \left(\frac{27}{8}\right)^x,$$

and use the x-intercept method, as shown in Figure 22. The x-intercept is .5, confirming the analytic solution.

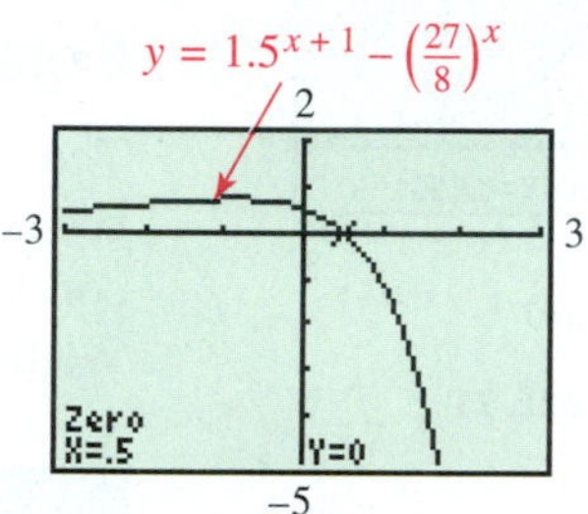

FIGURE 22

EXAMPLE 8 **Using a Graph to Solve Exponential Inequalities**

Use the graph in Figure 22 to solve each inequality.

(a) $1.5^{x+1} - \left(\frac{27}{8}\right)^x > 0$ **(b)** $1.5^{x+1} - \left(\frac{27}{8}\right)^x < 0$

Solution In Figure 22, the graph of $y = 1.5^{x+1} - \left(\frac{27}{8}\right)^x$ lies *above* the x-axis for values of x less than .5 and *below* the x-axis for values of x greater than .5. Tracing left and right will support this observation. The solution set for part (a) is $(-\infty, .5)$, and the solution set for part (b) is $(.5, \infty)$.

Compound Interest

The formula for *compound interest* (interest paid on both principal and interest) is an important application of exponential functions. The formula for simple interest is $I = Prt$, where P is the principal, or initial amount of money; r is the rate of interest expressed as a decimal; and t is time in years that the principal earns interest. Suppose $t = 1$ year. Then, at the end of the year, the amount has grown to

$$P + Pr = P(1 + r),$$

the original principal plus the interest. If this amount is left at the same interest rate for another year, the total amount becomes

$$[P(1 + r)] + [P(1 + r)]r = [P(1 + r)](1 + r) \qquad \text{Factor.}$$

$$= P(1 + r)^2.$$

After the third year, this will grow to

$$[P(1+r)^2] + [P(1+r)^2]r = [P(1+r)^2](1+r) \quad \text{Factor.}$$
$$= P(1+r)^3.$$

Continuing in this way produces a general formula for interest compounded n times per year.

Compound Interest Formula

Suppose that a principal of P dollars is invested at an annual interest rate r (in decimal form), compounded n times per year. Then the amount A accumulated after t years is given by the formula

$$A = P\left(1 + \frac{r}{n}\right)^{nt}.$$

EXAMPLE 9 **Using the Compound Interest Formula**

Suppose that \$1000 is invested at an annual rate of 8%, compounded quarterly (four times per year). Find the total amount in the account after 10 years if no withdrawals are made.

Analytic Solution

Use the compound interest formula with $P = 1000$, $r = .08$, $n = 4$, and $t = 10$.

$$A = P\left(1 + \frac{r}{n}\right)^{nt}$$

$$A = 1000\left(1 + \frac{.08}{4}\right)^{4\cdot 10} \quad \text{Substitute.}$$

$$A = 1000(1.02)^{40} \quad \text{Simplify.}$$

$$A \approx 2208.039664 \quad \text{Use a calculator.}$$

To the nearest cent, there will be \$2208.04 in the account after 10 years. This means that

$$\$2208.04 - \$1000 = \$1208.04$$

interest was earned. (Note that there are four quarters per year, so 10 years = 40 quarters.)

Graphing Calculator Solution

The function $Y_1 = 1000\left(1 + \frac{.08}{4}\right)^{4X}$ is graphed in Figure 23 and evaluated for $X = 10$. The Y-value at the bottom of the screen is the amount in the account. (Tracing to the right on this graph gives new meaning to the phrase "watching your money grow.")

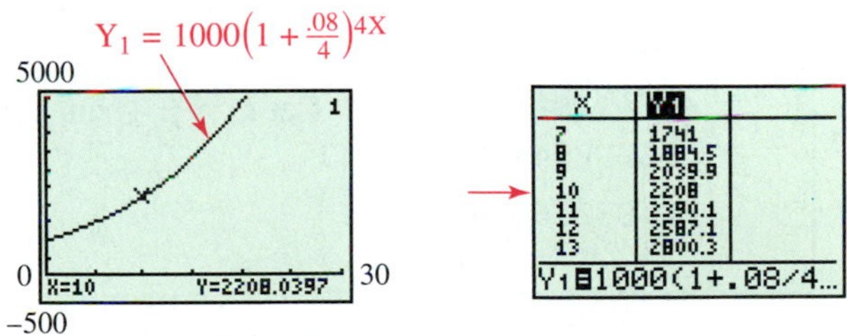

FIGURE 23

The table in Figure 23 can be used to numerically determine the amount in the account. Compare the value of Y_1 when $X = 10$ with the value in the analytic solution and the Y-value displayed at the bottom of the calculator graph. ■

k	$\left(1+\frac{1}{k}\right)^k$ (rounded)
1	2
2	2.25
5	2.48832
10	2.59374
100	2.70481
1000	2.71692
10,000	2.71815
1,000,000	2.71828

The Number e and Continuous Compounding

The more often interest is compounded within a given period, the more interest will be earned. Surprisingly, however, there is a limit on the amount of interest, no matter how often it is compounded. To see this, suppose that \$1 is invested at 100% interest per year, compounded k times per year. Then the interest rate (in decimal form) is 1.00 and the interest rate per period is $\frac{1}{k}$. According to the formula (with $P = 1$), the compound amount at the end of 1 year will be

$$A = \left(1 + \frac{1}{k}\right)^k.$$

A calculator gives the results in the margin for various values of k. The table suggests that as k increases, the value of $\left(1 + \frac{1}{k}\right)^k$ gets closer and closer to some fixed number. This is indeed the case. That fixed number is called e.

Value of e

To nine decimal places,

$$e \approx 2.718281828.$$

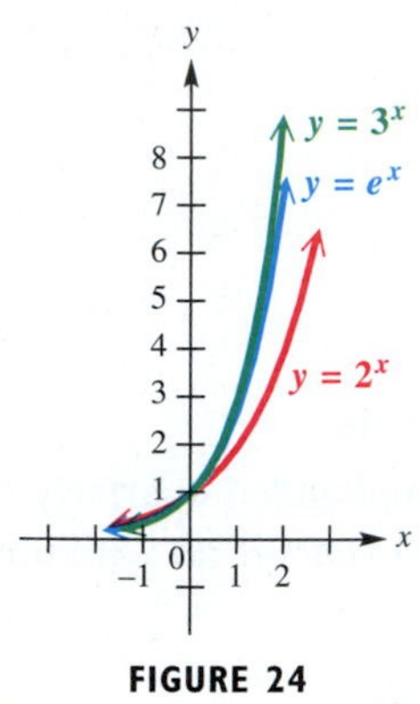

FIGURE 24

Figure 24 shows the functions defined by $y = 2^x$, $y = 3^x$, and $y = e^x$. Notice that because $2 < e < 3$, the graph of $y = e^x$ lies "between" the other two graphs.

As mentioned, the amount of interest earned increases with the frequency of compounding, but the value of the expression $\left(1 + \frac{1}{k}\right)^k$ approaches e as k gets larger. Consequently, the formula for compound interest approaches a limit as well, called the compound amount from *continuous compounding*.

To derive the formula for continuous compounding, we begin with the compound interest formula,

$$A = P\left(1 + \frac{r}{n}\right)^{nt}.$$

Let $k = \frac{n}{r}$. Then, $n = rk$, and with these substitutions, the formula becomes

$$A = P\left(1 + \frac{1}{k}\right)^{rkt} = P\left[\left(1 + \frac{1}{k}\right)^k\right]^{rt}.$$

If $n \to \infty$, $k \to \infty$ as well, and the expression $\left(1 + \frac{1}{k}\right)^k \to e$, as discussed earlier. This leads to the formula $A = Pe^{rt}$.

Continuous Compounding Formula

If an amount of P dollars is deposited at a rate of interest r (in decimal form) compounded continuously for t years, then the final amount in dollars is

$$A = Pe^{rt}.$$

TECHNOLOGY NOTE

Because e is such an important base for the exponential function, calculators are programmed to find powers of e.

```
e^(1)
       2.718281828
e^(-1)
       .3678794412
e^(2)
       7.389056099
```

Looking Ahead to Calculus

In calculus, the derivative allows us to determine the slope of a line tangent to the graph of a function. For the function $f(x) = e^x$, the derivative is the function f itself: $f'(x) = e^x$. Therefore, in calculus the exponential function with base e is much easier to work with than exponential functions having other bases.

EXAMPLE 10 Solving a Continuous Compounding Problem

Suppose \$5000 is deposited in an account paying 8% compounded continuously for 5 years. Find the total amount on deposit at the end of the 5 years.

Analytic Solution

Let $P = 5000$, $r = .08$, and $t = 5$ in the continuous compounding formula $A = Pe^{rt}$. Then

$$A = 5000e^{.08(5)} = 5000e^{.4}.$$

Using a calculator, we find, to the nearest cent, that

$$A = 5000e^{.4} \approx 7459.12.$$

There will be \$7459.12 in the account after 5 years. (This means that

$$\$7459.12 - \$5000 = \$2459.12$$

interest was earned.)

Graphing Calculator Solution

As Figure 25(a) shows, the graph of $Y_1 = 5000e^{.08X}$ indicates that when $X = 5$, $Y \approx 7459.12$.

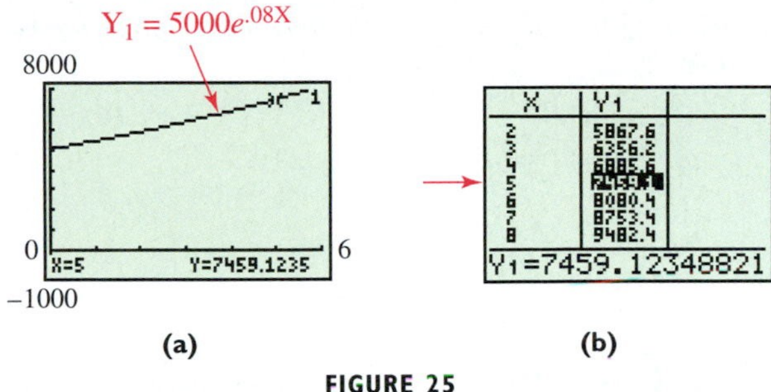

(a) (b)

FIGURE 25

The analytic result is supported numerically by the table in Figure 25(b), which shows values of $Y_1 = 5000e^{.08X}$ for several values of X. When $X = 5$, the table gives $Y_1 \approx 7459.12$. ■

An Application of Exponential Functions

As the U.S. population ages, there has been increased concern about the incidence of older people dying of influenza or pneumonia. Inoculating those over 55 years of age has significantly reduced the risk of older people dying from these illnesses.

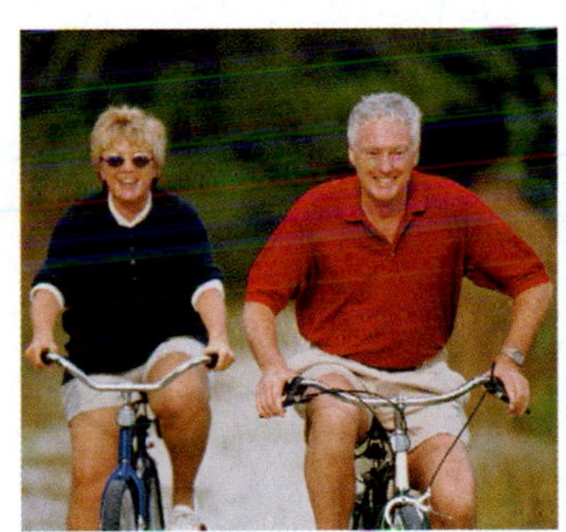

EXAMPLE 11 Modeling the Risk of Dying of Influenza or Pneumonia

The chances of dying of influenza or pneumonia increase exponentially after age 55 according to the function defined by

$$f(x) = r(1.082)^x,$$

where r is the risk (in decimal form) at age 55 and x is the number of years greater than 55. (*Source:* U.S. Census Bureau.) Compare the risk at age 75 with the risk at age 55.

Solution Let $x = 75 - 55 = 20$ and write $f(x)$ in terms of r.

$$f(x) = r(1.082)^{20} \approx 4.84r \quad \text{Use a calculator.}$$

Thus, the risk is almost five times as great at age 75 as at age 55. ■

5.2 Exercises

Use a calculator to find an approximation for each power. Give the maximum number of decimal places that your calculator displays.

1. $2^{\sqrt{10}}$
2. $3^{\sqrt{11}}$
3. $\left(\frac{1}{2}\right)^{\sqrt{2}}$
4. $\left(\frac{1}{3}\right)^{\sqrt{6}}$
5. $4.1^{-\sqrt{3}}$
6. $6.4^{-\sqrt{3}}$
7. $\sqrt{7}^{\sqrt{7}}$
8. $\sqrt{13}^{-\sqrt{13}}$

Use a calculator graph of each exponential function to graphically support the result found in the specified exercise.

9. $y = 2^x$ (Exercise 1)

10. $y = 3^x$ (Exercise 2)

11. $y = \left(\frac{1}{2}\right)^x$ (Exercise 3)

12. $y = \left(\frac{1}{3}\right)^x$ (Exercise 4)

Concept Check *In the figure, the graphs of $y = a^x$ for $a = 1.8, 2.3, 3.2, .4, .75,$ and $.31$ are given. They are identified by letter, but not necessarily in the same order as the values of a just given. Use your knowledge of how the exponential function behaves for various powers of a to identify each lettered graph.*

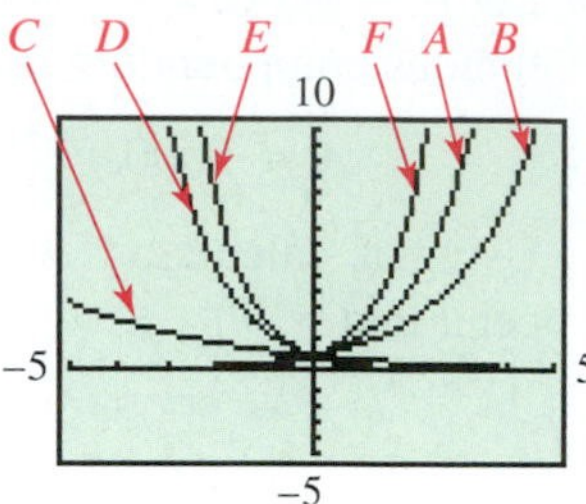

13. *A*

14. *B*

15. *C*

16. *D*

17. *E*

18. *F*

Relating Concepts

For individual or group investigation (Exercises 19–24)

In Exercises 19–24, assume that $f(x) = a^x$, where $a > 1$. ***Work these exercises in order.***

19. Is f a one-to-one function? If so, based on Section 5.1, what kind of related function exists for f?

20. If f has an inverse function f^{-1}, sketch f and f^{-1} on the same axes.

21. If f^{-1} exists, find an equation for $y = f^{-1}(x)$, using the method described in Section 5.1. You need not solve for y.

22. If $a = 10$, what is the equation for $y = f^{-1}(x)$? (You need not solve for y.)

23. If $a = e$, what is the equation for $y = f^{-1}(x)$? (You need not solve for y.)

24. If the point (p, q) is on the graph of f, then the point __________ is on the graph of f^{-1}.

Graph each function by hand and support your sketch with a calculator graph. Give the domain, range, and equation of the asymptote.

25. $f(x) = 3^x$

26. $f(x) = 4^x$

27. $f(x) = \left(\frac{1}{3}\right)^x$

28. $f(x) = \left(\frac{1}{4}\right)^x$

29. $f(x) = \left(\frac{3}{2}\right)^x$

30. $f(x) = \left(\frac{2}{3}\right)^x$

31. $f(x) = e^x$

32. $f(x) = -e^x$

33. $f(x) = e^{x+1}$

34. $f(x) = e^x - 1$

35. $f(x) = 10^x$

36. $f(x) = 10^{-x}$

37. $f(x) = 4^{-x}$

38. $f(x) = 6^{-x}$

Concept Check *In Exercises 39 and 40, refer to the graphs at the right and follow the directions in parts (a)–(f).*

39. **(a)** Is $a > 1$ or is $0 < a < 1$?

(b) Give the domain and range of f, and the equation of the asymptote.

(c) Sketch the graph of $g(x) = -a^x$.

(d) Give the domain and range of g, and the equation of the asymptote.

(e) Sketch the graph of $h(x) = a^{-x}$.

(f) Give the domain and range of h, and the equation of the asymptote.

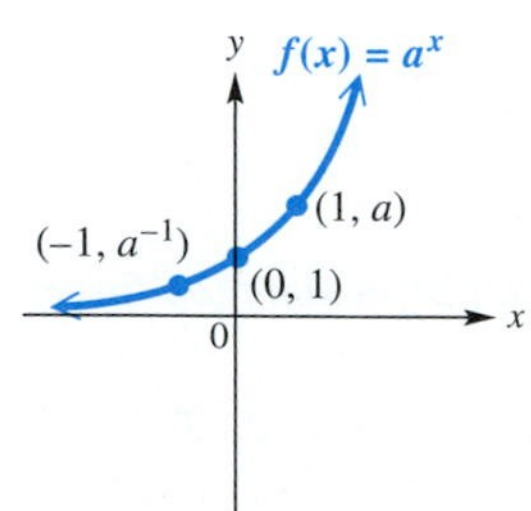

40. **(a)** Is $a > 1$ or is $0 < a < 1$?
(b) Give the domain and range of f, and the equation of the asymptote.
(c) Sketch the graph of $g(x) = a^x + 2$.
(d) Give the domain and range of g, and the equation of the asymptote.
(e) Sketch the graph of $h(x) = a^{x+2}$.
(f) Give the domain and range of h, and the equation of the asymptote.

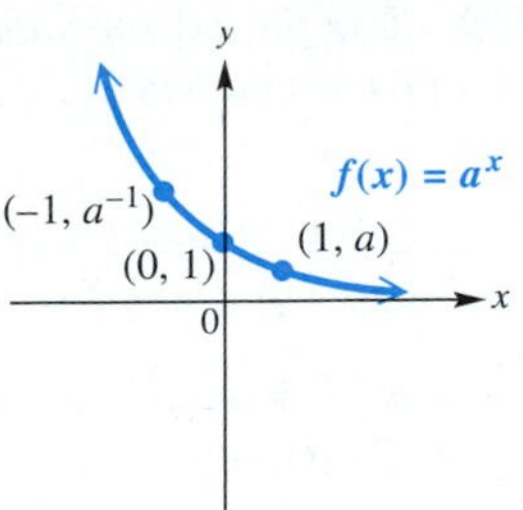

Sketch the graph of $f(x) = 2^x$. Then refer to it and use the techniques of Chapter 2 to graph each function as defined.

41. $f(x) = 2^x + 1$ **42.** $f(x) = 2^x - 4$ **43.** $f(x) = 2^{x+1}$ **44.** $f(x) = 2^{x-4}$

Sketch the graph of $f(x) = \left(\frac{1}{3}\right)^x$. Then refer to it and use the techniques of Chapter 2 to graph each function as defined.

45. $f(x) = \left(\frac{1}{3}\right)^x - 2$ **46.** $f(x) = \left(\frac{1}{3}\right)^x + 4$ **47.** $f(x) = \left(\frac{1}{3}\right)^{x+2}$ **48.** $f(x) = \left(\frac{1}{3}\right)^{x-4}$

Solve each equation.

49. $4^x = 2$ **50.** $125^x = 5$ **51.** $\left(\frac{1}{2}\right)^x = 4$ **52.** $\left(\frac{2}{3}\right)^x = \frac{9}{4}$

53. $2^{3-x} = 8$ **54.** $5^{2x+1} = 25$ **55.** $12^{x-3} = 1$ **56.** $(-3)^{5-x} = 1$

57. $e^{4x-1} = (e^2)^x$ **58.** $e^{3-x} = (e^3)^{-x}$ **59.** $27^{4x} = 9^{x+1}$ **60.** $32^x = 16^{1-x}$

61. $4^{x-2} = 2^{3x+3}$ **62.** $\left(\frac{1}{2}\right)^{3x-6} = 8^{x+1}$ **63.** $(\sqrt{2})^{x+4} = 4^x$ **64.** $(\sqrt[3]{5})^{-x} = \left(\frac{1}{5}\right)^{x+2}$

Solve each equation in part (a) analytically. Support your answer with a calculator graph. Then use the graph to solve the associated inequalities in parts (b) and (c).

65. **(a)** $2^{x+1} = 8$
(b) $2^{x+1} > 8$
(c) $2^{x+1} < 8$

66. **(a)** $3^{2-x} = 9$
(b) $3^{2-x} > 9$
(c) $3^{2-x} < 9$

67. **(a)** $27^{4x} = 9^{x+1}$
(b) $27^{4x} > 9^{x+1}$
(c) $27^{4x} < 9^{x+1}$

68. **(a)** $32^x = 16^{1-x}$
(b) $32^x > 16^{1-x}$
(c) $32^x < 16^{1-x}$

69. **(a)** $\left(\frac{1}{2}\right)^{-x} = \left(\frac{1}{4}\right)^{x+1}$
(b) $\left(\frac{1}{2}\right)^{-x} \geq \left(\frac{1}{4}\right)^{x+1}$
(c) $\left(\frac{1}{2}\right)^{-x} \leq \left(\frac{1}{4}\right)^{x+1}$

70. **(a)** $\left(\frac{2}{3}\right)^{x-1} = \left(\frac{81}{16}\right)^{x+1}$
(b) $\left(\frac{2}{3}\right)^{x-1} \leq \left(\frac{81}{16}\right)^{x+1}$
(c) $\left(\frac{2}{3}\right)^{x-1} \geq \left(\frac{81}{16}\right)^{x+1}$

71. ***Concept Check*** If $f(x) = a^x$ and $f(3) = 27$, find each value of $f(x)$.
(a) $f(1)$ **(b)** $f(-1)$
(c) $f(2)$ **(d)** $f(0)$

Concept Check *Give an equation of the form $f(x) = a^x$ to define the exponential function whose graph contains the given point.*

72. $(3, 8)$ **73.** $(-3, 64)$

Concept Check *Use properties of exponents to write each function in the form $f(t) = ka^t$, where k is a constant. (Hint: Recall that $a^{x+y} = a^x \cdot a^y$.)*

74. $f(t) = 3^{2t+3}$ **75.** $f(t) = \left(\frac{1}{3}\right)^{1-2t}$

76. The graph of $y = e^{x-3}$ can be obtained by translating the graph of $y = e^x$ to the right 3 units. Find a constant C such that the graph of $y = Ce^x$ is the same as the graph of $y = e^{x-3}$. Verify your result by graphing both functions.

Compound Amount *Use the appropriate compound interest formula to find the amount that will be in each account, given the stated conditions.*

77. \$20,000 invested at 3% annual interest for 4 years compounded **(a)** annually; **(b)** semiannually

78. \$35,000 invested at 4.2% annual interest for 3 years compounded **(a)** annually; **(b)** quarterly

79. \$27,500 invested at 3.95% annual interest for 5 years compounded **(a)** daily ($n = 365$); **(b)** continuously

80. \$15,800 invested at 4.6% annual interest for 6.5 years compounded **(a)** quarterly; **(b)** continuously

Comparing Investment Plans *In Exercises 81 and 82, decide which of the two plans will provide a better yield. (Interest rates stated are annual rates.)*

81. Plan A: \$40,000 invested for 3 years at 4.5%, compounded quarterly

Plan B: \$40,000 invested for 3 years at 4.4%, compounded continuously

82. Plan A: \$50,000 invested for 10 years at 4.75%, compounded daily ($n = 365$)

Plan B: \$50,000 invested for 10 years at 4.7%, compounded continuously

Use the table capabilities of your calculator to work Exercises 83 and 84.

83. ***Comparison of Two Accounts*** You have the choice of investing \$1000.00 at an annual rate of 5%, compounded either annually or monthly. Let Y_1 represent the investment compounded annually, and let Y_2 represent the investment compounded monthly. Graph both Y_1 and Y_2, and observe the slight difference in each curve. Then use a table to compare the graphs numerically. What is the difference between the returns on each investment after 1 year, 2 years, 5 years, 10 years, 20 years, 30 years, and 40 years?

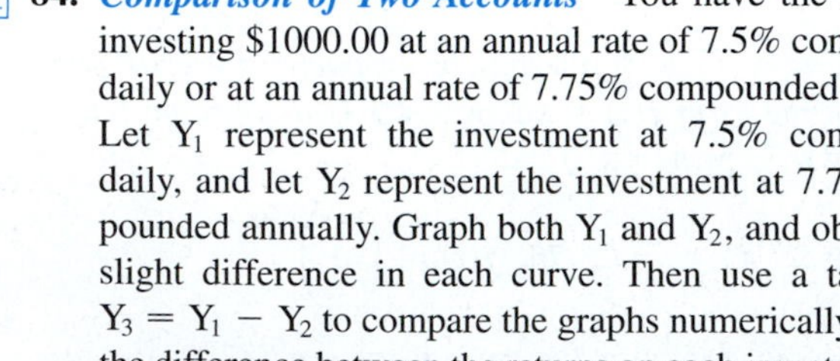

84. ***Comparison of Two Accounts*** You have the choice of investing \$1000.00 at an annual rate of 7.5% compounded daily or at an annual rate of 7.75% compounded annually. Let Y_1 represent the investment at 7.5% compounded daily, and let Y_2 represent the investment at 7.75% compounded annually. Graph both Y_1 and Y_2, and observe the slight difference in each curve. Then use a table with $Y_3 = Y_1 - Y_2$ to compare the graphs numerically. What is the difference between the returns on each investment after 1 year, 2 years, 5 years, 10 years, 20 years, 30 years, and 40 years? Why does the lower interest rate yield the greater return over time?

(Modeling) *Solve each problem.*

85. ***Atmospheric Pressure*** The atmospheric pressure (in millibars) at a given altitude (in meters) is shown in the table.

Altitude	Pressure	Altitude	Pressure
0	1013	6000	472
1000	899	7000	411
2000	795	8000	357
3000	701	9000	308
4000	617	10,000	265
5000	541		

Source: Miller, A. and J. Thompson, *Elements of Meteorology,* Fourth Edition, Charles E. Merrill Publishing Company, Columbus, Ohio, 1993.

(a) Use a graphing calculator to make a scatter diagram of the data for atmospheric pressure P at altitude x.

(b) Would a linear or exponential function fit the data better?

(c) The function defined by

$$P(x) = 1013e^{-.0001341x}$$

approximates the data. Use a graphing calculator to graph P and the data on the same coordinate axes.

(d) Use P to predict the pressures at 1500 m and 11,000 m, and compare them with the actual values of 846 millibars and 227 millibars, respectively.

86. ***World Population Growth*** Since 1990, world population in millions closely fits the exponential function defined by

$$y = 5282e^{.01405x},$$

where x is the number of years since 1990. (*Source:* U.S. Census Bureau.)

(a) The world population was about 6080 million in 2000. How closely does the function approximate this value?

(b) Use this model to approximate the population in 2005.

(c) Use the model to predict the population in 2010.

(d) Explain why the model may not be accurate for 2010.

87. ***Traffic Flow*** At an intersection, cars arrive randomly at an average rate of 30 cars per hour. Using the function defined by

$$f(x) = 1 - e^{-.5x},$$

highway engineers estimate the likelihood or probability that at least one car will enter the intersection within a period of x minutes. (*Source:* Mannering, F. and W.

Kilareski, *Principles of Highway Engineering and Traffic Analysis,* Second Edition, John Wiley and Sons, 1998.)

(a) Evaluate $f(2)$ and interpret the answer.

(b) Graph f for $0 \le x \le 60$. What happens to the likelihood that at least one car enters the intersection during a 60-minute period?

88. ***Growth of* E. coli *Bacteria*** A type of bacteria that inhabits the intestines of animals is named *E. coli* (*Escherichia coli*). These bacteria are capable of rapid growth and can be dangerous to humans—especially children. In one study, *E. coli* bacteria were capable of doubling in number every 49.5 minutes. Their number after x minutes can be modeled by the function defined by

$$N(x) = N_0 e^{.014x}.$$

(*Source:* Stent, G. S., *Molecular Biology of Bacterial Viruses,* W. H. Freeman, 1963.) Suppose $N_0 = 500{,}000$ is the initial number of bacteria per milliliter.

(a) Make a conjecture about the number of bacteria per milliliter after 99 minutes. Verify your conjecture.

(b) Determine graphically the time elapsed from when there were 25 million bacteria per milliliter.

5.3 Logarithms and Their Properties

Definition of Logarithm ■ Common Logarithms ■ Natural Logarithms ■ Properties of Logarithms ■ Change-of-Base Rule

Definition of Logarithm

In the exponential equation $2^3 = 8$, 3 is the exponent to which 2 must be raised in order to obtain 8. We call 3 the *logarithm* to the base 2 of 8, abbreviated $3 = \log_2 8$. ***A logarithm is an exponent*** and thus has the same properties as exponents.

Logarithm

For all positive numbers a, where $a \neq 1$,

$$a^y = x \quad \text{is equivalent to} \quad y = \log_a x.$$

A **logarithm** is an exponent, and $\log_a x$ is the exponent to which a must be raised in order to obtain x. The number a is called the **base** of the logarithm, and x is called the **argument** of the expression $\log_a x$. ***The value of x will always be positive.***

The table shows several pairs of equivalent statements written in both exponential and logarithmic forms.

Exponent

Exponential form: $a^y = x$

Base

Exponent

Logarithmic form: $y = \log_a x$

Base

Exponential Form	Logarithmic Form
$2^3 = 8$	$\log_2 8 = 3$
$\left(\frac{1}{2}\right)^{-4} = 16$	$\log_{1/2} 16 = -4$
$10^5 = 100{,}000$	$\log_{10} 100{,}000 = 5$
$3^{-4} = \frac{1}{81}$	$\log_3\left(\frac{1}{81}\right) = -4$
$5^1 = 5$	$\log_5 5 = 1$
$\left(\frac{3}{4}\right)^0 = 1$	$\log_{3/4} 1 = 0$

EXAMPLE 1 Solving Logarithmic Equations

Solve each equation by first rewriting it in exponential form.

(a) $x = \log_8 4$ **(b)** $\log_x 16 = 4$ **(c)** $\log_4 x = \frac{3}{2}$

Solution

(a)

$$x = \log_8 4$$

$$8^x = 4 \qquad \text{Write in exponential form.}$$

$$(2^3)^x = 2^2 \qquad \text{Write as a power of the same base, 2.}$$

$$2^{3x} = 2^2 \qquad (a^m)^n = a^{mn}$$

$$3x = 2 \qquad \text{Set exponents equal.}$$

$$x = \frac{2}{3} \qquad \text{Divide by 3.}$$

Since $8^{2/3} = \left(\sqrt[3]{8}\right)^2 = 2^2 = 4$, the solution set is $\left\{\frac{2}{3}\right\}$.

(b)

$$\log_x 16 = 4$$

$$x^4 = 16 \qquad \text{Write in exponential form.}$$

$$x = \pm\sqrt[4]{16} \qquad \text{Take fourth roots.}$$

$$x = \pm 2 \qquad \text{Simplify.}$$

Remember both the positive and negative roots.

Since the base x must be positive, reject -2. The solution set is $\{2\}$.

(c)

$$\log_4 x = \frac{3}{2}$$

$$x = 4^{3/2} \qquad \text{Write in exponential form.}$$

$$x = 8 \qquad 4^{3/2} = \left(\sqrt{4}\right)^3 = 8$$

Check that the solution set is $\{8\}$. ■

Common Logarithms

The two most important bases for logarithms are 10 and e. Base 10 logarithms are called **common logarithms.** The common logarithm of x is written $\mathbf{\log x}$, where the base is understood to be 10.

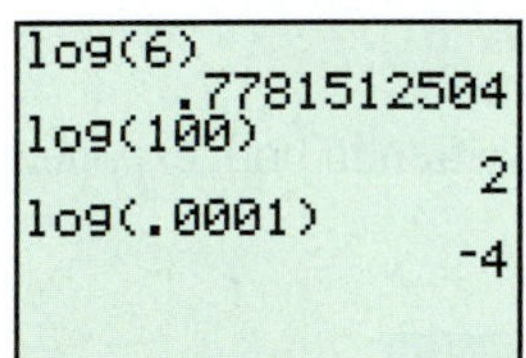

(a)

10^(log(6))
6
10^(log(100))
100
10^(log(.0001))
1E-4

(b)

FIGURE 26

Common Logarithm

For all positive numbers x, $\mathbf{\log x = \log_{10} x}$.

Remember that the argument of a common logarithm (and any base logarithm, for that matter) must be a positive number.

Figure 26(a) shows a typical graphing calculator screen with several common logarithms evaluated. The first display indicates that .7781512504 is (approximately) the exponent to which 10 must be raised in order to obtain 6. The second says that 2 is the exponent to which 10 must be raised in order to obtain 100. This is correct, since $100 = 10^2$. The third display indicates that -4 is the exponent to which 10 must be raised in order to obtain .0001. Again, this is correct, since $.0001 = 10^{-4}$.

Figure 26(b) shows how the definition of a common logarithm can be applied: ***Raising 10 to the power log x gives x as a result.***

```
log(12)
      1.079181246
log(.1)
               -1
log(3/5)
     -.2218487496
```

These are the results for Example 2.

EXAMPLE 2 **Evaluating Common Logarithms**

Evaluate each expression.

(a) $\log 12$ **(b)** $\log .1$ **(c)** $\log \frac{3}{5}$

Solution Use a calculator.

(a) $\log 12 \approx 1.079181246$ **(b)** $\log .1 = -1$ **(c)** $\log \frac{3}{5} \approx -.2218487496$ ■

TECHNOLOGY NOTE

An attempt to find the common logarithm of a nonpositive number results in an error message when a graphing calculator is in real number mode.

In chemistry, the **pH** of a solution is defined as

$$\mathbf{pH = -\log[H_3O^+]},$$

where $[H_3O^+]$ is the hydronium ion concentration in moles per liter.* The pH value is a measure of the acidity or alkalinity of a solution. Pure water has pH 7.0, substances with pH values greater than 7.0 are alkaline, and substances with pH values less than 7.0 are acidic. It is customary to round pH values to the nearest tenth.

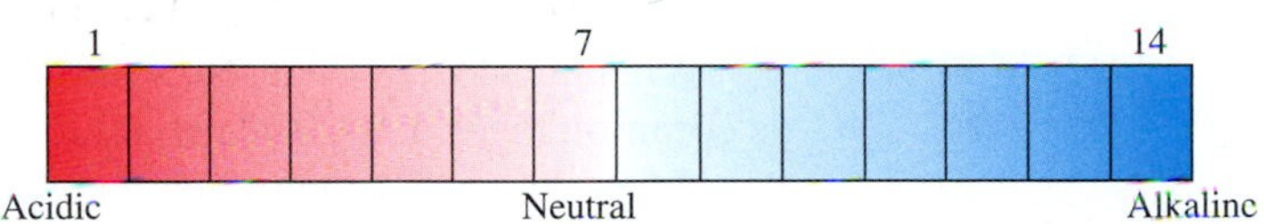

EXAMPLE 3 **Finding pH and $[H_3O^+]$**

(a) Find the pH of a solution with $[H_3O^+] = 2.5 \times 10^{-4}$. Is the solution acidic or alkaline?

(b) Find the hydronium ion concentration of a solution with $\text{pH} = 7.1$.

```
-log(2.5*10^(-4))
      3.602059991
10^(-7.1)
   7.943282347E-8
```

These are the calculations required in Example 3.

Solution

(a)
$$\text{pH} = -\log[H_3O^+]$$
$$\text{pH} = -\log(2.5 \times 10^{-4}) \quad \text{Substitute.}$$
$$\text{pH} \approx 3.6 \quad \text{Use a calculator.}$$

The solution is acidic because the pH is less than 7.0.

(b)
$$\text{pH} = -\log[H_3O^+]$$
$$7.1 = -\log[H_3O^+] \quad \text{Substitute.}$$
$$-7.1 = \log[H_3O^+] \quad \text{Multiply by } -1.$$
$$[H_3O^+] = 10^{-7.1} \quad \text{Write in exponential form.}$$
$$[H_3O^+] \approx 7.9 \times 10^{-8} \quad \text{Evaluate } 10^{-7.1} \text{ with a calculator.}$$ ■

*A *mole* is the amount of substance that contains the same number of molecules as the number of atoms in 12 grams of carbon 12.

Looking Ahead to Calculus

The natural logarithmic function, defined by $f(x) = \ln x$, and the reciprocal function, defined by $g(x) = \frac{1}{x}$, have an important relationship in calculus. The derivative of the natural logarithmic function is the reciprocal function.

Natural Logarithms

In many practical applications of logarithms, the number e is used as the base. Logarithms to base e are called **natural logarithms,** since they occur in the life sciences and economics in natural situations that involve growth and decay. The natural logarithm of a positive number x is written $\mathbf{\ln x}$ (read **"el en x"**).

Natural Logarithm

For all positive numbers x, $\quad \mathbf{\ln x = \log_e x}$.

Figure 27 shows traditional and calculator graphs of the natural logarithmic function.

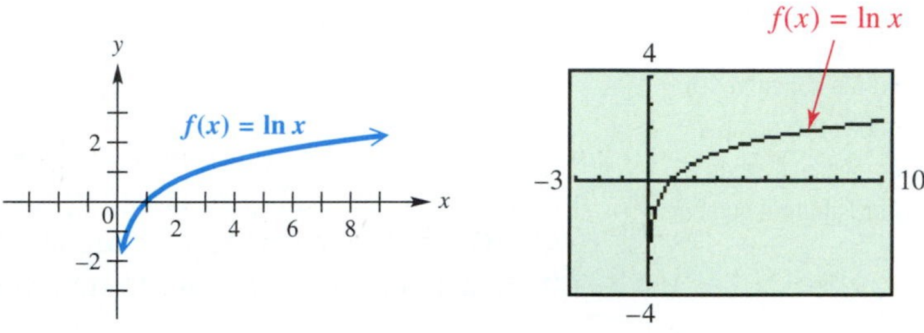

FIGURE 27

When a calculator is used, natural logarithms of numbers are found the same way as common logarithms. The natural logarithm key is usually found in conjunction with the e^x key. Figure 28 shows how the natural logarithmic function and the base e exponential function can be applied. For example, the first display in Figure 28(a) indicates that 2.079441542 is (approximately) the exponent to which e must be raised in order to obtain 8.

```
ln(8)
          2.079441542
ln(1)
                    0
ln(e^(4))
                    4
```

(a)

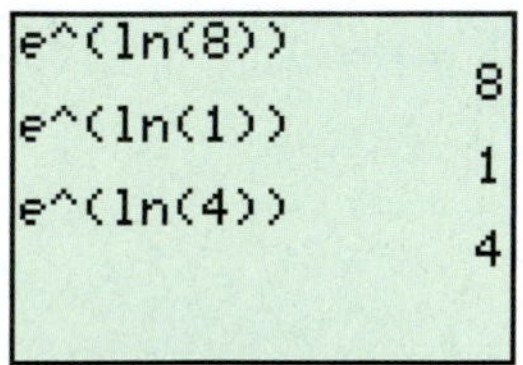

(b)

FIGURE 28

FOR DISCUSSION

1. The second display in Figure 28(a) indicates that $\ln 1 = 0$. Use your calculator to find log 1. Discuss your results. If 1 is the argument, does it matter what base is used? Why or why not?
2. The third display in Figure 28(a) indicates that $\ln(e^4) = 4$. Use the same expression, but replace 4 with the number of letters in your last name. What is the result? Can you make a generalization?
3. Does your generalization apply to common logarithms as well?

EXAMPLE 4 Evaluating Natural Logarithms

Evaluate each expression.

(a) $\ln 12$ **(b)** $\ln e^{10}$ **(c)** $\ln \sqrt{e}$

Solution Use a calculator. Refer to Item 2 in the "For Discussion" box for parts (b) and (c).

(a) $\ln 12 \approx 2.48490665$ **(b)** $\ln e^{10} = 10$ **(c)** $\ln \sqrt{e} = .5$ ■

EXAMPLE 5 **Solving a Continuous Compounding Problem**

The continuous compounding formula $A = Pe^{rt}$ was given in Section 5.2. Suppose that \$1000 is invested at 3% annual interest compounded continuously. How long will it take for the amount to grow to \$1500?

Analytic Solution

Let $P = 1000$, $A = 1500$, and $r = .03$ in the formula.

$$1500 = 1000e^{.03t} \qquad A = Pe^{rt}$$

$$1.5 = e^{.03t} \qquad \text{Divide by 1000.}$$

Because $.03t$ is the exponent to which e must be raised in order to obtain 1.5, it must equal the natural logarithm of 1.5, or $\ln 1.5$.

$$.03t = \ln 1.5 \qquad \text{Write in logarithmic form.}$$

$$t = \frac{\ln 1.5}{.03} \qquad \text{Divide by .03.}$$

$$t \approx 13.5 \qquad \text{Use a calculator.}$$

It will take about 13.5 years to grow to \$1500.

Graphing Calculator Solution

Graph $Y_1 = 1000e^{.03X}$ and $Y_2 = 1500$. Then find the point of intersection, as shown in Figure 29. The table in Figure 30 gives the same result. When time X is 13.5 years, the amount Y_1 is $1499.3 \approx 1500$ dollars.

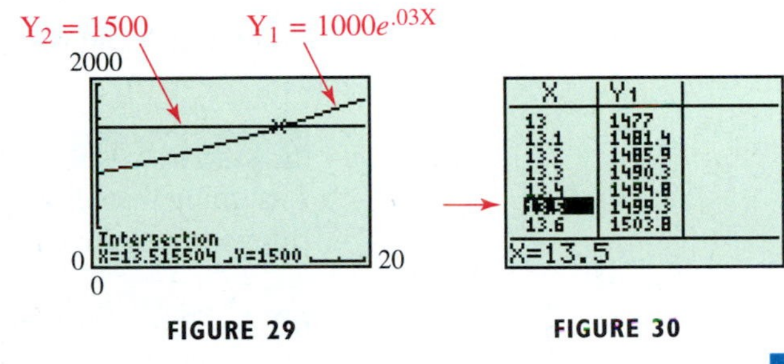

FIGURE 29

FIGURE 30

Properties of Logarithms

We now formally state several properties of logarithms.

Properties of Logarithms

For $a > 0$, $a \neq 1$, and any real number k,

1. $\log_a 1 = 0$. **2.** $\log_a a^k = k$. **3.** $a^{\log_a k} = k, \quad k > 0$.

For $x > 0$, $y > 0$, $a > 0$, $a \neq 1$, and any real number r,

Product Rule **4.** $\log_a xy = \log_a x + \log_a y$.
(The logarithm of the product of two numbers is equal to the sum of the logarithms of the numbers.)

Quotient Rule **5.** $\log_a \frac{x}{y} = \log_a x - \log_a y$.
(The logarithm of the quotient of two numbers is equal to the difference of the logarithms of the numbers.)

Power Rule **6.** $\log_a x^r = r \log_a x$.
(The logarithm of a number raised to a power is equal to the exponent multiplied by the logarithm of the number.)

Property 1 is true because $a^0 = 1$ for any nonzero value of a. Property 2 is verified by writing the equation in exponential form, giving the identity $a^k = a^k$. Property 3 is

Looking Ahead to Calculus

The product rule for logarithms (as well as the quotient and power rules) is proved in calculus by a different method than the one labeled "Proof" here. The typical calculus proof involves the derivative of the base e logarithmic function.

justified by the fact that $\log_a k$ is the exponent to which a must be raised in order to obtain k. Therefore, by the definition, $a^{\log_a k}$ must equal k. The proof of Property 4, the product rule, follows.

Proof Let $m = \log_a x$ and $n = \log_a y$.

Then, $a^m = x$ and $a^n = y$ — Definition of logarithm

$$a^m \cdot a^n = xy \quad \text{Multiply.}$$

$$a^{m+n} = xy \quad \text{Add exponents.}$$

$$\log_a xy = m + n \quad \text{Definition of logarithm}$$

$$\log_a xy = \log_a x + \log_a y. \quad \text{Substitute.}$$

Properties 5 and 6, the quotient and power rules, are proved in a similar way.

EXAMPLE 6 Using the Properties of Logarithms

Assuming that all variables represent positive real numbers, use the properties of logarithms to rewrite each expression.

(a) $\log 8x$ **(b)** $\log_9 \frac{15}{7}$ **(c)** $\log_5 \sqrt{8}$ **(d)** $\log .1$

(e) $2^{\log_2 3}$ **(f)** $\log_a \frac{x}{yz}$ **(g)** $\log_a \sqrt[3]{m^2}$ **(h)** $\log_b \sqrt[n]{\frac{x^3y^5}{z^m}}$

Solution

(a) $\log 8x = \log 8 + \log x$ — Product rule

(b) $\log_9 \frac{15}{7} = \log_9 15 - \log_9 7$ — Quotient rule

(c) $\log_5 \sqrt{8} = \log_5 8^{1/2} = \frac{1}{2}\log_5 8$ — Power rule

(d) $\log .1 = \log \frac{1}{10} = \log 10^{-1} = -1$ — Property 2

(e) $2^{\log_2 3} = 3$ — Property 3

(f) $\log_a \frac{x}{yz} = \log_a x - (\log_a y + \log_a z)$

$= \log_a x - \log_a y - \log_a z$ — Distributive property

Be careful with signs.

(g) $\log_a \sqrt[3]{m^2} = \log_a m^{2/3} = \frac{2}{3}\log_a m$ — Power rule

(h) $\log_b \sqrt[n]{\frac{x^3y^5}{z^m}} = \frac{1}{n}\log_b \frac{x^3y^5}{z^m}$ — Power rule

$= \frac{1}{n}(\log_b x^3 + \log_b y^5 - \log_b z^m)$ — Product and quotient rules

$= \frac{1}{n}(3\log_b x + 5\log_b y - m\log_b z)$ — Power rule

Use parentheses to avoid errors.

$= \frac{3}{n}\log_b x + \frac{5}{n}\log_b y - \frac{m}{n}\log_b z$ — Distributive property

■

EXAMPLE 7 **Using the Properties of Logarithms**

Use the properties of logarithms to write each expression as a single logarithm with coefficient 1. Assume that all variables represent positive real numbers.

(a) $\log_3(x + 2) + \log_3 x - \log_3 2$ **(b)** $2\log_a m - 3\log_a n$

(c) $\frac{1}{2}\log_b m + \frac{3}{2}\log_b 2n - \log_b m^2 n$

Solution

(a) $\log_3(x + 2) + \log_3 x - \log_3 2 = \log_3 \frac{(x + 2)x}{2}$ Product and quotient rules

(b) $2\log_a m - 3\log_a n = \log_a m^2 - \log_a n^3$ Power rule

$= \log_a \frac{m^2}{n^3}$ Quotient rule

(c) $\frac{1}{2}\log_b m + \frac{3}{2}\log_b 2n - \log_b m^2 n$

$= \log_b m^{1/2} + \log_b(2n)^{3/2} - \log_b m^2 n$ Power rule

$= \log_b \frac{m^{1/2}(2n)^{3/2}}{m^2 n}$ Product and quotient rules

$= \log_b \frac{2^{3/2}n^{1/2}}{m^{3/2}}$ Rules for exponents

$= \log_b \left(\frac{2^3 n}{m^3}\right)^{1/2}$ Rules for exponents

$= \log_b \sqrt{\frac{8n}{m^3}}$ Definition of $a^{1/n}$ ■

FOR DISCUSSION

1. Without using your calculator, determine exact values of $\log_3 9$ and $\log_3 27$. Then, use the fact that 16 is between 9 and 27 to determine between what two consecutive integers $\log_3 16$ must lie. Finally, use the change-of-base rule to support your answer.

2. Without using your calculator, determine exact values of $\log_5\left(\frac{1}{5}\right)$ and $\log_5 1$. Then, use the fact that .68 is between $\frac{1}{5}$ and 1 to determine between what two consecutive integers $\log_5 .68$ must lie. Finally, use the change-of-base rule to support your answer.

CAUTION ***There is no property of logarithms that allows us to rewrite a logarithm of a sum or difference.*** That is why, in Example 7(a), $\log_3(x + 2)$ was not written as $\log_3 x + \log_3 2$. Remember, $\log_3 x + \log_3 2 = \log_3(x \cdot 2)$. The distributive property does not apply here, since $\log_3(x + 2)$ is one term; "log" is a function name, *not* a factor.

Change-of-Base Rule

A natural question to ask at this point is "Can we use a calculator to find logarithms for bases other than 10 and e, such as $\log_5 17$?" The change-of-base rule shows that we can and how to do so.

Change-of-Base Rule

For any positive real numbers x, a, and b, where $a \neq 1$ and $b \neq 1$,

$$\log_a x = \frac{\log_b x}{\log_b a}.$$

Proof Let

$$
\begin{aligned}
y &= \log_a x. \\
a^y &= x && \text{Write in exponential form.} \\
\log_b a^y &= \log_b x && \text{Take logarithms of each side.} \\
y \log_b a &= \log_b x && \text{Power rule} \\
y &= \frac{\log_b x}{\log_b a} && \text{Divide each side by } \log_b a. \\
\log_a x &= \frac{\log_b x}{\log_b a} && \text{Substitute } \log_a x \text{ for } y.
\end{aligned}
$$

Any positive number other than 1 can be used for base b in the change-of-base rule, but usually the only practical bases are e and 10, since calculators typically give logarithms only for these two bases.

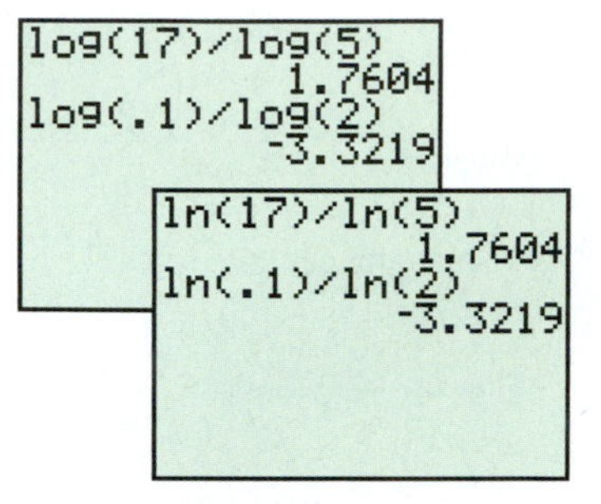

FIGURE 31

EXAMPLE 8 Using the Change-of-Base Rule

Use logarithms and the change-of-base rule to evaluate each expression. Round to four decimal places.

(a) $\log_5 17$ **(b)** $\log_2 .1$

Solution With the calculator set to four decimal places, Figure 31 shows how the results can be found by using either common or natural logarithms. The results are the same.

(a) $\log_5 17 = \dfrac{\log 17}{\log 5} = \dfrac{\ln 17}{\ln 5} \approx 1.7604$ **(b)** $\log_2 .1 \approx -3.3219$ ■

EXAMPLE 9 Modeling Diversity of Species

One measure of the diversity of species in an ecological community is the *index of diversity*

$$H = -(P_1 \log_2 P_1 + P_2 \log_2 P_2 + \cdots + P_n \log_2 P_n),$$

where $P_1, P_2, \ldots, P_n$ are the proportions of a sample belonging to each of n species found in the sample. Find the index of diversity in a community with two species, one with 90 members and the other with 10.

Solution Since there are a total of 100 members in the community, $P_1 = \frac{90}{100} = .9$ and $P_2 = \frac{10}{100} = .1$. First we use the change-of-base rule to find $\log_2 .9$ and $\log_2 .1$. (We use natural logarithms, but common logarithms could be used instead.)

$$\log_2 .9 = \left(\frac{\ln .9}{\ln 2}\right) \approx -.152 \quad \text{and} \quad \log_2 .1 = \left(\frac{\ln .1}{\ln 2}\right) \approx -3.32$$

Now we substitute to find H.

$$
\begin{aligned}
H &= -(.9 \log_2 .9 + .1 \log_2 .1) \\
&\approx -.9(-.152) - .1(-3.32) \\
&\approx .469
\end{aligned}
$$

The interpretation of this index varies with the number of species involved. For the case with two species, if they are equally distributed, the measure of diversity is 1. If there is little diversity, H is close to 0. In this example, since $H \approx .5$, there is neither great nor little diversity. ■

5.3 Exercises

Concept Check In Exercises 1 and 2, match the logarithm in Column I with its value in Column II. Remember that $\log_a x$ *is the exponent to which a must be raised in order to obtain x.*

I	II
1. (a) $\log_2 16$	**A.** 0
(b) $\log_3 1$	**B.** $\frac{1}{2}$
(c) $\log_{10} .1$	**C.** 4
(d) $\log_2 \sqrt{2}$	**D.** -3
(e) $\log_e \frac{1}{e^2}$	**E.** -1
(f) $\log_{1/2} 8$	**F.** -2

I	II
2. (a) $\log_3 81$	**A.** -2
(b) $\log_3 \frac{1}{3}$	**B.** -1
(c) $\log_{10} .01$	**C.** 0
(d) $\log_6 \sqrt{6}$	**D.** $\frac{1}{2}$
(e) $\log_e 1$	**E.** $\frac{9}{2}$
(f) $\log_3 27^{3/2}$	**F.** 4

For each statement, write an equivalent statement in logarithmic form.

3. $3^4 = 81$

4. $2^5 = 32$

5. $\left(\frac{1}{2}\right)^{-4} = 16$

6. $\left(\frac{2}{3}\right)^{-3} = \frac{27}{8}$

7. $10^{-4} = .0001$

8. $\left(\frac{1}{100}\right)^{-2} = 10{,}000$

9. $e^0 = 1$

10. $e^{1/3} = \sqrt[3]{e}$

For each statement, write an equivalent statement in exponential form.

11. $\log_6 36 = 2$

12. $\log_5 5 = 1$

13. $\log_{\sqrt{3}} 81 = 8$

14. $\log_4 \frac{1}{64} = -3$

15. $\log_{10} .001 = -3$

16. $\log_3 \sqrt[3]{9} = \frac{2}{3}$

17. $\log \sqrt{10} = .5$

18. $\ln e^6 = 6$

Solve each equation.

19. $\log_5 125 = x$

20. $\log_3 81 = x$

21. $\log_x 3^{12} = 24$

22. $\log_x 25 = 6$

23. $\log_6 x = -3$

24. $\log_4 x = -\frac{1}{6}$

25. $\log_x 16 = \frac{4}{3}$

26. $\log_{16/25} x = -\frac{3}{2}$

27. $\log_x .001 = -3$

28. $\log_3(x - 1) = 2$

29. $\log_9 \frac{\sqrt[4]{27}}{3} = x$

30. $\log_{1/4} \frac{16^2}{2^{-3}} = x$

Concept Check Simplify each expression.

31. (a) $3^{\log_3 7}$ **(b)** $4^{\log_4 9}$
(c) $12^{\log_{12} 4}$
(d) $a^{\log_a k}$ $(k > 0, a > 0, a \neq 1)$

32. (a) $\log_3 3^{19}$ **(b)** $\log_4 4^{17}$
(c) $\log_{12} 12^{1/3}$
(d) $\log_a \sqrt{a}$ $(a > 0, a \neq 1)$

33. (a) $\log_3 1$ **(b)** $\log_4 1$
(c) $\log_{12} 1$
(d) $\log_a 1$ $(a > 0, a \neq 1)$

34. **(a)** Explain in your own words the meaning of $\log_a x$. **(b)** In the expression $\log_a x$, why must x be nonnegative?

Concept Check *Evaluate each expression. Do not use a calculator.*

35. $\log 10^{1.5}$ **36.** $\log 10^{4.3}$ **37.** $\log 10^{2/5}$ **38.** $\log 10^{\sqrt{3}}$ **39.** $\ln e^{2/3}$

40. $\ln e^{.5}$ **41.** $\ln e^{\pi}$ **42.** $\ln e^{\sqrt{6}}$ **43.** $3 \ln e^{1.8}$ **44.** $\sqrt{2} \ln e^{\sqrt{2}}$

Use a calculator to find a decimal approximation for each common or natural logarithm.

45. $\log 43$ **46.** $\log 1247$ **47.** $\log .783$ **48.** $\log .014$

49. $\log 28^3$ **50.** $\log(47 \times 93)$ **51.** $\ln 43$ **52.** $\ln 1247$

53. $\ln .783$ **54.** $\ln .014$ **55.** $\ln 28^3$ **56.** $\ln(47 \times 93)$

Refer to Example 3. For each substance, find the pH from the given hydronium ion $[H_3O^+]$ *concentration.*

57. Grapefruit, 6.3×10^{-4} **58.** Limes, 1.6×10^{-2}

59. Crackers, 3.9×10^{-9} **60.** Sodium hydroxide (lye), 3.2×10^{-14}

Refer to Example 3. Find the hydronium ion $[H_3O^+]$ *concentration for each substance with the given pH.*

61. Soda pop, 2.7 **62.** Wine, 3.4 **63.** Beer, 4.8 **64.** Drinking water, 6.5

Suppose that \$2500 is invested in an account that pays interest compounded continuously. Find the amount of time that it would take for the account to grow to the given amount at the given rate of interest.

65. \$3000 at 3.75% **66.** \$3500 at 4.25% **67.** \$5000 at 5% **68.** \$5000 at 6%

Use the properties of logarithms to rewrite each logarithm if possible. Assume that all variables represent positive real numbers.

69. $\log_3 \frac{2}{5}$ **70.** $\log_4 \frac{6}{7}$ **71.** $\log_2 \frac{6x}{y}$ **72.** $\log_3 \frac{4p}{q}$

73. $\log_5 \frac{5\sqrt{7}}{3m}$ **74.** $\log_2 \frac{2\sqrt{3}}{5p}$ **75.** $\log_4(2x + 5y)$ **76.** $\log_6(7m + 3q)$

77. $\log_k \frac{pq^2}{m}$ **78.** $\log_z \frac{x^5y^3}{3}$ **79.** $\log_m \sqrt{\frac{r^3}{5z^5}}$ **80.** $\log_p \sqrt[3]{\frac{m^5}{kt^2}}$

Use the properties of logarithms to rewrite each expression as a single logarithm. Assume that all variables represent positive real numbers.

81. $\log_a x + \log_a y - \log_a m$ **82.** $(\log_b k - \log_b m) - \log_b a$ **83.** $2 \log_m a - 3 \log_m b^2$

84. $\frac{1}{2} \log_y p^3q^4 - \frac{2}{3} \log_y p^4q^3$ **85.** $2 \log_a(z - 1) + \log_a(3z + 2), z > 1$ **86.** $\log_b(2y + 5) - \frac{1}{2} \log_b(y + 3)$

87. $-\frac{2}{3} \log_5 5m^2 + \frac{1}{2} \log_5 25m^2$ **88.** $-\frac{3}{4} \log_3 16p^4 - \frac{2}{3} \log_3 8p^3$

Use the change-of-base rule to find an approximation for each logarithm.

89. $\log_5 10$ **90.** $\log_9 12$ **91.** $\log_{15} 5$ **92.** $\log_{1/2} 3$

93. $\log_{100} 83$ **94.** $\log_{200} 175$ **95.** $\log_{2.9} 7.5$ **96.** $\log_{5.8} 12.7$

Relating Concepts

For individual or group investigation (Exercises 97–102)

Work Exercises 97–102 in order.

97. Use the terminology of Chapter 2 to explain how the graph of $y = -3^x + 7$ can be obtained from the graph of $y = 3^x$.

98. Graph $y_1 = 3^x$ and $y_2 = -3^x + 7$ in the window $[-5, 5]$ by $[-10, 10]$ to support your answer in Exercise 97.

99. Use the capabilities of your calculator to find an approximation for the x-intercept of the graph of y_2 in Exercise 98.

100. Solve $0 = -3^x + 7$ for x, expressing x in terms of base 3 logarithms.

101. Use the change-of-base rule to find an approximation for the solution of the equation in Exercise 100.

102. Compare your results in Exercises 99 and 101.

(Modeling) *Solve each problem.*

103. ***Diversity of Species*** Suppose a sample of a small community shows two species with 50 individuals each. Find the index of diversity

$$H = -(P_1 \log_2 P_1 + P_2 \log_2 P_2 + \cdots + P_n \log_2 P_n).$$

104. ***Diversity of Species*** A virgin forest in northwestern Pennsylvania has four species of large trees with the following proportions of each: hemlock, .521; beech, .324; birch, .081; maple, .074. Find the index of diversity H, using the formula in Exercise 103.

105. ***Diversity of Species*** The number of species in a sample is approximated by

$$S(n) = a \ln\left(1 + \frac{n}{a}\right),$$

where n is the number of individuals in the sample and a is a constant that indicates the diversity of species in the community. If $a = .36$, find $S(n)$ for each value of n. (*Hint:* $S(n)$ must be a whole number.)

(a) 100 **(b)** 200 **(c)** 150 **(d)** 10

106. **(a)** Prove the quotient rule of logarithms.

(b) Prove the power rule of logarithms.

Reviewing Basic Concepts (Sections 5.1–5.3)

1. Is the function defined by the table a one-to-one function? Explain your answer.

x	-2	2	4	6	8
y	4	4	8	12	16

2. Give the inverse of each function defined.

(a)

x	7	8	9	10
y	12	21	32	45

(b) $f(x) = \dfrac{x + 5}{4}$

3. Graph $f(x) = 2x + 3$ and $f^{-1}(x)$ by hand on the same axes.

4. Graph $f(x) = 3^{-x}$ by hand.

5. Solve $4^{2x} = 8$ analytically. Use a calculator to support your answer.

6. Find the interest earned on \$600 at 4% compounded quarterly for 3 years.

7. Evaluate each logarithm without using a calculator.

(a) $\log \dfrac{1}{\sqrt{10}}$ **(b)** $2 \ln e^{1.5}$ **(c)** $\log_2 4$

8. Use the properties of logarithms to rewrite $\log \dfrac{3x^2}{5y}$.

9. Use the properties of logarithms to write

$$\ln 4 + \ln x - 3 \ln 2$$

as a single logarithm.

10. How long will it take for \$1500 deposited at 4.5% compounded continuously to grow to \$1650? Give the answer to the nearest tenth of a year.

5.4 Logarithmic Functions

Graphs of Logarithmic Functions ■ Applying Earlier Work to Logarithmic Functions ■ A Logarithmic Model

The exponential function defined by $f(x) = a^x$, $a > 1$, is increasing on its entire domain. If $0 < a < 1$, the function is decreasing on its entire domain. Therefore, for all allowable bases a of $f(x) = a^x$, the graph passes the horizontal line test and is one-to-one, so f has an inverse.

We can find the rule for f^{-1} analytically, using the steps described in Section 5.1.

$f(x) = a^x$	Exponential function
$y = a^x$	Replace $f(x)$ with y.
$x = a^y$	Interchange x and y.
$y = \log_a x$	Write in logarithmic form.
$f^{-1}(x) = \log_a x$	Replace y with $f^{-1}(x)$.

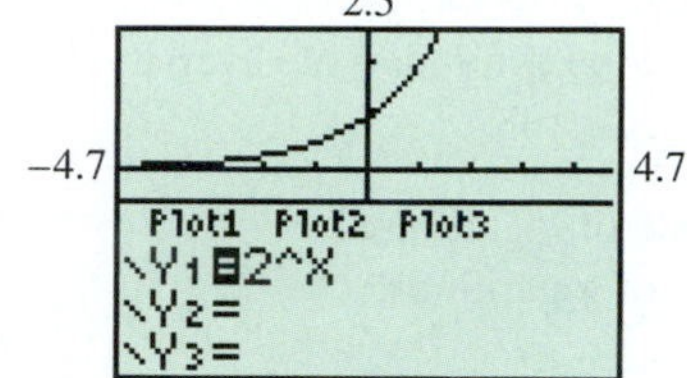

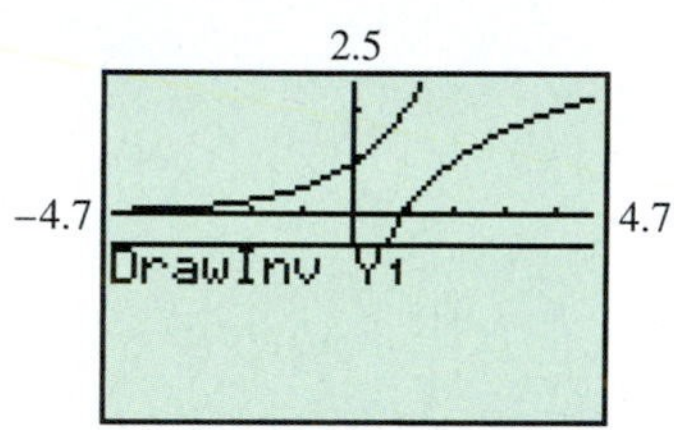

When directed to draw the inverse of $Y_1 = 2^X$, the calculator graphs $y = \log_2 x$.

The last equation indicates that the **logarithmic function** with base a is the inverse of the exponential function with base a. To confirm this, use results from Section 5.3 to show that $(f \circ f^{-1})(x) = x$ and $(f^{-1} \circ f)(x) = x$.

$$(f \circ f^{-1})(x) = f[f^{-1}(x)] = a^{\log_a x} = x$$

$$(f^{-1} \circ f)(x) = f^{-1}[f(x)] = \log_a a^x = x$$

Thus, the functions defined by $\boldsymbol{f(x) = a^x}$ and $\boldsymbol{g(x) = \log_a x}$ are inverse functions.

Graphs of Logarithmic Functions

Recall from Section 5.1 that the graph of the inverse of a one-to-one function can be obtained by reflecting the graph of the function across the line $y = x$. In Examples 1 and 2 of Section 5.2, we graphed $f(x) = 2^x$ and $g(x) = \left(\frac{1}{2}\right)^x$. We can now graph their inverses, $F(x) = \log_2 x$ and $G(x) = \log_{1/2} x$.

EXAMPLE 1 Graphing Logarithmic Functions

Graph each logarithmic function, and determine its domain and range.

(a) $F(x) = \log_2 x$ **(b)** $G(x) = \log_{1/2} x$

Solution

(a) Interchange the x- and y-coordinates in the table for $f(x) = 2^x$ from Section 5.2, and plot the new data to obtain the graph of $F(x) = \log_2 x$. The domain of F is $(0, \infty)$, which is the range of f. Similarly, the range of F is $(-\infty, \infty)$, which is the domain of f. See Figure 32.

(b) Use the table for $g(x) = \left(\frac{1}{2}\right)^x$ from Section 5.2 and a similar procedure to get the graph of $G(x) = \log_{1/2} x$. The domain of G is $(0, \infty)$ and the range is $(-\infty, \infty)$. See Figure 33.

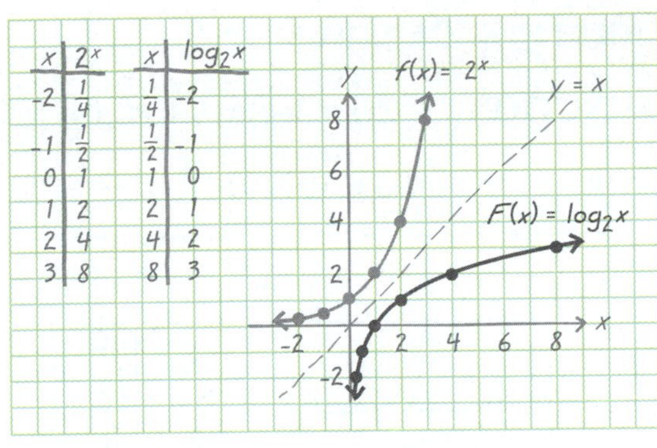

FIGURE 32

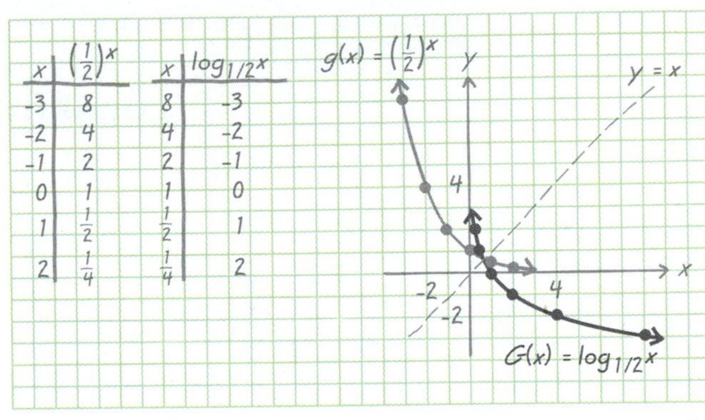

FIGURE 33

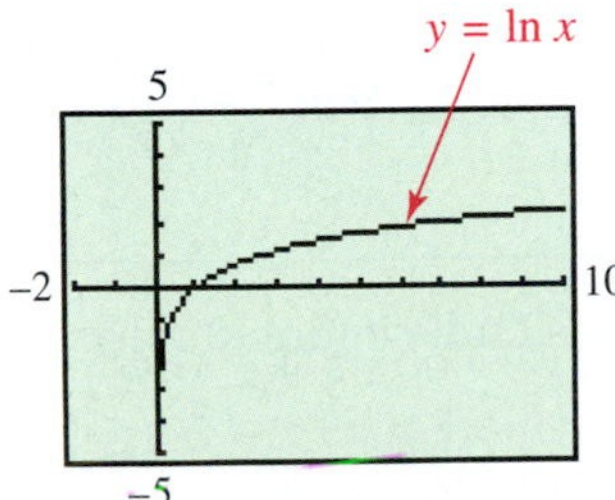

FIGURE 34

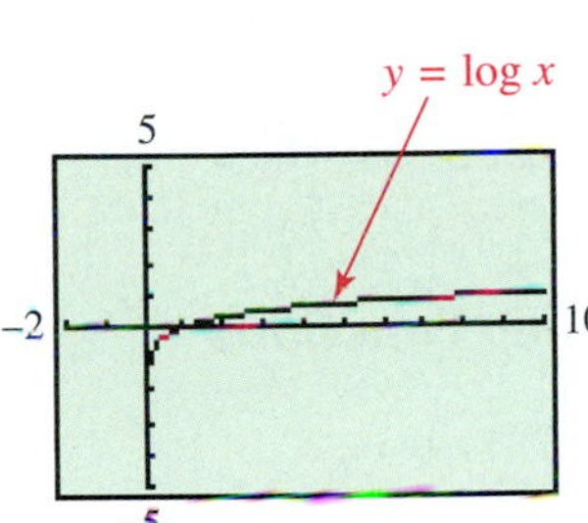

FIGURE 35

NOTE Another way to graph a logarithmic function, such as $y = \log_2 x$, is to write it in equivalent exponential form, $2^y = x$. Then substitute values for y and solve for x. ***Be careful to plot the ordered pairs correctly when using this method.***

The most important logarithmic functions are $y = \ln x$ (base e) and $y = \log x$ (base 10), which are graphed on a graphing calculator in Figures 34 and 35.

Because e and 10 are the only logarithmic bases on most graphing calculators, we must use the change-of-base rule to graph a logarithmic function for some other base. For example, to graph $y = \log_2 x$, we graph either

$$y = \frac{\log x}{\log 2} \quad \text{or} \quad y = \frac{\ln x}{\ln 2}.$$

We now summarize information about the graphs of logarithmic functions.

GCM TECHNOLOGY NOTE

In Section 5.2 we saw that it is difficult to interpret from a calculator screen the behavior of the graph of an exponential function as it approaches its horizontal asymptote. A similar problem occurs for graphs of logarithmic functions near their vertical asymptotes. In Figures 34 and 35, be aware that there are no endpoints and that as $x \to 0$ from the right, $y \to -\infty$.

FUNCTION CAPSULE

LOGARITHMIC FUNCTION $f(x) = \log_a x, \quad a > 1$

Domain: $(0, \infty)$ Range: $(-\infty, \infty)$

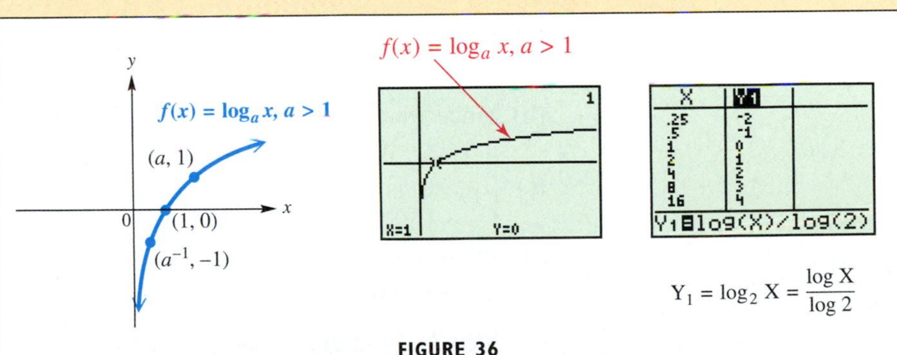

FIGURE 36

- $f(x) = \log_a x$, $a > 1$, is increasing and continuous on its entire domain, $(0, \infty)$.
- The y-axis is the vertical asymptote as $x \to 0$ from the right.
- The graph passes through the points $(a^{-1}, -1)$, $(1, 0)$, and $(a, 1)$.

FUNCTION CAPSULE

LOGARITHMIC FUNCTION $f(x) = \log_a x, \quad 0 < a < 1$

Domain: $(0, \infty)$ Range: $(-\infty, \infty)$

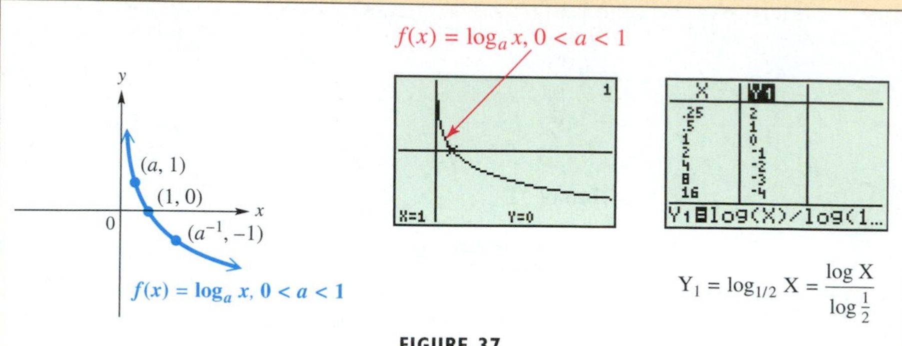

FIGURE 37

- $f(x) = \log_a x$, $0 < a < 1$, is decreasing and continuous on its entire domain, $(0, \infty)$.
- The y-axis is the vertical asymptote as $x \to 0$ from the right.
- The graph passes through the points $(a, 1)$, $(1, 0)$, and $(a^{-1}, -1)$.

A function of the form $y = \log_a f(x)$ is defined only for values for which $f(x) > 0$.

EXAMPLE 2 **Determining Domains of Logarithmic Functions**

Find the domain of each function.

(a) $f(x) = \log_2(x - 1)$ **(b)** $f(x) = (\log_3 x) - 1$

(c) $f(x) = \log_3|x|$ **(d)** $f(x) = \ln(x^2 - 4)$

Solution

(a) Since the argument of a logarithm must be positive, $x - 1 > 0$; thus, $x > 1$. The domain is $(1, \infty)$.

(b) Since we are interested in the logarithm to the base 3 of x (not $x - 1$), we must have $x > 0$. The domain is $(0, \infty)$.

(c) $|x| > 0$ is true for all real numbers x except 0. Therefore, the domain is $(-\infty, 0) \cup (0, \infty)$.

(d) We use the steps first introduced in Section 3.3 to solve the quadratic inequality $x^2 - 4 > 0$.

Step 1 Find the values of x that satisfy $x^2 - 4 = 0$.

$$x^2 - 4 = 0 \quad \text{Corresponding quadratic equation}$$

$$(x + 2)(x - 2) = 0 \quad \text{Factor.}$$

$$x + 2 = 0 \quad \text{or} \quad x - 2 = 0 \quad \text{Zero-product property}$$

$$x = -2 \quad \text{or} \quad x = 2$$

Step 2 The two numbers -2 and 2 divide a number line into the three intervals shown in Figure 38.

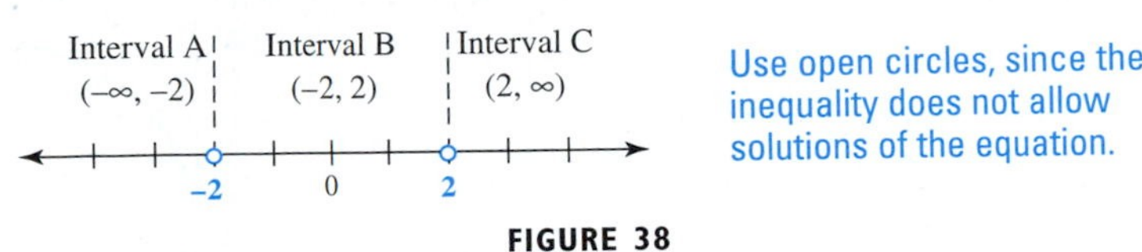

Use open circles, since the inequality does not allow solutions of the equation.

FIGURE 38

Step 3 Choose a test value in each interval to see if it satisfies the original inequality $x^2 - 4 > 0$.

Interval	Test Value	Is $x^2 - 4 > 0$ True or False?
A: $(-\infty, -2)$	-3	$(-3)^2 - 4 > 0$? $5 > 0$ True
B: $(-2, 2)$	0	$0^2 - 4 > 0$? $-4 > 0$ False
C: $(2, \infty)$	3	$3^2 - 4 > 0$? $5 > 0$ True

Since the values in Intervals A and C make the inequality true, the domain of $f(x) = \ln(x^2 - 4)$ is the union of those two intervals,

$$(-\infty, -2) \cup (2, \infty).$$ ■

EXAMPLE 3 **Graphing Translated Logarithmic Functions**

Graph each function. Give the domain, range, asymptote, and x-intercept, and tell whether the function is increasing or decreasing on its domain.

(a) $y = \log_2(x - 1)$ **(b)** $y = (\log_3 x) - 1$

Solution

X | Y1
-.5 | ERROR
0 | ERROR
.999 | ERROR
1 | ERROR
1.001 | -9.966
2 | 0
3 | 1

Y1 = log(X–1)/log…

This table numerically supports the result in Example 3(a): the domain of $y = \log_2(x - 1)$, entered here as $Y_1 = \log(X - 1)/\log 2$, consists of real numbers greater than 1.

(a) Because the argument is $x - 1$, the graph of $y = \log_2(x - 1)$ is the graph of $y = \log_2 x$ shifted 1 unit to the right. The vertical asymptote also moves 1 unit to the right, so its equation is $x = 1$. The x-intercept is 2. The domain is $(1, \infty)$, as found in Example 2(a), and the range is $(-\infty, \infty)$. The function is always increasing on its domain. See Figure 39 on the next page.

This graph can also be plotted by using the equivalent exponential form.

$$y = \log_2(x - 1)$$

$$x - 1 = 2^y \quad \text{Write in exponential form.}$$

$$x = 2^y + 1 \quad \text{Add 1.}$$

Using this equation, we choose values for y and then calculate each of the corresponding x-values.

x	y
$\frac{5}{4}$	-2
$\frac{3}{2}$	-1
2	0
3	1
5	2

In the equation $x = 2^y + 1$, choose values for y and calculate x.

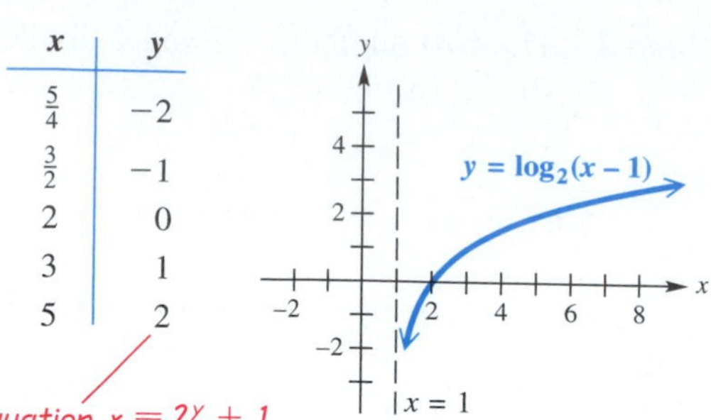

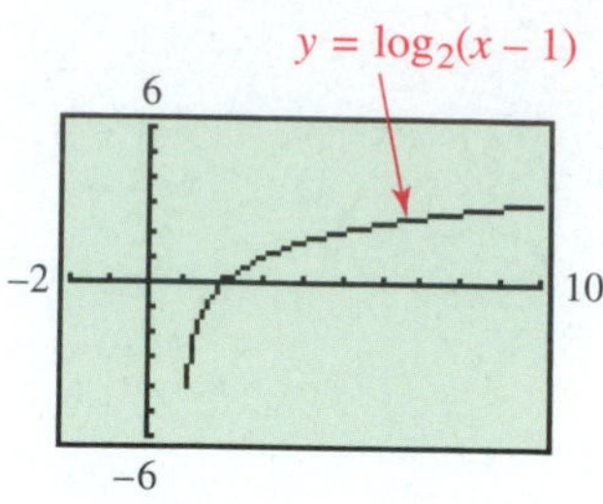

This calculator graph was obtained by using
$$y = \frac{\log(x-1)}{\log 2}.$$

FIGURE 39

(b) Here, 1 is subtracted from $\log_3 x$, so the graph of $y = \log_3 x$ is shifted 1 unit downward. The vertical asymptote—the y-axis, or $x = 0$—is not affected. The x-intercept of $y = (\log_3 x) - 1$ is 3. The domain is $(0, \infty)$, from Example 2(b), and the range is $(-\infty, \infty)$. The function is always increasing. See Figure 40.

The alternative method described in part (a) can also be used.

$$y = (\log_3 x) - 1$$
$$y + 1 = \log_3 x \qquad \text{Add 1.}$$
$$x = 3^{y+1} \qquad \text{Write in exponential form.}$$

Again, choose y-values and calculate the corresponding x-values.

x	y
$\frac{1}{3}$	-2
1	-1
3	0
9	1

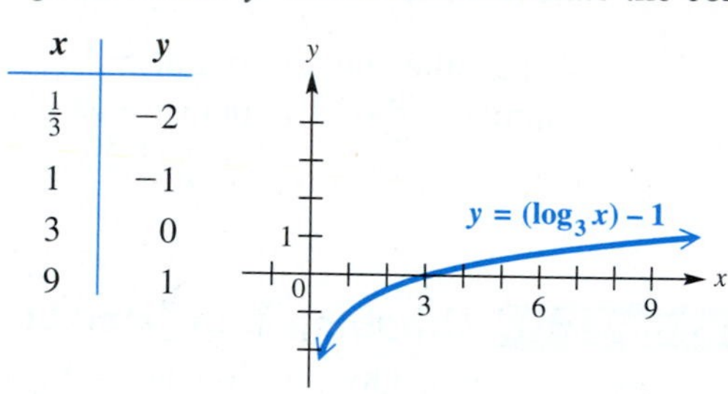

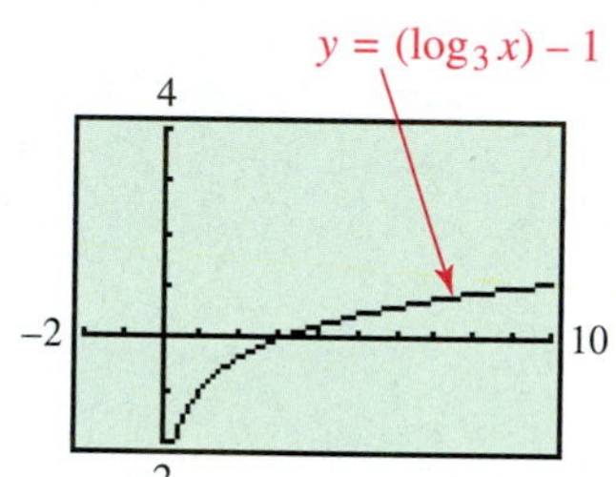

This calculator graph was obtained by using
$$y = \frac{\log x}{\log 3} - 1.$$

FIGURE 40

EXAMPLE 4 Determining Symmetry

Show that the graph of each function is symmetric with respect to the y-axis.

(a) $f(x) = \log_3|x|$ **(b)** $f(x) = \ln(x^2 - 4)$

Solution

(a) For all real numbers x, $|-x| = |x|$. Therefore,

$$f(-x) = \log_3|-x| = \log_3|x| = f(x).$$

Because $f(-x) = f(x)$, the graph is symmetric with respect to the y-axis, which serves as both vertical asymptote and axis of symmetry. See Figure 41. As we saw in Example 2(c), the domain is $(-\infty, 0) \cup (0, \infty)$.

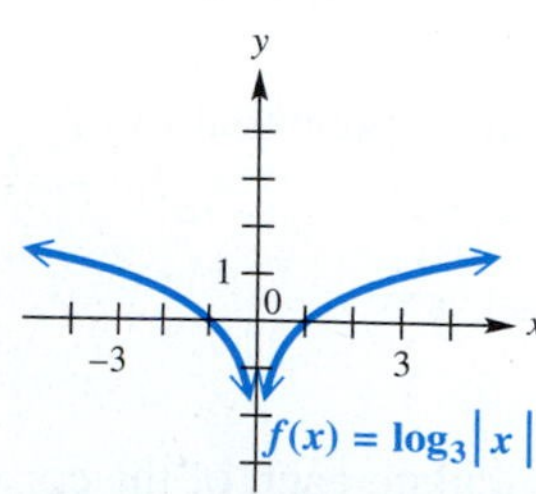

FIGURE 41

FOR DISCUSSION

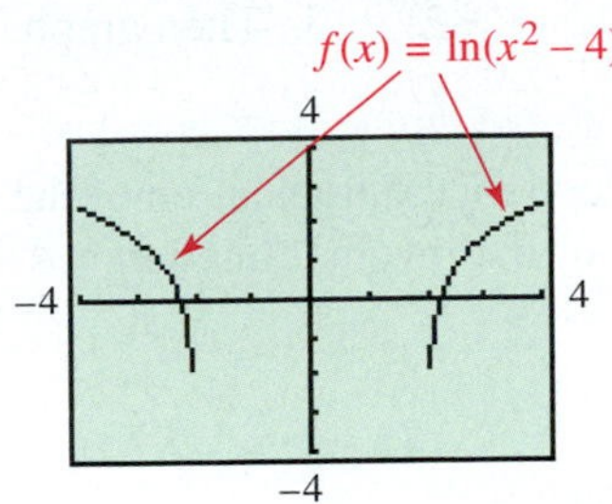

This calculator graph of $f(x) = \ln(x^2 - 4)$, discussed in Example 4(b), could easily be misinterpreted by someone not familiar with the underlying mathematical concepts. The lines $x = -2$ and $x = 2$ are vertical asymptotes. To suggest this conclusion, use a calculator and do the following:

1. Evaluate $f(2.1)$, $f(2.01)$, $f(2.001)$, and $f(2.0001)$ for $f(x) = \ln(x^2 - 4)$. Describe what happens to $f(x)$ as $x \to 2$ from the right.
2. Repeat Item 1 for $f(-2.1)$, $f(-2.01)$, $f(-2.001)$, and $f(-2.0001)$. Describe what happens to $f(x)$ as $x \to -2$ from the left.

(b) Since $x^2 = (-x)^2$ for all x,

$$\ln[(-x)^2 - 4] = \ln(x^2 - 4).$$

Thus, $f(-x) = f(x)$, and the graph is symmetric with respect to the y-axis, which is the axis of symmetry of the graph. The domain is $(-\infty, -2) \cup (2, \infty)$ from Example 2(d). See Figure 42.

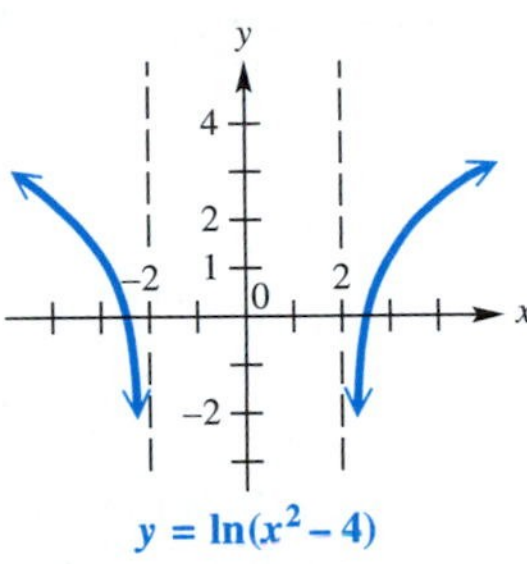

FIGURE 42

■

What Went Wrong?

A student wanted to support the power rule for logarithms (Property 6 from Section 5.3): $\log_a x^r = r\log_a x$. The student defined Y_1 and Y_2, as shown in the screen on the left, and expected the two graphs to be the same. However, the graph of Y_1 was different from the graph of Y_2, as shown.

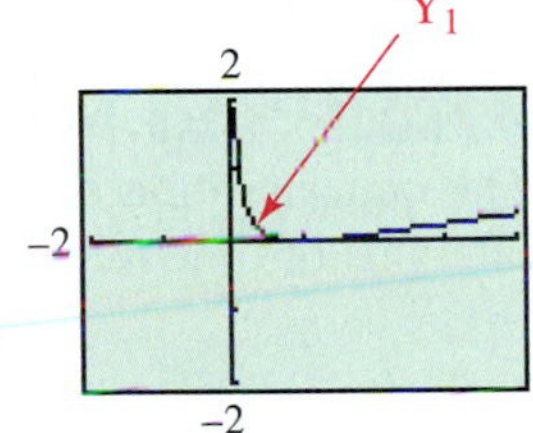

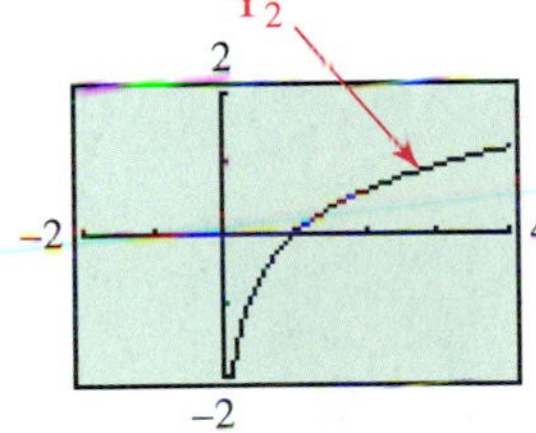

What Went Wrong? How can the student change the input for Y_1 to obtain the same graph as that of Y_2?

Applying Earlier Work to Logarithmic Functions

EXAMPLE 5 Using a Property of Logarithms to Translate a Graph

Describe how to graph $f(x) = \log 8x$ by translation.

Solution By the product rule for logarithms, $\log 8x = \log 8 + \log x$, so we can obtain the graph of $y_2 = \log 8x$ by shifting the graph of $y_1 = \log x$ upward $\log 8 \approx .90309$ unit. Figure 43 shows the graph. ■

$y_1 = \log x$ $y_2 = \log 8x$

The vertical distance between y_2 and y_1 is log 8.

FIGURE 43

Answer to What Went Wrong?

With some calculators, the first equation should be entered as $Y_1 = (\log(X^2))(X > 0)$.

EXAMPLE 6 Finding the Inverse of an Exponential Function

Find analytically the inverse of the function defined by $f(x) = -2^x + 3$. Then graph, and discuss the relationship between the two graphs.

Solution The function is one-to-one because it is a reflection and translation of the one-to-one function defined by $y = 2^x$. Find the equation of the inverse function.

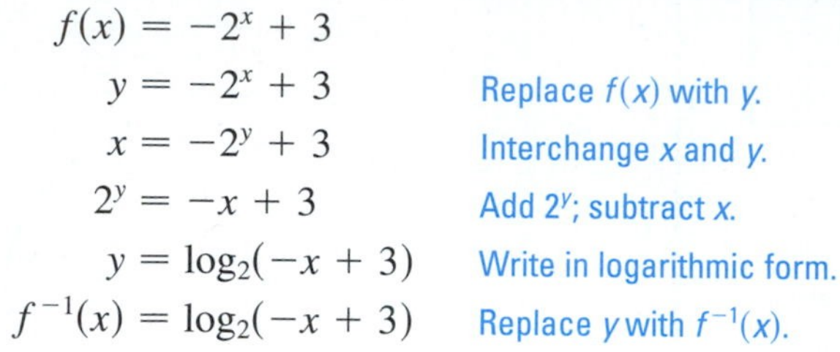

$$f(x) = -2^x + 3$$

$$y = -2^x + 3 \quad \text{Replace } f(x) \text{ with } y.$$

$$x = -2^y + 3 \quad \text{Interchange } x \text{ and } y.$$

$$2^y = -x + 3 \quad \text{Add } 2^y\text{; subtract } x.$$

$$y = \log_2(-x + 3) \quad \text{Write in logarithmic form.}$$

$$f^{-1}(x) = \log_2(-x + 3) \quad \text{Replace } y \text{ with } f^{-1}(x).$$

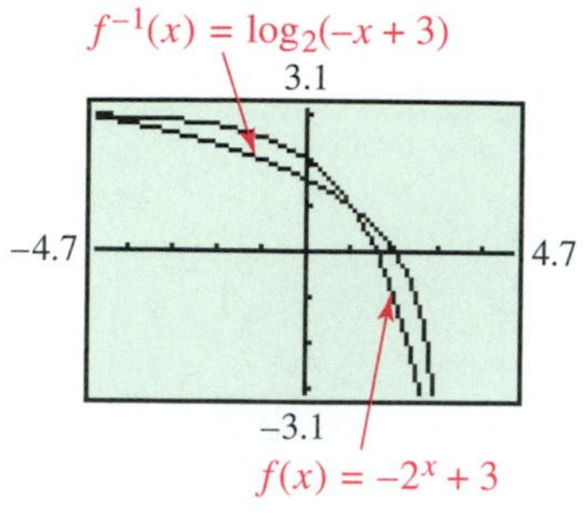

FIGURE 44

Figure 44 shows the graphs of both f and f^{-1}. The table compares some of the features of these inverses. Notice how the roles of x and y are reversed in f and f^{-1}.

Function	Domain	Range	x-intercept	y-intercept	Asymptote
$f(x) = -2^x + 3$	$(-\infty, \infty)$	$(-\infty, 3)$	$\log_2 3 \approx 1.58$	2	Horizontal: $y = 3$
$f^{-1}(x) = \log_2(-x + 3)$	$(-\infty, 3)$	$(-\infty, \infty)$	2	$\log_2 3 \approx 1.58$	Vertical: $x = 3$

■

A Logarithmic Model

EXAMPLE 7 Modeling Drug Concentration

The concentration of a drug injected into the bloodstream decreases with time. The intervals of time in hours when the drug should be administered are given by

$$T = \frac{1}{k} \ln \frac{C_2}{C_1},$$

where k is a constant determined by the drug in use, C_2 is the concentration at which the drug is harmful, and C_1 is the concentration below which the drug is ineffective. (*Source:* Horelick, B. and S. Koont, "Applications of Calculus to Medicine: Prescribing Safe and Effective Dosage," *UMAP Module 202,* 1977.) Thus, if $T = 4$, the drug should be administered every 4 hours.

For a certain drug, $k = \frac{1}{3}$, $C_2 = 5$, and $C_1 = 2$. How often should the drug be administered?

Solution Substitute the given values in the equation of the function.

$$T = \frac{1}{k} \ln \frac{C_2}{C_1} = \frac{1}{\frac{1}{3}} \ln \frac{5}{2} \approx 2.75$$

The drug should be given about every $2\frac{3}{4}$ hours. ■

5.4 Exercises

The graph of an exponential function f is given, with three points labeled. Sketch the graph of f^{-1} by hand, labeling three points on the graph. For f^{-1}, also state the domain, the range, whether it increases or decreases on its domain, and the equation of its vertical asymptote.

1.
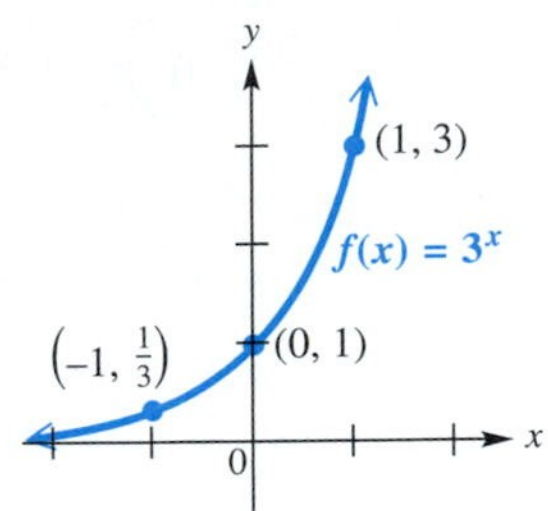

2.
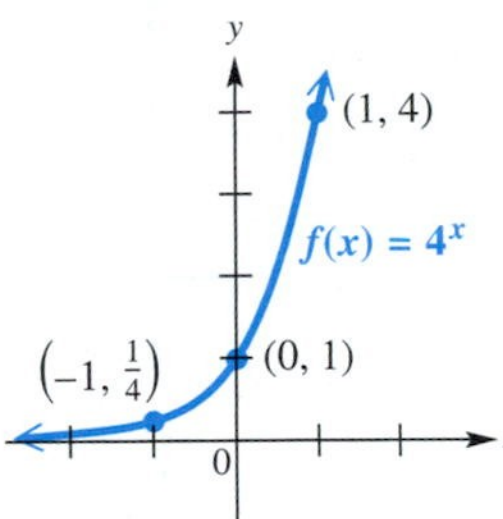

3.
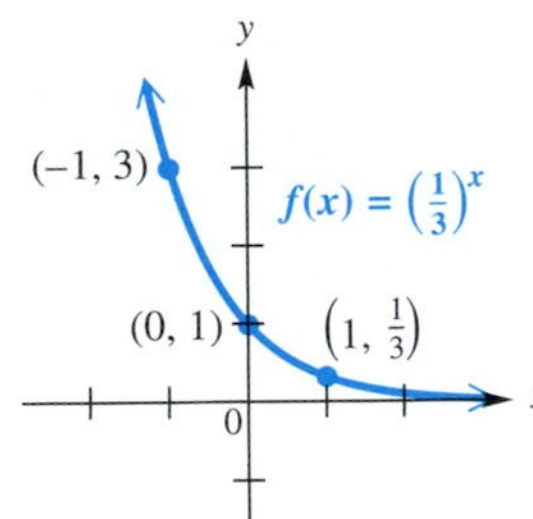

4.
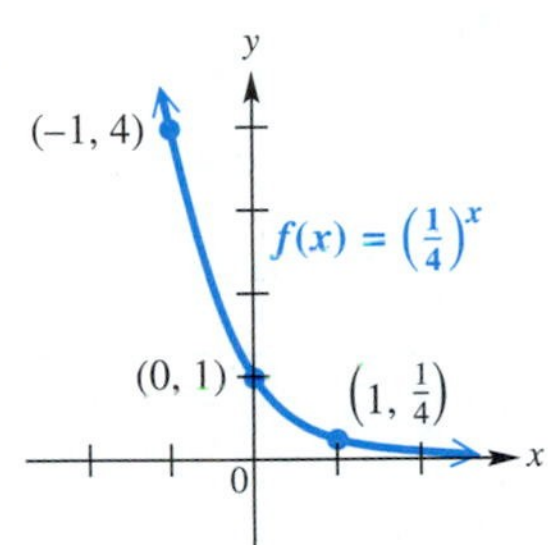

5.
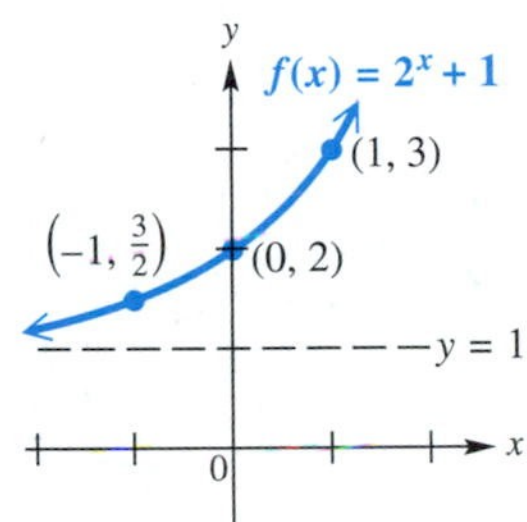

6.
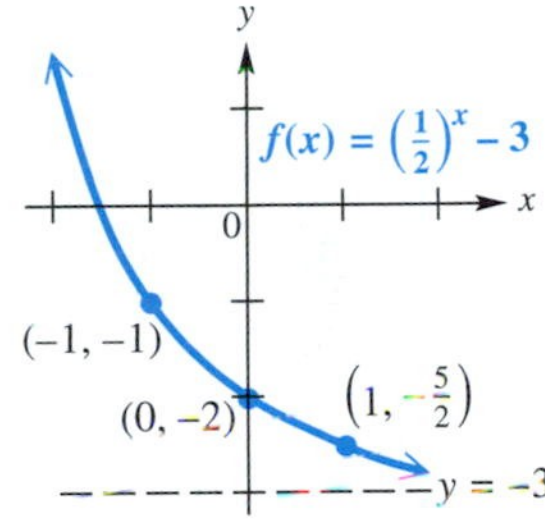

7. In Exercises 1–6, each function f is an exponential function. Therefore, each function f^{-1} is a(n) ________ function.

8. Compare the characteristics of the graph of $f(x) = \log_a x$ with those of the graph of $f(x) = a^x$ in Section 5.2. Make a list of characteristics that reinforce the idea that these are inverse functions.

Find the domain of each logarithmic function analytically. You may wish to support your answer graphically.

9. $y = \log 2x$

10. $y = \log \dfrac{1}{2}x$

11. $y = \log(-x)$

12. $y = \log\left(-\dfrac{1}{2}x\right)$

13. $y = \ln(x^2 + 7)$

14. $y = \ln(x^4 + 8)$

15. $\ln(-x^2 + 4)$

16. $y = \ln(-x^2 + 16)$

17. $y = \log_4(x^2 - 4x - 21)$

18. $y = \log_6(2x^2 - 7x - 4)$

19. $y = \log(x^3 - x)$

20. $y = \log(x^3 - 81x)$

21. $y = \log\left(\dfrac{x + 3}{x - 4}\right)$

22. $y = \log\left(\dfrac{x + 1}{x - 5}\right)$

23. $y = \log|3x - 7|$

24. $y = \log|6x + 6|$

Sketch the graph of $f(x) = \log_2 x$. Then refer to it and use the techniques of Chapter 2 to graph each function.

25. $f(x) = (\log_2 x) + 3$

26. $f(x) = \log_2(x + 3)$

27. $f(x) = |\log_2(x + 3)|$

Sketch the graph of $f(x) = \log_{1/2} x$. Then refer to it and use the techniques of Chapter 2 to graph each function.

28. $f(x) = (\log_{1/2} x) - 2$

29. $f(x) = \log_{1/2}(x - 2)$

30. $f(x) = |\log_{1/2}(x - 2)|$

Concept Check *In Exercises 31–38, match the correct graph in choices A–H to each equation.*

31. $y = e^x + 3$

32. $y = e^x - 3$

33. $y = e^{x+3}$

34. $y = e^{x-3}$

35. $y = \ln x + 3$

36. $y = \ln x - 3$

37. $y = \ln(x - 3)$

38. $y = \ln(x + 3)$

A.

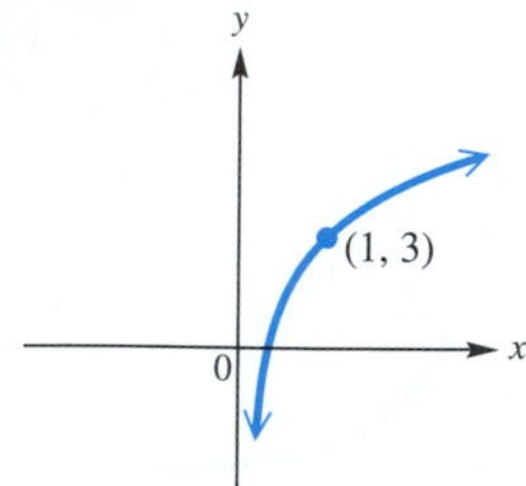

B.

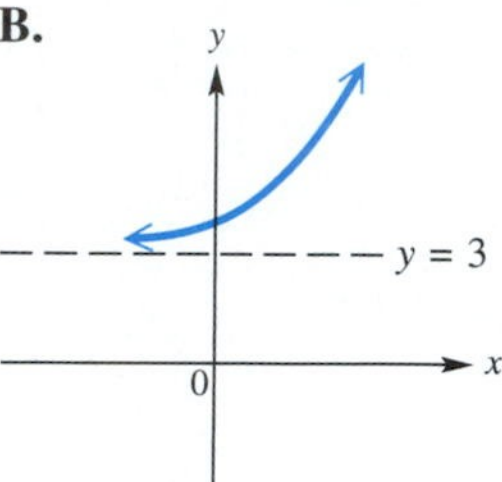

C.

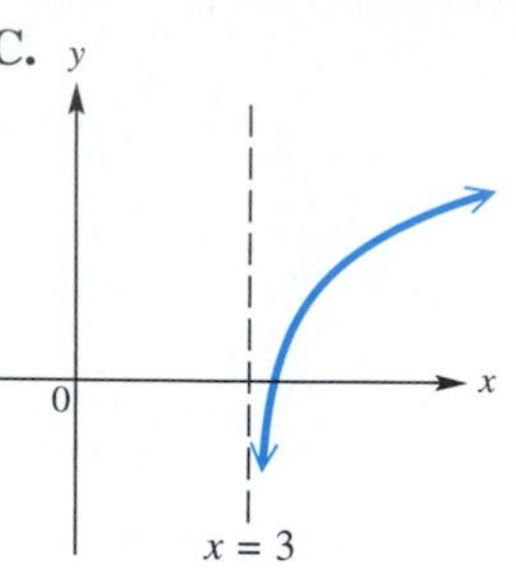

D.

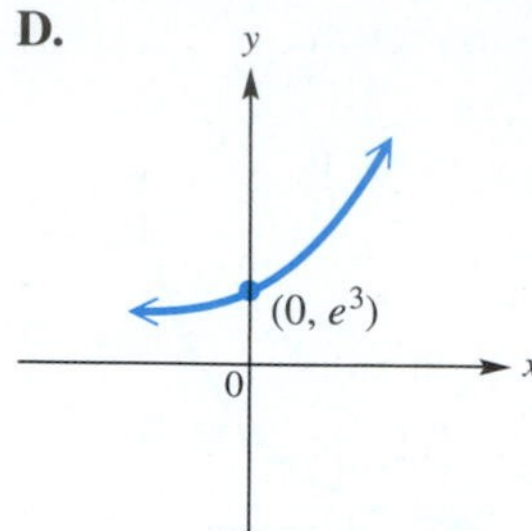

E.

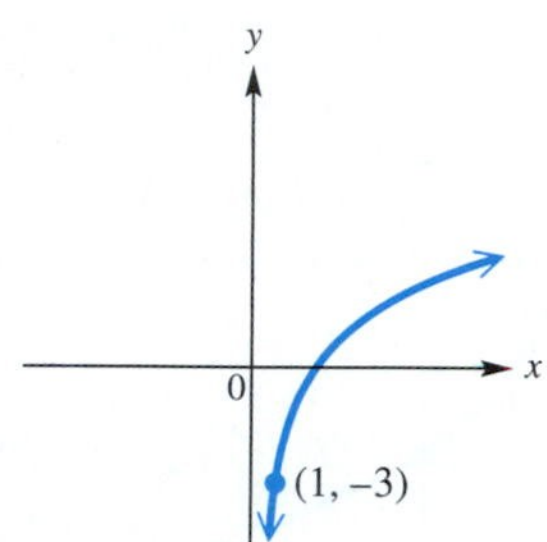

F.

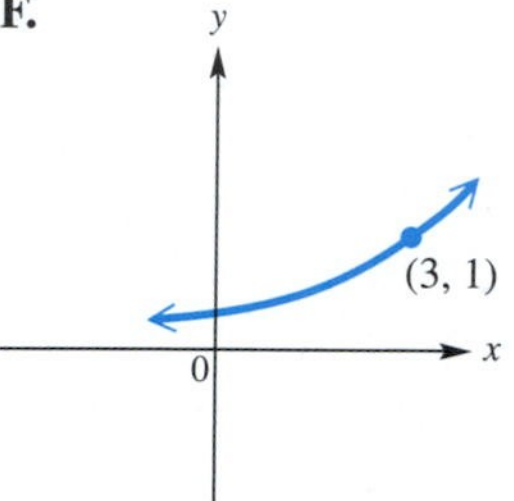

G.

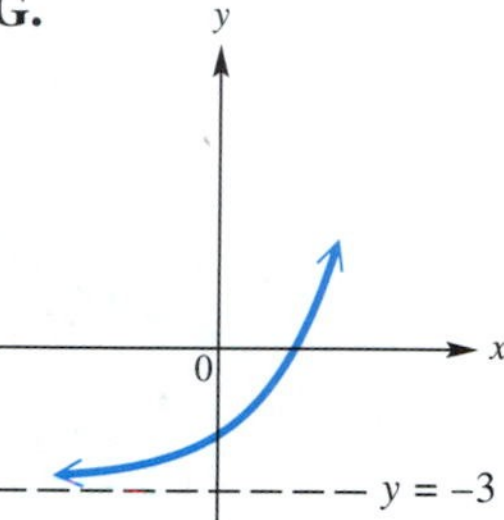

H.

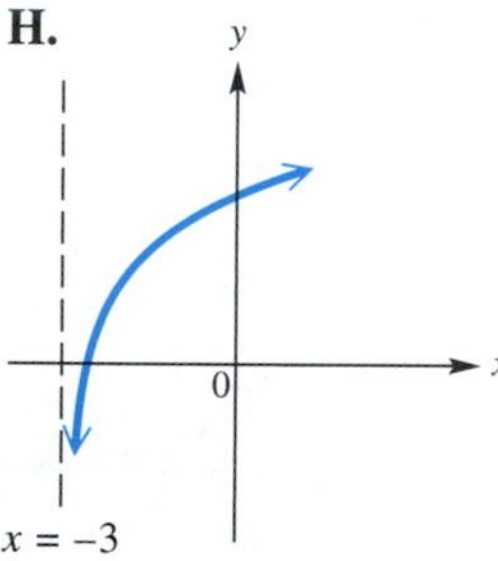

Graph each function.

39. $f(x) = \log_5 x$

40. $f(x) = \log_{10} x$

41. $f(x) = \log_{1/2}(1 - x)$

42. $f(x) = \log_{1/3}(3 - x)$

43. $f(x) = \log_3(x - 1)$

44. $f(x) = \log_2(x^2)$

In Exercises 45–50, **(a)** *explain how the graph of the given function can be obtained from the graph of* $y = \log_2 x$ *and* **(b)** *graph the function.*

45. $y = \log_2(x + 4)$

46. $y = \log_2(x - 6)$

47. $y = 3 \log_2 x + 1$

48. $y = -4 \log_2 x - 8$

49. $y = \log_2(-x) + 1$

50. $y = -\log_2(-x)$

 **51.** Graph $y = \log x^2$ and $y = 2 \log x$ on separate viewing screens. It would seem, at first glance, that by applying the power rule for logarithms, these graphs should be the same. Are they? If not, why not? (*Hint:* Consider the domain in each case.)

 **52.** Graph $f(x) = \log_3|x|$ in the window $[-4, 4]$ by $[-4, 4]$, and compare $f(x)$ with the traditional graph in Figure 41. How might one easily misinterpret the domain of the function simply by observing the calculator graph? What is the domain of this function?

Evaluate each logarithm in three ways: **(a)** *Use the definition of logarithm in Section 5.3 to find the exact value analytically.* **(b)** *Support the result of part (a) by using the change-of-base rule and common logarithms on your calculator.* **(c)** *Support the result of part (a) by locating the appropriate point on the graph of the function* $y = \log_a x$.

53. $\log_9 27$

54. $\log_4\left(\frac{1}{8}\right)$

55. $\log_{16}\left(\frac{1}{8}\right)$

56. $\log_2 \sqrt{8}$

For each exponential function f, find f^{-1} *analytically and graph both f and* f^{-1} *in the same viewing window.*

57. $f(x) = 4^x - 3$

58. $f(x) = \left(\frac{1}{2}\right)^x - 5$

59. $f(x) = -10^x + 4$

60. $f(x) = -e^x + 6$

61. ***Concept Check*** Use the graph to estimate each logarithm to the nearest tenth.

(a) $\log_3 .3$ **(b)** $\log_3 .8$

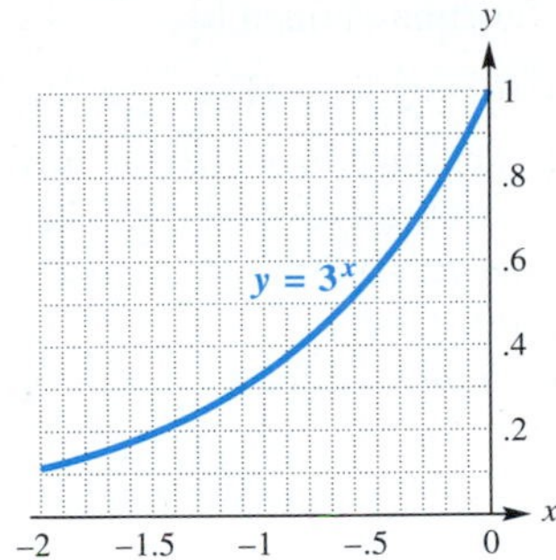

62. ***Concept Check*** Suppose $f(x) = \log_a x$ and $f(3) = 2$. Determine each function value.

(a) $f\left(\frac{1}{9}\right)$ **(b)** $f(27)$ **(c)** $f^{-1}(-2)$ **(d)** $f^{-1}(0)$

63. ***Concept Check*** If (5, 4) is on the graph of the logarithmic function with base a, which is true, $5 = \log_a 4$ or $4 = \log_a 5$?

Use a graphing calculator to solve each equation. Give solutions to the nearest hundredth.

64. $\log_{10} x = x - 2$

65. $2^{-x} = \log_{10} x$

66. $e^x = x^2$

(Modeling) *Solve each problem.*

67. ***Height of the Eiffel Tower*** The right side of Paris's Eiffel Tower has a shape that can be approximated by the graph of the function defined by

$$f(x) = -301 \ln \frac{x}{207}.$$

(*Source:* Banks, Robert B., *Towing Icebergs, Falling Dominoes, and Other Adventures in Applied Mathematics,* Princeton University Press, 1998.)

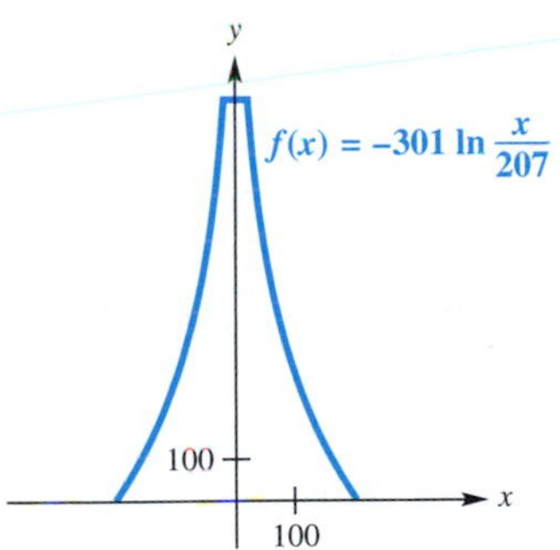

(a) Explain why the shape of the left side of the Eiffel Tower has the formula given by $f(-x)$.

(b) The short horizontal line at the top of the figure has length 15.7488 feet. Approximately how tall is the Eiffel Tower?

(c) Approximately how far from the center of the tower is the point on the right side that is 500 feet above the ground?

68. ***Age of a Whale*** The age in years of a female blue whale is approximated by

$$t = -2.57 \ln\left(\frac{87 - L}{63}\right),$$

where L is its length in feet.

(a) How old is a female blue whale that measures 80 feet?

(b) What is the length of a female blue whale that is 4 years old?

(c) The equation that defines t has domain $24 < L < 87$. Explain why.

69. ***Barometric Pressure*** The function defined by

$$f(x) = 27 + 1.105 \log(x + 1)$$

approximates the barometric pressure in inches of mercury at a distance of x miles from the eye of a typical hurricane. (*Source:* Miller, A. and R. Anthes, *Meteorology,* Fifth Edition, Charles E. Merrill, 1985.)

(a) Approximate the pressure 9 miles from the eye of the hurricane.

(b) The ordered pair $(99, 29.21)$ belongs to this function. What information does it convey?

70. ***Race Speed*** At the IAAF Grand Prix final races held in Paris on September 14, 2002, American sprinter Tim Montgomery ran the 100-meter race in 9.78 seconds. His speed in meters per second after t seconds is closely modeled by the function defined by

$$f(t) = 11.65(1 - e^{-t/1.27}).$$

(*Source:* Banks, Robert B., *Towing Icebergs, Falling Dominoes, and Other Adventures in Applied Mathematics,* Princeton University Press, 1998; *USA Today.*)

(a) How fast was he running as he crossed the finish line?

(b) After how many seconds was he running at the rate of 10 meters per second?

5.5 Exponential and Logarithmic Equations and Inequalities

Exponential Equations and Inequalities (Type 2) ■ Logarithmic Equations and Inequalities ■ Equations and Inequalities Involving Both Exponentials and Logarithms ■ Formulas Involving Exponentials and Logarithms

General methods for solving exponential and logarithmic equations depend on the following properties, which are based on the fact that exponential and logarithmic functions are, in general, one-to-one. Property 1 below was used in Section 5.2 to solve Type 1 exponential equations.

Properties of Logarithmic and Exponential Functions

For $b > 0$ and $b \neq 1$,

1. $b^x = b^y$ **if and only if** $x = y$.
2. If $x > 0$ and $y > 0$, then $\log_b x = \log_b y$ **if and only if** $x = y$.

Exponential Equations and Inequalities (Type 2)

A **Type 2 exponential equation** or **inequality** is one in which the exponential expressions *cannot* easily be written as powers of the same base. Examples of Type 2 equations are

$$7^x = 12 \quad \text{and} \quad 2^{3x+1} = 3^{4-x}.$$

The general strategy in solving these equations is to use Property 2 by taking the same base logarithm of each side (usually either common or natural) and then applying the power rule for logarithms to eliminate the variable exponents. Then, the equation is solved using algebraic techniques.

EXAMPLE 1 **Solving a Type 2 Exponential Equation**

Solve $7^x = 12$.

Analytic Solution

Property 1 cannot be used to solve this equation, so we apply Property 2. While any appropriate base b can be used to apply Property 2, the most practical is base 10 or base e.

$$7^x = 12$$

$$\ln 7^x = \ln 12 \quad \text{Take base } e \text{ logarithms.}$$

$$x \ln 7 = \ln 12 \quad \text{Power rule}$$

$$x = \frac{\ln 12}{\ln 7} \quad \text{Divide by ln 7.}$$

The expression $\frac{\ln 12}{\ln 7}$ is the *exact* solution of $7^x = 12$. Had we used common logarithms, the solution would have the form $\frac{\log 12}{\log 7}$. In either case, we use a calculator to find a decimal approximation.

$$\frac{\ln 12}{\ln 7} = \frac{\log 12}{\log 7} \approx 1.277 \quad \text{Nearest thousandth}$$

The exact solution set is $\left\{\frac{\ln 12}{\ln 7}\right\}$ or $\left\{\frac{\log 12}{\log 7}\right\}$, while $\{1.277\}$ is an approximate solution set.

Graphing Calculator Solution

Using the intersection-of-graphs method, we graph $y_1 = 7^x$ and $y_2 = 12$. The x-coordinate of the point of intersection is approximately 1.277, as seen in Figure 45(a). Figure 45(b) illustrates the x-intercept method of solution. The x-intercept of $y = 7^x - 12$ is also approximately 1.277.

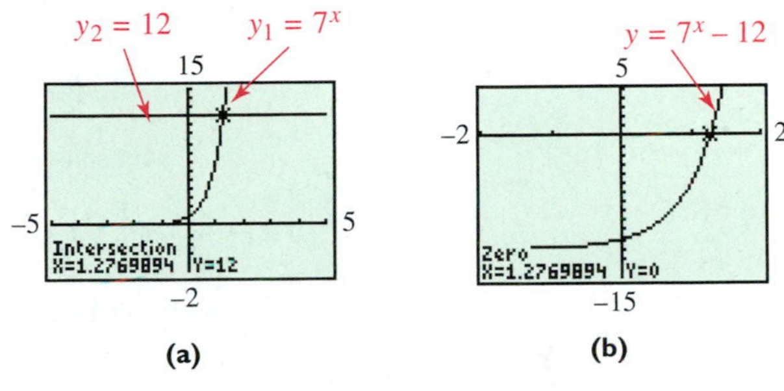

FIGURE 45

To check the analytic solution, we can also raise 7 to either of the powers $\frac{\ln 12}{\ln 7}$ or $\frac{\log 12}{\log 7}$ to obtain a result of 12. See Figure 46.

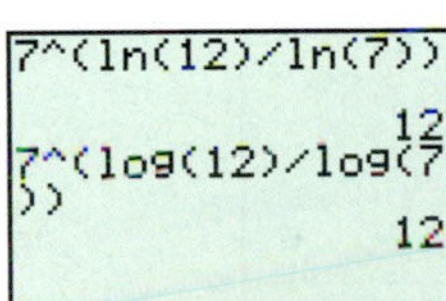

FIGURE 46

EXAMPLE 2 **Solving Type 2 Exponential Inequalities**

(a) Use Figure 45(a) to solve $7^x < 12$. **(b)** Use Figure 45(b) to solve $7^x > 12$.

Solution

(a) Because the graph of $y_1 = 7^x$ is below the graph of $y_2 = 12$ for all x-values less than $\frac{\ln 12}{\ln 7} \approx 1.277$, the solution set is

$$\left(-\infty, \frac{\ln 12}{\ln 7}\right) \quad \text{or} \quad (-\infty, 1.277).$$

(b) This inequality is equivalent to $7^x - 12 > 0$. Because the graph of $y = 7^x - 12$ is above the x-axis for all values of x greater than $\frac{\ln 12}{\ln 7} \approx 1.277$, the solution set is

$$\left(\frac{\ln 12}{\ln 7}, \infty\right) \quad \text{or} \quad (1.277, \infty).$$

EXAMPLE 3 **Solving a Type 2 Exponential Equation**

Solve $2^{3x+1} = 3^{4-x}$.

Solution

$$\begin{aligned}
2^{3x+1} &= 3^{4-x} \\
\log 2^{3x+1} &= \log 3^{4-x} && \text{Take common logarithms.} \\
(3x+1)\log 2 &= (4-x)\log 3 && \text{Power rule} \\
3x \log 2 + \log 2 &= 4 \log 3 - x \log 3 && \text{Distributive property} \\
3x \log 2 + x \log 3 &= 4 \log 3 - \log 2 && \text{Move all } x\text{-terms to one side.} \\
x(3 \log 2 + \log 3) &= 4 \log 3 - \log 2 && \text{Factor out } x \text{ on the left.} \\
\text{Exact solution} \longrightarrow x &= \frac{4 \log 3 - \log 2}{3 \log 2 + \log 3} && \text{Divide by } 3 \log 2 + \log 3. \\
\text{Approximate solution} \longrightarrow x &\approx 1.165 && \text{Nearest thousandth}
\end{aligned}$$

The exact solution can be written in a more compact form.

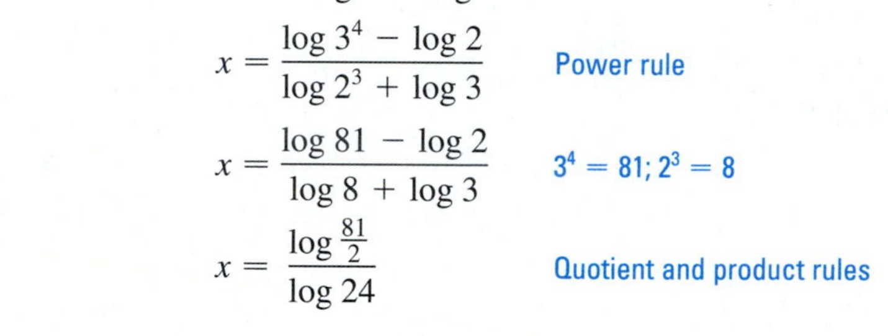

$$\begin{aligned}
x &= \frac{4 \log 3 - \log 2}{3 \log 2 + \log 3} && \text{Exact solution} \\
x &= \frac{\log 3^4 - \log 2}{\log 2^3 + \log 3} && \text{Power rule} \\
x &= \frac{\log 81 - \log 2}{\log 8 + \log 3} && 3^4 = 81;\ 2^3 = 8 \\
x &= \frac{\log \frac{81}{2}}{\log 24} && \text{Quotient and product rules}
\end{aligned}$$

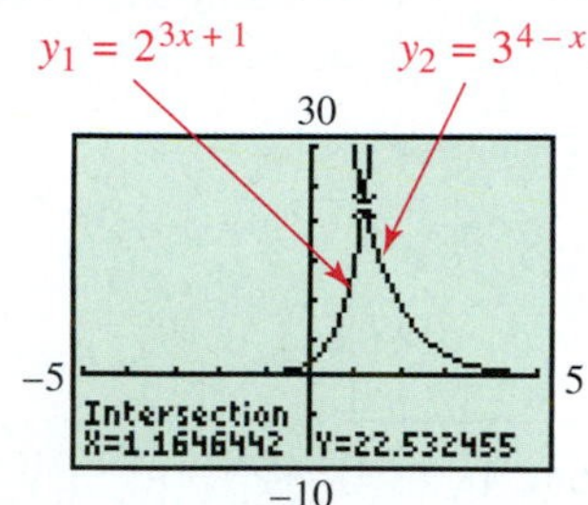

FIGURE 47

The exact solution set is $\left\{\frac{\log(81/2)}{\log 24}\right\}$. A calculator approximation of the solution set is $\{1.165\}$. As seen in Figure 47, the x-coordinate is approximately 1.165, supporting the solution. ■

FOR DISCUSSION

1. Use Figure 47 to determine the solution sets of

$$2^{3x+1} < 3^{4-x} \quad \text{and} \quad 2^{3x+1} > 3^{4-x}.$$

2. Discuss the difference between the exact solution and an approximate solution of an exponential equation. In general, can you find an exact solution from a graph?

Logarithmic Equations and Inequalities

When solving logarithmic equations, begin by noticing the domain of the variable. In some cases, the solution process leads to extraneous values. We use the properties of logarithms from Section 5.3 and Property 2 from this section to solve logarithmic equations.

EXAMPLE 4 **Solving a Logarithmic Equation**

Solve $\log_3(x+6) - \log_3(x+2) = \log_3 x$.

Analytic Solution

The domain must satisfy $x + 6 > 0$, $x + 2 > 0$, and $x > 0$. The intersection of their solution sets gives the domain $(0, \infty)$.

$$\log_3(x+6) - \log_3(x+2) = \log_3 x$$

$$\log_3 \frac{x+6}{x+2} = \log_3 x \qquad \text{Quotient property}$$

$$\frac{x+6}{x+2} = x \qquad \text{Property 2}$$

$$x + 6 = x(x+2) \qquad \text{Multiply by } x + 2.$$

$$x + 6 = x^2 + 2x \qquad \text{Distributive property}$$

$$x^2 + x - 6 = 0 \qquad \text{Standard form}$$

$$(x+3)(x-2) = 0 \qquad \text{Factor.}$$

$$x = -3 \quad \text{or} \quad x = 2 \qquad \text{Zero-product property}$$

The value $x = -3$ is not in the domain of $\log_3 x$ in the original equation, so the only valid solution is 2, giving the solution set $\{2\}$.

Check: $\log_3(2+6) - \log_3(2+2) = \log_3 2 \quad ? \qquad$ Let $x = 2$.

$$\log_3 \frac{8}{4} = \log_3 2 \quad ?$$

$$\log_3 2 = \log_3 2 \qquad \text{True}$$

Graphing Calculator Solution

Figure 48 shows that the x-coordinate of the point of intersection of the graphs of

$$y_1 = \log_3(x+6) - \log_3(x+2)$$

and

$$y_2 = \log_3 x$$

is 2, which agrees with our analytic solution. Notice that the graphs do not intersect when $x = -3$, supporting the conclusion in the analytic solution that -3 is an extraneous value.

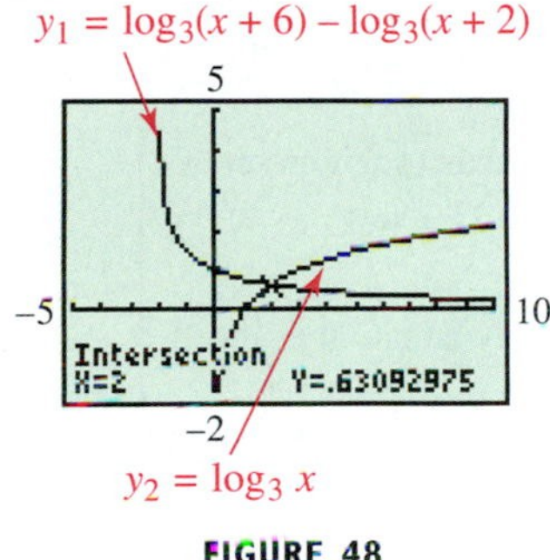

FIGURE 48

NOTE The equation in Example 4 could also have been solved by transforming so that all logarithmic terms appeared on one side. Then, by applying the properties of logarithms from Section 5.3 and rewriting in exponential form, the same solution set would result. This would also allow for solving graphically by the x-intercept method.

EXAMPLE 5 **Solving a Logarithmic Equation**

Solve $\log(3x+2) + \log(x-1) = 1$.

Analytic Solution

Find the domain of each logarithm. For $3x + 2 > 0$, $x > -\frac{2}{3}$ and for $x - 1 > 0$, $x > 1$, so the domain is $(1, \infty)$. Recall from Section 5.3 that **$\log x$ *means* $\log_{10} x$.**

$$\log(3x+2) + \log(x-1) = 1$$

$$\log[(3x+2)(x-1)] = 1 \qquad \text{Product rule}$$

$$(3x+2)(x-1) = 10^1 \qquad \text{Write in exponential form.}$$

$$3x^2 - x - 2 = 10 \qquad \text{Multiply.}$$

$$3x^2 - x - 12 = 0 \qquad \text{Standard form}$$

Graphing Calculator Solution

We arbitrarily choose to use the x-intercept method of solution. As seen in Figure 49 on the next page, the x-intercept is approximately 2.174, and since $\frac{1+\sqrt{145}}{6} \approx 2.174$, this confirms the analytic result.

(continued)

The equation $3x^2 - x - 12 = 0$ cannot be solved by factoring. Use the quadratic formula.

$$x = \frac{-b \pm \sqrt{b^2 - 4ac}}{2a} = \frac{1 \pm \sqrt{1 + 144}}{6} = \frac{1 \pm \sqrt{145}}{6}$$

$a = 3, b = -1, c = -12$

Since $\frac{1 - \sqrt{145}}{6}$ is less than 1, it is not in the domain and must be rejected, giving the solution set $\left\{\frac{1 + \sqrt{145}}{6}\right\}$.

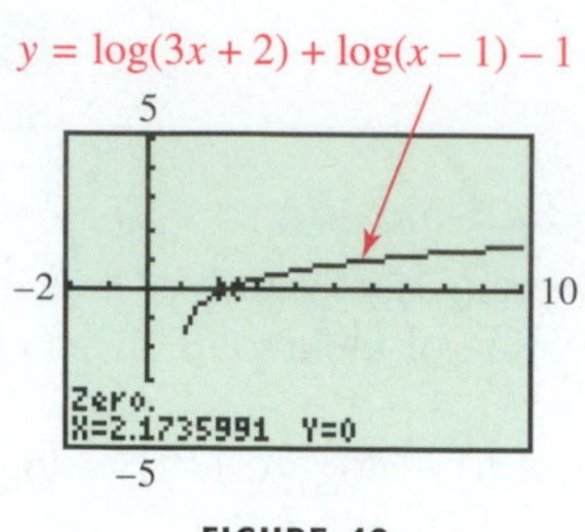

FIGURE 49

FOR DISCUSSION

Use Figure 49 from Example 5 to answer each item.

1. Give the exact solution set of $\log(3x + 2) + \log(x - 1) - 1 \geq 0$.
2. Give the exact solution set of $\log(3x + 2) + \log(x - 1) - 1 \leq 0$. (*Hint:* Pay attention to the domain.)

Equations and Inequalities Involving Both Exponentials and Logarithms

EXAMPLE 6 **Solving a More Complicated Exponential Equation**

Solve $e^{-2 \ln x} = \frac{1}{16}$.

Solution

$$e^{-2 \ln x} = \frac{1}{16} \qquad \text{The domain is } (0, \infty).$$

$$e^{\ln x^{-2}} = \frac{1}{16} \qquad \text{Power rule}$$

$$x^{-2} = \frac{1}{16} \qquad a^{\log_a x} = x$$

$$x^{-2} = 4^{-2} \qquad \frac{1}{16} = \frac{1}{4^2} = 4^{-2};\ -4 \text{ is not a solution, since } -4 < 0 \text{ and } x > 0.$$

$$x = 4 \qquad \text{Properties of exponents}$$

An analytic check shows that 4 is indeed a solution. The solution set is $\{4\}$.

EXAMPLE 7 **Solving a More Complicated Logarithmic Equation**

Solve $\ln e^{\ln x} - \ln(x - 3) = \ln 2$.

Solution

$$\ln e^{\ln x} - \ln(x - 3) = \ln 2 \qquad \text{The domain is } (3, \infty).$$

$$\ln x - \ln(x - 3) = \ln 2 \qquad e^{\ln x} = x$$

$$\ln \frac{x}{x - 3} = \ln 2 \qquad \text{Quotient rule}$$

$$\frac{x}{x - 3} = 2 \qquad \text{Property 2}$$

$$x = 2(x - 3) \qquad \text{Multiply by } x - 3.$$

$$x = 2x - 6 \qquad \text{Distributive property}$$

$$6 = x \qquad \text{Subtract } x\text{; add 6.}$$

Check:

$$
\begin{aligned}
\ln e^{\ln x} - \ln(x-3) &= \ln 2 && \text{Original equation}\\
\ln e^{\ln 6} - \ln(6-3) &= \ln 2 \quad ? && \text{Let } x = 6.\\
\ln 6 - \ln 3 &= \ln 2 \quad ? \\
\ln \frac{6}{3} &= \ln 2 \quad ? && \text{Quotient rule}\\
\ln 2 &= \ln 2 && \text{True}
\end{aligned}
$$

The solution set is $\{6\}$. ■

Solving Exponential and Logarithmic Equations

An exponential or logarithmic equation can be solved by changing the equation into one of the following forms, where a and b are real numbers, $a > 0$, and $a \neq 1$:

1. $\boldsymbol{a^{f(x)} = b}$
   Solve by taking logarithms of each side. (Natural logarithms are the best choice if $a = e$.)
2. $\boldsymbol{\log_a f(x) = \log_a g(x)}$
   From the given equation, $f(x) = g(x)$, which is solved analytically.
3. $\boldsymbol{\log_a f(x) = b}$
   Solve by changing to exponential form, $f(x) = a^b$.

Formulas Involving Exponentials and Logarithms

EXAMPLE 8 Solving an Exponential Formula from Psychology

The strength of a habit is a function of the number of times the habit is repeated. If N is the number of repetitions and H is the strength of the habit, then, according to psychologist C. L. Hull,

$$H = 1000(1 - e^{-kN}),$$

where k is a constant. Solve this formula for k.

Solution

$$
\begin{aligned}
H &= 1000(1 - e^{-kN})\\
\frac{H}{1000} &= 1 - e^{-kN} && \text{Divide by 1000.}\\
\frac{H}{1000} - 1 &= -e^{-kN} && \text{Subtract 1.}\\
e^{-kN} &= 1 - \frac{H}{1000} && \text{Multiply by } -1\text{; rewrite.}\\
\ln e^{-kN} &= \ln\left(1 - \frac{H}{1000}\right) && \text{Take logarithms on each side.}\\
-kN &= \ln\left(1 - \frac{H}{1000}\right) && \ln e^x = x\\
k &= -\frac{1}{N}\ln\left(1 - \frac{H}{1000}\right) && \text{Multiply by } -\tfrac{1}{N}.
\end{aligned}
$$

■

EXAMPLE 9 Solving a Logarithmic Formula from Biology

The number of species in a sample is given by the formula

$$S = a\ln\left(1 + \frac{n}{a}\right),$$

where n is the number of individuals in the sample and a is a constant indicating the diversity of species in the community. Solve the formula for n.

Solution

$$S = a\ln\left(1 + \frac{n}{a}\right)$$

$$\frac{S}{a} = \ln\left(1 + \frac{n}{a}\right) \qquad \text{Divide by } a.$$

$$e^{S/a} = 1 + \frac{n}{a} \qquad \text{Write in exponential form.}$$

$$e^{S/a} - 1 = \frac{n}{a} \qquad \text{Subtract 1.}$$

$$n = a(e^{S/a} - 1) \qquad \text{Multiply by } a\text{; rewrite.}$$

■

EXAMPLE 10 Modeling the Life Span of a Robin

A study monitored the life spans of 129 robins over a 4-year period. The equation

$$y = \frac{2 - \log(100 - x)}{.42}$$

was developed to calculate the number of years y it takes for x percent of the robin population to die. (*Source:* Lack, D., *The Life of a Robin,* Collins, 1965.) How many robins had died after 6 months?

Solution Let $y = \frac{6}{12} = .5$ year, and solve the equation for x.

$$y = \frac{2 - \log(100 - x)}{.42}$$

$$.5 = \frac{2 - \log(100 - x)}{.42} \qquad \text{Let } y = .5.$$

$$.21 = 2 - \log(100 - x) \qquad \text{Multiply by .42.}$$

$$\log(100 - x) = 1.79 \qquad \text{Add } \log(100 - x)\text{; subtract .21.}$$

$$10^{1.79} = 100 - x \qquad \text{Write in exponential form.}$$

$$x = 100 - 10^{1.79} \qquad \text{Add } x\text{; subtract } 10^{1.79}.$$

$$x \approx 38.3 \qquad \text{Use a calculator.}$$

About 38% of the robins had died after 6 months. ■

5.5 Exercises

In Exercises 1–4, use the graph of f to find the solution set of **(a)** $f(x) = 0$ *and* **(b)** $f(x) > 0$.

1.

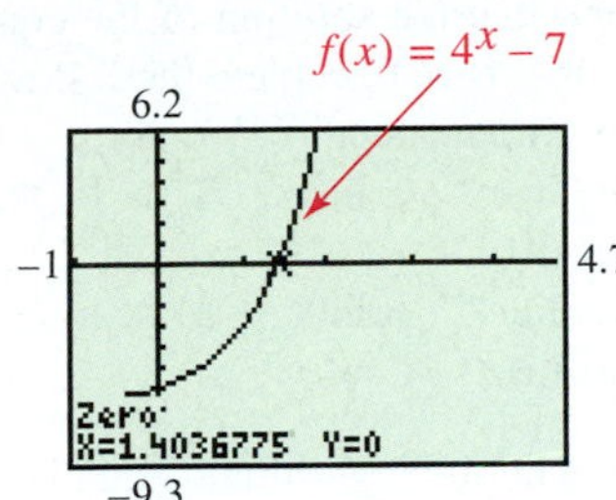

2.

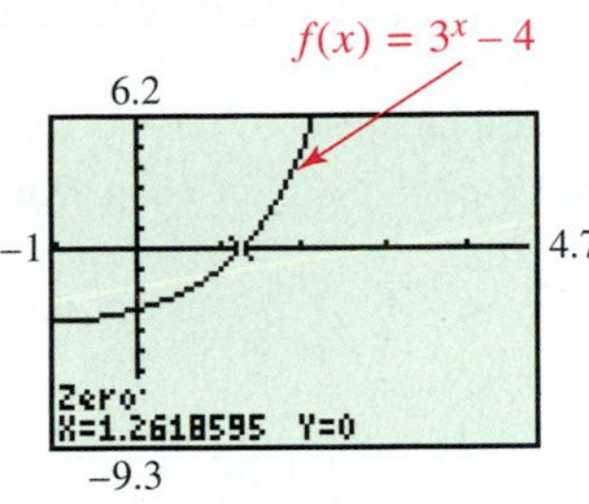

3.

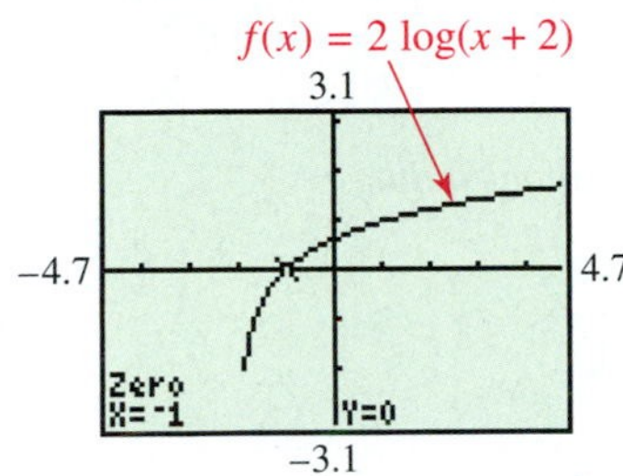

4.

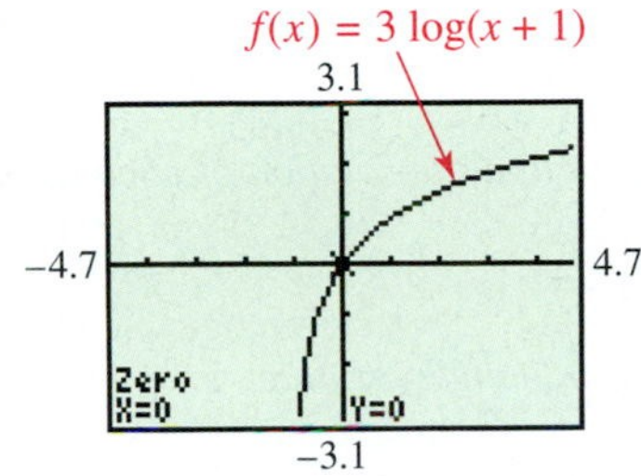

Solve each exponential equation. Express the solution set so that **(a)** *solutions are in exact form and, if irrational,* **(b)** *solutions are approximated to the nearest thousandth. Support your solutions by using a calculator.*

5. $3^x = 7$

6. $5^x = 13$

7. $\left(\frac{1}{2}\right)^x = 5$

8. $\left(\frac{1}{3}\right)^x = 6$

9. $.8^x = 4$

10. $.6^x = 3$

11. $4^{x-1} = 3^{2x}$

12. $2^{x+3} = 5^x$

13. $6^{x+1} = 4^{2x-1}$

14. $3^{x-4} = 7^{2x+5}$

15. $2^x = -5$

16. $3^x = -4$

17. $e^{x-3} = 2^{3x}$

18. $e^{.5x} = 3^{1-2x}$

19. $\left(\frac{1}{3}\right)^x = -3$

20. $\left(\frac{1}{9}\right)^x = -9$

21. $.05(1.15)^x = 5$

22. $1.2(.9)^x = .6$

23. $3(2)^{x-2} + 1 = 100$

24. $5(1.2)^{3x-2} + 1 = 7$

25. $2(1.05)^x + 3 = 10$

26. $3(1.4)^x - 4 = 60$

27. $5(1.015)^{x-1980} = 8$

28. $30 - 3(.75)^{x-1} = 29$

Solve each logarithmic equation. Express all solutions in exact form. Support your solutions by using a calculator.

29. $5 \ln x = 10$

30. $3 \log x = 2$

31. $\ln(4x) = 1.5$

32. $\ln(2x) = 5$

33. $\log(2 - x) = .5$

34. $\ln(1 - x) = \frac{1}{2}$

35. $\log_6(2x + 4) = 2$

36. $\log_5(8 - 3x) = 3$

37. $\log_4(x^3 + 37) = 3$

38. $\log_7(x^3 + 65) = 0$

39. $\ln x + \ln x^2 = 3$

40. $\log x + \log x^2 = 3$

41. $\log x + \log(x - 21) = 2$

42. $\log x + \log(3x - 13) = 1$

43. $\ln(4x - 2) - \ln 4 = -\ln(x - 2)$

44. $\ln(5 + 4x) - \ln(3 + x) - \ln 3 = 0$

45. $\log_5(x + 2) + \log_5(x - 2) = 1$

46. $\log_2(x - 7) + \log_2 x = 3$

47. $\log_7(4x) - \log_7(x + 3) = \log_7 x$

48. $\log_2(2x) + \log_2(x + 2) = \log_2 16$

49. $\ln e^x - 2 \ln e = \ln e^4$

50. $\log_2(\log_2 x) = 1$

51. $\log x = \sqrt{\log x}$

52. $\ln(\ln x) = 0$

53. In Example 6, we found that the solution of the equation $e^{-2 \ln x} = \frac{1}{16}$ is 4. Complete the following:
(a) Solve the equation graphically.
(b) Solve the inequality $e^{-2 \ln x} < \frac{1}{16}$, using the graph from part (a).
(c) Solve the inequality $e^{-2 \ln x} > \frac{1}{16}$, using the graph from part (a).

54. In Example 7, we found that the solution of the equation $\ln e^{\ln x} - \ln(x - 3) = \ln 2$ is 6. Complete the following:
(a) Solve the equation graphically.
(b) Solve the inequality $\ln e^{\ln x} - \ln(x - 3) > \ln 2$, using the graph from part (a).
(c) Solve the inequality $\ln e^{\ln x} - \ln(x - 3) < \ln 2$, using the graph from part (a).

55. A student told a friend, "You must reject any negative solution of an equation involving logarithms." Is this correct? Write an explanation of your answer.

56. Use a graph to explain why the logarithmic equation

$$\ln x - \ln(x + 1) = \ln 5$$

has no solution.

Use a graphing calculator to find the solution of each equation. Round your result to the nearest thousandth.

57. $1.5^{\log x} = e^{.5}$

58. $1.5^{\ln x} = 10^{.5}$

Solve each formula for the indicated variable.

59. $r = p - k \ln t$, for t

60. $p = a + \dfrac{k}{\ln x}$, for x

61. $T = T_0 + (T_1 - T_0)10^{-kt}$, for t

62. $A = \dfrac{Pi}{1 - (1 + i)^{-n}}$, for n

63. $A = T_0 + Ce^{-kt}$, for k

64. $y = \dfrac{K}{1 + ae^{-bx}}$, for b

65. $y = A + B(1 - e^{-Cx})$, for x

66. $m = 6 - 2.5 \log\left(\dfrac{M}{M_0}\right)$, for M

67. $\log A = \log B - C \log x$, for A

68. $d = 10 \log\left(\dfrac{I}{I_0}\right)$, for I

69. $A = P\left(1 + \dfrac{r}{n}\right)^{nt}$, for t

70. $D = 160 + 10 \log x$, for x

Relating Concepts

For individual or group investigation (Exercises 71–76)

The methods of solving quadratic equations from Chapter 3 can be applied to equations that are quadratic in form. Consider the equation $e^{2x} - 4e^x + 3 = 0$. ***Work Exercises 71–76 in order.***

71. The expression e^{2x} is equivalent to $(e^x)^2$. Explain why.

72. The given equation is equivalent to the equation $(e^x)^2 - 4e^x + 3 = 0$. Factor the left side.

73. Solve the equation in Exercise 72 by the zero-product property. Give exact values.

74. Solve the equation in Exercise 72 by using a calculator graph of $y = e^{2x} - 4e^x + 3$.

75. Use the graph from Exercise 74 to solve the inequality $e^{2x} - 4e^x + 3 > 0$.

76. Use the graph from Exercise 74 to solve the inequality $e^{2x} - 4e^x + 3 < 0$.

In general, it is not possible to find exact solutions analytically for equations that involve exponential or logarithmic functions together with polynomial, radical, and rational functions. Solve each equation using a graphical method, and express solutions to the nearest thousandth if an approximation is appropriate.

77. $x^2 = 2^x$

78. $x^2 - 4 = e^{x-4} + 4$

79. $\log x = x^2 - 8x + 14$

80. $\ln x = -\sqrt[3]{x+3}$

81. $e^x = \dfrac{1}{x+2}$

82. $3^{-x} = \sqrt{x+5}$

Use any method (analytic or graphical) to solve each equation.

83. $\log_2 \sqrt{2x^2} - 1 = .5$

84. $\log x^2 = (\log x)^2$

85. $\ln(\ln e^{-x}) = \ln 3$

86. $e^{x+\ln 3} = 4e^x$

(Modeling) Life Span of Robins *Use the equation*

$$y = \frac{2 - \log(100 - x)}{.42}$$

from Example 10 for Exercises 87 and 88.

87. Estimate analytically the percentage of robins that died after 2 years.

88. Estimate graphically the number of years elapsed for 75% of the robins to die.

(Modeling) Salinity *The salinity of the oceans changes with latitude and depth. In the tropics, the salinity increases on the surface of the ocean due to rapid evaporation. In the higher latitudes, there is less evaporation and rainfall causes the salinity to be less on the surface than at lower depths. The function given by*

$$f(x) = 31.5 + 1.1 \log(x + 1)$$

models salinity to depths of 1000 *meters at a latitude of* 57.5°N. *The variable x is the depth in meters, and $f(x)$ is in grams of salt per kilogram of seawater. (Source: Hartman, D.,* Global Physical Climatology, *Academic Press, 1994.)*

89. Approximate analytically the depth where the salinity equals 33.

90. Estimate graphically the salinity at a depth of 500 meters.

Reviewing Basic Concepts (Sections 5.4 and 5.5)

1. Fill in the blanks: If $f(x) = 3^x$ and $g(x) = \log_3 x$, then functions f and g are __________ functions, and their graphs are __________ with respect to the line with equation __________. The domain of f is the __________ of g and vice versa.

Let $f(x) = 2 - \log_2(x - 1)$ in Exercises 2–5.

2. Graph $f(x)$.

3. Give the equation of the asymptote of the graph of $f(x)$ and any intercepts.

4. Compare the graph of $f(x)$ with the graph of $g(x) = \log_2 x$.

5. Find the inverse of the function defined by $f(x)$.

Solve each equation.

6. $3^{2x-1} = 4^x$

7. $\ln 5x - \ln(x + 2) = \ln 3$

8. $10^{5 \log x} = 32$

9. $H = 1000(1 - e^{-kN})$, for N

10. ***(Modeling) Caloric Intake*** The function defined by

$$f(x) = 280 \ln(x + 1) + 1925$$

models the number of calories consumed daily by a person owning x acres of land in a developing country. (*Source:* Grigg, D., *The World Food Problem,* Blackwell Publishers, 1993.) Estimate the number of acres owned for the average intake to be 2300 calories per day.

5.6 Further Applications and Modeling with Exponential and Logarithmic Functions

Physical Science Applications ■ Financial Applications ■ Biological and Medical Applications ■ Modeling Data with Exponential and Logarithmic Functions

Looking Ahead to Calculus

The exponential growth and decay function formulas are studied in calculus in conjunction with the topic known as *differential equations.*

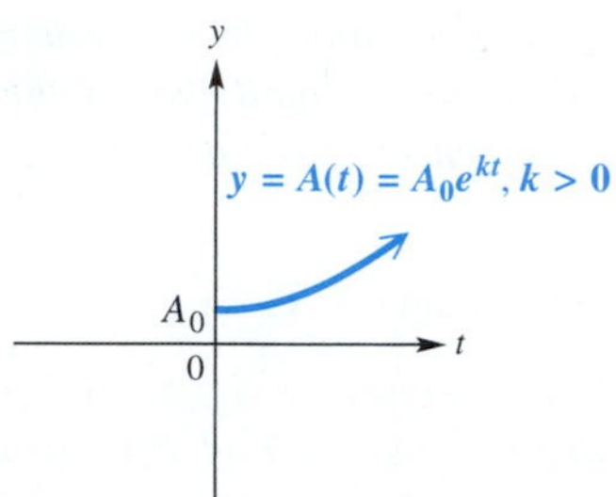

Exponential growth function

FIGURE 50

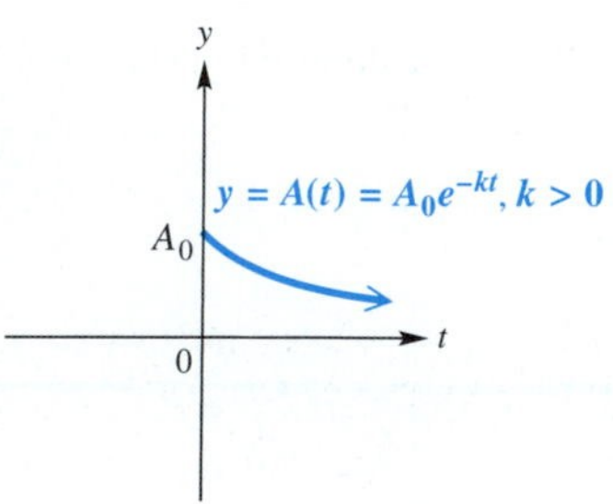

Exponential decay function

FIGURE 51

Physical Science Applications

The formula $A = Pe^{rt}$ for continuous compounding of interest is an example of an exponential growth function. A function of the form

$$A(t) = A_0e^{kt}, \qquad \text{Exponential growth function}$$

where A_0 represents the initial quantity present, t represents time elapsed, $k > 0$ represents the growth constant associated with the quantity, and $A(t)$ represents the amount present at time t, is called an **exponential growth function.** This is an increasing function, because $e > 1$ and $k > 0$. On the other hand, for $k > 0$, a function of the form

$$A(t) = A_0e^{-kt} \qquad \text{Exponential decay function}$$

is an **exponential decay function.** It is a decreasing function because

$$e^{-k} = (e^{-1})^k = \left(\frac{1}{e}\right)^k$$

and $0 < \frac{1}{e} < 1$. In both cases, we usually restrict t to be nonnegative, giving domain $[0, \infty)$. (Why?) Figures 50 and 51 respectively show graphs of typical growth and decay functions.

If a quantity decays exponentially, the amount of time that it takes to reach one-half its original amount is called the **half-life.**

EXAMPLE 1 Analyzing an Exponential Decay Function Involving Radioactive Isotopes

Nuclear energy derived from radioactive isotopes can be used to supply power to space vehicles. Suppose that the output of the radioactive power supply for a certain satellite is given by the function defined by

$$y = 40e^{-.004t},$$

where y is measured in watts and t is the time in days.

(a) What is the initial output of the power supply?

(b) After how many days will the output be reduced to 35 watts?

(c) After how many days will the output be half of its initial amount? (That is, what is its half-life?) Support this result with a graph.

Solution

(a) $y = 40e^{-.004t} = 40e^{-.004(0)} = 40e^0 = 40 \qquad e^0 = 1$

The initial output is 40 watts.

(b) Let $y = 35$, and solve for t.

$$y = 40e^{-.004t}$$

$$35 = 40e^{-.004t}$$

$$\frac{35}{40} = e^{-.004t} \quad \text{Divide by 40.}$$

$$\ln\frac{35}{40} = \ln e^{-.004t} \quad \text{Take logarithms of each side.}$$

$$\ln\frac{35}{40} = -.004t \quad \ln e^k = k$$

$$t = \frac{\ln\frac{35}{40}}{-.004} \quad \text{Divide by } -.004.$$

$$t \approx 33.4 \quad \text{Use a calculator.}$$

The output will be reduced to 35 watts in about 33.4 days.

(c) Because the initial amount is 40, we must find the value of t for which $y = \frac{1}{2}(40) = 20$.

$$y = 40e^{-.004t}$$

$$20 = 40e^{-.004t}$$

$$.5 = e^{-.004t} \quad \text{Divide by 40.}$$

$$\ln .5 = \ln e^{-.004t} \quad \text{Take logarithms of each side.}$$

$$\ln .5 = -.004t \quad \ln e^k = k$$

$$t \approx 173 \quad \text{Divide by } -.004\text{; simplify.}$$

The half-life is approximately 173 days. The graph in Figure 52 also shows that when $x = t = 173$, $y \approx 20 = \frac{1}{2}(40)$. ■

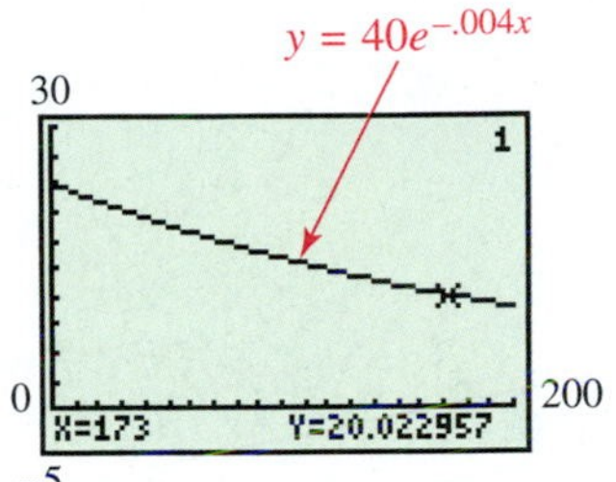

FIGURE 52

EXAMPLE 2 Using an Exponential Function to Find the Age of a Fossil

Carbon 14 is a radioactive form of carbon that is found in all living plants and animals. After a plant or animal dies, the radiocarbon disintegrates. Using a technique called *carbon dating,* scientists determine the age of the remains by comparing the amount of carbon 14 present with the amount found in living plants and animals. The amount of carbon 14 present after t years is given by the exponential function defined by $A(t) = A_0e^{-kt}$, with $k = \frac{\ln 2}{5700}$. Find the half-life of carbon 14.

Solution

$$A(t) = A_0e^{-kt}$$

$$\frac{1}{2}A_0 = A_0e^{-\frac{\ln 2}{5700}t} \quad \text{Let } A(t) = \tfrac{1}{2}A_0 \text{ and } k = \tfrac{\ln 2}{5700}.$$

$$\frac{1}{2} = e^{-\frac{\ln 2}{5700}t} \quad \text{Divide by } A_0.$$

$$\ln\frac{1}{2} = \ln e^{-\frac{\ln 2}{5700}t} \quad \text{Take logarithms of each side.}$$

$$\ln\frac{1}{2} = -\frac{\ln 2}{5700}t \qquad \ln e^x = x$$

$$-\frac{5700}{\ln 2}\ln\frac{1}{2} = t \qquad \text{Multiply by } -\tfrac{5700}{\ln 2}.$$

$$-\frac{5700}{\ln 2}(\ln 1 - \ln 2) = t \qquad \text{Quotient rule}$$

$$-\frac{5700}{\ln 2}(-\ln 2) = t \qquad \ln 1 = 0$$

$$5700 = t \qquad \text{Simplify.}$$

The half-life is 5700 years. ■

EXAMPLE 3 Finding an Exponential Decay Function Given Half-life

Radium 226, which decays according to the function defined by $A(t) = A_0e^{-kt}$, where t is time in years, has a half-life of about 1612 years. Find k. Then find how long it takes a 10-gram sample to decay to 6 grams.

Solution The half-life tells us that when $t = 1612$, $A(t) = \frac{1}{2}A_0$.

$$A(t) = A_0e^{-kt}$$

$$\frac{1}{2}A_0 = A_0e^{-k(1612)} \qquad \text{Substitute.}$$

$$\frac{1}{2} = e^{-1612k} \qquad \text{Divide by } A_0.$$

$$\ln\frac{1}{2} = \ln e^{-1612k} \qquad \text{Take logarithms of each side.}$$

$$\ln\frac{1}{2} = -1612k \qquad \ln e^x = x$$

$$k = \frac{\ln\frac{1}{2}}{-1612} \approx .00043 \qquad \text{Divide by } -1612\text{; approximate.}$$

Thus, radium 226 decays according to the equation

$$A(t) = A_0e^{-.00043t}.$$

Now we let $A(t) = 6$ and $A_0 = 10$ to find t.

$$6 = 10e^{-.00043t} \qquad \text{Substitute.}$$

$$.6 = e^{-.00043t} \qquad \text{Divide by 10.}$$

$$\ln .6 = \ln e^{-.00043t} \qquad \text{Take logarithms of each side.}$$

$$\ln .6 = -.00043t \qquad \ln e^x = x$$

$$t = \frac{\ln .6}{-.00043} \qquad \text{Divide by } -.00043.$$

$$t \approx 1188 \qquad \text{Use a calculator.}$$

The 10-gram sample will decay to 6 grams in about 1188 years. ■

EXAMPLE 4 **Measuring Sound Intensity**

The loudness of sounds is measured in a unit called a *decibel.* To measure with this unit, we first assign an intensity of I_0 to a very faint sound, called the *threshold sound.* If a particular sound has intensity I, then the decibel rating of this louder sound is

$$d = 10 \log \frac{I}{I_0}.$$

In the movie *The Lord of the Rings: The Two Towers,* action sequences reached levels of $10^{9.5}I_0$. (*Source:* Canadian Broadcast Company News.) Find the decibel rating of these sequences, and compare it with the average rating of 80 decibels for a ringing telephone.

Solution

$$d = 10 \log \frac{10^{9.5}I_0}{I_0} \qquad \text{Substitute } I = 10^{9.5}I_0 \text{ in the equation.}$$

$$d = 10 \log 10^{9.5}$$

$$d = 10(9.5) \qquad \log 10^{9.5} = 9.5$$

$$d = 95$$

The action sequences reached 95 decibels. Since 95 decibels is equivalent to $10^{9.5}I_0$ and 80 decibels is equivalent to $10^{8.0}I_0$, their ratio is

$$\frac{10^{9.5}I_0}{10^{8.0}I_0} = 10^{1.5} \approx 31.6.$$

The action sequences are 31.6 times the intensity of a ringing telephone. ■

Financial Applications

The compound interest formulas

$$A = P\left(1 + \frac{r}{n}\right)^{nt} \quad \text{and} \quad A = Pe^{rt}$$

were introduced in Section 5.2. The first applies to interest compounded n times per year, the second to interest compounded continuously. We can use logarithms to determine how long it takes for an investment to grow to a desired amount.

If a quantity grows exponentially, the amount of time that it takes to become twice its original amount is called the **doubling time.** This is analogous to half-life for quantities that decay exponentially. In both cases, the initial amount present does not affect either the doubling time or half-life.

EXAMPLE 5 **Solving Compound Interest Formulas for *t***

(a) How long will it take \$1000 invested at 6% interest compounded quarterly to grow to \$2700?

(b) How long will it take for the money in an account that is compounded continuously at 8% interest to double?

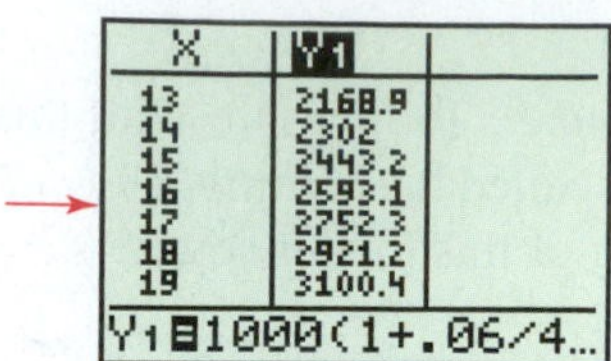

This table numerically supports the result of Example 5(a). Notice that $Y_1 = 2700$ for some X-value between 16 and 17, closer to 17. We found analytically that this value is approximately 16.678.

Solution

(a)

$$A = P\left(1 + \frac{r}{n}\right)^{nt}$$

$$2700 = 1000\left(1 + \frac{.06}{4}\right)^{4t} \qquad A = 2700,\ P = 1000,\ r = .06,\ n = 4$$

$$2700 = 1000(1.015)^{4t} \qquad \text{Solve for } t.$$

$$2.7 = 1.015^{4t} \qquad \text{Divide by 1000.}$$

$$\log 2.7 = \log 1.015^{4t} \qquad \text{Take logarithms of each side.}$$

$$\log 2.7 = 4t \log 1.015 \qquad \text{Power rule}$$

$$\frac{\log 2.7}{4 \log 1.015} = t \qquad \text{Divide by } 4 \log 1.015.$$

$$t \approx 16.678 \qquad \text{Use a calculator.}$$

Since interest is compounded quarterly, it will take about $16\frac{3}{4}$ years for the initial amount to grow to \$2700.

(b) Use the formula for continuous compounding, $A = Pe^{rt}$, to find the time t that makes $A = 2P$. Substitute $2P$ for A and .08 for r; then solve for t.

$$A = Pe^{rt}$$

$$2P = Pe^{.08t} \qquad A = 2P,\ r = .08$$

$$2 = e^{.08t} \qquad \text{Divide by } P.$$

$$\ln 2 = \ln e^{.08t} \qquad \text{Take logarithms of each side.}$$

$$\ln 2 = .08t \qquad \ln e^x = x$$

$$\frac{\ln 2}{.08} = t \qquad \text{Divide by .08.}$$

$$t \approx 8.664 \qquad \text{Use a calculator.}$$

The amount will double in about $8\frac{2}{3}$ years. ■

GCM TECHNOLOGY NOTE

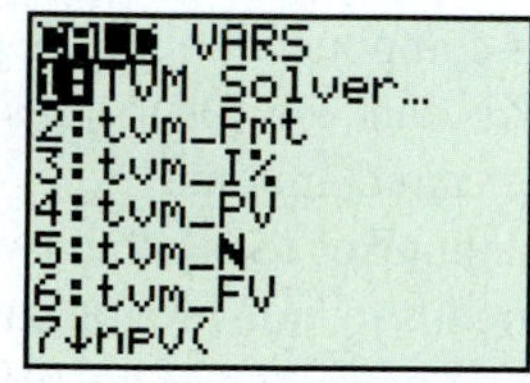

Graphing calculators such as the TI-83/84 Plus are capable of financial calculations. Consult the graphing calculator manual that accompanies this text or your owner's manual.

A loan is **amortized** if both the principal and interest are paid by a sequence of equal periodic payments.

Amortization Payments

A loan of P dollars at interest rate i per period (as a decimal) may be amortized in n equal periodic payments of R dollars made at the end of each period, where

$$R = \frac{P}{\left[\dfrac{1 - (1 + i)^{-n}}{i}\right]}.$$

The total interest I that will be paid during the term of the loan is

$$I = nR - P.$$

GCM **EXAMPLE 6** **Using Amortization to Finance an Automobile**

Following the traditional American process of haggling, you agree to purchase a used SUV for \$24,000. After a down payment of \$4000, the balance will be paid off in 36 equal monthly payments at 8.5% interest per year. Find the amount of each payment. How much interest will you pay over the life of the loan?

Solution Use the amortization formula.

$$R = \frac{20{,}000}{\left[\frac{1 - (1 + .007083)^{-36}}{.007083}\right]} \approx 631.35 \qquad P = 24{,}000 - 4000 = 20{,}000, \; i = \frac{.085}{12} \approx .007083, \; n = 36$$

The monthly payment will be \$631.35. The total interest paid will be

$$36(\$631.35) - \$20{,}000 = \$2728.60.$$

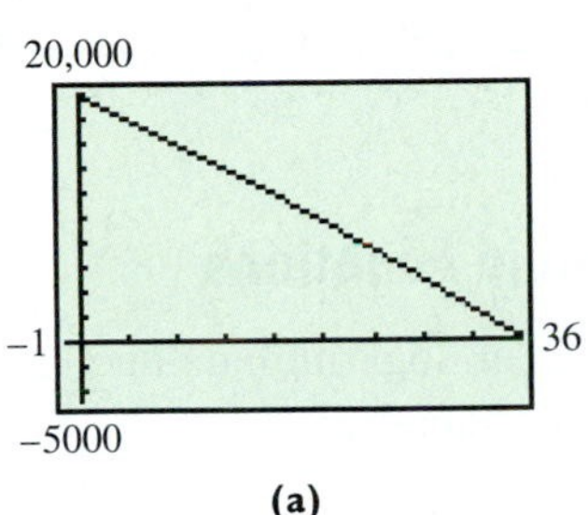

(a)

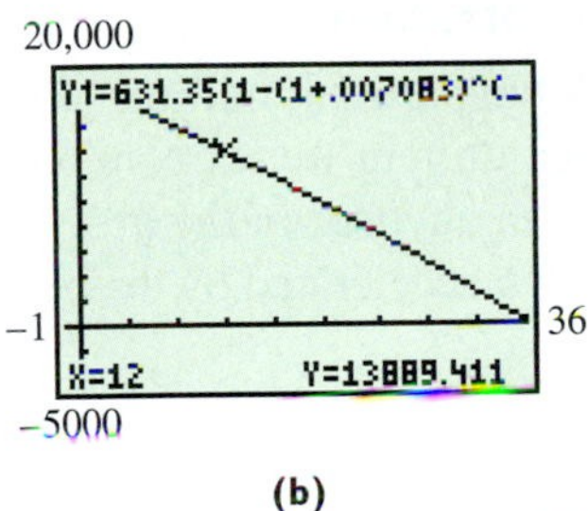

(b)

FIGURE 53

In Example 6, the approximate unpaid balance y after x payments is given by

$$y = R\left[\frac{1 - (1 + i)^{-(n-x)}}{i}\right].$$

For example, the unpaid balance after 1 payment is

$$y = \$631.35\left[\frac{1 - (1 + .007083)^{-35}}{.007083}\right] \approx \$19{,}510.40.$$

A graph of y as a function of x is shown in Figure 53(a). We can find the unpaid balance after any number of payments, x, by finding the y-value that corresponds to x. For example, the remaining balance after 1 year, or 12 payments, is shown at the bottom of the screen in Figure 53(b). You may be surprised that the remaining balance on a \$20,000 loan after 12 payments is as large as \$13,889.41. This is because most of each early payment on a loan goes toward interest.

Biological and Medical Applications

EXAMPLE 7 **Determining the Amount of Medication**

When physicians prescribe medication, they must consider how the drug's effectiveness decreases over time. If, each hour, a drug is only 90% as effective as the previous hour, at some point the patient will not be receiving enough medication and must receive another dose. For example, if the initial dose was 200 milligrams and the drug was administered 3 hours ago, the exponential expression $200(.90)^2$ represents the amount of effective medication still available. Thus, $200(.90)^2 = 162$ milligrams are still in the system. (The exponent is the number of hours since the drug was administered, less 1.)

How long will it take for this initial dose to reach the dangerously low level of 50 milligrams?

Analytic Solution

We must solve the equation $200(.90)^x = 50$.

$$200(.90)^x = 50$$
$$(.90)^x = .25 \quad \text{Divide by 200.}$$
$$\log(.90)^x = \log .25 \quad \text{Take logarithms.}$$
$$x \log .90 = \log .25 \quad \text{Power rule}$$
$$x = \frac{\log .25}{\log .90} \approx 13.16 \quad \text{Divide by log .90.}$$

Since x represents one *less than* the number of hours from when the drug was administered, the drug will reach a level of 50 milligrams in about 14 hours.

Graphing Calculator Solution

The graph in Figure 54 confirms the result that $y = 50$ when $x \approx 13.16$.

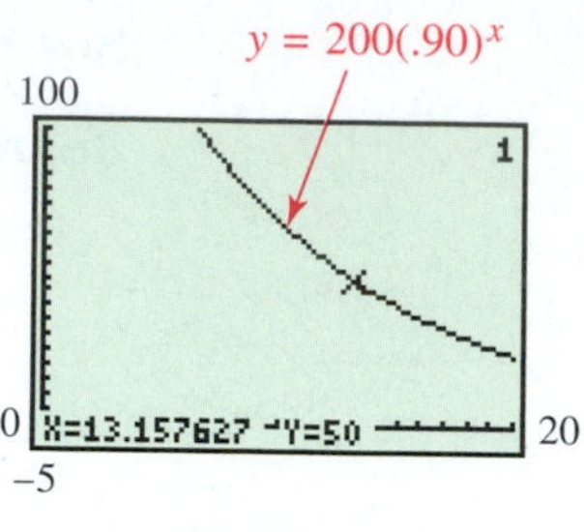

FIGURE 54

Modeling Data with Exponential and Logarithmic Functions

Some data can be modeled quite well with exponential or logarithmic functions. Graphing calculators can determine such models for appropriate data.

GCM **EXAMPLE 8** **Modeling Atmospheric CO_2 Concentrations**

Future concentrations of atmospheric carbon dioxide (CO_2) in parts per million (ppm) are shown in the table. (These concentrations assume that current trends continue.) CO_2 levels in the year 2000 were greater than they had been anytime in the previous 160,000 years. The increase in concentration of CO_2 has been accelerated by the burning of fossil fuels and by deforestation.

Year	2000	2050	2100	2150	2200
CO_2 (ppm)	364	467	600	769	987

Source: Turner, R., *Environmental Economics, an Elementary Approach,* Johns Hopkins University Press, 1993.

(a) Let $x = 0$ correspond to 2000 and $x = 200$ to 2200. Find values for C and a so that $f(x) = Ca^x$ models these data.

(b) Use a graphing calculator with regression capability to find an exponential function g that models all the data, and graph it with the data points. How does function g compare with function f from part (a)?

(c) Estimate CO_2 concentrations for the year 2025.

Solution

(a) The concentration is 364 when $x = 0$, so $C = 364$. This gives

$$f(x) = Ca^x = 364a^x.$$

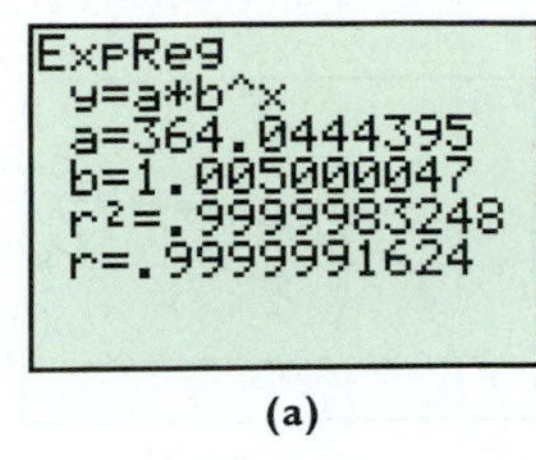

(a)

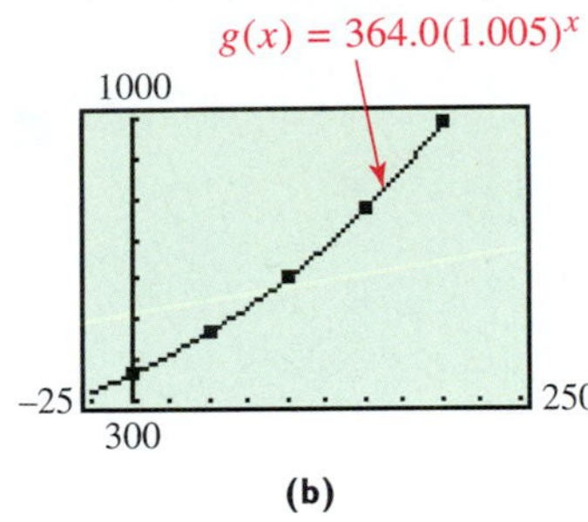

(b)

FIGURE 55

One possibility for determining a is to require that the graph of f pass through the point (2200, 987), where $x = 200$. Thus, $f(200) = 987$.

$$364a^x = f(x)$$

$$364 \cdot a^{200} = 987 \qquad f(200) = 987$$

$$a^{200} = \frac{987}{364} \qquad \text{Divide by 364.}$$

$$(a^{200})^{1/200} = \left(\frac{987}{364}\right)^{1/200} \qquad \text{Raise to the } \tfrac{1}{200}\text{th power.}$$

$$a = \left(\frac{987}{364}\right)^{1/200} \qquad \text{Property of exponents}$$

$$a \approx 1.005 \qquad \text{Use a calculator.}$$

Hence, $f(x) = 364(1.005)^x$. (Answers may vary slightly.)

(b) Figure 55(a) shows the coefficients given by the calculator for exponential regression. With these values, $g(x) \approx 364.0(1.005)^x$, which is essentially the same as $f(x)$ from part (a). The data points and the function are graphed in Figure 55(b).

(c) Since 2025 corresponds to $x = 25$, evaluate $f(25)$.

$$f(25) = 364(1.005)^{25} \approx 412.$$

The concentration of carbon dioxide could reach 412 ppm by 2025. ■

EXAMPLE 9 Modeling Interest Rates

The table lists the interest rates for certificates of deposit during January 2005. Rates are for deposits of $10,000 or more.

Time	1 year	2 years	3 years	4 years	5 years
Yield (%)	2.326	2.685	3.148	3.354	3.665

Source: Union Bank of California.

(a) Make a scatter diagram of the data. What type of function might model these data?

(b) Use least-squares regression to obtain a formula $f(x) = a + b \ln x$ that models the data.

(c) Graph f and the data in the same viewing window.

Solution

(a) Enter the data points (1, 2.326), (2, 2.685), (3, 3.148), (4, 3.354), and (5, 3.665) into a calculator. A scatter diagram of the data is shown in Figure 56. A logarithmic function model may be appropriate.

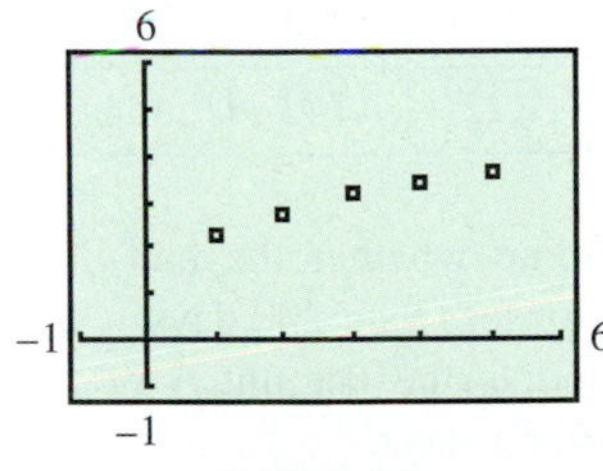

FIGURE 56

(b) In Figures 57(a) and 57(b) on the next page, least-squares regression gives a logarithmic function defined (approximately) by

$$f(x) = 2.244 + .8263 \ln x.$$

(c) A graph of f and the data are shown in Figure 57(c).

(a)

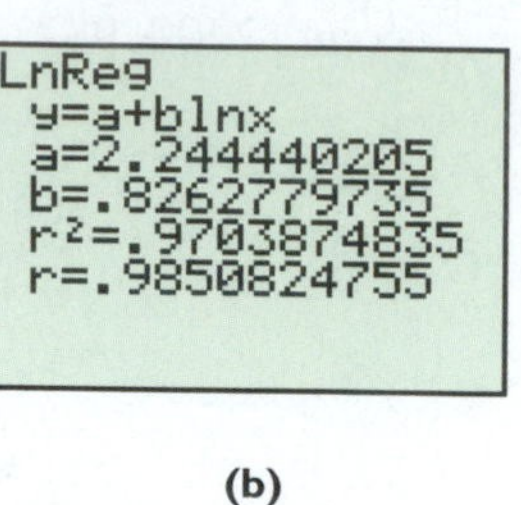

(b)

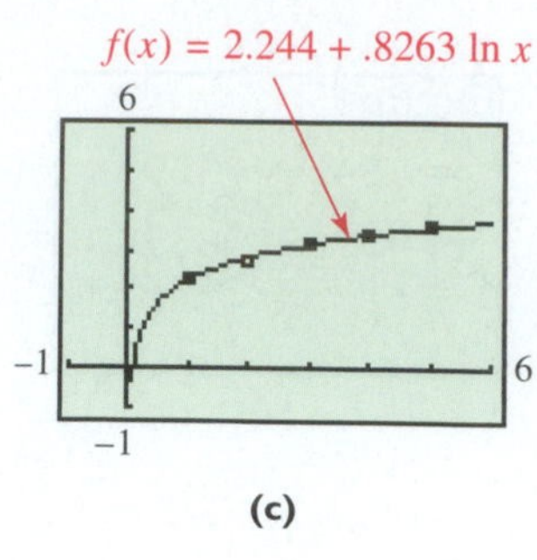

(c)

FIGURE 57

5.6 Exercises

(Modeling) The exercises in this set are grouped according to discipline; they involve exponential or logarithmic models.

Physical Sciences *(Exercises 1–20)*

The information in Example 2 allows us to use the function defined by $A(t) = A_0e^{-.0001216t}$ *to approximate the amount of carbon 14 remaining in a sample, where t is in years. Use this function in Exercises 1–4.* (*Note*: $-.0001216 \approx -\frac{\ln 2}{5700}$.)

1. **Carbon 14 Dating** Suppose an Egyptian mummy is discovered in which the amount of carbon 14 present is only about one-third the amount found in the atmosphere. About how long ago did the Egyptian die?

2. **Carbon 14 Dating** A sample from a refuse deposit near the Strait of Magellan had 60% of the carbon 14 of a contemporary living sample. How old was the sample?

3. **Carbon 14 Dating** Paint from the Lascaux caves of France contains 15% of the normal amount of carbon 14. Estimate the age of the paint.

4. **Carbon 14 Dating** Estimate the age of a specimen that contains 20% of the carbon 14 of a comparable living specimen.

5. **Radioactive Lead 210** The half-life of radioactive lead 210 is 21.7 years.
   (a) Find an exponential decay model for lead 210.
   (b) How long will it take a sample of 500 grams to decay to 400 grams?
   (c) How much of the sample of 500 grams will remain after 10 years?

6. **Radioactive Cesium 137** Radioactive cesium 137 was emitted in large amounts in the Chernobyl nuclear power station accident in Russia on April 26, 1986. The amount of cesium 137 remaining after x years in an initial sample of 100 milligrams can be described by

$$A(x) = 100e^{-.02295x}.$$

(*Source:* Mason, C., *Biology of Freshwater Pollution*, John Wiley and Sons, 1991.)
   (a) How much is remaining after 50 years? Is the half-life of cesium 137 greater or less than 50 years?
   (b) Estimate graphically the half-life of cesium 137.

7. **Radioactive Polonium 210** The table shows the amount y of polonium 210 remaining after x days from an initial sample of 2 milligrams.

x (days)	0	100	200	300
y (milligrams)	2	1.22	.743	.453

   (a) Use the table to determine whether the half-life of polonium 210 is greater or less than 200 days.
   (b) Find a formula that models the amount A of polonium 210 in the table after x days.
   (c) Estimate graphically the half-life of polonium 210.

8. *Sound Intensity* Use the formula

$$d = 10 \log \frac{I}{I_0}$$

to find the average decibel level for each noise source with the given intensity I. For comparison, conversational speech has a sound level of about 60 decibels.

(a) Jackhammer: $31{,}620{,}000{,}000 I_0$

(b) Movie: *Harry Potter and the Chamber of Secrets:* $1.995 \times 10^9 I_0$

(c) Rock singer screaming into microphone: $10^{14} I_0$

The intensity of an earthquake, measured on the Richter scale, is given by $\log \frac{I}{I_0}$, *where* I_0 *is the intensity of an earthquake of a certain small size. Use this information in Exercises 9 and 10.*

9. *Earthquake Intensity* On November 3, 2002, Denali Park, Alaska, was shaken by an earthquake that measured 7.9 on the Richter scale.

(a) Express this reading in terms of I_0.

(b) In March of the same year, a quake measuring 6.1 on the Richter scale killed about 1000 people in the Hindu Kush region of Afghanistan. Express the intensity of a 6.1 reading in terms of I_0.

(c) Compare this 6.1 reading with the force of an earthquake with measure 6.0.

10. *Earthquake Intensity* The earthquake off the coast of Northern Sumatra on Dec. 26, 2004, had a Richter scale rating of 9.0.

(a) Express the intensity of this earthquake in terms of I_0.

(b) On Dec. 14, 2004, the Cayman Islands experienced an earthquake with a Richter scale rating of 6.8. Express the intensity of this earthquake in terms of I_0.

(c) Compare the intensities of the Sumatran and Cayman Islands earthquakes.

11. *Rock Sample Age* Use the function defined by

$$t = T \frac{\ln\left[1 + 8.33\left(\frac{A}{K}\right)\right]}{\ln 2}$$

to estimate the age t in years of a rock sample if tests show that $\frac{A}{K}$ is .103 for the sample. Let $T = 1.26 \times 10^9$.

12. *Air Pressure* The air pressure in pounds per square inch h feet above sea level is given by

$$P(h) = 14.7e^{-.0000385h}.$$

At approximately what height is the pressure 10% of the pressure at sea level?

13. *Magnitude of a Star* The magnitude of a star is defined by the equation

$$M = 6 - \frac{5}{2} \log \frac{I}{I_0},$$

where I_0 is the measure of a just-visible star and I is the actual intensity of the star being measured. The dimmest stars are of magnitude 6, and the brightest are of magnitude 1. Determine the ratio of light intensities between a star of magnitude 1 and a star of magnitude 3.

Newton's law of cooling says that the rate at which a body cools is proportional to the difference C in temperature between the body and the environment around it. The temperature $f(t)$ *of the body at time t in appropriate units after being introduced into an environment with a constant temperature* T_0 *is*

$$f(t) = T_0 + Ce^{-kt},$$

where C and k are constants. Use this result in Exercises 14–16.

14. *Newton's Law of Cooling* Boiling water at 100°C is placed in a freezer at 0°C. The temperature of the water is 50°C after 24 minutes. Find the temperature of the water after 96 minutes.

15. *Newton's Law of Cooling* A pot of coffee with a temperature of 100°C is set down in a room with a temperature of 20°C. The coffee cools to 60°C after 1 hour.

(a) Write an equation to model the data.

(b) Find the temperature after a half hour.

(c) How long will it take for the coffee to cool to 50°C? Support your answer graphically.

16. *Newton's Law of Cooling* A piece of metal is heated to 300°C and then placed in a cooling liquid at 50°C. After 4 minutes, the metal has cooled to 175°C. Find its temperature after 12 minutes.

17. *Greenhouse Gases* Chlorofluorocarbons (CFCs) are gases that increase the greenhouse effect and damage the ozone layer. CFC 12 is one type of chlorofluorocarbon used in refrigeration, air conditioning, and foam insulation. The table lists future concentrations of CFC 12 in parts per billion (ppb) if current trends continue.

Year	2000	2005	2010	2015	2020
CFC 12 (ppb)	.72	.88	1.07	1.31	1.60

Source: Turner, R., *Environmental Economics, an Elementary Approach,* Johns Hopkins University Press, 1993.

(a) Let $x = 0$ correspond to 2000 and $x = 20$ to 2020. Find values for C and a so that $f(x) = Ca^x$ models these data.
(b) Estimate the CFC 12 concentration in 2013.

18. *Global Warming* Greenhouse gases such as carbon dioxide trap heat from the sun. Presently, the net incoming solar radiation reaching Earth's surface is approximately 240 watts per square meter (w/m^2). Any portion of this amount that is due to greenhouse gases is called *radiative forcing*. The table lists the estimated increase in radiative forcing R over the levels in 1750.

x (year)	1800	1850	1900	1950	2000
$R(x)$ (w/m^2)	.2	.4	.6	1.2	2.4

Source: Nilsson, A., *Greenhouse Earth*, John Wiley and Sons, 1992.

(a) Estimate constants C and k so that $R(x) = Ce^{kx}$ models the data. Let $x = 0$ correspond to 1800.
(b) Estimate the year when the additional radiative forcing could reach 3 w/m^2.

19. *Global Warming* See Exercise 18. The relationship between radiative forcing R and the increase in average global temperature T in degrees Fahrenheit can be modeled by

$$T(R) = 1.03R.$$

For example, $T(2) = 1.03(2) = 2.06$ means that if Earth's atmosphere traps an additional 2 watts per square meter, then the average global temperature may increase by 2.06°F if all other factors remain constant. (*Source:* Clime, W., *The Economics of Global Warming,* Institute for International Economics, 1992.)

(a) Use $R(x)$ from Exercise 18 to write an equation for $(T \circ R)(x)$, where x is the year and $x = 0$ corresponds to the year 1800.
(b) Evaluate $(T \circ R)(100)$, and interpret its meaning.

20. *Reducing Carbon Emissions* Governments could reduce carbon emissions by placing a tax on fossil fuels. The cost–benefit equation

$$\ln(1 - P) = -.0034 - .0053x$$

estimates the relationship between a tax of x dollars per ton of carbon and the percent P reduction in emissions of carbon, where P is in decimal form. Determine P when $x = 60$. Interpret the result. (*Source:* Clime, W., *The Economics of Global Warming,* Institute for International Economics, 1992.)

Finance (*Exercises 21–38*)

21. *Interest on an Account* How long will it take for $1000 to grow to $5000 at an interest rate of 3.5% if interest is compounded **(a)** quarterly **(b)** continuously?

22. *Interest on an Account* How long will it take for $5000 to grow to $8400 at an interest rate of 6% if interest is compounded **(a)** semiannually **(b)** continuously?

23. *Interest on an Account* Tom Tupper wants to buy a $30,000 car. He has saved $27,000. Find the number of years (to the nearest tenth) it will take for his $27,000 to grow to $30,000 at 6% interest compounded quarterly.

24. *Doubling Time* Find the doubling time of an investment earning 2.5% interest if interest is compounded **(a)** quarterly **(b)** continuously?

25. *Comparison of Investment* Faye Korn, who is self-employed, wants to invest $60,000 in a pension plan. One investment offers 7% compounded quarterly. Another offers 6.75% compounded continuously. Which investment will earn more interest in 5 years? How much more will the better plan earn?

26. *Growth of an Account* See Exercise 25. If Faye chooses the plan with continuous compounding, how long will it take for her $60,000 to grow to $80,000?

The interest rate stated by a financial institution is sometimes called the ***nominal rate.*** *If interest is compounded, the actual rate is, in general, higher than the nominal rate, and is called the* ***effective rate.*** *If r is the nominal rate and n is the number of times interest is compounded annually, then*

$$R = \left(1 + \frac{r}{n}\right)^n - 1$$

is the effective rate. Here, R represents the annual rate that the investment would earn if simple interest were paid. Use this formula in Exercises 27 and 28.

27. *Effective Rate* Find the effective rate if the nominal rate is 6% and interest is compounded quarterly.

28. *Effective Rate* Find the effective rate if the nominal rate is 4.5% and interest is compounded daily ($n = 365$).

In the formula $A = P\left(1 + \frac{r}{n}\right)^{nt}$, we can interpret P as the ***present value*** *of A dollars t years from now, earning annual interest r compounded n times per year. In this context, A is called the* ***future value.*** *If we solve the formula for P, we obtain*

$$P = A\left(1 + \frac{r}{n}\right)^{-nt}.$$

Use this formula in Exercises 29–32.

29. *Present Value* Find the present value of $10,000 5 years from now if interest is compounded semiannually at 12%.

30. *Present Value* Find the present value of $25,000 2.75 years from now if interest is compounded quarterly at 6%.

31. *Future Value* Find the interest rate necessary for a present value of $25,000 to grow to a future value of $31,360 if interest is compounded annually for 2 years.

32. *Future Value* Find the interest rate necessary for a present value of $1200 to grow to a future value of $1780 if interest is compounded quarterly for 5 years.

In Exercises 33–35, use the amortization formulas given in this section to find **(a)** *the monthly payment on a loan with the given conditions and* **(b)** *the total interest that will be paid during the term of the loan.*

33. *Amortization* $8500 is amortized over 4 years with an interest rate of 7.5%.

34. *Amortization* $9600 is amortized over 5 years with an interest rate of 9.2%.

35. *Amortization* $125,000 is amortized over 30 years with an interest rate of 7.25%.

36. *Amortization Payments* Use the formula

$$y = R\left[\frac{1 - (1 + i)^{-(n-x)}}{i}\right],$$

where y is the unpaid balance after x payments have been made on a loan with n payments of R dollars. Find the balance after 120 payments have been made on the loan in Exercise 35.

37. *Growth of an Account* Use the table feature of your graphing calculator to work parts (a) and (b).

(a) Find how long it will take $1500 invested at 5.75%, compounded daily, to triple in value. Locate the solution by systematically decreasing ΔTbl. Find the answer to the nearest day. (Find your answer to the nearest day by eventually letting $\Delta\text{Tbl} = \frac{1}{365}$. The decimal part of the solution can be multiplied by 365 to determine the number of days greater than the nearest year. For example, if the solution is determined to be 16.2027 years, then multiply .2027 by 365 to get 73.9855. The solution is then, to the nearest day, 16 years and 74 days.) Confirm your answer analytically.

(b) Find how long it will take $2000 invested at 8%, compounded daily, to be worth $5000.

38. *Growth of an Account* If interest is compounded continuously and the interest rate is tripled, what effect will this have on the time required for an investment to double?

Biology and Medicine (Exercises 39–46)

39. *Growth of Bacteria* *Escherichia coli* is a strain of bacteria that occurs naturally in many organisms. Under certain conditions, the number of bacteria present in a colony is approximated by

$$A(t) = A_0 e^{.023t},$$

where t is in minutes. If $A_0 = 2{,}400{,}000$, find the number of bacteria at each time.

(a) 5 minutes **(b)** 10 minutes **(c)** 60 minutes

40. *Growth of Bacteria* The growth of bacteria in food products makes it necessary to date some products (such as milk) so that they will be sold and consumed before the bacteria count becomes too high. Suppose that, under certain storage conditions, the number of bacteria present in a product is given by

$$f(t) = 500e^{.1t},$$

where t is time in days after packing of the product and the value of $f(t)$ is in millions.

(a) If the product cannot be safely eaten after the bacteria count reaches 3,000,000,000, how long will this take?

(b) If $t = 0$ corresponds to January 1, what date should be placed on the product?

A large cloud of radioactive debris from a nuclear explosion has floated over the Pacific Northwest, contaminating much of the hay supply. Farmers in the area are concerned that the cows which eat this hay will give contaminated milk. (The tolerance level for radioactive iodine in milk is 0.) The percent of the initial amount of radioactive iodine still present in the hay after t days is approximated by $P(t) = 100e^{-.1t}$.

Use this information in Exercises 41 and 42.

41. *Radioactive Iodine Level* Some scientists feel that the hay is safe after the level of radioactive iodine has declined to 10% of the original level. Find the number of days before the hay can be used.

42. *Radioactive Iodine Level* Other scientists believe that the hay is not safe until the level of radioactive iodine has declined to only 1% of the original level. Find the number of days this would take.

For Exercises 43 and 44, refer to Example 7.

43. *Drug Level* If 250 milligrams of a drug are administered and the drug is only 75% as effective each subsequent hour, how much effective medicine will remain in the person's system after 6 hours?

44. *Drug Level* A new drug has been introduced that is 80% as effective each hour as the previous hour. A minimum of 20 milligrams must remain in the patient's bloodstream during the course of treatment. If 100 milligrams are administered, how many hours may elapse before another dose is necessary?

45. *Epidemics* In 1666 the village of Eyam, located in England, experienced an outbreak of the Great Plague. Out of 261 people in the community, only 83 survived. The table shows a function f that computes the number of people who had not (yet) been infected after x days.

x	0	15	30	45
$f(x)$	254	240	204	150
x	60	75	90	125
$f(x)$	125	103	97	83

Source: Raggett, G., "Modeling the Eyam Plague," *The Institute of Mathematics and Its Applications* 18.

(a) Use a table to represent a function g that computes the number of people in Eyam who were infected after x days.

(b) Write an equation that shows the relationship between $f(x)$ and $g(x)$.

(c) Use graphing to decide which equation represents $g(x)$ better,

$$y_1 = \frac{171}{1 + 18.6e^{-.0747x}} \quad \text{or} \quad y_2 = 18.3(1.024)^x.$$

(d) Use your results from parts (b) and (c) to find a formula for $f(x)$.

46. *Heart Disease Death Rates* The table lists heart disease death rates per 100,000 people in 2001 for selected ages.

Age	30	40	50	60	70
Death Rate	8.0	29.6	92.9	246.9	635.1

Source: Center for Disease Control.

(a) Make a scatter diagram of the data in the window $[25, 75]$ by $[-100, 700]$.

(b) Find an exponential function f that models the data.

(c) Estimate the heart disease death rate for people who are 80 years old.

Social Sciences (*Exercises 47–52*)

47. *Social Security* If major reform occurs in the Social Security system, individuals may be able to invest some of their contributions into individual accounts. Many of these accounts would be managed by financial firms that charge fees. The table lists the amount in billions of dollars that may be collected if fees are .93% of assets each year.

Year	2005	2010	2015	2020
Fees (in billions)	20	41	80	136

Source: Social Security Advisory Council.

(a) Use exponential regression to find a and b so that $f(x) = ab^x$ models the data. Let $x = 5$ correspond to 2005, $x = 10$ to 2010, and so on.

(b) Graph f and the data.

(c) Estimate the fees in 2013.

48. *Telecommuting* Some workers use technology such as fax machines, e-mail, and computers to work at home rather than in the office. The table lists the expected telecommuters in millions from 2001 to 2006.

Year	2001	2002	2003	2004
Telecommuters	19.7	32.7	41.3	44.4
Year	2005	2006		
Telecommuters	49.5	52.8		

Source: International Telework Association and Council (ITAC).

Find values for a and b so that $f(x) = a + b \ln x$ models the data, where $x = 1$ corresponds to 2001, $x = 2$ to 2002, and so on.

49. *Tribal Gaming Revenues* Revenues in the United States from all forms of tribal gaming increased between 1999 and 2004. The function represented by

$$f(x) = 9.8e^{.1345x}$$

models these revenues in billions of dollars. In this model, x represents the year, where $x = 0$ corresponds to 1999. (*Source:* National Indian Gaming Commission.)

(a) Estimate tribal gaming revenues in 2004.

(b) Determine analytically the year when these revenues reached \$14.7 billion. (Round to the nearest whole number.)

(c) Support your result in part (a) graphically.

50. ***Population Growth*** In 2000 India's population reached 1.003 billion. It is projected to be 1.362 billion in 2025. (*Source:* U.S. Census Bureau.)

(a) Find values for C and a so that $f(x) = Ca^{x-2000}$ models the population of India in year x.

(b) Estimate India's population in 2020.

(c) Use f to determine the year when India's population might reach 1.5 billion.

51. ***Midair Near Collisions*** The table shows the number of aircraft midair near collisions y in year x.

x	2000	2001	2002	2003
y	239	205	180	162

Source: Federal Aviation Administration.

(a) Use exponential regression to approximate constants C and a so that $f(x) = Ca^{x-2000}$ models the data.

(b) Support your answer by graphing f and the data.

52. ***Population Growth*** The population of Tennessee in millions is given by

$$P(x) = 5.33e^{.0133x}$$

where $x = 0$ corresponds to 1995, $x = 5$ to 2000, and so on. (*Source:* U.S. Census Bureau.)

(a) Determine analytically the year when the population of Tennessee was 5.9 million.

(b) Solve part (a) graphically.

(c) According to this model, when will the population reach 6.25 million?

(Modeling) In real life, populations of bacteria, insects, and animals do not continue to grow indefinitely. Initially, population growth may be slow. Then, as their numbers increase, so does the rate of growth. After a region has become heavily populated or saturated, the population usually levels off because of limited resources. This type of growth may be modeled by a ***logistic function*** *represented by*

$$f(x) = \frac{c}{1 + ae^{-bx}},$$

where a, b, and c are positive constants.

Refer to this information for Exercises 53 and 54.

53. ***Heart Disease*** As age increases, so does the likelihood of coronary heart disease (CHD). The fraction of people x years old with some CHD is modeled by

$$f(x) = \frac{.9}{1 + 271e^{-.122x}}.$$

(*Source:* Hosmer, D. and S. Lemeshow, *Applied Logistic Regression,* John Wiley and Sons, 1989.)

(a) Evaluate $f(25)$ and $f(65)$. Interpret the results.

(b) At what age does this likelihood equal 50%.

54. ***Tree Growth*** The height of a tree in feet after x years is modeled by

$$f(x) = \frac{50}{1 + 47.5e^{-.22x}}.$$

(a) Make a table for f starting at $x = 10$ and incrementing by 10. What appears to be the maximum height of the tree?

(b) Graph f and identify the horizontal asymptote. Explain its significance.

(c) After how long was the tree 30 feet tall?

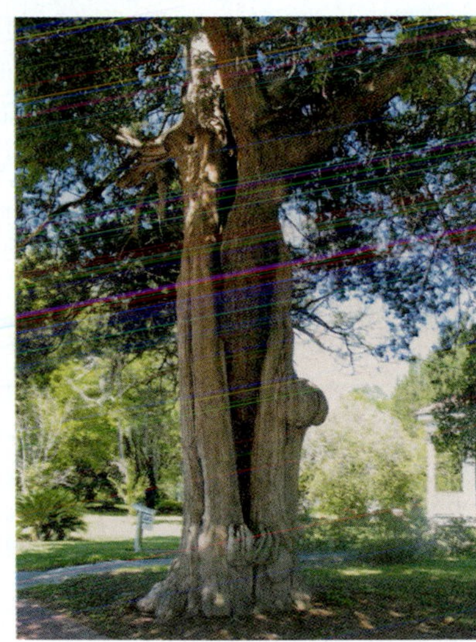

CHAPTER 5 SUMMARY

KEY TERMS & SYMBOLS	KEY CONCEPTS
5.1 Inverse Functions one-to-one function inverse function f^{-1} (f-inverse)	**ONE-TO-ONE FUNCTION** A function f is a one-to-one function if, for elements a and b from the domain of f, $a \neq b$ implies $f(a) \neq f(b)$. **HORIZONTAL LINE TEST** If every horizontal line intersects the graph of a function at no more than one point, then the function is one-to-one. **INVERSE FUNCTION** Let f be a one-to-one function. Then g is the inverse function of f and f is the inverse function of g if $(f \circ g)(x) = x$ for every x in the domain of g, and $(g \circ f)(x) = x$ for every x in the domain of f. **FINDING THE EQUATION OF THE INVERSE OF $y = f(x)$** For a one-to-one function f defined by an equation $y = f(x)$, find the defining equation of the inverse as follows: ***Step 1*** Interchange x and y. ***Step 2*** Solve for y. ***Step 3*** Replace y with $f^{-1}(x)$. (Any restrictions on x and y should be considered.) **IMPORTANT FACTS ABOUT INVERSES** **1.** If f is one-to-one, then f^{-1} exists. **2.** The domain of f is equal to the range of f^{-1}, and the range of f is equal to the domain of f^{-1}. **3.** If the point (a, b) lies on the graph of f, then (b, a) lies on the graph of f^{-1}. The graphs of f and f^{-1} are reflections of each other across the line $y = x$.
5.2 Exponential Functions exponential function the number e	**EXPONENTIAL FUNCTION** If $a > 0$, $a \neq 1$, then $f(x) = a^x$ is the exponential function with base a. **CHARACTERISTICS OF EXPONENTIAL FUNCTIONS** $f(x) = a^x, \quad a > 1$ ■ This function is increasing and continuous on its entire domain, $(-\infty, \infty)$. ■ The x-axis is the horizontal asymptote as $x \to -\infty$. ■ The graph passes through the points $(-1, a^{-1})$, $(0, 1)$, and $(1, a)$. $f(x) = a^x, \quad 0 < a < 1$ ■ This function is decreasing and continuous on its entire domain, $(-\infty, \infty)$. ■ The x-axis is the horizontal asymptote as $x \to \infty$. ■ The graph passes through the points $(-1, a^{-1})$, $(0, 1)$, and $(1, a)$.

(continued)

KEY TERMS & SYMBOLS	KEY CONCEPTS
	SOLVING EXPONENTIAL EQUATIONS (TYPE 1) To solve a Type 1 exponential equation such as $4^x = 8^{2x-3}$, write each base as a power of the same rational number base, apply the power rule for exponents, set exponents equal, and solve the resulting equation. **COMPOUNDING FORMULAS** Compounded n times per year: $A = P\left(1 + \frac{r}{n}\right)^{nt}$ Compounded continuously: $A = Pe^{rt}$
5.3 Logarithms and Their Properties logarithm, $\log_a x$ base argument common logarithm, $\log x$ pH natural logarithm, $\ln x$	**LOGARITHM** For all positive numbers a, where $a \neq 1$, $a^y = x$ is equivalent to $y = \log_a x$. A logarithm is an exponent, and $\log_a x$ is the exponent to which a must be raised in order to obtain x. The number a is called the base of the logarithm, and x is called the argument of the expression $\log_a x$. The value of x will always be positive. **COMMON LOGARITHM** For all positive numbers x, $\log x = \log_{10} x$. **NATURAL LOGARITHM** For all positive numbers x, $\ln x = \log_e x$. **PROPERTIES OF LOGARITHMS** For $a > 0$, $a \neq 1$, and any real number k, **1.** $\log_a 1 = 0$. **2.** $\log_a a^k = k$. **3.** $a^{\log_a k} = k$, $k > 0$. For $x > 0$, $y > 0$, $a > 0$, $a \neq 1$, and any real number r, **Product Rule** **4.** $\log_a xy = \log_a x + \log_a y$. (The logarithm of the product of two numbers is equal to the sum of the logarithms of the numbers.) **Quotient Rule** **5.** $\log_a \frac{x}{y} = \log_a x - \log_a y$. (The logarithm of the quotient of two numbers is equal to the difference of the logarithms of the numbers.) **Power Rule** **6.** $\log_a x^r = r \log_a x$. (The logarithm of a number raised to a power is equal to the exponent multiplied by the logarithm of the number.) **CHANGE-OF-BASE RULE** For any positive real numbers x, a, and b, where $a \neq 1$ and $b \neq 1$, $\log_a x = \frac{\log_b x}{\log_b a}$. *(continued)*

KEY TERMS & SYMBOLS	KEY CONCEPTS
5.4 Logarithmic Functions logarithmic function	The functions defined by $f(x) = a^x$ and $g(x) = \log_a x$ are inverses. **CHARACTERISTICS OF LOGARITHMIC FUNCTIONS** $f(x) = \log_a x, \quad a > 1$ ■ This function is increasing and continuous on its entire domain, $(0, \infty)$. ■ The y-axis is the vertical asymptote as $x \to 0$ from the right. ■ The graph passes through the points $(a^{-1}, -1)$, $(1, 0)$, and $(a, 1)$. $f(x) = \log_a x, \quad 0 < a < 1$ ■ This function is decreasing and continuous on its entire domain, $(0, \infty)$. ■ The y-axis is the vertical asymptote as $x \to 0$ from the right. ■ The graph passes through the points $(a, 1)$, $(1, 0)$, and $(a^{-1}, -1)$.
5.5 Exponential and Logarithmic Equations and Inequalities	**PROPERTIES OF LOGARITHMIC AND EXPONENTIAL FUNCTIONS** For $b > 0$ and $b \neq 1$, 1. $b^x = b^y$ **if and only if** $x = y$. 2. If $x > 0$ and $y > 0$, then $\log_b x = \log_b y$ **if and only if** $x = y$. **SOLVING EXPONENTIAL AND LOGARITHMIC EQUATIONS** To solve a Type 2 exponential equation such as $2^x = 3^{x+1}$, take the same base logarithm on each side, apply the power rule for logarithms so that the variables are no longer in the exponents, and solve the resulting equation. An exponential or logarithmic equation can be solved by changing the equation into one of the following forms, where a and b are real numbers, $a > 0$, and $a \neq 1$: 1. $a^{f(x)} = b$ Solve by taking logarithms of each side. (Natural logarithms are the best choice if $a = e$.) 2. $\log_a f(x) = \log_a g(x)$ From the given equation, $f(x) = g(x)$, which is solved analytically. 3. $\log_a f(x) = b$ Solve by changing to exponential form, $f(x) = a^b$.
5.6 Further Applications and Modeling with Exponential and Logarithmic Functions exponential growth function exponential decay function half-life doubling time amortized	**GROWTH** $A(t) = A_0 e^{kt}, \quad k > 0$ **DECAY** $A(t) = A_0 e^{-kt}, \quad k > 0$ **SOUND INTENSITY** $d = 10 \log \frac{I}{I_0}$ **AMORTIZATION PAYMENTS** $R = \dfrac{P}{\left[\dfrac{1 - (1 + i)^{-n}}{i}\right]}$

CHAPTER 5 Review Exercises

Determine whether each function is one-to-one. Assume that the graphs shown are comprehensive.

1.

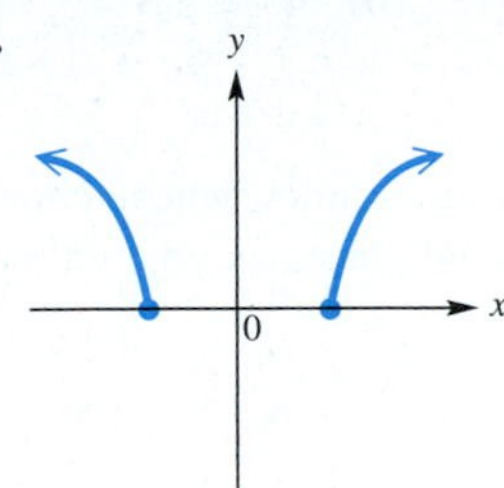

2.

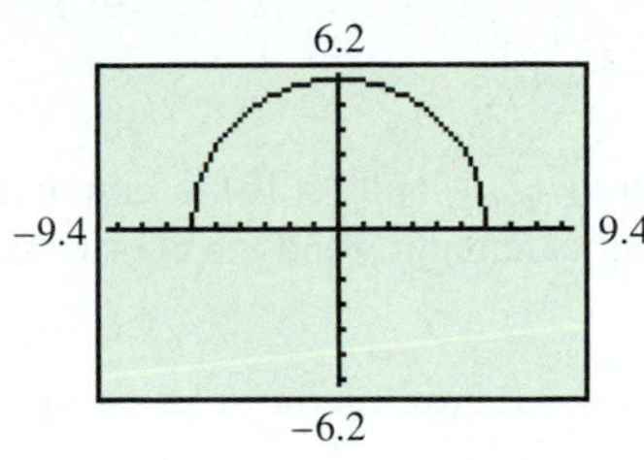

3.

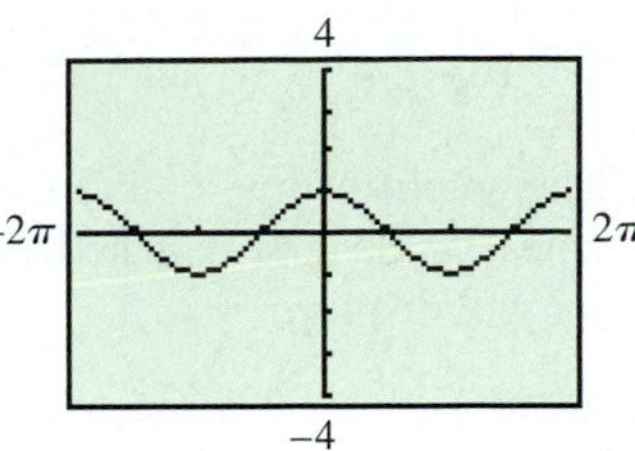

4. $f(x) = \sqrt{3x + 2}$

For Exercises 5–10, consider the function defined by $f(x) = \sqrt[3]{2x - 7}$.

5. What is the domain of f?

6. What is the range of f?

7. Explain why f^{-1} exists.

8. Find a formula for $f^{-1}(x)$.

9. Graph both f and f^{-1} in a square viewing window, along with the line $y = x$. Describe how the graphs of f and f^{-1} are related.

10. Verify analytically that

$$(f \circ f^{-1})(x) = x \quad \text{and} \quad (f^{-1} \circ f)(x) = x.$$

Concept Check *Match each equation with the graph that most closely resembles its graph. Assume that* $a > 1$.

11. $y = a^{x+2}$

12. $y = a^x + 2$

13. $y = -a^x + 2$

14. $y = a^{-x} + 2$

A.

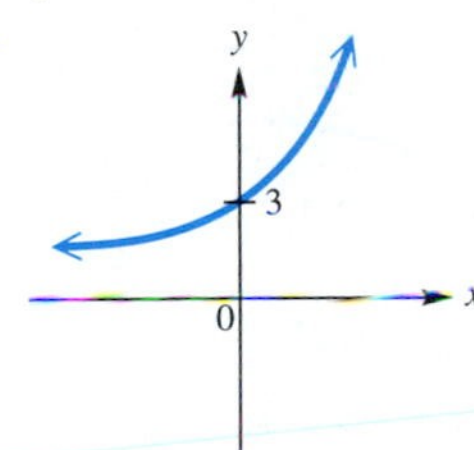

B.

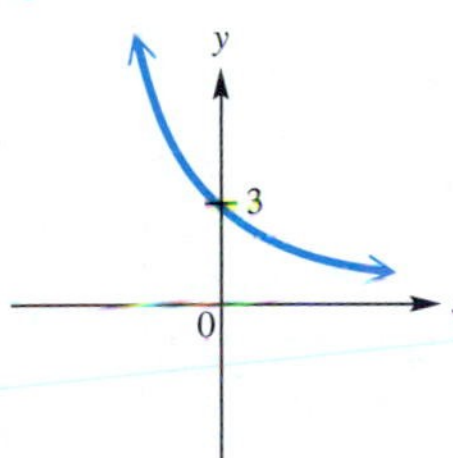

C.

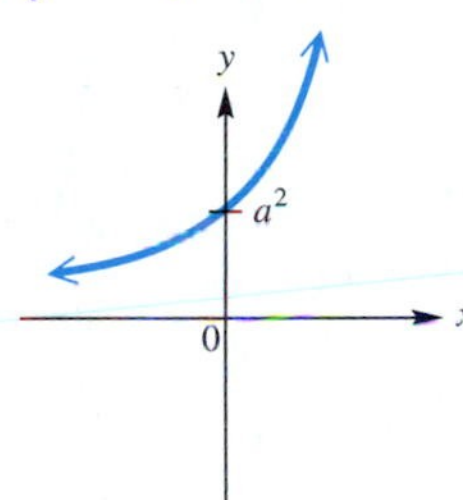

D.

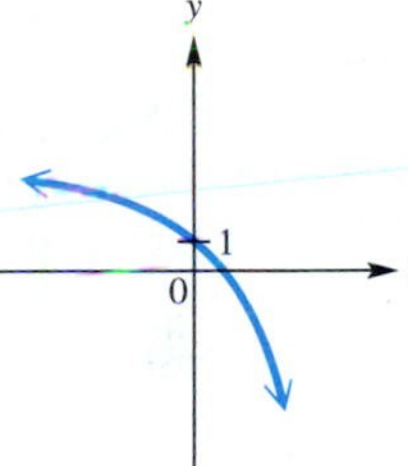

Concept Check *Consider the exponential function defined by* $y = f(x) = a^x$, *graphed at the right. Answer the following based on the graph.*

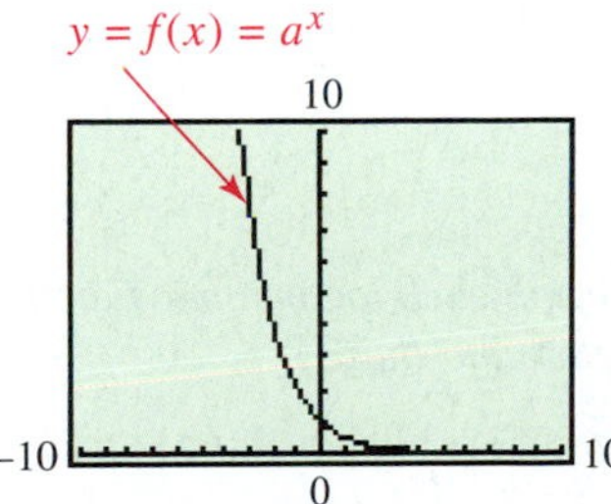

15. What is true about the value of a in comparison to 0 and 1?

16. What is the domain of f?

17. What is the range of f?

18. What is the value of $f(0)$?

19. Sketch the graph of $y = f^{-1}(x)$ by hand.

20. What is the expression that defines $f^{-1}(x)$?

Graph each exponential function. Give the domain and range.

21. $f(x) = \left(\dfrac{1}{2}\right)^{x-1}$

22. $f(x) = e^x + 2$

23. $f(x) = -4^x$

24. *Concept Check* If $f(x) = a^{-x}$ and $0 < a < 1$, is f increasing or decreasing on its domain?

Solve the equation in part (a) analytically. Then use a graph to solve the inequality in part (b).

25. **(a)** $\left(\frac{1}{8}\right)^{-2x} = 2^{x+3}$
(b) $\left(\frac{1}{8}\right)^{-2x} \geq 2^{x+3}$

26. **(a)** $3^{-x} = \left(\frac{1}{27}\right)^{1-2x}$
(b) $3^{-x} < \left(\frac{1}{27}\right)^{1-2x}$

27. **(a)** $.5^{-x} = .25^{x+1}$
(b) $.5^{-x} > .25^{x+1}$

28. **(a)** $.4^{x} = 2.5^{1-x}$
(b) $.4^{x} < 2.5^{1-x}$

29. The graphs of $y = x^2$ and $y = 2^x$ have the points $(2, 4)$ and $(4, 16)$ in common. There is a third point in common whose coordinates can be approximated by using a graphing calculator. Find the coordinates, giving as many decimal places as your calculator will display.

30. Solve $3^x = \pi$ by the intersection-of-graphs method. Express the solution to the nearest thousandth.

Use a calculator to find an approximation for each logarithm to four decimal places.

31. $\log 58.3$

32. $\log .00233$

33. $\ln 58.3$

34. $\log_2 .00233$

Evaluate each expression, giving an exact or approximate value as directed. In the case of approximations, give four decimal places.

35. $\log_{13} 1$ (exact)

36. $\ln e^{\sqrt{6}}$ (exact)

37. $\log_5 5^{12}$ (exact)

38. $7^{\log_7 13}$ (exact)

39. $3^{\log_3 5}$ (exact)

40. $\log_4 9$ (approximate)

Concept Check *In Exercises 41–46, identify the corresponding graph in choices A–F for each function.*

41. $f(x) = \log_2 x$

42. $f(x) = \log_2(2x)$

43. $f(x) = \log_2\left(\frac{1}{x}\right)$

44. $f(x) = \log_2\left(\frac{x}{2}\right)$

45. $f(x) = \log_2(x - 1)$

46. $f(x) = \log_2(-x)$

A.

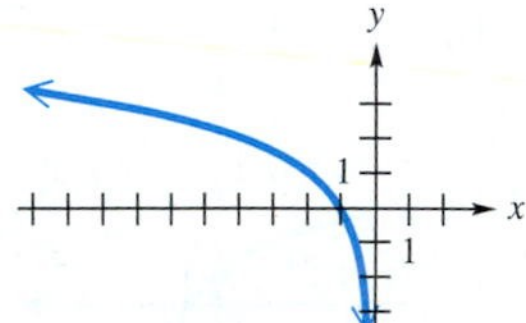

B.

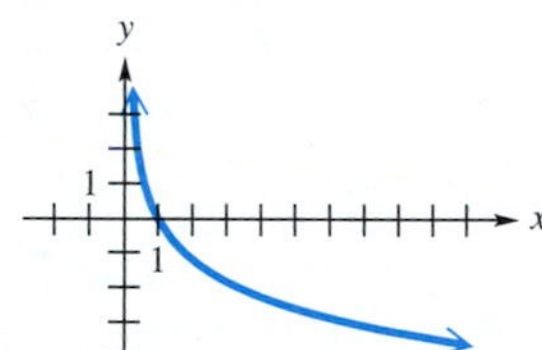

C.

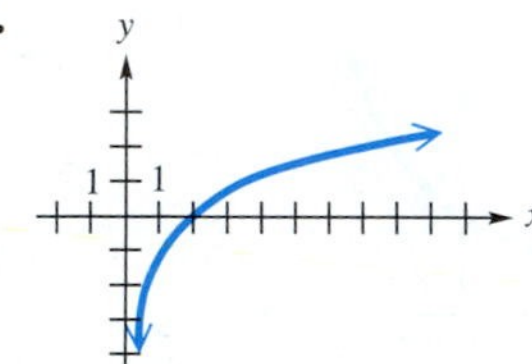

D.

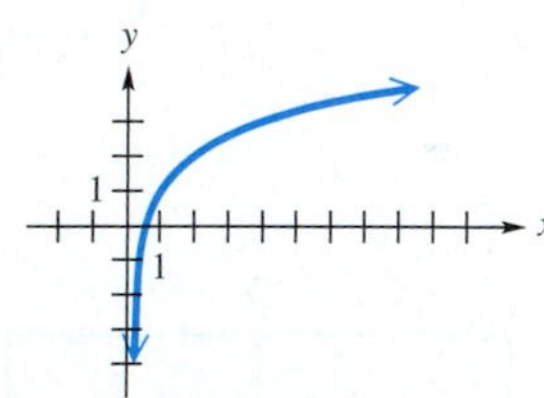

E.

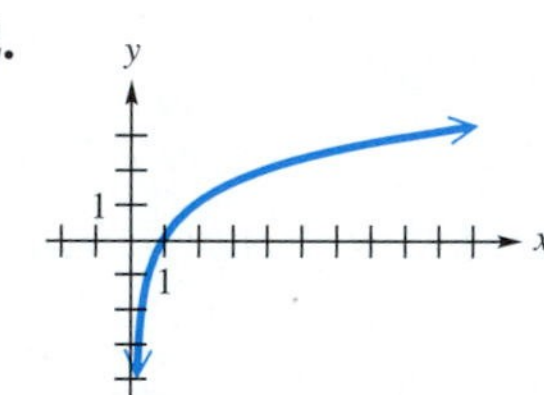

F.

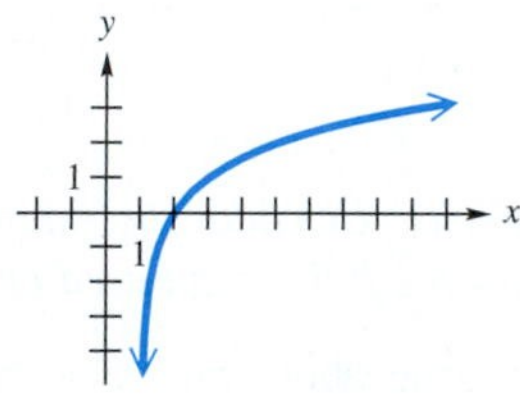

Graph each logarithmic function. Give the domain and range. (Hint: See Exercises 21–23, as those exponential functions are closely related to these.)

47. $g(x) = 1 + \log_{1/2} x$

48. $g(x) = \ln(x - 2)$

49. $g(x) = \log_4(-x)$

50. **Concept Check** How do the functions in Exercises 47–49 relate to those in Exercises 21–23?

51. **Concept Check** What is the base of the logarithmic function whose graph contains the point $(81, 4)$?

52. **Concept Check** What is the base of the exponential function whose graph contains the point $\left(-4, \frac{1}{16}\right)$?

Write each logarithm as a sum, difference, or product of logarithms. Assume variables represent positive numbers.

53. $\log_3 \frac{mn}{5r}$

54. $\log_2 \frac{\sqrt{7}}{15}$

55. $\log_5\left(x^2 y^4 \sqrt[5]{m^3 p}\right)$

56. $\log_7(7k + 5r^2)$

Solve the equation in part (a) analytically. Then use a graph to solve the inequality in part (b).

57. **(a)** $\log(x + 3) + \log x = 1$
(b) $\log(x + 3) + \log x > 1$

58. **(a)** $\ln e^{\ln x} - \ln(x - 4) = \ln 3$
(b) $\ln e^{\ln x} - \ln(x - 4) \leq \ln 3$

59. **(a)** $\ln e^{\ln 2} - \ln(x - 1) = \ln 5$
(b) $\ln e^{\ln 2} - \ln(x - 1) \geq \ln 5$

Solve each equation. Express the solution set so that **(a)** *solutions are in exact form and, if irrational,* **(b)** *solutions are approximated to the nearest thousandth. Support your solutions by using a calculator.*

60. $8^x = 32$

61. $\frac{8}{27} = x^{-3}$

62. $10^{2x-3} = 17$

63. $e^{x+1} = 10$

64. $\log_{64} x = \frac{1}{3}$

65. $\ln(6x) - \ln(x + 1) = \ln 4$

66. $\log_{12}(2x) + \log_{12}(x - 1) = 1$

67. $\log_{16}\sqrt{x + 1} = \frac{1}{4}$

68. $\ln x + 3\ln 2 = \ln\left(\frac{2}{x}\right)$

69. $\ln[\ln(e^{-x})] = \ln 3$

70. $\ln e^x - \ln e^3 = \ln e^5$

71. $2^x = -3$

Solve each formula for the indicated variable.

72. $N = a + b\ln\left(\frac{c}{d}\right)$, for c

73. $y = y_0 e^{-kt}$, for t

Use the x-intercept method to estimate the solution(s) of each equation. Round to the nearest thousandth.

74. $\log_{10} x = x - 2$

75. $2^{-x} = \log_{10} x$

76. $x^2 - 3 = \log x$

Solve each application.

77. ***Interest Rate*** To the nearest tenth, what interest rate, compounded annually, will produce \$8780 if \$3500 is left at interest for 10 years?

78. ***Growth of an Account*** Find the number of years (to the nearest tenth) needed for \$48,000 to become \$58,344 at 5% interest compounded semiannually.

79. ***Growth of an Account*** Lateisha Shaw deposits \$12,000 for 8 years in an account paying 5% compounded annually. She then leaves the money alone, with no further deposits, at 6% compounded annually for an additional 6 years. Find the total amount on deposit after the entire 14-year period.

80. ***Growth of an Account*** Suppose that \$2000 is invested in an account that pays 3% annually and is left for 5 years.

(a) How much will be in the account if interest is compounded quarterly (4 times per year)?

(b) How much will be in the account if interest is compounded continuously?

(c) To the nearest tenth of a year, how long will it take the \$2000 to triple if interest is compounded continuously?

(Modeling) Solve each problem.

81. ***Runway Length*** There is a mathematical relationship between an airplane's weight x and the runway length required at takeoff. For some airplanes, the minimum runway length in thousands of feet may be modeled by

$$L(x) = 3\log x,$$

where x is measured in thousands of pounds. (*Source:* Haefner, L., *Introduction to Transportation Systems,* Holt, Rinehart and Winston, 1986.)

(a) Graph L in the window $[0, 50]$ by $[0, 6]$. Interpret the graph.

(b) If the weight of an airplane increases tenfold from 10,000 to 100,000 pounds, does the length of the required runway also increase by a factor of 10? Explain.

82. ***Caloric Intake*** The function defined by

$$f(x) = 280\ln(x + 1) + 1925$$

models the number of calories consumed daily by a person owning x acres of land in a developing country. Find the number of acres owned by someone whose average calorie intake is 2200 calories per day. (*Source:* Grigg, D., *The World Food Problem,* Blackwell Publishers, 1993.)

83. *Pollutant Concentration* The concentration of a pollutant, in grams per liter, in the east fork of the Big Weasel River is approximated by

$$P(x) = .04e^{-4x},$$

where x is the number of miles downstream from a paper mill where the measurement is taken. Find the following:

(a) $P(.5)$ **(b)** $P(1)$

(c) The concentration of the pollutant 2 miles downstream

(d) The number of miles downstream where the concentration of the pollutant is .002 gram per liter.

84. *Repetitive Skills* A person learning certain skills involving repetition tends to learn quickly at first. Then, learning tapers off and approaches some upper limit. Suppose the number of symbols per minute that a textbook typesetter can produce is given by

$$p(x) = 250 - 120(2.8)^{-.5x},$$

where x is the number of months the typesetter has been in training. Find the following:

(a) $p(2)$ **(b)** $p(10)$.

(c) Graph $y = p(x)$ in the window $[0, 10]$ by $[0, 300]$, and support the answer to part (a).

85. *Free Fall of a Skydiver* A skydiver in free fall travels at a speed of

$$v(t) = 176(1 - e^{-.18t})$$

feet per second after t seconds. How long will it take for the skydiver to attain a speed of 147 feet per second (100 mph)?

CHAPTER 5 Test

1. Match each equation in parts (a)–(d) with its graph.

(a) $y = \log_{1/2} x$ **(b)** $y = e^x$

(c) $y = \ln x$ **(d)** $y = \left(\frac{1}{2}\right)^x$

(e) Which pairs of functions in parts (a)–(d) are inverses?

A.

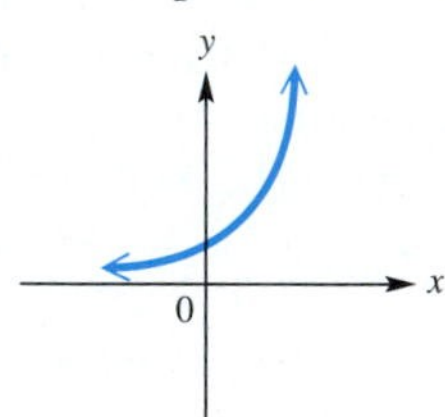

B.

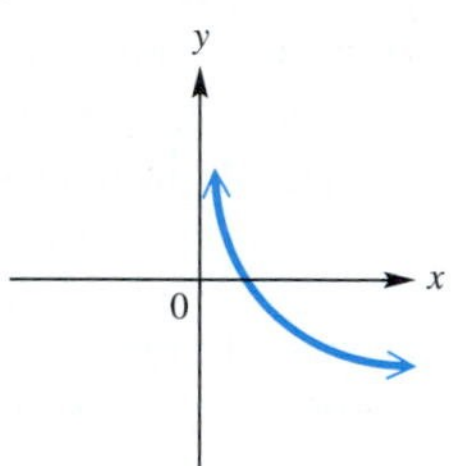

C.

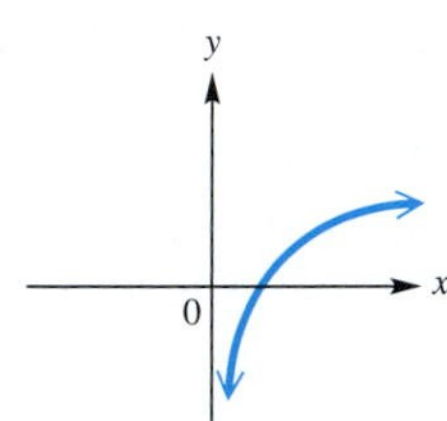

D.

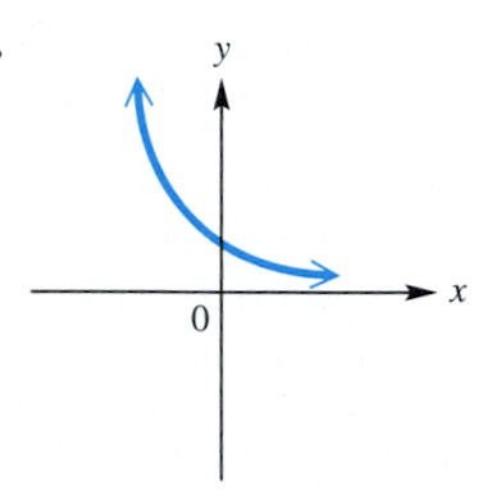

2. Consider the function defined by $f(x) = -2^{x-1} + 8$.

(a) Graph f in the standard viewing window of your calculator.

(b) Give the domain and range of f.

(c) Does the graph have an asymptote? If so, is it vertical or horizontal, and what is its equation?

(d) Find the x- and y-intercepts analytically, and use the graph from part (a) to support your answers graphically.

(e) Find $f^{-1}(x)$.

3. Solve the equation $\left(\frac{1}{8}\right)^{2x-3} = 16^{x+1}$ analytically.

4. *Growth of an Account* Suppose that \$10,000 is invested at 5.5% for 4 years. Find the total amount present at the end of that period if interest is compounded **(a)** quarterly and **(b)** continuously.

5. One of your friends is taking another mathematics course and tells you, "I have no idea what an expression like $\log_5 27$ really means." Write an explanation of what it means, and tell how you can find an approximation for it with a calculator.

6. Use a calculator to find an approximation of each logarithm to the nearest thousandth.

(a) $\log 45.6$ **(b)** $\ln 470$ **(c)** $\log_3 769$

7. Use the properties of logarithms to write $\log \frac{m^3n}{\sqrt{y}}$ as an equivalent expression.

8. Solve $A = Pe^{rt}$ for t.

9. Consider the equation $\log_2 x + \log_2(x + 2) = 3$.

(a) Solve the equation analytically. If there is an extraneous value, what is it?

(b) To support the solution in part (a), we can graph $y_1 = \log_2 x + \log_2(x + 2) - 3$ and find the x-intercept. Write an expression for y_1, using the change-of-base rule with base 10, and graph the function.

(c) Use the graph to solve the logarithmic inequality $\log_2 x + \log_2(x + 2) > 3$.

Solve each equation. Give the solution set **(a)** *with an exact value and then* **(b)** *with an approximation to the nearest thousandth.*

10. $2e^{5x+2} = 8$

11. $6^{2-x} = 2^{3x+1}$

12. $\log(\ln x) = 1$

13. *(Modeling) Population Growth* A population is increasing according to the equation

$$y = 2e^{.02t},$$

where y is in millions and t is in years. Match each question in parts (a)–(d) with one of the solution methods A–D.

(a) How long will it take for the population to triple?
(b) When will the population reach 3 million?
(c) How large will the population be in 3 years?
(d) How large will the population be in 4 months?

A. Evaluate $2e^{.02(1/3)}$.
B. Solve $2e^{.02t} = 3 \cdot 2$ for t.
C. Evaluate $2e^{.02(3)}$.
D. Solve $2e^{.02t} = 3$ for t.

14. *(Modeling) Drug Level in the Bloodstream* After a medical drug is injected into the bloodstream, it is gradually eliminated from the body. Graph each function in parts (a)–(d) on the interval $[0, 10]$. Use $[0, 500]$ as the range of $A(t)$. Use a graphing calculator to determine the function that best models the amount $A(t)$ (in milligrams) of a drug remaining in the body after t hours if 350 milligrams were initially injected.

(a) $A(t) = t^2 - t + 350$
(b) $A(t) = 350 \log(t + 1)$
(c) $A(t) = 350(.75)^t$
(d) $A(t) = 100(.95)^t$

15. *(Modeling) Decay of Radium* Radium 226 has a half-life of about 1600 years. An initial sample weighs 2 grams.

(a) Find a formula for the decay function.
(b) Find the amount left after 9600 years.
(c) Find the time for the initial amount to decay to .5 gram.

CHAPTER 5 PROJECT

Modeling Motor Vehicle Sales in the United States (with a lesson about the careless use of mathematical models)

During the 1990s, both the numbers of new motor vehicles sold and the numbers of new trucks sold in the United States increased significantly. The table gives approximations for these numbers in 1992, 1996, and 1999. Note that new motor vehicle sales *include* new truck sales.

Year	New Motor Vehicle Sales	New Truck Sales
1992	13,100,000	4,900,000
1996	15,500,000	6,900,000
1999	17,400,000	8,700,000

Source: U.S. Bureau of Economic Analysis.

Growth in sales can sometimes be modeled by a function having the form $f(x) = a(b)^x$, where a and b are constants. The constant a represents the sales during an initial year, and the variable x corresponds to the number of years that have passed since the initial year. Once a rule for $f(x)$ is found, the constant b can be used to determine how fast sales increased. For example, if $b = 1.03$, then sales increased by 3% per year.

Activities

1. Find a function defined by $f(x) = a(b)^x$ that models new motor vehicle sales (in millions) in 1992 and in 1999. Let $x = 0$ correspond to 1992 and $x = 7$ to 1999. (*Hint:* To determine values for a and b, let $f(0) = 13.1$ and $f(7) = 17.4$.)
2. Now, similar to the way $f(x)$ was found, find a function defined by $g(x) = c(d)^x$ that models new truck sales (in millions) in 1992 and in 1999.
3. Use $f(x)$ and $g(x)$ to estimate, to the nearest tenth of a percent, the annual increases in sales of new motor vehicles and in sales of new trucks from 1992 to 1999.
4. Estimate sales of new motor vehicles and sales of new trucks in 1996. Compare your results with the actual values shown in the table.
5. Use $f(x)$ and $g(x)$ to predict new motor vehicle sales and new truck sales in 2020. Discuss your results.
6. Graph f and g together in the window $[0, 33]$ by $[0, 80]$. Do the functions f and g give realistic sales figures from 1992 to 2025? Explain. Comment on the following statement: "A model is best when it is used to estimate values between given data points."

6

Analytic Geometry

ASTEROIDS CAN BE likened to interplanetary garbage. These large rocks hurtle through space, following an orbit around the sun. When an asteroid strikes a planet or moon, the destruction can be devastating.

In June 2004, astronomers working at Kitt Peak National Observatory near Tucson, Arizona, discovered an asteroid one-quarter mile in diameter that appeared to be heading toward Earth. In December, the asteroid, named 2004 MN4, rose from level 2 to level 4 on the Torino probability-of-impact scale, making it the first near-Earth asteroid to reach the latter level. The projected path showed that 2004 MN4 could hit Earth on April 13, 2029, releasing energy equivalent to a 1000-to-1500-megaton bomb. After exhaustive research and refined calculations of the asteroid's orbit, astronomers determined in January 2005 that 2004 MN4 would not hit Earth, but would likely pass within 41,000 miles of it.

Paths of planets, asteroids, comets, and objects launched into space can be modeled by equations and graphs of *conic sections*. We study conic sections in this chapter.

Source: http://impact.arc.nasa.gov

Chapter Outline

6.1 Circles and Parabolas

Conic Sections ■ Equations and Graphs of Circles ■ Equations and Graphs of Parabolas ■ Translations of Parabolas ■ An Application of Parabolas

Conic Sections

Parabolas, *circles*, *ellipses*, and *hyperbolas* form a group of curves known as the **conic sections,** because they are the result of intersecting a cone with a plane. Figure 1 illustrates these curves, which all can be defined mathematically by using the distance formula from Section 1.1,

$$d(P, R) = \sqrt{(x_2 - x_1)^2 + (y_2 - y_1)^2}. \quad \text{Distance formula}$$

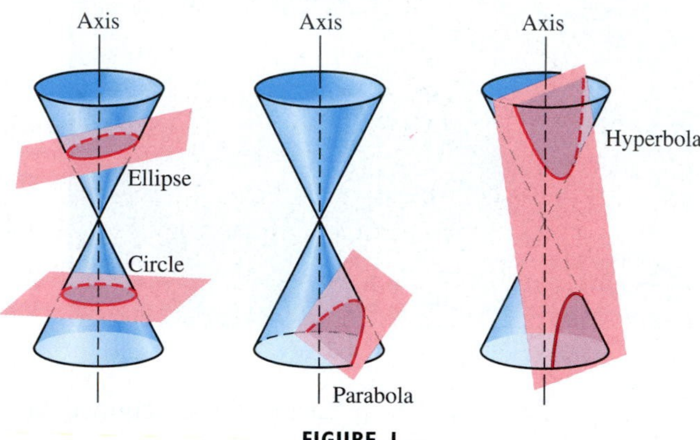

FIGURE 1

If the plane intersects the cone at the vertex, the intersections illustrated in Figure 1 are reduced respectively to a point, a line, and two intersecting lines. These are called *degenerate conic sections.*

Equations and Graphs of Circles

> **Circle**
>
> A **circle** is a set of points in a plane that are equidistant from a fixed point. The distance is called the **radius** of the circle, and the fixed point is called the **center.**

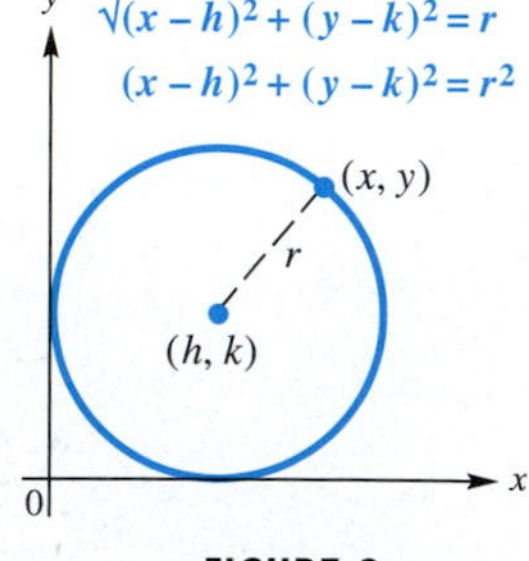

FIGURE 2

Suppose a circle has center (h, k) and radius $r > 0$. See Figure 2. Then the distance between the center (h, k) and any point (x, y) on the circle must equal r. Thus, an equation of the circle is

$$\sqrt{(x - h)^2 + (y - k)^2} = r \quad \text{Distance formula}$$

or

$$(x - h)^2 + (y - k)^2 = r^2. \quad \text{Square both sides.}$$

> **Center–Radius Form of the Equation of a Circle**
>
> The **center–radius form** of the equation of a circle with center (h, k) and radius r is
>
> $$(x - h)^2 + (y - k)^2 = r^2.$$

Notice that a circle is the graph of a relation that is not a function, since it does not pass the vertical line test.

EXAMPLE 1 Finding the Equation of a Circle

Find the center–radius form of the equation of a circle with radius 6 and center $(-3, 4)$. Graph the circle by hand. Give the domain and range of the relation.

Solution Using the center–radius form with $h = -3$, $k = 4$, and $r = 6$, we find that the equation of the circle is

$$[x - (-3)]^2 + (y - 4)^2 = 6^2 \quad \text{or} \quad (x + 3)^2 + (y - 4)^2 = 36.$$

The graph is shown in Figure 3. Because the center is $(-3, 4)$ and the radius is 6, the circle must pass through the four points $(-3 \pm 6, 4)$ and $(-3, 4 \pm 6)$, as illustrated in the table. Using these four points, we see that the domain is $[-9, 3]$ and the range is $[-2, 10]$.

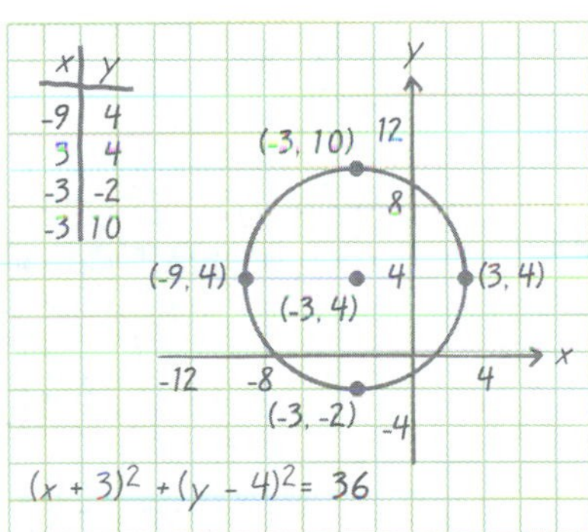

FIGURE 3

If a circle has center at the origin $(0, 0)$, then its equation is found by using $h = 0$ and $k = 0$ in the center–radius form.

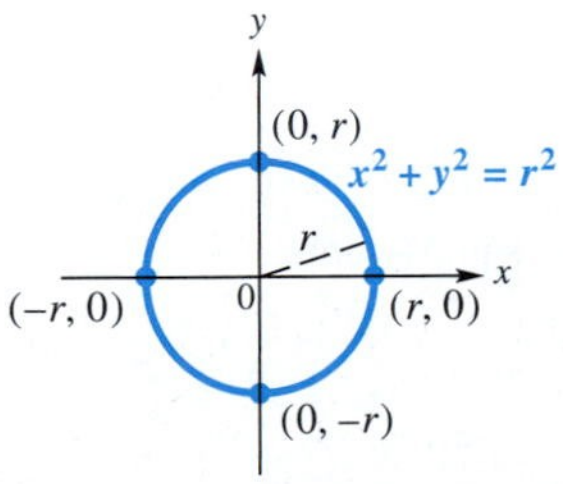

FIGURE 4

> **Equation of a Circle with Center at the Origin**
>
> A circle with center $(0, 0)$ and radius r (see Figure 4) has equation
>
> $$x^2 + y^2 = r^2.$$

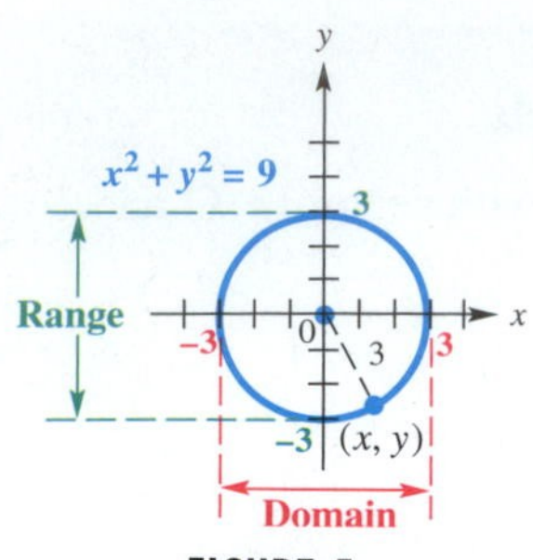

FIGURE 5

TECHNOLOGY NOTE

To obtain an undistorted graph of a circle in function mode on a graphing calculator screen, a *square viewing window* must be used. For example, the graph of $x^2 + y^2 = 9$ in a rectangular window looks like an ellipse.

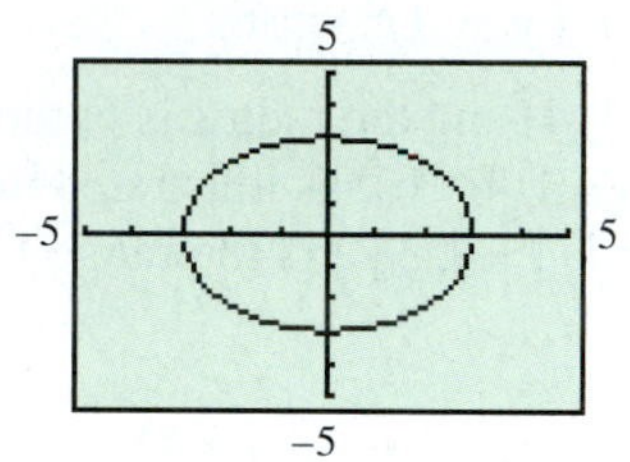

EXAMPLE 2 Finding the Equation of a Circle with Center at the Origin

Find the equation of a circle with center at the origin and radius 3. Graph the relation, and state the domain and range.

Solution Using the form $x^2 + y^2 = r^2$ with $r = 3$, we find that the equation of the circle is $x^2 + y^2 = 9$. See Figure 5. The domain and range are $[-3, 3]$. ■

As mentioned in Section 4.4, a graphing calculator in function mode *cannot* directly graph a circle. We must solve the equation of the circle for y, obtaining two functions y_1 and y_2. The union of these two graphs is the graph of the circle.

GCM

EXAMPLE 3 Graphing Circles

Use a graphing calculator to graph each circle in a square viewing window.

(a) $x^2 + y^2 = 9$ **(b)** $(x + 3)^2 + (y - 4)^2 = 36$

Solution In both cases, first solve for y. Recall that if $k > 0$, $y^2 = k$ has *two* real solutions: $\sqrt{k}$ and $-\sqrt{k}$.

(a)

$$x^2 + y^2 = 9$$

$$y^2 = 9 - x^2 \quad \text{Subtract } x^2.$$

$$y = \pm\sqrt{9 - x^2} \quad \text{Take square roots.}$$

Remember both the positive and negative square roots.

Graph $y_1 = \sqrt{9 - x^2}$ and $y_2 = -\sqrt{9 - x^2}$. See Figure 6, and compare with Figure 5.

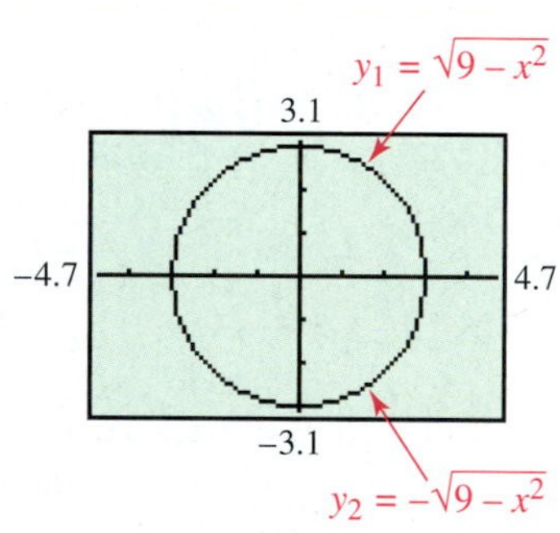

FIGURE 6

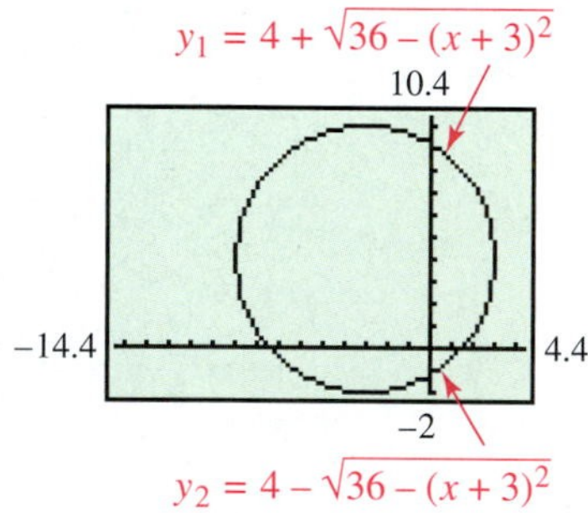

FIGURE 7

TECHNOLOGY NOTE

If we graph the circle in Figure 6 without using a decimal window, we may get the figure shown below, where the two graphs do not meet at the points $(-3, 0)$ and $(3, 0)$. In some cases, a decimal window does *not* correct this problem. Although the technology is deceiving, mathematically this is a complete circle.

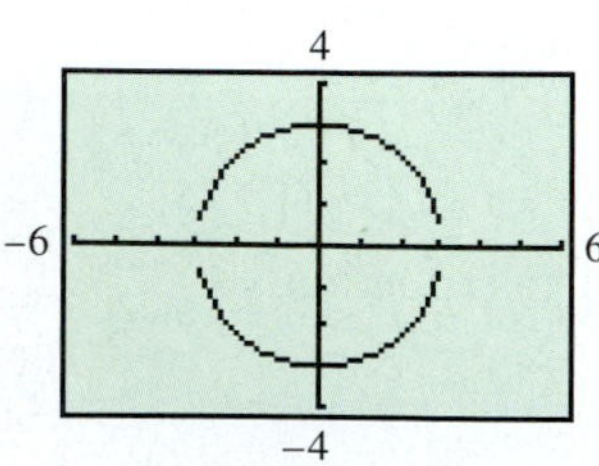

(b)

$$(x + 3)^2 + (y - 4)^2 = 36$$

$$(y - 4)^2 = 36 - (x + 3)^2 \quad \text{Subtract } (x + 3)^2.$$

$$y - 4 = \pm\sqrt{36 - (x + 3)^2} \quad \text{Take square roots.}$$

$$y = 4 \pm \sqrt{36 - (x + 3)^2} \quad \text{Add 4.}$$

The two functions graphed in Figure 7 are

$$y_1 = 4 + \sqrt{36 - (x + 3)^2} \quad \text{and} \quad y_2 = 4 - \sqrt{36 - (x + 3)^2}.$$ ■

Starting with $(x - h)^2 + (y - k)^2 = r^2$ and squaring $x - h$ and $y - k$ gives the **general form of the equation of a circle,**

$$x^2 + y^2 + cx + dy + e = 0, \quad (*)$$

where c, d, and e are real numbers. Given an equation in this form, we can complete the square to get an equation of the form

$$(x - h)^2 + (y - k)^2 = m, \qquad \text{for some number } m.$$

Looking Ahead to Calculus

The circle $x^2 + y^2 = 1$ is called the *unit circle*. It is important in interpreting the *trigonometric* or *circular* functions in calculus.

There are three possibilities for the graph, based on the value of m:

1. If $m > 0$, then $r^2 = m$, and the equation represents a circle with radius $\sqrt{m}$.
2. If $m = 0$, then the equation represents the single point (h, k).
3. If $m < 0$, then no points satisfy the equation.

EXAMPLE 4 **Finding the Center and Radius by Completing the Square**

Decide whether each equation has a circle as its graph.

(a) $x^2 - 6x + y^2 + 10y + 25 = 0$ **(b)** $x^2 + 10x + y^2 - 4y + 33 = 0$

(c) $2x^2 + 2y^2 - 6x + 10y = 1$

Solution

(a) Since $x^2 - 6x + y^2 + 10y + 25 = 0$ has the form of equation $(*)$, it represents a circle, a single point, or no points at all. To decide, complete the square on x and y separately.

$$(x^2 - 6x \quad\ ) + (y^2 + 10y \quad\ ) = -25 \qquad \text{Add } -25.$$

$$\left[\frac{1}{2}(-6)\right]^2 = (-3)^2 = 9 \quad \text{and} \quad \left[\frac{1}{2}(10)\right]^2 = 5^2 = 25$$

Add 9 and 25 on the left to complete the two squares, and to compensate, add 9 and 25 on the right.

$$(x^2 - 6x + 9) + (y^2 + 10y + 25) = -25 + 9 + 25 \qquad \text{Complete the square.}$$

Add 9 and 25 on both sides.

$$(x - 3)^2 + (y + 5)^2 = 9 \qquad \text{Factor; add.}$$

Since $9 > 0$, the equation represents a circle with center at $(3, -5)$ and radius 3. See Figure 8.

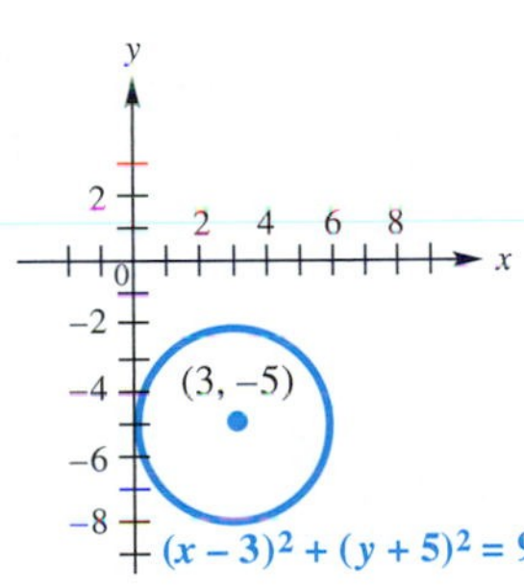

FIGURE 8

(b)
$$x^2 + 10x + y^2 - 4y + 33 = 0$$

$$(x^2 + 10x \quad\ ) + (y^2 - 4y \quad\ ) = -33 \qquad \text{Add } -33.$$

$$\left[\frac{1}{2}(10)\right]^2 = 25 \quad \text{and} \quad \left[\frac{1}{2}(-4)\right]^2 = 4$$

$$(x^2 + 10x + 25) + (y^2 - 4y + 4) = -33 + 25 + 4 \qquad \text{Complete the square.}$$

$$(x + 5)^2 + (y - 2)^2 = -4 \qquad \text{Factor; add.}$$

Because $-4 < 0$, there are no ordered pairs (x, y), with x and y both real numbers, satisfying the equation. The graph contains no points.

(c) $2x^2 + 2y^2 - 6x + 10y = 1$

To complete the square, make the coefficients of the x^2- and y^2-terms 1.

$$2(x^2 - 3x) + 2(y^2 + 5y) = 1 \qquad \text{Group the terms: factor out 2.}$$

Add $2\left(\frac{9}{4}\right)$ and $2\left(\frac{25}{4}\right)$ on both sides.

$$2\left(x^2 - 3x + \frac{9}{4}\right) + 2\left(y^2 + 5y + \frac{25}{4}\right) = 1 + 2\left(\frac{9}{4}\right) + 2\left(\frac{25}{4}\right) \qquad \text{Complete the square.}$$

$$2\left(x - \frac{3}{2}\right)^2 + 2\left(y + \frac{5}{2}\right)^2 = 18 \qquad \text{Factor; simplify.}$$

$$\left(x - \frac{3}{2}\right)^2 + \left(y + \frac{5}{2}\right)^2 = 9 \qquad \text{Divide each side by 2.}$$

$(x - \frac{3}{2})^2 + (y + \frac{5}{2})^2 = 9$

$(\frac{3}{2}, -\frac{5}{2})$

FIGURE 9

The equation has a circle with center at $\left(\frac{3}{2}, -\frac{5}{2}\right)$ and radius 3 as its graph. See Figure 9. ■

Equations and Graphs of Parabolas

The definition of a parabola is also based on distance.

> **Parabola**
>
> A **parabola** is a set of points in a plane equidistant from a fixed point and a fixed line. The fixed point is called the **focus,** and the fixed line the **directrix,** of the parabola.

As shown in Figure 10, the axis of symmetry of a parabola passes through the focus and is perpendicular to the directrix. The vertex is the midpoint of the line segment joining the focus and directrix on the axis.

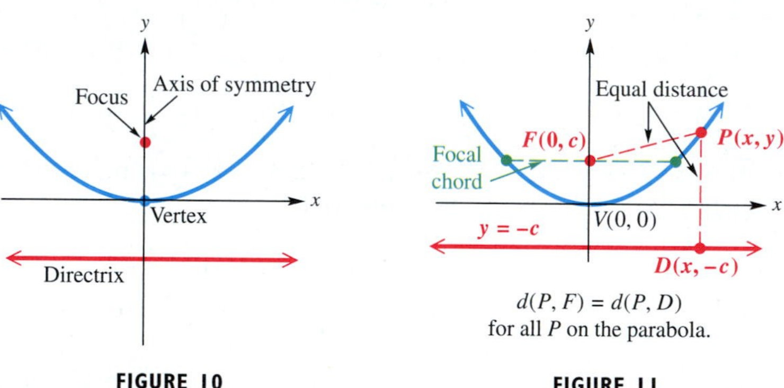

FIGURE 10

FIGURE 11

We can find an equation of a parabola from the preceding definition. Let the directrix be the line $y = -c$ and the focus be the point F with coordinates $(0, c)$. See Figure 11. To find the equation of the set of points that are the same distance from the line $y = -c$ and the point $(0, c)$, choose one such point P and give it coordinates (x, y). Since $d(P, F)$ and $d(P, D)$ must be equal, the distance formula gives the following.

$$d(P,F) = d(P,D) \quad \text{See Figure 11.}$$

$$\sqrt{(x-0)^2 + (y-c)^2} = \sqrt{(x-x)^2 + [y-(-c)]^2} \quad \text{Distance formula}$$

$$\sqrt{x^2 + (y-c)^2} = \sqrt{(y+c)^2} \quad \text{Simplify.}$$

$$x^2 + (y-c)^2 = (y+c)^2 \quad \text{Square both sides.}$$

$$x^2 + y^2 - 2yc + c^2 = y^2 + 2yc + c^2 \quad \text{Square both binomials.}$$

$$x^2 = 4cy \quad \text{Subtract } y^2 \text{ and } c^2\text{; add } 2yc.$$

Remember the middle term when squaring a binomial.

This discussion is summarized as follows.

Parabola with a Vertical Axis

The parabola with focus $(0, c)$ and directrix $y = -c$ has equation

$$\mathbf{x^2 = 4cy.}$$

The parabola has vertical axis $x = 0$ and opens upward if $c > 0$ or downward if $c < 0$.

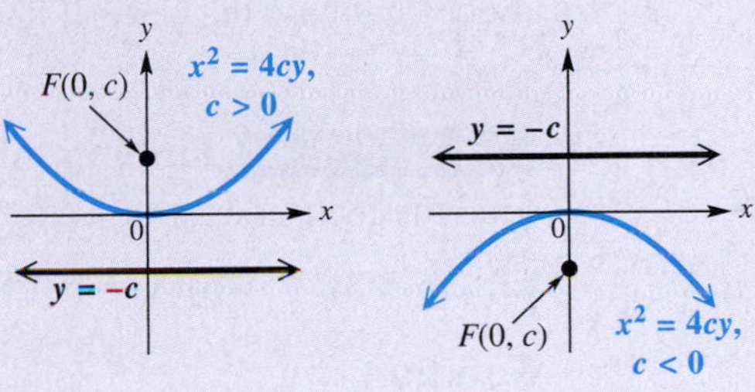

NOTE When graphing a parabola, remember that the focus is located "inside" the parabola. If the parabola opens upward, the focus is above the vertex; if the parabola opens downward, the focus is below the vertex.

The **focal chord** through the focus and perpendicular to the axis of symmetry of a parabola has length $|4c|$. To see this, note in Figure 11 that the endpoints of the chord are $(-x, c)$ and (x, c). Let $y = c$ in the equation of the parabola and solve for x.

$$x^2 = 4cy$$

$$x^2 = 4c^2 \quad \text{Substitute } c \text{ for } y.$$

$$x = |2c| \quad \text{Take the positive square root.}$$

The length of half the focal chord is $|2c|$, so its full length is $|4c|$. This fact is useful when graphing a parabola because it means that the points with coordinates $(-2c, c)$ and $(2c, c)$ lie on the parabola.

If the directrix is the line $x = -c$ and the focus is $(c, 0)$, then the definition of a parabola and the distance formula leads to the equation of a parabola with a horizontal axis. (See Exercise 113.)

Parabola with a Horizontal Axis

The parabola with focus $(c, 0)$ and directrix $x = -c$ has equation

$$y^2 = 4cx.$$

The parabola has horizontal axis $y = 0$ and opens to the right if $c > 0$ or to the left if $c < 0$.

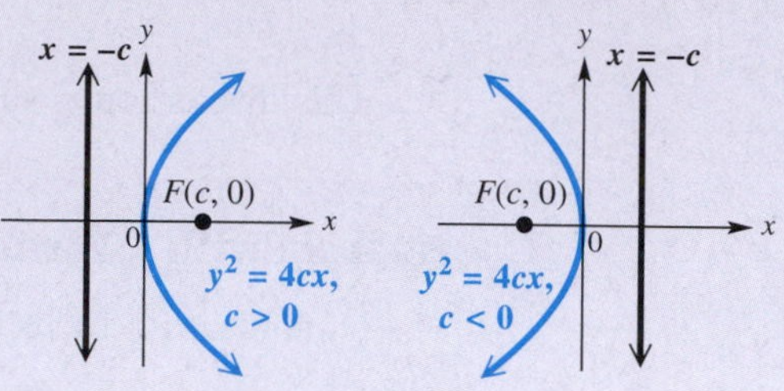

Notice that ***the graph of a parabola with a horizontal axis is not a function,*** since it does not pass the vertical line test.

EXAMPLE 5 **Finding Information about Parabolas from Equations**

Find the focus, directrix, vertex, and axis of each parabola. Then graph each parabola.

(a) $x^2 = 8y$ **(b)** $y^2 = -28x$

Solution

(a) The equation $x^2 = 8y$ has the form $x^2 = 4cy$, so set $4c = 8$, from which $c = 2$. Since the x-term is squared, the parabola is vertical, with focus $(0, c) = (0, 2)$ and directrix $y = -2$. The vertex is $(0, 0)$, and the axis of the parabola is the y-axis. See Figure 12, which also shows the endpoints of the focal chord.

TECHNOLOGY NOTE

To graph $y^2 = -28x$ with a calculator, solve for y to obtain

$$y = \pm\sqrt{-28x}.$$

Graphing each equation results in the entire parabola being shown.

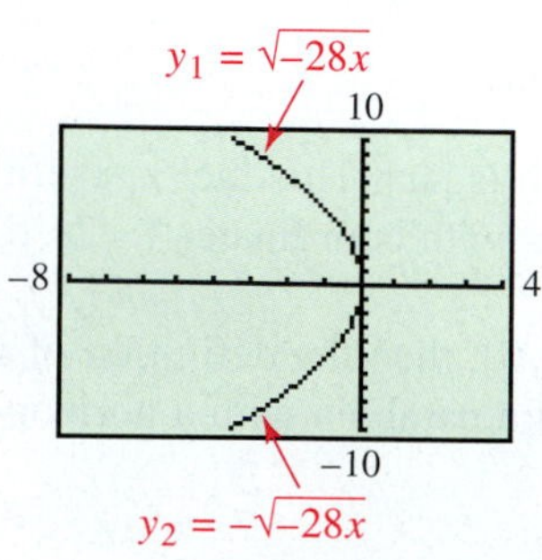

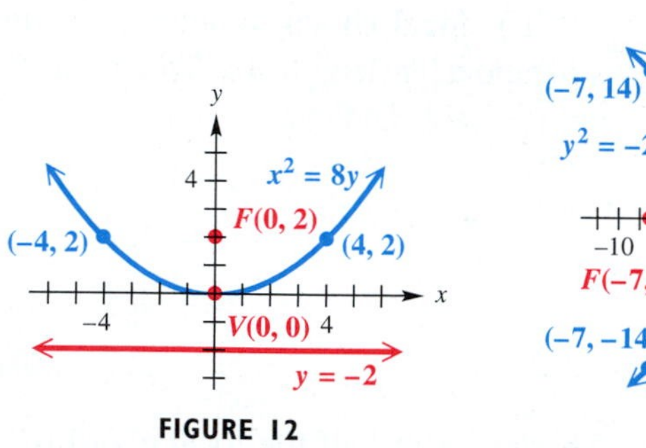

FIGURE 12

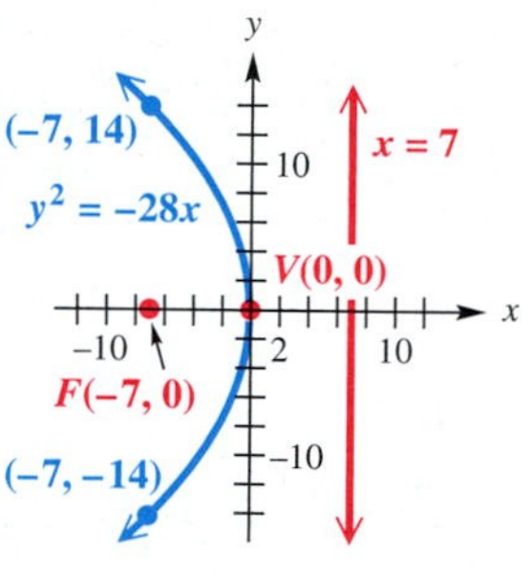

FIGURE 13

(b) The equation $y^2 = -28x$ has the form $y^2 = 4cx$, with $4c = -28$, so $c = -7$. The parabola is horizontal, with focus $(-7, 0)$, directrix $x = 7$, vertex $(0, 0)$, and x-axis as axis of the parabola. The focal chord has endpoints $(-7, -14)$ and $(-7, 14)$. Since c is negative, the graph opens to the left, as shown in Figure 13. ■

EXAMPLE 6 **Writing Equations of Parabolas**

Write an equation for each parabola.

(a) Focus $\left(\frac{2}{3}, 0\right)$ and vertex at the origin

(b) Vertical axis, vertex at the origin, through the point $(-2, 12)$

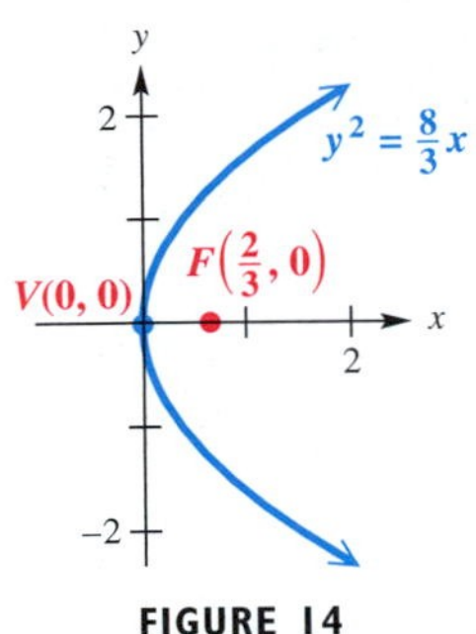

FIGURE 14

Solution

(a) Since the focus $\left(\frac{2}{3}, 0\right)$ and the vertex $(0, 0)$ are both on the x-axis, the parabola is horizontal. It opens to the right because $c = \frac{2}{3}$ is positive. See Figure 14. The equation, which will have the form $y^2 = 4cx$, is

$$y^2 = 4\left(\frac{2}{3}\right)x, \quad \text{or} \quad y^2 = \frac{8}{3}x.$$

(b) The parabola will have an equation of the form $x^2 = 4cy$ because the axis is vertical and the vertex is $(0, 0)$. Since the point $(-2, 12)$ is on the graph, it must satisfy the equation.

$$x^2 = 4cy$$

$$(-2)^2 = 4c(12) \qquad \text{Let } x = -2 \text{ and } y = 12.$$

$$4 = 48c$$

$$c = \frac{1}{12} \qquad \text{Solve for } c.$$

Thus, $\quad x^2 = 4cy = 4\left(\frac{1}{12}\right)y, \quad$ which gives $\quad x^2 = \frac{1}{3}y \quad$ or $\quad y = 3x^2.$ ■

Translations of Parabolas

The equations $x^2 = 4cy$ and $y^2 = 4cx$ can be extended to parabolas having vertex (h, k) by replacing x and y with $x - h$ and $y - k$, respectively.

Equation Forms for Translated Parabolas

A parabola with vertex (h, k) has an equation of the form

$$(x - h)^2 = 4c(y - k) \qquad \text{Vertical axis}$$

or

$$(y - k)^2 = 4c(x - h), \qquad \text{Horizontal axis}$$

where the focus is a distance $|c|$ from the vertex.

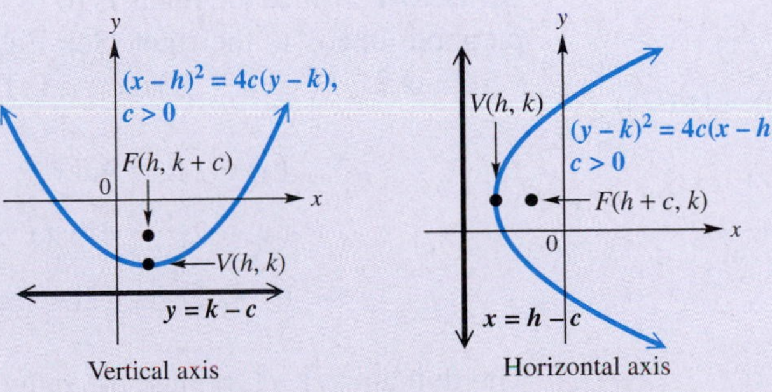

EXAMPLE 7 **Graphing a Parabola with Vertex (h, k)**

Graph the parabola given by the equation $x = -\frac{1}{8}(y + 3)^2 + 2$. Label the vertex, focus, and directrix.

Solution Rewrite the equation in the form $(y - k)^2 = 4c(x - h)$.

$$x = -\frac{1}{8}(y + 3)^2 + 2$$

$$x - 2 = -\frac{1}{8}(y + 3)^2 \qquad \text{Subtract 2.}$$

$$-8(x - 2) = (y + 3)^2 \qquad \text{Multiply by } -8.$$

$$(y + 3)^2 = -8(x - 2) \qquad \text{Rewrite.}$$

The vertex is $(2, -3)$. Since $4c = -8$, we have $c = -2$, and the parabola opens to the left. The focus is 2 units to the left of the vertex, and the directrix is 2 units to the right of the vertex. Thus, the focus is $(0, -3)$ and the directrix is $x = 4$. See Figure 15.

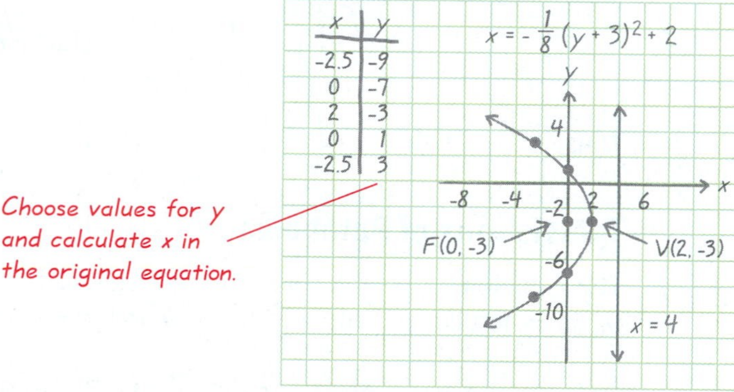

FIGURE 15

EXAMPLE 8 **Writing an Equation of a Parabola**

Write an equation for the parabola with focus $(2, 2)$ and vertex $(-1, 2)$, and graph it. Give the domain and range.

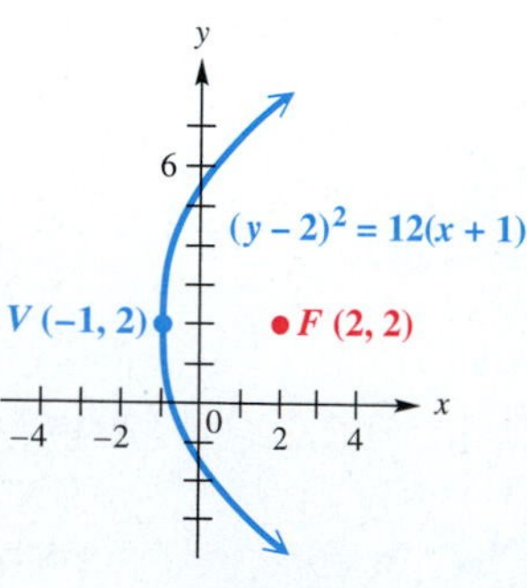

FIGURE 16

Solution Since the focus is to the right of the vertex, the axis is horizontal and the parabola opens to the right. See Figure 16. The distance between the focus and the vertex is $2 - (-1) = 3$, so $c = 3$. The equation of the parabola is

$$(y - k)^2 = 4c(x - h) \qquad \text{Parabola with horizontal axis}$$

$$(y - 2)^2 = 4(3)(x + 1) \qquad \text{Substitute for } c, h, \text{ and } k.$$

$$(y - 2)^2 = 12(x + 1). \qquad \text{Multiply.}$$

The domain is $[-1, \infty)$ and the range is $(-\infty, \infty)$.

EXAMPLE 9 **Completing the Square to Graph a Horizontal Parabola**

Graph $x = -2y^2 + 4y - 3$. Give the domain and range.

Analytic Solution

$$x = -2y^2 + 4y - 3$$

$$x + 3 = -2y^2 + 4y \quad \text{Add 3.}$$

$$x + 3 = -2(y^2 - 2y) \quad \text{Factor out } -2.$$

$$x + 3 - 2 = -2(y^2 - 2y + 1) \quad \text{Complete the square.}$$

$$x + 1 = -2(y - 1)^2 \quad \text{Add; factor the trinomial.}$$

$$(y - 1)^2 = -\frac{1}{2}(x + 1) \quad (*) \quad \text{Multiply by } -\tfrac{1}{2}\text{; rewrite.}$$

The parabola opens to the left, is narrower than the graph of $x = y^2$, and has vertex $(-1, 1)$. The domain is $(-\infty, -1]$ and the range is $(-\infty, \infty)$. See Figure 17.

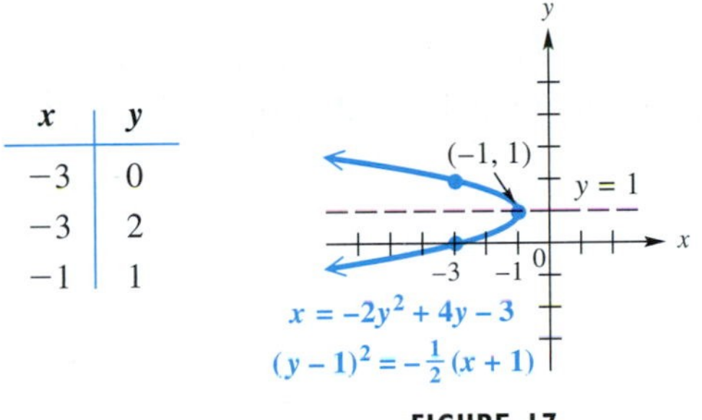

x	y
-3	0
-3	2
-1	1

FIGURE 17

Graphing Calculator Solution

To obtain a calculator graph, solve the equation $(y - 1)^2 = -\frac{1}{2}(x + 1)$ $(*)$ from the analytic solution for y.

$$y - 1 = \pm\sqrt{\frac{x + 1}{-2}} \quad \text{Take square roots.}$$

$$y = 1 \pm \sqrt{\frac{x + 1}{-2}} \quad \text{Add 1.}$$

Graphing these equations as y_1 and y_2 results in Figure 18.

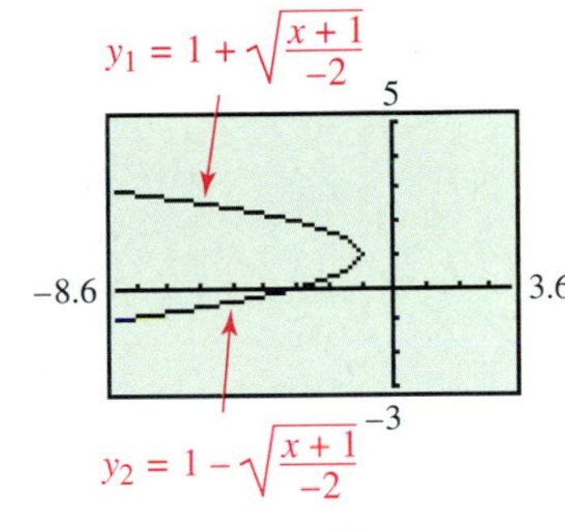

FIGURE 18

An Application of Parabolas

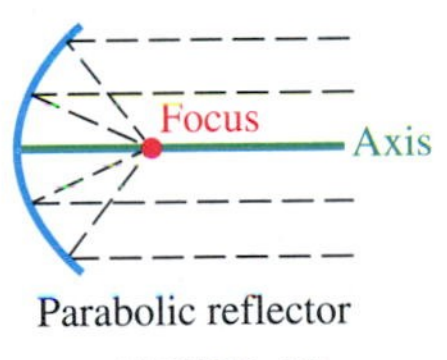

Parabolic reflector

FIGURE 19

The geometric properties of parabolas lead to many practical applications. For example, if a light source is placed at the focus of a parabolic reflector, as in Figure 19, light rays reflect parallel to the axis, making a spotlight or flashlight. The process also works in reverse. Light rays from a distant source come in parallel to the axis and are reflected to a point at the focus. This use of parabolic reflection is seen in the satellite dishes used to pick up signals from communications satellites.

EXAMPLE 10 **Modeling the Reflective Property of Parabolas**

The Parkes radio telescope has a parabolic dish shape with diameter 210 feet and depth 32 feet. Because of this parabolic shape, distant rays hitting the dish are reflected directly toward the focus. A cross section of the dish is shown in Figure 20. (*Source:* Mar, J. and H. Liebowitz, *Structure Technology for Large Radio and Radar Telescope Systems,* MIT Press, 1969.)

(a) Determine an equation describing this cross section by placing the vertex at the origin with the parabola opening upward.

(b) The receiver must be placed at the focus of the parabola. How far from the vertex of the parabolic dish should the receiver be located?

Focus

210 ft

32 ft

FIGURE 20

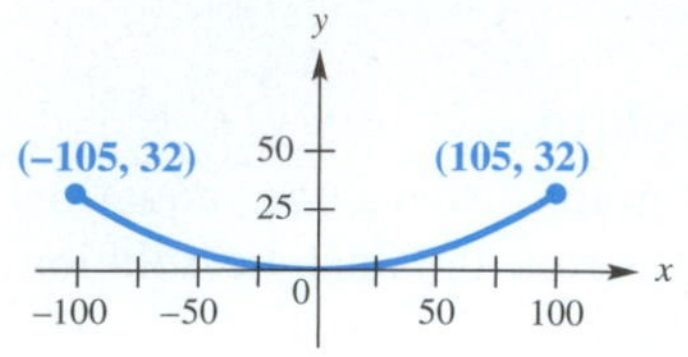

y axis, x axis; points (−105, 32) and (105, 32); ticks −100, −50, 0, 50, 100 and 25, 50

FIGURE 21

Solution

(a) Locate the vertex at the origin, as shown in Figure 21. The form of the parabola is $x^2 = 4cy$. The parabola must pass through the point $\left(\frac{210}{2}, 32\right) = (105, 32)$. Thus,

$$x^2 = 4cy \quad \text{Vertical parabola}$$
$$(105)^2 = 4c(32) \quad \text{Let } x = 105 \text{ and } y = 32.$$
$$11{,}025 = 128c$$
$$c = \frac{11{,}025}{128}. \quad \text{Solve for } c.$$

The cross section can be modeled by the equation

$$x^2 = 4cy$$
$$x^2 = 4\left(\frac{11{,}025}{128}\right)y \quad \text{Substitute for } c.$$
$$x^2 = \frac{11{,}025}{32}y.$$

(b) The distance between the vertex and the focus is c. In part (a), we found that $c = \frac{11{,}025}{128} \approx 86.1$, so the receiver should be located at (0, 86.1), or 86.1 ft above the vertex. ■

6.1 Exercises

 **1.** Describe the graph of the equation

$$(x-3)^2 + (y-3)^2 = 0.$$

 **2.** Describe the graph of the equation

$$(x-3)^2 + (y-3)^2 = -1.$$

Match each equation in Column I with the appropriate description in Column II.

I	II
3. $x = 2y^2$	**A.** Circle; center $(3, -4)$; radius 5
4. $y = 2x^2$	**B.** Parabola; opens left
5. $x^2 = -3y$	**C.** Parabola; opens upward
6. $y^2 = -3x$	**D.** Circle; center $(-3, 4)$; radius 5
7. $x^2 + y^2 = 5$	**E.** Parabola; opens right
8. $(x-3)^2 + (y+4)^2 = 25$	**F.** Circle; center $(0, 0)$; radius $\sqrt{5}$
9. $(x+3)^2 + (y-4)^2 = 25$	**G.** No points on its graph
10. $x^2 + y^2 = -4$	**H.** Parabola; opens downward

Find the center–radius form for each circle satisfying the given conditions.

11. Center $(1, 4)$; radius 3

12. Center $(-2, 5)$; radius 4

13. Center $(0, 0)$; radius 1

14. Center $(0, 0)$; radius 5

15. Center $\left(\frac{2}{3}, -\frac{4}{5}\right)$; radius $\frac{3}{7}$

16. Center $\left(-\frac{1}{2}, -\frac{1}{4}\right)$; radius $\frac{12}{5}$

17. Center $(-1, 2)$; passing through $(2, 6)$

18. Center $(2, -7)$; passing through $(-2, -4)$

19. Center $(-3, -2)$; tangent to the x-axis (*Hint: Tangent to* means touching at one point.)

20. Center $(5, -1)$; tangent to the y-axis

Relating Concepts

For individual or group investigation (Exercises 21–24)

The figure shows a circle and a diameter of the circle. The endpoints of the diameter have coordinates $(-1, 3)$ *and* $(5, -9)$. ***Work Exercises 21–24 in order.***

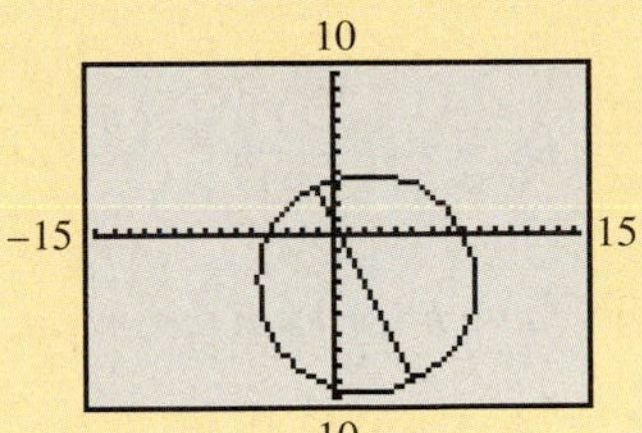

21. Find the coordinates of the center of the circle.

22. Find the radius of the circle.

23. Give the center–radius form of the equation of the circle.

24. Give the center–radius form of the equation of the circle with endpoints of a diameter having coordinates $(3, -5)$ and $(-7, 3)$.

Graph each circle by hand if possible. Give the domain and range.

25. $x^2 + y^2 = 4$

26. $x^2 + y^2 = 36$

27. $x^2 + y^2 = 0$

28. $x^2 + y^2 = -9$

29. $(x - 2)^2 + y^2 = 36$

30. $(x + 2)^2 + (y - 5)^2 = 16$

31. $(x - 5)^2 + (y + 4)^2 = 49$

32. $(x - 4)^2 + (y - 3)^2 = 25$

33. $(x + 3)^2 + (y + 2)^2 = 36$

34. $x^2 + (y - 2)^2 = 9$

35. $(x - 1)^2 + (y + 2)^2 = 16$

36. $(x + 1)^2 + y^2 + 2 = 0$

Graph each circle using a graphing calculator. Use a square viewing window. Give the domain and range.

37. $x^2 + y^2 = 81$

38. $x^2 + (y + 3)^2 = 49$

39. $(x - 3)^2 + (y - 2)^2 = 25$

40. $(x + 2)^2 + (y + 3)^2 = 36$

Decide whether each equation has a circle as its graph. If it does, give the center and radius.

41. $x^2 + 6x + y^2 + 8y + 9 = 0$

42. $x^2 + 8x + y^2 - 6y + 16 = 0$

43. $x^2 - 4x + y^2 + 12y = -4$

44. $x^2 - 12x + y^2 + 10y = -25$

45. $4x^2 + 4x + 4y^2 - 16y - 19 = 0$

46. $9x^2 + 12x + 9y^2 - 18y - 23 = 0$

47. $x^2 + 2x + y^2 - 6y + 14 = 0$

48. $x^2 + 4x + y^2 - 8y + 32 = 0$

49. $x^2 - 2x + y^2 + 4y + 9 = 0$

50. $4x^2 + 4x + 4y^2 - 4y - 3 = 0$

51. $9x^2 + 36x + 9y^2 = -32$

52. $9x^2 + 9y^2 + 54y = -72$

Concept Check *Each equation in Exercises 53–60 defines a parabola. Without actually graphing, match the equation in Column I with its description in Column II.*

I

53. $(x - 4)^2 = y + 2$

54. $(x - 2)^2 = y + 4$

55. $y + 2 = -(x - 4)^2$

56. $y = -(x - 2)^2 - 4$

57. $(y - 4)^2 = x + 2$

58. $(y - 2)^2 = x + 4$

59. $x + 2 = -(y - 4)^2$

60. $x = -(y - 2)^2 - 4$

II

A. Vertex $(2, -4)$; opens downward
B. Vertex $(2, -4)$; opens upward
C. Vertex $(4, -2)$; opens downward
D. Vertex $(4, -2)$; opens upward
E. Vertex $(-2, 4)$; opens left
F. Vertex $(-2, 4)$; opens right
G. Vertex $(-4, 2)$; opens left
H. Vertex $(-4, 2)$; opens right

61. *Concept Check* For the graph of $(x - h)^2 = 4c(y - k)$, in what quadrant is the vertex if:
(a) $h < 0, k < 0$; **(b)** $h < 0, k > 0$; **(c)** $h > 0, k < 0$; **(d)** $h > 0, k > 0$?

62. *Concept Check* Repeat parts (a)–(d) of Exercise 61 for the graph of $(y - k)^2 = 4c(x - h)$.

Give the focus, directrix, and axis of each parabola.

63. $x^2 = 16y$ **64.** $x^2 = 4y$ **65.** $x^2 = -\frac{1}{2}y$ **66.** $x^2 = \frac{1}{9}y$

67. $y^2 = \frac{1}{16}x$ **68.** $y^2 = -\frac{1}{32}x$ **69.** $y^2 = -16x$ **70.** $y^2 = -4x$

Write an equation for each parabola with vertex at the origin.

71. Focus $(0, -2)$ **72.** Focus $(5, 0)$ **73.** Focus $\left(-\frac{1}{2}, 0\right)$ **74.** Focus $\left(0, \frac{1}{4}\right)$

75. Through $(2, -2\sqrt{2})$; opening to the right

76. Through $(\sqrt{3}, 3)$; opening upward

77. Through $(\sqrt{10}, -5)$; opening downward

78. Through $(-3, 3)$; opening to the left

79. Through $(2, -4)$; symmetric with respect to the y-axis

80. Through $(3, 2)$; symmetric with respect to the x-axis

Find an equation of a parabola that satisfies the given conditions.

81. Focus $(0, 2)$; vertex $(0, 1)$ **82.** Focus $(-1, 2)$; vertex $(3, 2)$ **83.** Focus $(0, 0)$; directrix $x = -2$

84. Focus $(2, 1)$; directrix $x = -1$ **85.** Focus $(-1, 3)$; directrix $y = 7$ **86.** Focus $(1, 2)$; directrix $y = 4$

87. Horizontal axis; vertex $(-2, 3)$; passing through $(-4, 0)$

88. Horizontal axis; vertex $(-1, 2)$; passing through $(2, 3)$

Graph each parabola by hand, and check using a graphing calculator. Give the vertex, axis, domain, and range.

89. $y = (x + 3)^2 - 4$ **90.** $y = (x - 5)^2 - 4$ **91.** $y = -2(x + 3)^2 + 2$ **92.** $(x - 2)^2 = \frac{3}{2}(y + 1)$

93. $y = x^2 - 2x + 3$ **94.** $y = x^2 + 6x + 5$ **95.** $y = 2x^2 - 4x + 5$ **96.** $y = -3x^2 + 24x - 46$

97. $x = y^2 + 2$ **98.** $x = (y + 1)^2$ **99.** $x = (y - 3)^2$ **100.** $(y + 2)^2 = x + 1$

101. $x = (y - 4)^2 + 2$ **102.** $x = -2(y + 3)^2$ **103.** $x = \frac{2}{3}y^2 - 4y + 8$ **104.** $x = y^2 + 2y - 8$

105. $x = -4y^2 - 4y - 3$ **106.** $x = 2y^2 - 4y + 6$ **107.** $x = -2y^2 + 2y - 3$ **108.** $2x = y^2 - 4y + 6$

109. $2x = y^2 - 2y + 9$ **110.** $x = -3y^2 + 6y - 1$ **111.** $y^2 - 4y + 4 = 4x + 4$ **112.** $y^2 + 2y + 1 = -2x + 4$

113. Prove that the parabola with focus $(c, 0)$ and directrix $x = -c$ has equation $y^2 = 4cx$.

(Modeling) *Solve each problem.*

114. *Path of an Object on a Planet* When an object moves under the influence of a constant force (without air resistance), its path is parabolic. This is the path of a ball thrown near the surface of a planet or other celestial object. Suppose two balls are simultaneously thrown upward at a 45° angle on two different planets. If their initial velocities are both 30 mph, then their xy-coordinates in feet at time x in seconds can be expressed by the equation

$$y = x - \frac{g}{1922}x^2,$$

where g is the acceleration due to gravity. The value of g will vary with the mass and size of the planet. (*Source:* Zeilik, M., S. Gregory, and E. Smith, *Introductory Astronomy and Astrophysics,* Saunders College Publishers, 1992.)

(a) On Earth, $g = 32.2$; on Mars, $g = 12.6$. Find the two equations, and use the same screen of a graphing calculator to graph the paths of the two balls thrown on Earth and Mars. Use the window $[0, 180]$ by $[0, 120]$. (*Hint:* If possible, set the mode on your graphing calculator to simultaneous.)

(b) Determine the difference in the horizontal distances traveled by the two balls.

115. *Path of an Object on a Planet* Refer to Exercise 114. Suppose the two balls are now thrown upward at a 60° angle on Mars and the moon. If their initial velocities are 60 mph, then their *xy*-coordinates in feet at time x in seconds can be expressed by the equation

$$y = \frac{19}{11}x - \frac{g}{3872}x^2.$$

(*Source:* Zeilik, M., S. Gregory, and E. Smith, *Introductory Astronomy and Astrophysics,* Saunders College Publishers, 1992.)

(a) On the same screen, graph the paths of the balls if $g = 5.2$ for the moon.

(b) Determine the maximum height of each ball to the nearest foot.

116. *Design of a Radio Telescope* The U.S. Naval Research Laboratory designed a giant radio telescope weighing 3450 tons. Its parabolic dish had a diameter of 300 feet, with a focal length (the distance from the focus to the parabolic surface) of 128.5 feet. Determine the maximum depth of the 300-foot dish. (*Source:* Mar, J. and H. Liebowitz, *Structure Technology for Large Radio and Radar Telescope Systems,* MIT Press, 1969.)

117. *Path of an Alpha Particle* When an alpha particle is moving in a horizontal path along the positive x-axis and passes between charged plates, it is deflected in a parabolic path. If the plate is charged with 2000 volts and is .4 meter long, then an alpha particle's path can be described by the equation

$$y = -\frac{k}{2v_0}x^2,$$

where $k = 5 \times 10^{-9}$ is constant and v_0 is the initial velocity of the particle. If $v_0 = 10^7$ meters per second, what is the deflection of the alpha particle's path in the y-direction when $x = .4$ meter? (*Source:* Semat, H. and J. Albright, *Introduction to Atomic and Nuclear Physics,* Holt, Rinehart and Winston, 1972.)

118. *Height of a Bridge's Cable Supports* The cable in the center portion of a bridge is supported as shown in the figure to form a parabola. The center support is 10 feet high, the tallest supports are 210 feet high, and the distance between the two tallest supports is 400 feet. Find the height of the remaining supports if the supports are evenly spaced. (Ignore the width of the supports.)

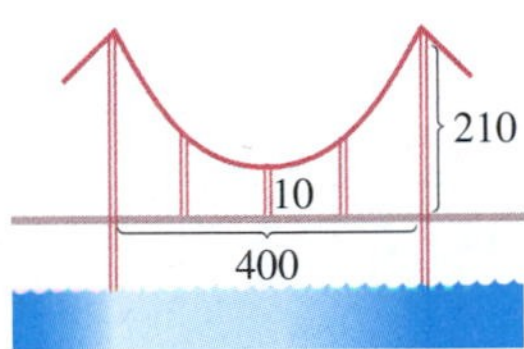

119. *Parabolic Arch* An arch in the shape of a parabola has the dimensions shown in the figure. How wide is the arch 9 feet up?

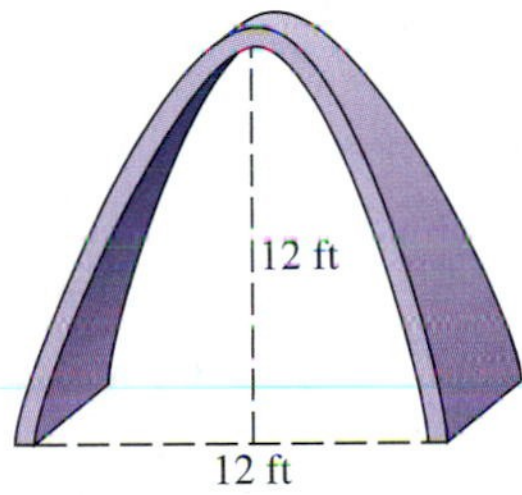

120. *Headlight* A headlight is being constructed in the shape of a paraboloid with depth 4 inches and diameter 5 inches, as illustrated in the figure. Determine the distance d that the bulb should be from the vertex in order to have the beam of light shine straight ahead.

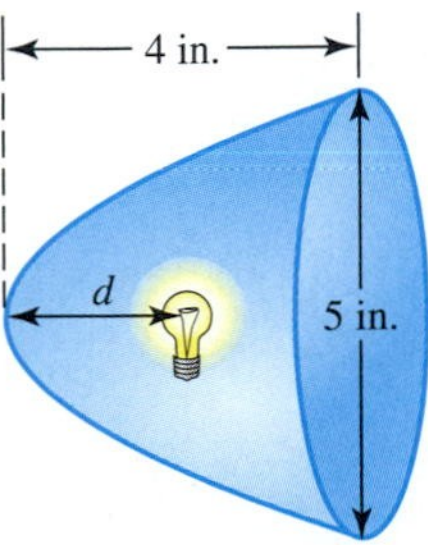

6.2 Ellipses and Hyperbolas

Equations and Graphs of Ellipses ■ Translations of Ellipses ■ An Application of Ellipses ■ Equations and Graphs of Hyperbolas ■ Translations of Hyperbolas

Equations and Graphs of Ellipses

The *ellipse* is another relation whose equation is based on the distance formula.

Ellipse

An **ellipse** is the set of all points in a plane, the sum of whose distances from two fixed points is constant. Each fixed point is called a **focus** (plural, **foci**) of the ellipse.

FIGURE 22

As shown in Figure 22, an ellipse has two axes of symmetry: the **major axis** (the longer one) and the **minor axis** (the shorter one). ***The foci are always located on the major axis.*** The midpoint of the major axis is the **center** of the ellipse, and the endpoints of the major axis are the **vertices** of the ellipse. ***The graph of an ellipse is not the graph of a function.*** (Why?)

FIGURE 23

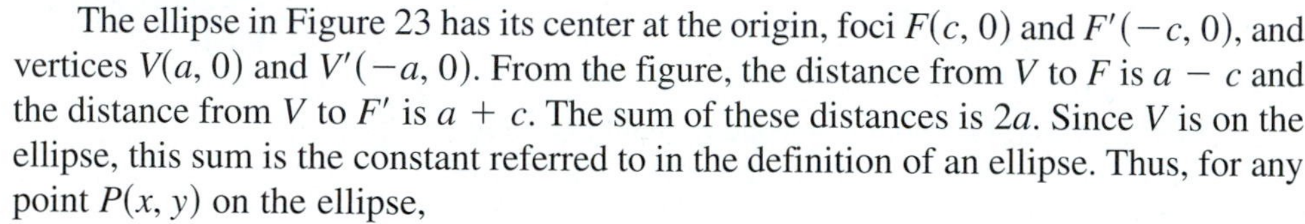

The ellipse in Figure 23 has its center at the origin, foci $F(c, 0)$ and $F'(-c, 0)$, and vertices $V(a, 0)$ and $V'(-a, 0)$. From the figure, the distance from V to F is $a - c$ and the distance from V to F' is $a + c$. The sum of these distances is $2a$. Since V is on the ellipse, this sum is the constant referred to in the definition of an ellipse. Thus, for any point $P(x, y)$ on the ellipse,

$$d(P, F) + d(P, F') = 2a.$$

By the distance formula,

$$d(P, F) = \sqrt{(x - c)^2 + y^2}, \quad \text{and} \quad d(P, F') = \sqrt{[x - (-c)]^2 + y^2} = \sqrt{(x + c)^2 + y^2}.$$

Thus,

$$\sqrt{(x - c)^2 + y^2} + \sqrt{(x + c)^2 + y^2} = 2a$$

$$\sqrt{(x - c)^2 + y^2} = 2a - \sqrt{(x + c)^2 + y^2} \qquad \text{Isolate } \sqrt{(x - c)^2 + y^2}.$$

$$(x - c)^2 + y^2 = 4a^2 - 4a\sqrt{(x + c)^2 + y^2} + (x + c)^2 + y^2 \qquad \text{Square both sides.}$$

$$x^2 - 2cx + c^2 + y^2 = 4a^2 - 4a\sqrt{(x + c)^2 + y^2} + x^2 + 2cx + c^2 + y^2 \qquad \text{Square } x - c\text{; square } x + c.$$

$$4a\sqrt{(x + c)^2 + y^2} = 4a^2 + 4cx \qquad \text{Isolate } 4a\sqrt{(x + c)^2 + y^2}.$$

$$a\sqrt{(x + c)^2 + y^2} = a^2 + cx \qquad \text{Divide by 4.}$$

$$a^2(x^2 + 2cx + c^2 + y^2) = a^4 + 2ca^2x + c^2x^2 \qquad \text{Square both sides.}$$

$$a^2x^2 + 2ca^2x + a^2c^2 + a^2y^2 = a^4 + 2ca^2x + c^2x^2 \qquad \text{Distributive property}$$

$$a^2x^2 + a^2c^2 + a^2y^2 = a^4 + c^2x^2 \qquad \text{Subtract } 2ca^2x.$$

$$a^2x^2 - c^2x^2 + a^2y^2 = a^4 - a^2c^2 \qquad \text{Rearrange terms.}$$

$$(a^2 - c^2)x^2 + a^2y^2 = a^2(a^2 - c^2) \qquad \text{Factor.}$$

$$\frac{x^2}{a^2} + \frac{y^2}{a^2 - c^2} = 1. \quad (*) \qquad \text{Divide by } a^2(a^2 - c^2).$$

Since $B(0, b)$ is on the ellipse in Figure 23, we have

$$d(B, F) + d(B, F') = 2a$$

$$\sqrt{(-c)^2 + b^2} + \sqrt{c^2 + b^2} = 2a$$

$$2\sqrt{c^2 + b^2} = 2a \qquad \text{Combine terms.}$$

$$\sqrt{c^2 + b^2} = a \qquad \text{Divide by 2.}$$

$$c^2 + b^2 = a^2 \qquad \text{Square both sides.}$$

$$b^2 = a^2 - c^2. \qquad \text{Subtract } c^2.$$

Replacing $a^2 - c^2$ with b^2 in equation (*) gives

$$\frac{x^2}{a^2} + \frac{y^2}{b^2} = 1,$$

the standard form of the equation of an ellipse centered at the origin and with foci on the x-axis. If the vertices and foci were on the y-axis, an almost identical derivation could be used to get the standard form

$$\frac{x^2}{b^2} + \frac{y^2}{a^2} = 1.$$

Looking Ahead to Calculus

Methods of calculus can be used to solve problems involving ellipses. For example, differentiation is used to find the slope of the tangent line at a point on the ellipse, and integration is used to find the length of any arc of the ellipse.

Standard Forms of Equations for Ellipses

The ellipse with center at the origin and equation

$$\frac{x^2}{a^2} + \frac{y^2}{b^2} = 1 \quad (a > b > 0)$$

has vertices $(\pm a, 0)$, endpoints of the minor axis $(0, \pm b)$, and foci $(\pm c, 0)$, where $c^2 = a^2 - b^2$.

Major axis on x-axis

The ellipse with center at the origin and equation

$$\frac{x^2}{b^2} + \frac{y^2}{a^2} = 1 \quad (a > b > 0)$$

has vertices $(0, \pm a)$, endpoints of the minor axis $(\pm b, 0)$, and foci $(0, \pm c)$, where $c^2 = a^2 - b^2$.

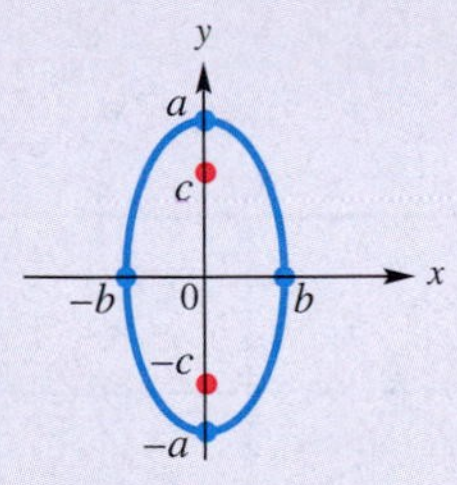

Major axis on y-axis

Do not be confused by the two standard forms—in one case a^2 is associated with x^2; in the other case a^2 is associated with y^2. In practice, we need only find the intercepts of the graph—if the positive x-intercept is larger than the positive y-intercept, then the major axis is horizontal; otherwise, it is vertical. When using $a^2 - c^2 = b^2$, or $a^2 - b^2 = c^2$, choose a^2 and b^2 so that $a^2 > b^2$.

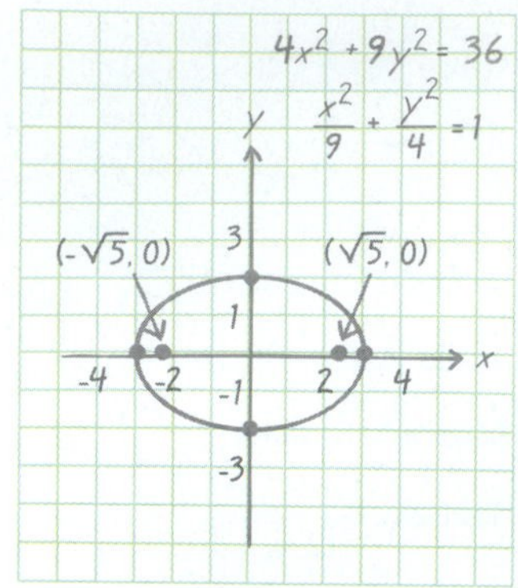

FIGURE 24

EXAMPLE 1 Graphing an Ellipse Centered at the Origin

Graph each ellipse, and find the foci. Give the domain and range.

(a) $4x^2 + 9y^2 = 36$ **(b)** $4x^2 + y^2 = 64$

Solution

(a)

$$4x^2 + 9y^2 = 36$$

$$\frac{x^2}{9} + \frac{y^2}{4} = 1 \qquad \text{Divide by 36 to write in standard form.}$$

Be sure to divide *each* term by 36.

Thus, the x-intercepts are ± 3, and the y-intercepts are ± 2. See Figure 24. Since $9 > 4$, we find the foci by letting $a^2 = 9$ and $b^2 = 4$ in $c^2 = a^2 - b^2$.

$$c^2 = 9 - 4 = 5, \quad \text{so} \quad c = \sqrt{5}.$$

(By definition, $c > 0$. See Figure 23.) The major axis is along the x-axis, so the foci have coordinates $(-\sqrt{5}, 0)$ and $(\sqrt{5}, 0)$. The domain of this relation is $[-3, 3]$, and the range is $[-2, 2]$.

(b)

$$4x^2 + y^2 = 64$$

$$\frac{x^2}{16} + \frac{y^2}{64} = 1 \qquad \text{Divide by 64 to write in standard form.}$$

Be sure to divide *each* term by 64.

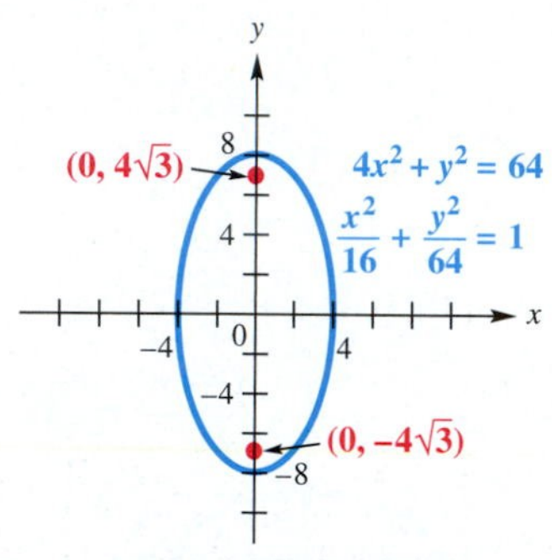

FIGURE 25

The x-intercepts are ± 4; the y-intercepts are ± 8. See Figure 25. Here $64 > 16$, so $a^2 = 64$ and $b^2 = 16$. Thus,

$$c^2 = 64 - 16 = 48, \quad \text{so} \quad c = \sqrt{48} = \sqrt{16 \cdot 3} = 4\sqrt{3}.$$

The major axis is on the y-axis, so the coordinates of the foci are $(0, -4\sqrt{3})$ and $(0, 4\sqrt{3})$. The domain of the relation is $[-4, 4]$; the range is $[-8, 8]$. ■

The graph of an ellipse is *not* the graph of a function. To graph the ellipse in Example 1(a) with a graphing calculator, solve for y to get equations of two functions.

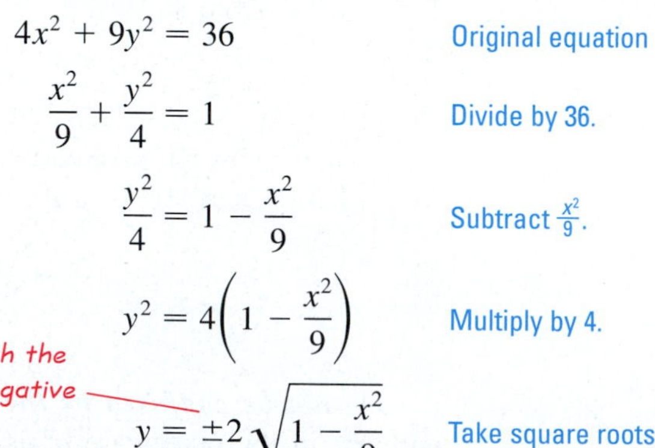

$$4x^2 + 9y^2 = 36 \qquad \text{Original equation}$$

$$\frac{x^2}{9} + \frac{y^2}{4} = 1 \qquad \text{Divide by 36.}$$

$$\frac{y^2}{4} = 1 - \frac{x^2}{9} \qquad \text{Subtract } \tfrac{x^2}{9}.$$

$$y^2 = 4\left(1 - \frac{x^2}{9}\right) \qquad \text{Multiply by 4.}$$

$$y = \pm 2\sqrt{1 - \frac{x^2}{9}} \qquad \text{Take square roots.}$$

Remember both the positive and negative square roots.

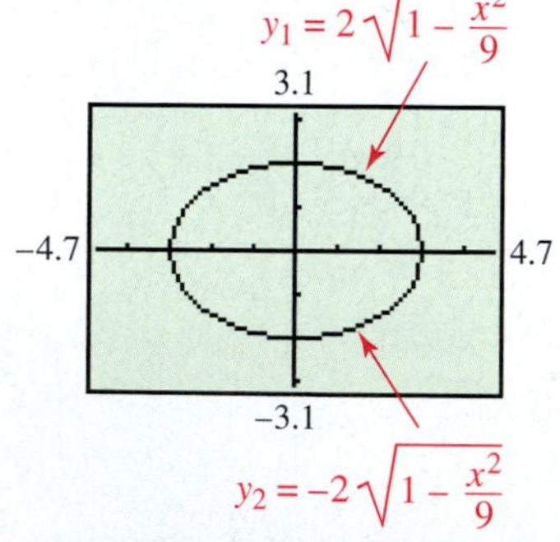

Since $y_2 = -y_1$, we can enter $-y_1$ to get the portion of the graph below the x-axis.

FIGURE 26

The ellipse is obtained by graphing both equations, as shown in Figure 26.

EXAMPLE 2 **Writing the Equation of an Ellipse**

Write the equation of the ellipse having center at the origin, foci at $(0, 3)$ and $(0, -3)$, and major axis 8 units long.

Solution Since the major axis is 8 units long, $2a = 8$ and $a = 4$. To find b^2, use the relationship $a^2 - b^2 = c^2$, with $a = 4$ and $c = 3$.

$$a^2 - b^2 = c^2$$

$$4^2 - b^2 = 3^2 \quad \text{Substitute for } a \text{ and } c.$$

$$16 - b^2 = 9$$

$$b^2 = 7$$

Since the foci are on the y-axis, we use the larger intercept, a, to find the denominator for y^2, giving the equation in standard form as

$$\frac{x^2}{7} + \frac{y^2}{16} = 1.$$

A graph of this ellipse is shown in Figure 27. The domain of this relation is $\left[-\sqrt{7}, \sqrt{7}\right]$, and the range is $[-4, 4]$.

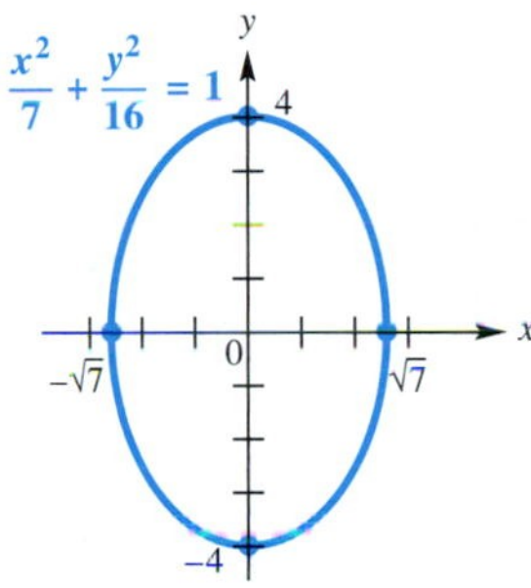

FIGURE 27

Translations of Ellipses

Just as a circle need not have its center at the origin, an ellipse also may have its center translated from the origin.

Standard Forms for Ellipses Centered at (h, k)

An ellipse with center at (h, k) and either a horizontal or vertical major axis satisfies one of the following equations, where $a > b > 0$ and $c^2 = a^2 - b^2$ with $c > 0$:

$$\frac{(x-h)^2}{a^2} + \frac{(y-k)^2}{b^2} = 1$$ Major axis: horizontal; foci: $(h \pm c, k)$; vertices: $(h \pm a, k)$

$$\frac{(x-h)^2}{b^2} + \frac{(y-k)^2}{a^2} = 1.$$ Major axis: vertical; foci: $(h, k \pm c)$; vertices: $(h, k \pm a)$

EXAMPLE 3 **Graphing an Ellipse Translated from the Origin**

Graph $\dfrac{(x-2)^2}{9}+\dfrac{(y+1)^2}{16}=1$. Give the domain and range.

Analytic Solution

The graph of this equation is an ellipse centered at $(2,-1)$. For this ellipse, $a=4$ and $b=3$. Since $a=4$ is associated with y^2, the vertices of the ellipse are on the vertical line through $(2,-1)$. Find the vertices by locating two points on the vertical line through $(2,-1)$, one 4 units up from $(2,-1)$ and one 4 units down. The vertices are $(2,3)$ and $(2,-5)$. Locate two other points on the ellipse by locating points on the horizontal line through $(2,-1)$, one 3 units to the right and one 3 units to the left. The graph is shown in Figure 28. The domain is $[-1,5]$, and the range is $[-5,3]$.

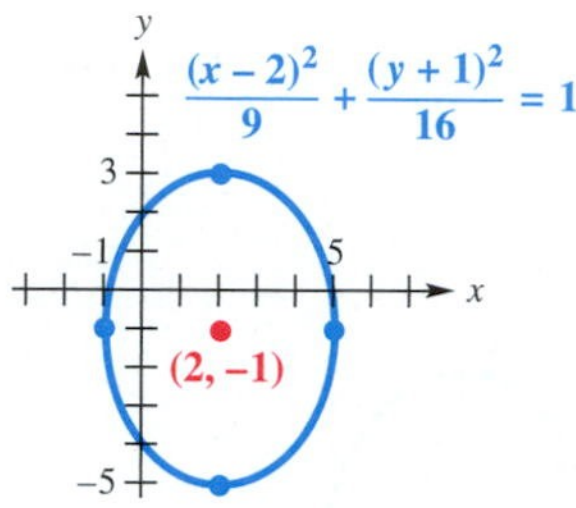

FIGURE 28

Graphing Calculator Solution

Solve for y in the equation of the ellipse to obtain

$$y=-1\pm 4\sqrt{1-\frac{(x-2)^2}{9}}.$$

The + sign yields the top half of the ellipse, while the − sign yields the bottom half. See Figure 29.

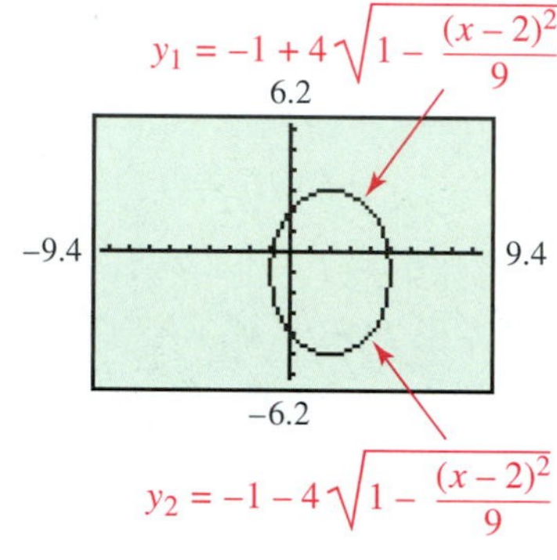

FIGURE 29

NOTE An ellipse is symmetric with respect to its major axis, its minor axis, and its center. ***If a = b in the equation of an ellipse, the graph is that of a circle.***

EXAMPLE 4 **Finding the Standard Form for an Ellipse**

Write $4x^2-16x+9y^2+54y+61=0$ in the standard form for an ellipse centered at (h,k). Identify the center and vertices.

Solution

$$4x^2-16x+9y^2+54y+61=0$$

$$4(x^2-4x+\quad)+9(y^2+6y+\quad)=-61 \qquad \text{Distributive property; add } -61.$$

$$4(x^2-4x+4)+9(y^2+6y+9)=-61+4(4)+9(9) \qquad \text{Complete the square.}$$

$$4(x-2)^2+9(y+3)^2=36 \qquad \text{Factor; add.}$$

$$\frac{(x-2)^2}{9}+\frac{(y+3)^2}{4}=1 \qquad \text{Divide each side by 36.}$$

The center is $(2,-3)$. Because $a=3$ and the major axis is horizontal, the vertices of the ellipse are $(2\pm 3,-3)$, or $(5,-3)$ and $(-1,-3)$.

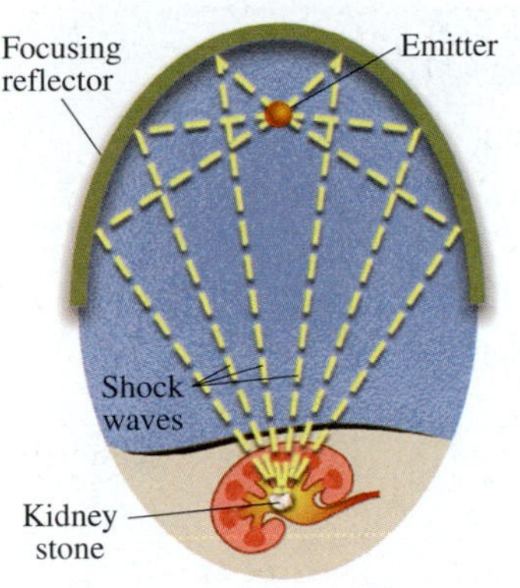

The top of an ellipse is illustrated in this depiction of how a lithotripter crushes a kidney stone.

FIGURE 30

An Application of Ellipses

Ellipses have many useful applications. As Earth makes its yearlong journey around the sun, it traces an ellipse. Spacecraft travel around Earth in elliptical orbits, and planets make elliptical orbits around the sun.

An application from medicine is a lithotripter, a machine used to crush kidney stones via shock waves. The patient is placed in an elliptical tub with the kidney stone at one focus of the ellipse. A beam is projected from the other focus to the tub so that it reflects to hit the kidney stone. See Figure 30.

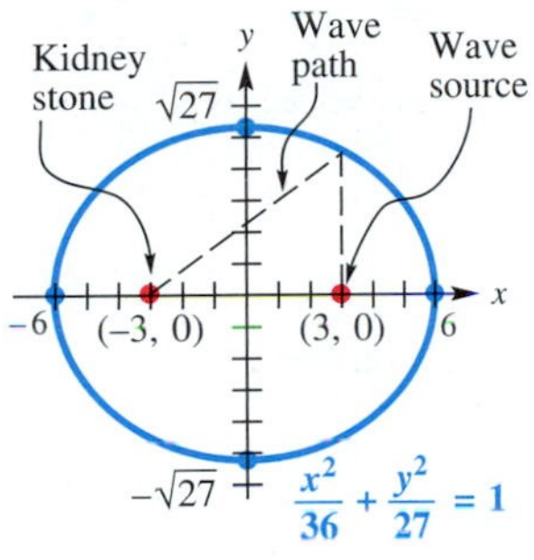

FIGURE 31

EXAMPLE 5 Modeling the Reflective Property of Ellipses

If a lithotripter is based on the ellipse

$$\frac{x^2}{36}+\frac{y^2}{27}=1,$$

determine how many units the kidney stone and the wave source must be placed from the center of the ellipse.

Solution The kidney stone and the source of the beam must be placed at the foci, $(c, 0)$ and $(-c, 0)$. Here $a^2 = 36$ and $b^2 = 27$, so

$$c=\sqrt{a^2-b^2}=\sqrt{36-27}=\sqrt{9}=3.$$

Thus, the foci are $(-3, 0)$ and $(3, 0)$, so the kidney stone and the source both must be placed on a line 3 units from the center. See Figure 31. ■

Equations and Graphs of Hyperbolas

An ellipse was defined as the set of all points in a plane, the *sum* of whose distances from two fixed points is constant. A *hyperbola* is defined similarly.

> **Hyperbola**
>
> A **hyperbola** is the set of all points in a plane such that the absolute value of the *difference* of the distances from two fixed points is constant. The two fixed points are called the **foci** of the hyperbola.

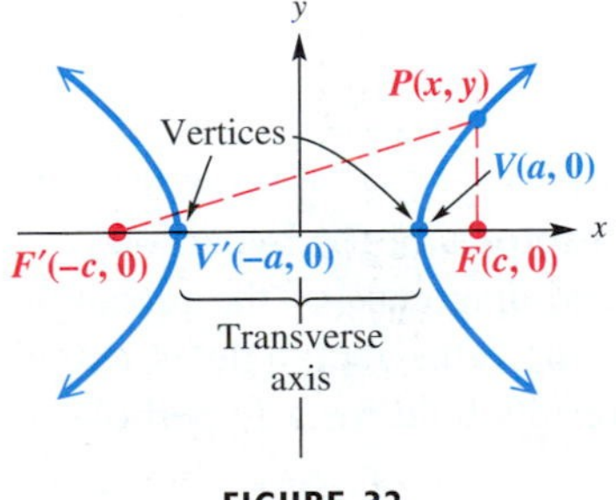

FIGURE 32

Suppose a hyperbola has center at the origin and foci at $F'(-c, 0)$ and $F(c, 0)$. See Figure 32. The midpoint of the segment $F'F$ is the center of the hyperbola, and the points $V'(-a, 0)$ and $V(a, 0)$ are the **vertices** of the hyperbola. The line segment $V'V$ is the **transverse axis** of the hyperbola.

Using the distance formula in conjunction with the definition of a hyperbola, we can verify the following standard forms of the equations for hyperbolas centered at the origin. (See Exercise 95.)

Standard Forms of Equations for Hyperbolas

The hyperbola with center at the origin and equation

$$\frac{x^2}{a^2} - \frac{y^2}{b^2} = 1$$

has vertices $(\pm a, 0)$, asymptotes $y = \pm\frac{b}{a}x$, and foci $(\pm c, 0)$, where $c^2 = a^2 + b^2$.

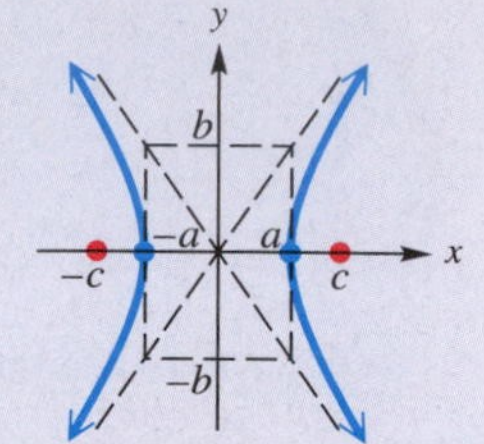

Transverse axis on x-axis

The hyperbola with center at the origin and equation

$$\frac{y^2}{a^2} - \frac{x^2}{b^2} = 1$$

has vertices $(0, \pm a)$, asymptotes $y = \pm\frac{a}{b}x$, and foci $(0, \pm c)$, where $c^2 = a^2 + b^2$.

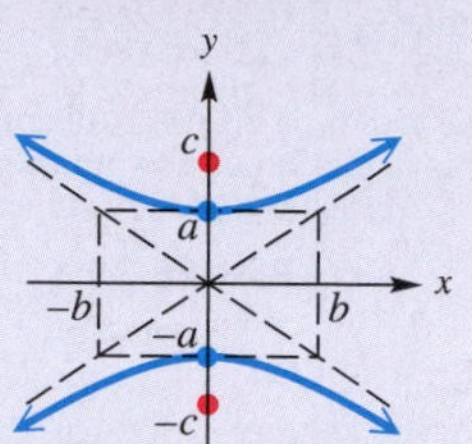

Transverse axis on y-axis

Starting with the first equation for a hyperbola and solving for y gives

$$\frac{x^2}{a^2} - \frac{y^2}{b^2} = 1$$

$$\frac{x^2}{a^2} - 1 = \frac{y^2}{b^2} \qquad \text{Subtract 1; add } \tfrac{y^2}{b^2}.$$

$$\frac{x^2 - a^2}{a^2} = \frac{y^2}{b^2} \qquad \text{Write the left side as a single fraction.}$$

$$y = \pm\frac{b}{a}\sqrt{x^2 - a^2}. \quad (*) \qquad \text{Take square roots; multiply by } b.$$

Remember both the positive and negative square roots.

If x^2 is very large in comparison to a^2, the difference $x^2 - a^2$ would be relatively close to x^2. If this happens, then the points satisfying equation (*) would approach one of the lines

$$y = \pm\frac{b}{a}x.$$

Thus, as $|x|$ gets larger and larger, the points of the hyperbola $\frac{x^2}{a^2} - \frac{y^2}{b^2} = 1$ become closer to the lines $y = \pm\frac{b}{a}x$. These lines, called the **asymptotes** of the hyperbola, are helpful in sketching the graph. The lines are the extended diagonals of the rectangle whose vertices are (a, b), $(-a, b)$, $(a, -b)$, and $(-a, -b)$. This rectangle is called the **fundamental rectangle** of the hyperbola. Similar results hold for a hyperbola of the form $\frac{y^2}{a^2} - \frac{x^2}{b^2} = 1$.

EXAMPLE 6 Using Asymptotes to Graph a Hyperbola

Sketch the asymptotes and graph the hyperbola $\frac{x^2}{25} - \frac{y^2}{49} = 1$. Give the foci, domain, and range.

Analytic Solution

For this hyperbola, $a = 5$ and $b = 7$. With these values, $y = \pm\frac{b}{a}x$ becomes $y = \pm\frac{7}{5}x$. Choosing $x = 5$ gives $y = \pm 7$. Choosing $x = -5$ also gives $y = \pm 7$. These four points, $(\pm 5, \pm 7)$, are the corners of the fundamental rectangle shown in Figure 33.

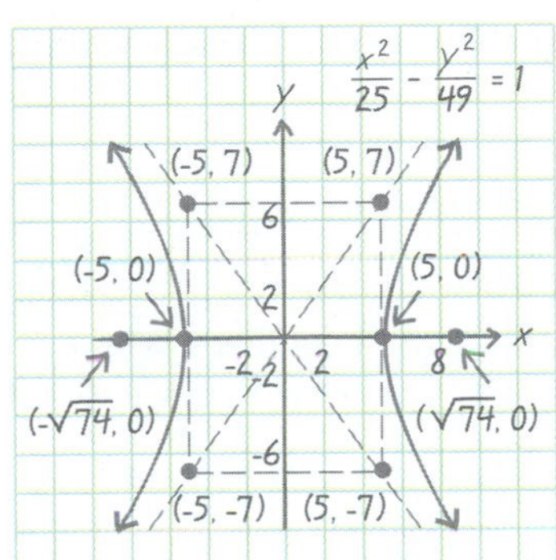

FIGURE 33

The extended diagonals of this rectangle are the asymptotes of the hyperbola. See Figure 33. Note that the hyperbola has vertices $(-5, 0)$ and $(5, 0)$. We find the foci by letting

$$c^2 = a^2 + b^2 = 25 + 49 = 74, \quad \text{so} \quad c = \sqrt{74}.$$

Therefore, the foci are $(-\sqrt{74}, 0)$ and $(\sqrt{74}, 0)$. The domain of this relation is $(-\infty, -5] \cup [5, \infty)$, and the range is $(-\infty, \infty)$.

Graphing Calculator Solution

Solve the given equation for y.

$$\frac{x^2}{25} - \frac{y^2}{49} = 1$$

$$-\frac{y^2}{49} = 1 - \frac{x^2}{25} \quad \text{Subtract } \frac{x^2}{25}.$$

$$\frac{y^2}{49} = \frac{x^2}{25} - 1 \quad \text{Multiply by } -1.$$

$$\frac{y}{7} = \pm\sqrt{\frac{x^2}{25} - 1} \quad \text{Take square roots.}$$

$$y = \pm 7\sqrt{\frac{x^2}{25} - 1} \quad \text{Multiply by 7.}$$

The graphs of the hyperbola and its asymptotes are shown in Figure 34. The graph of y_1 creates the upper half of the hyperbola, while the graph of y_2 creates the lower half.

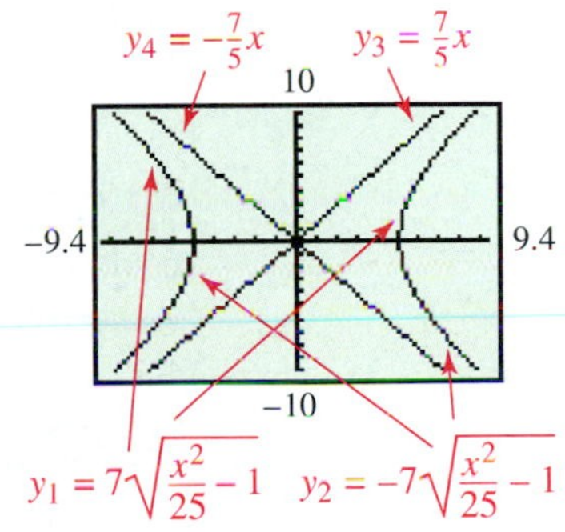

FIGURE 34

EXAMPLE 7 Graphing a Hyperbola Centered at the Origin

Graph $25y^2 - 4x^2 = 9$. Give the domain and range.

Solution

$$25y^2 - 4x^2 = 9$$

$$\frac{25y^2}{9} - \frac{4x^2}{9} = 1 \quad \text{Divide by 9.}$$

Be sure to divide *each* term by 9.

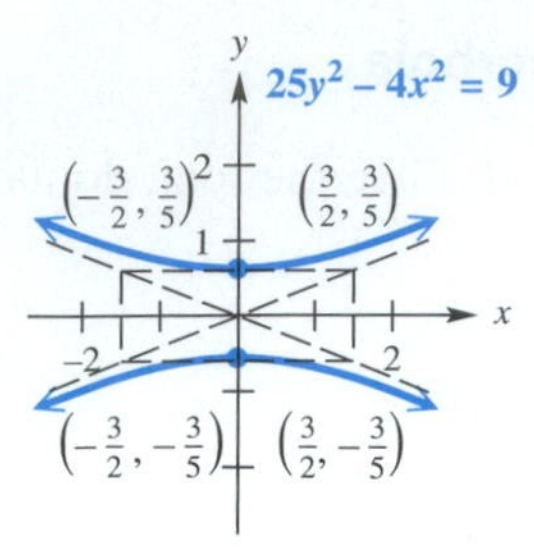

FIGURE 35

To determine the values of a and b, write the equation $\frac{25y^2}{9} - \frac{4x^2}{9} = 1$ as

$$\frac{y^2}{\frac{9}{25}} - \frac{x^2}{\frac{9}{4}} = 1. \qquad \frac{a}{b} = \frac{1}{\frac{b}{a}}$$

This hyperbola is centered at the origin, has foci on the y-axis, and has y-intercepts $-\frac{3}{5}$ and $\frac{3}{5}$. Use the four points $\left(\pm\frac{3}{2}, \pm\frac{3}{5}\right)$ to get the fundamental rectangle shown in Figure 35. Use the diagonals of this rectangle to determine the asymptotes for the graph. The domain is $(-\infty, \infty)$, and the range is $\left(-\infty, -\frac{3}{5}\right] \cup \left[\frac{3}{5}, \infty\right)$. ■

Translations of Hyperbolas

Just like other conics, hyperbolas may have their centers translated from the origin.

Standard Forms for Hyperbolas Centered at (h, k)

A hyperbola with center (h, k) and either a horizontal or vertical transverse axis satisfies one of the following equations, where $c^2 = a^2 + b^2$:

$$\frac{(x-h)^2}{a^2} - \frac{(y-k)^2}{b^2} = 1$$

Transverse axes: horizontal; vertices: $(h \pm a, k)$; foci: $(h \pm c, k)$; asymptotes: $y = \pm\frac{b}{a}(x-h) + k$

$$\frac{(y-k)^2}{a^2} - \frac{(x-h)^2}{b^2} = 1.$$

Transverse axes: vertical; vertices: $(h, k \pm a)$; foci: $(h, k \pm c)$; asymptotes: $y = \pm\frac{a}{b}(x-h) + k$

NOTE The asymptotes for a hyperbola *always* pass through the center (h, k). By the point–slope form of a line, the equation of any asymptote is $y = m(x-h) + k$. If the transverse axis is horizontal, then $m = \pm\frac{b}{a}$; if it is vertical, then $m = \pm\frac{a}{b}$.

EXAMPLE 8 Graphing a Hyperbola Translated from the Origin

Graph $\frac{(y+2)^2}{9} - \frac{(x+3)^2}{4} = 1$, and identify the center. Give the domain and range.

Solution This equation represents a hyperbola centered at $(-3, -2)$. For this vertical hyperbola, $a = 3$ and $b = 2$. The x-values of the vertices are -3. Locate the y-values of the vertices by taking the y-value of the center, -2, and adding and subtracting 3. Thus, the vertices are $(-3, 1)$ and $(-3, -5)$. The asymptotes have slopes $\pm\frac{3}{2}$ and pass through the center $(-3, -2)$. The equations of the asymptotes,

$$y = \pm\frac{3}{2}(x+3) - 2,$$

can be found by using the point–slope form of the equation of a line. The graph is shown in Figure 36. The domain of the relation is $(-\infty, \infty)$, and the range is $(-\infty, -5] \cup [1, \infty)$. ■

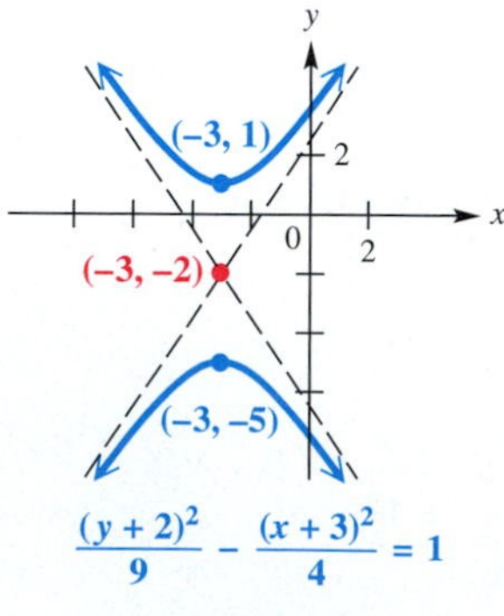

FIGURE 36

FOR DISCUSSION

How would you graph the hyperbola in Example 8,

$$\frac{(y+2)^2}{9} - \frac{(x+3)^2}{4} = 1,$$

on your graphing calculator? Duplicate the calculator graph of that hyperbola shown in Figure 37. The two branches of a hyperbola are reflections about two different axes and also about a point. What are the axes and the point for the hyperbola shown in the figure?

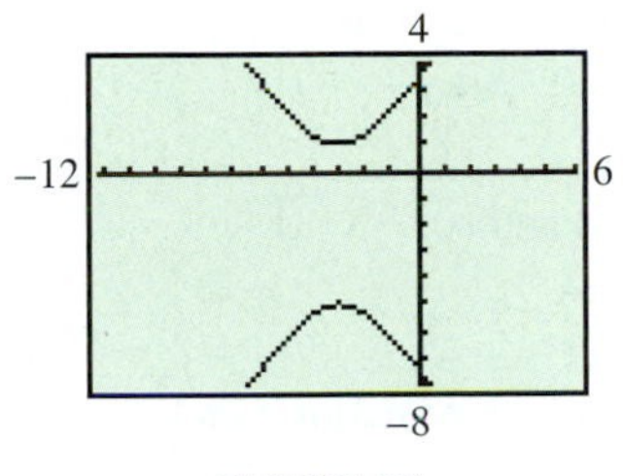

FIGURE 37

EXAMPLE 9 Writing the Equation of a Hyperbola

Write an equation of the hyperbola having center $(1, 2)$, focus $(6, 2)$, and vertex $(4, 2)$.

Solution Because the center, focus, and vertex lie on the line $y = 2$, the hyperbola has a horizontal transverse axis. The center is $(1, 2)$, so the standard form of the equation is

$$\frac{(x-1)^2}{a^2} - \frac{(y-2)^2}{b^2} = 1,$$

with a and b to be determined. The distance from the focus to the center is $6 - 1 = 5$, and the distance from the vertex to the center is $4 - 1 = 3$, so $c = 5$ and $a = 3$. Because $a^2 + b^2 = c^2$, it follows that $b = 4$. Thus, the equation of the hyperbola is

$$\frac{(x-1)^2}{25} - \frac{(y-2)^2}{16} = 1.$$ ■

EXAMPLE 10 Finding the Standard Form for a Hyperbola

Write $9x^2 - 18x - 4y^2 - 16y = 43$ in the standard form for a hyperbola centered at (h, k). Identify the center and vertices.

Solution

$$9x^2 - 18x - 4y^2 - 16y = 43$$

$$9(x^2 - 2x + \quad) - 4(y^2 + 4y + \quad) = 43 \qquad \text{Distributive property}$$

$$9(x^2 - 2x + 1) - 4(y^2 + 4y + 4) = 43 + 9(1) - 4(4) \qquad \text{Complete the square.}$$

$$9(x-1)^2 - 4(y+2)^2 = 36 \qquad \text{Factor; add.}$$

$$\frac{(x-1)^2}{4} - \frac{(y+2)^2}{9} = 1 \qquad \text{Divide each side by 36.}$$

The center is $(1, -2)$. Because $a = 2$ and the transverse axis is horizontal, the vertices of the hyperbola are $(1 \pm 2, -2)$, or $(3, -2)$ and $(-1, -2)$. ■

6.2 Exercises

Concept Check *Match each equation in Column I with the appropriate description in Column II.*

I

1. $\dfrac{x^2}{4} + \dfrac{y^2}{16} = 1$
2. $\dfrac{x^2}{16} + \dfrac{y^2}{4} = 1$
3. $\dfrac{x^2}{4} - \dfrac{y^2}{16} = 1$
4. $\dfrac{y^2}{4} - \dfrac{x^2}{16} = 1$
5. $\dfrac{(x+2)^2}{9} + \dfrac{(y-4)^2}{25} = 1$
6. $\dfrac{(x-2)^2}{9} + \dfrac{(y+4)^2}{25} = 1$
7. $\dfrac{(x+2)^2}{9} - \dfrac{(y-4)^2}{25} = 1$
8. $\dfrac{(x-2)^2}{9} - \dfrac{(y-4)^2}{25} = 1$

II

A. Hyperbola; center $(2, 4)$
B. Ellipse; foci $(\pm 2\sqrt{3}, 0)$
C. Hyperbola; foci $(0, \pm 2\sqrt{5})$
D. Hyperbola; center $(-2, 4)$
E. Ellipse; center $(-2, 4)$
F. Center $(0, 0)$; horizontal transverse axis
G. Ellipse; foci $(0, \pm 2\sqrt{3})$
H. Vertical major axis; center $(2, -4)$

9. Explain how a circle can be interpreted as a special case of an ellipse.

10. *Concept Check* If an ellipse has endpoints of the minor axis and vertices at $(-3,0)$, $(3,0)$, $(0,5)$, and $(0,-5)$, what is its domain? What is its range?

Graph each ellipse by hand. Give the domain and range. Give the foci in Exercises 11–14.

11. $\frac{x^2}{9}+\frac{y^2}{4}=1$

12. $\frac{x^2}{16}+\frac{y^2}{36}=1$

13. $9x^2+6y^2=54$

14. $12x^2+8y^2=96$

15. $\frac{25y^2}{36}+\frac{64x^2}{9}=1$

16. $\frac{16y^2}{9}+\frac{121x^2}{25}=1$

Graph each ellipse and label its center. Give the domain and range.

17. $\frac{(x-1)^2}{9}+\frac{(y+3)^2}{25}=1$

18. $\frac{(x+3)^2}{16}+\frac{(y-2)^2}{36}=1$

19. $\frac{(x-2)^2}{16}+\frac{(y-1)^2}{9}=1$

20. $\frac{(x+3)^2}{25}+\frac{(y+2)^2}{36}=1$

21. $\frac{(x+1)^2}{64}+\frac{(y-2)^2}{49}=1$

22. $\frac{(x-4)^2}{9}+\frac{(y+2)^2}{4}=1$

Find an equation for each ellipse.

23. x-intercepts ± 4; foci $(-2,0)$ and $(2,0)$

24. y-intercepts ± 3; foci $(0,\sqrt{3})$ and $(0,-\sqrt{3})$

25. Endpoints of major axis at $(6, 0)$ and $(-6, 0)$; $c = 4$

26. Vertices $(0,5)$ and $(0,-5)$; $b = 2$

27. Center $(3,-2)$; $a = 5$; $c = 3$; major axis vertical

28. Center $(2,0)$; minor axis of length 6; major axis horizontal and of length 9

29. Major axis of length 6; foci $(0, 2)$ and $(0, -2)$

30. Minor axis of length 4; foci $(-5, 0)$ and $(5, 0)$

31. Center $(5, 2)$; minor axis vertical, with length 8; $c = 3$

32. Center $(-3, 6)$; major axis vertical, with length 10; $c = 2$

33. Vertices $(4, 9)$ and $(4, 1)$; minor axis of length 6

34. Foci at $(-3, -3)$ and $(7, -3)$; the point $(2, 1)$ on ellipse

Write the equation in standard form for an ellipse centered at (h, k). Identify the center and vertices.

35. $9x^2+18x+4y^2-8y-23=0$

36. $9x^2-36x+16y^2-64y-44=0$

37. $4x^2+8x+y^2+2y+1=0$

38. $x^2-6x+9y^2=0$

39. $4x^2+16x+5y^2-10y+1=0$

40. $2x^2+4x+3y^2-18y+23=0$

41. $16x^2-16x+4y^2+12y=51$

42. $16x^2+48x+4y^2-20y+57=0$

Graph each hyberbola by hand. Give the domain and range. Give the center in Exercises 51–56.

43. $\frac{x^2}{16}-\frac{y^2}{9}=1$

44. $\frac{y^2}{9}-\frac{x^2}{9}=1$

45. $49y^2-36x^2=1764$

46. $144x^2-49y^2=7056$

47. $\frac{4x^2}{9}-\frac{25y^2}{16}=1$

48. $x^2-y^2=1$

49. $9x^2-4y^2=1$

50. $25y^2-9x^2=1$

51. $\frac{(x-1)^2}{9}-\frac{(y+3)^2}{25}=1$

52. $\frac{(x+3)^2}{16}-\frac{(y-2)^2}{36}=1$

53. $\frac{(y-5)^2}{4}-\frac{(x+1)^2}{9}=1$

54. $\frac{(y+1)^2}{25}-\frac{(x-3)^2}{36}=1$

55. $16(x+5)^2-(y-3)^2=1$

56. $4(x+9)^2-25(y+6)^2=100$

Find an equation for each hyperbola.

57. x-intercepts ± 3; foci $(-4, 0)$ and $(4, 0)$

58. y-intercepts ± 5; foci $(0, 3\sqrt{3})$ and $(0, -3\sqrt{3})$

59. Asymptotes $y = \pm \frac{3}{5}x$; y-intercepts 3 and -3

60. y-intercept -2; center at origin; passing through $(2, 3)$

61. Vertices $(0, 6)$ and $(0, -6)$; asymptotes $y = \pm \frac{1}{2}x$

62. Vertices $(-10, 0)$ and $(10, 0)$; asymptotes $y = \pm 5x$

63. Vertices $(-3, 0)$ and $(3, 0)$; passing through $(6, 1)$

64. Vertices $(0, 5)$ and $(0, -5)$; passing through $(3, 10)$

65. Foci $(0, \sqrt{13})$ and $(0, -\sqrt{13})$; asymptotes $y = \pm 5x$

66. Foci $(-3\sqrt{5}, 0)$ and $(3\sqrt{5}, 0)$; asymptotes $y = \pm 2x$

67. Vertices $(4, 5)$ and $(4, 1)$; asymptotes $y = \pm 7(x - 4) + 3$

68. Vertices $(5, -2)$ and $(1, -2)$; asymptotes $y = \pm \frac{3}{2}(x - 3) - 2$

69. Center $(1, -2)$; focus $(4, -2)$; vertex $(3, -2)$

70. Center $(9, -7)$; focus $(9, 3)$; vertex $(9, -1)$

Write the equation in standard form for a hyperbola centered at (h, k). Identify the center and vertices.

71. $x^2 - 2x - y^2 + 2y = 4$

72. $y^2 + 4y - x^2 + 2x = 6$

73. $3y^2 + 24y - 2x^2 + 12x + 24 = 0$

74. $4x^2 + 16x - 9y^2 + 18y = 29$

75. $x^2 - 6x - 2y^2 + 7 = 0$

76. $y^2 + 8y - 3x^2 + 13 = 0$

77. $4y^2 + 32y - 5x^2 - 10x + 39 = 0$

78. $5x^2 + 10x - 7y^2 + 28y = 58$

Relating Concepts

For individual or group investigation (Exercises 79–84)

Consider the ellipse and hyperbola defined by

$$\frac{x^2}{16} + \frac{y^2}{12} = 1 \quad \textit{and} \quad \frac{x^2}{4} - \frac{y^2}{12} = 1,$$

respectively. ***Work Exercises 79–84 in order.***

79. Find the foci of the ellipse. Call them F_1 and F_2.

80. Graph the ellipse with your calculator, and trace to find the coordinates of several points on the ellipse.

81. For each of the points P, verify that

$[\text{distance of } P \text{ from } F_1] + [\text{distance of } P \text{ from } F_2] = 8.$

82. Repeat Exercises 79 and 80 for the hyperbola.

83. For each of the points P from Exercise 82, verify that

$|[\text{distance of } P \text{ from } F_1] - [\text{distance of } P \text{ from } F_2]| = 4.$

84. How do Exercises 81 and 83 relate to the definitions of the ellipse and the hyperbola given in this section?

(Modeling) Solve each problem.

85. *Shape of a Lithotripter* A patient's kidney stone is placed 12 units away from the source of the shock waves of a lithotripter. The lithotripter is based on an ellipse with a minor axis that measures 16 units. Find an equation of an ellipse that would satisfy this situation.

86. *Orbit of Venus* The orbit of Venus is an ellipse, with the sun at one focus. An approximate equation for the orbit is

$$\frac{x^2}{5013} + \frac{y^2}{4970} = 1,$$

where x and y are measured in millions of miles.
(a) Find the length of the major axis.
(b) Find the length of the minor axis.

87. *The Roman Colosseum* The Roman Colosseum is an ellipse with major axis 620 feet and minor axis 513 feet. Find the distance between the foci of this ellipse.

88. *The Roman Colosseum* A formula for the approximate circumference of an ellipse is

$$C \approx 2\pi \sqrt{\frac{a^2 + b^2}{2}},$$

where a and b are the lengths shown in the figure. Use this formula to find the approximate circumference of the Roman Colosseum. (See Exercise 87.)

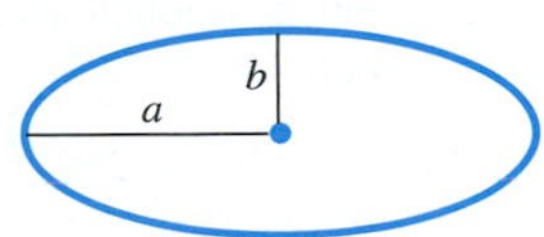

89. *Height of an Overpass* A one-way road passes under an overpass in the form of half of an ellipse 15 feet high at the center and 20 feet wide. Assuming that a truck is 12 feet wide, what is the height of the tallest truck that can pass under the overpass?

90. *Location-Finding System* Ships and planes often use a location-finding system called LORAN. With this system, a radio transmitter at M on the figure sends out a series of pulses. When each pulse is received at transmitter S, it then sends out a pulse. A ship at P receives pulses from both M and S. A receiver on the ship measures the difference in the arrival times of the pulses. The navigator then consults a special map, showing certain curves according to the differences in arrival times. In this way, the ship can be located as lying on a portion of what type of curve?

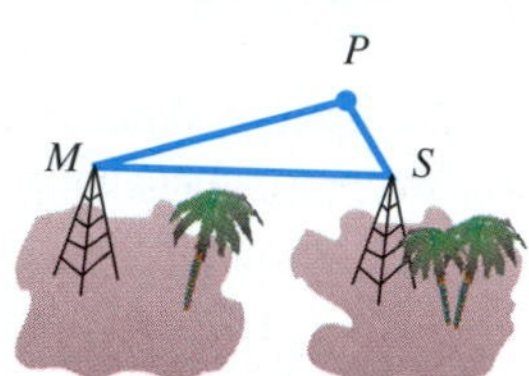

91. *Orbit of a Satellite* The coordinates in miles for the orbit of the artificial satellite *Explorer VII* can be described by the equation

$$\frac{x^2}{a^2} + \frac{y^2}{b^2} = 1,$$

where $a = 4465$ and $b = 4462$. Earth's center is located at one focus of its elliptical orbit. (*Source:* Loh, W., *Dynamics and Thermodynamics of Planetary Entry,* Prentice-Hall, 1963; Thomson, W., *Introduction to Space Dynamics,* John Wiley and Sons, 1961.)

(a) Graph both the orbit of *Explorer VII* and Earth on the same coordinate axes if the average radius of Earth is 3960 miles. Use the window $[-6750, 6750]$ by $[-4500, 4500]$.

(b) Determine the maximum and minimum heights of the satellite above Earth's surface.

92. *Design of a Sports Complex* Two buildings in a sports complex are shaped and positioned like a portion of the branches of the hyperbola

$$400x^2 - 625y^2 = 250{,}000.$$

In this equation, x and y are in meters.

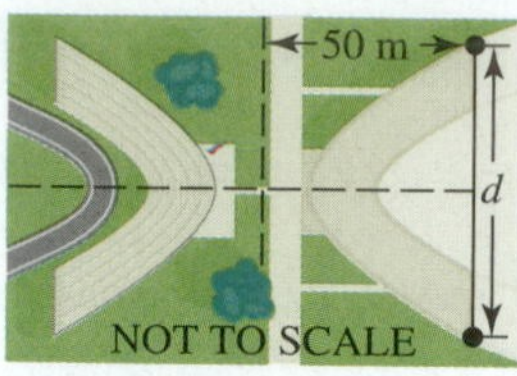

(a) How far apart are the buildings at their closest point?

(b) Find the distance d in the figure.

93. *Structure of an Atom* In 1911, Ernest Rutherford discovered the basic structure of the atom by "shooting" positively charged alpha particles with a speed of 10^7 meters per second at a piece of gold foil 6×10^{-7} meter thick. Only a small percentage of the alpha particles struck a gold nucleus head-on and were deflected directly back toward their source. The rest of the particles often followed a hyperbolic trajectory because they were repelled by positively charged gold nuclei. Thus, Rutherford proposed that the atom was composed of mostly empty space and a small, dense nucleus. The figure shows an alpha particle A initially approaching a gold nucleus N and being deflected at an angle $\theta = 90°$. N is located at a focus of the hyperbola, and the trajectory of A passes through a vertex of the hyperbola. (*Source:* Semat, H. and J. Albright, *Introduction to Atomic and Nuclear Physics,* Holt, Rinehart and Winston, 1972.)

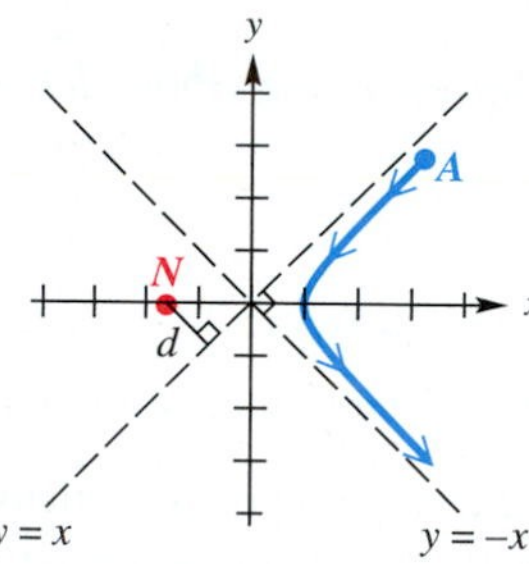

(a) Determine the equation of the trajectory of the alpha particle if $d = 5 \times 10^{-14}$ meter.

(b) What was the minimum distance between the centers of the alpha particle and the gold nucleus?

94. *Sound Detection* Microphones are placed at points $(-c, 0)$ and $(c, 0)$. An explosion occurs at point $P(x, y)$ having positive x-coordinate.

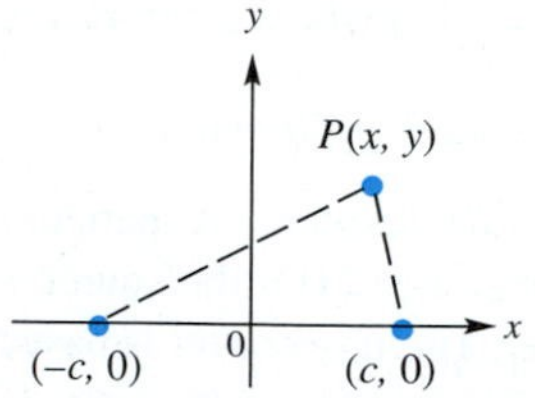

The sound is detected at the closer microphone t seconds before being detected at the farther microphone. Assume that sound travels at a speed of 330 meters per second, and show that P must be on the hyperbola

$$\frac{x^2}{330^2t^2} - \frac{y^2}{4c^2 - 330^2t^2} = \frac{1}{4}.$$

95. Suppose a hyperbola has center at the origin, foci at $F'(-c, 0)$ and $F(c, 0)$, and equation $d(P, F') - d(P, F) = 2a$. Let $b^2 = c^2 - a^2$, and show that an equation of the hyperbola is

$$\frac{x^2}{a^2} - \frac{y^2}{b^2} = 1.$$

96. Use the definition of an ellipse to find an equation of an ellipse with foci $(3, 0)$ and $(-3, 0)$ where the sum of the distances from any point of the ellipse to the two foci is 10.

97. Use the definition of a hyperbola to find an equation of a hyperbola with center at the origin, foci $(-2, 0)$ and $(2, 0)$, and the absolute value of the difference of the distances from any point of the hyperbola to the two foci equal to 2.

Reviewing Basic Concepts (Sections 6.1 and 6.2)

1. Match each definition in A–D with the appropriate conic section.
(a) Circle **(b)** Parabola **(c)** Ellipse **(d)** Hyperbola

A. A set of points in a plane such that the sum of their distances from two fixed points is constant
B. A set of points in a plane that are equidistant from a fixed point
C. A set of points in a plane such that the absolute value of the difference of their distances from two fixed points is constant
D. A set of points in a plane equidistant from a fixed point and a fixed line

Graph each conic section by hand and check with your calculator.

2. $12x^2 - 4y^2 = 48$

3. $y = 2x^2 + 3x - 1$

4. $x^2 + y^2 - 2x + 2y - 2 = 0$

5. $4x^2 + 9y^2 = 36$

6. Given the two vertices and two foci of a conic section, how can you tell whether it is an ellipse or a hyperbola?

Write an equation for each conic section.

7. Center $(2, -1)$; radius 3

8. Foci $(\pm 4, 0)$; major axis length 12

9. Vertices $(0, \pm 2)$; foci $(0, \pm 4)$

10. Single focus $\left(0, \frac{1}{2}\right)$; vertex at the origin

6.3 Summary of the Conic Sections

Characteristics ■ Identifying Conic Sections ■ Eccentricity

Characteristics

The conic sections presented in this chapter have equations that can be written in the form

$$Ax^2 + Dx + Cy^2 + Ey + F = 0,$$

where either A or C must be nonzero.

The special characteristics of each conic section $Ax^2 + Dx + Cy^2 + Ey + F = 0$ are summarized in the following table.

Conic Section	Characteristic	Example
Parabola	Either $A = 0$ or $C = 0$, but not both	$y = x^2$ $x = 3y^2 + 2y - 4$
Circle	$A = C \neq 0$	$x^2 + y^2 = 16$
Ellipse	$A \neq C$, $AC > 0$	$\frac{x^2}{16} + \frac{y^2}{25} = 1$
Hyperbola	$AC < 0$	$x^2 - y^2 = 1$

The chart summarizes our work with conic sections.

Equation	Graph	Description	Identification
$(x - h)^2 = 4c(y - k)$	y, x, 0, (h, k), Parabola	Opens up if $c > 0$, down if $c < 0$. Vertex is (h, k).	x^2-term is present. y is not squared.
$(y - k)^2 = 4c(x - h)$	y, x, 0, (h, k), Parabola	Opens to the right if $c > 0$, to the left if $c < 0$. Vertex is (h, k).	y^2-term is present. x is not squared.
$(x - h)^2 + (y - k)^2 = r^2$	y, x, 0, (h, k), r, Circle	Center is (h, k), radius is r.	x^2- and y^2-terms have the same positive coefficient.
$\frac{x^2}{a^2} + \frac{y^2}{b^2} = 1 \quad (a > b > 0)$	y, x, 0, (0, b), (−a, 0), (a, 0), (0, −b), Ellipse	x-intercepts are a and $-a$. y-intercepts are b and $-b$.	x^2- and y^2-terms have different positive coefficients.

(continued)

Equation	Graph	Description	Identification
$\frac{x^2}{b^2} + \frac{y^2}{a^2} = 1 \quad (a > b > 0)$	(0, *a*) (−*b*, 0) (*b*, 0) 0 (0, −*a*) Ellipse	*x*-intercepts are *b* and −*b*. *y*-intercepts are *a* and −*a*.	x^2- and y^2-terms have different positive coefficients.
$\frac{x^2}{a^2} - \frac{y^2}{b^2} = 1$	(−*a*, 0) (*a*, 0) 0 Hyperbola	*x*-intercepts are *a* and −*a*. Asymptotes are found from the four points: $(\pm a, \pm b)$.	x^2-term has a positive coefficient. y^2-term has a negative coefficient.
$\frac{y^2}{a^2} - \frac{x^2}{b^2} = 1$	(0, *a*) 0 (0, −*a*) Hyperbola	*y*-intercepts are *a* and −*a*. Asymptotes are found from the four points: $(\pm b, \pm a)$.	y^2-term has a positive coefficient. x^2-term has a negative coefficient.

NOTE To find the value of c for an ellipse, use the equation $c^2 = a^2 - b^2$; for a hyperbola, use the equation $c^2 = a^2 + b^2$. If the center is $(0, 0)$, then the foci of these conic sections are located either at $(\pm c, 0)$ or at $(0, \pm c)$.

Identifying Conic Sections

To recognize the type of graph that a given conic section has, we sometimes need to transform the equation into a more familiar form.

EXAMPLE 1 Determining Types of Conic Sections from Equations

Decide on the type of conic section represented by each equation, and give each graph.

(a) $x^2 = 25 + 5y^2$ **(b)** $x^2 - 8x + y^2 + 10y = -41$

(c) $4x^2 - 16x + 9y^2 + 54y = -61$ **(d)** $x^2 - 6x + 8y - 7 = 0$

Solution

(a)

$$x^2 = 25 + 5y^2$$

$$x^2 - 5y^2 = 25 \qquad \text{Subtract } 5y^2.$$

$$\frac{x^2}{25} - \frac{y^2}{5} = 1 \qquad \text{Divide by 25.}$$

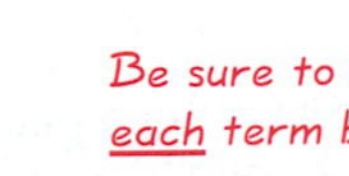

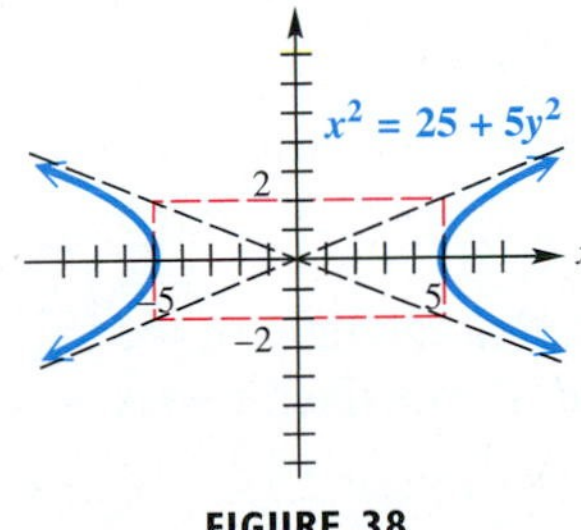

FIGURE 38

The equation represents a hyperbola centered at the origin and with asymptotes $y = \pm\frac{\sqrt{5}}{5}x$. The vertices are $(-5, 0)$ and $(5, 0)$. See Figure 38.

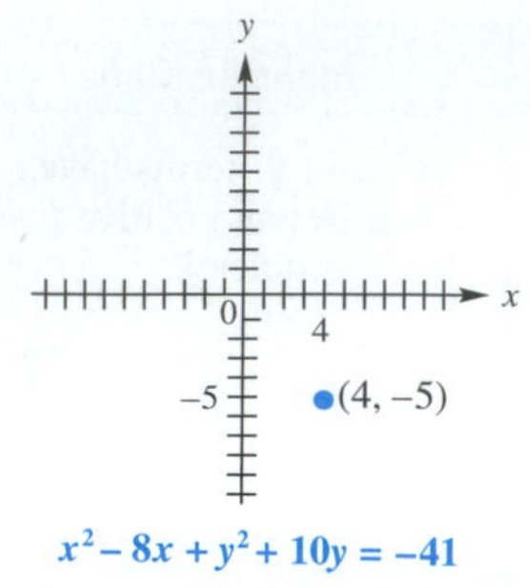

$x^2 - 8x + y^2 + 10y = -41$

FIGURE 39

(b)

$$x^2 - 8x + y^2 + 10y = -41$$

$$(x^2 - 8x + 16) + (y^2 + 10y + 25) = -41 + 16 + 25 \quad \text{Complete the square on both } x \text{ and } y.$$

$$(x - 4)^2 + (y + 5)^2 = 0 \quad \text{Factor; add.}$$

This result shows that the equation is that of a circle with radius 0—that is, the point $(4, -5)$. See Figure 39. Had a negative number (instead of 0) been obtained on the right, the equation would have represented no points at all, and there would be no points on its graph.

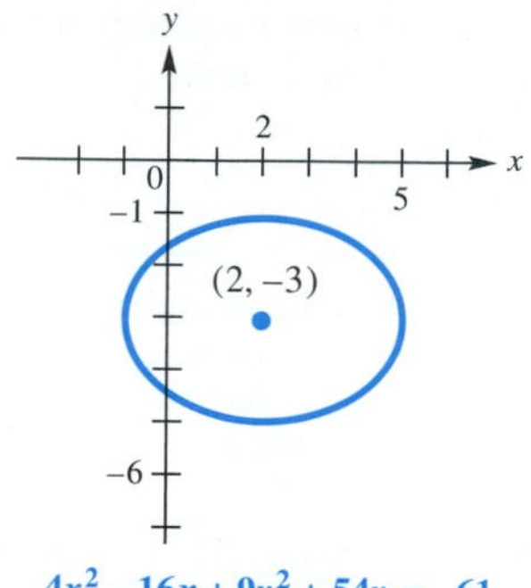

$4x^2 - 16x + 9y^2 + 54y = -61$

FIGURE 40

(c) Since the coefficients of the x^2- and y^2-terms of $4x^2 - 16x + 9y^2 + 54y = -61$ are unequal and both positive, this equation may represent an ellipse, but not a circle. (It may also represent a single point or no points at all.)

$$4x^2 - 16x + 9y^2 + 54y = -61$$

$$4(x^2 - 4x \quad) + 9(y^2 + 6y \quad) = -61 \quad \text{Factor out 4 and 9.}$$

$$4(x^2 - 4x + 4) + 9(y^2 + 6y + 9) = -61 + 4(4) + 9(9) \quad \text{Complete the square.}$$

$$4(x - 2)^2 + 9(y + 3)^2 = 36 \quad \text{Factor; simplify.}$$

$$\frac{(x - 2)^2}{9} + \frac{(y + 3)^2}{4} = 1 \quad \text{Divide by 36.}$$

This equation represents an ellipse having center $(2, -3)$. See Figure 40.

(d) Since only one variable of $x^2 - 6x + 8y - 7 = 0$ is squared (x, and not y), the equation represents a parabola. Rearrange the terms so that the term $8y$ (with the variable that is not squared) is alone on one side.

$$x^2 - 6x + 8y - 7 = 0$$

$$8y = -x^2 + 6x + 7 \quad \text{Isolate the term with } y.$$

$$8y = -(x^2 - 6x \quad) + 7 \quad \text{Regroup; factor out } -1.$$

$$8y = -(x^2 - 6x + 9) + 7 + 9 \quad \text{Complete the square.}$$

$$8y = -(x - 3)^2 + 16 \quad \text{Factor; add.}$$

$$(x - 3)^2 = -8y + 16 \quad \text{Add } (x-3)^2\text{; add } -8y.$$

$$(x - 3)^2 = -8(y - 2) \quad \text{Factor.}$$

Be careful with signs.

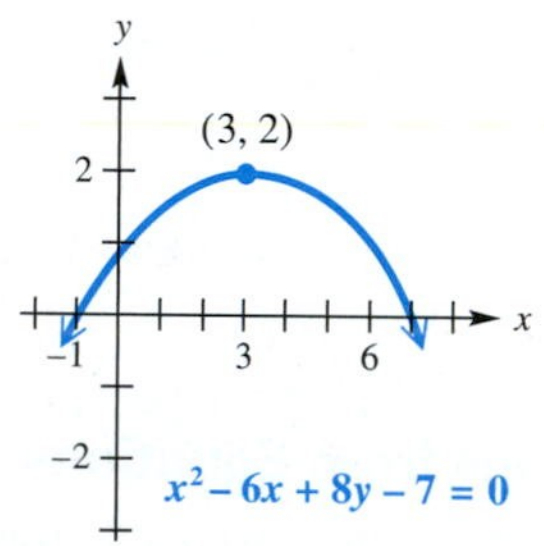

$x^2 - 6x + 8y - 7 = 0$

FIGURE 41

The parabola has vertex $(3, 2)$ and opens downward, as shown in Figure 41. ■

Eccentricity

In Sections 6.1 and 6.2, we introduced definitions of the conic sections. The conic sections (or *conics*) can all be characterized by one general definition.

Conic

A **conic** is the set of all points $P(x, y)$ in a plane such that the ratio of the distance from P to a fixed point and the distance from P to a fixed line is constant.

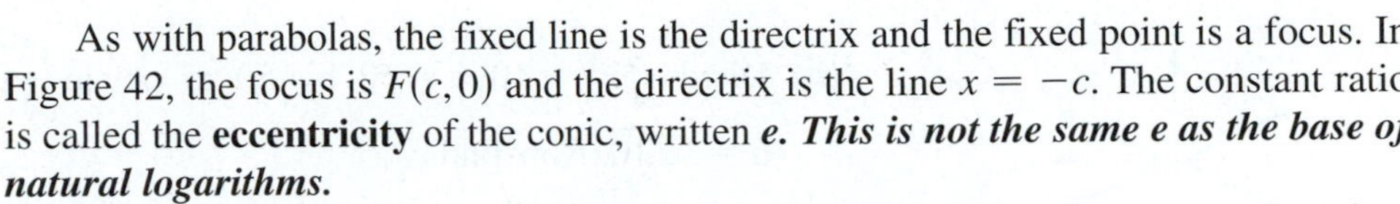

FIGURE 42

As with parabolas, the fixed line is the directrix and the fixed point is a focus. In Figure 42, the focus is $F(c, 0)$ and the directrix is the line $x = -c$. The constant ratio is called the **eccentricity** of the conic, written e. ***This is not the same e as the base of natural logarithms.***

If the conic is a parabola, then by definition, the distances $d(P, F)$ and $d(P, D)$ in Figure 42 are equal. Thus, every parabola has eccentricity 1.

For an ellipse, eccentricity is a measure of its "roundness." The constant ratio in the definition is $e = \frac{c}{a}$, where (as before) c is the distance from the center of the figure to a focus and a is the distance from the center to a vertex. By the definition of an ellipse, $a^2 > b^2$ and $c = \sqrt{a^2 - b^2}$. Thus, for the ellipse,

$$0 < c < a$$

$$0 < \frac{c}{a} < 1 \qquad \text{Divide by } a.$$

$$0 < e < 1. \qquad e = \tfrac{c}{a}$$

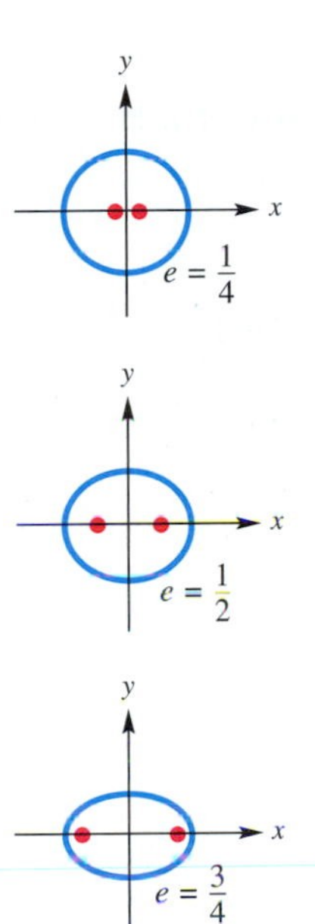

FIGURE 43

If a is constant, letting c approach 0 would force the ratio $\frac{c}{a}$ to approach 0, which also forces b to approach a (so that $\sqrt{a^2 - b^2} = c$ would approach 0). Since b determines the endpoints of the minor axis, the lengths of the major and minor axes are almost the same, producing an ellipse very close in shape to a circle when e is very close to 0. In a similar manner, if e approaches 1, then b will approach 0.

The path of Earth around the sun is an ellipse that is very nearly circular. In fact, for this ellipse, $e \approx .017$. By contrast, the path of Halley's comet is a very flat ellipse, with $e \approx .97$. Figure 43 compares ellipses with different eccentricities. The locations of the foci are shown in each case.

The equation of a circle can be written

$$(x - h)^2 + (y - k)^2 = r^2$$

$$\frac{(x - h)^2}{r^2} + \frac{(y - k)^2}{r^2} = 1.$$

In a circle, the foci coincide with the center, so $a = b$, $c = \sqrt{a^2 - b^2} = 0$, and thus $e = 0$.

EXAMPLE 2 Finding Eccentricity from Equations of Ellipses

Find the eccentricity of each ellipse.

(a) $\frac{x^2}{9} + \frac{y^2}{16} = 1$ **(b)** $5x^2 + 10y^2 = 50$

Solution

(a) Since $16 > 9$, let $a^2 = 16$, which gives $a = 4$. Also,

$$c = \sqrt{a^2 - b^2} = \sqrt{16 - 9} = \sqrt{7}.$$

Finally, $e = \frac{c}{a} = \frac{\sqrt{7}}{4} \approx .66.$

(b) Divide $5x^2 + 10y^2 = 50$ by 50 to obtain $\frac{x^2}{10} + \frac{y^2}{5} = 1$. Here, $a^2 = 10$, with $a = \sqrt{10}$. Now find c.

$$c = \sqrt{10 - 5} = \sqrt{5} \quad \text{and} \quad e = \frac{\sqrt{5}}{\sqrt{10}} = \frac{1}{\sqrt{2}} \approx .71.$$

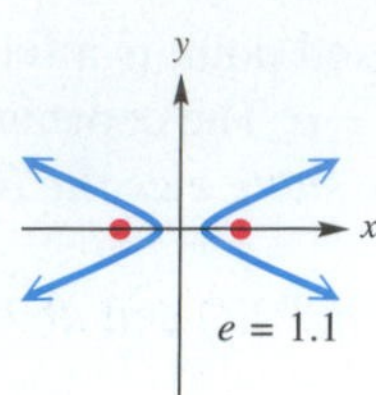

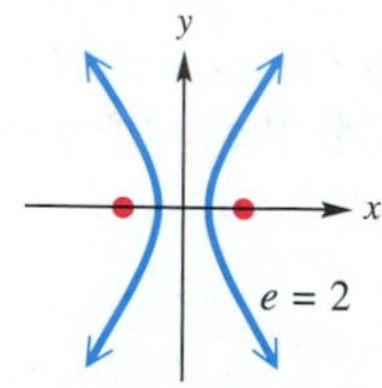

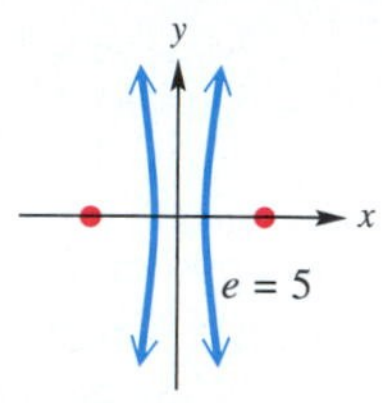

FIGURE 44

The hyperbola in standard form

$$\frac{x^2}{a^2} - \frac{y^2}{b^2} = 1 \quad \text{or} \quad \frac{y^2}{a^2} - \frac{x^2}{b^2} = 1,$$

where $c = \sqrt{a^2 + b^2}$, also has eccentricity $e = \frac{c}{a}$. By definition, $c = \sqrt{a^2 + b^2} > a$, so $\frac{c}{a} > 1$, and for a hyperbola, $e > 1$. Narrow hyperbolas have e near 1, and wide hyperbolas have large e. See Figure 44.

EXAMPLE 3 Finding Eccentricity from the Equation of a Hyperbola

Find the eccentricity of the hyperbola $\frac{x^2}{9} - \frac{y^2}{4} = 1$.

Solution Here $a^2 = 9$; thus, $a = 3$, $c = \sqrt{9 + 4} = \sqrt{13}$, and

$$e = \frac{c}{a} = \frac{\sqrt{13}}{3} \approx 1.2.$$

The table summarizes this discussion of eccentricity.

Conic	Eccentricity
Circle	$e = 0$
Parabola	$e = 1$
Ellipse	$e = \frac{c}{a}$ and $0 < e < 1$
Hyperbola	$e = \frac{c}{a}$ and $e > 1$

EXAMPLE 4 Finding Equations of Conics by Using Eccentricity

Find an equation for each conic with center at the origin.

(a) Focus $(3, 0)$; eccentricity 2 **(b)** Vertex $(0, -8)$; $e = \frac{1}{2}$

Solution

(a) Since $e = 2$, which is greater than 1, the conic is a hyperbola with $c = 3$.

$$e = \frac{c}{a}$$

$$2 = \frac{3}{a} \qquad \text{Let } e = 2 \text{ and } c = 3.$$

$$a = \frac{3}{2} \qquad \text{Solve for } a.$$

Now we find b.

$$b^2 = c^2 - a^2 = 9 - \frac{9}{4} = \frac{27}{4} \qquad c^2 = 3^2 = 9;\ a^2 = \left(\tfrac{3}{2}\right)^2 = \tfrac{9}{4}$$

The given focus is on the x-axis, so the x^2-term is positive, and the equation is

$$\frac{x^2}{\frac{9}{4}} - \frac{y^2}{\frac{27}{4}} = 1 \quad \text{or} \quad \frac{4x^2}{9} - \frac{4y^2}{27} = 1.$$

(b) The graph of the conic is an ellipse because $e = \frac{1}{2} < 1$. From the given vertex $(0, -8)$, we know that the vertices are on the y-axis and $a = 8$.

$$e = \frac{c}{a}$$

$$\frac{1}{2} = \frac{c}{8} \qquad \text{Let } e = \tfrac{1}{2} \text{ and } a = 8.$$

$$c = 4 \qquad \text{Solve for } c.$$

Since $b^2 = a^2 - c^2 = 64 - 16 = 48$, the equation is

$$\frac{x^2}{48} + \frac{y^2}{64} = 1.$$ ■

EXAMPLE 5 Applying an Ellipse to the Orbit of a Planet

The orbit of the planet Mars is an ellipse with the sun at one focus. The eccentricity of the ellipse is .0935, and the closest Mars comes to the sun is 128.5 million miles. (*Source: The World Almanac and Book of Facts,* 2001.) Find the maximum distance of Mars from the sun.

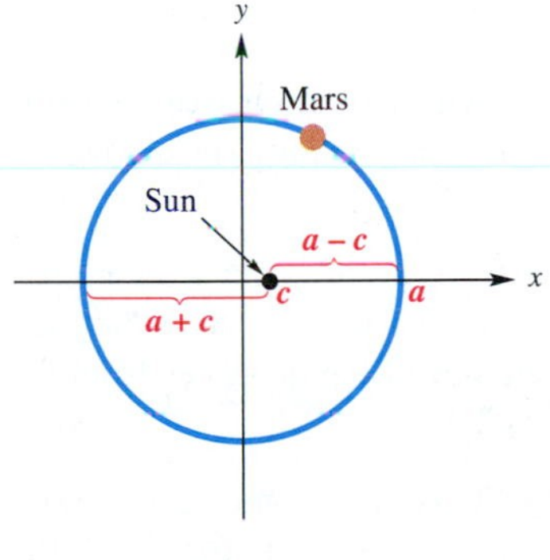

Not to scale

FIGURE 45

Solution Figure 45 shows the orbit of Mars with the origin at the center of the ellipse and the sun at one focus. Mars is closest to the sun when Mars is at the right endpoint of the major axis and farthest from the sun when Mars is at the left endpoint. Therefore, the smallest distance is $a - c$, and the greatest distance is $a + c$. Since $a - c = 128.5$, $c = a - 128.5$. Using this value of c and $e = .0935$, we can find a.

$$e = \frac{c}{a}$$

$$.0935 = \frac{a - 128.5}{a} \qquad \text{Substitute for } e \text{ and } c.$$

$$.0935a = a - 128.5 \qquad \text{Multiply by } a.$$

$$-.9065a = -128.5 \qquad \text{Subtract } a.$$

$$a \approx 141.8 \qquad \text{Divide by } -.9065.$$

Then $\quad c = 141.8 - 128.5 = 13.3$

and $\quad a + c = 141.8 + 13.3 = 155.1.$

Thus, the maximum distance of Mars from the sun is about 155.1 million miles. ■

6.3 Exercises

Without graphing, identify the type of graph that each equation has.

1. $x^2 + y^2 = 144$
2. $(x-2)^2 + (y+3)^2 = 25$
3. $y = 2x^2 + 3x - 4$
4. $x = 3y^2 + 5y - 6$
5. $x = -3(y-4)^2 + 1$
6. $\frac{x^2}{25} + \frac{y^2}{36} = 1$
7. $\frac{x^2}{49} + \frac{y^2}{100} = 1$
8. $x^2 - y^2 = 1$
9. $\frac{x^2}{4} - \frac{y^2}{16} = 1$
10. $\frac{(x+2)^2}{9} + \frac{(y-4)^2}{16} = 1$
11. $\frac{x^2}{25} - \frac{y^2}{25} = 1$
12. $y = 4(x+3)^2 - 7$

Identify the type of graph for each equation. It may be necessary to transform the equation.

13. $\frac{x^2}{4} = 1 - \frac{y^2}{9}$
14. $\frac{x^2}{4} = 1 + \frac{y^2}{9}$
15. $\frac{x^2}{4} + \frac{y^2}{4} = -1$
16. $x^2 = 25 + y^2$
17. $x^2 = 25 - y^2$
18. $9x^2 + 36y^2 = 36$
19. $x^2 = 4y - 8$
20. $\frac{(x+3)^2}{16} + \frac{(y-2)^2}{16} = 1$
21. $\frac{(x-4)^2}{8} + \frac{(y+1)^2}{2} = 0$
22. $y^2 - 4y = x + 4$
23. $(x+7)^2 + (y-5)^2 + 4 = 0$
24. $4(x-3)^2 + 3(y+4)^2 = 0$
25. $3x^2 + 6x + 3y^2 - 12y = 12$
26. $2x^2 - 8x + 2y^2 + 20y = 12$
27. $x^2 - 6x + y = 0$
28. $x - 4y^2 - 8y = 0$
29. $4x^2 - 8x - y^2 - 6y = 6$
30. $x^2 + 2x = x^2 - 4y - 2$
31. $4x^2 - 8x + 9y^2 + 54y = -84$
32. $3x^2 + 12x + 3y^2 = -11$
33. $6x^2 - 12x + 6y^2 - 18y + 25 = 0$
34. $4x^2 - 24x + 5y^2 + 10y + 41 = 0$
35. **Concept Check** Suppose that both A and C are 0 in the equation $Ax^2 + Bx + Cy^2 + Dy + E = 0$. What kind of graph does this equation have?

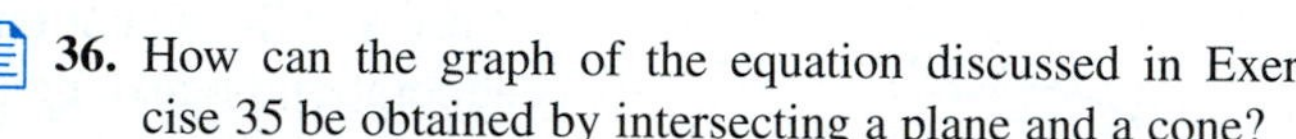

36. How can the graph of the equation discussed in Exercise 35 be obtained by intersecting a plane and a cone?
37. **Concept Check** Identify the type of conic section consisting of the set of all points in the plane for which the sum of the distances from the points $(5, 0)$ and $(-5, 0)$ is 14.
38. **Concept Check** Identify the type of conic section consisting of the set of all points in the plane for which the absolute value of the difference of the distances from the points $(3, 0)$ and $(-3, 0)$ is 2.
39. **Concept Check** Identify the type of conic section consisting of the set of all points in the plane for which the distance from the point $(3, 0)$ is one and one-half times the distance from the line $x = \frac{4}{3}$.
40. **Concept Check** Identify the type of conic section consisting of the set of all points in the plane for which the distance from the point $(2, 0)$ is one-third the distance from the line $x = 10$.

Find the eccentricity of each ellipse or hyperbola.

41. $12x^2 + 9y^2 = 36$
42. $8x^2 - y^2 = 16$
43. $x^2 - y^2 = 4$
44. $x^2 + 2y^2 = 8$
45. $4x^2 + 7y^2 = 28$
46. $9x^2 - y^2 = 1$
47. $x^2 - 9y^2 = 18$
48. $x^2 + 10y^2 = 10$

Write an equation for each conic. Each parabola has vertex at the origin, and each ellipse or hyperbola is centered at the origin.

49. Focus $(0, 8)$; $e = 1$
50. Focus $(-2, 0)$; $e = 1$
51. Focus $(3, 0)$; $e = \frac{1}{2}$

52. Focus $(0, -2)$; $e = \frac{2}{3}$

53. Vertex $(-6, 0)$; $e = 2$

54. Vertex $(0, 4)$; $e = \frac{5}{3}$

55. Focus $(0, -1)$; $e = 1$

56. Focus $(2, 0)$; $e = \frac{6}{5}$

57. Vertical major axis of length 6; $e = \frac{4}{5}$

58. y-intercepts -4 and 4; $e = \frac{7}{3}$

59. ***Concept Check*** Calculator graphs are shown in Figures A–D. Arrange the figures in order so that the first in the list has the smallest eccentricity and the rest have eccentricities in increasing order.

A.

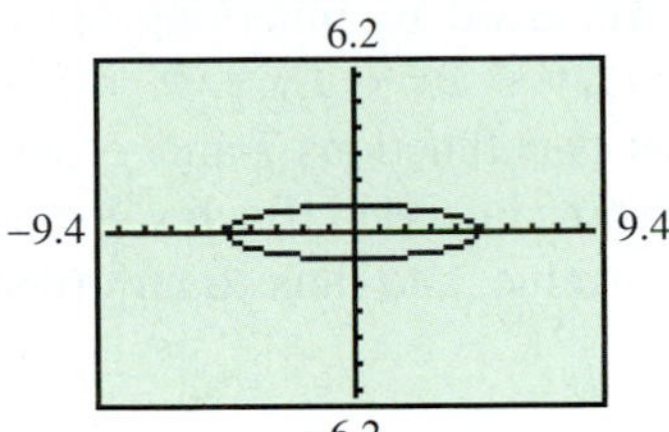

B.

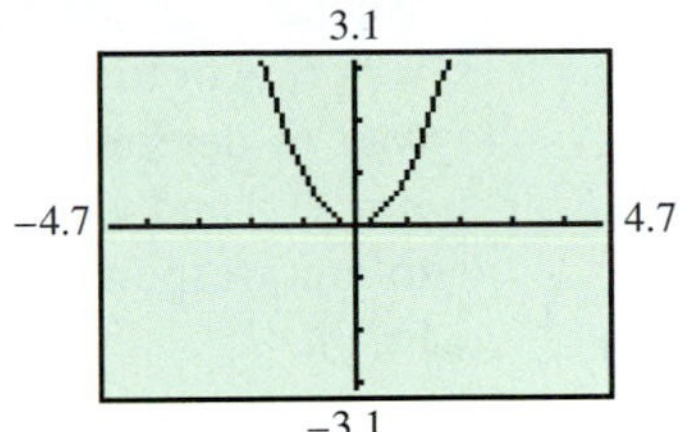

C.

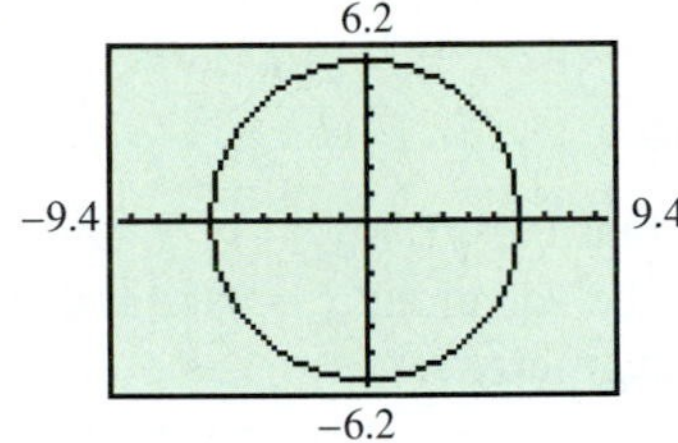

D.

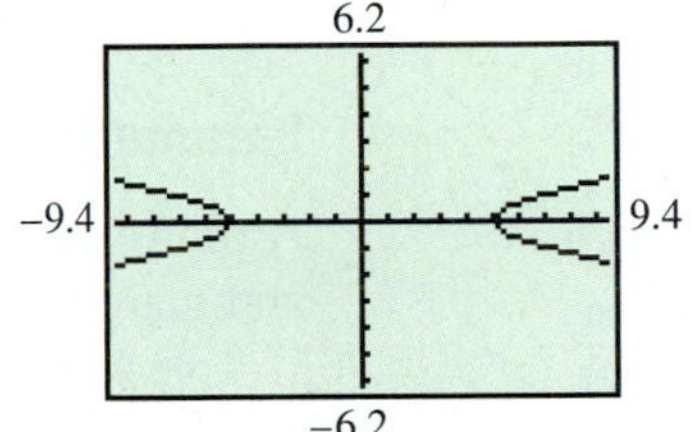

(Modeling) *Solve each application.*

60. ***Orbit of Mars*** The orbit of Mars around the sun is an ellipse with equation

$$\frac{x^2}{5013} + \frac{y^2}{4970} = 1,$$

where x and y are measured in millions of miles. Find the eccentricity of this ellipse.

61. ***Orbits of Neptune and Pluto*** Neptune and Pluto both have elliptical orbits with the sun at one focus. Neptune's orbit has $a = 30.1$ astronomical units (AU) and eccentricity $e = .009$, whereas Pluto's orbit has $a = 39.4$ and $e = .249$. (1 AU is equal to the average distance from Earth to the sun and is approximately 149,600,000 kilometers.) (*Source:* Zeilik, M., S. Gregory, and E. Smith, *Introductory Astronomy and Astrophysics,* Saunders College Publishers, 1992.)

(a) Position the sun at the origin, and determine an equation for each orbit.

(b) Graph both equations on the same coordinate axes. Use the window $[-60, 60]$ by $[-40, 40]$.

62. ***Velocity of a Planet in Orbit*** The maximum and minimum velocities in kilometers per second of a planet moving in an elliptical orbit can be calculated with the equations

$$v_{\text{max}} = \frac{2\pi a}{P}\sqrt{\frac{1+e}{1-e}} \quad \text{and} \quad v_{\text{min}} = \frac{2\pi a}{P}\sqrt{\frac{1-e}{1+e}},$$

where a is in kilometers, P is its orbital period in seconds, and e is the eccentricity of the orbit. (*Source:* Zeilik, M., S. Gregory, and E. Smith, *Introductory Astronomy and Astrophysics,* Saunders College Publishers, 1992.)

(a) Calculate v_{max} and v_{min} for Earth if $a = 1.496 \times 10^8$ kilometers and $e = .0167$.

(b) If an object has a circular orbit, what can be said about its orbital velocity?

(c) Kepler showed that the sun is located at a focus of a planet's elliptical orbit. He also showed that a planet has minimum velocity when its distance from the sun is maximum, and a planet has maximum velocity when its distance from the sun is minimum. Where do the maximum and minimum velocities occur in an elliptical orbit?

63. ***Distance between Halley's Comet and the Sun*** The famous Halley's comet last passed Earth in February 1986 and will next return in 2062. Halley's comet has an elliptical orbit of eccentricity .9673 with the sun at one of the foci. The greatest distance of the comet from the sun is 3281 million miles. (*Source: The World Almanac and Book of Facts,* 2001.) Find the least distance between Halley's comet and the sun.

64. ***Orbit of Earth*** The orbit of Earth is an ellipse with the sun at one focus. The distance between Earth and the sun ranges from 91.4 to 94.6 million miles. (*Source: The World Almanac and Book of Facts,* 2001.) Find the eccentricity of Earth's orbit.

6.4 Parametric Equations

Graphs of Parametric Equations and Their Rectangular Equivalents ■ Alternative Forms of Parametric Equations ■ An Application of Parametric Equations

We have graphed sets of ordered pairs that correspond to functions of the form $y = f(x)$ or to relations of the form $Ax^2 + Bxy + Cy^2 + Dx + Ey + F = 0$. Another way to determine a set of ordered pairs involves two functions f and g defined by $x = f(t)$ and $y = g(t)$, where t is a real number in some interval I. Each value of t leads to a corresponding x-value and a corresponding y-value, and thus to an ordered pair (x, y).

GCM TECHNOLOGY NOTE

In addition to graphing rectangular equations, graphing calculators are capable of graphing plane curves defined by parametric equations. The calculator must be set in parametric mode, and the window requires intervals for the parameter t, as well as for x and y. Consult the graphing calculator manual that accompanies this text or your owner's manual.

Parametric Equations of a Plane Curve

A **plane curve** is a set of points (x, y) such that $x = f(t)$, $y = g(t)$, and f and g are both continuous on an interval I. The equations $x = f(t)$ and $y = g(t)$ are **parametric equations** with **parameter t.**

Graphs of Parametric Equations and Their Rectangular Equivalents

GCM **EXAMPLE 1 Graphing a Plane Curve Defined Parametrically**

Graph the plane curve defined by the parametric equations

$$x = t^2, \quad y = 2t + 3, \quad \text{for } t \text{ in } [-3, 3],$$

and then find an equivalent rectangular equation.

Analytic Solution

Make a table of corresponding values of t, x, and y over the domain of t. Then plot the points. The graph is a portion of a parabola with horizontal axis $y = 3$. See Figure 46. The arrowheads indicate the direction the curve traces as t increases.

t	x	y
−3	9	−3
−2	4	−1
−1	1	1
0	0	3
1	1	5
2	4	7
3	9	9

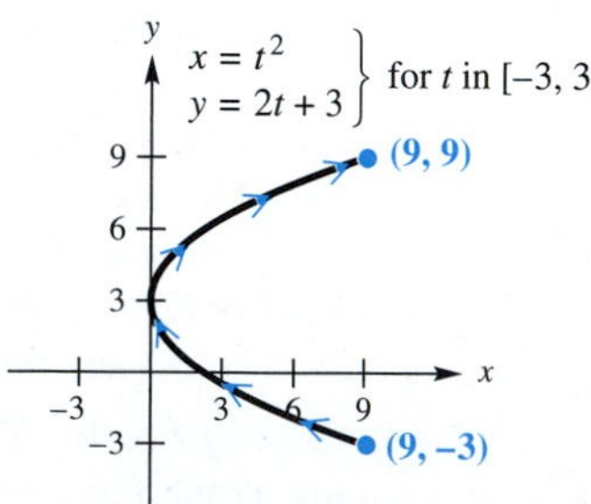

FIGURE 46

Graphing Calculator Solution

Figure 47(a) shows a calculator table of values.

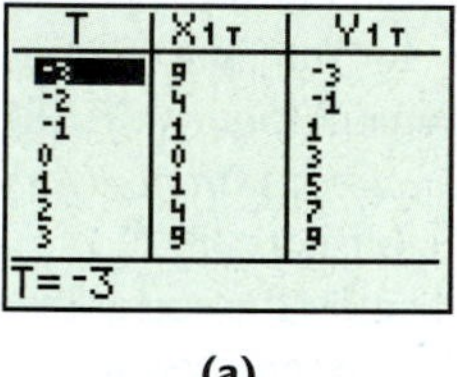

(a)

FIGURE 47

Figure 47(b) shows a graph. The formation of the curve as T increases can be seen by using the pause feature and carefully observing the path of the curve.

(continued)

To find an equivalent rectangular equation, we eliminate the parameter t. For this curve, we begin by solving for t in the second equation, $y = 2t + 3$, because it leads to a unique solution.

$$y = 2t + 3$$
$$2t = y - 3 \qquad \text{Subtract 3; rewrite.}$$
$$t = \frac{y-3}{2} \qquad \text{Divide by 2.}$$

Now we substitute the result in the first equation to get

$$x = t^2 = \left(\frac{y-3}{2}\right)^2 = \frac{(y-3)^2}{4} = \frac{1}{4}(y-3)^2.$$

This is indeed an equation of a horizontal parabola that opens to the right. Because t is in $[-3, 3]$, x is in $[0, 9]$ and y is in $[-3, 9]$. The rectangular equation must be given with its restricted domain as

$$x = \frac{1}{4}(y-3)^2, \quad \text{for } x \text{ in } [0, 9].$$

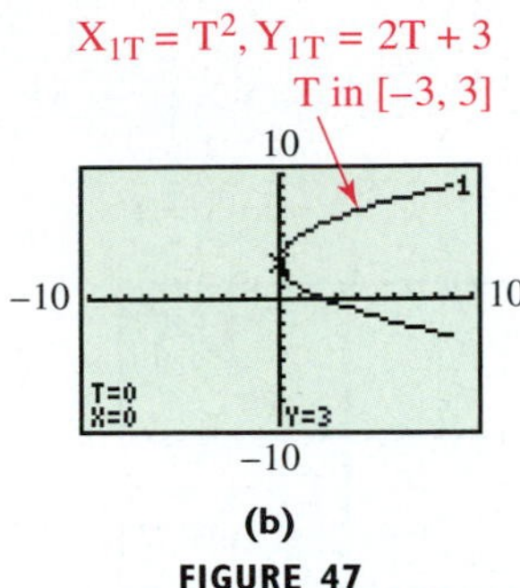

(b)

FIGURE 47

The displays at the bottom of Figures 47(a) and (b) indicate particular values of T, X, and Y. How do these relate to the parametric and rectangular forms of the equation of this parabola?

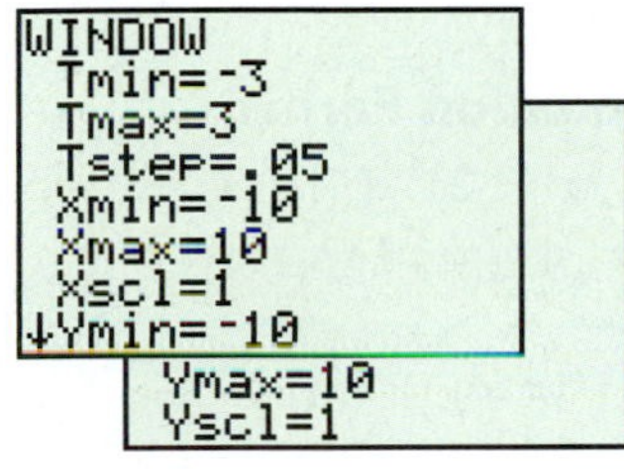

For the parametric equations in Example 1, the screens show the entries that produce the standard window shown in Figure 47.

EXAMPLE 2 Graphing a Plane Curve Defined Parametrically

Graph the plane curve defined by

$$x = 2t + 5, \quad y = \sqrt{4 - t^2}, \quad \text{for } t \text{ in } [-2, 2],$$

and then find an equivalent rectangular equation.

Solution A graph is shown in Figure 48. To get an equivalent rectangular equation, solve the first equation, which does not involve a radical, for t, and then substitute into the second equation. Since t is in $[-2, 2]$, x is in the interval $[1, 9]$.

$$2t + 5 = x, \quad \text{so} \quad t = \frac{x-5}{2}.$$

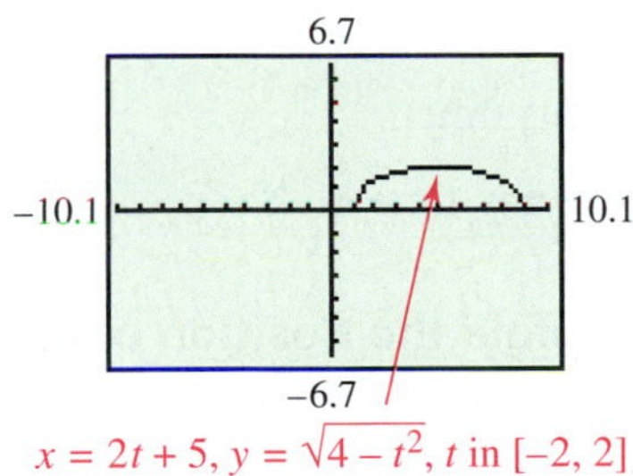

FIGURE 48

The rectangular equation is

$$y = \sqrt{4 - t^2}$$
$$y = \sqrt{4 - \left(\frac{x-5}{2}\right)^2} \qquad \text{Substitute.}$$
$$y^2 = 4 - \left(\frac{x-5}{2}\right)^2 \qquad \text{Square both sides.}$$
$$y^2 = 4 - \frac{(x-5)^2}{4} \qquad 2^2 = 4$$
$$4y^2 = 16 - (x-5)^2 \qquad \text{Multiply by 4.}$$
$$4y^2 + (x-5)^2 = 16 \qquad \text{Add } (x-5)^2.$$
$$\frac{y^2}{4} + \frac{(x-5)^2}{16} = 1, \qquad \text{Divide by 16.}$$

which represents the complete ellipse. (By definition, the parametric graph has $y \geq 0$, so it is only the upper half of the ellipse.)

EXAMPLE 3 Graphing a Line Defined Parametrically

Graph the plane curve defined by

$$x = t^2, \quad y = t^2,$$

and then find an equivalent rectangular equation.

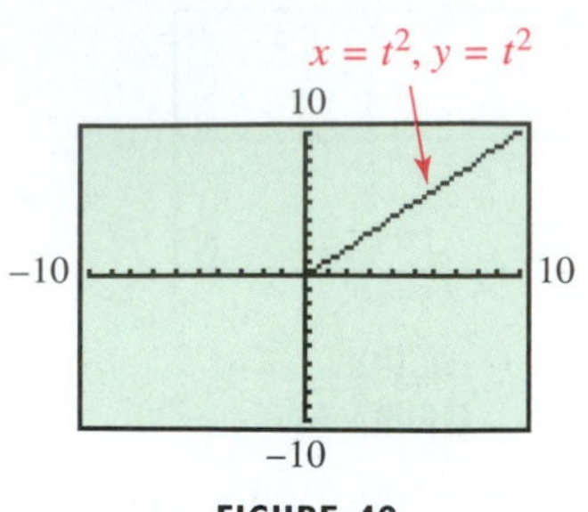

FIGURE 49

Solution Figure 49 shows the graph of the curve in a standard window with both x and y in $[-10, 10]$. The graph is a ray, because both x and y must be greater than or equal to 0. Since both x and y equal t^2, $y = x$. To be equivalent, however, the rectangular equation must be given as

$$y = x, \quad x \geq 0.$$

Alternative Forms of Parametric Equations

Parametric representations of a curve are not unique. In fact, there are infinitely many parametric representations of a given curve. If the curve can be described by a rectangular equation $y = f(x)$ with domain X, then one simple parametric representation is

$$x = t, \quad y = f(t), \quad \text{for } t \text{ in } X.$$

FOR DISCUSSION

Consider the rectangular equation

$$y - 1 = (x - 2)^2.$$

Why are the parametric equations

$$x = \sqrt{t} + 2, \quad y = t + 1,$$

for t in $[0, \infty)$ not equivalent to the rectangular equation?

EXAMPLE 4 Finding Alternative Parametric Equation Forms

Give two parametric representations for the parabola $y = (x - 2)^2 + 1$.

Solution The simplest choice is to let

$$x = t, \quad y = (t - 2)^2 + 1, \quad \text{for } t \text{ in } (-\infty, \infty).$$

Another choice that leads to a simpler equation for y is

$$x = t + 2, \quad y = t^2 + 1, \quad \text{for } t \text{ in } (-\infty, \infty).$$

An Application of Parametric Equations

One application of parametric equations is to determine the path of a moving object whose position is given by the function defined by

$$x = f(t), \quad y = g(t), \quad \text{where } t \text{ represents time.}$$

The parametric equations give the position of the object at any time t.

EXAMPLE 5 Using Parametric Equations to Define the Position of an Object in Motion

The motion of a projectile moving in a direction at an angle $\theta = 45°$ with the horizontal (neglecting air resistance) is given by

$$x = v_0 \frac{\sqrt{2}}{2} t, \quad y = v_0 \frac{\sqrt{2}}{2} t - 16t^2, \quad \text{for } t \text{ in } [0, k],$$

where t is time in seconds, v_0 is the initial speed of the projectile in feet per second, x and y are in feet, and k is a positive real number. See Figure 50 on the next page. Find the rectangular form of the equation.

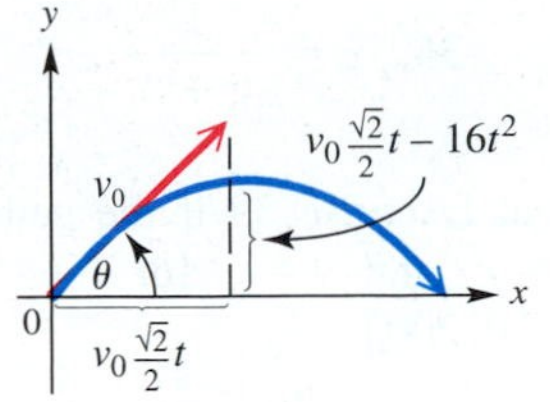

FIGURE 50

Solution Solving the first equation for t and substituting the result into the second equation gives (after simplification)

$$y = x - \frac{32}{v_0^2}x^2,$$

the equation of a vertical parabola opening downward, as shown in Figure 50. ■

6.4 Exercises

For each plane curve, use a graphing calculator to generate the curve over the interval for the parameter t, in the window specified. Then, find a rectangular equation for the curve.

1. $x = 2t,\ y = t + 1$, for t in $[-2, 3]$; window: $[-8, 8]$ by $[-8, 8]$

2. $x = t + 2,\ y = t^2$, for t in $[-1, 1]$; window: $[0, 4]$ by $[-2, 2]$

3. $x = \sqrt{t},\ y = 3t - 4$, for t in $[0, 4]$; window: $[-6, 6]$ by $[-6, 10]$

4. $x = t^2,\ y = \sqrt{t}$, for t in $[0, 4]$; window: $[-2, 20]$ by $[0, 4]$

5. $x = t^3 + 1,\ y = t^3 - 1$, for t in $[-3, 3]$; window: $[-30, 30]$ by $[-30, 30]$

6. $x = 2t - 1,\ y = t^2 + 2$, for t in $[-10, 10]$; window: $[-20, 20]$ by $[0, 120]$

7. $x = 2^t,\ y = \sqrt{3t - 1}$, for t in $\left[\frac{1}{3}, 4\right]$; window: $[-2, 30]$ by $[-2, 10]$

8. $x = \ln(t - 1),\ y = 2t - 1$, for t in $(1, 10]$; window: $[-5, 5]$ by $[-2, 20]$

9. $x = t + 2,\ y = -\frac{1}{2}\sqrt{9 - t^2}$, for t in $[-3, 3]$; window: $[-6, 6]$ by $[-4, 4]$

10. $x = t,\ y = \sqrt{4 - t^2}$, for t in $[-2, 2]$; window: $[-6, 6]$ by $[-4, 4]$

11. $x = t,\ y = \frac{1}{t}$, for t in $(-\infty, 0) \cup (0, \infty)$; window: $[-6, 6]$ by $[-4, 4]$

12. $x = 2t - 1,\ y = \frac{1}{t}$, for t in $(-\infty, 0) \cup (0, \infty)$; window: $[-6, 6]$ by $[-4, 4]$

For each plane curve, find a rectangular equation.

13. $x = 3t,\ y = t - 1$, for t in $(-\infty, \infty)$

14. $x = t + 3,\ y = 2t$, for t in $(-\infty, \infty)$

15. $x = 3t^2,\ y = t + 1$, for t in $(-\infty, \infty)$

16. $x = t - 2,\ y = \frac{1}{2}t^2 + 1$, for t in $(-\infty, \infty)$

17. $x = 3t^2,\ y = 4t^3$, for t in $(-\infty, \infty)$

18. $x = 2t^3,\ y = -t^2$, for t in $(-\infty, \infty)$

19. $x = t,\ y = \sqrt{t^2 + 2}$, for t in $(-\infty, \infty)$

20. $x = \sqrt{t},\ y = t^2 - 1$, for t in $[0, \infty)$

21. $x = e^t,\ y = e^{-t}$, for t in $(-\infty, \infty)$

22. $x = e^{2t},\ y = e^t$, for t in $(-\infty, \infty)$

23. $x = \dfrac{1}{\sqrt{t + 2}},\ y = \dfrac{t}{t + 2}$, for t in $(-2, \infty)$

24. $x = \dfrac{t}{t - 1},\ y = \dfrac{1}{\sqrt{t - 1}}$, for t in $(1, \infty)$

25. $x = t + 2,\ y = \dfrac{1}{t + 2}$, for $t \neq -2$

26. $x = t - 3,\ y = \dfrac{2}{t - 3}$, for $t \neq 3$

27. $x = t^2,\ y = 2\ln t$, for t in $(0, \infty)$

28. $x = \ln t,\ y = 3\ln t$, for t in $(0, \infty)$

Give two parametric representations for each plane curve. Use your calculator to verify your results.

29. $y = 2x + 3$

30. $y = \dfrac{3}{2}x - 4$

31. $y = \sqrt{3x + 2}$, x in $\left[-\frac{2}{3}, \infty\right)$

32. $y = (x + 1)^2 + 1$

33. $x = y^3 + 1$ **34.** $x = 2(y - 3)^2 - 4$ **35.** $x = \sqrt{y + 1}$ **36.** $x = \dfrac{1}{y + 1}$

(Modeling) Solve each application.

37. ***Firing a Projectile*** A projectile is fired with an initial velocity of 400 feet per second at an angle of 45° with the horizontal. (See Example 5.)
(a) Find the time when it strikes the ground.
(b) Find the range (horizontal distance covered).
(c) What is the maximum altitude?

38. ***Firing a Projectile*** If a projectile is fired at an angle of 30° with the horizontal, the parametric equations that describe its motion are

$$x = v_0\frac{\sqrt{3}}{2}t, \quad y = \frac{v_0}{2}t - 16t^2, \quad \text{for } t \text{ in } [0, \infty).$$

Repeat Exercise 37 if the projectile is fired at 800 feet per second.

39. ***Path of a Projectile*** A projectile moves so that its position at any time t is given by the equations $x = 60t$ and $y = 80t - 16t^2$. Graph the path of the projectile, and find the equivalent rectangular equation. Use the window $[0, 300]$ by $[0, 200]$.

40. ***Path of a Projectile*** Repeat Exercise 39 if the path is given by the equations $x = t^2$ and $y = -16t + 64\sqrt{t}$. Use the window $[0, 300]$ by $[0, 200]$.

41. Show that the rectangular equation for the curve defined by

$$x = v_0\frac{\sqrt{2}}{2}t, \quad y = v_0\frac{\sqrt{2}}{2}t - 16t^2, \quad \text{for } t \text{ in } [0, k],$$

is $y = x - \dfrac{32}{v_0^2}x^2$. (See Example 5.)

42. Find the vertex of the parabola given by the rectangular equation of Exercise 41.

43. Give two parametric representations of the line through the point (x_1, y_1) with slope m.

44. Give two parametric representations of the parabola $y = a(x - h)^2 + k$.

Reviewing Basic Concepts (Sections 6.3 and 6.4)

Write each equation in standard form, and name the type of conic section defined.

1. $3x^2 + y^2 - 6x + 6y = 0$ **2.** $y^2 - 2x^2 + 8y - 8x - 4 = 0$ **3.** $3y^2 + 12y + 5x = 3$

Find the eccentricity of each conic section.

4. $x^2 + 25y^2 = 25$ **5.** $8y^2 - 4x^2 = 8$

Find an equation of each conic. (Hint: *A sketch may be helpful.*)

6. Focus $(-2, 0)$; vertex $(0, 0)$; eccentricity 1 **7.** Foci $(\pm 3, 0)$; major axis length 10

8. Foci $(0, \pm 5)$; vertices $(0, \pm 4)$

Solve each problem.

9. The figure represents an elliptical stone arch with the dimensions (in feet) indicated. Find the height y of the arch 6 feet from the center of the base.

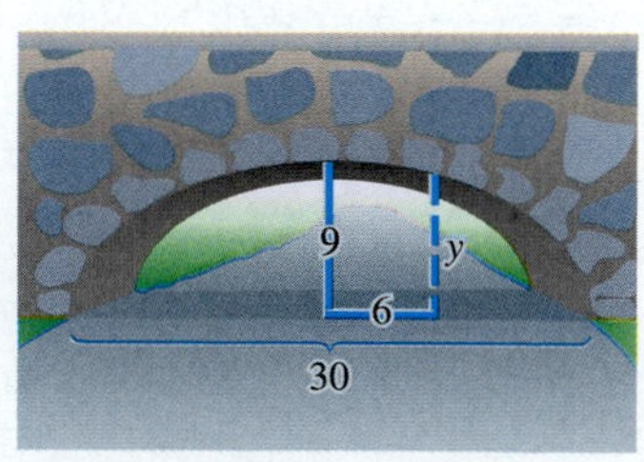

10. A plane curve is defined by

$$x = 2t, \quad y = \sqrt{t^2 + 1}, \quad \text{for } t \text{ in } (-\infty, \infty).$$

(a) Graph the curve by hand and support your graph with your calculator.
(b) Find an equivalent rectangular equation for the curve.

CHAPTER 6 SUMMARY

KEY TERMS & SYMBOLS	KEY CONCEPTS
6.1 Circles and Parabolas conic sections circle radius center parabola focus directrix focal chord	**CIRCLE** A circle is a set of points in a plane that are equidistant from a fixed point. The distance is called the radius of the circle, and the fixed point is called the center. **CENTER–RADIUS FORM OF THE EQUATION OF A CIRCLE** The circle with center (h, k) and radius r has equation $(x - h)^2 + (y - k)^2 = r^2$. A circle with center $(0, 0)$ and radius r has equation $x^2 + y^2 = r^2$. **GENERAL FORM OF THE EQUATION OF A CIRCLE** For real numbers c, d, and e, the general form of the equation of a circle is $$x^2 + y^2 + cx + dy + e = 0.$$ **PARABOLA** A parabola is a set of points in a plane equidistant from a fixed point and a fixed line. The fixed point is called the focus, and the fixed line the directrix, of the parabola. **PARABOLA WITH A VERTICAL AXIS** The parabola with focus $(0, c)$ and directrix $y = -c$ has equation $x^2 = 4cy$. The parabola has vertical axis $x = 0$ and opens upward if $c > 0$ or downward if $c < 0$. **PARABOLA WITH A HORIZONTAL AXIS** The parabola with focus $(c, 0)$ and directrix $x = -c$ has equation $y^2 = 4cx$. The parabola has horizontal axis $y = 0$ and opens to the right if $c > 0$ or to the left if $c < 0$. **TRANSLATION OF A PARABOLA** The parabola with vertex (h, k) and vertical line $x = h$ as axis has an equation of the form $(x - h)^2 = 4c(y - k)$. The parabola opens upward if $c > 0$ or downward if $c < 0$. The parabola with vertex (h, k) and horizontal line $y = k$ as axis has an equation of the form $(y - k)^2 = 4c(x - h)$. The parabola opens to the right if $c > 0$ or to the left if $c < 0$.
6.2 Ellipses and Hyperbolas ellipse focus (foci) major axis minor axis center vertices hyperbola transverse axis asymptotes fundamental rectangle	**ELLIPSE** An ellipse is the set of all points in a plane, the *sum* of whose distances from two fixed points is constant. Each fixed point is called a focus (plural *foci*) of the ellipse. **STANDARD FORMS OF EQUATIONS FOR ELLIPSES** The ellipse with center at the origin and equation $\frac{x^2}{a^2} + \frac{y^2}{b^2} = 1$ has vertices $(\pm a, 0)$, endpoints of the minor axis $(0, \pm b)$, and foci $(\pm c, 0)$, where $c^2 = a^2 - b^2$ and $a > b > 0$. The ellipse with center at the origin and equation $\frac{x^2}{b^2} + \frac{y^2}{a^2} = 1$ has vertices $(0, \pm a)$, endpoints of the minor axis $(\pm b, 0)$, and foci $(0, \pm c)$, where $c^2 = a^2 - b^2$ and $a > b > 0$. **TRANSLATED ELLIPSES** The preceding equations can be extended to ellipses having center (h, k) by replacing x and y with $x - h$ and $y - k$, respectively.

(continued)

KEY TERMS & SYMBOLS	KEY CONCEPTS
	HYPERBOLA A hyperbola is the set of all points in a plane such that the absolute value of the *difference* of the distances from two fixed points is constant. The two fixed points are called the foci of the hyperbola. **STANDARD FORMS OF EQUATIONS FOR HYPERBOLAS** The hyperbola with center at the origin and equation $\frac{x^2}{a^2} - \frac{y^2}{b^2} = 1$ has vertices $(\pm a, 0)$ and foci $(\pm c, 0)$, where $c^2 = a^2 + b^2$. The hyperbola with center at the origin and equation $\frac{y^2}{a^2} - \frac{x^2}{b^2} = 1$ has vertices $(0, \pm a)$ and foci $(0, \pm c)$, where $c^2 = a^2 + b^2$. **TRANSLATED HYPERBOLAS** The preceding equations can be extended to hyperbolas having center (h, k) by replacing x and y with $x - h$ and $y - k$, respectively.
6.3 Summary of the Conic Sections conic eccentricity, e	The conic sections in this chapter have equations that can be written in the form $$Ax^2 + Dx + Cy^2 + Ey + F = 0.$$
6.4 Parametric Equations plane curve parametric equations parameter	**PARAMETRIC EQUATIONS OF A PLANE CURVE** A plane curve is a set of points (x, y) such that $x = f(t)$, $y = g(t)$, and f and g are both continuous on an interval I. The equations $x = f(t)$ and $y = g(t)$ are parametric equations with parameter t.

Conic Section	Characteristic	Example
Parabola	Either $A = 0$ or $C = 0$, but not both	$y = x^2$ $x = 3y^2 + 2y - 4$
Circle	$A = C \neq 0$	$x^2 + y^2 = 16$
Ellipse	$A \neq C, AC > 0$	$\frac{x^2}{16} + \frac{y^2}{25} = 1$
Hyperbola	$AC < 0$	$x^2 - y^2 = 1$

CONIC

A conic is the set of all points $P(x, y)$ in a plane such that the ratio of the distance from P to a fixed point and the distance from P to a fixed line is constant. This constant ratio is called the eccentricity of the conic.

Conic	Eccentricity
Circle	$e = 0$
Parabola	$e = 1$
Ellipse	$e = \frac{c}{a}$ and $0 < e < 1$
Hyperbola	$e = \frac{c}{a}$ and $e > 1$

CHAPTER 6 Review Exercises

Write an equation for the circle satisfying the given conditions. Graph it by hand, and give the domain and range.

1. Center $(-2, 3)$; radius 5
2. Center $(\sqrt{5}, -\sqrt{7})$; radius $\sqrt{3}$
3. Center $(-8, 1)$; passing through $(0, 16)$
4. Center $(3, -6)$; tangent to the x-axis

Find the center and radius of each circle.

5. $x^2 - 4x + y^2 + 6y + 12 = 0$
6. $x^2 - 6x + y^2 - 10y + 30 = 0$
7. $2x^2 + 14x + 2y^2 + 6y = -2$
8. $3x^2 + 3y^2 + 33x - 15y = 0$
9. Describe the graph of $(x - 4)^2 + (y - 5)^2 = 0$.

Give the focus, directrix, and axis for each parabola, graph it by hand, and give the domain and range.

10. $y^2 = -\frac{2}{3}x$
11. $y^2 = 2x$
12. $3x^2 - y = 0$
13. $x^2 + 2y = 0$

Write an equation for each parabola with vertex at the origin.

14. Focus $(4, 0)$
15. Through $(2, 5)$; opening to the right
16. Through $(3, -4)$; opening downward

Write an equation for each parabola.

17. Vertex $(-5, 6)$; focus $(2, 6)$
18. Vertex $(4, 3)$; focus $(4, 5)$

Graph each ellipse or hyperbola by hand, and give the domain, range, and coordinates of the vertices.

19. $\frac{x^2}{5} + \frac{y^2}{9} = 1$
20. $\frac{x^2}{16} + \frac{y^2}{4} = 1$
21. $\frac{x^2}{64} - \frac{y^2}{36} = 1$
22. $\frac{y^2}{25} - \frac{x^2}{9} = 1$
23. $\frac{(x-3)^2}{4} + (y+1)^2 = 1$
24. $\frac{(x-2)^2}{9} + \frac{(y+3)^2}{4} = 1$
25. $\frac{(y+2)^2}{4} - \frac{(x+3)^2}{9} = 1$
26. $\frac{(x+1)^2}{16} - \frac{(y-2)^2}{4} = 1$

Write an equation for each conic section with center at the origin.

27. Ellipse: vertex $(0, 4)$; focus $(0, 2)$
28. Ellipse: x-intercept 6; focus $(-2, 0)$
29. Hyperbola: focus $(0, -5)$; y-intercepts -4 and 4
30. Hyperbola: y-intercept -2; passing through $(2, 3)$
31. Focus $(0, -3)$; $e = \frac{2}{3}$
32. Focus $(5, 0)$; $e = \frac{5}{2}$
33. Consider the circle with equation $x^2 + y^2 + 2x + 6y - 15 = 0$.
    (a) What are the coordinates of the center?
    (b) What is the radius?
    (c) What two functions must be graphed to graph this circle with your calculator in function mode?

Concept Check *Match each equation in Column I with the appropriate description in Column II.*

I	II
34. $4x^2 + y^2 = 36$	**A.** Circle; center $(1, -2)$; radius 6
35. $x = 2y^2 + 3$	**B.** Hyperbola; center $(2, 1)$
36. $(x - 1)^2 + (y + 2)^2 = 36$	**C.** Ellipse; major axis on x-axis
37. $\frac{x^2}{36} + \frac{y^2}{9} = 1$	**D.** Ellipse; major axis on y-axis
38. $(y - 1)^2 - (x - 2)^2 = 36$	**E.** Parabola; opens right
39. $y^2 = 36 + 4x^2$	**F.** Hyperbola; transverse axis on y-axis

Write the equation in standard form for an ellipse or a hyperbola centered at (h, k). Identify the center and the vertices.

40. $4x^2 + 8x + 25y^2 - 250y = -529$

41. $5x^2 + 20x + 2y^2 - 8y = -18$

42. $x^2 + 4x - 4y^2 + 24y = 36$

43. $4y^2 + 8y - 3x^2 + 6x = 11$

Find the eccentricity of each ellipse or hyperbola.

44. $9x^2 + 25y^2 = 225$

45. $4x^2 + 9y^2 = 36$

46. $9x^2 - y^2 = 9$

Write an equation for each conic section.

47. Parabola with vertex $(-3, 2)$ and y-intercepts 5 and -1

48. Hyperbola with foci $(0, 12)$ and $(0, -12)$; asymptotes $y = \pm x$

49. Ellipse consisting of all points in the plane, the sum of whose distances from $(0, 0)$ and $(4, 0)$ is 8

50. Hyperbola consisting of all points in the plane for which the absolute value of the difference of the distances from $(0, 0)$ and $(0, 4)$ is 2

Use a graphing calculator to graph each plane curve in the specified window.

51. $x = 4t - 3$, $y = t^2$, for t in $[-3, 4]$;
window: $[-20, 20]$ by $[-20, 20]$

52. $x = t^2$, $y = t^3$, for t in $[-2, 2]$;
window: $[-15, 15]$ by $[-10, 10]$

53. $x = t + \ln t$, $y = t + e^t$, for t in $(0, 2]$;
window: $[-5, 5]$ by $[0, 10]$

Find a rectangular equation for each plane curve.

54. $x = 3t + 2$, $y = t - 1$, for t in $[-5, 5]$

55. $x = \sqrt{t - 1}$, $y = \sqrt{t}$, for t in $[1, \infty)$

56. $x = \frac{1}{t + 3}$, $y = t + 3$, for $t \neq -3$

(Modeling) *Solve each application.*

57. ***Orbit of Venus*** The orbit of Venus is an ellipse with the sun at one focus. The eccentricity of the orbit is $e = .006775$, and the major axis has length 134.5 million miles. (*Source: The World Almanac and Book of Facts,* 2001.) Find the least and greatest distances of Venus from the sun.

58. ***Orbit of the Comet Swift–Tuttle*** Comet Swift–Tuttle has an elliptical orbit of eccentricity $e = .964$ with the sun at one focus. Find the equation of the comet, given that the closest it comes to the sun is 89 million miles.

Trajectory of a Satellite *When a satellite is near Earth, its orbital trajectory may trace out a hyperbola, a parabola, or an ellipse. The type of trajectory depends on the satellite's velocity V in meters per second. It will be hyperbolic if $V > k/\sqrt{D}$, parabolic if $V = k/\sqrt{D}$, and elliptic if $V < k/\sqrt{D}$, where $k = 2.82 \times 10^7$ is a constant and D is the distance in meters from the satellite to the center of Earth.* (*Source:* Loh, W., *Dynamics and Thermodynamics of Planetary Entry,* Prentice-Hall, 1963; Thomson, W., *Introduction to Space Dynamics,* John Wiley and Sons, 1961.)

Use this information in Exercises 59–61 on the next page.

59. When the artificial satellite *Explorer IV* was at a maximum distance D of 42.5×10^6 meters from Earth's center, it had a velocity V of 2090 meters per second. Determine the shape of its trajectory.

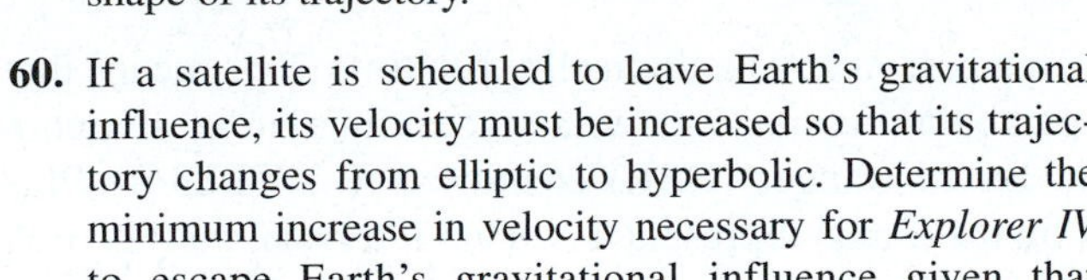

60. If a satellite is scheduled to leave Earth's gravitational influence, its velocity must be increased so that its trajectory changes from elliptic to hyperbolic. Determine the minimum increase in velocity necessary for *Explorer IV* to escape Earth's gravitational influence given that $D = 42.5 \times 10^6$ meters.

61. Explain why it is easier to change a satellite's trajectory from an ellipse to a hyperbola when D is maximum rather than minimum.

62. If $Ax^2 + Cy^2 + Dx + Ey + F = 0$ is the general equation of an ellipse, find its center point by completing the square.

CHAPTER 6 Test

1. Match each equation in Column I with the appropriate description in Column II.

I	II
(a) $\dfrac{(x+3)^2}{4} - \dfrac{(y+2)^2}{16} = 1$	**A.** Circle; center $(3, 2)$; radius 4
(b) $(x-3)^2 + (y-2)^2 = 16$	**B.** Hyperbola; center $(-3, -2)$
(c) $(x+3)^2 + (y-2)^2 = 16$	**C.** Ellipse; center $(-3, -2)$
(d) $(x+3)^2 = -(y-4)$	**D.** Circle; center $(-3, 2)$; radius 4
(e) $x - 4 = (y-2)^2$	**E.** Parabola; opens downward
(f) $\dfrac{(x+3)^2}{4} + \dfrac{(y+2)^2}{16} = 1$	**F.** Parabola; opens right

2. Give the coordinates of the focus, and the equation of the directrix, for the parabola with equation $y^2 = \frac{1}{8}x$.

3. Graph $y = -\sqrt{1 - \dfrac{x^2}{36}}$. Is it the graph of a function? Find the domain and range.

4. What two functions are used to graph $\dfrac{x^2}{25} - \dfrac{y^2}{49} = 1$ on a calculator in function mode?

Graph each relation by hand. Identify the graph, and give the radius, center, vertices, and foci, as applicable.

5. $\dfrac{y^2}{4} - \dfrac{x^2}{9} = 1$

6. $x^2 + 4y^2 + 2x - 16y + 17 = 0$

7. $y^2 - 8y - 2x + 22 = 0$

8. $x^2 + (y-4)^2 = 9$

9. $\dfrac{(x-3)^2}{49} + \dfrac{(y+1)^2}{16} = 1$

10. $x = 4t^2 - 4$, $y = t - 1$, for t in $[-1.5, 1.5]$

11. Write an equation for each conic.
(a) Center at the origin; focus $(0, -2)$; $e = 1$
(b) Center at the origin; vertical major axis of length 6; $e = \frac{5}{6}$

12. ***(Modeling) Height of a Bridge's Arch*** An arch of a bridge has the shape of the top half of an ellipse. The arch is 40 feet wide and 12 feet high at the center. Find the equation of the complete ellipse. Find the height of the arch 10 feet from the center at the bottom.

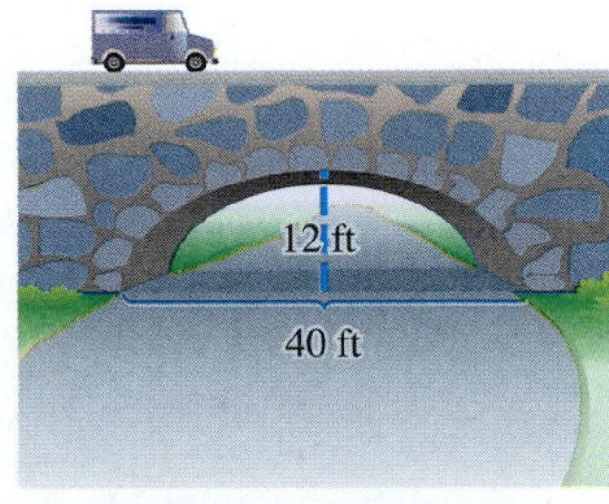

CHAPTER 6 PROJECT

Modeling the Path of a Bouncing Ball

The height of each bounce of a bouncing ball varies quadratically with time. This means that the graph of each bounce will be a parabola. Figure A shows a scatter diagram of a bouncing ball, captured with a Calculator Based Ranger (CBR) connected to a TI-83/84 Plus graphing calculator. The tick marks on the x-axis represent 1-second intervals, and the tick marks on the y-axis represent 1-foot intervals.

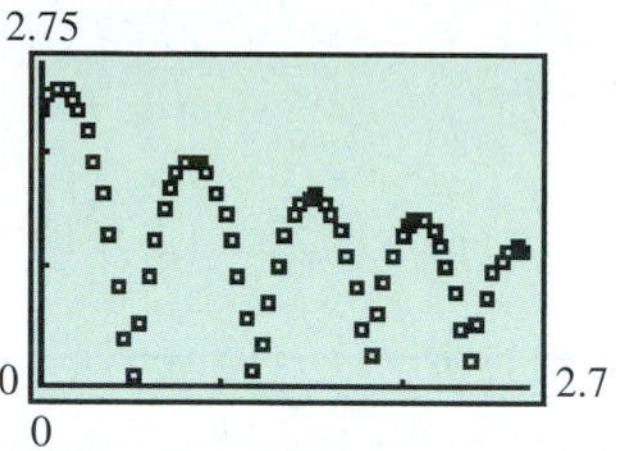

FIGURE A

Since each bounce can be modeled with a parabola, the entire path of the ball can be modeled with a piecewise-defined function. The graph consists of all or part of five bounces; therefore, the function will have five pieces. We can find this function by using the data lists that generated the scatter diagram.

Time	Height	Time	Height	Time	Height
0	2.37249	.934912	1.80222	1.869694	.60089
.042496	2.48284	.977404	1.66124	1.912184	.88197
.084992	2.536	1.019894	1.47971	1.954674	1.10494
.127488	2.53149	1.062384	1.23962	1.997164	1.27115
.169984	2.46122	1.104874	.92341	2.039654	1.37565
.21248	2.34186	1.147364	.55945	2.082144	1.41574
.254976	2.16168	1.189854	.12928	2.124634	1.40223
.297472	1.92384	1.232344	.37297	2.167124	1.33602
.339968	1.61799	1.274834	.73107	2.209614	1.20809
.382464	1.27476	1.317324	1.02881	2.252104	1.02476
.42496	.85089	1.359814	1.26845	2.294594	.78512
.467456	.37882	1.402304	1.44863	2.337084	.48648
.509952	.10045	1.444794	1.54592	2.379574	.20991
.552448	.53423	1.487284	1.60808	2.422064	.5108
.594944	.9153	1.529774	1.6225	2.464554	.76125
.63744	1.23286	1.572264	1.57295	2.507044	.9635
.679936	1.48782	1.614754	1.47836	2.549534	1.08467
.722432	1.6752	1.657244	1.32701	2.592024	1.1689
.764928	1.81754	1.699734	1.10854	2.634514	1.19683
.807424	1.89321	1.742224	.83332	2.677004	1.168
.84992	1.91889	1.784714	.49234	2.719494	1.0671
.892416	1.88826	1.827204	.2509	2.761984	.95134

First, we determine the endpoints of each parabola or partial parabola by looking through the data list. For example, to find the endpoints of the second parabola, look in the list for the first height that is close to 0. We find that the first near-0 height of .10045 occurs at time .509952, and the next near-0 height of .12928 occurs at 1.189854. So, the domain of the second parabola is the interval $(.51, 1.19)$. Its vertex is at time .84992, which corresponds to the largest height, 1.91889, in the list between these endpoints.

The equation of each parabola should be in the form $y = a(x - h)^2 + k$. Since the ball is a projectile in motion, a is half the acceleration due to gravity: -16 feet per second. The values of h and k are the coordinates of each vertex—that is, $(.84992, 1.91889)$ for the second parabola. The resolution of the calculator screen is such that rounding to hundredths will not result in loss of detail. Thus, the equation and domain of the second parabola are $y = -16(x - .85)^2 + 1.92$ and $(.51, 1.19)$, respectively.

Activities

1. Find the equations and domains of the four remaining parabolas. Write a piecewise-defined function for the five parabolas and graph it. Prepare a report that includes the piecewise-defined function, the graph, and a discussion comparing the graph with the scatter diagram in Figure A.
2. The TI-83/84 Plus calculator has a graph style called "Path" in which a circular cursor traces the leading edge of the graph and draws a path. If you are using this model, set the graph style to Path and watch the ball bounce. (*Note:* In reality, the ball bounces straight up and down.)
3. Working in groups of 3 or 4, gather your own original data. Write a piecewise-defined function to model the path of the ball. (Different groups should use different balls.)

7

Systems of Equations and Inequalities; Matrices

IMAGINE THAT WE are developing coffee blends for Starbucks Corporation, the leading roaster and retailer of specialty coffees in the world. Three varieties of coffee—Arabian Mocha Sanai, Organic Shade-Grown Mexico, and Guatemala Antigua—are combined and roasted, yielding 50-pound batches of coffee beans. To develop the perfect balance of flavor, boldness, and acidity, we combine twice as many pounds of Guatemala Antigua, which retails for \$10.19 per pound as Arabian Mocha Sanai, which sells for \$15.99 per pound. Organic Shade-Grown Mexico rounds out the blend at \$12.99 per pound. How many pounds of each variety must we use to create a blend that retails for \$12.37 per pound?

We can determine the number of pounds of each variety of coffee to use in the blend by setting up a *system of equations in three variables*,

$$\begin{aligned} a + b + c &= 50 \\ c &= 2a \\ 15.99a + 12.99b + 10.19c &= 12.37(50), \end{aligned}$$

where a represents the number of pounds of Arabian Mocha Sanai, b the number of pounds of Organic Shade-Grown Mexico, and c the number of pounds of Guatemala Antigua. In this chapter, we learn to write and solve systems of equations and inequalities in two, three, or four variables for many business and science applications.

Source: Starbucks Corp.

Chapter Outline

7.1 Systems of Equations

Linear Systems ■ Substitution Method ■ Elimination Method ■ Special Systems ■ Nonlinear Systems ■ Applications of Systems

Linear Systems

The definition of a linear equation given in Chapter 1 can be extended to more variables; any equation of the form

$$a_1x_1 + a_2x_2 + \cdots + a_nx_n = b,$$

for real numbers $a_1, a_2, \ldots, a_n$ (not all of which are 0) and b, is a **linear equation,** or a **first-degree equation in n unknowns.**

A set of equations is called a **system of equations.** The **solutions** of a system of equations must satisfy *every* equation in the system. If all the equations in a system are linear, the system is a **system of linear equations,** or a **linear system.**

The solution set of a linear equation in two variables is an infinite set of ordered pairs. Since the graph of such an equation is a straight line, there are three possible outcomes for the graph of a system of two linear equations in two variables, as shown in Figure 1. They are as follows:

1. The graphs of the two equations intersect in a single point whose coordinates give the only solution of the system. In this case, the system is **consistent,** and the equations are **independent.** This is the most common case. See Figure 1(a).
2. The graphs are distinct parallel lines. In this case, the system is said to be **inconsistent.** That is, there is no solution common to both equations. See Figure 1(b).
3. The graphs are the same line. In this case, the equations are said to be **dependent,** and any solution of one equation is also a solution of the other. Thus, there are infinitely many solutions. See Figure 1(c).

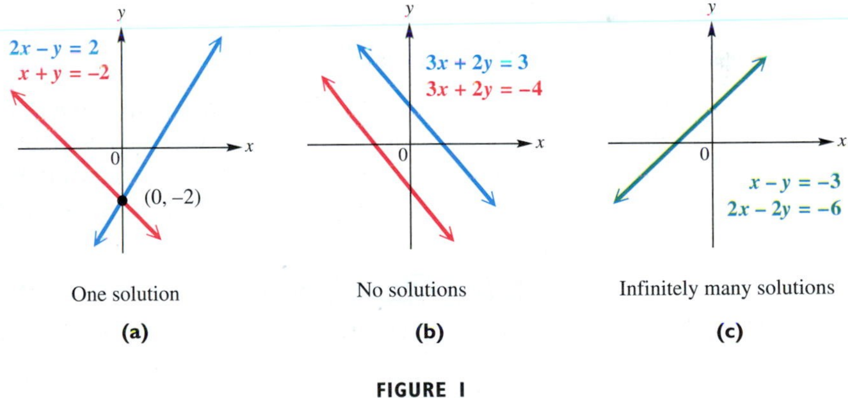

FIGURE 1

Substitution Method

Although the *number* of solutions of a linear system can usually be seen from the graph of the equations of the system, determining an exact solution from a graph is often difficult. In such cases, we use algebraic methods of solution.

In a system of two equations with two variables, the **substitution method** involves using one equation to find an expression for one variable in terms of the other, and then substituting into the other equation of the system.

GCM EXAMPLE 1 Solving a System by Substitution

Solve the system.

$$3x + 2y = 11 \quad (1)$$
$$-x + y = 3 \quad (2)$$

Analytic Solution

One way to solve this system is to solve equation (2) for y, getting $y = x + 3$. Then substitute $x + 3$ for y in equation (1) and solve for x.

$$3x + 2y = 11 \quad (1)$$
$$3x + 2(x + 3) = 11 \quad \text{Let } y = x + 3.$$
$$3x + 2x + 6 = 11 \quad \text{Distributive property}$$
$$5x + 6 = 11 \quad \text{Combine terms.}$$
$$5x = 5 \quad \text{Subtract 6.}$$
$$x = 1 \quad \text{Divide by 5.}$$

Replace x with 1 in $y = x + 3$ to find that $y = 1 + 3 = 4$, so the solution is $(1, 4)$. Check by substituting 1 for x and 4 for y in *both* equations of the original system. The solution set is $\{(1, 4)\}$.

Graphing Calculator Solution

To find the solution graphically, first solve equations (1) and (2) for y to get $Y_1 = -1.5X + 5.5$ and $Y_2 = X + 3$. Graph Y_1 and Y_2 in the standard viewing window to find that the point of intersection is $(1, 4)$, as seen in Figure 2. The table in Figure 3 shows numerically that when $X = 1$, both Y_1 and Y_2 are 4.

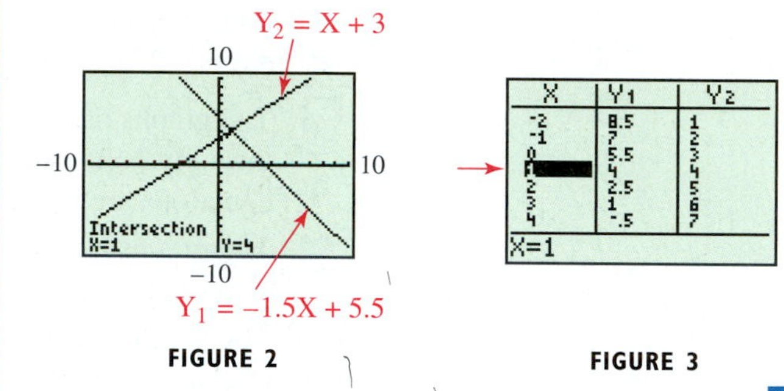

FIGURE 2 FIGURE 3

Elimination Method

Looking Ahead to Calculus

In calculus, the definite integral (see *Looking Ahead to Calculus,* p. 183),

$$\int_a^b [f(x) - g(x)]\,dx,$$

gives the area between the graphs of f and g from $x = a$ to $x = b$. A system of equations is used to find the x-values a and b where the two graphs intersect.

Another way to solve a system of two equations, the **elimination method,** uses multiplication and addition to eliminate a variable from one equation. In this process, the given system is replaced by new systems that have the same solution set as the original system. Systems that have the same solution set are called **equivalent systems.**

EXAMPLE 2 Solving a System by Elimination

Solve the system.

$$3x - 4y = 1 \quad (1)$$
$$2x + 3y = 12 \quad (2)$$

Solution To eliminate x, multiply each side of equation (1) by -2 and each side of equation (2) by 3 to get

$$-6x + 8y = -2 \quad \text{Multiply (1) by } -2. \quad (3)$$
$$6x + 9y = 36. \quad \text{Multiply (2) by 3.} \quad (4)$$

This new system will have the same solution set as the given system. We add equations (3) and (4) to eliminate x and solve the result for y.

$$\begin{aligned} -6x + 8y &= -2 && \text{(3)} \\ 6x + 9y &= 36 && \text{(4)} \\ \hline 17y &= 34 && \text{Add.} \\ y &= 2 && \text{Divide by 17.} \end{aligned}$$

Substitute 2 for y in equation (1) or (2).

$$\begin{aligned} 3x - 4(2) &= 1 && \text{(1)} \\ 3x &= 9 \\ x &= 3 \end{aligned}$$

Write the x-value first in the ordered pair.

The solution is (3, 2), which can be checked by substituting 3 for x and 2 for y in *both* equations of the original system.

Check:

$$\begin{aligned} 3x - 4y &= 1 && \text{(1)} \\ 3(3) - 4(2) &= 1 && \text{?} \\ 1 &= 1 && \text{True} \end{aligned}$$

$$\begin{aligned} 2x + 3y &= 12 && \text{(2)} \\ 2(3) + 3(2) &= 12 && \text{?} \\ 12 &= 12 && \text{True} \end{aligned}$$

Both check; the solution set is $\{(3, 2)\}$. ■

What Went Wrong?

To find the solution in Example 2 graphically, a student solved each equation for y, entered the functions shown on the left, and got the graph shown on the right.

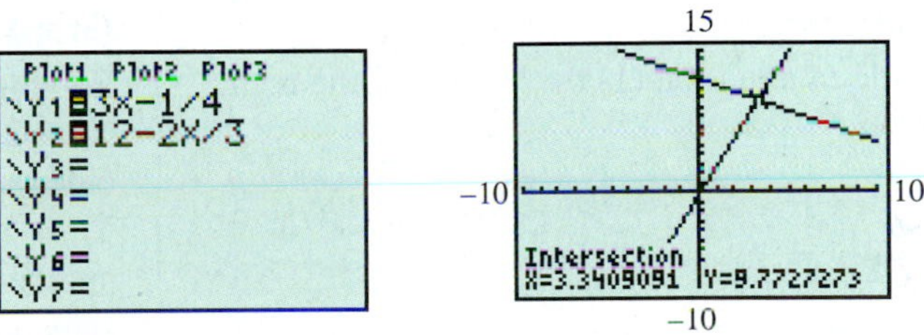

What Went Wrong? What can be done to correct the graph?

Special Systems

The examples presented so far in this section have all been consistent systems with a single solution. This is not always the case.

Answer to What Went Wrong?

The student neglected to insert parentheses as necessary. Y_1 should be entered as $(3X - 1)/4$ and Y_2 as $(12 - 2X)/3$.

EXAMPLE 3 Solving an Inconsistent System

Solve the system.

$$3x - 2y = 4 \quad (1)$$
$$-6x + 4y = 7 \quad (2)$$

Analytic Solution

Eliminate the variable x by multiplying each side of equation (1) by 2 and then adding the result to equation (2).

$$\begin{aligned} 6x - 4y &= 8 && \text{Multiply (1) by 2.} \\ -6x + 4y &= 7 && \text{(2)} \\ \hline 0 &= 15 && \text{False} \end{aligned}$$

Both variables were eliminated, leaving the false statement $0 = 15$. This contradiction indicates that these two equations have no solutions in common. The system is inconsistent, and the solution set is $\emptyset$.

Graphing Calculator Solution

In slope–intercept form, equation (1) is $y = 1.5x - 2$ and equation (2) is $y = 1.5x + 1.75$. Since the lines have the same slope, but different y-intercepts, they are parallel and have no point of intersection, which supports the analytic solution. See Figure 4.

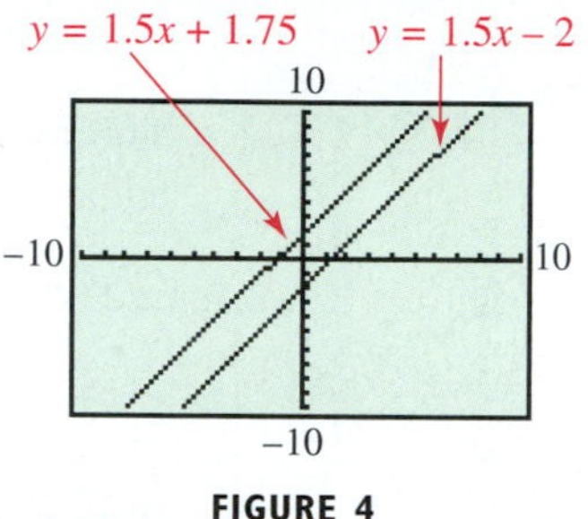

FIGURE 4

EXAMPLE 4 Solving a System with Dependent Equations

Solve the system.

$$8x - 2y = -4 \quad (1)$$
$$-4x + y = 2 \quad (2)$$

Analytic Solution

Divide each side of equation (1) by 2, and add the result to equation (2).

$$\begin{aligned} 4x - y &= -2 && \text{Divide (1) by 2.} \\ -4x + y &= 2 && \text{(2)} \\ \hline 0 &= 0 && \text{True} \end{aligned}$$

The result, $0 = 0$, is a true statement, which indicates that the equations of the original system are equivalent. Any ordered pair (x, y) that satisfies either equation will satisfy the system. From equation (2),

$$-4x + y = 2 \quad \text{or} \quad y = 2 + 4x.$$

The solutions of the system can be written as a set of ordered pairs $(x, 2 + 4x)$, for any real number x. Some ordered pairs in the solution set are $(0, 2 + 4 \cdot 0) = (0, 2)$, $(1, 2 + 4 \cdot 1) = (1, 6)$, $(3, 14)$, $\left(-\frac{1}{2}, 0\right)$, and $(-2, -6)$. The solution set is $\{(x, 2 + 4x)\}$.

Graphing Calculator Solution

Solving the equations for y gives $Y_1 = \frac{8X + 4}{2}$ and $Y_2 = 2 + 4X$. The graphs of the two equations coincide, as seen in Figure 5. The table indicates that $Y_1 = Y_2$ for arbitrarily selected values of X, providing another way to show that the two equations lead to the same graph.

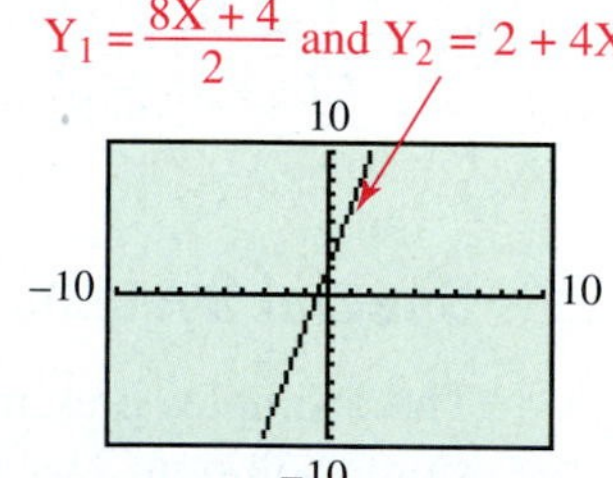

X	Y1	Y2
0	2	2
1	6	6
3	14	14
-.5	0	0
-2	-6	-6

Y1■(8X+4)/2

FIGURE 5

Refer to the analytic solution to see how the solution set can be written in terms of an arbitrary variable.

NOTE Alternatively, we could solve equation (2) in Example 4 for x and write the solution set with y arbitrary: $\left\{\left(\frac{y-2}{4}, y\right)\right\}$. Substitute values for y and verify that the same ordered pairs result.

Nonlinear Systems

A **nonlinear system of equations** is a system in which at least one of the equations is not a linear equation.

EXAMPLE 5 Solving a Nonlinear System by Substitution

Solve the system.

$$3x^2 - 2y = 5 \quad (1)$$
$$x + 3y = -4 \quad (2)$$

Analytic Solution

The graph of equation (1) is a parabola, and that of equation (2) is a line. There may be 0, 1, or 2 points of intersection of these graphs.

Although either equation could be solved for either variable, it is simpler here to solve the linear equation (2) for y, since x is squared in equation (1).

$$x + 3y = -4 \quad (2)$$
$$3y = -4 - x \quad \text{Subtract } x.$$
$$y = \frac{-4 - x}{3} \quad \text{Divide by 3.} \quad (3)$$

Substitute this value of y into equation (1).

$$3x^2 - 2\left(\frac{-4 - x}{3}\right) = 5 \quad (1)$$
$$9x^2 - 2(-4 - x) = 15 \quad \text{Multiply each term by 3.}$$
$$9x^2 + 8 + 2x = 15 \quad \text{Distributive property}$$
$$9x^2 + 2x - 7 = 0 \quad \text{Standard form}$$
$$(9x - 7)(x + 1) = 0 \quad \text{Factor.}$$
$$9x - 7 = 0 \quad \text{or} \quad x + 1 = 0 \quad \text{Zero-product property}$$
$$x = \frac{7}{9} \quad \text{or} \quad x = -1 \quad \text{Solve for } x.$$

Be careful with signs.

Substitute both values of x into equation (3) to find the corresponding y-values.

$$y = \frac{-4 - \frac{7}{9}}{3} = -\frac{43}{27} \qquad y = \frac{-4 - (-1)}{3} = -1$$

The solution set is $\left\{\left(\frac{7}{9}, -\frac{43}{27}\right), (-1, -1)\right\}$.

Graphing Calculator Solution

Graph equation (1) as $y_1 = 1.5x^2 - 2.5$ and equation (2) as $y_2 = -\frac{1}{3}x - \frac{4}{3}$. Figure 6 indicates the points of intersection, $(-1, -1)$ and $(.\overline{7}, 1.\overline{592})$, confirming the analytic solution.

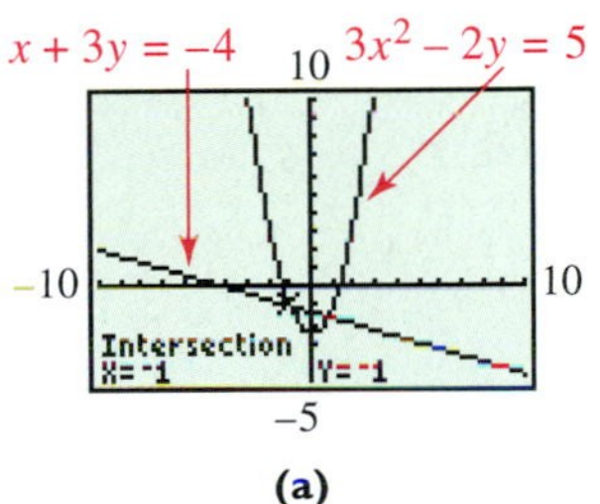

(a)

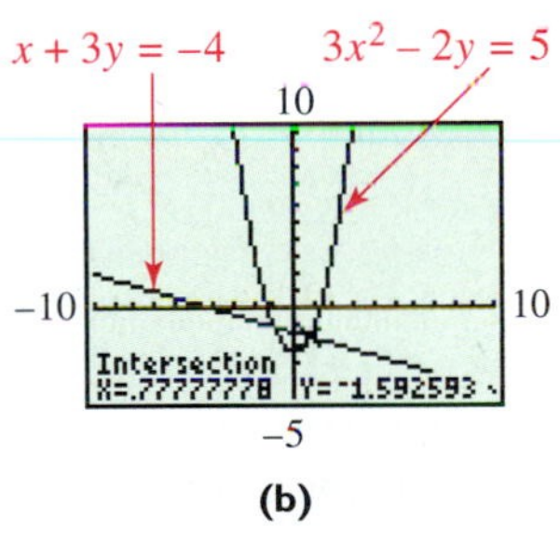

(b)

FIGURE 6

The decimal values for x and y displayed at the bottom of Figure 6(b) represent the repeating decimal forms for $\frac{7}{9}$ and $-\frac{43}{27}$, respectively.

GCM **EXAMPLE 6 Solving a Nonlinear System by Elimination**

Solve the system.

$$x^2 + y^2 = 4 \quad (1)$$
$$2x^2 - y^2 = 8 \quad (2)$$

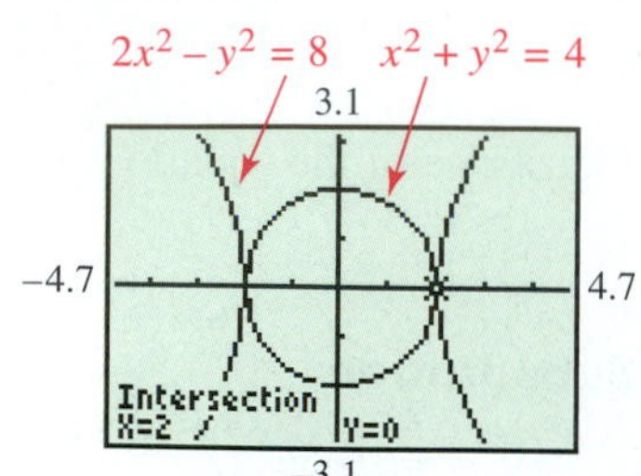

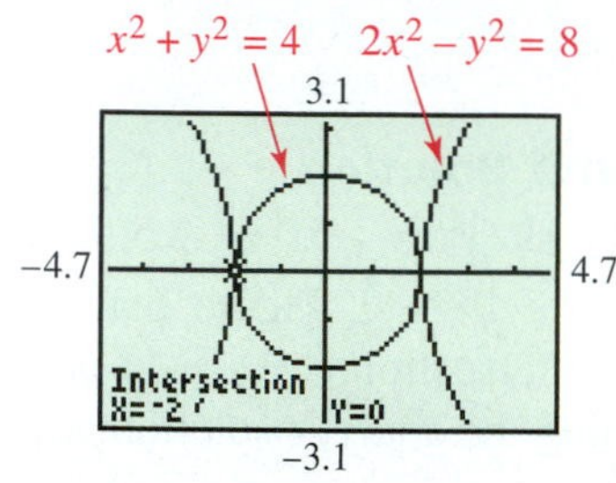

The two points of intersection are $(2, 0)$ and $(-2, 0)$.

FIGURE 7

Solution The graph of equation (1) is a circle and that of equation (2) is a hyperbola. Visualizing them suggests that there may be 0, 1, 2, 3, or 4 points of intersection.

Add the two equations to eliminate y^2.

$$x^2 + y^2 = 4 \quad (1)$$
$$2x^2 - y^2 = 8 \quad (2)$$
$$3x^2 \quad = 12 \quad \text{Add.}$$
$$x^2 = 4 \quad \text{Divide by 3.}$$
$$x = 2 \quad \text{or} \quad x = -2 \quad \text{Square root property}$$

Remember both the positive and negative square roots.

Substituting into equation (1) gives the corresponding values of y.

If $x = 2$, then

$$2^2 + y^2 = 4$$
$$y^2 = 0$$
$$y = 0.$$

If $x = -2$, then

$$(-2)^2 + y^2 = 4$$
$$y^2 = 0$$
$$y = 0.$$

Check that the solution set of the system is $\{(2, 0), (-2, 0)\}$. See Figure 7. ■

Many systems are difficult or even impossible to solve analytically. Graphing calculators allow us to solve such systems graphically.

EXAMPLE 7 Solving a Nonlinear System Graphically

Solve the system.

$$y = 2^x \quad (1)$$
$$|x + 2| - y = 0 \quad (2)$$

TECHNOLOGY NOTE

Example 7 illustrates why proficiency in setting various windows is important when studying algebra with a graphing calculator.

Solution Enter equation (1) as $y_1 = 2^x$ and equation (2) as $y_2 = |x + 2|$. As seen in Figure 8, the various windows indicate that there are three points of intersection of the graphs and thus three solutions. Using the capabilities of the calculator, we find that $(2, 4)$ is an exact solution and both $(-2.22, .22)$ and $(-1.69, .31)$ are approximate solutions. Therefore, the solution set is $\{(2, 4), (-2.22, .22), (-1.69, .31)\}$.

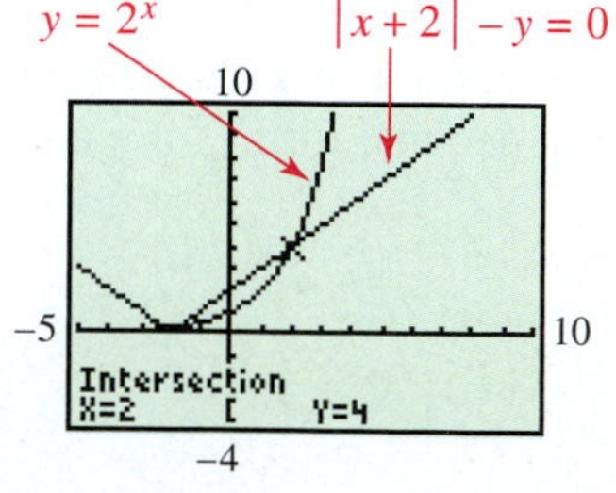

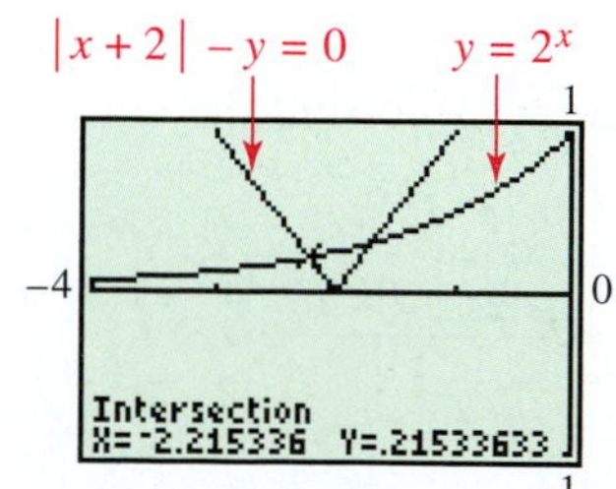

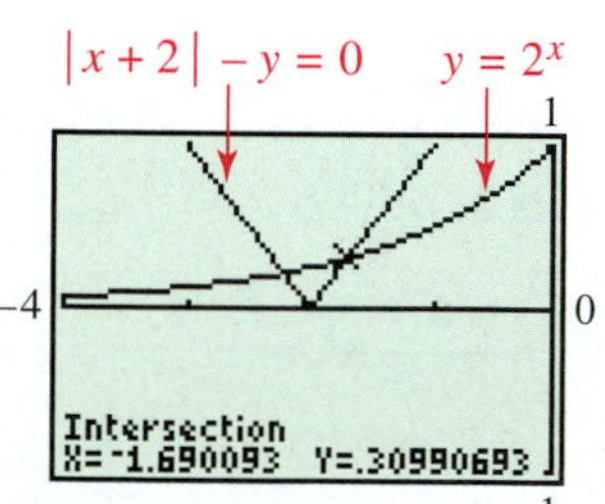

FIGURE 8

■

Applications of Systems

Many applied problems involve more than one unknown quantity. To solve such a problem with a system of equations, we can use the four-step process for solving application problems, presented in Section 1.6. This process is summarized as follows.

Solving Application Problems

Step 1 **Read the problem** carefully and assign different variables to represent each quantity.

Step 2 **Write a system of equations** involving these variables.

Step 3 **Solve the system of equations** and determine the solution.

Step 4 **Look back and check** your solution. Does it seem reasonable?

EXAMPLE 8 **Using a Linear System to Solve an Application**

Landmark Title IX legislation prohibits sex discrimination in sports programs. In a recent year, the national average spent on two varsity athletes, one female and one male, was \$6050 for Division I-A schools. However, average expenditures for a male athlete exceeded those for a female athlete by \$3900. Determine how much was spent per varsity athlete for each gender. (*Source: USA Today.*)

Solution

Step 1 Let x represent average expenditures per male athlete and y represent average expenditures per female athlete.

Step 2 Since the average amount spent on one female and one male athlete was \$6050, one equation is

$$\frac{x+y}{2} = 6050.$$

The expenditures for a male athlete exceeded those for a female athlete by \$3900. Thus, $x - y = 3900$, which gives the system of equations

$$\frac{x+y}{2} = 6050 \quad (1)$$

$$x - y = 3900. \quad (2)$$

Step 3 Multiply equation (1) by 2 and then add the resulting equations.

$$\begin{aligned} x + y &= 12{,}100 && \text{Multiply (1) by 2.} \\ x - y &= 3900 && \text{(2)} \\ \hline 2x \quad &= 16{,}000 && \text{Add.} \\ x &= 8000 && \text{Divide by 2.} \end{aligned}$$

The average expenditure per male athlete was \$8000. To determine y, substitute $x = 8000$ into equation (2).

$$x - y = 3900 \qquad (2)$$
$$8000 - y = 3900$$
$$y = 4100$$

The average amount spent on each female athlete was \$4100.

Step 4 The average of \$8000 and \$4100 is $\frac{\$8000 + \$4100}{2} = \$6050$, and $\$8000 - \$4100 = \$3900$. The solution checks. ■

EXAMPLE 9 Using a Nonlinear System to Find the Dimensions of a Box

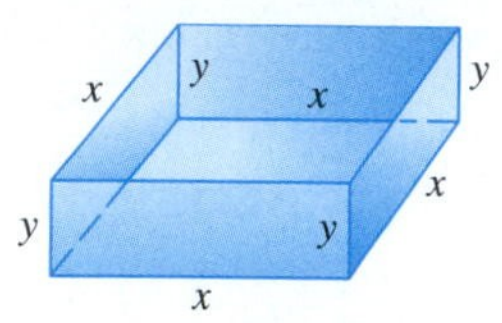

FIGURE 9

A box with an open top has a square base and four sides of equal height. The volume of the box is 75 cubic inches, and the surface area is 85 square inches. What are the dimensions of the box?

Solution Figure 9 shows a diagram of such a box. If each side of the square base measures x inches and the height measures y inches, then the volume is

$$x^2y = 75 \qquad \text{Volume formula} \quad (1)$$

and the surface area is

$$x^2 + 4xy = 85. \qquad \text{Sum of areas of base and sides} \quad (2)$$

We solve equation (1) for y to get $y = \frac{75}{x^2}$ and substitute this into equation (2).

$$x^2 + 4x\left(\frac{75}{x^2}\right) = 85 \qquad (2)$$
$$x^2 + \frac{300}{x} = 85 \qquad \text{Multiply.}$$
$$x^3 + 300 = 85x \qquad \text{Multiply by } x.$$
$$x^3 - 85x + 300 = 0 \qquad \text{Subtract } 85x.$$

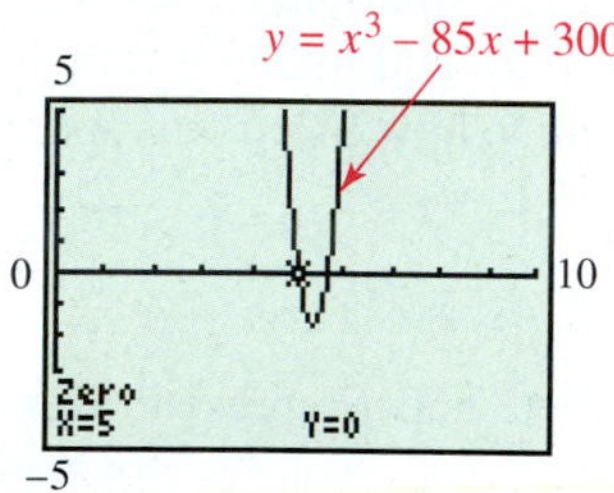

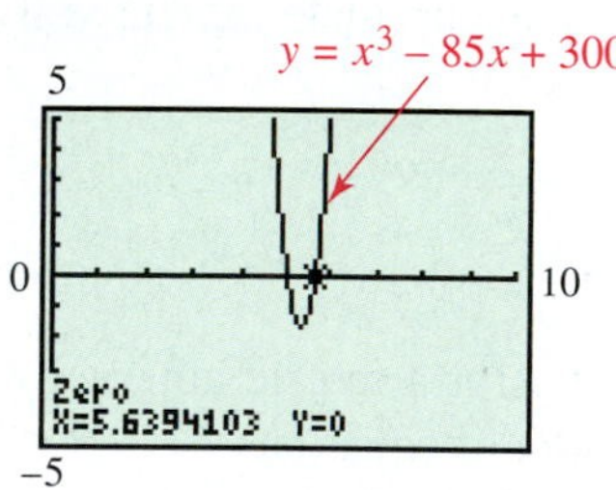

FIGURE 10

The solutions of this equation are x-intercepts of the graph of $y = x^3 - 85x + 300$. A comprehensive graph of this function indicates that there are three real solutions. However, one of them is negative and must be rejected. As Figure 10 indicates, one positive solution is 5 and the other positive solution is approximately 5.64. By substituting back into equation (1), we find that when $x = 5$, $y = 3$, and when $x \approx 5.64$, $y \approx 2.36$. Therefore, this problem has two solutions: The box may have base 5 inches by 5 inches and height 3 inches, or it may have base 5.64 inches by 5.64 inches and height 2.36 inches (approximately). ■

7.1 Exercises

Effects of NAFTA *In 1994, the North American Free Trade Agreement (NAFTA) made the United States, Mexico, and Canada the largest free-trade zone in the world. The figure at the right shows the projected levels of migration from Mexico to the United States with and without NAFTA.*

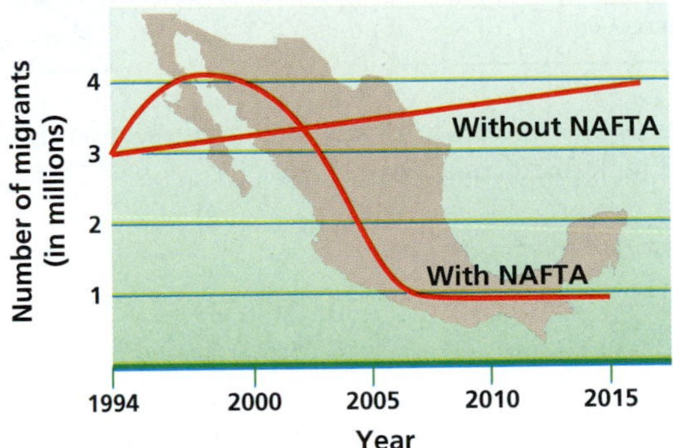

Source: *Population Bulletin*, June 1999.

1. In what year do both projections produce the same level of migration?
2. Refer to Exercise 1. In the year that the two projections produced the same level of migration, what was the level?
3. Express as an ordered pair the solution of the system containing the graphs of the two projections.
4. Use the terms *increasing* and *decreasing* to describe the trends for the "With NAFTA" graph.
5. If equations of the form $y = f(t)$ were found to model either of the two graphs, then t would represent __________ and y would represent ______________.

6. Explain why each graph is that of a function.

Use each graph to estimate the solution of the system of equations. Then solve the system analytically.

7. 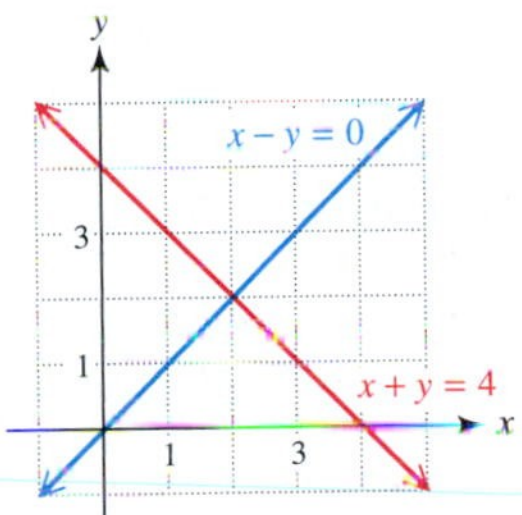

8. 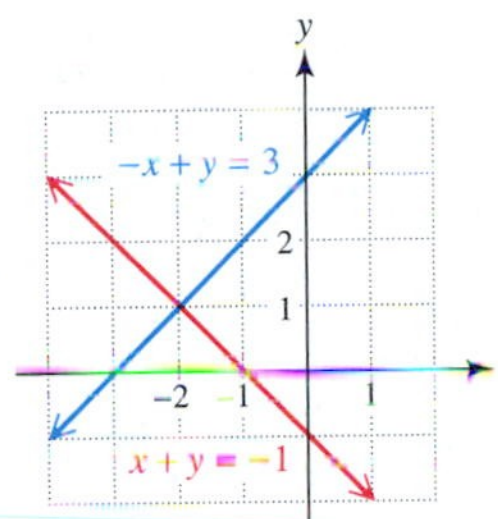

9. 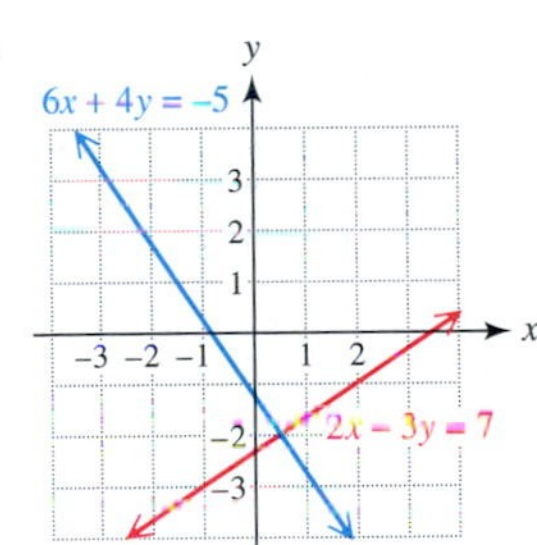

10. 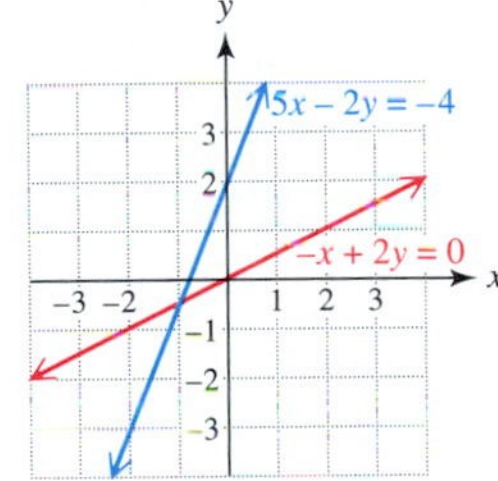

Solve each system by substitution.

11. $6x - y = 5$
$y = x$

12. $5x + y = 2$
$y = -3x$

13. $x + 2y = -1$
$2x + y = 4$

14. $2x + y = -11$
$x + 3y = -8$

15. $y = 2x + 3$
$3x + 4y = 78$

16. $y = 4x - 6$
$2x + 5y = -8$

17. $3x - 2y = 12$
$5x = 4 - 2y$

18. $8x + 3y = 2$
$5x = 17 + 6y$

19. $4x - 5y = -11$
$2x + y = 5$

20. $7x - y = -10$
$3y - x = 10$

21. $4x + 5y = 7$
$9y = 31 + 2x$

22. $-2x = 6y + 18$
$-29 = 5y - 3x$

23. $3x - 7y = 15$
$3x + 7y = 15$

24. $3y = 5x + 6$
$x + y = 2$

25. $2x - 7y = 8$
$-3x + \dfrac{21}{2}y = 5$

26. $.6x - .2y = 2$
$-1.2x + .4y = 3$

27. *Concept Check* Only one of the screens gives the correct graphical solution of the system in Exercise 19. Which one is it?

A.

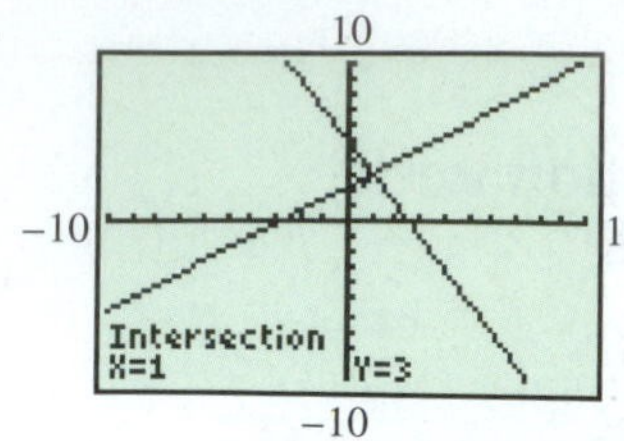

B.

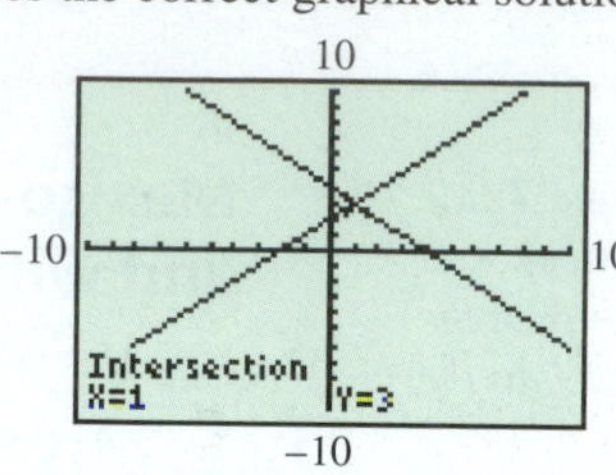

C.

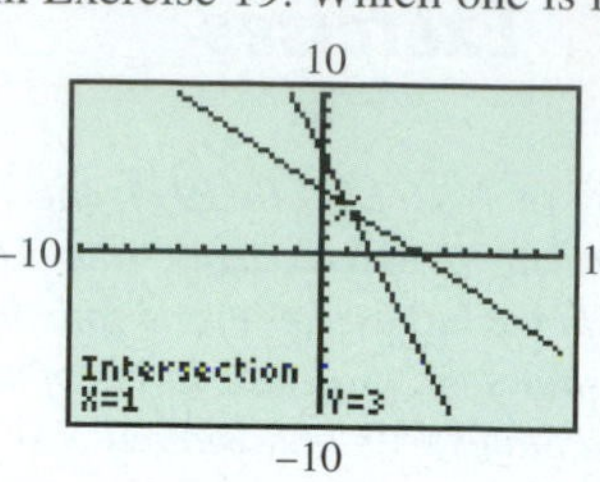

28. *Concept Check* Refer to Example 2 in this section. If we began solving the system by eliminating y, by what numbers might we have multiplied equations (1) and (2)?

Solve each system by elimination.

29. $3x - y = -4$, $x + 3y = 12$

30. $2x - 3y = -7$, $5x + 4y = 17$

31. $4x + 3y = -1$, $2x + 5y = 3$

32. $5x + 7y = 6$, $10x - 3y = 46$

33. $12x - 5y = 9$, $3x - 8y = -18$

34. $6x + 7y = -2$, $7x - 6y = 26$

35. $4x - y = 9$, $-8x + 2y = -18$

36. $x + y = 4$, $3x + 3y = 12$

37. $9x - 5y = 1$, $-18x + 10y = 1$

38. $3x + 2y = 5$, $6x + 4y = 8$

39. $3x + y = 6$, $6x + 2y = 1$

40. $3x + 5y = -2$, $9x + 15y = -6$

41. $\frac{x}{2} + \frac{y}{3} = 8$, $\frac{2x}{3} + \frac{3y}{2} = 17$

42. $\frac{x}{5} + 3y = 31$, $2x - \frac{y}{5} = 8$

43. $\frac{2x - 1}{3} + \frac{y + 2}{4} = 4$, $\frac{x + 3}{2} - \frac{x - y}{3} = 3$

44. $\frac{x + 6}{5} + \frac{2y - x}{10} = 1$, $\frac{x + 2}{4} + \frac{3y + 2}{5} = -3$

Use a graphing calculator to solve each system. Express solutions with approximations to the nearest thousandth.

45. $\sqrt{3}x - y = 5$, $100x + y = 9$

46. $\frac{11}{3}x + y = .5$, $.6x - y = 3$

47. $\sqrt{5}x + \sqrt[3]{6}\,y = 9$, $\sqrt{2}x + \sqrt[5]{9}\,y = 12$

48. $\pi x + ey = 3$, $ex + \pi y = 4$

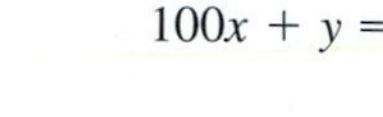 **49.** Explain how one can determine whether a system is inconsistent or has dependent equations when using the substitution or elimination method.

50. *Concept Check* For what value(s) of k will the following system of linear equations have no solution? infinitely many solutions?

$$x - 2y = 3$$
$$-2x + 4y = k$$

Draw a sketch of the two graphs described, with the indicated number of points of intersection. (There may be more than one way to do this.)

51. A line and a circle; no points

52. A line and a circle; one point

53. A line and a circle; two points

54. A line and a hyperbola; no points

55. A line and a hyperbola; one point

56. A line and a hyperbola; two points

57. A circle and an ellipse; four points

58. A parabola and an ellipse; one point

59. A parabola and a hyperbola; four points

60. A circle and a hyperbola; two points

Solve each nonlinear system of equations analytically.

61. $y = -x^2 + 2$, $x - y = 0$

62. $y = (x - 1)^2$, $x - 3y = -1$

63. $3x^2 + 2y^2 = 5$, $x - y = -2$

64. $x^2 + y^2 = 5$
$-3x + 4y = 2$

65. $x^2 + y^2 = 10$
$2x^2 - y^2 = 17$

66. $x^2 + y^2 = 4$
$2x^2 - 3y^2 = -12$

67. $x^2 + 2y^2 = 9$
$3x^2 - 4y^2 = 27$

68. $2x^2 + 3y^2 = 5$
$3x^2 - 4y^2 = -1$

69. $2x^2 + 2y^2 = 20$
$3x^2 + 3y^2 = 30$

70. $x^2 + y^2 = 4$
$5x^2 + 5y^2 = 28$

71. $y = |x - 1|$
$y = x^2 - 4$

72. $2x^2 - y^2 = 4$
$|x| = |y|$ (*Hint:* $|x| = |y|$ is equivalent to $x^2 = y^2$.)

Solve each system graphically. Give x- and y-coordinates correct to the nearest hundredth.

73. $y = \log(x + 5)$
$y = x^2$

74. $y = 5^x$
$xy = 1$

75. $y = e^{x+1}$
$2x + y = 3$

76. $y = \sqrt[3]{x - 4}$
$x^2 + y^2 = 6$

(Modeling) *Solve each problem.*

77. ***Robberies*** The total number of robberies during 2000 and 2001 was 831,000. From 2000 to 2001, the number of robberies declined by 15,000. (*Source*: FBI.)

(a) Write a system of equations whose solution represents the number of robberies committed in each of these years.

(b) Solve the system.

 **(c)** Interpret the solution.

78. ***Smallpox Vaccinations*** During the first four months of 2004, a total of 6767 people were vaccinated for smallpox in Florida and Texas. There were 273 more people vaccinated in Florida than in Texas. (*Source*: FBI.)

(a) Write a system of linear equations whose solution represents the number of vaccinations given in each state.

(b) Solve the system.

 **(c)** Interpret the solution.

79. ***Time on the Internet*** From 2001 to 2002, the average number of hours that a user spent on the Internet each week increased by 13%, amounting to 1.3 more hours. Find the average number of hours that a user spent on the Internet each week in 2001 and 2002. (*Source*: UCLA Center for Communications Policy.)

80. ***Dimensions of a Box*** A box has an open top, rectangular sides, and a square base. The volume of the box is 576 cubic inches, and the surface area of the outside of the box is 336 square inches. Find the dimensions of the box.

81. ***Dimensions of a Box*** A box has rectangular sides and a rectangular top and base that are twice as long as they are wide. The volume of the box is 588 cubic inches, and the surface area of the outside of the box is 448 square inches. Find the dimensions of the box.

82. ***Investments*** A student invests \$5000 at 5% and 7% annually. After 1 year, the student receives a total of \$325 in interest. How much did the student invest at each interest rate?

83. ***Heart Rate*** In one study, a group of conditioned athletes was exercised to exhaustion. Let x represent an athlete's heart rate 5 seconds after stopping exercise and y the rate after 10 seconds. It was found that the maximum heart rate H for these athletes satisfied the two equations

$$H = .491x + .468y + 11.2$$
$$H = -.981x + 1.872y + 26.4.$$

If an athlete had maximum heart rate $H = 180$, determine x and y graphically. Interpret your answer. (*Source:* Thomas, V., *Science and Sport,* Faber and Faber, 1970.)

84. ***Heart Rate*** Repeat Exercise 83 for an athlete with a maximum heart rate of 195.

85. ***Tuna and Shrimp Consumption*** The total amounts (in millions of pounds) of canned tuna and fresh shrimp available for U.S. consumption during the years 1990–1996 are modeled by the linear functions defined by

$$y = -6.393x + 894.9 \quad \text{Tuna}$$
$$y = 19.14x + 746.9. \quad \text{Shrimp}$$

In both equations, x is the number of years since 1990. (*Source:* U.S. National Oceanic and Atmospheric Administration.)

(a) Solve this system of equations analytically. Give answers to the nearest tenth.

(b) Interpret the solution found in part (a).
(c) Graphically support the analytic solution in part (a).

86. ***Populations of Minorities in the United States*** The current and estimated resident populations (in percent) of blacks and Hispanics in the United States for the years 1990–2050 are modeled by the linear functions defined by

$$y = .0515x + 12.3 \quad \text{Blacks}$$
$$y = .255x + 9.01. \quad \text{Hispanics}$$

In each case, x represents the number of years since 1990. (*Source:* U.S. Census Bureau.)

(a) Solve the system to find the year when these population percents will be equal.
(b) What percent of the U.S. resident population will be black or Hispanic in the year found in part (a)?
(c) Graphically support the analytic solution in part (a).
(d) Which population is increasing more rapidly? (*Hint:* Consider the slopes of the lines.)

87. ***Geometry*** Approximate graphically the radius and height of a cylindrical container with volume 50 cubic inches and lateral surface area 65 square inches.

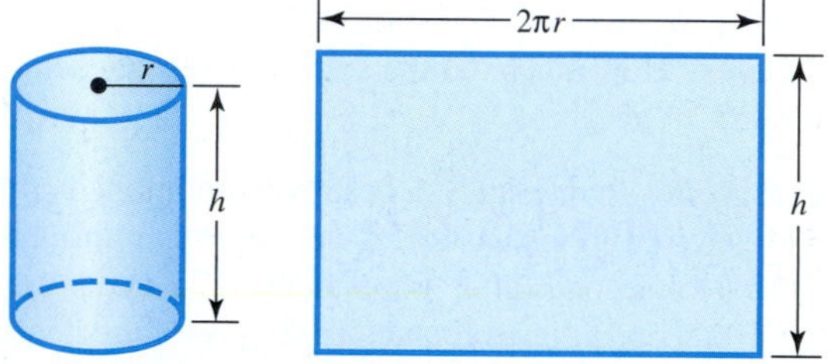

88. ***Geometry*** Determine graphically whether it is possible to construct a cylindrical container, *including* the top and bottom, with volume 38 cubic inches and surface area 38 square inches.

89. ***Roof Truss*** The forces or weights W_1 and W_2 exerted on each rafter for the roof truss shown in the figure are determined by the following system of linear equations. Solve the system.

$$W_1 + \sqrt{2}\,W_2 = 300$$
$$\sqrt{3}\,W_1 - \sqrt{2}\,W_2 = 0$$

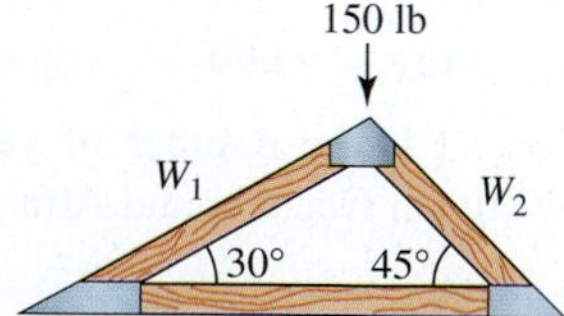

90. ***Height and Weight*** The relationship between a professional basketball player's height h in inches and weight w in pounds was modeled by using two samples of players. The resulting equations were

$$w = 7.46h - 374$$
$$w = 7.93h - 405.$$

Assume that $65 \le h \le 85$.

(a) Use each equation to predict the weight of a professional basketball player who is 6 feet 11 inches.
(b) Determine graphically the height at which the two models give the same weight.
(c) For each model, what change in weight is associated with a 1-inch increase in height?

91. ***Traffic Control*** The figure shows two intersections labeled A and B that involve one-way streets. The numbers and variables represent the average traffic flow rates measured in vehicles per hour. For example, an average of 500 vehicles per hour enter intersection A from the west, whereas 150 vehicles per hour enter this intersection from the north. A stoplight will control the unknown traffic flow denoted by the variables x and y. Use the fact that the number of vehicles entering an intersection must equal the number leaving to determine x and y.

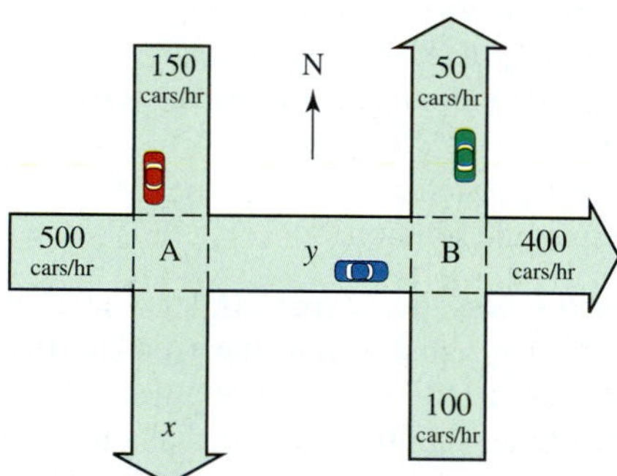

(Modeling) Break-Even Point *The break-even point for a company is the point where costs equal revenues. If both cost and revenue are expressed as linear equations, the break-even point is the solution of a linear system. In each exercise, C represents cost in dollars to produce x items, and R represents revenue in dollars from the sale of x items. Use the substitution method to find the break-even point in each case—that is, the point where*

$$C = R.$$

Then find the value of C and R at that point.

92. $C = 20x + 10{,}000$
$R = 30x - 11{,}000$

93. $C = 4x + 125$
$R = 9x - 200$

(Modeling) Equilibrium Supply and Demand *In each exercise, p is the price of an item in dollars, while q represents the supply in one equation and the demand in the other. Find the equilibrium price and the equilibrium supply/demand.*

94. $p = 80 - \frac{3}{5}q$

$p = \frac{2}{5}q$

95. $p = 630 - \frac{3}{4}q$

$p = \frac{3}{4}q$

96. Let the supply and demand equations for banana smoothies be

$$\text{supply: } p = \frac{3}{2}q \quad \text{and} \quad \text{demand: } p = 81 - \frac{3}{4}q.$$

(a) Graph these equations on the same axes. Let $x = q$ and $y = p$, and use the viewing window $[0, 120]$ by $[0, 100]$. Find the point of intersection.

(b) Find the equilibrium demand analytically.

(c) Find the equilibrium price analytically.

Relating Concepts

For individual or group investigation (Exercises 97–102)

Because variables appear in denominators, the system

$$\frac{5}{x} + \frac{15}{y} = 16$$
$$\frac{5}{x} + \frac{4}{y} = 5$$

is not a linear system. However, we can solve it in a manner similar to the method for solving a linear system by using a substitution-of-variable technique. Let $t = \frac{1}{x}$ *and let* $u = \frac{1}{y}$. ***Work Exercises 97–102 in order.***

97. Write a system of equations in t and u by making the appropriate substitutions.

98. Solve the system in Exercise 97 for t and u.

99. Solve the given system for x and y by using the equations relating t to x and u to y.

100. Refer to the first equation in the given system, and solve for y in terms of x to obtain a rational function.

101. Repeat Exercise 100 for the second equation in the given system.

102. Using a viewing window of $[0, 10]$ by $[0, 2]$, show that the point of intersection of the graphs of the functions in Exercises 100 and 101 has the same x- and y-values as found in Exercise 99.

Use the substitution-of-variable technique from the preceding Relating Concepts exercises to solve each system analytically.

103. $\frac{2}{x} + \frac{1}{y} = \frac{3}{2}$

$\frac{3}{x} - \frac{1}{y} = 1$

104. $\frac{2}{x} + \frac{1}{y} = 11$

$\frac{3}{x} - \frac{5}{y} = 10$

105. $\frac{2}{x} + \frac{3}{y} = 18$

$\frac{4}{x} - \frac{5}{y} = -8$

106. $\frac{1}{x} + \frac{3}{y} = \frac{16}{5}$

$\frac{5}{x} + \frac{4}{y} = 5$

7.2 Solution of Linear Systems in Three Variables

Geometric Considerations ■ Analytic Solution of Systems in Three Variables ■ Applications of Systems ■ Curve Fitting Using a System

Geometric Considerations

We can extend the ideas of systems of equations in two variables to linear equations of the form

$$Ax + By + Cz = D.$$

A solution of such an equation is called an **ordered triple** and is denoted (x, y, z). For example, $(1, 2, -4)$ is a solution of $2x + 5y - 3z = 24$. The solution set of such an equation is an infinite set of ordered triples.

In geometry, the graph of a linear equation in three variables is a plane in three-dimensional space. Considering the possible intersections of the planes representing three equations in three unknowns shows that the solution set of such a system may be either a single ordered triple (x, y, z), an infinite set of ordered triples (dependent equations), or the empty set (an inconsistent system). See Figure 11.

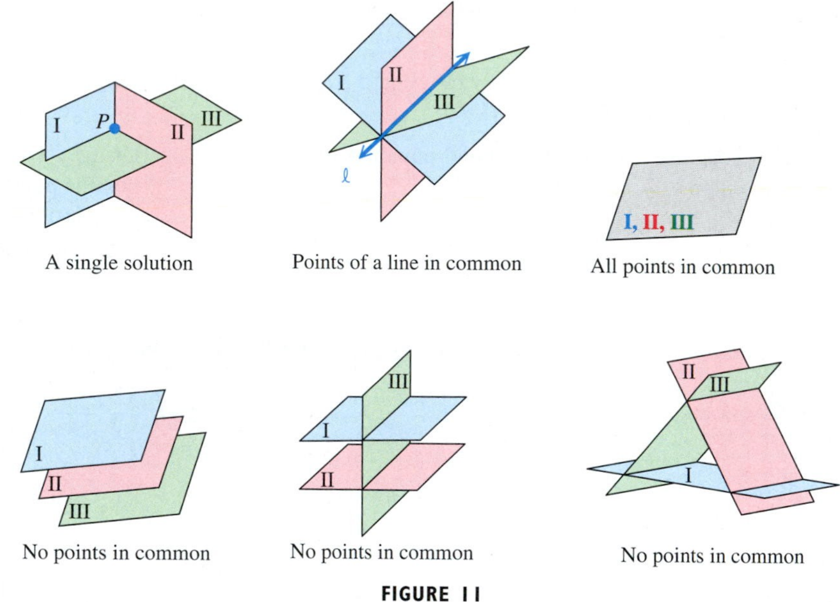

FIGURE 11

Analytic Solution of Systems in Three Variables

To solve a linear system with three unknowns, first eliminate a variable from any two of the equations. Then eliminate the *same variable* from a different pair of equations. Eliminate a second variable using the resulting two equations in two variables to get an equation with just one variable whose value you can now determine. Find the values of the remaining variables by substitution.

EXAMPLE 1 Solving a System of Three Equations in Three Variables

Solve the system.

$$\begin{aligned} 3x + 9y + 6z &= 3 \quad (1) \\ 2x + y - z &= 2 \quad (2) \\ x + y + z &= 2 \quad (3) \end{aligned}$$

Solution Eliminate z by adding equations (2) and (3) to get

$$3x + 2y = 4. \quad (4)$$

To eliminate z from another pair of equations, multiply both sides of equation (2) by 6 and add the result to equation (1).

$$\begin{aligned} 12x + 6y - 6z &= 12 \quad \text{Multiply (2) by 6.} \\ 3x + 9y + 6z &= 3 \quad (1) \\ \hline 15x + 15y \quad &= 15 \quad (5) \end{aligned}$$

To eliminate x from equations (4) and (5), multiply both sides of equation (4) by -5 and add the result to equation (5). Solve the resulting equation for y.

$$\begin{aligned} -15x - 10y &= -20 \quad \text{Multiply (4) by } -5. \\ 15x + 15y &= 15 \quad (5) \\ \hline 5y &= -5 \\ y &= -1 \end{aligned}$$

Using $y = -1$, find x from equation (4) by substitution.

$$\begin{aligned} 3x + 2(-1) &= 4 \quad \text{Let } y = -1 \text{ in (4).} \\ x &= 2 \end{aligned}$$

Substitute 2 for x and -1 for y in equation (3) to find z.

$$\begin{aligned} 2 + (-1) + z &= 2 \quad (3) \\ z &= 1 \end{aligned}$$

Verify that the ordered triple $(2, -1, 1)$ satisfies all three equations in the *original* system. The solution set is $\{(2, -1, 1)\}$. ■

CAUTION Be careful not to end up with two equations that still have *three* variables. ***Remember to eliminate the same variable from each pair of equations.***

EXAMPLE 2 Solving a System of Two Equations in Three Variables

Solve the system.

$$\begin{aligned} x + 2y + z &= 4 \quad (1) \\ 3x - y - 4z &= -9 \quad (2) \end{aligned}$$

Solution Geometrically, the solution is the intersection of the two planes given by equations (1) and (2). The intersection of two different nonparallel planes is a line. Thus, there will be infinitely many ordered triples in the solution set, representing the points on the line of intersection.

To eliminate x, multiply both sides of equation (1) by -3 and add the result to equation (2). (Either y or z could have been eliminated instead.)

$$\begin{aligned} -3x - 6y - 3z &= -12 & \text{Multiply (1) by } -3. \\ 3x - y - 4z &= -9 & (2) \\ \hline -7y - 7z &= -21 & (3) \end{aligned}$$

Now solve equation (3) for z.

$$\begin{aligned} -7y - 7z &= -21 & (3) \\ -7z &= 7y - 21 \\ z &= -y + 3 \end{aligned}$$

This gives z in terms of y. Express x also in terms of y by solving equation (1) for x and substituting $-y + 3$ for z in the result.

$$\begin{aligned} x + 2y + z &= 4 & (1) \\ x &= -2y - z + 4 & \text{Solve for } x. \\ x &= -2y - (-y + 3) + 4 & \text{Substitute } -y + 3 \text{ for } z. \\ x &= -y + 1 & \text{Simplify.} \end{aligned}$$

Be careful with signs.

The system has infinitely many solutions. For any value of y, the value of z is $-y + 3$ and the value of x is $-y + 1$. For example, if $y = 1$, then $x = -1 + 1 = 0$ and $z = -1 + 3 = 2$, giving the solution $(0, 1, 2)$. Verify that another solution is $(-1, 2, 1)$.

With y arbitrary, the solution set is of the form $\{(-y + 1, y, -y + 3)\}$. ■

NOTE Had we solved equation (3) in Example 2 for y instead of z, the solution would have had a different form, but would have led to the same set of solutions. In that case we would have z arbitrary, and the solution set would be of the form $\{(-2 + z, 3 - z, z)\}$. By choosing $z = 2$, one solution would be $(0, 1, 2)$, which was found in Example 2.

A system like the one in Example 2 occurs when two of the equations in a system of three equations in three variables are dependent. In such a case, there are really only two equations in three variables, and Example 2 illustrates the method of solution. On the other hand, an inconsistent system is indicated by a false statement at some point in the solution, as in Example 3 of Section 7.1.

Applications of Systems

EXAMPLE 3 Solving a System to Satisfy Feed Requirements

An animal feed is made from three ingredients: corn, soybeans, and cottonseed. One unit of each ingredient provides units of protein, fat, and fiber, as shown in the table. How many units of each ingredient should be used to make a feed that contains 22 units of protein, 28 units of fat, and 18 units of fiber?

	Corn	Soybeans	Cottonseed	Total
Protein	.25	.4	.2	22
Fat	.4	.2	.3	28
Fiber	.3	.2	.1	18

Solution Let x represent the number of units of corn, y the number of units of soybeans, and z the number of units of cottonseed that are required. Since the total amount of protein must be 22 units, the first row of the table yields

$$.25x + .4y + .2z = 22. \quad (1)$$

Also, for the 28 units of fat,

$$.4x + .2y + .3z = 28, \quad (2)$$

and for the 18 units of fiber,

$$.3x + .2y + .1z = 18. \quad (3)$$

To clear decimals, multiply each side of the first equation by 100 and each side of the second and third equations by 10, to get the system

$$\begin{aligned} 25x + 40y + 20z &= 2200 && (4) \\ 4x + 2y + 3z &= 280 && (5) \\ 3x + 2y + z &= 180. && (6) \end{aligned}$$

Using the method described in Example 1 shows that

$$x = 40, \quad y = 15, \quad \text{and} \quad z = 30.$$

The feed should contain 40 units of corn, 15 units of soybeans, and 30 units of cottonseed to fulfill the given requirements. ■

EXAMPLE 4 Solving a Feed Requirements Application with Fewer Equations than Variables

In Example 3, suppose that only the fat and fiber content of the feed are of interest. How would the solution be changed?

Solution We would need to solve the system

$$\begin{aligned} 4x + 2y + 3z &= 280 \\ 3x + 2y + z &= 180, \end{aligned}$$

which does not have a unique solution, since there are two equations with three variables. Following the procedure outlined in Example 2, we find that the solution set of this system is $\{(100 - 2z, 2.5z - 60, z)\}$. In this applied problem, however, all three variables must be nonnegative, so z must satisfy the conditions

$$100 - 2z \geq 0, \quad 2.5z - 60 \geq 0, \quad z \geq 0.$$

From the first inequality, $z \leq 50$; from the second inequality, $z \geq 24$; thus, $24 \leq z \leq 50$. Only solutions with z in this range are usable. ■

Curve Fitting Using a System

Recall from Chapter 3 that the graph of a quadratic function is a parabola with a vertical axis. Given three noncollinear points, we can find the equation of a parabola that passes through them.

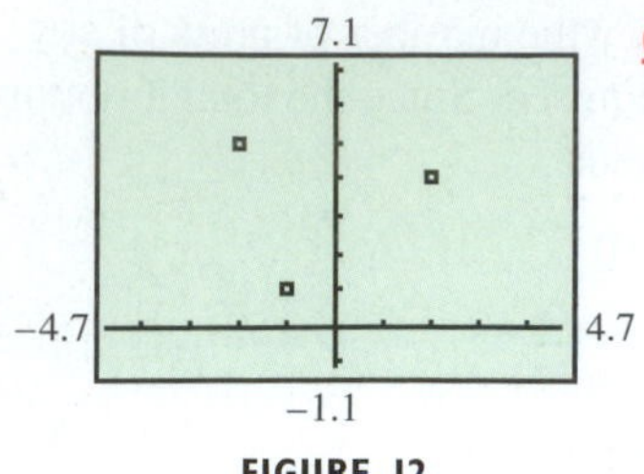

FIGURE 12

GCM EXAMPLE 5 Using a System to Fit a Parabola to Three Data Points

Figure 12 shows three data points: $(2, 4)$, $(-1, 1)$, and $(-2, 5)$. Find the equation of the parabola (with vertical axis) that passes through these points.

Solution Since the three points lie on the graph of the equation $y = ax^2 + bx + c$, they must satisfy the equation. Substituting each ordered pair into the equation gives three equations in three variables.

Use parentheses around substituted values to avoid errors.

$$4 = a(2)^2 + b(2) + c \quad \text{or} \quad 4 = 4a + 2b + c \qquad (1)$$
$$1 = a(-1)^2 + b(-1) + c \quad \text{or} \quad 1 = a - b + c \qquad (2)$$
$$5 = a(-2)^2 + b(-2) + c \quad \text{or} \quad 5 = 4a - 2b + c \qquad (3)$$

To solve this system, first eliminate c, using equations (1) and (2).

$$\begin{aligned} 4 &= 4a + 2b + c && (1) \\ -1 &= -a + b - c && \text{Multiply (2) by } -1. \\ \hline 3 &= 3a + 3b && (4) \end{aligned}$$

Now, use equations (2) and (3) to also eliminate c.

$$\begin{aligned} 1 &= a - b + c && (2) \\ -5 &= -4a + 2b - c && \text{Multiply (3) by } -1. \\ \hline -4 &= -3a + b && (5) \end{aligned}$$

Solve the system of equations (4) and (5) in two variables by eliminating a.

$$\begin{aligned} 3 &= 3a + 3b && (4) \\ -4 &= -3a + b && (5) \\ \hline -1 &= 4b && \text{Add.} \\ -\frac{1}{4} &= b && \text{Solve for } b. \end{aligned}$$

Find a by substituting $-\frac{1}{4}$ for b in transformed equation (4).

$$\begin{aligned} 1 &= a + b && \text{Equation (4) divided by 3} \\ 1 &= a - \frac{1}{4} && \text{Let } b = -\tfrac{1}{4}. \\ \frac{5}{4} &= a && \text{Solve for } a. \end{aligned}$$

Finally, find c by substituting $a = \frac{5}{4}$ and $b = -\frac{1}{4}$ in equation (2).

$$\begin{aligned} 1 &= a - b + c && (2) \\ 1 &= \frac{5}{4} - \left(-\frac{1}{4}\right) + c && \text{Let } a = \tfrac{5}{4},\ b = -\tfrac{1}{4}. \\ 1 &= \frac{6}{4} + c && \text{Simplify.} \\ -\frac{1}{2} &= c && \text{Subtract } \tfrac{6}{4}. \end{aligned}$$

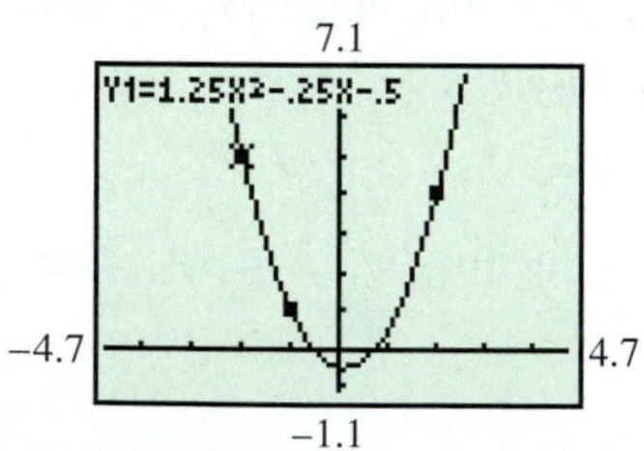

The figure shows that the parabola contains the three data points. A calculator with regression capability can also be used to find the quadratic function that exactly fits these data points.

An equation of the parabola is

$$y = \frac{5}{4}x^2 - \frac{1}{4}x - \frac{1}{2}, \quad \text{or equivalently,} \quad y = 1.25x^2 - .25x - .5.$$ ■

7.2 Exercises

 1. Describe the procedure that would be used to do the following in solving the system

$$\begin{aligned} x + 2y - 3z &= -10 \quad (1)\\ 3x - y + z &= 5 \quad (2)\\ 2x + y - 4z &= -18. \quad (3) \end{aligned}$$

(a) Use equations (1) and (2) to eliminate y by multiplying equation (2) by an appropriate constant.
(b) Use equations (1) and (3) to eliminate x by multiplying equation (1) by an appropriate constant.
(c) Use equations (2) and (3) to eliminate z by multiplying equation (2) by an appropriate constant.

Concept Check *Verify that the given ordered triple is a solution of the system.*

2. $(-3, 6, 1)$
$$\begin{aligned} 2x + y - z &= -1\\ x - y + 3z &= -6\\ -4x + y + z &= 19 \end{aligned}$$

3. $\left(\frac{1}{2}, -\frac{3}{4}, \frac{1}{6}\right)$
$$\begin{aligned} 2x + 8y - 6z &= -6\\ x + y + z &= -\frac{1}{12}\\ x + 3z &= 1 \end{aligned}$$

4. $(-.2, .4, .5)$
$$\begin{aligned} 5x - y + 2z &= -.4\\ x + 4z &= 1.8\\ -3y + z &= -.7 \end{aligned}$$

Solve each system analytically. (Hint: *In Exercises 25–28, let* $t = \frac{1}{x}$, $u = \frac{1}{y}$, *and* $v = \frac{1}{z}$. *Solve for t, u, and v, and then solve for x, y, and z.*)

5. $\begin{aligned} x + y + z &= 2\\ 2x + y - z &= 5\\ x - y + z &= -2 \end{aligned}$

6. $\begin{aligned} 2x + y + z &= 9\\ -x - y + z &= 1\\ 3x - y + z &= 9 \end{aligned}$

7. $\begin{aligned} x + 3y + 4z &= 14\\ 2x - 3y + 2z &= 10\\ 3x - y + z &= 9 \end{aligned}$

8. $\begin{aligned} 4x - y + 3z &= -2\\ 3x + 5y - z &= 15\\ -2x + y + 4z &= 14 \end{aligned}$

9. $\begin{aligned} x + 2y + 3z &= 8\\ 3x - y + 2z &= 5\\ -2x - 4y - 6z &= 5 \end{aligned}$

10. $\begin{aligned} 3x - 2y - 8z &= 1\\ 9x - 6y - 24z &= -2\\ x - y + z &= 1 \end{aligned}$

11. $\begin{aligned} x + 4y - z &= 6\\ 2x - y + z &= 3\\ 3x + 2y + 3z &= 16 \end{aligned}$

12. $\begin{aligned} 4x - 3y + z &= 9\\ 3x + 2y - 2z &= 4\\ x - y + 3z &= 5 \end{aligned}$

13. $\begin{aligned} 5x + y - 3z &= -6\\ 2x + 3y + z &= 5\\ -3x - 2y + 4z &= 3 \end{aligned}$

14. $\begin{aligned} 2x - 5y + 4z &= -35\\ 5x + 3y - z &= 1\\ x + y + z &= 1 \end{aligned}$

15. $\begin{aligned} x - 3y - 2z &= -3\\ 3x + 2y - z &= 12\\ -x - y + 4z &= 3 \end{aligned}$

16. $\begin{aligned} x + y + z &= 3\\ 3x - 3y - 4z &= -1\\ x + y + 3z &= 11 \end{aligned}$

17. $\begin{aligned} 2x + 6y - z &= 6\\ 4x - 3y + 5z &= -5\\ 6x + 9y - 2z &= 11 \end{aligned}$

18. $\begin{aligned} 8x - 3y + 6z &= -2\\ 4x + 9y + 4z &= 18\\ 12x - 3y + 8z &= -2 \end{aligned}$

19. $\begin{aligned} x + z &= 4\\ x + y &= 4\\ y + z &= 4 \end{aligned}$

20. $\begin{aligned} x - z &= 2\\ x + y &= -3\\ y - z &= 1 \end{aligned}$

21. $\begin{aligned} 2x + y - z &= -4\\ y + 2z &= 12\\ 2x - z &= -4 \end{aligned}$

22. $\begin{aligned} 3x + 2y - z &= -1\\ 3y + z &= 12\\ x - 3z &= -3 \end{aligned}$

23. $\begin{aligned} 2x + 3y + 4z &= 3\\ 6x + 3y + 8z &= 6\\ 6y - 4z &= 1 \end{aligned}$

24. $\begin{aligned} 10x + 2y - 3z &= 0\\ 5x + 4y + 6z &= -1\\ 6y + 3z &= 2 \end{aligned}$

25. $\begin{aligned} \frac{1}{x} + \frac{1}{y} - \frac{1}{z} &= \frac{1}{4}\\ \frac{2}{x} - \frac{1}{y} + \frac{3}{z} &= \frac{9}{4}\\ -\frac{1}{x} - \frac{2}{y} + \frac{4}{z} &= 1 \end{aligned}$

26. $\dfrac{3}{x} + \dfrac{2}{y} - \dfrac{1}{z} = \dfrac{11}{6}$
$\dfrac{1}{x} - \dfrac{1}{y} + \dfrac{3}{z} = -\dfrac{11}{12}$
$\dfrac{2}{x} + \dfrac{1}{y} + \dfrac{1}{z} = \dfrac{7}{12}$

27. $\dfrac{2}{x} - \dfrac{2}{y} + \dfrac{1}{z} = -1$
$\dfrac{4}{x} + \dfrac{1}{y} - \dfrac{2}{z} = -9$
$\dfrac{1}{x} + \dfrac{1}{y} - \dfrac{3}{z} = -9$

28. $\dfrac{5}{x} - \dfrac{1}{y} - \dfrac{2}{z} = -6$
$-\dfrac{1}{x} + \dfrac{3}{y} - \dfrac{3}{z} = -12$
$\dfrac{2}{x} - \dfrac{1}{y} - \dfrac{1}{z} = 6$

29. $x - 4y + 2z = -2$
$x + 2y - 2z = -3$
$x - y = 4$

30. $2x + y + 3z = 4$
$-3x - y - 4z = 5$
$x + y + 2z = 0$

31. $x + y + z = 0$
$x - y - z = 3$
$x + 3y + 3z = 5$

32. *Concept Check* Consider the linear equation in three variables $x + y + z = 4$. Find a pair of linear equations that, when considered together with the given equation, will form a system having **(a)** exactly one solution, **(b)** no solution, **(c)** infinitely many solutions.

33. *Concept Check* Refer to Example 2 in this section. Write the solution set with x arbitrary.

34. Using your immediate surroundings, give an example of three planes that intersect in a single point.

35. Using your immediate surroundings, give an example of three planes that intersect in a line.

Solve each system in terms of the arbitrary variable z.

36. $x - 2y + 3z = 6$
$2x - y + 2z = 5$

37. $3x + 4y - z = 13$
$x + y + 2z = 15$

38. $5x - 4y + z = 9$
$x + y = 15$

39. $3x - 5y - 4z = -7$
$y - z = -13$

40. $x - y + z = -6$
$4x + y + z = 7$

41. $3x - 2y + z = 15$
$x + 4y - z = 11$

42. $x + 3y + z = 6$
$3x + y - z = 6$
$x - y - z = 0$

43. $2x - y + 2z = 6$
$-x + y + z = 0$
$-x - 3z = -6$

44. $x + 2y + z = 0$
$3x + 2y - z = 4$
$-x + 2y + 3z = -4$

(Modeling) Solve each problem.

45. *Mixing Waters* A sparkling-water distributor wants to make up 300 gallons of sparkling water to sell for \$6.00 per gallon. She wishes to mix three grades of water selling for \$9.00, \$3.00, and \$4.50 per gallon, respectively. She must use twice as much of the \$4.50 water as the \$3.00 water. How many gallons of each should she use?

46. *Coin Collecting* A coin collection made up of pennies, nickels, and quarters contains a total of 29 coins. The number of quarters is 8 less than the number of pennies. The total face value of the coins is \$1.77. How many of each denomination are there?

47. *Pricing Concert Tickets* Frank Capek and his Generation Gap group sells three kinds of concert tickets: "up close," "middle," and "farther back." "Up close" tickets cost \$6 more than "middle" tickets, while "middle" tickets cost \$3 more than "farther back" tickets. Twice the cost of an "up close" ticket is \$3 more than 3 times the cost of a "farther back" seat. Find the price of each kind of ticket.

48. *Mixing Glue* A glue company needs to make some glue that it can sell for \$120 per barrel. It wants to use 150 barrels of glue worth \$100 per barrel, along with some glue worth \$150 per barrel and glue worth \$190 per barrel. It must use the same number of barrels of \$150 and \$190 glue. How much of the \$150 and \$190 glue will be needed? How many barrels of \$120 glue will be produced?

49. *Triangle Dimensions* The sum of the measures of the angles of any triangle is 180°. In a certain triangle, the largest angle measures 55° less than twice the medium angle, and the smallest measures 25° less than the medium angle. Find the measures of each of the three angles.

50. *Triangle Dimensions* The perimeter of a triangle is 59 inches. The longest side is 11 inches longer than the medium side, and the medium side is 3 inches more than the shortest side. Find the length of each side.

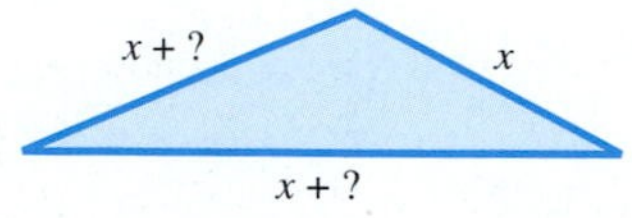

51. *Investment Decisions* Norma James invested \$10,000 received in an inheritance in three parts. With one part, she bought mutual funds that offered a return of 4% per year. The second part, which amounted to twice the first, was used to buy government bonds paying 4.5% per year. She put the rest into a savings account that paid 2.5% annual interest. During the first year, the total interest was \$415. How much did Norma invest at each rate?

52. *Investment Decisions* Tom Accardo won \$100,000 in the Louisiana state lottery. He invested part of the money in real estate with an annual return of 5% and another part in a money market account at 4.5% interest. He invested the rest, which amounted to \$20,000 less than the sum of the other two parts, in certificates of deposit that pay 3.75%. If the total annual interest on the money was \$4450, how much was invested at each rate?

53. *Scheduling Deliveries* Doctor Rug sells rug-cleaning machines. The EZ model weighs 10 pounds and comes in a 10-cubic-foot box. The compact model weighs 20 pounds and comes in an 8-cubic-foot box. The commercial model weighs 60 pounds and comes in a 28-cubic-foot box. Each of the company's delivery vans has 248 cubic feet of space and can hold a maximum of 440 pounds. In order for a van to be fully loaded, how many of each model should it carry?

54. *Scheduling Production* Ciolino's makes dining room furniture. A buffet requires 30 hours for construction and 10 hours for finishing, a chair 10 hours for construction and 10 hours for finishing, and a table 10 hours for construction and 30 hours for finishing. The construction department has 350 hours of labor and the finishing department has 150 hours of labor available each week. How many pieces of each type of furniture should be produced each week if the factory is to run at full capacity?

55. *Predicting Home Prices* The table shows the selling prices for three homes. Price P is given in thousands of dollars, age A in years, and home size S in thousands of square feet. These data may be modeled by the equation $P = a + bA + cS$.

Price (P)	Age (A)	Size (S)
190	20	2
320	5	3
50	40	1

(a) Write a system of linear equations whose solution gives a, b, and c.
(b) Solve this system of linear equations.
(c) Predict the price of a home that is 10 years old and has 2500 square feet.

Curve Fitting *Find the equation of the parabola (with vertical axis) that passes through the data points specified. Then support your answer with a graph.*

56. $(2, 9)$, $(-2, 1)$, $(-3, 4)$

57. $(1.5, 6.25)$, $(0, -2)$, $(-1.5, 3.25)$

58. $(2, 14)$, $(0, 0)$, $(-1, -1)$

59. *Curve Fitting* The table was generated using a function defined by $Y_1 = aX^2 + bX + c$. Use any three points from the table to find the equation that defines the function.

X	Y1
-3	5.48
-2	2.9
-1	1.26
0	.56
1	.8
2	1.98
3	4.1

X=-3

Curve Fitting *Given three noncollinear points, there is one and only one circle that passes through them. Knowing that the equation of a circle may be written in the form*

$$x^2 + y^2 + ax + by + c = 0,$$

find the equation of the circle passing through the given points.

60. $(-1, 3)$, $(6, 2)$, and $(-2, -4)$

61. $(-1, 5)$, $(6, 6)$, and $(7, -1)$

62. $(2, 1)$, $(-1, 0)$, and $(3, 3)$

63.

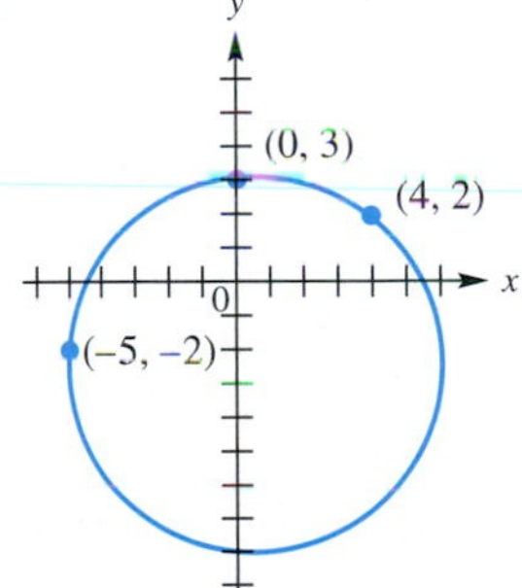

(Modeling) Position of a Particle *Suppose that the position of a particle moving along a straight line is given by*

$$s(t) = at^2 + bt + c,$$

where t is time in seconds and a, b, and c are real numbers.

64. If $s(0) = 5$, $s(1) = 23$, and $s(2) = 37$, find the equation that defines $s(t)$. Then find $s(8)$.

65. If $s(0) = -10$, $s(1) = 6$, and $s(2) = 30$, find the equation that defines $s(t)$. Then find $s(10)$.

7.3 Solution of Linear Systems by Row Transformations

Matrix Row Transformations ■ Row Echelon Method ■ Reduced Row Echelon Method ■ Special Cases ■ An Application of Matrices

Matrix Row Transformations

The concepts used to solve linear systems of equations in the previous sections can be streamlined by using *matrices* (singular: *matrix*). Consider a system of three equations and three unknowns.

$$\begin{aligned} a_1x + b_1y + c_1z &= d_1 \\ a_2x + b_2y + c_2z &= d_2 \\ a_3x + b_3y + c_3z &= d_3 \end{aligned} \quad \text{can be written as} \quad \begin{bmatrix} a_1 & b_1 & c_1 & d_1 \\ a_2 & b_2 & c_2 & d_2 \\ a_3 & b_3 & c_3 & d_3 \end{bmatrix}. \quad \text{Matrix}$$

GCM TECHNOLOGY NOTE

The manner in which matrices are entered and displayed in a calculator varies greatly among different manufacturers and even among the various models manufactured by the same corporation.

Such a rectangular array of numbers enclosed by brackets is called a **matrix.** Each number in the array is an **element** or **entry.** The constants in the last column of the matrix can be set apart from the coefficients of the variables with a vertical line, as shown in the following **augmented matrix.** (Because the matrix of coefficients has an extra column determined by the constants of the system, the coefficient matrix is *augmented.*)

$$\text{Rows} \rightarrow \left[\begin{array}{ccc|c} a_1 & b_1 & c_1 & d_1 \\ a_2 & b_2 & c_2 & d_2 \\ a_3 & b_3 & c_3 & d_3 \end{array}\right] \quad \text{Augmented matrix}$$

Columns

As an example, the system on the left has the augmented matrix shown on the right.

$$\begin{aligned} x + 3y + 2z &= 1 \\ 2x + y - z &= 2 \\ x + y + z &= 2 \end{aligned} \qquad \left[\begin{array}{ccc|c} 1 & 3 & 2 & 1 \\ 2 & 1 & -1 & 2 \\ 1 & 1 & 1 & 2 \end{array}\right] \quad \text{Augmented matrix}$$

Figure 13 shows how this matrix can be entered into a TI-83/84 Plus calculator. The matrix has 3 rows (horizontal) and 4 columns (vertical), so it is a 3×4 (read "three by four") matrix. To refer to a number in the matrix, use its row and column numbers. For example, the number 3 is in the first row, second column.

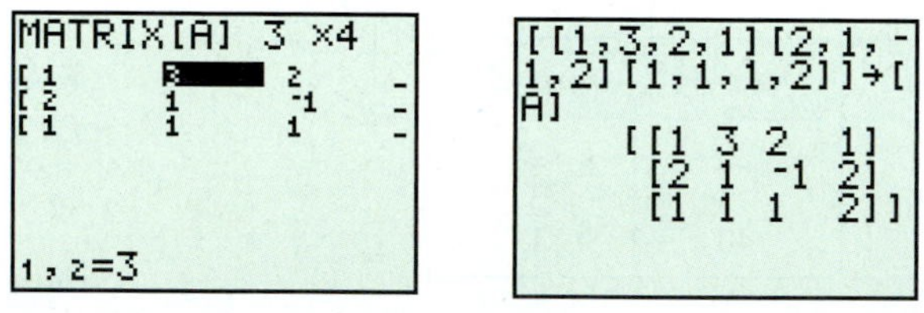

FIGURE 13

The rows of an augmented matrix can be treated just like the equations of the corresponding system of linear equations. Since the augmented matrix is nothing more than a short form of the system, any transformation of the matrix that results in an equivalent system of equations can be performed.

GCM TECHNOLOGY NOTE

Consult the graphing calculator manual that accompanies this text or your owner's manual to see how these transformations are accomplished with your model.

Matrix Row Transformations

For any augmented matrix of a system of linear equations, the following row transformations will result in the matrix of an equivalent system:

1. Any two rows may be interchanged.
2. The elements of any row may be multiplied by a nonzero real number.
3. Any row may be changed by adding to its elements a multiple of the corresponding elements of another row.

EXAMPLE 1 Using Row Transformations

Use matrix row transformations to transform

$$\left[\begin{array}{rrr|r} 1 & 3 & 5 & 2 \\ 0 & 1 & 2 & 3 \\ 1 & -1 & -2 & 4 \end{array}\right]$$

to each matrix.

(a) $\left[\begin{array}{rrr|r} 0 & 1 & 2 & 3 \\ 1 & 3 & 5 & 2 \\ 1 & -1 & -2 & 4 \end{array}\right]$ **(b)** $\left[\begin{array}{rrr|r} -2 & -6 & -10 & -4 \\ 0 & 1 & 2 & 3 \\ 1 & -1 & -2 & 4 \end{array}\right]$ **(c)** $\left[\begin{array}{rrr|r} 1 & 3 & 5 & 2 \\ 0 & 1 & 2 & 3 \\ 0 & -4 & -7 & 2 \end{array}\right]$

Analytic Solution

(a) Interchange the first two rows in the given matrix.

$$\left[\begin{array}{rrr|r} 0 & 1 & 2 & 3 \\ 1 & 3 & 5 & 2 \\ 1 & -1 & -2 & 4 \end{array}\right] \begin{array}{l} R_2 \rightarrow R_1 \\ R_1 \rightarrow R_2 \\ R_3 \text{ is unchanged.} \end{array}$$

(b) Multiply the elements of the first row by -2.

$$\left[\begin{array}{rrr|r} -2 & -6 & -10 & -4 \\ 0 & 1 & 2 & 3 \\ 1 & -1 & -2 & 4 \end{array}\right] \begin{array}{l} -2R_1 \\ R_2 \text{ and } R_3 \text{ are} \\ \text{unchanged.} \end{array}$$

(c) Multiply each element of the first row by -1, and add it to the corresponding element in the third row. For example, to obtain 0 in row 3, column 1, multiply 1 by -1 and add to 1: $1(-1) + 1 = 0$. The final result is

$$\left[\begin{array}{rrr|r} 1 & 3 & 5 & 2 \\ 0 & 1 & 2 & 3 \\ 0 & -4 & -7 & 2 \end{array}\right]. \begin{array}{l} R_1 \text{ and } R_2 \text{ are} \\ \text{unchanged.} \\ -1R_1 + R_3 \end{array}$$

Graphing Calculator Solution

Figure 14 shows the given matrix, designated [A]. The TI-83/84 Plus uses the commands indicated in Figure 15 to perform the transformations.

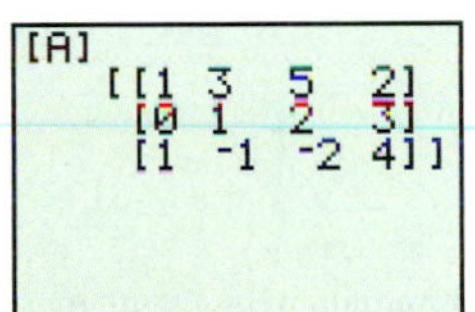

FIGURE 14

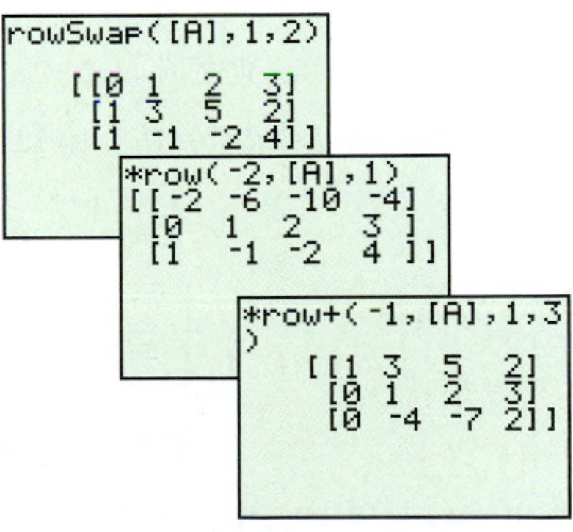

FIGURE 15

$$\left[\begin{array}{ccc|c} 1 & 5 & 3 & 7 \\ 0 & 1 & 2 & 9 \\ 0 & 0 & 1 & 4 \end{array}\right]$$

Echelon (triangular) form of an augmented matrix

Row Echelon Method

Matrix row transformations are used to transform the augmented matrix of a system into one that is in **echelon** (or **triangular**) **form.** In this form, the matrix will have 1s down the diagonal from upper left to lower right and 0s below each 1, as shown in the matrix in the margin. The 1s lie on the **main diagonal** of the matrix. Once a system of linear equations is in echelon form, **back-substitution** can be used to find the solution set. The **row echelon method** uses matrices to solve a system of linear equations, as shown in Examples 2 and 3.

GCM **EXAMPLE 2** **Solving a System by the Row Echelon Method**

Solve the system.

$$\begin{aligned} 5x + 2y &= 1 \\ 2x - y &= 4 \end{aligned}$$

Analytic Solution

$$\left[\begin{array}{cc|c} 5 & 2 & 1 \\ 2 & -1 & 4 \end{array}\right] \quad \text{Augmented matrix}$$

Transform the first row so that the first entry is 1. To do this, multiply row 1 by $\frac{1}{5}$.

$$\left[\begin{array}{cc|c} 1 & \frac{2}{5} & \frac{1}{5} \\ 2 & -1 & 4 \end{array}\right] \quad \tfrac{1}{5}R_1$$

Transform so that the entry below the main diagonal (that is, 2) is 0. Multiply row 1 by -2 and add to row 2.

$$\left[\begin{array}{cc|c} 1 & \frac{2}{5} & \frac{1}{5} \\ 0 & -\frac{9}{5} & \frac{18}{5} \end{array}\right] \quad \begin{array}{l} R_1 \text{ is unchanged.} \\ -2R_1 + R_2 \end{array}$$

Multiply row 2 by $-\frac{5}{9}$ to get 1 on the main diagonal.

$$\left[\begin{array}{cc|c} 1 & \frac{2}{5} & \frac{1}{5} \\ 0 & 1 & -2 \end{array}\right] \quad -\tfrac{5}{9}R_2$$

This matrix represents the system

$$x + \frac{2}{5}y = \frac{1}{5}$$
$$y = -2.$$

Since $y = -2$, use *back-substitution* to find x.

$$x + \frac{2}{5}(-2) = \frac{1}{5} \quad \text{Let } y = -2.$$

$$x - \frac{4}{5} = \frac{1}{5} \quad \text{Multiply.}$$

$$x = 1 \quad \text{Add } \tfrac{4}{5}.$$

Write the x-value first in the ordered pair.

The solution set is $\{(1, -2)\}$.

Graphing Calculator Solution

Figure 16 shows matrix [A].

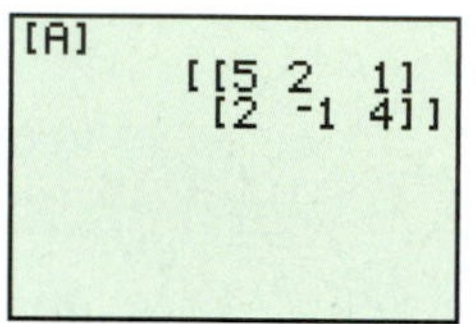

FIGURE 16

Using the ref(command on the TI-83/84 Plus, we can find the row echelon form. See Figure 17.

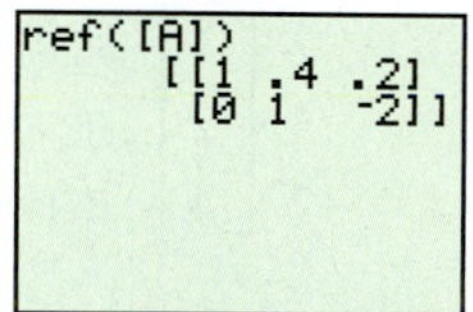

FIGURE 17

The decimal entries can be converted to fractions, as shown in Figure 18.

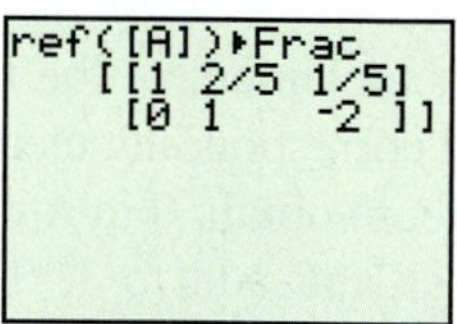

FIGURE 18

The augmented matrix in Figure 18 corresponds to the one in color in the analytic solution. The rest of the solution process is the same.

The row echelon method can be extended to larger systems. The final matrix will always have 0s below the diagonal of 1s to the left of the vertical bar. To transform the matrix, work column by column from upper left to lower right. For each column, first perform the step that gives the 1 on the main diagonal. Then obtain 0s below the 1 in that column.

EXAMPLE 3 Solving a System by the Row Echelon Method (Analytic)

Solve the system.

$$\begin{aligned} x - y + 5z &= -6 \\ 3x + 3y - z &= 10 \\ x + 3y + 2z &= 5 \end{aligned}$$

Solution

$$\left[\begin{array}{ccc|c} 1 & -1 & 5 & -6 \\ 3 & 3 & -1 & 10 \\ 1 & 3 & 2 & 5 \end{array}\right] \quad \text{Augmented matrix}$$

There is already a 1 in row 1, column 1. Next, we must obtain 0s in the rest of column 1. First, add to row 2 the results of multiplying row 1 by -3.

$$\left[\begin{array}{ccc|c} 1 & -1 & 5 & -6 \\ 0 & 6 & -16 & 28 \\ 1 & 3 & 2 & 5 \end{array}\right] \quad -3R_1 + R_2$$

Continuing this process, we transform the matrix into row echelon form as follows:

$$\left[\begin{array}{ccc|c} 1 & -1 & 5 & -6 \\ 0 & 6 & -16 & 28 \\ 0 & 4 & -3 & 11 \end{array}\right] \quad -1R_1 + R_3$$

$$\left[\begin{array}{ccc|c} 1 & -1 & 5 & -6 \\ 0 & 1 & -\frac{8}{3} & \frac{14}{3} \\ 0 & 4 & -3 & 11 \end{array}\right] \quad \tfrac{1}{6}R_2$$

$$\left[\begin{array}{ccc|c} 1 & -1 & 5 & -6 \\ 0 & 1 & -\frac{8}{3} & \frac{14}{3} \\ 0 & 0 & \frac{23}{3} & -\frac{23}{3} \end{array}\right] \quad -4R_2 + R_3$$

$$\left[\begin{array}{ccc|c} 1 & -1 & 5 & -6 \\ 0 & 1 & -\frac{8}{3} & \frac{14}{3} \\ 0 & 0 & 1 & -1 \end{array}\right]. \quad \begin{array}{l}\text{Row echelon form} \\ \tfrac{3}{23}R_3\end{array}$$

All integer elements can be obtained by multiplying row 2 by 3. However, the matrix is no longer in row echelon form.

$$\left[\begin{array}{ccc|c} 1 & -1 & 5 & -6 \\ 0 & 3 & -8 & 14 \\ 0 & 0 & 1 & -1 \end{array}\right] \quad 3R_2$$

$$\left[\begin{array}{rrr|r} 1 & -1 & 5 & -6 \\ 0 & 3 & -8 & 14 \\ 0 & 0 & 1 & -1 \end{array}\right]$$

The final matrix, repeated in the margin, corresponds to the system of equations

$$\begin{aligned} x - y + 5z &= -6 \quad (1) \\ 3y - 8z &= 14 \quad (2) \\ z &= -1. \quad (3) \end{aligned}$$

Since $z = -1$ from equation (3), use back-substitution into equation (2) to find that $y = 2$. Back-substitute $y = 2$ and $z = -1$ into equation (1) to find that $x = 1$. The solution set is $\{(1, 2, -1)\}$. ■

EXAMPLE 4 Solving a System by the Row Echelon Method (Graphical)

Solve the system of Example 3 using the row reduction capability of a graphing calculator.

$$\begin{aligned} x - y + 5z &= -6 \\ 3x + 3y - z &= 10 \\ x + 3y + 2z &= 5 \end{aligned}$$

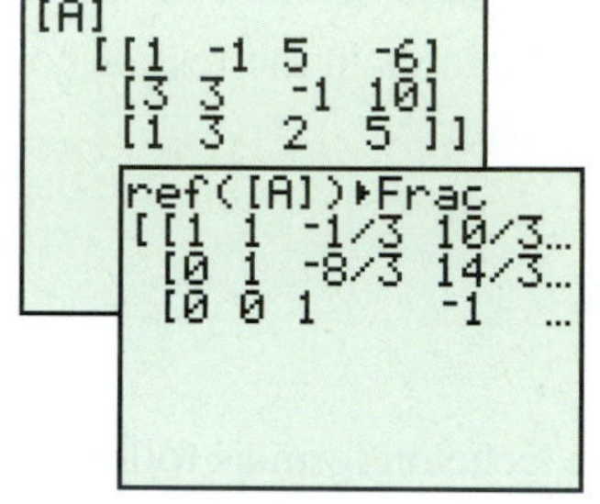

FIGURE 19

Solution Enter the augmented matrix of the system as matrix [A]. See the top screen in Figure 19. Then use the ref(command to obtain the row echelon form. The matrix in the bottom screen indicates that the row echelon form is

$$\begin{aligned} x + y - \frac{1}{3}z &= \frac{10}{3} \\ y - \frac{8}{3}z &= \frac{14}{3} \\ z &= -1. \end{aligned}$$

Using back-substitution, we find that the solution set is $\{(1, 2, -1)\}$. ■

NOTE In Example 4, the entries in row 1, columns 2, 3, and 4, are different from the corresponding entries in the row echelon form shown in Example 3, because the steps were performed in an alternative way. However, the solution set is the same.

Reduced Row Echelon Method

Another matrix method for solving systems is the **reduced row echelon method.** Earlier, we saw that the row echelon form of a matrix has 1s along the main diagonal and 0s below. The reduced row echelon form has 1s along the main diagonal and 0s both below *and above*. For example, the augmented matrix of the system

$$\begin{aligned} x + y + z &= 6 \\ 2x - y + z &= 5 \\ 3x + y - 2z &= 9 \end{aligned} \quad \text{is} \quad \left[\begin{array}{rrr|r} 1 & 1 & 1 & 6 \\ 2 & -1 & 1 & 5 \\ 3 & 1 & -2 & 9 \end{array}\right].$$

By using row transformations, this augmented matrix can be transformed to

$$\left[\begin{array}{rrr|r} 1 & 0 & 0 & 3 \\ 0 & 1 & 0 & 2 \\ 0 & 0 & 1 & 1 \end{array}\right], \quad \text{which represents the system} \quad \begin{aligned} x &= 3 \\ y &= 2 \\ z &= 1. \end{aligned}$$

The solution set is $\{(3, 2, 1)\}$.

GCM **EXAMPLE 5** **Solving a System by the Reduced Row Echelon Method**

Use a graphing calculator with reduced row echelon capability to solve the system.

$$\begin{aligned} x + 2y + z &= 30 \\ 3x + 2y + 2z &= 56 \\ 2x + 3y + z &= 44 \end{aligned}$$

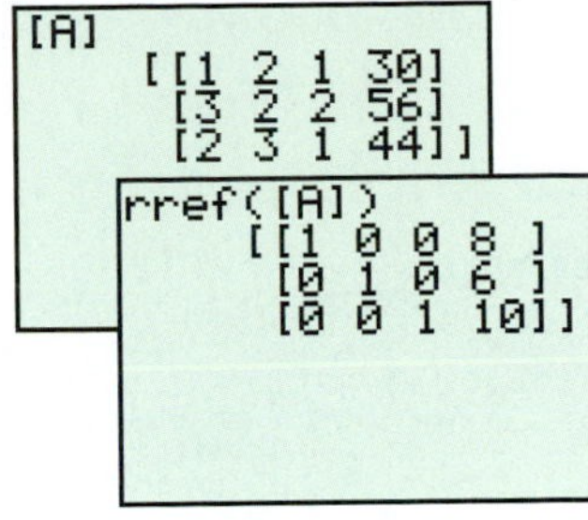

FIGURE 20

Solution The top screen in Figure 20 shows the augmented matrix of the system, and the bottom screen shows its reduced row echelon form, which represents the system

$$\begin{aligned} x &= 8 \\ y &= 6 \\ z &= 10, \end{aligned}$$

and has solution set $\{(8, 6, 10)\}$. ■

Special Cases

Whenever a row of the augmented matrix is of the form

$$0 \quad 0 \quad 0 \quad \ldots \quad |a, \quad \text{where } a \neq 0,$$

the system is inconsistent and there will be no solution, since this row corresponds to the equation $0 = a$. A row of the matrix of a linear system in the form

$$0 \quad 0 \quad 0 \quad \ldots \quad |0$$

indicates that the equations of the system are dependent.

EXAMPLE 6 **Solving a Special System (No Solution)**

Solve the system.

$$\begin{aligned} x + 2y + z &= 4 \\ x - 2y + 3z &= 1 \\ 2x + 4y + 2z &= 9 \end{aligned}$$

Analytic Solution

$$\left[\begin{array}{ccc|c} 1 & 2 & 1 & 4 \\ 1 & -2 & 3 & 1 \\ 2 & 4 & 2 & 9 \end{array}\right] \quad \text{Augmented matrix}$$

$$\left[\begin{array}{ccc|c} 1 & 2 & 1 & 4 \\ 1 & -2 & 3 & 1 \\ 0 & 0 & 0 & 1 \end{array}\right] \quad -2R_1 + R_3$$

The final row indicates that the system is inconsistent and has solution set $\emptyset$.

Graphing Calculator Solution

Augmented matrix [A] and the row echelon form are shown in Figure 21. The final row indicates an inconsistent system with solution set $\emptyset$. A similar conclusion can be made if the rref(command is used.

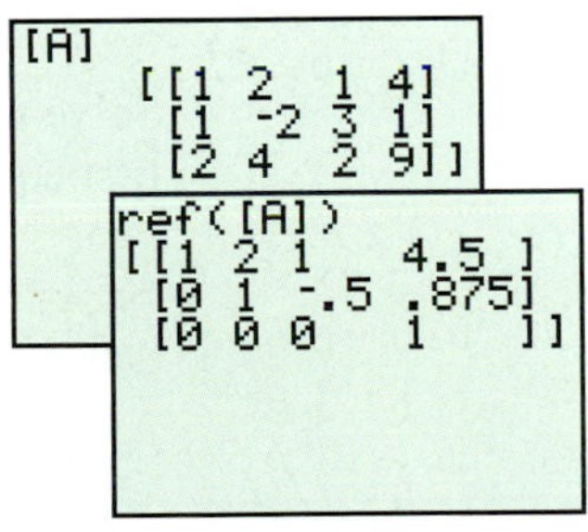

FIGURE 21

■

EXAMPLE 7 **Solving a Special System (Dependent Equations)**

Solve the system.

$$\begin{aligned} 2x - 5y + 3z &= 1 \\ x + 4y - 2z &= 8 \\ 4x - 10y + 6z &= 2 \end{aligned}$$

Solution

$$\left[\begin{array}{rrr|r} 2 & -5 & 3 & 1 \\ 1 & 4 & -2 & 8 \\ 4 & -10 & 6 & 2 \end{array}\right] \quad \text{Augmented matrix}$$

$$\left[\begin{array}{rrr|r} 2 & -5 & 3 & 1 \\ 1 & 4 & -2 & 8 \\ 0 & 0 & 0 & 0 \end{array}\right] \quad -2R_1 + R_3$$

The final row of 0s indicates that the system has dependent equations. The first two rows represent the system

$$\begin{aligned} 2x - 5y + 3z &= 1 \\ x + 4y - 2z &= 8. \end{aligned}$$

Recall from Section 7.2 that a system with two equations and three variables may have infinitely many solutions.

$$\left[\begin{array}{rrr|r} 2 & -5 & 3 & 1 \\ 1 & 4 & -2 & 8 \end{array}\right] \quad \text{Augmented matrix}$$

$$\left[\begin{array}{rrr|r} 1 & 4 & -2 & 8 \\ 2 & -5 & 3 & 1 \end{array}\right] \quad \begin{array}{l} R_2 \rightarrow R_1 \\ R_1 \rightarrow R_2 \end{array}$$

$$\left[\begin{array}{rrr|r} 1 & 4 & -2 & 8 \\ 0 & -13 & 7 & -15 \end{array}\right] \quad -2R_1 + R_2$$

$$\left[\begin{array}{rrr|r} 1 & 4 & -2 & 8 \\ 0 & 1 & -\frac{7}{13} & \frac{15}{13} \end{array}\right] \quad -\tfrac{1}{13}R_2$$

This is as far as we go with the row echelon method. The equations that correspond to the final matrix are

$$x + 4y - 2z = 8 \quad \text{and} \quad y - \frac{7}{13}z = \frac{15}{13}.$$

Solve the second equation for y: $y = \frac{15}{13} + \frac{7}{13}z$. Now substitute this result for y in the first equation and solve for x.

$$\begin{aligned} x + 4y - 2z &= 8 \\ x + 4\left(\frac{15}{13} + \frac{7}{13}z\right) - 2z &= 8 \qquad \text{Let } y = \tfrac{15}{13} + \tfrac{7}{13}z. \\ x + \frac{60}{13} + \frac{28}{13}z - 2z &= 8 \qquad \text{Distributive property} \end{aligned}$$

$$x + \frac{60}{13} + \frac{2}{13}z = 8 \qquad \frac{28}{13}z - 2z = \frac{2}{13}z$$

$$x = 8 - \frac{60}{13} - \frac{2}{13}z \qquad \text{Solve for } x.$$

$$x = \frac{44}{13} - \frac{2}{13}z \qquad \text{Simplify.}$$

The solution set written with z arbitrary is

$$\left\{\left(\frac{44 - 2z}{13}, \frac{15 + 7z}{13}, z\right)\right\}.$$

An Application of Matrices

EXAMPLE 8 Determining a Model Using Given Data

Three food shelters are operated by a charitable organization. Three different quantities are computed: monthly food costs F in dollars, number N of people served per month, and monthly charitable receipts R in dollars. The data are shown in the table.

Food Costs (F)	Number Served (N)	Charitable Receipts (R)
3000	2400	8000
4000	2600	10,000
8000	5900	14,000

Source: Sanders, D., *Statistics: A First Course*, McGraw-Hill, 1995.

(a) Model the data using the equation $F = aN + bR + c$, where a, b, and c are constants.

(b) Predict the food costs for a shelter that serves 4000 people and receives charitable receipts of \$12,000. Round your answer to the nearest hundred dollars.

Solution

(a) Since $F = aN + bR + c$, the constants a, b, and c satisfy the following three equations:

$$\begin{aligned} 3000 &= a(2400) + b(8000) + c \\ 4000 &= a(2600) + b(10{,}000) + c \\ 8000 &= a(5900) + b(14{,}000) + c. \end{aligned}$$

We can rewrite this system as

$$\begin{aligned} 2400a + 8000b + c &= 3000 \\ 2600a + 10{,}000b + c &= 4000 \\ 5900a + 14{,}000b + c &= 8000 \end{aligned} \quad \text{which has augmented matrix} \quad \left[\begin{array}{ccc|c} 2400 & 8000 & 1 & 3000 \\ 2600 & 10{,}000 & 1 & 4000 \\ 5900 & 14{,}000 & 1 & 8000 \end{array}\right].$$

Using the reduced row echelon method (see the "For Discussion" box), we find that $a \approx .6897$, $b \approx .4310$, and $c \approx -2103$; thus,

$$F = .6897N + .4310R - 2103.$$

FOR DISCUSSION

The three screens shown here depict matrix [A] from Example 8, the row echelon form of [A], and the reduced row echelon form of [A].

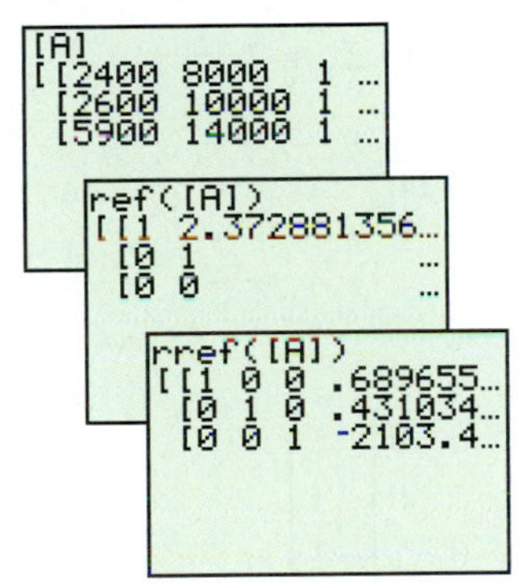

1. Use your calculator to re-create the screens shown here. (Note that only a portion of the matrix is visible.)

2. Which one of the two bottom screens provides an easier interpretation of the solution of the system? Why?

(b) To predict food costs for a shelter that serves 4000 people and receives charitable receipts of \$12,000, let $N = 4000$ and $R = 12{,}000$ and evaluate F.

$$F = .6897N + .4310R - 2103$$
$$F = .6897(4000) + .4310(12{,}000) - 2103 = 5827.8$$

This model predicts monthly food costs of approximately \$5800. ■

7.3 Exercises

Concept Check *Use the given row transformation to transform each matrix.*

1. $\begin{bmatrix} 2 & 4 \\ 4 & 7 \end{bmatrix}$ $-2R_1$

2. $\begin{bmatrix} -1 & 4 \\ 7 & 0 \end{bmatrix}$ $7R_1$

3. $\begin{bmatrix} 1 & 5 & 6 \\ -2 & 3 & -1 \\ 4 & 7 & 0 \end{bmatrix}$ $R_1 + R_2$

4. $\begin{bmatrix} 2 & 5 & 6 \\ 4 & -1 & 2 \\ 3 & 7 & 1 \end{bmatrix}$ $R_3 + R_1$

5. $\begin{bmatrix} -3 & 1 & -4 \\ 2 & 1 & 3 \\ -7 & 5 & 2 \end{bmatrix}$ $-5R_2 + R_3$

6. $\begin{bmatrix} 4 & 10 & -8 \\ 7 & 4 & 3 \\ -1 & 1 & 0 \end{bmatrix}$ $-4R_3 + R_2$

Write the augmented matrix for each system. Do not solve the system.

7. $2x + 3y = 11$
$x + 2y = 8$

8. $3x + 5y = -13$
$2x + 3y = -9$

9. $x + 5y = 6$
$x = 3$

10. $2x + 7y = 1$
$5x = -15$

11. $2x + y + z = 3$
$3x - 4y + 2z = -7$
$x + y + z = 2$

12. $4x - 2y + 3z = 4$
$3x + 5y + z = 7$
$5x - y + 4z = 7$

13. $x + y = 2$
$2y + z = -4$
$z = 2$

14. $x = 6$
$y + 2z = 2$
$x - 3z = 6$

Write the system of equations associated with each augmented matrix.

15. $\left[\begin{array}{cc|c} 2 & 1 & 1 \\ 3 & -2 & -9 \end{array}\right]$

16. $\left[\begin{array}{cc|c} 1 & -5 & -18 \\ 6 & 2 & 20 \end{array}\right]$

17. $\left[\begin{array}{ccc|c} 1 & 0 & 0 & 2 \\ 0 & 1 & 0 & 3 \\ 0 & 0 & 1 & -2 \end{array}\right]$

18. $\left[\begin{array}{ccc|c} 1 & 0 & 1 & 4 \\ 0 & 1 & 0 & 2 \\ 0 & 0 & 1 & 3 \end{array}\right]$

19.
```
[A]
 [[3  2  1 1 ]
  [0  2  4 22]
  [-1 -2 3 15]]
```

20.
```
[B]
 [[2 1  3  12 ]
  [4 -3 0  10 ]
  [5 0  -4 -11]]
```

Each augmented matrix is in row echelon form and represents a linear system. Use back-substitution to solve the system if possible.

21. $\left[\begin{array}{cc|c} 1 & 2 & 3 \\ 0 & 1 & -1 \end{array}\right]$

22. $\left[\begin{array}{cc|c} 1 & -5 & 6 \\ 0 & 0 & 0 \end{array}\right]$

23. $\left[\begin{array}{cc|c} 1 & -1 & 2 \\ 0 & 1 & 0 \end{array}\right]$

24. $\left[\begin{array}{cc|c} 1 & 4 & -2 \\ 0 & 1 & 3 \end{array}\right]$

25. $\left[\begin{array}{ccc|c} 1 & 1 & -1 & 4 \\ 0 & 1 & -1 & 2 \\ 0 & 0 & 1 & 1 \end{array}\right]$

26. $\left[\begin{array}{ccc|c} 1 & -2 & -1 & 0 \\ 0 & 1 & -3 & 1 \\ 0 & 0 & 1 & 2 \end{array}\right]$

27. $\left[\begin{array}{ccc|c} 1 & 2 & -1 & 5 \\ 0 & 1 & -2 & 1 \\ 0 & 0 & 0 & 0 \end{array}\right]$

28. $\left[\begin{array}{ccc|c} 1 & -1 & 2 & 8 \\ 0 & 1 & -4 & 2 \\ 0 & 0 & 1 & -1 \end{array}\right]$

29. $\left[\begin{array}{ccc|c} 1 & 2 & 1 & -3 \\ 0 & 1 & -3 & \frac{1}{2} \\ 0 & 0 & 0 & 4 \end{array}\right]$

30. $\left[\begin{array}{ccc|c} 1 & 0 & -4 & \frac{3}{4} \\ 0 & 1 & 2 & 1 \\ 0 & 0 & 0 & -3 \end{array}\right]$

Solve each system of equations.

31. $\begin{aligned} x + y &= 5 \\ x - y &= -1 \end{aligned}$

32. $\begin{aligned} x + 2y &= 5 \\ 2x + y &= -2 \end{aligned}$

33. $\begin{aligned} x + y &= -3 \\ 2x - 5y &= -6 \end{aligned}$

34. $\begin{aligned} 3x - 2y &= 4 \\ 3x + y &= -2 \end{aligned}$

35. $\begin{aligned} 2x - 3y &= 10 \\ 2x + 2y &= 5 \end{aligned}$

36. $\begin{aligned} 4x + y &= 5 \\ 2x + y &= 3 \end{aligned}$

37. $\begin{aligned} 2x - 3y &= 2 \\ 4x - 6y &= 1 \end{aligned}$

38. $\begin{aligned} x + 2y &= 1 \\ 2x + 4y &= 3 \end{aligned}$

39. $\begin{aligned} 6x - 3y &= 1 \\ -12x + 6y &= -2 \end{aligned}$

40. $\begin{aligned} x - y &= 1 \\ -x + y &= -1 \end{aligned}$

41. $\begin{aligned} x + y &= -1 \\ y + z &= 4 \\ x + z &= 1 \end{aligned}$

42. $\begin{aligned} x - z &= -3 \\ y + z &= 9 \\ x + z &= 7 \end{aligned}$

43. $\begin{aligned} x + y - z &= 6 \\ 2x - y + z &= -9 \\ x - 2y + 3z &= 1 \end{aligned}$

44. $\begin{aligned} x + 3y - 6z &= 7 \\ 2x - y + 2z &= 0 \\ x + y + 2z &= -1 \end{aligned}$

45. $\begin{aligned} -x + y &= -1 \\ y - z &= 6 \\ x + z &= -1 \end{aligned}$

46. $\begin{aligned} x + y &= 1 \\ 2x - z &= 0 \\ y + 2z &= -2 \end{aligned}$

47. $\begin{aligned} 2x - y + 3z &= 0 \\ x + 2y - z &= 5 \\ 2y + z &= 1 \end{aligned}$

48. $\begin{aligned} 4x + 2y - 3z &= 6 \\ x - 4y + z &= -4 \\ -x + 2z &= 2 \end{aligned}$

49. $\begin{aligned} 2.1x + .5y + 1.7z &= 4.9 \\ -2x + 1.5y - 1.7z &= 3.1 \\ 5.8x - 4.6y + .8z &= 9.3 \end{aligned}$

50. $\begin{aligned} .1x + .3y + 1.7z &= .6 \\ .6x + .1y - 3.1z &= 6.2 \\ 2.4y + .9z &= 3.5 \end{aligned}$

51. $\begin{aligned} 53x + 95y + 12z &= 108 \\ 81x - 57y - 24z &= -92 \\ -9x + 11y - 78z &= 21 \end{aligned}$

52. $\begin{aligned} 103x - 886y + 431z &= 1200 \\ -55x + 981y &= 1108 \\ -327x + 421y + 337z &= 99 \end{aligned}$

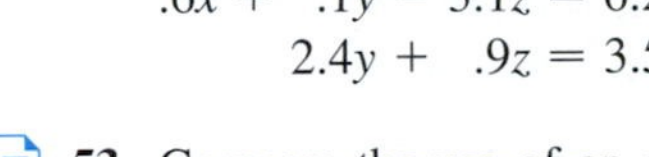

53. Compare the use of an augmented matrix as a short-hand way of writing a system of linear equations with the use of synthetic division as a shorthand way to divide polynomials.

54. Compare the use of the third row transformation on a matrix with the elimination method of solving a system of linear equations.

Solve each system. Let z be the arbitrary variable if necessary.

55. $\begin{aligned} x - 3y + 2z &= 10 \\ 2x - y - z &= 8 \end{aligned}$

56. $\begin{aligned} 3x + y - z &= 12 \\ x + 2y + z &= 10 \end{aligned}$

57. $\begin{aligned} x + 2y - z &= 0 \\ 3x - y + z &= 6 \\ -2x - 4y + 2z &= 0 \end{aligned}$

58. $\begin{aligned} 3x + 5y - z &= 0 \\ 4x - y + 2z &= 1 \\ -6x - 10y + 2z &= 0 \end{aligned}$

59. $\begin{aligned} x - 2y + z &= 5 \\ -2x + 4y - 2z &= 2 \\ 2x + y - z &= 2 \end{aligned}$

60. $\begin{aligned} 3x + 6y - 3z &= 12 \\ -x - 2y + z &= 16 \\ x + y - 2z &= 20 \end{aligned}$

Solve each system of four equations in four variables. Express the solutions in the form (x, y, z, w).

61. $\begin{aligned} x + 3y - 2z - w &= 9 \\ 4x + y + z + 2w &= 2 \\ -3x - y + z - w &= -5 \\ x - y - 3z - 2w &= 2 \end{aligned}$

62. $\begin{aligned} 3x + 2y - w &= 0 \\ 2x + z + 2w &= 5 \\ x + 2y - z &= -2 \\ 2x - y + z + w &= 2 \end{aligned}$

(Modeling) *Solve each application.*

63. *Food Shelter Costs* Three food shelters have monthly food costs F in dollars, number N of people served per month, and monthly charitable receipts R in dollars, as shown in the table.

Food Costs (F)	Number Served (N)	Charitable Receipts (R)
1300	1800	5000
5300	3200	12,000
6500	4500	13,000

(a) Model these data by using

$$F = aN + bR + c,$$

where a, b, and c are constants.

(b) Predict food costs for a shelter that serves 3500 people and receives charitable receipts of $12,500. Round your answer to the nearest hundred dollars.

64. *Paid Vacation for Employees* The average number y of paid days off for full-time workers at medium-to-large companies after x years is listed in the table.

x (years)	1	15	30
y (days)	9.4	18.8	21.9

Source: Bureau of Labor Statistics.

(a) Determine the coefficients for $f(x) = ax^2 + bx + c$ so that f models these data.
(b) Graph f with the data in the window $[-4, 32]$ by $[8, 23]$.
(c) Estimate the number of paid days off after 3 years of experience. Compare it with the actual value of 11.2 days.

65. *Scheduling Production* A company produces two models of bicycles: model A and model B. Model A requires 2 hours of assembly time, and model B requires 3 hours of assembly time. The parts for model A cost \$25 per bike; those for model B cost \$30 per bike. If the company has a total of 34 hours of assembly time and \$365 available per day for these two models, what is the maximum number of each model that can be made in a day and use all of the available resources?

66. *Scheduling Production* Caltek Computer Company makes two products: computer monitors and printers. Both require time on two machines: monitors, 1 hour on machine A and 2 hours on machine B; printers, 3 hours on machine A and 1 hour on machine B. Both machines operate 15 hours per day. What is the maximum number of each product that can be produced per day under these conditions?

67. *Financing Expansion* To get funds necessary for a planned expansion, a small company took out three loans totaling \$25,000. The company was able to borrow some of the money at 8%. It borrowed \$2000 more than $\frac{1}{2}$ the amount of the 8% loan at 10% and the rest at 9%. The total annual interest was \$2220. How much did the company borrow at each rate?

68. *Investment Decisions* Rick Pal deposits some money in a bank account paying 3% per year. He uses some additional money, amounting to $\frac{1}{3}$ the amount placed in the bank, to buy bonds paying 4% per year. With the balance of his funds, he buys a 4.5% certificate of deposit. The first year, his investments bring a return of \$400. If the total of the investments is \$10,000, how much did he invest at each rate?

(Modeling) Each set of data in Exercises 69 and 70 can be modeled by $f(x) = ax^2 + bx + c$, where x denotes the year.
(a) *Find a linear system whose solution represents values of a, b, and c.*
(b) *Find the solution by using a method from this section.*
(c) *Graph f and the data in the same viewing window.*
(d) *Make your own prediction using f.*

69. *Chronic Health Care* A large percentage of the U.S. population will require chronic health care in the coming decades. The average caregiving age is 50–64, while the typical person needing chronic care is 85 or older. The ratio of potential caregivers to those needing chronic health care will shrink in the coming years, as shown in the table.

Year	1990	2010	2030
Ratio	11	10	6

Source: Robert Wood Johnson Foundation, *Chronic Care in America: A 21st Century Challenge.*

70. *Carbon Dioxide Levels* Carbon dioxide (CO_2) is a greenhouse gas. Its concentration in parts per million (ppm) has been measured at Mauna Loa, Hawaii, during past years. The table lists measurements for three selected years.

Year	1958	1973	2003
CO_2 (ppm)	315	325	376

Source: Mauna Loa Observatory.

(Modeling) Traffic Flow Each figure in Exercises 71 and 72 shows three one-way streets with intersections A, B, and C. Numbers indicate the average traffic flow in vehicles per minute. The variables x, y, and z denote unknown traffic flows that need to be determined for timing of stoplights.
(a) *If the number of vehicles per minute entering an intersection must equal the number exiting an intersection, verify that the system of linear equations describes the traffic flow.*
(b) *Rewrite the system and solve.*
(c) *Interpret your solution.*

71. A: $x + 5 = y + 7$
B: $z + 6 = x + 3$
C: $y + 3 = z + 4$

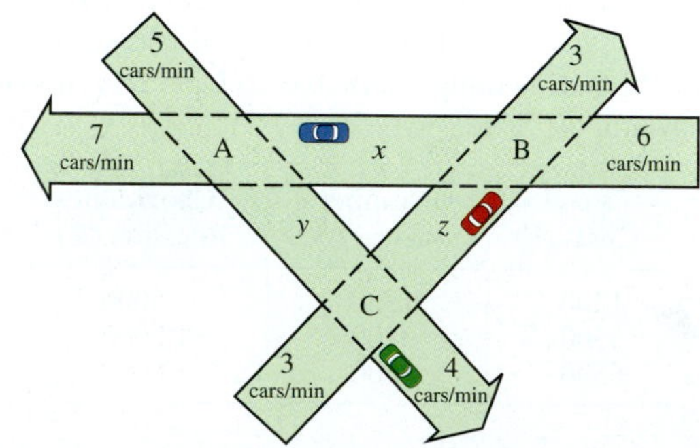

72. A: $x + 7 = y + 4$
B: $4 + 5 = x + z$
C: $y + 8 = 9 + 4$

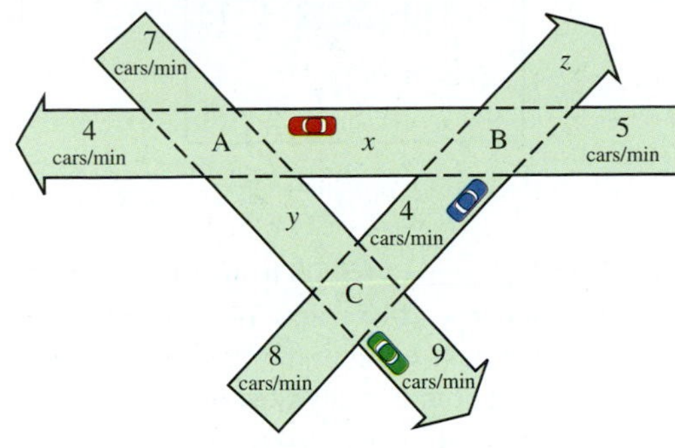

(Modeling) *Solve each problem.*

73. ***Fawn Population*** To model spring fawn count F from adult antelope population A, precipitation P, and severity of winter W, environmentalists have used the equation

$$F = a + bA + cP + dW,$$

where the coefficients a, b, c, and d are constants that must be determined before using the equation. The table lists the results of four different years.

Fawns	Adults	Precip. (in inches)	Winter Severity
239	871	11.5	3
234	847	12.2	2
192	685	10.6	5
343	969	14.2	1

Source: Brase, C. and C. Brase, *Understandable Statistics,* D.C. Heath and Company, 1995; Bureau of Land Management.

(a) Substitute the values for F, A, P, and W from the table for each of the four years into the given equation $F = a + bA + cP + dW$ to obtain four linear equations involving a, b, c, and d.

(b) Write an augmented matrix representing the system, and solve for a, b, c, and d.

(c) Write the equation for F, using the values from part (b) for the coefficients.

(d) If a winter has severity 3, adult antelope population 960, and precipitation 12.6 inches, predict the spring fawn count. (Compare this with the actual count of 320.)

74. ***Weight of a Black Bear*** The table shows weight W, neck size N, overall length L, and chest size C for four bears.

W (pounds)	N (inches)	L (inches)	C (inches)
125	19	57.5	32
316	26	65	42
436	30	72	48
514	30.5	75	54

Source: M. Triola, *Elementary Statistics*; Minitab, Inc.

(a) We can model these data with the equation

$$W = a + bN + cL + dC,$$

where a, b, c, and d are constants. To do so, represent a system of linear equations by a 4×5 augmented matrix whose solution gives values for a, b, c, and d.

(b) Solve the system. Round each value to the nearest thousandth.

 **(c)** Predict the weight of a bear with $N = 24$, $L = 63$, and $C = 39$. Interpret the result.

75. ***Blending Coffee Beans*** Solve the system of linear equations discussed in the introduction to this chapter.

$$a + b + c = 50$$
$$c = 2a$$
$$15.99a + 12.99b + 10.19c = 12.37(50)$$

 **76.** ***Blending Coffee Beans*** Rework Exercise 75 if Guatemala Antigua retails for \$12.49 per pound instead of \$10.19 per pound. Does your answer seem reasonable? Explain.

Reviewing Basic Concepts (Sections 7.1–7.3)

Solve each system, using the method indicated.

1. (Elimination)

$$2x - 3y = 18$$
$$5x + 2y = 7$$

2. (Graphical)

$$2x + y = -4$$
$$-x + 2y = 2$$

3. (Substitution)

$$5x + 10y = 10$$
$$x + 2y = 2$$

4. (Elimination)

$$x - y = 6$$
$$x - y = 4$$

5. (Analytic, with graphical support)

$$6x + 2y = 10$$
$$2x^2 - 3y = 11$$

6. (Row echelon method)

$$x + y + z = 1$$
$$-x + y + z = 5$$
$$y + 2z = 5$$

7. (Reduced row echelon method)

$$\begin{aligned} 2x + 4y + 4z &= 4 \\ x + 3y + z &= 4 \\ -x + 3y + 2z &= -1 \end{aligned}$$

8. Solve the system represented by the augmented matrix.

```
[A]
  [[2 1  2 10]
   [1 0  2 5 ]
   [1 -2 2 1 ]]
```

(Modeling) *Solve each problem.*

9. ***Television Sales*** In a recent year, about 32 million television sets were sold. For every 10 televisions sold with stereo sound, about 19 sold without stereo sound. (*Source: Consumer Electronics*.) About how many of each type of television were sold that year?

10. ***Investments*** A sum of \$5000 is invested in three mutual funds that pay 8%, 11%, and 14% interest rates. The amount of money invested in the fund paying 14% equals the total amount of money invested in the other two funds, and the total annual interest from all three funds is \$595. Find the amount invested at each rate.

7.4 Matrix Properties and Operations

Terminology of Matrices ■ Operations on Matrices ■ Applying Matrix Algebra

Terminology of Matrices

Suppose that you are the manager of an electronics store and one day you receive the following products from two distributors: from Wholesale Enterprises, 2 cell phones, 7 DVDs, and 5 video games; from Discount Distributors, 4 cell phones, 6 DVDs, and 9 video games. We can organize the information in a table.

Distributor	Product		
	Cell Phones	DVDs	Video Games
Wholesale Enterprises	2	7	5
Discount Distributors	4	6	9

As long as we remember what each number represents, we can remove all the labels and write the numbers in the table as a matrix.

$$\begin{bmatrix} 2 & 7 & 5 \\ 4 & 6 & 9 \end{bmatrix} \quad 2 \times 3 \text{ matrix}$$

Matrices are classified by their **dimension**—that is, by the number of rows and columns they contain. For example, the preceding matrix has 2 rows and 3 columns, with dimension 2×3; a matrix with m rows and n columns has dimension $m \times n$. ***The number of rows is always given first.***

Certain matrices have special names: an $n \times n$ matrix is a **square matrix** of order n. Also, a matrix with just one row is a **row matrix,** and a matrix with just one column is a **column matrix.**

TECHNOLOGY NOTE

You should familiarize yourself with the matrix capabilities of your particular calculator. Refer to the graphing calculator manual that accompanies this text or your owner's manual.

Two matrices are equal if they have the same dimension and if corresponding elements, position by position, are equal. Using this definition, the matrices

$$\begin{bmatrix} 2 & 1 \\ 3 & -5 \end{bmatrix} \quad \text{and} \quad \begin{bmatrix} 1 & 2 \\ -5 & 3 \end{bmatrix}$$

are *not* equal (even though they contain the same elements and are the same dimension), because the corresponding elements differ.

GCM EXAMPLE 1 Classifying Matrices by Dimension

Find the dimension of each matrix, and determine any special characteristics.

(a) $\begin{bmatrix} 6 & 5 \\ 3 & 4 \\ 5 & -1 \end{bmatrix}$ **(b)** $\begin{bmatrix} 3 \\ -5 \\ 0 \\ 2 \end{bmatrix}$ **(c)** $\begin{bmatrix} 1 & 6 & 5 & -2 & 5 \end{bmatrix}$

Solution

(a) This is a 3×2 matrix, because it has 3 rows and 2 columns.

(b) This matrix is a 4×1 *column matrix.*

(c) $\begin{bmatrix} 1 & 6 & 5 & -2 & 5 \end{bmatrix}$ is a 1×5 matrix. It is an example of a *row matrix.* ■

GCM EXAMPLE 2 Determining Equality of Matrices

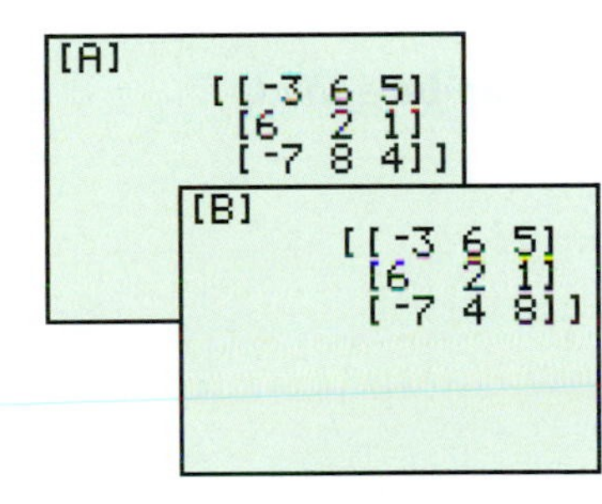

FIGURE 22

(a) If $A = \begin{bmatrix} 2 & 1 \\ p & q \end{bmatrix}$ and $B = \begin{bmatrix} x & y \\ -1 & 0 \end{bmatrix}$, find the values of x, y, p, and q such that $A = B$.

(b) Are matrices [A] and [B] in Figure 22 equal?

Solution

(a) From the definition of equality, the only way that the statement

$$\begin{bmatrix} 2 & 1 \\ p & q \end{bmatrix} = \begin{bmatrix} x & y \\ -1 & 0 \end{bmatrix}$$

can be true is if $x = 2$, $y = 1$, $p = -1$, and $q = 0$.

(b) The matrices are not equal. Corresponding entries in the third row are not equal. ■

Operations on Matrices

At the beginning of this section, we gave an example using the matrix

$$\begin{bmatrix} 2 & 7 & 5 \\ 4 & 6 & 9 \end{bmatrix},$$

where the columns represent the numbers of certain products (cell phones, DVDs, and video games, respectively) and the rows represent the two different distributors (Wholesale Enterprises and Discount Distributors, respectively). For example, the element 7 represents 7 DVDs received from Wholesale Enterprises.

TECHNOLOGY NOTE

Graphing calculators can perform operations on matrices, provided that the dimensions of the matrices are compatible for the operation. A dimension error message will occur if the operation cannot be performed.

On another day, the shipments from these two distributors are described with the matrix

$$\begin{bmatrix} 3 & 12 & 10 \\ 15 & 11 & 8 \end{bmatrix}.$$

Here, for example, 8 video games were received from Discount Distributors. The total number of products received from the distributors on the two days can be found by adding the corresponding elements of the two matrices.

$$\begin{bmatrix} 2 & 7 & 5 \\ 4 & 6 & 9 \end{bmatrix} + \begin{bmatrix} 3 & 12 & 10 \\ 15 & 11 & 8 \end{bmatrix} = \begin{bmatrix} 2+3 & 7+12 & 5+10 \\ 4+15 & 6+11 & 9+8 \end{bmatrix}$$

$$= \begin{bmatrix} 5 & 19 & 15 \\ 19 & 17 & 17 \end{bmatrix}$$

The 5 in the sum indicates that, on the two days, 5 cell phones were received from Wholesale Enterprises. Generalizing from this example leads to the definition of matrix addition.

Matrix Addition

The sum of two $m \times n$ matrices A and B is the $m \times n$ matrix $A + B$ in which each element is the sum of the corresponding elements of A and B.

CAUTION ***Only matrices with the same dimension can be added.***

GCM **EXAMPLE 3** **Adding Matrices**

Find each sum if possible.

(a) $\begin{bmatrix} 5 & -6 \\ 8 & 9 \end{bmatrix} + \begin{bmatrix} -4 & 6 \\ 8 & -3 \end{bmatrix}$ **(b)** $\begin{bmatrix} 2 \\ 5 \\ 8 \end{bmatrix} + \begin{bmatrix} -6 \\ 3 \\ 12 \end{bmatrix}$

(c) $A + B$ if $A = \begin{bmatrix} 5 & 8 \\ 6 & 2 \end{bmatrix}$ and $B = \begin{bmatrix} 3 & 9 & 1 \\ 4 & 2 & 5 \end{bmatrix}$

Analytic Solution

(a) $\begin{bmatrix} 5 & -6 \\ 8 & 9 \end{bmatrix} + \begin{bmatrix} -4 & 6 \\ 8 & -3 \end{bmatrix}$

$$= \begin{bmatrix} 5+(-4) & -6+6 \\ 8+8 & 9+(-3) \end{bmatrix}$$

$$= \begin{bmatrix} 1 & 0 \\ 16 & 6 \end{bmatrix}$$

(b) $\begin{bmatrix} 2 \\ 5 \\ 8 \end{bmatrix} + \begin{bmatrix} -6 \\ 3 \\ 12 \end{bmatrix} = \begin{bmatrix} -4 \\ 8 \\ 20 \end{bmatrix}$

Graphing Calculator Solution

(a) Figure 23(b) shows the sum of matrices A and B defined in Figure 23(a).

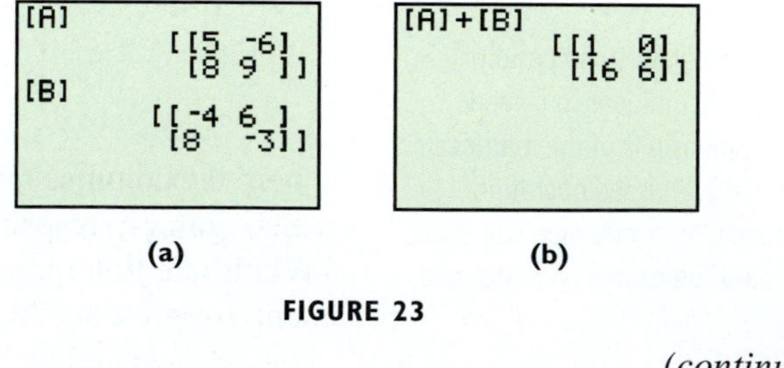

(a) (b)

FIGURE 23

(continued)

(c) The matrices

$$A = \begin{bmatrix} 5 & 8 \\ 6 & 2 \end{bmatrix} \quad 2 \times 2 \text{ matrix}$$

and $$B = \begin{bmatrix} 3 & 9 & 1 \\ 4 & 2 & 5 \end{bmatrix} \quad 2 \times 3 \text{ matrix}$$

have different dimensions and so cannot be added; the sum $A + B$ does not exist.

(b) The screen in Figure 24 shows how the sum of two column matrices entered directly on the home screen is displayed.

(c) A graphing calculator will return an ERROR message if it is directed to perform an operation on matrices that is not possible due to a dimension mismatch. See Figure 25.

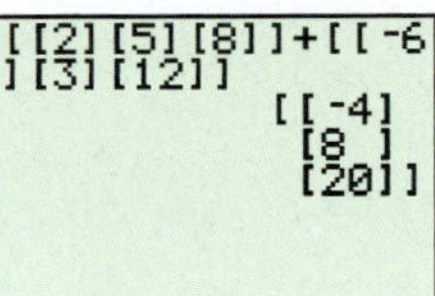

FIGURE 24

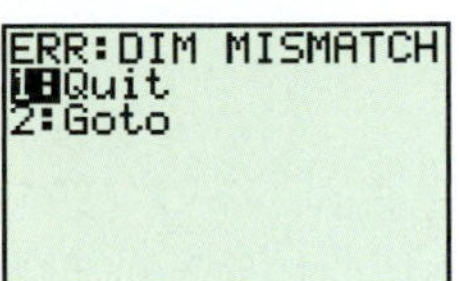

FIGURE 25

A matrix containing only zeros as elements is called a **zero matrix.**

$$[0 \quad 0 \quad 0] \quad 1 \times 3 \text{ zero matrix} \qquad \begin{bmatrix} 0 & 0 & 0 \\ 0 & 0 & 0 \end{bmatrix} \quad 2 \times 3 \text{ zero matrix}$$

The additive inverse of a real number a is the unique real number $-a$ such that $a + (-a) = 0$ and $-a + a = 0$. Given matrix A, the elements of matrix $-A$ are the additive inverses of the corresponding elements of A. (Remember that each element of A is a real number and thus has an additive inverse.) For example, if

$$A = \begin{bmatrix} -5 & 2 & -1 \\ 3 & 4 & -6 \end{bmatrix}, \quad \text{then} \quad -A = \begin{bmatrix} 5 & -2 & 1 \\ -3 & -4 & 6 \end{bmatrix}.$$

To check, test that $A + (-A)$ equals the zero matrix O.

$$A + (-A) = \begin{bmatrix} -5 & 2 & -1 \\ 3 & 4 & -6 \end{bmatrix} + \begin{bmatrix} 5 & -2 & 1 \\ -3 & -4 & 6 \end{bmatrix} = \begin{bmatrix} 0 & 0 & 0 \\ 0 & 0 & 0 \end{bmatrix} = O$$

Then, test that $-A + A$ is also O. Matrix $-A$ is the **additive inverse,** or **negative,** of matrix A. Every matrix has a unique additive inverse.

What Went Wrong?

A student entered [A] into his calculator as shown in Figure 26(a). He attempted to find the additive inverse of [A] and got the error message shown in Figure 26(b).

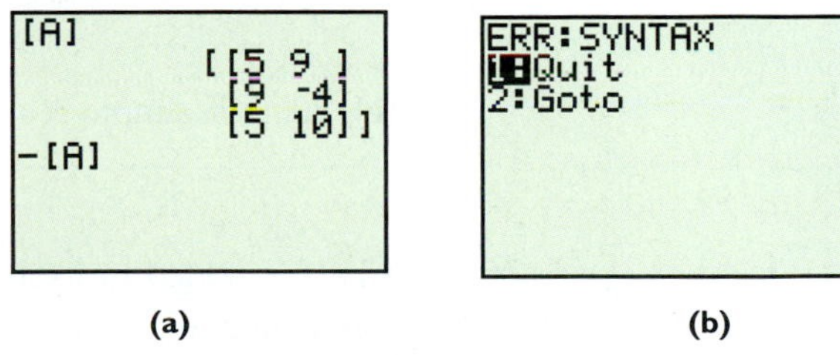

(a) (b)

FIGURE 26

What Went Wrong? How can he correct his error?

Answer to What Went Wrong?

The student entered the subtraction symbol, rather than the negative symbol, preceding [A]. He can obtain −[A] by using the correct symbol.

Just as subtraction of real numbers is defined in terms of the additive inverse, subtraction of matrices is defined in the same way.

Matrix Subtraction

If A and B are matrices with the same dimension, then

$$A - B = A + (-B).$$

GCM **EXAMPLE 4 Subtracting Matrices**

Find each difference.

(a) $\begin{bmatrix} -5 & 6 \\ 2 & 4 \end{bmatrix} - \begin{bmatrix} -3 & 2 \\ 5 & -8 \end{bmatrix}$ **(b)** $[8 \quad 6 \quad -4] - [3 \quad 5 \quad -8]$

(c) $A - B$ if $A = \begin{bmatrix} -2 & 5 \\ 0 & 1 \end{bmatrix}$ and $B = \begin{bmatrix} 3 \\ 5 \end{bmatrix}$

Analytic Solution

(a) $\begin{bmatrix} -5 & 6 \\ 2 & 4 \end{bmatrix} - \begin{bmatrix} -3 & 2 \\ 5 & -8 \end{bmatrix}$

$= \begin{bmatrix} -5 & 6 \\ 2 & 4 \end{bmatrix} + \begin{bmatrix} 3 & -2 \\ -5 & 8 \end{bmatrix}$

$= \begin{bmatrix} -2 & 4 \\ -3 & 12 \end{bmatrix}$

(b) $[8 \quad 6 \quad -4] - [3 \quad 5 \quad -8]$

$= [5 \quad 1 \quad 4]$

(c) Matrices A and B have different dimensions, so their difference does not exist.

Graphing Calculator Solution

(a) In Figure 27, the two matrices are defined as [C] and [D]; the difference [C] − [D] is displayed in Figure 28.

```
[C]
        [[-5 6]
         [2  4]]
[D]
        [[-3 2 ]
         [5  -8]]
```

FIGURE 27

```
[C]-[D]
        [[-2 4 ]
         [-3 12]]
```

FIGURE 28

(b) See Figure 29.

```
[E]
        [[8 6 -4]]
[F]
        [[3 5 -8]]
[E]-[F]
        [[5 1 4]]
```

FIGURE 29

(c) As in Example 3(c), the calculator returns a dimension error message.

If a matrix A is added to itself, each element in the sum is twice as large as the corresponding element of A. For example,

$$\begin{bmatrix} 2 & 5 \\ 1 & 3 \\ 4 & 6 \end{bmatrix} + \begin{bmatrix} 2 & 5 \\ 1 & 3 \\ 4 & 6 \end{bmatrix} = \begin{bmatrix} 4 & 10 \\ 2 & 6 \\ 8 & 12 \end{bmatrix} = 2\begin{bmatrix} 2 & 5 \\ 1 & 3 \\ 4 & 6 \end{bmatrix}.$$

In the last expression, the number 2 in front of the matrix is called a **scalar** to distinguish it from a matrix. A scalar is just a special name for a real number. This example suggests the definition of multiplication of a matrix by a scalar.

Multiplication of a Matrix by a Scalar

The product of a scalar k and a matrix A is the matrix kA, each of whose elements is k times the corresponding element of A.

GCM EXAMPLE 5 Multiplying Matrices by Scalars

Perform each multiplication.

(a) $5\begin{bmatrix} 2 & -3 \\ 0 & 4 \end{bmatrix}$ **(b)** $\frac{3}{4}\begin{bmatrix} 20 & 36 \\ 12 & -16 \end{bmatrix}$

Analytic Solution

(a) $5\begin{bmatrix} 2 & -3 \\ 0 & 4 \end{bmatrix} = \begin{bmatrix} 5(2) & 5(-3) \\ 5(0) & 5(4) \end{bmatrix}$ Multiply each element by 5.

$$= \begin{bmatrix} 10 & -15 \\ 0 & 20 \end{bmatrix}$$

(b) $\frac{3}{4}\begin{bmatrix} 20 & 36 \\ 12 & -16 \end{bmatrix} = \begin{bmatrix} \frac{3}{4}(20) & \frac{3}{4}(36) \\ \frac{3}{4}(12) & \frac{3}{4}(-16) \end{bmatrix}$

$$= \begin{bmatrix} 15 & 27 \\ 9 & -12 \end{bmatrix}$$

Graphing Calculator Solution

See Figures 30 and 31.

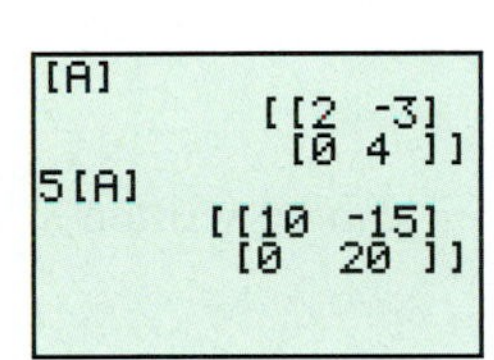

FIGURE 30

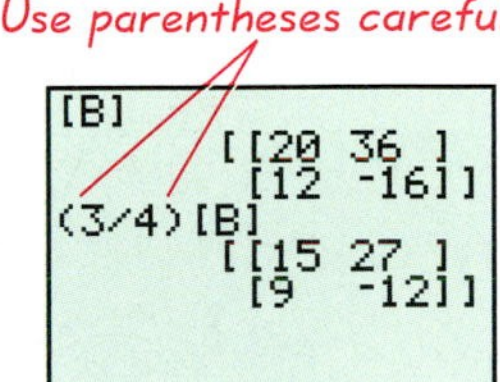

FIGURE 31

Returning to the electronics store example, recall the matrix for the number of each type of product received from the two distributors.

$$\begin{bmatrix} 2 & 7 & 5 \\ 4 & 6 & 9 \end{bmatrix}$$

Now suppose each cell phone costs the store \$120, each DVD costs \$18, and each video game costs \$9. To find the total cost of the products from Wholesale Enterprises, we multiply as follows.

Type of Product	Number of Items	Cost per Item	Total Cost
Cell phone	2	\$120	\$240
DVD	7	\$ 18	\$126
Video game	5	\$ 9	\$ 45
			\$411 ← Total from Wholesale Enterprises

The products from Wholesale Enterprises cost a total of \$411. This result is the sum of three products:

$$2(\$120) + 7(\$18) + 5(\$9) = \$411.$$

In the same way, using the second row of the matrix and the three product costs gives the total cost from Discount Distributors.

$$4(\$120) + 6(\$18) + 9(\$9) = \$669$$

The product costs can be written as a column matrix.

$$\begin{bmatrix} 120 \\ 18 \\ 9 \end{bmatrix}$$

The total costs from each distributor can also be written as a column matrix.

$$\begin{bmatrix} 411 \\ 669 \end{bmatrix}$$

The product of the matrices

$$\begin{bmatrix} 2 & 7 & 5 \\ 4 & 6 & 9 \end{bmatrix} \quad \text{and} \quad \begin{bmatrix} 120 \\ 18 \\ 9 \end{bmatrix}$$

can be written

$$\begin{bmatrix} 2 & 7 & 5 \\ 4 & 6 & 9 \end{bmatrix} \begin{bmatrix} 120 \\ 18 \\ 9 \end{bmatrix} = \begin{bmatrix} 2 \cdot 120 + 7 \cdot 18 + 5 \cdot 9 \\ 4 \cdot 120 + 6 \cdot 18 + 9 \cdot 9 \end{bmatrix} = \begin{bmatrix} 411 \\ 669 \end{bmatrix}.$$

Each element of the product was found by multiplying the elements of the *rows* of the matrix on the left and the corresponding elements of the *column* of the matrix on the right and then finding the sum of these products. Notice that the product of a 2×3 matrix and a 3×1 matrix is a 2×1 matrix.

Generalizing from this example gives the definition of matrix multiplication.

Matrix Multiplication

The product AB of an $m \times n$ matrix A and an $n \times k$ matrix B is found as follows:

To find the ith row, jth column element of AB, multiply each element in the ith row of A by the corresponding element in the jth column of B. The sum of these products will give the element of row i, column j of AB. The dimension of AB is $m \times k$.

The product AB can be found only if the number of columns of A is the same as the number of rows of B. The final product will have as many rows as A and as many columns as B.

EXAMPLE 6 Deciding whether Two Matrices Can Be Multiplied

Suppose matrix A has dimension 2×2, while matrix B has dimension 2×4. Can the product AB be calculated? If so, what is the dimension of the product? Can the product BA be calculated?

Solution The following diagram helps answer these questions.

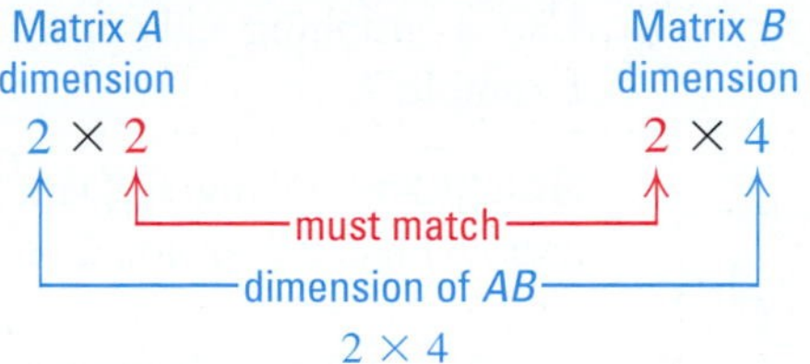

The product AB can be calculated because A has two columns and B has two rows. The dimension of the product is 2×4.

The product BA cannot be found. To see why, consider the following diagram. The number of columns in B is different from the number of rows in A.

■

EXAMPLE 7 Multiplying Matrices

Find the product AB of the two matrices $A = \begin{bmatrix} -3 & 4 & 2 \\ 5 & 0 & 4 \end{bmatrix}$ and $B = \begin{bmatrix} -6 & 4 \\ 2 & 3 \\ 3 & -2 \end{bmatrix}$.

Solution Since A has dimension 2×3 and B has dimension 3×2, they are compatible for multiplication. The product AB will have dimension 2×2.

$$\begin{matrix} & A & B \end{matrix}$$

Step 1 $\begin{bmatrix} -3 & 4 & 2 \\ 5 & 0 & 4 \end{bmatrix}\begin{bmatrix} -6 & 4 \\ 2 & 3 \\ 3 & -2 \end{bmatrix}$ $\quad -3(-6) + 4(2) + 2(3) = 32$

Step 2 $\begin{bmatrix} -3 & 4 & 2 \\ 5 & 0 & 4 \end{bmatrix}\begin{bmatrix} -6 & 4 \\ 2 & 3 \\ 3 & -2 \end{bmatrix}$ $\quad -3(4) + 4(3) + 2(-2) = -4$

Step 3 $\begin{bmatrix} -3 & 4 & 2 \\ 5 & 0 & 4 \end{bmatrix}\begin{bmatrix} -6 & 4 \\ 2 & 3 \\ 3 & -2 \end{bmatrix}$ $\quad 5(-6) + 0(2) + 4(3) = -18$

Step 4 $\begin{bmatrix} -3 & 4 & 2 \\ 5 & 0 & 4 \end{bmatrix}\begin{bmatrix} -6 & 4 \\ 2 & 3 \\ 3 & -2 \end{bmatrix}$ $\quad 5(4) + 0(3) + 4(-2) = 12$

Step 5 Write the product.

$$\begin{bmatrix} -3 & 4 & 2 \\ 5 & 0 & 4 \end{bmatrix}\begin{bmatrix} -6 & 4 \\ 2 & 3 \\ 3 & -2 \end{bmatrix} = \begin{bmatrix} 32 & -4 \\ -18 & 12 \end{bmatrix}$$

■

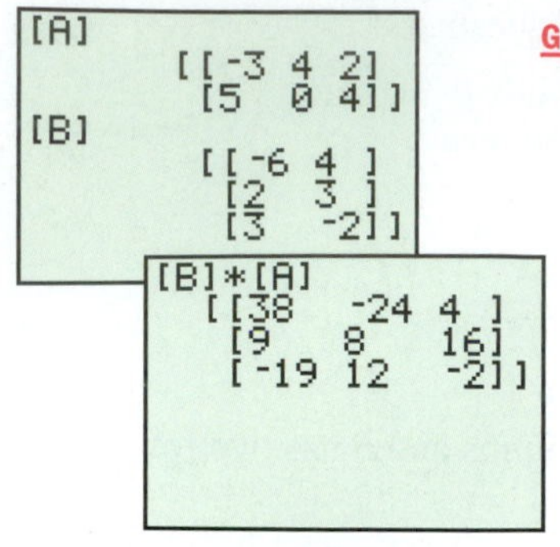

FIGURE 32

GCM **EXAMPLE 8** **Multiplying Matrices**

Use a graphing calculator to find the product BA of the two matrices given in Example 7.

Solution Define [A] and [B] as shown in Figure 32. Since [B] has dimension 3×2 and [A] has dimension 2×3, the dimension of the product BA is 3×3. ■

As shown in Examples 7 and 8,

$$AB \neq BA.$$

In general, ***matrix multiplication is not commutative.*** In fact, in some cases, one of the products may be defined while the other is not. See Example 6.

```
[A]
        [[1  3]
         [-2 4]]
[B]
        [[2  5 ]
         [10 -3]]
[A]*[B]
        [[32 -4 ]
         [36 -22]]
[B]*[A]
        [[-8 26]
         [16 18]]
```

FIGURE 33

Figure 33 shows matrices [A] and [B] and both products from Example 9.

GCM **EXAMPLE 9** **Multiplying Square Matrices**

Find AB and BA, given $A = \begin{bmatrix} 1 & 3 \\ -2 & 4 \end{bmatrix}$ and $B = \begin{bmatrix} 2 & 5 \\ 10 & -3 \end{bmatrix}$.

Solution

$$AB = \begin{bmatrix} 1 & 3 \\ -2 & 4 \end{bmatrix}\begin{bmatrix} 2 & 5 \\ 10 & -3 \end{bmatrix} = \begin{bmatrix} 2 + 30 & 5 - 9 \\ -4 + 40 & -10 - 12 \end{bmatrix} = \begin{bmatrix} 32 & -4 \\ 36 & -22 \end{bmatrix}$$

$$BA = \begin{bmatrix} 2 & 5 \\ 10 & -3 \end{bmatrix}\begin{bmatrix} 1 & 3 \\ -2 & 4 \end{bmatrix} = \begin{bmatrix} 2 - 10 & 6 + 20 \\ 10 + 6 & 30 - 12 \end{bmatrix} = \begin{bmatrix} -8 & 26 \\ 16 & 18 \end{bmatrix}$$ ■

Applying Matrix Algebra

EXAMPLE 10 **Using Matrix Multiplication to Model Plans for a Subdivision**

A contractor builds three kinds of houses, models X, Y, and Z, with a choice of two styles, colonial or ranch. Matrix A shows the number of each kind of house the contractor is planning to build for a new 100-home subdivision. The amounts for each of the main materials used depend on the style of the house. These amounts are shown in matrix B, while matrix C gives the cost in dollars for each kind of material. Concrete is measured here in cubic yards, lumber in 1000 board feet, brick in 1000s, and shingles in 100 square feet.

$$\begin{array}{c} \\ \text{Model X} \\ \text{Model Y} \\ \text{Model Z} \end{array} \begin{array}{c} \begin{array}{cc} \text{Colonial} & \text{Ranch} \end{array} \\ \begin{bmatrix} 0 & 30 \\ 10 & 20 \\ 20 & 20 \end{bmatrix} \end{array} = A$$

$$\begin{array}{c} \\ \text{Colonial} \\ \text{Ranch} \end{array} \begin{array}{c} \begin{array}{cccc} \text{Concrete} & \text{Lumber} & \text{Brick} & \text{Shingles} \end{array} \\ \begin{bmatrix} 10 & 2 & 0 & 2 \\ 50 & 1 & 20 & 2 \end{bmatrix} \end{array} = B \qquad \begin{array}{c} \text{Concrete} \\ \text{Lumber} \\ \text{Brick} \\ \text{Shingles} \end{array} \begin{array}{c} \text{Cost per Unit} \\ \begin{bmatrix} 20 \\ 180 \\ 60 \\ 25 \end{bmatrix} \end{array} = C$$

(a) What is the total cost of materials for all houses of each model?

(b) How much of each of the four kinds of material must be ordered?

(c) Use a graphing calculator to find the total cost of the materials.

Solution

(a) To find the cost of materials for each model, first find matrix AB, which will give the total amount of each material needed for all houses of each model.

$$AB = \begin{bmatrix} 0 & 30 \\ 10 & 20 \\ 20 & 20 \end{bmatrix} \begin{bmatrix} 10 & 2 & 0 & 2 \\ 50 & 1 & 20 & 2 \end{bmatrix} = \begin{array}{c} \begin{array}{cccc} \text{Concrete} & \text{Lumber} & \text{Brick} & \text{Shingles} \end{array} \\ \begin{bmatrix} 1500 & 30 & 600 & 60 \\ 1100 & 40 & 400 & 60 \\ 1200 & 60 & 400 & 80 \end{bmatrix} \end{array} \begin{array}{l} \text{Model X} \\ \text{Model Y} \\ \text{Model Z} \end{array}$$

Multiplying AB and the cost matrix C gives the total cost of materials for each model.

$$(AB)C = \begin{bmatrix} 1500 & 30 & 600 & 60 \\ 1100 & 40 & 400 & 60 \\ 1200 & 60 & 400 & 80 \end{bmatrix} \begin{bmatrix} 20 \\ 180 \\ 60 \\ 25 \end{bmatrix} = \begin{array}{c} \text{Cost} \\ \begin{bmatrix} 72{,}900 \\ 54{,}700 \\ 60{,}800 \end{bmatrix} \end{array} \begin{array}{l} \text{Model X} \\ \text{Model Y} \\ \text{Model Z} \end{array}$$

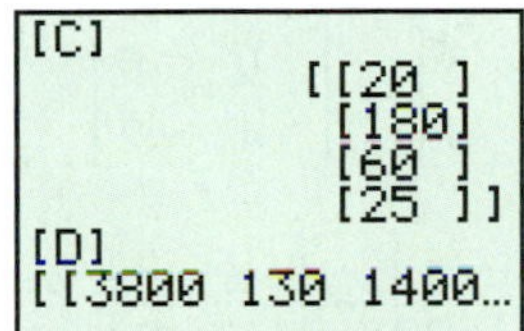

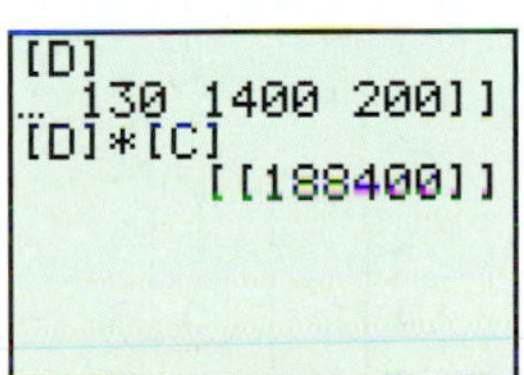

FIGURE 34

(b) The totals of the columns of matrix AB will give a matrix whose elements represent the amounts of each material needed for the subdivision. Call this matrix D, and write it as a row matrix.

$$D = [3800 \quad 130 \quad 1400 \quad 200]$$

(c) The total cost of all the materials is given by the product of matrix C, the cost matrix, and matrix D, the total amounts matrix. To multiply these and get a 1×1 matrix representing the total cost requires multiplying a 1×4 matrix and a 4×1 matrix. This is why in part (b) a row matrix was written rather than a column matrix. The total materials cost is given by DC. Figure 34 shows how a graphing calculator computes this product. The total cost of the materials is \$188,400. ■

To help keep track of the quantities a matrix represents, let matrix A from Example 10 represent models/styles, matrix B represent styles/materials, and matrix C represent materials/cost. In each case, the meaning of the rows is written first and that of the columns second. When the product AB was found in Example 10, the rows of the matrix represented models and the columns represented materials. Therefore, the matrix product AB represents models/materials.

7.4 Exercises

Find the dimension of each matrix. Identify any square, column, or row matrices.

1. $\begin{bmatrix} -3 & 6 \\ 7 & -4 \end{bmatrix}$

2. $\begin{bmatrix} 2 & -8 & 6 \\ 1 & 0 & -5 \\ 5 & -2 & 3 \end{bmatrix}$

3. $\begin{bmatrix} -6 & 8 & 0 & 0 \\ 4 & 1 & 9 & 2 \\ 3 & -5 & 7 & 1 \end{bmatrix}$

4. $\begin{bmatrix} -3 & 4 & 2 & 1 \\ 0 & 8 & 6 & 3 \end{bmatrix}$

5. $\begin{bmatrix} 2 \\ 4 \end{bmatrix}$ 6. $[4\ 9]$ 7. $[-9]$ 8. $\begin{bmatrix} 0 & 0 & 0 & 0 & 0 \\ 0 & 0 & 0 & 0 & 0 \end{bmatrix}$

Find the value of each variable.

9. $\begin{bmatrix} w & x \\ y & z \end{bmatrix} = \begin{bmatrix} 3 & 2 \\ -1 & 4 \end{bmatrix}$

10. $\begin{bmatrix} 2 & 5 & 6 \\ 1 & m & n \end{bmatrix} = \begin{bmatrix} z & y & w \\ 1 & 8 & -2 \end{bmatrix}$

11. $\begin{bmatrix} 0 & 5 & x \\ -1 & 3 & y+2 \\ 4 & 1 & z \end{bmatrix} = \begin{bmatrix} 0 & w+3 & 6 \\ -1 & 3 & 0 \\ 4 & 1 & 8 \end{bmatrix}$

12. $\begin{bmatrix} 3+x & 4 & t \\ 5 & 8-w & y+1 \\ -4 & 3 & 2r \end{bmatrix} = \begin{bmatrix} 9 & 4 & 6 \\ z+3 & w & 9 \\ p & q & r \end{bmatrix}$

13. $\begin{bmatrix} -7+z & 4r & 8s \\ 6p & 2 & 5 \end{bmatrix} + \begin{bmatrix} -9 & 8r & 3 \\ 2 & 5 & 4 \end{bmatrix} = \begin{bmatrix} 2 & 36 & 27 \\ 20 & 7 & 12a \end{bmatrix}$

14. $\begin{bmatrix} a+2 & 3z+1 & 5m \\ 8k & 0 & 3 \end{bmatrix} + \begin{bmatrix} 3a & 2z & 5m \\ 2k & 5 & 6 \end{bmatrix} = \begin{bmatrix} 10 & -14 & 80 \\ 10 & 5 & 9 \end{bmatrix}$

15. Your friend missed the lecture on adding matrices. In your own words, explain to her how to add two matrices.

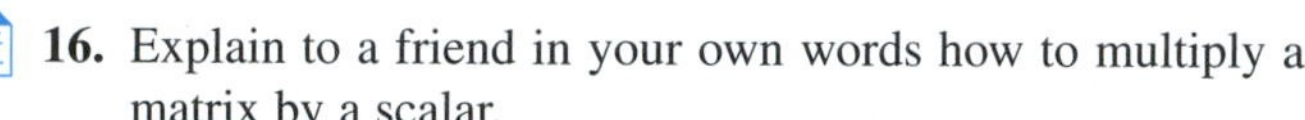
16. Explain to a friend in your own words how to multiply a matrix by a scalar.

Perform each operation if possible.

17. $\begin{bmatrix} 6 & -9 & 2 \\ 4 & 1 & 3 \end{bmatrix} + \begin{bmatrix} -8 & 2 & 5 \\ 6 & -3 & 4 \end{bmatrix}$

18. $\begin{bmatrix} 9 & 4 \\ -8 & 2 \end{bmatrix} + \begin{bmatrix} -3 & 2 \\ -4 & 7 \end{bmatrix}$

19. $\begin{bmatrix} -6 & 8 \\ 0 & 0 \end{bmatrix} - \begin{bmatrix} 0 & 0 \\ -4 & -2 \end{bmatrix}$

20. $\begin{bmatrix} 1 & -4 \\ 2 & -3 \\ -8 & 4 \end{bmatrix} - \begin{bmatrix} -6 & 9 \\ -2 & 5 \\ -7 & -12 \end{bmatrix}$

21. $\begin{bmatrix} 6 & -2 \\ 5 & 4 \end{bmatrix} + \begin{bmatrix} -1 & 7 \\ 7 & -4 \end{bmatrix}$

22. $\begin{bmatrix} 12 & -5 \\ 10 & 3 \end{bmatrix} - \begin{bmatrix} 6 & 9 \\ -2 & 0 \end{bmatrix}$

23. $\begin{bmatrix} -8 & 4 & 0 \\ 2 & 5 & 0 \end{bmatrix} + \begin{bmatrix} 6 & 3 \\ 8 & 9 \end{bmatrix}$

24. $\begin{bmatrix} 2 \\ 3 \end{bmatrix} - \begin{bmatrix} 8 & 1 \\ 9 & 4 \end{bmatrix}$

25. $\begin{bmatrix} 9 & 4 & 1 & -2 \\ 5 & -6 & 3 & 4 \\ 2 & -5 & 1 & 2 \end{bmatrix} - \begin{bmatrix} -2 & 5 & 1 & 3 \\ 0 & 1 & 0 & 2 \\ -8 & 3 & 2 & 1 \end{bmatrix} + \begin{bmatrix} 2 & 4 & 0 & 3 \\ 4 & -5 & 1 & 6 \\ 2 & -3 & 0 & 8 \end{bmatrix}$

26. $\begin{bmatrix} 6 & -2 & 4 \\ -2 & 5 & 8 \\ 1 & 0 & 2 \end{bmatrix} + \begin{bmatrix} 3 & 0 & 8 \\ 1 & -2 & 4 \\ 6 & 9 & -2 \end{bmatrix} - \begin{bmatrix} -4 & 2 & 1 \\ 0 & 3 & -2 \\ 4 & 2 & 0 \end{bmatrix}$

27. $2\begin{bmatrix} 2 & -1 \\ 5 & 1 \\ 0 & 3 \end{bmatrix} + \begin{bmatrix} 5 & 0 \\ 7 & -3 \\ 1 & 1 \end{bmatrix} - \begin{bmatrix} 9 & -4 \\ 4 & 4 \\ 1 & 6 \end{bmatrix}$

28. $-3\begin{bmatrix} 3 & 8 \\ -1 & -9 \end{bmatrix} + 5\begin{bmatrix} 4 & -8 \\ 1 & 6 \end{bmatrix}$

29. $2\begin{bmatrix} 2 & -1 & -1 \\ -1 & 2 & -1 \\ -1 & -1 & 2 \end{bmatrix} + 3\begin{bmatrix} 1 & 2 & 3 \\ 2 & 1 & 3 \\ 2 & 3 & 1 \end{bmatrix}$

30. $3\begin{bmatrix} 1 & 0 & 3 & -1 \\ 0 & 1 & 2 & -1 \\ 1 & 0 & -3 & 1 \end{bmatrix} - 4\begin{bmatrix} -1 & 0 & 0 & 4 \\ 0 & -1 & 3 & 2 \\ 2 & 0 & 1 & -1 \end{bmatrix}$

31. $3\begin{bmatrix} 6 & -1 & 4 \\ 2 & 8 & -3 \\ -4 & 5 & 6 \end{bmatrix} + 5\begin{bmatrix} -2 & -8 & -6 \\ 4 & 1 & 3 \\ 2 & -1 & 5 \end{bmatrix}$

32. $4\begin{bmatrix} 1 & -4 \\ 2 & -3 \\ -8 & 4 \end{bmatrix} - 3\begin{bmatrix} -6 & 9 \\ -2 & 5 \\ -7 & -12 \end{bmatrix}$

Matrices [A] *and* [B] *are shown on the screen. Find each matrix.*

33. 2[A] 34. −3[B]

35. 2[A] − [B] 36. −2[A] + [B]

37. 5[A] + .5[B] 38. −4[A] + 1.5[B]

```
[A]
        [[-2 4]
         [0  3]]
[B]
        [[-6 2]
         [4  0]]
```

39. *Concept Check* Based on the screen shown here, what is matrix [A]?

```
[B]
     [[4   6  -5]
      [-6  3   2 ]]
[A]+[B]
     [[6    12 0 ]
      [-10  -4 11]]
```

40. *Concept Check* Based on the screen shown here, what is matrix [B]?

```
[A]
     [[3   6  5]
      [-2  1  4]]
[A]-[B]
     [[9   0  -5]
      [-4  6  -3]]
```

The dimensions of matrices A and B are given. Find the dimensions of the product AB and of the product BA if the products are defined. If they are not defined, say so.

41. A is 4×2; B is 2×4.

42. A is 3×1; B is 1×3.

43. A is 3×5; B is 5×2.

44. A is 7×3; B is 2×7.

45. A is 4×3; B is 2×5.

46. A is 1×6; B is 2×4.

47. *Concept Check* The product MN of two matrices can be found only if the number of ________ of M equals the number of ________ of N.

48. *Concept Check* In finding the product AB of matrices A and B, the first row, second column, entry is found by multiplying the ________ elements in A and the ________ elements in B and then ________ these products.

Find each matrix product if possible.

49. $\begin{bmatrix} p & q \\ r & s \end{bmatrix}\begin{bmatrix} a & c \\ b & d \end{bmatrix}$

50. $\begin{bmatrix} a & b & c \\ d & e & f \\ g & h & i \end{bmatrix}\begin{bmatrix} x \\ y \\ z \end{bmatrix}$

51. $\begin{bmatrix} 3 & -4 & 1 \\ 5 & 0 & 2 \end{bmatrix}\begin{bmatrix} -1 \\ 4 \\ 2 \end{bmatrix}$

52. $\begin{bmatrix} -6 & 3 & 5 \\ 2 & 9 & 1 \end{bmatrix}\begin{bmatrix} -2 \\ 0 \\ 3 \end{bmatrix}$

53. $\begin{bmatrix} 5 & 2 \\ -1 & 4 \end{bmatrix}\begin{bmatrix} 3 & -2 \\ 1 & 0 \end{bmatrix}$

54. $\begin{bmatrix} -4 & 0 \\ 1 & 3 \end{bmatrix}\begin{bmatrix} -2 & 4 \\ 0 & 1 \end{bmatrix}$

55. $\begin{bmatrix} 2 & 2 & -1 \\ 3 & 0 & 1 \end{bmatrix}\begin{bmatrix} 0 & 2 \\ -1 & 4 \\ 0 & 2 \end{bmatrix}$

56. $\begin{bmatrix} -9 & 2 & 1 \\ 3 & 0 & 0 \end{bmatrix}\begin{bmatrix} 2 \\ -1 \\ 4 \end{bmatrix}$

57. $\begin{bmatrix} -2 & -3 & -4 \\ 2 & -1 & 0 \\ 4 & -2 & 3 \end{bmatrix}\begin{bmatrix} 0 & 1 & 4 \\ 1 & 2 & -1 \\ 3 & 2 & -2 \end{bmatrix}$

58. $\begin{bmatrix} -1 & 2 & 0 \\ 0 & 3 & 2 \\ 0 & 1 & 4 \end{bmatrix}\begin{bmatrix} 2 & -1 & 2 \\ 0 & 2 & 1 \\ 3 & 0 & -1 \end{bmatrix}$

59. $\begin{bmatrix} -2 & 4 & 1 \end{bmatrix}\begin{bmatrix} 3 & -2 & 4 \\ 2 & 1 & 0 \\ 0 & -1 & 4 \end{bmatrix}$

60. $\begin{bmatrix} 0 & 3 & -4 \end{bmatrix}\begin{bmatrix} -2 & 6 & 3 \\ 0 & 4 & 2 \\ -1 & 1 & 4 \end{bmatrix}$

Given $A = \begin{bmatrix} 4 & -2 \\ 3 & 1 \end{bmatrix}$, $B = \begin{bmatrix} 5 & 1 \\ 0 & -2 \\ 3 & 7 \end{bmatrix}$, *and* $C = \begin{bmatrix} -5 & 4 & 1 \\ 0 & 3 & 6 \end{bmatrix}$, *find each product if possible.*

61. BA

62. AC

63. BC

64. CB

65. AB

66. CA

67. A^2

68. A^3

69. *Concept Check* Compare the answers to Exercises 61 and 65, 63 and 64, and 62 and 66. Is matrix multiplication commutative?

70. *Concept Check* For any matrices P and Q, what must be true for both PQ and QP to exist?

Solve each problem.

71. ***Income from Yogurt*** Yagel's Yougurt sells three types of yogurt: nonfat, regular, and supercreamy, at three locations. Location I sells 50 gallons of nonfat, 100 gallons of regular, and 30 gallons of supercreamy each day. Location II sells 10 gallons of nonfat, and Location III sells 60 gallons of nonfat each day. Daily sales of regular yogurt are 90 gallons at Location II and 120 gallons at Location III. At Location II, 50 gallons of supercreamy are sold each day, and 40 gallons of supercreamy are sold each day at Location III.

(a) Write a 3×3 matrix that shows the sales figures for the three locations, with the rows representing the three locations.

(b) The incomes per gallon for nonfat, regular, and supercreamy are \$12, \$10, and \$15, respectively. Write a 1×3 or 3×1 matrix displaying the incomes.

(c) Find a matrix product that gives the daily income at each of the three locations.

(d) What is Yagel's Yogurt's total daily income from the three locations?

72. ***Purchasing Costs*** The Bread Box, a small neighborhood bakery, sells four main items: sweet rolls, bread, cakes, and pies. The amount of each ingredient (in cups, except for eggs) required for these items is given by matrix A.

	Eggs	Flour	Sugar	Shortening	Milk
Rolls (doz)	1	4	$\frac{1}{4}$	$\frac{1}{4}$	1
Bread (loaves)	0	3	0	$\frac{1}{4}$	0
Cakes	4	3	2	1	1
Pies (crust)	0	1	0	$\frac{1}{3}$	0

$= A$

The cost (in cents) for each ingredient when purchased in large lots or small lots is given in matrix B.

	Cost	
	Large lot	Small lot
Eggs	5	5
Flour	8	10
Sugar	10	12
Shortening	12	15
Milk	5	6

$= B$

(a) Use matrix multiplication to find a matrix giving the comparative cost per item for the two purchase options.

(b) Suppose a day's orders consist of 20 dozen sweet rolls, 200 loaves of bread, 50 cakes, and 60 pies. Write the orders as a 1×4 matrix, and using matrix multiplication, write as a matrix the amount of each ingredient needed to fill the day's orders.

(c) Use matrix multiplication to find a matrix giving the costs under the two purchase options to fill the day's orders.

73. ***(Modeling) Predator–Prey Relationship*** In certain parts of the Rocky Mountains, deer are the main food source for mountain lions. When the deer population d is large, the mountain lions (m) thrive. However, a large mountain lion population drives down the size of the deer population. Suppose the fluctuations of the two populations from year to year can be modeled with the matrix equation

$$\begin{bmatrix} m_{n+1} \\ d_{n+1} \end{bmatrix} = \begin{bmatrix} .51 & .4 \\ -.05 & 1.05 \end{bmatrix} \begin{bmatrix} m_n \\ d_n \end{bmatrix}.$$

The numbers in the column matrices give the numbers of animals in the two populations after n years and $n + 1$ years, where the number of deer is measured in hundreds.

(a) Give the equation for d_{n+1} obtained from the second row of the square matrix. Use this equation to determine the rate the deer population will grow from year to year if there are no mountain lions.

(b) Suppose we start with a mountain lion population of 2000 and a deer population of 500,000 (that is, 5000 hundred deer). How large would each population be after 1 year? 2 years?

(c) Consider part (b), but change the initial mountain lion population to 4000. Show that the populations would each grow at a steady annual rate of 1.01.

74. ***(Modeling) Northern Spotted Owl Population*** Several years ago, mathematical ecologists created a mathematical model to analyze population dynamics of the northern spotted owl. The ecologists divided the female owl population into three categories: juvenile (up to 1 year old), subadult (1 to 2 years old), and adult (over 2 years old). They concluded that the change in the makeup of the northern spotted owl population in successive years could be described by the following matrix equation:

$$\begin{bmatrix} j_{n+1} \\ s_{n+1} \\ a_{n+1} \end{bmatrix} = \begin{bmatrix} 0 & 0 & .33 \\ .18 & 0 & 0 \\ 0 & .71 & .94 \end{bmatrix} \begin{bmatrix} j_n \\ s_n \\ a_n \end{bmatrix}.$$

The numbers in the column matrices give the numbers of females in the three age groups after n years and $n + 1$ years. Multiplying the matrices yields

$j_{n+1} = .33a_n$ Each year, 33 juvenile females are born for each 100 adult females.

$s_{n+1} = .18j_n$ Each year, 18% of the juvenile females survive to become subadults.

$a_{n+1} = .71s_n + .94a_n.$ Each year, 71% of the subadults survive to become adults and 94% of the adults survive.

(*Source:* Lamberson, R. H., R. McKelvey, B. R. Noon, and C. Voss, "A Dynamic Analysis of Northern Spotted Owl Viability in a Fragmented Forest Landscape," *Conservation Biology*, Vol. 6, No. 4, December, 1992.)

(a) Suppose there are currently 3000 female northern spotted owls: 690 juveniles, 210 subadults, and 2100 adults. Use the preceding matrix equation to determine the total number of female owls for each of the next 5 years.

(b) Using advanced techniques from linear algebra, we can show that, in the long run,

$$\begin{bmatrix} j_{n+1} \\ s_{n+1} \\ a_{n+1} \end{bmatrix} \approx .98359 \begin{bmatrix} j_n \\ s_n \\ a_n \end{bmatrix}.$$

What can we conclude about the long-term fate of the northern spotted owl?

(c) In this model, the main impediment to the survival of the northern spotted owl is the number .18 in the second row of the 3×3 matrix. This number is low for two reasons: The first year of life is precarious for most animals living in the wild, and juvenile owls must eventually leave the nest and establish their own territory. If much of the forest near their original home has been cleared, then they are vulnerable to predators while searching for a new home. Suppose that, due to better forest management, the number .18 can be increased to .3. Rework part (a) under this new assumption.

Let

$$A = \begin{bmatrix} a_{11} & a_{12} \\ a_{21} & a_{22} \end{bmatrix}, \quad B = \begin{bmatrix} b_{11} & b_{12} \\ b_{21} & b_{22} \end{bmatrix}, \quad \textit{and} \quad C = \begin{bmatrix} c_{11} & c_{12} \\ c_{21} & c_{22} \end{bmatrix},$$

where all the elements are real numbers. Use these matrices to show that each statement is true for 2×2 matrices.

75. $A + B = B + A$ (commutative property)

76. $A + (B + C) = (A + B) + C$ (associative property)

77. $(AB)C = A(BC)$ (associative property)

78. $A(B + C) = AB + AC$ (distributive property)

79. $c(A + B) = cA + cB$, for any real number c

80. $(c + d)A = cA + dA$, for any real numbers c and d

81. $(cA)d = (cd)A$

82. $(cd)A = c(dA)$

7.5 Determinants and Cramer's Rule

Determinants of 2 × 2 Matrices ■ Determinants of Larger Matrices ■ Derivation of Cramer's Rule ■ Using Cramer's Rule to Solve Systems

For convenience, we use subscript notation to name the elements of a matrix, as in the following matrix A:

$$A = \begin{bmatrix} a_{11} & a_{12} & a_{13} & \cdots & a_{1n} \\ a_{21} & a_{22} & a_{23} & \cdots & a_{2n} \\ a_{31} & a_{32} & a_{33} & \cdots & a_{3n} \\ \vdots & \vdots & \vdots & & \vdots \\ a_{m1} & a_{m2} & a_{m3} & \cdots & a_{mn} \end{bmatrix}.$$

With this notation, the row 1, column 1 element is a_{11}; the row 2, column 3 element is a_{23}; and, in general, the row i, column j element is a_{ij}. We use subscript notation in this section to define determinants.

Determinants of 2 × 2 Matrices

Associated with every square matrix A is a real number called the **determinant** of A. The symbols $|A|$, $\delta(A)$, and $\det A$ all represent the determinant of A. In this text, we use **det A.**

Looking Ahead to Calculus

Determinants are utilized in calculus to find *vector cross products,* which are used to study the effect of forces in the plane or in space. They are also used to express the results of certain vector operations.

Determinant of a 2 × 2 Matrix

The **determinant of a 2 × 2 matrix A,** where

$$A = \begin{bmatrix} a_{11} & a_{12} \\ a_{21} & a_{22} \end{bmatrix},$$

is a real number defined as $\det A = a_{11}a_{22} - a_{21}a_{12}$.

NOTE You should be able to distinguish between a matrix and its determinant. A matrix is an array of numbers, while a determinant is a *single real number* associated with a square matrix.

GCM EXAMPLE 1 Evaluating the Determinant of a 2 × 2 Matrix

Find $\det A$ if $A = \begin{bmatrix} -3 & 4 \\ 6 & 8 \end{bmatrix}$.

Analytic Solution

Use the definition just given.

$$\det A = \det \begin{bmatrix} -3 & 4 \\ 6 & 8 \end{bmatrix}$$

$$= -3(8) - 6(4)$$

($-3 = a_{11}$, $8 = a_{22}$, $6 = a_{21}$, $4 = a_{12}$)

$$= -48$$

Graphing Calculator Solution

Graphing calculators with matrix capabilities can compute determinants. Figure 35 shows both [A] and det[A].

```
[A]
          [[-3 4]
           [6  8]]
det([A])
             -48
```

FIGURE 35

EXAMPLE 2 Solving an Equation Involving a Determinant

If $A = \begin{bmatrix} x & 3 \\ -1 & 5 \end{bmatrix}$ and $\det A = 33$, find the value of x.

Solution Since $\det A = 5x - (-1)(3)$, or $5x + 3$, we have

$$5x + 3 = 33$$

$$5x = 30 \quad \text{Subtract 3.}$$

$$x = 6. \quad \text{Divide by 5.}$$

Check to see that $\det \begin{bmatrix} 6 & 3 \\ -1 & 5 \end{bmatrix} = 33$.

Determinants of Larger Matrices

Determinant of a 3 × 3 Matrix

The **determinant of a 3 × 3 matrix** A**,** where

$$A = \begin{bmatrix} a_{11} & a_{12} & a_{13} \\ a_{21} & a_{22} & a_{23} \\ a_{31} & a_{32} & a_{33} \end{bmatrix},$$

is a real number defined as

$$\begin{aligned} \det A = &(a_{11}a_{22}a_{33} + a_{12}a_{23}a_{31} + a_{13}a_{21}a_{32}) \\ &-(a_{31}a_{22}a_{13} + a_{32}a_{23}a_{11} + a_{33}a_{21}a_{12}). \end{aligned}$$

NOTE The above definition is not usually memorized in its general form. The method of cofactors, which is described next, is used more often because it allows us to calculate the determinant of any square matrix.

A method for calculating 3 × 3 determinants is found by rearranging and factoring the terms given in the definition to get

$$\det \begin{bmatrix} a_{11} & a_{12} & a_{13} \\ a_{21} & a_{22} & a_{23} \\ a_{31} & a_{32} & a_{33} \end{bmatrix} = a_{11}(a_{22}a_{33} - a_{32}a_{23}) - a_{21}(a_{12}a_{33} - a_{32}a_{13}) + a_{31}(a_{12}a_{23} - a_{22}a_{13}).$$

Each quantity in parentheses represents the determinant of a 2 × 2 matrix that is the part of the 3 × 3 matrix remaining when the row and column of the multiplier are eliminated, as shown here.

$$a_{11}(a_{22}a_{33} - a_{32}a_{23}) \quad \begin{bmatrix} a_{11} & a_{12} & a_{13} \\ a_{21} & a_{22} & a_{23} \\ a_{31} & a_{32} & a_{33} \end{bmatrix}$$

$$a_{21}(a_{12}a_{33} - a_{32}a_{13}) \quad \begin{bmatrix} a_{11} & a_{12} & a_{13} \\ a_{21} & a_{22} & a_{23} \\ a_{31} & a_{32} & a_{33} \end{bmatrix}$$

$$a_{31}(a_{12}a_{23} - a_{22}a_{13}) \quad \begin{bmatrix} a_{11} & a_{12} & a_{13} \\ a_{21} & a_{22} & a_{23} \\ a_{31} & a_{32} & a_{33} \end{bmatrix}$$

The determinant of each 2 × 2 matrix is called a **minor** of the associated element in the 3 × 3 matrix. The symbol M_{ij} represents the minor that results when row i and column j are eliminated. The list on the next page gives some of the minors from the preceding matrix.

Element	Minor	Element	Minor
a_{11}	$M_{11} = \det\begin{bmatrix} a_{22} & a_{23} \\ a_{32} & a_{33} \end{bmatrix}$	a_{22}	$M_{22} = \det\begin{bmatrix} a_{11} & a_{13} \\ a_{31} & a_{33} \end{bmatrix}$
a_{21}	$M_{21} = \det\begin{bmatrix} a_{12} & a_{13} \\ a_{32} & a_{33} \end{bmatrix}$	a_{23}	$M_{23} = \det\begin{bmatrix} a_{11} & a_{12} \\ a_{31} & a_{32} \end{bmatrix}$
a_{31}	$M_{31} = \det\begin{bmatrix} a_{12} & a_{13} \\ a_{22} & a_{23} \end{bmatrix}$	a_{33}	$M_{33} = \det\begin{bmatrix} a_{11} & a_{12} \\ a_{21} & a_{22} \end{bmatrix}$

In a 4×4 matrix, the minors are determinants of 3×3 matrices. Similarly, an $n \times n$ matrix has minors that are determinants of $(n - 1) \times (n - 1)$ matrices.

To find the determinant of a 3×3 or larger square matrix, first choose any row or column. Then the minor of each element in that row or column must be multiplied by $+1$ or -1, depending on whether the sum of the row number and column number is even or odd. The product of a minor and the number $+1$ or -1 is called a *cofactor*.

Cofactor

Let M_{ij} be the minor for element a_{ij} in an $n \times n$ matrix. The **cofactor** of a_{ij}, written A_{ij}, is

$$A_{ij} = (-1)^{i+j} \cdot M_{ij}.$$

EXAMPLE 3 Finding the Cofactor of an Element

Find the cofactor of each given element of the matrix

$$\begin{bmatrix} 6 & 2 & 4 \\ 8 & 9 & 3 \\ 1 & 2 & 0 \end{bmatrix}.$$

(a) 6 **(b)** 3 **(c)** 8

Solution

(a) Since 6 is in the first row, first column of the matrix, $i = 1$ and $j = 1$. $M_{11} = \det\begin{bmatrix} 9 & 3 \\ 2 & 0 \end{bmatrix} = -6$. The cofactor is

$$(-1)^{1+1}(-6) = 1(-6) = -6.$$

(b) Here $i = 2$ and $j = 3$. $M_{23} = \det\begin{bmatrix} 6 & 2 \\ 1 & 2 \end{bmatrix} = 10$. The cofactor is

$$(-1)^{2+3}(10) = -1(10) = -10.$$

(c) We have $i = 2$ and $j = 1$. $M_{21} = \det\begin{bmatrix} 2 & 4 \\ 2 & 0 \end{bmatrix} = -8$. The cofactor is

$$(-1)^{2+1}(-8) = -1(-8) = 8.$$ ■

Finally, the determinant of an $n \times n$ matrix is found as follows.

Finding the Determinant of an $n \times n$ Matrix

Multiply each element in any row or column of the matrix by its cofactor. The sum of these products gives the value of the determinant.

The process of forming this sum of products is called **expansion by a row or column.**

GCM **EXAMPLE 4 Evaluating the Determinant of a 3 × 3 Matrix**

Evaluate $\det\begin{bmatrix} 2 & -3 & -2 \\ -1 & -4 & -3 \\ -1 & 0 & 2 \end{bmatrix}$, expanding by the second column.

In Example 1 we saw that the determinant of a 2 × 2 matrix can be found using a graphing calculator. Figure 36 shows how a calculator evaluates the 3 × 3 determinant in Example 4.

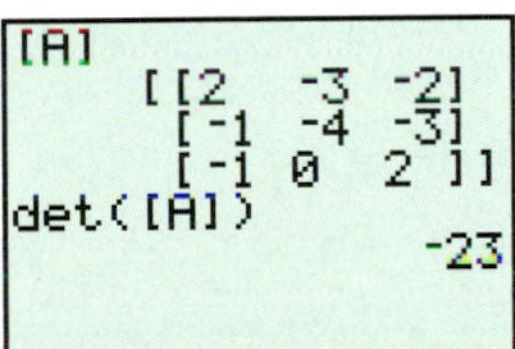

FIGURE 36

Solution First find the minor of each element in the second column.

$$M_{12} = \det\begin{bmatrix} -1 & -3 \\ -1 & 2 \end{bmatrix} = -1(2) - (-1)(-3) = -5$$

$$M_{22} = \det\begin{bmatrix} 2 & -2 \\ -1 & 2 \end{bmatrix} = 2(2) - (-1)(-2) = 2$$

$$M_{32} = \det\begin{bmatrix} 2 & -2 \\ -1 & -3 \end{bmatrix} = 2(-3) - (-1)(-2) = -8$$

Now find the cofactor of each of these minors.

$$A_{12} = (-1)^{1+2} \cdot M_{12} = (-1)^3(-5) = -1(-5) = 5$$

$$A_{22} = (-1)^{2+2} \cdot M_{22} = (-1)^4 \cdot 2 = 1 \cdot 2 = 2$$

$$A_{32} = (-1)^{3+2} \cdot M_{32} = (-1)^5(-8) = -1(-8) = 8$$

Find the determinant by multiplying each cofactor by its corresponding element in the matrix and finding the sum of the products.

$$\det\begin{bmatrix} 2 & -3 & -2 \\ -1 & -4 & -3 \\ -1 & 0 & 2 \end{bmatrix} = a_{12} \cdot A_{12} + a_{22} \cdot A_{22} + a_{32} \cdot A_{32}$$

$$= -3(5) + (-4)2 + 0(8)$$

$$= -15 + (-8) + 0 = -23$$ ■

CAUTION Keep track of all negative signs when evaluating determinants by hand. Write down each step as in the examples. Skipping steps leads to computational errors.

In Example 4, we would have found exactly the same answer by using any row or column of the matrix. One reason we used column 2 is that it contains a 0 element, so it was not really necessary to calculate M_{32} and A_{32}. One learns quickly that 0s are "friendly" in work with determinants.

For 3 × 3 Matrices

+	−	+
−	+	−
+	−	+

Instead of calculating $(-1)^{i+j}$ for a given element, the sign checkerboard in the margin can be used. The signs alternate for each row and column, beginning with + in the first row, first column, position. If we expand a 3×3 matrix about row 3, for example, the first minor would have a + sign associated with it, the second minor a − sign, and the third minor a + sign. This array of signs can be extended to determinants of 4×4 and larger matrices.

GCM **EXAMPLE 5** **Evaluating the Determinant of a 4 × 4 Matrix**

Evaluate.

$$\det\begin{bmatrix} -1 & -2 & 3 & 2 \\ 0 & 1 & 4 & -2 \\ 3 & -1 & 4 & 0 \\ 2 & 1 & 0 & 3 \end{bmatrix}$$

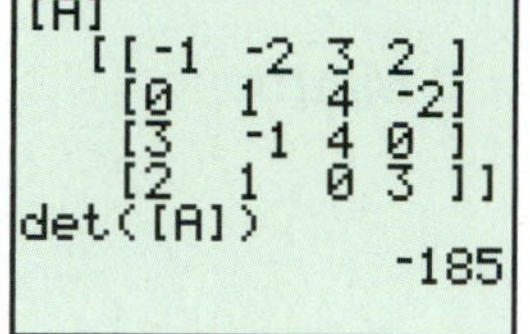

FIGURE 37

Solution The determinant of a 4×4 matrix can be found by using cofactors, but the computation is tedious. Using a graphing calculator with matrix capabilities is much easier. Figure 37 indicates that the desired determinant is −185. ■

Derivation of Cramer's Rule

Determinants can be used to solve a linear system of equations in the form

$$a_1x + b_1y = c_1 \quad (1)$$
$$a_2x + b_2y = c_2 \quad (2)$$

by elimination as follows:

$$a_1b_2x + b_1b_2y = c_1b_2 \quad \text{Multiply (1) by } b_2.$$
$$-a_2b_1x - b_1b_2y = -c_2b_1 \quad \text{Multiply (2) by } -b_1.$$
$$(a_1b_2 - a_2b_1)x = c_1b_2 - c_2b_1 \quad \text{Add.}$$
$$x = \frac{c_1b_2 - c_2b_1}{a_1b_2 - a_2b_1} \quad \text{if } a_1b_2 - a_2b_1 \neq 0.$$

Similarly,

$$-a_1a_2x - a_2b_1y = -a_2c_1 \quad \text{Multiply (1) by } -a_2.$$
$$a_1a_2x + a_1b_2y = a_1c_2 \quad \text{Multiply (2) by } a_1.$$
$$(a_1b_2 - a_2b_1)y = a_1c_2 - a_2c_1 \quad \text{Add.}$$
$$y = \frac{a_1c_2 - a_2c_1}{a_1b_2 - a_2b_1} \quad \text{if } a_1b_2 - a_2b_1 \neq 0.$$

Both numerators and the common denominator of these values for x and y can be written as determinants, since

$$c_1b_2 - c_2b_1 = \det\begin{bmatrix} c_1 & b_1 \\ c_2 & b_2 \end{bmatrix}, \quad a_1c_2 - a_2c_1 = \det\begin{bmatrix} a_1 & c_1 \\ a_2 & c_2 \end{bmatrix}, \quad \text{and} \quad a_1b_2 - a_2b_1 = \det\begin{bmatrix} a_1 & b_1 \\ a_2 & b_2 \end{bmatrix}.$$

With these determinants, the solutions for x and y become

$$x = \frac{\det\begin{bmatrix} c_1 & b_1 \\ c_2 & b_2 \end{bmatrix}}{\det\begin{bmatrix} a_1 & b_1 \\ a_2 & b_2 \end{bmatrix}} \quad \text{and} \quad y = \frac{\det\begin{bmatrix} a_1 & c_1 \\ a_2 & c_2 \end{bmatrix}}{\det\begin{bmatrix} a_1 & b_1 \\ a_2 & b_2 \end{bmatrix}} \quad \text{if } \det\begin{bmatrix} a_1 & b_1 \\ a_2 & b_2 \end{bmatrix} \neq 0.$$

These results are summarized as **Cramer's rule.**

Cramer's Rule for 2 × 2 Systems

The solution of the system

$$a_1x + b_1y = c_1$$
$$a_2x + b_2y = c_2$$

is given by $\quad x = \dfrac{D_x}{D} \quad$ and $\quad y = \dfrac{D_y}{D},$

where

$$D_x = \det\begin{bmatrix} c_1 & b_1 \\ c_2 & b_2 \end{bmatrix}, \quad D_y = \det\begin{bmatrix} a_1 & c_1 \\ a_2 & c_2 \end{bmatrix}, \quad \text{and} \quad D = \det\begin{bmatrix} a_1 & b_1 \\ a_2 & b_2 \end{bmatrix} \neq 0.$$

CAUTION ***Cramer's rule does not apply if $D = 0$.*** When $D = 0$, the system is inconsistent or has dependent equations. For this reason, evaluate D first.

Using Cramer's Rule to Solve Systems

EXAMPLE 6 Applying Cramer's Rule to a System with Two Equations

Use Cramer's rule to solve the system.

$$5x + 7y = -1$$
$$6x + 8y = 1$$

Analytic Solution

By Cramer's rule, $x = \dfrac{D_x}{D}$ and $y = \dfrac{D_y}{D}$. Find D first, because if $D = 0$, Cramer's rule does not apply. If $D \neq 0$, then find D_x and D_y.

$$D = \det\begin{bmatrix} 5 & 7 \\ 6 & 8 \end{bmatrix} = 5(8) - 6(7) = -2$$

$$D_x = \det\begin{bmatrix} -1 & 7 \\ 1 & 8 \end{bmatrix} = -1(8) - 1(7) = -15$$

$$D_y = \det\begin{bmatrix} 5 & -1 \\ 6 & 1 \end{bmatrix} = 5(1) - 6(-1) = 11$$

From Cramer's rule,

$$x = \frac{D_x}{D} = \frac{-15}{-2} = \frac{15}{2} \quad \text{and} \quad y = \frac{D_y}{D} = \frac{11}{-2} = -\frac{11}{2}.$$

The solution set is $\left\{\left(\frac{15}{2}, -\frac{11}{2}\right)\right\}$, as can be verified by substituting into the given system.

Graphing Calculator Solution

See Figure 38. Enter D, D_x, and D_y as [A], [B], and [C], respectively. Then find the desired quotients

$$x = \frac{\det([B])}{\det([A])} \quad \text{and} \quad y = \frac{\det([C])}{\det([A])}.$$

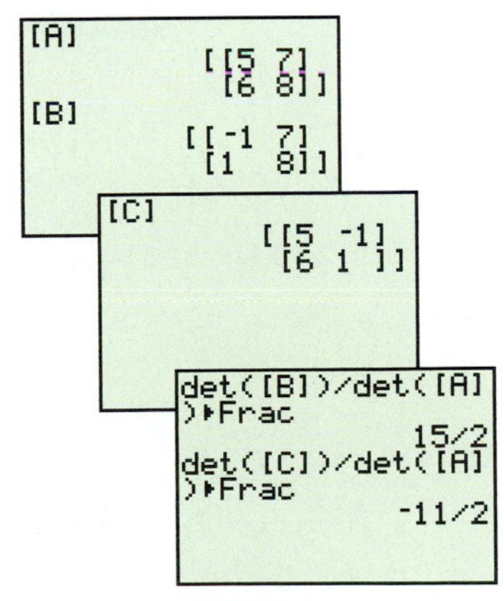

FIGURE 38

Cramer's rule can be generalized to systems of three equations in three variables (or n equations in n variables).

Cramer's Rule for 3 × 3 Systems

The solution of the system

$$\begin{aligned} a_1x + b_1y + c_1z &= d_1 \\ a_2x + b_2y + c_2z &= d_2 \\ a_3x + b_3y + c_3z &= d_3 \end{aligned}$$

is given by $$x = \frac{D_x}{D}, \quad y = \frac{D_y}{D}, \quad \text{and} \quad z = \frac{D_z}{D},$$

where $$D_x = \det\begin{bmatrix} d_1 & b_1 & c_1 \\ d_2 & b_2 & c_2 \\ d_3 & b_3 & c_3 \end{bmatrix}, \qquad D_y = \det\begin{bmatrix} a_1 & d_1 & c_1 \\ a_2 & d_2 & c_2 \\ a_3 & d_3 & c_3 \end{bmatrix},$$

$$D_z = \det\begin{bmatrix} a_1 & b_1 & d_1 \\ a_2 & b_2 & d_2 \\ a_3 & b_3 & d_3 \end{bmatrix}, \quad \text{and} \quad D = \det\begin{bmatrix} a_1 & b_1 & c_1 \\ a_2 & b_2 & c_2 \\ a_3 & b_3 & c_3 \end{bmatrix} \neq 0.$$

EXAMPLE 7 Applying Cramer's Rule to a System with Three Equations

Use Cramer's rule to solve the system.

$$\begin{aligned} x + y - z &= -2 \\ 2x - y + z &= -5 \\ x - 2y + 3z &= 4 \end{aligned}$$

Solution Verify that the required determinants are

$$D = \det\begin{bmatrix} 1 & 1 & -1 \\ 2 & -1 & 1 \\ 1 & -2 & 3 \end{bmatrix} = -3, \quad D_x = \det\begin{bmatrix} -2 & 1 & -1 \\ -5 & -1 & 1 \\ 4 & -2 & 3 \end{bmatrix} = 7,$$

$$D_y = \det\begin{bmatrix} 1 & -2 & -1 \\ 2 & -5 & 1 \\ 1 & 4 & 3 \end{bmatrix} = -22, \quad D_z = \det\begin{bmatrix} 1 & 1 & -2 \\ 2 & -1 & -5 \\ 1 & -2 & 4 \end{bmatrix} = -21.$$

Thus, $$x = \frac{D_x}{D} = \frac{7}{-3} = -\frac{7}{3}, \quad y = \frac{D_y}{D} = \frac{-22}{-3} = \frac{22}{3}, \quad z = \frac{D_z}{D} = \frac{-21}{-3} = 7,$$

so the solution set is $\left\{\left(-\frac{7}{3}, \frac{22}{3}, 7\right)\right\}$. ■

EXAMPLE 8 **Verifying That Cramer's Rule Does Not Apply**

Show why Cramer's rule does not apply to the system.

$$\begin{aligned} 2x - 3y + 4z &= 10 \\ 6x - 9y + 12z &= 24 \\ x + 2y - 3z &= 5 \end{aligned}$$

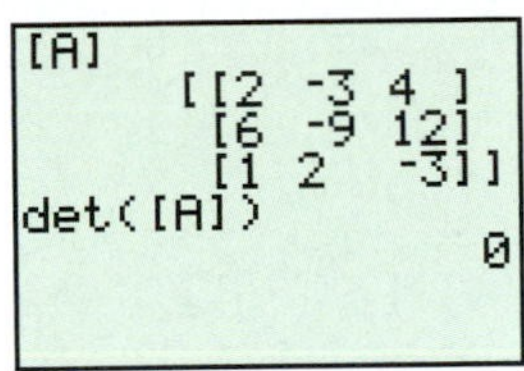

FIGURE 39

Solution We must show that $D = 0$. Figure 39 confirms this fact, with $D = \det([A])$. When $D = 0$, the system either is inconsistent or contains dependent equations. We could use the elimination or row echelon method to tell which is the case. Verify that this system is inconsistent. ■

7.5 Exercises

Find each determinant.

1. $\det\begin{bmatrix} -5 & 9 \\ 4 & -1 \end{bmatrix}$

2. $\det\begin{bmatrix} -1 & 3 \\ -2 & 9 \end{bmatrix}$

3. $\det\begin{bmatrix} -1 & -2 \\ 5 & 3 \end{bmatrix}$

4. $\det\begin{bmatrix} 6 & -4 \\ 0 & -1 \end{bmatrix}$

5. $\det\begin{bmatrix} 9 & 3 \\ -3 & -1 \end{bmatrix}$

6. $\det\begin{bmatrix} 0 & 2 \\ 1 & 5 \end{bmatrix}$

7. $\det\begin{bmatrix} 3 & 4 \\ 5 & -2 \end{bmatrix}$

8. $\det\begin{bmatrix} -9 & 7 \\ 2 & 6 \end{bmatrix}$

Find the cofactor of each element in the second row for each matrix.

9. $\begin{bmatrix} -2 & 0 & 1 \\ 1 & 2 & 0 \\ 4 & 2 & 1 \end{bmatrix}$

10. $\begin{bmatrix} 1 & -1 & 2 \\ 1 & 0 & 2 \\ 0 & -3 & 1 \end{bmatrix}$

11. $\begin{bmatrix} 1 & 2 & -1 \\ 2 & 3 & -2 \\ -1 & 4 & 1 \end{bmatrix}$

12. $\begin{bmatrix} 2 & -1 & 4 \\ 3 & 0 & 1 \\ -2 & 1 & 4 \end{bmatrix}$

Find each determinant.

13. $\det\begin{bmatrix} 4 & -7 & 8 \\ 2 & 1 & 3 \\ -6 & 3 & 0 \end{bmatrix}$

14. $\det\begin{bmatrix} 8 & -2 & -4 \\ 7 & 0 & 3 \\ 5 & -1 & 2 \end{bmatrix}$

15. $\det\begin{bmatrix} 1 & 2 & 0 \\ -1 & 2 & -1 \\ 0 & 1 & 4 \end{bmatrix}$

16. $\det\begin{bmatrix} 2 & 1 & -1 \\ 4 & 7 & -2 \\ 2 & 4 & 0 \end{bmatrix}$

17. $\det\begin{bmatrix} 10 & 2 & 1 \\ -1 & 4 & 3 \\ -3 & 8 & 10 \end{bmatrix}$

18. $\det\begin{bmatrix} 7 & -1 & 1 \\ 1 & -7 & 2 \\ -2 & 1 & 1 \end{bmatrix}$

19. $\det\begin{bmatrix} 1 & -2 & 3 \\ 0 & 0 & 0 \\ 1 & 10 & -12 \end{bmatrix}$

20. $\det\begin{bmatrix} 2 & 3 & 0 \\ 1 & 9 & 0 \\ -1 & -2 & 0 \end{bmatrix}$

21. $\det\begin{bmatrix} 3 & 3 & -1 \\ 2 & 6 & 0 \\ -6 & -6 & 2 \end{bmatrix}$

22. $\det\begin{bmatrix} 5 & -3 & 2 \\ -5 & 3 & -2 \\ 1 & 0 & 1 \end{bmatrix}$

23. $\det\begin{bmatrix} .4 & -.8 & .6 \\ .3 & .9 & .7 \\ 3.1 & 4.1 & -2.8 \end{bmatrix}$

24. $\det\begin{bmatrix} -.3 & -.1 & .9 \\ 2.5 & 4.9 & -3.2 \\ -.1 & .4 & .8 \end{bmatrix}$

Solve each determinant equation for x.

25. $\det\begin{bmatrix} 5 & x \\ -3 & 2 \end{bmatrix} = 6$

26. $\det\begin{bmatrix} -.5 & 2 \\ x & x \end{bmatrix} = 0$

27. $\det\begin{bmatrix} x & 3 \\ x & x \end{bmatrix} = 4$

28. $\det\begin{bmatrix} 2x & x \\ 11 & x \end{bmatrix} = 6$

29. $\det\begin{bmatrix} -2 & 0 & 1 \\ -1 & 3 & x \\ 5 & -2 & 0 \end{bmatrix} = 3$

30. $\det\begin{bmatrix} 4 & 3 & 0 \\ 2 & 0 & 1 \\ -3 & x & -1 \end{bmatrix} = 5$

31. $\det\begin{bmatrix} 5 & 3x & -3 \\ 0 & 2 & -1 \\ 4 & -1 & x \end{bmatrix} = -7$

32. $\det\begin{bmatrix} 2x & 1 & -1 \\ 0 & 4 & x \\ 3 & 0 & 2 \end{bmatrix} = x$

Area of a Triangle *A triangle with vertices at* (x_1, y_1), (x_2, y_2), *and* (x_3, y_3), *as shown in the figure, has area equal to the absolute value of D, where*

$$D = \frac{1}{2}\det\begin{bmatrix} x_1 & y_1 & 1 \\ x_2 & y_2 & 1 \\ x_3 & y_3 & 1 \end{bmatrix}.$$

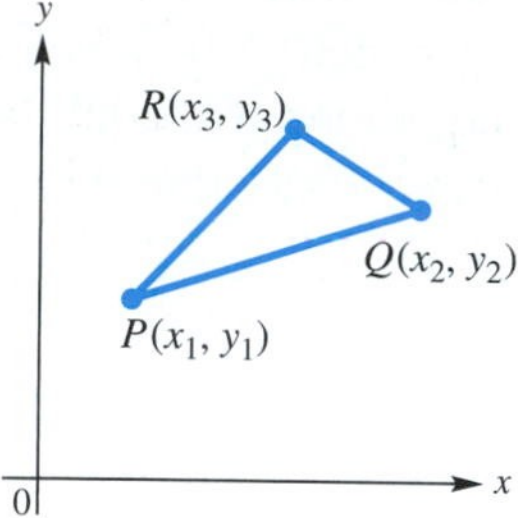

Use D to find the area of each triangle with coordinates as given.

33. $P(0,0), Q(0,2), R(1,4)$

34. $P(0,1), Q(2,0), R(1,5)$

35. $P(2,5), Q(-1,3), R(4,0)$

36. $P(2,-2), Q(0,0), R(-3,-4)$

37. $P(1,2), Q(4,3), R(3,5)$

38. ***Area of a Triangle*** Find the area of a triangular lot whose vertices have coordinates in feet of $(101.3, 52.7)$, $(117.2, 253.9)$, and $(313.1, 301.6)$. (*Source:* Al-Khafaji, A. and J. Tooley, *Numerical Methods in Engineering Practice,* Holt, Rinehart and Winston, 1995.)

Use the concept of the area of a triangle discussed in Exercises 33–38 to determine whether the three points are collinear.

39. $(1, 3), (-3, 11), (2, 1)$

40. $(3, 6), (-1, -6), (5, 11)$

41. $(-2, -5), (4, 4), (2, 3)$

42. $(4, -5), (-2, 10), (6, -10)$

Evaluate each determinant.

43. $\det\begin{bmatrix} 3 & -6 & 5 & -1 \\ 0 & 2 & -1 & 3 \\ -6 & 4 & 2 & 0 \\ -7 & 3 & 1 & 1 \end{bmatrix}$

44. $\det\begin{bmatrix} 4 & 5 & -1 & -1 \\ 2 & -3 & 1 & 0 \\ -5 & 1 & 3 & 9 \\ 0 & -2 & 1 & 5 \end{bmatrix}$

45. $\det\begin{bmatrix} 4 & 0 & 0 & 2 \\ -1 & 0 & 3 & 0 \\ 2 & 4 & 0 & 1 \\ 0 & 0 & 1 & 2 \end{bmatrix}$

46. $\det\begin{bmatrix} -2 & 0 & 4 & 2 \\ 3 & 6 & 0 & 4 \\ 0 & 0 & 0 & 3 \\ 9 & 0 & 2 & -1 \end{bmatrix}$

Several theorems are useful for calculating determinants. These theorems are true for square matrices of any dimension.

Determinant Theorems

1. If every element in a row (or column) of matrix A is 0, then $\det A = 0$.
2. If the rows of matrix A are the corresponding columns of matrix B, then $\det B = \det A$.
3. If any two rows (or columns) of matrix A are interchanged to form matrix B, then $\det B = -\det A$.
4. Suppose matrix B is formed by multiplying every element of a row (or column) of matrix A by the real number k. Then $\det B = k \cdot \det A$.
5. If two rows (or columns) of a matrix A are identical, then $\det A = 0$.
6. Changing a row (or column) of a matrix by adding a constant times another row (or column) to it does not change the determinant of the matrix.

Use the determinant theorems to find each determinant.

47. $\det\begin{bmatrix} 1 & 0 & 0 \\ 1 & 0 & 1 \\ 3 & 0 & 0 \end{bmatrix}$

48. $\det\begin{bmatrix} -1 & 2 & 4 \\ 4 & -8 & -16 \\ 3 & 0 & 5 \end{bmatrix}$

49. $\det\begin{bmatrix} 6 & 8 & -12 \\ -1 & 0 & 2 \\ 4 & 0 & -8 \end{bmatrix}$

50. $\det\begin{bmatrix} 4 & 8 & 0 \\ -1 & -2 & 1 \\ 2 & 4 & 3 \end{bmatrix}$

51. $\det\begin{bmatrix} -4 & 1 & 4 \\ 2 & 0 & 1 \\ 0 & 2 & 4 \end{bmatrix}$

52. $\det\begin{bmatrix} 6 & 3 & 2 \\ 1 & 0 & 2 \\ 5 & 7 & 3 \end{bmatrix}$

Relating Concepts

For individual or group investigation (Exercises 53–56)

Determinants can be used to find the equation of a line passing through two given points. ***Work Exercises 53–56 in order.***

53. Expand the determinant in the equation $\det\begin{bmatrix} x & y & 1 \\ 2 & 3 & 1 \\ -1 & 4 & 1 \end{bmatrix} = 0.$

54. Find the equation of the line through (2, 3) and (−1, 4). How does your answer compare with the answer to Exercise 53?

55. Write the equation of the line through the points (x_1, y_1) and (x_2, y_2), using the point–slope formula.

56. Expand the following determinant, and compare the resulting equation with your answer to Exercise 55. What do you find?

$$\det\begin{bmatrix} x & y & 1 \\ x_1 & y_1 & 1 \\ x_2 & y_2 & 1 \end{bmatrix} = 0$$

57. ***Concept Check*** For the following system, match each determinant defined by Cramer's rule in (a)–(d) with its equivalent from choices A–D.

$$\begin{aligned} 4x + 3y - 2z &= 1 \\ 7x - 4y + 3z &= 2 \\ -2x + y - 8z &= 0 \end{aligned}$$

(a) D **(b)** D_x **(c)** D_y **(d)** D_z

A. $\det\begin{bmatrix} 1 & 3 & -2 \\ 2 & -4 & 3 \\ 0 & 1 & -8 \end{bmatrix}$ **B.** $\det\begin{bmatrix} 4 & 3 & 1 \\ 7 & -4 & 2 \\ -2 & 1 & 0 \end{bmatrix}$ **C.** $\det\begin{bmatrix} 4 & 1 & -2 \\ 7 & 2 & 3 \\ -2 & 0 & -8 \end{bmatrix}$ **D.** $\det\begin{bmatrix} 4 & 3 & -2 \\ 7 & -4 & 3 \\ -2 & 1 & -8 \end{bmatrix}$

58. ***Concept Check*** For the following system, $D = -43$, $D_x = -43$, $D_y = 0$, and $D_z = 43$. Use Cramer's rule to find the solution set.

$$\begin{aligned} x + 3y - 6z &= 7 \\ 2x - y + z &= 1 \\ x + 2y + 2z &= -1 \end{aligned}$$

Use Cramer's rule to solve each system of equations. If $D = 0$, use another method to complete the solution.

59. $x + y = 4$
$2x - y = 2$

60. $3x + 2y = -4$
$2x - y = -5$

61. $4x + 3y = -7$
$2x + 3y = -11$

62. $4x - y = 0$
$2x + 3y = 14$

63. $3x + 2y = 4$
$6x + 4y = 8$

64. $1.5x + 3y = 5$
$2x + 4y = 3$

65. $2x - 3y = -5$
$x + 5y = 17$

66. $x + 9y = -15$
$3x + 2y = 5$

67. $\begin{aligned} 4x - y + 3z &= -3 \\ 3x + y + z &= 0 \\ 2x - y + 4z &= 0 \end{aligned}$

68. $\begin{aligned} 5x + 2y + z &= 15 \\ 2x - y + z &= 9 \\ 4x + 3y + 2z &= 13 \end{aligned}$

69. $\begin{aligned} 2x - y + 4z &= -2 \\ 3x + 2y - z &= -3 \\ x + 4y + 2z &= 17 \end{aligned}$

70. $\begin{aligned} x + y + z &= 4 \\ 2x - y + 3z &= 4 \\ 4x + 2y - z &= -15 \end{aligned}$

71. $\begin{aligned} 5x - y &= -4 \\ 3x + 2z &= 4 \\ 4y + 3z &= 22 \end{aligned}$

72. $\begin{aligned} 3x + 5y &= -7 \\ 2x + 7z &= 2 \\ 4y + 3z &= -8 \end{aligned}$

73. $\begin{aligned} 2x - y + 3z &= 1 \\ -2x + y - 3z &= 2 \\ 5x - y + z &= 2 \end{aligned}$

74. $\begin{aligned} -2x - 2y + 3z &= 4 \\ 5x + 7y - z &= 2 \\ 2x + 2y - 3z &= -4 \end{aligned}$

75. $\begin{aligned} 3x + 2y - w &= 0 \\ 2x + z + 2w &= 5 \\ x + 2y - z &= -2 \\ 2x - y + z + w &= 2 \end{aligned}$

76. $\begin{aligned} x + 2y - z + w &= 8 \\ 2x - y + 2w &= 8 \\ y + 3z &= 5 \\ x - z &= 4 \end{aligned}$

77. In your own words, explain what happens when you apply Cramer's rule if $D = 0$.

78. Describe D_x, D_y, and D_z in terms of the coefficients and constants in a given system of equations.

7.6 Solution of Linear Systems by Matrix Inverses

Identity Matrices ■ Multiplicative Inverses of Square Matrices ■ Using Determinants to Find Inverses ■ Solving Linear Systems Using Inverse Matrices ■ Curve Fitting Using a System

In Section 7.4, we saw several parallels between the set of real numbers and the set of matrices. Another similarity is that both sets have identity and inverse elements for multiplication.

Identity Matrices

By the identity property for real numbers, for any real number a,

$$a \cdot 1 = a \quad \text{and} \quad 1 \cdot a = a.$$

If there is to be a multiplicative **identity matrix** I, such that

$$AI = A \quad \text{and} \quad IA = A,$$

for any matrix A, then A and I must be square matrices with the same dimension.

Figure 40 shows how the identity matrix I_2 is displayed on a graphing calculator with matrix capabilities.

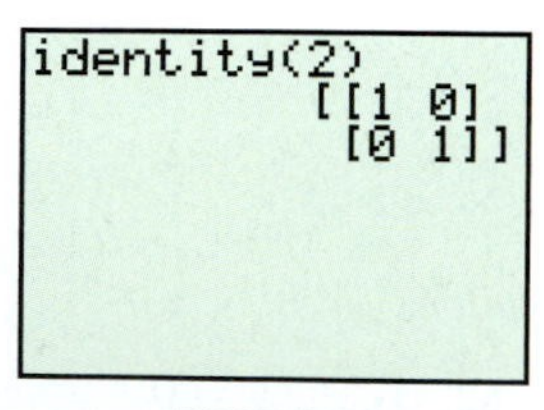

FIGURE 40

2 × 2 Identity Matrix

If I_2 represents the 2×2 identity matrix, then

$$I_2 = \begin{bmatrix} 1 & 0 \\ 0 & 1 \end{bmatrix}.$$

To verify that I_2 is the 2×2 identity matrix, we must show that $AI = A$ and $IA = A$ for any 2×2 matrix A. Let

$$A = \begin{bmatrix} x & y \\ z & w \end{bmatrix}.$$

Then

$$AI = \begin{bmatrix} x & y \\ z & w \end{bmatrix}\begin{bmatrix} 1 & 0 \\ 0 & 1 \end{bmatrix} = \begin{bmatrix} x \cdot 1 + y \cdot 0 & x \cdot 0 + y \cdot 1 \\ z \cdot 1 + w \cdot 0 & z \cdot 0 + w \cdot 1 \end{bmatrix} = \begin{bmatrix} x & y \\ z & w \end{bmatrix} = A,$$

and

$$IA = \begin{bmatrix} 1 & 0 \\ 0 & 1 \end{bmatrix}\begin{bmatrix} x & y \\ z & w \end{bmatrix} = \begin{bmatrix} 1 \cdot x + 0 \cdot z & 1 \cdot y + 0 \cdot w \\ 0 \cdot x + 1 \cdot z & 0 \cdot y + 1 \cdot w \end{bmatrix} = \begin{bmatrix} x & y \\ z & w \end{bmatrix} = A.$$

GCM EXAMPLE 1 Using the 2 × 2 Identity Matrix

Let $A = \begin{bmatrix} -2 & 6 \\ 3 & 5 \end{bmatrix}$. Show that $AI_2 = A$ and $I_2A = A$.

Analytic Solution

$$AI_2 = \begin{bmatrix} -2 & 6 \\ 3 & 5 \end{bmatrix}\begin{bmatrix} 1 & 0 \\ 0 & 1 \end{bmatrix}$$

$$= \begin{bmatrix} -2(1) + 6(0) & -2(0) + 6(1) \\ 3(1) + 5(0) & 3(0) + 5(1) \end{bmatrix}$$

$$= \begin{bmatrix} -2 & 6 \\ 3 & 5 \end{bmatrix} = A$$

$$I_2A = \begin{bmatrix} 1 & 0 \\ 0 & 1 \end{bmatrix}\begin{bmatrix} -2 & 6 \\ 3 & 5 \end{bmatrix}$$

$$= \begin{bmatrix} 1(-2) + 0(3) & 1(6) + 0(5) \\ 0(-2) + 1(3) & 0(6) + 1(5) \end{bmatrix}$$

$$= \begin{bmatrix} -2 & 6 \\ 3 & 5 \end{bmatrix} = A$$

Graphing Calculator Solution

Define [A] as shown in Figure 41. Then show that $AI_2 = A$ and $I_2A = A$.

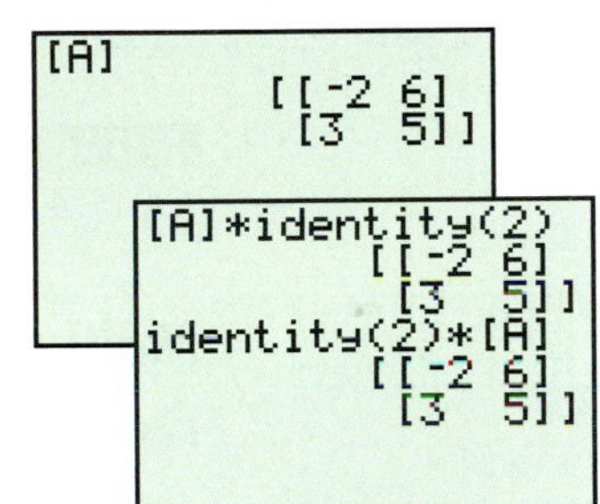

FIGURE 41

Figure 42 shows how a graphing calculator displays I_3 and I_4.

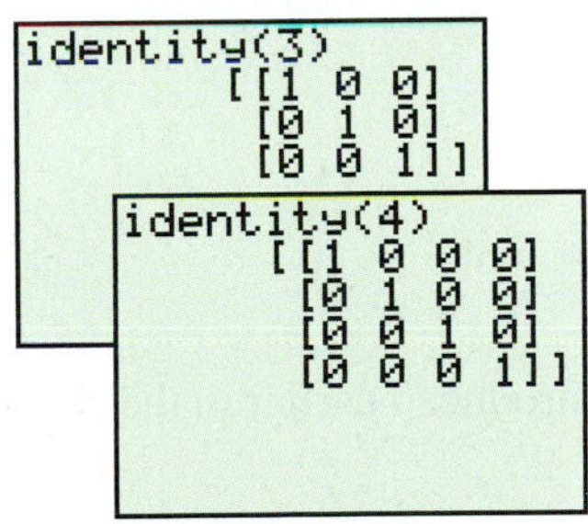

FIGURE 42

The 2 × 2 identity matrix I_2 suggests the following generalization.

$n \times n$ Identity Matrix

For any value of n, there is an $n \times n$ identity matrix having 1s down the main diagonal and 0s elsewhere. The $\boldsymbol{n \times n}$ **identity matrix** is given by

$$I_n = \begin{bmatrix} \mathbf{1} & \mathbf{0} & \cdots & \mathbf{0} \\ \mathbf{0} & \mathbf{1} & \cdots & \mathbf{0} \\ \vdots & \vdots & \boldsymbol{a_{ij}} & \vdots \\ \mathbf{0} & \mathbf{0} & \cdots & \mathbf{1} \end{bmatrix}.$$

Here, $a_{ij} = 1$ when $i = j$ (the diagonal elements), and $a_{ij} = 0$ otherwise.

EXAMPLE 2 **Using the 3 × 3 Identity Matrix**

Let $A = \begin{bmatrix} -2 & 4 & 0 \\ 3 & 5 & 9 \\ 0 & 8 & -6 \end{bmatrix}$. Show that $AI_3 = A$.

Analytic Solution

Using the 3×3 identity matrix and the definition of matrix multiplication gives

$$AI_3 = \begin{bmatrix} -2 & 4 & 0 \\ 3 & 5 & 9 \\ 0 & 8 & -6 \end{bmatrix}\begin{bmatrix} 1 & 0 & 0 \\ 0 & 1 & 0 \\ 0 & 0 & 1 \end{bmatrix}$$

$$= \begin{bmatrix} -2 & 4 & 0 \\ 3 & 5 & 9 \\ 0 & 8 & -6 \end{bmatrix} = A.$$

Graphing Calculator Solution

See Figure 43.

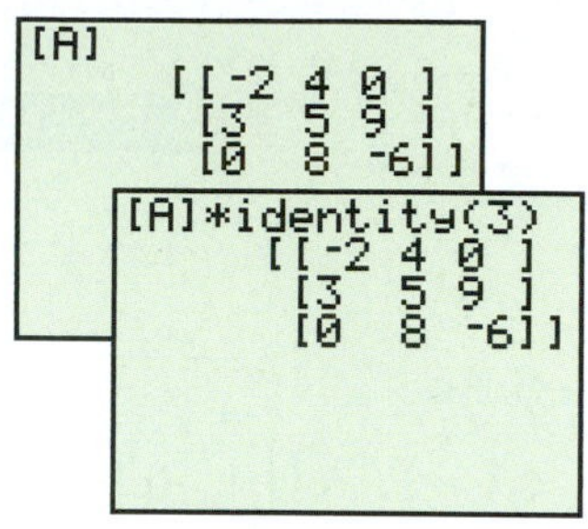

FIGURE 43

NOTE Because multiplication by an *identity* matrix is commutative, we can also show that $I_3A = A$ in Example 2.

Multiplicative Inverses of Square Matrices

For every nonzero real number a, there is a multiplicative inverse $\frac{1}{a}$ such that

$$a \cdot \frac{1}{a} = 1 \quad \text{and} \quad \frac{1}{a} \cdot a = 1.$$

(Recall: $\frac{1}{a}$ is also written a^{-1}.) In a similar way, if A is an $n \times n$ matrix, then its **multiplicative inverse,** written A^{-1}, must satisfy both

$$AA^{-1} = I_n \quad \text{and} \quad A^{-1}A = I_n.$$

This result means that only a square matrix can have a multiplicative inverse.

CAUTION Although $a^{-1} = \frac{1}{a}$ for any nonzero real number a, if A is a matrix, then

$$A^{-1} \neq \frac{1}{A}.$$

In fact, $\frac{1}{A}$ has no meaning, since 1 is a *number* and A is a *matrix.*

To find the matrix A^{-1}, we use row transformations, introduced earlier in the chapter. As an example, we find the inverse of

$$A = \begin{bmatrix} 2 & 4 \\ 1 & -1 \end{bmatrix}.$$

Let the unknown inverse matrix be

$$A^{-1} = \begin{bmatrix} x & y \\ z & w \end{bmatrix}.$$

By the definition of matrix inverse, $AA^{-1} = I_2$, or

$$AA^{-1} = \begin{bmatrix} 2 & 4 \\ 1 & -1 \end{bmatrix}\begin{bmatrix} x & y \\ z & w \end{bmatrix} = \begin{bmatrix} 1 & 0 \\ 0 & 1 \end{bmatrix}.$$

By matrix multiplication,

$$\begin{bmatrix} 2x + 4z & 2y + 4w \\ x - z & y - w \end{bmatrix} = \begin{bmatrix} 1 & 0 \\ 0 & 1 \end{bmatrix}.$$

Setting corresponding elements equal gives the system of equations

$$\begin{aligned} 2x + 4z &= 1 && (1) \\ 2y + 4w &= 0 && (2) \\ x - z &= 0 && (3) \\ y - w &= 1. && (4) \end{aligned}$$

Since equations (1) and (3) involve only x and z, while equations (2) and (4) involve only y and w, these four equations lead to two systems of equations,

$$\begin{aligned} 2x + 4z &= 1 \\ x - z &= 0 \end{aligned} \quad \text{and} \quad \begin{aligned} 2y + 4w &= 0 \\ y - w &= 1. \end{aligned}$$

Writing the two systems as augmented matrices gives

$$\left[\begin{array}{cc|c} 2 & 4 & 1 \\ 1 & -1 & 0 \end{array}\right] \quad \text{and} \quad \left[\begin{array}{cc|c} 2 & 4 & 0 \\ 1 & -1 & 1 \end{array}\right].$$

Each of these systems can be solved by the reduced row echelon method. However, since the elements to the left of the vertical bar are identical, the two systems can be combined into one matrix,

$$\left[\begin{array}{cc|cc} 2 & 4 & 1 & 0 \\ 1 & -1 & 0 & 1 \end{array}\right],$$

and solved simultaneously. We need to transform the numbers on the left of the vertical bar to the 2×2 identity matrix.

$$\left[\begin{array}{cc|cc} 1 & -1 & 0 & 1 \\ 2 & 4 & 1 & 0 \end{array}\right] \quad \text{Interchange } R_1 \text{ and } R_2 \text{ to get 1 in the upper left corner.}$$

$$\left[\begin{array}{cc|cc} 1 & -1 & 0 & 1 \\ 0 & 6 & 1 & -2 \end{array}\right] \quad -2R_1 + R_2$$

$$\left[\begin{array}{cc|cc} 1 & -1 & 0 & 1 \\ 0 & 1 & \frac{1}{6} & -\frac{1}{3} \end{array}\right] \quad \tfrac{1}{6}R_2$$

$$\left[\begin{array}{cc|cc} 1 & 0 & \frac{1}{6} & \frac{2}{3} \\ 0 & 1 & \frac{1}{6} & -\frac{1}{3} \end{array}\right] \quad R_2 + R_1$$

The numbers in the first column to the right of the vertical bar in the final matrix give the values of x and z. The second column gives the values of y and w. That is,

$$\left[\begin{array}{cc|cc} 1 & 0 & x & y \\ 0 & 1 & z & w \end{array}\right] = \left[\begin{array}{cc|cc} 1 & 0 & \frac{1}{6} & \frac{2}{3} \\ 0 & 1 & \frac{1}{6} & -\frac{1}{3} \end{array}\right]$$

so that

$$A^{-1} = \begin{bmatrix} x & y \\ z & w \end{bmatrix} = \begin{bmatrix} \frac{1}{6} & \frac{2}{3} \\ \frac{1}{6} & -\frac{1}{3} \end{bmatrix}.$$

To check, multiply A by A^{-1}. The result should be I_2.

$$AA^{-1} = \begin{bmatrix} 2 & 4 \\ 1 & -1 \end{bmatrix}\begin{bmatrix} \frac{1}{6} & \frac{2}{3} \\ \frac{1}{6} & -\frac{1}{3} \end{bmatrix} = \begin{bmatrix} \frac{1}{3}+\frac{2}{3} & \frac{4}{3}-\frac{4}{3} \\ \frac{1}{6}-\frac{1}{6} & \frac{2}{3}+\frac{1}{3} \end{bmatrix} = \begin{bmatrix} 1 & 0 \\ 0 & 1 \end{bmatrix} = I_2$$

Thus, $$A^{-1} = \begin{bmatrix} \frac{1}{6} & \frac{2}{3} \\ \frac{1}{6} & -\frac{1}{3} \end{bmatrix}.$$

If A^{-1} exists, then it is unique. If A^{-1} does not exist, then A is a **singular matrix.** The process for finding the multiplicative inverse A^{-1} for any $n \times n$ matrix A that has an inverse is summarized as follows.

Finding an Inverse Matrix

To obtain A^{-1} for any $n \times n$ matrix A for which A^{-1} exists, follow these steps:

Step 1 Form the augmented matrix $[A|I_n]$, where I_n is the $n \times n$ identity matrix.

Step 2 Perform row transformations on $[A|I_n]$, to obtain a matrix of the form $[I_n|B]$.

Step 3 Matrix B is A^{-1}.

NOTE To confirm that two $n \times n$ matrices A and B are inverses of each other, it is sufficient to show that $AB = I_n$. It is not necessary to show also that $BA = I_n$.

EXAMPLE 3 Finding the Inverse of a 3 × 3 Matrix

Find A^{-1} if $A = \begin{bmatrix} 1 & 0 & 1 \\ 2 & -2 & -1 \\ 3 & 0 & 0 \end{bmatrix}$.

Solution Use row transformations as follows.

Step 1 Write the augmented matrix $[A|I_3]$.

$$\left[\begin{array}{ccc|ccc} 1 & 0 & 1 & 1 & 0 & 0 \\ 2 & -2 & -1 & 0 & 1 & 0 \\ 3 & 0 & 0 & 0 & 0 & 1 \end{array}\right]$$

Step 2 Since 1 is already in the upper left-hand corner as desired, begin by using the row transformation that will result in 0 for the first element in the second row. Multiply the elements of the first row by -2, and add the result to the second row.

$$\left[\begin{array}{ccc|ccc} 1 & 0 & 1 & 1 & 0 & 0 \\ 0 & -2 & -3 & -2 & 1 & 0 \\ 3 & 0 & 0 & 0 & 0 & 1 \end{array}\right] \quad -2R_1 + R_2$$

To get 0 for the first element in the third row, multiply the elements of the first row by -3 and add the result to the third row.

$$\left[\begin{array}{rrr|rrr} 1 & 0 & 1 & 1 & 0 & 0 \\ 0 & -2 & -3 & -2 & 1 & 0 \\ 0 & 0 & -3 & -3 & 0 & 1 \end{array}\right] \quad -3R_1 + R_3$$

To get 1 for the second element in the second row, multiply the elements of the second row by $-\frac{1}{2}$.

$$\left[\begin{array}{rrr|rrr} 1 & 0 & 1 & 1 & 0 & 0 \\ 0 & 1 & \frac{3}{2} & 1 & -\frac{1}{2} & 0 \\ 0 & 0 & -3 & -3 & 0 & 1 \end{array}\right] \quad -\tfrac{1}{2}R_2$$

Continuing this process, we transform the matrix as follows.

$$\left[\begin{array}{rrr|rrr} 1 & 0 & 1 & 1 & 0 & 0 \\ 0 & 1 & \frac{3}{2} & 1 & -\frac{1}{2} & 0 \\ 0 & 0 & 1 & 1 & 0 & -\frac{1}{3} \end{array}\right] \quad -\tfrac{1}{3}R_3$$

$$\left[\begin{array}{rrr|rrr} 1 & 0 & 0 & 0 & 0 & \frac{1}{3} \\ 0 & 1 & \frac{3}{2} & 1 & -\frac{1}{2} & 0 \\ 0 & 0 & 1 & 1 & 0 & -\frac{1}{3} \end{array}\right] \quad -1R_3 + R_1$$

$$\left[\begin{array}{rrr|rrr} 1 & 0 & 0 & 0 & 0 & \frac{1}{3} \\ 0 & 1 & 0 & -\frac{1}{2} & -\frac{1}{2} & \frac{1}{2} \\ 0 & 0 & 1 & 1 & 0 & -\frac{1}{3} \end{array}\right] \quad -\tfrac{3}{2}R_3 + R_2$$

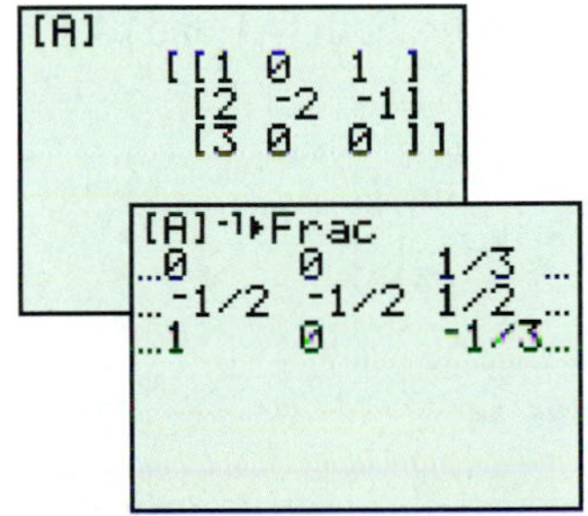

A graphing calculator can be used to find the inverse of a matrix. The screens support the result in Example 3. The elements of the inverse are expressed as fractions, so it is easier to compare with the inverse matrix found in the example.

Step 3 The last transformation shows that the inverse is

$$A^{-1} = \begin{bmatrix} 0 & 0 & \frac{1}{3} \\ -\frac{1}{2} & -\frac{1}{2} & \frac{1}{2} \\ 1 & 0 & -\frac{1}{3} \end{bmatrix}.$$

Confirm this result by forming the product $A^{-1}A$ or AA^{-1}, each of which equals the matrix I_3. ■

Using Determinants to Find Inverses

Determinants can be used to find the inverses of matrices. Suppose that $A = \begin{bmatrix} a & b \\ c & d \end{bmatrix}$ and that A has inverse $A^{-1} = \begin{bmatrix} x & y \\ z & w \end{bmatrix}$. Then it can be shown that the elements of A^{-1} are determined by the equations

$$x = \frac{d}{ad - cb}, \quad y = \frac{-b}{ad - cb}, \quad z = \frac{-c}{ad - cb}, \quad \text{and} \quad w = \frac{a}{ad - cb}.$$

Verify these results. This discussion is summarized in the box on the next page.

Inverse of a 2 × 2 Matrix

If $A = \begin{bmatrix} a & b \\ c & d \end{bmatrix}$ and $\det A \neq 0$, then

$$A^{-1} = \frac{1}{\det A}\begin{bmatrix} d & -b \\ -c & a \end{bmatrix} \quad \text{or} \quad A^{-1} = \begin{bmatrix} \frac{d}{ad - cb} & \frac{-b}{ad - cb} \\ \frac{-c}{ad - cb} & \frac{a}{ad - cb} \end{bmatrix}.$$

If $\det A = 0$, then A^{-1} does not exist and A is a singular matrix.

GCM **EXAMPLE 4** **Finding the Inverse of a 2 × 2 Matrix**

Find A^{-1}, if it exists, for each matrix.

(a) $A = \begin{bmatrix} 2 & 3 \\ 1 & -1 \end{bmatrix}$ **(b)** $A = \begin{bmatrix} 3 & -6 \\ 2 & -4 \end{bmatrix}$

Analytic Solution

(a) First we find $\det A$ and determine whether A^{-1} actually exists.

$$\det A = 2(-1) - 1(3)$$
$$= -5$$

Since $-5 \neq 0$, A^{-1} exists. Thus,

$$A^{-1} = \frac{1}{-5}\begin{bmatrix} -1 & -3 \\ -1 & 2 \end{bmatrix}$$
$$= \begin{bmatrix} \frac{1}{5} & \frac{3}{5} \\ \frac{1}{5} & -\frac{2}{5} \end{bmatrix}.$$

(b) Here, A^{-1} does not exist, since $\det A = 0$:

$$\det A = 3(-4) - 2(-6)$$
$$= 0.$$

Graphing Calculator Solution

(a) See Figure 44, which illustrates both [A] and $[A]^{-1}$.

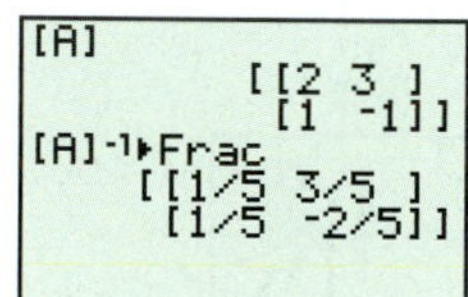

FIGURE 44

(b) The calculator returns a singular matrix error message when it is directed to find the inverse of a matrix whose determinant is 0. See Figure 45.

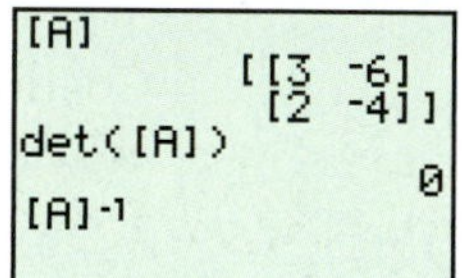

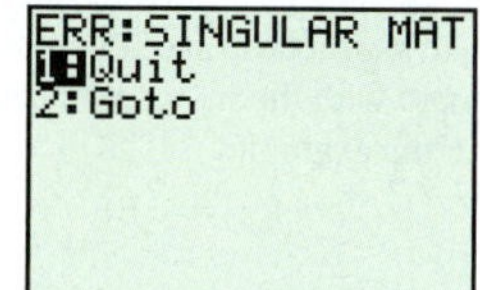

FIGURE 45

Solving Linear Systems Using Inverse Matrices

We used matrices to solve systems of linear equations by the row echelon method in Section 7.3. Another way to use matrices to solve linear systems is to write the system as a matrix equation $AX = B$, where A is the matrix of the coefficients of the variables of the system, X is the matrix of the variables, and B is the matrix of the constants. Matrix A is called the **coefficient matrix.**

To solve the matrix equation $AX = B$, first see if A^{-1} exists. Assuming that it does, use the facts that $A^{-1}A = I$ and $IX = X$ to get

$$\begin{aligned} AX &= B \\ A^{-1}(AX) &= A^{-1}B && \text{Multiply each side by } A^{-1}. \\ (A^{-1}A)X &= A^{-1}B && \text{Associative property} \\ IX &= A^{-1}B && \text{Multiplicative inverse property} \\ X &= A^{-1}B. && \text{Identity property} \end{aligned}$$

CAUTION ***When multiplying by matrices on each side of a matrix equation, be careful to multiply in the same order on each side.*** Multiplication of matrices is not commutative.

EXAMPLE 5 **Using a Matrix Inverse to Solve a System of Equations**

Use the inverse of the coefficient matrix to solve the system.

$$\begin{aligned} 2x + y + 3z &= 1 \\ x - 2y + z &= -3 \\ -3x + y - 2z &= -4 \end{aligned}$$

Analytic Solution

We find the determinant of the coefficient matrix to be sure that it is not 0. If

$$A = \begin{bmatrix} 2 & 1 & 3 \\ 1 & -2 & 1 \\ -3 & 1 & -2 \end{bmatrix},$$

then $\det A = -10 \neq 0$ and A^{-1} exists. We evaluate $A^{-1}B$ with

$$X = \begin{bmatrix} x \\ y \\ z \end{bmatrix} \quad \text{and} \quad B = \begin{bmatrix} 1 \\ -3 \\ -4 \end{bmatrix}.$$

$$A^{-1}B = \begin{bmatrix} 2 & 1 & 3 \\ 1 & -2 & 1 \\ -3 & 1 & -2 \end{bmatrix}^{-1} \begin{bmatrix} 1 \\ -3 \\ -4 \end{bmatrix}$$

$$= \begin{bmatrix} -\frac{3}{10} & -\frac{1}{2} & -\frac{7}{10} \\ \frac{1}{10} & -\frac{1}{2} & -\frac{1}{10} \\ \frac{1}{2} & \frac{1}{2} & \frac{1}{2} \end{bmatrix} \begin{bmatrix} 1 \\ -3 \\ -4 \end{bmatrix} = \begin{bmatrix} 4 \\ 2 \\ -3 \end{bmatrix}$$

Since $X = A^{-1}B$, we have $x = 4$, $y = 2$, and $z = -3$. The solution set is $\{(4, 2, -3)\}$. The elements of A^{-1} can be found by hand, as demonstrated in Example 3, or by using a calculator.

Graphing Calculator Solution

Entering the coefficient matrix as [A], we find that det([A]) = −10 and $[A]^{-1}$ exists. See Figure 46. Then we enter the column of constants as [B] and find the matrix product $[A]^{-1}[B]$. The display shows the column matrix representing $x = 4$, $y = 2$, and $z = -3$. The solution set is $\{(4, 2, -3)\}$.

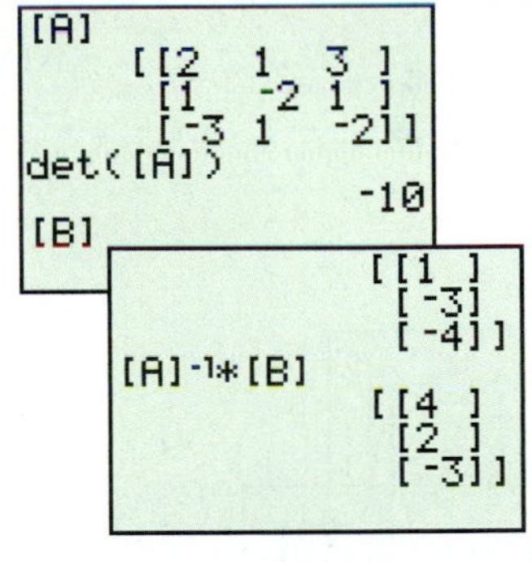

FIGURE 46

NOTE If the determinant of the coefficient matrix A is 0, the system either is inconsistent or has dependent equations and A^{-1} does not exist.

FOR DISCUSSION

1. Using the window $[-5, 5]$ by $[-50, 50]$, reproduce the graph in Figure 47, based on the result of Example 6.
2. If your calculator has *cubic regression*, use it to find the same results as in Example 6.

Curve Fitting Using a System

EXAMPLE 6 Using a System to Find the Equation of a Cubic Polynomial

Figure 47 shows four views of the graph of a polynomial function of the form $P(x) = ax^3 + bx^2 + cx + d$. Use the points indicated to write a system of four equations in the variables a, b, c, and d, and then use the inverse matrix method to solve the system. What is the equation that defines this graph?

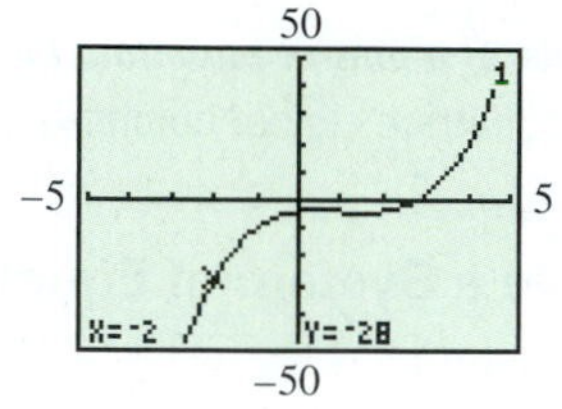

First view

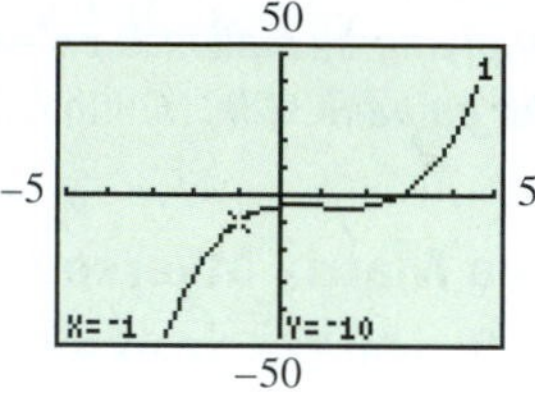

Second view

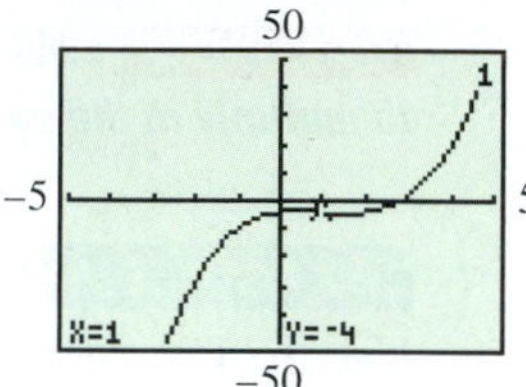

Third view

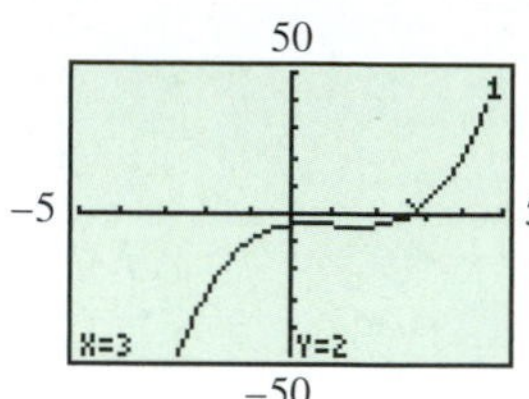

Fourth view

FIGURE 47

Solution We see from the graph that $P(-2) = -28$, $P(-1) = -10$, $P(1) = -4$, and $P(3) = 2$. From the first of these, we get

$$P(-2) = a(-2)^3 + b(-2)^2 + c(-2) + d = -28$$

or, equivalently,

$$-8a + 4b - 2c + d = -28.$$

Similarly, from the others, we find the following equations:

$$\begin{aligned} &\text{From } P(-1) = -10: & -a + b - c + d &= -10 \\ &\text{From } P(1) = -4: & a + b + c + d &= -4 \\ &\text{From } P(3) = 2: & 27a + 9b + 3c + d &= 2. \end{aligned}$$

Now we must solve the system formed by these four equations:

$$\begin{aligned} -8a + 4b - 2c + d &= -28 \\ -a + b - c + d &= -10 \\ a + b + c + d &= -4 \\ 27a + 9b + 3c + d &= 2. \end{aligned}$$

We use the inverse matrix method to solve the system, letting

$$A = \begin{bmatrix} -8 & 4 & -2 & 1 \\ -1 & 1 & -1 & 1 \\ 1 & 1 & 1 & 1 \\ 27 & 9 & 3 & 1 \end{bmatrix}, \quad X = \begin{bmatrix} a \\ b \\ c \\ d \end{bmatrix}, \quad \text{and} \quad B = \begin{bmatrix} -28 \\ -10 \\ -4 \\ 2 \end{bmatrix}.$$

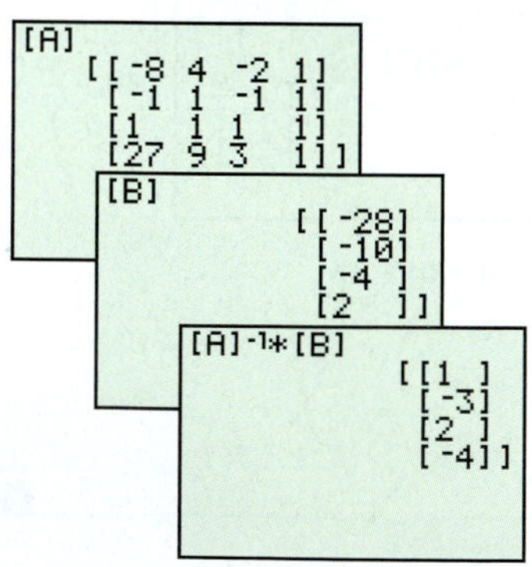

FIGURE 48

Because $\det A = 240 \neq 0$, a unique solution exists. On the basis of our earlier discussion, $X = A^{-1}B$. Figure 48 shows how the matrix inverse method yields the values $a = 1$, $b = -3$, $c = 2$, and $d = -4$. The polynomial is defined by $P(x) = x^3 - 3x^2 + 2x - 4$. ■

7.6 Exercises

Determine whether A and B are inverses by calculating AB and BA.

1. $A = \begin{bmatrix} 5 & 7 \\ 2 & 3 \end{bmatrix}; B = \begin{bmatrix} 3 & -7 \\ -2 & 5 \end{bmatrix}$

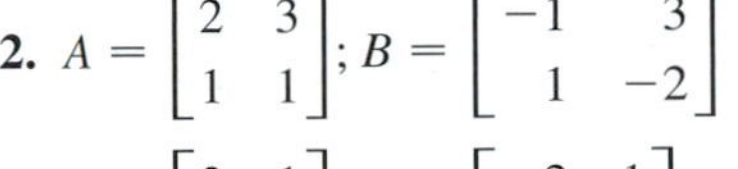

2. $A = \begin{bmatrix} 2 & 3 \\ 1 & 1 \end{bmatrix}; B = \begin{bmatrix} -1 & 3 \\ 1 & -2 \end{bmatrix}$

3. $A = \begin{bmatrix} -1 & 2 \\ 3 & -5 \end{bmatrix}; B = \begin{bmatrix} -5 & -2 \\ -3 & -1 \end{bmatrix}$

4. $A = \begin{bmatrix} 2 & 1 \\ 3 & 2 \end{bmatrix}; B = \begin{bmatrix} 2 & 1 \\ -3 & 2 \end{bmatrix}$

5. $A = \begin{bmatrix} 0 & 1 & 0 \\ 0 & 0 & -2 \\ 1 & -1 & 0 \end{bmatrix}; B = \begin{bmatrix} 1 & 0 & 1 \\ 1 & 0 & 0 \\ 0 & -1 & 0 \end{bmatrix}$

6. $A = \begin{bmatrix} 1 & 2 & 0 \\ 0 & 1 & 0 \\ 0 & 1 & 0 \end{bmatrix}; B = \begin{bmatrix} 1 & -2 & 0 \\ 0 & 1 & 0 \\ 0 & -1 & 1 \end{bmatrix}$

7. $A = \begin{bmatrix} -1 & -1 & -1 \\ 4 & 5 & 0 \\ 0 & 1 & -3 \end{bmatrix}; B = \begin{bmatrix} 15 & 4 & -5 \\ -12 & -3 & 4 \\ -4 & -1 & 1 \end{bmatrix}$

8. $A = \begin{bmatrix} 1 & 3 & 3 \\ 1 & 4 & 3 \\ 1 & 3 & 4 \end{bmatrix}; B = \begin{bmatrix} 7 & -3 & -3 \\ -1 & 1 & 0 \\ -1 & 0 & 1 \end{bmatrix}$

9. Under what condition will the inverse of a square matrix not exist?

10. Explain why a 2×2 matrix will not have an inverse if either a column or a row contains all 0s.

For each matrix, find A^{-1} by hand if it exists.

11. $A = \begin{bmatrix} 3 & 7 \\ 2 & 5 \end{bmatrix}$

12. $A = \begin{bmatrix} -5 & 3 \\ -8 & 5 \end{bmatrix}$

13. $A = \begin{bmatrix} -1 & -2 \\ 3 & 4 \end{bmatrix}$

14. $A = \begin{bmatrix} -1 & 2 \\ -2 & -1 \end{bmatrix}$

15. $A = \begin{bmatrix} -6 & 4 \\ -3 & 2 \end{bmatrix}$

16. $A = \begin{bmatrix} 5 & 10 \\ -3 & -6 \end{bmatrix}$

17. $A = \begin{bmatrix} .6 & .2 \\ .5 & .1 \end{bmatrix}$

18. $A = \begin{bmatrix} .8 & -.3 \\ .5 & -.2 \end{bmatrix}$

19. $A = \begin{bmatrix} 0 & 0 & 1 \\ 1 & 0 & 0 \\ 0 & 1 & 0 \end{bmatrix}$

20. $A = \begin{bmatrix} 1 & 0 & 0 \\ 1 & 1 & 0 \\ 0 & 1 & 1 \end{bmatrix}$

21. $A = \begin{bmatrix} 1 & 0 & 1 \\ 2 & 1 & 3 \\ -1 & 1 & 1 \end{bmatrix}$

22. $A = \begin{bmatrix} -2 & 1 & 0 \\ 1 & 0 & 1 \\ -1 & 1 & 0 \end{bmatrix}$

For each matrix, find A^{-1} if it exists.

23. $A = \begin{bmatrix} 1 & 0 & 0 \\ 0 & -1 & 0 \\ 1 & 0 & 1 \end{bmatrix}$

24. $A = \begin{bmatrix} 1 & 3 & 3 \\ 1 & 4 & 3 \\ 1 & 3 & 4 \end{bmatrix}$

25. $A = \begin{bmatrix} -1 & -1 & -1 \\ 4 & 5 & 0 \\ 0 & 1 & -3 \end{bmatrix}$

26. $A = \begin{bmatrix} 2 & 0 & 4 \\ 3 & 1 & 5 \\ -1 & 1 & -2 \end{bmatrix}$

27. $A = \begin{bmatrix} -.4 & .1 & .2 \\ 0 & .6 & .8 \\ .3 & 0 & -.2 \end{bmatrix}$

28. $A = \begin{bmatrix} .8 & .2 & .1 \\ -.2 & 0 & .3 \\ 0 & 0 & .5 \end{bmatrix}$

29. $A = \begin{bmatrix} 2 & 1 & 2 \\ 5 & 10 & 5 \\ 3 & 6 & 3 \end{bmatrix}$

30. $A = \begin{bmatrix} 5 & -3 & 2 \\ -5 & 3 & -2 \\ 1 & 0 & 1 \end{bmatrix}$

31. $A = \begin{bmatrix} \sqrt{2} & .5 \\ -17 & \frac{1}{2} \end{bmatrix}$

32. $A = \begin{bmatrix} \frac{2}{3} & .7 \\ 22 & \sqrt{3} \end{bmatrix}$

33. $A = \begin{bmatrix} 1.4 & .5 & .59 \\ .84 & 1.36 & .62 \\ .56 & .47 & 1.3 \end{bmatrix}$

34. $A = \begin{bmatrix} \frac{1}{2} & \frac{1}{4} & \frac{1}{3} \\ 0 & \frac{1}{4} & \frac{1}{3} \\ \frac{1}{2} & \frac{1}{2} & \frac{1}{3} \end{bmatrix}$

Solve each system by using the matrix inverse method.

35. $\begin{aligned} 2x - y &= -8 \\ 3x + y &= -2 \end{aligned}$

36. $\begin{aligned} x + 3y &= -12 \\ 2x - y &= 11 \end{aligned}$

37. $\begin{aligned} 2x + 3y &= -10 \\ 3x + 4y &= -12 \end{aligned}$

38. $\begin{aligned} 2x - 3y &= 10 \\ 2x + 2y &= 5 \end{aligned}$

39. $\begin{aligned} 2x - 5y &= 10 \\ 4x - 5y &= 15 \end{aligned}$

40. $\begin{aligned} 2x - 3y &= 2 \\ 4x - 6y &= 1 \end{aligned}$

41. $\begin{aligned} 2x + 4z &= 14 \\ 3x + y + 5z &= 19 \\ -x + y - 2z &= -7 \end{aligned}$

42. $\begin{aligned} 3x + 6y + 3z &= 12 \\ 6x + 4y - 2z &= -4 \\ y - z &= -3 \end{aligned}$

43. $\begin{aligned} x + 3y + z &= 2 \\ x - 2y + 3z &= -3 \\ 2x - 3y - z &= 34 \end{aligned}$

44. $\begin{aligned} x + y - z &= 6 \\ 2x - y + z &= -9 \\ x - 2y + 3z &= 1 \end{aligned}$

45. $\begin{aligned} x + 3y - 2z - w &= 9 \\ 4x + y + z + 2w &= 2 \\ -3x - y + z - w &= -5 \\ x - y - 3z - 2w &= 2 \end{aligned}$

46. $\begin{aligned} 3x + 2y - w &= 0 \\ 2x + z + 2w &= 5 \\ x + 2y - z &= -2 \\ 2x - y + z + w &= 2 \end{aligned}$

47. $\begin{aligned} x - \sqrt{2}y &= 2.6 \\ .75x + y &= -7 \end{aligned}$

48. $\begin{aligned} 2.1x + y &= \sqrt{5} \\ \sqrt{2}x - 2y &= 5 \end{aligned}$

49. $\begin{aligned} \pi x + ey + \sqrt{2}z &= 1 \\ ex + \pi y + \sqrt{2}z &= 2 \\ \sqrt{2}x + ey + \pi z &= 3 \end{aligned}$

50. $\begin{aligned} (\log 2)x + (\ln 3)y + (\ln 4)z &= 1 \\ (\ln 3)x + (\log 2)y + (\ln 8)z &= 5 \\ (\log 12)x + (\ln 4)y + (\ln 8)z &= 9 \end{aligned}$

Curve Fitting *In Exercises 51 and 52, use the method of Example 6 to find the cubic polynomial P(x) that defines the curve shown in the four figures.*

51.

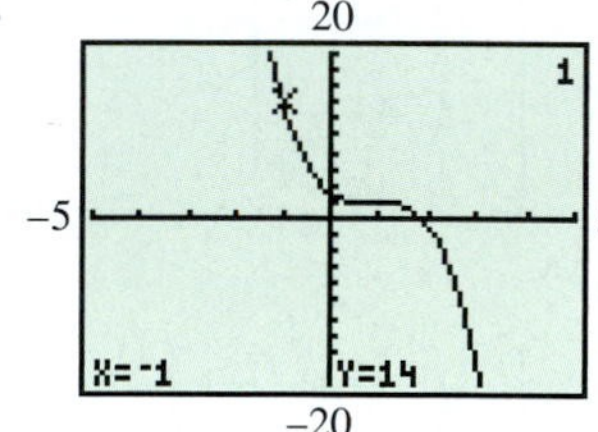

First view

Second view

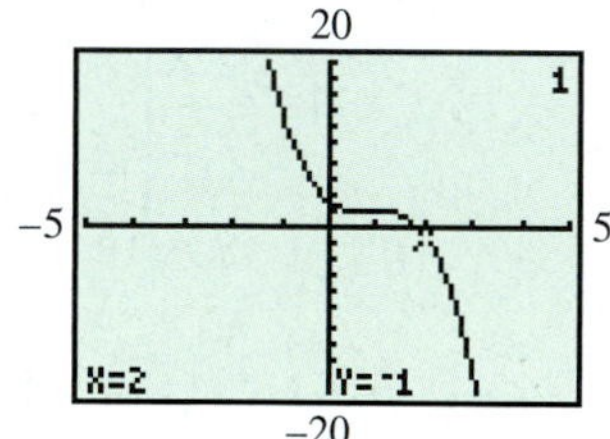

Third view

Fourth view

52.

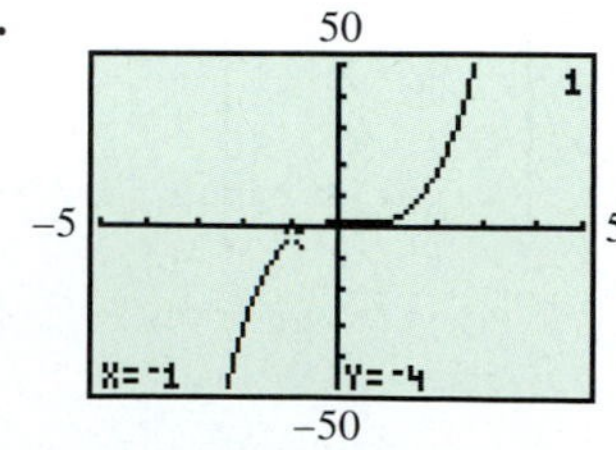

First view

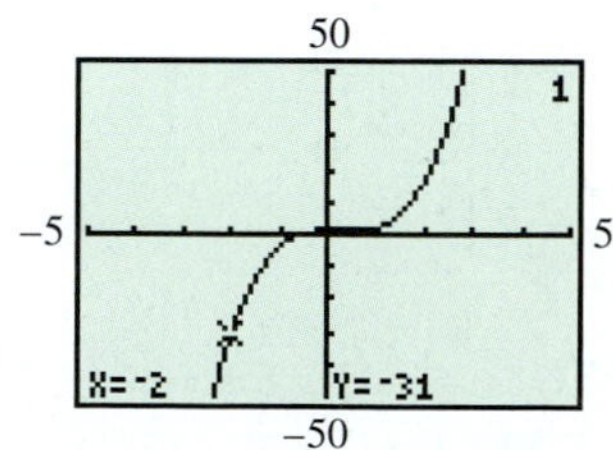

Second view

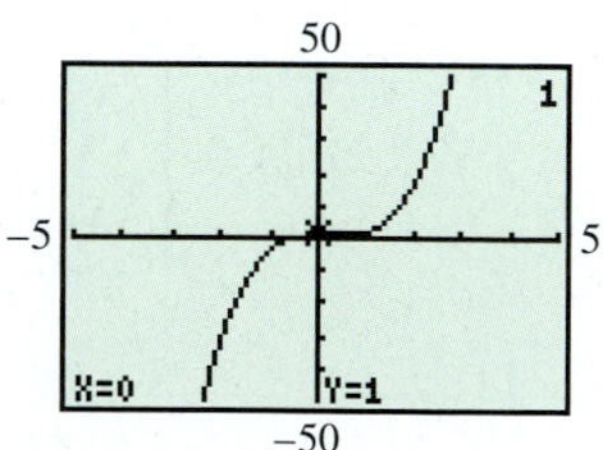

Third view

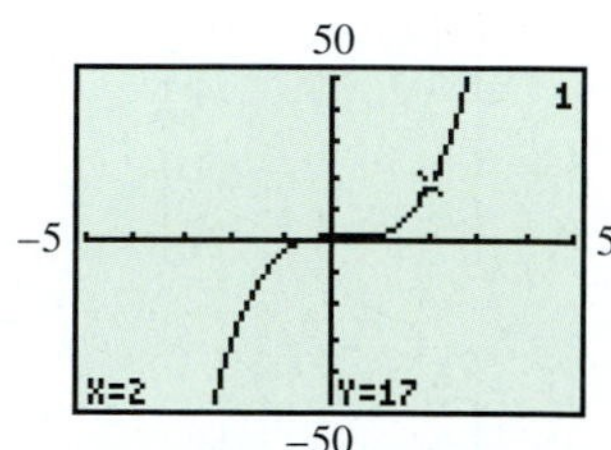

Fourth view

53. ***Curve Fitting*** Find the fourth-degree polynomial $P(x)$ satisfying the following conditions: $P(-2) = 13$, $P(-1) = 2$, $P(0) = -1$, $P(1) = 4$, and $P(2) = 41$.

54. ***Curve Fitting*** Find the fifth-degree polynomial $P(x)$ satisfying the following conditions: $P(-2) = -8$, $P(-1) = -1$, $P(0) = -4$, $P(1) = -5$, $P(2) = 8$, and $P(3) = 167$.

(Modeling) *Solve each problem.*

55. ***Determining Prices*** A group of students bought 3 soft drinks and 2 boxes of popcorn at a movie for \$8.50. A second group bought 4 soft drinks and 3 boxes of popcorn for \$12.

(a) Find a matrix equation $AX = B$ whose solution gives the individual prices of a soft drink and a box of popcorn. Solve this matrix equation by using A^{-1}.

(b) Could these prices be determined if both groups had bought 3 soft drinks and 2 boxes of popcorn for \$8.50? Try to calculate A^{-1} and explain your results.

56. ***Cost of CDs*** A music store has compact discs that sell for three prices, marked A, B, and C. The last column in the table shows the total cost of a purchase. Use this information to determine the cost of one CD of each type by setting up a matrix equation and solving it with an inverse.

A	B	C	Total
2	3	4	\$120.91
1	4	0	\$ 62.95
2	1	3	\$ 79.94

57. ***Tire Sales*** In a study, the relationship among annual tire sales T in thousands, automobile registrations A in millions, and personal disposable income I in millions of dollars was investigated. Representative data for three different years are shown in the table.

T	A	I
10,170	113	308
15,305	133	622
21,289	155	1937

Source: Jarrett, J., *Business Forecasting Methods,* Basil Blackwell, 1991.

The data were modeled by

$$T = aA + bI + c,$$

where a, b, and c are constants.

(a) Use the data to write a system of linear equations whose solution gives a, b, and c.

(b) Solve this linear system. Write a formula for T.

(c) If $A = 118$ and $I = 311$, predict T. (The actual value for T was 11,314.)

58. ***Plate Glass Sales*** The amount of plate glass G sold can be affected by the number of new building contracts B issued and automobiles A produced, since plate glass is used in buildings and cars. A plate glass company in California wanted to forecast sales. The table contains sales data for three consecutive years. All units are in millions.

G	A	B
603	5.54	37.1
657	6.93	41.3
779	7.64	45.6

Source: Makridakis, S. and S. Wheelwright, *Forecasting Methods for Management,* John Wiley and Sons, 1989.

The data were modeled by

$$G = aA + bB + c,$$

where a, b, and c are constants.

(a) Write a system of linear equations whose solution gives a, b, and c.

(b) Solve this linear system. Write a formula for G.

(c) For the following year, it was estimated that $A = 7.75$ and $B = 47.4$. Predict G. (The actual value for G was \$878 million.)

59. ***Home Prices*** Real-estate companies sometimes study how the selling price of a home is related to its size and condition. The table contains data for sales of three homes. Price P is measured in thousands of dollars, home size S is in square feet, and condition C is rated on a scale from 1 to 10, where 10 represents excellent condition. The variables were found to be related by the linear equation $P = a + bS + cC$.

P	S	C
122	1500	8
130	2000	5
158	2200	10

(a) Use the table to write a system of linear equations whose solution gives a, b, and c.

(b) Estimate the selling price of a home with 1800 square feet and in condition 7.

60. ***Traffic Flow*** Refer to Exercises 71 and 72 in Section 7.3. The figure shows four one-way streets with intersections A, B, C, and D. Numbers indicate the average traffic flow in vehicles per minute. The variables x_1, x_2, x_3, and x_4 denote unknown traffic flows.

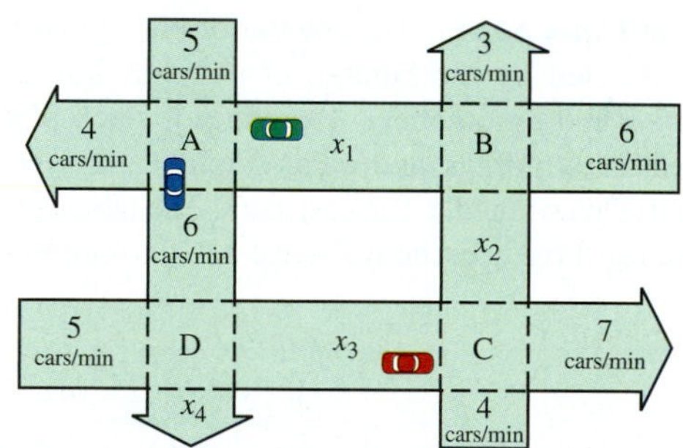

(a) The number of vehicles per minute entering an intersection equals the number exiting an intersection. Verify that the given system of linear equations describes the traffic flow.

$$\text{A: } x_1 + 5 = 4 + 6$$
$$\text{B: } x_2 + 6 = x_1 + 3$$
$$\text{C: } x_3 + 4 = x_2 + 7$$
$$\text{D: } 6 + 5 = x_3 + x_4$$

(b) Write the system as $AX = B$, and solve by using A^{-1}.

(c) Interpret your results.

Each graphing calculator screen shows $[A]^{-1}$ *for some matrix* $[A]$. *Find matrix* $[A]$.

61. 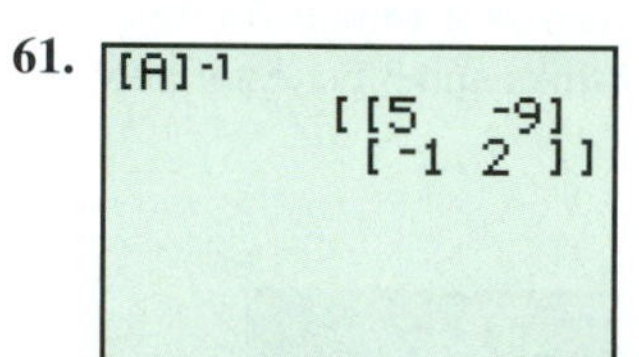

62. 

63. 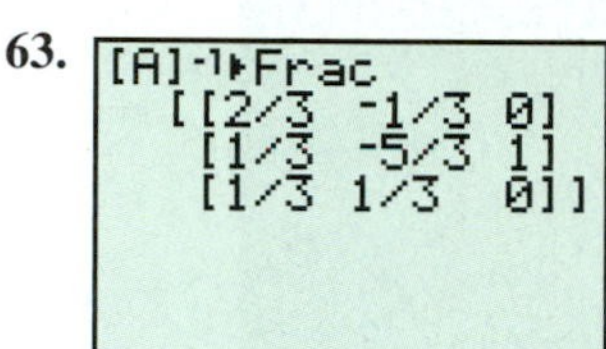

64. 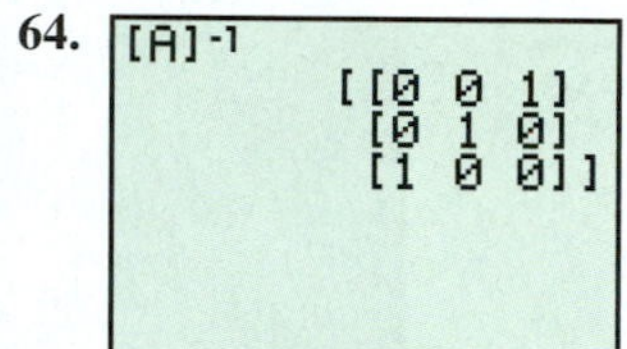

65. Let $A = \begin{bmatrix} a & 0 & 0 \\ 0 & b & 0 \\ 0 & 0 & c \end{bmatrix}$, where a, b, and c are nonzero real numbers. Find A^{-1}.

66. Let $A = \begin{bmatrix} 1 & 0 & 0 \\ 0 & 0 & -1 \\ 0 & 1 & -1 \end{bmatrix}$. Show that $A^3 = I_3$, and use this result to find the inverse of A.

Reviewing Basic Concepts (Sections 7.4–7.6)

Find the following, if possible, given matrices

$$A = \begin{bmatrix} -5 & 4 \\ 2 & -1 \end{bmatrix}, \quad B = \begin{bmatrix} 0 & -2 \\ 3 & -4 \end{bmatrix}, \quad C = \begin{bmatrix} 2 & -3 & 1 \\ -2 & 1 & 0 \\ 0 & -1 & 4 \end{bmatrix}, \quad D = \begin{bmatrix} 0 & 4 & 1 \\ 0 & 2 & 2 \\ 1 & 1 & 1 \end{bmatrix}.$$

1. $A - B$ **2.** $-3B$ **3.** A^2 **4.** CD

5. det A **6.** det C **7.** A^{-1} **8.** C^{-1}

9. ***Roof Trusses*** Linear systems occur in the design of roof trusses for new homes and buildings. The simplest type of roof truss is a triangle. The truss shown in the figure is used to frame roofs of small buildings. If a 100-pound force is applied at the peak of the truss, then the forces or weights W_1 and W_2 exerted parallel to each rafter of the truss are determined by the following linear system of equations:

$$\frac{\sqrt{3}}{2}(W_1 + W_2) = 100$$
$$W_1 - W_2 = 0.$$

Solve the system, using Cramer's rule to find W_1 and W_2. (*Source:* Hibbeler, R., *Structural Analysis,* Prentice-Hall, 1995.)

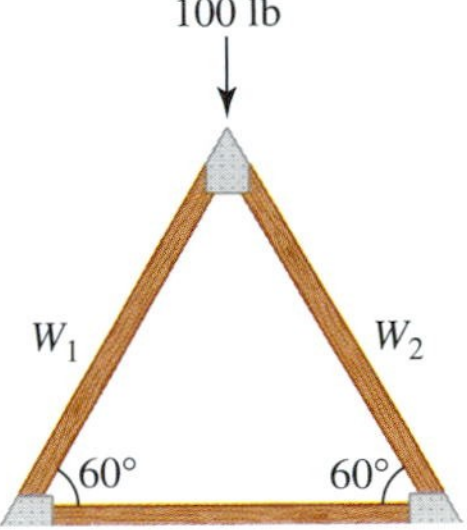

10. Solve the system, using the inverse matrix method.

$$\begin{aligned} 2x + y + 2z &= 10 \\ y + 2z &= 4 \\ x - 2y + 2z &= 1 \end{aligned}$$

7.7 Systems of Inequalities and Linear Programming

Solving Linear Inequalities ■ Solving Systems of Inequalities ■ Linear Programming

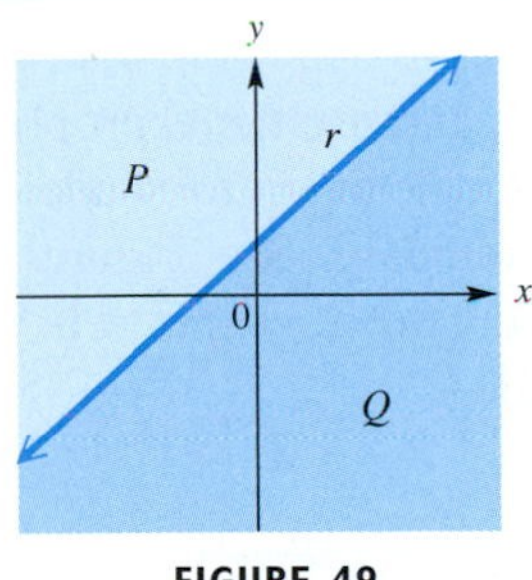

FIGURE 49

Many mathematical descriptions of real situations are best expressed as inequalities rather than equations. For example, a firm might be able to use a machine *no more* than 12 hours a day, while the production of *at least* 500 cases of a certain product might be required to meet a contract. The simplest way to see the solution of an inequality in two variables is to draw its graph.

A line divides a plane into three sets of points: the points of the line itself and the points belonging to the two regions determined by the line. Each of these two regions is called a **half plane.** In Figure 49, line r divides the plane into three different sets of points: line r, half plane P, and half plane Q. The points on r belong neither to P nor to Q. Line r is the **boundary** of each half plane.

Solving Linear Inequalities

A **linear inequality in two variables** is an inequality of the form

$$Ax + By \le C,$$

where A, B, and C are real numbers with A and B not both equal to 0. (The symbol $\le$ could be replaced with $\ge$, $<$, or $>$.) The graph of a linear inequality is a half plane, perhaps with its boundary.

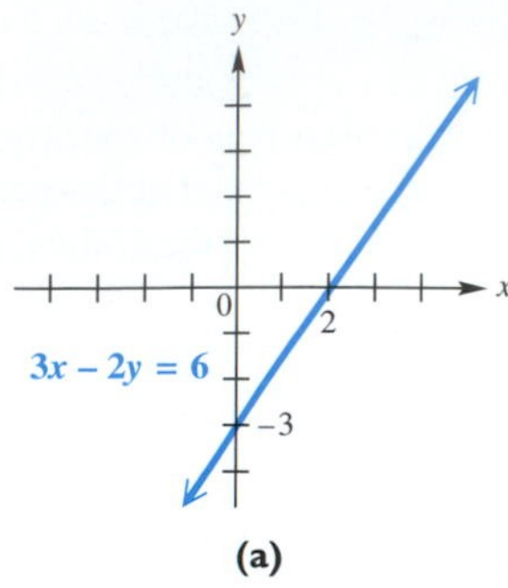

(a)

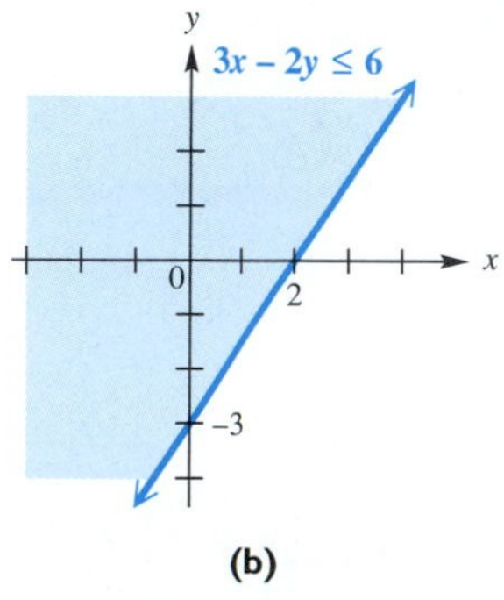

(b)

FIGURE 50

For example, to graph the linear inequality

$$3x - 2y \leq 6,$$

first graph the boundary, $3x - 2y = 6$, as shown in Figure 50(a). Since the points of the line $3x - 2y = 6$ satisfy $3x - 2y \leq 6$, this line is part of the solution. To decide which half plane is part of the solution, solve the original inequality for y.

$$3x - 2y \leq 6$$

$$-2y \leq -3x + 6 \quad \text{Subtract } 3x.$$

$$y \geq \frac{3}{2}x - 3 \quad \text{Multiply by } -\tfrac{1}{2}\text{; change } \leq \text{ to } \geq.$$

Reverse the inequality symbol when multiplying by a negative number.

For a particular value of x, the inequality will be satisfied by all values of y that are *greater than* or equal to $\frac{3}{2}x - 3$. Thus, the solution contains the half plane *above* the line, as shown in Figure 50(b).

CAUTION To determine whether to shade above or below a boundary line, solve the inequality for y.

1. If the symbol is $>$ or $\geq$, shade *above* the boundary.
2. If the symbol is $<$ or $\leq$, shade *below* the boundary.

GCM **EXAMPLE 1 Graphing a Linear Inequality**

Graph $x + 4y > 4$.

Analytic Solution

The boundary here is the line $x + 4y = 4$. Since the points on this line do not satisfy $x + 4y > 4$, make the line dashed, as in Figure 51. To decide which half plane represents the solution, solve for y.

$$x + 4y > 4$$

$$4y > -x + 4 \quad \text{Subtract } x.$$

$$y > -\frac{1}{4}x + 1 \quad \text{Divide by 4.}$$

Since y is *greater than* $-\frac{1}{4}x + 1$, the graph of the solution is the half plane *above* the boundary, as shown in the figure.

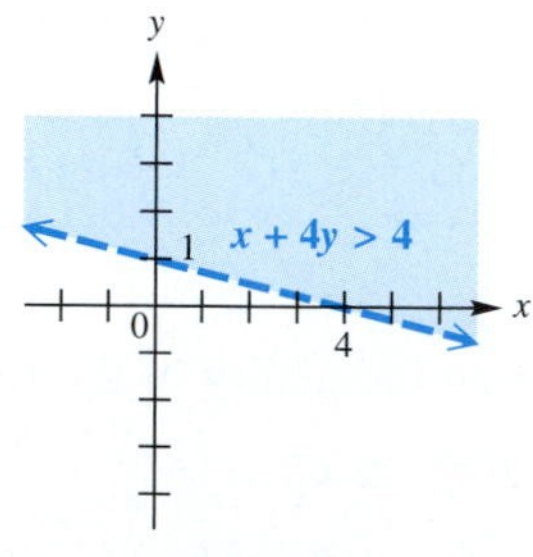

FIGURE 51

Graphing Calculator Solution

Solve the corresponding linear equation (the boundary) for y to get

$$y = -\frac{1}{4}x + 1.$$

Graph this boundary, and use the appropriate commands for your calculator to shade the region above the boundary. See Figure 52. (Notice that the calculator does not tell you which region to shade.)

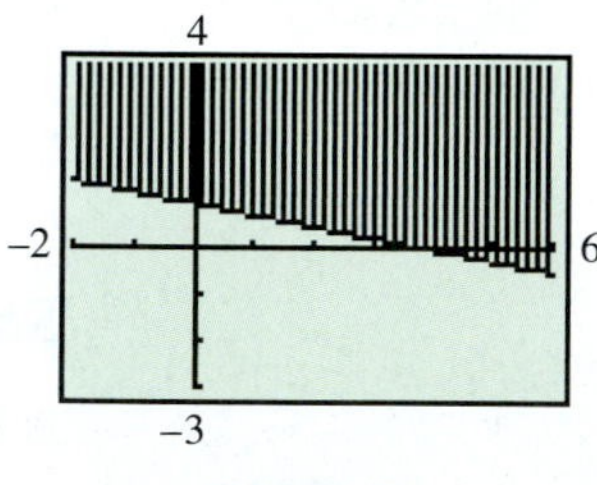

FIGURE 52

(continued)

Alternatively, or as a check, choose a point not on the boundary line and substitute into the inequality. The point $(0, 0)$ is a good choice if it does not lie on the boundary, since the substitution is easily done.

$$x + 4y > 4 \quad \text{Original inequality}$$
$$0 + 4(0) > 4 \quad \text{Use } (0, 0) \text{ as a test point.}$$
$$0 > 4 \quad \text{False}$$

Since the point $(0, 0)$ is below the boundary, the points that satisfy the inequality must be above the boundary, which agrees with the preceding result.

The calculator graph does *not* distinguish between solid boundary lines and dashed boundary lines. Whether the boundary is solid or dashed must be determined by looking at the inequality symbol. ***We must understand the mathematics to correctly interpret a calculator graph of an inequality.***

Graphing an Inequality

1. For a function f, the graph of $y < f(x)$ consists of all the points that are *below* the graph of $y = f(x)$; the graph of $y > f(x)$ consists of all the points that are *above* the graph of $y = f(x)$. (Similar statements can be made for $\leq$ and $\geq$, with boundaries included.)
2. If the inequality is not or cannot be solved for y, choose a test point not on the boundary. If the test point satisfies the inequality, the graph includes all points on the same side of the boundary as the test point. Otherwise, the graph includes all points on the other side of the boundary.

Solving Systems of Inequalities

The solution set of a **system of inequalities,** such as the system

$$x > 6 - 2y$$
$$x^2 < 2y,$$

is the intersection of the solution sets of its members. We find this intersection by graphing the solution sets of both inequalities on the same coordinate axes and identifying, by shading, the region common to both graphs.

GCM EXAMPLE 2 Graphing a System of Two Inequalities

Graph the solution set of the system.

$$x > 6 - 2y$$
$$x^2 < 2y$$

Analytic Solution

Figures 53(a) and (b) on the next page show the graphs of $x > 6 - 2y$ and $x^2 < 2y$. The methods of Section 7.1 can be used to show that the boundaries intersect at the points $(2, 2)$ and $\left(-3, \frac{9}{2}\right)$. The solution set of the system is shown in Figure 53(c). Since the points on the boundaries of $x > 6 - 2y$ and $x^2 < 2y$ do not belong to the graph of the solution, the boundaries are dashed.

Graphing Calculator Solution

As usual, solve each inequality for y first. Enter the boundary equations,

$$y_1 = \frac{6 - x}{2} \quad \text{and} \quad y_2 = \frac{x^2}{2}.$$

(continued)

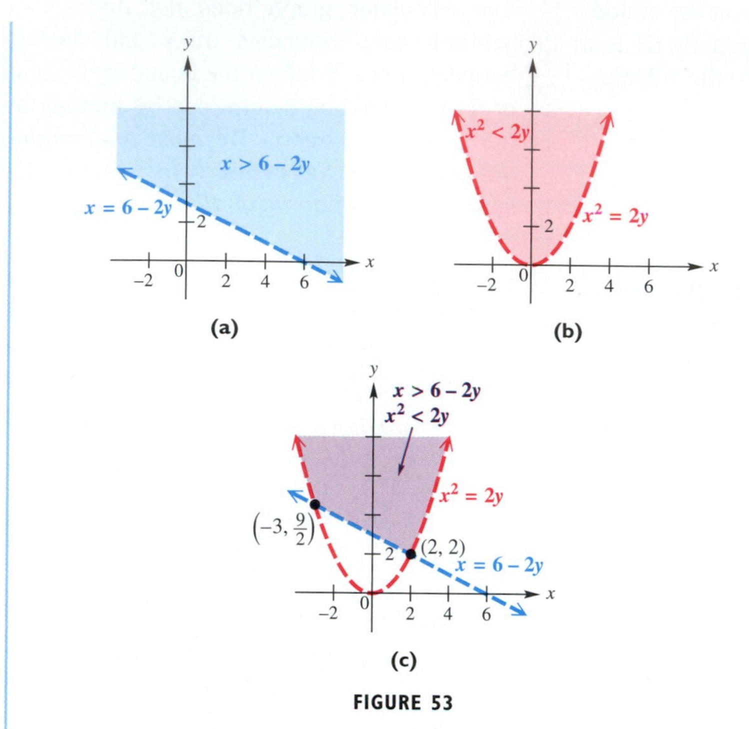

FIGURE 53

In both inequalities,

$$y_1 > \frac{6 - x}{2} \quad \text{and} \quad y_2 > \frac{x^2}{2},$$

y is *greater than* an expression involving x. Therefore, use the capability of your calculator to shade *above* each boundary. See Figure 54.

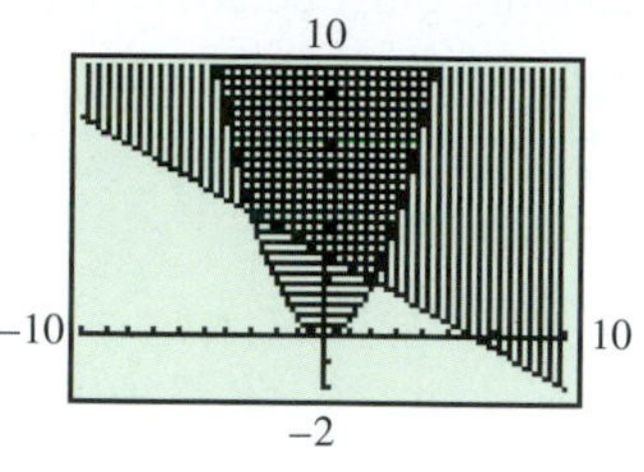

FIGURE 54

The crosshatched region supports the solution shown in Figure 53(c). ■

EXAMPLE 3 Graphing a System of Three Inequalities

Graph the solution set of the system.

$$y \geq 2^x$$
$$9x^2 + 4y^2 \leq 36$$
$$2x + y < 1$$

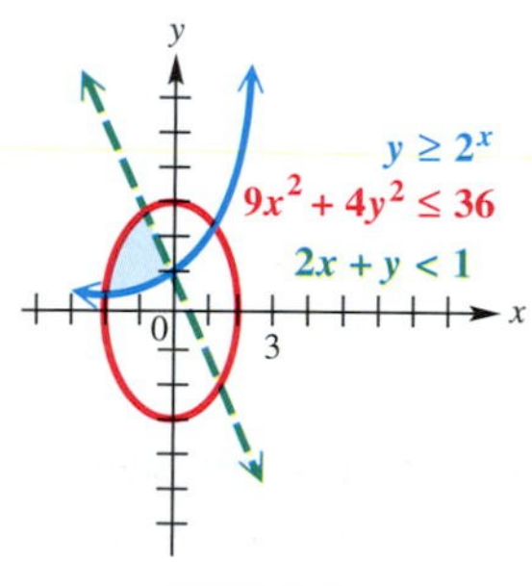

FIGURE 55

Solution Graph the three inequalities on the same axes and shade the region common to all three. See Figure 55. Two boundary lines are solid and one is dashed. ■

Linear Programming

An important application of mathematics to business and social science is called *linear programming*. We use **linear programming** to find an optimum value—for example, minimum cost or maximum profit. Linear programming was first developed to solve problems in allocating supplies for the U.S. Air Force during World War II.

EXAMPLE 4 Finding a Maximum Profit Model

The Charlson Company makes two products: VCRs and DVD players. Each VCR gives a profit of \$30, while each DVD player produces a \$70 profit. The company must manufacture at least 10 VCRs per day to satisfy one of its customers, but no more than 50, because of production restrictions. The number of DVD players produced cannot exceed 60 per day, and the number of VCRs cannot exceed the number of DVD players. How many of each should the company manufacture to obtain maximum profit?

Solution First, we translate the statement of the problem into symbols.

Let x = number of VCRs to be produced daily,

and y = number of DVD players to be produced daily.

The company must produce at least 10 VCRs (10 or more), so

$$x \geq 10.$$

Since no more than 50 VCRs may be produced,

$$x \leq 50.$$

No more than 60 DVD players may be made in one day, so

$$y \leq 60.$$

The number of VCRs may not exceed the number of DVD players translates as

$$x \leq y.$$

The numbers of VCRs and of DVD players cannot be negative, so

$$x \geq 0 \quad \text{and} \quad y \geq 0.$$

These restrictions, or **constraints,** form the system of inequalities

$$x \geq 10, \quad x \leq 50, \quad y \leq 60, \quad x \leq y, \quad x \geq 0, \quad y \geq 0.$$

Each VCR gives a profit of \$30, so the daily profit from production of x VCRs is $30x$ dollars. Also, the profit from production of y DVD players will be $70y$ dollars per day. The total daily profit is thus

$$\text{profit} = 30x + 70y.$$

This equation defines the function to be maximized, called the **objective function.**

To find the maximum possible profit that the company can make, subject to these constraints, we sketch the graph of each constraint. The only feasible values of x and y are those which satisfy all constraints—that is, the values that lie in the intersection of the graphs of the constraints. The intersection is shown in Figure 56. Any point lying inside the shaded region or on the boundary in the figure satisfies the restrictions as to the number of VCRs and DVD players that may be produced. (For practical purposes, however, only points with integer coefficients are useful.) This region is called the **region of feasible solutions.** The **vertices** (singular: **vertex**), or **corner points,** of the region of feasible solutions have coordinates

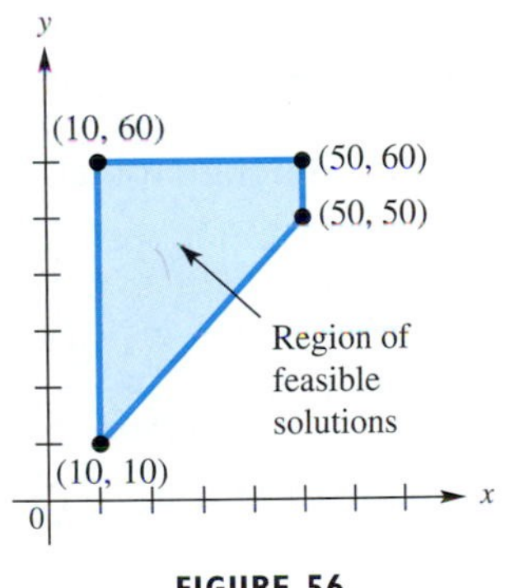

FIGURE 56

$$(10, 10), \quad (10, 60), \quad (50, 50), \quad \text{and} \quad (50, 60).$$

We must find the value of the objective function $30x + 70y$ for each vertex. We want the vertex that produces the maximum possible value of $30x + 70y$.

$$\begin{aligned}
(10, 10)&: \quad 30(10) + 70(10) = 1000 \\
(10, 60)&: \quad 30(10) + 70(60) = 4500 \\
(50, 50)&: \quad 30(50) + 70(50) = 5000 \\
(50, 60)&: \quad 30(50) + 70(60) = 5700 \longleftarrow \text{Maximum}
\end{aligned}$$

The maximum profit is obtained when 50 VCRs and 60 DVD players are produced each day. This maximum profit will be $30(50) + 70(60) = 5700$ dollars per day. ■

To justify the procedure used in Example 4, consider the following scenario. The Charlson Company needed to find values of x and y in the shaded region of Figure 56 that produce the maximum profit—that is, the maximum value of $30x + 70y$. To locate the point (x, y) that gives the maximum profit, add to the graph of that figure lines corresponding to arbitrarily chosen profits of \$0, \$1000, \$3000, and \$7000:

$$30x + 70y = 0, \quad 30x + 70y = 1000, \quad 30x + 70y = 3000, \quad 30x + 70y = 7000.$$

For instance, each point on the line $30x + 70y = 3000$ corresponds to production values that yield a profit of \$3000.

Figure 57 shows the region of feasible solutions, together with these lines. The lines are parallel, and the higher the line, the greater the profit. The line $30x + 70y = 7000$ yields the greatest profit, but does not contain any points of the region of feasible solutions. To find the feasible solution of greatest profit, lower the line $30x + 70y = 7000$ until it contains a feasible solution—that is, until it touches the region of feasible solutions at point A, a vertex of the region. See Figure 58.

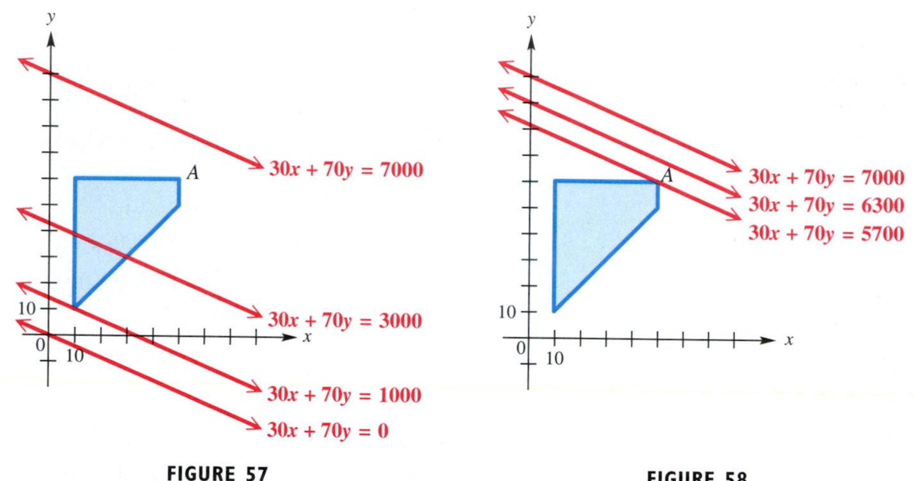

FIGURE 57

FIGURE 58

The result observed in Figure 58 holds for *every* linear programming problem.

Fundamental Theorem of Linear Programming

If the optimal value for a linear programming problem exists, it occurs at a vertex of the region of feasible solutions.

Solving a Linear Programming Problem

Step 1 Write the objective function and all necessary constraints.

Step 2 Graph the region of feasible solutions.

Step 3 Identify all vertices (corner points).

Step 4 Find the value of the objective function at each vertex.

Step 5 The solution is given by the vertex producing the optimal value of the objective function.

EXAMPLE 5 Finding a Minimum Cost Model

Robin takes vitamin pills each day. She wants at least 16 units of Vitamin A, at least 5 units of Vitamin B_1, and at least 20 units of Vitamin C. She can choose between red pills, costing 10¢ each, that contain 8 units of A, 1 of B_1, and 2 of C; and blue pills, costing 20¢ each, that contain 2 units of A, 1 of B_1, and 7 of C. How many of each pill should she buy to minimize her cost and yet fulfill her daily requirements?

Solution

Step 1 Let x represent the number of red pills to buy, and let y represent the number of blue pills to buy. Then the cost in pennies per day is given by

$$\text{cost} = 10x + 20y.$$

Robin buys x of the 10¢ pills and y of the 20¢ pills, and she gets 8 units of Vitamin A from each red pill and 2 units of Vitamin A from each blue pill. Thus, she gets $8x + 2y$ units of A per day. Since she wants at least 16 units,

$$8x + 2y \geq 16.$$

Each red pill and each blue pill supplies 1 unit of Vitamin B_1. Robin wants at least 5 units per day, so

$$x + y \geq 5.$$

For Vitamin C, the inequality is

$$2x + 7y \geq 20.$$

Also, $x \geq 0$ and $y \geq 0$, since Robin cannot buy negative numbers of the pills.

Step 2 The intersection of the graphs of

$$8x + 2y \geq 16, \quad x + y \geq 5, \quad 2x + 7y \geq 20, \quad x \geq 0, \quad \text{and} \quad y \geq 0$$

is given in Figure 59.

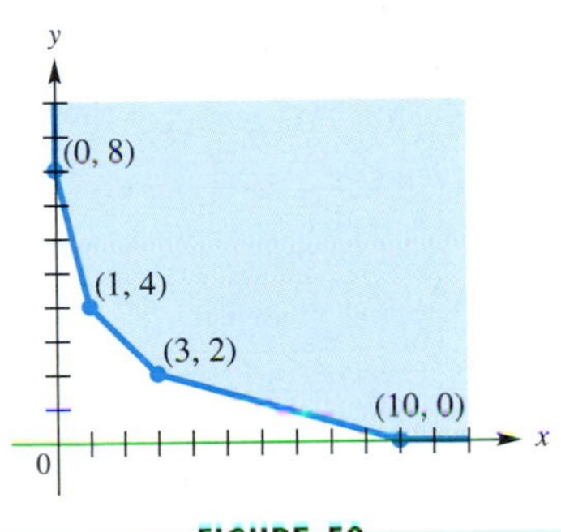

FIGURE 59

Step 3 The vertices are (0, 8), (1, 4), (3, 2), and (10, 0).

Steps 4 and 5 The minimum cost occurs at (3, 2). See the following table.

Point	Cost = $10x + 20y$
(0, 8)	$10(0) + 20(8) = 160$
(1, 4)	$10(1) + 20(4) = 90$
(3, 2)	$10(3) + 20(2) = 70$ ← Minimum
(10, 0)	$10(10) + 20(0) = 100$

Robin's best choice is to buy 3 red pills and 2 blue pills, for a total cost of 70¢ per day. She receives just the minimum amounts of Vitamins B_1 and C and an excess of Vitamin A. ■

7.7 Exercises

Graph each inequality.

1. $x \leq 3$
2. $y \leq -2$
3. $x + 2y \leq 6$
4. $x - y \geq 2$
5. $2x + 3y \geq 4$
6. $4y - 3x < 5$
7. $3x - 5y > 6$
8. $x < 3 + 2y$
9. $5x \leq 4y - 2$
10. $2x > 3 - 4y$
11. $y < 3x^2 + 2$
12. $y \leq x^2 - 4$
13. $y > (x - 1)^2 + 2$
14. $y > 2(x + 3)^2 - 1$
15. $x^2 + (y + 3)^2 \leq 16$
16. $(x - 4)^2 + (y + 3)^2 \leq 9$

17. In your own words, explain how to determine whether the boundary of the graph of an inequality is solid or dashed.

18. When graphing $y > 3x - 6$, would you shade above or below the line $y = 3x - 6$? Explain your answer.

Concept Check *Use the concepts of this section to work Exercises 19–22.*

19. For $Ax + By \geq C$, if $B > 0$, would you shade above or below the boundary line?

20. For $Ax + By \geq C$, if $B < 0$, would you shade above or below the boundary line?

21. Which one of the following is a description of the graph of the inequality $(x - 5)^2 + (y - 2)^2 < 4$?
    A. The region inside a circle with center $(-5, -2)$ and radius 2
    B. The region inside a circle with center $(5, 2)$ and radius 2
    C. The region inside a circle with center $(-5, -2)$ and radius 4
    D. The region outside a circle with center $(5, 2)$ and radius 4

22. Which one of the given inequalities satisfies the following description: the region outside an ellipse centered at the origin, with x-intercepts 4 and -4 and y-intercepts 9 and -9?
    A. $\frac{x^2}{4} + \frac{y^2}{9} > 1$
    B. $\frac{x^2}{16} - \frac{y^2}{81} > 1$
    C. $\frac{x^2}{16} + \frac{y^2}{81} > 0$
    D. $\frac{x^2}{16} + \frac{y^2}{81} > 1$

Concept Check *In Exercises 23–26, match the inequality with the appropriate calculator graph. Do not use your calculator; instead, use your knowledge of the concepts involved in graphing inequalities.*

23. $y \leq 3x - 6$
24. $y \geq 3x - 6$
25. $y \leq -3x - 6$
26. $y \geq -3x - 6$

A.

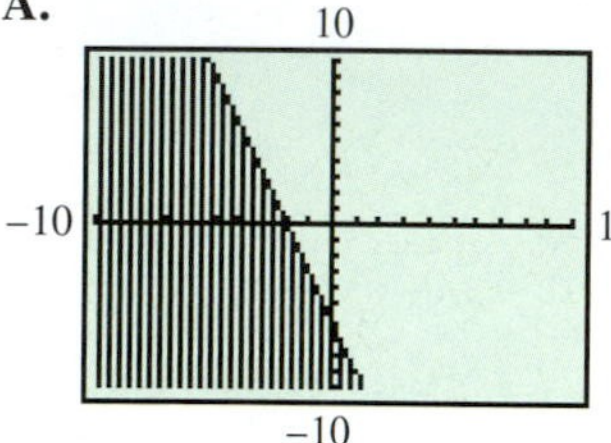

B.

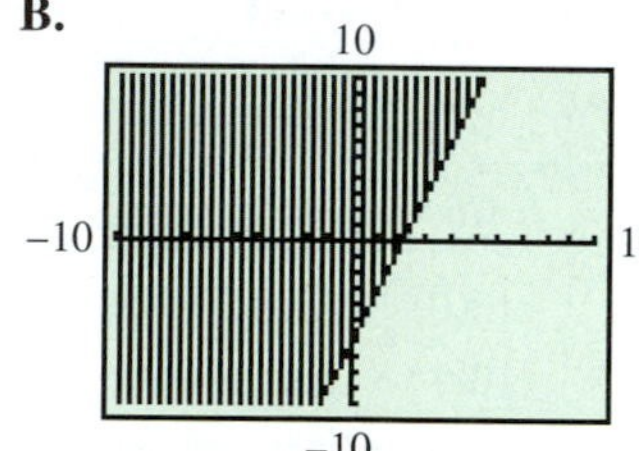

C.

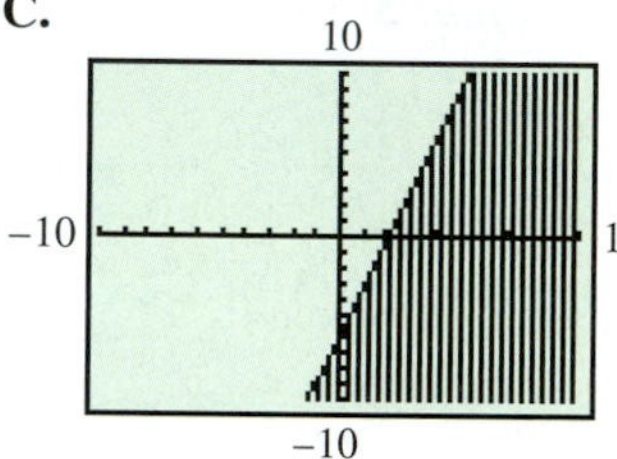

D.

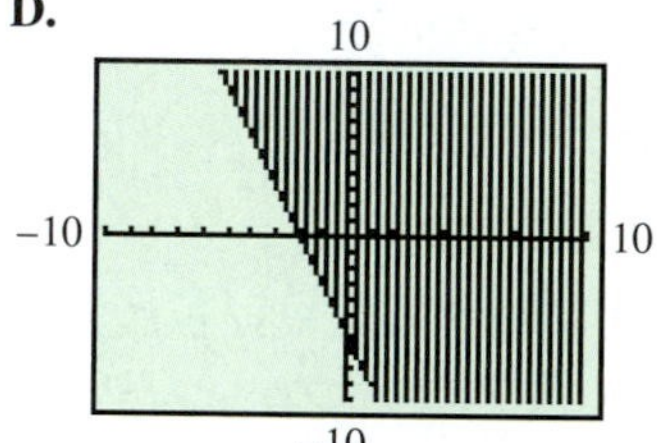

Graph the solution set of each system of inequalities by hand.

27. $x + y \geq 0$
    $2x - y \geq 3$
28. $x + y \leq 4$
    $x - 2y \geq 6$
29. $2x + y > 2$
    $x - 3y < 6$
30. $4x + 3y < 12$
    $y + 4x > -4$
31. $3x + 5y \leq 15$
    $x - 3y \geq 9$
32. $y \leq x$
    $x^2 + y^2 < 1$
33. $4x - 3y \leq 12$
    $y \leq x^2$
34. $y \leq -x^2$
    $y \geq x^2 - 6$
35. $x + y \leq 9$
    $x \leq -y^2$
36. $x + 2y \leq 4$
    $y \geq x^2 - 1$
37. $y \leq (x + 2)^2$
    $y \geq -2x^2$
38. $x - y < 1$
    $-1 < y < 1$

39. $x + y \le 36$
$-4 \le x \le 4$

40. $y > x^2 + 4x + 4$
$y < -x^2$

41. $y \ge (x - 2)^2 + 3$
$y \le -(x - 1)^2 + 6$

42. $x \ge 0$
$x + y \le 4$
$2x + y \le 5$

43. $3x - 2y \ge 6$
$x + y \le -5$
$y \le 4$

44. $-2 < x < 3$
$-1 \le y \le 5$
$2x + y < 6$

45. $-2 < x < 2$
$y > 1$
$x - y > 0$

46. $x + y \le 4$
$x - y \le 5$
$4x + y \le -4$

47. $x \le 4$
$x \ge 0$
$y \ge 0$
$x + 2y \ge 2$

48. $2y + x \ge -5$
$y \le 3 + x$
$x \le 0$
$y \le 0$

49. $2x + 3y \le 12$
$2x + 3y > -6$
$3x + y < 4$
$x \ge 0$
$y \ge 0$

50. $y \ge 3^x$
$y \ge 2$

51. $y \le \left(\frac{1}{2}\right)^x$
$y \ge 4$

52. $\ln x - y \ge 1$
$x^2 - 2x - y \le 1$

53. $y \le \log x$
$y \ge |x - 2|$

54. $e^{-x} - y \le 1$
$x - 2y \ge 4$

55. *Concept Check* Which one of the choices that follow is a description of the solution set of the following system?

$$x^2 + 4y^2 < 36$$
$$y < x$$

A. All points outside the ellipse $x^2 + 4y^2 = 36$ and above the line $y = x$
B. All points outside the ellipse $x^2 + 4y^2 = 36$ and below the line $y = x$
C. All points inside the ellipse $x^2 + 4y^2 = 36$ and above the line $y = x$
D. All points inside the ellipse $x^2 + 4y^2 = 36$ and below the line $y = x$

56. *Concept Check* Fill in each blank with the appropriate response. The graph of the system

$$y > x^2 + 2$$
$$x^2 + y^2 < 16$$
$$y < 7$$

consists of all points ________ (above/below) the parabola given by $y = x^2 + 2$, ________ (inside/outside) the circle $x^2 + y^2 = 16$, and ________ (above/below) the line $y = 7$.

Concept Check *In Exercises 57–60, match each system of inequalities with the appropriate calculator graph. Do not use your calculator; instead, use your knowledge of the concepts involved in graphing systems of inequalities.*

57. $y \ge x$
$y \le 2x - 3$

58. $y \ge x^2$
$y < 5$

59. $x^2 + y^2 \le 16$
$y \ge 0$

60. $y \le x$
$y \ge 2x - 3$

A.

B.

C.

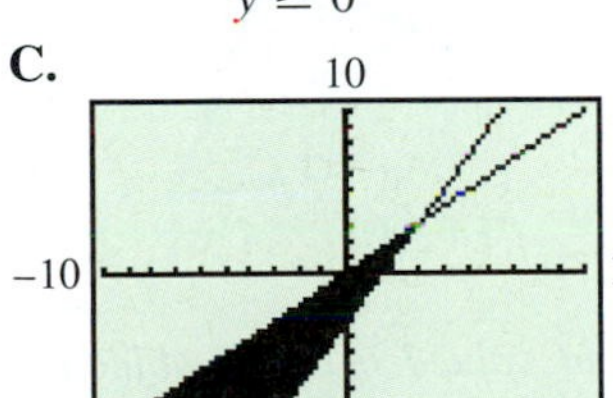

D.

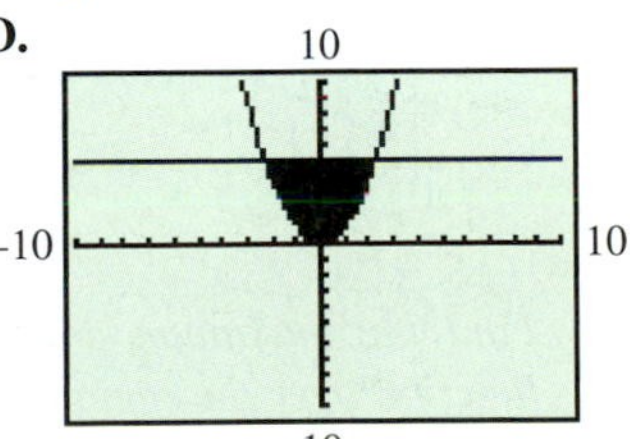

Use the shading capabilities of your graphing calculator to graph each inequality or system of inequalities.

61. $3x + 2y \ge 6$

62. $y \le x^2 + 5$

63. $x + y \ge 2$
$x + y \le 6$

64. $y \ge |x + 2|$
$y \le 6$

65. $y \ge 2^x$
$y \le 8$

66. $y \le x^3 + x^2 - 4x - 4$

67. ***Concept Check*** Find a system of linear inequalities for which the graph is the region in the first quadrant between and inclusive of the pair of lines $x + 2y - 8 = 0$ and $x + 2y = 12$.

68. ***Cost of Vitamins*** The figure shows the region of feasible solutions for the vitamin problem of Example 5 and the straight-line graph of all combinations of red and blue pills for which the cost is 40 cents.

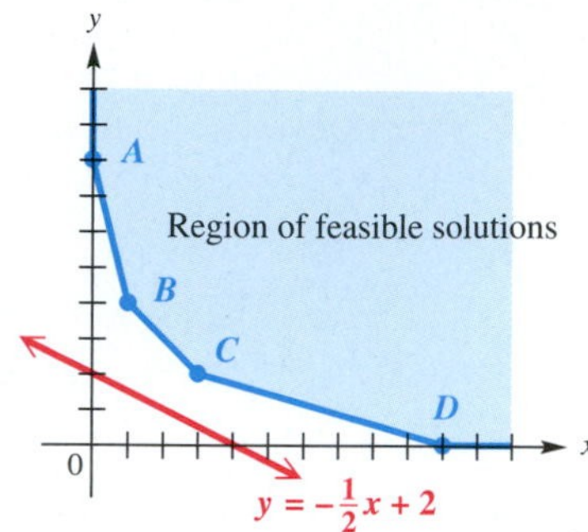

(a) The cost function is $c = 10x + 20y$. Give the linear equation (in slope–intercept form) of the line of constant cost c.

(b) As c increases, does the line of constant cost move up or down?

(c) By inspection, find the vertex of the region of feasible solutions that gives the optimal solution.

The graphs show regions of feasible solutions. Find the maximum and minimum values of each objective function.

69. $3x + 5y$

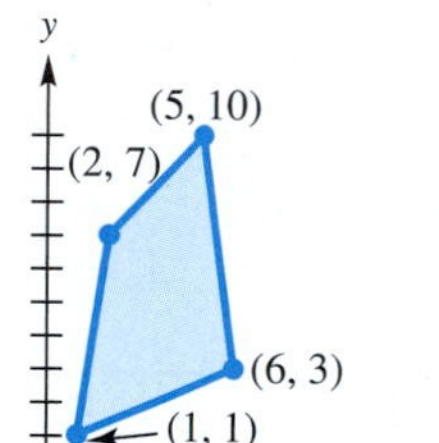

70. $6x + y$

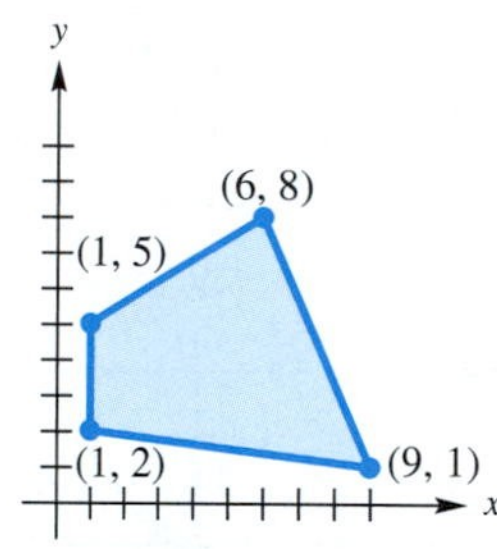

Find the maximum and minimum values of each objective function over the region of feasible solutions shown.

71. $3x + 5y$ **72.** $5x + 5y$

73. $10y$ **74.** $3x - y$

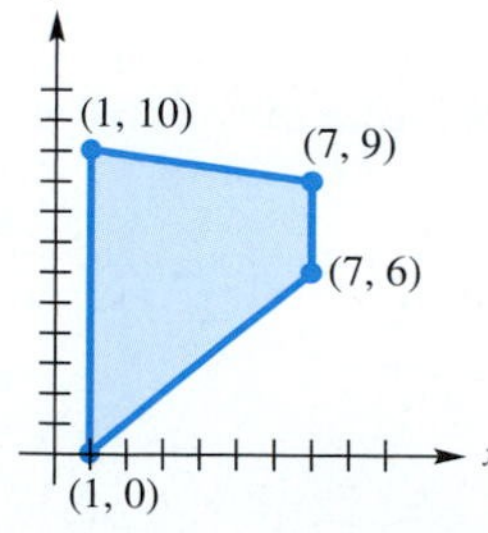

(Modeling) Solve each problem.

75. ***Inquiries about Displayed Products*** A wholesaler of party goods wishes to display her products at a convention of social secretaries in such a way that she gets the maximum number of inquiries about her whistles and hats. Her booth at the convention has 12 square meters of floor space to be used for display purposes. A display unit for hats requires 2 square meters, and one for whistles requires 4 square meters. Experience tells the wholesaler that she should never have more than a total of 5 units of whistles and hats on display at one time. If she receives three inquiries for each unit of hats and two inquiries for each unit of whistles on display, how many of each should she display in order to get the maximum number of inquiries? What is that maximum number?

76. ***Profit from Farm Animals*** Farmer Jones raises only pigs and geese. She wants to raise no more than 16 animals, with no more than 12 geese. She spends \$50 to raise a pig and \$20 to raise a goose. She has \$500 available for this purpose. Find the maximum profit she can make if she makes a profit of \$80 per goose and \$40 per pig. Indicate how many pigs and geese she should raise to achieve this maximum.

77. ***Shipment Costs*** A manufacturer of refrigerators must ship at least 100 refrigerators to its two West Coast warehouses. Each warehouse holds a maximum of 100 refrigerators. Warehouse A holds 25 refrigerators already, while warehouse B has 20 on hand. It costs \$12 to ship a refrigerator to warehouse A and \$10 to ship one to warehouse B. How many refrigerators should be shipped to each warehouse to minimize cost? What is the minimum cost?

78. ***Diet Requirements*** Theo, who is dieting, requires two food supplements: I and II. He can get these supplements from two different products *A* and *B*, as shown in the following table.

Supplement (grams/serving)	I	II
Product *A*	3	2
Product *B*	2	4

Theo's physician has recommended that he include at least 15 grams of each supplement in his daily diet. If product *A* costs 25¢ per serving and product *B* costs 40¢ per serving, how can he satisfy his requirements most economically?

79. ***Gasoline Revenues*** A manufacturing process requires that oil refineries manufacture at least 2 gallons of gasoline for each gallon of fuel oil. To meet winter demand for fuel oil, at least 3 million gallons a day must be produced. The demand for gasoline is no more than 6.4 million gallons per day. If the price of gasoline is $1.90 per gallon and the price of fuel oil is $1.50 per gallon, how much of each should be produced to maximize revenue?

80. ***Manufacturing Revenues*** A machine shop manufactures two types of bolts. Each can be made on any of three groups of machines, but the time required on each group differs, as shown in the following table.

Bolt	Machine Group		
	I	II	III
Type A	.1 min	.1 min	.1 min
Type B	.1 min	.4 min	.5 min

Production schedules are made up one day at a time. In a day, there are 240, 720, and 160 minutes available, respectively, on these machines. Type A bolts sells for 10¢ and Type B bolts for 12¢. How many of each type of bolt should be manufactured per day to maximize revenue? What is the maximum revenue?

81. ***Aid to Disaster Victims*** Earthquake victims in China need medical supplies and bottled water. Each medical kit measures 1 cubic foot and weighs 10 pounds. Each container of water is also 1 cubic foot, but weights 20 pounds. The plane can carry only 80,000 pounds, with total volume 6000 cubic feet. Each medical kit will aid 4 people, while each container of water will serve 10 people. How many of each should be sent in order to maximize the number of people aided? How many people will be aided?

82. ***Aid to Disaster Victims*** If each medical kit could aid 6 people instead of 4, how would the results in Exercise 81 change?

7.8 Partial Fractions

Decomposition of Rational Expressions ■ Distinct Linear Factors ■ Repeated Linear Factors ■ Distinct Linear and Quadratic Factors ■ Repeated Quadratic Factors

Decomposition of Rational Expressions

Add rational expressions →

$$\frac{2}{x+1} + \frac{3}{x} = \frac{5x+3}{x(x+1)}$$

← Partial fraction decomposition

Looking Ahead to Calculus

In calculus, partial fraction decomposition provides a powerful technique for determining integrals of rational functions.

The sums of rational expressions are found by combining two or more such expressions into one rational expression. Here, the reverse process is considered: Given one rational expression, express it as the sum of two or more rational expressions. A special type of sum of rational expressions is called the **partial fraction decomposition;** each term in the sum is a **partial fraction.** The technique of decomposing a rational expression into partial fractions is useful in calculus and other areas of mathematics.

To form a partial fraction decomposition of a rational expression, follow these steps.

Partial Fraction Decomposition of $\frac{f(x)}{g(x)}$

Step 1 If $\frac{f(x)}{g(x)}$ is not a proper fraction (a fraction whose numerator is of lesser degree than the denominator), divide $f(x)$ by $g(x)$. For example,

$$\frac{x^4 - 3x^3 + x^2 + 5x}{x^2 + 3} = x^2 - 3x - 2 + \frac{14x + 6}{x^2 + 3}.$$

Then apply the following steps to the remainder, which is a proper fraction.

Step 2 Factor $g(x)$ completely into factors of the form $(ax + b)^m$ or $(cx^2 + dx + e)^n$, where $cx^2 + dx + e$ is irreducible and m and n are integers.

(continued)

Partial Fraction Decomposition of $\frac{f(x)}{g(x)}$ *(continued)*

Step 3 **(a)** For each distinct linear factor $(ax + b)$, the decomposition must include the term $\frac{A}{ax+b}$.

(b) For each repeated linear factor $(ax + b)^m$, the decomposition must include the terms

$$\frac{A_1}{ax+b} + \frac{A_2}{(ax+b)^2} + \cdots + \frac{A_m}{(ax+b)^m}.$$

Step 4 **(a)** For each distinct quadratic factor $(cx^2 + dx + e)$, the decomposition must include the term $\frac{Bx+C}{cx^2+dx+e}$.

(b) For each repeated quadratic factor $(cx^2 + dx + e)^n$, the decomposition must include the terms

$$\frac{B_1x + C_1}{cx^2+dx+e} + \frac{B_2x+C_2}{(cx^2+dx+e)^2} + \cdots + \frac{B_nx+C_n}{(cx^2+dx+e)^n}.$$

Step 5 Use algebraic techniques to solve for the constants in the numerators of the decomposition.

To find the constants in Step 5, the goal is to get a system with as many equations as there are unknowns in the numerators. One method for finding these equations is to substitute values for x in the rational equation formed in Steps 3 or 4.

Distinct Linear Factors

EXAMPLE 1 Finding a Partial Fraction Decomposition

Find the partial fraction decomposition of $\dfrac{2x^4 - 8x^2 + 5x - 2}{x^3 - 4x}$.

Solution The numerator has greater degree than the denominator, so divide first.

$$\begin{array}{r} 2x \\ x^3 - 4x\overline{)2x^4 - 8x^2 + 5x - 2} \\ \underline{2x^4 - 8x^2} \\ 5x - 2 \end{array}$$

The quotient is $\dfrac{2x^4 - 8x^2 + 5x - 2}{x^3 - 4x} = 2x + \dfrac{5x-2}{x^3-4x}$. Now work with the remainder fraction. Factor the denominator as $x^3 - 4x = x(x+2)(x-2)$. Since the factors are distinct linear factors, use Step 3(a) to write the decomposition as

$$\frac{5x-2}{x^3-4x} = \frac{A}{x} + \frac{B}{x+2} + \frac{C}{x-2}, \qquad (1)$$

where A, B, and C are constants that need to be found. Multiply each side of equation (1) by $x(x+2)(x-2)$ to get

$$5x - 2 = A(x+2)(x-2) + Bx(x-2) + Cx(x+2). \qquad (2)$$

Equation (1) is an identity, since both sides represent the same rational expression. Thus, equation (2) is also an identity. Equation (1) holds for all values of x except 0, -2, and 2. However, equation (2) holds for all values of x. In particular, substituting 0 for x in equation (2) gives $-2 = -4A$, so $A = \frac{1}{2}$. Similarly, choosing $x = -2$ gives $-12 = 8B$, so $B = -\frac{3}{2}$. Finally, choosing $x = 2$ gives $8 = 8C$, so $C = 1$. The remainder rational expression can be written as the following sum of partial fractions:

$$\frac{5x - 2}{x^3 - 4x} = \frac{1}{2x} + \frac{-3}{2(x + 2)} + \frac{1}{x - 2}.$$

The given rational expression can be written as

$$\frac{2x^4 - 8x^2 + 5x - 2}{x^3 - 4x} = 2x + \frac{1}{2x} + \frac{-3}{2(x + 2)} + \frac{1}{x - 2}.$$

Check the work by combining the terms on the right. ■

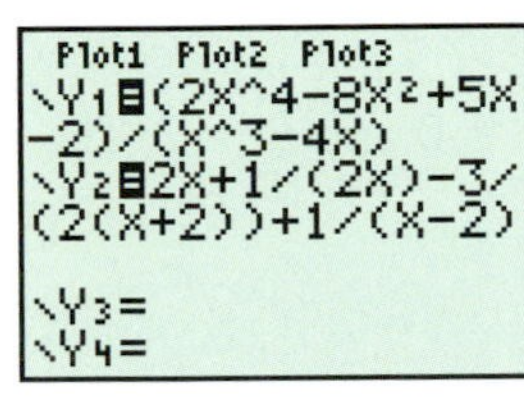

(a)

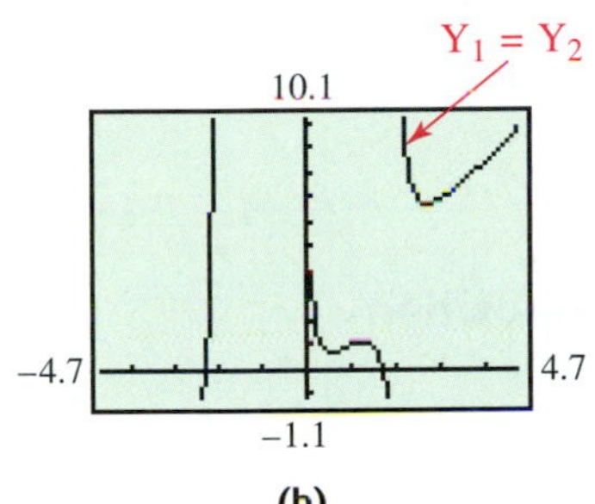

(b)

FIGURE 60

We can use a graphing calculator to *check* the work in Example 1. We define Y_1 using the original rational expression and Y_2 using the partial fraction decomposition:

$$Y_1 = \frac{2X^4 - 8X^2 + 5X - 2}{X^3 - 4X}, \quad Y_2 = 2X + \frac{1}{2X} + \frac{-3}{2(X + 2)} + \frac{1}{X - 2}.$$

See Figure 60(a). When we graph both Y_1 and Y_2 on the same screen, the graphs should be indistinguishable. See Figure 60(b).

NOTE The method of graphical support just described is not foolproof; however, it gives a fairly accurate portrayal of whether the partial fraction decomposition is correct. A table can be used for support as well.

Repeated Linear Factors

EXAMPLE 2 Finding a Partial Fraction Decomposition

Find the partial fraction decomposition of $\frac{2x}{(x - 1)^3}$.

Solution This is a proper fraction. The denominator is already factored with repeated linear factors. We write the decomposition as shown, using Step 3(b).

$$\frac{2x}{(x - 1)^3} = \frac{A}{x - 1} + \frac{B}{(x - 1)^2} + \frac{C}{(x - 1)^3}$$

We clear denominators by multiplying each side of this equation by $(x - 1)^3$.

$$2x = A(x - 1)^2 + B(x - 1) + C$$

Substituting 1 for x leads to $C = 2$, so

$$2x = A(x - 1)^2 + B(x - 1) + 2. \quad (1)$$

We substituted 1, the only root, and we still need to find values for A and B. However, *any* number can be substituted for x. For example, when we choose $x = -1$ (because it is easy to substitute), equation (1) becomes

$$\begin{aligned} 2(-1) &= A(-1-1)^2 + B(-1-1) + 2 && \text{Let } x = -1 \text{ in (1).} \\ -2 &= 4A - 2B + 2 && \\ -4 &= 4A - 2B && \text{Subtract 2.} \\ -2 &= 2A - B. && \text{Divide by 2.} \quad (2) \end{aligned}$$

Substituting 0 for x in equation (1) gives

$$\begin{aligned} 0 &= A - B + 2 \\ 2 &= -A + B. \quad (3) \end{aligned}$$

Now we solve the system of equations (2) and (3) to get $A = 0$ and $B = 2$. The partial fraction decomposition is

$$\frac{2x}{(x-1)^3} = \frac{2}{(x-1)^2} + \frac{2}{(x-1)^3}.$$

We needed three substitutions because there were three constants to evaluate: A, B, and C. To check this result, we could combine the terms on the right. ■

Distinct Linear and Quadratic Factors

EXAMPLE 3 Finding a Partial Fraction Decomposition

Find the partial fraction decomposition of $\dfrac{x^2 + 3x - 1}{(x+1)(x^2+2)}$.

Solution This denominator has distinct linear and quadratic factors, with neither repeated. Since $x^2 + 2$ cannot be factored, it is irreducible. The partial fraction decomposition is

$$\frac{x^2 + 3x - 1}{(x+1)(x^2+2)} = \frac{A}{x+1} + \frac{Bx + C}{x^2 + 2}.$$

Multiply each side by $(x+1)(x^2+2)$ to get

$$x^2 + 3x - 1 = A(x^2 + 2) + (Bx + C)(x + 1). \quad (1)$$

First, substitute -1 for x to get

Use parentheses around substituted values to avoid errors.

$$\begin{aligned} (-1)^2 + 3(-1) - 1 &= A[(-1)^2 + 2] + 0 \\ -3 &= 3A \\ A &= -1. \end{aligned}$$

Replace A with -1 in equation (1) and substitute any value for x. If $x = 0$, then

$$\begin{aligned} 0^2 + 3(0) - 1 &= -1(0^2 + 2) + (B \cdot 0 + C)(0 + 1) \\ -1 &= -2 + C \\ C &= 1. \end{aligned}$$

Now, letting $A = -1$ and $C = 1$, substitute again in equation (1), using another number for x. For $x = 1$,

$$\begin{aligned} 3 &= -3 + (B + 1)(2) \\ 6 &= 2B + 2 \\ B &= 2. \end{aligned}$$

With $A = -1$, $B = 2$, and $C = 1$, the partial fraction decomposition is

$$\frac{x^2 + 3x - 1}{(x + 1)(x^2 + 2)} = \frac{-1}{x + 1} + \frac{2x + 1}{x^2 + 2}.$$

Again, this work can be checked by combining terms on the right. ■

For fractions with denominators that have quadratic factors, another method is often more convenient. The system of equations is formed by equating coefficients of like terms on each side of the partial fraction decomposition. For instance, in Example 3, after each side was multiplied by the common denominator, the equation was

$$x^2 + 3x - 1 = A(x^2 + 2) + (Bx + C)(x + 1).$$

Multiplying on the right and collecting like terms, we have

$$\begin{aligned} x^2 + 3x - 1 &= Ax^2 + 2A + Bx^2 + Bx + Cx + C \\ x^2 + 3x - 1 &= (A + B)x^2 + (B + C)x + (C + 2A). \end{aligned}$$

Now, equating the coefficients of like powers of x gives the three equations

$$\begin{aligned} 1 &= A + B \\ 3 &= B + C \\ -1 &= C + 2A. \end{aligned}$$

Solving this system of equations for A, B, and C would give the partial fraction decomposition. The next example uses a combination of the two methods.

Repeated Quadratic Factors

EXAMPLE 4 Finding a Partial Fraction Decomposition

Find the partial fraction decomposition of $\dfrac{2x}{(x^2 + 1)^2(x - 1)}$.

Solution This expression has both a linear factor and a repeated quadratic factor. By Steps 3(a) and 4(b) from the beginning of this section,

$$\frac{2x}{(x^2 + 1)^2(x - 1)} = \frac{Ax + B}{x^2 + 1} + \frac{Cx + D}{(x^2 + 1)^2} + \frac{E}{x - 1}.$$

Multiplying each side by $(x^2 + 1)^2(x - 1)$ leads to

$$2x = (Ax + B)(x^2 + 1)(x - 1) + (Cx + D)(x - 1) + E(x^2 + 1)^2. \qquad (1)$$

If $x = 1$, equation (1) reduces to $2 = 4E$, or $E = \frac{1}{2}$. Substituting $\frac{1}{2}$ for E in equation (1) and combining terms on the right gives

$$2x = \left(A + \frac{1}{2}\right)x^4 + (-A + B)x^3 + (A - B + C + 1)x^2$$

$$+ (-A + B + D - C)x + \left(-B - D + \frac{1}{2}\right). \qquad (2)$$

To get additional equations involving the unknowns, equate the coefficients of like powers of x on each side of equation (2). Setting corresponding coefficients of x^4 equal yields $0 = A + \frac{1}{2}$, or $A = -\frac{1}{2}$. From the corresponding coefficients of x^3, $0 = -A + B$, which means that since $A = -\frac{1}{2}$, $B = -\frac{1}{2}$. Using the coefficients of x^2 gives $0 = A - B + C + 1$. Since $A = -\frac{1}{2}$ and $B = -\frac{1}{2}$, $C = -1$. Finally, from the coefficients of x, $2 = -A + B + D - C$. Substituting for A, B, and C gives $D = 1$. With $A = -\frac{1}{2}$, $B = -\frac{1}{2}$, $C = -1$, $D = 1$, and $E = \frac{1}{2}$, the given fraction has the partial fraction decomposition

$$\frac{2x}{(x^2 + 1)^2(x - 1)} = \frac{-\frac{1}{2}x - \frac{1}{2}}{x^2 + 1} + \frac{-x + 1}{(x^2 + 1)^2} + \frac{\frac{1}{2}}{x - 1},$$

or

$$\frac{2x}{(x^2 + 1)^2(x - 1)} = \frac{-(x + 1)}{2(x^2 + 1)} + \frac{-x + 1}{(x^2 + 1)^2} + \frac{1}{2(x - 1)}.$$

Simplify complex fractions.

In summary, to solve for the constants in the numerators of a partial fraction decomposition, use either of the following methods or a combination of the two.

Techniques for Decomposition into Partial Fractions

Method 1 For Linear Factors

Step 1 Multiply each side of the rational expression by the common denominator.

Step 2 Substitute the zero of each factor into the resulting equation. For repeated linear factors, substitute as many other numbers as is necessary to find all the constants in the numerators. The number of substitutions required will equal the number of constants A, B, ...

Method 2 For Quadratic Factors

Step 1 Multiply each side of the rational expression by the common denominator.

Step 2 Collect terms on the right side of the resulting equation.

Step 3 Equate the coefficients of like terms to get a system of equations.

Step 4 Solve the system to find the constants in the numerators.

7.8 Exercises

Find the partial fraction decomposition for each rational expression.

1. $\dfrac{5}{3x(2x + 1)}$

2. $\dfrac{3x - 1}{x(x + 1)}$

3. $\dfrac{4x + 2}{(x + 2)(2x - 1)}$

4. $\dfrac{x + 2}{(x + 1)(x - 1)}$

5. $\dfrac{x}{x^2 + 4x - 5}$

6. $\dfrac{5x - 3}{(x + 1)(x - 3)}$

7. $\dfrac{2x}{(x + 1)(x + 2)^2}$

8. $\dfrac{2}{x^2(x + 3)}$

9. $\dfrac{4}{x(1 - x)}$

10. $\dfrac{x + 1}{x^2(1 - x)}$

11. $\dfrac{4x^2 - x - 15}{x(x + 1)(x - 1)}$

12. $\dfrac{2x + 1}{(x + 2)^3}$

13. $\dfrac{x^2}{x^2 + 2x + 1}$

14. $\dfrac{3}{x^2 + 4x + 3}$

15. $\dfrac{2x^5 + 3x^4 - 3x^3 - 2x^2 + x}{2x^2 + 5x + 2}$

16. $\dfrac{6x^5 + 7x^4 - x^2 + 2x}{3x^2 + 2x - 1}$

17. $\dfrac{x^3 + 4}{9x^3 - 4x}$

18. $\dfrac{x^3 + 2}{x^3 - 3x^2 + 2x}$

19. $\dfrac{-3}{x^2(x^2 + 5)}$

20. $\dfrac{2x + 1}{(x + 1)(x^2 + 2)}$

21. $\dfrac{3x - 2}{(x + 4)(3x^2 + 1)}$

22. $\dfrac{3}{x(x + 1)(x^2 + 1)}$

23. $\dfrac{1}{x(2x + 1)(3x^2 + 4)}$

24. $\dfrac{x^4 + 1}{x(x^2 + 1)^2}$

25. $\dfrac{3x - 1}{x(2x^2 + 1)^2}$

26. $\dfrac{3x^4 + x^3 + 5x^2 - x + 4}{(x - 1)(x^2 + 1)^2}$

27. $\dfrac{-x^4 - 8x^2 + 3x - 10}{(x + 2)(x^2 + 4)^2}$

28. $\dfrac{x^2}{x^4 - 1}$

29. $\dfrac{5x^5 + 10x^4 - 15x^3 + 4x^2 + 13x - 9}{x^3 + 2x^2 - 3x}$

30. $\dfrac{3x^6 + 3x^4 + 3x}{x^4 + x^2}$

Determine whether each partial fraction decomposition is correct by graphing the left side and the right side of the equation on the same coordinate axes and observing whether the graphs coincide.

31. $\dfrac{4x^2 - 3x - 4}{x^3 + x^2 - 2x} = \dfrac{2}{x} + \dfrac{-1}{x - 1} + \dfrac{3}{x + 2}$

32. $\dfrac{1}{(x - 1)(x + 2)} = \dfrac{1}{x - 1} - \dfrac{1}{x + 2}$

33. $\dfrac{x^3 - 2x}{(x^2 + 2x + 2)^2} = \dfrac{x - 2}{x^2 + 2x + 2} + \dfrac{2}{(x^2 + 2x + 2)^2}$

34. $\dfrac{2x + 4}{x^2(x - 2)} = \dfrac{-2}{x} + \dfrac{-2}{x^2} + \dfrac{2}{x - 2}$

Reviewing Basic Concepts (Sections 7.7 and 7.8)

Graph the solution set of each inequality or system of inequalities.

1. $-2x - 3y \le 6$

2. $x - y < 5$
$x + y \ge 3$

3. $y \ge x^2 - 2$
$x + 2y \ge 4$

4. $x^2 + y^2 \le 25$
$x^2 + y^2 \ge 9$

5. Which one of the following systems is represented by the given graph?

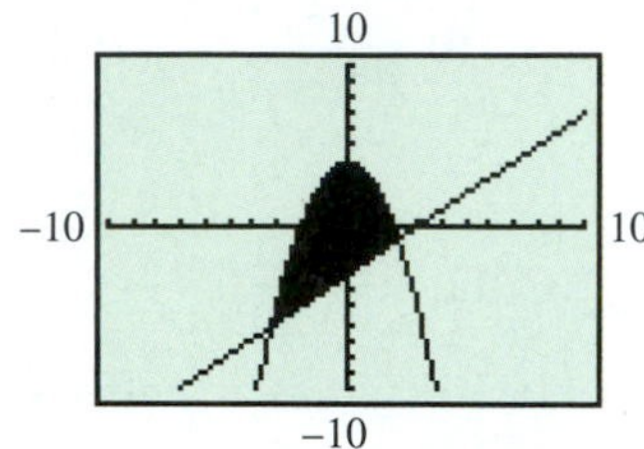

A. $y > x - 3$
$y < -x^2 + 4$

B. $y < x - 3$
$y < -x^2 + 4$

C. $y > x - 3$
$y > -x^2 + 4$

D. $y < x - 3$
$y > -x^2 + 4$

6. Find the minimum value of $2x + 3y$, subject to the following constraints:

$$x \ge 0$$
$$y \ge 0$$
$$x + y \ge 4$$
$$2x + y \le 8.$$

7. The graph shows a region of feasible solutions. Find the maximum and minimum values of $P = 3x + 5y$ over this region.

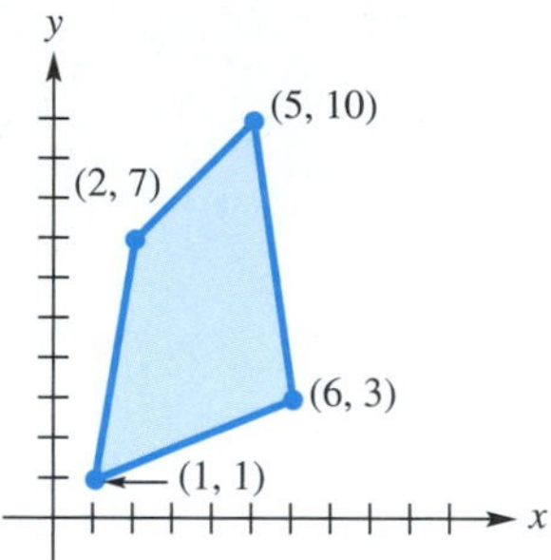

8. *(Modeling) Minimizing Cost* Two substances X and Y are found in pet food. Each substance contains the ingredients A and B. Substance X is 20% ingredient A and 50% ingredient B. Substance Y is 50% ingredient A and 30% ingredient B. The cost of substance X is $2 per pound and the cost of substance Y is $3 per pound. The pet store needs at least 251 pounds of ingredient A and at least 200 pounds of ingredient B. If cost is to be minimal, how many pounds of each substance should be ordered? Find the minimum cost.

Decompose each rational expression into partial fractions.

9. $\dfrac{10x + 13}{x^2 - x - 20}$

10. $\dfrac{2x^2 - 15x - 32}{(x - 1)(x^2 + 6x + 8)}$

CHAPTER 7 SUMMARY

KEY TERMS & SYMBOLS	KEY CONCEPTS
7.1 Systems of Equations linear equation (or first-degree equation) in n unknowns system of equations solutions	Systems of equations in two variables may be solved by the substitution method, the elimination method, or graphically, with the intersection-of-graphs method. **SUBSTITUTION METHOD** (for a system of two equations in two variables) Solve one equation for one variable in terms of the other. Substitute for that variable in the other equation, and solve for its value. Then substitute that value into the other equation to find the value of the remaining variable. *(continued)*

KEY TERMS & SYMBOLS	KEY CONCEPTS
system of linear equations (linear system) consistent system independent equations inconsistent system dependent equations substitution method elimination method equivalent systems nonlinear system of equations	**ELIMINATION METHOD** (for a system of two equations in two variables) Multiply one or both equations by appropriate numbers so that the sum of the coefficients of one of the variables is 0. Add the equations, and solve for the value of the remaining variable. Then substitute that value into either of the given equations to solve for the value of the other variable.
7.2 Solution of Linear Systems in Three Variables ordered triple (x, y, z)	**SOLVING A SYSTEM OF LINEAR EQUATIONS IN THREE VARIABLES** To solve a linear system with three unknowns, first eliminate a variable from any two of the equations. Then eliminate the *same variable* from a different pair of equations. Eliminate a second variable by using the resulting two equations in two variables to get an equation with just one variable whose value you can now determine. Find the values of the remaining variables by substitution.
7.3 Solution of Linear Systems by Row Transformations matrix (matrices) element (entry) augmented matrix echelon (triangular) form main diagonal back-substitution row echelon method reduced row echelon method	**MATRIX ROW TRANSFORMATIONS** For any augmented matrix of a system of linear equations, the following row transformations will result in the matrix of an equivalent system: **1.** Any two rows may be interchanged. **2.** The elements of any row may be multiplied by a nonzero real number. **3.** Any row may be changed by adding to its elements a multiple of the corresponding elements of another row.
7.4 Matrix Properties and Operations dimension square matrix row matrix column matrix zero matrix additive inverse (negative) scalar	**MATRIX ADDITION** The sum of two $m \times n$ matrices A and B is the $m \times n$ matrix $A + B$ in which each element is the sum of the corresponding elements of A and B. **MATRIX SUBTRACTION** If A and B are matrices with the same dimension, then $\mathbf{A - B = A + (-B)}$. **SCALAR MULTIPLICATION** If k is a scalar and A is a matrix, then kA is the matrix of the same dimension where each entry of A is multiplied by k. **MATRIX MULTIPLICATION** The product AB of an $m \times n$ matrix A and an $n \times k$ matrix B is found as follows: To find the ith row, jth column element of AB, multiply each element in the ith row of A by the corresponding element in the jth column of B. The sum of these products will give the element of row i, column j of AB. The dimension of AB is $m \times k$. *(continued)*

KEY TERMS & SYMBOLS	KEY CONCEPTS
7.5 Determinants and Cramer's Rule determinant det A minor cofactor expansion by a row or column	**DETERMINANT OF A 2 × 2 MATRIX** The determinant of a 2×2 matrix A, where $A = \begin{bmatrix} a_{11} & a_{12} \\ a_{21} & a_{22} \end{bmatrix}$, is a real number defined as $\det A = a_{11}a_{22} - a_{21}a_{12}$. **COFACTOR** Let M_{ij} be the minor for element a_{ij} in an $n \times n$ matrix. The cofactor of a_{ij} is $A_{ij} = (-1)^{i+j} \cdot M_{ij}$. **FINDING THE DETERMINANT OF AN $n \times n$ MATRIX** Multiply each element in any row or column of the matrix by its cofactor. The sum of these products gives the value of the determinant. **CRAMER'S RULE FOR 2 × 2 SYSTEMS** The solution of the system $$\begin{aligned} a_1x + b_1y &= c_1 \\ a_2x + b_2y &= c_2 \end{aligned}$$ is given by $x = \frac{D_x}{D}$ and $y = \frac{D_y}{D}$, where $D_x = \det\begin{bmatrix} c_1 & b_1 \\ c_2 & b_2 \end{bmatrix}$, $D_y = \det\begin{bmatrix} a_1 & c_1 \\ a_2 & c_2 \end{bmatrix}$, and $D = \det\begin{bmatrix} a_1 & b_1 \\ a_2 & b_2 \end{bmatrix} \neq 0$. Cramer's rule can be extended to 3×3 and larger systems.
7.6 Solution of Linear Systems by Matrix Inverses identity matrix I_n multiplicative inverse matrix singular matrix coefficient matrix	**INVERSE OF A 2 × 2 MATRIX** If $A = \begin{bmatrix} a & b \\ c & d \end{bmatrix}$ and $\det A \neq 0$, then $$A^{-1} = \frac{1}{\det A}\begin{bmatrix} d & -b \\ -c & a \end{bmatrix} = \begin{bmatrix} \frac{d}{ad - cb} & \frac{-b}{ad - cb} \\ \frac{-c}{ad - cb} & \frac{a}{ad - cb} \end{bmatrix}.$$ To find the multiplicative inverse of a 3×3 (or larger) matrix A, we often use a calculator. **FINDING AN INVERSE MATRIX** To obtain A^{-1} for any $n \times n$ matrix A for which A^{-1} exists, follow these steps: ***Step 1*** Form the augmented matrix $[A \mid I_n]$, where I_n is the $n \times n$ identity matrix. ***Step 2*** Perform row transformations on $[A \mid I_n]$ to obtain a matrix of the form $[I_n \mid B]$. ***Step 3*** Matrix B is A^{-1}.
7.7 Systems of Inequalities and Linear Programming	**GRAPHING AN INEQUALITY** **1.** For a function f, the graph of $y < f(x)$ consists of all the points that are *below* the graph of $y = f(x)$; the graph of $y > f(x)$ consists of all the points that are *above* the graph of $y = f(x)$. (Similar statements can be made for $\leq$ and $\geq$, with boundaries included.)

(continued)

KEY TERMS & SYMBOLS	KEY CONCEPTS
half plane boundary linear inequality in two variables system of inequalities linear programming constraints objective function region of feasible solutions vertex (corner point)	**2.** If the inequality is not or cannot be solved for y, choose a test point not on the boundary. If the test point satisfies the inequality, the graph includes all points on the same side of the boundary as the test point. Otherwise, the graph includes all points on the other side of the boundary. **SOLVING A LINEAR PROGRAMMING PROBLEM** ***Step 1*** Write the objective function and all necessary constraints. ***Step 2*** Graph the region of feasible solutions. ***Step 3*** Identify all vertices (corner points). ***Step 4*** Find the value of the objective function at each vertex. ***Step 5*** The solution is given by the vertex producing the optimal value of the objective function.
7.8 Partial Fractions partial fraction decomposition partial fraction	To solve for the constants in the numerators of a partial fraction decomposition, use either of the following methods or a combination of the two. **METHOD 1 FOR LINEAR FACTORS** ***Step 1*** Multiply each side of the rational expression by the common denominator. ***Step 2*** Substitute the zero of each factor into the resulting equation. For repeated linear factors, substitute as many other numbers as is necessary to find all the constants in the numerators. The number of substitutions required will equal the number of constants, $A, B, \ldots$. **METHOD 2 FOR QUADRATIC FACTORS** ***Step 1*** Multiply each side of the rational expression by the common denominator. ***Step 2*** Collect terms on the right side of the resulting equation. ***Step 3*** Equate the coefficients of like terms to get a system of equations. ***Step 4*** Solve the system to find the constants in the numerators.

CHAPTER 7 Review Exercises

Solve each system. Identify any systems with dependent equations and any inconsistent systems.

1. $4x - 3y = -1$
$3x + 5y = 50$

2. $\frac{x}{2} - \frac{y}{5} = \frac{11}{10}$
$2x - \frac{4y}{5} = \frac{22}{5}$

3. $4x + 5y = 5$
$3x + 7y = -6$

4. $y = x^2 - 1$
$x + y = 1$

5. $x^2 + y^2 = 2$
$3x + y = 4$

6. $x^2 + 2y^2 = 22$
$2x^2 - y^2 = -1$

7. $x^2 - 4y^2 = 19$
$x^2 + y^2 = 29$

8. $xy = 4$
$x - 6y = 2$

9. $x^2 - y^2 = 32$
$x = y + 4$

10. Use your calculator with window $[-18, 18]$ by $[-12, 12]$ to answer the following.
(a) Do the circle $x^2 + y^2 = 144$ and the line $x + 2y = 8$ have any points in common?
(b) Approximate any intersection points to the nearest tenth.
(c) Find the exact solution set of the system.

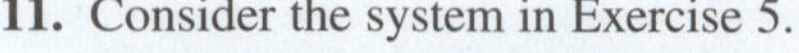
11. Consider the system in Exercise 5.
(a) To graph the first equation, what two functions must you enter into your calculator?
(b) To graph the second equation, what function must you enter?
(c) What would be an appropriate window in which to graph this system?

12. Can a system of two linear equations in two variables have exactly two solutions? Explain.

13. Can a system consisting of two equations in three variables have a unique solution? Explain.

Solve each system. Identify any systems with dependent equations and any inconsistent systems.

14. $\begin{aligned} 2x - 3y + z &= -5 \\ x + 4y + 2z &= 13 \\ 5x + 5y + 3z &= 14 \end{aligned}$

15. $\begin{aligned} x - 3y &= 12 \\ 2y + 5z &= 1 \\ 4x + z &= 25 \end{aligned}$

16. $\begin{aligned} x + y - z &= 5 \\ 2x + y + 3z &= 2 \\ 4x - y + 2z &= -1 \end{aligned}$

17. $\begin{aligned} 5x - 3y + 2z &= -5 \\ 2x + y - z &= 4 \\ -4x - 2y + 2z &= -1 \end{aligned}$

Use the reduced row echelon method to solve each system. Identify any systems with dependent equations and any inconsistent systems.

18. $\begin{aligned} 2x + 3y &= 10 \\ -3x + y &= 18 \end{aligned}$

19. $\begin{aligned} 3x + y &= -7 \\ x - y &= -5 \end{aligned}$

20. $\begin{aligned} x - z &= -3 \\ y + z &= 6 \\ 2x - 3z &= -9 \end{aligned}$

21. $\begin{aligned} 2x - y + 4z &= -1 \\ -3x + 5y - z &= 5 \\ 2x + 3y + 2z &= 3 \end{aligned}$

Perform each operation if possible. If not possible, say so.

22. $\begin{bmatrix} -5 & 4 & 9 \\ 2 & -1 & -2 \end{bmatrix} + \begin{bmatrix} 1 & -2 & 7 \\ 4 & -5 & -5 \end{bmatrix}$

23. $\begin{bmatrix} 3 \\ 2 \\ 5 \end{bmatrix} - \begin{bmatrix} 8 \\ -4 \\ 6 \end{bmatrix} + \begin{bmatrix} 1 \\ 0 \\ 2 \end{bmatrix}$

24. $\begin{bmatrix} 2 & 5 & 8 \\ 1 & 9 & 2 \end{bmatrix} - \begin{bmatrix} 3 & 4 \\ 7 & 1 \end{bmatrix}$

25. $3\begin{bmatrix} 2 & 4 \\ -1 & 4 \end{bmatrix} - 2\begin{bmatrix} 5 & 8 \\ 2 & -2 \end{bmatrix}$

26. $-1\begin{bmatrix} 3 & -5 & 2 \\ 1 & 7 & -4 \end{bmatrix} + 5\begin{bmatrix} 0 & 2 \\ -1 & 3 \end{bmatrix}$

27. $10\begin{bmatrix} 2x + 3y & 4x + y \\ x - 5y & 6x + 2y \end{bmatrix} + 2\begin{bmatrix} -3x - y & x + 6y \\ 4x + 2y & 5x - y \end{bmatrix}$

28. ***Concept Check*** Complete the following sentence: The sum of two $m \times n$ matrices A and B is found by ________.

Multiply if possible. If not possible, say so.

29. $\begin{bmatrix} -8 & 6 \\ 5 & 2 \end{bmatrix}\begin{bmatrix} 3 & -1 \\ 7 & 2 \end{bmatrix}$

30. $\begin{bmatrix} 3 & 2 & -1 \\ 4 & 0 & 6 \end{bmatrix}\begin{bmatrix} -2 & 0 \\ 0 & 2 \\ 3 & 1 \end{bmatrix}$

31. $\begin{bmatrix} 1 & -2 & 4 & 2 \\ 0 & 1 & -1 & 8 \end{bmatrix}\begin{bmatrix} -1 \\ 2 \\ 0 \\ 1 \end{bmatrix}$

32. $\begin{bmatrix} 1 & 2 & 5 \\ -3 & 4 & 7 \\ 0 & 2 & -1 \end{bmatrix}\begin{bmatrix} 4 & 2 & 3 \\ 10 & -5 & 6 \end{bmatrix}$

33. $\begin{bmatrix} 4 & 2 & 3 \\ 10 & -5 & 6 \end{bmatrix}\begin{bmatrix} 1 & 2 & 5 \\ -3 & 4 & 7 \\ 0 & 2 & -1 \end{bmatrix}$

34. $\begin{bmatrix} 3 & -1 & 0 \end{bmatrix}\begin{bmatrix} 1 & 3 & 2 \\ 2 & -4 & 0 \\ 5 & 7 & 3 \end{bmatrix}$

Find AB and BA to determine whether A and B are inverses.

35. $A = \begin{bmatrix} 3 & 2 \\ 13 & 9 \end{bmatrix}; B = \begin{bmatrix} 9 & -2 \\ -13 & 3 \end{bmatrix}$

36. $A = \begin{bmatrix} 1 & 0 \\ 2 & -3 \end{bmatrix}; B = \begin{bmatrix} 1 & 0 \\ \frac{2}{3} & -\frac{1}{3} \end{bmatrix}$

37. $A = \begin{bmatrix} 2 & 0 & 6 \\ 0 & 1 & 0 \\ 1 & 0 & 1 \end{bmatrix}; B = \begin{bmatrix} -1 & 0 & \frac{3}{2} \\ 0 & 1 & 0 \\ \frac{1}{4} & 0 & -1 \end{bmatrix}$

38. $A = \begin{bmatrix} 1 & 0 & 2 \\ 0 & 2 & 4 \\ 0 & 0 & 1 \end{bmatrix}; B = \begin{bmatrix} 1 & 0 & -2 \\ 0 & \frac{1}{2} & -2 \\ 0 & 0 & 1 \end{bmatrix}$

For each matrix, find A^{-1} if it exists.

39. $A = \begin{bmatrix} 6 & 3 \\ 10 & 5 \end{bmatrix}$

40. $A = \begin{bmatrix} -4 & 2 \\ 0 & 3 \end{bmatrix}$

41. $A = \begin{bmatrix} 2 & 0 \\ -1 & 5 \end{bmatrix}$

42. $A = \begin{bmatrix} 2 & 0 & 4 \\ 1 & -1 & 0 \\ 0 & 1 & -2 \end{bmatrix}$

43. $A = \begin{bmatrix} 2 & -1 & 0 \\ 1 & 0 & 1 \\ 1 & -2 & 0 \end{bmatrix}$

44. $A = \begin{bmatrix} 2 & 3 & 5 \\ -2 & -3 & -5 \\ 1 & 4 & 2 \end{bmatrix}$

Use the inverse matrix method to solve each system if possible. Identify any systems with dependent equations or any inconsistent systems.

45. $\begin{aligned} x + y &= 4 \\ 2x + 3y &= 10 \end{aligned}$

46. $\begin{aligned} 5x - 3y &= -2 \\ 2x + 7y &= -9 \end{aligned}$

47. $\begin{aligned} 2x + y &= 5 \\ 3x - 2y &= 4 \end{aligned}$

48. $\begin{aligned} x - 2y &= 7 \\ 3x + y &= 7 \end{aligned}$

49. $\begin{aligned} x + 2y &= -1 \\ 3y - z &= -5 \\ x + 2y - z &= -3 \end{aligned}$

50. $\begin{aligned} 3x - 2y + 4z &= 1 \\ 4x + y - 5z &= 2 \\ -6x + 4y - 8z &= -2 \end{aligned}$

51. $\begin{aligned} x + y + z &= 1 \\ 2x - y &= -2 \\ 3y + z &= 2 \end{aligned}$

52. $\begin{aligned} x &= -3 \\ y + z &= 6 \\ 2x - 3z &= -9 \end{aligned}$

Evaluate each determinant.

53. $\det\begin{bmatrix} -1 & 8 \\ 2 & 9 \end{bmatrix}$

54. $\det\begin{bmatrix} -2 & 4 \\ 0 & 3 \end{bmatrix}$

55. $\det\begin{bmatrix} -2 & 4 & 1 \\ 3 & 0 & 2 \\ -1 & 0 & 3 \end{bmatrix}$

56. $\det\begin{bmatrix} -1 & 2 & 3 \\ 4 & 0 & 3 \\ 5 & -1 & 2 \end{bmatrix}$

Solve each determinant equation for x.

57. $\det\begin{bmatrix} -3 & 2 \\ 1 & x \end{bmatrix} = 5$

58. $\det\begin{bmatrix} 3x & 7 \\ -x & 4 \end{bmatrix} = 8$

59. $\det\begin{bmatrix} 2 & 5 & 0 \\ 1 & 3x & -1 \\ 0 & 2 & 0 \end{bmatrix} = 4$

60. $\det\begin{bmatrix} 6x & 2 & 0 \\ 1 & 5 & 3 \\ x & 2 & -1 \end{bmatrix} = 2x$

Exercises 61 and 62 refer to the system $\begin{aligned} 3x - y &= 28 \\ 2x + y &= 2 \end{aligned}$.

61. Suppose you are asked to solve this system by using Cramer's rule.
(a) What is the value of D?
(b) What is the value of D_x?
(c) What is the value of D_y?
(d) Find x and y, using Cramer's rule.

62. Suppose you are asked to solve this system by using the inverse matrix method.
(a) What is A?
(b) What is B?
(c) Explain how you would use these matrices to go about solving the system.

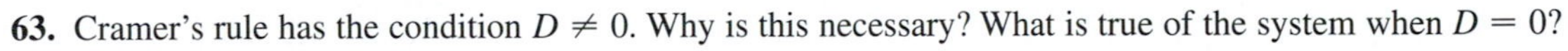

63. Cramer's rule has the condition $D \neq 0$. Why is this necessary? What is true of the system when $D = 0$?

Solve each system, if possible, by using Cramer's rule. Identify any systems with dependent equations or any inconsistent systems.

64. $\begin{aligned} 3x + y &= -1 \\ 5x + 4y &= 10 \end{aligned}$

65. $\begin{aligned} 3x + 7y &= 2 \\ 5x - y &= -22 \end{aligned}$

66. $\begin{aligned} 2x - 5y &= 8 \\ 3x + 4y &= 10 \end{aligned}$

67. $\begin{aligned} 3x + 2y + z &= 2 \\ 4x - y + 3z &= -16 \\ x + 3y - z &= 12 \end{aligned}$

68. $\begin{aligned} 5x - 2y - z &= 8 \\ -5x + 2y + z &= -8 \\ x - 4y - 2z &= 0 \end{aligned}$

69. $\begin{aligned} -x + 3y - 4z &= 2 \\ 2x + 4y + z &= 3 \\ 3x - z &= 9 \end{aligned}$

Solve each problem.

70. *Determining the Contents of a Meal* A cup of uncooked rice contains 15 grams of protein and 810 calories. A cup of uncooked soybeans contains 22.5 grams of protein and 270 calories. How many cups of each should be used for a meal containing 9.5 grams of protein and 324 calories?

71. *Determining Order Quantities* A company sells CDs for \$.40 each and diskettes for \$.30 each. The company receives \$38 for an order of 100 CDs and diskettes. However, the customer neglected to specify how many of each to send. Determine the number of CDs and the number of diskettes that should be sent.

72. *Mixing Teas* Three kinds of tea worth \$4.60, \$5.75, and \$6.50 per pound are to be mixed to get 20 pounds of tea worth \$5.25 per pound. The amount of \$4.60 tea used is to equal the amount of the other two kinds together. How many pounds of each tea should be used?

73. *Mixing Solutions of a Drug* A 5% solution of a drug is to be mixed with some 15% solution and some 10% solution to get 20 milliliters of 8% solution. The amount of 5% solution used must be 2 milliliters more than the sum of the other two solutions. How many milliliters of each solution should be used?

74. *(Modeling) Blood Pressure* In a study of adult males, it was believed that systolic blood pressure P was affected by both age A in years and weight W in pounds. This was modeled by $P = a + bA + cW$, where a, b, and c are constants. The table lists three individuals with representative blood pressures for the group. (*Source:* Brase, C.H. and C.P. Brase, *Understandable Statistics,* D.C. Heath and Company, 1995.)

P	A	W
113	39	142
138	53	181
152	65	191

(a) Use the table to approximate values for the constants a, b, and c.

(b) Estimate a typical systolic blood pressure for an individual who is 55 years old and weighs 175 pounds.

75. *Curve Fitting* Find the equation of the polynomial function of degree 3 whose graph passes through the points $(-2, 1)$, $(-1, 6)$, $(2, 9)$, and $(3, 26)$.

76. *Curve Fitting* Find the equation of the quadratic polynomial that defines the curve shown in the figure.

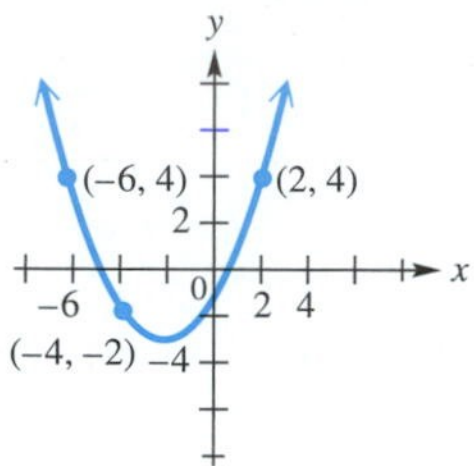

Graph the solution set of each system.

77. $x + y \leq 6$
$2x - y \geq 3$

78. $y \leq \frac{1}{3}x - 2$
$y^2 \leq 16 - x^2$

79. Find $x \geq 0$ and $y \geq 0$ such that

$$3x + 2y \leq 12$$
$$5x + y \geq 5$$

and $2x + 4y$ is maximized.

80. Find $x \geq 0$ and $y \geq 0$ such that

$$x + y \leq 50$$
$$2x + y \geq 20$$
$$x + 2y \geq 30$$

and $4x + 2y$ is minimized.

81. *(Modeling) Company Profit* A small company manufacturers two products: radios and DVD players. Each radio results in a profit of \$15, each DVD player a profit of \$35. Due to demand, the company must produce at least 5, and not more than, 25 radios per day. The number of radios cannot exceed the number of DVD players, and the number of DVD players cannot exceed 30. How many of each should the company manufacture to obtain maximum profit? What will that profit be?

Find the partial fraction decomposition of each rational expression.

82. $\dfrac{5x - 2}{x^2 - 4}$

83. $\dfrac{x + 2}{x^3 + 2x^2 + x}$

84. $\dfrac{x + 2}{x^3 - x^2 + 4x}$

CHAPTER 7 Test

1. Consider the system of equations.

$$x^2 - 4y^2 = -15$$
$$3x + y = 1$$

(a) What type of graph does each equation have?
(b) How many points of intersection of the two graphs are possible?
(c) Solve the system.
(d) Support the solutions, using a graphing calculator.

2. Solve the system.

$$\begin{aligned} 2x + y + z &= 3 \\ x + 2y - z &= 3 \\ 3x - y + z &= 5 \end{aligned}$$

3. Perform each matrix operation if possible.

(a) $3\begin{bmatrix} 2 & 3 \\ 1 & -4 \\ 5 & 9 \end{bmatrix} - \begin{bmatrix} -2 & 6 \\ 3 & -1 \\ 0 & 8 \end{bmatrix}$

(b) $\begin{bmatrix} 1 \\ 2 \end{bmatrix} + \begin{bmatrix} 4 \\ -6 \end{bmatrix} + \begin{bmatrix} 2 & 8 \\ -7 & 5 \end{bmatrix}$

(c) $\begin{bmatrix} 2 & 1 & -3 \\ 4 & 0 & 5 \end{bmatrix}\begin{bmatrix} 1 & 3 \\ 2 & 4 \\ 3 & -2 \end{bmatrix}$

4. Suppose A and B are both $n \times n$ matrices.
(a) Can AB be found?
(b) Can BA be found?

(c) Does $AB = BA$? Explain why or why not.
(d) If A is $n \times n$ and C is $m \times n$, can either AC or CA be found?

5. Evaluate each determinant.

(a) $\det\begin{bmatrix} 4 & 9 \\ -5 & -11 \end{bmatrix}$ (b) $\det\begin{bmatrix} 2 & 0 & 8 \\ -1 & 7 & 9 \\ 12 & 5 & -3 \end{bmatrix}$

6. Solve the system by using Cramer's rule.

$$\begin{aligned} 2x - 3y &= -33 \\ 4x + 5y &= 11 \end{aligned}$$

7. Consider the system of equations.

$$\begin{aligned} x + y - z &= -4 \\ 2x - 3y - z &= 5 \\ x + 2y + 2z &= 3 \end{aligned}$$

(a) Write the matrix of coefficients A, the matrix of variables X, and the matrix of constants B for this system.
(b) Find A^{-1}.
(c) Use the matrix inverse method to solve the system.
(d) If the first equation in the system is replaced by $.5x + y + z = 1.5$, the system cannot be solved with the matrix inverse method. Explain why.

8. ***(Modeling) Voter Turnout*** The table shows the percent y of voter turnout in the United States for three presidential elections in year x, where $x = 0$ corresponds to 1900. Find a quadratic function defined by $f(x) = ax^2 + bx + c$ that models these data. Graph f together with the data.

x	24	60	96
y	48.9	62.8	48.8

Source: Committee for the Study of the American Electorate.

9. The solution set of a system of inequalities is shown. Which system is it?

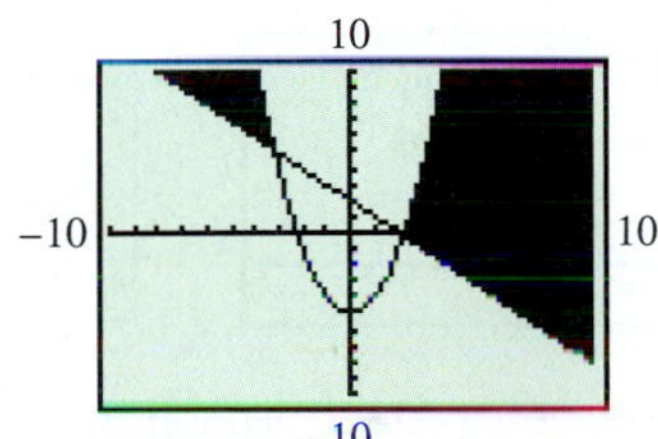

A. $y > 2 - x$, $y > x^2 - 5$
B. $y > 2 - x$, $y < x^2 - 5$
C. $y < 2 - x$, $y < x^2 - 5$
D. $y < 2 - x$, $y > x^2 - 5$

10. ***Storage Capacity*** An office manager wants to buy filing cabinets. Cabinet X costs \$100, requires 6 square feet of floor space, and holds 8 cubic feet. Cabinet Y costs \$200, requires 8 square feet of floor space, and holds 12 cubic feet. No more than \$1400 can be spent, and the office has room for no more than 72 square feet of cabinets. The office manager wants the maximum storage capacity within the limits imposed by funds and space. How many of each type of cabinet should be bought?

Find the partial fraction decomposition for each rational expression.

11. $\dfrac{7x - 1}{x^2 - x - 6}$

12. $\dfrac{x^2 - 11x + 6}{(x + 2)(x - 2)^2}$

CHAPTER 7 PROJECT

Finding a Polynomial Whose Graph Passes through Any Number of Given Points

In the Chapter 3 Project, we found a ninth-degree polynomial with zeros that depended on the digits of a Social Security number. In this project, we'll learn to find a polynomial that goes through a given set of points in the plane. Recall that three points define a second-degree polynomial, four points define a third-degree polynomial, five points define a fourth-degree polynomial, and so on. One important restriction on the points is that they must each have a different x-coordinate. Since polynomials define functions, no two distinct points can have the same x-coordinate.

In the following example, we use the same Social Security number (539-58-0954) that we used in the Chapter 3 Project to find an eighth-degree polynomial that lies on the nine points with x-coordinates from 1 to 9 and with y-coordinates that are the digits of the Social Security number. The nine ordered pairs are

$$(1,5), (2,3), (3,9), (4,5), (5,8), (6,0), (7,9), (8,5), \text{ and } (9,4).$$

These nine ordered pairs are entered in the lists shown in Figure A. To get the scatter diagram of the data in Figure C, we enter the window settings shown in Figure B.

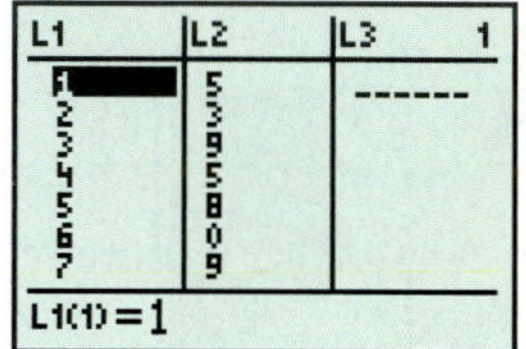

FIGURE A

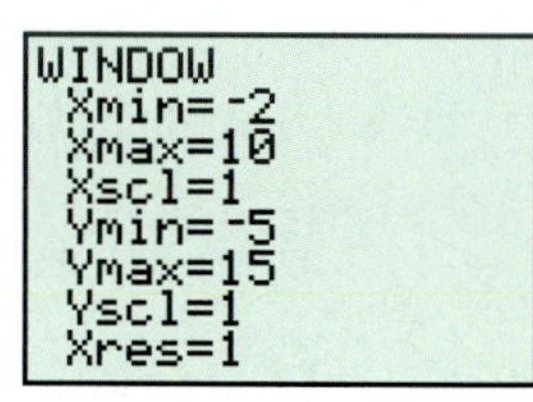

FIGURE B

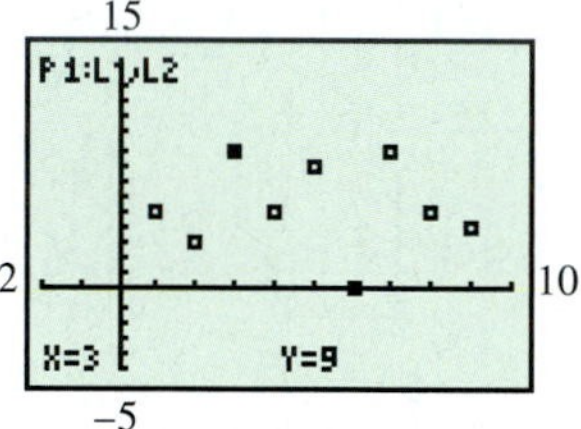

FIGURE C

We now use the nine ordered pairs to write a system of nine equations with nine unknowns by substituting each pair of x- and y-values into the general eighth-degree polynomial equation.

$$ax^8 + bx^7 + cx^6 + dx^5 + ex^4 + fx^3 + gx^2 + hx + i = y. \qquad (*)$$

$$a1^8 + b1^7 + c1^6 + d1^5 + e1^4 + f1^3 + g1^2 + h1^1 + i = 5$$
$$a2^8 + b2^7 + c2^6 + d2^5 + e2^4 + f2^3 + g2^2 + h2^1 + i = 3$$
$$a3^8 + b3^7 + c3^6 + d3^5 + e3^4 + f3^3 + g3^2 + h3^1 + i = 9$$
$$a4^8 + b4^7 + c4^6 + d4^5 + e4^4 + f4^3 + g4^2 + h4^1 + i = 5$$
$$a5^8 + b5^7 + c5^6 + d5^5 + e5^4 + f5^3 + g5^2 + h5^1 + i = 8$$
$$a6^8 + b6^7 + c6^6 + d6^5 + e6^4 + f6^3 + g6^2 + h6^1 + i = 0$$
$$a7^8 + b7^7 + c7^6 + d7^5 + e7^4 + f7^3 + g7^2 + h7^1 + i = 9$$
$$a8^8 + b8^7 + c8^6 + d8^5 + e8^4 + f8^3 + g8^2 + h8^1 + i = 5$$
$$a9^8 + b9^7 + c9^6 + d9^5 + e9^4 + f9^3 + g9^2 + h9^1 + i = 4$$

Next, we write an equivalent matrix equation $AX = B$.

$$\begin{bmatrix} 1 & 1 & 1 & 1 & 1 & 1 & 1 & 1 & 1 \\ 2^8 & 2^7 & 2^6 & 2^5 & 2^4 & 2^3 & 2^2 & 2 & 1 \\ 3^8 & 3^7 & 3^6 & 3^5 & 3^4 & 3^3 & 3^2 & 3 & 1 \\ 4^8 & 4^7 & 4^6 & 4^5 & 4^4 & 4^3 & 4^2 & 4 & 1 \\ 5^8 & 5^7 & 5^6 & 5^5 & 5^4 & 5^3 & 5^2 & 5 & 1 \\ 6^8 & 6^7 & 6^6 & 6^5 & 6^4 & 6^3 & 6^2 & 6 & 1 \\ 7^8 & 7^7 & 7^6 & 7^5 & 7^4 & 7^3 & 7^2 & 7 & 1 \\ 8^8 & 8^7 & 8^6 & 8^5 & 8^4 & 8^3 & 8^2 & 8 & 1 \\ 9^8 & 9^7 & 9^6 & 9^5 & 9^4 & 9^3 & 9^2 & 9 & 1 \end{bmatrix} \begin{bmatrix} a \\ b \\ c \\ d \\ e \\ f \\ g \\ h \\ i \end{bmatrix} \begin{matrix} \\ = \\ = \\ = \\ = \\ = \\ = \\ = \\ = \end{matrix} \begin{bmatrix} 5 \\ 3 \\ 9 \\ 5 \\ 8 \\ 0 \\ 9 \\ 5 \\ 4 \end{bmatrix}$$

To solve the matrix equation for the matrix of variables, enter matrices A and B into the matrix editor of your graphing calculator. (See Figures D and E.) Entries can be typed in exponential form; the calculator will evaluate each entry. Although the calculator displays large entries in scientific notation, it stores the exact value as shown in Figure D, for the sixth row, first column, entry of the matrix. (The screen is not large enough to display the entire matrix at one time.)

FIGURE D

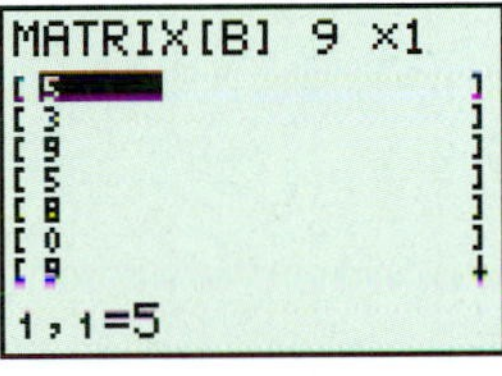

FIGURE E

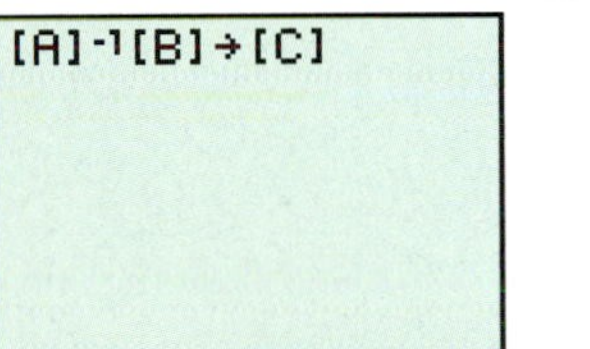

FIGURE F

FIGURE G

Solve the system by multiplying the inverse of matrix A by B, storing the solution as [C], as shown in Figures F and G. (You will need to scroll down to see the entire matrix [C].)

The entries in [C] are a, b, c, and so on, the coefficients of the polynomial equation (*). We can use the rows of [C] to enter the equation of the polynomial in the Y = menu, using the notation in Figure H. Be sure to use the matrix menu to enter [C] for each coordinate $[C]\,(m, n)$. Figure I shows the graph of the polynomial in the same window as the scatter diagram in Figure C. In Figure J, the window setting has been changed to show a graph that includes all local minima and maxima.

FIGURE H

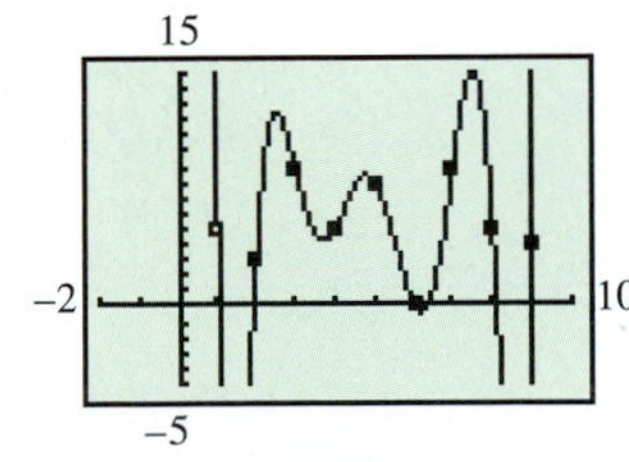

FIGURE I

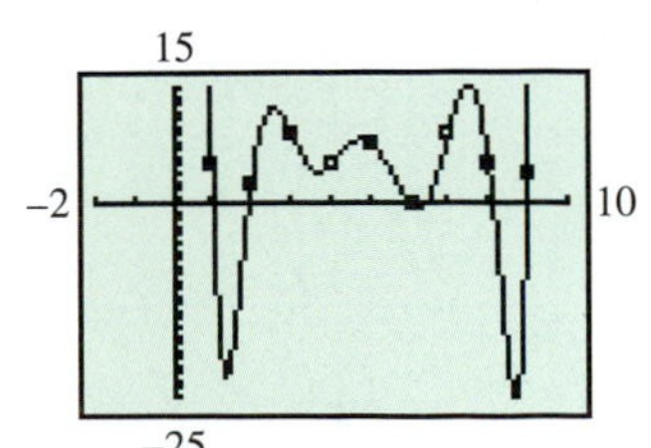

FIGURE J

Activities

1. Use the process just outlined to construct a second-degree polynomial with three ordered pairs. Use the digits 1, 2, and 3 as the x-coordinates and the first three digits of a Social Security number as the y-coordinates of the ordered pairs.
2. Construct the eighth-degree polynomial $f(x)$ that lies on the nine points with x-coordinates from 1 to 9 and y-coordinates that are the digits of a Social Security number.
3. Using your graphing calculator, find a comprehensive graph of $f(x)$, including all local extreme points, all intercepts, and clear end behavior. Sketch the graph on paper, or print it by using appropriate computer software. Label the extreme points with their coordinates.

8

Trigonometric Functions and Applications

MYSTERY SURROUNDS AN area of the Atlantic Ocean known as the Bermuda Triangle. According to the National Transportation Safety Board, more than 2000 boats and airplanes have disappeared while traversing this triangular-shaped region with vertices at Bermuda, Puerto Rico, and Miami.

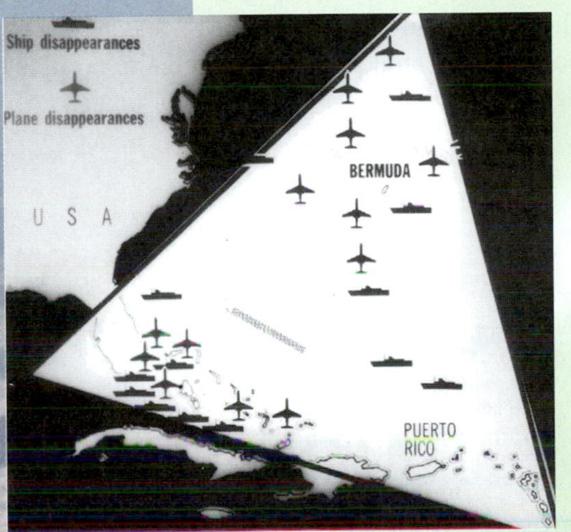

What could explain the disappearance of so many boats and planes on otherwise routine journeys? One theory researched by Graham Hawkes, a submarine explorer, postulates that plumes of methane gas released from breaks in the ocean floor could cause boats and ships to sink and airplanes to crash. Ascending bubbles violently displace water into the air and could force it over a boat, while simultaneously decreasing buoyancy under the boat. Clouds of the colorless, odorless gas rise above the water's surface as the bubbles burst. Airplane mechanics working with Hawkes and a team of scientists showed that with as little as a 1% concentration of methane gas in the air, a small airplane engine could stall, causing the plane to crash into the water.

In this chapter, we use trigonometric functions to solve problems related to navigation, periodic phenomena, and rotation. Navigators rely on trigonometry to guide ships and airplanes, safely avoid mountains or submerged objects, and perhaps prevent unintended entry into the Bermuda Triangle.

Source: www.bermuda-triangle.org; www.discovery.com.

Chapter Outline

8.1 Angles and Arcs

Basic Terminology ■ Degree Measure ■ Standard Position and Coterminal Angles ■ Radian Measure ■ Arc Lengths and Areas of Sectors ■ Angular and Linear Speed

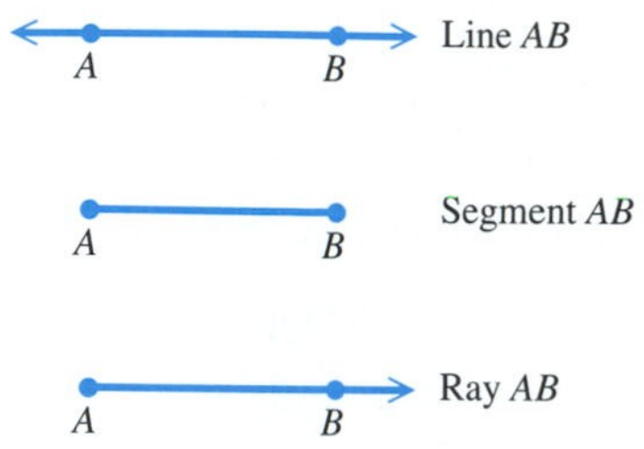

Line *AB*

Segment *AB*

Ray *AB*

FIGURE 1

Terminal side

Vertex *A*

Initial side

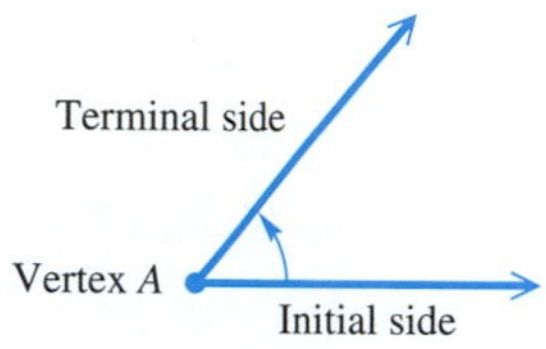

FIGURE 2

Basic Terminology

Two distinct points *A* and *B* determine a line called **line *AB*.** The portion of the line between *A* and *B*, including points *A* and *B*, is **segment *AB*.** The portion of line *AB* that starts at *A* and continues through *B* and on past *B* is called **ray *AB*.** Point *A* is the *endpoint of the ray.* See Figure 1.

An **angle** is formed by rotating a ray around its endpoint. The ray in its initial position is called the **initial side** of the angle, and the ray in its location after the rotation is the **terminal side** of the angle. The endpoint of the ray is the **vertex** of the angle. Figure 2 shows the initial and terminal sides of an angle with vertex *A*. If the rotation is counterclockwise, the angle is **positive.** If the rotation is clockwise, the angle is **negative.** See Figure 3.

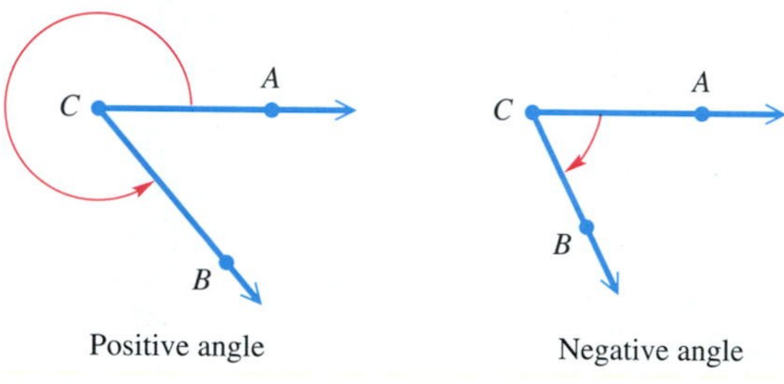

FIGURE 3

An angle can be named by using the name of its vertex. For example, either angle in Figure 3 can be called angle *C*. Alternatively, an angle can be named using three letters, with the vertex letter in the middle. Thus, either angle also could be named angle *ACB* or angle *BCA*.

Degree Measure

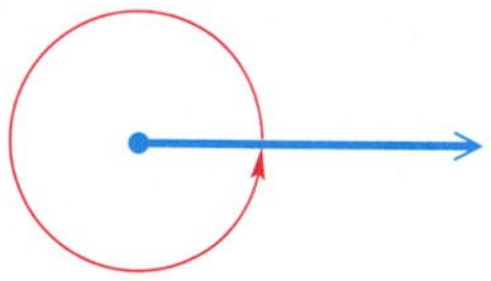

The most common unit used to measure the size of angles is the **degree.** Degree measure was developed by the Babylonians 4000 years ago.* To use degree measure, we assign 360 degrees to a complete rotation of a ray. In Figure 4, notice that the terminal side of the angle corresponds to its initial side when it makes a complete rotation. One degree, written **1°,** represents $\frac{1}{360}$ of a rotation. Therefore, 90° represents $\frac{90}{360} = \frac{1}{4}$ of a complete rotation, and 180° represents $\frac{180}{360} = \frac{1}{2}$ of a complete rotation.

A complete rotation of a ray gives an angle whose measure is 360°.

FIGURE 4

An angle measuring between 0° and 90° is an **acute angle.** An angle measuring exactly 90° is a **right angle.** An angle measuring more than 90° but less than 180° is an **obtuse angle,** and an angle of exactly 180° is a **straight angle.**

*The Babylonians were the first to subdivide the circumference of a circle into 360 parts. There are various theories as to why the number 360 was chosen. One is that it is approximately the number of days in a year, and it has many divisors, which makes it convenient to work with. Another involves a roundabout theory dealing with the length of a Babylonian mile.

In Figure 5, we use the Greek letter θ (theta)* to name each angle.

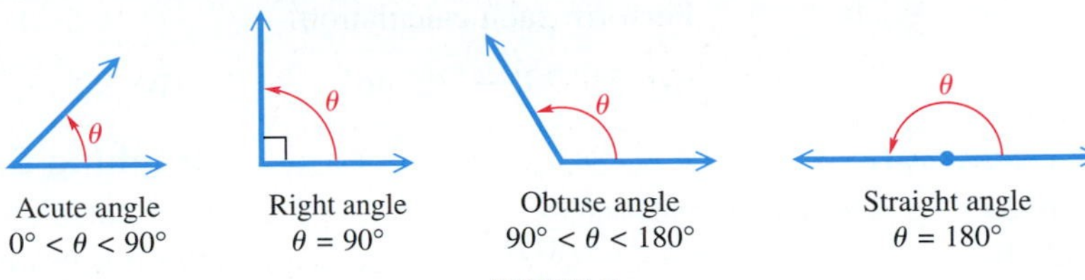

FIGURE 5

If the sum of the measures of two positive angles is 90°, the angles are called **complementary.** Two positive angles with measures whose sum is 180° are **supplementary.**

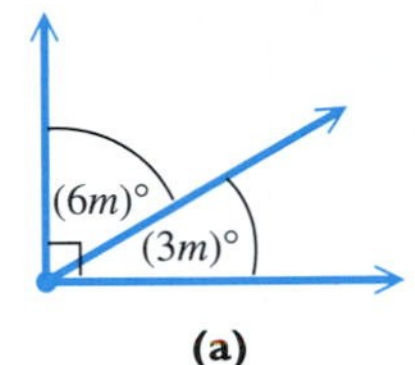

(a)

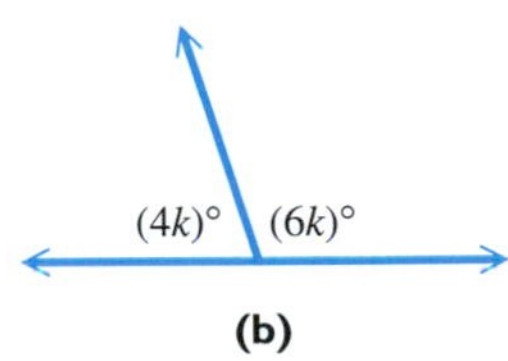

(b)

FIGURE 6

EXAMPLE 1 Finding Measures of Angles

Find the measure of each angle in Figure 6.

Solution

(a) In Figure 6(a), since the two angles form a right angle (as indicated by the ⌐ symbol), they are complementary angles.

$$6m + 3m = 90$$
$$9m = 90 \quad \text{Combine terms.}$$
$$m = 10 \quad \text{Divide by 9.}$$

The two angles have measures of $6m = 6(10) = 60°$ and $3m = 3(10) = 30°$. Be sure that you determine the measure of each angle by substituting 10 for m. Note that $60° + 30° = 90°$, so these two angles are complementary.

(b) The angles in Figure 6(b) are supplementary.

$$4k + 6k = 180$$
$$10k = 180 \quad \text{Combine terms.}$$
$$k = 18 \quad \text{Divide by 10.}$$

Remember to substitute for k.

These angle measures are $4(18) = 72°$ and $6(18) = 108°$. ■

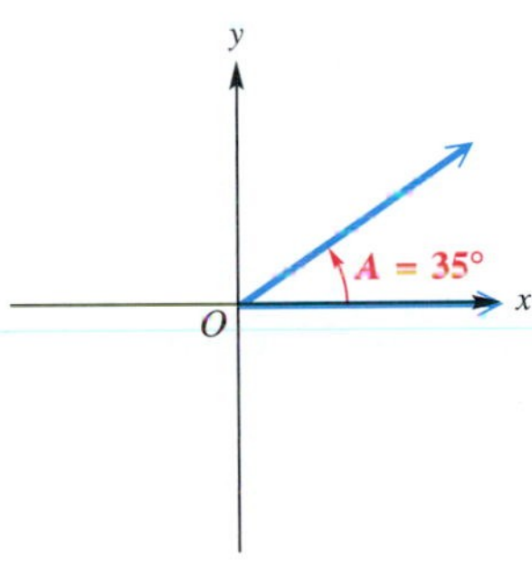

FIGURE 7

Do not confuse an angle with its measure. Angle A of Figure 7 is a rotation; the measure of the rotation is 35°. This measure is often expressed by saying that m(angle A) is 35°, where ***m*(angle *A*)** is read **"the measure of angle *A*."** It is convenient, however, to abbreviate $m(\text{angle } A) = 35°$ as simply angle $A = 35°$.

Traditionally, portions of a degree have been measured with minutes and seconds. One **minute,** written **1′**, is $\frac{1}{60}$ of a degree.

$$1' = \frac{1}{60}^{\circ} \quad \text{or} \quad 60' = 1°$$

One **second, 1″,** is $\frac{1}{60}$ of a minute.

$$1'' = \frac{1}{60}' = \frac{1}{3600}^{\circ} \quad \text{or} \quad 60'' = 1'$$

The measure 12° 42′ 38″ represents 12 degrees, 42 minutes, 38 seconds.

TECHNOLOGY NOTE

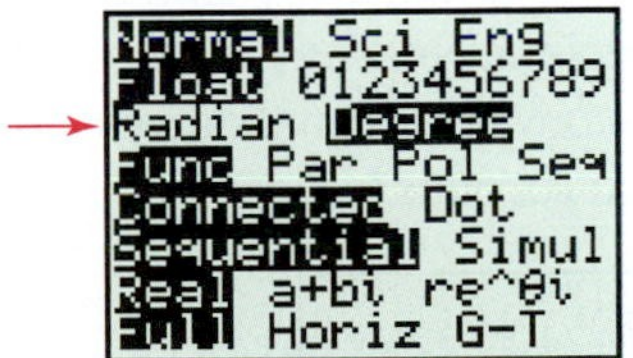

Graphing calculators can be set in either *radian mode* or *degree mode.* Be sure your calculator is in the correct mode before performing calculations involving angles.

*In addition to θ (theta), Greek letters such as α (alpha) and β (beta) are sometimes used to name angles.

GCM **EXAMPLE 2** **Calculating with Degrees and Minutes**

Perform each calculation.

(a) $51^\circ 29' + 32^\circ 46'$ **(b)** $90^\circ - 73^\circ 12'$

Analytic Solution

(a)

$$\begin{array}{r} 51^\circ 29' \\ +\ 32^\circ 46' \\ \hline 83^\circ 75' \end{array}$$

Add degrees and minutes separately.

We can write the sum $83^\circ 75'$ as

$$\begin{array}{r} 83^\circ \\ +\ \ 1^\circ 15' \\ \hline 84^\circ 15'. \end{array}$$

$75' = 60' + 15' = 1^\circ 15'$

(b)

$$\begin{array}{r} 89^\circ 60' \\ -\ 73^\circ 12' \\ \hline 16^\circ 48' \end{array}$$

Write 90° as $89^\circ 60'$.

Graphing Calculator Solution

Your calculator can be in either degree or radian mode. (We discuss *radian measure* later in this section.) Figure 8 shows the calculations.

```
51°29'+32°46'►DM
S
           84°15'0"
90°0'-73°12'►DMS
           16°48'0"
```

FIGURE 8

Angles are commonly measured in decimal degrees. For example,

$$12.4238^\circ = 12\frac{4238}{10{,}000}^\circ.$$

Decimal degrees are often more convenient to use than degrees, minutes, and seconds. For this reason it is important to be able to convert between these two angle measures. These conversions can be done by hand or with a calculator.

GCM **EXAMPLE 3** **Converting between Decimal Degrees and Degrees, Minutes, and Seconds**

(a) Convert $74^\circ 8' 14''$ to decimal degrees. Round to the nearest thousandth of a degree.

(b) Convert 34.817° to degrees, minutes, and seconds.

Analytic Solution

(a) $74^\circ 8' 14'' = 74^\circ + \frac{8}{60}^\circ + \frac{14}{3600}^\circ$ $\quad 1' = \frac{1}{60}^\circ$ and $1'' = \frac{1}{3600}^\circ$

$\approx 74^\circ + .1333^\circ + .0039^\circ$

$\approx 74.137^\circ$ $\quad$ Add; round.

(b) $34.817^\circ = 34^\circ + .817^\circ$

$= 34^\circ + .817(60')$ $\quad 1^\circ = 60'$

$= 34^\circ + 49.02'$

$= 34^\circ + 49' + .02'$

$= 34^\circ + 49' + .02(60'')$ $\quad 1' = 60''$

$= 34^\circ + 49' + 1.2''$

$= 34^\circ 49' 1.2''$

Graphing Calculator Solution

The first two results in Figure 9 show how $74^\circ 8' 14''$ is converted to decimal degrees. (The second result was obtained by fixing the display to three decimal places.) The final result shows how to convert decimal degrees to degrees, minutes, and seconds.

```
74°8'14"
        74.13722222
74°8'14"
             74.137
34.817►DMS
          34°49'1.2"
```

FIGURE 9

Standard Position and Coterminal Angles

An angle is in **standard position** if its vertex is at the origin and its initial side is along the positive x-axis. See Figures 10(a) and 10(b). An angle in standard position is said to lie in the quadrant in which its terminal side lies. For example, an acute angle is in quadrant I (Figure 10(a)) and an obtuse angle is in quadrant II (Figure 10(b)). Figure 10(c) shows ranges of angle measures for each quadrant when $0° < \theta < 360°$. Angles in standard position having their terminal sides along the x-axis or y-axis, such as angles with measures 90°, 180°, 270°, and so on, are called **quadrantal angles.**

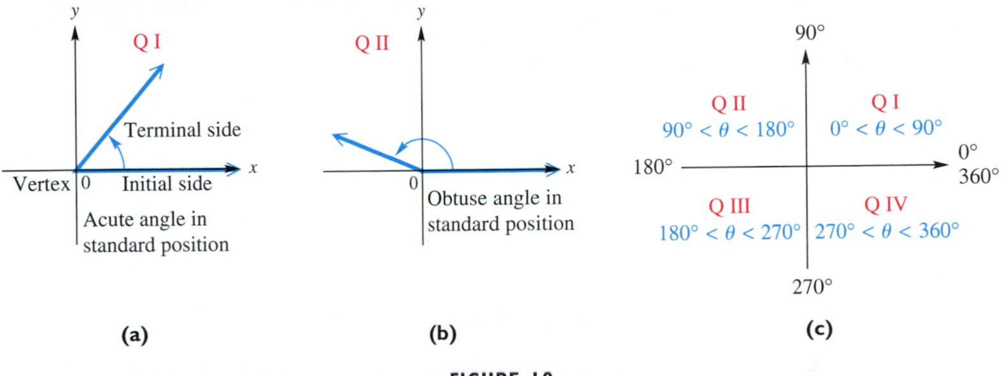

FIGURE 10

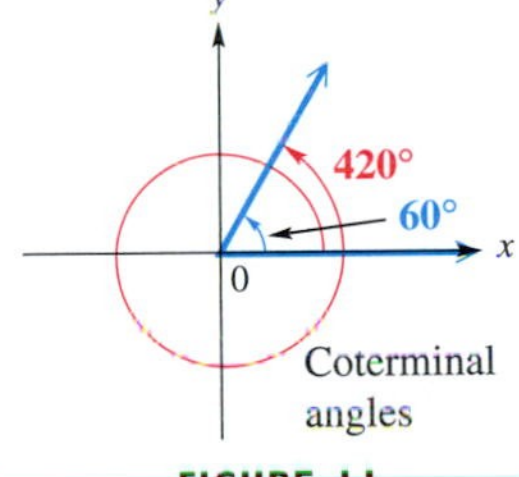

FIGURE 11

A complete rotation of a ray results in an angle measuring 360°. By continuing the rotation, angles of measure larger than 360° can be produced. The angles in Figure 11 with measures 60° and 420° have the same initial side and the same terminal side, but different amounts of rotation. Such angles are called **coterminal angles**—their measures differ by a multiple of 360°. The angles in Figure 12 are also coterminal.

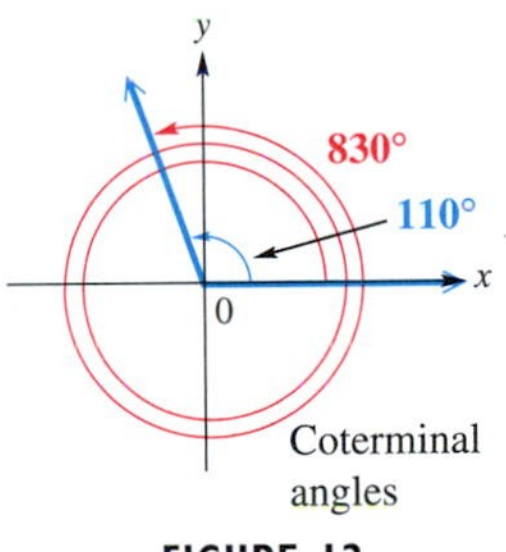

FIGURE 12

EXAMPLE 4 Finding Measures of Coterminal Angles

Find the angle of smallest possible positive measure coterminal with each angle.

(a) 908° **(b)** −75°

Solution

(a) Add or subtract 360° as many times as needed to obtain an angle with measure greater than 0° but less than 360°. Since

$$908° - 2 \cdot 360° = 908° - 720° = 188°,$$

an angle of 188° is coterminal with an angle of 908°. See Figure 13.

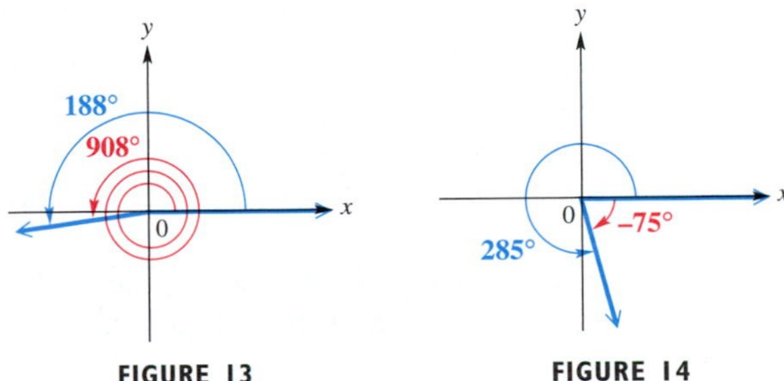

FIGURE 13 **FIGURE 14**

(b) Use a rotation of $360° + (-75°) = 285°$. See Figure 14. ■

Sometimes we want to find an expression that will generate all angles coterminal with a given angle. For example, any angle coterminal with 60° can be obtained by adding an appropriate integer multiple of 360° to 60°. If we let n represent any integer, then the expression

$$60° + n \cdot 360°$$

represents all such coterminal angles. The table shows a few possibilities.

Value of n	Angle Coterminal with 60°
2	$60° + 2 \cdot 360° = 780°$
1	$60° + 1 \cdot 360° = 420°$
0	$60° + 0 \cdot 360° = 60°$ (the angle itself)
−1	$60° + (-1) \cdot 360° = -300°$

Radian Measure

In work involving applications of trigonometry, angles are often measured in degrees. In more advanced work in mathematics, *radian measure* of angles is preferred.

Figure 15 shows an angle θ in standard position along with a circle of radius r. The vertex of θ is at the center of the circle. Angle θ intercepts an arc on the circle equal in length to the radius of the circle. So, angle θ is said to have a measure of 1 radian.

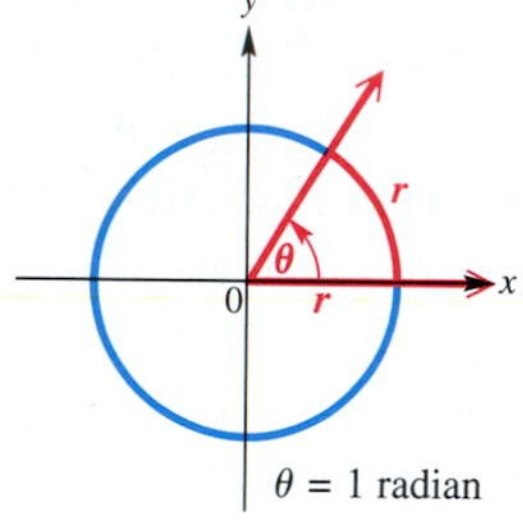

FIGURE 15

Radian

An angle with vertex at the center of a circle that intercepts an arc on the circle equal in length to the radius of the circle has measure **1 radian.**

It follows that an angle of measure 2 radians intercepts an arc equal in length to twice the radius of the circle, an angle of measure $\frac{1}{2}$ radian intercepts an arc equal in length to half the radius of the circle, and so on.

The **circumference** of a circle—the distance around the circle—is given by $C = 2\pi r$, where r is the radius of the circle. The formula $C = 2\pi r$ shows that the radius can be laid off 2π times around a circle. Therefore, an angle of 360°, which corresponds to a complete circle, intercepts an arc equal in length to 2π times the radius of the circle. Thus, an angle of 360° has measure 2π radians:

$$\mathbf{360° = 2\pi \text{ radians.}}$$

An angle of 180° is half the degree measure of an angle of 360°, so an angle of 180° has half the radian measure of an angle of 360°.

$$\mathbf{180° = \frac{1}{2}(2\pi) \text{ radians} = \pi \text{ radians}}$$ Degree–radian relationship

We can use the relationship $180° = \pi$ radians to develop a method for converting between degrees and radians as follows.

$$180° = \pi \text{ radians}$$

$$1° = \frac{\pi}{180} \text{ radian} \quad \text{Divide by 180.} \qquad \text{or} \qquad 1 \text{ radian} = \frac{180°}{\pi} \quad \text{Divide by } \pi.$$

Converting between Degrees and Radians

1. Multiply a degree measure by $\frac{\pi}{180}$ radian and simplify to convert to radians.
2. Multiply a radian measure by $\frac{180°}{\pi}$ and simplify to convert to degrees.

NOTE ***1 radian is equivalent to approximately* 57.3°.**

GCM **EXAMPLE 5 Converting Degrees to Radians**

Convert each degree measure to radians.

(a) 45° **(b)** 249.8°

Analytic Solution

(a) $45° = 45\left(\frac{\pi}{180} \text{ radian}\right)$ Multiply by $\frac{\pi}{180}$ radian.

$= \frac{\pi}{4}$ radian

(b) $249.8° = 249.8\left(\frac{\pi}{180} \text{ radian}\right)$

≈ 4.360 radians Round to nearest thousandth.

Graphing Calculator Solution

For Figure 16, the calculator is in radian mode. ***When exact values involving π are required, such as $\frac{\pi}{4}$ in part (a), calculator approximations are not acceptable.***

```
45°
          .7853981634
π/4
          .7853981634
249.8°
          4.359832471
```

FIGURE 16

GCM **EXAMPLE 6 Converting Radians to Degrees**

Convert each radian measure to degrees.

(a) $\frac{9\pi}{4}$ **(b)** 4.25 (Give the answer in decimal degrees.)

Analytic Solution

(a) $\frac{9\pi}{4} = \frac{9\pi}{4}\left(\frac{180°}{\pi}\right)$ Multiply by $\frac{180°}{\pi}$.

$= 405°$

(b) $4.25 = 4.25\left(\frac{180°}{\pi}\right)$

$\approx 243.5°$ Round to nearest tenth.

Graphing Calculator Solution

Figure 17 shows how a calculator in degree mode converts radian measures to decimal degrees.

```
(9π/4)ʳ
                  405
4.25ʳ
          243.5070629
```

FIGURE 17

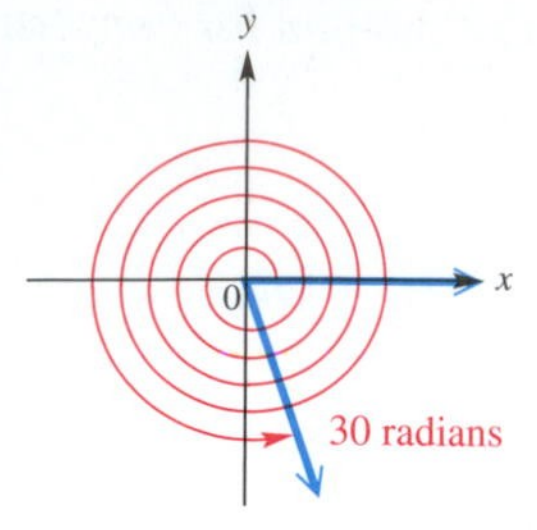

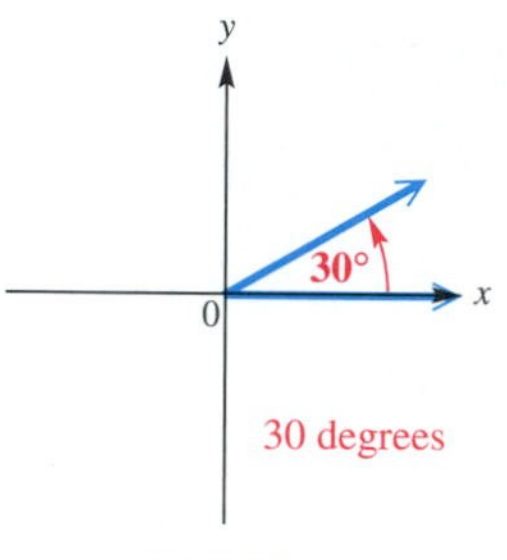

FIGURE 18

If no unit of measure is specified for an angle, radian measure is understood.

CAUTION Figure 18 shows angles measuring 30 radians and 30°. Be careful not to confuse them.

The table that follows and Figure 19 give some equivalent angles measured in degrees and radians. Keep in mind that **180° = π radians.**

Equivalent Angle Measures in Degrees and Radians

	Radians			Radians	
Degrees	**Exact**	**Approximate**	**Degrees**	**Exact**	**Approximate**
0°	0	0	90°	$\frac{\pi}{2}$	1.5708
30°	$\frac{\pi}{6}$	.5236	180°	π	3.1416
45°	$\frac{\pi}{4}$	.7854	270°	$\frac{3\pi}{2}$	4.7124
60°	$\frac{\pi}{3}$	1.0472	360°	2π	6.2832

Looking Ahead to Calculus

In calculus, radian measure is much easier to work with than degree measure. If x is measured in radians, then the derivative of $f(x) = \sin x$ is

$$f'(x) = \cos x.$$

However, if x is measured in degrees, then the derivative of $f(x) = \sin x$ is

$$f'(x) = \frac{\pi}{180} \cos x.$$

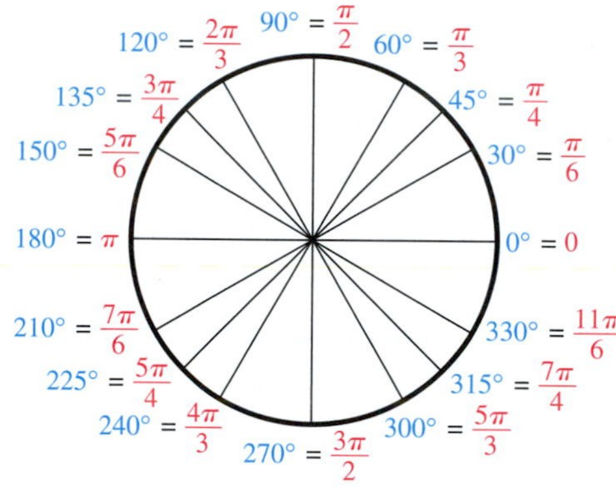

FIGURE 19

We use radian measure to simplify certain formulas, two of which follow. Both would be more complicated if expressed in degrees.

Arc Lengths and Areas of Sectors

The formula to find the length of an arc of a circle is derived from the fact (proved in geometry) that the length of an arc is proportional to the measure of its central angle.

In Figure 20, angle QOP has measure 1 radian and intercepts an arc of length r on the circle. Angle ROT has measure θ radians and intercepts an arc of length s on the circle. Since the lengths of the arcs are proportional to the measures of their central angles,

$$\frac{s}{r} = \frac{\theta}{1}$$

$$s = r\theta. \quad \text{Multiply each side by } r.$$

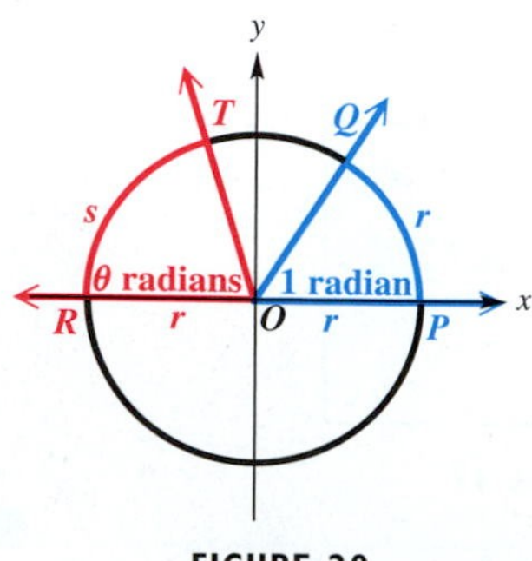

FIGURE 20

Arc Length

The length s of the arc intercepted on a circle of radius r by a central angle of measure θ radians is given by the product of the radius and the radian measure of the angle, or

$$s = r\theta, \quad \theta \text{ in radians.}$$

CAUTION In applying the formula $s = r\theta$, the value of θ ***must be expressed in radians.***

EXAMPLE 7 **Finding Arc Length by Using $s = r\theta$**

A circle has radius 18.2 centimeters. Find the length of an arc intercepted by a central angle having each of the following measures.

(a) $\frac{3\pi}{8}$ radians **(b)** 144°

Solution

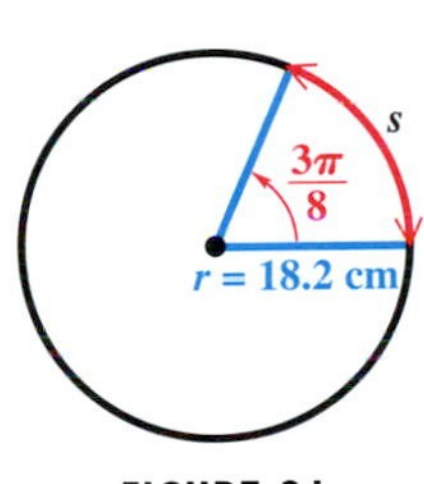

FIGURE 21

(a) As shown in Figure 21, $r = 18.2$ centimeters and $\theta = \frac{3\pi}{8}$.

$$s = r\theta = 18.2\left(\frac{3\pi}{8}\right) = \frac{54.6\pi}{8} \approx 21.4 \text{ centimeters}$$

(b) The formula $s = r\theta$ requires that θ be measured in radians. First, convert θ to radians by multiplying 144° by $\frac{\pi}{180}$ radian.

$$144° = 144\left(\frac{\pi}{180}\right) = \frac{4\pi}{5} \text{ radians} \qquad \text{Convert from degrees to radians.}$$

The length s is given by

$$s = r\theta = 18.2\left(\frac{4\pi}{5}\right) = \frac{72.8\pi}{5} \approx 45.7 \text{ centimeters.}$$

Be sure to use radians for θ in $s = r\theta$.

EXAMPLE 8 **Using Latitudes to Find Distance**

Reno, Nevada, is due north of Los Angeles. The latitude of Reno is 40° N, while that of Los Angeles is 34° N. (The N in 34° N means *north* of the equator.) If the radius of Earth is 6400 kilometers, find the north–south distance between the two cities.

Solution Latitude gives the measure of a central angle with vertex at Earth's center whose initial side goes through the equator and whose terminal side goes through the given location. As shown in Figure 22, the central angle between Reno and Los Angeles is $40° - 34° = 6°$. The distance between the two cities can be found by the formula $s = r\theta$, after 6° is first converted to radians.

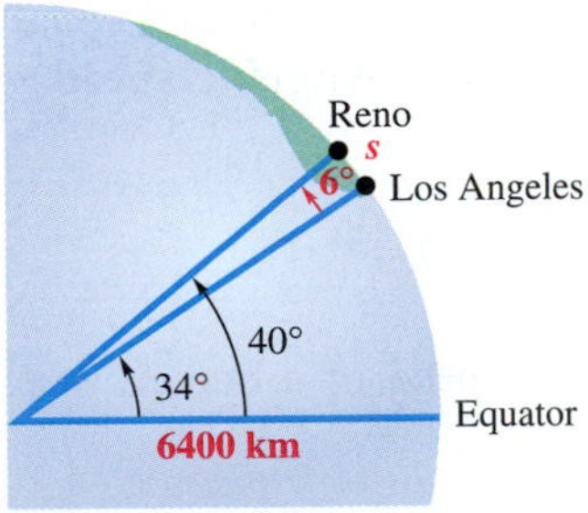

$$6° = 6\left(\frac{\pi}{180}\right) = \frac{\pi}{30} \text{ radian}$$

The distance between the two cities is

$$s = r\theta = 6400\left(\frac{\pi}{30}\right) \approx 670 \text{ kilometers.}$$

FIGURE 22

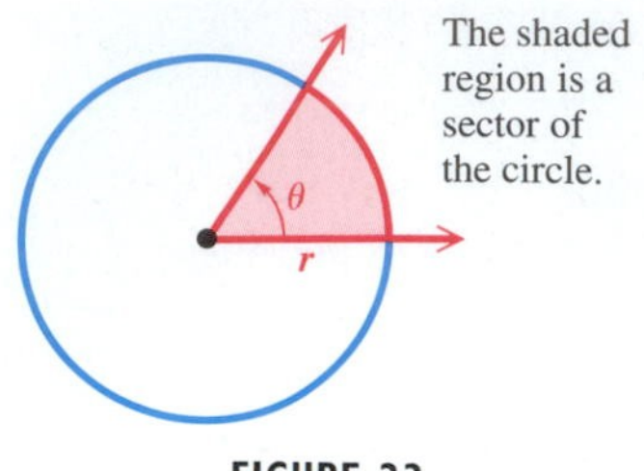

FIGURE 23

A **sector of a circle** is the portion of the interior of a circle intercepted by a central angle. Think of it as a "piece of pie." See Figure 23. The interior of a circle can be thought of as a sector with measure 2π radians. If a central angle for a sector has measure θ radians, then the sector makes up the fraction $\frac{\theta}{2\pi}$ of a complete circle. The area inside a circle with radius r is $A = \pi r^2$. Therefore, the area of the sector is given by the product of the fraction $\frac{\theta}{2\pi}$ and the total area, πr^2.

$$\text{area of sector} = \frac{\theta}{2\pi}(\pi r^2) = \frac{1}{2}r^2\theta, \quad \theta \text{ in radians}$$

This discussion is summarized as follows.

Area of a Sector

The area of a sector of a circle of radius r and central angle θ is given by

$$A = \frac{1}{2}r^2\theta, \quad \theta \text{ in radians.}$$

CAUTION As in the formula for arc length, the value of θ ***must be in radians*** when the formula for the area of a sector is used.

EXAMPLE 9 **Finding the Area of a Sector-Shaped Field**

Figure 24 shows a field in the shape of a sector of a circle. Find the area of the field.

FIGURE 24

Solution First, convert 15° to radians.

$$15° = 15\left(\frac{\pi}{180}\right) = \frac{\pi}{12} \text{ radian}$$

Now use the formula for the area of a sector.

$$A = \frac{1}{2}r^2\theta = \frac{1}{2}(321)^2\left(\frac{\pi}{12}\right) \approx 13{,}500 \text{ square meters}$$

Be sure to use radians for θ in $A = \frac{1}{2}r^2\theta$.

Angular and Linear Speed

The human joint that can be flexed the fastest is the wrist, which can rotate through 90°, or $\frac{\pi}{2}$ radians, in .045 second while holding a tennis racket. **Angular speed** ω (omega) measures the speed of rotation and is defined by

$$\omega = \frac{\theta}{t},$$

where θ is the angle of rotation and t is time. The angular speed of a human wrist swinging a tennis racket is

$$\omega = \frac{\theta}{t} = \frac{\frac{\pi}{2}}{.045} \approx 35 \text{ radians per second.}$$

The **linear speed** v at which the tip of the racket travels as a result of flexing the wrist is given by

$$v = r\omega,$$

where r is the radius (distance) from the tip of the racket to the wrist joint. Thus, if r equals 2 feet, then the speed at the tip of the racket is

$$v = r\omega \approx 2(35) = 70 \text{ feet per second,}$$

or about 48 mph. In a tennis serve, the arm rotates at the shoulder, so the final speed of the racket is considerably faster. (*Source:* Cooper, J. and R. Glassow, *Kinesiology*, 2nd ed., C.V. Mosby, 1968.)

EXAMPLE 10 Finding Angular Speed and Linear Speed

A belt runs a pulley of radius 6 centimeters at 80 revolutions per minute.

(a) Find the angular speed of the pulley in radians per second.

(b) Find the linear speed of the belt in centimeters per second.

Solution

(a) In 1 minute, the pulley makes 80 revolutions. Each revolution is 2π radians, for a total of

$$80(2\pi) = 160\pi \text{ radians per minute.}$$

Since there are 60 seconds in 1 minute, we find ω, the angular speed in radians per second, by dividing 160π by 60.

$$\omega = \frac{160\pi}{60} = \frac{8\pi}{3} \text{ radians per second}$$

(b) The linear speed of the belt will be the same as that of a point on the circumference of the pulley. Thus,

$$v = r\omega = 6\left(\frac{8\pi}{3}\right) = 16\pi \approx 50.3 \text{ centimeters per second.}$$ ■

Suppose that an object is moving at a constant speed. If it travels a distance s in time t, then its linear speed is given by

$$v = \frac{s}{t}.$$

If the object is traveling in a circle, then $s = r\theta$, where r is the radius of the circle and θ is the angle of rotation in radians. Thus, we can write

$$v = \frac{r\theta}{t}.$$

The formulas for angular and linear speed are summarized on the next page.

Angular Speed	Linear Speed
$\omega = \dfrac{\theta}{t}$ (ω in radians per unit time, θ in radians)	$v = \dfrac{s}{t}$ $v = \dfrac{r\theta}{t}$ $v = r\omega$

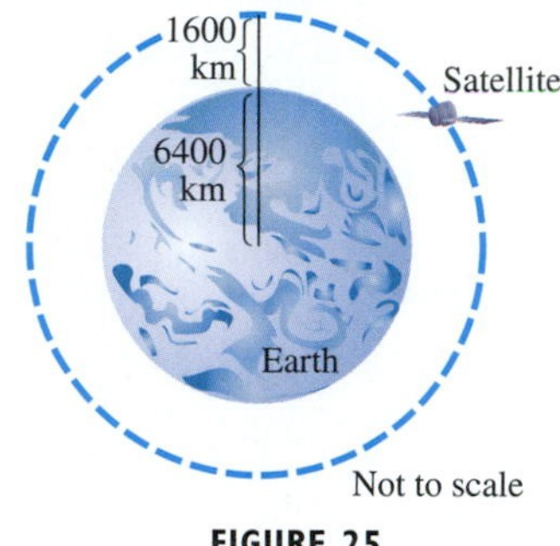

FIGURE 25

EXAMPLE 11 Finding Linear Speed and Distance Traveled by a Satellite

A satellite traveling in a circular orbit 1600 kilometers above the surface of Earth takes 2 hours to complete an orbit. The radius of Earth is 6400 kilometers. See Figure 25.

(a) Find the linear speed of the satellite.

(b) Find the distance the satellite travels in 4.5 hours.

Solution

(a) The distance of the satellite from the center of Earth is

$$r = 1600 + 6400 = 8000 \text{ kilometers.}$$

For one orbit, $\theta = 2\pi$, and

$$s = r\theta = 8000(2\pi) \text{ kilometers.}$$

Since it takes 2 hours to complete an orbit, the linear speed is

$$v = \frac{s}{t} = \frac{8000(2\pi)}{2} = 8000\pi \approx 25{,}000 \text{ kilometers per hour.}$$

(b) $s = vt = 8000\pi(4.5) = 36{,}000\pi \approx 110{,}000$ kilometers ■

8.1 Exercises

Find **(a)** *the complement and* **(b)** *the supplement of each angle.*

1. 30° **2.** 60° **3.** 45° **4.** $\dfrac{\pi}{3}$ **5.** $\dfrac{\pi}{4}$ **6.** $\dfrac{\pi}{12}$

7. *Concept Check* An angle measures x degrees.
(a) What is the measure of its complement?
(b) What is the measure of its supplement?

8. *Concept Check* An angle measures x radians.
(a) What is the measure of its complement?
(b) What is the measure of its supplement?

Find the degree measure of the smaller angle formed by the hands of a clock at the following times.

9.

10.

Find the measure of each angle in Exercises 11–16.

11.

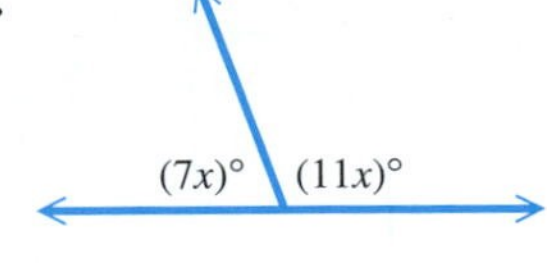

12.

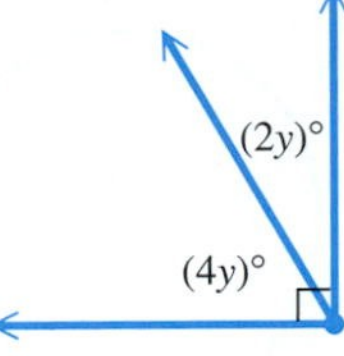

13.

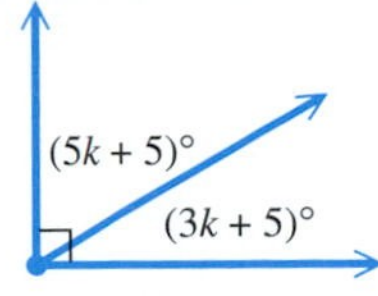

14.

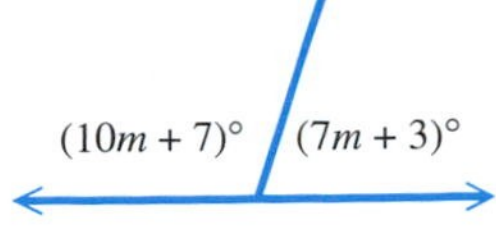

15. Supplementary angles with measures $6x - 4$ degrees and $8x - 12$ degrees

16. Complementary angles with measures $9z + 6$ degrees and $3z$ degrees

Perform each calculation.

17. $62° 18' + 21° 41'$ **18.** $75° 15' + 83° 32'$ **19.** $71° 18' - 47° 29'$

20. $47° 23' - 73° 48'$ **21.** $90° - 72° 58' 11''$ **22.** $90° - 36° 18' 47''$

Convert each angle measure to decimal degrees. Use a calculator, and round to the nearest thousandth of a degree if necessary.

23. $20° 54'$ **24.** $38° 42'$ **25.** $91° 35' 54''$ **26.** $34° 51' 35''$

Convert each angle measure to degrees, minutes, and seconds. Use a calculator as necessary.

27. $31.4296°$ **28.** $59.0854°$ **29.** $89.9004°$ **30.** $102.3771°$

Concept Check *Sketch each angle in standard position. Draw an arrow representing the correct amount of rotation. Find the measure of two other angles, one positive and one negative, that are coterminal with the given angle. Give the quadrant of each angle.*

31. $75°$ **32.** $89°$ **33.** $174°$ **34.** $234°$ **35.** $-61°$ **36.** $-159°$

Find the angle of smallest positive measure that is coterminal with the given angle.

37. $-40°$ **38.** $-98°$ **39.** $450°$ **40.** $539°$

41. $-\frac{\pi}{4}$ **42.** $-\frac{\pi}{3}$ **43.** $-\frac{3\pi}{2}$ **44.** $-\pi$

Give an expression that generates all angles coterminal with each angle. Let n represent any integer.

45. $30°$ **46.** $45°$ **47.** $-90°$ **48.** $-135°$

49. $\frac{\pi}{4}$ **50.** $\frac{\pi}{6}$ **51.** $-\frac{3\pi}{4}$ **52.** $-\frac{7\pi}{6}$

Convert each degree measure to radians. Leave answers as multiples of π.

53. $60°$ **54.** $90°$ **55.** $150°$ **56.** $270°$ **57.** $-45°$ **58.** $-210°$

Convert each radian measure to degrees.

59. $\frac{\pi}{3}$ **60.** $\frac{8\pi}{3}$ **61.** $\frac{7\pi}{4}$ **62.** $\frac{2\pi}{3}$ **63.** $\frac{11\pi}{6}$ **64.** $\frac{15\pi}{4}$

Convert each degree measure to radians. Round to the nearest hundredth.

65. $39°$ **66.** $74°$ **67.** $139° 10'$ **68.** $174° 50'$ **69.** $64.29°$ **70.** $122.62°$

Convert each radian measure to degrees. Write answers to the nearest minute.

71. 2 **72.** 5 **73.** 1.74 **74.** .3417 **75.** -1.3 **76.** -4

77. ***Concept Check*** The figure shows the same angles measured in both degrees and radians. Complete the missing measures.

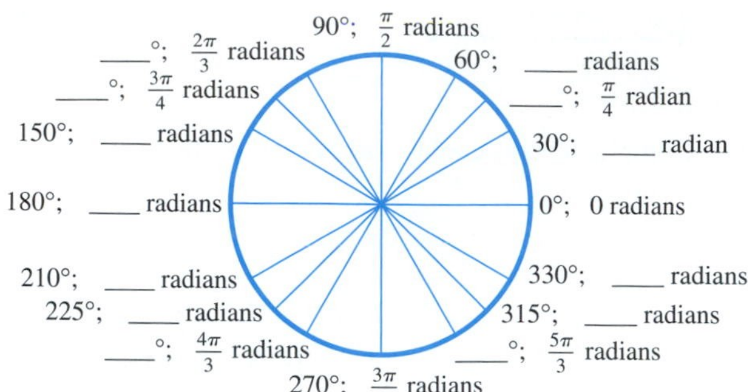

78. ***Railroad Engineering*** The term *grade* has several different meanings in construction work. Some engineers use the term to represent $\frac{1}{100}$ of a right angle and express it as a percent. For instance, an angle of .9° would be referred to as a 1% grade. (*Source:* Hay, W., *Railroad Engineering,* John Wiley and Sons, 1982.)

(a) By what number should you multiply a grade to convert it to radians?

(b) In a rapid-transit rail system, the maximum grade allowed between two stations is 3.5%. Express this angle in degrees and in radians.

Concept Check *Find the exact length of each arc intercepted by the given central angle.*

79.

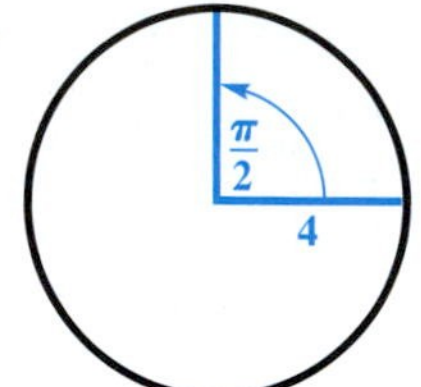

80.

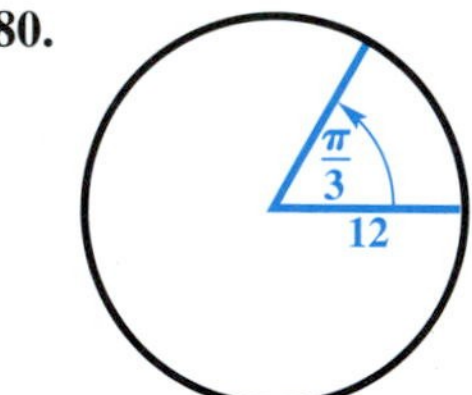

Concept Check *Find the radius of each circle.*

81.

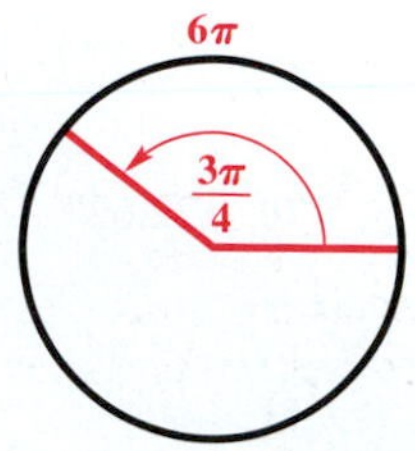

82.

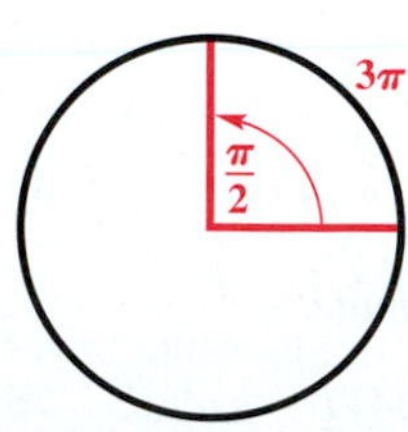

Concept Check *Find the measure of each central angle (in radians).*

83.

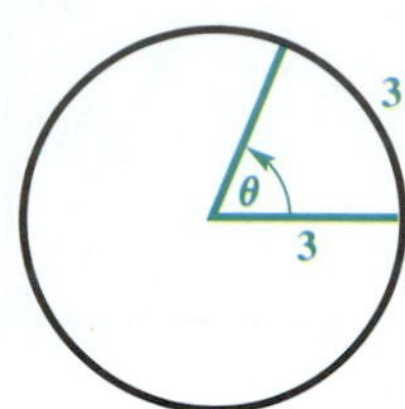

84.

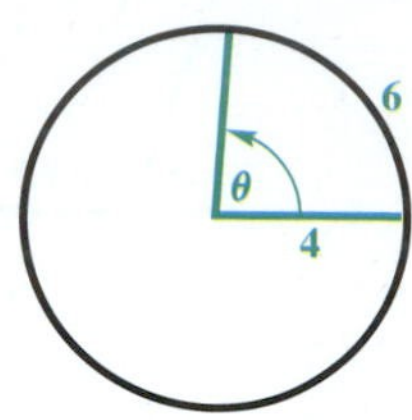

Find the length of each arc intercepted by a central angle θ in a circle of radius r.

85. $r = 12.3$ centimeters; $\theta = \frac{2\pi}{3}$ radians

86. $r = .892$ centimeter; $\theta = \frac{11\pi}{10}$ radians

87. $r = 4.82$ meters; $\theta = 60°$

88. $r = 71.9$ centimeters; $\theta = 135°$

Distance between Cities *Find the distance in kilometers between the pair of cities whose latitudes are given. Assume that the cities are on a north–south line and that the radius of Earth is 6400 kilometers. Round answers to the nearest hundred kilometers.*

89. Madison, South Dakota, 44° N, and Dallas, Texas, 33° N

90. Charleston, South Carolina, 33° N, and Toronto, Ontario, 43° N

91. New York City, New York, 41° N, and Lima, Peru, 12° S

92. Halifax, Nova Scotia, 45° N, and Buenos Aires, Argentina, 34° S

Use the formula $\omega = \frac{\theta}{t}$ to find the value of the missing variable.

93. $\theta = \frac{3\pi}{4}$ radians, $t = 8$ seconds

94. $\theta = \frac{2\pi}{5}$ radians, $t = 10$ seconds

95. $\theta = \frac{2\pi}{9}$ radian, $\omega = \frac{5\pi}{27}$ radian per minute

96. $\omega = .90674$ radian per minute, $t = 11.876$ minutes

The formula $\omega = \frac{\theta}{t}$ can be rewritten as $\theta = \omega t$. Substituting ωt for θ changes $s = r\theta$ to $s = r\omega t$. Use the formula $s = r\omega t$ to find the value of the missing variable.

97. $r = 6$ centimeters, $\omega = \frac{\pi}{3}$ radians per second, $t = 9$ seconds

98. $r = 9$ yards, $\omega = \frac{2\pi}{5}$ radians per second, $t = 12$ seconds

99. $s = 6\pi$ centimeters, $r = 2$ centimeters, $\omega = \frac{\pi}{4}$ radian per second

100. $s = \frac{3\pi}{4}$ kilometers, $r = 2$ kilometers, $t = 4$ seconds

Solve each problem.

101. *Pulley Raising a Weight* Refer to the figure and answer each question.

(a) How many inches will the weight rise if the pulley is rotated through an angle of $71° 50'$?

(b) Through what angle, to the nearest minute, must the pulley be rotated to raise the weight 6 inches?

102. *Pulley Raising a Weight* Find the radius of the pulley in the figure if a rotation of 51.6° raises the weight 11.4 centimeters.

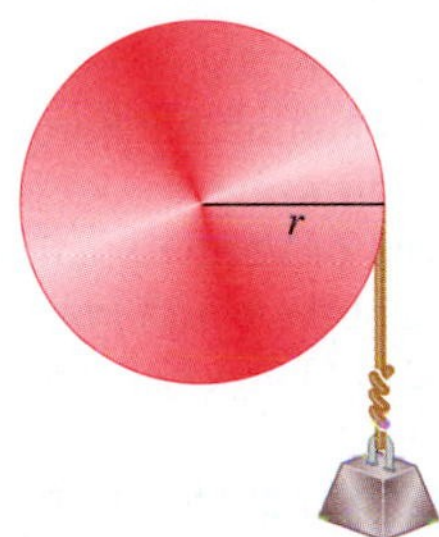

103. *Rotating Wheels* The rotation of the smaller wheel in the figure causes the larger wheel to rotate. Through how many degrees will the larger wheel rotate if the smaller one rotates through 60.0°?

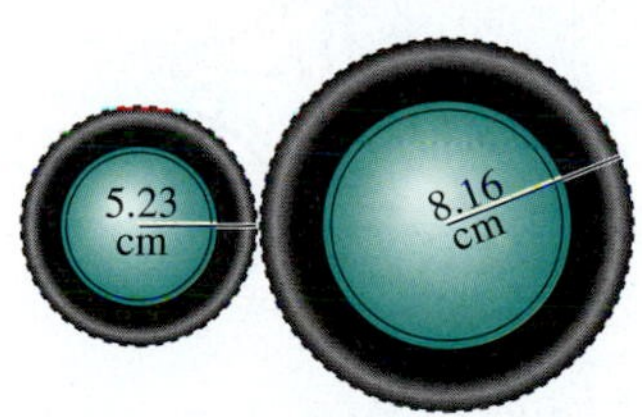

104. *Rotating Wheels* Find the radius of the larger wheel in the figure if the smaller wheel rotates 80.0° when the larger wheel rotates 50.0°.

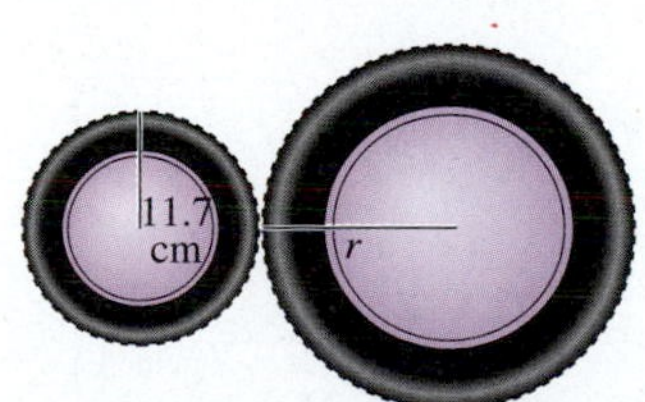

105. *Bicycle Chain Drive* The figure shows the chain drive of a bicycle. How far will the bicycle move if the pedals are rotated through 180°? Assume the radius of the bicycle wheel is 13.6 inches.

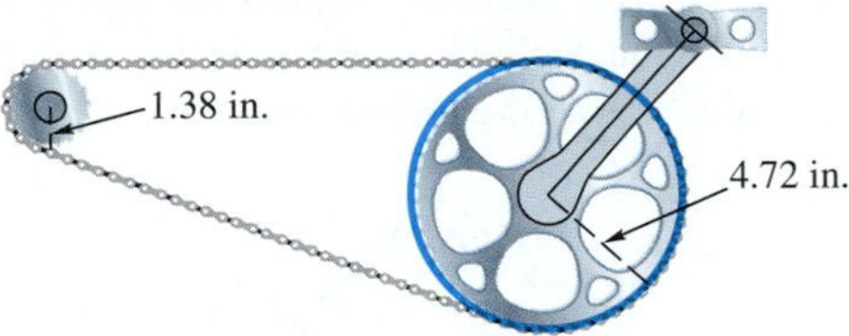

106. *Pickup Truck Speedometer* The speedometer of a pickup truck is designed to be accurate with tires of radius 14 inches.

(a) Find the number of rotations of a tire in 1 hour if the truck is driven 55 mph.

(b) Suppose that oversize tires of radius 16 inches are placed on the truck. If the truck is now driven for 1 hour with the speedometer reading 55 mph, how far has the truck gone? If the speed limit is 55 mph, does the driver deserve a speeding ticket?

Find the area of a sector of a circle having radius r and central angle θ.

107. $r = 29.2$ meters; $\theta = \frac{5\pi}{6}$ radians

108. $r = 59.8$ kilometers; $\theta = \frac{2\pi}{3}$ radians

109. $r = 12.7$ centimeters; $\theta = 81°$

110. $r = 18.3$ meters; $\theta = 125°$

Concept Check *Find the area of each sector. Express your answers in terms of π.*

111.

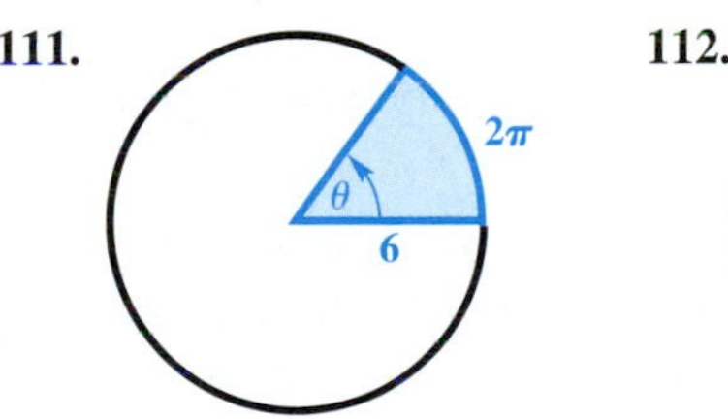

112.

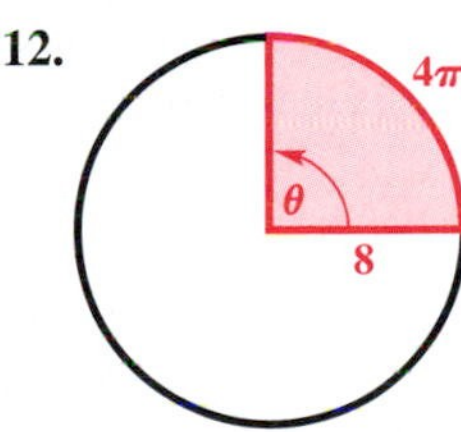

Solve each problem.

113. *Doppler Radar* Radar is used to identify severe weather. If Doppler radar can detect weather within a 240-mile radius and creates a new image every 48 seconds, find the area scanned by the radar in 1 second.

114. *Land Required for a Solar-Power Plant* A 300-megawatt solar-power plant needs approximately 950,000 square meters of land area to collect the required amount of energy from sunlight.

(a) If the land area is circular, what is its radius?

(b) If the land area is a 35° sector of a circle, what is its radius?

115. *Area Cleaned by a Windshield Wiper* The Ford Model A, built from 1928 to 1931, had a single windshield wiper on the driver's side. The total arm and blade was 10 inches long and rotated back and forth through an angle of 95°. The shaded region in the figure is the portion of the windshield cleaned by the 7-inch wiper blade. What is the area of the region cleaned?

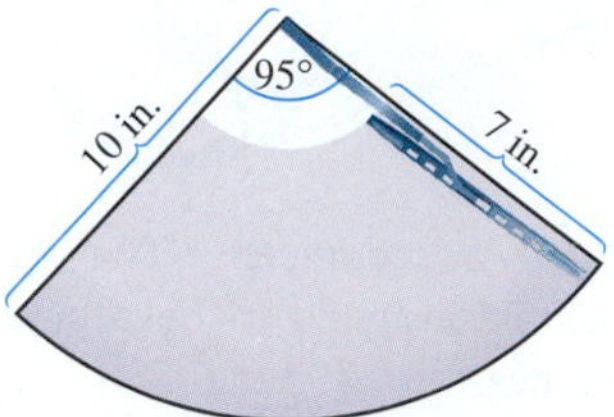

116. *Measures of a Structure* The figure shows Medicine Wheel, a Native American structure in northern Wyoming. This circular structure is perhaps 2500 years old. There are 27 spokes in the wheel, all equally spaced.

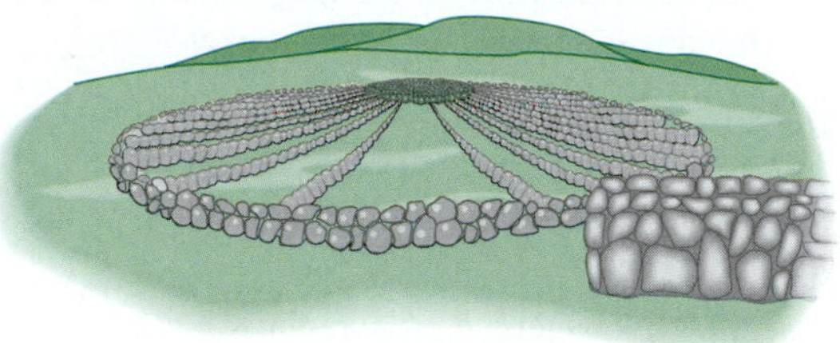

(a) Find the measure of each central angle in degrees and in radians.

(b) If the radius of the wheel is 76 feet, find the circumference.

(c) Find the length of each arc intercepted by consecutive pairs of spokes.

(d) Find the area of each sector formed by consecutive spokes.

117. *Size of a Corral* The unusual corral in the figure is separated into 26 areas, many of which approximate sectors of a circle. Assume that the corral has a diameter of 50 meters.

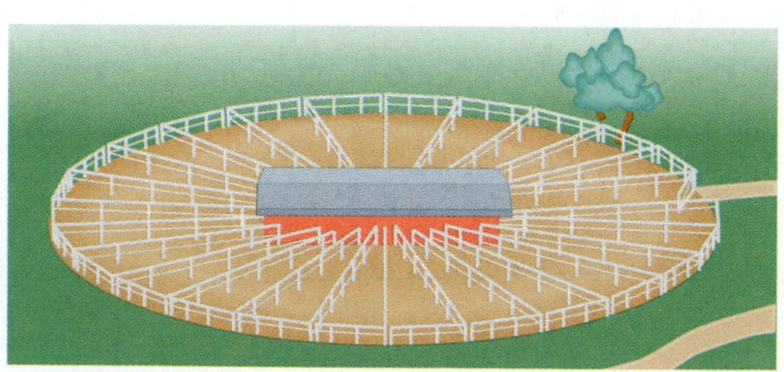

(a) Find the central angle for each region, assuming that the 26 regions are all equal sectors with the fences meeting at the center.

(b) What is the area of each sector?

118. *Area of a Lot* A frequent problem in surveying city lots and rural lands adjacent to curves of highways and railways is that of finding the area when one or more of the boundary lines is the arc of a circle. Find the area of the lot shown in the figure. (*Source:* Anderson, J. and E. Michael, *Introduction to Surveying*, McGraw-Hill, 1985.)

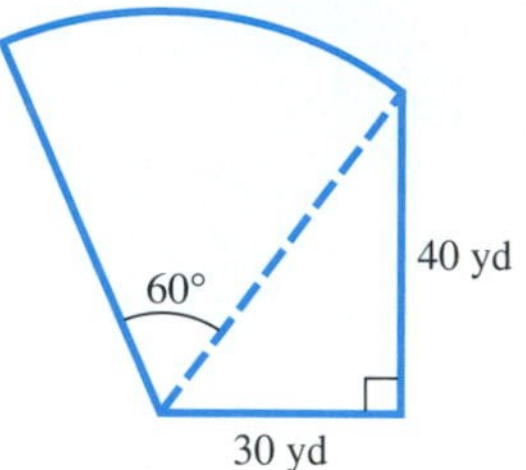

119. *Concept Check* Consider the area-of-a-sector formula $A = \frac{1}{2}r^2\theta$. What well-known formula corresponds to the special case $\theta = 2\pi$?

120. *Revolution of Earth* Earth revolves on its axis once every 24 hours. Assuming that Earth's radius is 6400 kilometers, find the following.

(a) Angular speed of Earth in radians per day

(b) Linear speed at the North Pole or South Pole

(c) Linear speed at Quito, Ecuador, a city on the equator

121. *Speed of a Bicycle* A bicycle has a tire 26 inches in diameter that is rotating at 15 radians per second. Approximate the speed of the bicycle in feet per second and in miles per hour.

122. *Angular Speed of a Pulley* The pulley shown has radius 12.96 centimeters. Suppose that it takes 18 seconds for 56 centimeters of belt to go around the pulley. Find the angular speed of the pulley in radians per second.

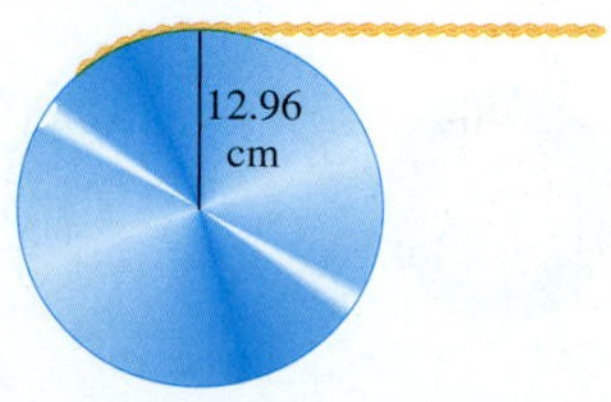

123. ***Angular Speed of Skateboard Wheels*** The wheels on a skateboard have diameter 2.25 inches. If a skateboarder is traveling downhill at 15 mph, determine the angular speed of the wheels in radians per second.

124. ***Angular Speed of Pulleys*** The two pulleys in the figure have radii 15 centimeters and 8 centimeters, respectively. The larger pulley rotates 25 times in 36 seconds. Find the angular speed of each pulley in radians per second.

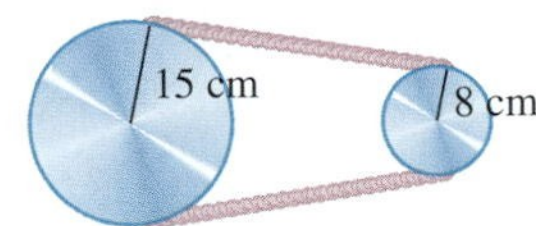

125. ***Speed of a Motor Propeller*** A 90-horsepower outboard motor at full throttle rotates its propeller 5000 revolutions per minute. Find the angular speed of the propeller in radians per second. What is the linear speed in inches per second of a point at the tip of the propeller if its diameter is 10 inches?

126. ***Speed of a Golf Clubhead*** The shoulder joint can rotate at about 25 radians per second. Assuming that a golfer's arm is straight and the distance from the shoulder to the clubhead is 5 feet, estimate the linear speed of the clubhead from the shoulder rotation. (*Source:* Cooper, J. and R. Glassow, *Kinesiology,* Second Edition, C.V. Mosby, 1968.)

127. ***Orbit of Earth*** Earth travels about the sun in an orbit that is almost circular. Assume that the orbit is a circle, with radius 93,000,000 miles. (See the figure at the top of the next column.)

(a) Assume that a year is 365 days, and find θ, the angle formed by Earth's movement in 1 day.

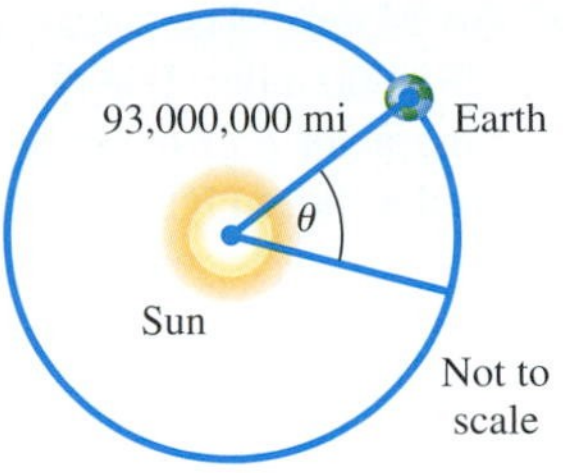

(b) Give the angular speed in radians per hour.

(c) Find the linear speed of Earth in miles per hour.

128. ***Nautical Miles*** *Nautical miles* are used by ships and airplanes. They are different from *statute miles,* which equal 5280 feet. A nautical mile is defined to be the arc length along the equator intercepted by a central angle *AOB* of 1 minute, as illustrated in the figure. If the equatorial radius of Earth is 3963 miles, use the arc length formula to approximate the number of statute miles in 1 nautical mile. Round your answer to two decimal places.

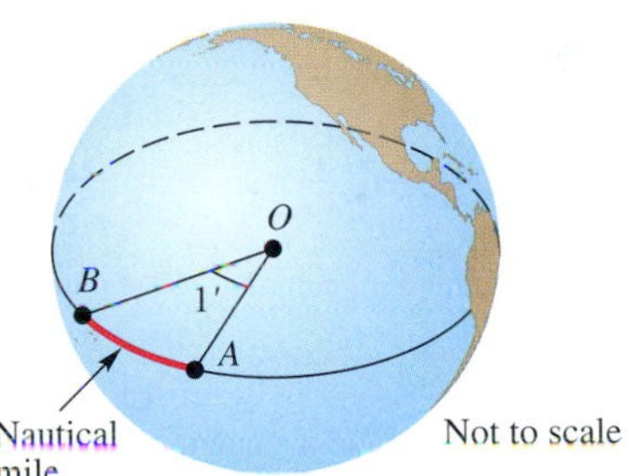

129. ***Circumference of Earth*** The first accurate estimate of the distance around Earth was done by the Greek astronomer Eratosthenes (276–195 B.C.), who noted that the noontime position of the sun at the summer solstice differed by 7° 12′ from the city of Syene to the city of Alexandria. (See the figure.) The distance between these two cities is 496 miles. Use the arc length formula to estimate the radius of Earth. Then find the circumference of Earth. (*Source:* Zeilik, M., *Introductory Astronomy and Astrophysics,* Third Edition, Saunders College Publishers, 1992.)

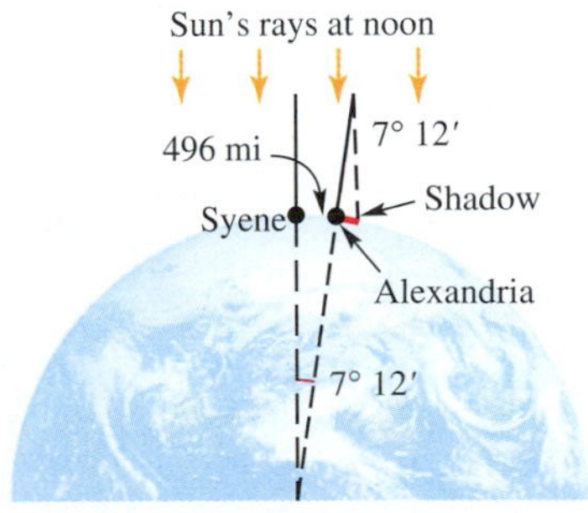

130. ***Diameter of the Moon*** The distance to the moon is approximately 238,900 miles. Use the arc length formula to *estimate* the diameter d of the moon if angle θ in the figure is measured to be .517°.

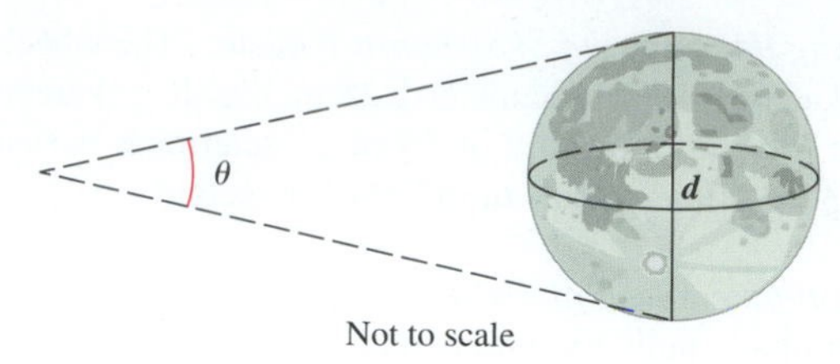

Not to scale

8.2 The Unit Circle

Trigonometric (Circular) Functions ■ Using a Calculator to Find Function Values ■ Exact Function Values for $\frac{\pi}{4}$, $\frac{\pi}{6}$, and $\frac{\pi}{3}$

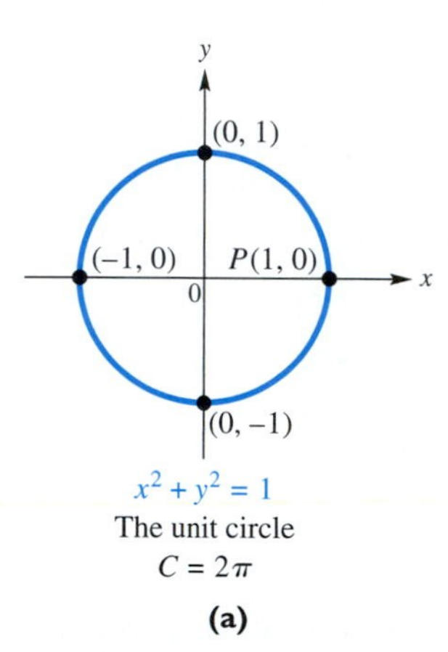

$x^2 + y^2 = 1$
The unit circle
$C = 2\pi$
(a)

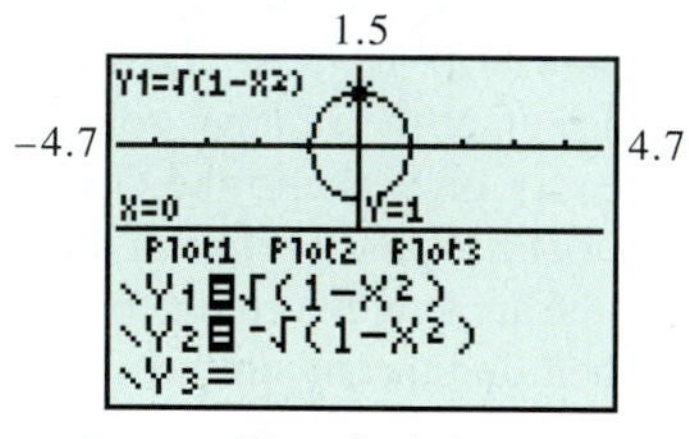

The unit circle
(b)

FIGURE 26

Trigonometric (Circular) Functions

As we saw in Chapter 6, the graph of the equation $\mathbf{x^2 + y^2 = 1}$ is a circle with center at the origin and radius 1. We refer to its graph as the **unit circle.** Because the radius of the unit circle is 1, its circumference is $C = 2\pi(1) = 2\pi$. See Figure 26(a).

Suppose that we consider the point $P(1, 0)$ on the unit circle as a starting point. We can move either clockwise or counterclockwise along an arc of the unit circle any distance s we wish. Let's agree that if s represents a real number and $s > 0$, then the movement is counterclockwise around the circle, while if $s < 0$, the movement is clockwise around the circle. (If $s = 0$, there is no movement.) Thus, every real number s is paired with one and only one point on the graph of the unit circle. For example, if $s = \frac{\pi}{2}$, then we start at $P(1, 0)$ and move $\frac{\pi}{2}$ units counterclockwise, ending at the point $(0, 1)$. (Note that the circumference of the circle is 2π, so a distance of $\frac{\pi}{2}$ equals one fourth of the circumference.) We say that the point $(0, 1)$ *corresponds* to $\frac{\pi}{2}$. See Figure 26(b).

Suppose that (x, y) is the point on the unit circle that corresponds to the real number s. We can define the six trigonometric functions of s as **sine, cosine, tangent, cosecant, secant,** and **cotangent,** which are abbreviated **sin, cos, tan, csc, sec,** and **cot.** (These are sometimes called *circular functions* because of their relationship to the unit circle.)

Trigonometric (Circular) Functions

If (x, y) is the point on the unit circle that corresponds to the real number s, then

$$\sin s = y \qquad \cos s = x \qquad \tan s = \frac{y}{x} \quad (x \neq 0)$$

$$\csc s = \frac{1}{y} \quad (y \neq 0) \qquad \sec s = \frac{1}{x} \quad (x \neq 0) \qquad \cot s = \frac{x}{y} \quad (y \neq 0).$$

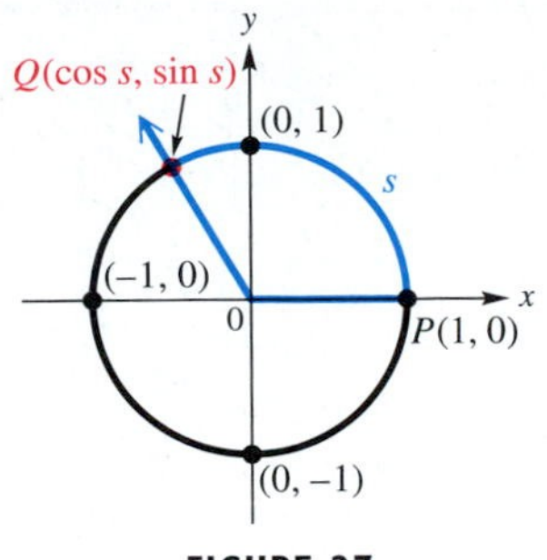

FIGURE 27

From the definitions, we see that there are three pairs of **reciprocal functions:**

sine and *cosecant,* *cosine* and *secant,* and *tangent* and *cotangent.*

Figure 27 indicates that, for the positive number s, the point Q corresponding to s has coordinates $(\cos s, \sin s)$.

Figure 28 justifies an important relationship between $\sin s$ and $\cos s$ for any real number s.* Since $\sin s = y$ and $\cos s = x$, we can replace x and y in $x^2 + y^2 = 1$ and obtain the Pythagorean relationship $(\cos s)^2 + (\sin s)^2 = 1$, or, as it is usually written,

$$\mathbf{\cos^2 s + \sin^2 s = 1.}$$

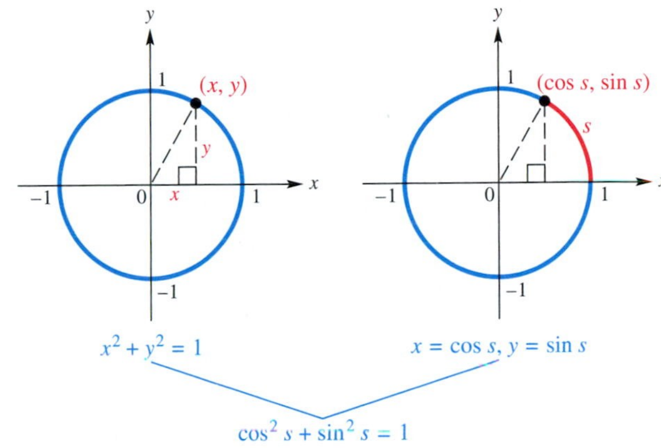

FIGURE 28

GCM TECHNOLOGY NOTE

Powers of trigonometric functions are not usually entered on calculators as they are written mathematically. For example, we write $\sin^2 s$ to represent the square of $\sin s$, but this would be entered as $(\sin s)^2$ on many calculators.

Since the ordered pair (x, y) represents a point on the unit circle,

$$-1 \le x \le 1 \quad \text{and} \quad -1 \le y \le 1,$$

so

$$-1 \le \cos s \le 1 \quad \text{and} \quad -1 \le \sin s \le 1.$$

For any value of s, both $\sin s$ and $\cos s$ exist, so the domain of these functions is the set of all real numbers. Note that $\tan s$ equals $\frac{y}{x}$, which is undefined when $x = 0$. Referring to the unit circle, we see that $x = 0$ when $s = \pm\frac{\pi}{2}, \pm\frac{3\pi}{2}$. Furthermore, if s is more than one revolution, then $x = 0$ when $s = \pm\frac{5\pi}{2}, \pm\frac{7\pi}{2}$, and so on. Therefore, the domain of the tangent function consists of real numbers s satisfying

$$s \neq (2n + 1)\frac{\pi}{2}, \quad n \text{ any integer.}$$

The definition of secant also has x in the denominator, so the domain of secant is the same as the domain of tangent. Both cotangent and cosecant are defined with denominator y. To guarantee that $y \neq 0$, the domain for each of these functions must be the set of all values of s satisfying

$$s \neq n\pi, \quad n \text{ any integer.}$$

*The authors wish to thank Professor Linda K. Schmidt of Greenville Technical College for her suggestion to include Figure 28.

Looking Ahead to Calculus

The domains of the trigonometric functions as defined in this section consist of sets of *real numbers*. Later in the chapter, we will see that the domains can also be sets of *angles*. Real number domains are used when studying trigonometric functions in calculus.

Domains

The domains of the six trigonometric (circular) functions are as follows. Assume that n is any integer and s is a real number.

Sine and Cosine Functions: $(-\infty, \infty)$

Tangent and Secant Functions: $\left\{ s \,\middle|\, s \neq (2n + 1)\frac{\pi}{2} \right\}$

Cotangent and Cosecant Functions: $\{s \mid s \neq n\pi\}$

From the definitions of the trigonometric functions, the following statements are true for the real number s, provided that the denominator is not 0:

$$\tan s = \frac{\sin s}{\cos s}, \qquad \cot s = \frac{\cos s}{\sin s}, \qquad \sec s = \frac{1}{\cos s}, \qquad \csc s = \frac{1}{\sin s}.$$

If $\cos s \neq 0$, then $\quad 1 + \tan^2 s = \sec^2 s,$

and if $\sin s \neq 0$, then $\quad 1 + \cot^2 s = \csc^2 s.$

Statements such as these are called *identities*. An **identity** is a statement that is true for all values in the domain of the variable. (We cover identities in more detail in Section 8.5 and in Chapter 9.)

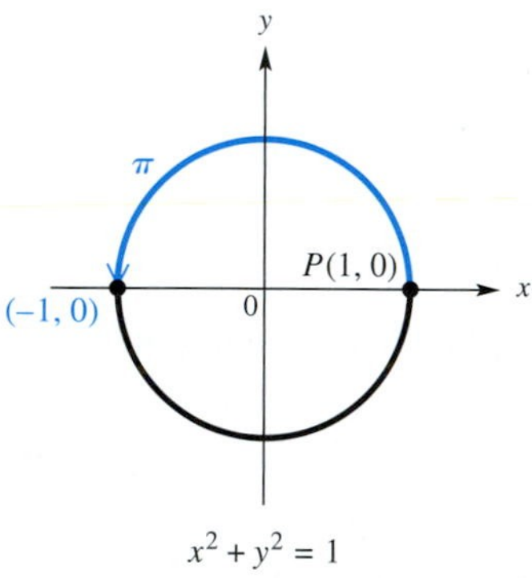

$x^2 + y^2 = 1$

(a)

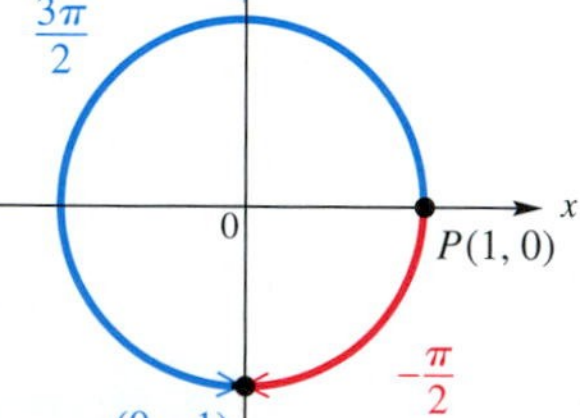

$x^2 + y^2 = 1$

(b)

FIGURE 29

EXAMPLE 1 Finding Function Values by Using the Unit Circle

For each number s, determine the point on the unit circle to which it corresponds and the trigonometric function values for s. State which functions are undefined for s.

(a) 0 **(b)** π **(c)** $\frac{3\pi}{2}$ **(d)** $-\frac{\pi}{2}$

Solution

(a) The starting point $P(1, 0)$, seen in Figure 29(a), corresponds to $s = 0$. Thus,

$$\sin 0 = 0 \qquad \cos 0 = 1 \qquad \tan 0 = \frac{0}{1} = 0$$

$$\csc 0 \text{ is undefined} \qquad \sec 0 = \frac{1}{1} = 1 \qquad \cot 0 \text{ is undefined.}$$

(b) Figure 29(a) shows that the point $(-1, 0)$ corresponds to $s = \pi$. (Since the circumference of the circle is 2π, a distance of π equals one half of the circumference.) Thus,

$$\sin \pi = 0 \qquad \cos \pi = -1 \qquad \tan \pi = \frac{0}{-1} = 0$$

$$\csc \pi \text{ is undefined} \qquad \sec \pi = \frac{1}{-1} = -1 \qquad \cot \pi \text{ is undefined.}$$

(c) The point $(0, -1)$ corresponds to $s = \frac{3\pi}{2}$, as seen in Figure 29(b), leading to

$$\sin\frac{3\pi}{2} = -1 \qquad \cos\frac{3\pi}{2} = 0 \qquad \tan\frac{3\pi}{2} \text{ is undefined}$$

$$\csc\frac{3\pi}{2} = \frac{1}{-1} = -1 \qquad \sec\frac{3\pi}{2} \text{ is undefined} \qquad \cot\frac{3\pi}{2} = \frac{0}{-1} = 0.$$

(d) The real number $s = -\frac{\pi}{2}$ also corresponds to $(0, -1)$, as seen in Figure 29(b). Thus, its trigonometric function values are the same as those of $\frac{3\pi}{2}$ in part (c). ■

The following table gives function values of multiples of $\frac{\pi}{2}$ from 0 to 2π.

Exact Function Values for 0, $\frac{\pi}{2}$, π, $\frac{3\pi}{2}$, and 2π

s	$\sin s$	$\cos s$	$\tan s$	$\cot s$	$\sec s$	$\csc s$
0	0	1	0	Undefined	1	Undefined
$\frac{\pi}{2}$	1	0	Undefined	0	Undefined	1
π	0	-1	0	Undefined	-1	Undefined
$\frac{3\pi}{2}$	-1	0	Undefined	0	Undefined	-1
2π	0	1	0	Undefined	1	Undefined

Using a Calculator to Find Function Values

Calculators are capable of determining trigonometric function values. ***When finding trigonometric function values based on the unit circle, the calculator must be in radian mode (as opposed to degree mode).*** See Figure 30.

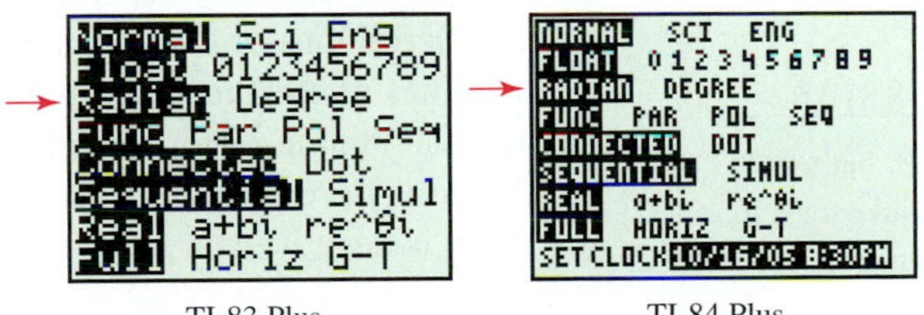

TI-83 Plus TI-84 Plus

Radian mode

FIGURE 30

EXAMPLE 2 Using a Calculator to Find Function Values

Use a calculator in radian mode to find the six trigonometric function values for π.

Solution These values were found in Example 1(b) by using the unit circle. In Figure 31(a), we show how a calculator displays $\cos\pi$, $\sin\pi$, and $\tan\pi$.

```
cos(π)
            -1
sin(π)
             0
tan(π)
             0
```

(a)

Figure 31

TECHNOLOGY NOTE

In general, calculators do not have keys designated for secant, cosecant, and cotangent. To find these values, we use the reciprocal relationships.

To find sec π, we use the fact that the secant function is the reciprocal of the cosine function, and we enter $1/\cos(\pi)$. The result is shown in Figure 31(b). The other two function values are undefined, and if we attempt to find csc π by entering the reciprocal of sin π, an error message occurs, as seen in Figure 31(c).

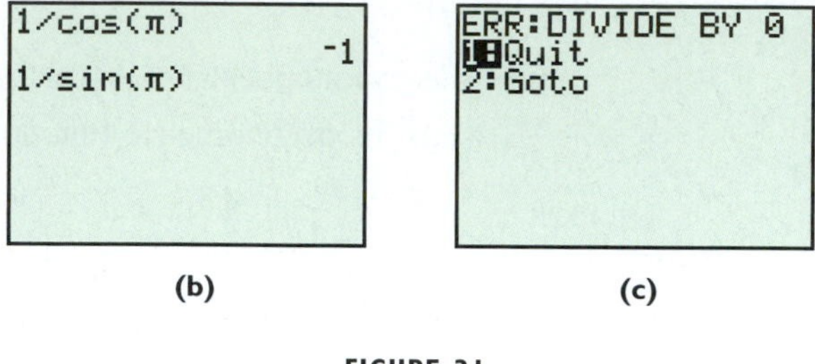

FIGURE 31

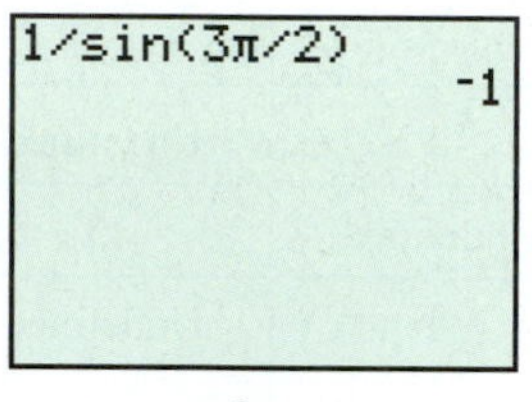

$\csc \frac{3\pi}{2} = -1$

FIGURE 32

CAUTION The calculator functions

$$\cos^{-1}, \quad \sin^{-1}, \quad \text{and} \quad \tan^{-1}$$

DO NOT represent reciprocals of the cosine, sine, and tangent. These are the *inverse trigonometric functions,* introduced later in the text. To find, for example, csc $\frac{3\pi}{2}$ with a calculator, enter $1/\sin(3\pi/2)$, as shown in Figure 32. Compare with the result of Example 1(c).

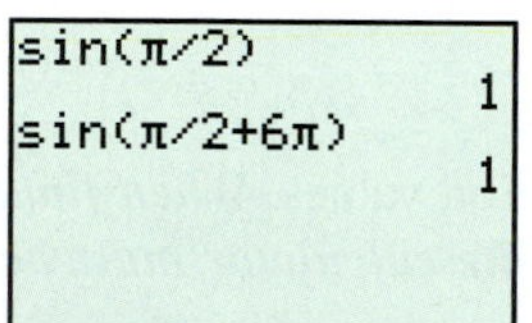

FIGURE 33

Because the circumference of the unit circle is 2π, any integer multiple of 2π added to s corresponds to the same point as s on the unit circle. Thus,

$$\cos(s + 2n\pi) = \cos s \quad \text{and} \quad \sin(s + 2n\pi) = \sin s$$

for any integer n. (The trigonometric functions are examples of a broader class of functions called *periodic functions,* whose values repeat in a fixed manner.) Figure 33 illustrates these facts for the sine function, with $s = \frac{\pi}{2}$ and $n = 3$.

The definitions of the trigonometric functions can be used to determine the signs of the values of the functions in each of the four quadrants. For example, if s corresponds to a point in quadrant I, then both x and y are positive, and all the trigonometric functions have positive values. In quadrant II, x is negative and y is positive, so $\cos s = x$ is negative, $\sin s = y$ is positive, $\tan s = \frac{y}{x}$ is negative, and so on.

FOR DISCUSSION

Duplicate Figure 33, but use different integer values for n to further support the equations. Then show that similar statements (such as $\cos(\pi + 4\pi) = \cos \pi$) are true for the cosine function.

Signs of Values of Trigonometric Functions

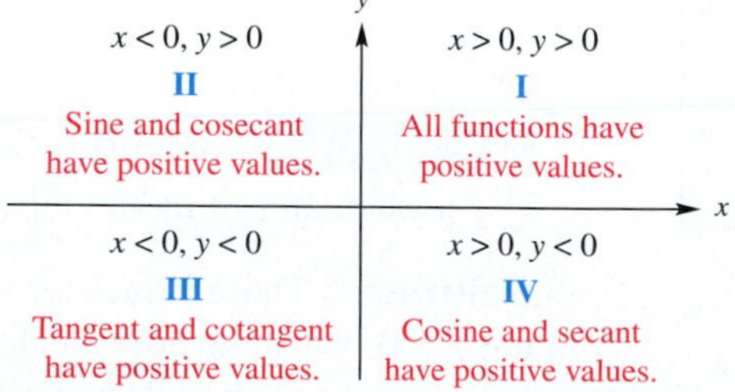

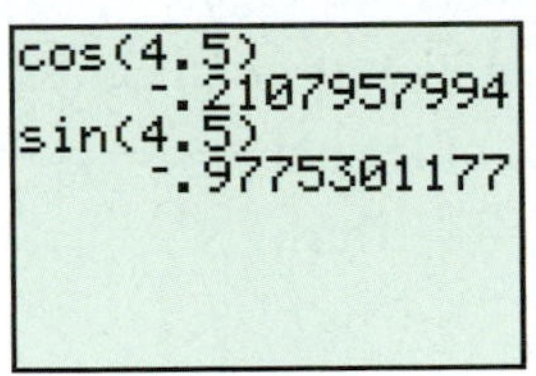

Radian mode

FIGURE 34

EXAMPLE 3 Finding the Cosine, Sine, and Quadrant

Use a calculator to find approximations for cos 4.5 and sin 4.5. Then determine the quadrant in which the point on the unit circle corresponding to 4.5 lies.

Solution Use a calculator as in Figure 34 to find the *approximations:*

$$\cos 4.5 \approx -.2107957994 \quad \text{and} \quad \sin 4.5 \approx -.9775301177.$$

Since $\cos 4.5 < 0$ and $\sin 4.5 < 0$, the point corresponding to 4.5 lies in quadrant III, where both x and y are negative. This is also reasonable, since $\pi < 4.5 < \frac{3\pi}{2}$. ■

Exact Function Values for $\frac{\pi}{4}$, $\frac{\pi}{6}$, and $\frac{\pi}{3}$

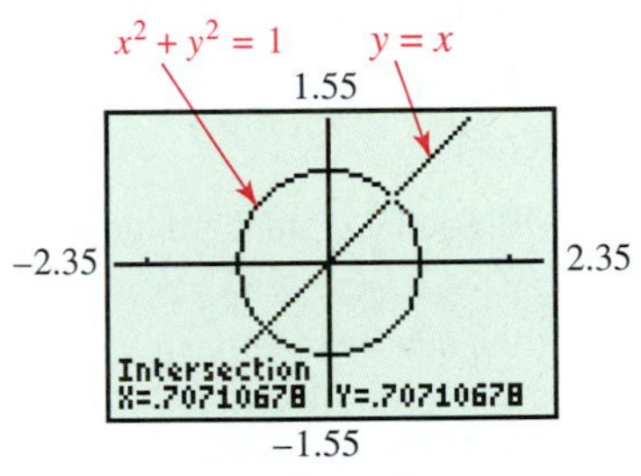

FIGURE 35

Using elementary algebra and geometry, we can find the *exact* sine and cosine values (and thus the other function values as well) for the real numbers $\frac{\pi}{4}$, $\frac{\pi}{6}$, and $\frac{\pi}{3}$. Because the unit circle is symmetric with respect to both axes and the origin, these results enable us to find all six trigonometric function values of integer multiples of these numbers as well.

The line $y = x$ intersects the quadrant I portion of the unit circle, and the point of intersection will give us exact values of $\cos \frac{\pi}{4}$ and $\sin \frac{\pi}{4}$. See Figure 35. (The display shows *approximations* of $x = \cos \frac{\pi}{4}$ and $y = \sin \frac{\pi}{4}$.) We must solve the system

$$x^2 + y^2 = 1 \qquad (1)$$
$$y = x. \qquad (2)$$

Substituting y for x in equation (1) gives

$$y^2 + y^2 = 1 \qquad \text{Let } y = x.$$
$$2y^2 = 1 \qquad \text{Combine terms.}$$
$$y^2 = \frac{1}{2} \qquad \text{Divide by 2.}$$
$$y = \sqrt{\frac{1}{2}} \qquad \text{Take square roots; } y > 0.$$
$$y = \frac{1}{\sqrt{2}} \cdot \frac{\sqrt{2}}{\sqrt{2}} = \frac{\sqrt{2}}{2}. \qquad \text{Rationalize the denominator.}$$

Since $y = x$, $x = \frac{\sqrt{2}}{2}$. Thus, the point $\left(\frac{\sqrt{2}}{2}, \frac{\sqrt{2}}{2}\right)$ corresponds to $\frac{\pi}{4}$, and

$$\sin \frac{\pi}{4} = \frac{\sqrt{2}}{2} \qquad \cos \frac{\pi}{4} = \frac{\sqrt{2}}{2} \qquad \tan \frac{\pi}{4} = \frac{\frac{\sqrt{2}}{2}}{\frac{\sqrt{2}}{2}} = 1$$

$$\csc \frac{\pi}{4} = \sqrt{2} \qquad \sec \frac{\pi}{4} = \sqrt{2} \qquad \cot \frac{\pi}{4} = \frac{\frac{\sqrt{2}}{2}}{\frac{\sqrt{2}}{2}} = 1.$$

FOR DISCUSSION

Use a calculator to find approximations for $\cos \frac{\pi}{4}$, $\sin \frac{\pi}{4}$, and $\frac{\sqrt{2}}{2}$. Show that all three expressions give the same approximation. Compare them with the x- and y-values shown at the bottom of Figure 35. They should be the same (except for perhaps the final digits, depending on how many digits the calculator displays).

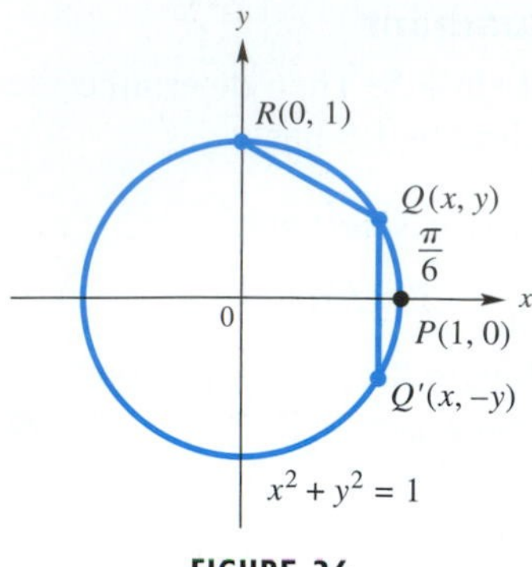

FIGURE 36

To find the point on the unit circle that corresponds to $\frac{\pi}{6}$, see Figure 36. Point $Q(x, y)$ corresponds to the real number $\frac{\pi}{6}$, and $Q'(x, -y)$ is its reflection across the x-axis, corresponding to $-\frac{\pi}{6}$. The chord QQ' cuts off an arc of length $\frac{\pi}{6} + |-\frac{\pi}{6}| = \frac{\pi}{3}$, as does the chord QR, since the length of arc QR is $\frac{\pi}{2} - \frac{\pi}{6} = \frac{\pi}{3}$. Thus,

$$d(Q, Q') = d(Q, R)$$

$$\sqrt{(x - x)^2 + [y - (-y)]^2} = \sqrt{(x - 0)^2 + (y - 1)^2} \quad \text{Distance formula}$$

$$\sqrt{(2y)^2} = \sqrt{x^2 + y^2 - 2y + 1} \quad \text{Simplify.}$$

Remember the middle term when squaring a binomial.

$$4y^2 = x^2 + y^2 - 2y + 1 \quad \text{Square each side.}$$

$$4y^2 = 1 - 2y + 1 \quad x^2 + y^2 = 1$$

$$4y^2 = 2 - 2y \quad \text{Add.}$$

$$2y^2 + y - 1 = 0 \quad \text{Divide by 2; standard form}$$

$$y = \frac{-1 \pm \sqrt{1 + 8}}{4} \quad \text{Let } a = 2, b = 1, c = -1 \text{ in the quadratic formula.}$$

$$y = \frac{-b \pm \sqrt{b^2 - 4ac}}{2a}$$

$$y = \frac{1}{2}. \quad \text{Simplify; } y > 0 \text{, so reject } y = -1.$$

To find x, we substitute $\frac{1}{2}$ for y in $x^2 + y^2 = 1$.

$$x^2 + \left(\frac{1}{2}\right)^2 = 1 \quad \text{Let } y = \tfrac{1}{2} \text{ in equation (1).}$$

$$x^2 = \frac{3}{4} \quad \left(\tfrac{1}{2}\right)^2 = \tfrac{1}{4}\text{; subtract } \tfrac{1}{4}.$$

$$x = \sqrt{\frac{3}{4}} \quad \text{Take square roots; } x > 0.$$

$$x = \frac{\sqrt{3}}{2} \quad \text{Simplify.}$$

Therefore, the real number $\frac{\pi}{6}$ corresponds to the point $\left(\frac{\sqrt{3}}{2}, \frac{1}{2}\right)$.

A similar argument shows that the real number $\frac{\pi}{3}$ corresponds to $\left(\frac{1}{2}, \frac{\sqrt{3}}{2}\right)$. These results lead to exact values of the trigonometric function values for $\frac{\pi}{6}$ and $\frac{\pi}{3}$, as summarized in the following table, along with the values for $\frac{\pi}{4}$ found earlier.

Exact Function Values for $\frac{\pi}{6}$, $\frac{\pi}{4}$, and $\frac{\pi}{3}$

s	$\sin s$	$\cos s$	$\tan s$	$\cot s$	$\sec s$	$\csc s$
$\frac{\pi}{6}$	$\frac{1}{2}$	$\frac{\sqrt{3}}{2}$	$\frac{\sqrt{3}}{3}$	$\sqrt{3}$	$\frac{2\sqrt{3}}{3}$	2
$\frac{\pi}{4}$	$\frac{\sqrt{2}}{2}$	$\frac{\sqrt{2}}{2}$	1	1	$\sqrt{2}$	$\sqrt{2}$
$\frac{\pi}{3}$	$\frac{\sqrt{3}}{2}$	$\frac{1}{2}$	$\sqrt{3}$	$\frac{\sqrt{3}}{3}$	2	$\frac{2\sqrt{3}}{3}$

The symmetry of the unit circle and the values just found lead to other function values, as seen in Figure 37.

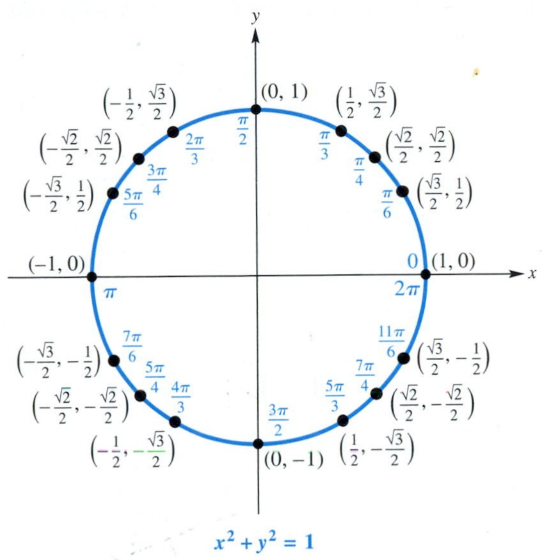

$x^2 + y^2 = 1$

FIGURE 37

EXAMPLE 4 Finding Exact Function Values

Use Figure 37 to find each value.

(a) $\sin \frac{5\pi}{6}$ **(b)** $\cos \frac{2\pi}{3}$ **(c)** $\tan \frac{4\pi}{3}$ **(d)** $\tan\left(-\frac{2\pi}{3}\right)$ **(e)** $\sec \frac{\pi}{3}$

Solution

(a) The point $\left(-\frac{\sqrt{3}}{2}, \frac{1}{2}\right)$ corresponds to $\frac{5\pi}{6}$ on the unit circle. Since the function is sine, we use the y-value of the point to obtain $\sin \frac{5\pi}{6} = \frac{1}{2}$.

(b) The point $\left(-\frac{1}{2}, \frac{\sqrt{3}}{2}\right)$ corresponds to $\frac{2\pi}{3}$. The cosine function requires that we use the x-value, so $\cos \frac{2\pi}{3} = -\frac{1}{2}$.

(c) By definition, $\tan s = \frac{y}{x}$. When $s = \frac{4\pi}{3}$, $(x, y) = \left(-\frac{1}{2}, -\frac{\sqrt{3}}{2}\right)$, so

$$\tan \frac{4\pi}{3} = \frac{-\frac{\sqrt{3}}{2}}{-\frac{1}{2}} = -\frac{\sqrt{3}}{2} \div \left(-\frac{1}{2}\right) = -\frac{\sqrt{3}}{2} \cdot \left(-\frac{2}{1}\right) = \sqrt{3}.$$

Be careful when simplifying a complex fraction.

(d) The point on the unit circle for $s = -\frac{2\pi}{3}$ is the same as that for $\frac{4\pi}{3}$. By the result of part (c), $\tan\left(-\frac{2\pi}{3}\right) = \sqrt{3}$.

(e) $\sec \frac{\pi}{3} = \frac{1}{\cos \frac{\pi}{3}} = \frac{1}{\frac{1}{2}} = 2$ ■

8.2 Exercises

Determine what fraction of the circumference of the unit circle each value of s represents. For example, $s = \pi$ represents half of the circumference of the unit circle.

1. $s = \frac{\pi}{3}$
2. $s = \frac{2\pi}{3}$
3. $s = \frac{\pi}{4}$
4. $s = \frac{3\pi}{4}$
5. $s = \frac{\pi}{6}$
6. $s = \frac{5\pi}{6}$
7. $s = 3\pi$
8. $s = \frac{5\pi}{2}$

Each figure shows an arc with length s along the circumference of a unit circle. Evaluate $\sin s$ *and* $\cos s$.

9.

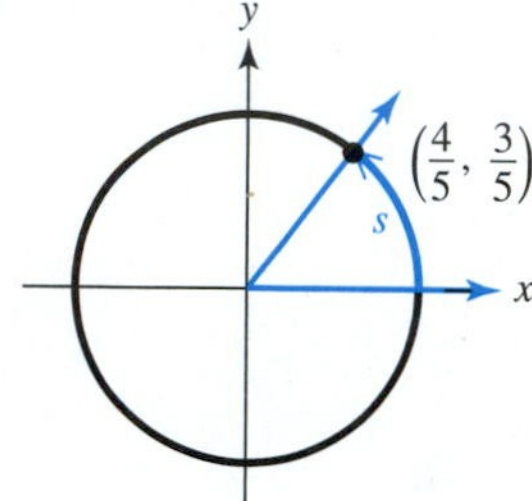

10.

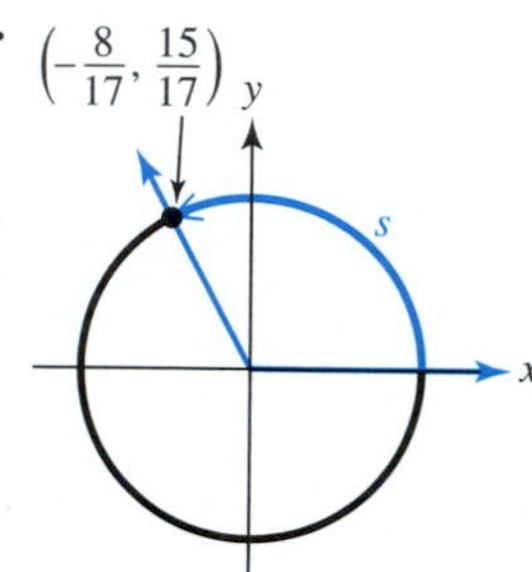

11.

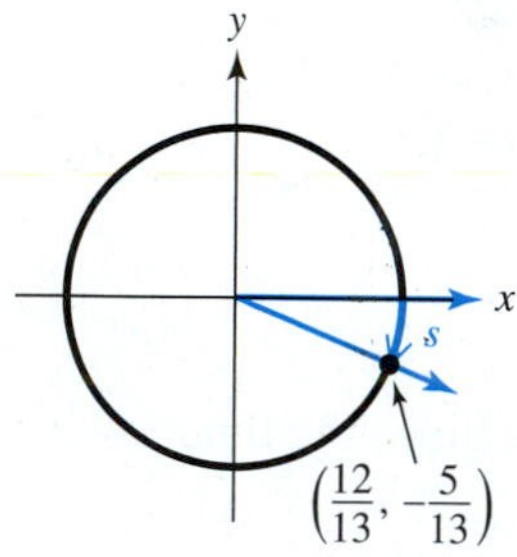

12.

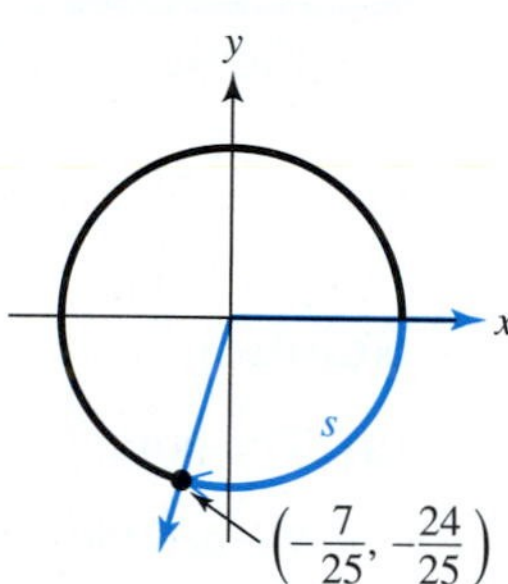

Refer to Figure 26(a). For each real number s, determine the point on the unit circle to which it corresponds. Determine $\cos s$ *and* $\sin s$. *State which functions are undefined for s.*>

13. $-\pi$
14. π
15. 2π
16. -2π
17. $\frac{3\pi}{2}$
18. $-\frac{3\pi}{2}$
19. $\frac{\pi}{2}$
20. $-\frac{\pi}{2}$

Refer to Figure 26(a). Find the trigonometric (circular) function values for each number s. State which functions are undefined for s. Use a calculator to confirm your results.

21. -5π
22. -3π
23. 6π
24. 8π
25. $-\frac{5\pi}{2}$
26. $-\frac{7\pi}{2}$
27. $\frac{5\pi}{2}$
28. $\frac{7\pi}{2}$

Use a calculator to find approximations for cos s *and* sin s *for each number* s*. Give as many decimal places as your calculator displays. (These are NOT exact values—they are only approximations.) Then determine the quadrant in which the point on the unit circle corresponding to* s *lies. Finally, find approximations for* tan s, cot s, sec s, *and* csc s.

29. .75 **30.** .95 **31.** -4.25 **32.** -3.75

33. -2.25 **34.** -2.75 **35.** 5.5 **36.** 5.75

Relating Concepts

For individual or group investigation (Exercises 37–42)

With your calculator in radian mode, ***work Exercises 37–42 in order.***

37. Let s represent the number of letters in your first name. Find an approximation for cos s.

38. Let n represent the number of letters in your last name. Find an approximation for $\cos(s + 2n\pi)$.

39. Compare your results from Exercises 37 and 38. Explain how and why they relate to each other as they do.

40. Let s represent the number of letters in your last name. Find an approximation for sin s.

41. Let n represent the number of letters in your first name. Find an approximation for $\sin(s + 2n\pi)$.

42. Compare your results from Exercises 40 and 41. Explain how and why they relate to each other as they do.

Use the figure and any appropriate relationship(s) among trigonometric functions to find each value.

43. $\sin \frac{7\pi}{6}$ **44.** $\cos \frac{5\pi}{3}$ **45.** $\tan \frac{3\pi}{4}$

46. $\cos \frac{7\pi}{6}$ **47.** $\sec \frac{2\pi}{3}$ **48.** $\csc \frac{11\pi}{6}$

49. $\cot \frac{5\pi}{6}$ **50.** $\cos\left(-\frac{4\pi}{3}\right)$ **51.** $\sin\left(-\frac{5\pi}{6}\right)$

52. $\tan \frac{17\pi}{3}$ **53.** $\sec \frac{23\pi}{6}$ **54.** $\csc \frac{13\pi}{3}$

55. $\cos\left(-\frac{13\pi}{6}\right)$ **56.** $\sin\left(-\frac{9\pi}{4}\right)$ **57.** $\tan\left(-\frac{13\pi}{4}\right)$

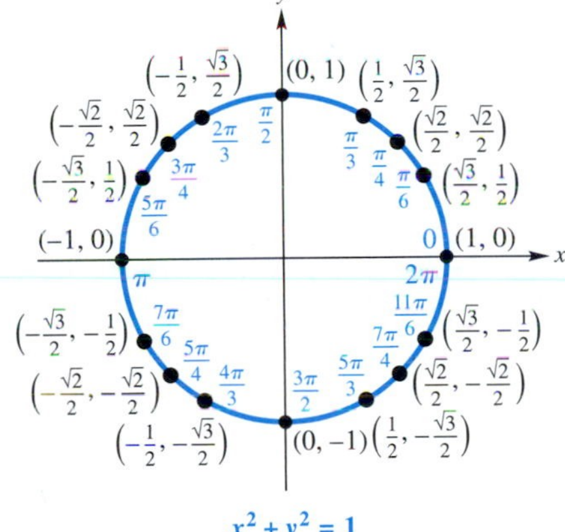

58. In the screen shown, the value 3 is stored in S. Then the value of $(\cos(S))^2 + (\sin(S))^2$ is shown to be 1. Duplicate this screen on your own calculator, but use several different values of S. Is the result always 1? Explain.

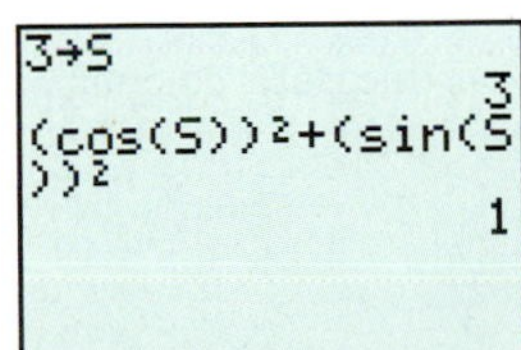

Use the identity $\cos^2 s + \sin^2 s = 1$ *to find the value of x or y, as appropriate. Then, assuming that s corresponds to the given point on the unit circle, find the six trigonometric function values for s.*

59. $\left(\frac{3}{5}, y\right), y > 0$ **60.** $\left(\frac{7}{25}, y\right), y > 0$ **61.** $\left(x, \frac{24}{25}\right), x < 0$

62. $\left(x, \frac{8}{17}\right), x < 0$

63. $\left(-\frac{1}{3}, y\right), y < 0$

64. $\left(-\frac{1}{4}, y\right), y < 0$

For each of the following, find tan *s*, cot *s*, sec *s*, *and* csc *s*. *(Do not use a calculator.)*

65. $\sin s = \frac{1}{2}, \cos s = \frac{\sqrt{3}}{2}$

66. $\sin s = \frac{3}{4}, \cos s = \frac{\sqrt{7}}{4}$

67. $\sin s = \frac{4}{5}, \cos s = -\frac{3}{5}$

68. $\sin s = -\frac{1}{2}, \cos s = -\frac{\sqrt{3}}{2}$

69. $\sin s = -\frac{\sqrt{3}}{2}, \cos s = \frac{1}{2}$

70. $\sin s = \frac{12}{13}, \cos s = \frac{5}{13}$

Decide in what quadrant the point corresponding to s must lie to satisfy the following conditions for s.

71. $\sin s > 0, \cos s < 0$

72. $\cos s > 0, \tan s > 0$

73. $\sec s < 0, \csc s < 0$

74. $\tan s > 0, \cos s < 0$

75. $\cos s > 0, \sin s < 0$

76. $\tan s < 0, \sin s > 0$

77. Prove that if $\cos s \neq 0$, then $1 + \tan^2 s = \sec^2 s$.

78. Prove that if $\sin s \neq 0$, then $1 + \cot^2 s = \csc^2 s$.

Let s be a real number corresponding to the point (a, b) *on the unit circle. Use this information to determine the sine and cosine of each real number.*

79. s

80. $s + 2\pi$

81. $s - 6\pi$

82. $s + \pi$

83. $-s$

84. $-s + \pi$

85. $s - \frac{\pi}{2}$

86. $s + \frac{\pi}{2}$

Reviewing Basic Concepts (Sections 8.1 and 8.2)

1. Give the complement and the supplement of each angle.

(a) 35° **(b)** $\frac{\pi}{4}$

2. Convert 32.25° to degrees, minutes, and seconds.

3. Convert 59° 35′ 30″ to decimal degrees.

4. Find the angle of smallest positive measure coterminal with each angle.

(a) 560° **(b)** $-\frac{2\pi}{3}$

5. Convert each angle measure as directed.

(a) 240° to radians **(b)** $\frac{3\pi}{4}$ radians to degrees

6. Suppose a circle has radius 3 centimeters.

(a) What is the exact length of an arc intercepted by a central angle of 120°?

(b) What is the exact area of the sector formed in part (a)?

7. Give the exact coordinates of the point on the unit circle that corresponds to the given value of *s*.

(a) -2π **(b)** $\frac{5\pi}{4}$ **(c)** $\frac{5\pi}{2}$

8. Give the values of the trigonometric functions of $-\frac{5\pi}{2}$. State which functions are undefined.

9. Give the exact values of the six trigonometric functions of each number.

(a) $\frac{7\pi}{6}$ **(b)** $-\frac{2\pi}{3}$

10. Give calculator approximations for the six trigonometric function values of 2.25.

8.3 Graphs of the Sine and Cosine Functions

Periodic Functions ■ Graph of the Sine Function ■ Graph of the Cosine Function ■ Graphing Techniques, Amplitude, and Period ■ Translations ■ Determining a Trigonometric Model Using Curve Fitting

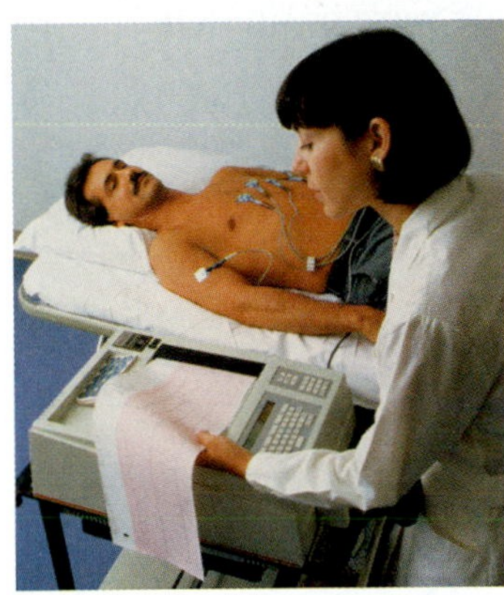

Periodic Functions

Many things in daily life repeat with a predictable pattern. In warm areas, electricity use goes up in summer and down in winter, the price of fresh fruit goes down in summer and up in winter, and attendance at amusement parks increases in spring and declines in autumn. Because the sine and cosine functions repeat their values over and over in a regular pattern, they are examples of *periodic functions*. Figure 38 shows a periodic graph that represents a normal heartbeat.

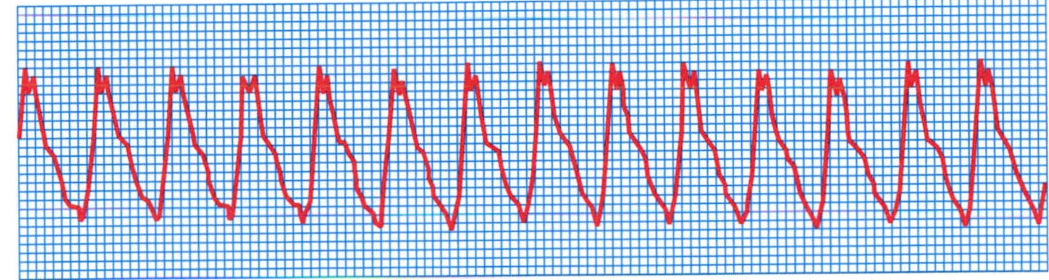

FIGURE 38

Looking Ahead to Calculus

Periodic functions are used throughout calculus, so you will need to know their characteristics. One use of these functions is to describe the location of a point in the plane by *polar coordinates*, an alternative to rectangular coordinates. (See Chapter 10.)

Periodic Function

A **periodic function** is a function f such that

$$f(x) = f(x + np),$$

for every real number x in the domain of f, every integer n, and some positive real number p. The smallest possible positive value of p is the **period** of the function.

The circumference of the unit circle is 2π, so the smallest value of p for which the sine and cosine functions repeat is 2π. ***Therefore, the sine and cosine functions are periodic functions with period 2π.***

Graph of the Sine Function

In the previous section, we saw that for a real number s, the point on the unit circle corresponding to s has coordinates $(\cos s, \sin s)$. See Figure 39, and trace along the circle to verify the results shown in the table.

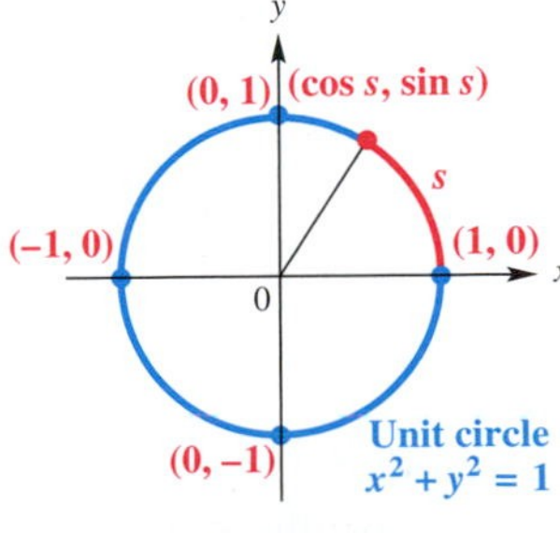

FIGURE 39

As s Increases from	$\sin s$	$\cos s$
0 to $\frac{\pi}{2}$	Increases from 0 to 1	Decreases from 1 to 0
$\frac{\pi}{2}$ to π	Decreases from 1 to 0	Decreases from 0 to -1
π to $\frac{3\pi}{2}$	Decreases from 0 to -1	Increases from -1 to 0
$\frac{3\pi}{2}$ to 2π	Increases from -1 to 0	Increases from 0 to 1

To avoid confusion when graphing the sine function, we use x rather than s; this corresponds to the letters in the xy-coordinate system. Selecting key values of x and finding the corresponding values of $\sin x$ leads to the tables in Figure 40.

To obtain the traditional graph of a portion of the sine function shown in Figure 40, we plot the points from the table of values and join them with a smooth curve. Since $y = \sin x$ is periodic and has $(-\infty, \infty)$ as its domain, the graph continues in the same pattern in both directions. This graph is called a **sine wave** or **sinusoid.**

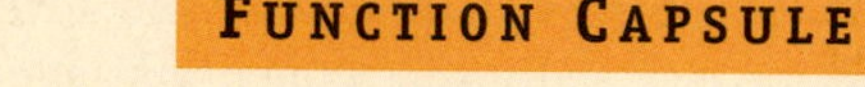

SINE FUNCTION $\quad f(x) = \sin x$

Domain: $(-\infty, \infty)$ $\quad$ Range: $[-1, 1]$

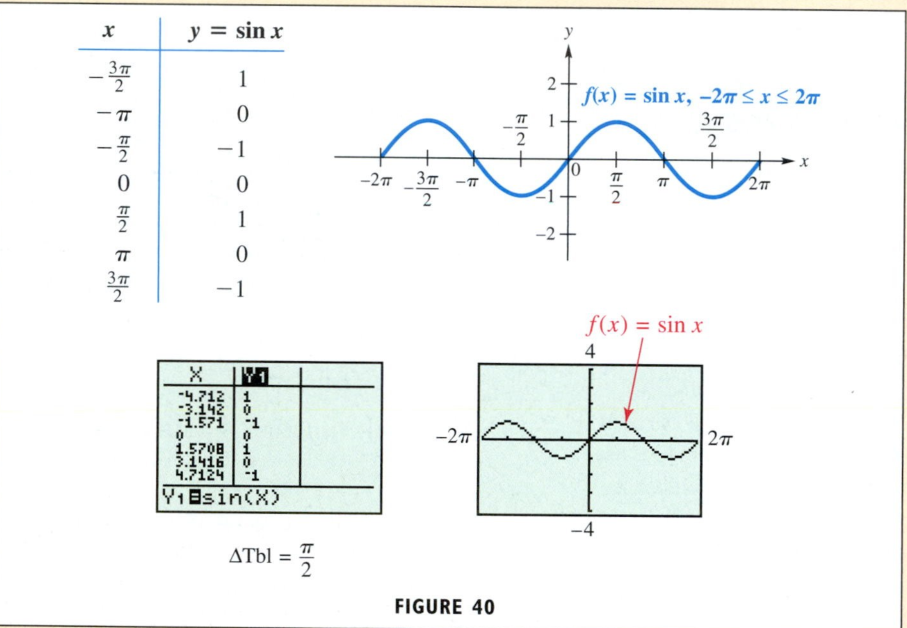

x	$y = \sin x$
$-\frac{3\pi}{2}$	1
$-\pi$	0
$-\frac{\pi}{2}$	-1
0	0
$\frac{\pi}{2}$	1
π	0
$\frac{3\pi}{2}$	-1

X	Y1
-4.712	1
-3.142	0
-1.571	-1
0	0
1.5708	1
3.1416	0
4.7124	-1

FIGURE 40

- The graph is continuous over its entire domain, $(-\infty, \infty)$.
- Its x-intercepts are of the form $n\pi$, where n is an integer.
- Its period is 2π.
- The graph is symmetric with respect to the origin, and it is an odd function. That is, for all x in the domain, $\sin(-x) = -\sin x$.

GCM TECHNOLOGY NOTE

Graphing calculators often have a window designated for graphing circular functions. We refer to the window $[-2\pi, 2\pi]$ by $[-4, 4]$ with Xscl $= \frac{\pi}{2}$ and Yscl $= 1$ as the **trig viewing window.** Your model may use a different "standard" viewing window for the graphs of trigonometric functions.

A comprehensive graph of a sinusoid consists of at least one period of the graph and shows the extreme points.

Graph of the Cosine Function

We find the graph of $y = \cos x$ in much the same way as the graph of $y = \sin x$. In the tables shown with Figure 41 on the next page for $y = \cos x$, we use the same values for x as before. Notice that the graph of $y = \cos x$ has the same shape as the graph of $y = \sin x$. It is, in fact, the graph of the sine function shifted, or translated, $\frac{\pi}{2}$ units to the left.

FUNCTION CAPSULE

COSINE FUNCTION $\quad f(x) = \cos x$

Domain: $(-\infty, \infty)$ $\quad$ Range: $[-1, 1]$

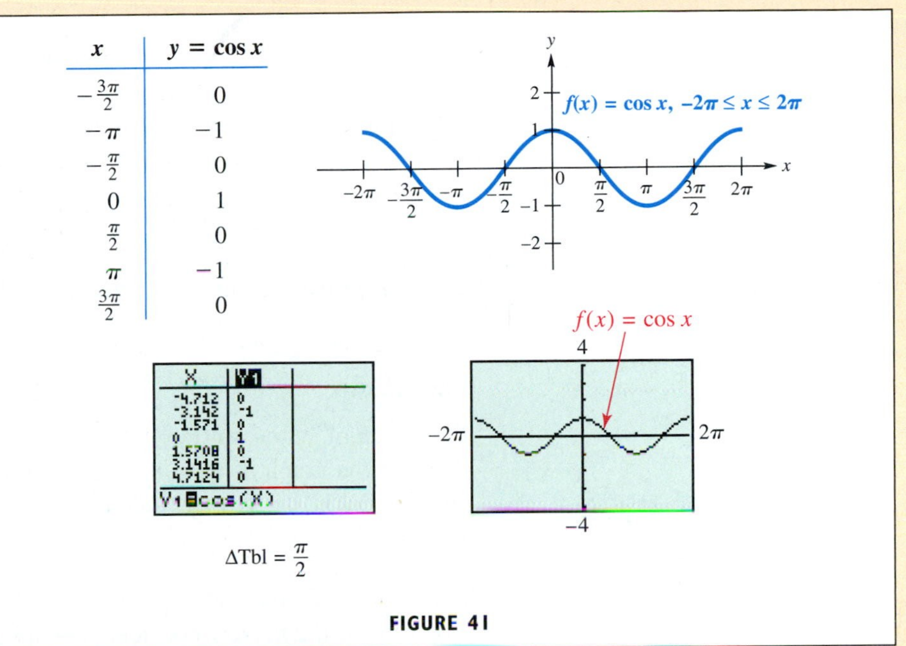

x	$y = \cos x$
$-\frac{3\pi}{2}$	0
$-\pi$	-1
$-\frac{\pi}{2}$	0
0	1
$\frac{\pi}{2}$	0
π	-1
$\frac{3\pi}{2}$	0

FIGURE 41

- The graph is continuous over its entire domain, $(-\infty, \infty)$.
- Its x-intercepts are of the form $(2n + 1)\frac{\pi}{2}$, where n is an integer.
- Its period is 2π.
- The graph is symmetric with respect to the y-axis, and it is an even function. That is, for all x in the domain, $\cos(-x) = \cos x$.

Looking Ahead to Calculus

The discussion of the derivative of a function in calculus shows that for the sine function, the slope of the tangent line at any point x is given by $\cos x$. For example, look at the graph of $y = \sin x$ and notice that a tangent line at $x = \pm\frac{\pi}{2}, \pm\frac{3\pi}{2}, \pm\frac{5\pi}{2}, \ldots$ will be horizontal and thus have slope 0. Now look at the graph of $y = \cos x$ and see that for these values, $\cos x = 0$.

Graphing Techniques, Amplitude, and Period

The examples that follow show graphs that are "stretched" or "shrunk" either vertically, horizontally, or both compared with the graphs of $y = \sin x$ or $y = \cos x$.

GCM **EXAMPLE 1** **Graphing $y = a \sin x$**

Graph $y = 2 \sin x$, and compare with the graph of $y = \sin x$.

Solution

For a given value of x, the value of y is twice as large as it would be for $y = \sin x$, as shown in the table of values on the next page. The only change in the graph is the range, which becomes $[-2, 2]$. See Figure 42 on the next page, which includes a graph of $y = \sin x$ for comparison.

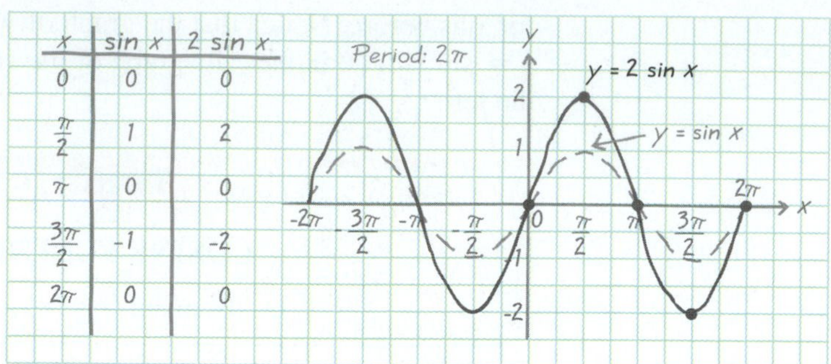

FIGURE 42

The **amplitude** of a *sinusoidal* function is half the difference between the maximum and minimum values. Thus, for the sine and cosine functions, the amplitude is

$$\frac{1}{2}[1-(-1)]=\frac{1}{2}(2)=1.$$

Generalizing from Example 1 gives the following.

Amplitude

The graph of $\boldsymbol{y = a \sin x}$ or $\boldsymbol{y = a \cos x}$, with $a \neq 0$, will have the same shape as the graph of $y = \sin x$ or $y = \cos x$, respectively, except with range $[-|a|, |a|]$. The amplitude is $|a|$.

No matter what the value of the amplitude, the periods of $y = a \sin x$ and $y = a \cos x$ are still 2π. Now consider $y = \sin 2x$. We can complete a table of values for the interval $[0, 2\pi]$.

x	0	$\frac{\pi}{4}$	$\frac{\pi}{2}$	$\frac{3\pi}{4}$	π	$\frac{5\pi}{4}$	$\frac{3\pi}{2}$	$\frac{7\pi}{4}$	2π
sin 2x	0	1	0	−1	0	1	0	−1	0

Note that one complete cycle occurs in π units, not 2π units. Therefore, the period here is π, which equals $\frac{2\pi}{2}$. What about $y = \sin 4x$? Look at the next table.

x	0	$\frac{\pi}{8}$	$\frac{\pi}{4}$	$\frac{3\pi}{8}$	$\frac{\pi}{2}$	$\frac{5\pi}{8}$	$\frac{3\pi}{4}$	$\frac{7\pi}{8}$	π
sin 4x	0	1	0	−1	0	1	0	−1	0

A complete cycle is achieved in $\frac{\pi}{2}$ or $\frac{2\pi}{4}$ units, which is reasonable, because

$$\sin\left(4 \cdot \frac{\pi}{2}\right) = \sin 2\pi = 0.$$

In general, the graph of a function of the form $y = \sin bx$ or $y = \cos bx$, for $b > 0$, will have a period different from 2π when $b \neq 1$. Since the values of $\sin bx$ or cos bx will take on all possible values as bx ranges from 0 to 2π, to find the period of either of these functions, we must solve the three-part inequality

$$0 \le bx \le 2\pi$$

$$0 \le x \le \frac{2\pi}{b}. \quad \text{Divide by the positive number } b.$$

Thus, the period is $\frac{2\pi}{b}$. By dividing the interval $\left[0, \frac{2\pi}{b}\right]$ into four equal parts, we obtain the values for which sin bx or cos bx is -1, 0, or 1. These values will give minimum points, x-intercepts, and maximum points on the graph. Once these points are determined, we can sketch the graph by joining the points with a smooth sinusoidal curve. (If a function has $b < 0$, then the identities of the next chapter can be used to write the function as one in which $b > 0$.)

NOTE To divide an interval into four equal parts, follow these steps.

Step 1 Find the midpoint of the interval by adding the x-values of the endpoints and dividing by 2.

Step 2 Find the two midpoints of the intervals found in Step 1, using the same procedure.

EXAMPLE 2 Graphing $y = \sin bx$

Graph $y = \sin 2x$, and compare with the graph of $y = \sin x$.

Solution For this function, $b = 2$, so the period is $\frac{2\pi}{2} = \pi$. Therefore, the graph will complete one period over the interval $[0, \pi]$. The endpoints are 0 and π, and the three points between the endpoints are

$$\frac{1}{2}\left(0 + \frac{\pi}{2}\right), \quad \frac{1}{2}(0 + \pi), \quad \text{and} \quad \frac{1}{2}\left(\frac{\pi}{2} + \pi\right),$$

which give the following x-values:

$0,$	$\frac{\pi}{4},$	$\frac{\pi}{2},$	$\frac{3\pi}{4},$	$\pi.$
↑	↑	↑	↑	↑
Left endpoint	First-quarter point	Midpoint	Third-quarter point	Right endpoint

We plot the points from the table of values given on the previous page and join them with a smooth sinusoidal curve. More of the graph can be sketched by repeating this cycle, as shown in Figure 43. The amplitude is not changed. The graph of $y = \sin x$ is included for comparison.

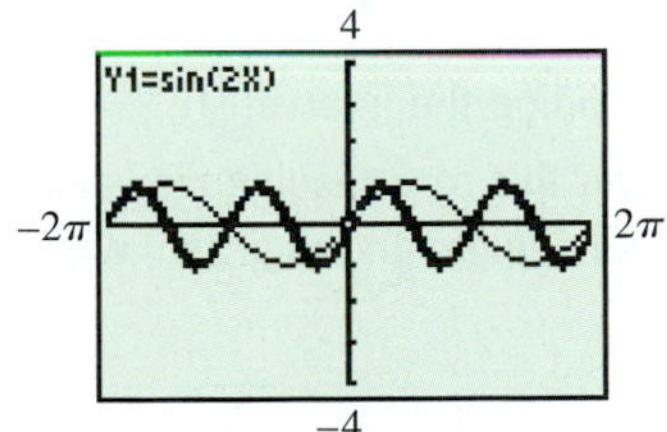

The thick graph of $y = \sin 2x$ oscillates twice as fast as the graph of $y = \sin x$.

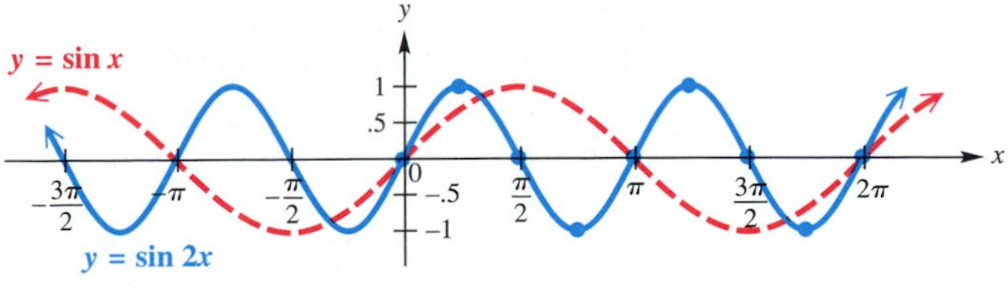

FIGURE 43

We can think of the graph of $y = \sin bx$ as a horizontal stretching of the graph of $y = \sin x$ when $0 < b < 1$ and a horizontal shrinking when $b > 1$. Generalizing from Example 2 leads to the following result.

Period

For $b > 0$, the graph of $\boldsymbol{y = \sin bx}$ will resemble that of $y = \sin x$, but with period $\frac{2\pi}{b}$. Also, the graph of $\boldsymbol{y = \cos bx}$ will resemble that of $y = \cos x$, but with period $\frac{2\pi}{b}$.

EXAMPLE 3 **Graphing $y = \cos bx$**

Graph $y = \cos \frac{2}{3}x$ over one period.

Solution The period is $\dfrac{2\pi}{\frac{2}{3}} = 3\pi$. We divide the interval $[0, 3\pi]$ into four equal parts to get the following x-values that yield minimum points, maximum points, and x-intercepts: $\mathbf{0}, \frac{3\pi}{4}, \frac{3\pi}{2}, \frac{9\pi}{4}$, and $\mathbf{3\pi}$. We use these values to obtain a table of key points for one period.

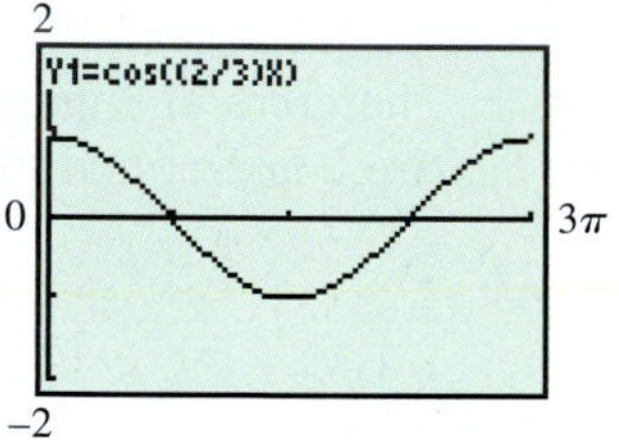

This screen shows a graph of the function in Example 3. By choosing Xscl = $\frac{3\pi}{4}$, the x-intercepts, maxima, and minima coincide with tick marks on the x-axis.

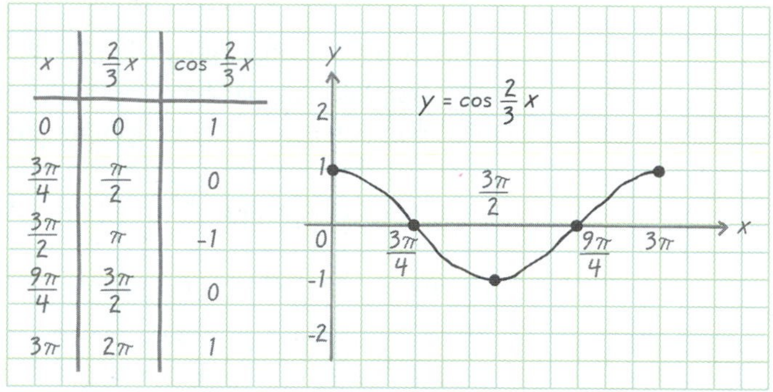

FIGURE 44

The amplitude is 1 because the maximum value is 1, the minimum value is -1, and half of $1 - (-1)$ is $\frac{1}{2}(2) = 1$. We plot these points and join them with a smooth curve, as shown in Figure 44. ■

NOTE Look at the middle row of the table in Example 3. Dividing the interval $\left[0, \frac{2\pi}{b}\right]$ into four equal parts will always give the values $0, \frac{\pi}{2}, \pi, \frac{3\pi}{2}$, and 2π for this row, resulting in values of -1, 0, or 1 for the sinusoidal function. These values lead to key points on the graph, which can then be easily sketched.

The method used in Examples 1–3 is summarized as follows.

Sketching Graphs of the Sine and Cosine Functions

To graph $y = a \sin bx$ or $y = a \cos bx$, with $b > 0$,

Step 1 Find the period, $\frac{2\pi}{b}$. Start at 0 on the x-axis, and lay off a distance of $\frac{2\pi}{b}$.

Step 2 Divide the interval into four equal parts. (See the note preceding Example 2.)

Step 3 Evaluate the function for each of the five x-values resulting from Step 2. The points will be maximum points, minimum points, and x-intercepts.

Step 4 Plot the points found in Step 3, and join them with a sinusoidal curve having amplitude $|a|$.

Step 5 Draw the graph over additional periods, to the right and to the left, as needed.

The function in Example 4 has both amplitude and period affected by constants.

EXAMPLE 4 **Graphing $y = a \sin bx$**

Graph $y = -2 \sin 3x$ using the preceding guidelines.

Solution

Step 1 For this function, $b = 3$, so the period is $\frac{2\pi}{3}$. The function will be graphed over the interval $\left[0, \frac{2\pi}{3}\right]$.

Step 2 Divide the interval $\left[0, \frac{2\pi}{3}\right]$ into four equal parts to get the x-values $0, \frac{\pi}{6}, \frac{\pi}{3}, \frac{\pi}{2}$, and $\frac{2\pi}{3}$.

Step 3 Make a table of values determined by the x-values from Step 2.

x	0	$\frac{\pi}{6}$	$\frac{\pi}{3}$	$\frac{\pi}{2}$	$\frac{2\pi}{3}$
$3x$	0	$\frac{\pi}{2}$	π	$\frac{3\pi}{2}$	2π
$\sin 3x$	0	1	0	-1	0
$-2 \sin 3x$	0	-2	0	2	0

Step 4 Plot the points $(0, 0)$, $\left(\frac{\pi}{6}, -2\right)$, $\left(\frac{\pi}{3}, 0\right)$, $\left(\frac{\pi}{2}, 2\right)$, and $\left(\frac{2\pi}{3}, 0\right)$, and join them with a sinusoidal curve having amplitude 2. See Figure 45.

Step 5 The graph can be extended by repeating the cycle.

Notice that when a is negative, the graph of $y = a \sin bx$ is a reflection across the x-axis of the graph of $y = |a| \sin bx$. ■

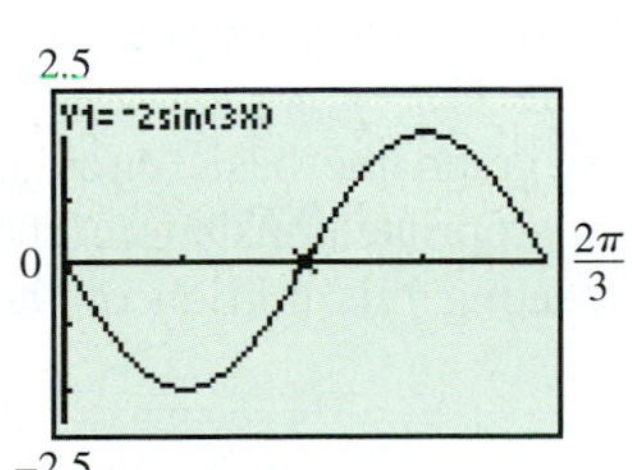

FIGURE 45

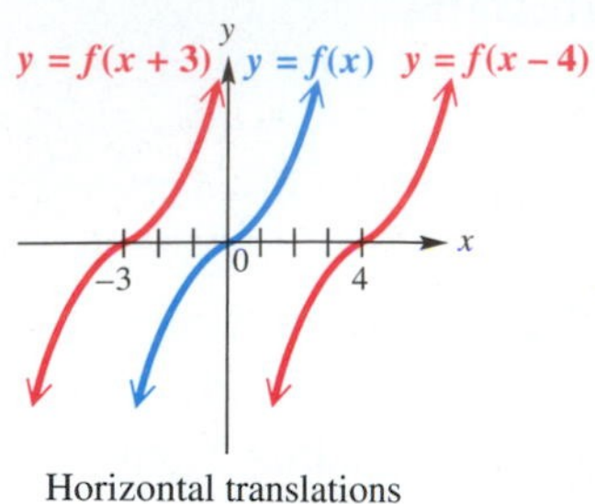

Horizontal translations of $y = f(x)$

FIGURE 46

Translations

In general, the graph of the function defined by $y = f(x - d)$ is translated *horizontally* compared with the graph of $y = f(x)$. The translation is d units to the right if $d > 0$ and $|d|$ units to the left if $d < 0$. See Figure 46. With circular functions, a horizontal translation is called a **phase shift.** In the function $y = f(x - d)$, the expression $x - d$ is called the **argument.**

In Example 5, we give two methods that can be used to sketch the graph of a circular function involving a phase shift.

EXAMPLE 5 Graphing $y = \sin(x - d)$

Graph $y = \sin\left(x - \frac{\pi}{3}\right)$.

Solution *Method 1* For the argument $x - \frac{\pi}{3}$ to result in all possible values throughout one period, it must take on all values between 0 and 2π, inclusive. Therefore, to find an interval of one period, we solve the three-part inequality

$$0 \leq x - \frac{\pi}{3} \leq 2\pi \quad \text{to obtain} \quad \frac{\pi}{3} \leq x \leq \frac{7\pi}{3}. \qquad \text{Add } \tfrac{\pi}{3} \text{ to each part.}$$

Divide the interval $\left[\frac{\pi}{3}, \frac{7\pi}{3}\right]$ into four equal parts to get the x-values $\frac{\pi}{3}, \frac{5\pi}{6}, \frac{4\pi}{3}, \frac{11\pi}{6}$, and $\frac{7\pi}{3}$. A table of values using these x-values follows.

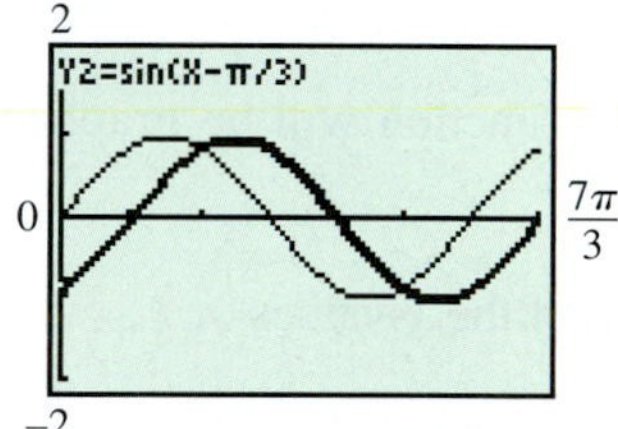

The screen shows the graph of $Y_1 = \sin X$ as a thin line and the graph of $Y_2 = \sin\left(X - \frac{\pi}{3}\right)$ as a thick line. Note the horizontal translation.

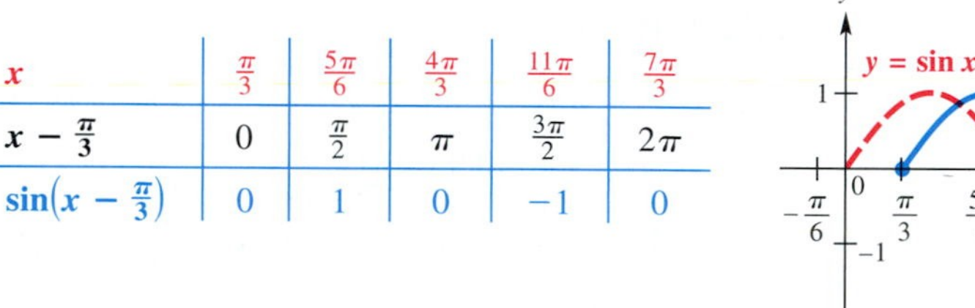

x	$\frac{\pi}{3}$	$\frac{5\pi}{6}$	$\frac{4\pi}{3}$	$\frac{11\pi}{6}$	$\frac{7\pi}{3}$
$x - \frac{\pi}{3}$	0	$\frac{\pi}{2}$	π	$\frac{3\pi}{2}$	2π
$\sin\left(x - \frac{\pi}{3}\right)$	0	1	0	−1	0

FIGURE 47

We join the corresponding points to get the graph shown in Figure 47. The period is 2π, and the amplitude is 1.

Method 2 We can also graph $y = \sin\left(x - \frac{\pi}{3}\right)$ by using a horizontal translation. The argument $x - \frac{\pi}{3}$ indicates that the graph will be translated $\frac{\pi}{3}$ units to the *right* (the phase shift) compared with the graph of $y = \sin x$. See Figure 47.

Therefore, to graph a function with this method, first graph the basic circular function, and then graph the desired function by using the appropriate translation. The graph can be extended through additional periods by repeating this portion of the graph as necessary. ■

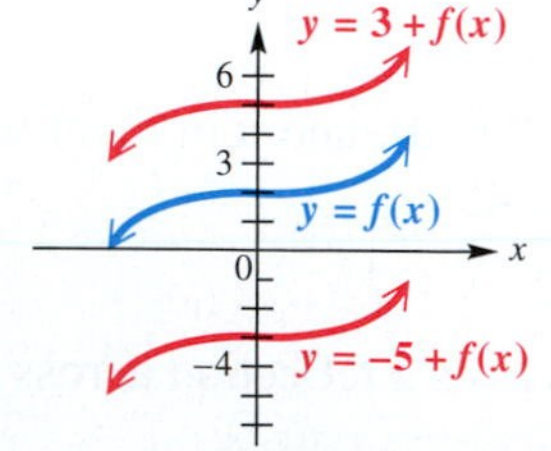

Vertical translations of $y = f(x)$

FIGURE 48

The graph of a function of the form $y = c + f(x)$ is translated *vertically* compared with the graph of $y = f(x)$. See Figure 48. The translation is c units up if $c > 0$ and $|c|$ units down if $c < 0$.

EXAMPLE 6 Graphing $y = c + a \cos bx$

Graph $y = 3 - 2 \cos 3x$.

Solution The values of y will be 3 greater than the corresponding values of y in $y = -2 \cos 3x$. This means that the graph of $y = 3 - 2 \cos 3x$ is the same as the graph of $y = -2 \cos 3x$, vertically translated 3 units up. Since the period of $y = -2 \cos 3x$ is $\frac{2\pi}{3}$, the key points have x-values $0, \frac{\pi}{6}, \frac{\pi}{3}, \frac{\pi}{2}$, and $\frac{2\pi}{3}$. Use these x-values to make a table of values.

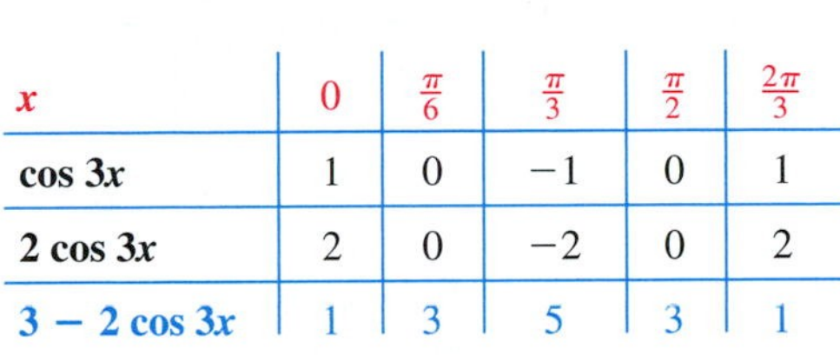

x	0	$\frac{\pi}{6}$	$\frac{\pi}{3}$	$\frac{\pi}{2}$	$\frac{2\pi}{3}$
$\cos 3x$	1	0	−1	0	1
$2 \cos 3x$	2	0	−2	0	2
$3 - 2 \cos 3x$	1	3	5	3	1

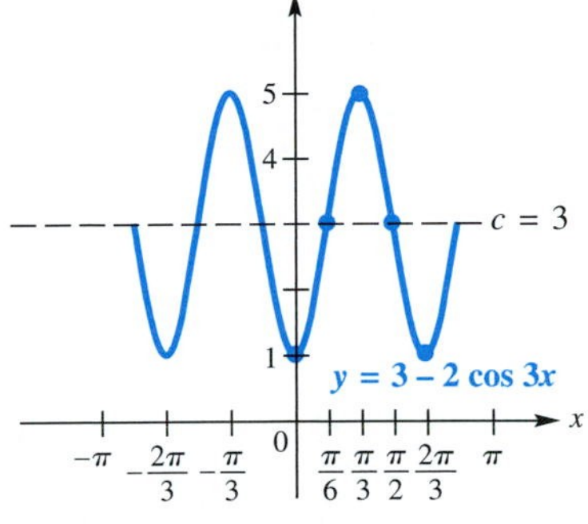

FIGURE 49

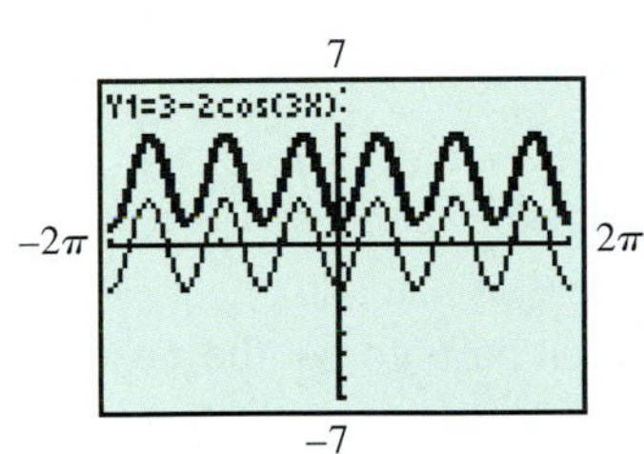

The function in Example 6 is shown using the thick graph style. Notice also the thin graph style for $y = -2 \cos 3x$.

The key points are shown on the graph in Figure 49, along with more of the graph, sketched using the fact that the function is periodic. ■

A function of the form

$$y = c + a \sin[b(x - d)] \quad \text{or} \quad y = c + a \cos[b(x - d)], \quad b > 0,$$

can be graphed according to the following guidelines.

Further Guidelines for Sketching Graphs of Sine and Cosine Functions

Method 1 Follow these steps.

Step 1 Find an interval whose length is one period $\frac{2\pi}{b}$ by solving the three-part inequality $0 \le b(x - d) \le 2\pi$.

Step 2 Divide the interval into four equal parts.

Step 3 Evaluate the function for each of the five x-values resulting from Step 2. The points will be maximum points, minimum points, and points that intersect the line $y = c$ ("middle" points of the wave).

Step 4 Plot the points found in Step 3, and join them with a sinusoidal curve having amplitude $|a|$.

Step 5 Draw the graph over additional periods, to the right and to the left, as needed.

Method 2 First graph the basic circular function. The amplitude of the function is $|a|$, and the period is $\frac{2\pi}{b}$. Then use translations to graph the desired function. The vertical translation is c units up if $c > 0$ and $|c|$ units down if $c < 0$. The horizontal translation (phase shift) is d units to the right if $d > 0$ and $|d|$ units to the left if $d < 0$.

EXAMPLE 7 Graphing $y = c + a\sin[b(x - d)]$

Graph $y = -1 + 2\sin(4x + \pi)$.

Solution We use the guidelines described by Method 1. First write the expression in the form $c + a\sin[b(x - d)]$ by rewriting $4x + \pi$ as $4\left(x + \frac{\pi}{4}\right)$:

$$y = -1 + 2\sin\left[4\left(x + \frac{\pi}{4}\right)\right]. \quad \text{Rewrite } 4x + \pi \text{ as } 4\left(x + \tfrac{\pi}{4}\right).$$

Step 1 Find an interval whose length is one period.

$$0 \le 4\left(x + \frac{\pi}{4}\right) \le 2\pi$$

$$0 \le x + \frac{\pi}{4} \le \frac{\pi}{2} \quad \text{Divide by 4.}$$

$$-\frac{\pi}{4} \le x \le \frac{\pi}{4} \quad \text{Subtract } \tfrac{\pi}{4}.$$

Step 2 Divide the interval given by $\left[-\frac{\pi}{4}, \frac{\pi}{4}\right]$ into four equal parts to get the x-values $-\frac{\pi}{4}, -\frac{\pi}{8}, 0, \frac{\pi}{8}$, and $\frac{\pi}{4}$.

Step 3 Make a table of values.

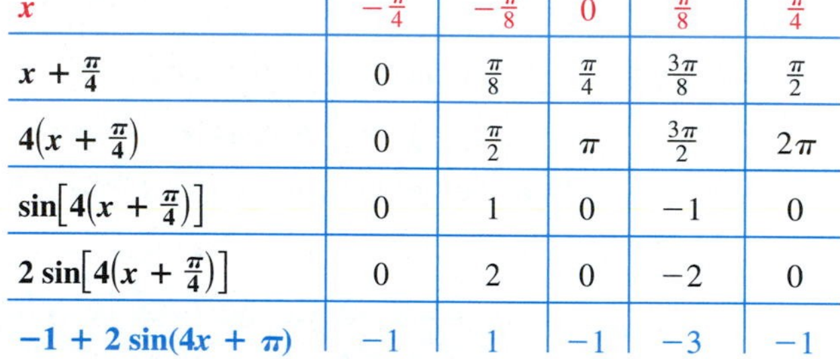

x	$-\frac{\pi}{4}$	$-\frac{\pi}{8}$	0	$\frac{\pi}{8}$	$\frac{\pi}{4}$
$x + \frac{\pi}{4}$	0	$\frac{\pi}{8}$	$\frac{\pi}{4}$	$\frac{3\pi}{8}$	$\frac{\pi}{2}$
$4\left(x + \frac{\pi}{4}\right)$	0	$\frac{\pi}{2}$	π	$\frac{3\pi}{2}$	2π
$\sin\left[4\left(x + \frac{\pi}{4}\right)\right]$	0	1	0	-1	0
$2\sin\left[4\left(x + \frac{\pi}{4}\right)\right]$	0	2	0	-2	0
$-1 + 2\sin(4x + \pi)$	-1	1	-1	-3	-1

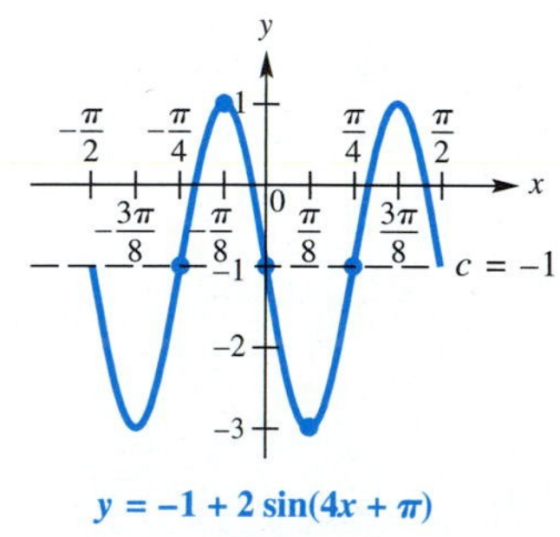

$y = -1 + 2\sin(4x + \pi)$

FIGURE 50

Steps 4 and 5 Plot the points found in the table and join them with a sinusoidal curve. Figure 50 shows the graph, extended to the right and left to include two full periods. ■

Determining a Trigonometric Model Using Curve Fitting

A sinusoidal function is often a good approximation of a set of periodic data points.

Month	°F	Month	°F
Jan	54	July	82
Feb	55	Aug	81
Mar	61	Sept	77
Apr	69	Oct	71
May	73	Nov	59
June	79	Dec	55

Source: Miller, A. and J. Thompson, *Elements of Meteorology*, Charles E. Merrill, 1975.

GCM **EXAMPLE 8 Modeling Temperature with a Sine Function**

The maximum monthly average temperature in New Orleans is 82°F and the minimum is 54°F. The table shows the monthly average temperatures. The scatter diagram for a 2-year interval in Figure 51 on the next page strongly suggests that the temperatures can be modeled with a sine curve.

(a) Using only the maximum and minimum temperatures, determine a function of the form $f(x) = a\sin[b(x - d)] + c$, where a, b, c, and d are constants, that models the monthly average temperature in New Orleans. Let x represent the month, with January corresponding to $x = 1$.

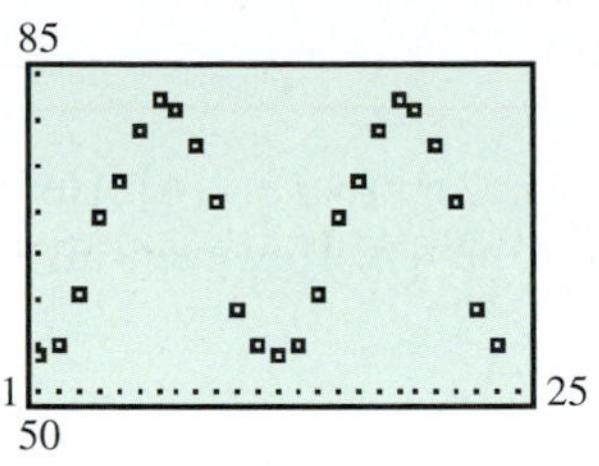

FIGURE 51

(b) On the same coordinate axes, graph f for a 2-year period, together with the actual data values found in the table.

(c) Use the *sine regression* feature of a graphing calculator to determine a second model for these data.

Solution

(a) We can use the maximum and minimum monthly average temperatures to find the amplitude a.

$$a = \frac{82 - 54}{2} = 14$$

The average of the maximum and minimum temperatures is a good choice for c.

$$c = \frac{82 + 54}{2} = 68.$$

Since temperatures repeat every 12 months, b is $\frac{2\pi}{12} = \frac{\pi}{6}$. The coldest month is January ($x = 1$), and the hottest month is July ($x = 7$), so we should choose d to be about 4. The table shows that temperatures are actually a little warmer after July than before, so we experiment with values just greater than 4 to find d. Trial and error with a calculator leads to $d = 4.2$. Thus,

$$f(x) = a \sin[b(x - d)] + c = 14 \sin\left[\frac{\pi}{6}(x - 4.2)\right] + 68.$$

(b) See Figure 52(a). The figure also shows the graph of $y = 14 \sin \frac{\pi}{6}x + 68$ for comparison. The horizontal translation of the model is fairly obvious here.

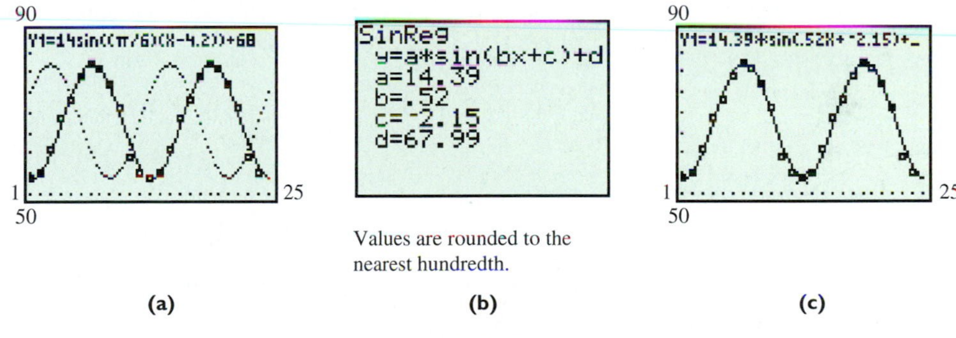

Values are rounded to the nearest hundredth.

(a) (b) (c)

FIGURE 52

(c) Using two years of the given data, the screen in Figure 52(b) shows the equation of the model, while Figure 52(c) shows its graph along with the data points. ■

NOTE If $f(x) = a \sin[b(x - d)] + c$ is used to model temperature data and if the greatest monthly average temperature is H and the least is L, then $a = \frac{1}{2}(H - L)$ and $c = \frac{1}{2}(H + L)$. If the period is 12 months, then $b = \frac{2\pi}{12}$ and if the maximum average temperature occurs in July, then $d \approx 4$.

What Went Wrong?

A student graphed $y = -1 + 3 \sin \frac{1}{2}x$ in the window $[-2\pi, 2\pi]$ by $[-4, 4]$. He knew that the amplitude should be 3 and the period should be 4π. However, his graph looked like this:

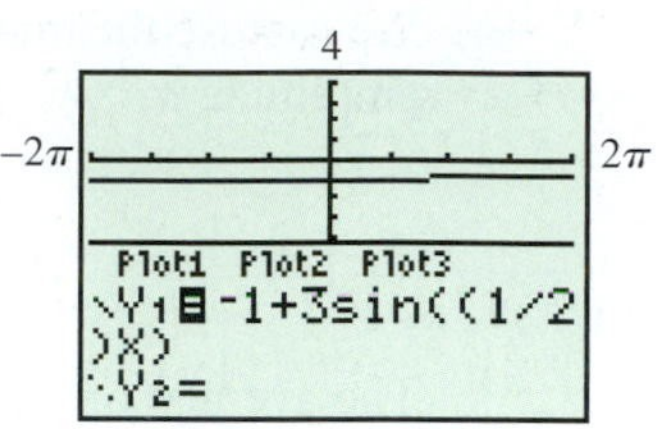

What Went Wrong? How can he obtain the correct graph?

Answer to What Went Wrong?

He graphed the function while the calculator was in *degree* mode. To obtain the correct graph, he should change the mode to *radian*.

8.3 Exercises

Concept Check *Without using a calculator, match each function defined in Exercises 1–8 with its graph in A–H.*

1. $y = \sin x$

2. $y = \cos x$

3. $y = -\sin x$

4. $y = -\cos x$

5. $y = \sin 2x$

6. $y = \cos 2x$

7. $y = 2 \sin x$

8. $y = 2 \cos x$

A.

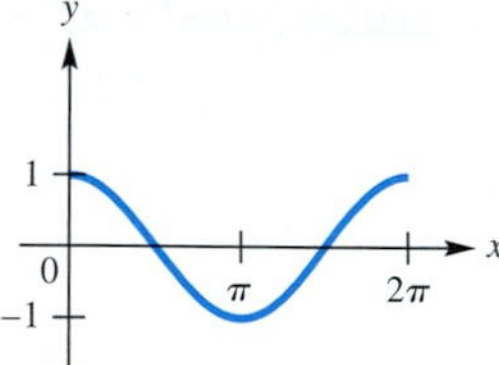

B.

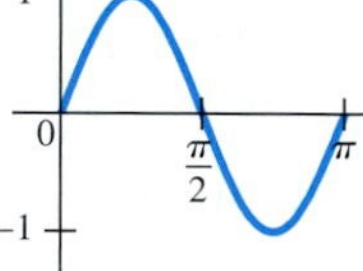

C.

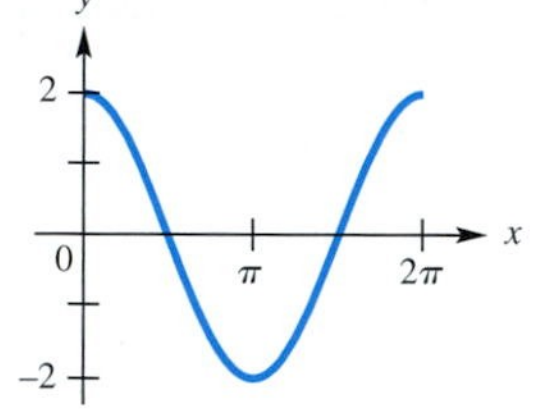

D.

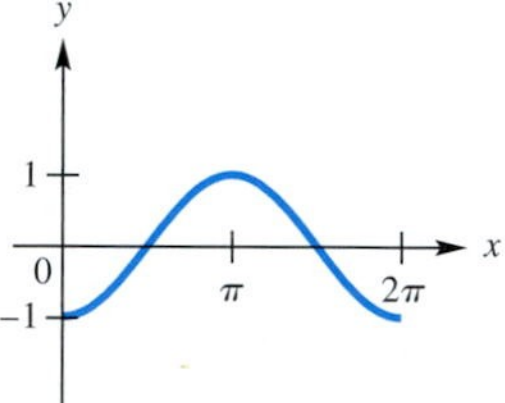

E.

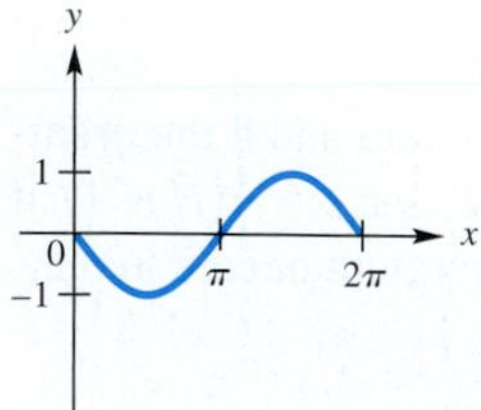

F.

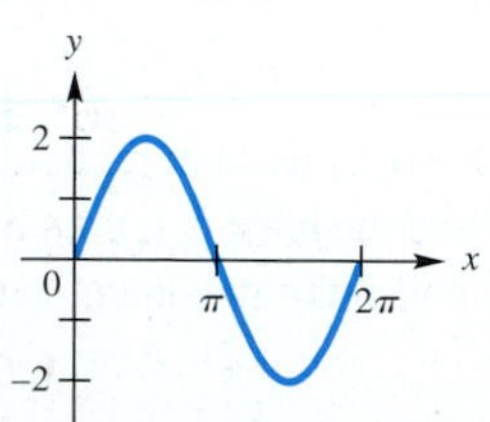

G.

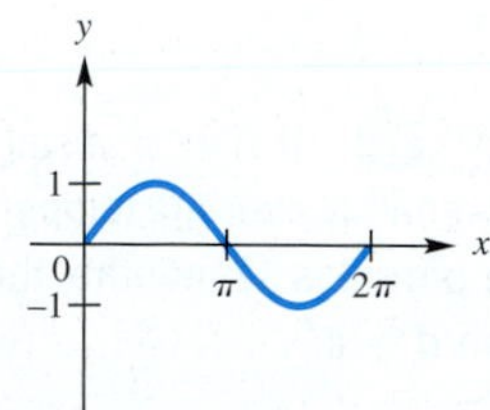

H.

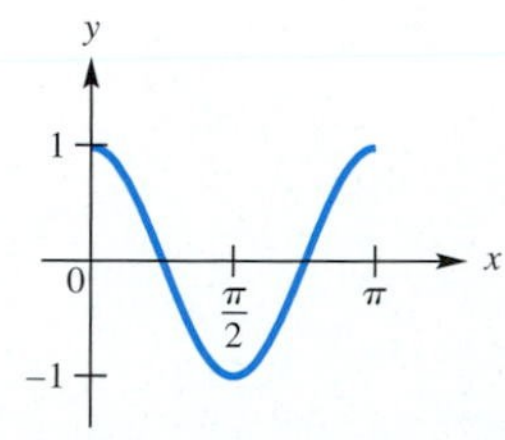

Concept Check *Without using a calculator, match each function defined in Exercises 9–16 with its graph in A–H.*

9. $y = \sin\left(x - \dfrac{\pi}{4}\right)$ **10.** $y = \sin\left(x + \dfrac{\pi}{4}\right)$ **11.** $y = \cos\left(x - \dfrac{\pi}{4}\right)$ **12.** $y = \cos\left(x + \dfrac{\pi}{4}\right)$

13. $y = 1 + \sin x$ **14.** $y = -1 + \sin x$ **15.** $y = 1 + \cos x$ **16.** $y = -1 + \cos x$

A.

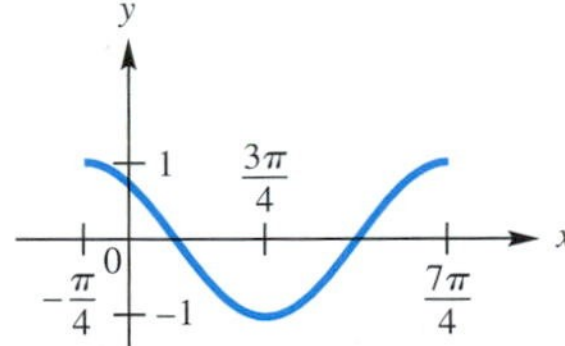

B.

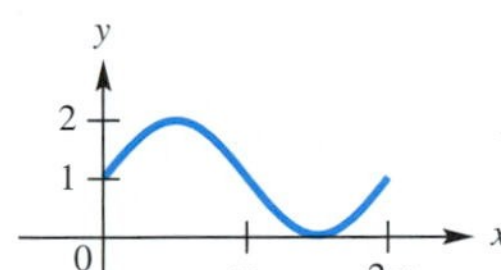

C.

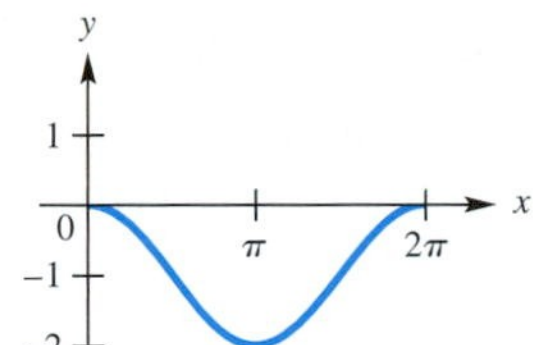

D.

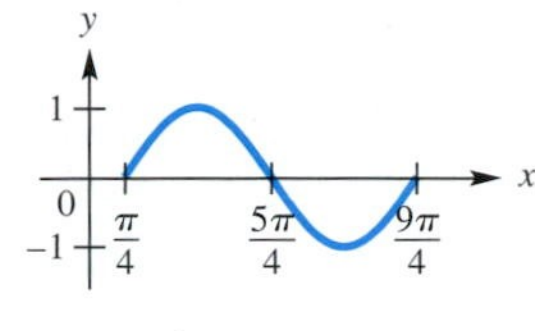

E.

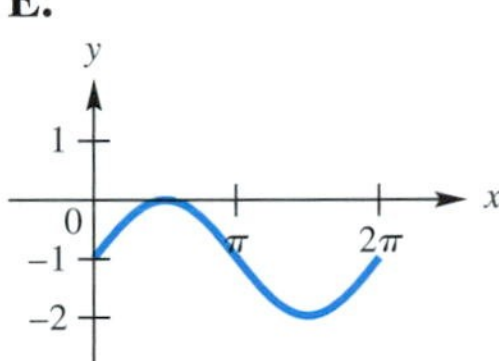

F.

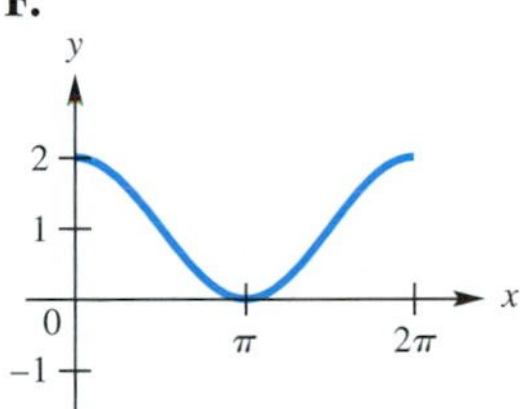

G.

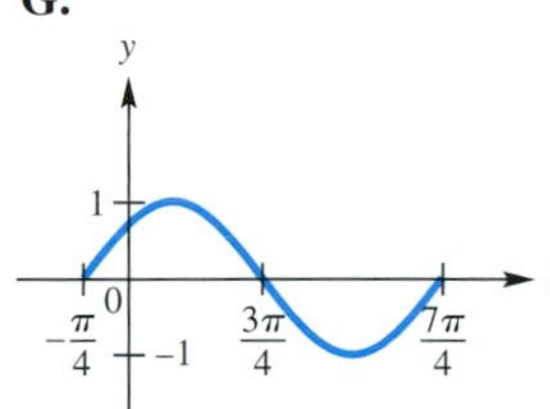

H.

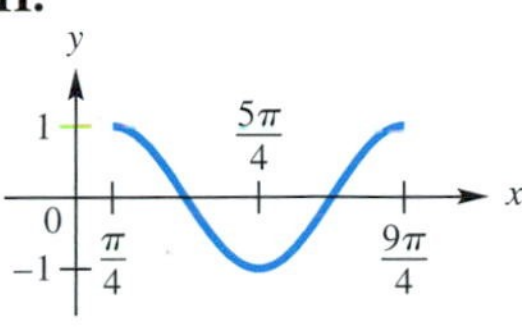

Concept Check *Match each function defined in Column I with the appropriate description in Column II.*

I	II
17. $y = 3\sin(2x - 4)$	**A.** Amplitude $= 2$; period $= \frac{\pi}{2}$; phase shift $= \frac{3}{4}$
18. $y = 2\sin(3x - 4)$	**B.** Amplitude $= 3$; period $= \pi$; phase shift $= 2$
19. $y = 4\sin(3x - 2)$	**C.** Amplitude $= 4$; period $= \frac{2\pi}{3}$; phase shift $= \frac{2}{3}$
20. $y = 2\sin(4x - 3)$	**D.** Amplitude $= 2$; period $= \frac{2\pi}{3}$; phase shift $= \frac{4}{3}$

Concept Check *In Exercises 21 and 22, give the equation of a sine function having the graph shown.*

21.

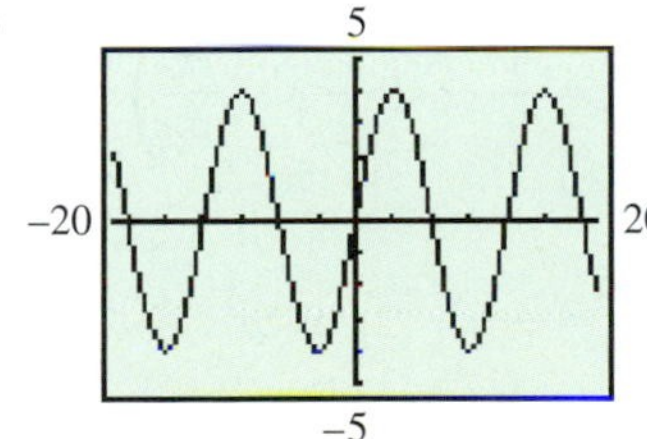


```
WINDOW
 Xmin=-20
 Xmax=20
 Xscl=3.1415926...
 Ymin=-5
 Ymax=5
 Yscl=1
 Xres=1
```

22.

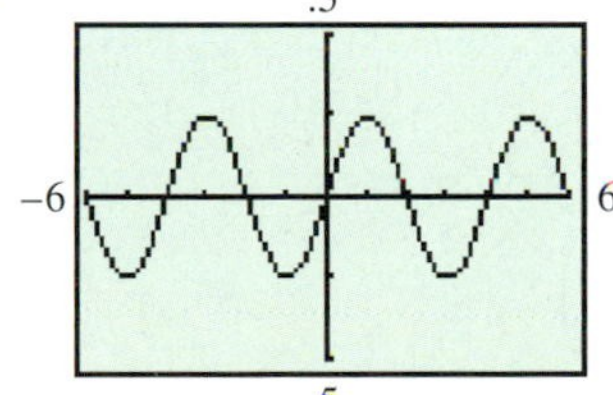

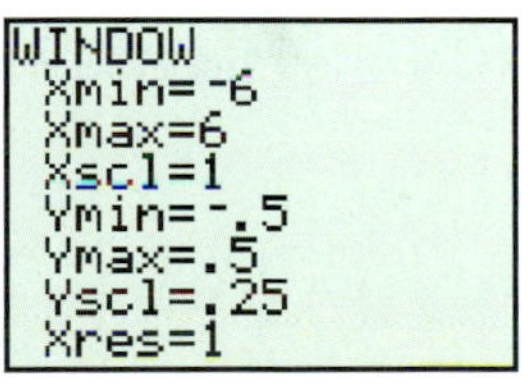

Relating Concepts

For individual or group investigation (Exercises 23–30)

Consider the function defined by $f(x) = -5 + 3\sin\left[2\left(x - \frac{\pi}{2}\right)\right]$. ***Answer Exercises 23–30 in order.***

23. Because the maximum value of the sine function is ________, the maximum value of $\sin\left[2\left(x - \frac{\pi}{2}\right)\right]$ is ________, the maximum value of $3\sin\left[2\left(x - \frac{\pi}{2}\right)\right]$ is ________, and thus the maximum value of $-5 + 3\sin\left[2\left(x - \frac{\pi}{2}\right)\right]$ is ________.

24. Because the minimum value of the sine function is ________, the minimum value of $\sin\left[2\left(x - \frac{\pi}{2}\right)\right]$ is ________, the minimum value of $3\sin\left[2\left(x - \frac{\pi}{2}\right)\right]$ is ________, and thus the minimum value of $-5 + 3\sin\left[2\left(x - \frac{\pi}{2}\right)\right]$ is ________.

(continued)

25. Based on the answers in Exercises 23 and 24, what is the range of f? Why will the trig viewing window as defined in the text not provide a comprehensive graph of f?

26. To obtain a comprehensive graph of f, Ymax must be at least ________ and Ymin must be at most ________.

27. Explain why using Xmin $= -2\pi$ and Xmax $= 2\pi$ will show exactly four periods of the graph of f.

28. Use your calculator to graph f in the window $[-2\pi, 2\pi]$ by $[-10, 5]$.

29. A "border" of empty space appears above and below the calculator graph from Exercise 28. To eliminate the border, what Ymin and Ymax values should you use to obtain a comprehensive graph?

30. Evaluate the approximate values of the function for $x = -2$ and for $x = -2 + \pi$. What is the value in each case? Why is this so?

Graph each function over the interval $[-2\pi, 2\pi]$. Give the amplitude.

31. $y = 2 \cos x$

32. $y = 3 \sin x$

33. $y = \frac{2}{3} \sin x$

34. $y = \frac{3}{4} \cos x$

35. $y = -\cos x$

36. $y = -\sin x$

37. $y = -2 \sin x$

38. $y = -3 \cos x$

Graph each function over a two-period interval. Give the period and amplitude.

39. $y = \sin \frac{1}{2}x$

40. $y = \sin \frac{2}{3}x$

41. $y = \cos 2x$

42. $y = \cos \frac{3}{4}x$

43. $y = 2 \sin \frac{1}{4}x$

44. $y = 3 \sin 2x$

45. $y = -2 \cos 3x$

46. $y = -5 \cos 2x$

Graph each function over a two-period interval. State the phase shift.

47. $y = \sin\left(x - \frac{\pi}{4}\right)$

48. $y = \cos\left(x - \frac{\pi}{3}\right)$

49. $y = 2 \cos\left(x - \frac{\pi}{3}\right)$

50. $y = 3 \sin\left(x - \frac{3\pi}{2}\right)$

Find the **(a)** *amplitude,* **(b)** *period,* **(c)** *phase shift (if any),* **(d)** *vertical translation (if any), and* **(e)** *range of each function. Then graph the function over at least one period.*

51. $y = -4 \sin(2x - \pi)$

52. $y = 3 \cos(4x + \pi)$

53. $y = \frac{1}{2} \cos\left(\frac{1}{2}x - \frac{\pi}{4}\right)$

54. $y = -\frac{1}{4} \sin\left(\frac{3}{4}x + \frac{\pi}{8}\right)$

55. $y = 1 - \frac{2}{3} \sin \frac{3}{4}x$

56. $y = -1 - 2 \cos 5x$

57. $y = 1 - 2 \cos \frac{1}{2}x$

58. $y = -3 + 3 \sin \frac{1}{2}x$

59. $y = -3 + 2 \sin\left(x + \frac{\pi}{2}\right)$

60. $y = 4 - 3 \cos(x - \pi)$

61. $y = \frac{1}{2} + \sin\left[2\left(x + \frac{\pi}{4}\right)\right]$

62. $y = -\frac{5}{2} + \cos\left[3\left(x - \frac{\pi}{6}\right)\right]$

63. $y = 2 \sin(x - \pi)$

64. $y = \frac{2}{3} \cos\left(x + \frac{\pi}{2}\right)$

65. $y = 4 \cos\left(\frac{1}{2}x + \frac{\pi}{2}\right)$

66. $y = -\cos\left[\frac{2}{3}\left(x - \frac{\pi}{3}\right)\right]$

67. $y = 2 - \sin\left(3x - \frac{\pi}{5}\right)$

68. $y = -1 + \frac{1}{2} \cos(2x - 3\pi)$

(Modeling) *Solve each problem.*

69. ***Average Temperatures*** The graph models the monthly average temperature y in degrees Fahrenheit for a city in Canada, where x is the month.

(a) Find the maximum and minimum monthly average temperatures.

(b) Find the amplitude and period. Interpret the results.

(c) Explain what the x-intercepts represent.

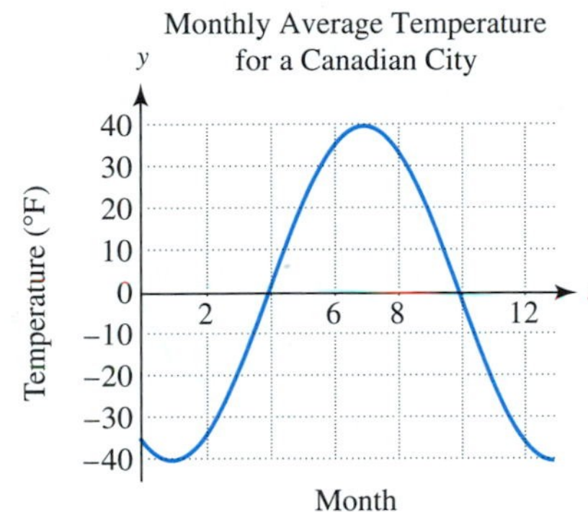

70. ***Average Temperatures*** The graph in Exercise 69 is given by $y = 40\cos\left[\frac{\pi}{6}(x - 7)\right]$. Modify this equation to model the following situations.

(a) The maximum monthly average temperature is 50°F and the minimum is −50°F.

(b) The maximum monthly average temperature is 60°F and the minimum is −20°F.

(c) The maximum monthly average temperature occurs in August and the minimum occurs in February.

71. ***Ocean Temperatures*** The graph models the Gulf of Mexico water temperatures in degrees Fahrenheit at St. Petersburg, Florida. (*Source:* J. Williams.)

(a) Estimate the maximum and minimum water temperatures. When do they occur?

(b) What would happen to the amplitude of the graph if the minimum water temperature decreased to 50°F? Make a sketch of this situation.

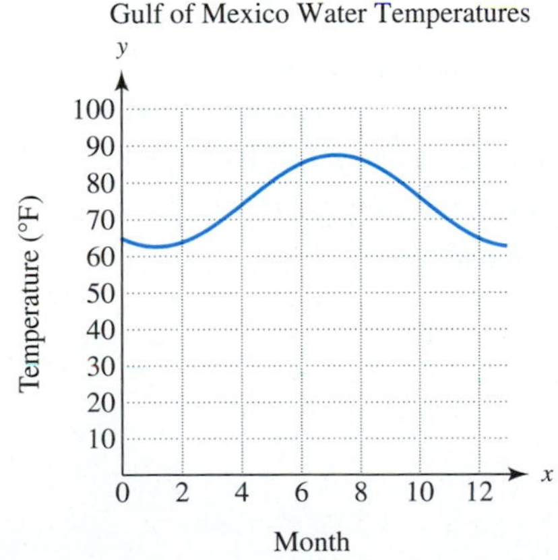

72. ***Ocean Temperatures*** The graph in Exercise 71 is described by the equation

$$y = 12.4\sin\left[\frac{\pi}{6}(x - 4.2)\right] + 75.$$

Modify this equation to model the following situations.

(a) The monthly average water temperatures vary between 60°F and 90°F.

(b) The monthly average water temperatures vary between 50°F and 70°F.

Tides for Kahului Harbor *The graph shows the tides for Kahului Harbor, on the island of Maui, Hawaii. To identify high and low tides and times for other Maui areas, the following adjustments must be made.*

Hana:	High, +40 minutes, +.1 foot;
	Low, +18 minutes, −.2 foot
Makena:	High, +1:21, −.5 foot;
	Low, +1:09, −.2 foot
Maalaea:	High, +1:52, −.1 foot;
	Low, +1:19, −.2 foot
Lahaina:	High, +1:18, −.2 foot;
	Low, +1:01, −.1 foot

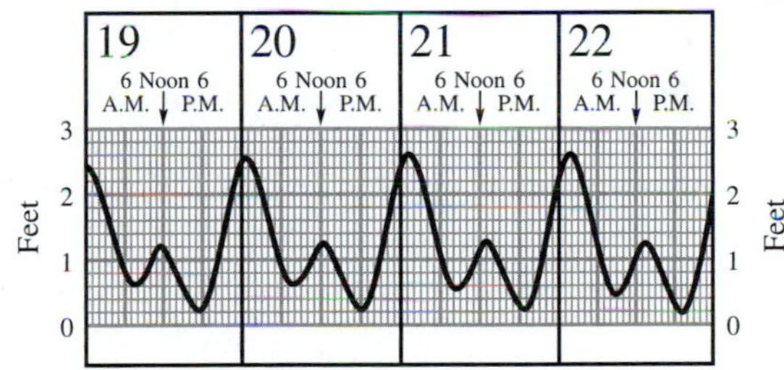

Source: Maui News. Original chart prepared by Edward K. Noda and Associates.

Use the graph to work Exercises 73–78.

73. The graph is an example of a periodic function. What is the period (in hours)?

74. What is the amplitude?

75. At what time on January 20 was low tide at Kahului? What was the height?

76. Repeat Exercise 75 for Maalaea.

77. At what time on January 22 was high tide at Kahului? What was the height?

78. Repeat Exercise 77 for Lahaina.

79. ***Annual Average Temperatures*** Scientists believe that the annual average temperatures in a given location make up a periodic function. The average temperature at a given place during a given season fluctuates as time goes on, from colder to warmer and back to colder. The graph shows an idealized description of the temperature (in °F) for the last few thousand years of a location at the same latitude as Anchorage, AK.

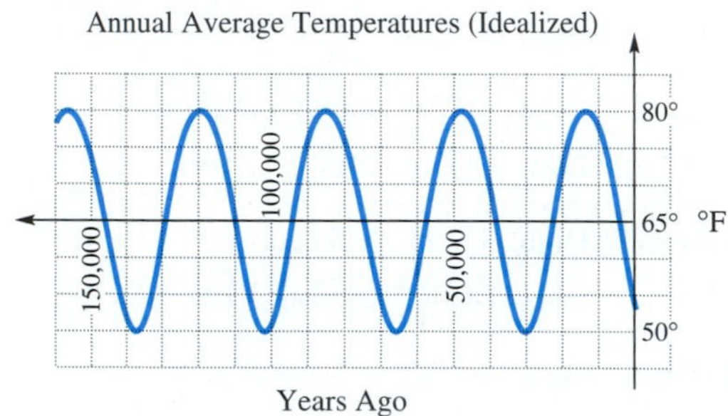

(a) Find the highest and lowest annual average temperatures recorded.

(b) Use these two numbers to find the amplitude.

(c) Find the period of the function.

(d) What is the trend of the temperature now?

80. ***Blood Pressure Variation*** The graph gives the variation in blood pressure for a typical person. Systolic and diastolic pressures are the upper and lower limits of the periodic changes in pressure that produce the pulse. The length of time between peaks is called the period of the pulse.

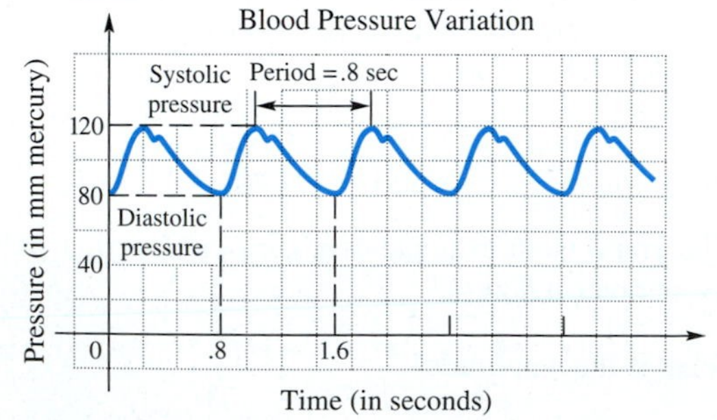

(a) Find the amplitude of the graph.

(b) Find the pulse rate (the number of pulse beats in one minute) for this person.

81. ***Activity of a Nocturnal Animal*** Many activities of living organisms are periodic. For example, the graph shows the times that a certain nocturnal animal begins its evening activity.

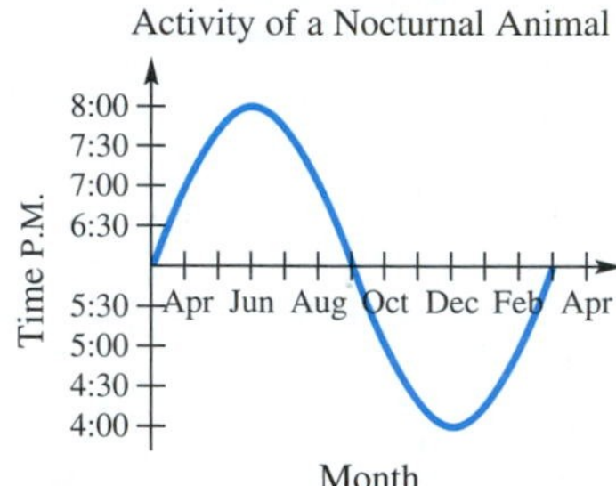

(a) Find the amplitude of this graph.

(b) Find the period.

82. ***Voltage*** The voltage E in an electrical circuit is given by the equation

$$E = 5 \cos 120\pi t,$$

where t is measured in seconds.

(a) Find the amplitude and the period.

(b) How many cycles are completed in 1 second? (The number of cycles (periods) completed in 1 second is the *frequency* of the function.)

(c) Find E when $t = 0, .03, .06, .09,$ and $.12$.

(d) Graph E for $0 \leq t \leq \frac{1}{30}$.

83. ***Voltage*** For another electrical circuit, the voltage E is modeled by

$$E = 3.8 \cos 40\pi t,$$

where t is time measured in seconds.

(a) Find the amplitude and the period.

(b) Find the frequency. See Exercise 82(b).

(c) Find E when $t = .02, .04, .08, .12,$ and $.14$.

(d) Graph one period of E.

84. ***Atmospheric Carbon Dioxide*** At Mauna Loa, Hawaii, atmospheric carbon dioxide levels in parts per million (ppm) have been measured regularly since 1958. The function defined by

$$L(x) = .022x^2 + .55x + 316 + 3.5\sin(2\pi x)$$

can be used to model these levels, where x is in years and $x = 0$ corresponds to 1960. (*Source:* Nilsson, A., *Greenhouse Earth,* John Wiley and Sons, 1992.)

(a) Graph L for $15 \leq x \leq 35$. (*Hint:* For the range, use $325 \leq y \leq 365$.)

(b) When do the seasonal maximum and minimum carbon dioxide levels occur?

(c) L is the sum of a quadratic function and a sine function. What is the significance of each of these functions? Discuss what physical phenomena may be responsible for each function.

85. ***Atmospheric Carbon Dioxide*** Refer to Exercise 84. The carbon dioxide content in the atmosphere at Barrow, Alaska, in parts per million (ppm) can be modeled with the function defined by

$$C(x) = .04x^2 + .6x + 330 + 7.5\sin(2\pi x),$$

where $x = 0$ corresponds to 1970. (*Source:* Zeilik, M., S. Gregory, and E. Smith, *Introductory Astronomy and Astrophysics,* Fourth Edition, Saunders College Publishers, 1998.)

(a) Graph C for $5 \leq x \leq 25$. (*Hint:* Use $320 \leq y \leq 380$.)

(b) Discuss possible reasons the amplitude of the oscillations in the graph of C is larger than the amplitude of the oscillations in the graph of L that models Hawaii in Exercise 84.

(c) Define a new function C that is valid if x represents the actual year, where $1970 \leq x \leq 1995$.

86. ***Monthly Average Temperatures*** The monthly average temperatures (in °F) in Vancouver, Canada, are shown in the table.

Month	°F	Month	°F
Jan	36	July	64
Feb	39	Aug	63
Mar	43	Sept	57
Apr	48	Oct	50
May	55	Nov	43
June	59	Dec	39

Source: Miller, A. and J. Thompson, *Elements of Meteorology*, Charles E. Merrill, 1975.

(a) Plot the monthly average temperatures over a 2-year period by letting $x = 1$ correspond to the month of January during the first year. Do the data seem to indicate a translated sine graph?

(b) The highest monthly average temperature is 64°F in July, and the lowest monthly average temperature is 36°F in January. Their average is 50°F. Graph the data together with the line $y = 50$. What does this line represent with regard to temperature in Vancouver?

(c) Approximate the amplitude, period, and phase shift of the translated sine wave indicated by the data.

(d) Determine a sinusoidal function of the form $f(x) = a\sin[b(x - d)] + c$, where a, b, c, and d are constants, that models the data.

(e) Graph f together with the data on the same coordinate axes. How well does f model the given data?

(f) Use the sine regression capability of a graphing calculator to find the equation of a sine curve that fits these data.

87. ***Monthly Average Temperatures*** The monthly average temperatures (in °F) in Phoenix, Arizona, are shown in the table.

Month	°F	Month	°F
Jan	51	July	90
Feb	55	Aug	90
Mar	63	Sept	84
Apr	67	Oct	71
May	77	Nov	59
June	86	Dec	52

Source: Miller, A. and J. Thompson, *Elements of Meteorology*, Charles E. Merrill, 1975.

(a) Predict the yearly average temperature and compare it with the actual value of 70°F.

(b) Plot the monthly average temperature over a 2-year period by letting $x = 1$ correspond to January of the first year.

(c) Determine a sinusoidal function of the form $f(x) = a\cos[b(x - d)] + c$, where a, b, c, and d are constants, that models the data.

(d) Graph f together with the data on the same coordinate axes. How well does f model the data?

(e) Use the sine regression capability of a graphing calculator to find the equation of a sine curve that fits these data.

*In an article entitled "I Found Sinusoids in My Gas Bill" (*Mathematics Teacher*, January 2000), Cathy G. Schloemer presents the following graph that accompanied her gas bill.*

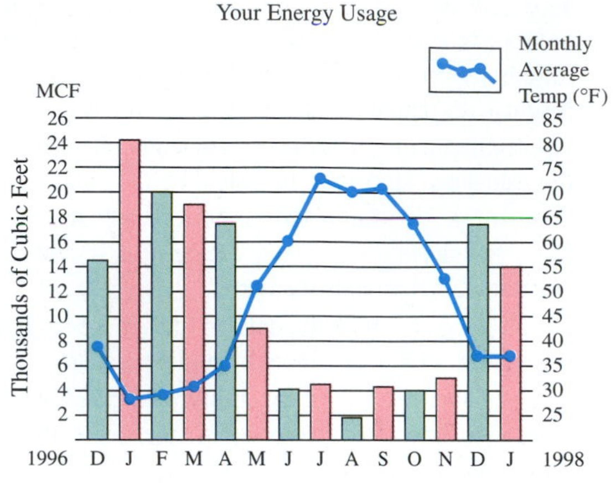

Notice that two sinusoids are suggested here: one for the behavior of the monthly average temperature and another for gas use in MCF (thousands of cubic feet).

Use the graph to answer Exercises 88 and 89.

88. If January 1997 is represented by $x = 1$, the data of estimated ordered pairs (month, temperature) are given in the lists shown on the two screens.

L1	L2	L3 1
1	28	------
2	29	
3	31	
4	35	
5	51	
6	60	
7	73	
L1(1) = 1		

L1	L2	L3 1
7	73	
8	70	
9	71	
10	64	
11	53	
12	37	

L1(13) =		

Use the sine regression feature of a graphing calculator to find a sine function that fits these data points. Then make a scatter diagram, and graph the function.

89. Again, if January 1997 is represented by $x = 1$, the data of estimated ordered pairs (month, gas use in MCF) are given in the lists shown on the two screens.

L1	L2	L3 1
1	24.2	------
2	20	
3	18.8	
4	17.5	
5	9.2	
6	4.2	
7	4.8	
L1(1) = 1		

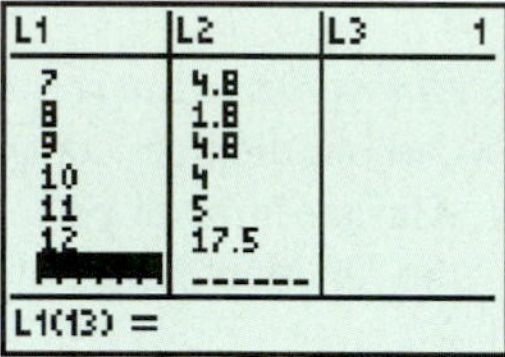

L1	L2	L3 1
7	4.8	
8	1.8	
9	4.8	
10	4	
11	5	
12	17.5	

L1(13) =		

Use the sine regression feature of a graphing calculator to find a sine function that fits these data points. Then make a scatter diagram, and graph the function.

8.4 Graphs of the Other Circular Functions

Graphs of the Cosecant and Secant Functions ■ Graphs of the Tangent and Cotangent Functions ■ Addition of Ordinates

Graphs of the Cosecant and Secant Functions

Since cosecant values are reciprocals of the corresponding sine values, the period of the function $y = \csc x$ is 2π, the same as for $y = \sin x$. When $\sin x = 1$, the value of $\csc x$ is also 1, and when $0 < \sin x < 1$, $\csc x > 1$. Also, if $-1 < \sin x < 0$, then $\csc x < -1$. (Verify these statements.) As $|x|$ approaches 0, $|\sin x|$ approaches 0, and $|\csc x|$ gets larger and larger. The graph of $y = \csc x$ approaches the vertical line $x = 0$ but never touches it, so the line $x = 0$ is a *vertical asymptote*. In fact, the lines $x = n\pi$, where n is any integer, are all vertical asymptotes.

Using the preceding information and plotting a few points shows that the graph takes the shape of the solid curve shown in Figure 53. To indicate how the two graphs are related, the graph of $y = \sin x$ is shown as a dashed curve.

GCM TECHNOLOGY NOTE

We use dot mode to graph the cosecant and secant functions on a graphing calculator to get an accurate picture. (If connected mode is used, the calculator will attempt to connect points that are actually separated by vertical asymptotes.) We enter csc x as $\frac{1}{\sin x}$ or $(\sin x)^{-1}$, and sec x as $\frac{1}{\cos x}$ or $(\cos x)^{-1}$.

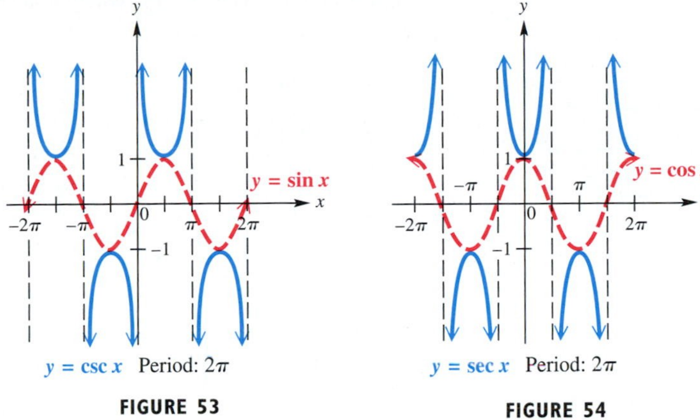

FIGURE 53

FIGURE 54

A similar analysis for the secant leads to the solid curve shown in Figure 54.

X	Y_1	Y_2
0	ERROR	ERROR
.3927	2.6131	-2.613
.7854	1.4142	-1.414
1.1781	1.0824	-1.082
1.5708	1	-1
1.9635	1.0824	-1.082
2.3562	1.4142	-1.414

X=0

For X in radians, the Y_1 and Y_2 columns give values of $Y_1 = \csc X$ and $Y_2 = \csc(-X)$ $\left(\Delta\text{Tbl} = \frac{\pi}{8}\right)$. The results suggest that, like the sine function, $\csc(-X) = -\csc X$.

FUNCTION CAPSULE

COSECANT FUNCTION $\quad f(x) = \csc x$

Domain: $\{x \mid x \neq n\pi$, where n is an integer$\}$ $\quad$ Range: $(-\infty, -1] \cup [1, \infty)$

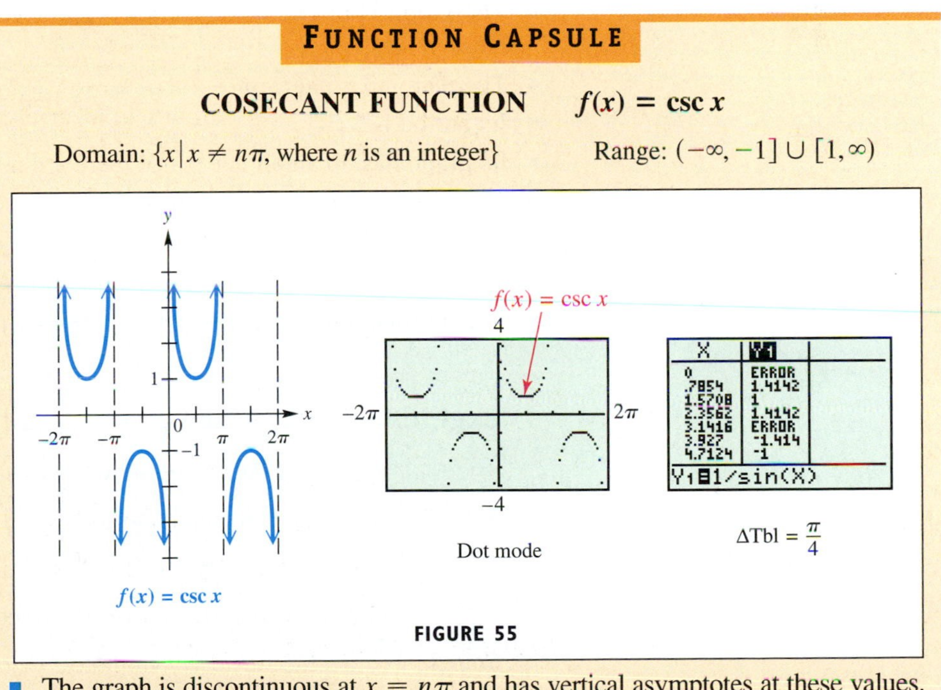

FIGURE 55

- The graph is discontinuous at $x = n\pi$ and has vertical asymptotes at these values.
- There are no x-intercepts.
- Its period is 2π.
- Its graph has no amplitude, since there are no maximum or minimum values.
- The graph is symmetric with respect to the origin, so it is an odd function. That is, for all x in the domain, $\csc(-x) = -\csc x$.

X	Y_1	Y_2
0	1	1
.3927	1.0824	1.0824
.7854	1.4142	1.4142
1.1781	2.6131	2.6131
1.5708	ERROR	ERROR
1.9635	−2.613	−2.613
2.3562	−1.414	−1.414

X=0

The table shows values of $Y_1 = \sec X$ and $Y_2 = \sec(-X)$. These values suggest that, like the cosine function, $\sec(-X) = \sec X$.

FUNCTION CAPSULE

SECANT FUNCTION $\quad f(x) = \sec x$

Domain: $\{x \mid x \neq (2n + 1)\frac{\pi}{2}$, where n is an integer$\}$ $\qquad$ Range: $(-\infty, -1] \cup [1, \infty)$

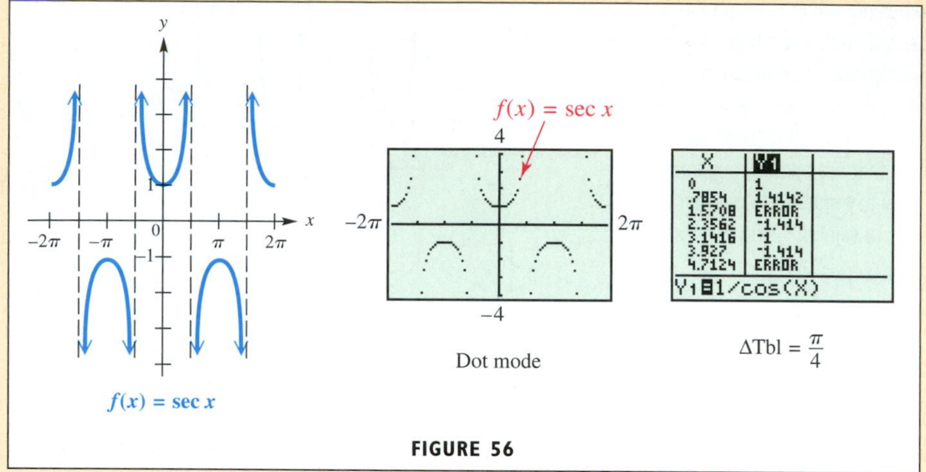

FIGURE 56

- The graph is discontinuous at $x = (2n + 1)\frac{\pi}{2}$ and has vertical asymptotes at these values.
- There are no x-intercepts.
- Its period is 2π.
- Its graph has no amplitude, since there are no maximum or minimum values.
- The graph is symmetric with respect to the y-axis, so it is an even function. That is, for all x in the domain, $\sec(-x) = \sec x$.

Use the following guidelines when graphing the cosecant and secant functions.

Sketching Graphs of the Cosecant and Secant Functions

To graph $\boldsymbol{y = a \csc bx}$ or $\boldsymbol{y = a \sec bx}$, with $b > 0$,

Step 1 Graph the corresponding reciprocal function as a guide.

For $y = a \csc bx$, $\quad$ use $y = a \sin bx$ as a guide.

For $y = a \sec bx$, $\quad$ use $y = a \cos bx$ as a guide.

Step 2 Sketch the vertical asymptotes, whose equations have the form $x = k$, where k is an x-intercept of the graph of the guide function.

Step 3 Sketch the graph of the desired function by drawing the typical U-shaped branches between the adjacent asymptotes. The branches will be above the graph of the guide function when the guide function values are positive and below the graph of the guide function when the guide function values are negative. The graph will resemble those in Figures 55 and 56.

Like graphs of the sine and cosine functions, graphs of the cosecant and secant functions may be translated vertically and horizontally.

EXAMPLE 1 Graphing $y = a \sec bx$

Graph $y = 2 \sec \frac{1}{2}x$.

Solution

Step 1 This function involves the secant, so the corresponding reciprocal function involves the cosine. The guide function to graph is

$$y = 2 \cos \frac{1}{2}x.$$

Using the guidelines of Section 8.3, we find that one period of the graph lies along the interval that satisfies the inequality

$$0 \leq \frac{1}{2}x \leq 2\pi, \quad \text{or} \quad 0 \leq x \leq 4\pi. \qquad \text{Multiply by 2.}$$

Dividing this interval into four equal parts gives the key points

$$(\mathbf{0,2}), \quad (\boldsymbol{\pi},\mathbf{0}), \quad (\mathbf{2}\boldsymbol{\pi},\mathbf{-2}), \quad (\mathbf{3}\boldsymbol{\pi},\mathbf{0}), \quad \text{and} \quad (\mathbf{4}\boldsymbol{\pi},\mathbf{2}),$$

which are joined with a smooth dashed curve to indicate that this graph is only a guide. An additional period is graphed, as seen in Figure 57(a).

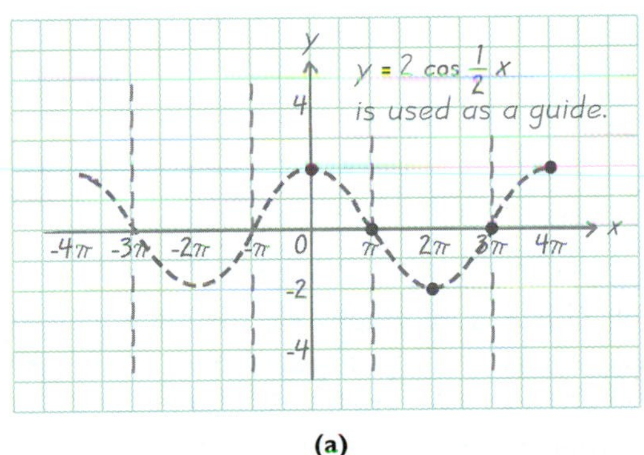

(a)

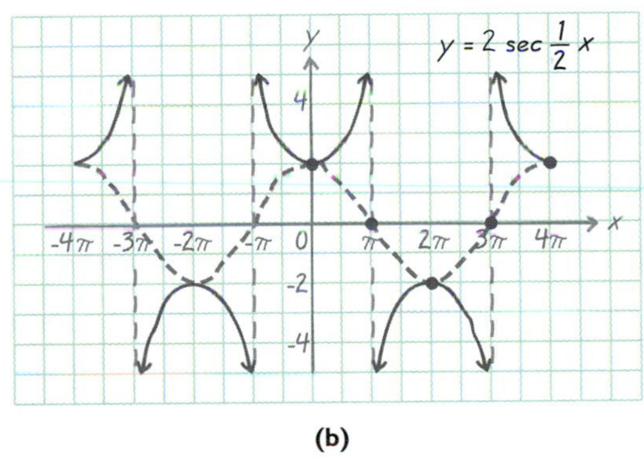

(b)

FIGURE 57

6

Y1=2/cos((1/2)X)

-4π 4π

-6

Dot mode

Compare this graph to the dark graph in Figure 57(b).

Step 2 We sketch the vertical asymptotes as shown in Figure 57(a). These occur at x-values for which the guide function equals 0, such as

$$\boldsymbol{x = -3\pi, \quad x = -\pi, \quad x = \pi,} \quad \text{and} \quad \boldsymbol{x = 3\pi.}$$

Step 3 We sketch the graph of $y = 2 \sec \frac{1}{2}x$ by drawing the typical U-shaped branches, approaching the asymptotes. See Figure 57(b). ■

NOTE In Figure 57(b), the vertical asymptotes and the dashed curve of $y = 2 \cos \frac{1}{2}x$ are ***not*** part of the graph of $y = 2 \sec \frac{1}{2}x$. They are aids that can be used when graphing by hand. The actual graph of $y = 2 \sec \frac{1}{2}x$ consists of only the solid curve.

EXAMPLE 2 Graphing $y = a\csc(x - d)$

Graph $y = \frac{3}{2}\csc\left(x - \frac{\pi}{2}\right)$.

Analytic Solution

First we graph the corresponding guide function

$$y = \frac{3}{2}\sin\left(x - \frac{\pi}{2}\right).$$

Compared with the graph of $y = \sin x$, the graph of $y = \frac{3}{2}\sin\left(x - \frac{\pi}{2}\right)$ has amplitude $\frac{3}{2}$ and phase shift $\frac{\pi}{2}$ units to the right. Its graph is shown as a dashed curve in Figure 58. The x-intercepts of the sinusoidal graph of $y = \frac{3}{2}\sin\left(x - \frac{\pi}{2}\right)$ correspond to the vertical asymptotes found in the graph of $y = \frac{3}{2}\csc\left(x - \frac{\pi}{2}\right)$. These asymptotes are shown as vertical dashed lines. We sketch the graph of $y = \frac{3}{2}\csc\left(x - \frac{\pi}{2}\right)$ by using the sinusoidal graph of $y = \frac{3}{2}\sin\left(x - \frac{\pi}{2}\right)$ as a guide. Two periods are shown in Figure 58.

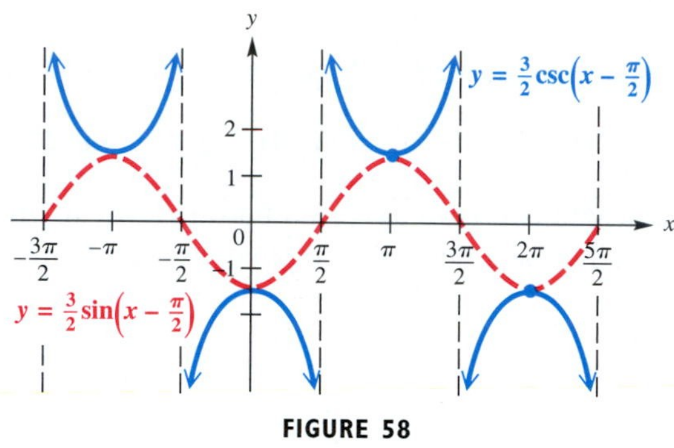

FIGURE 58

Graphing Calculator Solution

Figure 59 shows graphs of

$$Y_1 = \frac{3}{2}\left[1/\sin\left(X - \frac{\pi}{2}\right)\right].$$

Connected mode draws vertical lines appearing between the portions of the graph, while dot mode does not.

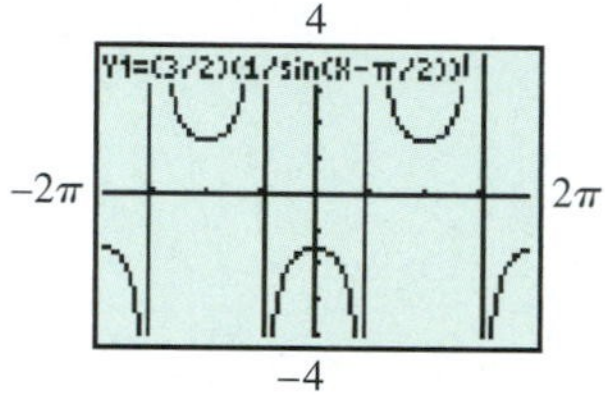

Connected mode

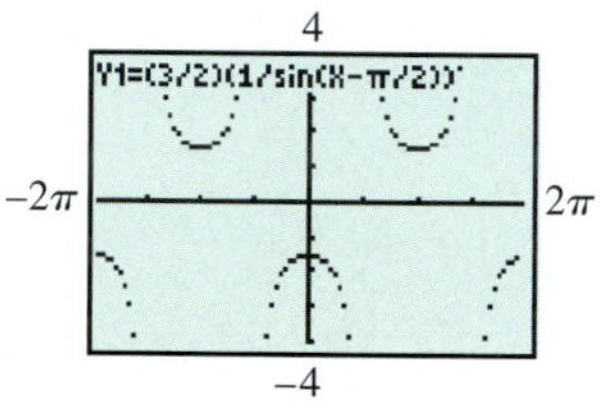

Dot mode

FIGURE 59

Graphs of the Tangent and Cotangent Functions

Unlike the four functions whose graphs we studied previously, the tangent function has period π. Because $\tan x = \frac{\sin x}{\cos x}$, tangent values are 0 when sine values are 0 and are undefined when cosine values are 0. (Verify this.) As x-values go from $-\frac{\pi}{2}$ to $\frac{\pi}{2}$, tangent values go from $-\infty$ to ∞ and increase throughout the interval. Those same values repeat as x goes from $\frac{\pi}{2}$ to $\frac{3\pi}{2}$, $\frac{3\pi}{2}$ to $\frac{5\pi}{2}$, and so on. The graph of $y = \tan x$ from $-\pi$ to $\frac{3\pi}{2}$ is shown in Figure 60.

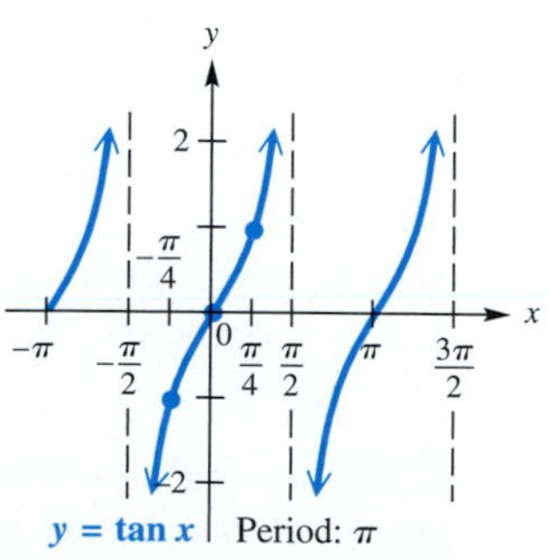

FIGURE 60

The cotangent function also has period π. Because $\cot x = \frac{\cos x}{\sin x}$, cotangent values are 0 when cosine values are 0 and are undefined when sine values are 0. As x-values go from 0 to π, cotangent values go from ∞ to $-\infty$ and decrease throughout the interval. Those same values repeat as x goes from π to 2π, 2π to 3π, and so on. The graph of $y = \cot x$ from $-\pi$ to π is shown in Figure 61.

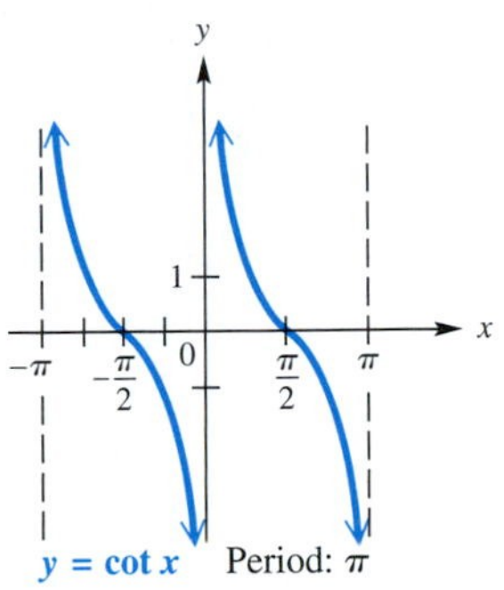

FIGURE 61

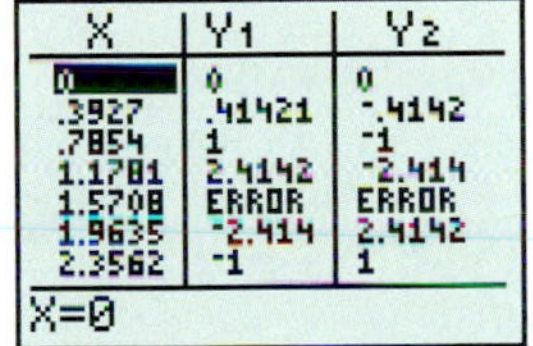

X	Y1	Y2
0	0	0
.3927	.41421	-.4142
.7854	1	-1
1.1781	2.4142	-2.414
1.5708	ERROR	ERROR
1.9635	-2.414	2.4142
2.3562	-1	1

X=0

The table shows values of $Y_1 = \tan X$ and $Y_2 = \tan(-X)$. These values suggest that $\tan(-X) = -\tan X$.

FUNCTION CAPSULE

TANGENT FUNCTION $f(x) = \tan x$

Domain: $\{x \mid x \neq (2n + 1)\frac{\pi}{2}$, where n is an integer$\}$ Range: $(-\infty, \infty)$

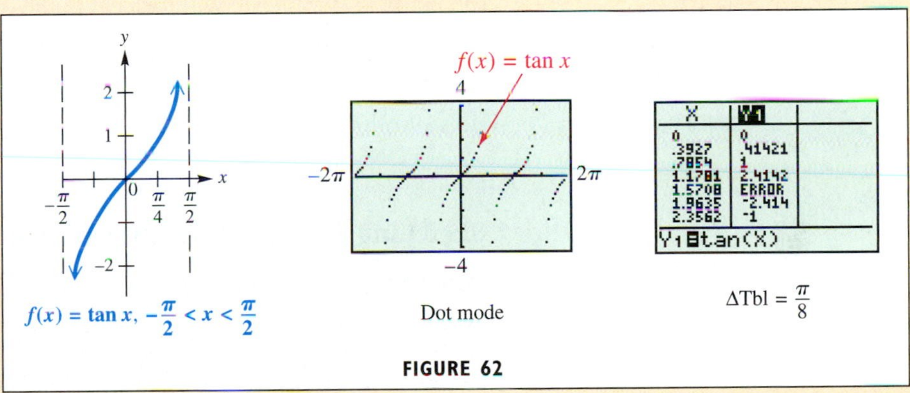

FIGURE 62

- The graph is discontinuous at $x = (2n + 1)\frac{\pi}{2}$ and has vertical asymptotes at these values.
- The x-intercepts are of the form $x = n\pi$.
- Its period is π.
- Its graph has no amplitude, since there are no minimum or maximum values.
- The graph is symmetric with respect to the origin, so it is an odd function. That is, for all x in the domain, $\tan(-x) = -\tan x$.

NOTE The sine, cosine, secant, and cosecant functions all have period 2π. Only the tangent and cotangent have period π.

X	Y1	Y2
0	ERROR	ERROR
.3927	2.4142	-2.414
.7854	1	-1
1.1781	.41421	-.4142
1.5708	0	0
1.9635	-.4142	.41421
2.3562	-1	1

X=0

In this table, $Y_1 = \cot X$ and $Y_2 = \cot(-X)$. As with the tangent function, the table suggests that $\cot(-X) = -\cot X$.

FUNCTION CAPSULE

COTANGENT FUNCTION $f(x) = \cot x$

Domain: $\{x \mid x \neq n\pi$, where n is an integer$\}$ Range: $(-\infty, \infty)$

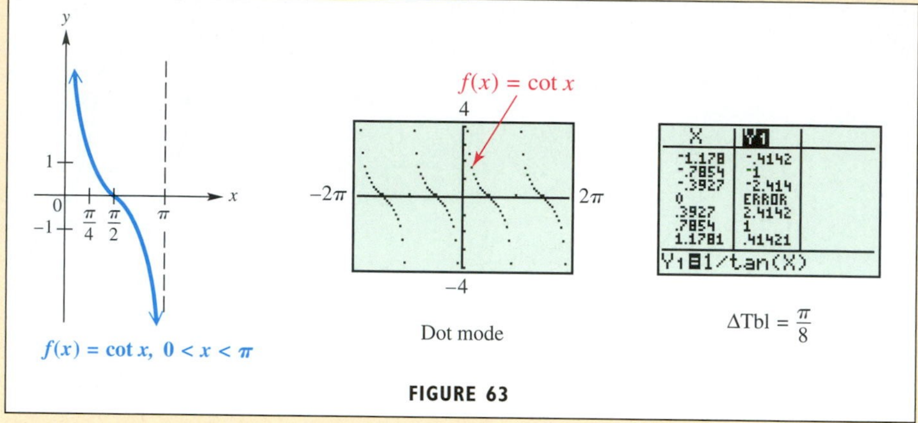

FIGURE 63

- The graph is discontinuous at $x = n\pi$ and has vertical asymptotes at these values.
- The x-intercepts are of the form $x = (2n + 1)\frac{\pi}{2}$.
- Its period is π.
- Its graph has no amplitude, since there are no minimum or maximum values.
- The graph is symmetric with respect to the origin, so it is an odd function. That is, for all x in the domain, $\cot(-x) = -\cot x$.

Use the following guidelines when graphing the tangent and cotangent functions.

Sketching Graphs of the Tangent and Cotangent Functions

To graph $\boldsymbol{y = a \tan bx}$ or $\boldsymbol{y = a \cot bx}$, with $b > 0$,

Step 1 The period is $\frac{\pi}{b}$. To locate two adjacent vertical asymptotes, solve the following equations for x:

For $y = a \tan bx$: $bx = -\frac{\pi}{2}$ and $bx = \frac{\pi}{2}$.

For $y = a \cot bx$: $bx = 0$ and $bx = \pi$.

Step 2 Sketch the two vertical asymptotes found in Step 1.

Step 3 Divide the interval formed by the vertical asymptotes into four equal parts.

Step 4 Evaluate the function for the first-quarter point, midpoint, and third-quarter point, using the x-values found in Step 3.

Step 5 Join the points with a smooth curve, approaching the vertical asymptotes. Indicate additional asymptotes and periods as necessary. The graph will resemble those in Figures 62 and 63.

EXAMPLE 3 Graphing $y = \tan bx$

Graph $y = \tan 2x$ using the preceding guidelines.

Solution

Step 1 The period of this function is $\frac{\pi}{2}$. To locate two adjacent vertical asymptotes, solve $2x = -\frac{\pi}{2}$ and $2x = \frac{\pi}{2}$ (since this is a tangent function). The two asymptotes have equations $x = -\frac{\pi}{4}$ and $x = \frac{\pi}{4}$.

Step 2 Sketch the two vertical asymptotes $x = \pm\frac{\pi}{4}$, as shown in Figure 64.

Step 3 Divide the interval $\left(-\frac{\pi}{4}, \frac{\pi}{4}\right)$ into four equal parts to obtain key x-values:

first-quarter value: $\mathbf{-\frac{\pi}{8}}$, middle value: **0**, third-quarter value: $\mathbf{\frac{\pi}{8}}$.

Step 4 Evaluate the function for the x-values found in Step 3.

x	$-\frac{\pi}{8}$	0	$\frac{\pi}{8}$
$2x$	$-\frac{\pi}{4}$	0	$\frac{\pi}{4}$
$\tan 2x$	-1	0	1

Step 5 Join these points with a smooth curve, approaching the vertical asymptotes. See Figure 64. Another period has been graphed, one half-period to the left and one half-period to the right.

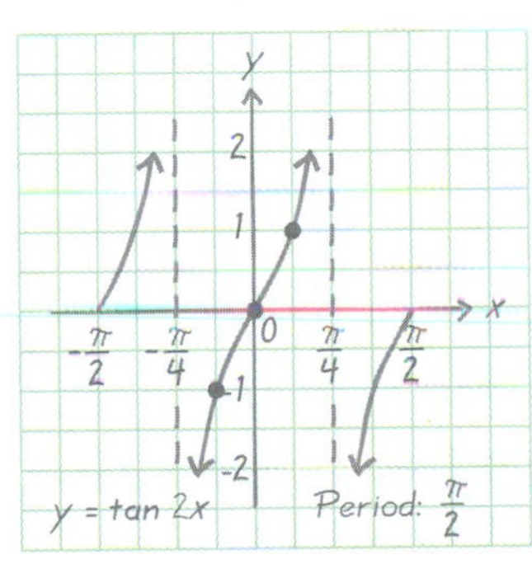

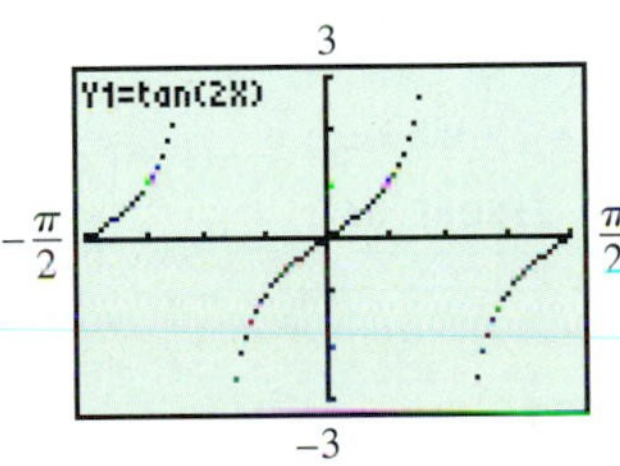

Dot mode

FIGURE 64

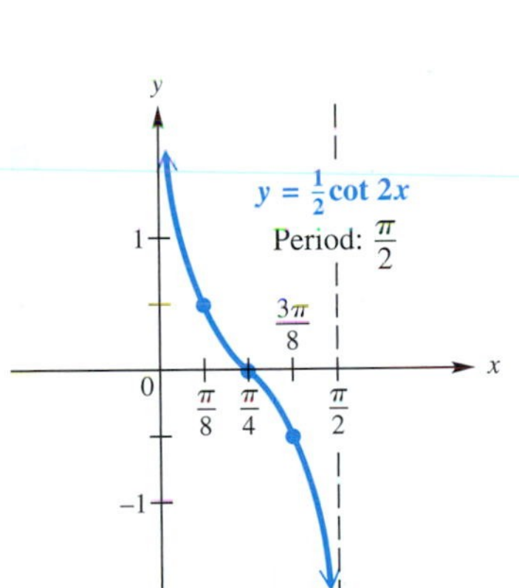

3

Y1=(1/2)(1/tan(2X))

0 π

−3

Dot mode

FIGURE 65

EXAMPLE 4 Graphing $y = a \cot bx$

Graph $y = \dfrac{1}{2} \cot 2x$.

Solution Because this function involves the cotangent, we can locate two adjacent asymptotes by solving the equations $2x = 0$ and $2x = \pi$. The lines $x = 0$ (the y-axis) and $x = \frac{\pi}{2}$ are two such asymptotes. We divide the interval $0 < x < \frac{\pi}{2}$ into four equal parts, getting key x-values of $\frac{\pi}{8}$, $\frac{\pi}{4}$, and $\frac{3\pi}{8}$. Evaluating the function at these x-values gives the key points $\left(\frac{\pi}{8}, \frac{1}{2}\right)$, $\left(\frac{\pi}{4}, 0\right)$, and $\left(\frac{3\pi}{8}, -\frac{1}{2}\right)$. Joining these points with a smooth curve approaching the asymptotes gives the graph shown in Figure 65. ■

As with other functions, the graphs of the tangent and cotangent functions may be translated horizontally as well as vertically.

EXAMPLE 5 Graphing a Tangent Function with a Vertical Translation

Graph $y = 2 + \tan x$.

Analytic Solution

Every value of y for this function will be 2 units more than the corresponding value of y in $y = \tan x$, causing the graph of $y = 2 + \tan x$ to be translated upward 2 units compared with the graph of $y = \tan x$. See Figure 66.

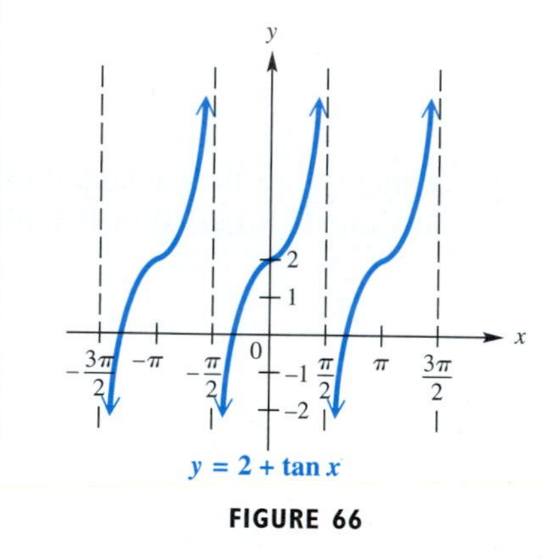

FIGURE 66

Graphing Calculator Solution

To see the vertical translation, observe the coordinates displayed at the bottoms of the screens in Figures 67 and 68. For $x = \frac{\pi}{4} \approx .78539816$,

$$Y_1 = \tan X = 1,$$

while for the same x-value,

$$Y_2 = 2 + \tan X = 2 + 1 = 3.$$

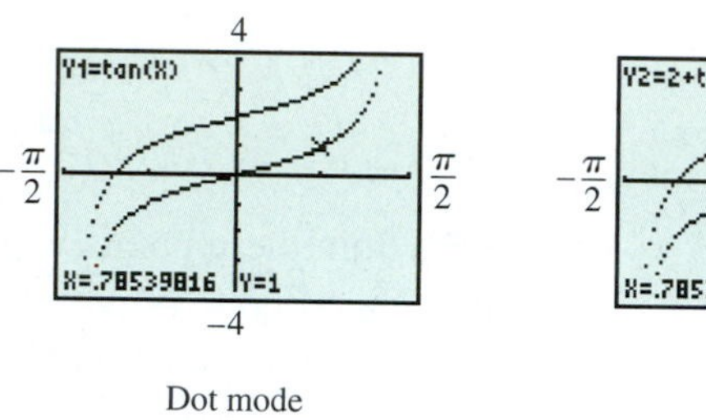

Dot mode

FIGURE 67

Dot mode

FIGURE 68

■

EXAMPLE 6 Graphing a Cotangent Function with Vertical and Horizontal Translations

Graph $y = -2 - \cot\left(x - \frac{\pi}{4}\right)$.

Solution Here, $b = 1$, so the period is π. The graph will be translated downward 2 units (because $c = -2$), will be reflected across the x-axis (because of the negative sign in front of the cotangent), and will have a phase shift to the right $\frac{\pi}{4}$ unit $\left(\text{because of the argument } \left(x - \frac{\pi}{4}\right)\right)$. To locate adjacent asymptotes, since this function involves the cotangent, we solve the following equations:

$$x - \frac{\pi}{4} = 0, \quad \text{so } x = \frac{\pi}{4} \qquad \text{and} \qquad x - \frac{\pi}{4} = \pi, \quad \text{so } x = \frac{5\pi}{4}.$$

Dividing the interval $\frac{\pi}{4} < x < \frac{5\pi}{4}$ into four equal parts and evaluating the function at the three key x-values within the interval gives the points $\left(\frac{\pi}{2}, -3\right)$, $\left(\frac{3\pi}{4}, -2\right)$, and $(\pi, -1)$. Join these points with a smooth curve. This period of the graph, along with the one in the domain interval $\left(-\frac{3\pi}{4}, \frac{\pi}{4}\right)$, is shown in Figure 69. ■

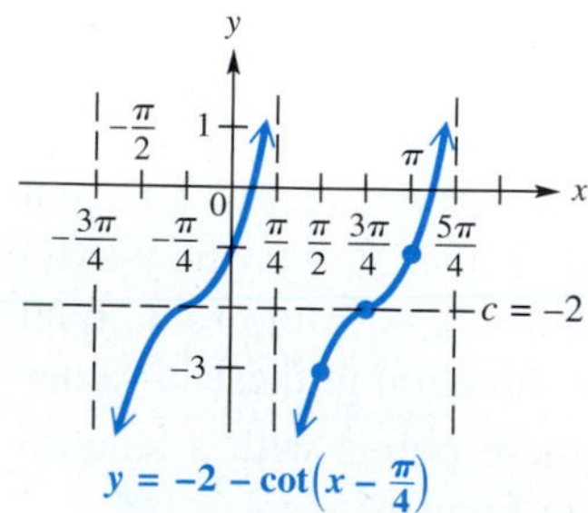

FIGURE 69

Addition of Ordinates

New functions can be formed by adding or subtracting other functions. A function formed by combining two other functions, such as

$$y = \cos x + \sin x,$$

has historically been graphed with a method known as *addition of ordinates*. (The x-value of a point is called its *abscissa*, while its y-value is called its *ordinate*.) To apply that method to this function, we would graph the functions $y = \cos x$ and $y = \sin x$. Then, for selected values of x, we would add $\cos x$ and $\sin x$, and plot the points $(x, \cos x + \sin x)$. Connecting the selected points with a sinusoidal curve gives the graph of the desired function. While this method illustrates some valuable concepts involving the arithmetic of functions, it is time consuming.

The technique is easily illustrated with graphing calculators. Let $Y_1 = \cos X$, $Y_2 = \sin X$, and $Y_3 = Y_1 + Y_2$. Figure 70 shows the result when Y_1 and Y_2 are graphed with thin lines and $Y_3 = \cos X + \sin X$ is graphed with a thick line. Notice that for $X = \frac{\pi}{6} \approx .52359878$, $Y_1 + Y_2 = Y_3$.

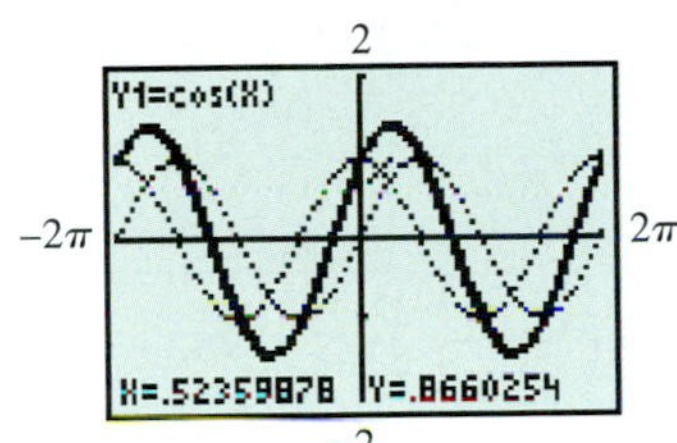

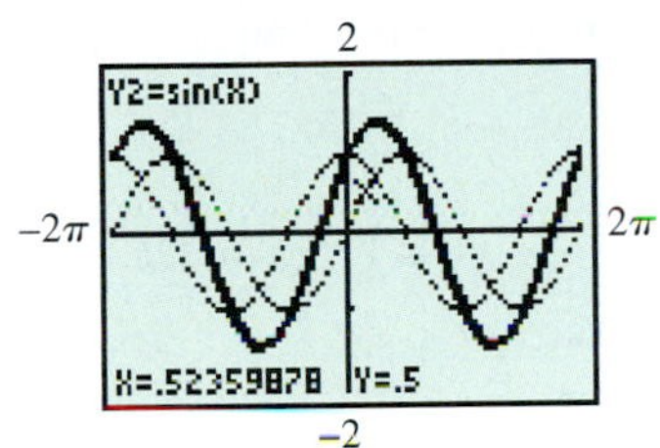

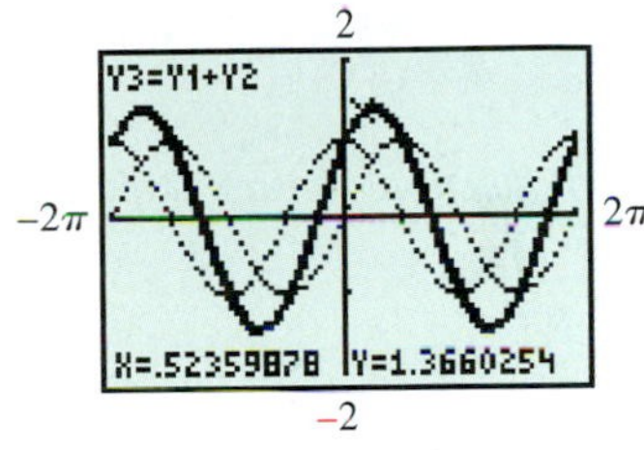

FIGURE 70

8.4 Exercises

Concept Check *Without using a calculator, match each function defined in Exercises 1–6 with its graph in A–F.*

1. $y = -\csc x$

2. $y = -\sec x$

3. $y = -\tan x$

4. $y = -\cot x$

5. $y = \tan\left(x - \frac{\pi}{4}\right)$

6. $y = \cot\left(x - \frac{\pi}{4}\right)$

A.

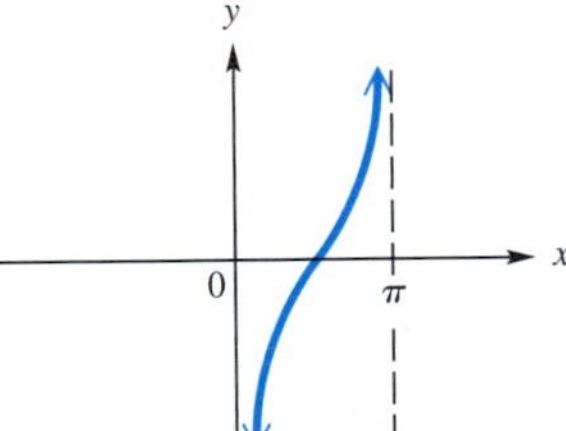

B.

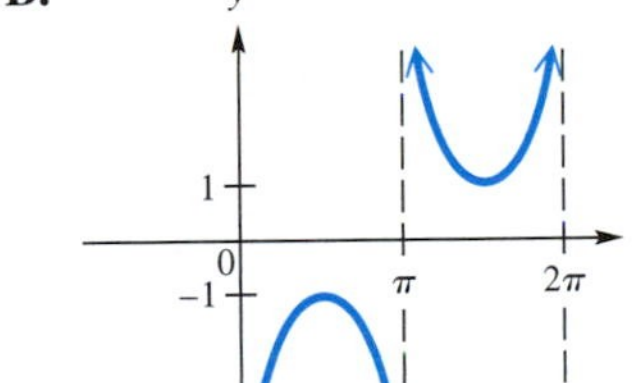

C.

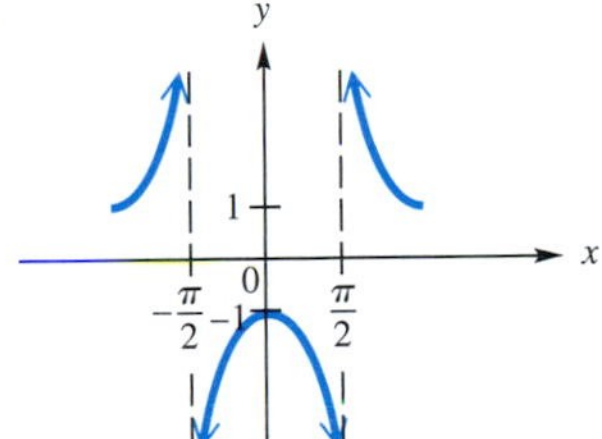

D.

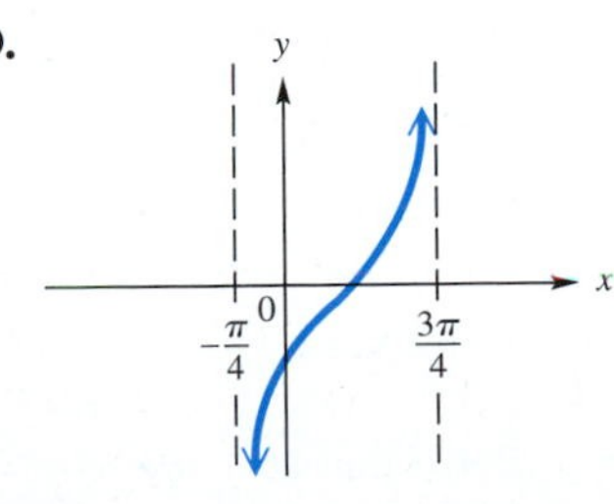

E.

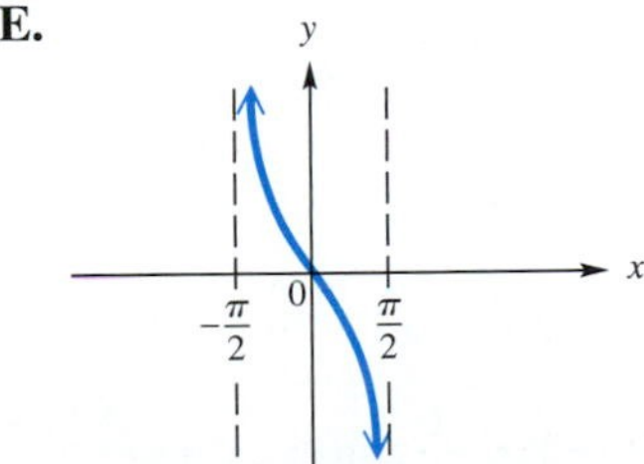

F.

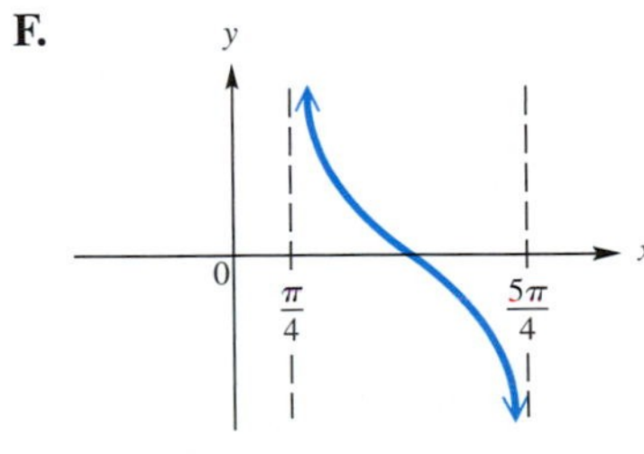

Concept Check *Tell whether each statement is* true *or* false. *If* false, *tell why.*

7. The smallest positive number k for which $x = k$ is an asymptote for the tangent function is $\frac{\pi}{2}$.

8. The smallest positive number k for which $x = k$ is an asymptote for the cotangent function is $\frac{\pi}{2}$.

9. The tangent and secant functions are undefined for the same values.

10. The secant and cosecant functions are undefined for the same values.

11. The graph of $y = \tan x$ in Figure 62 suggests that $\tan(-x) = \tan x$ for all x in the domain of $\tan x$.

12. The graph of $y = \sec x$ in Figure 56 suggests that $\sec(-x) = \sec x$ for all x in the domain of $\sec x$.

Find the **(a)** *period,* **(b)** *phase shift (if any), and* **(c)** *range of each function.*

13. $y = 2 \csc \frac{1}{2}x$

14. $y = 3 \csc 2x$

15. $y = -2 \sec\left(x + \frac{\pi}{2}\right)$

16. $y = -\frac{3}{2} \sec(x - \pi)$

17. $y = \frac{5}{2} \cot\left[\frac{1}{3}\left(x - \frac{\pi}{2}\right)\right]$

18. $y = -3 \tan\left[\frac{1}{2}\left(x + \frac{\pi}{4}\right)\right]$

19. $f(x) = \frac{1}{2} \sec(2x + \pi)$

20. $f(x) = -\frac{1}{3} \csc\left(\frac{1}{2}x - \frac{\pi}{2}\right)$

21. $y = -1 - \tan\left(x + \frac{\pi}{4}\right)$

22. $y = 2 + \cot\left(2x - \frac{\pi}{3}\right)$

Graph each function over a one-period interval.

23. $y = 3 \sec \frac{1}{4}x$

24. $y = -2 \sec \frac{1}{2}x$

25. $y = -\frac{1}{2} \csc\left(x + \frac{\pi}{2}\right)$

26. $y = \frac{1}{2} \csc\left(x - \frac{\pi}{2}\right)$

27. $y = \csc\left(x - \frac{\pi}{4}\right)$

28. $y = \sec\left(x + \frac{3\pi}{4}\right)$

29. $y = \sec\left(x + \frac{\pi}{4}\right)$

30. $y = \csc\left(x + \frac{\pi}{3}\right)$

31. $y = \sec\left(\frac{1}{2}x + \frac{\pi}{3}\right)$

32. $y = \csc\left(\frac{1}{2}x - \frac{\pi}{4}\right)$

33. $y = 2 + 3 \sec(2x - \pi)$

34. $y = 1 - 2 \csc\left(x + \frac{\pi}{2}\right)$

35. $y = 1 - \frac{1}{2} \csc\left(x - \frac{3\pi}{4}\right)$

36. $y = 2 + \frac{1}{4} \sec\left(\frac{1}{2}x - \pi\right)$

Graph each function over a one-period interval.

37. $y = \tan 4x$

38. $y = \tan \frac{1}{2}x$

39. $y = 2 \tan x$

40. $y = 2 \cot x$

41. $y = 2 \tan \frac{1}{4}x$

42. $y = \frac{1}{2} \cot x$

43. $y = \cot 3x$

44. $y = -\cot \frac{1}{2}x$

45. $y = -2 \tan \frac{1}{4}x$

46. $y = 3 \tan \frac{1}{2}x$

47. $y = \frac{1}{2} \cot 4x$

48. $y = -\frac{1}{2} \cot 2x$

Graph each function over a two-period interval.

49. $y = \tan(2x - \pi)$

50. $y = \tan\left(\frac{x}{2} + \pi\right)$

51. $y = \cot\left(3x + \frac{\pi}{4}\right)$

52. $y = \cot\left(2x - \frac{3\pi}{2}\right)$

53. $y = 1 + \tan x$

54. $y = -2 + \tan x$

55. $y = 1 - \cot x$

56. $y = -2 - \cot x$

57. $y = -1 + 2 \tan x$

58. $y = 3 + \frac{1}{2} \tan x$

59. $y = -1 + \frac{1}{2} \cot(2x - 3\pi)$

60. $y = -2 + 3 \tan(4x + \pi)$

61. *Concept Check* If c is any number, then how many solutions does the equation $c = \tan x$ have in the interval $(-2\pi, 2\pi]$?

62. *Concept Check* If c is any number such that $-1 < c < 1$, then how many solutions does the equation $c = \sec x$ have over the entire domain of the secant function?

Solve each problem.

63. Graph $y = \tan x$ and $y = x$ simultaneously in the window $[-1, 1]$ by $[-1, 1]$. Write a sentence or two describing the relationship of $\tan x$ to x for small x-values.

64. Between each pair of successive asymptotes, a portion of the graph of $y = \sec x$ or $y = \csc x$ resembles a parabola. Can each of these portions actually be a parabola? Explain.

65. *(Modeling) Distance of a Rotating Beacon* A rotating beacon is located at point A next to a long wall. See the figure at the top of the next column. The beacon is 4 meters from the wall. The distance d is given by

$$d = 4 \tan 2\pi t,$$

where t is time measured in seconds since the beacon started rotating. (When $t = 0$, the beacon is aimed at point R. When the beacon is aimed to the right of R, the value of d is positive; d is negative if the beacon is aimed to the left of R.)

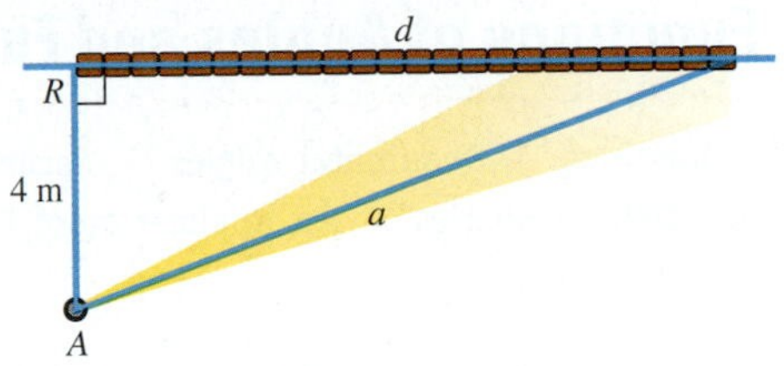

Find d for each time.

(a) $t = 0$ **(b)** $t = .2$

(c) $t = .8$ **(d)** $t = 1.2$

(e) Why is .25 a meaningless value for t?

66. *(Modeling) Distance of a Rotating Beacon* In the figure for Exercise 65, the distance a is given by

$$a = 4|\sec 2\pi t|.$$

Find a for each time.

(a) $t = 0$ **(b)** $t = .86$ **(c)** $t = 1.24$

Addition of Ordinates *Use a graphing calculator to graph Y_1, Y_2, and $Y_1 + Y_2$ on the same screen. Evaluate each of the three functions at $x = \frac{\pi}{6}$, and verify that* $Y_1(\frac{\pi}{6}) + Y_2(\frac{\pi}{6}) = (Y_1 + Y_2)(\frac{\pi}{6})$.

67. $Y_1 = \sin X$; $Y_2 = \sin 2X$

68. $Y_1 = \cos X$; $Y_2 = \cos 2X$

69. $Y_1 = \sin 2X$; $Y_2 = \cos \frac{1}{2}X$

70. $Y_1 = \tan X$; $Y_2 = \tan 2X$

Reviewing Basic Concepts (Sections 8.3 and 8.4)

1. Graph one period of $y = -\cos x$. State the period and the amplitude.

2. Graph $y = 3 \sin(\pi x + \pi)$ on the interval $-2 \le x \le 2$. State the amplitude, period, and phase shift.

3. *(Modeling) Daylight Hours* The graph of

$$f(x) = 6.5 \sin\left[\frac{\pi}{6}(x - 3.65)\right] + 12.4$$

models the daylight hours at 60°N latitude, where $x = 1$ corresponds to January 1, $x = 2$ to February 1, and so on.

(a) Estimate the maximum and minimum number of daylight hours.

(b) Interpret the amplitude and period.

Graph one period of each function. State the period, phase shift, domain, and range.

4. $y = \csc\left(x - \frac{\pi}{4}\right)$

5. $y = \sec\left(x + \frac{\pi}{4}\right)$

Graph one period of each function. State the period, domain, and range.

6. $y = 2 \tan x$

7. $y = 2 \cot x$

8.5 Functions of Angles and Fundamental Identities

Trigonometric Functions ■ Quadrantal Angles ■ Reciprocal Identities ■ Signs and Ranges of Function Values ■ Pythagorean Identities ■ Quotient Identities ■ An Application of Trigonometric Functions

Trigonometric Functions

Earlier in this chapter, we defined the trigonometric functions with real number domains. We now view them from a different perspective. We start with an angle θ in standard position and choose any point P having coordinates (x, y) on the terminal side of angle θ. (Point P must *not* be the vertex of the angle.) See Figure 71.

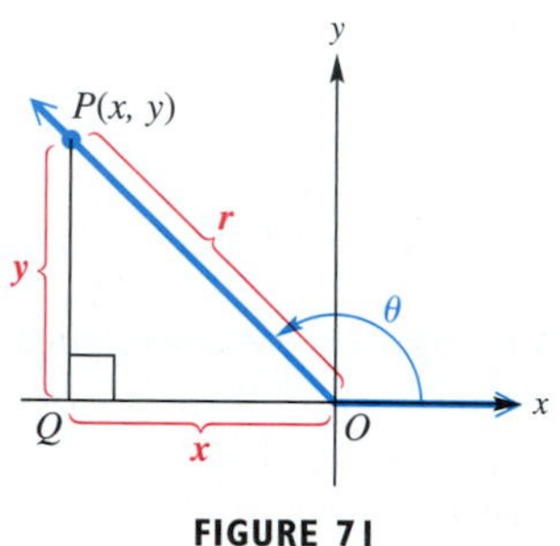

FIGURE 71

A perpendicular from P to the x-axis at point Q determines a right triangle having vertices at O, P, and Q. The distance r from $P(x, y)$ to the origin, $(0, 0)$, can be found from the distance formula.

$$r = \sqrt{(x - 0)^2 + (y - 0)^2} = \sqrt{x^2 + y^2}$$ $r > 0$, since distance is never negative.

We can define the six trigonometric functions using the set of all angles in standard position as their domains.

Trigonometric Functions

Let (x, y) be a point other than the origin on the terminal side of an angle θ in standard position. The distance from the point to the origin is $r = \sqrt{x^2 + y^2}$. The six trigonometric functions of θ are as follows.

$$\sin \theta = \frac{y}{r} \qquad \cos \theta = \frac{x}{r} \qquad \tan \theta = \frac{y}{x} \quad (x \neq 0)$$

$$\csc \theta = \frac{r}{y} \quad (y \neq 0) \qquad \sec \theta = \frac{r}{x} \quad (x \neq 0) \qquad \cot \theta = \frac{x}{y} \quad (y \neq 0)$$

NOTE Although Figure 71 shows a second-quadrant angle, these definitions apply to any angle θ. Because of the restrictions on the denominators in the definitions of tangent, cotangent, secant, and cosecant, some angles have undefined function values.

EXAMPLE 1 Finding Function Values of an Angle

The terminal side of an angle θ in standard position passes through the point $(8, 15)$. Find the values of the six trigonometric functions of angle θ.

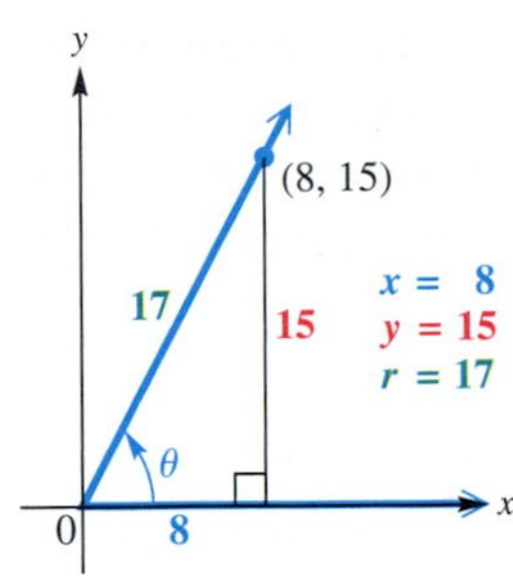

FIGURE 72

Solution Figure 72 shows angle θ and the triangle formed by dropping a perpendicular from the point $(8, 15)$ to the x-axis. The point $(8, 15)$ is 8 units to the right of the y-axis and 15 units above the x-axis, so $x = 8$ and $y = 15$. The value of r is

$$r = \sqrt{x^2 + y^2} = \sqrt{8^2 + 15^2} = \sqrt{64 + 225} = \sqrt{289} = 17.$$

We can now find the values of the six trigonometric functions of angle θ.

$$\sin \theta = \frac{y}{r} = \frac{15}{17} \qquad \cos \theta = \frac{x}{r} = \frac{8}{17} \qquad \tan \theta = \frac{y}{x} = \frac{15}{8}$$

$$\csc \theta = \frac{r}{y} = \frac{17}{15} \qquad \sec \theta = \frac{r}{x} = \frac{17}{8} \qquad \cot \theta = \frac{x}{y} = \frac{8}{15}$$

■

EXAMPLE 2 Finding Function Values of an Angle

The terminal side of an angle θ in standard position passes through $(-3, -4)$. Find the values of the six trigonometric functions of angle θ.

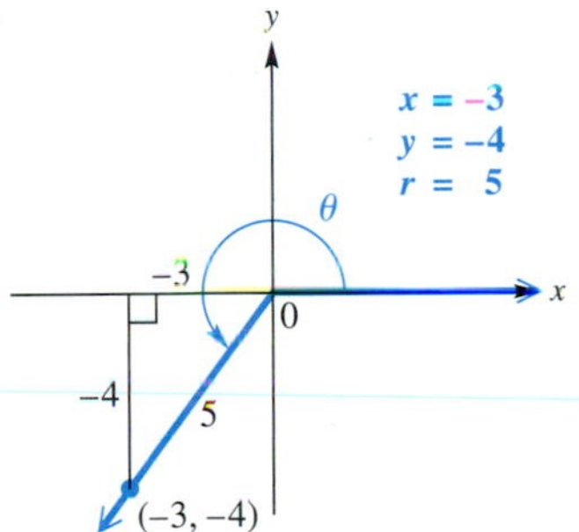

FIGURE 73

Solution As shown in Figure 73, $x = -3$ and $y = -4$. The value of r is

$$r = \sqrt{(-3)^2 + (-4)^2} = \sqrt{25} = 5.$$

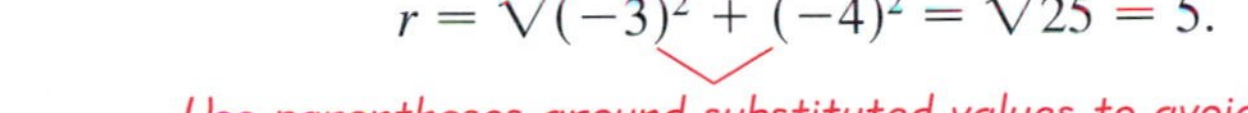

(Remember that $r > 0$.) By the definitions of the trigonometric functions,

$$\sin \theta = \frac{-4}{5} = -\frac{4}{5} \qquad \cos \theta = \frac{-3}{5} = -\frac{3}{5} \qquad \tan \theta = \frac{-4}{-3} = \frac{4}{3}$$

$$\csc \theta = \frac{5}{-4} = -\frac{5}{4} \qquad \sec \theta = \frac{5}{-3} = -\frac{5}{3} \qquad \cot \theta = \frac{-3}{-4} = \frac{3}{4}.$$

■

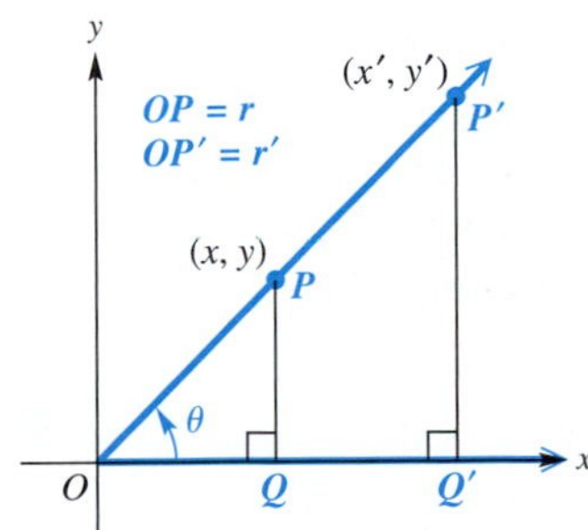

FIGURE 74

We can find the six trigonometric functions by using *any* point other than the origin on the terminal side of an angle. Refer to Figure 74, which shows an angle θ and two distinct points on its terminal side. Point P has coordinates (x, y), and point P' (read **"P-prime"**) has coordinates (x', y'). Let r be the length of the hypotenuse of triangle OPQ, and let r' be the length of the hypotenuse of triangle $OP'Q'$. Since corresponding sides of similar triangles are in proportion,

$$\frac{y}{r} = \frac{y'}{r'},$$

so $\sin \theta = \frac{y}{r}$ is the same no matter which point is used to find it. Similar results hold for the other five functions.

We can also find the trigonometric function values of an angle if we know the equation of the line coinciding with the terminal ray. The graph of the equation

$$Ax + By = 0$$

is a line that passes through the origin. If we restrict x to have only nonpositive or only nonnegative values, we obtain as the graph a ray with endpoint at the origin. For example, the graph of $x + 2y = 0, x \geq 0$, shown in Figure 75, is a ray that can serve as the terminal side of an angle in standard position. By choosing any point on the ray, we can find the trigonometric function values of the angle.

FIGURE 75

EXAMPLE 3 Finding Function Values of an Angle

GCM

Find the six trigonometric function values of the angle θ in standard position if the terminal side of θ is defined by $x + 2y = 0, x \geq 0$.

Analytic Solution

The angle is shown in Figure 76. We can use *any* point except $(0, 0)$ on the terminal side of θ to find the trigonometric function values. We choose $x = 2$ and find the corresponding y-value.

$$x + 2y = 0, x \geq 0$$

$$2 + 2y = 0 \quad \text{Let } x = 2.$$

$$2y = -2 \quad \text{Subtract 2.}$$

$$y = -1 \quad \text{Divide by 2.}$$

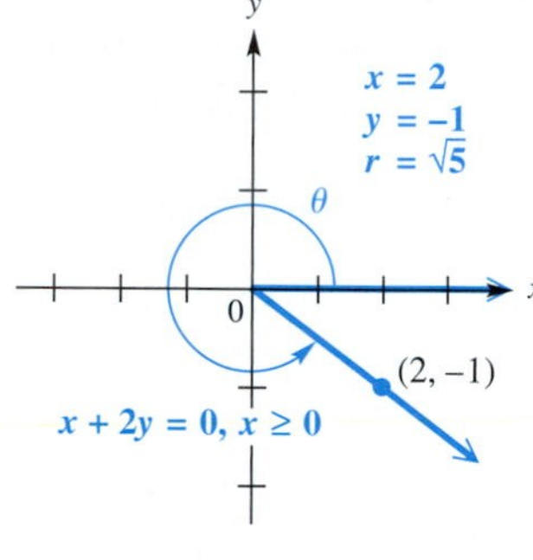

FIGURE 76

The point $(2, -1)$ lies on the terminal side, and the corresponding value of r is $r = \sqrt{2^2 + (-1)^2} = \sqrt{5}$. Thus,

$$\sin\theta = \frac{y}{r} = \frac{-1}{\sqrt{5}} = \frac{-1}{\sqrt{5}} \cdot \frac{\sqrt{5}}{\sqrt{5}} = -\frac{\sqrt{5}}{5}$$

$$\cos\theta = \frac{x}{r} = \frac{2}{\sqrt{5}} = \frac{2}{\sqrt{5}} \cdot \frac{\sqrt{5}}{\sqrt{5}} = \frac{2\sqrt{5}}{5}$$

Rationalize denominators.

$$\tan\theta = \frac{y}{x} = -\frac{1}{2} \qquad \csc\theta = \frac{r}{y} = \frac{\sqrt{5}}{-1} = -\sqrt{5}$$

$$\sec\theta = \frac{r}{x} = \frac{\sqrt{5}}{2} \qquad \cot\theta = \frac{x}{y} = \frac{2}{-1} = -2.$$

Graphing Calculator Solution

Figure 77 shows the graph of

$$x + 2y = 0, \quad x \geq 0.$$

We can graph the ray by entering the equation as

$$Y_1 = (-1/2)X \quad \text{or} \quad Y_1 = -.5X,$$

with the restriction $X \geq 0$. We used the capability of the calculator to find the y-value for $X = 2$, which produced the point and values of X and Y shown on the screen.

2

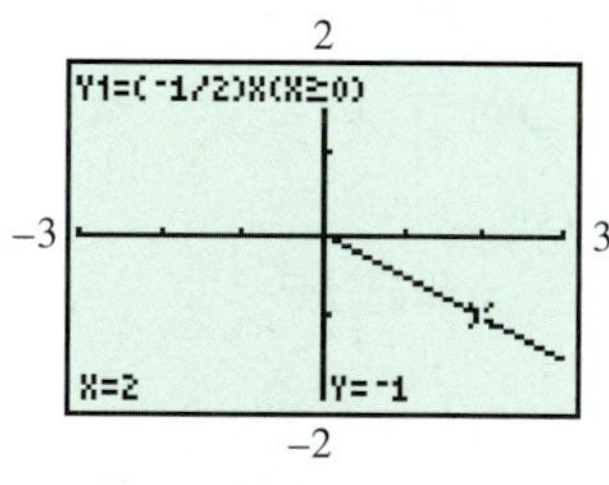

−3 3

−2

FIGURE 77

We can now use these values of x and y and the definitions of the trigonometric functions to find the six function values of θ, as shown in the analytic solution.

Recall that when the equation of a line is written using slope–intercept form $y = mx + b$, the coefficient of x is the slope of the line. In Example 3, $x + 2y = 0$ can be written as $y = -\frac{1}{2}x$, so the slope is $-\frac{1}{2}$. Notice that $\tan\theta = -\frac{1}{2}$. ***In general, it is true that $m = \tan\theta$.***

FOR DISCUSSION

Either individually or with a group of students, rework Example 3, using a different value for x. Find the corresponding y-value, and then show that the six trigonometric function values you obtain are the same as the previous ones.

NOTE The trigonometric function values we found in Examples 1–3 are *exact.* If we were to use a calculator to approximate these values, the decimal results would not be acceptable if exact values were required.

Quadrantal Angles

If the terminal side of an angle in standard position lies along the y-axis, any point on this terminal side has x-coordinate 0. Similarly, an angle with terminal side on the x-axis has y-coordinate 0 for any point on the terminal side. Since the values of x and y appear in the denominators of some trigonometric functions, some trigonometric function values of quadrantal angles (i.e., those with terminal side on an axis) are undefined.

GCM **EXAMPLE 4** **Finding Function Values of Quadrantal Angles**

Find the values of the six trigonometric functions for each angle.

(a) An angle of 90°

(b) An angle θ in standard position with terminal side through $(-3, 0)$

Analytic Solution

(a) We select any point on the terminal side of a 90° angle, such as $(0, 1)$ shown in Figure 78. Here, $x = 0$ and $y = 1$, so $r = 1$.

$$\sin 90^\circ = \frac{y}{r} = \frac{1}{1} = 1 \qquad \csc 90^\circ = \frac{r}{y} = \frac{1}{1} = 1$$

$$\cos 90^\circ = \frac{x}{r} = \frac{0}{1} = 0 \qquad \sec 90^\circ = \frac{r}{x} = \frac{1}{0} \quad \text{(undefined)}$$

$$\tan 90^\circ = \frac{y}{x} = \frac{1}{0} \quad \text{(undefined)} \qquad \cot 90^\circ = \frac{x}{y} = \frac{0}{1} = 0$$

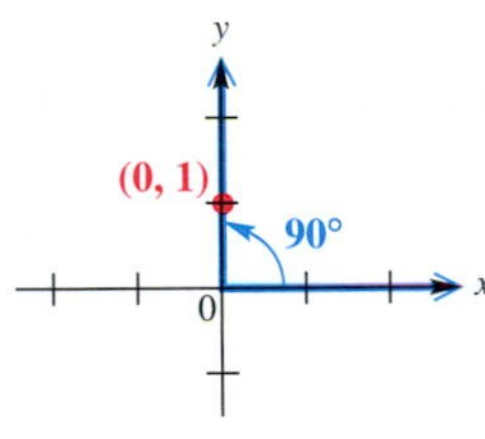

FIGURE 78

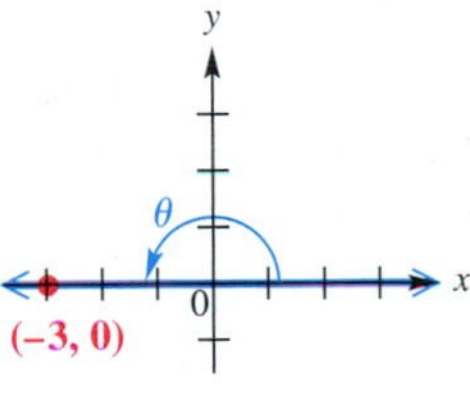

FIGURE 79

(b) Figure 79 shows the angle. Here, $x = -3$, $y = 0$, and $r = 3$, so

$$\sin \theta = \frac{0}{3} = 0 \qquad \csc \theta = \frac{3}{0} \quad \text{(undefined)}$$

$$\cos \theta = \frac{-3}{3} = -1 \qquad \sec \theta = \frac{3}{-3} = -1$$

$$\tan \theta = \frac{0}{-3} = 0 \qquad \cot \theta = \frac{-3}{0} \quad \text{(undefined).}$$

Graphing Calculator Solution

A calculator set in degree mode returns the correct values for sin 90° and cos 90°. See Figure 80. The bottom screen shows an ERROR message for tan 90°, because 90° is not in the domain of the tangent function.

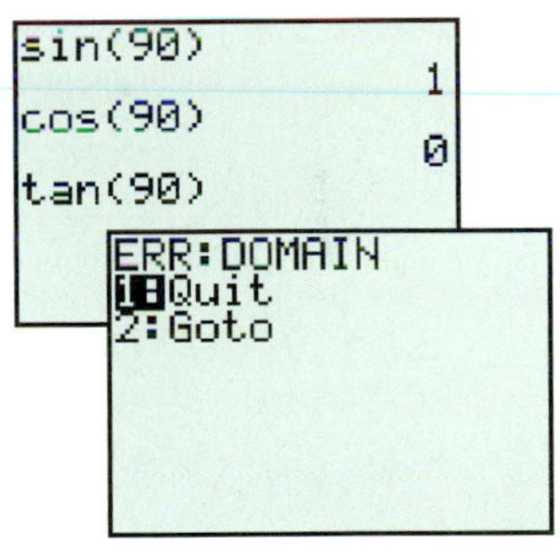

Degree mode

FIGURE 80

There are no calculator keys for finding the function values of cotangent, secant, or cosecant. Later in this section, we discuss how to find these function values with a calculator.

The conditions under which the trigonometric function values of quadrantal angles are undefined are summarized here.

Undefined Function Values

If the terminal side of a quadrantal angle lies along the y-axis, the tangent and secant functions are undefined. If it lies along the x-axis, the cotangent and cosecant functions are undefined.

Function values of the most commonly used quadrantal angles, 0°, 90°, 180°, 270°, and 360°, are summarized in the following table. We saw a similar table based on the unit circle interpretation in Section 8.2.

θ	$\sin\theta$	$\cos\theta$	$\tan\theta$	$\cot\theta$	$\sec\theta$	$\csc\theta$
0°	0	1	0	Undefined	1	Undefined
90°	1	0	Undefined	0	Undefined	1
180°	0	-1	0	Undefined	-1	Undefined
270°	-1	0	Undefined	0	Undefined	-1
360°	0	1	0	Undefined	1	Undefined

Values given in this table can also be found with a calculator that has trigonometric function keys. ***Make sure the calculator is set in degree mode.***

CAUTION ***One of the most common errors involving calculators in trigonometry occurs when the calculator is set for the wrong mode.*** Be sure that you know how to set your calculator in either radian or degree mode.

Reciprocal Identities

In Section 8.2, we introduced several statements that are true for all real numbers for which the functions are defined. For example, $\sin^2 s + \cos^2 s = 1$ and $\tan s = \frac{1}{\cot s}$ are two of these equations, called *identities.* They hold for angles as well.

The definitions of the trigonometric functions on page 622 were written so that functions in the same column are reciprocals of each other. Since $\sin\theta = \frac{y}{r}$ and $\csc\theta = \frac{r}{y}$, we have $\sin\theta = \frac{1}{\csc\theta}$ and $\csc\theta = \frac{1}{\sin\theta}$, provided that $\sin\theta \neq 0$. Also, $\cos\theta$ and $\sec\theta$ are reciprocals, as are $\tan\theta$ and $\cot\theta$. In summary, we have the **reciprocal identities,** which hold for any angle θ that does not lead to a 0 denominator.

Reciprocal Identities

$$\sin\theta = \frac{1}{\csc\theta} \qquad \cos\theta = \frac{1}{\sec\theta} \qquad \tan\theta = \frac{1}{\cot\theta}$$

$$\csc\theta = \frac{1}{\sin\theta} \qquad \sec\theta = \frac{1}{\cos\theta} \qquad \cot\theta = \frac{1}{\tan\theta}$$

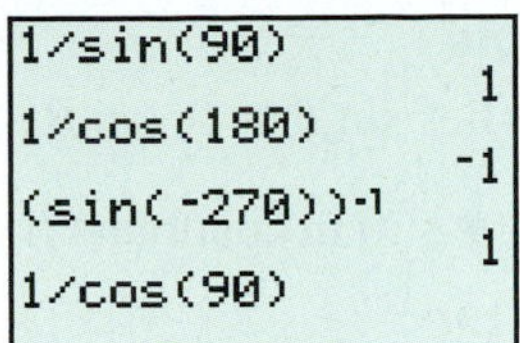

Degree mode

(a)

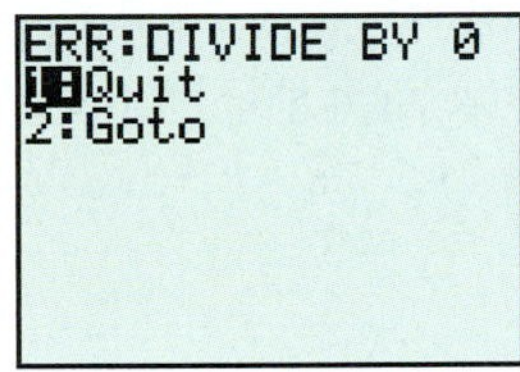

(b)

FIGURE 81

Figure 81(a) shows how csc 90°, sec 180°, and csc(−270°) are found by using the reciprocal identities and the reciprocal key of a graphing calculator in degree mode. ***Be sure not to use the inverse trigonometric function keys to find the reciprocal function values.*** Attempting to find sec 90° by entering $\frac{1}{\cos 90°}$ produces an ERROR message, indicating that the reciprocal is undefined. See Figure 81(b). Compare these results with those in the table of quadrantal angle function values.

NOTE Identities can be written in different forms. For example,

$$\sin\theta = \frac{1}{\csc\theta} \quad \text{can also be written} \quad \csc\theta = \frac{1}{\sin\theta} \quad \text{and} \quad (\sin\theta)(\csc\theta) = 1.$$

EXAMPLE 5 Using the Reciprocal Identities

Find each function value.

(a) $\cos\theta$ if $\sec\theta = \frac{5}{3}$ **(b)** $\sin\theta$ if $\csc\theta = -\frac{\sqrt{12}}{2}$

Solution

(a) Since $\cos\theta$ is the reciprocal of $\sec\theta$,

$$\cos\theta = \frac{1}{\sec\theta} = \frac{1}{\frac{5}{3}} = 1 \div \frac{5}{3} = 1 \cdot \frac{3}{5} = \frac{3}{5}.$$

Be careful when simplifying a complex fraction.

(b) Since $\sin\theta = \frac{1}{\csc\theta}$,

$$\sin\theta = \frac{1}{-\frac{\sqrt{12}}{2}} = \frac{-2}{\sqrt{12}} = \frac{-2}{2\sqrt{3}} = \frac{-1}{\sqrt{3}} = \frac{-1}{\sqrt{3}} \cdot \frac{\sqrt{3}}{\sqrt{3}} = -\frac{\sqrt{3}}{3}.$$ ■

Remember to rationalize the denominator.

Signs and Ranges of Function Values

In the definitions of the trigonometric functions, r is the distance from the origin to the point (x, y), so $r > 0$. If we choose a point (x, y) in quadrant I, then both x and y will be positive, so the values of all six functions will be positive in quadrant I.

A point (x, y) in quadrant II has $x < 0$ and $y > 0$. This makes the values of sine and cosecant positive for quadrant II angles, while the other four functions take on negative values. Similar results can be obtained for the other quadrants, as summarized here. (Compare this to the summary in Section 8.2.)

Signs of Function Values

θ in Quadrant	$\sin\theta$	$\cos\theta$	$\tan\theta$	$\cot\theta$	$\sec\theta$	$\csc\theta$
I	+	+	+	+	+	+
II	+	−	−	−	−	+
III	−	−	+	+	−	−
IV	−	+	−	−	+	−

$x < 0, y > 0, r > 0$
II
Sine and cosecant positive

$x > 0, y > 0, r > 0$
I
All functions positive

$x < 0, y < 0, r > 0$
III
Tangent and cotangent positive

$x > 0, y < 0, r > 0$
IV
Cosine and secant positive

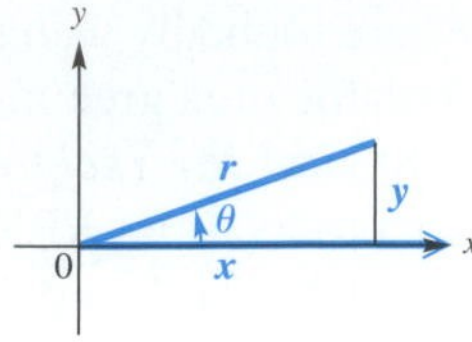

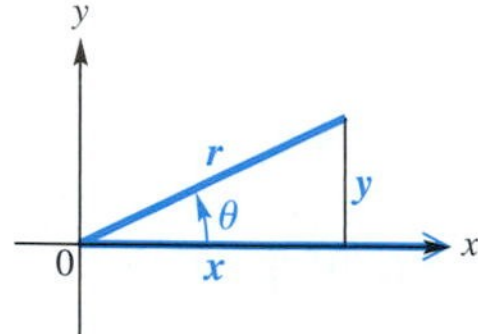

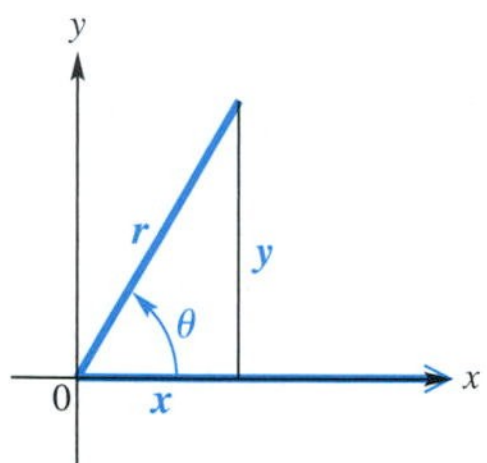

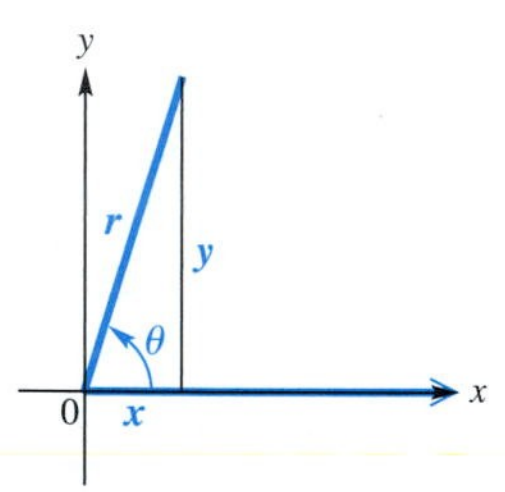

FIGURE 82

EXAMPLE 6 Identifying the Quadrant of an Angle

Identify the quadrant (or quadrants) of any angle θ that satisfies $\sin\theta > 0$, $\tan\theta < 0$.

Solution Since $\sin\theta > 0$ in quadrants I and II, while $\tan\theta < 0$ in quadrants II and IV, both conditions are met only in quadrant II. ■

In Sections 8.3 and 8.4, we saw how the ranges of trigonometric functions were determined using graphs of functions. The ranges can also be justified using angle-based definitions. Figure 82 shows an angle θ as it increases in measure from near 0° toward 90°. In each case, the value of r is the same. As the measure of the angle increases, y increases, but never exceeds r, so $y \leq r$. Dividing each side by the positive number r gives $\frac{y}{r} \leq 1$.

In a similar way, angles in quadrant IV suggest that $-1 \leq \frac{y}{r}$, so

$$-1 \leq \frac{y}{r} \leq 1$$

$$-1 \leq \sin\theta \leq 1. \qquad \frac{y}{r} = \sin\theta \text{ for any angle } \theta.$$

Similarly,

$$-1 \leq \cos\theta \leq 1.$$

The tangent of an angle is defined as $\frac{y}{x}$. It is possible that $x < y$, $x = y$, or $x > y$. For this reason, $\frac{y}{x}$ can take any value, so $\tan\theta$ can be any real number, as can $\cot\theta$.

The functions $\sec\theta$ and $\csc\theta$ are reciprocals of the functions $\cos\theta$ and $\sin\theta$, respectively, making

$$\sec\theta \leq -1 \text{ or } \sec\theta \geq 1 \qquad \text{and} \qquad \csc\theta \leq -1 \text{ or } \csc\theta \geq 1.$$

In summary, the ranges of the trigonometric functions are as follows.

Ranges of Trigonometric Functions

For any angle θ for which the indicated functions exist,

1. $-1 \leq \sin\theta \leq 1$ and $-1 \leq \cos\theta \leq 1$;
2. $\tan\theta$ and $\cot\theta$ may equal any real number;
3. $\sec\theta \leq -1$ or $\sec\theta \geq 1$ and $\csc\theta \leq -1$ or $\csc\theta \geq 1$.

(Notice that sec θ and csc θ are never between −1 and 1, exclusively.)

EXAMPLE 7 Deciding whether a Value Is in the Range of a Function

Decide whether each statement is *possible* or *impossible*.

(a) $\sin\theta = \sqrt{8}$ **(b)** $\tan\theta = 110.47$ **(c)** $\sec\theta = .6$

Solution

(a) For any value of θ, $-1 \leq \sin\theta \leq 1$. Since $\sqrt{8} > 1$, it is impossible to find a value of θ with $\sin\theta = \sqrt{8}$.

(b) The tangent function can take on any real number, so $\tan\theta = 110.47$ is possible.

(c) Since $\sec\theta \leq -1$ or $\sec\theta \geq 1$, the statement $\sec\theta = .6$ is impossible. ■

Pythagorean Identities

Recall from Section 8.2 that for any real number s, $\sin^2 s + \cos^2 s = 1$. This is true because the point $(\cos s, \sin s)$ on the unit circle satisfies $x^2 + y^2 = 1$. We can derive this identity for angles using the relationship $x^2 + y^2 = r^2$.

$$\frac{x^2}{r^2} + \frac{y^2}{r^2} = \frac{r^2}{r^2} \qquad \text{Divide by } r^2.$$

$$\left(\frac{x}{r}\right)^2 + \left(\frac{y}{r}\right)^2 = 1 \qquad \text{Power rule for exponents}$$

$$(\cos\theta)^2 + (\sin\theta)^2 = 1 \qquad \cos\theta = \tfrac{x}{r},\ \sin\theta = \tfrac{y}{r}$$

or

$$\sin^2\theta + \cos^2\theta = 1 \qquad \text{Rewrite.}$$

Starting again with $x^2 + y^2 = r^2$ and dividing through by x^2 gives

$$\frac{x^2}{x^2} + \frac{y^2}{x^2} = \frac{r^2}{x^2} \qquad \text{Divide by } x^2.$$

$$1 + \left(\frac{y}{x}\right)^2 = \left(\frac{r}{x}\right)^2 \qquad \text{Power rule for exponents}$$

$$1 + (\tan\theta)^2 = (\sec\theta)^2 \qquad \tan\theta = \tfrac{y}{x},\ \sec\theta = \tfrac{r}{x}$$

or

$$1 + \tan^2\theta = \sec^2\theta. \qquad \text{Rewrite.}$$

On the other hand, dividing through by y^2 leads to

$$1 + \cot^2\theta = \csc^2\theta.$$

GCM TECHNOLOGY NOTE

Powers of trigonometric functions are not usually entered on calculators as they are written mathematically. For example, we write $\sin^2\theta$ to represent the square of $\sin\theta$, but this would be entered as $(\sin\theta)^2$ on many calculators.

These three identities are called the **Pythagorean identities,** since the original equation that led to them, $x^2 + y^2 = r^2$, comes from the Pythagorean theorem.

Pythagorean Identities

$$\sin^2\theta + \cos^2\theta = 1 \qquad 1 + \tan^2\theta = \sec^2\theta \qquad 1 + \cot^2\theta = \csc^2\theta$$

As before, we have given only one form of each identity. However, algebraic transformations produce equivalent identities. For example, by subtracting $\sin^2\theta$ from each side of $\sin^2\theta + \cos^2\theta = 1$, we get the equivalent identity

$$\cos^2\theta = 1 - \sin^2\theta.$$

You should be able to transform these identities and recognize equivalent forms.

Quotient Identities

Recall that $\sin\theta = \frac{y}{r}$ and $\cos\theta = \frac{x}{r}$. Consider the quotient of $\sin\theta$ and $\cos\theta$.

$$\frac{\sin\theta}{\cos\theta} = \frac{\frac{y}{r}}{\frac{x}{r}} = \frac{y}{r} \div \frac{x}{r} = \frac{y}{r} \cdot \frac{r}{x} = \frac{y}{x} = \tan\theta, \quad \cos\theta \neq 0$$

Similarly, $\frac{\cos\theta}{\sin\theta} = \cot\theta$, for $\sin\theta \neq 0$. Thus, we have the **quotient identities.**

Looking Ahead to Calculus

The reciprocal, Pythagorean, and quotient identities are used repeatedly in calculus to find limits, derivatives, and integrals of trigonometric functions. These identities are also used to rewrite expressions in a form that permits simplifying a square root. For example, if $a \geq 0$ and $x = a \sin \theta$,

$$\begin{aligned}\sqrt{a^2 - x^2} &= \sqrt{a^2 - a^2 \sin^2 \theta} \\ &= \sqrt{a^2(1 - \sin^2 \theta)} \\ &= \sqrt{a^2 \cos^2 \theta} \\ &= a|\cos \theta|.\end{aligned}$$

Quotient Identities

$$\frac{\sin \theta}{\cos \theta} = \tan \theta \qquad \frac{\cos \theta}{\sin \theta} = \cot \theta$$

EXAMPLE 8 Finding Function Values, Given One Value and the Quadrant

Find $\sin \theta$ and $\cos \theta$ if $\tan \theta = \frac{4}{3}$ and θ is in quadrant III.

Solution Since θ is in quadrant III, $\sin \theta$ and $\cos \theta$ will both be negative. It is tempting to say that since $\tan \theta = \frac{\sin \theta}{\cos \theta}$ and $\tan \theta = \frac{4}{3}$, then $\sin \theta = -4$ and $\cos \theta = -3$. *This is incorrect*—both $\sin \theta$ and $\cos \theta$ must be in the interval $[-1, 1]$.

We use the Pythagorean identity $1 + \tan^2 \theta = \sec^2 \theta$ to find $\sec \theta$ and then the reciprocal identity $\cos \theta = \frac{1}{\sec \theta}$ to find $\cos \theta$.

$$1 + \tan^2 \theta = \sec^2 \theta \qquad \text{Pythagorean identity}$$

$$1 + \left(\frac{4}{3}\right)^2 = \sec^2 \theta \qquad \tan \theta = \tfrac{4}{3}$$

$$1 + \frac{16}{9} = \sec^2 \theta$$

$$\frac{25}{9} = \sec^2 \theta$$

$$-\frac{5}{3} = \sec \theta \qquad \text{Choose the negative square root, since } \sec \theta \text{ is negative when } \theta \text{ is in quadrant III.}$$

$$-\frac{3}{5} = \cos \theta \qquad \text{Secant and cosine are reciprocals.}$$

Since $\sin^2 \theta = 1 - \cos^2 \theta$,

$$\sin^2 \theta = 1 - \left(-\frac{3}{5}\right)^2 \qquad \cos \theta = -\tfrac{3}{5}$$

$$\sin^2 \theta = 1 - \frac{9}{25}$$

$$\sin^2 \theta = \frac{16}{25}$$

$$\sin \theta = -\frac{4}{5}. \qquad \text{Choose the negative square root.}$$

Therefore, $\sin \theta = -\frac{4}{5}$ and $\cos \theta = -\frac{3}{5}$. ■

NOTE Example 8 can also be worked by drawing θ in standard position in quadrant III, finding r to be 5, and then using the definitions of $\sin \theta$ and $\cos \theta$ in terms of x, y, and r.

EXAMPLE 9 **Using Identities to Find a Trigonometric Expression**

If θ is in quadrant IV, find an expression for $\sec\theta$ in terms of $\sin\theta$.

Solution

$$\sin^2\theta + \cos^2\theta = 1 \quad \text{Pythagorean identity}$$

$$\cos^2\theta = 1 - \sin^2\theta \quad \text{Subtract } \sin^2\theta.$$

$$\cos\theta = \pm\sqrt{1 - \sin^2\theta} \quad \text{Take square roots.}$$

$$\sec\theta = \pm\frac{1}{\sqrt{1 - \sin^2\theta}} \quad \sec\theta = \frac{1}{\cos\theta}$$

Angle θ is in quadrant IV, so $\sec\theta > 0$. Thus, we choose the positive expression.

$$\sec\theta = \frac{1}{\sqrt{1 - \sin^2\theta}} \quad \sec\theta > 0$$

$$= \frac{\sqrt{1 - \sin^2\theta}}{1 - \sin^2\theta} \quad \text{Rationalize the denominator.}$$

An Application of Trigonometric Functions

Grade or slope is a measure of steepness and indicates whether a highway is uphill or downhill. A 5% grade indicates that a road is increasing 5 vertical feet for each 100-foot increase in horizontal distance. *Grade resistance* is the gravitational force acting on a vehicle and is given by

$$R = W\sin\theta,$$

where W is the weight of the vehicle and θ is the angle associated with the grade. See Figure 83. For an uphill grade $\theta > 0$ and for a downhill grade $\theta < 0$. (*Source:* Mannering, F. and W. Kilareski, *Principles of Highway Engineering and Traffic Analysis,* Second Edition, John Wiley and Sons, 1998.)

FIGURE 83

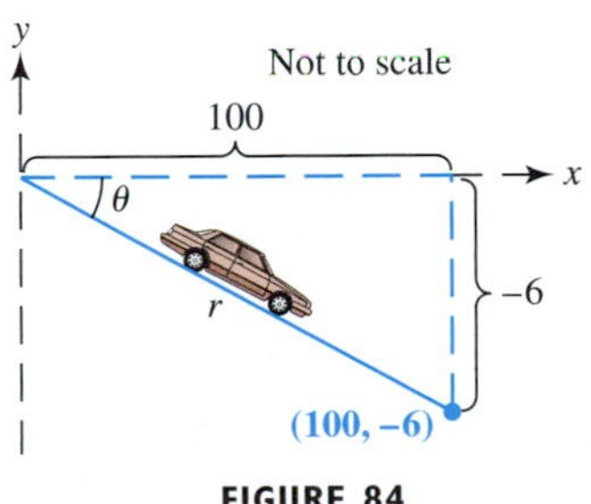

FIGURE 84

EXAMPLE 10 **Calculating Grade Resistance**

A downhill highway grade is modeled by the line $y = -.06x$ in quadrant IV.

(a) Find the grade of the road.

(b) Determine the grade resistance for a 3000-pound car. Interpret the result.

Solution

(a) The slope of the line is $-.06$, so when x *increases* by 100 feet, y *decreases* by 6 feet. See Figure 84. Thus, this road has a grade of -6%.

FIGURE 84 (repeated)

(b) First we must find $\sin \theta$. From Figure 84, we see that the point $(100, -6)$ lies on the terminal side of θ. Thus,

$$r = \sqrt{x^2 + y^2} = \sqrt{100^2 + (-6)^2} = \sqrt{10{,}036},$$

and $$\sin \theta = \frac{y}{r} = \frac{-6}{\sqrt{10{,}036}}.$$

The grade resistance is

$$R = W \sin \theta = 3000\left(\frac{-6}{\sqrt{10{,}036}}\right) \approx -179.7 \text{ pounds.}$$

On this stretch of highway, gravity would pull a 3000-pound vehicle *downhill* with a force of about 180 pounds. Note that a downhill grade results in a negative grade resistance. ■

8.5 Exercises

Sketch an angle θ in standard position such that θ has the smallest possible positive measure and the given point is on the terminal side of θ.

1. $(5, -12)$

2. $(-12, -5)$

Find the values of the six trigonometric functions for the angles in standard position having the following points on their terminal sides. Rationalize denominators when applicable.

3. $(-3, 4)$

4. $(-4, -3)$

5. $(0, 2)$

6. $(-4, 0)$

7. $(1, \sqrt{3})$

8. $(-2\sqrt{3}, -2)$

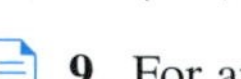 **9.** For any nonquadrantal angle θ, $\sin \theta$ and $\csc \theta$ will have the same sign. Explain why.

10. *Concept Check* If the terminal side of an angle θ is in quadrant III, what is the sign of each of the trigonometric function values of θ?

Concept Check *Suppose that the point (x, y) is in the indicated quadrant. Decide whether the given ratio is positive or negative.* (Hint: *Drawing a sketch may help.*)

11. II, $\frac{x}{r}$

12. III, $\frac{y}{r}$

13. IV, $\frac{y}{x}$

14. III, $\frac{x}{y}$

In Exercises 15–18, an equation with a restriction on x is given. This is an equation of the terminal side of an angle θ in standard position. Sketch the smallest positive such angle θ, and find the values of the six trigonometric functions of θ.

15. $2x + y = 0, x \geq 0$

16. $3x + 5y = 0, x \geq 0$

17. $-6x - y = 0, x \leq 0$

18. $-5x - 3y = 0, x \leq 0$

19. Find the six trigonometric function values of the quadrantal angle $\theta = 540°$.

Use the trigonometric function values of quadrantal angles given in this section to evaluate each expression. An expression such as $\cot^2 90°$ *means* $(\cot 90°)^2$, *which is equal to* $0^2 = 0$.

20. $3 \sec 180° - 5 \tan 360°$

21. $4 \csc 270° + 3 \cos 180°$

22. $\tan 360° + 4 \sin 180° + 5 \cos^2 180°$

23. $2 \sec 0° + 4 \cot^2 90° + \cos 360°$

24. $\sin^2 180° + \cos^2 180°$

25. $\sin^2 360° + \cos^2 360°$

If n is an integer, $n \cdot 180°$ *represents an integer multiple of* 180°, $(2n + 1) \cdot 90°$ *represents an odd integer multiple of* 90°, *and so on. Decide whether each expression is equal to* 0, 1, or -1 *or is* undefined.

26. $\cos[(2n + 1) \cdot 90°]$

27. $\tan(n \cdot 180°)$

28. $\cos[(2n + 1) \cdot 180°]$

29. $\sin(n \cdot 180°)$

Provide conjectures in Exercises 30–33.

30. The angles 15° and 75° are complementary. With your calculator, find sin 15° and cos 75°. Make a conjecture about the sines and cosines of complementary angles, and test your hypothesis on other pairs of complementary angles. (*Note:* This relationship will be discussed in detail in the next section.)

31. The angles 25° and 65° are complementary. With your calculator, find tan 25° and cot 65°. Make a conjecture about the tangents and cotangents of complementary angles, and test your hypothesis on other pairs of complementary angles. (*Note:* This relationship will be discussed in detail in the next section.)

32. With your calculator, find sin 10° and sin(−10°). Make a conjecture about the sine of an angle and its negative, and test your hypothesis on other angles. Also, use a geometric argument with the definition of $\sin \theta$ to justify your hypothesis. (*Note:* This relationship will be discussed in detail in Section 9.1.)

33. With your calculator, find cos 20° and cos(−20°). Make a conjecture about the cosine of an angle and its negative, and test your hypothesis on other angles. Also, use a geometric argument with the definition of $\cos \theta$ to justify your hypothesis. (*Note:* This relationship will be discussed in detail in Section 9.1.)

In Exercises 34–39, set your graphing calculator in parametric and degree modes. Set the window and functions as shown here, and graph. A circle of radius 1 will appear on the screen. Trace a short distance around the circle. In the screen, the point on the circle corresponds to an angle T = 25°. *Since* $r = 1$, cos 25° *is* X ≈ .90630779, *and* sin 25° *is* Y ≈ .42261826.

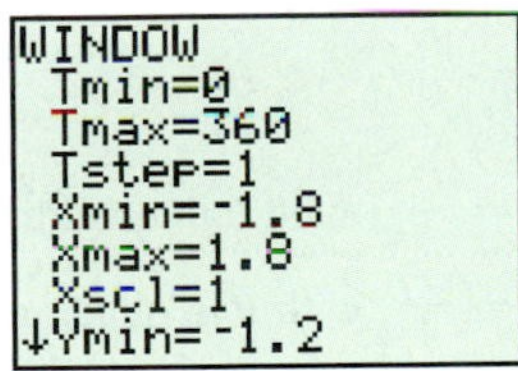

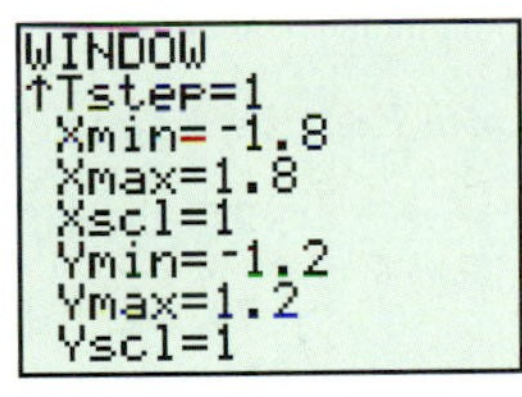

This screen is a continuation of the previous one.

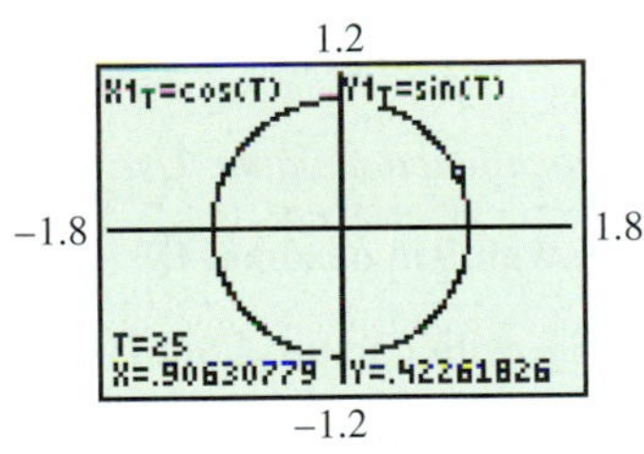

34. Use the right- and left-arrow keys to move to the point corresponding to 20°. What are cos 20° and sin 20°?

35. For what angle T, 0° ≤ T ≤ 90°, is cos T ≈ .766?

36. For what angle T, 0° ≤ T ≤ 90°, is sin T ≈ .574?

37. For what angle T, 0° ≤ T ≤ 90°, does cos T = sin T?

38. As T increases from 0° to 90°, does the cosine increase or decrease? What about the sine?

39. As T increases from 90° to 180°, does the cosine increase or decrease? What about the sine?

40. ***Concept Check*** What positive number a is its own reciprocal? Find a value of θ for which $\sin \theta = \csc \theta = a$.

41. ***Concept Check*** What negative number a is its own reciprocal? Find a value of θ for which $\cos \theta = \sec \theta = a$.

Use the appropriate reciprocal identity to find each function value. Rationalize denominators when applicable. In Exercises 46 and 47, use a calculator.

42. $\cos \theta$ if $\sec \theta = -2.5$

43. $\cot \theta$ if $\tan \theta = -\frac{1}{5}$

44. $\sin \theta$ if $\csc \theta = \sqrt{15}$

45. $\tan \theta$ if $\cot \theta = -\frac{\sqrt{5}}{3}$

46. $\sin \theta$ if $\csc \theta = 1.42716321$

47. $\cos \theta$ if $\sec \theta = 9.80425133$

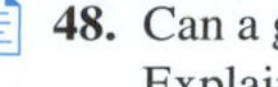

48. Can a given angle θ satisfy both $\sin \theta > 0$ and $\csc \theta < 0$? Explain.

49. Explain what is wrong with the following item that appears on a trigonometry test: Find $\sec \theta$, given that $\cos \theta = \frac{3}{2}$.

Find the tangent of each angle.

50. $\cot \theta = -3$

51. $\cot \theta = \dfrac{\sqrt{3}}{3}$

52. $\cot \theta = .4$

Find the smallest positive value of θ (in degrees) that makes the statement true.

53. $\tan(3\theta - 4°) = \dfrac{1}{\cot(5\theta - 8°)}$

54. $\sec(2\theta + 6°) \cos(5\theta + 3°) = 1$

Identify the quadrant or quadrants for the angle satisfying the given conditions.

55. $\sin \theta > 0$; $\cos \theta < 0$

56. $\cos \theta > 0$; $\tan \theta > 0$

57. $\tan \theta > 0$; $\cot \theta > 0$

58. $\tan \theta < 0$; $\cot \theta < 0$

Give the signs of the sine, cosine, and tangent functions for each angle.

59. 129°

60. 183°

61. 298°

62. 412°

63. −82°

64. −121°

Concept Check *Without using a calculator, decide which is greater.*

65. sin 30° or tan 30°

66. sin 20° or sin 21°

67. sin 33° or sec 33°

Decide whether each statement is possible *or* impossible *for an angle θ.*

68. $\sin \theta = 2$

69. $\cos \theta = -1.001$

70. $\tan \theta = .92$

71. $\cot \theta = -12.1$

72. $\sec \theta = 1$

73. $\tan \theta = 1$

74. $\sin \theta = \frac{1}{2}$ and $\csc \theta = 2$

75. $\tan \theta = 2$ and $\cot \theta = -2$

Use identities to find each function value. Use a calculator in Exercises 82 and 83.

76. $\tan \theta$ if $\sec \theta = 3$, with θ in quadrant IV

77. $\sin \theta$ if $\cos \theta = -\frac{1}{4}$, with θ in quadrant II

78. $\csc \theta$ if $\cot \theta = -\frac{1}{2}$, with θ in quadrant IV

79. $\sec \theta$ if $\tan \theta = \frac{\sqrt{7}}{3}$, with θ in quadrant III

80. $\cos \theta$ if $\csc \theta = -4$, with θ in quadrant III

81. $\sin \theta$ if $\sec \theta = 2$, with θ in quadrant IV

82. $\cot \theta$ if $\csc \theta = -3.5891420$, with θ in quadrant III

83. $\tan \theta$ if $\sin \theta = .49268329$, with θ in quadrant II

In Exercises 84 and 85, each graphing calculator screen is obtained for a particular stored value of X. What will the screen display for the value of the expression in the final line of the display?

84.

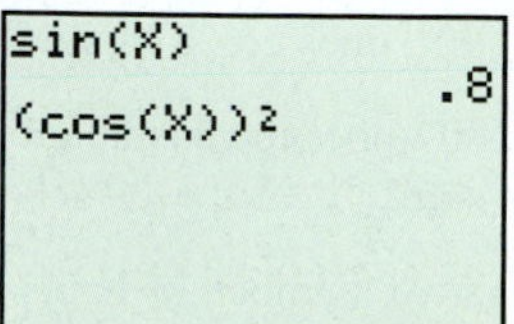

85.

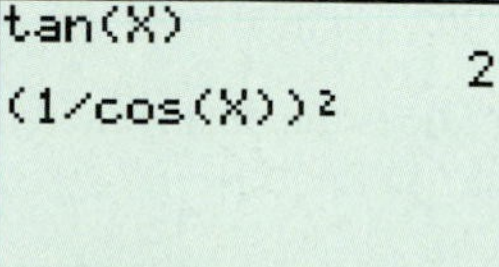

86. *Concept Check* Is there an angle θ with $\cos \theta = -.6$ and $\sin \theta = .8$?

Find all trigonometric function values for each angle. Use a calculator in Exercises 91 and 92.

87. $\tan \theta = -\frac{15}{8}$, with θ in quadrant II

88. $\cos \theta = -\frac{3}{5}$, with θ in quadrant III

89. $\tan \theta = \sqrt{3}$, with θ in quadrant III

90. $\sin \theta = \frac{\sqrt{5}}{7}$, with $\tan \theta > 0$

91. $\cot \theta = -1.49586$, with θ in quadrant IV

92. $\sin \theta = .164215$, with θ in quadrant II

Use fundamental identities to find each expression.

93. Write $\cos \theta$ in terms of $\sin \theta$ if θ is acute.

94. Write $\sec \theta$ in terms of $\cos \theta$.

95. Write $\sin \theta$ in terms of $\cot \theta$ if θ is in quadrant III.

96. Write $\tan \theta$ in terms of $\cos \theta$ if θ is in quadrant IV.

97. Write $\tan \theta$ in terms of $\sin \theta$ if θ is in quadrant I or IV.

98. Write $\sin \theta$ in terms of $\sec \theta$ if θ is in quadrant I or II.

Work each problem.

99. Derive the identity $1 + \cot^2 \theta = \csc^2 \theta$ by dividing $x^2 + y^2 = r^2$ by y^2.

100. Using a method similar to the one given in this section showing that $\frac{\sin \theta}{\cos \theta} = \tan \theta$, show that $\frac{\cos \theta}{\sin \theta} = \cot \theta$.

101. *Concept Check* *True* or *false?* For all angles θ, $\sin \theta + \cos \theta = 1$. If false, give an example showing why.

102. *Concept Check* *True* or *false?* Since $\cot \theta = \frac{\cos \theta}{\sin \theta}$, if $\cot \theta = \frac{1}{2}$ with θ in quadrant I, then $\cos \theta = 1$ and $\sin \theta = 2$. If false, explain why.

103. *Highway Grade* Suppose the uphill grade of a highway can be modeled by the line $y = .03x$ in quadrant I.
(a) Find the grade of the hill.
(b) Determine the grade resistance for a gravel truck weighing 25,000 pounds.

104. *Highway Grade* Suppose the downhill grade of a highway can be modeled by the line $y = -.09x$ in quadrant IV.
(a) Find the grade of the hill.
(b) Determine the grade resistance for a car weighing 3800 pounds.

8.6 Evaluating Trigonometric Functions

Definitions of the Trigonometric Functions ■ Trigonometric Function Values of Special Angles ■ Cofunction Identities ■ Reference Angles ■ Special Angles as Reference Angles ■ Finding Function Values with a Calculator ■ Finding Angle Measures

Definitions of the Trigonometric Functions

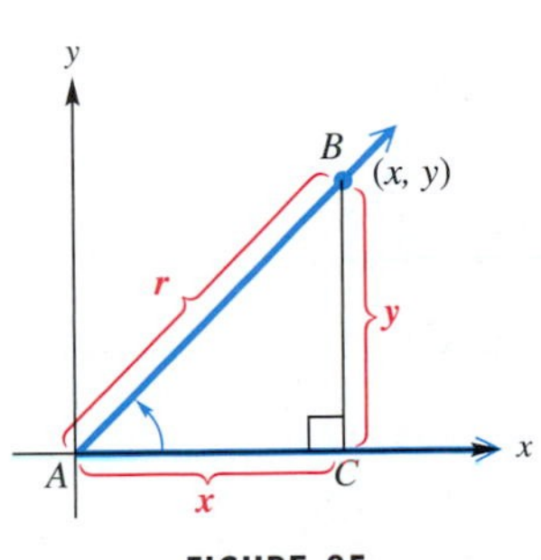

FIGURE 85

In Section 8.5, we used angles in standard position to define the trigonometric functions. There is another way to approach them: as ratios of the sides of right triangles. Figure 85 shows an acute angle A in standard position. The definitions of the trigonometric function values of angle A require x, y, and r. In Figure 85, x and y are the lengths of the two legs of the right triangle ABC; r is the length of the hypotenuse.

The side of length y is called the **side opposite** angle A, and the side of length x is called the **side adjacent** to angle A. We use the lengths of these sides to replace x and y in the definitions of the trigonometric functions, and the length of the hypotenuse to replace r, to get the right-triangle-based definitions on the next page.

Right-Triangle-Based Definitions of Trigonometric Functions

For any acute angle A in standard position,

$$\sin A = \frac{y}{r} = \frac{\text{side opposite}}{\text{hypotenuse}} \qquad \csc A = \frac{r}{y} = \frac{\text{hypotenuse}}{\text{side opposite}}$$

$$\cos A = \frac{x}{r} = \frac{\text{side adjacent}}{\text{hypotenuse}} \qquad \sec A = \frac{r}{x} = \frac{\text{hypotenuse}}{\text{side adjacent}}$$

$$\tan A = \frac{y}{x} = \frac{\text{side opposite}}{\text{side adjacent}} \qquad \cot A = \frac{x}{y} = \frac{\text{side adjacent}}{\text{side opposite}}.$$

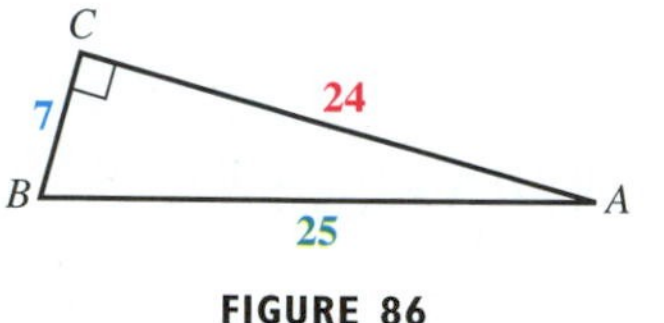

FIGURE 86

EXAMPLE 1 Finding Trigonometric Function Values of an Acute Angle

Find the values of sin A, cos A, and tan A in the right triangle in Figure 86.

Solution The length of the side opposite angle A is 7, the length of the side adjacent to angle A is 24, and the length of the hypotenuse is 25. Thus,

$$\sin A = \frac{\text{side opposite}}{\text{hypotenuse}} = \frac{7}{25} \qquad \cos A = \frac{\text{side adjacent}}{\text{hypotenuse}} = \frac{24}{25} \qquad \tan A = \frac{\text{side opposite}}{\text{side adjacent}} = \frac{7}{24}.$$

NOTE Because the cosecant, secant, and cotangent ratios are the reciprocals of the sine, cosine, and tangent values, $\csc A = \frac{25}{7}$, $\sec A = \frac{25}{24}$, and $\cot A = \frac{24}{7}$ in Example 1.

Trigonometric Function Values of Special Angles

Previously, we found exact function values for the real numbers $\frac{\pi}{4}$, $\frac{\pi}{6}$, and $\frac{\pi}{3}$. These real numbers correspond to radian measured angles, which in turn measure 45°, 30°, and 60°. The same values can be found using angles in triangles. We start with an equilateral triangle—a triangle with all sides of equal length. Each angle of such a triangle measures 60°. While the results we will obtain are independent of the length, for convenience we choose the length of each side to be 2 units. See Figure 87(a).

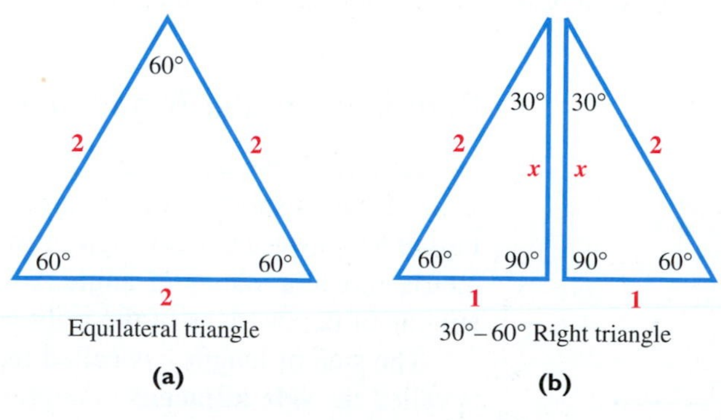

FIGURE 87

Bisecting one angle of this equilateral triangle leads to two right triangles, each of which has angles of 30°, 60°, and 90°, as shown in Figure 87(b) on the preceding page. Since the hypotenuse of one of these right triangles has length 2, the shortest side will have length 1. (Why?) If x represents the length of the medium side, then

$$2^2 = 1^2 + x^2 \quad \text{Pythagorean theorem}$$
$$4 = 1 + x^2$$
$$3 = x^2 \quad \text{Subtract 1.}$$
$$\sqrt{3} = x. \quad \text{Choose the positive root. (Why?)}$$

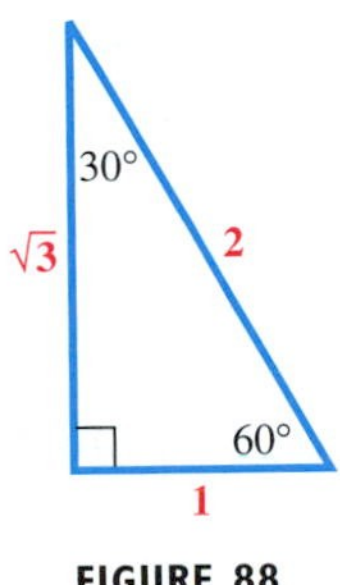

FIGURE 88

Figure 88 summarizes our results, using a 30°–60° right triangle. As shown in the figure, the side opposite the 30° angle has length 1; that is, for the 30° angle,

$$\text{hypotenuse} = 2, \quad \text{side opposite} = 1, \quad \text{side adjacent} = \sqrt{3}.$$

Now we use the definitions of the trigonometric functions.

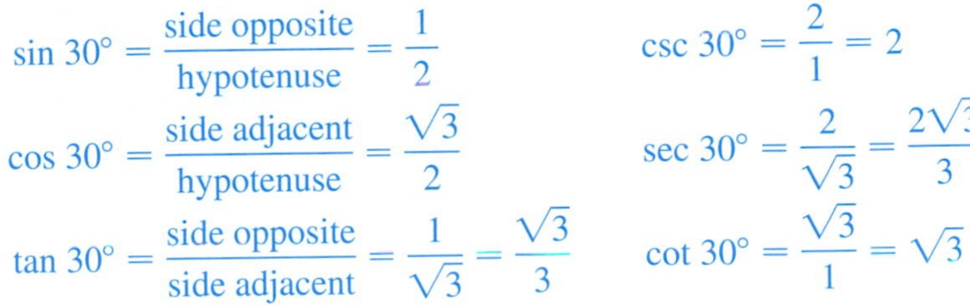

$$\sin 30° = \frac{\text{side opposite}}{\text{hypotenuse}} = \frac{1}{2} \qquad \csc 30° = \frac{2}{1} = 2$$
$$\cos 30° = \frac{\text{side adjacent}}{\text{hypotenuse}} = \frac{\sqrt{3}}{2} \qquad \sec 30° = \frac{2}{\sqrt{3}} = \frac{2\sqrt{3}}{3}$$
$$\tan 30° = \frac{\text{side opposite}}{\text{side adjacent}} = \frac{1}{\sqrt{3}} = \frac{\sqrt{3}}{3} \qquad \cot 30° = \frac{\sqrt{3}}{1} = \sqrt{3}$$

EXAMPLE 2 Finding Trigonometric Function Values for 60°

Find the six trigonometric function values for a 60° angle.

Solution Refer to Figure 88 to find the following ratios.

$$\sin 60° = \frac{\sqrt{3}}{2} \qquad \cos 60° = \frac{1}{2} \qquad \tan 60° = \sqrt{3}$$
$$\csc 60° = \frac{2\sqrt{3}}{3} \qquad \sec 60° = 2 \qquad \cot 60° = \frac{\sqrt{3}}{3}$$

■

We find the values of the trigonometric functions for 45° by starting with a 45°–45° right triangle, as shown in Figure 89. This triangle is isosceles, and for convenience, we choose the lengths of the equal sides to be 1 unit. (As before, the results are independent of the length of the equal sides of the right triangle.) Since the shorter sides each have length 1, if r represents the length of the hypotenuse, then

$$1^2 + 1^2 = r^2 \quad \text{Pythagorean theorem}$$
$$2 = r^2$$
$$\sqrt{2} = r. \quad \text{Choose the positive root.}$$

45° 1 $r = \sqrt{2}$ 45° 1

45°–45° Right triangle

FIGURE 89

Now we use the measures indicated on the 45°–45° right triangle in Figure 89.

$$\sin 45° = \frac{1}{\sqrt{2}} = \frac{\sqrt{2}}{2} \qquad \cos 45° = \frac{1}{\sqrt{2}} = \frac{\sqrt{2}}{2} \qquad \tan 45° = \frac{1}{1} = 1$$
$$\csc 45° = \frac{\sqrt{2}}{1} = \sqrt{2} \qquad \sec 45° = \frac{\sqrt{2}}{1} = \sqrt{2} \qquad \cot 45° = \frac{1}{1} = 1$$

GCM TECHNOLOGY NOTE

Using a graphing calculator to find the decimal value for cos 30° gives a decimal approximation for $\frac{\sqrt{3}}{2}$. The decimal is an approximation, while $\frac{\sqrt{3}}{2}$ is exact. In general, unless a trigonometric function value is a rational number, a graphing calculator will provide only a decimal approximation for the irrational function value. (The TI-89 is an exception.)

Function values for 30° or $\frac{\pi}{6}$, 45° or $\frac{\pi}{4}$, and 60° or $\frac{\pi}{3}$ are summarized in the table that follows. (Compare to the similar table in Section 8.2.)

Function Values for Special Angles

θ	$\sin\theta$	$\cos\theta$	$\tan\theta$	$\cot\theta$	$\sec\theta$	$\csc\theta$
30° or $\frac{\pi}{6}$	$\frac{1}{2}$	$\frac{\sqrt{3}}{2}$	$\frac{\sqrt{3}}{3}$	$\sqrt{3}$	$\frac{2\sqrt{3}}{3}$	2
45° or $\frac{\pi}{4}$	$\frac{\sqrt{2}}{2}$	$\frac{\sqrt{2}}{2}$	1	1	$\sqrt{2}$	$\sqrt{2}$
60° or $\frac{\pi}{3}$	$\frac{\sqrt{3}}{2}$	$\frac{1}{2}$	$\sqrt{3}$	$\frac{\sqrt{3}}{3}$	2	$\frac{2\sqrt{3}}{3}$

Cofunction Identities

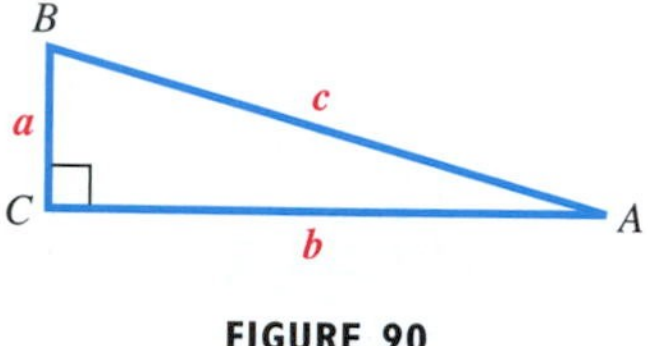

FIGURE 90

In a right triangle ABC with right angle C, the acute angles A and B are complementary. See Figure 90. The length of the side opposite angle A is a, and the length of the side opposite angle B is b. The length of the hypotenuse is c. In this triangle, $\sin A = \frac{a}{c}$. Since $\cos B$ is also equal to $\frac{a}{c}$,

$$\sin A = \frac{a}{c} = \cos B.$$

Similarly, $\tan A = \frac{a}{b} = \cot B$ and $\sec A = \frac{c}{b} = \csc B.$

Since angles A and B are *complementary* and $\sin A = \cos B$, the functions *sine* and *cosine* are called **cofunctions.** Also, tangent and cotangent are cofunctions, as are secant and cosecant. Since angles A and B are complementary, $A + B = 90°$, or $B = 90° - A$, giving $\sin A = \cos B = \cos(90° - A)$.

Similar results, called the **cofunction identities,** are true for the other trigonometric functions.

Cofunction Identities

If A is an acute angle measured in degrees, then

$$\sin A = \cos(90° - A) \quad \csc A = \sec(90° - A) \quad \tan A = \cot(90° - A)$$
$$\cos A = \sin(90° - A) \quad \sec A = \csc(90° - A) \quad \cot A = \tan(90° - A).$$

If A is an acute angle measured in radians, then

$$\sin A = \cos\left(\frac{\pi}{2} - A\right) \qquad \csc A = \sec\left(\frac{\pi}{2} - A\right)$$
$$\cos A = \sin\left(\frac{\pi}{2} - A\right) \qquad \sec A = \csc\left(\frac{\pi}{2} - A\right)$$
$$\tan A = \cot\left(\frac{\pi}{2} - A\right) \qquad \cot A = \tan\left(\frac{\pi}{2} - A\right).$$

NOTE These identities actually apply to all angles (not just acute angles). However, for our present discussion, we will need them only for acute angles.

EXAMPLE 3 **Writing Functions in Terms of Cofunctions**

Write each expression in terms of its cofunction.

(a) $\cos 52° 16'$ **(b)** $\tan \dfrac{\pi}{6}$

Solution

(a) Since $\cos A = \sin(90° - A)$,

$$\cos 52° 16' = \sin(90° - 52° 16') = \sin 37° 44'.$$

(b) Since $\tan A = \cot\left(\frac{\pi}{2} - A\right)$,

$$\tan \frac{\pi}{6} = \cot\left(\frac{\pi}{2} - \frac{\pi}{6}\right) = \cot \frac{\pi}{3}.$$

Reference Angles

Associated with every nonquadrantal angle in standard position is a positive acute angle called its *reference angle*. A **reference angle** for an angle θ, written $\boldsymbol{\theta'}$, is the positive acute angle made by the terminal side of angle θ and the x-axis. Figure 91 shows several angles θ (each less than one complete counterclockwise revolution) in quadrants II, III, and IV, with the reference angle θ' also shown. In quadrant I, if θ is acute, then θ and θ' are the same. If an angle θ is negative or has measure greater than 360° its reference angle is found by first finding its coterminal angle that is between 0° and 360° and then using the diagrams in Figure 91.

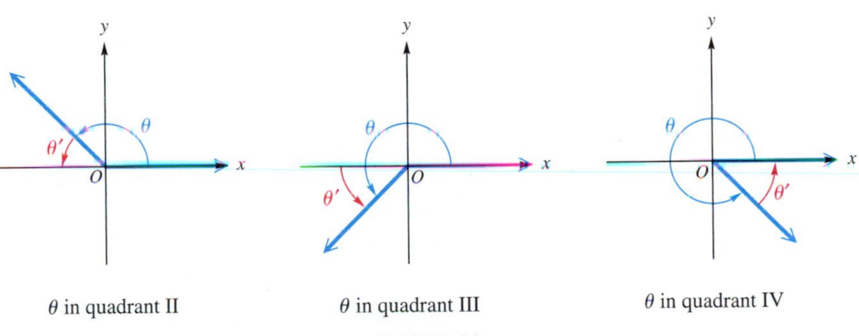

FIGURE 91

CAUTION A common error is to find the reference angle by using the terminal side of θ and the y-axis. ***The reference angle is always found with reference to the x-axis.***

EXAMPLE 4 **Finding Reference Angles**

Find the reference angle for each angle.

(a) 218° **(b)** 1387° **(c)** $\dfrac{5\pi}{6}$

Solution

(a) As shown in Figure 92, the positive acute angle made by the terminal side of this angle and the x-axis is $218° - 180° = 38°$. For $\theta = 218°$, the reference angle is $\theta' = 38°$.

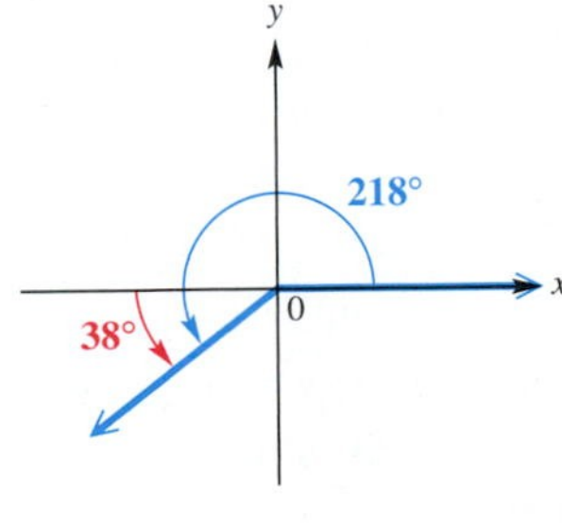

FIGURE 92

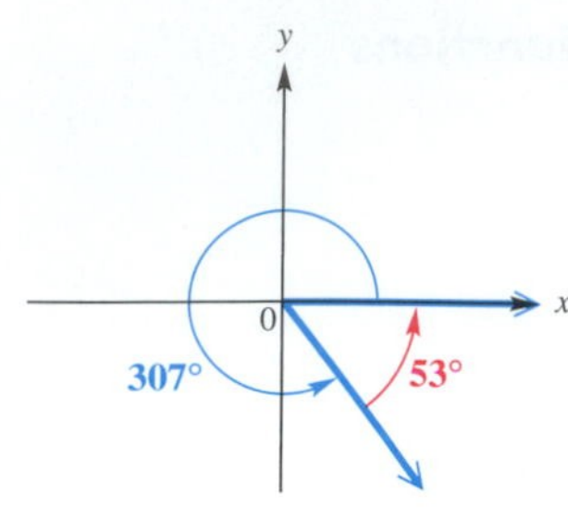

FIGURE 93

(b) First find a coterminal angle to 1387° between 0° and 360°. Divide 1387° by 360° to get a quotient of about 3.9. Begin by subtracting 360° three times (because of the 3 in 3.9): $1387° - 3 \cdot 360° = 307°$. The reference angle for 307° (and thus for 1387°) is $360° - 307° = 53°$. See Figure 93.

(c) The reference angle is $\pi - \frac{5\pi}{6} = \frac{\pi}{6}$. See Figure 94. ■

Special Angles as Reference Angles

We can now find exact trigonometric function values of angles with reference angles of 30°, 60°, or 45°.

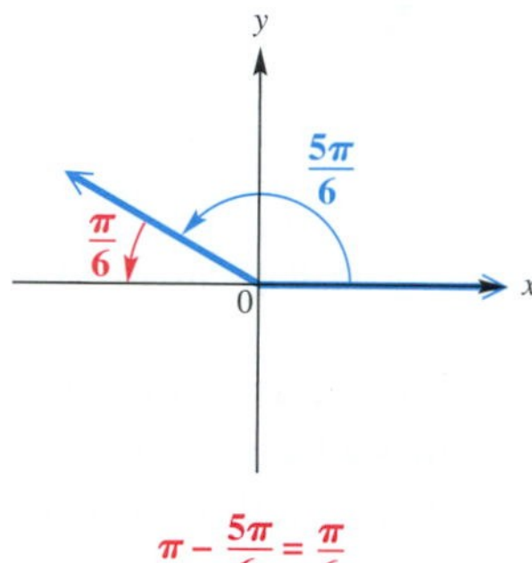

FIGURE 94

EXAMPLE 5 Finding Trigonometric Function Values of a Quadrant III Angle

Find the values of the trigonometric functions for 210°.

Solution An angle of 210° is shown in Figure 95. The reference angle is $210° - 180° = 30°$. To find the trigonometric function values of 210°, choose point P on the terminal side of the angle so that the distance from the origin O to P is 2. By the results from 30°–60° right triangles, the coordinates of point P become $(-\sqrt{3}, -1)$, with $x = -\sqrt{3}$, $y = -1$, and $r = 2$. Then, by the definitions of the trigonometric functions,

$$\sin 210° = -\frac{1}{2} \qquad \cos 210° = -\frac{\sqrt{3}}{2} \qquad \tan 210° = \frac{\sqrt{3}}{3}$$

$$\csc 210° = -2 \qquad \sec 210° = -\frac{2\sqrt{3}}{3} \qquad \cot 210° = \sqrt{3}.$$ ■

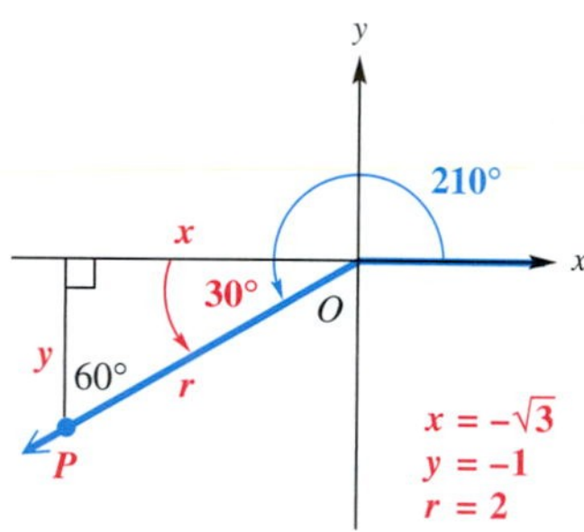

FIGURE 95

Notice in Example 5 that the trigonometric function values of 210° correspond in absolute value to those of its reference angle 30°. The signs are different for the sine, cosine, secant, and cosecant functions because 210° is a quadrant III angle. These results suggest a shortcut for finding the trigonometric function values of a nonacute angle, using the reference angle. In Example 5, the reference angle for 210° is 30°. Using the trigonometric function values of 30° and choosing the correct signs for a quadrant III angle, we obtain the same results found in Example 5.

Finding Trigonometric Function Values for Any Nonquadrantal Angle θ

Step 1 If $\theta > 360°$, or if $\theta < 0°$, then find a coterminal angle by adding or subtracting 360° as many times as needed to obtain an angle greater than 0° but less than 360°.

Step 2 Find the reference angle θ'.

Step 3 Find the trigonometric function values for reference angle θ'.

Step 4 Determine the correct signs for the values found in Step 3. (Use the table of signs in Section 8.5 if necessary.) This gives the values of the trigonometric functions for angle θ.

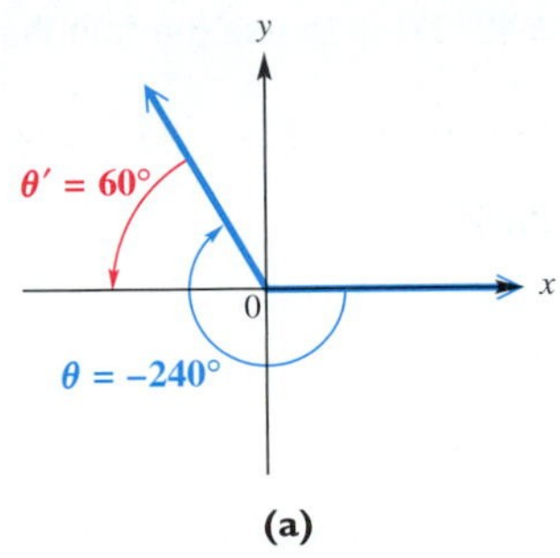

(a)

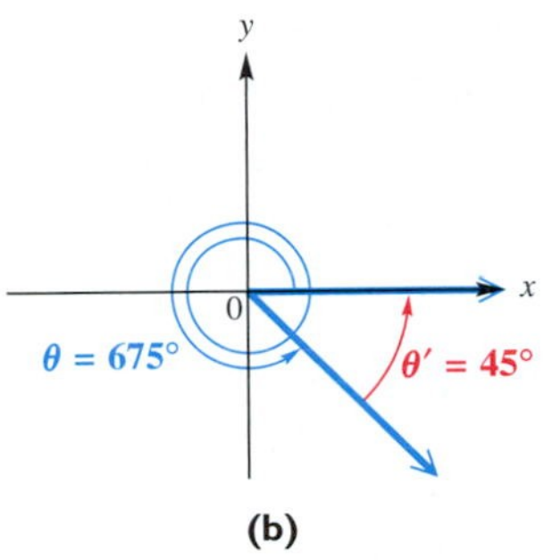

(b)

FIGURE 96

EXAMPLE 6 Finding Trigonometric Function Values by Using Reference Angles

Find the exact value of each expression.

(a) $\cos(-240°)$ **(b)** $\tan 675°$

Solution

(a) Since an angle of $-240°$ is coterminal with an angle of

$$-240° + 360° = 120°,$$

the reference angle is $180° - 120° = 60°$, as shown in Figure 96(a). Since the cosine is negative in quadrant II,

$$\cos(-240°) = \cos 120° = -\cos 60° = -\frac{1}{2}.$$

(b) Begin by subtracting 360° to get a coterminal angle between 0° and 360°.

$$675° - 360° = 315°$$

As shown in Figure 96(b), the reference angle is $360° - 315° = 45°$. An angle of 315° is in quadrant IV, so the tangent will be negative, and

$$\tan 675° = \tan 315° = -\tan 45° = -1.$$ ■

NOTE To determine the reference angle for an angle θ, always find the acute angle between the terminal side of θ and the x-axis. ***Do not use the y-axis when finding reference angles.***

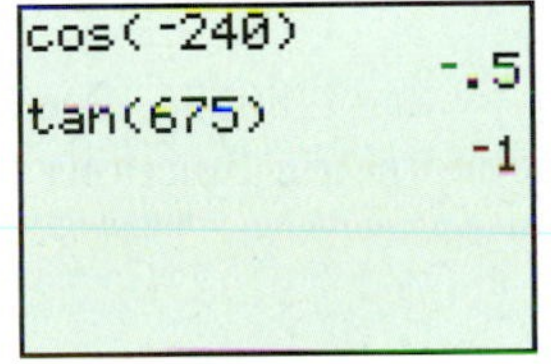

Degree mode

FIGURE 97

Finding Function Values with a Calculator

We saw in Section 8.2 that calculators are capable of finding trigonometric function values for real numbers. The same is true for angles. For example, the values of $\cos(-240°)$ and $\tan 675°$, found analytically in Example 6, are found with a calculator in Figure 97.

CAUTION We know that angles can be measured in either degrees or radians. When evaluating trigonometric functions of angles given in degrees, remember that the calculator must be set in *degree mode.*

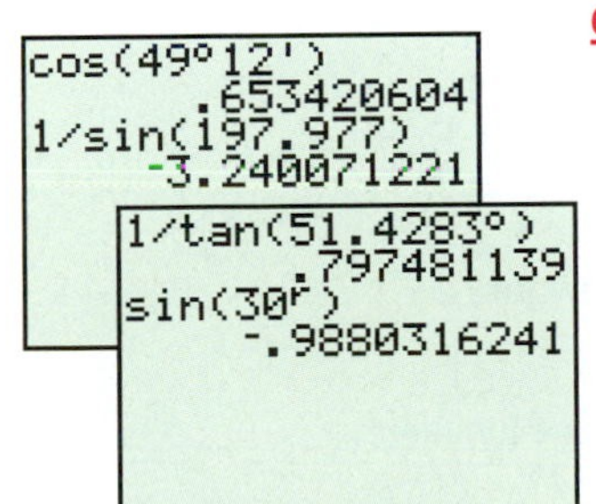

FIGURE 98

GCM

EXAMPLE 7 Approximating Trigonometric Function Values with a Calculator

Approximate the value of each expression.

(a) $\cos 49° 12'$ **(b)** $\csc 197.977°$ **(c)** $\cot 51.4283°$ **(d)** $\sin 30$

Solution The top screen in Figure 98 shows how to approximate the expressions in parts (a) and (b). The calculator is in degree mode. Notice that to find a cosecant value, we take the reciprocal of the sine value.

The bottom screen in Figure 98 shows how to approximate the expressions in parts (c) and (d). Here, we use the degree symbol (°) and the radian symbol (^{r}), which overrides the mode of the calculator. ■

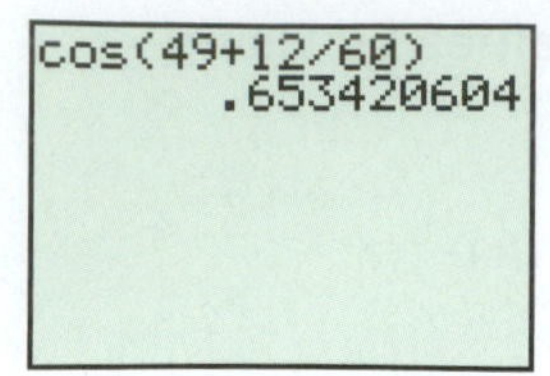

Degree mode

FIGURE 99

NOTE An alternative method of finding a value such as cos 49° 12′ is to use the fact that 60′ = 1°. Thus, as shown in Figure 99,

$$\cos 49° 12' = \cos\left(49 + \frac{12}{60}\right)^{\circ} \approx .653420604.$$

Finding Angle Measures

Sometimes we need to find the measure of an angle having a certain trigonometric function value. Graphing calculators have three *inverse functions* (denoted $\sin^{-1}$, $\cos^{-1}$, and $\tan^{-1}$) which do just that. If x is an appropriate number, then $\sin^{-1}x$, $\cos^{-1}x$, and $\tan^{-1}x$ give the measure of an angle whose sine, cosine, and tangent, respectively, is x. For the applications in this section, these functions will return measures of angles in quadrant I. (A more complete discussion of inverse functions follows in Chapter 9.)

GCM TECHNOLOGY NOTE

Students often confuse the symbols for the inverse trigonometric functions with the reciprocal functions. For example, $\sin^{-1} x$ represents an angle whose sine is x, *not* the reciprocal of $\sin x$ (which is $\csc x$). To find reciprocal function values, use the function together with the reciprocal function of the calculator.

GCM **EXAMPLE 8 Using Inverse Trigonometric Functions**

(a) Use a calculator to find an angle θ in degrees that satisfies $\sin \theta \approx .9677091705$.

(b) Use a calculator to find an angle θ in radians that satisfies $\tan \theta \approx .25$.

Solution

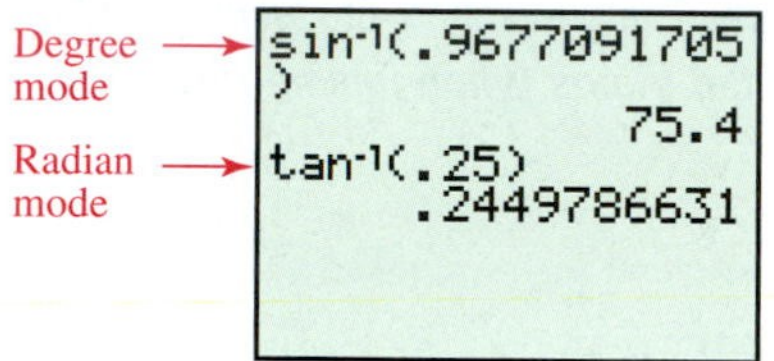

FIGURE 100

(a) With the calculator in *degree mode,* we find that an angle θ having sine value .9677091705 is 75.4°. (While there are infinitely many such angles, the calculator gives only this one.) We write this result as $\sin^{-1} .9677091705 \approx 75.4°$. See the first calculation in Figure 100.

(b) With the calculator in *radian mode,* we find $\tan^{-1} .25 \approx .2449786631$. See the second calculation in Figure 100. Note that inverse trigonometric functions can calculate angles in degrees or radians. ■

Grade resistance is the force F that causes a car to roll down a hill. It can be calculated by $F = W \sin \theta$, where θ represents the angle of the grade and W represents the weight of the vehicle. (See Figure 83 in Section 8.5.)

EXAMPLE 9 Calculating Highway Grade

Find the angle θ for which a 3000-pound car has a grade resistance of 500 pounds.

Solution

$$F = W \sin \theta$$

$$500 = 3000 \sin \theta \qquad \text{Let } F = 500 \text{ and } W = 3000.$$

$$\sin \theta = \frac{1}{6} \qquad \text{Divide by 3000; rewrite.}$$

$$\theta = \sin^{-1} \frac{1}{6} \qquad \text{Use the inverse sine function.}$$

$$\theta \approx 9.6° \qquad \text{Approximate.}$$

Use degree mode here.

Thus, if a road is inclined at approximately 9.6°, a 3000-pound car would be acted on by a force of 500 pounds pulling downhill. ■

EXAMPLE 10 Finding Angle Measures

Find all values of θ if θ is in the interval $[0°, 360°)$ and $\cos\theta = -\frac{\sqrt{2}}{2}$.

Analytic Solution

Since cosine is negative, θ must lie in quadrant II or III. Because the absolute value of $\cos\theta$ is $\frac{\sqrt{2}}{2}$, the reference angle θ' must be $\cos^{-1}\frac{\sqrt{2}}{2} = 45°$. The two possible angles θ are sketched in Figure 101.

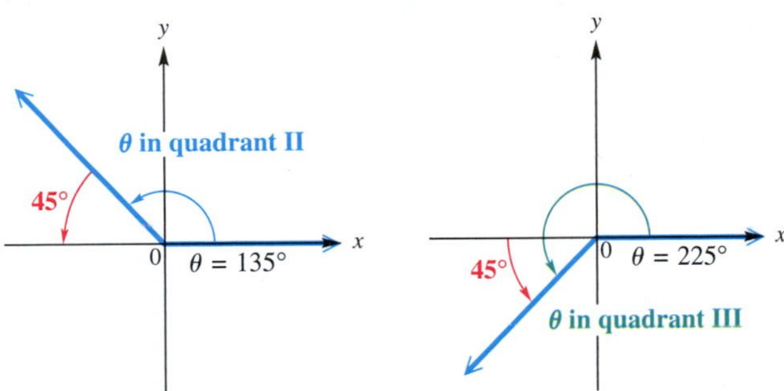

FIGURE 101

The quadrant II angle θ must equal $180° - 45° = 135°$, and the quadrant III angle θ must equal $180° + 45° = 225°$.

Graphing Calculator Solution

The screen in Figure 102 shows how the inverse cosine function is used to find the two values in $[0°, 360°)$ for which $\cos\theta = -\frac{\sqrt{2}}{2}$. Notice that $\cos^{-1}\left(-\frac{\sqrt{2}}{2}\right)$ yields only one value: 135°; to find the other value, we use the reference angle.

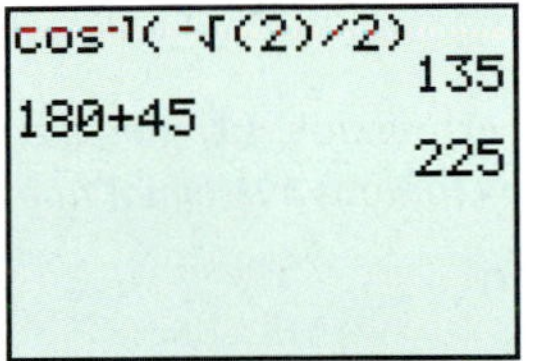

Degree mode

FIGURE 102

EXAMPLE 11 Finding Angle Measures

Find two angles in the interval $[0, 2\pi)$ that satisfy $\cos\theta \approx .3623577545$.

Solution With the calculator in radian mode, we find that one such θ is

$$\theta = \cos^{-1} .3623577545 \approx 1.2.$$

Since $1.2 < \frac{\pi}{2}$, θ is in quadrant I. The other value of θ will have its reference angle θ' equal to 1.2 and must be in quadrant IV, since the angle given by the calculator is in quadrant I and the cosine is also positive in quadrant IV. The other value of θ is

$$\theta = 2\pi - 1.2 \approx 5.083185307.$$

What Went Wrong?

A student was asked to find cot 20°. He set his calculator in degree mode and, knowing that cotangent is the reciprocal of tangent, produced the screen shown here. The correct answer is $\cot 20° \approx 2.747477419$.

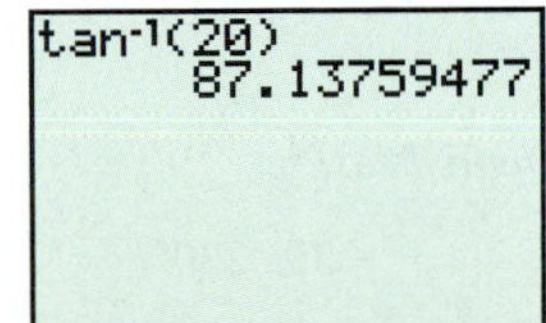

What Went Wrong? How can he obtain the correct answer?

Answer to What Went Wrong?

The student used the *inverse tangent function* ($\tan^{-1}$) rather than finding the *reciprocal* of the tangent of 20°. If he enters $1/\tan(20°)$, he will obtain the correct answer.

8.6 Exercises

Find exact values or expressions for the six trigonometric functions of angle A.

1.

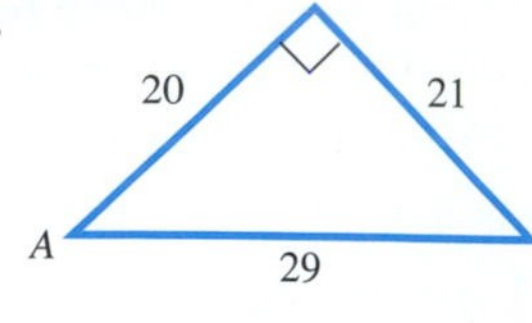

2.

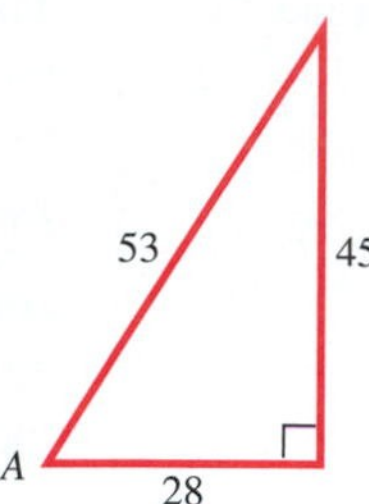

3.

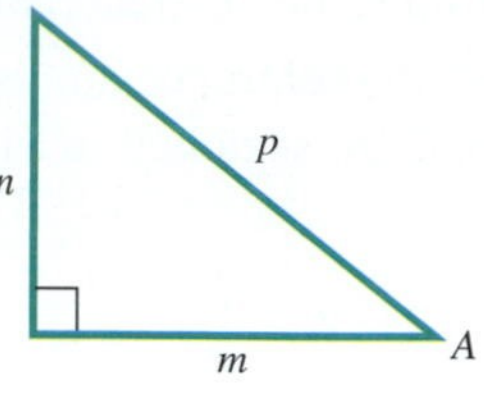

4.

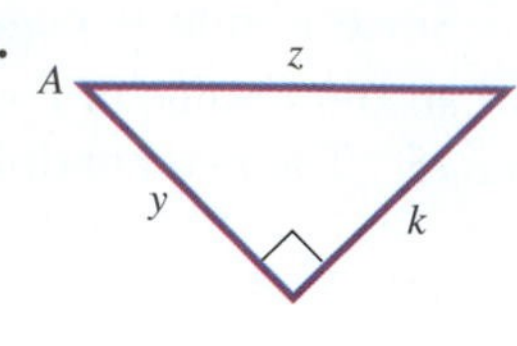

For each expression, **(a)** *give the exact value and* **(b)** *if the exact value is irrational, use your calculator to support your answer in part (a) by finding a decimal approximation.*

5. tan 30°

6. cot 30°

7. sin 30°

8. cos 30°

9. sec 30°

10. csc 30°

11. csc 45°

12. sec 45°

13. cos 45°

14. cot 45°

15. $\sin \frac{\pi}{3}$

16. $\cos \frac{\pi}{3}$

17. $\tan \frac{\pi}{3}$

18. $\cot \frac{\pi}{3}$

19. $\sec \frac{\pi}{3}$

20. $\csc \frac{\pi}{3}$

 **21.** A student was asked to give the exact value of sin 45°. Using his calculator, he gave the answer .7071067812. The teacher did not give him credit. Why?

 **22.** A student was asked to give an approximate value of sin 45. With her calculator in degree mode, she gave the value .7071067812. The teacher did not give her credit. What was her error?

Write each expression in terms of its cofunction.

23. cot 73°

24. sec 39°

25. sin 38°

26. cos 19°

27. tan 25° 43′

28. sin 38° 29′

29. $\cos \frac{\pi}{5}$

30. $\sin \frac{\pi}{3}$

31. tan .5

32. csc .3

Give the reference angle for each angle measure.

33. 98°

34. 212°

35. 230°

36. 130°

37. −135°

38. −60°

39. 750°

40. 480°

41. $\frac{4\pi}{3}$

42. $\frac{7\pi}{6}$

43. $-\frac{4\pi}{3}$

44. $-\frac{7\pi}{6}$

45. ***Concept Check*** In Example 5, why was 2 a good choice for r? Could any other positive number be used?

 **46.** Explain how the reference angle is used to find values of the trigonometric functions for an angle in quadrant III.

Complete the table with exact *trigonometric function values. Do* not *use a calculator.*

	θ	$\sin\theta$	$\cos\theta$	$\tan\theta$	$\cot\theta$	$\sec\theta$	$\csc\theta$
47.	**30°**	$\frac{1}{2}$	$\frac{\sqrt{3}}{2}$			$\frac{2\sqrt{3}}{3}$	2
48.	**45°**			1	1		
49.	**60°**		$\frac{1}{2}$	$\sqrt{3}$		2	
50.	**120°**	$\frac{\sqrt{3}}{2}$		$-\sqrt{3}$			$\frac{2\sqrt{3}}{3}$
51.	**135°**	$\frac{\sqrt{2}}{2}$	$-\frac{\sqrt{2}}{2}$			$-\sqrt{2}$	$\sqrt{2}$
52.	**150°**		$-\frac{\sqrt{3}}{2}$	$-\frac{\sqrt{3}}{3}$			2
53.	**210°**	$-\frac{1}{2}$		$\frac{\sqrt{3}}{3}$	$\sqrt{3}$		-2
54.	**240°**	$-\frac{\sqrt{3}}{2}$	$-\frac{1}{2}$			-2	$-\frac{2\sqrt{3}}{3}$

Find exact values of the six trigonometric functions for each angle by hand.

55. 300°

56. 315°

57. 405°

58. 420°

59. $\frac{11\pi}{6}$

60. $\frac{5\pi}{3}$

61. $-\frac{7\pi}{4}$

62. $-\frac{4\pi}{3}$

Use a calculator to find a decimal approximation for each value. Give as many digits as your calculator displays.

63. tan 29°

64. sin 38°

65. cot 41° 24′

66. csc 145° 45′

67. sec 183° 48′

68. cos 421° 30′

69. tan(−80° 6′)

70. sin(−317° 36′)

71. sin 2.5

72. cos 3.8

73. tan 5

74. sec 10

For each expression, **(a)** *write the function in terms of a function of the reference angle,* **(b)** *give the exact value, and* **(c)** *use a calculator to show that the decimal value or approximation for the given function is the same as the decimal value or approximation for your answer in part (b).*

75. $\sin\frac{7\pi}{6}$

76. $\cos\frac{5\pi}{3}$

77. $\tan\frac{3\pi}{4}$

78. $\sin\frac{5\pi}{3}$

79. $\cos\frac{7\pi}{6}$

80. $\tan\frac{4\pi}{3}$

Find all values of θ if θ is in the interval $[0°, 360°)$ and has the given function value.

81. $\sin\theta = \frac{1}{2}$

82. $\cos\theta = \frac{\sqrt{3}}{2}$

83. $\tan\theta = -\sqrt{3}$

84. $\sec\theta = -\sqrt{2}$

85. $\cot\theta = -\frac{\sqrt{3}}{3}$

86. $\cos\theta = \frac{\sqrt{2}}{2}$

Find all values of θ if θ is in the interval $[0°, 360°)$ and has the given function value. Give approximations to as many decimal places as your calculator displays.

87. $\cos\theta \approx .68716510$

88. $\cos\theta \approx .96476120$

89. $\sin\theta \approx .41298643$

90. $\sin\theta \approx .63898531$

91. $\tan\theta \approx .87692035$

92. $\tan\theta \approx 1.2841996$

Find two angles in the interval $[0, 2\pi)$ that satisfy the given equation. Give calculator approximations to as many digits as your calculator displays.

93. $\tan\theta \approx .21264138$

94. $\cos\theta \approx .78269876$

95. $\cot\theta \approx .29949853$

96. $\csc\theta \approx 1.0219553$

Relating Concepts

For individual or group investigation (Exercises 97–108)

In a square window of your calculator that gives a good picture of the first quadrant, graph the line $y = \sqrt{3}x$ with $x \geq 0$. Then, trace to any point on the line. See the figure. What we see is a simulated view of an angle in standard position, with terminal side in quadrant I. Store the values of x and y in convenient memory locations, and call them x_1 and y_1. ***Work Exercises 97–108 in order.***

97. Find the value of $\sqrt{x_1^2 + y_1^2}$ and store it in a convenient memory location. (Call it r.) What does this number mean geometrically?

98. With your calculator in degree mode, find $\tan^{-1}\left(\frac{y_1}{x_1}\right)$.

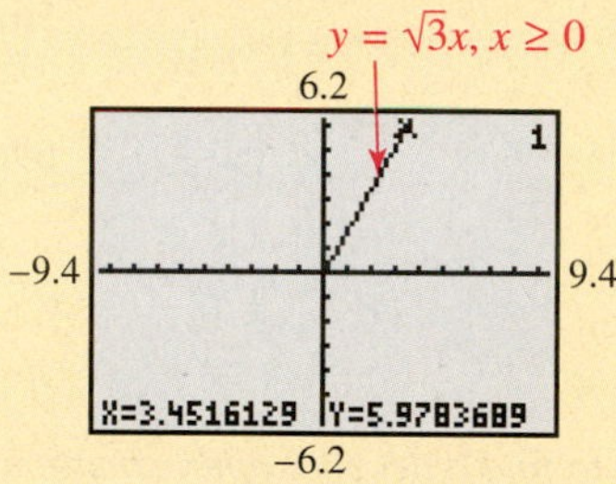

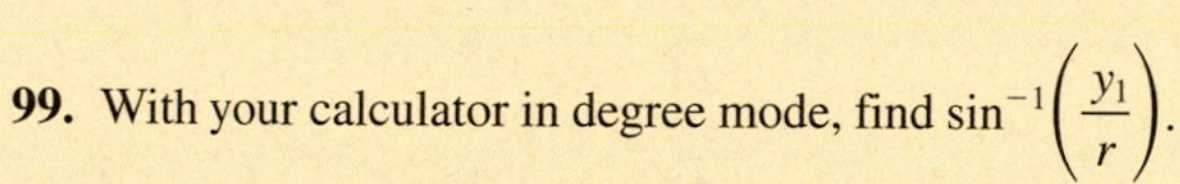

99. With your calculator in degree mode, find $\sin^{-1}\left(\frac{y_1}{r}\right)$.

100. With your calculator in degree mode, find $\cos^{-1}\left(\frac{x_1}{r}\right)$.

101. Your answers in Exercises 98–100 should all be the same. How does this answer relate to the angle formed between the positive x-axis and the line?

102. Find the value of $\frac{y_1}{x_1}$. Now square it. What do you get? What is the exact value of $\frac{y_1}{x_1}$?

103. Look at the equation of the line you graphed, and make a conjecture: The ________ of a line passing through the origin is equal to the ________ of the angle it forms with the positive x-axis.

104. Find the value of $\left(\frac{x_1}{r}\right)^2 + \left(\frac{y_1}{r}\right)^2$. What identity does this illustrate?

105. Find $\csc 60°$. Then find the value of $\frac{r}{y_1}$. Do they agree?

106. Graph $y_2 = \sqrt{1 - x^2}$ as a second curve in the same viewing window. This is one-half of a circle centered at the origin with radius 1. Now use the intersection feature of your calculator to determine the x- and y-coordinates of the points of intersection of the two graphs. What are they?

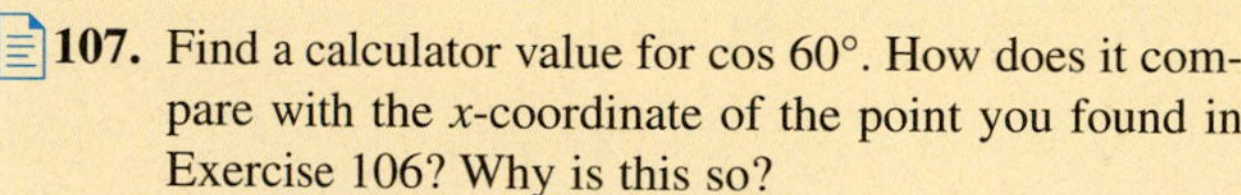

107. Find a calculator value for $\cos 60°$. How does it compare with the x-coordinate of the point you found in Exercise 106? Why is this so?

108. Find a calculator approximation for $\sin 60°$. How does it compare with the y-coordinate of the point you found in Exercise 106? Why is this so?

Highway Grade *Exercises 109 and 110 refer to Example 9. Use the formula $F = W \sin \theta$ to find the angle θ for which a car with weight W has grade resistance F.*

109. $W = 5000$ lb, $F = 400$ lb

110. $W = 3500$ lb, $F = 130$ lb

(Modeling) Speed of Light *When a light ray travels from one medium, such as air, to another medium, such as water or glass, the speed of the light and the direction in which the ray is traveling change. (This is why a fish under water is in a different position than it appears to be.) The changes are given by Snell's law,*

$$\frac{c_1}{c_2} = \frac{\sin \theta_1}{\sin \theta_2},$$

where c_1 is the speed of light in the first medium, c_2 is the speed of light in the second medium, and θ_1 and θ_2 are the angles shown in the figure. (Source: The Physics Classroom, www.glenbrook.k12.il.us*) In Exercises 111–116, assume that $c_1 = 3 \times 10^8$ meters per second.*

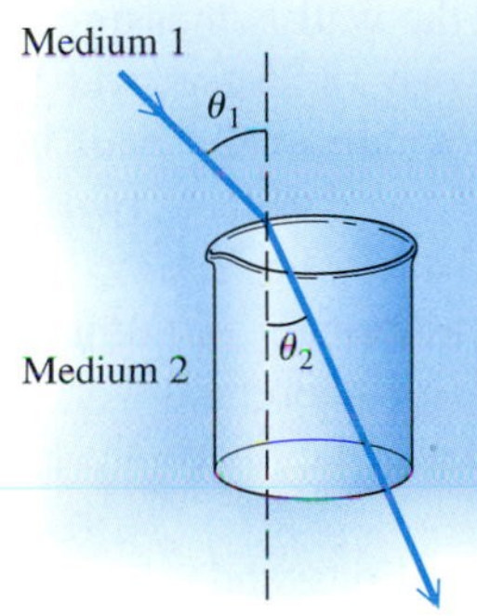

Find the speed of light in the second medium.

111. $\theta_1 = 46°$; $\theta_2 = 31°$

112. $\theta_1 = 39°$; $\theta_2 = 28°$

Find θ_2 for the following values of θ_1 and c_2. Round to the nearest degree.

113. $\theta_1 = 40°$; $c_2 = 1.5 \times 10^8$ meters per second

114. $\theta_1 = 62°$; $c_2 = 2.6 \times 10^8$ meters per second

(Modeling) Fish's View of the World *The figure at the top of the next column shows a fish's view of the world above the surface of the water. (Source:* Walker, Jearl, "The Amateur Scientist," *Scientific American,* March 1984. Illustration by Michael Goodman.*) Suppose that a light ray comes from the horizon, enters the water, and strikes the fish's eye.*

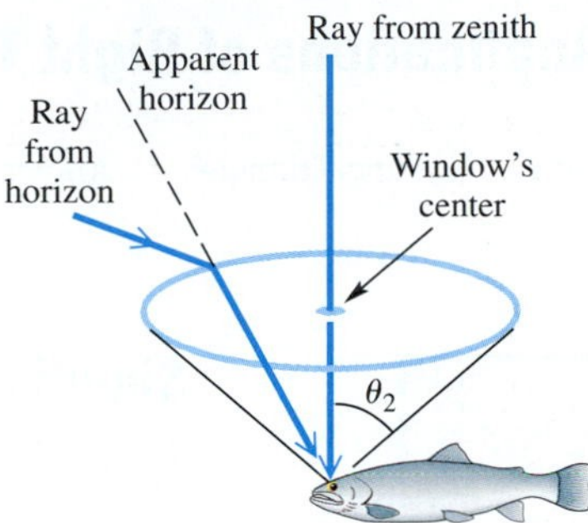

115. Assume that this ray gives a value of 90° for angle θ_1 in the formula for Snell's law. (In a practical situation, the angle would probably be a little less than 90°.) The speed of light in water is about 2.254×10^8 meters per second. Find angle θ_2.

116. Suppose an object is located at a true angle of 29.6° above the horizon. Find the apparent angle above the horizon to a fish.

117. ***(Modeling) Braking Distance*** If air resistance is ignored, the braking distance D (in feet) for an automobile to change its velocity from V_1 to V_2 (feet per second) can be modeled by the equation

$$D = \frac{1.05(V_1^2 - V_2^2)}{64.4(K_1 + K_2 + \sin \theta)}.$$

K_1 is a constant determined by the efficiency of the brakes and tires, K_2 is a constant determined by the rolling resistance of the automobile, and θ is the grade of the highway. (*Source:* Mannering, F. and W. Kilareski, *Principles of Highway Engineering and Traffic Analysis,* Second Edition, John Wiley and Sons, 1998.)

(a) Compute the number of feet required to slow a car from 55 to 30 mph while traveling uphill on a grade of $\theta = 3.5°$. Let $K_1 = .4$ and $K_2 = .02$. (*Hint:* Change miles per hour to feet per second.)

(b) Repeat part (a) with $\theta = -2°$.

(c) How is the braking distance affected by the grade θ? Does this agree with your driving experience?

118. ***(Modeling) Car's Speed at Collision*** Refer to Exercise 117. An automobile is traveling at 90 mph on a highway with a downhill grade of $\theta = -3.5°$. The driver sees a stalled truck in the road 200 feet away and immediately applies the brakes. Assuming that a collision cannot be avoided, how fast (in miles per hour) is the car traveling when it hits the truck? (Use the same values for K_1 and K_2 as in Exercise 117.)

8.7 Applications of Right Triangles

Significant Digits ■ Solving Triangles ■ Angles of Elevation or Depression ■ Bearing ■ Further Applications of Trigonometric Functions

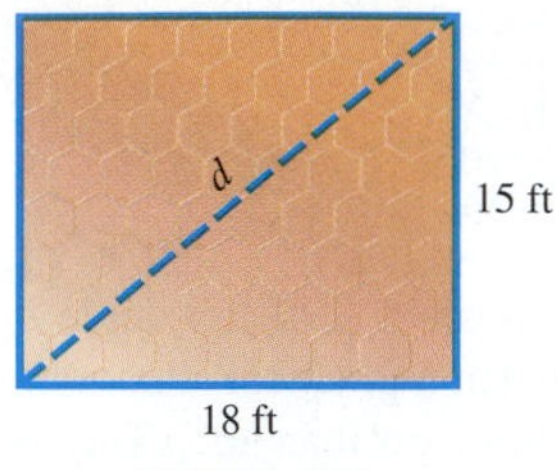

FIGURE 103

Significant Digits

Suppose we quickly measure a room as 15 feet by 18 feet. See Figure 103. To calculate the length of a diagonal of the room, we use the Pythagorean theorem.

$$d^2 = 15^2 + 18^2$$
$$d^2 = 549$$
$$d = \sqrt{549} \approx 23.430749$$

Should this answer be given as the length of the diagonal of the room? Of course not. The number 23.430749 contains six decimal places, while the original data of 15 feet and 18 feet are accurate only to the nearest foot. Since the result of a problem can be no more accurate than the least accurate number in any calculation, we really should say that the diagonal of the 15- by 18-foot room is about 23 feet.

If a wall measured to the nearest foot is 18 feet long, this actually means that the wall has length between 17.5 feet and 18.5 feet. If the wall is measured more accurately as 18.3 feet long, then its length is really between 18.25 feet and 18.35 feet. A measurement of 18.00 feet would indicate that the length of the wall is between 17.995 feet and 18.005 feet. The measurement 18 feet is said to have two *significant digits* of accuracy; 18.0 has three significant digits, and 18.00 has four.

A **significant digit** is a digit obtained by actual measurement. A number that represents the result of counting, or a number that results from theoretical work and is not the result of a measurement, is an **exact number.** For example, there are 50 states in the United States, so 50 is an exact number.

Most values of trigonometric functions are approximations, and all measurements are approximations. When performing calculations involving approximate numbers, start by determining the number that has the least number of significant digits. Round your final answer to the same number of significant digits as this number. ***Remember that your answer is no more accurate than the least accurate number in your calculation.***

Use the following table to determine the significant digits in angle measure. This table assumes that the degree portion of the measurement contains two significant digits.

Significant Digits for Angles

Angle Measure to Nearest	Number of Significant Digits
Degree	2
Ten minutes, or nearest tenth of a degree	3
Minute, or nearest hundredth of a degree	4
Tenth of a minute, or nearest thousandth of a degree	5

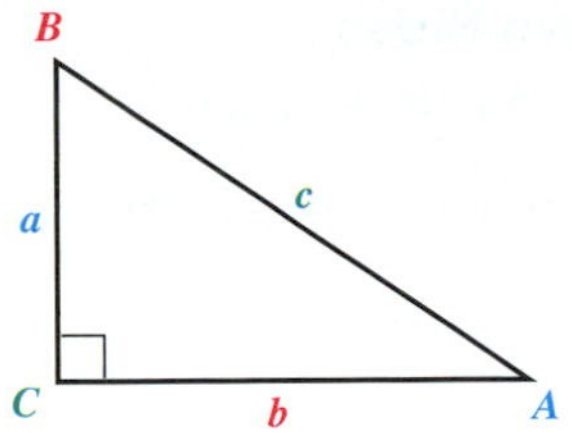

In solving triangles, a labeled sketch is an important aid.

FIGURE 104

For example, an angle measuring $52° 30'$ has three significant digits (assuming that $30'$ is measured to the nearest 10 minutes).

Solving Triangles

To *solve a triangle* means to find the measures of all the angles and sides of the triangle. As shown in Figure 104, we use a to represent the length of the side opposite angle A, b for the length of the side opposite angle B, and so on. In a right triangle, the letter c is reserved for the hypotenuse.

EXAMPLE 1 Solving a Right Triangle, Given an Angle and a Side

Solve right triangle ABC, with $A = 34° 30'$ and $c = 12.7$ inches. See Figure 105.

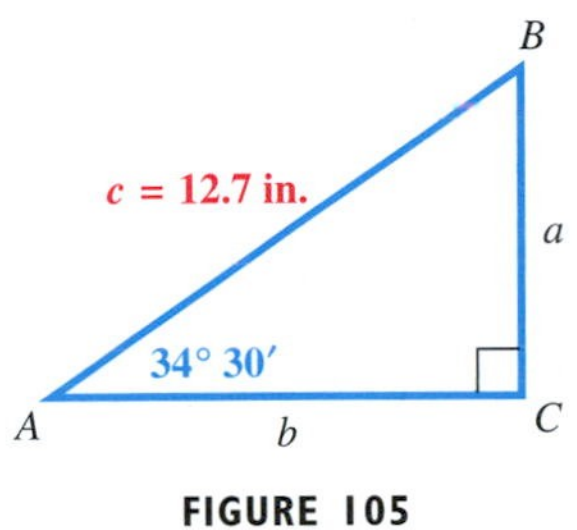

FIGURE 105

Solution To solve the triangle, find the measures of the remaining sides and angles. To find the value of a, use a trigonometric function involving the known values of angle A and side c. Since the sine of angle A is given by the quotient of the side opposite A and the hypotenuse, use $\sin A$.

$$\sin A = \frac{a}{c} \qquad \sin A = \frac{\text{side opposite}}{\text{hypotenuse}}$$

$$\sin 34° 30' = \frac{a}{12.7} \qquad A = 34° 30',\ c = 12.7$$

$$a = 12.7 \sin 34° 30' \qquad \text{Multiply by 12.7; rewrite.}$$

$$a \approx 12.7(.56640624) \qquad \text{Use a calculator.}$$

$$a \approx 7.19 \text{ inches} \qquad \text{Three significant digits}$$

To find the value of b, we could use the Pythagorean theorem. It is better, however, to use the information given in the problem rather than a result just calculated. If a mistake is made in finding a, then b also would be incorrect. Moreover, rounding more than once may cause the result to be less accurate. Use $\cos A$.

$$\cos A = \frac{b}{c} \qquad \cos A = \frac{\text{side adjacent}}{\text{hypotenuse}}$$

$$\cos 34° 30' = \frac{b}{12.7}$$

$$b = 12.7 \cos 34° 30'$$

$$b \approx 10.5 \text{ inches} \qquad \text{Three significant digits}$$

Once b is found, the Pythagorean theorem can be used as a check. All that remains to solve triangle ABC is to find the measure of angle B. Since $A + B = 90°$,

$$B = 90° - A = 89° 60' - 34° 30' = 55° 30'.$$

NOTE In Example 1, we could have found the measure of angle B first and then used the trigonometric function values of B to find the unknown sides. The process of solving a right triangle can usually be done in several ways. ***To maintain accuracy, always use given information as much as possible, and avoid rounding off in intermediate steps.***

Looking Ahead to Calculus

The derivatives of the parametric equations $x = f(t)$ and $y = g(t)$ often represent the rates of change of physical quantities, such as velocities. In such cases, the derivatives are called *related rates* because a change in one causes a related change in the other. Determining these rates in calculus often requires solving right triangles. Many problems that require the maximum or minimum value of some quantity also involve solving a right triangle.

GCM TECHNOLOGY NOTE

Once you have mastered the material on solving right triangles, you may wish to write a program that will accomplish this goal. You will need to consider the various cases of what is given and what must be found.

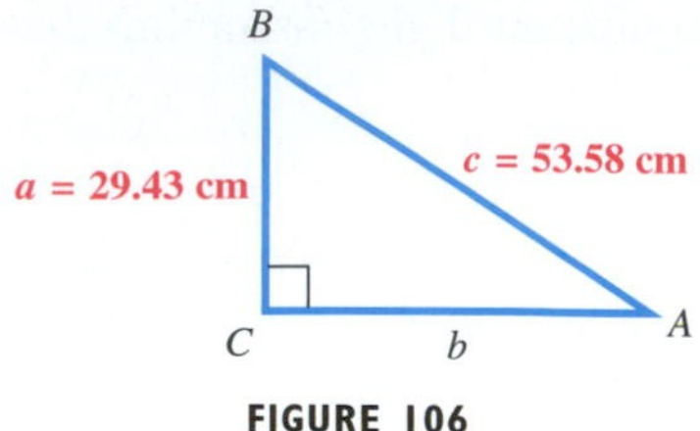

FIGURE 106

EXAMPLE 2 Solving a Right Triangle, Given Two Sides

Solve right triangle ABC if $a = 29.43$ centimeters and $c = 53.58$ centimeters.

Solution We draw a sketch showing the given information, as in Figure 106. One way to begin is to find angle A by using the sine function.

$$\sin A = \frac{\text{side opposite}}{\text{hypotenuse}} = \frac{29.43}{53.58}$$

Using a calculator, we find that $A = \sin^{-1} \frac{29.43}{53.58} \approx 33.32°$. The measure of B is approximately $90° - 33.32° = 56.68°$. We now find b from the Pythagorean theorem.

$$b^2 = c^2 - a^2 \qquad \text{Pythagorean theorem solved for } b^2$$

$$b^2 = 53.58^2 - 29.43^2 \qquad c = 53.58,\ a = 29.43$$

$$b \approx 44.77 \text{ centimeters} \qquad \text{Subtract; take square roots.}$$

■

Angles of Elevation or Depression

Y
Angle of elevation
X
Horizontal

(a)

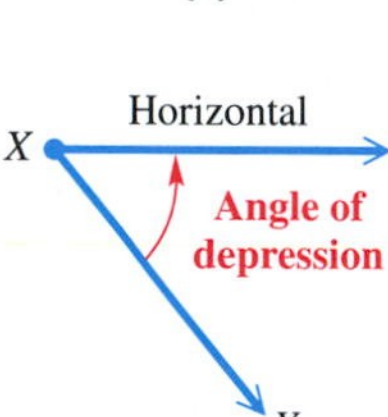

(b)

FIGURE 107

Many applications of right triangles involve angles of elevation or depression. The **angle of elevation** from point X to point Y (above X) is the acute angle formed by ray XY and a horizontal ray with endpoint at X. See Figure 107(a). The **angle of depression** from point X to point Y (below X) is the acute angle formed by ray XY and a horizontal ray with endpoint X. See Figure 107(b).

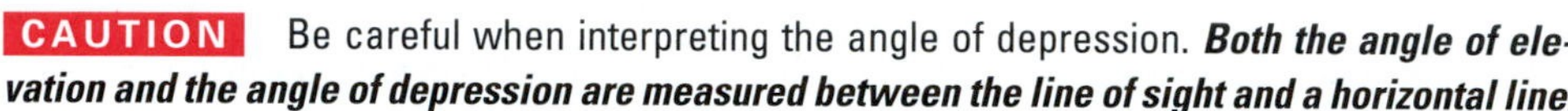

CAUTION Be careful when interpreting the angle of depression. ***Both the angle of elevation and the angle of depression are measured between the line of sight and a horizontal line.***

To solve applied trigonometry problems, we often solve triangles.

Solving an Applied Trigonometry Problem

Step 1 Draw a sketch, and label it with the given information. Label the quantity to be found with a variable.

Step 2 Use the sketch to write an equation relating the given quantities to the variable.

Step 3 Solve the equation, and check that your answer makes sense.

EXAMPLE 3 Finding the Angle of Elevation when Lengths Are Known

The length of the shadow of a building 34.09 meters tall is 37.62 meters. Find the angle of elevation of the sun.

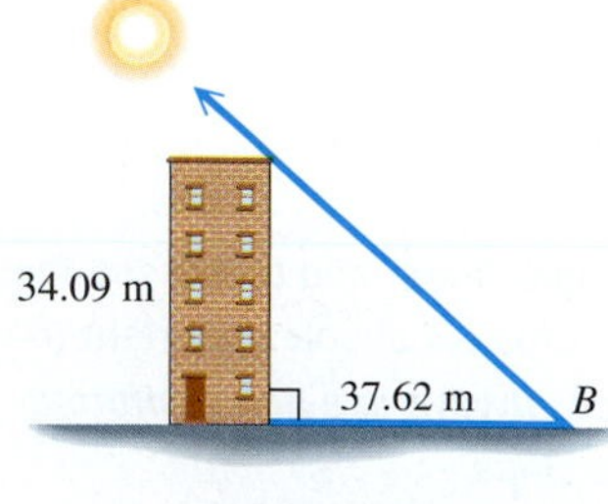

FIGURE 108

Solution As shown in Figure 108, the angle of elevation of the sun is angle B. Since the side opposite B and the side adjacent to B are known, use the tangent ratio to find B.

$$\tan B = \frac{34.09}{37.62}, \quad \text{so} \quad B = \tan^{-1} \frac{34.09}{37.62} \approx 42.18°$$

The angle of elevation of the sun is 42.18°. ■

Bearing

Other applications of right triangles involve **bearing,** an important idea in navigation. There are two methods for expressing bearing. When a single angle is given, such as 164°, it is understood that the bearing is measured in a clockwise direction from due north. Several sample bearings using this first method are shown in Figure 109.

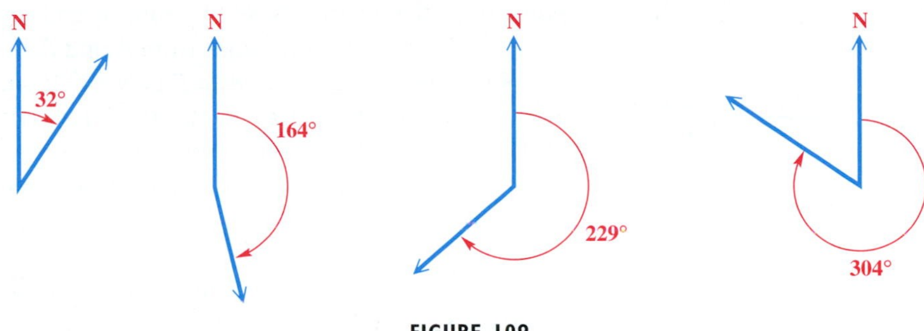

FIGURE 109

EXAMPLE 4 Solving a Problem Involving Bearing (First Method)

Radar stations A and B are on an east–west line, with A west of B, 3.7 kilometers apart. Station A detects a plane at C, on a bearing of 61°. Station B simultaneously detects the same plane, on a bearing of 331°. Find the distance from A to C.

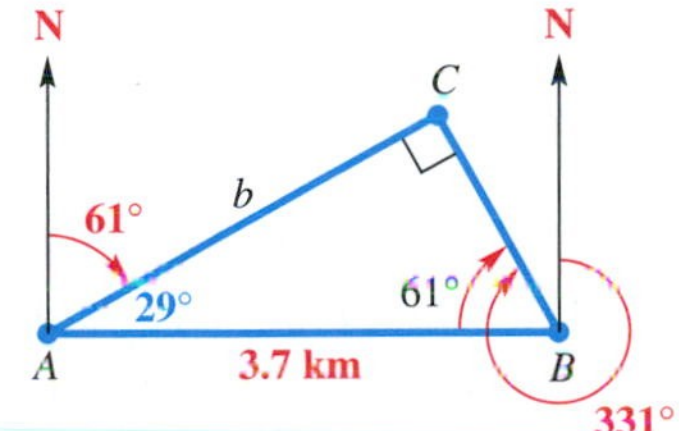

FIGURE 110

Solution Draw a sketch showing the given information, as in Figure 110. Since a line drawn due north is perpendicular to an east–west line, right angles are formed at A and B, so angles CAB and CBA can be found. Angle C is a right angle because angles CAB and CBA are complementary. Find distance b by using the cosine function.

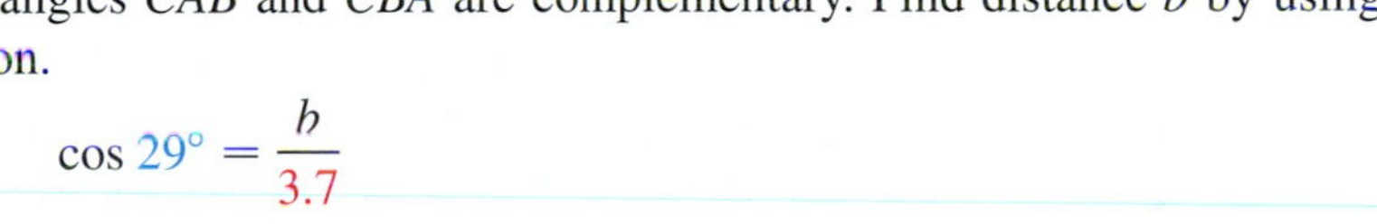

$$\cos 29° = \frac{b}{3.7}$$

$$3.7 \cos 29° = b \qquad \text{Multiply by 3.7.}$$

$$b \approx 3.2 \text{ kilometers} \qquad \text{Use a calculator; round to the nearest tenth.}$$

CAUTION ***The importance of a correctly labeled sketch when solving applications like that in Example 4 cannot be overemphasized.*** Some of the necessary information is often not directly stated in the problem and can be determined only from the sketch.

The second method for expressing bearing starts with a north–south line and uses an acute angle to show the direction, either east or west, from this line. Figure 111 shows several sample bearings using this system. Either N or S always comes first, followed by an acute angle, and then E or W.

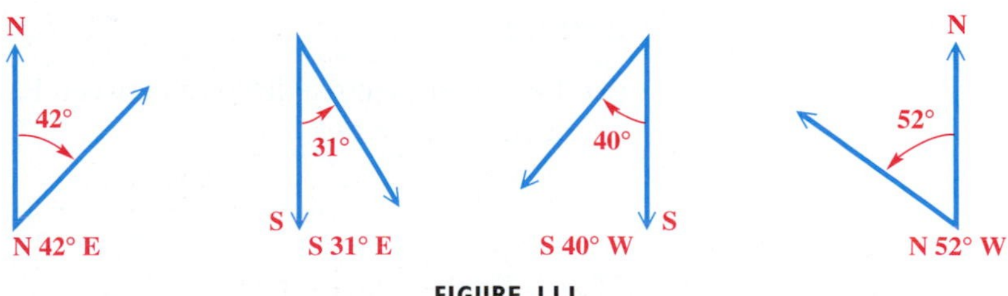

FIGURE 111

EXAMPLE 5 **Solving a Problem Involving Bearing (Second Method)**

The bearing from A to C is S 52° E. The bearing from A to B is N 84° E. The bearing from B to C is S 38° W. A plane flying at 250 mph takes 2.4 hours to go from A to B. Find the distance from A to C.

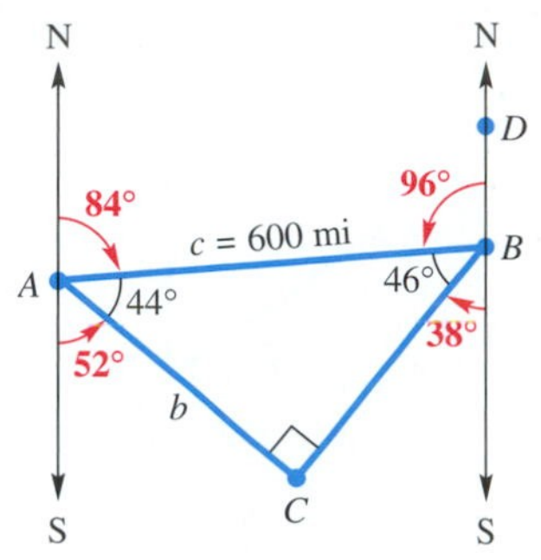

FIGURE 112

Solution Make a sketch. First draw the two bearings from point A. Then choose a point B on the bearing N 84° E from A, and draw the bearing to C. Point C will be located where the bearing lines from A and B intersect, as shown in Figure 112.

Since the bearing from A to B is N 84° E, angle ABD is $180° - 84° = 96°$. Thus, angle ABC is 46°. Also, angle BAC is $180° - (84° + 52°) = 44°$. Angle C is $180° - (44° + 46°) = 90°$. From the statement of the problem, a plane flying at 250 mph takes 2.4 hours to go from A to B. The distance from A to B is

$$c = \text{rate} \times \text{time} = 250(2.4) = 600 \text{ miles.}$$

To find b, the distance from A to C, use the sine function. (The cosine function could also be used.)

$$\sin 46° = \frac{b}{c}$$

$$\sin 46° = \frac{b}{600} \qquad \text{Let } c = 600.$$

$$600 \sin 46° = b \qquad \text{Multiply by 600.}$$

$$b \approx 430 \text{ miles} \qquad \text{Two significant digits}$$

Further Applications of Trigonometric Functions

For centuries, astronomers wanted to know how far it was to the stars. Not until 1838 did the astronomer Friedrich Bessel determine the distance to a star called 61 Cygni. He used a *parallax** method that relied on the measurement of small angles. See Figure 113. As Earth revolves around the sun, the observed parallax of 61 Cygni is $\theta \approx .0000811°$. Because stars are so distant, parallax angles are very small. (*Source:* Freebury, H., *A History of Mathematics,* MacMillan, 1961; Zeilik, M. et al., *Introductory Astronomy and Astrophysics,* Third Edition, Saunders College Publishers, 1992.)

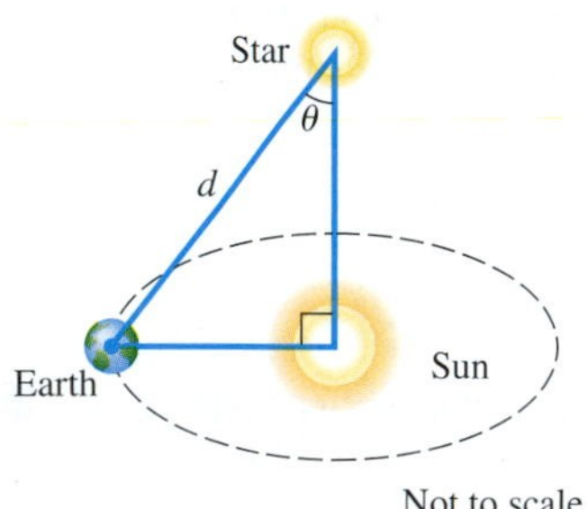

FIGURE 113

EXAMPLE 6 **Calculating the Distance to a Star**

One of the nearest stars is Alpha Centauri, which has parallax $\theta \approx .000212°$.

(a) If the Earth–Sun distance is 93,000,000 miles, calculate the distance to Alpha Centauri.

(b) A light-year is defined to be the distance that light travels in 1 year and equals about 5.9 trillion miles. Find the distance to Alpha Centauri in light-years.

Solution

(a) Let d represent the distance between Earth and Alpha Centauri. From Figure 113, it can be seen that

$$\sin \theta = \frac{93{,}000{,}000}{d} \quad \text{or} \quad d = \frac{93{,}000{,}000}{\sin \theta}.$$

*You observe parallax when you ride in a car and see a nearby object apparently moving backward with respect to more distant objects.

Substituting .000212° for θ gives

$$d = \frac{93{,}000{,}000}{\sin .000212^\circ} \approx 2.51 \times 10^{13} \text{ miles.}$$

(b) This distance equals $\dfrac{2.51 \times 10^{13}}{5.9 \times 10^{12}} \approx 4.3$ light-years. ■

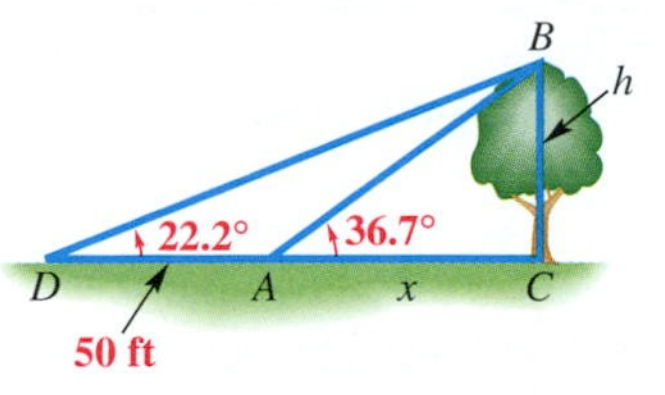

FIGURE 114

EXAMPLE 7 Solving a Problem Involving the Angle of Elevation

Francisco needs to know the height of a tree. From a given point on the ground, he finds that the angle of elevation to the top of the tree is 36.7°. He then moves back 50 feet. From the second point, the angle of elevation to the top of the tree is 22.2°. See Figure 114. Find the height of the tree.

Analytic Solution

Figure 114 shows two unknowns: x, the distance from the center of the trunk of the tree at ground level to the point where the first observation was made, and h, the height of the tree. Since nothing is given about the length of the hypotenuse of either triangle ABC or triangle BCD, we use a ratio that does not involve the hypotenuse: the tangent. (See Figure 115 in the Graphing Calculator Solution.)

In triangle ABC, $\tan 36.7^\circ = \dfrac{h}{x}$, or $h = x \tan 36.7^\circ$.

In triangle BCD, $\tan 22.2^\circ = \dfrac{h}{50 + x}$, or $h = (50 + x) \tan 22.2^\circ$.

Each of these expressions equals h, so the expressions must be equal.

$$x \tan 36.7^\circ = (50 + x) \tan 22.2^\circ$$

Now we solve for x.

$$x \tan 36.7^\circ = 50 \tan 22.2^\circ + x \tan 22.2^\circ \quad \text{Distributive property}$$

$$x \tan 36.7^\circ - x \tan 22.2^\circ = 50 \tan 22.2^\circ \quad \text{Write } x \text{ terms on one side.}$$

$$x(\tan 36.7^\circ - \tan 22.2^\circ) = 50 \tan 22.2^\circ \quad \text{Factor out } x \text{ on the left.}$$

$$x = \frac{50 \tan 22.2^\circ}{\tan 36.7^\circ - \tan 22.2^\circ} \quad \text{Divide by the coefficient of } x.$$

From before, $h = x \tan 36.7^\circ$. Substituting for x gives

$$h = \left(\frac{50 \tan 22.2^\circ}{\tan 36.7^\circ - \tan 22.2^\circ}\right) \tan 36.7^\circ.$$

Graphing Calculator Solution*

In Figure 115, we superimposed Figure 114 on coordinate axes with the origin at D. The tangent of the angle between the x-axis and the graph of a line with equation $y = mx + b$ is the slope of the line, m. For line DB, $m = \tan 22.2^\circ$. Since $b = 0$ here, the equation of line DB is $y_1 = (\tan 22.2^\circ)x$. The equation of line AB is $y_2 = (\tan 36.7^\circ)x + b$. Since $b \neq 0$ here, we use the point $A(50, 0)$ to find the equation.

$$y_2 - y_1 = m(x - x_1) \quad \text{Point–slope form}$$

$$y_2 - 0 = m(x - 50) \quad x_1 = 50,\ y_1 = 0$$

$$y_2 = \tan 36.7^\circ(x - 50)$$

Lines y_1 and y_2 are graphed in Figure 116 on the next page. The y-coordinate of the point of intersection of the graphs of these two lines gives the length of BC, or h. Thus, $h \approx 45$.

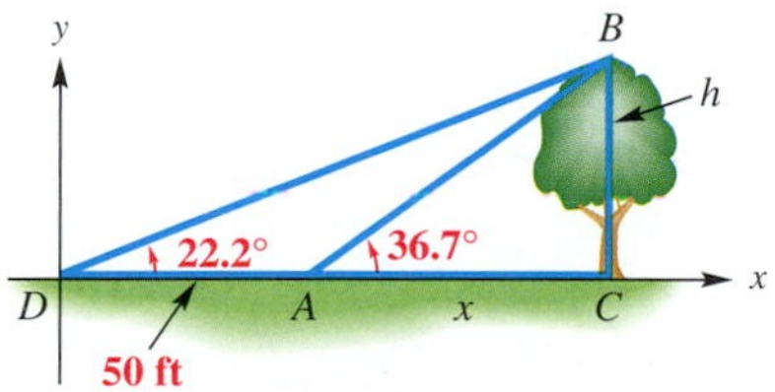

FIGURE 115

(continued)

Source: Adapted with permission from "Letter to the Editor," by Robert Ruzich (*Mathematics Teacher*, Volume 88, Number 1). Copyright © 1995 by the National Council of Teachers of Mathematics.

From a calculator,

$$\tan 36.7° \approx .74537703 \quad \text{and} \quad \tan 22.2° \approx .40809244$$

so $\tan 36.7° - \tan 22.2° \approx .74537703 - .40809244 = .33728459$

and $$h \approx \left(\frac{50(.40809244)}{.33728459}\right).74537703 \approx 45.$$

The height of the tree is approximately 45 feet.

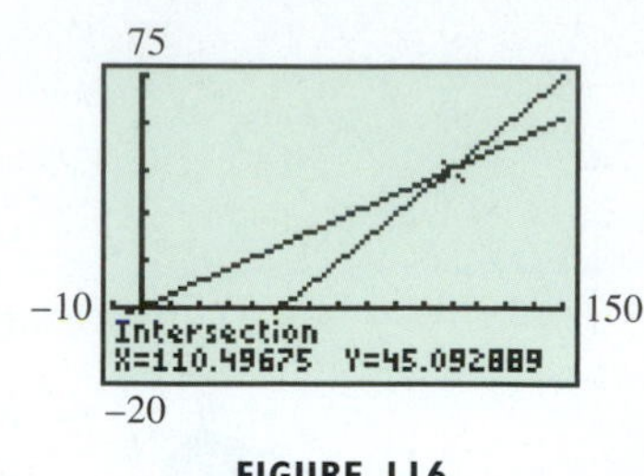

FIGURE 116

NOTE In practice, we usually do not write down the intermediate calculator approximation steps. However, we have done this in Example 7 so that you may follow the steps more easily.

8.7 Exercises

Solve each right triangle.

1.
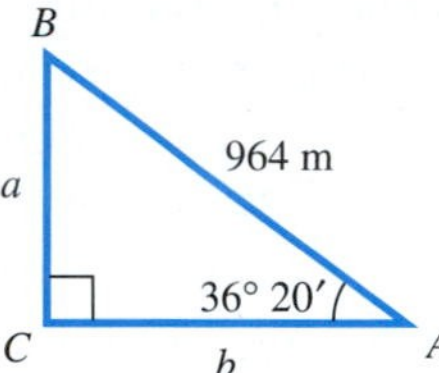

2.
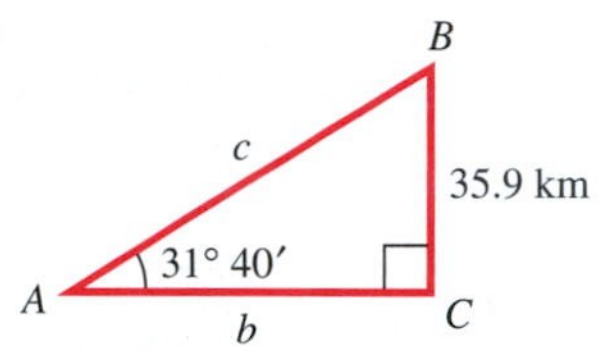

3.
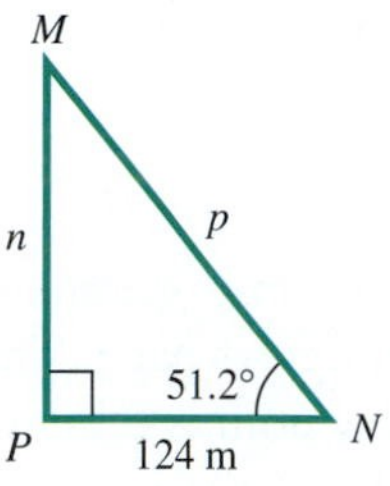

4.
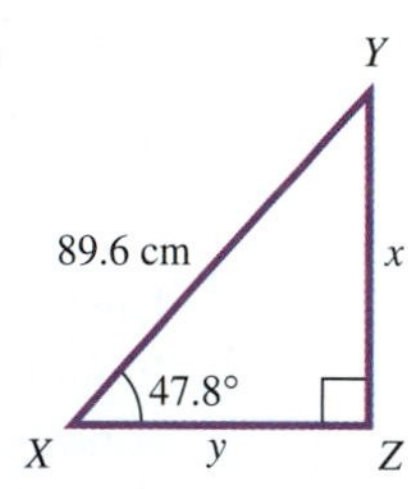

5.
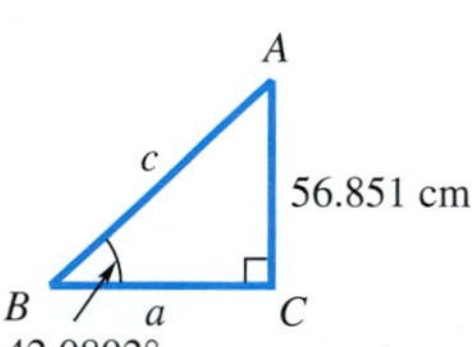

6.
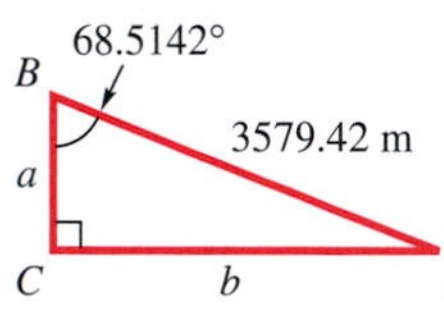

7.
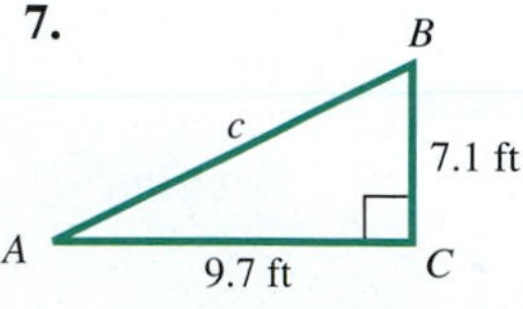

8.
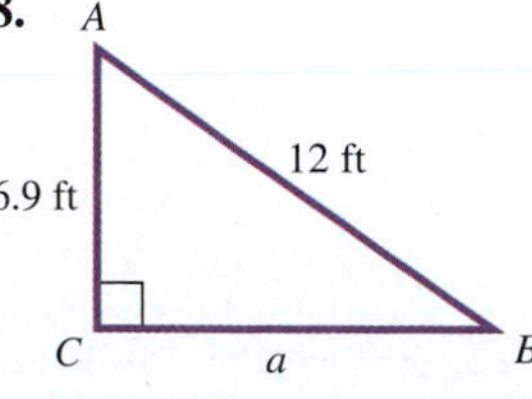

9.
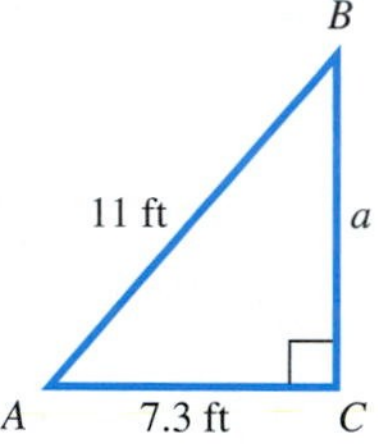

10.
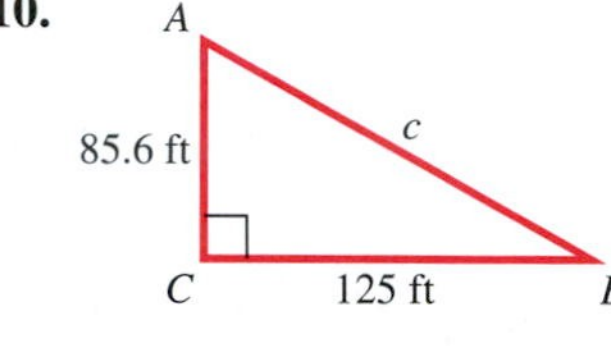

Solve each right triangle. In each case, $C = 90°$. If angle information is given in degrees and minutes, give answers in the same way. If given in decimal degrees, do likewise in answers. When two sides are given, give angles in degrees and minutes.

11. $A = 28.00°$; $c = 17.4$ feet

12. $B = 46.00°$; $c = 29.7$ meters

13. $B = 73.00°$; $b = 128$ inches

14. $A = 61°\ 00'$; $b = 39.2$ centimeters

15. $a = 76.4$ yards; $b = 39.3$ yards

16. $a = 958$ meters; $b = 489$ meters

17. *Concept Check* If we are given an acute angle and a side of a right triangle, what unknown part of the triangle requires the least work to find?

18. Can a right triangle be solved if we are given measures of its two acute angles and no side lengths? Explain.

Find the measure of the angle formed by the line passing through the origin and the positive part of the x-axis. Use the values displayed at the bottom of the screen.

19.

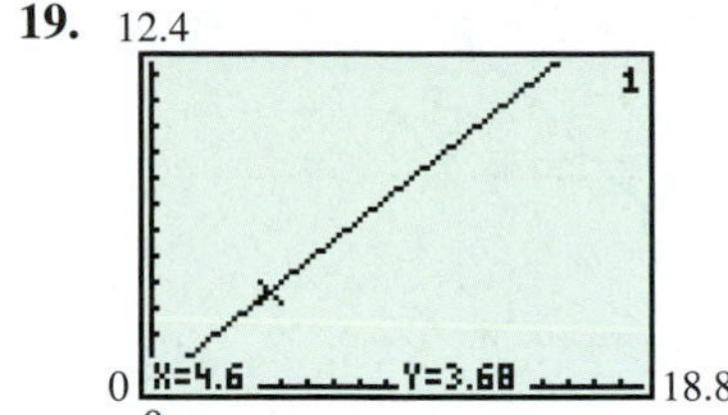

20.

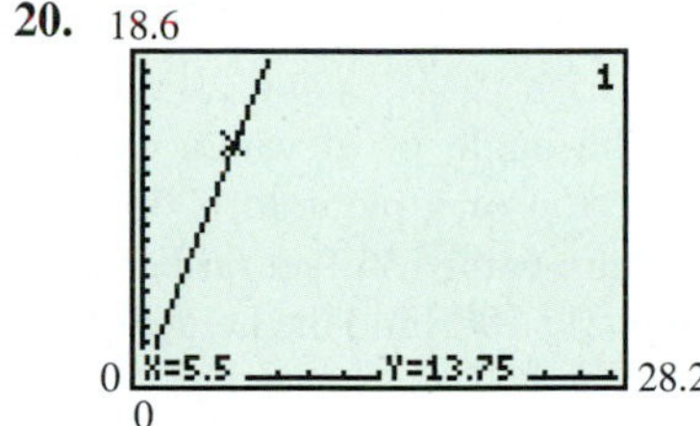

21. Explain why the angle of depression *DAB* has the same measure as the angle of elevation *ABC* in the figure. (Assume that line *AD* is parallel to line *CB*.)

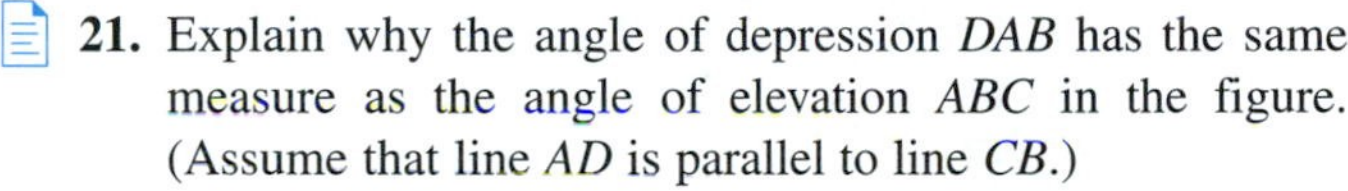

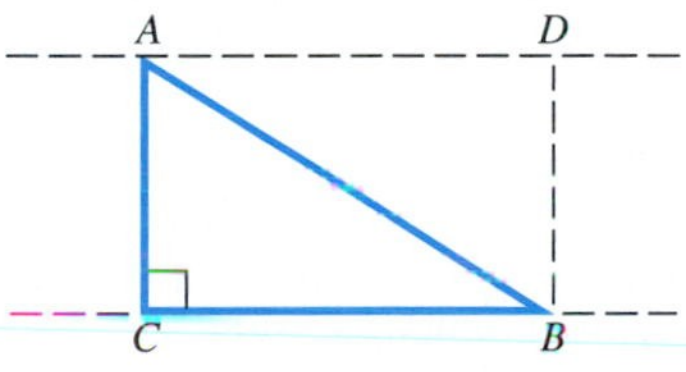

AD is parallel to *BC*.

22. Why is angle *CAB* *not* an angle of depression in the figure for Exercise 21? Why is angle *CAB* *not* an angle of elevation in the figure?

23. ***Concept Check*** When bearing is given as a single angle measure, how is the angle represented in a sketch?

24. ***Concept Check*** When bearing is given as N (or S), then the angle measure, then E (or W), how is the angle represented in a sketch?

Solve each problem.

25. ***Height of a Ladder on a Wall*** A 13.5-meter fire-truck ladder is leaning against a wall. (See the figure at the top of the next column.) Find the distance *d* the ladder goes up the wall (above the top of the fire truck) if the ladder makes an angle of 43° 50′ with the horizontal.

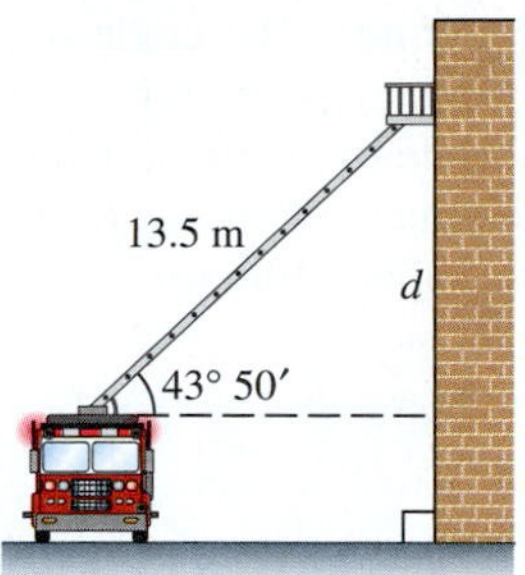

26. ***Length of a Guy Wire*** A weather tower used to measure wind speed has a guy wire attached to it 175 feet above the ground. The angle between the wire and the vertical tower is 57.0°. Approximate the length of the guy wire.

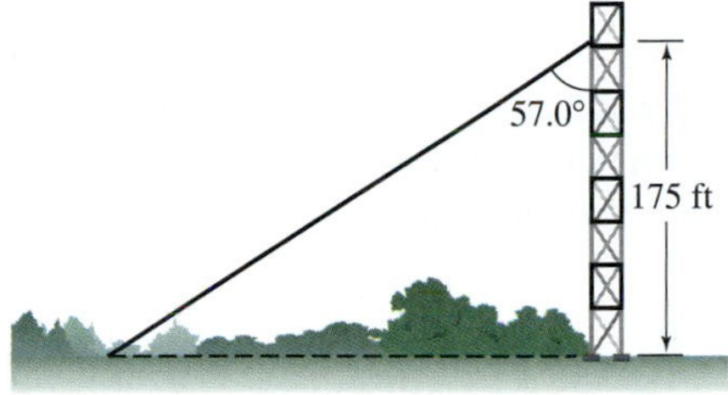

27. ***Length of a Shadow*** Suppose that the angle of elevation of the sun is 23.4°. Find the length of the shadow cast by Diane Carr, who is 5.75 feet tall.

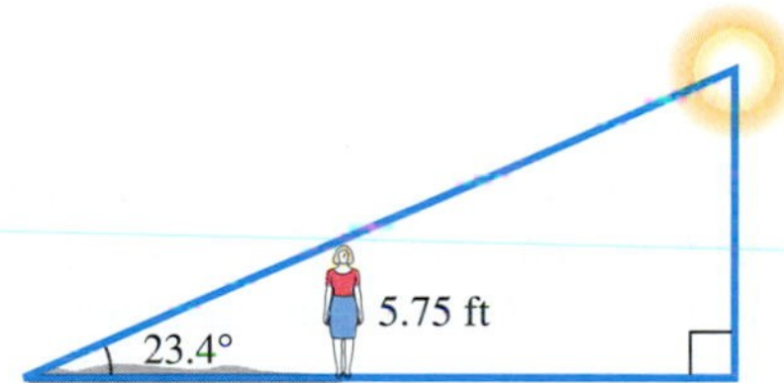

28. ***Height of a Tower*** The shadow of a vertical tower is 40.6 meters long when the angle of elevation of the sun is 34.6°. Find the height of the tower.

29. ***Height of a Building*** From a window 30 feet above the street, the angle of elevation to the top of the building across the street is 50.0° and the angle of depression to the base of this building is 20.0°. Find the height of the building across the street.

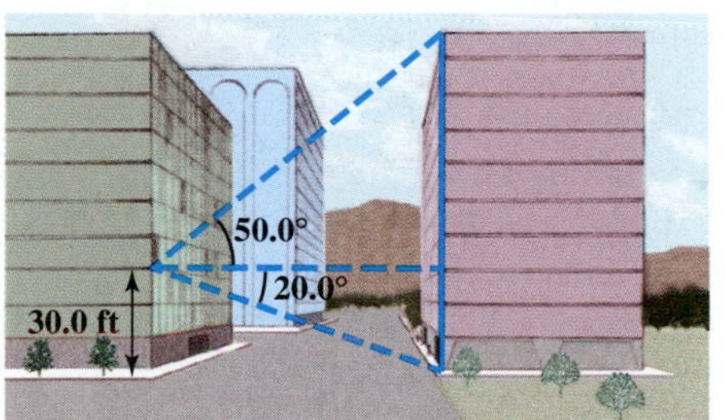

30. ***Height of a Building*** The angle of elevation from the top of a small building to the top of a nearby taller building is $46° 40'$, while the angle of depression to the bottom is $14° 10'$. If the smaller building is 28.0 meters high, find the height of the taller building.

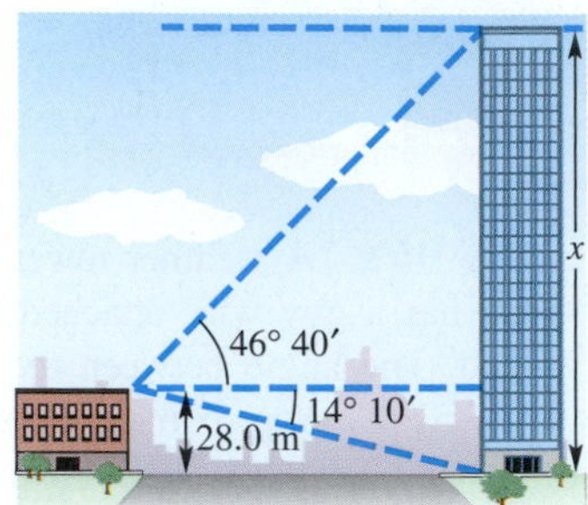

31. ***Angle of a Television Lens*** A security camera is to be mounted on a bank wall so as to have a good view of the head teller. Find the angle of depression of the lens.

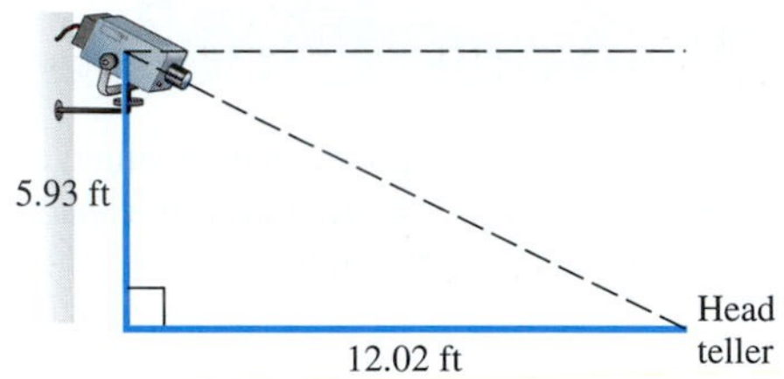

32. ***Angle of Depression of a Floodlight*** A company safety committee has recommended that a floodlight be mounted in a parking lot so as to illuminate the employee exit. See the figure. Find the angle of depression of the light.

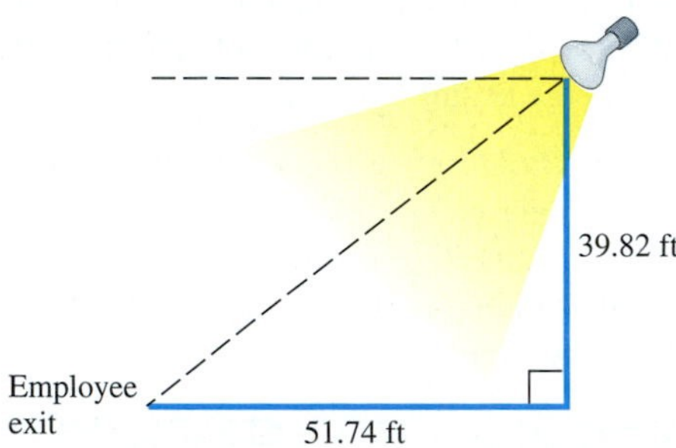

33. ***Distance through a Tunnel*** A tunnel is to be dug from A to B. Both A and B are visible from C. (See the figure at the top of the next column.) If AC is 1.4923 miles, BC is 1.0837 miles, and C is 90°, find the measures of angles A and B.

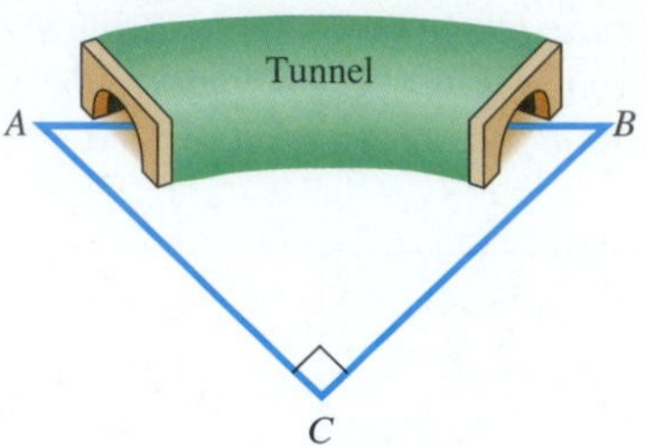

34. ***Angle of Depression from a Plane*** An airplane is flying 10,500 feet above the level ground. The angle of depression from the plane to the base of a tree is $13° 50'$. How far horizontally must the plane fly to be directly over the tree?

35. ***Height of a Pyramid*** The angle of elevation from a point on the ground to the top of a pyramid is $35° 30'$. The angle of elevation from a point 135 feet farther back to the top of the pyramid is $21° 10'$. Find the height of the pyramid.

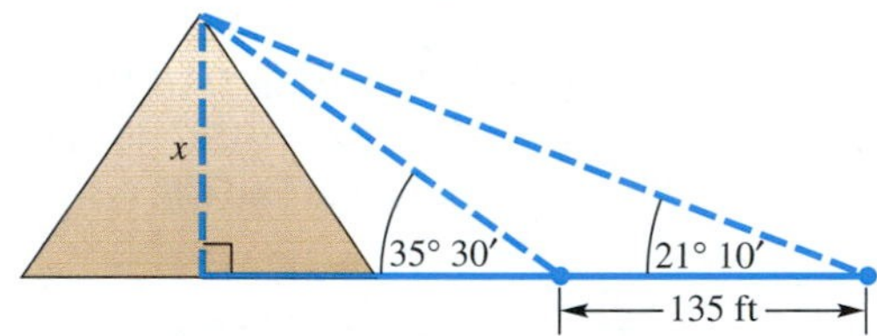

36. ***Distance between a Whale and a Lighthouse*** Debbie Glockner-Ferrari, a humpback whale researcher, is watching a whale approach directly toward her as she observes from the top of a lighthouse. When she first begins watching, the angle of depression of the whale is $15° 50'$. Just as the whale turns away from the lighthouse, the angle of depression is $35° 40'$. If the height of the lighthouse is 68.7 meters, find the distance x traveled by the whale as it approaches the lighthouse.

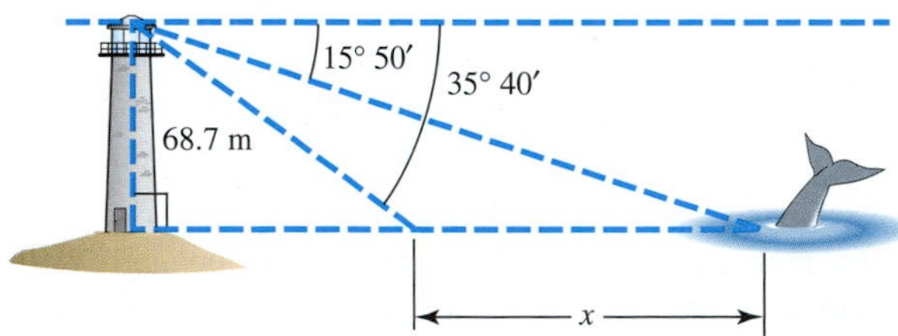

37. ***Height of an Antenna*** An antenna is on top of the center of a house. The angle of elevation from a point on the ground 28.0 meters from the center of the house to the top of the antenna is $27° 10'$, and the angle of elevation to the bottom of the antenna is $18° 10'$. Find the height of the antenna.

38. ***Height of Mt. Whitney*** The angle of elevation from Lone Pine to the top of Mt. Whitney is $10° 50'$. Van Dong Le, traveling 7.00 kilometers from Lone Pine along a straight, level road toward Mt. Whitney, finds the angle of elevation to be $22° 40'$. Find the height of the top of Mt. Whitney above the level of the road.

39. ***Length of a Side of a Lot*** A piece of land has the shape shown in the figure. Find the value of x.

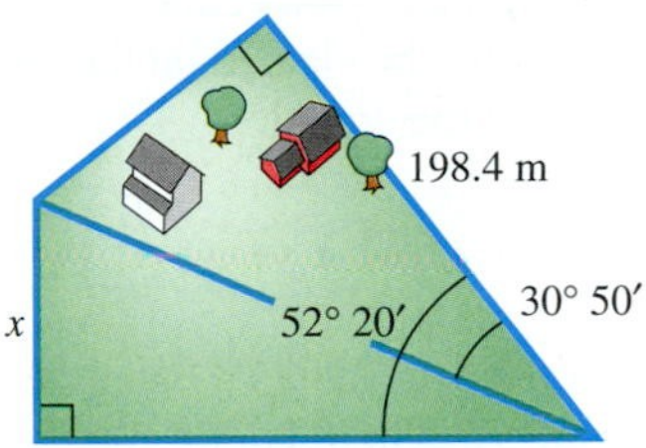

40. Find the value of x in the figure.

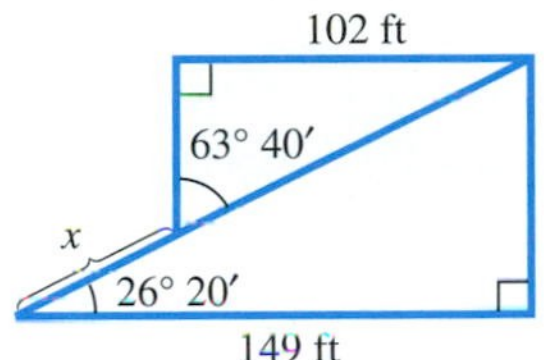

41. ***Distance between Two Ships*** A ship leaves port and sails on a bearing of N $28° 10'$ E. Another ship leaves the same port at the same time and sails on a bearing of S $61° 50'$ E. If the first ship sails at 24.0 mph and the second sails at 28.0 mph, find the distance x between the two ships after 4 hours.

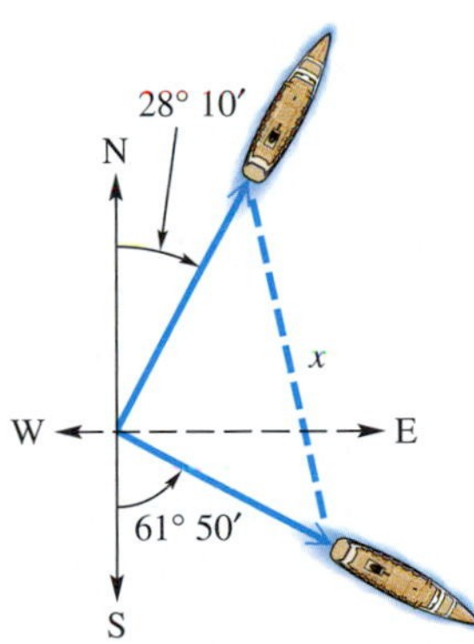

42. ***Distance between Transmitters*** Radio direction finders are set up at two points A and B, which are 2.50 miles apart on an east–west line. From A, it is found that the bearing of a signal from a radio transmitter is N $36° 20'$ E, while the bearing of the same signal from B is N $53° 40'$ W. Find the distance of the transmitter from B.

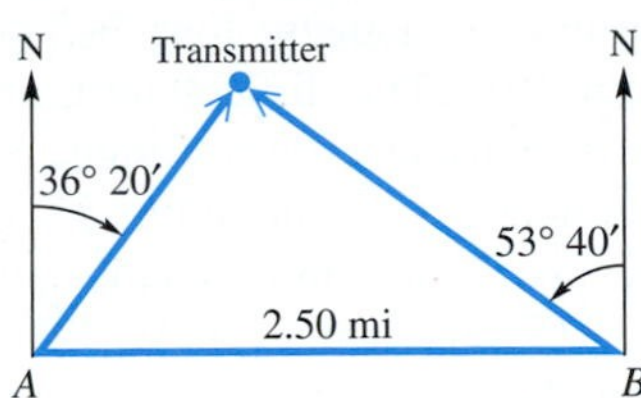

43. ***Transmitter Distance*** In Exercise 42, find the distance of the transmitter from A.

44. ***Distance Traveled by a Plane*** A plane flies 1.5 hours at 110 mph on a bearing of 40°. It then turns and flies 1.3 hours at the same speed on a bearing of 130°. How far is the plane from its starting point?

45. ***Distance Traveled by a Ship*** A ship travels 50 kilometers on a bearing of 27° and then travels on a bearing of 117° for 140 kilometers. Find the distance x between the starting point and the ending point.

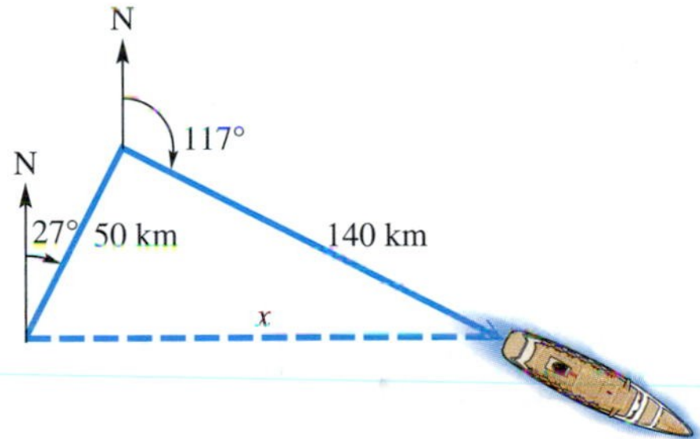

46. ***Distance between Two Ships*** Two ships leave a port at the same time. The first ship sails on a bearing of 40° at 18 knots (nautical miles per hour) and the second at a bearing of 130° at 26 knots. How far apart are they after 1.5 hours?

47. ***Distance Across a Lake*** To find the distance RS across a lake, a surveyor lays off $RT = 53.1$ meters, with angle $T = 32° 10'$ and angle $S = 57° 50'$. Find length RS.

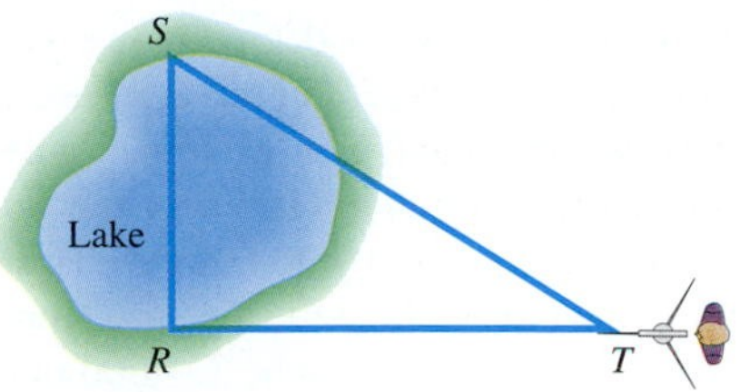

48. ***Height of Mt. Everest*** The highest mountain peak in the world is Mt. Everest, located in the Himalayas. The height of this enormous mountain was determined in 1856 by surveyors using trigonometry long before the peak was first climbed in 1953. This difficult measurement had to be done from a great distance. At an altitude of 14,545 feet on a different mountain, the straight line distance to the peak of Mt. Everest is 27.0134 miles and its angle of elevation is $\theta = 5.82°$. (*Source:* Dunham, W., *The Mathematical Universe,* John Wiley and Sons, 1994.)

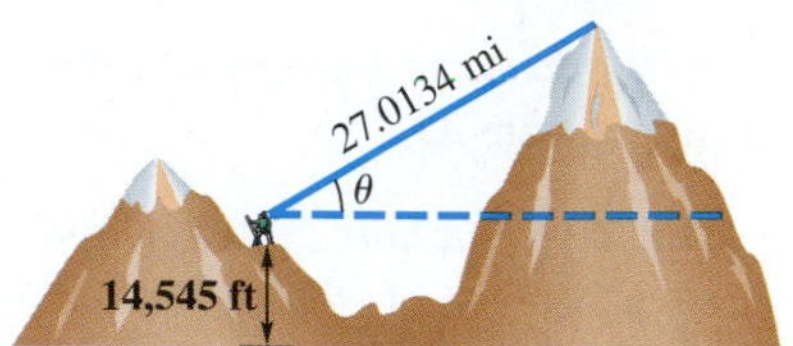

(a) Approximate the height (in feet) of Mt. Everest.

(b) In the actual measurement, Mt. Everest was over 100 miles away and the curvature of Earth had to be taken into account. Would the curvature of Earth make the peak appear taller or shorter than it actually is?

49. ***Cloud Ceiling*** The U.S. Weather Bureau defines a *cloud ceiling* as the altitude of the lowest clouds that cover more than half the sky. To determine a cloud ceiling, a powerful searchlight projects a circle of light vertically onto the bottom of the cloud. An observer sights the circle of light in the crosshairs of a tube called a *clinometer*. A pendant hanging vertically from the tube and resting on a protractor gives the angle of elevation. Find the cloud ceiling if the searchlight is located 1000 feet from the observer and the angle of elevation is 30.0° as measured with a clinometer at eye height 6 feet. (Assume three significant digits.)

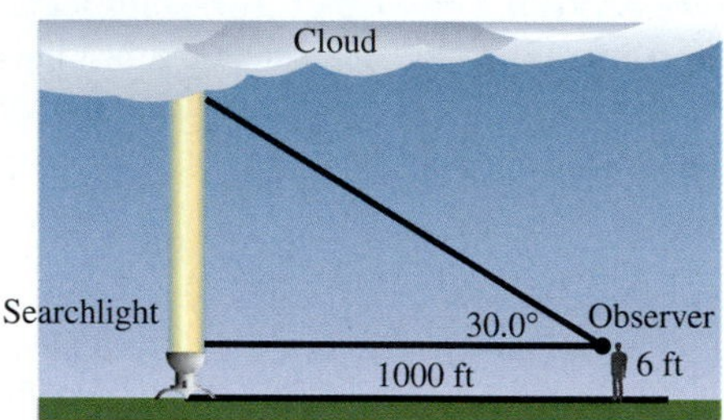

50. ***Error in Measurement*** A degree may seem like a very small unit, but an error of 1 degree in measuring an angle may be very significant. For example, suppose a laser beam directed toward the visible center of the moon misses its assigned target by 30 seconds. How far is it (in miles) from the target? Take the distance from the surface of Earth to that of the moon to be 234,000 miles. (*Source: A Sourcebook of Applications of School Mathematics* by Donald Bushaw et al. Copyright © 1980 by The Mathematical Association of America.)

51. ***(Modeling) Stopping Distance on a Curve*** When an automobile travels along a circular curve, objects like trees and buildings situated on the inside of the curve can obstruct a driver's vision. These obstructions prevent the driver from seeing sufficiently far down the highway to ensure a safe stopping distance. In the figure, the *minimum* distance d that should be cleared on the inside of the highway is modeled by the equation

$$d = R\left(1 - \cos\frac{\beta}{2}\right).$$

(*Source:* Mannering, F. and W. Kilareski, *Principles of Highway Engineering and Traffic Analysis,* Second Edition, John Wiley and Sons, 1998.)

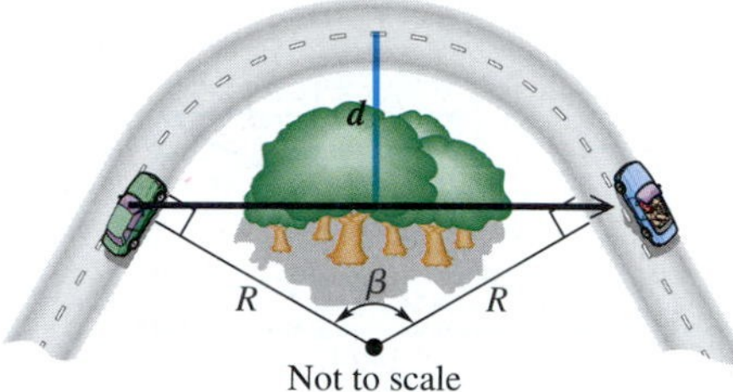

Not to scale

(a) It can be shown that if β is measured in degrees, then $\beta \approx \frac{57.3S}{R}$, where S is the safe stopping distance for the given speed limit. Compute d for a 55-mph speed limit if $S = 336$ feet and $R = 600$ feet.

(b) Compute d for a 65-mph speed limit if $S = 485$ feet and $R = 600$ feet.

(c) How does the speed limit affect the amount of land that should be cleared on the inside of the curve?

52. ***(Modeling) Highway Curve Design*** Highway curves are sometimes banked so that the outside of the curve is slightly elevated or inclined above the inside of the curve, as shown in the figure on the next page. This inclination is called the *superelevation*. It is important that both the curve's radius and superelevation are correct for a given speed limit. The relationship between a car's velocity v in feet per second, the safe radius r of the curve in feet, and the superelevation θ in degrees is modeled by

$$r = \frac{v^2}{4.5 + 32.2 \tan\theta}.$$

(*Source:* Mannering, F. and W. Kilareski, *Principles of Highway Engineering and Traffic Analysis,* Second Edition, John Wiley and Sons, 1998.)

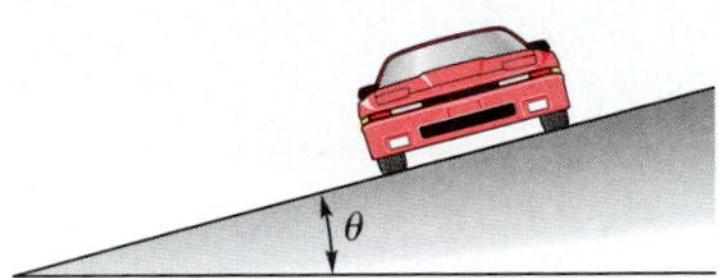

(a) A curve has a speed limit of 66 feet per second (45 mph) and a superelevation of $\theta = 3°$. Approximate the safe radius r.

(b) Find r if $\theta = 5°$ and $v = 66$.

(c) Make a conjecture about how increasing θ with $v = 66$ affects the safe radius r.

(d) Find v if $r = 1150$ and $\theta = 2.1°$.

53. *(Modeling) Distance between the Sun and a Star* Suppose that a star forms an angle θ with respect to Earth and the sun. Let the coordinates of Earth be (x, y), of the star be $(0, 0)$, and of the sun be $(x, 0)$. See the figure.

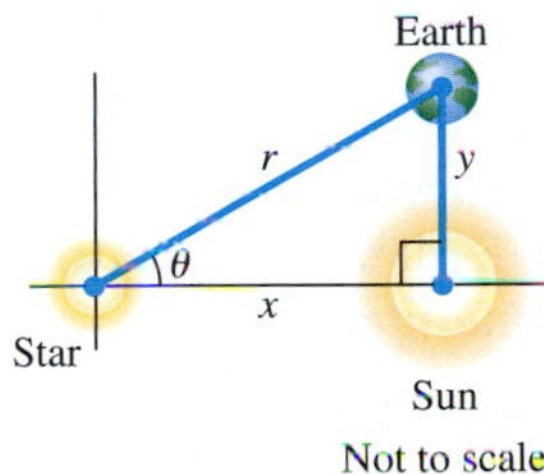

Find an equation for x, the distance between the sun and the star, as follows.

(a) Write an equation involving a trigonometric function that relates x, y, and θ.

(b) Solve your equation for x.

54. *Area of a Solar Cell* A solar cell converts the energy of sunlight directly into electrical energy. The amount of energy a cell produces depends on its area. Suppose a solar cell is hexagonal, as shown in the figure. Express its area in terms of $\sin \theta$ and any side x. (*Hint:* Consider one of the six equilateral triangles from the hexagon. See the figure.) (*Source:* Kastner, B., *Space Mathematics,* NASA, 1985.)

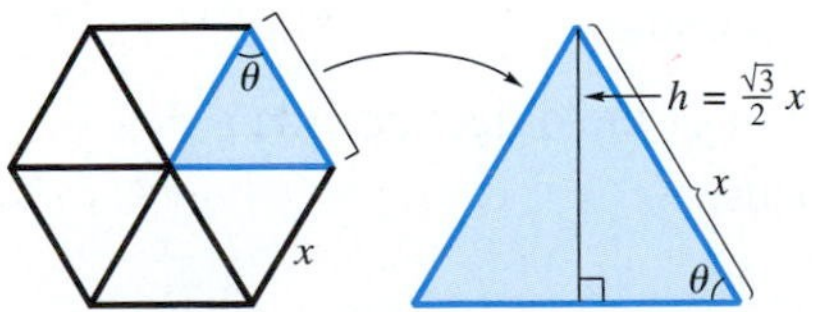

Find the exact value of each part labeled with a variable in each figure.

55.

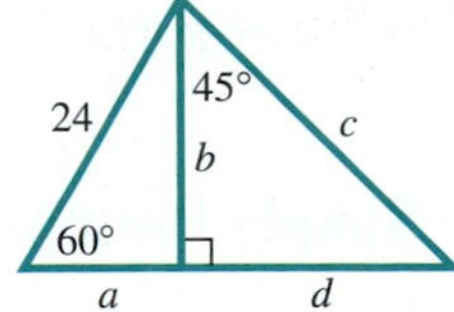

56.

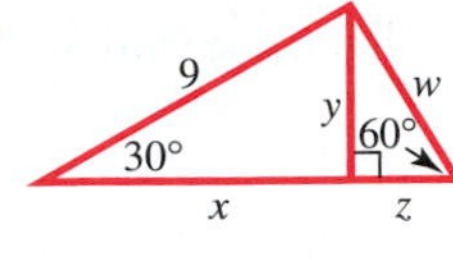

57.

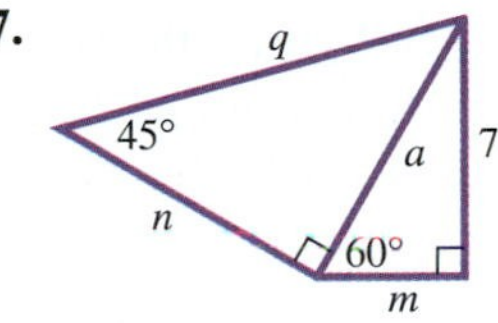

58.

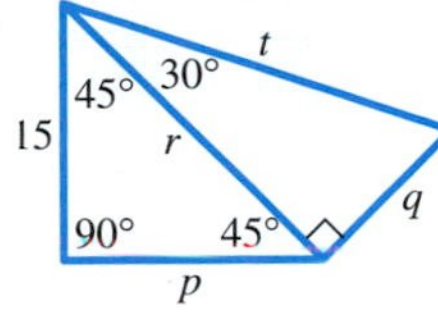

Find a formula for the area of each figure in terms of s.

59.

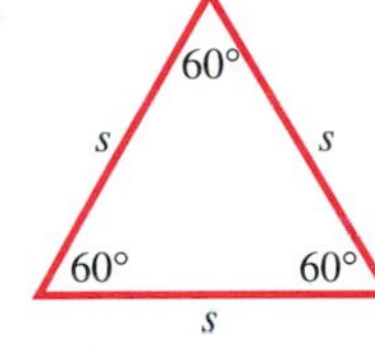

60.

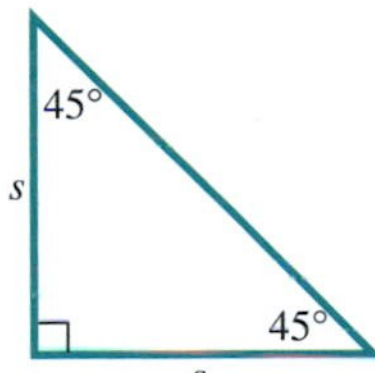

61. *Concept Check* The figure shows a vertical stick with length 1 casting a shadow on the ground. The angle of elevation is θ. Which trigonometric function of θ describes the length of the shadow?

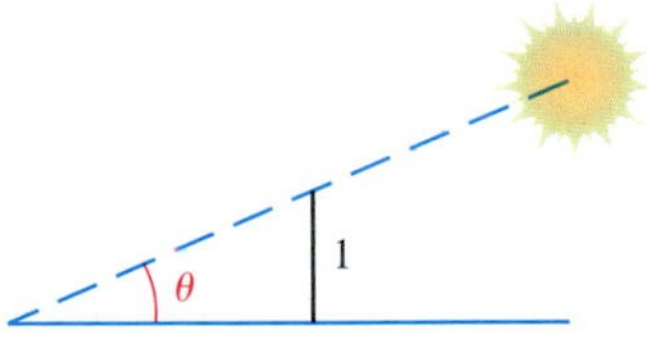

62. *Concept Check* The figure shows a horizontal stick with length 1 casting a shadow on a wall. Which trigonometric function of θ describes the length of the shadow?

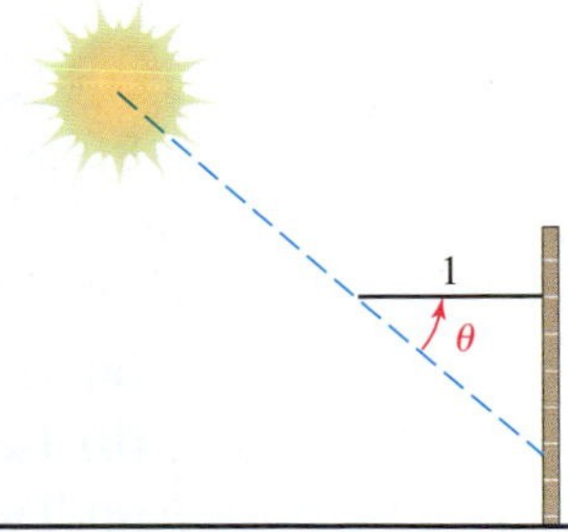

8.8 Harmonic Motion

Simple Harmonic Motion ■ Damped Oscillatory Motion

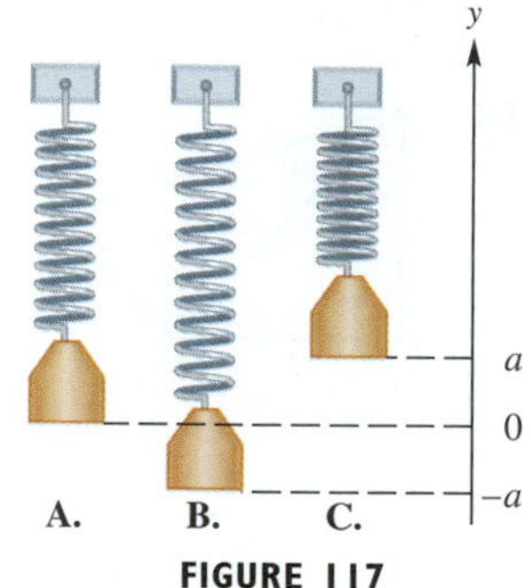

FIGURE 117

Simple Harmonic Motion

In part A of Figure 117, a spring with a weight attached to its free end is in its equilibrium (or rest) position. If the weight is pulled down a units and released (part B of the figure), the spring's elasticity causes the weight to rise a units $(a > 0)$ above the equilibrium position, as seen in part C, and then oscillate about the equilibrium position. If friction is neglected, this oscillatory motion is described mathematically by a sinusoid. Other applications of this type of motion include sound, electric current, and electromagnetic waves.

To develop a general equation for such motion, consider Figure 118. Suppose the point $P(x, y)$ moves around the circle counterclockwise at a uniform angular speed ω. Assume that at time $t = 0$, P is at $(a, 0)$. The angle swept out by ray OP at time t is given by $\theta = \omega t$. The coordinates of point P at time t are

$$x = a \cos \theta = a \cos \omega t \quad \text{and} \quad y = a \sin \theta = a \sin \omega t.$$

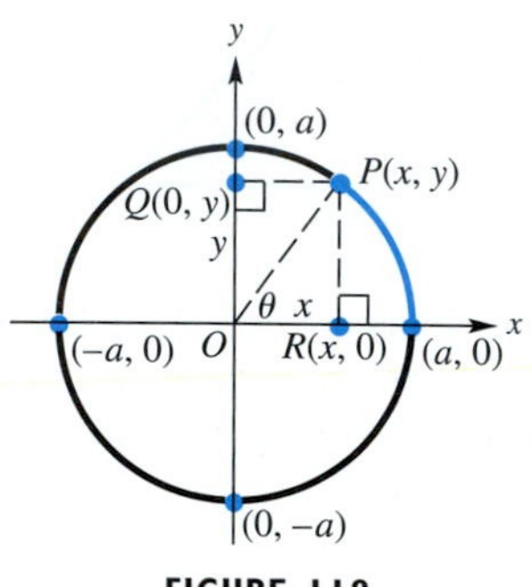

FIGURE 118

As P moves around the circle in Figure 118 from the point $(a, 0)$, the point $Q(0, y)$ oscillates back and forth along the y-axis between the points $(0, a)$ and $(0, -a)$. Similarly, the point $R(x, 0)$ oscillates back and forth between $(a, 0)$ and $(-a, 0)$. This oscillatory motion is called **simple harmonic motion.**

The amplitude of the motion is $|a|$, and the period $\frac{2\pi}{\omega}$. The moving points P and Q or P and R complete one oscillation or cycle per period. The number of cycles per unit of time, called the **frequency,** is the reciprocal of the period, $\frac{\omega}{2\pi}$, where $\omega > 0$.

Simple Harmonic Motion

The position of a point oscillating about an equilibrium position at time t is modeled by either

$$s(t) = a \cos \omega t \quad \text{or} \quad s(t) = a \sin \omega t,$$

where a and ω are constants, with $\omega > 0$. The amplitude of the motion is $|a|$, the period is $\frac{2\pi}{\omega}$, and the frequency is $\frac{\omega}{2\pi}$.

EXAMPLE 1 Modeling the Motion of a Spring

Suppose that an object is attached to a coiled spring such as the one in Figure 117. It is pulled down a distance of 5 units from its equilibrium position and then released. The time for one complete oscillation is 4 seconds.

(a) Give an equation that models the position of the object at time t.

(b) Determine the position at $t = 1.5$ seconds.

(c) Find the frequency.

Solution

(a) When the object is released at $t = 0$, the distance of the object from the equilibrium position is 5 inches below equilibrium. If $s(t)$ is to model the motion, then $s(0)$ must equal -5. We use

$$s(t) = a \cos \omega t,$$

with $a = -5$. We choose the cosine function because $\cos \omega(0) = \cos 0 = 1$, and $-5 \cdot 1 = -5$. (Had we chosen the sine function, a phase shift would have been required.) The period is 4, so

$$\frac{2\pi}{\omega} = 4, \quad \text{or} \quad \omega = \frac{\pi}{2}. \qquad \text{Solve for } \omega.$$

Thus, the motion is modeled by

$$s(t) = -5 \cos \frac{\pi}{2} t.$$

(b) After 1.5 seconds, the position is

$$s(1.5) = -5 \cos\left[\frac{\pi}{2}(1.5)\right] \approx 3.54 \text{ inches.}$$

Since $3.54 > 0$, the object is above the equilibrium position.

(c) The frequency is the reciprocal of the period, or $\frac{1}{4}$ cycle per second. ■

EXAMPLE 2 Analyzing Harmonic Motion

Suppose that an object oscillates according to the model

$$s(t) = 8 \sin 3t,$$

where t is in seconds and $s(t)$ is in feet. Analyze the motion.

Solution The motion is harmonic because the model is of the form $s(t) = a \sin \omega t$. Since $a = 8$, the object oscillates 8 feet in either direction from its starting point. The period $\frac{2\pi}{3} \approx 2.1$ is the time, in seconds, it takes for one complete oscillation. The frequency is the reciprocal of the period, so the object completes $\frac{3}{2\pi} \approx .48$ oscillation per second. ■

Damped Oscillatory Motion

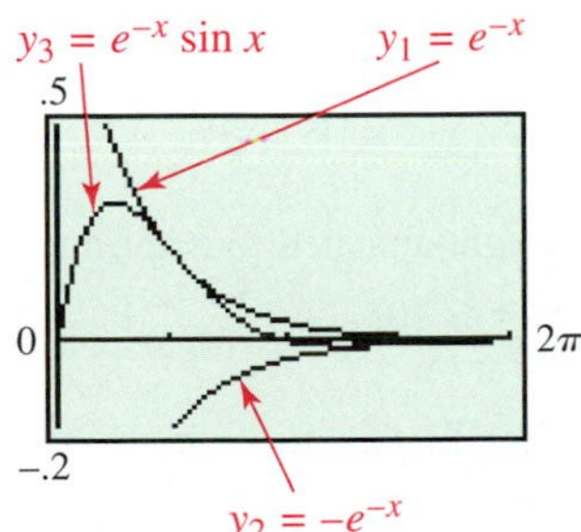

FIGURE 119

In the example of the stretched spring, we disregarded the effect of friction. Friction causes the amplitude of the motion to diminish gradually until the weight comes to rest. In this situation, we say that the motion has been *damped* by the force of friction. Most oscillatory motions are damped, and the decrease in amplitude follows the pattern of exponential decay. A typical example of **damped oscillatory motion** is provided by the function defined by

$$s(t) = e^{-t} \sin t.$$

Figure 119 shows how the graph of $y_3 = e^{-x} \sin x$ is bounded above by the graph of $y_1 = e^{-x}$ and below by the graph of $y_2 = -e^{-x}$. The damped-motion curve dips

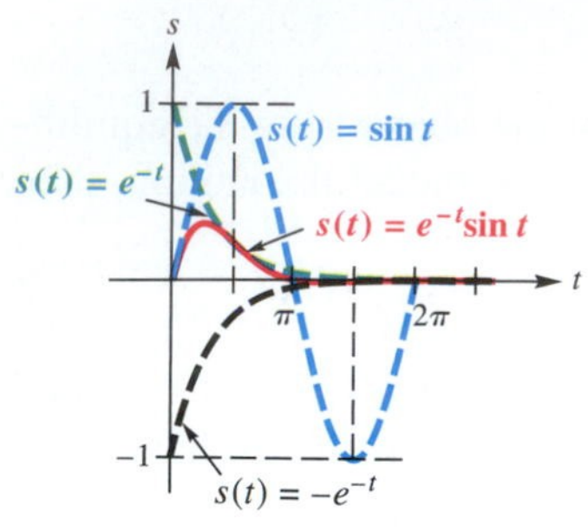

FIGURE 120

below the x-axis at $x = \pi$ but stays above the graph of y_2. Figure 120 shows a traditional graph, along with the graph of $s = \sin t$.

Shock absorbers are put on an automobile in order to damp oscillatory motion. Instead of oscillating up and down for a long while after hitting a bump or pothole, the oscillations of the car are quickly damped out for a smoother ride.

8.8 Exercises

(Modeling) Springs *Suppose that a weight on a spring has initial position $s(0)$ and period P.*

(a) *Find a function s given by $s(t) = a \cos \omega t$ that models the displacement of the weight.*

(b) *Evaluate $s(1)$. Is the weight moving upward, downward, or neither when $t = 1$? Support your results graphically.*

1. $s(0) = 2$ inches; $P = .5$ second

2. $s(0) = 5$ inches; $P = 1.5$ seconds

3. $s(0) = -3$ inches; $P = .8$ second

4. $s(0) = -4$ inches; $P = 1.2$ seconds

(Modeling) Music *A note on the piano has frequency F. Suppose the maximum displacement at the center of the piano wire is given by $s(0)$. Find constants a and ω so that the equation $s(t) = a \cos \omega t$ models this displacement. Graph s in the viewing window $[0, .05]$ by $[-.3, .3]$.*

5. $F = 27.5$; $s(0) = .21$

6. $F = 110$; $s(0) = .11$

7. $F = 55$; $s(0) = .14$

8. $F = 220$; $s(0) = .06$

(Modeling) *Solve each problem.*

9. *Particle Movement* Write the equation, and then determine the amplitude, period, and frequency, of the simple harmonic motion of a particle moving uniformly around a circle of radius 2 units, with angular speed **(a)** 2 radians per second and **(b)** 4 radians per second.

10. *Pendulum* What are the period P and frequency T of oscillation of a pendulum of length $\frac{1}{2}$ foot? (*Hint:* $P = 2\pi\sqrt{\frac{L}{32}}$, where L is the length of the pendulum in feet and P is in seconds.)

11. *Pendulum* How long should the pendulum be to have period 1 second? (See Exercise 10.)

12. *Spring* The formula for the up-and-down motion of a weight on a spring is given by

$$s(t) = a \sin \sqrt{\frac{k}{m}}\,t.$$

If the spring constant k is 4, what mass m must be used to produce a period of 1 second?

13. *Spring* A spring with spring constant $k = 2$ and a 1-unit mass m attached to it is stretched and then allowed to come to rest.

(a) If the spring is stretched $\frac{1}{2}$ foot and released, what are the amplitude, period, and frequency of the resulting oscillatory motion?

(b) What is the equation of the motion?

14. *Spring* The position of a weight attached to a spring is

$$s(t) = -5 \cos 4\pi t$$

inches after t seconds.

(a) What is the maximum height that the weight rises above the equilibrium position?

(b) What are the frequency and period?

(c) When does the weight first reach its maximum height?

(d) Calculate and interpret $s(1.3)$.

15. *Spring* The position of a weight attached to a spring is

$$s(t) = -4 \cos 10t$$

inches after t seconds.

(a) What is the maximum height that the weight rises above the equilibrium position?

(b) What are the frequency and period?

(c) When does the weight first reach its maximum height?

(d) Calculate and interpret $s(1.466)$.

16. *Spring* A weight attached to a spring is pulled down 3 inches below the equilibrium position.

(a) Assuming that the frequency is $\frac{6}{\pi}$ cycles per second, find a trigonometric model that gives the position of the weight at time t seconds.

(b) What is the period?

17. *Spring* A weight attached to a spring is pulled down 2 inches below the equilibrium position.

(a) Assuming that the period is $\frac{1}{3}$ second, find a trigonometric model that gives the position of the weight at time t seconds.

(b) What is the frequency?

 **18.** Use a graphing calculator to graph $y_1 = e^{-t}\sin t$, $y_2 = e^{-t}$, and $y_3 = -e^{-t}$ in the viewing window $[0, \pi]$ by $[-.5, .5]$.

(a) Find the x-intercepts of the graph of y_1. Explain the relationship of these x-intercepts to those of the graph of $y = \sin x$.

(b) Find the x-coordinate of any points of intersection of y_1 and y_2 or y_1 and y_3. How are these values related to the graph of $y = \sin x$?

19. *Damped Motion* A weight attached to a spring is submerged in water. Suppose that its displacement from its equilibrium position can be modeled by

$$D(t) = 2e^{-2t}\cos 2\pi t,$$

where D is in inches and $t \geq 0$ is in seconds. Note that when $D < 0$ the spring is stretched and when $D > 0$ the spring is compressed.

(a) What is the initial displacement of the weight when $t = 0$?

(b) What is the frequency of its oscillations?

(c) Determine graphically the times when the spring is *stretched* $\frac{1}{e}$ inch.

20. *Electrical Circuit* The voltage in a circuit with alternating current is given by

$$V(t) = 160e^{-20t}\cos 120\pi t,$$

where V is in volts and $t \geq 0$ is in seconds.

(a) What is the initial voltage when $t = 0$?

(b) What is the frequency of the voltage?

(c) Solve $V(t) = 100$ graphically.

Reviewing Basic Concepts (Sections 8.5–8.8)

1. An angle θ is in standard position and its terminal side passes through the point $(-2, 5)$. Find the exact value of each of the six trigonometric functions of θ.

2. Find the exact values of the trigonometric functions of a 270° angle.

3. Decide whether the following statements are *possible* or *impossible*.

(a) $\cos\theta = \dfrac{3}{2}$ (b) $\tan\theta = 300$

(c) $\csc\theta = 5$

4. If θ is a quadrant III angle with $\sin\theta = -\frac{2}{3}$, use identities to find the other trigonometric function values of θ.

5. Find the six trigonometric function values of angle A in the triangle shown.

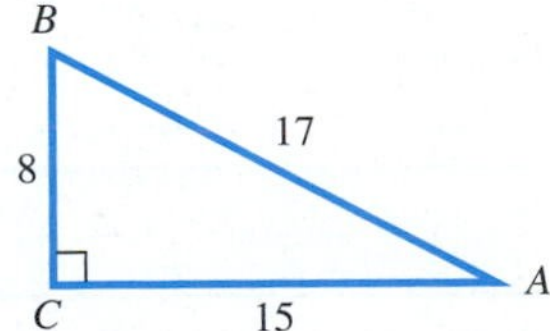

6. Complete the following table of exact function values.

θ	sin θ	cos θ	tan θ	cot θ	sec θ	csc θ
30° or $\frac{\pi}{6}$						
45° or $\frac{\pi}{4}$						
60° or $\frac{\pi}{3}$						

7. Write each expression using its cofunction.

(a) $\sin 27°$ (b) $\tan\dfrac{\pi}{5}$

8. Give the reference angle for each angle measure.

(a) 100° (b) −365° (c) $\dfrac{8\pi}{3}$

9. Find the exact values for the six trigonometric functions of 315°.

10. Use a calculator to find an approximation for each function value.

(a) $\sin 46°\,30'$ (b) $\tan(-100°)$ (c) $\csc 4$

11. Find all values of θ if θ is in the interval $[0°, 360°)$ and $\tan\theta = -\frac{\sqrt{3}}{3}$.

12. Find two angles in the interval $[0, 2\pi)$ that satisfy $\sin\theta = .68163876$.

13. ***Height of Mt. Kilimanjaro*** From a point A, the angle of elevation of Mount Kilimanjaro in Africa is 13.7°, and from a point B directly behind A, the angle of elevation is 10.4°. If the distance between A and B is 5 miles, approximate the height of Mount Kilimanjaro to the nearest hundred feet.

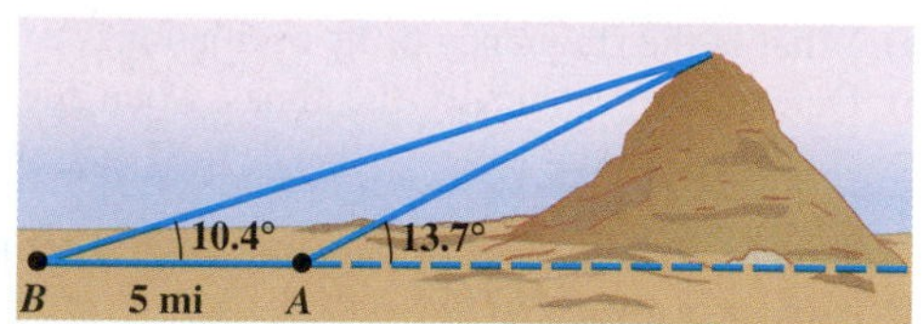

14. ***(Modeling) Spring*** The height of a weight attached to a spring is

$$s(t) = -4\cos 8\pi t$$

inches after t seconds.

(a) Find the maximum height that the weight rises above the equilibrium position $y = 0$.

(b) When does the weight first reach its maximum height if $t \geq 0$?

(c) What are the frequency and period?

CHAPTER 8 SUMMARY

KEY TERMS & SYMBOLS	KEY CONCEPTS
8.1 Angles and Arcs line segment ray angle initial side terminal side vertex positive angle negative angle degree (°) acute angle right angle obtuse angle straight angle complementary angles supplementary angles m(angle A) minute (′) second (″) angle in standard position quadrantal angle coterminal angle radian circumference of a circle sector of a circle angular speed, ω linear speed, v	An angle with vertex at the center of a circle and that intercepts an arc on the circle equal in length to the radius of the circle has measure **1 radian.** **DEGREE/RADIAN RELATIONSHIP** $180° = \pi$ **radians** **CONVERTING BETWEEN DEGREES AND RADIANS** **1.** Multiply a degree measure by $\frac{\pi}{180}$ radian and simplify to convert to radians. **2.** Multiply a radian measure by $\frac{180°}{\pi}$ and simplify to convert to degrees. **ARC LENGTH** The length s of the arc intercepted on a circle of radius r by a central angle of measure θ radians is given by the product of the radius and the radian measure of the angle, or $s = r\theta$, θ **in radians.** **AREA OF A SECTOR** The area of a sector of a circle of radius r and central angle θ is given by $A = \frac{1}{2}r^2\theta$, θ **in radians.** **ANGULAR AND LINEAR SPEED** **Angular Speed:** $\omega = \frac{\theta}{t}$ (ω in radians per unit time, θ in radians) **Linear Speed:** $v = \frac{s}{t}$; $v = \frac{r\theta}{t}$; $v = r\omega$

(continued)

KEY TERMS & SYMBOLS	KEY CONCEPTS
8.2 The Unit Circle unit circle circular functions sine cosine tangent cosecant secant cotangent	**TRIGONOMETRIC FUNCTIONS (UNIT CIRCLE APPROACH)** For s, x, and y as shown in the figure, the circular functions are $\sin s = y \qquad \csc s = \frac{1}{y} \quad (y \neq 0)$ $\cos s = x \qquad \sec s = \frac{1}{x} \quad (x \neq 0)$ $\tan s = \frac{y}{x} \quad (x \neq 0) \qquad \cot s = \frac{x}{y} \quad (y \neq 0).$ 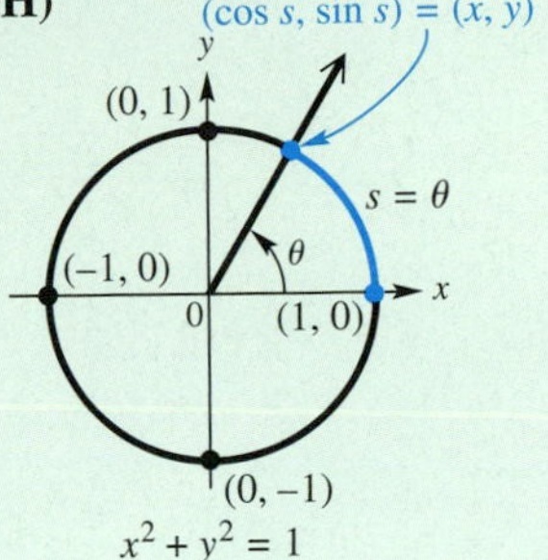
8.3 Graphs of the Sine and Cosine Functions periodic function period sine wave (sinusoid) sinusoidal function amplitude phase shift argument	**SINE AND COSINE FUNCTIONS** 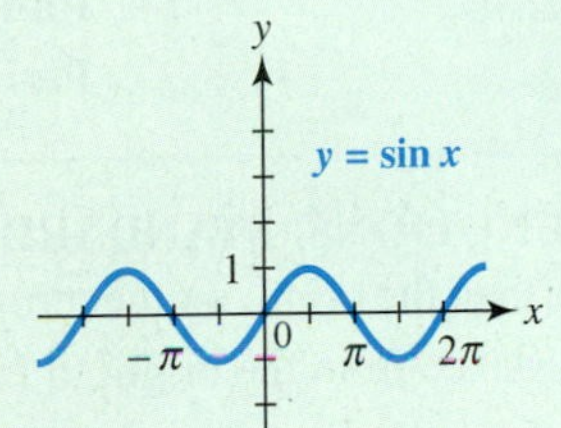 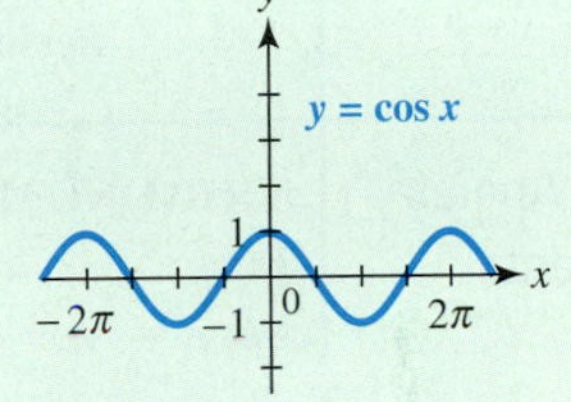 **Domain:** $(-\infty, \infty)$; **Range:** $[-1, 1]$ **Amplitude:** 1; **Period:** 2π **Domain:** $(-\infty, \infty)$; **Range:** $[-1, 1]$ **Amplitude:** 1; **Period:** 2π The graph of $y = c + a \sin[b(x - d)]$ or $y = c + a \cos[b(x - d)]$, $b > 0$, has **1.** amplitude $\lvert a \rvert$, **2.** period $\frac{2\pi}{b}$, **3.** vertical translation c units upward if $c > 0$ or $\lvert c \rvert$ units downward if $c < 0$, **4.** phase shift d units to the right if $d > 0$ or $\lvert d \rvert$ units to the left if $d < 0$.
8.4 Graphs of the Other Circular Functions	**COSECANT AND SECANT FUNCTIONS** 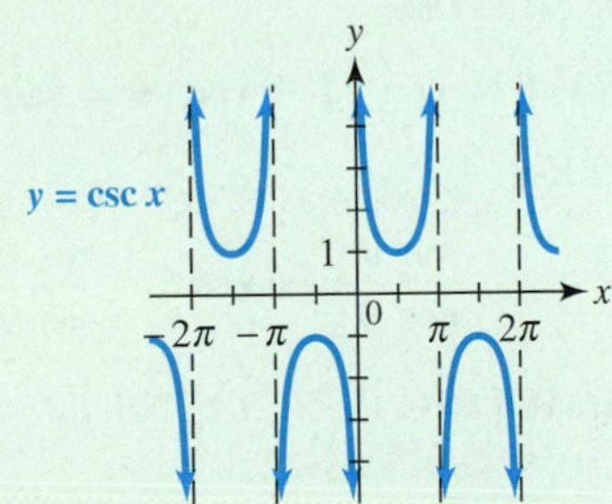 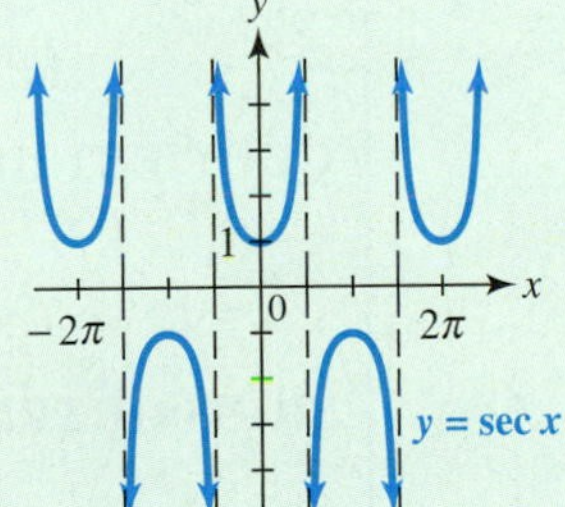 **Domain:** $\{x \mid x \neq n\pi$, where n is an integer$\}$ **Range:** $(-\infty, -1] \cup [1, \infty)$ **Period:** 2π **Domain:** $\{x \mid x \neq (2n + 1)\frac{\pi}{2}$, where n is an integer$\}$ **Range:** $(-\infty, -1] \cup [1, \infty)$ **Period:** 2π

(continued)

KEY TERMS & SYMBOLS	KEY CONCEPTS
	TANGENT AND COTANGENT FUNCTIONS (graph of $y = \tan x$) **Domain:** $\{x \mid x \neq (2n+1)\frac{\pi}{2}$, where n is an integer$\}$ **Range:** $(-\infty, \infty)$ **Period:** π (graph of $y = \cot x$) **Domain:** $\{x \mid x \neq n\pi$, where n is an integer$\}$ **Range:** $(-\infty, \infty)$ **Period:** π
8.5 Functions of Angles and Fundamental Identities	**TRIGONOMETRIC FUNCTIONS (STANDARD POSITION ANGLE APPROACH)** Let (x, y) be a point other than the origin on the terminal side of an angle θ in standard position. Let $r = \sqrt{x^2 + y^2}$, the distance from the origin to (x, y). Then $\sin\theta = \frac{y}{r}$ $\cos\theta = \frac{x}{r}$ $\tan\theta = \frac{y}{x}$ $(x \neq 0)$ $\csc\theta = \frac{r}{y}$ $(y \neq 0)$ $\sec\theta = \frac{r}{x}$ $(x \neq 0)$ $\cot\theta = \frac{x}{y}$ $(y \neq 0)$. **RECIPROCAL IDENTITIES** $\sin\theta = \frac{1}{\csc\theta}$ $\cos\theta = \frac{1}{\sec\theta}$ $\tan\theta = \frac{1}{\cot\theta}$ $\csc\theta = \frac{1}{\sin\theta}$ $\sec\theta = \frac{1}{\cos\theta}$ $\cot\theta = \frac{1}{\tan\theta}$ **PYTHAGOREAN IDENTITIES** $\sin^2\theta + \cos^2\theta = 1$ $1 + \tan^2\theta = \sec^2\theta$ $1 + \cot^2\theta = \csc^2\theta$ **QUOTIENT IDENTITIES** $\frac{\sin\theta}{\cos\theta} = \tan\theta$ $\frac{\cos\theta}{\sin\theta} = \cot\theta$ **SIGNS OF TRIGONOMETRIC FUNCTIONS** II: $x < 0, y > 0, r > 0$ — Sine and cosecant positive I: $x > 0, y > 0, r > 0$ — All functions positive III: $x < 0, y < 0, r > 0$ — Tangent and cotangent positive IV: $x > 0, y < 0, r > 0$ — Cosine and secant positive

(continued)

KEY TERMS & SYMBOLS	KEY CONCEPTS

8.6 Evaluating Trigonometric Functions

side opposite
side adjacent
reference angle

TRIGONOMETRIC FUNCTIONS (RIGHT TRIANGLE APPROACH)

For any acute angle A in standard position,

$$\sin A = \frac{y}{r} = \frac{\text{side opposite}}{\text{hypotenuse}} \qquad \csc A = \frac{r}{y} = \frac{\text{hypotenuse}}{\text{side opposite}}$$

$$\cos A = \frac{x}{r} = \frac{\text{side adjacent}}{\text{hypotenuse}} \qquad \sec A = \frac{r}{x} = \frac{\text{hypotenuse}}{\text{side adjacent}}$$

$$\tan A = \frac{y}{x} = \frac{\text{side opposite}}{\text{side adjacent}} \qquad \cot A = \frac{x}{y} = \frac{\text{side adjacent}}{\text{side opposite}}.$$

FUNCTION VALUES FOR SPECIAL ANGLES

θ	$\sin\theta$	$\cos\theta$	$\tan\theta$	$\cot\theta$	$\sec\theta$	$\csc\theta$
$30°$ or $\frac{\pi}{6}$	$\frac{1}{2}$	$\frac{\sqrt{3}}{2}$	$\frac{\sqrt{3}}{3}$	$\sqrt{3}$	$\frac{2\sqrt{3}}{3}$	2
$45°$ or $\frac{\pi}{4}$	$\frac{\sqrt{2}}{2}$	$\frac{\sqrt{2}}{2}$	1	1	$\sqrt{2}$	$\sqrt{2}$
$60°$ or $\frac{\pi}{3}$	$\frac{\sqrt{3}}{2}$	$\frac{1}{2}$	$\sqrt{3}$	$\frac{\sqrt{3}}{3}$	2	$\frac{2\sqrt{3}}{3}$

COFUNCTION IDENTITIES

If A is an acute angle measured in degrees, then

$$\sin A = \cos(90° - A) \qquad \csc A = \sec(90° - A)$$

$$\cos A = \sin(90° - A) \qquad \sec A = \csc(90° - A)$$

$$\tan A = \cot(90° - A) \qquad \cot A = \tan(90° - A).$$

If A is measured in radians, replace 90° with $\frac{\pi}{2}$.

REFERENCE ANGLES

θ' is the reference angle for angle θ. (If θ is acute, then $\theta' = \theta$.)

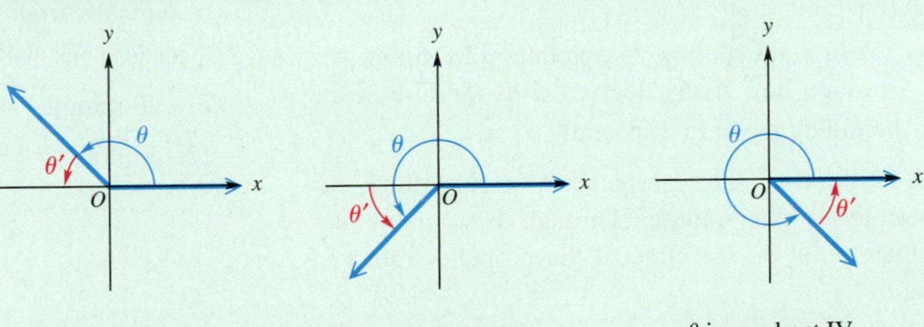

θ in quadrant II $\qquad$ θ in quadrant III $\qquad$ θ in quadrant IV

FINDING TRIGONOMETRIC FUNCTION VALUES FOR ANY NONQUADRANTAL ANGLE θ

Step 1 Add or subtract 360° as many times as needed to get an angle of at least 0° but less than 360°.

Step 2 Find the reference angle θ'.

Step 3 Find the trigonometric function values for θ'.

Step 4 Determine the correct signs for the values found in Step 3.

(continued)

KEY TERMS & SYMBOLS	KEY CONCEPTS
8.7 Applications of Right Triangles significant digit exact number angle of elevation angle of depression bearing	**SOLVING AN APPLIED TRIGONOMETRY PROBLEM** *Step 1* Draw a sketch, and label it with the given information. Label the quantity to be found with a variable. *Step 2* Use the sketch to write an equation relating the given quantities to the variable. *Step 3* Solve the equation, and check that your answer makes sense.
8.8 Harmonic Motion simple harmonic motion frequency damped oscillatory motion	**SIMPLE HARMONIC MOTION** The position of a point oscillating about an equilibrium position at time t is modeled by either $s(t) = a \cos \omega t$ or $s(t) = a \sin \omega t$, where a and ω are constants, with $\omega > 0$. The amplitude of the motion is $\lvert a \rvert$, the period is $\frac{2\pi}{\omega}$, and the frequency is $\frac{\omega}{2\pi}$.

CHAPTER 8 Review Exercises

1. Find the angle of smallest possible positive measure coterminal with an angle of $-174°$.

2. Let n represent any integer, and write an expression for all angles coterminal with an angle of $270°$.

Solve each problem.

3. ***Rotating Pulley*** A pulley is rotating 320 times per minute. Through how many degrees does a point on the edge of the pulley move in $\frac{2}{3}$ second?

4. ***Rotating Propeller*** The propeller of a speedboat rotates 650 times per minute. Through how many degrees will a point on the edge of the propeller rotate in 2.4 seconds?

5. Which is larger, an angle of $1°$ or an angle of 1 radian? Justify your answer.

6. Consider each angle in standard position having the given radian measure. In what quadrant does the terminal side lie?

(a) 3 **(b)** 4 **(c)** -2 **(d)** 7

Convert each degree measure to radians. Leave answers as multiples of π.

7. $120°$

8. $800°$

Convert each radian measure to degrees.

9. $\dfrac{5\pi}{4}$

10. $-\dfrac{6\pi}{5}$

Concept Check *Suppose the tip of the minute hand of a clock is 2 inches from the center of the clock. For each duration, determine the distance traveled by the tip of the minute hand.*

11. 20 minutes

12. 3 hours

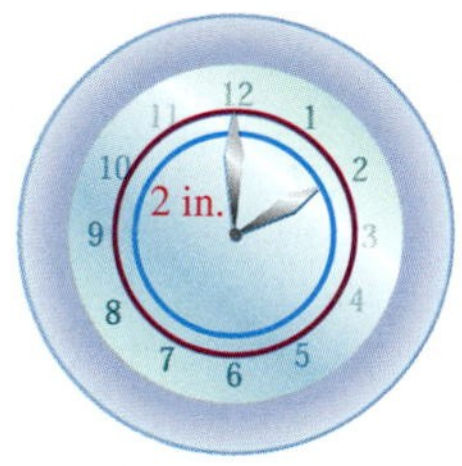

Solve each problem. Use a calculator as necessary.

13. ***Arc Length*** The radius of a circle is 15.2 cm. Find the length of an arc of the circle intercepted by a central angle of $\frac{3\pi}{4}$ radians.

14. ***Area of a Sector*** A central angle of $\frac{7\pi}{4}$ radians forms a sector of a circle. Find the area of the sector if the radius of the circle is 28.69 inches.

15. ***Height of a Tree*** A tree 2000 yards away subtends an angle of $1° 10'$. Find the height of the tree to two significant digits.

16. ***Concept Check*** Find the measure of the central angle θ (in radians) and the area of the sector.

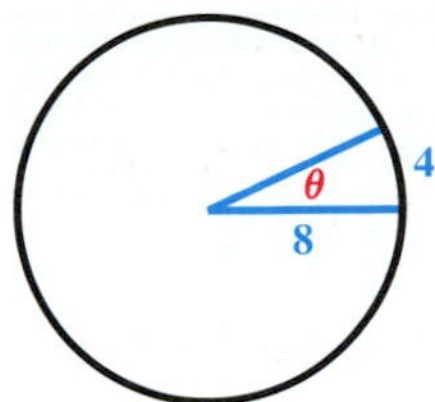

Find the exact function value. Do not use a calculator.

17. $\cos \frac{2\pi}{3}$

18. $\tan\left(-\frac{7\pi}{3}\right)$

19. $\csc\left(-\frac{11\pi}{6}\right)$

Approximate each circular function value to four decimal places.

20. $\cos(-.2443)$

21. $\cot 3.0543$

22. Use a calculator to approximate s in the interval $\left[0, \frac{\pi}{2}\right]$ if $\sin s = .49244294$.

23. ***Concept Check*** The screen shows a point on the unit circle. What is the length of the shortest arc of the circle from $(1, 0)$ to the point?

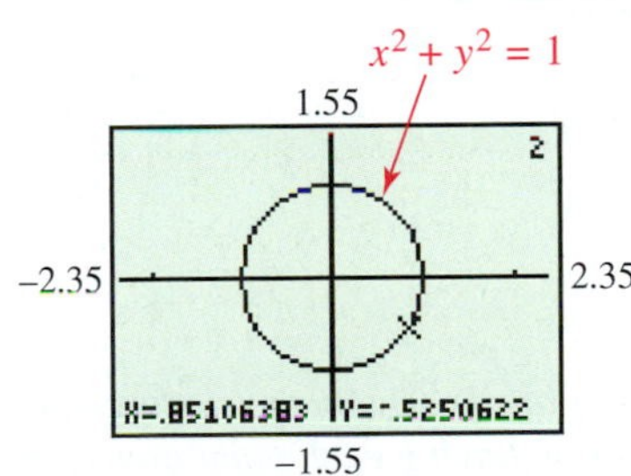

24. ***Concept Check*** Which one of the following is true about the graph of $y = 4 \sin 2x$?

A. It has amplitude 2 and period $\frac{\pi}{2}$.
B. It has amplitude 4 and period π.
C. Its range is $[0, 4]$.
D. Its range is $[-4, 0]$.

For each defined function, give the amplitude, period, vertical translation, and phase shift, as applicable. Then graph the function over at least two periods.

25. $y = 2 \sin x$

26. $y = \tan 3x$

27. $y = -\frac{1}{2} \cos 3x$

28. $y = 2 \sin 5x$

29. $y = 1 + 2 \sin \frac{1}{4}x$

30. $y = 3 - \frac{1}{4} \cos \frac{2}{3}x$

31. $y = 3 \cos\left(x + \frac{\pi}{2}\right)$

32. $y = -\sin\left(x - \frac{3\pi}{4}\right)$

33. $y = \frac{1}{2} \csc\left(2x - \frac{\pi}{4}\right)$

34. $y = \cot \frac{1}{2}x$

35. $y = -2 \sec 2x$

36. $y = \csc(2x - \pi)$

37. Suppose that f is a sinusoidal function with period π and $f\left(\frac{6\pi}{5}\right) = 1$. Explain why $f\left(-\frac{4\pi}{5}\right) = 1$.

38. Suppose that f is a periodic function such that $f(0) = 0$ and $f(2\pi) = 0$. Does it follow that the period of f is 2π? Explain.

In Exercises 39–44, graph each defined function over a one-period interval.

39. $y = 3 \cos 2x$

40. $y = \frac{1}{2} \cot 3x$

41. $y = \cos\left(x - \frac{\pi}{4}\right)$

42. $y = \tan\left(x - \frac{\pi}{2}\right)$

43. $y = 1 + 2 \cos 3x$

44. $y = -1 - 3 \sin 2x$

Concept Check *Identify one circular function that satisfies the description.*

45. Period is π, x-intercepts are of the form $n\pi$, where n is an integer

46. Period is 2π, graph passes through the origin

47. Period is 2π, graph passes through the point $\left(\frac{\pi}{2}, 0\right)$

48. Period is 2π, domain is $\{x \mid x \neq n\pi$, where n is an integer$\}$

49. Period is π, function is decreasing on the interval $(0, \pi)$

50. Period is 2π, has vertical asymptotes of the form $x = (2n + 1)\frac{\pi}{2}$, where n is an integer

For each graph, give the equation of a sine function having that graph.

51.

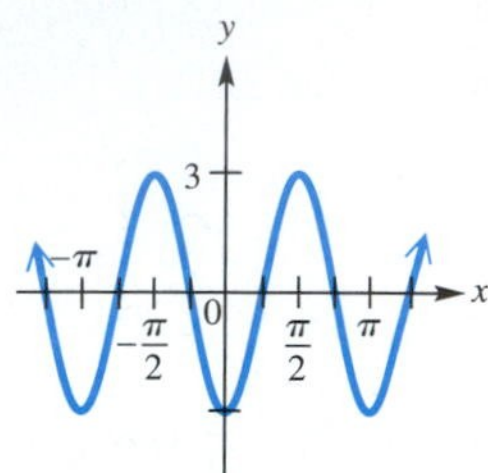

52.

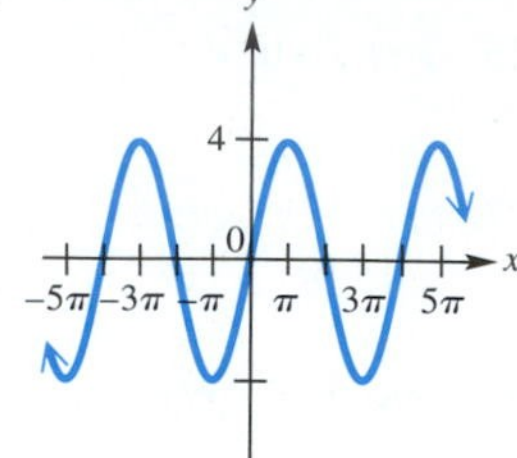

53.

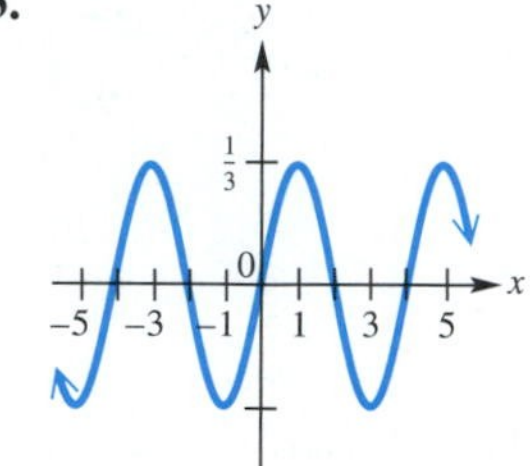

54.

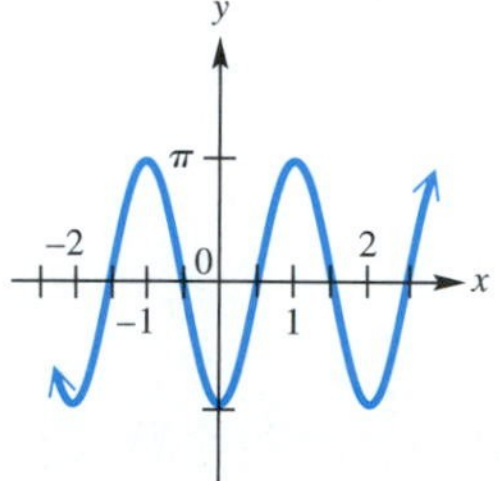

Solve each problem.

55. ***(Modeling) Monthly Average Temperature*** The monthly average temperature (in °F) in Chicago, Illinois, is shown in the table.

Month	°F	Month	°F
Jan	25	July	74
Feb	28	Aug	75
Mar	36	Sept	66
Apr	48	Oct	55
May	61	Nov	39
June	72	Dec	28

Source: Miller, A. and J. Thompson, *Elements of Meteorology*, Charles E. Merrill, 1975.

(a) Plot the monthly average temperature over a 2-year period by letting $x = 1$ correspond to January of the first year.

(b) Determine a modeling function of the form

$$f(x) = a \sin[b(x - d)] + c,$$

where a, b, c, and d are constants.

(c) Explain the significance of each constant.

(d) Graph f, together with the data, on the same coordinate axes. How well does f model the data?

(e) Use the sine regression capability of a graphing calculator to find the equation of a sine curve that fits these data.

56. ***(Modeling) Pollution Trends*** The amount of pollution in the air fluctuates with the seasons. It is lower after heavy spring rains and higher after periods of little rain. Additionally, the long-term trend is upward. An idealized graph of this situation is shown in the figure.

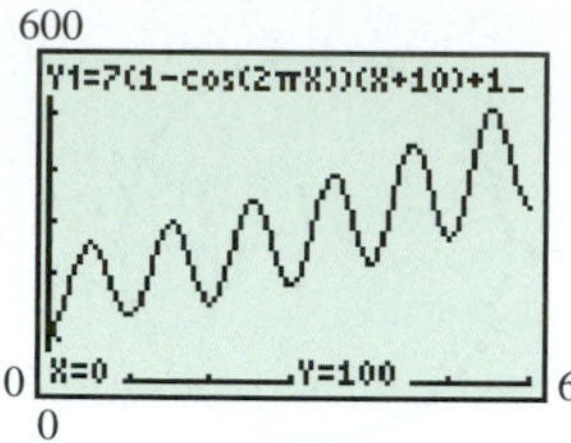

Circular functions can be used to model the fluctuating part of the pollution levels. Exponential functions can be used to model long-term growth. The pollution level in a certain area might be given by

$$y = 7(1 - \cos 2\pi x)(x + 10) + 100e^{.2x},$$

where x is time in years, with $x = 0$ representing January 1 of the base year. July 1 of the same year would be represented by $x = .5$, October 1 of the following year would be represented by $x = 1.75$, and so on. Find the pollution level on each date.

(a) January 1, base year

(b) July 1, base year

(c) January 1, following year

(d) July 1, following year

Find the trigonometric function values of each angle. If a value is undefined, say so.

57. **58.**

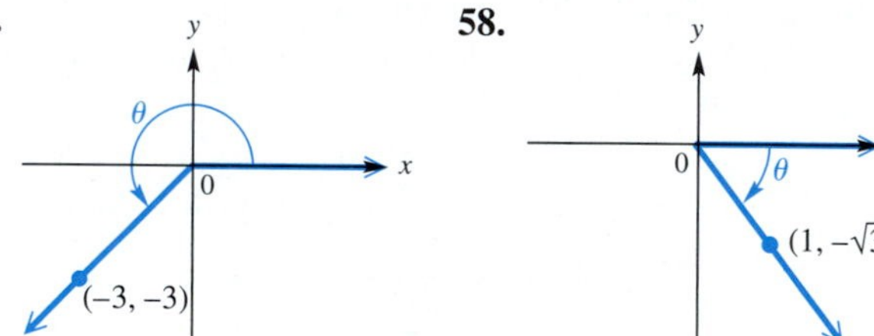

59. 180°

Find the values of all six trigonometric functions for an angle in standard position having each point on its terminal side.

60. $(3, -4)$

61. $(9, -2)$

62. $(-2\sqrt{2}, 2\sqrt{2})$

63. ***Concept Check*** If the terminal side of a quadrantal angle lies along the y-axis, which of its trigonometric functions are undefined?

Decide whether each statement is possible *or* impossible.

64. $\sec \theta = -\dfrac{2}{3}$

65. $\tan \theta = 1.4$

Find all six trigonometric function values for each angle. Rationalize denominators when applicable.

66. $\sin \theta = \frac{\sqrt{3}}{5}$ and $\cos \theta < 0$

67. $\cos \theta = -\frac{5}{8}$, with θ in quadrant III

68. If, for some particular angle θ, $\sin \theta < 0$ and $\cos \theta > 0$, in what quadrant must θ lie? What is the sign of $\tan \theta$?

69. Explain how you would use a calculator to find the cotangent of an angle θ whose tangent is 1.6778490. Then find $\cot \theta$.

Find the values of the six trigonometric functions for each angle A.

70.

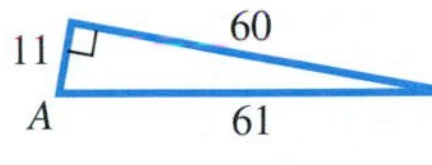

71.

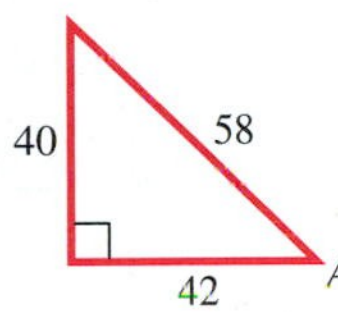

72. Explain why, in the figure, the cosine of angle A is equal to the sine of angle B.

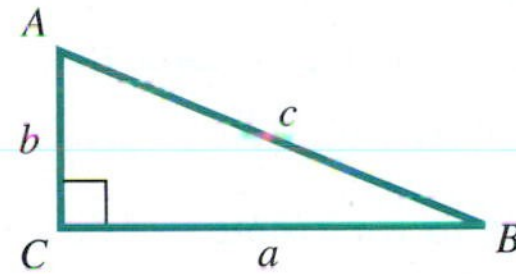

Find the values of the six trigonometric functions for each angle. Give exact values. Do not use a calculator. Rationalize denominators when applicable.

73. 300°

74. −225°

75. −390°

Use a calculator to find each value.

76. sin 72° 30′

77. sec 222° 30′

78. cot 305.6°

79. tan 11.7689°

80. ***Concept Check*** Which one of the following cannot be *exactly* determined with the methods of this chapter?

A. cos 135° **B.** cot(−45°)
C. sin 300° **D.** tan 140°

Approximate to the nearest tenth of a degree each value of θ, where θ is in the interval $[0°, 90°)$.

81. $\sin \theta = .82584121$

82. $\cot \theta = 1.1249386$

83. A student wants to use a calculator to find the value of cot 25°. However, instead of entering $\frac{1}{\tan 25}$, he enters $\tan^{-1} 25$. Assuming that the calculator is in degree mode, will this produce the correct answer? Explain.

84. For $\theta = 1997°$, use a calculator to find $\cos \theta$ and $\sin \theta$. From your results, decide what quadrant the angle lies in.

Solve each right triangle.

85.

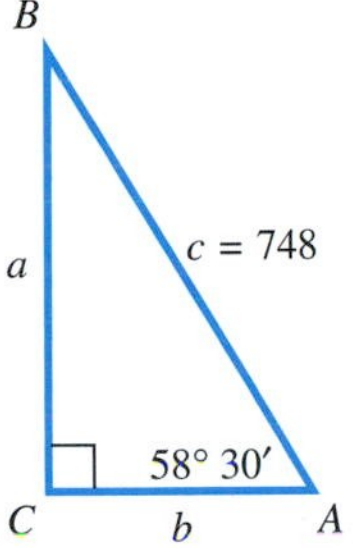

86. $A = 39.72°$; $b = 38.97$ m

Solve each problem.

87. ***Distance between Two Points*** The bearing of B from C is 254°. The bearing of A from C is 344°. The bearing of A from B is 32°. The distance from A to C is 780 meters. Find the distance from A to B.

88. ***Distance a Ship Sails*** The bearing from point A to point B is S 55° E and from point B to point C is N 35° E. If a ship sails from A to B, a distance of 80 kilometers, and then from B to C, a distance of 74 kilometers, how far is it from A to C?

89. ***Distance between Two Cars*** Two cars leave an intersection at the same time. One heads due south at 55 mph. The other travels due west. After two hours, the bearing of the car headed west from the car headed south is 324°. How far apart are they at that time?

90. ***Distance Traveled by a Sailboat*** From the top of a building that overlooks an ocean, an observer watches a boat sailing directly toward the building. If the observer is 150 feet above sea level, and if the angle of depression of the boat changes from 27° to 39° during the period of observation, approximate the distance the boat travels.

91. ***Height of a Tower*** The angle of elevation from a point 93.2 feet from the base of a tower to the top of the tower is $38° 20'$. Find the height of the tower.

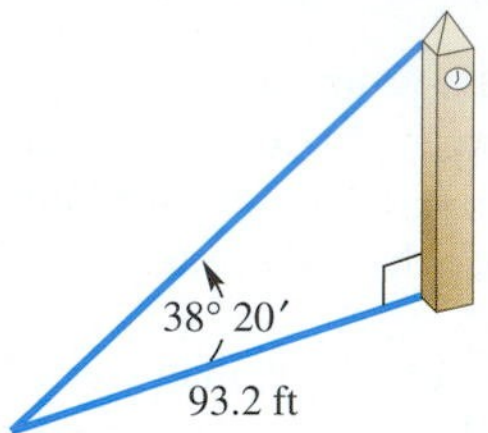

92. ***Height of a Tower*** The angle of depression of a television tower to a point on the ground 36.0 meters from the bottom of the tower is $29.5°$. Find the height of the tower.

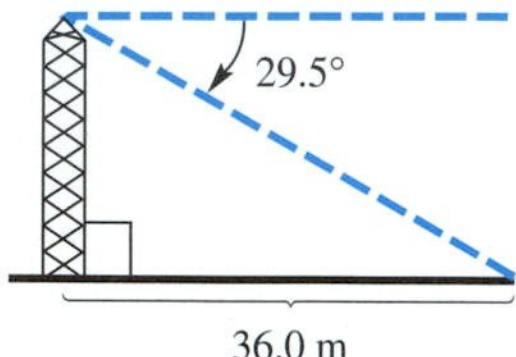

93. ***Height of a Triangle*** Find the measure of h to the nearest unit.

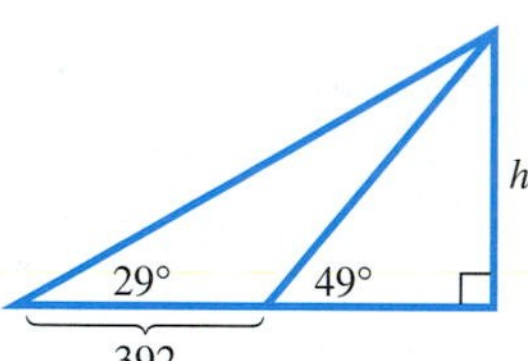

94. ***(Modeling) Fundamental Surveying Problem*** The first fundamental problem of surveying is to determine the coordinates of a point Q, given the coordinates of a point P; the distance between P and Q; and the bearing θ from P to Q. See the figure. (*Source:* Mueller, I. and K. Ramsayer, *Introduction to Surveying,* Frederick Ungar Publishing, 1979.)

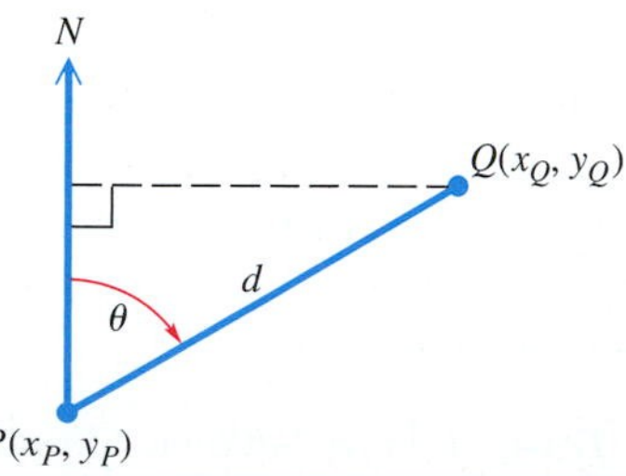

(a) Find a formula for the coordinates (x_Q, y_Q) of the point Q, given θ; the coordinates (x_P, y_P) of P; and the distance d between P and Q.

(b) Use your formula to determine the coordinates of the point (x_Q, y_Q) if $(x_P, y_P) = (123.62, 337.95)$, $\theta = 17° 19' 22''$, and $d = 193.86$ feet.

95. ***Atmospheric Effect on Sunlight*** The shortest path for the sun's rays through Earth's atmosphere occurs when the sun is directly overhead. Disregarding the curvature of Earth, as the sun moves lower on the horizon, the distance that sunlight passes through the atmosphere increases by a factor of $\csc \theta$, where θ is the angle of elevation of the sun. This increased distance reduces both the intensity of the sunlight and the amount of ultraviolet light that reaches Earth's surface. See the figure. (*Source:* Winter, C.-J., R. Sizmann, and L. L.Vant-Hunt (Editors), *Solar Power Plants,* Springer-Verlag, 1991.)

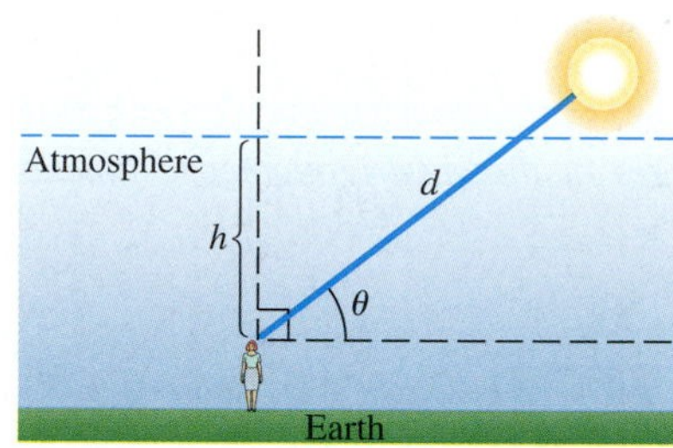

(a) Verify that $d = h \csc \theta$.

(b) Determine θ when $d = 2h$.

(c) The atmosphere filters out the ultraviolet light that causes skin to burn. Compare the difference between sunbathing when $\theta = \frac{\pi}{2}$ and $\theta = \frac{\pi}{3}$. Which measure gives less ultraviolet light?

96. ***Viewing Angle to an Object*** Let a person h_1 feet tall stand d feet from an object h_2 feet tall, where $h_2 > h_1$. Let θ be the angle of elevation to the top of the object. See the figure.

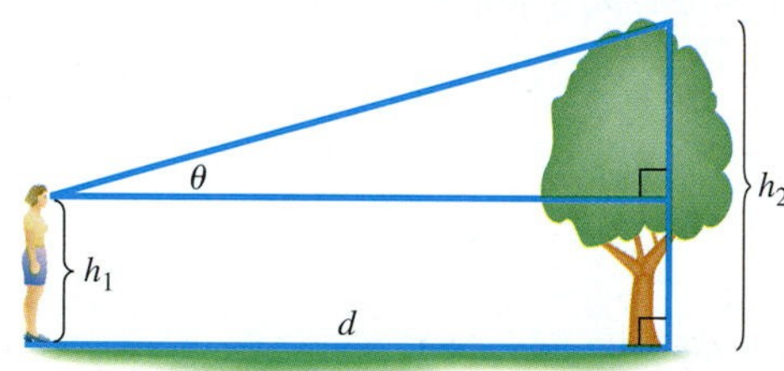

(a) Show that

$$d = (h_2 - h_1)\cot \theta.$$

(b) Let $h_2 = 55$ and $h_1 = 5$. Graph d for the interval $0 < \theta \leq \frac{\pi}{2}$.

97. Find the measure (in both degrees and radians) of the angle θ formed on the screen by the line passing through the origin and the positive part of the x-axis. Use the displayed values of x and y at the bottom of the screen.

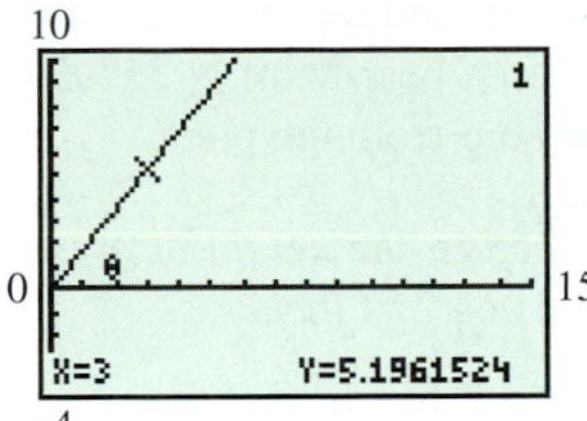

(Modeling) Harmonic Motion *An object in simple harmonic motion has position function s inches from an initial point, and t is the time in seconds. Find the amplitude, period, and frequency for each function.*

98. $s(t) = 3 \cos 2t$ **99.** $s(t) = 4 \sin \pi t$

100. In Exercise 98, what does the period represent? What does the amplitude represent?

101. In Exercise 99, what does the frequency represent? Find the position of the object from the initial point at 1.5 seconds, 2 seconds, and 3.25 seconds.

CHAPTER 8 Test

1. Find the angle of smallest positive measure coterminal with $-157°$.

2. *Rotating Tire* A tire rotates 450 times per minute. Through how many degrees does a point on the edge of the tire move in 1 second?

3. **(a)** Convert 120° to radians.
(b) Convert $\frac{9\pi}{10}$ radians to degrees.

4. A central angle of a circle with radius 150 centimeters cuts off an arc of 200 centimeters. Find each measure.
(a) The radian measure of the angle
(b) The area of a sector with that central angle

5. Find the length of the arc intercepted by an angle of 36° in a circle with radius 12 inches.

Graph each defined function over a two-period interval. Identify asymptotes when applicable.

6. $y = -1 + 2 \sin(x + \pi)$ **7.** $y = -\cos 2x$

8. $y = \tan\left(x - \dfrac{\pi}{2}\right)$

9. *(Modeling) Monthly Average Temperatures* The monthly average temperature (in °F) in Austin, Texas, can be modeled with the trigonometric function defined by

$$f(x) = 17.5 \sin\left[\frac{\pi}{6}(x - 4)\right] + 67.5,$$

where x is the month and $x = 1$ corresponds to January. (*Source:* Miller, A. and J. Thompson, *Elements of Meteorology,* Charles E. Merrill, 1975.)

(a) Graph f over the interval $1 \le x \le 25$.
(b) Find the amplitude, period, phase shift, and vertical translation of f.
(c) What is the monthly average temperature for the month of December?
(d) Find the maximum and minimum monthly average temperatures and the months when they occur.
(e) What would be an approximation for the *yearly* average temperature in Austin? How is this related to the vertical translation of the sine function in the formula for f?

10. Use the figure to find the following.

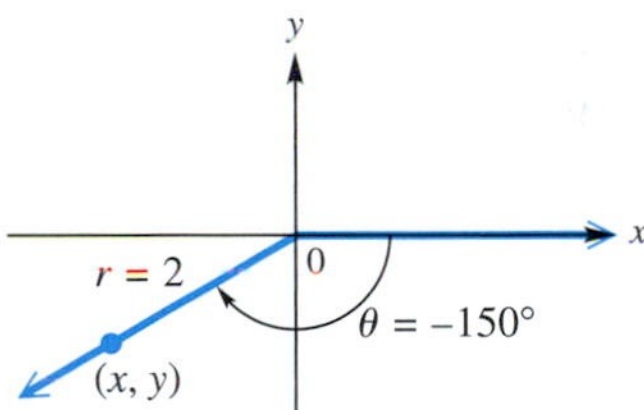

(a) The smallest positive angle coterminal with θ
(b) The values of x and y
(c) The values of the six trigonometric functions of θ
(d) The radian measure of θ

11. If $\cos \theta < 0$ and $\cot \theta > 0$, in what quadrant does θ lie?

12. If $(2, -5)$ is on the terminal side of an angle θ in standard position, find $\sin \theta$, $\cos \theta$, and $\tan \theta$.

13. If $\cos \theta = \frac{4}{5}$ and θ is in quadrant IV, find the values of the other trigonometric functions of θ.

14. Find the exact value of each variable in the figure.

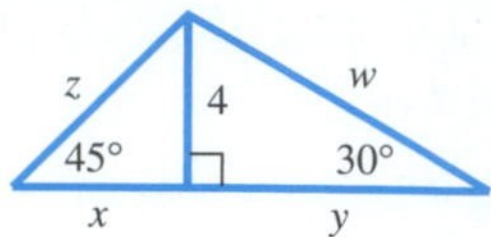

15. Find the exact value of $\cot(-750°)$.

16. Use a calculator to approximate the following.
(a) $\sin 78° 21'$ **(b)** $\tan 11.7689°$ **(c)** $\sec 58.9041°$

17. Use a calculator to approximate s in the interval $\left[0, \frac{\pi}{2}\right]$ if $\sin s = .82584121$.

18. Solve the triangle.

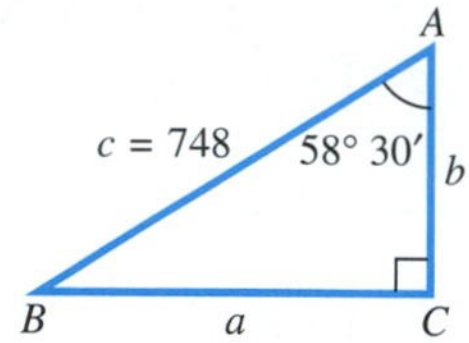

19. *Height of a Flagpole* To measure the height of a flagpole, Amado Carillo found that the angle of elevation from a point 24.7 feet from the base to the top is $32° 10'$. What is the height of the flagpole?

20. *Distance of a Ship from a Pier* A ship leaves a pier on a bearing of S 55° E and travels 80 kilometers. It then turns and continues on a bearing of N 35° E for 74 kilometers. How far is the ship from the pier?

Spring *After t seconds, the height of a weight attached to a spring is $s(t) = -3 \cos 2\pi t$ inches.*

21. Find the maximum height that the weight rises above the equilibrium position $y = 0$.

22. When is the first time that the weight reaches its maximum height if $t \geq 0$?

CHAPTER 8 PROJECT

Modeling Sunset Times

Sunset times at a location depend on its latitude and longitude and on the time of year. For example, see the table for sunset times at a location of 40° N on the longitude of Greenwich, England. (*Note:* Times are for the first day of each month.)

Month	Sunset	Month	Sunset
Jan	4:46 P.M.	July	7:33 P.M.
Feb	5:19 P.M.	Aug	7:14 P.M.
Mar	5:52 P.M.	Sept	6:32 P.M.
Apr	6:24 P.M.	Oct	5:42 P.M.
May	6:55 P.M.	Nov	4:58 P.M.
June	7:23 P.M.	Dec	4:35 P.M.

Source: World Almanac and Book of Facts, 2000.

Since the pattern for sunset time repeats itself every year, the data are periodic (with period 1 year, or 12 months). This means that there exists a transformation of a sine function of the form $y = a \sin[b(x - d)] + c$ that fits the data closely.

Activities

To find the function that fits the given data, follow these steps.

1. Use your graphing calculator to make a scatter diagram of the data. Plot two years of sunsets to display two full periods. For the independent variable, months, use the number of each month, 1 through 24 for two periods. For the dependent variable, sunset, the

times must be converted from hours and minutes to decimal hours. For example, 4:46 becomes $4 + \frac{46}{60}$, or 4.77 hours. Round all times to the nearest hundredth.

2. Find the amplitude of the function. Which of the constants in the form $y = a \sin[b(x - d)] + c$ gives the amplitude?

3. Find the period. Use the fact that the smallest positive number p for which $f(x) = f(x + np)$ is the period. Use the period to find the value of b.

4. Find the phase shift. Consider $\sin 0 = 0$. The corresponding point of the period of the data would be the spring equinox, March 21. Since the given sunset data are for the first day of each month, March 21 corresponds to $3 + \frac{21}{31} \approx 3.68$.

5. Find the vertical shift—the average of the maximum and minimum times.

6. Use the results of Items 2–5 to give the function that models the times for sunset. Verify your answer by graphing the function and the scatter diagram in the same window.

7. Use sine regression to find a function that models the data. The form of this function will be different from the one you found earlier. Are the two forms nearly equivalent?

9

Trigonometric Identities and Equations

ON THE NIGHT of October 20, 1955, a new sound exploded from a stage at Brooklyn High School in Cleveland, Ohio. The concert showcased rising stars in the music industry, including Elvis Presley and Bill Haley, and ushered in a new era of popular music—*rock and roll.* The distinctive melodic phrases, or *riffs,* generated by the musicians' electric guitars defined the new genre. Riffs were created as the musicians plucked or strummed the steel guitar strings with one hand, while pressing combinations of strings against the guitars' necks with the other. Vibrations caused by the strings' movements were amplified and broadcast to the crowd at the concert that night.

Sound waves, such as those initiated by musical instruments, travel through gases, liquids, and solids in sinusoidal patterns that can be graphed as sine or cosine functions and described by trigonometric equations. Rapid changes in pressure on the eardrum caused by sound waves send electrical impulses to the brain that are interpreted by humans and animals as noise. When sound waves are combined and organized to have rhythm, melody, harmony, and dynamics, the brain interprets them as music. The Beach Boys, shown above, have written some of the most beautiful harmonies of the rock and roll era.

In this chapter, we solve problems involving trigonometric equations and learn identities that enable us to relate trigonometric functions to each other.

Source: Rock and Roll Hall of Fame.

Chapter Outline

9.1 Trigonometric Identities

Fundamental Identities ■ Using the Fundamental Identities ■ Verifying Identities

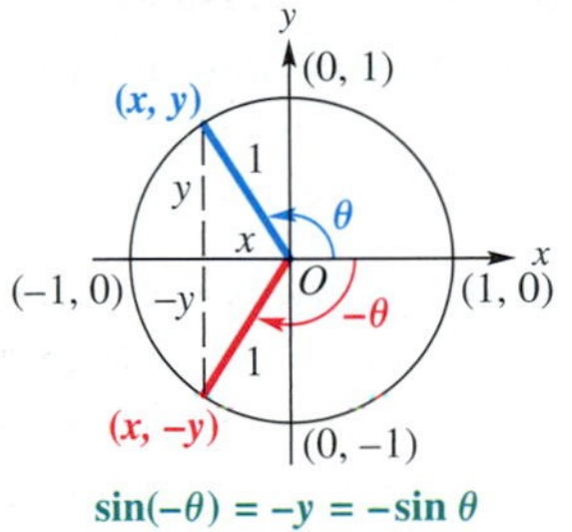

$\sin(-\theta) = -y = -\sin\theta$

FIGURE 1

Fundamental Identities

As suggested by the unit circle in Figure 1, an angle θ (in radians) determines an arc of length θ terminating at a point (x, y). Angle $-\theta$ determines a corresponding arc of length $-\theta$ terminating at the point $(x, -y)$. Therefore,

$$\sin(-\theta) = -y \quad \text{and} \quad \sin\theta = y,$$

so $\sin(-\theta)$ and $\sin\theta$ are negatives of each other, or

$$\sin(-\theta) = -\sin\theta.$$

Figure 1 shows an arc θ in quadrant II, but the same result holds for θ in any quadrant. Also, by definition,

$$\cos(-\theta) = x \quad \text{and} \quad \cos\theta = x,$$

so

$$\cos(-\theta) = \cos\theta.$$

We use these identities for $\sin(-\theta)$ and $\cos(-\theta)$ to find $\tan(-\theta)$ in terms of $\tan\theta$:

$$\tan(-\theta) = \frac{\sin(-\theta)}{\cos(-\theta)} = \frac{-\sin\theta}{\cos\theta} = -\frac{\sin\theta}{\cos\theta} \quad \text{or} \quad \tan(-\theta) = -\tan\theta.$$

Similar reasoning gives the remaining **negative-number** or **negative-angle identities,** which, together with the reciprocal, quotient, and Pythagorean identities from Chapter 8, are called the **fundamental identities.**

Looking Ahead to Calculus

Much of our work with identities in this chapter is preparation for calculus. Some calculus problems are simplified by making an appropriate trigonometric substitution. For example, if $x = 3\tan\theta$, then

$$\begin{aligned}\sqrt{9 + x^2} &= \sqrt{9 + (3\tan\theta)^2}\\ &= \sqrt{9 + 9\tan^2\theta}\\ &= \sqrt{9(1 + \tan^2\theta)}\\ &= 3\sqrt{1 + \tan^2\theta}\\ &= 3\sqrt{\sec^2\theta}.\end{aligned}$$

In the interval $\left(0, \frac{\pi}{2}\right)$, the value of $\sec\theta$ is positive, giving

$$\sqrt{9 + x^2} = 3\sec\theta.$$

Fundamental Identities

Reciprocal Identities

$$\cot\theta = \frac{1}{\tan\theta} \qquad \sec\theta = \frac{1}{\cos\theta} \qquad \csc\theta = \frac{1}{\sin\theta}$$

Quotient Identities

$$\tan\theta = \frac{\sin\theta}{\cos\theta} \qquad \cot\theta = \frac{\cos\theta}{\sin\theta}$$

Pythagorean Identities

$$\sin^2\theta + \cos^2\theta = 1 \qquad 1 + \tan^2\theta = \sec^2\theta \qquad 1 + \cot^2\theta = \csc^2\theta$$

Negative-Number Identities

$$\sin(-\theta) = -\sin\theta \qquad \cos(-\theta) = \cos\theta \qquad \tan(-\theta) = -\tan\theta$$
$$\csc(-\theta) = -\csc\theta \qquad \sec(-\theta) = \sec\theta \qquad \cot(-\theta) = -\cot\theta$$

NOTE The most commonly recognized forms of the fundamental identities are given in the preceding box. You must also recognize alternative forms of these identities. For example, two other forms of $\sin^2\theta + \cos^2\theta = 1$ are

$$\sin^2\theta = 1 - \cos^2\theta \quad \text{and} \quad \cos^2\theta = 1 - \sin^2\theta.$$

Using the Fundamental Identities

Any trigonometric function of a number or an angle can be expressed in terms of any other function.

EXAMPLE 1 **Expressing One Function in Terms of Another**

Express $\cos x$ in terms of $\tan x$.

Solution Since $\sec x$ is related to both $\cos x$ and $\tan x$ by identities, start with $1 + \tan^2 x = \sec^2 x$.

$$\frac{1}{1+\tan^2 x} = \frac{1}{\sec^2 x} \quad \text{Take reciprocals.}$$

$$\frac{1}{1+\tan^2 x} = \cos^2 x \quad \text{Reciprocal identity}$$

Remember both the positive and negative square roots.

$$\pm\sqrt{\frac{1}{1+\tan^2 x}} = \cos x \quad \text{Take the square root of each side.}$$

$$\cos x = \frac{\pm 1}{\sqrt{1+\tan^2 x}} \quad \text{Quotient rule; rewrite.}$$

$$\cos x = \frac{\pm\sqrt{1+\tan^2 x}}{1+\tan^2 x} \quad \text{Rationalize the denominator.}$$

Choose the + sign or the − sign, depending on the quadrant of x. ■

Each of the functions $\tan\theta$, $\cot\theta$, $\sec\theta$, and $\csc\theta$ can easily be expressed in terms of $\sin\theta$ and/or $\cos\theta$. We often make such substitutions in an expression to simplify it.

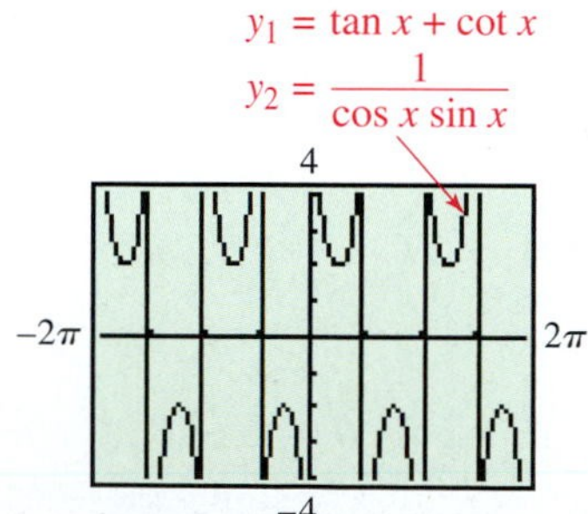

The graph supports the result in Example 2. The graphs of y_1 and y_2 appear to be identical.

EXAMPLE 2 **Rewriting an Expression in Terms of Sine and Cosine**

Write $\tan\theta + \cot\theta$ in terms of $\sin\theta$ and $\cos\theta$, and then simplify the expression.

Solution

$$\tan\theta + \cot\theta = \frac{\sin\theta}{\cos\theta} + \frac{\cos\theta}{\sin\theta} \quad \text{Quotient identities}$$

$$= \frac{\sin^2\theta}{\cos\theta\sin\theta} + \frac{\cos^2\theta}{\cos\theta\sin\theta} \quad \text{Write each fraction with the LCD, } \cos\theta\sin\theta.$$

$$= \frac{\sin^2\theta + \cos^2\theta}{\cos\theta\sin\theta} \quad \text{Add fractions.}$$

$$\tan\theta + \cot\theta = \frac{1}{\cos\theta\sin\theta} \quad \text{Pythagorean identity}$$ ■

CAUTION ***In working with trigonometric expressions and identities, be sure to write the argument of the function.*** For example, we would *not* write $\sin^2 + \cos^2 = 1$; an argument such as θ is necessary in this identity.

Verifying Identities

Looking Ahead to Calculus

Trigonometric identities are used in calculus to simplify trigonometric expressions, determine derivatives of trigonometric functions, and change the form of some integrals.

One of the skills required for more advanced work in mathematics (and especially in calculus) is the ability to use identities to write expressions in alternative forms. We develop this skill by using the fundamental identities to verify that an equation is an identity (for those values of the variable for which it is defined). Here are some hints to help you get started.

Hints for Verifying Identities

1. Learn the fundamental identities. Whenever you see either side of a fundamental identity, the other side should come to mind. Also, be aware of equivalent forms of the fundamental identities. For example, $\sin^2 \theta = 1 - \cos^2 \theta$ is an alternative form of $\sin^2 \theta + \cos^2 \theta = 1$.
2. Try to rewrite the more complicated side of the equation so that it is identical to the simpler side.
3. It is sometimes helpful to express all functions in the equation in terms of sine and cosine and then simplify the result.
4. Usually, any factoring or indicated algebraic operations should be performed. For example, the expression $\sin^2 x + 2 \sin x + 1$ can be factored as $(\sin x + 1)^2$. The sum or difference of two expressions, such as $\frac{1}{\sin \theta} + \frac{1}{\cos \theta}$, can be added or subtracted in the same way as any other rational expressions.

$$\frac{1}{\sin \theta} + \frac{1}{\cos \theta} = \frac{\cos \theta}{\sin \theta \cos \theta} + \frac{\sin \theta}{\sin \theta \cos \theta} \quad \text{Write with the LCD.}$$

$$= \frac{\cos \theta + \sin \theta}{\sin \theta \cos \theta} \quad \frac{a}{c} + \frac{b}{c} = \frac{a+b}{c}$$

5. As you select substitutions, keep in mind the side you are *not* changing, because it represents your goal. For example, to verify the identity

$$\tan^2 x + 1 = \frac{1}{\cos^2 x},$$

try to think of an identity that relates $\tan x$ to $\cos x$. Since $\sec x = \frac{1}{\cos x}$ and $\sec^2 x = \tan^2 x + 1$, the secant function is the best link between the two sides.

6. If an expression contains $1 + \sin x$, multiplying both numerator and denominator by $1 - \sin x$ would give $1 - \sin^2 x$, which could be replaced with $\cos^2 x$. Similar results for $1 - \sin x$, $1 + \cos x$, and $1 - \cos x$ may be useful.

TECHNOLOGY NOTE

To confirm the "coinciding" of two graphs like the ones indicated in Figure 2, trace to any X-value of Y_1 and then move the tracing cursor to Y_2. There should be no change in the value of Y for that particular value of X.

CAUTION ***Verifying identities is not the same as solving equations.*** Techniques used in solving equations, such as adding the same term to each side, or multiplying each side by the same term, are not useful when working with identities, because you are starting with a statement (to be verified) that may not be true.

To avoid the temptation to use algebraic properties of equations to verify identities, ***one strategy is to work with only one side and rewrite it to match the other side.***

GCM **EXAMPLE 3** **Verifying an Identity (Working with One Side)**

Verify that the given equation is an identity.

$$\cot x + 1 = \csc x(\cos x + \sin x)$$

Analytic Solution

We use fundamental identities to rewrite one side of the equation so that it is identical to the other side. Since the right side is more complicated, we work with it, using the third hint to change all functions to sine or cosine.

$$\cot x + 1 = \csc x(\cos x + \sin x) \qquad \text{Given equation}$$

Steps	Reasons
Right side of given equation: $\csc x(\cos x + \sin x) = \dfrac{1}{\sin x}(\cos x + \sin x)$	$\csc x = \frac{1}{\sin x}$
$= \dfrac{\cos x}{\sin x} + \dfrac{\sin x}{\sin x}$	Distributive property
$= \cot x + 1$ (Left side of given equation)	$\frac{\cos x}{\sin x} = \cot x$; $\frac{\sin x}{\sin x} = 1$

The given equation is an identity because the right side equals the left side.

Graphing Calculator Support*

We graph the two functions in the same window, with

$$Y_1 = \cot X + 1 = \frac{1}{\tan X} + 1$$

and

$$Y_2 = \csc X(\cos X + \sin X) = \frac{\cos X + \sin X}{\sin X}.$$

Each expression is in a form that can be entered in the calculator. The graphs coincide, as shown in Figure 2, which supports, but does not *verify*, the analytic result.

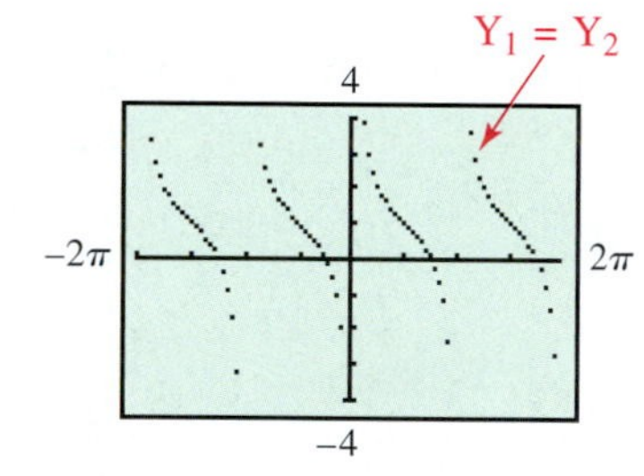

Dot mode

FIGURE 2

FOR DISCUSSION

We know that $\cot \frac{\pi}{2} + 1 = 0 + 1 = 1$ and $\csc \frac{\pi}{2}\left(\cos \frac{\pi}{2} + \sin \frac{\pi}{2}\right) = 1(0 + 1) = 1$. In the table in Figure 3, where Y_1 and Y_2 are defined as in Example 3, why don't all pairs of Y-values agree? How does the answer reinforce the warning that we must understand the mathematics to interpret the calculator results?

X	Y1	Y2
0	ERROR	ERROR
.3927	3.4142	3.4142
.7854	2	2
1.1781	1.4142	1.4142
1.5708	ERROR	1
1.9635	.58579	.58579
2.3562	0	0

X=1.570796326795

FIGURE 3

*To verify an identity, we must provide an analytical solution. ***A graph can only support, not prove, an identity,*** since we can graph only a finite interval of the domain.

EXAMPLE 4 Verifying an Identity (Working with One Side)

Verify that the given equation is an identity.

$$\frac{\tan t - \cot t}{\sin t \cos t} = \sec^2 t - \csc^2 t$$

Analytic Solution

We transform the more complicated left side to match the right side.

Left side of given equation

$$\frac{\tan t - \cot t}{\sin t \cos t} = \frac{\tan t}{\sin t \cos t} - \frac{\cot t}{\sin t \cos t} \qquad \frac{a-b}{c} = \frac{a}{c} - \frac{b}{c}$$

$$= \tan t \cdot \frac{1}{\sin t \cos t} - \cot t \cdot \frac{1}{\sin t \cos t} \qquad \frac{a}{b} = a \cdot \frac{1}{b}$$

$$= \frac{\sin t}{\cos t} \cdot \frac{1}{\sin t \cos t} - \frac{\cos t}{\sin t} \cdot \frac{1}{\sin t \cos t} \qquad \tan t = \frac{\sin t}{\cos t}, \ \cot t = \frac{\cos t}{\sin t}$$

$$= \frac{1}{\cos^2 t} - \frac{1}{\sin^2 t} \qquad \text{Multiply.}$$

$$= \sec^2 t - \csc^2 t \qquad \frac{1}{\cos^2 t} = \sec^2 t; \ \frac{1}{\sin^2 t} = \csc^2 t$$

Right side of given equation

The third hint about writing all trigonometric functions in terms of sine and cosine was used in the third line of the solution. Since the left side is identical to the right side, the given equation is an identity.

Graphing Calculator Support

Graphs of the two functions coincide. See Figure 4. The table shows that for selected values of X, the function values agree, further supporting the identity.

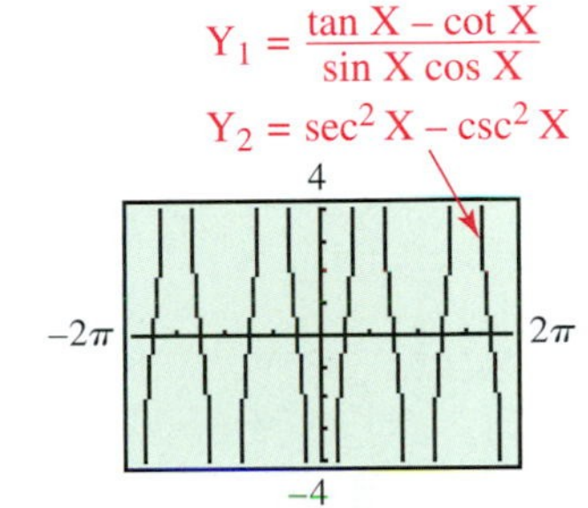

X	Y_1	Y_2
[illegible]	5.6569	5.6569
-.9817	1.7934	1.7934
-.7854	0	0
-.589	-1.793	-1.793
-.3927	-5.657	-5.657
-.1963	-25.23	-25.23
0	ERROR	ERROR

X=-1.1780972451

FIGURE 4

EXAMPLE 5 Verifying an Identity (Working with One Side)

Verify that the given equation is an identity.

$$\frac{\cos x}{1 - \sin x} = \frac{1 + \sin x}{\cos x}$$

Solution We work on the right side, using the last hint in the list given earlier.

$$\frac{1 + \sin x}{\cos x} = \frac{(1 + \sin x)(1 - \sin x)}{\cos x(1 - \sin x)} \qquad \text{Multiply by } \frac{1 - \sin x}{1 - \sin x} = 1.$$

$$= \frac{1 - \sin^2 x}{\cos x(1 - \sin x)} \qquad (x + y)(x - y) = x^2 - y^2$$

$$= \frac{\cos^2 x}{\cos x(1 - \sin x)} \qquad 1 - \sin^2 x = \cos^2 x$$

$$= \frac{\cos x}{1 - \sin x} \qquad \text{Lowest terms}$$

We could have obtained a similar result by working on the left side of the given equation and multiplying the numerator and denominator by $1 + \sin x$. ■

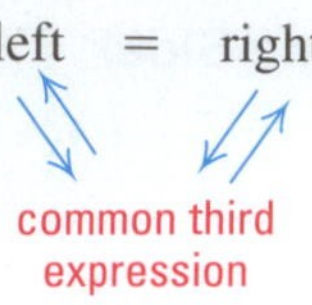

If both sides of an identity appear to be equally complicated, the identity can be verified if we work independently on the left side and on the right side, until each side is changed into some common third result. ***Each step, on each side, must be reversible.*** With all steps reversible, the procedure is as shown in the margin. The left side leads to the third expression, which leads back to the right side. This procedure is just a shortcut for the procedure used in Examples 3–5: One side is changed into the other side, but by going through an intermediate step.

EXAMPLE 6 Verifying an Identity (Working with Both Sides)

Verify that the given equation is an identity.

$$\frac{\sec \alpha + \tan \alpha}{\sec \alpha - \tan \alpha} = \frac{1 + 2 \sin \alpha + \sin^2 \alpha}{\cos^2 \alpha}$$

Solution Both sides appear equally complicated, so we verify the identity by changing each side into a common third expression. We work first on the left.

$$\frac{\sec \alpha + \tan \alpha}{\sec \alpha - \tan \alpha} = \frac{(\sec \alpha + \tan \alpha)\cos \alpha}{(\sec \alpha - \tan \alpha)\cos \alpha} \quad \text{Multiply by 1 in the form } \frac{\cos \alpha}{\cos \alpha}.$$

$$= \frac{\sec \alpha \cos \alpha + \tan \alpha \cos \alpha}{\sec \alpha \cos \alpha - \tan \alpha \cos \alpha} \quad \text{Distributive property}$$

$$= \frac{1 + \tan \alpha \cos \alpha}{1 - \tan \alpha \cos \alpha} \quad \sec \alpha \cos \alpha = 1$$

$$= \frac{1 + \frac{\sin \alpha}{\cos \alpha} \cdot \cos \alpha}{1 - \frac{\sin \alpha}{\cos \alpha} \cdot \cos \alpha} \quad \tan \alpha = \frac{\sin \alpha}{\cos \alpha}$$

$$= \frac{1 + \sin \alpha}{1 - \sin \alpha} \quad \text{Simplify.}$$

Now we work on the right side of the given equation. Begin by factoring.

$$\frac{1 + 2 \sin \alpha + \sin^2 \alpha}{\cos^2 \alpha} = \frac{(1 + \sin \alpha)^2}{\cos^2 \alpha} \quad a^2 + 2ab + b^2 = (a + b)^2$$

$$= \frac{(1 + \sin \alpha)^2}{1 - \sin^2 \alpha} \quad \cos^2 \alpha = 1 - \sin^2 \alpha$$

$$= \frac{(1 + \sin \alpha)^2}{(1 + \sin \alpha)(1 - \sin \alpha)} \quad \text{Factor } 1 - \sin^2 \alpha \text{ as the difference of squares.}$$

$$= \frac{1 + \sin \alpha}{1 - \sin \alpha} \quad \text{Lowest terms}$$

$y_1 = \frac{\sec x + \tan x}{\sec x - \tan x}$

$y_2 = \frac{1 + 2 \sin x + \sin^2 x}{\cos^2 x}$

4, −2π, 2π, −4

The graph confirms the result in Example 6.

We have verified that the given equation is an identity by showing that

$$\underbrace{\frac{\sec \alpha + \tan \alpha}{\sec \alpha - \tan \alpha}}_{\text{Left side of given equation}} = \underbrace{\frac{1 + \sin \alpha}{1 - \sin \alpha}}_{\text{Common third expression}} = \underbrace{\frac{1 + 2 \sin \alpha + \sin^2 \alpha}{\cos^2 \alpha}}_{\text{Right side of given equation}}.$$

■

FOR DISCUSSION

In Chapter 8, we introduced the cofunction relationships. For example, the functions sine and cosine are cofunctions. The equation

$$\frac{\cot x}{\csc x} = \cos x$$

represents an identity.

1. Replace each function in this identity with its cofunction.
2. Graph the two sides of the new equation from Item 1 as functions y_1 and y_2.
3. You should have found that $y_1 = y_2$. Verify this analytically.
4. Use cofunctions to rewrite the identity in Example 3.
5. Verify this new identity analytically and graphically.

There are usually several ways to verify a given identity. For instance, another way to begin verifying the identity in Example 6 is to work on the left.

$$\frac{\sec \alpha + \tan \alpha}{\sec \alpha - \tan \alpha} = \frac{\frac{1}{\cos \alpha} + \frac{\sin \alpha}{\cos \alpha}}{\frac{1}{\cos \alpha} - \frac{\sin \alpha}{\cos \alpha}} \qquad \text{Fundamental identities}$$

$$= \frac{\frac{1 + \sin \alpha}{\cos \alpha}}{\frac{1 - \sin \alpha}{\cos \alpha}} \qquad \text{Add and subtract fractions.}$$

$$= \frac{1 + \sin \alpha}{\cos \alpha} \div \frac{1 - \sin \alpha}{\cos \alpha} \qquad \text{Simplify the complex fraction.}$$

$$= \frac{1 + \sin \alpha}{\cos \alpha} \cdot \frac{\cos \alpha}{1 - \sin \alpha}$$

$$= \frac{1 + \sin \alpha}{1 - \sin \alpha}$$

Compare this with the result shown in Example 6 for the right side to see that the two sides indeed agree.

EXAMPLE 7 Applying a Pythagorean Identity to Radios

Tuners in radios select a radio station by adjusting the frequency. A tuner may contain an inductor L and a capacitor C, as illustrated in Figure 5. The energy stored in the inductor at time t is given by

$$L(t) = k \sin^2 2\pi Ft$$

An inductor and a capacitor

FIGURE 5

and the energy stored in the capacitor is given by

$$C(t) = k \cos^2 2\pi Ft,$$

where F is the frequency of the radio station and k is a constant. The total energy E in the circuit is given by

$$E(t) = L(t) + C(t).$$

Show that E is a constant function. (*Source:* Weidner, R. and R. Sells, *Elementary Classical Physics,* Vol. 2, Allyn & Bacon, 1973.)

Solution

$$\begin{aligned} E(t) &= L(t) + C(t) && \text{Given equation} \\ &= k \sin^2 2\pi Ft + k \cos^2 2\pi Ft && \text{Substitute.} \\ &= k(\sin^2 2\pi Ft + \cos^2 2\pi Ft) && \text{Factor.} \\ &= k(1) && \sin^2 \theta + \cos^2 \theta = 1. \text{ (Here, } \theta = 2\pi Ft.) \\ &= k && k \text{ is constant.} \end{aligned}$$

■

9.1 Exercises

Concept Check *Use the negative-number identities to decide whether each function is even or odd.*

1. $\sin x$ **2.** $\cos x$ **3.** $\tan x$ **4.** $\cot x$ **5.** $\sec x$ **6.** $\csc x$

Concept Check *For each expression in Column I, choose the expression from Column II that completes a fundamental identity.*

I	II
7. $\dfrac{\cos x}{\sin x} =$ ______	**A.** $\sin^2 x + \cos^2 x$
8. $\tan x =$ ______	**B.** $\cot x$
9. $\cos(-x) =$ ______	**C.** $\sec^2 x$
10. $\tan^2 x + 1 =$ ______	**D.** $\dfrac{\sin x}{\cos x}$
11. $1 =$ ______	**E.** $\cos x$

Concept Check *For each expression in Column I, choose the expression from Column II that completes an identity. You may have to rewrite one or both expressions.*

I	II
12. $-\tan x \cos x =$ ______	**A.** $\dfrac{\sin^2 x}{\cos^2 x}$
13. $\sec^2 x - 1 =$ ______	**B.** $\dfrac{1}{\sec^2 x}$
14. $\dfrac{\sec x}{\csc x} =$ ______	**C.** $\sin(-x)$
15. $1 + \sin^2 x =$ ______	**D.** $\csc^2 x - \cot^2 x + \sin^2 x$
16. $\cos^2 x =$ ______	**E.** $\tan x$

Write each expression as a trigonometric function of a positive number. (For example, $\sin(-3.4) = -\sin 3.4$.*)*

17. $\cos(-4.38)$ **18.** $\cos(-5.46)$ **19.** $\sin(-.5)$

20. $\sin(-2.5)$ **21.** $\tan\left(-\dfrac{\pi}{7}\right)$ **22.** $\cot\left(-\dfrac{4\pi}{7}\right)$

23. A student writes "$1 + \cot^2 = \csc^2$." Comment on this student's work.

24. Another student makes the following claim: "Since $\sin^2 \theta + \cos^2 \theta = 1$, I should also be able to say that $\sin \theta + \cos \theta = 1$ if I take the square root of both sides." Comment on this student's statement.

Complete this table so that each function in Exercises 25–30 is expressed in terms of the functions given across the top.

	$\sin \theta$	$\cos \theta$	$\tan \theta$	$\cot \theta$	$\sec \theta$	$\csc \theta$
25. $\sin \theta$	$\sin \theta$	$\pm\sqrt{1 - \cos^2 \theta}$	$\dfrac{\pm\tan \theta\sqrt{1 + \tan^2 \theta}}{1 + \tan^2 \theta}$			$\dfrac{1}{\csc \theta}$
26. $\cos \theta$		$\cos \theta$	$\dfrac{\pm\sqrt{\tan^2 \theta + 1}}{\tan^2 \theta + 1}$		$\dfrac{1}{\sec \theta}$	
27. $\tan \theta$			$\tan \theta$	$\dfrac{1}{\cot \theta}$		
28. $\cot \theta$			$\dfrac{1}{\tan \theta}$	$\cot \theta$	$\dfrac{\pm\sqrt{\sec^2 \theta - 1}}{\sec^2 \theta - 1}$	
29. $\sec \theta$		$\dfrac{1}{\cos \theta}$			$\sec \theta$	
30. $\csc \theta$	$\dfrac{1}{\sin \theta}$					$\csc \theta$

Write each expression as a constant, a single trigonometric function, or a power of a trigonometric function. (You may wish to use a graph to support your result.)

31. $\tan \theta \cos \theta$

32. $\cot \alpha \sin \alpha$

33. $\dfrac{\sin \beta \tan \beta}{\cos \beta}$

34. $\dfrac{\csc \theta \sec \theta}{\cot \theta}$

35. $\sec^2 x - 1$

36. $\csc^2 t - 1$

37. $\dfrac{\sin^2 x}{\cos^2 x} + \sin x \csc x$

38. $\dfrac{1}{\tan^2 \alpha} + \cot \alpha \tan \alpha$

Write each expression in terms of sine and cosine, and simplify it. (The final expression does not necessarily have to be in terms of sine and cosine.)

39. $\cot \theta \sin \theta$

40. $\sec \theta \cot \theta \sin \theta$

41. $\cos \theta \csc \theta$

42. $\cot^2 \theta(1 + \tan^2 \theta)$

43. $\sin^2 \theta(\csc^2 \theta - 1)$

44. $(\sec \theta - 1)(\sec \theta + 1)$

45. $(1 - \cos \theta)(1 + \sec \theta)$

46. $\dfrac{\cos \theta + \sin \theta}{\sin \theta}$

47. $\dfrac{\cos^2 \theta - \sin^2 \theta}{\sin \theta \cos \theta}$

48. $\dfrac{1 - \sin^2 \theta}{1 + \cot^2 \theta}$

49. $\sec \theta - \cos \theta$

50. $\dfrac{1 + \tan^2 \theta}{1 + \cot^2 \theta}$

51. $\sin \theta(\csc \theta - \sin \theta)$

52. $(\sec \theta + \csc \theta)(\cos \theta - \sin \theta)$

Perform each indicated operation and simplify the result.

53. $\cot \theta + \dfrac{1}{\cot \theta}$

54. $\dfrac{\sec x}{\csc x} + \dfrac{\csc x}{\sec x}$

55. $\tan s(\cot s + \csc s)$

56. $\cos \beta(\sec \beta + \csc \beta)$

57. $\dfrac{1}{\csc^2 \theta} + \dfrac{1}{\sec^2 \theta}$

58. $\dfrac{1}{\sin \alpha - 1} - \dfrac{1}{\sin \alpha + 1}$

59. $\dfrac{\cos x}{\sec x} + \dfrac{\sin x}{\csc x}$

60. $\dfrac{\cos \theta}{\sin \theta} + \dfrac{\sin \theta}{1 + \cos \theta}$

61. $(1 + \sin t)^2 + \cos^2 t$

62. $(1 + \tan s)^2 - 2 \tan s$

63. $\dfrac{1}{1 + \cos x} - \dfrac{1}{1 - \cos x}$

64. $(\sin \alpha - \cos \alpha)^2$

Relating Concepts

For individual or group investigation (Exercises 65–70)

Work Exercises 65–70 in order.

65. Graph $y = (\sec x + \tan x)(1 - \sin x)$.

66. The graph should look like the graph of a trigonometric function. Identify the trigonometric function.

67. Graph the function you identified in Exercise 66.

68. Do your graphs in Exercises 65 and 67 suggest an identity? If so, what is it?

69. Verify your answer to Exercise 68 analytically.

70. Repeat Exercises 65–69, using the function

$$y = \frac{\cos x + 1}{\sin x + \tan x}.$$

What is the identity?

In Exercises 71–88, verify that each equation is an identity.

71. $\dfrac{\cot \theta}{\csc \theta} = \cos \theta$

72. $\dfrac{\tan \theta}{\sec \theta} = \sin \theta$

73. $\cos^2 \theta (\tan^2 \theta + 1) = 1$

74. $\dfrac{\cos \theta + 1}{\tan^2 \theta} = \dfrac{\cos \theta}{\sec \theta - 1}$

75. $\dfrac{\tan^2 \gamma + 1}{\sec \gamma} = \sec \gamma$

76. $\sin^2 \beta(1 + \cot^2 \beta) = 1$

77. $\sin^2 \alpha + \tan^2 \alpha + \cos^2 \alpha = \sec^2 \alpha$

78. $\cot s + \tan s = \sec s \csc s$

79. $\dfrac{\sin^2 \gamma}{\cos \gamma} = \sec \gamma - \cos \gamma$

80. $\dfrac{\cos \alpha}{\sec \alpha} + \dfrac{\sin \alpha}{\csc \alpha} = \sec^2 \alpha - \tan^2 \alpha$

81. $\dfrac{\cos \theta}{\sin \theta \cot \theta} = 1$

82. $\sin^4 \theta - \cos^4 \theta = 2 \sin^2 \theta - 1$

83. $\tan^2 \gamma \sin^2 \gamma = \tan^2 \gamma + \cos^2 \gamma - 1$

84. $(1 - \cos^2 \alpha)(1 + \cos^2 \alpha) = 2 \sin^2 \alpha - \sin^4 \alpha$

85. $\dfrac{(\sec \theta - \tan \theta)^2 + 1}{\sec \theta \csc \theta - \tan \theta \csc \theta} = 2 \tan \theta$

86. $\dfrac{1}{1 - \sin \theta} + \dfrac{1}{1 + \sin \theta} = 2 \sec^2 \theta$

87. $\dfrac{1}{\tan \alpha - \sec \alpha} + \dfrac{1}{\tan \alpha + \sec \alpha} = -2 \tan \alpha$

88. $\dfrac{\csc \theta + \cot \theta}{\tan \theta + \sin \theta} = \cot \theta \csc \theta$

89. A student claims that the equation

$$\cos \theta + \sin \theta = 1$$

is an identity, since by letting $\theta = \frac{\pi}{2}$ we get $0 + 1 = 1$, a true statement. Comment on this student's reasoning.

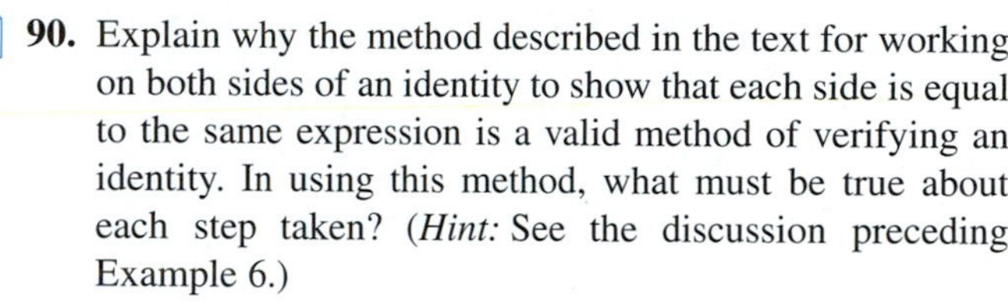

90. Explain why the method described in the text for working on both sides of an identity to show that each side is equal to the same expression is a valid method of verifying an identity. In using this method, what must be true about each step taken? (*Hint:* See the discussion preceding Example 6.)

(Modeling) *Work each problem.*

91. *Intensity of a Lamp* According to Lambert's law, the intensity of light from a single source on a flat surface at point P is given by

$$I = k \cos^2 \theta,$$

where k is a constant. See the figure. (*Source:* Winter, C., *Solar Power Plants,* Springer-Verlag, 1991.)

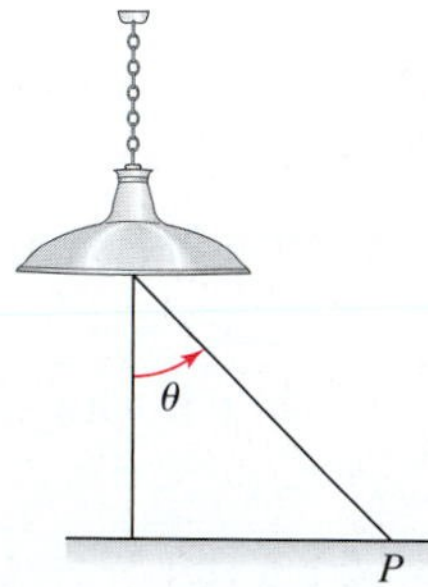

(a) Write I in terms of the sine function.

(b) Explain why the maximum value of I occurs when $\theta = 0$.

92. *Oscillating Spring* The distance or displacement y of a weight attached to an oscillating spring from its natural position is modeled by

$$y = 4 \cos 2\pi t,$$

where t is time in seconds. Potential energy is the energy of position and is given by

$$P = ky^2,$$

where k is a constant. The weight has the greatest potential energy when the spring is stretched the most. See the figure. (*Source:* Weidner, R., and R. Sells, *Elementary Classical Physics,* Vol. 1, Allyn & Bacon, 1973.)

(a) Write an expression for P that involves the cosine function.

(b) Use a fundamental identity to write P in terms of $\sin 2\pi t$.

93. ***Energy in an Oscillating Spring*** Refer to Exercise 92. Two types of mechanical energy are kinetic energy and potential energy. Kinetic energy is the energy of motion, and potential energy is the energy of position. A stretched spring has potential energy, which is converted to kinetic energy when the spring is released. If the potential energy of a weight attached to a spring is

$$P(t) = k\cos^2 4\pi t,$$

where k is a constant and t is time in seconds, then its kinetic energy is given by

$$K(t) = k\sin^2 4\pi t.$$

The total mechanical energy E is given by the equation $E(t) = P(t) + K(t)$.

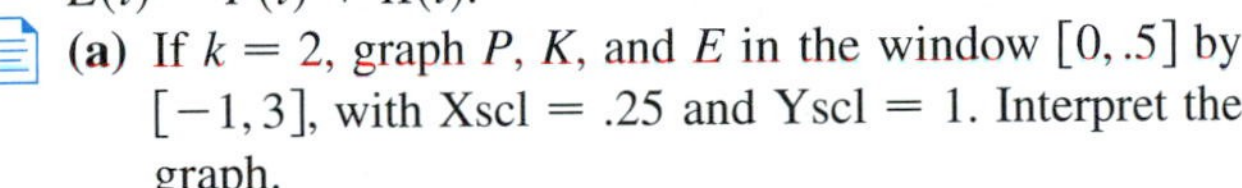

(a) If $k = 2$, graph P, K, and E in the window $[0, .5]$ by $[-1, 3]$, with Xscl $= .25$ and Yscl $= 1$. Interpret the graph.

(b) Make a table of K, P, and E, starting at $t = 0$ and incrementing by .05. Interpret the results.

(c) Use a fundamental identity to derive a simplified expression for $E(t)$.

94. ***Radio Tuners*** Refer to Example 7. Let the energy stored in the inductor be given by

$$L(t) = 3\cos^2 6{,}000{,}000t$$

and the energy in the capacitor be given by

$$C(t) = 3\sin^2 6{,}000{,}000t,$$

where t is time in seconds. The total energy E in the circuit is given by $E(t) = L(t) + C(t)$.

(a) Graph L, C, and E in the window $[0, 10^{-6}]$ by $[-1, 4]$, with Xscl $= 10^{-7}$ and Yscl $= 1$. Interpret the graph.

(b) Make a table of L, C, and E, starting at $t = 0$ and incrementing by 10^{-7}. Interpret your results.

(c) Use a fundamental identity to derive a simplified expression for $E(t)$.

9.2 Sum and Difference Identities

Cosine Sum and Difference Identities ■ Sine and Tangent Sum and Difference Identities

As mentioned earlier, the identities presented in this chapter hold true whether the arguments represent real numbers or degrees.

Cosine Sum and Difference Identities

FOR DISCUSSION

Use several pairs of different values for A and B to investigate whether the given statement is true in general.

1. $\cos(A - B) \stackrel{?}{=} \cos A - \cos B$ **2.** $\cos(A + B) \stackrel{?}{=} \cos A + \cos B$

The results of the preceding "For Discussion" box should convince you that $\cos(A - B)$ *does not equal* $\cos A - \cos B$ and that $\cos(A + B)$ *does not equal* $\cos A + \cos B$. For example, if $A = \frac{\pi}{2}$ and $B = 0$, then

$$\cos(A - B) = \cos\left(\frac{\pi}{2} - 0\right) = \cos\frac{\pi}{2} = 0,$$

while $$\cos A - \cos B = \cos\frac{\pi}{2} - \cos 0 = 0 - 1 = -1.$$

Similarly, when $A = \frac{\pi}{2}$ and $B = 0$, $\cos(A + B) = 0$ but $\cos A + \cos B = 1$.

There are identities for $\cos(A - B)$ and $\cos(A + B)$. To derive an identity for $\cos(A - B)$, we start by locating angles A and B in standard position on a unit circle, with $B < A$. Let S and Q be the points where the terminal sides of angles A and B, respectively, intersect the circle. Locate point R on the unit circle so that angle POR equals the difference $A - B$. See Figure 6.

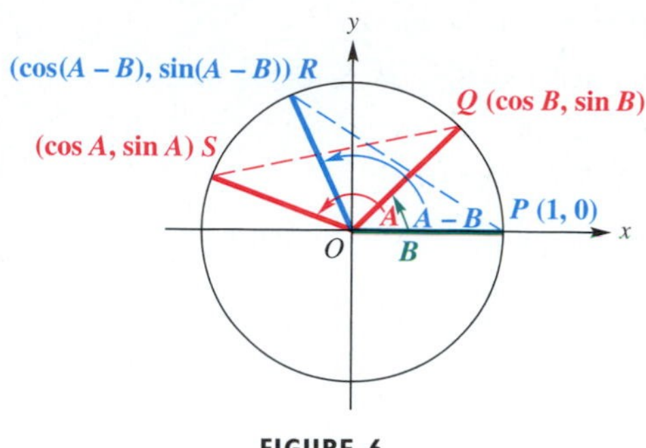

FIGURE 6

Point Q is on the unit circle, so by the work with circular functions in Chapter 8, the x-coordinate of Q is given by the cosine of angle B, while the y-coordinate of Q is given by the sine of angle B.

$$Q \text{ has coordinates } (\cos B, \sin B).$$

In the same way,

$$S \text{ has coordinates } (\cos A, \sin A)$$

and

$$R \text{ has coordinates } (\cos(A - B), \sin(A - B)).$$

Angle SOQ also equals $A - B$. Since the central angles SOQ and POR are equal, chords PR and SQ are equal. By the distance formula, since $PR = SQ$,

$$\sqrt{[\cos(A - B) - 1]^2 + [\sin(A - B) - 0]^2} = \sqrt{(\cos A - \cos B)^2 + (\sin A - \sin B)^2}.$$

Squaring both sides and clearing parentheses gives

$$\cos^2(A - B) - 2\cos(A - B) + 1 + \sin^2(A - B)$$
$$= \cos^2 A - 2\cos A \cos B + \cos^2 B + \sin^2 A - 2\sin A \sin B + \sin^2 B.$$

Because $\sin^2 x + \cos^2 x = 1$ for any value of x, rewrite the equation as

$$2 - 2\cos(A - B) = 2 - 2\cos A \cos B - 2\sin A \sin B$$
$$\cos(A - B) = \cos A \cos B + \sin A \sin B. \qquad \text{Subtract 2; divide by } -2.$$

Although Figure 6 shows angles A and B in the second and first quadrants, respectively, this result is the same for any values of these angles.

To find a similar expression for $\cos(A + B)$, rewrite $A + B$ as $A - (-B)$ and use the identity for $\cos(A - B)$.

$$\cos(A + B) = \cos[A - (-B)]$$
$$= \cos A \cos(-B) + \sin A \sin(-B) \qquad \text{Cosine difference idenity}$$
$$= \cos A \cos B + \sin A(-\sin B) \qquad \text{Negative-number identities}$$
$$\cos(A + B) = \cos A \cos B - \sin A \sin B$$

Cosine of a Sum or Difference

$$\cos(A - B) = \cos A \cos B + \sin A \sin B$$
$$\cos(A + B) = \cos A \cos B - \sin A \sin B$$

These identities are important in calculus and other areas of mathematics and are useful in some applications. Although a calculator can be used to find an approximation for cos 15°, for example, the method shown in Example 1 can be used to get an exact value by letting $A = 45°$ and $B = 30°$ and using the cosine difference identity.

GCM **EXAMPLE 1 Finding Exact Cosine Function Values**

Find the *exact* value of the following.

(a) $\cos 15°$ **(b)** $\cos \frac{5\pi}{12}$ **(c)** $\cos 87° \cos 93° - \sin 87° \sin 93°$

Solution

(a) To find cos 15°, we write 15° as the sum or difference of two angles with known function values, such as 45° and 30°, since

$$15° = 45° - 30°.$$

(We could also use $60° - 45°$.) Then we use the identity for the cosine of the difference of two angles.

$$\begin{aligned}\cos 15° &= \cos(45° - 30°) && 15° = 45° - 30° \\ &= \cos 45° \cos 30° + \sin 45° \sin 30° && \text{Cosine difference identity} \\ &= \frac{\sqrt{2}}{2} \cdot \frac{\sqrt{3}}{2} + \frac{\sqrt{2}}{2} \cdot \frac{1}{2} && \text{Substitute known values.} \\ &= \frac{\sqrt{6} + \sqrt{2}}{4} && \text{Multiply; add fractions.}\end{aligned}$$

```
cos(5π/12)
      .2588190451
(√(6)-√(2))/4
      .2588190451
```

Radian mode

This screen supports the result in Example 1(b).

(b)
$$\begin{aligned}\cos \frac{5\pi}{12} &= \cos\left(\frac{\pi}{6} + \frac{\pi}{4}\right) && \frac{\pi}{6} = \frac{2\pi}{12};\ \frac{\pi}{4} = \frac{3\pi}{12} \\ &= \cos\frac{\pi}{6}\cos\frac{\pi}{4} - \sin\frac{\pi}{6}\sin\frac{\pi}{4} && \text{Cosine sum identity} \\ &= \frac{\sqrt{3}}{2} \cdot \frac{\sqrt{2}}{2} - \frac{1}{2} \cdot \frac{\sqrt{2}}{2} && \text{Substitute known values.} \\ &= \frac{\sqrt{6} - \sqrt{2}}{4} && \text{Multiply; subtract fractions.}\end{aligned}$$

(c)
$$\begin{aligned}\cos 87° \cos 93° - \sin 87° \sin 93° &= \cos(87° + 93°) && \text{Cosine sum identity} \\ &= \cos 180° && \text{Add.} \\ &= -1\end{aligned}$$

■

NOTE In Example 1(b), we used the fact that $\frac{5\pi}{12} = \frac{\pi}{6} + \frac{\pi}{4}$. At first glance, this sum may not be obvious. Think of the values $\frac{\pi}{6}$ and $\frac{\pi}{4}$ in terms of fractions with denominator 12: $\frac{\pi}{6} = \frac{2\pi}{12}$ and $\frac{\pi}{4} = \frac{3\pi}{12}$. The following list may help you with problems of this type:

$$\frac{\pi}{3} = \frac{4\pi}{12}, \quad \frac{\pi}{4} = \frac{3\pi}{12}, \quad \frac{\pi}{6} = \frac{2\pi}{12}.$$

Using this list, we see, for example, that $\frac{\pi}{12} = \frac{\pi}{3} - \frac{\pi}{4}$ $\left(\text{or } \frac{\pi}{4} - \frac{\pi}{6}\right)$.

If either angle A or B in the identities for $\cos(A + B)$ and $\cos(A - B)$ is a quadrantal angle, then these identities allow us to write the expressions in terms of a single function of A or B.

EXAMPLE 2 **Reducing $\cos(A - B)$ to a Function of a Single Variable**

Write $\cos(180° - \theta)$ as a trigonometric function of θ.

Solution Replace A with 180° and B with θ in the cosine difference identity.

$$\begin{aligned} \cos(180° - \theta) &= \cos 180° \cos\theta + \sin 180° \sin\theta \\ &= (-1)\cos\theta + (0)\sin\theta \\ &= -\cos\theta \end{aligned}$$

Sine and Tangent Sum and Difference Identities

We can use the cosine sum and difference identities to derive similar identities for sine and tangent. Using the cofunction identity $\sin\theta = \cos\left(\frac{\pi}{2} - \theta\right)$ from Section 8.6 and replacing θ with $A + B$, we have

$$\begin{aligned} \sin(A + B) &= \cos\left[\frac{\pi}{2} - (A + B)\right] \\ &= \cos\left[\left(\frac{\pi}{2} - A\right) - B\right] \\ &= \cos\left(\frac{\pi}{2} - A\right)\cos B + \sin\left(\frac{\pi}{2} - A\right)\sin B && \text{Cosine difference identity} \\ \sin(A + B) &= \sin A \cos B + \cos A \sin B. && \text{Cofunction identities} \end{aligned}$$

Now we write $\sin(A - B)$ as $\sin[A + (-B)]$ and use the identity for $\sin(A + B)$.

$$\begin{aligned} \sin(A - B) &= \sin[A + (-B)] \\ &= \sin A \cos(-B) + \cos A \sin(-B) && \text{Sine sum identity} \\ \sin(A - B) &= \sin A \cos B - \cos A \sin B && \text{Negative-number identities} \end{aligned}$$

Sine of a Sum or Difference

$$\sin(A + B) = \sin A \cos B + \cos A \sin B$$
$$\sin(A - B) = \sin A \cos B - \cos A \sin B$$

To derive the identity for $\tan(A + B)$, we start with

$$\tan(A + B) = \frac{\sin(A + B)}{\cos(A + B)} \quad \text{Fundamental identity}$$

$$= \frac{\sin A \cos B + \cos A \sin B}{\cos A \cos B - \sin A \sin B}. \quad \text{Sum identities}$$

We express this result in terms of the tangent function.

$$\tan(A + B) = \frac{\dfrac{\sin A \cos B + \cos A \sin B}{1}}{\dfrac{\cos A \cos B - \sin A \sin B}{1}} \cdot \frac{\dfrac{1}{\cos A \cos B}}{\dfrac{1}{\cos A \cos B}} \quad \text{Multiply numerator and denominator by } \frac{1}{\cos A \cos B}.$$

$$= \frac{\dfrac{\sin A \cos B}{\cos A \cos B} + \dfrac{\cos A \sin B}{\cos A \cos B}}{\dfrac{\cos A \cos B}{\cos A \cos B} - \dfrac{\sin A \sin B}{\cos A \cos B}} \quad \text{Multiply numerators; multiply denominators.}$$

$$= \frac{\dfrac{\sin A}{\cos A} + \dfrac{\sin B}{\cos B}}{1 - \dfrac{\sin A}{\cos A} \cdot \dfrac{\sin B}{\cos B}} \quad \text{Simplify.}$$

$$\tan(A + B) = \frac{\tan A + \tan B}{1 - \tan A \tan B} \quad \tan\theta = \frac{\sin\theta}{\cos\theta}$$

Replacing B with $-B$ and using the fact that $\tan(-B) = -\tan B$ gives the identity for the tangent of the difference of two numbers.

Tangent of a Sum or Difference

$$\tan(A + B) = \frac{\tan A + \tan B}{1 - \tan A \tan B} \qquad \tan(A - B) = \frac{\tan A - \tan B}{1 + \tan A \tan B}$$

EXAMPLE 3 Finding Exact Sine and Tangent Function Values

Find the *exact* value of the following.

(a) $\sin 75°$ **(b)** $\tan \frac{7\pi}{12}$ **(c)** $\sin 40° \cos 160° - \cos 40° \sin 160°$

Solution

(a) $\sin 75° = \sin(45° + 30°)$ — $75° = 45° + 30°$

$= \sin 45° \cos 30° + \cos 45° \sin 30°$ — Sine sum identity

$= \frac{\sqrt{2}}{2} \cdot \frac{\sqrt{3}}{2} + \frac{\sqrt{2}}{2} \cdot \frac{1}{2}$ — Substitute known values.

$= \frac{\sqrt{6} + \sqrt{2}}{4}$ — Multiply; add fractions.

This screen indicates that the point (1.8325957, −3.732051), which approximates $\left(\frac{7\pi}{12}, -2-\sqrt{3}\right)$, lies on the graph of $y = \tan x$, supporting the result of Example 3(b):

$$\tan\frac{7\pi}{12} = -2-\sqrt{3}.$$

(b) $\tan\frac{7\pi}{12} = \tan\left(\frac{\pi}{3}+\frac{\pi}{4}\right)$ $\quad \frac{\pi}{3}=\frac{4\pi}{12}; \frac{\pi}{4}=\frac{3\pi}{12}$

$= \dfrac{\tan\frac{\pi}{3}+\tan\frac{\pi}{4}}{1-\tan\frac{\pi}{3}\tan\frac{\pi}{4}}$ Tangent sum identity

$= \dfrac{\sqrt{3}+1}{1-\sqrt{3}\cdot 1}$ Substitute known values.

$= \dfrac{\sqrt{3}+1}{1-\sqrt{3}}\cdot\dfrac{1+\sqrt{3}}{1+\sqrt{3}}$ Rationalize the denominator.

$= \dfrac{\sqrt{3}+3+1+\sqrt{3}}{1-3}$ Multiply.

$= \dfrac{4+2\sqrt{3}}{-2}$ Combine terms.

$= \dfrac{2(2+\sqrt{3})}{2(-1)}$ Factor out 2. (Factor first. Then divide out the common factor.)

$= -2-\sqrt{3}$ Lowest terms

(c) $\sin 40^\circ \cos 160^\circ - \cos 40^\circ \sin 160^\circ = \sin(40^\circ - 160^\circ)$ Sine difference identity

$= \sin(-120^\circ)$ Subtract.

$= -\sin 120^\circ$ Negative-number identity

$= -\dfrac{\sqrt{3}}{2}$ ■

EXAMPLE 4 Using Sum and Difference Identities

Write each function as an expression involving functions of θ alone.

(a) $\sin(30^\circ+\theta)$ **(b)** $\tan(45^\circ-\theta)$ **(c)** $\sin(180^\circ+\theta)$

Solution

(a) $\sin(30^\circ+\theta) = \sin 30^\circ \cos\theta + \cos 30^\circ \sin\theta$ Sine sum identity

$= \dfrac{1}{2}\cos\theta + \dfrac{\sqrt{3}}{2}\sin\theta$

(b) $\tan(45^\circ-\theta) = \dfrac{\tan 45^\circ - \tan\theta}{1+\tan 45^\circ \tan\theta}$ Tangent difference identity

$= \dfrac{1-\tan\theta}{1+\tan\theta}$

(c) $\sin(180^\circ+\theta) = \sin 180^\circ \cos\theta + \cos 180^\circ \sin\theta$ Sine sum identity

$= 0\cdot\cos\theta + (-1)\sin\theta$

$= -\sin\theta$ ■

An identity like the one derived in Example 4(c) is sometimes called a *reduction formula,* because the expression on the left is reduced to a function of θ alone.

EXAMPLE 5 Finding Function Values and the Quadrant of $A + B$

Suppose that A and B are angles in standard position, with $\sin A = \frac{4}{5}$, $\frac{\pi}{2} < A < \pi$, and $\cos B = -\frac{5}{13}$, $\pi < B < \frac{3\pi}{2}$. Find the following.

(a) $\sin(A + B)$ **(b)** $\tan(A + B)$ **(c)** The quadrant of $A + B$

Solution

(a) The identity for $\sin(A + B)$ requires $\sin A$, $\cos A$, $\sin B$, and $\cos B$. We are given values of $\sin A$ and $\cos B$. We must find values of $\cos A$ and $\sin B$.

$$\sin^2 A + \cos^2 A = 1 \qquad \text{Fundamental identity}$$

$$\frac{16}{25} + \cos^2 A = 1 \qquad \sin A = \frac{4}{5}$$

$$\cos^2 A = \frac{9}{25} \qquad \text{Subtract } \frac{16}{25}.$$

$$\cos A = -\frac{3}{5} \qquad \text{Since } A \text{ is in quadrant II, } \cos A < 0.$$

A similar procedure yields $\sin B = -\frac{12}{13}$. Now use the formula for $\sin(A + B)$.

$$\sin(A + B) = \frac{4}{5}\left(-\frac{5}{13}\right) + \left(-\frac{3}{5}\right)\left(-\frac{12}{13}\right) = -\frac{20}{65} + \frac{36}{65} = \frac{16}{65}$$

(b) Use the values of sine and cosine from part (a) to get $\tan A = -\frac{4}{3}$ and $\tan B = \frac{12}{5}$.

$$\tan(A + B) = \frac{-\frac{4}{3} + \frac{12}{5}}{1 - \left(-\frac{4}{3}\right)\left(\frac{12}{5}\right)} = \frac{\frac{16}{15}}{1 + \frac{48}{15}} = \frac{\frac{16}{15}}{\frac{63}{15}} = \frac{16}{15} \div \frac{63}{15} = \frac{16}{15} \cdot \frac{15}{63} = \frac{16}{63}$$

(c) From the results of parts (a) and (b), $\sin(A + B) = \frac{16}{65}$ and $\tan(A + B) = \frac{16}{63}$, both positive. Therefore, $A + B$ must be in quadrant I, since it is the only quadrant in which sine and tangent are both positive. ■

EXAMPLE 6 Applying the Cosine Difference Identity to Voltage

Common household electric current is called *alternating current* because the current alternates direction within the wires. The voltage V in a 115-volt outlet can be expressed by the equation $V = 163 \sin \omega t$, where ω is the angular velocity (in radians per second) of the rotating generator at the electrical plant and t is time in seconds. (*Source:* Bell, D., *Fundamentals of Electric Circuits,* Fourth Edition, Prentice-Hall, 1988.)

(a) Electric generators must rotate at 60 cycles per second so that household appliances and computers function properly. Determine ω for these electric generators.

(b) Graph V on the interval $0 \le t \le .05$.

(c) For what value of ϕ will the graph of $V = 163 \cos(\omega t - \phi)$ be the same as the graph of $V = 163 \sin \omega t$?

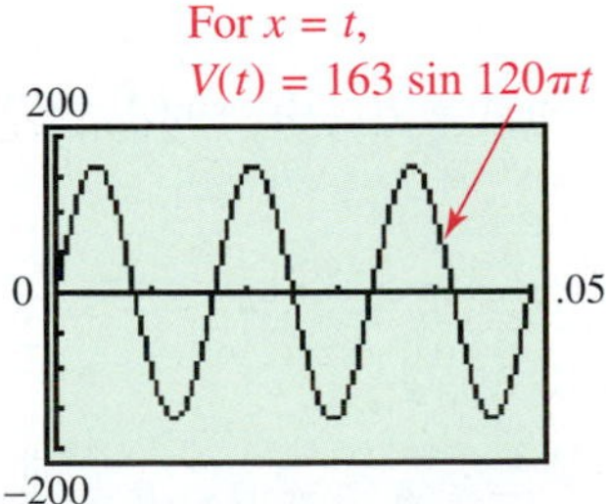

FIGURE 7

Solution

(a) Each cycle is 2π radians and at 60 cycles per second, $\omega = 60(2\pi) = 120\pi$ radians per second.

(b) $V = 163 \sin \omega t = 163 \sin 120\pi t$. Because the amplitude is 163 here, we choose $-200 \le V \le 200$ for the range, as shown in Figure 7.

(c) Since $\cos\left(x - \frac{\pi}{2}\right) = \cos\left(\frac{\pi}{2} - x\right) = \sin x$, we choose $\phi = \frac{\pi}{2}$. ■

9.2 Exercises

Concept Check *Match each expression in Column I with the correct expression in Column II to form an identity.*

I	II
1. $\cos(x + y) =$ ______	A. $\cos x \cos y + \sin x \sin y$
2. $\cos(x - y) =$ ______	B. $\sin x \sin y - \cos x \cos y$
3. $\sin(x + y) =$ ______	C. $\sin x \cos y + \cos x \sin y$
4. $\sin(x - y) =$ ______	D. $\sin x \cos y - \cos x \sin y$
	E. $\cos x \sin y - \sin x \cos y$
	F. $\cos x \cos y - \sin x \sin y$

Use identities to find the exact value of each expression.

5. $\sin \frac{\pi}{12}$
6. $\tan \frac{\pi}{12}$
7. $\sin\left(-\frac{5\pi}{12}\right)$
8. $\tan\left(-\frac{5\pi}{12}\right)$
9. $\sin\left(\frac{13\pi}{12}\right)$
10. $\cos\left(\frac{13\pi}{12}\right)$
11. $\cos 75°$
12. $\sin 105°$
13. $\tan 105°$
14. $\sin(-15°)$
15. $\cos(-15°)$
16. $\tan(-75°)$
17. $\cos \frac{\pi}{3} \cos \frac{2\pi}{3} - \sin \frac{\pi}{3} \sin \frac{2\pi}{3}$
18. $\cos \frac{7\pi}{8} \cos \frac{\pi}{8} + \sin \frac{7\pi}{8} \sin \frac{\pi}{8}$
19. $\sin 76° \cos 31° - \cos 76° \sin 31°$
20. $\sin 40° \cos 50° + \cos 40° \sin 50°$
21. $\dfrac{\tan 80° + \tan 55°}{1 - \tan 80° \tan 55°}$
22. $\dfrac{\tan 80° - \tan(-55°)}{1 + \tan 80° \tan(-55°)}$

Use identities to write each expression as a function of x alone.

23. $\sin(180° - x)$
24. $\sin(270° + x)$
25. $\cos(180° + x)$
26. $\cos(270° - x)$
27. $\sin(x - 90°)$
28. $\sin(x + 90°)$
29. $\tan(180° - x)$
30. $\tan(360° - x)$
31. $\cos\left(\frac{\pi}{2} - x\right)$
32. $\cos(\pi - x)$
33. $\cos\left(\frac{3\pi}{2} + x\right)$
34. $\sin(\pi - x)$
35. $\sin(\pi + x)$
36. $\tan(2\pi - x)$
37. $\cos(135° - x)$
38. $\sin(45° + x)$
39. $\tan(45° + x)$
40. $\tan(\pi + x)$
41. $\tan(\pi - x)$
42. $\sin\left(\frac{3\pi}{2} - x\right)$

Suppose that A and B are angles in standard position. Use the given information to find **(a)** $\sin(A + B)$, **(b)** $\sin(A - B)$, **(c)** $\tan(A + B)$, **(d)** $\tan(A - B)$, **(e)** *the quadrant of* $A + B$, *and* **(f)** *the quadrant of* $A - B$.

43. $\cos A = \frac{3}{5}$, $\sin B = \frac{5}{13}$, $0 < A < \frac{\pi}{2}, 0 < B < \frac{\pi}{2}$
44. $\sin A = \frac{3}{5}$, $\sin B = -\frac{12}{13}$, $0 < A < \frac{\pi}{2}, \pi < B < \frac{3\pi}{2}$
45. $\cos A = -\frac{8}{17}$, $\cos B = -\frac{3}{5}$, $\pi < A < \frac{3\pi}{2}, \pi < B < \frac{3\pi}{2}$
46. $\cos A = -\frac{15}{17}$, $\sin B = \frac{4}{5}$, $\frac{\pi}{2} < A < \pi, 0 < B < \frac{\pi}{2}$

Verify that each equation is an identity.

47. $\sin(x + y) + \sin(x - y) = 2 \sin x \cos y$

48. $\tan(x - y) - \tan(y - x) = \dfrac{2(\tan x - \tan y)}{1 + \tan x \tan y}$

49. $\dfrac{\cos(A - B)}{\cos A \sin B} = \tan A + \cot B$

50. $\dfrac{\sin(A + B)}{\cos A \cos B} = \tan A + \tan B$

51. $\dfrac{\sin(A - B)}{\sin B} + \dfrac{\cos(A - B)}{\cos B} = \dfrac{\sin A}{\sin B \cos B}$

52. $\dfrac{\tan(A + B) - \tan B}{1 + \tan(A + B) \tan B} = \tan A$

53. $\dfrac{\sin(x - y)}{\sin(x + y)} = \dfrac{\tan x - \tan y}{\tan x + \tan y}$

54. $\dfrac{\cos(A - B)}{\sin(A + B)} = \dfrac{1 + \cot A \cot B}{\cot A + \cot B}$

Exercise 55 and 56 refer to Example 6.

55. How many times does the current oscillate in .05 second?

56. What are the maximum and minimum voltages in this outlet? Is the voltage always equal to 115 volts?

(Modeling) *Solve each problem.*

57. *Back Stress* If a person bends at the waist with a straight back making an angle of θ degrees with the horizontal, then the force F exerted on the back muscles can be modeled by the equation

$$F = \frac{.6W \sin(\theta + 90^\circ)}{\sin 12^\circ},$$

where W is the weight of the person. (*Source:* Metcalf, H., *Topics in Classical Biophysics,* Prentice-Hall, 1980.)

(a) Calculate F when $W = 170$ pounds and $\theta = 30^\circ$.
(b) Use an identity to show that F is approximately equal to $2.9W \cos \theta$.
(c) For what value of θ is F maximum?

58. *Back Stress* Refer to Exercise 57.
(a) Suppose a 200-pound person bends at the waist so that $\theta = 45^\circ$. Estimate the force exerted by the person's back muscles.
(b) Approximate graphically the value of θ that results in the back muscles exerting a force of 400 pounds.

59. *Sound Waves* Sound is a result of waves applying pressure to a person's eardrum. For a pure sound wave radiating outward in a spherical shape, the trigonometric function

$$P = \frac{a}{r} \cos\left(\frac{2\pi r}{\lambda} - ct\right)$$

can be used to model the sound pressure P at a radius of r feet from the source; t is time in seconds, λ is length of the sound wave in feet, c is speed of sound in feet per second, and a is maximum sound pressure at the source measured in pounds per square foot. (*Source:* Beranek, L., *Noise and Vibration Control,* Institute of Noise Control Engineering, Washington, DC, 1988.) Let $\lambda = 4.9$ feet and $c = 1026$ feet per second.
(a) Let $a = .4$ pound per square foot. Graph the sound pressure at a distance $r = 10$ feet from its source over the interval $0 \le t \le .05$. Describe P at this distance.
(b) Now let $a = 3$ and $t = 10$. Graph the sound pressure for $0 \le r \le 20$. What happens to pressure P as radius r increases?
(c) Suppose a person stands at a radius r so that $r = n\lambda$, where n is a positive integer. Use the difference identity for cosine to simplify P in this situation.

60. *Voltage of a Circuit* When the two voltages

$$V_1 = 30 \sin 120\pi t \quad \text{and} \quad V_2 = 40 \cos 120\pi t$$

are applied to the same circuit, the resulting voltage V will equal their sum. (*Source:* Bell, D., *Fundamentals of Electric Circuits,* Second Edition, Reston Publishing Company, 1981.)
(a) Graph $V = V_1 + V_2$ over the interval $0 \le t \le .05$.
(b) Use the graph to estimate values for a and ϕ so that $V = a \sin(120\pi t + \phi)$.
(c) Use identities to verify that your expression for V is valid.

Reviewing Basic Concepts (Sections 9.1 and 9.2)

1. Write $\frac{\csc x}{\cot x} - \frac{\cot x}{\csc x}$ in terms of $\sin x$ and $\cos x$.

2. Use a sum or difference identity to find an exact value of $\tan\left(-\frac{\pi}{12}\right)$.

3. Find an exact value of

$$\cos 18^\circ \cos 108^\circ + \sin 18^\circ \sin 108^\circ.$$

4. Use a sum or difference identity to write $\sin\left(x - \frac{\pi}{4}\right)$ as a function of x alone.

5. Given $\sin A = \frac{2}{3}$ and $\cos B = -\frac{1}{2}$, where A is in quadrant II and B is in quadrant III, find $\sin(A + B)$, $\cos(A - B)$, and $\tan(A - B)$.

Verify that each equation is an identity.

6. $\csc^2 \theta - \cot^2 \theta = 1$

7. $\frac{\sin t}{1 - \cos t} = \frac{1 + \cos t}{\sin t}$

8. $\frac{\cot A - \tan A}{\csc A \sec A} = \cos^2 A - \sin^2 A$

9. $\frac{\sin(x - y)}{\sin x \sin y} = \cot y - \cot x$

10. ***(Modeling) Voltage*** A coil of wire rotating in a magnetic field induces a voltage given by

$$e = 20 \sin\left(\frac{\pi t}{4} - \frac{\pi}{2}\right).$$

Use an identity to express the right side of this equation in terms of $\cos \frac{\pi t}{4}$.

9.3 Further Identities

Double-Number Identities ■ Product-to-Sum and Sum-to-Product Identities ■ Half-Number Identities

Double-Number Identities

The **double-number identities** result from the sum identities when $A = B$ so that $A + B = A + A = 2A$. For example, if $B = A$, then the cosine sum identity $\cos(A + B) = \cos A \cos B - \sin A \sin B$ gives an identity for $\cos 2A$.

$$\cos 2A = \cos(A + A)$$
$$= \cos A \cos A - \sin A \sin A \quad \text{Cosine sum identity}$$
$$\cos 2A = \cos^2 A - \sin^2 A$$

Two other useful forms of this identity are obtained by substituting either $\cos^2 A = 1 - \sin^2 A$ or $\sin^2 A = 1 - \cos^2 A$. Replace $\cos^2 A$ with $1 - \sin^2 A$ to obtain

$$\cos 2A = \cos^2 A - \sin^2 A$$
$$= (1 - \sin^2 A) - \sin^2 A \quad \text{Fundamental identity}$$
$$\cos 2A = 1 - 2\sin^2 A,$$

or replace $\sin^2 A$ with $1 - \cos^2 A$ to obtain

$$\cos 2A = \cos^2 A - \sin^2 A$$
$$= \cos^2 A - (1 - \cos^2 A) \quad \text{Fundamental identity}$$
$$= \cos^2 A - 1 + \cos^2 A$$
$$\cos 2A = 2\cos^2 A - 1.$$

We find sin 2A with the identity for $\sin(A + B)$, again letting $B = A$.

$$\sin 2A = \sin(A + A)$$
$$= \sin A \cos A + \cos A \sin A \quad \text{Sine sum identity}$$
$$\sin 2A = 2 \sin A \cos A$$

Similarly, we use the identity for $\tan(A + B)$ to find tan 2A.

$$\tan 2A = \tan(A + A)$$
$$= \frac{\tan A + \tan A}{1 - \tan A \tan A} \quad \text{Tangent sum identity}$$
$$\tan 2A = \frac{2 \tan A}{1 - \tan^2 A}$$

Double-Number Identities

$$\cos 2A = \cos^2 A - \sin^2 A \qquad \cos 2A = 1 - 2\sin^2 A$$
$$\cos 2A = 2\cos^2 A - 1 \qquad \sin 2A = 2 \sin A \cos A$$
$$\tan 2A = \frac{2 \tan A}{1 - \tan^2 A}$$

EXAMPLE 1 Finding Function Values of 2θ

Given $\cos \theta = \frac{3}{5}$ and $\sin \theta < 0$, find $\sin 2\theta$, $\cos 2\theta$, and $\tan 2\theta$.

Solution To find $\sin 2\theta$, we must first find the value of $\sin \theta$.

$$\sin^2 \theta + \left(\frac{3}{5}\right)^2 = 1 \quad \sin^2\theta + \cos^2\theta = 1;\ \cos\theta = \tfrac{3}{5}$$
$$\sin^2 \theta = \frac{16}{25} \quad \text{Subtract } \tfrac{9}{25}.$$
$$\sin \theta = -\frac{4}{5} \quad \text{Take square roots; choose the negative square root, since } \sin\theta < 0.$$

Using the double-number identity for sine and the first form of the identity for $\cos 2\theta$ yields

$$\sin 2\theta = 2 \sin \theta \cos \theta = 2\left(-\frac{4}{5}\right)\left(\frac{3}{5}\right) = -\frac{24}{25}, \quad \sin\theta = -\tfrac{4}{5};\ \cos\theta = \tfrac{3}{5}$$

and

$$\cos 2\theta = \cos^2 \theta - \sin^2 \theta = \frac{9}{25} - \frac{16}{25} = -\frac{7}{25}.$$

The value of $\tan 2\theta$ can be found in two ways. One way is to use the double-number identity and the fact that $\tan\theta = \frac{\sin\theta}{\cos\theta} = \frac{-\frac{4}{5}}{\frac{3}{5}} = -\frac{4}{5} \div \frac{3}{5} = -\frac{4}{5}\cdot\frac{5}{3} = -\frac{4}{3}$.

$$\tan 2\theta = \frac{2\tan\theta}{1-\tan^2\theta} = \frac{2\left(-\frac{4}{3}\right)}{1-\left(-\frac{4}{3}\right)^2} = \frac{-\frac{8}{3}}{-\frac{7}{9}} = \frac{24}{7}$$

Be careful when simplifying a complex fraction.

Alternatively, we can find $\tan 2\theta$ by finding the quotient of $\sin 2\theta$ and $\cos 2\theta$.

$$\tan 2\theta = \frac{\sin 2\theta}{\cos 2\theta} = \frac{-\frac{24}{25}}{-\frac{7}{25}} = \frac{24}{7}$$

EXAMPLE 2 Verifying a Double-Number Identity

Verify that the given equation is an identity.

$$\cot x \sin 2x = 1 + \cos 2x$$

Analytic Solution

We start by working on the left side, using the hint from Section 9.1 about writing all functions in terms of sine and cosine.

$$\begin{aligned}
\cot x \sin 2x &= \frac{\cos x}{\sin x}\cdot \sin 2x && \cot x = \tfrac{\cos x}{\sin x}\\
&= \frac{\cos x}{\sin x}(2\sin x\cos x) && \sin 2x = 2\sin x\cos x\\
&= 2\cos^2 x && \text{Multiply.}\\
&= 1 + \cos 2x && 2\cos^2 x - 1 = \cos 2x
\end{aligned}$$

The final step illustrates the importance of being able to recognize alternative forms of identities.

Graphing Calculator Support

To support our analytic work, we show that the graphs of

$$y_1 = \cot x \sin 2x \quad \text{and} \quad y_2 = 1 + \cos 2x$$

coincide. Notice that the graph of y_2 is obtained from that of $y = \cos x$ with period changed to $\frac{2\pi}{2} = \pi$, shifted 1 unit upward. See Figure 8.

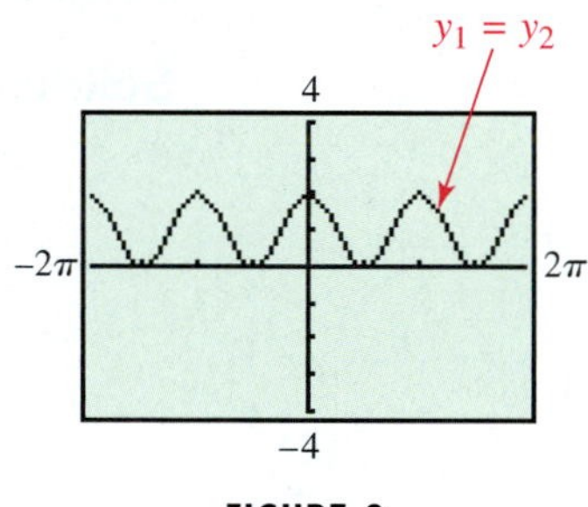

FIGURE 8

EXAMPLE 3 Simplifying Expressions by Using Double-Number Identities

Simplify each expression.

(a) $\cos^2 7x - \sin^2 7x$ **(b)** $\sin 15^\circ \cos 15^\circ$

Solution

(a) The expression $\cos^2 7x - \sin^2 7x$ suggests one of the double-number identities for cosine: $\cos 2A = \cos^2 A - \sin^2 A$. Substituting $7x$ for A gives

$$\cos^2 7x - \sin^2 7x = \cos 2(7x) = \cos 14x.$$

(b) If the given expression sin 15° cos 15° were 2 sin 15° cos 15°, we could apply the identity for sin 2*A* directly, since sin 2*A* = 2 sin *A* cos *A*.

$$\sin 15^\circ \cos 15^\circ = \frac{1}{2}(2) \sin 15^\circ \cos 15^\circ \qquad \text{Multiply by 1 in the form } \tfrac{1}{2}(2).$$

$$= \frac{1}{2}(2 \sin 15^\circ \cos 15^\circ) \qquad \text{Associative property}$$

$$= \frac{1}{2} \sin(2 \cdot 15^\circ) \qquad 2 \sin A \cos A = \sin 2A, \text{ with } A = 15^\circ$$

$$= \frac{1}{2} \sin 30^\circ \qquad \text{Multiply.}$$

$$= \frac{1}{2} \cdot \frac{1}{2} = \frac{1}{4} \qquad \sin 30^\circ = \tfrac{1}{2}$$

■

Identities involving larger multiples of the variable can be derived by repeated use of the double-number identities and the Pythagorean identities.

GCM EXAMPLE 4 Deriving a Multiple-Number Identity

Write sin 3*x* in terms of sin *x*.

Analytic Solution

$$\sin 3x = \sin(2x + x)$$

$$= \sin 2x \cos x + \cos 2x \sin x \qquad \text{Sine sum identity}$$

$$= (2 \sin x \cos x) \cos x + (\cos^2 x - \sin^2 x) \sin x \qquad \text{Double-number identities}$$

$$= 2 \sin x \cos^2 x + \cos^2 x \sin x - \sin^3 x \qquad \text{Multiply.}$$

$$= 2 \sin x(1 - \sin^2 x) + (1 - \sin^2 x) \sin x - \sin^3 x \qquad \cos^2 x = 1 - \sin^2 x$$

$$= 2 \sin x - 2 \sin^3 x + \sin x - \sin^3 x - \sin^3 x \qquad \text{Distributive property}$$

$$= 3 \sin x - 4 \sin^3 x \qquad \text{Combine terms.}$$

Graphing Calculator Support

The table in Figure 9 numerically supports the analytic solution. Here,

$$Y_2 = 3 \sin X - 4 \sin^3 X.$$

The table does not verify the identity because it does not list every possible X-value.

X	Y1	Y2
-1.178	.38268	.38268
-.7854	-.7071	-.7071
-.3927	-.9239	-.9239
0	0	0
.3927	.92388	.92388
.7854	.70711	.70711
1.1781	-.3827	-.3827

Y1=sin(3X)

FIGURE 9

■

EXAMPLE 5 Determining Wattage Consumption

If a toaster is plugged into a common household outlet, the wattage consumed is not constant. Instead, it varies at a high frequency according to the model

$$W = \frac{V^2}{R},$$

where *V* is the voltage and *R* is a constant that measures the resistance of the toaster in ohms. (*Source:* Bell, D., *Fundamentals of Electric Circuits,* Fourth Edition, Prentice-Hall, 1998.) Graph the wattage *W* consumed by a typical toaster with $R = 15$ and $V = 163 \sin 120\pi t$ over the interval $0 \le t \le .05$. How many oscillations are there?

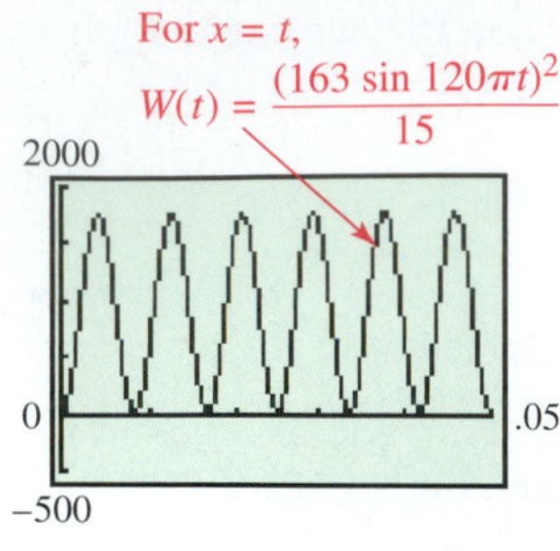

FIGURE 10

Solution Substituting the given values into the wattage equation gives

$$W = \frac{V^2}{R} = \frac{(163 \sin 120\pi t)^2}{15}.$$

To determine the range for W, we note that $\sin 120\pi t$ has maximum value 1, so the expression for W has maximum value $\frac{163^2}{15} \approx 1771$. The minimum value is 0. The graph in Figure 10 shows that there are six oscillations. ■

What Went Wrong?

To verify the proposed identity $\cos x = \cos 4x$, a student graphed both equations on the same screen, with the result shown on the left. He then compared the functions, using the table feature of his calculator, and got the result shown on the right. He was confused, because the graph indicates that the equation is not an identity, while the table indicates that it might be one.

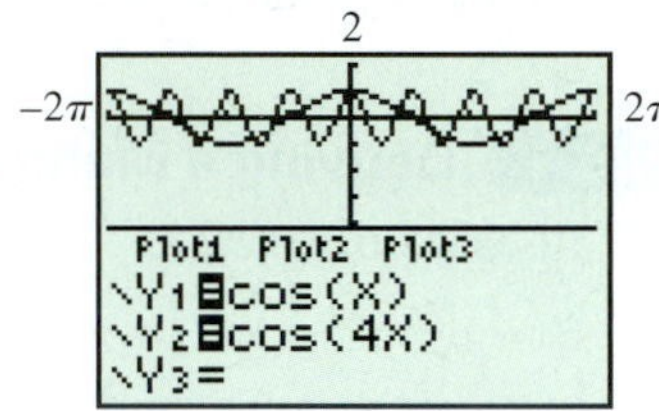

X	Y1	Y2
0	1	1
2.0944	-.5	-.5
4.1888	-.5	-.5
6.2832	1	1
8.3776	-.5	-.5
10.472	-.5	-.5
12.566	1	1

X=0

$\Delta\text{Tbl} = \frac{2\pi}{3} \approx 2.0944$
Tblmin = 0

What Went Wrong? What should the student conclude? Why?

Product-to-Sum and Sum-to-Product Identities

The identities for $\cos(A + B)$ and $\cos(A - B)$ are used to derive a group of identities useful in calculus.

Adding the identities for $\cos(A + B)$ and $\cos(A - B)$ gives

$$\begin{aligned} \cos(A + B) &= \cos A \cos B - \sin A \sin B \\ \cos(A - B) &= \cos A \cos B + \sin A \sin B \\ \hline \cos(A + B) + \cos(A - B) &= 2 \cos A \cos B \end{aligned}$$

or

$$\cos A \cos B = \frac{1}{2}[\cos(A + B) + \cos(A - B)].$$

Looking Ahead to Calculus

The product-to-sum identities are used in calculus to find *integrals* of functions that are products of trigonometric functions. One classic calculus text by Earl Swokowski includes the following example:

Evaluate $\int \cos 5x \cos 3x \, dx$.

The first solution line reads: "We may write $\cos 5x \cos 3x = \frac{1}{2}(\cos 8x + \cos 2x)$."

Answer to What Went Wrong?

He should conclude that $\cos x = \cos 4x$ is *not* an identity, based on the graph. The problem with the table is that a ΔTbl of $\frac{2\pi}{3}$ produces values where the graphs of the two functions intersect. Choosing a different interval, such as $\frac{\pi}{2}$, would show different function values at multiples of $\frac{\pi}{2}$.

Similarly, subtracting $\cos(A + B)$ from $\cos(A - B)$ gives

$$\sin A \sin B = \frac{1}{2}[\cos(A - B) - \cos(A + B)].$$

Using the identities for $\sin(A + B)$ and $\sin(A - B)$ in the same way, we get two more identities. Those and the previous ones are now summarized.

Product-to-Sum Identities

$$\cos A \cos B = \frac{1}{2}[\cos(A + B) + \cos(A - B)]$$

$$\sin A \sin B = \frac{1}{2}[\cos(A - B) - \cos(A + B)]$$

$$\sin A \cos B = \frac{1}{2}[\sin(A + B) + \sin(A - B)]$$

$$\cos A \sin B = \frac{1}{2}[\sin(A + B) - \sin(A - B)]$$

EXAMPLE 6 Using a Product-to-Sum Identity

Rewrite $\cos 2\theta \sin \theta$ as the sum or difference of two functions.

Solution Use the identity for $\cos A \sin B$, with $2\theta = A$ and $\theta = B$.

$$\cos 2\theta \sin \theta = \frac{1}{2}[\sin(2\theta + \theta) - \sin(2\theta - \theta)]$$

$$= \frac{1}{2}\sin 3\theta - \frac{1}{2}\sin \theta$$

From these new identities, we can derive another group of identities that are used to rewrite sums of trigonometric functions as products.

Sum-to-Product Identities

$$\sin A + \sin B = 2 \sin\left(\frac{A + B}{2}\right) \cos\left(\frac{A - B}{2}\right)$$

$$\sin A - \sin B = 2 \cos\left(\frac{A + B}{2}\right) \sin\left(\frac{A - B}{2}\right)$$

$$\cos A + \cos B = 2 \cos\left(\frac{A + B}{2}\right) \cos\left(\frac{A - B}{2}\right)$$

$$\cos A - \cos B = -2 \sin\left(\frac{A + B}{2}\right) \sin\left(\frac{A - B}{2}\right)$$

EXAMPLE 7 Using a Sum-to-Product Identity

Write $\sin 2\gamma - \sin 4\gamma$ as a product of two functions.

Solution Use the identity for $\sin A - \sin B$, with $2\gamma = A$ and $4\gamma = B$.

$$\begin{aligned}\sin 2\gamma - \sin 4\gamma &= 2\cos\left(\frac{2\gamma + 4\gamma}{2}\right)\sin\left(\frac{2\gamma - 4\gamma}{2}\right)\\ &= 2\cos\frac{6\gamma}{2}\sin\left(\frac{-2\gamma}{2}\right)\\ &= 2\cos 3\gamma\sin(-\gamma)\\ &= -2\cos 3\gamma\sin\gamma \qquad \sin(-\gamma) = -\sin\gamma\end{aligned}$$

Half-Number Identities

Looking Ahead to Calculus

Half-number identities for sine and cosine are used in calculus to eliminate the *xy*-term from an equation in the form $Ax^2 + Bxy + Cy^2 + Dx + Ey + F = 0$, so that the type of conic section it represents can be determined.

From the alternative forms of the identity for $\cos 2A$, we derive identities for $\sin \frac{A}{2}$, $\cos \frac{A}{2}$, and $\tan \frac{A}{2}$. These are known as **half-number identities.**

To derive the identity for $\sin \frac{A}{2}$, start with the following double-number identity for cosine and solve for $\sin x$.

$$\begin{aligned}\cos 2x &= 1 - 2\sin^2 x\\ 2\sin^2 x &= 1 - \cos 2x \qquad \text{Add } 2\sin^2 x\text{; subtract } \cos 2x.\\ \sin x &= \pm\sqrt{\frac{1 - \cos 2x}{2}} \qquad \text{Divide by 2; take square roots.}\\ \sin\frac{A}{2} &= \pm\sqrt{\frac{1 - \cos A}{2}} \qquad \text{Let } 2x = A \text{ so that } x = \tfrac{A}{2}.\end{aligned}$$

Remember both the positive and negative square roots.

The $\pm$ sign in this identity indicates that the appropriate sign is chosen depending on the quadrant of $\frac{A}{2}$. For example, if $\frac{A}{2}$ is a third-quadrant number on the unit circle, we choose the negative sign, since the sine function is negative in quadrant III.

We derive the identity for $\cos \frac{A}{2}$ by using the double-number identity for $\cos 2x$.

$$\begin{aligned}\cos 2x &= 2\cos^2 x - 1\\ 1 + \cos 2x &= 2\cos^2 x \qquad \text{Add 1.}\\ \cos^2 x &= \frac{1 + \cos 2x}{2} \qquad \text{Rewrite; divide by 2.}\\ \cos x &= \pm\sqrt{\frac{1 + \cos 2x}{2}} \qquad \text{Take square roots.}\\ \cos\frac{A}{2} &= \pm\sqrt{\frac{1 + \cos A}{2}} \qquad \text{Replace } x \text{ with } \tfrac{A}{2}.\end{aligned}$$

The $\pm$ sign is chosen as previously described.

An identity for $\tan \frac{A}{2}$ comes from the half-number identities for sine and cosine.

$$\tan\frac{A}{2} = \frac{\sin\frac{A}{2}}{\cos\frac{A}{2}} = \frac{\pm\sqrt{\frac{1 - \cos A}{2}}}{\pm\sqrt{\frac{1 + \cos A}{2}}} = \pm\sqrt{\frac{1 - \cos A}{1 + \cos A}}$$

We derive an alternative identity for $\tan \frac{A}{2}$ by using double-number identities.

$$\tan \frac{A}{2} = \frac{\sin \frac{A}{2}}{\cos \frac{A}{2}}$$

$$= \frac{2 \sin \frac{A}{2} \cos \frac{A}{2}}{2 \cos^2 \frac{A}{2}}$$ Multiply by $2 \cos \frac{A}{2}$ in numerator and denominator.

$$= \frac{\sin\left[2\left(\frac{A}{2}\right)\right]}{1 + \cos\left[2\left(\frac{A}{2}\right)\right]}$$ Double-number identities

$$\tan \frac{A}{2} = \frac{\sin A}{1 + \cos A}$$ Simplify.

From this identity for $\tan \frac{A}{2}$, we can also derive

$$\tan \frac{A}{2} = \frac{1 - \cos A}{\sin A}.$$

These last two identities for $\tan \frac{A}{2}$ do not require a choice of sign, as the first one does.

Half-Number Identities

$$\cos \frac{A}{2} = \pm\sqrt{\frac{1 + \cos A}{2}} \qquad \sin \frac{A}{2} = \pm\sqrt{\frac{1 - \cos A}{2}}$$

$$\tan \frac{A}{2} = \pm\sqrt{\frac{1 - \cos A}{1 + \cos A}} \qquad \tan \frac{A}{2} = \frac{\sin A}{1 + \cos A} \qquad \tan \frac{A}{2} = \frac{1 - \cos A}{\sin A}$$

In Example 1 of Section 9.2, we showed that the exact value of $\cos 15°$ or equivalently, $\cos \frac{\pi}{12}$, is $\frac{\sqrt{6} + \sqrt{2}}{4}$. We did this by using the identity for $\cos(A - B)$. Another form of the exact value of $\cos \frac{\pi}{12}$ can be found by using the identity for $\cos \frac{A}{2}$.

EXAMPLE 8 **Using a Half-Number Identity to Find an Exact Value**

Find the exact value of $\cos \frac{\pi}{12}$.

Solution

$$\cos \frac{\pi}{12} = \cos \frac{\frac{\pi}{6}}{2} = \sqrt{\frac{1 + \cos \frac{\pi}{6}}{2}}$$ Cosine half-number identity

$$= \sqrt{\frac{1 + \frac{\sqrt{3}}{2}}{2}} = \sqrt{\frac{\left(1 + \frac{\sqrt{3}}{2}\right) \cdot 2}{2 \cdot 2}}$$

$$= \frac{\sqrt{2 + \sqrt{3}}}{2}$$ $\sqrt{\frac{a}{b}} = \frac{\sqrt{a}}{\sqrt{b}}$

Verify that the final expression, $\frac{\sqrt{2 + \sqrt{3}}}{2}$, has approximation .9659258263, the same as that for $\frac{\sqrt{6} + \sqrt{2}}{4}$. ■

EXAMPLE 9 Finding Function Values of $\frac{x}{2}$

Given $\cos x = \frac{2}{3}$, with $\frac{3\pi}{2} < x < 2\pi$, find $\cos \frac{x}{2}$, $\sin \frac{x}{2}$, and $\tan \frac{x}{2}$.

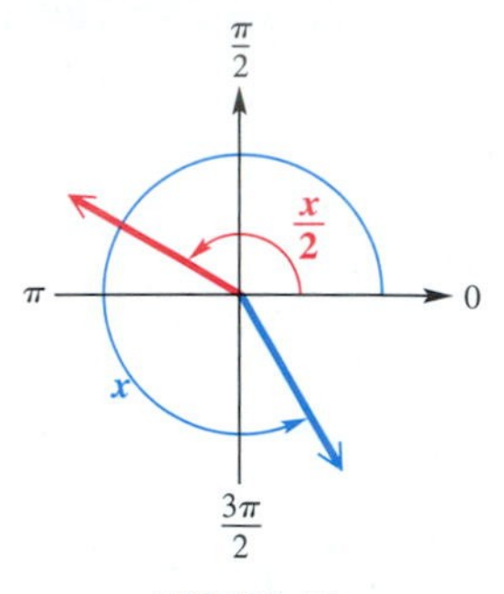

FIGURE 11

Solution The angle associated with $\frac{x}{2}$ terminates in quadrant II, since

$$\frac{3\pi}{2} < x < 2\pi$$

implies

$$\frac{3\pi}{4} < \frac{x}{2} < \pi. \quad \text{Divide by 2.}$$

See Figure 11. In quadrant II, the value of $\cos \frac{x}{2}$ is negative and the value of $\sin \frac{x}{2}$ is positive. Now use the appropriate half-number identities and simplify the radicals.

$$\cos \frac{x}{2} = -\sqrt{\frac{1 + \frac{2}{3}}{2}} = -\sqrt{\frac{5}{6}} = -\frac{\sqrt{5}}{\sqrt{6}} \cdot \frac{\sqrt{6}}{\sqrt{6}} = -\frac{\sqrt{30}}{6}$$

$$\sin \frac{x}{2} = \sqrt{\frac{1 - \frac{2}{3}}{2}} = \sqrt{\frac{1}{6}} = \frac{1}{\sqrt{6}} \cdot \frac{\sqrt{6}}{\sqrt{6}} = \frac{\sqrt{6}}{6} \quad \text{Rationalize all denominators.}$$

$$\tan \frac{x}{2} = \frac{\sin \frac{x}{2}}{\cos \frac{x}{2}} = \frac{\frac{\sqrt{6}}{6}}{-\frac{\sqrt{30}}{6}} = \frac{\sqrt{6}}{6} \cdot \left(-\frac{6}{\sqrt{30}}\right) = -\sqrt{\frac{6}{30}} = -\sqrt{\frac{1}{5}} = -\frac{\sqrt{5}}{5}$$

Be careful when simplifying a complex fraction.

Notice that it is not necessary to use a half-number identity for $\tan \frac{x}{2}$ once we find $\sin \frac{x}{2}$ and $\cos \frac{x}{2}$. (We could also use the quotient identity for tangent.) However, using this identity would provide an excellent check. ■

EXAMPLE 10 Simplifying Expressions by Using Half-Number Identities

Simplify each expression.

(a) $\pm\sqrt{\frac{1 + \cos 12x}{2}}$ **(b)** $\frac{1 - \cos 5\alpha}{\sin 5\alpha}$

Solution

(a) This matches part of the identity for $\cos \frac{A}{2}$. Replace A with $12x$ to get

$$\cos \frac{A}{2} = \pm\sqrt{\frac{1 + \cos A}{2}} = \pm\sqrt{\frac{1 + \cos 12x}{2}} = \cos \frac{12x}{2} = \cos 6x.$$

(b) Use the third identity for $\tan \frac{A}{2}$ with $5\alpha = A$ to obtain

$$\frac{1 - \cos 5\alpha}{\sin 5\alpha} = \tan \frac{5\alpha}{2}.$$ ■

FOR DISCUSSION

Refer to the solution of Example 10(b). Enter the left side of the identity as Y_1 and the right side as Y_2 in your calculator. (Use X for α.) In radian mode, enter several values to support the identity. Explain why an error message is returned for the value $\frac{\pi}{5}$.

9.3 Exercises

Use identities to find **(a)** $\sin 2\theta$ *and* **(b)** $\cos 2\theta$.

1. $\sin \theta = \frac{2}{5}$ and $\cos \theta < 0$

2. $\cos \theta = -\frac{12}{13}$ and $\sin \theta > 0$

3. $\tan \theta = 2$ and $\cos \theta > 0$

4. $\tan \theta = \frac{5}{3}$ and $\sin \theta < 0$

5. $\sin \theta = -\frac{\sqrt{5}}{7}$ and $\cos \theta > 0$

6. $\cos \theta = \frac{\sqrt{3}}{5}$ and $\sin \theta > 0$

Use an identity to write each expression as a single trigonometric function or as a single number.

7. $\cos^2 15° - \sin^2 15°$

8. $\frac{2 \tan 15°}{1 - \tan^2 15°}$

9. $1 - 2 \sin^2 15°$

10. $1 - 2 \sin^2 22.5°$

11. $2 \cos^2 67.5° - 1$

12. $\cos^2 \frac{\pi}{8} - \frac{1}{2}$

13. $\frac{\tan 51°}{1 - \tan^2 51°}$

14. $\frac{\tan 34°}{2(1 - \tan^2 34°)}$

15. $\frac{1}{4} - \frac{1}{2} \sin^2 47.1°$

16. $\frac{1}{8} \sin 29.5° \cos 29.5°$

17. $4 \sin 15° \cos 15°$

18. $2 - 4 \sin^2 15°$

Graph each expression and use the graph to conjecture an identity. Then verify your conjecture.

19. $\cos^4 x - \sin^4 x$

20. $\frac{4 \tan x \cos^2 x - 2 \tan x}{1 - \tan^2 x}$

Use the method of Example 4 to do the following. Then, support your result graphically, using the trig viewing window of your calculator.

21. Express $\cos 3x$ in terms of $\cos x$.

22. Express $\tan 3x$ in terms of $\tan x$.

23. Express $\tan 4x$ in terms of $\tan x$.

24. Express $\cos 4x$ in terms of $\cos x$.

Use a half-number identity to find an exact value for each trigonometric function.

25. $\sin \frac{\pi}{12}$

26. $\cos \frac{\pi}{8}$

27. $\tan\left(-\frac{\pi}{8}\right)$

28. $\cos 67.5°$

29. $\sin 67.5°$

30. $\tan 195°$

Use a half-number identity to find an exact value for each function, given the information about x.

31. $\cos \frac{x}{2}$, given $\cos x = \frac{1}{4}$ and $0 < x < \frac{\pi}{2}$

32. $\sin \frac{x}{2}$, given $\cos x = -\frac{5}{8}$ and $\frac{\pi}{2} < x < \pi$

33. $\tan \frac{x}{2}$, given $\sin x = \frac{3}{5}$ and $\frac{\pi}{2} < x < \pi$

34. $\cos \frac{x}{2}$, given $\sin x = -\frac{4}{5}$ and $\frac{3\pi}{2} < x < 2\pi$

35. $\tan \frac{x}{2}$, given $\tan x = \frac{\sqrt{7}}{3}$ and $\pi < x < \frac{3\pi}{2}$

36. $\tan \frac{x}{2}$, given $\tan x = -\frac{\sqrt{5}}{2}$ and $\frac{\pi}{2} < x < \pi$

37. Use the identity

$$\tan \frac{A}{2} = \frac{\sin A}{1 + \cos A}$$

to derive the equivalent identity

$$\tan \frac{A}{2} = \frac{1 - \cos A}{\sin A}$$

by multiplying the right-hand side by $\frac{1 - \cos A}{1 - \cos A}$.

38. The identity

$$\tan \frac{A}{2} = \pm\sqrt{\frac{1 - \cos A}{1 + \cos A}}$$

can be used to show that $\tan 22.5° = \sqrt{3 - 2\sqrt{2}}$, and the identity

$$\tan \frac{A}{2} = \frac{\sin A}{1 + \cos A}$$

can be used to show that $\tan 22.5° = \sqrt{2} - 1$. Verify that these answers are the same, without using a calculator. (*Hint:* If $a > 0$ and $b > 0$ and $a^2 = b^2$, then $a = b$.)

39. Consider the expression $\tan\left(\frac{\pi}{2} + x\right)$.

(a) Why can't we use the identity for $\tan(A + B)$ to express it as a function of x alone?

(b) Use the identity $\tan \theta = \frac{\sin \theta}{\cos \theta}$ to rewrite the expression in terms of sine and cosine.

(c) Use the result of part (b) to show that

$$\tan\left(\frac{\pi}{2} + x\right) = -\cot x.$$

40. Recall the identity for the tangent of a sum.

(a) Use the identity for $\tan(A + B)$ to write an identity for $\cot(A + B)$ in terms of $\cot A$ and $\cot B$.

(b) Consider the expression $\cot(\pi + x)$. Why can't we use the result of part (a) to rewrite that expression as a function of x alone?

(c) Use the ideas of parts (b) and (c) of Exercise 39 to show that $\cot(\pi + x) = \cot x$.

Use an identity to write each expression as a single trigonometric function.

41. $\sqrt{\frac{1 - \cos 40°}{2}}$

42. $\sqrt{\frac{1 + \cos 76°}{2}}$

43. $\sqrt{\frac{1 - \cos 147°}{1 + \cos 147°}}$

44. $\sqrt{\frac{1 + \cos 165°}{1 - \cos 165°}}$

45. $\frac{1 - \cos 59.74°}{\sin 59.74°}$

46. $\frac{\sin 158.2°}{1 + \cos 158.2°}$

Verify that each equation is an identity.

47. $\sin 4\alpha = 4 \sin \alpha \cos \alpha \cos 2\alpha$

48. $\frac{1 + \cos 2x}{\sin 2x} = \cot x$

49. $\frac{2 \cos 2\alpha}{\sin 2\alpha} = \cot \alpha - \tan \alpha$

50. $\sin 4\gamma = 4 \sin \gamma \cos \gamma - 8 \sin^3 \gamma \cos \gamma$

51. $\sin 2\alpha \cos 2\alpha = \sin 2\alpha - 4 \sin^3 \alpha \cos \alpha$

52. $\cos 2x = \frac{1 - \tan^2 x}{1 + \tan^2 x}$

53. $\tan s + \cot s = 2 \csc 2s$

54. $\frac{\cot \alpha - \tan \alpha}{\cot \alpha + \tan \alpha} = \cos 2\alpha$

55. $\sec^2 \frac{x}{2} = \frac{2}{1 + \cos x}$

56. $\cot^2 \frac{x}{2} = \frac{(1 + \cos x)^2}{\sin^2 x}$

Write each expression as a sum or difference of trigonometric functions.

57. $2 \sin 58° \cos 102°$

58. $5 \cos 3x \cos 2x$

59. $2 \cos 85° \sin 140°$

60. $\sin 4x \sin 5x$

Write each expression as a product of trigonometric functions.

61. $\cos 4x - \cos 2x$

62. $\cos 5t + \cos 8t$

63. $\sin 25° + \sin(-48°)$

64. $\sin 102° - \sin 95°$

65. $\cos 4x + \cos 8x$

66. $\sin 9B - \sin 3B$

(Modeling) Solve each problem.

67. ***Railroad Curves*** In the United States, circular railroad curves are designated by their *degree of curvature,* which is the central angle subtended by a chord of 100 feet. (*Source:* Hay, W., *Railroad Engineering,* John Wiley and Sons, 1982.)

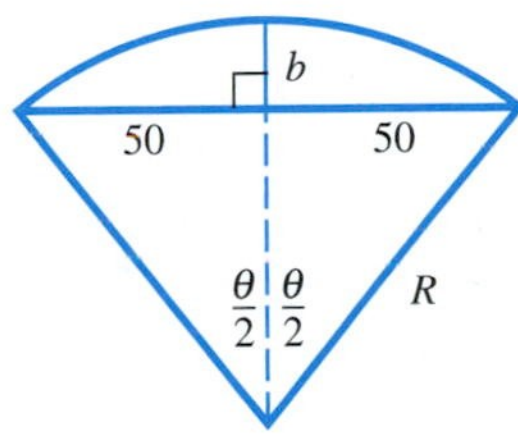

(a) Use the figure to write an expression for $\cos \frac{\theta}{2}$.

(b) Use the result of part (a) and the third half-number identity for tangent to write an expression for $\tan \frac{\theta}{4}$.

(c) If $b = 12$, what is the measure of angle θ to the nearest degree?

68. ***Distance Traveled by a Stone*** The distance D of an object thrown (or projected) from height h (feet) at angle θ with initial velocity v (feet per second) is modeled by the formula

$$D = \frac{v^2 \sin \theta \cos \theta + v \cos \theta \sqrt{(v \sin \theta)^2 + 64h}}{32}.$$

(*Source:* Kreighbaum, E. and K. Barthels, *Biomechanics,* Allyn & Bacon, 1996.)

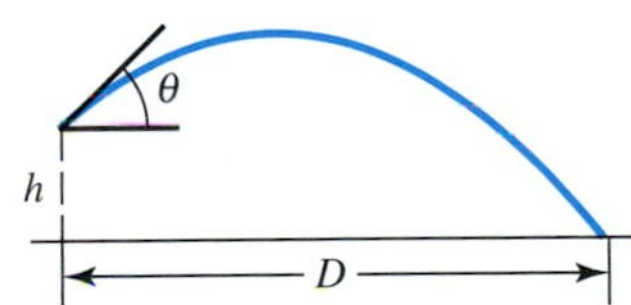

(a) Find D when $h = 0$; that is, when the object is projected from the ground.

(b) Suppose a car driving over loose gravel kicks up a small stone at a velocity of 36 feet per second (about 25 mph) and an angle $\theta = 30°$. How far will the stone travel?

(c) Repeat part (c) for a velocity of 40 feet per second and an angle of 32°.

69. ***Determining Wattage*** Amperage is a measure of the amount of electricity that is moving through a circuit, while voltage is a measure of the force pushing the electricity. The wattage W consumed by an electrical device can be determined by calculating the product of amperage I and voltage V. (*Source:* Wilcox, G. and C. Hesselberth, *Electricity for Engineering Technology,* Allyn & Bacon, 1970.)

(a) A household circuit has voltage

$$V = 163 \sin 120\pi t$$

when an incandescent light bulb is turned on with amperage

$$I = 1.23 \sin 120\pi t.$$

Graph the wattage $W = VI$ consumed by the light bulb over the interval $0 \le t \le .05$.

(b) Determine the maximum and minimum wattages used by the light bulb.

(c) Use identities to determine values for a, c, and ω so that

$$W = a \cos \omega t + c.$$

(d) Check your answer in part (c) by graphing both expressions for W on the same coordinate axes.

(e) Use the graph from part (a) to estimate the average wattage used by the light. How many watts do you think this incandescent light bulb is rated for?

70. ***Relating Voltage and Wattage*** Refer to Exercise 69. Suppose that voltage for an electric heater is given by

$$V = a \sin 2\pi\omega t$$

and amperage by

$$I = b \sin 2\pi\omega t,$$

where t is time in seconds.

(a) Find the period of the graph for the voltage.

(b) Show that the graph of the wattage $W = VI$ will have half the period of the voltage. Interpret this result.

Relating Concepts

For individual or group investigation (Exercises 71–90)

Verify that each equation is an identity by using any of the identities introduced in Sections 9.1–9.3.

71. $\tan \theta + \cot \theta = \sec \theta \csc \theta$

72. $\csc \theta \cos^2 \theta + \sin \theta = \csc \theta$

73. $\tan \dfrac{x}{2} = \csc x - \cot x$

74. $\sec(\pi - x) = -\sec x$

75. $\dfrac{\sin t}{1 + \cos t} = \dfrac{1 - \cos t}{\sin t}$

76. $\dfrac{1 - \sin t}{\cos t} = \dfrac{1}{\sec t + \tan t}$

77. $\sin 2\theta = \dfrac{2 \tan \theta}{1 + \tan^2 \theta}$

78. $\dfrac{2}{1 + \cos x} - \tan^2 \dfrac{x}{2} = 1$

79. $\cot \theta - \tan \theta = \dfrac{2 \cos^2 \theta - 1}{\sin \theta \cos \theta}$

80. $\dfrac{1}{\sec t - 1} + \dfrac{1}{\sec t + 1} = 2 \cot t \csc t$

81. $\dfrac{\sin(x + y)}{\cos(x - y)} = \dfrac{\cot x + \cot y}{1 + \cot x \cot y}$

82. $1 - \tan^2 \dfrac{\theta}{2} = \dfrac{2 \cos \theta}{1 + \cos \theta}$

83. $\dfrac{\sin \theta + \tan \theta}{1 + \cos \theta} = \tan \theta$

84. $\csc^4 x - \cot^4 x = \dfrac{1 + \cos^2 x}{1 - \cos^2 x}$

85. $\cos x = \dfrac{1 - \tan^2 \frac{x}{2}}{1 + \tan^2 \frac{x}{2}}$

86. $\cos 2x = \dfrac{2 - \sec^2 x}{\sec^2 x}$

87. $\dfrac{\tan^2 t + 1}{\tan t \csc^2 t} = \tan t$

88. $\dfrac{\sin s}{1 + \cos s} + \dfrac{1 + \cos s}{\sin s} = 2 \csc s$

89. $\tan 4\theta = \dfrac{2 \tan 2\theta}{2 - \sec^2 2\theta}$

90. $\tan\left(\dfrac{x}{2} + \dfrac{\pi}{4}\right) = \sec x + \tan x$

9.4 The Inverse Circular Functions

Review of Inverse Functions ■ Inverse Sine Function ■ Inverse Cosine Function ■ Inverse Tangent Function ■ Remaining Inverse Trigonometric Functions ■ Inverse Function Values

Review of Inverse Functions

We first discussed inverse functions in Section 5.1. We give a quick review here for a pair of inverse functions f and f^{-1}.

1. If a function f is one-to-one, then f has an inverse function f^{-1}.
2. In a one-to-one function, each x-value corresponds to only one y-value and each y-value corresponds to only one x-value.
3. The domain of f is the range of f^{-1}, and the range of f is the domain of f^{-1}.

4. The graphs of f and f^{-1} are reflections of each other about the line $y = x$.
5. To find $f^{-1}(x)$ from $f(x)$, follow these steps:

Step 1 Replace $f(x)$ with y and interchange x and y.

Step 2 Solve for y.

Step 3 Replace y with $f^{-1}(x)$.

In the remainder of this section, we use these facts to develop the inverse circular (trigonometric) functions.

Inverse Sine Function

Looking Ahead to Calculus

The inverse functions covered in this section are used in calculus to express the solutions of trigonometric equations and integrate certain types of functions.

From Figure 12 and the horizontal line test, we see that $y = \sin x$ does not define a one-to-one function. If we restrict the domain to the interval $\left[-\frac{\pi}{2}, \frac{\pi}{2}\right]$, which is the part of the graph in Figure 12 shown in color, this restricted function is one-to-one and has an inverse function. The range of $y = \sin x$ is $[-1, 1]$, so the domain of the inverse function will be $[-1, 1]$, and its range will be $\left[-\frac{\pi}{2}, \frac{\pi}{2}\right]$.

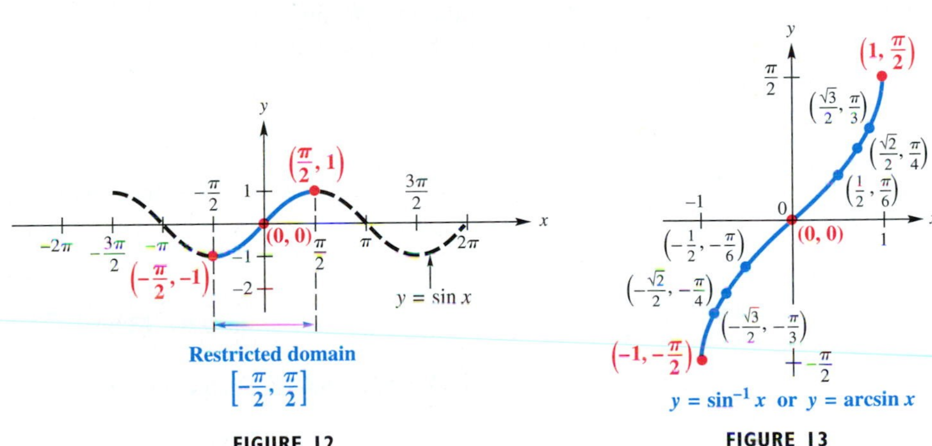

FIGURE 12 **FIGURE 13**

Reflecting the graph of $y = \sin x$ on the restricted domain across the line $y = x$ gives the graph of the inverse function, shown in Figure 13. The equation of the inverse of $y = \sin x$ is found by interchanging x and y to get $x = \sin y$. This equation is solved for y by writing $\boldsymbol{y = \sin^{-1} x}$ (read **"inverse sine of x"**). As Figure 13 shows, the domain of $y = \sin^{-1} x$ is $[-1, 1]$, while the restricted domain of $y = \sin x$, $\left[-\frac{\pi}{2}, \frac{\pi}{2}\right]$, is the range of $y = \sin^{-1} x$. An alternative notation for $\sin^{-1} x$ is $\arcsin x$.

Inverse Sine Function

$\boldsymbol{y = \sin^{-1} x}$ or $\boldsymbol{y = \arcsin x}$ means that $x = \sin y$, for $-\frac{\pi}{2} \le y \le \frac{\pi}{2}$.

We can think of $y = \sin^{-1} x$ or $y = \arcsin x$ as **"y is the number in the interval $\left[-\frac{\pi}{2}, \frac{\pi}{2}\right]$ whose sine is x."**

GCM EXAMPLE 1 Finding Inverse Sine Values

Find y in each equation.

(a) $y = \arcsin \frac{1}{2}$ **(b)** $y = \sin^{-1}(-1)$ **(c)** $y = \sin^{-1}(-2)$

Analytic Solution

(a) The graph of the function defined by $y = \arcsin x$ (Figure 13) includes the point $\left(\frac{1}{2}, \frac{\pi}{6}\right)$. Therefore,

$$\arcsin \frac{1}{2} = \frac{\pi}{6}.$$

Alternatively, we can think of $y = \arcsin \frac{1}{2}$ as "y is the number in $\left[-\frac{\pi}{2}, \frac{\pi}{2}\right]$ whose sine is $\frac{1}{2}$." Then we can rewrite the equation as $\sin y = \frac{1}{2}$. Since $\sin \frac{\pi}{6} = \frac{1}{2}$ and $\frac{\pi}{6}$ is in the range of the arcsin function, $y = \frac{\pi}{6}$.

(b) Writing the alternative equation, $\sin y = -1$, shows that $y = -\frac{\pi}{2}$. This can be verified by noticing that the point $\left(-1, -\frac{\pi}{2}\right)$ is on the graph of $y = \sin^{-1} x$.

(c) Because -2 is not in the domain of the inverse sine function, $\sin^{-1}(-2)$ does not exist. (Note that the sine of a number cannot be -2.)

Graphing Calculator Solution

Graph $Y_1 = \sin^{-1} X$ and locate the points with X-values $\frac{1}{2}$ and -1. Figure 14(a) shows that when $X = \frac{1}{2}$, $Y = \frac{\pi}{6} \approx .52359878$. Similarly, Figure 14(b) shows that when $X = -1$, $Y = -\frac{\pi}{2} \approx -1.570796$.

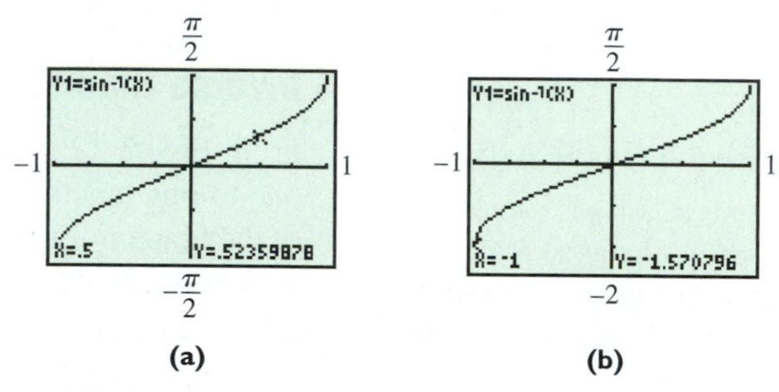

FIGURE 14

A calculator will give an error message for $\sin^{-1}(-2)$.

CAUTION In Example 1(b), it is tempting to give the value of $\sin^{-1}(-1)$ as $\frac{3\pi}{2}$, since $\sin \frac{3\pi}{2} = -1$. Notice, however, that $\frac{3\pi}{2}$ is not in the range of the inverse sine function.

FUNCTION CAPSULE

INVERSE SINE FUNCTION $y = \sin^{-1} x$ or $y = \arcsin x$

Domain: $[-1, 1]$ Range: $\left[-\frac{\pi}{2}, \frac{\pi}{2}\right]$

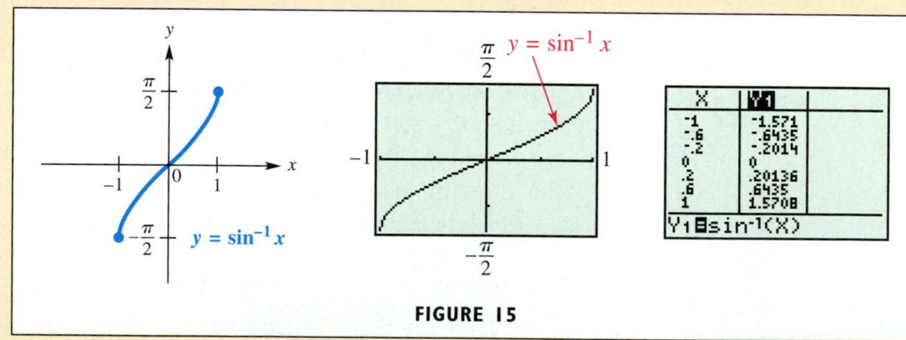

FIGURE 15

- The inverse sine function is increasing and continuous on its domain, $[-1, 1]$.
- Its x-intercept is 0, and its y-intercept is 0.
- Its graph is symmetric with respect to the origin, so it is an odd function. That is, for all x in the domain, $\sin^{-1}(-x) = -\sin^{-1} x$.

FOR DISCUSSION

Earlier in this section, we stated that $y = \sin^{-1} x$ means "y is the number in the interval $\left[-\frac{\pi}{2}, \frac{\pi}{2}\right]$ whose sine is x." Make a similar statement for

$$y = \cos^{-1} x.$$

Inverse Cosine Function

The function $y = \cos^{-1} x$ (or $y = \arccos x$) is defined by restricting the domain of the function $y = \cos x$ to the interval $[0, \pi]$, as in Figure 16, and then interchanging the roles of x and y. The graph of $y = \cos^{-1} x$ is shown in Figure 17.

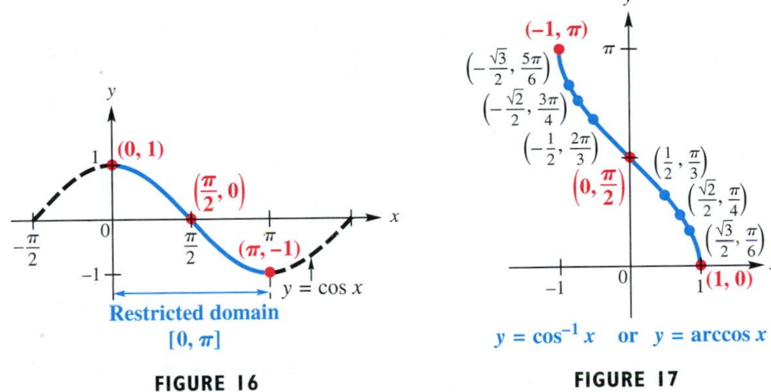

FIGURE 16

FIGURE 17

Inverse Cosine Function

$y = \cos^{-1} x$ or $y = \arccos x$ means that $x = \cos y$, for $0 \leq y \leq \pi$.

EXAMPLE 2 **Finding Inverse Cosine Values**

Find y in each equation.

(a) $y = \arccos 1$ **(b)** $y = \cos^{-1}\left(-\frac{\sqrt{2}}{2}\right)$

Analytic Solution

(a) Since the point $(1,0)$ lies on the graph of $y = \arccos x$ in Figure 17, the value of y is 0. Alternatively, we can think of $y = \arccos 1$ as "y is the number in $[0, \pi]$ whose cosine is 1," or $\cos y = 1$. Then $y = 0$, since $\cos 0 = 1$ and 0 is in the range of the arccosine function.

(b) We must find the value of y that satisfies $\cos y = -\frac{\sqrt{2}}{2}$, where y is in the interval $[0, \pi]$, the range of the function $y = \cos^{-1} x$. The only value for y that satisfies these conditions is $\frac{3\pi}{4}$. Again, this can be verified from the graph in Figure 17.

Graphing Calculator Solution

Figure 18(a) shows that, on the graph of $Y_1 = \cos^{-1} X$, when $X = 1$, $Y = 0$. Similarly, Figure 18(b) shows that when $X = -\frac{\sqrt{2}}{2} \approx -.7071068$, $Y = \frac{3\pi}{4} \approx 2.3561945$.

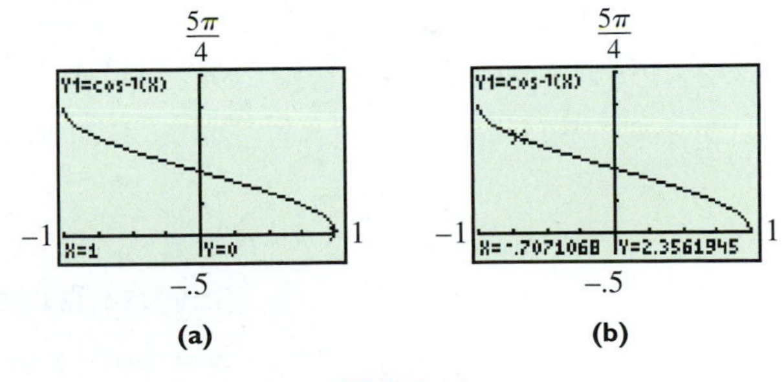

FIGURE 18

INVERSE COSINE FUNCTION $\quad y = \cos^{-1} x \quad$ or $\quad y = \arccos x$

Domain: $[-1, 1]$ $\qquad$ Range: $[0, \pi]$

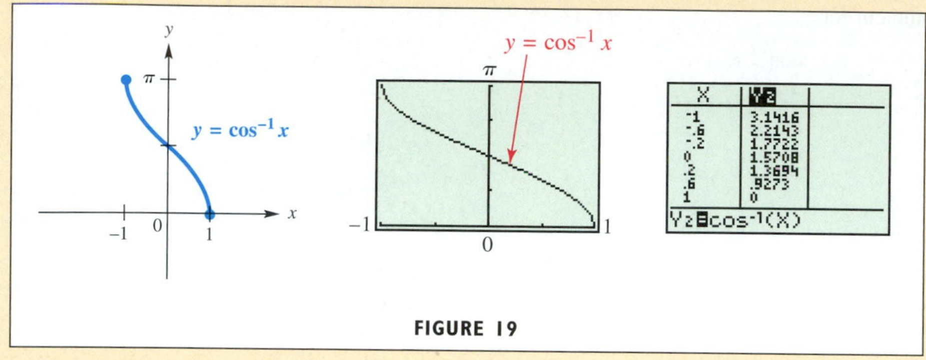

FIGURE 19

- The inverse cosine function is decreasing and continuous on its domain, $[-1, 1]$.
- Its x-intercept is 1, and its y-intercept is $\frac{\pi}{2}$.
- Its graph is neither symmetric with respect to the y-axis nor the origin.

Inverse Tangent Function

Restricting the domain of the function $y = \tan x$ to the open interval $\left(-\frac{\pi}{2}, \frac{\pi}{2}\right)$ yields a one-to-one function. By interchanging the roles of x and y, we obtain the inverse tangent function, given by $y = \tan^{-1} x$ or $y = \arctan x$. Figure 20 shows the graph of the restricted tangent function. Figure 21 gives the graph of $y = \tan^{-1} x$.

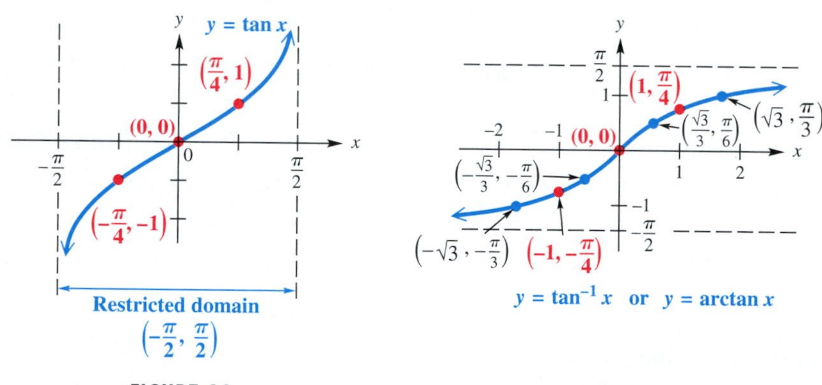

FIGURE 20 $\qquad$ FIGURE 21

Inverse Tangent Function

$y = \tan^{-1} x$ or $y = \arctan x$ means $x = \tan y$, for $-\frac{\pi}{2} < y < \frac{\pi}{2}$.

FUNCTION CAPSULE

INVERSE TANGENT FUNCTION $\quad y = \tan^{-1} x$ or $y = \arctan x$

Domain: $(-\infty, \infty)$ $\quad$ Range: $\left(-\frac{\pi}{2}, \frac{\pi}{2}\right)$

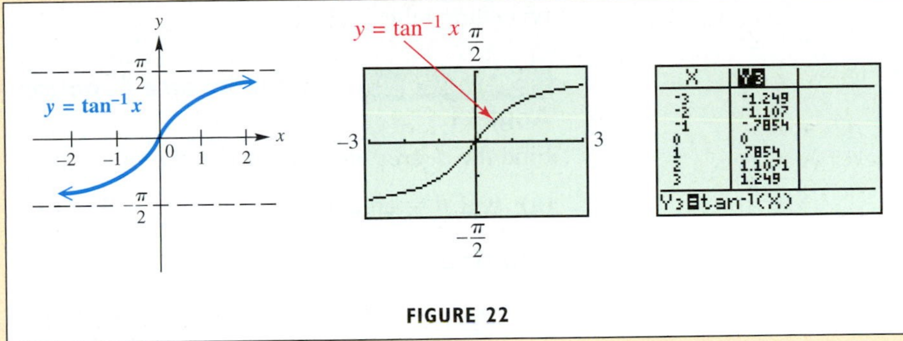

FIGURE 22

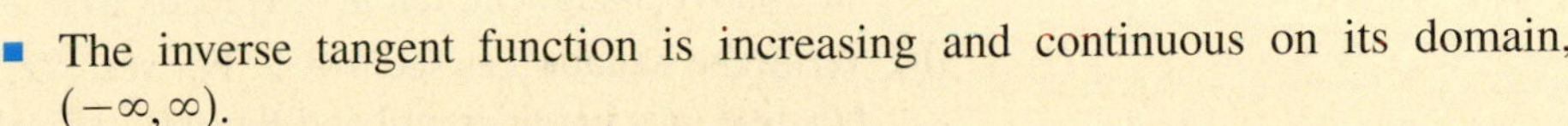

- The inverse tangent function is increasing and continuous on its domain, $(-\infty, \infty)$.
- Its x-intercept is 0, and its y-intercept is 0.
- Its graph is symmetric with respect to the origin, so it is an odd function. That is, for all x in the domain, $\tan^{-1}(-x) = -\tan^{-1} x$.
- The lines $y = \frac{\pi}{2}$ and $y = -\frac{\pi}{2}$ are horizontal asymptotes.

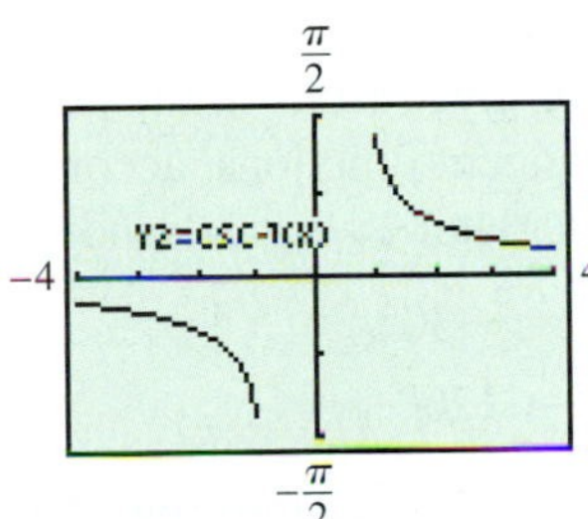

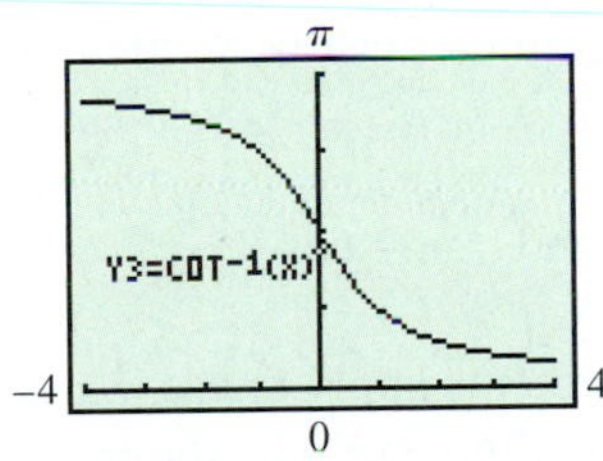

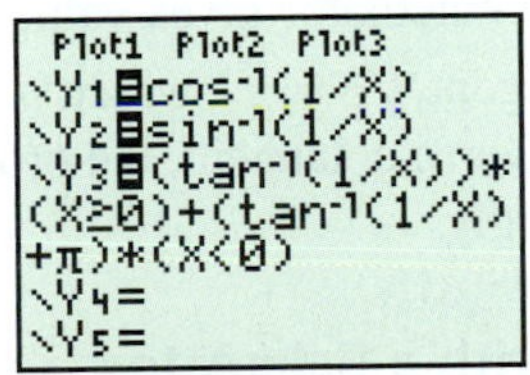


```
Plot1 Plot2 Plot3
\Y1=cos⁻¹(1/X)
\Y2=sin⁻¹(1/X)
\Y3=(tan⁻¹(1/X))*
(X≥0)+(tan⁻¹(1/X)
+π)*(X<0)
\Y4=
\Y5=
```

The first three screens show the graphs of the three remaining inverse circular functions. The last screen shows how they are defined.

FIGURE 23

Remaining Inverse Trigonometric Functions

The remaining three inverse trigonometric functions are defined similarly; their graphs are shown in Figure 23. All six inverse trigonometric functions with their domains and ranges are given in the following table.

Function	Domain	Range	
		Interval	Quadrants of the Unit Circle
$y = \sin^{-1} x$	$[-1, 1]$	$\left[-\frac{\pi}{2}, \frac{\pi}{2}\right]$	I and IV
$y = \cos^{-1} x$	$[-1, 1]$	$[0, \pi]$	I and II
$y = \tan^{-1} x$	$(-\infty, \infty)$	$\left(-\frac{\pi}{2}, \frac{\pi}{2}\right)$	I and IV
$y = \cot^{-1} x$	$(-\infty, \infty)$	$(0, \pi)$	I and II
$y = \sec^{-1} x$	$(-\infty, -1] \cup [1, \infty)$	$\left[0, \frac{\pi}{2}\right) \cup \left(\frac{\pi}{2}, \pi\right]$*	I and II
$y = \csc^{-1} x$	$(-\infty, -1] \cup [1, \infty)$	$\left[-\frac{\pi}{2}, 0\right) \cup \left(0, \frac{\pi}{2}\right]$*	I and IV

*The inverse secant and inverse cosecant functions are sometimes defined with different ranges. We use intervals that match their reciprocal functions (except for one missing point).

Inverse Function Values

The inverse circular functions are formally defined with real number ranges. However, there are times when it may be convenient to find degree-measured angles equivalent to these real number values. It is also often convenient to think in terms of the unit circle and choose the inverse function values on the basis of the quadrants given in the preceding table.

EXAMPLE 3 Finding Inverse Function Values (Degree-Measured Angles)

Find the *degree measure* of θ in the following.

(a) θ, if $\theta = \arctan 1$ **(b)** θ, if $\theta = \sec^{-1} 2$

Solution

(a) Here, θ must be in $(-90°, 90°)$, but since $1 > 0$, θ must be in quadrant I. The alternative statement, $\tan \theta = 1$, leads to $\theta = 45°$.

(b) Write the equation as $\sec \theta = 2$. For $\sec^{-1} x$, θ is in quadrant I or II. Because 2 is positive, θ is in quadrant I and $\theta = 60°$, since $\sec 60° = 2$. Note that $\frac{\pi}{3}$ (the radian equivalent of 60°) is in the range of the inverse secant function. ■

The inverse trigonometric function keys on a calculator give results in the proper quadrant for the inverse sine, inverse cosine, and inverse tangent functions, according to the definitions of these functions. For example, on a calculator in degree mode,

$$\sin^{-1} .5 = 30°, \qquad \sin^{-1}(-.5) = -30°,$$
$$\tan^{-1}(-1) = -45°, \quad \text{and} \quad \cos^{-1}(-.5) = 120°.$$

Finding $\cot^{-1} x$, $\sec^{-1} x$, and $\csc^{-1} x$ with a calculator is not as straightforward, because these functions must be expressed in terms of $\tan^{-1} x$, $\cos^{-1} x$, and $\sin^{-1} x$, respectively. If $y = \sec^{-1} x$, for example, then $\sec y = x$, which must be written as a cosine function as follows:

$$\text{If } \sec y = x, \text{ then } \frac{1}{\cos y} = x, \quad \text{or} \quad \cos y = \frac{1}{x}, \quad \text{and} \quad y = \cos^{-1} \frac{1}{x}.$$

In summary, to find $\sec^{-1} x$, we find $\cos^{-1} \frac{1}{x}$. Similar statements apply to $\csc^{-1} x$ and $\cot^{-1} x$. There is one additional consideration with $\cot^{-1} x$. Since we take the inverse tangent of the reciprocal to find the inverse cotangent, the calculator gives values of the inverse cotangent with the same range as the inverse tangent, $\left(-\frac{\pi}{2}, \frac{\pi}{2}\right)$, which is not the correct range for the inverse cotangent. Thus, for the inverse cotangent, the proper range must be considered and the results adjusted accordingly.

GCM **EXAMPLE 4 Finding Inverse Function Values with a Calculator**

(a) Find y in radians if $y = \csc^{-1}(-3)$.

(b) Find θ in degrees if $\theta = \text{arccot}(-.3541)$.

FIGURE 24

Solution

(a) With the calculator in radian mode, enter $\csc^{-1}(-3)$ as $\sin^{-1}\left(-\frac{1}{3}\right)$ to obtain $y \approx -.3398369095$. See Figure 24.

(b) Set the calculator in degree mode. A calculator gives the inverse tangent value of a negative number as a quadrant IV angle. The restriction on the range of arccotangent means that θ must be in quadrant II, so enter arccot$(-.3541)$ as $\tan^{-1}\left(\frac{1}{-.3541}\right) + 180°$. As shown in Figure 24, $\theta \approx 109.4990544°$. ■

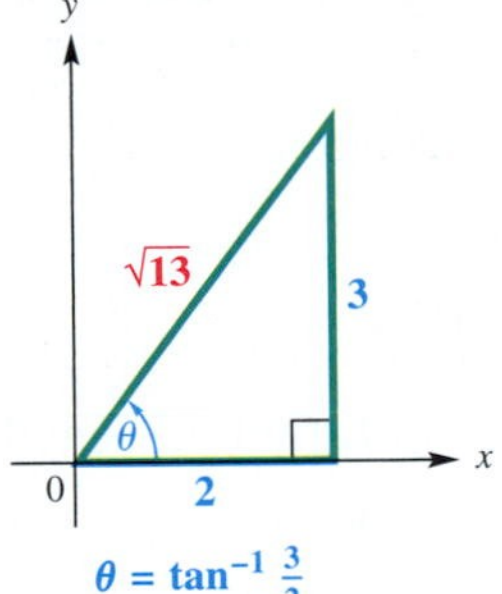

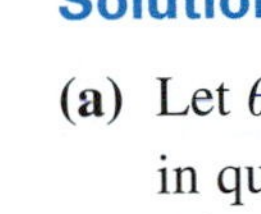

$\theta = \tan^{-1} \frac{3}{2}$

FIGURE 25

EXAMPLE 5 Finding Function Values by Using Definitions of the Trigonometric Functions

Evaluate each expression without using a calculator.

(a) $\sin\left(\tan^{-1} \frac{3}{2}\right)$ **(b)** $\tan\left(\cos^{-1}\left(-\frac{5}{13}\right)\right)$

Solution

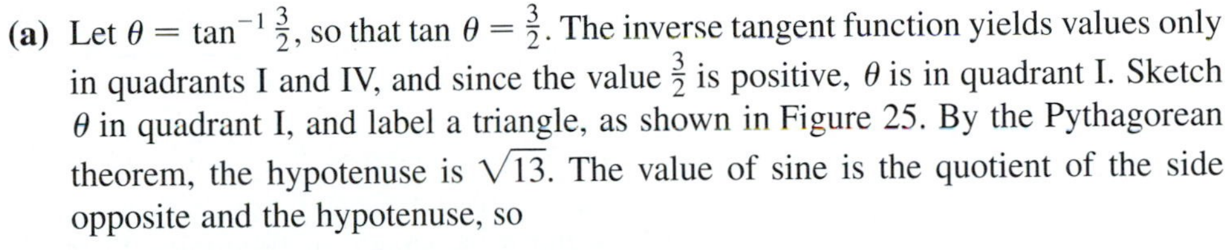

(a) Let $\theta = \tan^{-1} \frac{3}{2}$, so that $\tan \theta = \frac{3}{2}$. The inverse tangent function yields values only in quadrants I and IV, and since the value $\frac{3}{2}$ is positive, θ is in quadrant I. Sketch θ in quadrant I, and label a triangle, as shown in Figure 25. By the Pythagorean theorem, the hypotenuse is $\sqrt{13}$. The value of sine is the quotient of the side opposite and the hypotenuse, so

$$\sin\left(\tan^{-1} \frac{3}{2}\right) = \sin \theta = \frac{3}{\sqrt{13}} = \frac{3}{\sqrt{13}} \cdot \frac{\sqrt{13}}{\sqrt{13}} = \frac{3\sqrt{13}}{13}.$$

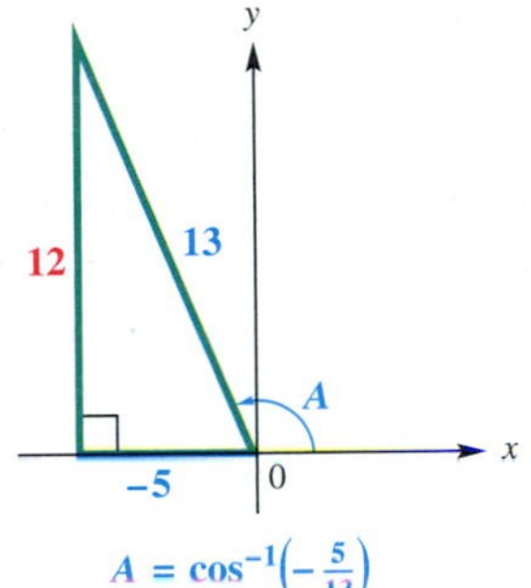

$A = \cos^{-1}\left(-\frac{5}{13}\right)$

FIGURE 26

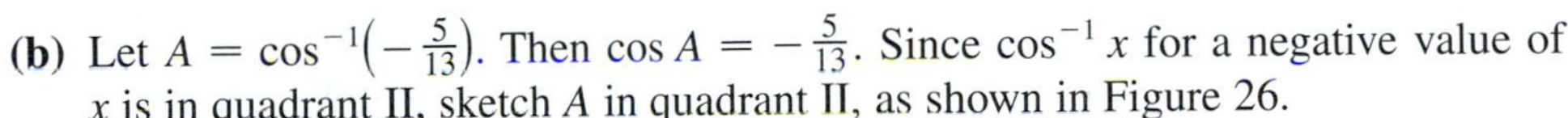

(b) Let $A = \cos^{-1}\left(-\frac{5}{13}\right)$. Then $\cos A = -\frac{5}{13}$. Since $\cos^{-1} x$ for a negative value of x is in quadrant II, sketch A in quadrant II, as shown in Figure 26.

$$\tan\left(\cos^{-1}\left(-\frac{5}{13}\right)\right) = \tan A = -\frac{12}{5}$$ ■

EXAMPLE 6 Finding Function Values by Using Identities

Evaluate each expression without using a calculator.

(a) $\cos\left(\arctan \sqrt{3} + \arcsin \frac{1}{3}\right)$ **(b)** $\tan\left(2 \arcsin \frac{2}{5}\right)$

Solution

(a) Let $A = \arctan \sqrt{3}$ and $B = \arcsin \frac{1}{3}$, so that $\tan A = \sqrt{3}$ and $\sin B = \frac{1}{3}$. Sketch both A and B in quadrant I. See Figure 27. Use the identity for $\cos(A + B)$.

$$\cos(A + B) = \cos A \cos B - \sin A \sin B$$

$$\cos\left(\arctan \sqrt{3} + \arcsin \frac{1}{3}\right) = \cos\left(\arctan \sqrt{3}\right) \cos\left(\arcsin \frac{1}{3}\right) - \sin\left(\arctan \sqrt{3}\right) \sin\left(\arcsin \frac{1}{3}\right) \quad (1)$$

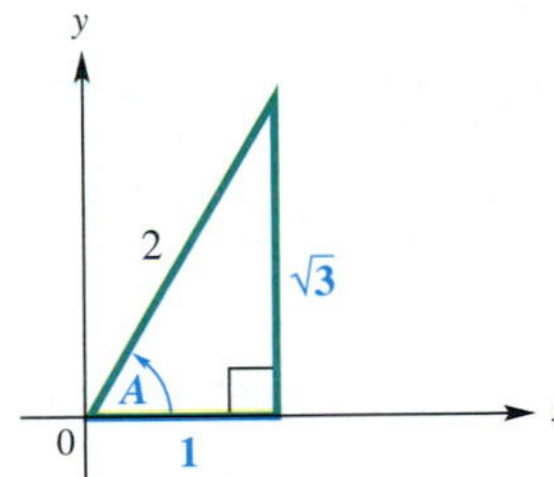

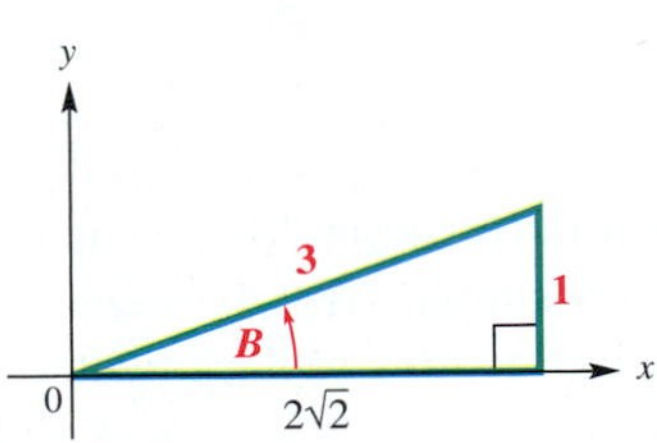

FIGURE 27

From Figure 27,

$$\cos\left(\arctan \sqrt{3}\right) = \cos A = \frac{1}{2}, \quad \cos\left(\arcsin \frac{1}{3}\right) = \cos B = \frac{2\sqrt{2}}{3},$$

$$\sin\left(\arctan \sqrt{3}\right) = \sin A = \frac{\sqrt{3}}{2}, \quad \sin\left(\arcsin \frac{1}{3}\right) = \sin B = \frac{1}{3}.$$

Substitute these values into equation (1) to get

$$\cos\left(\arctan\sqrt{3} + \arcsin\frac{1}{3}\right) = \frac{1}{2}\cdot\frac{2\sqrt{2}}{3} - \frac{\sqrt{3}}{2}\cdot\frac{1}{3} = \frac{2\sqrt{2}-\sqrt{3}}{6}.$$

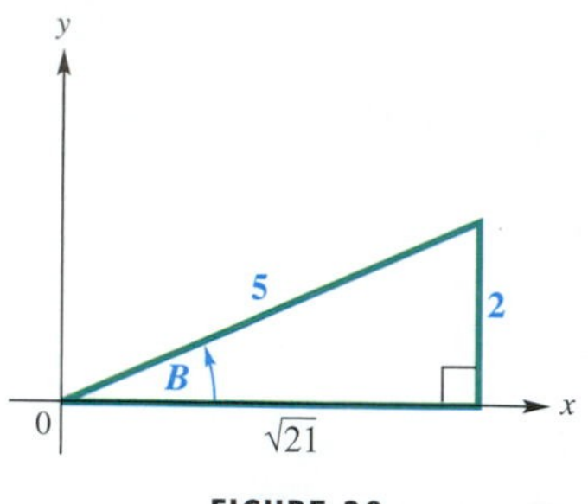

FIGURE 28

(b) Let $\arcsin\frac{2}{5} = B$. Then, from the double-number tangent identity,

$$\tan\left(2\arcsin\frac{2}{5}\right) = \tan 2B = \frac{2\tan B}{1-\tan^2 B}.$$

Since $\arcsin\frac{2}{5} = B$, $\sin B = \frac{2}{5}$. Sketch a triangle in quadrant I, find the length of the third side, and find tan B. From the triangle in Figure 28, $\tan B = \frac{2}{\sqrt{21}}$, and

$$\tan\left(2\arcsin\frac{2}{5}\right) = \frac{2\left(\frac{2}{\sqrt{21}}\right)}{1-\left(\frac{2}{\sqrt{21}}\right)^2} = \frac{\frac{4}{\sqrt{21}}}{1-\frac{4}{21}} = \frac{\frac{4}{\sqrt{21}}}{\frac{17}{21}} = \frac{4\sqrt{21}}{17}.$$ ■

Be careful when simplifying a complex fraction.

While the work shown in Examples 5 and 6 does not rely on a calculator, we can support our analytic work with one. By entering $\cos\left(\arctan\sqrt{3} + \arcsin\frac{1}{3}\right)$ from Example 6(a) into a calculator, we get the approximation .1827293862, the same approximation as when we enter $\frac{2\sqrt{2}-\sqrt{3}}{6}$ (the exact value obtained analytically). Similarly, we obtain the same approximation when we evaluate $\tan\left(2\arcsin\frac{2}{5}\right)$ and $\frac{4\sqrt{21}}{17}$, supporting our answer in Example 6(b).

EXAMPLE 7 Writing Function Values in Terms of *u*

Write each expression as an algebraic expression in u.

(a) $\sin(\tan^{-1} u)$ **(b)** $\cos(2\sin^{-1} u)$

Solution

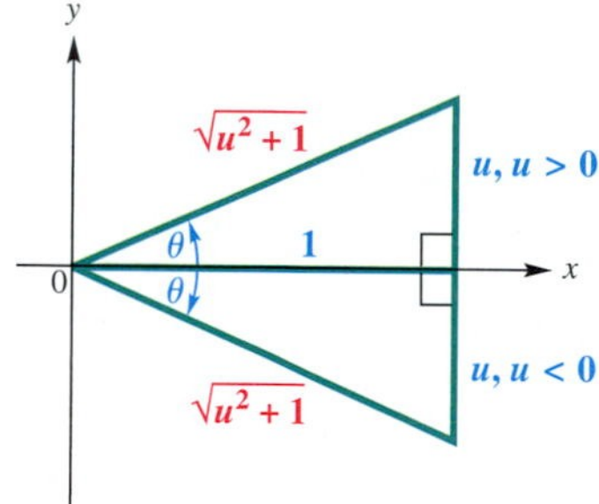

FIGURE 29

(a) Let $\theta = \tan^{-1} u$, so that $\tan\theta = u$. Here, u may be positive or negative. Since $-\frac{\pi}{2} < \tan^{-1} u < \frac{\pi}{2}$, sketch θ in quadrants I and IV and label two triangles. See Figure 29. Sine is given by the quotient of the side opposite and the hypotenuse, so

$$\sin(\tan^{-1} u) = \sin\theta = \frac{u}{\sqrt{u^2+1}} = \frac{u}{\sqrt{u^2+1}}\cdot\frac{\sqrt{u^2+1}}{\sqrt{u^2+1}} = \frac{u\sqrt{u^2+1}}{u^2+1}.$$

The result is positive when u is positive and negative when u is negative.

(b) Let $\theta = \sin^{-1} u$, so that $\sin\theta = u$. Use the identity $\cos 2\theta = 1 - 2\sin^2\theta$.

$$\cos(2\sin^{-1} u) = \cos 2\theta = 1 - 2\sin^2\theta = 1 - 2u^2$$ ■

FOR DISCUSSION

In Example 7(a), u can take on any real number value, but in Example 7(b), u can only take on values in $[-1, 1]$. Look at the final result in Example 7(b): $1 - 2u^2$. Suppose we let $u = 2$. How can we justify that this is not a valid replacement, recalling the range of the cosine function?

What Went Wrong?

A student found sin 1.74 on her calculator (set for radians). She then found the inverse sine of the answer, but got 1.401592654 instead of 1.74, the number she was expecting.

What Went Wrong (if anything)?

EXAMPLE 8 **Finding the Optimal Angle of Elevation of a Shot**

The optimal angle of elevation θ a shot-putter should aim for to throw the greatest distance depends on the velocity v of the put and the initial height h of the shot. See Figure 30. One model for θ that achieves this greatest distance is

$$\theta = \arcsin\left(\sqrt{\frac{v^2}{2v^2 + 64h}}\right).$$

(*Source:* Townend, M. Stewart, *Mathematics in Sport,* Chichester, Ellis Horwood Limited, 1984.)

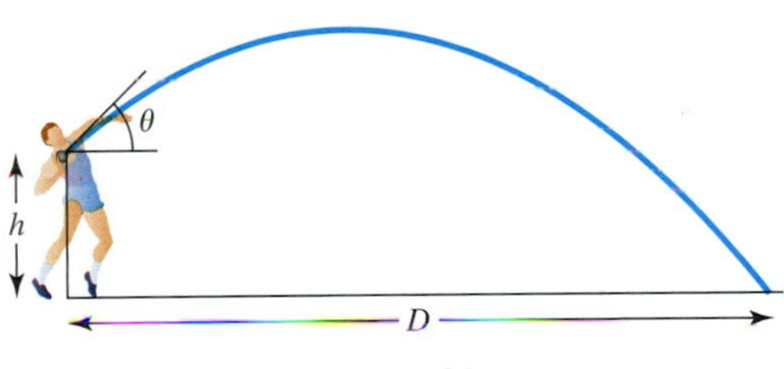

FIGURE 30

Suppose a shot-putter can consistently put the steel shot with $h = 7.6$ feet and $v = 42$ feet per second. At what angle should he put the shot to maximize distance?

Solution To find this angle, substitute and use a calculator in degree mode.

$$\theta = \arcsin\left(\sqrt{\frac{42^2}{2(42^2) + 64(7.6)}}\right) \quad h = 7.6,\ v = 42$$

$$\approx 41.5°$$

The athlete should put the shot at an angle of about 41.5° to maximize distance. ■

Answer to What Went Wrong?

Her thinking was incorrect. 1.74 is not in the range $\left[-\frac{\pi}{2}, \frac{\pi}{2}\right]$ of the inverse sine function. Her answer is actually $\pi - 1.74$, which places the result in the correct interval. Note that $\frac{\pi}{2} \approx 1.57$.

9.4 Exercises

Concept Check *In Exercises 1–4, write short answers and fill in the blanks.*

1. Consider the inverse sine function, defined by $y = \sin^{-1} x$, or $y = \arcsin x$.
  (a) What is its domain?
  (b) What is its range?
  (c) For this function, as x increases, y increases. Therefore, it is a(n) ______ (decreasing/increasing) function.
  (d) Why is $\arcsin(-2)$ not defined?

2. Consider the inverse cosine function, defined by $y = \cos^{-1} x$, or $y = \arccos x$.
  (a) What is its domain?
  (b) What is its range?
  (c) For this function, as x increases, y decreases. Therefore, it is a(n) ______ (decreasing/increasing) function.
  (d) $\text{Arccos}\left(-\frac{1}{2}\right) = \frac{2\pi}{3}$ is a true statement. Why is $\arccos\left(-\frac{1}{2}\right)$ not equal to $-\frac{4\pi}{3}$?

3. Consider the inverse tangent function, defined by $y = \tan^{-1} x$, or $y = \arctan x$.
  (a) What is its domain?
  (b) What is its range?
  (c) For this function, as x increases, y increases. Therefore, it is a(n) ______ (decreasing/increasing) function.
  (d) Is there any real number x for which $\arctan x$ is not defined? If so, what is it (or what are they)?

4. Consider the three other inverse trigonometric functions, as defined in this section.
  (a) Give the domain and range of the inverse cosecant function.
  (b) Give the domain and range of the inverse secant function.
  (c) Give the domain and range of the inverse cotangent function.

5. **Concept Check** Is $\sec^{-1} a$ calculated as $\cos^{-1} \frac{1}{a}$ or as $\frac{1}{\cos^{-1} a}$?

6. **Concept Check** For positive values of a, $\cot^{-1} a$ is calculated as $\tan^{-1} \frac{1}{a}$. How is $\cot^{-1} a$ calculated for negative values of a?

Find the exact value of each real number y. Do not use a calculator.

7. $y = \tan^{-1} 1$

8. $y = \sin^{-1} 0$

9. $y = \cos^{-1}(-1)$

10. $y = \arctan(-1)$

11. $y = \sin^{-1}(-1)$

12. $y = \cos^{-1} \frac{1}{2}$

13. $y = \arctan 0$

14. $y = \arcsin\left(-\frac{\sqrt{3}}{2}\right)$

15. $y = \arccos 0$

16. $y = \tan^{-1}(-1)$

17. $y = \sin^{-1} \frac{\sqrt{2}}{2}$

18. $y = \cos^{-1}\left(-\frac{1}{2}\right)$

19. $y = \arccos\left(-\frac{\sqrt{3}}{2}\right)$

20. $y = \arcsin\left(-\frac{\sqrt{2}}{2}\right)$

21. $y = \cot^{-1}(-1)$

22. $y = \sec^{-1}(-\sqrt{2})$

23. $y = \csc^{-1}(-2)$

24. $y = \text{arccot}(-\sqrt{3})$

25. $y = \text{arcsec} \frac{2\sqrt{3}}{3}$

26. $y = \csc^{-1} \sqrt{2}$

Give the degree measure of θ. Do not use a calculator.

27. $\theta = \arctan(-1)$

28. $\theta = \arccos\left(-\frac{1}{2}\right)$

29. $\theta = \arcsin\left(-\frac{\sqrt{3}}{2}\right)$

30. $\theta = \arcsin\left(-\frac{\sqrt{2}}{2}\right)$

31. $\theta = \cot^{-1}\left(-\frac{\sqrt{3}}{3}\right)$

32. $\theta = \sec^{-1}(-2)$

33. $\theta = \csc^{-1}(-2)$

34. $\theta = \csc^{-1}(-1)$

Use a calculator to give each value of θ in decimal degrees.

35. $\theta = \sin^{-1}(-.13349122)$

36. $\theta = \cos^{-1}(-.13348816)$

37. $\theta = \arccos(-.39876459)$

38. $\theta = \arcsin .77900016$

39. $\theta = \csc^{-1} 1.9422833$

40. $\theta = \cot^{-1} 1.7670492$

Use a calculator to give each real number value of y. (Be sure the calculator is in radian mode.)

41. $y = \arctan 1.1111111$

42. $y = \arcsin .81926439$

43. $y = \cot^{-1}(-.92170128)$

44. $y = \sec^{-1}(-1.2871684)$

45. $y = \arcsin .92837781$

46. $y = \arccos .44624593$

Draw by hand the graph of each inverse function as defined in the text.

47. $y = \cot^{-1} x$

48. $y = \csc^{-1} x$

49. $y = \sec^{-1} x$

50. $y = \operatorname{arccsc} 2x$

51. $y = \operatorname{arcsec} \frac{1}{2}x$

52. Explain why attempting to find $\sin^{-1} 1.003$ on your calculator will result in an error message.

53. Explain why you are able to find $\tan^{-1} 1.003$ on your calculator. Why is this situation different from the one described in Exercise 52?

Relating Concepts

*For individual or group discussion (Exercises 54–56)**

54. Consider the function defined by $f(x) = 3x - 2$ and its inverse $f^{-1}(x) = \frac{x+2}{3}$. Simplify $f[f^{-1}(x)]$ and $f^{-1}[f(x)]$. What do you notice in each case? What would the graph of each composition look like?

55. Use a graphing calculator in radian mode to graph $y = \tan(\tan^{-1} x)$ in the standard viewing window. How does this compare with the graph you described in Exercise 54?

56. Use a graphing calculator in radian and dot modes to graph $y = \tan^{-1}(\tan x)$ in the standard viewing window. Why does this graph not agree with the graph you found in Exercise 55?

Give the exact real number value of each expression without using a calculator.

57. $\sin\left(\sin^{-1}\frac{1}{2}\right)$

58. $\cos\left(\cos^{-1}\frac{\sqrt{3}}{2}\right)$

59. $\sin^{-1}\left(\sin\frac{4\pi}{3}\right)$

60. $\cos^{-1}\left(\cos\left(-\frac{\pi}{6}\right)\right)$

61. $\cos^{-1}\left(\cos\frac{3\pi}{2}\right)$

62. $\sin^{-1}\left(\sin\frac{3\pi}{2}\right)$

63. $\tan(\tan^{-1} 5)$

64. $\tan(\tan^{-1}(-4))$

65. $\sec(\sec^{-1} 2)$

66. $\csc(\csc^{-1} 3)$

67. $\tan^{-1}\left(\tan\frac{5\pi}{6}\right)$

68. $\tan^{-1}\left(\tan\frac{3\pi}{4}\right)$

69. $\tan\left(\arccos\frac{3}{4}\right)$

70. $\sin\left(\arccos\frac{1}{4}\right)$

71. $\cos(\tan^{-1}(-2))$

72. $\sec\left(\sin^{-1}\left(-\frac{1}{5}\right)\right)$

73. $\sin\left(2\tan^{-1}\frac{12}{5}\right)$

74. $\cos\left(2\sin^{-1}\frac{1}{4}\right)$

75. $\cos\left(2\arctan\frac{4}{3}\right)$

76. $\tan\left(2\cos^{-1}\frac{1}{4}\right)$

77. $\sin\left(2\cos^{-1}\frac{1}{5}\right)$

*The authors wish to thank Carol Walker of Hinds Community College for making a suggestion on which these exercises are based.

78. $\cos(2\tan^{-1}(-2))$

79. $\cos\left(\sin^{-1}\frac{3}{5} - \cos^{-1}\frac{12}{13}\right)$

80. $\cos\left(\cos^{-1}\frac{3}{5} + \sin^{-1}\frac{5}{13}\right)$

81. $\tan\left(\tan^{-1}\frac{3}{4} + \tan^{-1}\frac{12}{5}\right)$

82. $\tan\left(\sin^{-1}\frac{8}{17} + \tan^{-1}\frac{4}{3}\right)$

83. $\cos\left(\tan^{-1}\frac{5}{12} - \tan^{-1}\frac{3}{4}\right)$

84. $\cos\left(\sin^{-1}\frac{3}{5} + \cos^{-1}\frac{5}{13}\right)$

85. $\sin\left(\sin^{-1}\frac{1}{2} + \tan^{-1}(-3)\right)$

86. $\tan\left(\cos^{-1}\frac{\sqrt{3}}{2} - \sin^{-1}\left(-\frac{3}{5}\right)\right)$

Use a calculator to find each value. Give answers as real numbers.

87. $\cos(\tan^{-1} .5)$

88. $\sin(\cos^{-1} .25)$

89. $\tan(\arcsin .12251014)$

90. $\cot(\arccos .58236841)$

Write each expression as an algebraic expression in u, $u > 0$.

91. $\sec(\cos^{-1} u)$

92. $\cot(\tan^{-1} u)$

93. $\sin(\arccos u)$

94. $\tan(\arccos u)$

95. $\cot(\arcsin u)$

96. $\cos(\arcsin u)$

97. $\sin\left(\sec^{-1}\frac{u}{2}\right)$

98. $\cos\left(\tan^{-1}\frac{3}{u}\right)$

99. $\tan\left(\sin^{-1}\frac{u}{\sqrt{u^2+2}}\right)$

100. $\sec\left(\cos^{-1}\frac{u}{\sqrt{u^2+5}}\right)$

101. $\sec\left(\operatorname{arccot}\frac{\sqrt{4-u^2}}{u}\right)$

102. $\csc\left(\arctan\frac{\sqrt{9-u^2}}{u}\right)$

(Modeling) Solve each problem.

103. ***Angle of Elevation of a Shot*** Refer to Example 8.

(a) What is the optimal angle when $h = 0$?

(b) Fix h at 6 feet and regard θ as a function of v. As v gets larger and larger, the graph approaches a horizontal asymptote. Find the equation of that asymptote.

104. ***Angle of Elevation of a Plane*** Suppose an airplane flying faster than sound goes directly over you. Assume that the plane is flying at a constant altitude. At the instant you feel the sonic boom from the plane, the angle of elevation to the plane is given by

$$\alpha = 2 \arcsin \frac{1}{m},$$

where m is the *Mach number* of the plane's speed. (The Mach number is the ratio of the speed of the plane to the speed of sound.) Find α to the nearest degree for each value of m.

(a) $m = 1.2$ **(b)** $m = 1.5$

(c) $m = 2$ **(d)** $m = 2.5$

105. ***Viewing Angle of an Observer*** A painting 3 feet high and 6 feet from the floor will cut off an angle

$$\theta = \tan^{-1}\left(\frac{3x}{x^2 + 4}\right)$$

to an observer. Assume that the observer is x feet from the wall where the painting is displayed and that the eyes of the observer are 5 feet above the ground.

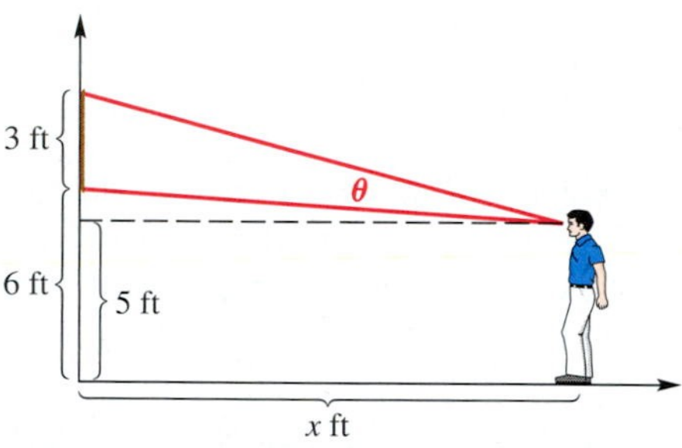

Find the value of θ for each value of x. Round to the nearest degree.

(a) $x = 3$ **(b)** $x = 6$ **(c)** $x = 9$

(d) Derive the given formula for θ. (*Hint:* Use right triangles and the identity for $\tan(\theta + \alpha)$.)

(e) Graph the function for θ with a graphing calculator, and determine the distance that maximizes the angle.

106. ***Communications Satellite Coverage*** The figure shows a stationary communications satellite positioned 20,000 miles above the equator. What percentage of the equator can be seen from the satellite? The diameter of Earth is 7927 miles at the equator.

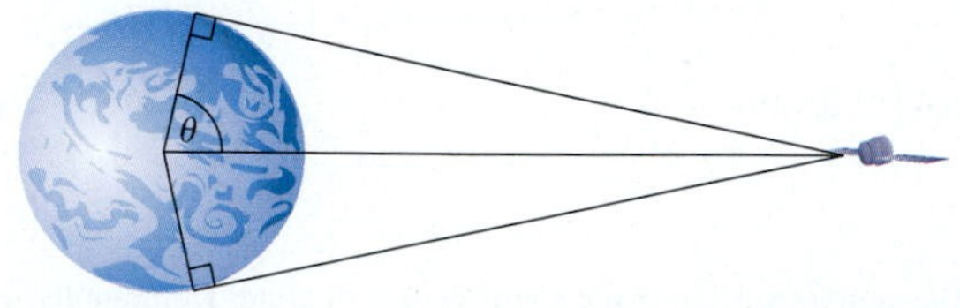

107. ***Landscaping Formula*** A shrub is planted in a 100-foot-wide space between buildings measuring 75 feet and 150 feet tall. The location of the shrub determines how much sun it receives each day. Show that if θ is the angle in the figure and x is the distance of the shrub from the taller building, then the value of θ (in radians) is given by

$$\theta = \pi - \arctan\left(\frac{75}{100 - x}\right) - \arctan\left(\frac{150}{x}\right).$$

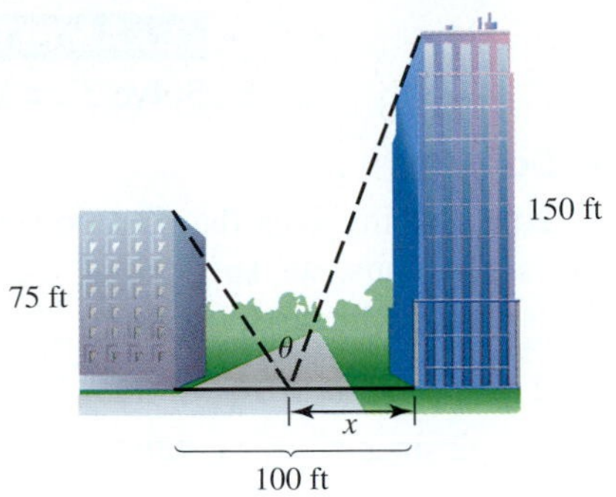

Reviewing Basic Concepts (Sections 9.3 and 9.4)

Use identities to work Exercises 1–5.

1. Find $\tan x$, given $\cos 2x = -\frac{5}{12}$ and $\frac{\pi}{2} < x < \pi$.

2. Find $\sin 2\theta$, $\cos 2\theta$, and $\tan 2\theta$ if $\sin \theta = -\frac{1}{3}$ and θ is in quadrant III.

3. Find the exact value of $\sin 75°$.

4. Write $2 \sin 25° \cos 150°$ as a sum or difference of trigonometric functions.

5. Verify each identity.

(a) $\sin^2 \frac{x}{2} = \frac{\tan x - \sin x}{2 \tan x}$

(b) $\frac{\sin 2x}{2 \sin x} = \cos^2 \frac{x}{2} - \sin^2 \frac{x}{2}$

6. Find the exact value of each real number y.

(a) $y = \arccos \frac{\sqrt{3}}{2}$ **(b)** $y = \sin^{-1}\left(-\frac{\sqrt{2}}{2}\right)$

7. Find the exact value of each angle θ measured in degrees.

(a) $\theta = \arccos .5$ **(b)** $\theta = \cot^{-1}(-1)$

8. Graph $y = 2 \csc^{-1} x$.

9. Find the exact value of each expression.

(a) $\cot\left(\arcsin\left(-\frac{2}{3}\right)\right)$

(b) $\cos\left(\tan^{-1} \frac{5}{12} - \sin^{-1} \frac{3}{5}\right)$

10. Write $\sin(\text{arccot } u)$ as an algebraic expression in u, $u > 0$.

9.5 Trigonometric Equations and Inequalities (I)

Equations Solvable by Linear Methods ■ Equations Solvable by Factoring ■ Equations Solvable by the Quadratic Formula ■ Using Trigonometric Identities to Solve Equations

Looking Ahead to Calculus

There are many instances in calculus where it is necessary to solve trigonometric equations. Examples include solving related-rate problems and optimization problems.

Earlier in this chapter, we studied trigonometric equations that were identities. We now consider trigonometric equations that are *conditional*—that is, equations that are satisfied by some values but not others.

Equations Solvable by Linear Methods

Conditional equations with trigonometric (or circular) functions can often be solved by analytic methods and trigonometric identities.

EXAMPLE 1 Solving a Trigonometric Equation by a Linear Method

Solve $2 \sin x - 1 = 0$ over the interval $[0, 2\pi)$.

Analytic Solution

Since this equation involves the first power of $\sin x$, it is linear in $\sin x$. Thus, we solve it by using the usual method for solving a linear equation.

$$2 \sin x - 1 = 0$$

$$2 \sin x = 1 \quad \text{Add 1.}$$

$$\sin x = \frac{1}{2} \quad \text{Divide by 2.}$$

The two values of x in the interval $[0, 2\pi)$ whose sine is $\frac{1}{2}$ are $\frac{\pi}{6}$ and $\frac{5\pi}{6}$. These are obtained by finding

$$\sin^{-1}\frac{1}{2} \quad \text{and} \quad \pi - \sin^{-1}\frac{1}{2}.$$

We also get these values by using the unit circle or the reference angle analysis discussed in Chapter 8. Therefore, the solution set in the specified interval is $\left\{\frac{\pi}{6}, \frac{5\pi}{6}\right\}$.

Graphing Calculator Solution

We graph $y = 2 \sin x - 1$ over the desired interval and find that the x-intercepts have the same decimal approximations as $\frac{\pi}{6}$ and $\frac{5\pi}{6}$. See Figure 31.

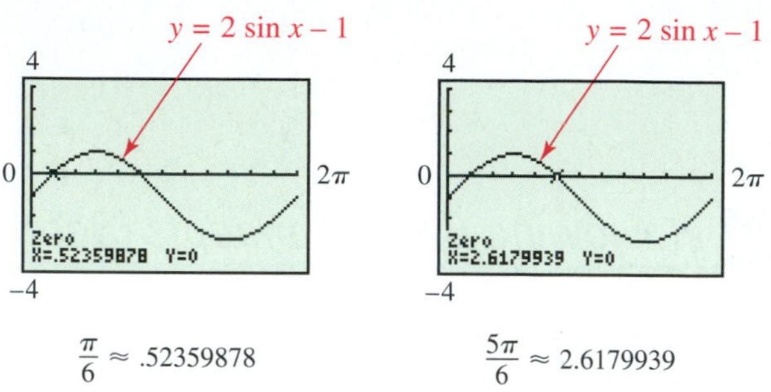

$\frac{\pi}{6} \approx .52359878$ $\qquad$ $\frac{5\pi}{6} \approx 2.6179939$

FIGURE 31

We used Xscl $= \frac{\pi}{6}$ in Figure 31, so we can see that the graph intersects the x-axis at the first and fifth tick marks, representing $\frac{\pi}{6}$ and $\frac{5\pi}{6}$, respectively. This allows us to observe exact values of solutions from the graph. ■

EXAMPLE 2 Solving Trigonometric Inequalities

Solve **(a)** $2 \sin x - 1 > 0$ and **(b)** $2 \sin x - 1 < 0$ over the interval $[0, 2\pi)$.

Solution To solve the inequality in part (a), we must identify the x-values in the interval $[0, 2\pi)$ for which the graph is *above* the x-axis. Similarly, to solve the inequality in part (b), we must identify the x-values for which the graph is *below* the x-axis. The graph in Figure 31 indicates the following solutions in $[0, 2\pi)$.

(a) The solution set of $2 \sin x - 1 > 0$ is $\left(\frac{\pi}{6}, \frac{5\pi}{6}\right)$.

(b) The solution set of $2 \sin x - 1 < 0$ is $\left[0, \frac{\pi}{6}\right) \cup \left(\frac{5\pi}{6}, 2\pi\right)$. ■

Equations Solvable by Factoring

EXAMPLE 3 Solving a Trigonometric Equation by Factoring

Solve $\sin x \tan x = \sin x$ over the interval $[0°, 360°)$.

Solution

$$\sin x \tan x = \sin x$$

$$\sin x \tan x - \sin x = 0 \quad \text{Subtract } \sin x.$$

$$\sin x(\tan x - 1) = 0 \quad \text{Factor.}$$

$$\sin x = 0 \quad \text{or} \quad \tan x - 1 = 0 \quad \text{Zero-product property}$$

$$\tan x = 1$$

$$x = 0° \text{ or } x = 180° \qquad x = 45° \text{ or } x = 225°$$

The solution set is $\{0°, 45°, 180°, 225°\}$. ■

CAUTION There are four solutions in Example 3. Trying to solve the equation by dividing each side by $\sin x$ would give just $\tan x = 1$, which would give $x = 45°$ or $x = 225°$. The other two solutions would not appear. The missing solutions are the ones that make the divisor, $\sin x$, equal 0. ***For this reason, we must avoid dividing by a variable expression.***

EXAMPLE 4 Solving a Trigonometric Equation and an Associated Inequality

Solve each equation over the interval $[0, 2\pi)$.

(a) $\tan^2 x + \tan x - 2 = 0$ **(b)** $\tan^2 x + \tan x - 2 > 0$

Solution

(a) This equation is quadratic in the term $\tan x$ and can be solved by factoring.

$$\tan^2 x + \tan x - 2 = 0$$

$$(\tan x - 1)(\tan x + 2) = 0 \quad \text{Factor.}$$

$$\tan x - 1 = 0 \quad \text{or} \quad \tan x + 2 = 0 \quad \text{Zero-product property}$$

$$\tan x = 1 \quad \text{or} \quad \tan x = -2$$

The solutions for $\tan x = 1$ in the interval $[0, 2\pi)$ are $\frac{\pi}{4}$ and $\frac{5\pi}{4}$.

To solve $\tan x = -2$ in the interval, we use a calculator set in *radian* mode. We find that

$$\tan^{-1}(-2) \approx -1.107148718.$$

However, due to the way the calculator determines this value, it is not in the desired interval. Because the period of the tangent function is π, we add π and then 2π to $\tan^{-1}(-2)$ to obtain the solutions in the desired interval.

$$x = \tan^{-1}(-2) + \pi \approx 2.034443936$$

$$x = \tan^{-1}(-2) + 2\pi \approx 5.176036589$$

The solution set is

$$\Big\{\underbrace{\frac{\pi}{4}, \frac{5\pi}{4}}_{\text{Exact values}}, \underbrace{2.03, 5.18}_{\text{Approximate values to the nearest hundredth}}\Big\}.$$

FOR DISCUSSION

What is the solution set of $\tan^2 x + \tan x - 2 < 0$ over the interval $[0, 2\pi)$? (*Hint:* Refer to Figure 32.)

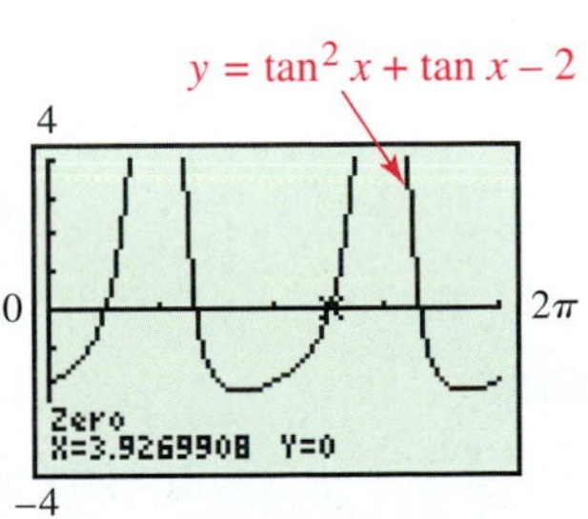

FIGURE 32

(b) From the graph of $y = \tan^2 x + \tan x - 2$ as seen in Figure 32, there are four x-intercepts in the interval $[0, 2\pi)$. (The figure shows the solution $\frac{5\pi}{4}$, since a decimal approximation of this number is 3.9269908. The other three solutions can be found similarly.) The graph lies *above* the x-axis for the following subset of the interval $[0, 2\pi)$:

$$\left(\frac{\pi}{4}, \frac{\pi}{2}\right) \cup \left(\frac{\pi}{2}, 2.03\right) \cup \left(\frac{5\pi}{4}, \frac{3\pi}{2}\right) \cup \left(\frac{3\pi}{2}, 5.18\right).$$

This is the solution set of the inequality. Notice that tangent is not defined when $x = \frac{\pi}{2}$ and when $x = \frac{3\pi}{2}$, so these values must be excluded from the solution set. ■

Equations Solvable by the Quadratic Formula

EXAMPLE 5 Solving a Trigonometric Equation by Using the Quadratic Formula

Solve $\cot x(\cot x + 3) = 1$ over the interval $[0, 2\pi)$.

Solution We multiply the factors on the left and subtract 1.

$$\cot^2 x + 3 \cot x - 1 = 0 \quad \text{Standard quadratic form}$$

Since this equation cannot be solved by factoring, we use the quadratic formula, with $a = 1$, $b = 3$, $c = -1$, and $\cot x$ as the variable.

$$\cot x = \frac{-b \pm \sqrt{b^2 - 4ac}}{2a} = \frac{-3 \pm \sqrt{9 + 4}}{2} = \frac{-3 \pm \sqrt{13}}{2}$$

$$\cot x \approx -3.302775638 \quad \text{or} \quad \cot x \approx .3027756377 \quad \text{Use a calculator.}$$

Since we cannot find inverse cotangent values directly on a calculator, we use the fact that $\cot x = \frac{1}{\tan x}$ and take reciprocals to get

$$\tan x \approx -.3027756377 \quad \text{or} \quad \tan x \approx 3.302775638$$
$$x \approx -.2940013018 \quad \text{or} \quad x \approx 1.276795025.$$

The first of these, $-.2940013018$, is not in the desired interval. Since the period of the cotangent function is π, we add π and then 2π to $-.2940013018$ to get 2.847591352 and 5.989184005, respectively.

The second value, 1.276795025, is in the desired interval. We add π to it to get another solution in the interval: $1.276795025 + \pi$. Rounding to the nearest hundredth gives 1.28, 2.85, 4.42, and 5.99 for the four solutions in the interval, and the solution set is $\{1.28, 2.85, 4.42, 5.99\}$.

The graph of $y = \cot^2 x + 3 \cot x - 1$, shown in Figure 33, indicates the solution 1.276795025. The others can be found similarly.

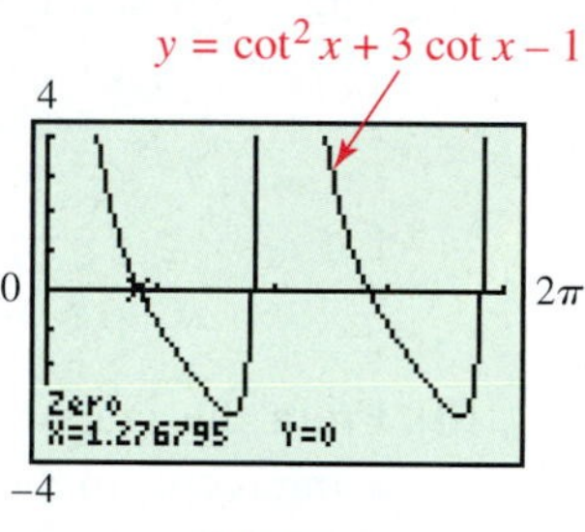

FIGURE 33

Using Trigonometric Identities to Solve Equations

Recall that squaring each side of an equation such as $\sqrt{x + 4} = x + 2$ will yield all solutions but may also give extraneous values. (Verify that in this equation 0 is a solution, while -3 is extraneous.) The same situation may occur when trigonometric equations are solved in this manner.

EXAMPLE 6 **Solving a Trigonometric Equation by Squaring**

Solve $\tan x + \sqrt{3} = \sec x$ over the interval $[0, 2\pi)$.

Analytic Solution

Since the tangent and secant functions are related by the identity $1 + \tan^2 x = \sec^2 x$, square both sides and write $\sec^2 x$ in terms of $\tan^2 x$.

$$(\tan x + \sqrt{3})^2 = (\sec x)^2 \quad \text{Square each side.}$$

$$\tan^2 x + 2\sqrt{3}\tan x + 3 = \sec^2 x \quad (x + y)^2 = x^2 + 2xy + y^2$$

$$\tan^2 x + 2\sqrt{3}\tan x + 3 = 1 + \tan^2 x \quad \text{Pythagorean identity}$$

$$2\sqrt{3}\tan x = -2 \quad \text{Subtract } 3 + \tan^2 x.$$

$$\tan x = -\frac{1}{\sqrt{3}} = -\frac{\sqrt{3}}{3} \quad \text{Divide by } 2\sqrt{3}\text{; rationalize the denominator.}$$

The possible solutions in the given interval are $\frac{5\pi}{6}$ and $\frac{11\pi}{6}$. Check $\frac{5\pi}{6}$ first.

Left side: $\tan x + \sqrt{3} = \tan\frac{5\pi}{6} + \sqrt{3} = -\frac{\sqrt{3}}{3} + \sqrt{3} = \frac{2\sqrt{3}}{3}$

Right side: $\sec x = \sec\frac{5\pi}{6} = -\frac{2\sqrt{3}}{3}$ ← Not equal

The check shows that $\frac{5\pi}{6}$ is not a solution. Now check $\frac{11\pi}{6}$.

Left side: $\tan\frac{11\pi}{6} + \sqrt{3} = -\frac{\sqrt{3}}{3} + \sqrt{3} = \frac{2\sqrt{3}}{3}$

Right side: $\sec\frac{11\pi}{6} = \frac{2\sqrt{3}}{3}$ ← Equal

This solution satisfies the equation, so the solution set is $\left\{\frac{11\pi}{6}\right\}$.

Graphing Calculator Solution

We will use the x-intercept method. Graph

$$y = \tan x + \sqrt{3} - \sec x$$

over the desired interval, as in Figure 34. The graph shows that the only x-intercept on the interval $[0, 2\pi)$ is 5.7595865, which is an approximation for $\frac{11\pi}{6}$, the solution found analytically.

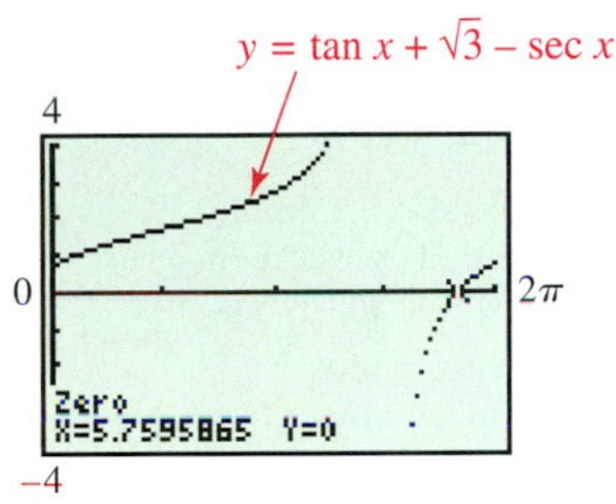

Dot mode; radian mode

FIGURE 34

The methods for solving trigonometric equations can be summarized as follows.

Solving a Trigonometric Equation Analytically

1. Decide whether the equation is linear or quadratic, so you can determine the solution method.
2. If only one trigonometric function is present, first solve the equation for that function.
3. If more than one trigonometric function is present, rearrange the equation so that one side equals 0. Then try to factor and set each factor equal to 0 to solve.
4. If the equation is quadratic in form, but not factorable, use the quadratic formula. Check that solutions are in the desired interval.
5. Try using identities to change the form of the equation. It may be helpful to square each side of the equation first. If this is done, check for extraneous values.

Solving a Trigonometric Equation Graphically

1. For an equation of the form $f(x) = g(x)$, use the intersection-of-graphs method.
2. For an equation of the form $f(x) = 0$, use the x-intercept method.

9.5 Exercises

Solve each equation for solutions over the interval $[0, 2\pi)$ by first solving for the trigonometric function.

1. $2 \cos x + 1 = 0$

2. $2 \sin x + 1 = 0$

3. $5 \sin x - 5 = 0$

4. $3 \cos x + 3 = 0$

5. $2 \tan x + 1 = -1$

6. $2 \cot x + 1 = -1$

7. $2 \cos x + 5 = 6$

8. $2 \sin x + 3 = 4$

9. $2 \csc x + 4 = \csc x + 6$

10. $2 \sec x + 1 = \sec x + 3$

11. $(\cot x - 1)(\sqrt{3} \cot x + 1) = 0$

12. $(\csc x + 2)(\csc x - \sqrt{2}) = 0$

13. $\cos x \cot x = \cos x$

14. $\sin x \cot x = \sin x$

15. $\sin^2 x - 2 \sin x + 1 = 0$

16. $\cos^2 x + 2 \cos x + 1 = 0$

17. $4(1 + \sin x)(1 - \sin x) = 3$

18. $(\cot x - \sqrt{3})(2 \sin x + \sqrt{3}) = 0$

19. $\tan x + 1 = \sqrt{3} + \sqrt{3} \cot x$

20. $\tan x - \cot x = 0$

21. $2 \sin x - 1 = \csc x$

22. $\cos^2 x = \sin^2 x$

23. $\cos^2 x - \sin^2 x = 1$

24. $\csc^2 x = 2 \cot x$

In Exercises 25–33, solve **(a)** $f(x) = 0$, **(b)** $f(x) > 0$, *and* **(c)** $f(x) < 0$ *over the interval* $[0, 2\pi)$.

25. $f(x) = -2 \cos x + 1$

26. $f(x) = 2 \sin x + 1$

27. $f(x) = \tan^2 x - 3$

28. $f(x) = \sec^2 x - 1$

29. $f(x) = 2 \cos^2 x - \sqrt{3} \cos x$

30. $f(x) = 2 \sin^2 x + 3 \sin x + 1$

31. $f(x) = \sin^2 x \cos x - \cos x$

32. $f(x) = \sin^2 x \cos^2 x$

33. $f(x) = 2 \tan^2 x \sin x - \tan^2 x$

Relating Concepts

For individual or group investigation (Exercises 34–38)

Work Exercises 34–38 in order.

34. Write the equation $\tan^3 x = 3 \tan x$ with 0 on one side.

35. Solve the equation you wrote in Exercise 34 over $[0, 2\pi)$ by factoring.

36. Verify that the solutions you found in Exercise 35 satisfy the equation.

37. Solve $\tan^3 x = 3 \tan x$ by dividing each side by $\tan x$.

38. Do your answers in Exercises 35 and 37 agree? Explain.

In Exercises 39–44, give solutions over the interval $[0, 2\pi)$ *as approximations to the nearest hundredth when exact values cannot be determined. You will need to use the quadratic formula. Give answers to Exercises 45–47 in degrees.*

39. $3\sin^2 x - \sin x = 2$

40. $9\sin^2 x = 6\sin x + 1$

41. $\tan^2 x + 4\tan x + 2 = 0$

42. $3\cot^2 x - 3\cot x = 1$

43. $2\cos^2 x + 2\cos x = 1$

44. $\sin^2 x - 2\sin x + 3 = 0$

45. $\sec^2 \theta = 2\tan\theta + 4$

46. $\cot\theta + 2\csc\theta = 3$

47. $2\sin\theta = 1 - 2\cos\theta$

Solve each equation graphically over the interval $[0, 2\pi)$*. Express solutions to the nearest hundredth. (*Hint: *In Exercise 51, the equation has three solutions.)*

48. $\cot x + 2\csc x = 3$

49. $2\sin x = 1 - 2\cos x$

50. $\sin^3 x + \sin x = 1$

51. $2\cos^3 x + \sin x = -1$

52. $e^x = \sin x + 3$

53. $\ln x = \cos x$

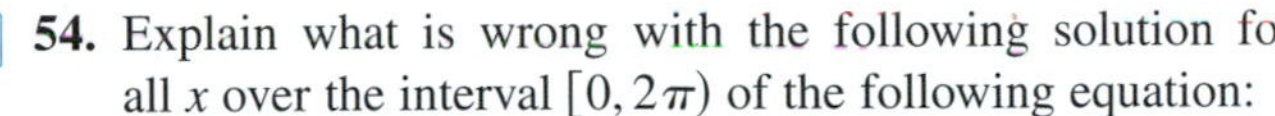

54. Explain what is wrong with the following solution for all x over the interval $[0, 2\pi)$ of the following equation:

$$\sin^2 x - \sin x = 0$$

$$\sin x - 1 = 0 \quad \text{Divide by } \sin x.$$

$$\sin x = 1 \quad \text{Add 1.}$$

$$x = \frac{\pi}{2}.$$

(Modeling) *Solve each problem.*

55. ***Daylight Hours in New Orleans*** The seasonal variation in length of daylight can be modeled by a sine function. For example, the daily number of hours of daylight in New Orleans is given by

$$h = \frac{35}{3} + \frac{7}{3}\sin\frac{2\pi x}{365},$$

where x is the number of days after March 21 (disregarding leap year). (*Source:* Bushaw, Donald et al., *A Sourcebook of Applications of School Mathematics.* Copyright © 1980 by The Mathematical Association of America.)

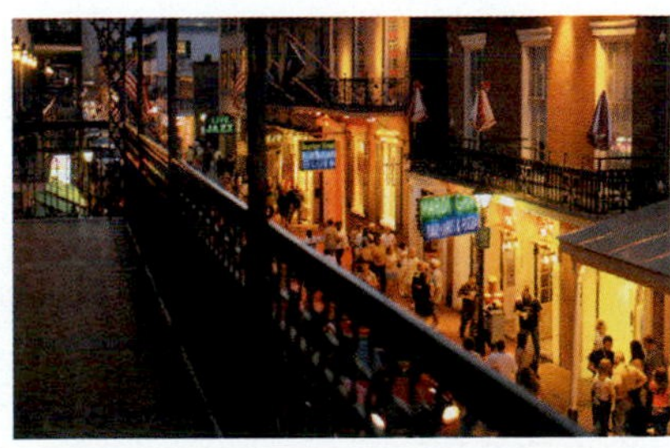

(a) On what date will there be about 14 hours of daylight?
(b) What date has the least number of hours of daylight?
(c) When will there be about 10 hours of daylight?

56. ***Mach Number for a Plane*** An airplane flying faster than sound sends out sound waves that form a cone. The cone intersects the ground to form a hyperbola. As this hyperbola passes over a particular point on the ground, a sonic boom is heard at that point. If α is the angle at the vertex of the cone, then

$$\sin\frac{\alpha}{2} = \frac{1}{m},$$

where m is the Mach number for the speed of the plane. (See Section 9.4, Exercise 104.) We assume that $m > 1$. Find the measure of α, in degrees, if $m = 1.5$.

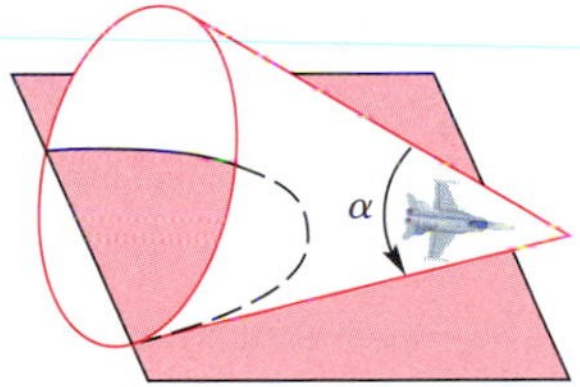

57. ***Accident Reconstruction*** The equation

$$.342D\cos\theta + h\cos^2\theta = \frac{16D^2}{V^2}$$

is used to reconstruct accidents in which a vehicle vaults into the air after hitting an obstruction. V is the velocity in feet per second of the vehicle when it hits the obstruction, D is the distance (in feet) from the obstruction to the vehicle's landing point, and h is the difference in height (in feet) between the landing point and the takeoff point. Angle θ is the takeoff angle—the angle between the horizontal and the path of the vehicle. Find θ to the nearest degree if $V = 60$, $D = 80$, and $h = 2$.

58. ***Maximum Viewing Angle*** The bottom of a 10-foot-high movie screen is located 2 feet above the eyes of the viewers, all of whom are sitting at the same level. A viewer seated 5 feet from the screen has the maximum viewing angle, given by x in the equation

$$\frac{\tan x + .4}{1 - .4 \tan x} = 2.4.$$

Find the maximum viewing angle (in degrees).

Distance of a Particle from a Starting Point *A particle moves along a straight line. The distance of the particle from a starting point at time t is given by*

$$s(t) = \sin t + 2 \cos t.$$

Find a value of t in $\left[0, \frac{\pi}{2}\right)$ *that satisfies each equation.*

59. $s(t) = \dfrac{2 + \sqrt{3}}{2}$

60. $s(t) = \dfrac{3\sqrt{2}}{2}$

9.6 Trigonometric Equations and Inequalities (II)

Equations and Inequalities Involving Multiple-Number Identities ■ Equations and Inequalities Involving Half-Number Identities ■ An Application of Trigonometric Equations

Equations and Inequalities Involving Multiple-Number Identities

EXAMPLE 1 Solving an Equation by Using a Double-Number Identity

Solve $\cos 2x = \cos x$ over the interval $[0, 2\pi)$.

Analytic Solution

First change $\cos 2x$ to a trigonometric function of x alone. Use the identity $\cos 2x = 2\cos^2 x - 1$ so that the equation involves only the cosine of x.

$$\cos 2x = \cos x$$

$$2\cos^2 x - 1 = \cos x \quad \text{Double-number identity}$$

$$2\cos^2 x - \cos x - 1 = 0 \quad \text{Standard form}$$

$$(2\cos x + 1)(\cos x - 1) = 0 \quad \text{Factor.}$$

$$2\cos x + 1 = 0 \quad \text{or} \quad \cos x - 1 = 0 \quad \text{Zero-product property}$$

$$\cos x = -\frac{1}{2} \qquad \cos x = 1$$

$$x = \frac{2\pi}{3} \quad \text{or} \quad x = \frac{4\pi}{3} \qquad x = 0$$

The solution set is $\left\{0, \frac{2\pi}{3}, \frac{4\pi}{3}\right\}$ over the interval $[0, 2\pi)$.

Graphing Calculator Solution

Graph $y = \cos 2x - \cos x$ in an appropriate window, and find the x-intercepts. The display in Figure 35 shows that one x-intercept is 2.0943951, which is an approximation for $\frac{2\pi}{3}$. The other two x-intercepts correspond to 0 and $\frac{4\pi}{3}$, respectively.

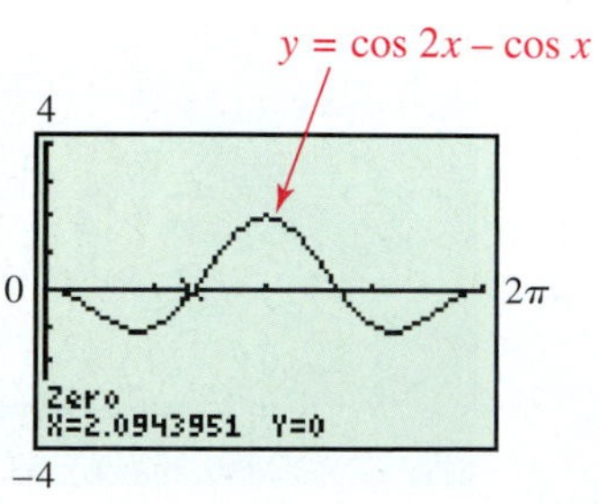

Radian mode

FIGURE 35

CAUTION In the analytic solution in Example 1, cos $2x$ cannot be changed to cos x by dividing by 2, since 2 is *not* a factor of cos $2x$.

$$\frac{\cos 2x}{2} \neq \cancel{\cos x}$$

To change cos $2x$ to a trigonometric function of x alone, use one of the identities for cos $2x$.

EXAMPLE 2 Solving Inequalities Involving a Function of 2x

Refer to Example 1 and Figure 35 to solve each inequality over the interval $[0, 2\pi)$.

(a) $\cos 2x < \cos x$ **(b)** $\cos 2x > \cos x$

FOR DISCUSSION

Refer to Example 2 and discuss how the solution sets would be affected with the following minor changes.

1. Solve $\cos 2x \leq \cos x$.
2. Solve $\cos 2x \geq \cos x$ over the interval $[0, 4\pi)$.

Solution

(a) The given inequality is equivalent to $\cos 2x - \cos x < 0$. We determine the solution set of this inequality by finding the x-values where the graph of the equation $y = \cos 2x - \cos x$ is *below* the x-axis. From Figure 35, we see that this occurs in the interval $[0, 2\pi)$ when x is in the interval $\left(0, \frac{2\pi}{3}\right) \cup \left(\frac{4\pi}{3}, 2\pi\right)$.

(b) Here, we want the x-values where the graph of $y = \cos 2x - \cos x$ is *above* the x-axis, so the solution set is $\left(\frac{2\pi}{3}, \frac{4\pi}{3}\right)$. ■

Some equations require an additional step to adjust the solution interval.

EXAMPLE 3 Solving an Equation by Using a Double-Number Identity

Solve $4 \sin x \cos x = \sqrt{3}$ over the interval $[0, 2\pi)$.

Analytic Solution

The identity $2 \sin x \cos x = \sin 2x$ is useful here.

$$4 \sin x \cos x = \sqrt{3}$$

$$2(2 \sin x \cos x) = \sqrt{3} \qquad 4 = 2 \cdot 2$$

$$2 \sin 2x = \sqrt{3} \qquad 2 \sin x \cos x = \sin 2x$$

$$\sin 2x = \frac{\sqrt{3}}{2} \qquad \text{Divide by 2.}$$

From the given domain, $0 \leq x < 2\pi$, the domain for $2x$ is $0 \leq 2x < 4\pi$. We list all solutions in this interval.

$$2x = \frac{\pi}{3}, \frac{2\pi}{3}, \frac{7\pi}{3}, \frac{8\pi}{3}$$

$$x = \frac{\pi}{6}, \frac{\pi}{3}, \frac{7\pi}{6}, \frac{4\pi}{3} \qquad \text{Divide by 2.}$$

We found the final two solutions for $2x$ by adding 2π to $\frac{\pi}{3}$ and $\frac{2\pi}{3}$, respectively, giving the solution set $\left\{\frac{\pi}{6}, \frac{\pi}{3}, \frac{7\pi}{6}, \frac{4\pi}{3}\right\}$ for x.

Graphing Calculator Solution

We use the intersection-of-graphs method. Figure 36 indicates that the x-coordinate of one point of intersection is .52359878, an approximation for $\frac{\pi}{6}$. The other three intersection points correspond to the remaining solutions, $\frac{\pi}{3}$, $\frac{7\pi}{6}$, and $\frac{4\pi}{3}$.

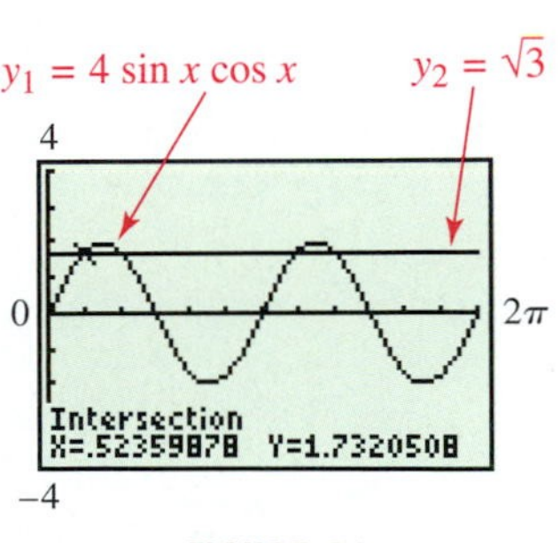

FIGURE 36 ■

EXAMPLE 4 **Solving an Equation That Involves Squaring Each Side**

Solve $\tan 3x + \sec 3x = 2$ over the interval $[0, 2\pi)$.

Solution Since the tangent and secant functions are related by the identity $1 + \tan^2 \theta = \sec^2 \theta$, we begin by expressing everything in terms of the secant.

$$\tan 3x + \sec 3x = 2$$

$$\tan 3x = 2 - \sec 3x \quad \text{Subtract } \sec 3x.$$

$$\tan^2 3x = 4 - 4 \sec 3x + \sec^2 3x \quad \text{Square each side.}$$

Remember the middle term when squaring a binomial.

$$\sec^2 3x - 1 = 4 - 4 \sec 3x + \sec^2 3x \quad \tan^2 3x = \sec^2 3x - 1$$

$$0 = 5 - 4 \sec 3x \quad \text{Subtract } \sec^2 3x\text{; add 1.}$$

$$4 \sec 3x = 5 \quad \text{Add } 4 \sec 3x.$$

$$\sec 3x = \frac{5}{4} \quad \text{Divide by 4.}$$

$$\frac{1}{\cos 3x} = \frac{5}{4} \quad \sec \theta = \frac{1}{\cos \theta}$$

$$\cos 3x = \frac{4}{5} \quad \text{Use reciprocals.}$$

We multiply the inequality $0 \le x < 2\pi$ by 3 to find that the interval for $3x$ is $[0, 6\pi)$. Using a calculator and the fact that the cosine is positive in quadrants I and IV, we get

$$3x \approx .64350111, 5.6396842, 6.9266864, 11.922870, 13.209872, 18.206055.$$

Dividing by 3 gives

$$x \approx .21450037, 1.8798947, 2.3088955, 3.9742898, 4.4032906, 6.0686849.$$

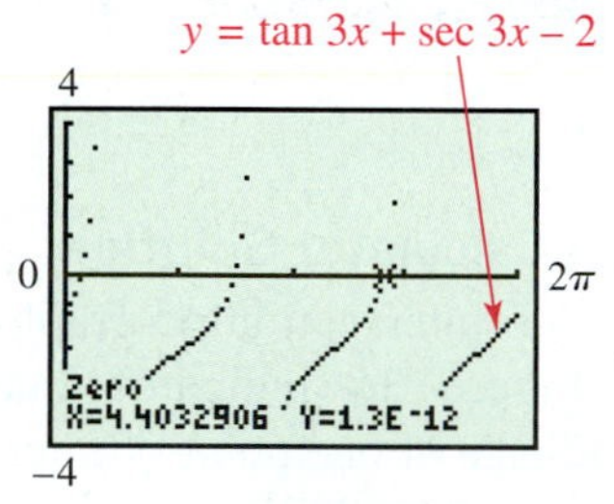

Dot mode

FIGURE 37

Recall from Section 4.5 that when each side of an equation is squared, there may be extraneous values. From the graph of $y = \tan 3x + \sec 3x - 2$ in Figure 37, we see that in the interval $[0, 2\pi)$ there are only three x-intercepts. One of these is approximately 4.4032906, which is one of the six possible solutions found. (See the display in the figure.) The other two can also be verified by a calculator. They are .21450037 and 2.3088955. To the nearest thousandth, the solution set over the interval $[0, 2\pi)$ is $\{.215, 2.309, 4.403\}$. ■

Equations and Inequalities Involving Half-Number Identities

EXAMPLE 5 **Solving an Equation by Using a Half-Number Identity**

Solve $2 \sin \frac{x}{2} = 1$ over the interval $[0, 2\pi)$.

Solution Write the interval $[0, 2\pi)$ as the inequality

$$0 \le x < 2\pi$$

$$0 \le \frac{x}{2} < \pi. \quad \text{Divide by 2.}$$

To find all values of $\frac{x}{2}$ in the interval $[0, \pi)$ that satisfy the given equation, first solve for $\sin \frac{x}{2}$.

$$2 \sin \frac{x}{2} = 1$$

$$\sin \frac{x}{2} = \frac{1}{2} \quad \text{Divide by 2.}$$

The two numbers in the interval $[0, \pi)$ whose sine is $\frac{1}{2}$ are $\frac{\pi}{6}$ and $\frac{5\pi}{6}$, so

$$\frac{x}{2} = \frac{\pi}{6} \quad \text{or} \quad \frac{x}{2} = \frac{5\pi}{6}$$

$$x = \frac{\pi}{3} \quad \text{or} \quad x = \frac{5\pi}{3}. \quad \text{Multiply by 2.}$$

The solution set in the given interval is $\left\{\frac{\pi}{3}, \frac{5\pi}{3}\right\}$. ■

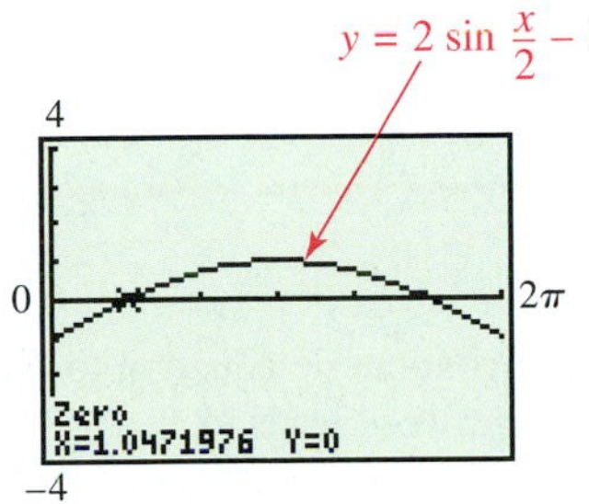

FIGURE 38

EXAMPLE 6 Solving an Inequality Involving a Function of $\frac{x}{2}$

Use a graph to solve $2 \sin \frac{x}{2} > 1$ over the interval $[0, 2\pi)$.

Solution Referring to Figure 38, where the x-scale is $\frac{\pi}{3}$, we see that the graph of $y = 2 \sin \frac{x}{2} - 1$ lies *above* the x-axis between $\frac{\pi}{3}$ and $\frac{5\pi}{3}$. Thus, the solution set is the open interval $\left(\frac{\pi}{3}, \frac{5\pi}{3}\right)$. ■

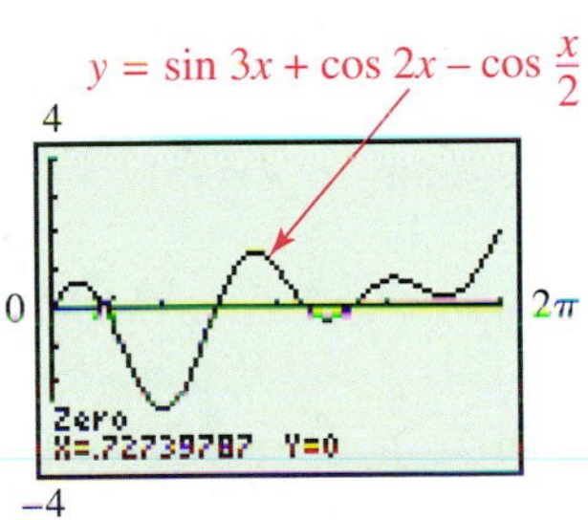

FIGURE 39

EXAMPLE 7 Solving an Equation by Using Only a Graphical Approach

Use a graph to solve $\sin 3x + \cos 2x = \cos \frac{x}{2}$ over the interval $[0, 2\pi)$.

Solution We solve this equation with the x-intercept method. The graph of $y = \sin 3x + \cos 2x - \cos \frac{x}{2}$ is shown in Figure 39. The least positive solution, .72739787, is indicated on the screen. The solution 0 can be verified by direct substitution. The solution set over the interval $[0, 2\pi)$ is $\{0, .727, 2.288, 3.524, 4.189\}$. ■

An Application of Trigonometric Equations

EXAMPLE 8 Describing a Musical Tone from a Graph

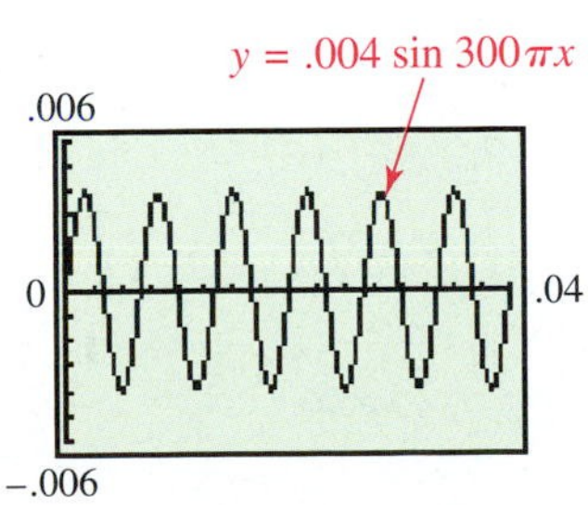

FIGURE 40

A basic component of music is a pure tone. The graph in Figure 40 models the sinusoidal pressure $y = P$ in pounds per square foot from a pure tone at time $x = t$ in seconds.

(a) The frequency of a pure tone is often measured in hertz. One *hertz* is equal to 1 cycle per second and is abbreviated Hz. What is the frequency f in hertz of the pure tone shown in the graph?

(b) The time for the tone to produce one complete cycle is called the *period.* Approximate the period T of the pure tone, in seconds.

(c) An equation for the graph is $y = .004 \sin 300\pi x$. Use a calculator to estimate all solutions of the equation $y = .004$ on the interval $[0, .02]$.

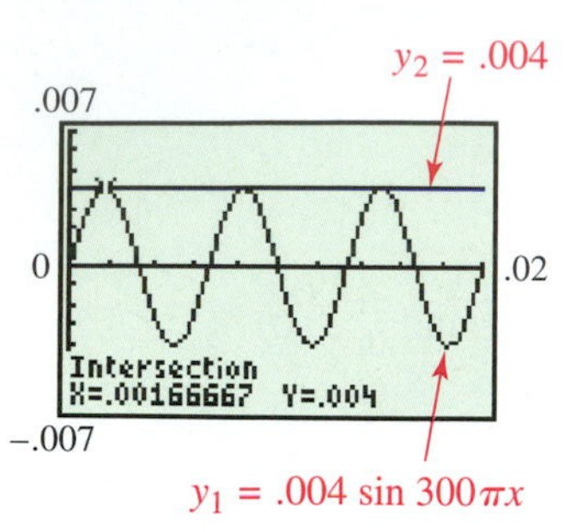

FIGURE 41

Solution

(a) From the graph in Figure 40 on the previous page, we see that there are 6 cycles in .04 second. This is equivalent to $\frac{6}{.04} = 150$ cycles per second. The pure tone has a frequency of $f = 150$ Hz.

(b) Six periods cover a time of .04 second. Therefore, the period would be equal to $T = \frac{.04}{6} = \frac{1}{150}$, or $.00\overline{6}$ second.

(c) If we reproduce the graph in Figure 40 on a calculator as y_1 and also graph the second function as $y_2 = .004$, we can determine that the approximate values of x at the points of intersection of the graphs on the interval $[0, .02]$ are

$$.0017, \quad .0083, \quad \text{and} \quad .015.$$

See Figure 41. ■

9.6 Exercises

Concept Check *Answer each question.*

1. Suppose you are solving a trigonometric equation for solutions in $[0, 2\pi)$ and your work leads to

$$2x = \frac{2\pi}{3}, 2\pi, \frac{8\pi}{3}.$$

What are the corresponding values of x?

2. Suppose you are solving a trigonometric equation to find solutions in $[0°, 360°)$ and your work leads to

$$\frac{1}{3}\theta = 45°, 60°, 75°, 90°.$$

What are the corresponding values of θ?

Solve each equation in part (a) analytically over the interval $[0, 2\pi)$. Then use a graph to solve each inequality in part (b).

3. (a) $\cos 2x = \frac{\sqrt{3}}{2}$
(b) $\cos 2x > \frac{\sqrt{3}}{2}$

4. (a) $\cos 2x = -\frac{1}{2}$
(b) $\cos 2x > -\frac{1}{2}$

5. (a) $\sin 3x = -1$
(b) $\sin 3x < -1$

6. (a) $\sin 3x = 0$
(b) $\sin 3x < 0$

7. (a) $\sqrt{2} \cos 2x = -1$
(b) $\sqrt{2} \cos 2x \le -1$

8. (a) $2\sqrt{3} \sin 2x = \sqrt{3}$
(b) $2\sqrt{3} \sin 2x \le \sqrt{3}$

9. (a) $\sin \frac{x}{2} = \sqrt{2} - \sin \frac{x}{2}$
(b) $\sin \frac{x}{2} > \sqrt{2} - \sin \frac{x}{2}$

10. (a) $\sin x = \sin 2x$
(b) $\sin x > \sin 2x$

Solve each equation over the interval $[0, 2\pi)$.

11. $\sin \frac{x}{2} = \cos \frac{x}{2}$

12. $\sec \frac{x}{2} = \cos \frac{x}{2}$

13. $\sin^2 \frac{x}{2} - 1 = 0$

14. $\sin x \cos x = \frac{1}{4}$

15. $\sin 2x = 2 \cos^2 x$

16. $\csc^2 \frac{x}{2} = 2 \sec x$

17. $\cos x - 1 = \cos 2x$

18. $1 - \sin x = \cos 2x$

Solve each equation over the interval $[0°, 360°)$.

19. $\sqrt{2} \sin 3\theta - 1 = 0$

20. $-2 \cos 2\theta = \sqrt{3}$

21. $\cos \frac{\theta}{2} = 1$

22. $\sin \frac{\theta}{2} = 1$

23. $2\sqrt{3} \sin \frac{\theta}{2} = 3$

24. $2\sqrt{3} \cos \frac{\theta}{2} = -3$

25. $2 \sin \theta = 2 \cos 2\theta$

26. $\cos \theta = \sin^2 \frac{\theta}{2}$

27. $2 - \sin 2\theta = 4 \sin 2\theta$

28. $4 \cos 2\theta = 8 \sin \theta \cos \theta$

29. $2 \cos^2 2\theta = 1 - \cos 2\theta$

30. $\sin \theta - \sin 2\theta = 0$

Use a graphical method to solve each equation over the interval $[0, 2\pi)$. Round values to the nearest thousandth.

31. $\sin x + \sin 3x = \cos x$

32. $\sin 3x - \sin x = 0$

33. $\cos 2x + \cos x = 0$

34. $\sin 4x + \sin 2x = 2 \cos x$

35. $\cos \frac{x}{2} = 2 \sin 2x$

36. $\sin \frac{x}{2} + \cos 3x = 0$

37. What is wrong with the following solution? Solve the equation $\tan 2\theta = 2$ over the interval $[0, 2\pi)$.

$$\tan 2\theta = 2$$

$$\frac{\tan 2\theta}{2} = \frac{2}{2}$$

$$\tan \theta = 1$$

$$\theta = \frac{\pi}{4} \quad \text{or} \quad \theta = \frac{5\pi}{4}$$

The solutions are $\frac{\pi}{4}$ and $\frac{5\pi}{4}$.

38. The equation $\cot \frac{x}{2} - \csc \frac{x}{2} - 1 = 0$ has no solution over the interval $[0, 2\pi)$. From this information, what can be said about the graph of

$$y = \cot \frac{x}{2} - \csc \frac{x}{2} - 1$$

over this interval? Confirm your answer by actually graphing the function over the interval.

(Modeling) *Solve each problem.*

39. ***Inducing Voltage*** A coil of wire rotating in a magnetic field induces a voltage given by

$$e = 20 \sin\left(\frac{\pi t}{4} - \frac{\pi}{2}\right),$$

where t is time in seconds. Find the smallest positive time to produce each voltage.

(a) 0 **(b)** $10\sqrt{3}$

40. ***Nautical Mile*** The British nautical mile is defined as the length of a minute of arc of a meridian. Since Earth is flat at its poles, the nautical mile, in feet, is given by

$$L = 6077 - 31 \cos 2\theta,$$

where θ is the latitude in degrees. (See the figure.) (*Source:* Bushaw, Donald et al., *A Sourcebook of Applications of School Mathematics.* Copyright © 1980 by The Mathematical Association of America. Reprinted by permission.)

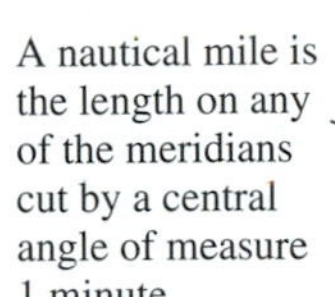

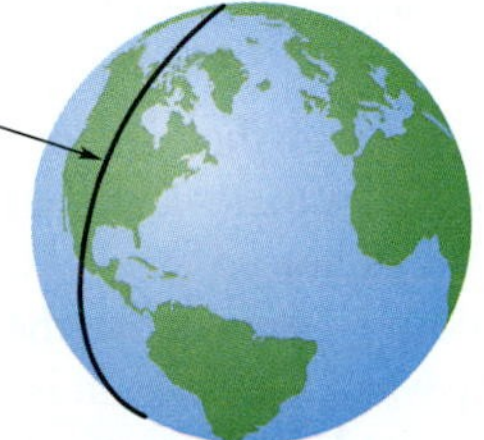

(a) Find the latitude between 0° and 90° at which the nautical mile is 6074 feet.

(b) At what latitude between 0° and 90° is the nautical mile 6108 feet?

(c) In the United States, the nautical mile is defined everywhere as 6080.2 feet. At what latitude between 0° and 90° does this agree with the British nautical mile?

41. ***Ear Pressure from a Pure Tone*** No musical instrument can generate a true pure tone, which has a unique, constant frequency and amplitude that sounds rather dull and uninteresting. The pressures caused by pure tones on the eardrum are sinusoidal. The change in pressure P in pounds per square feet on a person's eardrum from a pure tone at time t in seconds can be modeled by the equation

$$P = A \sin(2\pi f t + \phi),$$

where f is the frequency in cycles per second and ϕ is the phase angle. When P is positive, there is an increase in pressure and the eardrum is pushed inward; when P is negative, there is a decrease in pressure and the eardrum is pushed outward. (*Source:* Roederer, J., *Introduction to the Physics and Psychophysics of Music,* The English Universities Press, 1973.)

(a) Middle C has frequency 261.63 cycles per second. Graph this tone with $A = .004$ and $\phi = \frac{\pi}{7}$ in the window $[0, .005]$ by $[-.005, .005]$.

(b) Determine analytically the values of t for which $P = 0$ on $[0, .005]$, and support your answers graphically.

(c) Determine graphically when $P < 0$ on $[0, .005]$.

(d) Would an eardrum hearing this tone be vibrating outward or inward when $P < 0$?

42. ***Ear Pressure from a Vibrating String*** If a string with a fundamental frequency of 110 Hz is plucked in the middle, it will vibrate at the odd harmonics of 110, 330, 550, ... Hz but not at the even harmonics of 220, 440, 660, ... Hz. The resulting pressure P caused by the string can be approximated with the equation

$$P = .003 \sin 220\pi t + \frac{.003}{3} \sin 660\pi t + \frac{.003}{5} \sin 1100\pi t + \frac{.003}{7} \sin 1540\pi t.$$

(*Source:* Benade, A., *Fundamentals of Musical Acoustics,* Oxford University Press, 1976; Roederer, J., *Introduction to the Physics and Psychophysics of Music*, The English Universities Press, 1973.)

(a) Graph P in the window $[0, .03]$ by $[-.005, .005]$.

(b) Use the graph to describe the shape of the sound wave that is produced.

(c) At lower frequencies, the inner ear will hear a tone only when the eardrum is moving outward. (See Exercise 41.) Determine the times on the interval $[0, .03]$ when this will occur.

43. ***Hearing Beats in Music*** Musicians sometimes tune instruments by playing the same tone on two different instruments and listening for a phenomenon known as *beats*. Beats occur when two tones vary in frequency by only a few hertz. When the two instruments are in tune, the beats disappear. The ear hears beats because the pressure slowly rises and falls as a result of the slight variation in the frequency. This phenomenon can be seen on a graphing calculator. (*Source:* Pierce, J., *The Science of Musical Sound,* Scientific American Books, 1992.)

(a) Consider two tones with frequencies of 220 and 223 Hz and pressures

$$P_1 = .005 \sin 440\pi t \quad \text{and} \quad P_2 = .005 \sin 446\pi t,$$

respectively. Graph the pressure $P = P_1 + P_2$ felt by an eardrum over the 1-second interval $[.15, 1.15]$. How many beats are there in 1 second?

(b) Repeat part (a) with frequencies of 220 and 216.

(c) Determine a simple way to find the number of beats per second if the frequency of each tone is given.

44. ***Hearing Different Tones*** Small speakers like those found in older radios and telephones often cannot vibrate slower than 200 Hz—yet 35 keys on a piano have frequencies below 200 Hz. When a musical instrument creates a tone of 110 Hz, it also creates tones at 220, 330, 440, 550, 660, ... Hz. A small speaker cannot reproduce the 110-Hz vibration, but it can reproduce the higher frequencies, which are called the upper harmonics. The low tones can still be heard, because the speaker produces *difference tones* of the upper harmonics. The difference between consecutive frequencies is 110 Hz, and this difference tone will be heard by a listener. We can model this phenomenon with a graphing calculator. (*Source:* Benade, A., *Fundamentals of Musical Acoustics,* Oxford University Press, 1976.)

(a) In the window $[0, .03]$ by $[-2, 2]$, graph the upper harmonics represented by the pressure

$$P = \frac{1}{2} \sin[2\pi(220)t] + \frac{1}{3} \sin[2\pi(330)t] + \frac{1}{4} \sin[2\pi(440)t].$$

(b) Estimate all t-coordinates where P is maximum.

(c) What does a person hear in addition to the frequencies of 220, 330, and 440 Hz?

(d) Graph the pressure produced by a speaker that can vibrate at 110 Hz and above.

Reviewing Basic Concepts (Sections 9.5 and 9.6)

Solve each equation over the interval $[0, 2\pi)$.

1. $\cos 2x = \frac{\sqrt{3}}{2}$

2. $2 \sin x + 1 = 0$

3. $(\tan x - 1)(\cos x - 1) = 0$

4. $2 \cos^2 x = \sqrt{3} \cos x$

Solve each equation over the interval $[0°, 360°)$. Round values to the nearest tenth when exact values cannot be determined.

5. $3 \cot^2 \theta - 3 \cot \theta = 1$

6. $4 \cos^2 \theta + 4 \cos \theta - 1 = 0$

7. $2 \sin \theta - 1 = \csc \theta$

8. $\sec^2 \dfrac{\theta}{2} = 2$

Solve each equation graphically over the interval $[0, 2\pi)$.

9. $x^2 + \sin x - x^3 - \cos x = 0$

10. $x^3 - \cos^2 x = \dfrac{1}{2}x - 1$

CHAPTER 9 SUMMARY

KEY TERMS & SYMBOLS	KEY CONCEPTS
9.1 Trigonometric Identities	**FUNDAMENTAL IDENTITIES** **Reciprocal Identities** $\cot \theta = \dfrac{1}{\tan \theta}$ $\sec \theta = \dfrac{1}{\cos \theta}$ $\csc \theta = \dfrac{1}{\sin \theta}$ **Quotient Identities** $\tan \theta = \dfrac{\sin \theta}{\cos \theta}$ $\cot \theta = \dfrac{\cos \theta}{\sin \theta}$ **Pythagorean Identities** $\sin^2 \theta + \cos^2 \theta = 1$ $1 + \tan^2 \theta = \sec^2 \theta$ $1 + \cot^2 \theta = \csc^2 \theta$ **Negative-Number Identities** $\sin(-\theta) = -\sin \theta$ $\cos(-\theta) = \cos \theta$ $\tan(-\theta) = -\tan \theta$ $\csc(-\theta) = -\csc \theta$ $\sec(-\theta) = \sec \theta$ $\cot(-\theta) = -\cot \theta$
9.2 Sum and Difference Identities	**SUM AND DIFFERENCE IDENTITIES** $\cos(A - B) = \cos A \cos B + \sin A \sin B$ $\cos(A + B) = \cos A \cos B - \sin A \sin B$ $\sin(A + B) = \sin A \cos B + \cos A \sin B$ $\sin(A - B) = \sin A \cos B - \cos A \sin B$ $\tan(A + B) = \dfrac{\tan A + \tan B}{1 - \tan A \tan B}$ $\tan(A - B) = \dfrac{\tan A - \tan B}{1 + \tan A \tan B}$
9.3 Further Identities	**DOUBLE-NUMBER IDENTITIES** $\cos 2A = \cos^2 A - \sin^2 A$ $\cos 2A = 1 - 2 \sin^2 A$ $\cos 2A = 2 \cos^2 A - 1$ $\sin 2A = 2 \sin A \cos A$ $\tan 2A = \dfrac{2 \tan A}{1 - \tan^2 A}$ *(continued)*

KEY TERMS & SYMBOLS	KEY CONCEPTS

PRODUCT-TO-SUM IDENTITIES

$$\cos A \cos B = \frac{1}{2}[\cos(A+B) + \cos(A-B)]$$

$$\sin A \sin B = \frac{1}{2}[\cos(A-B) - \cos(A+B)]$$

$$\sin A \cos B = \frac{1}{2}[\sin(A+B) + \sin(A-B)]$$

$$\cos A \sin B = \frac{1}{2}[\sin(A+B) - \sin(A-B)]$$

SUM-TO-PRODUCT IDENTITIES

$$\sin A + \sin B = 2\sin\left(\frac{A+B}{2}\right)\cos\left(\frac{A-B}{2}\right)$$

$$\sin A - \sin B = 2\cos\left(\frac{A+B}{2}\right)\sin\left(\frac{A-B}{2}\right)$$

$$\cos A + \cos B = 2\cos\left(\frac{A+B}{2}\right)\cos\left(\frac{A-B}{2}\right)$$

$$\cos A - \cos B = -2\sin\left(\frac{A+B}{2}\right)\sin\left(\frac{A-B}{2}\right)$$

HALF-NUMBER IDENTITIES

$$\cos\frac{A}{2} = \pm\sqrt{\frac{1+\cos A}{2}} \qquad \sin\frac{A}{2} = \pm\sqrt{\frac{1-\cos A}{2}}$$

$$\tan\frac{A}{2} = \pm\sqrt{\frac{1-\cos A}{1+\cos A}} \qquad \tan\frac{A}{2} = \frac{\sin A}{1+\cos A} \qquad \tan\frac{A}{2} = \frac{1-\cos A}{\sin A}$$

(The sign is chosen based on the quadrant of $\frac{A}{2}$.)

9.4 The Inverse Circular Functions

$\sin^{-1} x$ or $\arcsin x$
$\cos^{-1} x$ or $\arccos x$
$\tan^{-1} x$ or $\arctan x$
$\cot^{-1} x$ or $\text{arccot}\, x$
$\sec^{-1} x$ or $\text{arcsec}\, x$
$\csc^{-1} x$ or $\text{arccsc}\, x$

INVERSE CIRCULAR FUNCTIONS

Function	Domain	Range: Interval	Range: Quadrants of the Unit Circle
$y = \sin^{-1} x$	$[-1, 1]$	$\left[-\frac{\pi}{2}, \frac{\pi}{2}\right]$	I and IV
$y = \cos^{-1} x$	$[-1, 1]$	$[0, \pi]$	I and II
$y = \tan^{-1} x$	$(-\infty, \infty)$	$\left(-\frac{\pi}{2}, \frac{\pi}{2}\right)$	I and IV
$y = \cot^{-1} x$	$(-\infty, \infty)$	$(0, \pi)$	I and II
$y = \sec^{-1} x$	$(-\infty, -1] \cup [1, \infty)$	$\left[0, \frac{\pi}{2}\right) \cup \left(\frac{\pi}{2}, \pi\right]$	I and II
$y = \csc^{-1} x$	$(-\infty, -1] \cup [1, \infty)$	$\left[-\frac{\pi}{2}, 0\right) \cup \left(0, \frac{\pi}{2}\right]$	I and IV

KEY TERMS & SYMBOLS	KEY CONCEPTS
9.5 Trigonometric Equations and Inequalities (I)	**SOLVING A TRIGONOMETRIC EQUATION ANALYTICALLY** **1.** Decide whether the equation is linear or quadratic, so you can determine the solution method. **2.** If only one trigonometric function is present, first solve the equation for that function. **3.** If more than one trigonometric function is present, rearrange the equation so that one side equals 0. Then try to factor and set each factor equal to 0 to solve. **4.** If the equation is quadratic in form, but not factorable, use the quadratic formula. Check that solutions are in the desired interval. **5.** Try using identities to change the form of the equation. It may be helpful to square each side of the equation first. If this is done, check for extraneous values. **SOLVING A TRIGONOMETRIC EQUATION GRAPHICALLY** **1.** For an equation of the form $f(x) = g(x)$, use the intersection-of-graphs method. **2.** For an equation of the form $f(x) = 0$, use the x-intercept method.
9.6 Trigonometric Equations and Inequalities (II)	Use the multiple-number and half-number identities to rewrite the equations. Then use the steps given in Section 9.5.

CHAPTER 9 Review Exercises

1. Give all the trigonometric functions f that satisfy the condition $f(-x) = -f(x)$.

2. Give all the trigonometric functions f that satisfy the condition $f(-x) = f(x)$.

Use the negative-number identities to write each expression as a function of a positive number.

3. $\cos(-3)$ **4.** $\sin(-3)$ **5.** $\tan(-3)$ **6.** $\sec(-3)$ **7.** $\csc(-3)$ **8.** $\cot(-3)$

Concept Check *For each expression in Group I, give the letter of the expression in Group II that completes an identity.*

I		II	
9. $\sec x =$ ______	**10.** $\tan x =$ ______	**A.** $\frac{1}{\sin x}$	**B.** $\frac{1}{\cos x}$
11. $\cot x =$ ______	**12.** $\tan^2 x + 1 =$ ______	**C.** $\frac{\sin x}{\cos x}$	**D.** $\frac{1}{\cot^2 x}$
13. $\tan^2 x =$ ______	**14.** $\csc x =$ ______	**E.** $\frac{1}{\cos^2 x}$	**F.** $\frac{\cos x}{\sin x}$

Use identities to write each expression in terms of $\sin\theta$ *and* $\cos\theta$, *and simplify.*

15. $\dfrac{\cot\theta}{\sec\theta}$ **16.** $\tan^2\theta(1 + \cot^2\theta)$ **17.** $\csc\theta + \cot\theta$

18. Use the trigonometric identities to find the remaining five trigonometric functions of x, given that $\cos x = \frac{3}{5}$ and x is in quadrant IV.

19. Given that $\tan x = -\frac{5}{4}$, where $\frac{\pi}{2} < x < \pi$, use the trigonometric identities to find the other trigonometric functions of x.

Concept Check *For each expression in Group I, choose the expression from Group II with the same value.*

I

20. $\cos 210°$
21. $\sin 35°$
22. $\tan(-35°)$
23. $-\sin 35°$
24. $\cos 35°$
25. $\cos 75°$
26. $\sin 75°$
27. $\sin 300°$
28. $\cos 300°$
29. $\tan(-55°)$

II

A. $\sin(-35°)$
B. $\cos 55°$
C. $\sqrt{\dfrac{1 + \cos 150°}{2}}$
D. $2 \sin 150° \cos 150°$
E. $\cos 150° \cos 60° - \sin 150° \sin 60°$
F. $\cot(-35°)$
G. $\cos^2 150° - \sin^2 150°$
H. $\sin 15° \cos 60° + \cos 15° \sin 60°$
I. $\cos(-35°)$
J. $\cot 125°$

Verify analytically that each equation is an identity.

30. $\sin^2 x - \sin^2 y = \cos^2 y - \cos^2 x$

31. $2 \cos^3 x - \cos x = \dfrac{\cos^2 x - \sin^2 x}{\sec x}$

32. $-\cot \dfrac{x}{2} = \dfrac{\sin 2x + \sin x}{\cos 2x - \cos x}$

33. $\dfrac{\sin^2 x}{2 - 2 \cos x} = \cos^2 \dfrac{x}{2}$

34. $\dfrac{\sin 2x}{\sin x} = \dfrac{2}{\sec x}$

35. $2 \cos A - \sec A = \cos A - \dfrac{\tan A}{\csc A}$

36. $\dfrac{2 \tan B}{\sin 2B} = \sec^2 B$

37. $1 + \tan^2 \alpha = 2 \tan \alpha \csc 2\alpha$

38. $\dfrac{\sin t}{1 - \cos t} = \cot \dfrac{t}{2}$

39. $\dfrac{2 \cot x}{\tan 2x} = \csc^2 x - 2$

40. $\tan \theta \sin 2\theta = 2 - 2 \cos^2 \theta$

41. $2 \tan x \csc 2x - \tan^2 x = 1$

Give the exact real number value of each expression without using a calculator.

42. $\sin^{-1} \dfrac{\sqrt{2}}{2}$

43. $\arccos\left(-\dfrac{1}{2}\right)$

44. $\arctan \dfrac{\sqrt{3}}{3}$

45. $\sec^{-1}(-2)$

46. $\text{arccsc}\, \dfrac{2\sqrt{3}}{3}$

47. $\cot^{-1}(-1)$

Give the degree measure of θ. Do not use a calculator.

48. $\theta = \arccos \dfrac{1}{2}$

49. $\theta = \arcsin\left(-\dfrac{\sqrt{3}}{2}\right)$

50. $\theta = \tan^{-1} 0$

Use a calculator to give the degree of measure of θ.

51. $\theta = \arcsin(-.656059029)$

52. $\theta = \arccos .7095707365$

53. $\theta = \arctan(-.1227845609)$

54. $\theta = \cot^{-1} 4.704630109$

55. $\theta = \sec^{-1} 28.65370835$

56. $\theta = \csc^{-1} 19.10732261$

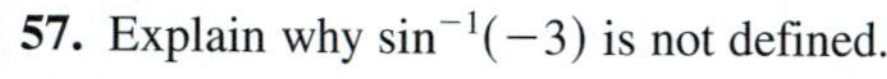

57. Explain why $\sin^{-1}(-3)$ is not defined.

58. Even though $\sin^{-1}(-3)$ is not defined, $\tan^{-1}(-3)$ is defined. Explain why this is so.

Evaluate each expression without a calculator.

59. $\sin\left(\sin^{-1} \dfrac{1}{2}\right)$

60. $\sin\left(\cos^{-1} \dfrac{3}{4}\right)$

61. $\cos(\arctan 3)$

62. $\sec\left(2 \sin^{-1}\left(-\dfrac{1}{3}\right)\right)$

63. $\cos^{-1}\left(\cos \dfrac{3\pi}{2}\right)$

64. $\tan\left(\sin^{-1} \dfrac{3}{5} + \cos^{-1} \dfrac{5}{7}\right)$

Write each expression as an algebraic expression in u, $u > 0$.

65. $\sin(\tan^{-1} u)$

66. $\cos\left(\arctan \frac{u}{\sqrt{1-u^2}}\right)$

67. $\tan\left(\arccos \frac{u}{\sqrt{u^2+1}}\right)$

Solve each equation for solutions in the interval $[0, 2\pi)$.

68. $\sin^2 x = 1$

69. $2 \tan x - 1 = 0$

70. $3 \sin^2 x - 5 \sin x + 2 = 0$

71. $\tan x = \cot x$

72. $5 \cot^2 x + 3 \cot x = 2$

73. $\sec \frac{x}{2} = \cos \frac{x}{2}$

74. $\sin 2x = \cos 2x + 1$

75. $2 \sin 2x = 1$

76. $\sin 2x + \sin 4x = 0$

77. $\cos x - \cos 2x = 2 \cos x$

78. $\tan 2x = \sqrt{3}$

79. $\cos^2 \frac{x}{2} - 2 \cos \frac{x}{2} + 1 = 0$

80. Use the results of Exercise 74 and a calculator graph of

$$f(x) = \sin 2x - \cos 2x - 1$$

to find the solution set of each inequality over the interval $[0, 2\pi)$.

(a) $\sin 2x > \cos 2x + 1$ **(b)** $\sin 2x < \cos 2x + 1$

(Modeling) *Solve each problem.*

81. ***Viewing Angle of an Observer*** A student is sitting in a desk 5 feet to the left of a 10-foot-wide blackboard, as shown in the figure. If the student is sitting x feet from the front of the classroom and has a viewing angle of θ radians, do the following.

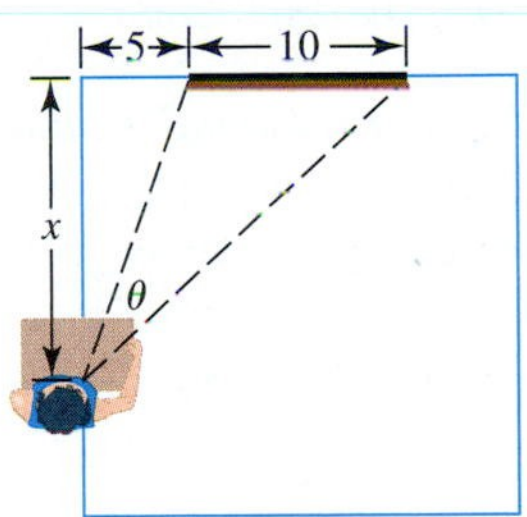

(a) Show that the value of θ is given by the function defined by $\theta = f(x)$, where

$$f(x) = \arctan\left(\frac{15}{x}\right) - \arctan\left(\frac{5}{x}\right).$$

(b) Graph this function with a graphing calculator to estimate the value of x that maximizes the viewing angle.

82. ***Motion Formed by a Moving Arm*** Suppose that the following equation describes the motion formed by a rhythmically moving arm:

$$y = \frac{1}{3} \sin \frac{4\pi t}{3}.$$

Here, t is time (in seconds) and y represents the measure of the angle formed by the upper arm in radians and a vertical line. See the figure.

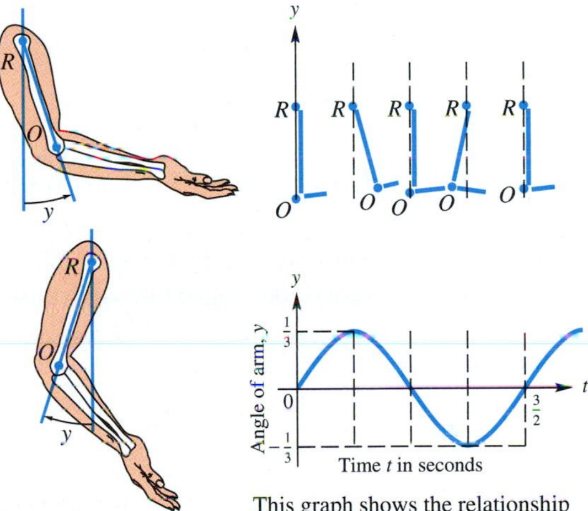

This graph shows the relationship between angle y and time t in seconds.

Source: De Sapio, Rodolfo. *Calculus for the Life Sciences.* Copyright © 1978 by W.H. Freeman and Company. Reprinted by permission.

(a) Solve the equation for t.

(b) At what time does the arm form an angle of .3 radian?

Alternating Electric Current *The study of alternating electric current requires the solutions of equations of the form* $i = I_{max} \sin 2\pi ft$, *for time t in seconds, where i is instantaneous current in amperes,* I_{max} *is maximum current in amperes, and f is the number of cycles per second. Find the smallest positive value of t, given the following data.*

83. $i = 40, I_{max} = 100, f = 60$

84. $i = 50, I_{max} = 100, f = 120$

85. ***Speed of Light in Different Mediums*** Snell's law states that

$$\frac{c_1}{c_2} = \frac{\sin \theta_1}{\sin \theta_2},$$

where c_1 is the speed of light in one medium, c_2 is the speed of light in a second medium, and θ_1 and θ_2 are the angles shown in the figure. Suppose a light is shining up through water into the air as in the figure. As θ_1 increases, θ_2 approaches 90°, at which point no light will emerge from the water. Assume that the ratio $\frac{c_1}{c_2}$ in this case is .752. For what value of θ_1 does $\theta_2 = 90°$? This value of θ_1 is called the *critical angle* for water.

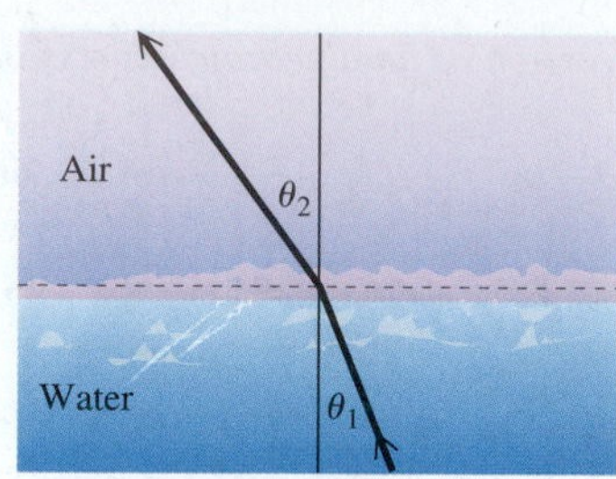

86. ***Speed of Light in Different Mediums*** Refer to Exercise 85. What happens when θ_1 is greater than the critical angle?

CHAPTER 9 Test

Remember to set your calculator for radian or degree measure as required.

Given $\sin y = -\frac{2}{3}$, $\cos x = -\frac{1}{5}$, $\frac{\pi}{2} < x < \pi$, *and* $\pi < y < \frac{3\pi}{2}$, *find the exact values for the following.*

1. $\sin(x + y)$

2. $\cos(x - y)$

3. $\tan \frac{y}{2}$

4. $\cos 2x$

5. Express $\tan^2 x - \sec^2 x$ in terms of $\sin x$ and $\cos x$, and simplify.

6. Graph $y = \sec x - \sin x \tan x$, and use the graph to conjecture an identity. Verify your conjecture analytically.

Verify that each equation is an identity.

7. $\sec^2 B = \dfrac{1}{1 - \sin^2 B}$

8. $\cos 2A = \dfrac{\cot A - \tan A}{\csc A \sec A}$

Use an identity to write each expression as a trigonometric function of θ alone.

9. $\cos(270° - \theta)$

10. $\sin(\pi + \theta)$

11. Consider the function defined by $f(x) = -2 \sin^{-1} x$.

(a) Sketch the graph.

(b) Give the domain and range.

(c) Explain why $f(2)$ is not defined.

Find the exact real number value of each expression.

12. $\arccos\left(-\frac{1}{2}\right)$

13. $\tan^{-1} 0$

14. $\csc^{-1} \dfrac{2\sqrt{3}}{3}$

15. $\sin^{-1}\left(\sin \dfrac{5\pi}{6}\right)$

16. $\cos\left(\arcsin \dfrac{2}{3}\right)$

17. $\sin\left(2 \cos^{-1} \dfrac{1}{3}\right)$

Write each expression as an algebraic expression in u, u > 0.

18. $\sec(\cos^{-1} u)$

19. $\tan(\arcsin u)$

Solve each equation or inequality over the indicated interval.

20. $\sin^2 \theta = \cos^2 \theta + 1$, $[0, 2\pi)$

21. $\csc^2 \theta - 2 \cot \theta = 4$, $[0°, 360°)$

22. $\cos x = \cos 2x$, $[0, 2\pi)$

23. $2\sqrt{3} \sin \dfrac{\theta}{2} = 3$, $[0°, 360°)$

24. $2 \sin x - 1 \le 0$, $[0, 2\pi)$

25. ***(Modeling) Migratory Animal Count*** The number of migratory animals (in hundreds) seen in one year and counted at a certain checkpoint is given by

$$T(t) = 50 + 50 \cos\left(\frac{\pi}{6} t\right),$$

where t is time in months, with $t = 0$ corresponding to July.

(a) Give the domain and range of T.

(b) Graph $T(t)$ over its domain.

(c) Use the graph to determine the maximum and minimum values of T and when they occur.

(d) Find $T(3)$ analytically, and support your result graphically. Use symmetry to find $T(9)$.

(e) When does the count reach 75 hundred animals?

(f) Write the equation

$$T = 50 + 50 \cos\left(\frac{\pi}{6} t\right)$$

as an equation involving arccosine by solving for t.

CHAPTER 9 PROJECT

Modeling a Damped Pendulum

A damped pendulum oscillates back and forth. The period (time for one back and forth oscillation) remains constant, but the amplitude diminishes at a constant rate. Therefore, the composition of an exponential function multiplied by a sine function provides a mathematical formula that models the pendulum's behavior. While several forms are possible for such a function, we use

$$y = r^x a \sin[b(x - d)] + c,$$

where x represents time, a denotes amplitude, b designates $\frac{2\pi}{\text{period}}$, c stands for vertical shift, d symbolizes horizontal phase shift, and r signifies the rate at which the amplitude diminishes.

The data in the table are for an object oscillating for 10 seconds. The data were gathered with a CBR data collector.

Time (seconds)	Distance (from probe in feet)	Time (seconds)	Distance (from probe in feet)	Time (seconds)	Distance (from probe in feet)	Time (seconds)	Distance (from probe in feet)	Time (seconds)	Distance (from probe in feet)
.1	2.11	2.1	2.77	4.1	3.20	6.1	3.40	8.1	3.35
.2	2.34	2.2	2.97	4.2	3.37	**6.2**	**3.43**	8.2	3.28
.3	2.62	2.3	3.23	4.3	3.47	6.3	3.41	8.3	3.17
.4	2.96	2.4	3.44	**4.4**	**3.52**	6.4	3.35	8.4	3.04
.5	3.28	2.5	3.56	4.5	3.50	6.5	3.21	8.5	2.90
.6	3.54	**2.6**	**3.61**	4.6	3.42	6.6	3.07	8.6	2.77
.7	3.69	2.7	3.59	4.7	3.29	6.7	2.91	8.7	2.66
.8	**3.81**	2.8	3.51	4.8	3.11	6.8	2.75	8.8	2.59
.9	3.78	2.9	3.37	4.9	2.92	6.9	2.62	**8.9**	**2.56**
1.0	3.65	3.0	3.15	5.0	2.74	7.0	2.54	9.0	2.59
1.1	3.48	3.1	2.93	5.1	2.57	**7.1**	**2.50**	9.1	2.65
1.2	3.24	3.2	2.69	5.2	2.46	7.2	2.54	9.2	2.76
1.3	2.95	3.3	2.59	**5.3**	**2.42**	7.3	2.60	9.3	2.88
1.4	2.64	3.4	2.37	5.4	2.44	7.4	2.72	9.4	3.01
1.5	2.41	**3.5**	**2.33**	5.5	2.55	7.5	2.87	9.5	3.13
1.6	2.28	3.6	2.34	5.6	2.66	7.6	3.01	9.6	3.23
1.7	**2.17**	3.7	2.43	5.7	2.83	7.7	3.15	9.7	3.30
1.8	2.20	3.8	2.58	5.8	3.00	7.8	3.27	**9.8**	**3.32**
1.9	2.37	3.9	2.77	5.9	3.17	7.9	3.35	9.9	3.30
2.0	2.46	4.0	2.99	6.0	3.31	**8.0**	**3.37**	10.0	3.23

Figure A is a scatter diagram of the data.

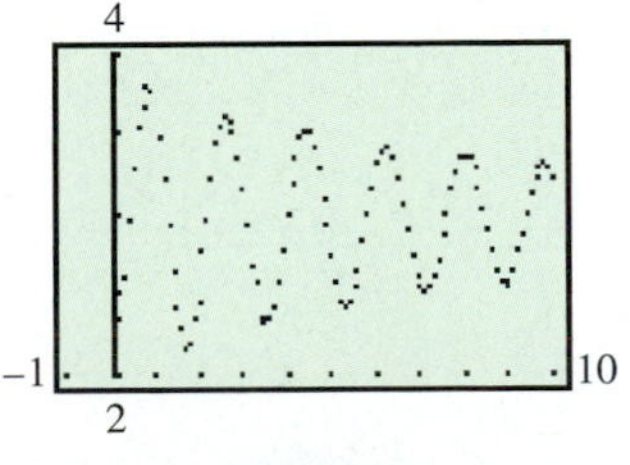

FIGURE A

Activities

1. Reproduce the scatter diagram in Figure A, but to avoid entering so many data points into your calculator by hand, enter only those that correspond to the maximum and minimum distances (printed in boldface type). Use the window $[0, 10]$ by $[-1, 5]$ with $\text{Xscl} = \text{Yscl} = 1$.
2. Find a function that models the data by determining the values of a, b, c, d, and r in the equation given earlier. (*Hints:* **(a)** Find the amplitude by calculating the distance between the extreme heights and dividing by 2. **(b)** Find the period by calculating the time difference between successive maxima (or minima). **(c)** Find the horizontal phase shift by finding the time of the first data point at the mean height. **(d)** Find the rate at which the amplitude diminishes by calculating the ratios of the first two consecutive maxima.)
3. Graph the function from Activity 2 together with your scatter diagram. Is it a good fit?
4. If you have access to a CBL or CBR data collector, work in small groups to gather your own data. For the pendulum, use any object that swings or vibrates. The data given earlier were gathered by clamping a meterstick to a secure beam and then pulling and releasing the free end of the stick to cause it to vibrate.

10

Applications of Trigonometry; Vectors

WORLD TECHNOLOGY TOOK a giant step forward when the *Global Positioning System (GPS),* a network of more than 24 communications satellites, achieved full operational capability on July 17, 1995. GPS satellites transmit signals that are picked up by GPS receivers on Earth. Computers in the receivers use the transmitted signals from satellites to calculate exact latitude (north–south) and longitude (east–west) coordinates of the receiver, yielding accuracy within 3 feet.

GPS satellites trace nearly circular orbits eleven hours, 58 minutes, in duration on six different paths around the globe. The number of satellites and diversity of paths allow visibility of 4 to 10 satellites from any point in the world at any given time. The satellites transmit signals composed of two carrier frequencies, two digital codes, and a navigation message, from an altitude 12,552 miles above Earth's surface. GPS receivers on Earth determine their location from signal data received simultaneously from any four of the satellites.

The magnitude and direction of four arrows, or *vectors,* representing signals from four GPS satellites to a receiver on Earth, are used to calculate location on the globe. Navigators use a similar calculation process with three vectors, called *triangulation,* to determine the location of a ship or an airplane with reference to three known points on Earth's surface.

In this chapter, we use vectors to solve problems involving navigation, weight and force balancing, and distance and angle determination, among others.

Source: A. El-Rabbany, *Introduction to GPS, The Global Positioning System.*

Chapter Outline

10.1 The Law of Sines

The concepts developed in Chapter 8 for right triangles can be extended to *all* triangles. Every triangle has three sides and three angles. If any three of the six measures of a triangle are known (provided that at least one measure is the length of a side), then the other three measures can be found. This is called *solving the triangle.*

Congruency and Oblique Triangles

The following axioms from geometry allow us to prove that two triangles are congruent (that is, their corresponding sides and angles are equal).

Congruence Axioms

Side–Angle–Side (SAS)	If two sides and the included angle of one triangle are equal, respectively, to two sides and the included angle of a second triangle, then the triangles are congruent.
Angle–Side–Angle (ASA)	If two angles and the included side of one triangle are equal, respectively, to two angles and the included side of a second triangle, then the triangles are congruent.
Side–Side–Side (SSS)	If three sides of one triangle are equal, respectively, to three sides of a second triangle, then the triangles are congruent.

Keep in mind that whenever SAS, ASA, or SSS is given, the triangle is unique. We continue to label triangles as we did earlier with right triangles: side a opposite angle A, side b opposite angle B, and side c opposite angle C.

A triangle that is not a right triangle is called an **oblique triangle.** The measures of the three sides and the three angles of a triangle can be found if at least one side and any other two measures are known. There are four possible cases.

Data Required for Solving Oblique Triangles

Case 1 One side and two angles are known (SAA or ASA).

Case 2 Two sides and one angle not included between the two sides are known (SSA). This case may lead to more than one triangle.

Case 3 Two sides and the angle included between the two sides are known (SAS).

Case 4 Three sides are known (SSS).

NOTE ***If we know three angles of a triangle, we cannot find unique side lengths, since AAA assures us only of similarity, not congruence.*** For example, there are infinitely many triangles ABC with $A = 35°$, $B = 65°$, and $C = 80°$.

Cases 1 and 2, discussed in this section, require the *law of sines*. Cases 3 and 4, discussed in the next section, require the *law of cosines*.

Derivation of the Law of Sines

To derive the law of sines, we start with an oblique triangle, such as the *acute triangle* in Figure 1(a) or the *obtuse triangle* in Figure 1(b). This discussion applies to both triangles. First, construct the perpendicular from B to side AC (or its extension). Let h be the length of this perpendicular. Then c is the hypotenuse of right triangle ADB, and a is the hypotenuse of right triangle BDC. By results from Chapter 8,

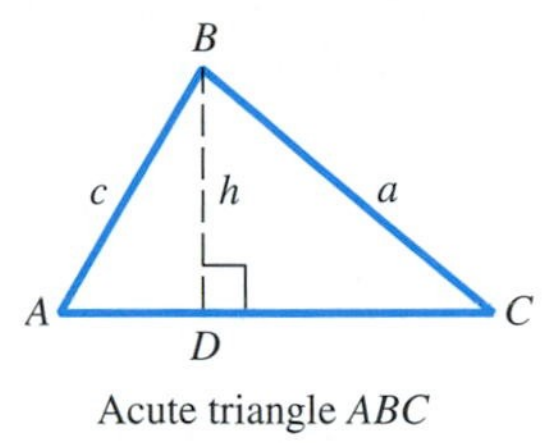

Acute triangle ABC

(a)

Obtuse triangle ABC

(b)

FIGURE 1

$$\text{in triangle } ADB, \qquad \sin A = \frac{h}{c}, \quad \text{or} \quad h = c \sin A;$$

$$\text{in triangle } BDC, \qquad \sin C = \frac{h}{a}, \quad \text{or} \quad h = a \sin C.$$

Since $h = c \sin A$ and $h = a \sin C$,

$$a \sin C = c \sin A$$

$$\frac{a}{\sin A} = \frac{c}{\sin C}. \qquad \text{Divide each side by } \sin A \sin C.$$

In a similar way, by constructing perpendicular lines from the other vertices, it can be shown that

$$\frac{a}{\sin A} = \frac{b}{\sin B} \quad \text{and} \quad \frac{b}{\sin B} = \frac{c}{\sin C}.$$

This discussion proves the following theorem.

Law of Sines

In any triangle ABC with sides a, b, and c,

$$\frac{a}{\sin A} = \frac{b}{\sin B}, \qquad \frac{a}{\sin A} = \frac{c}{\sin C}, \qquad \text{and} \qquad \frac{b}{\sin B} = \frac{c}{\sin C}.$$

This can be written in compact form as

$$\frac{a}{\sin A} = \frac{b}{\sin B} = \frac{c}{\sin C}.$$

When solving for an angle, we use an alternative form of the law of sines:

$$\frac{\sin A}{a} = \frac{\sin B}{b} = \frac{\sin C}{c}. \qquad \text{Alternative form}$$

NOTE In using the law of sines, a good strategy is to select a form so that the unknown variable is in the numerator with all other variables known.

Applications of Triangles

If two angles and a side are known (Case 1, SAA or ASA), then the law of sines can be used to solve the triangle.

EXAMPLE 1 **Using the Law of Sines to Solve a Triangle (SAA)**

Solve triangle ABC if $A = 32.0°$, $B = 81.8°$, and $a = 42.9$ centimeters.

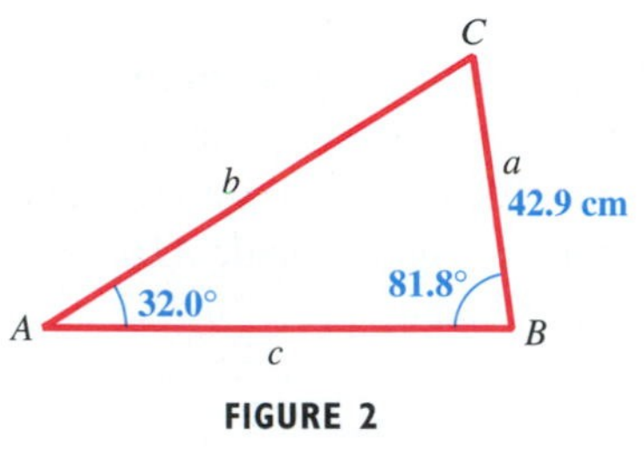

FIGURE 2

Solution Draw a triangle, roughly to scale, and label the given parts as in Figure 2. Since the values of A, B, and a are known, we use the form of the law of sines that involves these variables and then solve for b.

Choose a form that has the unknown variable in the numerator.

$$\frac{a}{\sin A} = \frac{b}{\sin B} \quad \text{Law of sines}$$

$$\frac{42.9}{\sin 32.0°} = \frac{b}{\sin 81.8°} \quad \text{Substitute the given values.}$$

$$b = \frac{42.9 \sin 81.8°}{\sin 32.0°} \quad \text{Multiply by } \sin 81.8°\text{; rewrite.}$$

$$b = 80.1 \text{ centimeters} \quad \text{Approximate with a calculator.}$$

To find C, use the fact that the sum of the angles of any triangle is 180°.

$$A + B + C = 180°$$

$$C = 180° - A - B$$

$$C = 180° - 32.0° - 81.8° = 66.2°$$

Use the law of sines to find c. (Why does the Pythagorean theorem not apply?)

$$\frac{a}{\sin A} = \frac{c}{\sin C} \quad \text{Law of sines}$$

$$\frac{42.9}{\sin 32.0°} = \frac{c}{\sin 66.2°} \quad \text{Substitute.}$$

$$c = \frac{42.9 \sin 66.2°}{\sin 32.0°} \quad \text{Multiply by } \sin 66.2°\text{; rewrite.}$$

$$c = 74.1 \text{ centimeters} \quad \text{Approximate with a calculator.}$$

■

NOTE In solving triangles in this chapter, we will often use the equality symbol, =, despite the fact that the values shown are actually approximations.

EXAMPLE 2 **Using the Law of Sines in an Application (ASA)**

A surveyor wishes to measure the distance across the Big Muddy River. See Figure 3. The surveyor determines that $C = 112.90°$, $A = 31.10°$, and $b = 347.6$ feet. Find the distance a.

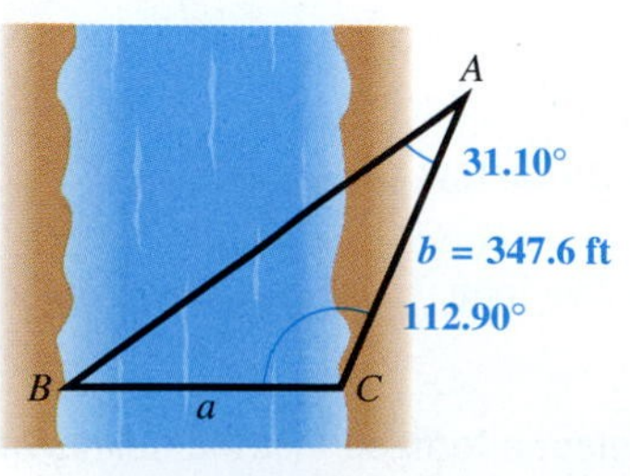

FIGURE 3

Solution To use the law of sines, one side and the angle opposite it must be known. Since b is the only side whose length is given, angle B must be found first.

$$B = 180° - A - C$$

$$B = 180° - 31.10° - 112.90° = 36.00°$$

GCM TECHNOLOGY NOTE

Programs can be written to solve triangles using the methods of this chapter. Triangle *ABC* of Example 2 is solved below, with values to the nearest tenth, using a program based on ASA.

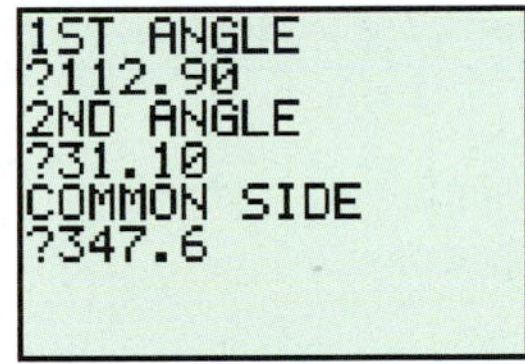

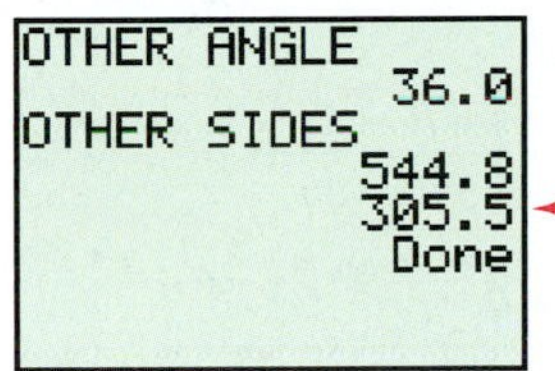

Now use the form of the law of sines involving A, B, and b to find a.

$$\frac{a}{\sin A} = \frac{b}{\sin B} \qquad \text{Law of sines}$$

$$\frac{a}{\sin 31.10^\circ} = \frac{347.6}{\sin 36.00^\circ} \qquad \text{Substitute.}$$

$$a = \frac{347.6 \sin 31.10^\circ}{\sin 36.00^\circ} \qquad \text{Multiply by } \sin 31.10^\circ.$$

$$a = 305.5 \text{ feet} \qquad \text{Use a calculator to approximate.}$$

The next example involves the concept of bearing, first discussed in Chapter 8.

EXAMPLE 3 Using the Law of Sines in an Application (ASA)

Two ranger stations are on an east–west line 110 mi apart. A forest fire is located on a bearing of N 42° E from the western station at A and a bearing of N 15° E from the eastern station at B. How far is the fire from the western station?

Solution Figure 4 shows the two stations at points A and B and the fire at point C. Angle $BAC = 90^\circ - 42^\circ = 48^\circ$, the obtuse angle at B equals $90^\circ + 15^\circ = 105^\circ$, and the third angle, C, equals $180^\circ - 105^\circ - 48^\circ = 27^\circ$. We use the law of sines to find side b.

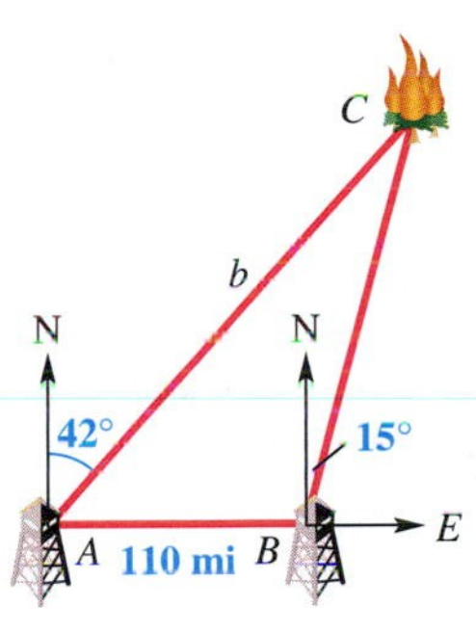

FIGURE 4

$$\frac{b}{\sin B} = \frac{c}{\sin C} \qquad \text{Law of sines}$$

$$\frac{b}{\sin 105^\circ} = \frac{110}{\sin 27^\circ} \qquad \text{Substitute.}$$

$$b = \frac{110 \sin 105^\circ}{\sin 27^\circ} \qquad \text{Multiply by } \sin 105^\circ.$$

$$b = 234 \text{ mi} \qquad \text{Approximation}$$

Ambiguous Case

If we are given the lengths of two sides and the angle opposite one of them (Case 2, SSA), it is possible that zero, one, or two such triangles exist. (Recall that there is no SSA congruence theorem.)

To illustrate, suppose we know the measure of acute angle A of triangle ABC, the length of side a, and the length of side b. We draw angle A having adjacent side of length b, as shown in Figure 5. Now we must draw the side of length a opposite angle A. The table on the next page shows possible outcomes. This situation (SSA) is called the **ambiguous case** of the law of sines.

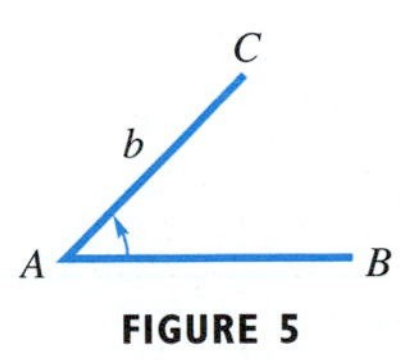

FIGURE 5

We can apply the law of sines to the values of a, b, and A and use some basic properties of geometry and trigonometry to determine which situation applies. The following basic facts should be kept in mind:

1. For any angle θ of a triangle, $0 < \sin \theta \leq 1$. If $\sin \theta = 1$, then $\theta = 90^\circ$ and the triangle is a right triangle.

2. $\sin\theta = \sin(180^\circ - \theta)$. (That is, supplementary angles have the same sine value.)
3. The smallest angle is opposite the shortest side, the largest angle is opposite the longest side, and the middle-valued angle is opposite the intermediate side (assuming that the triangle has sides that are all of different lengths).

If angle A is acute, there are four possible outcomes.

Number of Possible Triangles	Sketch	Conditions Necessary for Case to Hold
0	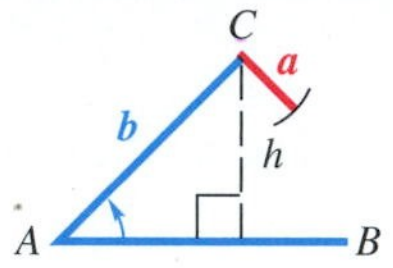	$\sin B > 1$, $a < h < b$
1	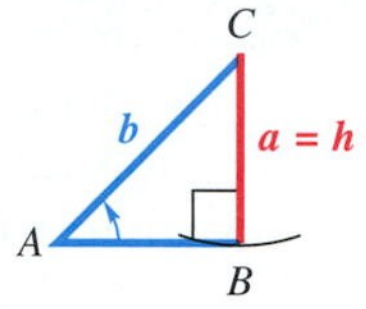	$\sin B = 1$, $a = h < b$
1	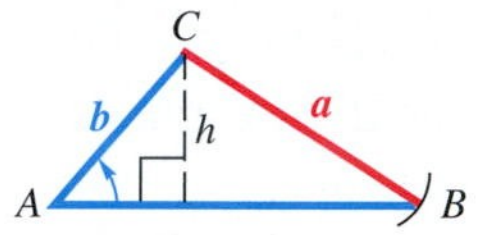	$0 < \sin B < 1$, $a \geq b$
2	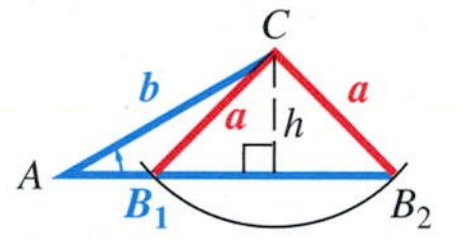	$0 < \sin B_2 < 1$, $a < b$

If angle A is obtuse, there are two possible outcomes.

Number of Possible Triangles	Sketch	Conditions Necessary for Case to Hold
0	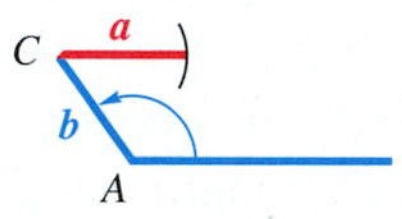	$\sin B \geq 1$, $a \leq b$
1	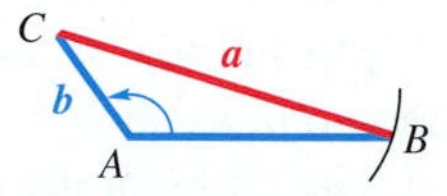	$0 < \sin B < 1$, $a > b$

The following two examples illustrate the cases of no triangle and two triangles.

EXAMPLE 4 Solving the Ambiguous Case (No Such Triangle)

Solve triangle ABC if $B = 55° 40'$, $b = 8.94$ meters, and $a = 25.1$ meters.

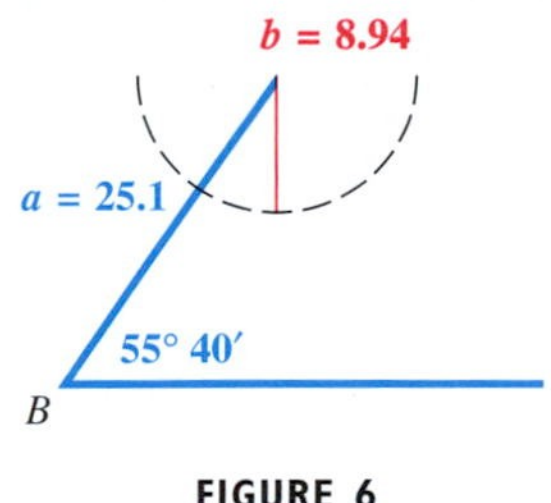

FIGURE 6

Solution Since we are given B, b, and a, we use the law of sines to find A.

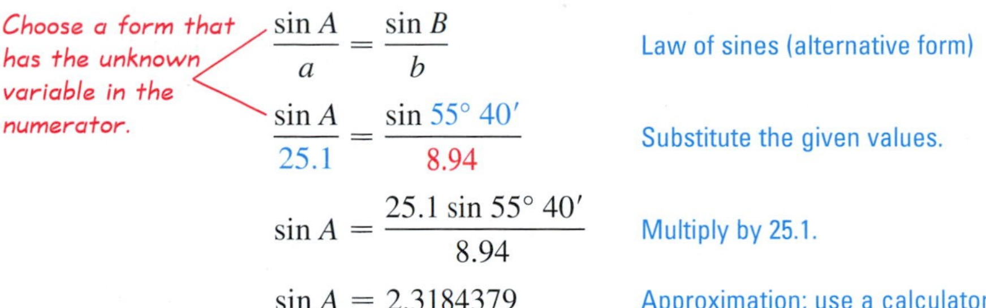
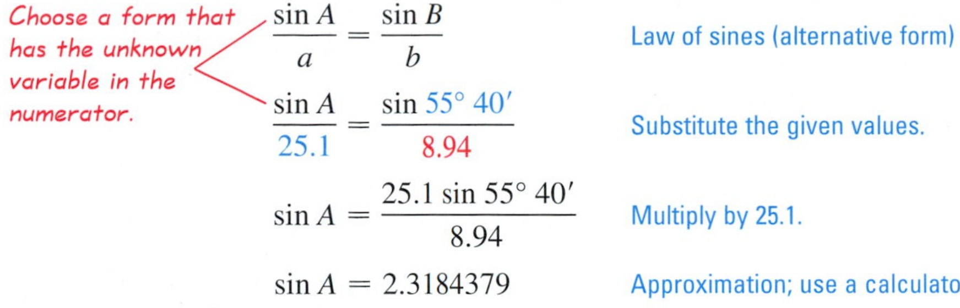

$$\frac{\sin A}{a} = \frac{\sin B}{b} \quad \text{Law of sines (alternative form)}$$

$$\frac{\sin A}{25.1} = \frac{\sin 55° 40'}{8.94} \quad \text{Substitute the given values.}$$

$$\sin A = \frac{25.1 \sin 55° 40'}{8.94} \quad \text{Multiply by 25.1.}$$

$$\sin A = 2.3184379 \quad \text{Approximation; use a calculator.}$$

Since $\sin A$ cannot be greater than 1, there can be no such angle A and thus no triangle with the given information. An attempt to sketch such a triangle leads to the situation shown in Figure 6. ■

TECHNOLOGY NOTE

Because there are 60′ in 1°, the expression sin 55° 40′ in Example 4 can be evaluated by entering sin(55 + 40/60) into your calculator. Be sure to use degree mode.

NOTE In the ambiguous case, we are given SSA—that is, two sides and an angle opposite one of the sides. For example, suppose that b, c, and angle C are given. This situation represents the ambiguous case because angle C is opposite side c.

EXAMPLE 5 Solving the Ambiguous Case (Two Triangles)

Solve triangle ABC if $A = 55.3°$, $a = 22.8$ feet, and $b = 24.9$ feet.

Solution To begin, use the law of sines to find angle B.

$$\frac{\sin A}{a} = \frac{\sin B}{b}$$

$$\frac{\sin 55.3°}{22.8} = \frac{\sin B}{24.9}$$

$$\sin B = \frac{24.9 \sin 55.3°}{22.8}$$

$$\sin B = .8978678 \quad \text{Approximation}$$

Since $\sin B \approx .8978678$, one value of B, to the nearest tenth, is

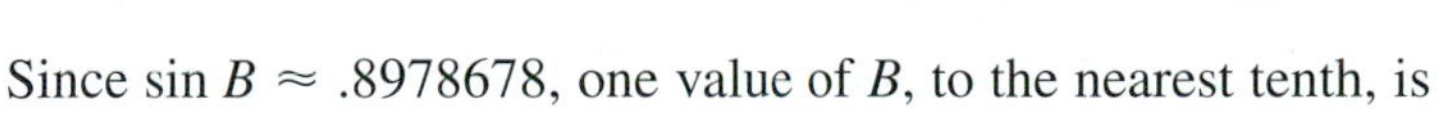

$$B_1 = 63.9°. \quad \text{Use the inverse sine function.}$$

Supplementary angles have the same sine value, so another *possible* value of B is

$$B_2 = 180° - 63.9° = 116.1°.$$

To see if $B_2 = 116.1°$ is a valid possibility, simply add 116.1° to the measure of A, 55.3°. Since $116.1° + 55.3° = 171.4°$, and this sum is less than 180°, it is a valid angle measure for this triangle.

Now separately solve triangles AB_1C_1 and AB_2C_2 shown in Figure 7. Begin with AB_1C_1. Find C_1 first.

$$C_1 = 180° - A - B_1 = 180° - 55.3° - 63.9° = 60.8°$$

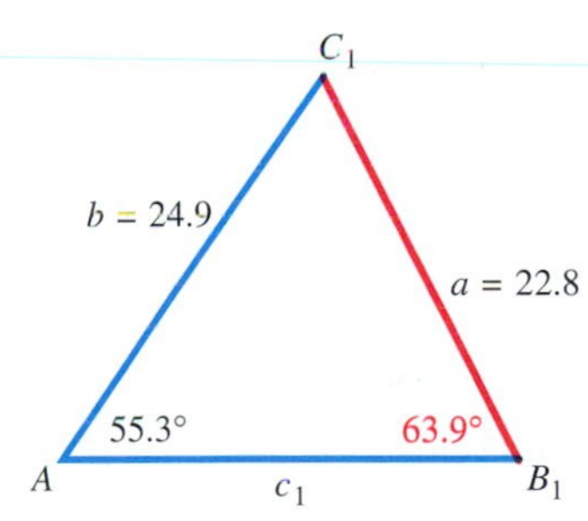

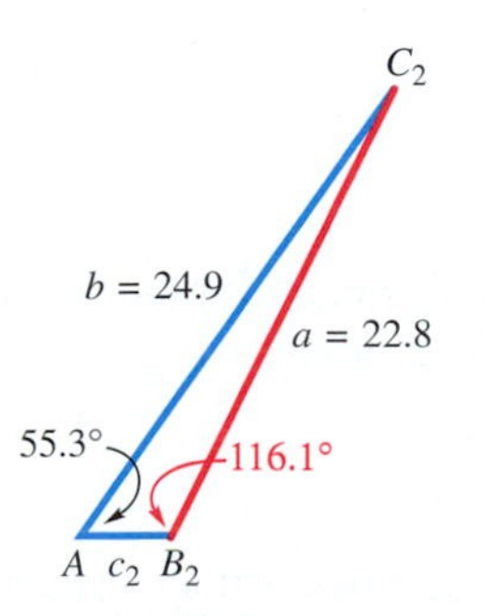

FIGURE 7

Use the law of sines to find c_1.

$$\frac{a}{\sin A} = \frac{c_1}{\sin C_1}$$

$$\frac{22.8}{\sin 55.3°} = \frac{c_1}{\sin 60.8°}$$

$$c_1 = \frac{22.8 \sin 60.8°}{\sin 55.3°}$$

$$c_1 = 24.2 \text{ feet} \qquad \text{Approximation}$$

To solve triangle AB_2C_2, first find C_2.

$$C_2 = 180° - A - B_2 = 180° - 55.3° - 116.1° = 8.6°$$

Use the law of sines to find c_2.

$$\frac{a}{\sin A} = \frac{c_2}{\sin C_2}$$

$$\frac{22.8}{\sin 55.3°} = \frac{c_2}{\sin 8.6°}$$

$$c_2 = \frac{22.8 \sin 8.6°}{\sin 55.3°}$$

$$c_2 = 4.15 \text{ feet} \qquad \text{Approximation}$$

The ambiguous case results in zero, one, or two triangles. The following guidelines can be used to determine how many triangles there are.

FOR DISCUSSION

Discuss the steps that you would take to solve each triangle. Do not actually solve the triangle.

1. $B = 85°, C = 40°, b = 26$
2. $A = 55°, B = 65°, c = 4$
3. $a = 17, b = 22, B = 32°$
4. $A = 20°, b = 7, a = 9$
5. $c = 225, A = 103.2°, B = 62.5°$

Number of Triangles Satisfying the Ambiguous Case (SSA)

Let sides a and b and angle A be given in triangle ABC. (The law of sines can be used to calculate the value of $\sin B$.)

1. If $\sin B > 1$, then *no triangle* satisfies the given conditions.
2. If $\sin B = 1$, then *one triangle* satisfies the given conditions and $B = 90°$.
3. If $0 < \sin B < 1$, then either *one or two triangles* satisfy the given conditions.
   - **(a)** If $\sin B = k$, then let $B_1 = \sin^{-1} k$ and use B_1 for B in the first triangle.
   - **(b)** Let $B_2 = 180° - B_1$. If $A + B_2 < 180°$, then a second triangle exists. In this case, use B_2 for B in the second triangle.

NOTE The preceding guidelines can be applied whenever two sides and an angle opposite one of the sides are given.

EXAMPLE 6 **Solving the Ambiguous Case (One Triangle)**

Solve triangle ABC, given that $A = 43.5°$, $a = 10.7$ inches, and $c = 7.2$ inches.

Solution To find angle C, use the law of sines.

$$\frac{\sin C}{c} = \frac{\sin A}{a}$$

$$\frac{\sin C}{7.2} = \frac{\sin 43.5°}{10.7}$$

$$\sin C = \frac{7.2 \sin 43.5°}{10.7} = .46319186$$

$$C = 27.6° \qquad \text{Use the inverse sine function.}$$

This is the acute angle. Another value of C that has $\sin C = .46319186$ is

$$C = 180° - 27.6° = 152.4°.$$

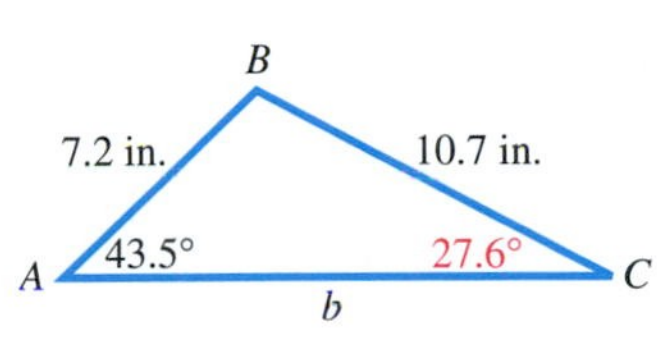

FIGURE 8

Notice in the given information that $c < a$, meaning that angle C must have measure *less than* that of angle A. Notice also that when we add this possible obtuse angle to the given angle $A = 43.5°$, we obtain

$$152.4° + 43.5° = 195.9°,$$

which is greater than 180°. So there can be only one triangle. See Figure 8. Then

$$B = 180° - 27.6° - 43.5° = 108.9°,$$

and we can find side b with the law of sines.

$$\frac{b}{\sin B} = \frac{a}{\sin A}$$

$$\frac{b}{\sin 108.9°} = \frac{10.7}{\sin 43.5°}$$

$$b = \frac{10.7 \sin 108.9°}{\sin 43.5°}$$

$$b = 14.7 \text{ inches} \qquad \text{Approximation}$$

■

EXAMPLE 7 **Analyzing Data Involving an Obtuse Angle**

Without using the law of sines, explain why $A = 104°$, $a = 26.8$ meters, and $b = 31.3$ meters cannot be valid for a triangle ABC.

Solution Since A is an obtuse angle, it is the largest angle. So the largest side of the triangle must be a. However, we are given $b > a$; thus, $B > A$, which is impossible if A is obtuse. Therefore, no such triangle ABC exists. ■

If the law of sines had actually been applied in Example 7, we would have obtained $\sin B > 1$, which has no solution. Verify this on your calculator.

What Went Wrong?

A student used the law of sines to solve for angle B in a triangle ABC, where $b = 23$, $c = 14$, and $C = 52°$. He entered the calculation shown on the first screen and got the results in the second screen.

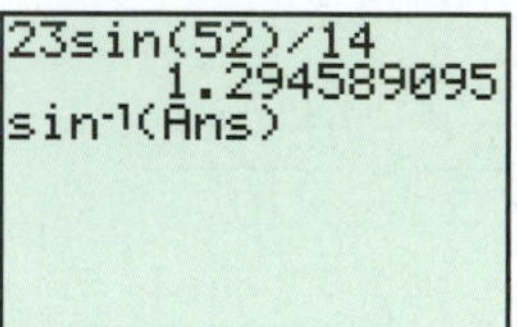

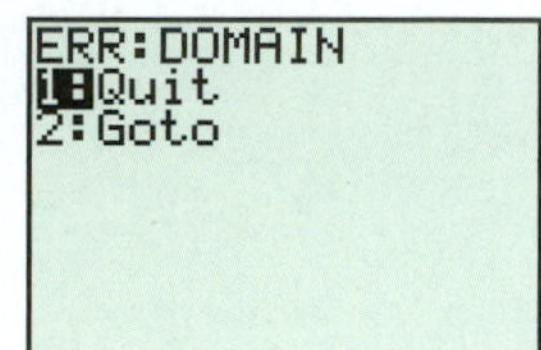

What Went Wrong? What is the solution, if any, for angle B?

Answer to What Went Wrong?

Since $\sin B > 1$, there is no solution for angle B. Triangle ABC cannot exist.

10.1 Exercises

1. *Concept Check* Consider this oblique triangle. Which one of the following proportions is *not* valid?

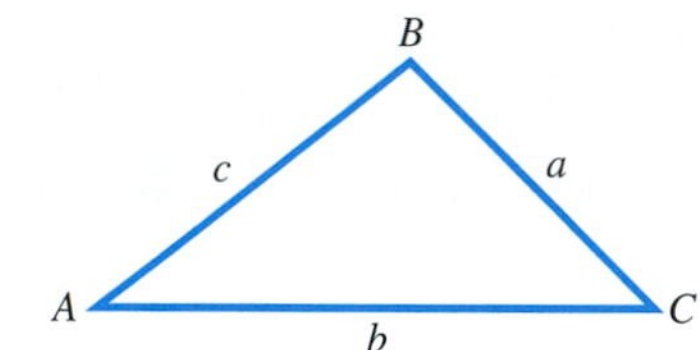

A. $\dfrac{a}{b} = \dfrac{\sin A}{\sin B}$ B. $\dfrac{a}{\sin A} = \dfrac{b}{\sin B}$

C. $\dfrac{\sin A}{a} = \dfrac{b}{\sin B}$ D. $\dfrac{\sin A}{a} = \dfrac{\sin B}{b}$

2. *Concept Check* Which two of the following situations do not provide sufficient information for solving a triangle by the law of sines?

   A. You are given two angles and the side included between them.

   B. You are given two angles and a side opposite one of them.

   C. You are given two sides and the angle included between them.

   D. You are given three sides.

Concept Check *Given the following angles and sides, decide whether solving triangle ABC results in the ambiguous case.*

3. A, B, and a
4. A, C, and c
5. a, b, and c
6. A, a, and b
7. B, b, and c
8. A, b, and c
9. C, a, and c
10. A, B, and b

Find the exact length of each side a. Do not use a calculator.

11.

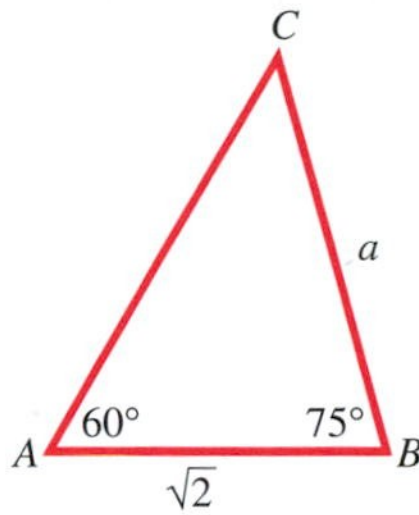

12.

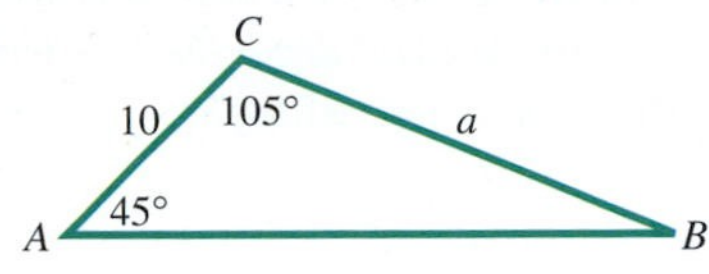

To the Student: Calculator Considerations

When making approximations, do not round off during intermediate steps—wait until the final step to give the appropriate approximation.

Solve each triangle.

13.

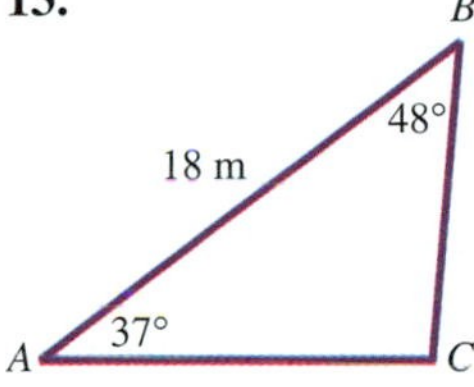

14.

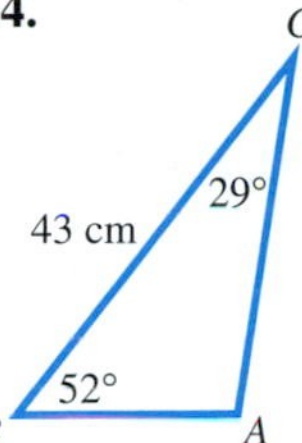

15.

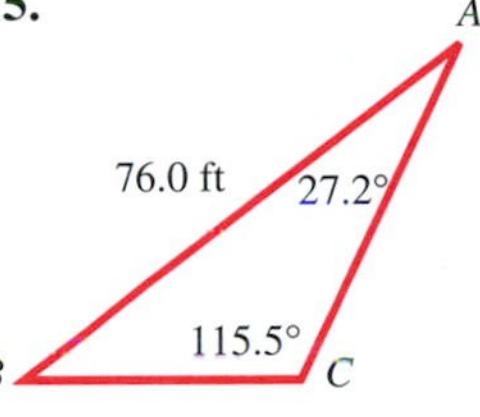

16.

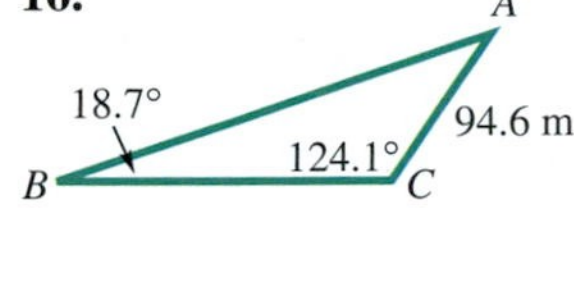

17. $A = 37°$, $C = 95°$, $c = 18$ meters

18. $B = 52°$, $C = 29°$, $a = 43$ centimeters

19. $C = 74.08°$, $B = 69.38°$, $c = 45.38$ meters

20. $A = 87.2°$, $b = 75.9$ yards, $C = 74.3°$

21. $B = 38° \ 40'$, $a = 19.7$ centimeters, $C = 91° \ 40'$

22. $B = 20° \ 50'$, $C = 103° \ 10'$, $b = 132$ feet

23. $A = 35.3°$, $B = 52.8°$, $b = 675$ feet

24. $A = 68.41°$, $B = 54.23°$, $a = 12.75$ feet

25. $A = 39.70°$, $C = 30.35°$, $b = 39.74$ meters

26. $C = 71.83°$, $B = 42.57°$, $a = 2.614$ centimeters

27. $B = 42.88°$, $C = 102.40°$, $b = 3974$ feet

28. $A = 18.75°$, $B = 51.53°$, $c = 2798$ yards

29. ***Concept Check*** Which one of the following sets of data does not determine a unique triangle?

A. $A = 40°$, $B = 60°$, $C = 80°$ **B.** $a = 5$, $b = 12$, $c = 13$

C. $a = 3$, $b = 7$, $C = 50°$ **D.** $a = 2$, $b = 2$, $c = 2$

30. ***Concept Check*** Which one of the following sets of data determines a unique triangle?

A. $A = 50°$, $B = 50°$, $C = 80°$ **B.** $a = 3$, $b = 5$, $c = 20$

C. $A = 40°$, $B = 20°$, $C = 30°$ **D.** $a = 7$, $b = 24$, $c = 25$

Find the unknown angles in each triangle that exists.

31. $A = 29.7°$, $b = 41.5$ feet, $a = 27.2$ feet

32. $B = 48.2°$, $a = 890$ centimeters, $b = 697$ centimeters

33. $B = 74.3°$, $a = 859$ meters, $b = 783$ meters

34. $C = 82.2°$, $a = 10.9$ kilometers, $c = 7.62$ kilometers

35. $A = 142.13°$, $b = 5.432$ feet, $a = 7.297$ feet

36. $B = 113.72°$, $a = 189.6$ yards, $b = 243.8$ yards

Concept Check *In each figure, a line of length h is to be drawn from the given point to the positive x-axis in order to form a triangle. For what values(s) of h can you draw the following?*

(a) *two triangles* **(b)** *exactly one triangle* **(c)** *no triangle*

37.

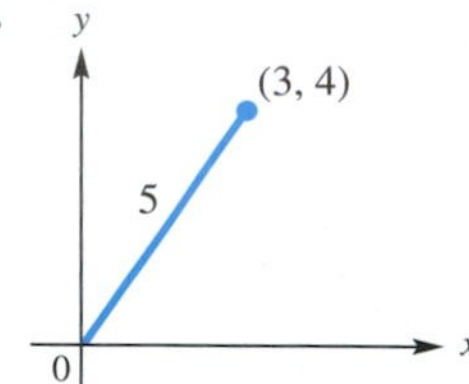

38.

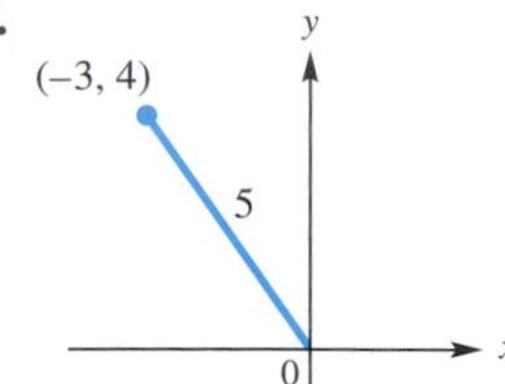
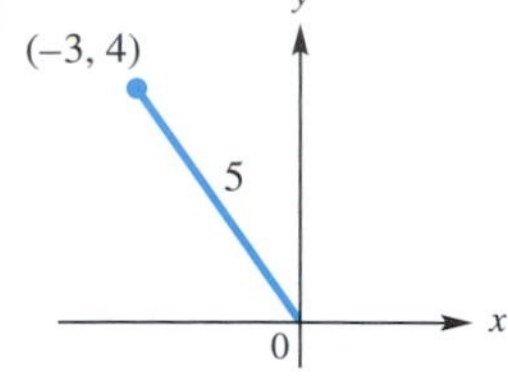

Determine the number of triangles ABC possible with the given parts.

39. $a = 31, b = 26, B = 48°$

40. $a = 35, b = 30, A = 40°$

41. $a = 50, b = 61, A = 58°$

42. $B = 54°, c = 28, b = 23$

Solve each triangle.

43. $A = 42.5°$, $a = 15.6$ feet, $b = 8.14$ feet

44. $C = 52.3°$, $a = 32.5$ yards, $c = 59.8$ yards

45. $B = 72.2°$, $b = 78.3$ meters, $c = 145$ meters

46. $C = 68.5°$, $c = 258$ centimeters, $b = 386$ centimeters

47. $A = 38° 40'$, $a = 9.72$ kilometers, $b = 11.8$ kilometers

48. $C = 29° 50'$, $a = 8.61$ meters, $c = 5.21$ meters

49. $B = 32° 50'$, $a = 7540$ centimeters, $b = 5180$ centimeters

50. $C = 22° 50'$, $b = 159$ millimeters, $c = 132$ millimeters

Solve each problem.

51. In Example 1, we ask the question "Why does the Pythagorean theorem not apply?" Answer this question.

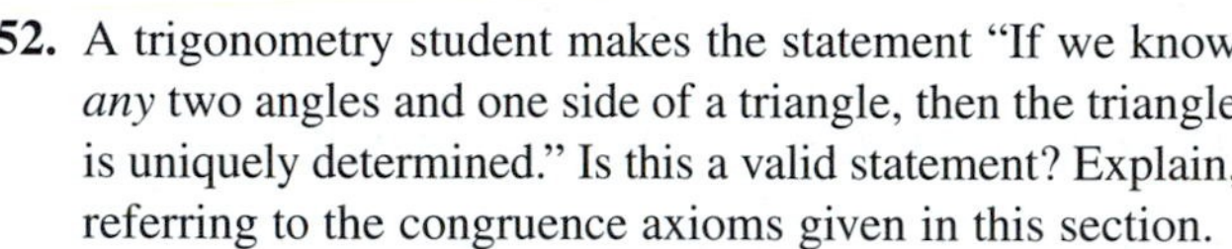

52. A trigonometry student makes the statement "If we know *any* two angles and one side of a triangle, then the triangle is uniquely determined." Is this a valid statement? Explain, referring to the congruence axioms given in this section.

53. Explain why the law of sines cannot be used to solve a triangle if we are given the lengths of the three sides of the triangle.

54. If a is twice as long as b, is angle A twice as large as angle B? Explain.

55. ***Distance across a River*** To find the distance AB across a river, a distance $BC = 354$ meters is laid off on one side of the river. It is found that $B = 112° 10'$ and $C = 15° 20'$. Find AB.

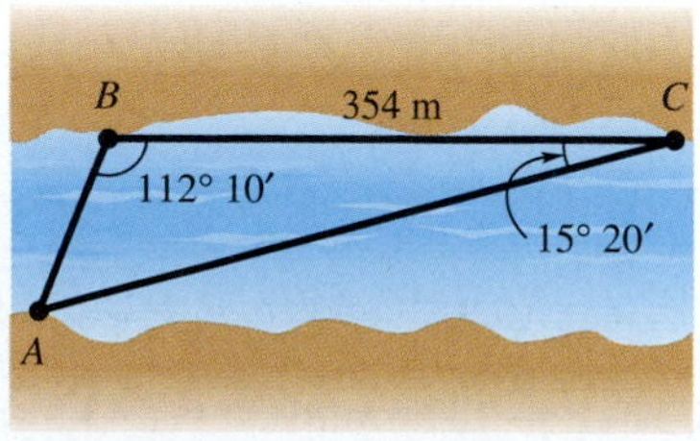

56. ***Distance across a Canyon*** To determine the distance RS across a deep canyon, Joanna lays off a distance $TR = 582$ yards. She then finds that $T = 32° 50'$ and $R = 102° 20'$. Find RS.

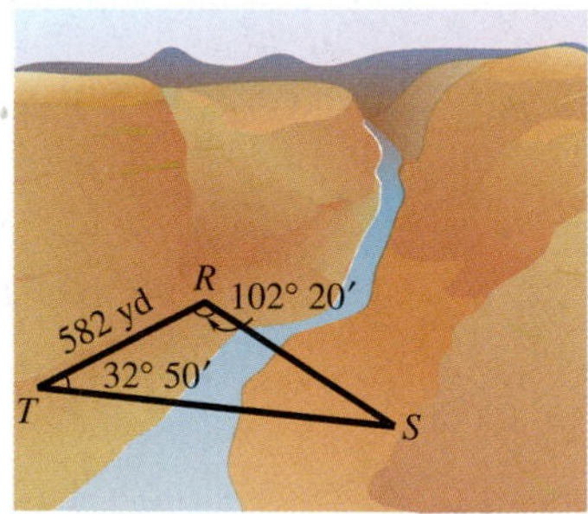

57. ***Distance between Radio Direction Finders*** Radio direction finders are at points A and B, which are 3.46 miles apart on an east–west line, with A west of B. From A, the bearing of a certain radio transmitter is 47.7°; from B, the bearing is 302.5°. Find the distance of the transmitter from A.

58. ***Distance a Ship Travels*** A ship is sailing due north. At a certain point, the bearing of a lighthouse 12.5 kilometers away is N 38.8° E. Later on, the captain notices that the bearing of the lighthouse has become S 44.2° E. How far did the ship travel between the two observations of the lighthouse?

59. ***Measurement of a Folding Chair*** A folding chair is to have a seat 12.0 inches deep with angles as shown in the figure. How far down from the seat should the crossing legs be joined? (*Hint:* Find x in the figure.)

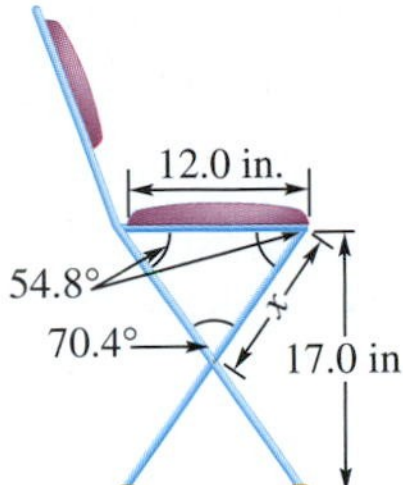

60. ***Distance across a River*** Mark notices that the bearing of a tree on the opposite bank of a river flowing north is 115.45°. Lisa is on the same bank as Mark, but 428.3 meters away. She notices that the bearing of the tree is 45.47°. The two banks are parallel. What is the distance across the river?

61. ***Angle Formed by Centers of Gears*** Three gears are arranged as shown in the figure. Find angle θ.

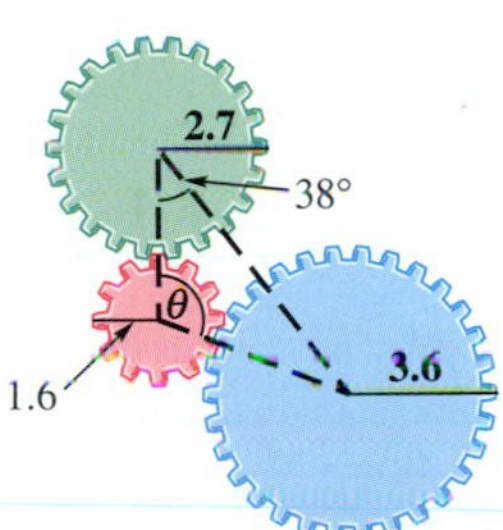

62. ***Distance between Atoms*** Three atoms with radii 2.0, 3.0, and 4.5 are arranged as in the figure. Find the distance between the centers of atoms *A* and *C*.

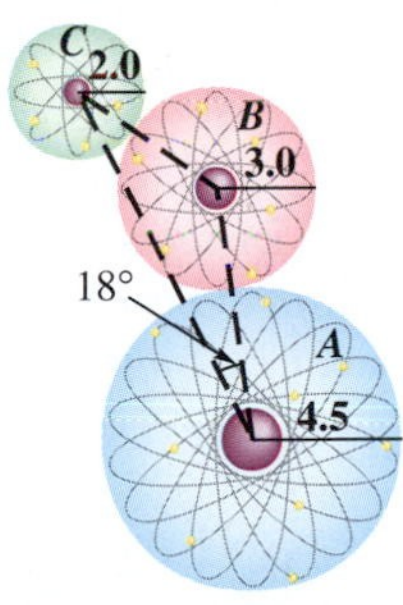

63. ***Distance between a Ship and a Lighthouse*** The bearing of a lighthouse from a ship was found to be N 37° E. After the ship sailed 2.5 miles due south, the new bearing was N 25° E. Find the distance between the ship and the lighthouse at each location.

64. ***Height of a Balloon*** A balloonist is directly above a straight road 1.5 miles long that joins two towns. She finds that the town closer to her is at an angle of depression of 35° and the farther town is at an angle of depression of 31°. How high above the ground is the balloon?

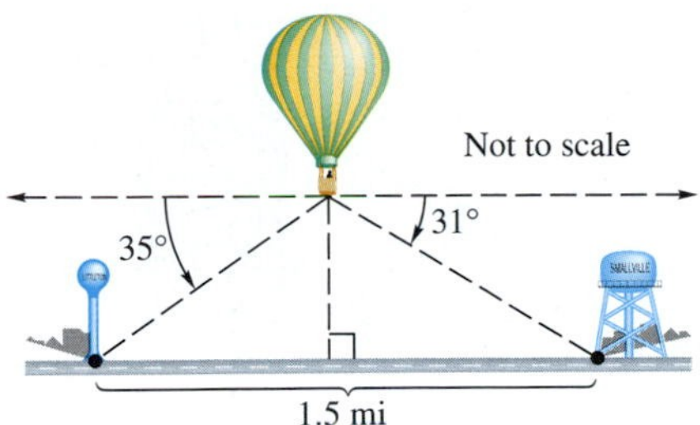

65. ***Distance from Ship to Shore*** From shore station A, a ship C is observed in the direction N 22.4° E. The same ship is observed to be in the direction N 10.6° W from shore station B, located a distance of 25.5 kilometers exactly southeast of A. Find the distance of the ship from station A.

66. ***Height of a Helicopter*** A helicopter is sighted at the same time by two ground observers who are 3 miles apart on the same side of the helicopter. (See the figure.) They report angles of elevation of 20.5° and 27.8°. How high is the helicopter?

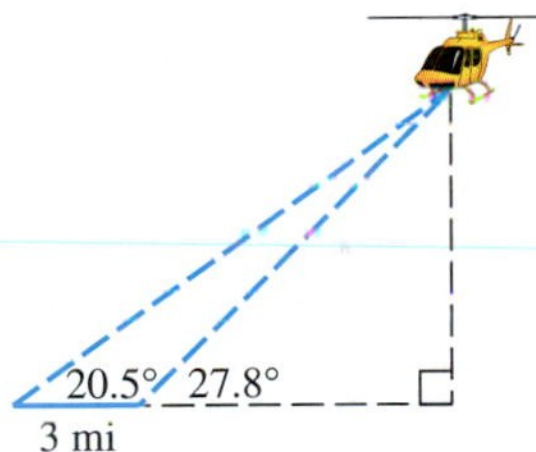

67. ***Distance from a Rocket to a Radar Station*** A rocket-tracking facility has two radar stations T_1 and T_2, placed 1.73 kilometers apart, that lock onto the rocket and continuously transmit the angles of elevation to a computer. Find the distance to the rocket from T_1 at the moment when the angles of elevation are 28.1° and 79.5°, as shown in the figure.

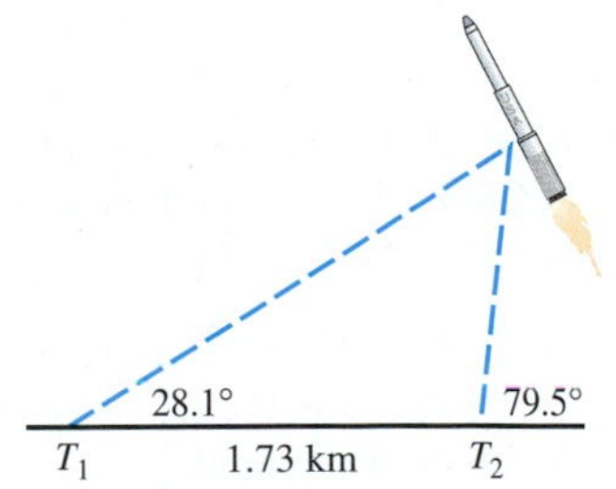

68. ***Height of a Clock Tower*** A surveyor standing 48.0 meters from the base of a building measures the angle to the top of the building to be 37.4°. The surveyor then measures the angle to the top of a clock tower on the building to be 45.6°. Find the height of the clock tower.

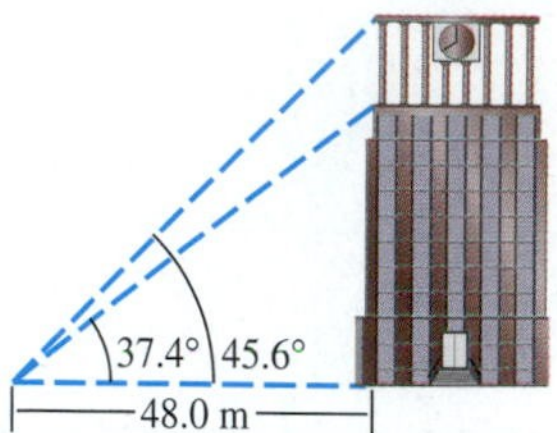

69. ***Distance between a Pin and a Rod*** A slider crank mechanism is shown in the figure. Find the distance between the wrist pin W and the connecting rod center C.

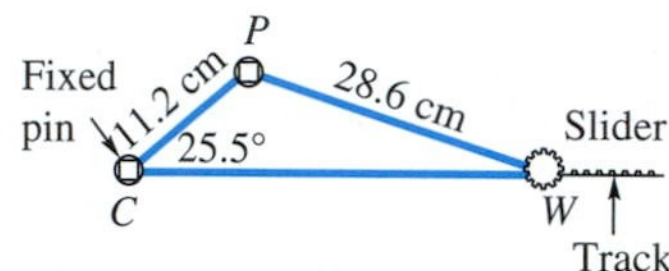

70. ***Path of a Satellite*** A satellite is traveling in a circular orbit 1600 kilometers above Earth. It will pass directly over a tracking station at noon. The satellite takes 2 hours to make a complete orbit. Assume that the radius of Earth is 6400 kilometers. The tracking antenna is aimed 30° above the horizon. See the figure. At what time (before noon) will the satellite pass through the beam of the antenna? (*Source: Space Mathematics* by B. Kastner, Ph.D. Copyright © 1972 by the National Aeronautics and Space Administration. Courtesy of NASA.)

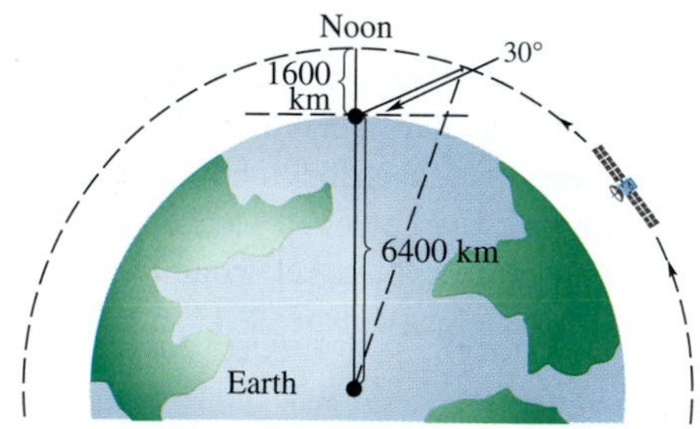

71. Apply the law of sines to the following: $a = \sqrt{5}$, $c = 2\sqrt{5}$, $A = 30°$. What is the value of sin C? What is the measure of C? Based on its angle measures, what kind of triangle is triangle ABC?

72. Explain the condition that must exist to determine that there is no triangle satisfying the given values of a, b, and B once the value of sin A is found.

73. Without using the law of sines, explain why no triangle ABC exists satisfying $A = 103° 20'$, $a = 14.6$ feet, and $b = 20.4$ feet.

74. Apply the law of sines to the data given in Example 7. Describe in your own words what happens when you try to find the measure of angle B, using a calculator.

75. ***Survey of Property*** A surveyor reported the following data about a piece of property: "The property is triangular in shape, with dimensions shown in the figure." Use the law of sines to see whether such a piece of property could exist.

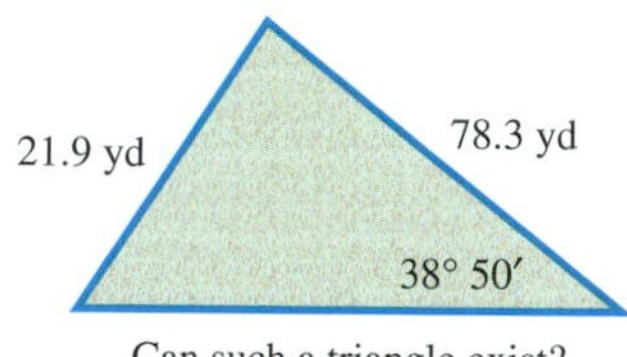

Can such a triangle exist?

76. ***Survey of Property*** The surveyor tries again: "A second triangular piece of property has dimensions shown in the figure." This time it turns out that the surveyor did not consider every possible case. Use the law of sines to show why.

77. ***Distance to the Moon*** Since the moon is a relatively close celestial object, its distance from Earth can be measured directly by taking two photographs of the moon at precisely the same time from two different locations on Earth. The moon will have a different angle of elevation at each location. On April 29, 1976, at 11:35 A.M., the lunar angles of elevation during a partial solar eclipse at Bochum in upper Germany and at Donaueschingen in lower Germany were measured as 52.6997° and 52.7430°, respectively. (See the figure.) The two cities are 398 kilometers apart. Calculate the distance to the moon from Bochum on this day, and compare it with the actual value of 406,000 kilometers. Disregard the curvature of Earth in this calculation. (*Source:* Scholosser, W., T. Schmidt-Kaler, and E. Milone, *Challenges of Astronomy,* Springer-Verlag, 1991.)

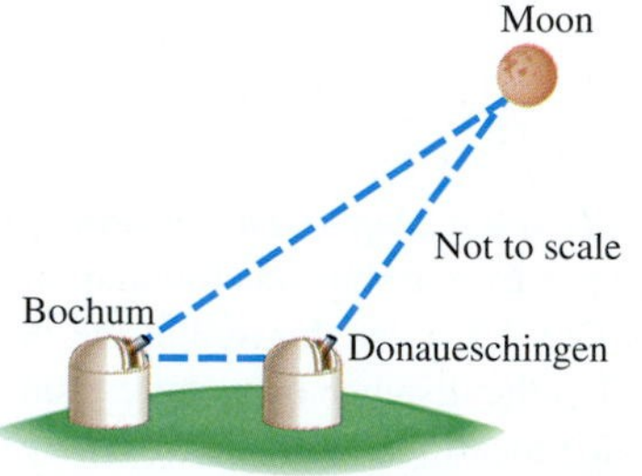

78. ***Ground Distances Measured by Aerial Photography*** The distance shown in an aerial photograph is determined by both the focal length of the lens and the tilt of the camera from the perpendicular to the ground. A camera lens with a 12-inch focal length has an angular coverage of 60°. If an aerial photograph is taken with this camera tilted 35° at an altitude of 5000 feet, calculate the distance d in miles that will be shown in the photograph. (See the figure.) (*Sources:* Brooks, R., and D. Johannes, *Photoarchaeology,* Dioscorides Press, 1990; Moffitt, F., *Photogrammetry,* International Textbook Company, 1967.)

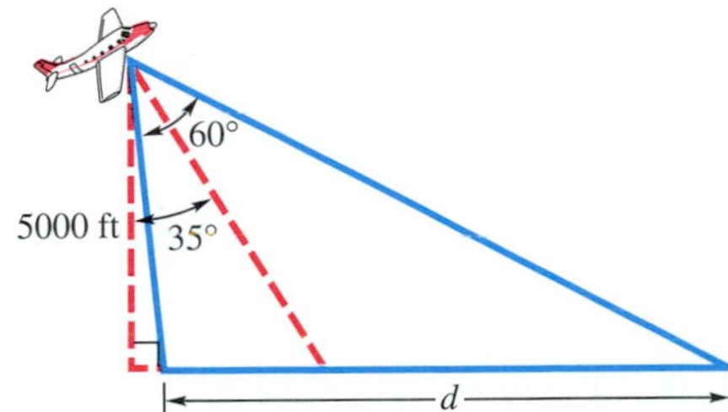

79. ***U.S. Flag*** The flag of the United States includes the colors red, white, and blue. Which color is predominant? Clearly, the answer is either red or white. (It can be shown that only 18.73% of the total area is blue.) (*Source: Slicing Pizzas, Racing Turtles, and Further Adventures in Applied Mathematics,* Banks, R., Princeton University Press, 1999; Schneider, D.)

(a) Let R denote the radius of the circumscribing circle of a five-pointed star appearing on the American flag. The star can be decomposed into 10 congruent triangles. In the figure at the top of the next column, r is the radius of the circumscribing circle of the pentagon in the interior of the star. Show that the area of the star is

$$A = \left[5\frac{\sin A \sin B}{\sin(A + B)}\right]R^2.$$

(*Hint:* $\sin C = \sin[180° - (A + B)] = \sin(A + B)$.)

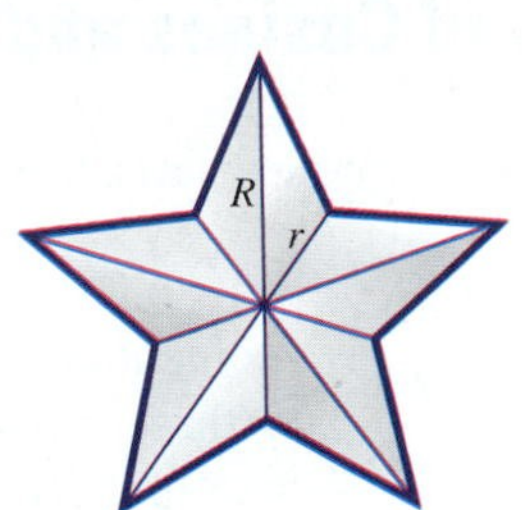

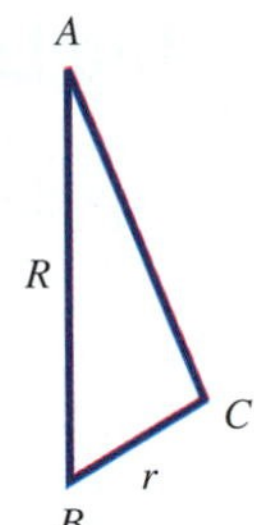

(b) Angles A and B have values 18° and 36°, respectively. Express the area of a star in terms of its radius R.

(c) To determine whether red or white is predominant, we must know the measurements of the flag. Consider a flag of width 10 inches, length 19 inches, length of each upper stripe 11.4 inches, and radius R of the circumscribing circle of each star .308 inch. The 13 stripes consist of six matching pairs of red and white stripes and one additional red, upper stripe. Therefore, we must compare the area of a red, upper stripe with the total area of the 50 white stars.

(i) Compute the area of the red, upper stripe.

(ii) Compute the total area of the 50 white stars.

(iii) Which color occupies the greatest area on the flag?

Relating Concepts

For individual or group investigation (Exercises 80–84)

In any triangle, the longest side is opposite the largest angle. This result from geometry can be proven with trigonometry. To prove it for acute triangles, ***work Exercises 80–84 in order.*** *(The case for obtuse triangles will be considered in the Section 10.2 Exercises.)*

80. Is the graph of the function $y = \sin x$ increasing or decreasing over the interval $\left[0, \frac{\pi}{2}\right]$?

81. Suppose angle A is the largest angle of an acute triangle, and let B be an angle smaller than A. Explain why $\frac{\sin B}{\sin A} < 1$.

82. Solve for b in the first form of the law of sines.

83. Use the results in Exercises 81 and 82 to show that $b < a$.

84. Use the result proved in Exercises 80–83 to explain why no triangle ABC satisfies $A = 83°$, $a = 14$, and $b = 20$.

10.2 The Law of Cosines and Area Formulas

Derivation of the Law of Cosines ■ Applications of Triangles ■ Area Formulas

As mentioned in Section 10.1, if we are given two sides and the included angle (Case 3) or three sides (Case 4) of a triangle, then a unique triangle is formed. These are the SAS and SSS cases, respectively. Both cases require using the *law of cosines.* Remember the following property of triangles when applying the law of cosines.

Triangle Side Length Restriction

In any triangle, the sum of the lengths of any two sides must be greater than the length of the remaining side.

For example, it would be impossible to construct a triangle with sides of lengths 3, 4, and 10. See Figure 9.

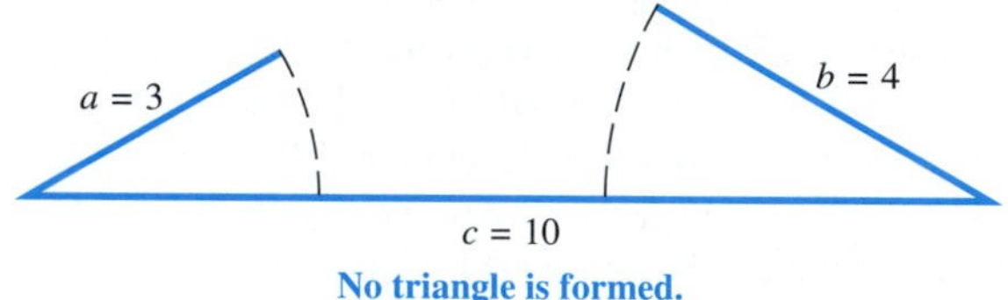

No triangle is formed.

FIGURE 9

Derivation of the Law of Cosines

To derive the law of cosines, let ABC be any oblique triangle. Choose a coordinate system so that vertex B is at the origin and side BC is along the positive x-axis. See Figure 10.

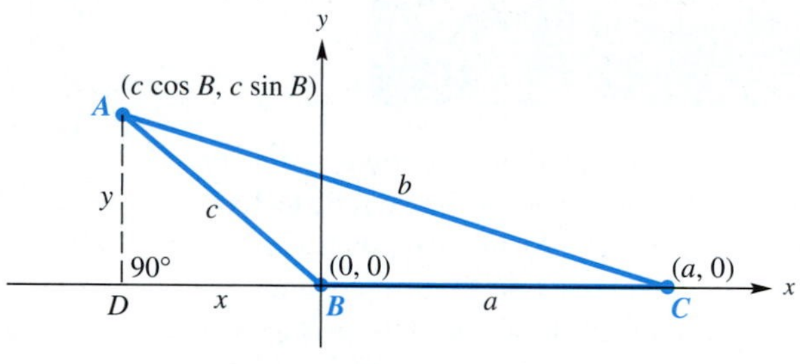

FIGURE 10

Let (x, y) be the coordinates of vertex A of the triangle. For angle B, whether obtuse or acute,

$$\sin B = \frac{y}{c} \quad \text{and} \quad \cos B = \frac{x}{c}$$

$$y = c \sin B \quad \text{and} \quad x = c \cos B. \quad \text{(Here } x \text{ is negative if } B \text{ is obtuse.)}$$

Thus, the coordinates of point A become $(c \cos B, c \sin B)$.

Point C has coordinates $(a, 0)$, and AC has length b. By the distance formula,

$$b = \sqrt{(c \cos B - a)^2 + (c \sin B - 0)^2}$$

$$b^2 = (c \cos B - a)^2 + (c \sin B)^2 \qquad \text{Square each side.}$$

$$= c^2 \cos^2 B - 2ac \cos B + a^2 + c^2 \sin^2 B \qquad \text{Multiply.}$$

$(x - y)^2 = x^2 - 2xy + y^2$

$$= a^2 + c^2(\cos^2 B + \sin^2 B) - 2ac \cos B \qquad \text{Properties of real numbers}$$

$$= a^2 + c^2(1) - 2ac \cos B \qquad \text{Fundamental identity}$$

$$b^2 = a^2 + c^2 - 2ac \cos B.$$

This result is one form of the law of cosines. In our work, if we had placed A or C at the origin, we would have obtained the same result, but with the variables rearranged. Two other forms appear in the following box.

Law of Cosines

In any triangle ABC with sides a, b, and c,

$$a^2 = b^2 + c^2 - 2bc \cos A,$$
$$b^2 = a^2 + c^2 - 2ac \cos B,$$
$$c^2 = a^2 + b^2 - 2ab \cos C.$$

NOTE If we let $C = 90°$ in the third form of the law of cosines, we have $\cos C = \cos 90° = 0$, and the formula becomes

$$c^2 = a^2 + b^2,$$

the familiar equation of the Pythagorean theorem. ***The Pythagorean theorem is a special case of the law of cosines.***

Applications of Triangles

EXAMPLE 1 Using the Law of Cosines in an Application (SAS)

A surveyor wishes to find the distance between two points A and B on opposite sides of a lake. While standing at point C, she finds that $AC = 259$ meters, $BC = 423$ meters, and angle ACB measures $132° 40'$. See Figure 11. Find the distance AB.

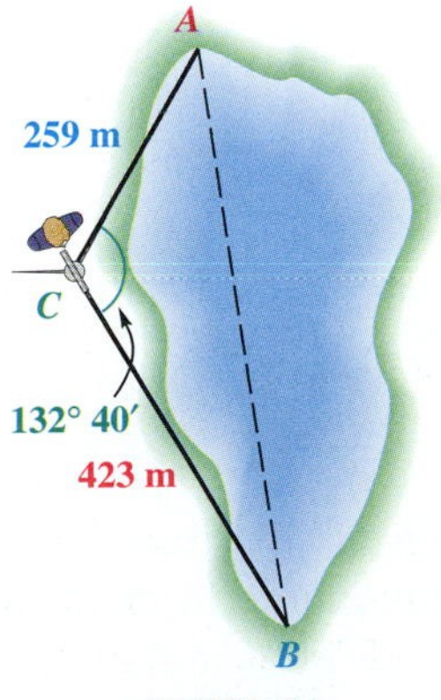

FIGURE 11

Solution The law of cosines can be used here, since we know the lengths of two sides of the triangle and the measure of the included angle.

$$AB^2 = 259^2 + 423^2 - 2(259)(423) \cos 132° 40' \qquad \text{Substitute.}$$

$$AB^2 = 394{,}510.6 \qquad \text{Use a calculator.}$$

$$AB = 628 \qquad \text{Approximate the square root.}$$

The distance between the points is approximately 628 meters. ■

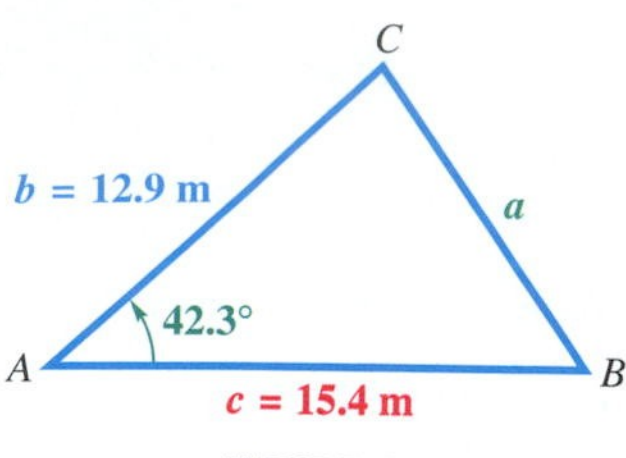

FIGURE 12

EXAMPLE 2 Using the Law of Cosines to Solve a Triangle (SAS)

Solve triangle ABC if $A = 42.3°$, $b = 12.9$ meters, and $c = 15.4$ meters.

Solution See Figure 12. We start by finding a with the law of cosines.

$$a^2 = b^2 + c^2 - 2bc \cos A \quad \text{Law of cosines}$$
$$a^2 = 12.9^2 + 15.4^2 - 2(12.9)(15.4) \cos 42.3° \quad \text{Substitute.}$$
$$a^2 = 109.7 \quad \text{Use a calculator.}$$
$$a = 10.47 \text{ meters} \quad \text{Take square roots.}$$

Of the two remaining angles, B and C, B must be the smaller, since it is opposite the shorter of the two sides b and c. Therefore, B cannot be obtuse.

$$\frac{\sin A}{a} = \frac{\sin B}{b} \quad \text{Law of sines (alternative form)}$$
$$\frac{\sin 42.3°}{10.47} = \frac{\sin B}{12.9} \quad \text{Substitute.}$$
$$\sin B = \frac{12.9 \sin 42.3°}{10.47} \quad \text{Multiply by 10.47; rewrite.}$$
$$B = 56.0° \quad \text{Use the inverse sine function.}$$

The easiest way to find C is to subtract the measures of A and B from 180°.

$$C = 180° - A - B = 180° - 42.3° - 56.0° = 81.7°$$ ■

CAUTION Had we chosen to use the law of sines to find C rather than B in Example 2, we would not have known whether C equals 81.7° or its supplement, 98.3°.

GCM TECHNOLOGY NOTE

A calculator program based on SSS can be used to support the result of Example 3.

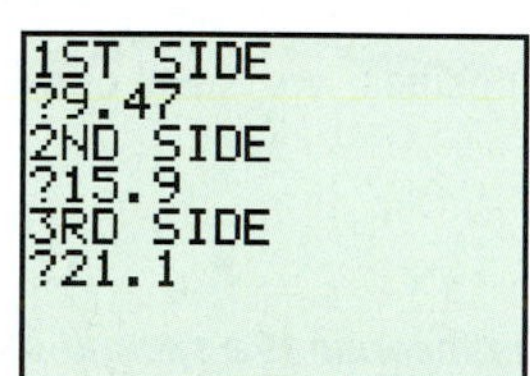

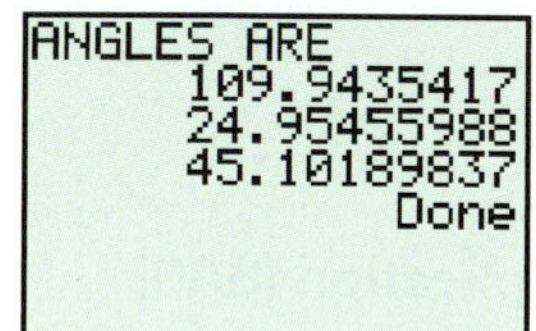

EXAMPLE 3 Using the Law of Cosines to Solve a Triangle (SSS)

Solve triangle ABC if $a = 9.47$ feet, $b = 15.9$ feet, and $c = 21.1$ feet.

Solution Given the lengths of three sides of the triangle, we can use the law of cosines to solve for any angle of the triangle. We solve for C, the largest angle. We will know that C is obtuse if $\cos C < 0$.

$$c^2 = a^2 + b^2 - 2ab \cos C \quad \text{Law of cosines}$$
$$\cos C = \frac{a^2 + b^2 - c^2}{2ab} \quad \text{Solve for } \cos C.$$
$$\cos C = \frac{9.47^2 + 15.9^2 - 21.1^2}{2(9.47)(15.9)} \quad \text{Substitute.}$$
$$\cos C = -.34109402 \quad \text{Use a calculator.}$$
$$C = 109.9° \quad \text{Use the inverse cosine function.}$$

We can use either the law of sines or the law of cosines to find $B = 45.1°$. (Verify this.) Since $A = 180° - B - C$, we obtain $A = 25.0°$. ■

FIGURE 13

Trusses are frequently used to support roofs on buildings, as illustrated in Figure 13. The simplest type of roof truss is a triangle, as shown in Figure 14 on the next page. One basic task in constructing a roof truss is to cut the ends of the rafters so that the roof has the correct slope. (*Source:* Riley, W., L. Sturges, and D. Morris, *Statics and Mechanics of Materials,* John Wiley and Sons, 1995.)

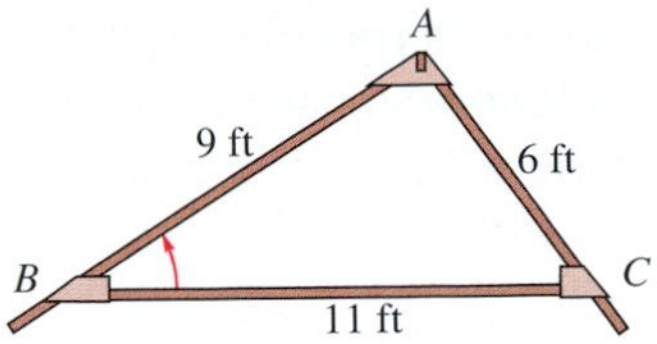

A

9 ft

6 ft

B

C

11 ft

FIGURE 14

EXAMPLE 4 **Designing a Roof Truss (SSS)**

Find angle B for the truss shown in Figure 14.

Solution Let $a = 11$, $b = 6$, and $c = 9$ in the law of cosines.

$$b^2 = a^2 + c^2 - 2ac \cos B \quad \text{Law of cosines}$$

$$2ac \cos B = a^2 + c^2 - b^2 \quad \text{Add } 2ac \cos B\text{; subtract } b^2.$$

$$\cos B = \frac{a^2 + c^2 - b^2}{2ac} \quad \text{Solve for } \cos B.$$

$$\cos B = \frac{11^2 + 9^2 - 6^2}{2(11)(9)} \quad \text{Substitute.}$$

$$\cos B = .8384 \quad \text{Use a calculator.}$$

$$B = 33° \quad \text{Use the inverse cosine function.}$$

Four cases can occur in solving an oblique triangle. These cases are summarized in the following table, along with a suggested procedure for solving each one. Other procedures will work too, but we give the one that is usually most efficient. In all four cases, it is assumed that the given information actually produces a triangle.

Oblique Triangle	Suggested Procedure for Solving
Case 1: One side and two angles are known. **(SAA or ASA)**	***Step 1*** Use the angle sum formula ($A + B + C = 180°$) to find the remaining angle. ***Step 2*** Use the law of sines to find the remaining sides.
Case 2: Two sides and one angle (not included between the two sides) are known. **(SSA)**	*This is the ambiguous case; there may be no triangle, one triangle, or two triangles.* ***Step 1*** Use the law of sines to find an angle. ***Step 2*** Use the angle sum formula to find the remaining angle. ***Step 3*** Use the law of sines to find the remaining side. *If two triangles exist, repeat Steps 2 and 3.*
Case 3: Two sides and the included angle are known. **(SAS)**	***Step 1*** Use the law of cosines to find the third side. ***Step 2*** Use the law of sines to find the smaller of the two remaining angles. ***Step 3*** Use the angle sum formula to find the remaining angle.
Case 4: Three sides are known. **(SSS)**	***Step 1*** Use the law of cosines to find the largest angle. ***Step 2*** Use the law of sines to find either remaining angle. ***Step 3*** Use the angle sum formula to find the remaining angle.

FOR DISCUSSION

Discuss the steps that you would take to solve each triangle. Do not actually solve the triangle.

1. $A = 32°, B = 67°, a = 20$
2. $b = 22, c = 15, A = 39°$
3. $a = 5, c = 7, C = 22°$
4. $b = 13, c = 15, B = 22°$
5. $a = 5, b = 7, c = 9$

Area Formulas

The law of cosines can be used to derive a formula for the area of a triangle, given the lengths of the three sides. This formula is known as **Heron's formula,** named after the Greek mathematician Heron of Alexandria, who lived around A.D. 75. It is found in his work *Metrica.* Heron's formula can be used for the case SSS.

Heron of Alexandria

Heron's Area Formula (SSS)

If a triangle has sides of lengths a, b, and c, and if the **semiperimeter** is

$$s = \frac{1}{2}(a + b + c),$$

then the area of the triangle is

$$\mathcal{A} = \sqrt{s(s - a)(s - b)(s - c)}.$$

EXAMPLE 5 Using Heron's Formula to Find an Area

The distance "as the crow flies" from Los Angeles to New York is 2451 miles, from New York to Montreal is 331 miles, and from Montreal to Los Angeles is 2427 miles. What is the area of the triangular region having these three cities as vertices? (Ignore the curvature of Earth.)

Montreal
c = 2427 mi
b = 331 mi
Los Angeles a = 2451 mi New York
Not to scale

FIGURE 15

Solution In Figure 15, we let $a = 2451$, $b = 331$, and $c = 2427$. Then the semiperimeter is

$$s = \frac{1}{2}(2451 + 331 + 2427) = 2604.5.$$

From Heron's formula, the area is

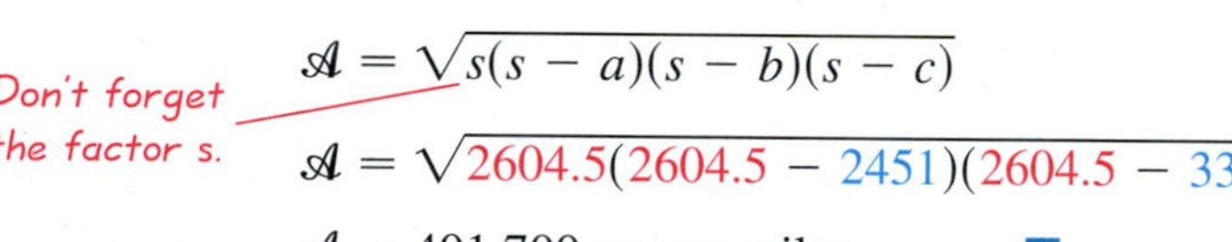

$$\mathcal{A} = \sqrt{s(s - a)(s - b)(s - c)}$$

$$\mathcal{A} = \sqrt{2604.5(2604.5 - 2451)(2604.5 - 331)(2604.5 - 2427)}$$

$$\mathcal{A} = 401{,}700 \text{ square miles.}$$ ■

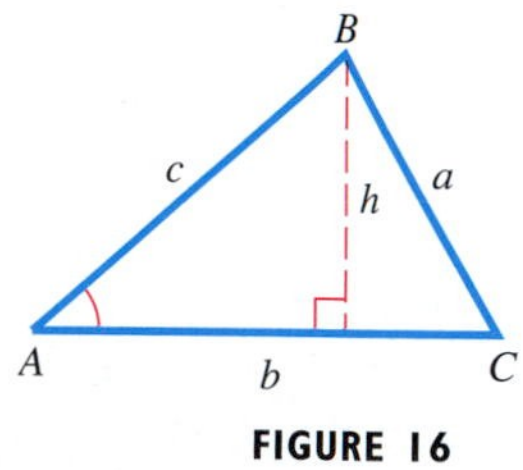

FIGURE 16

If we know the measures of two sides of a triangle and the angle included between them, we can find the area $\mathcal{A}$ of the triangle. This area is given by $\mathcal{A} = \frac{1}{2}bh$, where b is the base of the triangle and h is the height. Using trigonometry, we can find a formula for the area of the triangle shown in Figure 16.

$$\sin A = \frac{h}{c} \quad \text{or} \quad h = c \sin A$$

Thus, the area equals

$$\mathcal{A} = \frac{1}{2}bh = \frac{1}{2}bc \sin A.$$

Since the labels for the vertices in triangle ABC could be rearranged, three area formulas can be written. Notice that these formulas can be applied when SAS is given.

Area of a Triangle (SAS)

In any triangle ABC, the area $\mathcal{A}$ is given by the following formulas:

$$\mathcal{A} = \frac{1}{2}bc \sin A, \quad \mathcal{A} = \frac{1}{2}ab \sin C, \quad \text{and} \quad \mathcal{A} = \frac{1}{2}ac \sin B.$$

That is, the area is half the product of the lengths of two sides and the sine of the angle included between them.

EXAMPLE 6 Finding the Area of a Triangle (SAS)

Find the area of triangle ABC in Figure 17.

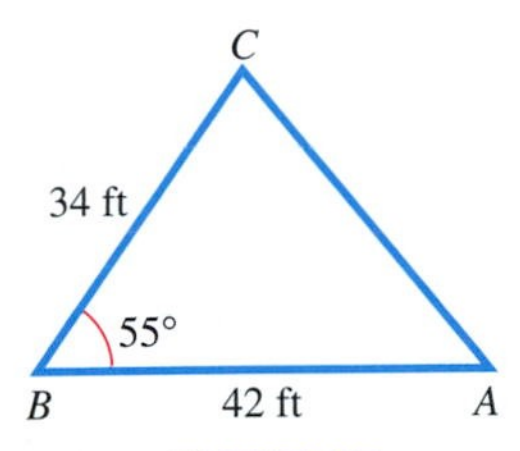

FIGURE 17

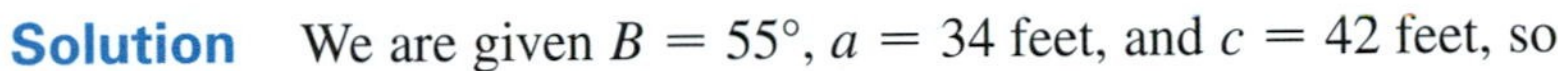

Solution We are given $B = 55°$, $a = 34$ feet, and $c = 42$ feet, so

$$\begin{aligned}\mathcal{A} &= \frac{1}{2}ac \sin B \\ &= \frac{1}{2}(34)(42)\sin 55° \\ &= 585 \text{ square feet.}\end{aligned}$$

EXAMPLE 7 Finding the Area of a Triangle (ASA)

Find the area of triangle ABC if $A = 24°\,40'$, $b = 27.3$ centimeters, and $C = 52°\,40'$.

Solution Before the area formula can be used, we must find either a or c. Since the sum of the measures of the angles of any triangle is 180°,

$$B = 180° - 24°\,40' - 52°\,40' = 102°\,40'.$$

We can use the law of sines to find a.

$$\begin{aligned}\frac{a}{\sin A} &= \frac{b}{\sin B} \\ \frac{a}{\sin 24°\,40'} &= \frac{27.3}{\sin 102°\,40'} \\ a &= 11.7 \text{ centimeters}\end{aligned}$$

Now, we find the area.

$$\begin{aligned}\mathcal{A} &= \frac{1}{2}ab \sin C \\ &= \frac{1}{2}(11.7)(27.3)\sin 52°\,40' \\ &= 127 \text{ square centimeters}\end{aligned}$$

What Went Wrong?

A student evaluates the expression in Example 7 for calculating the area of a triangle. The calculator screen shows the result, which does not equal 127 square centimeters.

What Went Wrong? What should the student do to correct the situation?

Answer to What Went Wrong?

The calculator should be set to degree mode rather than radian mode.

10.2 Exercises

Concept Check *Assume triangle ABC has standard labeling and complete the following.*

(a) *Determine whether* SAA, ASA, SSA, SAS, *or* SSS *is given.*

(b) *Decide whether the law of sines or the law of cosines should be used to begin solving the triangle.*

1. a, b, and C **2.** A, C, and c **3.** a, b, and A **4.** a, b, and c

5. A, B, and c **6.** a, c, and A **7.** a, B, and C **8.** b, c, and A

Find the length of the remaining side of each triangle. Do not use a calculator.

9.

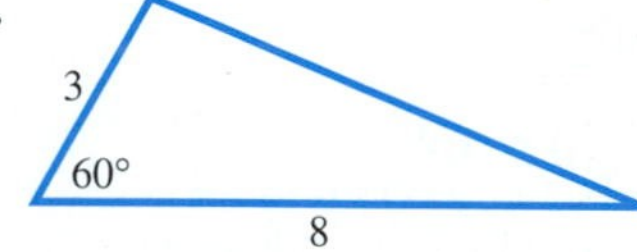

10.

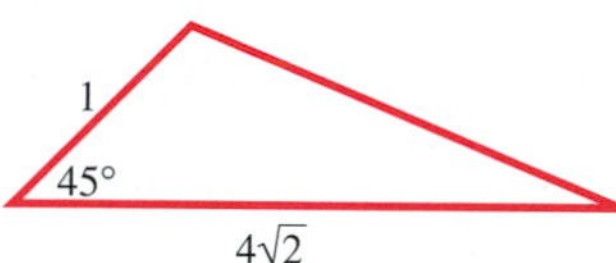

Find the value of θ in each triangle. Do not use a calculator.

11.

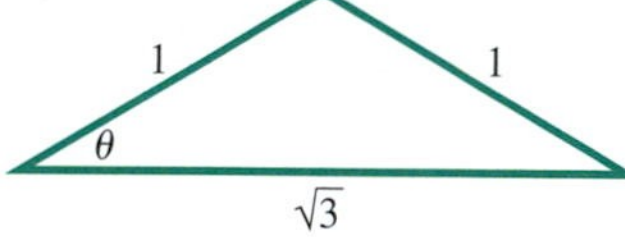

12.

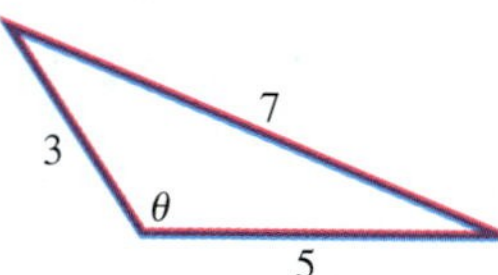

Solve each triangle. Approximate values to the nearest tenth.

13.

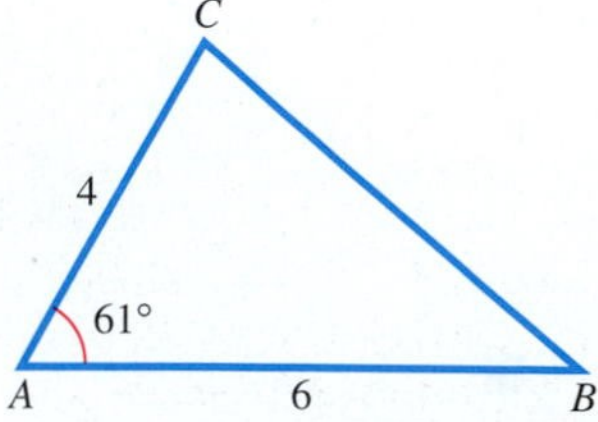

14.

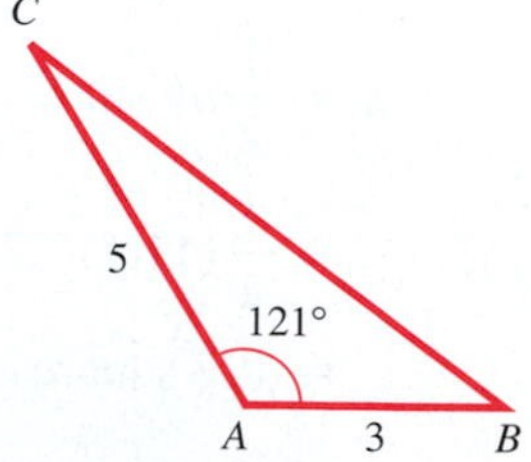

15.

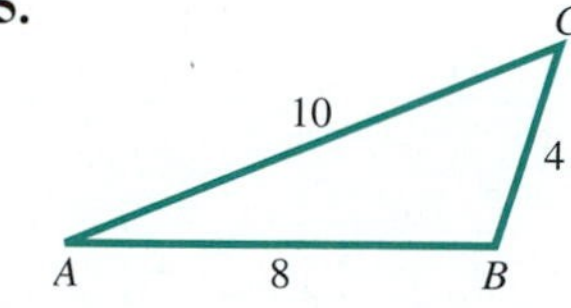

16. 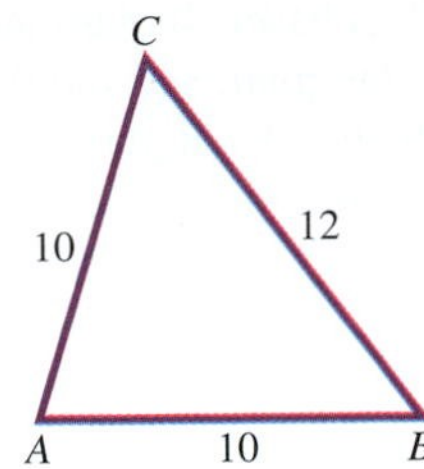

17. 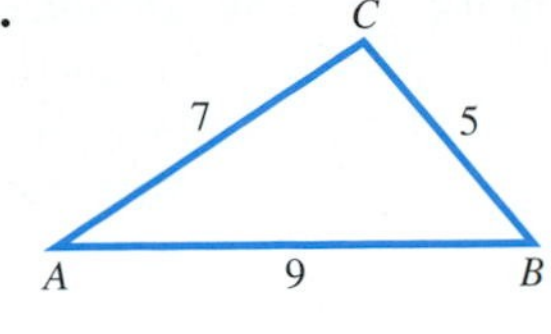

18. 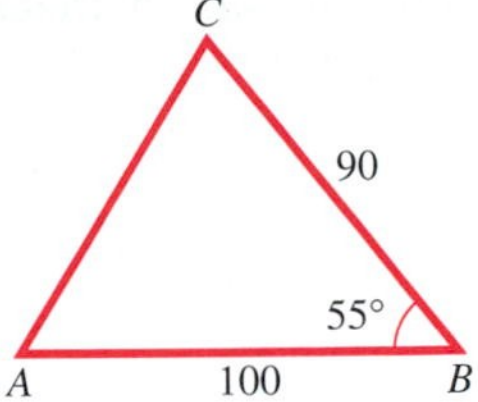

Solve each triangle.

19. $C = 28.3°$, $b = 5.71$ inches, $a = 4.21$ inches

20. $A = 41.4°$, $b = 2.78$ yards, $c = 3.92$ yards

21. $C = 45.6°$, $b = 8.94$ meters, $a = 7.23$ meters

22. $A = 67.3°$, $b = 37.9$ kilometers, $c = 40.8$ kilometers

23. $a = 9.3$ centimeters, $b = 5.7$ centimeters, $c = 8.2$ centimeters

24. $a = 28$ feet, $b = 47$ feet, $c = 58$ feet

25. $a = 42.9$ meters, $b = 37.6$ meters, $c = 62.7$ meters

26. $a = 189$ yards, $b = 214$ yards, $c = 325$ yards

27. $AB = 1240$ feet, $AC = 876$ feet, $BC = 965$ feet

28. $AB = 298$ meters, $AC = 421$ meters, $BC = 324$ meters

29. $A = 80° \, 40'$, $b = 143$ centimeters, $c = 89.6$ centimeters

30. $C = 72° \, 40'$, $a = 327$ feet, $b = 251$ feet

31. $B = 74.80°$, $a = 8.919$ inches, $c = 6.427$ inches

32. $C = 59.70°$, $a = 3.725$ miles, $b = 4.698$ miles

33. $A = 112.8°$, $b = 6.28$ meters, $c = 12.2$ meters

34. $B = 168.2°$, $a = 15.1$ centimeters, $c = 19.2$ centimeters

35. $a = 3.0$ feet, $b = 5.0$ feet, $c = 6.0$ feet

36. $a = 4.0$ feet, $b = 5.0$ feet, $c = 8.0$ feet

Solve each problem.

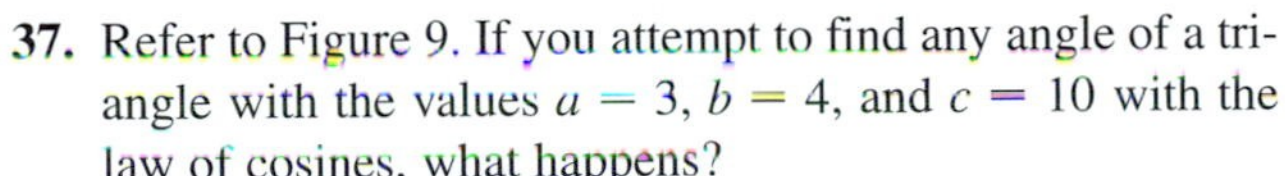

37. Refer to Figure 9. If you attempt to find any angle of a triangle with the values $a = 3$, $b = 4$, and $c = 10$ with the law of cosines, what happens?

38. "The shortest distance between two points is a straight line." Explain how this relates to the geometric property which states that the sum of the lengths of any two sides of a triangle must be greater than the remaining side.

39. ***Distance across a Lake*** Points A and B are on opposite sides of Lake Yankee. From a third point, C, the angle between the lines of sight to A and B is 46.3°. If AC is 350 meters long and BC is 286 meters long, find AB.

40. ***Diagonals of a Parallelogram*** The sides of a parallelogram are 4.0 centimeters and 6.0 centimeters. One angle is 58° while another is 122°. Find the lengths of the diagonals of the parallelogram.

41. ***Flight Distance*** Airports A and B are 450 kilometers apart, on an east–west line. Tom flies in a northeast direction from A to airport C. From C, he flies 359 kilometers on a heading of 128° 40′ to B. How far is C from A?

42. ***Distance between Two Ships*** Two ships leave a harbor together, traveling on courses that have an angle of 135° 40′ between them. If they each travel 402 miles, how far apart are they?

43. ***Distance between a Ship and a Rock*** A ship is sailing east. At one point, the bearing of a submerged rock is 45° 20′. After the ship has sailed 15.2 miles, the bearing of the rock has become 308° 40′. Find the distance of the ship from the rock at the latter point.

44. ***Distance between a Ship and a Submarine*** From an airplane flying over the ocean, the angle of depression to a submarine lying just under the surface is 24° 10′. At the same moment, the angle of depression from the airplane to a battleship is 17° 30′. (See the figure.) The distance from the airplane to the battleship is 5120 feet. Find the distance between the battleship and the submarine. (Assume that the airplane, submarine, and battleship are in a vertical plane.)

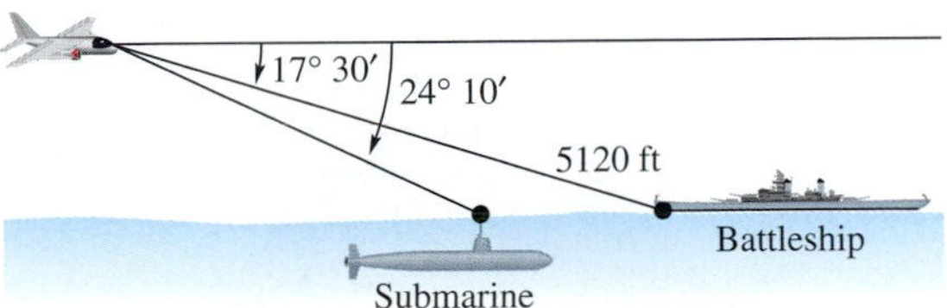

45. ***Distance between Two Boats*** Two boats leave a dock together. Each travels in a straight line. The angle between their courses measures 54° 10′. One boat travels 36.2 kilometers per hour and the other 45.6 kilometers per hour. How far apart will they be after 3 hours?

46. ***Truss Construction*** A triangular truss is shown in the figure. Find angle θ.

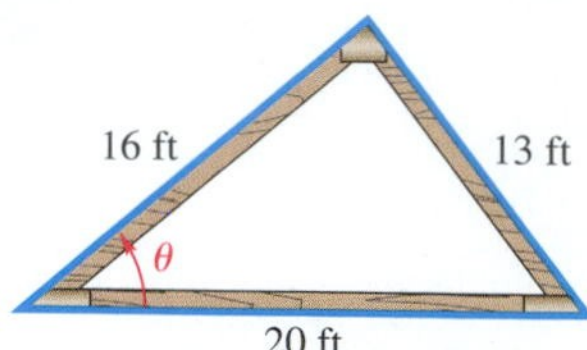

47. ***Distance between Points on a Crane*** A crane with a counterweight is shown in the figure. Find the horizontal distance between points A and B.

48. ***Distance between a Beam and Cables*** A weight is supported by cables attached to both ends of a balance beam, as shown in the figure. What angles are formed between the beam and the cables?

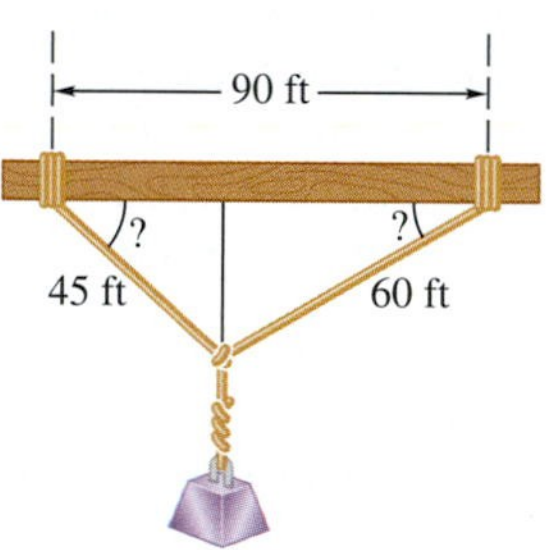

49. ***Length of a Tunnel*** To measure the distance through a mountain for a proposed tunnel, a point C is chosen that can be reached from each end of the tunnel. (See the figure.) If $AC = 3800$ meters, $BC = 2900$ meters, and angle $C = 110°$, find the length of the tunnel.

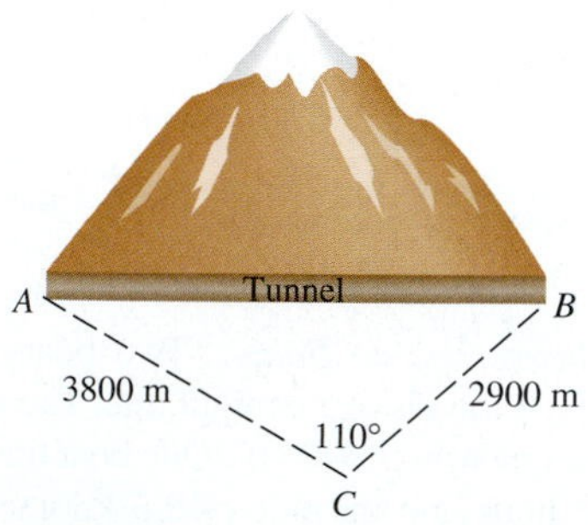

50. ***Distance on a Baseball Diamond*** A baseball diamond is a square 90 feet on a side, with home plate and the three bases as vertices. The pitcher's rubber is located 60.5 feet from home plate. Find the distance from the pitcher's rubber to each of the bases.

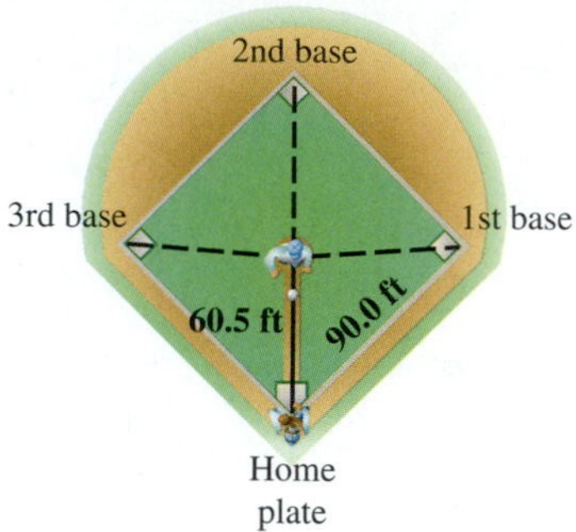

51. ***Distance between Ends of the Vietnam Memorial*** The Vietnam Veterans Memorial in Washington, DC, is V-shaped with equal sides of length 246.75 feet, and the angle between these sides measuring 125° 12′. Find the distance between the ends of the two sides. (*Source:* Pamphlet obtained at Vietnam Veterans Memorial.)

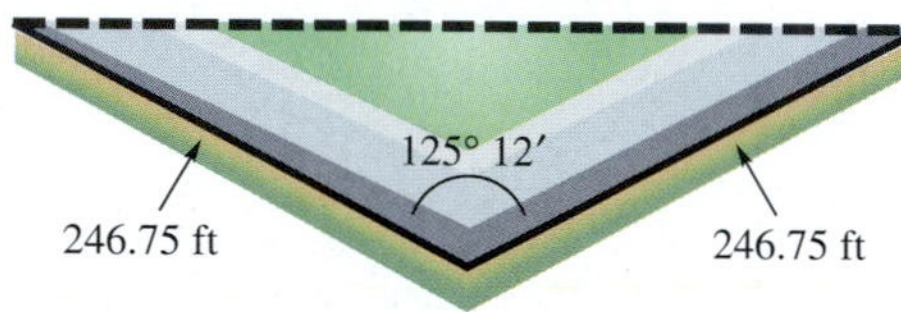

52. ***Distance between a Ship and a Point*** Starting at point A, a ship sails 18.5 kilometers on a bearing of 189°, then turns and sails 47.8 kilometers on a bearing of 317°. Find the distance of the ship from point A.

53. ***Bearing of One Town to Another*** Two towns 21 miles apart are separated by a dense forest. (See the figure.) To travel from town A to town B, a person must go 17 miles on a bearing of 325°, then turn and continue for 9 miles to reach town B. Find the bearing of B from A.

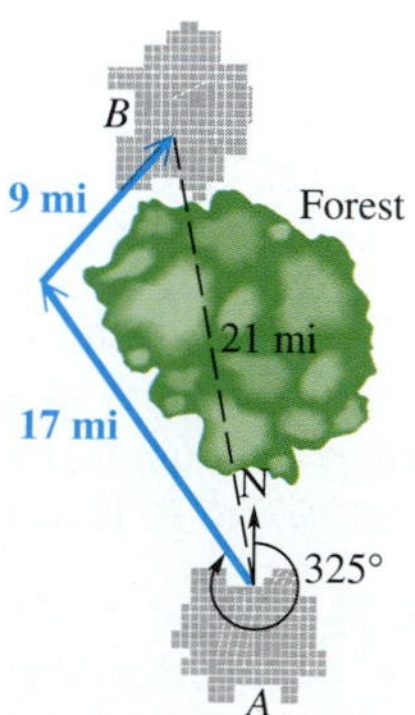

54. ***Distance between Two Factories*** Two factories blow their whistles at exactly 5:00. A man hears the two blasts at 3 seconds and 6 seconds after 5:00, respectively. The angle between his lines of sight to the two factories is 42.2°. If sound travels 344 meters per second, how far apart are the factories?

55. ***Distance between a Satellite and a Tracking Station*** A satellite traveling in a circular orbit 1600 kilometers above Earth is due to pass directly over a tracking station at noon. (See the figure.) Assume that the satellite takes 2 hours to make an orbit and that the radius of Earth is 6400 kilometers. Find the distance between the satellite and the tracking station at 12:03 P.M. (*Source: Space Mathematics* by B. Kastner, Ph.D. Copyright © 1972 by the National Aeronautics and Space Administration. Courtesy of NASA.)

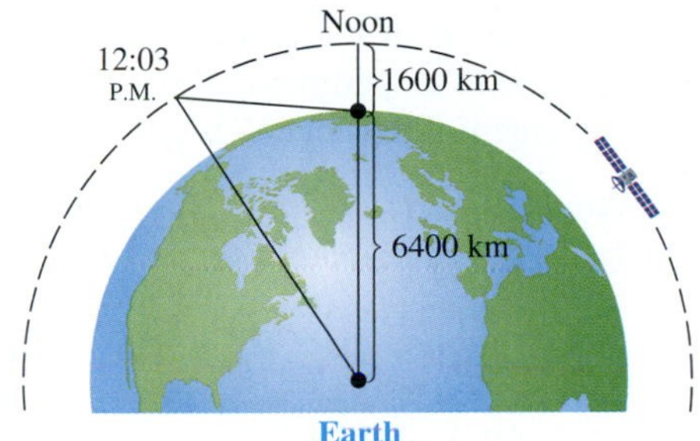

56. ***Path of a Ship*** A ship sailing due east in the North Atlantic has been warned to change course to avoid icebergs. The captain turns and sails on a bearing of 62°, then changes course again to a bearing of 115° until the ship reaches its original course. (See the figure.) How much farther did the ship have to travel to avoid the icebergs?

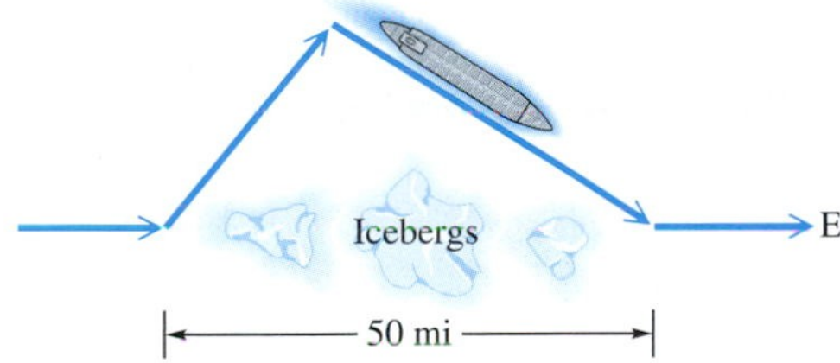

57. ***Angle in a Parallelogram*** A parallelogram has sides of length 25.9 centimeters and 32.5 centimeters. The longer diagonal has length 57.8 centimeters. Find the angle opposite the longer diagonal.

58. ***Distance between an Airplane and a Mountain*** A person in a plane flying straight north observes a mountain at a bearing of 24.1°. At the time, the plane is 7.92 kilometers from the mountain. A short time later, the bearing to the mountain becomes 32.7°. How far is the airplane from the mountain when the second bearing is taken?

59. ***Layout for a Playhouse*** The layout for a child's playhouse has the dimensions given in the figure. Find x.

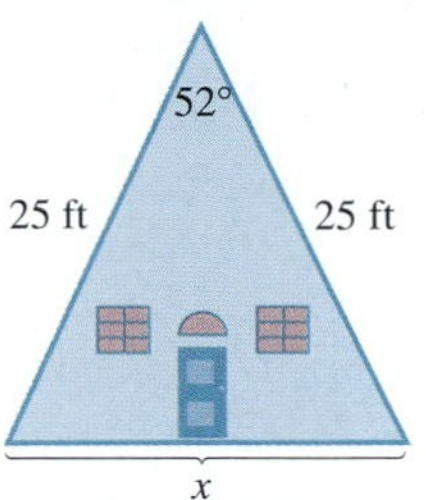

60. ***Distance between Two Towns*** To find the distance between two small towns, an electronic distance-measuring (EDM) instrument is placed on a hill from which both towns are visible. The distance to each town from the EDM and the angle between the two lines of sight are measured. (See the figure.) Find the distance between the towns.

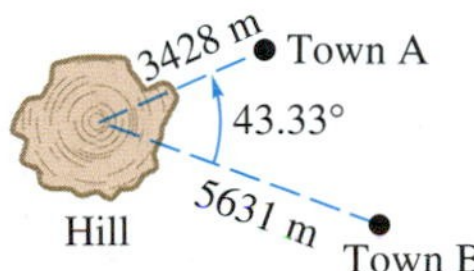

Relating Concepts

For individual or group investigation (Exercises 61–64)

In any triangle, the longest side is opposite the largest angle. This result from geometry was proven for acute triangles in Exercises 80–84 in Section 10.1. To prove it for obtuse triangles, ***work Exercises 61–64 in order.***

61. Suppose angle A is the largest angle of an obtuse triangle. Why is $\cos A$ negative?

62. Consider the law of cosines expression for a^2, and show that $a^2 > b^2 + c^2$.

63. Use the result of Exercise 62 to show that $a > b$ and $a > c$.

64. Use the result of Exercise 63 to explain why no triangle ABC satisfies $A = 103°$, $a = 25$, and $c = 30$.

Find the area of each triangle.

65. $A = 42.5°$, $b = 13.6$ meters, $c = 10.1$ meters

66. $B = 124.5°$, $a = 30.4$ centimeters, $c = 28.4$ centimeters

67. $a = 12$ meters, $b = 16$ meters, $c = 25$ meters

68. $a = 154$ centimeters, $b = 179$ centimeters, $c = 183$ centimeters

69. $a = 76.3$ feet, $b = 109$ feet, $c = 98.8$ feet

70. $a = 22$ inches, $b = 45$ inches, $c = 31$ inches

71. $a = 25.4$ yards, $b = 38.2$ yards, $c = 19.8$ yards

72. $a = 15.89$ inches, $b = 21.74$ inches, $c = 10.92$ inches

Solve each problem.

73. ***Cans of Paint Needed for a Job*** A painter needs to cover a triangular region 75 meters by 68 meters by 85 meters. A can of paint covers 75 square meters of area. How many cans (to the next higher number of cans) will be needed?

74. ***Area of a Triangular Lot*** A real estate agent wants to find the area of a triangular lot. A surveyor takes measurements and finds that two sides are 52.1 meters and 21.3 meters, and the angle between them is 42.2°. What is the area of the lot?

75. ***Area of a Metal Plate*** A painter is going to apply a special coating to a triangular metal plate on a new building. Two sides measure 16.1 meters and 15.2 meters. She knows that the angle between these sides is 125°. What is the area of the surface she plans to cover with the coating?

76. ***Area of the Bermuda Triangle*** Find the area of the Bermuda Triangle if the sides of the triangle have approximate lengths of 850 miles, 925 miles, and 1300 miles.

77. ***Perfect Triangles*** A *perfect triangle* is a triangle whose sides have whole-number lengths and whose area is numerically equal to its perimeter. Show that the triangle with sides of lengths 9, 10, and 17 is perfect.

78. ***Heron Triangles*** A *Heron triangle* is a triangle having integer sides and integer area. Show that each of the following is a Heron triangle.

(a) $a = 11, b = 13, c = 20$
(b) $a = 13, b = 14, c = 15$
(c) $a = 7, b = 15, c = 20$

10.3 Vectors and Their Applications

Basic Terminology ■ Algebraic Interpretation of Vectors ■ Operations with Vectors ■ Dot Product and the Angle between Vectors ■ Applications of Vectors

Basic Terminology

Many quantities involve magnitudes, such as 45 pounds or 60 mph. These quantities are called **scalars** and can be represented by real numbers. Other quantities, called **vector quantities,** involve both magnitude and direction. Typical vector quantities are velocity, acceleration, and force. For example, traveling 50 mph *east* represents a vector quantity.

A vector quantity is often represented with a directed line segment, called a **vector.** The length of the vector represents the **magnitude** of the vector quantity. The direction of the vector, indicated with an arrow, represents the direction of the quantity. For example, the vector in Figure 18 represents a force of 10 pounds applied at an angle of 30° from the horizontal.

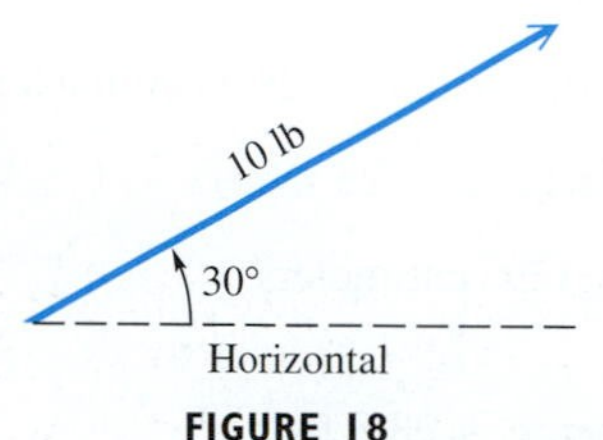

FIGURE 18

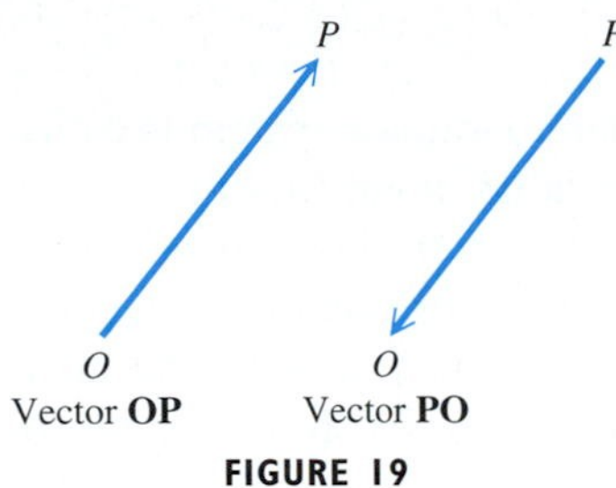

FIGURE 19

The symbol for a vector is often printed in boldface type. To write vectors by hand, it is customary to use an arrow over the letter or letters. Thus, **OP** and $\overrightarrow{OP}$ both represent vector **OP**. Vectors may be named with either one lowercase or uppercase letter, or two uppercase letters. When two letters are used, the first indicates the **initial point** and the second indicates the **terminal point** of the vector. Knowing these points gives the direction of the vector. For example, vectors **OP** and **PO** in Figure 19 are *not* the same vectors. They have the same magnitude, but *opposite directions.* The magnitude of vector **OP** is written $|\mathbf{OP}|$.

Two vectors are equal if and only if they both have the same direction and the same magnitude. In Figure 20, vectors **A** and **B** are equal, as are vectors **C** and **D**.

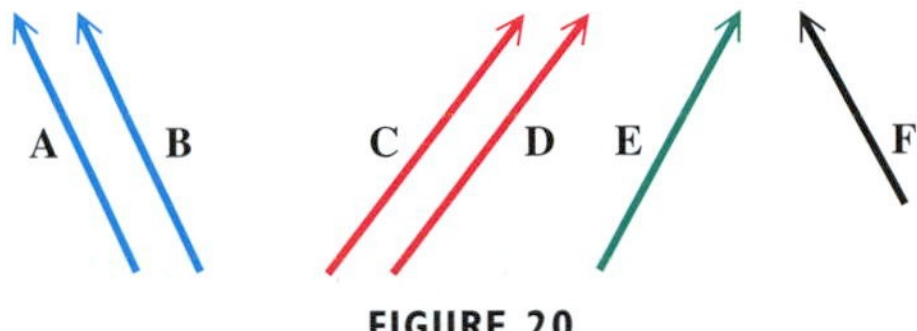

FIGURE 20

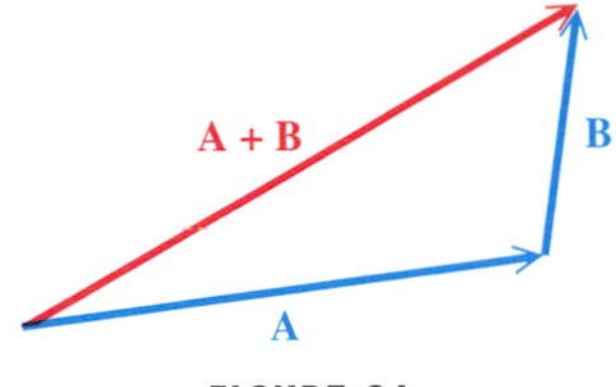

FIGURE 21

To find the sum of two vectors **A** and **B**, we place the initial point of vector **B** at the terminal point of vector **A**, as shown in Figure 21. The vector with the same initial point as **A** and the same terminal point as **B** is the sum **A** + **B**. The sum of two vectors is also a vector.

Another way to find the sum of two vectors is to use the **parallelogram rule.** Place vectors **A** and **B** so that their initial points coincide. Then, complete a parallelogram that has **A** and **B** as two sides. The diagonal of the parallelogram with the same initial point as **A** and **B** is the sum **A** + **B** found by the definition. Compare Figures 21 and 22. Parallelograms can be used to show that vector **B** + **A** is the same as vector **A** + **B** or that **A** + **B** = **B** + **A**, so vector addition is commutative. The vector sum **A** + **B** is called the **resultant** of vectors **A** and **B**.

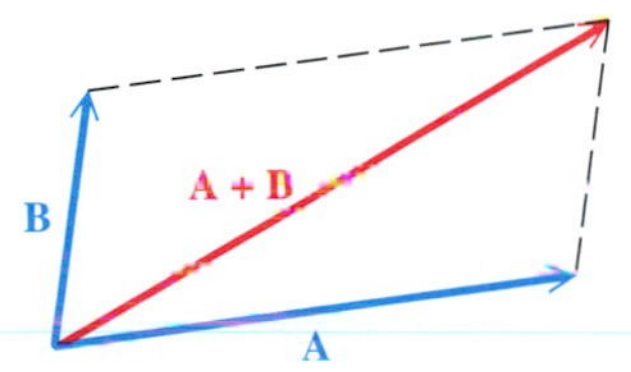

FIGURE 22

For every vector **v**, there is a vector −**v** that has the same magnitude as **v** but opposite direction. Vector −**v** is called the **opposite** of **v**. See Figure 23. The sum of **v** and −**v** has magnitude 0 and is called the **zero vector.** As with real numbers, to subtract vector **B** from vector **A**, find the vector sum **A** + (−**B**). See Figure 24.

v
−v

Vectors **v** and −**v** are opposites.

FIGURE 23

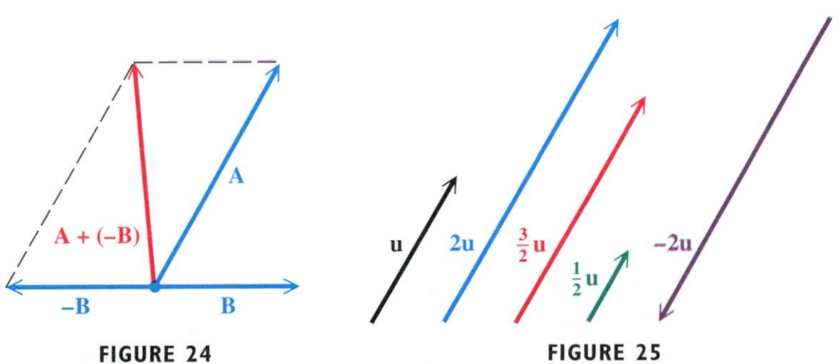

FIGURE 24 **FIGURE 25**

The **scalar product** of a real number (or scalar) k and a vector **u** is the vector $k \cdot \mathbf{u}$, which has magnitude $|k|$ times the magnitude of **u**. As suggested by Figure 25, the vector $k \cdot \mathbf{u}$ has the same direction as **u** if $k > 0$ and opposite direction if $k < 0$.

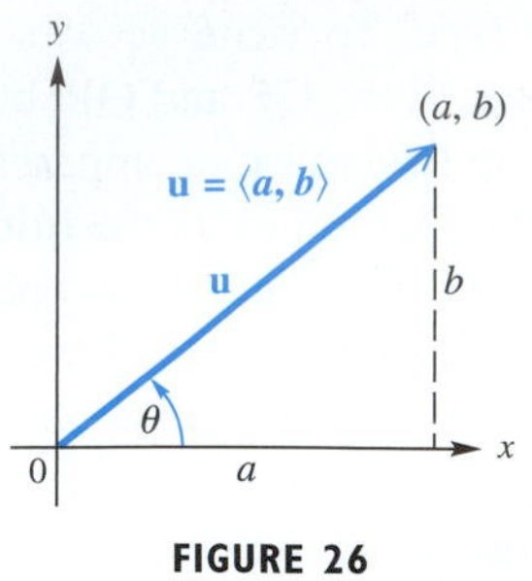

FIGURE 26

Looking Ahead to Calculus
In addition to two-dimensional vectors in a plane, calculus courses introduce three-dimensional vectors in space. (See Appendix C.) The magnitude of the two-dimensional vector $\langle a, b\rangle$ is given by $\sqrt{a^2 + b^2}$. If we extend this to the three-dimensional vector $\langle a, b, c\rangle$, the expression becomes $\sqrt{a^2 + b^2 + c^2}$. Similar extensions are made for other concepts.

Algebraic Interpretation of Vectors

A vector with its initial point at the origin in a rectangular coordinate system is called a **position vector.** A position vector **u** with its endpoint at the point (a, b) is written $\langle a, b\rangle$, so $\mathbf{u} = \langle a, b\rangle$. This means that every vector in the real plane corresponds to an ordered pair of real numbers. Thus, geometrically, a vector is a directed line segment; algebraically, it is an ordered pair. The numbers a and b are, respectively, the **horizontal component** and **vertical component** of vector **u**. Figure 26 shows the vector $\mathbf{u} = \langle a, b\rangle$. The positive angle between the x-axis and a position vector is called the **direction angle** for the vector. In Figure 26, θ is the direction angle for vector **u**.

From Figure 26, we can see that the magnitude and direction of a vector are related to its horizontal and vertical components.

Magnitude and Direction Angle of a Vector $\langle a, b\rangle$

The magnitude (length) of vector $\mathbf{u} = \langle a, b\rangle$ is given by

$$|\mathbf{u}| = \sqrt{a^2 + b^2}.$$

The direction angle θ satisfies $\tan\theta = \frac{b}{a}$, where $a \neq 0$.

GCM **EXAMPLE 1** **Finding Magnitude and Direction Angle**

Find the magnitude and direction angle of $\mathbf{u} = \langle 3, -2\rangle$.

Analytic Solution

The magnitude of vector **u** is

$$|\mathbf{u}| = \sqrt{3^2 + (-2)^2} = \sqrt{13}.$$

To find the direction angle θ, start with $\tan\theta = \frac{b}{a} = \frac{-2}{3} = -\frac{2}{3}$. Vector **u** has a positive horizontal component and a negative vertical component, placing the position vector in quadrant IV. A calculator gives $\tan^{-1}\left(-\frac{2}{3}\right) \approx -33.7°$. Adding 360° yields the direction angle $\theta = 326.3°$. See Figure 27.

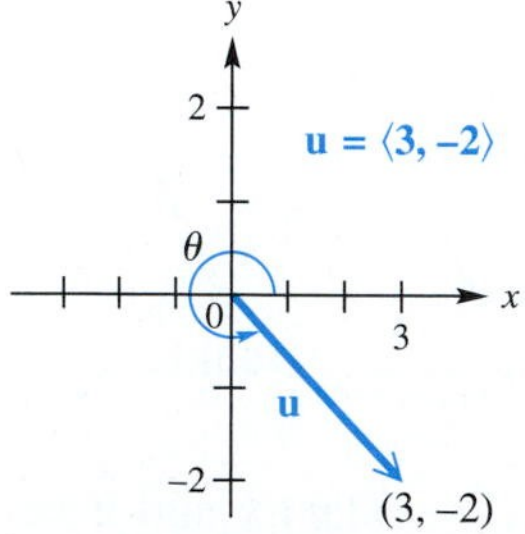

FIGURE 27

Graphing Calculator Solution

A calculator returns the magnitude and direction angle, given the horizontal and vertical components. An approximation for $\sqrt{13}$ is given, and the direction angle is returned as a measure with the smallest possible absolute value. We must add 360° to the value of θ to obtain the least positive direction angle. See Figure 28.

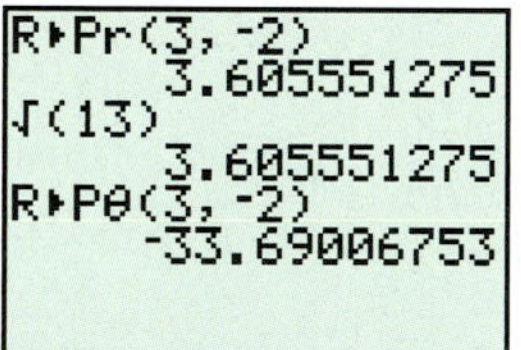

FIGURE 28

The TI-83/84 Plus calculator will always return an angle θ in the interval $(-180°, 180°]$ (or $(-\pi, \pi]$) for a conversion such as this.

Horizontal and Vertical Components

The horizontal and vertical components, respectively, of a vector **u** having magnitude $|\mathbf{u}|$ and direction angle θ are given by

$$a = |\mathbf{u}| \cos \theta \quad \text{and} \quad b = |\mathbf{u}| \sin \theta.$$

That is, $\mathbf{u} = \langle a, b \rangle = \langle |\mathbf{u}| \cos \theta, |\mathbf{u}| \sin \theta \rangle$.

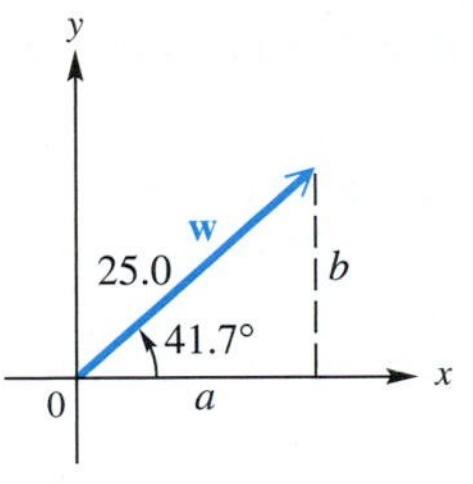

FIGURE 29

GCM **EXAMPLE 2** **Finding Horizontal and Vertical Components**

Vector **w** in Figure 29 has magnitude 25.0 and direction angle 41.7°. Find the horizontal and vertical components.

Analytic Solution

Use the two formulas in the box, with $|\mathbf{w}| = 25.0$ and $\theta = 41.7°$.

$$a = 25.0 \cos 41.7° \qquad b = 25.0 \sin 41.7°$$
$$a = 18.7 \qquad b = 16.6$$

Therefore, $\mathbf{w} = \langle 18.7, 16.6 \rangle$. The horizontal component is 18.7, and the vertical component is 16.6 (rounded to the nearest tenth).

Graphing Calculator Solution

See Figure 30. The calculator is in degree mode.

```
P▸Rx(25.0,41.7)
             18.7
P▸Ry(25.0,41.7)
             16.6
```

FIGURE 30

EXAMPLE 3 **Writing Vectors in the Form $\langle a, b \rangle$**

Write each vector in Figure 31 in the form $\langle a, b \rangle$.

FIGURE 31

Solution

$$\mathbf{u} = \langle 5 \cos 60°, 5 \sin 60° \rangle = \left\langle 5 \cdot \tfrac{1}{2}, 5 \cdot \tfrac{\sqrt{3}}{2} \right\rangle = \left\langle \tfrac{5}{2}, \tfrac{5\sqrt{3}}{2} \right\rangle$$

$$\mathbf{v} = \langle 2 \cos 180°, 2 \sin 180° \rangle = \langle 2(-1), 2(0) \rangle = \langle -2, 0 \rangle$$

$$\mathbf{w} = \langle 6 \cos 280°, 6 \sin 280° \rangle \approx \langle 1.0419, -5.9088 \rangle \qquad \text{Use a calculator.}$$

EXAMPLE 4 **Finding the Magnitude of a Resultant**

Two forces of 15 and 22 newtons act on a point in the plane. (A *newton* is a unit of force that equals .225 pound.) If the angle between the forces is 100°, find the magnitude of the resultant force.

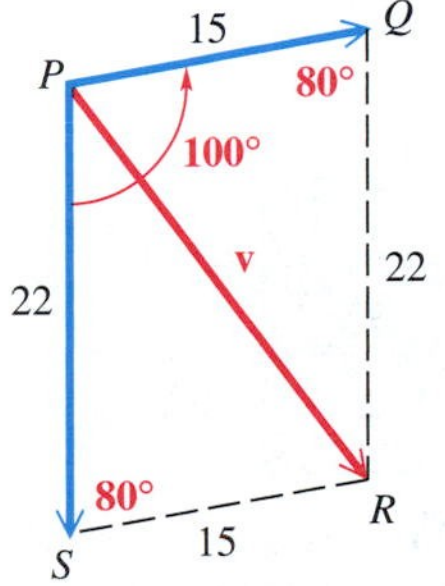

FIGURE 32

Solution As shown in Figure 32, a parallelogram that has the forces as adjacent sides can be formed. The angles of the parallelogram adjacent to angle P measure 80°, since adjacent angles of a parallelogram are supplementary. Opposite sides of the parallelogram are equal in length. The resultant force divides the parallelogram into two triangles. Use the law of cosines with either triangle to obtain

$$|\mathbf{v}|^2 = 15^2 + 22^2 - 2(15)(22) \cos 80° \qquad \text{Law of cosines}$$
$$|\mathbf{v}|^2 = 594 \qquad \text{Simplify.}$$
$$|\mathbf{v}| = 24. \qquad \text{Take the square root.}$$

To the nearest unit, the magnitude of the resultant force is 24 newtons.

In Example 4 we used a property of parallelograms. The following properties of parallelograms are helpful in studying vectors:

1. A parallelogram is a quadrilateral whose opposite sides are parallel.
2. The opposite sides and opposite angles of a parallelogram are equal, and adjacent angles of a parallelogram are supplementary.
3. The diagonals of a parallelogram bisect each other, but do not necessarily bisect the angles of the parallelogram.

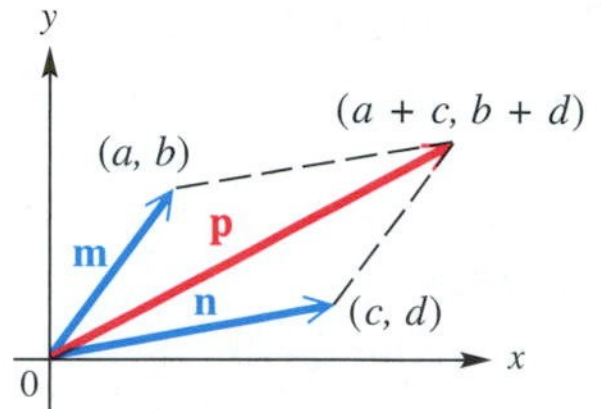

FIGURE 33

Operations with Vectors

In Figure 33, $\mathbf{m} = \langle a, b \rangle$, $\mathbf{n} = \langle c, d \rangle$, and $\mathbf{p} = \langle a + c, b + d \rangle$. Using geometry, we can show that the endpoints of the three vectors and their origin form a parallelogram. Since a diagonal of this parallelogram gives the resultant of **m** and **n**, we have $\mathbf{p} = \mathbf{m} + \mathbf{n}$, or

$$\langle a + c, b + d \rangle = \langle a, b \rangle + \langle c, d \rangle.$$

Similarly, we could verify the following vector operations.

Vector Operations

For any real numbers a, b, c, d, and k,

$$\langle a, b \rangle + \langle c, d \rangle = \langle a + c, b + d \rangle$$
$$k \cdot \langle a, b \rangle = \langle ka, kb \rangle.$$
$$\text{If } \mathbf{a} = \langle a_1, a_2 \rangle, \text{ then } -\mathbf{a} = \langle -a_1, -a_2 \rangle.$$
$$\langle a, b \rangle - \langle c, d \rangle = \langle a, b \rangle + -\langle c, d \rangle = \langle a - c, b - d \rangle$$

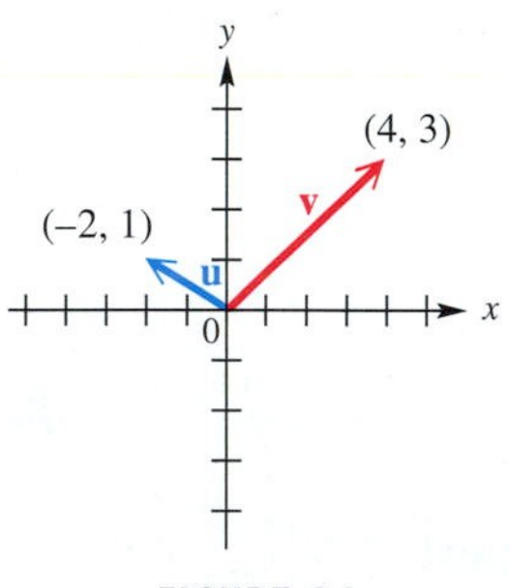

FIGURE 34

GCM **EXAMPLE 5** **Performing Vector Operations**

Let $\mathbf{u} = \langle -2, 1 \rangle$ and $\mathbf{v} = \langle 4, 3 \rangle$. Find **(a)** $\mathbf{u} + \mathbf{v}$, **(b)** $-2\mathbf{u}$, and **(c)** $4\mathbf{u} - 3\mathbf{v}$. See Figure 34.

Analytic Solution

(a) $\mathbf{u} + \mathbf{v} = \langle -2, 1 \rangle + \langle 4, 3 \rangle$
$= \langle -2 + 4, 1 + 3 \rangle$
$= \langle 2, 4 \rangle$

(b) $-2\mathbf{u} = -2 \cdot \langle -2, 1 \rangle$
$= \langle -2(-2), -2(1) \rangle$
$= \langle 4, -2 \rangle$

(c) $4\mathbf{u} - 3\mathbf{v} = 4 \cdot \langle -2, 1 \rangle - 3 \cdot \langle 4, 3 \rangle$
$= \langle -8, 4 \rangle - \langle 12, 9 \rangle$
$= \langle -8 - 12, 4 - 9 \rangle$
$= \langle -20, -5 \rangle$

Graphing Calculator Solution

Vector arithmetic can be performed with a graphing calculator, as shown in Figure 35.

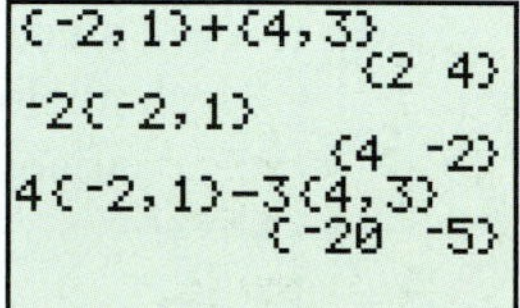

FIGURE 35

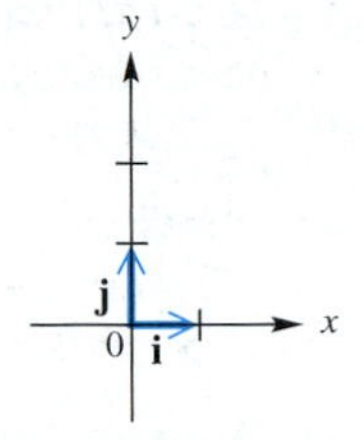

FIGURE 36

A **unit vector** is a vector that has magnitude 1. Two very useful unit vectors are defined as follows and shown in Figure 36.

$$\mathbf{i} = \langle 1,0 \rangle \qquad \mathbf{j} = \langle 0,1 \rangle$$

With the unit vectors **i** and **j**, we can express any other vector $\langle a,b \rangle$ in the form $a\mathbf{i} + b\mathbf{j}$, as shown in Figure 37, where $\langle 3,4 \rangle = 3\mathbf{i} + 4\mathbf{j}$. The vector operations previously given can be restated in $a\mathbf{i} + b\mathbf{j}$ notation.

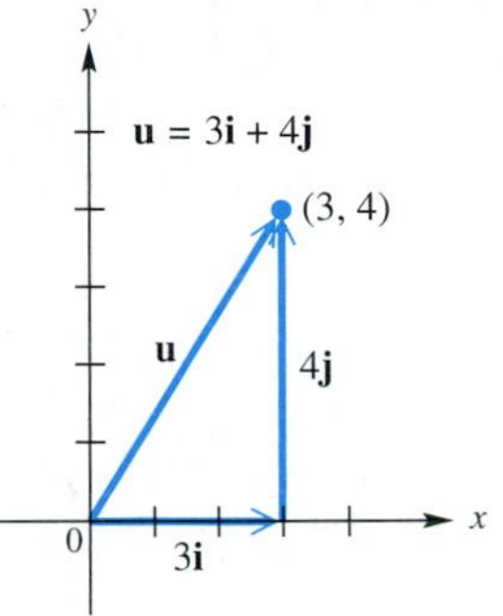

FIGURE 37

i, j Form for Vectors

If $\mathbf{v} = \langle a,b \rangle$, then $\mathbf{v} = a\mathbf{i} + b\mathbf{j}$.

Dot Product and the Angle between Vectors

The *dot product* of two vectors is a real number, not a vector. It is also known as the *inner product*. Dot products are used to determine the angle between two vectors, derive geometric theorems, and solve physics problems.

Dot Product

The **dot product** of the two vectors $\mathbf{u} = \langle a,b \rangle$ and $\mathbf{v} = \langle c,d \rangle$ is denoted $\mathbf{u} \cdot \mathbf{v}$, read "**u** dot **v**," and given by

$$\mathbf{u} \cdot \mathbf{v} = ac + bd.$$

GCM **EXAMPLE 6** **Finding the Dot Product**

Find each dot product.

(a) $\langle 2,3 \rangle \cdot \langle 4,-1 \rangle$ **(b)** $(6\mathbf{i} + 4\mathbf{j}) \cdot (-2\mathbf{i} + 3\mathbf{j})$

Solution

(a) $\langle 2,3 \rangle \cdot \langle 4,-1 \rangle = 2(4) + 3(-1) = 5$

(b) $(6\mathbf{i} + 4\mathbf{j}) \cdot (-2\mathbf{i} + 3\mathbf{j}) = \langle 6,4 \rangle \cdot \langle -2,3 \rangle = 6(-2) + 4(3) = 0$ ■

The following properties of dot products are easily verified.

Properties of the Dot Product

For all vectors **u**, **v**, and **w** and real numbers k,

(a) $\mathbf{u} \cdot \mathbf{v} = \mathbf{v} \cdot \mathbf{u}$ **(b)** $\mathbf{u} \cdot (\mathbf{v} + \mathbf{w}) = \mathbf{u} \cdot \mathbf{v} + \mathbf{u} \cdot \mathbf{w}$

(c) $(\mathbf{u} + \mathbf{v}) \cdot \mathbf{w} = \mathbf{u} \cdot \mathbf{w} + \mathbf{v} \cdot \mathbf{w}$ **(d)** $(k\mathbf{u}) \cdot \mathbf{v} = k(\mathbf{u} \cdot \mathbf{v}) = \mathbf{u} \cdot (k\mathbf{v})$

(e) $\mathbf{0} \cdot \mathbf{u} = 0$ **(f)** $\mathbf{u} \cdot \mathbf{u} = |\mathbf{u}|^2$.

To prove the first part of property (d), we let $\mathbf{u} = \langle a, b \rangle$ and $\mathbf{v} = \langle c, d \rangle$. Then

$$\begin{aligned}(k\mathbf{u}) \cdot \mathbf{v} &= (k\langle a, b \rangle) \cdot \langle c, d \rangle = \langle ka, kb \rangle \cdot \langle c, d \rangle \\ &= kac + kbd = k(ac + bd) \\ &= k(\langle a, b \rangle \cdot \langle c, d \rangle) = k(\mathbf{u} \cdot \mathbf{v}).\end{aligned}$$

The proofs of the remaining properties are similar.

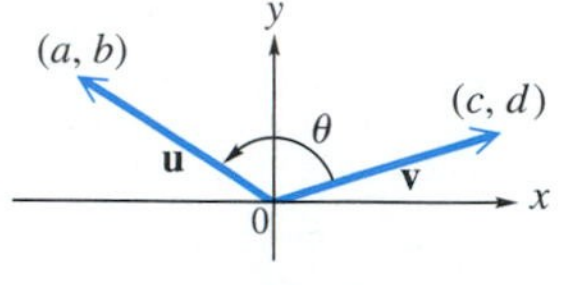

FIGURE 38

The dot product of two vectors can be positive, 0, or negative. A geometric interpretation of the dot product explains when each of these cases occurs. This interpretation involves the angle between the two vectors. Consider the vectors $\mathbf{u} = \langle a, b \rangle$ and $\mathbf{v} = \langle c, d \rangle$, as shown in Figure 38. The **angle θ between u and v** is defined to be the angle having the two vectors as its sides for which $0° \leq \theta \leq 180°$. The following theorem relates the dot product to the angle between the vectors. Its proof is outlined in Exercise 110.

Geometric Interpretation of Dot Product

If θ is the angle between the two nonzero vectors **u** and **v**, where $0° \leq \theta \leq 180°$, then

$$\mathbf{u} \cdot \mathbf{v} = |\mathbf{u}||\mathbf{v}| \cos \theta, \quad \text{or equivalently,} \quad \cos \theta = \frac{\mathbf{u} \cdot \mathbf{v}}{|\mathbf{u}||\mathbf{v}|}.$$

EXAMPLE 7 Finding the Angle between Two Vectors

Find the angle θ between the two vectors $\mathbf{u} = \langle 3, 4 \rangle$ and $\mathbf{v} = \langle 2, 1 \rangle$.

Solution Use the preceding geometric interpretation.

$$\begin{aligned}\cos \theta &= \frac{\boldsymbol{u} \cdot \boldsymbol{v}}{|\boldsymbol{u}||\boldsymbol{v}|} = \frac{\langle 3, 4 \rangle \cdot \langle 2, 1 \rangle}{|\langle 3, 4 \rangle||\langle 2, 1 \rangle|} \\ &= \frac{3(2) + 4(1)}{\sqrt{9 + 16} \cdot \sqrt{4 + 1}} = \frac{10}{5\sqrt{5}} = .894427191 \\ \theta &= \cos^{-1} .894427191 = 26.57°\end{aligned}$$

■

For angles θ between 0° and 180°, $\cos \theta$ is positive, 0, or negative when θ is less than, equal to, or greater than 90°, respectively. Therefore, the dot product is positive, 0, or negative according to this table.

Dot Product	Angle between Vectors
Positive	Acute
0	Right
Negative	Obtuse

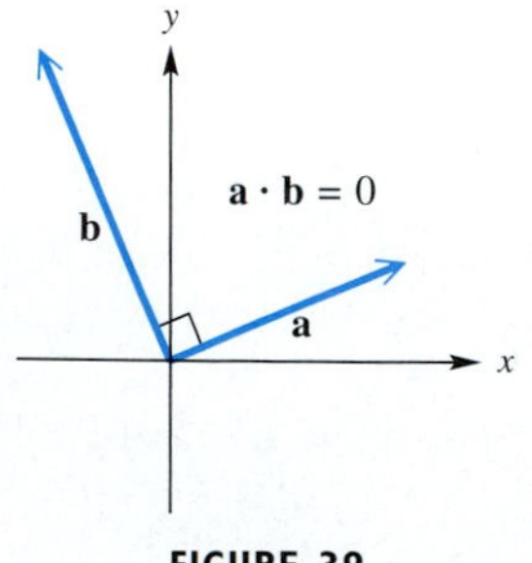

FIGURE 39

NOTE If $\mathbf{a} \cdot \mathbf{b} = 0$ for two nonzero vectors **a** and **b**, then $\cos \theta = 0$ and $\theta = 90°$. Thus, **a** and **b** are perpendicular or **orthogonal vectors.** See Figure 39.

Applications of Vectors

The law of sines and the law of cosines are often used to solve applied problems involving vectors.

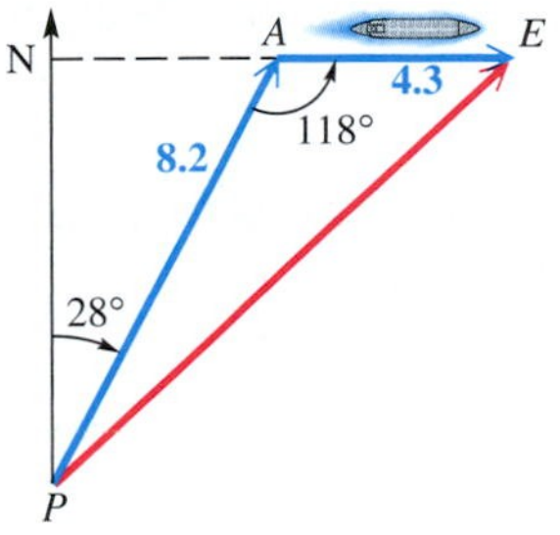

FIGURE 40

EXAMPLE 8 Applying Vectors to a Navigation Problem

A ship leaves port on a bearing of 28° and travels 8.2 miles. The ship then turns due east and travels 4.3 miles. How far is the ship from port? What is its bearing from port?

Solution In Figure 40, vectors **PA** and **AE** represent the ship's path. The magnitude and bearing of the resultant **PE** can be found as follows. Triangle *PNA* is a right triangle, so angle $NAP = 90° - 28° = 62°$. Then angle $PAE = 180° - 62° = 118°$. Use the law of cosines to find $|\mathbf{PE}|$, the magnitude of vector **PE**.

$$|\mathbf{PE}|^2 = 8.2^2 + 4.3^2 - 2(8.2)(4.3)\cos 118° \quad \text{Law of cosines}$$
$$|\mathbf{PE}|^2 = 118.84 \quad \text{Approximate.}$$
$$|\mathbf{PE}| = 10.9, \quad \text{or} \quad 11 \text{ miles} \quad \text{Take the square root; nearest integer}$$

To find the bearing of the ship from port, first find angle *APE*. Use the law of sines, along with the value of $|\mathbf{PE}|$, before rounding to 11.

$$\frac{\sin APE}{4.3} = \frac{\sin 118°}{10.9} \quad \text{Law of sines}$$
$$\sin APE = \frac{4.3 \sin 118°}{10.9} \quad \text{Multiply by 4.3.}$$
$$APE = 20° \quad \text{Use the inverse sine function.}$$

The ship is 11 miles from port at a bearing of $28° + 20° = 48°$. ■

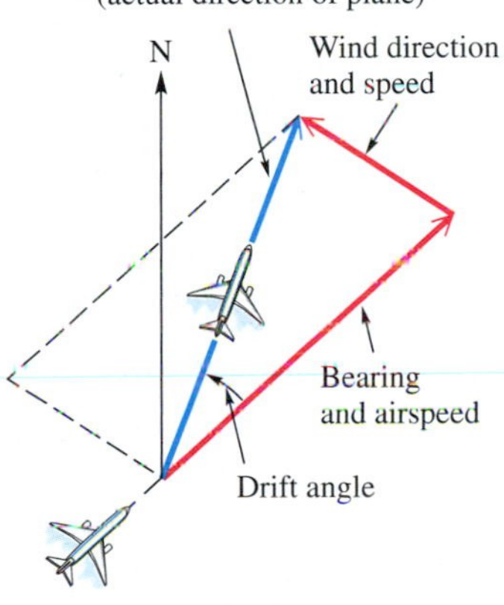

FIGURE 41

In air navigation, the **airspeed** of a plane is its speed relative to the air, while the **ground speed** is its speed relative to the ground. Because of wind, these two speeds are usually different. The ground speed of the plane is represented by the vector sum of the airspeed and wind speed vectors. See Figure 41.

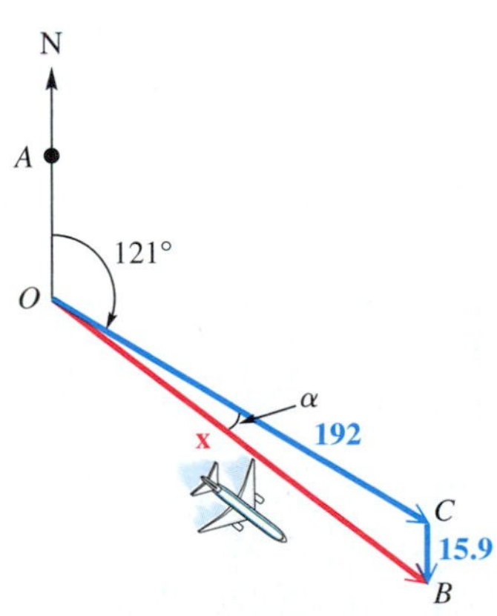

FIGURE 42

EXAMPLE 9 Applying Vectors to a Navigation Problem

A plane with an airspeed of 192 mph is headed on a bearing of 121°. A north wind is blowing (from north to south) at 15.9 mph. Find the ground speed and the actual bearing of the plane.

Solution In Figure 42, the ground speed is represented by $|\mathbf{x}|$. We must find angle α to find the bearing, which will be $121° + \alpha$. From Figure 42, angle *BCO* equals angle *AOC*, which equals 121°. We find $|\mathbf{x}|$ by using the law of cosines.

$$|\mathbf{x}|^2 = 192^2 + 15.9^2 - 2(192)(15.9)\cos 121°$$
$$|\mathbf{x}|^2 = 40{,}261$$
$$|\mathbf{x}| = 200.7, \quad \text{or} \quad 201 \text{ mph}$$

We find α by using the law of sines and the value of $|\mathbf{x}|$ before rounding.

$$\frac{\sin \alpha}{15.9} = \frac{\sin 121^\circ}{200.7}$$

$$\sin \alpha = .06792320$$

$$\alpha = 3.89^\circ \quad \text{Approximate.}$$

To the nearest degree, α is 4°. The ground speed is about 201 mph on a bearing of approximately $121^\circ + 4^\circ = 125^\circ$. ■

In the next example, we use vectors to solve an inclined plane problem.

EXAMPLE 10 Finding a Required Force

Find the force required to pull a wagon weighing 50 pounds up a ramp inclined 20° to the horizontal. (Assume that there is no friction.)

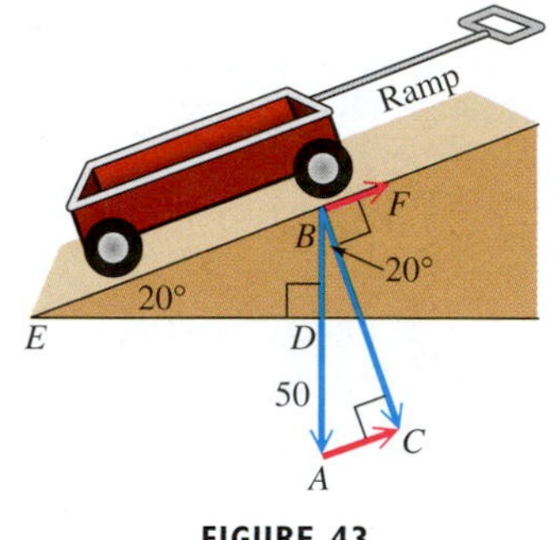

FIGURE 43

Solution In Figure 43, the vertical 50-pound force **BA** represents the force of gravity. It is the sum of the vectors **BC** and **−AC**. The vector **BC** represents the force with which the wagon pushes against the ramp. The vector **BF** represents the force that would pull the wagon up the ramp. Since vectors **BF** and **AC** are equal, $|\mathbf{AC}|$ gives the magnitude of the required force.

Vectors **BF** and **AC** are parallel, so angle *EBD* equals angle *A*. Since angle *BDE* and angle *C* are right angles, triangles *CBA* and *DEB* have two corresponding angles equal and so are similar triangles. Therefore, angle *ABC* equals angle *E*, which is 20°. From right triangle *ABC*,

$$\sin 20^\circ = \frac{|\mathbf{AC}|}{50}$$

$$|\mathbf{AC}| = 50 \sin 20^\circ = 17.$$

To the nearest pound, a 17-pound force will be required to pull the wagon up the ramp. ■

10.3 Exercises

Concept Check *Refer to vectors* **m** *through* **t** *at the right.*

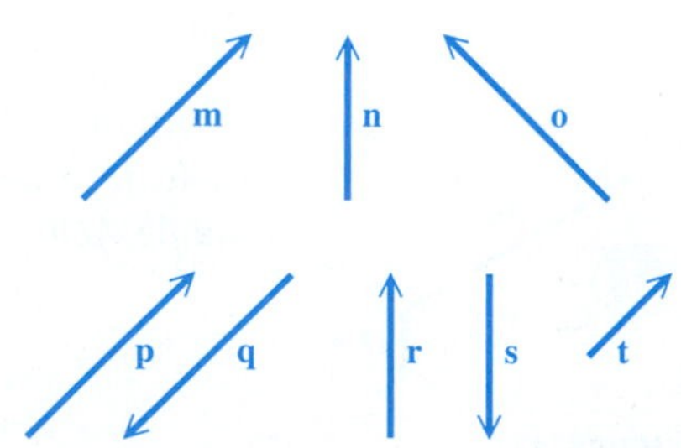

1. Name all pairs of vectors that appear to be equal.
2. Name all pairs of vectors that are opposites.
3. Name all pairs of vectors such that the first is a scalar multiple of the other, with the scalar positive.
4. Name all pairs of vectors such that the first is a scalar multiple of the other, with the scalar negative.

Refer to vectors **a** *through* **h**. *Make a copy or a sketch of each vector, and then draw a sketch to represent each vector in Exercises 5–16. For example, find* **a** + **e** *by placing* **a** *and* **e** *so that their initial points coincide. Then, use the parallelogram rule to find the resultant, shown in the figure at the right.*

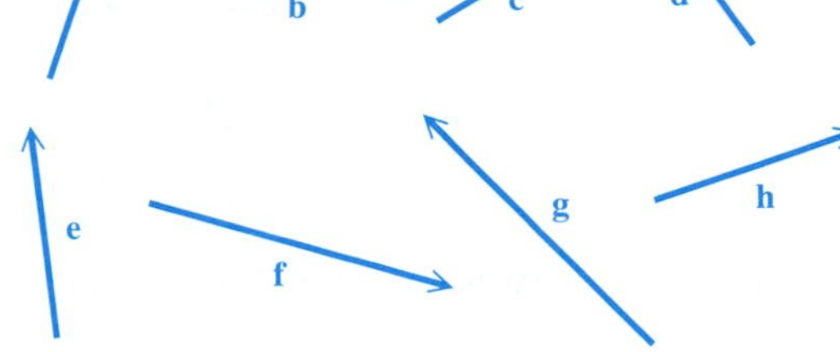

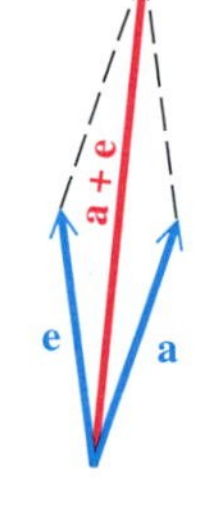

5. $-\mathbf{b}$ **6.** $-\mathbf{g}$

7. $3\mathbf{a}$ **8.** $2\mathbf{h}$

9. $\mathbf{a} + \mathbf{b}$ **10.** $\mathbf{h} + \mathbf{g}$

11. $\mathbf{a} - \mathbf{c}$ **12.** $\mathbf{d} - \mathbf{e}$

13. $\mathbf{a} + (\mathbf{b} + \mathbf{c})$ **14.** $(\mathbf{a} + \mathbf{b}) + \mathbf{c}$

15. $\mathbf{c} + \mathbf{d}$ **16.** $\mathbf{d} + \mathbf{c}$

In Exercises 17–22, use the figure to find each vector: **(a)** $\mathbf{a} + \mathbf{b}$ **(b)** $\mathbf{a} - \mathbf{b}$ **(c)** $-\mathbf{a}$.

17.

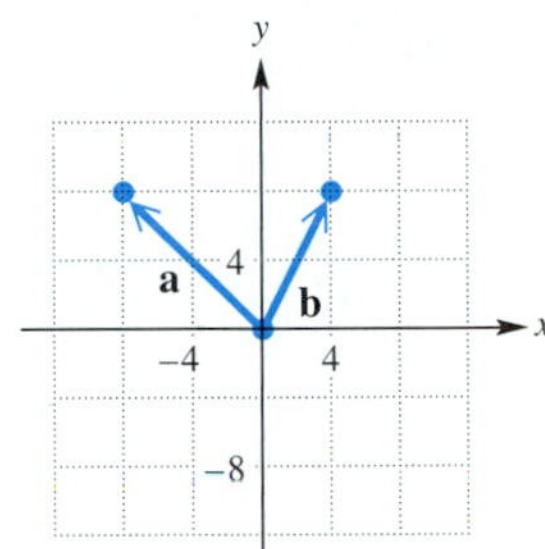

18.

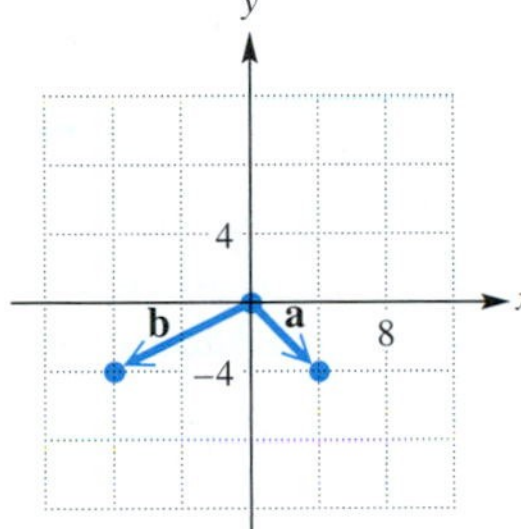

19.

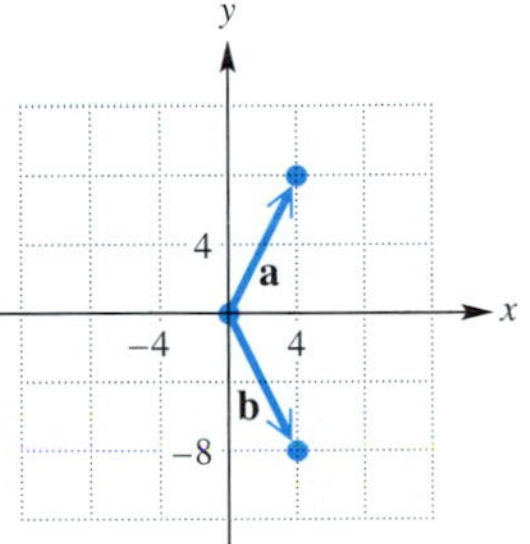

20.

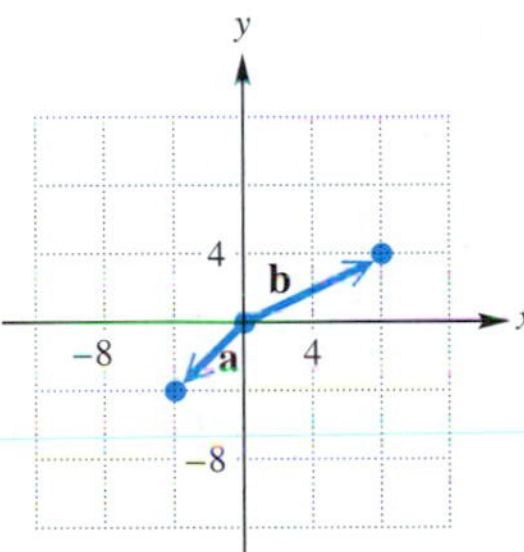

21.

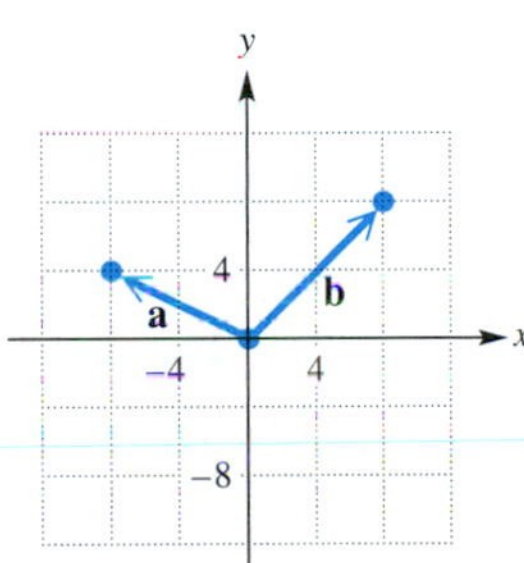

22.

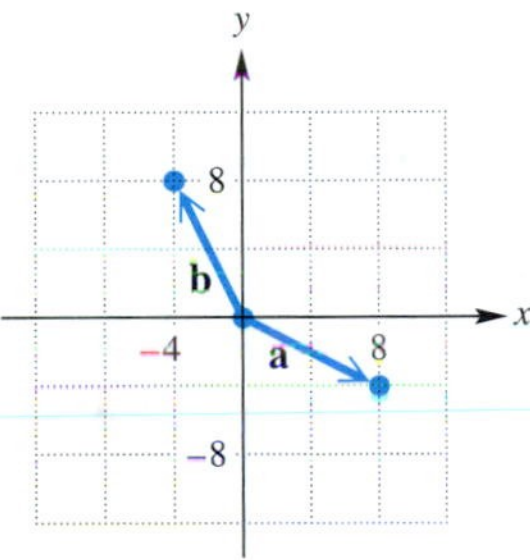

Given vectors **a** *and* **b**, *find* **(a)** $2\mathbf{a}$ **(b)** $2\mathbf{a} + 3\mathbf{b}$ **(c)** $\mathbf{b} - 3\mathbf{a}$.

23. $\mathbf{a} = 2\mathbf{i}$, $\mathbf{b} = \mathbf{i} + \mathbf{j}$ **24.** $\mathbf{a} = -\mathbf{i} + 2\mathbf{j}$, $\mathbf{b} = \mathbf{i} - \mathbf{j}$

25. $\mathbf{a} = \langle -1, 2 \rangle$, $\mathbf{b} = \langle 3, 0 \rangle$ **26.** $\mathbf{a} = \langle -2, -1 \rangle$, $\mathbf{b} = \langle -3, 2 \rangle$

Given $\mathbf{u} = \langle -2, 5 \rangle$ *and* $\mathbf{v} = \langle 4, 3 \rangle$, *find each vector.*

27. $\mathbf{u} + \mathbf{v}$ **28.** $\mathbf{u} - \mathbf{v}$ **29.** $\mathbf{v} - \mathbf{u}$ **30.** $5\mathbf{v}$ **31.** $-5\mathbf{v}$ **32.** $3\mathbf{u} + 6\mathbf{v}$

Write each vector **v** *in the form* $\langle a, b \rangle$. *In Exercises 33–36, give exact values. In Exercises 37–40, give approximations to the nearest hundredth.*

33.

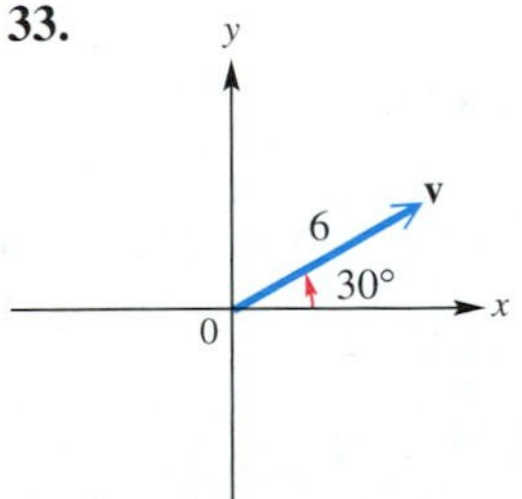

34.

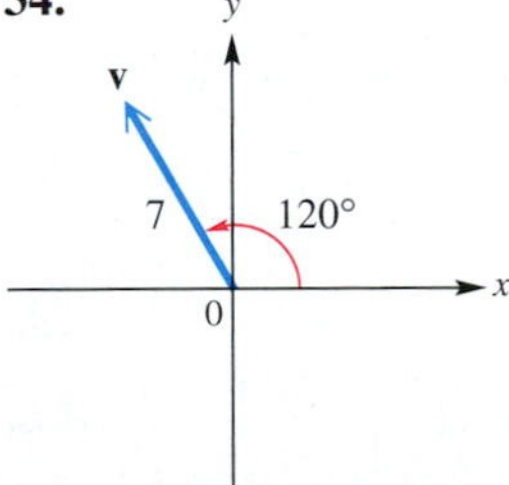

35.

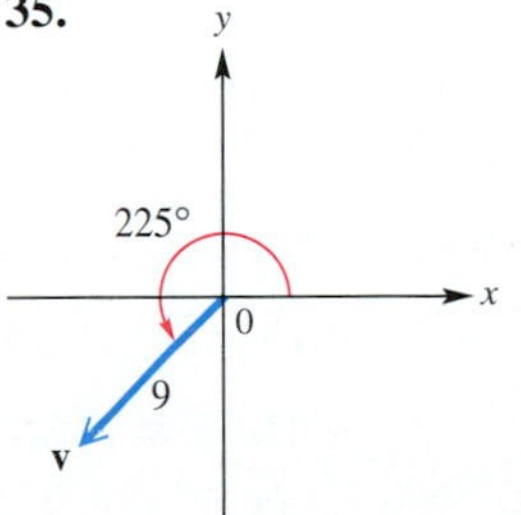

36.

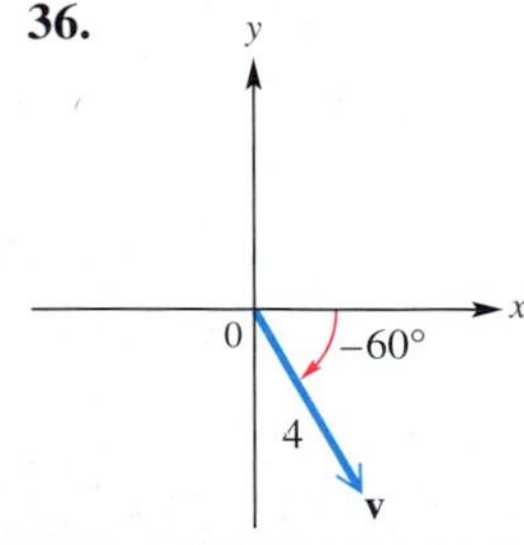

37.

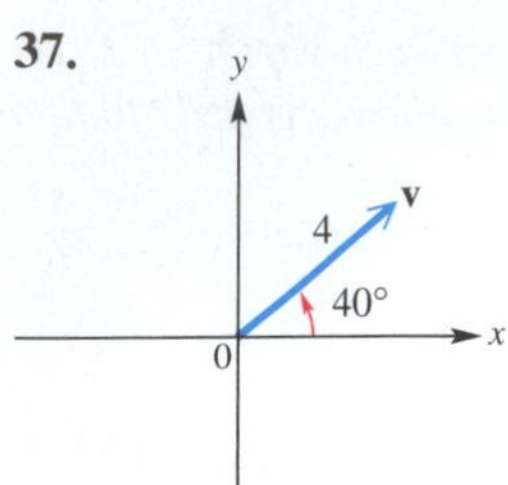

38.

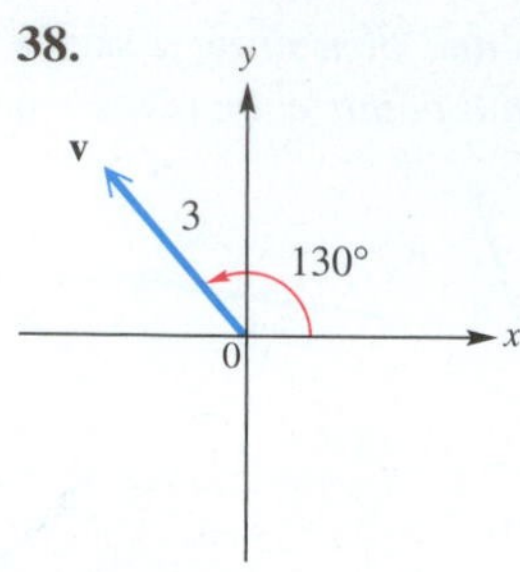

39.

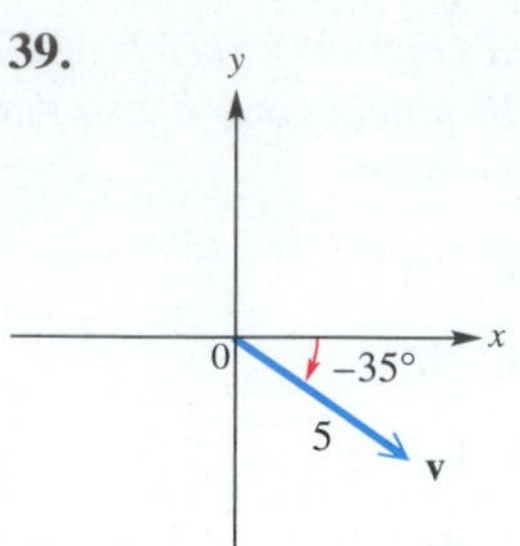

40.

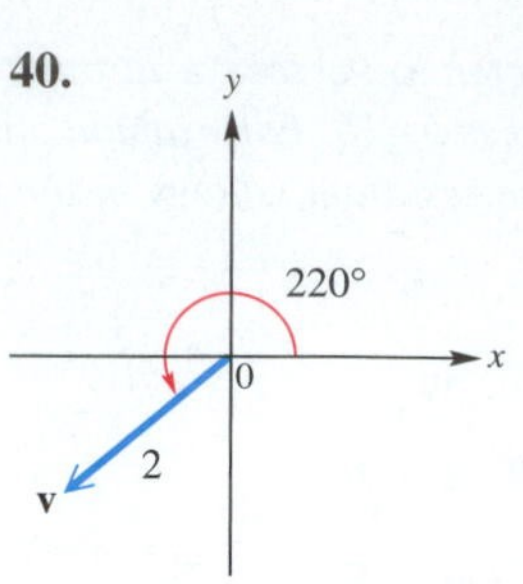

Use the parallelogram rule to find the magnitude of the resultant force of the two forces shown in each figure.

41.

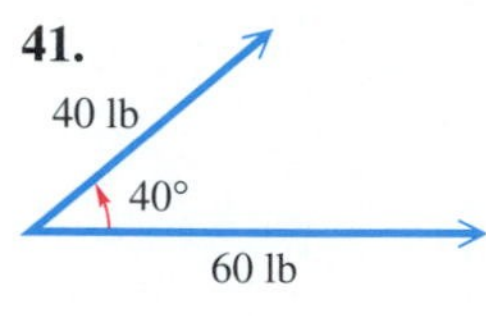

42.

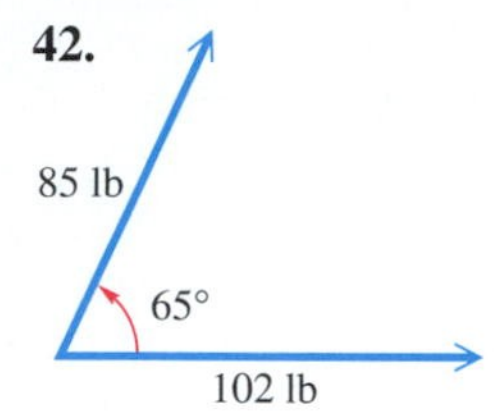

43.

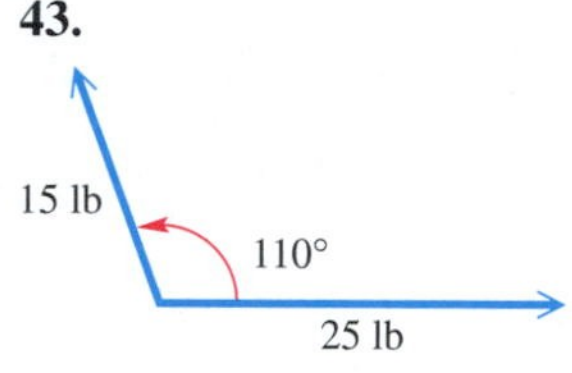

44.

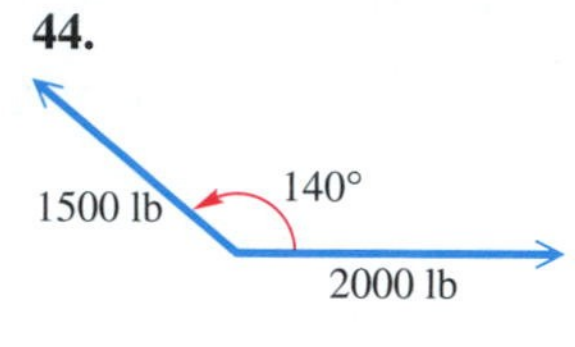

Write each vector in the form $a\mathbf{i} + b\mathbf{j}$. Round a and b to the nearest hundredth, if necessary.

45. $\langle -5, 8 \rangle$ **46.** $\langle 6, -3 \rangle$ **47.** $\langle 2, 0 \rangle$ **48.** $\langle 0, -4 \rangle$

49. Direction angle 45°, magnitude 8 **50.** Direction angle 210°, magnitude 3

51. Direction angle 115°, magnitude .6 **52.** Direction angle 208°, magnitude .9

Find the magnitude and direction angle (to the nearest tenth) for each vector. Give the measure of the direction angle as an angle in $[0, 360°)$.

53. $\langle 1, 1 \rangle$ **54.** $\langle -4, 4\sqrt{3} \rangle$ **55.** $\langle 8\sqrt{2}, -8\sqrt{2} \rangle$ **56.** $\langle \sqrt{3}, -1 \rangle$

57. $\langle 15, -8 \rangle$ **58.** $\langle -7, 24 \rangle$ **59.** $\langle -6, 0 \rangle$ **60.** $\langle 0, -12 \rangle$

Find the dot product of each pair of vectors.

61. $\langle 6, -1 \rangle, \langle 2, 5 \rangle$ **62.** $\langle -3, 8 \rangle, \langle 7, -5 \rangle$ **63.** $\langle 2, -3 \rangle, \langle 6, 5 \rangle$

64. $\langle 1, 2 \rangle, \langle 3, -1 \rangle$ **65.** $4\mathbf{i}, 5\mathbf{i} - 9\mathbf{j}$ **66.** $2\mathbf{i} + 4\mathbf{j}, -\mathbf{j}$

Find the angle between each pair of vectors.

67. $\langle 2, 1 \rangle, \langle -3, 1 \rangle$ **68.** $\langle 1, 7 \rangle, \langle 1, 1 \rangle$ **69.** $\langle 1, 2 \rangle, \langle -6, 3 \rangle$

70. $\langle 4, 0 \rangle, \langle 2, 2 \rangle$ **71.** $3\mathbf{i} + 4\mathbf{j}, \mathbf{j}$ **72.** $-5\mathbf{i} + 12\mathbf{j}, 3\mathbf{i} + 2\mathbf{j}$

Let $\mathbf{u} = \langle -2, 1 \rangle$, $\mathbf{v} = \langle 3, 4 \rangle$, and $\mathbf{w} = \langle -5, 12 \rangle$. Evaluate each expression.

73. $(3\mathbf{u}) \cdot \mathbf{v}$ **74.** $\mathbf{u} \cdot (\mathbf{v} - \mathbf{w})$ **75.** $\mathbf{u} \cdot \mathbf{v} - \mathbf{u} \cdot \mathbf{w}$ **76.** $\mathbf{u} \cdot (3\mathbf{v})$

Determine whether each pair of vectors is orthogonal.

77. $\langle 1, 2 \rangle, \langle -6, 3 \rangle$ **78.** $\langle 3, 4 \rangle, \langle 6, 8 \rangle$ **79.** $\langle 1, 0 \rangle, \langle \sqrt{2}, 0 \rangle$

80. $\langle 1, 1 \rangle, \langle 1, -1 \rangle$ **81.** $\sqrt{5}\mathbf{i} - 2\mathbf{j}, -5\mathbf{i} + 2\sqrt{5}\mathbf{j}$ **82.** $-4\mathbf{i} + 3\mathbf{j}, 8\mathbf{i} - 6\mathbf{j}$

Solve each problem.

83. ***Magnitudes of Forces*** A force of 176 pounds makes an angle of 78° 50′ with a second force. The resultant of the two forces makes an angle of 41° 10′ with the first force. Find the magnitude of the second force and of the resultant.

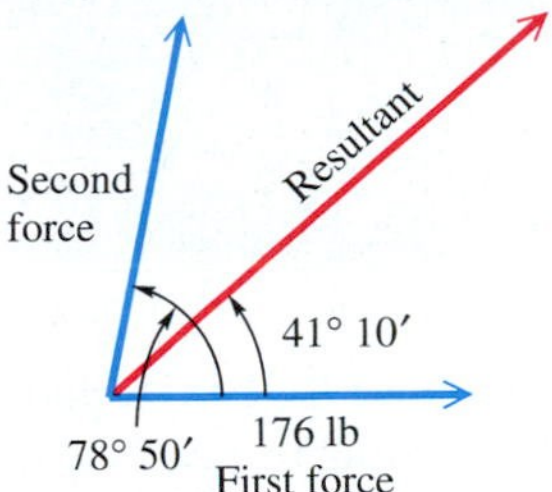

84. ***Magnitudes of Forces*** A force of 28.7 pounds makes an angle of 42° 10′ with a second force. The resultant of the two forces makes an angle of 32° 40′ with the first force. Find the magnitude of the second force and of the resultant.

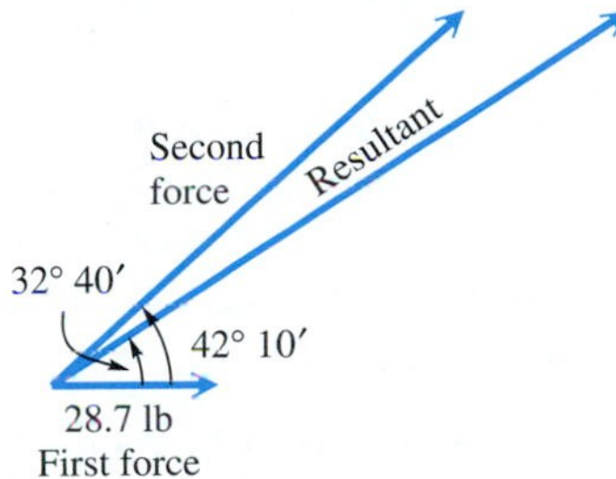

85. ***Angle of a Hill Slope*** A force of 25 pounds is required to hold an 80-pound crate from rolling on a hill. What angle does the hill make with the horizontal?

86. ***Force Needed to Keep a Car Parked*** Find the force required to keep a 3000-pound car parked on a hill that makes an angle of 15° with the horizontal.

87. ***Force Needed for a Monolith*** To build the pyramids in Egypt, it is believed that giant causeways were constructed to transport the building materials to the site. One such causeway is said to have been 3000 feet long, with a slope of about 2.3°. How much force would be required to hold a 60-ton monolith on this causeway?

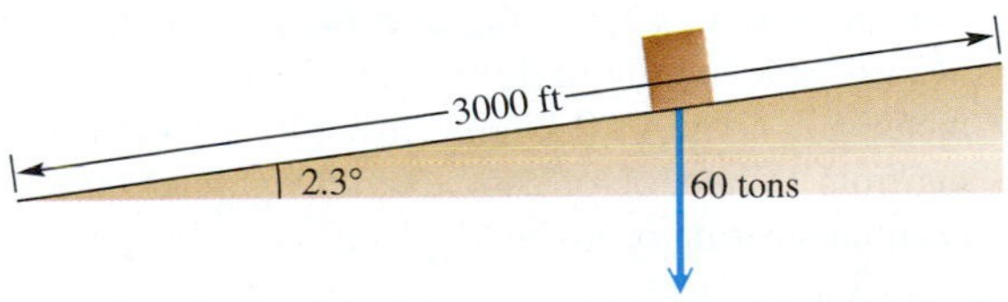

88. ***Angles between Forces*** Three forces acting at a point are in equilibrium. The forces are 980 pounds, 760 pounds, and 1220 pounds. Find the angles between the directions of the forces. (*Hint:* Arrange the forces to form the sides of a triangle.)

89. ***Incline Angle*** A force of 18 pounds is required to hold a 60-pound stump grinder on an incline. What angle does the incline make with the horizontal?

90. ***Incline Angle*** A force of 30 pounds is required to hold an 80-pound pressure washer on an incline. What angle does the incline make with the horizontal?

91. ***Weight of a Crate and Tension of a Rope*** A crate is supported by two ropes. One rope makes an angle of 46° 20′ with the horizontal and has a tension of 89.6 pounds on it. The other rope is horizontal. Find the weight of the crate and the tension in the horizontal rope.

92. ***Weight of a Box*** Two people are carrying a box. One person exerts a force of 150 pounds at an angle of 62.4° with the horizontal. The other person exerts a force of 114 pounds at an angle of 54.9°. Find the weight of the box.

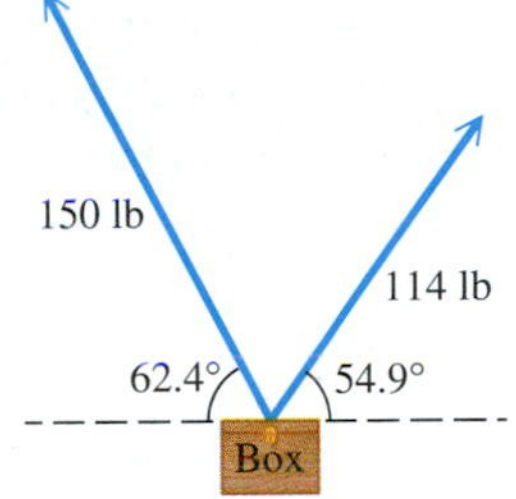

93. ***Distance and Bearing of a Ship*** A ship leaves port on a bearing of 34.0° and travels 10.4 miles. The ship then turns due east and travels 4.6 miles. How far is the ship from port, and what is its bearing from port?

94. ***Distance and Bearing of a Luxury Liner*** A luxury liner leaves port on a bearing of 110.0° and travels 8.8 miles. It then turns due west and travels 2.4 miles. How far is the liner from port, and what is its bearing from port?

95. *Distance of a Ship from Its Starting Point* Starting at point A, a ship sails 18.5 kilometers on a bearing of 189°, then turns and sails 47.8 kilometers on a bearing of 317°. Find the distance of the ship from point A.

96. *Distance of a Ship from Its Starting Point* Starting at point X, a ship sails 15.5 kilometers on a bearing of 200°, then turns and sails 2.4 kilometers on a bearing of 320°. Find the distance of the ship from point X.

97. *Distance and Direction of a Motorboat* A motorboat sets out in the direction N 80° E. The speed of the boat in still water is 20.0 mph. If the current is flowing directly south and the actual direction of the motorboat is due east, find the speed of the current and the actual speed of the motorboat.

98. *Path Traveled by a Plane* The aircraft carrier *Tallahassee* is traveling at sea on a steady course with a bearing of 30° at 32 mph. Patrol planes on the carrier have enough fuel for 2.6 hours of flight when traveling at a speed of 520 mph. One of the pilots takes off on a bearing of 338° and then turns and heads in a straight line, so as to be able to catch the carrier and land on the deck at the exact instant that his fuel runs out. If the pilot left at 2 P.M., at what time did he turn to head for the carrier?

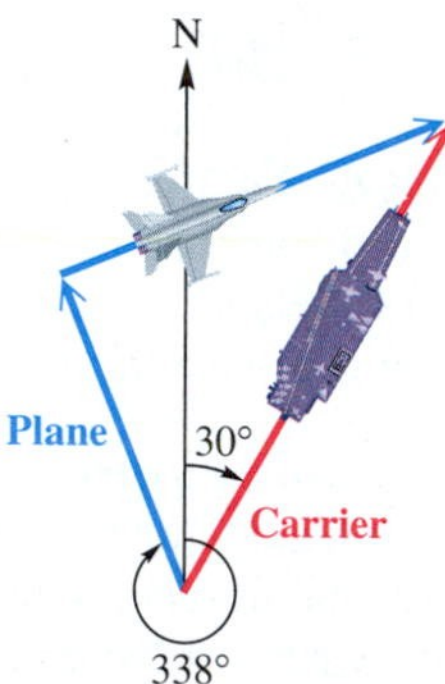

99. *Bearing and Ground Speed of a Plane* An airline route from San Francisco to Honolulu is on a bearing of 233.0°. A jet flying at 450 mph on that bearing runs into a wind blowing at 39.0 mph from a direction of 114.0°. Find the resulting bearing and ground speed of the plane.

100. *Movement of a Motorboat* Suppose you would like to cross a 132-foot-wide river in a motorboat. Assume that the motorboat can travel at 7 mph relative to the water and that the current is flowing west at the rate of 3 mph. The bearing θ is chosen so that the motorboat will land at a point exactly across from the starting point.

(a) At what speed will the motorboat be traveling relative to the banks?

(b) How long will it take for the motorboat to make the crossing?

(c) What is the measure of angle θ?

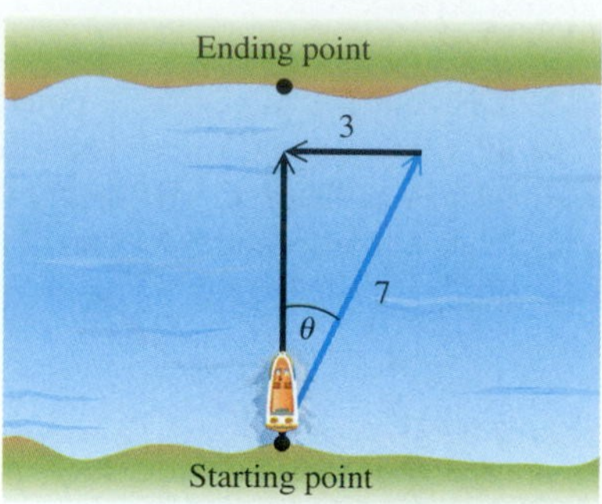

101. *Airspeed and Ground Speed* A pilot wants to fly on a bearing of 74.9°. Flying due east, he finds that a 42-mph wind, blowing from the south, puts him on course. Find the airspeed and the ground speed.

102. *Bearing of a Plane* A plane flies 650 mph on a bearing of 175.3°. A 25-mph wind from a direction of 266.6° blows against the plane. Find the resulting bearing of the plane.

103. *Bearing and Ground Speed of a Plane* A pilot is flying at 190 mph. He wants his flight path to be on a bearing of 64° 30′. A wind is blowing from the south at 35.0 mph. Find the bearing he should fly, and find the plane's ground speed.

104. *Bearing and Ground Speed of a Plane* A pilot is flying at 168 mph. She wants her flight path to be on a bearing of 57° 40′. A wind is blowing from the south at 27.1 mph. Find the bearing the pilot should fly, and find the plane's ground speed.

105. *Bearing and Airspeed of a Plane* What bearing and airspeed are required for a plane to fly 400 miles due north in 2.5 hours if the wind is blowing from a direction of 328° at 11 mph?

106. *Ground Speed and Bearing of a Plane* A plane is headed due south with an airspeed of 192 mph. A wind from a direction of 78° is blowing at 23 mph. Find the ground speed and resulting bearing of the plane.

107. *Ground Speed and Bearing of a Plane* An airplane is headed on a bearing of 174° at an airspeed of 240 kilometers per hour. A 30-kilometer-per-hour wind is blowing from a direction of 245°. Find the ground speed and resulting bearing of the plane.

108. *Velocity of a Star* The space velocity $\mathbf{v}$ of a star relative to the sun can be expressed as the resultant vector of two perpendicular vectors: the radial velocity $\mathbf{v}_r$, and the tangential velocity $\mathbf{v}_t$, where $\mathbf{v} = \mathbf{v}_r + \mathbf{v}_t$. If a star is located near the sun and its space velocity is large, then its motion across the sky will also be large. Barnard's Star is a relatively close star, 35 trillion miles from the sun. It moves across the sky through an angle of 10.34″ per year, the largest angle of any known star. Its radial velocity is $\mathbf{v}_r = 67$ miles per second toward the sun. (*Source:* Zeilik, M., S. Gregory, and E. Smith, *Introductory Astronomy and Astrophysics,* Second Edition, Saunders College Publishing, 1998; Acker, A. and C. Jaschek, *Astronomical Methods and Calculations,* John Wiley & Sons, 1986.)

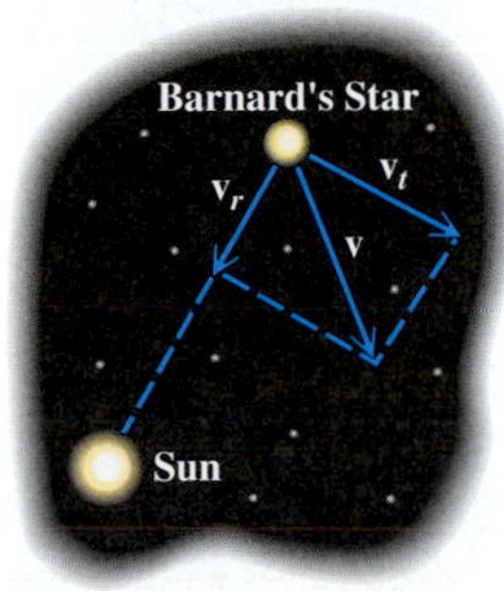

Not to scale

(a) Approximate the tangential velocity $\mathbf{v}_t$ of Barnard's Star. (*Hint:* Use the arc length formula $s = r\theta$.)
(b) Compute the magnitude of **v**.

109. ***(Modeling) Measuring Rainfall*** Suppose that vector **R** models the amount of rainfall in inches and the direction it falls, and vector **A** models the area in square inches and orientation of the opening of a rain gauge, as illustrated in the figure in the next column. The total volume V of water collected in the gauge is given by $V = |\mathbf{R} \cdot \mathbf{A}|$. This formula calculates the volume of water collected even if the wind is blowing the rain in a slanted direction or the rain gauge is not exactly vertical. Let $\mathbf{R} = \mathbf{i} - 2\mathbf{j}$ and $\mathbf{A} = .5\mathbf{i} + \mathbf{j}$.

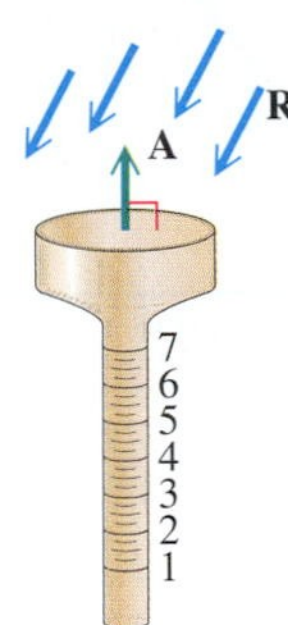

(a) Find $|\mathbf{R}|$ and $|\mathbf{A}|$. Interpret your results.
(b) Calculate V and interpret this result.
(c) For the rain gauge to collect the maximum amount of water, what should be true about vectors **R** and **A**?

110. ***The Dot Product*** In the figure, $\mathbf{a} = \langle a_1, a_2 \rangle$, $\mathbf{b} = \langle b_1, b_2 \rangle$, and $\mathbf{a} - \mathbf{b} = \langle a_1 - b_1, a_2 - b_2 \rangle$. Apply the law of cosines to the triangle and derive the equation $\mathbf{a} \cdot \mathbf{b} = |\mathbf{a}|\,|\mathbf{b}| \cos \theta$.

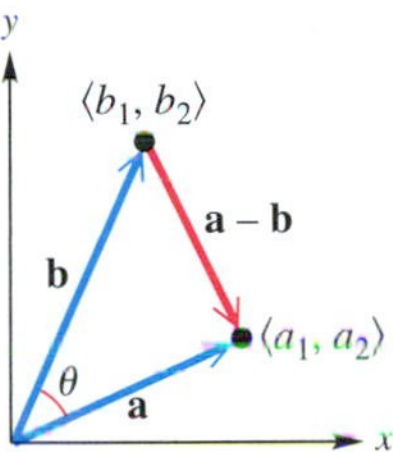

Reviewing Basic Concepts (Sections 10.1–10.3)

Round values to the nearest tenth in Exercises 1–5.

1. Solve triangle ABC if $A = 44°$, $C = 62°$, and $a = 12$.

2. Solve triangle ABC if $A = 32°$, $a = 6$, and $b = 8$. How many solutions are there?

3. Solve triangle ABC if $C = 41°$, $c = 12$, and $a = 7$. How many solutions are there?

4. Use the law of cosines to solve each triangle.

(a)

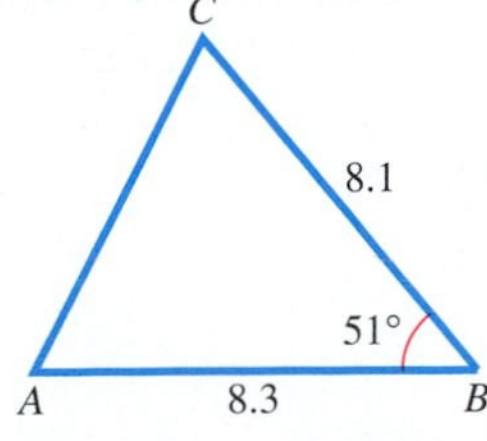

(b)

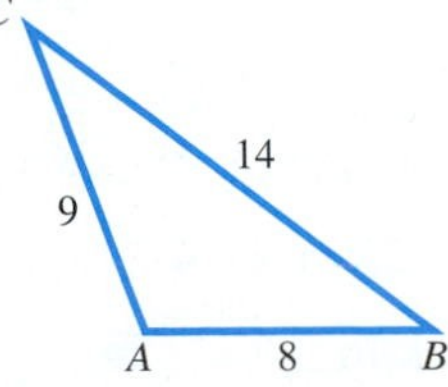

5. Find the area of triangle ABC if $a = 4.5$, $b = 5.2$, and $C = 55°$.

6. Find the area of triangle ABC if $a = 6$, $b = 7$, and $c = 9$.

7. Let $\mathbf{v} = 2\mathbf{i} - \mathbf{j}$ and $\mathbf{u} = -3\mathbf{i} + 2\mathbf{j}$. Find each expression.
(a) $2\mathbf{v} + \mathbf{u}$
(b) $2\mathbf{v}$
(c) $\mathbf{v} - 3\mathbf{u}$

8. Let $\mathbf{a} = \langle 3, -2 \rangle$ and $\mathbf{b} = \langle -1, 3 \rangle$. Find $\mathbf{a} \cdot \mathbf{b}$ and the angle between **a** and **b**, rounded to the nearest tenth of a degree.

9. ***Resultant Force*** Find the magnitude of the resultant force of the two forces shown in the figure.

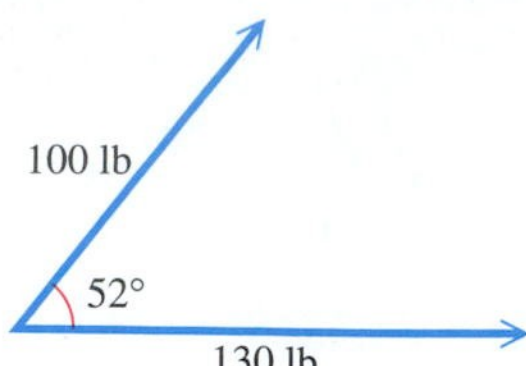

10. ***Height of an Airplane*** Two observation points A and B are 950 feet apart. From these points, the angles of elevation of an airplane are 52° and 57°. (See the figure.) Find the height of the airplane.

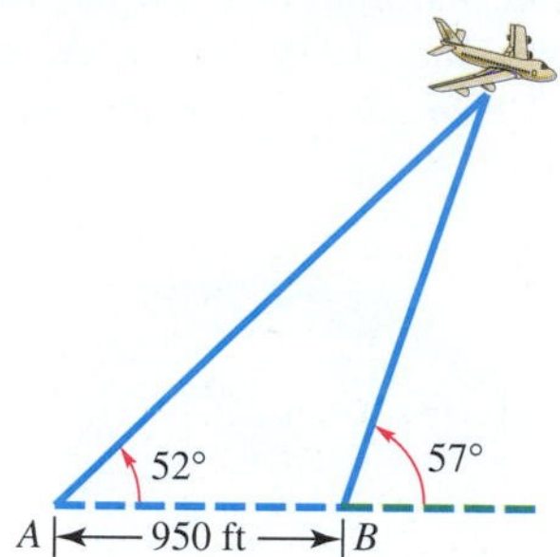

10.4 Trigonometric (Polar) Form of Complex Numbers

The Complex Plane and Vector Representation ■ Trigonometric (Polar) Form ■ Products of Complex Numbers in Trigonometric Form ■ Quotients of Complex Numbers in Trigonometric Form

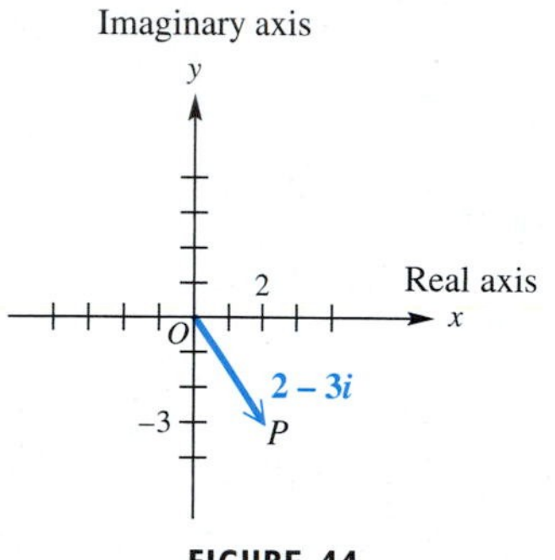

FIGURE 44

The Complex Plane and Vector Representation

Unlike real numbers, complex numbers cannot be ordered. One way to organize and illustrate them is by using a graph. To graph a complex number such as $2 - 3i$, we modify the familiar rectangular coordinate system by calling the horizontal axis the **real axis** and the vertical axis the **imaginary axis.** Complex numbers can be graphed in this **complex plane,** as shown in Figure 44 for the complex number $2 - 3i$. Each complex number $a + bi$ determines a unique position vector with initial point $(0, 0)$ and terminal point (a, b). This relationship shows the close connection between vectors and complex numbers.

NOTE This geometric representation is the reason that $a + bi$ is called the **rectangular form** of a complex number. (*Rectangular form* is also called *standard form.*)

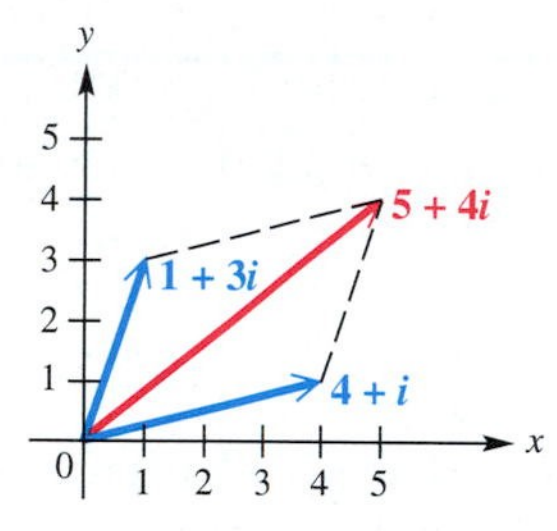

FIGURE 45

Recall that the sum of the two complex numbers $4 + i$ and $1 + 3i$ is

$$(4 + i) + (1 + 3i) = 5 + 4i. \quad \text{Add real parts; add imaginary parts.}$$

Graphically, the sum of two complex numbers is represented by the vector that is the resultant of the vectors corresponding to the two numbers, as shown in Figure 45.

EXAMPLE 1 Expressing the Sum of Complex Numbers Graphically

Find the sum of $6 - 2i$ and $-4 - 3i$. Graph both complex numbers and their resultant.

Solution The sum is found by adding the two numbers.

$$(6 - 2i) + (-4 - 3i) = 2 - 5i$$

The graphs are shown in Figure 46. ■

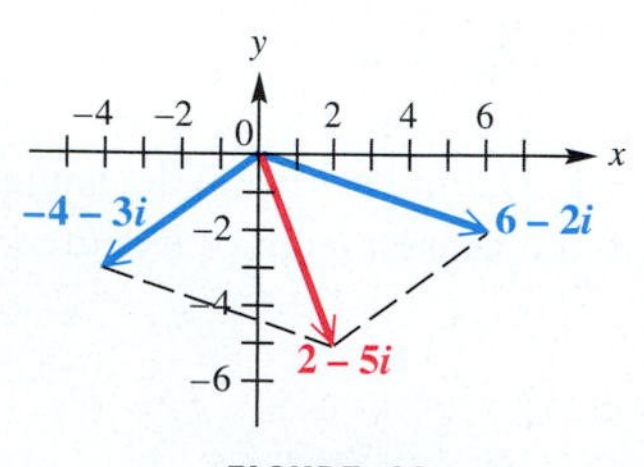

FIGURE 46

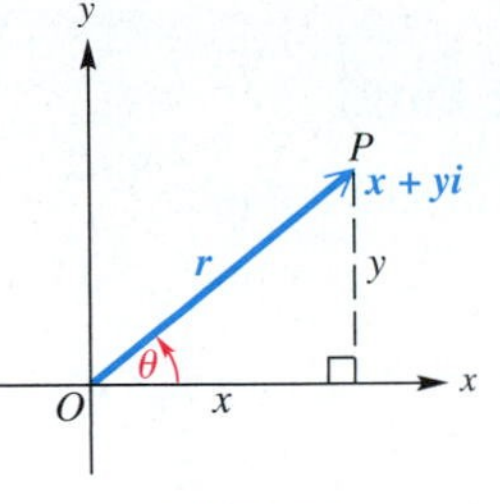

FIGURE 47

Trigonometric (Polar) Form

Figure 47 shows the complex number $x + yi$ that corresponds to a vector **OP** with direction angle θ and magnitude r. The following relationships among x, y, r, and θ can be verified from Figure 47.

Relationships among x, y, r, and θ

$$x = r\cos\theta \qquad y = r\sin\theta$$

$$r = \sqrt{x^2 + y^2} \qquad \tan\theta = \frac{y}{x}, \quad \text{if } x \neq 0$$

Substituting $x = r\cos\theta$ and $y = r\sin\theta$ into $x + yi$ gives

$$\begin{aligned} x + yi &= r\cos\theta + (r\sin\theta)i \\ &= r(\cos\theta + i\sin\theta). \end{aligned}$$

Trigonometric (Polar) Form of a Complex Number

The expression

$$r(\cos\theta + i\sin\theta)$$

is called the **trigonometric form** (or **polar form**) of the complex number $x + yi$. The expression $\cos\theta + i\sin\theta$ is sometimes abbreviated **cis θ.** Using this notation, $r(\cos\theta + i\sin\theta)$ is written **r cis θ.**

The number r is the **modulus** or **absolute value** of $x + yi$, and θ is the **argument** of $x + yi$. In this section, we often choose the value of θ in the interval $[0°, 360°)$. However, any angle coterminal with θ would work as the argument.

GCM **EXAMPLE 2 Converting from Trigonometric to Rectangular Form**

Express $2(\cos 300° + i\sin 300°)$ in rectangular form.

Analytic Solution

As discussed in Chapter 8, we know that $\cos 300° = \frac{1}{2}$ and $\sin 300° = -\frac{\sqrt{3}}{2}$, so

$$\begin{aligned} 2(\cos 300° + i\sin 300°) &= 2\left(\frac{1}{2} - i\frac{\sqrt{3}}{2}\right) \\ &= 1 - i\sqrt{3}. \end{aligned}$$

Notice that the real part is positive and the imaginary part is negative; this is consistent with 300° being a quadrant IV angle.

Graphing Calculator Solution

Figure 48 confirms the analytic solution.

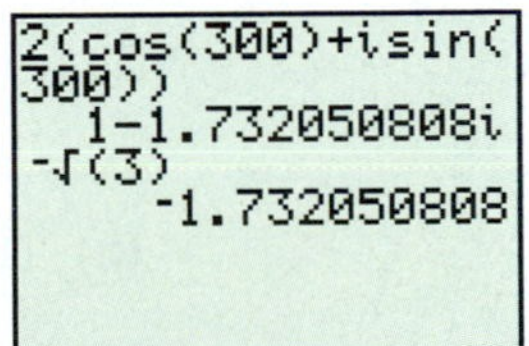

The imaginary part is an approximation for $-\sqrt{3}$.

FIGURE 48

Converting from Rectangular to Trigonometric Form

Step 1 Sketch a graph of the number $x + yi$ in the complex plane.

Step 2 Find r by using the equation $r = \sqrt{x^2 + y^2}$.

Step 3 Find θ by using the equation $\tan\theta = \frac{y}{x}$, $x \neq 0$, choosing the quadrant indicated in Step 1.

CAUTION Errors often occur in Step 3. ***Be sure to choose the correct quadrant for θ by referring to the graph sketched in Step 1.***

GCM **EXAMPLE 3** **Converting from Rectangular to Trigonometric Form**

Write each complex number in trigonometric form.

(a) $-\sqrt{3} + i$ (Use radian measure.) **(b)** $-3i$ (Use degree measure.)

TECHNOLOGY NOTE

The top screen shows how to convert from rectangular (x, y) form to trigonometric form. The calculator is in radian mode. The results agree with our analytic results in Example 3(a).

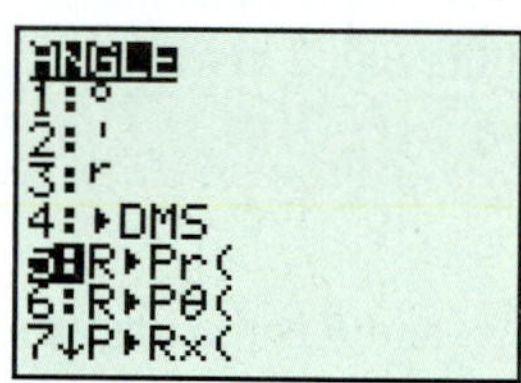


```
R▸Pr(-√(3),1)
                2
R▸Pθ(-√(3),1)
      2.617993878
5π/6
      2.617993878
```

Solution

(a) We start by sketching the graph of $-\sqrt{3} + i$ in the complex plane, as shown in Figure 49. Next, we find r and then we find θ.

$$r = \sqrt{x^2 + y^2} = \sqrt{(-\sqrt{3})^2 + 1^2} = \sqrt{3 + 1} = 2 \qquad x = -\sqrt{3},\ y = 1$$

$$\tan\theta = \frac{y}{x} = \frac{1}{-\sqrt{3}} = \frac{1}{-\sqrt{3}} \cdot \frac{\sqrt{3}}{\sqrt{3}} = -\frac{\sqrt{3}}{3}$$

Since $\tan\theta = -\frac{\sqrt{3}}{3}$, the reference angle for θ in radians is $\frac{\pi}{6}$. From the graph, we see that θ is in quadrant II, so $\theta = \pi - \frac{\pi}{6} = \frac{5\pi}{6}$. In trigonometric form,

Be sure to choose the correct quadrant.

$$-\sqrt{3} + i = 2\left(\cos\frac{5\pi}{6} + i\sin\frac{5\pi}{6}\right) = 2\ \text{cis}\ \frac{5\pi}{6}.$$

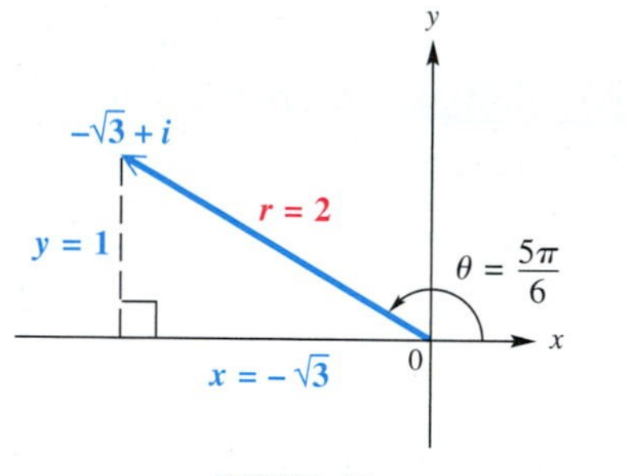

FIGURE 49

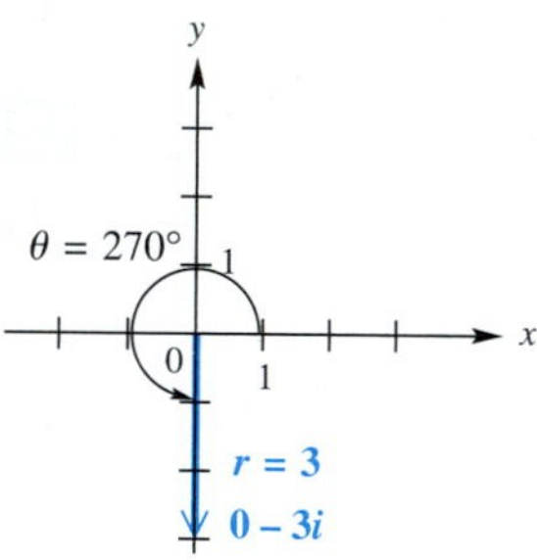

FIGURE 50

(b) The sketch of $-3i$ is shown in Figure 50. Since $-3i = 0 - 3i$, we have $x = 0$ and $y = -3$. We find r as follows.

$$r = \sqrt{0^2 + (-3)^2} = \sqrt{0 + 9} = \sqrt{9} = 3$$

We cannot find θ by using $\tan\theta = \frac{y}{x}$, because $x = 0$. From the graph, a value for θ is 270°. In trigonometric form,

When the angle is quadrantal, pay close attention to the graph.

$$-3i = 3(\cos 270° + i\sin 270°) = 3\ \text{cis}\ 270°.$$ ■

Source: Figure from Crownover, R., *Introduction to Fractals and Chaos.* Copyright © 1995. Boston: Jones and Bartlett Publishers. Reprinted with permission.

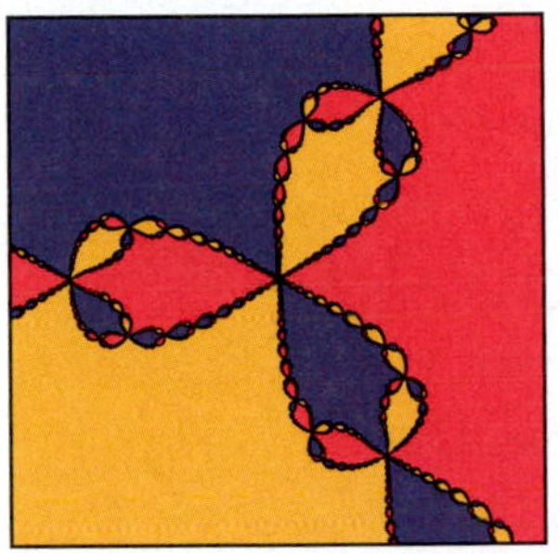

Source: Kincaid, D. R. and Cheney, E.W., *Numerical Analysis: Mathematics of Scientific Computing,* 1st edition, ©1991. Reprinted with permission of Brooks/Cole, an imprint of the Wadsworth Group, a division of Thomson Learning.

NOTE In Example 3, we gave answers in both forms: $r(\cos\theta + i\sin\theta)$ and $r\text{ cis }\theta$. These forms are used interchangeably throughout the rest of this chapter.

High-resolution computer graphics and complex numbers make it possible to produce beautiful shapes called *fractals.* Benoit B. Mandelbrot first used the term *fractal* in 1975. At its basic level, a fractal is a geometric figure with an endless self-similarity property. A fractal image repeats itself infinitely with ever-decreasing dimensions. Although most current applications of fractals are related to creating fascinating images and pictures, fractals do have tremendous potential in applied science. Two examples of fractals are shown in the margin. The top figure is the *Mandelbrot set,* while the bottom figure is an amazing graphical solution to a difficult problem first presented by Sir Arthur Cayley in 1879. It is called *Newton's basins of attraction for the cube roots of unity.*

EXAMPLE 4 Deciding whether a Complex Number Is in the Julia Set

The fractal called the *Julia set* is shown in Figure 51. To determine whether a complex number $z = a + bi$ is in this Julia set, perform the following sequence of calculations. Repeatedly compute the values of $z^2 - 1$, $(z^2 - 1)^2 - 1$, $[(z^2 - 1)^2 - 1]^2 - 1, \ldots.$ If the moduli or absolute values of any of the resulting complex numbers exceed 2, then the complex number z is not in the Julia set. Otherwise z is part of this set and the point (a, b) should be shaded in the graph.

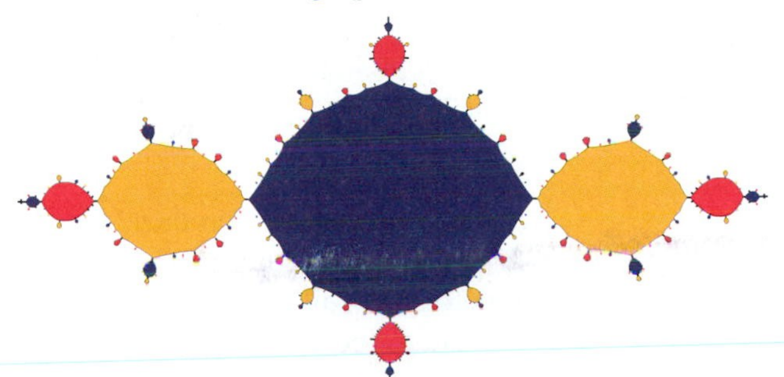

Source: Figure from Crownover, R., *Introduction to Fractals and Chaos.* Copyright © 1995. Boston: Jones and Bartlett Publishers. Reprinted with permission.

FIGURE 51

Determine whether each number belongs to the Julia set.

(a) $z = 0 + 0i$ **(b)** $z = 1 + 1i$

Solution

(a) Here,

$$\begin{aligned} z &= 0 + 0i = 0, \\ z^2 - 1 &= 0^2 - 1 = -1, \\ (z^2 - 1)^2 - 1 &= (-1)^2 - 1 = 0, \\ [(z^2 - 1)^2 - 1]^2 - 1 &= 0^2 - 1 = -1, \end{aligned}$$

and so on. We see that the calculations repeat as $0, -1, 0, -1$, and so on. The moduli are either 0 or 1, which do not exceed 2, so $0 + 0i$ is in the Julia set and the point $(0, 0)$ is part of the graph.

(b) We have $z^2 - 1 = (1 + i)^2 - 1 = (1 + 2i + i^2) - 1 = -1 + 2i$. The modulus is $\sqrt{(-1)^2 + 2^2} = \sqrt{5}$. Since $\sqrt{5}$ is greater than 2, $1 + 1i$ is not in the Julia set and $(1, 1)$ is not part of the graph. ■

GCM TECHNOLOGY NOTE

With the calculator in complex and degree modes, the math menu can be used to find the angle and the magnitude (absolute value) of the vector that corresponds to a given complex number.

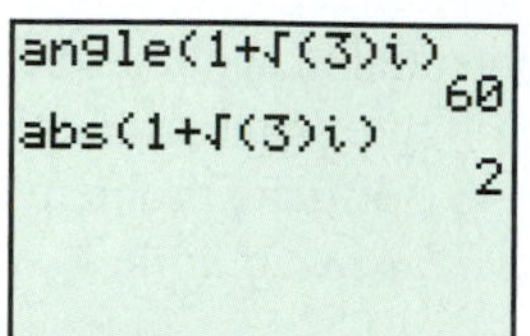

Products of Complex Numbers in Trigonometric Form

Using the FOIL method to multiply complex numbers in rectangular form, we find the product of $1 + i\sqrt{3}$ and $-2\sqrt{3} + 2i$ as follows.

$$\begin{aligned}(1 + i\sqrt{3})(-2\sqrt{3} + 2i) &= -2\sqrt{3} + 2i - 2i(3) + 2i^2\sqrt{3} && \text{FOIL}\\ &= -2\sqrt{3} + 2i - 6i - 2\sqrt{3} && i^2 = -1\\ &= -4\sqrt{3} - 4i && \text{Combine terms.}\end{aligned}$$

We can also find this same product by first converting the complex numbers $1 + i\sqrt{3}$ and $-2\sqrt{3} + 2i$ to trigonometric form as explained earlier in this section.

$$1 + i\sqrt{3} = 2(\cos 60° + i \sin 60°)$$
$$-2\sqrt{3} + 2i = 4(\cos 150° + i \sin 150°)$$

If we multiply the trigonometric forms, and if we use the identities for the cosine and the sine of the sum of two angles, then the result is

$$\begin{aligned}&[2(\cos 60° + i \sin 60°)][4(\cos 150° + i \sin 150°)]\\ &= 2 \cdot 4(\cos 60° \cdot \cos 150° + i \sin 60° \cdot \cos 150°\\ &\quad + i \cos 60° \cdot \sin 150° + i^2 \sin 60° \cdot \sin 150°)\\ &= 8[(\cos 60° \cdot \cos 150° - \sin 60° \cdot \sin 150°)\\ &\quad + i(\sin 60° \cdot \cos 150° + \cos 60° \cdot \sin 150°)]\\ &= 8[\cos(60° + 150°) + i \sin(60° + 150°)]\\ &= 8(\cos 210° + i \sin 210°).\end{aligned}$$

The modulus of the product, 8, is equal to the product of the moduli of the factors, $2 \cdot 4$, and the argument of the product, 210°, is the sum of the arguments of the factors, $60° + 150°$.

As expected, the product obtained by multiplying by the first method is the rectangular form of the product obtained by multiplying by the second method.

$$8(\cos 210° + i \sin 210°) = 8\left(-\frac{\sqrt{3}}{2} - \frac{1}{2}i\right) = -4\sqrt{3} - 4i$$

We can generalize this work in the following *product theorem.*

Product Theorem

If $r_1(\cos \theta_1 + i \sin \theta_1)$ and $r_2(\cos \theta_2 + i \sin \theta_2)$ are any two complex numbers, then

$$\begin{aligned}&[r_1(\cos \theta_1 + i \sin \theta_1)] \cdot [r_2(\cos \theta_2 + i \sin \theta_2)]\\ &= r_1 r_2[\cos(\theta_1 + \theta_2) + i \sin(\theta_1 + \theta_2)].\end{aligned}$$

In compact form, this is written

$$(r_1 \text{ cis } \theta_1)(r_2 \text{ cis } \theta_2) = r_1 r_2 \text{ cis}(\theta_1 + \theta_2).$$

GCM **EXAMPLE 5** **Using the Product Theorem**

Find the product of $3(\cos 45° + i \sin 45°)$ and $2(\cos 135° + i \sin 135°)$.

Solution

$$[3(\cos 45° + i \sin 45°)][2(\cos 135° + i \sin 135°)]$$
$$= 3 \cdot 2[\cos(45° + 135°) + i \sin(45° + 135°)] \quad \text{Product theorem}$$
$$= 6(\cos 180° + i \sin 180°) \quad \text{Multiply; add.}$$
$$= 6(-1 + i \cdot 0) = 6(-1) = -6$$

■

Quotients of Complex Numbers in Trigonometric Form

The rectangular form of the quotient of the complex numbers $1 + i\sqrt{3}$ and $-2\sqrt{3} + 2i$ is

$$\frac{1 + i\sqrt{3}}{-2\sqrt{3} + 2i} = \frac{(1 + i\sqrt{3})(-2\sqrt{3} - 2i)}{(-2\sqrt{3} + 2i)(-2\sqrt{3} - 2i)} \quad \text{Multiply by the conjugate of the denominator.}$$
$$= \frac{-2\sqrt{3} - 2i - 6i - 2i^2\sqrt{3}}{12 - 4i^2} \quad \text{FOIL; } (x + y)(x - y) = x^2 - y^2$$
$$= \frac{-8i}{16} = -\frac{1}{2}i. \quad \text{Simplify.}$$

Writing $1 + i\sqrt{3}$, $-2\sqrt{3} + 2i$, and $-\frac{1}{2}i$ in trigonometric form gives

$$1 + i\sqrt{3} = 2(\cos 60° + i \sin 60°),$$
$$-2\sqrt{3} + 2i = 4(\cos 150° + i \sin 150°),$$

and

$$-\frac{1}{2}i = \frac{1}{2}[\cos(-90°) + i \sin(-90°)].$$

The modulus of the quotient, $\frac{1}{2}$, is the quotient of the two moduli, $\frac{2}{4} = \frac{1}{2}$. The argument of the quotient, $-90°$, is the difference of the two arguments, $60° - 150° = -90°$. Generalizing from this example leads to the *quotient theorem.*

Quotient Theorem

If $r_1(\cos \theta_1 + i \sin \theta_1)$ and $r_2(\cos \theta_2 + i \sin \theta_2)$ are complex numbers, where $r_2(\cos \theta_2 + i \sin \theta_2) \neq 0$, then

$$\frac{r_1(\cos \theta_1 + i \sin \theta_1)}{r_2(\cos \theta_2 + i \sin \theta_2)} = \frac{r_1}{r_2}[\cos(\theta_1 - \theta_2) + i \sin(\theta_1 - \theta_2)].$$

In compact form, this is written

$$\frac{r_1 \text{ cis } \theta_1}{r_2 \text{ cis } \theta_2} = \frac{r_1}{r_2} \text{cis}(\theta_1 - \theta_2).$$

FOR DISCUSSION

In Example 6, the complex number

$$10 \text{ cis}(-60°) = 5 - 5i\sqrt{3}$$

was divided by the complex number

$$5 \text{ cis}(150°) = -\frac{5\sqrt{3}}{2} + \frac{5}{2}i.$$

Would you rather perform this division (by hand) in trigonometric form or in rectangular form? Explain your reasoning.

EXAMPLE 6 Using the Quotient Theorem

Find the quotient $\frac{10 \text{ cis}(-60°)}{5 \text{ cis } 150°}$. Write the result in rectangular form.

Solution

$$\frac{10 \text{ cis}(-60°)}{5 \text{ cis } 150°} = \frac{10}{5} \text{cis}(-60° - 150°) \quad \text{Quotient theorem}$$

$$= 2 \text{ cis}(-210°) \quad \text{Divide; subtract.}$$

$$= 2[\cos(-210°) + i \sin(-210°)] \quad \text{Rewrite.}$$

$$= 2\left[-\frac{\sqrt{3}}{2} + i\left(\frac{1}{2}\right)\right] \quad \cos(-210°) = -\frac{\sqrt{3}}{2};\ \sin(-210°) = \frac{1}{2}$$

$$= -\sqrt{3} + i \quad \text{Rectangular form}$$

■

10.4 Exercises

1. *Concept Check* The modulus of a complex number represents the ______________ of the vector representing it in the complex plane.

2. *Concept Check* What is the geometric interpretation of the argument of a complex number?

Graph each complex number as a vector in the complex plane.

3. $6 - 5i$
4. $2 - 2i\sqrt{3}$
5. $-4i$
6. $3i$
7. -8
8. 2

Give the rectangular form of the complex number represented in each graph.

9.

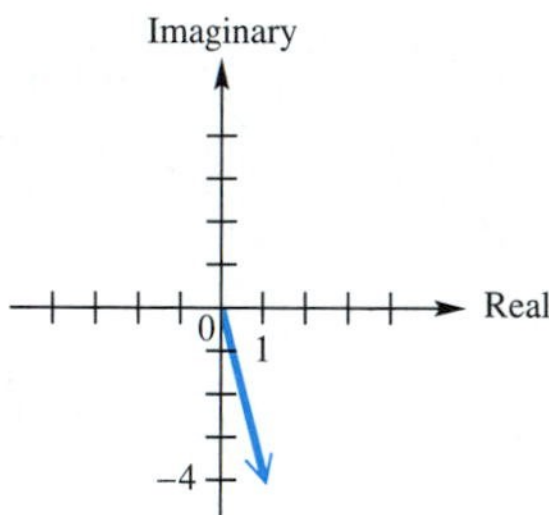

10.

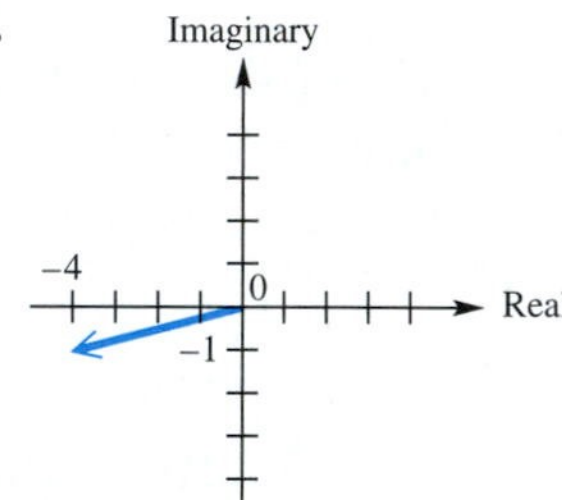

11. *Concept Check* What must be true for a complex number to also be a real number?

12. *Concept Check* If a real number is graphed in the complex plane, on what axis does the vector lie?

13. *Concept Check* A complex number of the form $a + bi$ will have its corresponding vector lying on the y-axis, provided that $a =$ ___________.

14. Are the notations $a + bi$ and $a\mathbf{i} + b\mathbf{j}$ equivalent? Explain.

Find the resultant (sum) of each pair of complex numbers. Express $a + bi$ in rectangular form.

15. $4 - 3i, -1 + 2i$
16. $2 + 3i, -4 - i$
17. $-3, 3i$
18. $6, -2i$
19. $2 + 6i, -2i$
20. $-5 - 8i, -1$

Find the modulus r of the number.

21. $1 + i$ **22.** $3 - 4i$ **23.** $12 - 5i$ **24.** $-24 + 7i$

25. -6 **26.** $15i$ **27.** $2 - 3i$ **28.** $11 - 60i$

Write each complex number in rectangular form. Give exact values for the real and imaginary parts.

29. $2(\cos 45° + i \sin 45°)$ **30.** $4(\cos 60° + i \sin 60°)$ **31.** $10 \text{ cis } 90°$

32. $8 \text{ cis } 270°$ **33.** $4(\cos 240° + i \sin 240°)$ **34.** $2(\cos 330° + i \sin 330°)$

35. $\cos \frac{\pi}{6} + i \sin \frac{\pi}{6}$ **36.** $3\left(\cos \frac{5\pi}{6} + i \sin \frac{5\pi}{6}\right)$ **37.** $5 \text{ cis}\left(-\frac{\pi}{6}\right)$

38. $6 \text{ cis } \frac{3\pi}{4}$ **39.** $\sqrt{2} \text{ cis } \pi$ **40.** $\sqrt{3} \text{ cis } \frac{3\pi}{2}$

Write each complex number in the trigonometric form $r(\cos \theta + i \sin \theta)$, where r is exact and $-180° < \theta \leq 180°$. Use your calculator to support your result.

41. $3 - 3i$ **42.** $-2 + 2i\sqrt{3}$ **43.** $1 + i\sqrt{3}$

44. $-3 - 3i\sqrt{3}$ **45.** $-2i$ **46.** 7

Write each complex number in trigonometric form, where r is exact and $-\pi < \theta \leq \pi$.

47. $4\sqrt{3} + 4i$ **48.** $\sqrt{3} - i$ **49.** $-\sqrt{2} + i\sqrt{2}$

50. $-5 - 5i$ **51.** -4 **52.** $5i$

Find each product in rectangular form, using exact values.

53. $[3(\cos 60° + i \sin 60°)][2(\cos 90° + i \sin 90°)]$ **54.** $[4(\cos 30° + i \sin 30°)][5(\cos 120° + i \sin 120°)]$

55. $[2(\cos 45° + i \sin 45°)][2(\cos 225° + i \sin 225°)]$ **56.** $[8(\cos 300° + i \sin 300°)][5(\cos 120° + i \sin 120°)]$

57. $\left[5 \text{ cis } \frac{\pi}{2}\right]\left[3 \text{ cis } \frac{\pi}{4}\right]$ **58.** $\left[6 \text{ cis } \frac{2\pi}{3}\right]\left[5 \text{ cis}\left(-\frac{\pi}{6}\right)\right]$

59. $\left[\sqrt{3} \text{ cis } \frac{\pi}{4}\right]\left[\sqrt{3} \text{ cis } \frac{5\pi}{4}\right]$ **60.** $\left[\sqrt{2} \text{ cis } \frac{5\pi}{6}\right]\left[\sqrt{2} \text{ cis } \frac{3\pi}{2}\right]$

Find each quotient in rectangular form, using exact values.

61. $\dfrac{4(\cos 120° + i \sin 120°)}{2(\cos 150° + i \sin 150°)}$ **62.** $\dfrac{16(\cos 300° + i \sin 300°)}{8(\cos 60° + i \sin 60°)}$ **63.** $\dfrac{10\left(\cos \frac{5\pi}{4} + i \sin \frac{5\pi}{4}\right)}{5\left(\cos \frac{\pi}{4} + i \sin \frac{\pi}{4}\right)}$

64. $\dfrac{24\left(\cos \frac{5\pi}{6} + i \sin \frac{5\pi}{6}\right)}{2\left(\cos \frac{\pi}{6} + i \sin \frac{\pi}{6}\right)}$ **65.** $\dfrac{3 \text{ cis } \frac{61\pi}{36}}{9 \text{ cis } \frac{13\pi}{36}}$ **66.** $\dfrac{12 \text{ cis } 293°}{6 \text{ cis } 23°}$

Solve each problem. Refer to Example 4 for Exercises 70 and 71.

67. *(Modeling) Alternating Current* The alternating current in amps in an electric inductor is given by

$$I = \frac{E}{Z},$$

where E is the voltage and $Z = R + X_L i$ is the impedance. If $E = 8(\cos 20° + i \sin 20°)$, $R = 6$, and $X_L = 3$, find the current. Give the answer in rectangular form.

68. *(Modeling) Electric Current* The current in a circuit with voltage E, resistance R, capacitive reactance X_c, and inductive resistance X_L is given by

$$I = \frac{E}{R + (X_L - X_c)i}.$$

Find I if $E = 12(\cos 25° + i \sin 25°)$, $R = 3$, $X_L = 4$, and $X_c = 6$. Give the answer in rectangular form.

69. *(Modeling) Impedance* In the parallel electric circuit shown in the figure, the impedance Z can be calculated by using the equation

$$Z = \frac{1}{\frac{1}{Z_1} + \frac{1}{Z_2}},$$

where Z_1 and Z_2 are the impedances for the branches of the circuit. If $Z_1 = 50 + 25i$ and $Z_2 = 60 + 20i$, calculate Z.

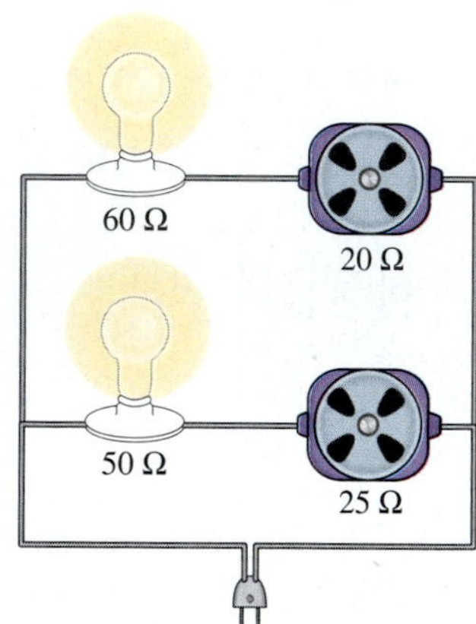

70. Is $z = -.2i$ in the Julia set?

71. The graph of the Julia set in Figure 51 appears to be symmetric with respect to both the x-axis and y-axis. Complete the following to show that this is true.

(a) Show that complex conjugates have the same modulus.

(b) Compute $z_1^2 - 1$ and $z_2^2 - 1$, where $z_1 = a + bi$ and $z_2 = a - bi$.

(c) Discuss why, if (a, b) is in the Julia set, then so is $(a, -b)$.

(d) Can we conclude that the graph of the Julia set must be symmetric with respect to the x-axis?

(e) Using a similar argument, show that the Julia set must also be symmetric with respect to the y-axis.

72. *Concept Check* Without actually performing the operations, state why the products

$$[2(\cos 45° + i \sin 45°)][5(\cos 90° + i \sin 90°)]$$

and

$$[2(\cos(-315°) + i \sin(-315°))] \cdot [5(\cos(-270°) + i \sin(-270°))]$$

are the same.

73. Consider the equation $(r \text{ cis } \theta)^2 = (r \text{ cis } \theta)(r \text{ cis } \theta) = r^2 \text{ cis}(\theta + \theta) = r^2 \text{ cis } 2\theta$. State in your own words how we can square a complex number in trigonometric form. (In the next section, we will develop this idea further.)

74. *Concept Check* Under what conditions is the difference between two nonreal complex numbers $a + bi$ and $c + di$ a real number?

75. *Concept Check* Give the smallest *positive* radian measure of θ if $r > 0$ and **(a)** $r \text{ cis } \theta$ has real part equal to 0, **(b)** $r \text{ cis } \theta$ has imaginary part equal to 0.

76. Use your calculator in radian mode to find the trigonometric form of $3 + 5i$. Approximate values to the nearest hundredth.

10.5 Powers and Roots of Complex Numbers

Powers of Complex Numbers (De Moivre's Theorem) ■ Roots of Complex Numbers

Powers of Complex Numbers (De Moivre's Theorem)

Because raising a number to a positive integer power is a repeated application of the product rule, it would seem likely that a theorem for finding powers of complex numbers exists. For example,

$$\begin{aligned}[r(\cos\theta + i\sin\theta)]^2 &= [r(\cos\theta + i\sin\theta)][r(\cos\theta + i\sin\theta)]\\ &= r\cdot r[\cos(\theta+\theta) + i\sin(\theta+\theta)]\\ &= r^2(\cos 2\theta + i\sin 2\theta).\end{aligned}$$

In the same way, $[r(\cos\theta + i\sin\theta)]^3 = r^3(\cos 3\theta + i\sin 3\theta).$

These results suggest the following theorem for positive integer values of n. Although the theorem is stated and can be proved for all n, we use it only for positive integer values of n and their reciprocals.

Abraham De Moivre (1667–1754)

De Moivre's Theorem

If $r(\cos\theta + i\sin\theta)$ is a complex number, and if n is any real number, then

$$[r(\cos\theta + i\sin\theta)]^n = r^n(\cos n\theta + i\sin n\theta).$$

In compact form, this is written

$$[r \text{ cis } \theta]^n = r^n(\text{cis } n\theta).$$

This theorem is named after the French expatriate friend of Isaac Newton, Abraham De Moivre, although he never explicitly stated it.

EXAMPLE 1 Finding a Power of a Complex Number

Find $(1 + i\sqrt{3})^8$ and express the result in rectangular form.

Solution Convert $1 + i\sqrt{3}$ into trigonometric form as in Section 10.4.

$$1 + i\sqrt{3} = 2(\cos 60^\circ + i\sin 60^\circ)$$

Now apply De Moivre's theorem.

$$\begin{aligned}(1 + i\sqrt{3})^8 &= [2(\cos 60^\circ + i\sin 60^\circ)]^8\\ &= 2^8[\cos(8\cdot 60^\circ) + i\sin(8\cdot 60^\circ)] && \text{De Moivre's theorem}\\ &= 256(\cos 480^\circ + i\sin 480^\circ)\\ &= 256(\cos 120^\circ + i\sin 120^\circ) && 480^\circ \text{ and } 120^\circ \text{ are coterminal.}\\ &= 256\left(-\frac{1}{2} + i\frac{\sqrt{3}}{2}\right) && \cos 120^\circ = -\tfrac{1}{2};\ \sin 120^\circ = \tfrac{\sqrt{3}}{2}\\ &= -128 + 128i\sqrt{3} && \text{Rectangular form}\end{aligned}$$

■

Roots of Complex Numbers

Every nonzero complex number has exactly n distinct complex nth roots. De Moivre's theorem can be extended to find all nth roots of a complex number.

*n*th Root

For a positive integer n, the complex number $a + bi$ is an ***n*th root** of the complex number $x + yi$ if

$$(a + bi)^n = x + yi.$$

To find the three complex cube roots of $8(\cos 135° + i \sin 135°)$, for example, look for a complex number, say, $r(\cos \alpha + i \sin \alpha)$, that will satisfy

$$[r(\cos \alpha + i \sin \alpha)]^3 = 8(\cos 135° + i \sin 135°).$$

By De Moivre's theorem, this equation becomes

$$r^3(\cos 3\alpha + i \sin 3\alpha) = 8(\cos 135° + i \sin 135°).$$

To satisfy this equation, set $r^3 = 8$ and $\cos 3\alpha + i \sin 3\alpha = \cos 135° + i \sin 135°$. The first of these conditions implies that $r = 2$, and the second implies that

$$\cos 3\alpha = \cos 135° \quad \text{and} \quad \sin 3\alpha = \sin 135°.$$

For these equations to be satisfied, 3α must represent an angle that is coterminal with 135°. Therefore, we must have

$$3\alpha = 135° + 360° \cdot k, \quad k \text{ any integer,}$$

or

$$\alpha = \frac{135° + 360° \cdot k}{3}, \quad k \text{ any integer.}$$

Now, let k take on the integer values 0, 1, and 2.

$$\text{If } k = 0, \text{ then} \quad \alpha = \frac{135° + 0°}{3} = 45°.$$

$$\text{If } k = 1, \text{ then} \quad \alpha = \frac{135° + 360°}{3} = \frac{495°}{3} = 165°.$$

$$\text{If } k = 2, \text{ then} \quad \alpha = \frac{135° + 720°}{3} = \frac{855°}{3} = 285°.$$

In the same way, $\alpha = 405°$ when $k = 3$. But note that $405° = 45° + 360°$, so $\sin 405° = \sin 45°$ and $\cos 405° = \cos 45°$. Similarly, if $k = 4$, $\alpha = 525°$, which has the same sine and cosine values as 165°. To continue with larger values of k would just be repeating solutions already found. Therefore, all of the cube roots (three of them) can be found by letting $k = 0$, 1, and 2, in turn.

When $k = 0$, the root is $2(\cos 45° + i \sin 45°)$.
When $k = 1$, the root is $2(\cos 165° + i \sin 165°)$.
When $k = 2$, the root is $2(\cos 285° + i \sin 285°)$.

In summary, we see that $2(\cos 45° + i \sin 45°)$, $2(\cos 165° + i \sin 165°)$, and $2(\cos 285° + i \sin 285°)$ are the three cube roots of $8(\cos 135° + i \sin 135°)$.

Generalizing our results, we state the following theorem.

nth Root Theorem

If n is any positive integer, r is a positive real number, and θ is in degrees, then the nonzero complex number $r(\cos \theta + i \sin \theta)$ has exactly n distinct nth roots, given by

$$\sqrt[n]{r}(\cos \alpha + i \sin \alpha),$$

where

$$\alpha = \frac{\theta + 360° \cdot k}{n} \quad \text{or} \quad \alpha = \frac{\theta}{n} + \frac{360° \cdot k}{n}, \qquad k = 0, 1, 2, \ldots, n - 1.$$

NOTE In the statement of the nth root theorem, if θ is in radians, then

$$\alpha = \frac{\theta + 2\pi k}{n} \quad \text{or} \quad \alpha = \frac{\theta}{n} + \frac{2\pi k}{n}.$$

EXAMPLE 2 Finding Complex Roots

Find the two square roots of $4i$. Write the roots in rectangular form, and check your results directly with a calculator.

Solution First write $4i$ in trigonometric form as

$$4i = 4\left(\cos \frac{\pi}{2} + i \sin \frac{\pi}{2}\right).$$

Here, $r = 4$ and $\theta = \frac{\pi}{2}$. The square roots have modulus $\sqrt{4} = 2$ and arguments as follows.

$$\alpha = \frac{\frac{\pi}{2}}{2} + \frac{2\pi k}{2} = \frac{\pi}{4} + \pi k \qquad \text{Simplify.}$$

Be careful when simplifying a complex fraction.

Since there are two square roots, let $k = 0$ and 1, in turn.

$$\text{If } k = 0, \text{ then} \quad \alpha = \frac{\pi}{4} + \pi \cdot 0 = \frac{\pi}{4}.$$

$$\text{If } k = 1, \text{ then} \quad \alpha = \frac{\pi}{4} + \pi \cdot 1 = \frac{5\pi}{4}.$$

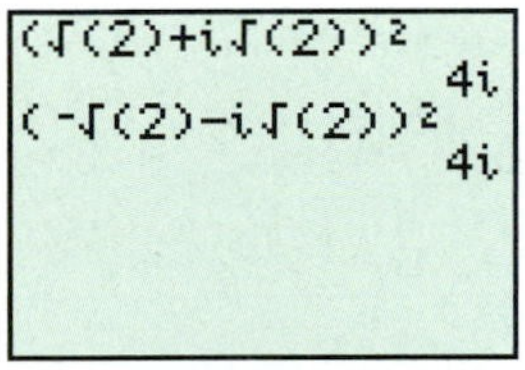

With these values for α, the square roots are $2 \text{ cis } \frac{\pi}{4}$ and $2 \text{ cis } \frac{5\pi}{4}$, which can be written in rectangular form as

$$\sqrt{2} + i\sqrt{2} \quad \text{and} \quad -\sqrt{2} - i\sqrt{2}.$$

Check the results by squaring each answer, as shown in Figure 52. ■

```
(√(2)+i√(2))²
                 4i
(-√(2)-i√(2))²
                 4i
```

FIGURE 52

TECHNOLOGY NOTE

In Example 3, $-8 + 8i\sqrt{3}$ is converted into the form $16(\cos 120° + i \sin 120°)$. In the screen below, a graphing calculator performs this conversion.

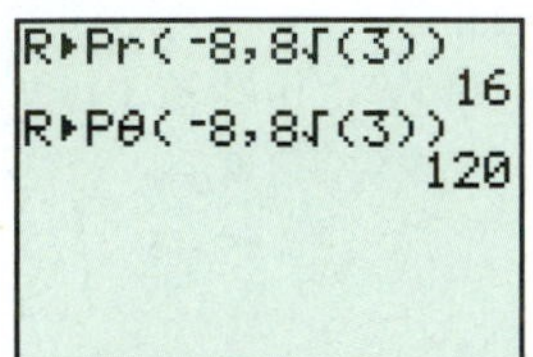

Degree mode

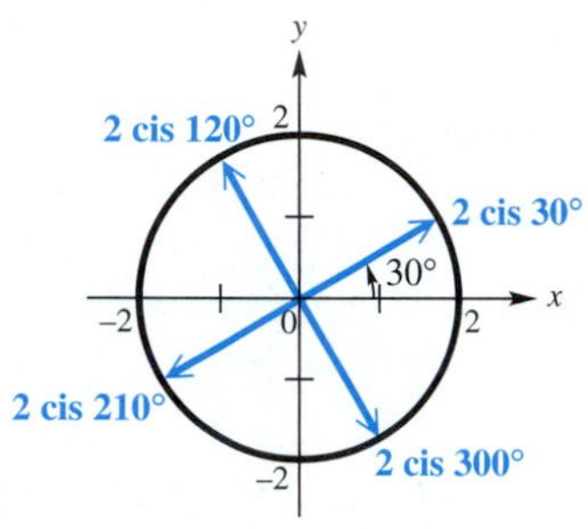

FIGURE 53

EXAMPLE 3 Finding Complex Roots

Find all fourth roots of $-8 + 8i\sqrt{3}$. Write the roots in rectangular form.

Solution First write $-8 + 8i\sqrt{3}$ in trigonometric form as

$$-8 + 8i\sqrt{3} = 16 \text{ cis } 120°.$$

Here, $r = 16$ and $\theta = 120°$. The fourth roots of this number have modulus $\sqrt[4]{16} = 2$ and arguments as follows.

$$\alpha = \frac{120°}{4} + \frac{360° \cdot k}{4} = 30° + 90° \cdot k$$

Since there are four fourth roots, let $k = 0, 1, 2$, and 3, in turn.

$$\begin{aligned} &\text{If } k = 0, \text{ then} \quad \alpha = 30° + 90° \cdot 0 = 30°. \\ &\text{If } k = 1, \text{ then} \quad \alpha = 30° + 90° \cdot 1 = 120°. \\ &\text{If } k = 2, \text{ then} \quad \alpha = 30° + 90° \cdot 2 = 210°. \\ &\text{If } k = 3, \text{ then} \quad \alpha = 30° + 90° \cdot 3 = 300°. \end{aligned}$$

Using these angles, we find that the fourth roots are

$$2 \text{ cis } 30°, \quad 2 \text{ cis } 120°, \quad 2 \text{ cis } 210°, \quad \text{and} \quad 2 \text{ cis } 300°.$$

These four roots can be written in rectangular form as

$$\sqrt{3} + i, \quad -1 + i\sqrt{3}, \quad -\sqrt{3} - i, \quad \text{and} \quad 1 - i\sqrt{3}.$$

The graphs of these roots are on a circle that has center at the origin and radius 2, as shown in Figure 53. The roots are equally spaced about the circle, 90° apart. ■

EXAMPLE 4 Solving an Equation by Finding Complex Roots

Find all complex number solutions of $x^5 - 1 = 0$. Graph them as vectors in the complex plane.

Solution Write the equation as

$$x^5 - 1 = 0, \quad \text{or} \quad x^5 = 1.$$

While there is only one real number solution, 1, there are five complex number solutions. To find these solutions, first write 1 in trigonometric form as

$$1 = 1 + 0i = 1(\cos 0° + i \sin 0°).$$

The modulus of the fifth roots is $\sqrt[5]{1} = 1$, and the arguments are given by

$$0° + 72° \cdot k, \quad k = 0, 1, 2, 3, 4.$$

With these arguments, the fifth roots are

$$\begin{aligned} &1(\cos 0° + i \sin 0°), && k = 0 \\ &1(\cos 72° + i \sin 72°), && k = 1 \\ &1(\cos 144° + i \sin 144°), && k = 2 \\ &1(\cos 216° + i \sin 216°), && k = 3 \\ \text{and} \quad &1(\cos 288° + i \sin 288°). && k = 4 \end{aligned}$$

The solution set of the equation is {cis 0°, cis 72°, cis 144°, cis 216°, cis 288°}. The first of these roots equals 1; the others cannot easily be expressed in rectangular form but can be approximated with a calculator. The five fifth roots all lie on a unit circle, equally spaced around it every 72°. See Figure 54.

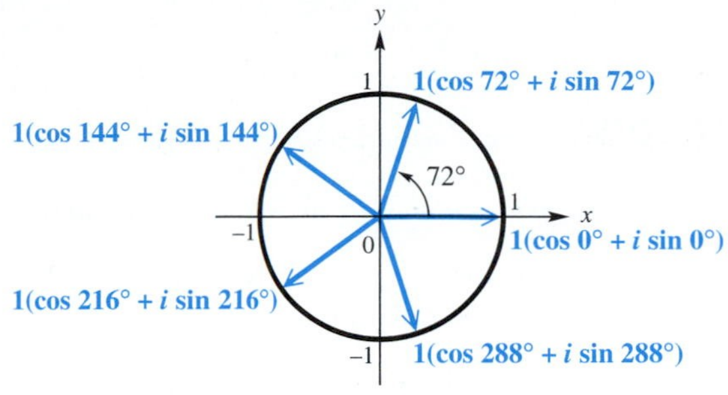

FIGURE 54

10.5 Exercises

Find each power. Write the answer in rectangular form.

1. $[3(\cos 30° + i \sin 30°)]^3$

2. $[2(\cos 135° + i \sin 135°)]^4$

3. $\left(\cos \frac{\pi}{4} + i \sin \frac{\pi}{4}\right)^8$

4. $\left(\cos \frac{\pi}{5} + i \sin \frac{\pi}{5}\right)^5$

5. $\left[2\left(\cos \frac{2\pi}{3} + i \sin \frac{2\pi}{3}\right)\right]^3$

6. $\left[3\left(\cos \frac{3\pi}{4} + i \sin \frac{3\pi}{4}\right)\right]^4$

7. $[3 \text{ cis } 100°]^3$

8. $[3 \text{ cis } 40°]^3$

9. $(\sqrt{3} + i)^3$

10. $(1 + \sqrt{3}i)^4$

11. $(2\sqrt{2} - 2i\sqrt{2})^6$

12. $(2 - 2i\sqrt{3})^4$

13. $\left(-\frac{\sqrt{2}}{2} + \frac{\sqrt{2}}{2}i\right)^4$

14. $\left(\frac{\sqrt{2}}{2} - \frac{\sqrt{2}}{2}i\right)^8$

15. $(1 - i)^6$

16. $(-1 + i)^7$

17. $(-2 - 2i)^5$

18. $(-3 - 3i)^3$

Find the cube roots of each complex number. Leave the answer in trigonometric form. Then graph each cube root as a vector in the complex plane.

19. 1

20. i

21. $8(\cos 60° + i \sin 60°)$

22. $27(\cos 300° + i \sin 300°)$

23. $-8i$

24. $27i$

25. -64

26. 27

27. $1 + i\sqrt{3}$

28. $2 - 2i\sqrt{3}$

29. $-2\sqrt{3} + 2i$

30. $\sqrt{3} - i$

Find all indicated roots and express them in rectangular form. Check your results with a calculator.

31. The square roots of $4(\cos 120° + i \sin 120°)$

32. The cube roots of $27(\cos 180° + i \sin 180°)$

33. The cube roots of $\cos 180° + i \sin 180°$

34. The fourth roots of $16(\cos 240° + i \sin 240°)$

35. The square roots of i

36. The square roots of $-4i$

37. The cube roots of $64i$

38. The fourth roots of -1

39. The fourth roots of 81

40. The square roots of $-1 + i\sqrt{3}$

41. For the real number 1, find and graph all indicated roots. Give answers in rectangular form.
(a) Fourth **(b)** Sixth

42. For the complex number i, find and graph all indicated roots. Give answers in trigonometric form.
(a) Square **(b)** Fourth

43. Explain why a positive real number must have a positive real nth root.

44. *Concept Check* *True* or *false:* **(a)** Every real number must have two real square roots. **(b)** Some real numbers have three real cube roots.

Relating Concepts

For individual or group investigation (Exercises 45–50)

We examine how the three complex cube roots of -8 *can be found in two different ways.* ***Work Exercises 45–50 in order.***

45. All complex roots of the equation $x^3 + 8 = 0$ are cube roots of -8. Factor $x^3 + 8$ as the sum of two cubes.

46. One of the factors found in Exercise 45 is linear. Set it equal to 0, solve, and determine the real cube root of -8.

47. One of the factors found in Exercise 45 is quadratic. Set it equal to 0, solve, and determine the rectangular forms of the other two cube roots of -8.

48. Use the method described in this section to find the three complex cube roots of -8. Give them in trigonometric form.

49. Convert the trigonometric forms found in Exercise 48 to rectangular form.

50. Compare your results in Exercises 46 and 47 with those in Exercise 49. What do you notice?

Find all complex solutions for each equation. Leave your answers in trigonometric form.

51. $x^4 + 1 = 0$ **52.** $x^4 + 16 = 0$ **53.** $x^5 - i = 0$ **54.** $x^4 - i = 0$

55. $x^3 + 1 = 0$ **56.** $x^3 + i = 0$ **57.** $x^3 - 8 = 0$ **58.** $x^5 + 1 = 0$

59. *(Modeling) Basins of Attraction* The second fractal shown in the margin in Section 10.4 is the solution to Cayley's problem of determining the basins of attraction for the cube roots of unity (1). The three cube roots of unity are

$$w_1 = 1, \quad w_2 = -\frac{1}{2} + \frac{\sqrt{3}}{2}i,$$

and
$$w_3 = -\frac{1}{2} - \frac{\sqrt{3}}{2}i.$$

This fractal can be generated by repeatedly evaluating the function defined by

$$f(z) = \frac{2z^3 + 1}{3z^2},$$

where z is a complex number. We begin by picking $z_1 = a + bi$ and then successively computing

$$z_2 = f(z_1), \quad z_3 = f(z_2), \quad z_4 = f(z_3), \quad \ldots .$$

If the resulting values of $f(z)$ approach w_1, color the pixel at (a, b) red. If they approach w_2, color it blue, and if they approach w_3, color it yellow. If this process continues for a large number of different z_1, the fractal in the figure will appear. Determine the appropriate color of the pixel for each value of z_1. (*Source:* Crownover, R., *Introduction to Fractals and Chaos,* Jones and Bartlett, 1995.)

(a) $z_1 = i$ **(b)** $z_1 = 2 + i$
(c) $z_1 = -1 - i$

60. *(Modeling)* The screens illustrate how a pentagon can be graphed with parametric equations. Note that a pentagon has five sides, and the T-step is $\frac{360}{5} = 72$. The display at the bottom of the graph screen indicates that one fifth root of 1 is $1 + 0i = 1$.

```
WINDOW
 Tmin=0
 Tmax=360
 Tstep=72
 Xmin=-1.8
 Xmax=1.8
 Xscl=1
↓Ymin=-1.2
```

```
WINDOW
↑Tstep=72
 Xmin=-1.8
 Xmax=1.8
 Xscl=1
 Ymin=-1.2
 Ymax=1.2
 Yscl=1
```

This is a continuation of the previous screen.

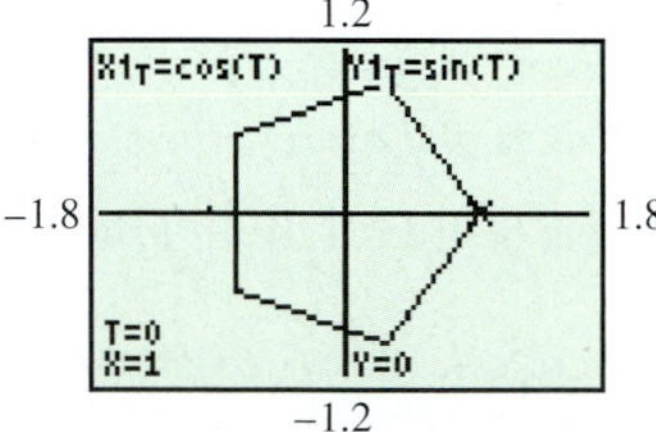

The calculator is in parametric, degree, and connected graph modes.

(a) Use this technique to find all fifth roots of 1, and express the real and imaginary parts in decimal form.
(b) Use this method to find the first three 10th roots of 1.

Reviewing Basic Concepts (Sections 10.4 and 10.5)

1. Write the complex number $2(\cos 60° + i \sin 60°)$ in rectangular form.

2. Find the modulus of the complex number $3 - 4i$.

3. Write $-\sqrt{2} + i\sqrt{2}$ in trigonometric form, where $0° \le \theta < 360°$.

For $z_1 = 4(\cos 135° + i \sin 135°)$

and $z_2 = 2(\cos 45° + i \sin 45°)$,

find each product or quotient and write your answers in both trigonometric and rectangular forms.

4. $z_1 \cdot z_2$

5. $\dfrac{z_1}{z_2}$

6. Evaluate the power $(4 \text{ cis } 17°)^3$, and write your answer in trigonometric form.

7. Evaluate the power $(2 - 2i)^4$, and write your answer in rectangular form.

8. Find the cube roots of -64, and express them in rectangular form.

9. Find the square roots of $2i$, and express them in rectangular form.

10. Find all complex solutions of the equation $x^3 = -1$. Leave your answers in trigonometric form.

10.6 Polar Equations and Graphs

Polar Coordinate System ■ Graphs of Polar Equations ■ Classifying Polar Equations ■ Converting Equations

Polar Coordinate System

Pole

Polar axis

FIGURE 55

We have been using the rectangular coordinate system to graph equations. The **polar coordinate system** is based on a point, called the **pole,** and a ray, called the **polar axis.** The polar axis is usually drawn in the direction of the positive x-axis. See Figure 55.

In Figure 56, the pole has been placed at the origin of a rectangular coordinate system, so that the polar axis coincides with the positive x-axis. Point P has rectangular coordinates (x, y). Point P can also be located by giving the directed angle θ from the positive x-axis to ray OP and the directed distance r from the pole to point P. The ordered pair (r, θ) gives **polar coordinates** of point P. If $r > 0$, point P lies on the terminal side of θ and if $r < 0$, point P lies on the ray pointing in the opposite direction of the terminal side of θ, a distance $|r|$ from the origin. See Figure 57.

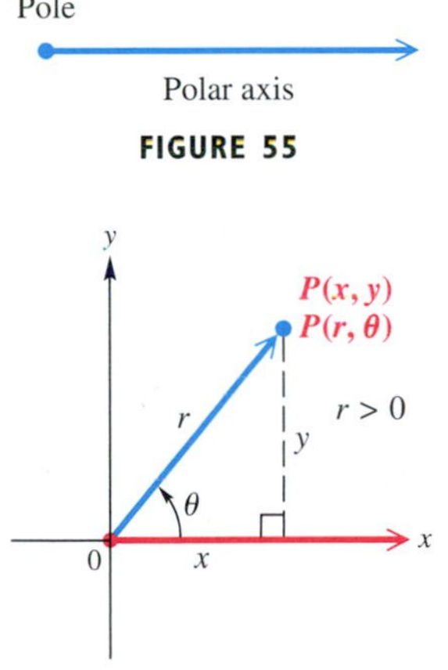

FIGURE 56

Using Figure 56 and trigonometry, we can establish the following relationships between rectangular and polar coordinates.

y

(r, θ)

r > 0

θ

x

O

(−r, θ)

FIGURE 57

Rectangular and Polar Coordinates

If a point has rectangular coordinates (x, y) and polar coordinates (r, θ), then these coordinates are related as follows.

$$x = r\cos\theta \qquad y = r\sin\theta$$

$$r^2 = x^2 + y^2 \qquad \tan\theta = \frac{y}{x}, \quad \text{if } x \neq 0$$

GCM **EXAMPLE 1 Plotting Points with Polar Coordinates**

Plot each point by hand in the polar coordinate system. Then determine the rectangular coordinates of each point.

(a) $P(2, 30°)$ **(b)** $Q\left(-4, \frac{2\pi}{3}\right)$ **(c)** $R\left(5, -\frac{\pi}{4}\right)$

Solution

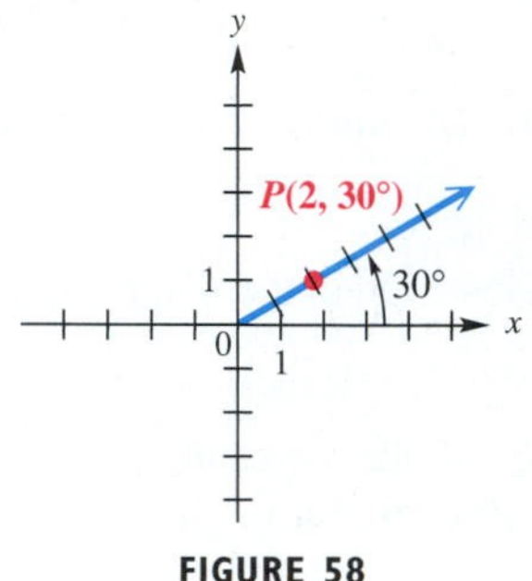

FIGURE 58

(a) In this case, $r = 2$ and $\theta = 30°$, so the point P is located 2 units from the origin in the positive direction on a ray making a 30° angle with the polar axis, as shown in Figure 58.

Using the conversion equations, we find the rectangular coordinates as follows.

$$x = r\cos\theta = 2\cos 30° = 2\left(\frac{\sqrt{3}}{2}\right) = \sqrt{3}$$

$$y = r\sin\theta = 2\sin 30° = 2\left(\frac{1}{2}\right) = 1$$

The rectangular coordinates are $\left(\sqrt{3}, 1\right)$.

(b) Since r is negative, Q is 4 units in the opposite direction from the pole on an extension of the $\frac{2\pi}{3}$ ray. See Figure 59. The rectangular coordinates are

$$x = -4\cos\frac{2\pi}{3} = -4\left(-\frac{1}{2}\right) = 2$$

and

$$y = -4\sin\frac{2\pi}{3} = -4\left(\frac{\sqrt{3}}{2}\right) = -2\sqrt{3}.$$

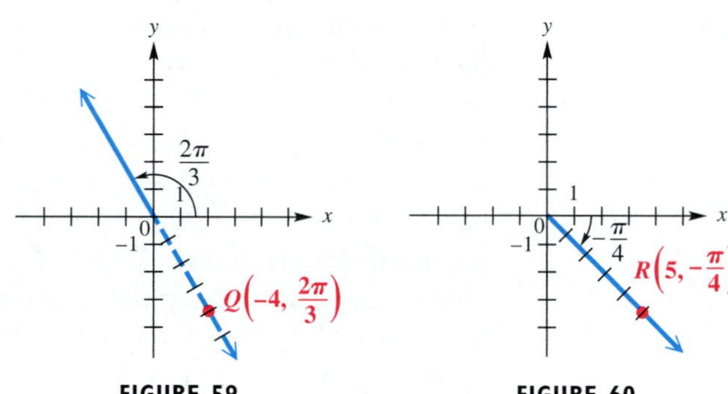

FIGURE 59 FIGURE 60

(c) Point R is shown in Figure 60. Since θ is negative, the angle is measured in the clockwise direction. Furthermore, we have

$$x = 5\cos\left(-\frac{\pi}{4}\right) = \frac{5\sqrt{2}}{2} \quad \text{and} \quad y = 5\sin\left(-\frac{\pi}{4}\right) = -\frac{5\sqrt{2}}{2}.$$ ■

Looking Ahead to Calculus

Techniques studied in calculus and associated with derivatives and integrals provide methods of finding slopes of tangent lines to polar curves, areas bounded by such curves, and lengths of their arcs.

One important difference between rectangular coordinates and polar coordinates is that while a given point in the plane can have only one pair of rectangular coordinates, this same point can have an infinite number of pairs of polar coordinates. For example, $(2, 30°)$ locates the same point as $(2, 390°)$, or $(2, -330°)$, or $(-2, 210°)$.

GCM **EXAMPLE 2 Giving Alternative Forms for Coordinates of a Point**

(a) Give three other pairs of polar coordinates for the point $P(3, 140°)$.

(b) Determine two pairs of polar coordinates, with $r > 0$, for the point with rectangular coordinates $(-1, 1)$.

Solution

(a) Three pairs that could be used for the point are $(3, -220°)$, $(-3, 320°)$, and $(-3, -40°)$. See Figure 61.

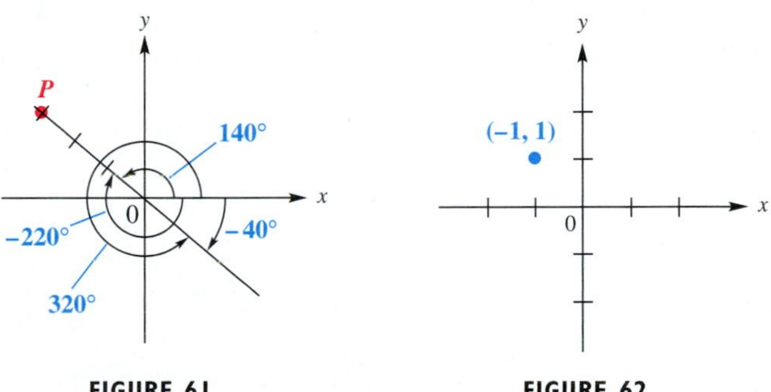

FIGURE 61 FIGURE 62

(b) As shown in Figure 62, the point $(-1, 1)$ lies in the second quadrant. Since $\tan \theta = \frac{1}{-1} = -1$, one possible value for θ is 135°. Also,

$$r = \sqrt{x^2 + y^2} = \sqrt{(-1)^2 + 1^2} = \sqrt{2}.$$

Therefore, two pairs of polar coordinates are $(\sqrt{2}, 135°)$ and $(\sqrt{2}, -225°)$. (Any angle coterminal with 135° could have been used for the second angle.) ■

Graphs of Polar Equations

Equations in x and y are called **rectangular** (or **Cartesian**) **equations.** An equation in which r and θ are the variables is a **polar equation.**

$$r = 3 \sin \theta, \qquad r = 2 + \cos \theta, \qquad r = \theta \qquad \text{Polar equations}$$

While the rectangular forms of lines and circles are the ones most often encountered, they can also be defined in terms of polar coordinates. The polar equation of the line $ax + by = c$ can be derived as follows.

$$ax + by = c \qquad \text{Rectangular equation}$$

$$a(r \cos \theta) + b(r \sin \theta) = c \qquad \text{Convert to polar coordinates.}$$

$$r(a \cos \theta + b \sin \theta) = c \qquad \text{Factor out } r.$$

$$r = \frac{c}{a \cos \theta + b \sin \theta} \qquad \text{Polar equation of a line}$$

For the circle $x^2 + y^2 = a^2$, we have

$$x^2 + y^2 = a^2$$

$$(\sqrt{x^2 + y^2})^2 = a^2 \qquad (\sqrt{k})^2 = k$$

$$r^2 = a^2 \qquad \sqrt{x^2 + y^2} = r$$

$$r = \pm a. \qquad \text{Polar equation of a circle; } r \text{ can be negative in polar coordinates.}$$

EXAMPLE 3 Examining Polar and Rectangular Equations of Lines and Circles

For each rectangular equation, give its equivalent polar equation and sketch its graph.

(a) $y = x - 3$ **(b)** $x^2 + y^2 = 4$

Solution

(a) This is the equation of a line. Rewrite $y = x - 3$ as

$$x - y = 3. \quad \text{Standard form: } ax + by = c$$

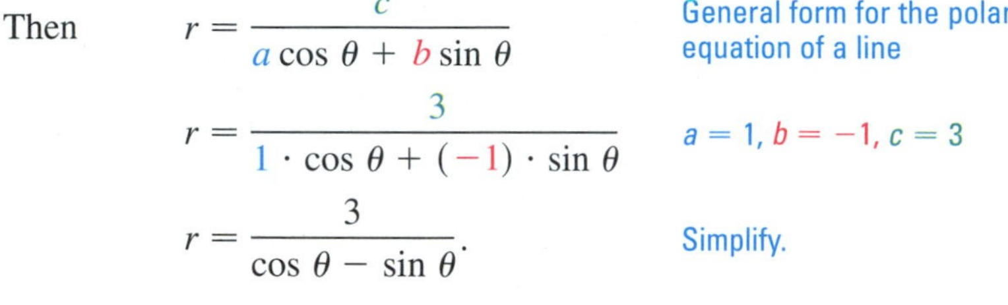

Then

$$r = \frac{c}{a\cos\theta + b\sin\theta} \quad \text{General form for the polar equation of a line}$$

$$r = \frac{3}{1\cdot\cos\theta + (-1)\cdot\sin\theta} \quad a = 1,\ b = -1,\ c = 3$$

$$r = \frac{3}{\cos\theta - \sin\theta}. \quad \text{Simplify.}$$

Its graph is shown in Figure 63.

$y = x - 3$ (rectangular)
$r = \dfrac{3}{\cos\theta - \sin\theta}$ (polar)

FIGURE 63

(b) The graph of $x^2 + y^2 = 4$ is a circle with center at the origin and radius 2.

$$x^2 + y^2 = 4$$

$$\left(\sqrt{x^2 + y^2}\right)^2 = (\pm 2)^2$$

$$\sqrt{x^2 + y^2} = 2 \quad \text{or} \quad \sqrt{x^2 + y^2} = -2$$

$$r = 2 \quad \text{or} \quad r = -2$$

(In polar coordinates, we may have $r < 0$.) The graphs of $r = 2$ and $r = -2$ coincide. The resulting graph is shown in Figure 64. ■

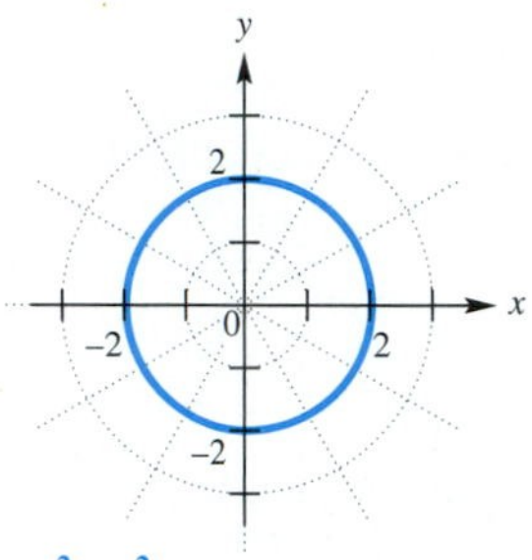

$x^2 + y^2 = 4$ (rectangular)
$r = 2$ or $r = -2$ (polar)

FIGURE 64

The tables in Examples 4–7 that follow include approximations for cosine values and r.

GCM EXAMPLE 4 Graphing a Polar Equation (Cardioid)

Graph $r = 1 + \cos\theta$.

Analytic Solution

To graph this equation, find some ordered pairs (as in the table). Once the pattern of values of r becomes clear, it is not necessary to find more ordered pairs. That is why we stopped with the ordered pair $(1.9, 330°)$.

θ	$\cos\theta$	$r = 1 + \cos\theta$	θ	$\cos\theta$	$r = 1 + \cos\theta$
0°	1	2	135°	−.7	.3
30°	.9	1.9	150°	−.9	.1
45°	.7	1.7	180°	−1	0
60°	.5	1.5	270°	0	1
90°	0	1	315°	.7	1.7
120°	−.5	.5	330°	.9	1.9

Graphing Calculator Solution

We choose degree mode and graph values of θ in the interval $[0°, 360°]$. The screen below and the continuation at the top of the next page show the choices needed to generate the graph.

```
WINDOW
 θmin=0
 θmax=360
 θstep=5
 Xmin=-2.35
 Xmax=2.35
 Xscl=1
↓Ymin=-1.55
```

(continued)

Connect the points in order—from $(2, 0°)$, to $(1.9, 30°)$, to $(1.7, 45°)$, and so on. See Figure 65. This curve is called a **cardioid** because of its heart shape. It has been graphed on a **polar grid.**

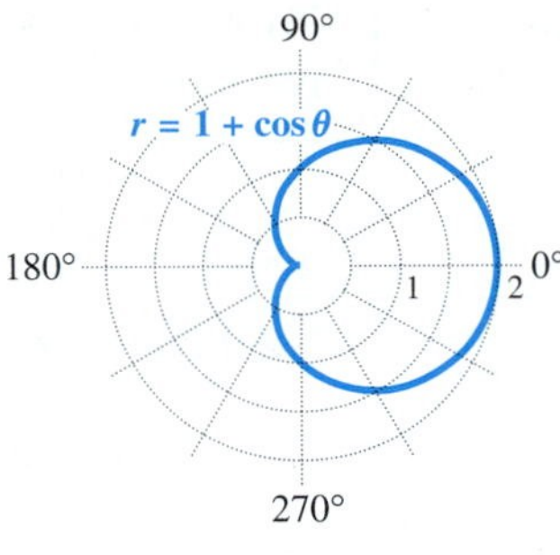

FIGURE 65

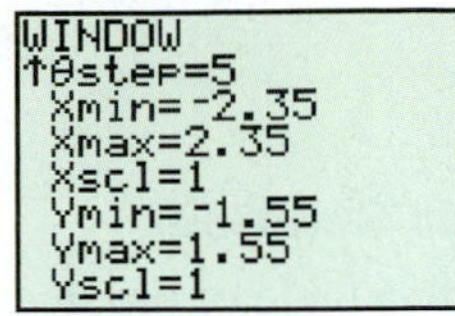

This is a continuation of the previous screen.

The calculator graph is shown in Figure 66.

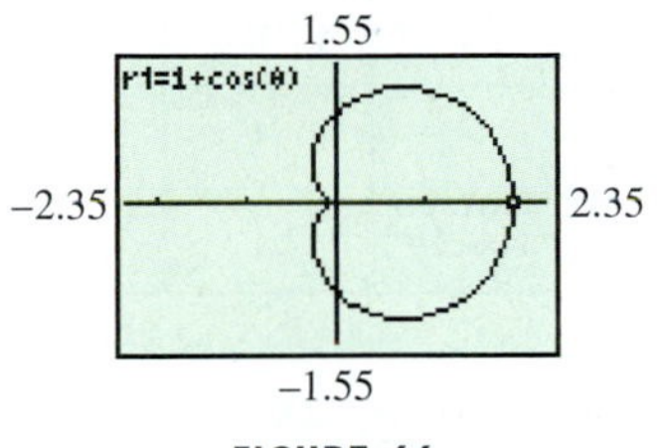

FIGURE 66

GCM EXAMPLE 5 Graphing a Polar Equation (Rose)

Graph $r = 3 \cos 2\theta$.

Solution Because of the argument 2θ, the graph requires a larger number of points than when the argument is θ. A few ordered pairs are given in the table. Complete the table similarly through the first 180°, so that 2θ has values up to 360°.

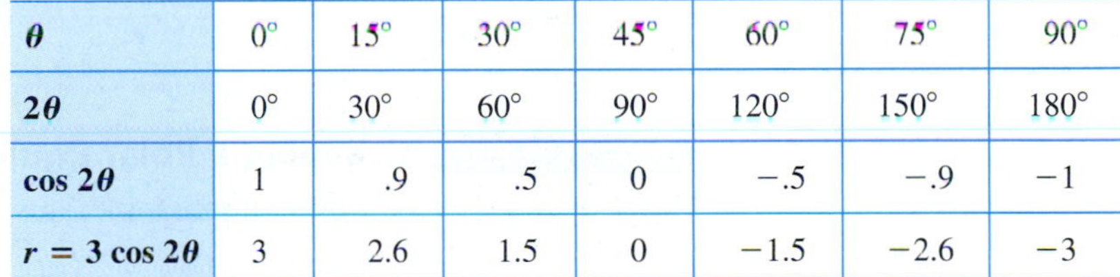

θ	0°	15°	30°	45°	60°	75°	90°
2θ	0°	30°	60°	90°	120°	150°	180°
$\cos 2\theta$	1	.9	.5	0	−.5	−.9	−1
$r = 3\cos 2\theta$	3	2.6	1.5	0	−1.5	−2.6	−3

90°
$r = 3 \cos 2\theta$
180° 0° Start Finish 3
270°

FIGURE 67

Plotting these points in order gives the graph, called a **four-leaved rose.** Notice in Figure 67 how the graph is developed with a continuous curve, beginning with the upper half of the right horizontal leaf and ending with the lower half of that leaf. As the graph is traced, the curve goes through the pole four times. ■

3.1
r1=3cos(2θ)
−4.7 4.7
−3.1

To graph the equation $r = 3 \cos 2\theta$ of Example 5 on a calculator, use polar graphing mode, degree mode, $\theta\text{min} = 0$, $\theta\text{max} = 360°$, $\theta\text{step} = 15°$, and window settings as shown.

FOR DISCUSSION

Trace the rose curve of Example 5 and watch the cursor to verify that the numbering that appears in Figure 67 is indeed correct.

The equation $r = 3 \cos 2\theta$ in Example 5 has a graph that belongs to a family of curves called **roses.** The graphs of $\boldsymbol{r = a \sin n\theta}$ and $\boldsymbol{r = a \cos n\theta}$ are roses, with n leaves if n is odd and $2n$ leaves if n is even. The value of a determines the length of the leaves.

GCM EXAMPLE 6 Graphing a Polar Equation (Lemniscate)

Graph $r^2 = \cos 2\theta$.

Analytic Solution

Complete a table of ordered pairs, and sketch the graph, as in Figure 68. The point $(-1, 0°)$, with r negative, may be plotted as $(1, 180°)$. Also, $(-.7, 30°)$ may be plotted as $(.7, 210°)$, and so on.

θ	0°	30°	45°	135°	150°	180°
2θ	0°	60°	90°	270°	300°	360°
$\cos 2\theta$	1	.5	0	0	.5	1
$r = \pm\sqrt{\cos 2\theta}$	±1	±.7	0	0	±.7	±1

Values of θ for $45° < \theta < 135°$ are not included in the table, because the corresponding values of $\cos 2\theta$ are negative (quadrants II and III) and so do not have real square roots. Values of θ larger than 180° give 2θ larger than 360° and would repeat the points already found. This curve is called a **lemniscate.**

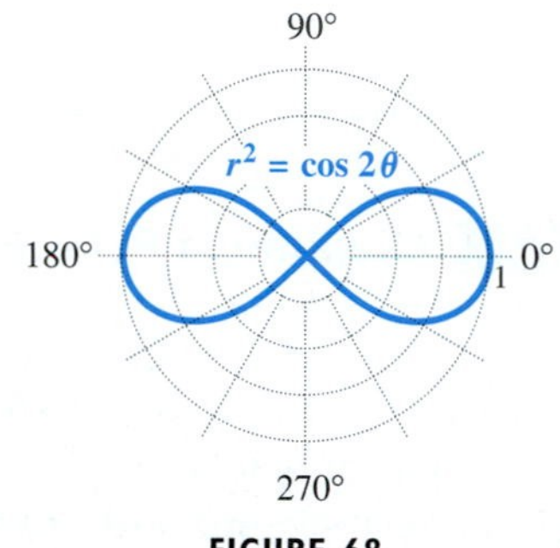

FIGURE 68

Graphing Calculator Solution

To graph $r^2 = \cos 2\theta$ with a graphing calculator, let

$$r_1 = \sqrt{\cos 2\theta}$$

and $$r_2 = -\sqrt{\cos 2\theta}.$$

See Figure 69.

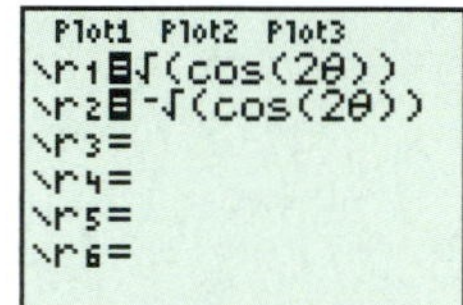

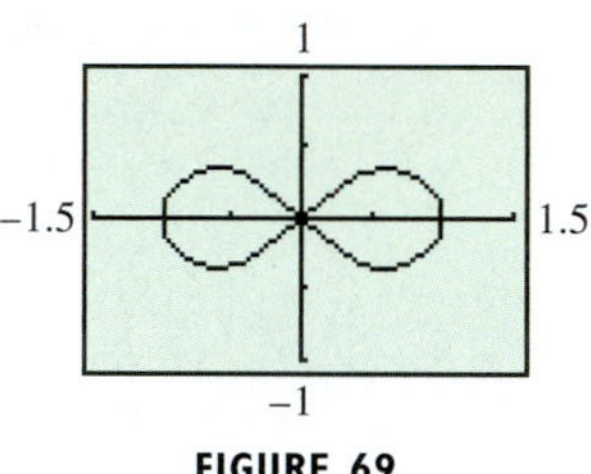

FIGURE 69

EXAMPLE 7 Graphing a Polar Equation (Spiral of Archimedes)

Graph $r = 2\theta$ (θ measured in radians).

Solution Some ordered pairs are shown in the table. Since $r = 2\theta$, rather than a trigonometric function of θ, we must also consider negative values of θ. Radian measures have been rounded. The graph in Figure 70 is called a **spiral of Archimedes.**

θ (radians)	$r = 2\theta$	θ (radians)	$r = 2\theta$
$-\pi$	−6.3	$\frac{\pi}{3}$	2.1
$-\frac{\pi}{2}$	−3.1	$\frac{\pi}{2}$	3.1
$-\frac{\pi}{4}$	−1.6	π	6.3
0	0	$\frac{3\pi}{2}$	9.4
$\frac{\pi}{6}$	1	2π	12.6

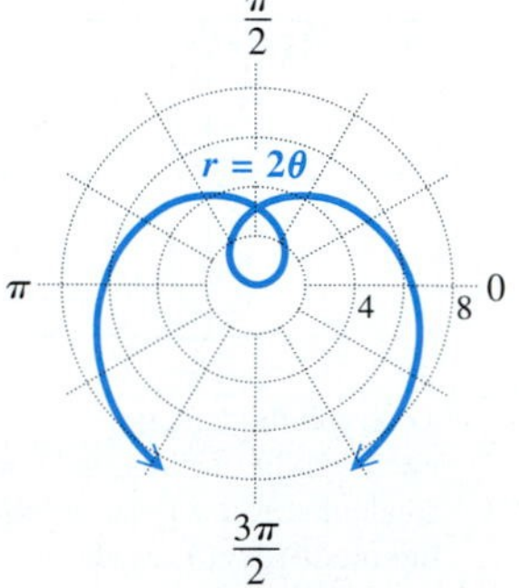

FIGURE 70

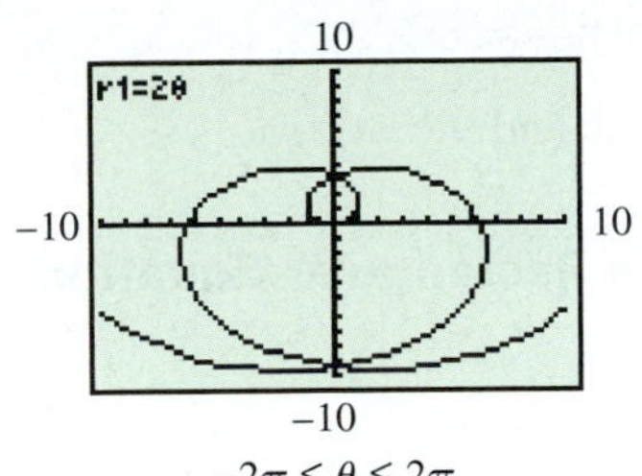

$-2\pi \leq \theta \leq 2\pi$

More of the spiral can be seen in this calculator graph of the spiral in Example 7 than is shown in Figure 70.

FOR DISCUSSION

Refer to the calculator graph in the margin. Experiment with various minimum and maximum values of θ for $r = 2\theta$. What behaviors do you observe? (As you experiment with various maximums and minimums for θ, you will also need to enlarge the window to accommodate the increased size of the graph.)

Classifying Polar Equations

The following table summarizes some common polar graphs and forms of their equations. (In addition to circles, lemniscates, and roses, which were just presented, we include **limaçons.** Cardioids are the special case of limaçons where $\left|\frac{a}{b}\right| = 1$.)

Circles and Lemniscates

Circles		Lemniscates	
$r = a \cos \theta$	$r = a \sin \theta$	$r^2 = a^2 \sin 2\theta$	$r^2 = a^2 \cos 2\theta$

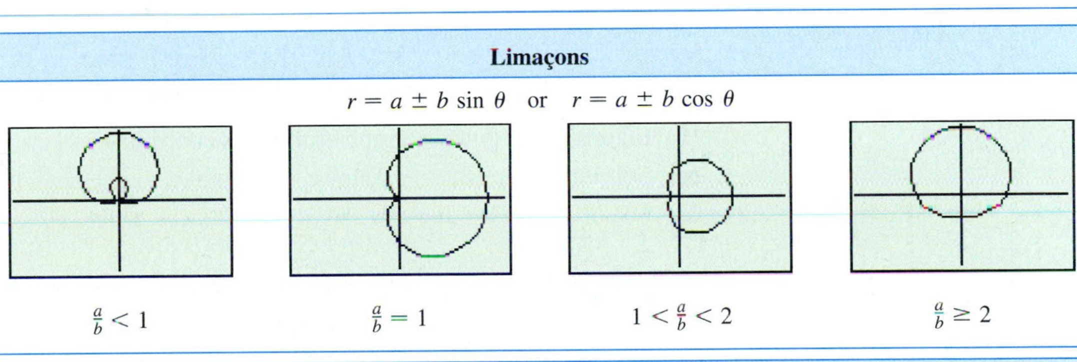

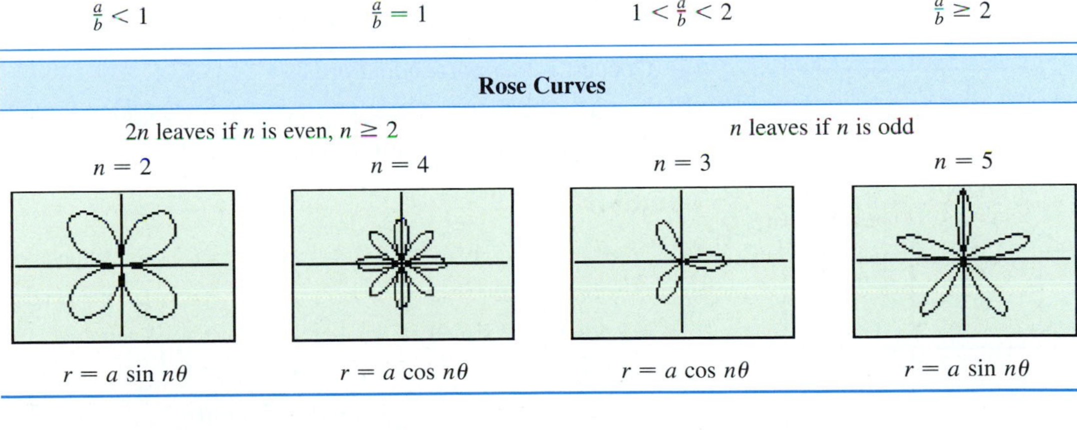

NOTE Some other polar curves are the *cissoid,* the *kappa curve,* the *conchoid,* the *trisectrix,* the *cruciform,* the *strophoid,* and the *lituus.* You may wish to refer to older textbooks on analytic geometry to investigate them.

Converting Equations

We can transform polar equations to rectangular equations and vice versa.

EXAMPLE 8 **Converting a Polar Equation to a Rectangular Equation**

Consider the polar equation $r = \dfrac{4}{1 + \sin \theta}$.

(a) Convert it to a rectangular equation.

(b) Use a graphing calculator to graph the polar equation for $0 \le \theta \le 2\pi$.

(c) Use a graphing calculator to graph the rectangular equation.

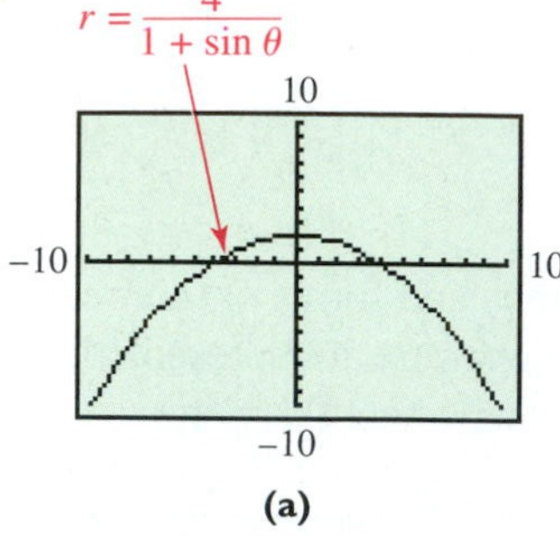

(a)

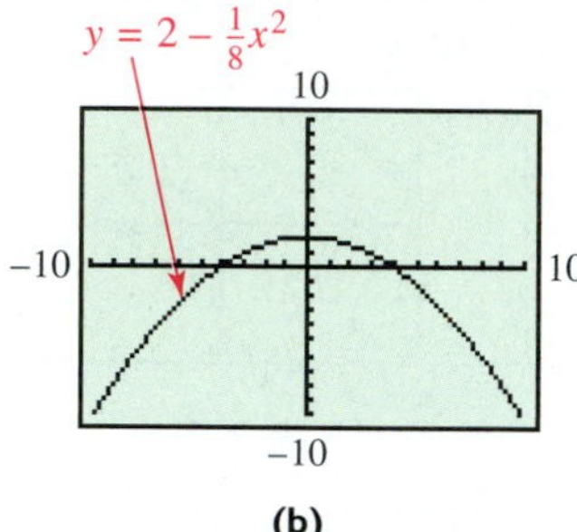

(b)

FIGURE 71

Solution

(a) Multiply each side of the equation by the denominator on the right, to clear the fraction.

$$r = \frac{4}{1 + \sin \theta} \qquad \text{Polar equation}$$

$$r + r \sin \theta = 4 \qquad \text{Multiply by } 1 + \sin \theta.$$

$$\sqrt{x^2 + y^2} + y = 4 \qquad \text{Let } r = \sqrt{x^2 + y^2} \text{ and } y = r \sin \theta.$$

$$\sqrt{x^2 + y^2} = 4 - y \qquad \text{Subtract } y.$$

$$x^2 + y^2 = (4 - y)^2 \qquad \text{Square each side.}$$

$$x^2 + y^2 = 16 - 8y + y^2 \qquad \text{Expand the right side.}$$

$$x^2 = -8y + 16 \qquad \text{Subtract } y^2.$$

$$x^2 = -8(y - 2) \qquad \text{Rectangular equation}$$

(b) Figure 71(a) shows a graph with polar coordinates.

(c) Solving $x^2 = -8(y - 2)$ for y, we obtain $y = 2 - \frac{1}{8}x^2$. Its graph is shown in Figure 71(b). Notice that the two graphs are the same parabola. ■

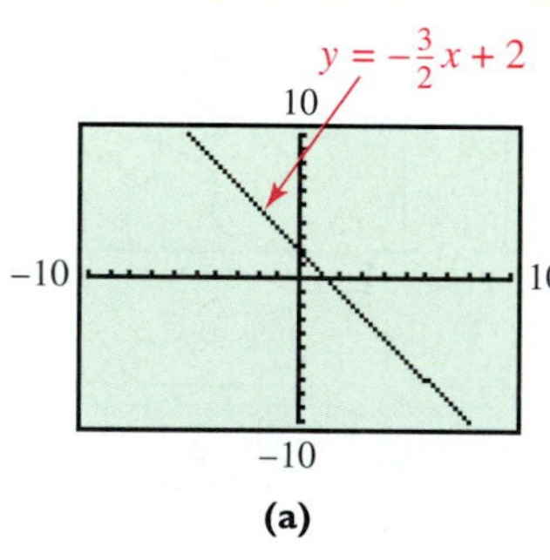

(a)

EXAMPLE 9 **Converting a Rectangular Equation to a Polar Equation**

Consider the rectangular equation $3x + 2y = 4$.

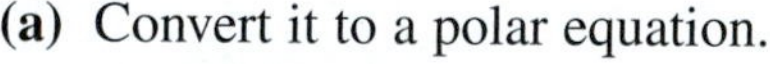

(a) Convert it to a polar equation.

(b) Use a graphing calculator to graph the rectangular equation.

(c) Use a graphing calculator to graph the polar equation for $0° \le \theta \le 360°$.

Solution

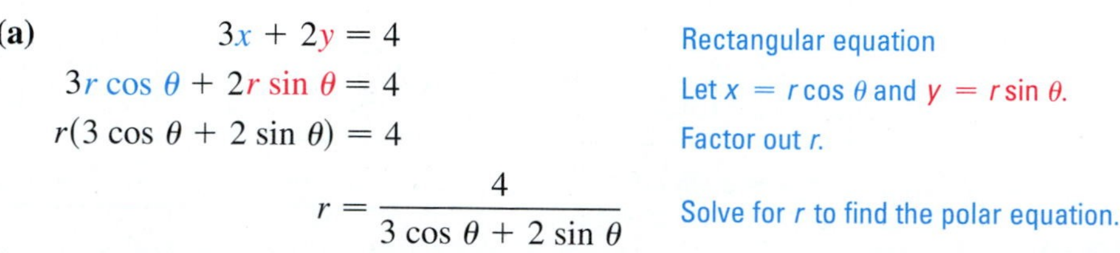

(a)

$$3x + 2y = 4 \qquad \text{Rectangular equation}$$

$$3r \cos \theta + 2r \sin \theta = 4 \qquad \text{Let } x = r \cos \theta \text{ and } y = r \sin \theta.$$

$$r(3 \cos \theta + 2 \sin \theta) = 4 \qquad \text{Factor out } r.$$

$$r = \frac{4}{3 \cos \theta + 2 \sin \theta} \qquad \text{Solve for } r \text{ to find the polar equation.}$$

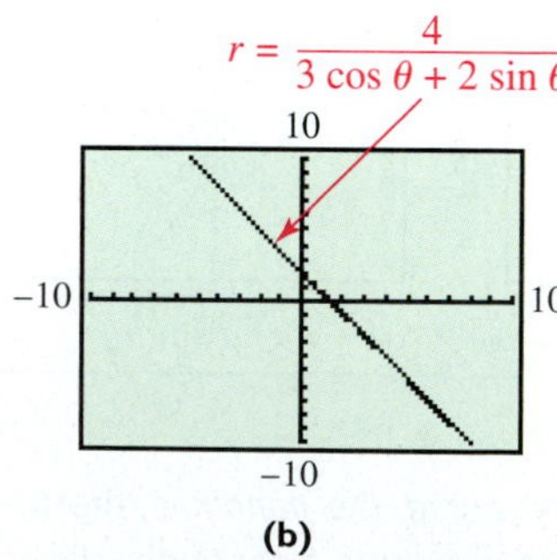

(b)

FIGURE 72

(b) We solve the given rectangular equation for y to obtain $y = -\frac{3}{2}x + 2$. Its graph is shown in Figure 72(a).

(c) The polar equation found in part (a) is graphed in Figure 72(b). ■

10.6 Exercises

1. *Concept Check* For each point given in polar coordinates, state the quadrant in which the point lies if it is graphed in a rectangular coordinate system.
(a) $(5, 135°)$ (b) $(2, 60°)$
(c) $(6, -30°)$ (d) $(4.6, 213°)$

2. *Concept Check* For each point given in polar coordinates, state the axis on which the point lies if it is graphed in a rectangular coordinate system. Also, state whether it is on the positive portion or the negative portion of the axis. (For example, $(5, 0°)$ lies on the positive x-axis.)
(a) $(7, 360°)$ (b) $(4, 180°)$
(c) $(2, -90°)$ (d) $(8, 450°)$

Plot each point, given its polar coordinates. Give two other pairs of polar coordinates for each point.

3. $(1, 45°)$
4. $(3, 120°)$
5. $(-2, 135°)$
6. $(-4, 27°)$
7. $(5, -60°)$
8. $(2, -45°)$
9. $(-3, -210°)$
10. $(-1, -120°)$
11. $(3, 300°)$
12. $(4, 270°)$

Plot the point whose rectangular coordinates are given. Then determine two pairs of polar coordinates for the point with $0° \leq \theta < 360°$.

13. $(-1, 1)$
14. $(1, 1)$
15. $(0, 3)$
16. $(0, -3)$
17. $(\sqrt{2}, \sqrt{2})$
18. $(-\sqrt{2}, \sqrt{2})$
19. $\left(\frac{\sqrt{3}}{2}, \frac{3}{2}\right)$
20. $\left(-\frac{\sqrt{3}}{2}, -\frac{1}{2}\right)$
21. $(3, 0)$
22. $(-2, 0)$

For each rectangular equation, give its equivalent polar equation and sketch its graph.

23. $x - y = 4$
24. $x + y = -7$
25. $x^2 + y^2 = 16$
26. $x^2 + y^2 = 9$
27. $2x + y = 5$
28. $3x - 2y = 6$

Concept Check *In Exercises 29–32, match each polar graph to its corresponding equation from choices A–D.*

29. $r = 3$
30. $r = \cos 3\theta$
31. $r = \cos 2\theta$
32. $r = \dfrac{2}{\cos \theta + \sin \theta}$

A.
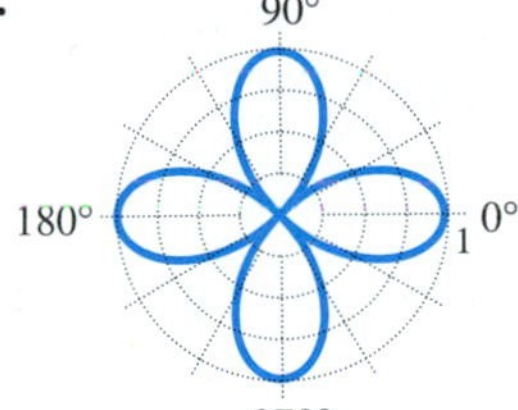

B.
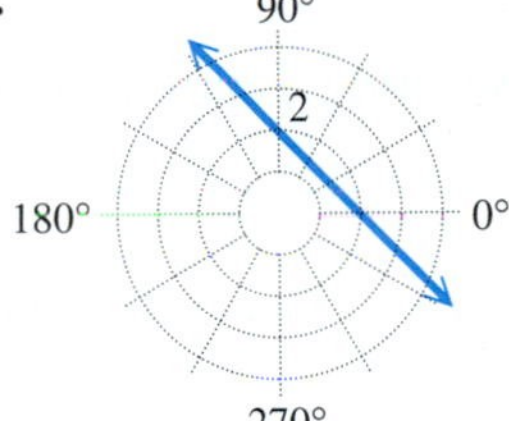

C.
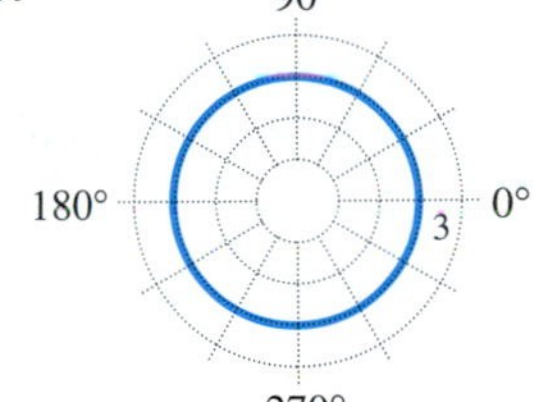

D.
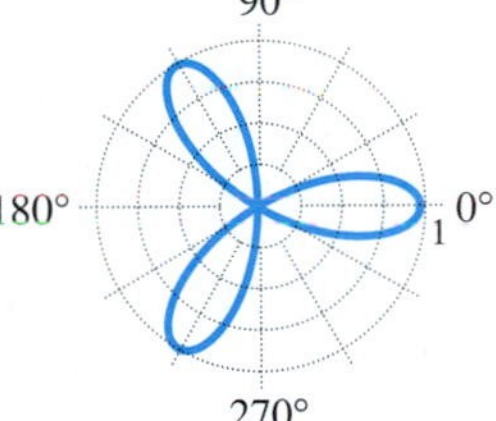

Graph each polar equation for θ in $[0°, 360°)$. In Exercises 33–42 identify the type of polar graph.

33. $r = 2 + 2\cos\theta$
34. $r = 8 + 6\cos\theta$
35. $r = 3 + \cos\theta$
36. $r = 2 - \cos\theta$
37. $r = 4\cos 2\theta$
38. $r = 3\cos 5\theta$
39. $r^2 = 4\cos 2\theta$
40. $r^2 = 4\sin 2\theta$
41. $r = 4 - 4\cos\theta$
42. $r = 6 - 3\cos\theta$
43. $r = 2\sin\theta\tan\theta$
(This is a *cissoid*.)
44. $r = \dfrac{\cos 2\theta}{\cos\theta}$
(This is a *cissoid* with a loop.)

Concept Check *The polar graphs in this section exhibit symmetry. Visualize an xy-plane superimposed on the polar coordinate system, with the pole at the origin and the polar axis on the positive x-axis. Then a polar graph may be symmetric with respect to the x-axis (the polar axis), the y-axis* $\left(\text{the line } \theta = \frac{\pi}{2}\right)$*, or the origin (the pole). Use this informatin to work Exercises 45 and 46.*

45. Complete the missing ordered pairs in the graphs.

(a)

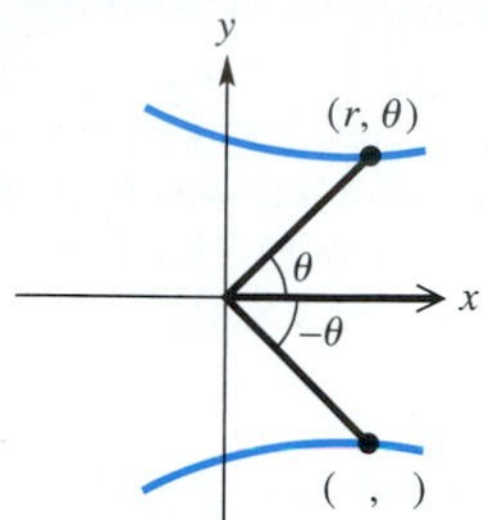

(b)

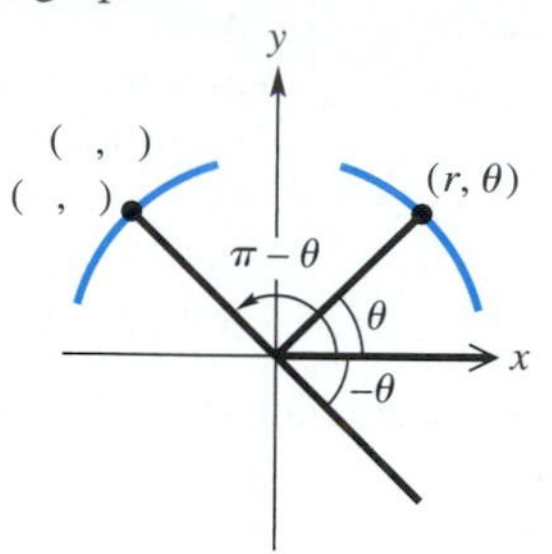

(c)

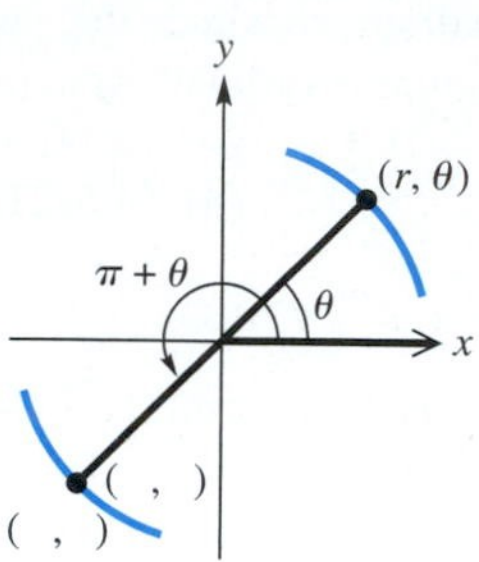

46. On the basis of your results in Exercise 45, fill in the blanks with the correct responses.

(a) The graph of $r = f(\theta)$ is symmetric with respect to the polar axis if substitution of __________ for θ leads to an equivalent equation.

(b) The graph of $r = f(\theta)$ is symmetric with respect to the vertical line $\theta = \frac{\pi}{2}$ if substitution of __________ for θ leads to an equivalent equation.

(c) Alternatively, the graph of $r = f(\theta)$ is symmetric with respect to the vertical line $\theta = \frac{\pi}{2}$ if substitution of __________ for r and __________ for θ leads to an equivalent equation.

(d) The graph of $r = f(\theta)$ is symmetric with respect to the pole if substitution of __________ for r leads to an equivalent equation.

(e) Alternatively, the graph of $r = f(\theta)$ is symmetric with respect to the pole if substitution of __________ for θ leads to an equivalent equation.

(f) In general, the completed statements in parts (a)–(e) mean that graphs of polar equations of the form $r = a \pm b \cos \theta$ (where a may be 0) are symmetric with respect to __________.

(g) In general, the completed statements in parts (a)–(e) mean that graphs of polar equations of the form $r = a \pm b \sin \theta$ (where a may be 0) are symmetric with respect to __________.

47. Explain the method you would use to graph (r, θ) by hand if $r < 0$.

48. Explain why, if $r > 0$, the points (r, θ) and $(-r, \theta + 180°)$ have the same graph.

Concept Check *Answer each question.*

49. If a point lies on an axis in the rectangular plane, then what kind of angle must θ be if (r, θ) represents the point in polar coordinates?

50. What will the graph of $r = k$ be, for $k > 0$?

51. How would the graph of Figure 65 change if the equation were $r = 1 - \cos \theta$?

52. How would the graph of Figure 65 change if the equation were $r = 1 + \sin \theta$?

53. The graphs of rose curves have equations of the form $r = a \sin n\theta$ or $r = a \cos n\theta$. What does the value of a determine? What does the value of n determine?

54. In Exercise 53, if $n = 1$, what will the graph be? What will the value of a determine?

For each equation, find an equivalent equation in rectangular coordinates. Then graph the equation.

55. $r = 2 \sin \theta$

56. $r = 2 \cos \theta$

57. $r = \dfrac{2}{1 - \cos \theta}$

58. $r = \dfrac{3}{1 - \sin \theta}$

59. $r = -2 \cos \theta - 2 \sin \theta$

60. $r = \dfrac{3}{4 \cos \theta - \sin \theta}$

61. $r = 2 \sec \theta$

62. $r = -5 \csc \theta$

63. $r = \dfrac{2}{\cos \theta + \sin \theta}$

64. $r = \dfrac{2}{2 \cos \theta + \sin \theta}$

The graph of $r = a\theta$ in polar coordinates is an example of the spiral of Archimedes. With your calculator set to radian mode, use the given value of a and interval of θ to graph the spiral in the window specified.

65. $a = 1, 0 \le \theta \le 4\pi, [-15, 15]$ by $[-15, 15]$

66. $a = 2, -4\pi \le \theta \le 4\pi, [-30, 30]$ by $[-30, 30]$

67. $a = 1.5, -4\pi \le \theta \le 4\pi, [-20, 20]$ by $[-20, 20]$

68. $a = -1, 0 \le \theta \le 12\pi, [-40, 40]$ by $[-40, 40]$

Find the polar coordinates of the points of intersection of the given curves for the specified interval of θ.

69. $r = 4 \sin \theta, r = 1 + 2 \sin \theta; 0 \le \theta < 2\pi$

70. $r = 3, r = 2 + 2 \cos \theta; 0° \le \theta < 360°$

71. $r = 2 + \sin \theta, r = 2 + \cos \theta; 0 \le \theta < 2\pi$

72. $r = \sin 2\theta, r = \sqrt{2} \cos \theta; 0 \le \theta < \pi$

(Modeling) *Solve each problem.*

73. ***Planetary Orbits*** The polar equation

$$r = \frac{a(1 - e^2)}{1 + e \cos \theta},$$

where a is the average distance in astronomical units from the sun and e is a constant called the eccentricity, can be used to graph the orbits of the planets. (See Section 6.3.) The sun will be located at the pole. The table lists a and e for the planets.

Planet	a	e
Mercury	.39	.206
Venus	.78	.007
Earth	1.00	.017
Mars	1.52	.093
Jupiter	5.20	.048
Saturn	9.54	.056
Uranus	19.20	.047
Neptune	30.10	.009
Pluto	39.40	.249

Source: Karttunen, H., P. Kröger, H. Oja, M. Putannen, and K. Donners (Editors), *Fundamental Astronomy*, Springer-Verlag, 1994; Zeilik, M., S. Gregory, and E. Smith, *Introductory Astronomy and Astrophysics*, Saunders College Publishers, 1992.

(a) Graph the orbits of the four closest planets on the same polar grid. Choose a viewing window that results in a graph with nearly circular orbits.

(b) Plot the orbits of Earth, Jupiter, Uranus, and Pluto on the same polar grid. How does Earth's distance from the sun compare with the distance from the sun to these planets?

(c) Use graphing to determine whether Pluto is always the farthest planet from the sun.

74. ***Radio Towers and Broadcasting Patterns*** Many times, radio stations do not broadcast in all directions with the same intensity. To avoid interference with an existing station to the north, a new station may be licensed to broadcast only east and west. To create an east–west signal, two radio towers are sometimes used, as illustrated in the figure. Locations where the radio signal is received correspond to the interior of the curve defined by

$$r^2 = 40{,}000 \cos 2\theta,$$

where the polar axis (or positive x-axis) points east.

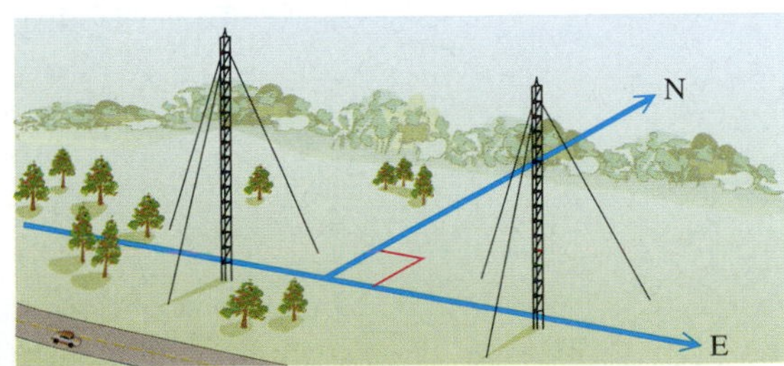

(a) Graph $r^2 = 40{,}000 \cos 2\theta$ for $0^\circ \leq \theta \leq 360^\circ$, with units in miles. Assuming that the radio towers are located near the pole, use the graph to describe the regions where the signal can be received and where the signal cannot be received.

(b) Suppose a radio signal pattern is given by

$$r^2 = 22{,}500 \sin 2\theta.$$

Graph this pattern and interpret the results.

10.7 More Parametric Equations

Parametric Equations with Trigonomeric Functions ■ The Cycloid ■ Applications of Parametric Equations

Parametric Equations with Trigonometric Functions

In Section 6.4, parametric equations were introduced. However, these equations did not include trigonometric functions. If we use trigonometric functions in parametric equations, many interesting curves can be drawn, as shown in Figure 73.

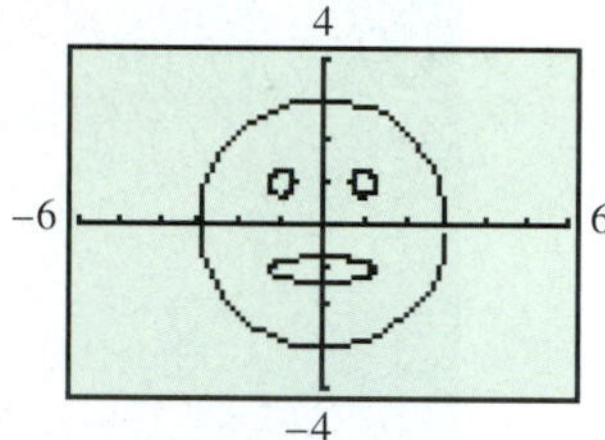

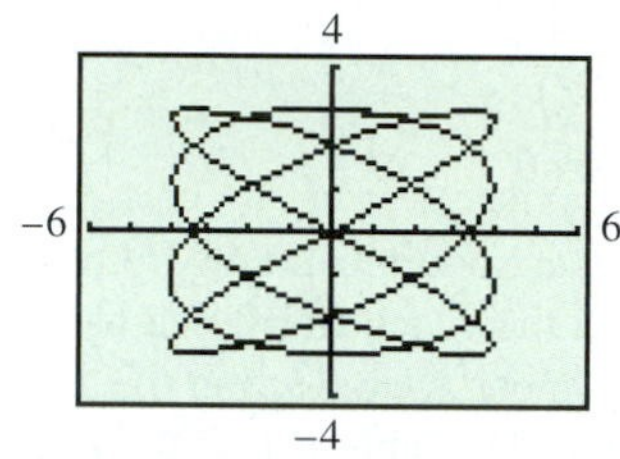

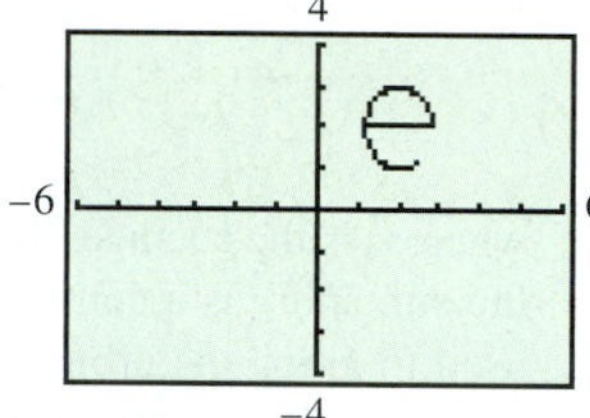

FIGURE 73

EXAMPLE 1 Graphing a Circle with Parametric Equations

Graph $x = 2 \cos t$ and $y = 2 \sin t$, for $0 \leq \theta \leq 2\pi$. Find an equivalent equation by using rectangular coordinates.

TECHNOLOGY NOTE

When graphing parametric equations, be sure that your calculator is set in parametric mode. A square window is necessary for the curve in Example 1 to appear circular rather than elliptical.

Solution Let $X_{1T} = 2\cos(T)$ and $Y_{1T} = 2\sin(T)$, and graph these parametric equations, as shown in Figure 74.

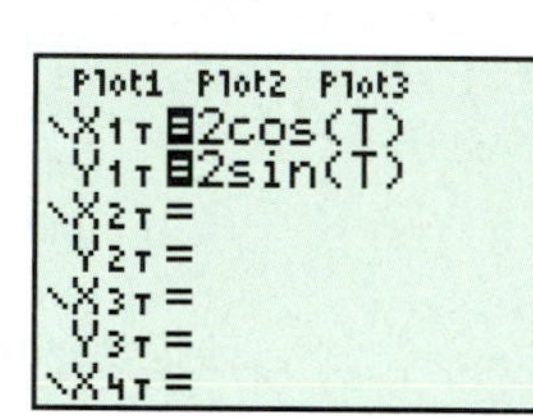

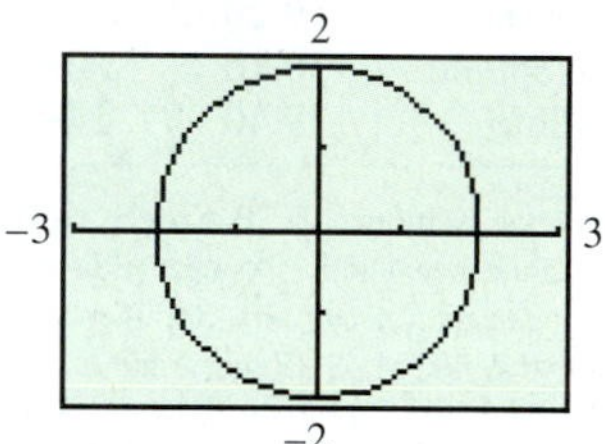

FIGURE 74

To verify that this graph is a circle, consider the following.

$$\begin{aligned} x^2 + y^2 &= (2 \cos t)^2 + (2 \sin t)^2 && x = 2 \cos t,\ y = 2 \sin t \\ &= 4 \cos^2 t + 4 \sin^2 t && \text{Properties of exponents} \\ &= 4(\cos^2 t + \sin^2 t) && \text{Distributive property} \\ &= 4 && \cos^2 t + \sin^2 t = 1 \end{aligned}$$

The parametric equations are equivalent to $x^2 + y^2 = 4$, which is the equation of a circle with center $(0, 0)$ and radius 2. ■

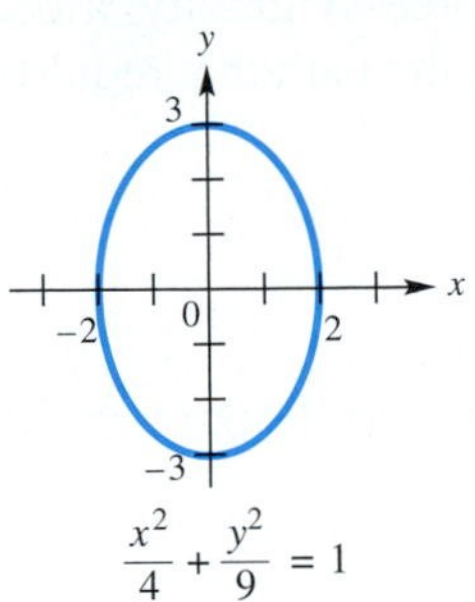

$\frac{x^2}{4} + \frac{y^2}{9} = 1$

FIGURE 75

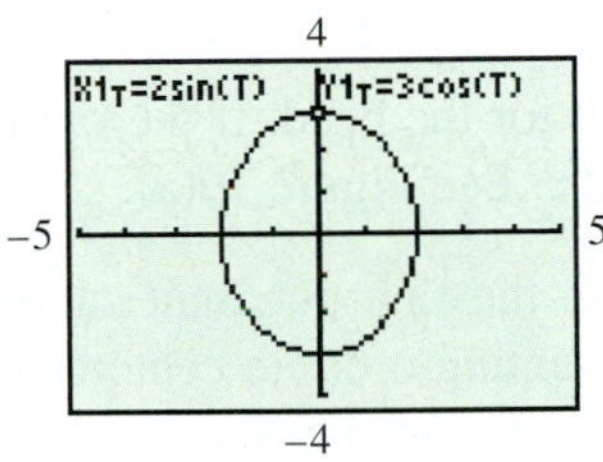

FIGURE 76

EXAMPLE 2 Graphing an Ellipse with Parametric Equations

Graph the plane curve defined by $x = 2 \sin t$, $y = 3 \cos t$, for t in $[0, 2\pi]$.

Solution To convert to a rectangular equation, square both sides of each equation; solve one for $\sin^2 t$, the other for $\cos^2 t$.

$$x = 2 \sin t \qquad\qquad y = 3 \cos t$$
$$x^2 = 4 \sin^2 t \qquad\qquad y^2 = 9 \cos^2 t$$
$$\frac{x^2}{4} = \sin^2 t \qquad\qquad \frac{y^2}{9} = \cos^2 t$$

Now add corresponding sides of the two equations.

$$\frac{x^2}{4} + \frac{y^2}{9} = \sin^2 t + \cos^2 t$$
$$\frac{x^2}{4} + \frac{y^2}{9} = 1$$

This is the equation of an ellipse, as shown in Figure 75. The ellipse can be graphed directly with a calculator in parametric mode. See Figure 76. ■

The Cycloid

Looking Ahead to Calculus
The cycloid is a special case of a curve traced out by a point at a given distance from the center of a circle as the circle rolls along a straight line. See Figure 77. Such a curve is called a *trochoid.* Parametrically defined curves are studied in calculus.

The path traced by a fixed point on the circumference of a circle rolling along a line is called a *cycloid.* A **cycloid** is defined by

$$x = at - a \sin t, \quad y = a - a \cos t, \quad \text{for } t \text{ in } (-\infty, \infty).$$

EXAMPLE 3 Graphing a Cycloid

Graph the cycloid with $a = 1$ for t in $[0, 2\pi]$.

Solution There is no simple way to find a rectangular equation for the cycloid from its parametric equations. Instead, begin with a table of values. (Some are approximations.)

t	0	$\frac{\pi}{4}$	$\frac{\pi}{2}$	π	$\frac{3\pi}{2}$	2π
x	0	.08	.6	π	5.7	2π
y	0	.3	1	2	1	0

Plotting the ordered pairs (x, y) from the table of values leads to the portion of the graph from 0 to 2π in Figure 77.

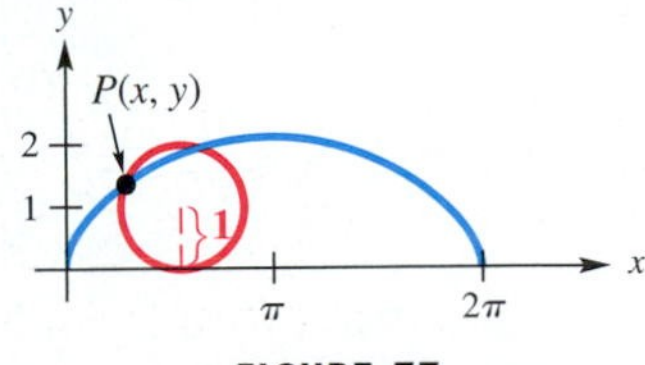

FIGURE 77 ■

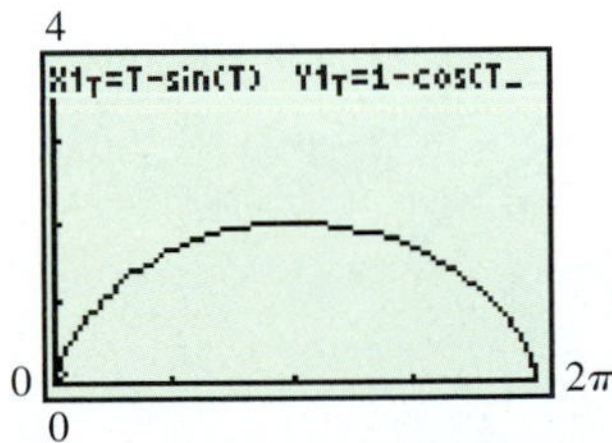

It is easier to graph the cycloid of Example 3 with a graphing calculator in parametric mode than with traditional methods.

FIGURE 78

The cycloid has an interesting physical property. If a flexible cord or wire goes through points P and Q as in Figure 78, and, due to the force of gravity, a bead is allowed to slide without friction along this path from P to Q, the path that requires the least time takes the shape of an inverted cycloid.

Applications of Parametric Equations

Parametric equations are used frequently in computer graphics to design a variety of figures, letters, and shapes.

GCM **EXAMPLE 4** **Creating a Drawing with Parametric Equations**

Graph a smiley face using parametric equations.

Solution ***Head*** We use a circle centered at the origin for the head. If the radius is 2, then we let $x = 2 \cos t$ and $y = 2 \sin t$ for $0 \leq t \leq 2\pi$. See Figure 79(a).

Eyes For the eyes, we use two small circles. The eye in the first quadrant can be modeled by $x = 1 + .3 \cos t$ and $y = 1 + .3 \sin t$, representing a circle centered at $(1, 1)$ with radius .3. The eye in the second quadrant can be modeled by $x = -1 + .3 \cos t$ and $y = 1 + .3 \sin t$ for $0 \leq t \leq 2\pi$, which is a circle centered at $(-1, 1)$ with radius .3. See Figure 79(b).

Mouth For the smile, we can use the lower half of a circle. Using trial and error, we arrive at $x = .5 \cos \frac{1}{2}t$ and $y = -.5 - .5 \sin \frac{1}{2}t$, a semicircle centered at $(0, -.5)$ with radius .5. Since we are letting $0 \leq t \leq 2\pi$, the term $\frac{1}{2}t$ ensures that only half a circle (semicircle) is drawn. The minus sign before $.5 \sin \frac{1}{2}t$ in the y-equation results in the lower half of the semicircle being drawn rather than the upper half. The final result is shown in Figure 79(c). We can add the pupils by plotting the points $(1, 1)$ and $(-1, 1)$. Turning off the coordinate axes rids the face of lines. ■

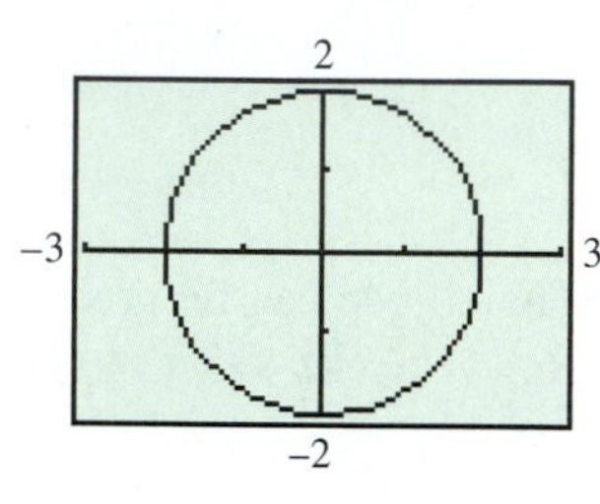

(a)

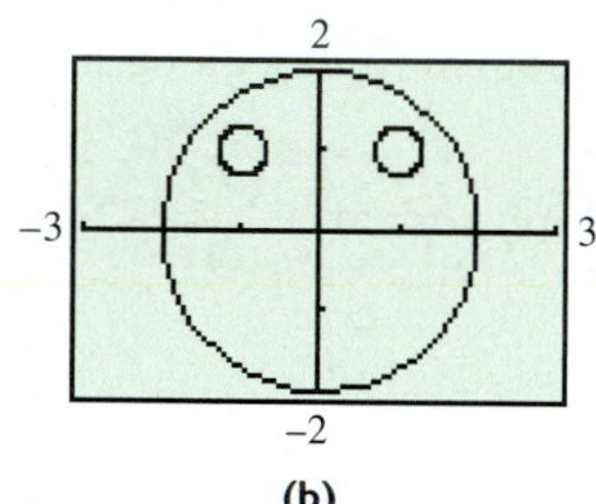

(b)

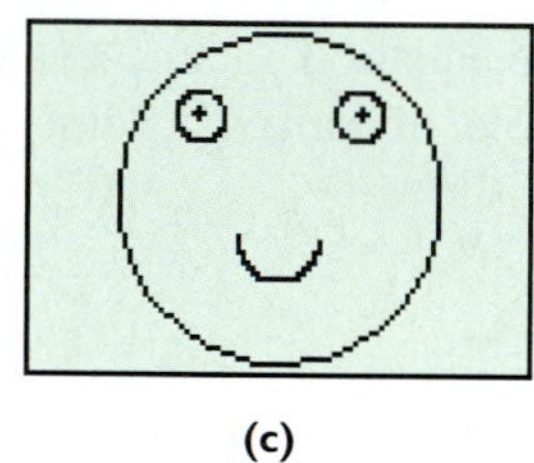
(c)

FIGURE 79

Parametric equations are used to simulate motion. If a ball is thrown with a velocity of v feet per second at an angle θ with the horizontal, its flight can be modeled by the parametric equations

$$x = (v \cos \theta)t \quad \text{and} \quad y = (v \sin \theta)t - 16t^2 + h,$$

where t is in seconds and h is the ball's initial height in feet above the ground. The term $-16t^2$ occurs because gravity is pulling downward. See Figure 80. These equations ignore air resistance.

FOR DISCUSSION

Modify the face in Example 4 so that it is frowning. Try to find a way to make the right eye shut rather than be open.

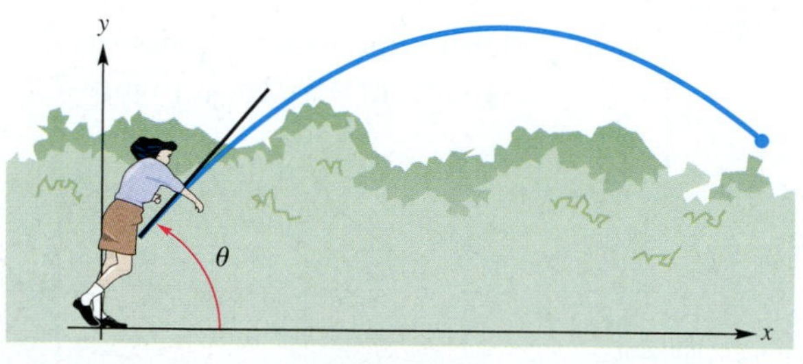

FIGURE 80

GCM EXAMPLE 5 Simulating Motion with Parametric Equations

Three golf balls are hit simultaneously into the air at 132 feet per second (90 mph) at angles of 30°, 50°, and 70° with the horizontal.

(a) Assuming that the ground is level, determine graphically which ball travels the farthest. Estimate this distance.

(b) Which ball reaches the greatest height? Estimate this height.

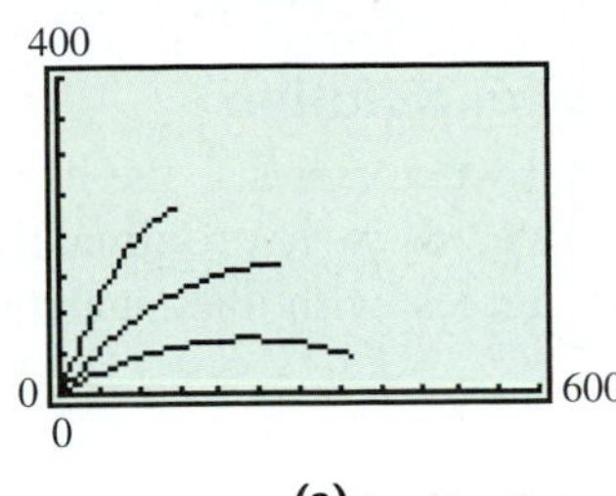

(a)

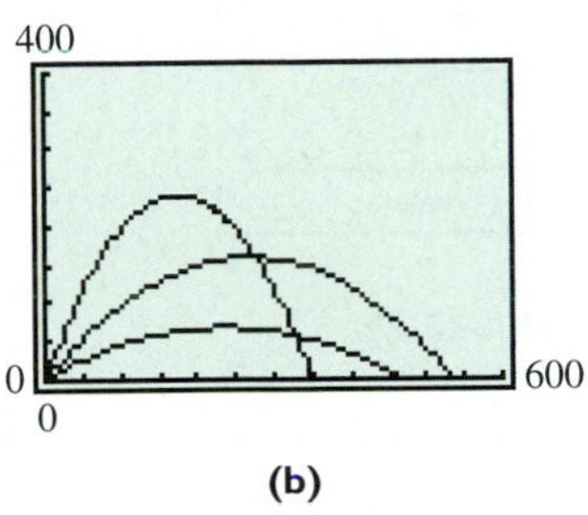

(b)

FIGURE 81

Solution

(a) The three sets of parametric equations determined by the three golf balls are as follows, since $h = 0$.

$$x_1 = (132 \cos 30°)t, \qquad y_1 = (132 \sin 30°)t - 16t^2$$
$$x_2 = (132 \cos 50°)t, \qquad y_2 = (132 \sin 50°)t - 16t^2$$
$$x_3 = (132 \cos 70°)t, \qquad y_3 = (132 \sin 70°)t - 16t^2$$

Graphs are shown in Figure 81(a), where $0 \leq t \leq 9$. We used a graphing calculator in simultaneous mode so that we could view all three balls in flight at the same time. From the graph in Figure 81(b), we can see that the ball hit at 50° travels the farthest distance. Using the trace feature, we estimate this distance to be about 540 feet.

(b) The ball hit at 70° reaches the greatest height, about 240 feet. ■

EXAMPLE 6 Examining Parametric Equations of Flight

A small rocket is launched from a table that is 3.36 feet above the ground. Its initial velocity is 64 feet per second, and it is launched at an angle of 30° with respect to the ground. Find the rectangular equation that models this path. What type of path does the rocket follow?

Solution Its path is defined by the parametric equations

$$x = (64 \cos 30°)t \quad \text{and} \quad y = (64 \sin 30°)t - 16t^2 + 3.36$$

or, equivalently,

$$x = 32\sqrt{3}t \quad \text{and} \quad y = -16t^2 + 32t + 3.36.$$

Solving $x = 32\sqrt{3}t$ for t, we obtain $t = \frac{x}{32\sqrt{3}}$. Substituting for t in the other parametric equation yields

$$y = -16\left(\frac{x}{32\sqrt{3}}\right)^2 + 32\left(\frac{x}{32\sqrt{3}}\right) + 3.36$$
$$y = -\frac{1}{192}x^2 + \frac{\sqrt{3}}{3}x + 3.36. \qquad \text{Simplify.}$$

Because this equation defines a parabola, the rocket follows a parabolic path. ■

FOR DISCUSSION

If a golf ball is hit at 88 feet per second (60 mph), use trial and error to find the angle θ that results in a maximum distance for the ball.

GCM **EXAMPLE 7** **Analyzing the Path of a Projectile**

Determine the total flight time and the horizontal distance traveled by the rocket in Example 6.

Analytic Solution

The equation $y = -16t^2 + 32t + 3.36$ gives the vertical position of the rocket at time t. We need to determine those values of t for which $y = 0$, since these values correspond to the rocket at ground level. This yields

$$0 = -16t^2 + 32t + 3.36.$$

From the quadratic formula, the solutions are $t = -.1$ or $t = 2.1$. Since t represents time, $t = -.1$ is an unacceptable answer. Therefore, the flight time is 2.1 seconds.

The rocket was in the air for 2.1 seconds, so we can use $t = 2.1$ and the parametric equation that models the horizontal position, $x = 32\sqrt{3}t$, to get

$$x = 32\sqrt{3}(2.1) \approx 116.4 \text{ feet}.$$

Graphing Calculator Solution

Figure 82 shows that when $t = 2.1$, the horizontal distance covered is approximately 116.4 feet, which agrees with the analytic solution.

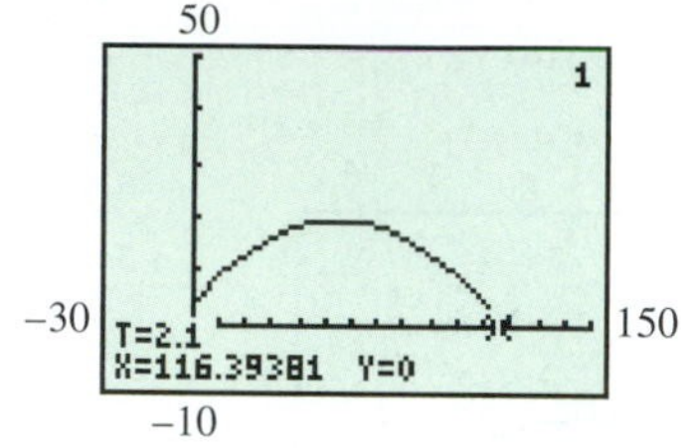

FIGURE 82

What Went Wrong?

A student graphed $x = \cos \frac{1}{2}t$, $y = \sin \frac{1}{2}t$, which should be a circle with radius 1 because $x^2 + y^2 = \cos^2 \frac{1}{2}t + \sin^2 \frac{1}{2}t = 1$. However, the following screen shows the graph obtained by the student.

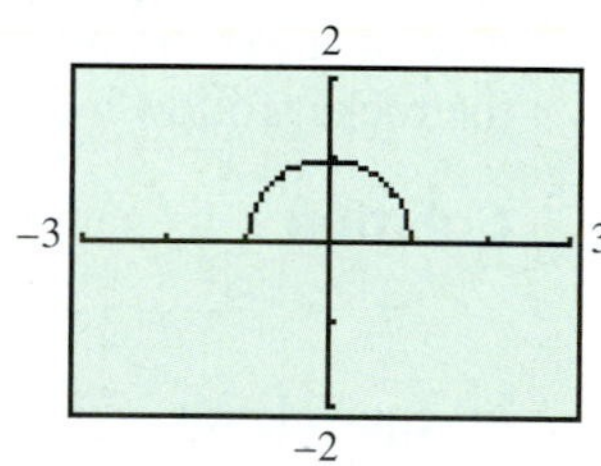

What Went Wrong? What should the student do to change the graph from a semicircle to a circle?

Answer to What Went Wrong?

The periods for $y = \cos \frac{1}{2}t$ and $y = \sin \frac{1}{2}t$ are both 4π. The interval for t should be $0 \le t \le 4\pi$, rather than $0 \le t \le 2\pi$, which is the interval the student used.

10.7 *Exercises*

Find a rectangular equation for each curve and describe the curve.

1. $x = 3 \sin t$, $y = 3 \cos t$; for t in $[-\pi, \pi]$

2. $x = 2 \sin t$, $y = 2 \cos t$; for t in $[0, 2\pi]$

3. $x = 2 \cos^2 t$, $y = 2 \sin^2 t$; for t in $\left[0, \frac{\pi}{2}\right]$

4. $x = \sqrt{5} \sin t$, $y = \sqrt{3} \cos t$; for t in $[0, 2\pi]$

5. $x = 3 \tan t, y = 2 \sec t$; for t in $\left(-\frac{\pi}{2}, \frac{\pi}{2}\right)$

6. $x = \cot t, y = \csc t$; for t in $(0, \pi)$

Graph each pair of parametric equations for $0 \leq t \leq 2\pi$. Describe any differences in the two graphs.

7. **(a)** $x = 3 \cos t, \quad y = 3 \sin t$
(b) $x = 3 \cos 2t, \quad y = 3 \sin 2t$

8. **(a)** $x = 2 \cos t, \quad y = 2 \sin t$
(b) $x = 2 \cos t, \quad y = -2 \sin t$

9. **(a)** $x = 3 \cos t, \quad y = 3 \sin t$
(b) $x = 3 \sin t, \quad y = 3 \cos t$

10. **(a)** $x = -1 + \cos t, \quad y = 2 + \sin t$
(b) $x = 1 + \cos t, \quad y = 2 + \sin t$

Find a rectangular equation for each curve and graph the curve.

11. $x = \sin t, y = \csc t$; for t in $(0, \pi)$

12. $x = \tan t, y = \cot t$; for t in $\left(0, \frac{\pi}{2}\right)$

13. $x = 2 + \sin t, y = 1 + \cos t$; for t in $[0, 2\pi]$

14. $x = 1 + 2 \sin t, y = 2 + 3 \cos t$; for t in $[0, 2\pi]$

Graph each pair of parametric equations.

15. $x = 2 + \cos t, y = \sin t - 1; 0 \leq t \leq 2\pi$

16. $x = -2 + \cos t, y = \sin t + 1; 0 \leq t \leq 2\pi$

17. $x = \cos^3 t, y = \sin^3 t; 0 \leq t \leq 2\pi$

18. $x = \cos^5 t, y = \sin^5 t; 0 \leq t \leq 2\pi$

19. $x = |3 \sin t|, y = |3 \cos t|; 0 \leq t \leq \pi$

20. $x = 3 \sin 2t, y = 3 \cos t; 0 \leq t \leq 2\pi$

Graph each cycloid for t in the specified interval.

21. $x = t - \sin t, y = 1 - \cos t$; for t in $[0, 4\pi]$

22. $x = 2t - 2 \sin t, y = 2 - 2 \cos t$; for t in $[0, 8\pi]$

Graph each set of parametric equations for $0 \leq t \leq 2\pi$ in the window $[0, 6]$ by $[0, 4]$. Identify the letter of the alphabet that is being graphed.

23. $x_1 = 1, \quad y_1 = 1 + t/\pi$
$x_2 = 1 + t/(3\pi), \quad y_2 = 2$
$x_3 = 1 + t/(2\pi), \quad y_3 = 3$

24. $x_1 = 1, \quad y_1 = 1 + t/\pi$
$x_2 = 1 + t/(3\pi), \quad y_2 = 2$
$x_3 = 1 + t/(2\pi), \quad y_3 = 3$
$x_4 = 1 + t/(2\pi), \quad y_4 = 1$

25. $x_1 = 1, \quad y_1 = 1 + t/\pi$
$x_2 = 1 + 1.3 \sin .5t, \quad y_2 = 2 + \cos .5t$

26. $x_1 = 2 + .8 \cos .85t, \quad y_1 = 2 + \sin .85t$
$x_2 = 1.2 + t/(1.3\pi), \quad y_2 = 2$

(Modeling) Designing Letters *Find a set of parametric equations that results in a letter similar to the one shown in each figure. Use the window $[-4.7, 4.7]$ by $[-3.1, 3.1]$, and turn off the coordinate axes. Answers may vary.*

27. 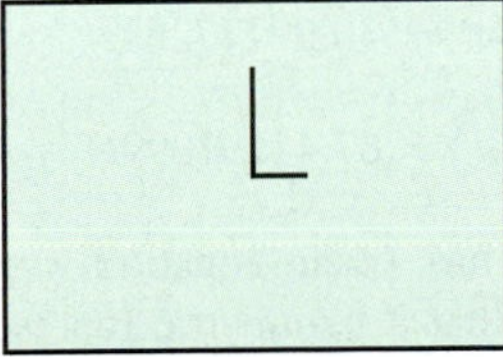

28. 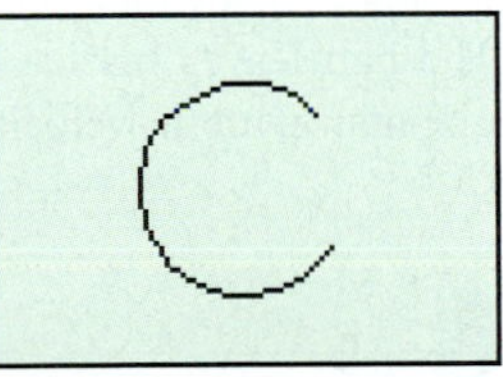

29. 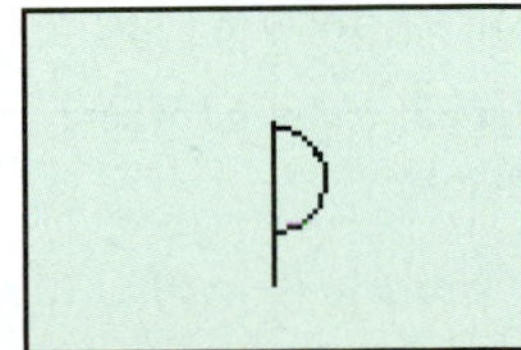

30. 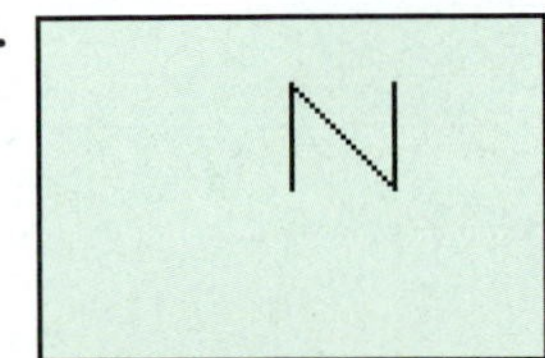

31. ***(Modeling) Designing a Face*** Refer to Example 4. Use parametric equations to create your own smiley face. This face should have a head, a mouth, and eyes.

32. ***(Modeling) Designing a Face*** Add a nose to the face that you designed in Exercise 31.

Lissajous Figures *The middle screen in Figure 73 is an example of a Lissajous figure. Lissajous figures occur in electronics and may be used to find the frequency of an unknown voltage. Graph each Lissajous figure for* $0 \le t \le 6.5$ *in the window* $[-6, 6]$ *by* $[-4, 4]$.

33. $x = 2 \cos t$, $y = 3 \sin 2t$

34. $x = 3 \cos 2t$, $y = 3 \sin 3t$

35. $x = 3 \sin 4t$, $y = 3 \cos 3t$

36. $x = 4 \sin 4t$, $y = 3 \sin 5t$

(Modeling) *In Exercises 37–40, do the following.*

(a) *Determine the parametric equations that model the path of the projectile.*

(b) *Determine the rectangular equation that models the path of the projectile.*

(c) *Determine the time the projectile is in flight and the horizontal distance covered.*

37. ***Flight of a Model Rocket*** A model rocket is launched from the ground with a velocity of 48 feet per second at an angle of 60° with respect to the ground.

38. ***Flight of a Golf Ball*** Tyler McGinnis is playing golf. He hits a golf ball from the ground at an angle of 60° with respect to the ground at a velocity of 150 feet per second.

39. ***Flight of a Softball*** Sally hits a softball when it is 2 feet above the ground. The ball leaves her bat at an angle of 20° with respect to the ground at a velocity of 88 feet per second.

40. ***Flight of a Baseball*** Pronk hits a baseball when it is 2.5 feet above the ground. The ball leaves his bat at an angle of 29° from the horizontal with a velocity of 136 feet per second.

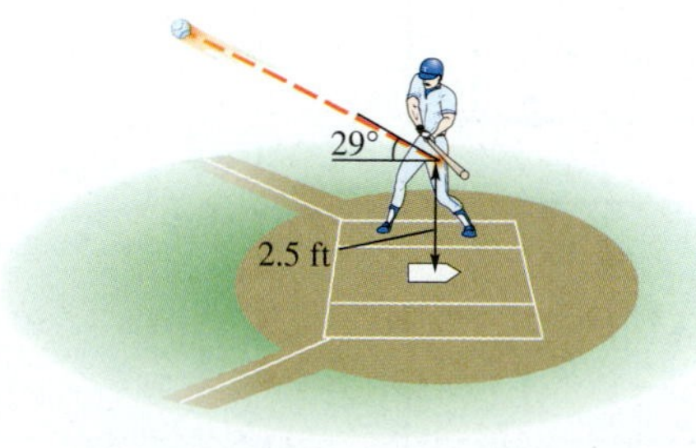

41. ***(Modeling) Simulating Gravity on the Moon*** If an object is thrown on the moon, then the parametric equations of flight are

$$x = (v \cos \theta)t \quad \text{and} \quad y = (v \sin \theta)t - 2.66t^2 + h.$$

Estimate the distance that a golf ball hit at 88 feet per second (60 mph) at an angle of 45° with the horizontal travels on the moon if the moon's surface is level.

42. ***(Modeling) Flight of a Baseball*** A baseball is hit from a height of 3 feet at a 60° angle above the horizontal. Its initial velocity is 64 feet per second.

(a) Write parametric equations that model the flight of the baseball.

(b) Determine the horizontal distance traveled by the ball in the air. Assume that the ground is level.

(c) What is the maximum height of the baseball? At that time, how far has the ball traveled horizontally?

(d) Would the ball clear a 5-foot-high fence that is 100 feet from the batter?

(Modeling) Path of a Projectile *In Exercises 43 and 44, a projectile has been launched from the ground with an initial velocity of 88 feet per second. You are given parametric equations that model the path of the projectile.*

(a) *Graph the parametric equations.*

(b) *Approximate* θ, *the angle the projectile makes with the horizontal at launch, to the nearest tenth of a degree.*

(c) *On the basis of your answer to part (b), write parametric equations for the projectile, using the cosine and sine functions.*

43. $x = 82.69265063t$, $y = -16t^2 + 30.09777261t$

44. $x = 56.56530965t$, $y = -16t^2 + 67.41191099t$

45. The spiral of Archimedes has polar equation $r = a\theta$, where $r^2 = x^2 + y^2$. Show that a parametric representation of the spiral of Archimedes is

$$x = a\theta \cos \theta, \quad y = a\theta \sin \theta, \quad \text{for } \theta \text{ in } (-\infty, \infty).$$

46. Show that the hyperbolic spiral $r\theta = a$, where $r^2 = x^2 + y^2$, is given parametrically by

$$x = \frac{a \cos \theta}{\theta}, \quad y = \frac{a \sin \theta}{\theta}, \quad \text{for } \theta \text{ in } (-\infty, 0) \cup (0, \infty).$$

Reviewing Basic Concepts (Sections 10.6 and 10.7)

1. For the point with polar coordinates $(-2, 130°)$, state the quadrant in which the point lies if it is graphed in a rectangular coordinate system.

2. Let point P have rectangular coordinates $(-2, 2)$. Determine two pairs of equivalent polar coordinates for point P.

3. Graph the polar equation $r = 2 - 2\cos\theta$. Identify the type of polar graph.

4. Find an equivalent equation in rectangular coordinates for the polar equation $r = 2\cos\theta$.

5. Find an equivalent equation in polar coordinates for the equation $x + y = 6$.

6. Find a rectangular equation for the curve described by $x = 2\cos t$, $y = 4\sin t$, $0 \le t < 2\pi$.

7. Graph the parametric equations

$$x = 2 - \sin t, \quad y = \cos t - 1,$$

for $0 \le t < 2\pi$.

8. *(Modeling) Flight of a Golf Ball* Vijay hits a golf ball with a 45° angle of elevation from the top of a ridge that is 50 feet above an area of level ground. The initial velocity of the ball is 88 feet per second, or 60 mph. Find the horizontal distance traveled by the ball in the air.

CHAPTER 10 SUMMARY

KEY TERMS & SYMBOLS	KEY CONCEPTS
10.1 The Law of Sines side–angle–side (SAS) angle–side–angle (ASA) side–side–side (SSS) side–angle–angle (SAA) side–side–angle (SSA) oblique triangle ambiguous case	**LAW OF SINES** In any triangle ABC with sides a, b, and c, $\dfrac{a}{\sin A} = \dfrac{b}{\sin B}$, $\dfrac{a}{\sin A} = \dfrac{c}{\sin C}$, and $\dfrac{b}{\sin B} = \dfrac{c}{\sin C}$, or, in compact form, $\dfrac{a}{\sin A} = \dfrac{b}{\sin B} = \dfrac{c}{\sin C}$.
10.2 The Law of Cosines and Area Formulas semiperimeter	**LAW OF COSINES** In any triangle ABC with sides a, b, and c, $a^2 = b^2 + c^2 - 2bc\cos A$, $b^2 = a^2 + c^2 - 2ac\cos B$, and $c^2 = a^2 + b^2 - 2ab\cos C$. **HERON'S AREA FORMULA (SSS)** If a triangle has sides of lengths a, b, and c, and if the semiperimeter is $s = \frac{1}{2}(a + b + c)$, then the area of the triangle is $\mathcal{A} = \sqrt{s(s-a)(s-b)(s-c)}$. *(continued)*

KEY TERMS & SYMBOLS	KEY CONCEPTS
	AREA OF A TRIANGLE (SAS) In any triangle ABC, the area $\mathscr{A}$ is given by the following formulas: $\mathscr{A} = \frac{1}{2}bc \sin A, \quad \mathscr{A} = \frac{1}{2}ab \sin C, \quad \text{and} \quad \mathscr{A} = \frac{1}{2}ac \sin B.$ That is, the area is half the product of the lengths of two sides and the sine of the angle included between them.
10.3 Vectors and Their Applications scalars vector quantities vector, **v**, **OP**, or $\overrightarrow{OP}$ initial point terminal point magnitude, $\|\mathbf{OP}\|$ parallelogram rule resultant vector opposite of **v** zero vector scalar product position vector, $\langle a, b\rangle$ horizontal component vertical component direction angle unit vectors, **i**, **j** dot product angle between two vectors orthogonal vectors airspeed ground speed	**RESULTANT VECTOR** The resultant (or sum) **A** + **B** of vectors **A** and **B** is shown in the figure. A + B B A **MAGNITUDE AND DIRECTION ANGLE OF A VECTOR** The magnitude (length) of vector $\mathbf{u} = \langle a, b\rangle$ is given by $\|\mathbf{u}\| = \sqrt{a^2 + b^2}$. The direction angle θ satisfies $\tan \theta = \frac{b}{a}$, where $a \neq 0$. **VECTOR OPERATIONS** For any real numbers a, b, c, d, and k, $\langle a, b\rangle + \langle c, d\rangle = \langle a + c, b + d\rangle$ $k \cdot \langle a, b\rangle = \langle ka, kb\rangle$ If $\mathbf{a} = \langle a_1, a_2\rangle$, then $-\mathbf{a} = \langle -a_1, -a_2\rangle$. $\langle a, b\rangle - \langle c, d\rangle = \langle a, b\rangle + -\langle c, d\rangle = \langle a - c, b - d\rangle$ If $\mathbf{u} = \langle x, y\rangle$ has direction angle θ, then $\mathbf{u} = \langle \|\mathbf{u}\| \cos \theta, \|\mathbf{u}\| \sin \theta\rangle$. **i, j FORM FOR VECTORS** If $\mathbf{v} = \langle a, b\rangle$, then $\mathbf{v} = a\mathbf{i} + b\mathbf{j}$, where $\mathbf{i} = \langle 1, 0\rangle$ and $\mathbf{j} = \langle 0, 1\rangle$. **DOT PRODUCT** The dot product of the two vectors $\mathbf{u} = \langle a, b\rangle$ and $\mathbf{v} = \langle c, d\rangle$, denoted $\mathbf{u} \cdot \mathbf{v}$, is given by $\mathbf{u} \cdot \mathbf{v} = ac + bd.$ If θ is the angle between **u** and **v**, where $0° \leq \theta \leq 180°$, then $\mathbf{u} \cdot \mathbf{v} = \|\mathbf{u}\|\|\mathbf{v}\| \cos \theta$. **PROPERTIES OF THE DOT PRODUCT** For all vectors **u**, **v**, and **w** and real numbers k, (a) $\mathbf{u} \cdot \mathbf{v} = \mathbf{v} \cdot \mathbf{u}$ (b) $\mathbf{u} \cdot (\mathbf{v} + \mathbf{w}) = \mathbf{u} \cdot \mathbf{v} + \mathbf{u} \cdot \mathbf{w}$ (c) $(\mathbf{u} + \mathbf{v}) \cdot \mathbf{w} = \mathbf{u} \cdot \mathbf{w} + \mathbf{v} \cdot \mathbf{w}$ (d) $(k\mathbf{u}) \cdot \mathbf{v} = k(\mathbf{u} \cdot \mathbf{v}) = \mathbf{u} \cdot (k\mathbf{v})$ (e) $\mathbf{0} \cdot \mathbf{u} = 0$ (f) $\mathbf{u} \cdot \mathbf{u} = \|\mathbf{u}\|^2$. *(continued)*

KEY TERMS & SYMBOLS	KEY CONCEPTS

10.4 Trigonometric (Polar) Form of Complex Numbers
real axis
imaginary axis
complex plane
rectangular form
trigonometric (polar) form
modulus (absolute value)
argument
	TRIGONOMETRIC (POLAR) FORM OF A COMPLEX NUMBER
	The expression $\boldsymbol{r(\cos\theta + i\sin\theta)}$ is called the trigonometric form (or polar form) of the complex number $x + yi$. The expression $\cos\theta + i\sin\theta$ is sometimes abbreviated cis θ. In this notation, $\boldsymbol{r(\cos\theta + i\sin\theta)}$ is written $\boldsymbol{r \text{ cis } \theta}$.
	
	PRODUCT THEOREM
	If $r_1(\cos\theta_1 + i\sin\theta_1)$ and $r_2(\cos\theta_2 + i\sin\theta_2)$ are any two complex numbers, then
	$$[r_1(\cos\theta_1 + i\sin\theta_1)]\cdot[r_2(\cos\theta_2 + i\sin\theta_2)] = r_1r_2[\cos(\theta_1+\theta_2) + i\sin(\theta_1+\theta_2)],$$
	or, in compact form, $(r_1 \text{ cis } \theta_1)(r_2 \text{ cis } \theta_2) = r_1r_2 \text{ cis}(\theta_1+\theta_2)$.
	
	QUOTIENT THEOREM
	If $r_1(\cos\theta_1 + i\sin\theta_1)$ and $r_2(\cos\theta_2 + i\sin\theta_2)$ are any two complex numbers, where $r_2(\cos\theta_2 + i\sin\theta_2) \neq 0$, then
	$$\frac{r_1(\cos\theta_1 + i\sin\theta_1)}{r_2(\cos\theta_2 + i\sin\theta_2)} = \frac{r_1}{r_2}[\cos(\theta_1-\theta_2) + i\sin(\theta_1-\theta_2)],$$
	or, in compact form, $\dfrac{r_1 \text{ cis } \theta_1}{r_2 \text{ cis } \theta_2} = \dfrac{r_1}{r_2}\text{cis}(\theta_1-\theta_2)$.

10.5 Powers and Roots of Complex Numbers *n*th root of a complex number	**DE MOIVRE'S THEOREM** If $r(\cos\theta + i\sin\theta)$ is a complex number, and if n is any real number, then $$[r(\cos\theta + i\sin\theta)]^n = r^n(\cos n\theta + i\sin n\theta),$$ or, in compact form, $[r \text{ cis } \theta]^n = r^n(\text{cis } n\theta)$. ***n*TH ROOT** For a positive integer n, the complex number $a + bi$ is an nth root of the complex number $x + yi$ if $(a + bi)^n = x + yi$. ***n*TH ROOT THEOREM** If n is any positive integer, r is a positive real number, and θ is in degrees, then the nonzero complex number $r(\cos\theta + i\sin\theta)$ has exactly n distinct nth roots, given by $$\sqrt[n]{r}(\cos\alpha + i\sin\alpha),$$ where $\alpha = \dfrac{\theta + 360°\cdot k}{n}$ or $\alpha = \dfrac{\theta}{n} + \dfrac{360°\cdot k}{n}$, $k = 0, 1, 2, \ldots, n-1$.
10.6 Polar Equations and Graphs polar coordinate system pole polar axis polar coordinates	**RECTANGULAR AND POLAR COORDINATES** The following relationships hold between the point (x, y) in the rectangular coordinate plane and the point (r, θ) in the polar coordinate plane. $$x = r\cos\theta, \quad y = r\sin\theta, \quad r^2 = x^2 + y^2, \quad \tan\theta = \frac{y}{x}, \text{ if } x \neq 0$$ *(continued)*

KEY TERMS & SYMBOLS	KEY CONCEPTS
polar equation cardioid polar grid rose curve (four-leaved rose) lemniscate spiral of Archimedes limaçon	Polar coordinates determine a point by locating it θ degrees relative to the polar axis (that is, the positive x-axis) and r units from the pole (or origin). **POLAR GRAPHS** Some of the most common polar graphs are defined by the equations that follow. $r = a\cos\theta$, $r = a\sin\theta$ Circles; $r^2 = a^2\sin 2\theta$, $r^2 = a^2\cos 2\theta$ Lemniscates $r = a \pm b\sin\theta$, $r = a \pm b\cos\theta$ Limaçons; $r = a\sin n\theta$, $r = a\cos n\theta$ Rose curves
10.7 More Parametric Equations cycloid	**PARAMETRIC EQUATIONS** Parametric equations with trigonometric functions can be used to describe many different types of curves. Examples include circles, ellipses, and cycloids. **HEIGHT OF AN OBJECT** If an object has an initial velocity v and initial height h, and travels so that its initial angle of elevation is θ, then its flight after t seconds is modeled by the parametric equations $x = (v\cos\theta)t$ and $y = (v\sin\theta)t - 16t^2 + h.$

CHAPTER 10 Review Exercises

Use the law of sines or the law of cosines to find the indicated part of each triangle ABC.

1. $C = 74° 10'$, $c = 96.3$ meters, $B = 39° 30'$; find b
2. $a = 165$ meters, $A = 100.2°$, $B = 25.0°$; find b
3. $a = 86.14$ inches, $b = 253.2$ inches, $c = 241.9$ inches; find A
4. $a = 14.8$ feet, $b = 19.7$ feet, $c = 31.8$ feet; find B
5. $A = 129° 40'$, $a = 127$ feet, $b = 69.8$ feet; find B
6. $B = 39° 50'$, $b = 268$ centimeters, $a = 340$ centimeters; find A
7. $B = 120.7°$, $a = 127$ feet, $c = 69.8$ feet; find b
8. $A = 46.2°$, $b = 184$ centimeters, $c = 192$ centimeters; find a

Find the area of each triangle ABC with the given information.

9. $b = 840.6$ meters, $c = 715.9$ meters, $A = 149.3°$
10. $a = 6.90$ feet, $b = 10.2$ feet, $C = 35° 10'$
11. $a = .913$ kilometer, $b = .816$ kilometer, $c = .582$ kilometer
12. $a = 43$ meters, $b = 32$ meters, $c = 51$ meters

Solve each problem.

13. ***Height of a Balloon*** The angles of elevation of a balloon from two points A and B on level ground are $24° 50'$ and $47° 20'$, respectively. As shown in the figure, points A and B are in the same vertical plane and are 8.4 miles apart. Approximate the height of the balloon above the ground to the nearest tenth of a mile.

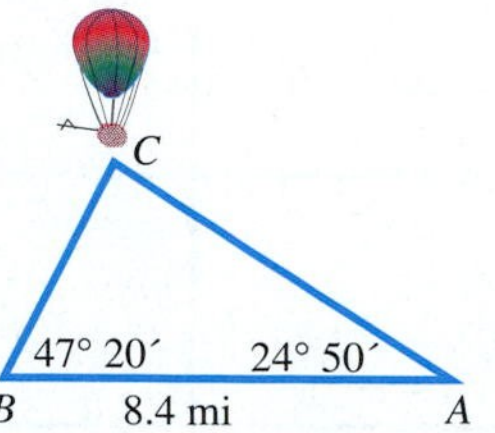

14. ***Distance in a Boat Race*** The course for a boat race starts at point X and goes in the direction S 48° W to point Y. It then turns and goes S 36° E to point Z and finally returns back to point X. If point Z lies 10 kilometers directly south of point X, find the distance from Y to Z to the nearest kilometer.

15. ***Radio Direction Finders*** Radio direction finders are placed at points A and B, which are 3.46 miles apart on an east–west line, with A west of B. From A, the bearing of an illegal radio transmitter is 48°; from B, the bearing is 302°. Find the distance between the transmitter and A.

16. ***Distance across a Canyon*** To measure the distance AB across a canyon for a power line, a surveyor measures angles B and C and the distance BC. (See the figure.) What is the distance from A to B?

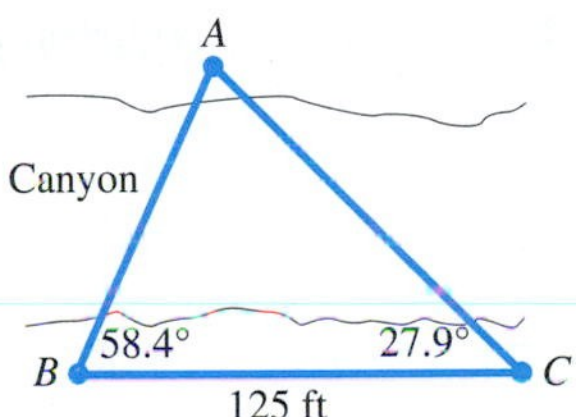

17. ***Length of a Brace*** A banner on an 8.0-foot pole is to be mounted on a building at an angle of 115°, as shown in the figure. Find the length of the brace.

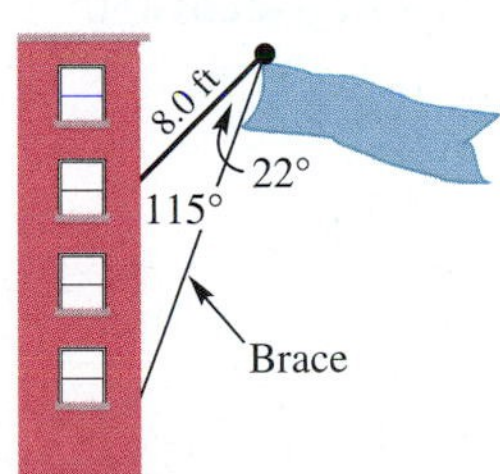

18. ***Hanging Sculpture*** A hanging sculpture in an art gallery is to be hung with two wires of lengths 15.0 feet and 12.2 feet so that the angle between them is 70.3°. How far apart should the ends of the wire be placed on the ceiling?

19. ***Pipeline Position*** A pipeline is to run between points A and B, which are separated by a protected wetlands area. To avoid the wetlands, the pipe will run from point A to C and then to B. The distances involved are $AB = 150$ kilometers, $AC = 102$ kilometers, and $BC = 135$ kilometers. What angle should be used at point C?

20. If we are given a, A, and C in a triangle ABC, does the possibility of the ambiguous case exist? If not, explain why.

21. Can a triangle ABC exist if $a = 4.7$, $b = 2.3$, and $c = 7.0$? If not, explain why. Answer this question without using trigonometry.

22. ***Concept Check*** Given that $a = 10$ and $B = 30°$ in triangle ABC, determine the values of b for which A has
(a) exactly one value,
(b) two values,
(c) no value.

23. ***Concept Check*** If angle C of a triangle ABC measures 90°, what does the law of cosines become?

24. ***Concept Check*** When applying the law of cosines to find an angle, how can we tell if the angle is acute or obtuse before evaluating the inverse cosine value?

25. Use the vectors shown here to sketch $\mathbf{a} + 3\mathbf{c}$.

Find the magnitude and direction angle for **u**, *rounded to the nearest tenth of a degree.*

26. $\mathbf{u} = \langle 21, -20 \rangle$

27. $\mathbf{u} = \langle -9, 12 \rangle$

Vector **v** *has the given magnitude and direction angle. Write* **v** *in the form* $\langle a, b \rangle$.

28. $|\mathbf{v}| = 50$, $\theta = 45°$ (give exact values)

29. $|\mathbf{v}| = 69.2$, $\theta = 75°$

30. $|\mathbf{v}| = 964$, $\theta = 154° 20'$

Find **(a)** *the dot product and* **(b)** *the angle between the pairs of vectors.*

31. $\mathbf{u} = \langle 6, 2 \rangle$, $\mathbf{v} = \langle 3, -2 \rangle$

32. $\mathbf{u} = \langle 2\sqrt{3}, 2 \rangle$, $\mathbf{v} = \langle 5, 5\sqrt{3} \rangle$

33. Are the vectors $\mathbf{u} = \langle 5, -1 \rangle$ and $\mathbf{v} = \langle -2, -10 \rangle$ orthogonal? Explain.

Solve each problem.

34. ***Weight of a Sled and Child*** Paula and Steve are pulling their daughter Jessie on a sled as shown in the figure. Steve pulls with a force of 18 pounds at an angle of 10°. Paula pulls with a force of 12 pounds at an angle of 15°. Find the magnitude of the resultant force.

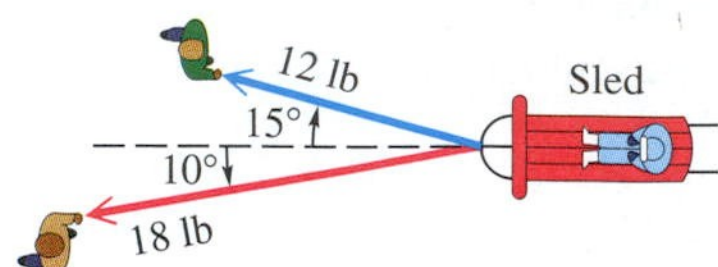

35. ***Barge Movement*** One rope pulls a barge directly east with a force of 1000 newtons. Another rope pulls the barge to the northeast with a force of 2000 newtons. Find the resultant force acting on the barge, and find the angle between the resultant and the first rope.

36. ***Direction and Airspeed*** A plane has an airspeed of 520 mph. The pilot wishes to fly on a course of 310°. A wind of 37 mph is blowing from a bearing of 212°. On what bearing should the pilot fly, and what will be her actual speed?

37. ***Bearing and Speed*** A long-distance swimmer starts out swimming a steady 3.2 mph due north. A 5.1-mph current is flowing on a bearing of 12°. What is the swimmer's resulting bearing and speed?

38. ***(Modeling) Wind and Vectors*** A wind can be described by $\mathbf{v} = 6\mathbf{i} + 8\mathbf{j}$, where vector $\mathbf{j}$ points north and represents a south wind of 1 mph.

(a) What is the speed of the wind?

(b) Find $3\mathbf{v}$. Interpret the result.

(c) Interpret the wind if it switches to $\mathbf{u} = -8\mathbf{i} + 8\mathbf{j}$.

Relating Concepts

For individual or group investigation (Exercises 39–44)

Consider the two vectors **v** *and* **u** *shown. Assume all values are exact.* ***Work Exercises 39–44 in order.***

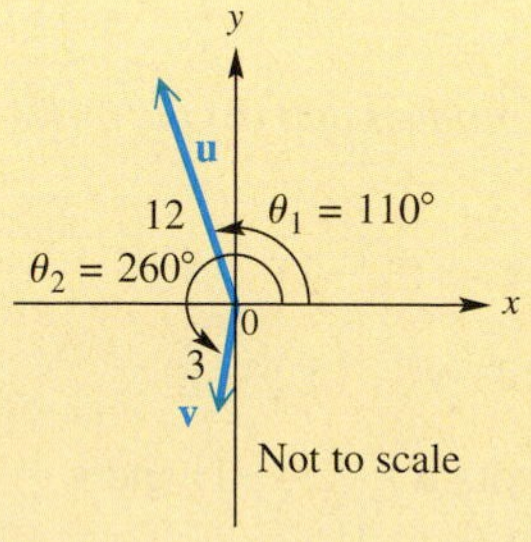

39. Use trigonometry alone (without vector notation) to find the magnitude and direction angle of $\mathbf{u} + \mathbf{v}$. Use the law of cosines and the law of sines in your work.

40. Find the horizontal and vertical components of $\mathbf{u}$, using your calculator.

41. Find the horizontal and vertical components of $\mathbf{v}$, using your calculator.

42. Find the horizontal and vertical components of $\mathbf{u} + \mathbf{v}$ by adding the results you obtained in Exercises 40 and 41.

43. Use your calculator to find the magnitude and direction angle of the vector $\mathbf{u} + \mathbf{v}$.

44. Compare your answers in Exercises 39 and 43. What do you notice? Which method of solution do you prefer?

Graph each complex number as a vector in the complex plane.

45. $5i$

46. $-4 + 2i$

Find the resultant of each pair of complex numbers.

47. $7 + 3i$ and $-2 + i$

48. $2 - 4i$ and $5 + i$

Perform each conversion in Exercises 49–52.

	Rectangular Form	Trigonometric Form
49.	$-2 + 2i$	________
50.	________	$3(\cos 90° + i \sin 90°)$
51.	________	$2 \text{ cis } 225°$
52.	$-4 + 4i\sqrt{3}$	________

Perform each indicated operation. Give answers in rectangular form.

53. $[5(\cos 90° + i \sin 90°)][6(\cos 180° + i \sin 180°)]$

54. $[3 \text{ cis } 135°][2 \text{ cis } 105°]$

55. $\dfrac{2(\cos 60° + i \sin 60°)}{8(\cos 300° + i \sin 300°)}$

56. $\dfrac{4 \text{ cis } 270°}{2 \text{ cis } 90°}$

57. $(\sqrt{3} + i)^3$

58. $(2 - 2i)^5$

59. $(\cos 100° + i \sin 100°)^6$

60. $(\text{cis } 20°)^3$

Find the indicated roots and graph as vectors in the complex plane. Leave answers in polar form.

61. The cube roots of $-27i$

62. The fourth roots of $16i$

63. The fifth roots of 32

64. Solve the equation $x^4 + i = 0$. Leave solutions in polar form.

Convert to rectangular coordinates. Give exact values.

65. $(12, 225°)$

66. $\left(-8, -\frac{\pi}{3}\right)$

Convert to polar coordinates, with $-180° < \theta \leq 180°$. Give exact values.

67. $(-6, 6)$

68. $(0, -5)$

Use a graphing calculator to graph each polar equation for $0° \leq \theta \leq 360°$. Use a square window.

69. $r = 4 \cos \theta$ (circle)

70. $r = 1 - 2 \sin \theta$ (limaçon with a loop)

71. $r = 2 \sin 4\theta$ (eight-leaved rose)

72. Sketch a traditional graph of

$$r = 1 + 2 \sin \theta \quad \text{(limaçon with a loop).}$$

Find an equivalent equation in rectangular coordinates.

73. $r = \dfrac{3}{1 + \cos \theta}$

74. $r = \dfrac{4}{2 \sin \theta - \cos \theta}$

75. $r = \sin \theta + \cos \theta$

76. $r = 2$

Find an equivalent equation in polar coordinates.

77. $x = -3$

78. $y = x$

79. $y = x^2$

80. $x = y^2$

Find a rectangular equation for each plane curve with the given parametric equations.

81. $x = \cos 2t$, $y = \sin t$, for t in $(-\pi, \pi)$

82. $x = 5 \tan t$, $y = 3 \sec t$, for t in $\left(-\frac{\pi}{2}, \frac{\pi}{2}\right)$

83. Graph the curve defined by the following parametric equations: $x = t + \cos t$, $y = \sin t$, for t in $[0, 2\pi]$.

84. ***(Modeling) Flight of a Baseball*** A baseball is hit when it is 3.2 feet above the ground. It leaves the bat with a velocity of 118 feet per second at an angle of 27° with respect to the ground. Find the horizontal distance the baseball travels in the air.

CHAPTER 10 Test

1. Find the indicated part of each triangle.

(a) $A = 25.2°$, $a = 6.92$ yards, $b = 4.82$ yards; find C

(b) $C = 118°$, $b = 132$ kilometers, $a = 75.1$ kilometers; find c

(c) $a = 17.3$ feet, $b = 22.6$ feet, $c = 29.8$ feet; find B

2. Given that $a = 10$ and $B = 150°$ in triangle ABC, determine the values of b for which A has

(a) exactly one value **(b)** two values

(c) no value.

3. What conditions determine whether three positive numbers can represent the sides of a triangle?

4. Solve each problem.

(a) ***Distance between Boats*** Two boats leave a dock together. Each travels in a straight line. The angle between their courses measures 54.2°. One boat travels 36.2 kilometers per hour, and the other travels 45.6 kilometers per hour. How far apart will they be after 3 hours?

(b) ***Distance between Points on a Softball Field*** The pitcher's mound on a regulation softball field is 46 feet from home plate. The distance between the bases is 60 feet, as shown in the figure. How far from third base (point T) is the pitcher's mound (point M)? Give your answer to the nearest foot.

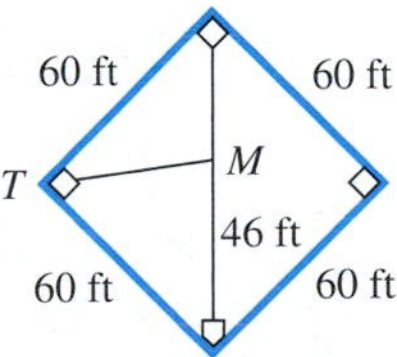

(c) ***Horizontal and Vertical Components*** Find the horizontal and vertical components of the vector with magnitude 569 that is inclined 127.5° from the horizontal.

(d) *Magnitude of a Resultant* Find the magnitude of the resultant of forces of 475 pounds and 586 pounds that form an angle of 78.2°.

5. For the vectors $\mathbf{u} = \langle -1, 3 \rangle$ and $\mathbf{v} = \langle 2, -6 \rangle$ find the following.

(a) $\mathbf{u} + \mathbf{v}$ (b) $-3\mathbf{v}$

(c) $\mathbf{u} \cdot \mathbf{v}$

(d) The angle between $\mathbf{u}$ and $\mathbf{v}$

6. Find the following for the complex numbers

$$4 \text{ cis } 240° \quad \text{and} \quad -4 + 4i\sqrt{3}.$$

(a) The rectangular form of 4 cis 240°

(b) The trigonometric form of $-4 + 4i\sqrt{3}$

(c) Their resultant in the form $a + bi$

7. Perform each indicated operation. Give the answer in rectangular form.

(a) $3(\cos 30° + i \sin 30°) \cdot 5(\cos 90° + i \sin 90°)$

(b) $\dfrac{2 \text{ cis } 315°}{4 \text{ cis } 45°}$

(c) $\left(1 - i\sqrt{3}\right)^5$

(d) Find the fourth roots of $\sqrt{3} + i$. Leave the answers in trigonometric form.

8. Do the following for the polar equation

$$r = 4 \cos \theta.$$

(a) Graph the equation for $0° \le \theta \le 360°$.

(b) Find an equivalent equation in rectangular coordinates.

(c) Is the graph from part (a) what you would expect for the graph of the equation from part (b)? Explain.

9. Find an equivalent equation in polar coordinates for $-x + 2y = 4$ in the form $r = f(\theta)$.

10. Graph the curve defined by the following parametric equations: $x = 2 \cos 2t$, $y = 2 \sin 2t$, for t in $[0, 2\pi]$.

CHAPTER 10 PROJECT

When Is a Circle Really a Polygon?

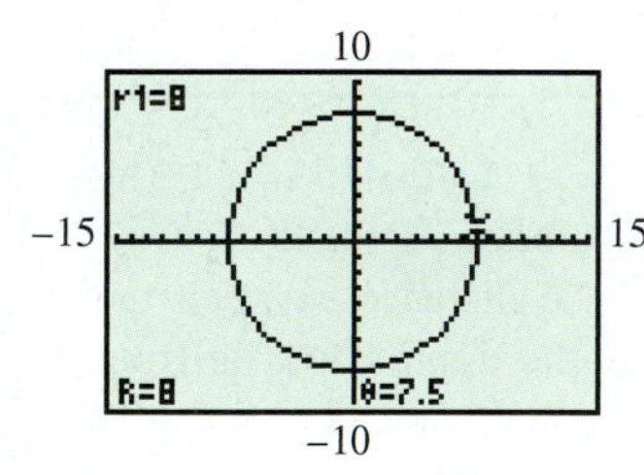

FIGURE A

When we draw a line or circle with pencil and paper, the pencil point is drawing an infinite number of points on a continuous curve. A graphing calculator plots a series of points and connects the points with line segments. This means that the polar circle $r = 8$, graphed in degree mode with the default setting for θ-step (7.5°), is not really a circle, but a polygon with vertices plotted every 7.5° around the origin and connected by line segments. Since $\frac{360}{7.5} = 48$, the "circle" is really a polygon with 48 vertices (a 48-gon).

Thus, to draw a 48-gon (with 48 sides), set your calculator in degree and polar modes and set the format menu for polar graphing. Set θmin $= 0$, θmax $= 360$, and θ-step $= 7.5$, with x in $[-15, 15]$ and y in $[-10, 10]$. Enter $r_1 = 8$ in the Y-menu. The resulting 48-gon is shown in Figure A.

Activities

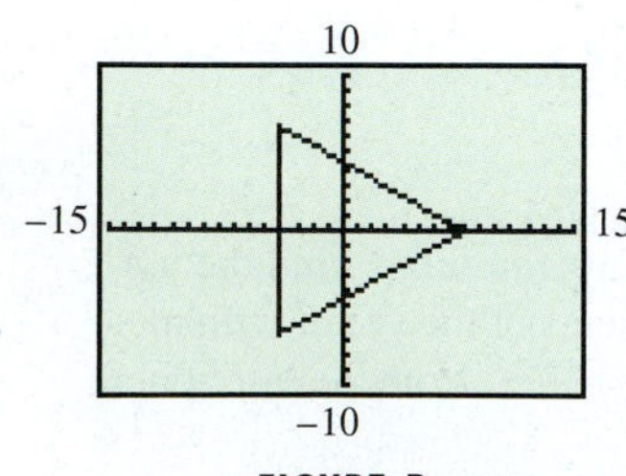

FIGURE B

1. A smaller θ-step makes a polygon with more sides, producing a more accurate-looking circle, but taking longer to draw. By changing θ-step appropriately, draw a 360-gon.
2. To construct any regular polygon, only the θ-step needs to be changed. For a triangle, a three-sided polygon, using θ-step $= 120$ produces the graph in Figure B.

   Changing only θ-step, construct the following regular polygons: a square, a pentagon, a hexagon, a heptagon, an octagon, a nonagon, and a decagon (figures with 4, 5, 6, 7, 8, 9, and 10 sides, respectively). We show the hexagon in Figure C on the next page. Which θ-step works for each polygon? (*Hint:* How did we choose θ-step $= 120$ for the triangle? It involves θmax as well as the number of sides.)

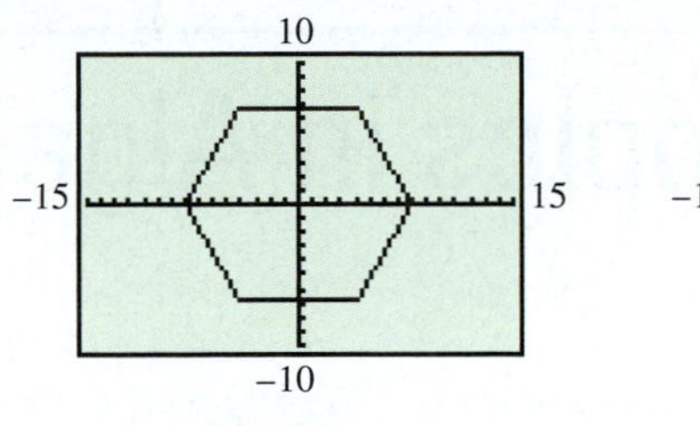

FIGURE C

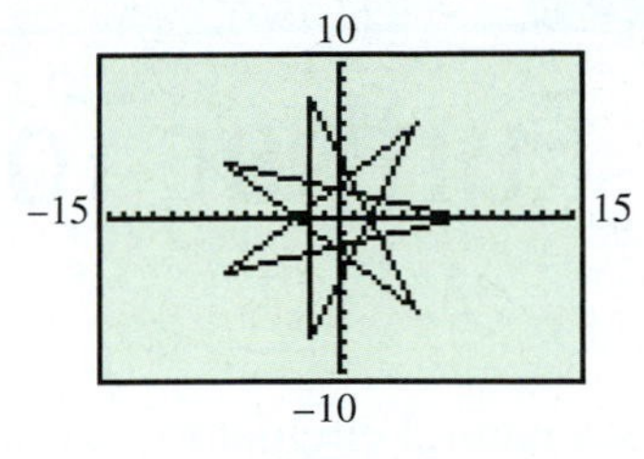

FIGURE D

3. Star polygons can be constructed similarly, changing only θmax and θ-step. We show the star polygon with 7 points in Figure D, where we used $\theta\text{-step} = \left(\frac{360}{7}\right) \cdot 3 = 154.29$ and $\theta\text{max} = 360 \cdot 3 = 1080$. Construct star polygons with 5, 9, and 11 points.

This investigation illustrates the importance of plotting enough points for an accurate representation of a curve. As demonstrated here, with "selected sampling," we can severely distort the expected results.

11

Further Topics in Algebra

EARTH HAS A natural circulatory system called the *Great Ocean Conveyor Belt*. This system circulates warm water from the Gulf Stream to northern latitudes and cool water from the North Atlantic to southern latitudes, greatly influencing and moderating temperatures throughout the world.

Global warming disrupts the normal flow of the Conveyor Belt. When greenhouse gases raise the average yearly temperature of the atmosphere, heavier-than-normal precipitation falls over the North Atlantic. Runoff from the land and polar ice melt add more fresh water, which lowers the ocean's salinity level and decreases its density. With lowered density, the Conveyor Belt slows or stops. As circulation slows, climate changes occur around the globe, including drought in some regions and torrential rainfall in others. When the Conveyor Belt stops, winter temperatures in northern latitudes get progressively colder without the warming waters of the Gulf Stream to moderate them, and a new Ice Age begins.

Topics in this chapter such as *sequences* and *probability* allow scientists to model and predict the results of continued global warming on Earth's climate.

Source: Woods Hole Oceanographic Institute; United Nations Environment Programme.

Great Ocean Conveyor Belt

Chapter Outline

11.1 Sequences and Series

Sequences ■ Series and Summation Notation ■ Summation Properties

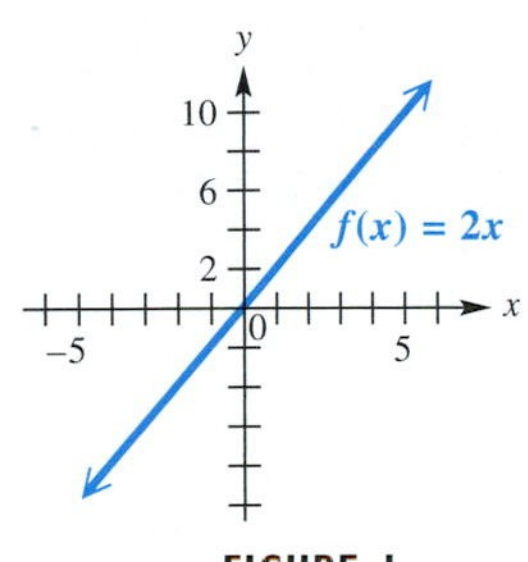

FIGURE 1

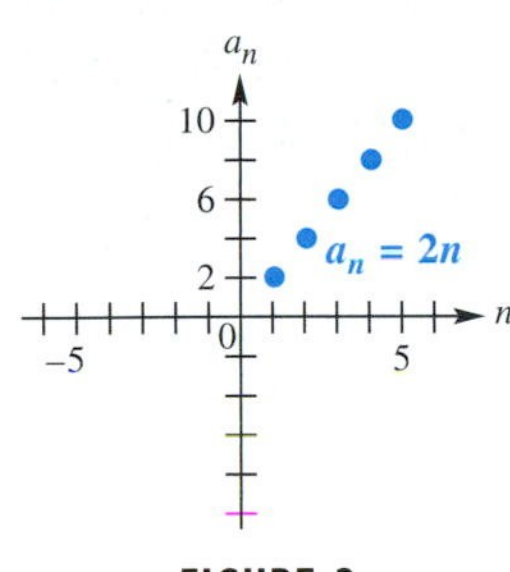

FIGURE 2

Sequences

A *sequence* is a function that computes an ordered list. For example, the average person in the United States uses 100 gallons of water each day. The function defined by $f(n) = 100n$ generates the terms of the sequence

$$100, \ 200, \ 300, \ 400, \ 500, \ 600, \ 700, \ldots,$$

when $n = 1, 2, 3, 4, 5, 6, 7, \ldots$. This function represents the number of gallons of water used by the average person after n days.

As another example, if \$100 is deposited into a savings account paying 5% interest compounded annually, then the function defined by $g(n) = 100(1.05)^n$ calculates the account balance after n years. The terms of the sequence are $g(1), g(2), g(3), g(4), g(5), g(6), g(7), \ldots,$ and can be approximated as

$$105, \ 110.25, \ 115.76, \ 121.55, \ 127.63, \ 134.01, \ 140.71, \ldots.$$

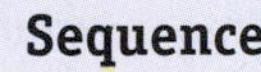

Sequence

A **sequence** is a function that has a set of natural numbers as its domain.

Instead of using $f(x)$ notation to indicate a sequence, it is customary to use $\boldsymbol{a_n}$, where $\boldsymbol{a_n = f(n)}$. ***The letter n is used instead of x as a reminder that n represents a natural number.*** The elements in the range of a sequence, called the **terms** of the sequence, are $a_1, a_2, a_3, \ldots$. The elements of both the domain and the range of a sequence are *ordered*. The first term is found by letting $n = 1$, the second term by letting $n = 2$, and so on. The **general term,** or ***n*th term,** of the sequence is a_n.

Figure 1 shows the graph of $f(x) = 2x$, and Figure 2 shows that of $a_n = 2n$. Notice that $f(x)$ is a continuous function and a_n is discontinuous. To graph a_n, we plot points of the form $(n, 2n)$ for $n = 1, 2, 3, \ldots$.

GCM EXAMPLE 1 Finding Terms of Sequences

Write the first five terms of each sequence.

(a) $a_n = \dfrac{n+1}{n+2}$ **(b)** $a_n = (-1)^n \cdot n$ **(c)** $a_n = \dfrac{(-1)^n}{2^n}$

Solution

(a) Replacing n in $a_n = \frac{n+1}{n+2}$ with 1, 2, 3, 4, and 5 gives

$$n = 1: \quad a_1 = \frac{1+1}{1+2} = \frac{2}{3}; \qquad n = 2: \quad a_2 = \frac{2+1}{2+2} = \frac{3}{4};$$

$$n = 3: \quad a_3 = \frac{3+1}{3+2} = \frac{4}{5}; \qquad n = 4: \quad a_4 = \frac{4+1}{4+2} = \frac{5}{6};$$

$$n = 5: \quad a_5 = \frac{5+1}{5+2} = \frac{6}{7}.$$

(b) Replace n in $a_n = (-1)^n \cdot n$ with 1, 2, 3, 4, and 5 to obtain

$$n = 1: \quad a_1 = (-1)^1 \cdot 1 = -1; \qquad n = 2: \quad a_2 = (-1)^2 \cdot 2 = 2;$$
$$n = 3: \quad a_3 = (-1)^3 \cdot 3 = -3; \qquad n = 4: \quad a_4 = (-1)^4 \cdot 4 = 4;$$
$$n = 5: \quad a_5 = (-1)^5 \cdot 5 = -5.$$

(c) For $a_n = \frac{(-1)^n}{2^n}$, we have $a_1 = -\frac{1}{2}$, $a_2 = \frac{1}{4}$, $a_3 = -\frac{1}{8}$, $a_4 = \frac{1}{16}$, and $a_5 = -\frac{1}{32}$. ■

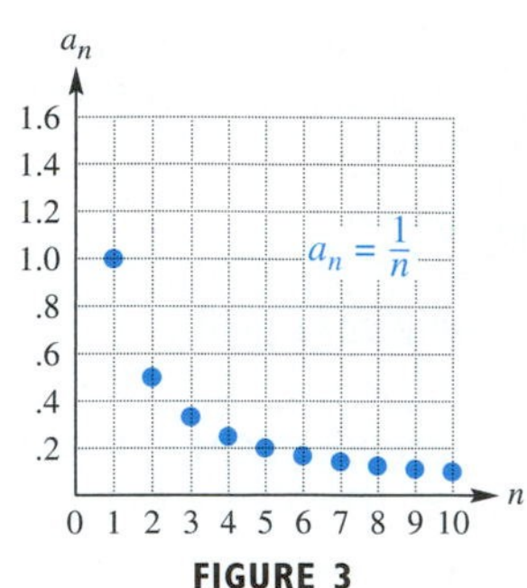

FIGURE 3

A sequence is a **finite sequence** if the domain is the set $\{1, 2, 3, 4, \ldots, n\}$, where n is a natural number. An **infinite sequence** has the set of *all* natural numbers as its domain. For example, the sequence of natural number multiples of 2,

$$2, 4, 6, 8, 10, 12, 14, \ldots, \qquad \text{is infinite,}$$

but the sequence of days in June,

$$1, 2, 3, 4, \ldots, 29, 30, \qquad \text{is finite.}$$

If the terms of an infinite sequence get closer and closer to some real number, the sequence is said to be **convergent** and to **converge** to that real number. For example, the sequence defined by $a_n = \frac{1}{n}$ approaches 0 as n becomes large. Thus, a_n is a convergent sequence that converges to 0. A graph of this sequence for $n = 1, 2, 3, \ldots, 10$ is shown in Figure 3. The terms of a_n approach the horizontal axis.

A sequence that does not converge to some number is **divergent.** The terms of the sequence $a_n = n^2$ are

$$1,\ 4,\ 9,\ 16,\ 25,\ 36,\ 49,\ 64,\ 81, \ldots.$$

This sequence is divergent because as n becomes large, the values for a_n do not approach a fixed number; rather, they increase without bound.

Some sequences are defined by a **recursive definition,** a definition in which each term is defined as an expression involving the previous term or terms. By contrast, the sequences in Example 1 were defined *explicitly,* with a formula for a_n that does not depend on a previous term.

GCM **TECHNOLOGY NOTE**

Some graphing calculators have a designated sequence mode to investigate and graph sequences defined in terms of n, where n is a natural number.

Using sequence mode on the TI-83/84 Plus to list the first 10 terms of the sequence with general term $a_n = n + \frac{1}{n}$ produces the result shown in the top figure. Additional terms can be seen by scrolling to the right. The bottom figure shows a calculator graph of $a_n = n + \frac{1}{n}$. Notice that for $n = 5$, the term is $5 + \frac{1}{5} = 5.2$.

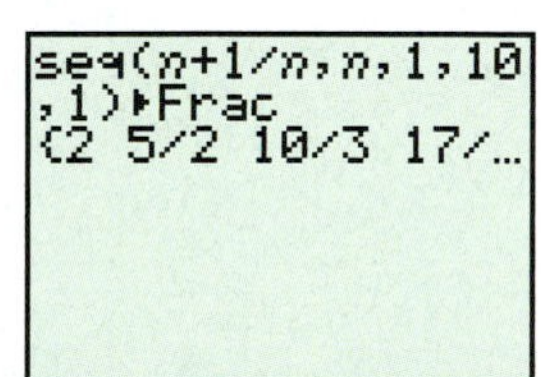

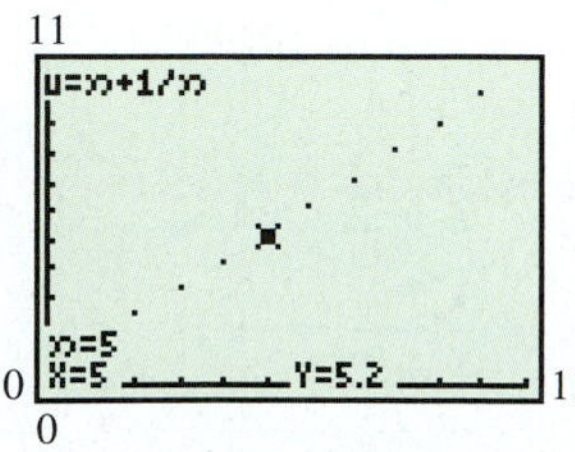

The fifth term is 5.2.

GCM **EXAMPLE 2** **Using a Recursion Formula**

Find the first four terms of each sequence.

(a) $a_1 = 4$; for $n > 1$, $a_n = 2 \cdot a_{n-1} + 1$

(b) $a_1 = 2$; for $n > 1$, $a_n = a_{n-1} + n - 1$

Solution

(a) This is a recursive definition. We know that $a_1 = 4$. Since $a_n = 2 \cdot a_{n-1} + 1$,

$$a_2 = 2 \cdot a_1 + 1 = 2 \cdot 4 + 1 = 9$$
$$a_3 = 2 \cdot a_2 + 1 = 2 \cdot 9 + 1 = 19$$
$$a_4 = 2 \cdot a_3 + 1 = 2 \cdot 19 + 1 = 39.$$

(b) $a_1 = 2$

$$a_2 = a_1 + 2 - 1 = 2 + 1 = 3$$
$$a_3 = a_2 + 3 - 1 = 3 + 2 = 5$$
$$a_4 = a_3 + 4 - 1 = 5 + 3 = 8 \quad ■$$

Leonardo of Pisa (Fibonacci)
(1170–1250)

FOR DISCUSSION

One of the most famous sequences in mathematics is the **Fibonacci sequence,**

$$1, 1, 2, 3, 5, 8, 13, 21, 34, 55, \ldots,$$

named for the Italian mathematician Leonardo of Pisa, who was also known as Fibonacci. The Fibonacci sequence is found in numerous places in nature. For example, male honeybees hatch from eggs that have not been fertilized, so a male bee has only one parent, a female. On the other hand, female honeybees hatch from fertilized eggs, so a female has two parents, one male and one female. The number of ancestors in consecutive generations of bees follows the Fibonacci sequence. Successive terms in the sequence also appear in plants, such as the daisy head, pineapple, and pinecone.

1. Try to discover the pattern in the Fibonacci sequence.
2. Using the given description, write a recursive definition that calculates the number of ancestors of a male bee in each generation.

EXAMPLE 3 Modeling Insect Population Growth

Frequently, the population of a particular insect grows rapidly at first and then levels off because of competition for limited resources. In one study, the behavior of the winter moth was modeled with a sequence similar to the following, where a_n represents the population density in thousands per acre during year n. (*Source:* Varley, G. and G. Gradwell, "Population models for the winter moth," Symposium of the Royal Entomological Society of London 4.)

$$a_1 = 1$$
$$a_n = 2.85a_{n-1} - .19a_{n-1}^{\,2}, \quad \text{for } n \geq 2$$

(a) Give a table of values for $n = 1, 2, 3, \ldots, 10$.

(b) Graph the sequence. Describe what happens to the population density of the winter moth.

Solution

(a) Evaluate $a_1, a_2, a_3, \ldots, a_{10}$ recursively. Since $a_1 = 1$,

$$a_2 = 2.85a_1 - .19a_1^{\,2} = 2.85(1) - .19(1)^2 = 2.66,$$

and $$a_3 = 2.85a_2 - .19a_2^{\,2} = 2.85(2.66) - .19(2.66)^2 \approx 6.24.$$

Approximate values for other terms are given in the table that follows. Figure 4 shows the computation of the sequence, denoted by $u(n)$ rather than a_n, with a calculator.

n	u(n)
1	1
2	2.66
3	6.2366
4	10.384
5	9.1069
6	10.197
7	9.3056

n=1

FIGURE 4

n	1	2	3	4	5	6	7	8	9	10
a_n	1	2.66	6.24	10.4	9.11	10.2	9.31	10.1	9.43	9.98

(b) The graph of a sequence is a set of discrete points. Plot the points

$$(1, 1), \quad (2, 2.66), \quad (3, 6.24), \quad \ldots, \quad (10, 9.98),$$

as shown in Figure 5(a). At first, the insect population increases rapidly and then oscillates about the line $y = 9.7$. (See the "For Discussion" in the margin.) The oscillations become smaller as n increases, indicating that the population density may stabilize near 9.7 thousand per acre. In Figure 5(b), the first 20 terms have been plotted with a calculator.

FOR DISCUSSION

In Example 3, the insect population stabilizes near the value $k = 9.7$ thousand. This value of k can be found by solving the quadratic equation $k = 2.85k - .19k^2$. Explain why.

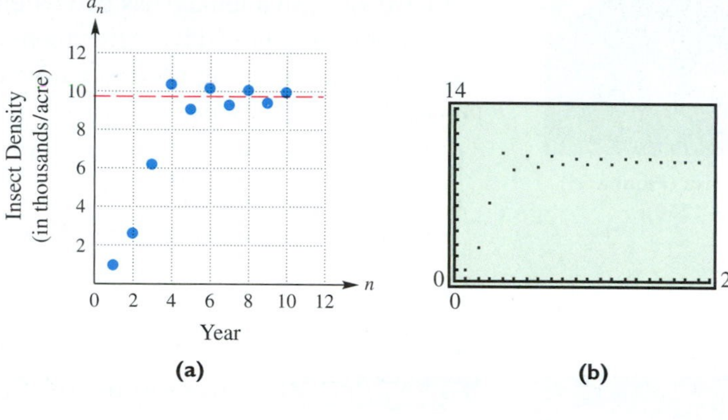

FIGURE 5

Series and Summation Notation

Suppose a person has a starting salary of \$30,000 and receives a \$2000 raise each year. Then

$$30{,}000, \; 32{,}000, \; 34{,}000, \; 36{,}000, \; 38{,}000$$

are terms of the sequence that describe this person's salaries over a 5-year period. The total earned is given by the *finite series*

$$30{,}000 + 32{,}000 + 34{,}000 + 36{,}000 + 38{,}000,$$

whose sum is \$170,000. Any sequence can be used to define a series. For example, the infinite sequence

$$1, \frac{1}{3}, \frac{1}{9}, \frac{1}{27}, \frac{1}{81}, \frac{1}{243}, \ldots$$

defines the terms of the *infinite series*

$$1 + \frac{1}{3} + \frac{1}{9} + \frac{1}{27} + \frac{1}{81} + \frac{1}{243} + \cdots.$$

If a sequence has terms $a_1, a_2, a_3, \ldots$, then S_n is defined as the sum of the first n terms. That is,

$$S_n = a_1 + a_2 + a_3 + \cdots + a_n.$$

The sum of the terms of a sequence, called a **series,** is written using **summation notation.** The symbol **Σ,** the Greek capital letter *sigma,* is used to indicate a sum.

Looking Ahead to Calculus

Sigma (summation) notation, Σ, is introduced in a first calculus course in conjunction with the *definite integral,* symbolized with an elongated S ($\int$) to which limits of integration are applied. The definition of a definite integral is

$$\int_a^b f(x)\,dx = \lim_{n \to \infty} \sum_{i=1}^{n} f(x_i)\,\Delta x_i.$$

The definite integral can be used to calculate the area between two curves.

Series

A **finite series** is an expression of the form

$$S_n = a_1 + a_2 + a_3 + \cdots + a_n = \sum_{i=1}^{n} a_i,$$

and an **infinite series** is an expression of the form

$$S_\infty = a_1 + a_2 + a_3 + \cdots + a_n + \cdots = \sum_{i=1}^{\infty} a_i.$$

The letter i is called the **index of summation.**

CAUTION ***Do not confuse this use of i with the use of i to represent the imaginary unit.*** Other letters, such as k and j, also may be used for the index of summation.

GCM **EXAMPLE 4** **Using Summation Notation**

Evaluate the series $\sum_{k=1}^{6} (2^k + 1)$.

Analytic Solution

Write each of the six terms. Then evaluate the sum.

$$\begin{aligned}\sum_{k=1}^{6} (2^k + 1) &= (2^1 + 1) + (2^2 + 1) + (2^3 + 1) \\ &\quad + (2^4 + 1) + (2^5 + 1) + (2^6 + 1) \\ &= (2 + 1) + (4 + 1) + (8 + 1) \\ &\quad + (16 + 1) + (32 + 1) + (64 + 1) \\ &= 3 + 5 + 9 + 17 + 33 + 65 \\ &= 132\end{aligned}$$

Graphing Calculator Solution

A graphing calculator can store the sequence into a list L_1 and then compute the sum of the six terms in the list. The screen in Figure 6 confirms the analytic result.

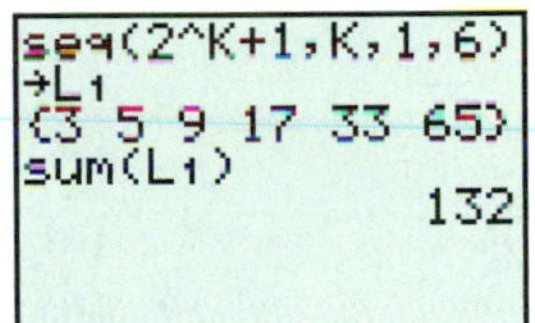

FIGURE 6

EXAMPLE 5 **Using Summation Notation with Subscripts**

Write the terms for each series. Evaluate each sum if possible.

(a) $\sum_{j=3}^{6} a_j$ **(b)** $\sum_{i=1}^{3} (6x_i - 2)$ if $x_1 = 2$, $x_2 = 4$, and $x_3 = 6$

(c) $\sum_{i=1}^{4} f(x_i)\,\Delta x$ if $f(x) = x^2$, $x_1 = 0$, $x_2 = 2$, $x_3 = 4$, $x_4 = 6$, and $\Delta x = 2$

Looking Ahead to Calculus

Sums like the one in Example 5(c) are frequently used in calculus.

Solution

(a) $\sum_{j=3}^{6} a_j = a_3 + a_4 + a_5 + a_6$

(b) Let $i = 1, 2$, and 3, respectively, with $x_1 = 2$, $x_2 = 4$, and $x_3 = 6$.

$$\sum_{i=1}^{3}(6x_i - 2) = (6x_1 - 2) + (6x_2 - 2) + (6x_3 - 2)$$

$$= (6 \cdot 2 - 2) + (6 \cdot 4 - 2) + (6 \cdot 6 - 2) \qquad \text{Substitute the given values for } x_1, x_2, \text{ and } x_3.$$

Use the order of operations.

$$= 10 + 22 + 34 \qquad \text{Simplify.}$$

$$= 66 \qquad \text{Add.}$$

(c) Let $i = 1, 2, 3$, and 4, respectively, with $f(x) = x^2$, $x_1 = 0$, $x_2 = 2$, $x_3 = 4$, and $x_4 = 6$.

$$\sum_{i=1}^{4} f(x_i)\,\Delta x = f(x_1)\,\Delta x + f(x_2)\,\Delta x + f(x_3)\,\Delta x + f(x_4)\,\Delta x$$

$$= x_1^{\,2}\,\Delta x + x_2^{\,2}\,\Delta x + x_3^{\,2}\,\Delta x + x_4^{\,2}\,\Delta x$$

$$= 0^2(2) + 2^2(2) + 4^2(2) + 6^2(2) \qquad f(x) = x^2;\ \Delta x = 2$$

$$= 0 + 8 + 32 + 72 \qquad \text{Simplify.}$$

$$= 112 \qquad \text{Add.}$$

■

EXAMPLE 6 Estimating π with a Series

The infinite series given by

$$\frac{\pi^4}{90} = \frac{1}{1^4} + \frac{1}{2^4} + \frac{1}{3^4} + \frac{1}{4^4} + \frac{1}{5^4} + \cdots + \frac{1}{n^4} + \cdots$$

can be used to estimate π.

(a) Approximate π by finding the sum of the first four terms.

(b) Use a calculator to approximate π by summing the first 50 terms. Compare the result with the decimal value of π.

Solution

(a) Summing the first four terms gives

$$\frac{\pi^4}{90} \approx \frac{1}{1^4} + \frac{1}{2^4} + \frac{1}{3^4} + \frac{1}{4^4} \approx 1.078751929.$$

This approximation can be solved for π by multiplying by 90 and then taking the fourth root. Thus,

$$\pi \approx \sqrt[4]{90(1.078751929)} \approx 3.139.$$

FIGURE 7

(b) As shown in Figure 7, the first 50 terms of the series provide the approximation $\pi \approx 3.141590776$. This computation matches the value of π for the first five decimal places. ■

Summation Properties

Properties of summation provide useful shortcuts for evaluating series.

Summation Properties

If $a_1, a_2, a_3, \ldots, a_n$ and $b_1, b_2, b_3, \ldots, b_n$ are two sequences and c is a constant, then, for every positive integer n,

(a) $\sum_{i=1}^{n} c = nc$ **(b)** $\sum_{i=1}^{n} ca_i = c\sum_{i=1}^{n} a_i$

(c) $\sum_{i=1}^{n} (a_i + b_i) = \sum_{i=1}^{n} a_i + \sum_{i=1}^{n} b_i$ **(d)** $\sum_{i=1}^{n} (a_i - b_i) = \sum_{i=1}^{n} a_i - \sum_{i=1}^{n} b_i$.

To prove Property (a), expand the series to get

$$c + c + c + c + \cdots + c,$$

where there are n terms of c, so the sum is nc.

Property (c) also can be proved by first expanding the series:

$$\sum_{i=1}^{n} (a_i + b_i) = (a_1 + b_1) + (a_2 + b_2) + \cdots + (a_n + b_n)$$

$$= (a_1 + a_2 + \cdots + a_n) + (b_1 + b_2 + \cdots + b_n)$$

Commutative and associative properties

$$= \sum_{i=1}^{n} a_i + \sum_{i=1}^{n} b_i.$$

Proofs of the other two properties are similar.

The following results about summations can be proved by mathematical induction. (See Section 11.5.)

Summation Rules

$$\sum_{i=1}^{n} i = 1 + 2 + \cdots + n = \frac{n(n+1)}{2}$$

$$\sum_{i=1}^{n} i^2 = 1^2 + 2^2 + \cdots + n^2 = \frac{n(n+1)(2n+1)}{6}$$

$$\sum_{i=1}^{n} i^3 = 1^3 + 2^3 + \cdots + n^3 = \frac{n^2(n+1)^2}{4}$$

EXAMPLE 7 **Using the Summation Properties**

Use the summation properties to find each sum.

(a) $\sum_{i=1}^{40} 5$ **(b)** $\sum_{i=1}^{22} 2i$ **(c)** $\sum_{i=1}^{14} (2i^2 - 3)$

Solution

(a) $\sum_{i=1}^{40} 5 = 40(5) = 200$ Property (a) with $n = 40$ and $c = 5$

(b) $\sum_{i=1}^{22} 2i = 2\sum_{i=1}^{22} i$ Property (b) with $c = 2$ and $a_i = i$

$= 2 \cdot \frac{22(22 + 1)}{2}$ Summation rules

$= 506$ Simplify.

(c) $\sum_{i=1}^{14} (2i^2 - 3) = \sum_{i=1}^{14} 2i^2 - \sum_{i=1}^{14} 3$ Property (d) with $a_i = 2i^2$ and $b_i = 3$

$= 2\sum_{i=1}^{14} i^2 - \sum_{i=1}^{14} 3$ Property (b) with $c = 2$ and $a_i = i^2$

$= 2 \cdot \frac{14(14 + 1)(2 \cdot 14 + 1)}{6} - 14(3)$ Summation rules; Property (a)

$= 1988$ Simplify. ■

TECHNOLOGY NOTE

In Figures 6 and 8, the calculator was set in function mode in order to use the variables *K* and *I* instead of *n*, used in Figure 7.

EXAMPLE 8 **Using the Summation Properties**

Evaluate $\sum_{i=1}^{6} (i^2 + 3i + 5)$.

Analytic Solution

$\sum_{i=1}^{6} (i^2 + 3i + 5) = \sum_{i=1}^{6} i^2 + \sum_{i=1}^{6} 3i + \sum_{i=1}^{6} 5$ Property (c)

$= \sum_{i=1}^{6} i^2 + 3\sum_{i=1}^{6} i + \sum_{i=1}^{6} 5$ Property (b)

$= \sum_{i=1}^{6} i^2 + 3\sum_{i=1}^{6} i + 6(5)$ Property (a)

$= \frac{6(6 + 1)(2 \cdot 6 + 1)}{6} + 3\left[\frac{6(6 + 1)}{2}\right] + 6(5)$ Summation rules

$= 91 + 3(21) + 6(5)$ Simplify.

$= 184$

Graphing Calculator Solution

Figure 8 confirms the analytic result. We use two commands on one line, to define the sequence and then compute its sum. (This approach is more efficient than using two separate commands, as seen in Figure 6 of Example 4.)

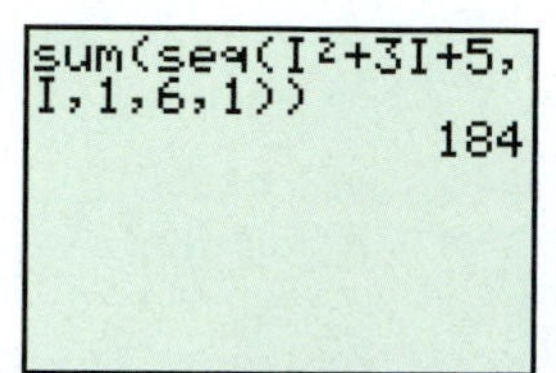

FIGURE 8 ■

11.1 Exercises

Write the first five terms of each sequence.

1. $a_n = 4n + 10$

2. $a_n = 6n - 3$

3. $a_n = 2^{n-1}$

4. $a_n = -3^n$

5. $a_n = \left(\frac{1}{3}\right)^n (n - 1)$

6. $a_n = (-2)^n(n)$

7. $a_n = (-1)^n(2n)$

8. $a_n = (-1)^{n-1}(n + 1)$

9. $a_n = \dfrac{4n - 1}{n^2 + 2}$

10. $a_n = \dfrac{n^2 - 1}{n^2 + 1}$

11. Your friend does not understand what is meant by the nth term, or general term, of a sequence. How would you explain this idea?

12. How are sequences related to functions?

Decide whether each sequence is finite or infinite.

13. The sequence of days of the week

14. The sequence of dates in the month of November

15. $1, 2, 3, 4$

16. $-1, -2, -3, -4$

17. $1, 2, 3, 4, \ldots$

18. $-1, -2, -3, -4, \ldots$

19. $a_1 = 3$; for $2 \le n \le 10$, $a_n = 3 \cdot a_{n-1}$

20. $a_1 = 1$; $a_2 = 3$; for $n \ge 3$, $a_n = a_{n-1} + a_{n-2}$

Find the first four terms of each sequence.

21. $a_1 = -2$, $a_n = a_{n-1} + 3$, for $n > 1$

22. $a_1 = -1$, $a_n = a_{n-1} - 4$, for $n > 1$

23. $a_1 = 1$, $a_2 = 1$, $a_n = a_{n-1} + a_{n-2}$, for $n \ge 3$ (the Fibonacci sequence)

24. $a_1 = 2$, $a_n = n \cdot a_{n-1}$, for $n > 1$

Evaluate each series.

25. $\sum_{i=1}^{5} (2i + 1)$

26. $\sum_{i=1}^{6} (3i - 2)$

27. $\sum_{j=1}^{4} \frac{1}{j}$

28. $\sum_{i=1}^{5} (i + 1)^{-1}$

29. $\sum_{i=1}^{4} i^i$

30. $\sum_{k=1}^{4} (k + 1)^2$

31. $\sum_{k=1}^{6} (-1)^k \cdot k$

32. $\sum_{i=1}^{7} (-1)^{i+1} \cdot i^2$

33. $\sum_{i=2}^{5} (6 - 3i)$

34. $\sum_{i=3}^{7} (5i + 2)$

35. $\sum_{i=-2}^{3} 2(3)^i$

36. $\sum_{i=-1}^{2} 5(2)^i$

37. $\sum_{i=-1}^{5} (i^2 - 2i)$

38. $\sum_{i=3}^{6} (2i^2 + 1)$

39. $\sum_{i=1}^{5} (3^i - 4)$

40. $\sum_{i=1}^{4} [(-2)^i - 3]$

Evaluate the terms of each sum, where $x_1 = -2$, $x_2 = -1$, $x_3 = 0$, $x_4 = 1$, and $x_5 = 2$.

41. $\sum_{i=1}^{5} x_i$

42. $\sum_{i=1}^{5} -x_i$

43. $\sum_{i=1}^{5} (2x_i + 3)$

44. $\sum_{i=1}^{4} x_i^2$

45. $\sum_{i=1}^{3} (3x_i - x_i^2)$

46. $\sum_{i=1}^{3} (x_i^2 + 1)$

47. $\sum_{i=2}^{5} \frac{x_i + 1}{x_i + 2}$

48. $\sum_{i=1}^{5} \frac{x_i}{x_i + 3}$

Evaluate the terms of $\sum_{i=1}^{4} f(x_i)\,\Delta x$, with $x_1 = 0$, $x_2 = 2$, $x_3 = 4$, $x_4 = 6$, and $\Delta x = .5$, for each function.

49. $f(x) = 4x - 7$

50. $f(x) = 6 + 2x$

51. $f(x) = 2x^2$

52. $f(x) = x^2 - 1$

53. $f(x) = \dfrac{-2}{x + 1}$

54. $f(x) = \dfrac{5}{2x - 1}$

Evaluate each series.

55. $\sum_{i=1}^{100} 6$

56. $\sum_{i=1}^{20} 5i$

57. $\sum_{i=1}^{15} i^2$

58. $\sum_{i=1}^{50} 2i^3$

59. $\sum_{i=1}^{5} (5i + 3)$

60. $\sum_{i=1}^{5} (8i - 1)$

61. $\sum_{i=1}^{5} (4i^2 - 2i + 6)$

62. $\sum_{i=1}^{6} (2 + i - i^2)$

63. $\sum_{i=1}^{4} (3i^3 + 2i - 4)$

64. $\sum_{i=1}^{6} (i^2 + 2i^3)$

Use a graphing calculator to graph the first 10 terms of each sequence. Make a conjecture as to whether the sequence converges or diverges. If you think it converges, determine the number to which it converges.

65. $a_n = \dfrac{n + 4}{2n}$

66. $a_n = \dfrac{1 + 4n}{2n}$

67. $a_n = 2e^n$

68. $a_n = n(n + 2)$

69. $a_n = \left(1 + \dfrac{1}{n}\right)^n$

70. $a_n = 5 - \dfrac{1}{n}$

Solve each problem.

71. ***Estimating π*** Find the sum of the first six terms of the series

$$\frac{\pi^4}{90} = \frac{1}{1^4} + \frac{1}{2^4} + \frac{1}{3^4} + \frac{1}{4^4} + \frac{1}{5^4} + \cdots + \frac{1}{n^4} + \cdots$$

presented in Example 6. Use your result to estimate π. Compare your answer with the actual value of π.

72. ***(Modeling) Insect Population*** Suppose an insect population density in thousands per acre during year n can be modeled by the recursively defined sequence

$$a_1 = 8$$
$$a_n = 2.9a_{n-1} - .2a_{n-1}^{\,2}, \quad \text{for } n > 1.$$

(a) Find the population for $n = 1, 2, 3$.

(b) Graph the sequence for $n = 1, 2, 3, \ldots, 20$. Use the window $[0, 21]$ by $[0, 14]$. Interpret the graph.

73. ***(Modeling) Bacteria Growth*** If certain bacteria are cultured in a medium with sufficient nutrients, they will double in size and then divide every 40 minutes. Let N_1 be the initial number of bacteria cells, N_2 the number after 40 minutes, N_3 the number after 80 minutes, and N_j the number after $40(j - 1)$ minutes. (*Source:* Hoppensteadt, F. and C. Peskin, *Mathematics in Medicine and the Life Sciences*, Springer-Verlag, 1992.)

(a) Write N_{j+1} in terms of N_j for $j \geq 1$.

(b) Determine the number of bacteria after two hours if $N_1 = 230$.

(c) Graph the sequence N_j for $j = 1, 2, 3, \ldots, 7$. Use the window $[0, 10]$ by $[0, 15{,}000]$.

(d) Describe the growth of these bacteria when there are unlimited nutrients.

74. ***(Modeling) Verhulst's Model for Bacteria Growth*** Refer to Exercise 73. If the bacteria are not cultured in a medium with sufficient nutrients, competition will ensue and the growth will slow. According to Verhulst's model, the number of bacteria N_j at time $40(j - 1)$ minutes can be determined by the sequence

$$N_{j+1} = \left[\frac{2}{1 + (N_j/K)}\right]N_j,$$

where K is a constant and $j \geq 1$. (*Source:* Hoppensteadt, F. and C. Peskin, *Mathematics in Medicine and the Life Sciences*, Springer-Verlag, 1992.)

(a) If $N_1 = 230$ and $K = 5000$, make a table of N_j for $j = 1, 2, 3, \ldots, 20$. Round values in the table to the nearest integer.

(b) Graph the sequence N_j for $j = 1, 2, 3, \ldots, 20$. Use the window $[0, 20]$ by $[0, 6000]$.

(c) Describe the growth of these bacteria when there are limited nutrients.

(d) Make a conjecture as to why K is called the *saturation constant*. Test your conjecture by changing the value of K in the given formula.

75. ***Estimating Powers of e*** The series

$$e^a \approx 1 + a + \frac{a^2}{2!} + \frac{a^3}{3!} + \cdots + \frac{a^n}{n!},$$

where $n! = 1 \cdot 2 \cdot 3 \cdot 4 \cdot \cdots \cdot n$, can be used to estimate the value of e^a for any real number a. Use the first eight terms of this series to approximate each expression. Compare this estimate with the actual value. Give values to six decimal places.

(a) e **(b)** e^{-1} **(c)** $\sqrt{e}$

76. ***Estimating Square Roots*** The sequence defined recursively by

$$a_1 = k$$
$$a_n = \frac{1}{2}\left(a_{n-1} + \frac{k}{a_{n-1}}\right), \quad \text{for } n > 1$$

can be used to compute $\sqrt{k}$ for any positive number k. This sequence was known to Sumerian mathematicians 4000 years ago, but it is still used today. Use the sequence to approximate the given square root by finding a_6. Compare your result with the actual value. (*Source:* Heinz-Otto, P., *Chaos and Fractals,* Springer-Verlag, 1993.)

(a) $\sqrt{2}$ **(b)** $\sqrt{8}$ **(c)** $\sqrt{11}$

11.2 Arithmetic Sequences and Series

Arithmetic Sequences ■ Arithmetic Series

Arithmetic Sequences

A sequence in which each term after the first is obtained by adding a fixed number to the previous term is an **arithmetic sequence** (or **arithmetic progression**). The fixed number that is added is the **common difference.** The sequence

$$5, 9, 13, 17, 21, \ldots$$

is an arithmetic sequence because each term after the first is obtained by adding 4 to the previous term. That is,

$$9 = 5 + 4, \quad 13 = 9 + 4, \quad 17 = 13 + 4, \quad 21 = 17 + 4,$$

and so on. The common difference is 4.

If the common difference of an arithmetic sequence is d, then, for every positive integer n in its domain,

$$d = a_{n+1} - a_n. \quad \text{Common difference } d$$

EXAMPLE 1 Finding the Common Difference

Find the common difference d of the arithmetic sequence $-9, -7, -5, -3, -1, \ldots$.

Solution We find d by choosing any two consecutive terms and subtracting the first from the second. Choosing -7 and -5 gives

$$d = -5 - (-7) = 2.$$

Be careful when subtracting a negative number.

Choosing -9 and -7 gives $d = -7 - (-9) = 2$, the same result. ■

EXAMPLE 2 Finding Terms Given a_1 and d

Write the first five terms of each arithmetic sequence.

(a) The first term is 7 and the common difference is -3.

(b) $a_1 = -12, d = 5$

Solution

(a) $a_1 = 7$

$a_2 = 7 + (-3) = 4$ $\quad$ $a_1 = 7,\ d = -3$

$a_3 = 4 + (-3) = 1$

$a_4 = 1 + (-3) = -2$

$a_5 = -2 + (-3) = -5$

(b) $a_1 = -12$ $\quad$ Starting with a_1, add d to each term to get the next term.

$a_2 = -12 + 5 = -7$

$a_3 = -7 + 5 = -2$

$a_4 = -2 + 5 = 3$

$a_5 = 3 + 5 = 8$ ■

If a_1 is the first term of an arithmetic sequence and d is the common difference, then the terms of the sequence are given by

$$a_1 = a_1$$
$$a_2 = a_1 + d$$
$$a_3 = a_2 + d = a_1 + d + d = a_1 + 2d$$
$$a_4 = a_3 + d = a_1 + 2d + d = a_1 + 3d$$
$$a_5 = a_4 + d = a_1 + 3d + d = a_1 + 4d$$
$$a_6 = a_5 + d = a_1 + 4d + d = a_1 + 5d,$$

and, in general, $\quad a_n = a_1 + (n - 1)d.$

nth Term of an Arithmetic Sequence

In an arithmetic sequence with first term a_1 and common difference d, the nth term is given by

$$\mathbf{a_n = a_1 + (n - 1)d.}$$

EXAMPLE 3 Finding Terms of an Arithmetic Sequence

Find a_{13} and a_n for the arithmetic sequence $-3, 1, 5, 9, \ldots$.

Solution Here, $a_1 = -3$ and $d = 1 - (-3) = 4$. First find a_{13}.

$a_n = a_1 + (n - 1)d$ $\quad$ Formula for the nth term

$a_{13} = a_1 + (13 - 1)d$ $\quad$ $n = 13$

$a_{13} = -3 + (12)4$ $\quad$ $a_1 = -3,\ d = 4$ $\quad$ *Work inside the parentheses first.*

$a_{13} = 45$ $\quad$ Simplify.

Find a_n by substituting values for a_1 and d in the formula for a_n.

$a_n = -3 + (n - 1) \cdot 4$ $\quad$ $a_1 = -3,\ d = 4$

$a_n = -3 + 4n - 4$ $\quad$ Distributive property

$a_n = 4n - 7$ $\quad$ Simplify. ■

EXAMPLE 4 **Finding Terms of an Arithmetic Sequence**

Find a_{18} and a_n for the arithmetic sequence having $a_2 = 9$ and $a_3 = 15$.

Solution Find d first; $d = a_3 - a_2 = 15 - 9 = 6$.

Since	$a_2 = a_1 + d,$	
	$9 = a_1 + 6$	$a_2 = 9, d = 6$
	$a_1 = 3.$	Simplify.
Then,	$a_{18} = 3 + (18 - 1)6$	Formula for a_n; $a_1 = 3, n = 18, d = 6$
	$a_{18} = 105,$	Simplify.
and	$a_n = 3 + (n - 1)6$	
	$a_n = 3 + 6n - 6$	Distributive property
	$a_n = 6n - 3.$	Simplify. ■

EXAMPLE 5 **Finding the First Term of an Arithmetic Sequence**

Suppose that an arithmetic sequence has $a_8 = -16$ and $a_{16} = -40$. Find a_1.

Solution Since $a_{16} = a_8 + 8d$,

$$8d = a_{16} - a_8 = -40 - (-16) = -24,$$

so $d = -3$. To find a_1, use the equation $a_8 = a_1 + 7d$.

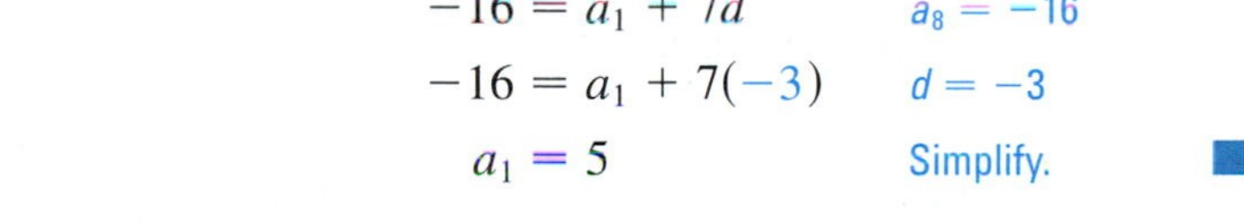

$-16 = a_1 + 7d$	$a_8 = -16$
$-16 = a_1 + 7(-3)$	$d = -3$
$a_1 = 5$	Simplify. ■

The graph of any sequence is a scatter diagram. To determine the characteristics of the graph of an arithmetic sequence, start by rewriting the formula for the nth term.

$a_n = a_1 + (n - 1)d$	Formula for the nth term
$= a_1 + nd - d$	Distributive property
$= dn + (a_1 - d)$	Commutative and associative properties
$= dn + c$	$c = a_1 - d$

The points on the graph of an arithmetic sequence are determined by $f(n) = dn + c$, where n is a natural number. Thus, the points on the graph of f must lie on the *line*

$$y = dx + c.$$

Slope ↑ (d) y-intercept ↑ (c)

For example, the sequence a_n shown in Figure 9(a) is an arithmetic sequence because the points that make up its graph are collinear (lie on a line). The slope determined by these points is 2, so the common difference d equals 2. On the other hand, the sequence b_n shown in Figure 9(b) is not an arithmetic sequence, because the points are *not* collinear.

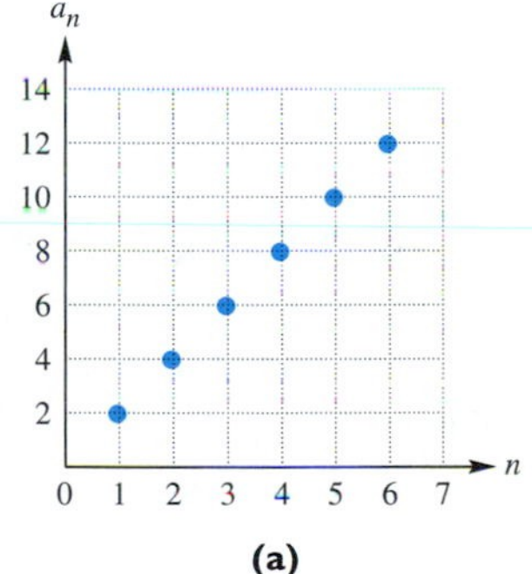

(a)

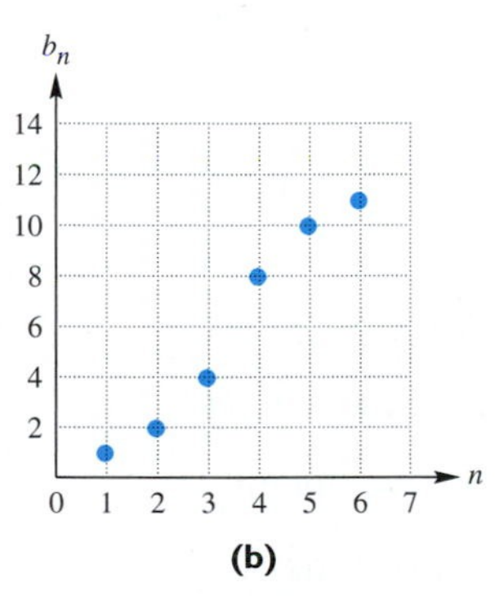

(b)

FIGURE 9

EXAMPLE 6 Finding the *n*th Term from a Graph

Find a formula for the nth term of the sequence a_n shown in Figure 10. What are the domain and range of this sequence?

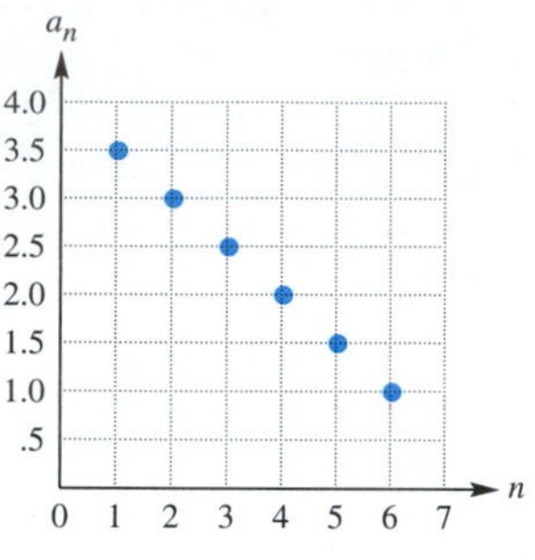

FIGURE 10

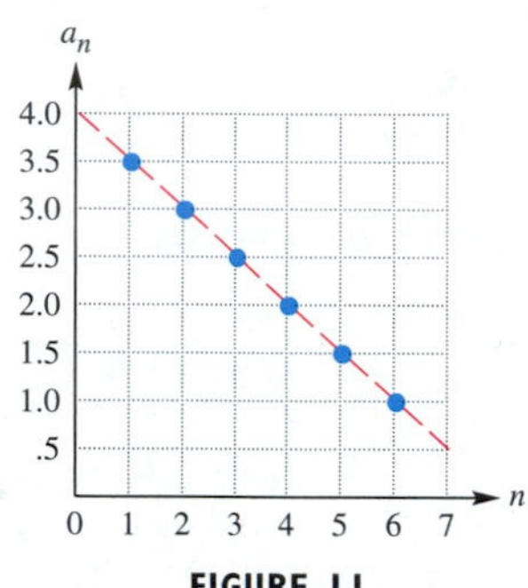

FIGURE 11

Solution The points in Figure 10 lie on a line, so the sequence is arithmetic. The equation of the dashed line shown in Figure 11 is $y = -.5x + 4$, so the nth term of this sequence is determined by

$$a_n = -.5n + 4.$$

The sequence is composed of the points $(1, 3.5)$, $(2, 3)$, $(3, 2.5)$, $(4, 2)$, $(5, 1.5)$, and $(6, 1)$. Therefore, the domain of the sequence is $\{1, 2, 3, 4, 5, 6\}$, and the range is $\{1, 1.5, 2, 2.5, 3, 3.5\}$. ■

Arithmetic Series

The sum of the terms of an arithmetic sequence is an **arithmetic series.** To illustrate, suppose that a person borrows \$3000 and agrees to pay \$100 per month plus interest of 1% per month on the unpaid balance until the loan is paid off. The first month, \$100 is paid to reduce the loan, plus interest of $(.01)3000 = 30$ dollars. The second month, another \$100 is paid toward the loan, and $(.01)2900 = 29$ dollars is paid for interest. Since the loan is reduced by \$100 each month, interest payments decrease by $(.01)100 = 1$ dollar each month, forming the arithmetic sequence

$$30,\ 29,\ 28, \ldots,\ 3,\ 2,\ 1.$$

The total interest paid is given by the sum of the terms of this sequence. Now we develop a formula to find that sum without adding all 30 numbers directly. Since the sequence is arithmetic, we can write the sum of the first n terms as

$$S_n = a_1 + [a_1 + d] + [a_1 + 2d] + \cdots + [a_1 + (n - 1)d].$$

We used the formula for the general term in the last expression. Now we write the same sum in reverse order, beginning with a_n and *subtracting d.*

$$S_n = a_n + [a_n - d] + [a_n - 2d] + \cdots + [a_n - (n - 1)d]$$

Adding respective sides of these two equations term by term, we obtain

$$S_n + S_n = (a_1 + a_n) + (a_1 + a_n) + \cdots + (a_1 + a_n)$$

$$2S_n = n(a_1 + a_n). \quad \text{There are } n \text{ terms of } (a_1 + a_n).$$

Now solve for S_n.

$$2S_n = n(a_1 + a_n)$$

$$S_n = \frac{n}{2}(a_1 + a_n) \quad \text{Divide by 2.}$$

$$S_n = \frac{n}{2}[a_1 + a_1 + (n-1)d] \quad a_n = a_1 + (n-1)d$$

$$S_n = \frac{n}{2}[2a_1 + (n-1)d] \quad \text{Alternative form}$$

FOR DISCUSSION

Explain why there is no formula for the sum of the terms of an infinite arithmetic sequence.

Sum of the First n Terms of an Arithmetic Sequence

If an arithmetic sequence has first term a_1 and common difference d, then the sum of the first n terms is given by

$$S_n = \frac{n}{2}(a_1 + a_n) \quad \text{or} \quad S_n = \frac{n}{2}[2a_1 + (n-1)d].$$

The first formula is used when the first and last terms are known; otherwise, the second formula is used. For example, in the sequence of interest payments given earlier,

$$S_n = \frac{n}{2}(a_1 + a_n) \quad \text{First formula for } S_n$$

gives

$$S_{30} = \frac{30}{2}(30 + 1) = 15(31) = 465, \quad n = 30,\ a_1 = 30,\ a_n = 1$$

so a total of $465 interest will be paid over the 30 months.

EXAMPLE 7 Using the Sum Formulas

(a) Evaluate S_{12} for the arithmetic sequence $-9, -5, -1, 3, 7, \ldots$.

(b) Use a formula for S_n to evaluate the sum of the first 60 positive integers.

Solution

(a) We want the sum of the first 12 terms.

$$S_n = \frac{n}{2}[2a_1 + (n-1)d] \quad \text{Second formula for } S_n$$

$$S_{12} = \frac{12}{2}[2(-9) + 11(4)] = 156 \quad n = 12,\ a_1 = -9,\ d = 4$$

(b)

$$S_n = \frac{n}{2}(a_1 + a_n) \quad \text{First formula for } S_n$$

$$S_{60} = \frac{60}{2}(1 + 60) = 1830 \quad n = 60,\ a_1 = 1,\ a_{60} = 60$$

■

EXAMPLE 8 Using the Sum Formulas

The sum of the first 17 terms of an arithmetic sequence is 187. If $a_{17} = -13$, find a_1 and d.

Solution

$$S_{17} = \frac{17}{2}(a_1 + a_{17}) \qquad \text{Use the first formula for } S_n, \text{ with } n = 17.$$

$$187 = \frac{17}{2}(a_1 - 13) \qquad S_{17} = 187,\ a_{17} = -13$$

$$22 = a_1 - 13 \qquad \text{Multiply by } \tfrac{2}{17}.$$

$$a_1 = 35 \qquad \text{Simplify.}$$

Since $a_{17} = a_1 + (17 - 1)d$,

$$-13 = 35 + 16d \qquad a_{17} = -13,\ a_1 = 35$$

$$-48 = 16d \qquad \text{Subtract 35.}$$

$$d = -3. \qquad \text{Divide by 16; rewrite.}$$

Any sum of the form

$$\sum_{i=1}^{n} (di + p),$$

where d and p are real numbers, represents the sum of the terms of an arithmetic sequence having first term $a_1 = d(1) + p = d + p$ and common difference d. These sums can be evaluated with the formulas in this section.

EXAMPLE 9 Using Summation Notation

Evaluate each sum.

(a) $\sum_{i=1}^{10} (4i + 8)$ **(b)** $\sum_{k=3}^{9} (4 - 3k)$

Analytic Solution

(a) This sum contains the first 10 terms of an arithmetic sequence.

$$a_1 = 4 \cdot 1 + 8 = 12 \qquad \text{First term}$$

$$a_{10} = 4 \cdot 10 + 8 = 48 \qquad \text{Last term}$$

Thus, $$\sum_{i=1}^{10} (4i + 8) = S_{10} = \frac{10}{2}(12 + 48) = 300.$$

(b) The first few terms are

$$[4 - 3(3)] + [4 - 3(4)] + [4 - 3(5)] + \cdots = -5 + (-8) + (-11) + \cdots.$$

Thus $a_1 = -5$ and $d = -3$. If the sequence started with $k = 1$, there would be nine terms. Since it starts at 3, two of those terms are missing, so there are seven terms and $n = 7$. Use the second formula for S_n.

$$\sum_{k=3}^{9} (4 - 3k) = \frac{7}{2}[2(-5) + 6(-3)] = -98$$

Graphing Calculator Solution

The screen in Figure 12 shows the sums for the series in parts (a) and (b). These results agree with the analytic solutions. Remember that a series is the sum of the terms of a sequence.

```
sum(seq(4I+8,I,1
,10,1))
                300
sum(seq(4-3K,K,3
,9,1))
                -98
```

FIGURE 12

11.2 Exercises

Find the common difference d for each arithmetic sequence.

1. $2, 5, 8, 11, \ldots$

2. $4, 10, 16, 22, \ldots$

3. $3, -2, -7, -12, \ldots$

4. $-8, -12, -16, -20, \ldots$

5. $x + 3y, 2x + 5y, 3x + 7y, \ldots$

6. $t^2 + q, -4t^2 + 2q, -9t^2 + 3q, \ldots$

Write the first five terms of each arithmetic sequence.

7. The first term is 8, and the common difference is 6.

8. The first term is -2, and the common difference is 12.

9. $a_1 = 5, d = -2$

10. $a_1 = 4, d = 3$

11. $a_3 = 10, d = -2$

12. $a_1 = 3 - \sqrt{2}, a_2 = 3$

Find a_8 and a_n for each arithmetic sequence.

13. $a_1 = 5, d = 2$

14. $a_1 = -3, d = -4$

15. $a_3 = 2, d = 1$

16. $a_4 = 5, d = -2$

17. $a_1 = 8, a_2 = 6$

18. $a_1 = 6, a_2 = 3$

19. $a_{10} = 6, a_{12} = 15$

20. $a_{15} = 8, a_{17} = 2$

21. $a_1 = x, a_2 = x + 3$

22. $a_2 = y + 1, d = -3$

Find a_1 for each arithmetic sequence.

23. $a_5 = 27, a_{15} = 87$

24. $a_{12} = 60, a_{20} = 84$

25. $S_{16} = -160, a_{16} = -25$

26. $S_{28} = 2926, a_{28} = 199$

Find the sum of the first 10 terms of each arithmetic sequence.

27. $a_1 = 8, d = 3$

28. $a_1 = -9, d = 4$

29. $a_3 = 5, a_4 = 8$

30. $a_2 = 9, a_4 = 13$

31. $5, 9, 13, \ldots$

32. $8, 6, 4, \ldots$

33. $a_1 = 10, a_{10} = 5.5$

34. $a_1 = -8, a_{10} = -1.25$

Find a_1 and d for each arithmetic sequence.

35. $S_{20} = 1090, a_{20} = 102$

36. $S_{31} = 5580, a_{31} = 360$

37. $S_{12} = -108, a_{12} = -19$

38. $S_{25} = 650, a_{25} = 62$

Find a formula for the nth term of the arithmetic sequence shown in each graph. Then state the domain and range of the sequence.

39.

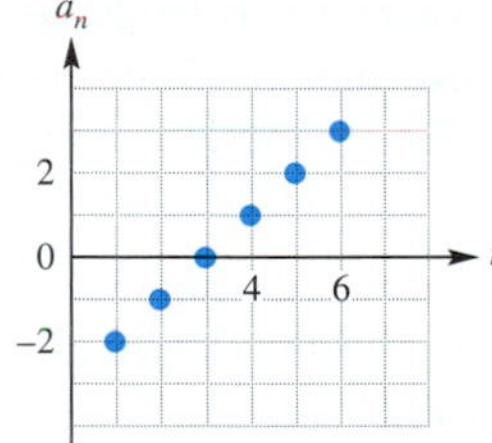

40.

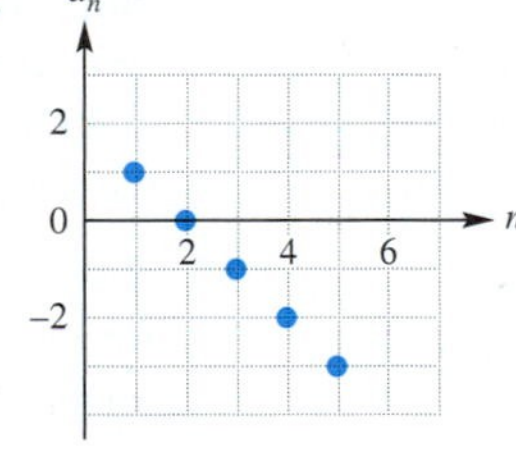

41.

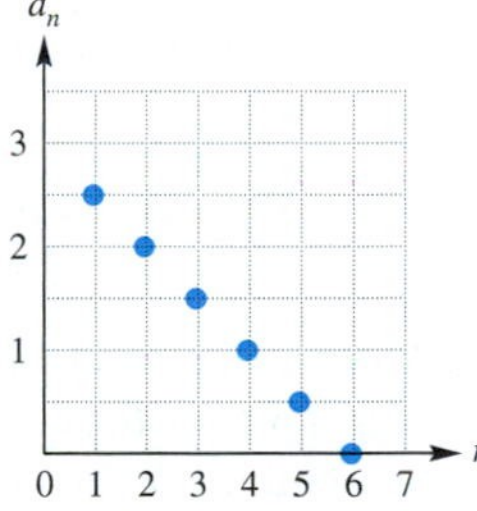

42.

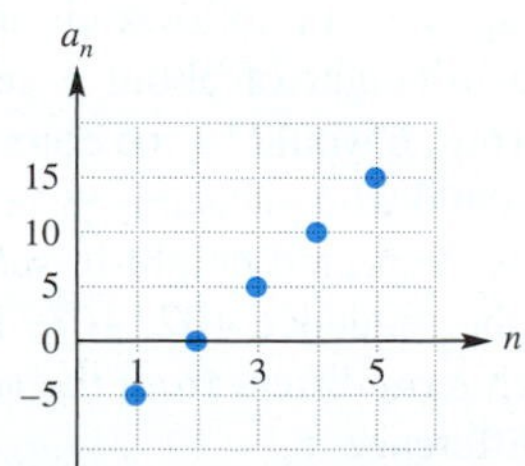

43.

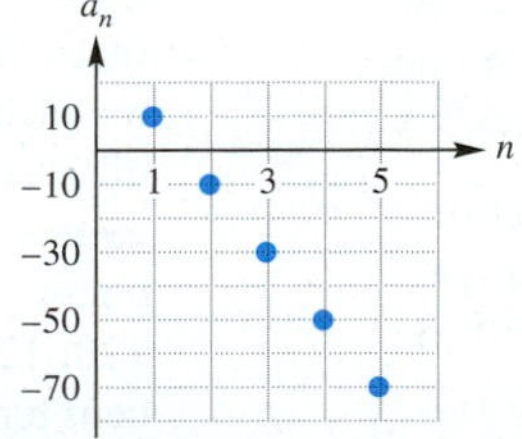

44.

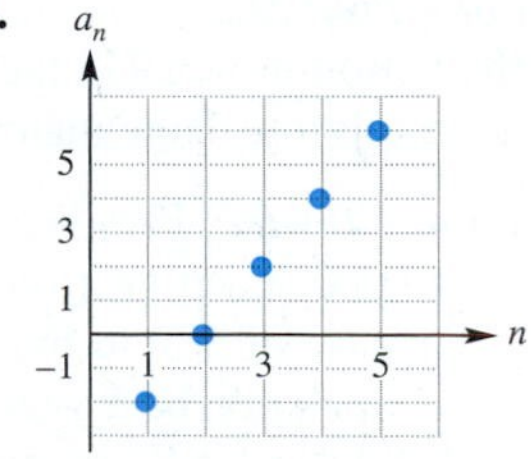

Use a formula to find the sum of each arithmetic series.

45. $3 + 5 + 7 + 9 + 11 + 13 + 15 + 17$

46. $7.5 + 6 + 4.5 + 3 + 1.5 + 0 + (-1.5)$

47. $1 + 2 + 3 + 4 + \cdots + 50$

48. $1 + 3 + 5 + 7 + \cdots + 97$

49. $-7 + (-4) + (-1) + 2 + 5 + \cdots + 98 + 101$

50. $89 + 84 + 79 + 74 + \cdots + 9 + 4$

51. The first 40 terms of the series given by $a_n = 5n$

52. The first 50 terms of the series given by $a_n = 1 - 3n$

Evaluate each sum.

53. $\sum_{i=1}^{3} (i + 4)$

54. $\sum_{i=1}^{5} (i - 8)$

55. $\sum_{j=1}^{10} (2j + 3)$

56. $\sum_{j=1}^{15} (5j - 9)$

57. $\sum_{i=1}^{12} (-5 - 8i)$

58. $\sum_{k=1}^{19} (-3 - 4k)$

59. $\sum_{i=1}^{1000} i$

60. $\sum_{k=1}^{2000} k$

Relating Concepts

For individual or group investigation (Exercises 61–64)

Let $f(x) = mx + b$. ***Work Exercises 61–64 in order.***

61. Find $f(1)$, $f(2)$, and $f(3)$.

62. Consider the sequence $f(1), f(2), f(3), \ldots.$ Is it an arithmetic sequence?

63. If the sequence is arithmetic, what is the common difference?

64. What is a_n for the sequence described in Exercise 62?

Use the sequence feature of a graphing calculator to evaluate the sum of the first 10 terms of the arithmetic sequence. In Exercises 67 and 68, round to the nearest thousandth.

65. $a_n = 4.2n + 9.73$

66. $a_n = 8.42n + 36.18$

67. $a_n = \sqrt{8}n + \sqrt{3}$

68. $a_n = -\sqrt[3]{4}n + \sqrt{7}$

Solve each problem.

69. ***Integer Sum*** Find the sum of all the integers from 51 to 71.

70. ***Integer Sum*** Find the sum of all the integers from -8 to 30.

71. ***Clock Chimes*** If a clock strikes the proper number of chimes each hour on the hour, how many times will it chime in a month of 30 days?

72. ***Telephone Pole Stack*** A stack of telephone poles has 30 in the bottom row, 29 in the next, and so on, with one pole in the top row. How many poles are in the stack?

73. ***Population Growth*** Five years ago, the population of a city was 49,000. Each year, the zoning commission permits an increase of 580 in the population. What will the maximum population be 5 years from now?

74. ***Supports on a Slide*** A slide of uniform slope is to be built on a level piece of land. There are to be 20 equally spaced supports, with the longest support 15 meters long and the shortest 2 meters long. Find the total length of all the supports.

75. ***Rungs of a Ladder*** How much material would be needed for the rungs of a ladder of 31 rungs if the rungs taper uniformly from 18 inches to 28 inches?

76. ***(Modeling) Growth Pattern for Children*** The normal growth pattern for children aged 3–11 follows an arithmetic sequence. An increase in height of about 6 centimeters per year is expected. Thus, 6 would be the common difference of the sequence. A child who measures 96 centimeters at age 3 would have his expected height in subsequent years represented by the sequence 102, 108, 114, 120, 126, 132, 138, 144. Each term differs from the adjacent terms by the common difference, 6.

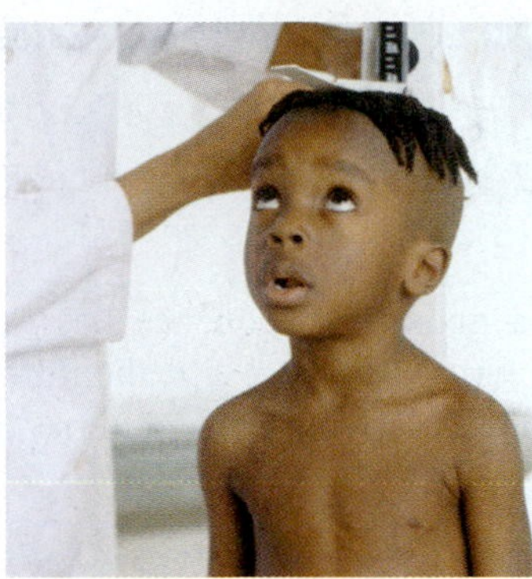

(a) If a child measures 98.2 centimeters at age 3 and 109.8 centimeters at age 5, what would be the common difference of the arithmetic sequence describing his yearly height?

(b) What would we expect his height to be at age 8?

77. *Concept Check* Find all arithmetic sequences $a_1, a_2, a_3, \ldots$ such that $a_1^2, a_2^2, a_3^2, \ldots$ is also an arithmetic sequence.

78. Suppose that $a_1, a_2, a_3, \ldots$ and $b_1, b_2, b_3, \ldots$ are arithmetic sequences. Let $d_n = a_n + c \cdot b_n$, for any real number c and every positive integer n. Show that $d_1, d_2, d_3, \ldots$ is an arithmetic sequence.

79. *Concept Check* Suppose that $a_1, a_2, a_3, a_4, a_5, \ldots$ is an arithmetic sequence. Is $a_1, a_3, a_5, \ldots$ an arithmetic sequence?

80. Explain why the sequence $\log 2, \log 4, \log 8, \log 16, \ldots$ is arithmetic.

11.3 Geometric Sequences and Series

Geometric Sequences ■ Geometric Series ■ Infinite Geometric Series ■ Annuities

Geometric Sequences

Suppose you agreed to work for 1¢ the first day, 2¢ the second day, 4¢ the third day, 8¢ the fourth day, and so on, with your wages doubling each day. How much will you earn on day 20? How much will you have earned altogether in 20 days? These questions will be answered in this section.

A **geometric sequence** (or **geometric progression**) is a sequence in which each term after the first is obtained by multiplying the preceding term by a constant nonzero real number, called the **common ratio.** The sequence of wages

$$1,\ 2,\ 4,\ 8,\ 16, \ldots$$

is an example of a geometric sequence in which the first term is 1 and the common ratio is 2. Notice that if we divide any term (except the first) by the preceding term, we obtain the common ratio $r = 2$.

$$\frac{a_2}{a_1} = \frac{2}{1} = 2; \qquad \frac{a_3}{a_2} = \frac{4}{2} = 2; \qquad \frac{a_4}{a_3} = \frac{8}{4} = 2; \qquad \frac{a_5}{a_4} = \frac{16}{8} = 2$$

If the common ratio of a geometric sequence is r, then, by the definition of a geometric sequence,

$$r = \frac{a_{n+1}}{a_n} \qquad \text{Common ratio } r$$

for every positive integer n. ***Therefore, the common ratio can be found by choosing any term except the first and dividing it by the preceding term.***

In the geometric sequence $2, 8, 32, 128, \ldots,$ we have $r = 4$. Notice that

$$8 = 2 \cdot 4$$
$$32 = 8 \cdot 4 = (2 \cdot 4) \cdot 4 = 2 \cdot 4^2$$
$$128 = 32 \cdot 4 = (2 \cdot 4^2) \cdot 4 = 2 \cdot 4^3.$$

To generalize this pattern, assume that a geometric sequence has first term a_1 and common ratio r. The second term can be written as $a_2 = a_1 r$, the third can be written as $a_3 = a_2 r = (a_1 r)r = a_1 r^2$, and so on. Following this pattern, the nth term is $a_n = a_1 r^{n-1}$.

nth Term of a Geometric Sequence

In the geometric sequence with first term a_1 and common ratio r, neither of which is zero, the nth term is given by

$$a_n = a_1 r^{n-1}.$$

EXAMPLE 1 Finding the *n*th Term of a Geometric Sequence of Wages

The formula for the nth term of a geometric sequence can be used to answer the first question posed at the beginning of this section. How much will be earned on day 20 if daily wages follow the sequence $1, 2, 4, 8, 16, \ldots$?

Solution To answer the question, let $a_1 = 1$ and $r = 2$, and find a_{20}.

$$a_{20} = a_1 r^{19} = 1(2)^{19} = 524{,}288 \text{ cents, or } \$5242.88$$ ■

EXAMPLE 2 Using the Formula for the *n*th Term

Find a_5 and a_n for the geometric sequence $4, -12, 36, -108, \ldots$.

Solution The first term, a_1, is 4. Find r by choosing any term except the first and dividing it by the preceding term. For example, $r = \frac{36}{-12} = -3$. Since $a_4 = -108$,

$$a_5 = -3(-108) = 324.$$

The fifth term also could be found by using the formula $a_n = a_1 r^{n-1}$ and replacing n with 5, r with -3, and a_1 with 4.

$$a_5 = 4 \cdot (-3)^{5-1} = 4 \cdot (-3)^4 = 324$$

By the formula, $a_n = 4 \cdot (-3)^{n-1}$. ■

EXAMPLE 3 Using the Formula for the *n*th Term

Find a_1 and r for the geometric sequence with third term 20 and sixth term 160.

Solution Use the formula for the nth term of a geometric sequence.

For $n = 3$, $a_3 = a_1 r^2 = 20$;

for $n = 6$, $a_6 = a_1 r^5 = 160$.

Because $a_1r^2 = 20$, $a_1 = \frac{20}{r^2}$. Substitute this value for a_1 in the second equation.

$$a_1r^5 = 160$$
$$\left(\frac{20}{r^2}\right)r^5 = 160$$
$$20r^3 = 160 \qquad \frac{r^5}{r^2} = r^{5-2} = r^3$$
$$r^3 = 8 \qquad \text{Divide by 20.}$$
$$r = 2 \qquad \text{Take cube roots.}$$

Since $a_1r^2 = 20$ and $r = 2$,

$$a_1(2)^2 = 20 \qquad \text{Substitute.}$$
$$4a_1 = 20$$
$$a_1 = 5. \qquad \text{Divide by 4.}$$

■

EXAMPLE 4 **Modeling a Population of Fruit Flies**

A population of fruit flies is growing in such a way that each generation is 1.5 times as large as the last generation. Suppose there were 100 insects in the first generation. How many would there be in the fourth generation?

Solution The population can be written as a geometric sequence with a_1 as the first-generation population, a_2 the second-generation population, and so on. Then the fourth-generation population is a_4. Use the formula for a_n.

$$a_4 = a_1r^3 = 100(1.5)^3 = 100(3.375) = 337.5 \qquad n = 4,\ a_1 = 100,\ r = 1.5$$

In the fourth generation, the population numbers about 338 insects. ■

Geometric Series

A **geometric series** is the sum of the terms of a geometric sequence. In applications, it may be necessary to find the sum of the terms of such a sequence. For example, a scientist might want to know the total number of insects in four generations of the population discussed in Example 4. This population would equal $a_1 + a_2 + a_3 + a_4$, or

$$100 + 100(1.5) + 100(1.5)^2 + 100(1.5)^3 = 812.5 \approx 813 \text{ insects.}$$

To find a formula for the sum S_n of the first n terms of a geometric sequence, first write the sum as

$$S_n = a_1 + a_2 + a_3 + \cdots + a_n$$

or

$$S_n = a_1 + a_1r + a_1r^2 + \cdots + a_1r^{n-1}. \qquad (1)$$

If $r = 1$, $S_n = na_1$, which is a correct formula for this case. If $r \neq 1$, multiply each side of equation (1) by r to obtain

$$rS_n = a_1r + a_1r^2 + a_1r^3 + \cdots + a_1r^n. \qquad (2)$$

If equation (2) is subtracted from equation (1),

$$S_n = a_1 + a_1 r + a_1 r^2 + \cdots + a_1 r^{n-1} \qquad (1)$$

$$rS_n = \qquad a_1 r + a_1 r^2 + \cdots + a_1 r^{n-1} + a_1 r^n \qquad (2)$$

$$S_n - rS_n = a_1 \qquad - a_1 r^n \qquad \text{Subtract.}$$

$$S_n(1 - r) = a_1(1 - r^n) \qquad \text{Factor.}$$

$$S_n = \frac{a_1(1 - r^n)}{1 - r}, \quad \text{where } r \neq 1. \qquad \text{Divide by } 1 - r.$$

Sum of the First n Terms of a Geometric Sequence

If a geometric sequence has first term a_1 and common ratio r, then the sum of the first n terms is given by

$$S_n = \frac{a_1(1 - r^n)}{1 - r}, \quad \text{where } r \neq 1.$$

We can use this formula to find the total fruit fly population in Example 4 over the four-generation period. With $n = 4$, $a_1 = 100$, and $r = 1.5$,

$$S_4 = \frac{100(1 - 1.5^4)}{1 - 1.5} = \frac{100(1 - 5.0625)}{-.5} = 812.5 \approx 813 \text{ insects.} \qquad \text{Same result}$$

EXAMPLE 5 Applying the Sum of the First n Terms

At the beginning of this section, we posed the following question: How much will you have earned altogether after 20 days?

Analytic Solution

We must find the total amount earned in 20 days with daily wages of 1, 2, 4, 8, ... cents. Since $a_1 = 1$ and $r = 2$,

$$S_{20} = \frac{1(1 - 2^{20})}{1 - 2} = 1{,}048{,}575 \text{ cents, or } \$10{,}485.75.$$

Not bad for 20 days of work!

Graphing Calculator Solution

See Figure 13, which confirms the analytic solution.

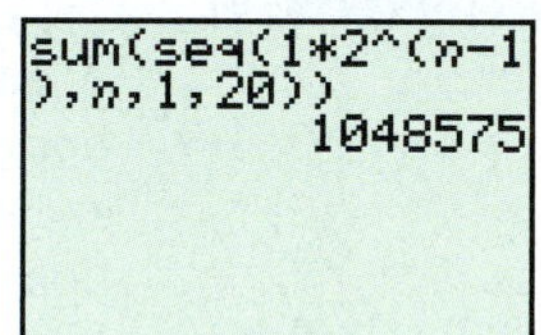

FIGURE 13

EXAMPLE 6 Finding the Sum of the First n Terms

Find $\sum_{i=1}^{6} 2 \cdot 3^i$.

Solution This series is the sum of the first six terms of a geometric sequence having $a_1 = 2 \cdot 3^1 = 6$ and $r = 3$. From the formula for S_n,

$$S_6 = \frac{6(1 - 3^6)}{1 - 3} = \frac{6(1 - 729)}{-2} = \frac{6(-728)}{-2} = 2184.$$

Infinite Geometric Series

We can extend our discussion of sums of sequences to include infinite geometric sequences such as

$$2,\ 1,\ \frac{1}{2},\ \frac{1}{4},\ \frac{1}{8},\ \frac{1}{16},\ldots,$$

with first term 2 and common ratio $\frac{1}{2}$. Using the formula for S_n gives the sequence of sums

$$S_1 = 2,\quad S_2 = 3,\quad S_3 = 3.5,\quad S_4 = 3.75,\quad S_5 = 3.875,\quad S_6 = 3.9375.$$

With a calculator in function mode, we define Y_1 as S_n.

$$Y_1 = \frac{2\left(1 - \left(\frac{1}{2}\right)^X\right)}{1 - \frac{1}{2}}$$

X	Y1
1	2
2	3
3	3.5
4	3.75
5	3.875
6	3.9375
7	3.9688

Y1=(2(1-(1/2)^X...

FIGURE 14

Using the table in Figure 14, we can observe the first few terms as X takes on the values $1, 2, 3, \ldots$. These terms seem to be getting closer and closer to the number 4. For no value of n is $S_n = 4$. However, if n is large enough, then S_n is as close to 4 as desired. As mentioned earlier, we say that the sequence converges to 4, expressed as

$$\lim_{n\to\infty} S_n = 4.$$

(Read: "The limit of S_n as n increases without bound is 4.") Since $\lim_{n\to\infty} S_n = 4$, the number 4 is called the *sum of the terms* of the infinite geometric sequence

$$2, 1, \frac{1}{2}, \frac{1}{4}, \ldots, \qquad \text{and} \qquad 2 + 1 + \frac{1}{2} + \frac{1}{4} + \cdots = 4.$$

Looking Ahead to Calculus

In calculus, different types of infinite series are studied. One important question to answer for each series is whether $\lim_{n\to\infty} S_n$ converges to a real number. In the discussion of

$$\lim_{n\to\infty} S_n = 4,$$

we used the phrases "large enough" and "as close as desired." This description is made more precise in a standard calculus course.

What Went Wrong?

The preceding discussion justified the statement that the sum of the terms $2, 1, \frac{1}{2}, \frac{1}{4}, \frac{1}{8}, \frac{1}{16}, \ldots$ gets closer and closer to 4 by adding more and more terms of the sequence. However, this sum will always differ from 4 by some small amount. The following figure is an extension of the table in Figure 14. According to the calculator, the sum *is* 4 when $X \geq 17$.

X	Y1
14	3.9998
15	3.9999
16	3.9999
17	4
18	4
19	4
20	4

Y1=(2(1-(1/2)^X...

What Went Wrong? Discuss the limitations of technology, basing your comments on this example.

Answers to What Went Wrong?

The calculator has room to display numbers to only four decimal places in the table. When the value of Y_1 reaches 3.99995 or greater, the calculator rounds up to 4. Try positioning the cursor on the 4 next to the 17 in the table. At the bottom of the screen a more accurate value of 3.99996948242 will appear for Y_1.

EXAMPLE 7 **Summing the Terms of an Infinite Geometric Sequence**

Find $1 + \frac{1}{3} + \frac{1}{9} + \frac{1}{27} + \cdots$.

Solution Use the formula for the sum of the first n terms of a geometric sequence to obtain

$$S_1 = 1, \quad S_2 = \frac{4}{3}, \quad S_3 = \frac{13}{9}, \quad S_4 = \frac{40}{27},$$

and, in general, $$S_n = \frac{1\left[1 - \left(\frac{1}{3}\right)^n\right]}{1 - \frac{1}{3}}. \qquad a_1 = 1, r = \frac{1}{3}$$

The following table shows the value of $\left(\frac{1}{3}\right)^n$ for larger and larger values of n.

n	1	10	100	200
$\left(\frac{1}{3}\right)^n$	$\frac{1}{3}$	1.69×10^{-5}	1.94×10^{-48}	3.76×10^{-96}

As n gets larger and larger, $\left(\frac{1}{3}\right)^n$ gets closer and closer to 0. That is,

$$\lim_{n \to \infty} \left(\frac{1}{3}\right)^n = 0,$$

making it reasonable that

$$\lim_{n \to \infty} S_n = \lim_{n \to \infty} \frac{1\left[1 - \left(\frac{1}{3}\right)^n\right]}{1 - \frac{1}{3}} = \frac{1(1 - 0)}{1 - \frac{1}{3}} = \frac{1}{\frac{2}{3}} = \frac{3}{2}.$$

Be careful simplifying a complex fraction.

Hence, $$1 + \frac{1}{3} + \frac{1}{9} + \frac{1}{27} + \cdots = \frac{3}{2}.$$ ■

2

0 6

0

This graph of the first six values of S_n in Example 7 shows its value approaching $\frac{3}{2}$. (The y-scale here is $\frac{1}{2}$.)

If a geometric sequence has first term a_1 and common ratio r, then

$$S_n = \frac{a_1(1 - r^n)}{1 - r} \qquad (r \neq 1)$$

for every positive integer n. If $-1 < r < 1$, then $\lim_{n \to \infty} r^n = 0$, and

$$\lim_{n \to \infty} S_n = \frac{a_1(1 - 0)}{1 - r} = \frac{a_1}{1 - r}.$$

This quotient, $\frac{a_1}{1 - r}$, is called the **sum of the terms of an infinite geometric sequence.** The limit $\lim_{n \to \infty} S_n$ is often expressed as S_∞ or $\sum_{i=1}^{\infty} a_i$.

Looking Ahead to Calculus

In calculus, you will find sums of the terms of infinite sequences that are not geometric.

Sum of the Terms of an Infinite Geometric Sequence

The sum of the terms of an infinite geometric sequence with first term a_1 and common ratio r, where $-1 < r < 1$, is given by

$$S_\infty = \frac{a_1}{1 - r}.$$

If $|r| > 1$, the terms get larger and larger in absolute value, so there is no limit as $n \to \infty$. Hence, the terms of the sequence will not have a sum.

EXAMPLE 8 Finding Sums of the Terms of Infinite Geometric Sequences

Find each sum.

(a) $\sum_{i=1}^{\infty} \left(-\frac{3}{4}\right)\left(-\frac{1}{2}\right)^{i-1}$ **(b)** $\sum_{i=1}^{\infty} \left(\frac{3}{5}\right)^{i}$

Solution

(a) Here, $a_1 = -\frac{3}{4}$ and $r = -\frac{1}{2}$. Since $-1 < r < 1$, the preceding formula applies.

$$S_\infty = \frac{a_1}{1 - r} = \frac{-\frac{3}{4}}{1 - \left(-\frac{1}{2}\right)} = \frac{-\frac{3}{4}}{\frac{3}{2}} = -\frac{3}{4} \div \frac{3}{2} = -\frac{3}{4} \cdot \frac{2}{3} = -\frac{1}{2}$$

Be careful simplifying a complex fraction.

(b) $\sum_{i=1}^{\infty} \left(\frac{3}{5}\right)^{i} = \frac{\frac{3}{5}}{1 - \frac{3}{5}} = \frac{\frac{3}{5}}{\frac{2}{5}} = \frac{3}{5} \div \frac{2}{5} = \frac{3}{5} \cdot \frac{5}{2} = \frac{3}{2}$ $\quad a_1 = \frac{3}{5}, r = \frac{3}{5}$ ■

Annuities

A sequence of equal payments made at equal intervals, such as car payments or house payments, is called an **annuity.** If the payments are accumulated in an account (with no withdrawals), the sum of the payments and interest on the payments is called the **future value** of the annuity. The formula that follows is derived from the formula for the sum of the terms of a geometric sequence.

Future Value of an Annuity

The formula for the future value of an annuity is

$$S = R\left[\frac{(1 + i)^n - 1}{i}\right],$$

where S is the future value, R is the payment at the end of each period, i is the interest rate in decimal form per period, and n is the number of periods.

EXAMPLE 9 Finding the Future Value of an Annuity

To save money for a trip, Paula Story deposited \$1000 at the *end* of each year for 4 years in an account paying 6% interest compounded annually. Find the future value of this annuity.

Solution Use the formula for S, with $R = 1000$, $i = .06$, and $n = 4$.

$$S = 1000\left[\frac{(1 + .06)^4 - 1}{.06}\right] \approx 4374.62$$

The future value of the annuity is \$4374.62. ■

11.3 Exercises

Write the terms of the geometric sequence that satisfies the given conditions.

1. $a_1 = \frac{5}{3}, r = 3, n = 4$ **2.** $a_1 = -\frac{3}{4}, r = \frac{2}{3}, n = 4$ **3.** $a_4 = 5, a_5 = 10, n = 5$ **4.** $a_3 = 16, a_4 = 8, n = 5$

Find a_5 and a_n for each geometric sequence.

5. $a_1 = 5, r = -2$ **6.** $a_1 = 8, r = -5$ **7.** $a_2 = -4, r = 3$

8. $a_3 = -2, r = 4$ **9.** $a_4 = 243, r = -3$ **10.** $a_4 = 18, r = 2$

11. $-4, -12, -36, -108, \ldots$ **12.** $-2, 6, -18, 54, \ldots$ **13.** $\frac{4}{5}, 2, 5, \frac{25}{2}, \ldots$

14. $\frac{1}{2}, \frac{2}{3}, \frac{8}{9}, \frac{32}{27}, \ldots$ **15.** $10, -5, \frac{5}{2}, -\frac{5}{4}, \ldots$ **16.** $3, -\frac{9}{4}, \frac{27}{16}, -\frac{81}{64}, \ldots$

Find a_1 and r for each geometric sequence.

17. $a_3 = 5, a_8 = \frac{1}{625}$ **18.** $a_2 = -6, a_7 = -192$ **19.** $a_4 = -\frac{1}{4}, a_9 = -\frac{1}{128}$ **20.** $a_3 = 50, a_7 = .005$

Use the formula for S_n to find the sum of the first five terms for each geometric sequence. Round the answers for Exercises 25 and 26 to the nearest hundredth.

21. $2, 8, 32, 128, \ldots$ **22.** $4, 16, 64, 256, \ldots$ **23.** $18, -9, \frac{9}{2}, -\frac{9}{4}, \ldots$

24. $12, -4, \frac{4}{3}, -\frac{4}{9}, \ldots$ **25.** $a_1 = 8.423, r = 2.859$ **26.** $a_1 = -3.772, r = -1.553$

Find each sum.

27. $\sum_{i=1}^{5} 3^i$ **28.** $\sum_{i=1}^{4} (-2)^i$ **29.** $\sum_{j=1}^{6} 48\left(\frac{1}{2}\right)^j$

30. $\sum_{j=1}^{5} 243\left(\frac{2}{3}\right)^j$ **31.** $\sum_{k=4}^{10} 2^k$ **32.** $\sum_{k=3}^{9} (-3)^k$

33. *Concept Check* Under what conditions does the sum of the terms of an infinite geometric sequence exist?

34. *Concept Check* The number .999... can be written as the sum of the terms of an infinite geometric sequence: $.9 + .09 + .009 + \cdots$. Here we have $a_1 = .9$ and $r = .1$. Use the formula for S_∞ to find this sum. Does your intuition indicate that your answer is correct?

Write the sum of each geometric series as a rational number. (See Exercise 34.)

35. $.8 + .08 + .008 + .0008 + \cdots$

36. $.7 + .07 + .007 + .0007 + \cdots$

37. $.45 + .0045 + .000045 + \cdots$

38. $.36 + .0036 + .000036 + \cdots$

Find r for each infinite geometric sequence. Identify any whose sum does not converge.

39. $12, 24, 48, 96, \ldots$

40. $625, 125, 25, 5, \ldots$

41. $-48, -24, -12, -6, \ldots$

42. $2, -10, 50, -250, \ldots$

Find each sum that converges.

43. $16 + 2 + \frac{1}{4} + \frac{1}{32} + \cdots$

44. $18 + 6 + 2 + \frac{2}{3} + \cdots$

45. $100 + 10 + 1 + \cdots$

46. $128 + 64 + 32 + \cdots$

47. $\frac{4}{3} + \frac{2}{3} + \frac{1}{3} + \cdots$

48. $\frac{1}{4} - \frac{1}{6} + \frac{1}{9} - \frac{2}{27} + \cdots$

49. $\sum_{i=1}^{\infty} 3\left(\frac{1}{4}\right)^{i-1}$

50. $\sum_{i=1}^{\infty} 5\left(-\frac{1}{4}\right)^{i-1}$

51. $\sum_{k=1}^{\infty} (.3)^k$

52. $\sum_{k=1}^{\infty} 10^{-k}$

Relating Concepts

For individual or group investigation (Exercises 53–56)

Let $g(x) = ab^x$. ***Work Exercises 53–56 in order.***

53. Find $g(1)$, $g(2)$, and $g(3)$.

54. Consider the sequence $g(1), g(2), g(3), \ldots$. Is it a geometric sequence? If so, what is the common ratio?

55. What is the general term of the sequence in Exercise 54?

56. Explain how geometric sequences are related to exponential functions.

Use a graphing calculator to evaluate each sum. Round to the nearest thousandth.

57. $\sum_{i=1}^{10} (1.4)^i$

58. $\sum_{j=1}^{6} -(3.6)^j$

59. $\sum_{j=3}^{8} 2(.4)^j$

60. $\sum_{i=4}^{9} 3(.25)^i$

Annuity Values *Find the future value of each annuity.*

61. Payments of \$1000 at the end of each year for 9 years at 4% interest compounded annually

62. Payments of \$800 at the end of each year for 12 years at 3% interest compounded annually

63. Payments of \$2430 at the end of each year for 10 years at 2.5% interest compounded annually

64. Payments of \$1500 at the end of each year for 6 years at 1.5% interest compounded annually

Solve each problem.

65. *(Modeling) Investment for Retirement* According to T. Rowe Price Associates, a person who has a moderate investment strategy and n years to retirement should have accumulated savings of a_n percent of his or her annual salary. The geometric sequence defined by

$$a_n = 1276(.916)^n$$

gives the appropriate percent for each year n.

(a) Find a_1 and r.

(b) Find and interpret the terms a_{10} and a_{20}.

66. ***(Modeling) Investment for Retirement*** Refer to Exercise 65. For someone who has a conservative investment strategy with n years to retirement, the geometric sequence is

$$a_n = 1278(.935)^n.$$

(*Source:* T. Rowe Price Associates.)

(a) Repeat part (a) of Exercise 65.

(b) Repeat part (b) of Exercise 65.

(c) Why are the answers in parts (a) and (b) larger than in Exercise 65?

67. ***(Modeling) Bacteria Growth*** The strain of bacteria described in Exercise 73 in the first section of this chapter will double in size and then divide every 40 minutes. Let a_1 be the initial number of bacteria cells, a_2 the number after 40 minutes, and a_n the number after $40(n - 1)$ minutes. (*Source:* Hoppensteadt, F. and C. Peskin, *Mathematics in Medicine and the Life Sciences*, Springer-Verlag, 1992.)

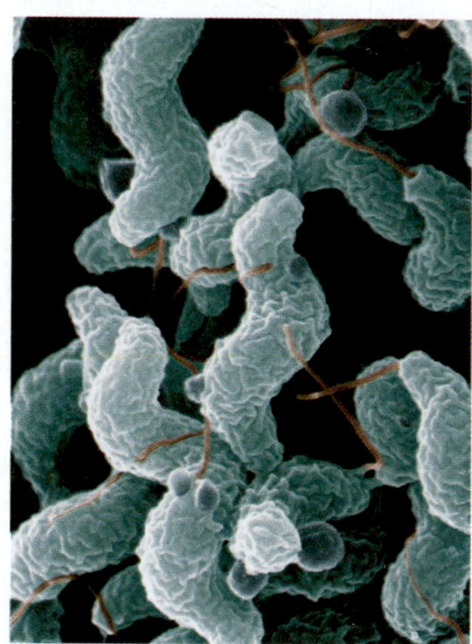

(a) Write a formula for the nth term a_n of the geometric sequence $a_1, a_2, a_3, \ldots, a_n, \ldots$.

(b) Determine the first value for n where $a_n > 1{,}000{,}000$ if $a_1 = 100$.

(c) How long does it take for the number of bacteria to exceed 1 million?

68. ***Photo Processing*** The final step in processing a black-and-white photographic print is to immerse the print in a chemical fixer. The print is then washed in running water. Under certain conditions, 98% of the fixer will be removed with 15 minutes of washing. How much of the original fixer would be left after 1 hour of washing?

69. ***Chemical Mixture*** A scientist has a vat containing 100 liters of a pure chemical. Twenty liters are drained and replaced with water. After complete mixing, 20 liters of the mixture are again drained and replaced with water. What will be the strength of the mixture after nine such drainings?

70. ***Half-Life of a Radioactive Substance*** The half-life of a radioactive substance is the time it takes for half the substance to decay. Suppose the half-life of a substance is 3 years and 10^{15} molecules of the substance are present initially. How many molecules will be present after 15 years?

71. ***Depreciation in Value*** Each year a machine loses 20% of the value it had at the beginning of the year. Find the value of the machine at the end of 6 years if it cost \$100,000 new.

72. ***Sugar Processing*** A sugar factory receives an order for 1000 units of sugar. The production manager thus orders the production of 1000 units of sugar. He forgets, however, that the production of sugar requires some sugar (to prime the machines, for example), so he ends up with only 900 units of sugar. He then orders an additional 100 units, and receives only 90 units. A further order for 10 units produces 9 units. Finally seeing that he is wrong, the manager decides to try mathematics. He views the production process as an infinite geometric progression with $a_1 = 1000$ and $r = .1$. Find the number of units of sugar that he should have ordered originally.

73. ***Swing of a Pendulum*** A pendulum bob swings through an arc 40 centimeters long on its first swing. Each swing thereafter, it swings only 80% as far as on the previous swing. How far will it swing altogether before coming to a complete stop?

74. ***Height of a Dropped Ball*** Penelope drops a ball from a height of 10 meters and notices that on each bounce the ball returns to about $\frac{3}{4}$ of its previous height. About how far will the ball travel before it comes to rest? (*Hint:* Consider the sum of two sequences.)

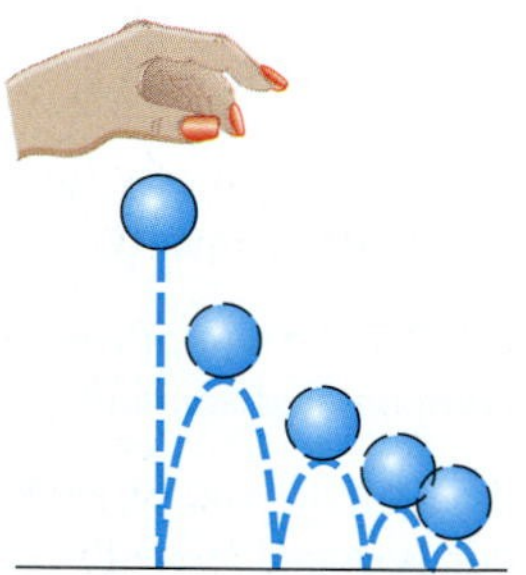

75. ***Number of Ancestors*** Each person has two parents, four grandparents, eight great-grandparents, and so on. What is the total number of ancestors a person has, going back five generations? ten generations?

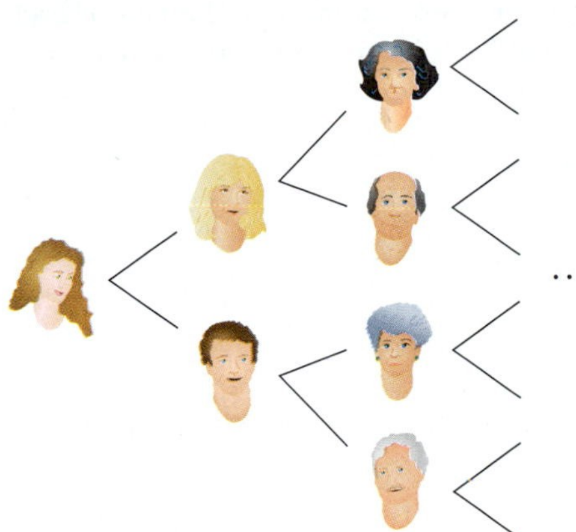

76. ***(Modeling) Drug Dosage*** Certain medical conditions are treated with a fixed dose of a drug administered at regular intervals. Suppose that a person is given 2 milligrams of a drug each day and that during each 24-hour period the body utilizes 40% of the amount of drug that was present at the beginning of the period.

(a) Show that the amount of the drug present in the body at the end of n days is $\sum_{i=1}^{n} 2(.6)^i$.

(b) What will be the approximate quantity of the drug in the body at the end of each day after the treatment has been administered over a long period?

77. ***Salaries*** You are offered a 6-week summer job and are asked to select one of the following salary options:

Option 1: \$5000 for the first day, with a \$10,000 raise each day for the remaining 29 days (that is, \$15,000 for day 2, \$25,000 for day 3, and so on)

Option 2: 1¢ for the first day, with the pay doubled each day thereafter (that is, 2¢ for day 2, 4¢ for day 3, and so on)

Which option would you choose?

78. ***Number of Ancestors*** Suppose a genealogical Web site allows you to identify all your ancestors that lived during the last 300 years. Assuming that each generation spans about 25 years, guess the number of ancestors that would be found during the 12 generations. Then use the formula for a geometric series to find the actual value.

79. ***Side Length of a Triangle*** A sequence of equilateral triangles is constructed. The first triangle has sides 2 meters in length. To get the second triangle, midpoints of the sides of the original triangle are connected. What is the length of the side of the eighth such triangle? See the figure.

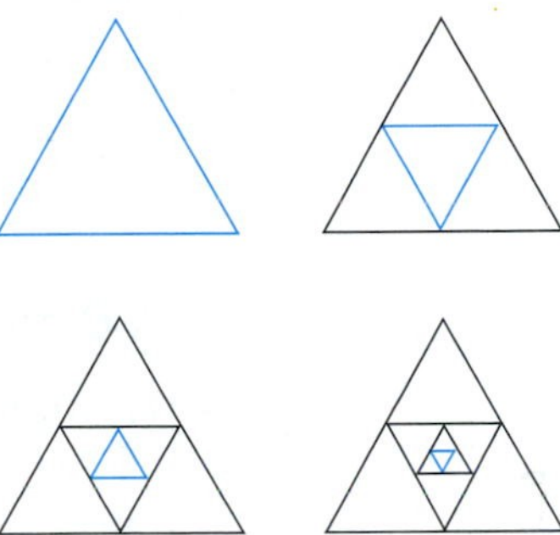

80. ***Perimeter and Area of Triangles***

(a) In Exercise 79, if the process could be continued indefinitely, what would be the total perimeter of all the triangles?

(b) What would be the total area of all the triangles, disregarding the overlapping?

81. ***Concept Check*** Let $a_1, a_2, a_3, \ldots$ and $b_1, b_2, b_3, \ldots$ be geometric sequences. Let $d_n = c \cdot a_n \cdot b_n$ for any real number c and every positive integer n. Show that $d_1, d_2, d_3, \ldots$ is a geometric sequence.

82. Explain why the sequence

$$\log 6, \log 36, \log 1296, \log 1{,}679{,}616, \ldots$$

is geometric.

Reviewing Basic Concepts (Sections 11.1–11.3)

1. Find the first five terms of the sequence $a_n = (-1)^{n-1}(4n)$.

2. Evaluate the series $\sum_{i=1}^{5} (3i + 1)$.

3. Decide whether the sequence $a_n = 1 - \frac{2}{n}$ converges or diverges. If it converges, determine the number to which it converges.

4. Write the first five terms of the arithmetic sequence with $a_1 = 8$ and $d = -2$.

5. Find a_1 for the arithmetic sequence with $a_5 = 5$ and $a_8 = 17$.

6. Find the sum of the first 10 terms of the arithmetic sequence with $a_1 = 2$ and $d = 5$.

7. Find a_3 and a_n for the geometric sequence with $a_1 = -2$ and $r = -3$.

8. Is the series $5 + 3 + \frac{9}{5} + \cdots + \frac{243}{625}$ arithmetic or geometric? Find the sum of the terms of this series.

9. Find the sum of the infinite geometric series $\sum_{k=1}^{\infty} 3\left(\frac{2}{3}\right)^k$.

10. ***Annuity Value*** Find the future value of an annuity if payments of \$500 are made at the end of each year for 13 years at 3.5% interest compounded annually.

11.4 The Binomial Theorem

A Binomial Expansion Pattern ■ Pascal's Triangle ■ n-Factorial ■ Binomial Coefficients ■ The Binomial Theorem ■ rth Term of a Binomial Expansion

Looking Ahead to Calculus

Students taking calculus study the binomial series, which follows from Isaac Newton's extension to the case where the exponent is no longer a positive integer. His result led to a series for $(1 + x)^k$, where k is a real number and $|x| < 1$.

A Binomial Expansion Pattern

The formula for writing the terms of $(x + y)^n$, where n is a natural number, is called the *binomial theorem.* Some expansions of $(x + y)^n$, for various nonnegative integer values of n, are as follows:

$$(x + y)^0 = 1$$
$$(x + y)^1 = x + y$$
$$(x + y)^2 = x^2 + 2xy + y^2$$
$$(x + y)^3 = x^3 + 3x^2y + 3xy^2 + y^3$$
$$(x + y)^4 = x^4 + 4x^3y + 6x^2y^2 + 4xy^3 + y^4$$
$$(x + y)^5 = x^5 + 5x^4y + 10x^3y^2 + 10x^2y^3 + 5xy^4 + y^5.$$

Notice that after the special case $(x + y)^0 = 1$, each expression begins with x raised to the same power as the binomial itself. That is, the expansion of $(x + y)^1$ has a first term of x^1, $(x + y)^2$ has a first term of x^2, $(x + y)^3$ has a first term of x^3, and so on. Also, the last term in each expansion is y to the same power as the binomial. Thus, the expansion of $(x + y)^n$ should begin with the term x^n and end with the term y^n.

Also, the exponent on x decreases by 1 in each term after the first, while the exponent on y, beginning with y in the second term, increases by 1 in each succeeding term. That is, the *variables* in the terms of the expansion of $(x + y)^n$ have the following pattern:

$$x^n,\ x^{n-1}y,\ x^{n-2}y^2,\ x^{n-3}y^3,\ \ldots,\ xy^{n-1},\ y^n.$$

This pattern suggests that the sum of the exponents on x and y in each term is n. For example, the third term in the preceding list is $x^{n-2}y^2$, and the sum of the exponents is $n - 2 + 2 = n$.

Pascal's Triangle

Writing just the *coefficients* in the terms of the expansion of $(x + y)^n$ gives the pattern shown on the next page.

Blaise Pascal (1623–1662)

Pascal's Triangle

											Row
					1						0
				1		1					1
			1		2		1				2
		1		3		3		1			3
	1		4		6		4		1		4
1		5		10		10		5		1	5

With the coefficients arranged in this way, each number in the triangle is the sum of the two numbers directly above it (one to the right and one to the left). For example, in row four, 1 is the sum of 1 (the only number above it), 4 is the sum of 1 and 3, 6 is the sum of 3 and 3, and so on. This triangular array of numbers is called **Pascal's triangle,** in honor of the 17th-century mathematician Blaise Pascal.

To find the coefficients for $(x + y)^6$, we need to include row six in Pascal's triangle. Adding adjacent numbers, we find that row six is

$$1 \quad 6 \quad 15 \quad 20 \quad 15 \quad 6 \quad 1.$$

Using these coefficients, we obtain the expansion of $(x + y)^6$:

$$(x + y)^6 = x^6 + 6x^5y + 15x^4y^2 + 20x^3y^3 + 15x^2y^4 + 6xy^5 + y^6.$$

GCM TECHNOLOGY NOTE

A calculator with a 10-digit display will give exact values of $n!$ for $n \leq 13$ and approximate values of $n!$ for $14 \leq n \leq 69$. The two screens confirm the results shown in the discussion.

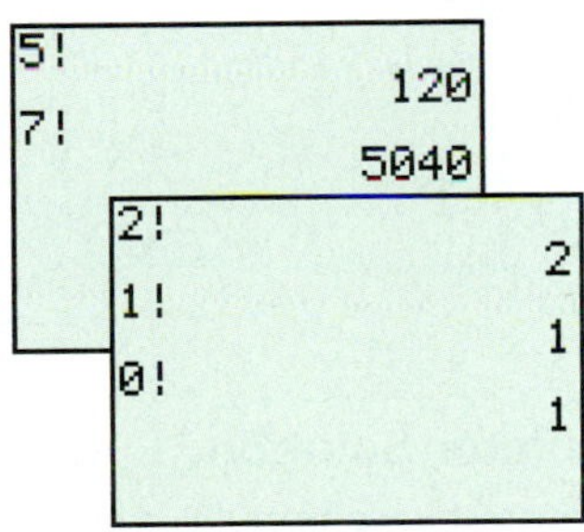

n-Factorial

Using Pascal's triangle to find the coefficients of $(x + y)^n$ becomes impractical for large values of n, because of the need to write all the preceding rows. A more efficient way of finding these coefficients uses **factorial notation.** The number $n!$ (read **"*n*-factorial"**) is defined as follows.

n-Factorial

For any positive integer n,

$$n! = n(n - 1)(n - 2) \cdots (3)(2)(1), \quad \text{and} \quad 0! = 1.$$

For example,

$$5! = 5 \cdot 4 \cdot 3 \cdot 2 \cdot 1 = 120,$$
$$7! = 7 \cdot 6 \cdot 5 \cdot 4 \cdot 3 \cdot 2 \cdot 1 = 5040,$$
$$2! = 2 \cdot 1 = 2, \quad 1! = 1, \quad \text{and} \quad 0! = 1.$$

Binomial Coefficients

Now, look at the coefficients of the expression

$$(x + y)^5 = x^5 + 5x^4y + 10x^3y^2 + 10x^2y^3 + 5xy^4 + y^5.$$

The coefficient of the second term, $5x^4y$, is 5, and the exponents on the variables are 4 and 1. Note that

$$5 = \frac{5!}{4!\,1!}.$$

The coefficient of the third term is 10, with exponents 3 and 2, and

$$10 = \frac{5!}{3!2!}.$$

The last term (the sixth term) can be written $y^5 = 1x^0y^5$, with coefficient 1 and exponents 0 and 5. Since $0! = 1$,

$$1 = \frac{5!}{0!5!}.$$

Generalizing from these examples, we find that the coefficient for the term of the expansion of $(x + y)^n$ in which the variable part is $x^r y^{n-r}$ (where $r \leq n$) is

$$\frac{n!}{r!(n-r)!}.$$

This number, called a **binomial coefficient,** is often written as $\binom{n}{r}$ (read **"*n* choose *r*"**) or ${}_nC_r$.

Binomial Coefficient

For nonnegative integers n and r, with $r \leq n$,

$${}_nC_r = \binom{n}{r} = \frac{n!}{r!(n-r)!}.$$

The binomial coefficients are numbers from Pascal's triangle. For example, $\binom{3}{0}$ is the first number in row three, and $\binom{7}{4}$ is the fifth number in row seven.

GCM EXAMPLE 1 Evaluating Binomial Coefficients

Evaluate each binomial coefficient.

(a) $\binom{6}{2}$ **(b)** $\binom{8}{0}$ **(c)** $\binom{10}{10}$

Analytic Solution

(a) $\binom{6}{2} = \frac{6!}{2!(6-2)!} = \frac{6!}{2!4!} = \frac{6 \cdot 5 \cdot 4 \cdot 3 \cdot 2 \cdot 1}{2 \cdot 1 \cdot 4 \cdot 3 \cdot 2 \cdot 1} = 15$

(b) $\binom{8}{0} = \frac{8!}{0!(8-0)!} = \frac{8!}{0!8!} = \frac{8!}{1 \cdot 8!} = 1$ $\quad 0! = 1$

(c) $\binom{10}{10} = \frac{10!}{10!(10-10)!} = \frac{10!}{10!0!} = 1$ $\quad 0! = 1$

Graphing Calculator Solution

Graphing calculators can compute binomial coefficients with the *combinations* function, denoted nCr. See Figure 15.

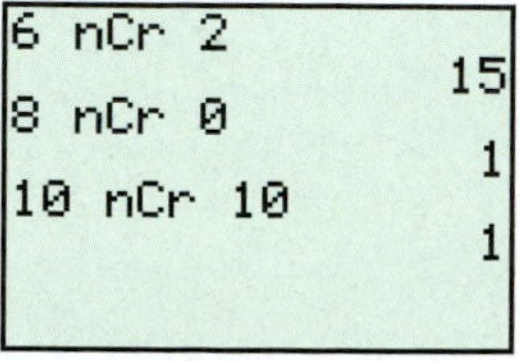

FIGURE 15

Refer again to Pascal's triangle. Notice the symmetry in each row, which suggests that the binomial coefficients should satisfy

$$\binom{n}{r} = \binom{n}{n-r},$$

since $$\binom{n}{r} = \frac{n!}{r!\,(n-r)!} \quad \text{and} \quad \binom{n}{n-r} = \frac{n!}{(n-r)!\,r!}.$$

The Binomial Theorem

Our observations about the expansion of $(x + y)^n$ are summarized as follows:

1. There are $n + 1$ terms in the expansion.
2. The first term is x^n and the last term is y^n.
3. In each succeeding term, the exponent on x decreases by 1 and the exponent on y increases by 1.
4. The sum of the exponents on x and y in any term is n.
5. The coefficient of the term with $x^r y^{n-r}$ or $x^{n-r} y^r$ is $\binom{n}{r}$.

These observations about the expansion of $(x + y)^n$ for any positive integer value of n suggest the **binomial theorem.** A proof is given in the next section.

TECHNOLOGY NOTE

The two screens illustrate how the sequence and table capabilities of the TI-83/84 Plus graphing calculator can generate rows of Pascal's triangle.

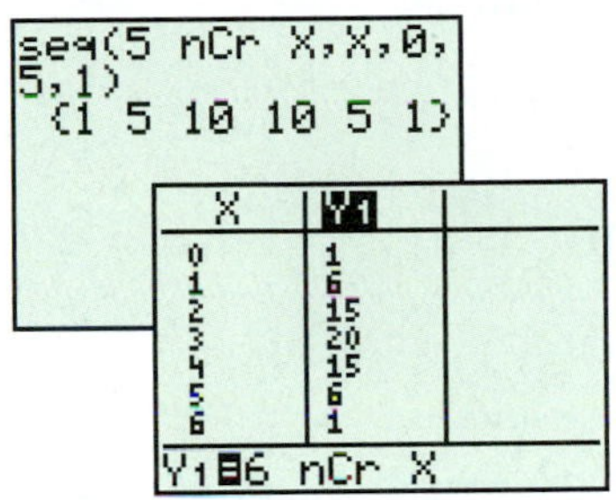

Binomial Theorem

For any positive integer n,

$$(x+y)^n = x^n + \binom{n}{1}x^{n-1}y + \binom{n}{2}x^{n-2}y^2 + \binom{n}{3}x^{n-3}y^3 + \cdots + \binom{n}{r}x^{n-r}y^r + \cdots + \binom{n}{n-1}xy^{n-1} + y^n.$$

NOTE The binomial theorem looks much more manageable written in summation notation. The theorem can be summarized as

$$(x+y)^n = \sum_{r=0}^{n} \binom{n}{r} x^{n-r} y^r.$$

Looking Ahead to Calculus

The binomial theorem is used to show that the derivative of $f(x) = x^n$ is given by the term nx^{n-1}. This fact is used extensively throughout the study of calculus.

FOR DISCUSSION

Consider the expansion of $(x + y)^5$:

$$(x+y)^5 = x^5 + 5x^4y + 10x^3y^2 + 10x^2y^3 + 5xy^4 + y^5.$$

Position ⟶ ① ② ③ ④ ⑤ ⑥

Choose any term except the sixth term, and multiply its numerical coefficient by the exponent on x. Then divide by the position number of the term. Compare your answer to the coefficient of the next term in the expansion. What do you notice? Repeat this procedure several times with this expansion and others, and make a conjecture.

EXAMPLE 2 Applying the Binomial Theorem

Write the binomial expansion of $(x + y)^9$.

Solution Use the binomial theorem.

$$\begin{aligned}(x+y)^9 &= x^9 + \binom{9}{1}x^8y + \binom{9}{2}x^7y^2 + \binom{9}{3}x^6y^3 + \binom{9}{4}x^5y^4 + \binom{9}{5}x^4y^5 \\ &\quad + \binom{9}{6}x^3y^6 + \binom{9}{7}x^2y^7 + \binom{9}{8}xy^8 + y^9 \\ &= x^9 + \frac{9!}{1!8!}x^8y + \frac{9!}{2!7!}x^7y^2 + \frac{9!}{3!6!}x^6y^3 + \frac{9!}{4!5!}x^5y^4 \\ &\quad + \frac{9!}{5!\,4!}x^4y^5 + \frac{9!}{6!3!}x^3y^6 + \frac{9!}{7!2!}x^2y^7 + \frac{9!}{8!1!}xy^8 + y^9 \\ &= x^9 + 9x^8y + 36x^7y^2 + 84x^6y^3 + 126x^5y^4 + 126x^4y^5 \\ &\quad + 84x^3y^6 + 36x^2y^7 + 9xy^8 + y^9\end{aligned}$$

EXAMPLE 3 Applying the Binomial Theorem

Expand $\left(a - \frac{b}{2}\right)^5$.

Solution Write the binomial as $\left(a - \frac{b}{2}\right)^5 = \left(a + \left(-\frac{b}{2}\right)\right)^5$. Then use the binomial theorem with $x = a$, $y = -\frac{b}{2}$, and $n = 5$.

$$\begin{aligned}\left(a - \frac{b}{2}\right)^5 &= a^5 + \binom{5}{1}a^4\left(-\frac{b}{2}\right) + \binom{5}{2}a^3\left(-\frac{b}{2}\right)^2 + \binom{5}{3}a^2\left(-\frac{b}{2}\right)^3 + \binom{5}{4}a\left(-\frac{b}{2}\right)^4 + \left(-\frac{b}{2}\right)^5 \\ &= a^5 + 5a^4\left(-\frac{b}{2}\right) + 10a^3\left(-\frac{b}{2}\right)^2 + 10a^2\left(-\frac{b}{2}\right)^3 + 5a\left(-\frac{b}{2}\right)^4 + \left(-\frac{b}{2}\right)^5 \\ &= a^5 - \frac{5}{2}a^4b + \frac{5}{2}a^3b^2 - \frac{5}{4}a^2b^3 + \frac{5}{16}ab^4 - \frac{1}{32}b^5\end{aligned}$$

EXAMPLE 4 Applying the Binomial Theorem

Expand $\left(\frac{3}{m^2} - 2\sqrt{m}\right)^4$. (Assume that $m > 0$.)

Solution Use the binomial theorem.

$$\begin{aligned}\left(\frac{3}{m^2} - 2\sqrt{m}\right)^4 &= \left(\frac{3}{m^2}\right)^4 + \binom{4}{1}\left(\frac{3}{m^2}\right)^3(-2\sqrt{m}) + \binom{4}{2}\left(\frac{3}{m^2}\right)^2(-2\sqrt{m})^2 + \binom{4}{3}\left(\frac{3}{m^2}\right)(-2\sqrt{m})^3 + (-2\sqrt{m})^4 \\ &= \frac{81}{m^8} + 4\left(\frac{27}{m^6}\right)(-2m^{1/2}) + 6\left(\frac{9}{m^4}\right)(4m) + 4\left(\frac{3}{m^2}\right)(-8m^{3/2}) + 16m^2 \qquad \sqrt{m} = m^{1/2} \\ &= \frac{81}{m^8} - \frac{216}{m^{11/2}} + \frac{216}{m^3} - \frac{96}{m^{1/2}} + 16m^2\end{aligned}$$

NOTE ***Any binomial expansion of the difference of two terms has alternating signs.***

*r*th Term of a Binomial Expansion

Any term in a binomial expansion can be determined without writing out the entire expansion. For example, the seventh term of $(x + y)^9$ has y raised to the sixth power (since y has the power 1 in the second term, the power 2 in the third term, and so on). The exponents on x and y in each term must have a sum of 9, so the exponent on x in the seventh term is $9 - 6 = 3$. Thus, writing the coefficient as given in the binomial theorem, we find the seventh term to be

$$\frac{9!}{6!\,3!}x^3y^6, \quad \text{or} \quad 84x^3y^6.$$

This is in fact the seventh term of $(x + y)^9$ found in Example 2. The discussion suggests the next theorem.

*r*th Term of the Binomial Expansion

The rth term of the binomial expansion of $(x + y)^n$, where $n \geq r - 1$, is

$$\binom{n}{r-1}x^{n-(r-1)}y^{r-1}.$$

EXAMPLE 5 **Finding a Specific Term of a Binomial Expansion**

Find the fourth term of the expansion of $(a + 2b)^{10}$.

Solution In the fourth term, $2b$ has exponent 3 and a has exponent $10 - 3 = 7$. Using the preceding formula, we find that the fourth term is

$$\binom{10}{3}a^7(2b)^3 = 120a^7(8b^3) = 960a^7b^3. \qquad n = 10,\ r = 4,\ x = a,\ y = 2b$$

11.4 Exercises

Evaluate the following. In Exercises 11 and 12, express the answer in terms of n.

1. $\frac{6!}{3!\,3!}$ **2.** $\frac{5!}{2!\,3!}$ **3.** $\frac{7!}{3!\,4!}$ **4.** $\frac{8!}{5!\,3!}$ **5.** $\binom{8}{3}$ **6.** $\binom{7}{4}$

7. $\binom{10}{8}$ **8.** $\binom{9}{6}$ **9.** $\binom{13}{13}$ **10.** $\binom{12}{12}$ **11.** $\binom{n}{n-1}$ **12.** $\binom{n}{n-2}$

13. ${}_8C_3$ **14.** ${}_9C_7$ **15.** ${}_{100}C_2$ **16.** ${}_{20}C_{15}$ **17.** ${}_5C_0$ **18.** ${}_6C_0$

19. *Concept Check* How many terms are there in the expansion of $(x + y)^8$?

20. *Concept Check* How many terms are there in the expansion of $(x + y)^{10}$?

21. *Concept Check* What are the first and last terms in the expansion of $(2x + 3y)^4$?

22. Describe in your own words how you would determine the binomial coefficient for the fifth term in the expansion of $(x + y)^8$.

Write the binomial expansion for each expression.

23. $(x + y)^6$
24. $(m + n)^4$
25. $(p - q)^5$
26. $(a - b)^7$
27. $(r^2 + s)^5$
28. $(m + n^2)^4$
29. $(p + 2q)^4$
30. $(3r - s)^6$
31. $(7p + 2q)^4$
32. $(4a - 5b)^5$
33. $(3x - 2y)^6$
34. $(7k - 9j)^4$
35. $\left(\frac{m}{2} - 1\right)^6$
36. $\left(3 + \frac{y}{3}\right)^5$
37. $\left(\sqrt{2}r + \frac{1}{m}\right)^4$
38. $\left(\frac{1}{k} - \sqrt{3}p\right)^3$

Write the indicated term of each binomial expansion.

39. Sixth term of $(4h - j)^8$
40. Eighth term of $(2c - 3d)^{14}$
41. Fifteenth term of $(a^2 + b)^{22}$
42. Twelfth term of $(2x + y^2)^{16}$
43. Fifteenth term of $(x - y^3)^{20}$
44. Tenth term of $(a^3 + 3b)^{11}$

Concept Check *Work Exercises 45–48.*

45. Find the middle term of $(3x^7 + 2y^3)^8$.

46. Find the two middle terms of $(-2m^{-1} + 3n^{-2})^{11}$.

47. Find the value of n for which the coefficients of the fifth and eighth terms in the expansion of $(x + y)^n$ are the same.

48. Find the term in the expansion of $\left(3 + \sqrt{x}\right)^{11}$ that contains x^4.

Relating Concepts

For individual or group investigation (Exercises 49–52)

The factorial of a positive integer n can be computed as a product: $n! = 1 \cdot 2 \cdot 3 \cdot \cdots \cdot n$. *Calculators and computers can evaluate factorials quickly. Before the days of technology, mathematicians developed a formula, called* ***Stirling's formula,*** *for approximating large factorials. Interestingly enough, it involves the irrational numbers* π *and* e.

$$n! \approx \sqrt{2\pi n} \cdot n^n \cdot e^{-n}$$

As an example, the exact value of 5! *is* 120, *and Stirling's formula gives the approximation as* 118.019168 *with a graphing calculator. This is "off" by less than* 2, *an error of only* 1.65%. ***Work Exercises 49–52 in order.***

49. Use a calculator and Stirling's formula to find the exact value of 10! and its approximation.

50. Subtract the smaller value from the larger value in Exercise 49. Divide it by 10! and convert to a percent. What is the percent error?

51. Repeat Exercises 49 and 50 for $n = 12$.

52. Repeat Exercises 49 and 50 for $n = 13$. What seems to happen as n increases?

In later courses, it is shown that

$$(1 + x)^n = 1 + nx + \frac{n(n-1)}{2!}x^2 + \frac{n(n-1)(n-2)}{3!}x^3 + \cdots$$

for any real number n (not just positive integer values) and any real number x, where $|x| < 1$. *Use this result to approximate each quantity in Exercises 53–56 to the nearest thousandth.*

53. $(1.02)^{-3}$
54. $\frac{1}{1.04^5}$
55. $(1.01)^{3/2}$
56. $(1.03)^{.2}$

11.5 Mathematical Induction

Proof by Mathematical Induction ■ Proving Statements ■ Generalized Principle of Mathematical Induction ■ Proof of the Binomial Theorem

Proof by Mathematical Induction

Many results in mathematics are claimed to hold for every positive integer. These results could be checked for $n = 1, n = 2, n = 3$, and so on, but since the set of positive integers is infinite, it would be impossible to check every possible case. For example, let S_n represent the statement that the sum of the first n positive integers is equal to $\frac{n(n+1)}{2}$. That is,

$$S_n: \quad 1 + 2 + 3 + \cdots + n = \frac{n(n+1)}{2}.$$

The truth of this statement can be checked quickly for the first few values of n.

If $n = 1$, then S_1 is $\quad 1 = \frac{1(1+1)}{2}$, which is true.

If $n = 2$, then S_2 is $\quad 1 + 2 = \frac{2(2+1)}{2}$, which is true.

If $n = 3$, then S_3 is $\quad 1 + 2 + 3 = \frac{3(3+1)}{2}$, which is true.

If $n = 4$, then S_4 is $\quad 1 + 2 + 3 + 4 = \frac{4(4+1)}{2}$, which is true.

Continuing in this way for any amount of time would still not prove that S_n is true for *every* positive integer value of n. To prove that such statements are true for every positive integer value of n, the following principle is often used.

Principle of Mathematical Induction

Let S_n be a statement concerning the positive integer n. Suppose that

1. S_1 is true.
2. For any positive integer k, $k \leq n$, if S_k is true, then S_{k+1} is also true.

Then S_n is true for every positive integer n.

A proof by mathematical induction can be explained as follows. By assumption (1), the statement is true when $n = 1$. By assumption (2), the fact that the statement is true for $n = 1$ implies that it is true for $n = 1 + 1 = 2$. Using (2) again, the statement is thus true for $2 + 1 = 3$, for $3 + 1 = 4$, for $4 + 1 = 5$, and so on. Continuing in this way shows that the statement must be true for *every* positive integer.

The situation is similar to that of an infinite number of dominoes lined up, as suggested in Figure 16. If the first domino is pushed over, it pushes the next, which pushes the next, and so on, indefinitely.

FIGURE 16

Another analogy to the principle of mathematical induction is an infinite ladder. Suppose the rungs are spaced so that, whenever you are on a rung, you know you can move to the next rung. Then *if* you can get to the first rung, you can go as high up the ladder as you wish.

Two separate steps are required for a proof by mathematical induction.

Proof by Mathematical Induction

Step 1 Prove that the statement is true for $n = 1$.

Step 2 Show that for any positive integer k, if S_k is true, then S_{k+1} is also true.

Proving Statements

EXAMPLE 1 Proving an Equality Statement

Let S_n represent the statement discussed at the beginning of this section,

$$1 + 2 + 3 + \cdots + n = \frac{n(n+1)}{2}.$$

Prove that S_n is true for every positive integer n.

Solution The proof by mathematical induction is as follows.

Step 1 Show that the statement is true when $n = 1$. If $n = 1$, S_1 becomes

$$1 = \frac{1(1+1)}{2}, \qquad \text{which is true.}$$

Step 2 Show that if S_k is true, then S_{k+1} is also true, where S_k is the statement

$$1 + 2 + 3 + \cdots + k = \frac{k(k+1)}{2},$$

and S_{k+1} is the statement

$$1 + 2 + 3 + \cdots + k + (k+1) = \frac{(k+1)[(k+1)+1]}{2}.$$

Start with S_k, and assume that it is true.

$$1 + 2 + 3 + \cdots + k = \frac{k(k+1)}{2}$$

Add $k + 1$ to each side of this equation to obtain S_{k+1}.

$$1 + 2 + 3 + \cdots + k + (k+1) = \frac{k(k+1)}{2} + (k+1)$$

$$= (k+1)\left(\frac{k}{2} + 1\right) \qquad \text{Factor out } k + 1.$$

$$= (k+1)\left(\frac{k+2}{2}\right) \qquad \text{Add inside the parentheses.}$$

$$1 + 2 + 3 + \cdots + k + (k+1) = \frac{(k+1)[(k+1)+1]}{2} \qquad \text{Multiply; } k + 2 = (k+1) + 1.$$

This final result is the statement for $n = k + 1$; it has been shown that if S_k is true, then S_{k+1} is also true. The two steps required for a proof by mathematical induction have been completed, so the statement S_n is true for every positive integer n. ■

EXAMPLE 2 Proving an Inequality Statement

Prove that if x is a real number between 0 and 1, then, for every positive integer n,

$$0 < x^n < 1.$$

Solution

Step 1 Here, S_1 is the statement

$$\text{if } 0 < x < 1, \quad \text{then} \quad 0 < x^1 < 1,$$

which is true.

Step 2 S_k is the statement

$$\text{if } 0 < x < 1, \quad \text{then} \quad 0 < x^k < 1.$$

To show that this statement implies that S_{k+1} is true, multiply all three parts of $0 < x^k < 1$ by x to get

$$x \cdot 0 < x \cdot x^k < x \cdot 1.$$

(Here the fact that $0 < x$ is used.) Simplify to obtain

$$0 < x^{k+1} < x.$$

Since $x < 1$,

$$0 < x^{k+1} < 1. \qquad \text{Replace } x \text{ with } 1.$$

By this work, if S_k is true, then S_{k+1} is true. Since both steps for a proof by mathematical induction have been completed, the given statement is true for every positive integer n. ■

Generalized Principle of Mathematical Induction

Some statements S_n are not true for the first few values of n, but are true for all values of n that are greater than or equal to some fixed integer j. The following slightly generalized form of the principle of mathematical induction takes care of these cases.

Generalized Principle of Mathematical Induction

Let S_n be a statement concerning the positive integer n. Let j be a fixed positive integer. Suppose that

Step 1 S_j is true.

Step 2 For any positive integer k, $k \geq j$, S_k implies S_{k+1}.

Then S_n is true for all positive integers n, where $n \geq j$.

EXAMPLE 3 Using the Generalized Principle

Let S_n represent the statement $2^n > 2n + 1$. Show that S_n is true for all values of n such that $n \geq 3$.

Solution (Check that S_n is false for $n = 1$ and $n = 2$.)

Step 1 Show that S_n is true for $n = 3$. If $n = 3$, S_n is

$$2^3 > 2 \cdot 3 + 1, \quad \text{or} \quad 8 > 7.$$

Thus, S_3 is true.

Step 2 Now show that S_k implies S_{k+1}, for $k \geq 3$, where

$$S_k \quad \text{is} \quad 2^k > 2k + 1$$

and

$$S_{k+1} \text{ is } 2^{k+1} > 2(k + 1) + 1.$$

Multiply each side of $2^k > 2k + 1$ by 2, obtaining

$$2 \cdot 2^k > 2(2k + 1)$$
$$2^{k+1} > 4k + 2$$
$$2^{k+1} > 2(k + 1) + 2k. \quad \text{Rewrite } 4k + 2 \text{ as } 2(k + 1) + 2k.$$

Since k is a positive integer greater than 3,

$$2k > 1.$$

It follows that

$$2^{k+1} > 2(k + 1) + 2k > 2(k + 1) + 1,$$

or

$$2^{k+1} > 2(k + 1) + 1, \quad \text{as required.}$$

Thus, S_k implies S_{k+1}, and this, together with the fact that S_3 is true, shows that S_n is true for every positive integer n greater than or equal to 3. ■

Proof of the Binomial Theorem

The binomial theorem, introduced in the previous section, can be proved by mathematical induction. That is, for any positive integer n and any numbers x and y,

$$(x + y)^n = x^n + \binom{n}{1}x^{n-1}y + \binom{n}{2}x^{n-2}y^2 + \binom{n}{3}x^{n-3}y^3 + \cdots + \binom{n}{r}x^{n-r}y^r + \cdots + \binom{n}{n-1}xy^{n-1} + y^n. \quad (1)$$

Proof Let S_n be statement (1). Begin by verifying S_n for $n = 1$,

$$S_1\text{:} \quad (x + y)^1 = x^1 + y^1, \quad \text{which is true.}$$

Now, assume that S_n is true for the positive integer k. Statement S_k becomes

$$S_k\text{:} \quad (x + y)^k = x^k + \frac{k!}{1!(k - 1)!}x^{k-1}y + \frac{k!}{2!(k - 2)!}x^{k-2}y^2 + \cdots + \frac{k!}{(k - 1)!\,1!}xy^{k-1} + y^k. \quad (2)$$

Definition of the binomial coefficient

Multiply the left side of equation (2) by $x + y$ and apply statement S_k.

$$\begin{aligned}(x + y)^k \cdot (x + y) \\ &= x(x + y)^k + y(x + y)^k \quad \text{Distributive property} \\ &= \left[x \cdot x^k + \frac{k!}{1!\,(k-1)!}x^k y + \frac{k!}{2!\,(k-2)!}x^{k-1}y^2 + \cdots + \frac{k!}{(k-1)!\,1!}x^2y^{k-1} + xy^k\right] \\ &\quad + \left[x^k \cdot y + \frac{k!}{1!\,(k-1)!}x^{k-1}y^2 + \cdots + \frac{k!}{(k-1)!\,1!}xy^k + y \cdot y^k\right]\end{aligned}$$

Rearrange terms to get

$$(x + y)^{k+1} = x^{k+1} + \left[\frac{k!}{1!(k-1)!} + 1\right]x^k y + \left[\frac{k!}{2!(k-2)!} + \frac{k!}{1!(k-1)!}\right]x^{k-1}y^2$$
$$+ \cdots + \left[1 + \frac{k!}{(k-1)!\,1!}\right]xy^k + y^{k+1}. \tag{3}$$

The first expression in brackets in equation (3) simplifies to $\binom{k+1}{1}$. To see this, note that

$$\binom{k+1}{1} = \frac{(k+1)(k)(k-1)(k-2)\cdots 1}{1 \cdot (k)(k-1)(k-2)\cdots 1} = k + 1.$$

Also, $$\frac{k!}{1!(k-1)!} + 1 = \frac{k(k-1)!}{1(k-1)!} + 1 = k + 1.$$

The second expression becomes $\binom{k+1}{2}$, the last $\binom{k+1}{k}$, and so on. The result of equation (3) is just equation (2) with every k replaced by $k + 1$. Thus, the truth of S_n when $n = k$ implies the truth of S_n for $n = k + 1$, which completes the proof of the theorem by mathematical induction.

11.5 Exercises

1. *Concept Check* When using mathematical induction to prove a statement, the domain of the variable must be all ____________.

2. Suppose that Step 2 in a proof by mathematical induction can be satisfied, but Step 1 cannot. May we conclude that the proof is complete? Explain.

3. *Concept Check* For which positive integers is the statement $2^n > 2n$ *not true*?

4. Write out in full and verify the statements S_1, S_2, S_3, S_4, and S_5 for the following formula. Then use mathematical induction to prove that the statement is true for every positive integer n.

$$2 + 4 + 6 + \cdots + 2n = n(n + 1)$$

Use mathematical induction to prove each statement. Assume that n is a positive integer.

5. $3 + 6 + 9 + \cdots + 3n = \dfrac{3n(n+1)}{2}$

6. $1 + 3 + 5 + \cdots + (2n - 1) = n^2$

7. $5 + 10 + 15 + \cdots + 5n = \dfrac{5n(n+1)}{2}$

8. $4 + 7 + 10 + \cdots + (3n + 1) = \dfrac{n(3n+5)}{2}$

9. $3 + 3^2 + 3^3 + \cdots + 3^n = \dfrac{3(3^n - 1)}{2}$

10. $1^2 + 2^2 + 3^2 + \cdots + n^2 = \dfrac{n(n + 1)(2n + 1)}{6}$

11. $1^3 + 2^3 + 3^3 + \cdots + n^3 = \dfrac{n^2(n + 1)^2}{4}$

12. $5 \cdot 6 + 5 \cdot 6^2 + 5 \cdot 6^3 + \cdots + 5 \cdot 6^n = 6(6^n - 1)$

13. $\dfrac{1}{1 \cdot 2} + \dfrac{1}{2 \cdot 3} + \dfrac{1}{3 \cdot 4} + \cdots + \dfrac{1}{n(n + 1)} = \dfrac{n}{n + 1}$

14. $7 \cdot 8 + 7 \cdot 8^2 + 7 \cdot 8^3 + \cdots + 7 \cdot 8^n = 8(8^n - 1)$

15. $\dfrac{4}{5} + \dfrac{4}{5^2} + \dfrac{4}{5^3} + \cdots + \dfrac{4}{5^n} = 1 - \dfrac{1}{5^n}$

16. $\dfrac{1}{2} + \dfrac{1}{2^2} + \dfrac{1}{2^3} + \cdots + \dfrac{1}{2^n} = 1 - \dfrac{1}{2^n}$

17. $\dfrac{1}{1 \cdot 4} + \dfrac{1}{4 \cdot 7} + \dfrac{1}{7 \cdot 10} + \cdots + \dfrac{1}{(3n - 2)(3n + 1)} = \dfrac{n}{3n + 1}$

18. $x^{2n} + x^{2n-1}y + \cdots + xy^{2n-1} + y^{2n} = \dfrac{x^{2n+1} - y^{2n+1}}{x - y}$

Find all positive integers n for which the given statement is not *true.*

19. $3^n > 6n$

20. $3^n > 2n + 1$

21. $2^n > n^2$

22. $n! > 2n$

Prove each statement by mathematical induction.

23. $(a^m)^n = a^{mn}$ (Assume that a and m are constant.)

24. $(ab)^n = a^n b^n$ (Assume that a and b are constant.)

25. $2^n > 2n$, if $n \geq 3$

26. $3^n > 2n + 1$, if $n \geq 2$

27. If $a > 1$, then $a^n > 1$.

28. If $a > 1$, then $a^n > a^{n-1}$.

29. If $0 < a < 1$, then $a^n < a^{n-1}$.

30. $2^n > n^2$, for $n > 4$

31. If $n \geq 4$, then $n! > 2^n$, where $n! = n(n - 1)(n - 2) \cdots (3)(2)(1)$.

32. $4^n > n^4$, for $n \geq 5$

Solve each problem.

33. ***Number of Handshakes*** Suppose that each of the n $(n \geq 2)$ people in a room shakes hands with everyone else, but not with himself. Show that the number of handshakes is $\frac{n^2 - n}{2}$.

34. ***Sides of a Polygon*** A series of sketches starts with an equilateral triangle having sides of length 1. In the steps that follow, equilateral triangles are constructed on each side of the previous figure. The length of the sides of each new triangle is $\frac{1}{3}$ the length of the sides of the previous triangles. Develop a formula for the number of sides of the nth figure. Use mathematical induction to prove your answer.

35. ***Perimeter*** Find the perimeter of the nth figure in Exercise 34.

36. ***Area*** Show that the area of the nth figure in Exercise 34 is

$$\sqrt{3}\left[\frac{2}{5} - \frac{3}{20}\left(\frac{4}{9}\right)^{n-1}\right].$$

37. ***Tower of Hanoi*** A pile of n rings, each smaller than the one below it, is on a peg on a board. Two other pegs are attached to the board. In the game called the *Tower of Hanoi* puzzle, all the rings must be moved, one at a time, to a different peg with no ring ever placed on top of a smaller ring. Find the least number of moves that would be required. Prove your result by mathematical induction.

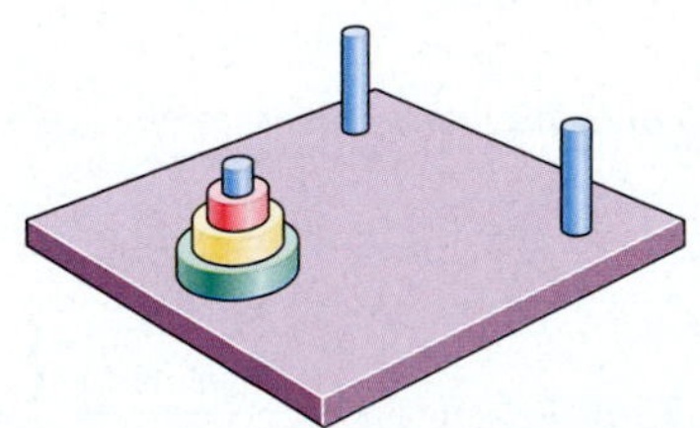

Reviewing Basic Concepts (Sections 11.4 and 11.5)

1. Calculate ${}_5C_3$.

2. Calculate $\binom{6}{2}$.

3. How many terms are there in the expansion of $(x + y)^n$?

4. Write the binomial expansion of $(a + 2b)^4$.

5. Write the binomial expansion of $\left(\frac{1}{m} - n^2\right)^3$.

6. Find the third term of $(x - 2y)^6$.

7. Use mathematical induction to prove that

$$4 + 8 + 12 + 16 + \cdots + 4n = 2n(n + 1).$$

8. Use mathematical induction to prove that $n^2 \le 2^n$ for $n \ge 4$.

11.6 Counting Theory

Fundamental Principle of Counting ■ Permutations ■ Combinations ■ Distinguishing between Permutations and Combinations

Fundamental Principle of Counting

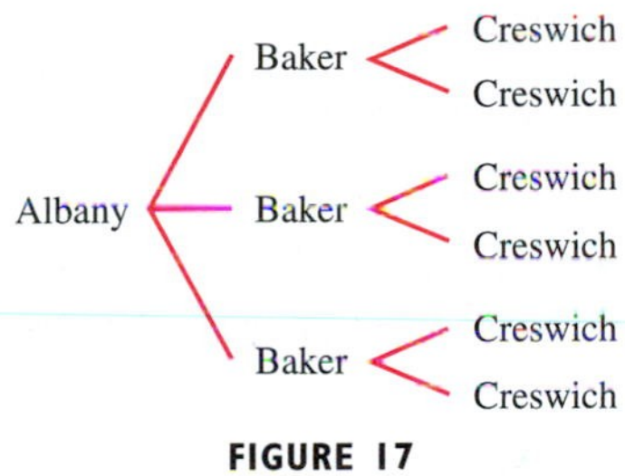

FIGURE 17

If there are 3 roads from Albany to Baker and 2 roads from Baker to Creswich, in how many ways can one travel from Albany to Creswich by way of Baker? For each of the 3 roads from Albany to Baker, there are 2 different roads from Baker to Creswich. Hence, there are $3 \cdot 2 = 6$ different ways to make the trip, as shown in the **tree diagram** in Figure 17.

Two events are **independent events** if neither influences the outcome of the other. The opening example illustrates the fundamental principle of counting with independent events.

Fundamental Principle of Counting

If n independent events occur, with

m_1 ways for event 1 to occur,

m_2 ways for event 2 to occur,

⋮

and m_n ways for event n to occur,

then there are

$$m_1 \cdot m_2 \cdot \cdots \cdot m_n$$

different ways for all n events to occur.

EXAMPLE 1 **Using the Fundamental Principle of Counting**

A restaurant offers a choice of 3 salads, 5 main dishes, and 2 desserts. Use the fundamental principle of counting to find the number of different 3-course meals that can be selected.

Solution Three events are involved: selecting a salad, selecting a main dish, and selecting a dessert. The first event can occur in 3 ways, the second in 5 ways, and the third in 2 ways; thus, there are

$$3 \cdot 5 \cdot 2 = 30 \text{ possible meals.}$$ ■

EXAMPLE 2 **Using the Fundamental Principle of Counting**

A teacher has 5 different books that she wishes to arrange in a row. How many different arrangements are possible?

Solution Five events are involved: selecting a book for the first spot, selecting a book for the second spot, and so on. For the first spot, the teacher has 5 choices. After a choice has been made, the teacher has 4 choices for the second spot. Continuing in this manner, we get 3 choices for the third spot, 2 for the fourth spot, and 1 for the fifth spot. By the fundamental principle of counting, there are

$$5 \cdot 4 \cdot 3 \cdot 2 \cdot 1 = 120 \text{ arrangements.}$$ ■

EXAMPLE 3 **Arranging *r* of *n* Items ($r < n$)**

Suppose the teacher in Example 2 wishes to place only 3 of the 5 books in a row. How many arrangements of 3 books are possible?

Solution The teacher still has 5 ways to fill the first spot, 4 ways to fill the second spot, and 3 ways to fill the third. Since only 3 books will be used, there are only 3 spots to be filled (3 events) instead of 5, with

$$5 \cdot 4 \cdot 3 = 60 \text{ arrangements.}$$ ■

Permutations

Refer to Example 3. Since each ordering of 3 books is considered a different *arrangement,* the number 60 in that example is called the number of *permutations* of 5 things taken 3 at a time, written $P(5, 3) = 60$. The number of ways of arranging 5 elements from a set of 5 elements, written $P(5, 5) = 120$, was found in Example 2.

A **permutation** of n elements taken r at a time is one of the *arrangements* of r elements from a set of n elements. Generalizing from the preceding examples, we find that the number of permutations of n elements taken r at a time, denoted by $P(n, r)$, is

$$\begin{aligned} P(n, r) &= n(n-1)(n-2)\cdots(n-r+1) \\ &= \frac{n(n-1)(n-2)\cdots(n-r+1)(n-r)(n-r-1)\cdots(2)(1)}{(n-r)(n-r-1)\cdots(2)(1)} \\ &= \frac{n!}{(n-r)!}. \end{aligned}$$

Permutations of n Elements Taken r at a Time

If $P(n,r)$ denotes the number of permutations of n elements taken r at a time, with $r \leq n$, then

$$P(n,r) = \frac{n!}{(n-r)!}.$$

Alternative notations for $P(n,r)$ are P_r^n and ${}_nP_r$.

GCM **EXAMPLE 4** **Using the Permutations Formula**

Find the following.

(a) The number of permutations of the letters L, M, and N

(b) The number of permutations of 2 of the 3 letters L, M, and N

Analytic Solution

(a) By the formula for $P(n,r)$ with $n = 3$ and $r = 3$,

$$P(3,3) = \frac{3!}{(3-3)!} = \frac{3!}{0!} = \frac{3!}{1} = 3 \cdot 2 \cdot 1 = 6. \quad 0! = 1$$

As shown in the tree diagram in Figure 18, the 6 permutations are

LMN, LNM, MLN, MNL, NLM, NML.

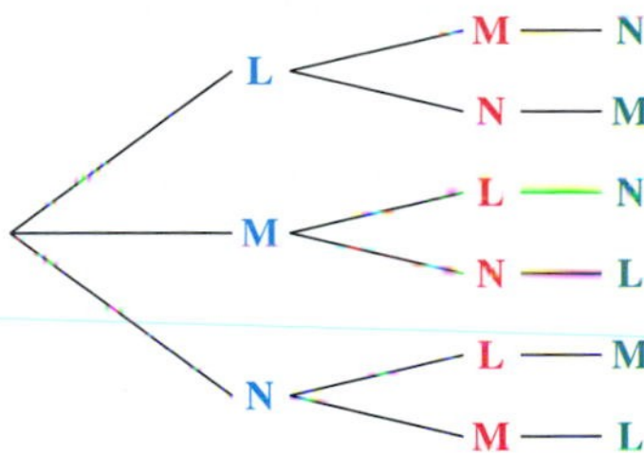

FIGURE 18

(b) $P(3,2) = \frac{3!}{(3-2)!} = \frac{3!}{1!} = \frac{6}{1} = 6$

This result is the same as the answer in part (a), because after the first two choices are made, the third choice is already determined, since only one letter is left.

Graphing Calculator Solution

The TI-83/84 Plus uses the notation nPr for evaluating permutations. This function, like the combinations function, is found in the MATH menu. The screen in Figure 19 confirms the analytic results in parts (a) and (b).

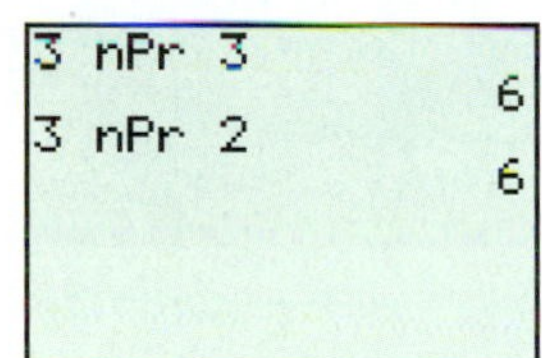

FIGURE 19

This example illustrates the general rule

$$P(n, n) = P(n, n-1).$$

■

EXAMPLE 5 **Using the Permutations Formula**

Suppose 8 people enter an event in a swim meet. In how many ways could the gold, silver, and bronze prizes be awarded?

Solution From the fundamental principle of counting, there are 3 events, resulting in $8 \cdot 7 \cdot 6 = 336$ choices. The formula for $P(n,r)$ gives the same result.

$$P(8,3) = \frac{8!}{5!} = \frac{8 \cdot 7 \cdot 6 \cdot 5 \cdot 4 \cdot 3 \cdot 2 \cdot 1}{5 \cdot 4 \cdot 3 \cdot 2 \cdot 1} = 336$$ ■

EXAMPLE 6 **Using the Permutations Formula**

In how many ways can 5 students be seated in a row of 5 desks?

Solution Use $P(n, n)$ with $n = 5$.

$$P(5, 5) = 5! = 5 \cdot 4 \cdot 3 \cdot 2 \cdot 1 = 120$$

Combinations

In Example 3, we saw that there are 60 ways that a teacher can arrange 3 of 5 different books in a row. That is, there are 60 permutations of 5 things taken 3 at a time. Suppose that the teacher does not wish to arrange the books in a row, but rather wishes to choose, *without regard to order*, any 3 of the 5 books to donate to a book sale. In how many ways can this be done?

At first glance, we might say 60 again, but this is incorrect. The number 60 counts all possible *arrangements* of 3 books chosen from 5. The following 6 arrangements, however, would all lead to the same set of 3 books being given to the book sale.

mystery, biography, textbook
mystery, textbook, biography
biography, mystery, textbook
biography, textbook, mystery
textbook, biography, mystery
textbook, mystery, biography

The list shows 6 different *arrangements* of 3 books, but only one *set* of 3 books. A subset of items selected *without regard to order* is called a **combination.** The number of combinations of 5 things taken 3 at a time is written $\binom{5}{3}$, ${}_5C_3$, or $C(5, 3)$.

NOTE This notation also represents the binomial coefficient. That is, binomial coefficients are combinations of n elements chosen r at a time.

To evaluate $\binom{5}{3}$ or $C(5, 3)$, start with the $5 \cdot 4 \cdot 3$ *permutations* of 5 things taken 3 at a time. Since order doesn't matter, and each subset of 3 items from the set of 5 items can have its elements rearranged in $3 \cdot 2 \cdot 1 = 3!$ ways, we find $\binom{5}{3}$ by dividing the number of permutations by 3!, or

$$\binom{5}{3} = \frac{5 \cdot 4 \cdot 3}{3!} = \frac{5 \cdot 4 \cdot 3}{3 \cdot 2 \cdot 1} = 10.$$

There are 10 ways that the teacher can choose 3 books for the book sale.

Generalizing this discussion gives the formula for the number of combinations of n elements taken r at a time:

$$C(n, r) = \binom{n}{r} = \frac{P(n, r)}{r!}.$$

Another version of the formula, found as follows, is most useful for calculation and is the one we used earlier to calculate binomial coefficients.

$$C(n, r) = \binom{n}{r} = \frac{P(n, r)}{r!} = \frac{n!}{(n-r)!} \cdot \frac{1}{r!} = \frac{n!}{(n-r)!\,r!}$$

Combinations of n Elements Taken r at a Time

If $C(n, r)$ or $\binom{n}{r}$ represents the number of combinations of n things taken r at a time, with $r \leq n$, then

$$C(n, r) = \binom{n}{r} = \frac{n!}{(n-r)!\,r!}.$$

NOTE $C(n, r) = {}_nC_r = \binom{n}{r}$

EXAMPLE 7 Using the Combinations Formula

How many different committees of 3 people can be chosen from a group of 8?

Analytic Solution

A committee is an unordered set. Use combinations.

$$\begin{aligned} C(8, 3) &= \binom{8}{3} \\ &= \frac{8!}{5!\,3!} \\ &= \frac{8 \cdot 7 \cdot 6 \cdot 5 \cdot 4 \cdot 3 \cdot 2 \cdot 1}{5 \cdot 4 \cdot 3 \cdot 2 \cdot 1 \cdot 3 \cdot 2 \cdot 1} \\ &= 56 \end{aligned}$$

Graphing Calculator Solution

The notation nCr is used by the TI-83/84 Plus to find combinations. The screen in Figure 20 agrees with the analytic result.

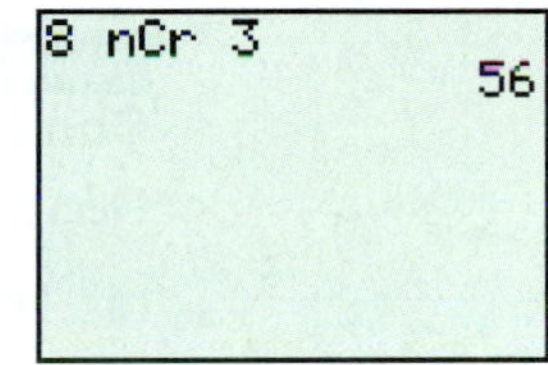

FIGURE 20

EXAMPLE 8 Using the Combinations Formula

Three lawyers are to be selected from a group of 30 to work on a special project.

(a) In how many different ways can the lawyers be selected?

(b) In how many ways can the group of 3 be selected if a certain lawyer must work on the project?

Solution

(a) Here we wish to know the number of 3-element combinations that can be formed from a set of 30 elements. (We want combinations, not permutations, since order within the group does not matter.)

$$C(30, 3) = \binom{30}{3} = \frac{30!}{27!\,3!} = 4060 \text{ ways}$$

(b) Since 1 lawyer has already been selected for the project, the problem is reduced to selecting 2 more from the remaining 29 lawyers.

$$C(29, 2) = \binom{29}{2} = \frac{29!}{27!\,2!} = 406 \text{ ways}$$

Distinguishing between Permutations and Combinations

Students often have difficulty determining whether to use permutations or combinations in solving problems. The following chart lists some of the similarities and differences between these two concepts.

Permutations	Combinations
Number of ways of selecting r items out of n items	
Repetitions are not allowed.	
Order is important.	Order is not important.
Arrangements of r items from a set of n items	Subsets of r items from a set of n items
$P(n, r) = \frac{n!}{(n-r)!}$	$C(n, r) = \binom{n}{r} = \frac{n!}{(n-r)!r!}$
Clue words: arrangement, schedule, order	Clue words: group, committee, sample, selection

EXAMPLE 9 Distinguishing between Permutations and Combinations

Decide whether permutations or combinations should be used to solve each problem.

(a) How many 4-digit codes are possible if no digits are repeated?

(b) A sample of 3 light bulbs is randomly selected from a batch of 15. How many different samples are possible?

(c) In a basketball tournament with 8 teams, how many games must be played so that each team plays every other team exactly once?

(d) In how many ways can 4 stockbrokers be assigned to 6 offices so that each broker has a private office?

Solution

(a) Since changing the order of the 4 digits results in a different code, permutations should be used.

(b) The order in which the 3 light bulbs are selected is not important. The sample is unchanged if the items are rearranged, so combinations should be used.

(c) The selection of 2 teams for a game is an *unordered* subset of 2 from the set of 8 teams. Use combinations.

(d) The office assignments are an *ordered* selection of 4 offices from the 6 offices. Exchanging the offices of any 2 brokers within a selection of 4 offices gives a different assignment, so permutations should be used. ■

To illustrate the differences between permutations and combinations in another way, suppose 2 cans of soup are to be selected from 4 cans on a shelf: noodle (N), bean (B), mushroom (M), and tomato (T). Then, as shown in Figure 21(a) on the next page, there are 12 ways to select 2 cans from the 4 cans if the order matters (if noodle first and bean second is considered different from bean, then noodle, for example).

On the other hand, if order is unimportant, then there are 6 ways to choose 2 cans of soup from the 4, as illustrated in Figure 21(b).

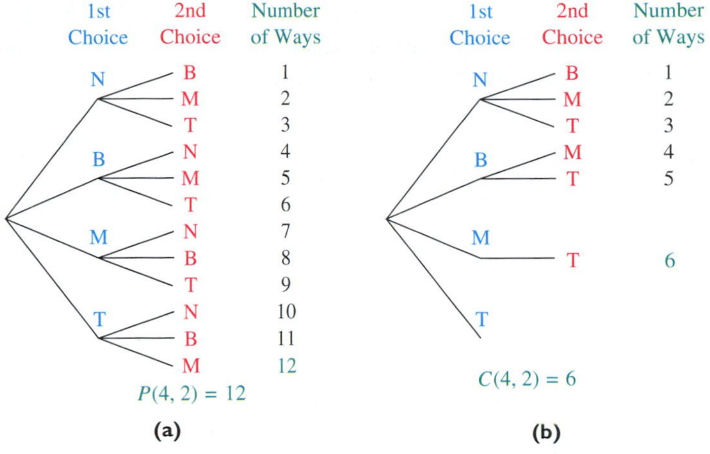

FIGURE 21

CAUTION Not all counting problems lend themselves to permutations or combinations. Whenever a tree diagram or the multiplication principle can be used directly, do so.

11.6 Exercises

Evaluate each expression.

1. $P(12, 8)$ **2.** $P(5, 5)$ **3.** $P(9, 2)$ **4.** $P(10, 9)$ **5.** $P(5, 1)$ **6.** $P(6, 0)$

7. $C(4, 2)$ **8.** $C(9, 3)$ **9.** $C(6, 0)$ **10.** $C(8, 1)$ **11.** $\binom{12}{4}$ **12.** $\binom{16}{3}$

Use a calculator to evaluate each expression.

13. ${}_{20}P_5$ **14.** ${}_{100}P_5$ **15.** ${}_{15}P_8$ **16.** ${}_{32}P_4$

17. ${}_{20}C_5$ **18.** ${}_{100}C_5$ **19.** $\binom{15}{8}$ **20.** $\binom{32}{4}$

21. ***Concept Check*** Decide whether the situation described involves a permutation or a combination of objects.
- **(a)** A telephone number
- **(b)** A Social Security number
- **(c)** A hand of cards in poker
- **(d)** A committee of politicians
- **(e)** The "combination" on a combination lock
- **(f)** A lottery choice of six numbers where the order does not matter
- **(g)** An automobile license plate

22. Explain the difference between a permutation and a combination. What should you look for in a problem to decide which of these is an appropriate method of solution?

Use the fundamental principle of counting or permutations to solve each problem.

23. ***Home Plan Choices*** How many different types of homes are available if a builder offers a choice of 5 basic plans, 3 roof styles, and 2 exterior finishes?

24. ***Auto Varieties*** An auto manufacturer produces 7 models, each available in 6 different colors, 4 different upholstery fabrics, and 5 interior colors. How many varieties of the auto are available?

25. ***Radio Station Call Letters*** How many different 4-letter radio station call letters can be made
- **(a)** if the first letter must be K or W and no letter may be repeated?

(b) if repeats are allowed (but the first letter is K or W)?
(c) How many of the 4-letter call letters (starting with K or W) with no repeats end in R?

26. *Meal Choices* A menu offers a choice of 3 salads, 8 main dishes, and 5 desserts. How many different 3-course meals (salad, main dish, dessert) are possible?

27. *Names for a Baby* A couple having a baby has narrowed down the choice of a name for the new baby to 3 first names and 5 middle names. How many different first- and middle-name pairings are possible?

28. *Concert Program Arrangement* A concert to raise money for an economics prize is to consist of 5 works: 2 overtures, 2 sonatas, and a piano concerto. In how many ways can a program with these 5 works be arranged?

29. *License Plates* For many years, the state of California used 3 letters followed by 3 digits on its automobile license plates.

(a) How many different license plates are possible with this arrangement?
(b) When the state ran out of new plates, the order was reversed to 3 digits followed by 3 letters. How many additional plates were then possible?
(c) Several years ago, the plates described in part (b) were also used up. The state then issued plates with 1 letter, followed by 3 digits, and then 3 letters. How many plates does this scheme provide?

30. *Telephone Numbers* How many 7-digit telephone numbers are possible if the first digit cannot be 0 and
(a) only odd digits may be used?
(b) the telephone number must be a multiple of 10 (that is, it must end in 0)?
(c) the telephone number must be a multiple of 100?
(d) the first 3 digits are 481?
(e) no repetitions are allowed?

31. *Seating People in a Row* In an experiment on social interaction, 6 people will sit in 6 seats in a row. In how many ways can this be done?

32. *Genetics Experiment* In how many ways can 7 of 10 monkeys be arranged in a row for a genetics experiment?

33. *Course Schedule Arrangement* A business school offers courses in keyboarding, spreadsheets, transcription, business English, technical writing, and accounting. In how many ways can a student arrange a schedule if 3 courses are taken?

34. *Course Schedule Arrangement* If your college offers 400 courses, 20 of which are in mathematics, and your counselor arranges your schedule of 4 courses by random selection, how many schedules are possible that do not include a math course?

35. *Club Officer Choices* In a club with 15 members, how many ways can a slate of 3 officers consisting of president, vice-president, and secretary/treasurer be chosen?

36. *Batting Orders* A baseball team has 20 players. How many 9-player batting orders are possible?

37. ***Basketball Positions*** In how many ways can 5 players be assigned to the 5 positions on a basketball team, assuming that any player can play any position? In how many ways can 10 players be assigned to the 5 positions?

38. ***Letter Arrangement*** How many ways can all the letters of the word TOUGH be arranged?

Solve each problem involving combinations.

39. ***Seminar Presenters*** A banker's association has 30 members. If 4 members are selected at random to present a seminar, how many different groups of 4 are possible?

40. ***Apple Samples*** How many different samples of 3 apples can be drawn from a crate of 25 apples?

41. ***Hamburger Choices*** Howard's Hamburger Heaven sells hamburgers with cheese, relish, lettuce, tomato, mustard, or ketchup. How many different hamburgers can be made that use any 3 of the extras?

42. ***Financial Planners*** Three financial planners are to be selected from a group of 12 to participate in a special program. In how many ways can this be done? In how many ways can the group that will not participate be selected?

43. ***Card Combinations*** Five cards marked respectively with the numbers 1, 2, 3, 4, and 5 are shuffled, and 2 cards are then drawn. How many different 2-card hands are possible?

44. ***Marble Samples*** If a bag contains 15 marbles, how many samples of 2 marbles can be drawn from it? how many samples of 4 marbles?

45. ***Marble Samples*** In Exercise 44, if the bag contains 3 yellow, 4 white, and 8 blue marbles, how many samples of 2 can be drawn in which both marbles are blue?

46. ***Apple Samples*** In Exercise 40, if it is known that there are 5 rotten apples in the crate,

(a) how many samples of 3 could be drawn in which all 3 are rotten?

(b) how many samples of 3 could be drawn in which there are 1 rotten apple and 2 good apples?

47. ***Convention Delegation Choices*** A city council is composed of 5 liberals and 4 conservatives. Three members are to be selected randomly as delegates to a convention.

(a) How many delegations are possible?

(b) How many delegations could have all liberals?

(c) How many delegations could have 2 liberals and 1 conservative?

(d) If 1 member of the council serves as mayor, how many delegations are possible that include the mayor?

48. ***Delegation Choices*** Seven workers decide to send a delegation of 2 to their supervisor to discuss their grievances.

(a) How many different delegations are possible?

(b) If it is decided that a certain employee must be in the delegation, how many different delegations are possible?

(c) If there are 2 women and 5 men in the group, how many delegations would include at least 1 woman?

Use any or all of the methods described in this section to solve each problem.

49. ***Course Schedule*** If Patrick McTernan has 8 courses to choose from, how many ways can he arrange his schedule if he must pick 4 of them?

50. ***Pineapple Samples*** How many samples of 3 pineapples can be drawn from a crate of 12?

51. ***Soup Ingredients*** Blanche Stowell specializes in making different vegetable soups with carrots, celery, beans, peas, mushrooms, and potatoes. How many different soups can she make with any 4 ingredients?

52. ***Secretary/Manager Assignments*** From a pool of 7 secretaries, 3 are selected to be assigned to 3 managers, 1 secretary to each manager. In how many ways can this be done?

53. ***Musical Chairs Seatings*** In a game of musical chairs, 12 children will sit in 11 chairs. One will be left out. How many seatings are possible?

54. ***Plant Samples*** In an experiment on plant hardiness, a researcher gathers 6 wheat plants, 3 barley plants, and 2 rye plants. She wishes to select 4 plants at random.

(a) In how many ways can this be done?

(b) In how many ways can it be done if exactly 2 wheat plants must be included?

55. ***Committee Choices*** In a club with 8 men and 11 women members, how many 5-member committees can be chosen that have the following?

(a) All men **(b)** All women

(c) 3 men and 2 women **(d)** No more than 3 women

56. ***Committee Choices*** From 10 names on a ballot, 4 will be elected to a political party committee. In how many ways can the committee of 4 be formed if each person will have a different responsibility?

57. ***Garage Door Openers*** The code for some garage door openers consists of 12 electrical switches that can be set to either 0 or 1 by the owner. With this type of opener, how many codes are possible? (*Source:* Promax.)

58. ***Combination Lock*** A typical combination for a padlock consists of 3 numbers from 0 to 39. Count the number of combinations that are possible with this type of lock if a number may be repeated.

59. ***Combination Lock*** A briefcase has 2 locks. The combination to each lock consists of a 3-digit number, where digits may be repeated. How many combinations are possible? (*Hint:* The word *combination* is a misnomer. Lock combinations are permutations because the arrangement of the numbers is important.)

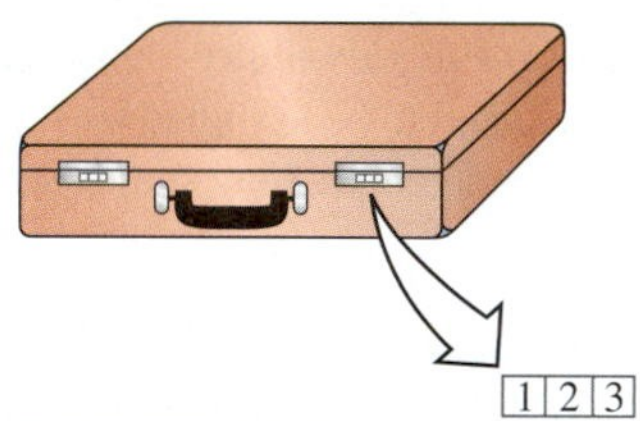

60. ***Lottery*** To win the jackpot in a lottery game, a person must pick 3 numbers from 0 to 9 in the correct order. If a number can be repeated, how many ways are there to play the game?

61. ***Keys*** How many distinguishable ways can 4 keys be put on a circular key ring? (*Hint*: Consider that clockwise and counter clockwise arrangements are not different.)

62. ***Sitting at a Round Table*** How many ways can 7 people sit at a round table? Assume that a different way means that at least 1 person is sitting next to someone different.

Prove each statement for positive integers n and r, with $r \leq n$. (Hint: Use the definitions of permutations and combinations.)

63. $P(n, n-1) = P(n, n)$

64. $P(n, 1) = n$

65. $P(n, 0) = 1$

66. $\binom{n}{n} = 1$

67. $\binom{n}{0} = 1$

68. $\binom{n}{1} = n$

69. $\binom{n}{n-1} = n$

70. $\binom{n}{n-r} = \binom{n}{r}$

71. Explain why the restriction $r \leq n$ is needed in the formula for $P(n, r)$.

72. If $P(n, r)$ is entered into a graphing calculator with noninteger values for n and r, what happens?

11.7 Probability

Basic Concepts ■ Complements and Venn Diagrams ■ Odds ■ Union of Two Events ■ Binomial Probability

Basic Concepts

Consider an experiment that has one or more **outcomes,** each of which is equally likely to occur. For example, the experiment of tossing a fair coin has two equally likely outcomes: heads (H) or tails (T). Also, the experiment of rolling a fair die has 6 equally likely outcomes: landing so the face that is up shows a 1, 2, 3, 4, 5, or 6.

The set S of all possible outcomes of a given experiment is called the **sample space** for the experiment. (In this text, all sample spaces are finite.) The sample space for the experiment of tossing a coin once consists of the outcomes H and T. This sample space can be written in set notation as

$$S = \{H, T\}.$$

Similarly, a sample space for the experiment of rolling a single die once is

$$S = \{1, 2, 3, 4, 5, 6\}.$$

Any subset of the sample space is called an **event.** In the experiment with the die, for example, "the number showing is a 3" is an event, say, $E_1 = \{3\}$. "The number showing is greater than 3" is also an event, say, $E_2 = \{4, 5, 6\}$. To represent the number of outcomes that belong to event E, the notation $n(E)$ is used. Then, $n(E_1) = 1$ and $n(E_2) = 3$.

The notation $P(E)$ is used for the *probability* of an event E. If the outcomes in the sample space for an experiment are equally likely, then the probability of event E occurring is found as follows.

Probability of Event *E*

In a sample space with equally likely outcomes, the **probability** of an event E, written $\boldsymbol{P(E)}$, is the ratio of the number of outcomes in sample space S that belong to event E, denoted $\boldsymbol{n(E)}$, to the total number of outcomes in sample space S, denoted $\boldsymbol{n(S)}$. That is,

$$P(E) = \frac{n(E)}{n(S)}.$$

To find the probability of event E_1 in the die experiment, start with the sample space $S = \{1, 2, 3, 4, 5, 6\}$, and the desired event $E_1 = \{3\}$. Since $n(E_1) = 1$ and $n(S) = 6$,

$$P(E_1) = \frac{n(E_1)}{n(S)} = \frac{1}{6}.$$

EXAMPLE 1 Finding Probabilities of Events

A single die is rolled. Write each event in set notation, and give the probability of the event.

(a) E_3: the number showing is even

(b) E_4: the number showing is greater than 4

(c) E_5: the number showing is less than 7

(d) E_6: the number showing is 7

Solution

(a) Since $E_3 = \{2, 4, 6\}$, $n(E_3) = 3$. As given earlier, $n(S) = 6$, so

$$P(E_3) = \frac{3}{6} = \frac{1}{2}.$$

(b) Again, $n(S) = 6$. Event $E_4 = \{5, 6\}$, with $n(E_4) = 2$, so

$$P(E_4) = \frac{2}{6} = \frac{1}{3}.$$

(c) $E_5 = \{1, 2, 3, 4, 5, 6\}$ and $P(E_5) = \frac{6}{6} = 1$.

(d) $E_6 = \emptyset$ and $P(E_6) = \frac{0}{6} = 0$. ■

In Example 1(c), $E_5 = S$. Therefore, the event E_5 is certain to occur every time the experiment is performed. ***An event that is certain to occur always has probability 1.*** On the other hand, $E_6 = \emptyset$ and $P(E_6) = 0$. ***The probability of an impossible event is always 0. For any event E, P(E) is between 0 and 1 inclusive.***

Complements and Venn Diagrams

The set of all outcomes in the sample space that do *not* belong to event E is called the **complement** of E, written $\boldsymbol{E'}$. In an experiment of drawing a single card from a standard deck of 52 cards, let E be the event "the card is an ace." Then E' is the event "the card is not an ace." From the definition of E', for an event E,

$$\boldsymbol{E \cup E' = S} \quad \text{and} \quad \boldsymbol{E \cap E' = \emptyset}.^*$$

FIGURE 22

Probability concepts can be illustrated with **Venn diagrams,** as shown in Figure 22. The rectangle in that figure represents the sample space in an experiment. The area inside the circle represents event E, and the area inside the rectangle, but outside the circle, represents event E'.

NOTE A standard deck of 52 cards has four suits: hearts ♥, diamonds ♦, spades ♠, and clubs ♣, with 13 cards of each suit. Each suit has a jack, a queen, and a king (sometimes called the "face cards"), an ace, and cards numbered from 2 to 10. The hearts and diamonds are red and the spades and clubs are black. We will refer to this standard deck of cards in this section.

EXAMPLE 2 Using the Complement

In the experiment of drawing a card from a well-shuffled deck, find the probability of event E, "the card is an ace," and event E'.

Solution There are 4 aces in the deck of 52 cards, so $n(E) = 4$ and $n(S) = 52$.

$$P(E) = \frac{n(E)}{n(S)} = \frac{4}{52} = \frac{1}{13}$$

Of the 52 cards, 48 are not aces.

$$P(E') = \frac{n(E')}{n(S)} = \frac{48}{52} = \frac{12}{13}$$

In Example 2, $P(E) + P(E') = \frac{1}{13} + \frac{12}{13} = 1$. This is always true for any event E and its complement E'. That is,

$$\boldsymbol{P(E) + P(E') = 1,}$$

which can be restated as

$$\boldsymbol{P(E) = 1 - P(E')} \quad \text{or} \quad \boldsymbol{P(E') = 1 - P(E).}$$

*The **union** of two sets A and B is the set $\boldsymbol{A \cup B}$ made up of all the elements that belong to either A or B or both. The **intersection** of sets A and B, written $\boldsymbol{A \cap B}$, is made up of all the elements that belong to both sets.

The last two equations, $P(E) = 1 - P(E')$ and $P(E') = 1 - P(E)$, suggest an alternative way to compute the probability of an event. For example, if $P(E) = \frac{1}{13}$, then

$$P(E') = 1 - \frac{1}{13} = \frac{12}{13}.$$

Odds

Sometimes probability statements are expressed in terms of *odds*, a comparison of $P(E)$ with $P(E')$. The **odds** in favor of an event E are expressed as the ratio of $P(E)$ to $P(E')$, or as the fraction $\frac{P(E)}{P(E')}$. For example, if the probability of rain can be established as $\frac{1}{3}$, the odds that it will rain are

$$P(\text{rain}) \text{ to } P(\text{no rain}) = \frac{1}{3} \text{ to } \frac{2}{3} = \frac{\frac{1}{3}}{\frac{2}{3}} = \frac{1}{3} \div \frac{2}{3} = \frac{1}{3} \cdot \frac{3}{2} = \frac{1}{2}, \quad \text{or} \quad 1 \text{ to } 2.$$

Be careful simplifying a complex fraction.

The odds that it will *not* rain are 2 to 1 $\left(\text{or } \frac{2}{3} \text{ to } \frac{1}{3}\right)$. If the odds in favor of an event are, say, 3 to 5, then the probability of the event is $\frac{3}{8}$, and the probability of the complement of the event is $\frac{5}{8}$. If the odds favoring event E are m to n, then

$$\mathbf{P(E) = \frac{m}{m + n}} \quad \text{and} \quad \mathbf{P(E') = \frac{n}{m + n}}.$$

EXAMPLE 3 Finding Odds in Favor of an Event

A shirt is selected at random from a dark closet containing 6 blue shirts and 4 shirts that are not blue. Find the odds in favor of a blue shirt being selected.

Solution Let E represent "a blue shirt is selected." Then

$$P(E) = \frac{6}{10}, \quad \text{or} \quad \frac{3}{5}, \quad \text{and} \quad P(E') = 1 - \frac{3}{5} = \frac{2}{5}.$$

Therefore, the odds in favor of a blue shirt being selected are

$$P(E) \text{ to } P(E') = \frac{3}{5} \text{ to } \frac{2}{5} = \frac{\frac{3}{5}}{\frac{2}{5}} = \frac{3}{5} \div \frac{2}{5} = \frac{3}{5} \cdot \frac{5}{2} = \frac{3}{2}, \quad \text{or} \quad 3 \text{ to } 2.$$

Be careful simplifying a complex fraction.

Union of Two Events

Since events are sets, we can use set operations to find the union of two events. Suppose a fair die is tossed. Let H be the event "the result is a 3" and K the event "the result is an even number." From the results earlier in this section,

$$H = \{3\} \qquad K = \{2, 4, 6\} \qquad H \cup K = \{2, 3, 4, 6\}$$

$$P(H) = \frac{1}{6} \qquad P(K) = \frac{3}{6} = \frac{1}{2} \qquad P(H \cup K) = \frac{4}{6} = \frac{2}{3}.$$

Notice that $P(H) + P(K) = P(H \cup K)$.

Before assuming that this relationship is true in general, consider another event G for this experiment: "The result is a 2."

$$G = \{2\} \qquad K = \{2, 4, 6\} \qquad G \cup K = \{2, 4, 6\}$$

$$P(G) = \frac{1}{6} \qquad P(K) = \frac{3}{6} = \frac{1}{2} \qquad P(G \cup K) = \frac{3}{6} = \frac{1}{2}$$

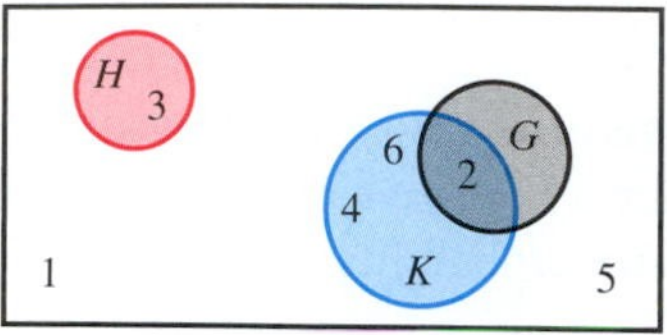

FIGURE 23

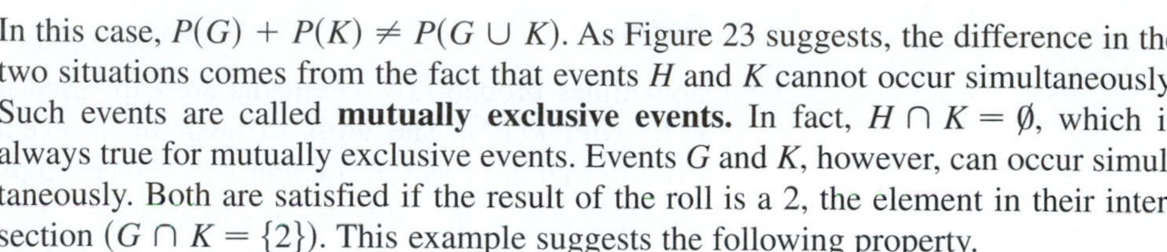

In this case, $P(G) + P(K) \neq P(G \cup K)$. As Figure 23 suggests, the difference in the two situations comes from the fact that events H and K cannot occur simultaneously. Such events are called **mutually exclusive events.** In fact, $H \cap K = \emptyset$, which is always true for mutually exclusive events. Events G and K, however, can occur simultaneously. Both are satisfied if the result of the roll is a 2, the element in their intersection ($G \cap K = \{2\}$). This example suggests the following property.

Probability of the Union of Two Events

For any events E and F,

$$P(E \text{ or } F) = P(E \cup F) = P(E) + P(F) - P(E \cap F).$$

EXAMPLE 4 Finding Probabilities of Unions

One card is drawn from a well-shuffled deck of 52 cards. What is the probability of each event?

(a) The card is an ace or a spade. **(b)** The card is a 3 or a king.

Solution

(a) The events "an ace is drawn" and "a spade is drawn" are not mutually exclusive, since it is possible to draw the ace of spades, an outcome satisfying both events.

$$P(\text{ace or spade}) = P(\text{ace}) + P(\text{spade}) - P(\text{ace and spade})$$

$$= \frac{4}{52} + \frac{13}{52} - \frac{1}{52} = \frac{16}{52} = \frac{4}{13}$$

(b) "A 3 is drawn" and "a king is drawn" are mutually exclusive events, because it is impossible to draw one card that is both a 3 and a king.

$$P(3 \text{ or } K) = P(3) + P(K) - P(3 \text{ and } K)$$

$$= \frac{4}{52} + \frac{4}{52} - 0 = \frac{8}{52} = \frac{2}{13}$$ ■

EXAMPLE 5 Finding Probabilities of Unions

Suppose two fair dice are rolled, one at a time. Find each probability.

(a) The first die shows a 2, or the sum of the two dice is 6 or 7.

(b) The sum of the two dice is at most 4.

Solution

(a) Think of the two dice as being distinguishable—the first one red and the second one green, for example. (Actually, the sample space is the same even if they are

not distinguishable.) A sample space with equally likely outcomes is shown in Figure 24, where $(1, 1)$ represents the event "the first (red) die shows a 1 and the second (green) die shows a 1," $(1, 2)$ represents "the first die shows a 1 and the second die shows a 2," and so on.

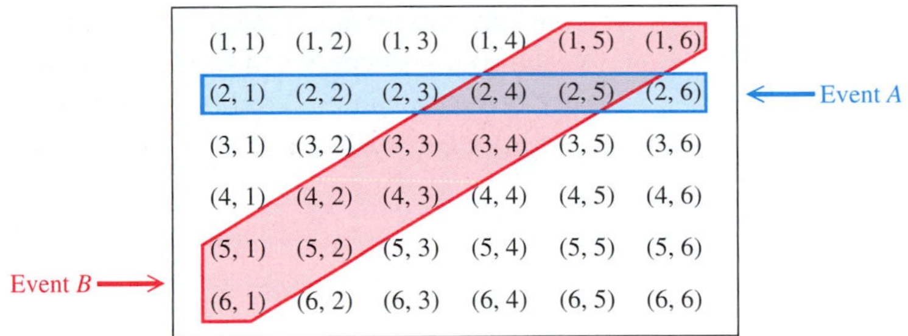

FIGURE 24

Let A represent the event "the first die shows a 2" and B represent the event "the sum of the results is 6 or 7." These events are indicated in the figure. From the diagram, event A has 6 elements, B has 11 elements, and the sample space has 36 elements. Thus,

$$P(A) = \frac{6}{36}, \quad P(B) = \frac{11}{36}, \quad \text{and} \quad P(A \cap B) = \frac{2}{36},$$

so

$$\begin{aligned} P(A \cup B) &= P(A) + P(B) - P(A \cap B) \\ &= \frac{6}{36} + \frac{11}{36} - \frac{2}{36} = \frac{15}{36} = \frac{5}{12}. \end{aligned}$$

(b) "At most 4" can be written as "2 or 3 or 4." (A sum of 1 is not possible.) Since the events represented by "2," "3," and "4" are mutually exclusive,

$$P(\text{at most } 4) = P(2 \text{ or } 3 \text{ or } 4) = P(2) + P(3) + P(4). \qquad (*)$$

The sample space for this experiment includes the 36 possible pairs of numbers shown in Figure 24. The pair $(1, 1)$ is the only one with a sum of 2, so $P(2) = \frac{1}{36}$. Also, $P(3) = \frac{2}{36}$, since both $(1, 2)$ and $(2, 1)$ give a sum of 3. The pairs $(1, 3)$, $(2, 2)$, and $(3, 1)$ have a sum of 4, so $P(4) = \frac{3}{36}$. Substituting into equation (*) gives

$$P(\text{at most } 4) = \frac{1}{36} + \frac{2}{36} + \frac{3}{36} = \frac{6}{36} = \frac{1}{6}.$$

The properties of probability are summarized as follows.

Properties of Probability

For any events E and F,

1. $0 \leq P(E) \leq 1$;
2. $P(\text{a certain event}) = 1$;
3. $P(\text{an impossible event}) = 0$;
4. $P(E') = 1 - P(E)$;
5. $P(E \text{ or } F) = P(E \cup F) = P(E) + P(F) - P(E \cap F)$.

CAUTION ***When finding the probability of a union, remember to subtract the probability of the intersection from the sum of the probabilities of the individual events.***

Binomial Probability

An experiment that consists of repeated independent trials with only two outcomes in each trial—success and failure—is called a **binomial experiment.** Let the probability of success in one trial be p. Then the probability of failure is $1 - p$, and the probability of r successes in n trials is given by

$$\binom{n}{r} p^r (1 - p)^{n-r}.$$

GCM **EXAMPLE 6** **Using a Binomial Experiment to Find Probabilities**

An experiment consists of rolling a die 10 times. Find each probability.

(a) Exactly 4 of the tosses result in a 3.

(b) In 9 of the 10 tosses, the result is not a 3.

Analytic Solution

(a) The probability of a 3 on one roll is $p = \frac{1}{6}$. Here, $n = 10$ and $r = 4$, so the required probability is

$$\binom{10}{4}\left(\frac{1}{6}\right)^4\left(1 - \frac{1}{6}\right)^{10-4} = 210\left(\frac{1}{6}\right)^4\left(\frac{5}{6}\right)^6$$
$$\approx .054.$$

(b) $\binom{10}{9}\left(\frac{5}{6}\right)^9\left(\frac{1}{6}\right)^1 \approx .323$ $\quad n = 10, r = 9, p = 1 - \frac{1}{6} = \frac{5}{6}$

Graphing Calculator Solution

Graphing calculators that have statistical distribution functions give binomial probabilities. In Figure 25, the numbers in parentheses separated by commas represent n, p, and r, respectively.

```
binompdf(10,(1/6
),4)
      .0542658759
binompdf(10,(5/6
),9)
      .3230111658
```

FIGURE 25

11.7 Exercises

State a sample space S with equally likely outcomes for each experiment.

1. A two-headed coin is tossed once.
2. Two ordinary coins are tossed.
3. Three ordinary coins are tossed.
4. Slips of paper, each of which is marked with the number 1, 2, 3, 4, or 5, are placed in a box. After mixing well, two slips are drawn, with the order not important.
5. The spinner shown here is spun twice.

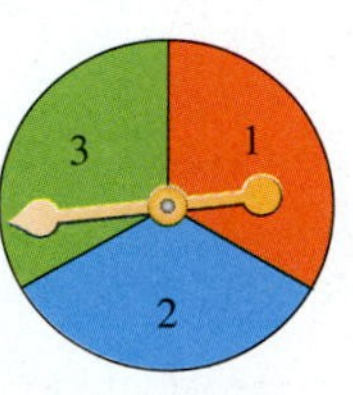

6. A die is rolled and then a coin is tossed.

Write each event in set notation. Give the probability of the event.

7. In Exercise 1,
(a) the result of the toss is heads.
(b) the result of the toss is tails.

8. In Exercise 2,
(a) both coins show the same face.
(b) at least one coin turns up heads.

9. In Exercise 5,
(a) the result is a repeated number.
(b) the second number is 1 or 3.
(c) the first number is even and the second number is odd.

10. In Exercise 4,
(a) both slips are marked with even numbers.
(b) both slips are marked with odd numbers.
(c) both slips are marked with the same number.
(d) one slip is marked with an odd number, the other with an even number.

11. A student gives the number to a probability problem as $\frac{6}{5}$. Explain why this answer must be incorrect.

12. *Concept Check* If the probability of an event is .857, what is the probability that the event will not occur?

Work each problem.

13. *Drawing a Marble* A marble is drawn at random from a box containing 3 yellow, 4 white, and 8 blue marbles. Find the probabilities in parts (a)–(c).
(a) A yellow marble is drawn.
(b) A black marble is drawn.
(c) The marble is yellow or white.
(d) What are the odds in favor of drawing a yellow marble?
(e) What are the odds against drawing a blue marble?

14. *Batting Average* A baseball player with a batting average of .300 comes to bat. What are the odds in favor of his getting a hit?

15. *Drawing Slips of Paper* In Exercise 4, what are the odds that the sum of the numbers on the two slips of paper is 5?

16. *Probability of Rain* If the odds that it will rain are 4 to 5, what is the probability of rain?

17. *Probability of a Candidate Losing* If the odds that a candidate will win an election are 3 to 2, what is the probability that the candidate will lose?

18. *Drawing a Card* A card is drawn from a well-shuffled deck of 52 cards. Find the probability that the card is
(a) a 9. **(b)** black. **(c)** a black 9.
(d) a heart. **(e)** a face card (K, Q, or J of any suit).
(f) red or a 3. **(g)** less than a 4 (consider aces as 1s).

19. *Guest Arrival at a Party* Mrs. Schmulen invites 10 relatives to a party: her mother, 2 uncles, 3 brothers, and 4 cousins. If the chances of any one guest arriving first are equally likely, find each probability.
(a) The first guest is an uncle or a brother.
(b) The first guest is a brother or a cousin.
(c) The first guest is a brother or her mother.

20. *Dice Rolls* Two dice are rolled. Find the probability of each event.
(a) The sum is at least 10.
(b) The sum is either 7 or at least 10.
(c) The sum is 2 or the dice both show the same number.

21. *Concept Check* Match each probability in parts (a)–(g) with one of the statements in A–F.
(a) $P(E) = -.1$ **(b)** $P(E) = .01$
(c) $P(E) = 1$ **(d)** $P(E) = 2$
(e) $P(E) = .99$ **(f)** $P(E) = 0$
(g) $P(E) = .5$
A. The event is certain to occur.
B. The event is impossible.
C. The event is very likely to occur.
D. The event is very unlikely to occur.
E. The event is just as likely to occur as not occur.
F. This probability cannot occur.

22. *Small-Business Loan* The probability that a bank with assets greater than or equal to $30 billion will make a loan to a small business is .002. What are the odds against such a bank making a small-business loan? (*Source: The Wall Street Journal* analysis of *CA1 Reports*.)

23. *Estimating Probability of Organ Transplants* In a recent year there were 51,277 people waiting for an organ transplant. The following table lists the number of patients waiting for the most common types of transplants.

Organ Transplant	Patients Waiting
Heart	3,774
Kidney	35,025
Liver	7,920
Lung	2,340

Source: Coalition on Organ and Tissue Donation.

Assuming that none of these people need two or more transplants, approximate the probability that a transplant patient chosen at random will need
(a) a kidney or a heart;
(b) neither a kidney nor a heart.

24. ***U.S. Population Origins*** Projected Hispanic and non-Hispanic U.S. populations (in thousands) for 2025 are given in the following table.

Type	Number
Hispanic	58,930
White	209,117
Black	43,511
Native American	2,744
Asian	20,748

Source: U.S. Census Bureau.

Assume that these projections are accurate. Estimate the probability that a U.S. resident selected at random in 2025 is

(a) Hispanic. **(b)** not white.
(c) Native American or black.
(d) What are the odds that a randomly selected U.S. resident is Asian?

25. ***Consumer Purchases*** The following table shows the probability that a customer at a department store will make a purchase in the indicated price range.

Cost	Probability
Below $5	.25
$5–$19.99	.37
$20–$39.99	.11
$40–$69.99	.09
$70–$99.99	.07
$100–$149.99	.08
$150 or more	.03

Find the probability that a customer makes a purchase that is

(a) less than $20. **(b)** $40 or more.
(c) more than $99.99. **(d)** less than $100.

26. ***State Lottery*** One game in a state lottery requires you to pick 1 heart, 1 club, 1 diamond, and 1 spade, in that order, from the 13 cards in each suit. What is the probability of getting all four picks correct and winning $5000?

27. ***State Lottery*** If three of the four selections in Exercise 26 are correct, the player wins $200. Find the probability of this outcome.

28. ***Partner Selection*** The law firm of Alam, Bartolini, Chinn, Dickinson, and Ellsberg has two senior partners: Alam and Bartolini. Two of the attorneys are to be selected to attend a conference. Assuming that all are equally likely to be selected, find each probability.

(a) Chinn is selected.
(b) Alam and Dickinson are selected.
(c) At least one senior partner is selected.

29. ***Opinion Survey*** The management of a firm surveys its workers, classified as follows for the purpose of an interview: 30% have worked for the company more than 5 years; 28% are female; 65% contribute to a voluntary retirement plan; and $\frac{1}{2}$ of the female workers contribute to the retirement plan. Find each probability.

(a) A male worker is selected.
(b) A worker is selected who has worked for the company 5 years or less.
(c) A worker is selected who contributes to the retirement plan or is female.

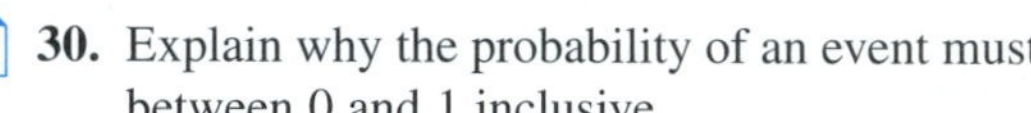

30. Explain why the probability of an event must be a number between 0 and 1 inclusive.

Gender Makeup of a Family *Suppose a family has 5 children. Suppose also that the probability of having a girl is $\frac{1}{2}$. Find the probability that the family has the following children.*

31. Exactly 2 girls and 3 boys

32. Exactly 3 girls and 2 boys

33. No girls

34. No boys

35. At least 3 boys

36. No more than 4 girls

College Student Smokers *The table gives the results of a survey of 14,000 college students who were cigarette smokers in a recent year.*

Number of Cigarettes per Day	Percent (as a Decimal)
Less than 1	.45
1 to 9	.24
10 to 19	.20
A pack of 20 or more	.11

Using the percents as probabilities, find the probability that, out of 10 of these student smokers selected at random, the following were true.

37. Four smoked less than 10 cigarettes per day.

38. Five smoked a pack or more per day.

39. Fewer than 2 smoked between 1 and 19 cigarettes per day.

40. No more than 3 smoked less than 1 cigarette per day.

A die is rolled 12 times. Find the probability of rolling the following.

41. Exactly 12 ones

42. Exactly 6 ones

43. No more than 3 ones

44. No more than 1 one

College Applications *The table gives the results of a survey of 282,549 freshmen from the class of 2006 at 437 of the nation's baccalaureate colleges and universities.*

Number of Colleges Applied to	1	2 or 3	4–6	7 or more
Percent (as a decimal)	.20	.29	.37	.14

Source: Higher Education Research Institute, UCLA, 2002.

Using the percents as probabilities, find the probability of each event for a randomly selected student.

45. The student applied to fewer than 4 colleges.

46. The student applied to at least 2 colleges.

47. The student applied to more than 3 colleges.

48. The student applied to no colleges.

49. ***Color-Blind Males*** The probability that a male will be color blind is .042. Find the probabilities that in a group of 53 men, the following are true.
(a) Exactly 5 are color blind.
(b) No more than 5 are color blind.
(c) At least 1 is color blind.

50. The screens in the next column illustrate how the table feature of a graphing calculator can be used to find the probabilities of having 0, 1, 2, 3, or 4 girls in a family of 4 children. (Note that 0 appears for X = 5 and X = 6. Why is this so?) Use this approach to answer the following.

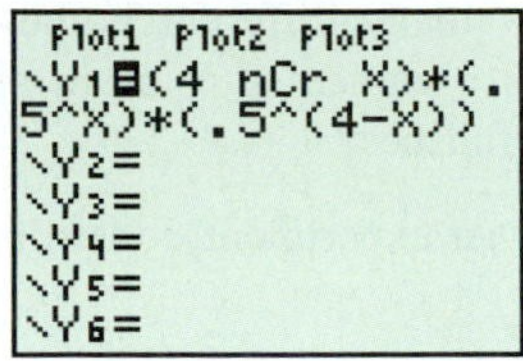

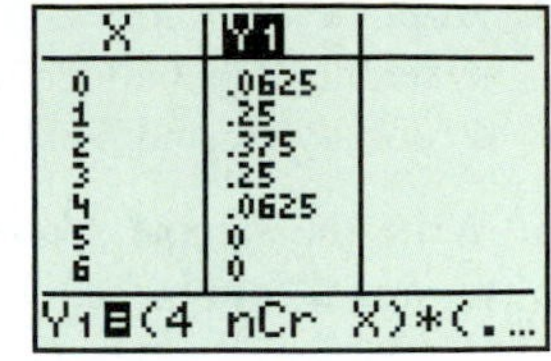

(a) Find the probabilities of having 0, 1, 2, or 3 boys in a family of 3 children.
(b) Find the probabilities of having 0, 1, 2, 3, 4, 5, or 6 girls in a family of 6 children.

51. ***(Modeling) Disease Infection*** What will happen when an infectious disease is introduced into a family? Suppose a family has I infected members and S members who are not infected, but are susceptible to contracting the disease. The probability P of k people not contracting the disease during a 1-week period can be calculated by the formula

$$P = \binom{S}{k} q^k (1 - q)^{S-k},$$

where $q = (1 - p)^I$ and p is the probability that a susceptible person contracts the disease from an infectious person. For example, if $p = .5$, then there is a 50% chance that a susceptible person exposed to one infectious person for 1 week will contract the disease. (*Source:* Hoppensteadt, F. and C. Peskin, *Mathematics in Medicine and the Life Sciences,* Springer-Verlag, 1992.)
(a) Compute the probability P of 3 family members not becoming infected within 1 week if there are currently 2 infected and 4 susceptible members. Assume that $p = .1$.
(b) A highly infectious disease can have $p = .5$. Repeat part (a) with this value of p.
(c) Determine the probability that everyone would become sick in a large family if initially $I = 1$, $S = 9$, and $p = .5$. Discuss the results.

52. ***(Modeling) Disease Infection*** Refer to Exercise 51. Suppose that, in a family, $I = 2$ and $S = 4$. If the probability P is .25 of there being $k = 2$ uninfected members after 1 week, estimate graphically the possible values of p. (*Hint:* Write P as a function of p.)

Reviewing Basic Concepts (Sections 11.6 and 11.7)

1. ***Book Arrangement*** A student has 4 different books that she wishes to arrange in a row. How many arrangements are there?

2. Calculate $P(7, 3)$.

3. ***Basketball Player Choices*** How many ways can 2 guards, 2 forwards, and 1 center be selected from a basketball team composed of 6 guards, 5 forwards, and 3 centers?

4. Calculate $\binom{10}{4}$.

5. ***Home Plan Choices*** How many different homes are available if a builder offers a choice of 9 basic plans, 4 roof styles, and 2 exterior finishes?

6. Write the sample space for an experiment consisting of tossing a coin twice.

7. ***Dice Roll*** Find the probability of rolling a sum of 11 with two dice.

8. ***Drawing Cards*** Find the probability of drawing 4 aces and 1 queen from a standard deck of 52 cards.

9. ***Probability of Rain*** If the odds that it will rain are 3 to 7, what is the probability of rain?

10. ***High School Graduates*** In a recent year there were 2.81 million high school graduates, of which 1.45 million were male. (*Source:* U.S. National Center for Education Statistics.) If one such high school graduate is selected at random, estimate the probability that this graduate is female.

CHAPTER 11 SUMMARY

KEY TERMS & SYMBOLS	KEY CONCEPTS
11.1 Sequences and Series sequence terms of a sequence general or nth term, a_n finite sequence infinite sequence convergent divergent recursive definition Fibonacci sequence series finite series infinite series summation (sigma) notation, $\sum_{n=1}^{\infty}$ index of summation, i	**SEQUENCE** A sequence is a function that has a set of natural numbers as its domain. **SERIES** A finite series is an expression of the form $S_n = a_1 + a_2 + a_3 + \cdots + a_n = \sum_{i=1}^{n} a_i,$ and an infinite series is an expression of the form $S_\infty = a_1 + a_2 + a_3 + \cdots + a_n + \cdots = \sum_{i=1}^{\infty} a_i.$ **SUMMATION PROPERTIES** If $a_1, a_2, a_3, \ldots, a_n$ and $b_1, b_2, b_3, \ldots, b_n$ are two sequences and c is a constant, then, for every positive integer n, (a) $\sum_{i=1}^{n} c = nc$ (b) $\sum_{i=1}^{n} ca_i = c\sum_{i=1}^{n} a_i$ (c) $\sum_{i=1}^{n} (a_i + b_i) = \sum_{i=1}^{n} a_i + \sum_{i=1}^{n} b_i$ (d) $\sum_{i=1}^{n} (a_i - b_i) = \sum_{i=1}^{n} a_i - \sum_{i=1}^{n} b_i.$
11.2 Arithmetic Sequences and Series arithmetic sequence (or progression) common difference, d arithmetic series	**ARITHMETIC SEQUENCE** A sequence in which each term after the first is obtained by adding a fixed number to the previous term is an arithmetic sequence. The fixed number that is added is the common difference. ***n*TH TERM OF AN ARITHMETIC SEQUENCE** In an arithmetic sequence with first term a_1 and common difference d, the nth term is given by $a_n = a_1 + (n - 1)d.$ *(continued)*

KEY TERMS & SYMBOLS	KEY CONCEPTS
	SUM OF THE FIRST *n* TERMS OF AN ARITHMETIC SEQUENCE If an arithmetic sequence has first term a_1 and common difference d, then the sum of the first n terms is given by $$S_n = \frac{n}{2}(a_1 + a_n) \quad \text{or} \quad S_n = \frac{n}{2}[2a_1 + (n-1)d].$$
11.3 Geometric Sequences and Series geometric sequence (or progression) common ratio, r geometric series annuity future value of an annuity	**GEOMETRIC SEQUENCE** A geometric sequence is a sequence in which each term after the first is obtained by multiplying the preceding term by a constant nonzero real number, called the common ratio. **nTH TERM OF A GEOMETRIC SEQUENCE** In the geometric sequence with first term a_1 and common ratio r, neither of which are zero, the nth term is given by $$a_n = a_1 r^{n-1}.$$ **SUM OF THE FIRST *n* TERMS OF A GEOMETRIC SEQUENCE** If a geometric sequence has first term a_1 and common ratio r, then the sum of the first n terms is given by $$S_n = \frac{a_1(1-r^n)}{1-r}, \quad \text{where } r \neq 1.$$ **SUM OF THE TERMS OF AN INFINITE GEOMETRIC SEQUENCE** The sum of the terms of an infinite geometric sequence with first term a_1 and common ratio r, where $-1 < r < 1$, is given by $$S_\infty = \frac{a_1}{1-r}.$$ If $r \leq -1$ or $r \geq 1$, the sum does not exist. **FUTURE VALUE OF AN ANNUITY** The formula for the future value of an annuity is $$S = R\left[\frac{(1+i)^n - 1}{i}\right],$$ where S is the future value, R is the payment at the end of each period, i is the interest rate in decimal form per period, and n is the number of periods.
11.4 The Binomial Theorem factorial notation n-factorial, $n!$ binomial coefficient binomial theorem	**PASCAL'S TRIANGLE** 1 — Row 0 1 1 — Row 1 1 2 1 — Row 2 1 3 3 1 — Row 3 1 4 6 4 1 — Row 4 1 5 10 10 5 1 — Row 5 *(continued)*

KEY TERMS & SYMBOLS	KEY CONCEPTS
	n-FACTORIAL For any positive integer n, $n! = n(n-1)(n-2)\cdots(3)(2)(1)$ and $0! = 1$. **BINOMIAL COEFFICIENT** For nonnegative integers n and r, with $r \le n$, ${}_nC_r = \binom{n}{r} = \frac{n!}{r!\,(n-r)!}$. **BINOMIAL THEOREM** For any positive integer n, $(x+y)^n = x^n + \binom{n}{1}x^{n-1}y + \binom{n}{2}x^{n-2}y^2 + \binom{n}{3}x^{n-3}y^3 + \cdots + \binom{n}{r}x^{n-r}y^r + \cdots + \binom{n}{n-1}xy^{n-1} + y^n$. **$r$TH TERM OF THE BINOMIAL EXPANSION** The rth term of the binomial expansion of $(x+y)^n$, where $n \ge r-1$, is $\binom{n}{r-1}x^{n-(r-1)}y^{r-1}$.
11.5 Mathematical Induction	**PROOF BY MATHEMATICAL INDUCTION** ***Step 1*** Prove that the statement is true for $n = 1$. ***Step 2*** Show that for any positive integer k, if S_k is true, then S_{k+1} is also true. **GENERALIZED PRINCIPLE OF MATHEMATICAL INDUCTION** Let S_n be a statement concerning the positive integer n. Let j be a fixed positive integer. Suppose that ***Step 1*** S_j is true. ***Step 2*** For any positive integer k, $k \ge j$, S_k implies S_{k+1}. Then S_n is true for all positive integers n, where $n \ge j$.
11.6 Counting Theory tree diagram independent events permutation, $P(n, r)$ combination, $C(n, r)$ or $\binom{n}{r}$	**FUNDAMENTAL PRINCIPLE OF COUNTING** If n independent events occur, with m_1 ways for event 1 to occur, m_2 ways for event 2 to occur, . . . and m_n ways for event n to occur, then there are $m_1 \cdot m_2 \cdot \cdots \cdot m_n$ different ways for all n events to occur. *(continued)*

KEY TERMS & SYMBOLS	KEY CONCEPTS

	PERMUTATIONS OF n ELEMENTS TAKEN r AT A TIME
	If $P(n, r)$ denotes the number of permutations of n elements taken r at a time, with $r \leq n$, then
	$$P(n, r) = \frac{n!}{(n - r)!}.$$
	COMBINATIONS OF n ELEMENTS TAKEN r AT A TIME
	If $C(n, r)$ or $\binom{n}{r}$ represents the number of combinations of n things taken r at a time, with $r \leq n$, then
	$$C(n, r) = \binom{n}{r} = \frac{n!}{(n - r)!r!}.$$

11.7 Probability outcomes sample space event probability complement, E' Venn diagram odds mutually exclusive events binomial experiment	**SAMPLE SPACE AND EVENT** The set S of all possible outcomes of a given experiment is called the sample space for the experiment. Any subset of S is called an event. **PROBABILITY OF EVENT E** In a sample space with equally likely outcomes, the probability of an event E, written $P(E)$, is the ratio of the number of outcomes in sample space S that belong to event E, denoted $n(E)$, to the total number of outcomes in sample space S, denoted $n(S)$. That is, $$P(E) = \frac{n(E)}{n(S)}.$$ **COMPLEMENT** The complement of an event E, written E', is the set of all outcomes not in E. Thus, $$P(E) + P(E') = 1.$$ **ODDS** The odds in favor of an event E are expressed as the ratio of $P(E)$ to $P(E')$, or $\frac{P(E)}{P(E')}$. **PROBABILITY OF THE UNION OF TWO EVENTS** For any events E and F, $$P(E \text{ or } F) = P(E \cup F) = P(E) + P(F) - P(E \cap F).$$ **PROPERTIES OF PROBABILITY** For any events E and F, **1.** $0 \leq P(E) \leq 1$; **2.** $P(\text{a certain event}) = 1$; **3.** $P(\text{an impossible event}) = 0$; **4.** $P(E') = 1 - P(E)$; **5.** $P(E \text{ or } F) = P(E \cup F) = P(E) + P(F) - P(E \cap F)$. **BINOMIAL PROBABILITY** An experiment that consists of repeated independent trials with only two outcomes in each trial—success and failure—is called a binomial experiment. Let the probability of success of one trial be p. Then the probability of failure is $1 - p$, and the probability of r successes in n trials is given by $$\binom{n}{r} p^r (1 - p)^{n-r}.$$

CHAPTER 11 Review Exercises

Write the first five terms for each sequence. State whether the sequence is arithmetic, geometric, *or* neither.

1. $a_n = \dfrac{n}{n+1}$

2. $a_n = (-2)^n$

3. $a_n = 2(n+3)$

4. $a_n = n(n+1)$

5. $a_1 = 5$; $a_n = a_{n-1} - 3$, for $n \geq 2$

In Exercises 6–9, write the first five terms of the sequence described.

6. Arithmetic, $a_2 = 10$, $d = -2$

7. Arithmetic, $a_3 = \pi$, $a_4 = 1$

8. Geometric, $a_1 = 6$, $r = 2$

9. Geometric, $a_1 = -5$, $a_2 = -1$

10. An arithmetic sequence has $a_5 = -3$ and $a_{15} = 17$. Find a_1 and a_n.

11. A geometric sequence has $a_1 = -8$ and $a_7 = -\frac{1}{8}$. Find a_4 and a_n.

Find a_8 for each arithmetic sequence.

12. $a_1 = 6$, $d = 2$

13. $a_1 = 6x - 9$, $a_2 = 5x + 1$

Find S_{12} for each arithmetic sequence.

14. $a_1 = 2$, $d = 3$

15. $a_2 = 6$, $d = 10$

Find a_5 for each geometric sequence.

16. $a_1 = -2$, $r = 3$

17. $a_3 = 4$, $r = \dfrac{1}{5}$

Find S_4 for each geometric sequence.

18. $a_1 = 3$, $r = 2$

19. $\dfrac{3}{4}, -\dfrac{1}{2}, \dfrac{1}{3}, \ldots$

20. ***Annuity Value*** Find the future value of an annuity that consists of payments of \$2000 at the end of each year for 5 years at 3% interest compounded annually.

Evaluate each sum that exists.

21. $\sum_{i=1}^{7} (-1)^{i-1}$

22. $\sum_{i=1}^{5} (i^2 + i)$

23. $\sum_{i=1}^{4} \dfrac{i+1}{i}$

24. $\sum_{j=1}^{10} (3j - 4)$

25. $\sum_{j=1}^{2500} j$

26. $\sum_{i=1}^{5} 4 \cdot 2^i$

27. $\sum_{i=1}^{\infty} \left(\dfrac{4}{7}\right)^i$

28. $\sum_{i=1}^{\infty} -2\left(\dfrac{6}{5}\right)^i$

Evaluate each sum that converges. If the series diverges, say so.

29. $24 + 8 + \dfrac{8}{3} + \dfrac{8}{9} + \cdots$

30. $-\dfrac{3}{4} + \dfrac{1}{2} - \dfrac{1}{3} + \dfrac{2}{9} - \cdots$

31. $\dfrac{1}{12} + \dfrac{1}{6} + \dfrac{1}{3} + \dfrac{2}{3} + \cdots$

32. $.9 + .09 + .009 + .0009 + \cdots$

Evaluate each sum, where $x_1 = 0$, $x_2 = 1$, $x_3 = 2$, $x_4 = 3$, $x_5 = 4$, and $x_6 = 5$.

33. $\sum_{i=1}^{4} (x_i^2 - 6)$

34. $\sum_{i=1}^{6} f(x_i)\,\Delta x$; $f(x) = (x-2)^3$, $\Delta x = .1$

Write each sum, using summation notation.

35. $4 - 1 - 6 - \cdots - 66$

36. $10 + 14 + 18 + \cdots + 86$

37. $4 + 12 + 36 + \cdots + 972$

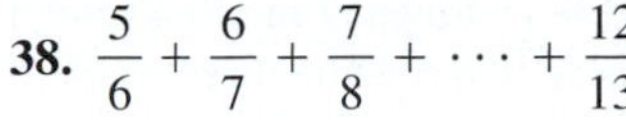

38. $\frac{5}{6} + \frac{6}{7} + \frac{7}{8} + \cdots + \frac{12}{13}$

Use the binomial theorem to expand each expression.

39. $(x + 2y)^4$

40. $(3z - 5w)^3$

41. $\left(3\sqrt{x} - \frac{1}{\sqrt{x}}\right)^5$

42. $(m^3 - m^{-2})^4$

Find the indicated term or terms for each expansion.

43. Sixth term of $(4x - y)^8$

44. Seventh term of $(m - 3n)^{14}$

45. First four terms of $(x + 2)^{12}$

46. Last three terms of $(2a + 5b)^{16}$

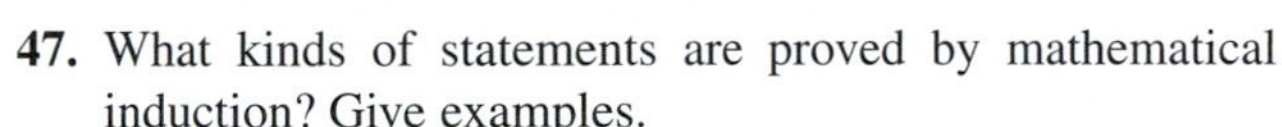

47. What kinds of statements are proved by mathematical induction? Give examples.

48. Describe a proof by mathematical induction.

Use mathematical induction to prove that each statement is true for every positive integer n.

49. $1 + 3 + 5 + 7 + \cdots + (2n - 1) = n^2$

50. $2 + 6 + 10 + 14 + \cdots + (4n - 2) = 2n^2$

51. $2 + 2^2 + 2^3 + \cdots + 2^n = 2(2^n - 1)$

52. $1^3 + 3^3 + 5^3 + \cdots + (2n - 1)^3 = n^2(2n^2 - 1)$

53. *Concept Check* Is a student identification number an example of a permutation or a combination?

Find the value of each expression.

54. $P(9, 2)$

55. $P(6, 0)$

56. $\binom{8}{3}$

57. $9!$

58. $C(10, 5)$

Solve each problem.

59. *Wedding Plans* Two people are planning their wedding. They can select from 2 different chapels, 4 soloists, 3 organists, and 2 ministers. How many different wedding arrangements are possible?

60. *Couch Styles* Bob Schiffer, who is furnishing his apartment, wants to buy a new couch. He can select from 5 different styles, each available in 3 different fabrics with 6 color choices. How many different couches are available?

61. *Summer Job Assignments* Four students are to be assigned to 4 different summer jobs. Each student is qualified for all 4 jobs. In how many ways can the jobs be assigned?

62. *Conference Delegations* A student body council consists of a president, vice president, secretary/treasurer, and 3 representatives at large. Three members are to be selected to attend a conference.
(a) How many different such delegations are possible?
(b) How many are possible if the president must attend?

63. *Tournament Outcomes* Nine football teams are competing for first-, second-, and third-place titles in a statewide tournament. In how many ways can the winners be determined?

64. *License Plates* How many different license plates can be formed with a letter followed by 3 digits and then 3 letters? How many such license plates have no repeats?

65. *Drawing a Marble* A marble is drawn at random from a box containing 4 green, 5 black, and 6 white marbles. Find each probability.
(a) A green marble is drawn.
(b) A marble that is not black is drawn.
(c) A blue marble is drawn.

66. *Drawing a Marble* Refer to Exercise 65, and answer each question.
(a) What are the odds in favor of drawing a green marble?
(b) What are the odds against drawing a white marble?
(c) What are the odds in favor of drawing a marble that is not white?

Drawing a Card *A card is drawn from a standard deck of 52 cards. Find the probability that the following is drawn.*

67. A black king

68. A face card or an ace

69. An ace or a diamond

70. A card that is not a diamond

Swimming Pool Filter Samples *A sample shipment of 5 swimming pool filters is chosen. The probability of exactly 0, 1, 2, 3, 4, or 5 filters being defective is given in the following table.*

Number Defective	0	1	2	3	4	5
Probability	.31	.25	.18	.12	.08	.06

Find the probability that the given number of filters are defective.

71. No more than 3

72. At least 2

73. More than 5

74. *Die Rolls* A die is rolled 12 times. Find the probability that exactly 2 of the rolls result in a 5.

75. *Coin Tosses* A coin is tossed 10 times. Find the probability that exactly 4 of the tosses result in a tail.

76. *Political Orientation* The table describes the political orientation of college freshmen in the class of 2006, as determined from a survey of 282,200 freshmen.

Political Orientation	**Number of Freshmen (in thousands)**
Far left	7.06
Liberal	71.48
Middle of the road	143.5
Conservative	56.51
Far right	3.673

Source: Higher Education Research Institute, UCLA, 2002.

(a) What is the probability that a randomly selected student from the class is in the conservative group?
(b) What is the probability that a randomly selected student from the class is on the far left or the far right politically?
(c) What is the probability of a randomly selected student from the class not being politically middle of the road?

CHAPTER 11 Test

1. Write the first five terms of each sequence. State whether the sequence is arithmetic, geometric, or neither.
(a) $a_n = (-1)^n(n + 2)$
(b) $a_n = -3 \cdot \left(\frac{1}{2}\right)^n$
(c) $a_1 = 2, a_2 = 3; a_n = a_{n-1} + 2a_{n-2}$, for $n \geq 3$

2. In each sequence described, find a_5.
(a) An arithmetic sequence with $a_1 = 1$ and $a_3 = 25$
(b) A geometric sequence with $a_1 = 81$ and $r = -\frac{2}{3}$

3. Find the sum of the first 10 terms of each sequence described.
(a) Arithmetic, with $a_1 = -43$ and $d = 12$
(b) Geometric, with $a_1 = 5$ and $r = -2$

4. Evaluate each sum that exists.
(a) $\sum_{i=1}^{30} (5i + 2)$
(b) $\sum_{i=1}^{5} (-3 \cdot 2^i)$
(c) $\sum_{i=1}^{\infty} (2^i) \cdot 4$
(d) $\sum_{i=1}^{\infty} 54\left(\frac{2}{9}\right)^i$

5. (a) Use the binomial theorem to expand $(2x - 3y)^4$.
(b) Find the third term in the expansion of $(w - 2y)^6$.

6. Evaluate the following.
(a) 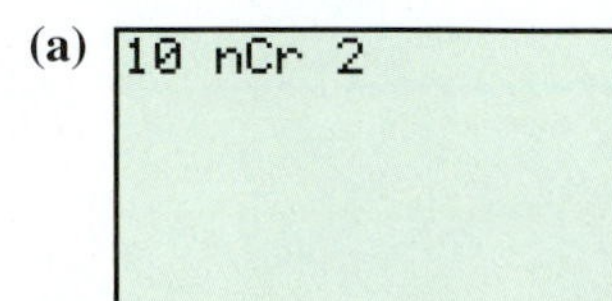

(b) $\binom{7}{3}$
(c) $7!$
(d) $P(11, 3)$

7. Use mathematical induction to prove that, for all positive integers n,

$$8 + 14 + 20 + 26 + \cdots + (6n + 2) = 3n^2 + 5n.$$

Solve each problem.

8. ***Athletic Shoe Choices*** A sports-shoe manufacturer makes athletic shoes in 4 different styles. Each style comes in 3 different colors, and each color comes in 2 different shades. How many different types of shoes can be made?

9. ***Club Officer Choices*** A club with 20 members plans to elect a president, a secretary, and a treasurer from its membership. If a member can hold at most one office, in how many ways can the three offices be filled?

10. ***Club Officer Choices*** Refer to Exercise 9. If there are 8 men and 12 women in the club, in how many ways can 2 men and 3 women be chosen to attend a conference?

11. ***Drawing a Card*** A single card is drawn from a standard deck of 52 cards.
    (a) Find the probability of drawing a red 3.
    (b) Find the probability of drawing a card that is not a face card.
    (c) Find the probability of drawing a king or a spade.
    (d) What are the odds in favor of drawing a face card?

12. ***Rolling a Die*** An experiment consists of rolling a die eight times. Find the probability of each event.
    (a) Exactly three rolls result in a 4.
    (b) All eight rolls result in a 6.

CHAPTER 11 PROJECT

Using Experimental Probabilities to Simulate Family Makeup

In the last section of this chapter, we calculated many theoretical probabilities. Even though the probability of obtaining a head on one toss of a coin is $\frac{1}{2}$, it does not mean that a head will appear on exactly one-half of the tosses. In the experiments that follow, we simulate the makeup of a family and experimentally determine the probability that a family of 4 children will consist of 2 boys and 2 girls. (Do you expect this probability to be $\frac{1}{2}$?)

Probabilities can be calculated in two ways. One way is *theoretically,* as described in the section on probability. The other is *experimentally,* by actually conducting the experiment and recording the successes compared with the total number of trials. The **law of large numbers** says that the larger the number of trials in an experiment, the closer the experimental probability will be to the theoretical probability.

Activities

1. Work with a partner. One member of the group should flip two coins simultaneously 20 times, and the other should tally the results in a table similar to the one shown here. The possible events are 2 heads and 0 tails, 1 head and 1 tail, and 0 heads and 2 tails.

Event	Tally	Relative Frequency	Experimental Probability (as a Decimal)
2 heads, 0 tails		$\frac{?}{20}$	
1 head, 1 tail		$\frac{?}{20}$	
0 heads, 2 tails		$\frac{?}{20}$	

(a) List the sample space and calculate the *theoretical* probability of each event.

(b) Explain why the three outcomes are not equally likely.

(c) Your experimental probabilities probably do not match the theoretical probabilities. Why not? How can you obtain better results?

2. With your partner, repeat the experiment, but this time flip four coins simultaneously 20 times and tally the results. The possible events are given in the following table.

Event	Tally	Relative Frequency	Experimental Probability (as a Decimal)
4 heads, 0 tails		$\frac{?}{20}$	
3 heads, 1 tail		$\frac{?}{20}$	
2 heads, 2 tails		$\frac{?}{20}$	
1 head, 3 tails		$\frac{?}{20}$	
0 heads, 4 tails		$\frac{?}{20}$	

(a) List the sample space and calculate the *theoretical* probability of each event.

(b) How frequently did you get 2 heads and 2 tails? Are you surprised?

(c) Your experimental probabilities likely do not match the theoretical probabilities. Why not? How can you obtain better results?

After working Activity 2, you probably agree that actual coin tossing is quite time consuming. To simulate coin tosses, we can use a random-integer generator on a graphing calculator, such as the TI-83/84 Plus. The first line in Figure A shows the syntax for generating a list of 4 integers on the closed interval $[0, 1]$; a toss of four coins can be simulated this way. Let us agree that 0 represents tails and 1 represents heads. See the six sample lines in Figure A.

By finding the sum of the entries in the list, we can determine the outcome. For example, a sum of 0 means that there are 4 tails, a sum of 1 means that there are 3 tails and 1 head, a sum of 2 means that there are 2 tails and 2 heads, and so on. The calculator can keep a tally of the sums, as shown in the short program on the home screen in Figure B.

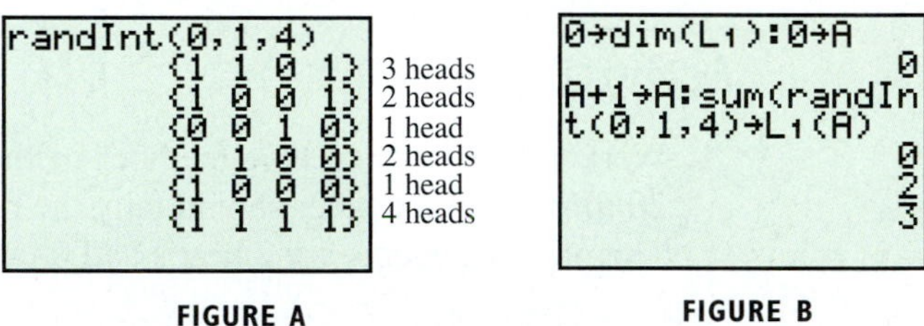

FIGURE A

FIGURE B

After running the program a large number of times, we can plot the results in a histogram, as seen in Figure C on the next page. The first display of Figure D shows how many times the experiment was run (150), the second display shows the experimental probability for a sum of 2 (meaning 2 tails and 2 heads), and the third display shows the theoretical probability for 2 heads and 2 tails.

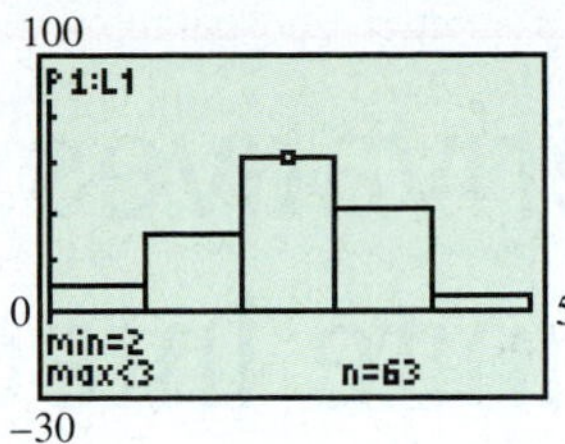

The sum 2 appears 63 times.

FIGURE C

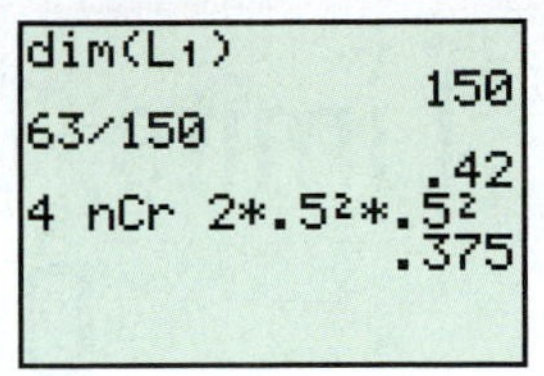

FIGURE D

3. With your partner, duplicate the procedure just described to determine an experimental probability of having 2 girls and 2 boys in a family of 4 children. Let 0 represent the outcome of having a girl and 1 represent the outcome of having a boy. Use the calculator to obtain 200 outcomes. How close is your result to the theoretical probability? It should be much closer than the one you found in Activity 2. Why is this so?

12

Limits, Derivatives, and Definite Integrals

IN 1665, CAMBRIDGE University was closed due to the plague, and Isaac Newton returned home. In two short years, he determined his three laws of motion. To accomplish this amazing feat, Newton invented calculus.

In 1905, Albert Einstein submitted an article that explained the theory of relativity. This theory overturned many of the traditional notions of space and time established by Newton. To develop this theory, Einstein used calculus.

Today, physicists are working on a new theory, called *string theory,* which holds that everything in the universe is composed of extremely small strings of energy, vibrating in 10 dimensions. Each vibrational pattern represents a different energy, and possibly, a different subatomic particle. Calculus is playing an important role in this new theory of matter.

Even though scientific models change over time, calculus remains timeless in its ability to adapt to new situations. This chapter introduces many important concepts found in calculus.

Source: Kaku, M., "Testing String Theory," *Discover,* August 2005.

Chapter Outline

12.1 An Introduction to Limits

Limit of a Function ■ Finding Limits of Various Types of Functions ■ Limits That Do Not Exist

In earlier chapters, we were concerned with static situations. We saw how to answer questions like the following:

How much would 100 items cost?
In how many years will the population reach 100,000?
How wide is an arch 10 meters from the ground?

With calculus, we are able to solve problems that involve changing situations:

How fast is the car moving after 20 seconds?
At what rate is the population growing?
What is the rate of change of profit when sales reach $1000?

Solving each of these problems requires a *derivative*. The definition of the derivative involves the idea of a *limit* of a function. The limit concept distinguishes calculus from other branches of mathematics, such as algebra and trigonometry.

Limit of a Function

Consider the rational function defined by

$$f(x) = \frac{x^2 - 4}{x - 2}.$$

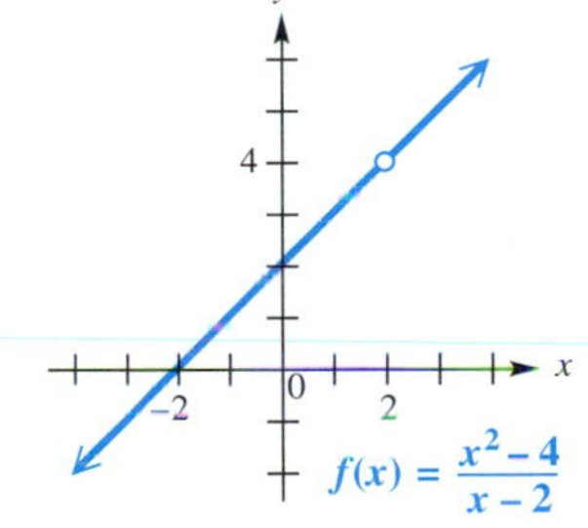

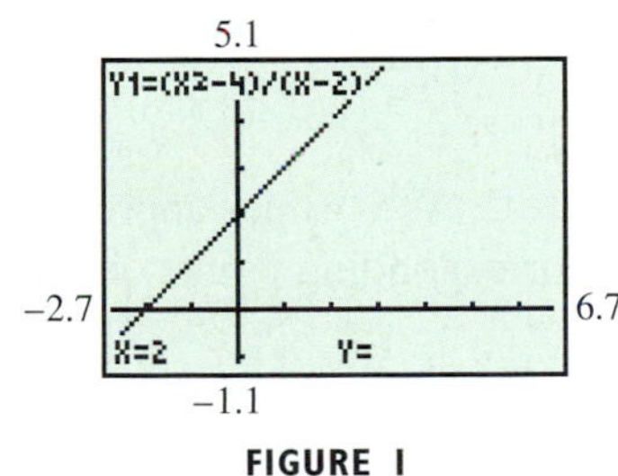

FIGURE 1

As we saw in Example 8 of Section 4.2, this function is not defined at $x = 2$, so its graph has a "hole" at $x = 2$. See Figure 1. However, we can calculate values of $f(x)$ for x near 2, as shown in the following table.

x approaches 2.

x	1.9	1.99	1.999 →	← 2.001	2.01	2.1
$f(x)$	3.9	3.99	3.999 →	← 4.001	4.01	4.1

$f(x)$ approaches 4.

From the graphs in Figure 1 and the table, we see that as x gets closer and closer to 2 (but never equals 2), the values of $f(x)$ get closer and closer to 4. We say, "The limit of $\frac{x^2 - 4}{x - 2}$ as x approaches 2 equals 4" and express it with the notation

$$\lim_{x \to 2} \frac{x^2 - 4}{x - 2} = 4.$$

In this example, the function was not defined at $x = 2$. However, it would be worthwhile to consider $\lim_{x \to 2} f(x)$ even if the function had been defined at $x = 2$. The following informal definition of limit applies to any function.

Looking Ahead to Calculus

The formal definition of a limit used in calculus is "$\lim_{x \to a} f(x) = L$ if for every $\epsilon > 0$ there exists a $\delta > 0$ such that $|f(x) - L| < \epsilon$ whenever $0 < |x - a| < \delta$."

Limit of a Function

Let f be a function and let a and L be real numbers. L is the **limit of $f(x)$ as x approaches a,** written

$$\lim_{x \to a} f(x) = L,$$

if the following conditions are met:

1. As x assumes values closer and closer (but not equal) to a on both sides of a, the corresponding values of $f(x)$ get closer and closer (and are perhaps equal) to L.
2. The value of $f(x)$ can be made as close to L as desired by taking values of x arbitrarily close to a.

This definition is informal because the expressions "closer and closer to" and "as close as desired" are not precise enough for formal mathematics.

NOTE ***The definition of a limit describes what happens to f(x) when x is near a.*** It is not affected by whether $f(a)$ is defined.

Finding Limits of Various Types of Functions

GCM **EXAMPLE 1** **Finding a Limit of a Polynomial Function**

Find $\lim_{x \to 1} (x^2 - 3x + 4)$.

Solution We can determine the behavior of the polynomial function defined by $f(x) = x^2 - 3x + 4$ near $x = 1$ by evaluating $f(x)$ for values of x that are closer and closer to 1 from both the left and the right. In the table, we have taken values of x whose distances from 1 are .1, .01, and .001 and evaluated $f(x)$ at each of them. Notice that the values of $f(x)$ get closer and closer to 2.

x approaches 1.

x	.9	.99	.999 →	← 1.001	1.01	1.1
$f(x)$	2.11	2.0101	2.001 →	← 1.999	1.9901	1.91

f(*x*) approaches 2.

The behavior of $f(x)$ is shown visually in the graph in Figure 2. As x-values approach 1 from both the right and the left, the y-coordinates of the corresponding points on the graph get closer and closer to 2.

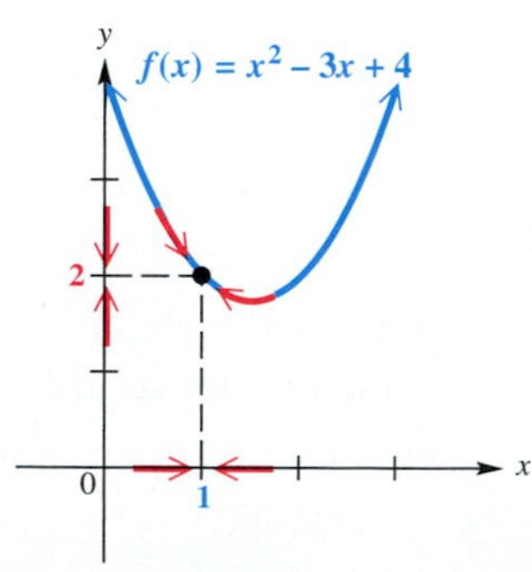

FIGURE 2

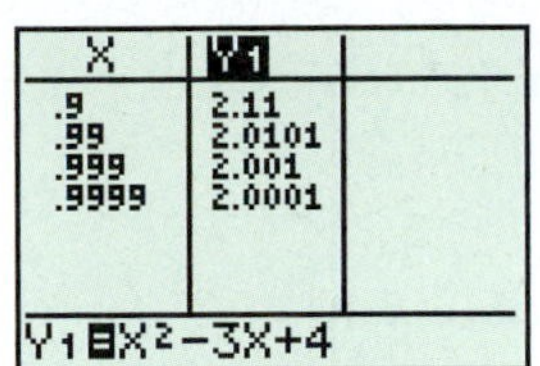

As $X \to 1$, $Y_1 \to 2$.

FIGURE 3

X	Y1
1.1	1.91
1.01	1.9901
1.001	1.999
1.0001	1.9999

Y1■X²-3X+4

As $X \to 1$, $Y_1 \to 2$.

FIGURE 4

From the preceding table and the graph shown in Figure 2, we see that

$$\lim_{x \to 1} (x^2 - 3x + 4) = 2.$$

The calculator tables in Figures 3 and 4 further support this conclusion. ■

CAUTION In Example 1, the function is defined at $x = 1$, and indeed, the value of the limit is $f(1)$. Although $\lim_{x \to a} f(x)$ will always equal $f(a)$ for polynomial functions, which are continuous, we cannot assume this to be true for all functions.

EXAMPLE 2 Finding a Limit of a Piecewise-Defined Function

Find $\lim_{x \to 3} f(x)$, where $f(x) = \begin{cases} 2x + 1 & \text{if } x \leq 3 \\ 4x - 5 & \text{if } x > 3 \end{cases}$.

Solution To determine the behavior of the function near $x = 3$, we can use a table and a graph, as in Example 1.

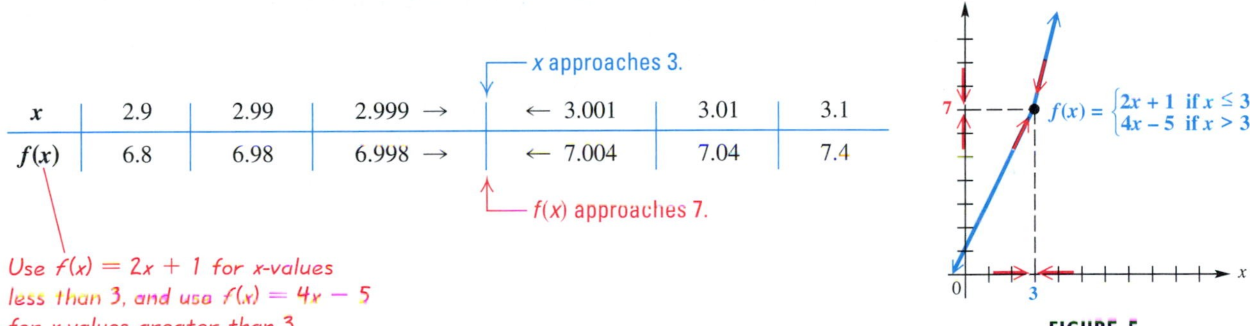

x	2.9	2.99	2.999 →	← 3.001	3.01	3.1
$f(x)$	6.8	6.98	6.998 →	← 7.004	7.04	7.4

FIGURE 5

From the table and the graph in Figure 5, we see that as x approaches 3 from either the left or the right, $f(x)$ approaches 7. Therefore,

$$\lim_{x \to 3} f(x) = 7.$$ ■

EXAMPLE 3 Finding a Limit of a Trigonometric Function

Find $\lim_{x \to 0} \frac{\sin x}{x}$, where x is measured in radians.

Solution In this example, 0, the number approached by x, is not in the domain of the function. However, since the limit as x approaches 0 is not affected by what happens at $x = 0$, we can continue as before. From the following table and the calculator table and graph in Figure 6 on the next page, we conclude that

$$\lim_{x \to 0} \frac{\sin x}{x} = 1.$$

x	−.1	−.01	−.001 →	← .001	.01	.1
$f(x)$	.9983342	.9999833	.9999998 →	← .9999998	.9999833	.9983342

x approaches 0.

f(*x*) approaches 1.

Looking Ahead to Calculus

The concept of the *derivative* of a function is central to calculus. The limit evaluated in Example 3 is used in calculus to show that the derivative of the sine function is the cosine function. Therefore, the value of the cosine function at any x-value gives a measure of the "steepness" of the graph of the sine function at that same x-value.

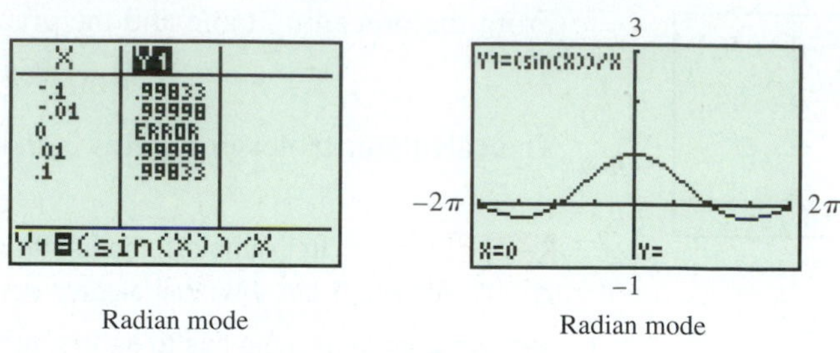

FIGURE 6

EXAMPLE 4 Finding a Limit as x Approaches a Negative Number

Find $\lim_{x \to -2} \dfrac{e^{x+2} + x + 1}{x + 2}$.

Solution Since x approaches a negative number, the table looks a little different from the previous tables. The numbers approaching -2 from the left are greater in absolute value than 2, and the numbers approaching -2 from the right are smaller in absolute value than 2. The table and the graph in Figure 7 indicate that

$$\lim_{x \to -2} \frac{e^{x+2} + x + 1}{x + 2} = 2.$$

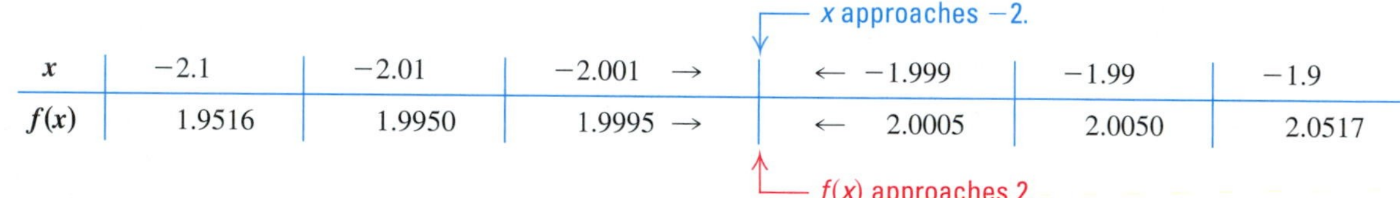

x approaches -2.

x	-2.1	-2.01	-2.001 →	← -1.999	-1.99	-1.9
$f(x)$	1.9516	1.9950	1.9995 →	← 2.0005	2.0050	2.0517

$f(x)$ approaches 2.

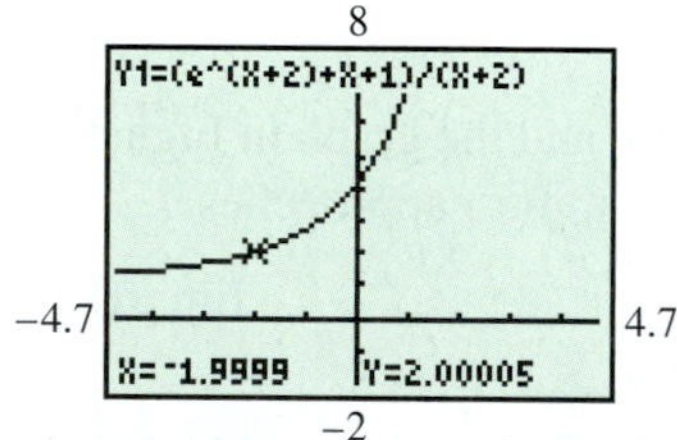

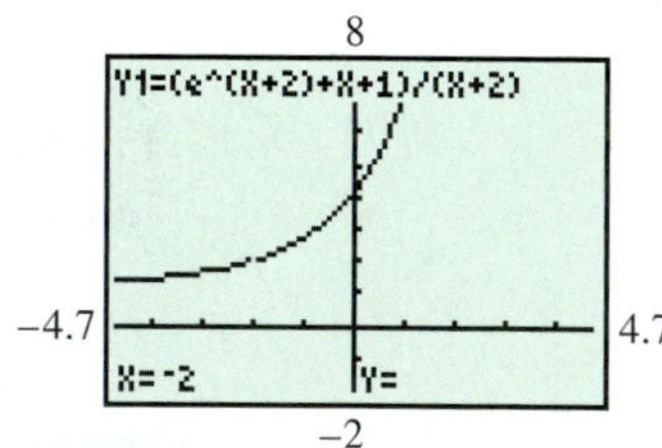

FIGURE 7

Limits That Do Not Exist

If there is no *single* value that is approached by $f(x)$ as x approaches a, we say that $f(x)$ does not have a limit as x approaches a, or $\lim_{x \to a} f(x)$ does not exist. The following examples illustrate three situations in which a limit can fail to exist.

EXAMPLE 5 Determining whether a Limit Exists

Find $\lim_{x \to 2} f(x)$, where $f(x) = \begin{cases} 4x - 5 & \text{if } x \le 2 \\ 3x - 5 & \text{if } x > 2 \end{cases}$.

Solution From the table and the graph in Figure 8, we see that as x gets closer and closer to 2 from the left, the values of $f(x)$ get closer and closer to 3. However, as x gets closer and closer to 2 from the right, the values of $f(x)$ get closer and closer to 1. Since $f(x)$ approaches different values depending on whether x approaches 2 from the left or from the right, $\lim_{x \to 2} f(x)$ *does not exist.*

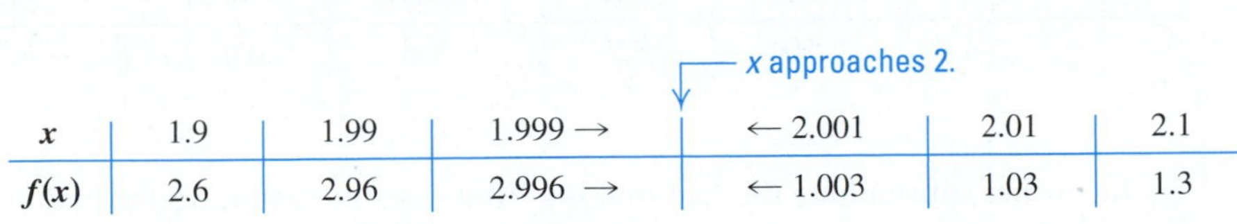

x	1.9	1.99	1.999 →	← 2.001	2.01	2.1
$f(x)$	2.6	2.96	2.996 →	← 1.003	1.03	1.3

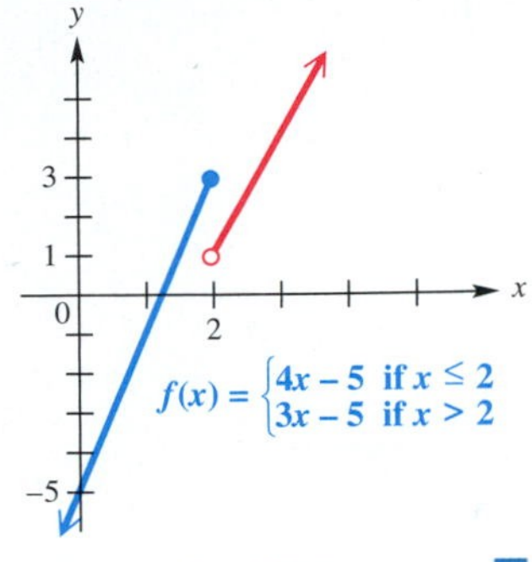

FIGURE 8

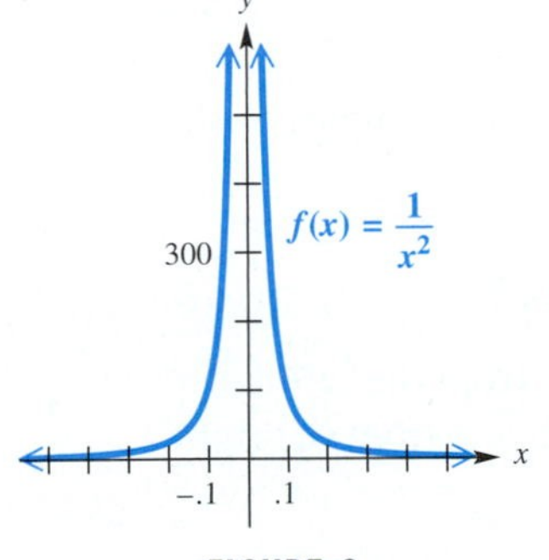

FIGURE 9

EXAMPLE 6 Determining whether a Limit Exists

Find $\lim_{x \to 0} f(x)$, where $f(x) = \dfrac{1}{x^2}$.

Solution The following table and the graph in Figure 9 illustrate that as the values of x approach 0, the corresponding values of $f(x)$ grow arbitrarily large. A limit L must be a (finite) real number. Therefore, we conclude that $\lim_{x \to 0} \dfrac{1}{x^2}$ does not exist.

x approaches 0.

x	−.1	−.01	−.001 →	← .001	.01	.1
$f(x)$	100	10,000	1,000,000 →	← 1,000,000	10,000	100

$f(x)$ becomes arbitrarily large.

EXAMPLE 7 Determining whether a Limit Exists

Find $\lim_{x \to 0} \sin \dfrac{1}{x}$.

Solution A graph of $f(x) = \sin \frac{1}{x}$ is shown in Figure 10.

TECHNOLOGY NOTE

The graph in Figure 10 is difficult to obtain using the typical low resolution of graphing calculators. Be sure to use radian mode.

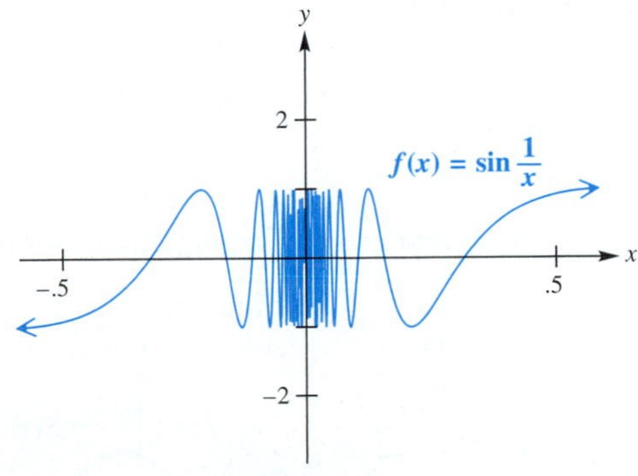

FIGURE 10

The graph oscillates faster and faster as x approaches 0, which leads us to believe that the limit does not exist. To help confirm our suspicion, we vary the table so that we can look at more values of x to the right of 0 than before. (Since $f(x)$ is an odd function, the corresponding values to the left of 0 will just be the negatives of these.)

x approaches 0.

x	.1	.01	.001	.0001	.00001	.000001
$f(x)$	$-.544$	$-.506$	.827	$-.306$	.036	$-.350$

$f(x)$ shows no direction.

To further substantiate the fact that the limit does not exist, notice that horizontal lines between $y = -1$ and $y = 1$ intersect the graph infinitely many times. Therefore, as $x \to 0$, the graph of $f(x)$ does not approach a single unique value. This limit does not exist. ■

The following summarizes the conclusions drawn from Examples 5–7.

Conditions under Which $\lim_{x \to a} f(x)$ Fails to Exist

1. $f(x)$ approaches a number L as x approaches a from the left and approaches a different number M as x approaches a from the right. (See Example 5.)
2. $f(x)$ becomes infinitely large in absolute value as x approaches a from either side. (See Example 6.)
3. $f(x)$ oscillates infinitely many times between two *fixed* values as x approaches a. (See Example 7.)

12.1 Exercises

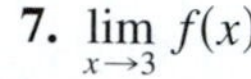

Concept Check *Tell whether each statement is* true *or* false. *If* false, *tell why.*

1. If a function f is defined at $x = a$, then $\lim_{x \to a} f(x) = f(a)$.

2. If $\lim_{x \to a} f(x)$ does not exist, then $f(x)$ necessarily approaches one value as x approaches a from the left and a different value as x approaches a from the right.

3. If $\lim_{x \to 1} f(x) = 5$, then 1 is in the domain of $f(x)$.

4. If $\lim_{x \to 1} f(x) = 5$, then 5 is in the range of $f(x)$.

5. If $\lim_{x \to a} f(x) = -5$, then $f(x)$ is between -5.001 and -4.999 for some value of x near a.

6. If $\lim_{x \to a} f(x) = b$, then $|f(x) - b| < .0001$ for some value of x near a.

Decide from the graph whether each limit exists. If a limit exists, find its value.

7. $\lim_{x \to 3} f(x)$

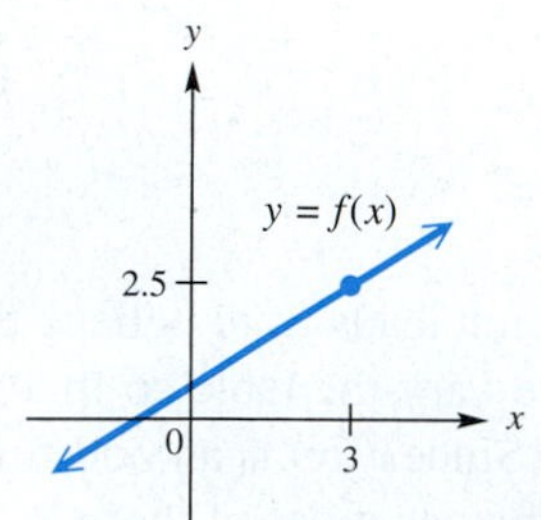

8. $\lim_{x \to 2} F(x)$

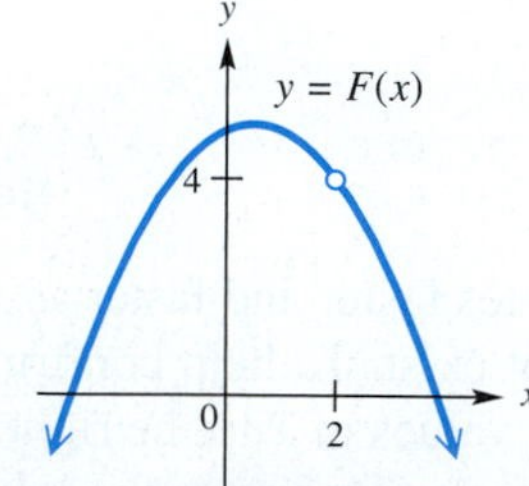

9. $\lim_{x \to 2} F(x)$

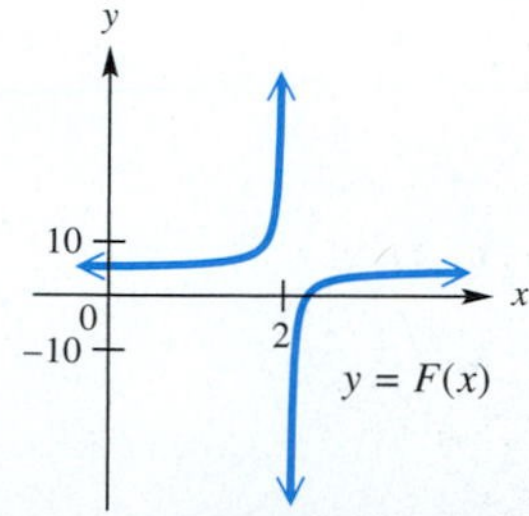

10. $\lim_{x\to 3} f(x)$

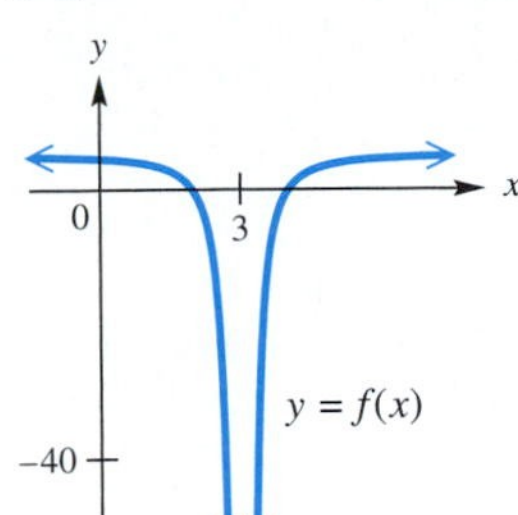

11. $\lim_{x\to 0} f(x)$

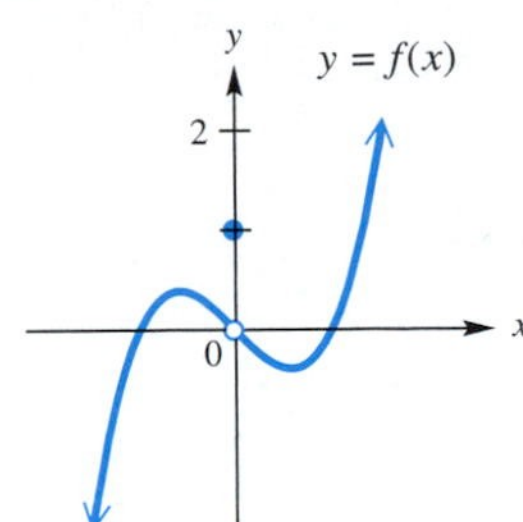

12. $\lim_{x\to 1} h(x)$

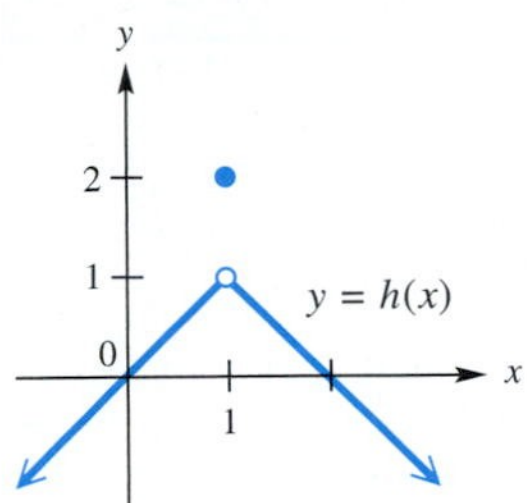

13. $\lim_{x\to 1} f(x)$

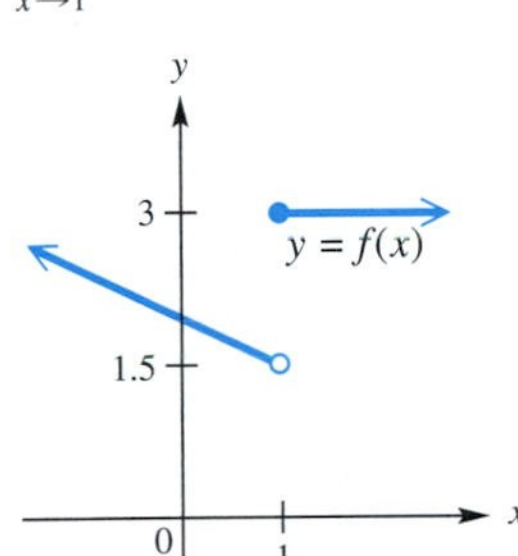

14. $\lim_{x\to 2} f(x)$

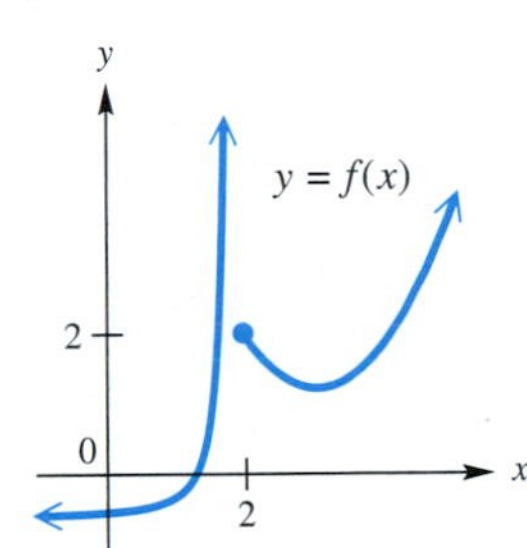

15. $\lim_{x\to 3} g(x)$

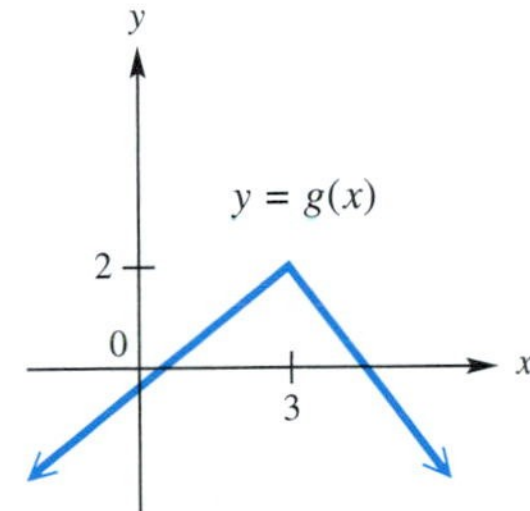

16. $\lim_{x\to -2} g(x)$

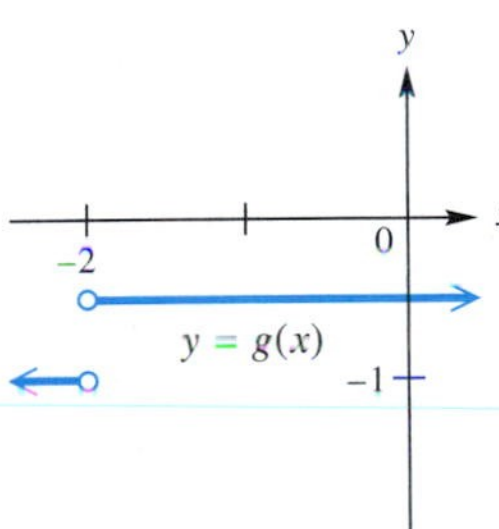

17. $\lim_{x\to .5} h(x)$

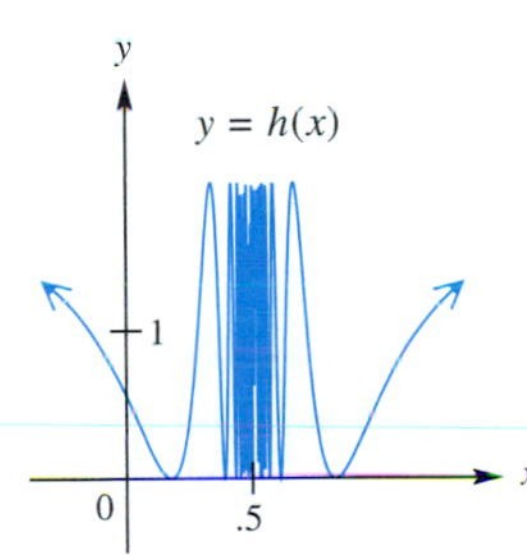

Infinitely many oscillations near .5

18. $\lim_{x\to 2} f(x)$

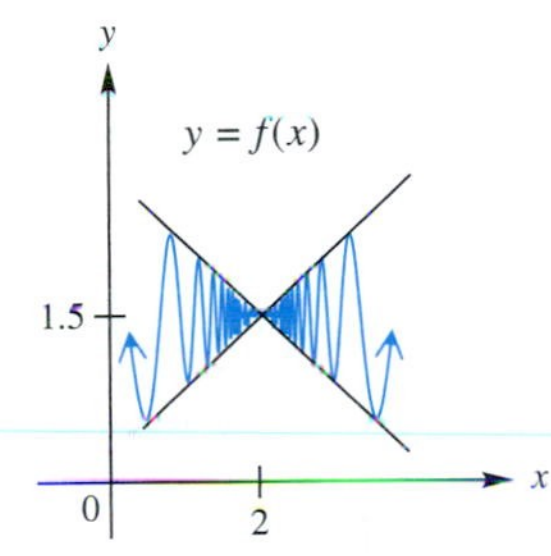

Infinitely many oscillations near 2

19. Use the table of values to estimate $\lim_{x\to 1} f(x)$.

x	.9	.99	.999	1.001	1.01	1.1
$f(x)$	1.9	1.99	1.999	2.001	2.01	2.1

20. Use the table of values to estimate $\lim_{x\to 2} f(x)$.

x	1.9	1.99	1.999	2.001	2.01	2.1
$f(x)$	−1.3	−1.05	−1.002	−.997	−.993	−.985

Complete each table and use the results to find the indicated limit.

21. If $f(x) = 2x^2 - 4x + 3$, find $\lim_{x\to 1} f(x)$.

x	.9	.99	.999	1.001	1.01	1.1
$f(x)$			1.000002	1.000002		

22. If $k(x) = \dfrac{x^3 - 2x - 4}{x - 2}$, find $\lim_{x\to 2} k(x)$.

x	1.9	1.99	1.999	2.001	2.01	2.1
$k(x)$		9.9401			10.0601	

23. If $f(x) = \dfrac{2x^3 + 3x^2 - 4x - 5}{x + 1}$, find $\lim_{x \to -1} f(x)$.

x	-1.1	-1.01	-1.001	$-.999$	$-.99$	$-.9$
$f(x)$	-3.68					-4.28

24. If $h(x) = \dfrac{\sqrt{x} - 2}{x - 1}$, find $\lim_{x \to 1} h(x)$.

x	.9	.99	.999	1.001	1.01	1.1
$h(x)$						

25. If $f(x) = \dfrac{\sqrt{x} - 3}{x - 3}$, find $\lim_{x \to 3} f(x)$.

x	2.9	2.99	2.999	3.001	3.01	3.1
$f(x)$						

26. If $f(x) = \dfrac{x^3 + 3x^2 + x + 3}{x + 3}$, find $\lim_{x \to -3} f(x)$.

x	-3.1	-3.01	-3.001	-2.999	-2.99	-2.9
$f(x)$						

27. If $f(x) = \dfrac{\sin 2x}{x}$, find $\lim_{x \to 0} f(x)$.

x	$-.1$	$-.01$	$-.001$	.001	.01	.1
$f(x)$						

28. If $f(x) = \dfrac{\sin 5x}{2x}$, find $\lim_{x \to 0} f(x)$.

x	$-.1$	$-.01$	$-.001$	.001	.01	.1
$f(x)$						

Use a table and/or graph to decide whether each limit exists. If a limit exists, find its value.

29. $\lim_{x \to 5} |2x - 4|$

30. $\lim_{x \to -1} \sqrt{5 + 3x}$

31. $\lim_{x \to 5} \dfrac{x^2 - 3x - 10}{x - 5}$

32. $\lim_{x \to 3} \dfrac{x + 1}{(x - 3)^2}$

33. $\lim_{x \to -2} \dfrac{x^2 + 2}{x + 2}$

34. $\lim_{x \to -4} \dfrac{x^2 + 5x + 4}{x + 4}$

35. $\lim_{x \to 2} \dfrac{x^2 - x - 2}{x - 2}$

36. $\lim_{x \to 1} \dfrac{4^x - x - 1}{x - 1}$

37. $\lim_{x \to 3} f(x)$, where

$$f(x) = \begin{cases} x + 7 & \text{if } x \le 3 \\ 5x - 5 & \text{if } x > 3 \end{cases}$$

38. $\lim_{x \to 1} f(x)$, where

$$f(x) = \begin{cases} 3x - 5 & \text{if } x \le 1 \\ 6 - 2x & \text{if } x > 1 \end{cases}$$

39. $\lim_{x \to 2} f(x)$, where

$$f(x) = \begin{cases} x^2 - 3 & \text{if } x < 2 \\ 5 - x^2 & \text{if } x > 2 \end{cases}$$

40. $\lim_{x \to -1} f(x)$, where

$$f(x) = \begin{cases} \sqrt{3 - x} & \text{if } x < -1 \\ \frac{4}{1 - x} & \text{if } x > -1 \end{cases}$$

41. $\lim_{x \to 1} f(x)$, where

$$f(x) = \begin{cases} e^x & \text{if } x \le 1 \\ \sqrt{x} & \text{if } x > 1 \end{cases}$$

42. $\lim_{x \to 1} \frac{\sqrt{x} - 1}{x - 1}$

43. $\lim_{x \to 0} \frac{x^3}{x - \sin x}$

44. $\lim_{x \to 0} \frac{\sin x}{\sin 2x}$

45. $\lim_{x \to 0} \frac{\cos x - 1}{x}$

46. $\lim_{x \to \pi} \frac{\tan^2 x}{1 + \sec x}$

47. $\lim_{x \to 0} \frac{e^{2x} - 1}{e^x - 1}$

48. $\lim_{x \to 1} \frac{\ln x}{x - 1}$

49. $\lim_{x \to 1} \frac{\ln x^2}{\ln x}$

50. $\lim_{x \to 0} \frac{e^{-x} - 1}{x}$

51. $\lim_{x \to 0} (x \sin x)$

52. $\lim_{x \to 0} \cos \frac{1}{x}$

53. A friend who is confused about limits wonders why you investigate the value of a function closer and closer to a point instead of just finding the value of the function at the point. How would you respond?

12.2 Techniques for Calculating Limits

Rules for Limits ■ Limits Involving Trigonometric Functions

In the previous section, we inferred the values of limits from tables and/or graphs. In this section, we develop algebraic techniques to evaluate limits.

Rules for Limits

The following rules are often helpful with evaluating limits. (Proofs of these rules require a formal definition of *limit.*)

Rules for Limits

1. Constant rule If k is a constant real number, then $\lim_{x \to a} k = k$.

2. Limit of x rule $\lim_{x \to a} x = a$

For the following rules, we assume that $\lim_{x \to a} f(x)$ and $\lim_{x \to a} g(x)$ both exist.

3. Sum and difference rules $\lim_{x \to a} [f(x) \pm g(x)] = \lim_{x \to a} f(x) \pm \lim_{x \to a} g(x)$

4. Product rule $\lim_{x \to a} [f(x) \cdot g(x)] = \lim_{x \to a} f(x) \cdot \lim_{x \to a} g(x)$

5. Quotient rule $\lim_{x \to a} \frac{f(x)}{g(x)} = \frac{\lim_{x \to a} f(x)}{\lim_{x \to a} g(x)}$, provided that $\lim_{x \to a} g(x) \neq 0$.

Looking Ahead to Calculus

The discussions at the right are *not* proofs of these rules. Proofs require the "epsilon-delta" definition of limit, discussed in the margin on page 898.

In the constant rule, the constant function defined by $f(x) = k$ always has the same value, k. Therefore, as x gets closer and closer to a, the values of $f(x)$ are always equal to k. (How much closer to k can you get?)

In the limit of x rule, the value of the identity $f(x)$ at x is the same as x. Therefore, as x gets closer and closer to a, the value of $f(x)$, which is the same as the value of x, gets closer and closer to a.

The sum and difference rules say that the limit of a sum or difference of two functions is the sum or difference of the limits. Suppose

$$\lim_{x \to a} f(x) = A \quad \text{and} \quad \lim_{x \to a} g(x) = B.$$

As x gets closer and closer to a, the values of $f(x)$ and $g(x)$ get closer and closer to A and B, respectively. Therefore, the sum $f(x) + g(x)$ gets closer and closer to the sum $A + B$; likewise the difference $f(x) - g(x)$ gets closer and closer to the difference $A - B$.

The product and quotient rules follow, since reasoning similar to that for the sum or difference applies for the product or quotient of two functions.

EXAMPLE 1 Finding a Limit of a Linear Function

Find $\lim_{x \to 4} (3 + 2x)$.

Solution

$$\begin{aligned} \lim_{x \to 4} (3 + 2x) &= \lim_{x \to 4} 3 + \lim_{x \to 4} 2x && \text{Sum rule} \\ &= 3 + \lim_{x \to 4} 2 \cdot \lim_{x \to 4} x && \text{Constant and product rules} \\ &= 3 + 2 \cdot 4 && \text{Constant and limit of } x \text{ rules} \\ &= 11 \end{aligned}$$

By the order of operations, first multiply, then add.

Letting $f(x) = 3 + 2x$, we see from Example 1 that $\lim_{x \to 4} f(x) = f(4)$. In fact, for *any* linear function defined by $f(x) = mx + b$,

$$\lim_{x \to a} f(x) = m \cdot a + b = f(a).$$

EXAMPLE 2 Finding a Limit of a Polynomial Function with One Term

Find $\lim_{x \to 5} 3x^2$.

Solution

$$\begin{aligned} \lim_{x \to 5} 3x^2 &= \lim_{x \to 5} 3 \cdot \lim_{x \to 5} x^2 && \text{Product rule} \\ &= 3 \cdot \lim_{x \to 5} x^2 && \text{Constant rule} \\ &= 3 \cdot \lim_{x \to 5} x \cdot \lim_{x \to 5} x && \text{Product rule} \\ &= 3 \cdot 5 \cdot 5 && \text{Limit of } x \text{ rule} \\ &= 75 \end{aligned}$$

Letting $f(x) = 3x^2$, we see from Example 2 that $\lim_{x \to 5} f(x) = 3 \cdot 5^2$. Actually, for *any* polynomial function in the form $f(x) = kx^n$,

$$\lim_{x \to a} f(x) = k \cdot a^n = f(a).$$

EXAMPLE 3 **Finding a Limit of a Polynomial Function**

Find $\lim_{x \to 2}(4x^3 - 6x + 1)$.

Solution

$$\begin{aligned}\lim_{x\to 2}(4x^3 - 6x + 1) &= \lim_{x\to 2} 4x^3 - \lim_{x\to 2} 6x + \lim_{x\to 2} 1 && \text{Difference and sum rules}\\ &= 4 \cdot 2^3 - 6 \cdot 2 + 1 && \text{Generalization following Example 2; constant rule}\\ &= 32 - 12 + 1 \\ &= 21\end{aligned}$$

Follow the order of operations.

Example 3 shows that $\lim_{x\to 2} f(x) = 4 \cdot 2^3 - 6 \cdot 2 + 1$ when $f(x) = 4x^3 - 6x + 1$. In fact, for *any* polynomial $p(x)$,

$$\lim_{x\to a} p(x) = p(a).$$

EXAMPLE 4 **Finding a Limit of a Rational Function**

Find $\lim_{x\to a} \frac{p(x)}{q(x)}$, where $p(x)$ and $q(x)$ are polynomials and $q(a) \neq 0$.

Solution Since $q(x)$ is a polynomial, $\lim_{x\to a} q(x) = q(a) \neq 0$. Therefore,

$$\lim_{x\to a} \frac{p(x)}{q(x)} = \frac{\lim_{x\to a} p(x)}{\lim_{x\to a} q(x)} = \frac{p(a)}{q(a)}. \qquad \text{Quotient rule; generalization following Example 3}$$

The generalization discussed after Example 3 and some additional rules are summarized as follows.

Rules for Limits (Continued)

For the following rules, we assume that $\lim_{x\to a} f(x)$ and $\lim_{x\to a} g(x)$ both exist.

6. **Polynomial rule** If $p(x)$ defines a polynomial function, then
$$\lim_{x\to a} p(x) = p(a).$$

7. **Rational function rule** If $f(x)$ defines a rational function $\frac{p(x)}{q(x)}$ with $q(a) \neq 0$, then
$$\lim_{x\to a} f(x) = f(a).$$

8. **Equal functions rule** If $f(x) = g(x)$ for all $x \neq a$, then
$$\lim_{x\to a} f(x) = \lim_{x\to a} g(x).$$

9. **Power rule** For any real number k,
$$\lim_{x\to a} [f(x)]^k = \left[\lim_{x\to a} f(x)\right]^k$$
provided that this limit exists.*

*The limit does not exist, for example, when $\lim_{x\to a} f(x) < 0$ and $k = \frac{1}{2}$, or when $\lim_{x\to a} f(x) = 0$ and $k \leq 0$.

Rules for Limits (Continued)

For the following rules, we assume that $\lim_{x \to a} f(x)$ exists.

10. Exponent rule For any real number $b > 0$,

$$\lim_{x \to a} b^{f(x)} = b^{\lim_{x \to a} f(x)}.$$

11. Logarithm rule For any real number $b > 0$ with $b \neq 1$,

$$\lim_{x \to a} [\log_b f(x)] = \log_b \left[\lim_{x \to a} f(x)\right]$$

provided that $\lim_{x \to a} f(x) > 0$.

The equal functions rule is reasonable because the limit of a function as x approaches a does not depend on the value of the function at a. In the power rule, k can be any real number for which the equation makes sense. For instance, if $k = \frac{1}{2}$, then the power rule applies to the square root of a function. The exponent and logarithm rules allow us to compute limits of generalized exponential and logarithmic functions.

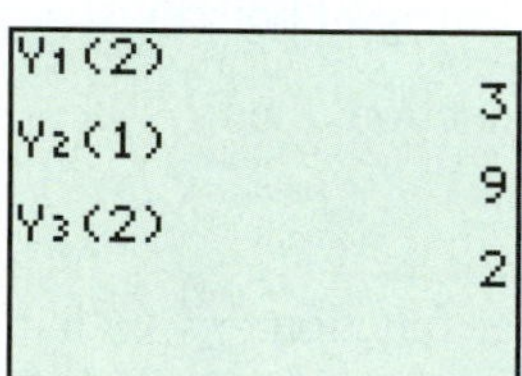

The screens show how a calculator evaluates the functions (and thus computes the limits) in Examples 5–7.

EXAMPLE 5 **Finding a Limit of a Square Root Function**

Find $\lim_{x \to 2} \sqrt{5x - 1}$.

Solution Apply the power rule with $f(x) = 5x - 1$ and $k = \frac{1}{2}$.

$$\begin{aligned} \lim_{x \to 2} \sqrt{5x - 1} &= \lim_{x \to 2} (5x - 1)^{1/2} \\ &= \left[\lim_{x \to 2} (5x - 1)\right]^{1/2} && \text{Power rule} \\ &= 9^{1/2} && \text{Polynomial rule} \\ &= 3 \end{aligned}$$

EXAMPLE 6 **Finding a Limit of an Exponential Function**

Find $\lim_{x \to 1} 3^{(x^2+1)}$.

Solution

$$\begin{aligned} \lim_{x \to 1} 3^{(x^2+1)} &= 3^{\lim_{x \to 1} (x^2+1)} && \text{Exponent rule} \\ &= 3^2 && \text{Polynomial rule} \\ &= 9 \end{aligned}$$

EXAMPLE 7 **Finding a Limit of a Logarithmic Function**

Find $\lim_{x \to 2} [\log(3x^5 + 4)]$.

Solution

$$\begin{aligned} \lim_{x \to 2} [\log(3x^5 + 4)] &= \log \left[\lim_{x \to 2} (3x^5 + 4)\right] && \text{Logarithm rule} \\ &= \log 100 && \text{Polynomial rule} \\ &= 2 \end{aligned}$$

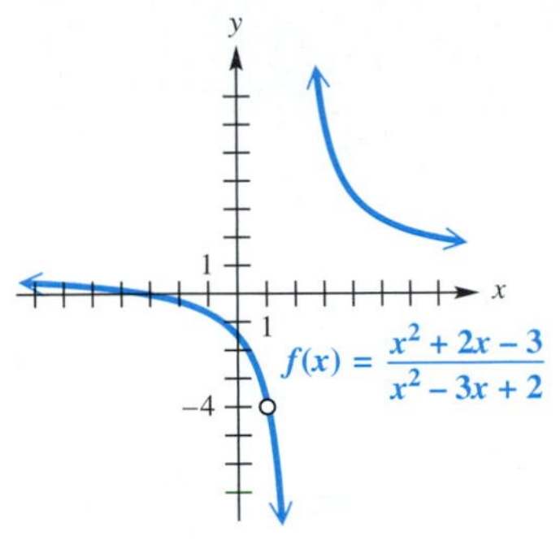

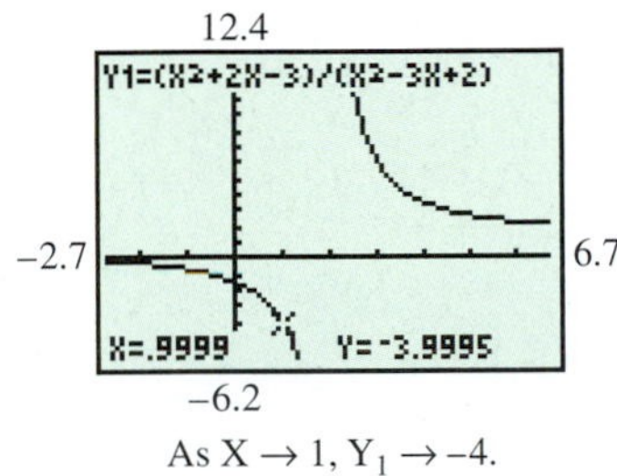

As $X \to 1$, $Y_1 \to -4$.

FIGURE 11

EXAMPLE 8 Finding a Limit of a Rational Function

Find $\lim_{x \to 1} \frac{x^2 + 2x - 3}{x^2 - 3x + 2}$.

Solution The rational function rule cannot be applied directly, since the value of the denominator is 0 when $x = 1$. However, we can find the limit after factoring the polynomials and dividing out the common factor.

$$\frac{x^2 + 2x - 3}{x^2 - 3x + 2} = \frac{(x + 3)(x - 1)}{(x - 2)(x - 1)} = \frac{x + 3}{x - 2}$$

Now we apply the equal functions rule with $f(x) = \frac{x^2 + 2x - 3}{x^2 - 3x + 2}$ and $g(x) = \frac{x + 3}{x - 2}$, two functions that are identical when $x \neq 1$.

$$\begin{aligned} \lim_{x \to 1} \frac{x^2 + 2x - 3}{x^2 - 3x + 2} &= \lim_{x \to 1} \frac{x + 3}{x - 2} && \text{Equal functions rule} \\ &= \frac{1 + 3}{1 - 2} && \text{Polynomial rule} \\ &= -4 && \text{Simplify.} \end{aligned}$$

See the graphs in Figure 11, which support this result. ■

NOTE In Example 8, we evaluated the limit by using *fraction reduction*. This technique will be used extensively in Section 12.4.

Limits Involving Trigonometric Functions

Many of the limits found in the previous section can be found algebraically by using the rules of limits presented in this section. However, some of the limits, such as $\lim_{x \to 0} \frac{\sin x}{x}$, can be obtained only with a geometric argument or by concepts covered in calculus courses. In Exercises 47–52, geometric arguments are used to derive

$$\lim_{x \to 0} \sin x = 0, \qquad \lim_{x \to 0} \cos x = 1, \qquad \text{and} \qquad \lim_{x \to 0} \frac{\sin x}{x} = 1.$$

Other limits, such as $\lim_{x \to 0} \frac{\cos x - 1}{x} = 0$, can be obtained algebraically from these limits.

EXAMPLE 9 Finding a Limit of a Trigonometric Function

Find $\lim_{x \to 0} \frac{\tan x}{x}$.

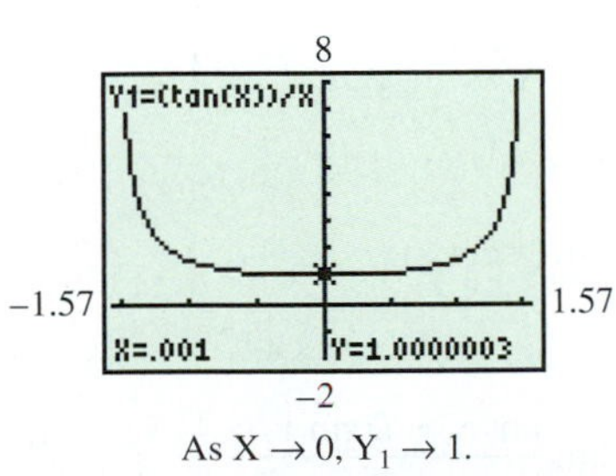

As $X \to 0$, $Y_1 \to 1$.

FIGURE 12

Solution The graph in Figure 12 *suggests* that this limit may equal 1; this is not, however, a proof. To show that 1 is the limit, we use trigonometric identities to rewrite the expression in terms of the sine and cosine functions.

$$\frac{\tan x}{x} = \frac{1}{x} \cdot \tan x = \frac{1}{x} \cdot \frac{\sin x}{\cos x} = \frac{\sin x}{x} \cdot \frac{1}{\cos x}.$$

Therefore,

$$\lim_{x \to 0} \frac{\tan x}{x} = \lim_{x \to 0} \left(\frac{\sin x}{x} \cdot \frac{1}{\cos x} \right) = \lim_{x \to 0} \frac{\sin x}{x} \cdot \lim_{x \to 0} \frac{1}{\cos x} = 1 \cdot 1 = 1.$$ ■

12.2 Exercises

Concept Check Let $\lim_{x \to 4} f(x) = 16$ and $\lim_{x \to 4} g(x) = 8$. Use the limit rules to find each limit.

1. $\lim_{x \to 4} [f(x) - g(x)]$

2. $\lim_{x \to 4} [g(x) \cdot f(x)]$

3. $\lim_{x \to 4} \dfrac{f(x)}{g(x)}$

4. $\lim_{x \to 4} [\log_2 f(x)]$

5. $\lim_{x \to 4} \sqrt{f(x)}$

6. $\lim_{x \to 4} [1 + f(x)]^2$

7. $\lim_{x \to 4} 2^{g(x)}$

8. $\lim_{x \to 4} \sqrt[3]{g(x)}$

9. $\lim_{x \to 4} \dfrac{f(x) + g(x)}{2g(x)}$

10. $\lim_{x \to 4} \dfrac{5g(x) + 2}{1 - f(x)}$

Determine each limit.

11. $\lim_{x \to -3} 7$

12. $\lim_{x \to 6} (-5)$

13. $\lim_{x \to \pi} x$

14. $\lim_{x \to -\sqrt{2}} x$

15. $\lim_{x \to 3} 4x^2$

16. $\lim_{x \to -2} (-3x^5)$

17. $\lim_{x \to -1} 4x^3$

18. $\lim_{x \to 1} (5x^8 - 3x^2 + 2)$

19. $\lim_{x \to 2} (x^3 + 4x^2 - 5)$

20. $\lim_{x \to 3} \dfrac{x^3 - 1}{x^2 + 1}$

21. $\lim_{x \to -1} \dfrac{2x + 3}{3x + 4}$

22. $\lim_{x \to 0} \dfrac{x^2 + 2x}{x}$

23. $\lim_{x \to 3} \dfrac{x^2 - 9}{x - 3}$

24. $\lim_{x \to -2} \dfrac{x^2 - 4}{x + 2}$

25. $\lim_{x \to -2} \dfrac{x^2 - x - 6}{x + 2}$

26. $\lim_{x \to 5} \dfrac{x^2 - 3x - 10}{x - 5}$

27. $\lim_{x \to 1} \dfrac{x^2 + x - 2}{x - 1}$

28. $\lim_{x \to 5} \dfrac{x^2 - 7x + 10}{x^2 - 25}$

29. $\lim_{x \to 3} \sqrt{6x - 2}$

30. $\lim_{x \to -4} (1 - 6x)^{3/2}$

31. $\lim_{x \to 1} 9^{1/(x+1)}$

32. $\lim_{x \to 3} 5^{\sqrt{x+1}}$

33. $\lim_{x \to 5} [\log_3(2x - 1)]$

34. $\lim_{x \to 4} \left[\log_2\left(14 + \sqrt{x}\right)\right]$

35. $\lim_{x \to 1} \left[\sqrt{x}(1 + x)\right]$

36. $\lim_{x \to 0} [2^{3x} - \ln(x + 1)]$

37. $\lim_{x \to 3} \dfrac{\sqrt{x + 1}}{\log_2(5x + 1)}$

38. $\lim_{x \to -1} \sqrt{x}$

39. $\lim_{x \to 1} \sqrt{3 - x}$

40. $\lim_{x \to 0} \sqrt{x}$

41. $\lim_{x \to 0} \dfrac{\sin x - 3x}{x}$

42. $\lim_{x \to 0} \dfrac{\sin x}{5x}$

43. $\lim_{x \to 0} (x \cot x)$

44. $\lim_{x \to 0} \dfrac{\sin^2 x}{x^2}$

45. $\lim_{x \to 0} \dfrac{\cos x - 1}{3x}$

46. $\lim_{x \to 0} \dfrac{\cos x + 2 \sin x - 1}{3x}$

Relating Concepts

For individual or group investigation (Exercises 47–52)

The definitions of the trigonometric functions imply that $AB = \sin x$, $OB = \cos x$, *and* $CD = \tan x$ *in the figure.* ***Work Exercises 47–52 in order.***

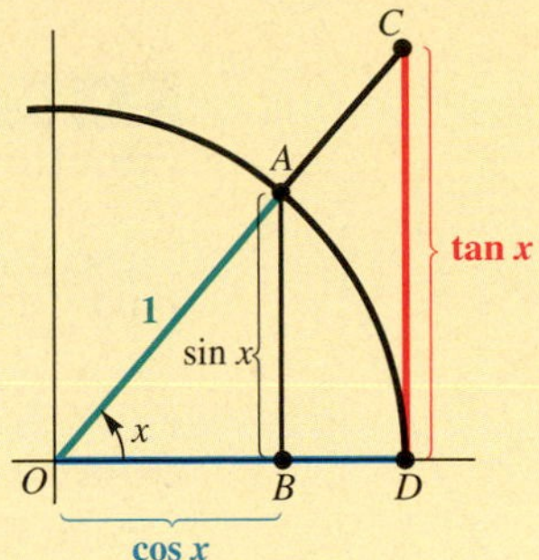

47. Use the figure to show that

$$\lim_{x\to 0} \sin x = 0 \quad \text{and} \quad \lim_{x\to 0} \cos x = 1.$$

48. Show that the area of triangle OAB is $\frac{1}{2}\sin x \cos x$.

49. Show that the area of triangle OCD is $\frac{1}{2}\tan x$.

50. Use the fact that the area of triangle $OAB <$ the area of sector $OAD <$ the area of triangle OCD to obtain the identity

$$\cos x < \frac{\sin x}{x} < \frac{1}{\cos x}.$$

(*Hint*: Recall that if $a < b < c$, then $\frac{1}{c} < \frac{1}{b} < \frac{1}{a}$.)

51. Why can we conclude that, for positive values of x, $\frac{\sin x}{x}$ approaches 1 as x approaches 0?

52. Use the result of Exercise 51 and the rules for limits to show that

$$\lim_{x\to 0} \frac{1-\cos x}{x} = 0.$$

(*Hint*: Multiply the numerator and denominator of $\frac{1-\cos x}{x}$ by the conjugate of the numerator, $1+\cos x$, and then factor out $\frac{\sin x}{x}$.)

12.3 One-Sided Limits; Limits Involving Infinity

Right- and Left-Hand Limits ■ Infinity as a Limit ■ Limits as x Approaches $\pm\infty$

Right- and Left-Hand Limits

The limits we have discussed so far are called **two-sided limits,** since the values of x get close to a from both the right and the left sides of a. Other types of limits, called **one-sided limits,** involve only values of x on one side of a.

The **right-hand limit,**

$$\lim_{x\to a^+} f(x) = L,$$

is read **"The limit of $f(x)$ as x approaches a from the right equals L."** It means that as x gets closer and closer to a from the right (that is, $x > a$), then the corresponding values of $f(x)$ get closer and closer to L. The value of $f(x)$ can be made as close to L as desired by taking values of x greater than a and close enough to a.

The **left-hand limit,**

$$\lim_{x\to a^-} f(x) = L,$$

read **"The limit of $f(x)$ as x approaches a from the left equals L,"** is defined similarly, but with x restricted to the *left* of a (that is, $x < a$).*

*Some textbooks refer to these limits as the "limit from the right" and the "limit from the left."

As with two-sided limits, limits from the right or left are not affected by whether $f(a)$ is defined; if it is, the value of $f(a)$ has no influence on the limit. The rules for limits presented in Section 12.2 are also valid for one-sided limits.

EXAMPLE 1 Finding One-Sided Limits of a Piecewise-Defined Function

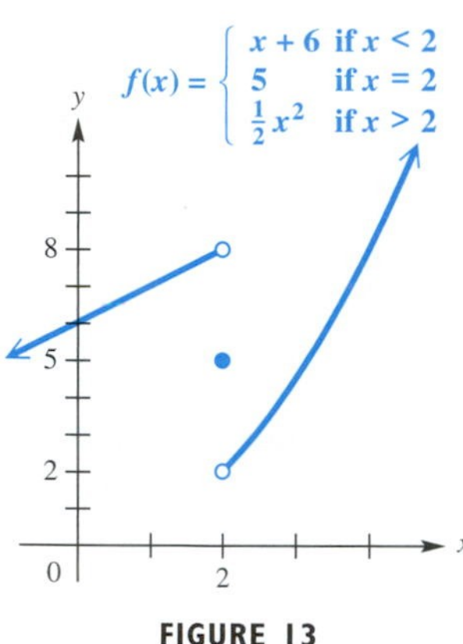

FIGURE 13

Find $\lim_{x \to 2^+} f(x)$ and $\lim_{x \to 2^-} f(x)$, where $f(x) = \begin{cases} x + 6 & \text{if } x < 2 \\ 5 & \text{if } x = 2 \\ \frac{1}{2}x^2 & \text{if } x > 2 \end{cases}$.

Solution Since $x > 2$ in the first limit, use the formula $f(x) = \frac{1}{2}x^2$. To find the second limit, where $x < 2$, use $f(x) = x + 6$.

$$\lim_{x \to 2^+} f(x) = \lim_{x \to 2^+} \frac{1}{2}x^2 = \frac{1}{2}(2^2) = 2$$

$$\lim_{x \to 2^-} f(x) = \lim_{x \to 2^-} (x + 6) = 2 + 6 = 8$$

These results are supported by the graph of f in Figure 13. ■

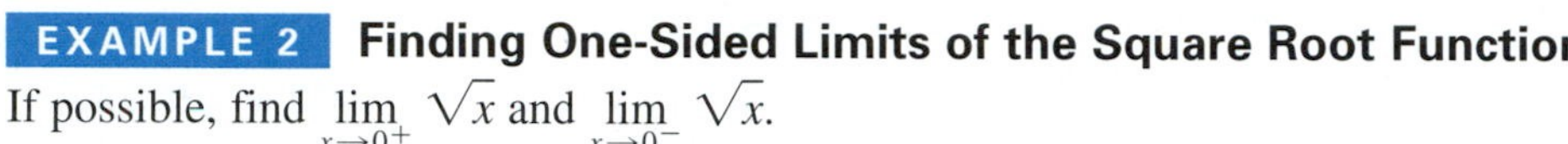

EXAMPLE 2 Finding One-Sided Limits of the Square Root Function

If possible, find $\lim_{x \to 0^+} \sqrt{x}$ and $\lim_{x \to 0^-} \sqrt{x}$.

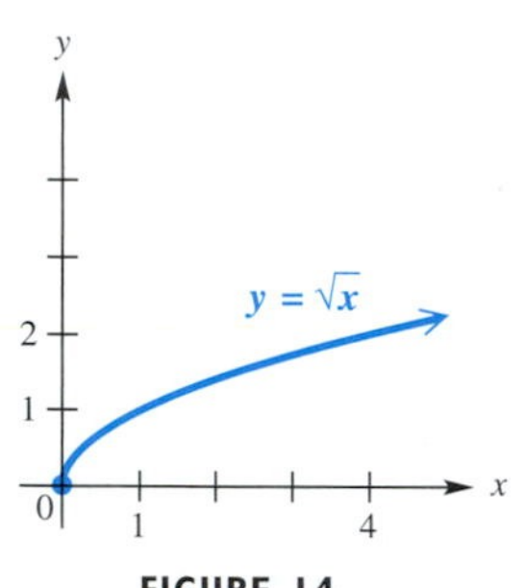

FIGURE 14

Solution The value of $\sqrt{x}$ is defined whenever $x \geq 0$. Thus, by the power function rule,

$$\lim_{x \to 0^+} \sqrt{x} = \sqrt{\lim_{x \to 0^+} x} = \sqrt{0} = 0.$$

However, the value of $\sqrt{x}$ is *undefined* whenever $x < 0$. Thus, $\lim_{x \to 0^-} \sqrt{x}$ does not exist. Notice in Figure 14 that the graph of $f(x) = \sqrt{x}$ does not exist to the left of $x = 0$. Therefore, a left-hand limit cannot exist at $x = 0$. ■

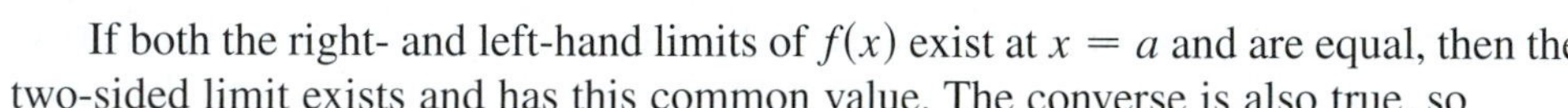

If both the right- and left-hand limits of $f(x)$ exist at $x = a$ and are equal, then the two-sided limit exists and has this common value. The converse is also true, so

$$\lim_{x \to a} f(x) = L \quad \textbf{if and only if} \quad \lim_{x \to a^+} f(x) = L \quad \textbf{and} \quad \lim_{x \to a^-} f(x) = L.$$

Infinity as a Limit

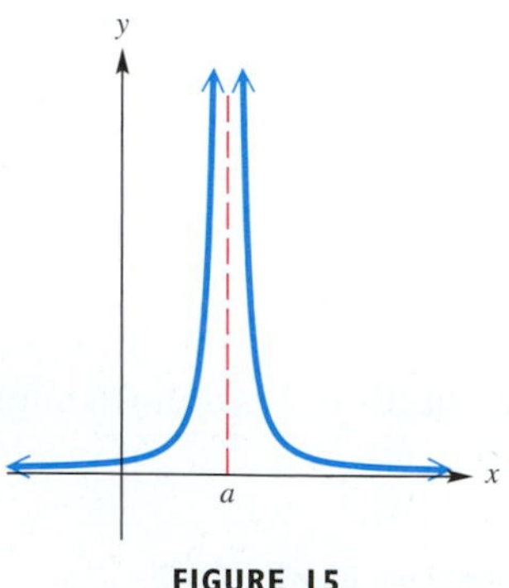

FIGURE 15

Figure 15 shows the graph of a function with a vertical asymptote. Neither the right-hand nor the left-hand limits exist for this function at $x = a$. However, as x approaches a from the right, the values of $f(x)$ increase without bound. We describe this behavior by writing

$$\lim_{x \to a^+} f(x) = \infty.$$

The infinity symbol does not represent a real number. Even though we say "The limit of $f(x)$ as x approaches a from the right equals infinity," the function does not have a right-hand limit. The notation $\lim_{x \to a^+} f(x) = \infty$ merely expresses the particular way in which the limit does not exist, namely, the function grows arbitrarily large as x approaches a from the right.

Similarly, we write

$$\lim_{x \to a^+} f(x) = -\infty$$

if the function values decrease without bound as x approaches a from the right. Analogous definitions can be made for

$$\lim_{x \to a^-} f(x) = \infty \quad \text{and} \quad \lim_{x \to a^-} f(x) = -\infty.$$

If $\lim_{x \to a^-} f(x) = \infty$ and $\lim_{x \to a^+} f(x) = \infty$, we write $\lim_{x \to a} f(x) = \infty$. Similarly, if $\lim_{x \to a^-} f(x) = -\infty$ and $\lim_{x \to a^+} f(x) = -\infty$, we write $\lim_{x \to a} f(x) = -\infty$.

The line given by $x = a$ is a vertical asymptote of $f(x)$ if at least one of the infinite behaviors discussed occurs as x approaches a. Figure 16 shows the four different types near vertical asymptotes that occur in the graphs of rational functions.

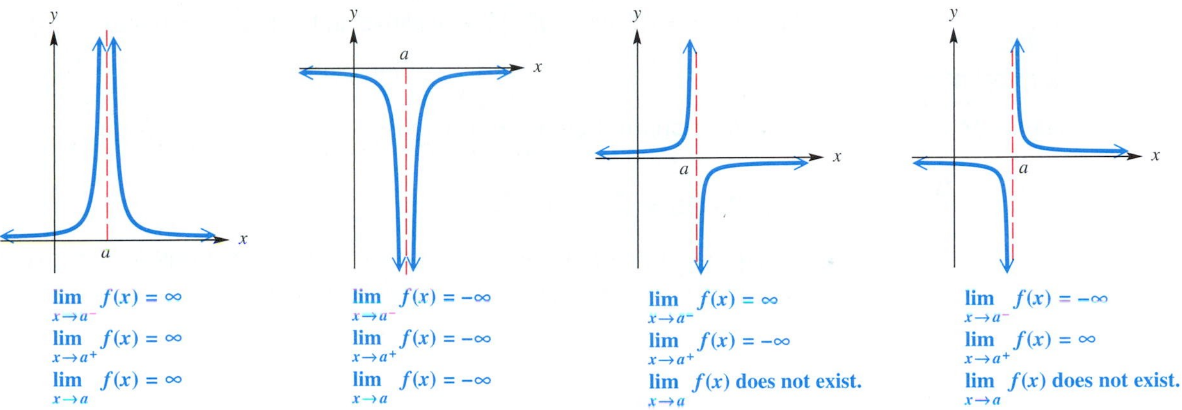

Vertical asymptotes

FIGURE 16

Two familiar functions with vertical asymptotes are shown in Figure 17. For the graph of the natural logarithmic function, $\lim_{x \to 0^+} f(x) = -\infty$. For the graph of the tangent function, $\lim_{x \to -\pi/2^+} f(x) = -\infty$ and $\lim_{x \to \pi/2^-} f(x) = \infty$.

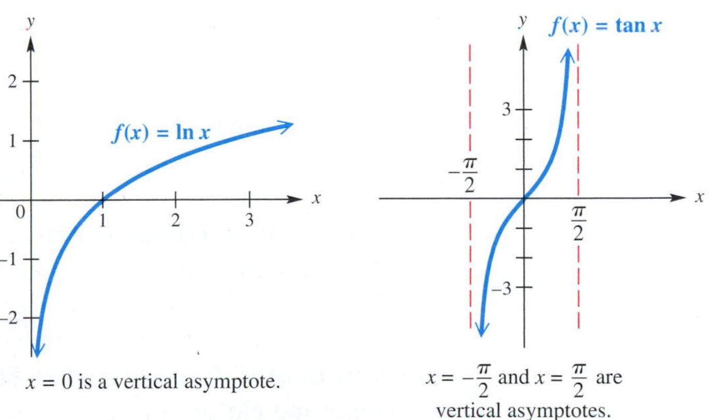

$x = 0$ is a vertical asymptote.

$x = -\frac{\pi}{2}$ and $x = \frac{\pi}{2}$ are vertical asymptotes.

FIGURE 17

EXAMPLE 3 Finding One-Sided Limits

Find $\lim_{x \to 2^+} f(x)$ and $\lim_{x \to 2^-} f(x)$, where $f(x) = \frac{1}{x-2}$.

Solution If x is slightly to the right of 2, say, $x = 2.001$, then

$$f(2.001) = \frac{1}{2.001 - 2} = \frac{1}{.001} = 1000.$$

As x moves closer and closer to 2, say, $x = 2.0001$, 2.00001, and 2.000001, the corresponding values of $f(x)$ are 10,000, 100,000, and 1,000,000. Continuing in this manner, we see that the values of $f(x)$ get arbitrarily large. Therefore,

$$\lim_{x \to 2^+} f(x) = \infty,$$

and the line $x = 2$ is a vertical asymptote.

A similar analysis from the left has x taking on values 1.999, 1.9999, 1.99999, and 1.999999 and $f(x)$ taking on values -1000, $-10{,}000$, $-100{,}000$, and $-1{,}000{,}000$. Since these values of $f(x)$ are negative and get arbitrarily large in absolute value,

$$\lim_{x \to 2^-} f(x) = -\infty.$$

See the graph in Figure 18. The table supports our conclusion. ■

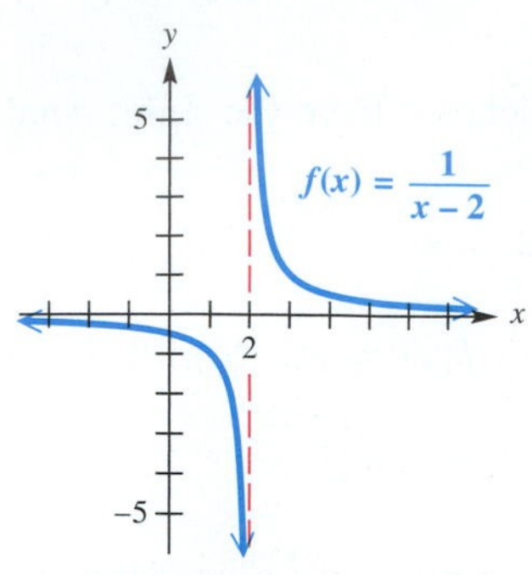

X	Y1
2.1	10
2.01	100
2.001	1000
1.9	-10
1.99	-100
1.999	-1000

Y1=1/(X-2)

FIGURE 18

Limits as x Approaches $\pm\infty$

Figure 19 shows portions of the graphs of two functions with horizontal asymptotes. We describe these asymptotic behaviors by writing

$$\lim_{x \to \infty} f(x) = 2 \quad \text{and} \quad \lim_{x \to -\infty} g(x) = 1.$$

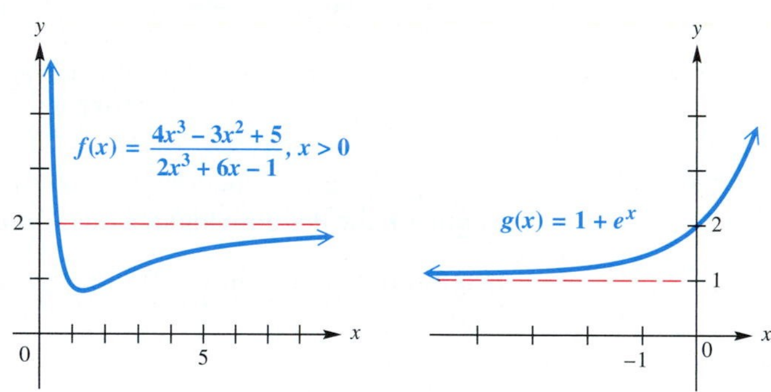

FIGURE 19

In general, the notation

$$\lim_{x \to \infty} f(x) = L,$$

read **"The limit of $f(x)$ as x approaches infinity equals L,"** means that $f(x)$ gets closer and closer to L as x gets larger and larger. Similarly, the notation

$$\lim_{x \to -\infty} f(x) = L,$$

read **"The limit of $f(x)$ as x approaches negative infinity equals L,"** means that $f(x)$ gets closer and closer to L as x assumes negative values of larger and larger magnitude. Limit Rules 1, 3–5, and 9–11 from the previous section are valid when $x \to a$ is replaced with either $x \to \infty$ or $x \to -\infty$.

GCM EXAMPLE 4 Finding Limits at Infinity

Find $\lim_{x \to \infty} f(x)$ and $\lim_{x \to -\infty} f(x)$, where $f(x) = 5 + \dfrac{10}{1 + e^{-.25x}}$.

Analytic Solution

The values of $e^{-.25x}$ become arbitrarily close to 0 as $x \to \infty$, so the expression $5 + \dfrac{10}{1 + e^{-.25x}}$ approaches $5 + \frac{10}{1 + 0} = 15$. Therefore,

$$\lim_{x \to \infty} f(x) = 15.$$

For negative values of x, $1 + e^{-.25x}$ gets arbitrarily large as x gets larger and larger in magnitude. Therefore,

$$5 + \frac{10}{1 + e^{-.25x}}$$

approaches $5 + 0 = 5$ as $x \to -\infty$. That is,

$$\lim_{x \to -\infty} f(x) = 5.$$

Graphing Calculator Solution

There are two ways to determine these limits with a graphing calculator: Look at the graph or analyze a table of function values. The graph in Figure 20 appears to approach the horizontal asymptote $y = 15$ as x moves to the right and to approach the horizontal asymptote $y = 5$ as x moves to the left. The tables in Figure 21 support these conclusions.

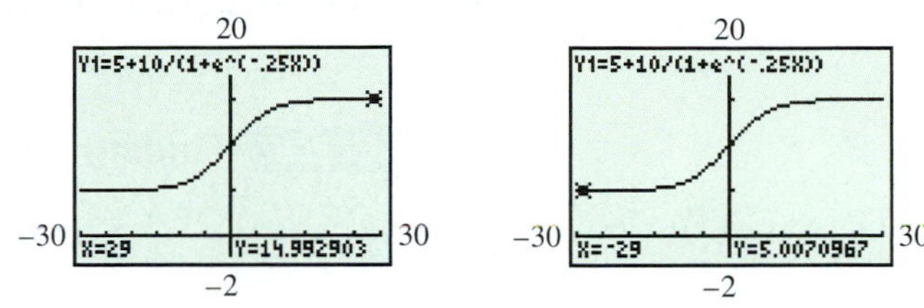

FIGURE 20

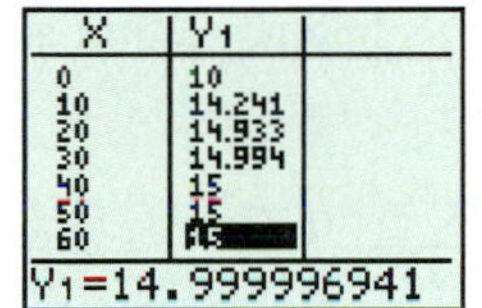

X	Y1
0	10
10	14.241
20	14.933
30	14.994
40	15
50	15
60	15

Y1=14.999996941

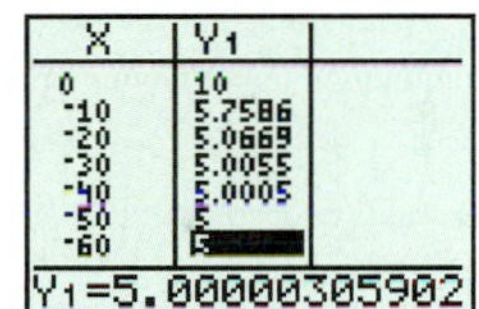

X	Y1
0	10
-10	5.7586
-20	5.0669
-30	5.0055
-40	5.0005
-50	5
-60	5

Y1=5.00000305902

FIGURE 21

NOTE A function can have two horizontal asymptotes as shown in Example 4.

EXAMPLE 5 Finding Limits at Infinity

Find $\lim_{x \to \infty} \dfrac{1}{x}$, $\lim_{x \to -\infty} \dfrac{1}{x}$, $\lim_{x \to \infty} \dfrac{1}{x^2}$, and $\lim_{x \to -\infty} \dfrac{1}{x^2}$.

Solution The following table, as well as the graphs of $f(x) = \frac{1}{x}$ and $g(x) = \frac{1}{x^2}$ shown in Figure 22, indicate that

$$\lim_{x \to \infty} \frac{1}{x} = 0, \quad \lim_{x \to -\infty} \frac{1}{x} = 0, \quad \lim_{x \to \infty} \frac{1}{x^2} = 0, \quad \text{and} \quad \lim_{x \to -\infty} \frac{1}{x^2} = 0.$$

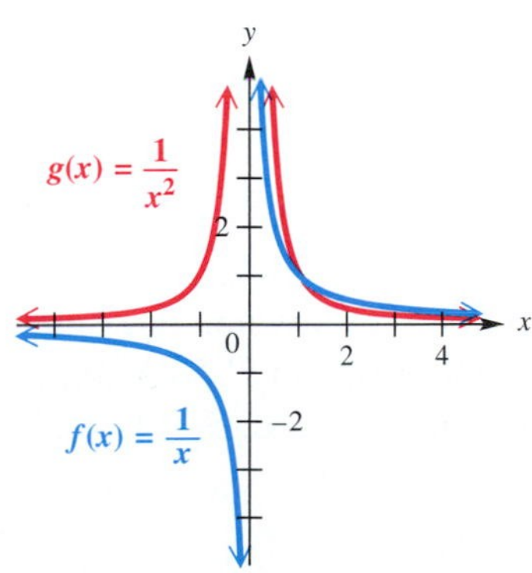

FIGURE 22

x	-100	-10	-1	1	10	100
$f(x) = \dfrac{1}{x}$	$-.01$	$-.1$	-1	1	.1	.01
$g(x) = \dfrac{1}{x^2}$	.0001	.01	1	1	.01	.0001

The results of Example 5 suggest the following rule.

Looking Ahead to Calculus

The concept of series discussed in Chapter 11 can be extended to infinite series. Functions of the form $\frac{1}{x^n}$ are often used to decide whether an infinite series is convergent. For example, if $0 \le f(x) \le \frac{1}{x^n}$, where $n > 1$, then $\sum_{i=0}^{\infty} f(i)$ is convergent.

Limits at Infinity of $\frac{1}{x^n}$

For any positive integer n,

$$\lim_{x \to \infty} \frac{1}{x^n} = 0 \quad \text{and} \quad \lim_{x \to -\infty} \frac{1}{x^n} = 0.$$

NOTE The first limit can be extended to include n as a positive real number. If x is positive, x^n exists for all positive real numbers n, so the first of the preceding two limits is valid.

EXAMPLE 6 Finding a Limit at Infinity

Find $\lim_{x \to \infty} \frac{5x^2 - 7x + 1}{2x^2 + x + 5}$.

Solution We can find limits for rational functions by dividing both numerator and denominator by the greatest power of x appearing in the denominator. (Recall that we used this technique to find asymptotes of rational functions in Chapter 4.)

$$\lim_{x \to \infty} \frac{5x^2 - 7x + 1}{2x^2 + x + 5} = \lim_{x \to \infty} \frac{5 - \frac{7}{x} + \frac{1}{x^2}}{2 + \frac{1}{x} + \frac{5}{x^2}}$$

Divide each term in the numerator and denominator by x^2.

$$= \frac{\lim_{x \to \infty}\left(5 - \frac{7}{x} + \frac{1}{x^2}\right)}{\lim_{x \to \infty}\left(2 + \frac{1}{x} + \frac{5}{x^2}\right)}$$

Quotient rule

$$= \frac{\lim_{x \to \infty} 5 - 7 \cdot \lim_{x \to \infty} \frac{1}{x} + \lim_{x \to \infty} \frac{1}{x^2}}{\lim_{x \to \infty} 2 + \lim_{x \to \infty} \frac{1}{x} + 5 \cdot \lim_{x \to \infty} \frac{1}{x^2}}$$

Constant, sum, difference, and product rules

$$= \frac{5 - 0 + 0}{2 + 0 + 0} = \frac{5}{2}$$

This result agrees with that indicated by the table and the graph in Figure 23. ■

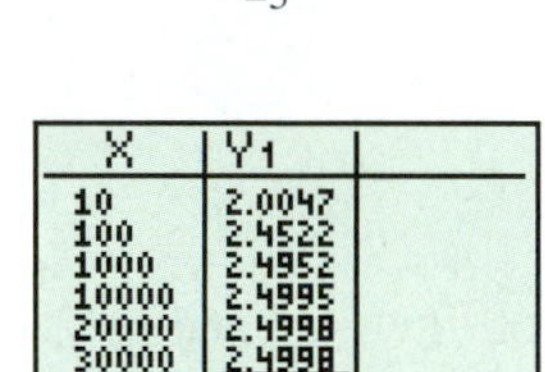

X	Y1
10	2.0047
100	2.4522
1000	2.4952
10000	2.4995
20000	2.4998
30000	2.4998
40000	2.4999

Y1=2.49988124789

FIGURE 23

One final limit behavior combines features of both limits at infinity and one-sided limits that get arbitrarily large. Functions such as x^n (where $n > 0$) and e^x do not have a limit as $x \to \infty$. They get larger and larger, increasing without bound. We describe this situation by writing

$$\lim_{x \to \infty} f(x) = \infty.$$

The limits described by $\lim_{x \to \infty} f(x) = -\infty$, $\lim_{x \to -\infty} f(x) = \infty$, and $\lim_{x \to -\infty} f(x) = -\infty$ have analogous meanings.

12.3 Exercises

Determine each limit.

1. **(a)** $\lim_{x\to 2^+} f(x)$ **(b)** $\lim_{x\to 2^-} f(x)$

where $f(x) = \begin{cases} x & \text{if } x < 2 \\ 3 & \text{if } x = 2 \\ 4 & \text{if } x > 2 \end{cases}$

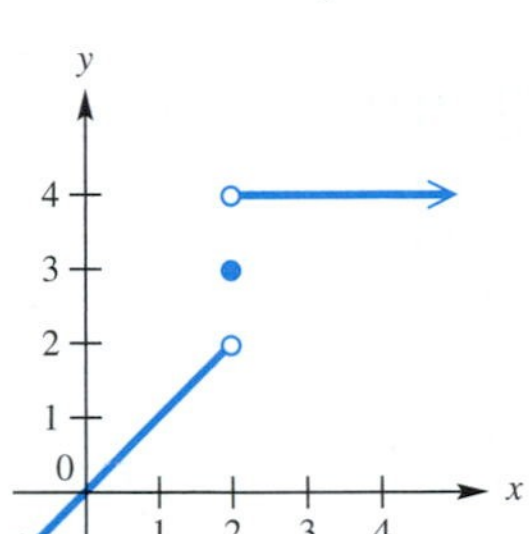

2. **(a)** $\lim_{x\to 4^+} f(x)$ **(b)** $\lim_{x\to 4^-} f(x)$

where $f(x) = \begin{cases} \frac{1}{2}x^2 & \text{if } x < 4 \\ 5 & \text{if } x = 4 \\ \sqrt{x} & \text{if } x > 4 \end{cases}$

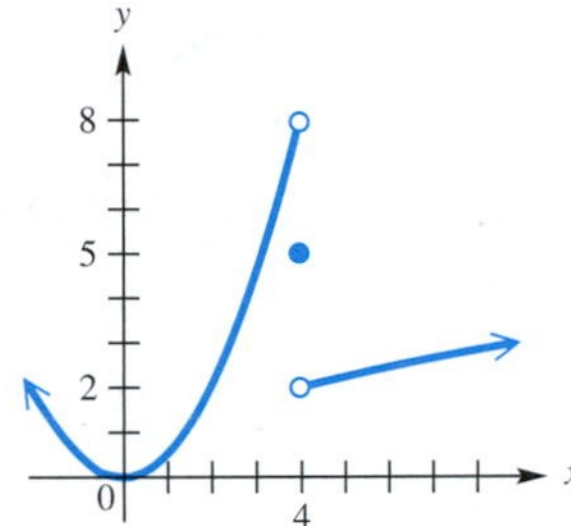

3. **(a)** $\lim_{x\to 3^+} f(x)$ **(b)** $\lim_{x\to 3^-} f(x)$

where $f(x) = \dfrac{x}{5(3 - x)^3}$

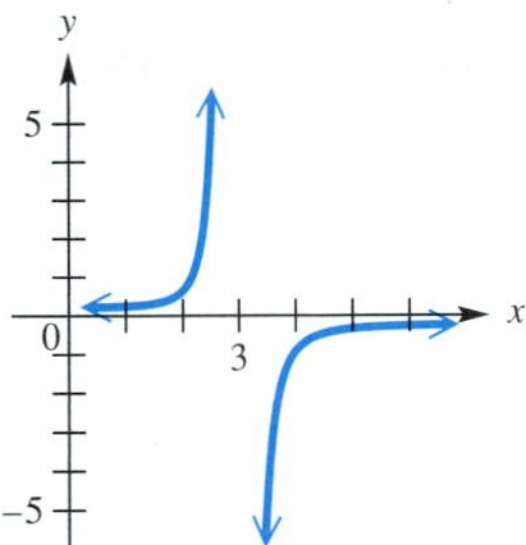

4. **(a)** $\lim_{x\to -2^+} f(x)$ **(b)** $\lim_{x\to -2^-} f(x)$

where $f(x) = \dfrac{x^2}{3(x + 2)}$

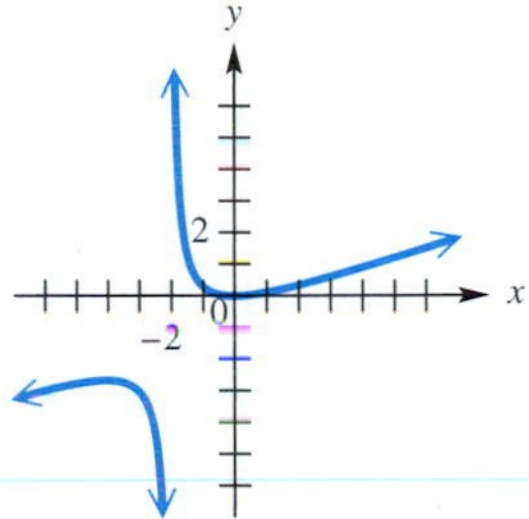

5. **(a)** $\lim_{x\to -1^+} f(x)$ **(b)** $\lim_{x\to -1^-} f(x)$

where $f(x) = \dfrac{x}{(x + 1)^2}$

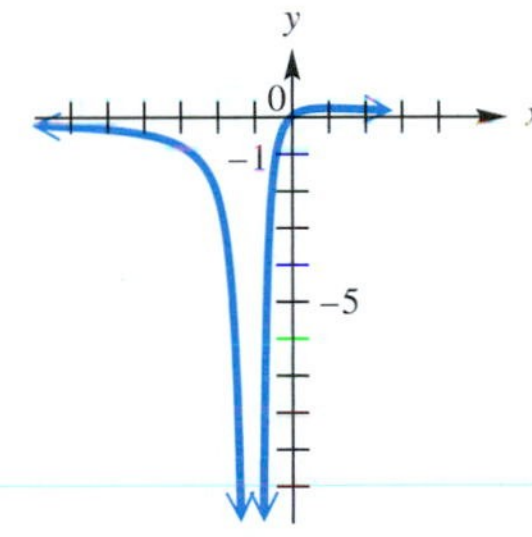

6. **(a)** $\lim_{x\to 1^+} f(x)$ **(b)** $\lim_{x\to 1^-} f(x)$

where $f(x) = \dfrac{2x}{(x - 1)^2}$

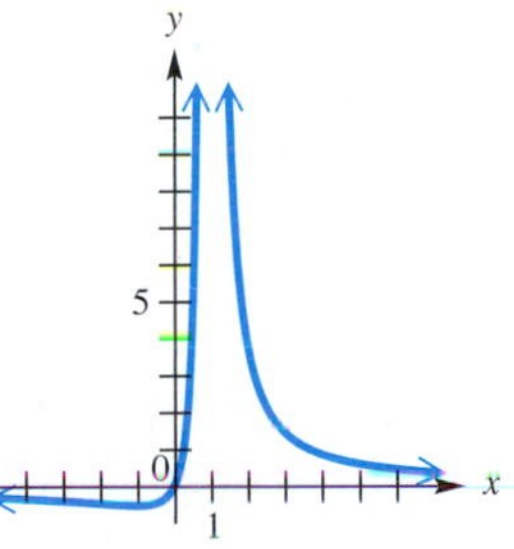

7. $\lim_{x\to 5^+} (3x - 5)$

8. $\lim_{x\to -4^-} x^3$

9. $\lim_{x\to 7^-} 100$

10. $\lim_{x\to 1^+} \sqrt{x - 1}$

11. $\lim_{x\to 2^-} \sqrt{2 - x}$

12. $\lim_{x\to -3^-} \sqrt{x + 3}$

13. $\lim_{x\to 0^-} \dfrac{|x|}{x}$

14. $\lim_{x\to 0^+} \dfrac{|x|}{x}$

15. $\lim_{x\to -3^-} \dfrac{|x + 3|}{x + 3}$

16. $\lim_{x\to -\infty} \dfrac{6x^2 + 1}{2x^2 + 3}$

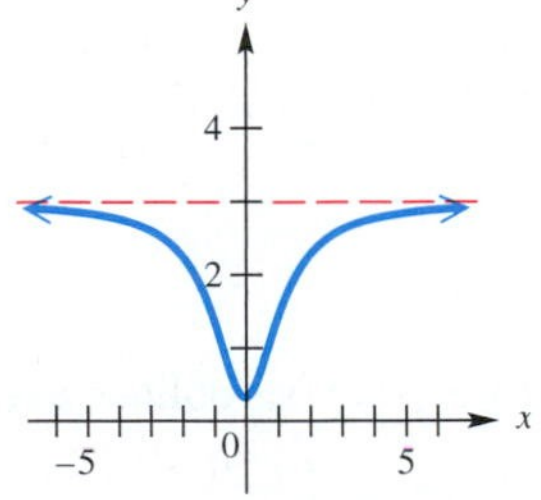

17. $\lim_{x\to \infty} (2 + e^{-x})$

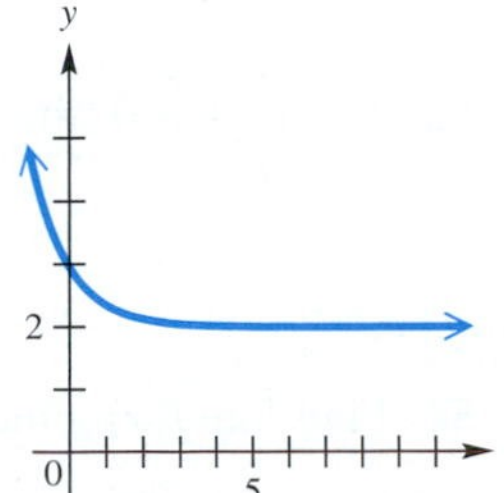

18. $\lim_{x\to \infty} (x \sin x)$

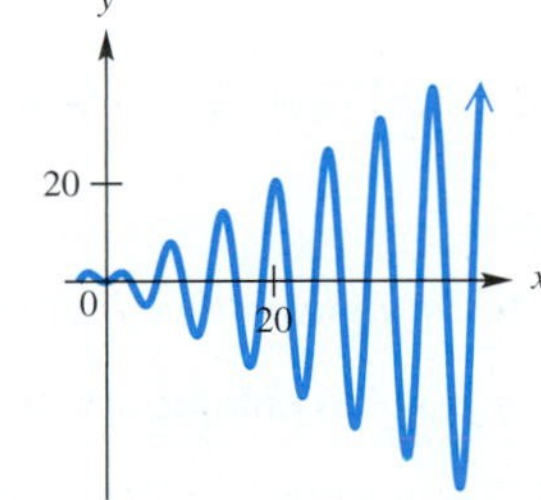

19. $\lim_{x\to \infty} \left(x + \dfrac{1}{x}\right)$

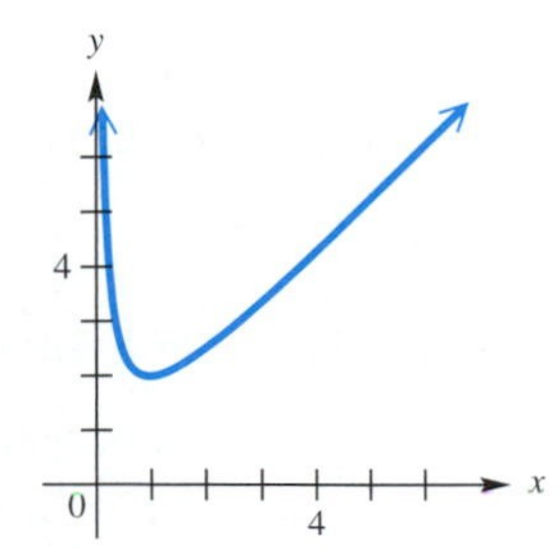

Use a table and/or graph to determine each limit.

20. $\lim_{x\to\infty}\left(x \sin \frac{5}{x}\right)$

21. $\lim_{x\to\infty}\left(x \sin \frac{1}{x^2}\right)$

22. $\lim_{x\to\infty}\left(\sqrt{x^2+x}-x\right)$

23. $\lim_{x\to\infty}\left(x-\sqrt{x^2+5}\right)$

Use a table and/or graph to find the asymptote(s) of each function.

24. $f(x)=\frac{e^x}{e^x-1}$

25. $f(x)=\tan^{-1}\frac{x}{x-1}$

26. $f(x)=\frac{x-\cos x}{x+\sin x}$

27. $f(x)=5-e^{-x}$

Determine each limit.

28. **(a)** $\lim_{x\to1^+} f(x)$ **(b)** $\lim_{x\to1^-} f(x)$

where $f(x)=\begin{cases}2x+3 & \text{if } x<1\\ 4 & \text{if } x=1\\ x^2 & \text{if } x>1\end{cases}$

29. **(a)** $\lim_{x\to2^+} f(x)$ **(b)** $\lim_{x\to2^-} f(x)$

where $f(x)=\begin{cases}7x & \text{if } x\le 2\\ x-1 & \text{if } x>2\end{cases}$

30. **(a)** $\lim_{x\to-1^+} f(x)$ **(b)** $\lim_{x\to-1^-} f(x)$

where $f(x)=\frac{1}{(1+x)^3}$

31. **(a)** $\lim_{x\to4^+} f(x)$ **(b)** $\lim_{x\to4^-} f(x)$

where $f(x)=\frac{x}{(4-x)^3}$

32. **(a)** $\lim_{x\to3^+} f(x)$ **(b)** $\lim_{x\to3^-} f(x)$

where $f(x)=\frac{1}{(x-3)^2}$

33. **(a)** $\lim_{x\to-3^+} f(x)$ **(b)** $\lim_{x\to-3^-} f(x)$

where $f(x)=\frac{x}{(x+3)^3}$

34. $\lim_{x\to\infty}\frac{3x}{5x-1}$

35. $\lim_{x\to\infty}\frac{5x}{3x-1}$

36. $\lim_{x\to-\infty}\frac{2x+3}{4x-7}$

37. $\lim_{x\to-\infty}\frac{8x+2}{2x-5}$

38. $\lim_{x\to\infty}\frac{x^2+2x}{2x^3-2x+1}$

39. $\lim_{x\to\infty}\frac{x^2+2x-5}{3x^2+2}$

40. $\lim_{x\to\infty}\frac{3x^3+2x-1}{2x^4-3x^3-2}$

41. $\lim_{x\to\infty}\frac{2x^2-1}{3x^4+2}$

42. $\lim_{x\to\infty}\frac{2x^3-x-3}{6x^2-x-1}$

43. $\lim_{x\to\infty}\frac{x^4-x^3-3x}{7x^2+9}$

44. $\lim_{x\to-\infty}\frac{-x^3-3x+1}{4x^3+5x^2-x}$

Concept Check *Write an expression for a function defined by $f(x)$ with the given features.*

45. $\lim_{x\to5^+} f(x)=\infty,\ \lim_{x\to5^-} f(x)=-\infty$

46. $f(x)$ is a quotient of two polynomials of degree greater than 2, $\lim_{x\to\infty} f(x)=0$

47. $f(x)$ is a polynomial, $\lim_{x\to\infty} f(x)=-\infty$

48. $\lim_{x\to5^+} f(x)=\infty,\ \lim_{x\to5^-} f(x)=\infty$

Concept Check *Find two functions defined by $f(x)$ and $g(x)$ with the given properties.*

49. $\lim_{x\to\infty} f(x)=\infty$, $\lim_{x\to\infty} g(x)=\infty$, and $\lim_{x\to\infty}[f(x)-g(x)]=\infty$

50. $\lim_{x\to\infty} f(x)=\infty$, $\lim_{x\to\infty} g(x)=\infty$, and $\lim_{x\to\infty}[f(x)-g(x)]=2$

51. $\lim_{x\to\infty} f(x)=0$, $\lim_{x\to\infty} g(x)=\infty$, and $\lim_{x\to\infty}[f(x)\cdot g(x)]=\infty$

52. $\lim_{x\to\infty} f(x)=0$, $\lim_{x\to\infty} g(x)=\infty$, and $\lim_{x\to\infty}[f(x)\cdot g(x)]=0$

Concept Check *In Exercises 53 and 54, assume that $f(x)$ has domain $[0,\infty)$.*

53. Find $\lim_{x\to\infty} f(x)$ if the graph of $y=f(x)$ has oblique asymptote $y=\frac{1}{2}x+3$.

54. Find $\lim_{x\to\infty} f(x)$ if the graph of $y=f(x)$ has oblique asymptote $y=-2x+3$.

Solve each problem.

55. Use a calculator to answer each question.

(a) From a graph of $y = xe^{-x}$, what do you think is the value of $\lim_{x \to \infty} (xe^{-x})$? Support your answer by evaluating the function for several large values of x.

(b) Repeat part (a), but this time using the graph of $y = x^2e^{-x}$.

(c) On the basis of your results from parts (a) and (b), what do you think is the value of $\lim_{x \to \infty} (x^n e^{-x})$ for other positive integers n?

56. Use a calculator to answer each question.

(a) From a graph of $y = \frac{\ln x}{x}$, what do you think is the value of $\lim_{x \to \infty} \frac{\ln x}{x}$? Support your answer by evaluating the function for several large values of x.

(b) Repeat part (a), but this time using the graph of $y = \frac{(\ln x)^2}{x}$.

(c) On the basis of your results from parts (a) and (b), what do you think is the value of $\lim_{x \to \infty} \frac{(\ln x)^n}{x}$, where n is a positive integer?

57. ***(Modeling) Quantity of a Drug*** Suppose a patient is given a .05-mg injection of a drug daily and each day 60% of the drug in the body is eliminated. The function defined by $y = f(t)$ in the figure shows the quantity of the drug in the body during the first two days of treatment. Interpret $\lim_{t \to 1^-} f(t)$ and $\lim_{t \to 1^+} f(t)$.

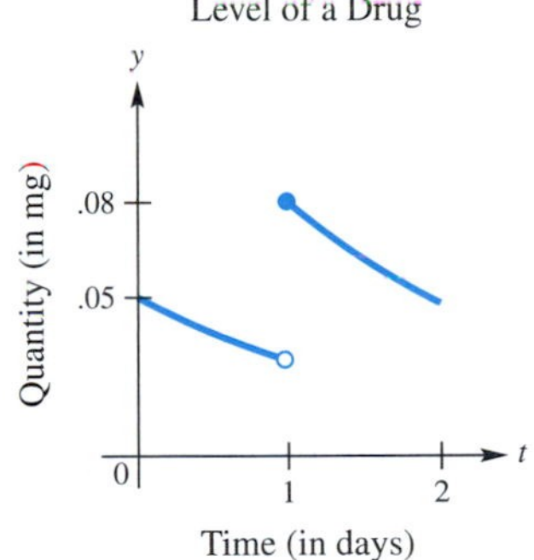

58. ***(Modeling) Speed of a Skydiver*** During free fall, the speed (in feet per second) of a skydiver after t seconds is given by

$$f(t) = 176(1 - e^{-.2t}).$$

Calculate $\lim_{t \to \infty} f(t)$ and give an interpretation of its value.

59. ***(Modeling) Evan's Price Adjustment Model*** If there is excess demand for a commodity, the price will rise rapidly at first and then more slowly according to what economists call the "Evan's price adjustment model." A typical function describing this behavior is given by

$$p(t) = 12 - 4e^{-.5t},$$

where $p(t)$ is the price in dollars after t days. Calculate $\lim_{t \to \infty} p(t)$ and give an interpretation of its value.

60. Explain why $\lim_{x \to \infty} \sin x$ does not exist.

Relating Concepts

For individual or group investigation (Exercises 61–64)

Recall from Section 3.5 that the end behavior of the graph of a polynomial function is determined by the degree of the polynomial and the sign of the leading coefficient. ***Work Exercises 61–64 in order,*** *to relate the concepts introduced in Section 3.5 with those of this chapter.*

61. Rewrite the polynomial function defined by

$$f(x) = a_nx^n + a_{n-1}x^{n-1} + \cdots + a_1x + a_0$$

as a product by factoring out the leading term, a_nx^n.

62. Use the rules for limits to show that

$$\lim_{x \to \infty} f(x) = a_n \cdot \lim_{x \to \infty} x^n \quad \text{and} \quad \lim_{x \to -\infty} f(x) = a_n \cdot \lim_{x \to -\infty} x^n.$$

63. Determine $\lim_{x \to \infty} f(x)$ and $\lim_{x \to -\infty} f(x)$ in each case.

(a) a_n positive, n even
(b) a_n negative, n even
(c) a_n positive, n odd
(d) a_n negative, n odd

64. Give the results of Exercise 63 in terms of end-behavior diagrams for graphs of polynomial functions.

Reviewing Basic Concepts (Sections 12.1–12.3)

1. Find each limit if it exists.

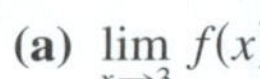
(a) $\lim_{x \to 3} f(x)$

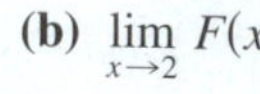
(b) $\lim_{x \to 2} F(x)$

(c) $\lim_{x \to 0} f(x)$

(d) $\lim_{x \to 3} g(x)$

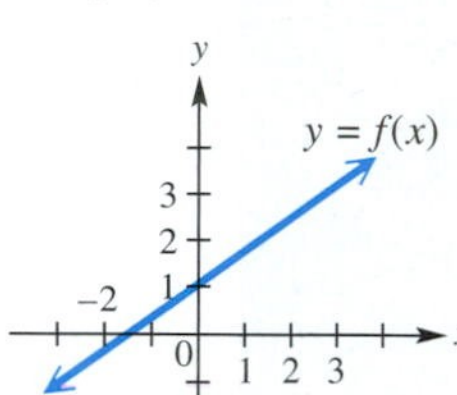

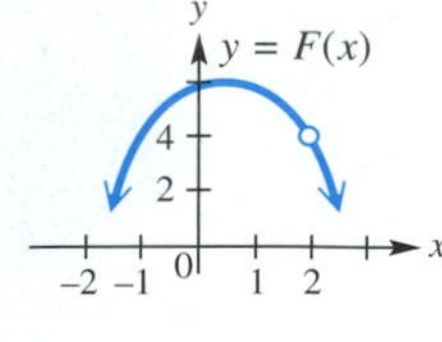

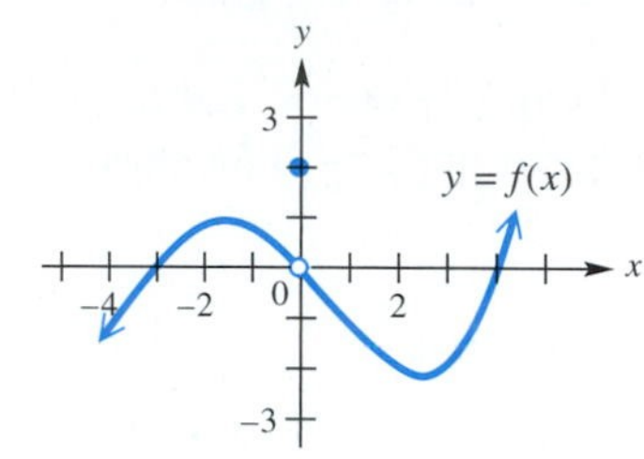

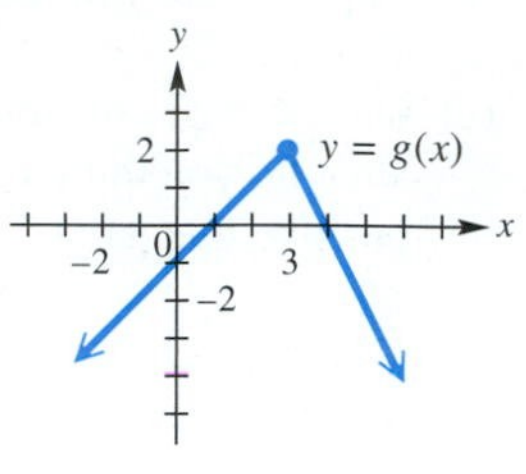

In Exercises 2 and 3, use the graph to find $\lim_{x \to a^-} f(x)$, $\lim_{x \to a^+} f(x)$, *and* $\lim_{x \to a} f(x)$, *if they exist.*

2. **(a)** $a = -2$ **(b)** $a = -1$

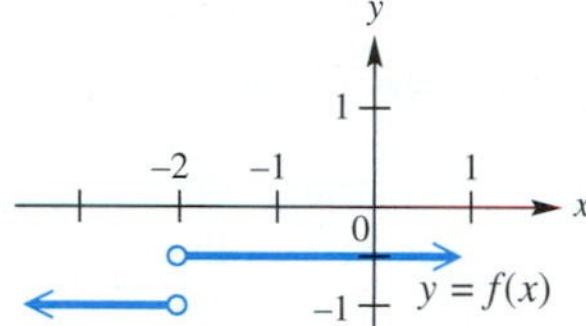

3. **(a)** $a = 1$ **(b)** $a = 2$

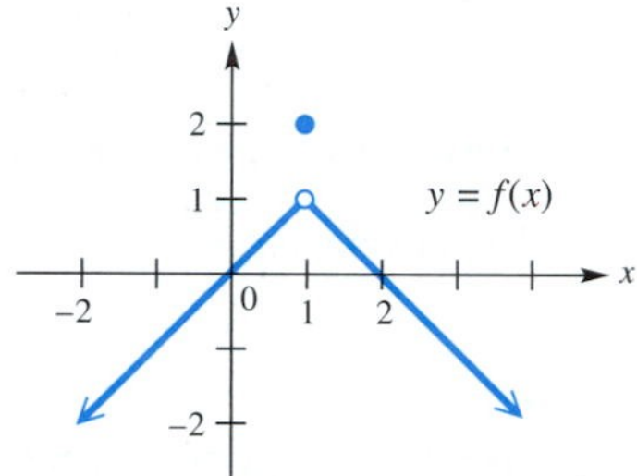

4. Find each limit if it exists.

(a) $\lim_{x \to \infty} f(x)$

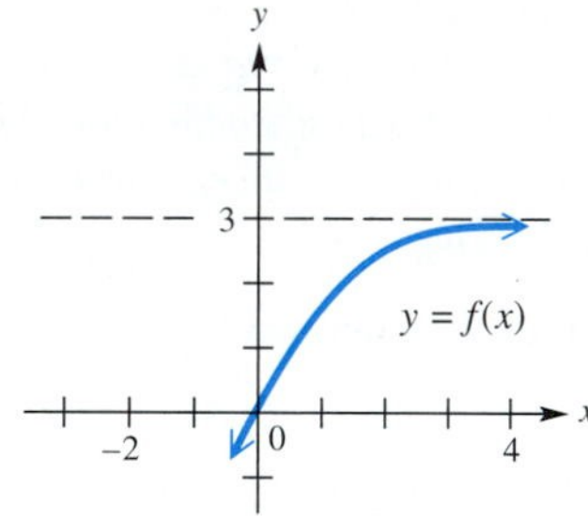

(b) $\lim_{x \to -\infty} g(x)$

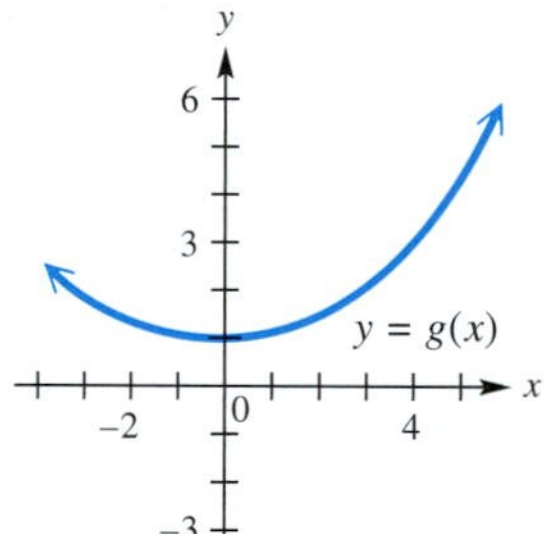

5. Find $\lim_{x \to 4} \dfrac{\sqrt{x} - 2}{x - 4}$, using a table of values near $x = 4$.

6. Let $\lim_{x \to 8} f(x) = 32$ and $\lim_{x \to 8} g(x) = 4$. Use the limit rules to find each limit.

(a) $\lim_{x \to 8} [f(x) - g(x)]$ **(b)** $\lim_{x \to 8} [g(x) \cdot f(x)]$ **(c)** $\lim_{x \to 8} \dfrac{f(x)}{g(x)}$ **(d)** $\lim_{x \to 8} [\log_2 f(x)]$

(e) $\lim_{x \to 8} \sqrt{f(x)}$ **(f)** $\lim_{x \to 8} \sqrt[3]{g(x)}$ **(g)** $\lim_{x \to 8} 2^{g(x)}$ **(h)** $\lim_{x \to 8} [1 + f(x)]^2$

(i) $\lim_{x \to 8} \dfrac{f(x) - g(x)}{4g(x)}$ **(j)** $\lim_{x \to 8} \dfrac{2g(x) + 3}{1 + f(x)}$

Find each limit if it exists.

7. $\lim_{x \to 3} \dfrac{x^2 - 9}{x - 3}$

8. $\lim_{x \to \infty} \dfrac{2x^2 - 1}{3x^4 + 5}$

9. $\lim_{x \to -\infty} \dfrac{2x^3 - x + 3}{6x^3 + 4x - 9}$

10. $\lim_{x \to 4} f(x)$, where $f(x) = \begin{cases} x + 2 & \text{if } x < 4 \\ x^2 - 6 & \text{if } x \geq 4 \end{cases}$

12.4 Tangent Lines and Derivatives

The Tangent Line as a Limit of Secant Lines ■ Derivative of a Function ■ Interpretation of the Derivative as a Rate of Change ■ Marginal Concept in Economics

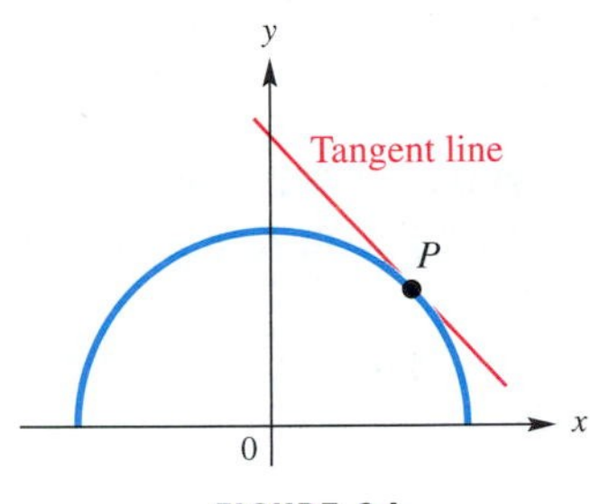

FIGURE 24

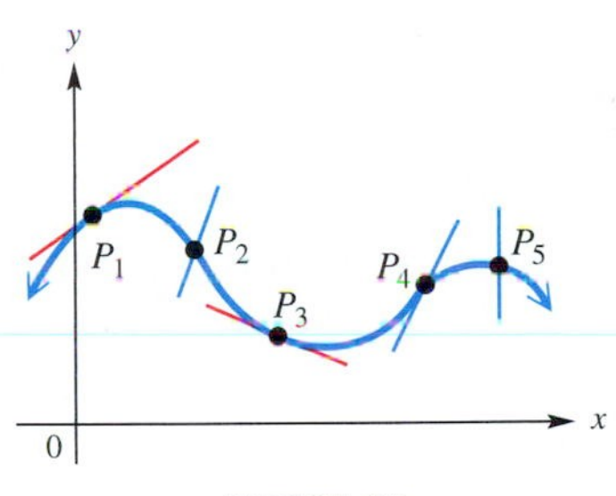

FIGURE 25

The Tangent Line as a Limit of Secant Lines

In geometry, a **tangent line** to a circle is defined as a line that touches the circle at only *one* point, as at the point P in Figure 24 (which shows the top half of a circle). If you think of this half circle as part of a curving road on which you are driving at night, then the tangent line indicates the direction of the light beam from your headlights as you pass through the point P.

The tangent line to an arbitrary curve at a point P on the curve should touch the curve at P, but not at any other nearby points, and should indicate the direction of the curve. In Figure 25, for example, the lines through P_1 and P_3 are tangent lines, while the lines through P_2 and P_5 are not. The tangent lines just touch the curve, while the other lines pass through it. To decide about the line at P_4, we need to define the idea of a tangent line to the graph of a function more carefully.

Let R be a fixed point with coordinates $(a, f(a))$ on the graph of a function $y = f(x)$, as in Figure 26. Choose a different point S on the graph, and draw the line through R and S; this line is called a **secant line.** If S has coordinates $(x, f(x))$, then, by the definition of slope, the slope of the secant line RS is given by

$$\frac{f(x) - f(a)}{x - a}. \quad \text{Slope of secant line}$$

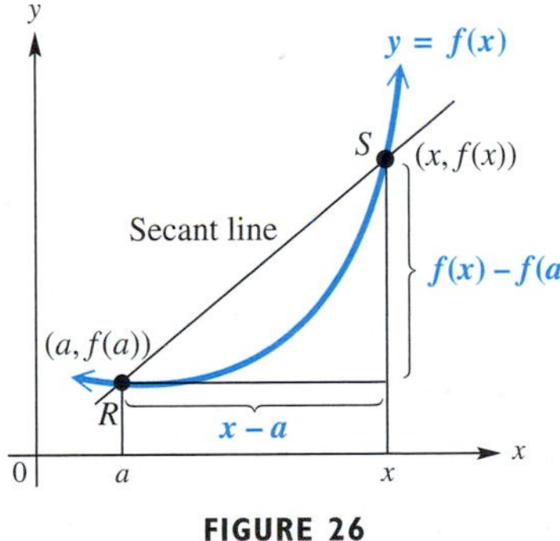

FIGURE 26

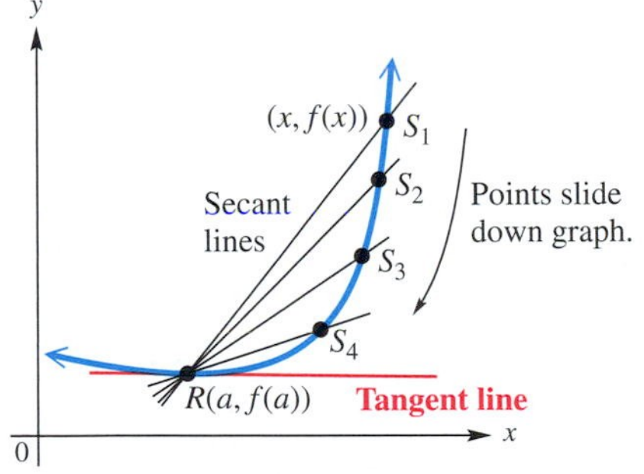

FIGURE 27

This slope corresponds to the average rate of change of y with respect to x over the interval from a to x. As x approaches a, point S will slide along the curve, getting closer and closer to the fixed point R. Figure 27 shows successive positions S_1,

S_2, S_3, and S_4 of the point S. Notice that the corresponding secant lines get closer and closer to the tangent line at R. If the slopes of the corresponding secant lines approach a limit as x approaches a, then the **slope of the tangent line** at point R is defined to be this limit.

Tangent Line

The **tangent line** of the graph of $y = f(x)$ at the point $(a, f(a))$ is the line through this point and having slope

$$m = \lim_{x \to a} \frac{f(x) - f(a)}{x - a},$$

provided that this limit exists. If the limit does not exist, then there is no tangent line at the point.

The slope of the tangent line at a point is also called the **slope of the curve** at the point and corresponds to the "instantaneous" rate of change of y with respect to x at the point.

In certain applications of mathematics, it is necessary to determine the equation of the line tangent to the graph of a function at a given point.

GCM EXAMPLE 1 Finding the Equation of a Tangent Line

Find the equation of the tangent line to the graph of the function defined by $f(x) = x^2 + 2$ when x has the value -1.

Analytic Solution

First find the slope of the tangent line at the given point by using the definition in the box, with $f(x) = x^2 + 2$ and $a = -1$.

$$\lim_{x \to a} \frac{f(x) - f(a)}{x - a} = \lim_{x \to -1} \frac{[x^2 + 2] - [(-1)^2 + 2]}{x - (-1)} \quad \text{Substitute.}$$

$$= \lim_{x \to -1} \frac{x^2 - 1}{x + 1} \quad \text{Simplify.}$$

$$= \lim_{x \to -1} \frac{(x + 1)(x - 1)}{(x + 1)} \quad \text{Factor.}$$

$$= \lim_{x \to -1} (x - 1) \quad \text{Simplify.}$$

$$= -2 \quad \text{Polynomial rule}$$

When $x = -1$, $f(x) = f(-1) = (-1)^2 + 2 = 3$. Now find the equation of the line through the point $(-1, 3)$ having slope -2.

$$y - y_1 = m(x - x_1) \quad \text{Point–slope form}$$

$$y - 3 = -2[x - (-1)] \quad y_1 = 3,\ m = -2,\ x_1 = -1$$

$$y - 3 = -2(x + 1) \quad \text{Work inside parentheses.}$$

$$y - 3 = -2x - 2 \quad \text{Distributive property}$$

$$y = -2x + 1 \quad \text{Add 3.}$$

Graphing Calculator Solution

Most calculators can draw a tangent line at a specified point on a graph. Figure 28 was generated with the Tangent command from the DRAW menu on the TI-83/84 Plus. At the bottom of the screen, the first line gives the x-coordinate of the point of "tangency" and the second line gives the equation of the tangent line.

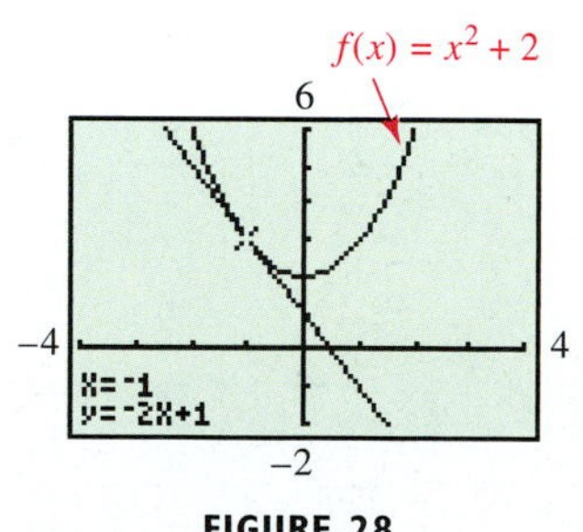

FIGURE 28

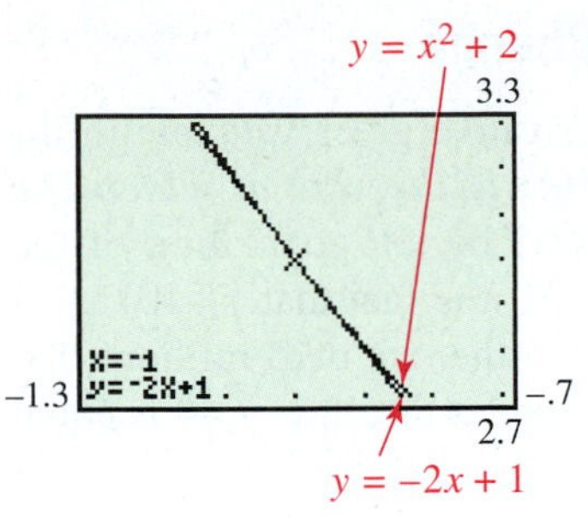

FIGURE 29

Figure 29 shows the result of using a calculator to zoom in on the point $(-1, 3)$ in Figure 28. Notice that in this close-up view, the graph and its tangent line appear virtually identical. This gives us another interpretation of the tangent line. Suppose, as we zoom in on a function, the graph appears to become a straight line. Then this line is the tangent line to the graph at that point. ***In other words, the tangent line captures the behavior of the function very close to the point under consideration.*** Consequently, another way to approximate the slope of the curve at a point is to zoom in on the curve near the point until it appears to be a straight line (the tangent line). Then we can estimate the slope by using any two points on that section of the curve.

Derivative of a Function

A function defined by $f(x)$ associates, with each number a in its domain, the height of the graph of $y = f(x)$ above the x-axis at the point where $x = a$. (Height is taken as negative for points below the x-axis.) Related to f is the function denoted $\mathbf{f'}$, read **"f prime,"** whose value at a is the *slope* of the curve $y = f(x)$ at the point where $x = a$. The function f' is called the *derivative of f.*

Looking Ahead to Calculus

In calculus, we derive symbolic techniques for converting the expression for $f(x)$ into an expression for $f'(x)$. It can be shown that if $f(x) = \sqrt{x}$, then $f'(x) = \frac{1}{2\sqrt{x}}$. Therefore, $f'(4) = \frac{1}{2\sqrt{4}} = \frac{1}{4}$, the same result obtained in Example 2.

Derivative

If a is in the domain of f, then the **derivative of f at a** is defined by

$$f'(a) = \lim_{x \to a} \frac{f(x) - f(a)}{x - a},$$

provided that this limit exists.

NOTE ***The derivative of f at a is equal to the slope of the tangent line to the graph of f at (a, f(a)).*** They are one and the same concept.

GCM **EXAMPLE 2** **Finding a Derivative**

Find $f'(4)$, where $f(x) = \sqrt{x}$.

Analytic Solution

$$f'(4) = \lim_{x \to 4} \frac{\sqrt{x} - \sqrt{4}}{x - 4} \quad \text{Definition of the derivative}$$

$$= \lim_{x \to 4} \frac{\sqrt{x} - 2}{x - 4} \quad \sqrt{4} = 2$$

$$= \lim_{x \to 4} \frac{\sqrt{x} - 2}{(\sqrt{x} - 2)(\sqrt{x} + 2)} \quad \text{Factor.}$$

$$= \lim_{x \to 4} \frac{1}{\sqrt{x} + 2} \quad \text{Simplify.}$$

$$= \frac{1}{\sqrt{4} + 2} = \frac{1}{4}$$

Therefore, $f'(4) = \frac{1}{4}$.

Graphing Calculator Solution

Calculators can approximate the value of $f'(a)$ with a single instruction. In Figure 30, $f'(4)$ is found with a TI-83/84 Plus. nDeriv(denotes "numerical approximation to the derivative," $\sqrt{(X)}$ is the function, X is the variable, and 4 is the value of a.

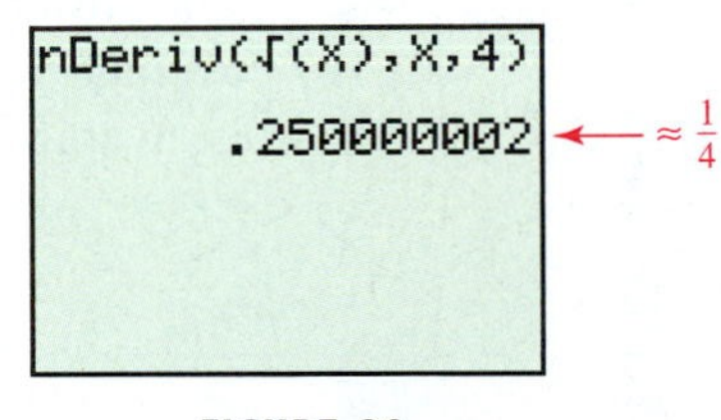

FIGURE 30

Interpretation of the Derivative as a Rate of Change

Consider a quantity that is changing with time. The derivative gives valuable information about the rate of the change. ***The value of $f'(a)$ is the rate at which the quantity is changing at time $t = a$.*** For instance, let $f(t)$ be the population of the United States (in millions) t years after 1800. In Figure 31, the fact that $f'(40) = .5$ tells us that in the year 1840 the population was growing at the rate of .5 million (that is, 500,000) people per year. Note that the slope of the tangent line at $t = 40$ is equal to the derivative $f'(40)$.

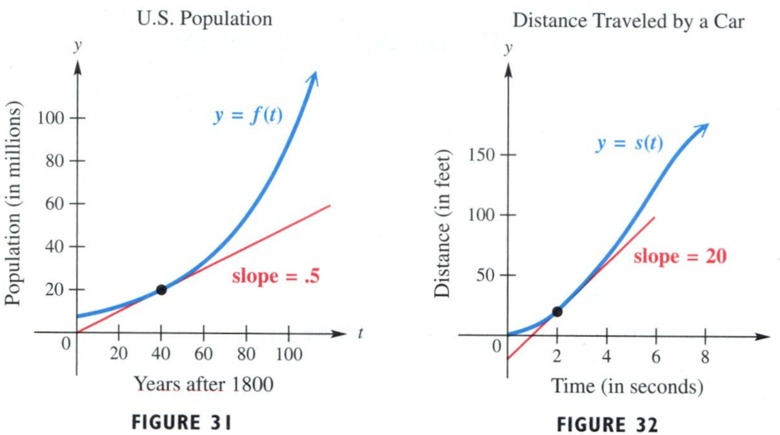

FIGURE 31 **FIGURE 32**

As another example, let $s(t)$ be the number of feet traveled by a car after t seconds. Figure 32 shows the distance traveled by a car on an 8-second test run. The fact that $s'(2) = 20$ tells us that after 2 seconds the car is traveling at a velocity of 20 feet per second. Note that the slope of the tangent line at $t = 2$ is equal to the derivative $s'(2)$.

We can analyze the function defined by $s(t)$ to see why $s'(2)$ is the car's velocity at time $t = 2$. (This value is sometimes called the *instantaneous velocity at time $t = 2$.*) Consider the time interval from time 2 seconds to 4 seconds. The average velocity during that interval is

$$\text{average velocity} = \frac{\text{distance traveled}}{\text{time elapsed}} = \frac{s(4) - s(2)}{4 - 2}.$$

This average velocity is an approximation to the (instantaneous) velocity at time 2 seconds. Actually, for any time $t > 2$, the average velocity during the time interval from 2 to t is

$$\text{average velocity} = \frac{\text{distance traveled}}{\text{time elapsed}} = \frac{s(t) - s(2)}{t - 2}.$$

The closer t is to 2, the better is the approximation. As t gets closer and closer to 2, the average velocity gets closer and closer to the (instantaneous) velocity at time 2. That is, average velocity approaches instantaneous velocity as $t \to 2$. Therefore,

$$\text{instantaneous velocity at time 2} = \lim_{t \to 2} \frac{s(t) - s(2)}{t - 2}.$$

A car traveling along a road is an example of motion along a straight line. The function defined by $s(t)$ is often called a **position function,** since it gives the location or position of the car on the straight line at time t. The velocity of the car tells how fast the position of the car changes and the direction of the car. The velocity at time t will be positive when the car is moving forward and negative when the car is backing up.

Looking Ahead to Calculus

The function $s'(t)$ is also denoted $v(t)$, since it gives the velocity of a car at any time. In calculus, we learn why $v'(t)$ tells how fast the car is accelerating at any time.

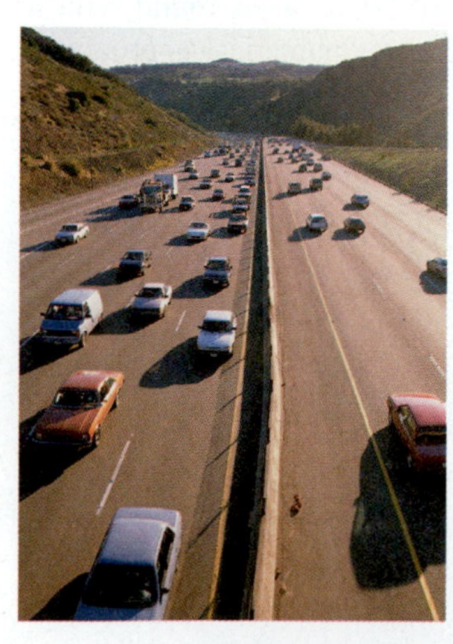

This discussion is summarized as follows.

Velocity as a Derivative

If an object is moving along a straight line and its position on the line at time t is $s(t)$, then the **velocity at time *a*** is

$$s'(a) = \lim_{t \to a} \frac{s(t) - s(a)}{t - a}.$$

Another example of an object moving along a straight line is a ball thrown straight up into the air. The ball initially rises (positive direction) and then falls (negative direction). In physics, it is shown that the height (in feet) of the ball above the ground after t seconds is

$$s(t) = -16t^2 + v_0 t + h_0,$$

where v_0 is the initial velocity and h_0 is the initial height. (This equation ignores air resistance.)

EXAMPLE 3 **Finding Information about a Ball Thrown into the Air**

Suppose a ball thrown straight up into the air has an initial height of 112 feet and an initial velocity of 96 feet per second.

(a) Give the equation for $s(t)$, the height of the ball after t seconds.

(b) How high above the ground is the ball at 5 seconds?

(c) What is the velocity of the ball at 5 seconds?

Solution

(a) Since $v_0 = 96$ and $h_0 = 112$,

$$s(t) = -16t^2 + 96t + 112.$$

(b) The height at 5 seconds is

$$s(5) = -16(5)^2 + 96(5) + 112 = 192 \text{ feet}.$$

(c) The velocity of the ball at 5 seconds is

$$s'(5) = \lim_{t \to 5} \frac{s(t) - s(5)}{t - 5}$$ Definition of the derivative

$$= \lim_{t \to 5} \frac{(-16t^2 + 96t + 112) - 192}{t - 5}$$ Substitute the results from parts (a) and (b).

$$= \lim_{t \to 5} \frac{-16t^2 + 96t - 80}{t - 5}$$ Simplify.

Be careful with signs.

$$= \lim_{t \to 5} \frac{-16(t^2 - 6t + 5)}{t - 5}$$ Factor out −16.

$$= \lim_{t \to 5} \frac{-16(t - 5)(t - 1)}{t - 5}$$ Factor the trinomial.

$$= \lim_{t \to 5} [-16(t - 1)]$$ Simplify.

$$= -64.$$

At 5 seconds, the ball is falling downward at a speed of 64 feet per second. ■

The interpretation of the derivative as an instantaneous rate of change applies to any quantity, even when the independent variable is not time.

The derivative $f'(a)$ measures the rate of change of $f(x)$ at $x = a$.

Marginal Concept in Economics

In economics, derivatives are often described by the adjective "marginal." For instance, if $C(x)$ is a cost function (the cost of producing x units of a commodity), then the value of the derivative $C'(a)$ is called the **marginal cost** at the production level of a units. Since the marginal cost is a derivative, its value gives the rate at which costs are increasing (or decreasing) with respect to the level of production, assuming that production is at level a.

For instance, if $C(1000) = 20{,}000$ and $C'(1000) = 15$, then the cost of producing 1000 units of goods is \$20,000 and the marginal cost when 1000 units of goods are produced is \$15. The marginal cost can be interpreted as the approximate cost of producing one additional unit of goods. That is, when $x = 1000$, the cost is increasing at the rate of 15 dollars per unit, so the cost of producing 1001 units will be about \$20,015.

If $R(x)$ and $P(x)$ define revenue and profit functions, then the values of $R'(a)$ and $P'(a)$ give the marginal revenue and the marginal profit, respectively, at $x = a$.

The following summarizes our work with derivatives of functions.

Summary of Ideas Related to $f'(x)$

1. **Geometric definition of $f'(a)$** $f'(a)$ is the slope of the tangent line to the graph of $f(x)$ at $x = a$.
2. **Algebraic definition of $f'(a)$** $f'(a) = \lim_{x \to a} \dfrac{f(x) - f(a)}{x - a}$
3. **Interpretation of $f'(a)$** $f'(a)$ represents the rate of change in $f(x)$ when $x = a$.

12.4 Exercises

Concept Check *Estimate the slope of the tangent line to each curve at the given point (x, y).*

1.

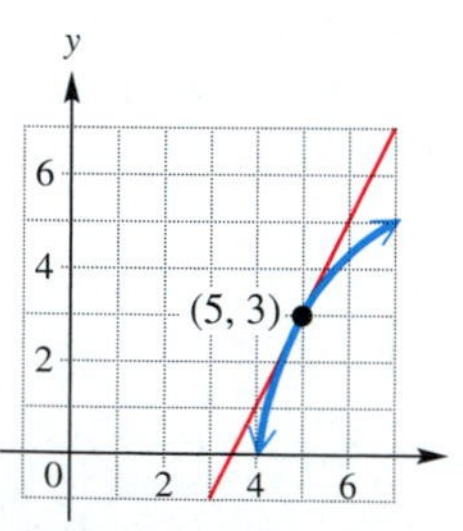

2.

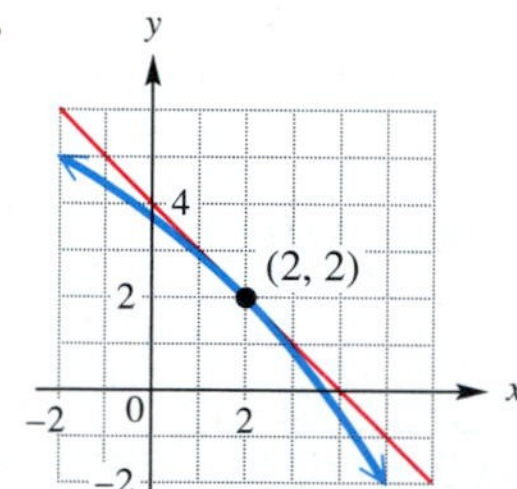

3.

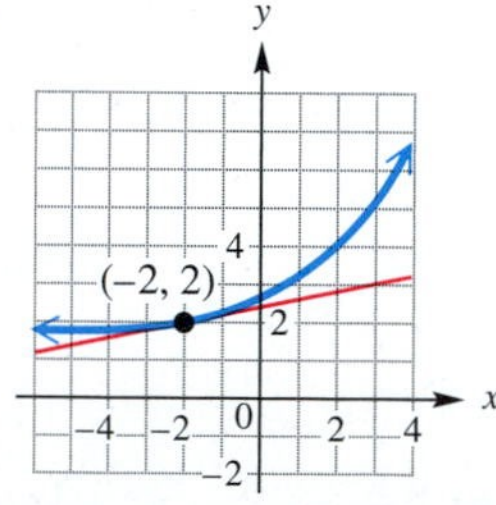

4.

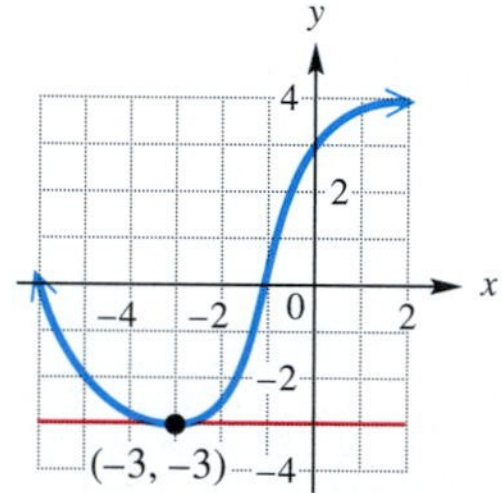

Find the slope of the tangent line to each curve when x has the given value.

5. $f(x) = x^2;\ x = 4$

6. $f(x) = x^2 + 5;\ x = 2$

7. $f(x) = -4x^2 + 11x;\ x = -2$

8. $f(x) = 6x^2 - 4x;\ x = -1$

9. $f(x) = -\dfrac{2}{x};\ x = 4$

10. $f(x) = \dfrac{6}{x};\ x = -1$

11. $f(x) = -3\sqrt{x};\ x = 1$

12. $f(x) = x^3;\ x = 1$

Find the equation of the tangent line to each curve when x has the given value. Verify your answer by graphing both $f(x)$ and the tangent line with a calculator.

13. $f(x) = x^2 + 2x;\ x = 3$

14. $f(x) = 6 - x^2;\ x = -1$

15. $f(x) = \dfrac{5}{x};\ x = 2$

16. $f(x) = \dfrac{-3}{x + 1};\ x = 1$

17. $f(x) = 4\sqrt{x};\ x = 9$

18. $f(x) = \sqrt{x};\ x = 25$

Concept Check *By considering the graph of the function but not calculating any limits, give the value of $f'(2)$ for each function.*

19. $f(x) = 5$

20. $f(x) = x$

21. $f(x) = -x$

22. $f(x) = 3x + 4$

Use a calculator to estimate $f'(a)$ for the given value of a.

23. $f(x) = e^x;\ a = 0$

24. $f(x) = \sin x;\ a = 0$

25. $f(x) = \dfrac{10x}{1 + .25x^2};\ a = 2$

26. $f(x) = \dfrac{1}{1 + x^2};\ a = 0$

27. $f(x) = x \cos x;\ a = \dfrac{\pi}{4}$

28. $f(x) = xe^x;\ a = 1$

Concept Check *In Exercises 29–32, use the figure to determine the derivative.*

29. The figure shows the graph of $f(x) = 1 + \frac{x}{x + 3}$, along with its tangent line at the point $(0, 1)$. What is $f'(0)$?

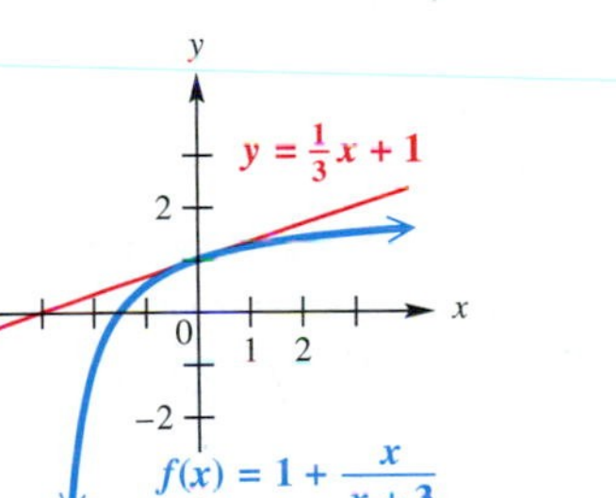

30. The figure shows the graph of $f(x) = \ln x$, along with its tangent line at the point $(1, 0)$. What is $f'(1)$?

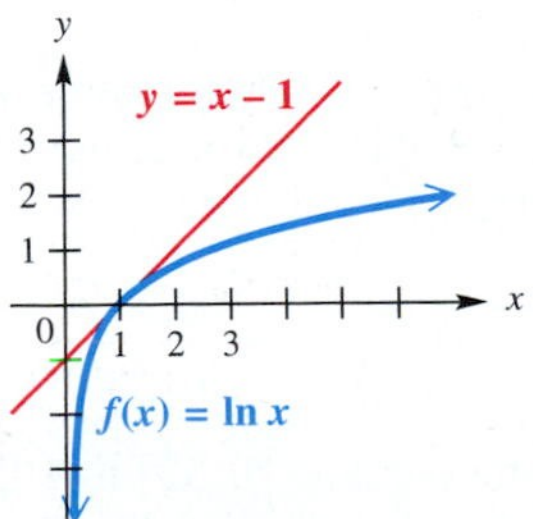

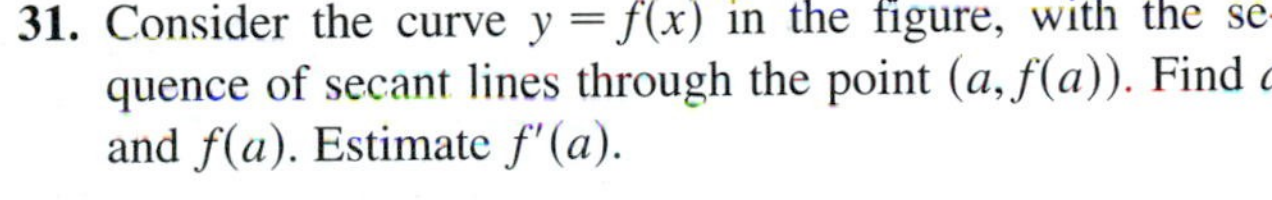

31. Consider the curve $y = f(x)$ in the figure, with the sequence of secant lines through the point $(a, f(a))$. Find a and $f(a)$. Estimate $f'(a)$.

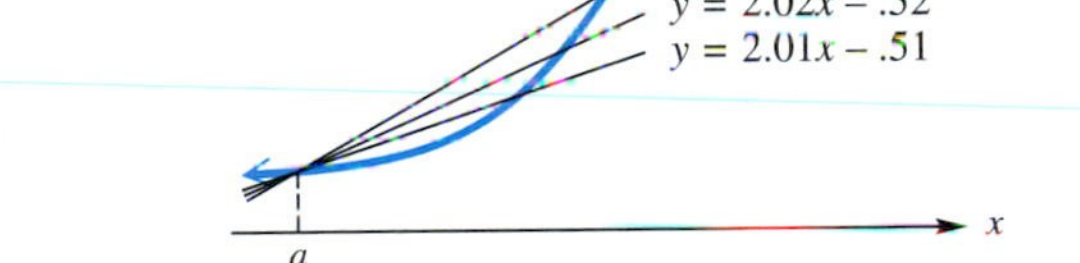

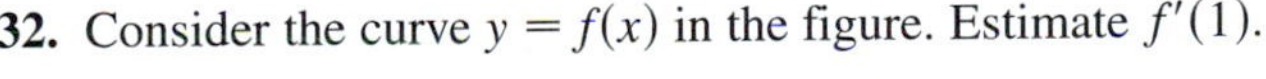

32. Consider the curve $y = f(x)$ in the figure. Estimate $f'(1)$.

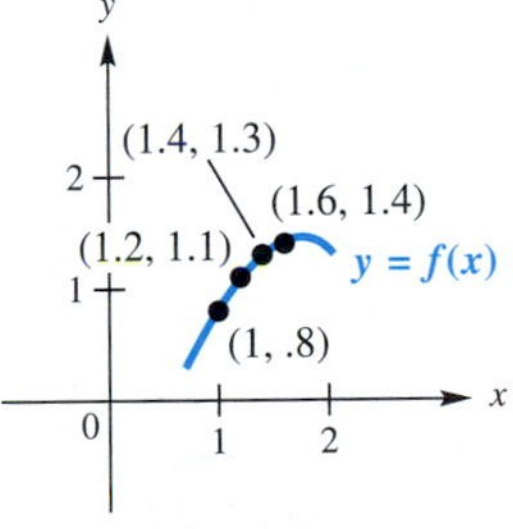

Concept Check *The position in feet of a car along a straight racetrack after t seconds is approximated by $s(t)$. Find the car's velocity in feet per second after 3 seconds.*

33. $s(t) = 9t^2$

34. $s(t) = 50\sqrt{t}$

35. $s(t) = 3t^3 - t^2$

36. $s(t) = 4t^2 + 5t + 1$

(Modeling) *Solve each problem.*

37. ***Interest Rates on Treasury Bonds*** In the figure, $f(t)$ is the interest rate (as a percent) on a 3-month United States treasury bond t years after January 1, 1979. The straight line is the tangent line to the graph of $y = f(t)$ when $t = 10$. How fast were interest rates rising on January 1, 1989?

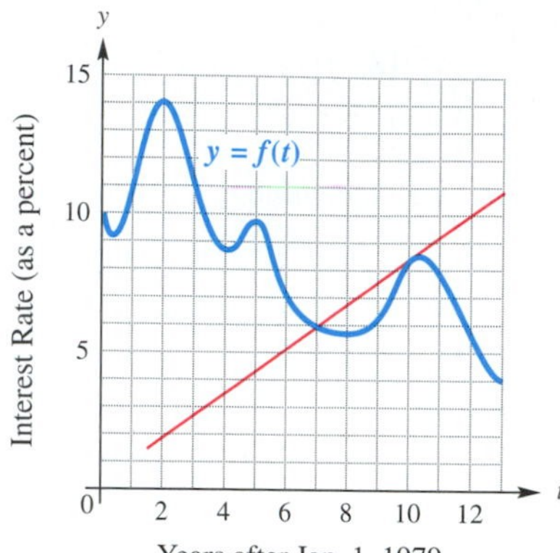

Source: Statistical Abstract of the United States.

38. ***Growth Rate in a Savings Account*** One hundred dollars is deposited in a savings account at 5% interest compounded continuously. The function defined by $f(t)$ shown in the figure gives the balance in the account after t years. At what rate (in dollars per year) is the balance growing after 14 years?

Balance in a Savings Account

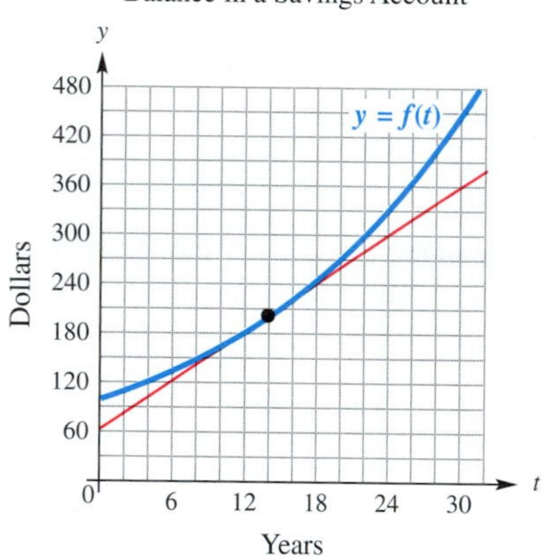

39. ***Weight of a Tumor*** The weight in grams of a cancerous tumor after t weeks is modeled by

$$W(t) = .1t^2.$$

At what rate is the tumor growing at time $t = 4$?

40. ***Velocity of a Ball*** A ball thrown straight up into the air has an initial height of 5 feet and an initial velocity of 128 feet per second. What is the velocity of the ball after 2 seconds?

41. ***Velocity of a Helicopter*** A helicopter is rising straight up in the air. Its distance from the ground t seconds after takeoff is $s(t)$ feet, where

$$s(t) = t^2 + t.$$

(a) How long will it take for the helicopter to rise 20 feet?

(b) Find the vertical velocity of the helicopter when it is 20 feet above the ground.

42. ***Marginal Profit*** Suppose that the total profit in hundreds of dollars from selling x items is given by

$$P(x) = 2x^2 - 5x + 6.$$

Find the marginal profit at $x = 2$.

43. ***Marginal Revenue*** The revenue (in thousands of dollars) from producing x units of an item is modeled by

$$R(x) = 10x - .002x^2.$$

Find the marginal revenue at $x = 1000$.

44. ***Spread of a Disease*** Epidemiologists estimate that t days after the flu begins to spread in a small town, the percent of the population infected by the flu is approximated by

$$p(t) = t^2 + 2t$$

for $0 \leq t \leq 5$. Find the instantaneous rate of change of the percent of the population infected at time $t = 3$.

(Modeling) *In Exercises 45 and 46, use a calculator to determine the derivative.*

45. ***Spread of a Rumor*** A rumor is spreading through a city. The number of people who have heard the rumor after t days is modeled by

$$f(t) = \frac{100{,}000}{1 + 9.134(.8)^t}.$$

Graph $f(t)$ in the window $[0, 32]$ by $[0, 100{,}000]$. Determine how fast the rumor is spreading after 8 days.

46. *Level of a Drug* When a drug is taken orally, the amount of the drug in the bloodstream after t hours is modeled by the function defined by

$$f(t) = 120(e^{-.2t} - e^{-t})$$

units. Graph the function in the window $[0, 16]$ by $[0, 70]$. How many units of the drug are in the bloodstream after 6 hours? At what rate is the level of the drug in the bloodstream decreasing after 6 hours?

(Modeling) Perhaps the largest single purchase you will make during your lifetime is the acquisition of a home. Since the amount of money required to buy a home is so large, you will probably borrow most of the money and pay off the loan in monthly payments for a specified period (normally 30 years) at a specified interest rate. Such a loan is called a mortgage.

Assuming that the amount of money to be borrowed and the payoff period have been decided, the size of the monthly payment will depend on the interest rate you are able to negotiate for the loan. If the interest rate is x%, then the monthly payment, f(x), is given by the formula

$$f(x) = \frac{P\left(\frac{x}{1200}\right)\left(1 + \frac{x}{1200}\right)^N}{\left(1 + \frac{x}{1200}\right)^N - 1},$$

where P is the amount of the loan (the principal*) and N is the duration of the loan in months.*

47. Find the monthly payment on a 30-year mortgage of \$200,000 at 8% interest. That is, find $f(x)$, where $P = 200{,}000$, $N = 360$, and $x = 8$.

48. Use a calculator to find $f'(8)$ for a 30-year \$200,000 mortgage.

49. Give the equation of the tangent line to $f(x)$ at $x = 8$.

50. Use the tangent line approximation to $f(x)$ in Exercise 49 to estimate how much your monthly payment will change if the interest rate
(a) rises to 8.25%; **(b)** falls to 7.5%.

12.5 Area and the Definite Integral

Areas by Approximation ■ The Definite Integral

Areas by Approximation

To calculate the areas of geometric figures such as rectangles, squares, triangles, and circles, we use specific formulas. In this section, we find the area of a figure or region that is bounded by general curves.

For example, under certain conditions, the area of a region can be thought of as a sum of parts. Figure 33 on the next page shows a region bounded by the positive y-axis, the positive x-axis, and the graph of $f(x) = \sqrt{4 - x^2}$. A very rough approximation of the area of this region can be found by using two circumscribed rectangles, as in Figure 34 on the next page. The height of the rectangle on the left is $f(0) = 2$, and the height of the rectangle on the right is $f(1) = \sqrt{3}$. The width of each rectangle is 1, making the total area of the two rectangles

$$1 \cdot f(0) + 1 \cdot f(1) = 2 + \sqrt{3} \approx 3.7321 \text{ square units.}$$

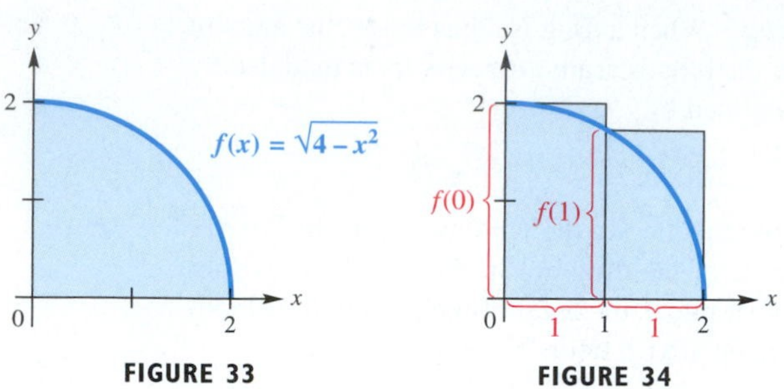

FIGURE 33

FIGURE 34

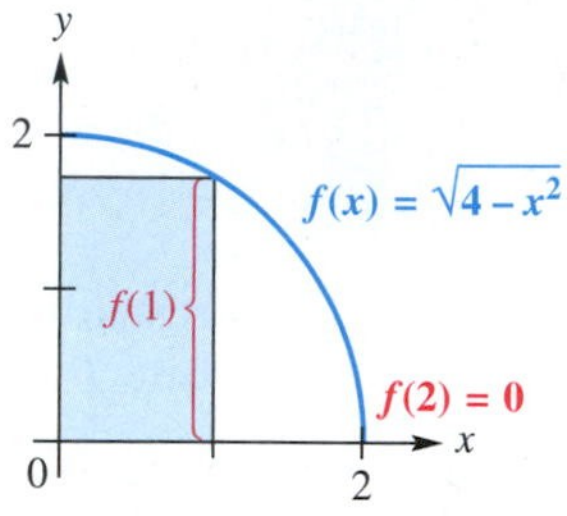

FIGURE 35

In this example, the function is decreasing, and we overestimate the area when we evaluate the function at the left endpoint to determine the height of the rectangle in that interval. If we use the right endpoint, however, the answer will be too small. For example, using the right endpoint 2, the area of the two rectangles is

$$1 \cdot f(1) + 1 \cdot f(2) = \sqrt{3} + 0 \approx 1.7321 \text{ square units.}$$

See Figure 35.

If the left endpoint gives an answer too big, and the right endpoint an answer too small, it seems reasonable to average the two answers. This produces the method called the **trapezoidal rule.** In this example, we get

$$\frac{3.7321 + 1.7321}{2} = 2.7321.$$

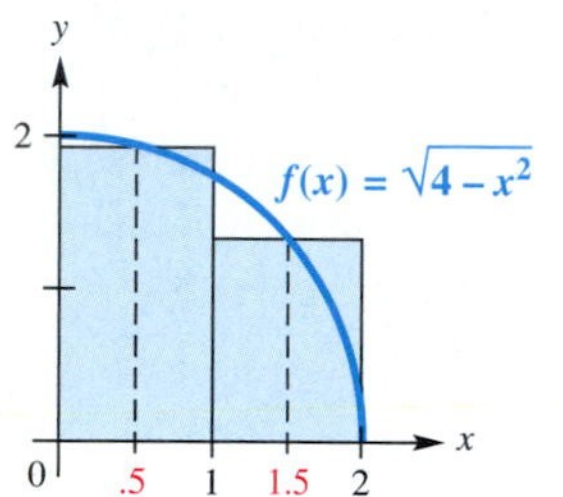

FIGURE 36

Another way to get an improved answer would be to use the midpoint of each interval, rather than the left endpoint or the right endpoint. This is called the **midpoint rule.** See Figure 36. In this example, that procedure leads to

$$\begin{aligned} 1 \cdot f(.5) + 1 \cdot f(1.5) &= \sqrt{3.75} + \sqrt{1.75} \\ &\approx 3.2594 \text{ square units.} \end{aligned}$$

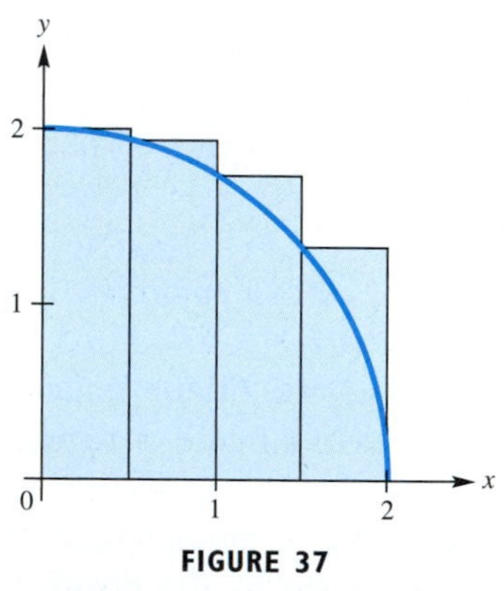

FIGURE 37

To improve the accuracy of all of the previous approximations, we could divide the interval from $x = 0$ to $x = 2$ into more parts. The result obtained by using the left endpoint again with four equal parts, each of width $\frac{1}{2}$, is shown in Figure 37. This approximation is greater than the actual area. As before, the height of each rectangle is given by the value of f at the left side of the rectangle, and its area is the width, $\frac{1}{2}$, multiplied by the height. The total area of the four rectangles is

$$\begin{aligned} &\frac{1}{2} \cdot f(0) + \frac{1}{2} \cdot f\left(\frac{1}{2}\right) + \frac{1}{2} \cdot f(1) + \frac{1}{2} \cdot f\left(\frac{3}{2}\right) \\ &= \frac{1}{2}(2) + \frac{1}{2}\left(\frac{\sqrt{15}}{2}\right) + \frac{1}{2}(\sqrt{3}) + \frac{1}{2}\left(\frac{\sqrt{7}}{2}\right) \\ &= 1 + \frac{\sqrt{15}}{4} + \frac{\sqrt{3}}{2} + \frac{\sqrt{7}}{4} \\ &\approx 3.4957 \text{ square units.} \end{aligned}$$

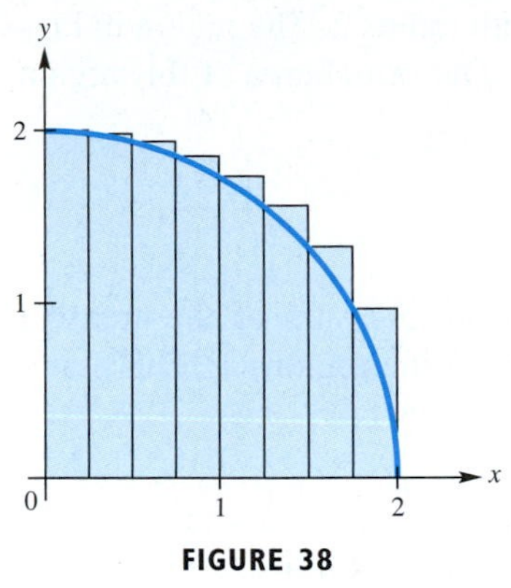

FIGURE 38

This approximation looks better, but it is still greater than the actual area. To improve the approximation, we divide the interval from $x = 0$ to $x = 2$ into eight parts with equal widths of $\frac{1}{4}$. See Figure 38. The total area of all these rectangles is

$$\frac{1}{4} \cdot f(0) + \frac{1}{4} \cdot f\left(\frac{1}{4}\right) + \frac{1}{4} \cdot f\left(\frac{1}{2}\right) + \frac{1}{4} \cdot f\left(\frac{3}{4}\right) + \frac{1}{4} \cdot f(1)$$

$$+ \frac{1}{4} \cdot f\left(\frac{5}{4}\right) + \frac{1}{4} \cdot f\left(\frac{3}{2}\right) + \frac{1}{4} \cdot f\left(\frac{7}{4}\right)$$

$$\approx 3.3398 \text{ square units.}$$

The process of approximating the area under a curve by using more and more rectangles to get a better and better approximation can be generalized. To do this, we divide the interval from $x = 0$ to $x = 2$ into n equal parts. Each of these n intervals has width

$$\frac{2 - 0}{n} = \frac{2}{n},$$

so each rectangle has width $\frac{2}{n}$ and height determined by the function value at the left side of the rectangle, the right side, or the midpoint. We could also average the left- and right-side values as before. Using a computer or graphing calculator to find approximations to the area for several values of n gives the results in the following table.

n	Left Sum	Right Sum	Trapezoidal Rule	Midpoint Rule
2	3.7321	1.7321	2.7321	3.2594
4	3.4957	2.4957	2.9957	3.1839
8	3.3398	2.8398	3.0898	3.1567
10	3.3045	2.9045	3.1045	3.1524
20	3.2285	3.0285	3.1285	3.1454
50	3.1783	3.0983	3.1383	3.1426
100	3.1604	3.1204	3.1404	3.1419
500	3.1455	3.1375	3.1415	3.1416

The numbers in the last four columns of this table represent approximations to the area under the curve, above the x-axis, and between the lines $x = 0$ and $x = 2$. As n becomes larger and larger, all four approximations become better and better, getting closer to the actual area. In this example, the exact area can be found by a formula from plane geometry. We write the given function as

$$y = \sqrt{4 - x^2}$$

to obtain

$$y^2 = 4 - x^2 \quad \text{Square each side.}$$

$$x^2 + y^2 = 4, \quad \text{Add } x^2.$$

which is the equation of a circle centered at the origin with radius 2. The region in Figure 33 is the quarter of this circle that lies in quadrant I. The actual area of this region is one-quarter of the area of the entire circle, or

$$\frac{1}{4}\pi(2)^2 = \pi \approx 3.1416.$$

As the number of rectangles increases without bound, the sum of the areas of these rectangles gets closer and closer to the actual area of the region, π. This statement can be written as

$$\lim_{n\to\infty} (\text{sum of areas of } n \text{ rectangles}) = \pi.$$

(The value of π was originally investigated by a process similar to this.)

Notice in this example that, for a particular value of n, the midpoint rule gave the best answer (the one closest to the true value of 3.1416), followed by the trapezoidal rule, followed by the left and right sums. In fact, the midpoint rule with $n = 20$ gives a value (3.1454) that is slightly more accurate than the left sum with $n = 500$ (3.1455).

The Definite Integral

Now we can generalize these concepts to get a method of finding the area bounded by the curve $y = f(x)$, the x-axis, and the vertical lines $x = a$ and $x = b$, as shown in Figure 39. To approximate this area, we could divide the region under the curve first into 10 rectangles (Figure 39(a)) and then into 20 rectangles (Figure 39(b)). The sums of the areas of the rectangles give approximations to the area under the curve.

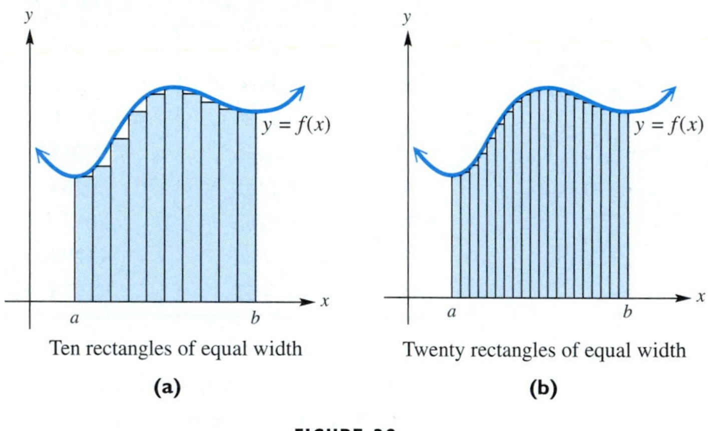

FIGURE 39

To develop a process that would yield the *exact* area, we begin by dividing the interval from a to b into n pieces of equal width, using each of these n pieces as the base of a rectangle. See Figure 40. Let x_1 be an arbitrary point in the first interval, x_2 be an arbitrary point in the second interval, and so on, up to the nth interval. In the graph of Figure 40 on the next page, the symbol $\boldsymbol{\Delta x}$ is used to represent the width of each of the intervals. The red rectangle is an arbitrary rectangle called the ith rectangle. Its area is the product of its length and width. Since the width of the ith rectangle is Δx and the length of the ith rectangle is given by the height $f(x_i)$,

$$\text{area of the } i\text{th rectangle} = f(x_i) \cdot \Delta x.$$

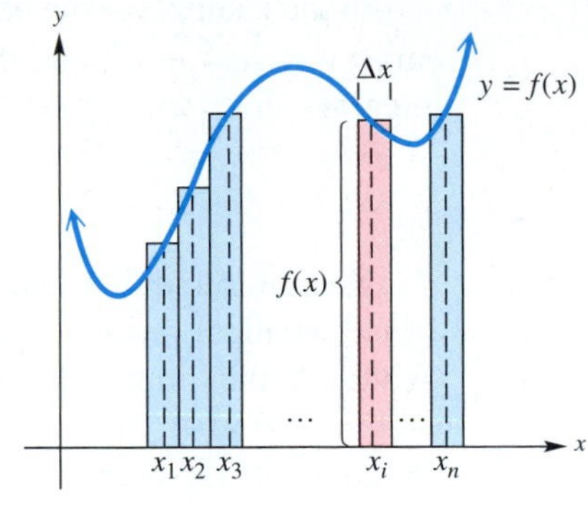

n rectangles of equal width

FIGURE 40

GCM TECHNOLOGY NOTE

Some calculators have a built-in function for evaluating the definite integral. The figure shows how a TI-83/84 Plus calculator computes

$$\int_0^2 \sqrt{4 - x^2}\,dx,$$

giving an answer of 3.141593074. This is an approximation with an error of approximately .0000004.

```
fnInt(√(4-X²),X,
0,2)
          3.141593074
```

The answer is an approximation for π.

The total area under the curve is approximated by the sum of the areas of all n of the rectangles. With sigma notation, the approximation to the total area becomes

$$\text{area of all } n \text{ rectangles} = \sum_{i=1}^{n} f(x_i) \cdot \Delta x.$$

The exact area is defined to be the limit of this sum (if the limit exists) as the number of rectangles increases without bound:

$$\text{exact area} = \lim_{n\to\infty} \sum_{i=1}^{n} f(x_i)\,\Delta x.$$

This limit, called the *definite integral* of $f(x)$ from a to b, is written as follows.

Definite Integral

If f is defined on the interval $[a, b]$, the **definite integral** of f from a to b is given by

$$\int_a^b f(x)\,dx = \lim_{n\to\infty} \sum_{i=1}^{n} f(x_i)\,\Delta x,$$

provided that the limit exists, where $\Delta x = \frac{b - a}{n}$ and x_i is *any* value of x in the ith interval.

The b above the integral sign in the definition is called the **upper limit of integration,** and the a is the **lower limit of integration.** This use of the word *limit* has nothing to do with the limit of the sum; it refers to the limits, or boundaries, on x. Note that the definite integral of a continuous function f will always exist.

The definite integral can be approximated by

$$\sum_{i=1}^{n} f(x_i)\,\Delta x.$$

If $f(x) \geq 0$ on the interval $[a, b]$, the definite integral gives the area under the curve between $x = a$ and $x = b$. In the midpoint rule, x_i is the midpoint of the ith interval. We may also let x_i be the left endpoint, the right endpoint, or any other point in the ith interval.

In the example at the beginning of this section, the area bounded by the x-axis, the curve $y = \sqrt{4 - x^2}$, and the lines $x = 0$ and $x = 2$ could be written as the definite integral

$$\int_0^2 \sqrt{4 - x^2}\,dx = \pi.$$

Keep in mind that finding the definite integral of a function can be thought of as a mathematical process that gives the sum of an infinite number of individual parts (within certain limits). The definite integral represents area only if the function is *non-negative* ($f(x) \geq 0$) at every x-value in the interval $[a, b]$. There are many other interpretations of the definite integral, and all of them involve this idea of approximation by appropriate sums.

EXAMPLE 1 Approximating Area

Approximate

$$\int_0^4 2x\,dx,$$

the area of the region under the graph of $f(x) = 2x$, above the x-axis, and between $x = 0$ and $x = 4$, by using four rectangles of equal width and whose heights are the values of the function at the midpoint of each rectangle.

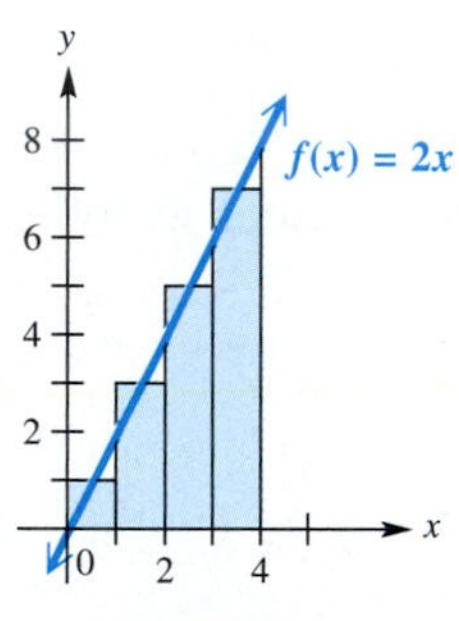

FIGURE 41

Solution We want to find the area of the shaded region in Figure 41. The heights of the four rectangles given by $f(x_i)$ for $i = 1, 2, 3$, and 4 are as follows.

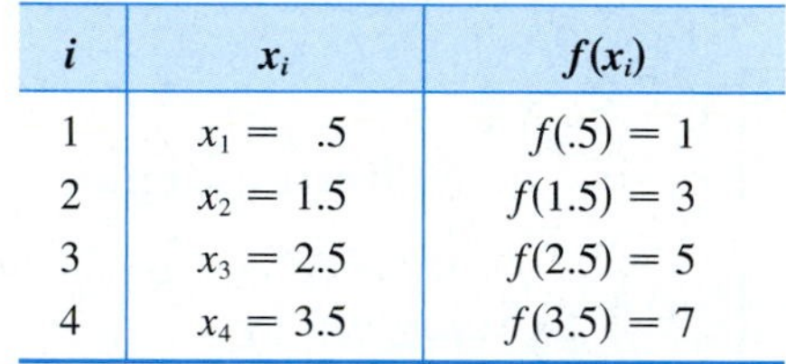

i	x_i	$f(x_i)$
1	$x_1 = .5$	$f(.5) = 1$
2	$x_2 = 1.5$	$f(1.5) = 3$
3	$x_3 = 2.5$	$f(2.5) = 5$
4	$x_4 = 3.5$	$f(3.5) = 7$

The width of each rectangle is $\Delta x = \frac{4 - 0}{4} = 1$. The sum of the areas of the four rectangles is

$$\begin{aligned}\sum_{i=1}^{4} f(x_i)\,\Delta x &= f(x_1)\,\Delta x + f(x_2)\,\Delta x + f(x_3)\,\Delta x + f(x_4)\,\Delta x \\ &= f(.5)\,\Delta x + f(1.5)\,\Delta x + f(2.5)\,\Delta x + f(3.5)\,\Delta x \\ &= 1(1) + 3(1) + 5(1) + 7(1) \\ &= 16.\end{aligned}$$

```
fnInt(2X,X,0,4)
                16
```

The figure shows how the TI-83/84 Plus confirms our answer in Example 1.

Using the formula for the area of a triangle, $A = \frac{1}{2}bh$, with b, the length of the base, equal to 4 and h, the height, equal to 8, gives

$$A = \frac{1}{2}bh = \frac{1}{2}(4)(8) = 16,$$

the exact value of the area. The approximation equals the exact area in this case because our use of the midpoints of each subinterval distributed the error evenly above and below the graph. ■

12.5 Exercises

1. Let $f(x) = 2x + 1$, $x_1 = 0$, $x_2 = 2$, $x_3 = 4$, $x_4 = 6$, and $\Delta x = 2$.
   (a) Find $\sum_{i=1}^{4} f(x_i)\,\Delta x$.
   (b) The sum in part (a) approximates a definite integral by using rectangles. The height of each rectangle is given by the value of the function at the left endpoint. Write the definite integral that the sum approximates.

2. Use the midpoint rule with $n = 4$ to approximate the area above the x-axis bounded by the graph of
$$f(x) = \sqrt{16 - x^2}$$
in the first quadrant.

In Exercises 3–12, approximate the area under the graph of $f(x)$ and above the x-axis, using each of the following methods with $n = 4$. **(a)** *Use left endpoints.* **(b)** *Use right endpoints.* **(c)** *Average the answers in parts (a) and (b).* **(d)** *Use midpoints.*

3. $f(x) = 3x + 2$ from $x = 1$ to $x = 5$

4. $f(x) = x + 5$ from $x = 2$ to $x = 4$

5. $f(x) = x + 2$ from $x = 0$ to $x = 4$

6. $f(x) = 3 + x$ from $x = 1$ to $x = 3$

7. $f(x) = x^2$ from $x = 1$ to $x = 5$

8. $f(x) = -x^2 + 4$ from $x = -2$ to $x = 2$

9. $f(x) = e^x - 1$ from $x = 0$ to $x = 4$

10. $f(x) = e^x + 1$ from $x = -2$ to $x = 2$

11. $f(x) = \dfrac{1}{x}$ from $x = 1$ to $x = 5$

12. $f(x) = \dfrac{2}{x}$ from $x = 1$ to $x = 9$

13. Consider the region below $f(x) = \frac{x}{2}$, above the x-axis, and between $x = 0$ and $x = 4$. Let x_i be the midpoint of the ith subinterval.
    (a) Approximate the area of the region, using four rectangles.
    (b) Find $\int_0^4 f(x)\,dx$ by using the formula for the area of a triangle.

14. Find $\int_0^5 (5 - x)\,dx$ by using the formula for the area of a triangle.

15. Find $\int_0^4 f(x)\,dx$ for each graph of $y = f(x)$.

(a)

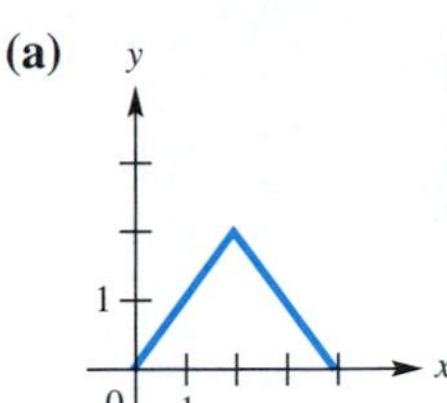

(b)

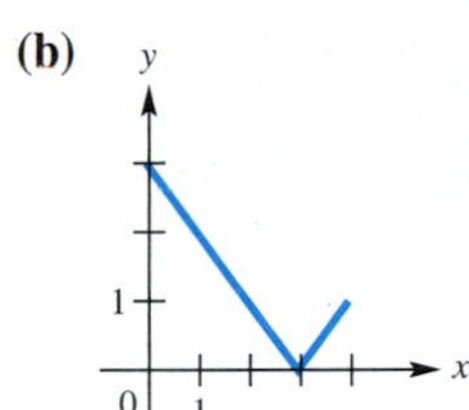

Find the exact value of each integral, using formulas from geometry.

16. $\displaystyle\int_{-3}^{3} \sqrt{9 - x^2}\,dx$

17. $\displaystyle\int_{-4}^{0} \sqrt{16 - x^2}\,dx$

18. $\displaystyle\int_{1}^{3} (5 - x)\,dx$

19. $\displaystyle\int_{2}^{5} (1 + 2x)\,dx$

20. ***(Modeling) Distance*** The curves on the next page show the velocity at t seconds after a car has accelerated from a dead stop. (*Source: Road & Track,* April–May 1978.) To find the total distance traveled by the car in reaching 100 mph, we must estimate the definite integral
$$\int_0^T v(t)\,dt,$$
where T represents the number of seconds it takes for the car to reach 100 mph.

Use the graphs to estimate this distance by adding the areas of rectangles with widths representing 5 seconds. (The last rectangle has a width representing 4 seconds or 3 seconds.) Use the midpoint rule. To adjust your answer to miles per hour, divide by 3600 (the number of seconds in an hour). You then have the number of miles that the car traveled in reaching 100 mph. Finally, multiply by 5280 feet per mile to convert the answer to feet.

(a) Estimate the distance traveled by the Porsche 928, using the graph.

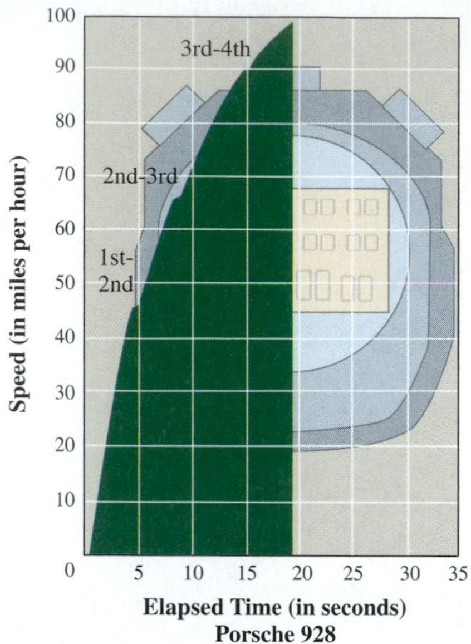

(b) Estimate the distance traveled by the BMW 733i, using the graph.

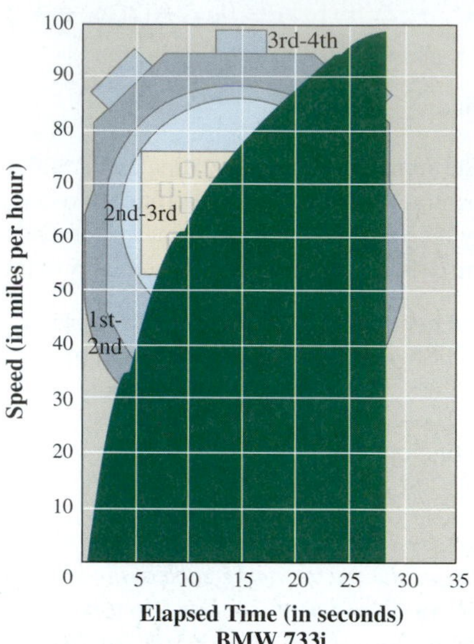

Reviewing Basic Concepts (Sections 12.4 and 12.5)

Estimate the slope of the tangent line to each curve at the given point (x, y).

1.

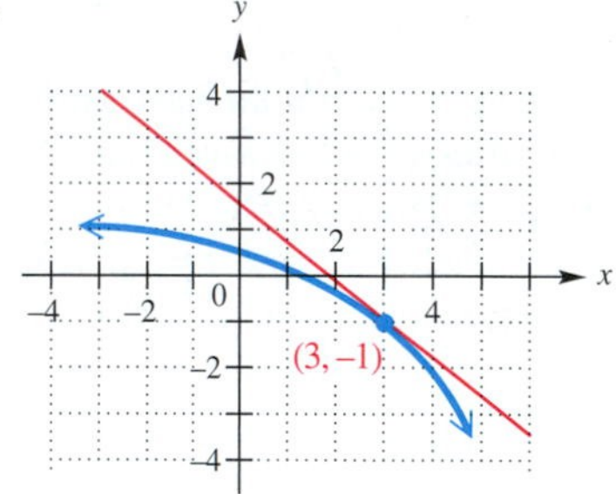

2.

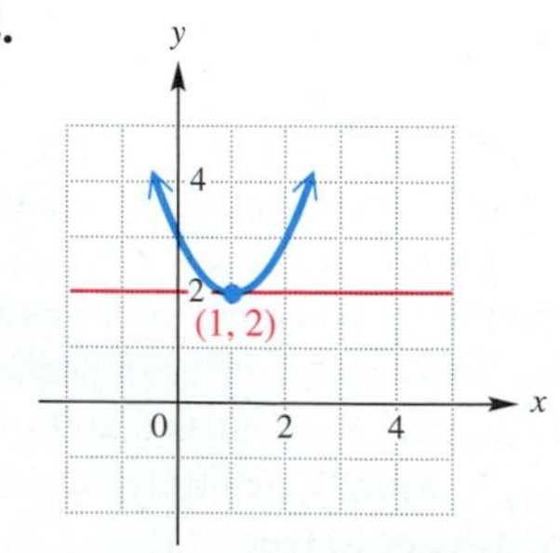

3.

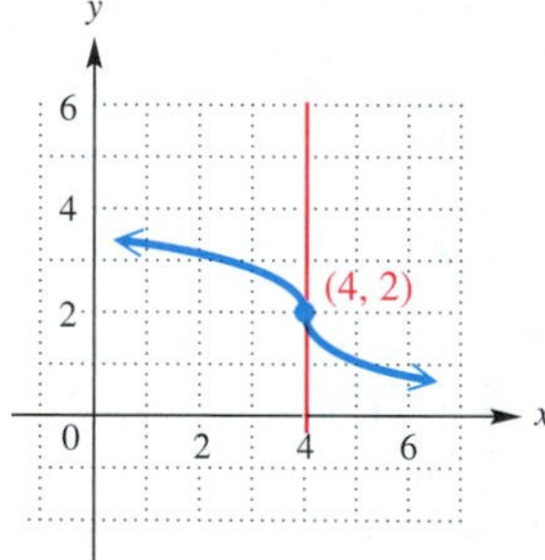

4. If $f(x) = k$, where k is a real number, what is the value of $f'(x)$?

5. ***(Modeling) Marginal Revenue*** The revenue in dollars from producing (and selling) x units of an item is modeled by

$$R(x) = -.0012x^2 + 3x.$$

Find the marginal revenue at a production level of 1800 units.

6. Find the equation of the tangent line to the graph of $f(x) = 6 - x^2$ when $x = -1$.

7. Find $f'(9)$, where $f(x) = \sqrt{x} + 2$.

8. Approximate the area under the graph of $f(x) = x^2$ and above the x-axis from $x = 0$ to $x = 5$ with $n = 5$, using midpoints.

9. Find the area bounded by the x-axis, the y-axis, and the graph of $y = 8 - x$, using the formula for the area of a triangle.

10. Use the formula for the area of a circle to find the exact value of

$$\int_0^3 \sqrt{9 - x^2}\, dx.$$

CHAPTER 12 SUMMARY

KEY TERMS & SYMBOLS	KEY CONCEPTS
12.1 An Introduction to Limits limit	**LIMIT OF A FUNCTION** Let f be a function and let a and L be real numbers. L is the limit of $f(x)$ as x approaches a, written $$\lim_{x \to a} f(x) = L,$$ if the following conditions are met: **1.** As x takes values closer and closer (but not equal) to a on both sides of a, the corresponding values of $f(x)$ get closer and closer (and perhaps equal) to L. **2.** The value of $f(x)$ can be made as close to L as desired by taking values of x arbitrarily close to a. **CONDITIONS UNDER WHICH $\lim_{x \to a} f(x)$ FAILS TO EXIST** **1.** $f(x)$ approaches a number L as x approaches a from the left and approaches a different number M as x approaches a from the right. **2.** $f(x)$ becomes infinitely large in absolute value as x approaches a from either side. **3.** $f(x)$ oscillates infinitely many times between two fixed values as x approaches a.
12.2 Techniques for Calculating Limits	**RULES FOR LIMITS** **1. Constant rule** If k is a constant real number, then $\lim_{x \to a} k = k$. **2. Limit of x rule** $\lim_{x \to a} x = a$ For the following rules, we assume that $\lim_{x \to a} f(x)$ and $\lim_{x \to a} g(x)$ both exist. **3. Sum and difference rules** $\lim_{x \to a}[f(x) \pm g(x)] = \lim_{x \to a} f(x) \pm \lim_{x \to a} g(x)$ **4. Product rule** $\lim_{x \to a}[f(x) \cdot g(x)] = \lim_{x \to a} f(x) \cdot \lim_{x \to a} g(x)$ **5. Quotient rule** $\lim_{x \to a} \dfrac{f(x)}{g(x)} = \dfrac{\lim_{x \to a} f(x)}{\lim_{x \to a} g(x)}$, provided that $\lim_{x \to a} g(x) \neq 0$ **6. Polynomial rule** If $p(x)$ defines a polynomial function, then $\lim_{x \to a} p(x) = p(a)$. **7. Rational function rule** If $f(x)$ defines a rational function $\frac{p(x)}{q(x)}$, with $q(a) \neq 0$, then $\lim_{x \to a} f(x) = f(a)$. *(continued)*

KEY TERMS & SYMBOLS	KEY CONCEPTS
	8. **Equal functions rule** If $f(x) = g(x)$ for all $x \neq a$, then $\lim_{x\to a} f(x) = \lim_{x\to a} g(x)$. 9. **Power rule** For any real number k, $\lim_{x\to a}[f(x)]^k = \left[\lim_{x\to a} f(x)\right]^k$, provided that this limit exists. 10. **Exponent rule** For any real number $b > 0$, $\lim_{x\to a} b^{f(x)} = b^{\lim_{x\to a} f(x)}$. 11. **Logarithm rule** For any real number $b > 0$ with $b \neq 1$, $\lim_{x\to a}[\log_b f(x)] = \log_b\left[\lim_{x\to a} f(x)\right]$, provided that $\lim_{x\to a} f(x) > 0$.
12.3 One-Sided Limits; Limits Involving Infinity two-sided limit one-sided limit right-hand limit left-hand limit limits as $x \to \pm\infty$	$\lim_{x\to a} f(x) = L$ **if and only if** $\lim_{x\to a^+} f(x) = L$ **and** $\lim_{x\to a^-} f(x) = L$ **LIMITS AT INFINITY OF** $\frac{1}{x^n}$ For any positive integer n, $\lim_{x\to\infty} \frac{1}{x^n} = 0$ and $\lim_{x\to-\infty} \frac{1}{x^n} = 0$.
12.4 Tangent Lines and Derivatives tangent line secant line slope of the tangent line, or slope of the curve derivative of f at a, $f'(a)$ position function velocity marginal cost	**TANGENT LINE** The tangent line of the graph of $y = f(x)$ at the point $(a, f(a))$ is the line through this point and having slope $$m = \lim_{x\to a} \frac{f(x) - f(a)}{x - a},$$ provided that this limit exists. If the limit does not exist, then there is no tangent line at the point. **DERIVATIVE** If a is in the domain of f, then the derivative of f at a is defined by $$f'(a) = \lim_{x\to a} \frac{f(x) - f(a)}{x - a},$$ provided that this limit exists. **VELOCITY AS A DERIVATIVE** If an object is moving along a straight line and its position on the line at time t is $s(t)$, then the velocity at time a is $$s'(a) = \lim_{t\to a} \frac{s(t) - s(a)}{t - a}.$$
12.5 Area and the Definite Integral trapezoidal rule midpoint rule definite integral, $\int_a^b f(x)\,dx$ limits of integration	**DEFINITE INTEGRAL** If f is defined on the interval $[a, b]$, the definite integral of f from a to b is given by $$\int_a^b f(x)\,dx = \lim_{n\to\infty} \sum_{i=1}^{n} f(x_i)\,\Delta x,$$ provided that the limit exists, where $\Delta x = \frac{b-a}{n}$ and x_i is *any* value of x in the ith interval.

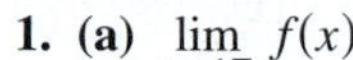

CHAPTER 12 Review Exercises

Decide whether the limits in Exercises 1–50 exist. If a limit exists, find its value. If the function approaches ∞ or $-\infty$, say so.

1. **(a)** $\lim_{x\to 1^-} f(x)$
(b) $\lim_{x\to 1^+} f(x)$
(c) $\lim_{x\to 1} f(x)$

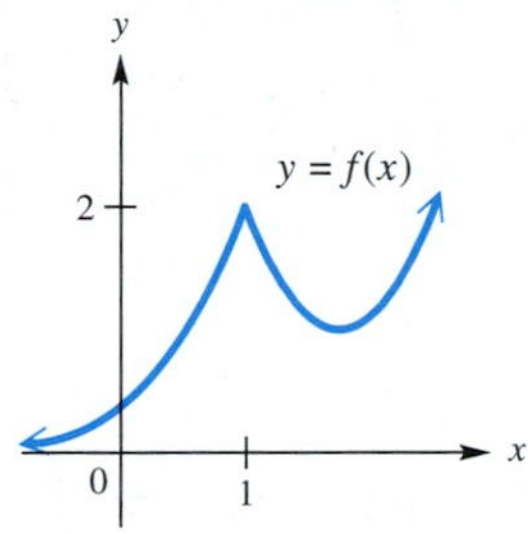

2. **(a)** $\lim_{x\to -1^-} f(x)$
(b) $\lim_{x\to -1^+} f(x)$
(c) $\lim_{x\to -1} f(x)$

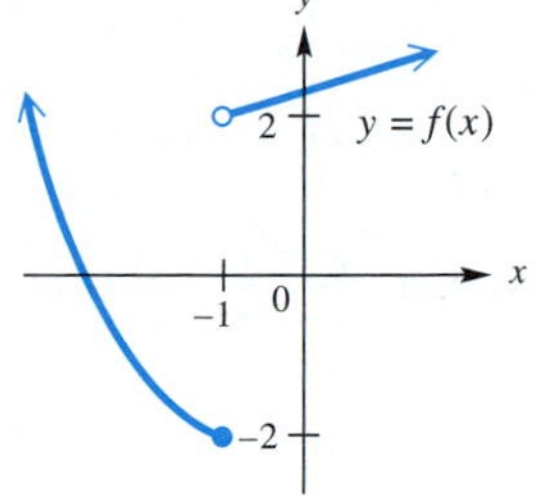

3. **(a)** $\lim_{x\to 4^-} f(x)$
(b) $\lim_{x\to 4^+} f(x)$
(c) $\lim_{x\to 4} f(x)$

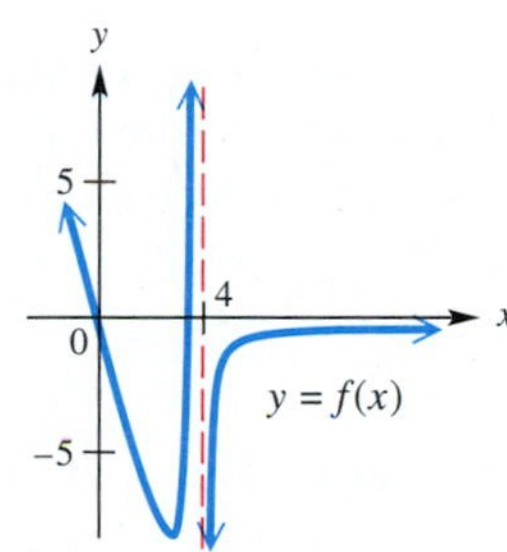

4. **(a)** $\lim_{x\to \infty} f(x)$
(b) $\lim_{x\to 0^-} f(x)$
(c) $\lim_{x\to 0^+} f(x)$

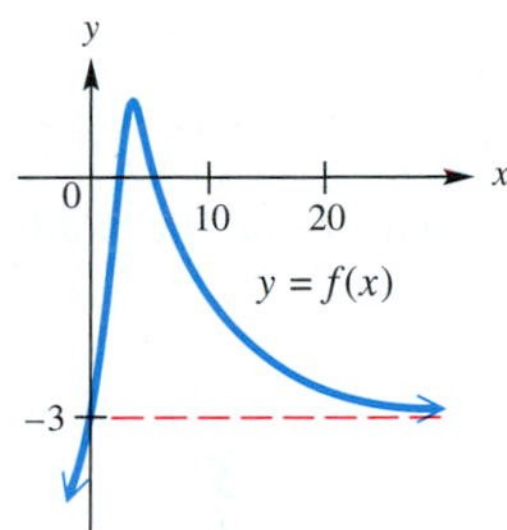

5. $\lim_{x\to 1} (2x^2 - 3x)$

6. $\lim_{x\to -1} (x - x^3)$

7. $\lim_{x\to 2} \frac{3x + 4}{x + 3}$

8. $\lim_{x\to 4} \frac{x - 4}{x^3 + 5x}$

9. $\lim_{x\to -1} \sqrt{5x + 21}$

10. $\lim_{x\to 2} 27^{(3-x)/(1+x)}$

11. $\lim_{x\to 1} [\log_2(5x + 3)]$

12. $\lim_{x\to -3} (5 - 9x)^{2/5}$

13. $\lim_{x\to 5} \frac{2x - 10}{5 - x}$

14. $\lim_{x\to 6} \frac{x^2 - 36}{x - 6}$

15. $\lim_{x\to -3} \frac{x^2 + 2x - 3}{x + 3}$

16. $\lim_{x\to 1} \frac{x^2 - 2x + 1}{x - 1}$

17. $\lim_{x\to 2} \frac{x^2 - x - 2}{x^2 - 5x + 6}$

18. $\lim_{x\to -2} \frac{x^3 + 3x^2 + 2x}{x^2 + 2x}$

19. $\lim_{x\to 1} \frac{x^2 + x}{x - 1}$

20. $\lim_{x\to 0} \frac{x^2 + 1}{x}$

21. $\lim_{x\to 0} \frac{\sin x}{3x}$

22. $\lim_{x\to 0} \frac{\sin \frac{1}{x}}{x}$

23. $\lim_{x\to 0} \frac{x \cos x - 1}{x^2}$

24. $\lim_{x\to 0} \frac{\tan x}{2x}$

25. $\lim_{x\to 2} f(x)$, where $f(x) = \begin{cases} 3x - 1 & \text{if } x \le 2 \\ x + 3 & \text{if } x > 2 \end{cases}$

26. $\lim_{x\to 1} f(x)$, where $f(x) = \begin{cases} x^2 & \text{if } x < 1 \\ 5 & \text{if } x = 1 \\ 3x - 2 & \text{if } x > 1 \end{cases}$

27. $\lim_{x\to 2^-} f(x)$, where $f(x) = \begin{cases} x^2 - 1 & \text{if } x < 2 \\ x^2 + 1 & \text{if } x \ge 2 \end{cases}$

28. $\lim_{x\to 0^+} f(x)$, where $f(x) = \begin{cases} \sqrt{x} & \text{if } x \ge 0 \\ \ln(-x) & \text{if } x < 0 \end{cases}$

29. $\lim_{x\to 0^-} f(x)$, where $f(x) = \begin{cases} \sqrt{x} & \text{if } x \ge 0 \\ \ln(-x) & \text{if } x < 0 \end{cases}$

30. $\lim_{x\to 0^-} f(x)$, where $f(x) = \begin{cases} \sin x & \text{if } x \ge 0 \\ \frac{\sin x}{x} & \text{if } x < 0 \end{cases}$

31. $\lim_{x \to 3^+} (\pi - 1)$

32. $\lim_{x \to -2^-} (-5x^3)$

33. $\lim_{x \to 7^-} \frac{1}{x - 7}$

34. $\lim_{x \to -2^+} \sqrt{x + 2}$

35. $\lim_{x \to 1^+} \sqrt{1 - x}$

36. $\lim_{x \to 2^-} (3x - 1)$

37. $\lim_{x \to 1^-} \frac{|x - 1|}{x - 1}$

38. $\lim_{x \to -2^+} \frac{x + 2}{|x + 2|}$

39. $\lim_{x \to 4^+} \frac{x}{(x - 4)^2}$

40. $\lim_{x \to 4^-} \frac{x}{(x - 4)^2}$

41. $\lim_{x \to 1^+} \frac{x^2}{(x - 1)^3}$

42. $\lim_{x \to 1^-} \frac{x^2}{(x - 1)^3}$

43. $\lim_{x \to \infty} \frac{5x + 1}{2x - 7}$

44. $\lim_{x \to \infty} \frac{4x^2 - 5x}{2x^2}$

45. $\lim_{x \to -\infty} \frac{x^3 + 1}{x^2 - 1}$

46. $\lim_{x \to -\infty} \frac{x^4 + x + 1}{x^5 - 2}$

47. $\lim_{x \to \infty} \left(5 + \frac{x}{1 + x^2}\right)$

48. $\lim_{x \to \infty} (e^{-2x} + 7)$

49. $\lim_{x \to \infty} \left(2 - \frac{3}{1 - e^{-x}}\right)$

50. $\lim_{x \to \infty} \left(\frac{5}{1 + 2^{-x}} - 3\right)$

Concept Check *In Exercises 51 and 52, write examples of functions defined by $f(x)$ and $g(x)$ with the given properties.*

51. Neither $\lim_{x \to 2} f(x)$ nor $\lim_{x \to 2} g(x)$ exists, but $\lim_{x \to 2} [f(x) + g(x)]$ exists.

52. $\lim_{x \to \infty} f(x) = \infty$, $\lim_{x \to \infty} g(x) = \infty$, and $\lim_{x \to \infty} \frac{f(x)}{g(x)} = \infty$

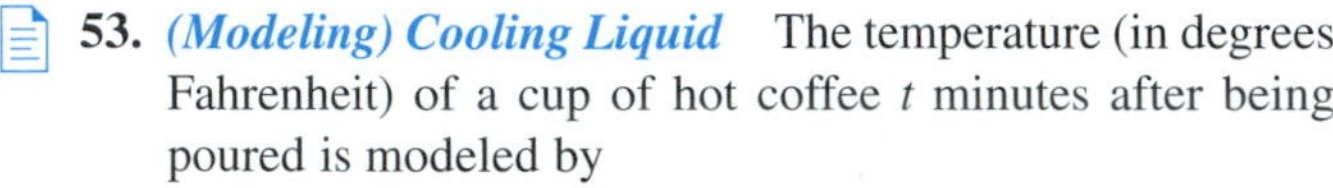

53. ***(Modeling) Cooling Liquid*** The temperature (in degrees Fahrenheit) of a cup of hot coffee t minutes after being poured is modeled by

$$f(t) = 70 + 110e^{-.25t}.$$

Find and interpret $\lim_{t \to \infty} f(t)$.

Find the slope of the tangent line to each curve when x has the given value.

54. $f(x) = 1 + x^2$; $x = 3$

55. $f(x) = \frac{3}{x}$; $x = 1$

Find the equation of the tangent line to each curve when x has the given value. Verify your answer by graphing both $f(x)$ and the tangent line with a calculator.

56. $f(x) = \frac{4}{x}$; $x = 2$

57. $f(x) = x^2 - x$; $x = 2$

58. ***Concept Check*** The figure shows the tangent line to the graph of $f(x) = \sqrt{x}$ at the point $(4, 2)$. What is

$$\lim_{x \to 4} \frac{\sqrt{x} - 2}{x - 4}?$$

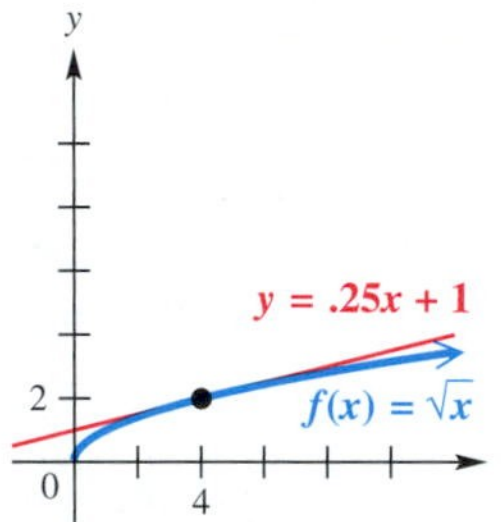

59. ***Concept Check*** The figure shows the tangent line to the graph of $f(x) = \sin x$ at the point $\left(\frac{\pi}{2}, 1\right)$. What is $f'\left(\frac{\pi}{2}\right)$?

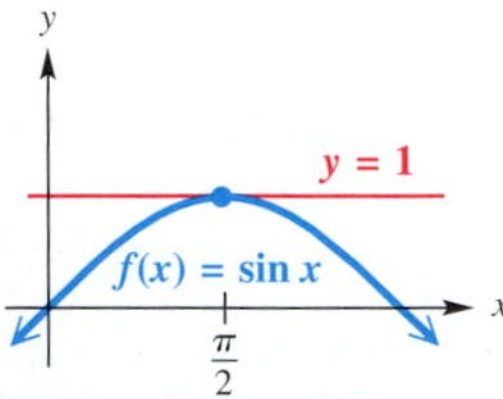

Use a calculator to estimate $f'(a)$ for the given value of a.

60. $f(x) = x \sin x$; $a = \pi$

61. $f(x) = e^x$; $a = 0$

Solve each problem.

62. ***(Modeling) Growth in Average Farm Size*** In the graph, $f(t)$ is the average size of a farm (in acres) in the United States t years after January 1, 1900. The straight line is the tangent line to the graph of $y = f(t)$ when $t = 50$. How fast was the average farm size increasing on January 1, 1950?

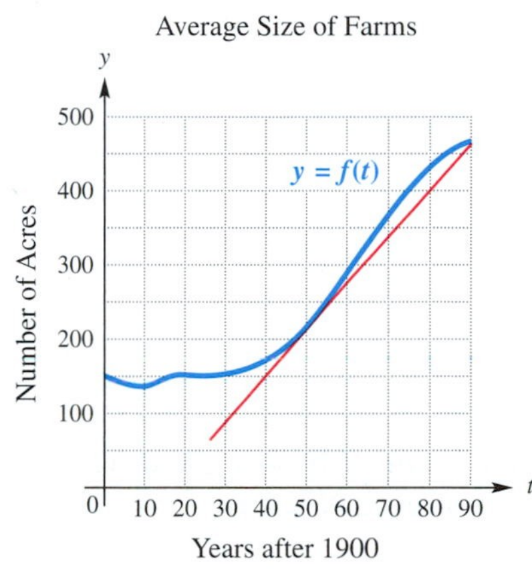

Source: Statistical Abstract of the United States.

63. ***Velocity of a Ball*** A ball thrown straight up into the air has an initial height of 4 feet and an initial velocity of 100 feet per second. What is the velocity of the ball after 3 seconds?

64. ***(Modeling) Marginal Cost*** The cost in hundreds of dollars to manufacture x units of goods is modeled by the function defined by

$$C(x) = .006x^3 - .7x^2 + 32x + 250.$$

Use a calculator to find the marginal cost at $x = 72$.

65. Let $f(x) = 3x + 1$, $x_1 = -1$, $x_2 = 0$, $x_3 = 1$, $x_4 = 2$, and $x_5 = 3$. Find

$$\sum_{i=1}^{5} f(x_i)\,\Delta x.$$

66. Find $\int_0^3 f(x)\,dx$ for the graph of $y = f(x)$.

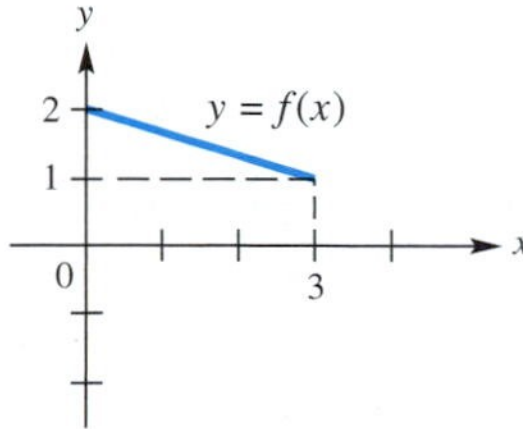

67. Using four rectangles, approximate the area under the graph of $f(x) = 2x + 3$ and above the x-axis from $x = 0$ to $x = 4$. Let the height of each rectangle be the function value on the left side.

68. Use the formula from geometry for the area of a trapezoid, $A = \frac{1}{2}h(B + b)$, to find

$$\int_0^4 (2x + 3)\,dx,$$

where B and b are the lengths of the parallel sides and h is the distance between them. Compare with Exercise 67.

69. Use the formula for the area of a triangle to find

$$\int_0^5 (5 - x)\,dx.$$

70. Use the formula for the area of a circle to find

$$\int_0^1 \sqrt{1 - x^2}\,dx.$$

CHAPTER 12 Test

Decide whether the limits in Exercises 1–18 exist. If a limit exists, find its value. If the function approaches ∞ or −∞, say so.

1. $\lim_{x \to 4^-} f(x)$

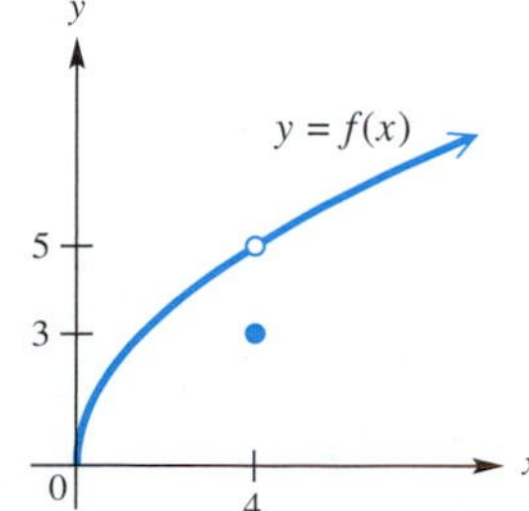

2. $\lim_{x \to \infty} f(x)$

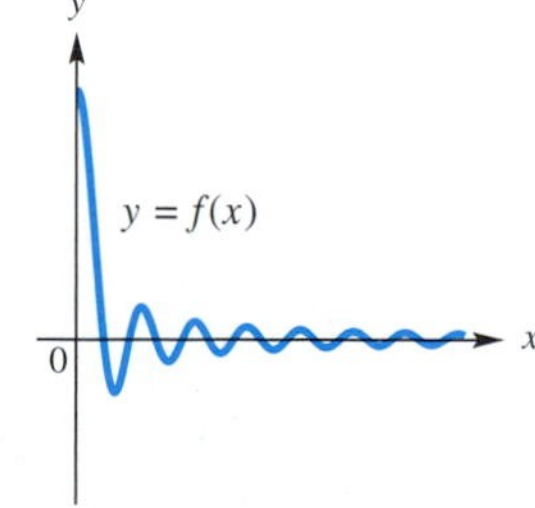

3. **(a)** $\lim_{x \to 3^+} f(x)$ **(b)** $\lim_{x \to 3^-} f(x)$

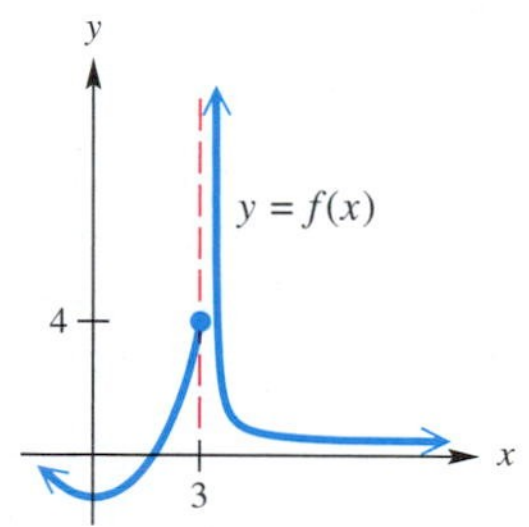

4. $\lim_{x \to 1} \frac{x^2 + x + 1}{x^2 + 1}$

5. $\lim_{x \to -2} \frac{x^2 + 2x}{x + 2}$

6. $\lim_{x \to 3} \frac{x^2 - 6x + 9}{x - 3}$

7. $\lim_{x \to 2} \frac{x^2 + x - 6}{x^2 - 4}$

8. $\lim_{x \to 3} \sqrt{x^2 + 7}$

9. $\lim_{x \to \infty} \frac{2x^2 - 3}{5x^2 + x + 1}$

10. $\lim_{x \to -\infty} \frac{3x - 4}{4x - 3}$

11. $\lim_{x \to 0} \frac{\sin x}{x \cos x}$

12. $\lim_{x \to 0} \frac{x^2 - 10}{x^3 + 1}$

13. $\lim_{x \to 1^+} f(x)$, where $f(x) = \begin{cases} x^2 + 1 & \text{if } x \leq 1 \\ 1 - 2x^2 & \text{if } x > 1 \end{cases}$

14. $\lim_{x \to 4^+} 55$

15. $\lim_{x \to 3^-} (4x^2 - 3x)$

16. $\lim_{x \to 2^+} \sqrt{2 - x}$

17. $\lim_{x \to 3^-} \frac{|3 - x|}{3 - x}$

18. $\lim_{x \to -1^+} \sqrt{x + 1}$

19. Find the slope of the tangent line to the graph of $f(x) = 2x^2 - 1$ at the point where $x = 1$.

20. Find the equation of the tangent line to the graph of $f(x) = -\frac{3}{x}$ at the point $(1, -3)$.

21. Use a calculator to estimate $f'(4)$, where $f(x) = \frac{1 + e^x}{x}$.

Solve each problem.

22. ***(Modeling) Rate of Radioactive Disintegration*** Five grams of radioactive cobalt are placed in a vault. The function defined by $f(t)$ shown in the graph gives the amount remaining after t years. At what rate (in grams per year) is the cobalt disintegrating after 14 years?

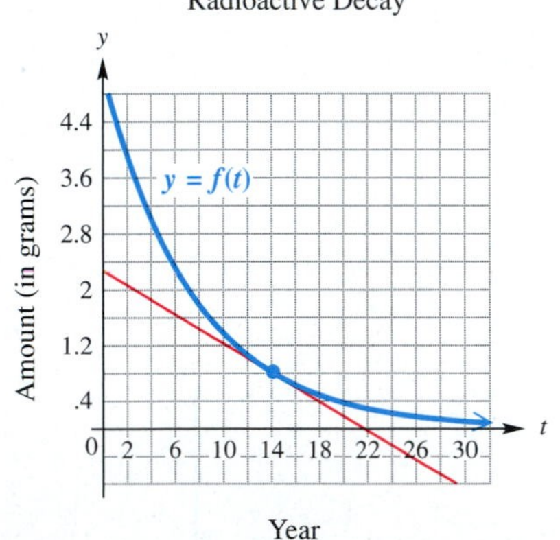

23. ***(Modeling) Social Diffusion*** The spread of a piece of information in a small town by mass media follows the model

$$f(t) = 40{,}000(1 - e^{-.25t}),$$

where $f(t)$ is the number of people who have heard the information after t days. Use a calculator to graph $y = f(t)$ for $0 \leq t \leq 32$, and determine $f'(3)$. What does $f'(3)$ tell about the situation?

24. Approximate the area under the graph of $f(x) = x^2$ from $x = 0$ to $x = 4$ and above the x-axis, using four rectangles. Let the height of each rectangle be the function value at the midpoint of the interval.

25. Use the formula for the area of a triangle to find

$$\int_0^6 (6 - x)\,dx.$$

CHAPTER 12 PROJECT

Instantaneous Rate of Change

The average rate of change of a function defined by $y = f(x)$ as x changes from a to b is given by the difference quotient

$$\frac{f(b) - f(a)}{b - a}.$$

An application of this expression comes from data that accompanied an article in the *New York Times* on December 28, 1993. It was found that the number of personal computers with CD-ROM drives (in millions) since 1990 could be modeled by the function $f(x) = .4525e^{1.556\sqrt{x}}$, where x represents the number of years since 1990. Using this function, we could have predicted that, over the first half of 1994, the average rate of change per year was

$$\frac{f(4.5) - f(4)}{4.5 - 4} = \frac{.4525e^{1.556\sqrt{4.5}} - .4525e^{1.556\sqrt{4}}}{.5}$$
$$\approx \frac{12.27797 - 10.16583}{.5}$$
$$= \frac{2.11214}{.5} = 4.224.$$

Therefore, the number of personal computers with CD-ROM drives increased during the first half of 1994 at a rate of 4.224 million computers per year.

The *instantaneous rate of change* of $y = f(x)$ is given by

$$\lim_{h \to 0} \frac{f(x + h) - f(x)}{h},$$

provided that this limit exists. In our example, we can determine the instantaneous rate of change at the beginning of 1994 ($x = 4$) in several ways by using a graphing calculator, two of which are described here. One way is to define Y_1 and Y_2, as shown in Figure A. (The calculator requires us to use X in place of h in the formula for the instantaneous rate of change.) By using smaller and smaller values of X, we can approximate the instantaneous rate of change in a table; the limit seems to be approximately 3.9545, as seen in Figure B.

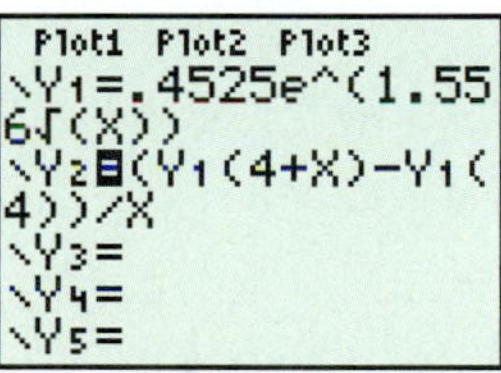

FIGURE A

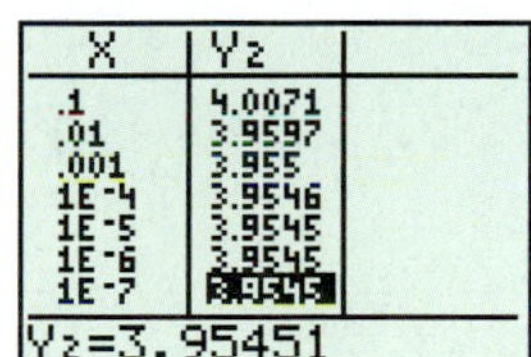

FIGURE B

The formula for the instantaneous rate of change is equivalent to that of the derivative. Another way to find the result obtained in the table of Figure B is to define Y_2 by using the numerical derivative feature, as seen in Figure C on the next page. Then we graph Y_2 and evaluate it for X = 4. The display at the bottom of Figure D agrees with our earlier result.

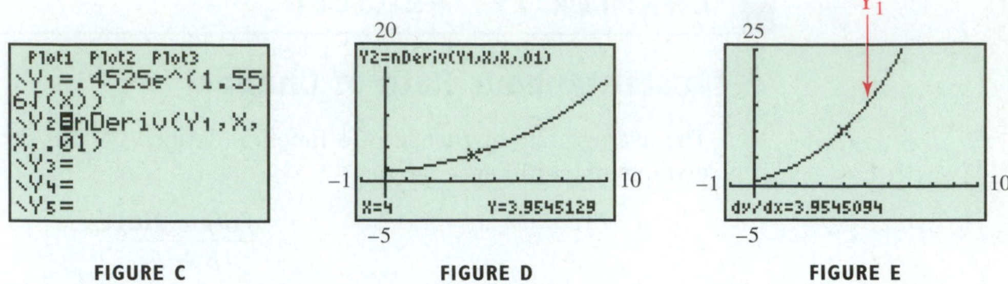

FIGURE C FIGURE D FIGURE E

Probably the easiest way to find the instantaneous rate of change is to graph Y_1 and use the $\frac{dy}{dx}$ capability of the calculator to find the value of the derivative at the desired value. Figure E shows the result, which closely corresponds to the value that we found earlier.

In conclusion, we can say that the instantaneous rate of change in the number of personal computers with CD-ROM drives at the beginning of 1994 was about 3.95 million computers per year.

Activities

Use one of the methods just described to answer the following.

1. Refer to Example 4 in Section 3.4 on pages 211–212. Enter the equation given by quadratic regression in part (e), and then use one of the methods described in this project to find the instantaneous rate of change in the year of your birth.

2. Refer to Example 7 in Section 3.8 on pages 255–256. Use one of the methods described in this project to find the instantaneous rate of change in 1998, first for the best-fitting cubic function in part (b) and then for the best-fitting quartic function in part (c).

3. Refer to Example 8 in Section 5.6 on pages 400–401. Enter the equation given by the exponential regression in part (b), and then use one of the methods described in this project to find the instantaneous rate of change in the year that is exactly 100 years after the year of your birth.

4. Refer to Exercise 48 in Section 5.6 on page 406. After finding the logarithmic regression equation, use one of the methods described in this project to find the instantaneous rate of change in the current year.

Reference: Basic Algebraic Concepts

IN THIS REFERENCE, we present a review of topics that are usually studied in beginning and intermediate algebra courses. You may wish to refer to the various sections of this reference from time to time if you need to refresh your memory on these basic concepts.

Chapter Outline

R.1 Review of Exponents and Polynomials

Rules for Exponents ■ Terminology for Polynomials ■ Adding and Subtracting Polynomials ■ Multiplying Polynomials

Rules for Exponents

By definition, the notation a^m (where m is a positive integer and a is a real number) means that a appears as a factor m times. In the same way, a^n (where n is a positive integer) means that a appears as a factor n times. In the product $a^m \cdot a^n$, the number a would appear $m + n$ times.

Product Rule

For all positive integers m and n and every real number a,

$$a^m \cdot a^n = a^{m+n}.$$

EXAMPLE 1 Using the Product Rule

Find each product.

(a) $y^4 \cdot y^7$ **(b)** $(6z^5)(9z^3)(2z^2)$

Solution

(a) $y^4 \cdot y^7 = y^{4+7} = y^{11}$ Keep the same base; add the exponents.

(b) $(6z^5)(9z^3)(2z^2) = (6 \cdot 9 \cdot 2)(z^5 z^3 z^2)$ Commutative and associative properties

$= 108z^{5+3+2}$ Product rule

$= 108z^{10}$ ■

Zero Exponent

For any nonzero real number a,

$$a^0 = 1.$$

Justification for this definition is given in Section R.4. The expression 0^0 ***is undefined.***

EXAMPLE 2 Using the Definition of a^0

Evaluate each power.

(a) 4^0 **(b)** $(-4)^0$ **(c)** -4^0 **(d)** $-(-4)^0$ **(e)** $(7r)^0, \quad r \neq 0$

Solution

(a) $4^0 = 1$ Base is 4.

(b) $(-4)^0 = 1$ Base is -4.

(c) $-4^0 = -(4^0) = -1$ Base is 4.

(d) $-(-4)^0 = -(1) = -1$ Base is -4.

(e) $(7r)^0 = 1, \quad r \neq 0$ Base is $7r$. ■

The expression $(2^5)^3$ can be written as

$$(2^5)^3 = 2^5 \cdot 2^5 \cdot 2^5.$$

By a generalization of the product rule for exponents, this product is

$$(2^5)^3 = 2^{5+5+5} = 2^{15}.$$

The same exponent could have been obtained by multiplying 3 and 5. This example suggests the first of the power rules given here. The others are found in a similar way.

Power Rules

For all positive integers m and n and all real numbers a and b,

$$(a^m)^n = a^{mn}, \quad (ab)^m = a^m b^m, \quad \left(\frac{a}{b}\right)^m = \frac{a^m}{b^m} \quad (b \neq 0).$$

EXAMPLE 3 Using the Power Rules

Simplify each expression.

(a) $(5^3)^2$ **(b)** $(3^4x^2)^3$ **(c)** $\left(\frac{2^5}{b^4}\right)^3, \quad b \neq 0$

Solution

(a) $(5^3)^2 = 5^{3(2)} = 5^6$ **(b)** $(3^4x^2)^3 = (3^4)^3(x^2)^3 = 3^{4(3)}x^{2(3)} = 3^{12}x^6$

(c) $\left(\frac{2^5}{b^4}\right)^3 = \frac{(2^5)^3}{(b^4)^3} = \frac{2^{15}}{b^{12}}, \quad b \neq 0$ ■

CAUTION Do not confuse expressions like mn^2 and $(mn)^2$, which are *not* equal. The second power rule given in the preceding box can be used only with the second expression: $(mn)^2 = m^2n^2$.

Terminology for Polynomials

An **algebraic expression** is the result of adding, subtracting, multiplying, dividing (except by 0), or finding roots or powers of any combination of variables and constants. Examples of algebraic expressions include

$$2x^2 - 3x, \quad \frac{5y}{2y-3}, \quad \sqrt{m^3 - 8}, \quad \text{and} \quad (3a + b)^4. \qquad \text{Algebraic expressions}$$

The simplest algebraic expressions, *polynomials,* are discussed in this section.

The product of a real number and one or more variables raised to powers is called a **term.** The real number is called the **numerical coefficient** or just the **coefficient.** The coefficient in $-3m^4$ is -3, while the coefficient in $-p^2$ is -1. **Like terms** are terms with the same variables, each raised to the same powers.

$$-13x^3, \quad 4x^3, \quad -x^3 \qquad \text{Like terms}$$

$$6y, \quad 6y^2, \quad 4y^3 \qquad \text{Unlike terms}$$

A **polynomial** is defined as a term or a finite sum of terms, with only positive or zero integer exponents permitted on the variables.

$$5x^3 - 8x^2 + 7x - 4, \quad 9p^5 - 3, \quad 8r^2, \quad 6 \qquad \text{Polynomials}$$

The terms of a polynomial cannot have variables in a denominator.

$$9x^2 - 4x + \frac{6}{x} \qquad \text{Not a polynomial}$$

The **degree of a term** with one variable is the exponent on the variable. For example, the degree of $2x^3$ is 3, the degree of $-x^4$ is 4, and the degree of $17x$ (that is, $17x^1$) is 1. The greatest degree of any term in a polynomial is called the **degree of the polynomial.** For example,

$$4x^3 - 2x^2 - 3x + 7 \text{ has degree } 3,$$

because the greatest degree of any term is 3 (the degree of $4x^3$). A nonzero constant such as -6, which can be written as $-6x^0$, has degree 0. (The polynomial 0 has undefined degree.)

A polynomial can have more than one variable. A term containing more than one variable has degree equal to the sum of all the exponents appearing on all the variables in the term. For example, $-3x^4y^3z^5$ has degree $4 + 3 + 5 = 12$. The degree of a polynomial in more than one variable is equal to the greatest degree of any term appearing in the polynomial. By this definition, the polynomial

$$2x^4y^3 - 3x^5y + x^6y^2 \text{ has degree } 8.$$

A polynomial containing exactly three terms is called a **trinomial;** one containing exactly two terms is a **binomial;** and a single-term polynomial is called a **monomial.** The following table shows several examples.

Polynomial	Degree	Type
$9p^7 - 4p^3 + 8p^2$	7	Trinomial
$29x^{11} + 8x^{15}$	15	Binomial
$-10r^6s^8$	14	Monomial
$5a^3b^7 - 3a^5b^5 + 4a^2b^9 - a^{10}$	11	None of these

Adding and Subtracting Polynomials

Since the variables used in polynomials represent real numbers, polynomials represent real numbers. This means that all properties of real numbers hold for polynomials. In particular, the distributive property holds, so

$$3m^5 - 7m^5 = (3 - 7)m^5 = -4m^5.$$

Thus, polynomials are added by adding coefficients of like terms; polynomials are subtracted by subtracting coefficients of like terms.

EXAMPLE 4 **Adding and Subtracting Polynomials**

Add or subtract, as indicated.

(a) $(2y^4 - 3y^2 + y) + (4y^4 + 7y^2 + 6y)$

(b) $(-3m^3 - 8m^2 + 4) - (m^3 + 7m^2 - 3)$

(c) $8m^4p^5 - 9m^3p^5 + (11m^4p^5 + 15m^3p^5)$

(d) $4(x^2 - 3x + 7) - 5(2x^2 - 8x - 4)$

Solution

(a) $(2y^4 - 3y^2 + y) + (4y^4 + 7y^2 + 6y)$

$= (2 + 4)y^4 + (-3 + 7)y^2 + (1 + 6)y$ Add coefficients of like terms.

$= 6y^4 + 4y^2 + 7y$

(b) $(-3m^3 - 8m^2 + 4) - (m^3 + 7m^2 - 3)$

$= (-3 - 1)m^3 + (-8 - 7)m^2 + [4 - (-3)]$ Subtract coefficients of like terms.

$= -4m^3 - 15m^2 + 7$

(c) $8m^4p^5 - 9m^3p^5 + (11m^4p^5 + 15m^3p^5) = 19m^4p^5 + 6m^3p^5$

(d) $4(x^2 - 3x + 7) - 5(2x^2 - 8x - 4)$

$= 4x^2 - 4(3x) + 4(7) - 5(2x^2) - 5(-8x) - 5(-4)$ Distributive property

$= 4x^2 - 12x + 28 - 10x^2 + 40x + 20$ Multiply.

$= -6x^2 + 28x + 48$ Add like terms. ■

As shown in parts (a), (b), and (d) of Example 4, polynomials in one variable are often written with their terms in **descending order** by degree, so the term of greatest degree is first, the one with the next greatest degree is next, and so on.

Multiplying Polynomials

We use the associative and distributive properties, together with the properties of exponents, to find the product of two polynomials. For example, to find the product of $3x - 4$ and $2x^2 - 3x + 5$, we treat $3x - 4$ as a single expression and use the distributive property as follows:

$$(3x - 4)(2x^2 - 3x + 5) = (3x - 4)(2x^2) + (3x - 4)(-3x) + (3x - 4)(5).$$

Now, we use the distributive property three separate times.

$$= 3x(2x^2) - 4(2x^2) + 3x(-3x) - 4(-3x) + 3x(5) - 4(5)$$
$$= 6x^3 - 8x^2 - 9x^2 + 12x + 15x - 20$$
$$= 6x^3 - 17x^2 + 27x - 20$$

Sometimes we find it more convenient to write such a product vertically.

$$\begin{array}{rrrrl} & 2x^2 & - \ 3x & + \ 5 & \\ & & 3x & - \ 4 & \\ \hline & -8x^2 & + \ 12x & - \ 20 & \leftarrow -4(2x^2 - 3x + 5) \\ 6x^3 & - \ 9x^2 & + \ 15x & & \leftarrow 3x(2x^2 - 3x + 5) \\ \hline 6x^3 & - \ 17x^2 & + \ 27x & - \ 20 & \text{Add in columns.} \end{array}$$

Place like terms in the same column.

EXAMPLE 5 **Multiplying Polynomials**

Multiply $(3p^2 - 4p + 1)(p^3 + 2p - 8)$.

Solution

$$
\begin{array}{rrrrrrl}
 & & & 3p^2 & -\ 4p & +\ 1 & \\
 & & & p^3 & +\ 2p & -\ 8 & \\
\hline
 & & & -24p^2 & +\ 32p & -\ 8 & \leftarrow -8(3p^2 - 4p + 1) \\
 & & 6p^3 & -\ 8p^2 & +\ 2p & & \leftarrow 2p(3p^2 - 4p + 1) \\
3p^5 & -\ 4p^4 & +\ p^3 & & & & \leftarrow p^3(3p^2 - 4p + 1) \\
\hline
3p^5 & -\ 4p^4 & +\ 7p^3 & -\ 32p^2 & +\ 34p & -\ 8 & \text{Add in columns.}
\end{array}
$$

The FOIL method is a convenient way to find the product of two binomials. The memory aid **FOIL** (for **F**irst, **O**utside, **I**nside, **L**ast) gives the pairs of terms to be multiplied to get the product, as shown in the next example.

EXAMPLE 6 **Using FOIL to Multiply Two Binomials**

Find each product.

(a) $(6m + 1)(4m - 3)$ **(b)** $(2x + 7)(2x - 7)$ **(c)** $r^2(3r + 2)(3r - 2)$

Solution

$$\begin{aligned}
\textbf{(a)}\ (6m + 1)(4m - 3) &= \overset{F}{6m(4m)} + \overset{O}{6m(-3)} + \overset{I}{1(4m)} + \overset{L}{1(-3)} \\
&= 24m^2 - 14m - 3 && -18m + 4m = -14m \\
\textbf{(b)}\ (2x + 7)(2x - 7) &= 4x^2 - 14x + 14x - 49 && \text{FOIL} \\
&= 4x^2 - 49 \\
\textbf{(c)}\ r^2(3r + 2)(3r - 2) &= r^2(9r^2 - 6r + 6r - 4) && \text{FOIL} \\
&= r^2(9r^2 - 4) && \text{Combine like terms.} \\
&= 9r^4 - 4r^2 && \text{Distributive property}
\end{aligned}$$

In part (a) of Example 6, the product of two binomials was a trinomial, while in parts (b) and (c), the product of two binomials was a binomial. The product of two binomials of the forms $x + y$ and $x - y$ is always a binomial. The squares of binomials of the forms $(x + y)^2$ and $(x - y)^2$ are also special products.

Special Products

Product of the Sum and Difference of Two Terms $(x + y)(x - y) = x^2 - y^2$

Square of a Binomial $(x + y)^2 = x^2 + 2xy + y^2$

$(x - y)^2 = x^2 - 2xy + y^2$

EXAMPLE 7 **Using the Special Products**

Find each product.

(a) $(3p + 11)(3p - 11)$ **(b)** $(5m^3 - 3)(5m^3 + 3)$

(c) $(9k - 11r^3)(9k + 11r^3)$ **(d)** $(2m + 5)^2$ **(e)** $(3x - 7y^4)^2$

Solution

(a) $(3p + 11)(3p - 11) = (3p)^2 - 11^2$ $\quad (x + y)(x - y) = x^2 - y^2$

$= 9p^2 - 121$

(b) $(5m^3 - 3)(5m^3 + 3) = (5m^3)^2 - 3^2$

$= 25m^6 - 9$

(c) $(9k - 11r^3)(9k + 11r^3) = (9k)^2 - (11r^3)^2$

$= 81k^2 - 121r^6$

(d) $(2m + 5)^2 = (2m)^2 + 2(2m)(5) + 5^2$ $\quad (x + y)^2 = x^2 + 2xy + y^2$

$= 4m^2 + 20m + 25$ Remember the middle term.

(e) $(3x - 7y^4)^2 = (3x)^2 - 2(3x)(7y^4) + (7y^4)^2$ $\quad (x - y)^2 = x^2 - 2xy + y^2$

$= 9x^2 - 42xy^4 + 49y^8$ ■

CAUTION As shown in Examples 7(d) and (e), the square of a binomial has *three* terms. Do not give $x^2 + y^2$ as the result of expanding $(x + y)^2$.

$$(x + y)^2 = x^2 + 2xy + y^2 \quad \text{Remember the middle term.}$$

Also, $$(x - y)^2 = x^2 - 2xy + y^2.$$

R.1 Exercises

Simplify each expression. Leave answers with exponents.

1. $(-4)^3 \cdot (-4)^2$
2. $(-5)^2 \cdot (-5)^6$
3. 2^0
4. -2^0
5. $(5m)^0, \quad m \neq 0$
6. $(-4z)^0, \quad z \neq 0$
7. $(2^2)^5$
8. $(6^4)^3$
9. $(2x^5y^4)^3$
10. $(-4m^3n^9)^2$
11. $-\left(\frac{p^4}{q}\right)^2$
12. $\left(\frac{r^8}{s^2}\right)^3$

Concept Check *Identify each expression as a* polynomial *or* not a polynomial. *For each polynomial, give the degree and identify it as a* monomial, binomial, trinomial, *or* none of these.

13. $-5x^{11}$
14. $9y^{12} + y^2$
15. $18p^5q + 6pq$
16. $2a^6 + 5a^2 + 4a$
17. $\sqrt{2}x^2 + \sqrt{3}x^6$
18. $-\sqrt{7}m^5n^2 + 2\sqrt{3}m^3n^2$
19. $\frac{1}{3}r^2s^2 - \frac{3}{5}r^4s^2 + rs^3$
20. $\frac{5}{p} + \frac{2}{p^2} + \frac{5}{p^3}$
21. $-5\sqrt{z} + 2\sqrt{z^3} - 5\sqrt{z^5}$

Find each sum or difference.

22. $(3x^2 - 4x + 5) + (-2x^2 + 3x - 2)$

23. $(4m^3 - 3m^2 + 5) + (-3m^3 - m^2 + 5)$

24. $(12y^2 - 8y + 6) - (3y^2 - 4y + 2)$

25. $(8p^2 - 5p) - (3p^2 - 2p + 4)$

26. $(6m^4 - 3m^2 + m) - (2m^3 + 5m^2 + 4m) + (m^2 - m)$

27. $-(8x^3 + x - 3) + (2x^3 + x^2) - (4x^2 + 3x - 1)$

Find each product.

28. $(4r - 1)(7r + 2)$

29. $(5m - 6)(3m + 4)$

30. $\left(3x - \frac{2}{3}\right)\left(5x + \frac{1}{3}\right)$

31. $\left(2m - \frac{1}{4}\right)\left(3m + \frac{1}{2}\right)$

32. $4x^2(3x^3 + 2x^2 - 5x + 1)$

33. $2b^3(b^2 - 4b + 3)$

34. $(2z - 1)(-z^2 + 3z - 4)$

35. $(m - n + k)(m + 2n - 3k)$

36. $(r - 3s + t)(2r - s + t)$

37. In words, state the formula for the square of a binomial.

38. In words, state the formula for the product of the sum and difference of two terms.

Find each product.

39. $(2m + 3)(2m - 3)$

40. $(8s - 3t)(8s + 3t)$

41. $(4m + 2n)^2$

42. $(a - 6b)^2$

43. $(5r + 3t^2)^2$

44. $(2z^4 - 3y)^2$

45. $[(2p - 3) + q]^2$

46. $[(4y - 1) + z]^2$

47. $[(3q + 5) - p][(3q + 5) + p]$

48. $[(9r - s) + 2][(9r - s) - 2]$

49. $[(3a + b) - 1]^2$

50. $[(2m + 7) - n]^2$

Perform the indicated operations.

51. $(6p + 5q)(3p - 7q)$

52. $(2p - 1)(3p^2 - 4p + 5)$

53. $(p^3 - 4p^2 + p) - (3p^2 + 2p + 7)$

54. $(6k - 3)^2$

55. $y(4x + 3y)(4x - 3y)$

56. $(r^5 - r^3 + r) + (3r^5 - 4r^4 + r^3 + 2r)$

57. $(2z + y)(3z - 4y)$

58. $(7m + 2n)(7m - 2n)$

59. $(3p + 5)^2$

60. $2(3r^2 + 4r + 2) - 3(-r^2 + 4r - 5)$

61. $p(4p - 6) + 2(3p - 8)$

62. $m(5m - 2) + 9(5 - m)$

63. $-y(y^2 - 4) + 6y^2(2y - 3)$

64. $-z^3(9 - z) + 4z(2 + 3z)$

R.2 Review of Factoring

Factoring Out the Greatest Common Factor ■ Factoring by Grouping ■ Factoring Trinomials ■ Factoring Special Products ■ Factoring by Substitution

The process of finding polynomials whose product equals a given polynomial is called **factoring.** For example, since $4x + 12 = 4(x + 3)$, both 4 and $x + 3$ are called **factors** of $4x + 12$. Also, $4(x + 3)$ is called the **factored form** of $4x + 12$. A polynomial that cannot be written as a product of two polynomials with integer coefficients is a **prime** or **irreducible polynomial.** A polynomial is **factored completely** when it is written as a product of prime polynomials with integer coefficients.

Factoring Out the Greatest Common Factor

Polynomials are factored by using the distributive property. For example, to factor $6x^2y^3 + 9xy^4 + 18y^5$, we look for the monomial that is the greatest common factor (GCF) of all the terms.

$$\begin{aligned} 6x^2y^3 + 9xy^4 + 18y^5 &= 3y^3(2x^2) + 3y^3(3xy) + 3y^3(6y^2) && \text{GCF} = 3y^3 \\ &= 3y^3(2x^2 + 3xy + 6y^2) && \text{Distributive property} \end{aligned}$$

EXAMPLE 1 Factoring Out the Greatest Common Factor

Factor out the greatest common factor from each polynomial.

(a) $9y^5 + y^2$ **(b)** $6x^2t + 8xt + 12t$

(c) $14m^4(m + 1) - 28m^3(m + 1) - 7m^2(m + 1)$

Solution

(a) $9y^5 + y^2 = y^2(9y^3) + y^2(1)$ GCF $= y^2$

$= y^2(9y^3 + 1)$ Distributive property

Remember to include the 1.

(b) $6x^2t + 8xt + 12t = 2t(3x^2 + 4x + 6)$ GCF $= 2t$

(c) The GCF is $7m^2(m + 1)$. Use the distributive property as follows:

$$\begin{aligned} 14m^4(m + 1) - 28m^3(m + 1) - 7m^2(m + 1) &= 7m^2(m + 1)[(2m^2 - 4m - 1)] \\ &= 7m^2(m + 1)(2m^2 - 4m - 1). \end{aligned}$$ ■

NOTE ***Factoring can always be checked by multiplying.***

Factoring by Grouping

When a polynomial has more than three terms, it can sometimes be factored using **factoring by grouping.** For example, to factor $ax + ay + 6x + 6y$, group the terms so that each group has a common factor.

$$\begin{aligned} ax + ay + 6x + 6y &= \underbrace{(ax + ay)}_{\text{Terms with common factor } a} + \underbrace{(6x + 6y)}_{\text{Terms with common factor } 6} \\ &= a(x + y) + 6(x + y) && \text{Factor each group.} \\ &= (x + y)(a + 6) && \text{Factor out } x + y. \end{aligned}$$

It is not always obvious which terms should be grouped. Experience and repeated trials are the most reliable tools when factoring.

EXAMPLE 2 Factoring by Grouping

Factor each polynomial by grouping.

(a) $mp^2 + 7m + 3p^2 + 21$ **(b)** $2y^2 + az - 2z - ay^2$

(c) $4x^3 + 2x^2 - 2x - 1$

Solution

(a) $mp^2 + 7m + 3p^2 + 21 = (mp^2 + 7m) + (3p^2 + 21)$ — Group the terms.

$= m(p^2 + 7) + 3(p^2 + 7)$ — Factor each group.

$= (p^2 + 7)(m + 3)$ — $p^2 + 7$ is a common factor.

Check: $(p^2 + 7)(m + 3) = mp^2 + 3p^2 + 7m + 21$ — FOIL

$= mp^2 + 7m + 3p^2 + 21$ — Commutative property

(b) $2y^2 + az - 2z - ay^2 = 2y^2 - 2z - ay^2 + az$ — Rearrange the terms.

$= (2y^2 - 2z) + (-ay^2 + az)$ — Group the terms.

$= 2(y^2 - z) + a(-y^2 + z)$ — Factor each group.

$= 2(y^2 - z) - a(y^2 - z)$ — Factor out $-a$ instead of a.

$= (y^2 - z)(2 - a)$ — Factor out $y^2 - z$.

(c) $4x^3 + 2x^2 - 2x - 1 = 2x^2(2x + 1) - 1(2x + 1)$ — Factor each group.

Be careful with signs.

$= (2x + 1)(2x^2 - 1)$ — Factor out $2x + 1$. ■

Factoring Trinomials

As shown here, factoring is the opposite of multiplication.

Multiplication →

$$(2x + 1)(3x - 4) = 6x^2 - 5x - 4$$

← Factoring

Since the product of two binomials is usually a trinomial, we can expect factorable trinomials (that have terms with no common factor) to have two binomial factors. Thus, factoring trinomials requires using FOIL in reverse.

EXAMPLE 3 Factoring Trinomials

Factor each trinomial.

(a) $4y^2 - 11y + 6$ **(b)** $6p^2 - 7p - 5$ **(c)** $16y^3 + 24y^2 - 16y$

Solution

(a) To factor this polynomial, we must find integers a, b, c, and d such that

$$4y^2 - 11y + 6 = (ay + b)(cy + d). \quad \text{FOIL}$$

Then $ac = 4$ and $bd = 6$. The positive factors of 4 are 4 and 1 or 2 and 2. Since the middle term has a negative coefficient, we consider only negative factors of 6. The possibilities are -2 and -3 or -1 and -6. We try various arrangements of these factors until we find one that gives the correct coefficient of y.

$(2y - 1)(2y - 6) = 4y^2 - 14y + 6$ — Incorrect

$(2y - 2)(2y - 3) = 4y^2 - 10y + 6$ — Incorrect

$(y - 2)(4y - 3) = 4y^2 - 11y + 6$ — Correct

The last trial gives the correct factorization.

(b) To factor $6p^2 - 7p - 5$, we try various possibilities. The positive factors of 6 could be 2 and 3 or 1 and 6. As factors of -5, we have -1 and 5 or -5 and 1.

$$(2p - 5)(3p + 1) = 6p^2 - 13p - 5 \qquad \text{Incorrect}$$
$$(3p - 5)(2p + 1) = 6p^2 - 7p - 5 \qquad \text{Correct}$$

Thus, $6p^2 - 7p - 5$ factors as $(3p - 5)(2p + 1)$.

(c) $16y^3 + 24y^2 - 16y = 8y(2y^2 + 3y - 2)$ Factor out the GCF, $8y$.

$= 8y(2y - 1)(y + 2)$ Factor the trinomial. ■

Factoring Special Products

Each of the special patterns for multiplication given in Section R.1 can be used in reverse to get a pattern for factoring. Perfect square trinomials can be factored as follows.

Perfect Square Trinomials

$$x^2 + 2xy + y^2 = (x + y)^2$$
$$x^2 - 2xy + y^2 = (x - y)^2$$

EXAMPLE 4 **Factoring Perfect Square Trinomials**

Factor each polynomial.

(a) $16p^2 - 40pq + 25q^2$ **(b)** $169x^2 + 104xy^2 + 16y^4$

Solution

(a) Since $16p^2 = (4p)^2$ and $25q^2 = (5q)^2$, use the second pattern shown in the box, with $4p$ replacing x and $5q$ replacing y.

$$16p^2 - 40pq + 25q^2 = (4p)^2 - 2(4p)(5q) + (5q)^2$$
$$= (4p - 5q)^2 \qquad \text{Answer}$$

Make sure that the middle term of the trinomial being factored, $-40pq$ here, is twice the product of the two terms in the binomial $4p - 5q$.

$$-40pq = 2(4p)(-5q)$$

(b) $169x^2 + 104xy^2 + 16y^4 = (13x + 4y^2)^2$, since $2(13x)(4y^2) = 104xy^2$. ■

The pattern for the product of the sum and difference of two terms gives the following factorization.

Difference of Squares

$$x^2 - y^2 = (x + y)(x - y)$$

EXAMPLE 5 Factoring Differences of Squares

Factor each polynomial.

(a) $4m^2 - 9$ **(b)** $256k^4 - 625m^4$

(c) $(a + 2b)^2 - 4c^2$ **(d)** $x^2 - 6x + 9 - y^4$

Solution

(a) $4m^2 - 9 = (2m)^2 - 3^2$ Write as the difference of squares.

$= (2m + 3)(2m - 3)$

(b) $256k^4 - 625m^4 = (16k^2)^2 - (25m^2)^2$ Write as the difference of squares.

$= (16k^2 + 25m^2)(16k^2 - 25m^2)$ Factor.

$= (16k^2 + 25m^2)(4k + 5m)(4k - 5m)$ Factor again.

Be sure to factor completely.

(c) $(a + 2b)^2 - 4c^2 = (a + 2b)^2 - (2c)^2$

$= [(a + 2b) + 2c][(a + 2b) - 2c]$

$= (a + 2b + 2c)(a + 2b - 2c)$

(d) $x^2 - 6x + 9 - y^4 = (x^2 - 6x + 9) - y^4$ Group terms.

$= (x - 3)^2 - (y^2)^2$ Write as the difference of squares.

$= [(x - 3) + y^2][(x - 3) - y^2]$ Factor.

$= (x - 3 + y^2)(x - 3 - y^2)$ ■

Difference and Sum of Cubes

Difference of cubes $x^3 - y^3 = (x - y)(x^2 + xy + y^2)$

Sum of cubes $x^3 + y^3 = (x + y)(x^2 - xy + y^2)$

EXAMPLE 6 Factoring Sums and Differences of Cubes

Factor each polynomial.

(a) $x^3 + 27$ **(b)** $m^3 - 64n^3$ **(c)** $8q^6 + 125p^9$

Solution

(a) $x^3 + 27 = x^3 + 3^3$ Write as a sum of cubes.

$= (x + 3)(x^2 - 3x + 3^2)$ Factor.

$= (x + 3)(x^2 - 3x + 9)$

(b) $m^3 - 64n^3 = m^3 - (4n)^3$ Write as a difference of cubes.

$= (m - 4n)[m^2 + m(4n) + (4n)^2]$ Factor.

$= (m - 4n)(m^2 + 4mn + 16n^2)$

(c) $8q^6 + 125p^9 = (2q^2)^3 + (5p^3)^3$

$= (2q^2 + 5p^3)[(2q^2)^2 - (2q^2)(5p^3) + (5p^3)^2]$

$= (2q^2 + 5p^3)(4q^4 - 10q^2p^3 + 25p^6)$ ■

Factoring by Substitution

Sometimes a polynomial can be factored by substituting one expression for another. Any variable *except* the original one can be used in the substitution.

EXAMPLE 7 Factoring by Substitution

Factor each polynomial.

(a) $6z^4 - 13z^2 - 5$ **(b)** $10(2a - 1)^2 - 19(2a - 1) - 15$ **(c)** $(2a - 1)^3 + 8$

Solution

(a) Replace z^2 with y, so $y^2 = (z^2)^2 = z^4$.

$$6z^4 - 13z^2 - 5 = 6y^2 - 13y - 5$$

$= (2y - 5)(3y + 1)$. Use FOIL to factor.

$= (2z^2 - 5)(3z^2 + 1)$ Replace y with z^2.

Remember to make the final substitution.

(Some students prefer to factor this type of trinomial directly, using trial and error with FOIL.)

(b) $10(2a - 1)^2 - 19(2a - 1) - 15$

$= 10m^2 - 19m - 15$ Replace $2a - 1$ with m.

$= (5m + 3)(2m - 5)$ Factor.

$= [5(2a - 1) + 3][2(2a - 1) - 5]$ Let $m = 2a - 1$.

$= (10a - 5 + 3)(4a - 2 - 5)$ Distributive property

$= (10a - 2)(4a - 7)$ Add.

$= 2(5a - 1)(4a - 7)$ Factor out the common factor.

(c) $(2a - 1)^3 + 8 = t^3 + 8$ Let $2a - 1 = t$.

$= t^3 + 2^3$ Write as a sum of cubes.

$= (t + 2)(t^2 - 2t + 4)$ Factor.

$= [(2a - 1) + 2][(2a - 1)^2 - 2(2a - 1) + 4]$ Let $t = 2a - 1$.

$= (2a + 1)(4a^2 - 4a + 1 - 4a + 2 + 4)$ Add; multiply.

$= (2a + 1)(4a^2 - 8a + 7)$ Combine like terms. ■

R.2 Exercises

1. *Concept Check* Match each polynomial in Column I with its factored form in Column II.

I	II
(a) $x^2 + 10xy + 25y^2$	**A.** $(x + 5y)(x - 5y)$
(b) $x^2 - 10xy + 25y^2$	**B.** $(x + 5y)^2$
(c) $x^2 - 25y^2$	**C.** $(x - 5y)^2$
(d) $25y^2 - x^2$	**D.** $(5y + x)(5y - x)$

2. *Concept Check* Match each polynomial in Column I with its factored form in Column II.

I	II
(a) $8x^3 - 27$	**A.** $(3 - 2x)(9 + 6x + 4x^2)$
(b) $8x^3 + 27$	**B.** $(2x - 3)(4x^2 + 6x + 9)$
(c) $27 - 8x^3$	**C.** $(2x + 3)(4x^2 - 6x + 9)$

Factor the greatest common factor from each polynomial.

3. $4k^2m^3 + 8k^4m^3 - 12k^2m^4$

4. $28r^4s^2 + 7r^3s - 35r^4s^3$

5. $2(a + b) + 4m(a + b)$

6. $4(y - 2)^2 + 3(y - 2)$

7. $(2y - 3)(y + 2) + (y + 5)(y + 2)$

8. $(6a - 1)(a + 2) + (6a - 1)(3a - 1)$

9. $(5r - 6)(r + 3) - (2r - 1)(r + 3)$

10. $(3z + 2)(z + 4) - (z + 6)(z + 4)$

11. $2(m - 1) - 3(m - 1)^2 + 2(m - 1)^3$

12. $5(a + 3)^3 - 2(a + 3) + (a + 3)^2$

Factor each polynomial by grouping.

13. $6st + 9t - 10s - 15$

14. $10ab - 6b + 35a - 21$

15. $10x^2 - 12y + 15x - 8xy$

16. $2m^4 + 6 - am^4 - 3a$

17. $t^3 + 2t^2 - 3t - 6$

18. $x^3 + 3x^2 - 5x - 15$

19. ***Concept Check*** Layla factored $16a^2 - 40a - 6a + 15$ by grouping and obtained $(8a - 3)(2a - 5)$. Jamal factored the same polynomial and got an answer of $(3 - 8a)(5 - 2a)$. Are both of these answers correct? If not, why not?

Factor each trinomial.

20. $6a^2 - 48a - 120$

21. $8h^2 - 24h - 320$

22. $3m^3 + 12m^2 + 9m$

23. $9y^4 - 54y^3 + 45y^2$

24. $6k^2 + 5kp - 6p^2$

25. $14m^2 + 11mr - 15r^2$

26. $5a^2 - 7ab - 6b^2$

27. $12s^2 + 11st - 5t^2$

28. $9x^2 - 6x^3 + x^4$

29. $30a^2 + am - m^2$

30. $24a^4 + 10a^3b - 4a^2b^2$

31. $18x^5 + 15x^4z - 75x^3z^2$

Factor each perfect square trinomial.

32. $9m^2 - 12m + 4$

33. $16p^2 - 40p + 25$

34. $32a^2 - 48ab + 18b^2$

35. $20p^2 - 100pq + 125q^2$

36. $4x^2y^2 + 28xy + 49$

37. $9m^2n^2 - 12mn + 4$

38. $(a - 3b)^2 - 6(a - 3b) + 9$

39. $(2p + q)^2 - 10(2p + q) + 25$

40. $(5r + 2s)^2 + 6(5r + 2s) + 9$

Factor each difference of squares.

41. $9a^2 - 16$

42. $16q^2 - 25$

43. $25s^4 - 9t^2$

44. $36z^2 - 81y^4$

45. $(a + b)^2 - 16$

46. $(p - 2q)^2 - 100$

47. $p^4 - 625$

48. $m^4 - 81$

49. ***Concept Check*** Which of the following is the correct complete factorization of $x^4 - 1$?
A. $(x^2 - 1)(x^2 + 1)$ **B.** $(x^2 + 1)(x + 1)(x - 1)$ **C.** $(x^2 - 1)^2$ **D.** $(x - 1)^2(x + 1)^2$

50. ***Concept Check*** Which of the following is the correct complete factorization of $x^3 + 8$?
A. $(x + 2)^3$ **B.** $(x + 2)(x^2 + 2x + 4)$ **C.** $(x + 2)(x^2 - 2x + 4)$ **D.** $(x + 2)(x^2 - 4x + 4)$

Factor each sum or difference of cubes.

51. $8 - a^3$

52. $r^3 + 27$

53. $125x^3 - 27$

54. $8m^3 - 27n^3$

55. $27y^9 + 125z^6$

56. $27z^3 + 729y^3$

57. $(r + 6)^3 - 216$

58. $(b + 3)^3 - 27$

59. $27 - (m + 2n)^3$

60. $125 - (4a - b)^3$

 **61.** Is the following factorization of $3a^4 + 14a^2 - 5$ correct? Explain. If it is incorrect, give the correct factors.

$$3a^4 + 14a^2 - 5 = 3u^2 + 14u - 5 \quad \text{Let } u = a^2.$$
$$= (3u - 1)(u + 5) \quad \text{Factor.}$$

Factor each polynomial by substitution.

62. $m^4 - 3m^2 - 10$

63. $a^4 - 2a^2 - 48$

64. $7(3k - 1)^2 + 26(3k - 1) - 8$

65. $6(4z - 3)^2 + 7(4z - 3) - 3$

66. $9(a - 4)^2 + 30(a - 4) + 25$

67. $20(4 - p)^2 - 3(4 - p) - 2$

Factor by any method.

68. $a^3(r + s) + b^2(r + s)$

69. $4b^2 + 4bc + c^2 - 16$

70. $(2y - 1)^2 - 4(2y - 1) + 4$

71. $x^2 + xy - 5x - 5y$

72. $8r^2 - 3rs + 10s^2$

73. $p^4(m - 2n) + q(m - 2n)$

74. $36a^2 + 60a + 25$

75. $4z^2 + 28z + 49$

76. $6p^4 + 7p^2 - 3$

77. $1000x^3 + 343y^3$

78. $b^2 + 8b + 16 - a^2$

79. $125m^6 - 216$

80. $q^2 + 6q + 9 - p^2$

81. $12m^2 + 16mn - 35n^2$

82. $216p^3 + 125q^3$

83. $4p^2 + 3p - 1$

84. $100r^2 - 169s^2$

85. $144z^2 + 121$

86. $(3a + 5)^2 - 18(3a + 5) + 81$

87. $(x + y)^2 - (x - y)^2$

88. $4z^4 - 7z^2 - 15$

R.3 Review of Rational Expressions

Domain of a Rational Expression ■ Lowest Terms of a Rational Expression ■ Multiplying and Dividing Rational Expressions ■ Adding and Subtracting Rational Expressions ■ Complex Fractions

An expression that is the quotient of two polynomials P and Q, with $Q \neq 0$, is a **rational expression.** Some examples of rational expressions are

$$\frac{x + 6}{x + 2}, \quad \frac{(x + 6)(x + 4)}{(x + 2)(x + 4)}, \quad \text{and} \quad \frac{2p^2 + 7p - 4}{5p^2 + 20p}. \qquad \text{Rational expressions}$$

Domain of a Rational Expression

The **domain** of a rational expression is the set of real numbers for which the expression is defined. Because the denominator cannot be 0, the domain consists of all real numbers except those that make the denominator 0. We find these numbers by setting the denominator equal to 0 and solving the resulting equation. For example, in the rational expression

$$\frac{x + 6}{x + 2},$$

the solution of the equation $x + 2 = 0$ is excluded from the domain. Since this solution is -2, the domain is the set of all real numbers x such that $x \neq -2$, written $\{x | x \neq -2\}$.

If the denominator of a rational expression contains a product, we find the domain by using the zero-product property (Section 3.3), which states that $ab = 0$ if and only if $a = 0$ or $b = 0$.

For example, to find the domain of

$$\frac{(x+6)(x+4)}{(x+2)(x+4)},$$

we solve as follows:

$$(x+2)(x+4)=0 \quad \text{Set denominator equal to 0.}$$
$$x+2=0 \quad \text{or} \quad x+4=0 \quad \text{Zero-product property}$$
$$x=-2 \quad \text{or} \quad x=-4. \quad \text{Solve each equation.}$$

The domain consists of the set of real numbers x such that $x \neq -2, -4$, written $\{x|x \neq -2, -4\}$.

Lowest Terms of a Rational Expression

A rational expression is in lowest terms when the greatest common factor of its numerator and denominator is 1. Just as the fraction $\frac{6}{8}$ is written in lowest terms as $\frac{3}{4}$, rational expressions can also be written in lowest terms. This is done with the fundamental principle of fractions.

Fundamental Principle of Fractions

$$\frac{ac}{bc}=\frac{a}{b} \quad (b \neq 0, c \neq 0)$$

EXAMPLE 1 Writing Rational Expressions in Lowest Terms

Write each rational expression in lowest terms.

(a) $\frac{2p^2+7p-4}{5p^2+20p}$ **(b)** $\frac{6-3k}{k^2-4}$

Solution

(a) $\frac{2p^2+7p-4}{5p^2+20p}=\frac{(2p-1)(p+4)}{5p(p+4)}$ Factor.

$=\frac{2p-1}{5p}$ Divide out the common factor.

To determine the domain, we find values of p that make the original denominator, $5p^2+20p$, equal to 0.

$$5p^2+20p=0$$
$$5p(p+4)=0 \quad \text{Factor.}$$
$$5p=0 \quad \text{or} \quad p+4=0 \quad \text{Set each factor equal to 0.}$$
$$p=0 \quad \text{or} \quad p=-4 \quad \text{Solve.}$$

The domain is $\{p|p \neq 0, -4\}$. ***From now on, we will assume such restrictions when writing rational expressions in lowest terms.***

(b) $\frac{6-3k}{k^2-4} = \frac{3(2-k)}{(k+2)(k-2)}$ Factor.

$= \frac{3(2-k)(-1)}{(k+2)(k-2)(-1)}$ $2-k$ and $k-2$ are opposites; multiply numerator and denominator by -1.

$= \frac{3(2-k)(-1)}{(k+2)(2-k)}$ $(k-2)(-1) = -k+2 = 2-k$

Be careful with signs.

$= \frac{-3}{k+2}$ Fundamental principle; lowest terms

Working in an alternative way would lead to the equivalent result $\frac{3}{-k-2}$. ■

CAUTION The fundamental principle requires a pair of common *factors,* one in the numerator and one in the denominator. ***Only after a rational expression has been factored can any common factors be divided out.*** For example,

$$\frac{2x+4}{6} = \frac{2(x+2)}{2 \cdot 3} = \frac{x+2}{3}.$$ Factor first; then divide.

Multiplying and Dividing Rational Expressions

Rational expressions are multiplied and divided by using definitions from earlier work with fractions.

Multiplying and Dividing Fractions

For fractions $\frac{a}{b}$ and $\frac{c}{d}$ $(b \neq 0, d \neq 0)$,

$$\frac{a}{b} \cdot \frac{c}{d} = \frac{ac}{bd} \quad \text{and} \quad \frac{a}{b} \div \frac{c}{d} = \frac{a}{b} \cdot \frac{d}{c}, \quad \text{if } \frac{c}{d} \neq 0.$$

EXAMPLE 2 **Multiplying and Dividing Rational Expressions**

Multiply or divide, as indicated.

(a) $\frac{2y^2}{9} \cdot \frac{27}{8y^5}$

(b) $\frac{3m^2-2m-8}{3m^2+14m+8} \cdot \frac{3m+2}{3m+4}$

(c) $\frac{5}{8m+16} \div \frac{7}{12m+24}$

(d) $\frac{3p^2+11p-4}{24p^3-8p^2} \div \frac{9p+36}{24p^4-36p^3}$

Solution

(a) $\frac{2y^2}{9} \cdot \frac{27}{8y^5} = \frac{2y^2 \cdot 27}{9 \cdot 8y^5}$ Multiply fractions.

$= \frac{2 \cdot 9 \cdot 3 \cdot y^2}{9 \cdot 2 \cdot 4 \cdot y^2 \cdot y^3}$ Factor.

$= \frac{3}{4y^3}$ Fundamental principle

(b) $$\frac{3m^2 - 2m - 8}{3m^2 + 14m + 8} \cdot \frac{3m + 2}{3m + 4} = \frac{(m - 2)(3m + 4)}{(m + 4)(3m + 2)} \cdot \frac{3m + 2}{3m + 4}$$ Factor.

$$= \frac{(m - 2)(3m + 4)(3m + 2)}{(m + 4)(3m + 2)(3m + 4)}$$ Multiply fractions.

$$= \frac{m - 2}{m + 4}$$ Fundamental principle

(c) $$\frac{5}{8m + 16} \div \frac{7}{12m + 24} = \frac{5}{8(m + 2)} \div \frac{7}{12(m + 2)}$$ Factor denominators.

$$= \frac{5}{8(m + 2)} \cdot \frac{12(m + 2)}{7}$$ Multiply by the reciprocal.

$$= \frac{5 \cdot 12(m + 2)}{8 \cdot 7(m + 2)}$$ Multiply fractions.

$$= \frac{15}{14}$$ Fundamental principle

(d) $$\frac{3p^2 + 11p - 4}{24p^3 - 8p^2} \div \frac{9p + 36}{24p^4 - 36p^3} = \frac{(p + 4)(3p - 1)}{8p^2(3p - 1)} \div \frac{9(p + 4)}{12p^3(2p - 3)}$$

$$= \frac{(p + 4)(3p - 1)(12p^3)(2p - 3)}{8p^2(3p - 1)(9)(p + 4)}$$

$$= \frac{12p^3(2p - 3)}{9 \cdot 8p^2}$$

$$= \frac{p(2p - 3)}{6}$$ ■

Adding and Subtracting Rational Expressions

Adding and subtracting rational expressions also depends on definitions from earlier work with fractions.

Adding and Subtracting Fractions

For fractions $\frac{a}{b}$ and $\frac{c}{d}$ $(b \neq 0, d \neq 0)$,

$$\frac{a}{b} + \frac{c}{d} = \frac{ad + bc}{bd} \quad \text{and} \quad \frac{a}{b} - \frac{c}{d} = \frac{ad - bc}{bd}.$$

In practice, we add or subtract rational expressions after rewriting them with a common denominator, preferably the least common denominator.

Finding the Least Common Denominator (LCD)

Step 1 Write each denominator as a product of prime factors.

Step 2 Form a product of all the different prime factors. Each factor should have as exponent the *greatest* exponent that appears on that factor.

EXAMPLE 3 Adding and Subtracting Rational Expressions

Add or subtract, as indicated.

(a) $\frac{5}{9x^2} + \frac{1}{6x}$ **(b)** $\frac{y+2}{y^2-y} - \frac{3y}{2y^2-4y+2}$

(c) $\frac{3}{(x-1)(x+2)} - \frac{1}{(x+3)(x-4)}$

Solution

(a) $\frac{5}{9x^2} + \frac{1}{6x}$

Step 1 Write each denominator as a product of prime factors.

$$9x^2 = 3^2 \cdot x^2$$
$$6x = 2^1 \cdot 3^1 \cdot x^1$$

Step 2 For the LCD, form the product of all the prime factors, with each factor having the greatest exponent that appears on it.

Greatest exponent on 3 is 2. Greatest exponent on x is 2.

$$\text{LCD} = 2^1 \cdot 3^2 \cdot x^2$$
$$= 18x^2$$

Write both of the given expressions with this denominator, and then add.

$$\frac{5}{9x^2} + \frac{1}{6x} = \frac{5 \cdot 2}{9x^2 \cdot 2} + \frac{1 \cdot 3x}{6x \cdot 3x} \quad \text{LCD} = 18x^2$$
$$= \frac{10}{18x^2} + \frac{3x}{18x^2}$$
$$= \frac{10+3x}{18x^2} \quad \text{Add the numerators.}$$

Always check to see that the answer is in lowest terms.

(b) Factor each denominator.

$$\frac{y+2}{y^2-y} - \frac{3y}{2y^2-4y+2} = \frac{y+2}{y(y-1)} - \frac{3y}{2(y-1)^2}$$

The LCD is $2y(y-1)^2$. Write each rational expression with this denominator.

$$\frac{y+2}{y(y-1)} - \frac{3y}{2(y-1)^2} = \frac{(y+2) \cdot 2(y-1)}{y(y-1) \cdot 2(y-1)} - \frac{3y \cdot y}{2(y-1)^2 \cdot y}$$
$$= \frac{2(y^2+y-2)}{2y(y-1)^2} - \frac{3y^2}{2y(y-1)^2} \quad \text{Multiply.}$$
$$= \frac{2y^2+2y-4-3y^2}{2y(y-1)^2} \quad \text{Multiply; subtract.}$$
$$= \frac{-y^2+2y-4}{2y(y-1)^2} \quad \text{Combine terms.}$$

(c) $\frac{3}{(x-1)(x+2)} - \frac{1}{(x+3)(x-4)}$

The LCD is $(x-1)(x+2)(x+3)(x-4)$.

$$= \frac{3(x+3)(x-4)}{(x-1)(x+2)(x+3)(x-4)} - \frac{(x-1)(x+2)}{(x+3)(x-4)(x-1)(x+2)}$$

$$= \frac{3(x^2-x-12)-(x^2+x-2)}{(x-1)(x+2)(x+3)(x-4)}$$ Multiply; subtract.

Be careful with signs.

$$= \frac{3x^2-3x-36-x^2-x+2}{(x-1)(x+2)(x+3)(x-4)}$$ Distributive property

$$= \frac{2x^2-4x-34}{(x-1)(x+2)(x+3)(x-4)}$$ Combine terms. ■

Complex Fractions

Any quotient of two rational expressions is called a **complex fraction.**

EXAMPLE 4 Simplifying Complex Fractions

Simplify each complex fraction.

(a) $\frac{6-\frac{5}{k}}{1+\frac{5}{k}}$ **(b)** $\frac{\frac{a}{a+1}+\frac{1}{a}}{\frac{1}{a}+\frac{1}{a+1}}$

Solution

(a) Multiply numerator and denominator by the LCD of all the fractions, k.

$$\frac{6-\frac{5}{k}}{1+\frac{5}{k}} = \frac{k\left(6-\frac{5}{k}\right)}{k\left(1+\frac{5}{k}\right)} = \frac{6k-k\left(\frac{5}{k}\right)}{k+k\left(\frac{5}{k}\right)} = \frac{6k-5}{k+5}$$

(b) Multiply numerator and denominator by the LCD of all the fractions.

$$\frac{\frac{a}{a+1}+\frac{1}{a}}{\frac{1}{a}+\frac{1}{a+1}} = \frac{\left(\frac{a}{a+1}+\frac{1}{a}\right)a(a+1)}{\left(\frac{1}{a}+\frac{1}{a+1}\right)a(a+1)}$$ LCD $= a(a+1)$

$$= \frac{\frac{a}{a+1}(a)(a+1)+\frac{1}{a}(a)(a+1)}{\frac{1}{a}(a)(a+1)+\frac{1}{a+1}(a)(a+1)}$$ Distributive property

$$= \frac{a^2+(a+1)}{(a+1)+a}$$ Multiply.

$$= \frac{a^2+a+1}{2a+1}$$ Combine terms.

Alternatively, first add the terms in the numerator and denominator, and then divide.

$$\frac{\frac{a}{a+1}+\frac{1}{a}}{\frac{1}{a}+\frac{1}{a+1}} = \frac{\frac{a^2+1(a+1)}{a(a+1)}}{\frac{1(a+1)+1(a)}{a(a+1)}}$$

Find the LCD; add terms in the numerator and denominator.

$$= \frac{\frac{a^2+a+1}{a(a+1)}}{\frac{2a+1}{a(a+1)}}$$

Combine terms in the numerator and denominator.

$$= \frac{a^2+a+1}{a(a+1)} \cdot \frac{a(a+1)}{2a+1}$$

Definition of division

$$= \frac{a^2+a+1}{2a+1}$$

Multiply fractions; write in lowest terms. ■

R.3 Exercises

Find the domain of each rational expression.

1. $\frac{x-2}{x+6}$
2. $\frac{x+5}{x-3}$
3. $\frac{2x}{5x-3}$
4. $\frac{6x}{2x-1}$
5. $\frac{-8}{x^2+1}$
6. $\frac{3x}{3x^2+7}$
7. $\frac{3x+7}{(4x+2)(x-1)}$
8. $\frac{9x+12}{(2x+3)(x-5)}$

Write each rational expression in lowest terms.

9. $\frac{25p^3}{10p^2}$
10. $\frac{14z^3}{6z^2}$
11. $\frac{8k+16}{9k+18}$
12. $\frac{20r+10}{30r+15}$
13. $\frac{3(t+5)}{(t+5)(t-3)}$
14. $\frac{-8(y-4)}{(y+2)(y-4)}$
15. $\frac{8x^2+16x}{4x^2}$
16. $\frac{36y^2+72y}{9y}$
17. $\frac{m^2-4m+4}{m^2+m-6}$
18. $\frac{r^2-r-6}{r^2+r-12}$
19. $\frac{8m^2+6m-9}{16m^2-9}$
20. $\frac{6y^2+11y+4}{3y^2+7y+4}$

Find each product or quotient.

21. $\frac{15p^3}{9p^2} \div \frac{6p}{10p^2}$
22. $\frac{3r^2}{9r^3} \div \frac{8r^3}{6r}$
23. $\frac{2k+8}{6} \div \frac{3k+12}{2}$
24. $\frac{5m+25}{10} \cdot \frac{12}{6m+30}$
25. $\frac{x^2+x}{5} \cdot \frac{25}{xy+y}$
26. $\frac{3m-15}{4m-20} \cdot \frac{m^2-10m+25}{12m-60}$
27. $\frac{4a+12}{2a-10} \div \frac{a^2-9}{a^2-a-20}$
28. $\frac{6r-18}{9r^2+6r-24} \cdot \frac{12r-16}{4r-12}$
29. $\frac{p^2-p-12}{p^2-2p-15} \cdot \frac{p^2-9p+20}{p^2-8p+16}$
30. $\frac{x^2+2x-15}{x^2+11x+30} \cdot \frac{x^2+2x-24}{x^2-8x+15}$
31. $\frac{m^2+3m+2}{m^2+5m+4} \div \frac{m^2+5m+6}{m^2+10m+24}$
32. $\frac{y^2+y-2}{y^2+3y-4} \div \frac{y^2+3y+2}{y^2+4y+3}$

33. $\dfrac{2m^2 - 5m - 12}{m^2 - 10m + 24} \div \dfrac{4m^2 - 9}{m^2 - 9m + 18}$

34. $\dfrac{6n^2 - 5n - 6}{6n^2 + 5n - 6} \cdot \dfrac{12n^2 - 17n + 6}{12n^2 - n - 6}$

35. $\dfrac{x^3 + y^3}{x^2 - y^2} \cdot \dfrac{x + y}{x^2 - xy + y^2}$

36. $\dfrac{8y^3 - 125}{4y^2 - 20y + 25} \cdot \dfrac{2y - 5}{y}$

37. $\dfrac{x^3 + y^3}{x^3 - y^3} \cdot \dfrac{x^2 - y^2}{x^2 + 2xy + y^2}$

38. $\dfrac{x^2 - y^2}{(x - y)^2} \cdot \dfrac{x^2 - xy + y^2}{x^2 - 2xy + y^2} \div \dfrac{x^3 + y^3}{(x - y)^4}$

39. ***Concept Check*** Which of these rational expressions equal -1? (Assume that all denominators are nonzero.)

A. $\dfrac{x - 4}{x + 4}$ **B.** $\dfrac{-x - 4}{x + 4}$ **C.** $\dfrac{x - 4}{4 - x}$ **D.** $\dfrac{x - 4}{-x - 4}$

 **40.** In your own words, explain how to find the least common denominator of two fractions.

Find each sum or difference.

41. $\dfrac{3}{2k} + \dfrac{5}{3k}$

42. $\dfrac{8}{5p} + \dfrac{3}{4p}$

43. $\dfrac{a + 1}{2} - \dfrac{a - 1}{2}$

44. $\dfrac{y + 6}{5} - \dfrac{y - 6}{5}$

45. $\dfrac{3}{p} + \dfrac{1}{2}$

46. $\dfrac{9}{r} - \dfrac{2}{3}$

47. $\dfrac{1}{6m} + \dfrac{2}{5m} + \dfrac{4}{m}$

48. $\dfrac{8}{3p} + \dfrac{5}{4p} + \dfrac{9}{2p}$

49. $\dfrac{1}{a + 1} - \dfrac{1}{a - 1}$

50. $\dfrac{1}{x + z} + \dfrac{1}{x - z}$

51. $\dfrac{m + 1}{m - 1} + \dfrac{m - 1}{m + 1}$

52. $\dfrac{2}{x - 1} + \dfrac{1}{1 - x}$

53. $\dfrac{3}{a - 2} - \dfrac{1}{2 - a}$

54. $\dfrac{q}{p - q} - \dfrac{q}{q - p}$

55. $\dfrac{x + y}{2x - y} - \dfrac{2x}{y - 2x}$

56. $\dfrac{m - 4}{3m - 4} + \dfrac{3m + 2}{4 - 3m}$

57. $\dfrac{1}{a^2 - 5a + 6} - \dfrac{1}{a^2 - 4}$

58. $\dfrac{-3}{m^2 - m - 2} - \dfrac{1}{m^2 + 3m + 2}$

59. $\dfrac{1}{x^2 + x - 12} - \dfrac{1}{x^2 - 7x + 12} + \dfrac{1}{x^2 - 16}$

60. $\dfrac{2}{2p^2 - 9p - 5} + \dfrac{p}{3p^2 - 17p + 10} - \dfrac{2p}{6p^2 - p - 2}$

61. $\dfrac{3a}{a^2 + 5a - 6} - \dfrac{2a}{a^2 + 7a + 6}$

62. $\dfrac{2k}{k^2 + 4k + 3} + \dfrac{3k}{k^2 + 5k + 6}$

Simplify each complex fraction.

63. $\dfrac{1 + \dfrac{1}{x}}{1 - \dfrac{1}{x}}$

64. $\dfrac{2 - \dfrac{2}{y}}{2 + \dfrac{2}{y}}$

65. $\dfrac{\dfrac{1}{x + 1} - \dfrac{1}{x}}{\dfrac{1}{x}}$

66. $\dfrac{\dfrac{1}{y + 3} - \dfrac{1}{y}}{\dfrac{1}{y}}$

67. $\dfrac{1 + \dfrac{1}{1 - b}}{1 - \dfrac{1}{1 + b}}$

68. $m - \dfrac{m}{m + \dfrac{1}{2}}$

69. $\dfrac{m - \dfrac{1}{m^2 - 4}}{\dfrac{1}{m + 2}}$

70. $\dfrac{\dfrac{3}{p^2 - 16} + p}{\dfrac{1}{p - 4}}$

R.4 Review of Negative and Rational Exponents

Negative Exponents and the Quotient Rule ■ Rational Exponents

Negative Exponents and the Quotient Rule

In the product rule for exponents, the exponents are *added* (that is, $a^m \cdot a^n = a^{m+n}$). Consider an expression such as $\frac{a^3}{a^7}$. If $a \neq 0$, then

$$\frac{a^3}{a^7} = \frac{a \cdot a \cdot a}{a \cdot a \cdot a \cdot a \cdot a \cdot a \cdot a} = \frac{1}{a \cdot a \cdot a \cdot a} = \frac{1}{a^4}.$$

This suggests that we should *subtract* exponents when dividing.

$$\frac{a^3}{a^7} = a^{3-7} = a^{-4}$$

The only way to keep these results consistent is to define a^{-4} as $\frac{1}{a^4}$. This example suggests the following definition.

Negative Exponent

If a is a nonzero real number and n is any integer, then

$$a^{-n} = \frac{1}{a^n}.$$

EXAMPLE 1 Using the Definition of a Negative Exponent

Evaluate each expression in parts (a)–(c). In parts (d) and (e), write the expression without negative exponents. Assume that all variables represent nonzero real numbers.

(a) 4^{-2} **(b)** -4^{-2} **(c)** $\left(\frac{2}{5}\right)^{-3}$ **(d)** x^{-5} **(e)** xy^{-3}

Solution

(a) $4^{-2} = \frac{1}{4^2} = \frac{1}{16}$

(b) $-4^{-2} = -\frac{1}{4^2} = -\frac{1}{16}$

(c) $\left(\frac{2}{5}\right)^{-3} = \frac{1}{\left(\frac{2}{5}\right)^3} = \frac{1}{\frac{8}{125}} = 1 \div \frac{8}{125} = 1 \cdot \frac{125}{8} = \frac{125}{8}$

(d) $x^{-5} = \frac{1}{x^5}$

(e) $xy^{-3} = x \cdot \frac{1}{y^3} = \frac{x}{y^3}$ ■

From part (c) of Example 1,

$$\left(\frac{2}{5}\right)^{-3} = \frac{125}{8} = \left(\frac{5}{2}\right)^3.$$

This result can be generalized: If $a \neq 0$ and $b \neq 0$, then, for any integer n,

$$\left(\frac{a}{b}\right)^{-n} = \left(\frac{b}{a}\right)^{n}.$$

The quotient rule for exponents follows from the definition of negative exponents.

Quotient Rule

For all integers m and n and all nonzero real numbers a,

$$\frac{a^m}{a^n} = a^{m-n}.$$

By the quotient rule, if $a \neq 0$,

$$\frac{a^m}{a^m} = a^{m-m} = a^0.$$

However, any nonzero quantity divided by itself equals 1. This is why we defined $a^0 = 1$ in Section R.1.

EXAMPLE 2 Using the Quotient Rule

Simplify each expression. Assume that all variables represent nonzero real numbers.

(a) $\frac{12^5}{12^2}$ **(b)** $\frac{a^5}{a^{-8}}$ **(c)** $\frac{16m^{-9}}{12m^{11}}$ **(d)** $\frac{25r^7z^5}{10r^9z}$

Solution

(a) $\frac{12^5}{12^2} = 12^{5-2} = 12^3$ **(b)** $\frac{a^5}{a^{-8}} = a^{5-(-8)} = a^{13}$

Use parentheses to avoid errors.

(c) $\frac{16m^{-9}}{12m^{11}} = \frac{16}{12} \cdot m^{-9-11} = \frac{4}{3}m^{-20} = \frac{4}{3} \cdot \frac{1}{m^{20}} = \frac{4}{3m^{20}}$

(d) $\frac{25r^7z^5}{10r^9z} = \frac{25}{10} \cdot \frac{r^7}{r^9} \cdot \frac{z^5}{z^1} = \frac{5}{2}r^{-2}z^4 = \frac{5z^4}{2r^2}$ ■

The rules for exponents stated in Section R.1 also apply to negative exponents.

EXAMPLE 3 Using the Rules for Exponents

Simplify each expression. Write answers without negative exponents. Assume that all variables represent nonzero real numbers.

(a) $3x^{-2}(4^{-1}x^{-5})^2$ **(b)** $\frac{5m^{-3}}{10m^{-5}}$ **(c)** $\frac{12p^3q^{-1}}{8p^{-2}q}$ **(d)** $\frac{(3x^2)^{-1}(3x^5)^{-2}}{(3^{-1}x^{-2})^2}$

Solution

(a)
$$\begin{aligned} 3x^{-2}(4^{-1}x^{-5})^2 &= 3x^{-2}(4^{-2}x^{-10}) && \text{Power rule} \\ &= 3 \cdot 4^{-2} \cdot x^{-2+(-10)} && \text{Rearrange factors; product rule} \\ &= 3 \cdot 4^{-2} \cdot x^{-12} \\ &= \frac{3}{16x^{12}} && \text{Write with positive exponents.} \end{aligned}$$

(b) $\dfrac{5m^{-3}}{10m^{-5}} = \dfrac{5}{10}m^{-3-(-5)}$ Quotient rule

$= \dfrac{1}{2}m^2$ or $\dfrac{m^2}{2}$

(c) $\dfrac{12p^3q^{-1}}{8p^{-2}q} = \dfrac{12}{8} \cdot \dfrac{p^3}{p^{-2}} \cdot \dfrac{q^{-1}}{q^1}$

$= \dfrac{3}{2} \cdot p^{3-(-2)}q^{-1-1}$ Quotient rule

$= \dfrac{3}{2}p^5q^{-2}$

$= \dfrac{3p^5}{2q^2}$ Write with positive exponents.

(d) $\dfrac{(3x^2)^{-1}(3x^5)^{-2}}{(3^{-1}x^{-2})^2} = \dfrac{3^{-1}x^{-2}3^{-2}x^{-10}}{3^{-2}x^{-4}}$ Power rule

$= \dfrac{3^{-1+(-2)}x^{-2+(-10)}}{3^{-2}x^{-4}} = \dfrac{3^{-3}x^{-12}}{3^{-2}x^{-4}}$ Product rule

$= 3^{-3-(-2)}x^{-12-(-4)} = 3^{-1}x^{-8}$ Quotient rule

$= \dfrac{1}{3x^8}$ Write with positive exponents. ■

CAUTION Notice the use of the power rule $(ab)^n = a^nb^n$ in Example 3(d): $(3x^2)^{-1} = 3^{-1}(x^2)^{-1} = 3^{-1}x^{-2}$. ***Remember to apply the exponent to a numerical coefficient.***

Rational Exponents

The definition of a^n can be extended to rational values of n by defining $a^{1/n}$ to be the nth root of a. By one of the power rules of exponents (extended to a rational exponent),

$$(a^{1/n})^n = a^{(1/n)n} = a^1 = a,$$

which suggests that $a^{1/n}$ is a number whose nth power is a.

The Expression $a^{1/n}$

***n* even** If n is an *even* positive integer, and if $a > 0$, then $a^{1/n}$ is the positive real number whose nth power is a. That is, $(a^{1/n})^n = a$. (In this case, $a^{1/n}$ is the principal nth root of a. See Section R.5.)

***n* odd** If n is an *odd* positive integer and *a is any real number,* then $a^{1/n}$ is the positive or negative real number whose nth power is a. That is, $(a^{1/n})^n = a$.

EXAMPLE 4 **Using the Definition of $a^{1/n}$**

Evaluate each expression.

(a) $36^{1/2}$ **(b)** $-100^{1/2}$ **(c)** $-(225)^{1/2}$ **(d)** $625^{1/4}$ **(e)** $(-1296)^{1/4}$

(f) $-1296^{1/4}$ **(g)** $(-27)^{1/3}$ **(h)** $-32^{1/5}$

Solution

(a) $36^{1/2} = 6$ because $6^2 = 36$.

(b) $-100^{1/2} = -10$

(c) $-(225)^{1/2} = -15$

(d) $625^{1/4} = 5$

(e) $(-1296)^{1/4}$ is not a real number. (Why?)

(f) $-1296^{1/4} = -6$

(g) $(-27)^{1/3} = -3$

(h) $-32^{1/5} = -2$ ■

The notation $a^{m/n}$ must be defined so that all the rules for exponents hold. For the power rule to hold, $(a^{1/n})^m$ must equal $a^{m/n}$. Therefore, $a^{m/n}$ is defined as follows.

Rational Exponent

For all integers m, all positive integers n, and all real numbers a for which $a^{1/n}$ is a real number,

$$a^{m/n} = (a^{1/n})^m.$$

EXAMPLE 5 Using the Definition of $a^{m/n}$

Evaluate each expression.

(a) $125^{2/3}$ **(b)** $32^{7/5}$ **(c)** $-81^{3/2}$

(d) $(-27)^{2/3}$ **(e)** $16^{-3/4}$ **(f)** $(-4)^{5/2}$

Solution

(a) $125^{2/3} = (125^{1/3})^2 = 5^2 = 25$

(b) $32^{7/5} = (32^{1/5})^7 = 2^7 = 128$

(c) $-81^{3/2} = -(81^{1/2})^3 = -(9)^3 = -729$

(d) $(-27)^{2/3} = [(-27)^{1/3}]^2 = (-3)^2 = 9$

(e) $16^{-3/4} = \frac{1}{16^{3/4}} = \frac{1}{(16^{1/4})^3} = \frac{1}{2^3} = \frac{1}{8}$

(f) $(-4)^{5/2}$ is not a real number because $(-4)^{1/2}$ is not a real number. ■

NOTE For all real numbers a, integers m, and positive integers n for which $a^{1/n}$ is a real number,

$$a^{m/n} = (a^{1/n})^m \quad \text{or} \quad a^{m/n} = (a^m)^{1/n}.$$

So $a^{m/n}$ can be evaluated either as $(a^{1/n})^m$ or as $(a^m)^{1/n}$. For example,

$$27^{4/3} = (27^{1/3})^4 = 3^4 = 81$$

or

$$27^{4/3} = (27^4)^{1/3} = 531{,}441^{1/3} = 81.$$

The first form, $(27^{1/3})^4$ (that is, $(a^{1/n})^m$), is easier to evaluate.

The earlier results concerning integer exponents also apply to rational exponents. These definitions and rules are now summarized.

Definitions and Rules for Exponents

Let r and s be rational numbers. The following results are valid whenever each expression is a real number:

$$a^r \cdot a^s = a^{r+s} \qquad (ab)^r = a^r \cdot b^r \qquad (a^r)^s = a^{rs}$$

$$\frac{a^r}{a^s} = a^{r-s} \qquad \left(\frac{a}{b}\right)^r = \frac{a^r}{b^r} \qquad a^{-r} = \frac{1}{a^r}.$$

EXAMPLE 6 **Using the Definitions and Rules for Exponents**

Simplify each expression. Assume that all variables represent positive real numbers.

(a) $\dfrac{27^{1/3} \cdot 27^{5/3}}{27^3}$ **(b)** $81^{5/4} \cdot 4^{-3/2}$ **(c)** $6y^{2/3} \cdot 2y^{1/2}$

(d) $\left(\dfrac{3m^{5/6}}{y^{3/4}}\right)^2 \cdot \left(\dfrac{8y^3}{m^6}\right)^{2/3}$ **(e)** $m^{2/3}(m^{7/3} + 2m^{1/3})$

Solution

(a) $\dfrac{27^{1/3} \cdot 27^{5/3}}{27^3} = \dfrac{27^{1/3+5/3}}{27^3}$ Product rule

$= \dfrac{27^2}{27^3} = 27^{2-3}$ Quotient rule

$= 27^{-1} = \dfrac{1}{27}$ Negative exponent

(b) $81^{5/4} \cdot 4^{-3/2} = (81^{1/4})^5(4^{1/2})^{-3} = 3^5 \cdot 2^{-3} = \dfrac{3^5}{2^3}$ or $\dfrac{243}{8}$

(c) $6y^{2/3} \cdot 2y^{1/2} = 12y^{2/3+1/2} = 12y^{7/6}$

(d) $\left(\dfrac{3m^{5/6}}{y^{3/4}}\right)^2 \cdot \left(\dfrac{8y^3}{m^6}\right)^{2/3} = \dfrac{9m^{5/3}}{y^{3/2}} \cdot \dfrac{4y^2}{m^4} = 36m^{5/3-4}y^{2-3/2} = \dfrac{36y^{1/2}}{m^{7/3}}$

(e) $m^{2/3}(m^{7/3} + 2m^{1/3}) = m^{2/3+7/3} + 2m^{2/3+1/3} = m^3 + 2m$ ■

EXAMPLE 7 **Factoring Expressions with Negative or Rational Exponents**

Factor out the least power of the variable. Assume that all variables represent positive real numbers.

(a) $12x^{-2} - 8x^{-3}$ **(b)** $4m^{1/2} + 3m^{3/2}$ **(c)** $y^{-1/3} + y^{2/3}$

Solution

(a) The least exponent on x here is -3. Since 4 is a common numerical factor, factor out $4x^{-3}$.

$$12x^{-2} - 8x^{-3} = 4x^{-3}(3x - 2)$$

Check by multiplying $4x^{-3}(3x - 2)$. The factored form can be written without negative exponents as $\frac{4(3x - 2)}{x^3}$.

(b) $4m^{1/2} + 3m^{3/2} = m^{1/2}(4 + 3m)$. To check, multiply $m^{1/2}$ by $4 + 3m$.

(c) $y^{-1/3} + y^{2/3} = y^{-1/3}(1 + y)$ or $\frac{1 + y}{y^{1/3}}$ ■

R.4 Exercises

Concept Check *Match each expression from Group I with the correct choice from Group II. Choices may be used once, more than once, or not at all.*

I		II	
1. $\left(\frac{4}{9}\right)^{3/2}$	**2.** $\left(\frac{4}{9}\right)^{-3/2}$	**A.** $\frac{9}{4}$	**B.** $-\frac{9}{4}$
3. $-\left(\frac{9}{4}\right)^{3/2}$	**4.** $-\left(\frac{4}{9}\right)^{-3/2}$	**C.** $-\frac{4}{9}$	**D.** $\frac{4}{9}$
5. $\left(\frac{8}{27}\right)^{2/3}$	**6.** $\left(\frac{8}{27}\right)^{-2/3}$	**E.** $\frac{8}{27}$	**F.** $-\frac{27}{8}$
7. $-\left(\frac{27}{8}\right)^{2/3}$	**8.** $-\left(\frac{27}{8}\right)^{-2/3}$	**G.** $\frac{27}{8}$	**H.** $-\frac{8}{27}$

Simplify each expression. Assume that all variables represent positive real numbers.

9. $(-4)^{-3}$

10. $(-5)^{-2}$

11. $\left(\frac{1}{2}\right)^{-3}$

12. $\left(\frac{2}{3}\right)^{-2}$

13. $-4^{1/2}$

14. $25^{1/2}$

15. $8^{2/3}$

16. $-81^{3/4}$

17. $27^{-2/3}$

18. $(-32)^{-4/5}$

19. $\left(\frac{27}{64}\right)^{-4/3}$

20. $\left(\frac{121}{100}\right)^{-3/2}$

21. $(16p^4)^{1/2}$

22. $(36r^6)^{1/2}$

23. $(27x^6)^{2/3}$

24. $(64a^{12})^{5/6}$

Perform the indicated operations. Write your answers with only positive exponents. Assume that all variables represent positive real numbers.

25. $2^{-3} \cdot 2^{-4}$

26. $5^{-2} \cdot 5^{-6}$

27. $27^{-2} \cdot 27^{-1}$

28. $9^{-4} \cdot 9^{-1}$

29. $\frac{4^{-2} \cdot 4^{-1}}{4^{-3}}$

30. $\frac{3^{-1} \cdot 3^{-4}}{3^2 \cdot 3^{-2}}$

31. $(m^{2/3})(m^{5/3})$

32. $(x^{4/5})(x^{2/5})$

33. $(1 + n)^{1/2}(1 + n)^{3/4}$

34. $(m + 7)^{-1/6}(m + 7)^{-2/3}$

35. $(2y^{3/4}z)(3y^{-2}z^{-1/3})$

36. $(4a^{-1}b^{2/3})(a^{3/2}b^{-3})$

37. $(4a^{-2}b^7)^{1/2} \cdot (2a^{1/4}b^3)^5$

38. $(x^{-2}y^{1/3})^5 \cdot (8x^2y^{-2})^{-1/3}$

39. $\left(\frac{r^{-2}}{s^{-5}}\right)^{-3}$

40. $\left(\frac{p^{-1}}{q^{-5}}\right)^{-2}$

41. $\left(\frac{-a}{b^{-3}}\right)^{-1}$

42. $\frac{7^{-1/3}7r^{-3}}{7^{2/3}r^{-2}}$

43. $\frac{12^{5/4}y^{-2}}{12^{-1}y^{-3}}$

44. $\frac{6k^{-4}(3k^{-1})^{-2}}{2^3k^{1/2}}$

45. $\frac{8p^{-3}(4p^2)^{-2}}{p^{-5}}$

46. $\frac{k^{-3/5}h^{-1/3}t^{2/5}}{k^{-1/5}h^{-2/3}t^{1/5}}$

47. $\frac{m^{7/3}n^{-2/5}p^{3/8}}{m^{-2/3}n^{3/5}p^{-5/8}}$

48. $\frac{m^{2/5}m^{3/5}m^{-4/5}}{m^{1/5}m^{-6/5}}$

49. $\frac{-4a^{-1}a^{2/3}}{a^{-2}}$

50. $\frac{8y^{2/3}y^{-1}}{2^{-1}y^{3/4}y^{-1/6}}$

51. $\frac{(k+5)^{1/2}(k+5)^{-1/4}}{(k+5)^{3/4}}$

52. $\frac{(x+y)^{-5/8}(x+y)^{3/8}}{(x+y)^{1/8}(x+y)^{-1/8}}$

Find each product. Assume that all variables represent positive real numbers.

53. $y^{5/8}(y^{3/8} - 10y^{11/8})$

54. $p^{11/5}(3p^{4/5} + 9p^{19/5})$

55. $-4k(k^{7/3} - 6k^{1/3})$

56. $-5y(3y^{9/10} + 4y^{3/10})$

57. $(x + x^{1/2})(x - x^{1/2})$

58. $(2z^{1/2} + z)(z^{1/2} - z)$

59. $(r^{1/2} - r^{-1/2})^2$

60. $(p^{1/2} - p^{-1/2})(p^{1/2} + p^{-1/2})$

Factor, using the given common factor. Assume that all variables represent positive real numbers.

61. $4k^{-1} + k^{-2};\quad k^{-2}$

62. $y^{-5} - 3y^{-3};\quad y^{-5}$

63. $9z^{-1/2} + 2z^{1/2};\quad z^{-1/2}$

64. $3m^{2/3} - 4m^{-1/3};\quad m^{-1/3}$

65. $p^{-3/4} - 2p^{-7/4};\quad p^{-7/4}$

66. $6r^{-2/3} - 5r^{-5/3};\quad r^{-5/3}$

67. $(p+4)^{-3/2} + (p+4)^{-1/2} + (p+4)^{1/2};\quad (p+4)^{-3/2}$

68. $(3r+1)^{-2/3} + (3r+1)^{1/3} + (3r+1)^{4/3};\quad (3r+1)^{-2/3}$

R.5 Review of Radicals

Radical Notation ■ Rules for Radicals ■ Simplifying Radicals ■ Operations with Radicals ■ Rationalizing Denominators

Radical Notation

In Section R.4, the notation $a^{1/n}$ was used for the nth root of a for appropriate values of a and n. An alternative (and more familiar) notation for $a^{1/n}$ uses *radical notation.*

Radical Notation for $a^{1/n}$

If a is a real number, n is a positive integer, and $a^{1/n}$ is a real number, then

$$\sqrt[n]{a} = a^{1/n}.$$

The symbol $\sqrt[n]{}$ is a **radical sign,** the number a is the **radicand,** and n is the **index** of the radical $\sqrt[n]{a}$. It is customary to use $\sqrt{a}$ instead of $\sqrt[2]{a}$ for the square root.

For even values of n (square roots, fourth roots, and so on) and $a > 0$, there are two nth roots, one positive and one negative. In such cases, the notation $\sqrt[n]{a}$ represents the positive root—the **principal nth root.** The negative root is written $-\sqrt[n]{a}$.

EXAMPLE 1 Evaluating Roots

Evaluate each root.

(a) $\sqrt[4]{16}$ **(b)** $-\sqrt[4]{16}$ **(c)** $\sqrt[4]{-16}$

(d) $\sqrt[5]{-32}$ **(e)** $\sqrt[3]{1000}$ **(f)** $\sqrt[6]{\frac{64}{729}}$

Solution

(a) $\sqrt[4]{16} = 16^{1/4} = (2^4)^{1/4} = 2$ **(b)** $-\sqrt[4]{16} = -16^{1/4} = -(2^4)^{1/4} = -2$

(c) $\sqrt[4]{-16}$ is not a real number. **(d)** $\sqrt[5]{-32} = [(-2)^5]^{1/5} = -2$

(e) $\sqrt[3]{1000} = 1000^{1/3} = 10$ **(f)** $\sqrt[6]{\frac{64}{729}} = \left(\frac{64}{729}\right)^{1/6} = \frac{2}{3}$ ■

With $a^{1/n}$ written as $\sqrt[n]{a}$, $a^{m/n}$ can also be written with radicals.

Radical Notation for $a^{m/n}$

If a is a real number, m is an integer, n is a positive integer, and $\sqrt[n]{a}$ is a real number, then

$$a^{m/n} = \left(\sqrt[n]{a}\right)^m = \sqrt[n]{a^m}.$$

EXAMPLE 2 Converting from Rational Exponents to Radicals

Write in radical form and simplify. Assume that all variable expressions represent positive real numbers.

(a) $8^{2/3}$ **(b)** $(-32)^{4/5}$ **(c)** $-16^{3/4}$ **(d)** $x^{5/6}$ **(e)** $3x^{2/3}$

(f) $2p^{-1/2}$ **(g)** $(3a + b)^{1/4}$

Solution

(a) $8^{2/3} = \left(\sqrt[3]{8}\right)^2 = 2^2 = 4$ **(b)** $(-32)^{4/5} = \left(\sqrt[5]{-32}\right)^4 = (-2)^4 = 16$

(c) $-16^{3/4} = -\left(\sqrt[4]{16}\right)^3 = -(2)^3 = -8$ **(d)** $x^{5/6} = \sqrt[6]{x^5}$

(e) $3x^{2/3} = 3\sqrt[3]{x^2}$ **(f)** $2p^{-1/2} = \frac{2}{p^{1/2}} = \frac{2}{\sqrt{p}}$

(g) $(3a + b)^{1/4} = \sqrt[4]{3a + b}$ ■

EXAMPLE 3 Converting from Radicals to Rational Exponents

Write each expression with rational exponents. Assume that all variable expressions represent positive real numbers.

(a) $\sqrt[4]{x^5}$ **(b)** $\sqrt{3y}$ **(c)** $10\left(\sqrt[5]{z}\right)^2$ **(d)** $5\sqrt[3]{(2x^4)^7}$ **(e)** $\sqrt{p^2 + q}$

Solution

(a) $\sqrt[4]{x^5} = x^{5/4}$ **(b)** $\sqrt{3y} = (3y)^{1/2} = 3^{1/2}y^{1/2}$

(c) $10\left(\sqrt[5]{z}\right)^2 = 10z^{2/5}$ **(d)** $5\sqrt[3]{(2x^4)^7} = 5(2x^4)^{7/3} = 5 \cdot 2^{7/3}x^{28/3}$

(e) $\sqrt{p^2 + q} = (p^2 + q)^{1/2}$ ■

By the definition of $\sqrt[n]{a}$, for any positive integer n, if $\sqrt[n]{a}$ is defined, then

$$\left(\sqrt[n]{a}\right)^n = a.$$

If a is positive, or if a is negative and n is an odd positive integer, then

$$\sqrt[n]{a^n} = a.$$

Because of the conditions just given, we *cannot* simply write $\sqrt{x^2} = x$. For example, if $x = -5$, then

$$\sqrt{x^2} = \sqrt{(-5)^2} = \sqrt{25} = 5 \neq x.$$

Since a negative value of x can produce a positive result, we use the absolute value. For any real number a,

$$\sqrt{a^2} = |a|.$$

For example,

$$\sqrt{(-9)^2} = |-9| = 9 \quad \text{and} \quad \sqrt{13^2} = |13| = 13.$$

This result can be generalized to any even nth root.

Evaluating $\sqrt[n]{a^n}$

If n is an even positive integer, then $\sqrt[n]{a^n} = |a|$.

If n is an odd positive integer, then $\sqrt[n]{a^n} = a$.

EXAMPLE 4 Using Absolute Value to Simplify Roots

Simplify each expression.

(a) $\sqrt{p^4}$ **(b)** $\sqrt[4]{p^4}$ **(c)** $\sqrt{16m^8r^6}$ **(d)** $\sqrt[6]{(-2)^6}$ **(e)** $\sqrt[5]{m^5}$

(f) $\sqrt{(2k+3)^2}$ **(g)** $\sqrt{x^2 - 4x + 4}$

Solution

(a) $\sqrt{p^4} = \sqrt{(p^2)^2} = |p^2| = p^2$ **(b)** $\sqrt[4]{p^4} = |p|$

(c) $\sqrt{16m^8r^6} = |4m^4r^3| = 4m^4|r^3|$

(d) $\sqrt[6]{(-2)^6} = |-2| = 2$

(e) $\sqrt[5]{m^5} = m$

(f) $\sqrt{(2k+3)^2} = |2k+3|$

(g) $\sqrt{x^2 - 4x + 4} = \sqrt{(x-2)^2} = |x-2|$ ■

NOTE When working with variable radicands, we usually assume that all expressions in radicands represent only nonnegative real numbers.

Rules for Radicals

The following three rules are the power rules for exponents written in radical notation.

Rules for Radicals

For all real numbers a and b, and positive integers m and n for which the indicated roots are real numbers,

$$\sqrt[n]{a} \cdot \sqrt[n]{b} = \sqrt[n]{ab} \qquad \sqrt[n]{\frac{a}{b}} = \frac{\sqrt[n]{a}}{\sqrt[n]{b}} \quad (b \neq 0) \qquad \sqrt[m]{\sqrt[n]{a}} = \sqrt[mn]{a}.$$

EXAMPLE 5 **Using the Rules for Radicals**

Simplify each expression. Assume that all variable expressions represent positive real numbers.

(a) $\sqrt{6} \cdot \sqrt{54}$ **(b)** $\sqrt[3]{m} \cdot \sqrt[3]{m^2}$ **(c)** $\sqrt{\frac{7}{64}}$

(d) $\sqrt[4]{\frac{a}{b^4}}$ **(e)** $\sqrt[7]{\sqrt[3]{2}}$ **(f)** $\sqrt[4]{\sqrt{3}}$

Solution

(a) $\sqrt{6} \cdot \sqrt{54} = \sqrt{6 \cdot 54} = \sqrt{324} = 18$

(b) $\sqrt[3]{m} \cdot \sqrt[3]{m^2} = \sqrt[3]{m^3} = m$

(c) $\sqrt{\frac{7}{64}} = \frac{\sqrt{7}}{\sqrt{64}} = \frac{\sqrt{7}}{8}$

(d) $\sqrt[4]{\frac{a}{b^4}} = \frac{\sqrt[4]{a}}{\sqrt[4]{b^4}} = \frac{\sqrt[4]{a}}{b}$

(e) $\sqrt[7]{\sqrt[3]{2}} = \sqrt[21]{2}$ Use the third rule.

(f) $\sqrt[4]{\sqrt{3}} = \sqrt[8]{3}$ ■

Simplifying Radicals

In work with fractions, it is customary to write a fraction in its simplest form. For example, $\frac{10}{2}$ is written as 5, $-\frac{9}{6}$ is written as $-\frac{3}{2}$, and $\frac{4}{16}$ is written as $\frac{1}{4}$. Similarly, expressions with radicals are often written in their simplest forms.

Simplified Radicals

An expression with radicals is simplified when the following conditions are satisfied.

1. The radicand has no factor raised to a power greater than or equal to the index.
2. The radicand has no fractions.
3. No denominator contains a radical.
4. Exponents in the radicand and the index of the radical have no common factor.
5. All indicated operations have been performed (if possible).

EXAMPLE 6 **Simplifying Radicals**

Simplify each radical.

(a) $\sqrt{175}$ **(b)** $-3\sqrt[5]{32}$ **(c)** $\sqrt{288m^5}$ **(d)** $\sqrt[3]{81x^5y^7z^6}$

Solution

(a) $\sqrt{175} = \sqrt{25 \cdot 7} = \sqrt{25} \cdot \sqrt{7} = 5\sqrt{7}$

(b) $-3\sqrt[5]{32} = -3\sqrt[5]{2^5} = -3 \cdot 2 = -6$

(c) $\sqrt{288m^5} = \sqrt{144m^4 \cdot 2m} = 12m^2\sqrt{2m}$

(d) $\sqrt[3]{81x^5y^7z^6} = \sqrt[3]{27 \cdot 3 \cdot x^3 \cdot x^2 \cdot y^6 \cdot y \cdot z^6}$ Factor.

$= \sqrt[3]{27x^3y^6z^6(3x^2y)}$ Group all perfect cubes.

$= 3xy^2z^2\sqrt[3]{3x^2y}$ Remove all perfect cubes from under the radical. ■

If the index of the radical and an exponent in the radicand have a common factor, the radical can be simplified by writing it in exponential form, simplifying the rational exponent, and then writing the result as a radical again.

EXAMPLE 7 **Simplifying Radicals by Writing Them with Rational Exponents**

Simplify each radical.

(a) $\sqrt[6]{3^2}$ **(b)** $\sqrt[6]{x^{12}y^3}$ $(y \geq 0)$ **(c)** $\sqrt[9]{\sqrt{6^3}}$

Solution

(a) $\sqrt[6]{3^2} = 3^{2/6} = 3^{1/3} = \sqrt[3]{3}$

(b) $\sqrt[6]{x^{12}y^3} = (x^{12}y^3)^{1/6} = x^2y^{3/6} = x^2y^{1/2} = x^2\sqrt{y}$

(c) $\sqrt[9]{\sqrt{6^3}} = \sqrt[9]{6^{3/2}} = (6^{3/2})^{1/9} = 6^{1/6} = \sqrt[6]{6}$ ■

In Example 7(a), we simplified $\sqrt[6]{3^2}$ to $\sqrt[3]{3}$. However, to simplify $\left(\sqrt[6]{x}\right)^2$, the variable x must be nonnegative. For example, suppose we write

$$(-8)^{2/6} = [(-8)^{1/6}]^2.$$

This result is not a real number, because $(-8)^{1/6}$ is not a real number. By contrast,

$$(-8)^{1/3} = -2.$$

Here, even though $\frac{2}{6} = \frac{1}{3}$,

$$\left(\sqrt[6]{x}\right)^2 \neq \sqrt[3]{x}.$$

If a is nonnegative, then $a^{m/n} = a^{mp/(np)}$ (for $p \neq 0$). Simplifying rational exponents on negative bases should be considered case by case.

Operations with Radicals

Radicals with the same radicand and the same index, such as $3\sqrt[4]{11pq}$ and $-7\sqrt[4]{11pq}$, are called **like radicals.** Examples of *unlike radicals* are

$2\sqrt{5}$ and $2\sqrt{3}$, Radicands are different.

as well as $2\sqrt{3}$ and $2\sqrt[3]{3}$. Indexes are different.

We add or subtract like radicals by using the distributive property. ***Only like radicals can be combined.*** Sometimes we need to simplify radicals before adding or subtracting.

EXAMPLE 8 Adding and Subtracting Like Radicals

Add or subtract, as indicated. Assume that all variables represent positive real numbers.

(a) $3\sqrt[4]{11pq} + \left(-7\sqrt[4]{11pq}\right)$ **(b)** $7\sqrt{2} - 8\sqrt{18} + 4\sqrt{72}$

(c) $\sqrt{98x^3y} + 3x\sqrt{32xy}$

Solution

(a) $3\sqrt[4]{11pq} + \left(-7\sqrt[4]{11pq}\right) = -4\sqrt[4]{11pq}$

(b)
$$\begin{aligned} 7\sqrt{2} - 8\sqrt{18} + 4\sqrt{72} &= 7\sqrt{2} - 8\sqrt{9 \cdot 2} + 4\sqrt{36 \cdot 2} && \text{Simplify radicals.} \\ &= 7\sqrt{2} - 8 \cdot 3\sqrt{2} + 4 \cdot 6\sqrt{2} \\ &= 7\sqrt{2} - 24\sqrt{2} + 24\sqrt{2} \\ &= 7\sqrt{2} && \text{Distributive property} \end{aligned}$$

(c)
$$\begin{aligned} \sqrt{98x^3y} + 3x\sqrt{32xy} &= \sqrt{49 \cdot 2 \cdot x^2 \cdot x \cdot y} + 3x\sqrt{16 \cdot 2 \cdot x \cdot y} \\ &= 7x\sqrt{2xy} + 3x(4)\sqrt{2xy} \\ &= 7x\sqrt{2xy} + 12x\sqrt{2xy} \\ &= 19x\sqrt{2xy} && \text{Distributive property} \end{aligned}$$

■

Multiplying radical expressions is much like multiplying polynomials.

EXAMPLE 9 Multiplying Radical Expressions

Find each product.

(a) $\left(\sqrt{2} + 3\right)\left(\sqrt{8} - 5\right)$ **(b)** $\left(\sqrt{7} - \sqrt{10}\right)\left(\sqrt{7} + \sqrt{10}\right)$

Solution

(a)
$$\begin{aligned} \left(\sqrt{2} + 3\right)\left(\sqrt{8} - 5\right) &= \sqrt{2}\left(\sqrt{8}\right) - \sqrt{2}(5) + 3\sqrt{8} - 3(5) && \text{FOIL} \\ &= \sqrt{16} - 5\sqrt{2} + 3\left(2\sqrt{2}\right) - 15 && \text{Multiply and simplify.} \\ &= 4 - 5\sqrt{2} + 6\sqrt{2} - 15 \\ &= -11 + \sqrt{2} && \text{Combine terms.} \end{aligned}$$

(b)
$$\begin{aligned} \left(\sqrt{7} - \sqrt{10}\right)\left(\sqrt{7} + \sqrt{10}\right) &= \left(\sqrt{7}\right)^2 - \left(\sqrt{10}\right)^2 && \text{Product of the sum and difference of two terms} \\ &= 7 - 10 \\ &= -3 \end{aligned}$$

■

Rationalizing Denominators

Condition 3 of the rules for simplifying radicals described earlier requires that no denominator contain a radical. The process of achieving this is called **rationalizing the denominator.** To rationalize a denominator, we multiply by a form of 1.

EXAMPLE 10 Rationalizing Denominators

Rationalize each denominator.

(a) $\dfrac{4}{\sqrt{3}}$ **(b)** $\dfrac{2}{\sqrt[3]{x}}$ $(x \neq 0)$

Solution

(a) $\dfrac{4}{\sqrt{3}} = \dfrac{4}{\sqrt{3}} \cdot \dfrac{\sqrt{3}}{\sqrt{3}} = \dfrac{4\sqrt{3}}{3}$ Multiply by $\dfrac{\sqrt{3}}{\sqrt{3}} = 1$.

(b) $\dfrac{2}{\sqrt[3]{x}} = \dfrac{2}{\sqrt[3]{x}} \cdot \dfrac{\sqrt[3]{x^2}}{\sqrt[3]{x^2}}$ $\dfrac{\sqrt[3]{x^2}}{\sqrt[3]{x^2}} = 1$

$= \dfrac{2\sqrt[3]{x^2}}{\sqrt[3]{x^3}} = \dfrac{2\sqrt[3]{x^2}}{x}$ $\sqrt[3]{x} \cdot \sqrt[3]{x^2} = \sqrt[3]{x^3} = x$ ■

In Example 9(b), we saw that

$$\left(\sqrt{7} - \sqrt{10}\right)\left(\sqrt{7} + \sqrt{10}\right) = -3,$$

a rational number. This result suggests a way to rationalize a denominator that is a binomial in which one or both terms is a radical. The expressions $a\sqrt{m} + b\sqrt{n}$ and $a\sqrt{m} - b\sqrt{n}$ are called **conjugates.**

EXAMPLE 11 Rationalizing a Binomial Denominator

Rationalize the denominator of $\dfrac{1}{1 - \sqrt{2}}$.

Solution The best approach is to multiply both numerator and denominator by the conjugate of the denominator—in this case, $1 + \sqrt{2}$.

$$\frac{1}{1 - \sqrt{2}} = \frac{1\left(1 + \sqrt{2}\right)}{\left(1 - \sqrt{2}\right)\left(1 + \sqrt{2}\right)} = \frac{1 + \sqrt{2}}{1 - 2} = -1 - \sqrt{2}$$ ■

R.5 Exercises

Concept Check *Match the rational exponent expression in Exercises 1–8 with the equivalent radical expression in A–H. Assume that $x \neq 0$.*

1. $(-3x)^{1/3}$ **2.** $-3x^{1/3}$ **3.** $(-3x)^{-1/3}$ **4.** $-3x^{-1/3}$ **5.** $(3x)^{1/3}$ **6.** $3x^{-1/3}$ **7.** $(3x)^{-1/3}$ **8.** $3x^{1/3}$

A. $\dfrac{3}{\sqrt[3]{x}}$ **B.** $-3\sqrt[3]{x}$ **C.** $\dfrac{1}{\sqrt[3]{3x}}$ **D.** $\dfrac{-3}{\sqrt[3]{x}}$ **E.** $3\sqrt[3]{x}$ **F.** $\sqrt[3]{-3x}$ **G.** $\sqrt[3]{3x}$ **H.** $\dfrac{1}{\sqrt[3]{-3x}}$

Write each expression in radical form. Assume that all variables represent positive real numbers.

9. $(-m)^{2/3}$ **10.** $p^{5/4}$ **11.** $(2m + p)^{2/3}$ **12.** $(5r + 3t)^{4/7}$

Write each expression with rational exponents. Assume that all variables represent nonnegative real numbers.

13. $\sqrt[5]{k^2}$ **14.** $-\sqrt[4]{z^5}$ **15.** $-3\sqrt{5p^3}$ **16.** $m\sqrt{2y^5}$

Concept Check *Answer each question.*

17. For which of the following cases is $\sqrt{ab} = \sqrt{a} \cdot \sqrt{b}$ a true statement?
A. a and b both positive **B.** a and b both negative

18. For what positive integers n greater than or equal to 2 is $\sqrt[n]{a^n} = a$ always a true statement?

19. For what values of x is $\sqrt{9ax^2} = 3x\sqrt{a}$ a true statement? Assume that $a \geq 0$.

20. Which of the following expressions is *not* simplified? Give the simplified form.

A. $\sqrt[3]{2y}$ **B.** $\frac{\sqrt{5}}{2}$ **C.** $\sqrt[4]{m^3}$ **D.** $\sqrt{\frac{3}{4}}$

Simplify each radical expression. Assume that all variables represent positive real numbers.

21. $\sqrt[3]{125}$ **22.** $\sqrt[4]{81}$ **23.** $\sqrt[5]{-3125}$ **24.** $\sqrt[3]{343}$ **25.** $\sqrt{50}$

26. $\sqrt{45}$ **27.** $\sqrt[3]{81}$ **28.** $\sqrt[3]{250}$ **29.** $-\sqrt[4]{32}$ **30.** $-\sqrt[4]{243}$

31. $-\sqrt{\frac{9}{5}}$ **32.** $-\sqrt[3]{\frac{3}{2}}$ **33.** $-\sqrt[3]{\frac{4}{5}}$ **34.** $\sqrt[4]{\frac{3}{2}}$ **35.** $\sqrt[3]{16(-2)^4(2)^8}$

36. $\sqrt[3]{25(3)^4(5)^3}$ **37.** $\sqrt{8x^5z^8}$ **38.** $\sqrt{24m^6n^5}$ **39.** $\sqrt[3]{16z^5x^8y^4}$ **40.** $-\sqrt[6]{64a^{12}b^8}$

41. $\sqrt[4]{m^2n^7p^8}$ **42.** $\sqrt[4]{x^8y^7z^9}$ **43.** $\sqrt[4]{x^4 + y^4}$ **44.** $\sqrt[3]{27 + a^3}$

45. $\sqrt{\frac{2}{3x}}$ **46.** $\sqrt{\frac{5}{3p}}$ **47.** $\sqrt{\frac{x^5y^3}{z^2}}$ **48.** $\sqrt{\frac{g^3h^5}{r^3}}$

49. $\sqrt[3]{\frac{8}{x^2}}$ **50.** $\sqrt[3]{\frac{9}{16p^4}}$ **51.** $\sqrt[4]{\frac{g^3h^5}{9r^6}}$ **52.** $\sqrt[4]{\frac{32x^5}{y^5}}$

53. $\frac{\sqrt[3]{mn} \cdot \sqrt[3]{m^2}}{\sqrt[3]{n^2}}$ **54.** $\frac{\sqrt[3]{8m^2n^3} \cdot \sqrt[3]{2m^2}}{\sqrt[3]{32m^4n^3}}$ **55.** $\frac{\sqrt[4]{32x^5y} \cdot \sqrt[4]{2xy^4}}{\sqrt[4]{4x^3y^2}}$ **56.** $\frac{\sqrt[4]{rs^2t^3} \cdot \sqrt[4]{r^3s^2t}}{\sqrt[4]{r^2t^3}}$

57. $\sqrt[3]{\sqrt{4}}$ **58.** $\sqrt[4]{\sqrt[3]{2}}$ **59.** $\sqrt[6]{\sqrt[3]{x}}$ **60.** $\sqrt[8]{\sqrt[4]{y}}$

Simplify each expression, assuming that all variables represent nonnegative numbers.

61. $4\sqrt{3} - 5\sqrt{12} + 3\sqrt{75}$ **62.** $2\sqrt{5} - 3\sqrt{20} + 2\sqrt{45}$ **63.** $3\sqrt{28p} - 4\sqrt{63p} + \sqrt{112p}$

64. $9\sqrt{8k} + 3\sqrt{18k} - \sqrt{32k}$ **65.** $2\sqrt[3]{3} + 4\sqrt[3]{24} - \sqrt[3]{81}$ **66.** $\sqrt[3]{32} - 5\sqrt[3]{4} + 2\sqrt[3]{108}$

67. $\frac{1}{\sqrt{3}} - \frac{2}{\sqrt{12}} + 2\sqrt{3}$ **68.** $\frac{1}{\sqrt{2}} + \frac{3}{\sqrt{8}} + \frac{1}{\sqrt{32}}$ **69.** $\frac{5}{\sqrt[3]{2}} - \frac{2}{\sqrt[3]{16}} + \frac{1}{\sqrt[3]{54}}$

70. $\frac{-4}{\sqrt[3]{3}} + \frac{1}{\sqrt[3]{24}} - \frac{2}{\sqrt[3]{81}}$ **71.** $(\sqrt{2} + 3)(\sqrt{2} - 3)$ **72.** $(\sqrt{5} + \sqrt{2})(\sqrt{5} - \sqrt{2})$

73. $(\sqrt[3]{11} - 1)(\sqrt[3]{11^2} + \sqrt[3]{11} + 1)$ **74.** $(\sqrt[3]{7} + 3)(\sqrt[3]{7^2} - 3\sqrt[3]{7} + 9)$ **75.** $(\sqrt{3} + \sqrt{8})^2$

76. $(\sqrt{2} - 1)^2$ **77.** $(3\sqrt{2} + \sqrt{3})(2\sqrt{3} - \sqrt{2})$ **78.** $(4\sqrt{5} - 1)(3\sqrt{5} + 2)$

Rationalize the denominator of each radical expression. Assume that all variables represent nonnegative numbers and that no denominators are 0.

79. $\dfrac{8}{\sqrt{5}}$ **80.** $\dfrac{3}{\sqrt{10}}$ **81.** $\dfrac{6}{\sqrt[3]{x^2}}$

82. $\dfrac{4}{\sqrt[3]{a^2}}$ **83.** $\dfrac{\sqrt{3}}{\sqrt{5}+\sqrt{3}}$ **84.** $\dfrac{\sqrt{7}}{\sqrt{3}-\sqrt{7}}$

85. $\dfrac{1+\sqrt{3}}{3\sqrt{5}+2\sqrt{3}}$ **86.** $\dfrac{\sqrt{7}-1}{2\sqrt{7}+4\sqrt{2}}$ **87.** $\dfrac{p}{\sqrt{p}+2}$

88. $\dfrac{\sqrt{r}}{3-\sqrt{r}}$ **89.** $\dfrac{a}{\sqrt{a+b}-1}$ **90.** $\dfrac{3m}{2+\sqrt{m+n}}$

CHAPTER R *Test*

1. Simplify each expression. Leave answers with exponents. Assume that all variables represent nonzero real numbers.

(a) $(-3)^4 \cdot (-3)^5$ **(b)** $(2x^3y)^2$ **(c)** $(-5z)^0$ **(d)** $\left(-\dfrac{4}{5}\right)^2$ **(e)** $-\left(\dfrac{m^2}{p}\right)^3$

2. Perform the indicated operations.

(a) $(x^2 - 3x + 2) - (x - 4x^2) + 3x(2x + 1)$ **(b)** $(6r - 5)^2$ **(c)** $(u + 2)(3u^2 - u + 4)$

(d) $(4x + 5)(4x - 5)$ **(e)** $[(5p - 1) + 4]^2$

3. Factor completely.

(a) $6x^2 - 17x + 7$ **(b)** $x^4 - 16$ **(c)** $z^2 - 6zk - 16k^2$ **(d)** $x^3y^2 - 9x^3 - 8y^2 + 72$

4. Write each rational expression in lowest terms, and give the domain.

(a) $\dfrac{16x^3}{4x^5}$ **(b)** $\dfrac{1+k}{k^2-1}$ **(c)** $\dfrac{x^2+x-2}{x^2+5x+6}$

5. Perform the indicated operations.

(a) $\dfrac{5x^2-9x-2}{30x^3+6x^2} \cdot \dfrac{2x^8+6x^7+4x^6}{x^4-3x^2-4}$ **(b)** $\dfrac{x}{x^2+3x+2} + \dfrac{2x}{2x^2-x-3}$

(c) $\dfrac{a+b}{2a-3} - \dfrac{a-b}{3-2a}$ **(d)** $\dfrac{g - \dfrac{2}{g}}{g - \dfrac{4}{g}}$

6. Simplify each expression. Assume that all variables represent positive real numbers.

(a) $(-7)^{-2}$ **(b)** $(16x^8)^{3/4}$ **(c)** $\left(\dfrac{64}{27}\right)^{-2/3}$

7. Perform the indicated operations. Write your answers with only positive exponents. Assume that all variables represent positive real numbers.

(a) $\dfrac{5^{-3} \cdot 5^{-1}}{5^{-2}}$ **(b)** $(x + 2)^{-1/5}(x + 2)^{-7/10}$ **(c)** $(m^{-1}n^{1/2})^4(4m^3n^{-2})^{-1/2}$

8. Simplify each expression. Assume that all variables represent positive real numbers.

(a) $\sqrt[4]{16}$ **(b)** $\sqrt[3]{\dfrac{2}{3}}$ **(c)** $\sqrt{18x^5y^8}$ **(d)** $\dfrac{\sqrt[4]{pq} \cdot \sqrt[4]{q^2}}{\sqrt[4]{p^3}}$

9. Simplify. Assume that all variables represent positive real numbers.

(a) $\sqrt{32x} + \sqrt{2x} - \sqrt{18x}$ **(b)** $(\sqrt[3]{2} + 4)(\sqrt[3]{4} + 1)$ **(c)** $(\sqrt{x} - \sqrt{y})(\sqrt{x} + \sqrt{y})$

10. Rationalize the denominator of $\dfrac{14}{\sqrt{11}-\sqrt{7}}$ and simplify.

Appendix A Geometry Formulas

The following table provides a summary of some important formulas from geometry.

Figure	Formula	Example
Square	Perimeter: $P = 4s$ Area: $A = s^2$	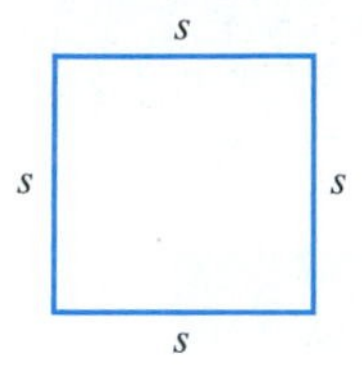
Rectangle	Perimeter: $P = 2L + 2W$ Area: $A = LW$	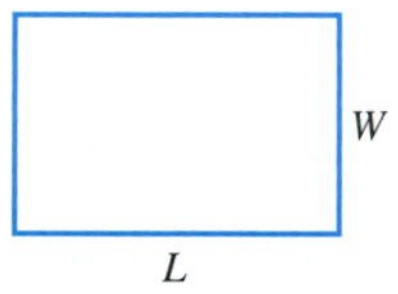
Triangle	Perimeter: $P = a + b + c$ Area: $A = \frac{1}{2}bh$	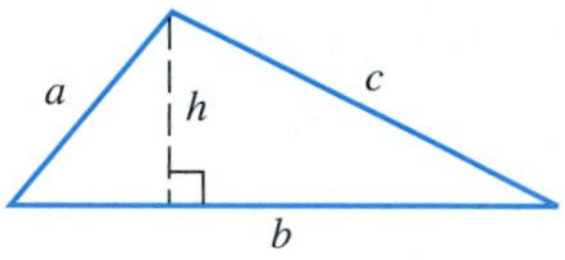
Pythagorean Theorem (for Right Triangles)	$c^2 = a^2 + b^2$	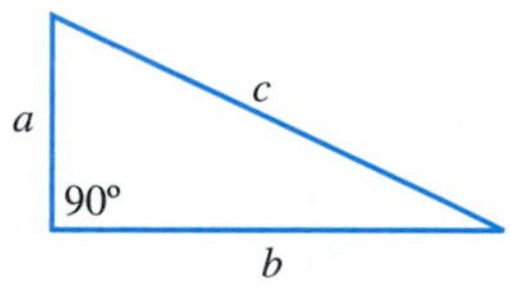
Sum of the Angles of a Triangle	$A + B + C = 180°$	
Circle	Diameter: $d = 2r$ Circumference: $C = 2\pi r = \pi d$ Area: $A = \pi r^2$	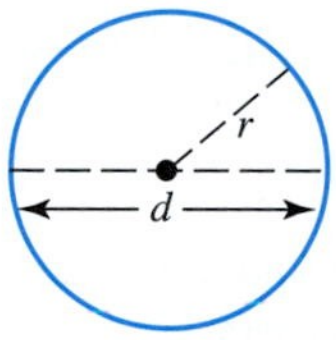
Parallelogram	Area: $A = bh$ Perimeter: $P = 2a + 2b$	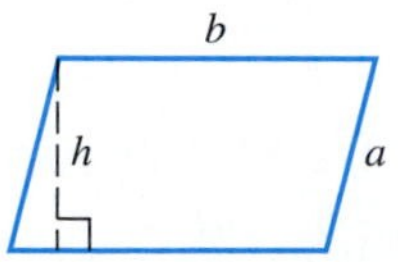

(continued)

Figure	Formula	Example
Trapezoid	Area: $A = \frac{1}{2}h(b_1 + b_2)$ Perimeter: $P = a + b_1 + c + b_2$	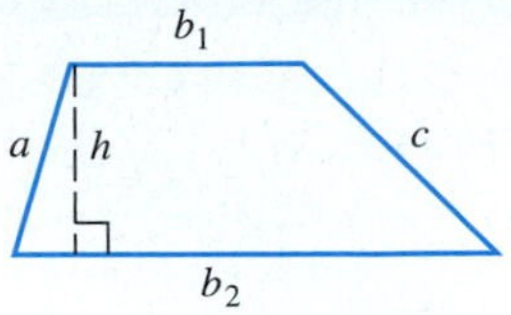 b_1, a, h, c, b_2
Sphere	Volume: $V = \frac{4}{3}\pi r^3$ Surface area: $S = 4\pi r^2$	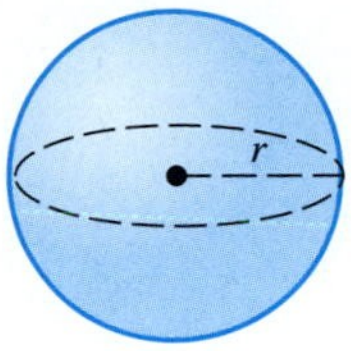 r
Cone	Volume: $V = \frac{1}{3}\pi r^2 h$ Surface area: $S = \pi r\sqrt{r^2 + h^2}$ (excludes the base)	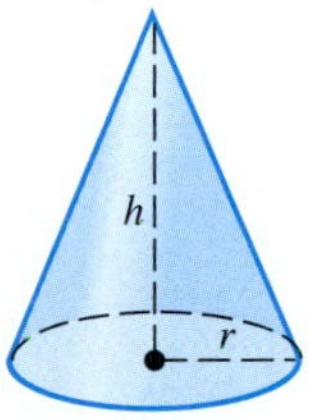 h, r
Cube	Volume: $V = e^3$ Surface area: $S = 6e^2$	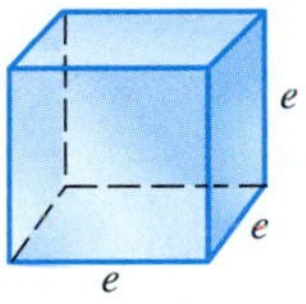 e, e, e
Rectangular Solid	Volume: $V = LWH$ Surface area: $S = 2HW + 2LW + 2LH$	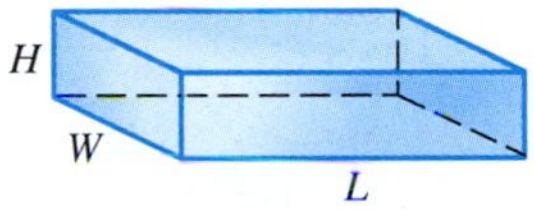 H, W, L
Right Circular Cylinder	Volume: $V = \pi r^2 h$ Surface area: $S = 2\pi rh + 2\pi r^2$ (includes top and bottom)	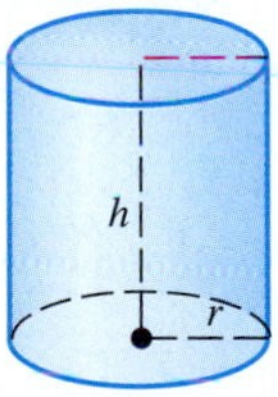 h, r
Right Pyramid	Volume: $V = \frac{1}{3}Bh$ B = area of the base	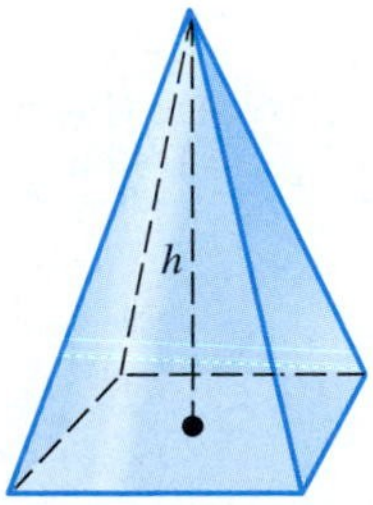 h

Appendix B Deciding Which Model Best Fits a Set of Data*

The Sum of the Squares of the Errors ■ Using a Graphing Calculator to Find the SSE ■ Choosing the Best Model from Different Types of Functions

In Chapters 1–5, we studied a variety of functions: linear, quadratic, polynomial, rational, power, root, exponential, logarithmic, and piecewise-defined. When presented with real data, it is unusual for the data to be modeled exactly by one of these functions. Earlier, we used the regression features of a graphing calculator to determine the best-fit model for a data set. But how do we know which type of function we should choose? And how will we know whether we have the best model from among the different types of functions available to us?

One way to find the best model for a set of data is to determine a measure that will allow us to compare one type of model with another. This can be done by using the **sum of the squares of the errors,** or **SSE.** By comparing the SSE of several potential models for a data set, we can determine which of the models is the best fit for the given data.

The Sum of the Squares of the Errors

When we find a function that models a set of real data, it is likely that the output values resulting from the model will not be exactly the same as the output values of the data set. The actual data point may be either higher or lower than the result produced by the algebraic model. The difference between the actual value and the model-predicted value is called the *error;* that is,

$$\text{Error} = \text{Actual data value} - \text{Model-predicted value.}$$

Error

If $M(x)$ is a model for a set of data and $M(x_i)$ is the output value at x_i, then the **error** produced by the model for input value x_i and actual output value a_i is given by

$$\mathbf{E_i = a_i - M(x_i).}$$

EXAMPLE 1 Finding the SSE

The average hourly wage for a skilled laborer since the year 1970 is given in the table. A linear model for these data is given by

$$M(x) = \frac{1}{2}x + 10,$$

where $x = 0$ represents the year 1960. Graph the actual data and the model-predicted data on the same set of coordinate axes. Then find the sum of the squares of errors for this model.

Year x	Average Hourly Wage in Dollars, $a(x)$
1970	15
1980	20
1990	24
2000	31

*The authors thank Sandee House at Georgia Perimeter Community College for the material in this appendix.

Solution Since $x = 0$ corresponds to the year 1960 and the first year of data is 1970, the first x value is 10, followed by 20, 30, and 40, as shown in the first column of the table below. The actual data values are given in the column headed $a(x)$. To find the model values, $M(x)$, use the given model,

$$M(x) = \frac{1}{2}x + 10.$$

For example, for $x = 10$, the model produces the output value of 15, since

$$M(10) = \frac{1}{2}(10) + 10 = 5 + 10 = 15.$$

To calculate the error produced by the model, find the difference between the actual output value $a(x)$ and the model value $M(x)$.

x	$a(x)$	Model Values $M(x)$	Error $a(x) - M(x)$	Comparison
10	15	15	$15 - 15 = 0$	Model produces the exact value.
20	20	20	$20 - 20 = 0$	Model produces the exact value.
30	24	25	$24 - 25 = -1$	Actual value is one less than the model-predicted value.
40	31	30	$31 - 30 = 1$	Actual value is one more than the model-predicted value.

The graph in Figure 1 shows the actual data and the model-predicted data.

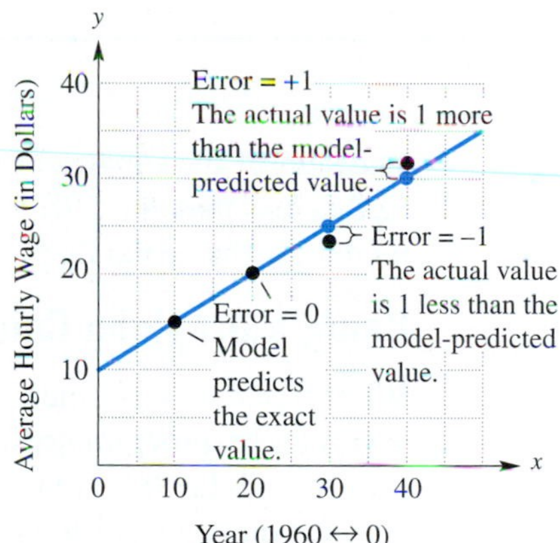

FIGURE 1

If we add the errors produced by the model, $0 + 0 + (-1) + 1$, our resulting value is 0. One might say that the model is a perfect match for the actual data. From the graph, however, this model does not always predict the exact output value. For that reason, simply adding up the errors does not give a good indication of the overall error of the model. To adjust for this effect, we find the *square* of each of the errors and then add to find the **sum of the squares of the errors.**

x	$a(x)$	Model Values $M(x) = \frac{1}{2}x + 10$	Error $a(x) - M(x)$	Error²
10	15	15	0	0
20	20	20	0	0
30	24	25	−1	1
40	31	30	1	1
		Sum	0	2 ← Sum of the squares of the errors

Therefore, the sum of the squares of the errors for this model is 2.

When using a graphing calculator to find the least-squares regression line, we are finding the model that minimizes the SSE, resulting in the line of best fit for the data—hence the name "least-squares regression line." ■

EXAMPLE 2 **Finding the SSE**

Use a graphing calculator to find the equation of the least-squares regression line for the data in Example 1. Find the SSE for the least-squares regression line, and compare the SSE with that of the model given in Example 1.

Solution Using a graphing calculator and entering the values for x and $a(x)$, we find that the equation of the least-squares regression line for these data is $y = .52x + 9.5$.

x	$a(x)$	$M(x) = .52x + 9.5$	Error	Error²
10	15	14.7	$15 - 14.7 = .3$	.09
20	20	19.9	$20 - 19.9 = .1$	.01
30	24	25.1	$24 - 25.1 = -1.1$	1.21
40	31	30.3	$31 - 30.3 = .7$	.49
				1.8 ← Sum of the squares of the errors

The sum of the squares of the errors for the least-squares regression line is 1.8. This SSE is less than the SSE for the model $M(x) = \frac{1}{2}x + 10$, meaning that this is a better model for the actual data. ■

GCM Using a Graphing Calculator to Find the SSE

With large data sets, finding the SSE can be tedious work. A graphing calculator, however, can be programmed to find the SSE. The steps for programming the TI-83/84 Plus to find the SSE are given in the graphing calculator manual that accompanies this text. If you have a different model, you may need to refer to the manual for that particular model, since there may be minor keystroke differences.

After programming is completed, the TI-83/84 Plus screen should look like the one in Figure 2.

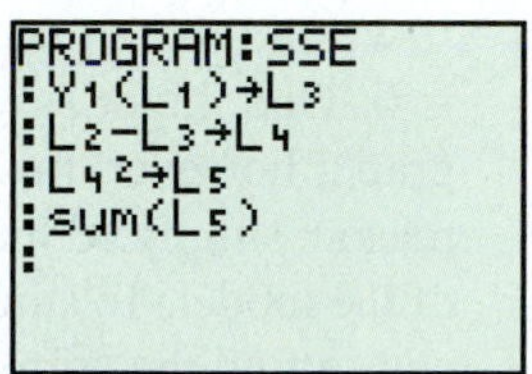

FIGURE 2

NOTE In order for this program to work properly, the following *must* be done *before* executing the program.

1. Your data *must* be stored in List 1 and List 2.
2. Your model *must* be stored in Y_1.

If the data or model is stored elsewhere, this program must be modified to work properly.

GCM **EXAMPLE 3** **Using the SSE Program**

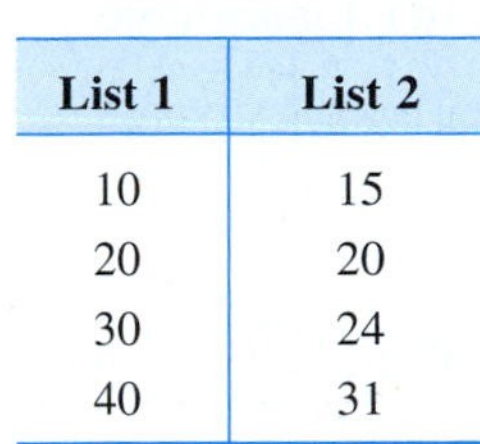

List 1	List 2
10	15
20	20
30	24
40	31

Use the SSE program to find the sum of the squares of the errors for the given model and the least-squares regression line from Examples 1 and 2.

Solution For the model $M(x) = \frac{1}{2}x + 10$, follow these steps.

Step 1 Verify that the data for x and $a(x)$ are stored in List 1 and List 2.

Step 2 Enter the model, $M(x) = \frac{1}{2}x + 10$, in Y_1.

Step 3 Execute the program. The screen should look like the one in Figure 3, indicating that the sum of the squares of the errors is 2, just as we found manually in Example 1.

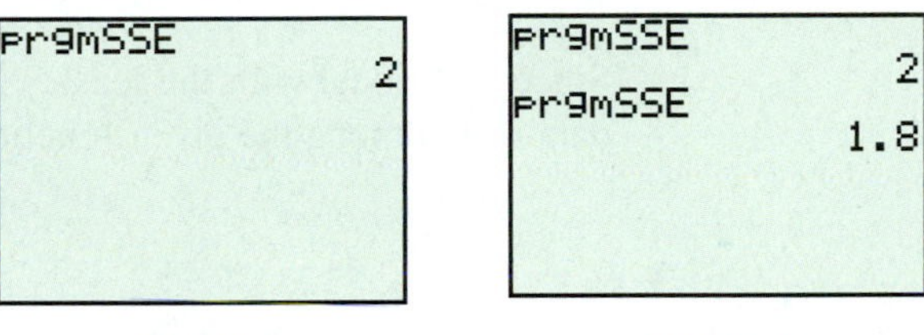

FIGURE 3 FIGURE 4

Follow the same procedure for the least-squares regression model found in Example 2, $M(x) = .52x + 9.5$.

Step 1 Verify that the data for x and $a(x)$ are stored in List 1 and List 2.

Step 2 Enter the new model, $M(x) = .52x + 9.5$, in Y_1.

Step 3 Execute the program. The screen should look like the one in Figure 4, indicating that the sum of the squares of the errors is 1.8, just as we found manually in Example 2. ■

Choosing the Best Model from Different Types of Functions

We often look at the graph of a set of data, such as those shown in Figure 5, and say, "It looks linear" or "It looks exponential." But how can we really tell which type of function best fits a set of data?

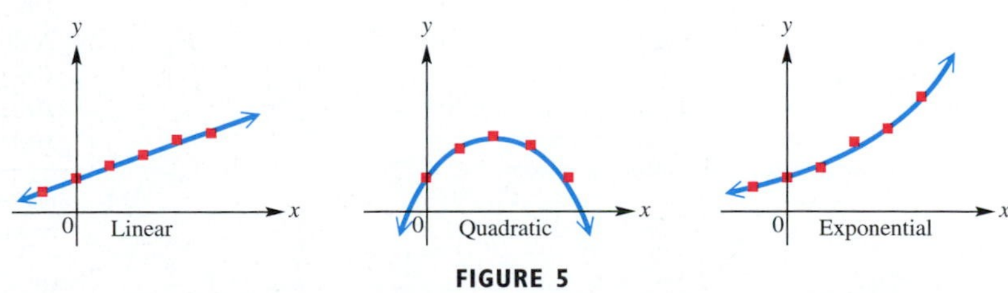

FIGURE 5

Year	Debt	Year	Debt
1995	4.974	2000	5.674
1996	5.225	2001	5.807
1997	5.413	2002	6.228
1998	5.526	2003	6.783
1999	5.656	2004	7.379

Source: World Almanac and Book of Facts 2005.

One way to determine the model of best fit is to find several different regression models and then compare the results of the SSE program for these models. The model with the least SSE is the best fit for that particular set of data.

EXAMPLE 4 Comparing SSE Values

The table shows the national debt (in trillions of dollars) from 1995 to 2004. Using $x = 0$ to represent 1990, find the following models for the data.

(a) linear **(b)** quadratic **(c)** exponential **(d)** logarithmic

Then find the SSE and determine which one is the best model for the given data.

Solution

	Model	Model Form	Rounded Model (to Six Decimal Places)	SSE
(a)	Linear	$y = ax + b$	$y = .227194x + 3.708158$	.614489
(b)	Quadratic	$y = ax^2 + bx + c$	$y = .027174x^2 - .289117x + 5.936445$	.224593
(c)	Exponential	$y = ab^x$	$y = 4.071018(1.038478)^x$	.505167
(d)	Logarithmic	$y = a + b \ln x$	$y = 1.645626 + 1.917431 \ln x$	1.007647

Since the model with the least SSE is the quadratic model, it is the best fit for these data from among the given functions. ■

Appendix B Exercises

For each data set, find the models as indicated and then use the SSE to answer the question.

1. ***Death Rate*** The number of deaths per 1000 people in the United States for the years 1993 to 2002 are given in the table. Using $x = 0$ to represent 1990, find linear and exponential regression models for the data. Based on the SSE for each model, which is the better fit?

Year	Number of Deaths Per 1000
1993	15.4
1994	15.0
1995	14.6
1996	14.4
1997	14.2
1998	14.3
1999	14.2
2000	14.4
2001	14.1
2002	13.9

Source: World Almanac and Book of Facts 2005.

2. ***Average Earnings*** The average hourly earnings of U.S. production workers for the years 1993 to 2003 are given in the table. Using $x = 0$ to represent 1990, find linear and natural logarithmic regression models for the data. Based on the SSE for each model, which is the better fit?

Year	Average Hourly Earnings
1993	\$11.03
1994	\$11.32
1995	\$11.64
1996	\$12.03
1997	\$12.49
1998	\$13.00
1999	\$13.47
2000	\$14.00
2001	\$14.53
2002	\$14.95
2003	\$15.35

Source: Bureau of Labor Statistics, U.S. Department of Labor.

3. ***Laboratory Cleanups*** The table shows the number of clandestine methamphetamine laboratory cleanups in Georgia from 1999 to 2004. Using $x = 0$ to represent 1995, find linear and quadratic regression models for the data. Based on the SSE for each model, which is the better fit?

Year	Lab Cleanups
1999	34
2000	86
2001	221
2002	392
2003	437
2004	489

Source: Atlanta Journal Constitution.

4. ***CNN Headline News Viewership*** The table shows the number of viewers in thousands for CNN Headline News for the years 1996 to 2000. Using $x = 0$ to represent 1995, find linear and natural logarithmic regression models for the data. Based on the SSE for each model, which is the better fit?

Year	Viewers
1996	189
1997	171
1998	154
1999	151
2000	141

Source: Atlanta Journal Constitution.

5. ***Women on Active Duty*** The table shows the percent of women on active duty in the U.S. Armed Forces in selected years from 1973 to 2003. Using $x = 0$ to represent 1970, find quadratic and natural logarithmic regression models for the data. Based on the SSE for each model, which is the better fit?

Year	Percent of Women in U.S. Armed Forces
1973	2.5
1975	4.6
1981	8.9
1987	10.2
1993	11.6
1997	13.6
2000	14.4
2003	14.9

Source: World Almanac and Book of Facts 2005.

Appendix C Vectors in Space

Rectangular Coordinates in Space ■ Vectors in Space ■ Vector Definitions and Operations ■ Direction Angles in Space ■ An Application of the Dot Product

Rectangular Coordinates in Space

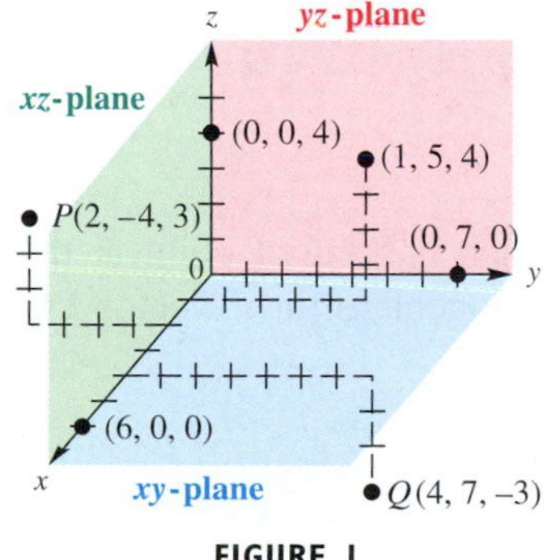

FIGURE 1

Vector quantities occur in space as well as in a plane, so properties for vectors in a two-dimensional plane can be extended to three-dimensional space. The plane determined by the x- and y-axes is called the ***xy*-plane.** A third axis is needed to locate points in space; the z-axis goes through the origin in the xy-plane and is perpendicular to both the x-axis and the y-axis. Figure 1 shows one way to orient the three axes, with the yz-plane as the plane of the page and the x-axis perpendicular to it.

We associate each point in space with an **ordered triple** (x, y, z). Figure 1 shows several ordered triples. The region of three-dimensional space where all coordinates are positive is called the **first octant.**

The distance formula given in Chapter 1 is a special case of the distance formula for points in space.

Distance Formula (Three-Dimensional Space)

If $P_1 = (x_1, y_1, z_1)$ and $P_2 = (x_2, y_2, z_2)$ are two points in a three-dimensional coordinate system, then the distance between P_1 and P_2 is given by

$$d(P_1, P_2) = \sqrt{(x_2 - x_1)^2 + (y_2 - y_1)^2 + (z_2 - z_1)^2}.$$

EXAMPLE 1 Finding the Distance between Two Points in Space

Find the distance between the points $P(2, -4, 3)$ and $Q(4, 7, -3)$ shown in Figure 1.

Solution By the distance formula,

$$d(P, Q) = \sqrt{(4-2)^2 + [7-(-4)]^2 + (-3-3)^2} = \sqrt{161}.$$ ■

Looking Ahead to Calculus

The concepts introduced in this section are extended in calculus to define a *vector-valued function:* a function whose domain is a set of real numbers and whose range is a set of vectors. For example,

$$\mathbf{R}(t) = f_1(t)\mathbf{i} + f_2(t)\mathbf{j} + f_3(t)\mathbf{k}$$

is a vector-valued function.

Vectors in Space

Extending the notation used to represent vectors in the plane, we denote a vector **v** in space with initial point at the origin as

$$\mathbf{v} = \langle a, b, c \rangle. \quad \text{Component form}$$

Using the unit vectors $\mathbf{i} = \langle 1, 0, 0 \rangle$, $\mathbf{j} = \langle 0, 1, 0 \rangle$, and $\mathbf{k} = \langle 0, 0, 1 \rangle$ in the directions of the positive x-axis, y-axis, and z-axis, respectively, we can also represent **v** as

$$\mathbf{v} = a\mathbf{i} + b\mathbf{j} + c\mathbf{k}, \quad \mathbf{i}, \mathbf{j}, \mathbf{k} \text{ form}$$

where a, b, and c are scalars. The scalars a, b, and c are the **components** of vector **v**.

EXAMPLE 2 Writing Vectors in i, j, k Form

Write the vectors **OP** and **OQ**, using the components for points P and Q defined in Example 1 and vectors **i**, **j**, and **k**.

Solution Since $P = (2, -4, 3)$, the components of **OP** are 2, -4, and 3. Similarly, the components of **OQ** are 4, 7, and -3. Then

$$\mathbf{OP} = 2\mathbf{i} - 4\mathbf{j} + 3\mathbf{k} \quad \text{and} \quad \mathbf{OQ} = 4\mathbf{i} + 7\mathbf{j} - 3\mathbf{k}.$$ ■

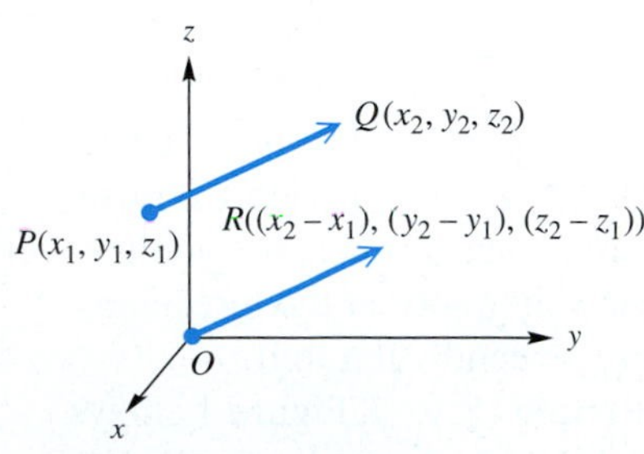

FIGURE 2

Vectors **OP** and **OQ** are position vectors because each has initial point at the origin. For $P = (x_1, y_1, z_1)$ and $Q = (x_2, y_2, z_2)$, the component form of vector **PQ** (which is not a position vector) is represented as

$$\mathbf{PQ} = \langle x_2 - x_1, y_2 - y_1, z_2 - z_1 \rangle.$$

As Figure 2 suggests, **PQ** is equal to the position vector

$$\mathbf{OR} = (x_2 - x_1)\mathbf{i} + (y_2 - y_1)\mathbf{j} + (z_2 - z_1)\mathbf{k}.$$

EXAMPLE 3 Finding a Position Vector That Corresponds to a Given Vector

Find the position vector corresponding to **PQ** with $P = (2, -4, 3)$ and $Q = (4, 7, -3)$.

Solution From the previous definition, the position vector is

$$\begin{aligned}\mathbf{v} &= (4 - 2)\mathbf{i} + [7 - (-4)]\mathbf{j} + (-3 - 3)\mathbf{k} \\ &= 2\mathbf{i} + 11\mathbf{j} - 6\mathbf{k}.\end{aligned}$$

Vector Definitions and Operations

The vector concepts discussed in Section 10.3 can be extended to vectors in space.

Vector Definitions and Operations

If $\mathbf{v} = a\mathbf{i} + b\mathbf{j} + c\mathbf{k}$ and $\mathbf{w} = d\mathbf{i} + e\mathbf{j} + f\mathbf{k}$ are vectors and g is a scalar, then

$$\mathbf{v} = \mathbf{w} \text{ if and only if } a = d,\ b = e, \text{ and } c = f,$$

$$\mathbf{v} + \mathbf{w} = (a + d)\mathbf{i} + (b + e)\mathbf{j} + (c + f)\mathbf{k},$$

$$\mathbf{v} - \mathbf{w} = (a - d)\mathbf{i} + (b - e)\mathbf{j} + (c - f)\mathbf{k},$$

$$g\mathbf{v} = ga\mathbf{i} + gb\mathbf{j} + gc\mathbf{k},$$

$$|\mathbf{v}| = \sqrt{a^2 + b^2 + c^2},$$

and

$$\mathbf{v} \cdot \mathbf{w} = ad + be + cf.$$

EXAMPLE 4 Performing Vector Operations

Find the following if $\mathbf{v} = 2\mathbf{i} + 6\mathbf{j} - 4\mathbf{k}$ and $\mathbf{w} = -\mathbf{i} + 5\mathbf{k}$.

(a) $\mathbf{v} + \mathbf{w}$ **(b)** $\mathbf{w} - \mathbf{v}$ **(c)** $-10\mathbf{w}$

(d) $|\mathbf{v}|$ **(e)** $\mathbf{v} \cdot \mathbf{w}$ **(f)** $\mathbf{w} \cdot \mathbf{v}$

Solution

(a) $\mathbf{v} + \mathbf{w} = (2 - 1)\mathbf{i} + (6 + 0)\mathbf{j} + (-4 + 5)\mathbf{k} = \mathbf{i} + 6\mathbf{j} + \mathbf{k}$

(b) $\mathbf{w} - \mathbf{v} = (-1 - 2)\mathbf{i} + (0 - 6)\mathbf{j} + [5 - (-4)]\mathbf{k} = -3\mathbf{i} - 6\mathbf{j} + 9\mathbf{k}$

(c) $-10\mathbf{w} = -10(-1)\mathbf{i} + (-10)5\mathbf{k} = 10\mathbf{i} - 50\mathbf{k}$

(d) $|\mathbf{v}| = \sqrt{2^2 + 6^2 + (-4)^2} = \sqrt{56} = 2\sqrt{14}$

(e) $\mathbf{v} \cdot \mathbf{w} = 2(-1) + 6(0) + (-4)5 = -22$

(f) $\mathbf{w} \cdot \mathbf{v} = -1(2) + 0(6) + 5(-4) = -22$

Properties of the dot product for vectors in two dimensions are extended to vectors in three dimensions, as is the geometric interpretation of the dot product.

If θ is the angle between two nonzero vectors **v** and **w** in space, where $0° \le \theta \le 180°$, then $\mathbf{v} \cdot \mathbf{w} = |\mathbf{v}||\mathbf{w}| \cos \theta$.

Dividing each side of the equation by $|\mathbf{v}||\mathbf{w}|$ gives the following result.

Angle between Two Vectors

If θ is the angle between two nonzero vectors $\mathbf{v}$ and $\mathbf{w}$, where $0° \leq \theta \leq 180°$, then

$$\cos \theta = \frac{\mathbf{v} \cdot \mathbf{w}}{|\mathbf{v}||\mathbf{w}|}.$$

EXAMPLE 5 Finding the Angle between Two Vectors

Find the angle between $\mathbf{v} = 7\mathbf{i} - 3\mathbf{j} - 5\mathbf{k}$ and $\mathbf{w} = 4\mathbf{i} + 6\mathbf{j} - 8\mathbf{k}$.

Solution First find $|\mathbf{v}|$ and $|\mathbf{w}|$.

$$|\mathbf{v}| = \sqrt{7^2 + (-3)^2 + (-5)^2} = \sqrt{83}$$

$$|\mathbf{w}| = \sqrt{4^2 + 6^2 + (-8)^2} = \sqrt{116} = 2\sqrt{29}$$

Now, use the formula for $\cos \theta$.

$$\cos \theta = \frac{\mathbf{v} \cdot \mathbf{w}}{|\mathbf{v}||\mathbf{w}|} = \frac{7(4) + (-3)6 + (-5)(-8)}{\sqrt{83} \cdot 2\sqrt{29}} \approx .5096$$

A calculator gives $\theta \approx 59.4°$. ■

Direction Angles in Space

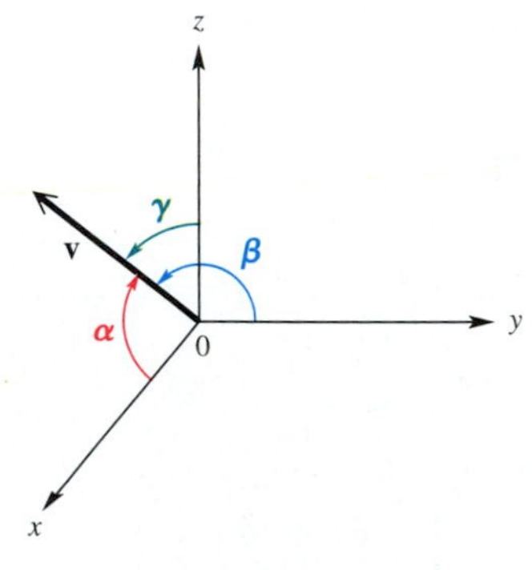

FIGURE 3

A vector in two dimensions is determined by its magnitude and direction angle. In three dimensions, a vector is determined by its magnitude and three direction angles—that is, the angles between the vector and each positive axis. As shown in Figure 3,

α is the direction angle between $\mathbf{v}$ and the positive x-axis,

β is the direction angle between $\mathbf{v}$ and the positive y-axis,

and γ is the direction angle between $\mathbf{v}$ and the positive z-axis.

We evaluate these angles using the expression for the cosine of the angle between two vectors. Recall that $\mathbf{i} = \langle 1, 0, 0 \rangle$, $\mathbf{j} = \langle 0, 1, 0 \rangle$, $\mathbf{k} = \langle 0, 0, 1 \rangle$, and each has magnitude 1. For $\mathbf{v} = a\mathbf{i} + b\mathbf{j} + c\mathbf{k}$, the three direction angles are determined as follows.

Direction Angles

$$\cos \alpha = \frac{\mathbf{v} \cdot \mathbf{i}}{|\mathbf{v}||\mathbf{i}|} = \frac{a}{|\mathbf{v}|}, \quad \cos \beta = \frac{\mathbf{v} \cdot \mathbf{j}}{|\mathbf{v}||\mathbf{j}|} = \frac{b}{|\mathbf{v}|}, \quad \cos \gamma = \frac{\mathbf{v} \cdot \mathbf{k}}{|\mathbf{v}||\mathbf{k}|} = \frac{c}{|\mathbf{v}|}.$$

The quantities $\cos \alpha$, $\cos \beta$, and $\cos \gamma$ are called **direction cosines.** In Exercise 49, you are asked to prove that

$$\cos^2 \alpha + \cos^2 \beta + \cos^2 \gamma = 1.$$

EXAMPLE 6 Finding Direction Angles of a Vector

Find the direction angles of $\mathbf{w} = 2\mathbf{i} - 3\mathbf{j} + 6\mathbf{k}$. Give answers in degrees.

Solution

$$|\mathbf{w}| = \sqrt{4 + 9 + 36} = 7$$

$$\cos \alpha = \frac{a}{|\mathbf{w}|} = \frac{2}{7} \qquad \cos \beta = \frac{b}{|\mathbf{w}|} = -\frac{3}{7} \qquad \cos \gamma = \frac{c}{|\mathbf{w}|} = \frac{6}{7}$$

$$\alpha \approx 73.4° \qquad \beta \approx 115.4° \qquad \gamma \approx 31.0°$$

EXAMPLE 7 Verifying the Sum of the Squares of the Direction Angle Cosines

Verify that $\cos^2 \alpha + \cos^2 \beta + \cos^2 \gamma = 1$ for the vector in Example 6.

Solution

$$\left(\frac{2}{7}\right)^2 + \left(-\frac{3}{7}\right)^2 + \left(\frac{6}{7}\right)^2 = \frac{4}{49} + \frac{9}{49} + \frac{36}{49} = 1$$

EXAMPLE 8 Finding a Direction Angle

Figure 4 shows that the angle between a vector **u** and the positive z-axis is 120°. The angle between **u** and the positive x-axis is 90°. What acute angle does **u** make with the positive y-axis?

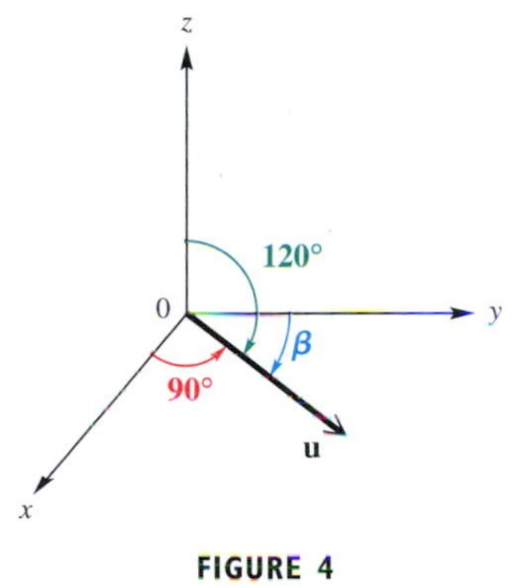

FIGURE 4

Solution Here, $\cos \gamma = \cos 120° = -\frac{1}{2}$ and $\cos \alpha = \cos 90° = 0$.

$$\cos^2 \alpha + \cos^2 \beta + \cos^2 \gamma = 1$$

$$\cos^2 \beta = 1 - \cos^2 \alpha - \cos^2 \gamma \quad \text{Subtract } \cos^2 \alpha \text{ and } \cos^2 \gamma.$$

$$\cos^2 \beta = 1 - 0 - \frac{1}{4} \quad \text{Substitute; } \left(-\tfrac{1}{2}\right)^2 = \tfrac{1}{4}.$$

$$\cos^2 \beta = \frac{3}{4} \quad \text{Subtract.}$$

$$\cos \beta = \frac{\sqrt{3}}{2} \quad \text{Take square roots; } 0° < \beta < 90°.$$

$$\beta = 30°$$

An Application of the Dot Product

The **work** done by a constant force **F** as it moves a particle from point P to point Q is defined as $\mathbf{F} \cdot \mathbf{PQ}$.

EXAMPLE 9 Applying the Dot Product to Work

Find the work (in units) done by a force $\mathbf{F} = \langle 3, 4, 2 \rangle$ that moves a particle from $P(1, 0, 5)$ to $Q(3, 3, 8)$.

Solution Since P is the initial point and Q is the terminal point,

$$\mathbf{PQ} = \langle 3 - 1, 3 - 0, 8 - 5 \rangle = \langle 2, 3, 3 \rangle.$$

Then the work done is

$$\mathbf{F} \cdot \mathbf{PQ} = \langle 3, 4, 2 \rangle \cdot \langle 2, 3, 3 \rangle$$
$$= 6 + 12 + 6 = 24 \text{ (work units).}$$

Appendix C *Exercises*

Concept Check *Fill in the blanks to complete each statement.*

1. The plane determined by the x-axis and the z-axis is called the __________.

2. A position vector has its __________ point at the origin.

3. The component form of the position vector with terminal point $(5, 3, -2)$ is __________.

4. The **i**, **j**, **k** form of the position vector with terminal point $(6, -1, -3)$ is __________.

Find the distance between the points P and Q.

5. $P = (0, 0, 0)$; $Q = (2, -2, 5)$

6. $P = (0, 0, 0)$; $Q = (7, 4, -1)$

7. $P = (10, 15, 9)$; $Q = (8, 3, -4)$

8. $P = (5, 4, -4)$; $Q = (3, 7, 2)$

9. $P = (20, 25, 16)$; $Q = (5, 5, 6)$

10. $P = (14, 10, 18)$; $Q = (-2, 4, 9)$

For each pair of points, find the position vector that corresponds to **PQ** *in component form and in* **i**, **j**, **k** *form.*

11. $P = (0, 0, 0)$; $Q = (2, -2, 5)$

12. $P = (0, 0, 0)$; $Q = (7, 4, -1)$

13. $P = (10, 15, 0)$; $Q = (8, 3, -4)$

14. $P = (0, 4, -4)$; $Q = (3, 7, 2)$

15. $P = (20, 25, 6)$; $Q = (5, 5, 16)$

16. $P = (14, 10, 18)$; $Q = (-2, 4, 9)$

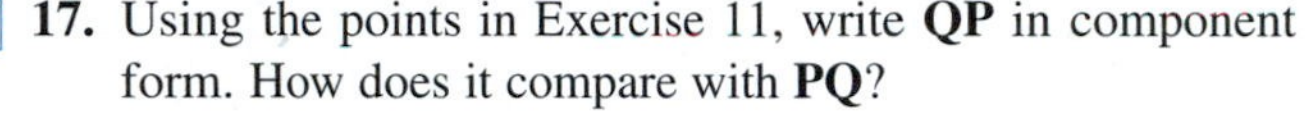

17. Using the points in Exercise 11, write **QP** in component form. How does it compare with **PQ**?

18. Compare the distance between points P and Q in three-dimensional space with the magnitude of vector **PQ**.

Given $\mathbf{u} = 2\mathbf{i} + 4\mathbf{j} + 7\mathbf{k}$, $\mathbf{v} = -3\mathbf{i} + 5\mathbf{j} + 2\mathbf{k}$, *and* $\mathbf{w} = 4\mathbf{i} - 3\mathbf{j} - 6\mathbf{k}$, *find the following.*

19. $\mathbf{u} - \mathbf{w}$

20. $\mathbf{v} + \mathbf{w}$

21. $4\mathbf{u} + 5\mathbf{v}$

22. $-\mathbf{v} + 3\mathbf{u}$

23. $|\mathbf{u}|$

24. $|\mathbf{w}|$

25. $|\mathbf{w} + \mathbf{u}|$

26. $|2\mathbf{v}|$

27. $\mathbf{v} \cdot \mathbf{w}$

28. $\mathbf{u} \cdot \mathbf{w}$

29. $\mathbf{v} \cdot \mathbf{v}$

30. $\mathbf{u} \cdot \mathbf{u}$

Find the angle between each pair of vectors.

31. $\langle 2, -2, 0 \rangle, \langle 5, -2, -1 \rangle$

32. $\langle 4, 0, 0 \rangle, \langle 5, 3, -2 \rangle$

33. $\langle 6, 0, 0 \rangle, \langle 8, 3, -4 \rangle$

34. $\langle -1, 2, -3 \rangle, \langle 0, -2, 1 \rangle$

35. $\langle 1, 0, 0 \rangle, \langle 0, 1, 0 \rangle$

36. $\langle 0, 0, 1 \rangle, \langle 0, 1, 0 \rangle$

Find the direction angles of each vector. Give answers in degrees. Check your work by showing that the sum of the squares of the cosines equals 1.

37. $\mathbf{u} = 2\mathbf{i} + 4\mathbf{j} + 7\mathbf{k}$

38. $\mathbf{v} = -3\mathbf{i} + 5\mathbf{j} + 2\mathbf{k}$

39. $\mathbf{w} = 4\mathbf{i} - 3\mathbf{j} - 6\mathbf{k}$

40. $\mathbf{y} = 2\mathbf{i} - 3\mathbf{j} + 4\mathbf{k}$

Work each problem.

41. The angle between vector **u** and the positive x-axis is 45°. The angle between **u** and the positive y-axis is 120°. Find the angle between **u** and the positive z-axis.

42. The direction angle between vector **v** and the y-axis is 135°. The direction angle between **v** and the z-axis is 90°. What is the direction angle between **v** and the x-axis?

43. What must be true for two vectors in space to be parallel?

44. Decide whether the vectors $\langle 3, 5, -1 \rangle$ and $\langle -12, -20, 4 \rangle$ are parallel. Explain your answer.

Work *Find the work (in units) done by a force* **F** *in moving a particle from P to Q.*

45. $\mathbf{F} = \langle 2, 0, 5 \rangle$; $P = (0, 0, 0)$; $Q = (1, 3, 2)$

46. $\mathbf{F} = \langle 3, 2, 0 \rangle$; $P = (0, 0, 0)$; $Q = (4, -2, 5)$

47. $\mathbf{F} = \mathbf{i} + 2\mathbf{j} - \mathbf{k}$; $P = (2, -1, 2)$; $Q = (5, 7, 8)$

48. $\mathbf{F} = 3\mathbf{i} + 2\mathbf{j}$; $P = (4, 7, 6)$; $Q = (10, 15, 12)$

49. Prove that, for direction cosines $\cos \alpha$, $\cos \beta$, and $\cos \gamma$, $\cos^2 \alpha + \cos^2 \beta + \cos^2 \gamma = 1$.

Appendix D Polar Form of Conic Sections

Up to this point, we have worked with equations of conic sections in rectangular form. If the focus of a conic section is at the pole, the polar form of its equation is

$$r = \frac{ep}{1 \pm e \cdot f(\theta)},$$

where f is the sine or cosine function.

Polar Forms of Conic Sections

A polar equation of the form

$$r = \frac{ep}{1 \pm e \cos \theta} \quad \text{or} \quad r = \frac{ep}{1 \pm e \sin \theta}$$

has a conic section as its graph. The eccentricity is e (where $e > 0$), and $|p|$ is the distance between the pole (focus) and the directrix.

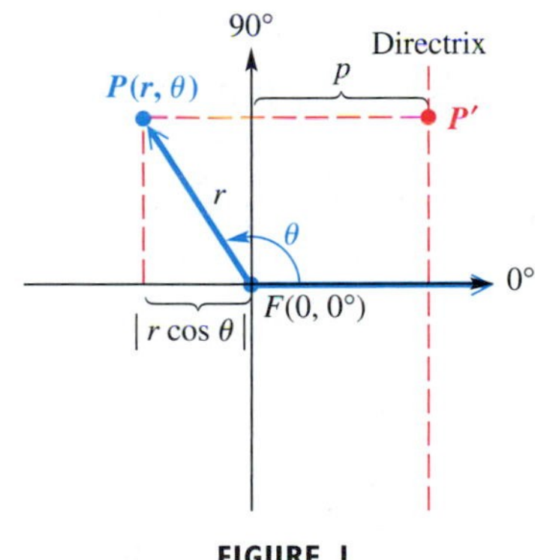

FIGURE 1

We can verify that $r = \frac{ep}{1 + e \cos \theta}$ does indeed satisfy the definition of a conic section. Consider Figure 1, where the directrix is vertical and $p > 0$ units to the right of the focus $F(0, 0°)$.

Let $P(r, \theta)$ be a point on the graph. Then the distance between P and the directrix is

$$PP' = |p - x|$$

$$= |p - r\cos\theta| \qquad x = r\cos\theta$$

$$= \left| p - \left(\frac{ep}{1 + e\cos\theta} \right) \cos\theta \right| \qquad \text{Use the equation for } r.$$

$$= \left| \frac{p(1 + e\cos\theta) - ep\cos\theta}{1 + e\cos\theta} \right| \qquad \text{Write with a common denominator.}$$

$$= \left| \frac{p + ep\cos\theta - ep\cos\theta}{1 + e\cos\theta} \right| \qquad \text{Distributive property}$$

$$PP' = \left| \frac{p}{1 + e\cos\theta} \right|. \qquad \text{Simplify.}$$

Since

$$r = \frac{ep}{1 + e\cos\theta},$$

we can multiply each side by $\frac{1}{e}$ to obtain

$$\frac{p}{1 + e\cos\theta} = \frac{r}{e}.$$

We substitute $\frac{r}{e}$ for the expression in the absolute value bars for PP'.

$$PP' = \left|\frac{p}{1 + e\cos\theta}\right| = \left|\frac{r}{e}\right| = \frac{|r|}{|e|} = \frac{|r|}{e}$$

The distance between the pole and P is $PF = |r|$, so the ratio of PF to PP' is

$$\frac{PF}{PP'} = \frac{|r|}{\frac{|r|}{e}} = |r| \div \left|\frac{r}{e}\right| = |r| \cdot \left|\frac{e}{r}\right| = e. \quad \text{Simplify the complex fraction; } |e| = e.$$

Thus, by the definition, the graph has eccentricity e and must be a conic.

In the preceding discussion, we assumed a vertical directrix to the right of the pole. There are three other possible situations, and all four are summarized in the table.

If the equation is:	then the directrix is:
$r = \frac{ep}{1 + e\cos\theta}$	*vertical*, p units to the *right* of the pole.
$r = \frac{ep}{1 - e\cos\theta}$	*vertical*, p units to the *left* of the pole.
$r = \frac{ep}{1 + e\sin\theta}$	*horizontal*, p units *above* the pole.
$r = \frac{ep}{1 - e\sin\theta}$	*horizontal*, p units *below* the pole.

EXAMPLE 1 Graphing a Conic Section with Equation in Polar Form

Graph $r = \frac{8}{4 + 4\sin\theta}$.

Analytic Solution

Divide both numerator and denominator by 4 to get

$$r = \frac{2}{1 + \sin\theta}.$$

From the preceding table, this is the equation of a conic with $ep = 2$ and $e = 1$. Thus, $p = 2$. Since $e = 1$, the graph is a parabola. The focus is at the pole, and the directrix is horizontal, 2 units *above* the pole. The vertex must have polar coordinates $(1, 90^\circ)$. Letting $\theta = 0^\circ$ and $\theta = 180^\circ$ gives the additional points $(2, 0^\circ)$ and $(2, 180^\circ)$. See Figure 2 on the next page.

Graphing Calculator Solution

Enter

$$r_1 = \frac{8}{4 + 4\sin\theta},$$

with the calculator in polar and degree modes. The first two screens in Figure 3 on the next page show the window settings, and the third screen shows the graph. Notice that the point $(1, 90^\circ)$ is indicated at the bottom of the third screen.

(continued)

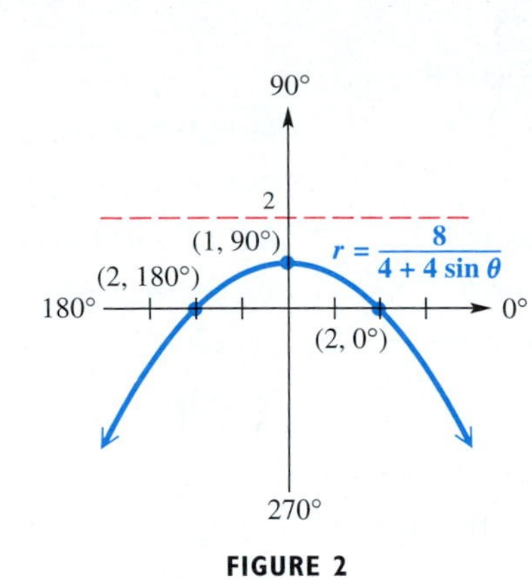

FIGURE 2

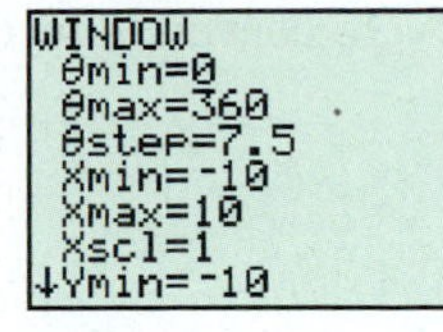

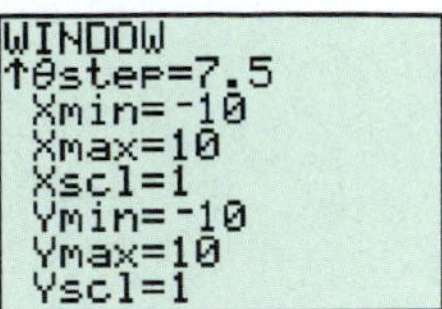

This is a continuation of the screen to the left.

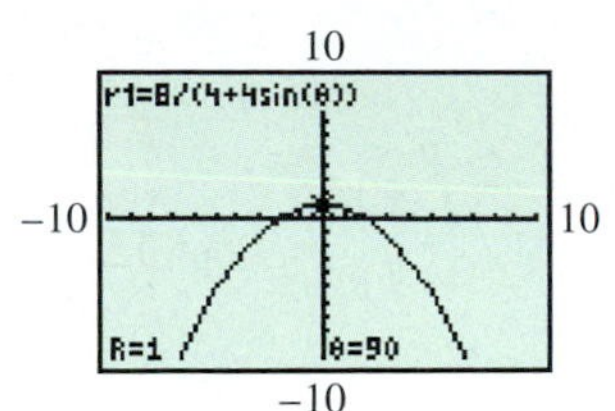

Degree mode

FIGURE 3

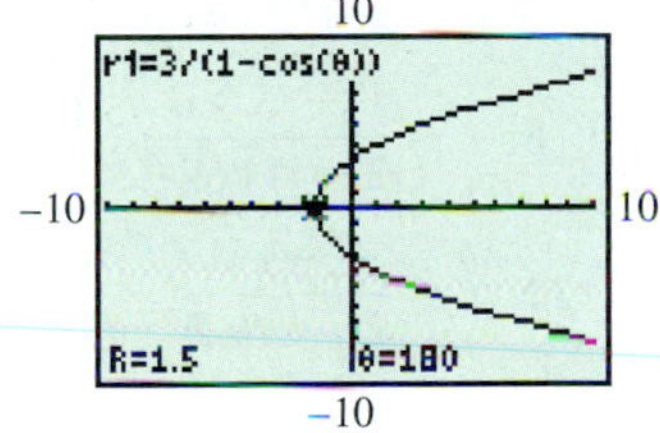

Degree mode

FIGURE 4

EXAMPLE 2 Finding a Polar Equation

Find the polar equation of a parabola with focus at the pole and vertical directrix 3 units to the left of the pole.

Solution The equation must be of the form

$$r = \frac{ep}{1 - e \cos \theta} = \frac{1 \cdot 3}{1 - 1 \cos \theta} = \frac{3}{1 - \cos \theta}. \qquad e = 1, p = 3$$

The calculator graph in Figure 4 supports our result. When $\theta = 180°$, $r = 1.5$. The distance from $F(0, 0°)$ to the directrix is $2r = 2(1.5) = 3$ units, as required. ■

EXAMPLE 3 Identifying and Converting from Polar to Rectangular Form

Identify the conic represented by the following equation. Convert it to rectangular form.

$$r = \frac{8}{2 - \cos \theta}$$

Solution To identify the type of conic, we divide both the numerator and the denominator on the right side by 2 to obtain

$$r = \frac{4}{1 - \frac{1}{2} \cos \theta}.$$

Be sure to divide each term in the numerator and denominator by 2.

This is a conic that has a vertical directrix with $e = \frac{1}{2}$; thus, it is an ellipse.

To convert to rectangular form, we start with the given equation.

$$r = \frac{8}{2 - \cos\theta}$$

$$r(2 - \cos\theta) = 8 \quad \text{Multiply by } 2 - \cos\theta.$$

$$2r - r\cos\theta = 8 \quad \text{Distributive property}$$

$$2r = r\cos\theta + 8 \quad \text{Add } r\cos\theta \text{ to each side.}$$

$$(2r)^2 = (r\cos\theta + 8)^2 \quad \text{Square each side.}$$

$$(2r)^2 = (x + 8)^2 \quad r\cos\theta = x$$

$$4r^2 = x^2 + 16x + 64 \quad \text{Multiply.}$$

Remember the middle term when squaring a binomial.

$$4(x^2 + y^2) = x^2 + 16x + 64 \quad r^2 = x^2 + y^2$$

$$4x^2 + 4y^2 = x^2 + 16x + 64 \quad \text{Distributive property}$$

$$3x^2 + 4y^2 - 16x - 64 = 0 \quad \text{General form}$$

The coefficients of x^2 and y^2 are both positive and are not equal, further supporting our assertion that the graph is an ellipse. ■

Appendix D Exercises

Graph each conic whose equation is given in polar form. Use a traditional or a calculator graph, as directed by your instructor.

1. $r = \frac{6}{3 + 3\sin\theta}$ **2.** $r = \frac{10}{5 + 5\sin\theta}$ **3.** $r = \frac{-4}{6 + 2\cos\theta}$ **4.** $r = \frac{-8}{4 + 2\cos\theta}$

5. $r = \frac{2}{2 - 4\sin\theta}$ **6.** $r = \frac{6}{2 - 4\sin\theta}$ **7.** $r = \frac{4}{2 - 4\cos\theta}$ **8.** $r = \frac{6}{2 - 4\cos\theta}$

9. $r = \frac{-1}{1 + 2\sin\theta}$ **10.** $r = \frac{-1}{1 - 2\sin\theta}$ **11.** $r = \frac{-1}{2 + \cos\theta}$ **12.** $r = \frac{-1}{2 - \cos\theta}$

Find a polar equation of the parabola with focus at the pole, satisfying the given conditions.

13. Vertical directrix 3 units to the right of the pole

14. Vertical directrix 4 units to the left of the pole

15. Horizontal directrix 5 units below the pole

16. Horizontal directrix 6 units above the pole

Find a polar equation for the conic with focus at the pole, satisfying the given conditions. Also, identify the type of conic represented.

17. $e = \frac{4}{5}$; vertical directrix 5 units to the right of the pole

18. $e = \frac{2}{3}$; vertical directrix 6 units to the left of the pole

19. $e = \frac{5}{4}$; horizontal directrix 8 units below the pole

20. $e = \frac{3}{2}$; horizontal directrix 4 units above the pole

Identify the type of conic represented, and convert the equation to rectangular form.

21. $r = \frac{6}{3 - \cos\theta}$ **22.** $r = \frac{8}{4 - \cos\theta}$ **23.** $r = \frac{-2}{1 + 2\cos\theta}$

24. $r = \frac{-3}{1 + 3\cos\theta}$ **25.** $r = \frac{-6}{4 + 2\sin\theta}$ **26.** $r = \frac{-12}{6 + 3\sin\theta}$

27. $r = \frac{10}{2 - 2\sin\theta}$ **28.** $r = \frac{12}{4 - 4\sin\theta}$

Appendix E Rotation of Axes

Derivation of Rotation Equations ■ Applying a Rotation Equation

Looking Ahead to Calculus

Rotation of axes is a topic traditionally covered in calculus texts, in conjunction with parametric equations and polar coordinates. The coverage in calculus is typically the same as that seen in this section.

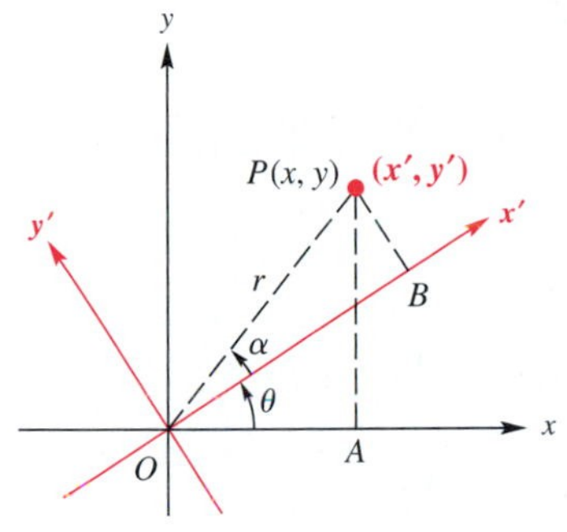

FIGURE 1

Derivation of Rotation Equations

If we begin with an xy-coordinate system having origin O and rotate the axes about O through an angle θ, the new coordinate system is called a **rotation** of the xy-system. Trigonometric identities are used to obtain equations for converting the coordinates of a point from the xy-system to the rotated $x'y'$-system. Let P be any point other than the origin, with coordinates (x, y) in the xy-system and (x', y') in the $x'y'$-system. See Figure 1. Let $OP = r$, and let α represent the angle made by OP and the x'-axis. As shown in Figure 1,

$$\cos(\theta + \alpha) = \frac{OA}{r} = \frac{x}{r},$$

$$\sin(\theta + \alpha) = \frac{AP}{r} = \frac{y}{r},$$

$$\cos \alpha = \frac{OB}{r} = \frac{x'}{r},$$

and

$$\sin \alpha = \frac{PB}{r} = \frac{y'}{r}.$$

These four statements can be written as

$$x = r\cos(\theta + \alpha), \quad y = r\sin(\theta + \alpha), \quad x' = r\cos\alpha, \quad y' = r\sin\alpha.$$

Using the trigonometric identity for the cosine of the sum of two angles gives

$$\begin{aligned} x &= r\cos(\theta + \alpha) \\ &= r(\cos\theta\cos\alpha - \sin\theta\sin\alpha) && \text{Cosine sum identity} \\ &= (r\cos\alpha)\cos\theta - (r\sin\alpha)\sin\theta && \text{Distributive property} \\ &= x'\cos\theta - y'\sin\theta. && \text{Substitute.} \end{aligned}$$

In the same way, the identity for the sine of the sum of two angles yields the equation

$$y = x'\sin\theta + y'\cos\theta.$$

This proves the following result.

Rotation Equations

If the rectangular coordinate axes are rotated about the origin through an angle θ, and if the coordinates of a point P are (x, y) and (x', y') in the xy-system and the $x'y'$-system, respectively, then the respective **rotation equations** are

$$\boldsymbol{x = x'\cos\theta - y'\sin\theta} \quad \text{and} \quad \boldsymbol{y = x'\sin\theta + y'\cos\theta}.$$

Applying a Rotation Equation

EXAMPLE 1 Finding an Equation after a Rotation

The equation of a curve is $x^2 + y^2 + 2xy + 2\sqrt{2}x - 2\sqrt{2}y = 0$. Find the resulting equation if the axes are rotated 45°. Graph the equation.

Solution If $\theta = 45°$, then $\sin\theta = \frac{\sqrt{2}}{2}$ and $\cos\theta = \frac{\sqrt{2}}{2}$, and the rotation equations become

$$x = \frac{\sqrt{2}}{2}x' - \frac{\sqrt{2}}{2}y' \quad \text{and} \quad y = \frac{\sqrt{2}}{2}x' + \frac{\sqrt{2}}{2}y'.$$

Substituting these values into the given equation yields

$$x^2 + y^2 + 2xy + 2\sqrt{2}x - 2\sqrt{2}y = 0$$

$$\left[\frac{\sqrt{2}}{2}x' - \frac{\sqrt{2}}{2}y'\right]^2 + \left[\frac{\sqrt{2}}{2}x' + \frac{\sqrt{2}}{2}y'\right]^2 + 2\left[\frac{\sqrt{2}}{2}x' - \frac{\sqrt{2}}{2}y'\right]\left[\frac{\sqrt{2}}{2}x' + \frac{\sqrt{2}}{2}y'\right]$$

$$+ 2\sqrt{2}\left[\frac{\sqrt{2}}{2}x' - \frac{\sqrt{2}}{2}y'\right] - 2\sqrt{2}\left[\frac{\sqrt{2}}{2}x' + \frac{\sqrt{2}}{2}y'\right] = 0.$$

Expanding these terms gives

$$\frac{1}{2}x'^2 - x'y' + \frac{1}{2}y'^2 + \frac{1}{2}x'^2 + x'y' + \frac{1}{2}y'^2 + x'^2 - y'^2$$

$$+ 2x' - 2y' - 2x' - 2y' = 0$$

$$2x'^2 - 4y' = 0 \qquad \text{Combine terms.}$$

$$x'^2 - 2y' = 0 \qquad \text{Divide by 2.}$$

$$x'^2 = 2y', \qquad \text{Add } 2y'.$$

the equation of a parabola. The graph is shown in Figure 2. ■

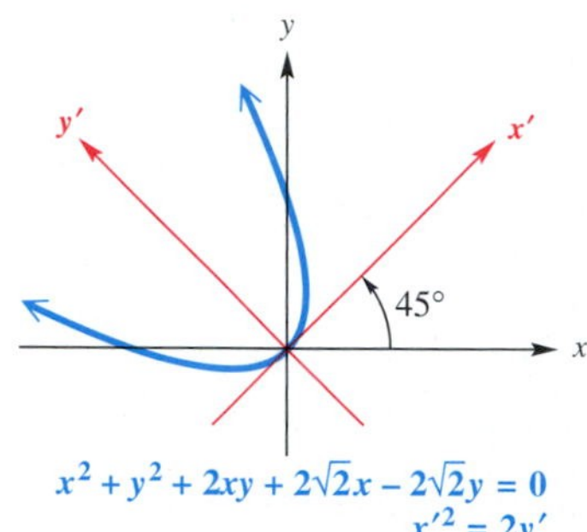

$x^2 + y^2 + 2xy + 2\sqrt{2}x - 2\sqrt{2}y = 0$
$x'^2 = 2y'$

FIGURE 2

We have graphed equations written in the general form

$$Ax^2 + Cy^2 + Dx + Ey + F = 0.$$

To graph an equation that has an xy-term by hand, it is necessary to find an appropriate **angle of rotation** to eliminate the xy-term.

Angle of Rotation

The xy-term is removed from the general equation

$$\mathbf{Ax^2 + Bxy + Cy^2 + Dx + Ey + F = 0}$$

by a rotation of the axes through an angle θ, $0° < \theta < 90°$, where

$$\cot 2\theta = \frac{A - C}{B}.$$

To find the rotation equations, first find $\sin\theta$ and $\cos\theta$. Example 2 illustrates a way to obtain $\sin\theta$ and $\cos\theta$ from $\cot 2\theta$ without first identifying angle θ.

EXAMPLE 2 Rotating and Graphing

Rotate the axes and graph $52x^2 - 72xy + 73y^2 = 200$.

Solution Here, $A = 52$, $B = -72$, and $C = 73$. By substitution,

$$\cot 2\theta = \frac{52 - 73}{-72} = \frac{-21}{-72} = \frac{7}{24}.$$

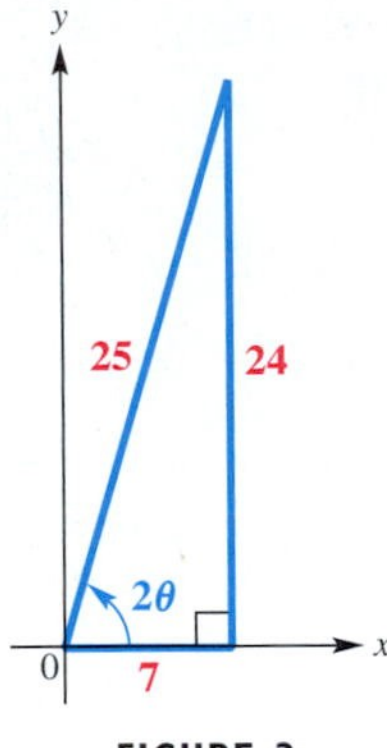

FIGURE 3

To find $\sin \theta$ and $\cos \theta$, use the trigonometric identities

$$\sin \theta = \sqrt{\frac{1 - \cos 2\theta}{2}} \quad \text{and} \quad \cos \theta = \sqrt{\frac{1 + \cos 2\theta}{2}}.$$

Sketch a right triangle and label it as in Figure 3, to see that $\cos 2\theta = \frac{7}{25}$. (Recall that in the two quadrants for which we are concerned, cosine and cotangent have the same sign.) Then

$$\sin \theta = \sqrt{\frac{1 - \frac{7}{25}}{2}} = \sqrt{\frac{9}{25}} = \frac{3}{5} \quad \text{and} \quad \cos \theta = \sqrt{\frac{1 + \frac{7}{25}}{2}} = \sqrt{\frac{16}{25}} = \frac{4}{5}.$$

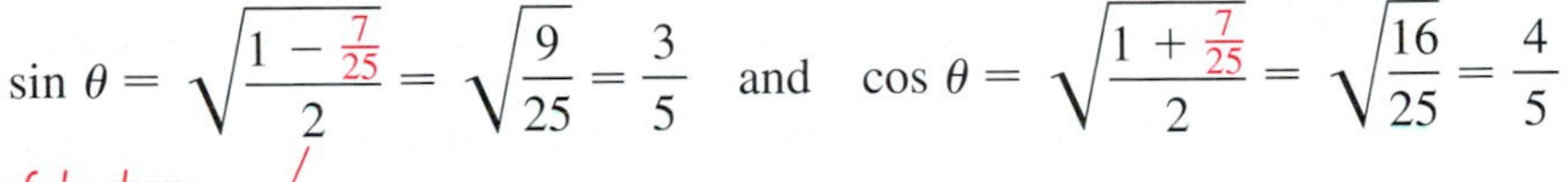

Be careful when simplifying a complex fraction.

Use these values for $\sin \theta$ and $\cos \theta$ to obtain

$$x = \frac{4}{5}x' - \frac{3}{5}y' \quad \text{and} \quad y = \frac{3}{5}x' + \frac{4}{5}y'.$$

Substituting these expressions for x and y into the original equation yields

$$52\left[\frac{4}{5}x' - \frac{3}{5}y'\right]^2 - 72\left[\frac{4}{5}x' - \frac{3}{5}y'\right]\left[\frac{3}{5}x' + \frac{4}{5}y'\right] + 73\left[\frac{3}{5}x' + \frac{4}{5}y'\right]^2 = 200$$

$$52\left[\frac{16}{25}x'^2 - \frac{24}{25}x'y' + \frac{9}{25}y'^2\right] - 72\left[\frac{12}{25}x'^2 + \frac{7}{25}x'y' - \frac{12}{25}y'^2\right] + 73\left[\frac{9}{25}x'^2 + \frac{24}{25}x'y' + \frac{16}{25}y'^2\right] = 200$$

$$25x'^2 + 100y'^2 = 200 \qquad \text{Combine terms.}$$

$$\frac{x'^2}{8} + \frac{y'^2}{2} = 1. \qquad \text{Divide by 200.}$$

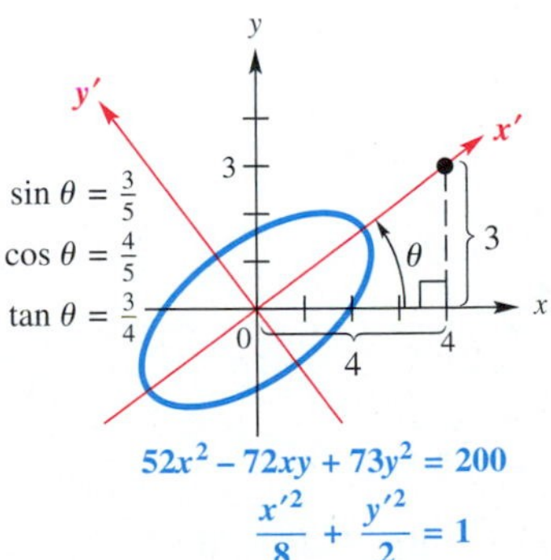

FIGURE 4

This is an equation of an ellipse having x'-intercepts $\pm 2\sqrt{2}$ and y'-intercepts $\pm\sqrt{2}$. The graph is shown in Figure 4. To find θ, use the fact that

$$\frac{\sin \theta}{\cos \theta} = \frac{\frac{3}{5}}{\frac{4}{5}} = \frac{3}{4} = \tan \theta, \quad \text{from which } \theta \approx 37°. \quad \blacksquare$$

Be careful when simplifying a complex fraction.

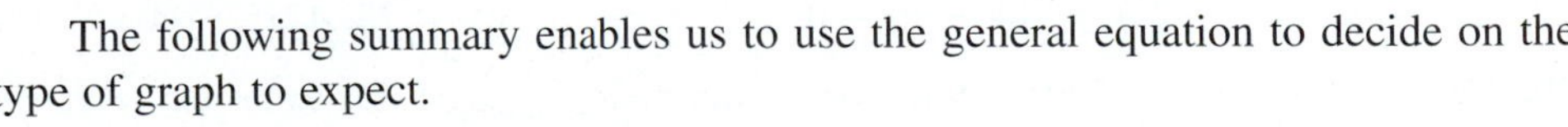

The following summary enables us to use the general equation to decide on the type of graph to expect.

Equations of Conics with xy-Term

If the general second-degree equation

$$Ax^2 + Bxy + Cy^2 + Dx + Ey + F = 0$$

has a graph, it will be one of the following:

(a) a circle or an ellipse (or a point) if $B^2 - 4AC < 0$;
(b) a parabola (or one line or two parallel lines) if $B^2 - 4AC = 0$;
(c) a hyperbola (or two intersecting lines) if $B^2 - 4AC > 0$;
(d) a straight line if $A = B = C = 0$, and $D \neq 0$ or $E \neq 0$.

Appendix E *Exercises*

Concept Check *Use the summary in this section to predict the graph of each second-degree equation.*

1. $4x^2 + 3y^2 + 2xy - 5x = 8$

2. $x^2 + 2xy - 3y^2 + 2y = 12$

3. $2x^2 + 3xy - 4y^2 = 0$

4. $x^2 - 2xy + y^2 + 4x - 8y = 0$

5. $4x^2 + 4xy + y^2 + 15 = 0$

6. $-x^2 + 2xy - y^2 + 16 = 0$

Find the angle of rotation θ that will remove the xy-term in each equation.

7. $2x^2 + \sqrt{3}xy + y^2 + x = 5$

8. $4\sqrt{3}x^2 + xy + 3\sqrt{3}y^2 = 10$

9. $3x^2 + \sqrt{3}xy + 4y^2 + 2x - 3y = 12$

10. $4x^2 + 2xy + 2y^2 + x - 7 = 0$

11. $x^2 - 4xy + 5y^2 = 18$

12. $3\sqrt{3}x^2 - 2xy + \sqrt{3}y^2 = 25$

Use the given angle of rotation to remove the xy-term, and graph each equation.

13. $x^2 - xy + y^2 = 6$; $\theta = 45°$

14. $2x^2 - xy + 2y^2 = 25$; $\theta = 45°$

15. $8x^2 - 4xy + 5y^2 = 36$; $\sin\theta = \dfrac{2}{\sqrt{5}}$

16. $5y^2 + 12xy = 10$; $\sin\theta = \dfrac{3}{\sqrt{13}}$

Remove the xy-term from each equation by performing a suitable rotation. Graph each equation.

17. $3x^2 - 2xy + 3y^2 = 8$

18. $x^2 + xy + y^2 = 3$

19. $x^2 - 4xy + y^2 = -5$

20. $x^2 + 2xy + y^2 + 4\sqrt{2}x - 4\sqrt{2}y = 0$

21. $7x^2 + 6\sqrt{3}xy + 13y^2 = 64$

22. $7x^2 + 2\sqrt{3}xy + 5y^2 = 24$

23. $3x^2 - 2\sqrt{3}xy + y^2 - 2x - 2\sqrt{3}y = 0$

24. $2x^2 + 2\sqrt{3}xy + 4y^2 = 5$

In each equation, remove the xy-term by rotation. Then translate the axes and sketch the graph.

25. $x^2 + 3xy + y^2 - 5\sqrt{2}y = 15$

26. $x^2 - \sqrt{3}xy + 2\sqrt{3}x - 3y - 3 = 0$

27. $4x^2 + 4xy + y^2 - 24x + 38y - 19 = 0$

28. $12x^2 + 24xy + 19y^2 - 12x - 40y + 31 = 0$

29. $16x^2 + 24xy + 9y^2 - 130x + 90y = 0$

30. $9x^2 - 6xy + y^2 - 12\sqrt{10}x - 36\sqrt{10}y = 0$

 **31.** Look at the box titled "Angle of Rotation." Explain why no rotation is applicable if the value of B is 0.

 **32.** Look at the equation involving cot 2θ in the box titled "Angle of Rotation." Explain why the angle of rotation must be 45° if the coefficients of x^2 and y^2 are equal and $B \neq 0$.

Answers to Selected Exercises

To The Student

If you need further help with algebra, you may want to obtain a copy of the *Student's Solution Manual* that goes with this book. It contains solutions to all the odd-numbered section and Chapter Review exercises and all the Relating Concepts, Reviewing Basic Concepts, and Chapter Test exercises. Your college bookstore either has the *Manual* or can order it for you.

In this section, we provide the answers that we think most students will obtain when they work the exercises by using the methods explained in the text. If your answer does not look exactly like the one given here, it is not necessarily wrong. In many cases there are equivalent forms of the answer. For example, if the answer section shows $\frac{3}{4}$ and your answer is .75, you have obtained the correct answer, but written it in a different (yet equivalent) form. Unless the directions specify otherwise, .75 is just as valid an answer as $\frac{3}{4}$. In general, if your answer does not agree with the one given in the text, see whether it can be transformed into the other form. If it can, then it is the correct answer. If you still have doubts, talk with your instructor.

CHAPTER 1 LINEAR FUNCTIONS, EQUATIONS, AND INEQUALITIES

1.1 Exercises *(pages 8–11)*

1. **(a)** 10 **(b)** 0, 10 **(c)** $-6, -\frac{12}{4}$ (or -3), 0, 10 **(d)** $-6, -\frac{12}{4}$ (or -3), $-\frac{5}{8}$, 0, .31, $.\overline{3}$, 10 **(e)** $-\sqrt{3}, 2\pi, \sqrt{17}$ **(f)** All are real numbers. **3.** **(a)** None are natural numbers. **(b)** None are whole numbers. **(c)** $-\sqrt{100}$ (or -10), -1 **(d)** $-\sqrt{100}$ (or -10), $-\frac{13}{6}$, -1, 5.23, $9.\overline{14}$, 3.14, $\frac{22}{7}$ **(e)** None are irrational numbers. **(f)** All are real numbers. **5.** natural number, integer, rational number, real number **7.** integer, rational number, real number **9.** rational number, real number **11.** real number **13.** natural numbers **15.** rational numbers **17.** integers

19. 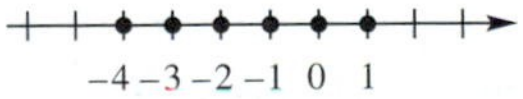

21. 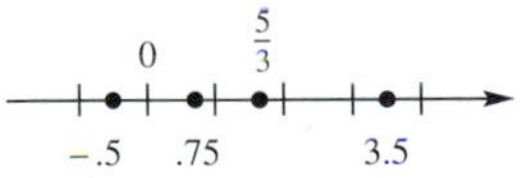

23. A rational number can be written as a fraction $\frac{p}{q}$, $q \neq 0$, where p and q are integers, whereas an irrational number cannot. **25.** I **27.** III **29.** none **31.** II **33.** none **35.** I or III **37.** II or IV **39.** y-axis

Graph for Exercises 25–33:

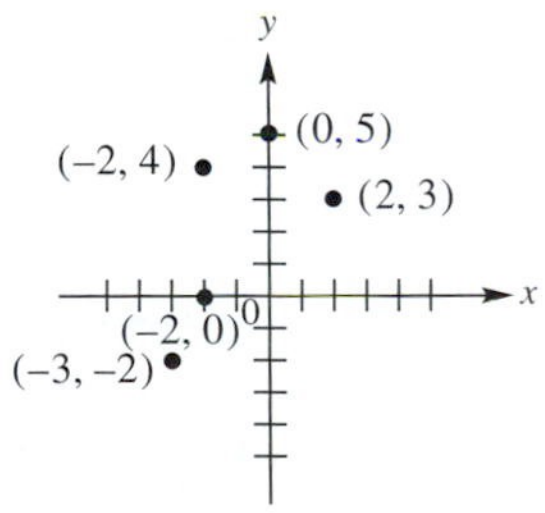

41. $[-5, 5]$ by $[-25, 25]$
43. $[-60, 60]$ by $[-100, 100]$
45. $[-500, 300]$ by $[-300, 500]$

47.

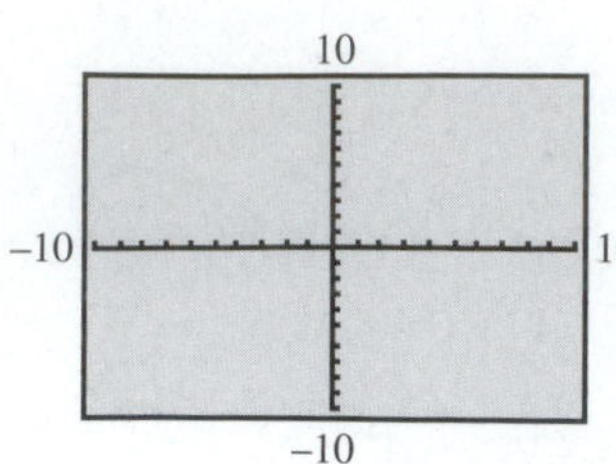

49.

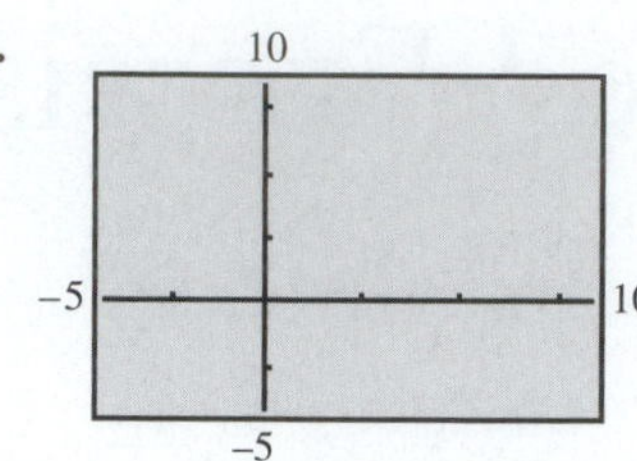

51.

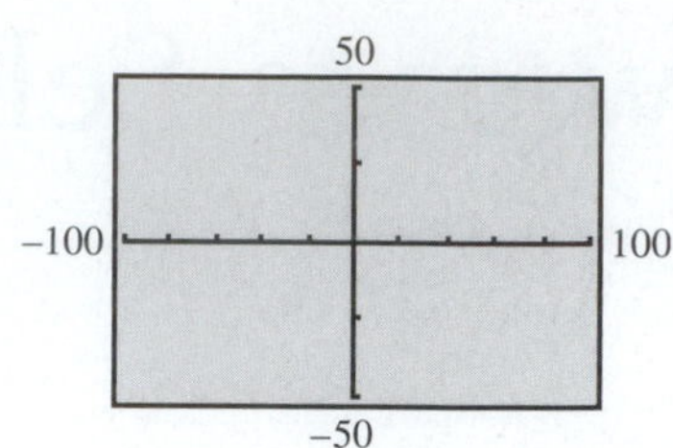

53. There are no tick marks. To set a screen with no tick marks on the axes, use Xscl = 0 and Yscl = 0. **55.** 7.616 **57.** 3.208 **59.** 3.045 **61.** 4.359 **63.** 311.987 **65.** .25 **67.** 5.66 **69.** 98.63 **71.** 8.25 **73.** .72 **75.** −2.82 **77.** $c = 17$ **79.** $b = 84$ **81.** $c = \sqrt{89}$ **83.** $a = 4$ **85.** **(a)** $\sqrt{40} = 2\sqrt{10}$ **(b)** $(-1, 4)$ **87.** **(a)** $\sqrt{205}$ **(b)** $\left(-\frac{1}{2}, 1\right)$ **89.** **(a)** 5 **(b)** $\left(\frac{7}{2}, 9\right)$ **91.** **(a)** $\sqrt{34}$ **(b)** $\left(-\frac{11}{2}, -\frac{7}{2}\right)$ **93.** **(a)** 5 **(b)** (7.7, 5.4) **95.** **(a)** $25x$ **(b)** $\left(\frac{19}{2}x, -11x\right)$ **97.** $Q(9, 14)$ **99.** $Q(-13.72, -5.01)$ **101.** $17,396 **103.** 1988: $12,471; 1998: $16,787 **105.** **(a)**

N
(0, 50)
d
Car A
Car B
(0, 0)
W
E
(−40, 0)
(0, −50)
S

(b) $d = \sqrt{4100} \approx 64.0$ mi **107.** $(a + b)^2 = a^2 + 2ab + b^2$; $c^2 + 4\left(\frac{1}{2}ab\right) = c^2 + 2ab$; $a^2 + 2ab + b^2 = c^2 + 2ab$, so subtract $2ab$ from each side to obtain $a^2 + b^2 = c^2$.

1.2 Exercises *(pages 20–22)*

1. $(-1, 4)$

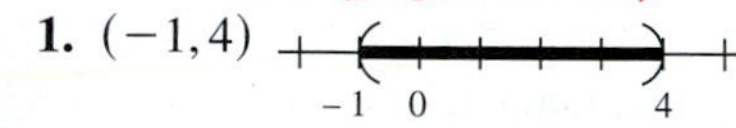

3. $(-\infty, 0)$

−1 0

5. $[1, 2)$

0 1 2

7. $\{x \mid -4 < x < 3\}$ **9.** $\{x \mid x \leq -1\}$ **11.** $\{x \mid -2 \leq x < 6\}$ **13.** $\{x \mid x \leq -4\}$ **15.** Use a parenthesis if the symbol is $<$ or $>$, and use a square bracket if the symbol is $\leq$ or $\geq$. **17.** domain: {5, 3, 4, 7}; range: {1, 2, 9, 6}; function **19.** domain: {1, 2, 3}; range: {6}; function **21.** domain: {4, 3, −2}; range: {1, −5, 3, 7}; not a function **23.** domain: {11, 12, 13, 14}; range: {−6, −7}; function **25.** domain: {0, 1, 2, 3, 4}; range: $\{\sqrt{2}, \sqrt{3}, \sqrt{5}, \sqrt{6}, \sqrt{7}\}$; function **27.** domain: $(-\infty, \infty)$; range: $(-\infty, \infty)$; function **29.** domain: $[-4, 4]$; range: $[-3, 3]$; not a function **31.** domain: $[2, \infty)$; range: $[0, \infty)$; function **33.** domain: $[-9, \infty)$; range: $(-\infty, \infty)$; not a function **35.** domain: {−5, −2, −1, −.5, 0, 1.75, 3.5}; range: {−1, 2, 3, 3.5, 4, 5.75, 7.5}; function **37.** domain: {2, 3, 5, 11, 17}; range: {1, 7, 20}; function **39.** 2 **41.** 7 **43.** −10 **45.** 4 **47.** −10 **49.** 5 **51.** 2 **53.** 11 **55.** $(-2, 3)$ **57.** 7; 8 **59.** **(a)** 0 **(b)** 4 **(c)** 2 **(d)** 4 **61.** **(a)** −3 **(b)** −2 **(c)** 0 **(d)** 2 **63.** **(a)–(d)** Answers will vary; see the definitions.

65.

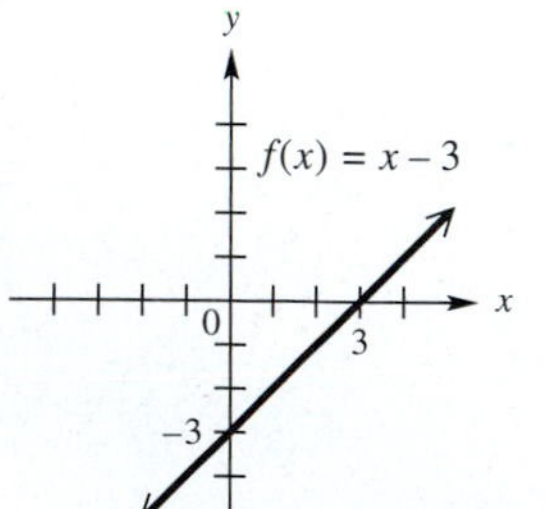

67.

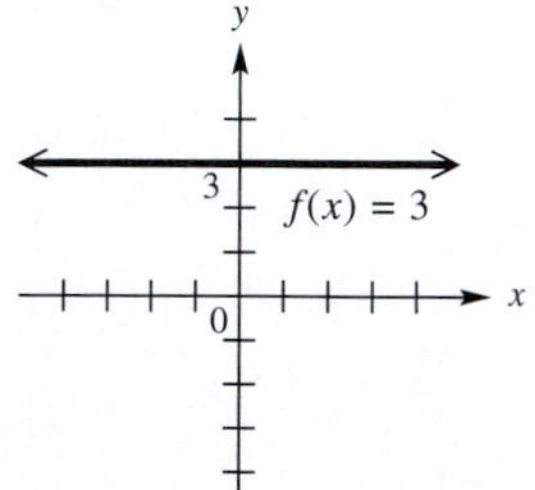

69.

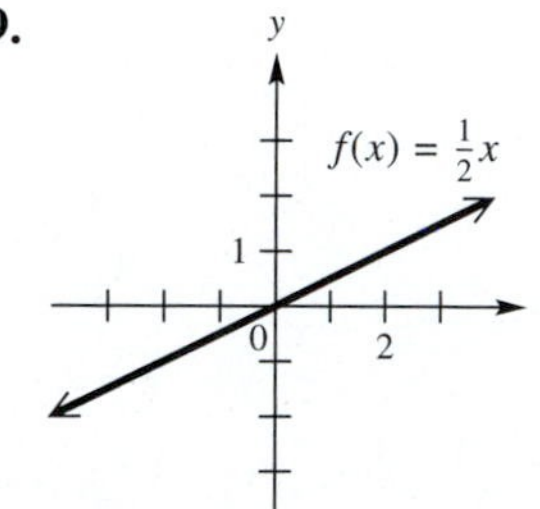

71.

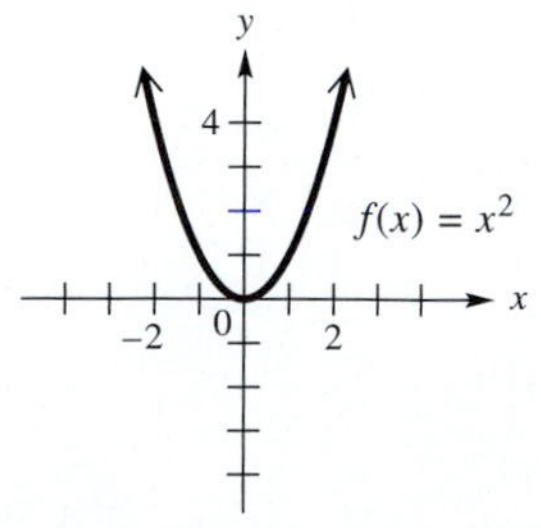

73. **(a)** $f(15) = 3$; If the delay is 15 sec, the lightning is 3 mi away. **(b)**

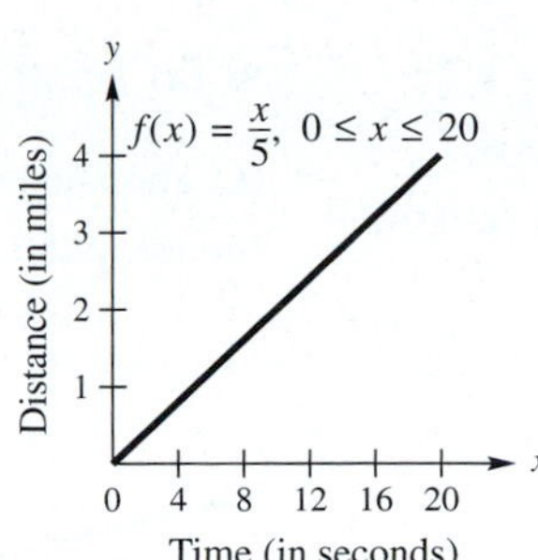

75. $f(x) = .075x; f(86) = \$6.45$ **77.** $f(x) = 92x + 75$; $1087

Reviewing Basic Concepts *(pages 22–23)*

1.

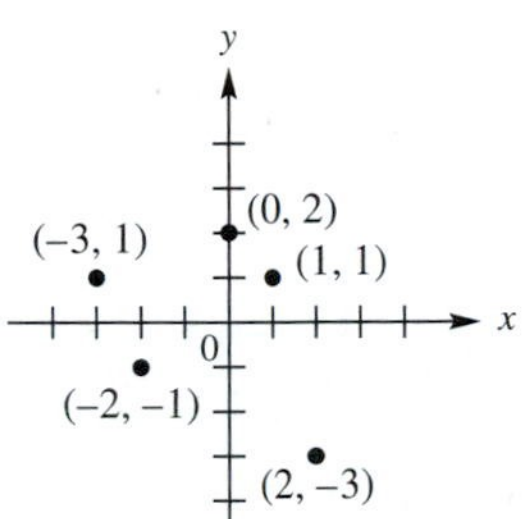

2. $d(P, Q) = \sqrt{149}; M = \left(1, \frac{3}{2}\right)$ **3.** 1.168 **4.** 60 in. **5.** $(-2, 5]; [4, \infty)$

6. not a function; domain: $[-2, 2]$; range: $[-3, 3]$ **7.** 23 **8.** $f(2) = 3$; $f(-1) = -3$

9.

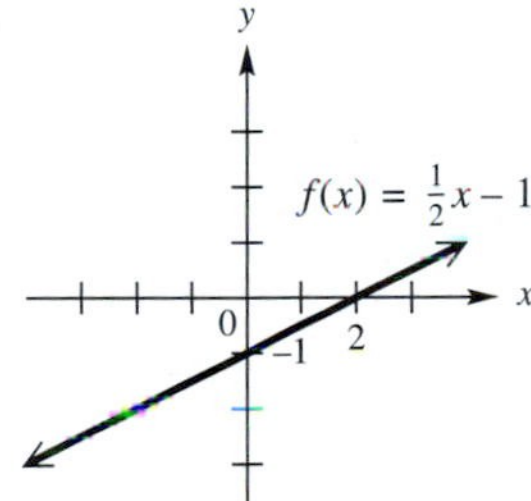

10. 34

1.3 Exercises *(pages 33–36)*

1. **(a)** $f(-2) = 0$; $f(4) = 6$ **(b)** -2 **(c)** The x-intercept is -2 and corresponds to the zero of f.

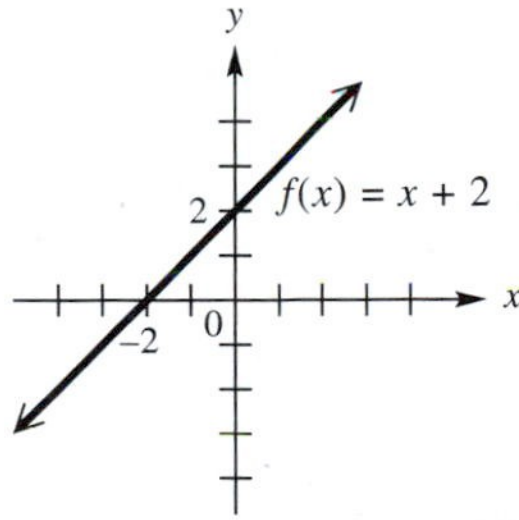

3. **(a)** $f(-2) = 3$; $f(4) = 0$ **(b)** 4 **(c)** The x-intercept is 4 and corresponds to the zero of f.

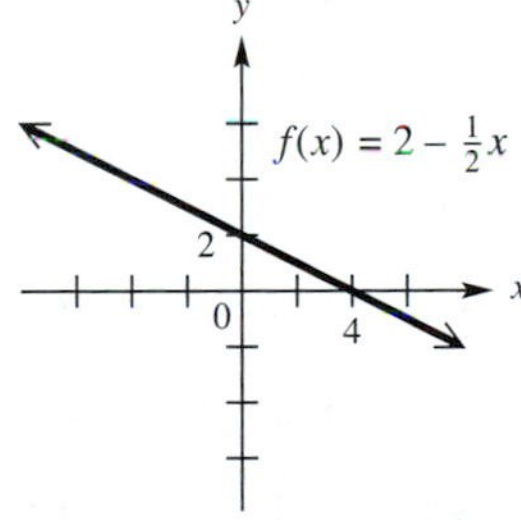

5. **(a)** $f(-2) = -\frac{2}{3}$; $f(4) = \frac{4}{3}$ **(b)** 0 **(c)** The x-intercept is 0 and corresponds to the zero of f.

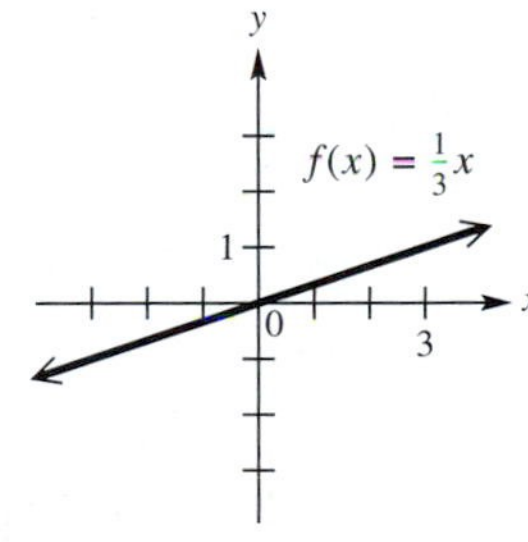

7. (a) $f(-2) = -.65$; $f(4) = 1.75$ **(b)** $-.375$ **(c)** The x-intercept is $-.375$ and corresponds to the zero of f.

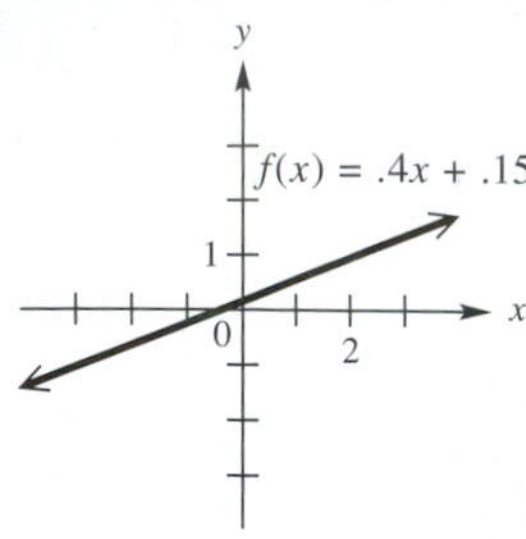

9. (a) $f(-2) = 1$; $f(4) = -\frac{1}{2}$ **(b)** 2 **(c)** The x-intercept is 2 and corresponds to the zero of f.

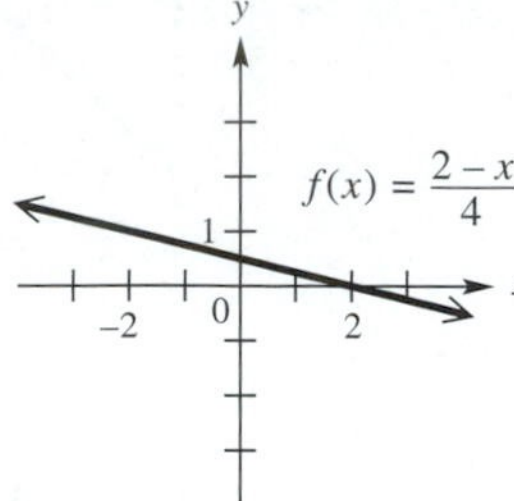

11. (a) $f(-2) = \frac{\pi - 6}{2}$; $f(4) = \frac{12 + \pi}{2}$ **(b)** $-\frac{\pi}{3}$ **(c)** The x-intercept is $-\frac{\pi}{3} \approx -1.047$ and corresponds to the zero of f.

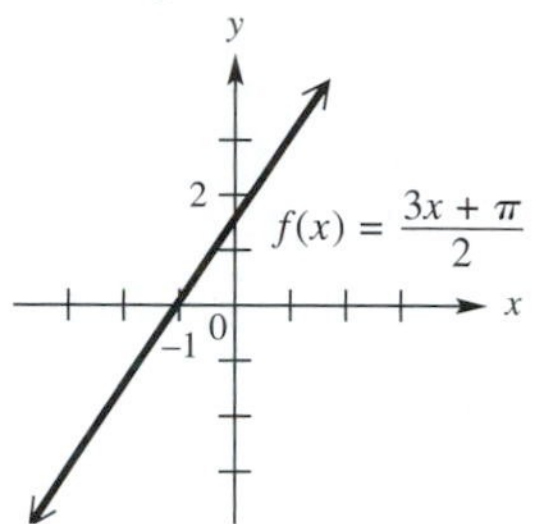

13. (a) 4 **(b)** -4 **(c)** $(-\infty, \infty)$ **(d)** $(-\infty, \infty)$ **(e)** 1

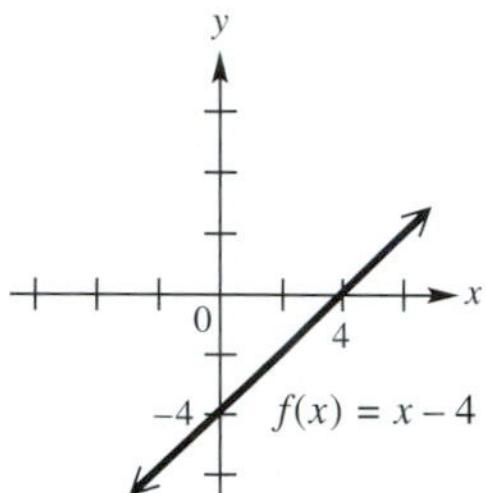

15. (a) 2 **(b)** -6 **(c)** $(-\infty, \infty)$ **(d)** $(-\infty, \infty)$ **(e)** 3

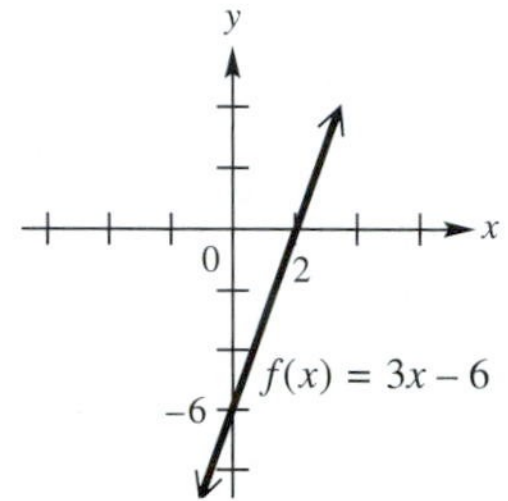

17. (a) 5 **(b)** 2 **(c)** $(-\infty, \infty)$ **(d)** $(-\infty, \infty)$ **(e)** $-\frac{2}{5}$

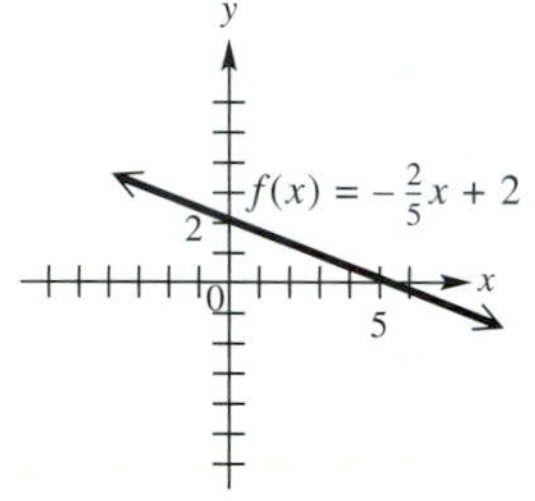

19. (a) 0 **(b)** 0 **(c)** $(-\infty, \infty)$ **(d)** $(-\infty, \infty)$ **(e)** 3

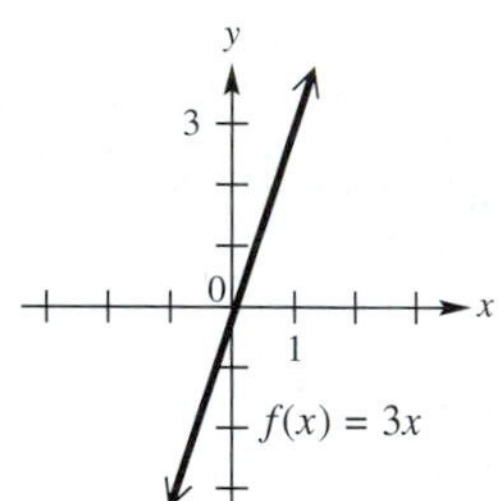

21. The point $(0, 0)$ must lie on the line.

23. (a) none **(b)** -3 **(c)** $(-\infty, \infty)$ **(d)** $\{-3\}$ **(e)** 0

25. (a) -1.5 **(b)** none **(c)** $\{-1.5\}$ **(d)** $(-\infty, \infty)$ **(e)** undefined

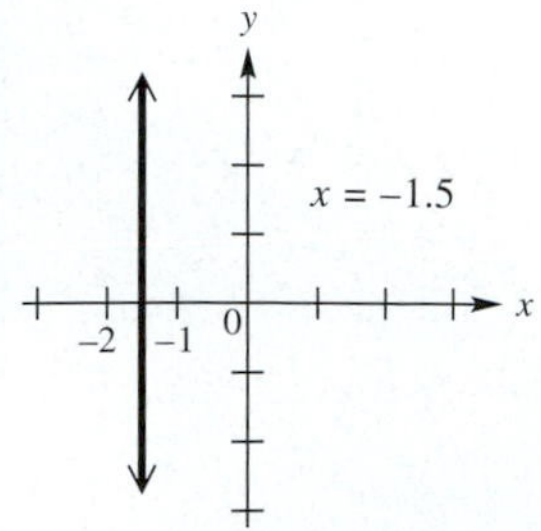

27. (a) 2 **(b)** none **(c)** $\{2\}$ **(d)** $(-\infty, \infty)$ **(e)** undefined

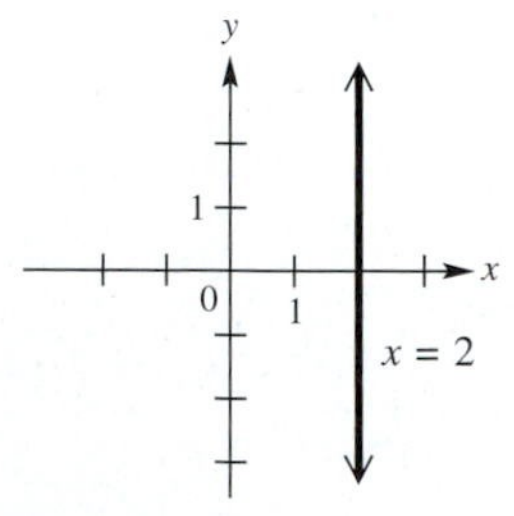

29. constant function **31.** $y = 3$ **33.** $x = 0$ **35.** window B **37.** window B **39.** 1 **41.** $\frac{7}{9}$ **43.** undefined **45.** 0 **47.** A **49.** C **51.** H **53.** B **55. (a)** -2; 1; $\frac{1}{2}$ **(b)** $f(x) = -2x + 1$ **(c)** $\frac{1}{2}$ **57. (a)** $-\frac{1}{3}$; 2; 6 **(b)** $f(x) = -\frac{1}{3}x + 2$ **(c)** 6 **59. (a)** -200; 300; $\frac{3}{2}$ **(b)** $f(x) = -200x + 300$ **(c)** $\frac{3}{2}$

61. 4; 2; $f(x) = 4x + 2$ **63.** -1.4; -3.1; $f(x) = -1.4x - 3.1$ **65.** A **67.** D

In Exercises 69–77, the two indicated points may vary.

69.

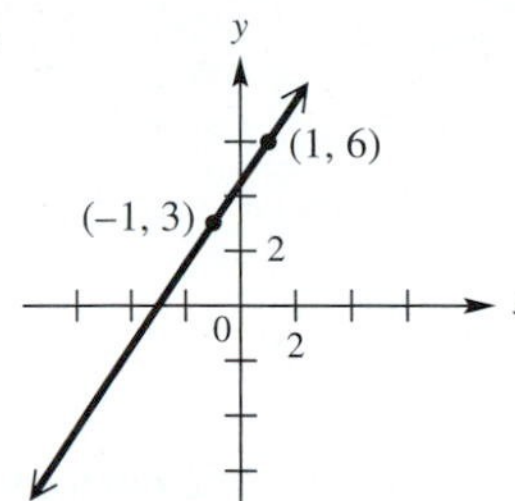

71.

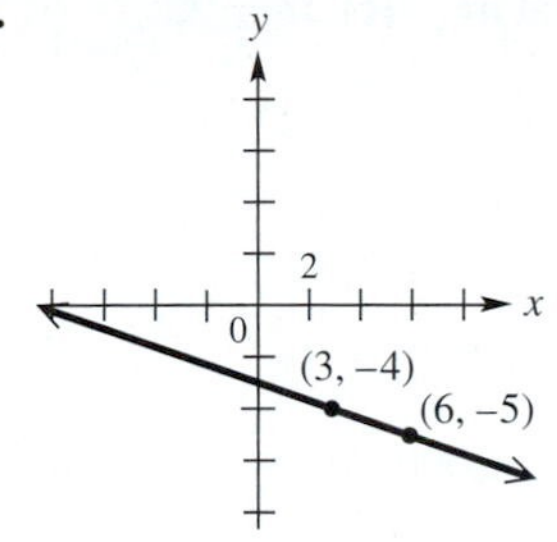

73.

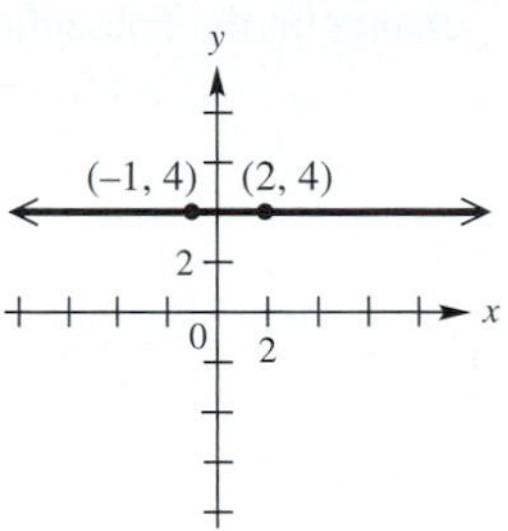

75.

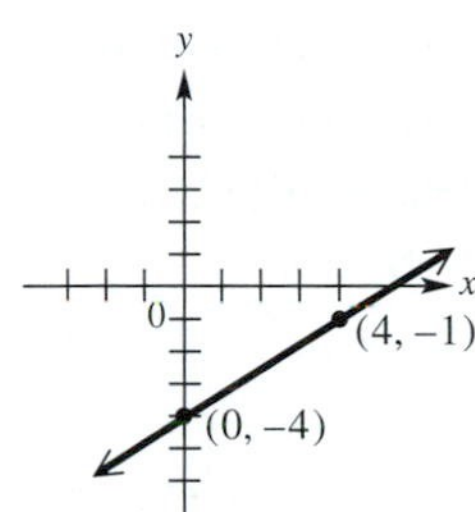

77.

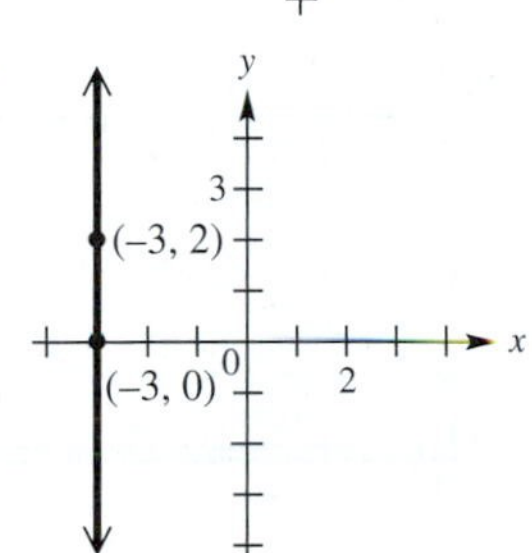

79. **(a)** $f(x) = 500x + 2000$ **(b)** Water is entering the pool at 500 gal per hr; the pool initially contains 2000 gal. **(c)** 5500 gal

81. **(a)** $f(x) = \frac{1}{4}x + 3$ **(b)** 3.625 in.

83. **(a)** $f(x) = -.096x + 14.8$ **(b)** $f(6) \approx 14.2$; This compares favorably to 14.1.

1.4 Exercises *(pages 45–50)*

1. $y = -2x + 5$ **3.** $y = 1.5x + 11.5$ **5.** $y = -.5x - 3$ **7.** $y = 2x - 5$ **9.** $y = \frac{1}{2}x + \frac{13}{24}$ **11.** $y = \frac{4}{5}x - \frac{14}{5}$

13. $y = -x - 4$ **15.** $y = x + 4$ **17.** $y = \frac{5}{2}x - \frac{31}{2}$ **19.** $y = -1.5x + 6.5$ **21.** $y = -\frac{1}{2}x + 5$

23. $y = 8x + 12$

Note: The graphs of the lines in Exercises 25–29 are the same as the graphs in Exercises 13–17 in Section 1.3. In the answers here, we give the x- and y-intercepts. Refer to the graphs given previously.

25. x: 4; y: -4 **27.** x: 2; y: -6 **29.** x: 5; y: 2 **31.** **33.**

We used $(0, 0)$ and $(1, 3)$.

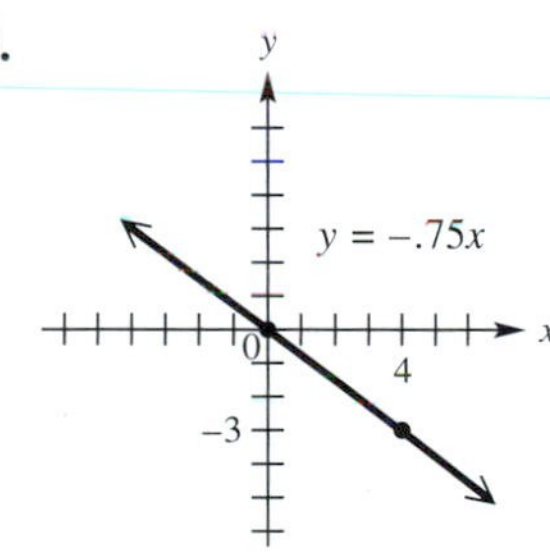

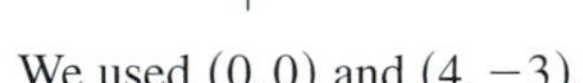
We used $(0, 0)$ and $(4, -3)$.

35. $y = -\frac{5}{3}x + 5$ **37.** $y = \frac{2}{7}x + \frac{4}{7}$ **39.** $y = -\frac{3}{4}x + \frac{25}{8}$ **41.** $y = -\frac{1}{3}x + \frac{11}{3}$ **43.** $y = \frac{5}{3}x + \frac{13}{3}$

45. $x = -5$ **47.** $y = -.2x + 7$ **49.** $y = \frac{1}{2}x$ **51.** **(a)** the Pythagorean theorem and its converse **(b)** $\sqrt{x_1^2 + m_1^2x_1^2}$ **(c)** $\sqrt{x_2^2 + m_2^2x_2^2}$ **(d)** $\sqrt{(x_2 - x_1)^2 + (m_2x_2 - m_1x_1)^2}$ **(f)** $-2x_1x_2(m_1m_2 + 1) = 0$ **(g)** Since $x_1 \neq 0$ and $x_2 \neq 0$, this implies $m_2m_1 + 1 = 0$, so $m_1m_2 = -1$. **(h)** The product of the slopes of perpendicular lines, neither of which is parallel to an axis, is -1. **53.** 3 **54.** 3 **55.** equal **56.** $\sqrt{10}$ **57.** $2\sqrt{10}$ **58.** $3\sqrt{10}$ **59.** The sum is $3\sqrt{10}$, which is equal to the answer in Exercise 58. **60.** B; C; A; C **61.** The midpoint is $(3, 3)$, which is the same as the middle entry in the table.

62. 7.5 **63.** **(a)** $y = 11x + 117$ **(b)** 11 mph **(c)** 117 mi **(d)** 130.75 mi **65.** **(a)** $y = 280x + 1480$ **(b)** 4000

67. **(a)** a linear relation

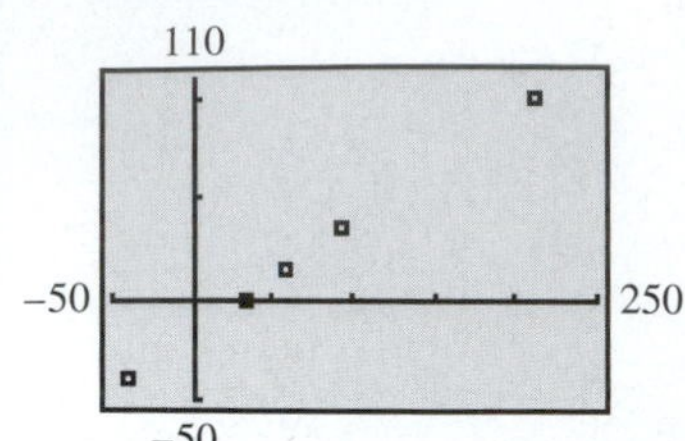

(b) $C(x) = \frac{5}{9}(x - 32)$; A slope of $\frac{5}{9}$ means the Celsius temperature changes 5° for every 9° change in the Fahrenheit temperature. **(c)** $28\frac{1}{3}$°C

69. **(a)** $y - 51.5 = \frac{81}{220}(x - 1980)$ **(b)** 1990: 55.2%; 1995: 57.0%; 2000: 58.9%; The value for 2000 is close, but the other two values are too low. **71.** **(a)** $y \approx 751.375x - 1{,}486{,}152.426$ **(b)**

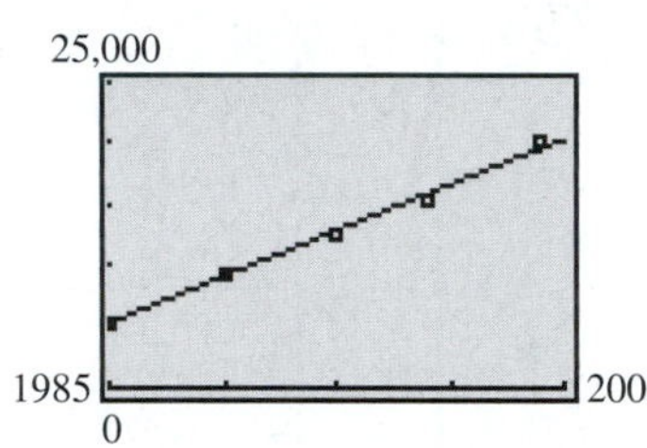

(c) It is about $10,587, which is quite close to the actual value of $10,498.

73. **(a)** $y \approx 14.68x + 277.82$ **(b)** about 2500 light-years **75.** $y \approx .101x + 11.6$; $r \approx .909$; strong, positive correlation since r is close to 1

Reviewing Basic Concepts *(page 50)*

1. $f(x) = 1.4x - 3.1; f(1.3) = -1.28$ **2.** x-intercept: $\frac{1}{2}$; y-intercept: 1; domain: $(-\infty, \infty)$; range: $(-\infty, \infty)$; slope: -2

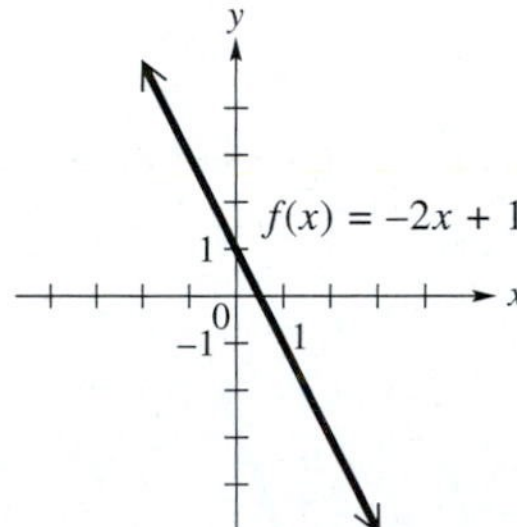

3. $\frac{2}{7}$ **4.** $x = -2$; $y = 10$ **5.**

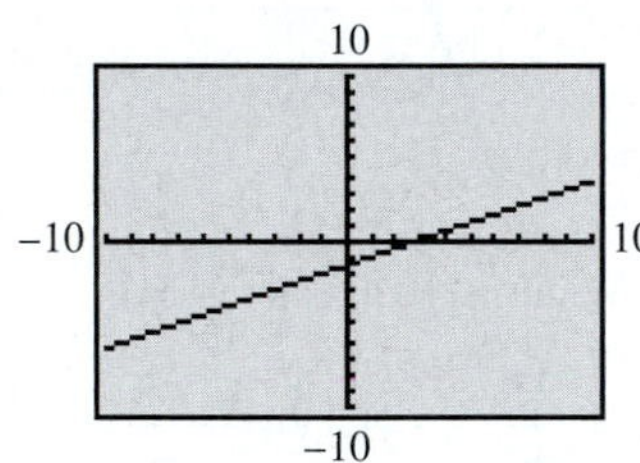

X	Y1
-3	-2.9
-2	-2.4
-1	-1.9
0	-1.4
1	-.9
2	-.4
3	.1

Y1=.5X−1.4

6. $y = 2x - 3$ **7.** $y = -\frac{2}{7}x + \frac{24}{7}$ **8.** $y = -\frac{2}{3}x + \frac{7}{3}$

9. (a)

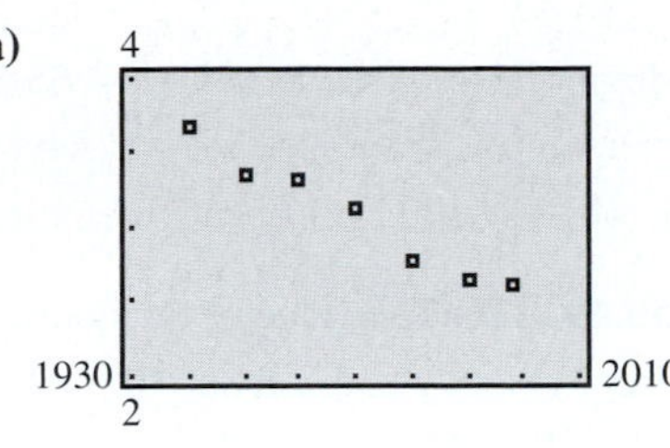

(b) negative **(c)** $y \approx -.01904x + 40.577$; $r \approx -.9792$
(d) 2.97; The estimate is close to the actual value of 2.94.

1.5 Exercises *(pages 62–66)*

1. -4 **3.** 0 **5.** $-\frac{23}{19}$ **7.** $\frac{8}{3}$ **9.** $-\frac{29}{6}$ **11.** $\{10\}$ **13.** $\{1\}$ **15.** $\{3\}$ **17.** When $x = 10$ is substituted into y_1 or y_2, the result is $y = 20$. **19.** $\emptyset$; contradiction **21.** $\{12\}$ **23.** $\{3\}$ **25.** $\{-2\}$ **27.** $\{7\}$ **29.** $\left\{\frac{17}{10}\right\}$ **31.** $\left\{\frac{5}{17}\right\}$ **33.** $\{75\}$ **35.** $\{-1\}$ **37.** $\{0\}$ **39.** $\left\{\frac{5}{4}\right\}$ **41.** $\{4\}$ **43.** $\{1.5\}$ **45.** $\{16.07\}$ **47.** $\{-1.46\}$ **49.** $\{-3.92\}$ **51.** identity; $(-\infty, \infty)$ **53.** identity; $(-\infty, \infty)$ **55.** contradiction; $\emptyset$ **57.** identity; $(-\infty, \infty)$ **59.** contradiction; $\emptyset$ **61.** $\{3\}$ **63.** $(3, \infty)$ **65.** $(-\infty, 3]$ **67.** $[3, \infty)$ **69.** $[3, \infty)$ **71. (a)** $(20, \infty)$ **(b)** $(-\infty, 20)$ **(c)** $[20, \infty)$ **(d)** $(-\infty, 20]$ **73. (a)** $\{4\}$ **(b)** $(4, \infty)$ **(c)** $(-\infty, 4)$ **75. (a)** $(8, \infty)$ **(b)** $(-\infty, 8]$ **77. (a)** $(5, \infty)$ **(b)** $(-\infty, 5]$ **79. (a)** $(-\infty, -3]$ **(b)** $(-3, \infty)$ **81. (a)** $\left(-\frac{3}{4}, \infty\right)$ **(b)** $\left(-\infty, -\frac{3}{4}\right]$ **83.** $(-\infty, 15]$ **85.** $(-6, \infty)$ **87.** $[-8, \infty)$ **89.** $(25, \infty)$ **91.** $\emptyset$ **93. (a)** It is moving away, since distance is increasing. **(b)** 1 hr; 3 hr **(c)** $[1, 3]$ **(d)** $(1, 6]$ **95.** $[1, 4]$ **97.** $(-6, -4)$ **99.** $(-16, 19]$ **101.** $\left[\frac{3\sqrt{2} - 1}{2}, \frac{3\sqrt{5} - 1}{2}\right]$ **103.** $\{3\}$; one; one

104. $(-\infty, 3)$; $(3, \infty)$; The value of $x = 3$ represents the boundary between the sets of real numbers given by $(-\infty, 3)$ and $(3, \infty)$.
105. $\{1.5\}$; $(1.5, \infty)$; $(-\infty, 1.5)$ **106. (a)** $\left\{-\frac{b}{a}\right\}$ **(b)** $\left(-\infty, -\frac{b}{a}\right)$; $\left(-\frac{b}{a}, \infty\right)$ **(c)** $\left(-\frac{b}{a}, \infty\right)$; $\left(-\infty, -\frac{b}{a}\right)$

107. (a) below .65 mi or $[0, .65)$ **(b)** $\left[0, \frac{15}{23.2}\right)$ **109.** $3.98\pi \le C \le 4.02\pi$

1.6 Exercises *(pages 73–78)*

1. 30 L **3.** A **5.** D **7.** 30 cm **9.** 7.6 cm **11.** 14 cm **13.** width: 21 in.; length: 28 in.; as a 35-in. screen **15.** 6 cm **17.** 612.86 yd **19.** $7\frac{1}{2}$ gal **21.** $2\frac{2}{3}$ gal **23.** 4 mL **25.** 8 L **27.** $266\frac{2}{3}$ gal

29. (a) $V(x) = 900x$ **(b)** $\frac{3}{50}x$ **(c)** $A = 2.4$ ach **(d)** $3.\overline{3}$ times **31. (a)** approximately .000021 for each individual **(b)** $C \approx .000021x$ **(c)** approximately 2 cases **(d)** approximately 413,000 cases

33. (a) $C(x) = .02x + 200$ **(b)** $R(x) = .04x$ **(c)** 10,000 **(d)** for $x < 10{,}000$, a loss; for $x > 10{,}000$, a profit

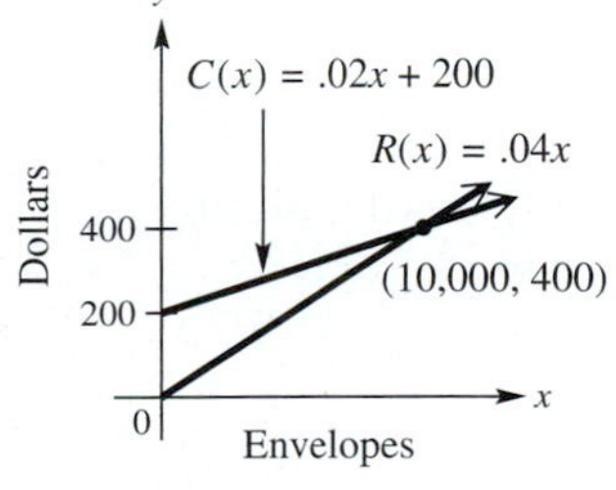

35. (a) $C(x) = 3.00x + 2300$ **(b)** $R(x) = 5.50x$ **(c)** 920 **(d)** for $x < 920$, a loss; for $x > 920$, a profit

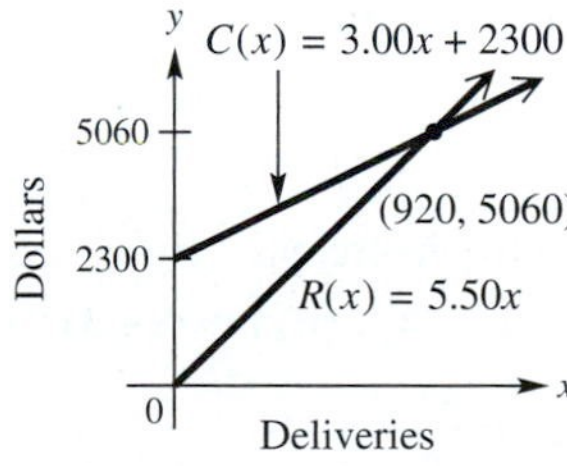

37. $k = 2.5$; $y = 20$ when $x = 8$ **39.** $k = .06$; $x = \$85$ when $y = \$5.10$ **41.** $30\frac{1}{3}$ lb per in.^{2} **43.** \$1048; $k = 65.5$

45. 46.7 ft^3 **47.** $51\frac{3}{7}$ ft **49.** $14\frac{1}{6}$ in. **51.** 12,500 **53.** **(a)** $y = 640x + 1100$ **(b)** \$17,100 **(c)** Locate the point (25, 17,100) on the graph of $y = 640x + 1100$. **55.** **(a)** $y = 2600x + 120{,}000$ **(b)** \$156,400 **(c)** The value of the house increased, on average, by \$2600 per year. **57.** **(a)** $A(x) = 806{,}400x$ **(b)** 24,192,000 gal **(c)** $P(x) = 40.32x$; about 40 pools **(d)** about 24.8 days **59.** **(a)** 500 cm^3 **(b)** 90°C **(c)** −273°C **61.** $P = \frac{I}{RT}$ **63.** $W = \frac{P - 2L}{2}$ or $W = \frac{P}{2} - L$

65. $h = \frac{2A}{b_1 + b_2}$ **67.** $H = \frac{S - 2LW}{2W + 2L}$ **69.** $h = \frac{3V}{\pi r^2}$ **71.** $n = \frac{2S}{a_1 + a_n}$ **73.** $g = \frac{2s}{t^2}$ **75.** short-term note: \$100,000; long-term note: \$140,000 **77.** \$10,000 at 2.5%; \$20,000 at 3% **79.** \$50,000 at 1.5%; \$90,000 at 4%

Reviewing Basic Concepts *(pages 78–79)*

1. $\{2\}$ **2.** $\{.757\}$ **3.** $\left\{-\frac{1}{4}\right\}$ **4.** **(a)** contradiction; $\emptyset$ **(b)** identity; $(-\infty, \infty)$ **(c)** conditional equation; $\left\{\frac{11}{4}\right\}$

5. $\left(-\infty, -\frac{5}{7}\right)$ **6.** $\left(-\frac{5}{2}, 3\right]$ **7.** **(a)** $\{2\}$ **(b)** $[2, \infty)$ **8.** 40.5 ft **9.** **(a)** $R(x) = 5.5x$ **(b)** $C(x) = 1.5x + 2500$ **(c)** 625 discs **10.** $h = \frac{V}{\pi r^2}$

Chapter 1 Review Exercises *(pages 82–86)*

1. $\sqrt{612} = 6\sqrt{17}$ **3.** −4 **5.** $-\frac{3}{4}$ **7.** 36 **9.** 11 **11.** **(a)** −3 **(b)** $y = -3x + 2$ **(c)** (.5, .5) **(d)** $\sqrt{90} = 3\sqrt{10}$

13. C **15.** A **17.** E **19.** domain: $[-6, 6]$; range: $[-6, 6]$ **21.** I **23.** B **25.** I **27.** O

29. $\left\{\frac{46}{7}\right\}$ **31.** $\left\{-\frac{5}{7}\right\}$ **33.** $\left\{\frac{7}{5}\right\}$ **35.** $(-\infty, 3)$ **37.** $\left(-\frac{10}{3}, \frac{46}{3}\right]$ **39.** $C(x) = 30x + 150$ **41.** 20

43. $f = \frac{AB(p + 1)}{24}$ **45.** **(a)** 41°F **(b)** about 21,000 ft **(c)** Graph $y = -3.52x + 58.6$. Then find the coordinates of the point where $x = 5$ to support the answer in **(a)**. Finally, find the coordinates of the point where $y = -15$ to support the answer in **(b)**.

47. **(a)** $y \approx .1199x - 234.4228$
(b) The model fits the data very well.

10
1978 2005
0

(c) The model yields \$5.26, which is \$.11 too high.

49. $y = 4x + 120$; 4 ft **51.** 1 hr, 8 min, 44 sec; This time is faster by 56 min, 11 sec. **53.** 7500 kg per m^2 **55.** **(a)** $f(x) = 1.2x - 1886.4$ **(b)** 517 **(c)** 523 **57.** $3\frac{3}{7}$ L **59.** $[300, \infty)$; 300 or more DVDs

Chapter 1 Test *(pages 87–88)*

1. **(a)** **(i)** $(-\infty, \infty)$ **(ii)** $[2, \infty)$ **(iii)** no x-intercepts **(iv)** 3 **(b)** **(i)** $(-\infty, \infty)$ **(ii)** $(-\infty, 0]$ **(iii)** 3 **(iv)** −3 **(c)** **(i)** $[-4, \infty)$ **(ii)** $[0, \infty)$ **(iii)** −4 **(iv)** 2 **2.** **(a)** $\{-4\}$ **(b)** $(-\infty, -4)$ **(c)** $[-4, \infty)$ **(d)** $\{-4\}$ **3.** **(a)** $\{5.5\}$ **(b)** $(-\infty, 5.5)$

(c) $(5.5, \infty)$ **(d)** $(-\infty, 5.5]$

4. (a) $\{-1\}$; The check leads to $-3 = -3$. **(b)** $(-1, \infty)$; The graph of $y_1 = f(x)$ is *above* the graph of $y_2 = g(x)$ for domain values greater than -1.

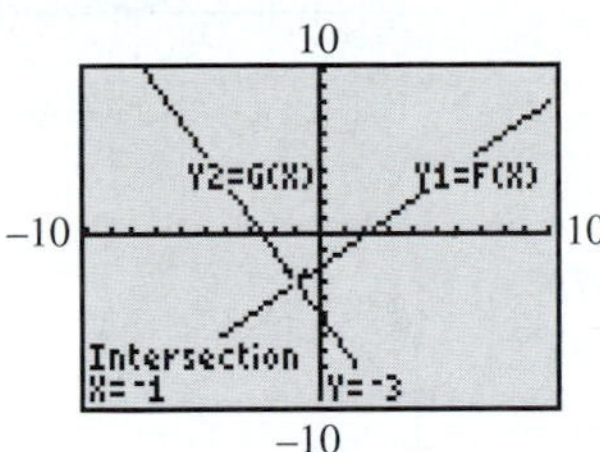

(c) $(-\infty, -1)$; The graph of $y_1 = f(x)$ is *below* the graph of $y_2 = g(x)$ for domain values less than -1.

5. (a) $\{-8\}$ **(b)** $[-8, \infty)$ **(c)** The x-intercept is -8, supporting the result of part (a). The graph of the linear function lies below or on the x-axis for domain values greater than or equal to -8, supporting the result of part (b).

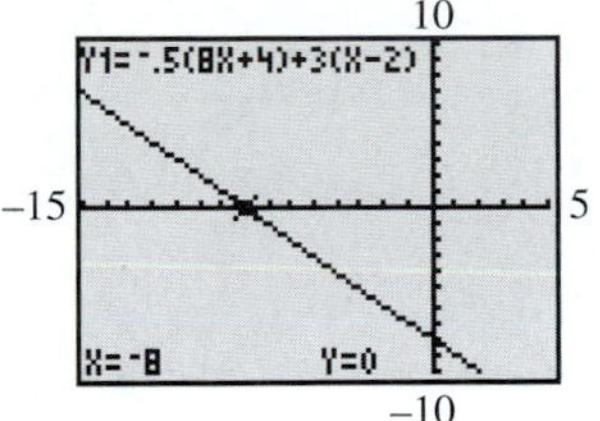

6. (a) \$21.51 **(b)** $m = 1.3205$; This means that during the years 1982 to 2002, the monthly rate increased, on the average, by about \$1.32 per year. **7. (a)** $y = -2x - 1$ **(b)** $y = -\frac{1}{2}x + \frac{7}{2}$ **8.** x-intercept: 2; y-intercept: $-\frac{3}{2}$; slope: $\frac{3}{4}$ **9.** horizontal: $y = 7$; vertical: $x = -3$ **10. (a)** $y \approx -.246x + 35.7$; $r \approx -.96$ **(b)** $y = -.246(40) + 35.7 = 25.9$°F, which is 1.1°F lower than the actual value. **11. (a)** $57{,}600\pi \approx 180{,}956$ in.^{3} **(b)** $g(x) = \frac{180{,}956}{231}x$ **(c)** 1958 gal **(d)** no; 2 downspouts

CHAPTER 2 ANALYSIS OF GRAPHS OF FUNCTIONS

2.1 Exercises *(pages 100–103)*

1. $(-\infty, \infty)$ **3.** $(0, 0)$ **5.** increases **7.** x-axis **9.** odd **11.** $(-\infty, \infty)$ **13.** $[0, \infty)$ **15.** $(-\infty, -3)$; $(-3, \infty)$ **17. (a)** $[3, \infty)$ **(b)** $(-\infty, 3]$ **(c)** none **(d)** $(-\infty, \infty)$ **(e)** $[0, \infty)$ **19. (a)** $(-\infty, 1]$ **(b)** $[4, \infty)$ **(c)** $[1, 4]$ **(d)** $(-\infty, \infty)$ **(e)** $(-\infty, 3]$ **21. (a)** none **(b)** $(-\infty, -2]$; $[3, \infty)$ **(c)** $(-2, 3)$ **(d)** $(-\infty, \infty)$ **(e)** $(-\infty, 1.5] \cup [2, \infty)$ **23.** increasing **25.** decreasing **27.** increasing **29.** decreasing **31.** decreasing **33.** increasing **35. (a)** no **(b)** yes **(c)** no **37. (a)** yes **(b)** no **(c)** no **39. (a)** yes **(b)** yes **(c)** yes **41. (a)** no **(b)** no **(c)** yes **43.** -12; -25; 21 **45. (a)** **(b)** **47.** even **61.** symmetric with respect to the origin **63.** symmetric with respect to the y-axis **65.** neither **67.** neither **69.** symmetric with respect to the y-axis **71.** symmetric with respect to the origin **73. (a)** Exercises 63, 64, 69, 70, and 72 **(b)** Exercises 61, 62, 68, and 71 **(c)** Exercises 65, 66, and 67

2.2 Exercises *(pages 108–112)*

1. $y = x^2 + 3$ **3.** $y = \sqrt{x} - 4$ **5.** $y = |x - 4|$ **7.** $y = (x + 7)^3$ **9.** To obtain the graph of g, shift the graph of f upward 4 units. **11.** B **13.** A **15.** B **17.** C **19.** B **21. (a)** $(-\infty, \infty)$ **(b)** $[-3, \infty)$ **23. (a)** $(-\infty, \infty)$

(b) $[-3, \infty)$ **25.** **(a)** $(-\infty, \infty)$ **(b)** $(-\infty, \infty)$ **27.** 4 **29.** -3 **31.**

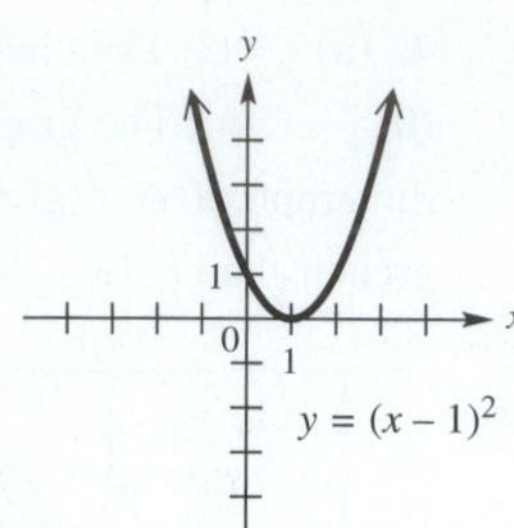

33.

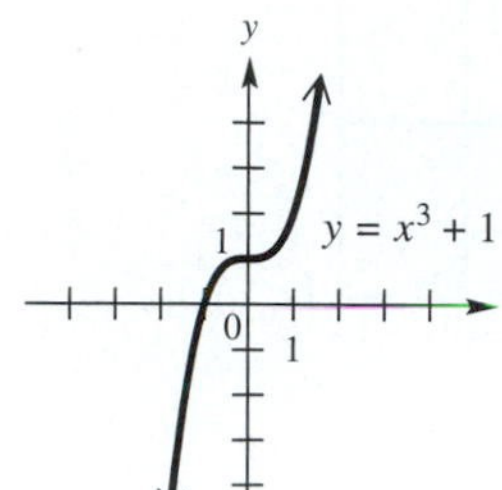

35.

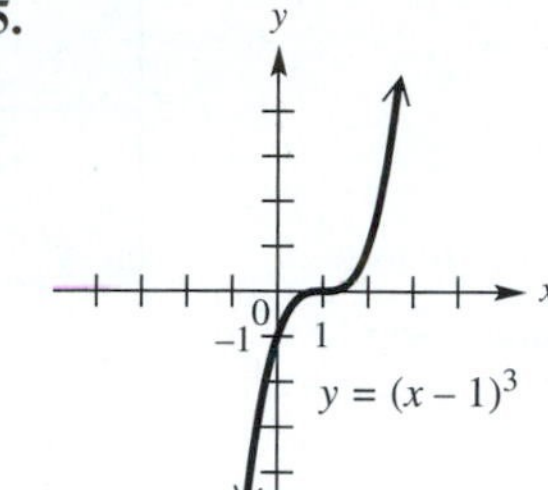

37.

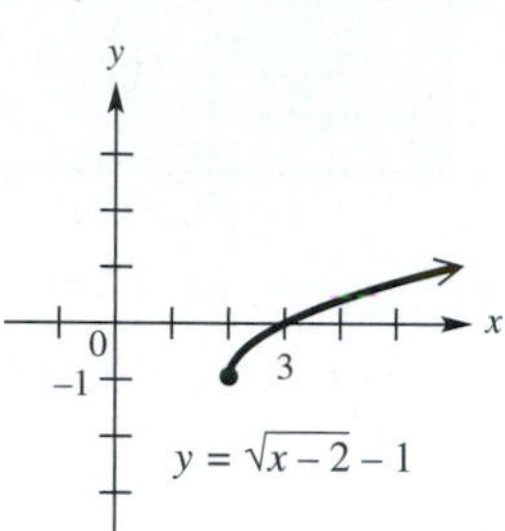

39.

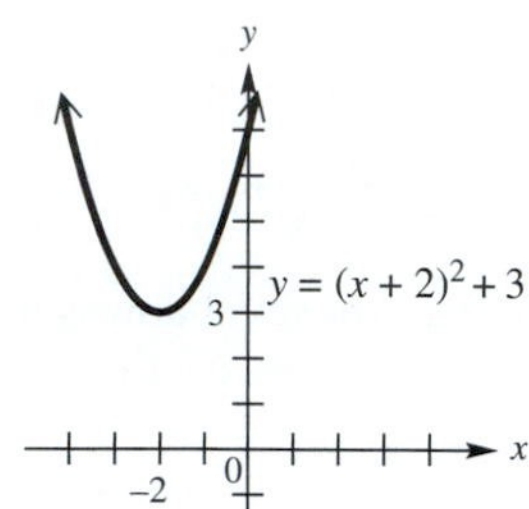

41.

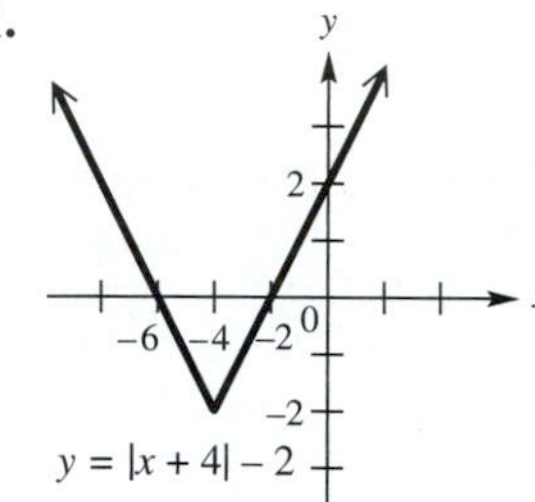

43. B **45.** A **47.**

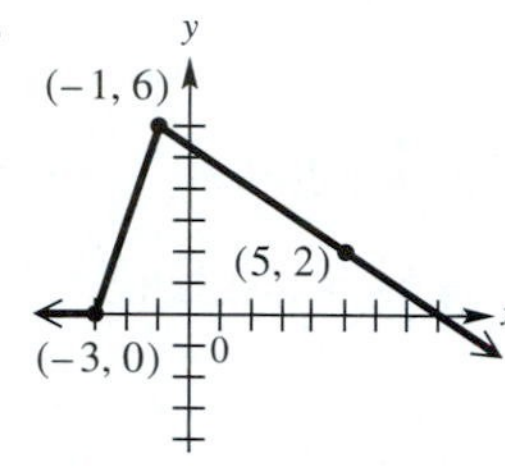

49.

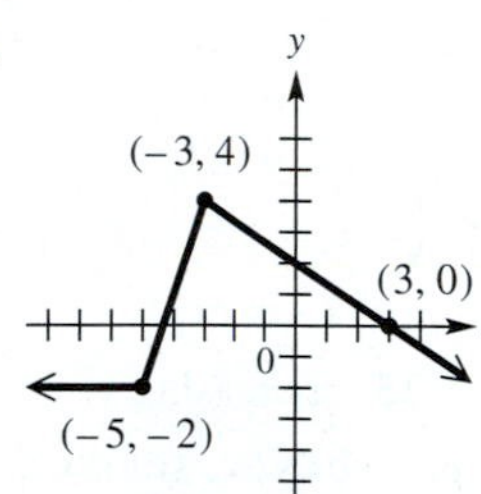

51. $y = (x + 4)^2 + 3$; **(a)** $[-4, \infty)$ **(b)** $(-\infty, -4]$ **53.** $y = x^3 - 5$; **(a)** $(-\infty, \infty)$ **(b)** none **55.** $y = \sqrt{x - 2} + 1$; **(a)** $[2, \infty)$ **(b)** none **57.** **(a)** $\{3, 4\}$ **(b)** $(-\infty, 3) \cup (4, \infty)$ **(c)** $(3, 4)$ **58.** **(a)** $\{\sqrt{2}\}$ **(b)** $(\sqrt{2}, \infty)$ **(c)** $(-\infty, \sqrt{2})$ **59.** **(a)** $\{-4, 5\}$ **(b)** $(-\infty, -4] \cup [5, \infty)$ **(c)** $[-4, 5]$ **60.** **(a)** $\emptyset$ **(b)** $[1, \infty)$ **(c)** $\emptyset$ **61.** $h = -3$; $k = 1$ **63.** **(a)** $y = 895.5(x - 1998) + 14{,}709$ **(b)** \$21,873 **65.** **(a)** $y \approx 190x + 2071.3$ **(b)** $y \approx 190(x - 1991) + 2071.3$ **(c)** about \$5300

67.

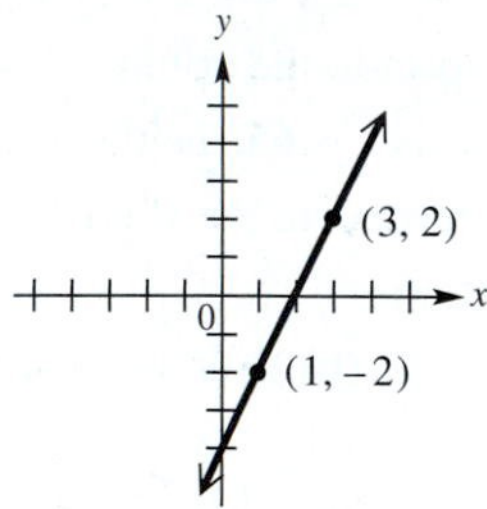

68. 2 **69.** $y_1 = 2x - 4$ **70.** $(1, 4)$ and $(3, 8)$ **71.** 2 **72.** $y_2 = 2x + 2$

73.

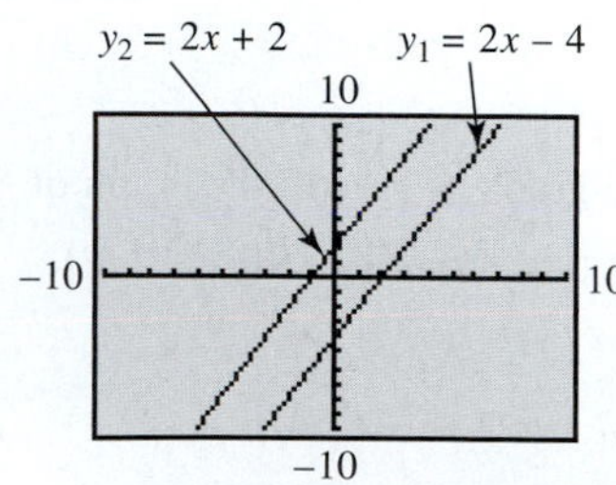

The graph of y_2 can be obtained by shifting the graph of y_1 upward 6 units. The constant, 6, comes from the 6 we added to each y-value in Exercise 70. **74.** c; c; the same as; c; upward (or positive vertical)

2.3 Exercises *(pages 121–125)*

1. $y = 2x^2$ **3.** $y = \sqrt{-x}$ **5.** $y = -3|x|$ **7.** $y = .25(-x)^3$ or $y = -.25x^3$

9.

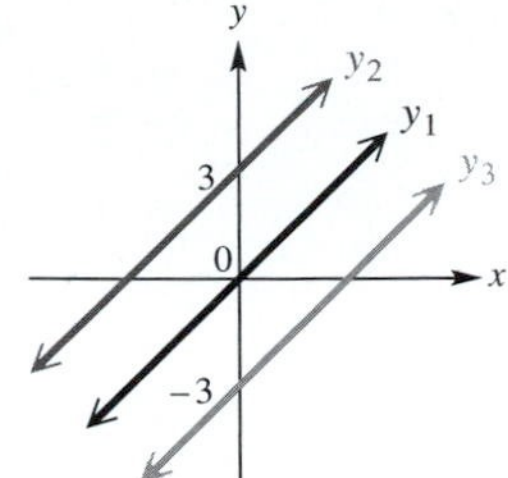

11.

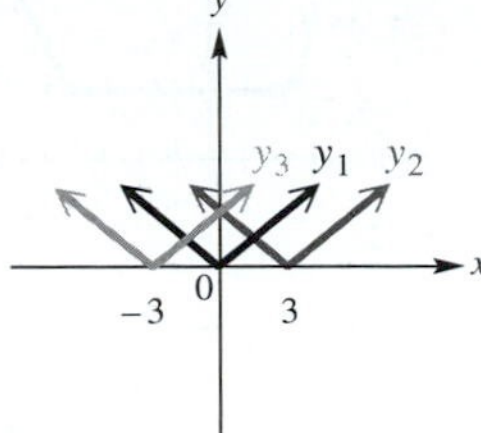

13.

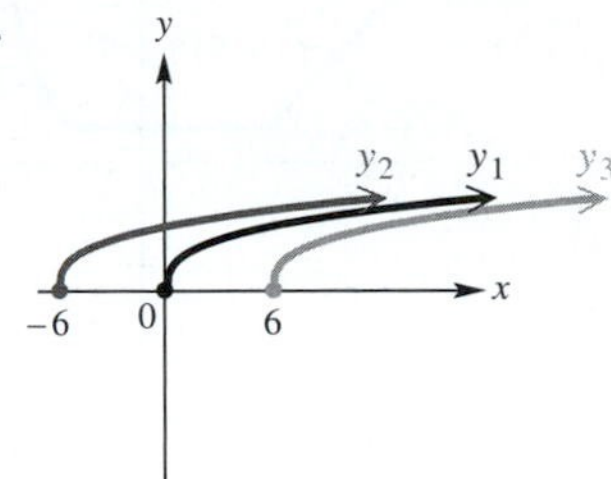

15.

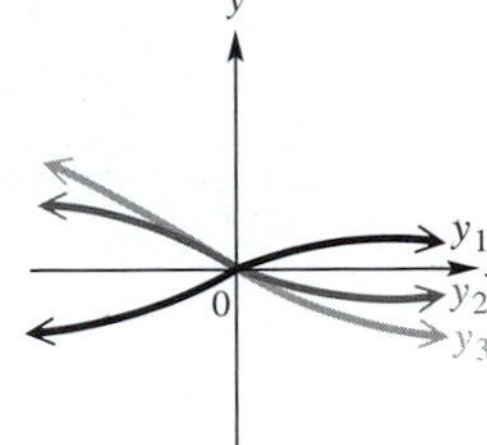

17.

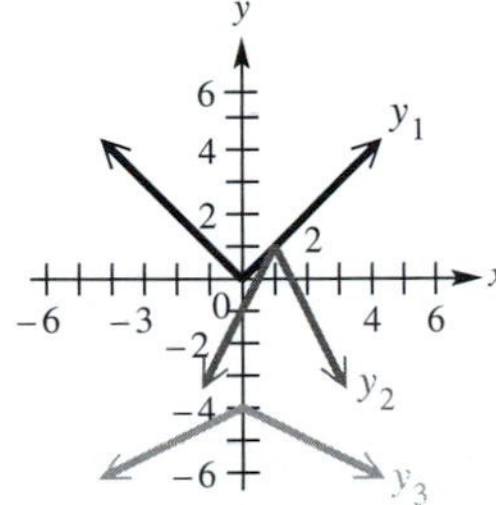

19.

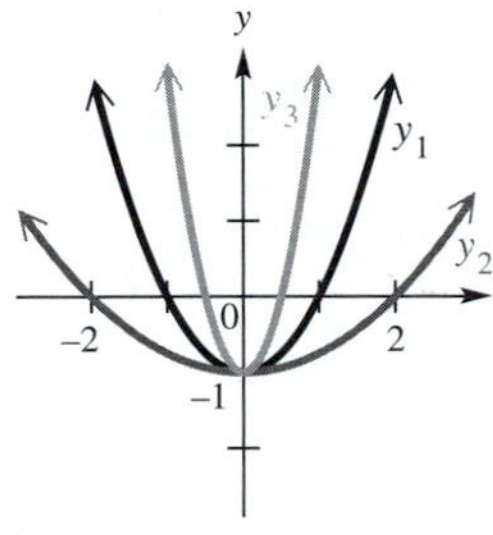

21.

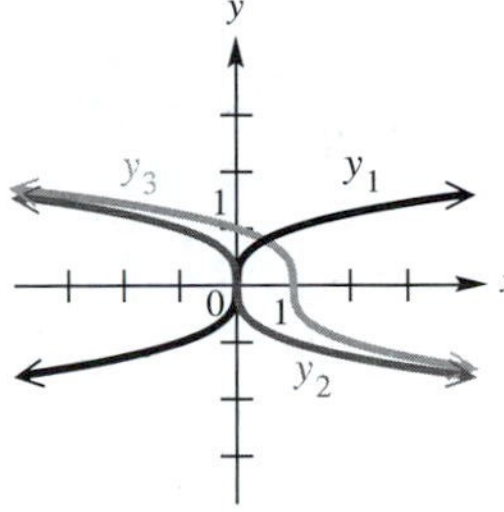

23. 4; x

25. 2; left; $\frac{1}{4}$; x; 3; downward (or negative) **27.** 3; right; 6

29. $y = \frac{1}{2}x^2 - 7$ **31.** $y = 4.5\sqrt{x - 3} - 6$ **33.** $g(x) = -(x - 5)^2 - 2$

35.

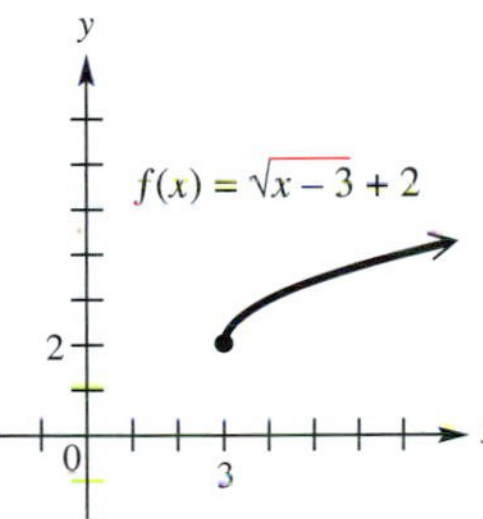

37.

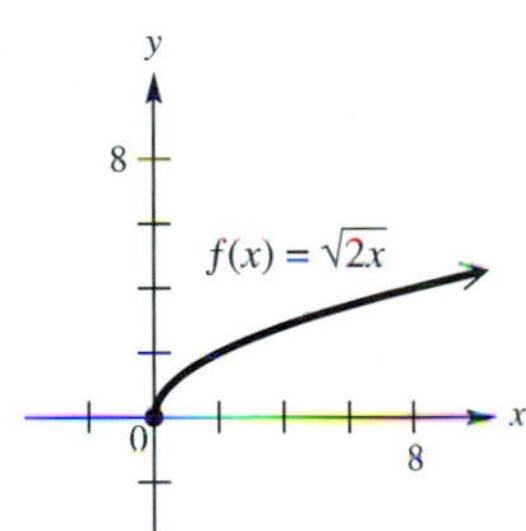

39.

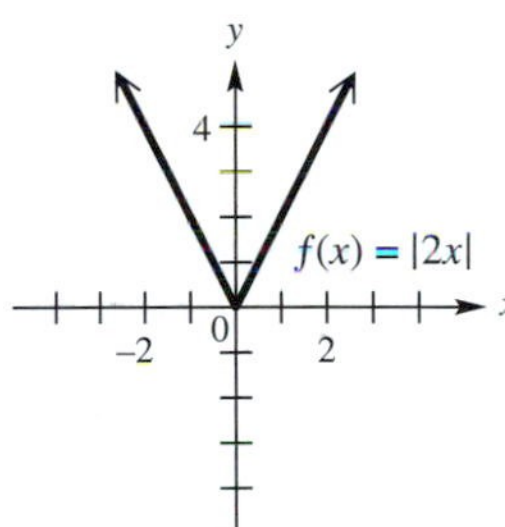

41.

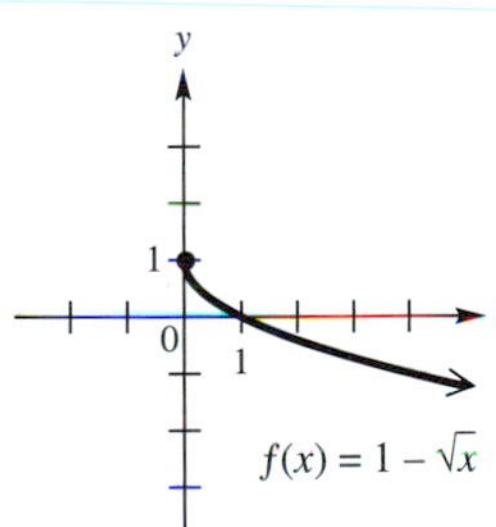

43.

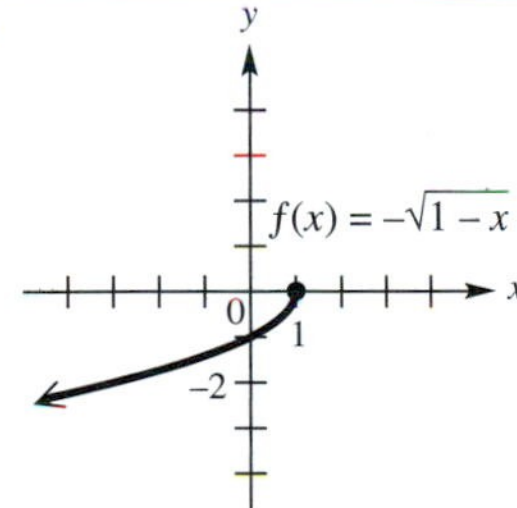

45.

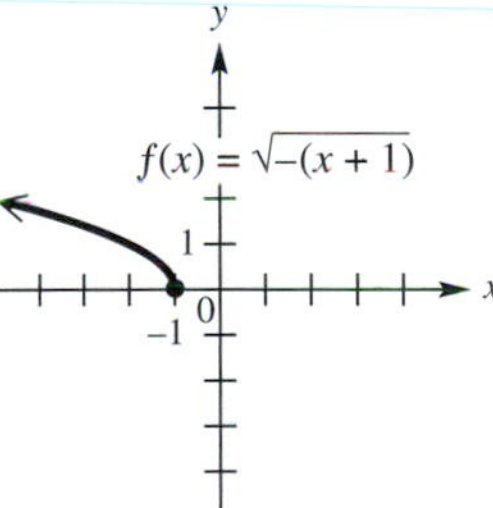

47.

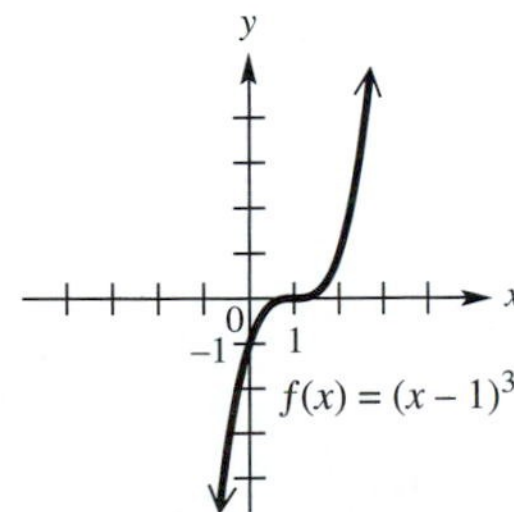

49.

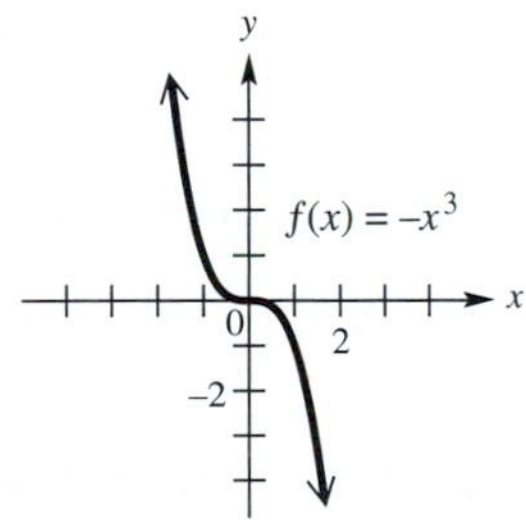

51. **(a)**

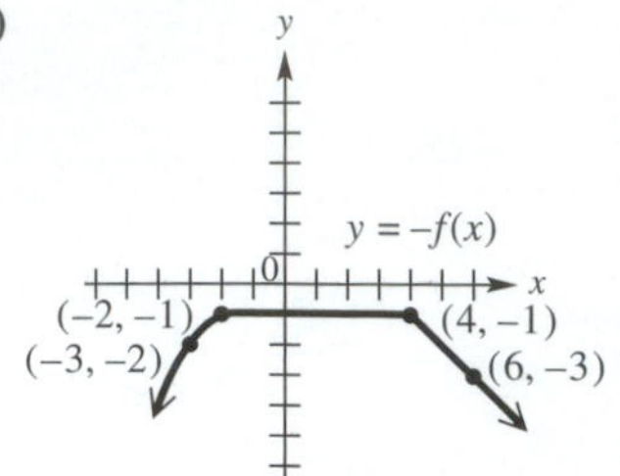

(b)

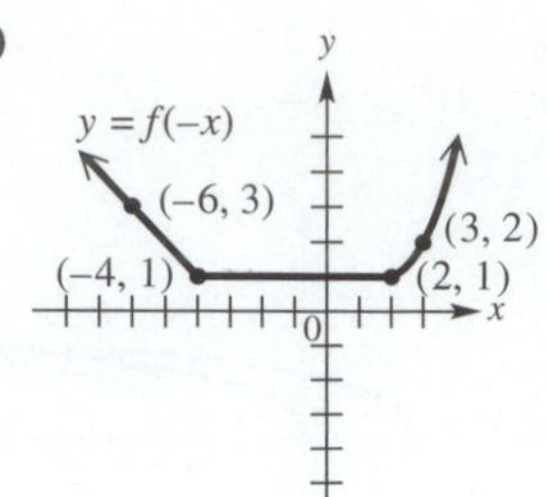

(c)

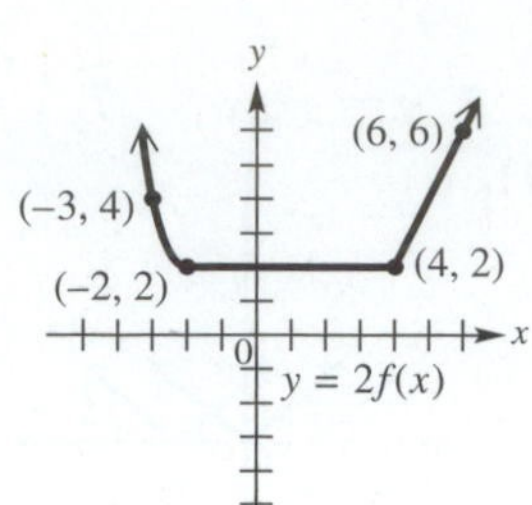

(d) $f(0) = 1$

53. **(a)**

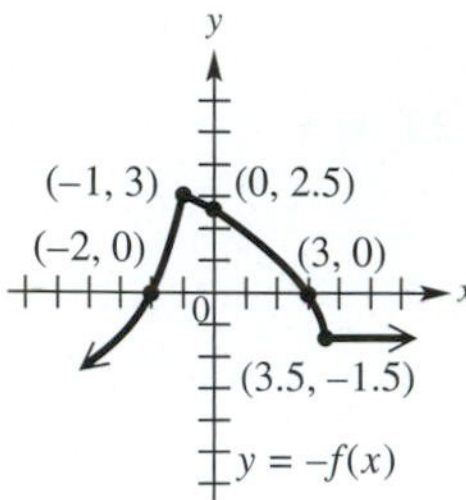

(b)

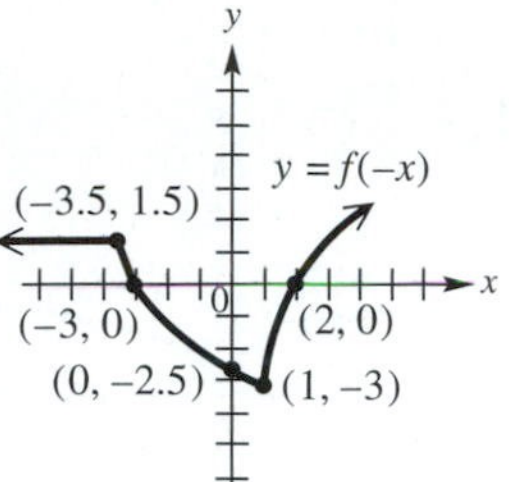

(c)

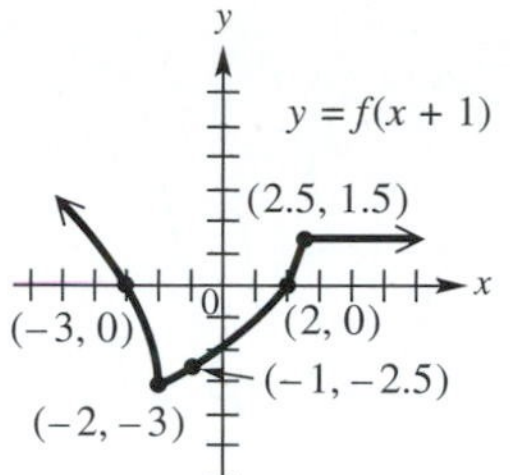

(d) -1 and 4

55. **(a)**

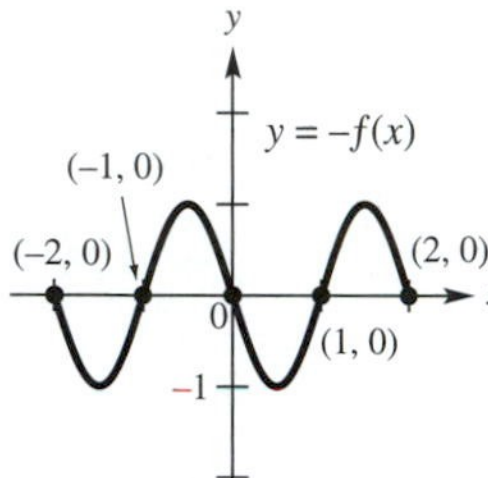

(b)

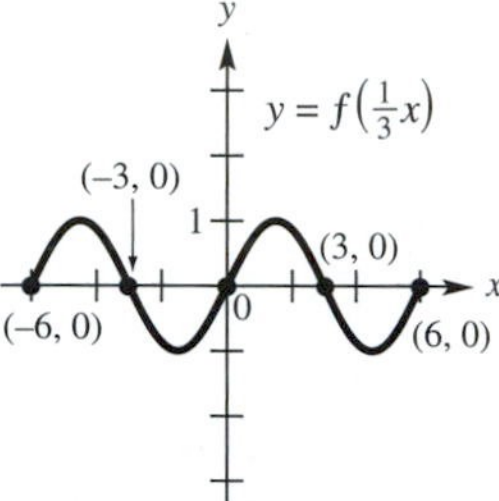

(c)

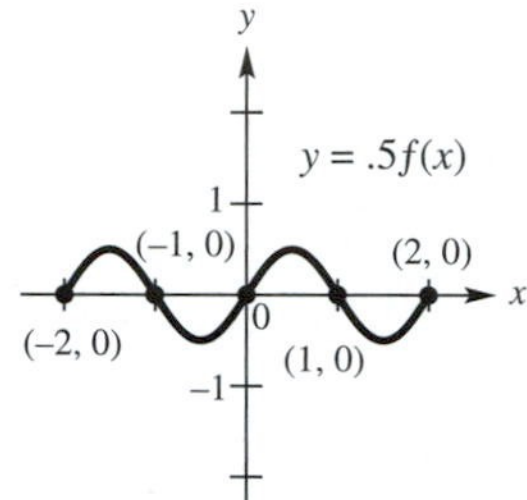

(d) symmetry with respect to the origin

57. **(a)**

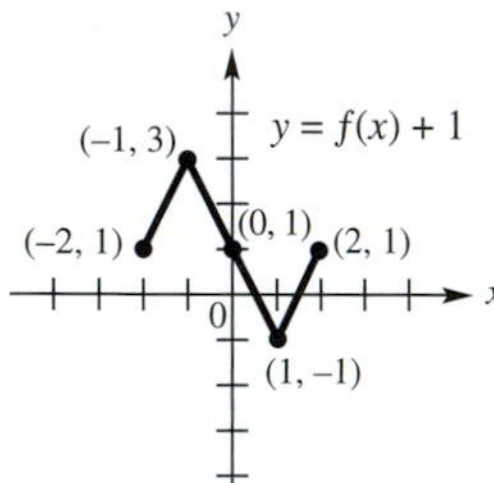

(b)

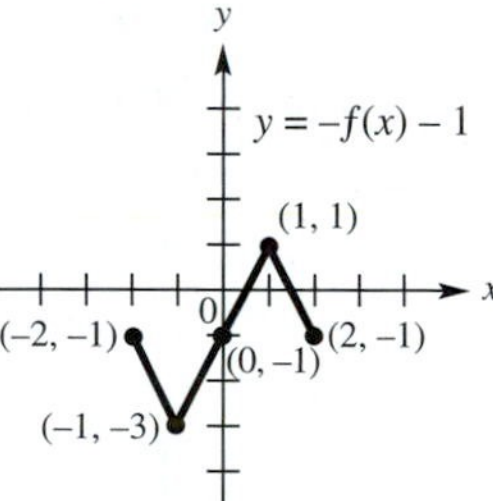

(c)

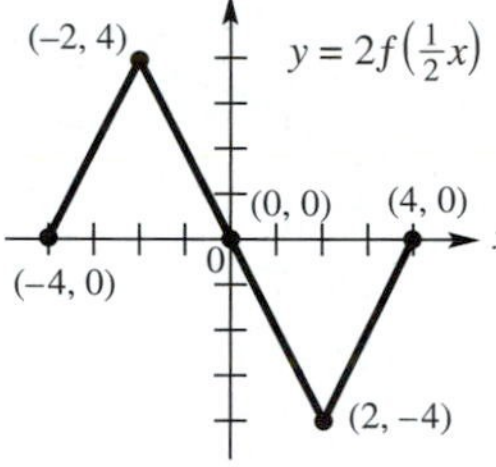

59. **(a)**

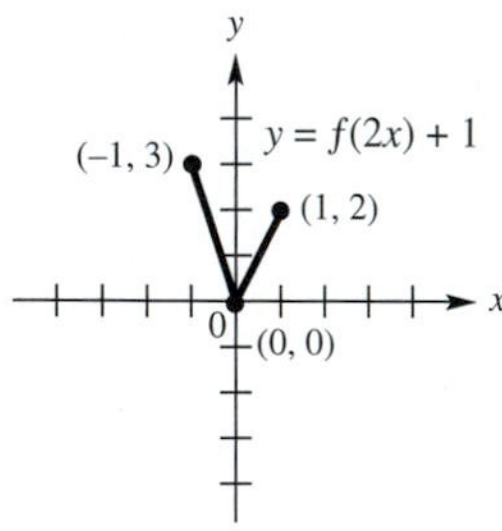

(b)

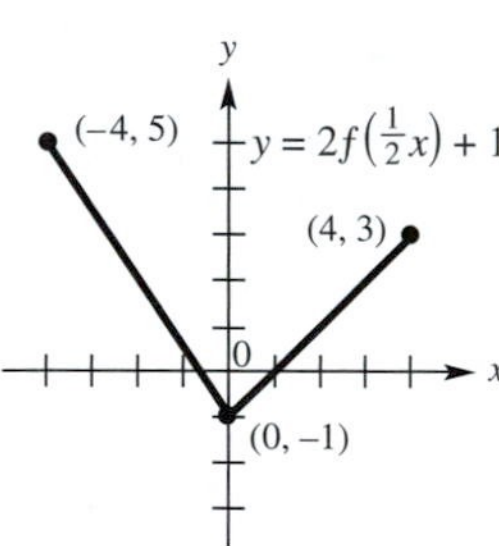

(c)

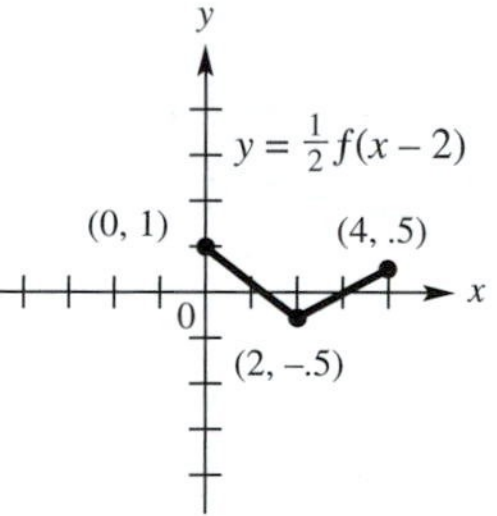

61. **(a)** r is an x-intercept. **(b)** $-r$ is an x-intercept. **(c)** $-r$ is an x-intercept. **63.** domain: $[1, 4]$; range: $[0, 3]$

65. domain: $[-1, 2]$; range: $[-3, 0]$ **67.** domain: $\left[-\frac{1}{2}, 1\right]$; range: $[0, 3]$ **69.** domain: $[-4, 8]$; range: $[0, 9]$

71. domain: $[-2, 1]$; range: $[0, 3]$ **73.** domain: $\left[-\frac{2}{3}, \frac{1}{3}\right]$; range: $[0, 3]$ **75.** domain: $[20, \infty)$; range: $[5, \infty)$

77. domain: $[-10, \infty)$; range: $(-\infty, 5]$ **79.** decreases **81.** increases **83. (a)** $[-1, 2]$ **(b)** $(-\infty, -1]$ **(c)** $[2, \infty)$
85. (a) $[1, \infty)$ **(b)** $[-2, 1]$ **(c)** $(-\infty, -2]$ **87.** $(8, 10)$ **89.** $y = 2|x + 2| - 1$ **91.** $y = -3|x| + 2$

Reviewing Basic Concepts *(pages 125–127)*

1. (a) $f(-3) = -6$ **(b)** $f(-3) = 6$ **(c)** $f(-3) = -6$ **(d)** $f(-3) = 6$ **2. (a)** B **(b)** D **(c)** E **(d)** A **(e)** C
3. (a) B **(b)** A **(c)** G **(d)** C **(e)** F **(f)** D **(g)** H **(h)** E
4. (a)

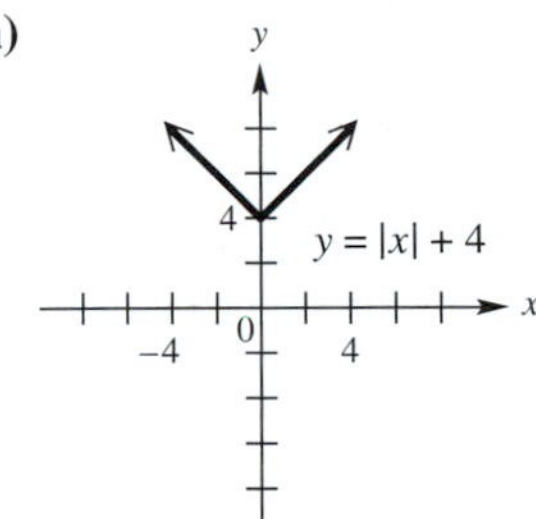

(b)

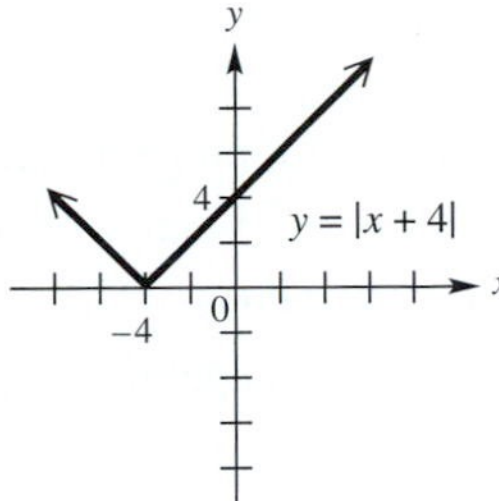

(c)

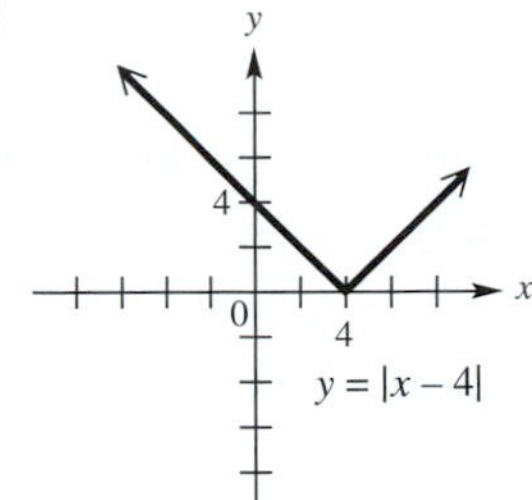

(d)

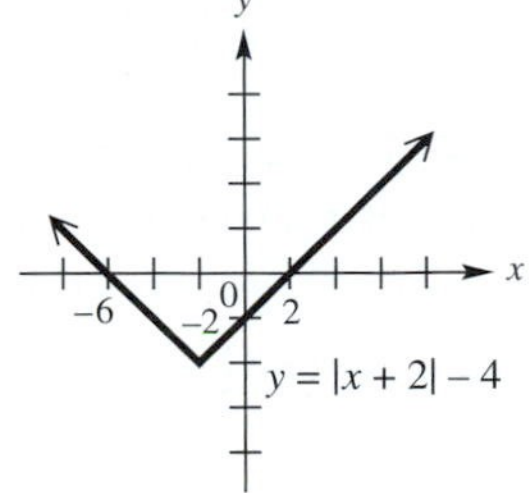

(e)

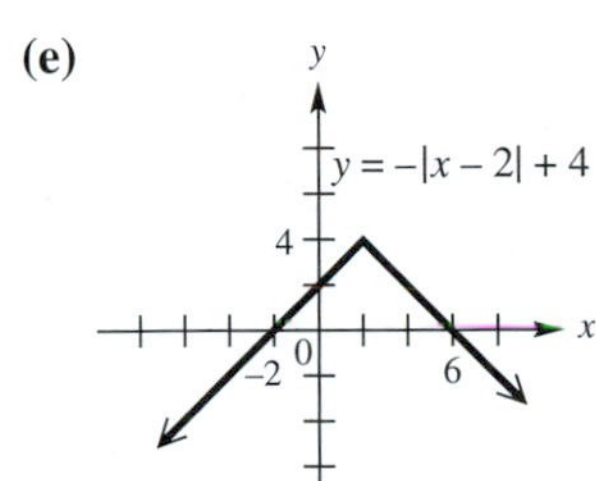

5. (a) It is the graph of $f(x) = |x|$ translated 1 unit to the left, reflected across the x-axis, and translated 3 units upward. The equation is $y = -|x + 1| + 3$. **(b)** It is the graph of $g(x) = \sqrt{x}$ translated 4 units to the left, reflected across the x-axis, and translated 2 units upward. The equation is $y = -\sqrt{x + 4} + 2$. **(c)** It is the graph of $g(x) = \sqrt{x}$ translated 4 units to the left, stretched vertically by a factor of 2, and translated 4 units downward. The equation is $y = 2\sqrt{x + 4} - 4$. **(d)** It is the graph of $f(x) = |x|$ translated 2 units to the right, shrunken vertically by a factor of $\frac{1}{2}$, and translated 1 unit downward. The equation is $y = \frac{1}{2}|x - 2| - 1$.

6. (a) 2 **(b)** 4 **7.** The graph of $y = F(x + h)$ is a horizontal translation of the graph of $y = F(x)$. The graph of $y = F(x) + h$ is not the same as the graph of $y = F(x + h)$ because the graph of $y = F(x) + h$ is a *vertical* translation of the graph of $y = F(x)$. **8.** The effect is either a (vertical) stretch or shrink, and perhaps a reflection across the x-axis. If $c > 0$, there is a stretch or shrink by a factor of c. If $c < 0$, there is a stretch or shrink by a factor of $|c|$, and a reflection across the x-axis. If $|c| > 1$, a stretch occurs; when $|c| < 1$, a shrink occurs.

9. (a)

x	$f(x)$
−3	4
−2	−6
−1	5
1	5
2	−6
3	4

(b)

x	$f(x)$
−3	4
−2	−6
−1	5
1	−5
2	6
3	−4

10. (a) $g(x) = -.279(x - 1992) + 5.532$
(b) $g(2006) = 1.626$ ppm

2.4 Exercises *(pages 133–138)*

1. 5 **3.** $[0, \infty)$ **5.** $[2, \infty)$ **7.**

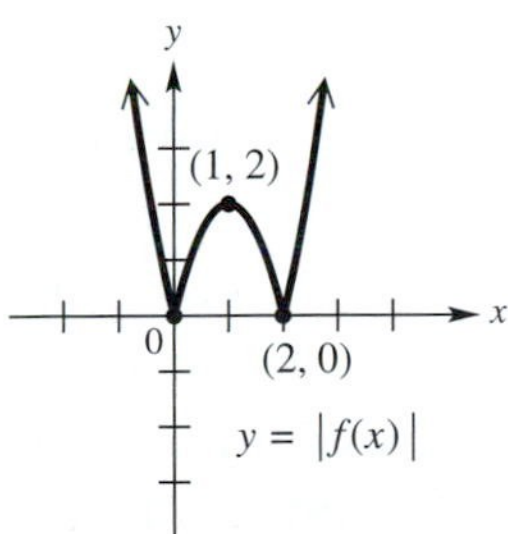

9.

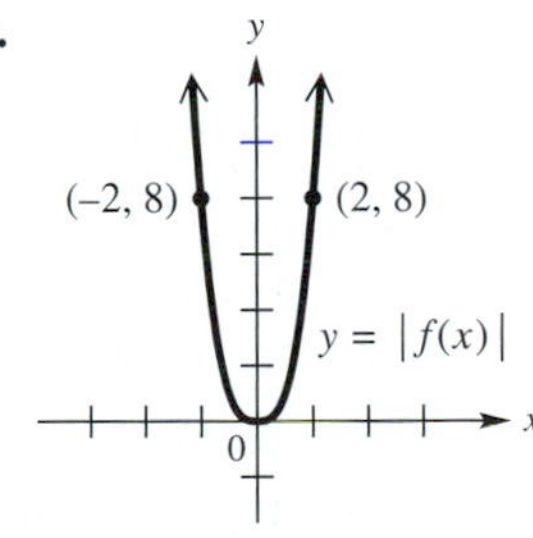

11. $y = |f(x)|$ has the same graph as $y = f(x)$. **13.**

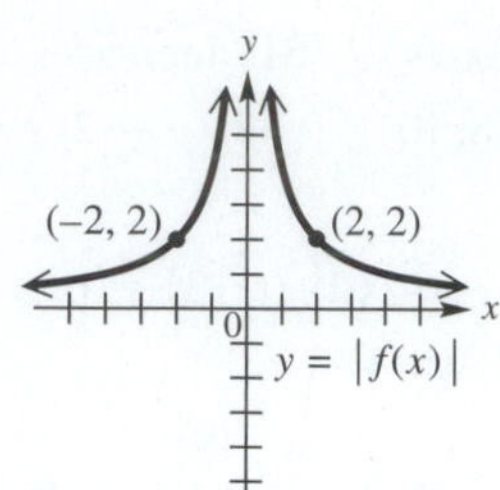

15.

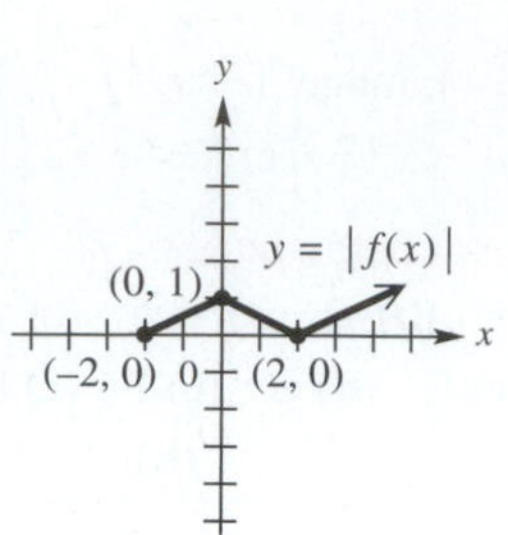

17. domain of $f(x)$: $(-\infty, \infty)$; range of $f(x)$: $[-2, \infty)$; domain of $|f(x)|$: $(-\infty, \infty)$; range of $|f(x)|$: $[0, \infty)$
19. domain of $f(x)$: $(-\infty, \infty)$; range of $f(x)$: $(-\infty, -1]$; domain of $|f(x)|$: $(-\infty, \infty)$; range of $|f(x)|$: $[1, \infty)$
21. domain of $f(x)$: $[-2, 3]$; range of $f(x)$: $[-2, 3]$; domain of $|f(x)|$: $[-2, 3]$; range of $|f(x)|$: $[0, 3]$
23. domain of $f(x)$: $[-2, 3]$; range of $f(x)$: $[-3, 1]$; domain of $|f(x)|$: $[-2, 3]$; range of $|f(x)|$: $[0, 3]$
25. (a)

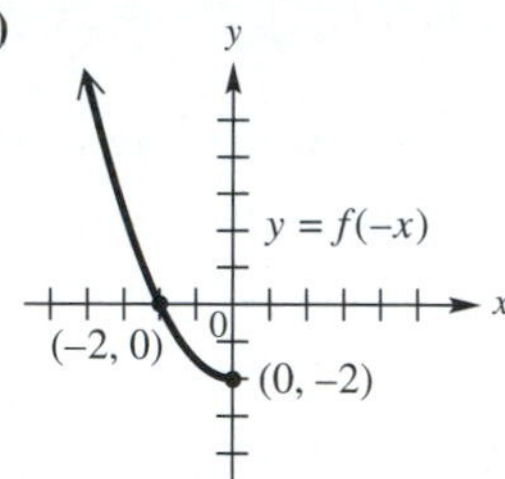

(b)

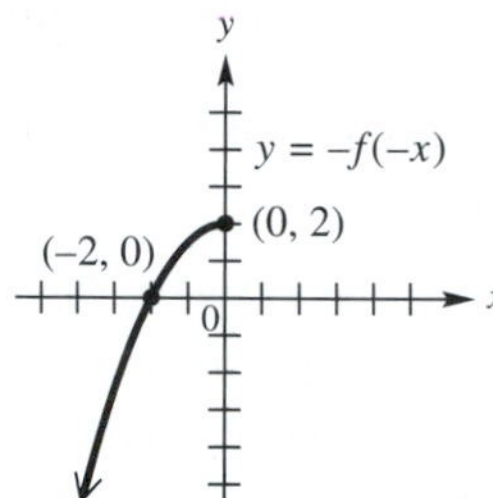

(c)

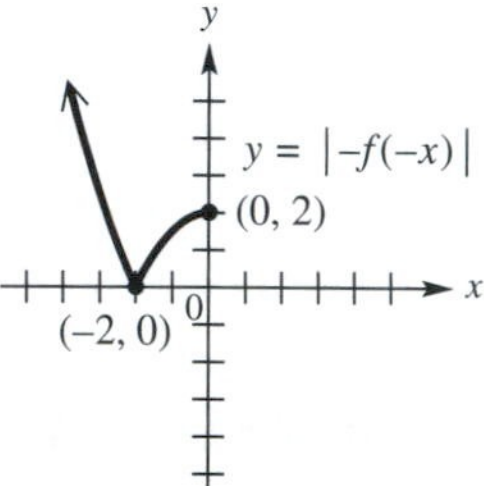

27. Figure A shows the graph of $y = f(x)$, while Figure B shows the graph of $y = |f(x)|$. **29. (a)** $\{-1, 6\}$ **(b)** $(-1, 6)$ **(c)** $(-\infty, -1) \cup (6, \infty)$ **31. (a)** $\{4\}$ **(b)** $\emptyset$ **(c)** $(-\infty, 4) \cup (4, \infty)$ **33.** The V-shaped graph is that of $f(x) = |.5x + 6|$, since this shape is typical of the graphs of absolute value functions of the form $f(x) = |ax + b|$. **34.** The straight line graph is that of $g(x) = 3x - 14$, which is a linear function. **35.** $\{8\}$ **36.** $(-\infty, 8)$ **37.** $(8, \infty)$ **38.** $\{8\}$
39. (a) $\{-13, 5\}$ **(b)** $(-\infty, -13) \cup (5, \infty)$ **(c)** $(-13, 5)$ **41. (a)** $\{2, 5\}$ **(b)** $(-\infty, 2] \cup [5, \infty)$ **(c)** $[2, 5]$
43. (a) $\left\{-\frac{3}{2}, \frac{1}{2}\right\}$ **(b)** $\left[-\frac{3}{2}, \frac{1}{2}\right]$ **(c)** $\left(-\infty, -\frac{3}{2}\right] \cup \left[\frac{1}{2}, \infty\right)$ **45. (a)** $\left\{\frac{5}{7}\right\}$ **(b)** $(-\infty, \infty)$ **(c)** $\left\{\frac{5}{7}\right\}$ **47. (a)** $\emptyset$ **(b)** $\emptyset$ **(c)** $(-\infty, \infty)$ **49.** $\left\{0, \frac{8}{3}\right\}$ **51.** $\left\{-\frac{1}{2}, 1\right\}$ **53.** $\{0, 1\}$ **55.** $(8, 22)$ **57.** $(-\infty, 1) \cup (2, \infty)$
59. $\left(-\infty, \frac{5}{3}\right] \cup \left[\frac{11}{3}, \infty\right)$ **61.** $(-\infty, 18) \cup (18, \infty)$ **63.** $\emptyset$ **65.** $(-\infty, \infty)$ **67. (a)** $\left\{-8, \frac{6}{5}\right\}$ **(b)** $(-\infty, -8) \cup \left(\frac{6}{5}, \infty\right)$ **(c)** $\left(-8, \frac{6}{5}\right)$ **69. (a)** $\left\{\frac{2}{3}, 8\right\}$ **(b)** $\left(-\infty, \frac{2}{3}\right) \cup (8, \infty)$ **(c)** $\left(\frac{2}{3}, 8\right)$ **71. (a)** $\left\{-3, \frac{5}{3}\right\}$ **(b)** $(-\infty, -3) \cup \left(\frac{5}{3}, \infty\right)$ **(c)** $\left(-3, \frac{5}{3}\right)$ **73. (a)** $\left\{-\frac{7}{8}\right\}$ **(b)** $\left(-\infty, -\frac{7}{8}\right)$ **(c)** $\left(-\frac{7}{8}, \infty\right)$ **75. (a)** $\{2, 8\}$ **(b)** $(2, 8)$ **(c)** $(-\infty, 2) \cup (8, \infty)$ **77.** $\{-3, 8\}$ **79.** $\{-2, 6\}$ **81. (a)** $19 \le T \le 67$ **(b)** The average monthly temperatures in Marquette vary between a low of 19°F and a high of 67°F. The monthly averages are always within 24° of 43°F.
83. (a) $28 \le T \le 72$ **(b)** The average monthly temperatures in Boston vary between a low of 28°F and a high of 72°F. The monthly averages are always within 22° of 50°F. **85. (a)** $49 \le T \le 74$ **(b)** The average monthly temperatures in Buenos Aires vary between a low of 49°F (possibly in July) and a high of 74°F (possibly in January). The monthly averages are always within 12.5° of 61.5°F. **87.** $[6.5, 9.5]$ **89. (a)** 9 **(b)** 113 or 147 **91.** $(-.05, .05)$ **93.** $(2.74975, 2.75025)$
95. 47°F **97.** 22°F **99.** $\{2\}$ **101.** $(-\infty, \infty)$ **103.** $\emptyset$

2.5 Exercises *(pages 144–148)*

1. (a) highest: 55 mph; lowest: 30 mph **(b)** 12 mi **(c)** $f(4) = 40$; $f(12) = 30$; $f(18) = 55$ **(d)** $x = 4, 6, 8, 12$, and 16. The speed limit changes at each discontinuity. **3. (a)** 50,000 gal; 30,000 gal **(b)** during the first and fourth days **(c)** $f(2) = 45$ thousand; $f(4) = 40$ thousand **(d)** 5000 gal per day **5. (a)** -10 **(b)** -2 **(c)** -1 **(d)** 2 **7. (a)** -3 **(b)** 1 **(c)** 0 **(d)** 9

9.

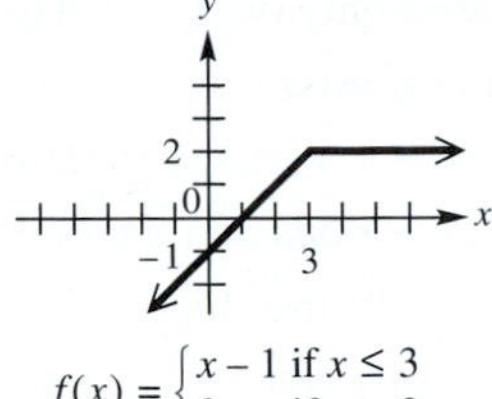

$f(x) = \begin{cases} x - 1 & \text{if } x \le 3 \\ 2 & \text{if } x > 3 \end{cases}$

11.

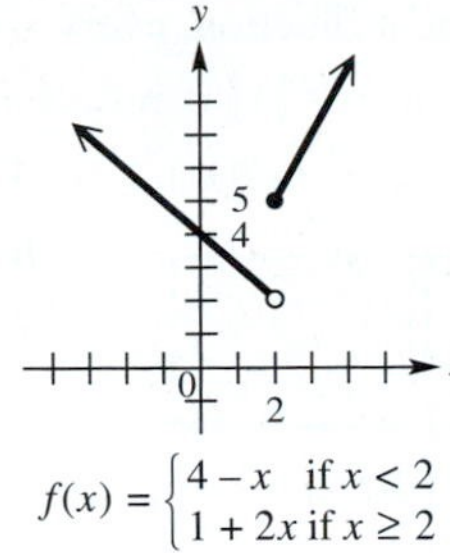

$f(x) = \begin{cases} 4 - x & \text{if } x < 2 \\ 1 + 2x & \text{if } x \ge 2 \end{cases}$

13.

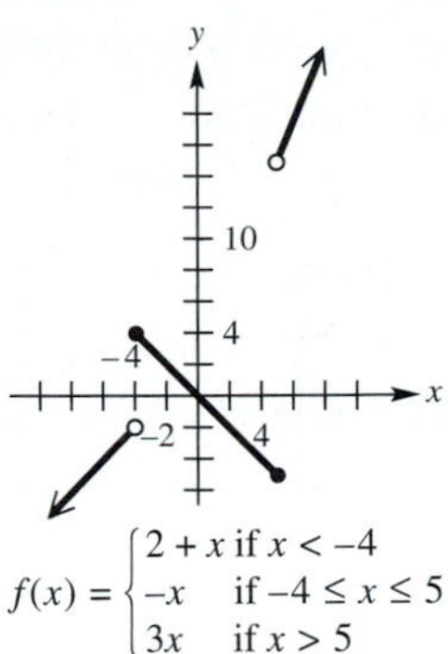

$f(x) = \begin{cases} 2 + x & \text{if } x < -4 \\ -x & \text{if } -4 \le x \le 5 \\ 3x & \text{if } x > 5 \end{cases}$

15.

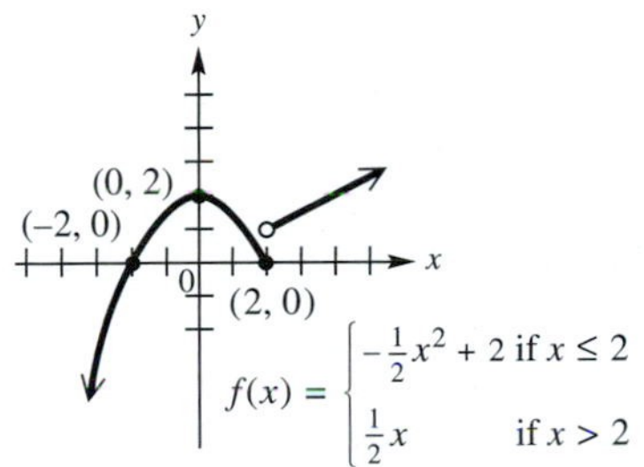

$f(x) = \begin{cases} -\frac{1}{2}x^2 + 2 & \text{if } x \le 2 \\ \frac{1}{2}x & \text{if } x > 2 \end{cases}$

17.

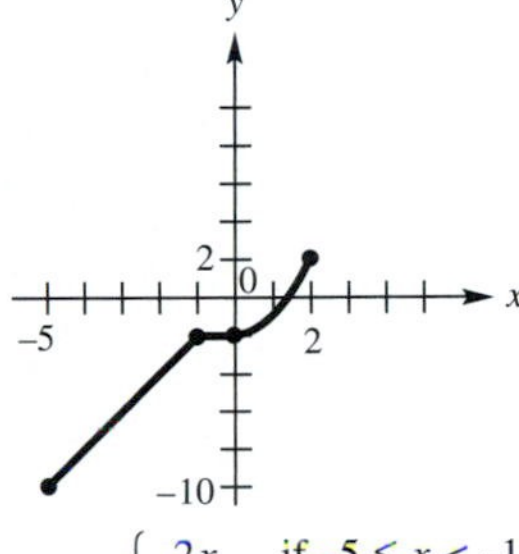

$f(x) = \begin{cases} 2x & \text{if } -5 \le x < -1 \\ -2 & \text{if } -1 \le x < 0 \\ x^2 - 2 & \text{if } 0 \le x \le 2 \end{cases}$

19.

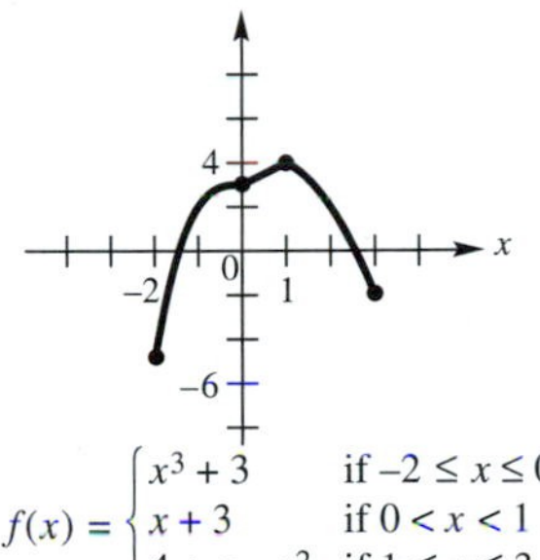

$f(x) = \begin{cases} x^3 + 3 & \text{if } -2 \le x \le 0 \\ x + 3 & \text{if } 0 < x < 1 \\ 4 + x - x^2 & \text{if } 1 \le x \le 3 \end{cases}$

21. B **23.** D **25.**

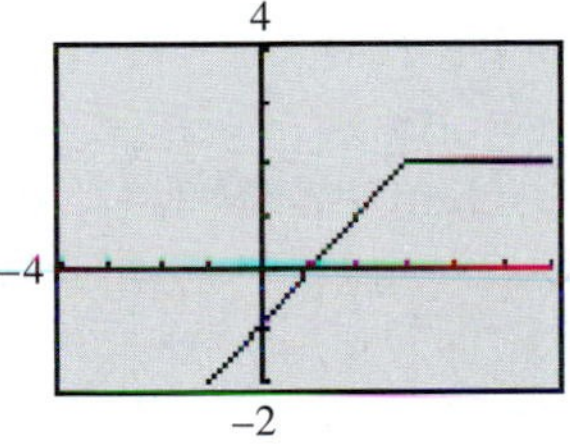

$f(x) = \begin{cases} x - 1 & \text{if } x \le 3 \\ 2 & \text{if } x > 3 \end{cases}$

27.

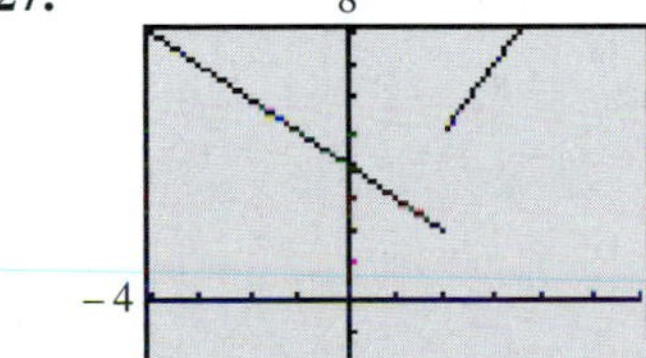

$f(x) = \begin{cases} 4 - x & \text{if } x < 2 \\ 1 + 2x & \text{if } x \ge 2 \end{cases}$

29.

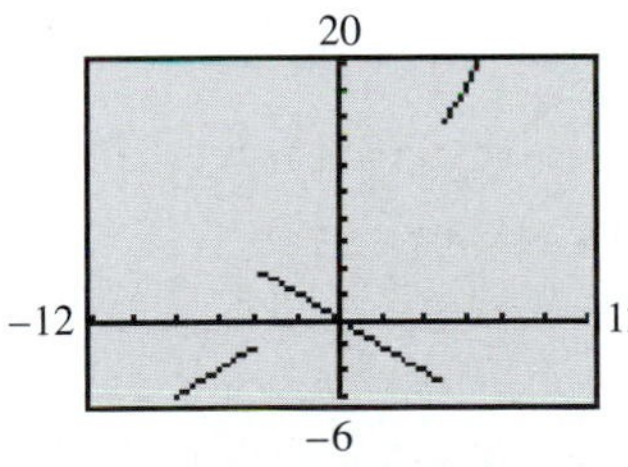

$f(x) = \begin{cases} 2 + x & \text{if } x < -4 \\ -x & \text{if } -4 \le x \le 5 \\ 3x & \text{if } x > 5 \end{cases}$

31.

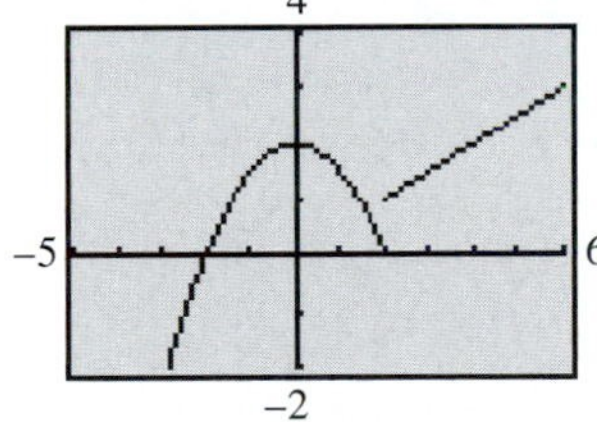

$f(x) = \begin{cases} -\frac{1}{2}x^2 + 2 & \text{if } x \le 2 \\ \frac{1}{2}x & \text{if } x > 2 \end{cases}$

33. $f(x) = \begin{cases} 2 & \text{if } x \le 0 \\ -1 & \text{if } x > 1 \end{cases}$; domain: $(-\infty, 0] \cup (1, \infty)$; range: $\{-1, 2\}$ **35.** $f(x) = \begin{cases} x & \text{if } x \le 0 \\ 2 & \text{if } x > 0 \end{cases}$; domain: $(-\infty, \infty)$; range: $(-\infty, 0] \cup \{2\}$ **37.** $f(x) = \begin{cases} \sqrt[3]{x} & \text{if } x < 1 \\ x + 1 & \text{if } x \ge 1 \end{cases}$; domain: $(-\infty, \infty)$; range: $(-\infty, 1) \cup [2, \infty)$ **39.** There is an *overlap* of intervals since the number 4 satisfies both conditions. To be a function, every x-value is used only once. **41.** The graph of $y = [\![x]\!]$ is shifted 1.5 units downward. **43.** The graph of $y = [\![x]\!]$ is reflected across the x-axis.

45.

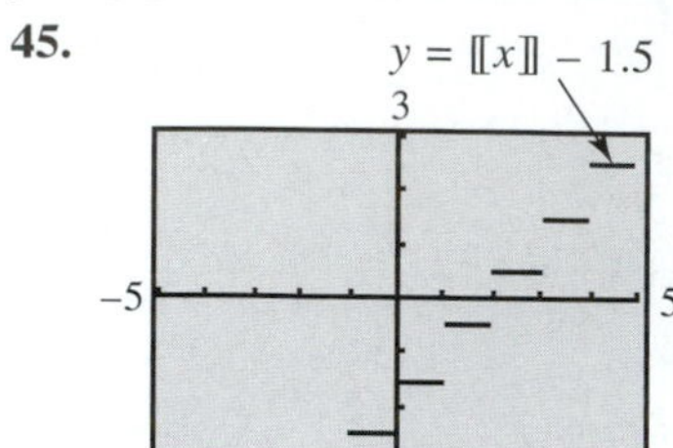

47.

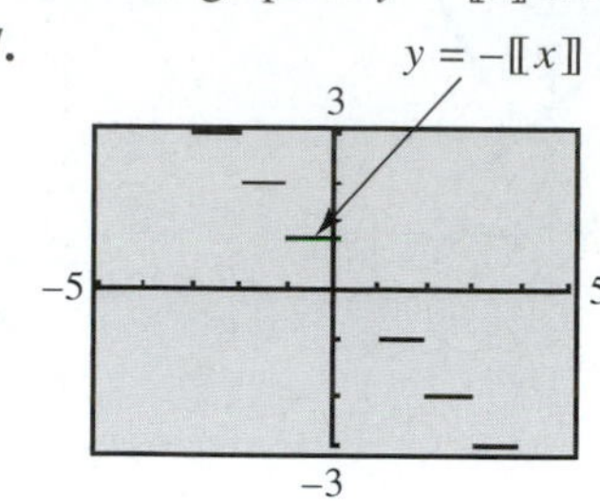

49. When $0 \le x \le 3$, the slope is 5, which means the inlet pipe is open and the outlet pipe is closed; when $3 < x \le 5$, the slope is 2, which means both pipes are open; when $5 < x \le 8$, the slope is 0, which means both pipes are closed; when $8 < x \le 10$, the slope is -3, which means the inlet pipe is closed and the outlet pipe is open.

51. **(a)** 140 **(b)** 220 **(c)** 220 **(d)** 220 **(e)** 220 **(f)** 60 **(g)** 60 **(h)**

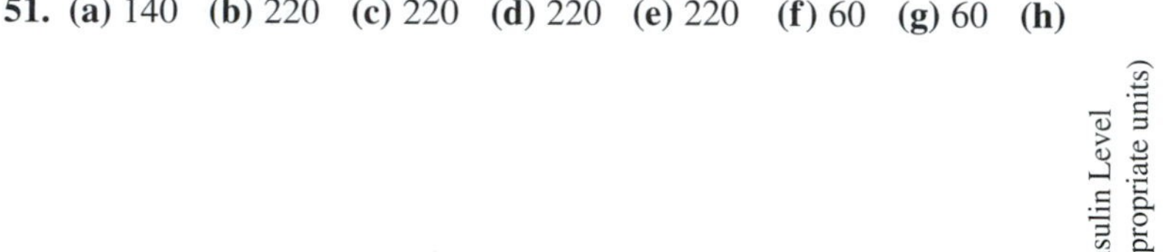

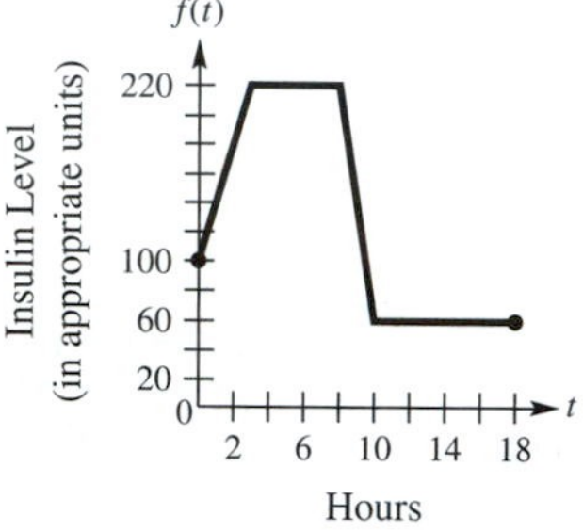

(i) If f were discontinuous, the insulin level would change instantaneously from one level to a second level. **53.** **(a)** From 1997 to 1999, there is a decrease of 700 cases per year. From 1999 to 2001, there is an increase of 200 cases per year.

(b) $f(x) = \begin{cases} -700x + 8100 & \text{if } 0 \le x \le 2 \\ 200x + 6300 & \text{if } 2 < x \le 4 \end{cases}$

55. **(a)**

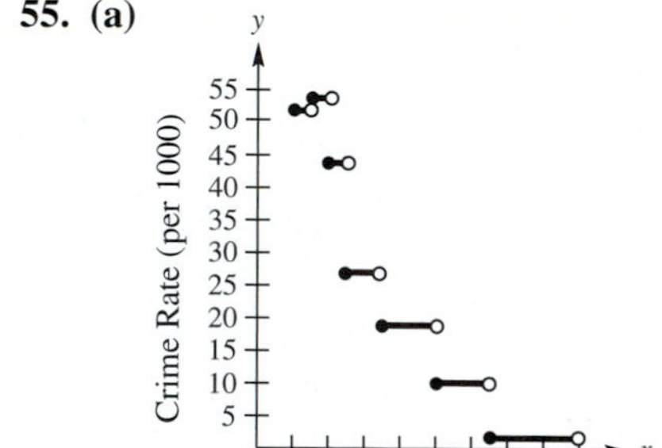

(b) The likelihood of being a victim peaks from age 16 up to age 20, then decreases.

57. **(a)** \$1.25 **(b)** $f(x) = \begin{cases} .50 & \text{if } 0 < x \le 1 \\ .50 - .25[\![1 - x]\!] & \text{if } 1 < x \le 5 \end{cases}$

59.

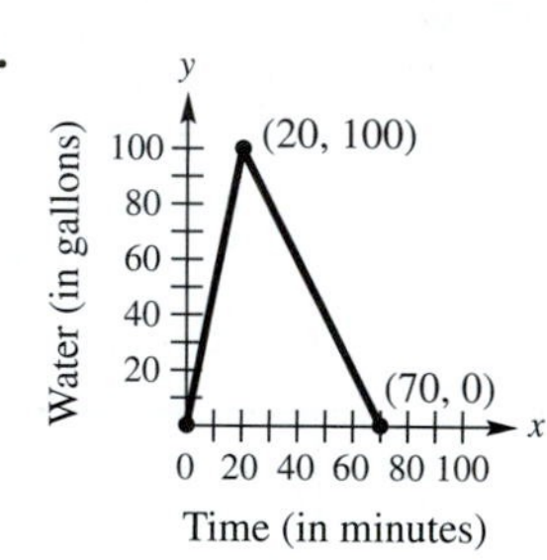

61. For x in the interval $(0, 2]$, $y = 25$. For x in the interval $(2, 3]$, $y = 25 + 3 = 28$. For x in the interval $(3, 4]$, $y = 28 + 3 = 31$, and so on. The graph is a step function. In this case, the first step has a different width.

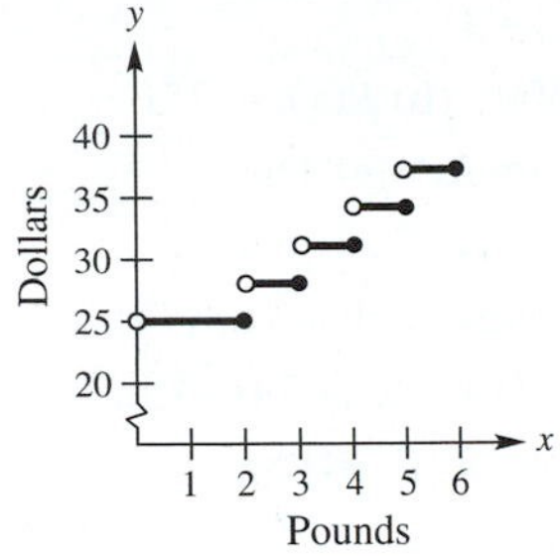

2.6 Exercises *(pages 156–161)*

1. E **3.** F **5.** A **7.** 2450 **9.** $256x^2 + 48x + 2$ **11.** 55 **13.** 1848 **15.** $-\frac{6}{7}$ **17.** 1122 **19.** 97

21. **(a)** $(f + g)(x) = 10x + 2$; $(f - g)(x) = -2x - 4$; $(fg)(x) = 24x^2 + 6x - 3$ **(b)** Domain is $(-\infty, \infty)$ in all cases. **(c)** $\left(\frac{f}{g}\right)(x) = \frac{4x - 1}{6x + 3}$; domain: $\left(-\infty, -\frac{1}{2}\right) \cup \left(-\frac{1}{2}, \infty\right)$ **(d)** $(f \circ g)(x) = 24x + 11$; domain: $(-\infty, \infty)$ **(e)** $(g \circ f)(x) = 24x - 3$; domain: $(-\infty, \infty)$ **23.** **(a)** $(f + g)(x) = |x + 3| + 2x$; $(f - g)(x) = |x + 3| - 2x$; $(fg)(x) = |x + 3|(2x)$ **(b)** Domain is $(-\infty, \infty)$ in all cases. **(c)** $\left(\frac{f}{g}\right)(x) = \frac{|x + 3|}{2x}$; domain: $(-\infty, 0) \cup (0, \infty)$ **(d)** $(f \circ g)(x) = |2x + 3|$; domain: $(-\infty, \infty)$ **(e)** $(g \circ f)(x) = 2|x + 3|$; domain: $(-\infty, \infty)$ **25.** **(a)** $(f + g)(x) = \sqrt[3]{x + 4} + x^3 + 5$; $(f - g)(x) = \sqrt[3]{x + 4} - x^3 - 5$; $(fg)(x) = \left(\sqrt[3]{x + 4}\right)(x^3 + 5)$ **(b)** Domain is $(-\infty, \infty)$ in all cases. **(c)** $\left(\frac{f}{g}\right)(x) = \frac{\sqrt[3]{x + 4}}{x^3 + 5}$; domain: $\left(-\infty, \sqrt[3]{-5}\right) \cup \left(\sqrt[3]{-5}, \infty\right)$ **(d)** $(f \circ g)(x) = \sqrt[3]{x^3 + 9}$; domain: $(-\infty, \infty)$ **(e)** $(g \circ f)(x) = x + 9$; domain: $(-\infty, \infty)$ **27.** **(a)** $(f + g)(x) = \sqrt{x^2 + 3} + x + 1$; $(f - g)(x) = \sqrt{x^2 + 3} - x - 1$; $(fg)(x) = \left(\sqrt{x^2 + 3}\right)(x + 1)$ **(b)** Domain is $(-\infty, \infty)$ in all cases. **(c)** $\left(\frac{f}{g}\right)(x) = \frac{\sqrt{x^2 + 3}}{x + 1}$; domain: $(-\infty, -1) \cup (-1, \infty)$ **(d)** $(f \circ g)(x) = \sqrt{x^2 + 2x + 4}$; domain: $(-\infty, \infty)$ (*Note:* To see that this is the domain, graph $y = x^2 + 2x + 4$ and note that $y > 0$ for all x.) **(e)** $(g \circ f)(x) = \sqrt{x^2 + 3} + 1$; domain: $(-\infty, \infty)$

29. **(a)** 2 **(b)** 4 **(c)** 0 **(d)** $-\frac{1}{3}$ **31.** **(a)** 3 **(b)** -5 **(c)** 2 **(d)** undefined **33.** **(a)** 5 **(b)** 5 **(c)** 0 **(d)** undefined

35.

x	$(f+g)(x)$	$(f-g)(x)$	$(fg)(x)$	$\left(\frac{f}{g}\right)(x)$
-2	6	-6	0	0
0	5	5	0	undefined
2	5	9	-14	-3.5
4	15	5	50	2

37. 7.7; 11.8; 19.5 **39.** 1991–1996 **41.** 6; It represents the dollars in billions spent for general science in 2000. **43.** space and other technologies; 1995–2000 **45.** **(a)** -4 **(b)** 2 **(c)** -4 **47.** **(a)** -3 **(b)** -2 **(c)** 0 **49.** **(a)** 5 **(b)** undefined **(c)** 4 **51.** 4; 2 **53.** -3 **55.** 93

57. **(a)** $(f \circ g)(x) = (x^2 + 3x - 1)^3$; domain: $(-\infty, \infty)$ **(b)** $(g \circ f)(x) = x^6 + 3x^3 - 1$; domain: $(-\infty, \infty)$ **(c)** $(f \circ f)(x) = x^9$; domain: $(-\infty, \infty)$ **59.** **(a)** $(f \circ g)(x) = 1 - x$; domain: $(-\infty, 1]$ **(b)** $(g \circ f)(x) = \sqrt{1 - x^2}$; domain: $[-1, 1]$ **(c)** $(f \circ f)(x) = x^4$; domain: $(-\infty, \infty)$ **61.** **(a)** $(f \circ g)(x) = \frac{1}{5x + 1}$; domain: $\left(-\infty, -\frac{1}{5}\right) \cup \left(-\frac{1}{5}, \infty\right)$ **(b)** $(g \circ f)(x) = \frac{5}{x + 1}$; domain: $(-\infty, -1) \cup (-1, \infty)$ **(c)** $(f \circ f)(x) = \frac{x + 1}{x + 2}$; domain: $(-\infty, -2) \cup (-2, -1) \cup (-1, \infty)$ **63.** **(a)** $(f \circ g)(x) = 8x^3 - 10x^2 + 1$; domain: $(-\infty, \infty)$ **(b)** $(g \circ f)(x) = 32x^3 + 28x^2 + 4x - 1$; domain: $(-\infty, \infty)$ **(c)** $(f \circ f)(x) = 4x + 3$; domain: $(-\infty, \infty)$

69. The graph of y_2 can be obtained by *reflecting* the graph of y_1 across the line $y_3 = x$.

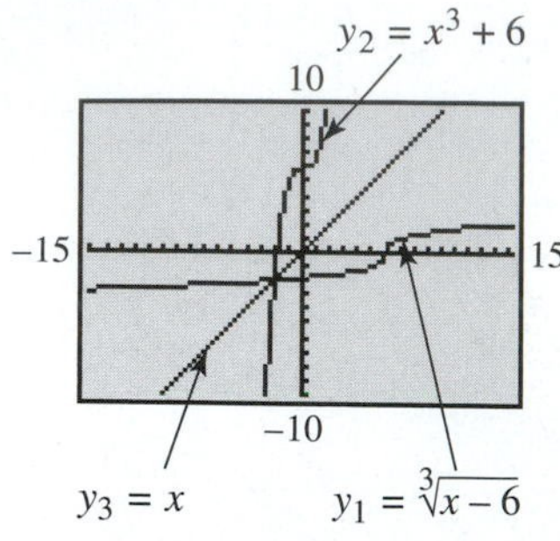

71. 4 **73.** $-12x - 1 - 6h$ **75.** $3x^2 + 3xh + h^2$

We give only one of many possible correct pairs of functions f and g in Exercises 77–81.

77. $f(x) = x^2$, $g(x) = 6x - 2$ **79.** $f(x) = \sqrt{x}$, $g(x) = x^2 - 1$

81. $f(x) = \sqrt{x} + 12$, $g(x) = 6x$ **83.** **(a)** $C(x) = 10x + 500$ **(b)** $R(x) = 35x$ **(c)** $P(x) = 35x - (10x + 500)$ or $P(x) = 25x - 500$ **(d)** 21 items **(e)** The smallest whole number for which $P(x) > 0$ is 21. Use a window of $[0, 30]$ by $[-1000, 500]$, for example. **85.** **(a)** $C(x) = 100x + 2700$ **(b)** $R(x) = 280x$ **(c)** $P(x) = 280x - (100x + 2700)$ or $P(x) = 180x - 2700$ **(d)** 16 items **(e)** The smallest whole number for which $P(x) > 0$ is 16. Use a window of $[0, 30]$ by $[-3000, 500]$, for example.

87. **(a)** $V(r) = \frac{4}{3}\pi(r + 3)^3 - \frac{4}{3}\pi r^3$ **(b)**

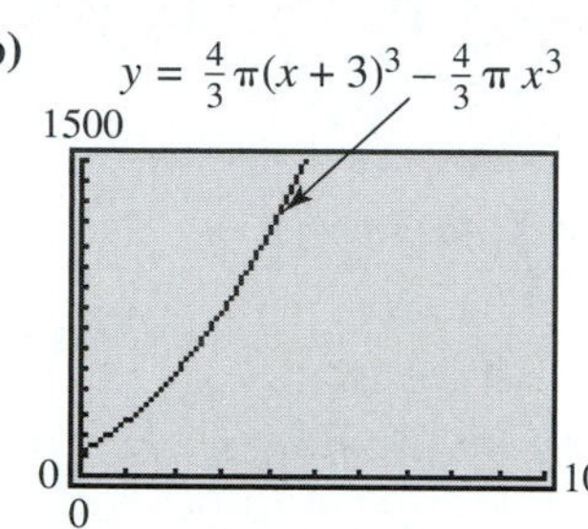

(c) 1168.67 in.^{3} **(d)** $V(4) = \frac{4}{3}\pi(7)^3 - \frac{4}{3}\pi(4)^3 \approx 1168.67$

89. **(a)** $P = 6x$; $P(x) = 6x$; This is a linear function.

(b) 4 represents the width of the rectangle and 24 represents the perimeter.

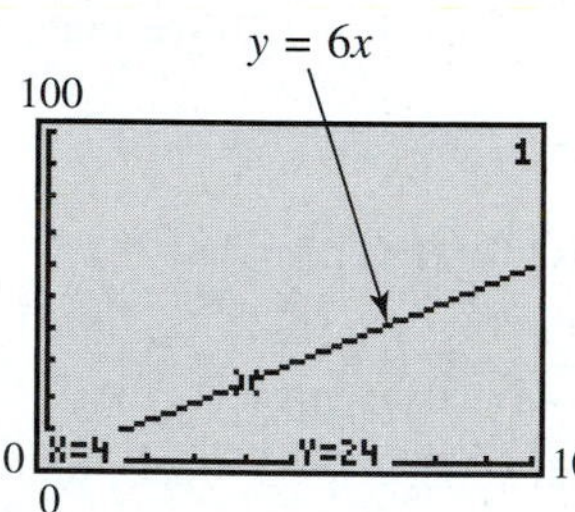

(c) (See graph for part (b).) perimeter = 24; This is the y-value shown on the screen for the integer x-value 4.

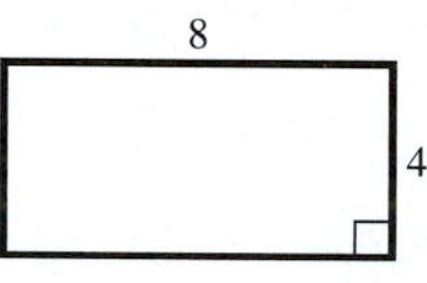

(d) (Answers may vary.) If the perimeter y of a rectangle satisfying the given conditions is 36, then the width x is 6.

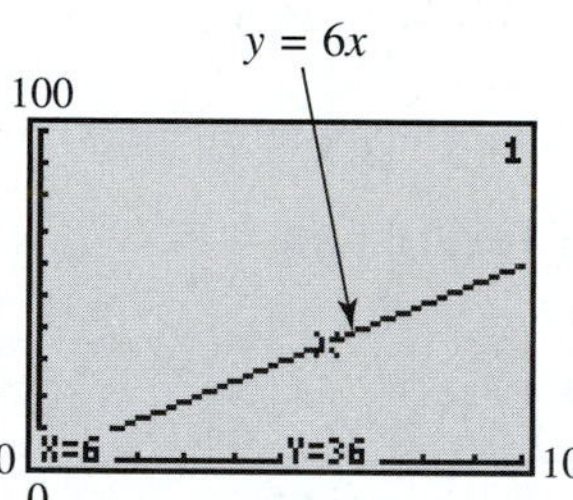

91. **(a)** $A(2x) = \sqrt{3}x^2$ **(b)** $A(16) = 64\sqrt{3}$ square units **(c)** On the graph of $y = \frac{\sqrt{3}}{4}x^2$, locate the point where $x = 16$ to find $y \approx 110.85$, an approximation for $64\sqrt{3}$. **93.** **(a)**

x	1999	2000	2001	2002	2003
$h(x)$	76	82	79	89	103

(b) $h(x) = f(x) + g(x)$

95. **(a)** 50 **(b)** $(f + g)(x)$ computes the total SO_2 emissions from burning coal and oil during year x.

(c)

x	1860	1900	1940	1970	2000
$(f + g)(x)$	2.4	12.8	26.5	50.0	78.0

97. **(a)** $h(x) = g(x) - f(x)$ **(b)** $h(1996) = 147$; $h(2006) = 153$ **(c)** $h(x) = .6(x - 1996) + 147$

Reviewing Basic Concepts *(page 162)*

1. **(a)** $\{-12, 4\}$ **(b)** $(-\infty, -12) \cup (4, \infty)$ **(c)** $[-12, 4]$

2.

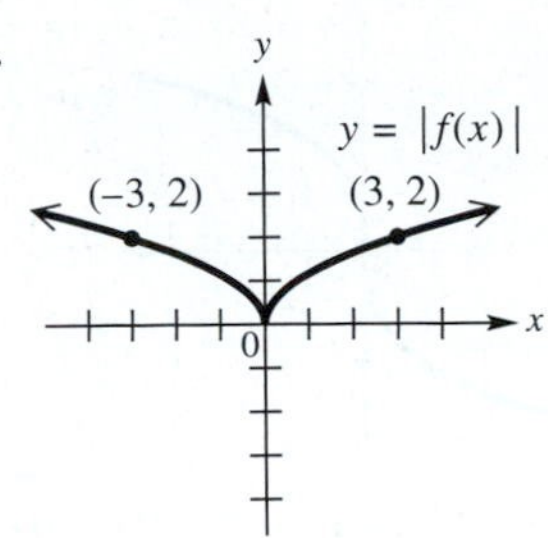

3. **(a)** $25.33 \le R_L \le 28.17$; $36.58 \le R_E \le 40.92$ **(b)** $5699.25 \le T_L \le 6338.25$; $8230.5 \le T_E \le 9207$

4. **(a)** -3 **(b)** 4 **(c)** 8

5. **(a)**

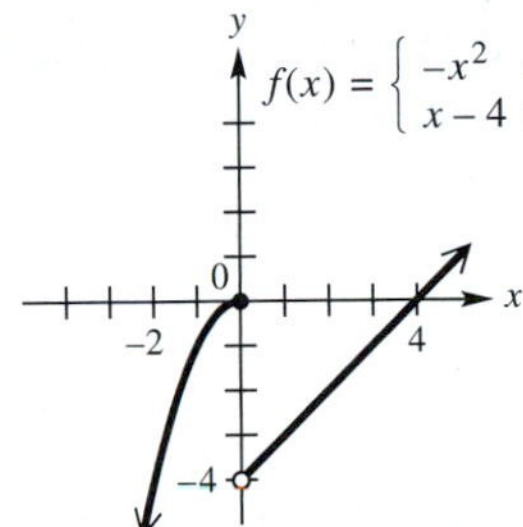

(b)

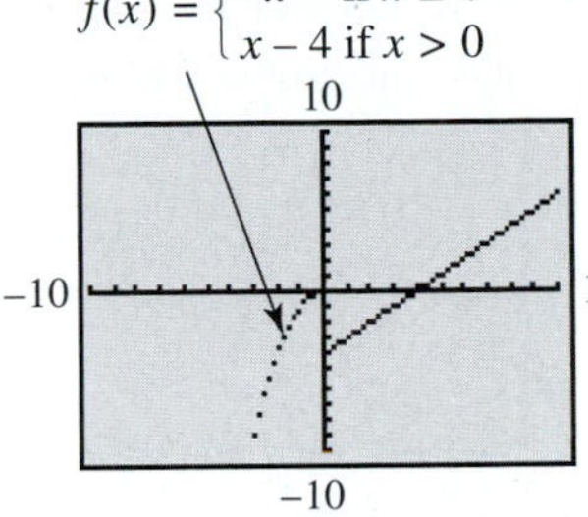

6. **(a)** -6 **(b)** -22 **(c)** 8 **(d)** $\frac{5}{9}$ **(e)** $-3x^2 - 4$ **(f)** $9x^2 + 24x + 16$

7. $f(x) = x^4$, $g(x) = x + 2$ (There are other possible choices for f and g.)

8. $-4x - 2h + 3$

9. **(a)** $y_1 = .04x$ **(b)** $y_2 = .025x + 12.5$ **(c)** It represents the total interest earned in both accounts for 1 year.

(d) $y_1 + y_2 = .04x + .025x + 12.5$ $28.75 is earned. **(e)** Evaluate $y_1 + y_2$ at $x = 250$ to get 28.75 (dollars).

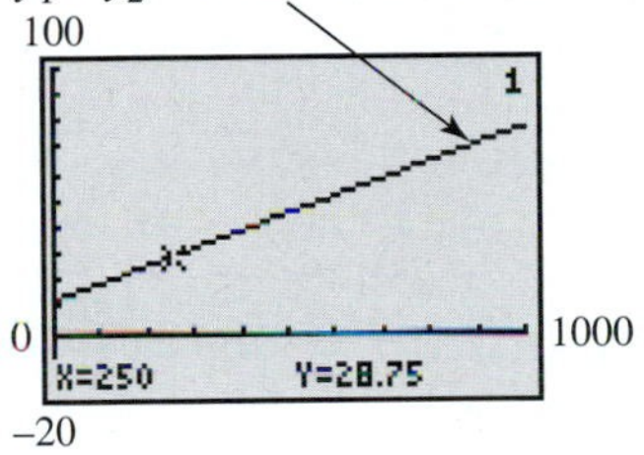

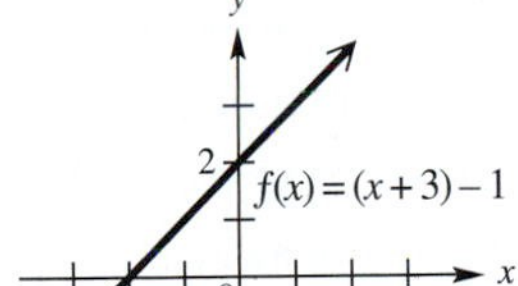

10. $S = \pi r^2\sqrt{5}$

Chapter 2 Review Exercises *(pages 166–169)*

1. true **3.** false; The domain of $f(x) = \sqrt{x}$ is $[0, \infty)$, while the domain of $f(x) = \sqrt[3]{x}$ is $(-\infty, \infty)$. **5.** true **7.** true

9. true **11.** $[0, \infty)$ **13.** $(-\infty, \infty)$ **15.** $(-\infty, \infty)$ **17.** $[0, \infty)$

19.

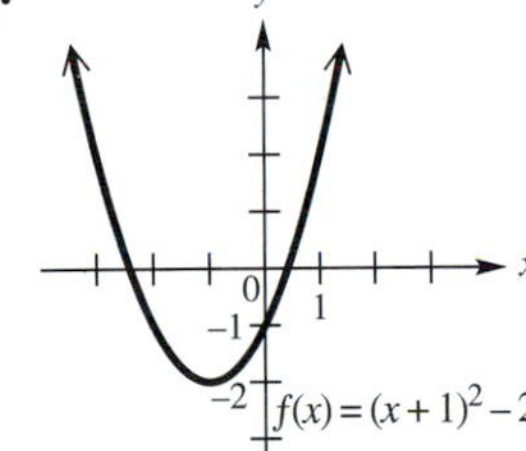

21.

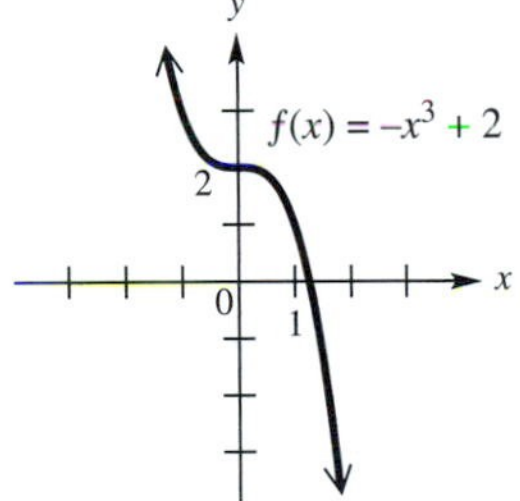

23.

25.
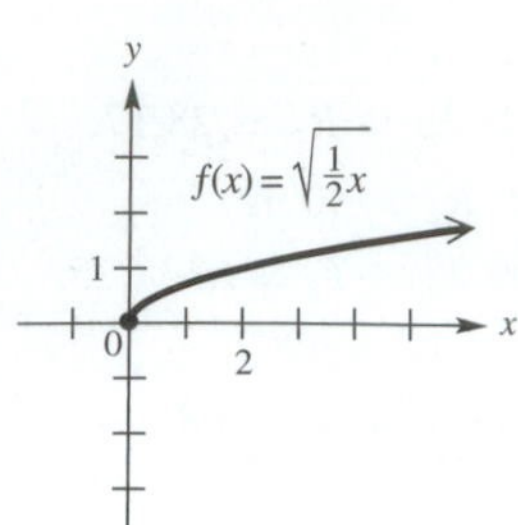

27.
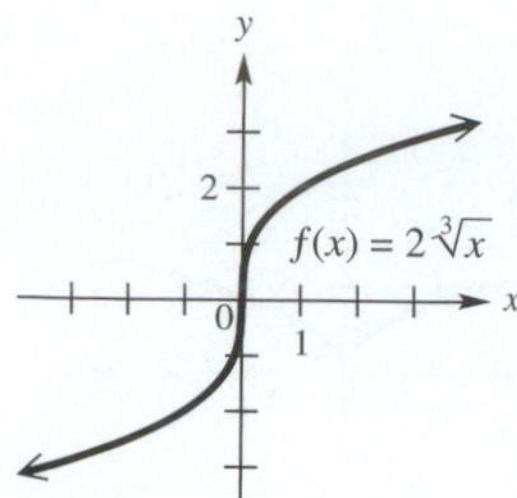

29.
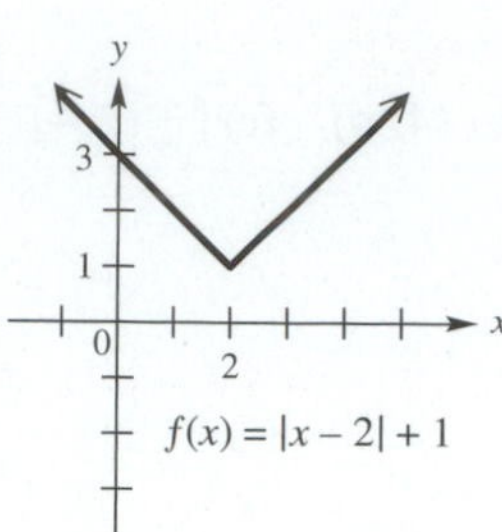

31. **(a)** $(-\infty,-2),[-2,1],(1,\infty)$ **(b)** $[-2,1]$ **(c)** $(-\infty,-2)$ **(d)** $(1,\infty)$ **(e)** $(-\infty,\infty)$ **(f)** $\{-2\}\cup[-1,1]\cup(2,\infty)$ **33.** x-axis symmetry, y-axis symmetry, origin symmetry; not a function **35.** y-axis symmetry; even function **37.** x-axis symmetry; not a function **39.** true **41.** true **43.** false; For example, $f(x) = x^3$ is odd, and $(2,8)$ is on the graph, but $(-2,8)$ is not. **45.** Start with the graph of $y = x^2$. Shift it 4 units to the left, stretch vertically by a factor of 3, reflect across the x-axis, and shift 8 units downward.

47.
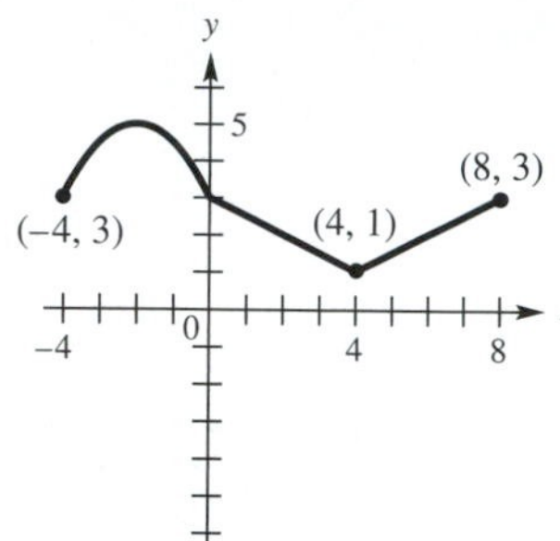

49.
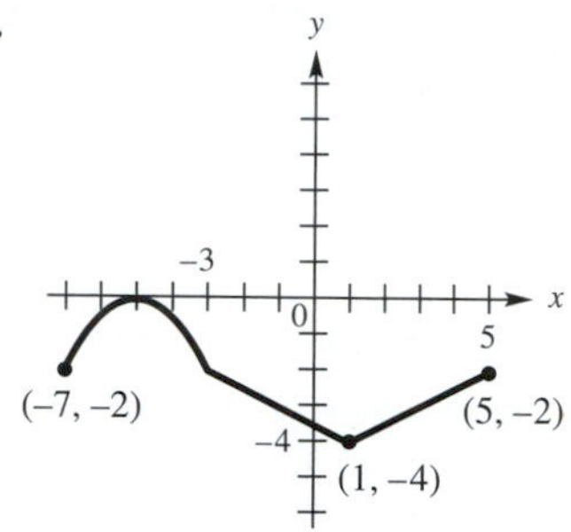

51.
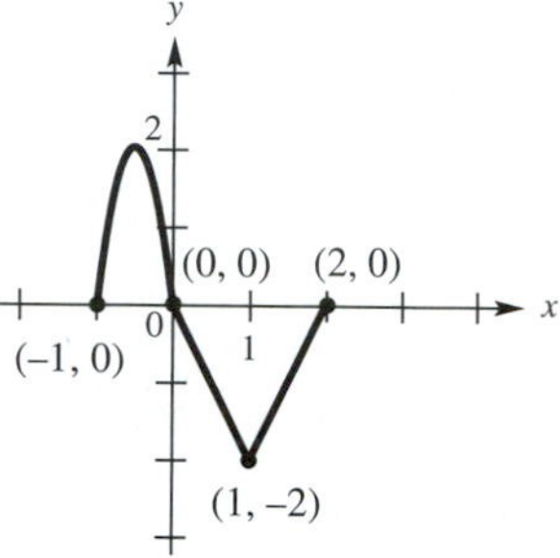

53. domain: $[-3, 4]$; range: $[2, 9]$ **55.** 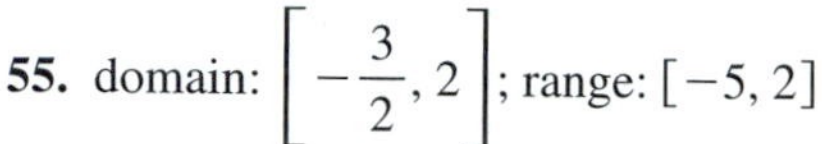 domain: $\left[-\frac{3}{2}, 2\right]$; range: $[-5, 2]$ **57.**
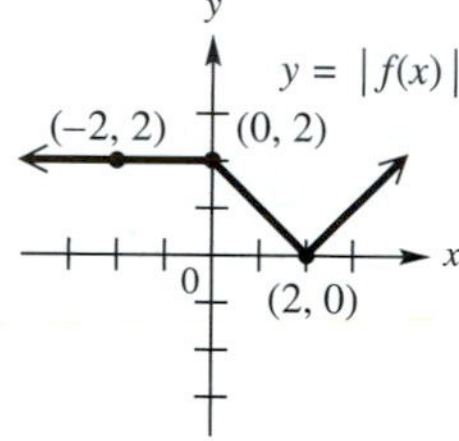

59. The graph of $y = |f(x)|$ is the same as that of $y = f(x)$. **61.** $\left\{-\frac{15}{4}, \frac{9}{4}\right\}$ **63.** $\left\{-\frac{7}{3}, 2\right\}$ **65.** $[-6, 1]$

67. $\left(-\infty, \frac{12}{5}\right) \cup \left(\frac{12}{5}, \infty\right)$ or $\left\{x \,\middle|\, x \neq \frac{12}{5}\right\}$ **69.** $\left\{-3, \frac{11}{3}\right\}$ **71.** The x-coordinates of the points of intersection of the graphs are -6 and 1. Thus, $\{-6, 1\}$ is the solution set of $y_1 = y_2$. The graph of y_1 lies on or below the graph of y_2 between -6 and 1, so the solution set of $y_1 \leq y_2$ is $[-6, 1]$. The graph of y_1 lies above the graph of y_2 everywhere else, so the solution set of $y_1 \geq y_2$ is $(-\infty, -6] \cup [1, \infty)$. **73.** Initially, the car is at home. After traveling 30 mph for 1 hr, the car is 30 mi away from home. During the second hour the car travels 20 mph until it is 50 mi away. During the third hour the car travels toward home at 30 mph until it is 20 mi away. During the fourth hour the car travels away from home at 40 mph until it is 60 mi away from home. During the last hour, the car travels 60 mi at 60 mph until it arrives home.

75.

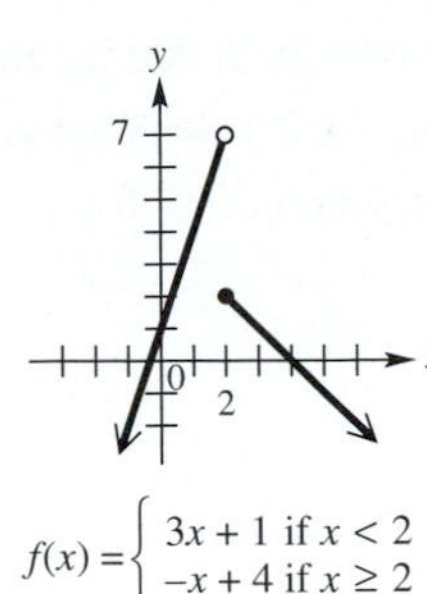

77.

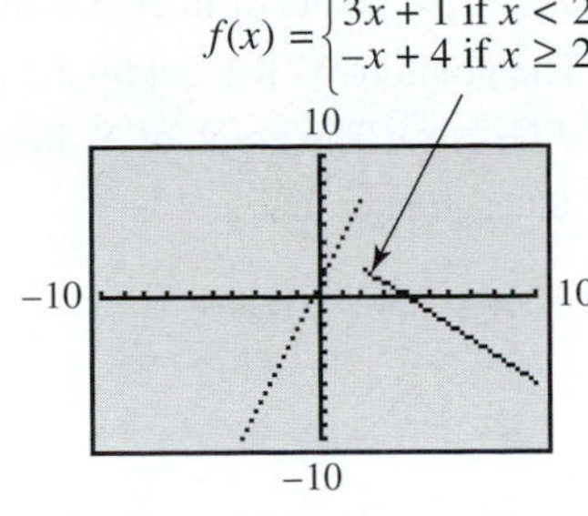

Dot mode

79. 5 **81.** 0 **83.** 3 **85.** 2 **87.** 8 **89.** −6 **91.** 2 **93.** 2 **95.** $f(x) = x^2$ and $g(x) = x^3 - 3x$ (There are other possible choices for f and g.)

97. $V(r) = \frac{4}{3}\pi(r + 4)^3 - \frac{4}{3}\pi r^3$ **99.** $f(x) = 36x$; $g(x) = 1760x$; $(f \circ g)(x) = 63{,}360x$

Chapter 2 Test *(pages 169–170)*

1. **(a)** D **(b)** D **(c)** C **(d)** B **(e)** C **(f)** C **(g)** C **(h)** D **(i)** D **(j)** C

2. **(a)**

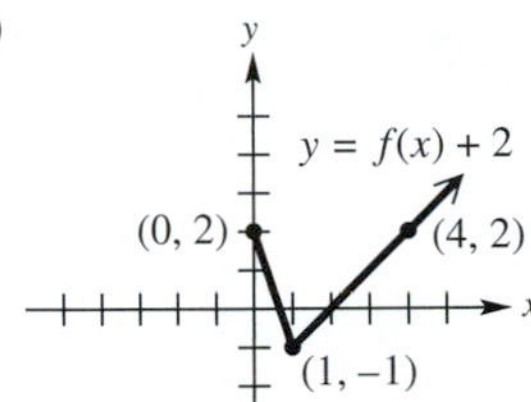

(b)

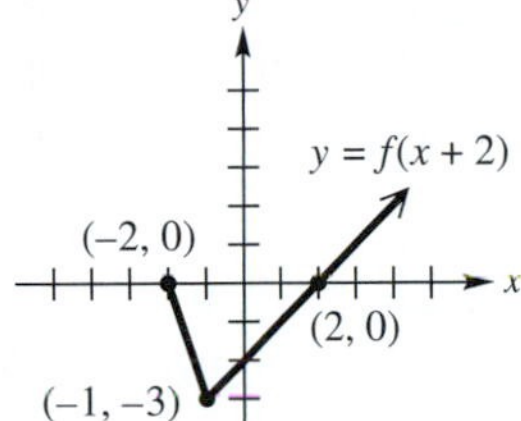

(c)

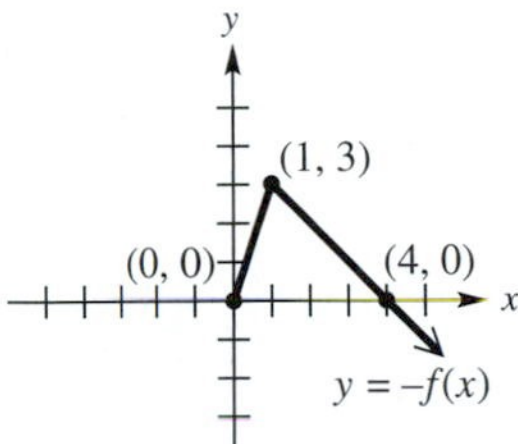

(d)

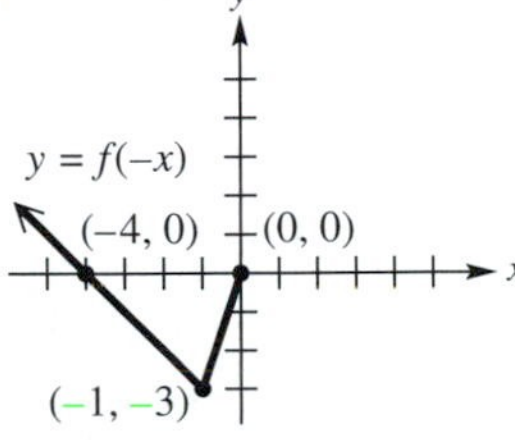

(e)

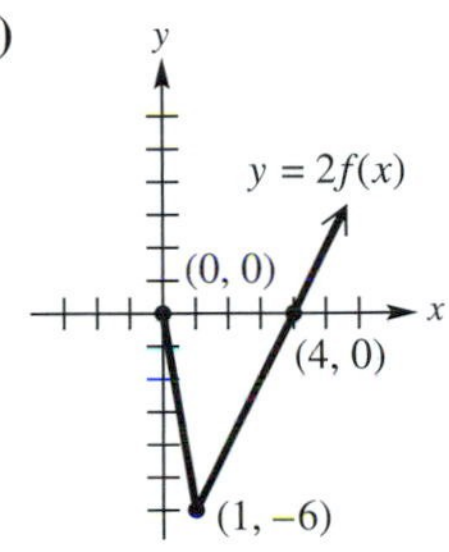

(f)

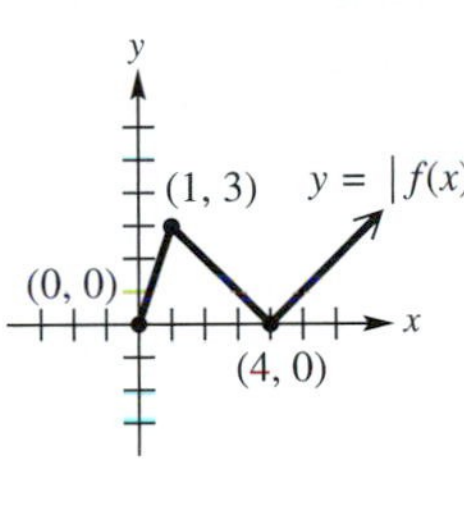

3. **(a)** $(-1, 4)$ **(b)** $(-4, 4)$

4. **(a)**

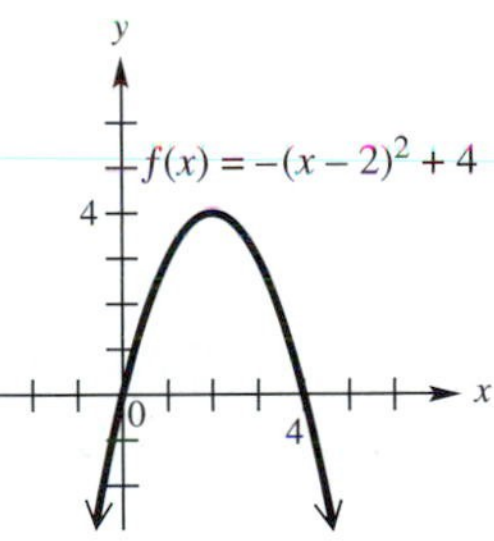

(b)

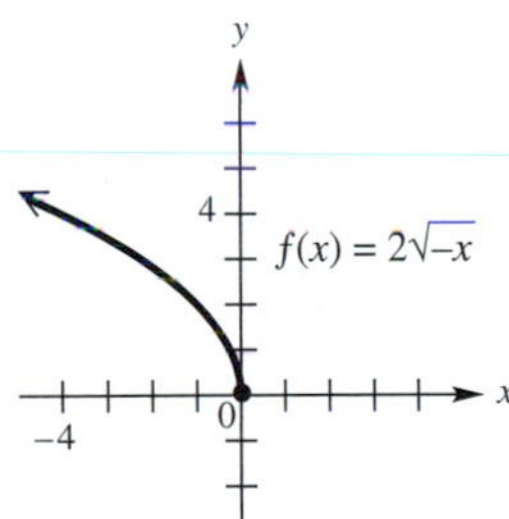

5. **(a)** $(-3, 6)$ **(b)** $(-3, -6)$

(c)

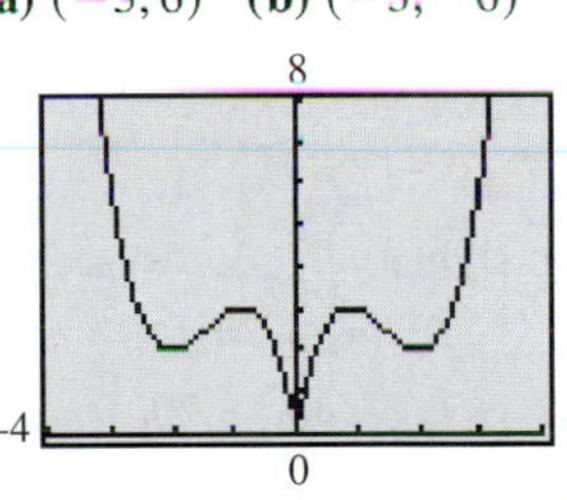

(We give an actual screen here. The drawing should resemble it.)

6. **(a)** Shift the graph of $y = \sqrt[3]{x}$ to the left 2 units, vertically stretch by a factor of 4, and shift 5 units downward.

(b)

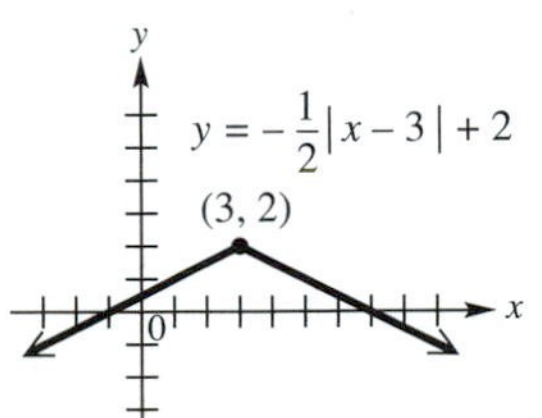

domain: $(-\infty, \infty)$; range: $(-\infty, 2]$

7. **(a)** $(-\infty, -3)$ **(b)** $(4, \infty)$ **(c)** $[-3, 4]$ **(d)** $(-\infty, -3), [-3, 4], (4, \infty)$ **(e)** $(-\infty, \infty)$ **(f)** $(-\infty, 2)$

8. $Y_1 = |4X + 8|$ $Y_2 = 4$

(a) $\{-3, -1\}$; The x-coordinates of the points of intersection of the graphs of Y_1 and Y_2 are -3 and -1. **(b)** $(-3, -1)$; The graph of Y_1 lies below the graph of Y_2 for x-values between -3 and -1. **(c)** $(-\infty, -3) \cup (-1, \infty)$; The graph of Y_1 lies above the graph of Y_2 for x-values less than -3 or for x-values greater than -1.

9. **(a)** $2x^2 - x + 1$ **(b)** $\dfrac{2x^2 - 3x + 2}{-2x + 1}$ **(c)** $\left(-\infty, \dfrac{1}{2}\right) \cup \left(\dfrac{1}{2}, \infty\right)$ **(d)** $8x^2 - 2x + 1$ **(e)** $4x + 2h - 3$

10. **(a)** $f(x) = \begin{cases} -x^2 + 3 & \text{if } x \le 1 \\ \sqrt[3]{x} + 2 & \text{if } x > 1 \end{cases}$ **(b)**

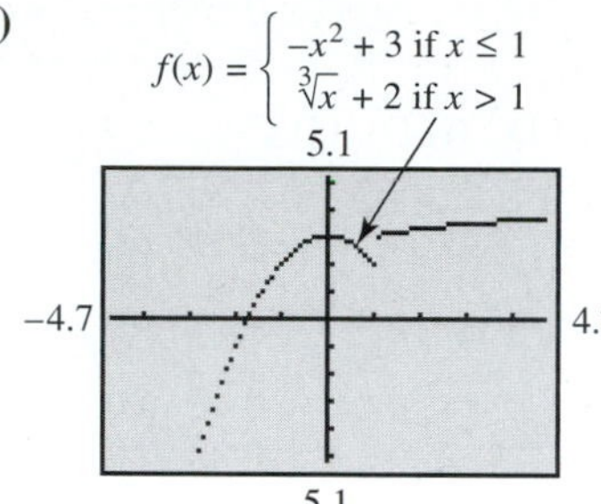

11. **(a)** Y1=.40int(X)+.75; X=5.5 Y=2.75

(b) \$2.75 is the cost for a 5.5-min call. See the display at the bottom of the screen.

12. **(a)** $C(x) = 3300 + 4.50x$ **(b)** $R(x) = 10.50x$ **(c)** $P(x) = R(x) - C(x) = 6.00x - 3300$ **(d)** 551

(e) Y1=6X−3300; X=550 Y=0

The first integer x-value for which $P(x) > 0$ is 551.

CHAPTER 3 POLYNOMIAL FUNCTIONS

3.1 Exercises *(pages 179–181)*

1. **(a)** 0 **(b)** -9 **(c)** pure imaginary **3.** **(a)** π **(b)** 0 **(c)** real **5.** **(a)** 3 **(b)** 7 **(c)** nonreal complex **7.** **(a)** 0 **(b)** $\sqrt{7}$ **(c)** pure imaginary **9.** **(a)** 0 **(b)** $\sqrt{7}$ **(c)** pure imaginary **11.** true **13.** true **15.** false; *Every* real number is a complex number. **17.** $10i$ **19.** $-20i$ **21.** $-i\sqrt{39}$ **23.** $5 + 2i$ **25.** $9 - 5i\sqrt{2}$ **27.** -3 **29.** -13 **31.** $-2\sqrt{6}$ **33.** $\sqrt{3}$ **35.** $i\sqrt{3}$ **37.** $\frac{1}{2}$ **39.** -2 **41.** $7 - i$ **43.** 2 **45.** $7 - 6i$ **47.** $1 - 10i$ **49.** 0 **51.** $8 - i$ **53.** $-14 + 2i$ **55.** $5 - 12i$ **57.** $-8 - 6i$ **59.** 13 **61.** 7 **63.** $25i$ **65.** $12 + 9i$ **67.** $20 + 15i$ **69.** i **71.** $-i$ **73.** 1 **75.** -1 **77.** $-i$ **79.** $-i$ **81.** 1 **83.** $\left(\frac{\sqrt{2}}{2} + \frac{\sqrt{2}}{2}i\right)^2 = i$ **85.** $5 + 3i$ **87.** $18i$ **89.** $-5 - i$ **91.** $-1 - 2i$ **93.** $2i$ **95.** $\frac{7}{25} - \frac{24}{25}i$ **97.** $\frac{7}{3}i$

99. We are multiplying by 1, the multiplicative identity.

3.2 Exercises *(pages 190–194)*

1. B **3.** D **5. (a)** $P(x) = (x-1)^2 - 16$ **(b)** $(1, -16)$ **(c)**

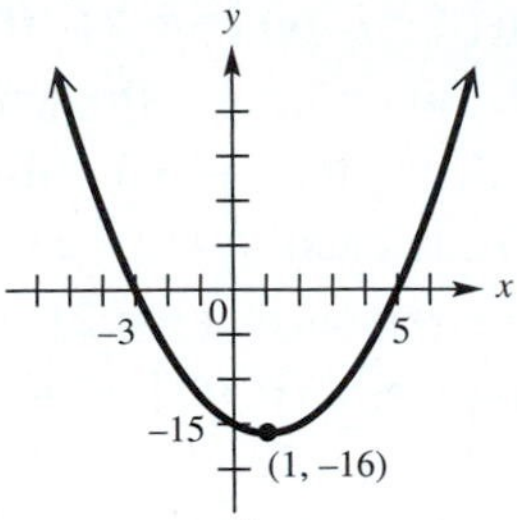

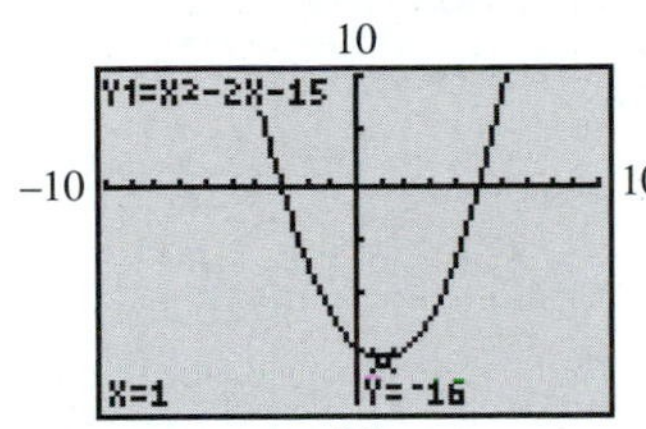

7. (a) $P(x) = -\left(x + \frac{3}{2}\right)^2 + \frac{49}{4}$ **(b)** $\left(-\frac{3}{2}, \frac{49}{4}\right)$

(c)

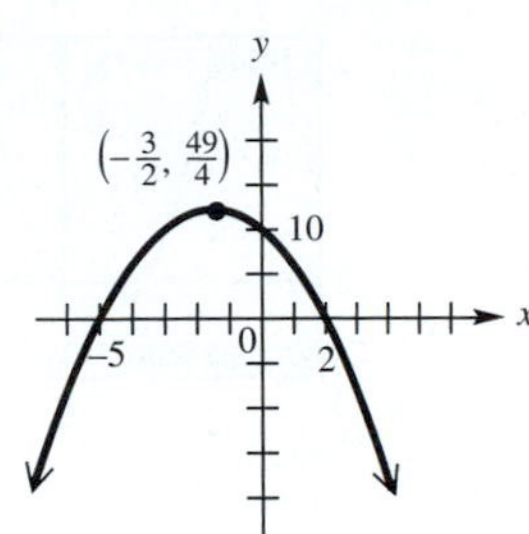

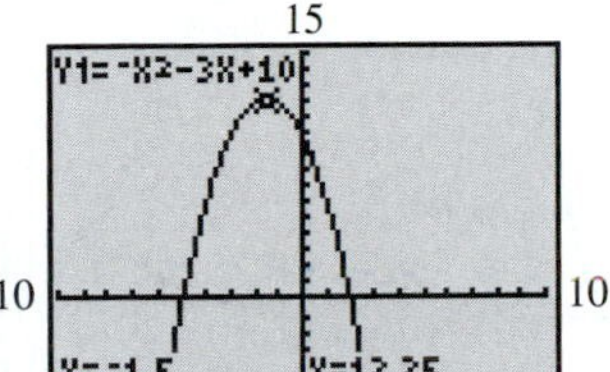

9. (a) $P(x) = (x-3)^2 - 9$ **(b)** $(3, -9)$

(c)

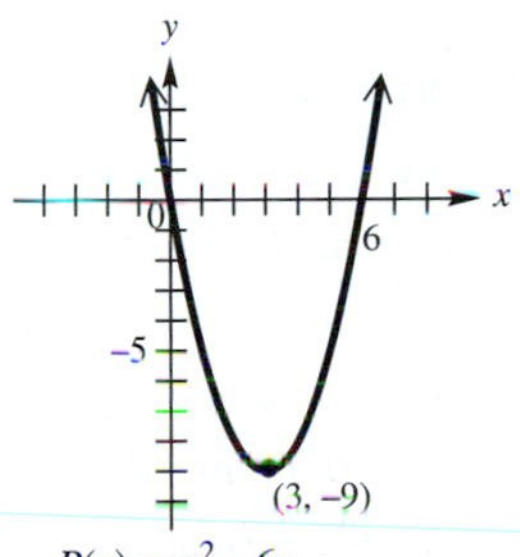

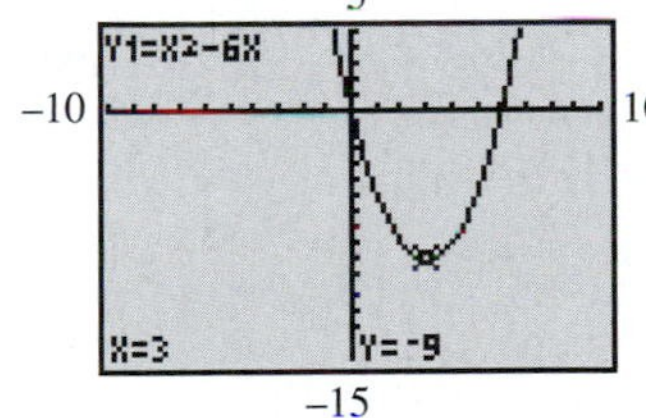

11. (a) $P(x) = 2\left(x - \frac{1}{2}\right)^2 - \frac{49}{2}$ **(b)** $\left(\frac{1}{2}, -\frac{49}{2}\right)$

(c)

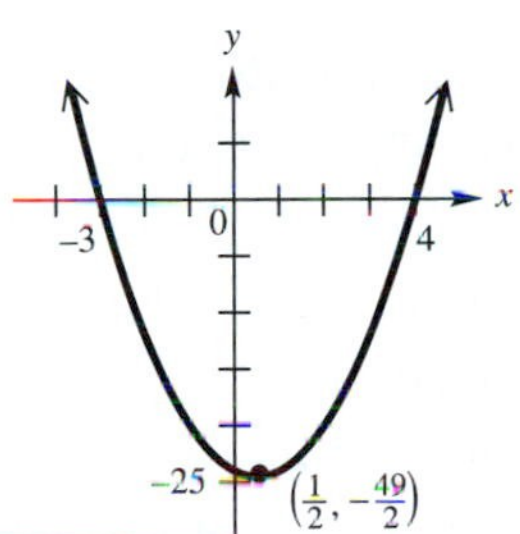

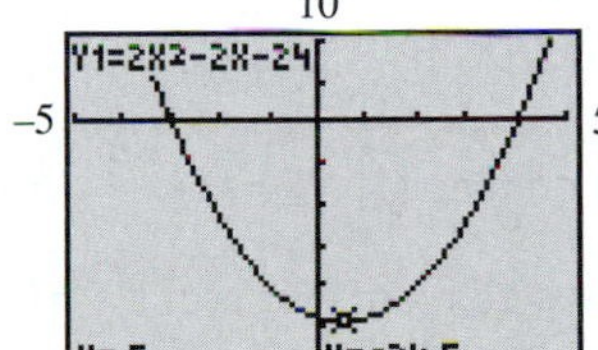

13. (a) $P(x) = -2\left(x - \frac{3}{2}\right)^2 + \frac{9}{2}$ **(b)** $\left(\frac{3}{2}, \frac{9}{2}\right)$ **(c)**

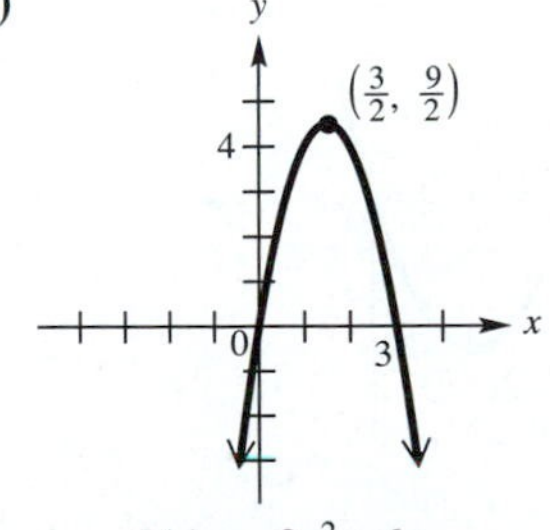

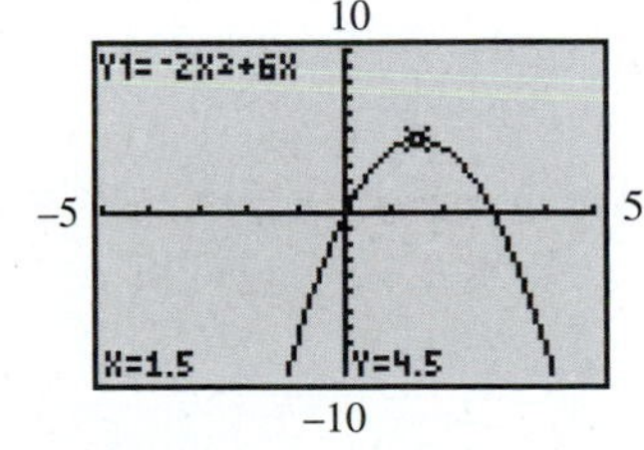

15. **(a)** $P(x) = 3\left(x + \frac{2}{3}\right)^2 - \frac{7}{3}$ **(b)** $\left(-\frac{2}{3}, -\frac{7}{3}\right)$

(c)

$\left(-\frac{2}{3}, -\frac{7}{3}\right)$

$P(x) = 3x^2 + 4x - 1$

Y1=3X²+4X−1

X=−.6666667 Y=−2.333333

17. **(a)** D **(b)** B **(c)** C **(d)** A **19.** **(a)** $(2, 0)$ **(b)** domain: $(-\infty, \infty)$; range: $[0, \infty)$ **(c)** $x = 2$ **(d)** $[2, \infty)$ **(e)** $(-\infty, 2]$ **(f)** minimum: 0 **21.** **(a)** $(-3, -4)$ **(b)** domain: $(-\infty, \infty)$; range: $[-4, \infty)$ **(c)** $x = -3$ **(d)** $[-3, \infty)$ **(e)** $(-\infty, -3]$ **(f)** minimum: -4 **23.** **(a)** $(-3, 2)$ **(b)** domain: $(-\infty, \infty)$; range: $(-\infty, 2]$ **(c)** $x = -3$ **(d)** $(-\infty, -3]$ **(e)** $[-3, \infty)$ **(f)** maximum: 2

25. **(a)** $(5, -4)$

(b)

$(5, -4)$

$P(x) = x^2 - 10x + 21$

27. **(a)** $(2, 2)$

(b)

$(2, 2)$

$y = -x^2 + 4x - 2$

29. **(a)** $(1, 3)$

(b)

$(1, 3)$

$P(x) = 2x^2 - 4x + 5$

31. **(a)** $(4, 2)$

(b)

$(4, 2)$

$P(x) = -3x^2 + 24x - 46$

33. **(a)** $(2.71, 5.20)$ **(b)** $-1.33, 6.74$ **35.** **(a)** $(1.12, .56)$ **(b)** none **37.** **(a)** $(.68, .57)$ **(b)** 0, 1.35 **39.** 3 **41.** none **43.** **(a)** $(4, -12)$ **(b)** minimum **(c)** -12 **(d)** $[-12, \infty)$ **45.** **(a)** $(1.5, 2)$ **(b)** maximum **(c)** 2 **(d)** $(-\infty, 2]$ **47.** quadratic; $a < 0$ **49.** quadratic; $a > 0$ **51.** linear; positive **53.** **(a)** quadratic; negative **(b)** $(8.8, 9.6)$ **(c)** In late-1998, the value of business mergers and acquisitions reached a maximum of \$9.6 trillion. **55.** **(a)** maximum **(b)** 1998; \$480.2 million **57.** **(a)** The value of t cannot be negative since t represents time elapsed from the throw. **(b)** Since the rock was projected from ground level, s_0, the initial height of the rock is 0. **(c)** $s(t) = -16t^2 + 90t$ **(d)** 99 ft **(e)** After 2.8125 sec, the maximum height, 126.5625 ft, is attained. Locate the vertex $(2.8125, 126.5625)$. **(f)** 5.625 sec

59. **(a)** The ball will not reach 355 ft because the graph of $y_1 = -16x^2 + 150x$ does not intersect the graph of $y_2 = 355$. **(b)** The ball reaches a height of 355 ft at $t \approx 1.4$ sec and $t \approx 14.2$ sec. **61.** $P(x) = 3x^2 + 6x - 1$ **63.** $P(x) = \frac{1}{2}x^2 - 8x + 35$

65. $P(x) = -\frac{2}{3}x^2 - \frac{16}{3}x - \frac{38}{3}$ **67.**

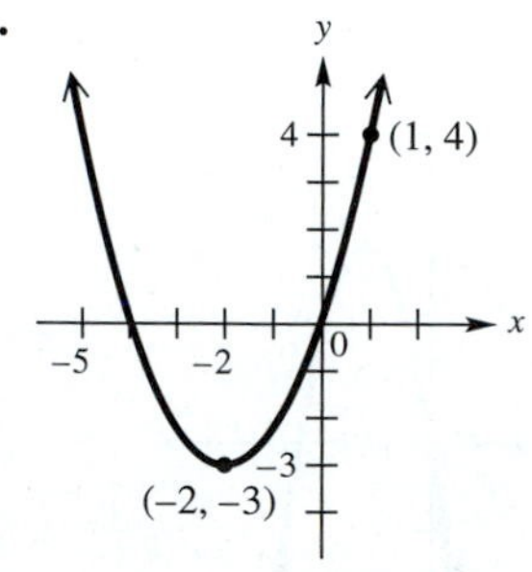

69.

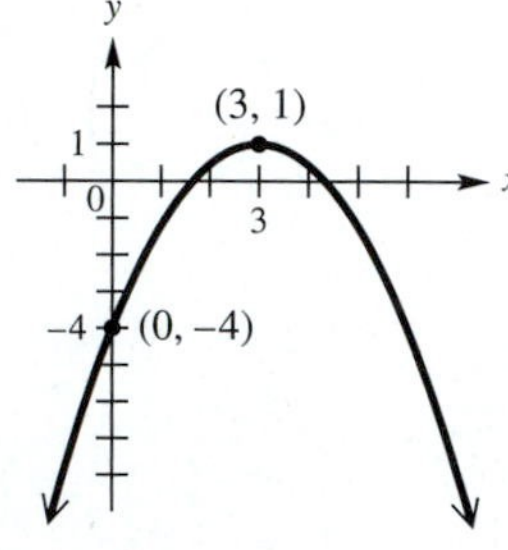

3.3 Exercises *(pages 205–208)*

1. G **3.** C **5.** H **7.** D **9.** D; $\left\{-\frac{1}{3}, 7\right\}$ **11.** C; $\{-4, 3\}$ **13.** $\{\pm 4\}$ **15.** $\{\pm 3\sqrt{5}\}$ **17.** $\{\pm 4i\}$

19. $\{\pm 3i\sqrt{2}\}$ **21.** $\left\{\frac{1 \pm 2\sqrt{3}}{3}\right\}$ **23.** $\left\{\frac{3}{5} \pm \frac{\sqrt{3}}{5}i\right\}$ **25.** $\{-4, 6\}$ **27.** $\left\{0, \frac{2}{3}\right\}$ **29.** $\left\{-\frac{1}{2}, \frac{3}{7}\right\}$

31. $\left\{\frac{1}{2}, 4\right\}$ **33.** $\{-8, 9\}$ **35.** $\{-1, 7\}$ **37.** $\{1 \pm \sqrt{5}\}$ **39.** $\left\{-\frac{1}{2} \pm \frac{1}{2}i\right\}$ **41.** $\left\{\frac{1 \pm \sqrt{5}}{2}\right\}$

43. $\{3 \pm \sqrt{2}\}$ **45.** $\left\{\frac{3}{2} \pm \frac{\sqrt{2}}{2}i\right\}$ **47.** $\left\{\frac{-3 \pm 3\sqrt{65}}{8}\right\}$ **49.** $\{-2, 8\}$ **51.** $\left\{\frac{2 \pm \sqrt{6}}{2}\right\}$ **53.** $\left\{-\frac{1}{2} \pm \frac{\sqrt{3}}{2}i\right\}$

55. $\left\{\frac{5}{2}\right\}$ **57.** $\left\{-\frac{2}{3}, 2\right\}$ **59.** $\{4 \pm 3i\sqrt{2}\}$ **61.** 0; one real solution; rational **63.** 84; two real solutions; irrational

65. -23; no real solutions **67.** $a = 1; b = -9; c = 20$ **69.** $a = 1; b = -2; c = -1$

In Exercises 71–75, answers may vary. We give one example for each exercise.

71. **73.** **75.** **77.** $\{2, 4\}$

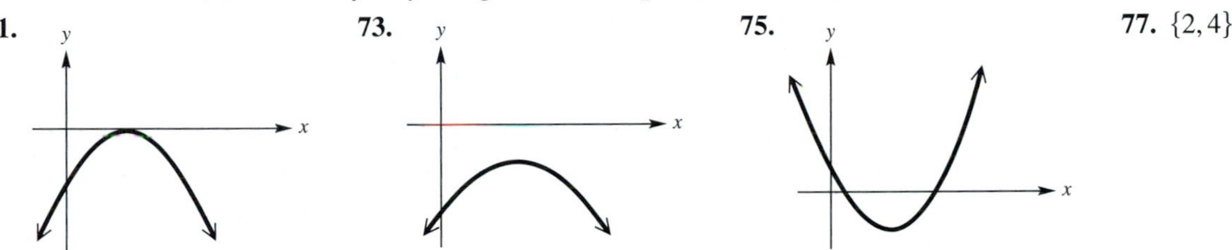

79. $(-\infty, 2) \cup (4, \infty)$ **81.** $(-\infty, 3) \cup (3, \infty)$ **83.** $(-\infty, \infty)$ **85.** no real solutions; two nonreal complex solutions

87. 3 **89.** yes; negative **91.** **(a)** $(-\infty, -3] \cup [-1, \infty)$ **(b)** $(-3, -1)$ **93.** **(a)** $\left(-\infty, \frac{1}{2}\right) \cup (4, \infty)$ **(b)** $\left[\frac{1}{2}, 4\right]$

95. **(a)** $(-\infty, -1] \cup [0, \infty)$ **(b)** $(-1, 0)$ **97.** **(a)** $\emptyset$ **(b)** $(-\infty, \infty)$ **99.** **(a)** $[1 - \sqrt{2}, 1 + \sqrt{2}]$

(b) $(-\infty, 1 - \sqrt{2}) \cup (1 + \sqrt{2}, \infty)$ **101.** $t = \frac{\pm\sqrt{2sg}}{g}$ **103.** $a = \pm\sqrt{c^2 - b^2}$ **105.** $r = \frac{\pm\sqrt{S\pi}}{2\pi}$ **107.** $e = \sqrt[3]{V}$

109. $v = \frac{\pm\sqrt[4]{Frk^3M^3}}{kM}$ **111.** $R = \frac{E^2 - 2Pr \pm E\sqrt{E^2 - 4Pr}}{2P}$ **113.** $x = -\frac{y}{2} \pm \frac{\sqrt{3}}{2}yi; y = -\frac{x}{2} \pm \frac{\sqrt{3}}{2}xi$

115. $x = \frac{2y \pm \sqrt{31y^2 + 9}}{9}; y = \frac{-2x \pm \sqrt{31x^2 - 3}}{3}$ **117.** 1991; 14,653

Reviewing Basic Concepts *(page 208)*

1. $3 + 7i$ **2.** $26i$ **3.** $-\frac{50}{13} + \frac{10}{13}i$ **4.**

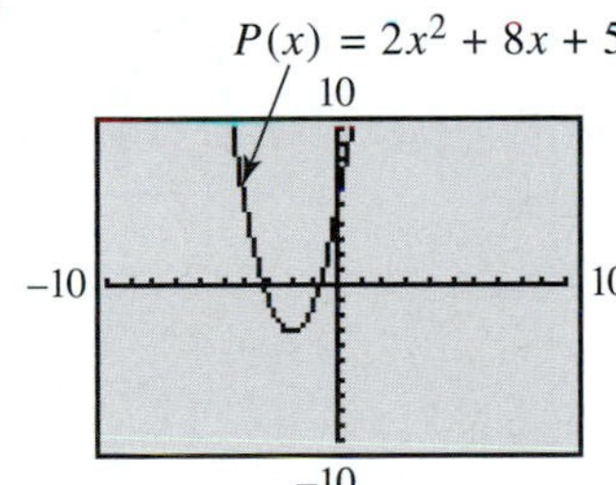

5. $(-2, -3)$; minimum **6.** $x = -2$

7. domain: $(-\infty, \infty)$; range: $[-3, \infty)$

8. $\left\{\pm\frac{5}{3}\right\}$ **9.** $\left\{-\frac{1}{3}, 2\right\}$

10. $\left\{\frac{1 \pm \sqrt{13}}{2}\right\}$ **11.** $\left[-\frac{1}{3}, 2\right]$

12. $\left(-\infty, \frac{1 - \sqrt{13}}{2}\right) \cup \left(\frac{1 + \sqrt{13}}{2}, \infty\right)$ **13.** $\emptyset$

3.4 Exercises *(pages 213–218)*

1. A **3. (a)** $30 - x$ **(b)** $0 < x < 30$ **(c)** $P(x) = -x^2 + 30x$ **(d)** 15 and 15; The maximum product is 225.
5. (a) $640 - 2x$ **(b)** $0 < x < 320$ **(c)** $A(x) = -2x^2 + 640x$ **(d)** between 57.04 ft and 85.17 ft or 234.83 ft and 262.96 ft **(e)** 160 ft by 320 ft; The maximum area is 51,200 ft^2. **7. (a)** $2x$ **(b)** length: $2x - 4$; width: $x - 4$; $x > 4$ **(c)** $V(x) = 4x^2 - 24x + 32$ **(d)** 8 in. by 20 in. **(e)** 13.0 to 14.2 in. **9.** 1.5 in. **11.** 20 ft **13.** 5 ft
15. a 17-ft ladder **17. (a)** $80 - x$ **(b)** $400 + 20x$ **(c)** $R(x) = -20x^2 + 1200x + 32{,}000$ **(d)** 5 or 55 **(e)** \$1000
19. (a) 23.32 ft per sec **(b)** yes **(c)** 12.88 ft **21. (a)** 3.5 ft **(b)** about .2 ft and 2.3 ft **(c)** 1.25 ft **(d)** about 3.78 ft
23. about 104.5 ft per sec; 71.25 mph **25. (a)** 19.2 hr **(b)** 84.3 ppm
27. (a)

120.2

3 15

39.8

(b) $f(x) = .6(x - 4)^2 + 50$ **(c)** There is a good fit.

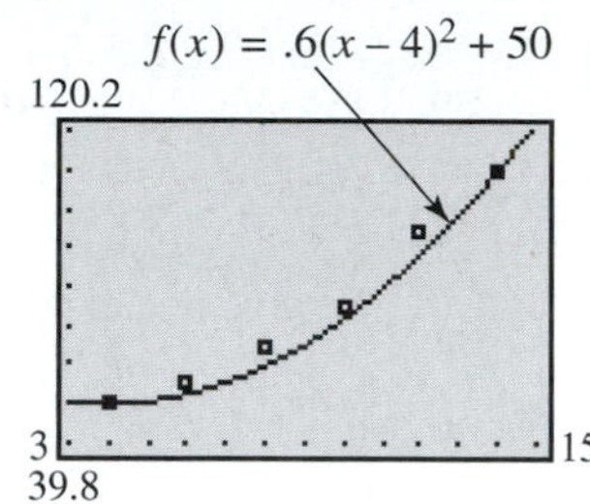

(d) $g(x) \approx .402x^2 - 1.175x + 48.343$ **(e) (i)** f: 136.4 thousand; g: 132.5 thousand **(ii)** f: 203.6 thousand; g: 185.6 thousand **(iii)** f: 314.6 thousand; g: 270.2 thousand **29. (a)**

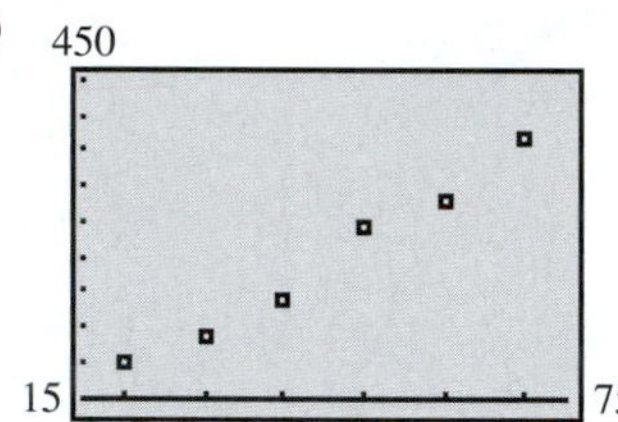

(b) $f(45) \approx 161.5$ ft is the stopping distance when the speed is 45 mph. **(c)** The model is quite good, although the stopping distances are a little low for the higher speeds.

3.5 Exercises *(pages 227–230)*

1. Shift the graph of $y = x^4$ to the left 3 units, stretch vertically by a factor of 2, and shift downward 7 units.

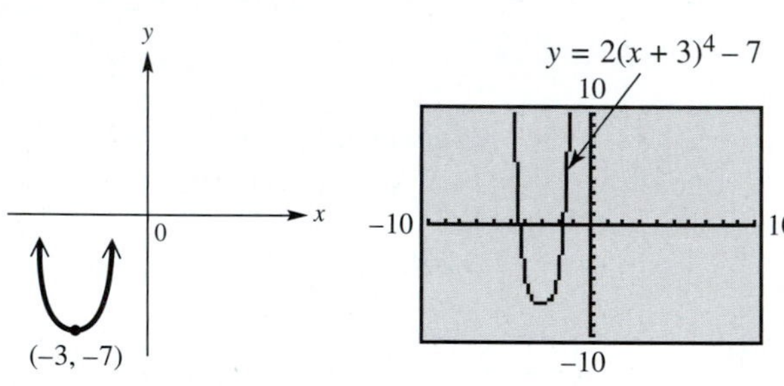

2. Shift the graph of $y = x^4$ to the left 1 unit, stretch vertically by a factor of 3, reflect across the x-axis, and shift upward 12 units.

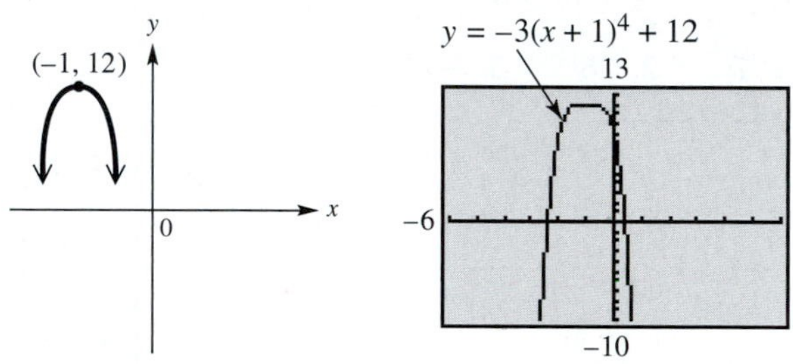

3. Shift the graph of $y = x^3$ to the right 1 unit, stretch vertically by a factor of 3, reflect across the x-axis, and shift upward 12 units.

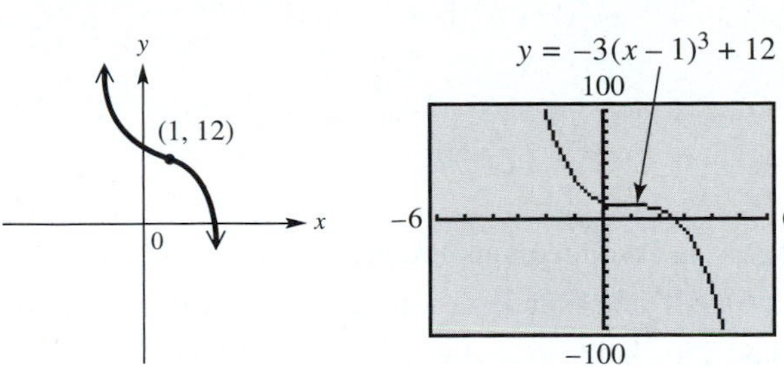

4. Shift the graph of $y = x^5$ to the right 1 unit, shrink vertically by a factor of .5, and shift 13 units upward.

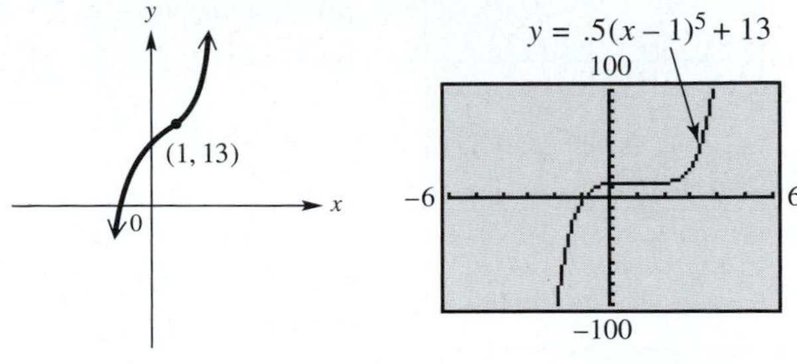

5. minimum degree: 4 **7.** (a, b) and (c, d), local maxima; (e, t), a local minimum **9.** (a, b), absolute maximum **11.** local maximum values of b and d; local minimum of t; absolute maximum of b **13.** The graph of $f(x) = x^n$ for $n \in \{\text{positive odd integers}\}$ will take the shape of the graph of $f(x) = x^3$, but gets steeper as n and x increase.

15. As the odd exponent n gets larger, the graph *flattens out* in the window $[-1, 1]$ by $[-1, 1]$.

$y = x^n$ for $n = 1, 3, 5, 7$

The graph of $y = x^7$ will be between $y = x^5$ and the x-axis in this window.

17. local maximum: $(2, 3.67)$; local minimum: $(3, 3.5)$ **19.** local maximum: $(-3.33, -1.85)$; local minimum: $(-4, -2)$ **21.** two: 2.10 and 2.15 **23.** none

25. 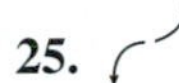 **27.** **29.** **31.** **33.** **35.**

37. D **39.** B **41.** A **43.** one **45.** B and D **47.** one **49.** B

51. false **53.** true **55.** true **57.** false

59.

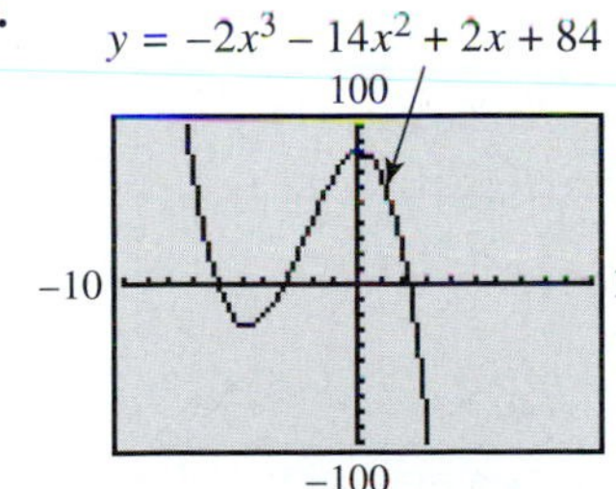

(a) $(-\infty, \infty)$ **(b)** $(-4.74, -27.03)$; not an absolute minimum point **(c)** $(.07, 84.07)$; not an absolute maximum point **(d)** $(-\infty, \infty)$ **(e)** x-intercepts: $-6, -3.19, 2.19$; y-intercept: 84 **(f)** $[-4.74, .07]$ **(g)** $(-\infty, -4.74]$; $[.07, \infty)$

61.

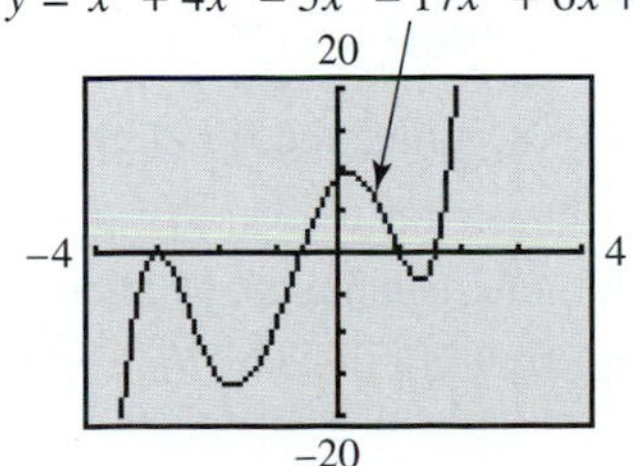

(a) $(-\infty, \infty)$ **(b)** $(-1.73, -16.39)$, $(1.35, -3.49)$; Neither is an absolute minimum point. **(c)** $(-3, 0)$, $(.17, 9.52)$; Neither is an absolute maximum point. **(d)** $(-\infty, \infty)$ **(e)** x-intercepts: $-3, -.62, 1, 1.62$; y-intercept: 9 **(f)** $(-\infty, -3]$; $[-1.73, .17]$; $[1.35, \infty)$ **(g)** $[-3, -1.73]$; $[.17, 1.35]$

63. $y = 2x^4 + 3x^3 - 17x^2 - 6x - 72$

(a) $(-\infty, \infty)$ **(b)** $(-2.63, -132.69)$ is an absolute minimum point; $(1.68, -99.90)$ **(c)** $(-.17, -71.48)$; not an absolute maximum point **(d)** $[-132.69, \infty)$ **(e)** x-intercepts: -4, 3; y-intercept: -72 **(f)** $[-2.63, -.17]$; $[1.68, \infty)$ **(g)** $(-\infty, -2.63]$; $[-.17, 1.68]$

65.

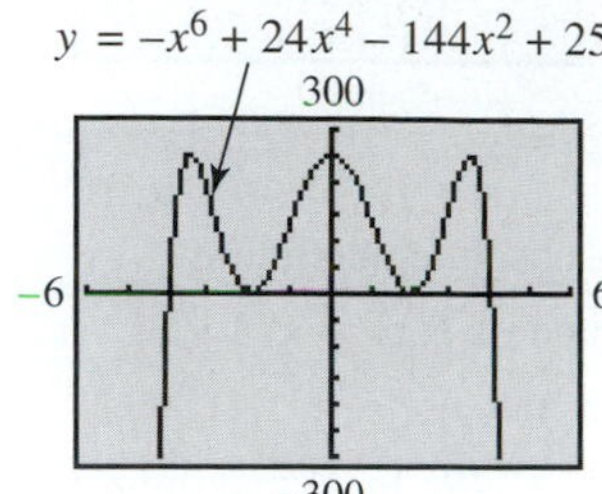

(a) $(-\infty, \infty)$ **(b)** $(-2, 0)$; $(2, 0)$; Neither is an absolute minimum point. **(c)** $(-3.46, 256)$, $(0, 256)$, and $(3.46, 256)$; All are absolute maximum points. **(d)** $(-\infty, 256]$ **(e)** x-intercepts: $-4, -2, 2, 4$; y-intercept: 256 **(f)** $(-\infty, -3.46]$; $[-2, 0]$; $[2, 3.46]$ **(g)** $[-3.46, -2]$; $[0, 2]$; $[3.46, \infty)$

There are many possible valid windows in Exercises 67–71. We give only one in each case.

67. $[-10, 10]$ by $[-40, 10]$ **69.** $[-10, 20]$ by $[-1500, 500]$ **71.** $[-10, 10]$ by $[-20, 500]$

73. (a)

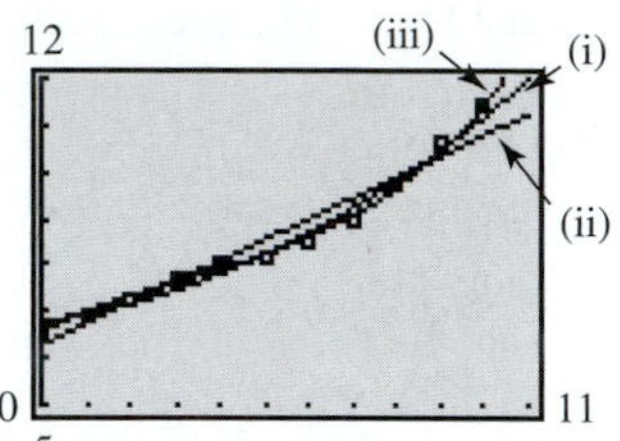

(b) All three model the data well, but (iii) models it best.

Reviewing Basic Concepts *(page 231)*

1. (a) The total length of the fence must satisfy $2L + 2x = 300$. From this equation, $L = 150 - x$. **(b)** $A(x) = x(150 - x)$ **(c)** $0 < x < 150$ **(d)** 50 m by 100 m **2. (a)**

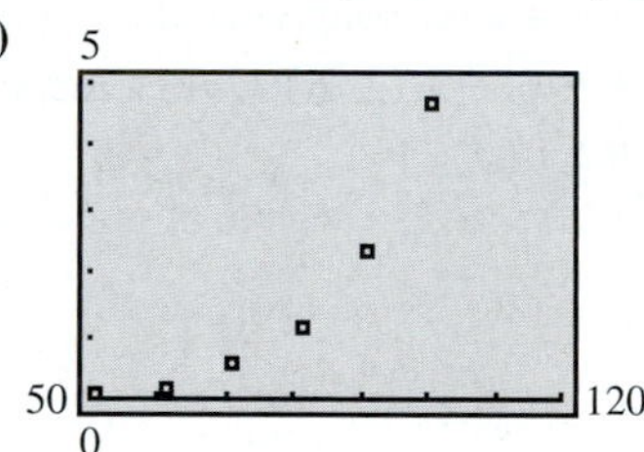

(b) $f(x) = .0018(x - 51)^2 + .1$
(c) $g(x) \approx .0026x^2 - .3139x + 9.426$

(d)

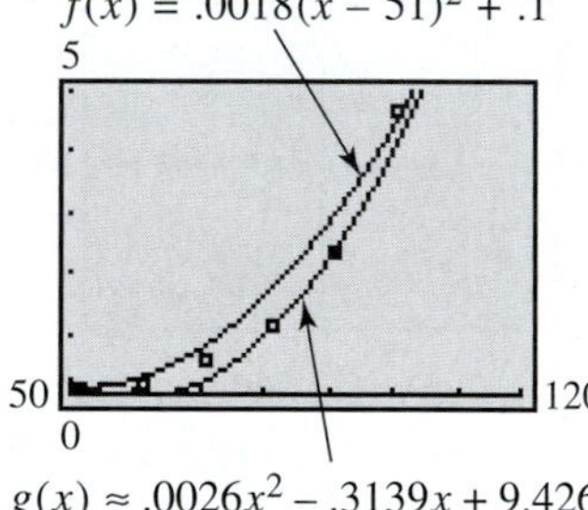

The regression function fits slightly better because it is closer to or passes through more data points. Neither function would fit the data for $x < 51$.

3. **4.** two; three **5.** **6.**

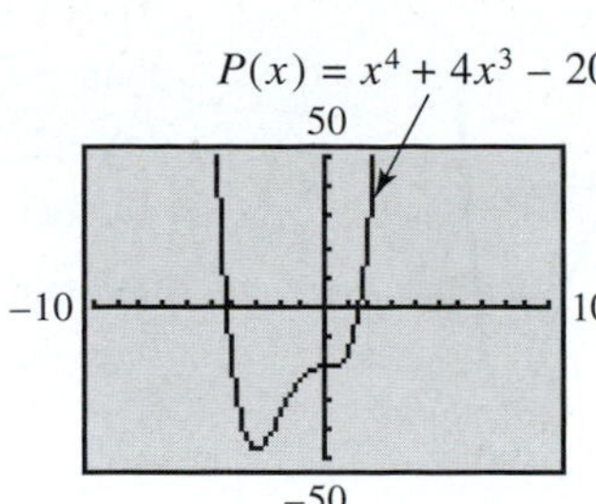

7. The only extreme point is $(-3, -47)$, an absolute minimum. **8.** x-intercepts: -4.26, 1.53; y-intercept: -20

3.6 Exercises (pages 239–240)

1. $P(1) = -5$ and $P(2) = 2$ differ in sign. The zero is approximately 1.79. **3.** $P(2) = 2$ and $P(2.5) = -.25$ differ in sign. The zero is approximately 2.39. **5.** $P(2) = 16$ and $P(1.5) = -.375$ differ in sign. The zero is approximately 1.52. **7.** $P(2.7) = .9219$ and $P(2.8) = -2.7616$ differ in sign. The zero is approximately 2.73. **9.** $P(-1.6) = -1.29376$ and $P(-1.5) = .15625$ differ in sign. The zero is approximately -1.51. **11.** There is at least one zero between 2 and 2.5. **13.** $x^2 - 3x - 2$ **15.** $3x^2 + 4x + \frac{3}{x-5}$ **17.** $x^3 - x^2 - 6x$ **19.** $x^2 + 3x + 3$ **21.** $-2x^2 + 2x - 3 + \frac{1}{x+1}$ **23.** $x^4 + x^3 + x^2 + x + 1$ **25.** 0 **27.** -25 **29.** -5 **31.** 3.625 **33.** -1 **35.** $\sqrt[3]{4}$ **37.** yes **39.** no **41.** yes **43.** no **45.** yes **47.** $x + 3, x - 1, x - 4$ **48.** $-3, 1, 4$ **49.** $-3, 1, 4$ **50.** $-10; -10$ **51.** $(-3, 1) \cup (4, \infty)$ **52.** $(-\infty, -3) \cup (1, 4)$ **53.** $-2, 1$ **55.** $\frac{-1-\sqrt{5}}{2}, \frac{-1+\sqrt{5}}{2}$ **57.** $\frac{1-\sqrt{13}}{6}, \frac{1+\sqrt{13}}{6}$ **59.** $-\sqrt{5}, \sqrt{5}$ **61.** $-\sqrt{3}, \sqrt{3}$ **63.** $P(x) = (x - 2)(2x - 5)(x + 3)$ **65.** $P(x) = (x + 4)(3x - 1)(2x + 1)$ **67.** $P(x) = (x + 3)(-3x + 1)(2x - 1)$ **69.** $P(x) = (x + 5)(x - \sqrt{3})(x + \sqrt{3})$ **71.** $P(x) = (x + 1)^2(x - 4)$

3.7 Exercises *(pages 248–250)*

Answers may vary in Exercises 1–5.

1. $P(x) = x^3 - 8x^2 + 21x - 20$ **3.** $P(x) = x^3 - 5x^2 + x - 5$ **5.** $P(x) = x^3 - 6x^2 + 10x$ **7.** $P(x) = -\frac{1}{6}x^3 + \frac{13}{6}x + 2$ **9.** $P(x) = -\frac{1}{2}x^3 - \frac{1}{2}x^2 + x$ **11.** $P(x) = -x^3 + 6x^2 - 10x + 8$ **13.** $-1 + i, -1 - i$ **15.** $-1 + \sqrt{2}, -1 - \sqrt{2}$ **17.** $-3i, \frac{1}{2} + \frac{\sqrt{3}}{2}i, \frac{1}{2} - \frac{\sqrt{3}}{2}i$

Answers may vary in Exercises 19–29.

19. $P(x) = x^2 - x - 20$ **21.** $P(x) = x^4 + x^3 - 5x^2 + x - 6$ **23.** $P(x) = x^3 - 3x^2 + 2$ **25.** $P(x) = x^4 - 7x^3 + 17x^2 - x - 26$ **27.** $P(x) = x^3 - 11x^2 + 33x + 45$ **29.** $P(x) = x^4 + 2x^3 - 10x^2 - 6x + 45$ **31.** Use synthetic division twice, with $k = -2$. Zeros are $-2, 3, -1$; $P(x) = (x + 2)^2(x - 3)(x + 1)$ **33.** 1, 3, or 5 **35.** **(a)** not possible **(b)** possible **(c)** not possible **(d)** possible

37.

$P(x) = 2x^3 - 5x^2 - x + 6$
$= (x + 1)(2x - 3)(x - 2)$

39.

$P(x) = x^4 - 18x^2 + 81$
$= (x - 3)^2(x + 3)^2$

41.

$P(x) = 2x^4 + x^3 - 6x^2 - 7x - 2$
$= (2x + 1)(x - 2)(x + 1)^2$

43.

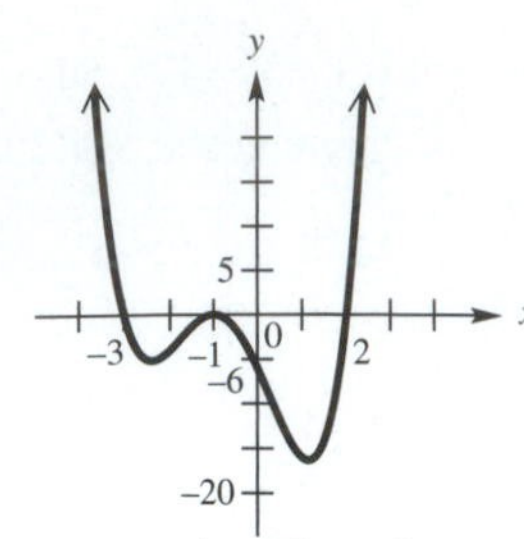

$P(x) = x^4 + 3x^3 - 3x^2 - 11x - 6$
$= (x + 3)(x + 1)^2(x - 2)$

45.

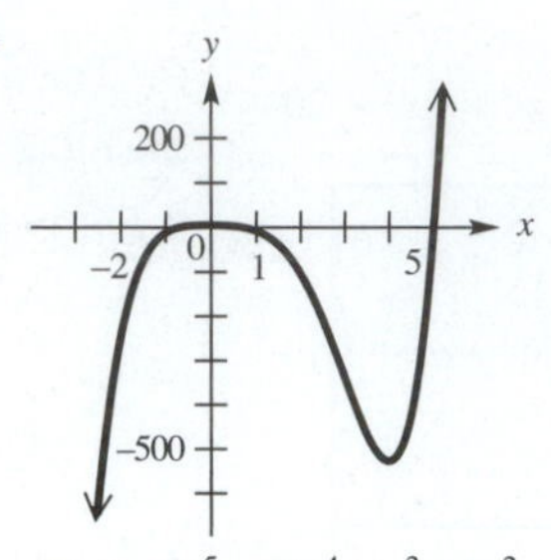

$P(x) = 2x^5 - 10x^4 + x^3 - 5x^2 - x + 5$
$= (x - 5)(x^2 + 1)(2x^2 - 1)$

47. **(a)** $\pm 1, \pm 2, \pm 5, \pm 10$ **(b)** Eliminate values less than -2 or greater than 5. **(c)** $-2, -1, 5$ **(d)** $P(x) = (x + 2)(x + 1)(x - 5)$ **49.** **(a)** $\pm 1, \pm 2, \pm 3, \pm 5, \pm 6, \pm 10, \pm 15, \pm 30$ **(b)** Eliminate values less than -5 or greater than 2. **(c)** $-5, -3, 2$ **(d)** $P(x) = (x + 5)(x + 3)(x - 2)$

51. **(a)** $\pm 1, \pm 2, \pm 3, \pm 4, \pm 6, \pm 12, \pm \frac{1}{2}, \pm \frac{3}{2}, \pm \frac{1}{3}, \pm \frac{2}{3}, \pm \frac{4}{3}, \pm \frac{1}{6}$ **(b)** Eliminate values less than -4 or greater than $\frac{3}{2}$. **(c)** $-4, -\frac{1}{3}, \frac{3}{2}$ **(d)** $P(x) = (x + 4)(3x + 1)(2x - 3)$

53. **(a)** $\pm 1, \pm 2, \pm 3, \pm 6, \pm \frac{1}{2}, \pm \frac{3}{2}, \pm \frac{1}{3}, \pm \frac{2}{3}, \pm \frac{1}{6}, \pm \frac{1}{12}, \pm \frac{1}{4}, \pm \frac{3}{4}$ **(b)** Eliminate values less than $-\frac{3}{2}$ or greater than $\frac{1}{2}$. **(c)** $-\frac{3}{2}, -\frac{2}{3}, \frac{1}{2}$ **(d)** $P(x) = (3x + 2)(2x + 3)(2x - 1)$ **55.** **(a)** $\pm 1, \pm 2, \pm 3, \pm 4, \pm 6, \pm 12, \pm \frac{1}{2}, \pm \frac{3}{2}, \pm \frac{1}{3}, \pm \frac{2}{3}, \pm \frac{4}{3}, \pm \frac{1}{4}, \pm \frac{3}{4}, \pm \frac{1}{6}, \pm \frac{1}{8}, \pm \frac{3}{8}, \pm \frac{1}{12}, \pm \frac{1}{24}$ **(b)** Eliminate values less than $-\frac{3}{2}$ or greater than $\frac{1}{2}$. **(c)** $-\frac{3}{2}, -\frac{2}{3}, \frac{1}{2}$ **(d)** $P(x) = 2(2x + 3)(3x + 2)(2x - 1)$ **57.** $-2, -1, \frac{5}{2}$ **59.** $\frac{3}{2}, 4$ **61.** possible: 0 or 2 positive real zeros, 1 negative real zero; actual: 0 positive, 1 negative **63.** possible: 1 positive real zero, 1 negative real zero; actual: 1 positive, 1 negative **65.** possible: 0 or 2 positive real zeros, 1 or 3 negative real zeros; actual: 0 positive, 1 negative

75. $P(x) = \frac{1}{2}(x + 6)(x - 2)(x - 5)$ or $P(x) = \frac{1}{2}x^3 - \frac{1}{2}x^2 - 16x + 30$ **77.** **(a)** positive zeros: 1; negative zeros: 3 or 1 **(b)** $\pm 1, \pm 2, \pm \frac{1}{2}$ **(c)** $-1, 1$ **(d)** no other real zeros **(e)** $-\frac{1}{4} + \frac{\sqrt{15}}{4}i, -\frac{1}{4} - \frac{\sqrt{15}}{4}i$ **(f)** $-1, 1$ **(g)** 2 **(h)** $P(4) = -570; (4, -570)$ **(i)** ⌒ **(j)**

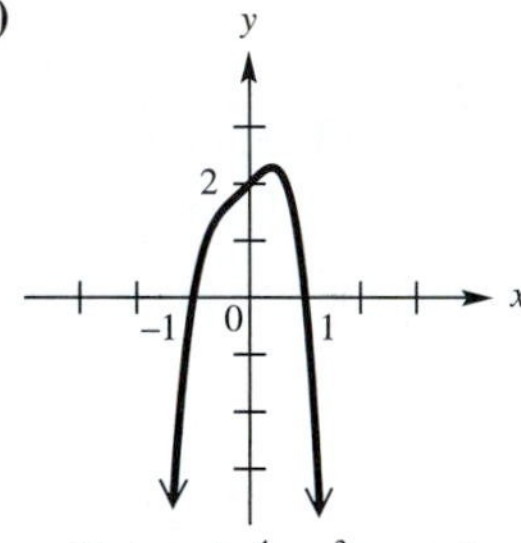

$P(x) = -2x^4 - x^3 + x + 2$

78. **(a)** positive zeros: 0; negative zeros: 4, 2 or 0 **(b)** $0, \pm 1, \pm 3, \pm 9, \pm 27, \pm \frac{1}{2}, \pm \frac{3}{2}, \pm \frac{9}{2}, \pm \frac{27}{2}, \pm \frac{1}{4}, \pm \frac{3}{4}, \pm \frac{9}{4}, \pm \frac{27}{4}$ **(c)** $0, -\frac{3}{2}$ (multiplicity 2) **(d)** no other real zeros **(e)** $\frac{1}{2} + \frac{\sqrt{11}}{2}i, \frac{1}{2} - \frac{\sqrt{11}}{2}i$ **(f)** $0, -\frac{3}{2}$ **(g)** 0 **(h)** $P(4) = 7260; (4, 7260)$ **(i)** **(j)**

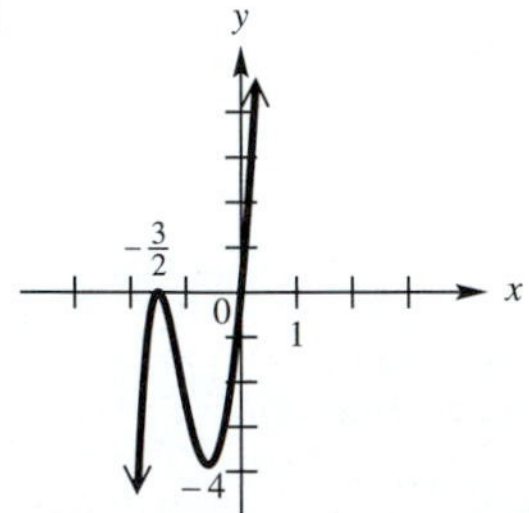

$P(x) = 4x^5 + 8x^4 + 9x^3 + 27x^2 + 27x$

79. **(a)** positive zeros: 1; negative zeros: 1
(b) $\pm 1, \pm 5, \pm \frac{1}{3}, \pm \frac{5}{3}$ **(c)** no rational zeros
(d) $-\sqrt{5}, \sqrt{5}$ **(e)** $-\frac{\sqrt{3}}{3} i, \frac{\sqrt{3}}{3} i$ **(f)** $-\sqrt{5}, \sqrt{5}$
(g) -5 **(h)** $P(4) = 539$; $(4, 539)$ **(i)**
(j)

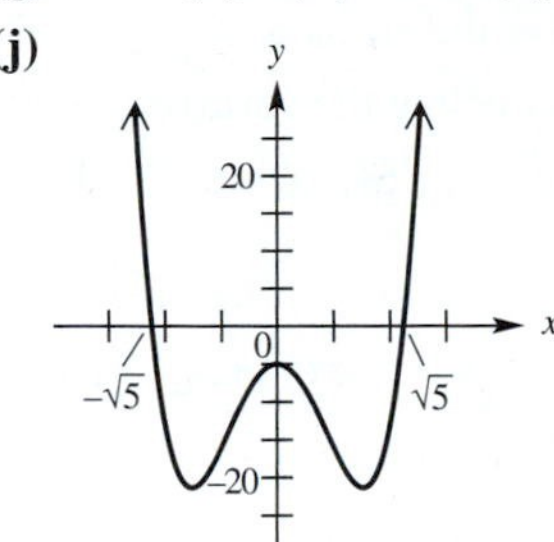

$P(x) = 3x^4 - 14x^2 - 5$

80. **(a)** positive zeros: 2 or 0; negative zeros: 3 or 1
(b) $\pm 1, \pm 3, \pm 9$ **(c)** $-3, -1$ (multiplicity 2), 1, 3
(d) no other real zeros **(e)** no other complex zeros
(f) $-3, -1, 1, 3$ **(g)** -9 **(h)** $P(4) = -525$; $(4, -525)$
(i) **(j)**

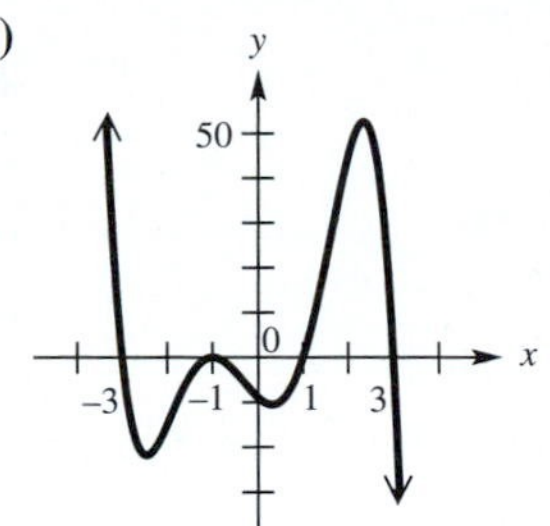

$P(x) = -x^5 - x^4 + 10x^3 + 10x^2 - 9x - 9$

81. **(a)** positive zeros: 4, 2 or 0; negative zeros: 0 **(b)** $\pm 1, \pm 2, \pm 3, \pm 4, \pm 6, \pm 12, \pm \frac{1}{3}, \pm \frac{2}{3}, \pm \frac{4}{3}$ **(c)** $\frac{1}{3}$, 2 (multiplicity 2), 3
(d) no other real zeros **(e)** no other complex zeros **(f)** $\frac{1}{3}$, 2, 3 **(g)** -12 **(h)** $P(4) = -44$; $(4, -44)$ **(i)**
(j)

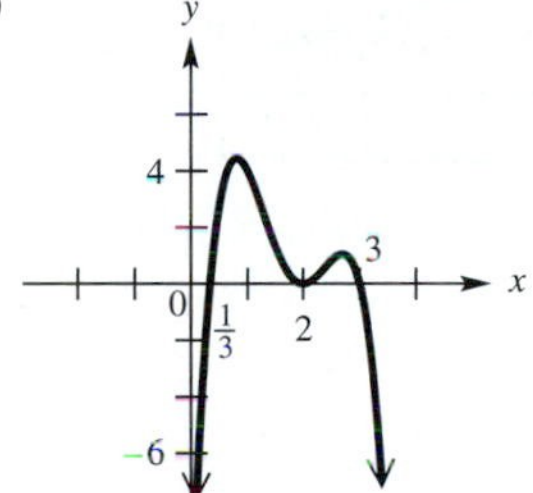

$P(x) = -3x^4 + 22x^3 - 55x^2 + 52x - 12$

82. For the function in Exercise 79: ± 2.236

3.8 Exercises *(pages 256–259)*

1. $\left\{0, \pm \frac{\sqrt{7}}{7} i\right\}$ **3.** $\left\{-\frac{2}{3}, \pm 1\right\}$ **5.** $\{\pm 1, \pm \sqrt{10}\}$ **7.** $\{-2, -1.5, 1.5, 2\}$ **9.** $\{-4, 4, -i, i\}$ **11.** $\{-8, 1, 8\}$
13. $\{-1.5, 0, 1\}$ **15.** $\{-1, \pm \sqrt{7}\}$ **17.** $\left\{0, \frac{-1 \pm \sqrt{73}}{6}\right\}$ **19.** $\left\{0, -\frac{1}{2} - \frac{\sqrt{3}}{2} i, -\frac{1}{2} + \frac{\sqrt{3}}{2} i\right\}$ **21.** $\{-4i, -i, i, 4i\}$
23. $\left\{-3, 2, -1 - i\sqrt{3}, -1 + i\sqrt{3}, \frac{3}{2} + \frac{3\sqrt{3}}{2} i, \frac{3}{2} - \frac{3\sqrt{3}}{2} i\right\}$

25.

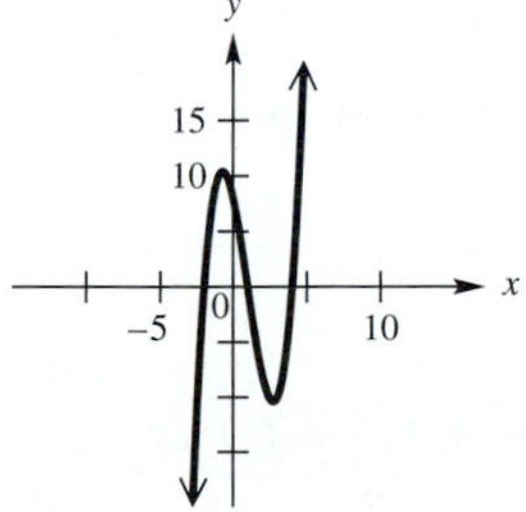

$P(x) = x^3 - 3x^2 - 6x + 8$
$= (x - 4)(x - 1)(x + 2)$

(a) $\{-2, 1, 4\}$ **(b)** $(-\infty, -2) \cup (1, 4)$
(c) $(-2, 1) \cup (4, \infty)$

27.

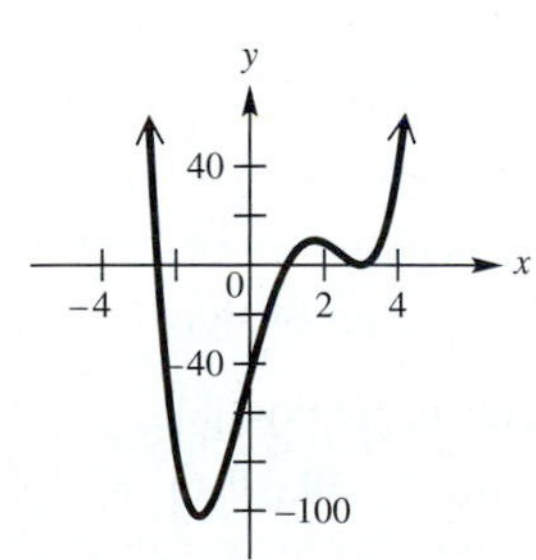

$P(x) = 2x^4 - 9x^3 - 5x^2 + 57x - 45$
$= (x - 3)^2(2x + 5)(x - 1)$

(a) $\{-2.5, 1, 3 \text{ (multiplicity 2)}\}$
(b) $(-2.5, 1)$ **(c)** $(-\infty, -2.5) \cup (1, 3) \cup (3, \infty)$

29.

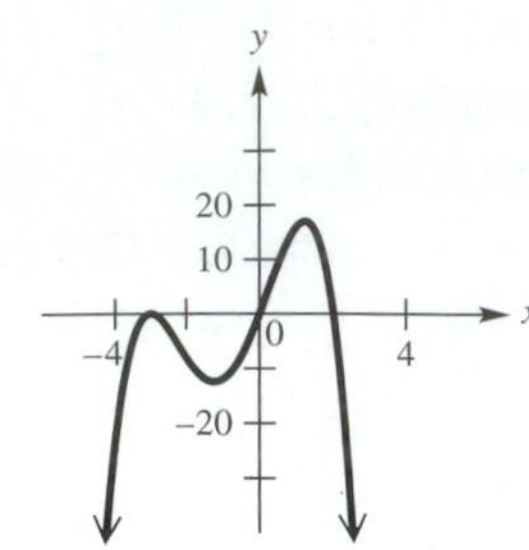

(a) $\{-3 \text{ (multiplicity 2)}, 0, 2\}$
(b) $\{-3\} \cup [0, 2]$ **(c)** $(-\infty, 0] \cup [2, \infty)$

31. $\{-.88, 2.12, 4.86\}$ **33.** $\{1.52\}$ **35.** $\{-.40, 2.02\}$ **37.** $\{-i, i\}$

39. $\left\{-1, \frac{1}{2} + \frac{\sqrt{3}}{2}i, \frac{1}{2} - \frac{\sqrt{3}}{2}i\right\}$

41. $\left\{3, -\frac{3}{2} - \frac{3\sqrt{3}}{2}i, -\frac{3}{2} + \frac{3\sqrt{3}}{2}i\right\}$ **43.** $\{-2, 2, -2i, 2i\}$

45. $\{-4, 2 + 2i\sqrt{3}, 2 - 2i\sqrt{3}\}$ **47.** $\{-3i\sqrt{2}, 3i\sqrt{2}\}$

49. **(a)** about 7.13 cm; The ball floats partly above the surface. **(b)** The sphere is more dense than water and sinks below the surface. **(c)** 10 cm; The balloon floats even with the surface. **51.** about 11.34 cm

53. **(a)** $0 < x < 6$ **(b)** $V(x) = 4x^3 - 60x^2 + 216x$ **(c)** about 2.35; about 228.16 in.^{3} **(d)** $.42 < x < 5$ **55.** about 2.61 in.

57. **(a)** $x - 1$ **(b)** $\sqrt{x^2 - (x - 1)^2}$ or $\sqrt{2x - 1}$ **(c)** $2x^3 - 5x^2 + 4x - 28{,}225 = 0$ **(d)** 7 in., 24 in., and 25 in.

59. **(a)** about 66.15 in.^{3} **(b)** .54 in. $< x <$ 2.92 in.

61. **(a)**

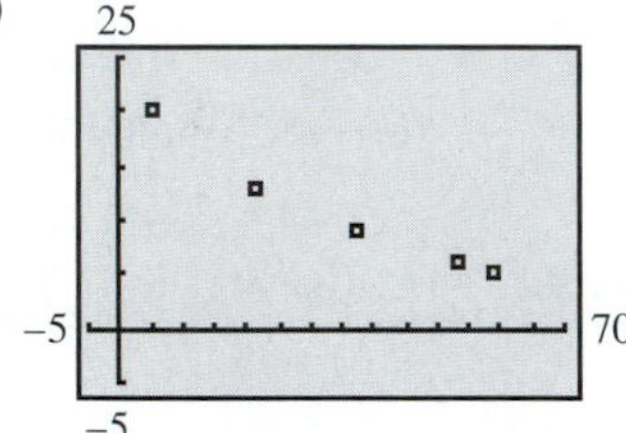

(b) $C(x) \approx .0035x^2 - .49x + 22$ **(c)** $C(x) \approx -.000068x^3 + .00987x^2 - .653x + 23$

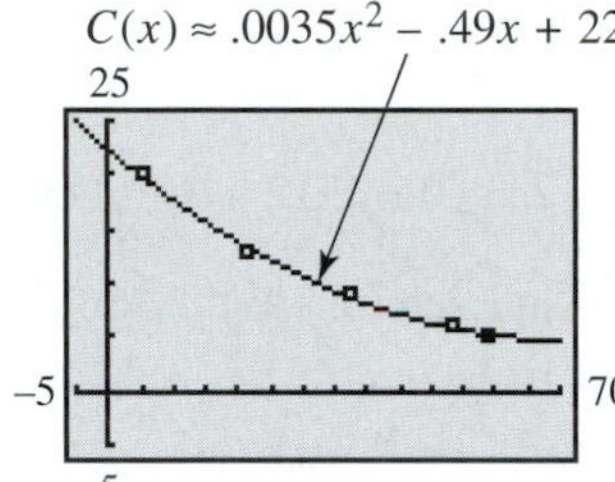

C(x) ≈ –.000068x³ + .00987x² – .653x + 23
25
–5
70
–5

(d) The cubic function is a slightly better fit. **(e)** $0 \le x < 31.92$ **63.** at approximately 3.4 sec

Reviewing Basic Concepts *(page 259)*

1. 30 **2.** no **3.** $P(x) = (x + 2)(2x - 3)(x - 2 - i)(x - 2 + i)$ **4.** $P(x) = x^3 - \frac{3}{2}x^2 + x - \frac{3}{2}$

5. $P(x) = x^4 + 6x^3 + 5x^2 + 8x + 80$ **6.** $-2, -1, \frac{5}{2}$ **7.** $\left\{\pm\sqrt{\frac{6 + \sqrt{33}}{3}}, \pm\sqrt{\frac{6 - \sqrt{33}}{3}}\right\}$

8. $P(x) \approx -1.055x^2 + 22.62x + 98.9$; $P(9) \approx 217$, which means that in 2007, about 217 million debit cards will be issued.

Chapter 3 Review Exercises *(pages 263–267)*

1. $18 - 4i$ **3.** $14 - 52i$ **5.** $\frac{1}{10} + \frac{3}{10}i$ **7.** $(-\infty, \infty)$ **9.** ∪ **11.** -8 **13.** $\left[\frac{3}{2}, \infty\right); \left(-\infty, \frac{3}{2}\right]$

15. The graph intersects the x-axis at -1 and 4, supporting the answer in (a). It lies above the x-axis when $x < -1$ or $x > 4$, supporting the answer in (b). It lies on or below the x-axis when x is between -1 and 4 inclusive, supporting the answer in (c).

17. Since the discriminant is greater than 0, there are two x-intercepts. **19.** $(1.04, 6.37)$ **21.** **(a)** 4 **(b)** 1 **(c)** two **(d)** none **23.** 25 sec **25.** 6.3 sec and 43.7 sec **27.** **(a)** $V(x) = 12x^2 - 128x + 256$ **(b)** 20 in. by 60 in. **(c)** One way is to graph $y_1 = V(x)$ and $y_2 = 2496$ and show that the graphs intersect at $x = 20$.

29. $P(-2) = 12$ and $P(-1) = -4$ differ in sign. **31.** -1 **33.** 28 **35.** $7 - 2i$

In Exercises 37 and 39, other answers are possible.

37. $P(x) = x^3 - 13x^2 + 46x - 48$ **39.** $P(x) = x^4 + 5x^3 + x^2 - 9x + 2$ **41.** yes **43.** $-3, 4, 1 - i, 1 + i$

45. $-2, \frac{1}{3}, 4$ **49.** $(x - 3)(x^2 + x - 1)$ **53.** $\{0, -1 + 2i, -1 - 2i\}$; 0 is the only x-intercept.

55. $Y_1 = (X + 4)\left(X - \left(\frac{5 - \sqrt{21}}{5}\right)\right)\left(X - \left(\frac{5 + \sqrt{21}}{5}\right)\right)$ **57.** even **59.** positive **61.** $(-\infty, a) \cup (b, c)$

63. $\{d, h\}$ **65.** Since $f(x)$ has three real zeros and a polynomial of degree 3 can have at most three zeros, there can be no other zeros, real or nonreal complex. **67.** true **69.** true **71.** false **73.** $(2, 0)$ **75.** $(-\infty, \infty)$

77. $\left\{-\sqrt{7}, -\frac{2}{3}, \sqrt{7}\right\}$ **79.** 4 in. by 4 in. by 4 in. **81.** 1387 thousand

Chapter 3 Test *(pages 267–268)*

1. (a) $20 - 9i$ **(b)** $4 - i$ **(c)** i **(d)** $12 + 16i$ **2. (a)** $(-1, 8)$ **(b)**

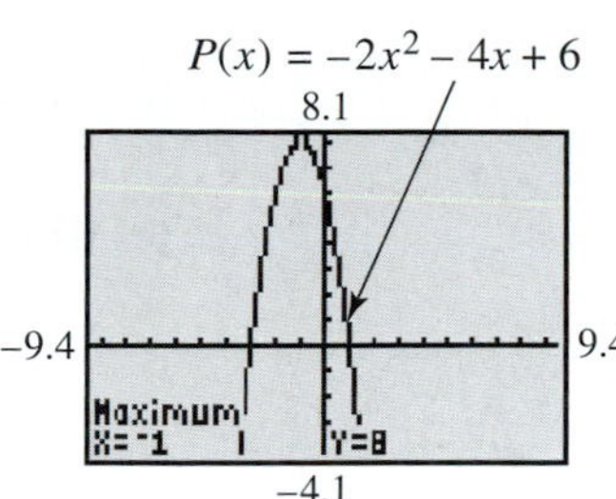

(c) $-3, 1$

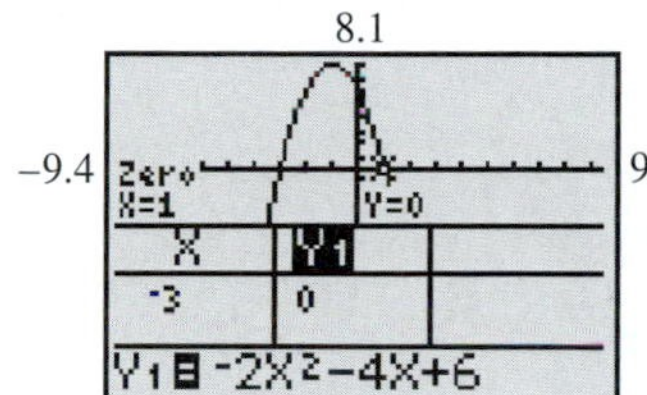

(d) 6 **(e)** domain: $(-\infty, \infty)$; range: $(-\infty, 8]$ **(f)** increasing: $(-\infty, -1]$; decreasing: $[-1, \infty)$

3. (a) $\left\{\frac{-3 \pm \sqrt{33}}{6}\right\}$ **(b)**

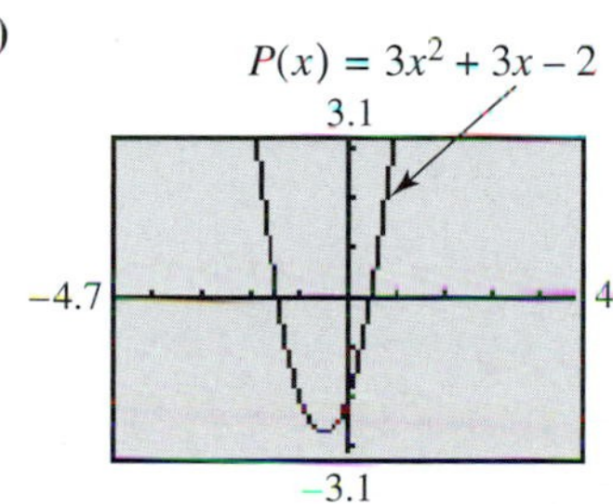

(i) $\left(\frac{-3 - \sqrt{33}}{6}, \frac{-3 + \sqrt{33}}{6}\right)$

(ii) $\left(-\infty, \frac{-3 - \sqrt{33}}{6}\right] \cup \left[\frac{-3 + \sqrt{33}}{6}, \infty\right)$

4. 15 in. by 12 in. by 4 in.

5. (a)

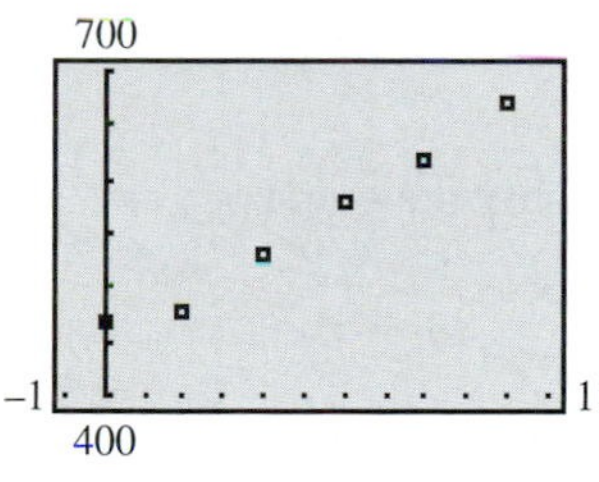

(b) $f(x) = 2x^2 + 470$ **(c)** $g(x) \approx .737x^2 + 13.8x + 461$; function g

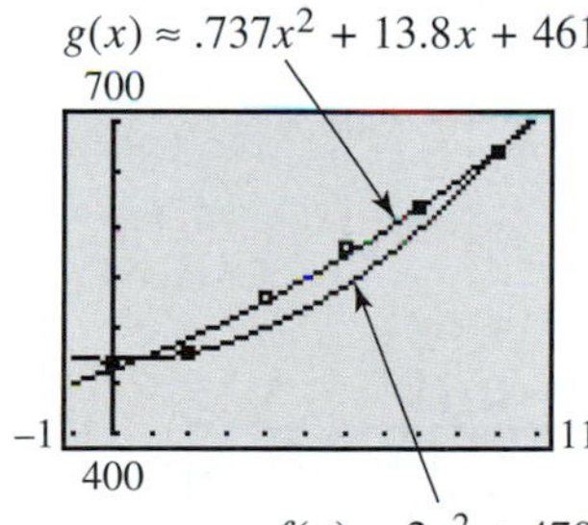

6. **(a)** $-1 + i\sqrt{3}, -1 - i\sqrt{3}$ **(b)** $\cup$

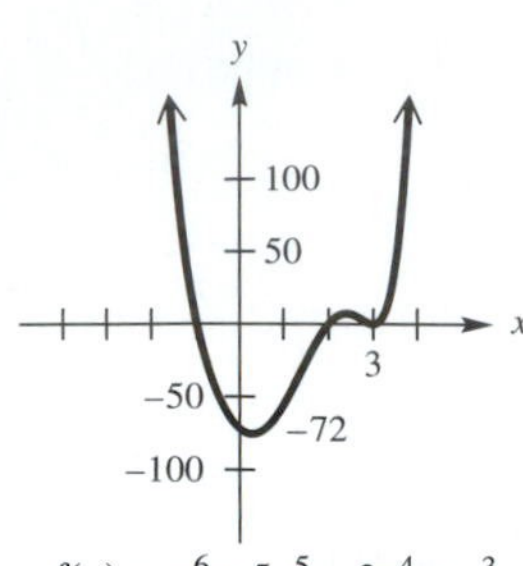

7. **(a)** $\left\{-\frac{5}{2}, \frac{5}{2}, -i, i\right\}$

(b)

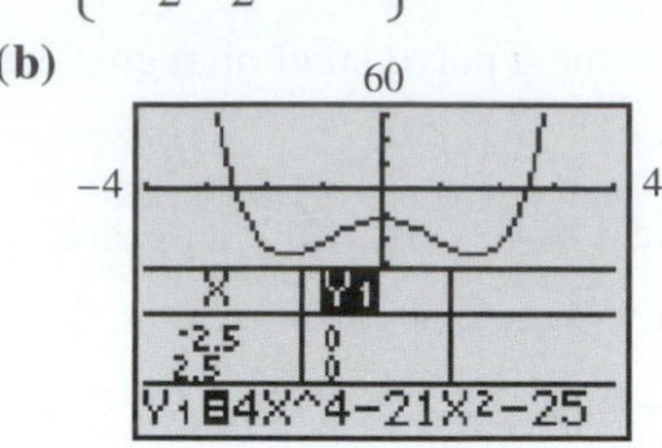

(c) It is symmetric with respect to the y-axis.

(d) **(i)** $\left(-\infty, -\frac{5}{2}\right] \cup \left[\frac{5}{2}, \infty\right)$ **(ii)** $\left(-\frac{5}{2}, \frac{5}{2}\right)$

8. **(a)** $\pm 1, \pm 2, \pm 11, \pm 22, \pm\frac{1}{3}, \pm\frac{2}{3}, \pm\frac{11}{3}, \pm\frac{22}{3}$ **(b)** $-2, \frac{1}{3}$ **(c)** $f(3) = -80$ and $f(4) = 330$ differ in sign. **(d)** positive zeros: 2 or 0; negative zeros: 2 or 0 **9.** **(a)** $\{.189, 1, 3.633\}$ **(b)** two **10.** **(a)** $f(x) \approx -.249x^3 + 4.47x^2 + .212x + 467$ **(b)** $g(x) \approx .0977x^4 - 2.20x^3 + 16.5x^2 - 22.1x + 470$ **(c)**

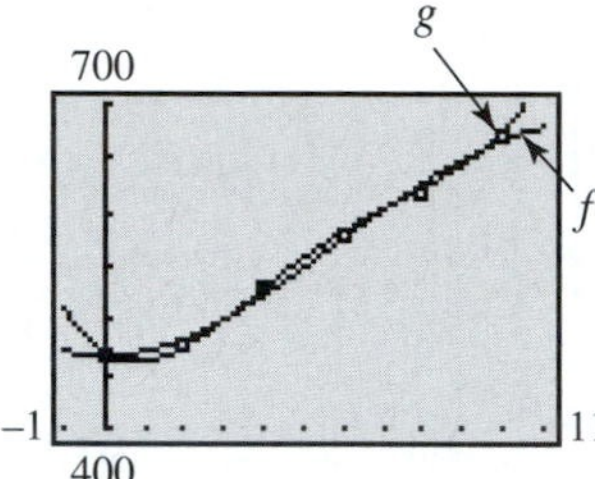

(d) cubic: 679 million; quartic: 726 million; The quartic function is a better estimate because it continues to increase, while the cubic function turns downward.

CHAPTER 4 RATIONAL, POWER, AND ROOT FUNCTIONS

4.1 Exercises *(pages 276–277)*

1. $(-\infty, 0) \cup (0, \infty)$; $(-\infty, 0) \cup (0, \infty)$ **3.** none; $(-\infty, 0) \cup (0, \infty)$; none **5.** $x = 3$; $y = 2$ **7.** even; symmetry with respect to the y-axis **9.** A, B, C **11.** A **13.** A **15.** A, C, D **17.** C

In Exercises 19–33, we give the domain and then the range below the traditional graph.

19. To obtain the graph of f, stretch the graph of $y = \frac{1}{x}$ vertically by a factor of 2.

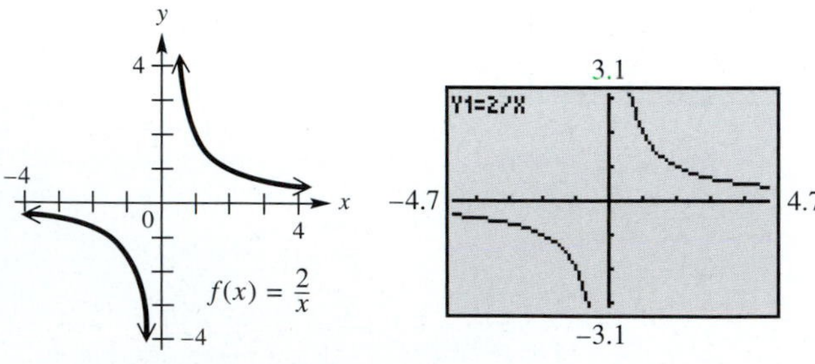

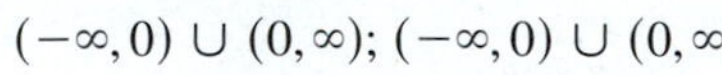

$(-\infty, 0) \cup (0, \infty)$; $(-\infty, 0) \cup (0, \infty)$

21. To obtain the graph of f, shift the graph of $y = \frac{1}{x}$ to the left 2 units.

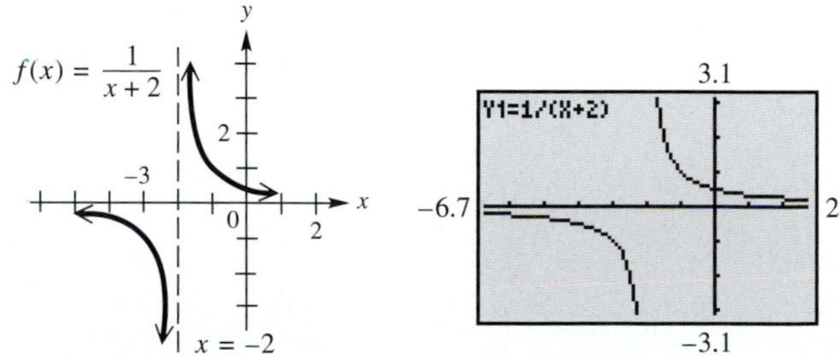

$(-\infty, -2) \cup (-2, \infty)$; $(-\infty, 0) \cup (0, \infty)$

23. To obtain the graph of f, shift the graph of $y = \dfrac{1}{x}$ upward 1 unit.

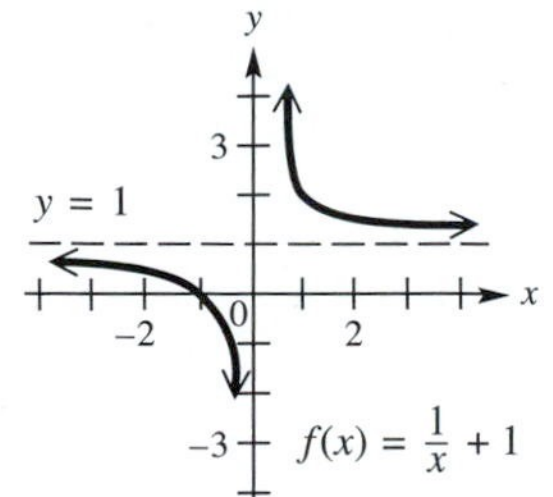

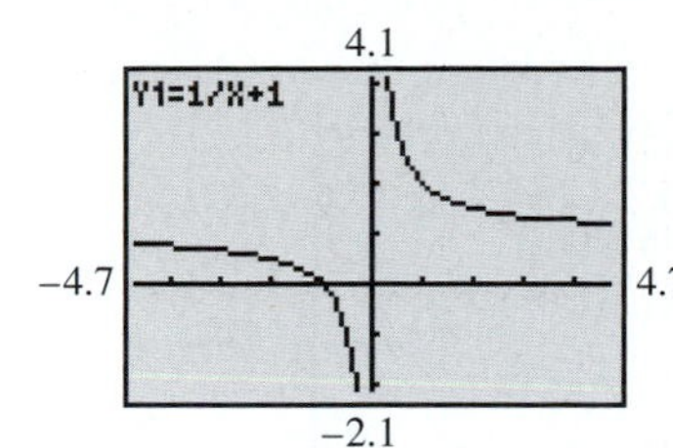

$(-\infty, 0) \cup (0, \infty)$; $(-\infty, 1) \cup (1, \infty)$

25. To obtain the graph of f, shift the graph of $y = \dfrac{1}{x}$ to the right 1 unit and upward 1 unit.

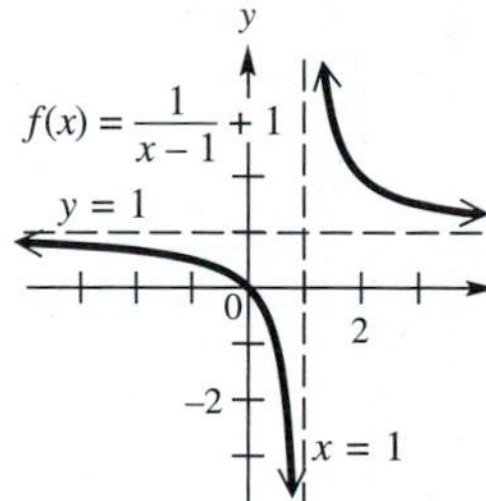

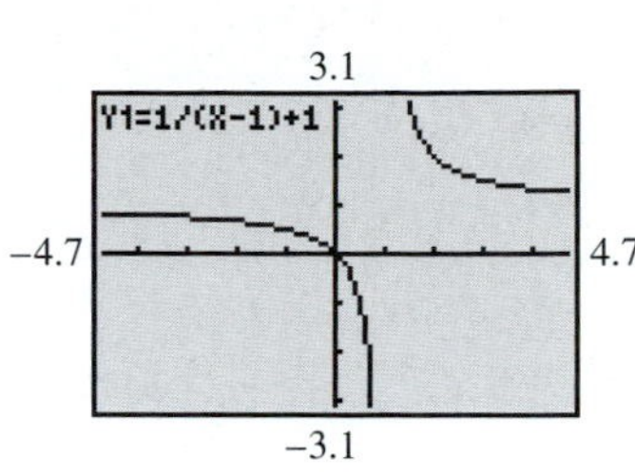

$(-\infty, 1) \cup (1, \infty)$; $(-\infty, 1) \cup (1, \infty)$

27. To obtain the graph of f, shift the graph of $y = \dfrac{1}{x^2}$ downward 2 units.

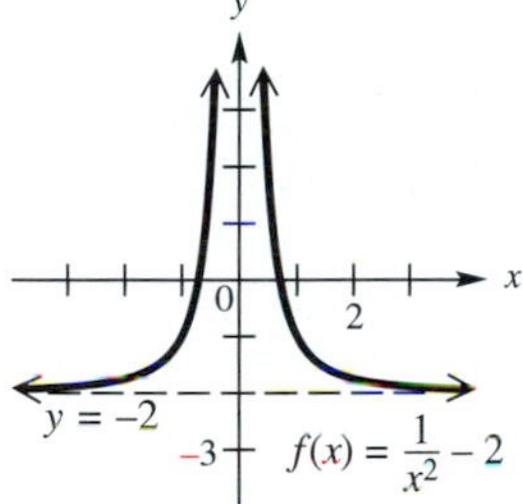

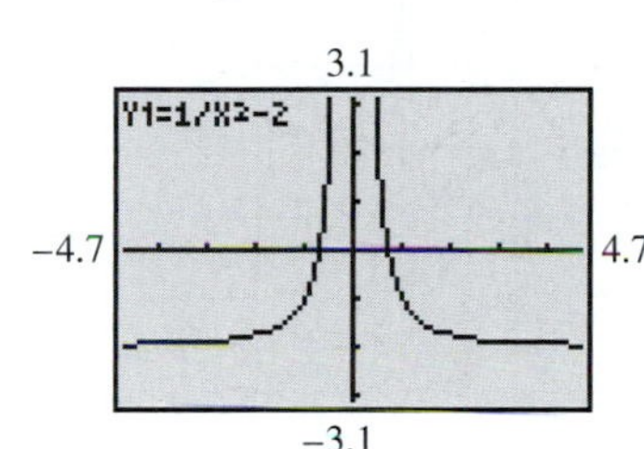

$(-\infty, 0) \cup (0, \infty)$; $(-2, \infty)$

29. To obtain the graph of f, stretch the graph of $y = \dfrac{1}{x^2}$ vertically by a factor of 2, and reflect across the x-axis.

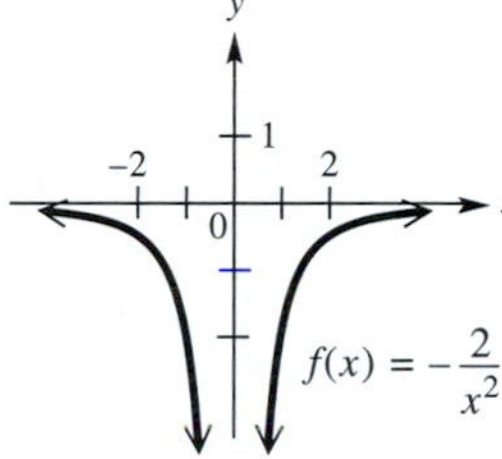

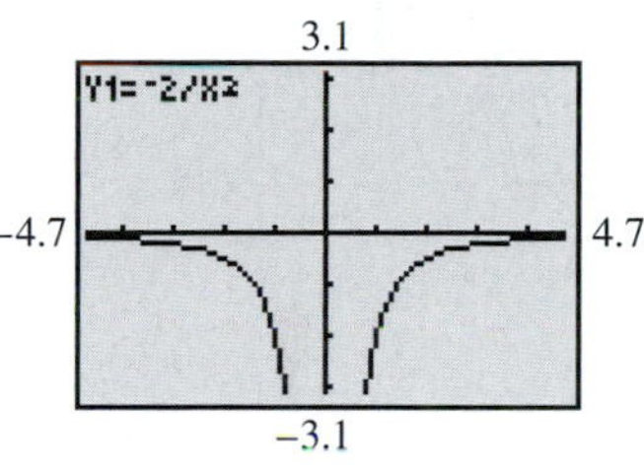

$(-\infty, 0) \cup (0, \infty)$; $(-\infty, 0)$

31. To obtain the graph of f, shift the graph of $y = \dfrac{1}{x^2}$ to the right 3 units.

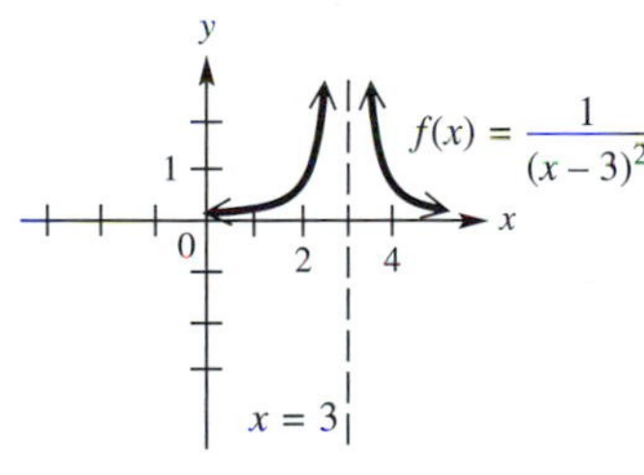

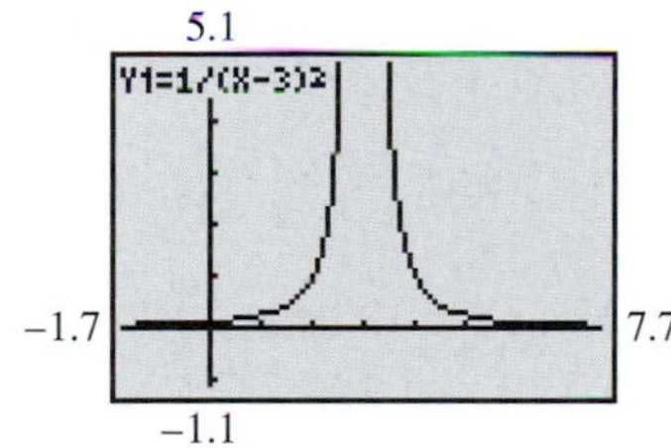

$(-\infty, 3) \cup (3, \infty)$; $(0, \infty)$

33. To obtain the graph of f, shift the graph of $y = \dfrac{1}{x^2}$ to the left 2 units, reflect across the x-axis, and shift 3 units downward.

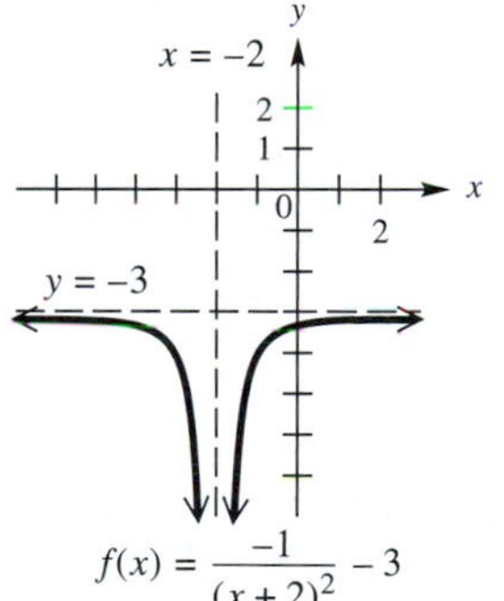

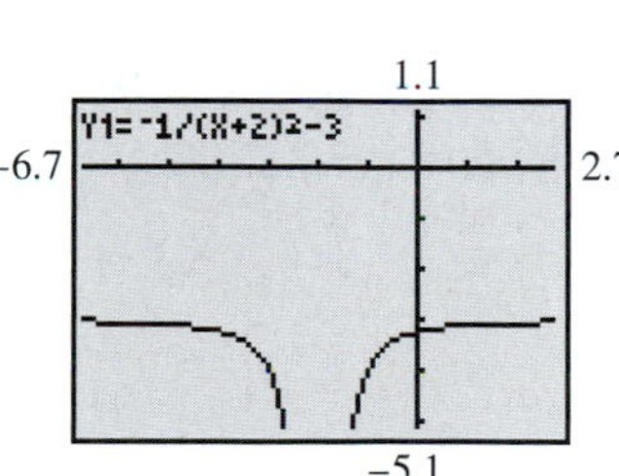

$(-\infty, -2) \cup (-2, \infty)$; $(-\infty, -3)$

35. C **37.** B **39.** vertical asymptote: $x = -1$; horizontal asymptote: $y = 1$; domain: $(-\infty, -1) \cup (-1, \infty)$; range: $(-\infty, 1) \cup (1, \infty)$ **41.** $f(x) = 1 + \dfrac{1}{x-2}$

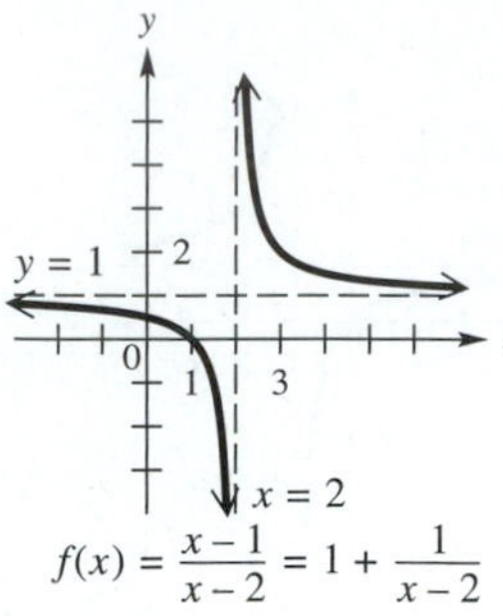

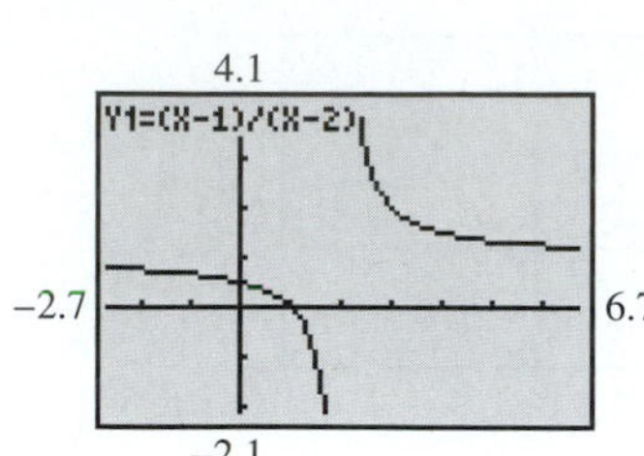

43. $f(x) = -2 + \dfrac{1}{x+3}$

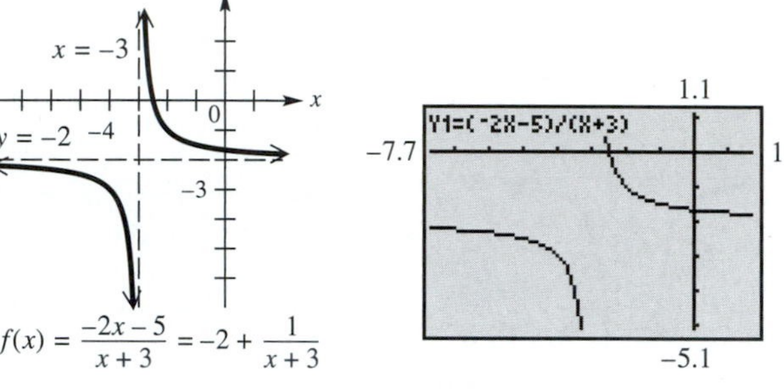

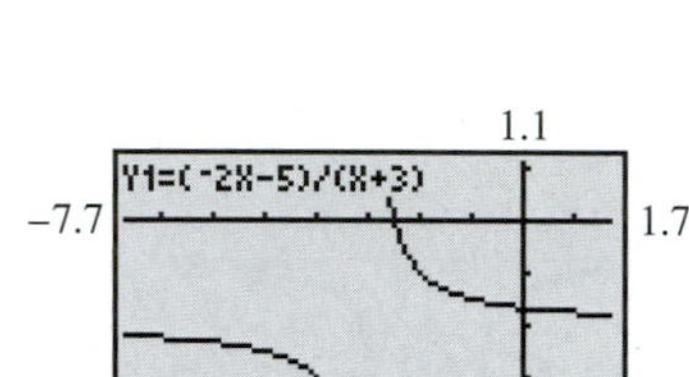

45. $f(x) = 2 + \dfrac{1}{x-3}$

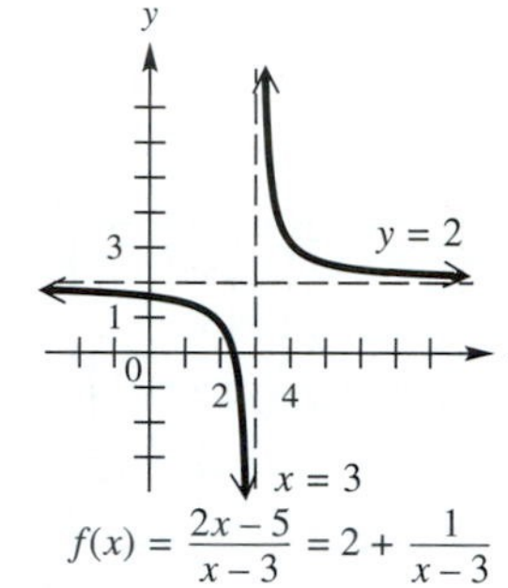

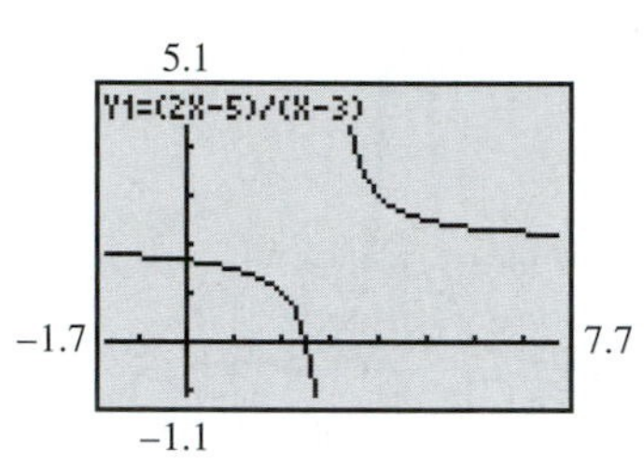

4.2 Exercises *(pages 287–290)*

1. D **3.** G **5.** E **7.** F

In Exercises 9–15 and 19–25, V.A. represents vertical asymptote, H.A. represents horizontal asymptote, and O.A. represents oblique asymptote.

9. V.A.: $x = 5$; H.A.: $y = 0$ **11.** V.A.: $x = -\dfrac{1}{2}$; H.A.: $y = -\dfrac{3}{2}$ **13.** V.A.: $x = -3$; O.A.: $y = x - 3$

15. V.A.: $x = -2$, $x = \dfrac{5}{2}$; H.A.: $y = \dfrac{1}{2}$ **17.** A **19.** V.A.: $x = 2$; H.A.: $y = 4$; $(-\infty, 2) \cup (2, \infty)$ **21.** V.A.: $x = \pm 2$; H.A.: $y = -4$; $(-\infty, -2) \cup (-2, 2) \cup (2, \infty)$ **23.** V.A.: none; H.A.: $y = 0$; $(-\infty, \infty)$ **25.** V.A.: $x = -1$; O.A.: $y = x - 1$; $(-\infty, -1) \cup (-1, \infty)$ **27.**

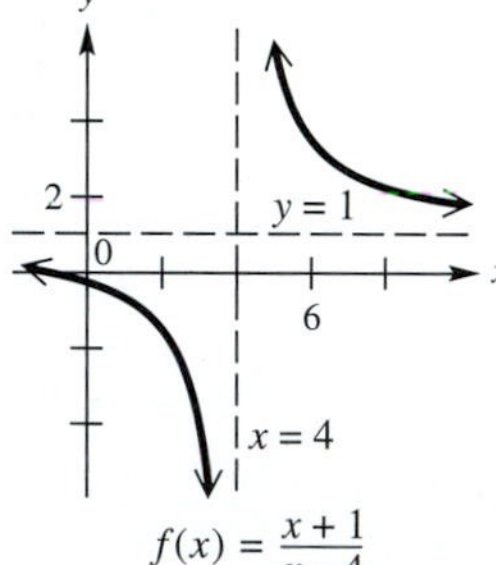

29.

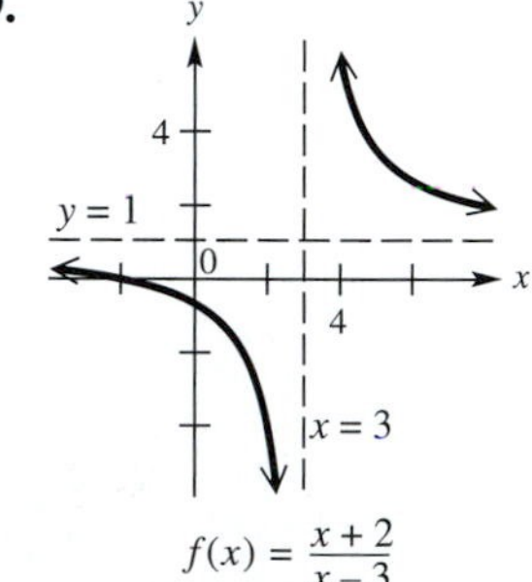

31.

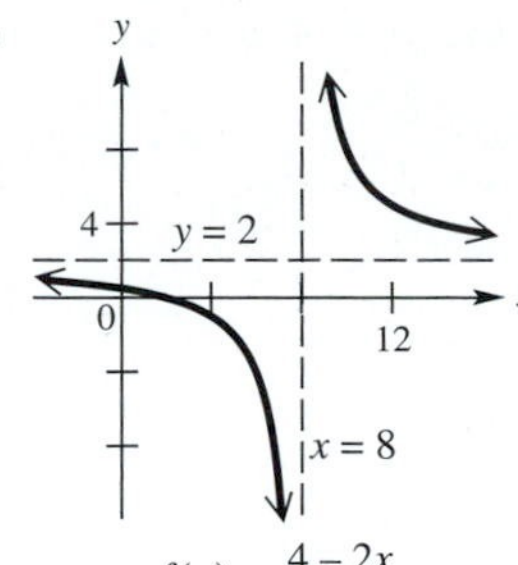

$f(x) = \dfrac{4 - 2x}{8 - x}$

33.

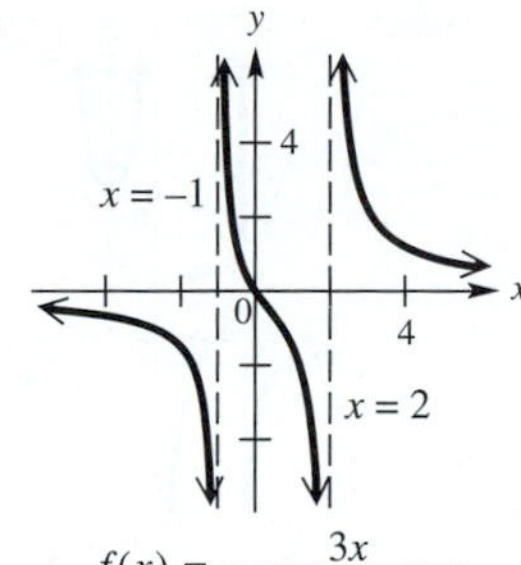

$f(x) = \dfrac{3x}{(x + 1)(x - 2)}$

35.

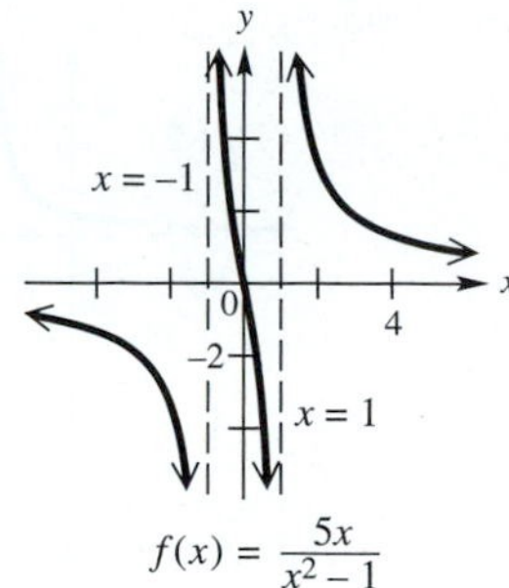

$f(x) = \dfrac{5x}{x^2 - 1}$

37.

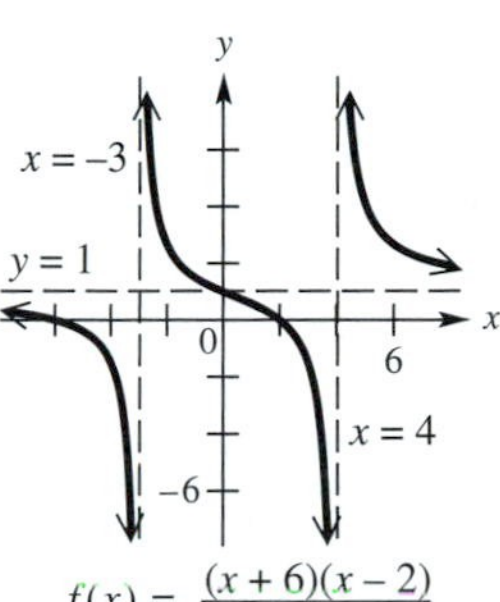

$f(x) = \dfrac{(x + 6)(x - 2)}{(x + 3)(x - 4)}$

39.

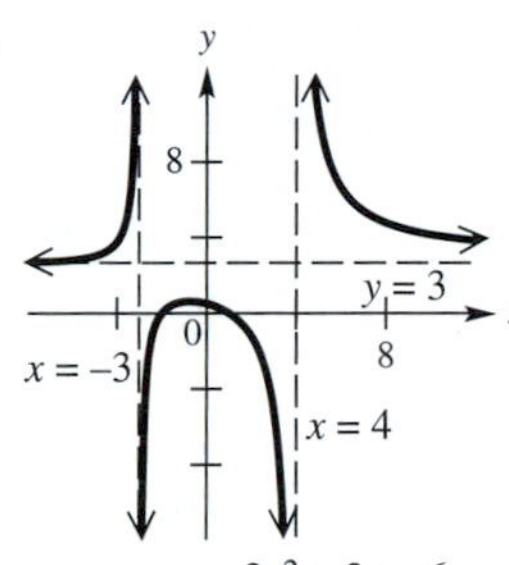

$f(x) = \dfrac{3x^2 + 3x - 6}{x^2 - x - 12}$

41.

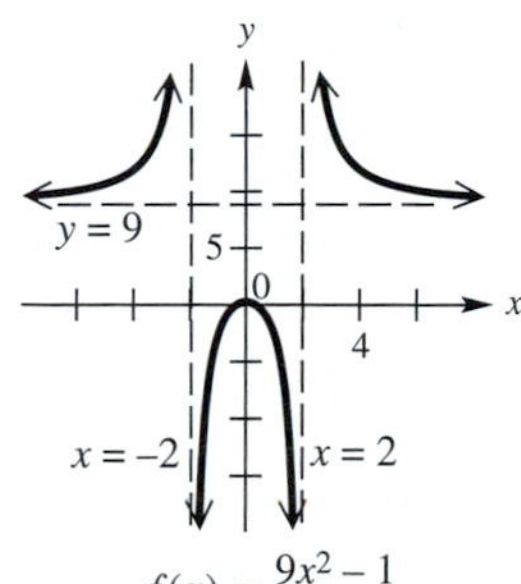

$f(x) = \dfrac{9x^2 - 1}{x^2 - 4}$

43.

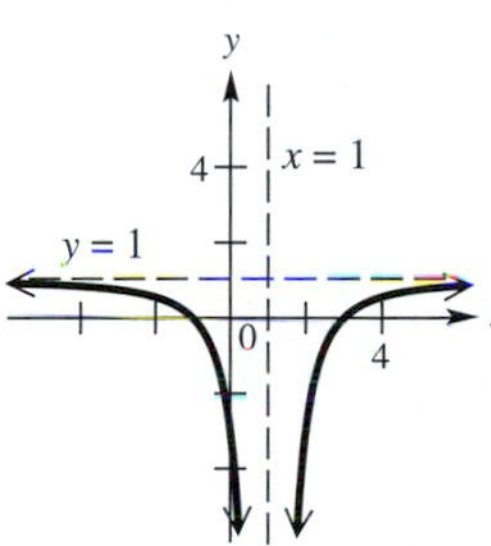

$f(x) = \dfrac{(x - 3)(x + 1)}{(x - 1)^2}$

45.

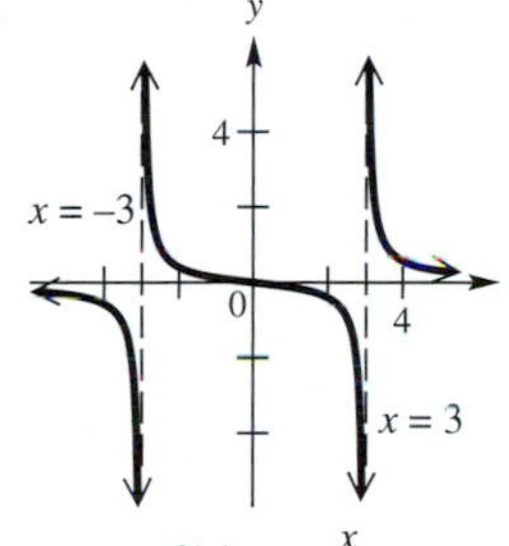

$f(x) = \dfrac{x}{x^2 - 9}$

47.

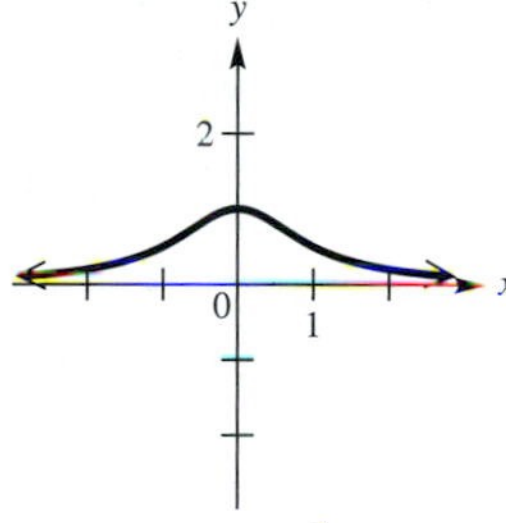

$f(x) = \dfrac{1}{x^2 + 1}$

49.

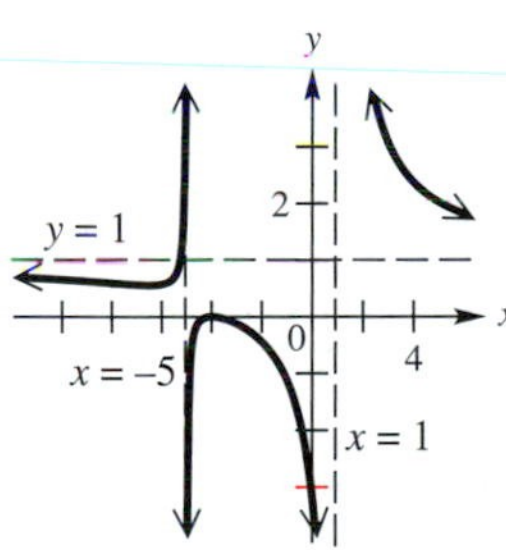

$f(x) = \dfrac{(x + 4)^2}{(x - 1)(x + 5)}$

51.

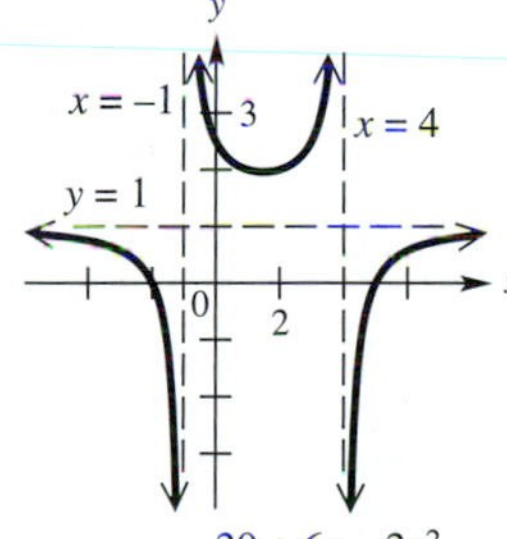

$f(x) = \dfrac{20 + 6x - 2x^2}{8 + 6x - 2x^2}$

53.

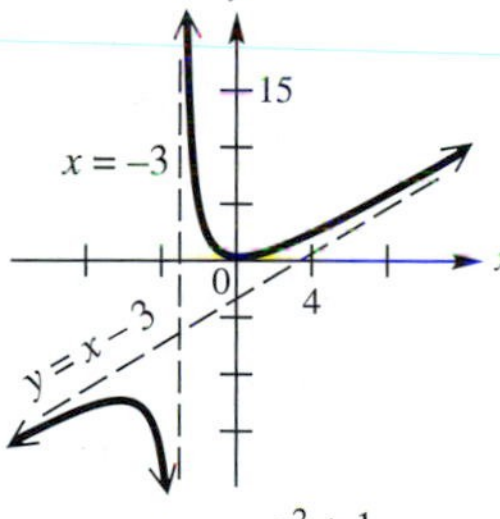

$f(x) = \dfrac{x^2 + 1}{x + 3}$

55.

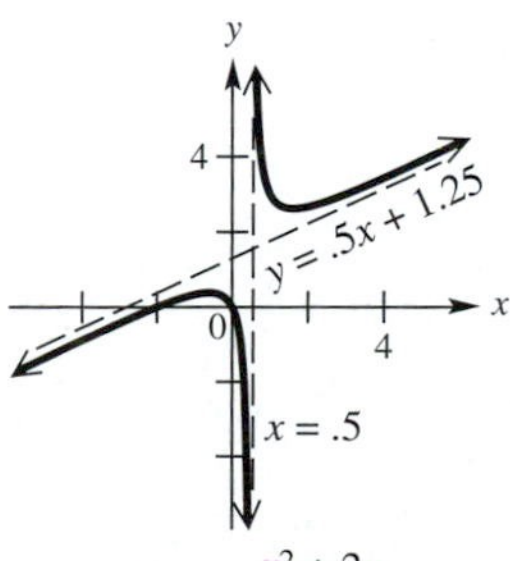

$f(x) = \dfrac{x^2 + 2x}{2x - 1}$

57.

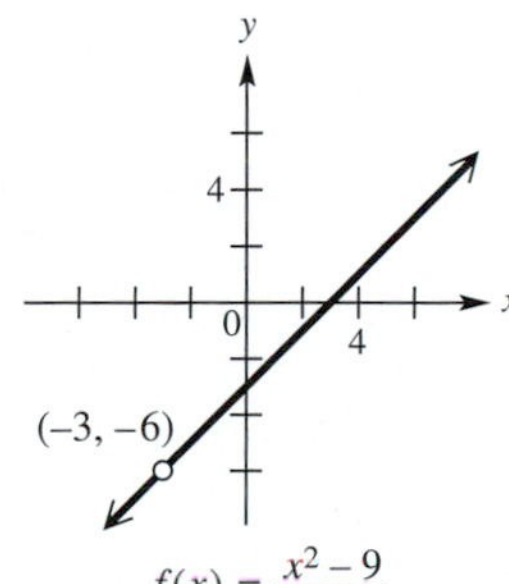

$f(x) = \dfrac{x^2 - 9}{x + 3}$

59.

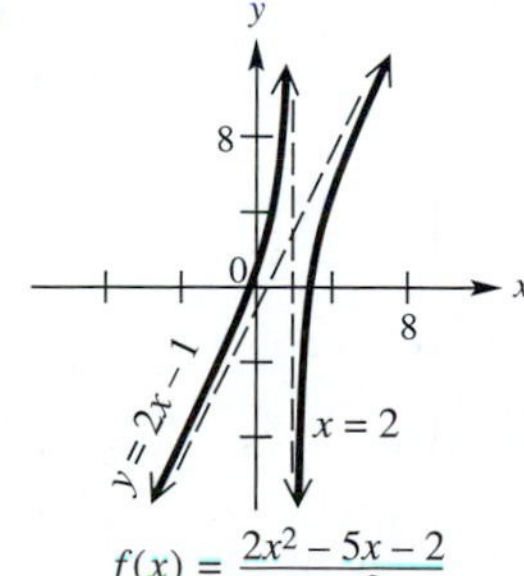

$f(x) = \dfrac{2x^2 - 5x - 2}{x - 2}$

61.

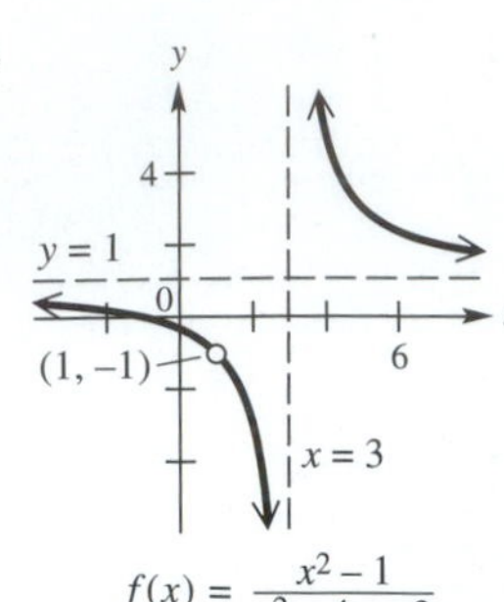

63.

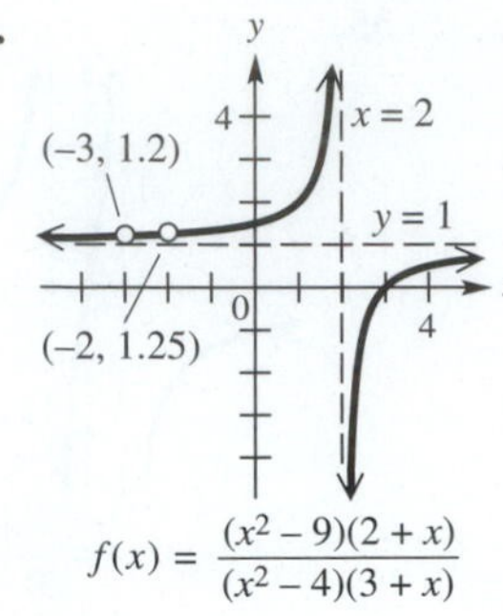

65.

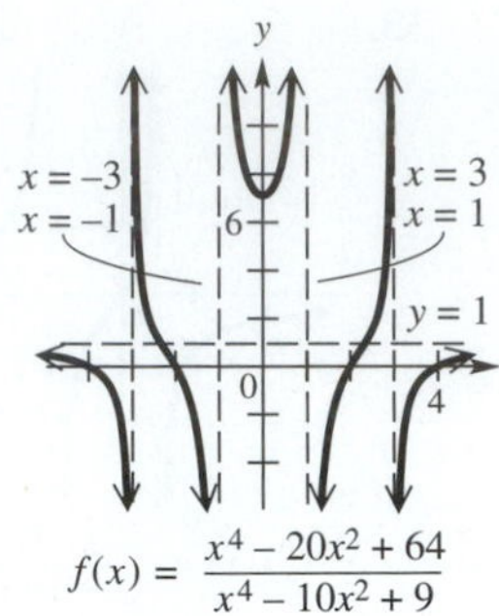

67. $p = 4$; $q = 2$

69. $p = -2$; $q = -1$

Answers may vary in Exercises 71 and 73.

71. $f(x) = \frac{(x - 3)(x + 2)}{(x - 2)(x + 2)} = \frac{x^2 - x - 6}{x^2 - 4}$ **73.** $f(x) = \frac{x - 2}{x(x - 4)} = \frac{x - 2}{x^2 - 4x}$ **75.** The complex solutions are $-\frac{1}{2} \pm \frac{\sqrt{15}}{2}i$. There are no real solutions and no vertical asymptotes.

77. (a)

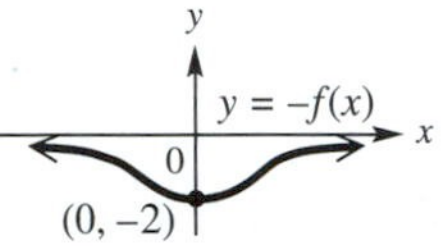

(b)

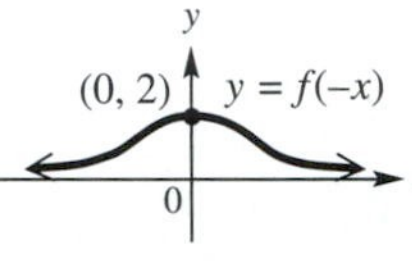

(Same as $y = f(x)$)

78. (a)

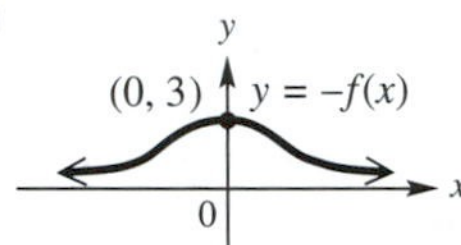

(b)

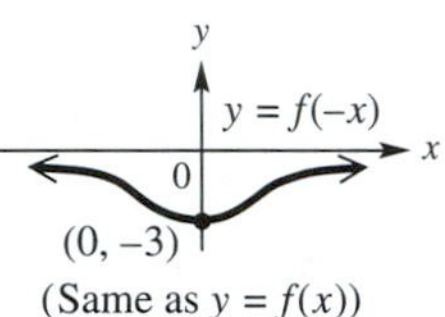

(Same as $y = f(x)$)

79. (a)

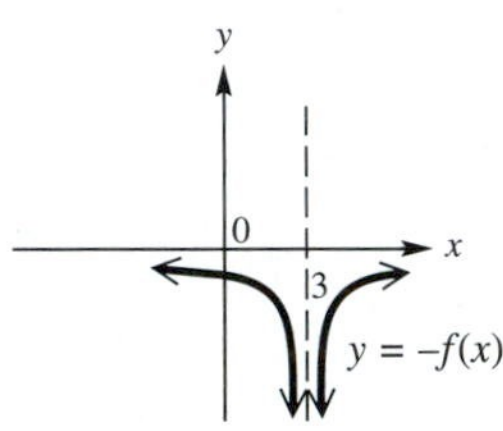

(b)

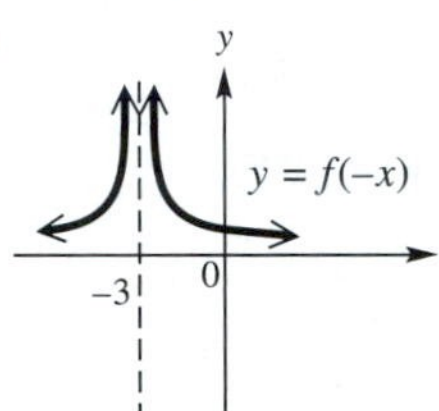

80. (a)

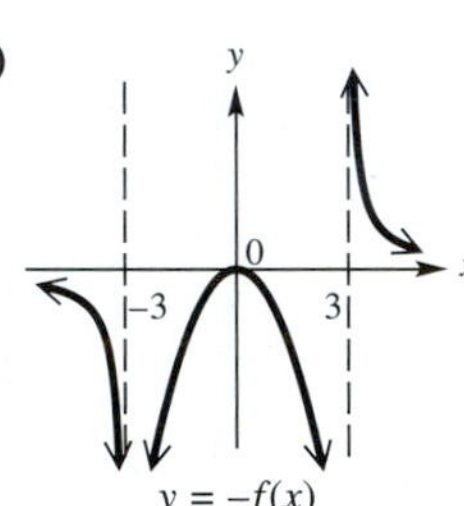

(b)

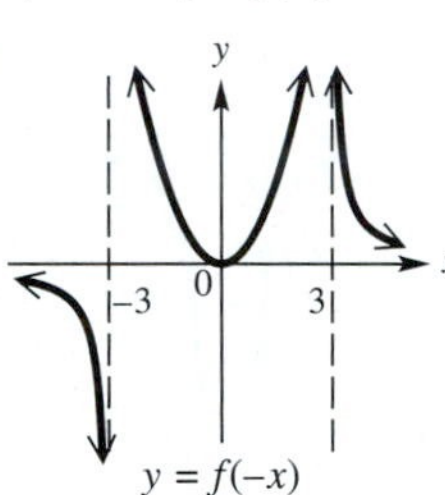

81. $y = -2x - 8$

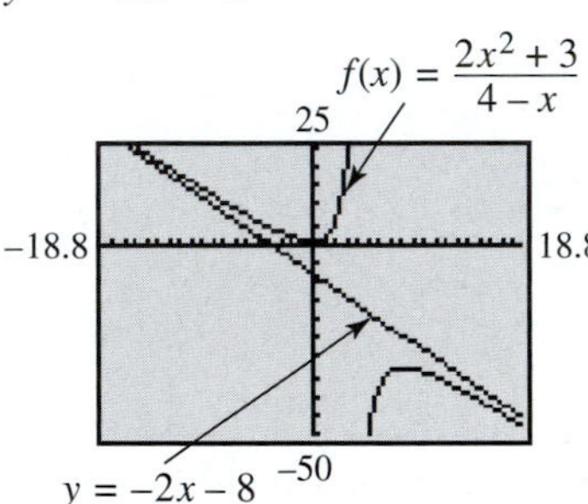

83. $y = -x + 3$

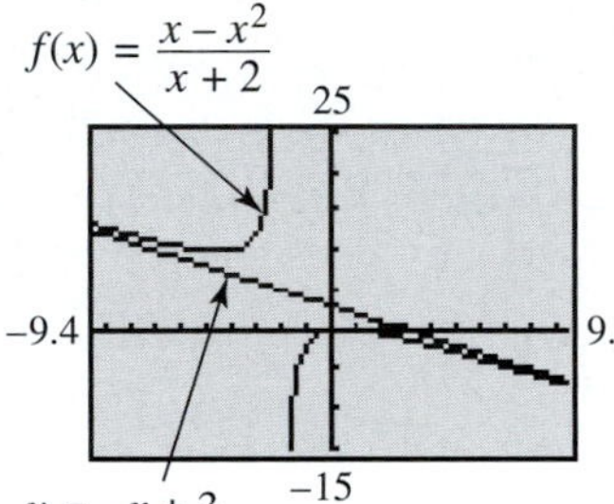

85. (a) $y = x + 1$ **(b)** at $x = 0$ and $x = 1$ **(c)** above

87.

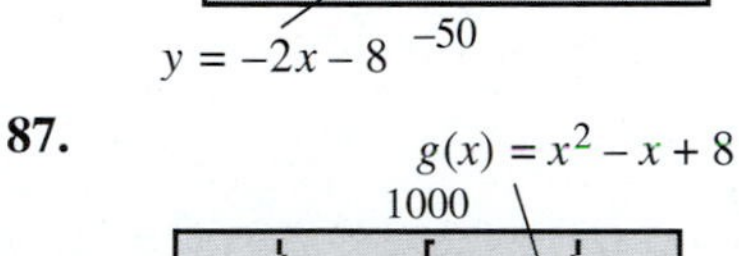

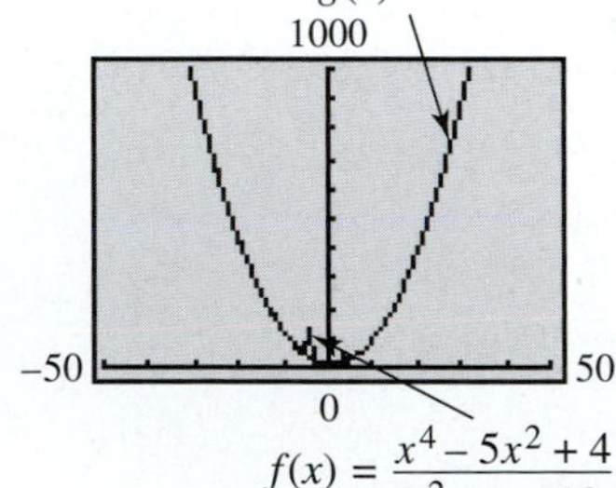

In this window, the two graphs seem to overlap (coincide), suggesting that as $|x| \to \infty$, the graph of f approaches the graph of g, giving an asymptotic effect.

89. **(a)** $g(x) = x - 3$ **(b)** f is undefined at X $= -3$, as indicated by the error message. **(c)** The "hole" is at $(-3, -6)$.

X	Y1	Y2
-6	-9	-9
-5	-8	-8
-4	-7	-7
-3	ERROR	-6
-2	-5	-5
-1	-4	-4
0	-3	-3

Y1=ERROR

3.1

−9.4 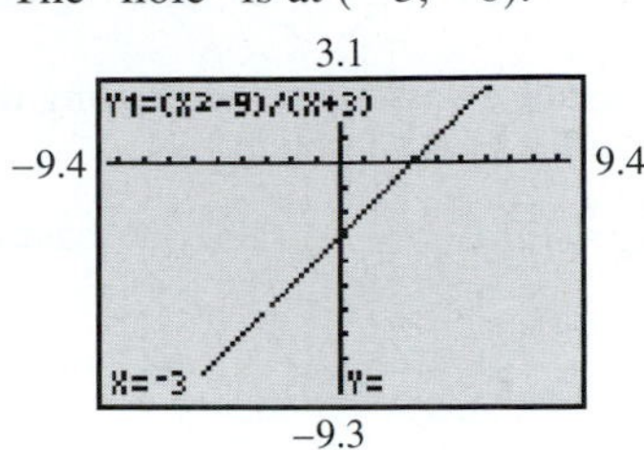 9.4

−9.3

91. **(a)** The horizontal asymptote has equation $y = 0$ (the x-axis). **(b)** The horizontal asymptote has equation $y = \frac{a}{b}$, where a is the leading coefficient of $p(x)$ and b is the leading coefficient of $q(x)$. **(c)** The oblique asymptote has equation $y = ax + b$, where $ax + b$ is the quotient (with remainder disregarded) found by dividing $p(x)$ by $q(x)$.

4.3 Exercises *(pages 300–305)*

1. **(a)** $\emptyset$ **(b)** $(-\infty, -2)$ **(c)** $(-2, \infty)$ **3.** **(a)** $\{-1\}$ **(b)** $(-1, 0)$ **(c)** $(-\infty, -1) \cup (0, \infty)$ **5.** **(a)** $\{0\}$ **(b)** $(-2, 0) \cup (2, \infty)$ **(c)** $(-\infty, -2) \cup (0, 2)$ **7.** **(a)** $\{0\}$ **(b)** $(-1, 0) \cup (0, 1)$ **(c)** $(-\infty, -1) \cup (1, \infty)$ **9.** **(a)** $\{0\}$ **(b)** $(1, 2) \cup (2, 3)$ **(c)** $(-\infty, -2) \cup (-2, 0) \cup (0, 1) \cup (3, \infty)$ **11.** **(a)** $\{2.5\}$ **(b)** $(2.5, 3)$ **(c)** $(-\infty, 2.5) \cup (3, \infty)$
13. His solution is not correct. If $x + 2 < 0$, it is necessary to reverse the direction of the inequality symbol. He should use test values, as explained in Example 3. The correct solution set is $\left(-2, \frac{1}{2}\right]$. **15.** **(a)** $\{3\}$ **(b)** $(-5, 3]$ **(c)** $(-\infty, -5) \cup [3, \infty)$
17. **(a)** $\emptyset$ **(b)** $(-\infty, -2)$ **(c)** $(-2, \infty)$ **19.** **(a)** $\left\{\frac{9}{5}\right\}$ **(b)** $(-\infty, 1) \cup \left(\frac{9}{5}, \infty\right)$ **(c)** $\left(1, \frac{9}{5}\right)$ **21.** **(a)** $\{-2\}$ **(b)** $(-\infty, -2] \cup (1, 2)$ **(c)** $[-2, 1) \cup (2, \infty)$ **23.** **(a)** $\emptyset$ **(b)** $\emptyset$ **(c)** $(-\infty, 2) \cup (2, \infty)$ **25.** **(a)** $\emptyset$ **(b)** $(-\infty, -1)$ **(c)** $(-1, \infty)$ **27.** $(-\infty, \infty)$ **29.** $\emptyset$ **31.** $\emptyset$ **33.** $(-\infty, \infty)$ **35.** $\{1\}$ **37.** $\{-3\}$ **39.** $\{-4\}$ **41.** $\{4, 9\}$
43. $\left\{\frac{-3 \pm \sqrt{29}}{2}\right\}$ **45.** $\left\{\frac{3 \pm \sqrt{3}}{3}\right\}$ **47.** $\left\{\pm\frac{1}{2}, \pm i\right\}$ **49.** $\emptyset$ **51.** $\{-10\}$ **53.** $\left\{\frac{27}{56}\right\}$ **55.** $(-\infty, 2)$
57. $(-\infty, -1) \cup \left(\frac{3}{2}, \infty\right)$ **59.** $(-\infty, -3) \cup (-1, 2)$ **61.** $(-1, 1) \cup \left[\frac{5}{2}, \infty\right)$ **63.** $(3, \infty)$
65. $(-\infty, 0) \cup \left(0, \frac{1}{2}\right] \cup [2, \infty)$ **67.** **(a)** $\{-3.54\}$ **(b)** $(-\infty, -3.54) \cup (1.20, \infty)$ **(c)** $(-3.54, 1.20)$

69. **(a)** $y = 10$

14

0 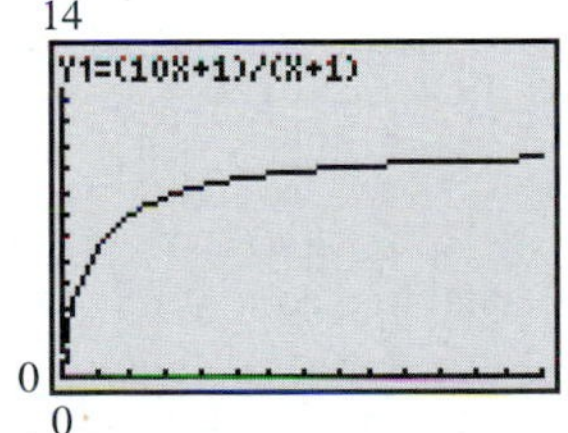 14

0

(b) When $x = 0$, there are 1,000,000 insects. **(c)** It starts to level off at 10,000,000. **(d)** The horizontal asymptote $y = 10$ represents the limiting population after a long time.

71. **(a)** about 12.4 cars per min **(b)** 3 **73.** There are two possible solutions: width $= 7$ in., length $= 14$ in., height $= 2$ in.; width ≈ 2.266 in., length ≈ 4.532 in., height ≈ 19.086 in.

75. **(a)** $f(400) = \frac{2540}{400} = 6.35$ in.; A curve designed for 60 mph with a radius of 400 ft should have the outer rail elevated 6.35 in.

(b) 50

0 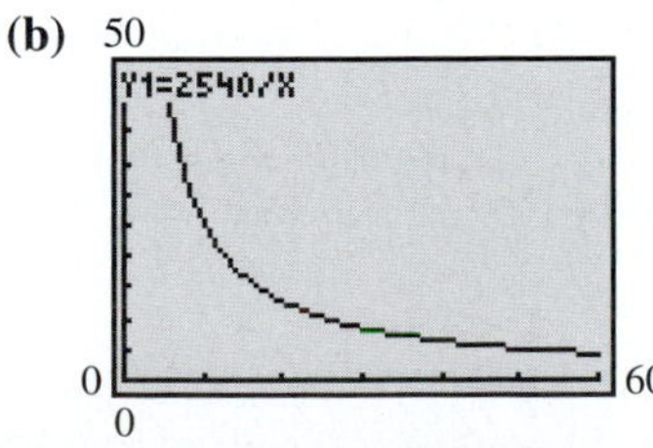 600

0

As the radius x of the curve increases, the elevation of the outer rail decreases. **(c)** The horizontal asymptote is $y = 0$. As the radius of the curve increases without bound ($x \rightarrow \infty$), the tracks become straight and no elevation or banking ($y \rightarrow 0$) is necessary. **(d)** 200 ft

77. **(a)** $D(.05) \approx 238$; The braking distance for a car traveling at 50 mph on a wet 5% uphill grade is about 238 ft. **(b)** As the uphill grade x increases, the braking distance decreases, which agrees with driving experience. **(c)** $x = \frac{13}{165} \approx .079$ or 7.9%

79. $\frac{32}{15}$ **81.** $\frac{18}{125}$ **83.** increases; decreases **85.** It becomes half as much. **87.** It becomes 27 times as much.

89. 21 **91.** $\frac{2}{9}$ ohm **93.** 17.8 lb **95.** 799.5 cm^3 **97.** 5.1 **99.** -1.4

Reviewing Basic Concepts *(pages 305–306)*

1.

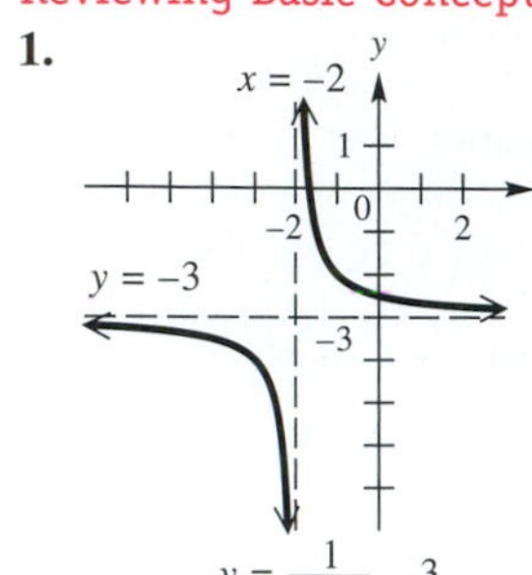

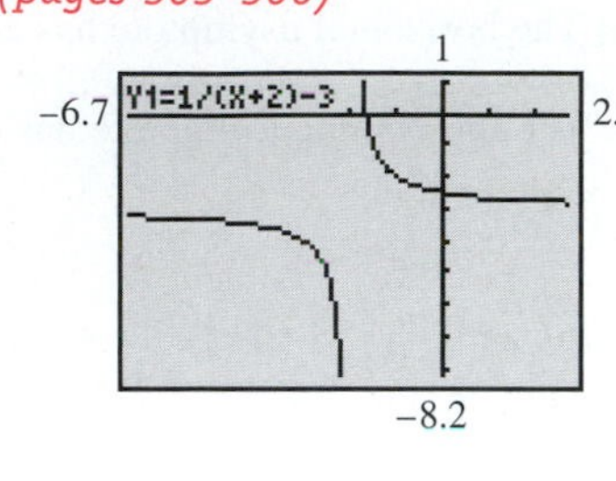

2. $(-\infty, -1) \cup (-1, 1) \cup (1, \infty)$ **3.** $x = 6$ **4.** $y = 1$ **5.** $y = x - 2$

6.

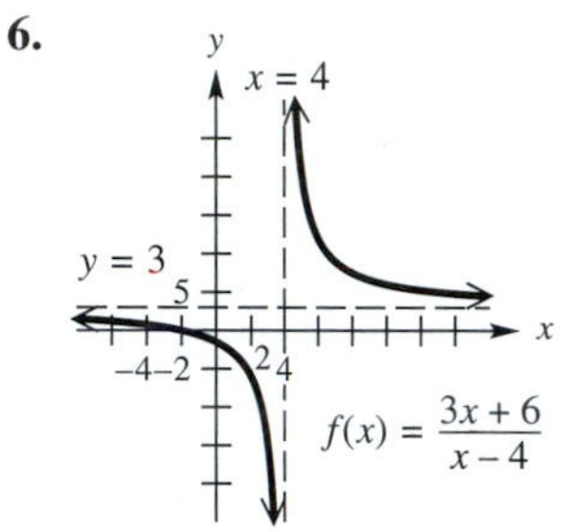

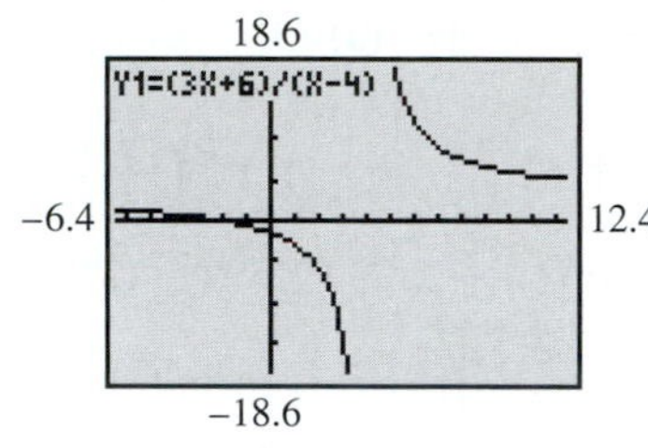

7. **(a)** $\{-4\}$ **(b)** $(-\infty, -4) \cup (2, \infty)$ **(c)** $(-4, 2)$

8. $\left(-\frac{1}{3}, \frac{3}{2}\right)$ **9.** inversely; height; 24

10. 4.3 vibrations per sec

4.4 Exercises *(pages 315–318)*

1. 13 **3.** -2 **5.** 729 **7.** $\frac{1}{25}$ **9.** 100 **11.** 4 **13.** $\frac{1}{8}$ **15.** -9 **17.** 2 **19.** 27 **21.** $(2x)^{1/3}$

23. $z^{5/3}$ **25.** $\frac{1}{y^{3/4}}$ **27.** $x^{5/6}$ **29.** $y^{3/4}$ **31.** -1.587401052 **33.** -5 **35.** -2.571281591 **37.** 1.464591888

39. .4252903703 **41.** 2 **43.** 1.267463962 **45.** .0322515344 **47.** $1.2^{1.62} \approx 1.34$ **49.** $50^{3/2} - 50^{1/2} \approx 346.48$

51. B **53.** **(a)** .125; $\frac{1}{8}$ **(b)** $\left(\sqrt[4]{16}\right)^{-3}$; $\sqrt[4]{16^{-3}}$ (There are other expressions.) Each is equal to .125. **(c)** Show that $.125 = \frac{1}{8}$.

55.

```
6 ˣ√81
            2.080083823
```

56.

```
81^(1/6)
            2.080083823
```

57.

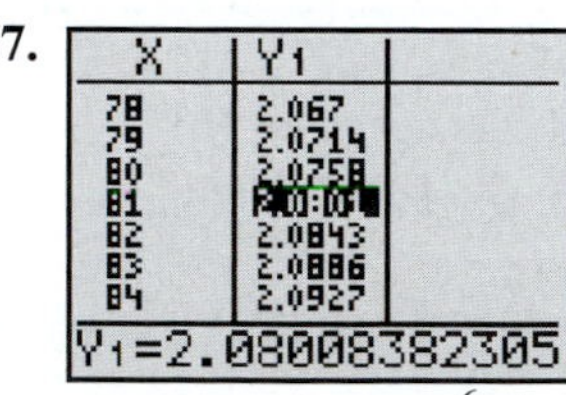

X	Y1
78	2.067
79	2.0714
80	2.0758
81	2.0801
82	2.0843
83	2.0886
84	2.0927

Y1=2.08008382305

In this table, $Y_1 = \sqrt[6]{X}$.

58.

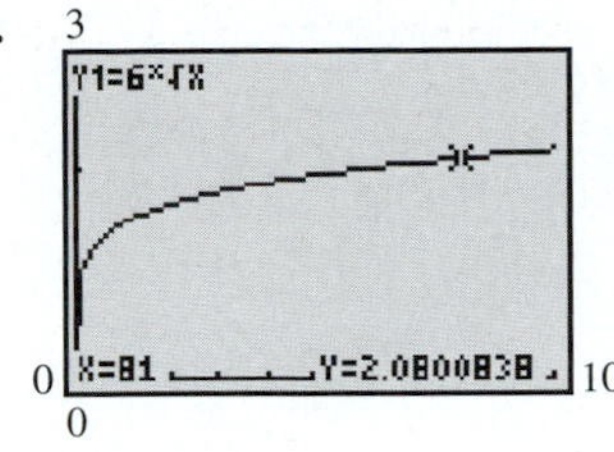

59. 3.2 ft^2 **61.** approximately 58.1 yr
63. **(a)** $a = 1960$ **(b)** $b \approx -1.2$ **(c)** $f(4) = 1960(4)^{-1.2} \approx 371$; If the zinc ion concentration reaches 371 mg per L, a rainbow trout will survive, on average, 4 min.
65. 1.06 g **67.** $a \approx 874.54$, $b \approx -.49789$ **69.** $\left[-\frac{5}{4}, \infty\right)$ **71.** $(-\infty, 6]$
73. $(-\infty, \infty)$ **75.** $[-7, 7]$ **77.** $[-1, 0] \cup [1, \infty)$

In Exercises 79–85, we give only the answers to parts (a)–(d). We do not show the graphs.

79. **(a)** $[0, \infty)$ **(b)** $\left[-\frac{5}{4}, \infty\right)$ **(c)** none **(d)** $\{-1.25\}$ **81.** **(a)** $(-\infty, 0]$ **(b)** $(-\infty, 6]$ **(c)** none **(d)** $\{6\}$
83. **(a)** $(-\infty, \infty)$ **(b)** $(-\infty, \infty)$ **(c)** none **(d)** $\{3\}$ **85.** **(a)** $[0, 7]$ **(b)** $[-7, 0]$ **(c)** $[0, 7]$ **(d)** $\{-7, 7\}$
87. Rewrite as $y = 3\sqrt{x + 3}$. Shift the graph of $y = \sqrt{x}$ to the left 3 units and stretch vertically by a factor of 3.
89. Rewrite as $y = 2\sqrt{x + 4} + 4$. Shift the graph of $y = \sqrt{x}$ to the left 4 units, stretch vertically by a factor of 2, and shift upward 4 units. **91.** Rewrite as $y = 3\sqrt[3]{x + 2} - 5$. Shift the graph of $y = \sqrt[3]{x}$ to the left 2 units, stretch vertically by a factor of 3, and shift downward 5 units.

93. The graph is a circle.
$y_1 = \sqrt{100 - x^2}$; $y_2 = -\sqrt{100 - x^2}$

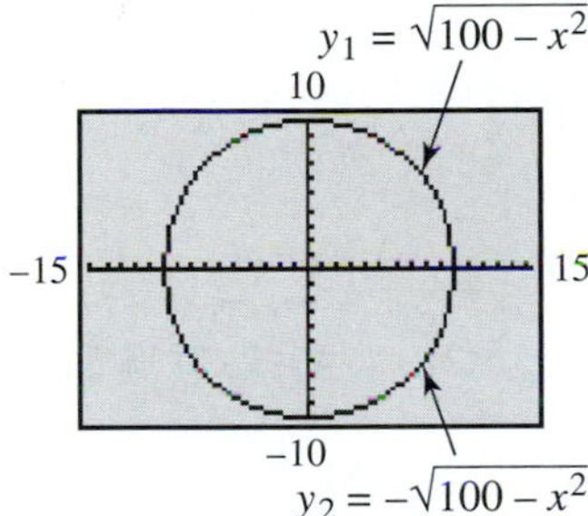

95. The graph is a (shifted) circle.
$y_1 = \sqrt{9 - (x - 2)^2}$; $y_2 = -\sqrt{9 - (x - 2)^2}$

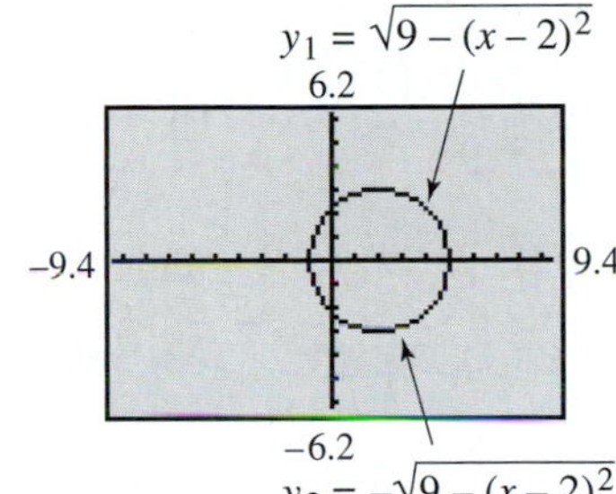

97. The graph is a horizontal parabola.
$y_1 = -3 + \sqrt{x}$; $y_2 = -3 - \sqrt{x}$

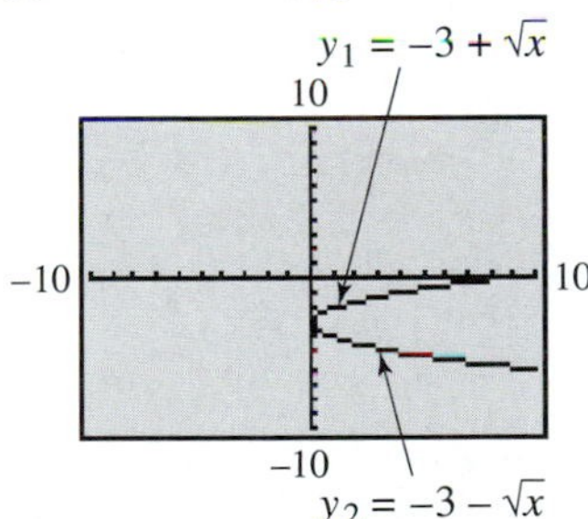

99. The graph is a horizontal parabola.
$y_1 = -2 + \sqrt{.5x + 3.5}$; $y_2 = -2 - \sqrt{.5x + 3.5}$

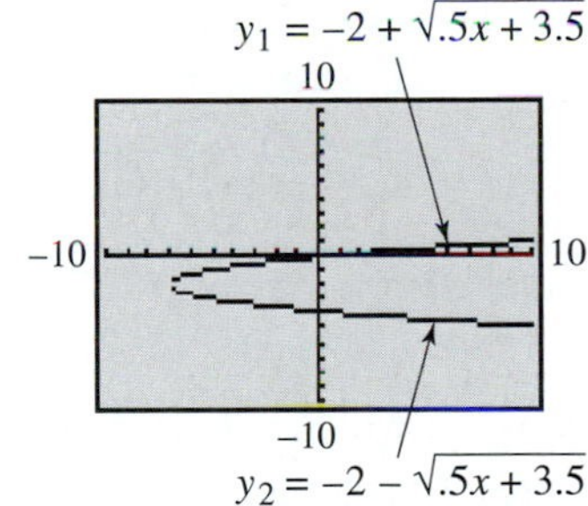

4.5 Exercises *(pages 324–328)*

1. **(a)** $\{4\}$ **(b)** $[4, \infty)$ **(c)** $[-5, 4]$ **3.** **(a)** $\{-3\}$ **(b)** $(-\infty, -3)$ **(c)** $(-3, \infty)$

5.

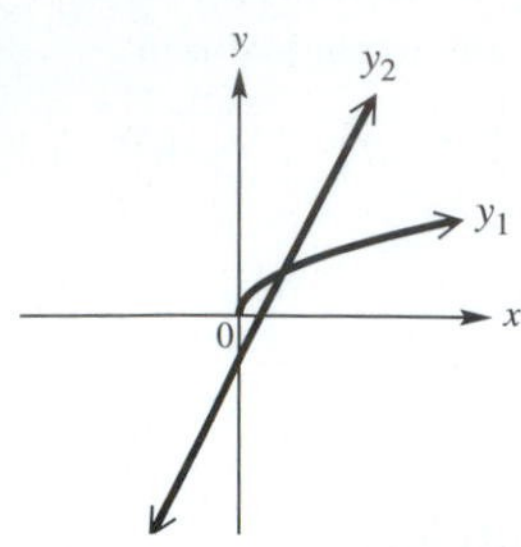

one real solution; $\{1\}$; $\frac{1}{4}$ is extraneous.

7.

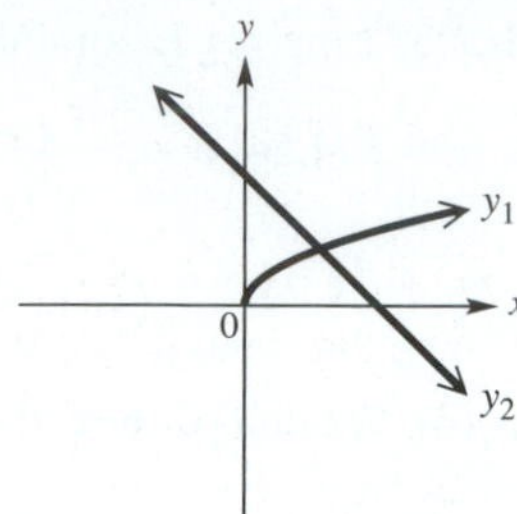

one real solution; $\left\{\frac{7 - \sqrt{13}}{2}\right\}$; $\frac{7 + \sqrt{13}}{2}$ is extraneous.

9.

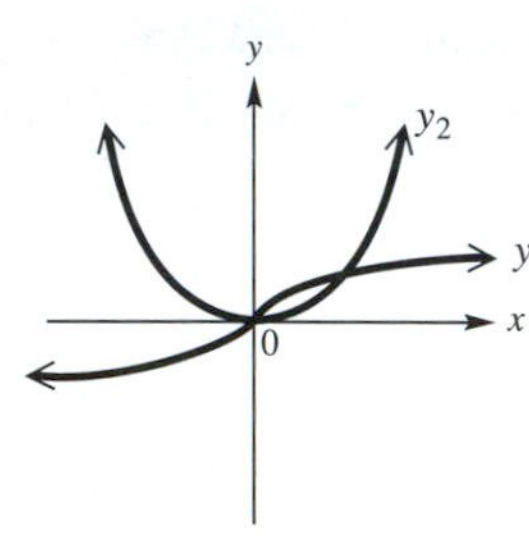

two real solutions; $\{0, 1\}$; no extraneous values

11. **(a)** $\{-1\}$ **(b)** $(-1, \infty)$ **(c)** $\left[-\frac{7}{3}, -1\right)$ **13.** **(a)** $\{3\}$ **(b)** $\left[-\frac{13}{4}, 3\right)$ **(c)** $(3, \infty)$ **15.** **(a)** $\{3\}$ **(b)** $\left[-\frac{1}{5}, 3\right)$ **(c)** $(3, \infty)$ **17.** **(a)** $\{2, 14\}$ **(b)** $(2, 14)$ **(c)** $(14, \infty)$ **19.** **(a)** $\{0, 3\}$ **(b)** $(-\infty, 0) \cup (3, \infty)$ **(c)** $(0, 3)$ **21.** **(a)** $\{0\}$ **(b)** $(0, \infty)$ **(c)** $\left[-\frac{1}{3}, 0\right)$ **23.** **(a)** $\{27\}$ **(b)** $[27, \infty)$ **(c)** $\left[\frac{5}{2}, 27\right]$ **25.** **(a)** $\{-8, 2\}$ **(b)** $(-\infty, -8) \cup (2, \infty)$ **(c)** $(-8, -6] \cup [0, 2)$ **27.** **(a)** $\left\{\frac{1}{4}, 1\right\}$ **(b)** $\left(-\infty, \frac{1}{4}\right) \cup (1, \infty)$ **(c)** $\left(\frac{1}{4}, 1\right)$

29. $\{2, 3\}$ **31.** $\{4\}$ **33.** $\{-1, 3\}$ **35.** $\{-28\}$ **37.** $\{2\}$ **39.** $\{65{,}538\}$ **41.** $\{-32, 32\}$ **43.** $\{27\}$ **45.** $\left\{-1, -\frac{1}{2}\right\}$ **47.** $\left\{-\frac{1}{4}, \frac{5}{7}\right\}$ **49.** $\{-8, 27\}$ **51.** $\{1\}$ **53.** $(4x - 4)^{1/3} = (x + 1)^{1/2}$ **54.** 6 **55.** $(4x - 4)^2 = (x + 1)^3$ **56.** $16x^2 - 32x + 16 = x^3 + 3x^2 + 3x + 1$, and thus, $x^3 - 13x^2 + 35x - 15 = 0$

57. three real roots

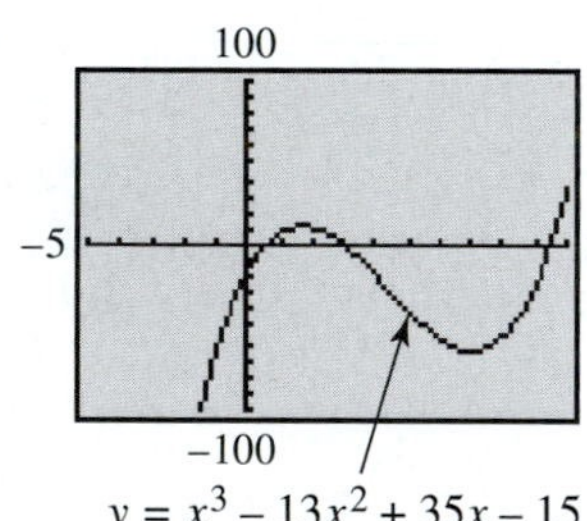

58.

$$\begin{array}{r|rrrr} 3 & 1 & -13 & 35 & -15 \\ & & 3 & -30 & 15 \\ \hline & 1 & -10 & 5 & 0 \end{array}$$

$P(3) = 0$

59. $P(x) = (x - 3)(x^2 - 10x + 5)$ **60.** $\{5 \pm 2\sqrt{5}\}$ **61.** 3; $5 + 2\sqrt{5}$; $5 - 2\sqrt{5}$

62. two real solutions

$y_3 = y_1 - y_2 = \sqrt[3]{4x - 4} - \sqrt{x + 1}$

.5
−2
20
−.5

63. $\{3, 5 + 2\sqrt{5}\}$; For the calculator solution, $5 + 2\sqrt{5} \approx 9.47$. **64.** The solution set of the original equation is a subset of the solution set of the equation in Exercise 56. The extraneous solution $5 - 2\sqrt{5}$ was obtained when each side of the original equation was raised to the sixth power. **65.** $\{0, 1\}$ **67.** $\left\{-\frac{2}{9}, 2\right\}$ **69.** $\{2, 18\}$ **71.** 4.5 km per sec **73.** 2.5 sec **75.** 91 mph

77. **(a)** $20 - x$ **(b)** x must be between 0 and 20. **(c)** $AP = \sqrt{x^2 + 12^2}$; $BP = \sqrt{(20 - x)^2 + 16^2}$
(d) $f(x) = \sqrt{x^2 + 12^2} + \sqrt{(20 - x)^2 + 16^2}$, $0 < x < 20$
(e)

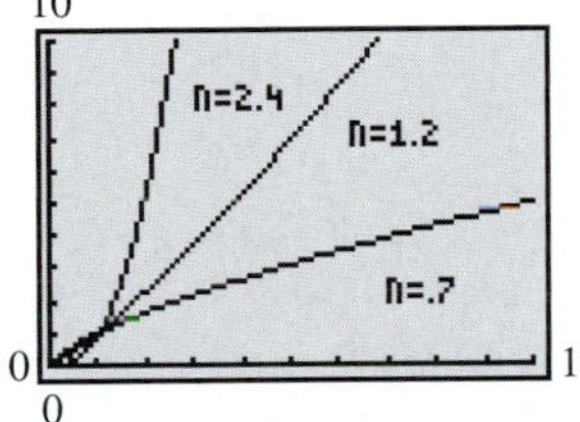

$f(4) \approx 35.28$; When the stake is 4 ft from the base of the 12-ft pole, approximately 35.28 ft of wire will be required. **(f)** When $x \approx 8.57$ ft, $f(x)$ is a minimum (approximately 34.41 ft).
(g) This problem has examined how the total amount of wire used can be expressed in terms of the distance from the stake at P to the base of the 12-ft pole. We find that the amount of wire used can be minimized when the stake is approximately 8.57 ft from the 12-ft pole.
79. Since $x \approx 1.31$, the hunter must travel $8 - x \approx 8 - 1.31 = 6.69$ mi along the river.
81. After 1.38 hr (1:23 P.M.), the ships are 33.28 mi apart.

Reviewing Basic Concepts *(page 328)*

1.

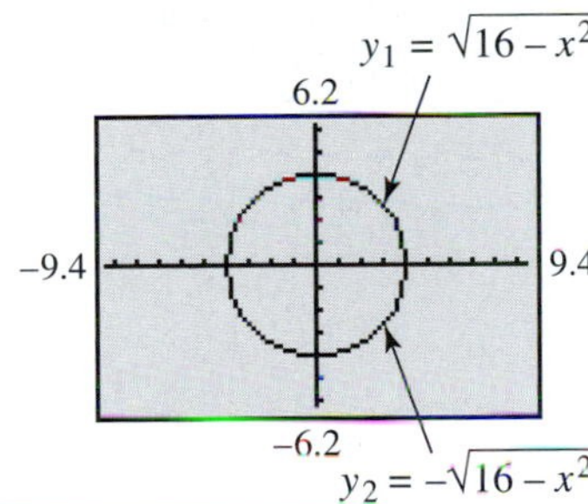

As the exponent increases in value, the curve rises more rapidly for $x \geq 1$. **2.** .24 m^2

3. $y_1 = \sqrt{16 - x^2}$; $y_2 = -\sqrt{16 - x^2}$

$y_1 = \sqrt{16 - x^2}$
6.2
−9.4
9.4
−6.2
$y_2 = -\sqrt{16 - x^2}$

4. $y_1 = -2 + \sqrt{x - 2}$; $y_2 = -2 - \sqrt{x - 2}$

$y_1 = -2 + \sqrt{x - 2}$
6.2
−9.4
9.4
−6.2
$y_2 = -2 - \sqrt{x - 2}$

5. $\{4\}$ **6.** $(4, \infty)$ **7.** $\left[-\frac{4}{3}, 4\right)$ **8.** $\{-1\}$ **9.** Any value less than 15 causes the radicand to be negative, and the calculator will not return square roots of negative numbers in the table. **10.** **(a)** dog: 148; person: 69 **(b)** 6.4 in.

Chapter 4 Review Exercises *(pages 331–334)*

1. **(a)** Reflect the graph of $y = \frac{1}{x}$ across the x-axis, and shift upward 6 units.
(b)

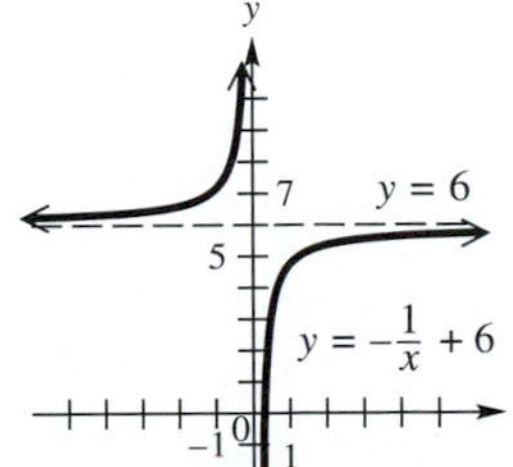

(c)

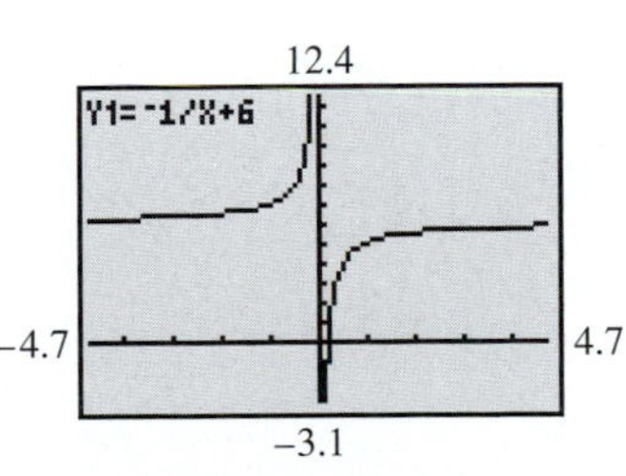

3. **(a)** Shift the graph of $y = \frac{1}{x^2}$ to the right 2 units, and reflect across the x-axis.
(b)

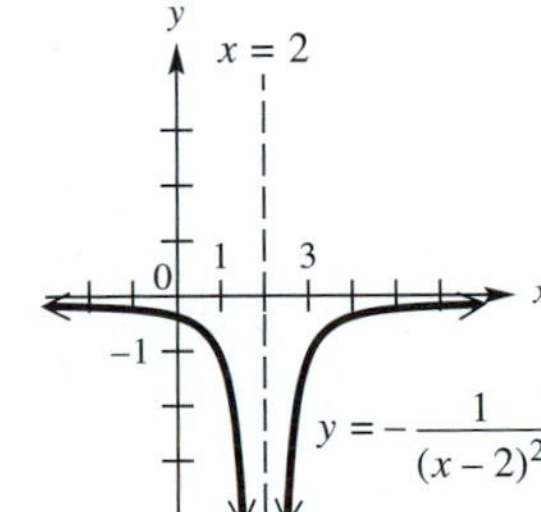

(c)

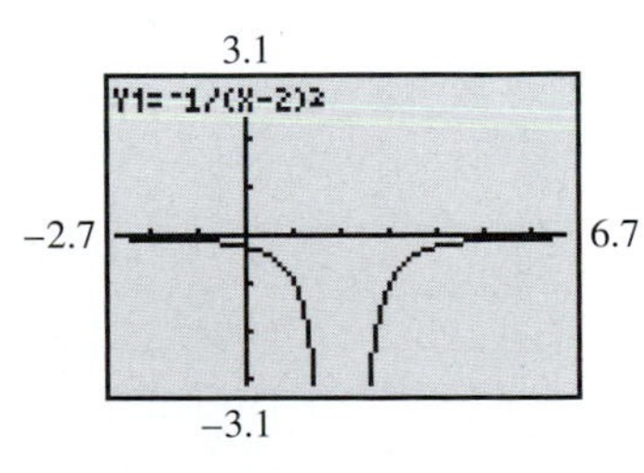

5. The degree of the numerator will be exactly 1 greater than the degree of the denominator.

7.

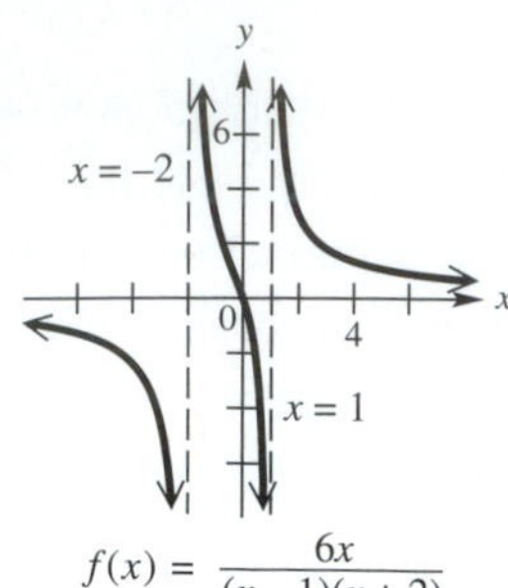

$f(x) = \frac{6x}{(x-1)(x+2)}$

9.

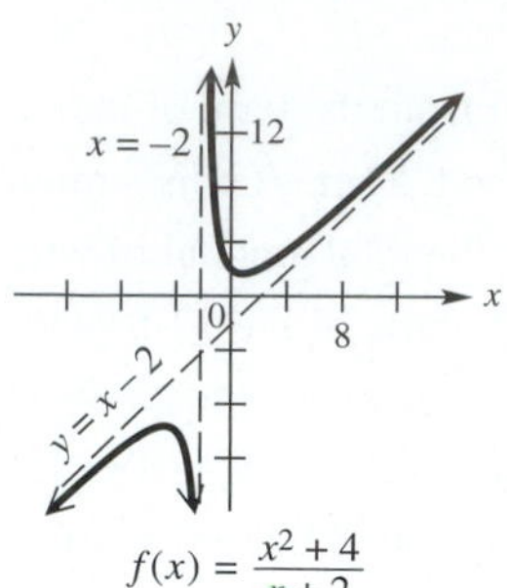

$f(x) = \frac{x^2+4}{x+2}$

11.

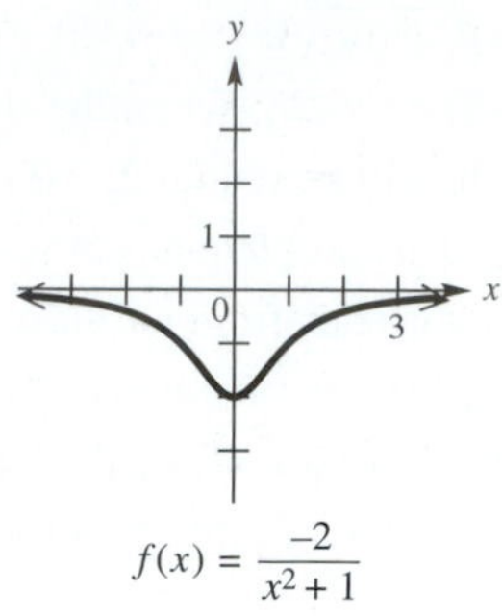

$f(x) = \frac{-2}{x^2+1}$

13.

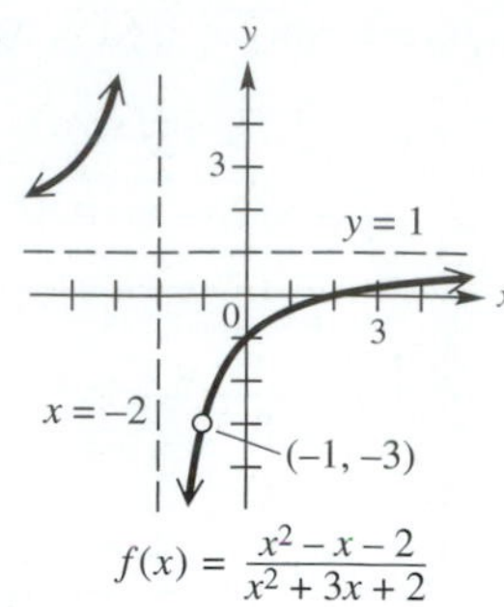

$f(x) = \frac{x^2-x-2}{x^2+3x+2}$

15. $f(x) = \frac{-3x+6}{x-1}$ **17.** **(a)** $\left\{\frac{2}{3}\right\}$ **(b)** $\left(-1, \frac{2}{3}\right)$ **(c)** $(-\infty, -1) \cup \left(\frac{2}{3}, \infty\right)$ **19.** **(a)** $\{0\}$ **(b)** $(-\infty, -1) \cup [0, 2)$ **(c)** $(-1, 0] \cup (2, \infty)$ **21.** $\{-2\}$ **23.** $(-2, -1)$ **25.** **(a)**

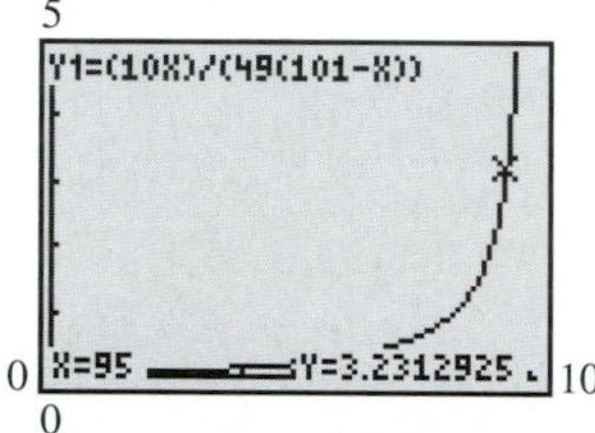

(b) approximately \$3231 (See the graph in part (a).)

27. **(a)** $[0, 36]$ **(b)** The average line length is less than or equal to 8 cars when the average arrival rate is 36 cars per hr or less.

29. 2 **31.** 847 **33.** 12.15 candela **35.** \$1375 **37.** $\frac{8}{9}$ metric ton **39.** $71\frac{1}{9}$ kg

41.

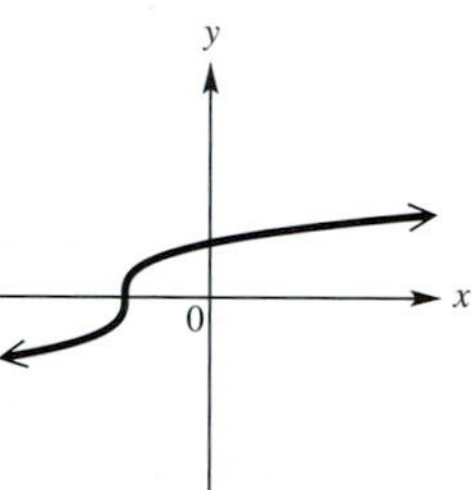

43.

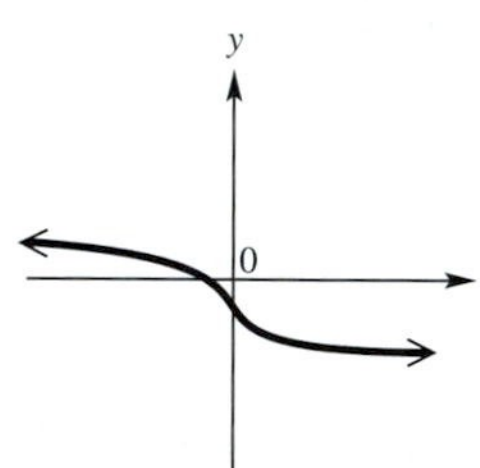

45. 12 **47.** $\frac{1}{3}$ **49.** $-\frac{2}{3}$ **51.** $\frac{1}{216}$ **53.** $\frac{1}{81}$ **55.** 2.42926041078 **57.** .0625 **59.** $[2, \infty)$ **61.** **(a)** none **(b)** $[2, \infty)$ **63.** **(a)** $\{2\}$ **(b)** $[-2.5, 2)$ **(c)** $(2, \infty)$ **65.** **(a)** $\{-1\}$ **(b)** $[-1, \infty)$ **(c)** $(-\infty, -1]$

67. no solutions

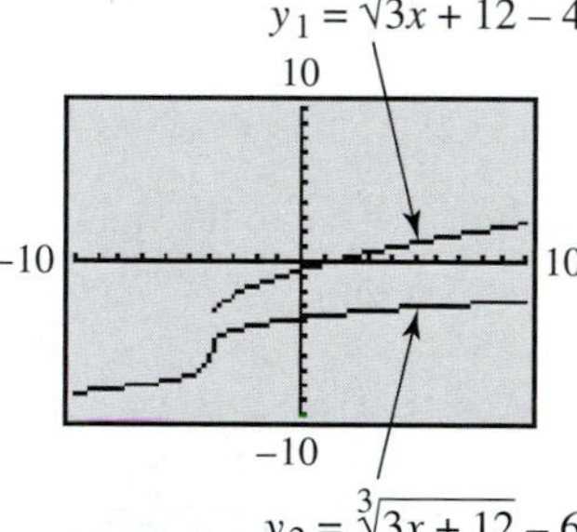

68. no x-intercepts

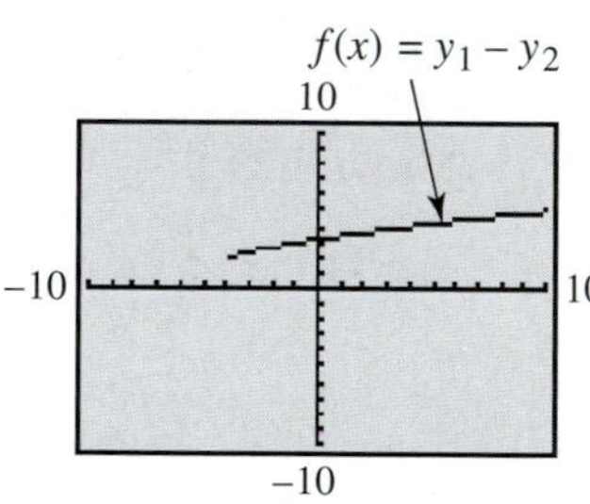

69. The graph of $-f(x) = y_2 - y_1$ is the reflection of the graph of $f(x) = y_1 - y_2$ across the x-axis.

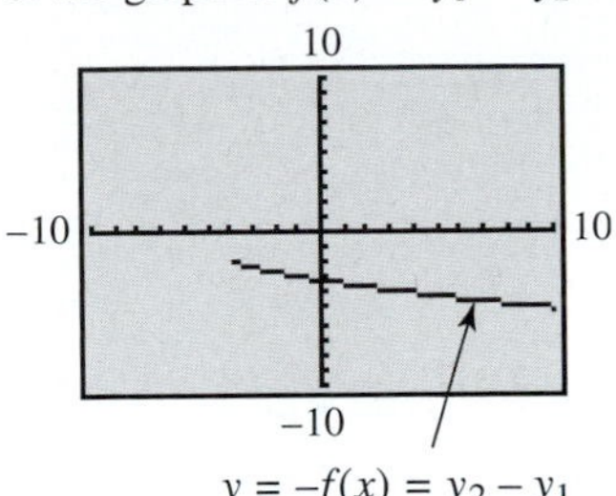

70. The graph of $y = f(-x)$ is the reflection of the graph of $y = f(x)$ across the y-axis.

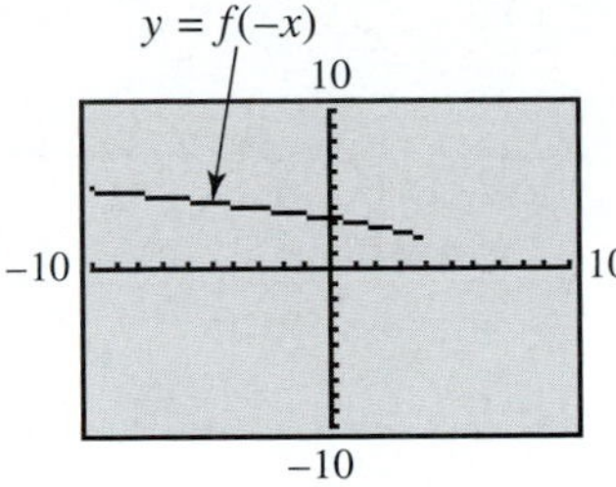

71. $\{4\}$ **73.** $\{6\}$ **75.** $\left\{\frac{81}{16}\right\}$ **77.** $\{15\}$ **79.** $\left\{-2, \frac{1}{3}\right\}$ **81.** $\{-1, 125\}$ **83.** $\{8\}$ **85.** **(a)** If the length L of the pendulum increases, so does the period of oscillation T. **(b)** There are a number of ways. One way is to realize that $k = \frac{L}{T^n}$ for some integer n. The ratio should be the constant k for each data point when the correct n is found. Another way is to use regression. **(c)** $k \approx .81$; $n = 2$ **(d)** 2.48 sec **(e)** T increases by a factor of $\sqrt{2} \approx 1.414$.

Chapter 4 Test *(pages 335–336)*

1. **(a)**

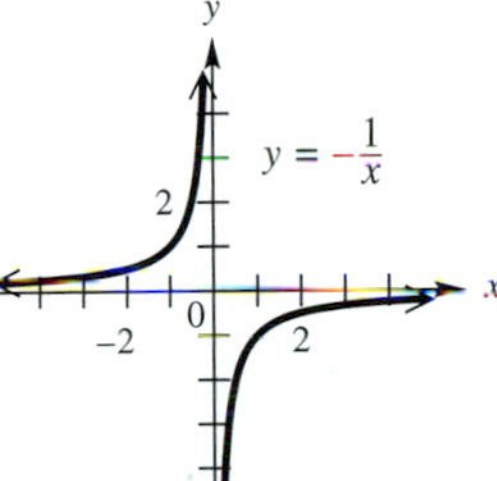

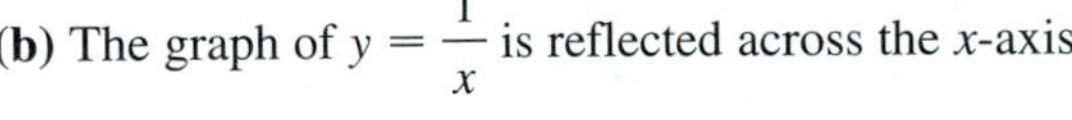

(b) The graph of $y = \frac{1}{x}$ is reflected across the x-axis or the y-axis.

(c)

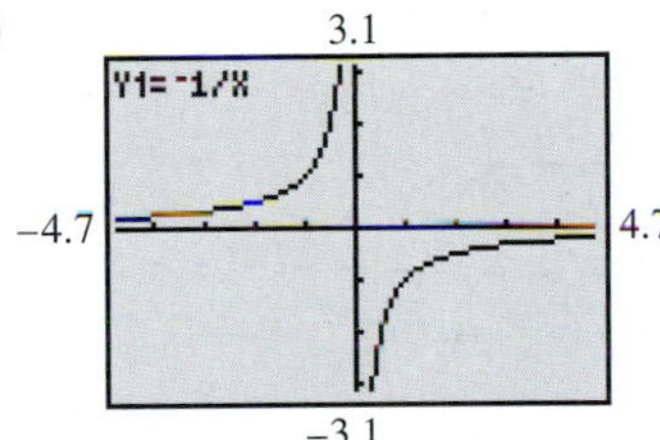

2. **(a)**

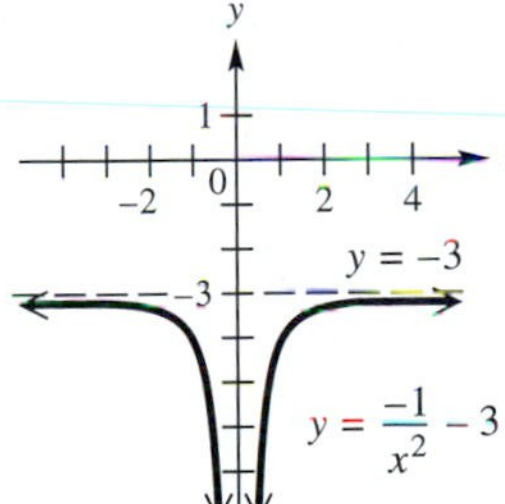

(b) The graph of $y = \frac{1}{x^2}$ is reflected across the x-axis and shifted 3 units downward.

(c)

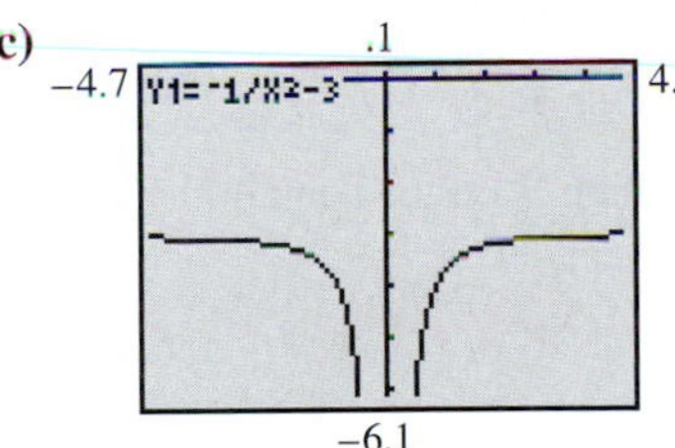

3. **(a)** $x = -1, x = 4$ **(b)** $y = 1$ **(c)** 1.5 **(d)** $-3, 2$ **(e)** $(.5, 1)$ **(f)**

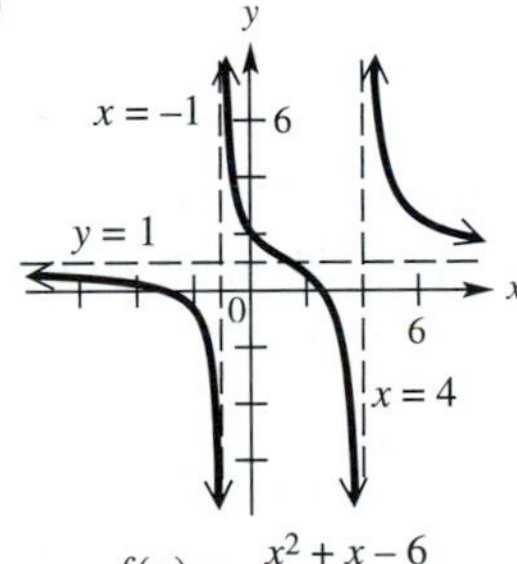

4. $y = 2x + 5$

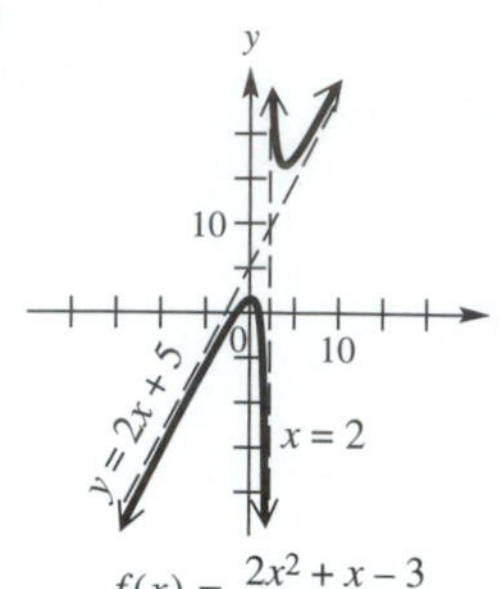

5. **(a)** -4 **(b)**

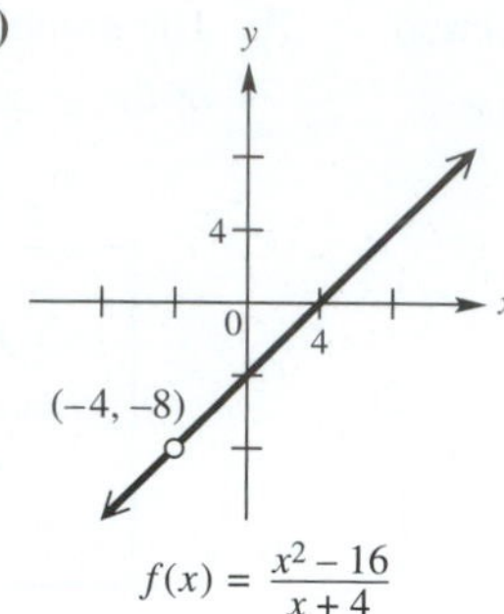

6. **(a)** $\{5\}$ **(b)** $(-\infty, -2) \cup (2, 5]$

7. **(a)** $W(30) = \frac{1}{10}$; $W(39) = 1$; $W(39.9) = 10$; When the rate is 30 vehicles per min, the average wait time is $\frac{1}{10}$ min (6 sec). The other results are interpreted similarly. **(b)**

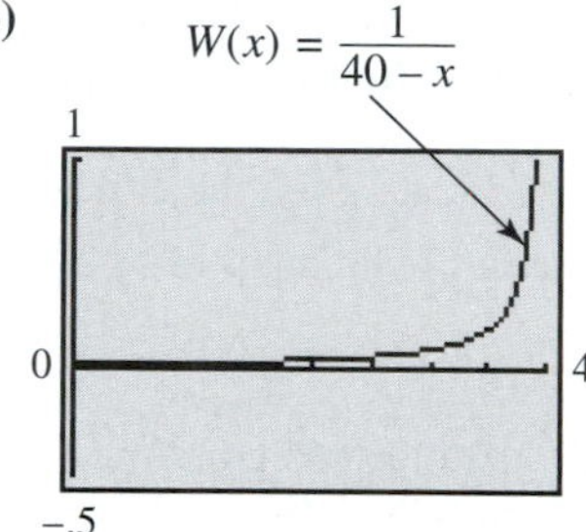

The vertical asymptote has equation $x = 40$. As x approaches 40, W gets larger and larger. **(c)** 39.8

8. 92; undernourished

9. radius: approximately 8.6 cm; amount: approximately 1394.9 cm^2

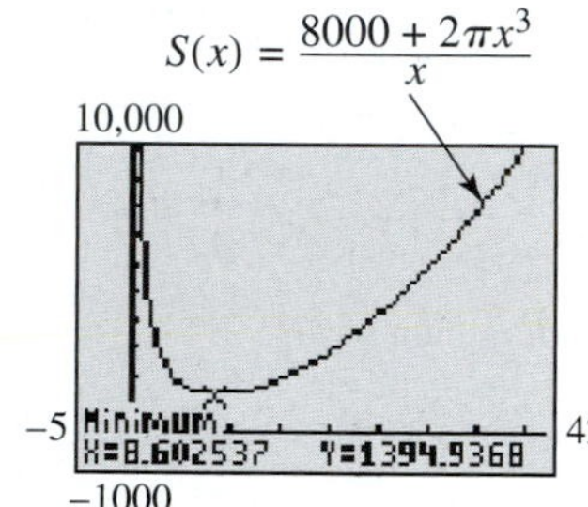

10.

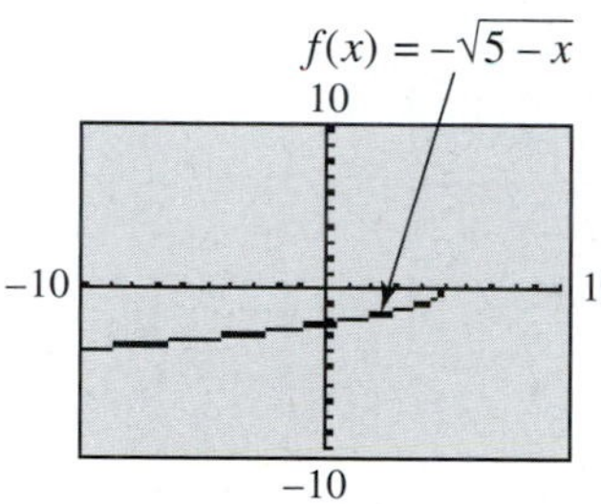

(a) $(-\infty, 5]$ **(b)** $(-\infty, 0]$ **(c)** increases **(d)** $\{5\}$ **(e)** $(-\infty, 5)$

11. **(a)** $\{0\}$

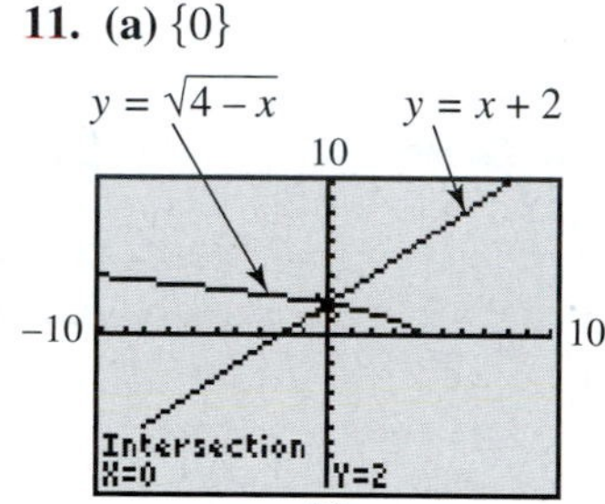

(b) $(-\infty, 0)$ **(c)** $[0, 4]$

12. The cable should be laid under water from P to a point S, which is on the bank 400 yd away from Q in the direction of R.

CHAPTER 5 INVERSE, EXPONENTIAL, AND LOGARITHMIC FUNCTIONS

5.1 Exercises *(pages 346–350)*

1. yes **3.** no **5.** no **7.** yes **9.** no **11.** yes **13.** yes **15.** no **17.** no **19.** yes **21.** yes **23.** yes **25.** The graph fails the horizontal line test because the end behavior is either ↖ ↗ or ↙ ↘. **27.** one-to-one **29.** x; $(g \circ f)(x)$ **31.** (b, a) **33.** $y = x$ **35.** does not; it is not one-to-one **43.** yes; $f^{-1} = \{(4, 10), (5, 20), (6, 30), (7, 40)\}$ **45.** no **47.** yes; $f^{-1} = \{(0^2, 0), (1^2, 1), (2^2, 2), (3^2, 3), (4^2, 4)\}$ **49.** untying your shoelaces **51.** leaving a room **53.** unwrapping a package

55. $f^{-1}(x) = \dfrac{x+4}{3}$

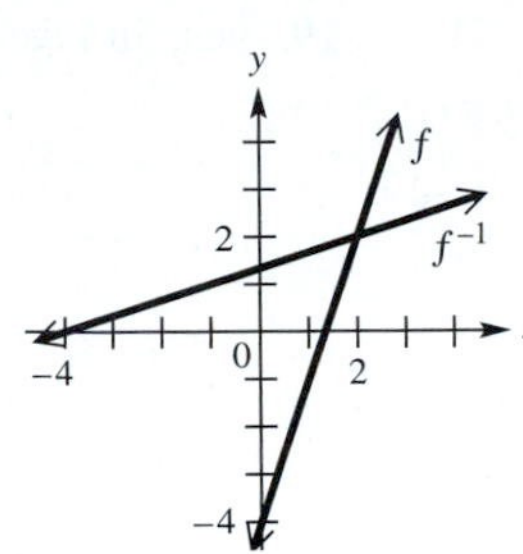

Domains and ranges of both f and f^{-1} are $(-\infty, \infty)$.

57. $f^{-1}(x) = \sqrt[3]{x-1}$

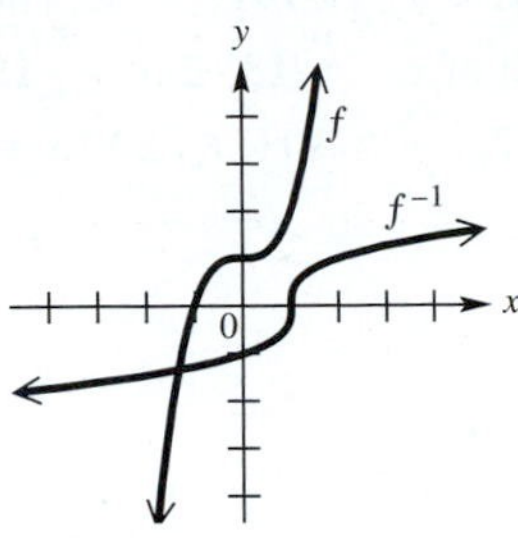

Domains and ranges of both f and f^{-1} are $(-\infty, \infty)$.

59. not one-to-one

61. $f^{-1}(x) = \dfrac{1}{x}$

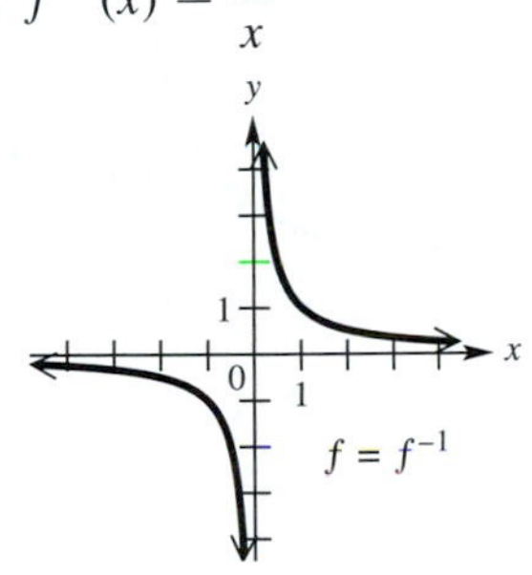

Domains and ranges of both f and f^{-1} are $(-\infty, 0) \cup (0, \infty)$.

63. $f^{-1}(x) = \dfrac{2-3x}{x}$

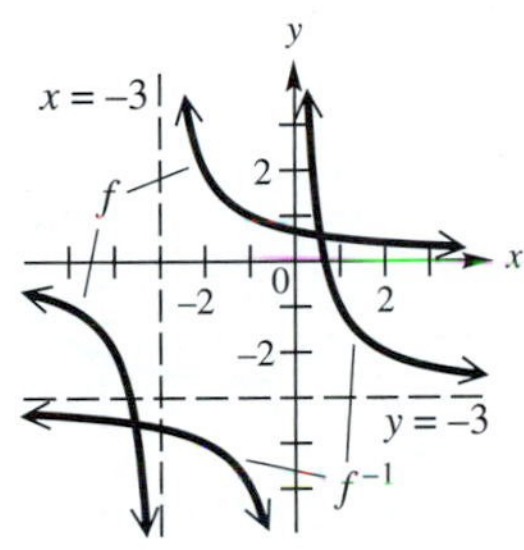

domain of f = range of $f^{-1} = (-\infty, -3) \cup (-3, \infty)$; domain of f^{-1} = range of $f = (-\infty, 0) \cup (0, \infty)$

65. $f^{-1}(x) = x^2 - 6, x \geq 0$

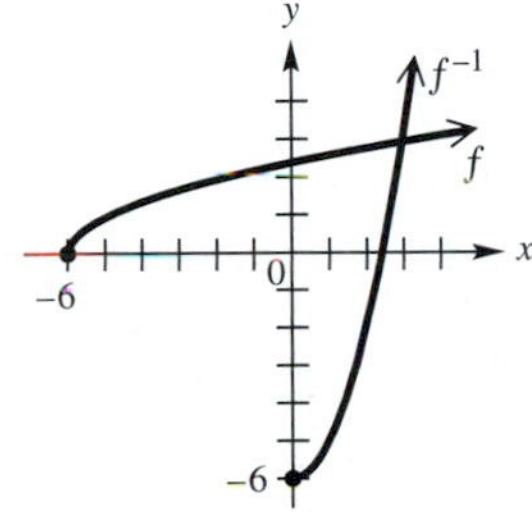

domain of f = range of $f^{-1} = [-6, \infty)$; domain of f^{-1} = range of $f = [0, \infty)$

67. 8 **69.** −8 **71.** 0 **73.** 4 **75.** 2 **77.** −2 **79.** yes **81.** no **83.** yes **85.** yes **87.** yes **89.** yes **91.** yes **93.**

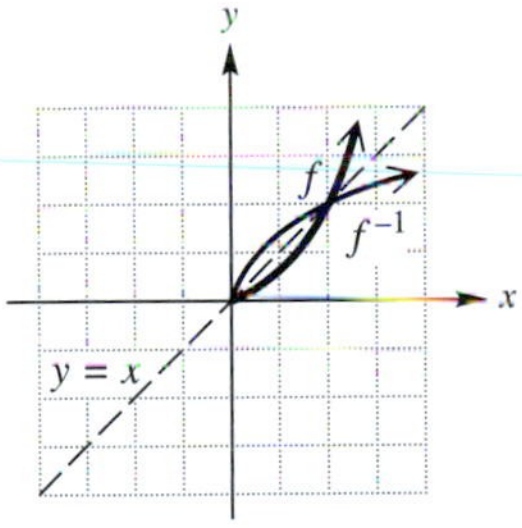

95.

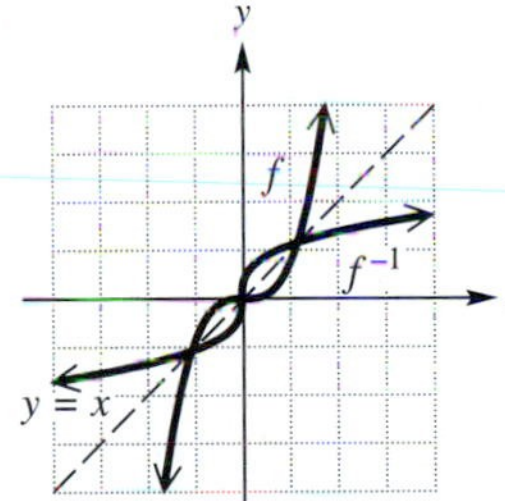

97.

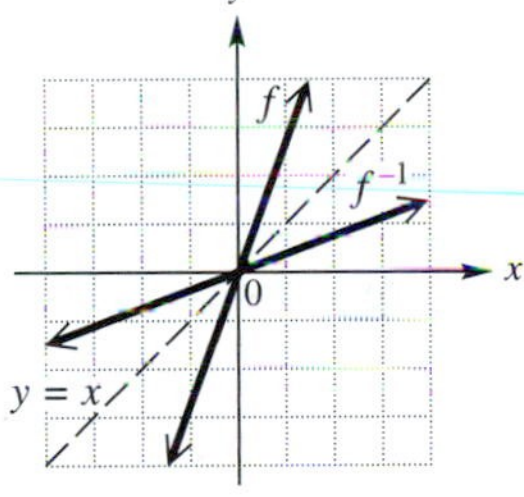

99. It represents the cost in dollars of building 1000 cars. **101.** $\dfrac{1}{a}$ **103.** not one-to-one **105.** one-to-one; $f^{-1}(x) = \dfrac{-5-3x}{x-1}$

In Exercises 107–111, we give only one of several possible choices for domain restriction.

107. $[0, \infty)$ **109.** $[6, \infty)$ **111.** $[0, \infty)$ **113.** $f^{-1}(x) = \sqrt{4-x}$ **115.** $f^{-1}(x) = x + 6, x \geq 0$

117. $f^{-1}(x) = \dfrac{x+1}{4}$; TREASURE HUNT IS ON **119.** $f^{-1}(x) = \sqrt[3]{x-1}$; 2745 3376 4097 5833 3376 9 1729 126 2198

5.2 Exercises *(pages 359–363)*

1. 8.952419619 **3.** .3752142272 **5.** .0868214883 **7.** 13.1207791 **9.** The point $(\sqrt{10}, 8.9524196)$ lies on the graph of y. **11.** The point $(\sqrt{2}, .37521423)$ lies on the graph of y. **13.** 2.3 **15.** .75 **17.** .31 **19.** yes; an inverse function **20.**

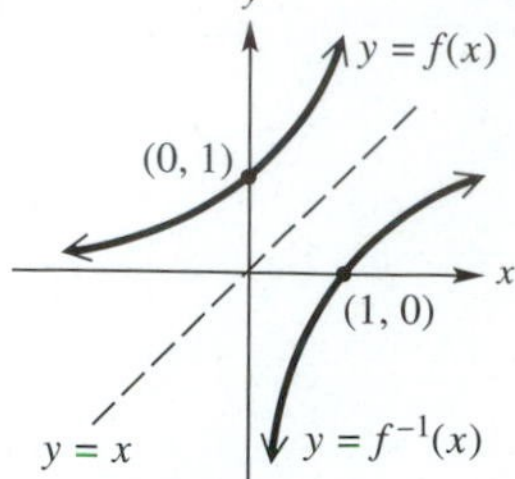

21. $x = a^y$ **22.** $x = 10^y$ **23.** $x = e^y$ **24.** (q, p)

In Exercises 25–37, we give domain, range, asymptote, and a traditional graph.

25. $(-\infty, \infty)$; $(0, \infty)$; $y = 0$

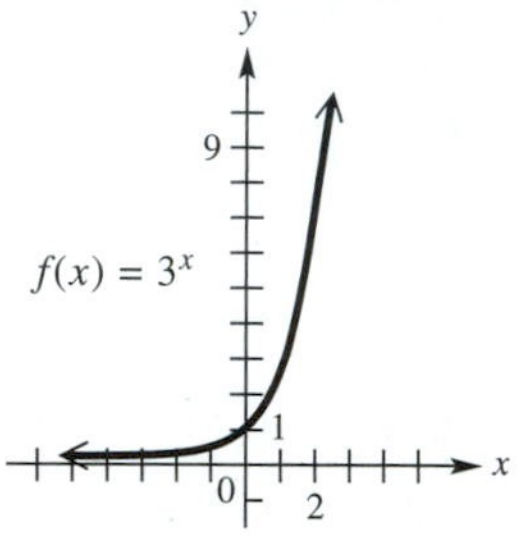

27. $(-\infty, \infty)$; $(0, \infty)$; $y = 0$

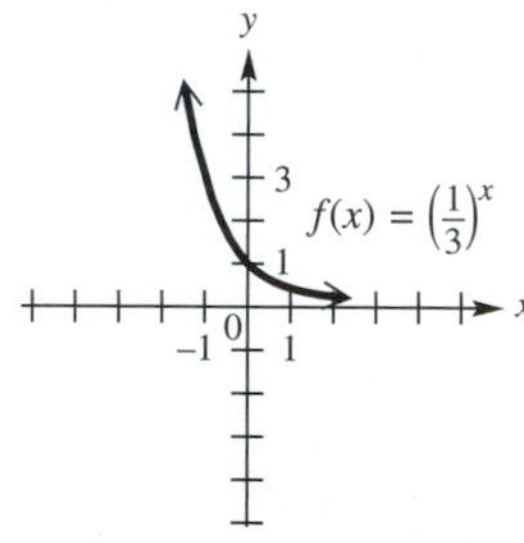

29. $(-\infty, \infty)$; $(0, \infty)$; $y = 0$

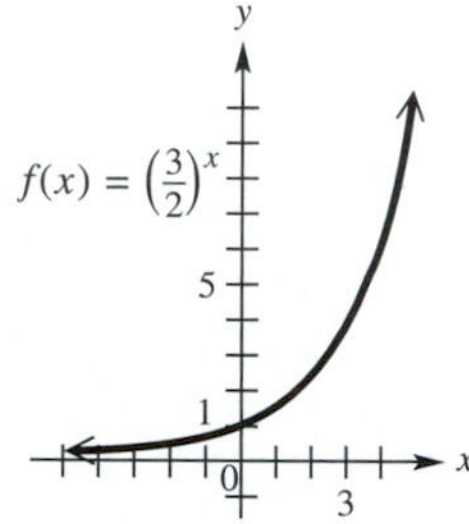

31. $(-\infty, \infty)$; $(0, \infty)$; $y = 0$

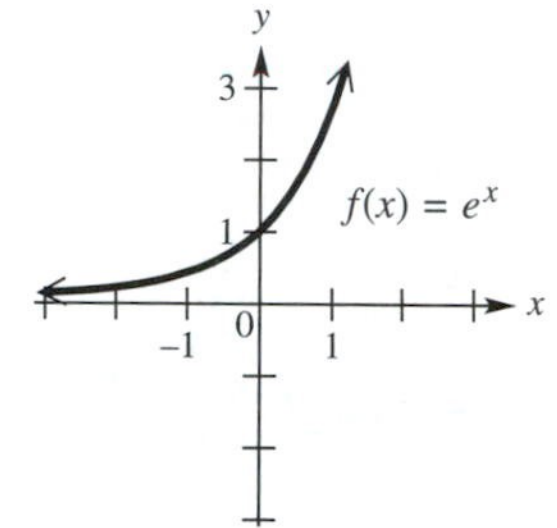

33. $(-\infty, \infty)$; $(0, \infty)$; $y = 0$

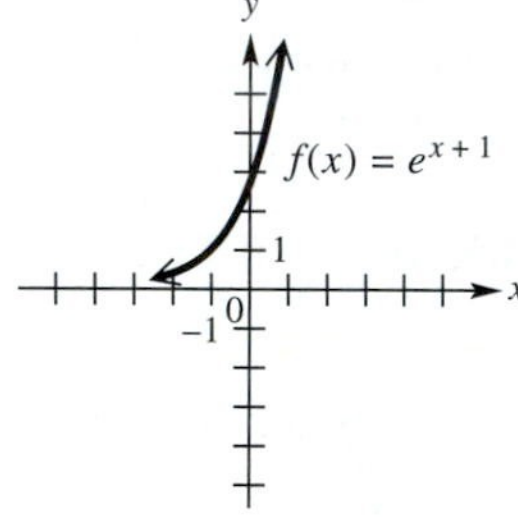

35. $(-\infty, \infty)$; $(0, \infty)$; $y = 0$

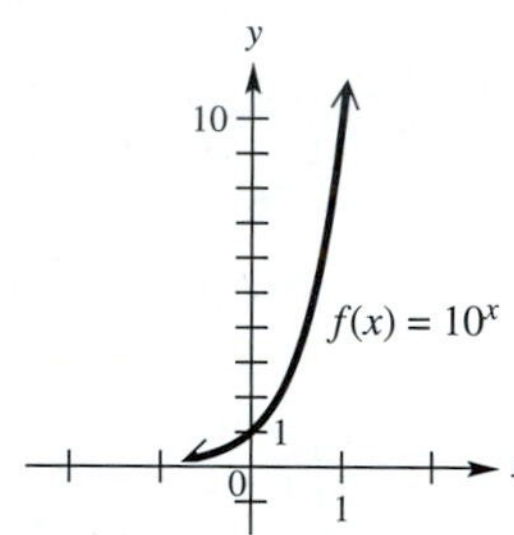

37. $(-\infty, \infty)$; $(0, \infty)$; $y = 0$

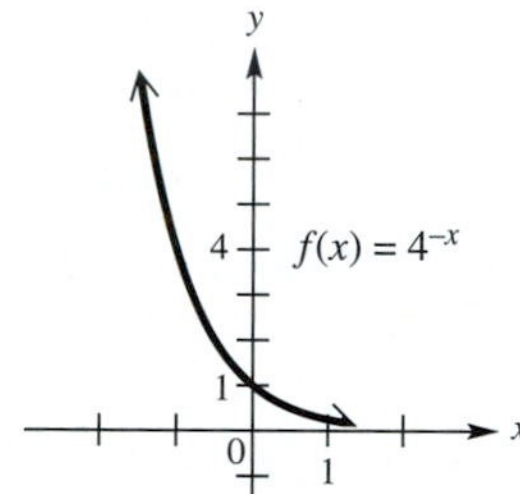

39. **(a)** $a > 1$ **(b)** $(-\infty, \infty)$; $(0, \infty)$; $y = 0$ **(c)**

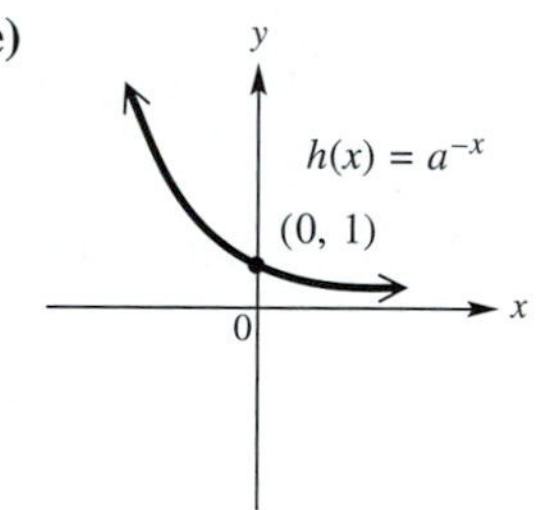

(d) $(-\infty, \infty)$; $(-\infty, 0)$; $y = 0$ **(e)** **(f)** $(-\infty, \infty)$; $(0, \infty)$; $y = 0$

41.

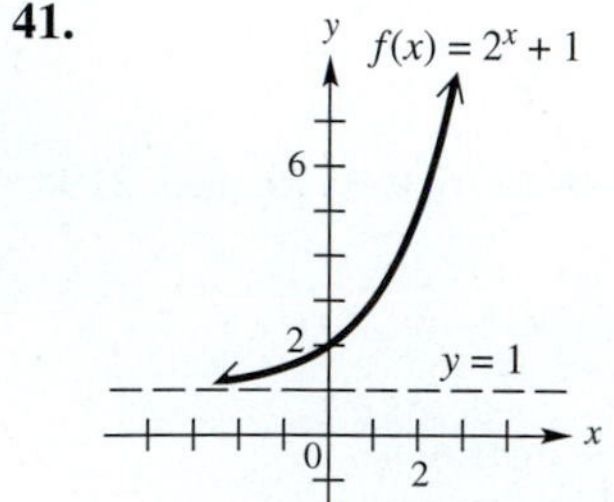

43. **45.** **47.**

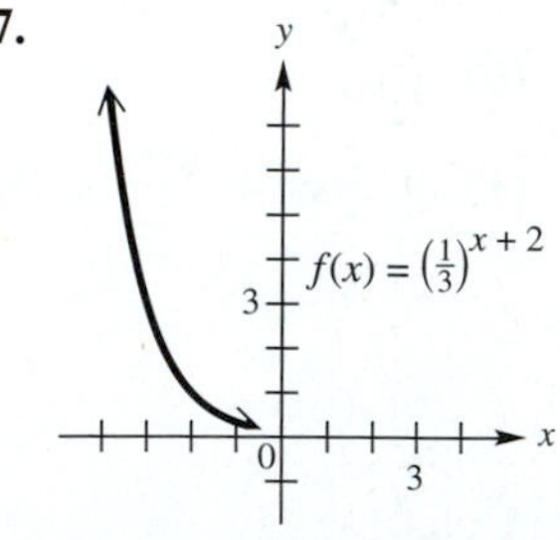

49. $\left\{\frac{1}{2}\right\}$ **51.** $\{-2\}$ **53.** $\{0\}$ **55.** $\{3\}$ **57.** $\left\{\frac{1}{2}\right\}$ **59.** $\left\{\frac{1}{5}\right\}$ **61.** $\{-7\}$ **63.** $\left\{\frac{4}{3}\right\}$ **65. (a)** $\{2\}$ **(b)** $(2, \infty)$ **(c)** $(-\infty, 2)$ **67. (a)** $\left\{\frac{1}{5}\right\}$ **(b)** $\left(\frac{1}{5}, \infty\right)$ **(c)** $\left(-\infty, \frac{1}{5}\right)$ **69. (a)** $\left\{-\frac{2}{3}\right\}$ **(b)** $\left[-\frac{2}{3}, \infty\right)$ **(c)** $\left(-\infty, -\frac{2}{3}\right]$ **71. (a)** 3 **(b)** $\frac{1}{3}$ **(c)** 9 **(d)** 1 **73.** $f(x) = \left(\frac{1}{4}\right)^x$ **75.** $f(t) = \left(\frac{1}{3}\right)9^t$ **77. (a)** \$22,510.18 **(b)** \$22,529.85 **79. (a)** \$33,504.35 **(b)** \$33,504.71 **81.** Plan A is better by \$102.65. **83.** \$1.16; \$2.44; \$7.08; \$18.11; \$59.34; \$145.80; \$318.43

85. (a)

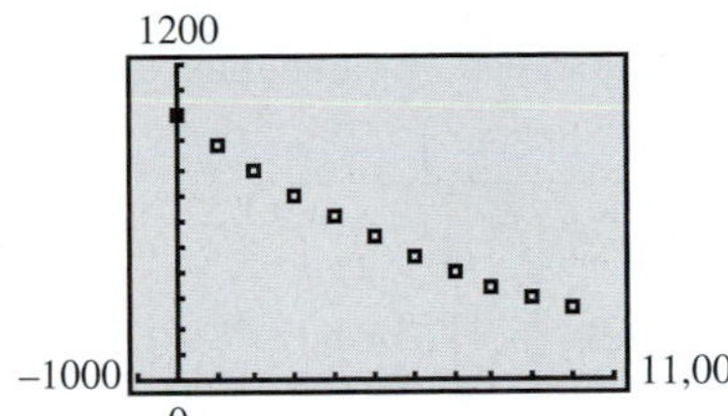

(b) exponential **(c)**

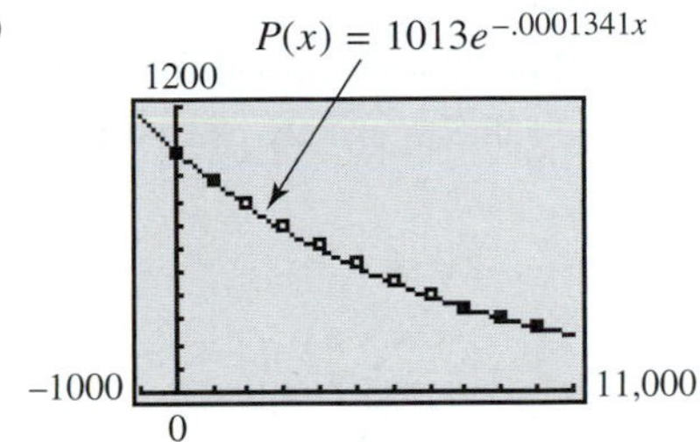

(d) $P(1500) \approx 828$ mb; $P(11{,}000) \approx 232$ mb **87. (a)** $f(2) = 1 - e^{-.5(2)} = 1 - e^{-1} \approx .63$; There is a 63% chance that at least one car will enter the intersection during a 2-min period. **(b)**

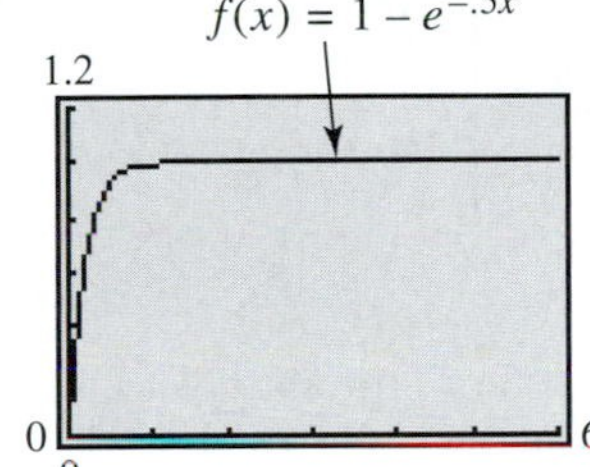

As time progresses, the probability increases and begins to approach 1. That is, it is almost certain that at least one car will enter the intersection during a 60-min period.

5.3 Exercises *(pages 371–373)*

1. (a) C **(b)** A **(c)** E **(d)** B **(e)** F **(f)** D **3.** $\log_3 81 = 4$ **5.** $\log_{1/2} 16 = -4$ **7.** $\log_{10} .0001 = -4$ or $\log .0001 = -4$ **9.** $\ln 1 = 0$ **11.** $6^2 = 36$ **13.** $(\sqrt{3})^8 = 81$ **15.** $10^{-3} = .001$ **17.** $10^{.5} = \sqrt{10}$ **19.** $\{3\}$ **21.** $\{\sqrt{3}\}$ or $\{3^{1/2}\}$ **23.** $\left\{\frac{1}{216}\right\}$ **25.** $\{8\}$ **27.** $\{10\}$ **29.** $\left\{-\frac{1}{8}\right\}$ **31. (a)** 7 **(b)** 9 **(c)** 4 **(d)** k **33. (a)** 0 **(b)** 0 **(c)** 0 **(d)** 0 **35.** 1.5 **37.** $\frac{2}{5}$ **39.** $\frac{2}{3}$ **41.** π **43.** 5.4 **45.** 1.633468456 **47.** −.1062382379 **49.** 4.341474094 **51.** 3.761200116 **53.** −.244622583 **55.** 9.996613531 **57.** 3.2 **59.** 8.4 **61.** 2×10^{-3} **63.** 1.6×10^{-5} **65.** 4.9 yr **67.** 13.9 yr **69.** $\log_3 2 - \log_3 5$ **71.** $\log_2 6 + \log_2 x - \log_2 y$ **73.** $1 + \frac{1}{2}\log_5 7 - \log_5 3 - \log_5 m$ **75.** cannot be rewritten **77.** $\log_k p + 2\log_k q - \log_k m$ **79.** $\frac{1}{2}(3\log_m r - \log_m 5 - 5\log_m z)$ **81.** $\log_a \frac{xy}{m}$ **83.** $\log_m \frac{a^2}{b^6}$ **85.** $\log_a[(z-1)^2(3z+2)]$ **87.** $\log_5(5^{1/3}m^{-1/3})$ or $\log_5 \frac{5^{1/3}}{m^{1/3}}$ **89.** 1.430676558 **91.** .5943161289 **93.** .9595390462 **95.** 1.892441722 **97.** Reflect the graph of $y = 3^x$ across the x-axis and shift 7 units upward.

98.

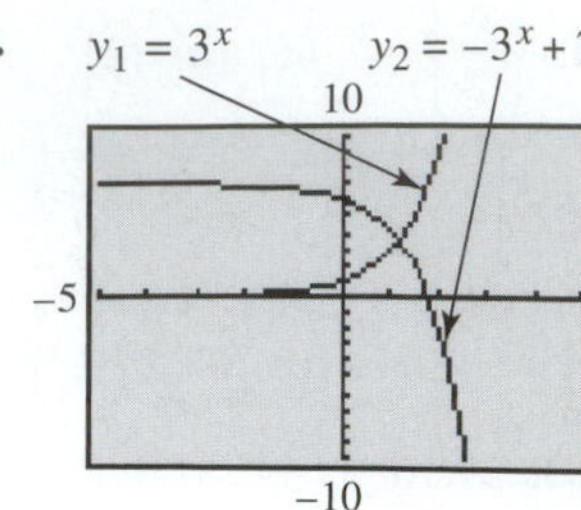

99. 1.7712437 **100.** $\{\log_3 7\}$ **101.** $\frac{\log 7}{\log 3} = \frac{\ln 7}{\ln 3} \approx 1.771243749$

102. The approximations are close enough to support the conclusion that the x-intercept is equal to $\log_3 7$. **103.** 1 **105.** **(a)** 2 **(b)** 2 **(c)** 2 **(d)** 1

Reviewing Basic Concepts *(page 373)*

1. No, because the x-values -2 and 2 both correspond to the y-value 4. In a one-to-one function, each y-value must correspond to exactly one x-value (and each x-value to exactly one y-value). **2.** **(a)**

x	12	21	32	45
y	7	8	9	10

(b) $f^{-1}(x) = 4x - 5$ **3.**

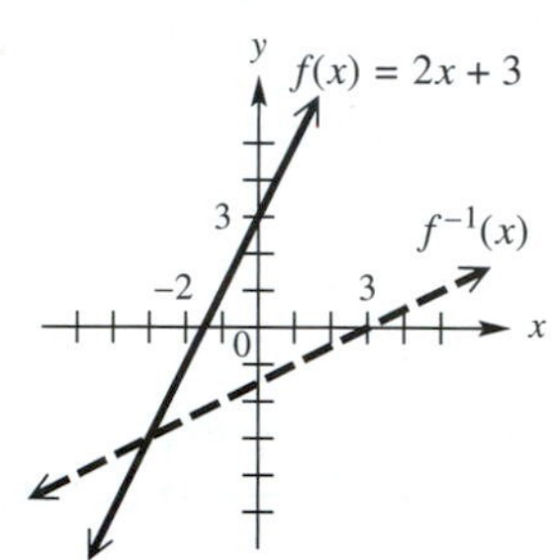

4.

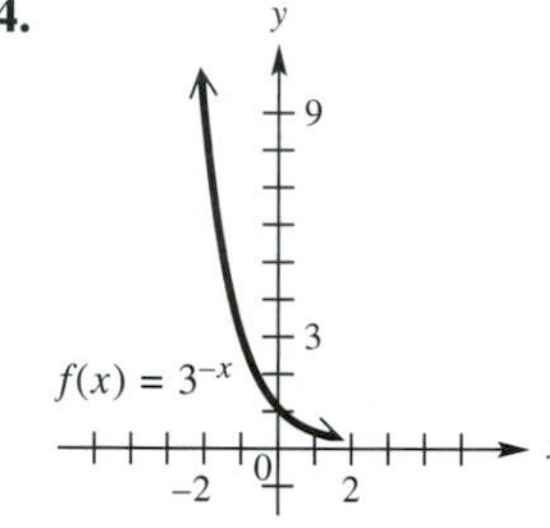

5. $\left\{\frac{3}{4}\right\}$ **6.** \$76.10

7. **(a)** $-\frac{1}{2}$ **(b)** 3 **(c)** 2

8. $\log 3 + 2 \log x - \log 5 - \log y$

9. $\ln \frac{x}{2}$ **10.** 2.1 yr

5.4 Exercises *(pages 381–384)*

1. $(0, \infty)$; $(-\infty, \infty)$; increases; $x = 0$ **3.** $(0, \infty)$; $(-\infty, \infty)$; decreases; $x = 0$ **5.** $(1, \infty)$; $(-\infty, \infty)$; increases; $x = 1$

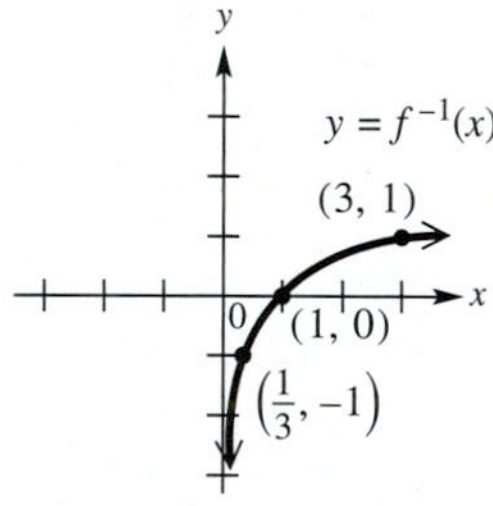

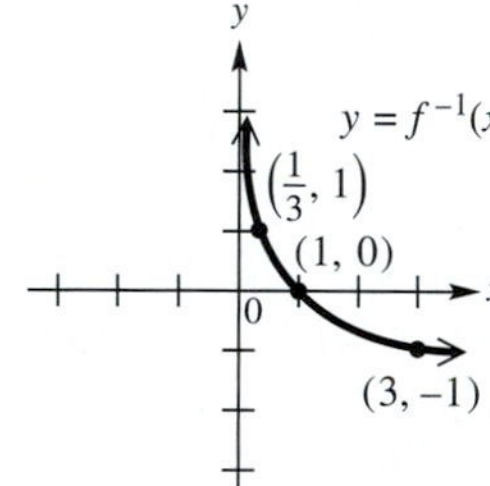

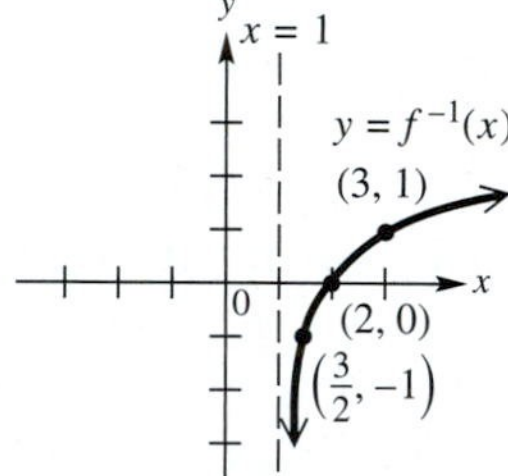

7. logarithmic **9.** $(0, \infty)$ **11.** $(-\infty, 0)$ **13.** $(-\infty, \infty)$ **15.** $(-2, 2)$ **17.** $(-\infty, -3) \cup (7, \infty)$

19. $(-1, 0) \cup (1, \infty)$ **21.** $(-\infty, -3) \cup (4, \infty)$ **23.** $\left(-\infty, \frac{7}{3}\right) \cup \left(\frac{7}{3}, \infty\right)$

25.

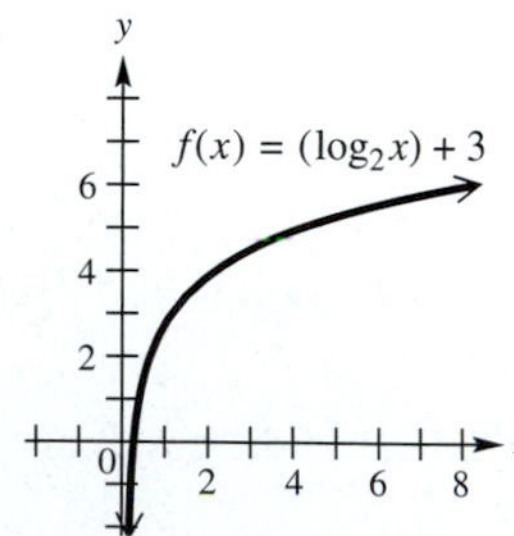

27.

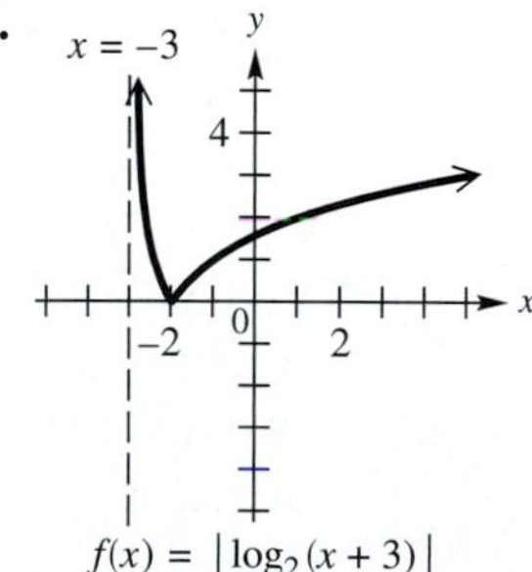

29.

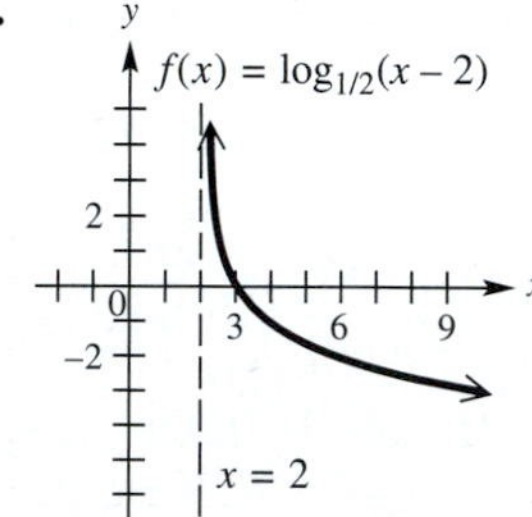
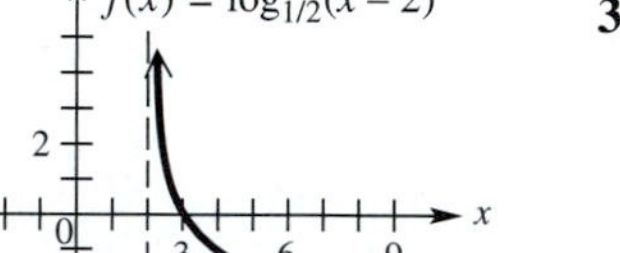

31. B **33.** D

35. A **37.** C

39.

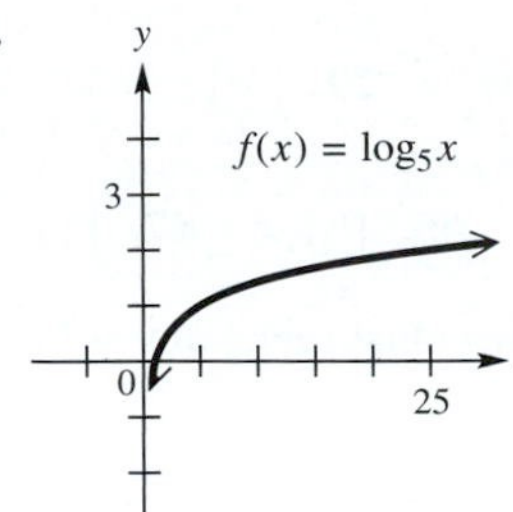

41.

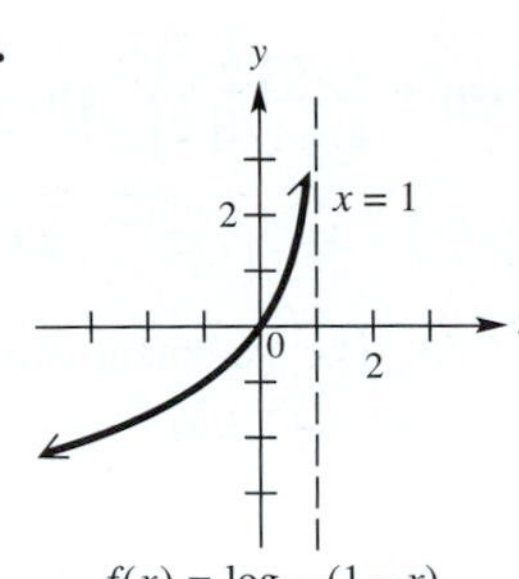

43.

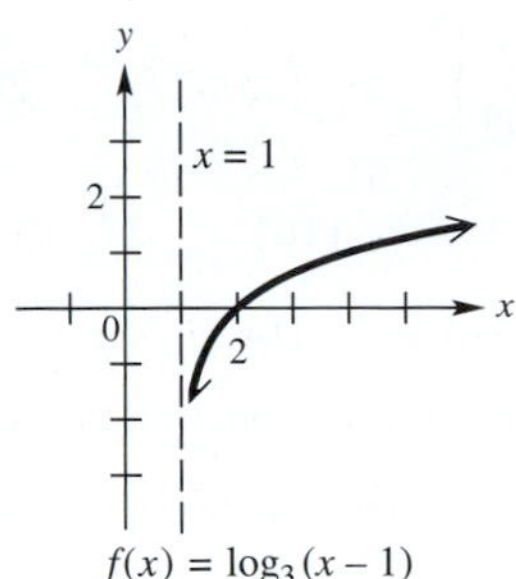

45. **(a)** The graph is shifted 4 units to the left.

(b)

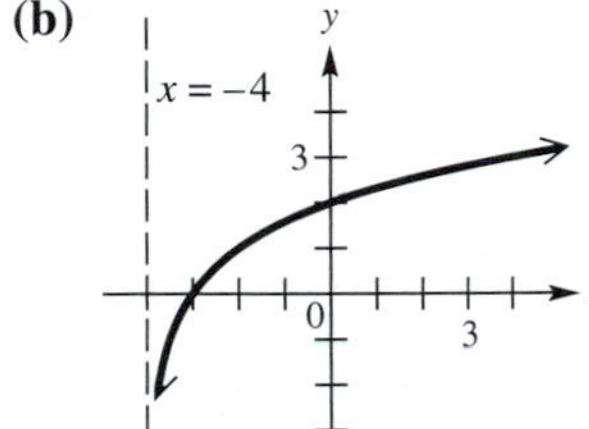

47. **(a)** The graph is stretched vertically by a factor of 3 and shifted 1 unit upward.

(b)

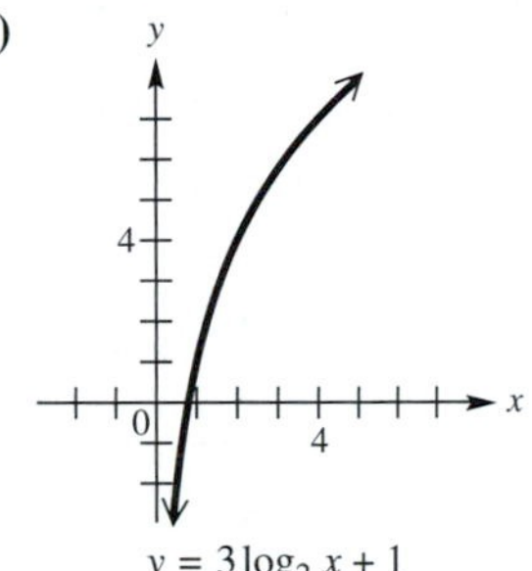

49. **(a)** The graph is reflected across the y-axis and shifted 1 unit upward.

(b)

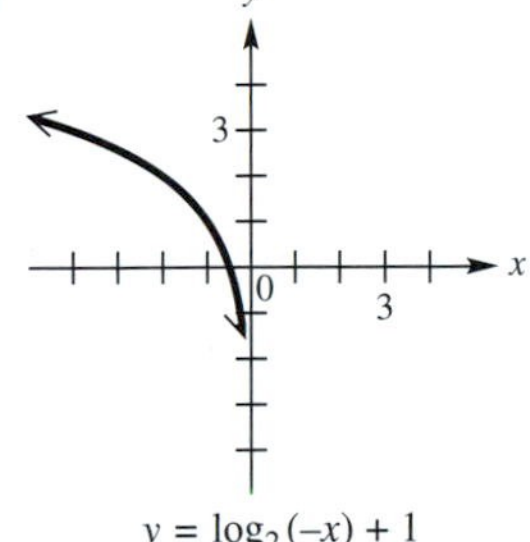

51. The graphs are not the same because the domain of $y = \log x^2$ is $(-\infty, 0) \cup (0, \infty)$, while the domain of $y = 2 \log x$ is $(0, \infty)$. The power rule does not apply if the argument is nonpositive. **53.** $\frac{3}{2}$ **55.** $-\frac{3}{4}$

57. $f^{-1}(x) = \log_4(x + 3)$

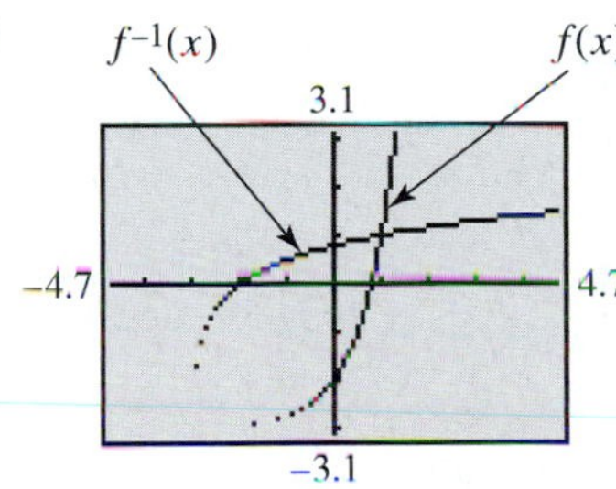

Dot mode

59. $f^{-1}(x) = \log(4 - x)$

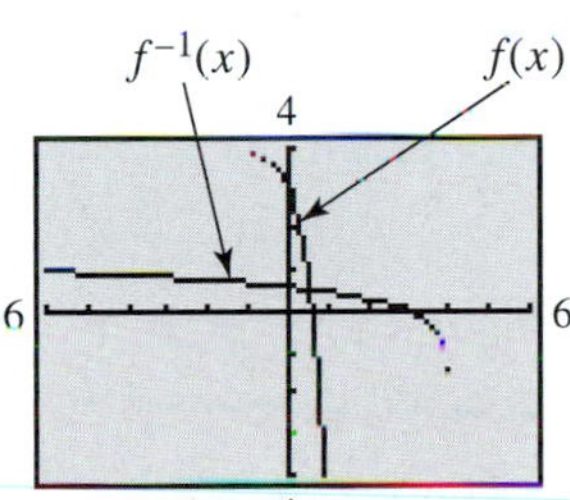

Dot mode

61. **(a)** -1.1 **(b)** $-.2$ **63.** $4 = \log_a 5$ **65.** $\{1.87\}$ **67.** **(a)** The left side is a reflection of the right side with respect to the axis of the tower. The graph of $f(-x)$ is the reflection of $f(x)$ with respect to the y-axis. **(b)** 984 ft **(c)** 39 ft
69. **(a)** 28.105 in. **(b)** It tells us that at 99 mi from the eye of a typical hurricane, the barometric pressure is 29.21 in.

5.5 Exercises *(pages 391–393)*

1. **(a)** $\{1.4036775\}$ **(b)** $(1.4036775, \infty)$ **3.** **(a)** $\{-1\}$ **(b)** $(-1, \infty)$

In Exercises 5–15 and 19–27, alternative forms involving natural logarithms are possible.

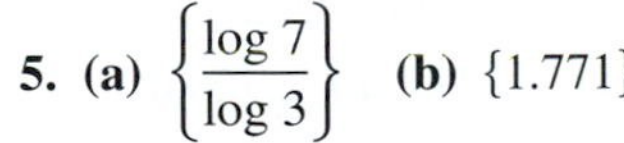

5. **(a)** $\left\{\frac{\log 7}{\log 3}\right\}$ **(b)** $\{1.771\}$ **7.** **(a)** $\left\{\frac{\log 5}{\log \frac{1}{2}}\right\}$ **(b)** $\{-2.322\}$ **9.** **(a)** $\left\{\frac{\log 4}{\log .8}\right\}$ **(b)** $\{-6.213\}$

11. **(a)** $\left\{\frac{\log 4}{\log 4 - 2\log 3}\right\}$ **(b)** $\{-1.710\}$ **13.** **(a)** $\left\{\frac{-\log 6 - \log 4}{\log 6 - 2\log 4}\right\}$ **(b)** $\{3.240\}$ **15.** **(a)** $\emptyset$

17. **(a)** $\left\{\frac{3}{1 - 3\ln 2}\right\}$ **(b)** $\{-2.779\}$ **19.** **(a)** $\emptyset$ **21.** **(a)** $\left\{\frac{2}{\log 1.15}\right\}$ **(b)** $\{32.950\}$ **23.** **(a)** $\left\{2 + \frac{\log 33}{\log 2}\right\}$

(b) {7.044} **25. (a)** $\left\{\dfrac{\log 3.5}{\log 1.05}\right\}$ **(b)** {25.677} **27. (a)** $\left\{1980 + \dfrac{\log \frac{8}{5}}{\log 1.015}\right\}$ **(b)** {2011.568} **29.** $\{e^2\}$

31. $\left\{\dfrac{e^{1.5}}{4}\right\}$ **33.** $\{2 - \sqrt{10}\}$ **35.** {16} **37.** {3} **39.** $\{e\}$ **41.** {25} **43.** {2.5} **45.** {3} **47.** {1}

49. {6} **51.** {1, 10} **53. (a)** {4} **(b)** $(4, \infty)$ **(c)** (0, 4) **55.** The statement is incorrect. We must reject any solution that is not in the domain of any logarithmic function in the equation. **57.** {17.106}

The answers in Exercises 59–69 may have other equivalent forms.

59. $t = e^{(p-r)/k}$ **61.** $t = -\dfrac{1}{k}\log\left(\dfrac{T - T_0}{T_1 - T_0}\right)$ **63.** $k = \dfrac{\ln\left(\frac{A - T_0}{C}\right)}{-t}$ **65.** $x = \dfrac{\ln\left(\frac{A + B - y}{B}\right)}{-C}$ **67.** $A = \dfrac{B}{x^C}$

69. $t = \dfrac{\log \frac{A}{P}}{n \log\left(1 + \frac{r}{n}\right)}$ **71.** It is true because $(a^m)^n = a^{mn}$, and so $(e^x)^2 = e^{x \cdot 2} = e^{2x}$. **72.** $(e^x - 3)(e^x - 1) = 0$

73. {0, ln 3} **74.**

$y = e^{2x} - 4e^x + 3$

The graph intersects the x-axis at 0 and ln 3 ≈ 1.099.

75. $(-\infty, 0) \cup (\ln 3, \infty)$ **76.** (0, ln 3) **77.** {−.767, 2, 4}

79. {2.454, 5.659} **81.** {−.443} **83.** {−2, 2} **85.** {−3}

87. 85.5% **89.** 22 m

Reviewing Basic Concepts *(page 393)*

1. inverse; symmetric; $y = x$; range **2.**

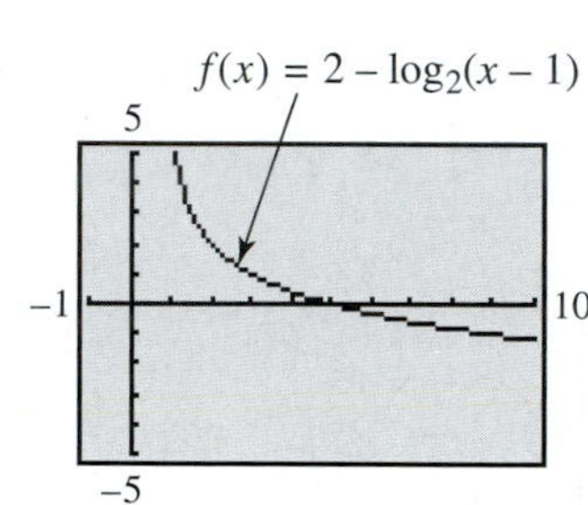

3. $x = 1$; x-intercept: 5, no y-intercept

4. The graph of $f(x)$ is the same as the graph of $g(x)$ reflected across the x-axis and shifted 1 unit to the right and 2 units upward.

5. $f^{-1}(x) = 1 + 2^{2-x}$

6. $\left\{\dfrac{\log 3}{2\log 3 - \log 4}\right\}$ or $\left\{\dfrac{\log 3}{\log \frac{9}{4}}\right\}$

7. {3} **8.** {2} **9.** $N = -\dfrac{1}{k}\ln\left(1 - \dfrac{H}{1000}\right)$ **10.** about 2.8 acres

5.6 Exercises *(pages 402–407)*

1. 9000 yr ago **3.** 16,000 yr old **5. (a)** $A = A_0e^{-.032t}$ **(b)** 6.97 yr **(c)** 363 g **7. (a)** less **(b)** $A = 2e^{-.005t}$ **(c)** about 140 days **9. (a)** $79{,}000{,}000I_0$ **(b)** $1{,}300{,}000I_0$ **(c)** The quake measuring 6.6 was about 1.26 times as strong. **11.** about 1.126 billion yr old **13.** Magnitude 1 is approximately 6.3 times as great as magnitude 3. **15. (a)** $f(t) = 20 + 80e^{-.693t}$ **(b)** 76.6ºC **(c)** 1.415 hr or about 1 hr and 25 min **17. (a)** $C \approx .72$, $a \approx 1.041$ **(b)** 1.21 ppb **19. (a)** $(T \circ R)(x) = .206e^{.0124x}$ **(b)** $(T \circ R)(100) = .7119$; In 1900 radiative forcing caused an approximately .7ºF increase in the average global temperature. **21. (a)** 46.2 yr **(b)** 46.0 yr **23.** 1.8 yr **25.** The better investment is 7% compounded quarterly; it will earn $800.31 more. **27.** 6.14% **29.** $5583.95 **31.** 12% **33. (a)** $205.52 **(b)** $1364.96 **35. (a)** $852.72 **(b)** $181,979.20 **37. (a)** 19 yr, 39 days **(b)** 11 yr, 166 days **39. (a)** 2,700,000 **(b)** 3,000,000 **(c)** 9,500,000 **41.** 23 days **43.** 59 mg

45. (a)

x	0	15	30	45	60	75	90	125
$g(x)$	7	21	57	111	136	158	164	178

(b) $g(x) = 261 - f(x)$ **(c)** y_1 models $g(x)$ better.

(d) $f(x) = 261 - \dfrac{171}{1 + 18.6e^{-.0747x}}$

47. **(a)** $f(x) \approx 10.98(1.14)^x$ **(b)**

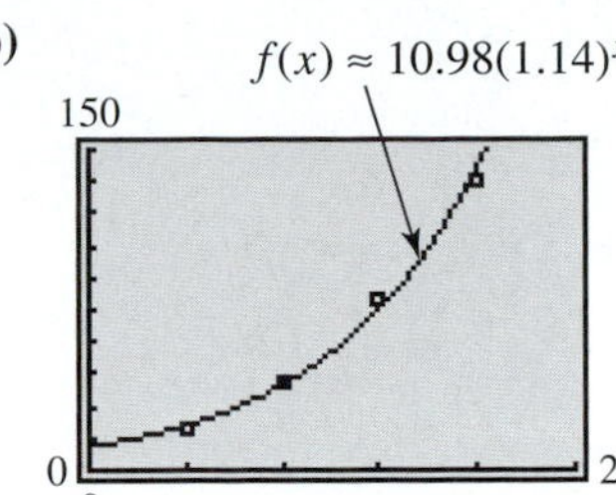

(c) about \$60 billion

49. **(a)** \$19.2 billion **(b)** 2002 **(c)**

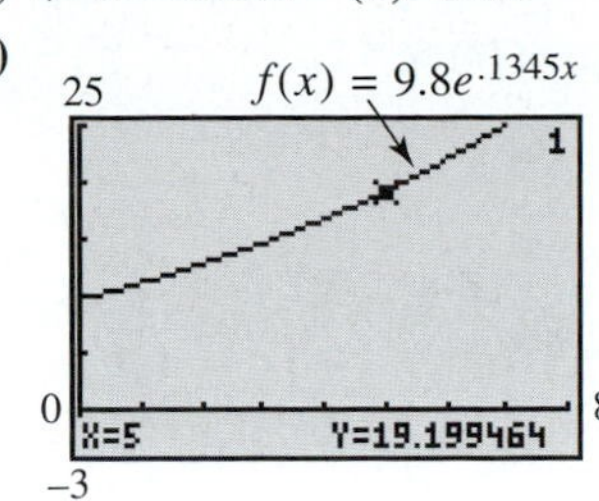

51. **(a)** $C \approx 236.16$; $a \approx .878$ **(b)**

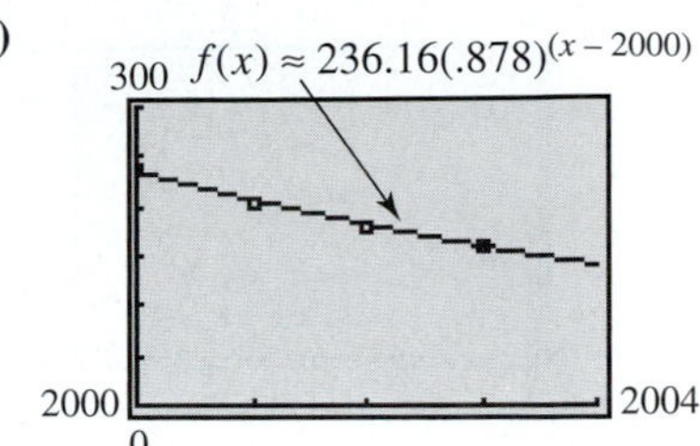

53. **(a)** .065; .82; Among people age 25, 6.5% have some CHD, while among people age 65, 82% have some CHD. **(b)** 48 yr

Chapter 5 Review Exercises *(pages 411–414)*

1. no **3.** no **5.** $(-\infty, \infty)$ **7.** Since f is one-to-one, it has an inverse. **9.** The graphs are reflections across the line $y = x$. (See the graph.) **11.** C **13.** D **15.** $0 < a < 1$ **17.** $(0, \infty)$ **19.**

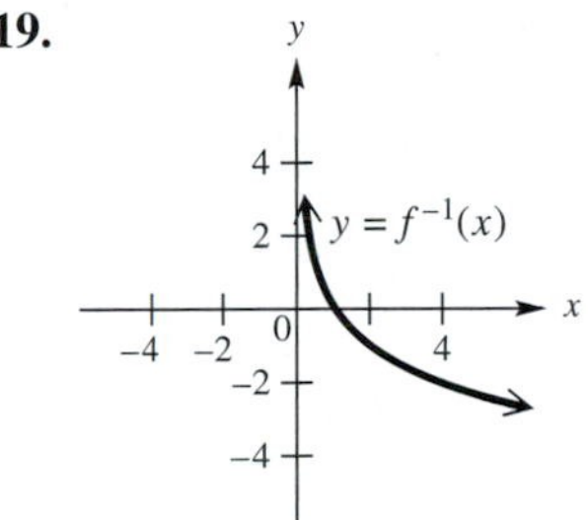

21. $(-\infty, \infty)$; $(0, \infty)$

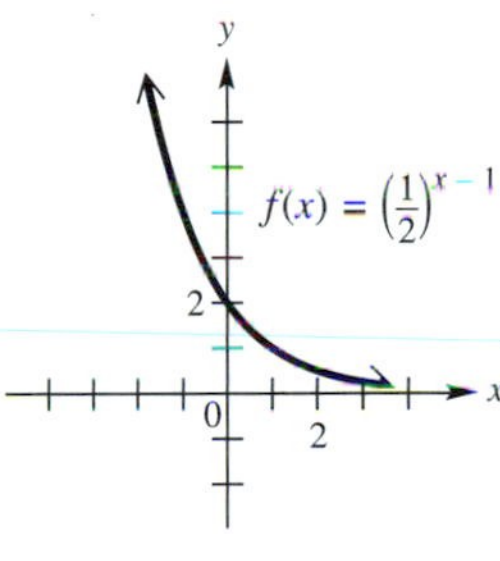

23. $(-\infty, \infty)$; $(-\infty, 0)$

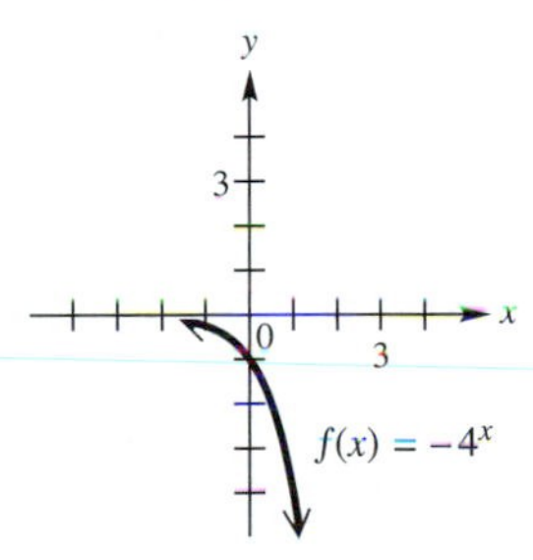

25. **(a)** $\left\{\frac{3}{5}\right\}$ **(b)** $\left[\frac{3}{5}, \infty\right)$ **27.** **(a)** $\left\{-\frac{2}{3}\right\}$ **(b)** $\left(-\frac{2}{3}, \infty\right)$ **29.** $(-.766664696, .58777475603)$ **31.** 1.7657 **33.** 4.0656 **35.** 0 **37.** 12 **39.** 5 **41.** E **43.** B **45.** F

47. $(0, \infty)$; $(-\infty, \infty)$

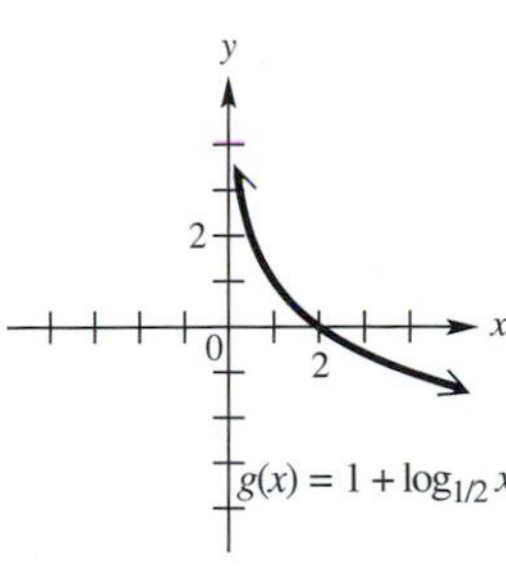

49. $(-\infty, 0)$; $(-\infty, \infty)$

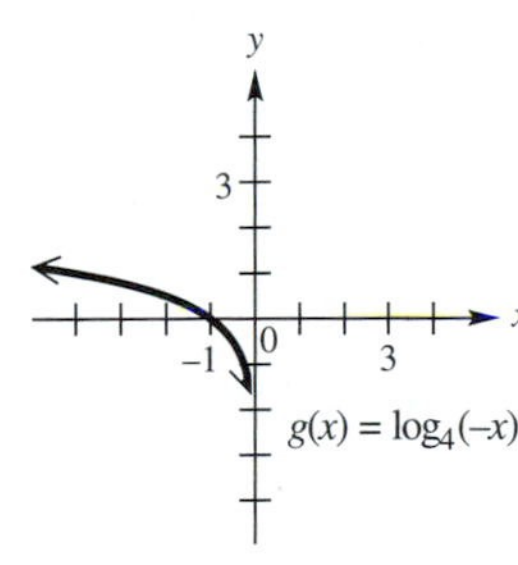

51. 3 **53.** $\log_3 m + \log_3 n - \log_3 5 - \log_3 r$ **55.** $2 \log_5 x + 4 \log_5 y + \frac{3}{5} \log_5 m + \frac{1}{5} \log_5 p$ **57.** **(a)** $\{2\}$ **(b)** $(2, \infty)$ **59.** **(a)** $\left\{\frac{7}{5}\right\}$ or $\{1.4\}$ **(b)** $(1, 1.4]$ **61.** **(a)** $\left\{\frac{3}{2}\right\}$ **63.** **(a)** $\{-1 + \ln 10\}$ **(b)** $\{1.303\}$ **65.** **(a)** $\{2\}$ **67.** **(a)** $\{3\}$

69. **(a)** $\{-3\}$ **71.** $\emptyset$ **73.** $t = \frac{\ln\left(\frac{y}{y_0}\right)}{-k}$ **75.** $\{1.874\}$ **77.** 9.6% **79.** \$25,149.59

81. **(a)**

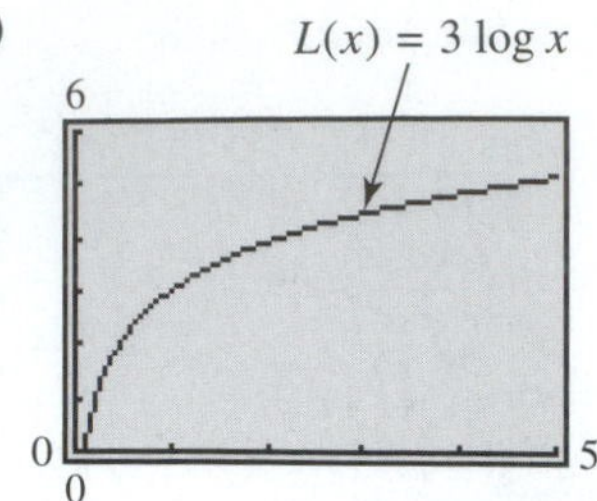

Since L is increasing, heavier planes require longer runways. **(b)** No, it increases by a factor of 2 to 6000 ft.

83. **(a)** .0054 g per L **(b)** .00073 g per L **(c)** .000013 g per L **(d)** .75 mi **85.** about 10 sec

Chapter 5 Test *(pages 414–415)*

1. **(a)** B **(b)** A **(c)** C **(d)** D **(e)** the equations of the functions in (a) and (d) and those in (b) and (c)

2. **(a)** $f(x) = -2^{x-1} + 8$ **(b)** domain: $(-\infty, \infty)$; range: $(-\infty, 8)$ **(c)** Yes, it has a horizontal asymptote with equation $y = 8$. **(d)** x-intercept: 4; y-intercept: 7.5 **(e)** $f^{-1}(x) = 1 + \frac{\log(8 - x)}{\log 2}$ (Any base logarithm can be used.) **3.** $\{.5\}$ **4.** **(a)** \$12,442.11 **(b)** \$12,460.77

5. The expression $\log_5 27$ is the exponent to which 5 must be raised in order to obtain 27. To find an approximation with a calculator, use the change-of-base rule. **6.** **(a)** 1.659 **(b)** 6.153 **(c)** 6.049 **7.** $3 \log m + \log n - \frac{1}{2} \log y$ **8.** $t = \frac{\ln A - \ln P}{r}$

9. **(a)** $\{2\}$; The extraneous value is -4. **(b)** $y_1 = \frac{\log x}{\log 2} + \frac{\log(x + 2)}{\log 2} - 3$

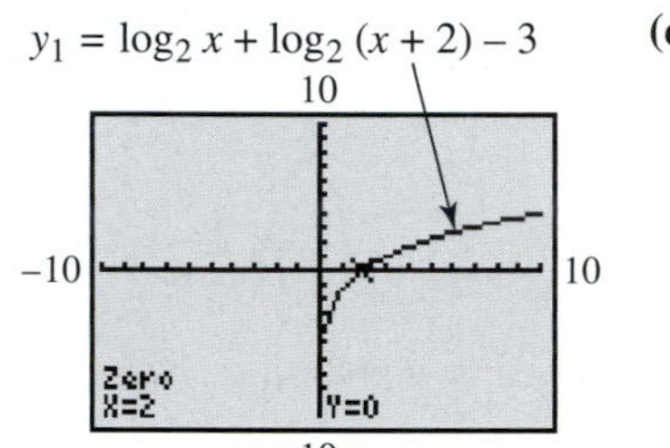

(c) $(2, \infty)$

10. **(a)** $\left\{\frac{-2 + \ln 4}{5}\right\}$ **(b)** $\{-.123\}$ **11.** **(a)** $\left\{\frac{\log 18}{\log 48}\right\}$ **(b)** $\{.747\}$ **12.** **(a)** $\{e^{10}\}$ **(b)** $\{22{,}026.466\}$

13. **(a)** B **(b)** D **(c)** C **(d)** A

14. **(a)** For $t = x$, $A(x) = x^2 - x + 350$ **(b)** For $t = x$, $A(x) = 350 \log(x + 1)$ **(c)**

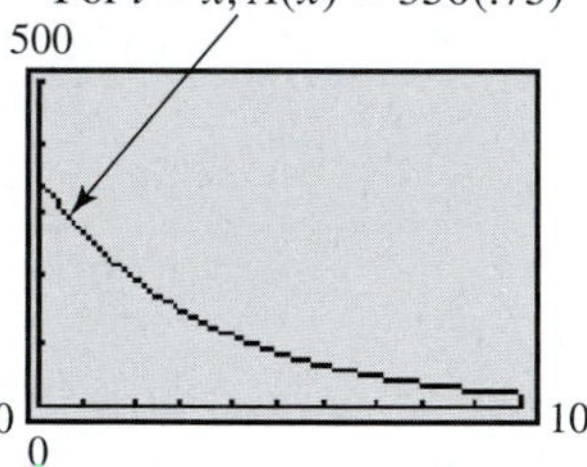

(d)

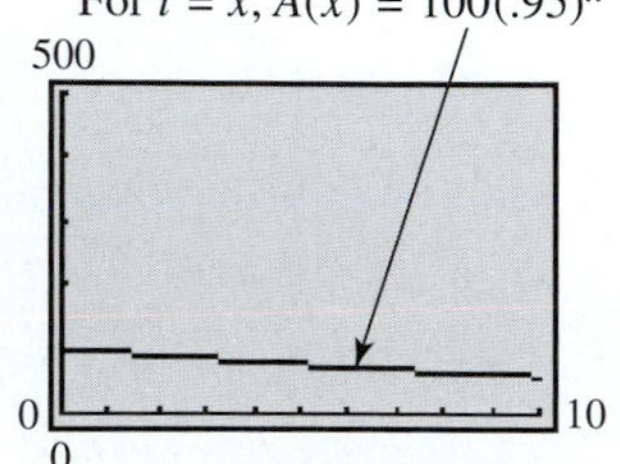

Function (c) best describes $A(t)$. **15.** **(a)** $A(t) = 2e^{-.000433t}$ **(b)** .03 g **(c)** about 3200 yr

CHAPTER 6 ANALYTIC GEOMETRY

6.1 Exercises *(pages 428–431)*

1. The graph is the point (3, 3). **3.** E **5.** H **7.** F **9.** D **11.** $(x - 1)^2 + (y - 4)^2 = 9$ **13.** $x^2 + y^2 = 1$
15. $\left(x - \frac{2}{3}\right)^2 + \left(y + \frac{4}{5}\right)^2 = \frac{9}{49}$ **17.** $(x + 1)^2 + (y - 2)^2 = 25$ **19.** $(x + 3)^2 + (y + 2)^2 = 4$ **21.** $(2, -3)$
22. $3\sqrt{5}$ **23.** $(x - 2)^2 + (y + 3)^2 = 45$ **24.** $(x + 2)^2 + (y + 1)^2 = 41$ **25.** domain: $[-2, 2]$; range: $[-2, 2]$

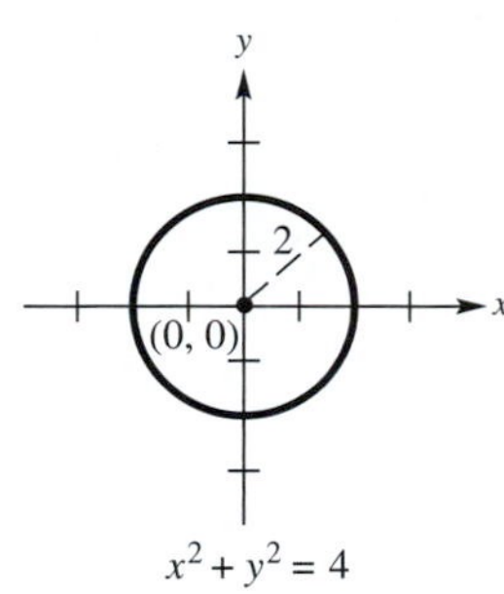

27. domain: $\{0\}$; range: $\{0\}$

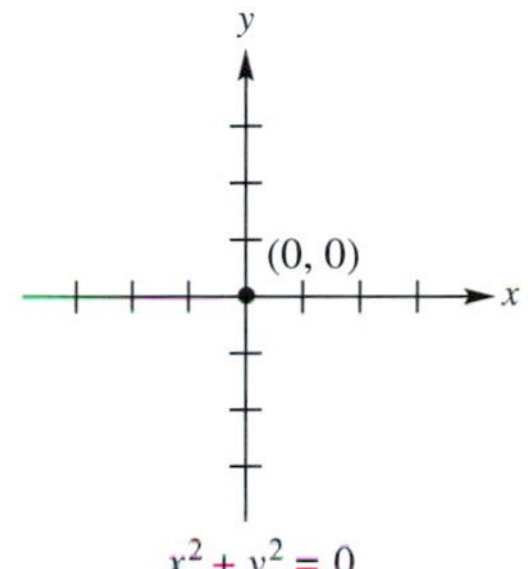
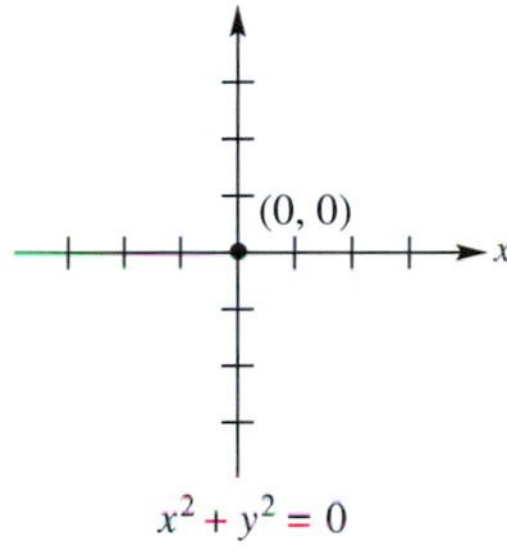

29. domain: $[-4, 8]$; range: $[-6, 6]$

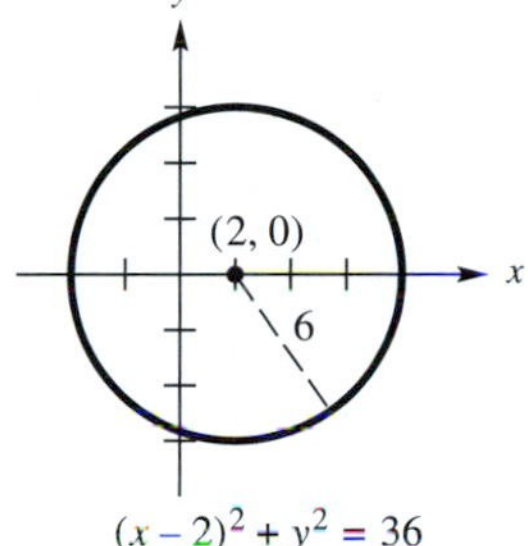

31. domain: $[-2, 12]$; range: $[-11, 3]$

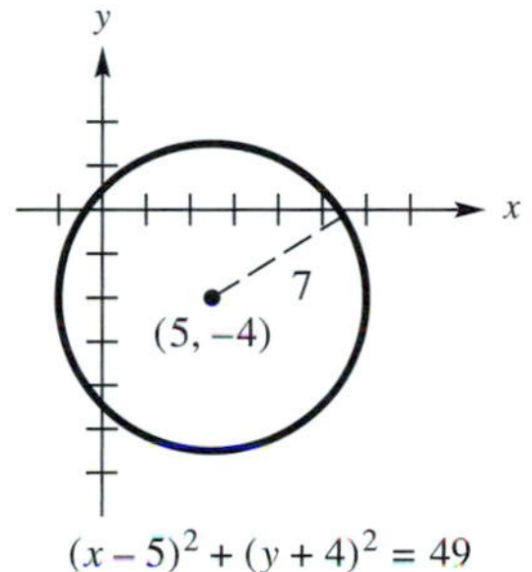

33. domain: $[-9, 3]$; range: $[-8, 4]$

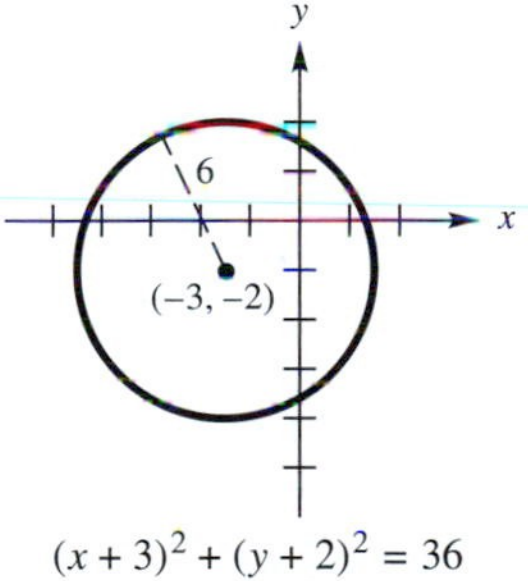

35. domain: $[-3, 5]$; range: $[-6, 2]$

y
x
4
(1, −2)
$(x - 1)^2 + (y + 2)^2 = 16$

37. domain: $[-9, 9]$; range: $[-9, 9]$

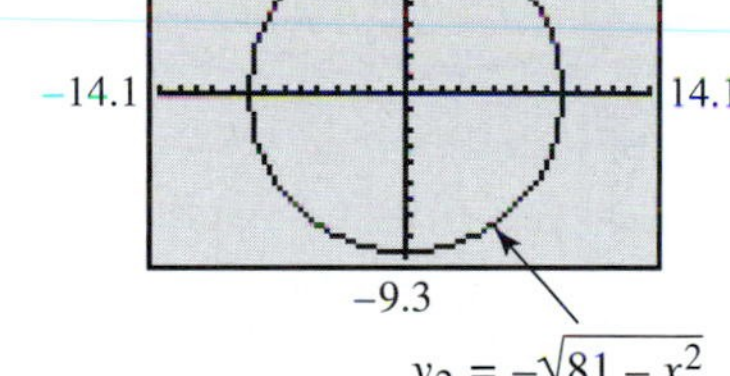

39. domain: $[-2, 8]$; range: $[-3, 7]$

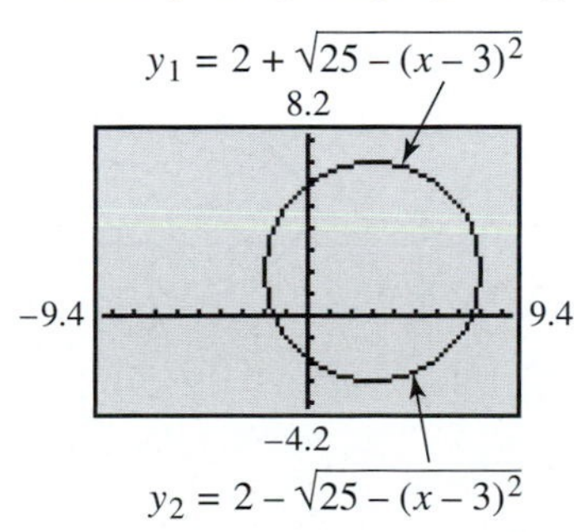

41. yes; center: $(-3, -4)$; radius: 4 **43.** yes; center: $(2, -6)$; radius: 6
45. yes; center: $\left(-\frac{1}{2}, 2\right)$; radius: 3 **47.** no **49.** no **51.** yes; center: $(-2, 0)$; radius: $\frac{2}{3}$ **53.** D **55.** C **57.** F **59.** E **61.** **(a)** III **(b)** II **(c)** IV **(d)** I
63. $(0, 4)$; $y = -4$; y-axis **65.** $\left(0, -\frac{1}{8}\right)$; $y = \frac{1}{8}$; y-axis **67.** $\left(\frac{1}{64}, 0\right)$; $x = -\frac{1}{64}$; x-axis **69.** $(-4, 0)$; $x = 4$; x-axis **71.** $x^2 = -8y$ **73.** $y^2 = -2x$
75. $y^2 = 4x$ **77.** $x^2 = -2y$ **79.** $x^2 = -y$ **81.** $x^2 = 4(y - 1)$
83. $y^2 = 4(x + 1)$ **85.** $(x + 1)^2 = -8(y - 5)$

87. $(y-3)^2 = -\frac{9}{2}(x+2)$

89. vertex: $(-3,-4)$; axis: $x=-3$; domain: $(-\infty,\infty)$; range: $[-4,\infty)$

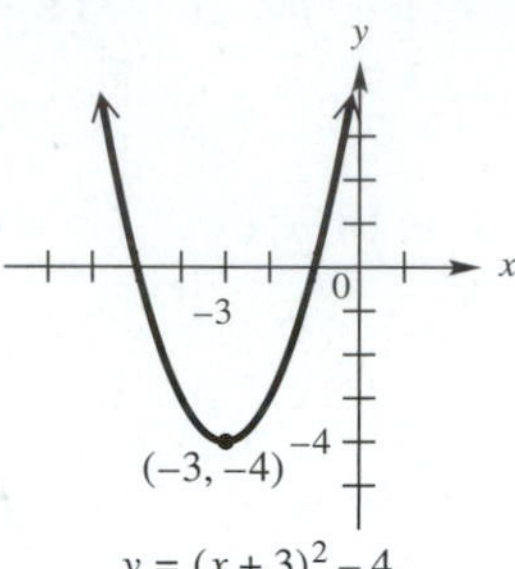

91. vertex: $(-3,2)$; axis: $x=-3$; domain: $(-\infty,\infty)$; range: $(-\infty,2]$

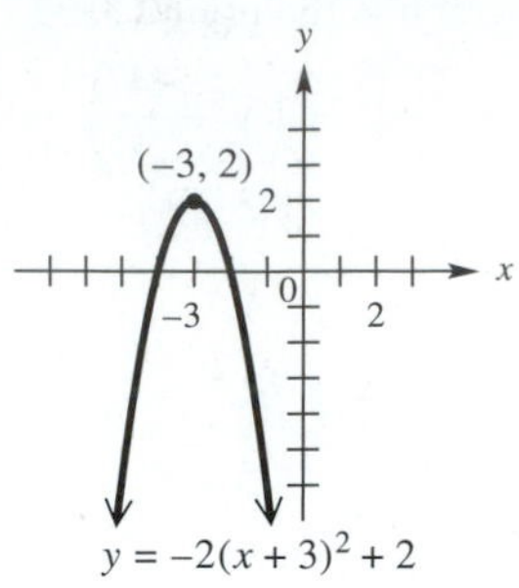

93. vertex: $(1,2)$; axis: $x=1$; domain: $(-\infty,\infty)$; range: $[2,\infty)$

$y = x^2 - 2x + 3$

95. vertex: $(1,3)$; axis: $x=1$; domain: $(-\infty,\infty)$; range: $[3,\infty)$

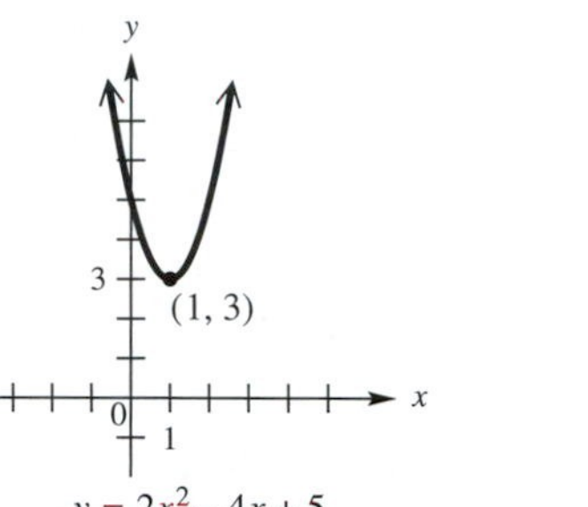

97. vertex: $(2,0)$; axis: $y=0$; domain: $[2,\infty)$; range: $(-\infty,\infty)$

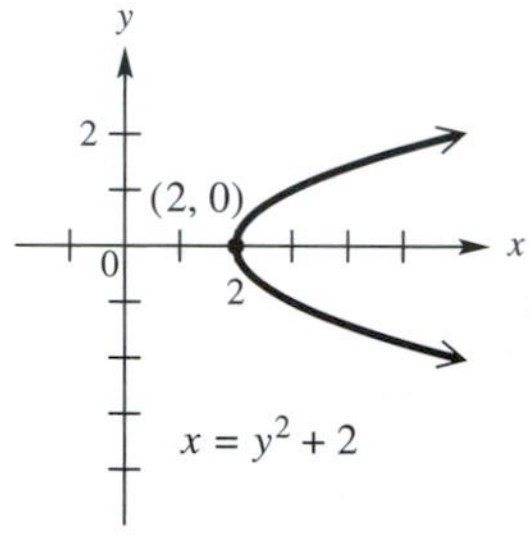

99. vertex: $(0,3)$; axis: $y=3$; domain: $[0,\infty)$; range: $(-\infty,\infty)$

$x = (y-3)^2$

101. vertex: $(2,4)$; axis: $y=4$; domain: $[2,\infty)$; range: $(-\infty,\infty)$

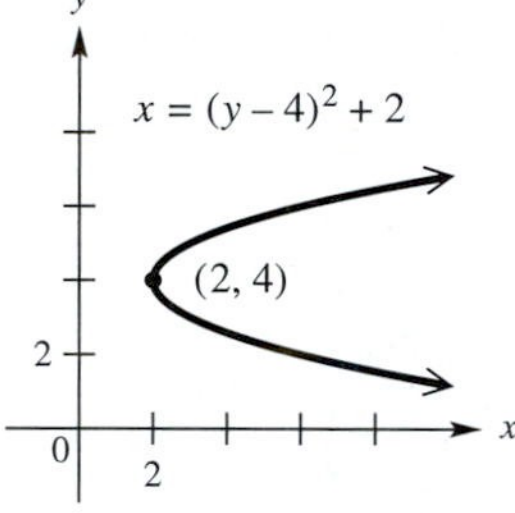

103. vertex: $(2,3)$; axis: $y=3$; domain: $[2,\infty)$; range: $(-\infty,\infty)$

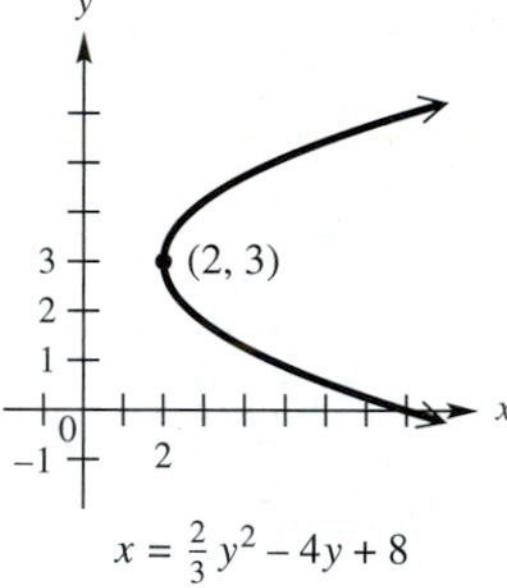

105. vertex: $\left(-2,-\frac{1}{2}\right)$; axis: $y=-\frac{1}{2}$; domain: $(-\infty,-2]$; range: $(-\infty,\infty)$

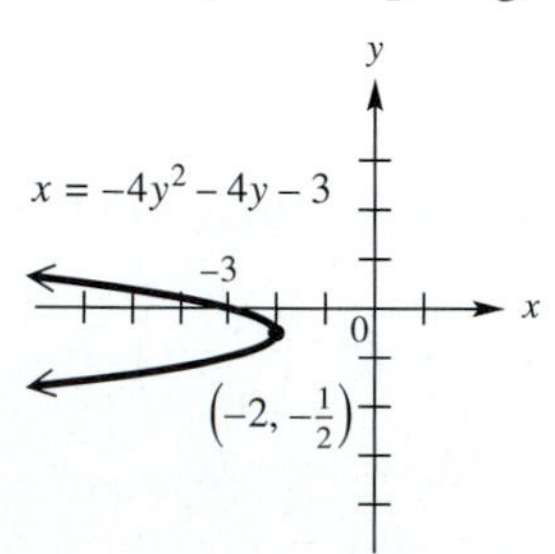

107. vertex: $\left(-\frac{5}{2},\frac{1}{2}\right)$; axis: $y=\frac{1}{2}$; domain: $\left(-\infty,-\frac{5}{2}\right]$; range: $(-\infty,\infty)$

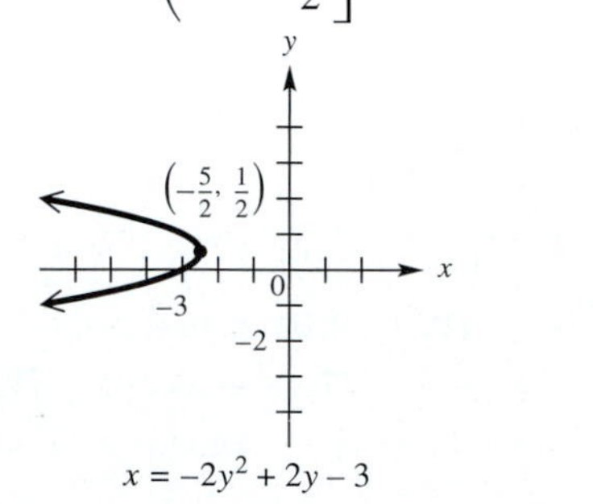

109. vertex: $(4,1)$; axis: $y=1$; domain: $[4,\infty)$; range: $(-\infty,\infty)$

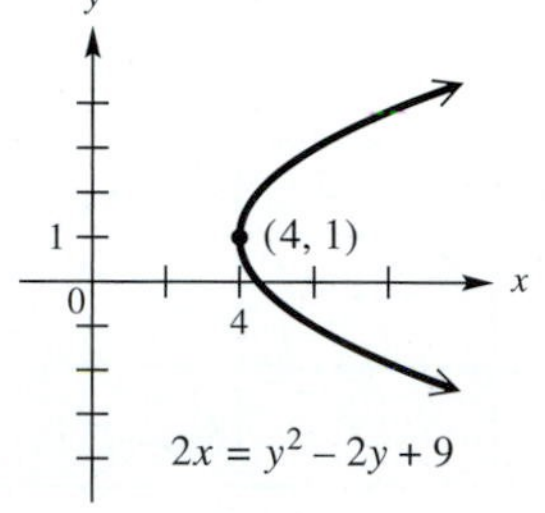

111. vertex: $(-1, 2)$; axis: $y = 2$; domain: $[-1, \infty)$; range: $(-\infty, \infty)$

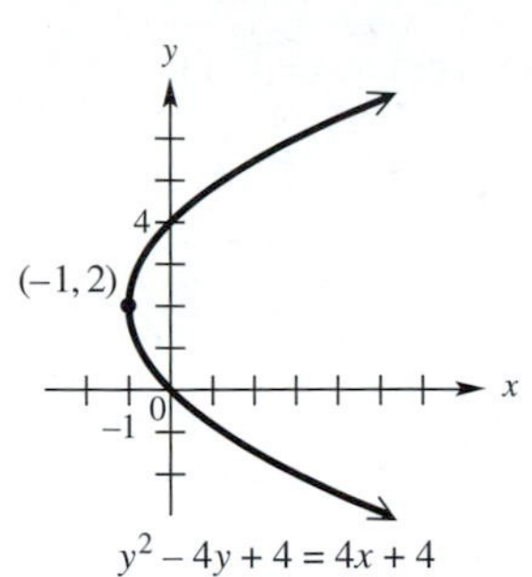

115. (a) Mars $y_2 = \frac{19}{11}x - \frac{12.6}{3872}x^2$; 1000; 0; 1500; 0; moon $y_1 = \frac{19}{11}x - \frac{5.2}{3872}x^2$

(b) Mars: approximately 229 ft; moon: approximately 555 ft **117.** 4×10^{-17} m downward **119.** 6 ft

6.2 Exercises *(pages 441–445)*

1. G **3.** F **5.** E **7.** D **9.** A circle can be interpreted as an ellipse whose two foci have the same coordinates; the "coinciding foci" give the center of the circle. **11.** domain: $[-3, 3]$; range: $[-2, 2]$; foci: $(-\sqrt{5}, 0)$, $(\sqrt{5}, 0)$

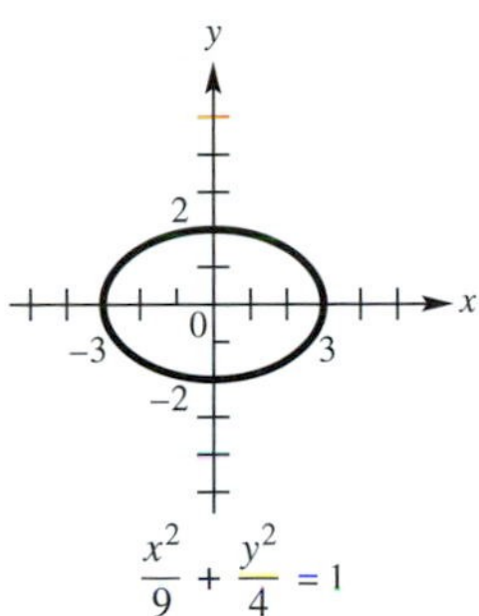

13. domain: $[-\sqrt{6}, \sqrt{6}]$; range: $[-3, 3]$; foci: $(0, -\sqrt{3})$, $(0, \sqrt{3})$

$9x^2 + 6y^2 = 54$

15. domain: $\left[-\frac{3}{8}, \frac{3}{8}\right]$; range: $\left[-\frac{6}{5}, \frac{6}{5}\right]$

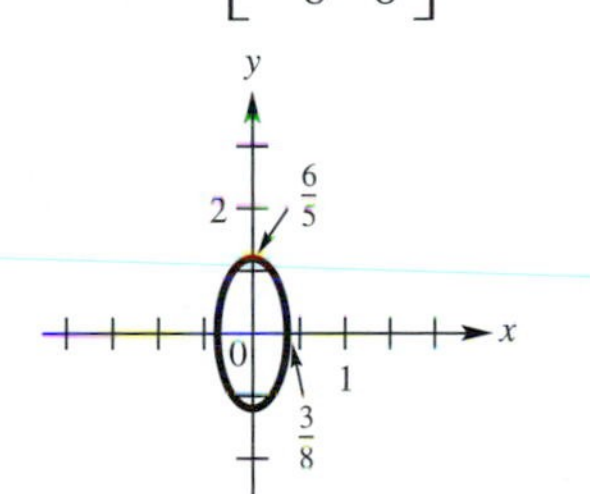

17. domain: $[-2, 4]$; range: $[-8, 2]$

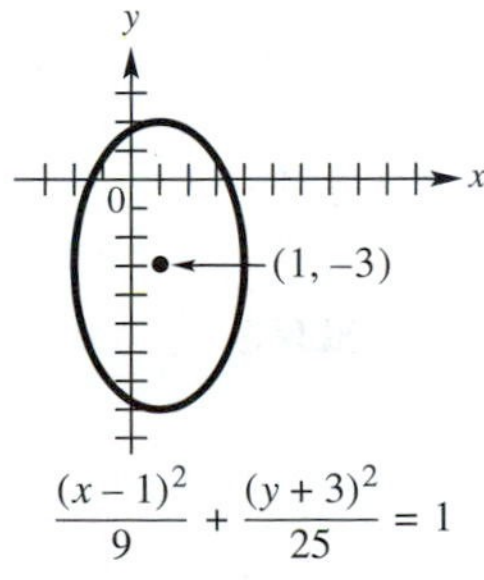

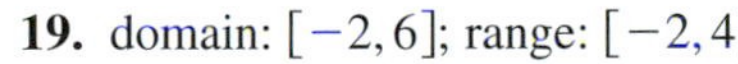

19. domain: $[-2, 6]$; range: $[-2, 4]$

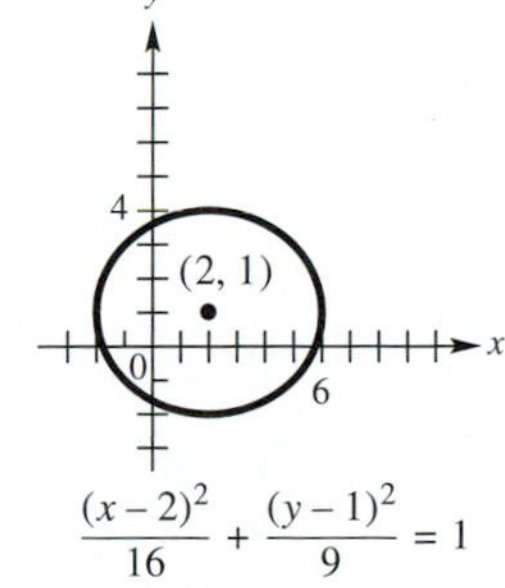

21. domain: $[-9, 7]$; range: $[-5, 9]$

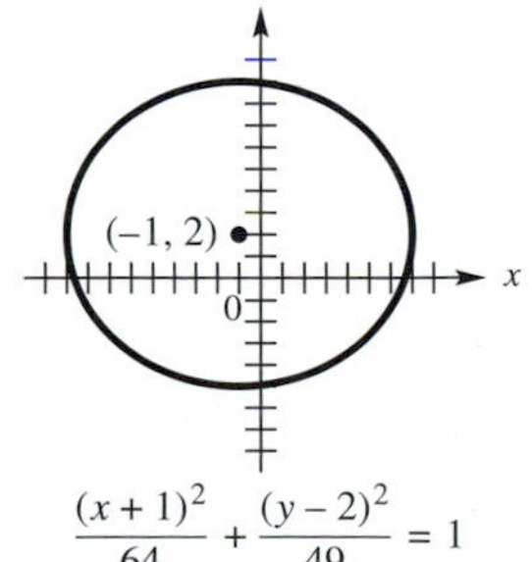

23. $\dfrac{x^2}{16} + \dfrac{y^2}{12} = 1$ **25.** $\dfrac{x^2}{36} + \dfrac{y^2}{20} = 1$ **27.** $\dfrac{(x-3)^2}{16} + \dfrac{(y+2)^2}{25} = 1$ **29.** $\dfrac{x^2}{5} + \dfrac{y^2}{9} = 1$

31. $\dfrac{(x-5)^2}{25} + \dfrac{(y-2)^2}{16} = 1$ **33.** $\dfrac{(x-4)^2}{9} + \dfrac{(y-5)^2}{16} = 1$ **35.** $\dfrac{(x+1)^2}{4} + \dfrac{(y-1)^2}{9} = 1$; center: $(-1, 1)$; vertices: $(-1, -2), (-1, 4)$ **37.** $\dfrac{(x+1)^2}{1} + \dfrac{(y+1)^2}{4} = 1$; center: $(-1, -1)$; vertices: $(-1, -3), (-1, 1)$

39. $\dfrac{(x+2)^2}{5} + \dfrac{(y-1)^2}{4} = 1$; center: $(-2, 1)$; vertices: $(-2-\sqrt{5}, 1), (-2+\sqrt{5}, 1)$ **41.** $\dfrac{(x-\frac{1}{2})^2}{4} + \dfrac{(y+\frac{3}{2})^2}{16} = 1$; center: $\left(\dfrac{1}{2}, -\dfrac{3}{2}\right)$; vertices: $\left(\dfrac{1}{2}, \dfrac{5}{2}\right), \left(\dfrac{1}{2}, -\dfrac{11}{2}\right)$

43. domain: $(-\infty, -4] \cup [4, \infty)$; range: $(-\infty, \infty)$

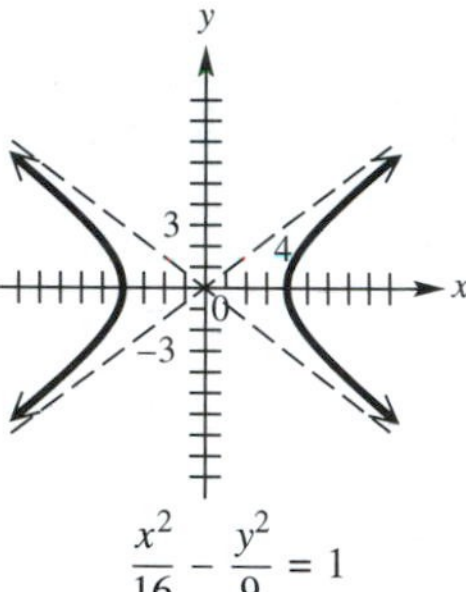

45. domain: $(-\infty, \infty)$; range: $(-\infty, -6] \cup [6, \infty)$

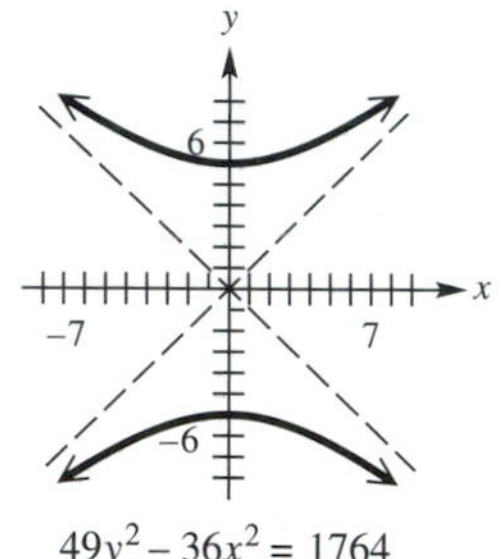

47. domain: $\left(-\infty, -\dfrac{3}{2}\right] \cup \left[\dfrac{3}{2}, \infty\right)$; range: $(-\infty, \infty)$

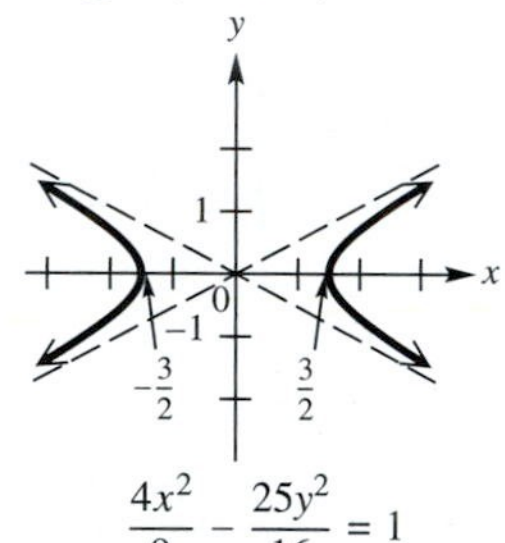

49. domain: $\left(-\infty, -\dfrac{1}{3}\right] \cup \left[\dfrac{1}{3}, \infty\right)$; range: $(-\infty, \infty)$

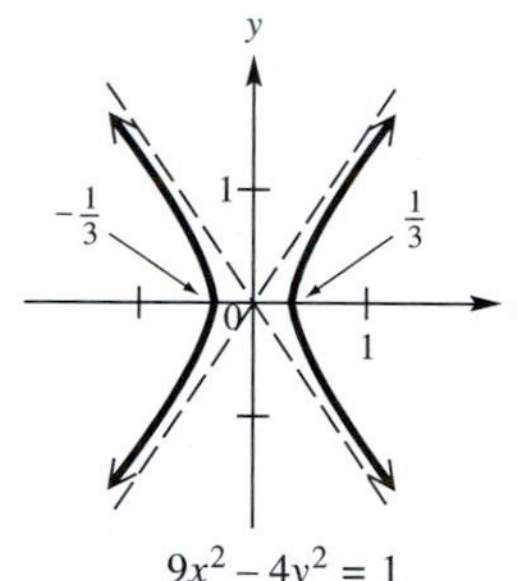

51. domain: $(-\infty, -2] \cup [4, \infty)$; range: $(-\infty, \infty)$; center: $(1, -3)$

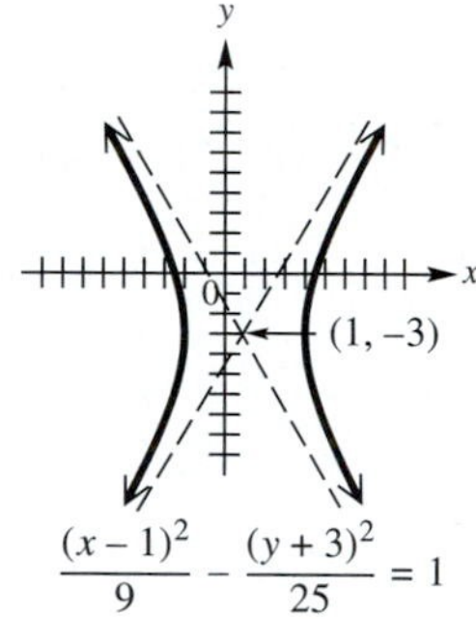

53. domain: $(-\infty, \infty)$; range: $(-\infty, 3] \cup [7, \infty)$; center: $(-1, 5)$

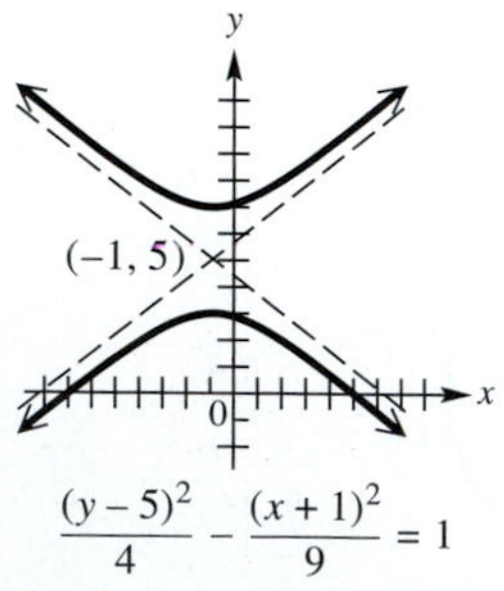

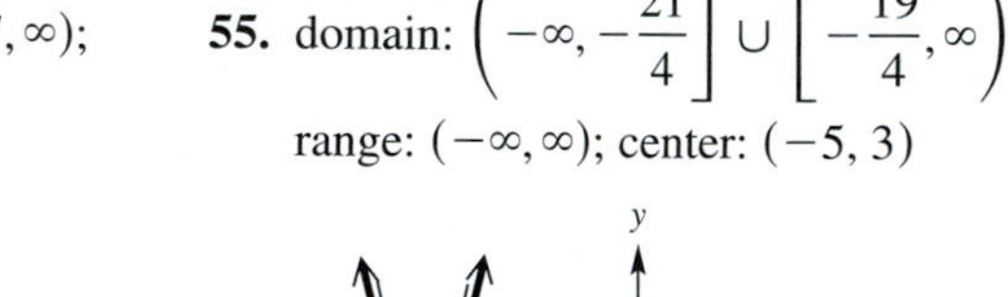

55. domain: $\left(-\infty, -\dfrac{21}{4}\right] \cup \left[-\dfrac{19}{4}, \infty\right)$; range: $(-\infty, \infty)$; center: $(-5, 3)$

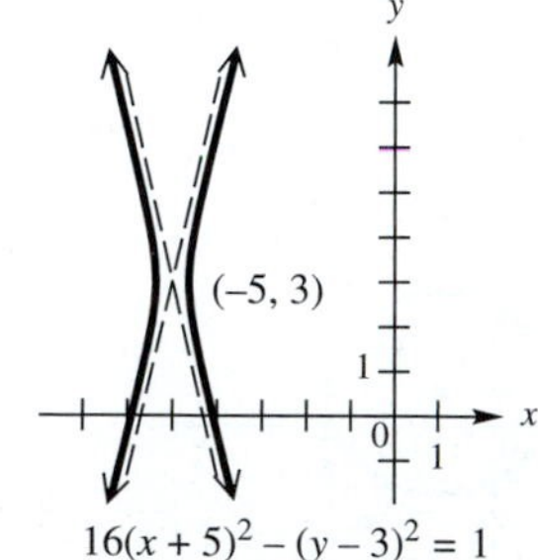

57. $\frac{x^2}{9} - \frac{y^2}{7} = 1$ **59.** $\frac{y^2}{9} - \frac{x^2}{25} = 1$ **61.** $\frac{y^2}{36} - \frac{x^2}{144} = 1$ **63.** $\frac{x^2}{9} - 3y^2 = 1$ **65.** $\frac{2y^2}{25} - 2x^2 = 1$

67. $\frac{(y-3)^2}{4} - \frac{49(x-4)^2}{4} = 1$ **69.** $\frac{(x-1)^2}{4} - \frac{(y+2)^2}{5} = 1$ **71.** $\frac{(x-1)^2}{4} - \frac{(y-1)^2}{4} = 1$; center: $(1, 1)$; vertices: $(-1, 1)$, $(3, 1)$ **73.** $\frac{(y+4)^2}{2} - \frac{(x-3)^2}{3} = 1$; center: $(3, -4)$; vertices: $(3, -4 - \sqrt{2})$, $(3, -4 + \sqrt{2})$

75. $\frac{(x-3)^2}{2} - \frac{(y-0)^2}{1} = 1$; center: $(3, 0)$; vertices: $(3 - \sqrt{2}, 0)$, $(3 + \sqrt{2}, 0)$ **77.** $\frac{(y+4)^2}{5} - \frac{(x+1)^2}{4} = 1$; center: $(-1, -4)$; vertices: $(-1, -4 - \sqrt{5})$, $(-1, -4 + \sqrt{5})$ **79.** $(-2, 0)$, $(2, 0)$

80.

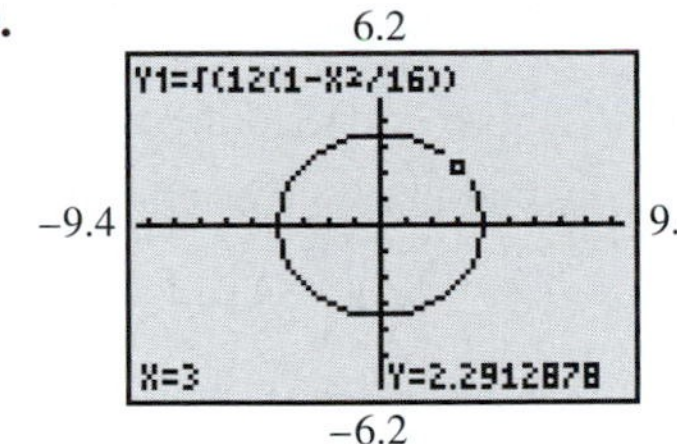

In addition to $(3, 2.2912878)$ shown on the screen, other points are $(0, 3.4641016)$ and $(-3, -2.291288)$. **81.** The points satisfy the equation.

82.

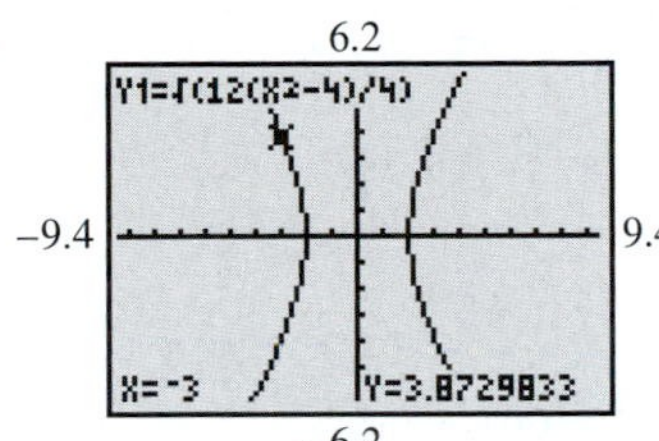

$(-4, 0)$, $(4, 0)$; In addition to the point shown on the screen, other points are $(-2, 0)$, $(2, 0)$, and $(4, 6)$. **83.** The points satisfy the equation. **84.** Exercise 81 demonstrates that the points on the graph satisfy the definition of the ellipse for that particular ellipse. Exercise 83 demonstrates similarly that the definition of the hyperbola is satisfied for that hyperbola.

85. $\frac{x^2}{100} + \frac{y^2}{64} = 1$ **87.** 348.2 ft **89.** just under 12 ft

91. (a)

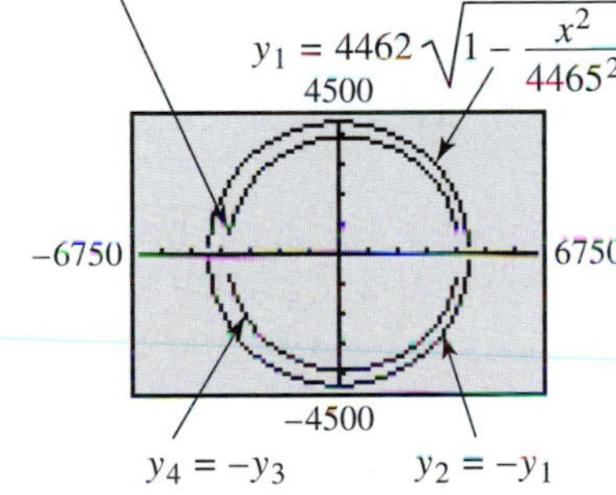

(b) maximum: approximately 669 mi; minimum: approximately 341 mi

93. (a) $x = \sqrt{y^2 + 2.5 \times 10^{-27}}$ **(b)** 1.2×10^{-13} m **97.** $x^2 - \frac{y^2}{3} = 1$

Reviewing Basic Concepts *(page 445)*

1. (a) B **(b)** D **(c)** A **(d)** C **2.**

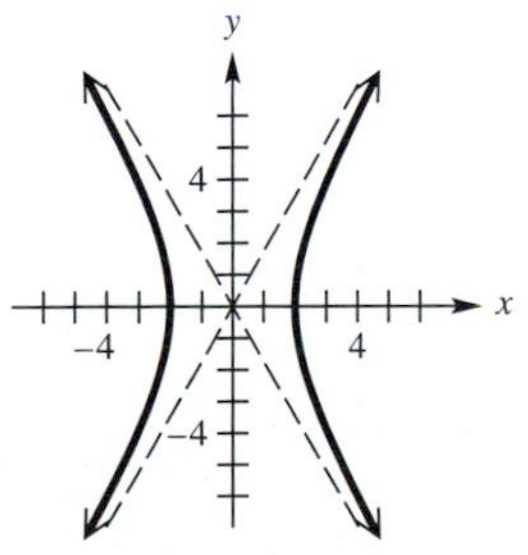

$12x^2 - 4y^2 = 48$

3.

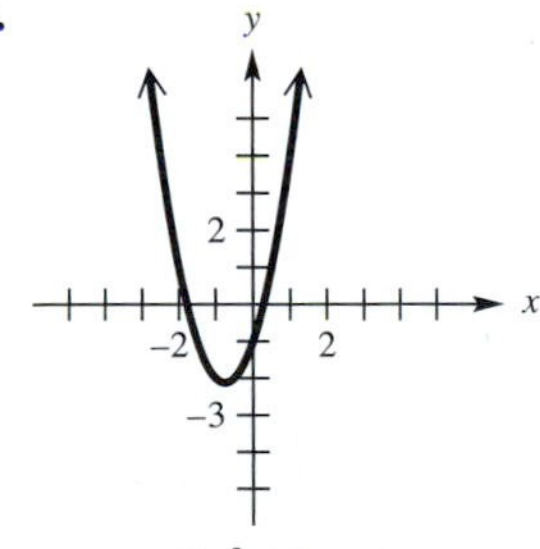

$y = 2x^2 + 3x - 1$

4.

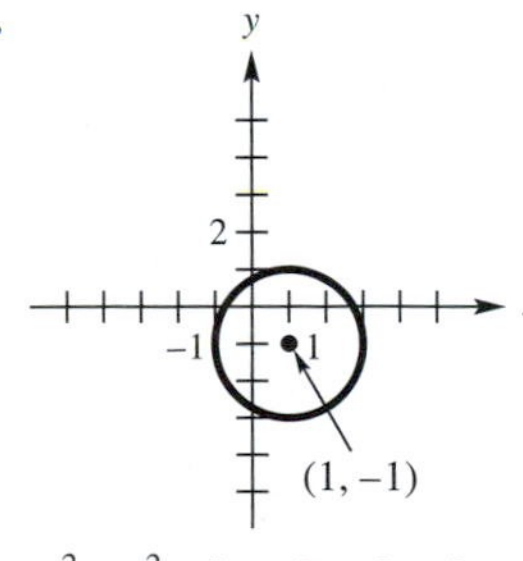

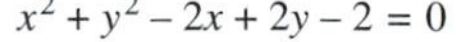

$x^2 + y^2 - 2x + 2y - 2 = 0$

5.

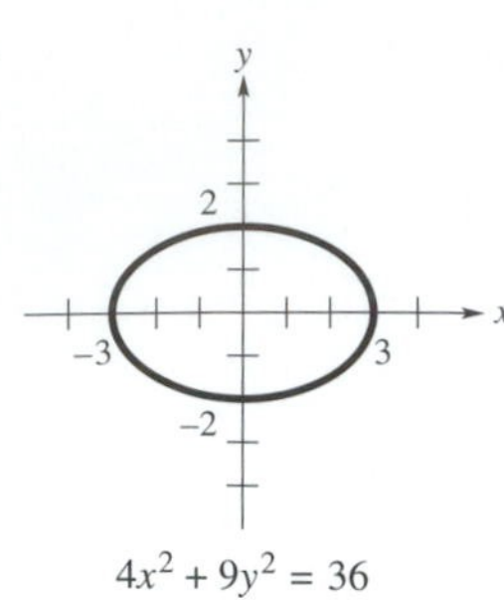

6. If $c < a$, it is an ellipse, and if $c > a$, it is a hyperbola. **7.** $(x - 2)^2 + (y + 1)^2 = 9$

8. $\frac{x^2}{36} + \frac{y^2}{20} = 1$ **9.** $\frac{y^2}{4} - \frac{x^2}{12} = 1$ **10.** $x^2 = 2y$

6.3 Exercises *(pages 452–453)*

1. circle **3.** parabola **5.** parabola **7.** ellipse **9.** hyperbola **11.** hyperbola **13.** ellipse **15.** no graph **17.** circle **19.** parabola **21.** point **23.** no graph **25.** circle **27.** parabola **29.** hyperbola **31.** ellipse **33.** no graph **35.** line **37.** ellipse **39.** hyperbola **41.** $\frac{1}{2}$ **43.** $\sqrt{2}$ **45.** $\frac{\sqrt{21}}{7}$ **47.** $\frac{\sqrt{10}}{3}$ **49.** $x^2 = 32y$ **51.** $\frac{x^2}{36} + \frac{y^2}{27} = 1$ **53.** $\frac{x^2}{36} - \frac{y^2}{108} = 1$ **55.** $x^2 = -4y$ **57.** $\frac{25x^2}{81} + \frac{y^2}{9} = 1$ **59.** C, A, B, D

61. **(a)** Neptune: $\frac{(x - .2709)^2}{30.1^2} + \frac{y^2}{30.1^2} = 1$; Pluto: $\frac{(x - 9.8106)^2}{39.4^2} + \frac{y^2}{38.16^2} = 1$ **(b)**

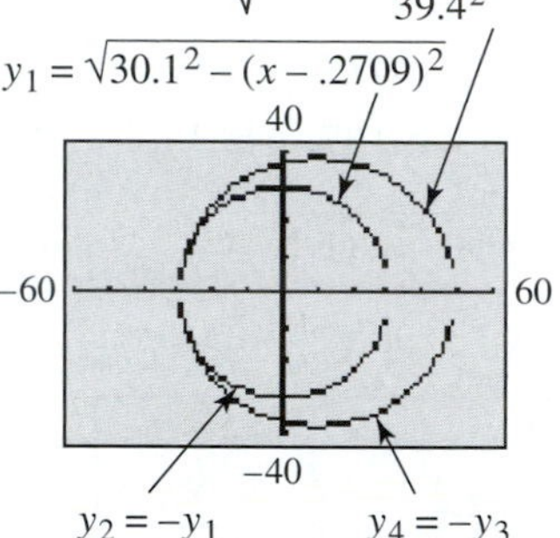

63. approximately 55 million mi

6.4 Exercises *(pages 457–458)*

1.

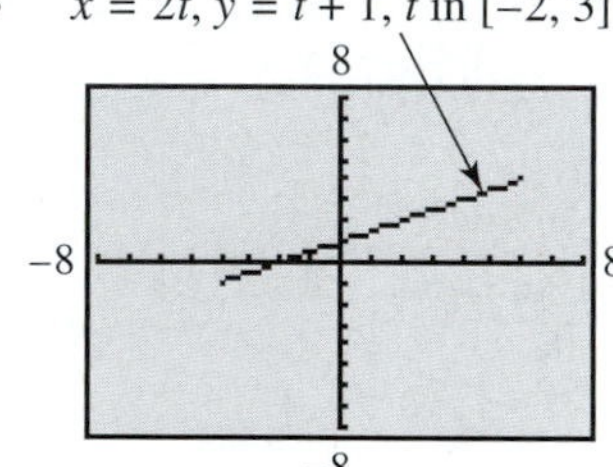

$y = \frac{1}{2}x + 1$, x in $[-4, 6]$

3.

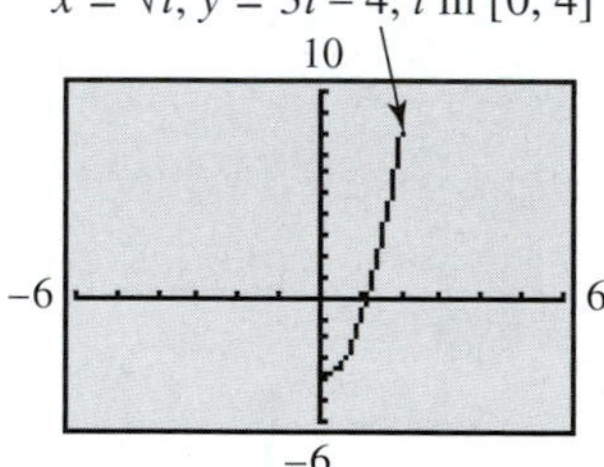

$y = 3x^2 - 4$, x in $[0, 2]$

5.

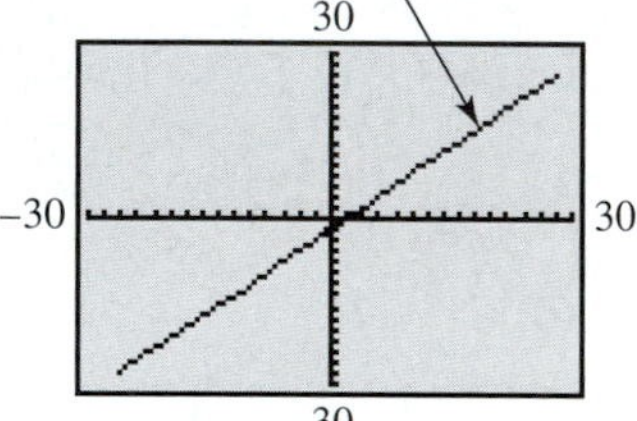

$y = x - 2$, x in $[-26, 28]$

7. $x = 2^t,\ y = \sqrt{3t-1},\ t \text{ in } \left[\frac{1}{3}, 4\right]$

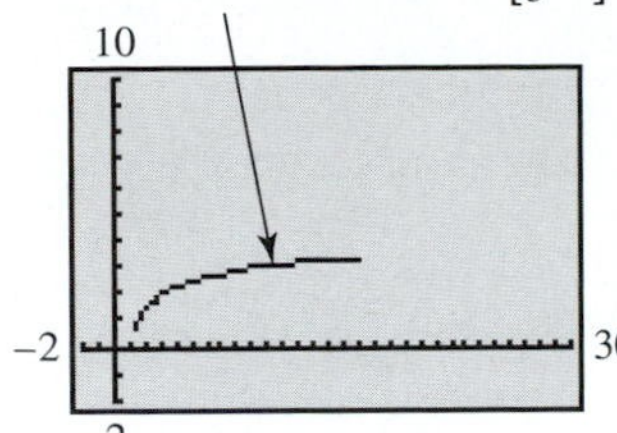

$x = 2^{(y^2+1)/3},\ y \text{ in } \left[0, \sqrt{11}\,\right]$

9. $x = t + 2,\ y = -\frac{1}{2}\sqrt{9 - t^2},$ $t \text{ in } [-3, 3]$

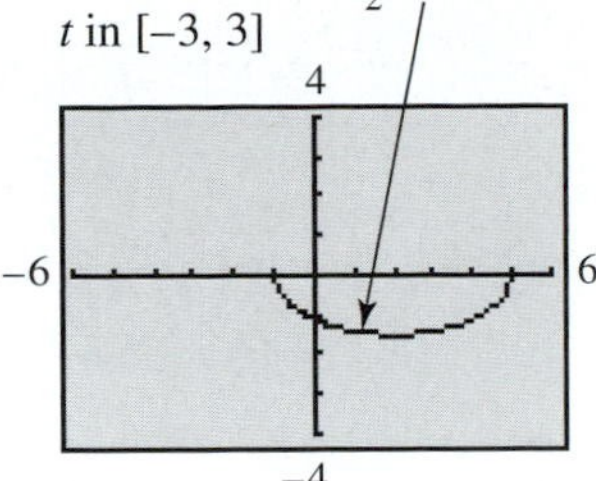

$y = -\dfrac{1}{2}\sqrt{9 - (x-2)^2},\ x \text{ in } [-1, 5]$

11. $x = t,\ y = \frac{1}{t},\ t \text{ in } (-\infty, 0) \cup (0, \infty)$

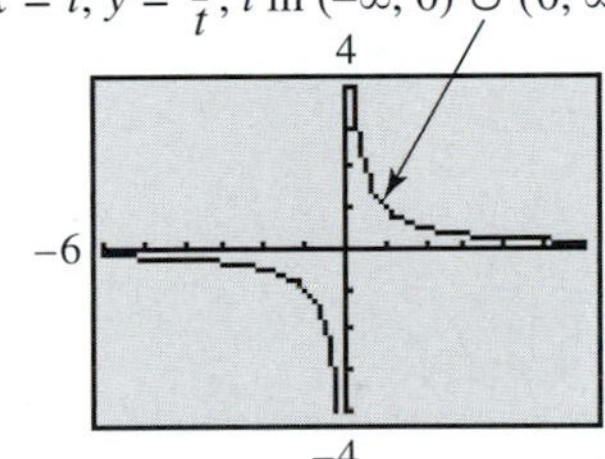

$y = \dfrac{1}{x},\ x \text{ in } (-\infty, 0) \cup (0, \infty)$

13. $y = \dfrac{1}{3}x - 1,\ x \text{ in } (-\infty, \infty)$ **15.** $x = 3(y-1)^2,\ y \text{ in } (-\infty, \infty)$

17. $x = 3\left(\dfrac{y}{4}\right)^{2/3},\ y \text{ in } (-\infty, \infty)$ **19.** $y = \sqrt{x^2 + 2},\ x \text{ in } (-\infty, \infty)$

21. $y = \dfrac{1}{x},\ x \text{ in } (0, \infty)$ **23.** $y = 1 - 2x^2,\ x \text{ in } (0, \infty)$ **25.** $y = \dfrac{1}{x},\ x \neq 0$

27. $y = \ln x,\ x \text{ in } (0, \infty)$

Other answers are possible for Exercises 29–35.

29. $x = \dfrac{1}{2}t,\ y = t + 3;\ x = \dfrac{t+3}{2},\ y = t + 6$ **31.** $x = \dfrac{1}{3}t,\ y = \sqrt{t+2},\ t \text{ in } [-2, \infty);\ x = \dfrac{t-2}{3},\ y = \sqrt{t},\ t \text{ in } [0, \infty)$

33. $x = t^3 + 1,\ y = t;\ x = t,\ y = \sqrt[3]{t-1}$ **35.** $x = \sqrt{t+1},\ y = t,\ t \text{ in } [-1, \infty);\ x = \sqrt{t},\ y = t - 1,\ t \text{ in } [0, \infty)$

37. **(a)** 17.7 sec **(b)** 5000 ft **(c)** 1250 ft **39.**

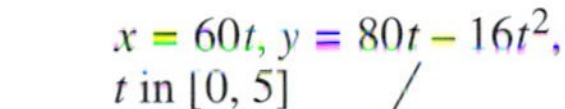

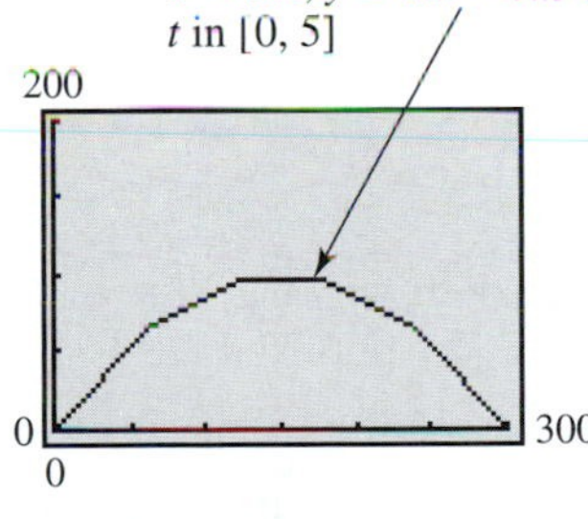

$y = \dfrac{4}{3}x - \dfrac{1}{225}x^2$

43. Many answers are possible, two of which are $x = t,\ y - y_1 = m(t - x_1)$ and $t = x - x_1,\ y = mt + y_1$.

Reviewing Basic Concepts *(page 458)*

1. $\dfrac{(x-1)^2}{4} + \dfrac{(y+3)^2}{12} = 1$; ellipse **2.** $\dfrac{(y+4)^2}{12} - \dfrac{(x+2)^2}{6} = 1$; hyperbola **3.** $(y+2)^2 = -\dfrac{5}{3}(x-3)$; parabola

4. $\dfrac{2\sqrt{6}}{5}$ **5.** $\sqrt{3}$ **6.** $y^2 = -8x$ **7.** $\dfrac{x^2}{25} + \dfrac{y^2}{16} = 1$ **8.** $\dfrac{y^2}{16} - \dfrac{x^2}{9} = 1$ **9.** $\dfrac{9\sqrt{21}}{5} \approx 8.25$ ft

10. (a)

$x = 2t$
$y = \sqrt{t^2 + 1}$
t in $(-\infty, \infty)$

(b) $y = \sqrt{\frac{x^2}{4} + 1}$

Chapter 6 Review Exercises *(pages 461–463)*

1. $(x + 2)^2 + (y - 3)^2 = 25$; domain: $[-7, 3]$; range: $[-2, 8]$

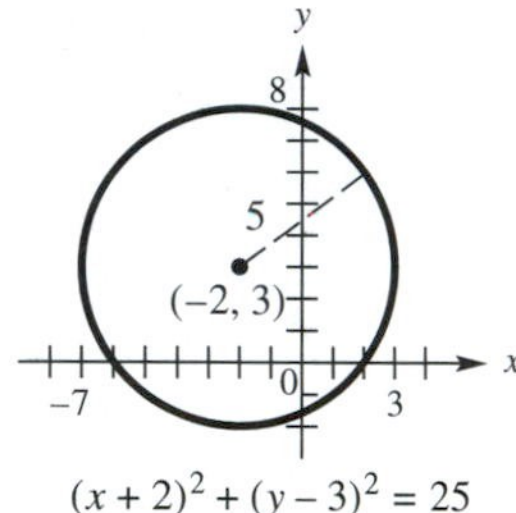

$(x + 2)^2 + (y - 3)^2 = 25$

3. $(x + 8)^2 + (y - 1)^2 = 289$; domain: $[-25, 9]$; range: $[-16, 18]$

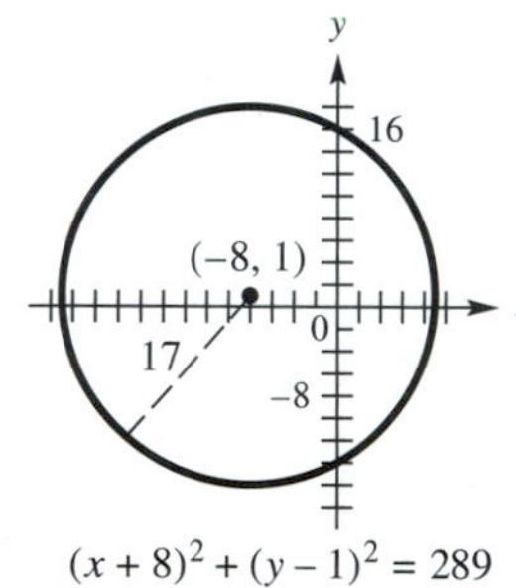

$(x + 8)^2 + (y - 1)^2 = 289$

5. $(2, -3)$; 1 **7.** $\left(-\frac{7}{2}, -\frac{3}{2}\right)$; $\frac{3\sqrt{6}}{2}$

9. The graph consists of the single point $(4, 5)$.

11.

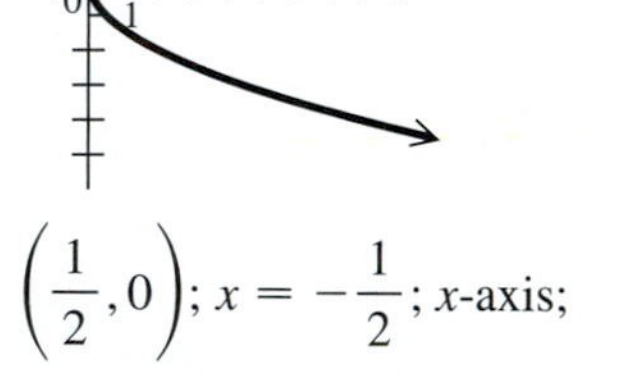

$\left(\frac{1}{2}, 0\right)$; $x = -\frac{1}{2}$; x-axis; domain: $[0, \infty)$; range: $(-\infty, \infty)$

13.

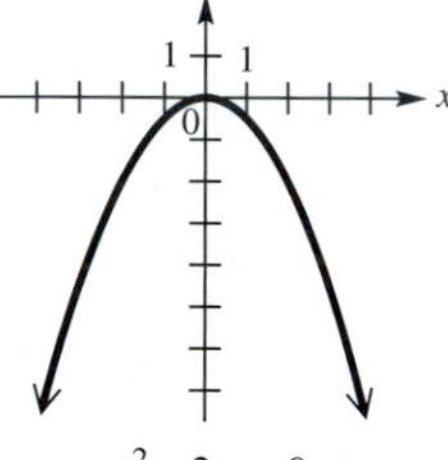

$x^2 + 2y = 0$

$\left(0, -\frac{1}{2}\right)$; $y = \frac{1}{2}$; y-axis; domain: $(-\infty, \infty)$; range: $(-\infty, 0]$

15. $y^2 = \frac{25}{2}x$ **17.** $(y - 6)^2 = 28(x + 5)$

19.

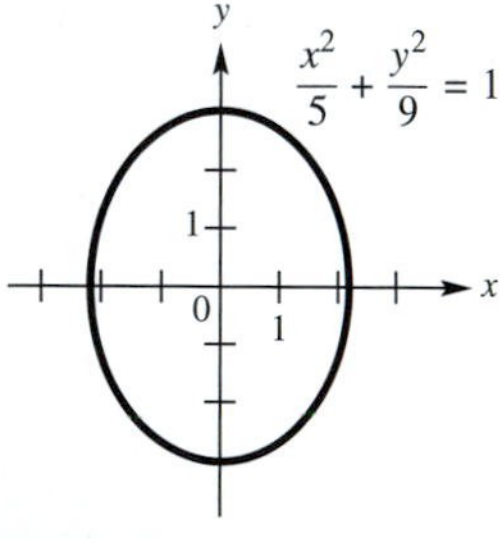

$[-\sqrt{5}, \sqrt{5}]$; $[-3, 3]$; $(0, -3)$, $(0, 3)$

21.

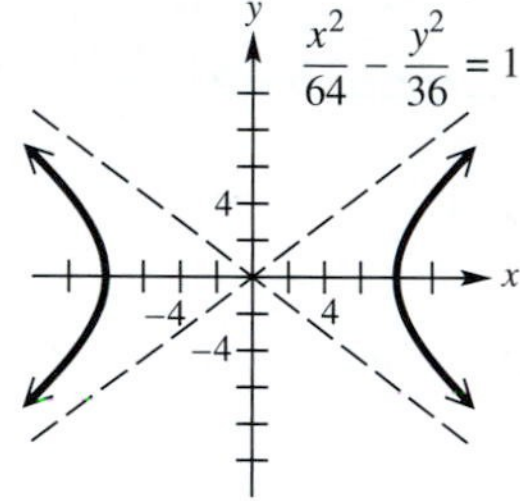

$(-\infty, -8] \cup [8, \infty)$; $(-\infty, \infty)$; $(-8, 0)$, $(8, 0)$

23.

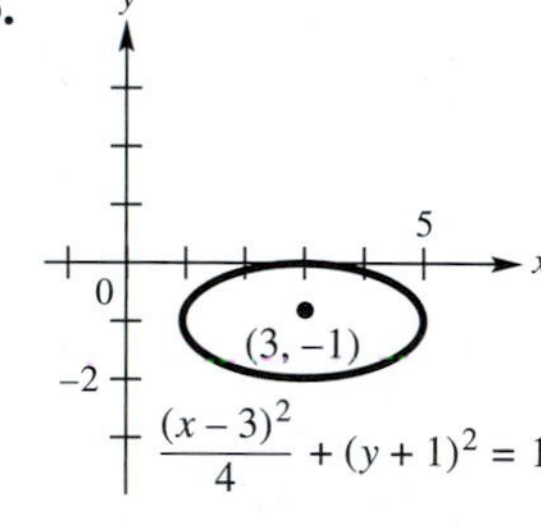

$[1, 5]$; $[-2, 0]$; $(1, -1)$, $(5, -1)$

25.

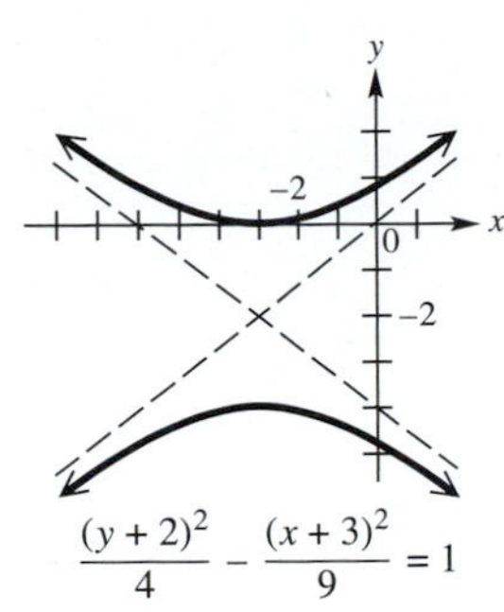

$(-\infty, \infty)$; $(-\infty, -4] \cup [0, \infty)$; $(-3, 0)$, $(-3, -4)$

27. $\frac{x^2}{12} + \frac{y^2}{16} = 1$ **29.** $\frac{y^2}{16} - \frac{x^2}{9} = 1$ **31.** $\frac{4x^2}{45} + \frac{4y^2}{81} = 1$

33. **(a)** $(-1, -3)$ **(b)** 5 **(c)** $y_1 = -3 + \sqrt{25 - (x+1)^2}$; $y_2 = -3 - \sqrt{25 - (x+1)^2}$ **35.** E **37.** C **39.** F

41. $\frac{(x+2)^2}{2} + \frac{(y-2)^2}{5} = 1$; center: $(-2, 2)$; vertices: $(-2, 2 - \sqrt{5})$, $(-2, 2 + \sqrt{5})$ **43.** $\frac{(y+1)^2}{3} - \frac{(x-1)^2}{4} = 1$; center: $(1, -1)$; vertices: $(1, -1 - \sqrt{3})$, $(1, -1 + \sqrt{3})$ **45.** $\frac{\sqrt{5}}{3}$ **47.** $(y-2)^2 = 3(x+3)$ **49.** $\frac{(x-2)^2}{16} + \frac{y^2}{12} = 1$

51.

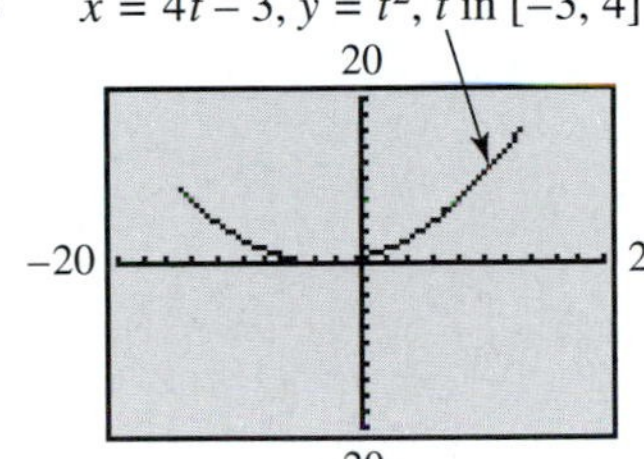

53.

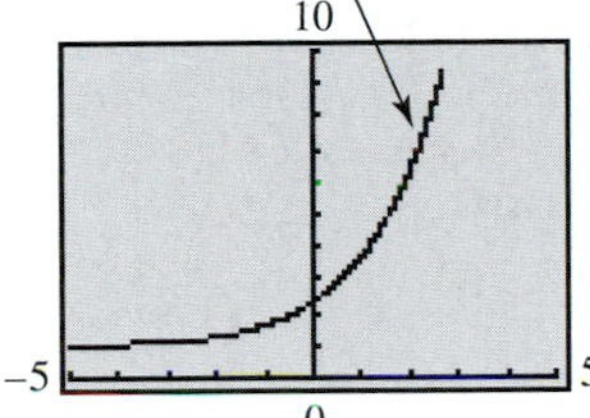

55. $y^2 - x^2 = 1$, x in $[0, \infty)$ **57.** 66.8 and 67.7 million mi **59.** elliptic **61.** The required increase in velocity is less when D is larger.

Chapter 6 Test *(page 463)*

1. **(a)** B **(b)** A **(c)** D **(d)** E **(e)** F **(f)** C **2.** $\left(\frac{1}{32}, 0\right)$; $x = -\frac{1}{32}$ **3.**

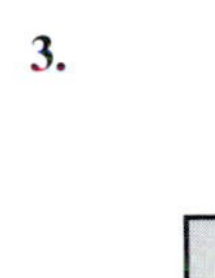

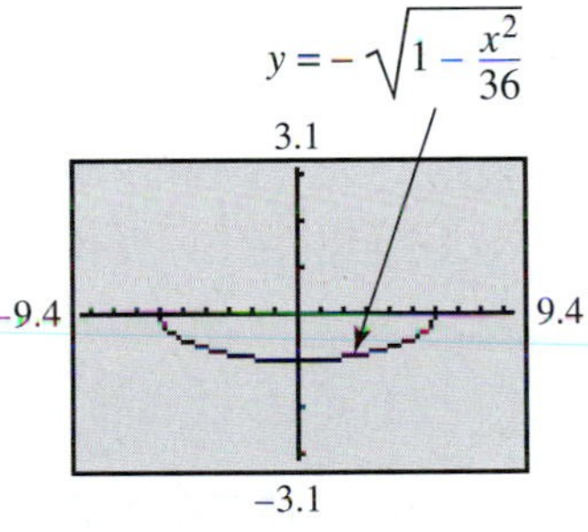

yes; domain: $[-6, 6]$; range: $[-1, 0]$

4. $y_1 = 7\sqrt{\frac{x^2}{25} - 1}$; $y_2 = -7\sqrt{\frac{x^2}{25} - 1}$ **5.**

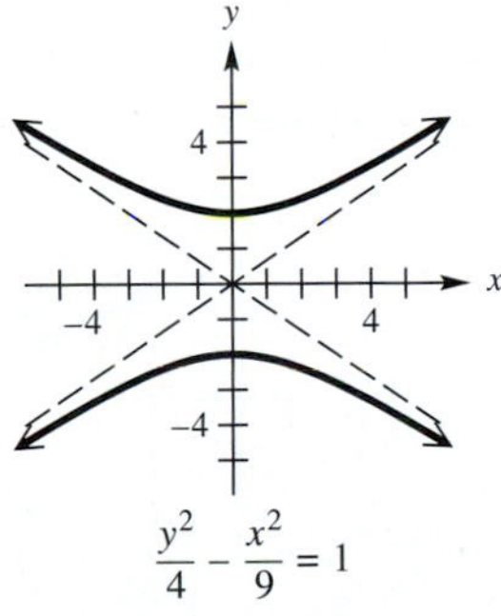

hyperbola; center: $(0, 0)$; vertices: $(0, -2)$, $(0, 2)$; foci: $(0, -\sqrt{13})$, $(0, \sqrt{13})$

6.

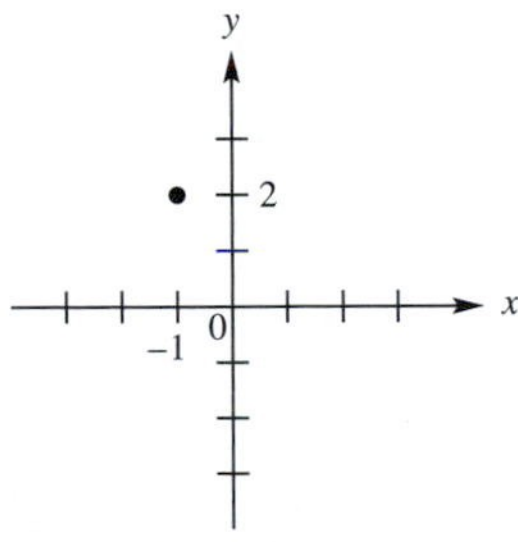

The graph is the point $(-1, 2)$.

7.

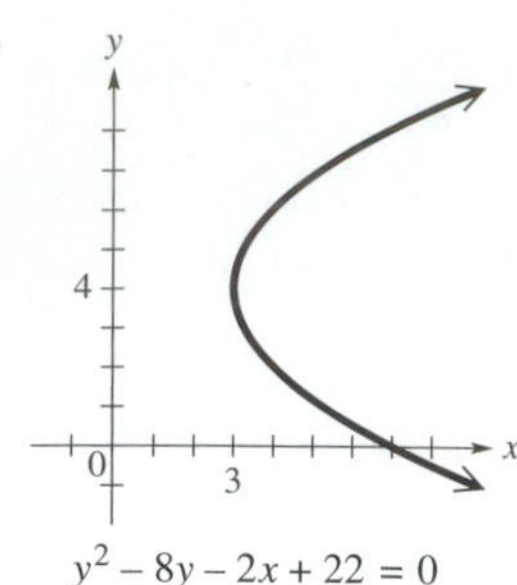

parabola; vertex: $(3, 4)$;
focus: $(3.5, 4)$

8.

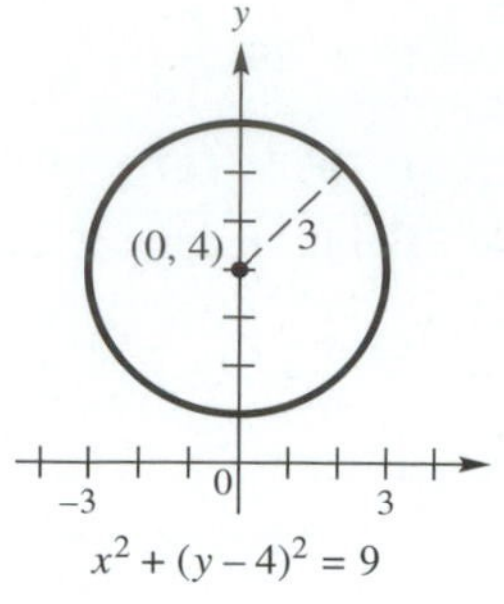

circle; center: $(0, 4)$;
radius: 3

9.

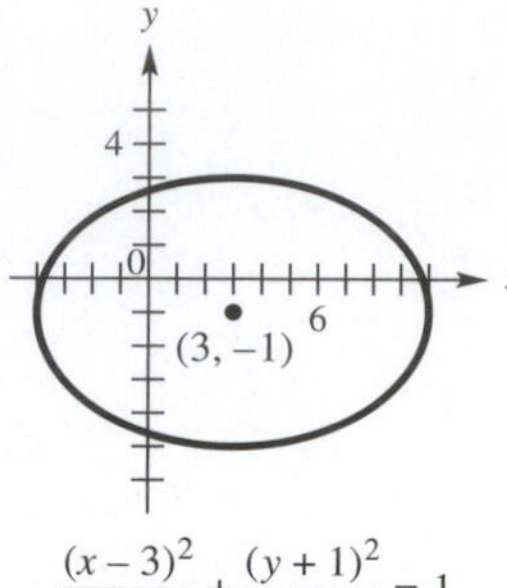

ellipse; center: $(3, -1)$;
vertices: $(-4, -1)$, $(10, -1)$;
foci: $\left(3 + \sqrt{33}, -1\right)$, $\left(3 - \sqrt{33}, -1\right)$

10.

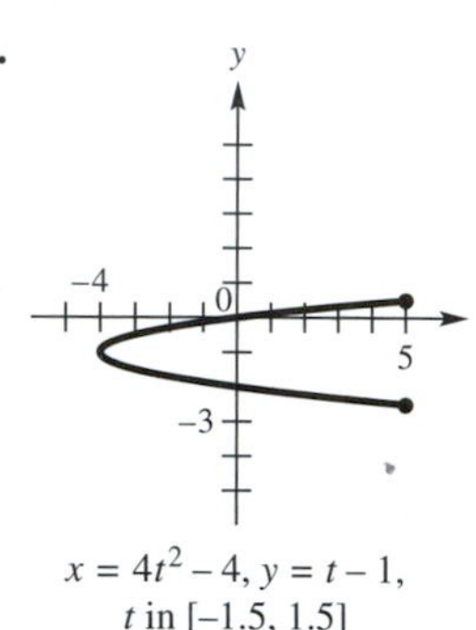

parabola; vertex: $(-4, -1)$;
focus: $\left(-\frac{63}{16}, -1\right)$

11. **(a)** $y = -\frac{1}{8}x^2$ **(b)** $\frac{x^2}{\frac{11}{4}} + \frac{y^2}{9} = 1$ or $\frac{4x^2}{11} + \frac{y^2}{9} = 1$

12. $\frac{x^2}{400} + \frac{y^2}{144} = 1$; approximately 10.39 ft

CHAPTER 7 SYSTEMS OF EQUATIONS AND INEQUALITIES; MATRICES

7.1 Exercises *(pages 475–479)*

1. approximately 2002 **3.** (2002, 3.1 million) **5.** the year; the number of migrants **7.** $\{(2, 2)\}$ **9.** $\left\{\left(\frac{1}{2}, -2\right)\right\}$ **11.** $\{(1, 1)\}$ **13.** $\{(3, -2)\}$ **15.** $\{(6, 15)\}$ **17.** $\{(2, -3)\}$ **19.** $\{(1, 3)\}$ **21.** $\{(-2, 3)\}$ **23.** $\{(5, 0)\}$ **25.** $\emptyset$ **27.** A **29.** $\{(0, 4)\}$ **31.** $\{(-1, 1)\}$ **33.** $\{(2, 3)\}$ **35.** $\left\{\left(\frac{y+9}{4}, y\right)\right\}$ or $\{(x, 4x - 9)\}$ **37.** $\emptyset$ **39.** $\emptyset$ **41.** $\{(12, 6)\}$ **43.** $\{(5, 2)\}$ **45.** $\{(.138, -4.762)\}$ **47.** $\{(-8.708, 15.668)\}$ **49.** An inconsistent system will conclude with no variables and a false statement, such as $0 = 1$. A system with dependent equations will conclude with no variables and a true statement, such as $0 = 0$. **51.**

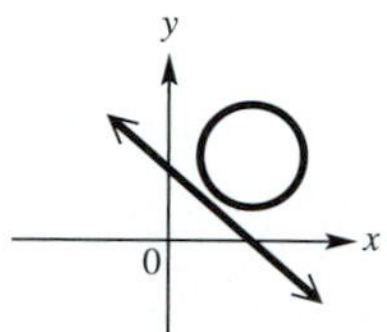

53.

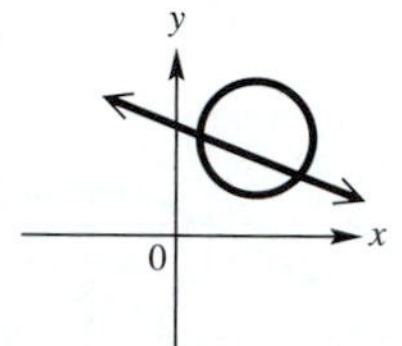

55.

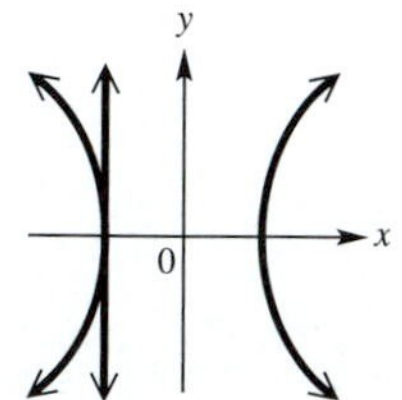

57. **59.**

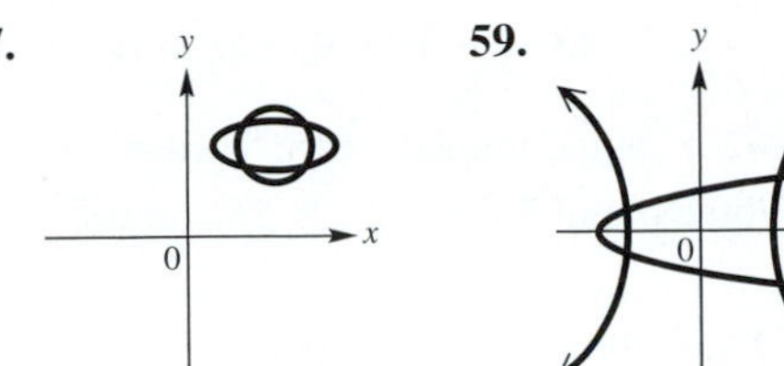

61. $\{(-2,-2),(1,1)\}$ **63.** $\left\{\left(-\frac{3}{5},\frac{7}{5}\right),(-1,1)\right\}$

65. $\{(3,1),(3,-1),(-3,1),(-3,-1)\}$ **67.** $\{(3,0),(-3,0)\}$

69. $\left\{\left(x,\pm\sqrt{10-x^2}\right)\right\}$

71. $\left\{\left(\frac{1+\sqrt{13}}{2},\frac{-1+\sqrt{13}}{2}\right),\left(\frac{-1-\sqrt{21}}{2},\frac{3+\sqrt{21}}{2}\right)\right\}$

73. $\{(-.79,.62),(.88,.77)\}$ **75.** $\{(.06,2.88)\}$

77. **(a)** Let x represent the number of robberies in 2000 and y the number in 2001.

$$x+y=831{,}000$$
$$x-y=15{,}000$$

(b) $\{(423{,}000, 408{,}000)\}$ **(c)** In 2000 there were 423,000 robberies, and in 2001 there were 408,000 robberies.

79. 10 hr in 2001, 11.3 hr in 2002 **81.** 14 by 7 by 6 in. or 5.17 by 10.34 by 11.00 in. **83.** $x\approx 177.1$, $y\approx 174.9$; If an athlete's maximum heart rate is 180 beats per minute (bpm), then it will be about 177 bpm after 5 sec and 175 bpm after 10 sec.

85. **(a)** $\{(5.8, 857.8)\}$ **(b)** During 1995, 857.8 million lb of canned tuna and fresh shrimp were available.

(c)

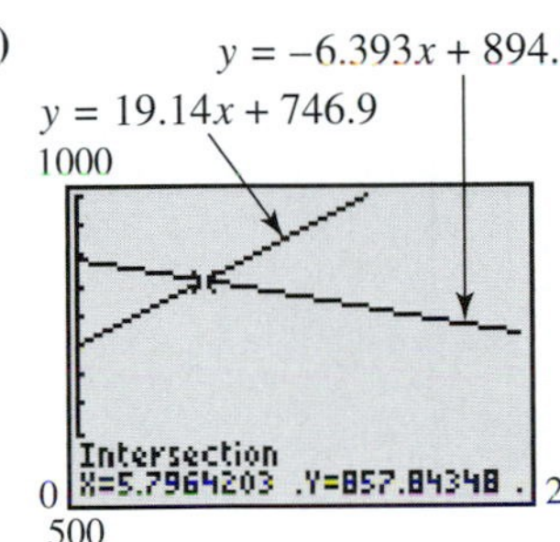

87. $r\approx 1.538$ in., $h\approx 6.724$ in. **89.** $W_1=\frac{300}{1+\sqrt{3}}\approx 109.8$ lb; $W_2=\frac{300\sqrt{3}}{\sqrt{6}+\sqrt{2}}\approx 134.5$ lb; $\{(109.8, 134.5)\}$ **91.** $x=300$; $y=350$

93. $x=65$; $R=C=\$385$ **95.** price: \$315; supply/demand: 420

97. $5t+15u=16$
$5t+\ \ 4u=5$ **98.** $t=\frac{1}{5}$; $u=1$ **99.** $x=5$; $y=1$ **100.** $y=\frac{-15x}{5-16x}$

101. $y=\frac{-4x}{5-5x}$ **102.**

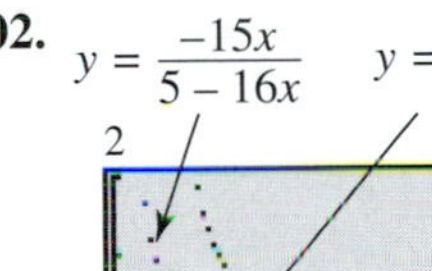

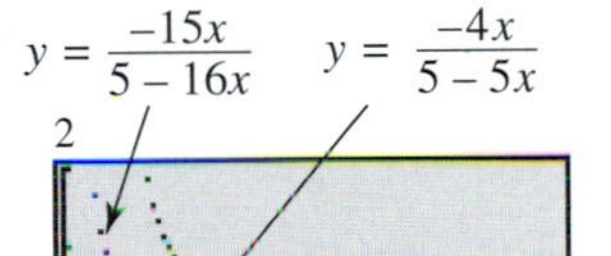

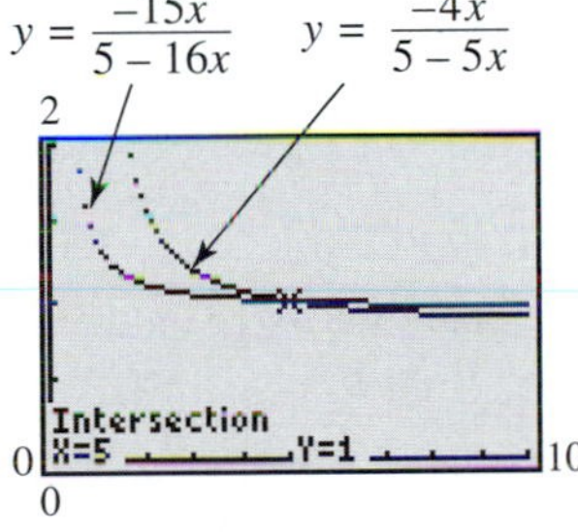

Dot mode

103. $\{(2,2)\}$ **105.**

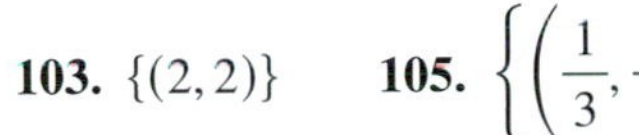
$\left\{\left(\frac{1}{3},\frac{1}{4}\right)\right\}$

7.2 Exercises *(pages 485–487)*

1. **(a)** Multiply equation (2) by 2 and add to get $7x-z=0$. **(b)** Multiply equation (1) by -2 and add to get $-3y+2z=2$. **(c)** Multiply equation (2) by 4 and add to get $14x-3y=2$. **3.** The first equation yields $-6=-6$, the second $-\frac{1}{12}=-\frac{1}{12}$, and the third $1=1$. **5.** $\{(1,2,-1)\}$ **7.** $\{(2,0,3)\}$ **9.** $\emptyset$ **11.** $\{(1,2,3)\}$ **13.** $\{(-1,2,1)\}$ **15.** $\{(4,1,2)\}$ **17.** $\left\{\left(\frac{1}{2},\frac{2}{3},-1\right)\right\}$ **19.** $\{(2,2,2)\}$ **21.** $\{(1,0,6)\}$ **23.** $\left\{\left(\frac{1}{2},\frac{1}{3},\frac{1}{4}\right)\right\}$ **25.** $\{(2,4,2)\}$ **27.** $\left\{\left(-1,1,\frac{1}{3}\right)\right\}$ **29.** $\emptyset$ **31.** $\emptyset$ **33.** $\{(x,1-x,x+2)\}$ **35.** For example, three pages of this book intersect at the spine.

37. $\{(47 - 9z, 7z - 32, z)\}$ **39.** $\{(3z - 24, z - 13, z)\}$ **41.** $\left\{\left(\frac{41 - z}{7}, \frac{2z + 9}{7}, z\right)\right\}$ **43.** $\{(-3z + 6, -4z + 6, z)\}$

45. \$9.00 water: 120 gal; \$3.00 water: 60 gal; \$4.50 water: 120 gal **47.** "up close": \$24; "in the middle": \$18; "farther back": \$15 **49.** 75°, 65°, 40° **51.** \$3000 at 4%; \$6000 at 4.5%; \$1000 at 2.5% **53.** three possibilities: 12 EZ, 16 compact, 0 commercial; 10 EZ, 8 compact, 3 commercial; 8 EZ, 0 compact, 6 commercial

55. **(a)** $\begin{aligned} a + 20b + 2c &= 190 \\ a + 5b + 3c &= 320 \\ a + 40b + c &= 50 \end{aligned}$ **(b)** $a = 30$, $b = -2$, and $c = 100$; that is, $P = 30 - 2A + 100A$ **(c)** \$260,000

57. $y = 3x^2 + x - 2$

7.6525

Y1=3X²+X-2

−1.8 1.8

−3.4025

59. $Y_1 = .47X^2 - .23X + .56$ **61.** $x^2 + y^2 - 6x - 4y - 12 = 0$

63. $x^2 + y^2 - 1.4x + 5.4y - 25.2 = 0$ **65.** $s(t) = 4t^2 + 12t - 10$; $s(10) = 510$

7.3 Exercises *(pages 496–499)*

1. $\begin{bmatrix} -4 & -8 \\ 4 & 7 \end{bmatrix}$ **3.** $\begin{bmatrix} 1 & 5 & 6 \\ -1 & 8 & 5 \\ 4 & 7 & 0 \end{bmatrix}$ **5.** $\begin{bmatrix} -3 & 1 & -4 \\ 2 & 1 & 3 \\ -17 & 0 & -13 \end{bmatrix}$ **7.** $\left[\begin{array}{cc|c} 2 & 3 & 11 \\ 1 & 2 & 8 \end{array}\right]$ **9.** $\left[\begin{array}{cc|c} 1 & 5 & 6 \\ 1 & 0 & 3 \end{array}\right]$

11. $\left[\begin{array}{ccc|c} 2 & 1 & 1 & 3 \\ 3 & -4 & 2 & -7 \\ 1 & 1 & 1 & 2 \end{array}\right]$ **13.** $\left[\begin{array}{ccc|c} 1 & 1 & 0 & 2 \\ 0 & 2 & 1 & -4 \\ 0 & 0 & 1 & 2 \end{array}\right]$ **15.** $\begin{aligned} 2x + y &= 1 \\ 3x - 2y &= -9 \end{aligned}$ **17.** $\begin{aligned} x &= 2 \\ y &= 3 \\ z &= -2 \end{aligned}$ **19.** $\begin{aligned} 3x + 2y + z &= 1 \\ 2y + 4z &= 22 \\ -x - 2y + 3z &= 15 \end{aligned}$

21. $\{(5, -1)\}$ **23.** $\{(2, 0)\}$ **25.** $\{(2, 3, 1)\}$ **27.** $\{(3 - 3z, 1 + 2z, z)\}$ **29.** $\emptyset$ **31.** $\{(2, 3)\}$ **33.** $\{(-3, 0)\}$

35. $\left\{\left(\frac{7}{2}, -1\right)\right\}$ **37.** $\emptyset$ **39.** $\left\{\left(x, \frac{6x - 1}{3}\right)\right\}$ or $\left\{\left(\frac{3y + 1}{6}, y\right)\right\}$ **41.** $\{(-2, 1, 3)\}$ **43.** $\{(-1, 23, 16)\}$

45. $\{(3, 2, -4)\}$ **47.** $\{(2, 1, -1)\}$ **49.** $\{(5.211, 3.739, -4.655)\}$ **51.** $\{(-.250, 1.284, -.059)\}$ **53.** In both cases, we simply write the coefficients and do not write the variables. This is possible because we agree on the order in which the variables appear (descending degree). **55.** $\left\{\left(\frac{5z + 14}{5}, \frac{5z - 12}{5}, z\right)\right\}$ **57.** $\left\{\left(\frac{12 - z}{7}, \frac{4z - 6}{7}, z\right)\right\}$ **59.** $\emptyset$

61. $\{(0, 2, -2, 1)\}$ **63.** **(a)** $F = .5714N + .4571R - 2014$ **(b)** \$5700 **65.** model A: 5 bicycles; model B: 8 bicycles

67. \$10,000 at 8%; \$7000 at 10%; \$8000 at 9%

69. **(a)** $\begin{aligned} 1990^2a + 1990b + c &= 11 \\ 2010^2a + 2010b + c &= 10 \\ 2030^2a + 2030b + c &= 6 \end{aligned}$ **(b)** $f(x) = -.00375x^2 + 14.95x - 14{,}889.125$ **(c)**

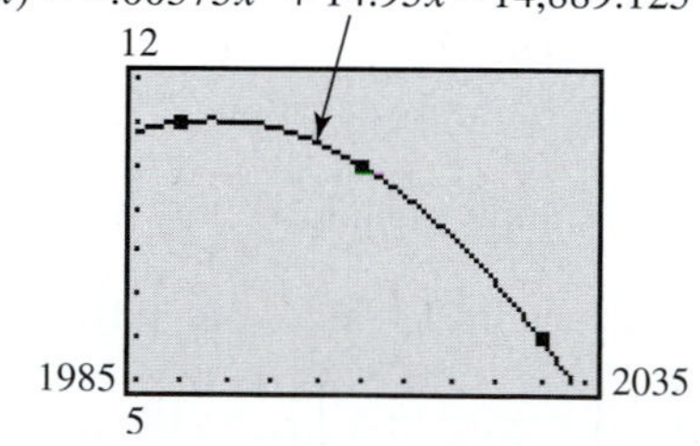

(d) Answers will vary. For example, in 2015 the predicted ratio is $f(2015) \approx 9.3$.

71. **(a)** At intersection A, incoming traffic is equal to $x + 5$. The outgoing traffic is given by $y + 7$. Therefore, $x + 5 = y + 7$. The other equations can be justified in a similar way.

(b) The three equations can be written as

$$x - y = 2$$
$$x - z = 3$$
$$y - z = 1.$$

The solution set can be written as $\{(z + 3, z + 1, z), \text{ where } z \geq 0\}$. **(c)** There are infinitely many solutions since some cars could be driving around the block continually.

73. **(a)** $a + 871b + 11.5c + 3d = 239$
$a + 847b + 12.2c + 2d = 234$
$a + 685b + 10.6c + 5d = 192$
$a + 969b + 14.2c + d = 343$
(b) $\left[\begin{array}{cccc|c} 1 & 871 & 11.5 & 3 & 239 \\ 1 & 847 & 12.2 & 2 & 234 \\ 1 & 685 & 10.6 & 5 & 192 \\ 1 & 969 & 14.2 & 1 & 343 \end{array}\right]$ The values are $a \approx -715.457$, $b \approx .34756$, $c \approx 48.6585$, and $d \approx 30.71951$.
(c) $F = -715.457 + .34756A + 48.6585P + 30.71951W$ **(d)** approximately 323, which is very close to 320

75. $a \approx 11.92$ lb, $b \approx 14.23$ lb, $c \approx 23.85$ lb

Reviewing Basic Concepts *(pages 499–500)*

1. $\{(3, -4)\}$ **2.** $\{(-2, 0)\}$

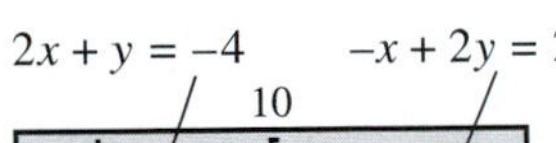

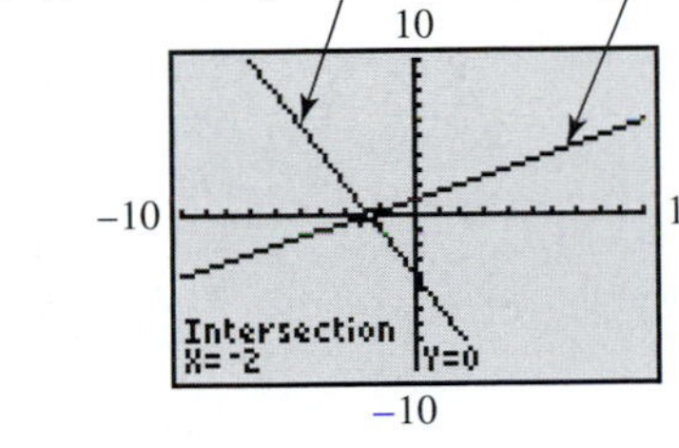

3. 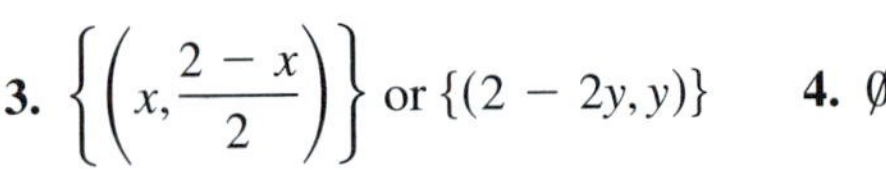 $\left\{\left(x, \frac{2 - x}{2}\right)\right\}$ or $\{(2 - 2y, y)\}$ **4.** $\emptyset$

5. $\{(2, -1), (-6.5, 24.5)\}$

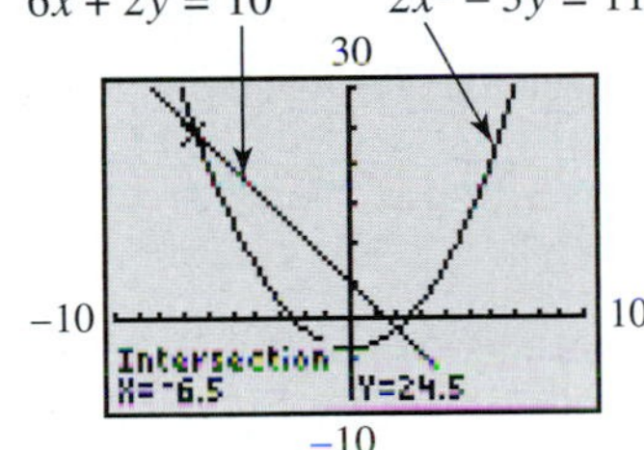

The other point of intersection is (2, −1).

6. $\{(-2, 1, 2)\}$ **7.** $\{(2, 1, -1)\}$ **8.** $\{(3, 2, 1)\}$ **9.** about 11.03 million with stereo sound; about 21 million without stereo sound **10.** \$1000 at 8%; \$1500 at 11%; \$2500 at 14%

7.4 Exercises *(pages 509–513)*

1. 2×2; square **3.** 3×4 **5.** 2×1; column **7.** 1×1; square, row, column **9.** $w = 3; x = 2; y = -1; z = 4$

11. $w = 2; x = 6; y = -2; z = 8$ **13.** $z = 18; r = 3; s = 3; p = 3; a = \frac{3}{4}$ **15.** The two matrices must have the same dimension. To find the sum, add the corresponding entries. The sum will be a matrix with the same dimension.

17. $\begin{bmatrix} -2 & -7 & 7 \\ 10 & -2 & 7 \end{bmatrix}$ **19.** $\begin{bmatrix} -6 & 8 \\ 4 & 2 \end{bmatrix}$ **21.** $\begin{bmatrix} 5 & 5 \\ 12 & 0 \end{bmatrix}$ **23.** cannot be added **25.** $\begin{bmatrix} 13 & 3 & 0 & -2 \\ 9 & -12 & 4 & 8 \\ 12 & -11 & -1 & 9 \end{bmatrix}$

27. $\begin{bmatrix} 0 & 2 \\ 13 & -5 \\ 0 & 1 \end{bmatrix}$ **29.** $\begin{bmatrix} 7 & 4 & 7 \\ 4 & 7 & 7 \\ 4 & 7 & 7 \end{bmatrix}$ **31.** $\begin{bmatrix} 8 & -43 & -18 \\ 26 & 29 & 6 \\ -2 & 10 & 43 \end{bmatrix}$ **33.** $\begin{bmatrix} -4 & 8 \\ 0 & 6 \end{bmatrix}$ **35.** $\begin{bmatrix} 2 & 6 \\ -4 & 6 \end{bmatrix}$ **37.** $\begin{bmatrix} -13 & 21 \\ 2 & 15 \end{bmatrix}$

39. $\begin{bmatrix} 2 & 6 & 5 \\ -4 & -7 & 9 \end{bmatrix}$ **41.** AB: 4×4; BA: 2×2 **43.** AB: 3×2; BA: not defined **45.** Neither AB nor BA is defined.

47. columns; rows **49.** $\begin{bmatrix} pa + qb & pc + qd \\ ra + sb & rc + sd \end{bmatrix}$ **51.** $\begin{bmatrix} -17 \\ -1 \end{bmatrix}$ **53.** $\begin{bmatrix} 17 & -10 \\ 1 & 2 \end{bmatrix}$ **55.** $\begin{bmatrix} -2 & 10 \\ 0 & 8 \end{bmatrix}$

57. $\begin{bmatrix} -15 & -16 & 3 \\ -1 & 0 & 9 \\ 7 & 6 & 12 \end{bmatrix}$ **59.** $[2 \quad 7 \quad -4]$ **61.** $\begin{bmatrix} 23 & -9 \\ -6 & -2 \\ 33 & 1 \end{bmatrix}$ **63.** $\begin{bmatrix} -25 & 23 & 11 \\ 0 & -6 & -12 \\ -15 & 33 & 45 \end{bmatrix}$ **65.** cannot be multiplied

67. $\begin{bmatrix} 10 & -10 \\ 15 & -5 \end{bmatrix}$ **69.** no **71. (a)** $\begin{bmatrix} 50 & 100 & 30 \\ 10 & 90 & 50 \\ 60 & 120 & 40 \end{bmatrix}$ **(b)** $\begin{bmatrix} 12 \\ 10 \\ 15 \end{bmatrix}$ **(c)** $\begin{bmatrix} 2050 \\ 1770 \\ 2520 \end{bmatrix}$ **(d)** \$6340

73. (a) $d_{n+1} = -.05m_n + 1.05d_n$; 1.05 **(b)** after 1 yr: 3020 mountain lions, 515,000 deer; after 2 yr: 3600 mountain lions, 525,700 deer

7.5 Exercises *(pages 521–524)*

1. −31 **3.** 7 **5.** 0 **7.** −26 **9.** 2, −6, 4 **11.** −6, 0, −6 **13.** 186 **15.** 17 **17.** 166 **19.** 0

21. 0 **23.** −5.5 **25.** $\left\{-\frac{4}{3}\right\}$ **27.** $\{-1, 4\}$ **29.** $\{-4\}$ **31.** $\{13\}$ **33.** 1 **35.** 9.5 **37.** 3.5 **39.** yes

41. no **43.** 298 **45.** −88 **47.** 0 **49.** 0 **51.** 16 **53.** $-x - 3y + 11 = 0$ or $x + 3y - 11 = 0$

54. $-x - 3y = -11$ or $x + 3y = 11$; They are equivalent. **55.** $y - y_1 = \frac{y_2 - y_1}{x_2 - x_1}(x - x_1)$

56. $y - y_1 = \frac{y_2 - y_1}{x_2 - x_1}(x - x_1)$; They are equivalent. **57. (a)** D **(b)** A **(c)** C **(d)** B **59.** $\{(2, 2)\}$ **61.** $\{(2, -5)\}$

63. $\left\{\left(\frac{4 - 2y}{3}, y\right)\right\}$ **65.** $\{(2, 3)\}$ **67.** $\{(-1, 2, 1)\}$ **69.** $\{(-3, 4, 2)\}$ **71.** $\{(0, 4, 2)\}$ **73.** $\emptyset$ **75.** $\{(-1, 2, 5, 1)\}$

77. If $D = 0$, Cramer's rule cannot be applied because there is no unique solution. There are either no solutions or infinitely many solutions.

7.6 Exercises *(pages 533–536)*

1. yes **3.** no **5.** no **7.** yes **9.** It will not exist if its determinant is equal to 0. **11.** $\begin{bmatrix} 5 & -7 \\ -2 & 3 \end{bmatrix}$

13. $\begin{bmatrix} 2 & 1 \\ -1.5 & -.5 \end{bmatrix}$ **15.** A^{-1} does not exist. **17.** $\begin{bmatrix} -2.5 & 5 \\ 12.5 & -15 \end{bmatrix}$ **19.** $\begin{bmatrix} 0 & 1 & 0 \\ 0 & 0 & 1 \\ 1 & 0 & 0 \end{bmatrix}$ **21.** $\begin{bmatrix} -2 & 1 & -1 \\ -5 & 2 & -1 \\ 3 & -1 & 1 \end{bmatrix}$

23. $\begin{bmatrix} 1 & 0 & 0 \\ 0 & -1 & 0 \\ -1 & 0 & 1 \end{bmatrix}$ **25.** $\begin{bmatrix} 15 & 4 & -5 \\ -12 & -3 & 4 \\ -4 & -1 & 1 \end{bmatrix}$ **27.** $\begin{bmatrix} -\frac{10}{3} & \frac{5}{9} & -\frac{10}{9} \\ \frac{20}{3} & \frac{5}{9} & \frac{80}{9} \\ -5 & \frac{5}{6} & -\frac{20}{3} \end{bmatrix}$ **29.** A^{-1} does not exist.

31. $\begin{bmatrix} .0543058761 & -.0543058761 \\ 1.846399787 & .153600213 \end{bmatrix}$ **33.** $\begin{bmatrix} .9987635516 & -.252092087 & -.3330564627 \\ -.5037783375 & 1.007556675 & -.2518891688 \\ -.2481013617 & -.2556769758 & 1.003768868 \end{bmatrix}$ **35.** $\{(-2, 4)\}$

37. $\{(4, -6)\}$ **39.** $\{(2.5, -1)\}$ **41.** $\{(3, 0, 2)\}$ **43.** $\left\{\left(12, -\frac{15}{11}, -\frac{65}{11}\right)\right\}$ **45.** $\{(0, 2, -2, 1)\}$

47. $\{(-3.542308934, -4.343268299)\}$ **49.** $\{(-.9704156959, 1.391914631, .1874077432)\}$

51. $P(x) = -2x^3 + 5x^2 - 4x + 3$ **53.** $P(x) = x^4 + 2x^3 + 3x^2 - x - 1$

55. (a) soft drink: \$1.50; popcorn: \$2.00 **(b)** No, A^{-1} does not exist.

57. **(a)**
$$\begin{aligned} 113a + 308b + c &= 10{,}170 \\ 133a + 622b + c &= 15{,}305 \\ 155a + 1937b + c &= 21{,}289 \end{aligned}$$
(b) $a \approx 251, b \approx .346, c \approx -18{,}300$; $T = 251A + .346I - 18{,}300$ **(c)** 11,426,000

59. **(a)** $\begin{aligned} a + 1500b + 8c &= 122 \\ a + 2000b + 5c &= 130 \\ a + 2200b + 10c &= 158 \end{aligned}$ or $\begin{bmatrix} 1 & 1500 & 8 \\ 1 & 2000 & 5 \\ 1 & 2200 & 10 \end{bmatrix}\begin{bmatrix} a \\ b \\ c \end{bmatrix} = \begin{bmatrix} 122 \\ 130 \\ 158 \end{bmatrix}$ **(b)** \$130,000 **61.** $\begin{bmatrix} 2 & 9 \\ 1 & 5 \end{bmatrix}$ **63.** $\begin{bmatrix} 1 & 0 & 1 \\ -1 & 0 & 2 \\ -2 & 1 & 3 \end{bmatrix}$

65. $\begin{bmatrix} \frac{1}{a} & 0 & 0 \\ 0 & \frac{1}{b} & 0 \\ 0 & 0 & \frac{1}{c} \end{bmatrix}$

Reviewing Basic Concepts *(pages 536–537)*

1. $\begin{bmatrix} -5 & 6 \\ -1 & 3 \end{bmatrix}$ **2.** $\begin{bmatrix} 0 & 6 \\ -9 & 12 \end{bmatrix}$ **3.** $\begin{bmatrix} 33 & -24 \\ -12 & 9 \end{bmatrix}$ **4.** $\begin{bmatrix} 1 & 3 & -3 \\ 0 & -6 & 0 \\ 4 & 2 & 2 \end{bmatrix}$ **5.** -3 **6.** -14 **7.** $\begin{bmatrix} \frac{1}{3} & \frac{4}{3} \\ \frac{2}{3} & \frac{5}{3} \end{bmatrix}$

8. $\begin{bmatrix} -\frac{2}{7} & -\frac{11}{14} & \frac{1}{14} \\ -\frac{4}{7} & -\frac{4}{7} & \frac{1}{7} \\ -\frac{1}{7} & -\frac{1}{7} & \frac{2}{7} \end{bmatrix}$ **9.** $W_1 = W_2 = \dfrac{100\sqrt{3}}{3} \approx 57.7$ lb **10.** $\{(3, 2, 1)\}$

7.7 Exercises *(pages 544–547)*

1.

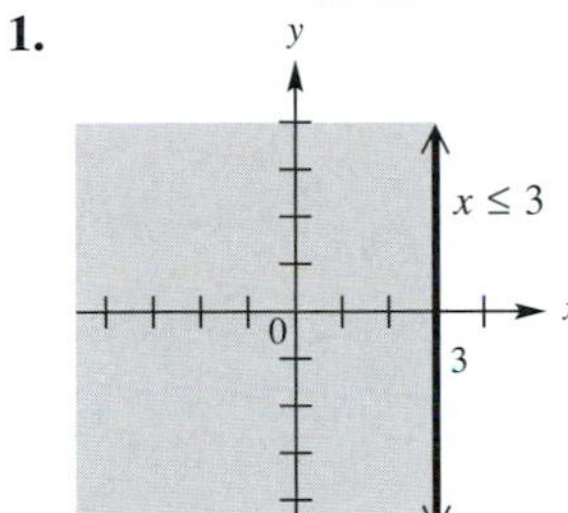

3.

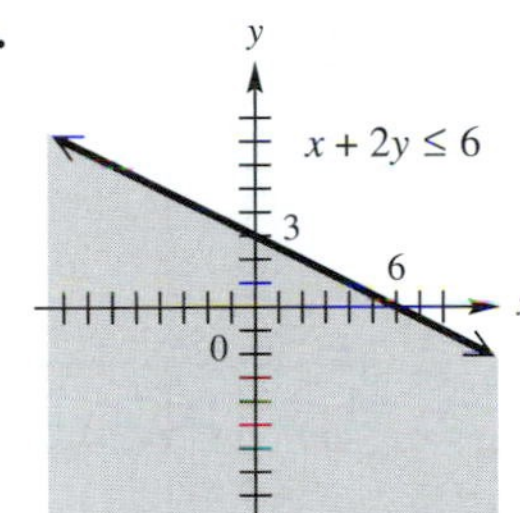

5.

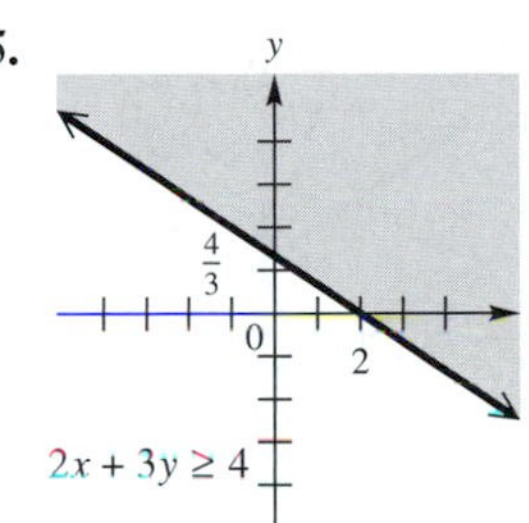

7.

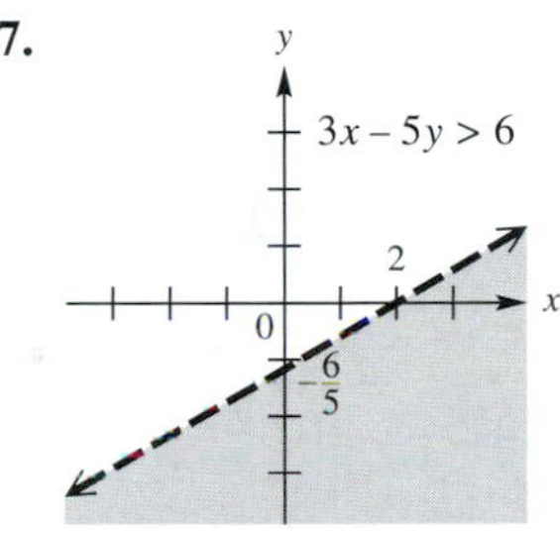

9.

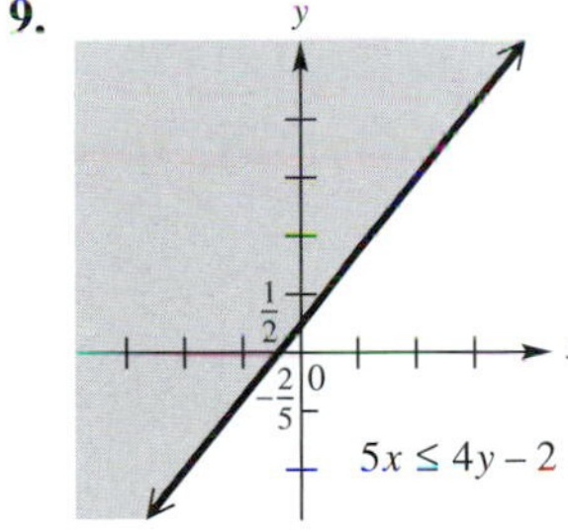

11.

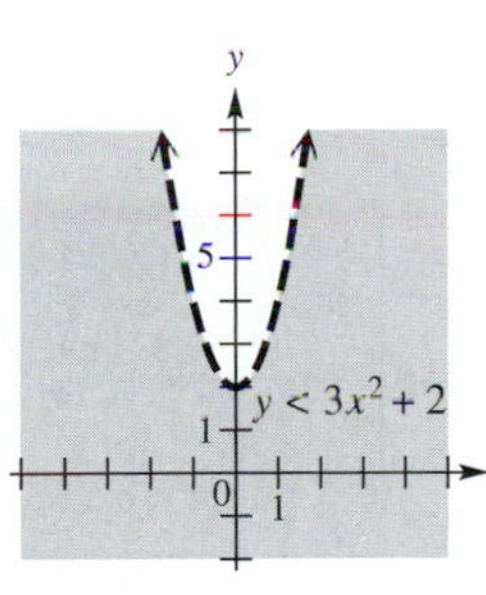

13.

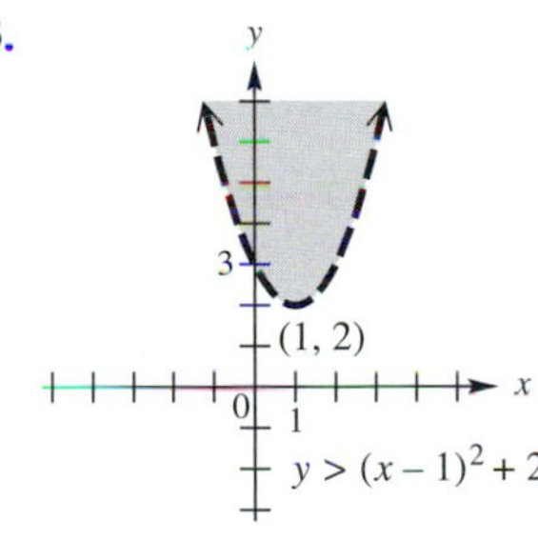

15.

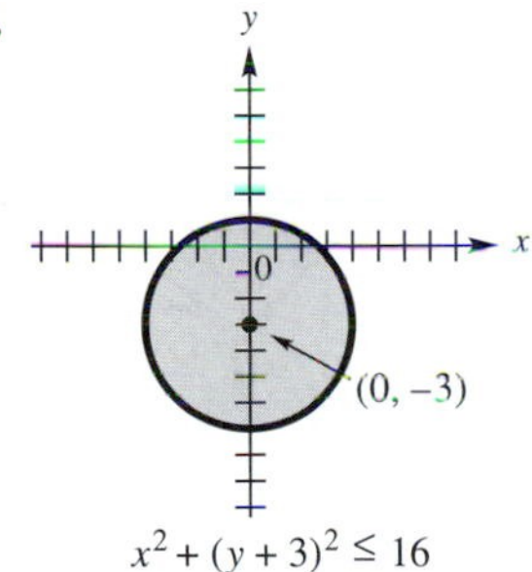

17. The boundary is solid if the symbol is $\geq$ or $\leq$ and dashed if the symbol is $>$ or $<$. **19.** above **21.** B **23.** C

25. A **27.**

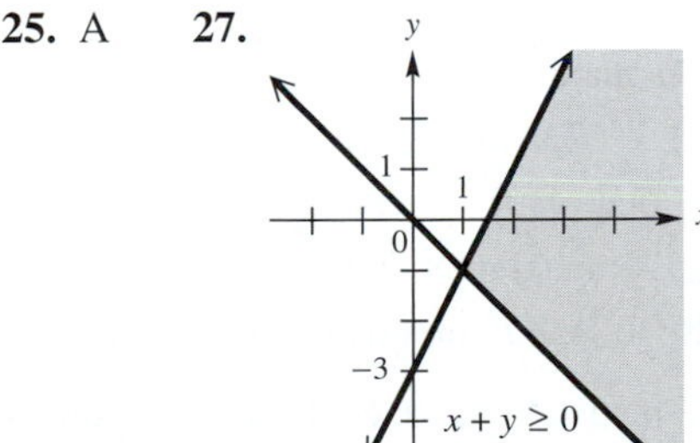

29.

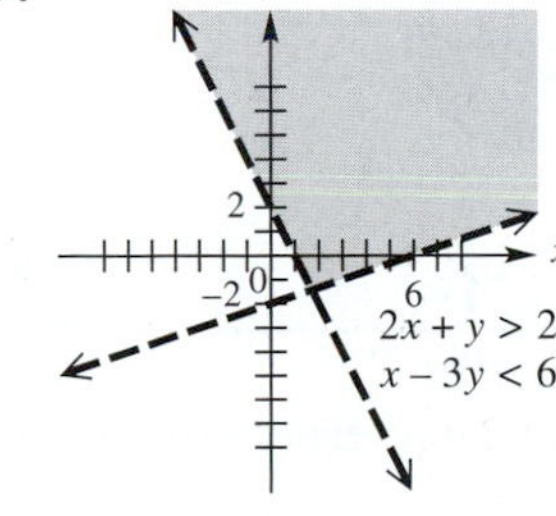

31.

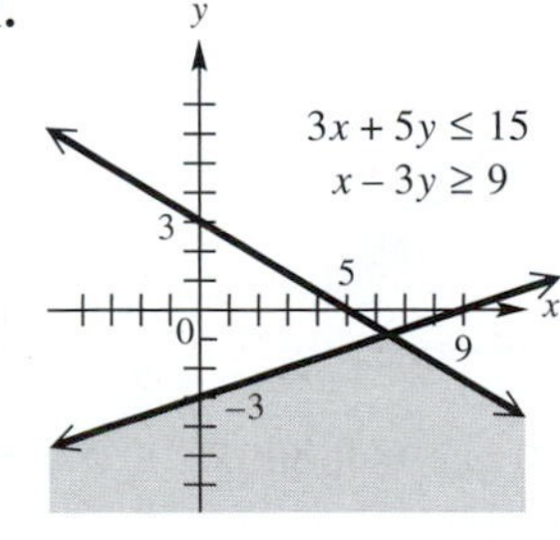

33.

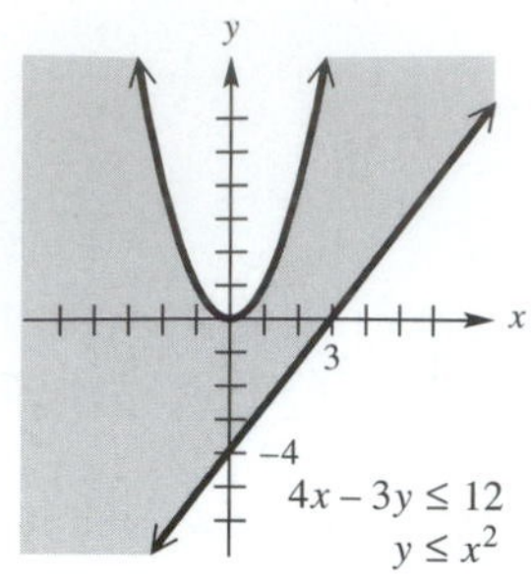

35.

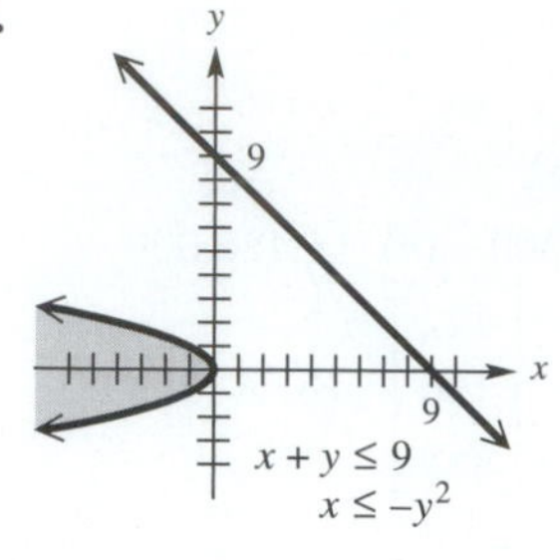

37.

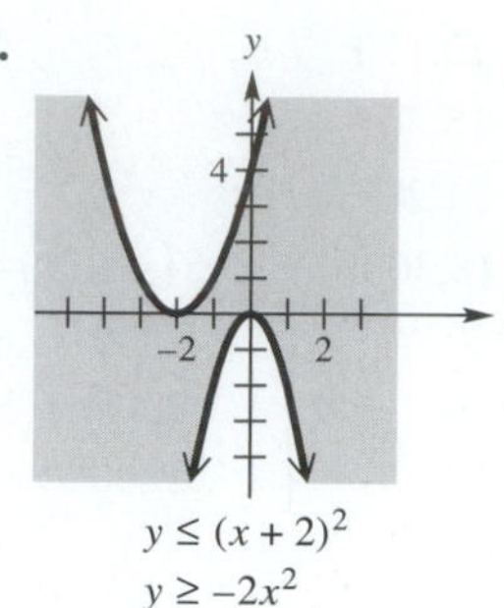

39.

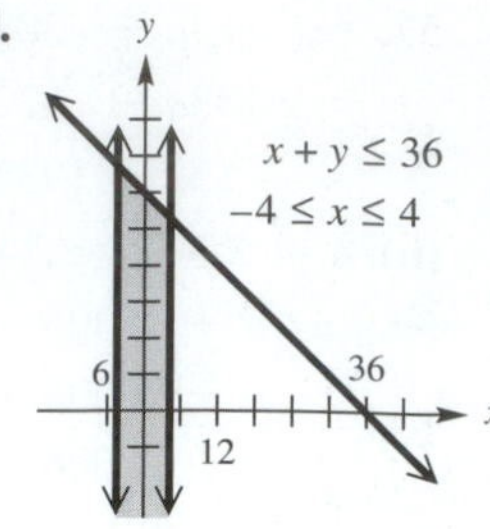

41.

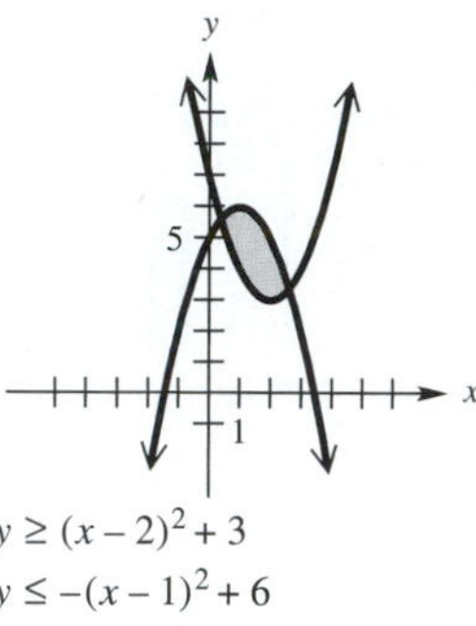

43.

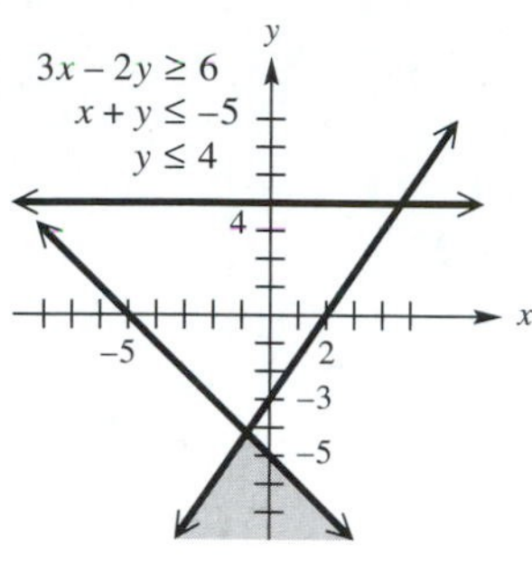

45.

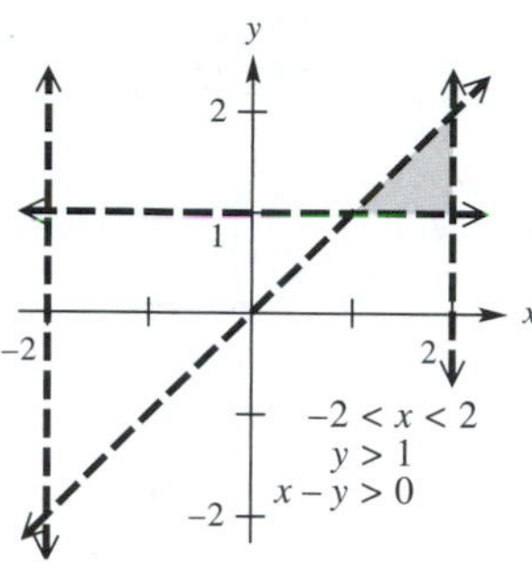

47.

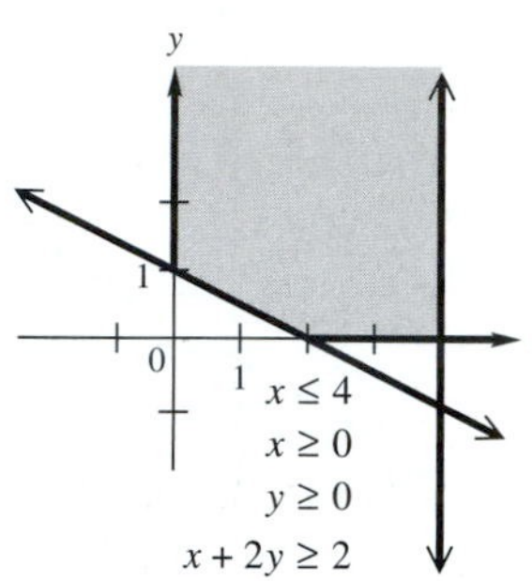

49.

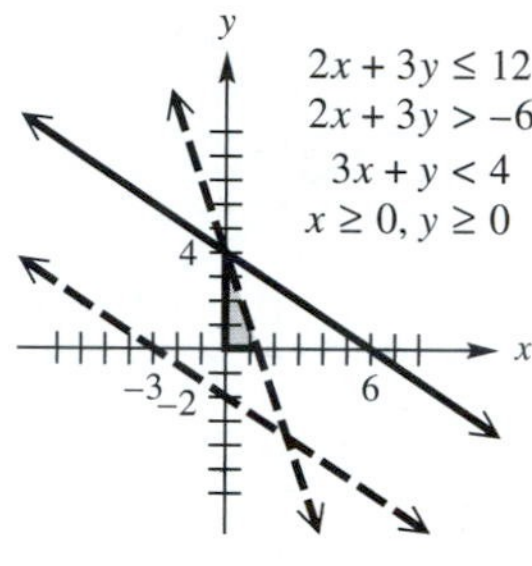

51.

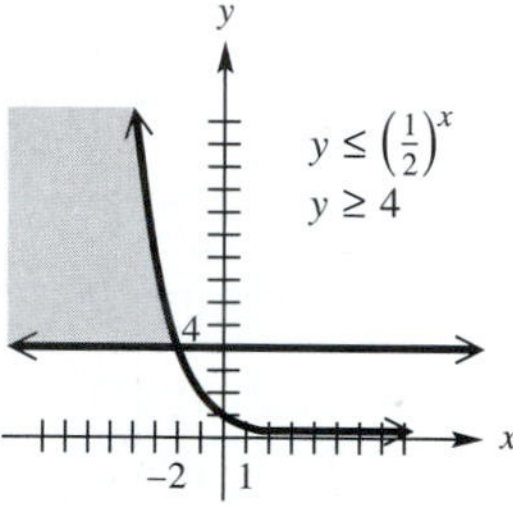

53.

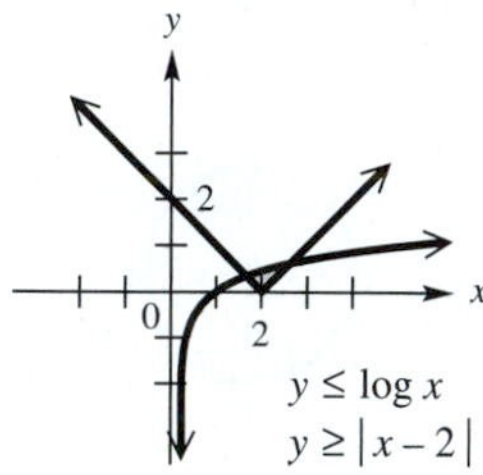

55. D **57.** A 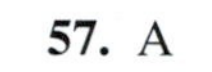 **59.** B **61.**

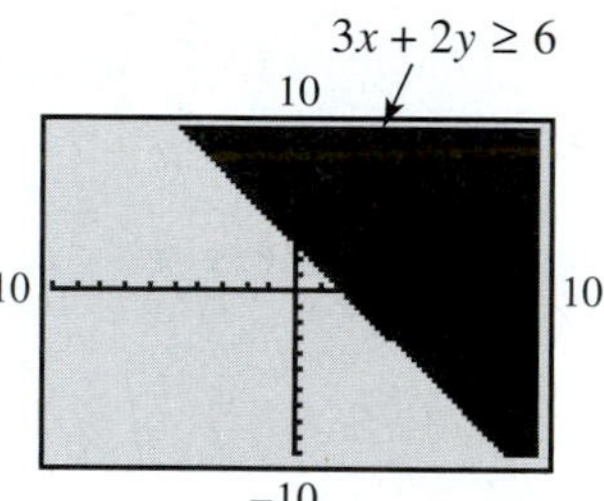

63.

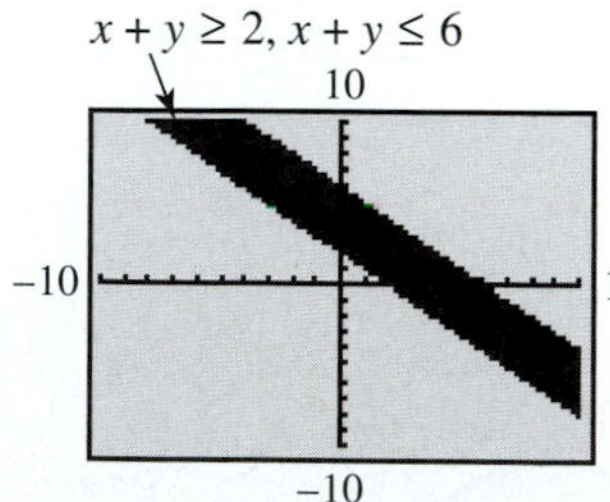

65.

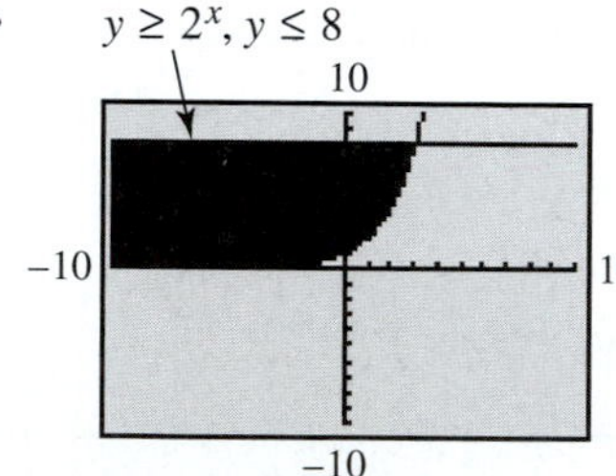

67. $x + 2y - 8 \geq 0, x + 2y \leq 12, x \geq 0, y \geq 0$ **69.** maximum of 65 at $(5, 10)$; minimum of 8 at $(1, 1)$ **71.** maximum of 66 at $(7, 9)$; minimum of 3 at $(1, 0)$ **73.** maximum of 100 at $(1, 10)$; minimum of 0 at $(1, 0)$ **75.** hat units: 5; whistle units: 0; maximum inquiries: 15 **77.** to A: 20; to B: 80; minimum cost: $1040

79. gasoline: 6,400,000 gal; fuel oil: 3,200,000 gal; maximum revenue: $16,960,000 **81.** medical kits: 0; containers of water: 4000; people aided: 40,000

7.8 Exercises *(page 553)*

1. $\frac{5}{3x} + \frac{-10}{3(2x+1)}$ **3.** $\frac{6}{5(x+2)} + \frac{8}{5(2x-1)}$ **5.** $\frac{5}{6(x+5)} + \frac{1}{6(x-1)}$ **7.** $\frac{-2}{x+1} + \frac{2}{x+2} + \frac{4}{(x+2)^2}$

9. $\frac{4}{x} + \frac{4}{1-x}$ **11.** $\frac{15}{x} + \frac{-5}{x+1} + \frac{-6}{x-1}$ **13.** $1 + \frac{-2}{x+1} + \frac{1}{(x+1)^2}$ **15.** $x^3 - x^2 + \frac{-1}{3(2x+1)} + \frac{2}{3(x+2)}$

17. $\frac{1}{9} + \frac{-1}{x} + \frac{25}{18(3x+2)} + \frac{29}{18(3x-2)}$ **19.** $\frac{-3}{5x^2} + \frac{3}{5(x^2+5)}$ **21.** $\frac{-2}{7(x+4)} + \frac{6x-3}{7(3x^2+1)}$

23. $\frac{1}{4x} + \frac{-8}{19(2x+1)} + \frac{-9x-24}{76(3x^2+4)}$ **25.** $\frac{-1}{x} + \frac{2x}{2x^2+1} + \frac{2x+3}{(2x^2+1)^2}$ **27.** $\frac{-1}{x+2} + \frac{3}{(x^2+4)^2}$

29. $5x^2 + \frac{3}{x} + \frac{-1}{x+3} + \frac{2}{x-1}$ **31.** coincide; correct **33.** do not coincide; not correct

Reviewing Basic Concepts *(page 554)*

1.

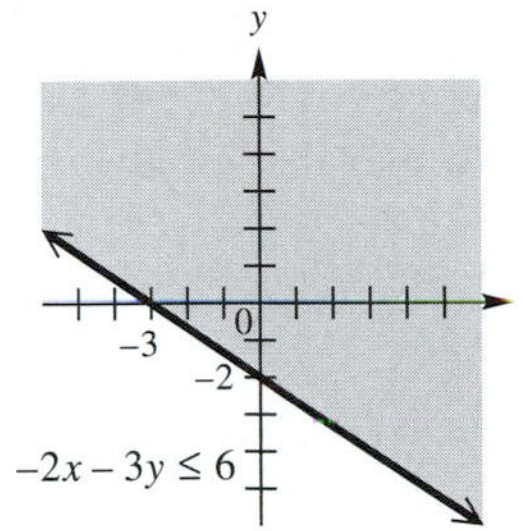

2.

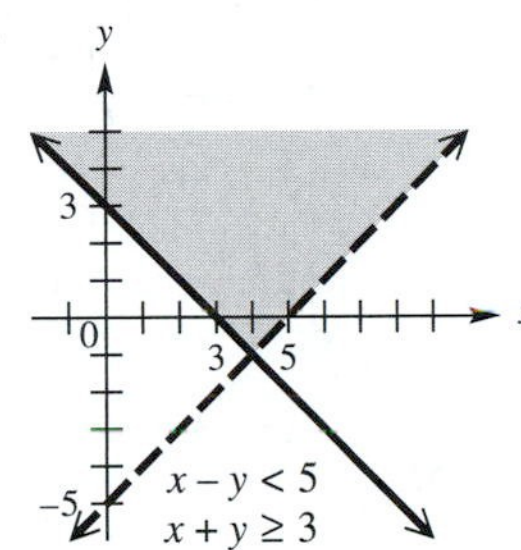

3.

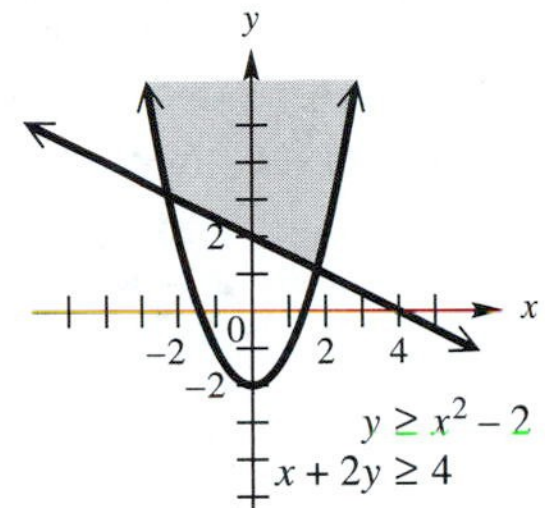

4.

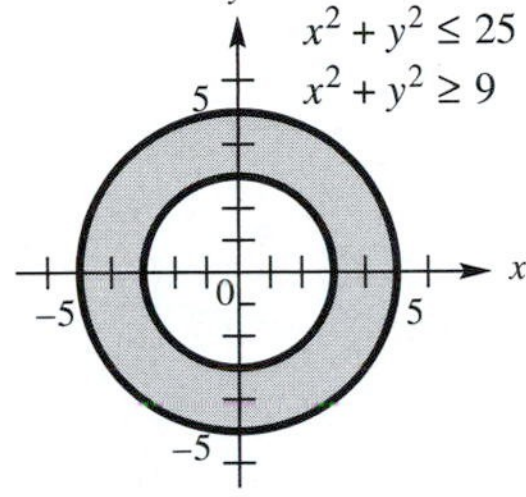

5. A **6.** Minimum value is 8 at $(4, 0)$. **7.** maximum: 65; minimum: 8 **8.** substance X: 130 lb; substance Y: 450 lb; minimum cost: \$1610 **9.** $\frac{7}{x-5} + \frac{3}{x+4}$ **10.** $\frac{-1}{x+2} + \frac{6}{x+4} + \frac{-3}{x-1}$

Chapter 7 Review Exercises *(pages 557–560)*

1. $\{(5, 7)\}$ **3.** $\{(5, -3)\}$ **5.** $\left\{\left(\frac{7}{5}, -\frac{1}{5}\right), (1, 1)\right\}$ **7.** $\{(-3\sqrt{3}, \sqrt{2}), (-3\sqrt{3}, -\sqrt{2}), (3\sqrt{3}, \sqrt{2}), (3\sqrt{3}, -\sqrt{2})\}$

9. $\{(6, 2)\}$ **11. (a)** $y_1 = \sqrt{2 - x^2}$, $y_2 = -\sqrt{2 - x^2}$ **(b)** $y_3 = -3x + 4$ **(c)** $[-3, 3]$ by $[-2, 2]$; Other windows are possible. **13.** No, a system consisting of two equations in three variables is represented by two planes in space. There will be no solutions or infinitely many solutions. **15.** $\{(6, -2, 1)\}$ **17.** $\emptyset$; inconsistent system **19.** $\{(-3, 2)\}$ **21.** $\{(0, 1, 0)\}$

23. $\begin{bmatrix} -4 \\ 6 \\ 1 \end{bmatrix}$ **25.** $\begin{bmatrix} -4 & -4 \\ -7 & 16 \end{bmatrix}$ **27.** $\begin{bmatrix} 14x + 28y & 42x + 22y \\ 18x - 46y & 70x + 18y \end{bmatrix}$ **29.** $\begin{bmatrix} 18 & 20 \\ 29 & -1 \end{bmatrix}$ **31.** $\begin{bmatrix} -3 \\ 10 \end{bmatrix}$

33. $\begin{bmatrix} -2 & 22 & 31 \\ 25 & 12 & 9 \end{bmatrix}$ **35.** yes **37.** no **39.** A^{-1} does not exist. **41.** $\begin{bmatrix} \frac{1}{2} & 0 \\ \frac{1}{10} & \frac{1}{5} \end{bmatrix}$ **43.** $\begin{bmatrix} \frac{2}{3} & 0 & -\frac{1}{3} \\ \frac{1}{3} & 0 & -\frac{2}{3} \\ -\frac{2}{3} & 1 & \frac{1}{3} \end{bmatrix}$

45. $\{(2, 2)\}$ **47.** $\{(2, 1)\}$ **49.** $\{(1, -1, 2)\}$ **51.** $\{(-1, 0, 2)\}$ **53.** -25 **55.** -44 **57.** $\left\{-\frac{7}{3}\right\}$

59. {all real numbers} **61. (a)** $D = 5$ **(b)** $D_x = 30$ **(c)** $D_y = -50$ **(d)** $x = 6$; $y = -10$; solution set: $\{(6, -10)\}$

63. If $D = 0$, there would be division by 0, which is undefined. The system will have no solutions or infinitely many solutions.

65. $\{(-4, 2)\}$ **67.** $\{(-4, 6, 2)\}$ **69.** $\left\{\left(\frac{172}{67}, -\frac{14}{67}, -\frac{87}{67}\right)\right\}$ **71.** CDs: 80; diskettes: 20 **73.** 5%: 11 mL; 15%: 3 mL; 10%: 6 mL **75.** $P(x) = x^3 - 2x + 5$

77.

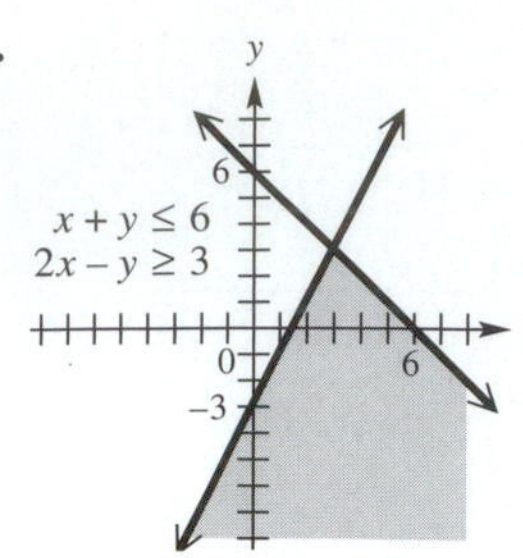

79. maximum value: 24, at $(0, 6)$ **81.** radios: 25; DVD players: 30; maximum profit: \$1425

83. $\frac{2}{x} + \frac{-2}{x+1} + \frac{-1}{(x+1)^2}$

Chapter 7 Test *(page 561)*

1. (a) first equation: a hyperbola; second equation: a line **(b)** 0, 1, or 2 **(c)** $\left\{(1, -2), \left(-\frac{11}{35}, \frac{68}{35}\right)\right\}$

(d)

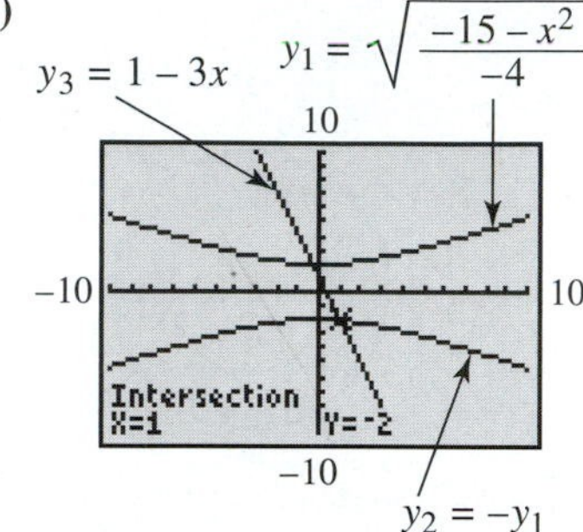

The other point of intersection is $\left(-\frac{11}{35}, \frac{68}{35}\right)$. **2.** $\{(2, 0, -1)\}$ **3. (a)** $\begin{bmatrix} 8 & 3 \\ 0 & -11 \\ 15 & 19 \end{bmatrix}$ **(b)** not possible **(c)** $\begin{bmatrix} -5 & 16 \\ 19 & 2 \end{bmatrix}$ **4. (a)** yes **(b)** yes **(c)** No. In general $AB \neq BA$ because matrix multiplication is not commutative. **(d)** AC cannot be found, but CA can.

5. (a) 1 **(b)** −844 **6.** $\{(-6, 7)\}$ **7. (a)** $A = \begin{bmatrix} 1 & 1 & -1 \\ 2 & -3 & -1 \\ 1 & 2 & 2 \end{bmatrix}, X = \begin{bmatrix} x \\ y \\ z \end{bmatrix}, B = \begin{bmatrix} -4 \\ 5 \\ 3 \end{bmatrix}$

(b) $A^{-1} = \begin{bmatrix} \frac{1}{4} & \frac{1}{4} & \frac{1}{4} \\ \frac{5}{16} & -\frac{3}{16} & \frac{1}{16} \\ -\frac{7}{16} & \frac{1}{16} & \frac{5}{16} \end{bmatrix}$ **(c)** $\{(1, -2, 3)\}$ **(d)** $A = \begin{bmatrix} .5 & 1 & 1 \\ 2 & -3 & -1 \\ 1 & 2 & 2 \end{bmatrix}$ and $\det A = 0$, so A^{-1} does not exist.

8. $f(x) \approx -.010764x^2 + 1.2903x + 24.133$

$f(x) \approx -.010764x^2 + 1.2903x + 24.133$

75

20 100

45

9. B **10.** cabinet X: 8; cabinet Y: 3

11. $\frac{4}{x-3} + \frac{3}{x+2}$

12. $\frac{-1}{x-2} + \frac{2}{x+2} + \frac{-3}{(x-2)^2}$

CHAPTER 8 TRIGONOMETRIC FUNCTIONS AND APPLICATIONS

8.1 Exercises *(pages 576–582)*

1. (a) 60° **(b)** 150° **3. (a)** 45° **(b)** 135° **5. (a)** $\frac{\pi}{4}$ **(b)** $\frac{3\pi}{4}$ **7. (a)** $(90 - x)°$ **(b)** $(180 - x)°$ **9.** 150°

11. 70°; 110° **13.** 55°; 35° **15.** 80°; 100° **17.** 83° 59′ **19.** 23° 49′ **21.** 17° 1′ 49″ **23.** 20.9°

25. 91.598° **27.** 31° 25′ 47″ **29.** 89° 54′ 1″

Angles other than those given are possible in Exercises 31–35.

31.

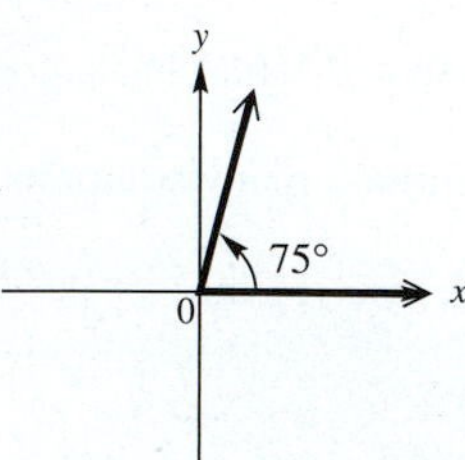

435°; −285°; quadrant I

33.

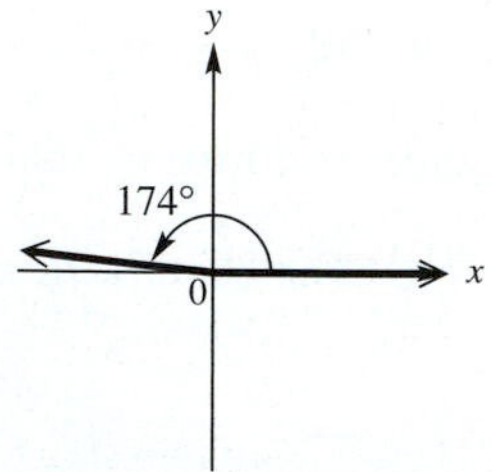

534°; −186°; quadrant II

35.

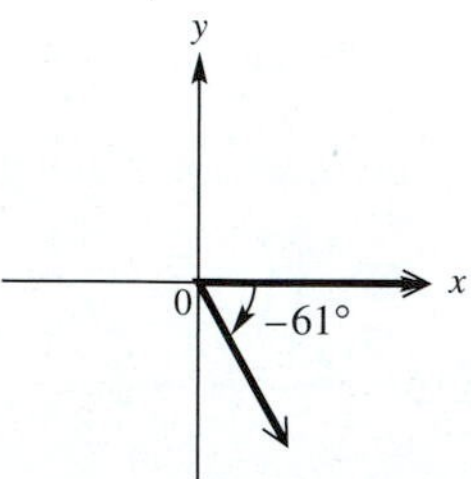

299°; −421°; quadrant IV

37. 320° **39.** 90° **41.** $\frac{7\pi}{4}$ **43.** $\frac{\pi}{2}$ **45.** $30° + n \cdot 360°$ **47.** $-90° + n \cdot 360°$ **49.** $\frac{\pi}{4} + 2n\pi$ **51.** $-\frac{3\pi}{4} + 2n\pi$ **53.** $\frac{\pi}{3}$ **55.** $\frac{5\pi}{6}$ **57.** $-\frac{\pi}{4}$ **59.** 60° **61.** 315° **63.** 330° **65.** .68 **67.** 2.43 **69.** 1.12 **71.** 114° 35′ **73.** 99° 42′ **75.** −74° 29′ **77.** We begin the answers with the blank next to 30°, and then proceed counterclockwise from there: $\frac{\pi}{6}$; 45; $\frac{\pi}{3}$; 120; 135; $\frac{5\pi}{6}$; π; $\frac{7\pi}{6}$; $\frac{5\pi}{4}$; 240; 300; $\frac{7\pi}{4}$; $\frac{11\pi}{6}$. **79.** 2π **81.** 8 **83.** 1 **85.** 25.8 cm **87.** 5.05 m **89.** 1200 km **91.** 5900 km **93.** $\frac{3\pi}{32}$ radian per sec **95.** $\frac{6}{5}$ min **97.** 18π cm **99.** 12 sec **101.** **(a)** 11.6 in. **(b)** 37° 5′ **103.** 38.5° **105.** 146 in. **107.** 1120 m^2 **109.** 114 cm^2 **111.** 6π **113.** approximately 3800 mi^2 **115.** 75.4 $in.^2$ **117.** **(a)** 13.85° **(b)** approximately 76 m^2 **119.** The area A of a circle of radius r is given by $A = \pi r^2$. **121.** $16\frac{1}{4}$ ft per sec, or about 11.1 mph **123.** 234.67 radians per sec **125.** $\frac{500\pi}{3}$ radians per sec; $\frac{2500\pi}{3}$ in. per sec **127.** **(a)** $\frac{2\pi}{365}$ radian **(b)** $\frac{\pi}{4380}$ radian per hr **(c)** approximately 66,700 mph **129.** radius: 3947 mi; circumference: 24,800 mi

8.2 Exercises *(pages 590–592)*

1. $\frac{1}{6}$ **3.** $\frac{1}{8}$ **5.** $\frac{1}{12}$ **7.** $\frac{3}{2}$ **9.** $\sin s = \frac{3}{5}$; $\cos s = \frac{4}{5}$ **11.** $\sin s = -\frac{5}{13}$; $\cos s = \frac{12}{13}$ **13.** $(-1, 0)$; -1; 0; cosecant and cotangent **15.** $(1, 0)$; 1; 0; cosecant and cotangent **17.** $(0, -1)$; 0; -1; secant and tangent **19.** $(0, 1)$; 0; 1; secant and tangent

In Exercises 21–27, we give, in order, sin *s*, cos *s*, tan *s*, cot *s*, sec *s*, and csc *s*.

21. 0, −1, 0, undefined, −1, undefined **23.** 0, 1, 0, undefined, 1, undefined **25.** −1, 0, undefined, 0, undefined, −1 **27.** 1, 0, undefined, 0, undefined, 1

In Exercises 29–35, we first give the quadrant for *s*. Then we give, in order, sin *s*, cos *s*, tan *s*, cot *s*, sec *s*, and csc *s*.

29. I; .68163876, .7316888689, .9315964599, 1.073426149, 1.366701125, 1.467052724 **31.** II; .8949893582, −.4460874899, −2.006309028, −.4984277029, −2.241712719, 1.117331721 **33.** III; −.77807361969, −.6281736227, 1.238627616, .8073451511, −1.591916572, −1.285226125 **35.** IV; −.7055403256, .7086697743, −.9955840522, −1.004435535, 1.411094471, −1.417353429

In Exercises 37–42, answers will vary, depending on the name used.

37. Suppose the first name is Shannon. Then $s = 7$, and $\cos 7 \approx .7539022543$.

38. Suppose the last name is Mulkey. Then $n = 6$, and $\cos(7 + 12\pi) \approx .7539022543$.

39. They are the same. The real numbers s and $s + 2n\pi$ correspond to the same point on the unit circle, because its circumference is 2π. **40.** Suppose the last name is Castellucio. Then $s = 11$, and $\sin 11 \approx -.9999902066$. **41.** Suppose the first name is Frankie. Then $n = 7$, and $\sin(11 + 14\pi) \approx -.9999902066$. **42.** See the answer to Exercise 39. **43.** $-\frac{1}{2}$ **45.** −1 **47.** −2 **49.** $-\sqrt{3}$ **51.** $-\frac{1}{2}$ **53.** $\frac{2\sqrt{3}}{3}$ **55.** $\frac{\sqrt{3}}{2}$ **57.** −1

In Exercises 59–63, we give, in order, the value of *x* or *y*; sin *s*, cos *s*, tan *s*, cot *s*, sec *s*, and csc *s*.

59. $y = \frac{4}{5}$; $\frac{4}{5}, \frac{3}{5}, \frac{4}{3}, \frac{3}{4}, \frac{5}{3}, \frac{5}{4}$ **61.** $x = -\frac{7}{25}$; $\frac{24}{25}, -\frac{7}{25}, -\frac{24}{7}, -\frac{7}{24}, -\frac{25}{7}, \frac{25}{24}$

63. $y = -\frac{2\sqrt{2}}{3}$; $-\frac{2\sqrt{2}}{3}, -\frac{1}{3}, 2\sqrt{2}, \frac{\sqrt{2}}{4}, -3, -\frac{3\sqrt{2}}{4}$

In Exercises 65–69, we give, in order, tan s, cot s, sec s, and csc s.

65. $\frac{\sqrt{3}}{3}, \sqrt{3}, \frac{2\sqrt{3}}{3}, 2$ **67.** $-\frac{4}{3}, -\frac{3}{4}, -\frac{5}{3}, \frac{5}{4}$ **69.** $-\sqrt{3}, -\frac{\sqrt{3}}{3}, 2, -\frac{2\sqrt{3}}{3}$ **71.** II **73.** III **75.** IV

79. $\sin s = b; \cos s = a$ **81.** $\sin(s - 6\pi) = b; \cos(s - 6\pi) = a$ **83.** $\sin(-s) = -b; \cos(-s) = a$

85. $\sin\left(s - \frac{\pi}{2}\right) = -a; \cos\left(s - \frac{\pi}{2}\right) = b$

Reviewing Basic Concepts *(page 592)*

1. (a) complement: 55°; supplement: 145° **(b)** complement: $\frac{\pi}{4}$; supplement: $\frac{3\pi}{4}$ **2.** $32°\ 15'\ 0''$ **3.** $59.591\overline{6}°$

4. (a) 200° **(b)** $\frac{4\pi}{3}$ **5. (a)** $\frac{4\pi}{3}$ **(b)** 135° **6. (a)** 2π cm **(b)** 3π cm^2 **7. (a)** $(1, 0)$ **(b)** $\left(-\frac{\sqrt{2}}{2}, -\frac{\sqrt{2}}{2}\right)$ **(c)** $(0, 1)$

In Exercises 8 and 9, we give, in order, sine, cosine, tangent, cotangent, secant, and cosecant.

8. -1; 0; undefined; 0; undefined; -1 **9. (a)** $-\frac{1}{2}; -\frac{\sqrt{3}}{2}; \frac{\sqrt{3}}{3}; \sqrt{3}; -\frac{2\sqrt{3}}{3}; -2$ **(b)** $-\frac{\sqrt{3}}{2}; -\frac{1}{2}; \sqrt{3}; \frac{\sqrt{3}}{3}; -2; -\frac{2\sqrt{3}}{3}$

10. $\cos 2.25 \approx -.6281736227$; $\sin 2.25 \approx .7780731969$; $\tan 2.25 \approx -1.238627616$; $\cot 2.25 \approx -.8073451511$; $\sec 2.25 \approx -1.591916572$; $\csc 2.25 \approx 1.285226125$

8.3 Exercises *(pages 604–610)*

1. G **3.** E **5.** B **7.** F **9.** D **11.** H **13.** B **15.** F **17.** B **19.** C **21.** $y = 4 \sin \frac{1}{2}x$ (There are other correct answers.) **23.** 1; 1; 3; -2 **24.** -1; -1; -3; -8 **25.** $[-8, -2]$; The trig viewing window as defined in the text has Ymin $= -4$ and Ymax $= 4$. Because the range of f includes values less than -4, the local minimum points will not appear in the trig viewing window. **26.** -2; -8 **27.** The period of f is $\frac{2\pi}{2} = \pi$, and the distance between -2π and 2π is 4π units. Since $\frac{4\pi}{\pi} = 4$, the interval $[-2\pi, 2\pi]$ will show 4 periods of the graph.

28. $f(x) = -5 + 3\sin\left[2\left(x - \frac{\pi}{2}\right)\right]$

5, -2π, 2π, -10

29. Ymin $= -8$; Ymax $= -2$ **30.** $f(-2) \approx -7.270407$; $f(-2 + \pi) \approx -7.270407$; Because the period of f is π, $f(p) = f(p + \pi)$ for any value of p. In this case, $p = -2$. **31.** 2

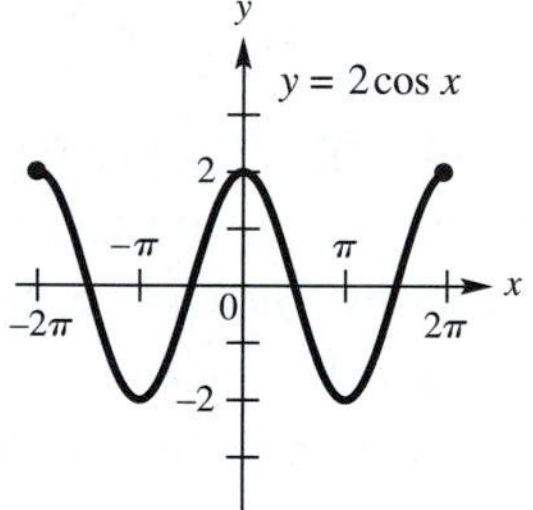

33. $\frac{2}{3}$

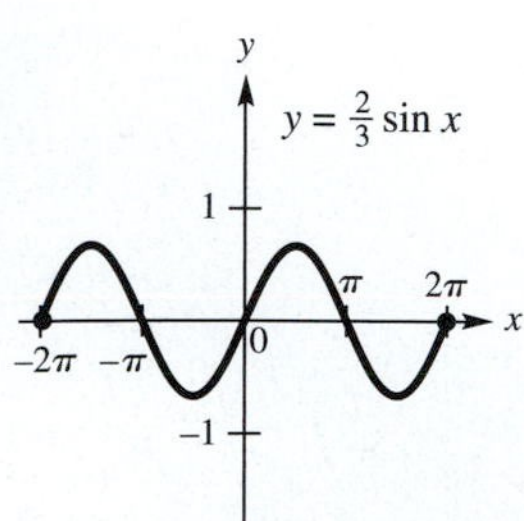

35. 1

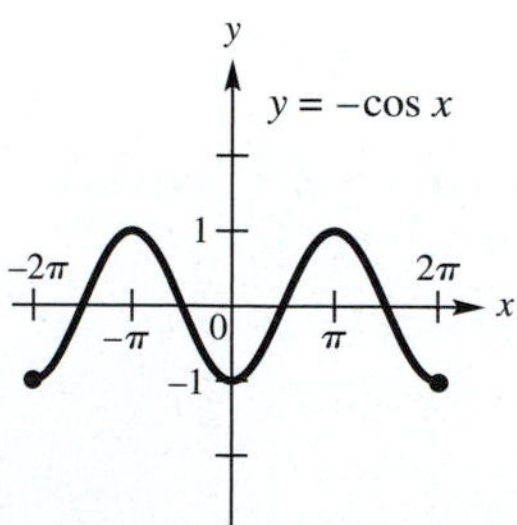

37. 2

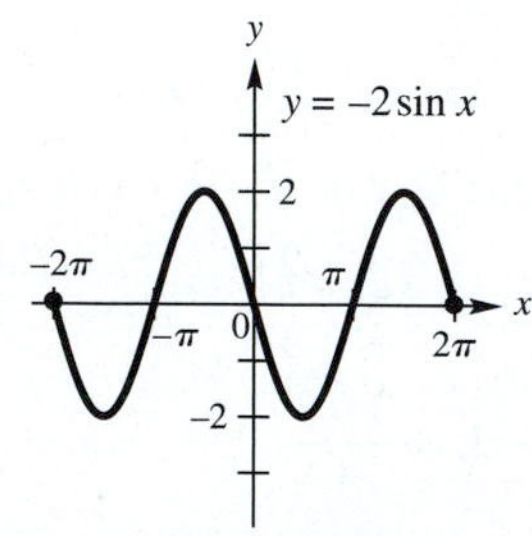

39. 4π; 1

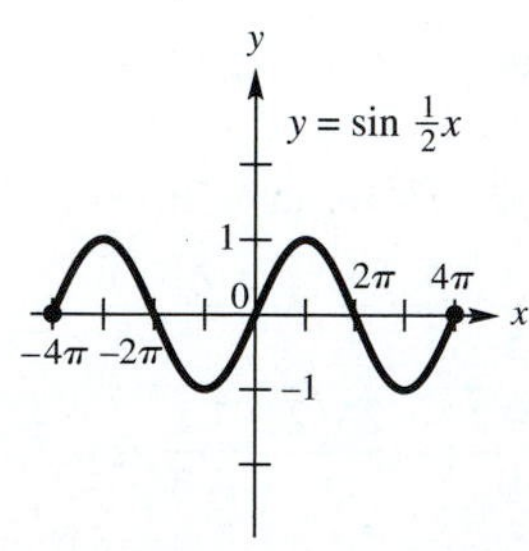

41. π; 1

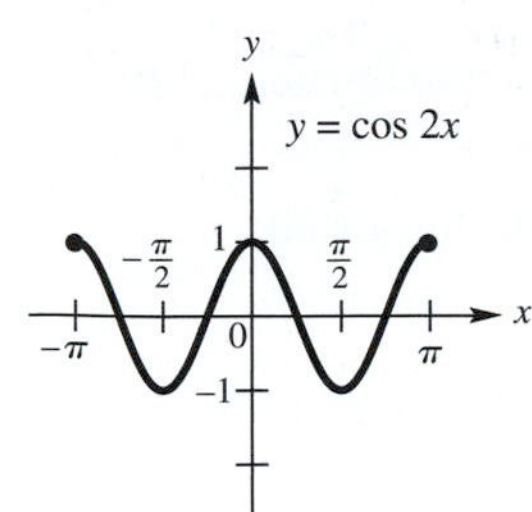

43. 8π; 2

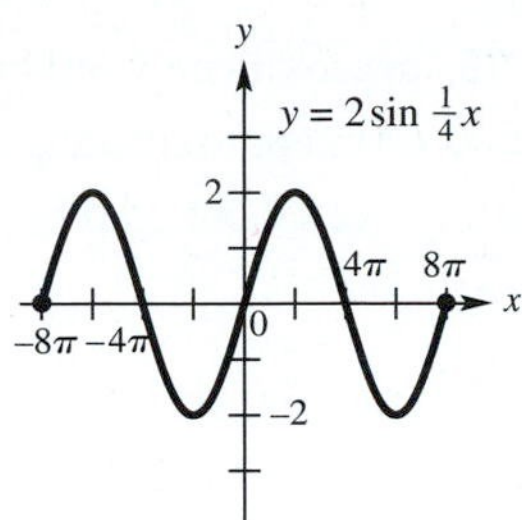

45. $\dfrac{2\pi}{3}$; 2

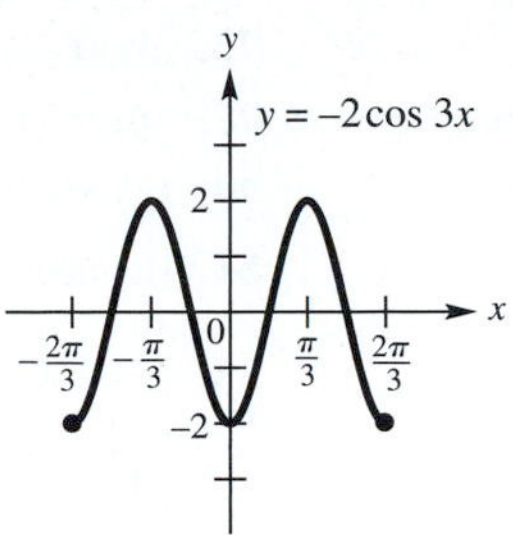

47. $\dfrac{\pi}{4}$

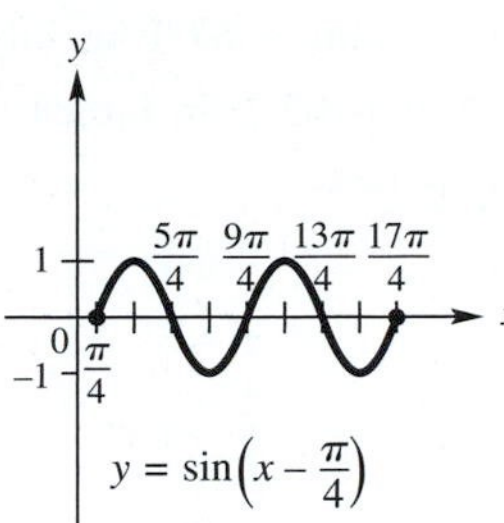

49. $\dfrac{\pi}{3}$

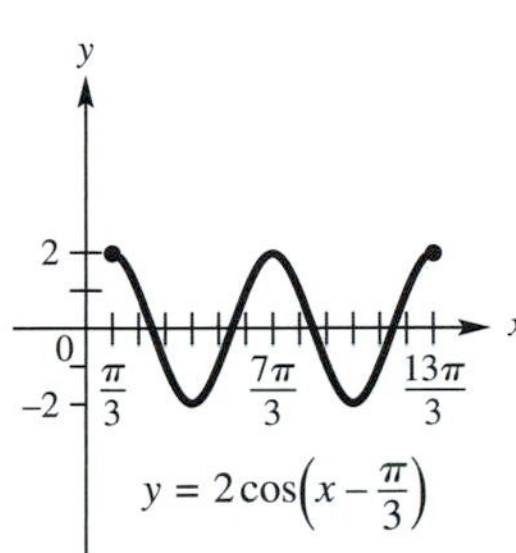

51. **(a)** 4 **(b)** π 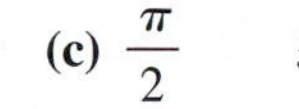 **(c)** $\dfrac{\pi}{2}$
(d) none **(e)** $[-4, 4]$

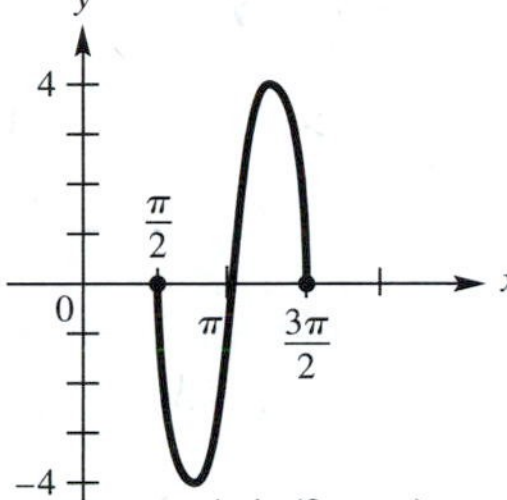

53. **(a)** $\dfrac{1}{2}$ **(b)** 4π **(c)** $\dfrac{\pi}{2}$
(d) none **(e)** $\left[-\dfrac{1}{2}, \dfrac{1}{2}\right]$

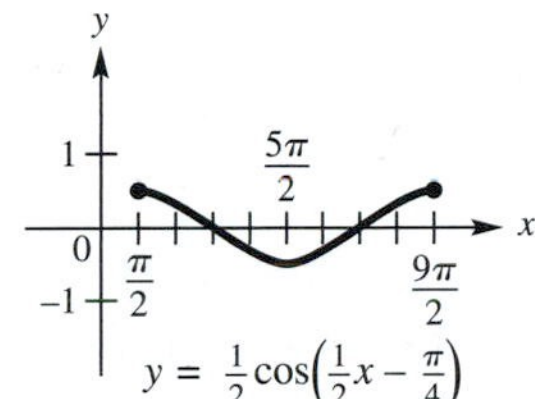

55. **(a)** $\dfrac{2}{3}$ **(b)** $\dfrac{8\pi}{3}$ **(c)** none
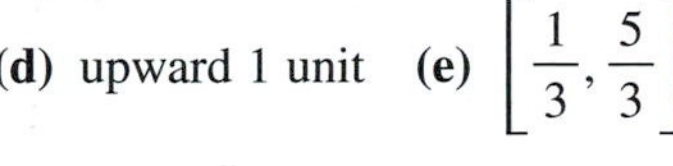
(d) upward 1 unit **(e)** $\left[\dfrac{1}{3}, \dfrac{5}{3}\right]$

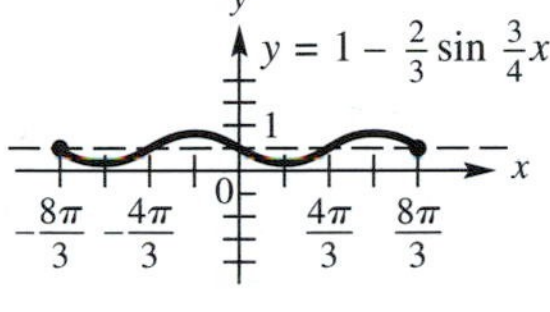

57. **(a)** 2 **(b)** 4π **(c)** none
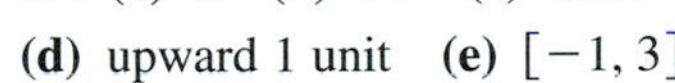
(d) upward 1 unit **(e)** $[-1, 3]$

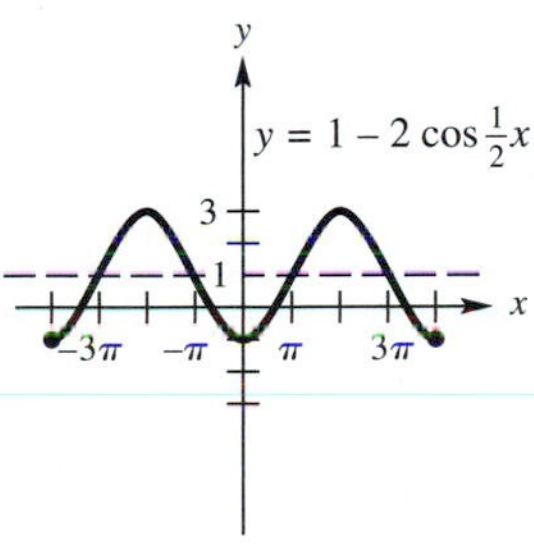

59. **(a)** 2 **(b)** 2π 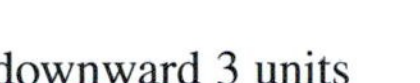 **(c)** $-\dfrac{\pi}{2}$
(d) downward 3 units
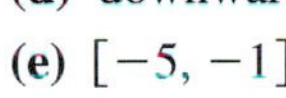
(e) $[-5, -1]$

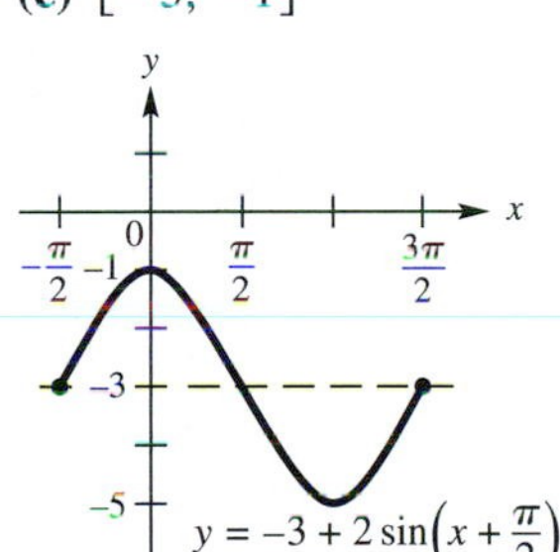

61. **(a)** 1 **(b)** π **(c)** $-\dfrac{\pi}{4}$
(d) upward $\dfrac{1}{2}$ unit **(e)** $\left[-\dfrac{1}{2}, \dfrac{3}{2}\right]$

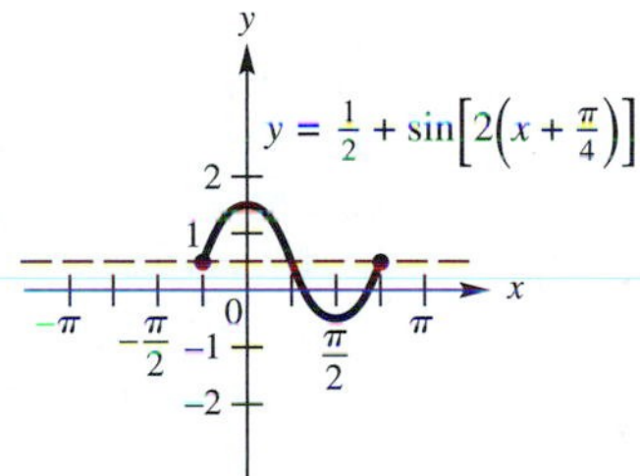

63. **(a)** 2 **(b)** 2π **(c)** π
(d) none **(e)** $[-2, 2]$

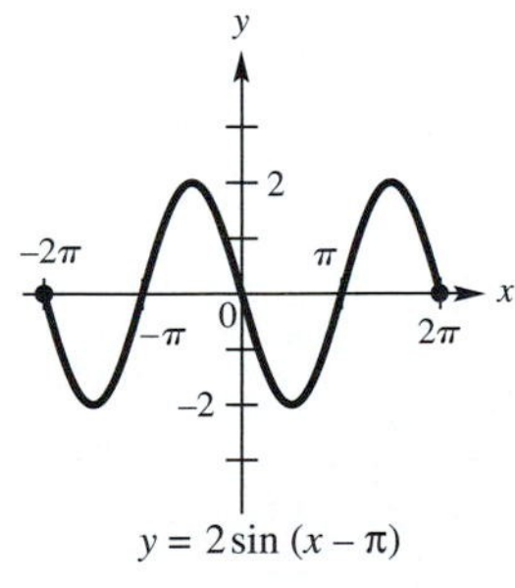

65. **(a)** 4 **(b)** 4π **(c)** $-\pi$
(d) none **(e)** $[-4, 4]$

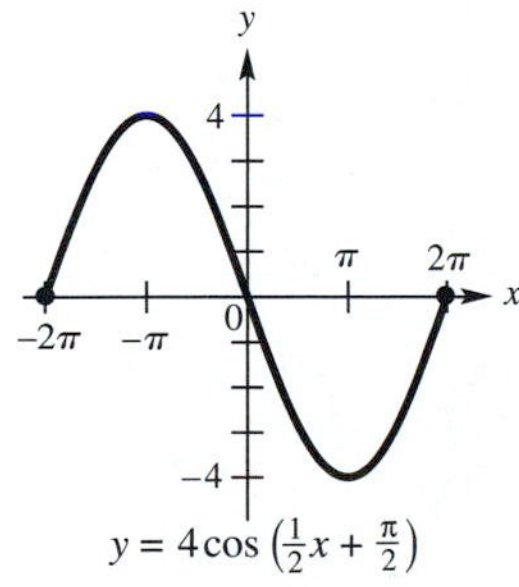

67. **(a)** 1 **(b)** $\dfrac{2\pi}{3}$ **(c)** $\dfrac{\pi}{15}$
(d) upward 2 units **(e)** $[1, 3]$

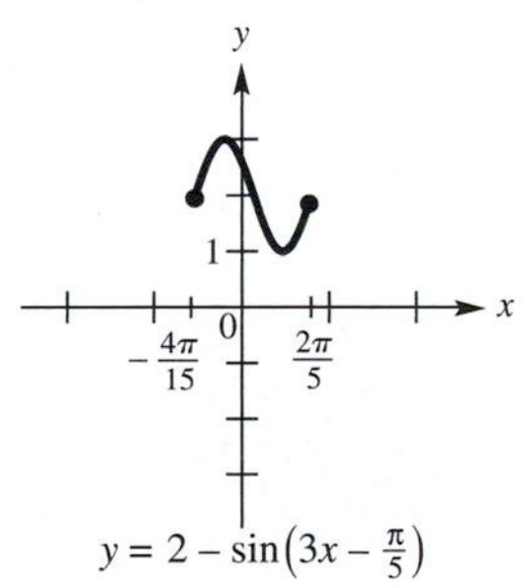

69. **(a)** 40°F; −40°F **(b)** amplitude: 40; period: 12; The monthly average temperatures vary by 80°F over a 12-month period. **(c)** The x-intercepts represent the months when the average temperature is 0°F.

71. **(a)** maximum ≈ 87°F in July; minimum ≈ 62°F in January or late December **(b)** increases

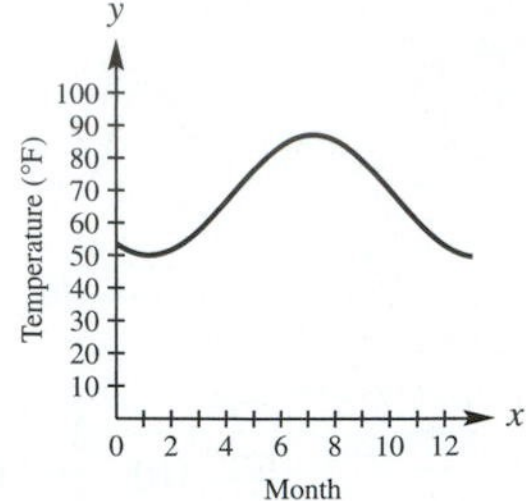

73. about 24 hr **75.** approximately 6:00 P.M.; approximately .2 ft

77. approximately 2:00 A.M.; approximately 2.6 ft

79. **(a)** 80°; 50° **(b)** 15° **(c)** about 35,000 yr **(d)** downward

81. **(a)** about 2 hr **(b)** 1 yr

83. **(a)** 3.8; $\frac{1}{20}$ **(b)** 20 **(c)** −3.074; 1.174; −3.074; −3.074; 1.174

(d)

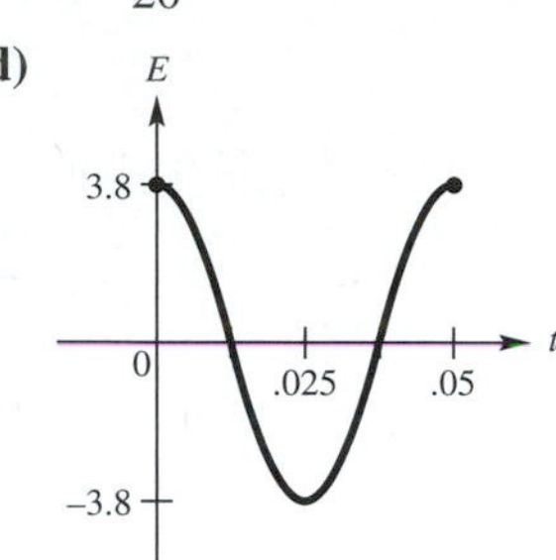

$E = 3.8 \cos 40\pi t$

85. **(a)** $C(x) = .04x^2 + .6x + 330 + 7.5 \sin(2\pi x)$

380, 320, 5, 25

(b) Answers will vary. The seasons are more dramatic in Alaska.

(c) $C(x) = .04(x - 1970)^2 + .6(x - 1970) + 330 + 7.5 \sin[2\pi(x - 1970)]$

87. **(a)** 70.4°; Answers may vary. **(b)**

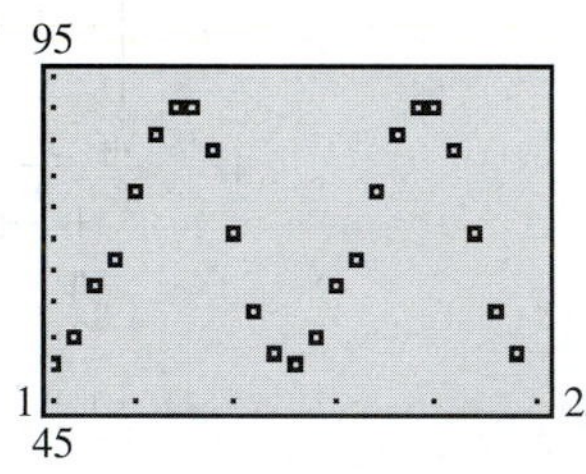

(c) $f(x) = 19.5 \cos\left[\frac{\pi}{6}(x - 7.2)\right] + 70.5$

(d) The function gives an excellent model for the data.

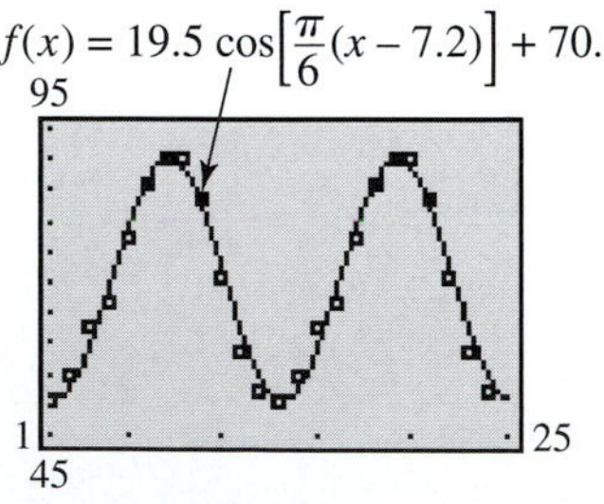

(e)

```
SinReg
 y=a*sin(bx+c)+d
 a=19.72
 b=.52
 c=-2.17
 d=70.47
```

TI-83/84 Plus fixed to the nearest hundredth

89.

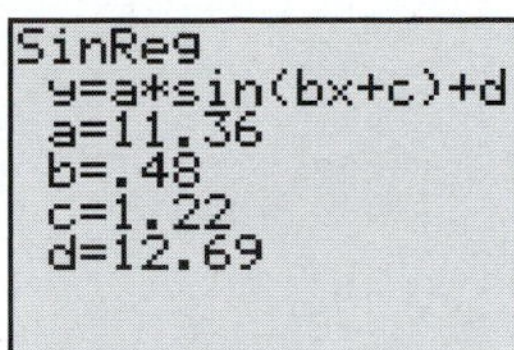

TI-83/84 Plus fixed to the nearest hundredth

30
Y1=11.36*sin(.48X+1.22)+1_
−.1 13
−2

8.4 Exercises *(pages 619–621)*

1. B **3.** E **5.** D **7.** true **9.** true **11.** false; $\tan(-x) = -\tan x$ for all x in the domain. **13.** **(a)** 4π **(b)** none **(c)** $(-\infty, -2] \cup [2, \infty)$ **15.** **(a)** 2π **(b)** $-\frac{\pi}{2}$ **(c)** $(-\infty, -2] \cup [2, \infty)$ **17.** **(a)** 3π **(b)** $\frac{\pi}{2}$ **(c)** $(-\infty, \infty)$ **19.** **(a)** π **(b)** $-\frac{\pi}{2}$ **(c)** $\left(-\infty, -\frac{1}{2}\right] \cup \left[\frac{1}{2}, \infty\right)$ **21.** **(a)** π **(b)** $-\frac{\pi}{4}$ **(c)** $(-\infty, \infty)$

23.

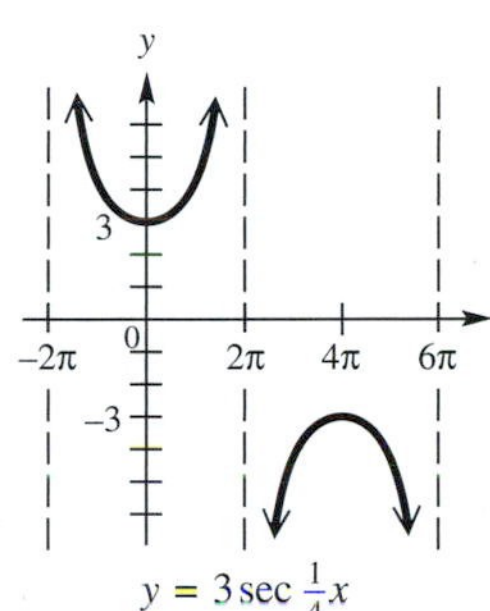

$y = 3 \sec \frac{1}{4}x$

25.

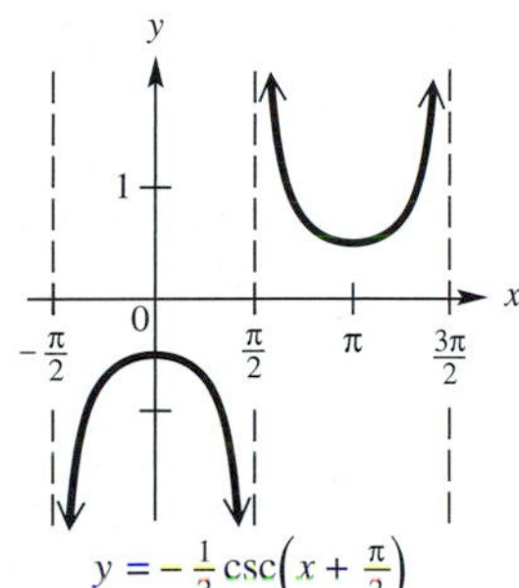

$y = -\frac{1}{2}\csc\left(x + \frac{\pi}{2}\right)$

27.

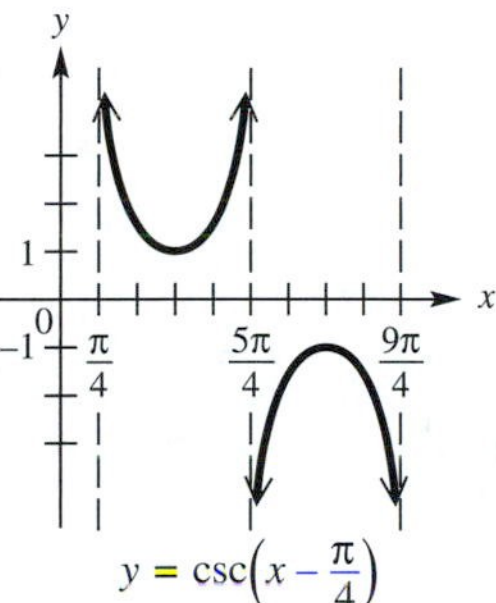

$y = \csc\left(x - \frac{\pi}{4}\right)$

29.

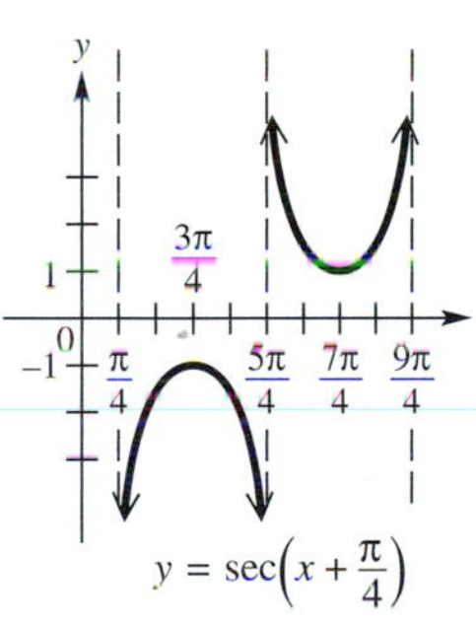

$y = \sec\left(x + \frac{\pi}{4}\right)$

31.

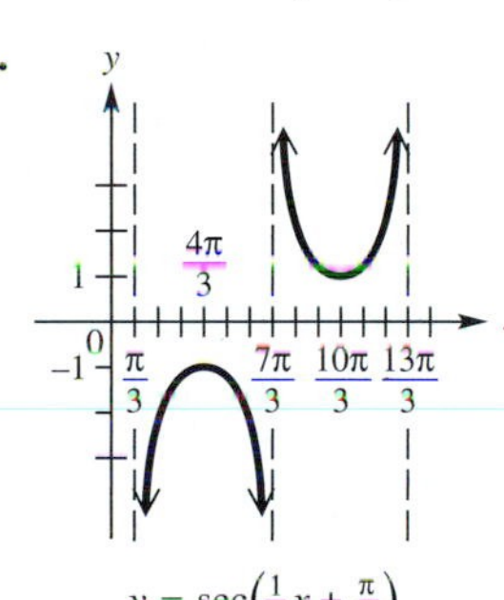

$y = \sec\left(\frac{1}{2}x + \frac{\pi}{3}\right)$

33.

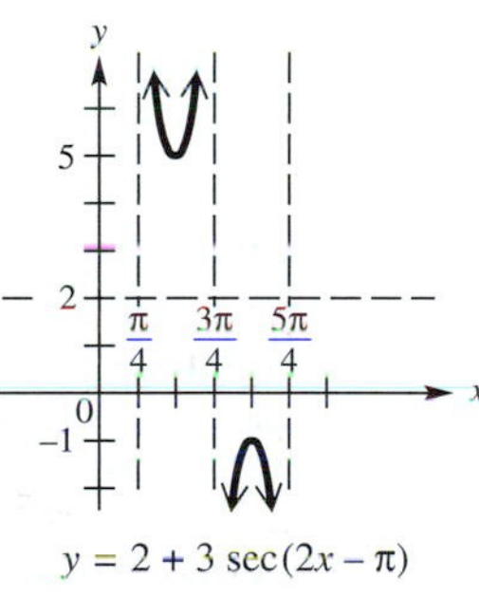

$y = 2 + 3\sec(2x - \pi)$

35.

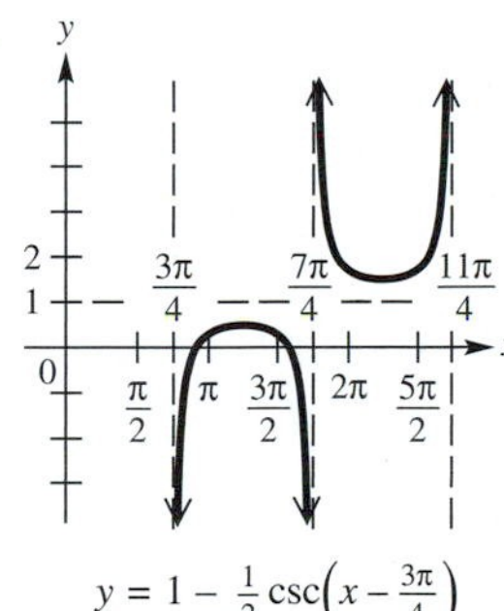

$y = 1 - \frac{1}{2}\csc\left(x - \frac{3\pi}{4}\right)$

37.

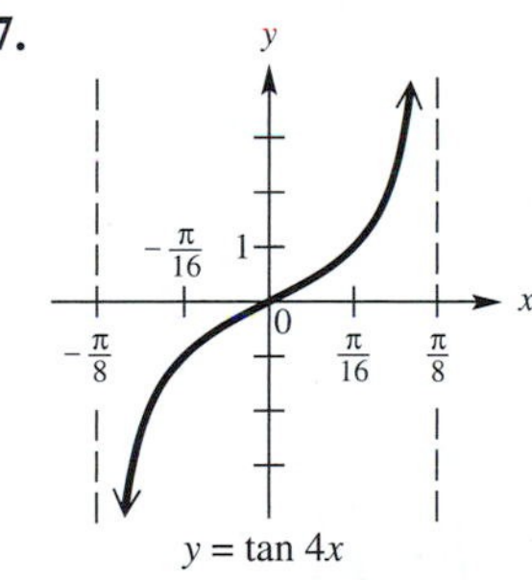

$y = \tan 4x$

39.

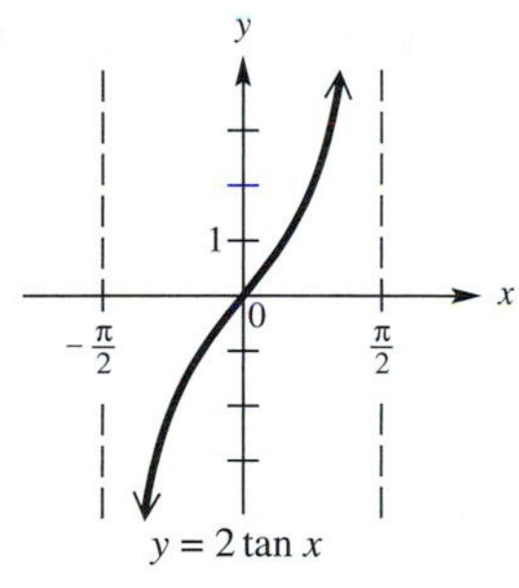

$y = 2\tan x$

41.

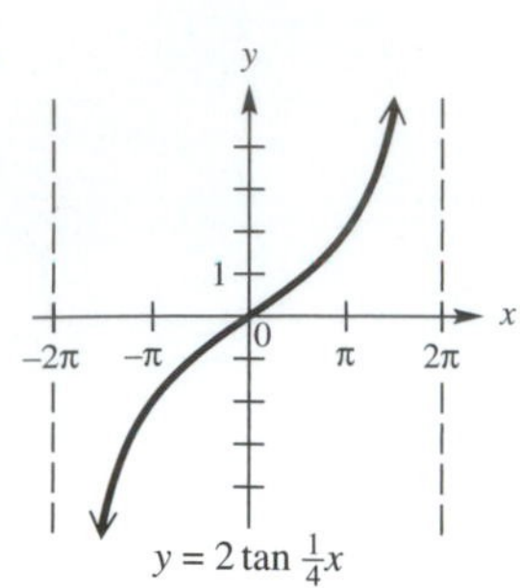

$y = 2\tan\frac{1}{4}x$

43.

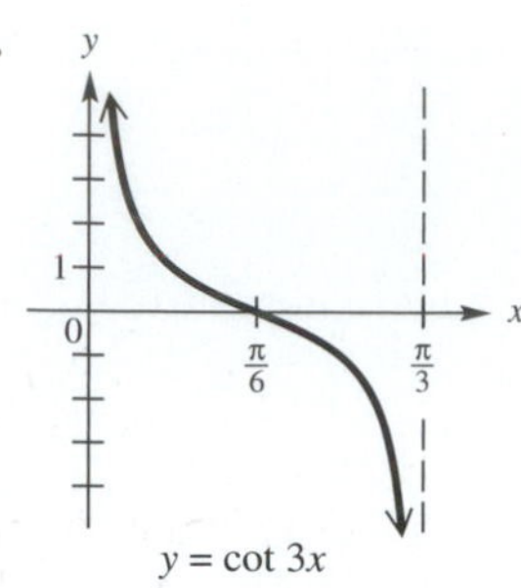

$y = \cot 3x$

45.

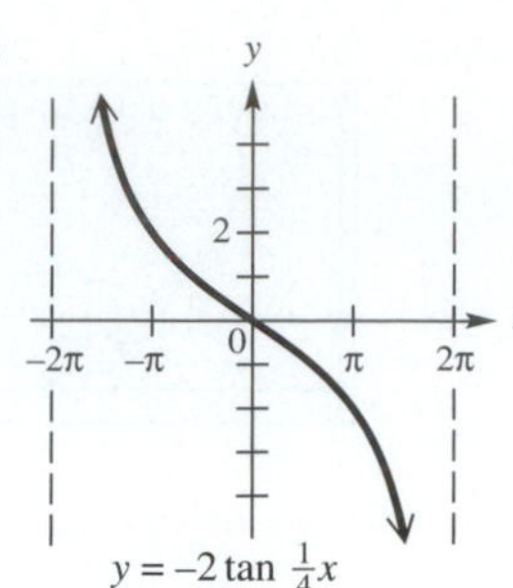

$y = -2\tan\frac{1}{4}x$

47.

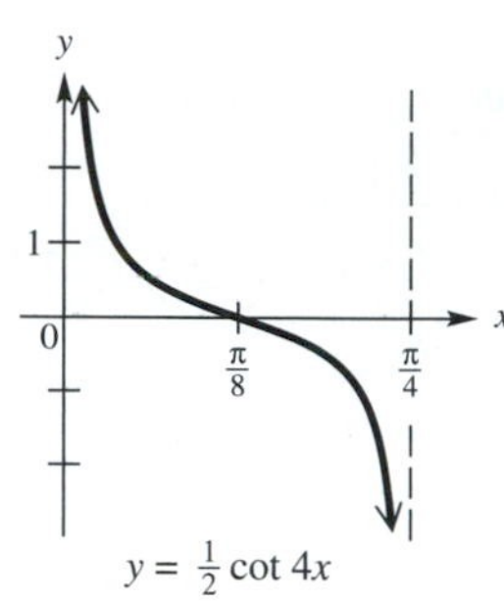

$y = \frac{1}{2}\cot 4x$

49.

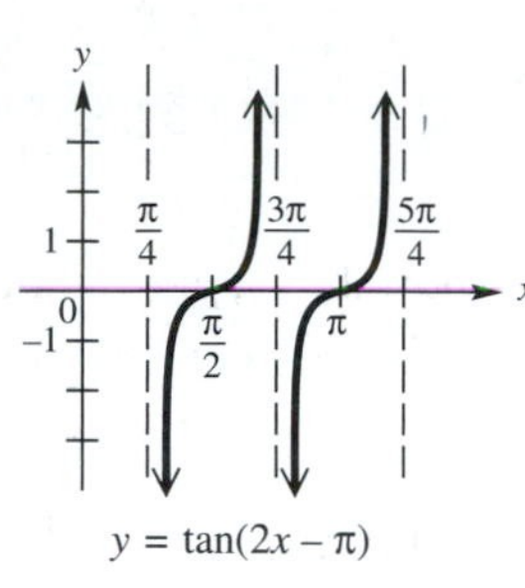

$y = \tan(2x - \pi)$

51.

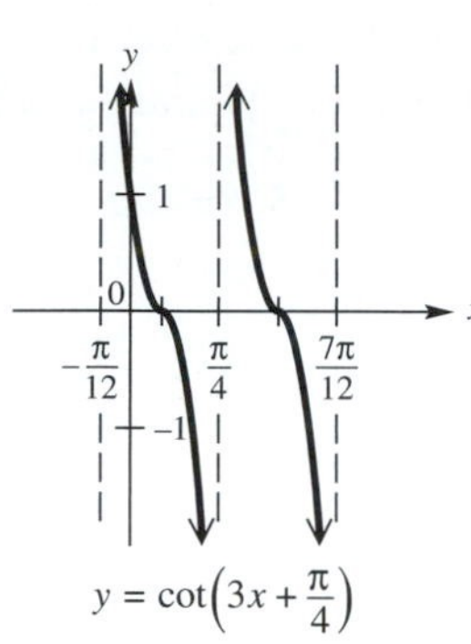

$y = \cot\left(3x + \frac{\pi}{4}\right)$

53.

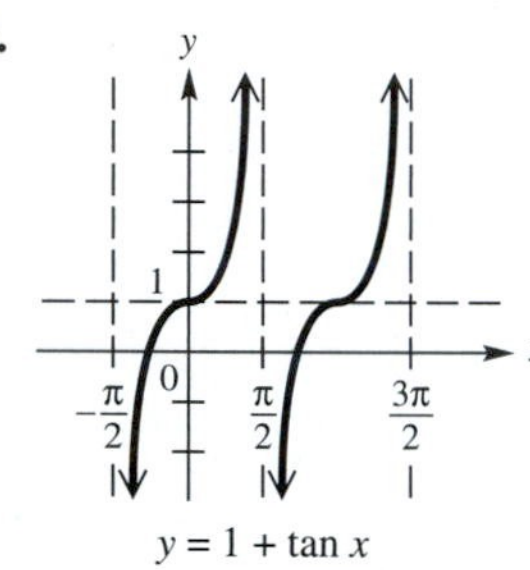

$y = 1 + \tan x$

55.

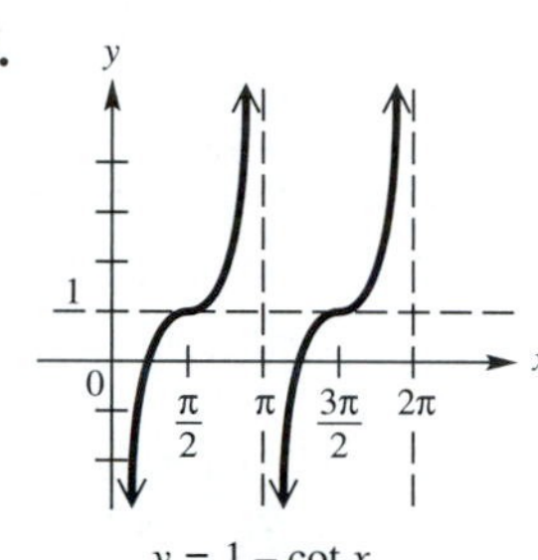

$y = 1 - \cot x$

57.

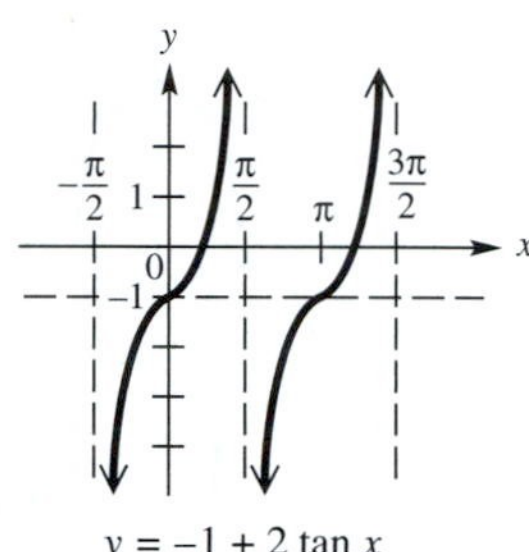

$y = -1 + 2\tan x$

59.

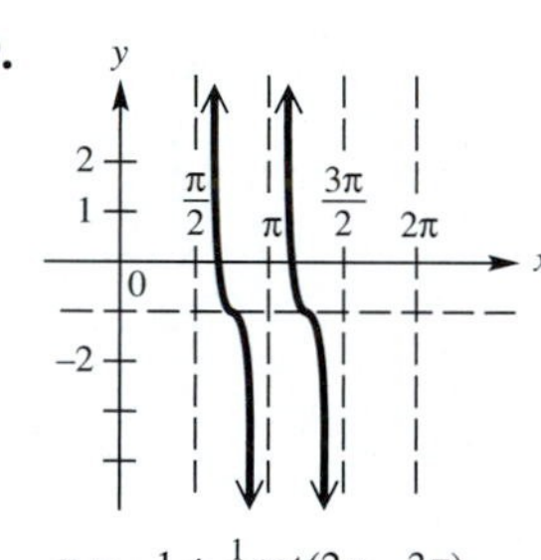

$y = -1 + \frac{1}{2}\cot(2x - 3\pi)$

61. 4 **63.** For values in the interval $-1 \leq x \leq 1$, x and $\tan x$ are approximately equal. As $x \rightarrow 0$, $\tan x$ approaches the value x. **65.** **(a)** 0 m **(b)** 12.3 m **(c)** −12.3 m **(d)** 12.3 m **(e)** It leads to $\tan\frac{\pi}{2}$, which is undefined. The beacon is shining parallel to the wall.

67. In both cases, the result is 1.366025404. **69.** In both cases, the result is 1.83195123.

Reviewing Basic Concepts *(page 621)*

1.

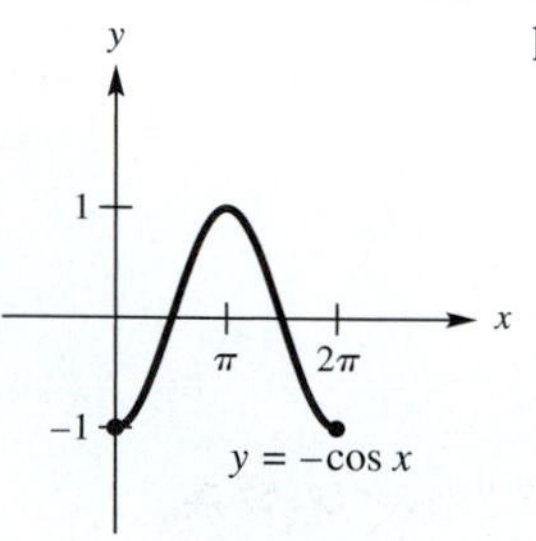

$y = -\cos x$

period: 2π; amplitude: 1

2.

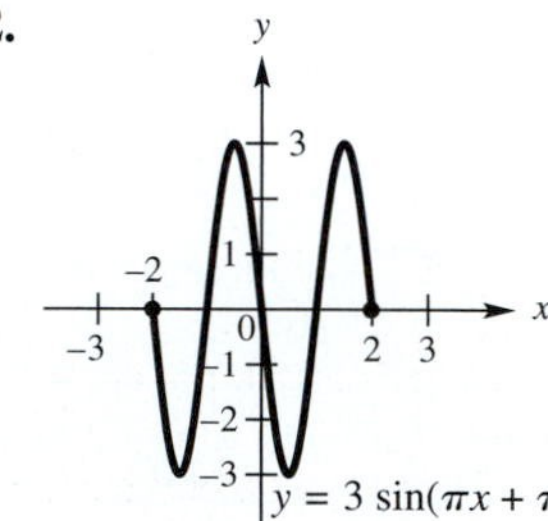

$y = 3\sin(\pi x + \pi)$

amplitude: 3; period: 2; phase shift: −1

3. **(a)** maximum: 18.9 hr; minimum: 5.9 hr **(b)** The amplitude represents half the difference in the daylight hours between the longest and shortest days. The period represents 12 months or one year.

4.

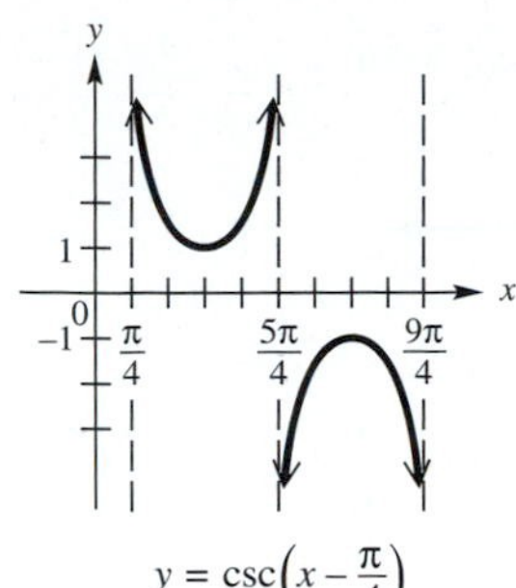

period: 2π; phase shift: $\frac{\pi}{4}$;

domain: $\left\{x \,\middle|\, x \neq \frac{\pi}{4} + n\pi, \text{ where } n \text{ is an integer}\right\}$;

range: $(-\infty, -1] \cup [1, \infty)$

5.

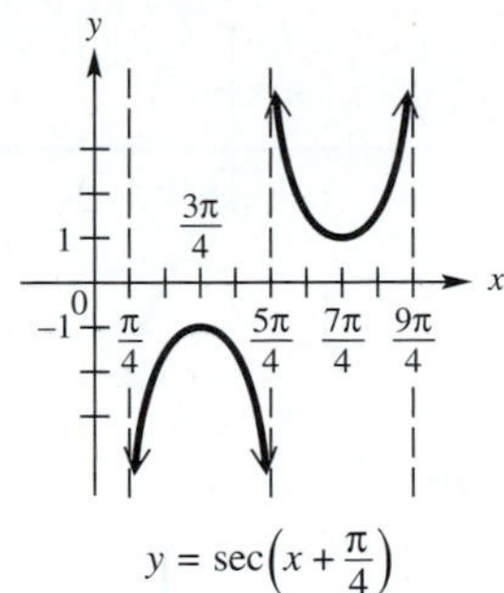

period: 2π; phase shift: $-\frac{\pi}{4}$;

domain: $\left\{x \,\middle|\, x \neq \frac{\pi}{4} + n\pi, \text{ where } n \text{ is an integer}\right\}$;

range: $(-\infty, -1] \cup [1, \infty)$

6.

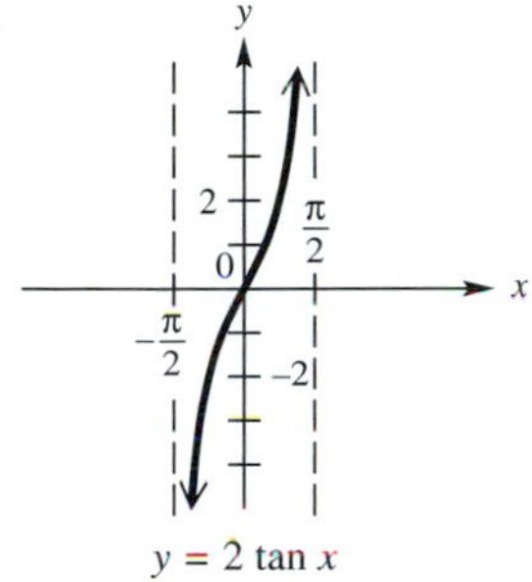

period: π; domain: $\left\{x \,\middle|\, x \neq \frac{\pi}{2} + n\pi, \text{ where } n \text{ is an integer}\right\}$;

range: $(-\infty, \infty)$

7.

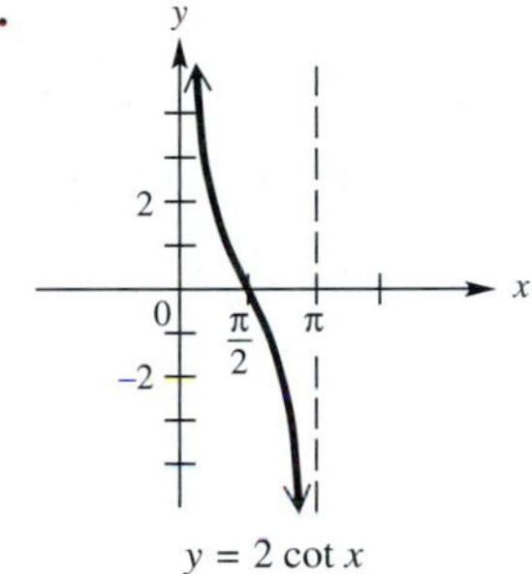

period: π; domain: $\{x \mid x \neq n\pi, \text{ where } n \text{ is an integer}\}$;
range: $(-\infty, \infty)$

8.5 Exercises *(pages 632–635)*

1.

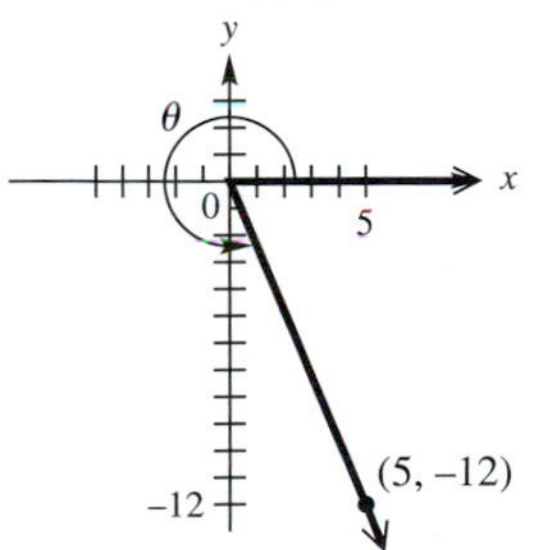

In Exercises 3–7, 15, and 17, we give, in order, sine, cosine, tangent, cotangent, secant, and cosecant.

3. $\frac{4}{5}; -\frac{3}{5}; -\frac{4}{3}; -\frac{3}{4}; -\frac{5}{3}; \frac{5}{4}$ **5.** 1; 0; undefined; 0; undefined; 1 **7.** $\frac{\sqrt{3}}{2}; \frac{1}{2}; \sqrt{3}; \frac{\sqrt{3}}{3}; 2; \frac{2\sqrt{3}}{3}$ **9.** The sine and cosecant functions are reciprocals, and reciprocals always have the same sign, since their product is 1, a positive number.

11. negative **13.** negative **15.**

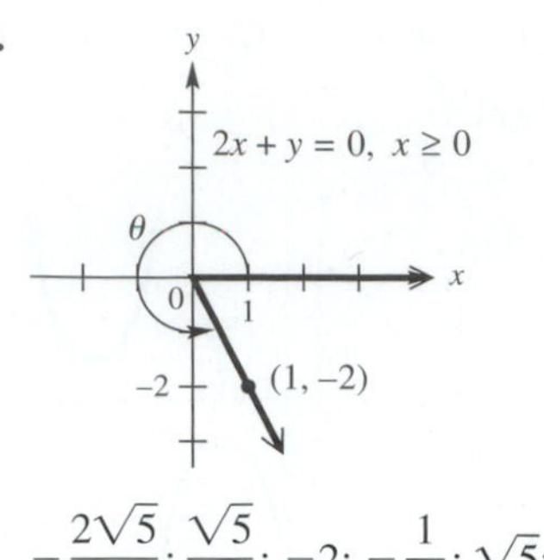

$-\frac{2\sqrt{5}}{5}; \frac{\sqrt{5}}{5}; -2; -\frac{1}{2}; \sqrt{5}; -\frac{\sqrt{5}}{2}$

17.

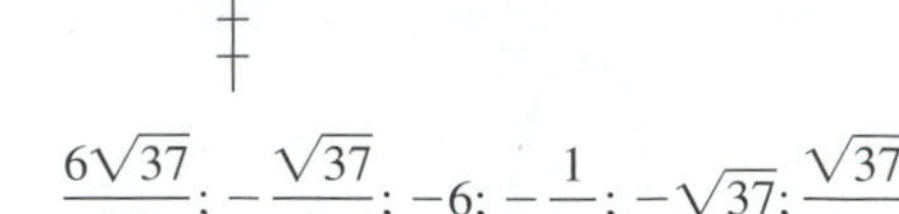

$\frac{6\sqrt{37}}{37}; -\frac{\sqrt{37}}{37}; -6; -\frac{1}{6}; -\sqrt{37}; \frac{\sqrt{37}}{6}$

19. $\sin 540° = 0$; $\cos 540° = -1$; $\tan 540° = 0$; csc 540° is undefined; $\sec 540° = -1$; cot 540° is undefined. **21.** -7 **23.** 3 **25.** 1 **27.** 0 **29.** 0 **31.** They are equal. **33.** They are equal. If $P(x, y)$ lies on the terminal side of θ, then $P(x, -y)$ lies on the terminal side of $-\theta$. Thus, $\cos\theta = x$ and $\cos(-\theta) = x$. **35.** 40° **37.** 45° **39.** decrease; decrease **41.** -1; $\theta = 180°$ **43.** -5 **45.** $-\frac{3\sqrt{5}}{5}$ **47.** .10199657 **49.** The cosine of any angle cannot be greater than 1 (or less than -1). **51.** $\sqrt{3}$ **53.** 2° **55.** II **57.** I or III **59.** +; −; − **61.** −; +; − **63.** −; +; − **65.** tan 30° **67.** sec 33° **69.** impossible **71.** possible **73.** possible **75.** impossible **77.** $\frac{\sqrt{15}}{4}$ **79.** $-\frac{4}{3}$ **81.** $-\frac{\sqrt{3}}{2}$ **83.** −.56616682 **85.** 5

In Exercises 87–91, we give, in order, sine, cosine, tangent, cotangent, secant, and cosecant.

87. $\frac{15}{17}; -\frac{8}{17}; -\frac{15}{8}; -\frac{8}{15}; -\frac{17}{8}; \frac{17}{15}$ **89.** $-\frac{\sqrt{3}}{2}; -\frac{1}{2}; \sqrt{3}; \frac{\sqrt{3}}{3}; -2; -\frac{2\sqrt{3}}{3}$ **91.** −.555762; .831342; −.668512; −1.49586; 1.20287; −1.79933 **93.** $\sqrt{1 - \sin^2\theta}$ **95.** $-\frac{1}{\sqrt{1 + \cot^2\theta}}$ or $-\frac{\sqrt{1 + \cot^2\theta}}{1 + \cot^2\theta}$ **97.** $\frac{\sin\theta}{\sqrt{1 - \sin^2\theta}}$ or $\frac{\sin\theta\sqrt{1 - \sin^2\theta}}{1 - \sin^2\theta}$ **101.** false; For example, $\sin 30° + \cos 30° \approx .5 + .8660 = 1.3660 \neq 1$. **103.** **(a)** 3% **(b)** $25{,}000\left(\frac{3}{\sqrt{10{,}009}}\right) \approx 750$ lb

8.6 Exercises *(pages 644–647)*

1. $\sin A = \frac{21}{29}$; $\cos A = \frac{20}{29}$; $\tan A = \frac{21}{20}$; $\cot A = \frac{20}{21}$; $\sec A = \frac{29}{20}$; $\csc A = \frac{29}{21}$

3. $\sin A = \frac{n}{p}$; $\cos A = \frac{m}{p}$; $\tan A = \frac{n}{m}$; $\cot A = \frac{m}{n}$; $\sec A = \frac{p}{m}$; $\csc A = \frac{p}{n}$

The number of digits in part (b) of Exercises 5–19 may vary.

5. **(a)** $\frac{\sqrt{3}}{3}$ **(b)** .5773502692 **7.** **(a)** $\frac{1}{2}$ **9.** **(a)** $\frac{2\sqrt{3}}{3}$ **(b)** 1.154700538 **11.** **(a)** $\sqrt{2}$ **(b)** 1.414213562 **13.** **(a)** $\frac{\sqrt{2}}{2}$ **(b)** .7071067812 **15.** **(a)** $\frac{\sqrt{3}}{2}$ **(b)** .8660254038 **17.** **(a)** $\sqrt{3}$ **(b)** 1.732050808 **19.** **(a)** 2 **21.** The value he gave was an *approximation,* not the exact value of sin 45°, which is $\frac{\sqrt{2}}{2}$. **23.** tan 17° **25.** cos 52° **27.** cot 64° 17′ **29.** $\sin\frac{3\pi}{10}$ **31.** $\cot\left(\frac{\pi}{2} - .5\right)$ **33.** 82° **35.** 50° **37.** 45° **39.** 30° **41.** $\frac{\pi}{3}$ **43.** $\frac{\pi}{3}$ **45.** It is easy to find one-half of 2, which is 1. This is, then, the measure of the side opposite the 30° angle, and the ratios are easily found. Yes, any positive number could have been used. **47.** $\frac{\sqrt{3}}{3}; \sqrt{3}$ **49.** $\frac{\sqrt{3}}{2}; \frac{\sqrt{3}}{3}; \frac{2\sqrt{3}}{3}$ **51.** $-1; -1$ **53.** $-\frac{\sqrt{3}}{2}; -\frac{2\sqrt{3}}{3}$

In Exercises 55–61, we give, in order, sine, cosine, tangent, cotangent, secant, and cosecant.

55. $-\frac{\sqrt{3}}{2}; \frac{1}{2}; -\sqrt{3}; -\frac{\sqrt{3}}{3}; 2; -\frac{2\sqrt{3}}{3}$ **57.** $\frac{\sqrt{2}}{2}; \frac{\sqrt{2}}{2}; 1; 1; \sqrt{2}; \sqrt{2}$ **59.** $-\frac{1}{2}; \frac{\sqrt{3}}{2}; -\frac{\sqrt{3}}{3}; -\sqrt{3}; \frac{2\sqrt{3}}{3}; -2$ **61.** $\frac{\sqrt{2}}{2}; \frac{\sqrt{2}}{2}; 1; 1; \sqrt{2}; \sqrt{2}$ **63.** .5543090515 **65.** 1.134277349 **67.** -1.002203376 **69.** -5.729741647 **71.** .5984721441 **73.** -3.380515006 **75.** **(a)** $-\sin\frac{\pi}{6}$ **(b)** $-\frac{1}{2}$ **(c)** $\sin\frac{7\pi}{6} = -\sin\frac{\pi}{6} = -.5$ **77.** **(a)** $-\tan\frac{\pi}{4}$ **(b)** -1 **(c)** $\tan\frac{3\pi}{4} = -\tan\frac{\pi}{4} = -1$ **79.** **(a)** $-\cos\frac{\pi}{6}$ **(b)** $-\frac{\sqrt{3}}{2}$ **(c)** $\cos\frac{7\pi}{6} = -\cos\frac{\pi}{6} \approx -.8660254038$ **81.** 30°; 150° **83.** 120°; 300° **85.** 120°; 300° **87.** 46.59388121°; 313.4061188° **89.** 24.39257624°; 155.6074238° **91.** 41.24818261°; 221.2481826° **93.** .2095206607; 3.351113314 **95.** 1.27979966; 4.421392314 **97.** It represents the distance from the point (x_1, y_1) to the origin. **98.** 60° **99.** 60° **100.** 60° **101.** It is a measure of the angle formed by the positive x-axis and the ray $y = \sqrt{3}x, x \geq 0$. **102.** $\left(\frac{y_1}{x_1}\right)^2 = 3$; exact value: $\sqrt{3}$ **103.** slope; tangent **104.** 1; It illustrates the identity $\cos^2\theta + \sin^2\theta = 1$. **105.** They agree, as they are both approximately 1.154700538. **106.** x-coordinate: .5; y-coordinate: .8660254 (approximately) **107.** $\cos 60° = .5$; This is the x-coordinate of the point found in Exercise 106. Because $r = 1$, here $\cos 60° = \frac{x}{1} = x = .5$. **108.** $\sin 60° \approx .86602540$; This is the y-coordinate of the point found in Exercise 106. Because $r = 1$, here $\sin 60° = \frac{y}{1} = y \approx .86602540$. **109.** $\theta = \sin^{-1}\frac{2}{25} \approx 4.6°$ **111.** 2×10^8 m per sec **113.** 19° **115.** 48.7° **117.** **(a)** approximately 155 ft **(b)** approximately 194 ft **(c)** As the grade decreases from uphill to downhill, the braking distance increases, which corresponds to driving experience.

8.7 Exercises *(pages 654–659)*

1. $B = 53° 40'$; $a = 571$ m; $b = 777$ m **3.** $M = 38.8°$; $n = 154$ m; $p = 198$ m **5.** $A = 47.9108°$; $c = 84.816$ cm; $a = 62.942$ cm **7.** $A = 36°$; $B = 54°$; $c = 12$ ft **9.** $A = 48°$; $B = 42°$; $a = 8.2$ ft **11.** $B = 62.00°$; $a = 8.17$ ft; $b = 15.4$ ft **13.** $A = 17.00°$; $a = 39.1$ in.; $c = 134$ in. **15.** $c = 85.9$ yd; $A = 62° 50'$; $B = 27° 10'$ **17.** The other acute angle requires the least work to find. **19.** 38.6598° **21.** Because AD and BC are parallel, angle DAB is congruent to angle ABC, as they are alternate interior angles of the transversal AB. (A theorem of elementary geometry assures us of this.) **23.** It is measured clockwise from the north. **25.** 9.35 m **27.** 13.3 ft **29.** 128 ft **31.** 26.3° or 26° 20′ **33.** $A = 35.987°$ or 35° 59′ 10″; $B = 54.013°$ or 54° 00′ 50″ **35.** 114 ft **37.** 5.18 m **39.** 84.7 m **41.** 148 mi **43.** 1.48 mi **45.** 150 km **47.** 33.4 m **49.** 583 ft **51.** **(a)** 23.4 ft **(b)** 48.3 ft **(c)** The faster the speed, the more land needs to be cleared inside the curve. **53.** **(a)** $\tan\theta = \frac{y}{x}$ **(b)** $x = \frac{y}{\tan\theta}$ **55.** $a = 12$; $b = 12\sqrt{3}$; $d = 12\sqrt{3}$; $c = 12\sqrt{6}$ **57.** $m = \frac{7\sqrt{3}}{3}$; $a = \frac{14\sqrt{3}}{3}$; $n = \frac{14\sqrt{3}}{3}$; $q = \frac{14\sqrt{6}}{3}$ **59.** $A = \frac{s^2\sqrt{3}}{4}$ **61.** $\cot\theta$

8.8 Exercises *(pages 662–663)*

1. **(a)** $s(t) = 2\cos 4\pi t$ **(b)** $s(1) = 2$; The weight is neither moving upward nor downward. At $t = 1$ the motion of the weight is changing from up to down. **3.** **(a)** $s(t) = -3\cos 2.5\pi t$ **(b)** $s(1) = 0$; upward

5. $s(t) = .21 \cos 55\pi t$

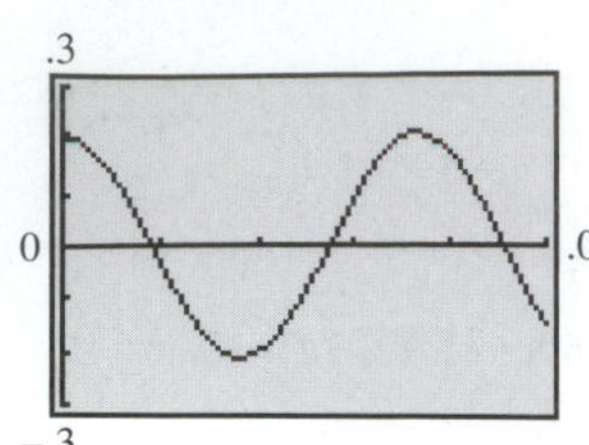

7. $s(t) = .14 \cos 110\pi t$

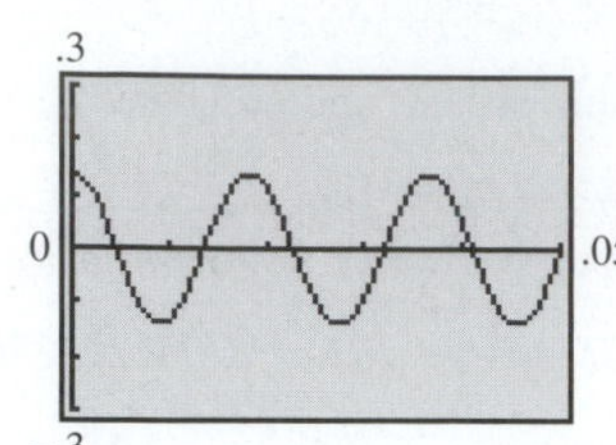

9. (a) $s(t) = 2 \sin 2t$; amplitude: 2; period: π; frequency: $\frac{1}{\pi}$ **(b)** $s(t) = 2 \sin 4t$; amplitude: 2; period: $\frac{\pi}{2}$; frequency: $\frac{2}{\pi}$ **11.** $\frac{8}{\pi^2}$ **13. (a)** amplitude: $\frac{1}{2}$; period: $\sqrt{2}\pi$; frequency: $\frac{\sqrt{2}}{2\pi}$ **(b)** $s(t) = \frac{1}{2} \sin \sqrt{2}t$

15. (a) 4 in. **(b)** $\frac{5}{\pi}$ cycles per sec; $\frac{\pi}{5}$ sec **(c)** after $\frac{\pi}{10}$ sec **(d)** approximately 2; After 1.466 sec, the weight is about 2 in. above the equilibrium position. **17. (a)** $s(t) = -2 \cos 6\pi t$ **(b)** 3 cycles per sec **19. (a)** The spring is compressed 2 inches. **(b)** 1 cycle per sec **(c)** approximately .305 sec and .636 sec

Reviewing Basic Concepts *(pages 663–664)*

1. $\sin\theta = \frac{5\sqrt{29}}{29}$; $\cos\theta = -\frac{2\sqrt{29}}{29}$; $\tan\theta = -\frac{5}{2}$; $\cot\theta = -\frac{2}{5}$; $\sec\theta = -\frac{\sqrt{29}}{2}$; $\csc\theta = \frac{\sqrt{29}}{5}$ **2.** $\sin 270° = -1$; $\cos 270° = 0$; $\tan 270°$ is undefined; $\cot 270° = 0$; $\sec 270°$ is undefined; $\csc 270° = -1$ **3. (a)** impossible **(b)** possible **(c)** possible **4.** $\cos\theta = -\frac{\sqrt{5}}{3}$; $\tan\theta = \frac{2\sqrt{5}}{5}$; $\cot\theta = \frac{\sqrt{5}}{2}$; $\sec\theta = -\frac{3\sqrt{5}}{5}$; $\csc\theta = -\frac{3}{2}$

5. $\sin A = \frac{8}{17}$; $\cos A = \frac{15}{17}$; $\tan A = \frac{8}{15}$; $\cot A = \frac{15}{8}$; $\sec A = \frac{17}{15}$; $\csc A = \frac{17}{8}$ **6.** Row 1: $\frac{1}{2}, \frac{\sqrt{3}}{2}, \frac{\sqrt{3}}{3}, \sqrt{3}, \frac{2\sqrt{3}}{3}, 2$; Row 2: $\frac{\sqrt{2}}{2}, \frac{\sqrt{2}}{2}, 1, 1, \sqrt{2}, \sqrt{2}$; Row 3: $\frac{\sqrt{3}}{2}, \frac{1}{2}, \sqrt{3}, \frac{\sqrt{3}}{3}, 2, \frac{2\sqrt{3}}{3}$ **7. (a)** $\cos 63°$ **(b)** $\cot \frac{3\pi}{10}$ **8. (a)** 80° **(b)** 5° **(c)** $\frac{\pi}{3}$

9. $\sin 315° = -\frac{\sqrt{2}}{2}$; $\cos 315° = \frac{\sqrt{2}}{2}$; $\tan 315° = -1$; $\cot 315° = -1$; $\sec 315° = \sqrt{2}$; $\csc 315° = -\sqrt{2}$

10. (a) .725374371 **(b)** 5.67128182 **(c)** −1.321348709 **11.** 150°; 330° **12.** .75; 2.391592654 **13.** 19,600 ft **14. (a)** 4 in. **(b)** after $\frac{1}{8}$ sec **(c)** 4 cycles per sec; $\frac{1}{4}$ sec

Chapter 8 Review Exercises *(pages 668–673)*

1. 186° **3.** 1280° **5.** 1 radian ≈ 57.3° > 1° **7.** $\frac{2\pi}{3}$ **9.** 225° **11.** $\frac{4\pi}{3}$ in. **13.** 35.8 cm **15.** 41 yd

17. $-\frac{1}{2}$ **19.** 2 **21.** −11.4266 **23.** .5528

25. 2; 2π; none; none

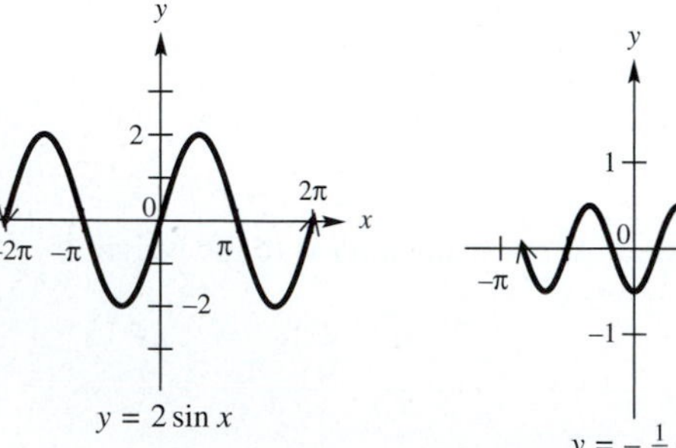

27. $\frac{1}{2}$; $\frac{2\pi}{3}$; none; none

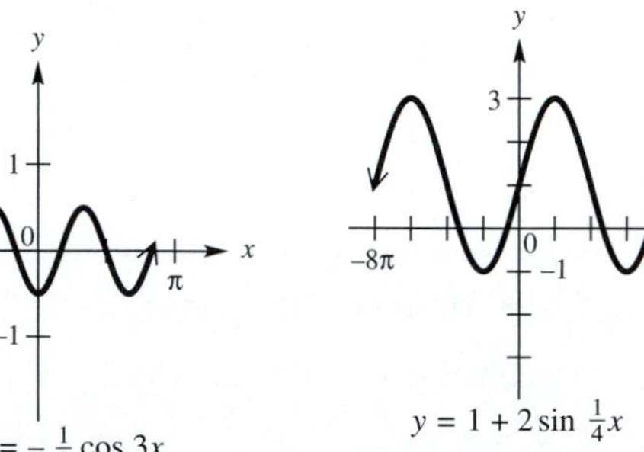

29. 2; 8π; upward 1 unit; none

y
3
x
−8π
0
−1
8π
y = 1 + 2 sin ¼x

31. 3; 2π; none; $-\frac{\pi}{2}$

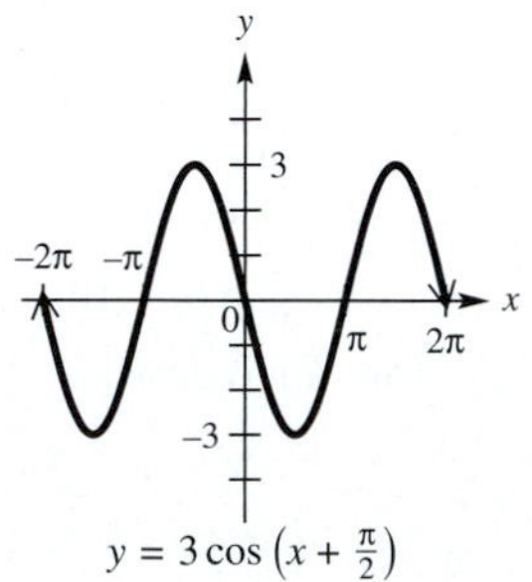

33. not applicable; π; none; $\dfrac{\pi}{8}$

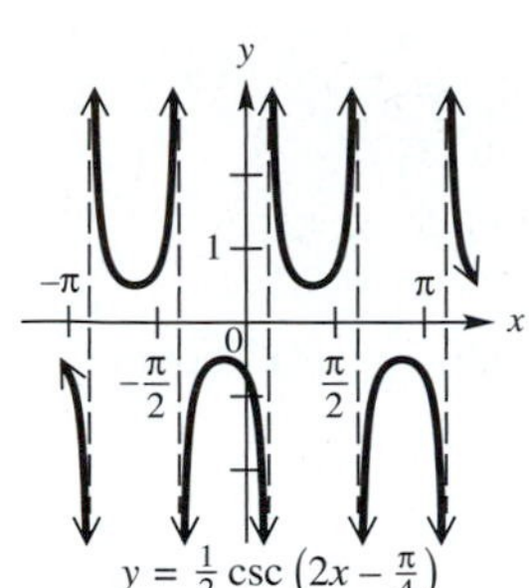

$y = \frac{1}{2}\csc\left(2x - \frac{\pi}{4}\right)$

35. not applicable; π; none; none

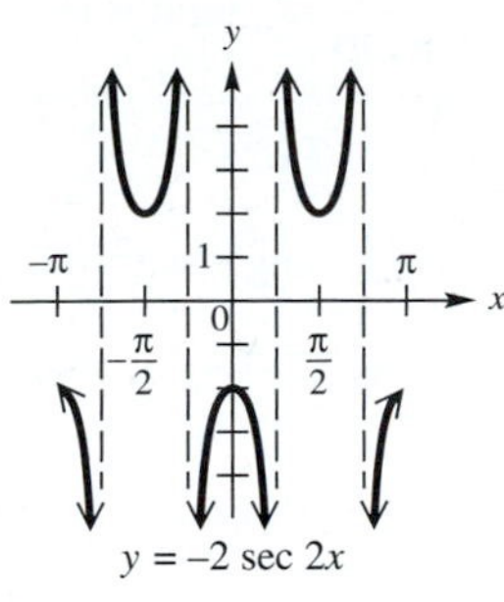

$y = -2 \sec 2x$

37. Because f has period π, $f(x) = f(x + n\pi)$ for integers n. Thus, $f\left(\dfrac{6\pi}{5}\right) = f\left(\dfrac{6\pi}{5} - 2\pi\right) = f\left(-\dfrac{4\pi}{5}\right)$, and it follows that $f\left(-\dfrac{4\pi}{5}\right) = 1$.

39.

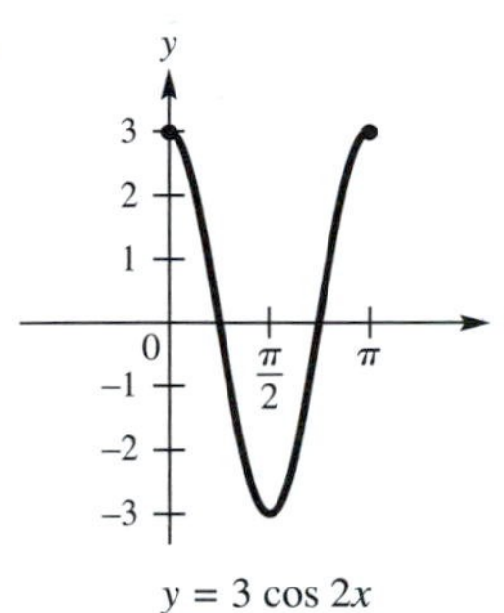

$y = 3 \cos 2x$

41.

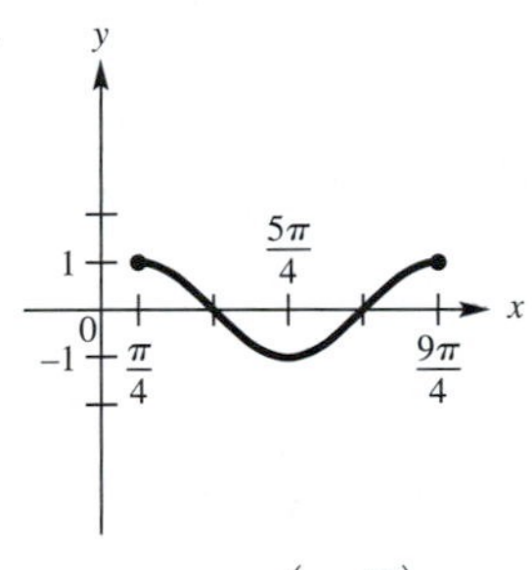

$y = \cos\left(x - \frac{\pi}{4}\right)$

43.

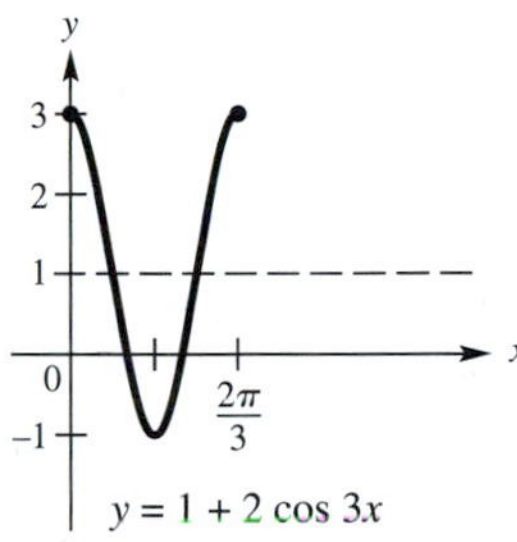

$y = 1 + 2 \cos 3x$

45. tangent **47.** cosine **49.** cotangent **51.** $y = 3 \sin\left[2\left(x - \dfrac{\pi}{4}\right)\right]$ **53.** $y = \dfrac{1}{3} \sin \dfrac{\pi}{2} x$

55. **(a)**

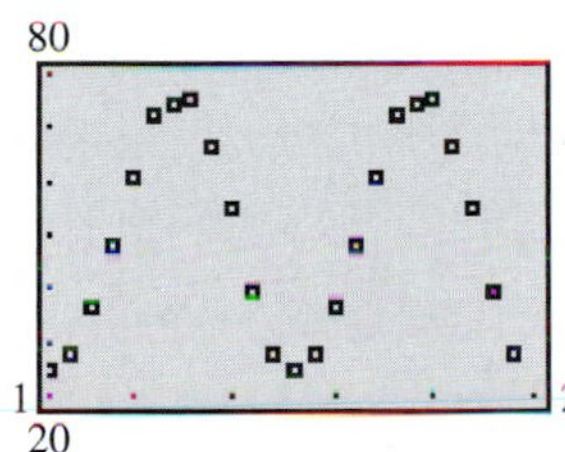

(b) $f(x) = 25 \sin\left[\dfrac{\pi}{6}(x - 4.2)\right] + 50$

(c) a represents the amplitude, $\dfrac{2\pi}{b}$ the period, c the vertical shift, and d the phase shift.

(d) The function gives an excellent model for the data.

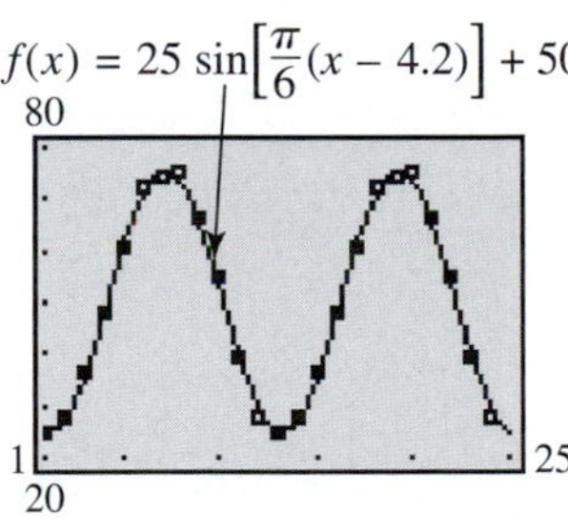

(e)

```
SinReg
 y=a*sin(bx+c)+d
 a=25.77
 b=.52
 c=-2.19
 d=50.57
```

In Exercises 57–61, we give, in order, sine, cosine, tangent, cotangent, secant, and cosecant.

57. $-\dfrac{\sqrt{2}}{2}$; $-\dfrac{\sqrt{2}}{2}$; 1; 1; $-\sqrt{2}$; $-\sqrt{2}$ **59.** 0; -1; 0; undefined; -1; undefined

61. $-\dfrac{2\sqrt{85}}{85}$; $\dfrac{9\sqrt{85}}{85}$; $-\dfrac{2}{9}$; $-\dfrac{9}{2}$; $\dfrac{\sqrt{85}}{9}$; $-\dfrac{\sqrt{85}}{2}$ **63.** tangent and secant **65.** possible

In Exercises 67 and 71–75, we give, in order, sine, cosine, tangent, cotangent, secant, and cosecant.

67. $-\frac{\sqrt{39}}{8}; -\frac{5}{8}; \frac{\sqrt{39}}{5}; \frac{5\sqrt{39}}{39}; -\frac{8}{5}; -\frac{8\sqrt{39}}{39}$ **69.** Evaluate $\frac{1}{1.6778490} \approx .5960011896$

71. $\frac{20}{29}; \frac{21}{29}; \frac{20}{21}; \frac{21}{20}; \frac{29}{21}; \frac{29}{20}$ **73.** $-\frac{\sqrt{3}}{2}; \frac{1}{2}; -\sqrt{3}; -\frac{\sqrt{3}}{3}; 2; -\frac{2\sqrt{3}}{3}$ **75.** $-\frac{1}{2}; \frac{\sqrt{3}}{2}; -\frac{\sqrt{3}}{3}; -\sqrt{3}; \frac{2\sqrt{3}}{3}; -2$

77. -1.3563417 **79.** .20834446 **81.** 55.7° **83.** No, it will give an angle whose tangent equals 25.

85. $B = 31° 30'$; $a = 638$; $b = 391$ **87.** 1200 m **89.** 140 mi **91.** 73.7 ft **93.** 419 **95.** **(b)** $\frac{\pi}{6}$ **(c)** There is less ultraviolet light when $\theta = \frac{\pi}{3}$. **97.** 60°; $\frac{\pi}{3}$ radians **99.** amplitude: 4; period: 2; frequency: $\frac{1}{2}$

101. The frequency is the number of cycles in one second; 4 in. below; at the initial point; $2\sqrt{2}$ in. below

Chapter 8 Test *(pages 673–674)*

1. 203° **2.** 2700° **3.** **(a)** $\frac{2\pi}{3}$ **(b)** 162° **4.** **(a)** $\frac{4}{3}$ **(b)** 15,000 cm^2 **5.** $\frac{12\pi}{5} \approx 7.54$ in.

6.

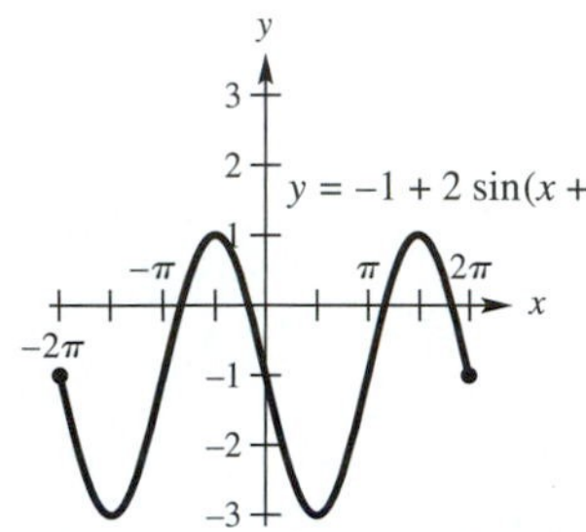

7.

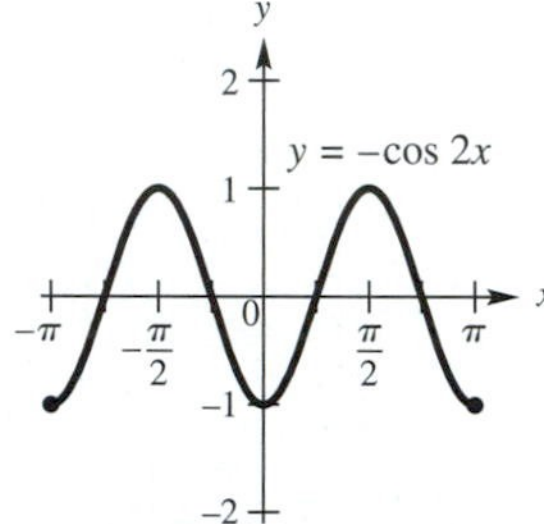

8.

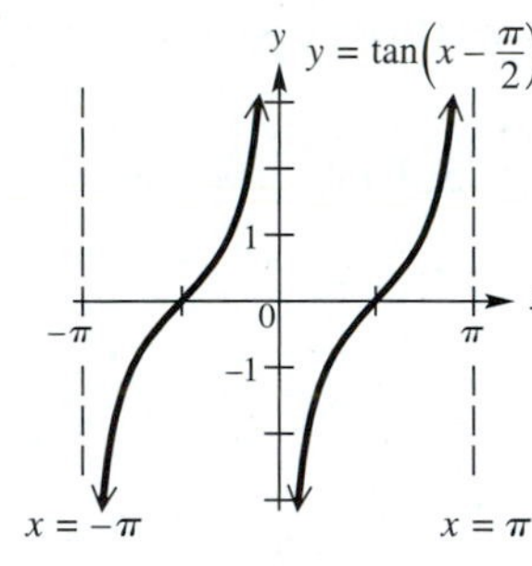

9. **(a)** $f(x) = 17.5 \sin\left[\frac{\pi}{6}(x - 4)\right] + 67.5$

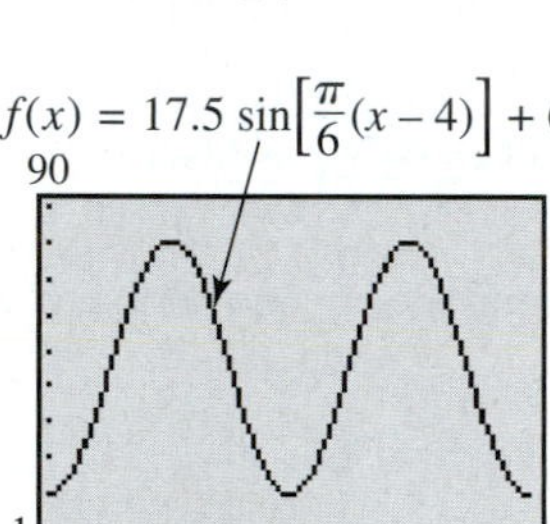

(b) 17.5; 12; 4; 67.5 upward
(c) approximately 52°F
(d) 50°F in January; 85°F in July
(e) approximately 67.5°; This is the vertical translation.

10. **(a)** 210° **(b)** $x = -\sqrt{3}$; $y = -1$ **(c)** $\sin(-150°) = -\frac{1}{2}$; $\cos(-150°) = -\frac{\sqrt{3}}{2}$; $\tan(-150°) = \frac{\sqrt{3}}{3}$; $\cot(-150°) = \sqrt{3}$; $\sec(-150°) = -\frac{2\sqrt{3}}{3}$; $\csc(-150°) = -2$ **(d)** $-\frac{5\pi}{6}$ **11.** III **12.** $\sin\theta = -\frac{5\sqrt{29}}{29}$; $\cos\theta = \frac{2\sqrt{29}}{29}$; $\tan\theta = -\frac{5}{2}$

13. $\sin\theta = -\frac{3}{5}$; $\tan\theta = -\frac{3}{4}$; $\cot\theta = -\frac{4}{3}$; $\sec\theta = \frac{5}{4}$; $\csc\theta = -\frac{5}{3}$ **14.** $x = 4$; $y = 4\sqrt{3}$; $z = 4\sqrt{2}$; $w = 8$

15. $-\sqrt{3}$ **16.** **(a)** .97939940 **(b)** .20834446 **(c)** 1.9362132 **17.** .97169234 **18.** $B = 31° 30'$; $a = 638$; $b = 391$

19. 15.5 ft **20.** 110 km **21.** 3 in. **22.** after $\frac{1}{2}$ sec

CHAPTER 9 TRIGONOMETRIC IDENTITIES AND EQUATIONS

9.1 Exercises *(pages 684–687)*

1. odd **3.** odd **5.** even **7.** B **9.** E **11.** A **13.** A **15.** D **17.** $\cos 4.38$ **19.** $-\sin .5$

21. $-\tan \frac{\pi}{7}$ **23.** The student has neglected to write the argument of the functions (for example, θ or t).

25. $\frac{\pm\sqrt{1+\cot^2\theta}}{1+\cot^2\theta}$; $\frac{\pm\sqrt{\sec^2\theta-1}}{\sec\theta}$ **27.** $\frac{\pm\sin\theta\sqrt{1-\sin^2\theta}}{1-\sin^2\theta}$; $\frac{\pm\sqrt{1-\cos^2\theta}}{\cos\theta}$; $\pm\sqrt{\sec^2\theta-1}$; $\frac{\pm\sqrt{\csc^2\theta-1}}{\csc^2\theta-1}$

29. $\frac{\pm\sqrt{1-\sin^2\theta}}{1-\sin^2\theta}$; $\pm\sqrt{\tan^2\theta+1}$; $\frac{\pm\sqrt{1+\cot^2\theta}}{\cot\theta}$; $\frac{\pm\csc\theta\sqrt{\csc^2\theta-1}}{\csc^2\theta-1}$ **31.** $\sin\theta$ **33.** $\tan^2\beta$ **35.** $\tan^2 x$

37. $\sec^2 x$ **39.** $\cos\theta$ **41.** $\cot\theta$ **43.** $\cos^2\theta$ **45.** $\sec\theta - \cos\theta$ **47.** $\cot\theta - \tan\theta$ **49.** $\frac{\sin^2\theta}{\cos\theta}$ **51.** $\cos^2\theta$

53. $\csc\theta\sec\theta$ or $\frac{1}{\sin\theta\cos\theta}$ **55.** $1+\sec s$ **57.** 1 **59.** 1 **61.** $2+2\sin t$ **63.** $-\frac{2\cos x}{\sin^2 x}$ or $-2\cot x\csc x$

65.

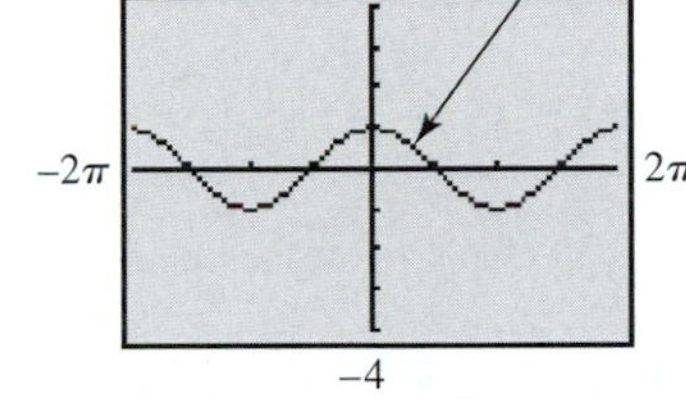

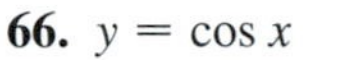

66. $y = \cos x$ **67.**

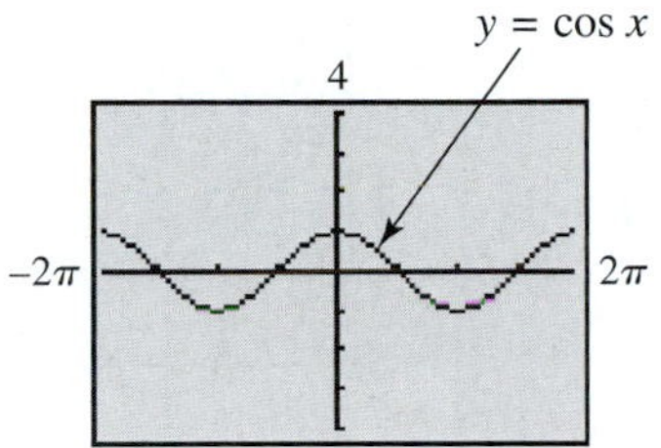

68. yes; $(\sec x + \tan x)(1 - \sin x) = \cos x$ **70.** $\frac{\cos x + 1}{\sin x + \tan x} = \cot x$ **89.** While the equation is true for the *particular* value $\theta = \frac{\pi}{2}$, it is not true in *general*. To be an identity, the equation must be true in *all* cases for which the functions involved are defined. **91.** **(a)** $I = k(1 - \sin^2\theta)$ **(b)** For $\theta = 2\pi n$ and all integers n, $\cos^2\theta = 1$, its maximum value, and I attains a maximum value of k.

93. **(a)** The total mechanical energy E is always 2. The spring has maximum potential energy when it is fully stretched but not moving. The spring has maximum kinetic energy when it is not stretched but is moving fastest.

3

E

K P

0 .5

–1

(b) Let $Y_1 = P(t)$, $Y_2 = K(t)$, and $Y_3 = E(t)$. $Y_3 = 2$ for all inputs. The spring is stretched the most (has greatest potential energy) when $t = .25, .5, .75, \ldots$. At these times kinetic energy is 0.

X	Y2	Y3
0	0	2
.05	.69098	2
.1	1.809	2
.15	1.809	2
.2	.69098	2
.25	0	2
.3	.69098	2

Y3 = Y1+Y2

(c) $E(t) = k$ or $E(t) = 2$

9.2 Exercises *(pages 694–695)*

1. F **3.** C **5.** $\frac{\sqrt{6}-\sqrt{2}}{4}$ **7.** $\frac{-\sqrt{2}-\sqrt{6}}{4}$ **9.** $\frac{-\sqrt{6}+\sqrt{2}}{4}$ **11.** $\frac{\sqrt{6}-\sqrt{2}}{4}$ **13.** $-\sqrt{3}-2$ **15.** $\frac{\sqrt{2}+\sqrt{6}}{4}$ **17.** -1 **19.** $\frac{\sqrt{2}}{2}$ **21.** -1 **23.** $\sin x$ **25.** $-\cos x$ **27.** $-\cos x$ **29.** $-\tan x$ **31.** $\sin x$ **33.** $\sin x$ **35.** $-\sin x$ **37.** $\frac{\sqrt{2}(\sin x - \cos x)}{2}$ **39.** $\frac{1+\tan x}{1-\tan x}$ **41.** $-\tan x$ **43.** **(a)** $\frac{63}{65}$ **(b)** $\frac{33}{65}$ **(c)** $\frac{63}{16}$ **(d)** $\frac{33}{56}$ **(e)** I **(f)** I **45.** **(a)** $\frac{77}{85}$ **(b)** $\frac{13}{85}$ **(c)** $-\frac{77}{36}$ **(d)** $\frac{13}{84}$ **(e)** II **(f)** I **55.** 3 **57.** **(a)** 425 lb **(c)** 0°

59. **(a)** The pressure P is oscillating.

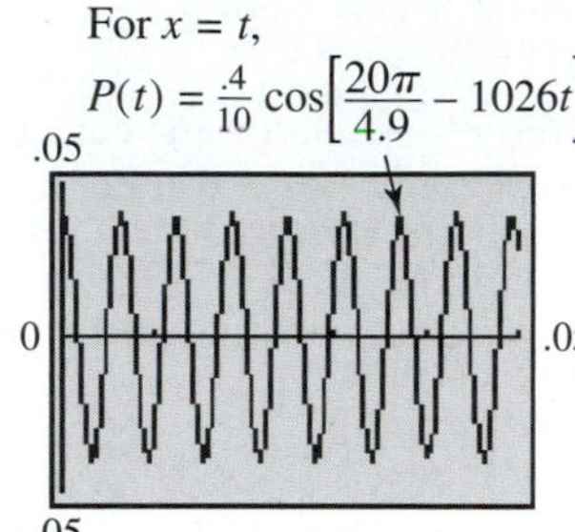

(b) The pressure oscillates and amplitude decreases as r increases.

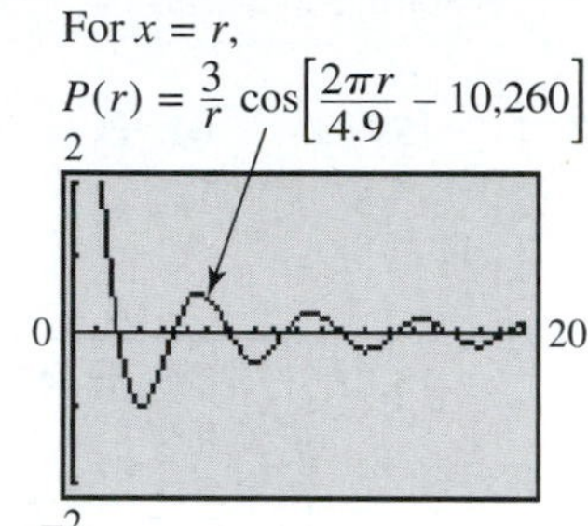

(c) $P = \frac{a}{n\lambda}\cos ct$

Reviewing Basic Concepts *(page 696)*

1. $\frac{\sin^2 x}{\cos x}$ **2.** $\frac{\sqrt{3}-3}{3+\sqrt{3}}$ or $\sqrt{3}-2$ **3.** 0 **4.** $\frac{\sqrt{2}}{2}(\sin x - \cos x)$ **5.** $\frac{-2+\sqrt{15}}{6}$; $\frac{\sqrt{5}-2\sqrt{3}}{6}$; $\frac{-2-\sqrt{15}}{\sqrt{5}-2\sqrt{3}}$ or $\frac{8\sqrt{5}+9\sqrt{3}}{7}$ **10.** $-20\cos\frac{\pi t}{4}$

9.3 Exercises *(pages 705–708)*

1. **(a)** $-\frac{4\sqrt{21}}{25}$ **(b)** $\frac{17}{25}$ **3.** **(a)** $\frac{4}{5}$ **(b)** $-\frac{3}{5}$ **5.** **(a)** $-\frac{4\sqrt{55}}{49}$ **(b)** $\frac{39}{49}$ **7.** $\frac{\sqrt{3}}{2}$ **9.** $\frac{\sqrt{3}}{2}$ **11.** $-\frac{\sqrt{2}}{2}$ **13.** $\frac{1}{2}\tan 102°$ **15.** $\frac{1}{4}\cos 94.2°$ **17.** 1 **19.** $\cos^4 x - \sin^4 x = \cos 2x$

Forms of answers may vary in Exercises 21 and 23.

21. $\cos 3x = 4\cos^3 x - 3\cos x$ **23.** $\tan 4x = \frac{4\tan x - 4\tan^3 x}{1 - 6\tan^2 x + \tan^4 x}$ **25.** $\frac{\sqrt{2-\sqrt{3}}}{2}$ **27.** $1-\sqrt{2}$ $\left(\text{or } -\sqrt{3-2\sqrt{2}}\right)$ **29.** $\frac{\sqrt{2+\sqrt{2}}}{2}$ **31.** $\frac{\sqrt{10}}{4}$ **33.** 3 **35.** $-\sqrt{7}$ **39.** **(a)** $\tan\frac{\pi}{2}$ is undefined, so it cannot be used. **(b)** $\tan\left(\frac{\pi}{2}+x\right) = \frac{\sin\left(\frac{\pi}{2}+x\right)}{\cos\left(\frac{\pi}{2}+x\right)}$ **41.** $\sin 20°$ **43.** $\tan 73.5°$ **45.** $\tan 29.87°$ **57.** $\sin 160° - \sin 44°$ **59.** $\sin 225° + \sin 55°$ **61.** $-2\sin 3x \sin x$ **63.** $-2\sin 11.5° \cos 36.5°$ **65.** $2\cos 6x \cos 2x$ **67.** **(a)** $\cos\frac{\theta}{2} = \frac{R-b}{R}$ **(b)** $\tan\frac{\theta}{4} = \frac{b}{50}$ **(c)** 54°

69. **(a)** For $x = t$, $W = VI =$ $(163 \sin 120\pi t)(1.23 \sin 120\pi t)$

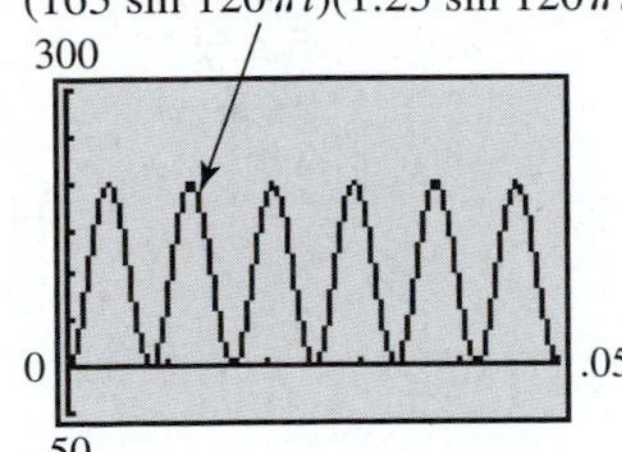

(b) maximum: 200.49 watts; minimum: 0 watts
(c) $a = -100.245$, $\omega = 240\pi$, $c = 100.245$
(e) 100.245 watts

9.4 Exercises *(pages 718–721)*

1. **(a)** $[-1, 1]$ **(b)** $\left[-\frac{\pi}{2}, \frac{\pi}{2}\right]$ **(c)** increasing **(d)** -2 is not in the domain. **3.** **(a)** $(-\infty, \infty)$ **(b)** $\left(-\frac{\pi}{2}, \frac{\pi}{2}\right)$ **(c)** increasing **(d)** no **5.** $\cos^{-1}\frac{1}{a}$ **7.** $\frac{\pi}{4}$ **9.** π **11.** $-\frac{\pi}{2}$ **13.** 0 **15.** $\frac{\pi}{2}$ **17.** $\frac{\pi}{4}$ **19.** $\frac{5\pi}{6}$ **21.** $\frac{3\pi}{4}$ **23.** $-\frac{\pi}{6}$ **25.** $\frac{\pi}{6}$ **27.** $-45°$ **29.** $-60°$ **31.** 120° **33.** $-30°$ **35.** $-7.6713835°$ **37.** 113.500970° **39.** 30.987961° **41.** .83798122 **43.** 2.3154725 **45.** 1.1900238

47.

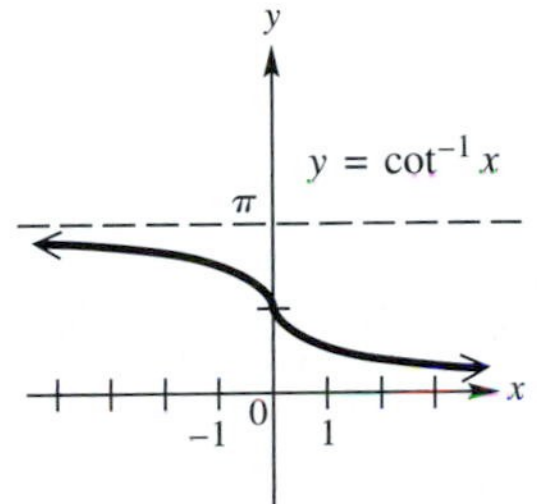

49.

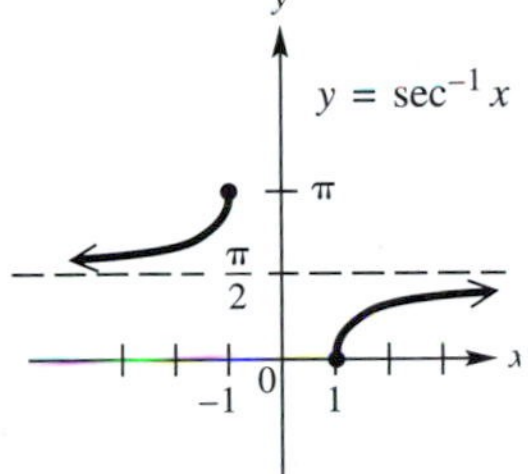

51.

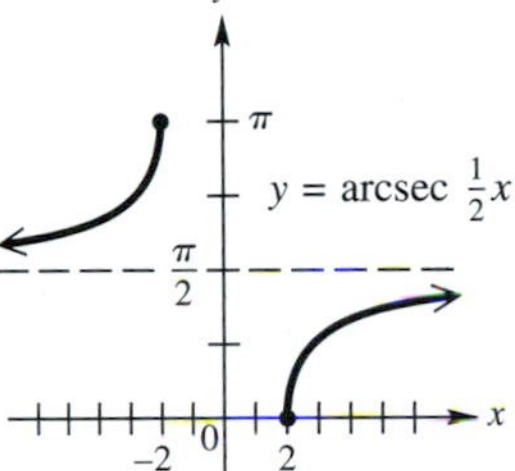

53. The domain of $y = \tan^{-1} x$ is $(-\infty, \infty)$. **54.** In both cases, the result is x. In each case, the graph is a straight line bisecting quadrants I and III (i.e., the line $y = x$).

55. It is the graph of $y = x$.

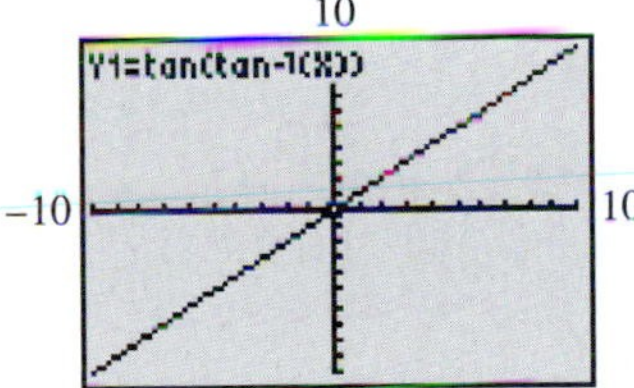

56. It does not agree because the range of the inverse tangent function is $\left(-\frac{\pi}{2}, \frac{\pi}{2}\right)$, not $(-\infty, \infty)$, as was the case in Exercise 55.

10
Y1=tan-1(tan(X))
−10 10
−10

57. $\frac{1}{2}$ **59.** $-\frac{\pi}{3}$ **61.** $\frac{\pi}{2}$ **63.** 5 **65.** 2 **67.** $-\frac{\pi}{6}$ **69.** $\frac{\sqrt{7}}{3}$ **71.** $\frac{\sqrt{5}}{5}$ **73.** $\frac{120}{169}$ **75.** $-\frac{7}{25}$ **77.** $\frac{4\sqrt{6}}{25}$ **79.** $\frac{63}{65}$ **81.** $-\frac{63}{16}$ **83.** $\frac{63}{65}$ **85.** $\frac{\sqrt{10} - 3\sqrt{30}}{20}$ **87.** .894427191 **89.** .1234399811 **91.** $\frac{1}{u}$ **93.** $\sqrt{1 - u^2}$ **95.** $\frac{\sqrt{1 - u^2}}{u}$ **97.** $\frac{\sqrt{u^2 - 4}}{u}$ **99.** $\frac{u\sqrt{2}}{2}$ **101.** $\frac{2\sqrt{4 - u^2}}{4 - u^2}$ **103.** **(a)** 45° **(b)** $\theta = 45°$

105. **(a)** 35° **(b)** 24° **(c)** 18° **(e)** about 2 ft

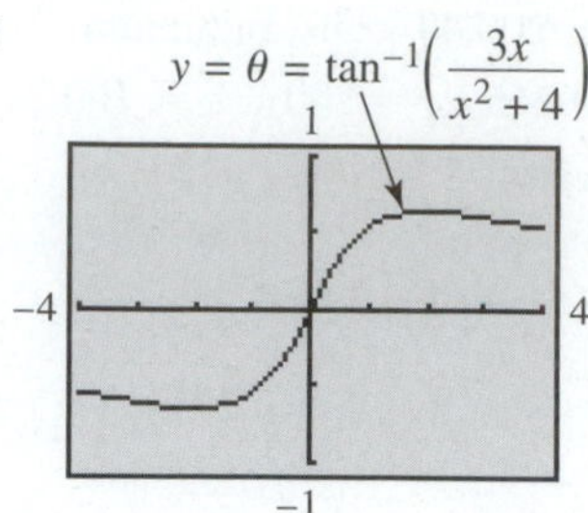

Reviewing Basic Concepts *(page 721)*

1. $-\frac{\sqrt{119}}{7}$ **2.** $\sin 2\theta = \frac{4\sqrt{2}}{9}$; $\cos 2\theta = \frac{7}{9}$; $\tan 2\theta = \frac{4\sqrt{2}}{7}$ **3.** $\frac{\sqrt{2}+\sqrt{6}}{4}$ **4.** $\sin 175° - \sin 125°$

6. **(a)** $\frac{\pi}{6}$ **(b)** $-\frac{\pi}{4}$ **7.** **(a)** 60° **(b)** 135° **8.**

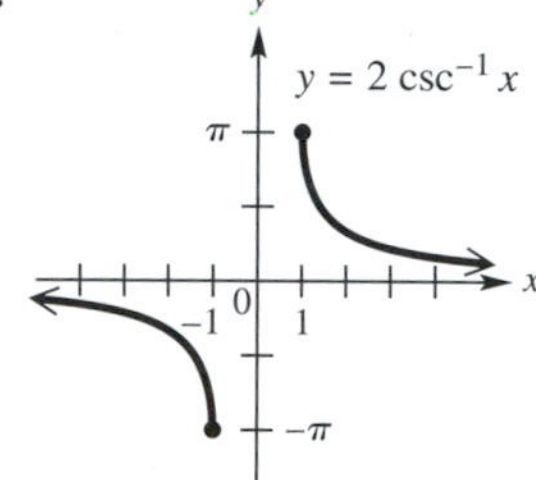

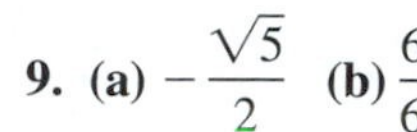

9. **(a)** $-\frac{\sqrt{5}}{2}$ **(b)** $\frac{63}{65}$ **10.** $\frac{\sqrt{u^2+1}}{u^2+1}$

9.5 Exercises *(pages 726–728)*

1. $\left\{\frac{2\pi}{3}, \frac{4\pi}{3}\right\}$ **3.** $\left\{\frac{\pi}{2}\right\}$ **5.** $\left\{\frac{3\pi}{4}, \frac{7\pi}{4}\right\}$ **7.** $\left\{\frac{\pi}{3}, \frac{5\pi}{3}\right\}$ **9.** $\left\{\frac{\pi}{6}, \frac{5\pi}{6}\right\}$ **11.** $\left\{\frac{\pi}{4}, \frac{2\pi}{3}, \frac{5\pi}{4}, \frac{5\pi}{3}\right\}$

13. $\left\{\frac{\pi}{4}, \frac{\pi}{2}, \frac{5\pi}{4}, \frac{3\pi}{2}\right\}$ **15.** $\left\{\frac{\pi}{2}\right\}$ **17.** $\left\{\frac{\pi}{6}, \frac{5\pi}{6}, \frac{7\pi}{6}, \frac{11\pi}{6}\right\}$ **19.** $\left\{\frac{\pi}{3}, \frac{3\pi}{4}, \frac{4\pi}{3}, \frac{7\pi}{4}\right\}$ **21.** $\left\{\frac{\pi}{2}, \frac{7\pi}{6}, \frac{11\pi}{6}\right\}$

23. $\{0, \pi\}$ **25.** **(a)** $\left\{\frac{\pi}{3}, \frac{5\pi}{3}\right\}$ **(b)** $\left(\frac{\pi}{3}, \frac{5\pi}{3}\right)$ **(c)** $\left[0, \frac{\pi}{3}\right) \cup \left(\frac{5\pi}{3}, 2\pi\right)$

27. **(a)** $\left\{\frac{\pi}{3}, \frac{2\pi}{3}, \frac{4\pi}{3}, \frac{5\pi}{3}\right\}$ **(b)** $\left(\frac{\pi}{3}, \frac{\pi}{2}\right) \cup \left(\frac{\pi}{2}, \frac{2\pi}{3}\right) \cup \left(\frac{4\pi}{3}, \frac{3\pi}{2}\right) \cup \left(\frac{3\pi}{2}, \frac{5\pi}{3}\right)$ **(c)** $\left[0, \frac{\pi}{3}\right) \cup \left(\frac{2\pi}{3}, \frac{4\pi}{3}\right) \cup \left(\frac{5\pi}{3}, 2\pi\right)$

29. **(a)** $\left\{\frac{\pi}{6}, \frac{\pi}{2}, \frac{3\pi}{2}, \frac{11\pi}{6}\right\}$ **(b)** $\left[0, \frac{\pi}{6}\right) \cup \left(\frac{\pi}{2}, \frac{3\pi}{2}\right) \cup \left(\frac{11\pi}{6}, 2\pi\right)$ **(c)** $\left(\frac{\pi}{6}, \frac{\pi}{2}\right) \cup \left(\frac{3\pi}{2}, \frac{11\pi}{6}\right)$ **31.** **(a)** $\left\{\frac{\pi}{2}, \frac{3\pi}{2}\right\}$

(b) $\left(\frac{\pi}{2}, \frac{3\pi}{2}\right)$ **(c)** $\left[0, \frac{\pi}{2}\right) \cup \left(\frac{3\pi}{2}, 2\pi\right)$ **33.** **(a)** $\left\{0, \frac{\pi}{6}, \frac{5\pi}{6}, \pi\right\}$ **(b)** $\left(\frac{\pi}{6}, \frac{\pi}{2}\right) \cup \left(\frac{\pi}{2}, \frac{5\pi}{6}\right)$

(c) $\left(0, \frac{\pi}{6}\right) \cup \left(\frac{5\pi}{6}, \pi\right) \cup \left(\pi, \frac{3\pi}{2}\right) \cup \left(\frac{3\pi}{2}, 2\pi\right)$ **34.** $\tan^3 x - 3\tan x = 0$ **35.** $\left\{0, \frac{\pi}{3}, \frac{2\pi}{3}, \pi, \frac{4\pi}{3}, \frac{5\pi}{3}\right\}$

36. They all satisfy the equation. **37.** $\left\{\frac{\pi}{3}, \frac{2\pi}{3}, \frac{4\pi}{3}, \frac{5\pi}{3}\right\}$ **38.** No. The solutions 0 and π were lost when dividing by $\tan x$, because both $\tan 0$ and $\tan \pi$ equal 0, and division by 0 is undefined. **39.** $\left\{\frac{\pi}{2}, 3.87, 5.55\right\}$ **41.** $\{1.86, 2.61, 5.00, 5.75\}$

43. $\{1.20, 5.09\}$ **45.** $\{135°, 315°, 71.6°, 251.6°\}$ **47.** $\{114.3°, 335.7°\}$ **49.** $\{1.99, 5.86\}$ **51.** $\{2.68, 4.46, 4.71\}$

53. $\{1.30\}$ **55. (a)** 91.3 days after March 21, on June 20 **(b)** 273.8 days after March 21, on December 19 **(c)** 228.7 days after March 21, on November 4, and again after 318.8 days, on February 2 **57.** $14°$ **59.** $\frac{\pi}{3}$

9.6 Exercises *(pages 732–734)*

1. $\frac{\pi}{3}, \pi, \frac{4\pi}{3}$ **3. (a)** $\left\{\frac{\pi}{12}, \frac{11\pi}{12}, \frac{13\pi}{12}, \frac{23\pi}{12}\right\}$ **(b)** $\left[0, \frac{\pi}{12}\right) \cup \left(\frac{11\pi}{12}, \frac{13\pi}{12}\right) \cup \left(\frac{23\pi}{12}, 2\pi\right)$ **5. (a)** $\left\{\frac{\pi}{2}, \frac{7\pi}{6}, \frac{11\pi}{6}\right\}$ **(b)** $\emptyset$ **7. (a)** $\left\{\frac{3\pi}{8}, \frac{5\pi}{8}, \frac{11\pi}{8}, \frac{13\pi}{8}\right\}$ **(b)** $\left[\frac{3\pi}{8}, \frac{5\pi}{8}\right] \cup \left[\frac{11\pi}{8}, \frac{13\pi}{8}\right]$ **9. (a)** $\left\{\frac{\pi}{2}, \frac{3\pi}{2}\right\}$ **(b)** $\left(\frac{\pi}{2}, \frac{3\pi}{2}\right)$ **11.** $\left\{\frac{\pi}{2}\right\}$ **13.** $\{\pi\}$ **15.** $\left\{\frac{\pi}{4}, \frac{\pi}{2}, \frac{5\pi}{4}, \frac{3\pi}{2}\right\}$ **17.** $\left\{\frac{\pi}{3}, \frac{\pi}{2}, \frac{3\pi}{2}, \frac{5\pi}{3}\right\}$ **19.** $\{15°, 45°, 135°, 165°, 255°, 285°\}$ **21.** $\{0°\}$ **23.** $\{120°, 240°\}$ **25.** $\{30°, 150°, 270°\}$ **27.** $\{11.8°, 78.2°, 191.8°, 258.2°\}$ **29.** $\{30°, 90°, 150°, 210°, 270°, 330°\}$ **31.** $\{.262, 1.309, 1.571, 3.403, 4.451, 4.712\}$ **33.** $\{1.047, 3.142, 5.236\}$ **35.** $\{.259, 1.372, 3.142, 4.911, 6.024\}$ **37.** $\frac{\tan 2\theta}{2} \neq \tan \theta$, because the 2 in 2θ is not a factor of the numerator. It is a factor in the argument of the tangent function. **39. (a)** 2 sec **(b)** $\frac{10}{3}$ sec

41. (a) For $x = t$,

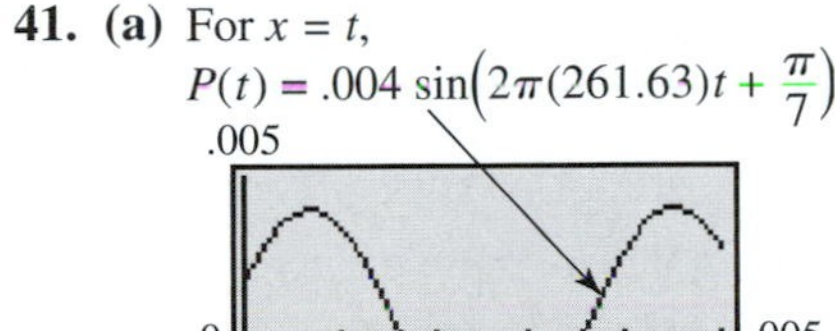

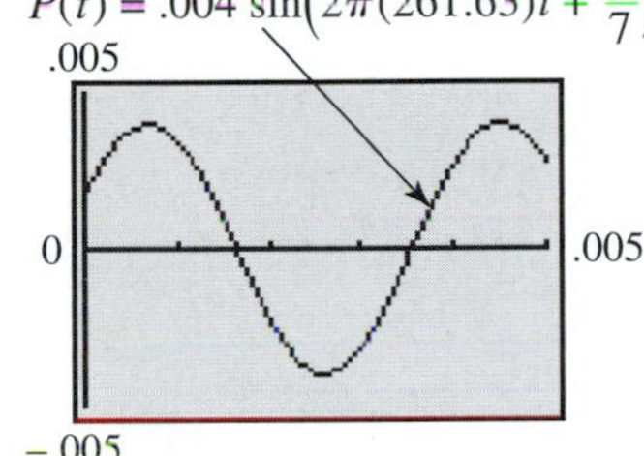

(b) .00164 and .00355
(c) $.00164 < t < .00355$ **(d)** outward

43. (a) 3 beats per sec

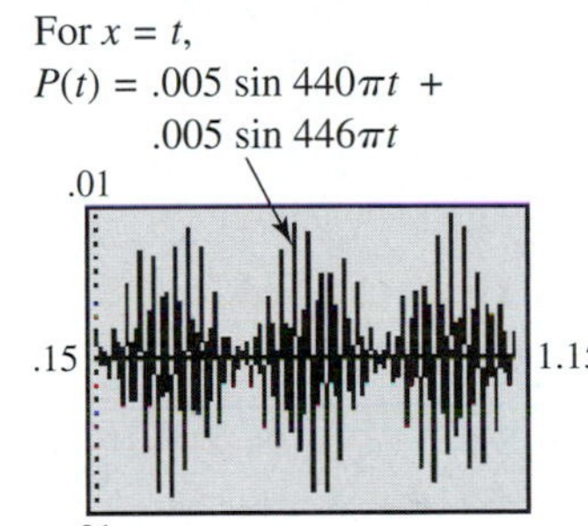

(b) 4 beats per sec

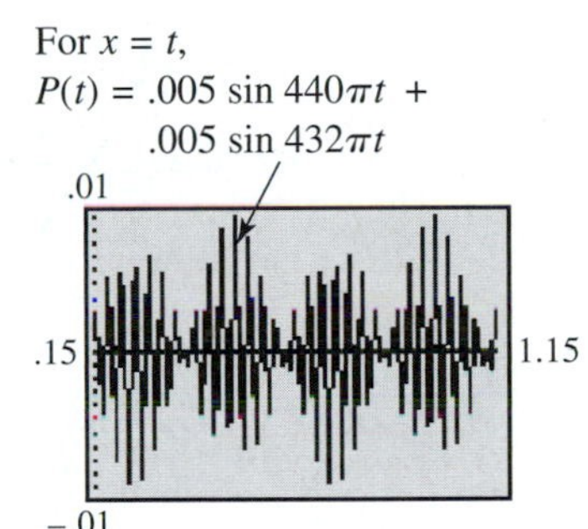

(c) The number of beats is equal to the absolute value of the difference in the frequencies of the two tones.

Reviewing Basic Concepts *(pages 734–735)*

1. $\left\{\frac{\pi}{12}, \frac{11\pi}{12}, \frac{13\pi}{12}, \frac{23\pi}{12}\right\}$ **2.** $\left\{\frac{7\pi}{6}, \frac{11\pi}{6}\right\}$ **3.** $\left\{0, \frac{\pi}{4}, \frac{5\pi}{4}\right\}$ **4.** $\left\{\frac{\pi}{6}, \frac{\pi}{2}, \frac{3\pi}{2}, \frac{11\pi}{6}\right\}$ **5.** $\{38.4°, 218.4°, 104.8°, 284.8°\}$ **6.** $\{78.0°, 282.0°\}$ **7.** $\{90°, 210°, 330°\}$ **8.** $\{90°, 270°\}$ **9.** $\{.68058878, 1.4158828\}$ **10.** $\{0, .37600772\}$

Chapter 9 Review Exercises *(pages 737–740)*

1. sine, tangent, cotangent, cosecant **3.** cos 3 **5.** $-\tan 3$ **7.** $-\csc 3$ **9.** B **11.** F **13.** D **15.** $\frac{\cos^2 \theta}{\sin \theta}$ **17.** $\frac{1 + \cos \theta}{\sin \theta}$ **19.** $\sin x = \frac{5\sqrt{41}}{41}$; $\cos x = -\frac{4\sqrt{41}}{41}$; $\cot x = -\frac{4}{5}$; $\sec x = -\frac{\sqrt{41}}{4}$; $\csc x = \frac{\sqrt{41}}{5}$ **21.** B **23.** A **25.** C **27.** D **29.** F **43.** $\frac{2\pi}{3}$ **45.** $\frac{2\pi}{3}$ **47.** $\frac{3\pi}{4}$ **49.** $-60°$ **51.** $-41°$ **53.** $-7°$ **55.** $88°$ **57.** There is no real number whose sine value is -3. The domain of inverse sine is $[-1, 1]$. **59.** $\frac{1}{2}$ **61.** $\frac{\sqrt{10}}{10}$ **63.** $\frac{\pi}{2}$

65. $\frac{u\sqrt{u^2+1}}{u^2+1}$ **67.** $\frac{1}{u}$ **69.** $\{.463647609, 3.605240263\}$ **71.** $\left\{\frac{\pi}{4}, \frac{3\pi}{4}, \frac{5\pi}{4}, \frac{7\pi}{4}\right\}$ **73.** $\{0\}$

75. $\left\{\frac{\pi}{12}, \frac{5\pi}{12}, \frac{13\pi}{12}, \frac{17\pi}{12}\right\}$ **77.** $\left\{\frac{\pi}{3}, \pi, \frac{5\pi}{3}\right\}$ **79.** $\{0\}$

81. **(b)** 8.6602567 ft; There may be a discrepancy in the final digits. **83.** .001 sec **85.** 48.8°

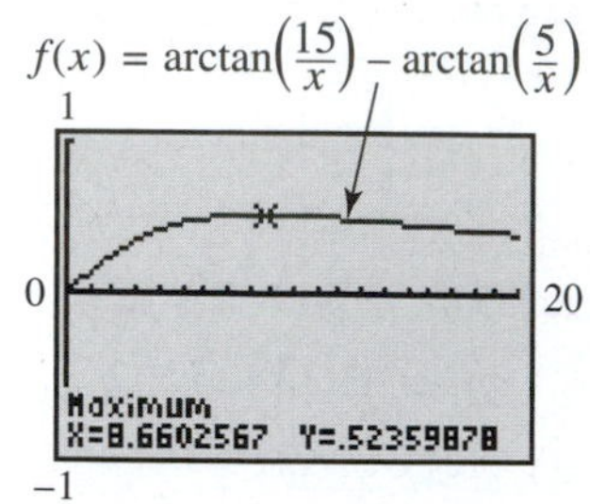

Chapter 9 Test *(page 740)*

1. $\frac{2-2\sqrt{30}}{15}$ **2.** $\frac{\sqrt{5}-4\sqrt{6}}{15}$ **3.** $\frac{-3-\sqrt{5}}{2}$ **4.** $-\frac{23}{25}$ **5.** -1 **6.** $\sec x - \sin x \tan x = \cos x$ **9.** $-\sin\theta$

10. $-\sin\theta$ **11.** **(a)**

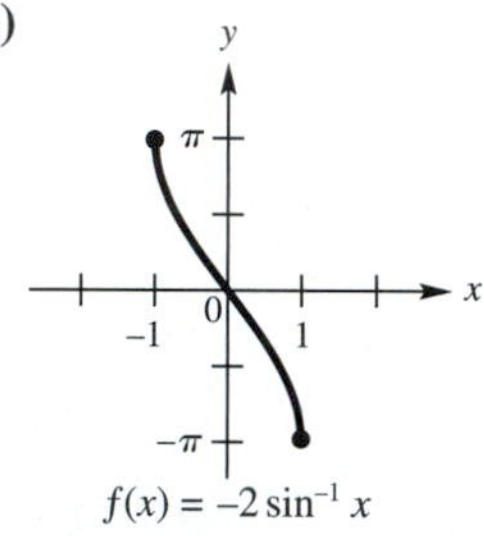

(b) domain: $[-1, 1]$; range: $[-\pi, \pi]$

(c) The domain of $\sin^{-1} x$ is $[-1, 1]$, and 2 is not in this interval. (No number has sine value 2.)

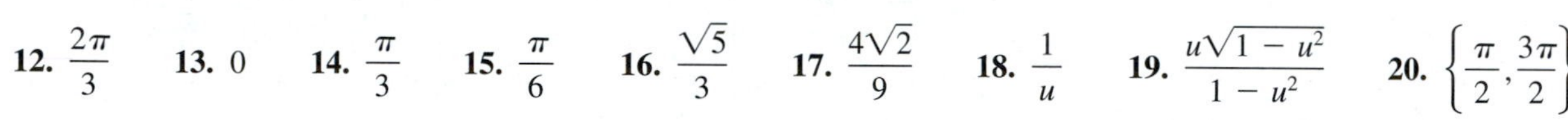

12. $\frac{2\pi}{3}$ **13.** 0 **14.** $\frac{\pi}{3}$ **15.** $\frac{\pi}{6}$ **16.** $\frac{\sqrt{5}}{3}$ **17.** $\frac{4\sqrt{2}}{9}$ **18.** $\frac{1}{u}$ **19.** $\frac{u\sqrt{1-u^2}}{1-u^2}$ **20.** $\left\{\frac{\pi}{2}, \frac{3\pi}{2}\right\}$

21. $\{18.4°, 135°, 198.4°, 315°\}$ **22.** $\left\{0, \frac{2\pi}{3}, \frac{4\pi}{3}\right\}$ **23.** $\{120°, 240°\}$ **24.** $\left[0, \frac{\pi}{6}\right] \cup \left[\frac{5\pi}{6}, 2\pi\right)$

25. **(a)** domain: $[0, 11]$; range: $[0, 100]$ (in hundreds) **(b)**

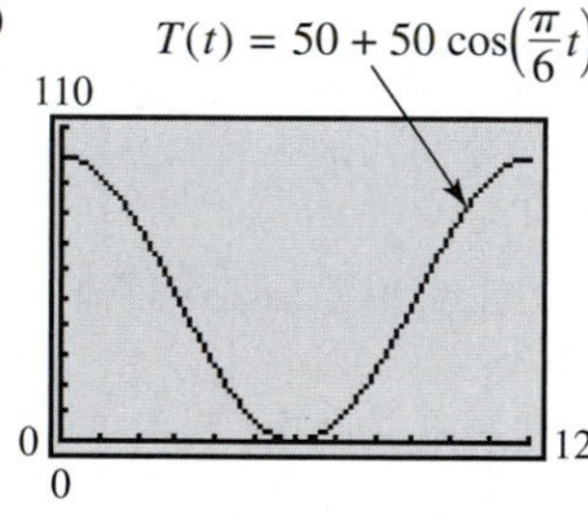

(c) The maximum of 10,000 animals occurs at 0 months (July). The minimum of 0 animals occurs at 6 months (January).

(d) There are 5000 animals at 3 months (October) and at 9 months (April).

(e) at 2 months (September) and 10 months (May)

(f) $t = \frac{6}{\pi}\arccos\left(\frac{T-50}{50}\right)$

CHAPTER 10 APPLICATIONS OF TRIGONOMETRY; VECTORS

10.1 Exercises *(pages 752–757)*

1. C **3.** no **5.** no **7.** yes **9.** yes **11.** $\sqrt{3}$ **13.** $C = 95°$, $b = 13$ m, $a = 11$ m **15.** $B = 37.3°$, $a = 38.5$ ft, $b = 51.0$ ft **17.** $B = 48°$, $a = 11$ m, $b = 13$ m **19.** $A = 36.54°$, $a = 28.10$ m, $b = 44.17$ m **21.** $A = 49° 40'$, $b = 16.1$ cm, $c = 25.8$ cm **23.** $C = 91.9°$, $a = 490$ ft, $c = 847$ ft **25.** $B = 110.0°$, $a = 27.01$ m,

$c = 21.37$ m **27.** $A = 34.72°$, $a = 3326$ ft, $c = 5704$ ft **29.** A **31.** $B_1 = 49.1°$, $C_1 = 101.2°$; $B_2 = 130.9°$, $C_2 = 19.4°$ **33.** no such triangle **35.** $B = 27.19°$, $C = 10.68°$ **37.** **(a)** $4 < h < 5$ **(b)** $h = 4$ or $h \geq 5$ **(c)** $h < 4$ **39.** 2 **41.** 0 **43.** $B = 20.6°$, $C = 116.9°$, $c = 20.6$ ft **45.** no such triangle **47.** $B_1 = 49° 20'$, $C_1 = 92° 00'$, $c_1 = 15.5$ km; $B_2 = 130° 40'$, $C_2 = 10° 40'$, $c_2 = 2.88$ km **49.** $A_1 = 52° 10'$, $C_1 = 95° 00'$, $c_1 = 9520$ cm; $A_2 = 127° 50'$, $C_2 = 19° 20'$, $c_2 = 3160$ cm **51.** The Pythagorean theorem only applies to right triangles. **53.** If we are given only three sides, then any equation from the law of sines will contain two unknowns. At least one angle measure must be given. **55.** 118 m **57.** 1.93 mi **59.** 10.4 in. **61.** 111° **63.** 5.1 mi, 7.2 mi **65.** 26.5 km **67.** 2.18 km **69.** 38.3 cm **71.** $\sin C = 1$; $C = 90°$; right triangle **73.** The longest side must be opposite the largest angle. Thus, B must be larger than A, which is impossible because A and B cannot both be obtuse. **75.** The property could not exist. **77.** about 419,000 km, which compares favorably to the actual value **79.** **(b)** $1.12257R^2$ **(c)** (i) 8.77 in.^{2} (ii) 5.32 in.^{2} (iii) red **80.** increasing **81.** If $B < A$, then $\sin B < \sin A$ because $y = \sin x$ is increasing on $\left[0, \frac{\pi}{2}\right]$. **82.** $b = \frac{a \sin B}{\sin A}$ **83.** $b = \frac{a \sin B}{\sin A} = a \cdot \frac{\sin B}{\sin A}$. Since $\frac{\sin B}{\sin A} < 1$, $b = a \cdot \frac{\sin B}{\sin A} < a \cdot 1 = a$, so $b < a$. **84.** If $B < A$ then $b < a$, but $b > a$ in triangle ABC.

10.2 Exercises *(pages 764–768)*

1. **(a)** SAS **(b)** law of cosines **3.** **(a)** SSA **(b)** law of sines **5.** **(a)** ASA **(b)** law of sines **7.** **(a)** ASA **(b)** law of sines **9.** 7 **11.** 30° **13.** $a = 5.4$, $B = 40.7°$, $C = 78.3°$ **15.** $A = 22.3°$, $B = 108.2°$, $C = 49.5°$ **17.** $A = 33.6°$, $B = 50.7°$, $C = 95.7°$ **19.** $c = 2.83$ in., $A = 44.9°$, $B = 106.8°$ **21.** $c = 6.46$ m, $A = 53.1°$, $B = 81.3°$ **23.** $A = 82°$, $B = 37°$, $C = 61°$ **25.** $C = 102° 10'$, $B = 35° 50'$, $A = 42° 00'$ **27.** $C = 84° 30'$, $B = 44° 40'$, $A = 50° 50'$ **29.** $a = 156$ cm, $B = 64° 50'$, $C = 34° 30'$ **31.** $b = 9.529$ in., $A = 64.59°$, $C = 40.61°$ **33.** $a = 15.7$ m, $B = 21.6°$, $C = 45.6°$ **35.** $A = 30°$, $B = 56°$, $C = 94°$ **37.** The value of $\cos \theta$ will be greater than 1; your calculator will give you an error message (or a complex number) when using the inverse cosine function. **39.** 257 m **41.** 281 km **43.** 10.8 mi **45.** 115 km **47.** 18 ft **49.** about 5500 m **51.** 438.14 ft **53.** 350° **55.** 2000 km **57.** 163.5° **59.** 22 ft **61.** Since A is obtuse, $90° < A < 180°$. The cosine of a quadrant II angle is negative. **62.** In $a^2 = b^2 + c^2 - 2bc \cos A$, $\cos A$ is negative, so $a^2 = b^2 + c^2 +$ (a positive quantity). Thus, $a^2 > b^2 + c^2$. **63.** $b^2 + c^2 > b^2$ and $b^2 + c^2 > c^2$. If $a^2 > b^2 + c^2$, then $a^2 > b^2$ and $a^2 > c^2$ from which $a > b$ and $a > c$ because a, b, and c are nonnegative. **64.** Because A is obtuse, it is the largest angle, so the longest side should be a, not c. **65.** 46.4 m^2 **67.** 78 m^2 **69.** 3650 ft^2 **71.** 228 yd^2 **73.** 33 cans **75.** 100 m^2 **77.** The area and perimeter are both 36.

10.3 Exercises *(pages 776–781)*

1. **m** and **p**; **n** and **r** **3.** **m** or **p** equal 2**t**, or **t** is one half **m** or **p**; also **m** = 1**p** and **n** = 1**r** **5.** **7.** **9.** **11.** **13.** **15.**

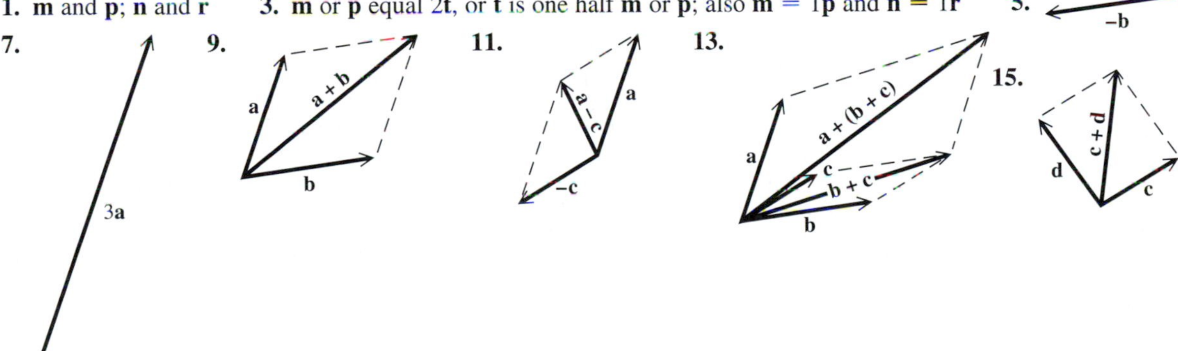

17. **(a)** $\langle -4, 16 \rangle$ **(b)** $\langle -12, 0 \rangle$ **(c)** $\langle 8, -8 \rangle$ **19.** **(a)** $\langle 8, 0 \rangle$ **(b)** $\langle 0, 16 \rangle$ **(c)** $\langle -4, -8 \rangle$ **21.** **(a)** $\langle 0, 12 \rangle$ **(b)** $\langle -16, -4 \rangle$ **(c)** $\langle 8, -4 \rangle$ **23.** **(a)** 4**i** **(b)** 7**i** + 3**j** **(c)** −5**i** + **j** **25.** **(a)** $\langle -2, 4 \rangle$ **(b)** $\langle 7, 4 \rangle$ **(c)** $\langle 6, -6 \rangle$ **27.** $\langle 2, 8 \rangle$ **29.** $\langle 6, -2 \rangle$

31. $\langle -20, -15\rangle$ **33.** $\langle 3\sqrt{3}, 3\rangle$ **35.** $\left\langle -\frac{9\sqrt{2}}{2}, -\frac{9\sqrt{2}}{2}\right\rangle$ **37.** $\langle 3.06, 2.57\rangle$ **39.** $\langle 4.10, -2.87\rangle$ **41.** 94.2 lb **43.** 24.4 lb **45.** $-5\mathbf{i} + 8\mathbf{j}$ **47.** $2\mathbf{i}$ **49.** $4\sqrt{2}\mathbf{i} + 4\sqrt{2}\mathbf{j}$ **51.** $-.25\mathbf{i} + .54\mathbf{j}$ **53.** $\sqrt{2}$; 45° **55.** 16; 315° **57.** 17; 331.9° **59.** 6; 180° **61.** 7 **63.** -3 **65.** 20 **67.** 135° **69.** 90° **71.** $\cos^{-1}\frac{4}{5} \approx 36.87°$ **73.** -6 **75.** -24 **77.** orthogonal **79.** not orthogonal **81.** not orthogonal **83.** 190 lb and 283 lb, respectively **85.** 18° **87.** 2.4 tons **89.** 17.5° **91.** weight: 64.8 lb; tension: 61.9 lb **93.** 13.5 mi; 50.4° **95.** 39.2 km **97.** current: 3.5 mph; motorboat: 19.7 mph **99.** bearing: 237°; ground speed: 470 mph **101.** ground speed: 161 mph; airspeed: 156 mph **103.** bearing: 74°; ground speed: 202 mph **105.** bearing: 358°; airspeed: 170 mph **107.** ground speed: 230 km per hr; bearing: 167° **109.** **(a)** $|\mathbf{R}| = \sqrt{5} \approx 2.2$, $|\mathbf{A}| = \sqrt{1.25} \approx 1.1$; About 2.2 in. of rain fell. The area of the opening of the rain gauge is about 1.1 in.^{2}. **(b)** $V = 1.5$; The volume of rain was 1.5 in.^{3}. **(c)** **R** and **A** should be parallel and point in opposite directions.

Reviewing Basic Concepts *(pages 781–782)*

1. $B = 74°$, $b = 16.6$, $c = 15.3$ **2.** two solutions: $B_1 = 45.0°$, $C_1 = 103°$, $c_1 = 11.0$ or $B_2 = 135.0°$, $C_2 = 13°$, $c_2 = 2.5$ **3.** one solution: $A = 22.5°$, $B = 116.5°$, $b = 16.4$ **4.** **(a)** $b = 7.1$, $A = 63.0°$, $C = 66.0°$ **(b)** $A = 110.7°$, $B = 37.0°$, $C = 32.3°$ **5.** 9.6 **6.** 21 **7.** **(a)** $\mathbf{i}$ **(b)** $4\mathbf{i} - 2\mathbf{j}$ **(c)** $11\mathbf{i} - 7\mathbf{j}$ **8.** -9; 142.1° **9.** 207 lb **10.** 7200 ft

10.4 Exercises *(pages 788–790)*

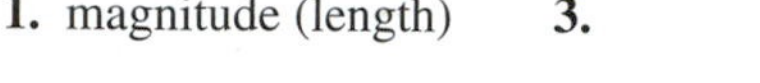

1. magnitude (length) **3.**

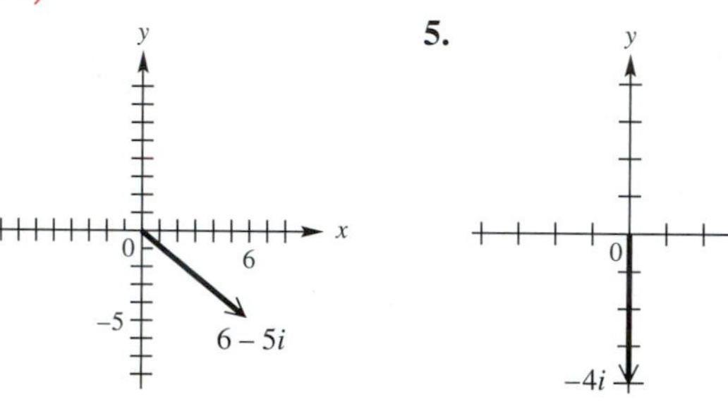

5.

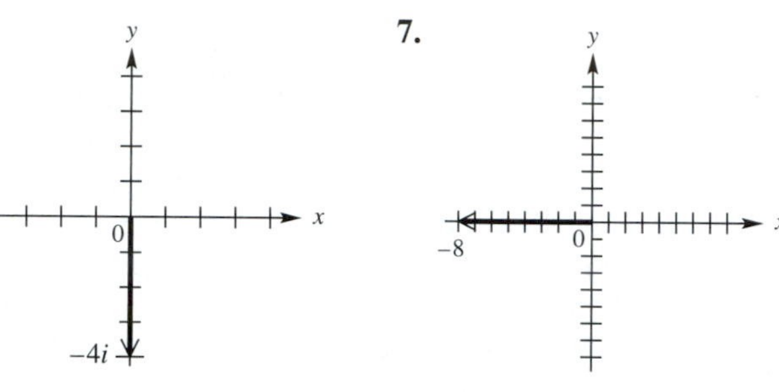

7.

9. $1 - 4i$ **11.** The imaginary part must be 0. **13.** 0 **15.** $3 - i$ **17.** $-3 + 3i$ **19.** $2 + 4i$ **21.** $\sqrt{2}$ **23.** 13 **25.** 6 **27.** $\sqrt{13}$ **29.** $\sqrt{2} + i\sqrt{2}$ **31.** $10i$ **33.** $-2 - 2i\sqrt{3}$ **35.** $\frac{\sqrt{3}}{2} + \frac{1}{2}i$ **37.** $\frac{5\sqrt{3}}{2} - \frac{5}{2}i$ **39.** $-\sqrt{2}$ **41.** $3\sqrt{2}[\cos(-45°) + i\sin(-45°)]$ **43.** $2(\cos 60° + i \sin 60°)$ **45.** $2[\cos(-90°) + i\sin(-90°)]$ **47.** $8\left(\cos\frac{\pi}{6} + i\sin\frac{\pi}{6}\right)$ **49.** $2\left(\cos\frac{3\pi}{4} + i\sin\frac{3\pi}{4}\right)$ **51.** $4(\cos \pi + i \sin \pi)$ **53.** $-3\sqrt{3} + 3i$ **55.** $-4i$ **57.** $-\frac{15\sqrt{2}}{2} + \frac{15\sqrt{2}}{2}i$ **59.** $-3i$ **61.** $\sqrt{3} - i$ **63.** -2 **65.** $-\frac{1}{6} - \frac{\sqrt{3}}{6}i$ **67.** $1.18 - .14i$ **69.** $27.43 + 11.5i$ **71.** **(b)** $z_1^2 - 1 = a^2 - b^2 - 1 + 2abi$; $z_2^2 - 1 = a^2 - b^2 - 1 - 2abi$ **(c)** If $z_1 = a + bi$ and $z_2 = a - bi$, then $z_1^2 - 1$ and $z_2^2 - 1$ are also conjugates with the same modulus. Therefore, if z_1 is in the Julia set, so is z_2. Thus, (a, b) in the Julia set implies $(a, -b)$ is also in the set. **(d)** yes **73.** To square a complex number in trigonometric form, square the modulus r and double the argument θ. **75.** **(a)** $\frac{\pi}{2}$ **(b)** π

10.5 Exercises *(pages 795–796)*

1. $27i$ **3.** 1 **5.** 8 **7.** $\frac{27}{2} - \frac{27\sqrt{3}}{2}i$ **9.** $8i$ **11.** $4096i$ **13.** -1 **15.** $8i$ **17.** $128 + 128i$

19. $\cos 0° + i \sin 0°$, $\cos 120° + i \sin 120°$, $\cos 240° + i \sin 240°$

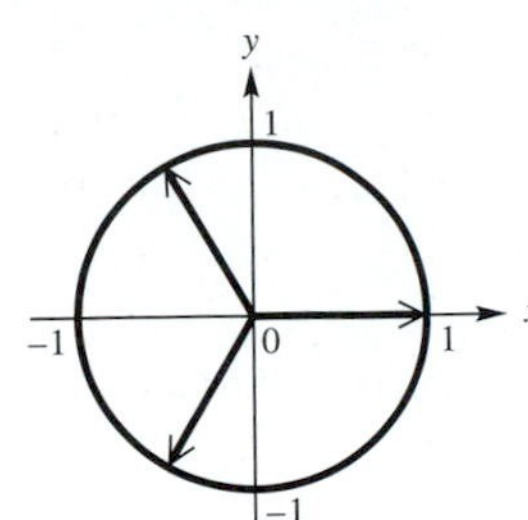

21. $2 \text{ cis } 20°$, $2 \text{ cis } 140°$, $2 \text{ cis } 260°$

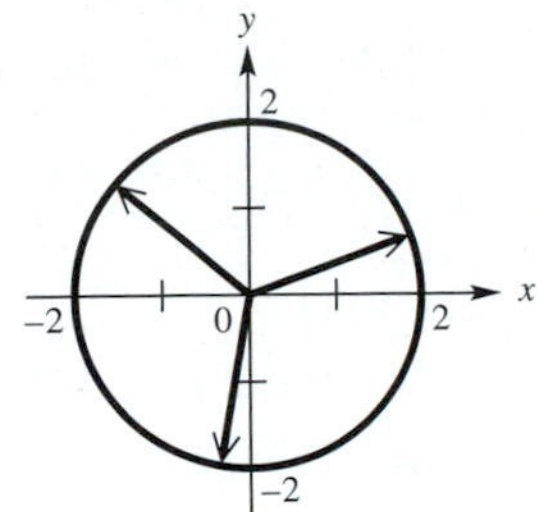

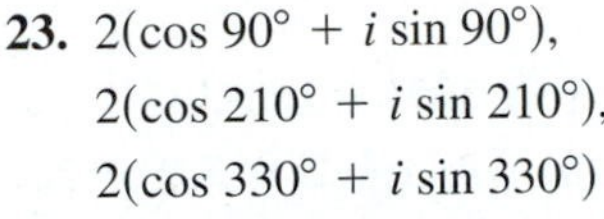

23. $2(\cos 90° + i \sin 90°)$, $2(\cos 210° + i \sin 210°)$, $2(\cos 330° + i \sin 330°)$

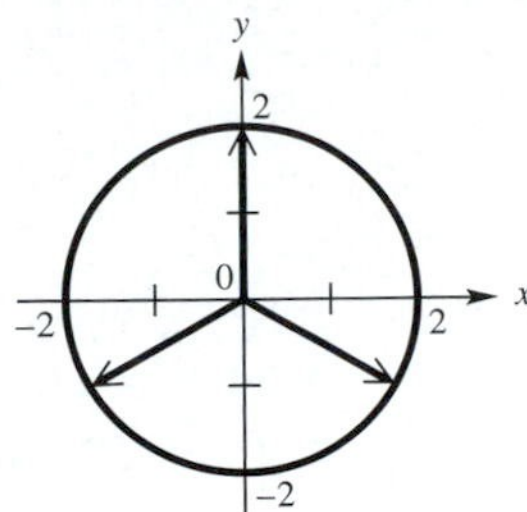

25. $4(\cos 60° + i \sin 60°)$, $4(\cos 180° + i \sin 180°)$, $4(\cos 300° + i \sin 300°)$

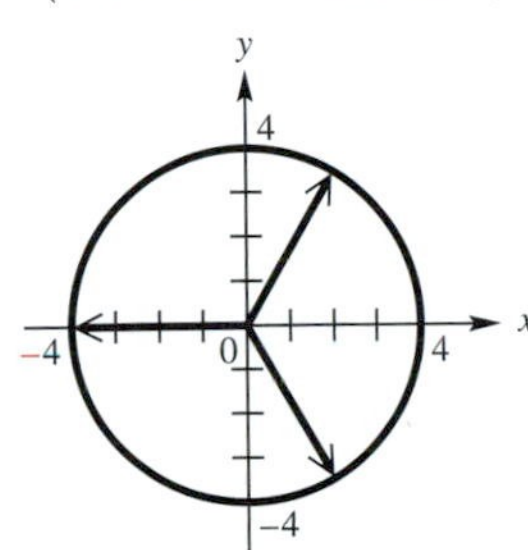

27. $\sqrt[3]{2}(\cos 20° + i \sin 20°)$, $\sqrt[3]{2}(\cos 140° + i \sin 140°)$, $\sqrt[3]{2}(\cos 260° + i \sin 260°)$

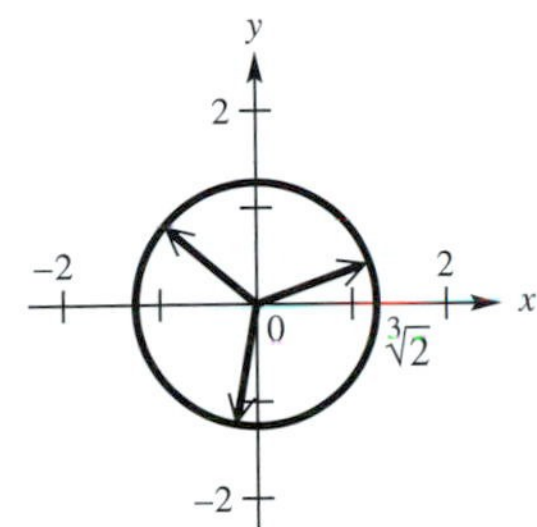

29. $\sqrt[3]{4}(\cos 50° + i \sin 50°)$, $\sqrt[3]{4}(\cos 170° + i \sin 170°)$, $\sqrt[3]{4}(\cos 290° + i \sin 290°)$

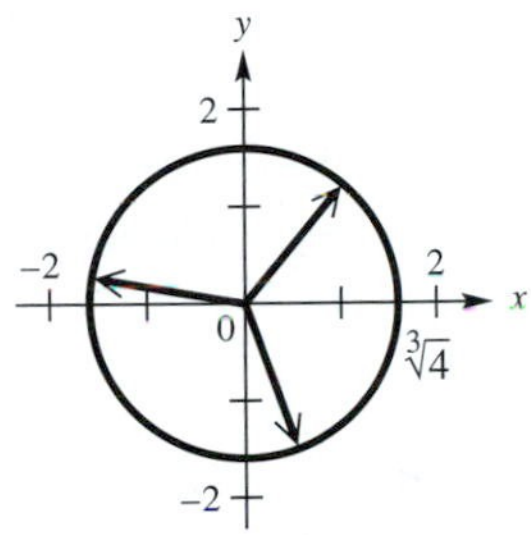

31. $1 + i\sqrt{3}, -1 - i\sqrt{3}$ **33.** $-1, \frac{1}{2} + \frac{\sqrt{3}}{2}i, \frac{1}{2} - \frac{\sqrt{3}}{2}i$ **35.** $\frac{\sqrt{2}}{2} + \frac{\sqrt{2}}{2}i, -\frac{\sqrt{2}}{2} - \frac{\sqrt{2}}{2}i$ **37.** $-4i, 2\sqrt{3} + 2i, -2\sqrt{3} + 2i$ **39.** $\pm 3, \pm 3i$

41. (a)

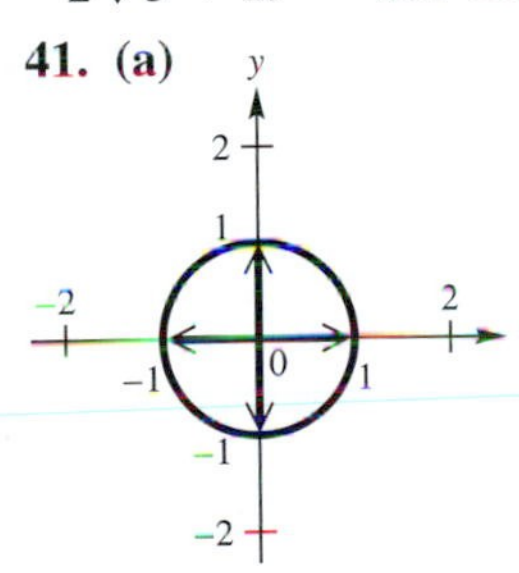

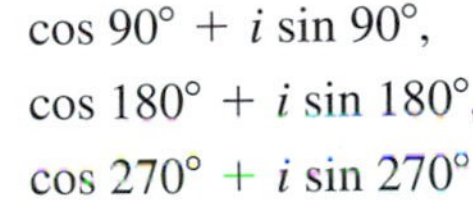

$\cos 0° + i \sin 0°$, $\cos 90° + i \sin 90°$, $\cos 180° + i \sin 180°$, $\cos 270° + i \sin 270°$

(b)

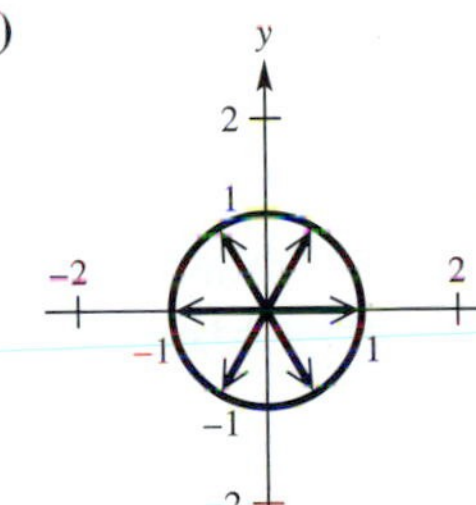

$\cos 0° + i \sin 0°$, $\cos 60° + i \sin 60°$, $\cos 120° + i \sin 120°$, $\cos 180° + i \sin 180°$, $\cos 240° + i \sin 240°$, $\cos 300° + i \sin 300°$

43. The argument for a positive real number is $\theta = 0°$. By the nth root theorem with $k = 0$, $\alpha = 0°$. Thus, one nth root has an argument of $0°$ and it must be real. **45.** $(x + 2)(x^2 - 2x + 4)$ **46.** $x + 2 = 0$ implies $x = -2$. **47.** $x^2 - 2x + 4 = 0$ implies $x = 1 + i\sqrt{3}$ or $x = 1 - i\sqrt{3}$. **48.** $2 \text{ cis } 60°, 2 \text{ cis } 180°, 2 \text{ cis } 300°$ **49.** $1 + i\sqrt{3}, -2, 1 - i\sqrt{3}$

50. They are the same.

51. $\{\cos 45° + i \sin 45°$, $\cos 135° + i \sin 135°$, $\cos 225° + i \sin 225°$, $\cos 315° + i \sin 315°\}$

53. $\{\cos 18° + i \sin 18°$, $\cos 90° + i \sin 90°$, $\cos 162° + i \sin 162°$, $\cos 234° + i \sin 234°$, $\cos 306° + i \sin 306°\}$

55. $\{\cos 60° + i \sin 60°$, $\cos 180° + i \sin 180°$, $\cos 300° + i \sin 300°\}$

57. $\{2(\cos 0° + i \sin 0°)$, $2(\cos 120° + i \sin 120°)$, $2(\cos 240° + i \sin 240°)\}$

59. (a) blue **(b)** red **(c)** yellow

Reviewing Basic Concepts (page 797)

1. $1 + i\sqrt{3}$ **2.** 5 **3.** $2(\cos 135° + i \sin 135°)$ **4.** $8(\cos 180° + i \sin 180°) = -8$ **5.** $2(\cos 90° + i \sin 90°) = 2i$ **6.** 64 cis 51° **7.** -64 **8.** $-4, 2 + 2i\sqrt{3}, 2 - 2i\sqrt{3}$ **9.** $1 + i, -1 - i$ **10.** {cis 60°, cis 180°, cis 300°}

10.6 Exercises (pages 805–808)

1. **(a)** II **(b)** I **(c)** IV **(d)** III

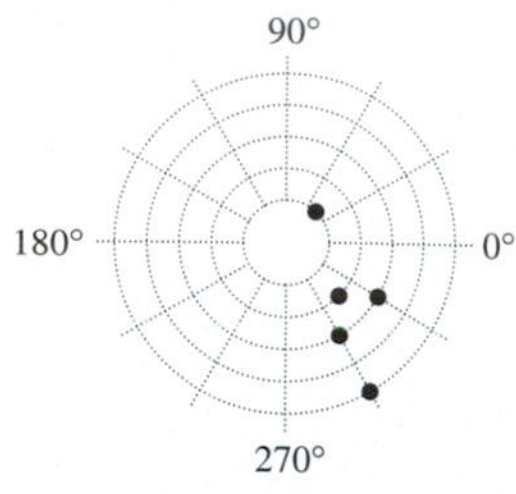

Graphs for Exercises 3, 5, 7, 9, 11

Answers may vary in Exercises 3–11.

3. $(1, 405°), (-1, 225°)$ **5.** $(-2, 495°), (2, 315°)$ **7.** $(5, 300°), (-5, 120°)$ **9.** $(-3, 150°), (3, -30°)$ **11.** $(3, 660°), (-3, 120°)$

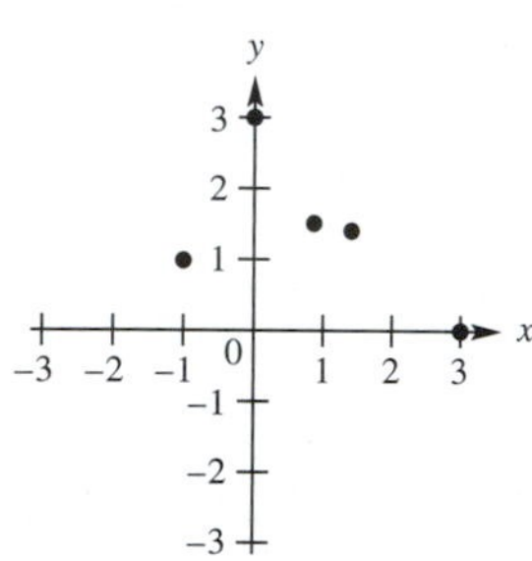

Graphs for Exercises 13, 15, 17, 19, 21

Answers may vary in Exercises 13–21.

13. $(\sqrt{2}, 135°), (-\sqrt{2}, 315°)$ **15.** $(3, 90°), (-3, 270°)$ **17.** $(2, 45°), (-2, 225°)$ **19.** $(\sqrt{3}, 60°), (-\sqrt{3}, 240°)$ **21.** $(3, 0°), (-3, 180°)$

23. $r = \dfrac{4}{\cos\theta - \sin\theta}$

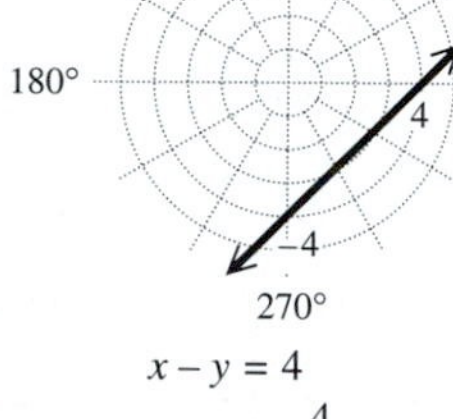

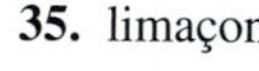

$x - y = 4$

$r = \dfrac{4}{\cos\theta - \sin\theta}$

25. $r = 4$ or $r = -4$

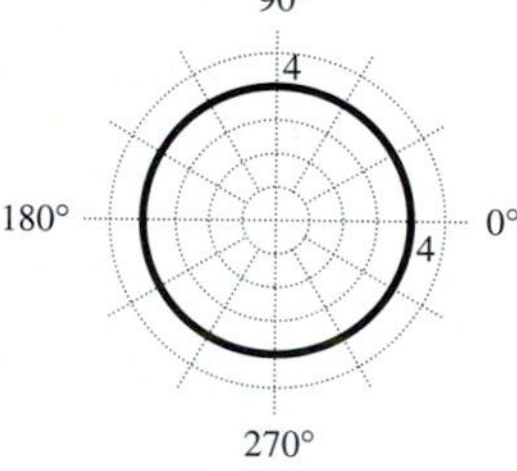
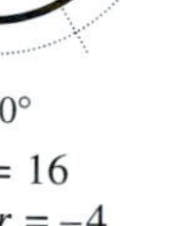

$x^2 + y^2 = 16$

$r = 4$ or $r = -4$

27. $r = \dfrac{5}{2\cos\theta + \sin\theta}$

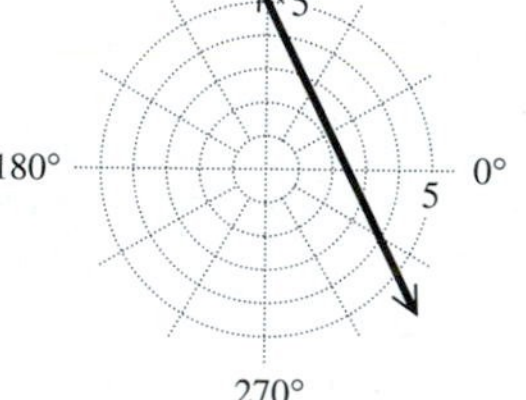

$2x + y = 5$

$r = \dfrac{5}{2\cos\theta + \sin\theta}$

29. C **31.** A

33. cardioid

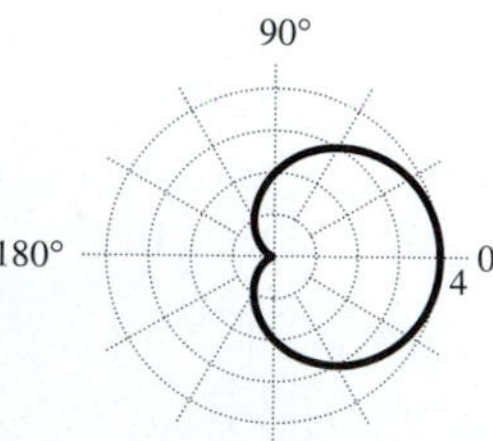

$r = 2 + 2\cos\theta$

35. limaçon

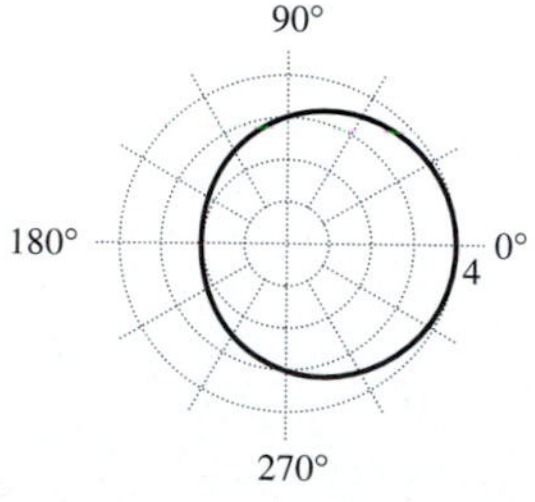

$r = 3 + \cos\theta$

37. four-leaved rose

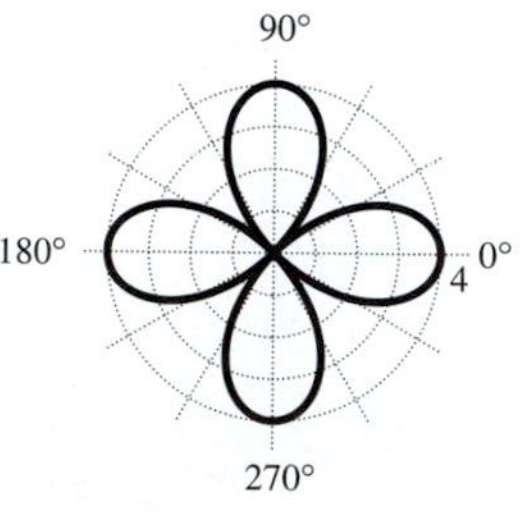

$r = 4\cos 2\theta$

39. lemniscate

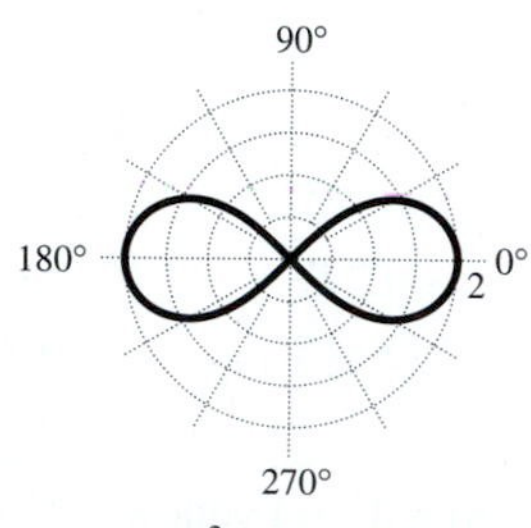

$r^2 = 4\cos 2\theta$

41. cardioid

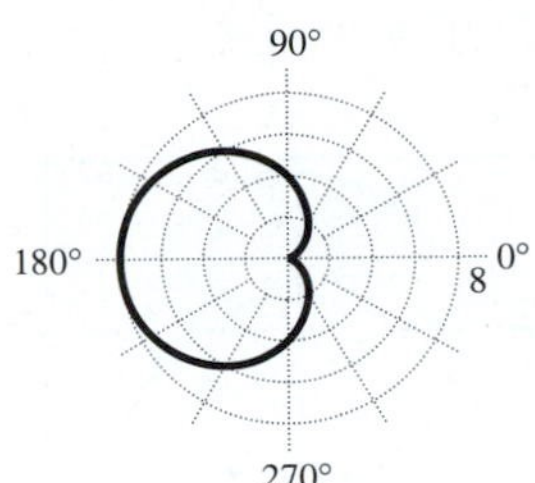

$r = 4 - 4\cos\theta$

43.

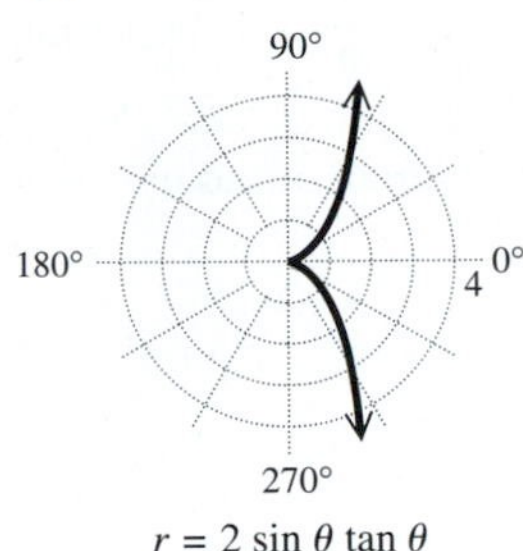

$r = 2\sin\theta\tan\theta$

45. **(a)** $(r, -\theta)$ **(b)** $(r, \pi - \theta)$ or $(-r, -\theta)$ **(c)** $(r, \pi + \theta)$ or $(-r, \theta)$ **47.** To graph (r, θ), $r < 0$, you could locate θ, add 180° to it, and move $|r|$ units along the terminal ray of $\theta + 180°$ in standard position. **49.** θ must be coterminal with 0°, 90°, 180°, or 270°; quadrantal **51.** reflected across the line $\theta = \frac{\pi}{2}$ (y-axis) **53.** The value of a determines the length of the leaves. The value of n determines the number of leaves—n leaves if n is odd and $2n$ leaves if n is even.

55. $x^2 + (y - 1)^2 = 1$

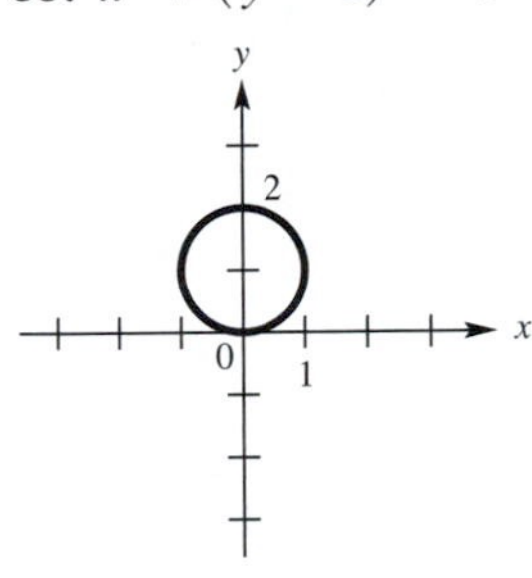

$r = 2\sin\theta$
$x^2 + (y - 1)^2 = 1$

57. $y^2 = 4(x + 1)$

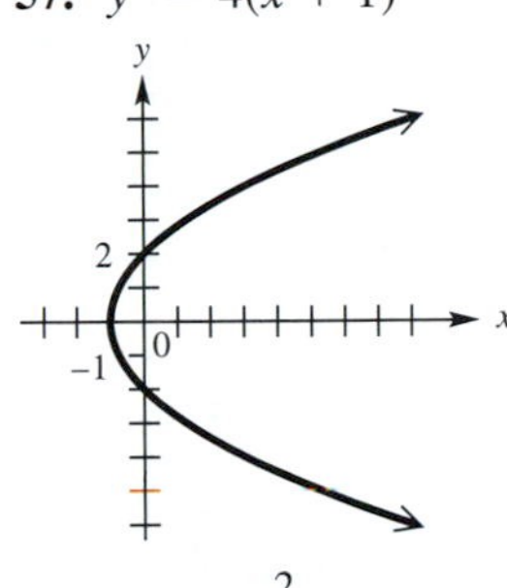

$r = \frac{2}{1 - \cos\theta}$
$y^2 = 4(x + 1)$

59. $(x + 1)^2 + (y + 1)^2 = 2$

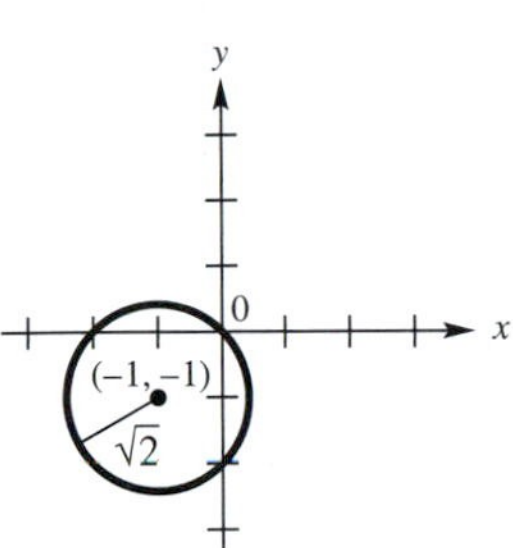

$r = -2\cos\theta - 2\sin\theta$
$(x + 1)^2 + (y + 1)^2 = 2$

61. $x = 2$

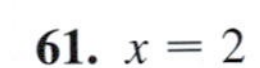

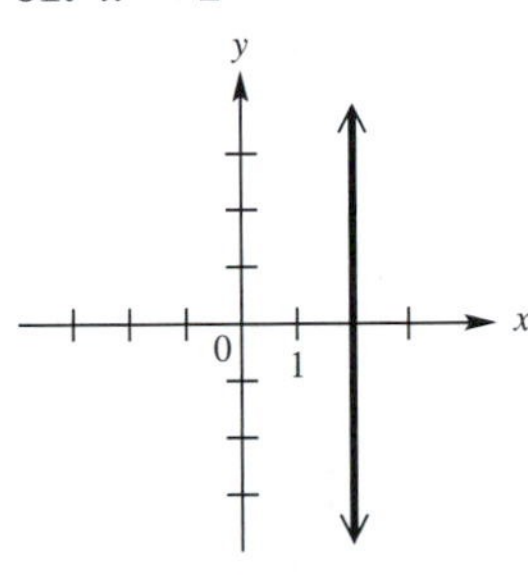

$r = 2\sec\theta$
$x = 2$

63. $x + y = 2$

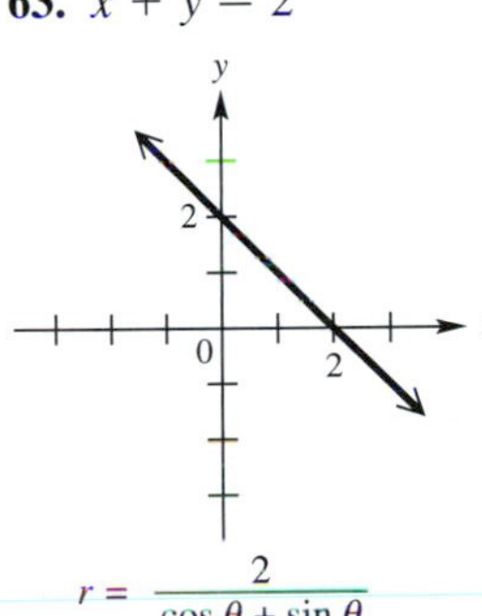

$r = \frac{2}{\cos\theta + \sin\theta}$
$x + y = 2$

65.

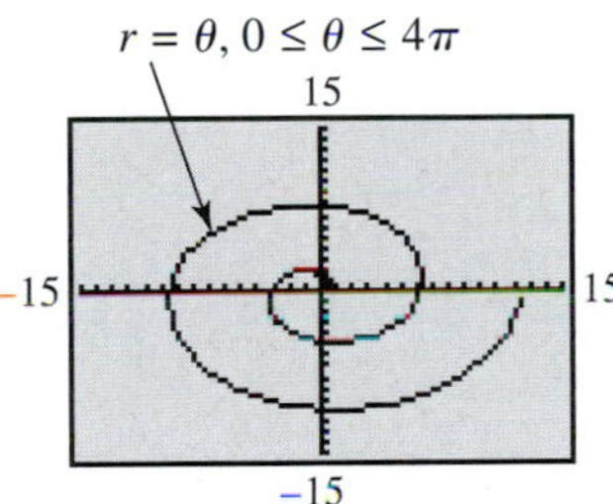

67.

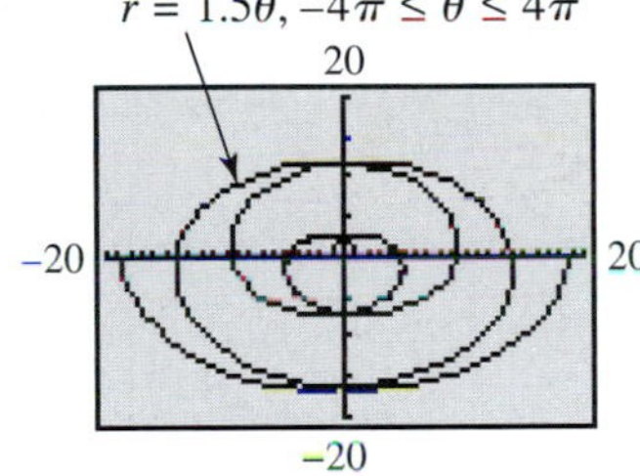

69. $\left(2, \frac{\pi}{6}\right), \left(2, \frac{5\pi}{6}\right)$ **71.** $\left(\frac{4 + \sqrt{2}}{2}, \frac{\pi}{4}\right), \left(\frac{4 - \sqrt{2}}{2}, \frac{5\pi}{4}\right)$

73. **(a)**

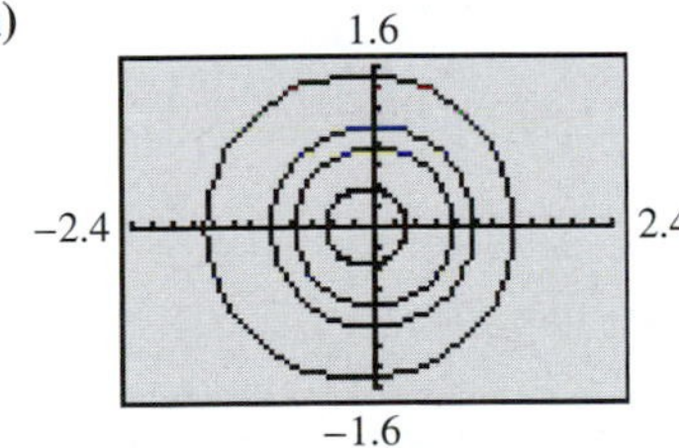

(b)

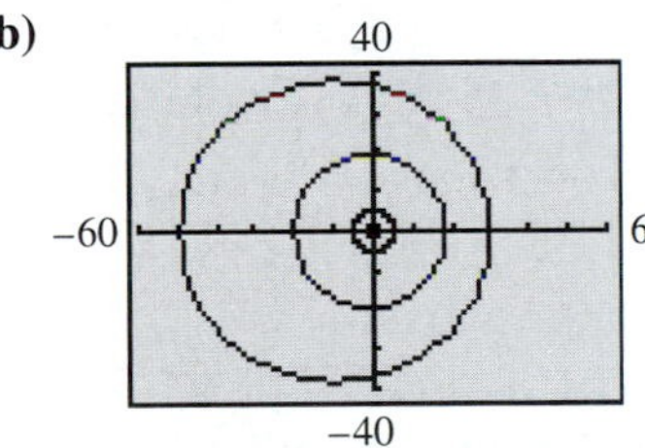

Earth is closest to the sun. **(c)** No, it is not.

10.7 Exercises *(pages 812–814)*

1. $x^2 + y^2 = 9$; circle with radius 3 **3.** $y = 2 - x$; line segment **5.** $\frac{y^2}{4} - \frac{x^2}{9} = 1$; hyperbola (upper branch)

The window in Exercises 7 and 9 is $[-4.7, 4.7]$ by $[-3.1, 3.1]$.

7. (a) traces a circle of radius 3 once

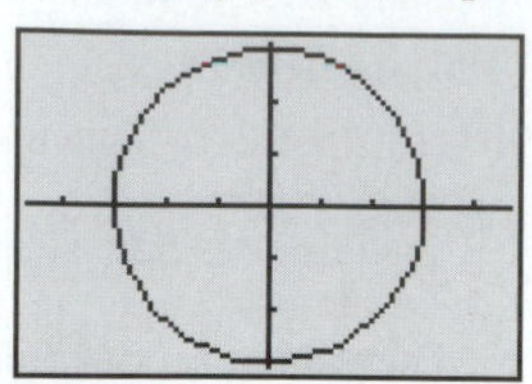

(b) traces a circle of radius 3 twice

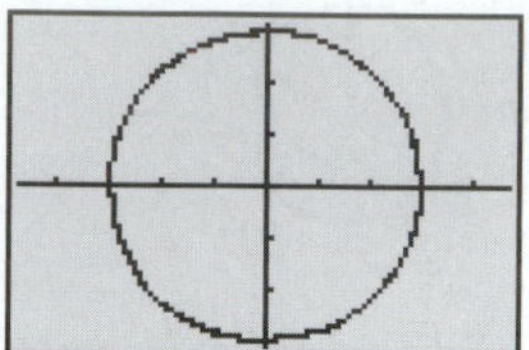

9. (a) traces a circle of radius 3 once counterclockwise, starting at (3, 0)

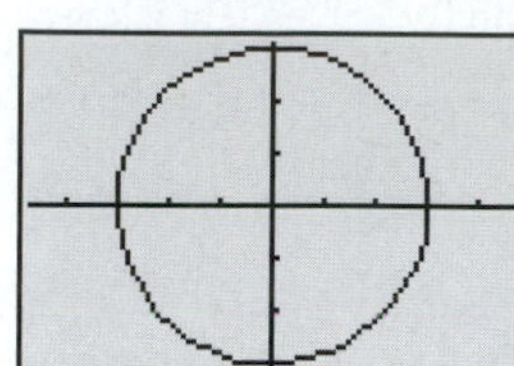

(b) traces a circle of radius 3 once clockwise starting at (0, 3)

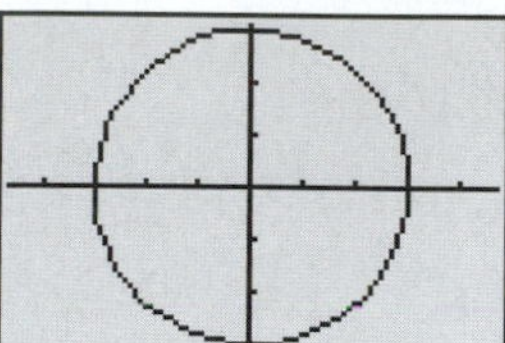

11.

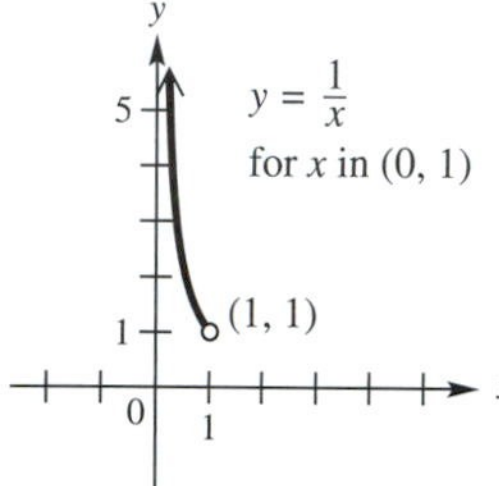

13.

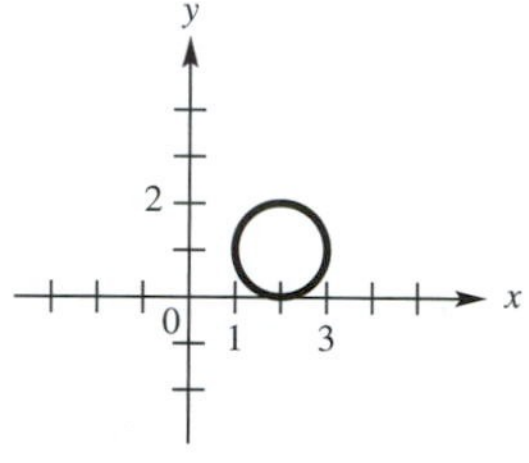

$(x - 2)^2 + (y - 1)^2 = 1$
for x in $[1, 3]$

15.

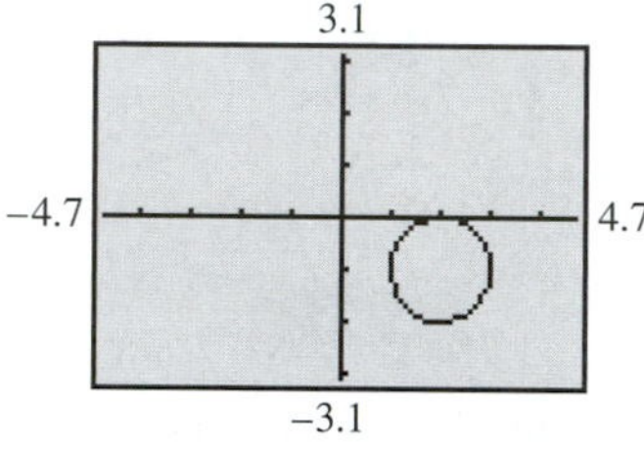

17.

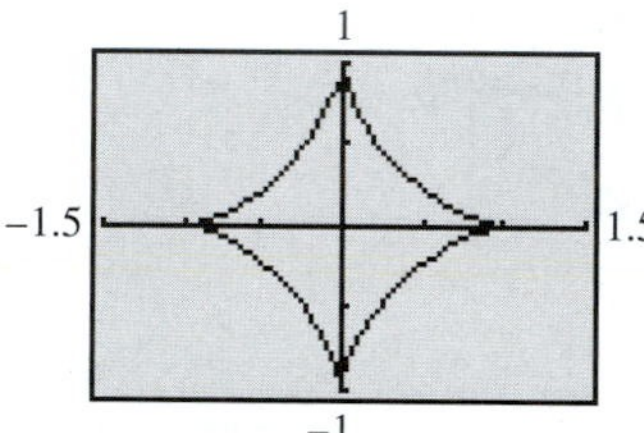

19.

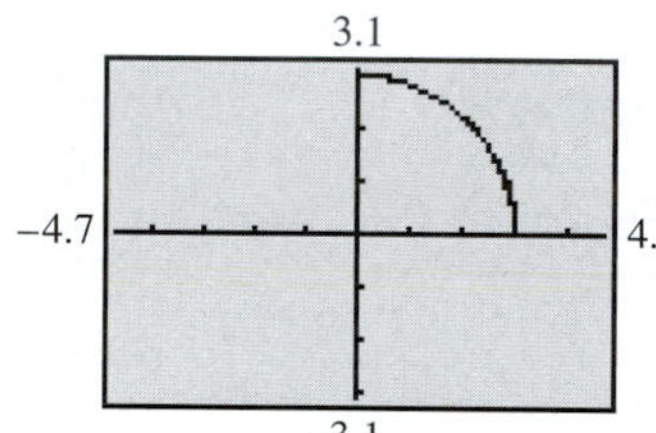

21.

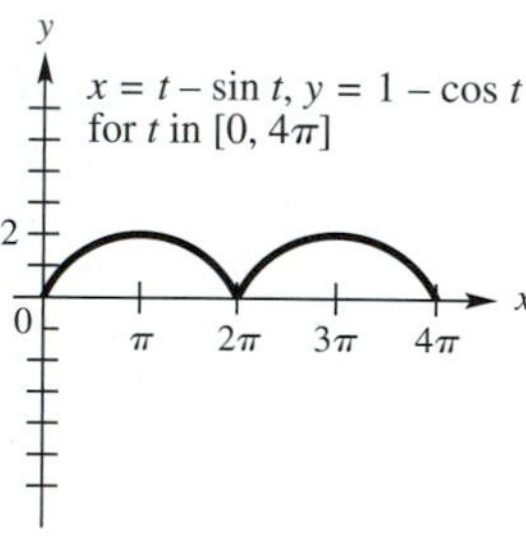

The window in Exercises 23 and 25 is $[0, 6]$ by $[0, 4]$.

23. F

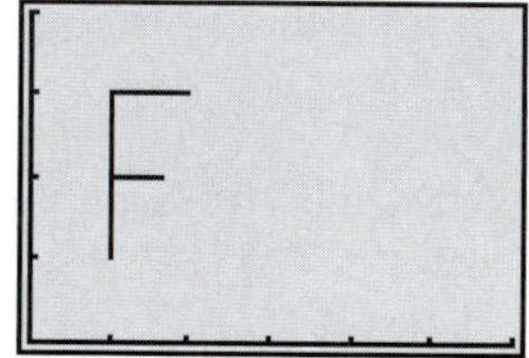

25. D

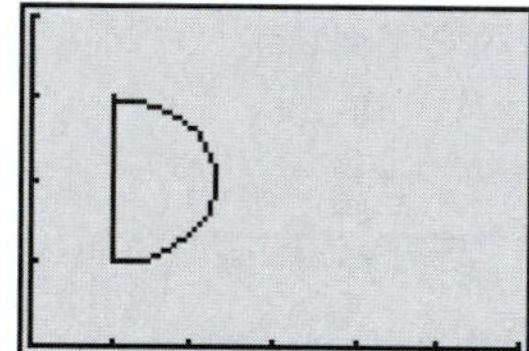

27. $x_1 = 0$, $y_1 = 2t$; $x_2 = t$, $y_2 = 0$; $0 \le t \le 1$; Answers may vary.

29. $x_1 = \sin t$, $y_1 = \cos t$; $x_2 = 0$, $y_2 = t - 2$; $0 \le t \le \pi$; answers may vary. **31.** Answers may vary.

The window in Exercises 33 and 35 is $[-6, 6]$ by $[-4, 4]$.

33.

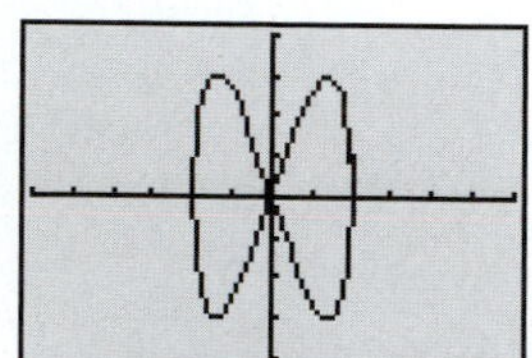

35.

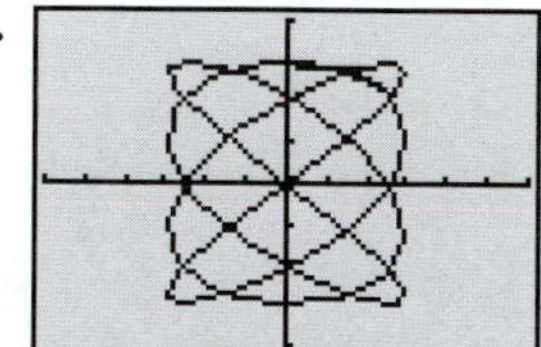

37. (a) $x = 24t$, $y = -16t^2 + 24\sqrt{3}t$

(b) $y = -\frac{1}{36}x^2 + \sqrt{3}x$ **(c)** 2.6 sec; 62 ft

39. **(a)** $x = (88 \cos 20°)t$, $y = 2 - 16t^2 + (88 \sin 20°)t$ **(b)** $y = 2 - \frac{x^2}{484 \cos^2 20°} + (\tan 20°)x$ **(c)** 1.9 sec; 161 ft

41. about 1456 ft **43.** **(a)**

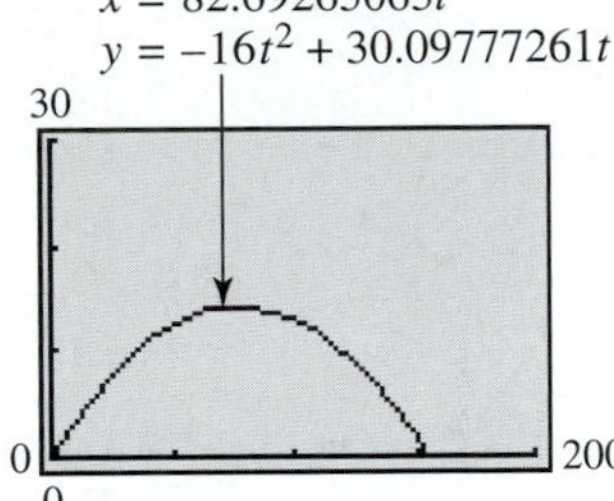

(b) 20.0° **(c)** $x = (88 \cos 20.0°)t$, $y = -16t^2 + (88 \sin 20.0°)t$

Reviewing Basic Concepts *(page 815)*

1. IV **2.** $(2\sqrt{2}, 135°)$, $(-2\sqrt{2}, -45°)$; Answers may vary.

3. cardioid

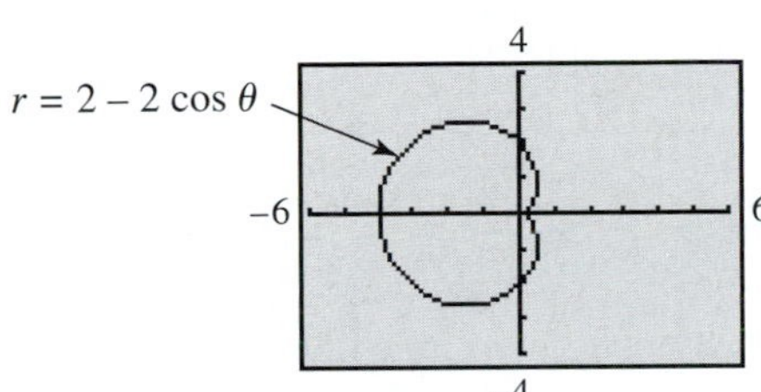

4. $(x - 1)^2 + y^2 = 1$ **5.** $r = \frac{6}{\cos \theta + \sin \theta}$

6. $\frac{x^2}{4} + \frac{y^2}{16} = 1$

7.

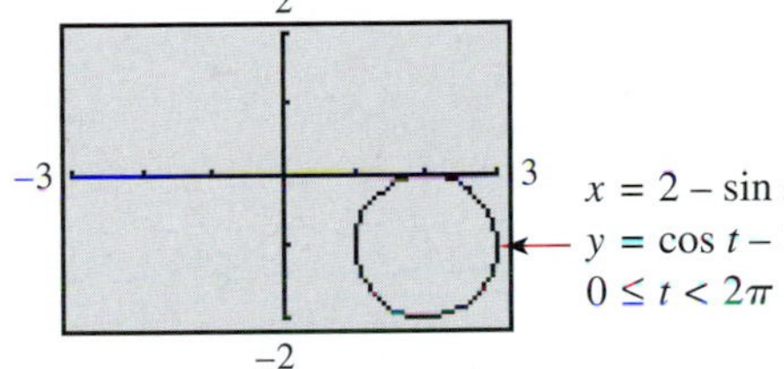

8. 285 ft

Chapter 10 Review Exercises *(pages 818–821)*

1. 63.7 m **3.** 19.87° or 19° 52′ **5.** 25° 00′ **7.** 173 ft **9.** 153,600 m^2 **11.** .234 km^2 **13.** 2.7 mi **15.** 1.91 mi **17.** 11 ft **19.** 77.1° **21.** No, since $a + b = c$. **23.** It becomes the Pythagorean theorem.

25.

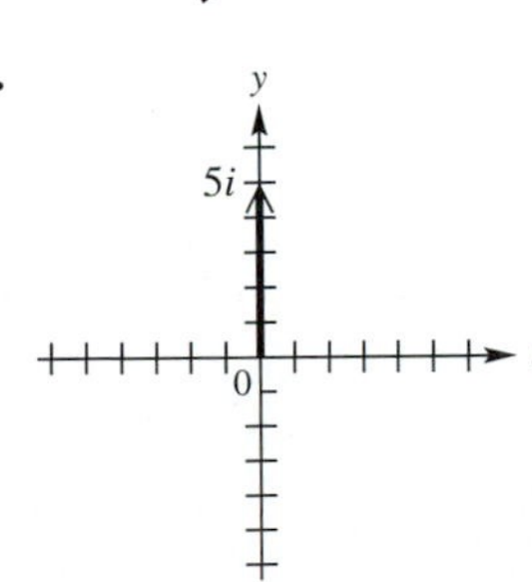

27. 15; 126.9° **29.** $\langle 17.9, 66.8 \rangle$ **31.** **(a)** 14 **(b)** 52.13° **33.** Yes they are, because $\mathbf{u} \cdot \mathbf{v} = 0$. **35.** 2800 newtons; 30.4° **37.** bearing: 7° 20′; speed: 8.3 mph **39.** magnitude: 9.52082827; direction angle: 119.0646784° **40.** $\langle -4.10424172, 11.27631145 \rangle$ **41.** $\langle -.520944533, -2.954423259 \rangle$ **42.** $\langle -4.625186253, 8.321888191 \rangle$ **43.** magnitude: 9.52082827; direction angle: 119.0646784° **44.** They are the same. Preference of method is an individual choice.

45.

47. $5 + 4i$ **49.** $2\sqrt{2}$ cis 135° **51.** $-\sqrt{2} - i\sqrt{2}$ **53.** $-30i$ **55.** $-\frac{1}{8} + \frac{\sqrt{3}}{8}i$ **57.** $8i$ **59.** $-\frac{1}{2} - \frac{\sqrt{3}}{2}i$

61. 3 cis 90°, 3 cis 210°, 3 cis 330°

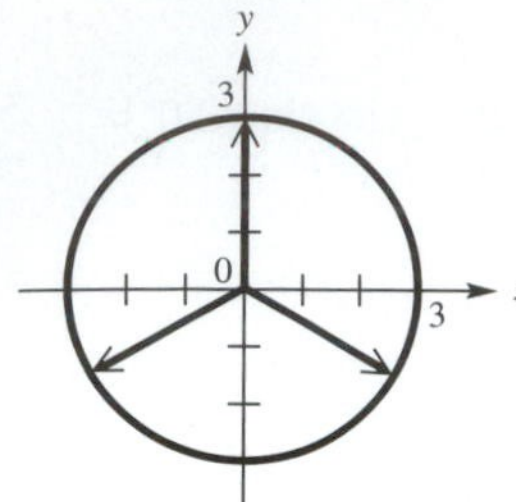

63. 2 cis 0°, 2 cis 72°, 2 cis 144°, 2 cis 216°, 2 cis 288°

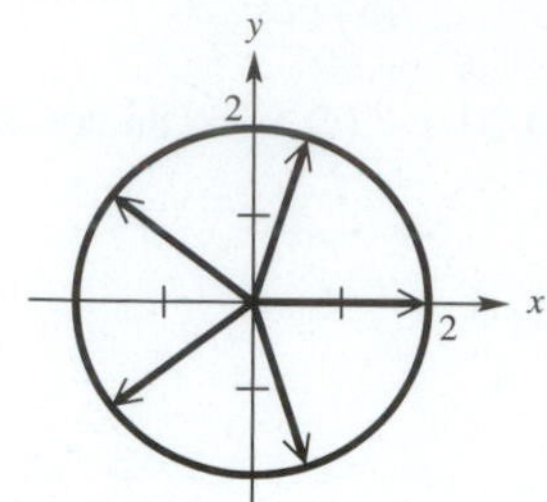

65. $(-6\sqrt{2}, -6\sqrt{2})$

67. $(6\sqrt{2}, 135°)$

69.

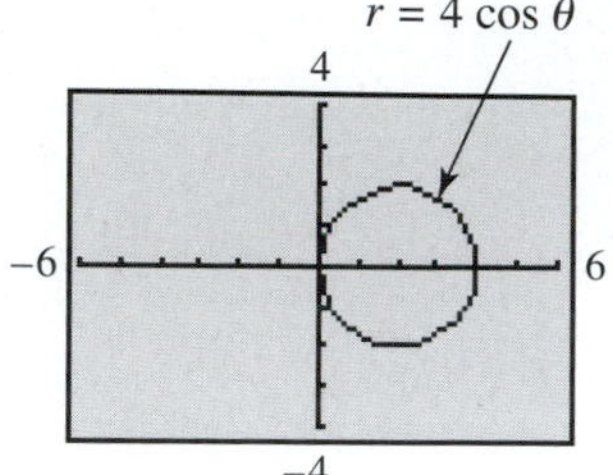

71.

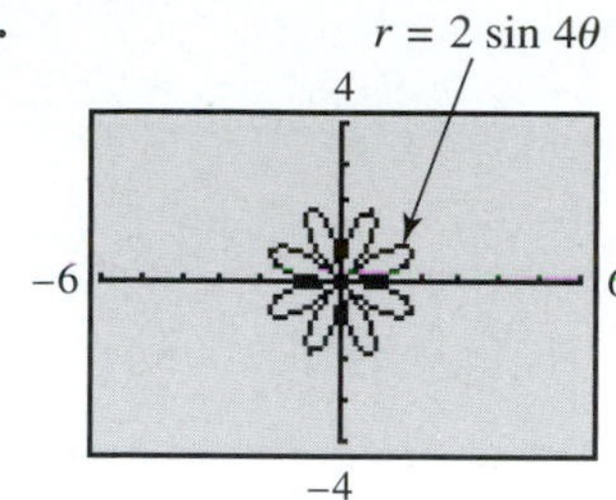

73. $y^2 + 6x - 9 = 0$ **75.** $x^2 - x + y^2 - y = 0$

77. $r = -\dfrac{3}{\cos\theta}$ **79.** $r = \dfrac{\tan\theta}{\cos\theta}$

81. $y^2 = -\dfrac{1}{2}(x - 1)$ or $2y^2 + x - 1 = 0$, for x in $[-1, 1]$

83.

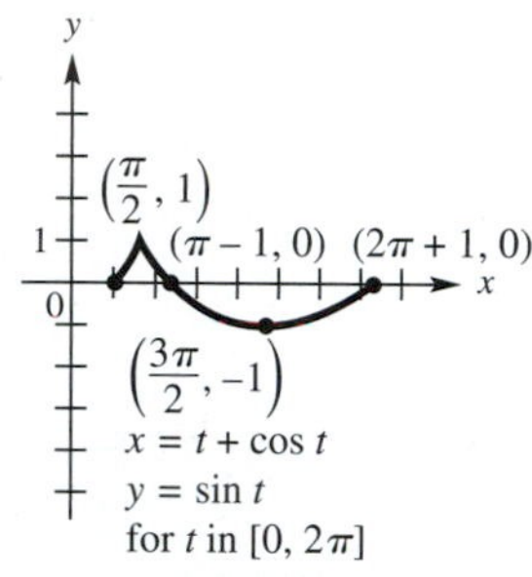

Chapter 10 Test *(pages 821–822)*

1. (a) 137.5° **(b)** 180 km **(c)** 49.0° **2. (a)** $b > 10$ **(b)** not possible **(c)** $b \le 10$ **3.** The sum of the two smaller numbers must be greater than the largest number. **4. (a)** 115 km **(b)** 43 ft **(c)** horizontal: −346; vertical: 451 **(d)** 826 lb **5. (a)** $\langle 1, -3\rangle$ **(b)** $\langle -6, 18\rangle$ **(c)** −20 **(d)** 180° **6. (a)** $-2 - 2i\sqrt{3}$ **(b)** 8 cis 120° **(c)** $-6 + 2i\sqrt{3}$

7. (a) $-\dfrac{15}{2} + \dfrac{15\sqrt{3}}{2}i$ **(b)** $-\dfrac{1}{2}i$ **(c)** $16 + 16i\sqrt{3}$ **(d)** $\sqrt[4]{2}$ cis 7.5°, $\sqrt[4]{2}$ cis 97.5°, $\sqrt[4]{2}$ cis 187.5°, $\sqrt[4]{2}$ cis 277.5°

8. (a)

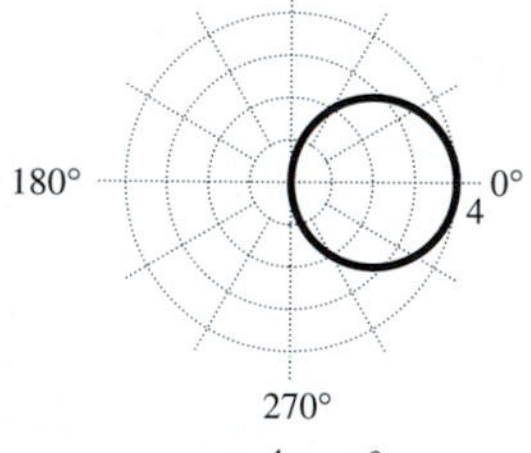

$r = 4\cos\theta$

(b) $x^2 - 4x + y^2 = 0$

(c) Yes, the graph of $x^2 - 4x + y^2 = 0$ is a circle with center $(2, 0)$ and radius 2.

9. $r = \dfrac{4}{2 \sin \theta - \cos \theta}$ **10.**

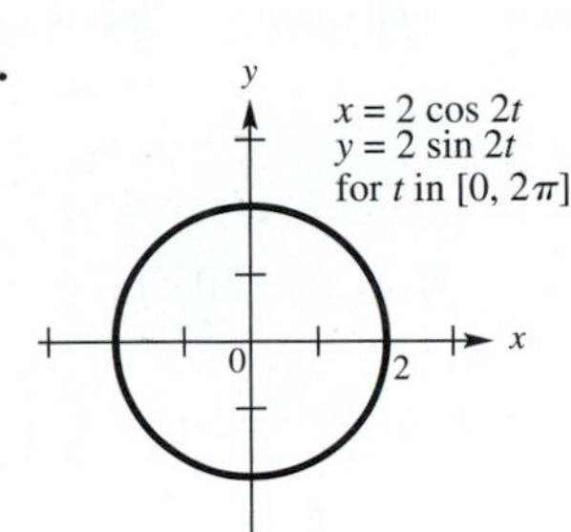

CHAPTER 11 FURTHER TOPICS IN ALGEBRA

11.1 Exercises *(pages 833–835)*

1. 14, 18, 22, 26, 30 **3.** 1, 2, 4, 8, 16 **5.** $0, \frac{1}{9}, \frac{2}{27}, \frac{1}{27}, \frac{4}{243}$ **7.** $-2, 4, -6, 8, -10$ **9.** $1, \frac{7}{6}, 1, \frac{5}{6}, \frac{19}{27}$

11. The *n*th term is the term that is in position *n*. For example, in the sequence 2, 4, 6, 8, . . . , the *third* term is 6, which is in position 3. The *n*th term here is given by $2n$. **13.** finite **15.** finite **17.** infinite **19.** finite **21.** $-2, 1, 4, 7$

23. 1, 1, 2, 3 **25.** 35 **27.** $\frac{25}{12}$ **29.** 288 **31.** 3 **33.** -18 **35.** $\frac{728}{9}$ **37.** 28 **39.** 343

41. $-2 - 1 + 0 + 1 + 2$ **43.** $-1 + 1 + 3 + 5 + 7$ **45.** $-10 - 4 + 0$ **47.** $0 + \frac{1}{2} + \frac{2}{3} + \frac{3}{4}$

49. $-3.5 + .5 + 4.5 + 8.5$ **51.** $0 + 4 + 16 + 36$ **53.** $-1 - \frac{1}{3} - \frac{1}{5} - \frac{1}{7}$ **55.** 600 **57.** 1240 **59.** 90

61. 220 **63.** 304 **65.** converges to $\frac{1}{2}$ **67.** diverges **69.** converges to $e \approx 2.71828$ **71.** 1.081123534; 3.1407; accurate to three decimal places rounded **73.** **(a)** $N_{j+1} = 2N_j$ for $j \geq 1$ **(b)** 1840 **(c)** 15,000

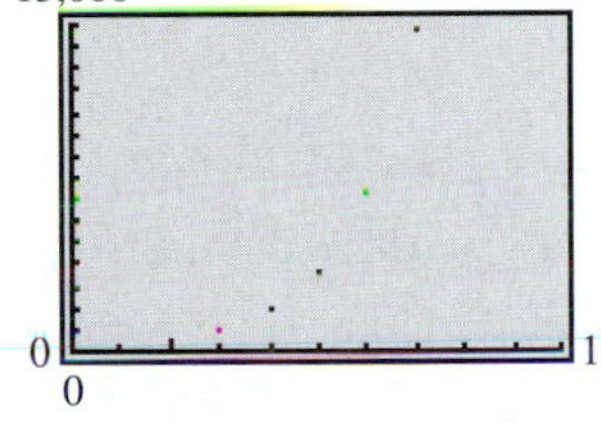

(d) The growth is very rapid. Since there is a doubling of the bacteria at equal intervals, their growth is exponential.

75. **(a)** 2.718254, $e \approx 2.718282$ **(b)** .367857, $e^{-1} \approx .367879$ **(c)** 1.648721, $\sqrt{e} = e^{1/2} \approx 1.648721$

11.2 Exercises *(pages 841–843)*

1. 3 **3.** -5 **5.** $x + 2y$ **7.** 8, 14, 20, 26, 32 **9.** $5, 3, 1, -1, -3$ **11.** 14, 12, 10, 8, 6

13. $a_8 = 19; a_n = 2n + 3$ **15.** $a_8 = 7; a_n = n - 1$ **17.** $a_8 = -6; a_n = -2n + 10$ **19.** $a_8 = -3; a_n = 4.5n - 39$

21. $a_8 = x + 21; a_n = x + 3n - 3$ **23.** 3 **25.** 5 **27.** 215 **29.** 125 **31.** 230 **33.** 77.5

35. $a_1 = 7; d = 5$ **37.** $a_1 = 1; d = -\frac{20}{11}$

In Exercises 39–43, *D* is the domain and *R* is the range.

39. $a_n = n - 3$; *D:* $\{1, 2, 3, 4, 5, 6\}$; *R:* $\{-2, -1, 0, 1, 2, 3\}$ **41.** $a_n = -\frac{1}{2}n + 3$; *D:* $\{1, 2, 3, 4, 5, 6\}$; *R:* $\{0, .5, 1, 1.5, 2, 2.5\}$

43. $a_n = -20n + 30$; *D:* $\{1, 2, 3, 4, 5\}$; *R:* $\{-70, -50, -30, -10, 10\}$ **45.** 80 **47.** 1275 **49.** 1739 **51.** 4100

53. 18 **55.** 140 **57.** -684 **59.** 500,500 **61.** $f(1) = m + b$; $f(2) = 2m + b$; $f(3) = 3m + b$ **62.** yes

63. m **64.** $a_n = mn + b$ **65.** 328.3 **67.** 172.884 **69.** 1281 **71.** 4680 **73.** 54,800 **75.** 713 in.
77. All terms are the same constant. **79.** yes

11.3 Exercises *(pages 850–853)*

1. $\frac{5}{3}$, 5, 15, 45 **3.** $\frac{5}{8}, \frac{5}{4}, \frac{5}{2}$, 5, 10 **5.** $a_5 = 80$; $a_n = 5(-2)^{n-1}$ **7.** $a_5 = -108$; $a_n = -\frac{4}{3}(3)^{n-1}$
9. $a_5 = -729$; $a_n = -9(-3)^{n-1}$ **11.** $a_5 = -324$; $a_n = -4(3)^{n-1}$ **13.** $a_5 = \frac{125}{4}$; $a_n = \frac{4}{5}\left(\frac{5}{2}\right)^{n-1}$
15. $a_5 = \frac{5}{8}$; $a_n = 10\left(-\frac{1}{2}\right)^{n-1}$ **17.** $a_1 = 125$; $r = \frac{1}{5}$ **19.** $a_1 = -2$; $r = \frac{1}{2}$ **21.** 682 **23.** $\frac{99}{8}$ **25.** 860.95
27. 363 **29.** $\frac{189}{4}$ **31.** 2032 **33.** The sum exists if $|r| < 1$. **35.** $\frac{8}{9}$ **37.** $\frac{5}{11}$ **39.** 2; The sum does not converge.
41. $\frac{1}{2}$ **43.** $\frac{128}{7}$ **45.** $\frac{1000}{9}$ **47.** $\frac{8}{3}$ **49.** 4 **51.** $\frac{3}{7}$ **53.** $g(1) = ab$; $g(2) = ab^2$; $g(3) = ab^3$
54. yes; The common ratio is b. **55.** $a_n = ab^n$ **56.** The independent variable is in the exponent. **57.** 97.739
59. .212 **61.** \$10,582.80 **63.** \$27,224.22 **65.** **(a)** $a_1 = 1169$; $r = .916$ **(b)** $a_{10} = 531$; $a_{20} = 221$; This means that a person who is 10 yr from retirement should have savings of 531% of his or her annual salary; a person 20 yr from retirement should have savings of 221% of his or her annual salary. **67.** **(a)** $a_n = a_1 \cdot 2^{n-1}$ **(b)** 15 (rounded from 14.28 since n must be a natural number) **(c)** 560 min, or 9 hr and 20 min **69.** 13.4% **71.** \$26,214.40 **73.** 200 cm **75.** 62; 2046
77. Option 2 pays better. **79.** $\frac{1}{64}$ m

Reviewing Basic Concepts *(pages 853–854)*

1. 4, −8, 12, −16, 20 **2.** 50 **3.** It converges to 1. **4.** 8, 6, 4, 2, 0 **5.** −11 **6.** 245 **7.** $a_3 = -18$; $a_n = -2(-3)^{n-1}$ **8.** geometric; $\frac{7448}{625}$ **9.** 6 **10.** \$8056.52

11.4 Exercises *(pages 859–860)*

1. 20 **3.** 35 **5.** 56 **7.** 45 **9.** 1 **11.** n **13.** 56 **15.** 4950 **17.** 1 **19.** 9 **21.** $16x^4$; $81y^4$
23. $x^6 + 6x^5y + 15x^4y^2 + 20x^3y^3 + 15x^2y^4 + 6xy^5 + y^6$ **25.** $p^5 - 5p^4q + 10p^3q^2 - 10p^2q^3 + 5pq^4 - q^5$
27. $r^{10} + 5r^8s + 10r^6s^2 + 10r^4s^3 + 5r^2s^4 + s^5$ **29.** $p^4 + 8p^3q + 24p^2q^2 + 32pq^3 + 16q^4$
31. $2401p^4 + 2744p^3q + 1176p^2q^2 + 224pq^3 + 16q^4$
33. $729x^6 - 2916x^5y + 4860x^4y^2 - 4320x^3y^3 + 2160x^2y^4 - 576xy^5 + 64y^6$
35. $\frac{m^6}{64} - \frac{3m^5}{16} + \frac{15m^4}{16} - \frac{5m^3}{2} + \frac{15m^2}{4} - 3m + 1$ **37.** $4r^4 + \frac{8\sqrt{2}r^3}{m} + \frac{12r^2}{m^2} + \frac{4\sqrt{2}r}{m^3} + \frac{1}{m^4}$ **39.** $-3584h^3j^5$
41. $319{,}770a^{16}b^{14}$ **43.** $38{,}760x^6y^{42}$ **45.** $90{,}720x^{28}y^{12}$ **47.** 11 **49.** exact: 3,628,800; approximate: 3,598,695.619
50. about .830% **51.** exact: 479,001,600; approximate: 475,687,486.5; about .692% **52.** exact: 6,227,020,800; approximate: 6,187,239,475; about .639%; As n increases, the percent error decreases. **53.** .942 **55.** 1.015

11.5 Exercises *(pages 865–866)*

1. positive integers (natural numbers) **3.** $n = 1$ and $n = 2$

Although we do not usually give proofs, the answers to Exercises 5 and 13 are shown here.

5. *Step 1:* $3(1) = 3$ and $\frac{3(1)(1+1)}{2} = \frac{6}{2} = 3$, so S is true for $n = 1$. *Step 2:* S_k: $3 + 6 + 9 + \cdots + 3k = \frac{3(k)(k+1)}{2}$
S_{k+1}: $3 + 6 + 9 + \cdots + 3(k+1) = \frac{3(k+1)((k+1)+1)}{2}$; Add $3(k+1)$ to each side of S_k and simplify until you obtain S_{k+1}. Since S is true for $n = 1$ and S is true for $n = k + 1$ when it is true for $n = k$, S is true for every positive integer n.

13. *Step 1:* $\frac{1}{1 \cdot 2} = \frac{1}{2}$ and $\frac{1}{1+1} = \frac{1}{2}$, so S is true for $n = 1$. *Step 2:* S_k: $\frac{1}{1 \cdot 2} + \frac{1}{2 \cdot 3} + \frac{1}{3 \cdot 4} + \cdots + \frac{1}{k(k+1)} = \frac{k}{k+1}$

S_{k+1}: $\frac{1}{1 \cdot 2} + \frac{1}{2 \cdot 3} + \cdots + \frac{1}{(k+1)[(k+1)+1]} = \frac{k+1}{(k+1)+1}$; Add $\frac{1}{(k+1)[(k+1)+1]}$ to each side of S_k and simplify until you obtain S_{k+1}. Since S is true for $n = 1$ and S is true for $n = k + 1$ when it is true for $n = k$, S is true for every positive integer n. **19.** $n = 1$ or 2 **21.** $n = 2, 3,$ or 4 **35.** $\frac{4^{n-1}}{3^{n-2}}$ or $3\left(\frac{4}{3}\right)^{n-1}$ **37.** $2^n - 1$

Reviewing Basic Concepts *(page 867)*

1. 10 **2.** 15 **3.** $n + 1$ **4.** $a^4 + 8a^3b + 24a^2b^2 + 32ab^3 + 16b^4$ **5.** $\frac{1}{m^3} - \frac{3n^2}{m^2} + \frac{3n^4}{m} - n^6$ **6.** $60x^4y^2$

11.6 Exercises *(pages 873–876)*

1. 19,958,400 **3.** 72 **5.** 5 **7.** 6 **9.** 1 **11.** 495 **13.** 1,860,480 **15.** 259,459,200 **17.** 15,504 **19.** 6435 **21. (a)** permutation **(b)** permutation **(c)** combination **(d)** combination **(e)** permutation **(f)** combination **(g)** permutation **23.** 30 **25. (a)** 27,600 **(b)** 35,152 **(c)** 1104 **27.** 15 **29. (a)** 17,576,000 **(b)** 17,576,000 **(c)** 456,976,000 **31.** 720 **33.** 120 **35.** 2730 **37.** 120; 30,240 **39.** 27,405 **41.** 20 **43.** 10 **45.** 28 **47. (a)** 84 **(b)** 10 **(c)** 40 **(d)** 28 **49.** 1680 **51.** 15 **53.** 479,001,600 **55. (a)** 56 **(b)** 462 **(c)** 3080 **(d)** 8526 **57.** 4096 **59.** 1,000,000 **61.** 6 **71.** If $r > n$, then $P(n, r) = \frac{n!}{(n-r)!}$ is undefined because $n - r < 0$, and thus $(n - r)!$ is undefined.

11.7 Exercises *(pages 882–885)*

1. $S = \{H\}$ **3.** $S = \{(H, H, H), (H, H, T), (H, T, T), (H, T, H), (T, T, T), (T, T, H), (T, H, T), (T, H, H)\}$ **5.** $S = \{(1, 1), (1, 2), (1, 3), (2, 1), (2, 2), (2, 3), (3, 1), (3, 2), (3, 3)\}$ **7. (a)** $\{H\}$; 1 **(b)** $\emptyset$; 0 **9. (a)** $\{(1, 1), (2, 2), (3, 3)\}$; $\frac{1}{3}$ **(b)** $\{(1, 1), (1, 3), (2, 1), (2, 3), (3, 1), (3, 3)\}$; $\frac{2}{3}$ **(c)** $\{(2, 1), (2, 3)\}$; $\frac{2}{9}$ **11.** $P = \frac{6}{5} > 1$ is impossible. **13. (a)** $\frac{1}{5}$ **(b)** 0 **(c)** $\frac{7}{15}$ **(d)** 1 to 4 **(e)** 7 to 8 **15.** 1 to 4 **17.** $\frac{2}{5}$ **19. (a)** $\frac{1}{2}$ **(b)** $\frac{7}{10}$ **(c)** $\frac{2}{5}$ **21. (a)** F **(b)** D **(c)** A **(d)** F **(e)** C **(f)** B **(g)** E **23. (a)** .76 **(b)** .24 **25. (a)** .62 **(b)** .27 **(c)** .11 **(d)** .89 **27.** $\frac{48}{28{,}561} \approx .001681$ **29. (a)** .72 **(b)** .70 **(c)** .79 **31.** $\frac{5}{16} = .3125$ **33.** $\frac{1}{32} = .03125$ **35.** $\frac{1}{2}$ **37.** .042246 **39.** .026864 **41.** approximately 4.6×10^{-10} **43.** approximately .875 **45.** .49 **47.** .51 **49. (a)** .047822 **(b)** .976710 **(c)** .897110 **51. (a)** approximately 40.4% **(b)** approximately 4.7% **(c)** approximately .2%; This means that in a large family or group of people, it is unlikely that everyone will become sick even though the disease is highly infectious.

Reviewing Basic Concepts *(pages 885–886)*

1. 24 **2.** 210 **3.** 450 **4.** 210 **5.** 72 **6.** $S = \{(H, H), (H, T), (T, H), (T, T)\}$ **7.** $\frac{2}{36} = \frac{1}{18}$ **8.** $\frac{4}{2{,}598{,}960} \approx .0000015$ **9.** $\frac{3}{10}$ **10.** approximately .484

Chapter 11 Review Exercises *(pages 890–892)*

1. $\frac{1}{2}, \frac{2}{3}, \frac{3}{4}, \frac{4}{5}, \frac{5}{6}$; neither **3.** 8, 10, 12, 14, 16; arithmetic **5.** 5, 2, −1, −4, −7; arithmetic **7.** $3\pi - 2, 2\pi - 1, \pi, 1, -\pi + 2$ **9.** $-5, -1, -\frac{1}{5}, -\frac{1}{25}, -\frac{1}{125}$ **11.** -1; $-8\left(\frac{1}{2}\right)^{n-1} = -\left(\frac{1}{2}\right)^{n-4}$ or 1; $-8\left(-\frac{1}{2}\right)^{n-1} = \left(-\frac{1}{2}\right)^{n-4}$ **13.** $-x + 61$ **15.** 612 **17.** $\frac{4}{25}$ **19.** $\frac{13}{36}$

21. 1 **23.** $\frac{73}{12}$ **25.** 3,126,250 **27.** $\frac{4}{3}$ **29.** 36 **31.** diverges **33.** -10

In Exercises 35 and 37, other answers are possible.

35. $\sum_{i=1}^{15}(-5i+9)$ **37.** $\sum_{i=1}^{6}4(3)^{i-1}$ **39.** $x^4+8x^3y+24x^2y^2+32xy^3+16y^4$

41. $243x^{5/2}-405x^{3/2}+270x^{1/2}-90x^{-1/2}+15x^{-3/2}-x^{-5/2}$ **43.** $-3584x^3y^5$ **45.** $x^{12}+24x^{11}+264x^{10}+1760x^9$

47. Statements containing the set of positive integers as their domains are proved by mathematical induction.

53. permutation **55.** 1 **57.** 362,880 **59.** 48 **61.** 24 **63.** 504 **65.** **(a)** $\frac{4}{15}$ **(b)** $\frac{2}{3}$ **(c)** 0 **67.** $\frac{1}{26}$

69. $\frac{4}{13}$ **71.** .86 **73.** 0 **75.** approximately .205

Chapter 11 Test *(pages 892–893)*

1. **(a)** $-3, 4, -5, 6, -7$; neither **(b)** $-\frac{3}{2}, -\frac{3}{4}, -\frac{3}{8}, -\frac{3}{16}, -\frac{3}{32}$; geometric **(c)** 2, 3, 7, 13, 27; neither

2. **(a)** $a_5=49$ **(b)** $a_5=16$ **3.** **(a)** 110 **(b)** -1705 **4.** **(a)** 2385 **(b)** -186 **(c)** The sum does not exist. **(d)** $\frac{108}{7}$

5. **(a)** $16x^4-96x^3y+216x^2y^2-216xy^3+81y^4$ **(b)** $60w^4y^2$ **6.** **(a)** 45 **(b)** 35 **(c)** 5040 **(d)** 990 **8.** 24

9. 6840 **10.** 6160 **11.** **(a)** $\frac{1}{26}$ **(b)** $\frac{10}{13}$ **(c)** $\frac{4}{13}$ **(d)** 3 to 10 **12.** **(a)** $\binom{8}{3}\left(\frac{1}{6}\right)^3\left(1-\frac{1}{6}\right)^5\approx .104$

(b) $\binom{8}{8}\left(\frac{1}{6}\right)^8\left(1-\frac{1}{6}\right)^0\approx .000000595$

CHAPTER 12 LIMITS, DERIVATIVES, AND DEFINITE INTEGRALS

12.1 Exercises *(pages 902–905)*

1. false; Consider $\lim_{x\to 0} f(x)$, where $f(x)=\begin{cases} x & \text{if } x\neq 0 \\ 2 & \text{if } x=0. \end{cases}$ **3.** false; In order for $\lim_{x\to a} f(x)$ to exist, $f(x)$ need only be defined *near a*, not *at a*. **5.** true **7.** 2.5 **9.** does not exist **11.** 0 **13.** does not exist **15.** 2 **17.** does not exist

19. 2 **21.**

x	.9	.99	.999	1.001	1.01	1.1
$f(x)$	1.02	1.0002	1.000002	1.000002	1.0002	1.02

; The limit is 1.

23.

x	-1.1	-1.01	-1.001	$-.999$	$-.99$	$-.9$
$f(x)$	-3.68	-3.9698	-3.997	-4.003	-4.0298	-4.28

; The limit is -4.

25.

x	2.9	2.99	2.999	3.001	3.01	3.1
$f(x)$	12.9706	127.084	1268.24	-1267.66	-126.506	-12.3932

; The limit does not exist.

27.

x	$-.1$	$-.01$	$-.001$	.001	.01	.1
$f(x)$	1.987	1.99987	1.9999987	1.9999987	1.99987	1.987

; The limit is 2.

29. 6 **31.** 7 **33.** does not exist **35.** 3 **37.** 10 **39.** 1 **41.** does not exist **43.** 6 **45.** 0 **47.** 2
49. 2 **51.** 0 **53.** There are situations where the limit of a function f at a cannot be evaluated as $f(a)$. An example is $\lim_{x\to 2}\frac{x^2-4}{x-2}$. This limit is 4, but $\frac{x^2-4}{x-2}$ cannot be evaluated at $x=2$, since 2 is not in the domain.

12.2 Exercises *(pages 910–911)*

1. 8 **3.** 2 **5.** 4 **7.** 256 **9.** $\frac{3}{2}$ **11.** 7 **13.** π **15.** 36 **17.** -4 **19.** 19 **21.** 1 **23.** 6 **25.** -5 **27.** 3 **29.** 4 **31.** 3 **33.** 2 **35.** 2 **37.** $\frac{1}{2}$ **39.** $\sqrt{2}$ **41.** -2 **43.** 1 **45.** 0

12.3 Exercises *(pages 917–919)*

1. **(a)** 4 **(b)** 2 **3.** **(a)** $-\infty$ **(b)** ∞ **5.** **(a)** $-\infty$ **(b)** $-\infty$ **7.** 10 **9.** 100 **11.** 0 **13.** -1 **15.** -1 **17.** 2 **19.** ∞ **21.** 0 **23.** 0 **25.** $y = \frac{\pi}{4}$ **27.** $y = 5$ **29.** **(a)** 1 **(b)** 14 **31.** **(a)** $-\infty$ **(b)** ∞ **33.** **(a)** $-\infty$ **(b)** ∞ **35.** $\frac{5}{3}$ **37.** 4 **39.** $\frac{1}{3}$ **41.** 0 **43.** ∞

In Exercises 45–51, other answers are possible.

45. $f(x) = \frac{1}{x-5}$ **47.** $f(x) = -x^3$ **49.** $f(x) = 2x$, $g(x) = x$ **51.** $f(x) = \frac{1}{x}$, $g(x) = x^2$ **53.** ∞ **55.** **(a)** 0 **(b)** 0 **(c)** 0 **57.** $\lim_{t \to 1^-} f(t)$ describes the amount of drug in the body just before the second injection is given, while $\lim_{t \to 1^+} f(t)$ describes the amount of drug in the body following that injection. **59.** 12; In the long run, the price will level off near \$12.

61. $f(x) = a_n x^n\left(1 + \frac{a_{n-1}}{a_n x} + \frac{a_{n-2}}{a_n x^2} + \cdots + \frac{a_1}{a_n x^{n-1}} + \frac{a_0}{a_n x^n}\right)$ **63.** **(a)** ∞, ∞ **(b)** $-\infty, -\infty$ **(c)** $\infty, -\infty$ **(d)** $-\infty, \infty$

64. **(a)** ∪ **(b)** ∩ **(c)** ↗ **(d)** ↘

Reviewing Basic Concepts *(pages 920–921)*

1. **(a)** 3 **(b)** 4 **(c)** 0 **(d)** 2 **2.** **(a)** $-1, -\frac{1}{2}$, does not exist **(b)** $-\frac{1}{2}, -\frac{1}{2}, -\frac{1}{2}$ **3.** **(a)** 1, 1, 1 **(b)** 0, 0, 0 **4.** **(a)** 3 **(b)** does not exist (∞) **5.** $\frac{1}{4}$ **6.** **(a)** 28 **(b)** 128 **(c)** 8 **(d)** 5 **(e)** $4\sqrt{2}$ **(f)** $\sqrt[3]{4}$ **(g)** 16 **(h)** 1089 **(i)** $\frac{7}{4}$ **(j)** $\frac{1}{3}$ **7.** 6 **8.** 0 **9.** $\frac{1}{3}$ **10.** does not exist

12.4 Exercises *(pages 926–929)*

1. 2 **3.** $\frac{1}{5}$ **5.** 8 **7.** 27 **9.** $\frac{1}{8}$ **11.** $-\frac{3}{2}$ **13.** $y = 8x - 9$ **15.** $y = -\frac{5}{4}x + 5$ **17.** $y = \frac{2}{3}x + 6$ **19.** 0 **21.** -1 **23.** 1 **25.** 0 **27.** .1517 **29.** $\frac{1}{3}$ **31.** 1, 1.5; 2 **33.** $s'(3) = 54$ ft per sec **35.** $s'(3) = 75$ ft per sec **37.** $\frac{4}{5}$% per yr **39.** .8 g per wk **41.** **(a)** 4 sec **(b)** 9 ft per sec **43.** \$6000 per unit **45.** about 5332 people per day **47.** \$1467.53 **49.** $y = 139.42x + 352.17$

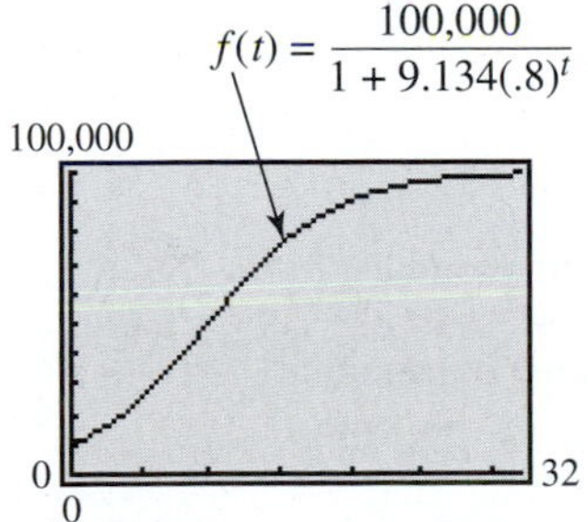

12.5 Exercises *(pages 935–936)*

1. **(a)** 56 **(b)** $\int_0^8 (2x + 1)\,dx$ **3.** **(a)** 38 **(b)** 50 **(c)** 44 **(d)** 44 **5.** **(a)** 14 **(b)** 18 **(c)** 16 **(d)** 16 **7.** **(a)** 30 **(b)** 54 **(c)** 42 **(d)** 41 **9.** **(a)** 27.19 **(b)** 80.79 **(c)** 53.99 **(d)** 47.43 **11.** **(a)** $\frac{25}{12}$ **(b)** $\frac{77}{60}$ **(c)** $\frac{101}{60}$ **(d)** $\frac{496}{315}$ **13.** **(a)** 4 **(b)** 4 **15.** **(a)** 4 **(b)** 5 **17.** 4π **19.** 24

Reviewing Basic Concepts *(pages 936–937)*

1. $-\frac{5}{6}$ **2.** 0 **3.** does not exist **4.** 0 **5.** $-\$1.32$ per unit **6.** $y = 2x + 7$ **7.** $\frac{1}{6}$ **8.** 41.25 **9.** 32 **10.** $\frac{9\pi}{4}$

Chapter 12 Review Exercises *(pages 939–941)*

1. **(a)** 2 **(b)** 2 **(c)** 2 **3.** **(a)** ∞ **(b)** $-\infty$ **(c)** does not exist **5.** -1 **7.** 2 **9.** 4 **11.** 3 **13.** -2 **15.** -4 **17.** -3 **19.** does not exist **21.** $\frac{1}{3}$ **23.** $-\infty$ **25.** 5 **27.** 3 **29.** $-\infty$ **31.** $\pi - 1$ **33.** $-\infty$ **35.** does not exist **37.** -1 **39.** ∞ **41.** ∞ **43.** $\frac{5}{2}$ **45.** $-\infty$ **47.** 5 **49.** -1

51. $f(x) = \frac{1}{x-2}$, $g(x) = -\frac{1}{x-2}$ (Other correct answers are possible.) **53.** 70; The coffee cools to 70°F. **55.** -3 **57.** $y = 3x - 4$ **59.** 0 **61.** 1 **63.** 4 ft per sec **65.** 20 **67.** 24 **69.** 12.5

Chapter 12 Test *(pages 941–942)*

1. 5 **2.** 0 **3.** **(a)** ∞ **(b)** 4 **4.** $\frac{3}{2}$ **5.** -2 **6.** 0 **7.** $\frac{5}{4}$ **8.** 4 **9.** $\frac{2}{5}$ **10.** $\frac{3}{4}$ **11.** 1 **12.** -10 **13.** -1 **14.** 55 **15.** 27 **16.** does not exist **17.** 1 **18.** 0 **19.** 4 **20.** $y = 3x - 6$ **21.** 10.1747 **22.** $\frac{1}{10}$ g per yr

23. 4723.67; After 3 days, the information is spreading at the rate of about 4724 people per day.

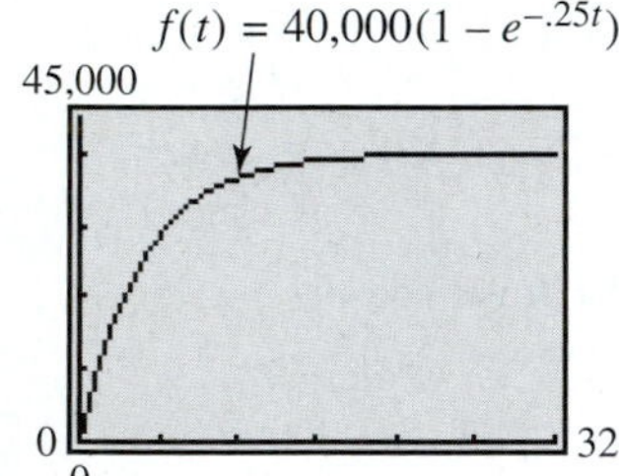

24. 21 **25.** 18

CHAPTER R REFERENCE: BASIC ALGEBRAIC CONCEPTS

R.1 Exercises *(pages 951–952)*

1. $(-4)^5$ **3.** 1 **5.** 1 **7.** 2^{10} **9.** $2^3x^{15}y^{12}$ **11.** $-\frac{p^8}{q^2}$ **13.** polynomial; degree 11; monomial **15.** polynomial; degree 6; binomial **17.** polynomial; degree 6; binomial **19.** polynomial; degree 6; trinomial **21.** not a polynomial **23.** $m^3 - 4m^2 + 10$ **25.** $5p^2 - 3p - 4$ **27.** $-6x^3 - 3x^2 - 4x + 4$ **29.** $15m^2 + 2m - 24$

31. $6m^2 + \frac{1}{4}m - \frac{1}{8}$ **33.** $2b^5 - 8b^4 + 6b^3$ **35.** $m^2 + mn - 2km - 2n^2 + 5kn - 3k^2$ **37.** Find the sum of the square of the first term, twice the product of the two terms, and the square of the last term. **39.** $4m^2 - 9$ **41.** $16m^2 + 16mn + 4n^2$ **43.** $25r^2 + 30rt^2 + 9t^4$ **45.** $4p^2 - 12p + 9 + 4pq - 6q + q^2$ **47.** $9q^2 + 30q + 25 - p^2$ **49.** $9a^2 + 6ab + b^2 - 6a - 2b + 1$ **51.** $18p^2 - 27pq - 35q^2$ **53.** $p^3 - 7p^2 - p - 7$ **55.** $16x^2y - 9y^3$ **57.** $6z^2 - 5yz - 4y^2$ **59.** $9p^2 + 30p + 25$ **61.** $4p^2 - 16$ **63.** $11y^3 - 18y^2 + 4y$

R.2 Exercises *(pages 957–959)*

1. (a) B **(b)** C **(c)** A **(d)** D **3.** $4k^2m^3(1 + 2k^2 - 3m)$ **5.** $2(a + b)(1 + 2m)$ **7.** $(y + 2)(3y + 2)$ **9.** $(r + 3)(3r - 5)$ **11.** $(m - 1)(2m^2 - 7m + 7)$ **13.** $(2s + 3)(3t - 5)$ **15.** $(2x + 3)(5x - 4y)$ **17.** $(t + 2)(t^2 - 3)$ **19.** Both are correct. **21.** $8(h - 8)(h + 5)$ **23.** $9y^2(y - 1)(y - 5)$ **25.** $(7m - 5r)(2m + 3r)$ **27.** $(3s - t)(4s + 5t)$ **29.** $(5a + m)(6a - m)$ **31.** $3x^3(2x + 5z)(3x - 5z)$ **33.** $(4p - 5)^2$ **35.** $5(2p - 5q)^2$ **37.** $(3mn - 2)^2$ **39.** $(2p + q - 5)^2$ **41.** $(3a + 4)(3a - 4)$ **43.** $(5s^2 + 3t)(5s^2 - 3t)$ **45.** $(a + b + 4)(a + b - 4)$ **47.** $(p^2 + 25)(p + 5)(p - 5)$ **49.** B **51.** $(2 - a)(4 + 2a + a^2)$ **53.** $(5x - 3)(25x^2 + 15x + 9)$ **55.** $(3y^3 + 5z^2)(9y^6 - 15y^3z^2 + 25z^4)$ **57.** $r(r^2 + 18r + 108)$ **59.** $(3 - m - 2n)(9 + 3m + 6n + m^2 + 4mn + 4n^2)$ **61.** It is incorrect because a^2 has not been substituted back for u; $(3a^2 - 1)(a^2 + 5)$ **63.** $(a^2 - 8)(a^2 + 6)$ **65.** $2(6z - 5)(8z - 3)$ **67.** $(18 - 5p)(17 - 4p)$ **69.** $(2b + c + 4)(2b + c - 4)$ **71.** $(x + y)(x - 5)$ **73.** $(m - 2n)(p^4 + q)$ **75.** $(2z + 7)^2$ **77.** $(10x + 7y)(100x^2 - 70xy + 49y^2)$ **79.** $(5m^2 - 6)(25m^4 + 30m^2 + 36)$ **81.** $(6m - 7n)(2m + 5n)$ **83.** $(4p - 1)(p + 1)$ **85.** prime **87.** $4xy$

R.3 Exercises *(pages 965–966)*

1. $\{x \mid x \neq -6\}$ **3.** $\left\{x \mid x \neq \frac{3}{5}\right\}$ **5.** $(-\infty, \infty)$ **7.** $\left\{x \mid x \neq -\frac{1}{2}, 1\right\}$ **9.** $\frac{5p}{2}$ **11.** $\frac{8}{9}$ **13.** $\frac{3}{t - 3}$ **15.** $\frac{2x + 4}{x}$ **17.** $\frac{m - 2}{m + 3}$ **19.** $\frac{2m + 3}{4m + 3}$ **21.** $\frac{25p^2}{9}$ **23.** $\frac{2}{9}$ **25.** $\frac{5x}{y}$ **27.** $\frac{2(a + 4)}{a - 3}$ **29.** 1 **31.** $\frac{m + 6}{m + 3}$ **33.** $\frac{m - 3}{2m - 3}$ **35.** $\frac{x + y}{x - y}$ **37.** $\frac{x^2 - xy + y^2}{x^2 + xy + y^2}$ **39.** B, C **41.** $\frac{19}{6k}$ **43.** 1 **45.** $\frac{6 + p}{2p}$ **47.** $\frac{137}{30m}$ **49.** $\frac{-2}{(a + 1)(a - 1)}$ **51.** $\frac{2m^2 + 2}{(m - 1)(m + 1)}$ **53.** $\frac{4}{a - 2}$ or $\frac{-4}{2 - a}$ **55.** $\frac{3x + y}{2x - y}$ or $\frac{-3x - y}{y - 2x}$ **57.** $\frac{5}{(a - 2)(a - 3)(a + 2)}$ **59.** $\frac{x - 11}{(x + 4)(x - 3)(x - 4)}$ **61.** $\frac{a^2 + 5a}{(a + 6)(a - 1)(a + 1)}$ **63.** $\frac{x + 1}{x - 1}$ **65.** $\frac{-1}{x + 1}$ **67.** $\frac{(2 - b)(1 + b)}{b(1 - b)}$ **69.** $\frac{m^3 - 4m - 1}{m - 2}$

R.4 Exercises *(pages 972–973)*

1. E **3.** F **5.** D **7.** B **9.** $-\frac{1}{64}$ **11.** 8 **13.** -2 **15.** 4 **17.** $\frac{1}{9}$ **19.** $\frac{256}{81}$ **21.** $4p^2$ **23.** $9x^4$ **25.** $\frac{1}{2^7}$ **27.** $\frac{1}{27^3}$ **29.** 1 **31.** $m^{7/3}$ **33.** $(1 + n)^{5/4}$ **35.** $\frac{6z^{2/3}}{y^{5/4}}$ **37.** $2^6a^{1/4}b^{37/2}$ **39.** $\frac{r^6}{s^{15}}$ **41.** $-\frac{1}{ab^3}$ **43.** $12^{9/4}y$ **45.** $\frac{1}{2p^2}$ **47.** $\frac{m^3p}{n}$ **49.** $-4a^{5/3}$ **51.** $\frac{1}{(k + 5)^{1/2}}$ **53.** $y - 10y^2$ **55.** $-4k^{10/3} + 24k^{4/3}$ **57.** $x^2 - x$ **59.** $r - 2 + r^{-1}$ or $r - 2 + \frac{1}{r}$ **61.** $k^{-2}(4k + 1)$ or $\frac{4k + 1}{k^2}$ **63.** $z^{-1/2}(9 + 2z)$ or $\frac{9 + 2z}{z^{1/2}}$ **65.** $p^{-7/4}(p - 2)$ or $\frac{p - 2}{p^{7/4}}$ **67.** $(p + 4)^{-3/2}(p^2 + 9p + 21)$ or $\frac{p^2 + 9p + 21}{(p + 4)^{3/2}}$

R.5 Exercises *(pages 979–981)*

1. F **3.** H **5.** G **7.** C **9.** $\sqrt[3]{(-m)^2}$ or $(\sqrt[3]{-m})^2$ **11.** $\sqrt[3]{(2m+p)^2}$ or $(\sqrt[3]{2m+p})^2$ **13.** $k^{2/5}$
15. $-3 \cdot 5^{1/2}p^{3/2}$ **17.** A **19.** $x \geq 0$ **21.** 5 **23.** -5 **25.** $5\sqrt{2}$ **27.** $3\sqrt[3]{3}$ **29.** $-2\sqrt[4]{2}$ **31.** $-\frac{3\sqrt{5}}{5}$
33. $-\frac{\sqrt[3]{100}}{5}$ **35.** $32\sqrt[3]{2}$ **37.** $2x^2z^4\sqrt{2x}$ **39.** $2zx^2y\sqrt[3]{2z^2x^2y}$ **41.** $np^2\sqrt[4]{m^2n^3}$ **43.** cannot be simplified further
45. $\frac{\sqrt{6x}}{3x}$ **47.** $\frac{x^2y\sqrt{xy}}{z}$ **49.** $\frac{2\sqrt[3]{x}}{x}$ **51.** $\frac{h\sqrt[4]{9g^3hr^2}}{3r^2}$ **53.** $\frac{m\sqrt[3]{n^2}}{n}$ **55.** $2\sqrt[4]{x^3y^3}$ **57.** $\sqrt[3]{2}$ **59.** $\sqrt[18]{x}$
61. $9\sqrt{3}$ **63.** $-2\sqrt{7p}$ **65.** $7\sqrt[3]{3}$ **67.** $2\sqrt{3}$ **69.** $\frac{13\sqrt[3]{4}}{6}$ **71.** -7 **73.** 10 **75.** $11+4\sqrt{6}$ **77.** $5\sqrt{6}$
79. $\frac{8\sqrt{5}}{5}$ **81.** $\frac{6\sqrt[3]{x}}{x}$ **83.** $\frac{\sqrt{15}-3}{2}$ **85.** $\frac{3\sqrt{5}-2\sqrt{3}+3\sqrt{15}-6}{33}$ **87.** $\frac{p(\sqrt{p}-2)}{p-4}$ **89.** $\frac{a(\sqrt{a+b}+1)}{a+b-1}$

Chapter R Test *(page 981)*

1. (a) $(-3)^9$ **(b)** $2^2x^6y^2$ **(c)** 1 **(d)** $\frac{4^2}{5^2}$ **(e)** $-\frac{m^6}{p^3}$ **2. (a)** $11x^2-x+2$ **(b)** $36r^2-60r+25$
(c) $3u^3+5u^2+2u+8$ **(d)** $16x^2-25$ **(e)** $25p^2+30p+9$ **3. (a)** $(3x-7)(2x-1)$ **(b)** $(x^2+4)(x+2)(x-2)$
(c) $(z+2k)(z-8k)$ **(d)** $(x-2)(x^2+2x+4)(y+3)(y-3)$ **4. (a)** $\frac{4}{x^2}$; $\{x \mid x \neq 0\}$ **(b)** $\frac{1}{k-1}$; $\{k \mid k \neq -1, 1\}$
(c) $\frac{x-1}{x+3}$; $\{x \mid x \neq -3, -2\}$ **5. (a)** $\frac{x^4(x+1)}{3(x^2+1)}$ **(b)** $\frac{x(4x+1)}{(x+2)(x+1)(2x-3)}$ **(c)** $\frac{2a}{2a-3}$ or $\frac{-2a}{3-2a}$ **(d)** $\frac{g^2-2}{(g+2)(g-2)}$
6. (a) $\frac{1}{49}$ **(b)** $8x^6$ **(c)** $\frac{9}{16}$ **7. (a)** $\frac{1}{5^2}$ **(b)** $\frac{1}{(x+2)^{9/10}}$ **(c)** $\frac{n^3}{2m^{11/2}}$ **8. (a)** 2 **(b)** $\frac{\sqrt[3]{18}}{3}$ **(c)** $3x^2y^4\sqrt{2x}$ **(d)** $\frac{\sqrt[4]{p^2q^3}}{p}$
9. (a) $2\sqrt{2x}$ **(b)** $6+\sqrt[3]{2}+4\sqrt[3]{4}$ **(c)** $x-y$ **10.** $\frac{7(\sqrt{11}+\sqrt{7})}{2}$

APPENDIX B DECIDING WHICH MODEL BEST FITS A SET OF DATA

Appendix B Exercises *(pages 988–989)*

1. linear model: $y = -.129091x + 15.418182$; SSE $= .430182$; exponential model: $y = 15.434482(.991194)^x$; SSE $= .418686$; The exponential model is the better of these two models. **3.** linear model: $y = 99.971429x - 373.314286$; SSE $= 7533.485714$; quadratic model: $y = -6.428571x^2 + 183.542857x - 626.171429$; SSE $= 5990.628571$; The quadratic model is the better of these two models. **5.** quadratic model: $y = -.008459x^2 + .688792x + 1.161391$; SSE $= 2.870134$; natural logarithmic model: $y = -3.454802 + 5.093917 \ln x$; SSE $= 2.261093$; The natural logarithmic model is the better of these two models.

APPENDIX C VECTORS IN SPACE

Appendix C Exercises *(pages 994–995)*

1. xz-plane **3.** $\langle 5, 3, -2\rangle$ **5.** $\sqrt{33}$ **7.** $\sqrt{317}$ **9.** $5\sqrt{29}$ **11.** $\langle 2, -2, 5\rangle$, $2\mathbf{i} - 2\mathbf{j} + 5\mathbf{k}$
13. $\langle -2, -12, -4\rangle$, $-2\mathbf{i} - 12\mathbf{j} - 4\mathbf{k}$ **15.** $\langle -15, -20, 10\rangle$, $-15\mathbf{i} - 20\mathbf{j} + 10\mathbf{k}$ **17.** $\langle -2, 2, -5\rangle$; $\mathbf{QP} = -\mathbf{PQ}$
19. $-2\mathbf{i} + 7\mathbf{j} + 13\mathbf{k}$ **21.** $-7\mathbf{i} + 41\mathbf{j} + 38\mathbf{k}$ **23.** $\sqrt{69}$ **25.** $\sqrt{38}$ **27.** -39 **29.** 38 **31.** 25.4°
33. 32.0° **35.** 90° **37.** $\alpha = 76.1°$; $\beta = 61.2°$; $\gamma = 32.6°$ **39.** $\alpha = 59.2°$; $\beta = 112.6°$; $\gamma = 140.2°$ **41.** 60°
43. Their position vectors are scalar multiples of each other. **45.** 12 work units **47.** 13 work units

APPENDIX D POLAR FORM OF CONIC SECTIONS

Appendix D Exercises *(page 998)*

In Exercises 1–11, we give only calculator graphs.

1.

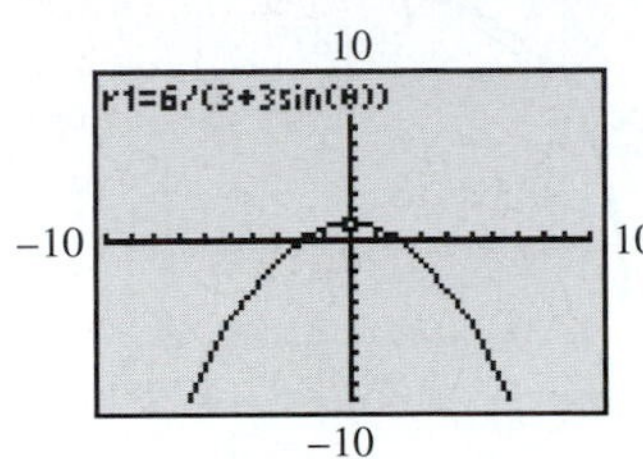

3.

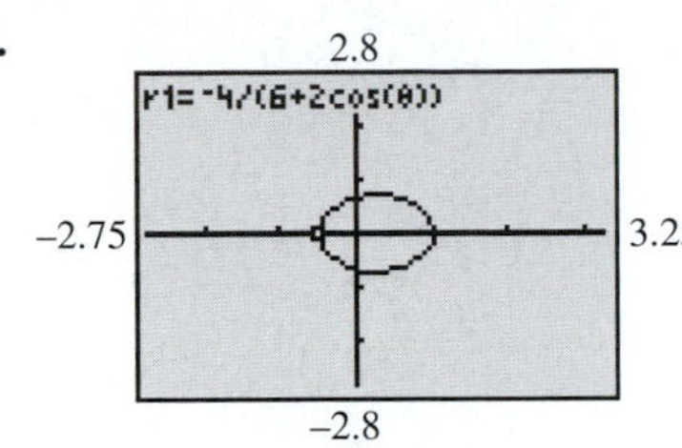

5.

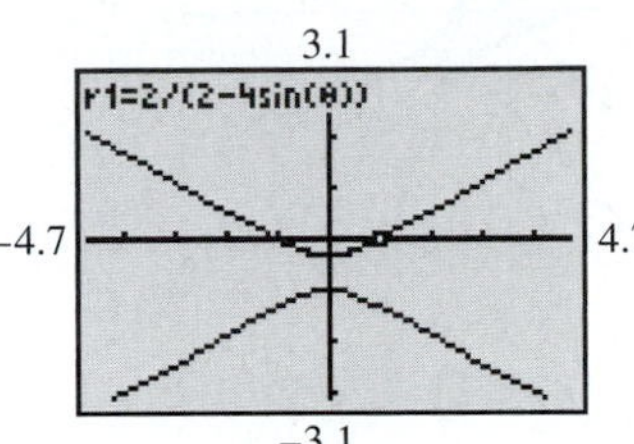

7.

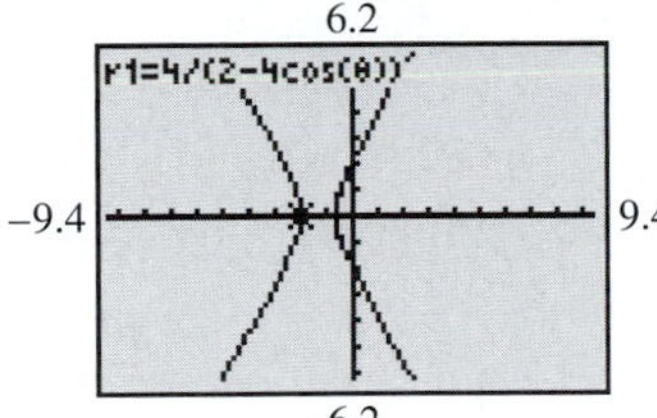

9.

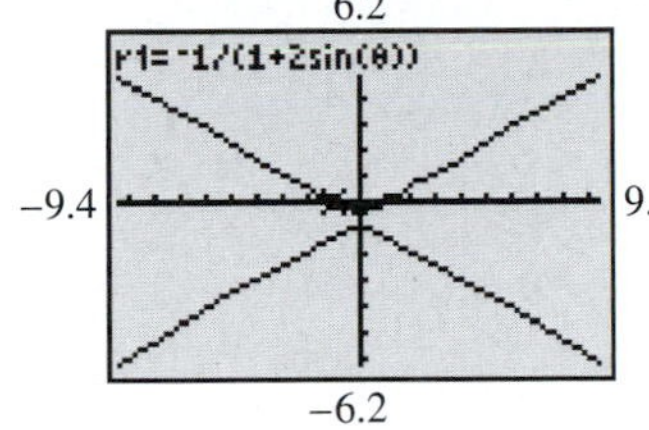

11.

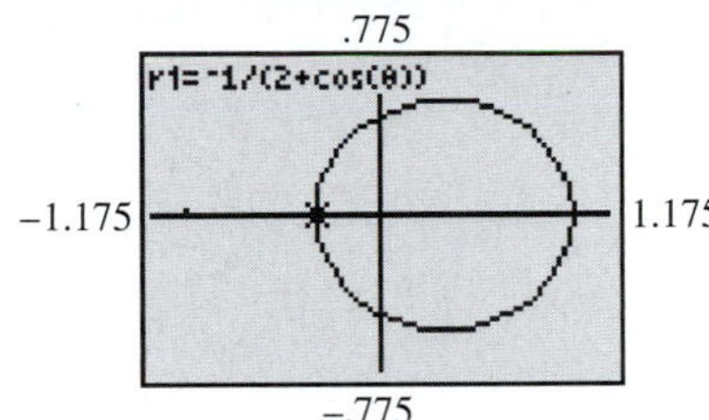

13. $r = \dfrac{3}{1 + \cos\theta}$ **15.** $r = \dfrac{5}{1 - \sin\theta}$ **17.** $r = \dfrac{20}{5 + 4\cos\theta}$; ellipse **19.** $r = \dfrac{40}{4 - 5\sin\theta}$; hyperbola

21. ellipse; $8x^2 + 9y^2 - 12x - 36 = 0$ **23.** hyperbola; $3x^2 - y^2 + 8x + 4 = 0$ **25.** ellipse; $4x^2 + 3y^2 - 6y - 9 = 0$

27. parabola; $x^2 - 10y - 25 = 0$

APPENDIX E ROTATION OF AXES

Appendix E Exercises *(pages 1002–1003)*

1. circle or ellipse or a point **3.** hyperbola or two intersecting lines **5.** parabola or one line or two parallel lines

7. 30° **9.** 60° **11.** 22.5°

13.

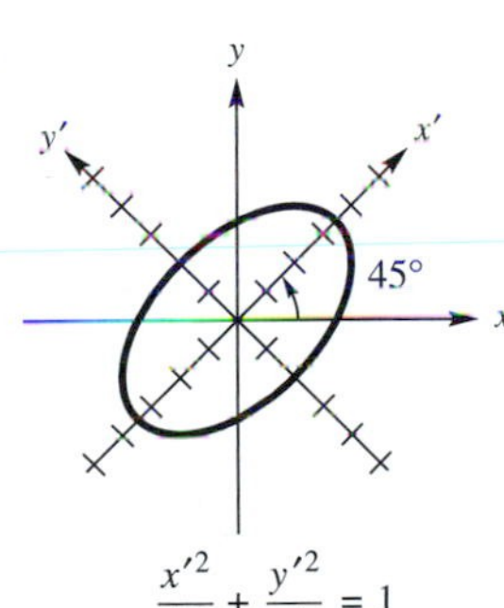

15.

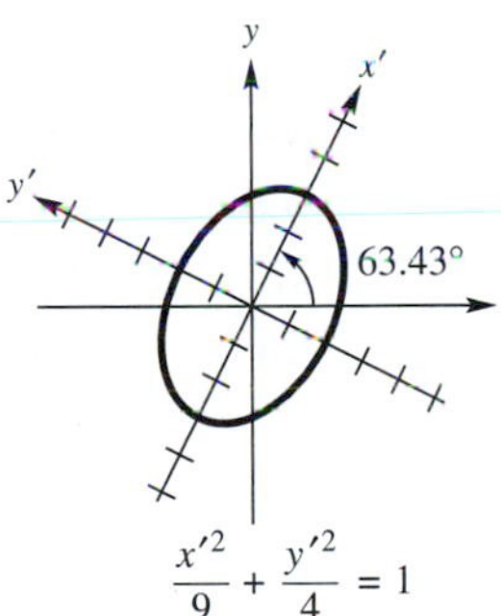

17.

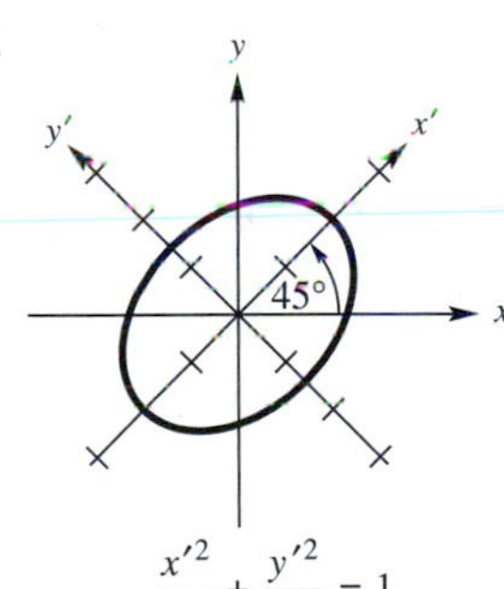

19.

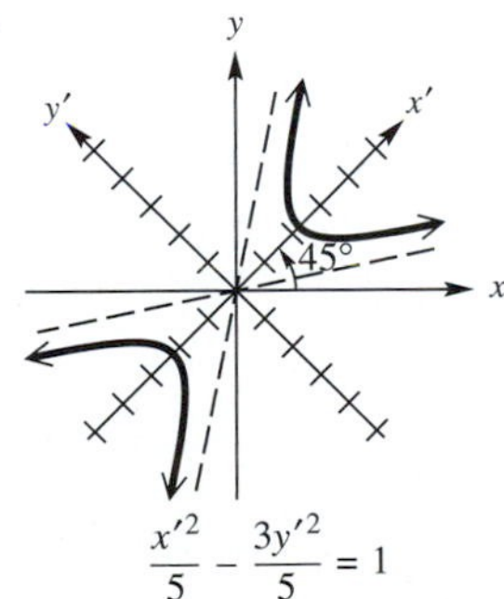

21.

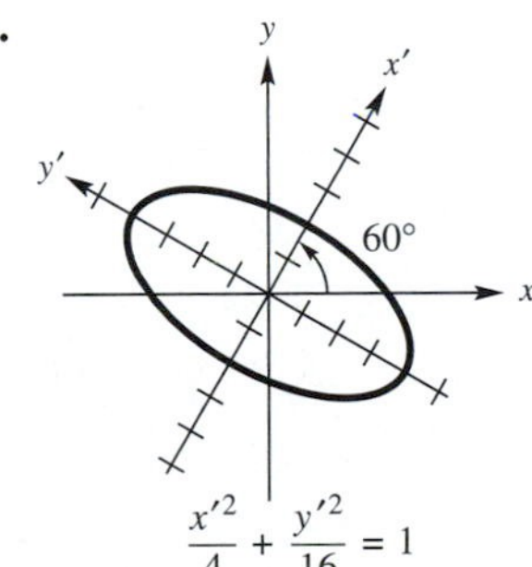

23.

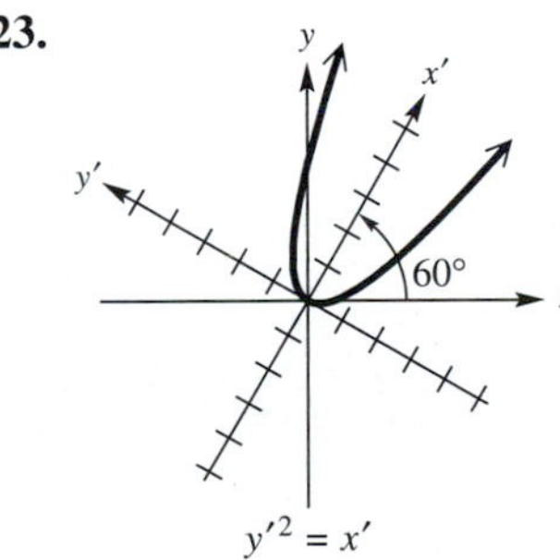

25.

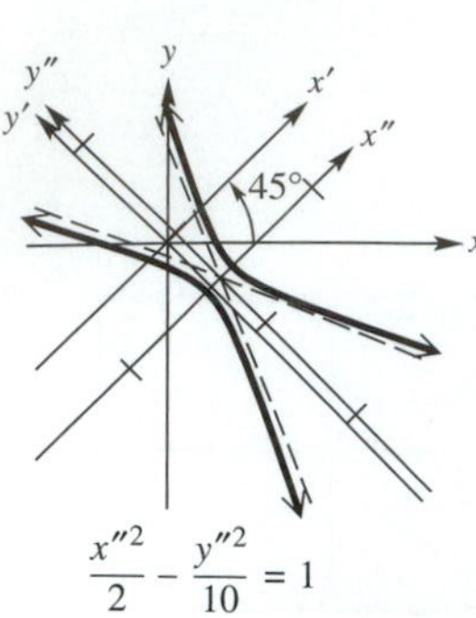

$\frac{x''^2}{2} - \frac{y''^2}{10} = 1$

27.

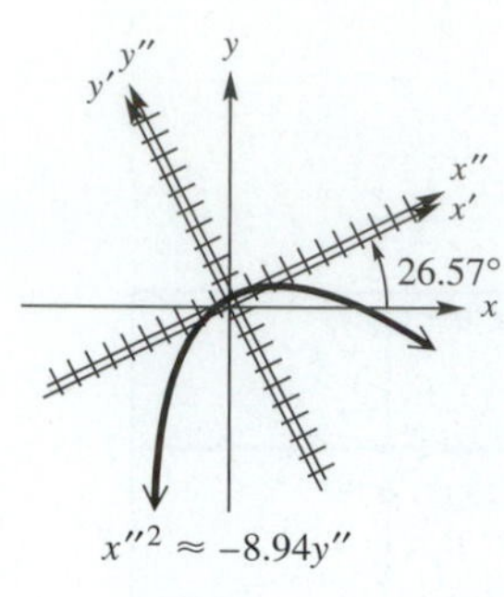

$x''^2 \approx -8.94y''$

29.

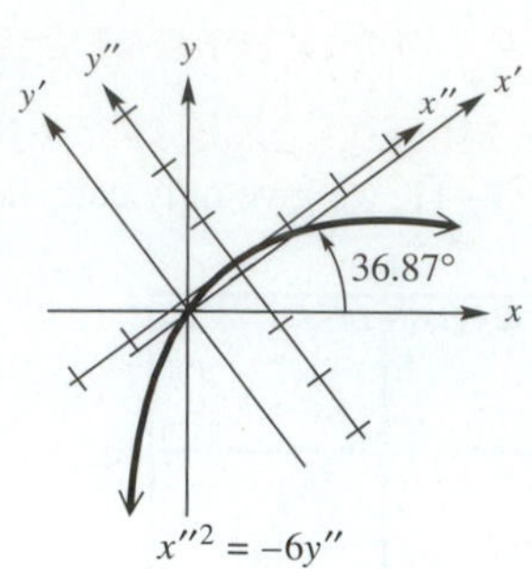

$x''^2 = -6y''$

31. If $B = 0$, $\cot 2\theta$ is undefined. The graph may be translated but is not rotated.

Index of Applications

Agriculture

Astronomy/Aerospace

Automotive

Biology/Life Sciences

Business

Chemistry

Construction

Consumer

Economics

Education

Energy

Engineering

Environment

Finance

General Interest

Geology

Geometry

Government

Labor

Health and Medicine

Physics

Probability

Sports/Entertainment

Statistics/Demographics

Technology

Transportation

Weather

Index

F

J

K

L

O

P

Q

R

T

U

V

W

X

Y

Z

Photo Credits: **p. 1** © Chris Rainier/CORBIS; **p. 1** L. Hobbs/PhotoLink/Getty Images; **p. 8** © Paul Kuroda/SuperStock; **p. 13** National Park Service Photo; **p. 23** © PhotoDisc; **p. 28** © Stockbyte/Getty Images; **p. 31** AP/Wide World Photos; **p. 45** © Najlah Feanny/Corbis; **p. 69** Color Day Production/Getty Images; **p. 73** Courtesy of Toshiba; **p. 74** Courtesy of Beth Anderson; **p. 76** © Joel W. Rogers/CORBIS; **p. 85** © Frank Polich/Reuters/Landov; **p. 88** AP/Wide World Photos; **p. 89** © Aya, www.starwheels.com; **p. 89** Artwork by Fred Casselman, Earth Echo Galleries,www.earthecho.com; **p. 89** Bee Smith/Getty Images; **p. 108** Thinkstock/Getty Images; **p. 111** David Mcglynn/Getty Images; **p. 137** © Royalty-Free/Corbis; **p. 143** U.S. Fish and Wildlife Service; **p. 148** © Tom Carter/Photo Edit; **p. 155** © Michael Newman/Photo Edit; **p. 160** Photodisc Collection/Getty Images; **p. 170** Laura Cavanaugh/UPI/Landov; **p. 173** © 2001 NBA Entertainment, photo by Walter Iooss, Jr.; **p. 173** © Paul J. Sutton/Duomo/Corbis; **p. 187** Jim Arbogast/Getty Images; **p. 193** © Reuters/CORBIS; **p. 193** © Roger Ressmeyer/CORBIS; **p. 205** Yellow Dog Productions/Getty Images; **p. 216** AP/Wide World Photos; **p. 255** Thinkstock/Getty Images; **p. 270** PhotoLink/Getty Images; **p. 295** © Getty Images; **p. 299** © Sam Diephuis/Getty Images; **p. 303** © PBNJ Productions/CORBIS; **p. 306** © Kraig Geiger/Contographer ®/Corbis; **p. 309** © Darrell Gulin/CORBIS; **p. 316** U.S. Fish and Wildlife Service; **p. 317** © Royalty-Free/Corbis; **p. 328** © Tom Brakefield/Getty Images; **p. 332** © CARDINALE STEPHANE/CORBIS SYGMA; **p. 338** U.S. Air Force Photo; **p. 338** Photodisc Collection/Getty Images; **p. 359** Stockbyte/Getty Images; **p. 370** U.S. Fish and Wildlife Service; **p. 380** Steve Cole/Getty Images; **p. 383** Dan Shapiro/NOAA; **p. 393** © Lightscapes Photography, Inc./CORBIS; **p. 397** © New Line/courtesy Everett Collection; **p. 402** © O. Alamany & E. Vicens/CORBIS; **p. 403** NASA Jet Propulsion Laboratory (NASA-JPL); **p. 407** Courtesy of B.K. Tomlin; **p. 415** Photodisc Collection/Getty Images; **p. 415** Photodisc Collection/Getty Images; **p. 417** NASA Goddard Space Flight Center (NASA-GSFC); **p. 417** NASA/JPL-Caltech/UMD; **p. 431** NASA and The Hubble Heritage Team (STScI/AURA); **p. 456** NASA Johnson Space Center (NASA-JSC); **p. 466** Don Farrall/Getty Images; **p. 466** Joel W. Rogers/CORBIS; **p. 473** Dominique Douic/Getty Images; **p. 477** Photodisc Collection/Getty Images; Comstock/Getty Images; **p. 500** Stewart Cohen/Getty Images; **p. 508** Photodisc Collection/Getty; **p. 512** U.S. Fish and Wildlife Service; **p. 535** Ryan McVay/Getty Images; **p. 565** © Topham/The Image Works; **p. 565** © S. Meltzer/PhotoLink/Getty Images; **p. 574** © Photodisc Collection/Getty Images; **p. 581** © Getty Images; **p. 593** © MacNeal Hospital/Stone; **p. 676** © Richard Cummins/SuperStock; **p. 676** © Neal Preston/CORBIS; **p. 695** © Corbis RF; **p. 699** © 2002 PhotoDisc; **p. 727** © Steve Allen/Getty Images; **p. 734** © Royalty-Free/Corbis; **p. 743** © Royalty-Free/Corbis; **p. 743** Courtesy of Garmin; **p. 757** © Rubberball/Getty Images; **p. 768** © Plush Studios/Getty Images; **p. 779** © Rob Melnychuk/Getty Images; **p. 791** University of St. Andrews, Scotland; **p. 807** Courtesy of NASA; **p. 815** © Reuters/CORBIS; **p. 819** © Flying Colours Ltd/Getty Images; **p. 824** Courtesy of Argonne National Laboratory; **p. 824** Courtesy of Argonne National Laboratory; **p. 827** USDA APHIS PPQ Archives, www.forestyimages.org; **p. 827** USDA Photo; **p. 843** USDA Photo; **p. 845** Comstock/Getty Images; **p. 852** Flying Colours Ltd./Getty Images; **p. 855** University of St. Andrews, Scotland; **p. 870** Courtesy of John Hornsby; **p. 874** Courtesy of Margaret Westmoreland; **p. 874** © PhotoDisc; **p. 878** Courtesy of Beth Anderson; **p. 884** Color Blind Images/Getty Images; **p. 891** Caroline Woodham/Getty Images; **p. 896** © Michael Banks/Getty Images; **p. 896** Kirk Weddle/Getty Images; **p. 919** © Digital Vision/Getty Images; **p. 924** © Photodisc Collection/Getty Images; **p. 928** © Jess Alford/Getty Images; **p. 929** © Ryan McVay/Getty Images; **p. 940** © Jonelle Weaver/Getty Images.

FUNCTION CAPSULES

IDENTITY FUNCTION $f(x) = x$

Domain: $(-\infty, \infty)$ Range: $(-\infty, \infty)$

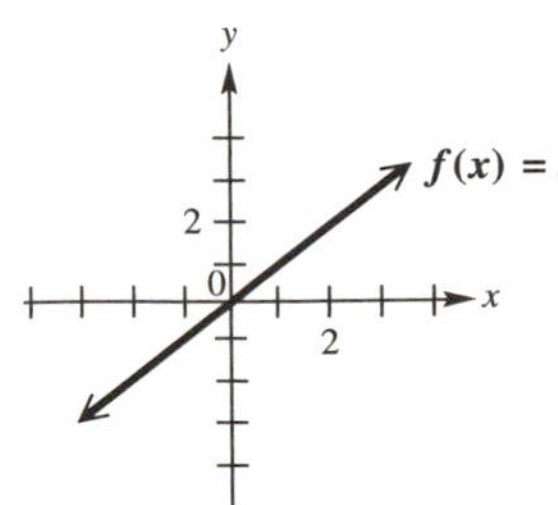

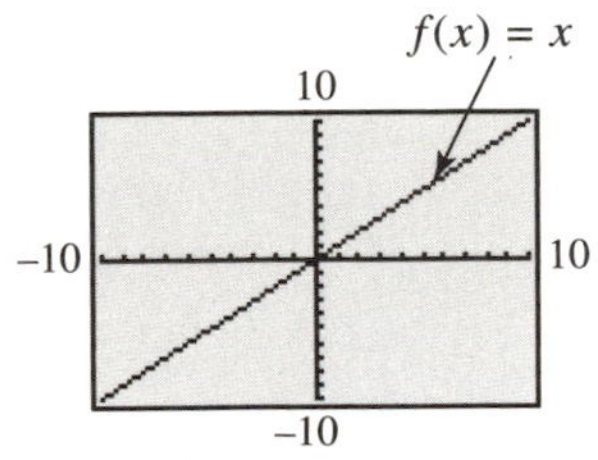

SQUARING FUNCTION $f(x) = x^2$

Domain: $(-\infty, \infty)$ Range: $[0, \infty)$

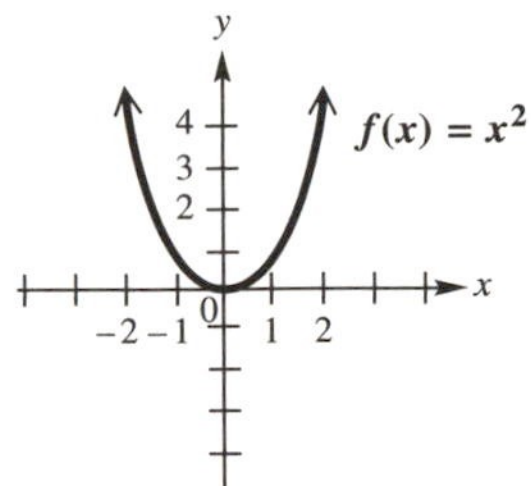

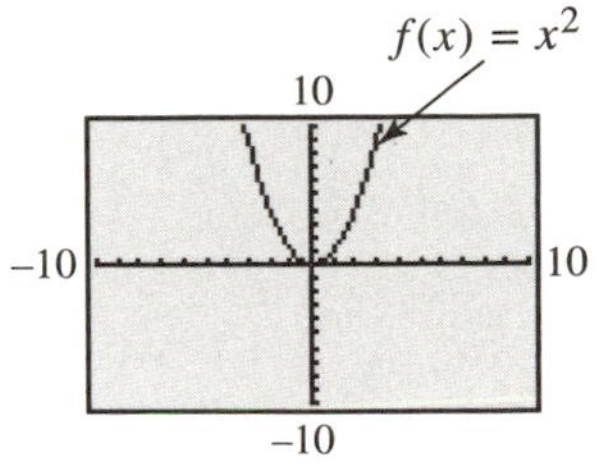

CUBING FUNCTION $f(x) = x^3$

Domain: $(-\infty, \infty)$ Range: $(-\infty, \infty)$

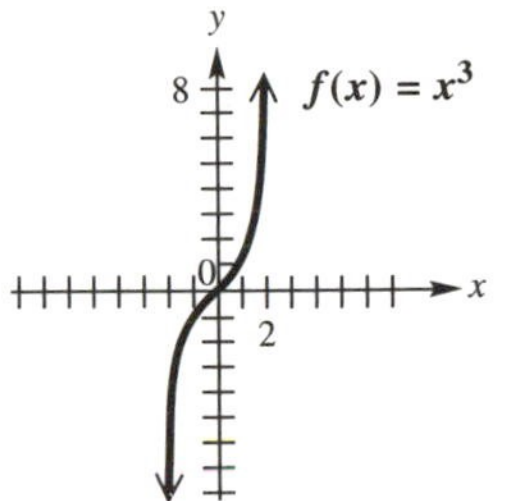

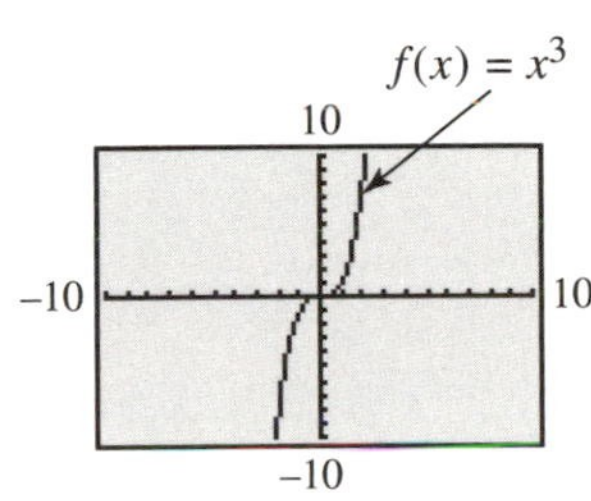

SQUARE ROOT FUNCTION $f(x) = \sqrt{x}$

Domain: $[0, \infty)$ Range: $[0, \infty)$

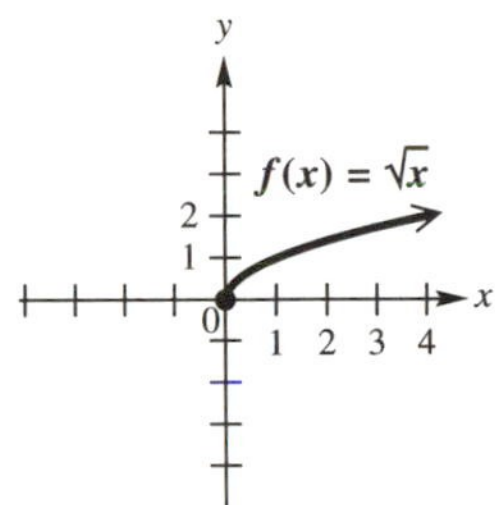

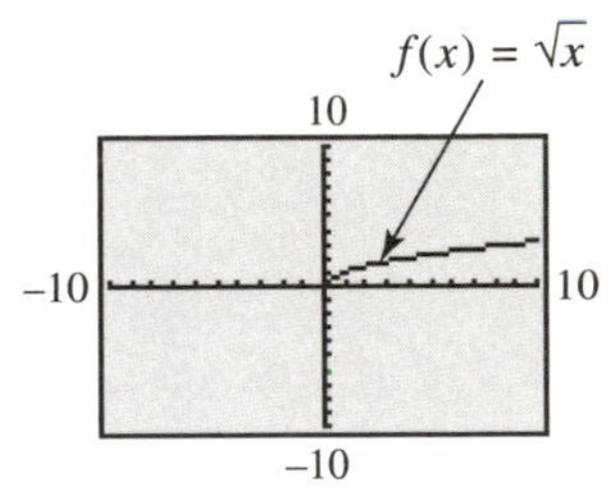

CUBE ROOT FUNCTION $f(x) = \sqrt[3]{x}$

Domain: $(-\infty, \infty)$ Range: $(-\infty, \infty)$

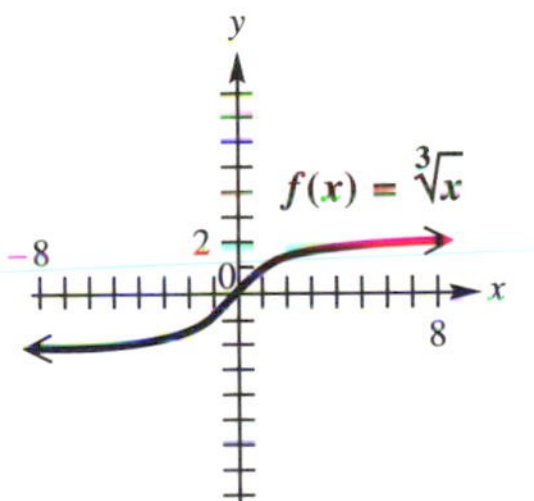

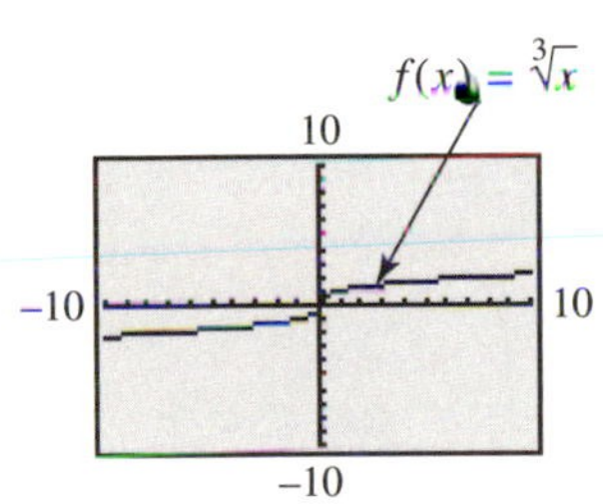

ABSOLUTE VALUE FUNCTION $f(x) = |x|$

Domain: $(-\infty, \infty)$ Range: $[0, \infty)$

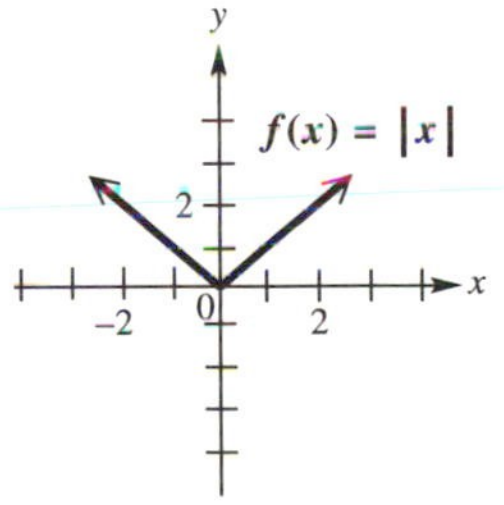

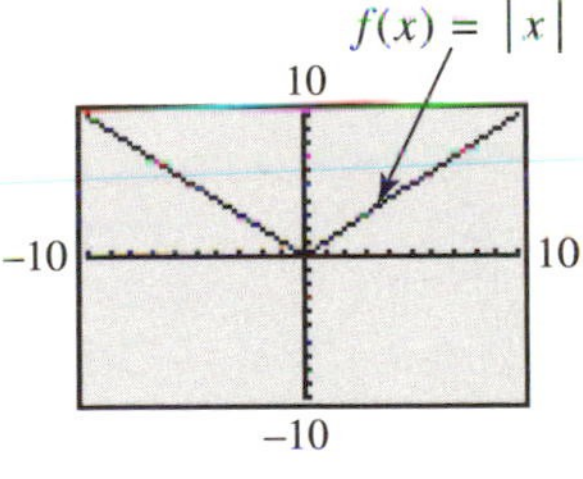

GREATEST INTEGER FUNCTION $f(x) = [\![x]\!]$

Domain: $(-\infty, \infty)$

Range: $\{x \mid x \text{ is an integer}\} = \{\ldots, -3, -2, -1, 0, 1, 2, 3, \ldots\}$

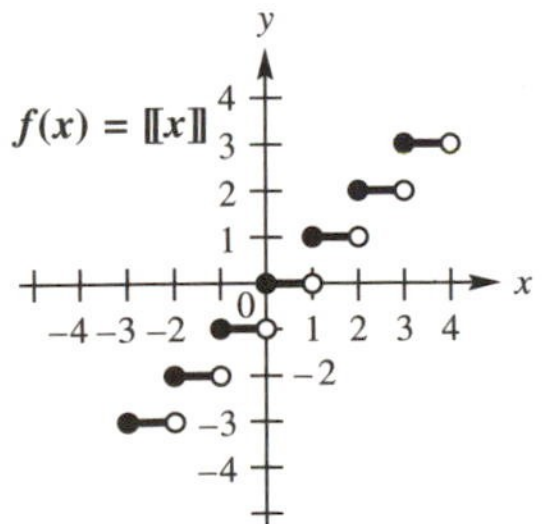

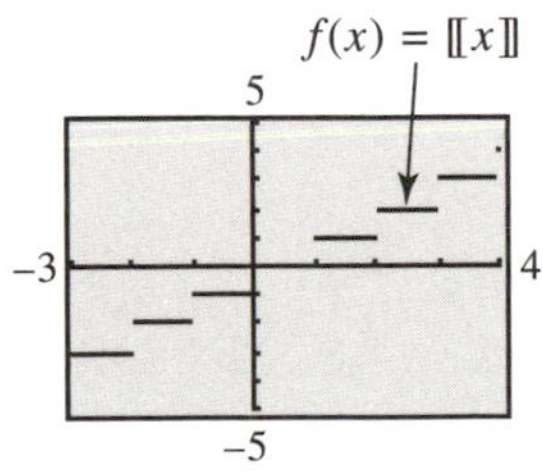

Dot mode

FUNCTION CAPSULES

RECIPROCAL FUNCTION $f(x) = \frac{1}{x}$

Domain: $(-\infty, 0) \cup (0, \infty)$ Range: $(-\infty, 0) \cup (0, \infty)$

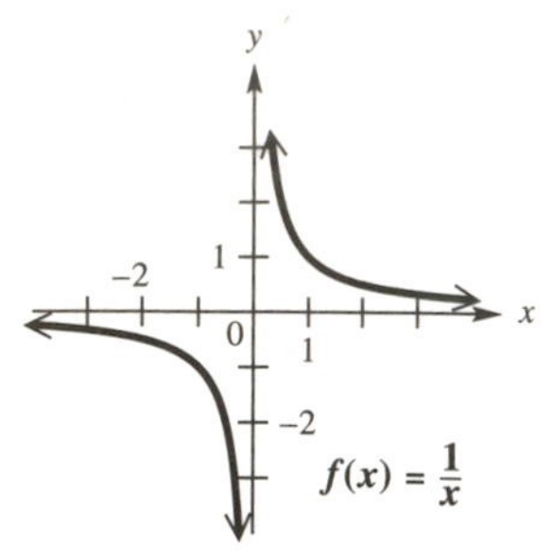

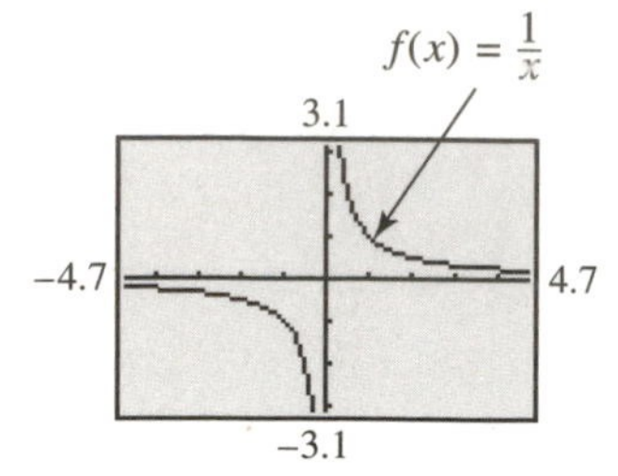

RATIONAL FUNCTION $f(x) = \frac{1}{x^2}$

Domain: $(-\infty, 0) \cup (0, \infty)$ Range: $(0, \infty)$

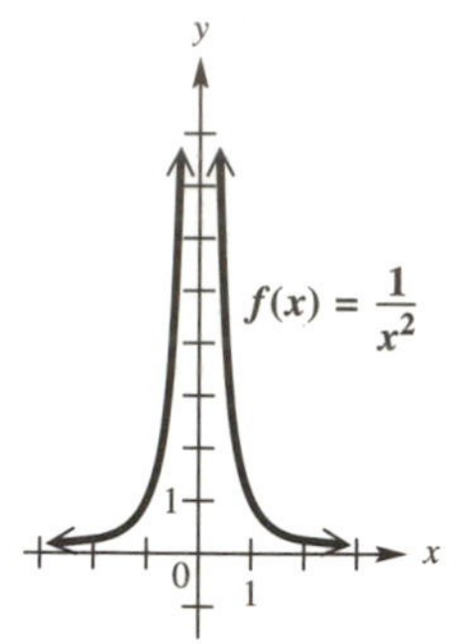

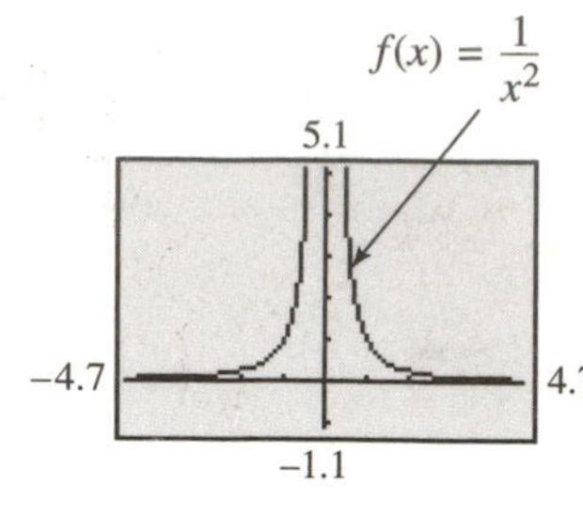

ROOT FUNCTION, n EVEN $f(x) = \sqrt[n]{x}$

Domain: $[0, \infty)$ Range: $[0, \infty)$

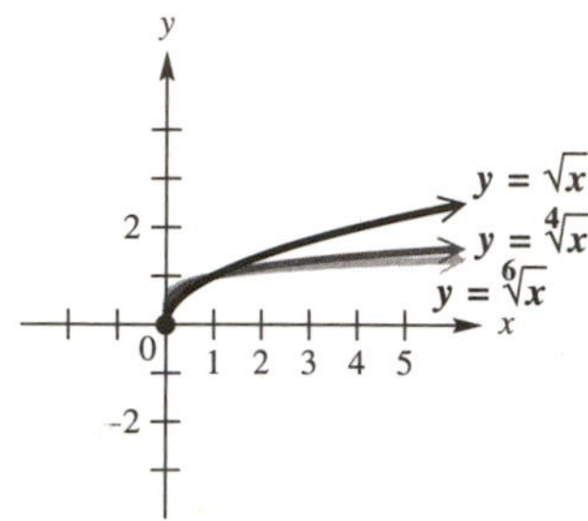

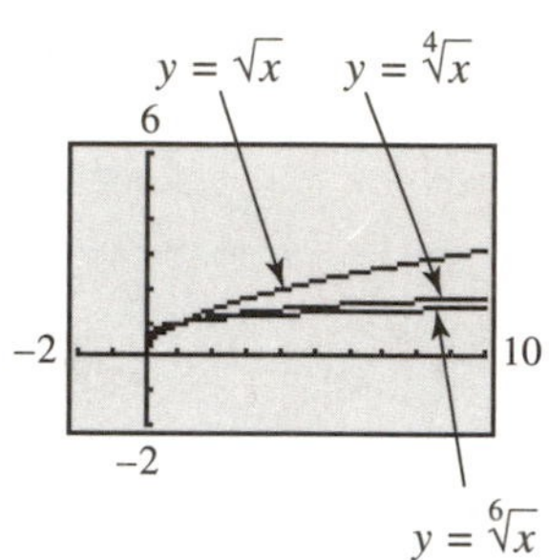

ROOT FUNCTION, n ODD $f(x) = \sqrt[n]{x}$

Domain: $(-\infty, \infty)$ Range: $(-\infty, \infty)$

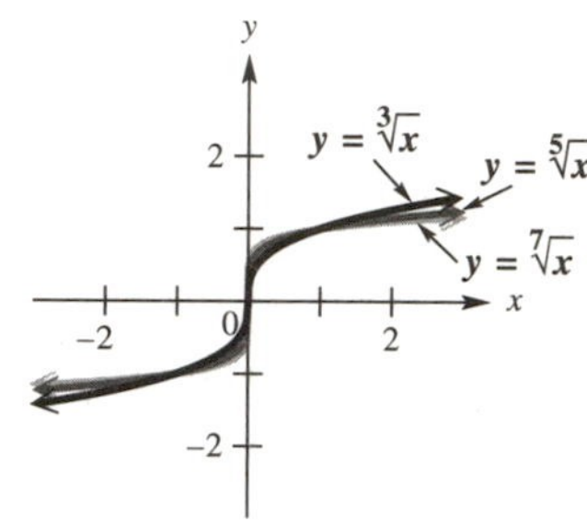

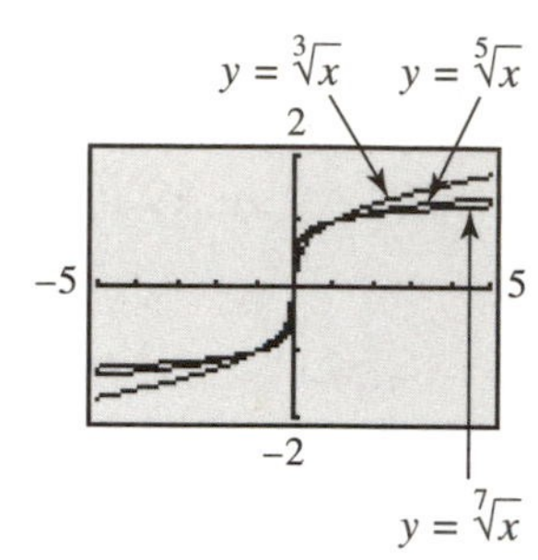

EXPONENTIAL FUNCTION $f(x) = a^x, \quad a > 1$

Domain: $(-\infty, \infty)$ Range: $(0, \infty)$

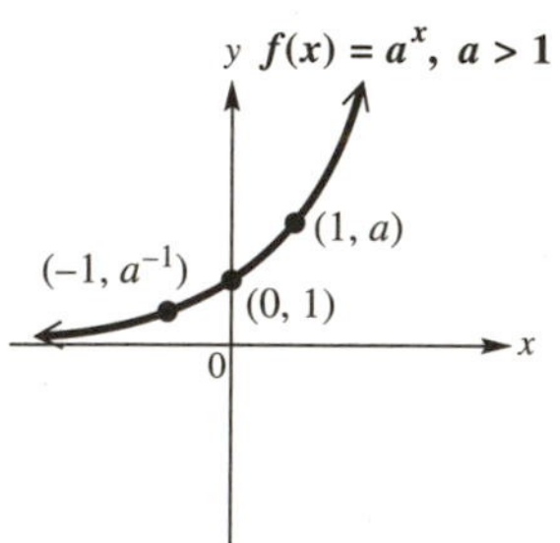

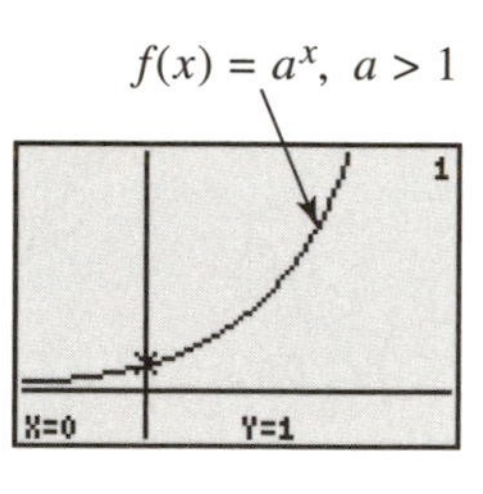

EXPONENTIAL FUNCTION $f(x) = a^x, \quad 0 < a < 1$

Domain: $(-\infty, \infty)$ Range: $(0, \infty)$

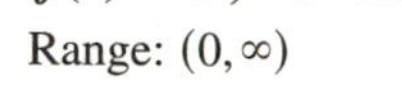

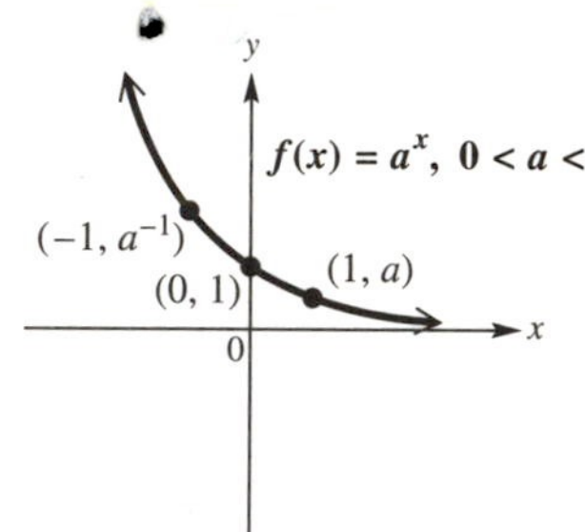

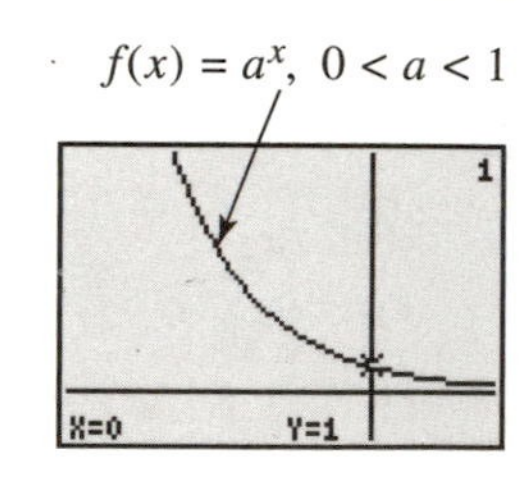

LOGARITHMIC FUNCTION $f(x) = \log_a x, \quad a > 1$

Domain: $(0, \infty)$ Range: $(-\infty, \infty)$

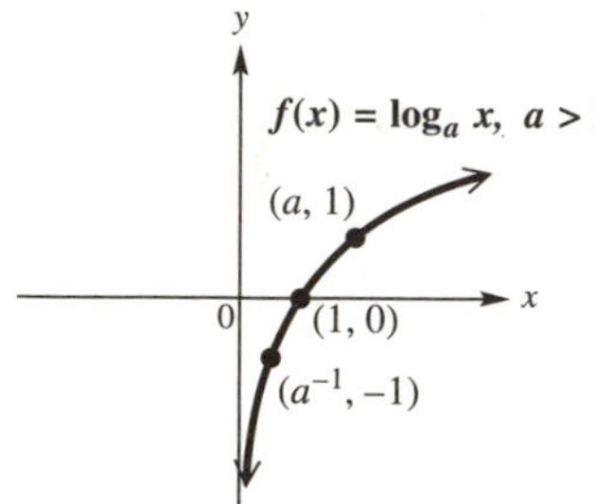

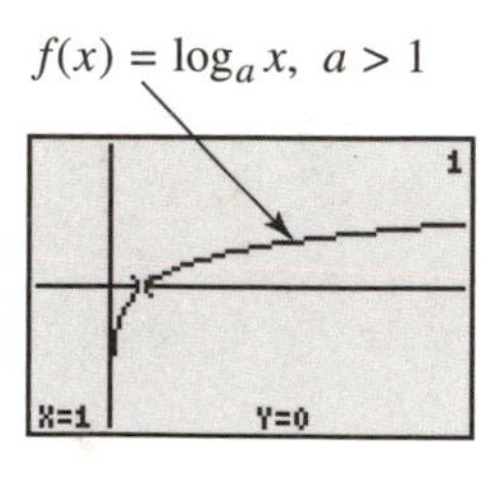

LOGARITHMIC FUNCTION $f(x) = \log_a x, \quad 0 < a < 1$

Domain: $(0, \infty)$ Range: $(-\infty, \infty)$

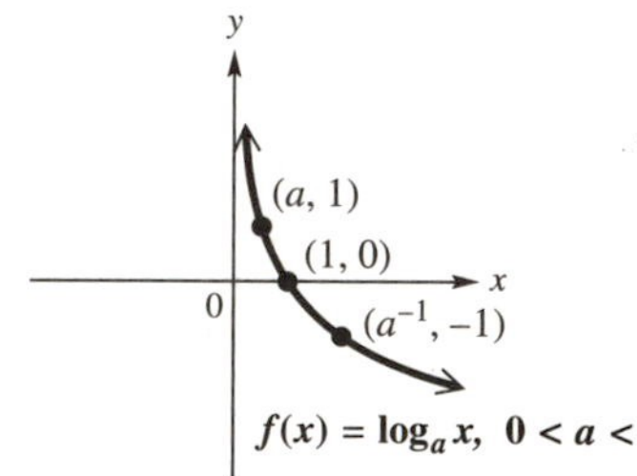

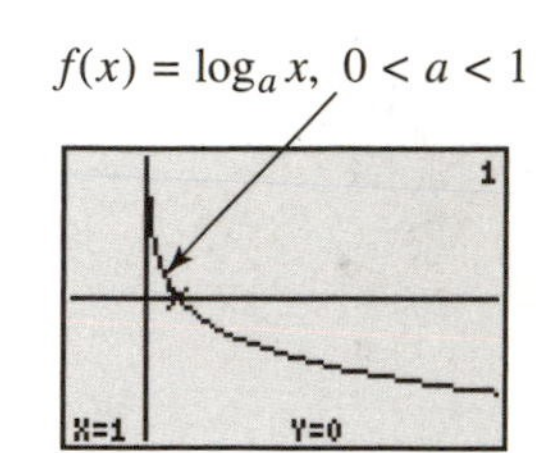